The Elements

Name	Symbol	Atomic Number	Relative Atomic Weight	Name	Symbol	Atomic Number	Relative Atomic Weight
Actinium	Ac	89	227.028	Mendelevium	Md	101	(258)
Aluminum	Al	13	26.9815	Mercury	Hg	80	200.59
Americium	Am	95	(243)	Molybdenum	Mo	42	95.94
Antimony	Sb	51	121.757	Neodymium	Nd	60	144.24
Argon	Ar	18	39.948	Neon	Ne	10	20.1797
Arsenic	As	33	74.9216	Neptunium	Np	93	237.048
Astatine	At	85	(210)	Nickel	Ni	28	58.693
Barium	Ba	56	137.327	Niobium	Nb	41	92.9064
Berkelium	Bk	97	(247)	Nitrogen	N	7	14.0067
Beryllium	Be	4	9.01218	Nobelium	No	102	(259)
Bismuth	Bi	83	208.980	Osmium	Os	76	190.23
Bohrium	Bh	107	(262)	Oxygen	O	8	15.9994
Boron	B	5	10.811	Palladium	Pd	46	106.42
Bromine	Br	35	79.904	Phosphorus	P	15	30.9738
Cadmium	Cd	48	112.411	Platinum	Pt	78	195.08
Calcium	Ca	20	40.078	Plutonium	Pu	94	(244)
Californium	Cf	98	(251)	Polonium	Po	84	(209)
Carbon	C	6	12.011	Potassium	K	19	39.0983
Cerium	Ce	58	140.115	Praseodymium	Pr	59	140.908
Cesium	Cs	55	132.905	Promethium	Pm	61	(145)
Chlorine	Cl	17	35.4527	Protactinium	Pa	91	231.036
Chromium	Cr	24	51.9961	Radium	Ra	88	226.025
Cobalt	Co	27	58.9332	Radon	Rn	86	(222)
Copper	Cu	29	63.546	Rhenium	Re	75	186.207
Curium	Cm	96	(247)	Rhodium	Rh	45	102.906
Dubnium	Db	105	(262)	Rubidium	Rb	37	85.4678
Dysprosium	Dy	66	162.50	Ruthenium	Ru	44	101.07
Einsteinium	Es	99	(252)	Rutherfordium	Rf	104	(261)
Erbium	Er	68	167.26	Samarium	Sm	62	150.36
Europium	Eu	63	151.965	Scandium	Sc	21	44.9559
Fermium	Fm	100	(257)	Seaborgium	Sg	106	(263)
Fluorine	F	9	18.9984	Selenium	Se	34	78.96
Francium	Fr	87	(223)	Silicon	Si	14	28.0855
Gadolinium	Gd	64	157.25	Silver	Ag	47	107.868
Gallium	Ga	31	69.723	Sodium	Na	11	22.9898
Germanium	Ge	32	72.61	Strontium	Sr	38	87.62
Gold	Au	79	196.967	Sulfur	S	16	32.066
Hafnium	Hf	72	178.49	Tantalum	Ta	73	180.948
Hassium	Hs	108	(265)	Technetium	Tc	43	(98)
Helium	He	2	4.00260	Tellurium	Te	52	127.60
Holmium	Ho	67	164.930	Terbium	Tb	65	158.925
Hydrogen	H	1	1.00794	Thallium	Tl	81	204.383
Indium	In	49	114.818	Thorium	Th	90	232.038
Iodine	I	53	126.904	Thulium	Tm	69	168.934
Iridium	Ir	77	192.22	Tin	Sn	50	118.710
Iron	Fe	26	55.847	Titanium	Ti	22	47.88
Krypton	Kr	36	83.80	Tungsten	W	74	183.84
Lanthanum	La	57	138.906	Uranium	U	92	238.029
Lawrencium	Lr	103	(260)	Vanadium	V	23	50.9415
Lead	Pb	82	207.2	Xenon	Xe	54	131.29
Lithium	Li	3	6.941	Ytterbium	Yb	70	173.04
Lutetium	Lu	71	174.967	Yttrium	Y	39	88.9059
Magnesium	Mg	12	24.3050	Zinc	Zn	30	65.39
Manganese	Mn	25	54.9381	Zirconium	Zr	40	91.224
Meitnerium	Mt	109	(266)				

Atomic masses in this table are relative to carbon-12 and limited to six significant figures, although some atomic masses are known more precisely. For certain radioactive elements the numbers listed (in parentheses) are the mass numbers of the most stable isotopes.

General Chemistry

General Chemistry

Principles and Modern Applications

Eighth Edition

Ralph H. Petrucci

California State University, San Bernardino

William S. Harwood

Indiana University, Bloomington

F. Geoffrey Herring

University of British Columbia

With contributions by Scott S. Perry, University of Houston

PRENTICE HALL

Upper Saddle River, New Jersey 07458

Library of Congress Cataloging-in-Publication Data

Petrucci , Ralph H.
  General chemistry: principles and modern applications.—8th ed. / Ralph H. Petrucci,
William S. Harwood, F. Goffrey Herring.
    p.   cm

  Includes index.
  ISBN 0-13-014329-4
  I. Chemistry. I. Harwood, William S. II. Herring, F. Geoffrey. III. Title.

QD31.3 .P47 2002
540—dc21   2001032331

Editor in Chief: John Challice
Development Editors: Deena Cloud; Karen Karlin
Production Editor: Debra A. Wechsler
Editorial Assistant: Eliana Ortiz
Marketing Assistant: Matthew Redstone
Executive Managing Editor: Kathleen Schiaparelli
Assistant Managing Editor: Beth Sturla
Project Manager: Kristen Kaiser
Media Editor: Paul Draper
Marketing Manager: Steve Sartori
Manufacturing Buyer: Michael Bell
Art Directors: Joseph Sengotta; Jonathan Boylan
Art Editor: Karen Branson; Shannon Sims
Interior Designer: Lisa A. Jones
Cover Designer: Jonathan Boylan
Photo Research: Beaura Ringrose
Assistant Managing Editor, Science Media: Alison Lorber
Vice President of Production and Manufacturing: David W. Riccardi
Editor in Chief, Development: Ray Mullaney
Manufacturing Manager: Trudy Pisciotti
Assistant to Art Director: John Christiana
Director of Creative Services: Paul Belfanti
Director of Design: Carole Anson
Managing Editor, Audio/Video Assets: Grace Hazeldine
Project Coordinator, Artworks: Connie Long
Illustrators, Artworks: Royce Copenhaver; Daniel Knopsnyder
Photo Research Administrator: Beth Boyd
Senior Manager, Artworks: Patty Burns
Production Manager, Artworks: Ronda Whitson
Manager, Production Technologies, Artworks: Matt Haas

Cover photos (from left to right): Computer graphics representation of the ATP-binding domain of phosphoglycerate kinase (Oxford Molecular Biophysics Laboratory/Science Photo Library/Photo Researchers, Inc.); chemical plant at dusk (Martin Bond/Science Photo Library/Photo Researchers, Inc.); liquid crystal color change (Richard Megna/ Fundamental Photographs); argon discharge globe (Ed Degginger/Color Pic, Inc.); computer graphics representation of one of the molecular orbitals of a probable reaction intermediary of benzine, C_6H_6 (Chemical Design/Science Photo Laboratory/Photo Researchers, Inc.); thermite reaction (Richard Price/Color Pic, Inc.); demonstration of magnetic levitation of yttrium-barium-copper oxide (David Parker/IMI/Univ of Birmingham/Science Photo Library/Photo Researchers, Inc.).

Printed in the United States of America
10 9 8 7 6 5 4 3 2

ISBN 0-13-014329-4

Pearson Education Ltd., *London*
Pearson Education Australia Pty. Limited, *Sydney*
Pearson Education Singapore, Pte. Ltd.
Pearson Education North Asia Ltd., *Hong Kong*
Pearson Education Canada Ltd., *Toronto*
Pearson Education de Mexico, S.A. de C.V.
Pearson Education Japan, *Tokyo*
Pearson Education Malaysia, Pte. Ltd.

Brief Contents

Contents

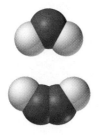

1 Matter—Its Properties and Measurement   1

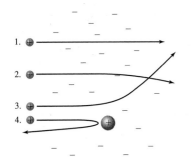

2 Atoms and the Atomic Theory   33

3 Chemical Compounds   65

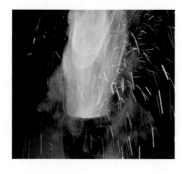

4   Chemical Reactions   107

5   Introduction to Reactions in Aqueous Solutions   139

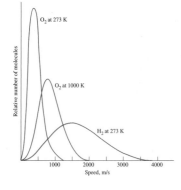

6   Gases   175

Translational

Rotational

Vibrational

Electrostatic
(Intermolecular attractions)

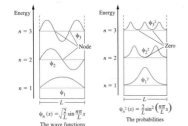

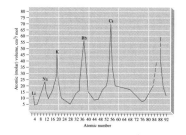

10 The Periodic Table and Some Atomic Properties 356

11 Chemical Bonding I: Basic Concepts 388

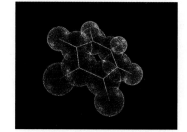

12 Chemical Bonding II: Additional Aspects 435

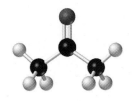

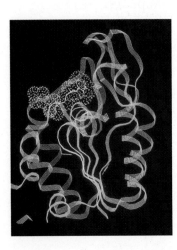

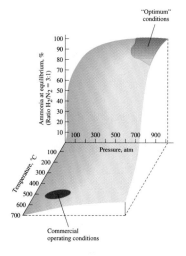

16 Principles of Chemical Equilibrium 626

17 Acids and Bases 665

18 Additional Aspects of Acid–Base Equilibria 710

19 Solubility and Complex-Ion Equilibria 749

20 Spontaneous Change: Entropy and Free Energy 782

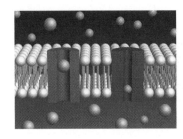

21 Electrochemistry 823

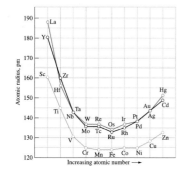

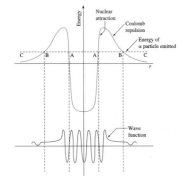

26   Nuclear Chemistry   1024

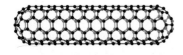

27   Organic Chemistry   1058

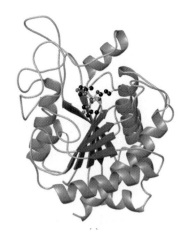

28 Chemistry of the Living State 1122

Appendixes

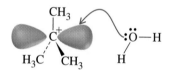

About the Authors

Ralph H. Petrucci

Ralph Petrucci received his B.S. in Chemistry from Union College and his Ph.D. from the University of Wisconsin–Madison. Following several years of teaching, research, consulting, and directing the NSF Institutes for Secondary School Science Teachers at Case Western Reserve University, Dr. Petrucci joined the planning staff of the new California State University campus at San Bernardino in 1964. There, in addition to his faculty appointment, he served as Chairman of the Natural Sciences Division and Dean of Academic Planning. Professor Petrucci, now retired from teaching, is the author of several books, including *General Chemistry* with John W. Hill.

William S. Harwood

Bill Harwood received his B.Sc. from the University of Massachusetts, Amherst and his Ph.D. in Inorganic Chemistry from Purdue University in 1986. He is currently a Professor of Science Education at Indiana University, Bloomington. Previously, Dr. Harwood was at the Department of Chemistry and Biochemistry at the University of Maryland, College Park. In his current role, Dr. Harwood continues to teach chemistry and conduct research in chemical education. He has received several awards for teaching excellence. Dr. Harwood is also active in the American Chemical Society and the Division of Chemical Education and was a consultant to AAAS project 2061. He is involved in the science reform efforts at both the pre-college and college levels. His research focuses on how best to use technology to improve learning in chemistry.

F. Geoffrey Herring

Geoff Herring received his B.Sc. and his Ph.D. in Physical Chemistry, both from the University of London. He is currently a Professor in the Department of Chemistry of the University of British Columbia, Vancouver. Dr. Herring has research interests in the area of biophysical chemistry and has published over 100 papers in the area of physical chemistry and chemical physics. Recently, Dr. Herring has undertaken studies in the use of information technology and interactive engagement methods in teaching general chemistry with a view to improving student comprehension and learning. Dr. Herring has taught chemistry from undergraduate to graduate levels for 30 years and has been the recipient of the Killam Prize for Excellence in Teaching.

Preface

"Know your audience." For this new edition, we have tried to follow this important advice to writers by attending more to the needs of those students who are taking a serious journey through the material. We also know that most general chemistry students have career interests not in chemistry, but in biology, medicine, engineering, environmental and agricultural sciences, and so on. And we understand that general chemistry will be the only college chemistry course for some students, and thus their only opportunity to learn some practical applications of chemistry. We have designed this book for all these students.

Students of this text should have already studied some chemistry. But those with no prior background and those who could use a refresher will find that the early chapters develop fundamental concepts from the most elementary ideas. Students who do plan to become professional chemists will also find opportunities in the text to pursue their own special interests.

The typical student may need help identifying and applying principles and visualizing their physical significance. The pedagogical features of this text are designed to provide this help. At the same time, we hope the text serves to sharpen student skills in problem solving and critical thinking. Thus, we have tried to strike the proper balances between principles and applications, qualitative and quantitative discussions, and rigor and simplification.

Throughout the text we provide real-world examples to enhance the discussion. Examples relevant to the biological sciences, engineering, and the environmental sciences will be found in numerous places. This should help to bring the chemistry alive for these students, and help them understand its relevance to their career interests. It also, in most cases, should help them master core concepts.

Organization

In this edition we retain the core organization of the sixth and seventh editions of this text, but with additional coverage of material, in depth and breadth, in a number of chapters. After a brief overview of core concepts in Chapter 1, we introduce atomic theory, including the periodic table, in Chapter 2. The periodic table is an extraordinarily useful tool, and presenting it early allows us to use the periodic table in new ways throughout the early chapters of the text. In Chapter 3 we introduce chemical compounds and their stoichiometry. Organic compounds are included in this presentation. The early introduction of organic compounds allows us to use organic examples throughout the book. Chapters 4 and 5 introduce chemical reactions. We discuss gases in Chapter 6, partly because they are familiar to students (which helps them build confidence), but also because some instructors prefer to cover this material early to better integrate their lecture and lab programs. Note that Chapter 6 can easily be deferred for coverage with the other states of matter, in Chapter 13. In Chapter 9 we delve more deeply into wave mechanics than in earlier editions, although we do so in a way that allows excision of this material at the instructor's discretion. As with previous editions, we have emphasized real-world chemistry in the final chapters that cover descriptive chemistry (Chapters 22–25), and we have tried to make this material easy to bring forward into earlier parts of the text. Moreover, many topics in these chapters can be covered selectively, without requiring the study of entire chapters. The text ends with heavily revised, comprehensive chapters on organic chemistry (Chapter 27) and biochemistry (Chapter 28).

Changes to This Edition

We have made a number of smaller organizational changes to improve the flow of information to the student and to reflect contemporary thoughts about how best to teach general chemistry. In Chapter 7 (Thermochemistry), the order in which heat and work are presented has been reversed from that of the 7th edition. Also, the concept of standard states is introduced earlier, so that most data in the chapter can be standard-state data. In Chapter 20 (Thermodynamics), the sections on entropy have been reorganized so that all of them precede the introduction to free energy.

▶ Increased level of detail and much more problem-solving pedagogy this edition.

Major changes in this edition have focused on increasing the depth of coverage and adding some more challenging end-of-chapter exercises. Specifically, in Chapter 3 (Chemical Compounds) we have added a section introducing organic compounds, including nomenclature. This allows more reference to organic compounds throughout the book and also suits those who chose to introduce organic chemistry earlier in the course. In Chapter 6 (Gases) there is greater emphasis on the use of SI units and more detail on the kinetic-molecular theory of gases. In Chapter 7 (Thermochemistry) the calculation of quantities of work and the discussion of state functions and path-dependent functions are more extensive than in the previous edition. Chapter 9 (Electrons in Atoms) has been significantly revised to include new sections dealing with wave mechanics and more information on the treatment of wave functions, quantum numbers, orbitals, and radial probability distributions. Chapter 10 (The Periodic Table and Some Atomic Properties) draws more on ideas from Chapter 9 than in earlier editions, permitting a fuller discussion of screening, penetration, and Z_{eff}. Chapter 11 (Basic Concepts of Chemical Bonding) has been rearranged to provide a clearer exposition of the general strategy for writing Lewis structures. In Chapter 12 (Additional Aspects of Chemical Bonding), the new ideas developed in Chapter 9 are applied to the hybridization of atomic orbitals and the treatment of molecular orbitals. Molecular orbital theory is extended to cover heteronuclear molecules.

Chapter 13 (Liquids, Solids, and Intermolecular Forces) features an expanded section on crystal structures. In Chapter 15 (Chemical Kinetics) the IUPAC-recommended definition of a general rate of reaction is used in the treatment of reaction rates. Also, reaction mechanisms and enzyme catalysis are presented in more detail.

A new feature in Chapter 17 (Acids and Bases) is a discussion of a general method for equilibrium calculations based on equilibrium constant expressions, material balances, and electroneutrality. In Chapter 20 (Thermodynamics), the concept of entropy is introduced in a new way, and the relationship between ΔG and $\Delta G°$ is developed and explained more fully. The chapters on descriptive inorganic chemistry (Chapters 22-24) have been updated and now include electrode potential (Latimer) diagrams. Chapter 27 (Organic Chemistry) now includes an introductory discussion of S_N1 and S_N2 reactions and other topics of interest to those covering more organic chemistry in this course. Discussions of metabolism and enzyme reactions have been added to Chapter 28 (Chemistry of the Living State).

In-Text Learning Aids for Students

As with previous editions, we have tried to create the most useful possible text for students. Here are some of the things that should make this so:

Important Expressions. The most significant equations, concepts, and rules are highlighted with colored panels so that students can readily find them.

Summary/Key Terms/Glossary. Each chapter concludes with a comprehensive verbal *Summary* of important concepts and factual information. The *Summary* is followed by a list of *Key Terms*—terms that appear in boldface type in the text and are defined again in the *Glossary* (Appendix E). Students can use *Key Terms* lists and the *Glossary* to help them master the terminology of general chemistry.

▶ Are You Wondering …?, probing questions asked by good students, are enhanced in this edition.

Are You Wondering …?   To help clarify matters that often puzzle students, we pose and then answer questions under this special heading. For obvious pedagogical reasons, these questions are cast in the form in which students typically ask them. Some are designed to help students avoid common misconceptions; others provide analogies or alternative explanations of a concept; still others address apparent inconsistencies in the material they are learning. In response to reviewer and student comments and suggestions, these have been expanded considerably in this edition. Specifically, the *Are You Wondering …?* format is used in a number of instances to introduce material directed at the better-prepared students. Some of these topics are pursued further in end-of-chapter exercises. These topics can be assigned or omitted at the discretion of the instructor.

▶ Keep in Mind margin notes are new to this edition.

Keep in Mind margin notes.   To help students appreciate the significance of earlier ideas, or to warn them about common pitfalls, we have added to this edition *Keep in Mind* margin notes. As the name suggests, these notes ask students to recall key information about concepts and problem-solving skills. At times, we use these in conjunction with worked examples to forewarn students about common mistakes.

▶ Detailed applications of chemistry are covered at the end of the chapter in a non-distracting way.

Focus On boxes.   We believe that relevant applications should be an integral part of the text and that asides should be limited to margin notes and *Are You Wondering …?* features. With this in mind, we have concluded the text of each chapter with a short essay on a practical topic appropriate to the chapter content. These essays, which may be considered optional reading, focus on ideas introduced in the chapter.

The Strongest Available Problem-Solving Focus

You probably won't become a better golfer just by watching Tiger Woods play; you have to get onto the course yourself, and often. To give students the support they need to develop strong problem-solving skills, we offer extensive in-text examples that cover all the key concepts introduced in the book, each accompanied by two practice examples. We also provide integrative problems as concluding in-text examples in each chapter and a very large selection of end-of-chapter exercises, including a set that integrate the student media:

▶ Many worked examples, carefully developed, step-by-step.

In-Text Illustrative Examples.   In each chapter, most concepts—especially those that students will be expected to apply in homework assignments and examinations—are illustrated with worked-out examples. To aid visual learners and to emphasize abstract concepts, in many cases a line drawing or photograph accompanies an example to help students visualize what is going on in the problem.

▶ **Two** practice examples after every in-text example.

Practice Examples.   These are designed to give students immediate practice in applying the principle(s) illustrated in the example. We offer two for every illustrative example. The first, Practice Example A, provides immediate practice in a problem very similar to the illustrative example. The second, Practice Example B, generally takes the student one step further than the illustrative example. This combination helps students to integrate and extend their knowledge and problem-solving skills. Answers to all Practice Examples are given in Appendix F. Complete solutions are given in the *Selected Solutions Manual*.

▶ Integrative Examples, designed to help students learn how to solve these more complicated problems, are new to this edition.

Integrative Examples.   The text includes a special category of problems that requires students to link various important problem types introduced in the chapter—with each other and with problem types from earlier chapters. These problems are meant to be challenging for students, and to help them learn how to solve such problems. Each chapter concludes with a multi-part *Integrative Example*, sometimes of a practical nature. In each case, the problem is broken down into parts, each part is solved, and intermediate results are combined into a final solution and answer.

▶ Integrative and Advanced Exercises are enhanced this edition.

▶ Feature Problems, the most challenging in the book, are expanded this edition.

▶ eMedia Exercises are new to this edition.

End-of-Chapter Exercises. Each chapter ends with five categories of exercises. *Review Questions* require straightforward application of principles introduced in the chapter, each generally involving a single concept, and either a numerical, symbolic, or short written (or verbal) answer. *Exercises* are grouped by categories related to the text sections, and they are of a broader nature than the *Review Questions*. The *Exercises* are paired, so that there are two problems of the same type. The *Integrative and Advanced Exercises* are not grouped by type. These are generally more difficult than those in the previous sections. They tend to integrate material from multiple sections, or multiple chapters, and they may introduce new ideas or pursue certain ideas further than is done in the text. *Feature Problems* are of special interest. These problems generally require the highest level of cognitive skill on the part of students to solve. Some of these problems retrace aspects of the history of chemistry; a few deal with classic experiments; others require students to interpret data or graphs; some present new material; some suggest alternative techniques for problem solving; and a few summarize main points of the chapter in a comprehensive manner. The *Feature Problems* are a resource that can be used in several ways: as discussion points in class, as assigned homework for individuals, or for collaborative group work. Finally, the *eMedia Exercises,* new to this edition, are questions that can only be solved using the interactive media accompanying this text. This permits the instructor to mandate the use of the media by simply assigning one or more of these problems.

Answers to all red-numbered problems are given in Appendix F. Full solutions to all red-numbered problems are found in the *Selected Solutions Manual.*

Supplements

For the Instructor

▶ New instructor's supplement.

Annotated Instructor's Edition (with Guide to Media Resources) (ISBN 0-13-017677-X). This special edition of the text provides marginal notes and information for instructors and TAs, including teaching tips, suggested lecture demonstrations, references to the chemical education literature, and icons identifying all art that appears on overhead transparencies and on the Media Portfolio CD-ROM for instructors.

▶ New instructor's supplement. Includes prebuilt MS Power-Point® slides.

Media Portfolio: Your Presentation Resource CD-ROM (dual platform; ISBN 0-13-017686-9). Specific to Petrucci/Harwood/Herring, this CD includes almost all art and photos from the book, over 61 short animations, 31 video demonstrations, and Java and Flash simulations from the Student iBook. All these pieces are presented in a thumbnail catalog format that allows easy porting of the files to presentation software such as MS PowerPoint®. Also included are electronic versions of suggested course outlines (which can be edited), a set of pre-built MS PowerPoint slides, and a special chemistry font that lets you quickly edit and add to the electronic files on the CD.

The *Instructor's Resource Manual (ISBN 0-13-017678-8)* by Michael L. Denniston, Georgia Perimeter College and Robert K. Wismer, Millersville University. Ideal for novice instructors or others using this text for the first time, this book integrates all ancillary material, offers Notes for the Instructor, lists key concepts, itemizes Chapter Objectives, and contains the solutions to the Advanced and Integrative Problems not found in Appendix F of the text.

▶ 50% more overhead transparencies this edition.

Transparencies (ISBN 0-13-017685-0) Includes over 250 full-color images from this text. Each of these is also provided in electronic form on the MediaPortfolio Instructor's CD.

Test Item File (ISBN 0-13-017679-6) by C. Alton Hassell, Baylor University. This hardcopy test bank includes over 2000 unique questions, each accuracy checked for this new edition and not available to students. These questions are also available in WebCT format for adopting institutions.

▶ New testing software

Prentice Hall Test Manager This newly updated testing software includes all 2000 questions from the Test Item File and permits easy creation and editing of quizzes. The software allows easy porting of quizzes into MS Word® format and also supports administration of quizzes over a LAN. Available in both Macintosh *(ISBN 0-13-017681-8)* and Windows *(ISBN 0-13-017670-2)* formats.

Solutions Manual (ISBN 0-13-017683-4) by Lucio Gelmini and Robert Hilts, both of Grant MacEwan College, and Robert K. Wismer, Millersville University. Contains completely revised, step-by-step solutions to all end-of-chapter (exercises except for eMedia exercises and the Advanced and Integrative Problems found in Appendix F of text). With instructor permission, these manuals may be made available to students.

▶ New to this edition! Three Course Management options.

Prentice Hall Course Management Solutions. Prentice Hall offers pre-built courses in a variety of Course Management systems, each of which lets you easily post your syllabus, communicate with students online or offline, administer quizzes, and record student results and track their progress.

Course Compass™ —the easiest way to get your course online! Three clicks and you're up. The course includes all media resources from the instructor CD, the student iBook, and over 5000 quiz questions.

BlackBoard® —for campuses that use the user-friendly system BlackBoard, consider a pre-built course that includes all media resources from the instructor CD, the student iBook, and over 5000 quiz questions.

WebCT® —for campuses that use the sophisticated course management tools of WebCT, the pre-built course offers everything above as well as over algorithmic questions developed in WebCT's calculation format.

Ask your Prentice Hall representative for details about any of these options.

For the Laboratory

Experiments in General Chemistry (ISBN 0-13-017688-5) by Gerald S. Weiss, Thomas G. Greco and Lyman H. Rickard, all at Millersville University. A comprehensive laboratory manual containing 37 experiments that parallel the text, including a final group of six experiments on qualitative cation analysis. There is an accompanying instructor's manual *(ISBN 0-13-017689-3)*.

For the Student

▶ Powerful new student supplement—an interactive electronic version of the text.

Student iBook (ISBN 0-13-017680-X) by Scott Perry, University of Houston. This interactive version of the text includes hundreds of animations, simulations, manipulable molecular models, movies, and interactive glossary terms, all integrated in-context within an electronic version of the text. Accessed easily using a web browser, this product allows students to see and discover chemistry in ways never before possible. Each interactive exercise is followed by a self-assessment question so students can make sure they understand the key points before moving on to the next topic. Organized exactly like the book, this product is available free with every new copy of the text.

▶ Enhanced student website features algorithmic questions.

The Petrucci/Harwood/Herring Companion Website **www.prenhall.com/petrucci** by Narayan S. Hosmane, Northern Illinois University. Now in its "second edition," this innovative online resource center is designed to specifically support and enhance students use of Petrucci/Harwood/Herring 8/e. It features

- A Problem Solving Center, where student have access to more than 2000 additional problems, including algorithmically generated questions and non-multiple-choice problems—all organized by chapter, with hints and specific feedback.

- A Visualization Center, where students can view hundreds of pre-built 3-D molecular models using Chime.
- Current Topics, where recent articles from the popular press are summarized and further questions are posted for students to answer on paper or online

Student Study Guide (ISBN 0-13-032567-8), by Dixie Goss of Hunter College and Robert K. Wismer of Millersville University, guides students through the text's coverage with discussion of chapter learning objectives, drill problems, self quizzes, and sample tests.

Student Solutions Manual (ISBN 0-13-017684-2), by Lucio Gelmini and Robert Hilts, both of Grant MacEwan College, and Robert Wismer of Millersville University. Contains full, step-by-step solutions to the red-numbered problems from the text (those answered in Appendix F of the textbook).

▶ New student supplement.

Math Review Toolkit (ISBN 0-13-032568-6), by Gary Long, Virginia Polytechnic Institute and State University. Contains a chapter-by-chapter review of the essential math skills required for each chapter as well as a brief review of writing in chemistry. Ideal for students for whom math is a major obstacle to success in the course. Available free with a new book; please see your Prentice Hall representative for details.

The New York Times/Prentice Hall Themes of the Times Supplement, in newspaper format, brings together a collection of recent chemistry-related articles from the pages of *The New York Times.* This free supplement, updated twice a year and available on request, encourages students to make connections between the chemistry they are learning in the classroom and the world around them. Available free with a new book; please see your Prentice Hall representative for details.

Prentice Hall Molecular Model Set for General and Organic Chemistry (ISBN 0-13-955444-0). This ball-and-stick model kit is designed for use in general chemistry and the student's next course in organic chemistry. It includes trigonal bipyramidal and octahedral atom centers as well as 14 carbon atoms.

Acknowledgments

Many people have given of their time, creativity, and support during the preparation of this edition. Numerous colleagues from many places have offered helpful suggestions through the several editions of this text. Some are cited specifically in these acknowledgments, but many others are not. To all, however, we are deeply grateful,

We are especially grateful for the patience and encouragement of our wives, Ruth Petrucci, Diana Harwood, and Jeanie Herring. They have been willing to give up precious family time so that we could produce this textbook. Without their love and support, this book would not have been possible.

We extend our sincere thanks and acknowledgements to those of our colleagues, in both the US and Canada, who gave us their advice and counsel during the preparation of this edition, either as commentators on the 7th edition, reviewers of the 8th edition manuscript, or as technical reviewers of 8th edition page proofs. We appreciate your time, thoughtfulness, and creativity:

Steven Adelman, Purdue University

Fakhrildeen Albahadily, University of Central Oklahoma

Margaret Asirvatham, University of Colorado

Alton J. Banks, North Carolina State University

Richard Bates, Georgetown University

Russel G. Baughman, Truman State University

Alexis O. Bawagan, Ottawa-Carleton Chemistry Institute

Azzedine Bensalem, Long Island University

Richard Bersohn, Columbia University

Joyce C. Brockwell, Northwestern University

Jim Byrd, California State University—Stanislaus

Lisheng Cai, The University of Illinois at Chicago

Rodney Cate, Midwestern State University

Thomas Chasteen, Sam Houston State University

Klaus Dichmann, Vanier College

Charles Drain, Hunter College

John Evans, New York University

Jan M. Fleisher, The College of New Jersey

Christopher G. Flinn, Memorial University of Newfoundland

Rene Fournier York University

Regina Frey, Washington University

Lucio Gelmini, Grant MacEwan College

Harry H. Gibson, Jr., Austin College

Marcia Gillette, University of Indiana, Kokomo

Jerry Goodisman, Syracuse University

C. Michael Greenlief, University of Missouri

Michael Hampton, York University

Sherman Henzel, Monroe Community College

Robert Hilts, Grant MacEwan College

Pamela Holt, Westmont College

Leonidas J. Jones III, St. Joseph College

George Kreishman, University of Cincinnati

Charles Kutal, University of Georgia

William LaCourse, University of Maryland, Baltimore County

Willem R. Leenstra, University of Vermont

N. Thornton Lipscomb, University of Louisville

John Maguire, Southern Methodist University—Dedman College

Albert Martin, Moravian College

Christie A. McDermott, University of Alberta

Wyatt R. Murphy, Seton Hall University

Allan Nishamura, Westmont College

Joseph Okoh, University of Maryland—Eastern Shore

Gren Patey, University of British Columbia

Bernard L. Powell, University of Texas—San Antonio

Vaughan Pultz, Truman State University

Mary Frances Richardson, Brock University

Darrin Richeson, University of Ottawa

Alan Storr, University of British Columbia

Iwao Teraoka, Polytechnic University—Brooklyn

Mark Thachuk, University of British Columbia

Robert Towery, Houston Baptist University

Maria Vogt, Bloomfield College

Harold Wilson, John Abbott College

We would also like to thank those who have reviewed previous editions of this text, including

J. Atherton, Memorial University of Newfoundland

Ronald M. Backus, American River College

Richard Bretz, The University of Michigan—Dearborn

Albert W. Burgstahler, The University of Kansas

Donald Campbell, University of Wisconsin—Eau Claire

Robert Crabtree, Yale University

Roberta Day, University of Massachusetts at Amherst

Bob Desiderato, University of North Texas

Daryl J. Doyle, Kettering University

John Forsberg, Saint Louis University

Frank Garland, The University of Michigan—Dearborn

Carter Gilmer, The University of Michigan—Dearborn

Peter L. Gold, The Pennsylvania State University

Stan Granda, University of Nevada, Las Vegas

Alton Hassell, Baylor University

Sherman Henzel, Monroe Community College

Andrew J. Holder, University of Missouri—Kansas City

Charles Keilin, Laney College

Donald Kleinfelter, University of Tennessee

Richard W. Kopp, East Tennessee State University

Edwin H. Lane, Williams Jewell College

Stacey Lowery-Bretz, The University of Michigan—Dearborn

John Maguire, Southern Methodist University

Patricia A. Metz, Texas Tech University

Quichee Mir, Yakima Valley College

Donald Newlin, Montgomery College

Robert Porod, Rock Valley College

Bernard L. Powell, University of Texas at San Antonio

Bruce Prall, Marian College

Paul Reinbold, Southern Nazarene University

George E. Shankle, Angelo State University

Maureen Scharberg, San Jose State University

Anil K. Sharma, Mississippi Valley State University

Julie Stewart, El Camino College

Tamar Y. Susskind, Oakland Community College

Duane Swank, Pacific Lutheran University

Carl A. von Frankenberg, University of Delaware

Garth W. Welch, Weber State University

Ronald Wikholm, University of Connecticut

Warren Yeakel, Henry Ford Community College.

We extend special thanks to the staff at Prentice Hall who took the last edition, paper, files, and sketches and turned them into this beautiful book. Eliana Ortiz, Editorial Assistant, was instrumental in coordinating the reviewing process. Debra Wechsler, our Production Editor, worked diligently and with great patience. We appreciate her careful attention to detail, her precision, her creative solutions to problems of layout, and her overall professionalism. Deena Cloud and Karen Karlin, our development editors, have done a fabulous job of keeping us on track, handling the myriad details, maintaining a sense of humor, and asking dozens of questions after the manner of the best students. Finally, we thank John Challice, our editor, for his unbounded enthusiasm, for coming to our aid in critical situations, and for marshalling all the forces required in this extensive revision.

Responding to feedback from our colleagues and students is the most important element in keeping this book on target from one revision to the next. Your comments are most welcome.

Ralph H. Petrucci
rhpetrucci@earthlink.net

William S. Harwood
wharwood@indiana.edu

F. Geoffrey Herring
fgh@chem.ubc.ca

WARNING: Many of the compounds described or pictured in this text are hazardous, as are many of the chemical reactions. Do not attempt any experiment pictured or implied in the text except with permission, in an authorized laboratory setting, and under adequate supervision.

Student's Guide to Using this Text

The next six pages walk you through some of the main features of this text and its integrated media resources. Using this text as designed will help you develop the essential knowledge and skills you need to succeed in chemistry. Good luck!

Keep in Mind Margin Notes ▶
These will help to remind you of ideas introduced earlier in the text that are important to understand what's currently being discussed.

KEEP IN MIND ▶
that if you know any four of the five quantities—q, m, specific heat, T_f, T_i—you can solve equation (7.5) for the remaining one.

Practice Example A: When 1.00 kg lead (specific heat = 100.0 °C is added to a quantity of water at 28.5 °C, the final tempera mixture is 35.2 °C. What is the mass of water present?

Practice Example B: A 100.0-g copper sample (specific hea at 100.0 °C is added to 50.0 g water at 26.5 °C. What is the fina copper–water mixture?

Significance of Specific-Heat Values
Table 7.1 lists the specific heats of several solid elements. The cific heat of aluminum compared with other metals helps to ac "miracle thaw" products designed to thaw frozen foods rapi cools only slowly as it transfers heat to the frozen food, and th

Student iBook icon
This icon tells you that there is an activity on the Student iBook related to what you're learning about. Look at the material on the iBook for a dynamic presentation.

the equilibrium vapor pressure P°, entropy increases by the amount ΔS in Figure 20-6, because the entropy of the ideal solution is greater than pure solvent, the entropy of the vapor produced by the vaporization of s the solution is also greater than the entropy of the vapor obtained fr solvent. For the vapor above the solution to have the higher entropy, it must have a greater number of accessible microscopic energy levels. vapor must be present in a larger volume, and hence, must be at a lo than the vapor coming from the pure solvent. This relationship cor Raoult's law: $P_A = x_A P_A^\circ$.

Absolute Entropies
To establish an *absolute* value of the entropy of a substance, we look tion in which the substance is in its lowest possible energy state, called th *energy*. We take this state to have an entropy of zero. Then we evalu changes as the substance is brought to other conditions of temperature a We add together these entropy changes and obtain a numerical valu solute entropy. The principle that permits this procedure is the **third l modynamics**, which can be stated as follows:

The entropy of a pure perfect crystal at 0 K is zero.

Figure 20-7 illustrates the method outlined in the preceding paragrap mining absolute entropy as a function of temperature. Where phase occur, equation (20.2) is used to evaluate the corresponding entropy ch temperature ranges in which there are no transitions, ΔS° values are ob measurements of specific heats as a function of temperature.
The absolute entropy of one mole of a substance in its standard sta the **standard molar entropy**, S°. Standard molar entropies of a num stances at 25 °C are tabulated in Appendix D. To use these values to c entropy change of a reaction, we use an equation with a familiar f equation 7.21).

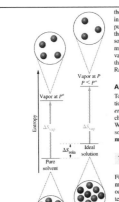

▲ FIGURE 20-6
An entropy-based rationale of Raoult's law

◀ Molecular Art ▼
It is sometimes difficult to visualize molecules and processes that can't be seen directly. To help you understand what's going on at the molecular level, carefully review and make sure you understand the molecular depictions provided in the text.

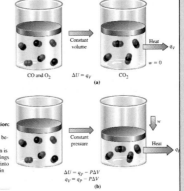

▶ FIGURE 7-13
Comparing heats of reaction at constant volume and constant pressure for the reaction:
$2\ CO(g) = O_2(g) \longrightarrow 2\ CO_2(g)$
(a) *No* work is performed at constant volume because the piston cannot move;
$q_V = \Delta U = -563.5\ \text{kJ}$ (b) When the reaction is carried out at constant pressure, the surroundings do work on the system as the system shrinks into a smaller volume. More heat is evolved than in the constant-volume reaction;
$q_P = \Delta H = -566.0\ \text{kJ}$.

Are You Wondering...

Why Na⁺(aq) does not act as an acid in aqueous solution?

Whether an aqueous solution of a metal ion is acidic depends on two principal factors. The first is the amount of charge on the cation; the second is the size of the ion. The greater the charge on the cation, the greater is the ability of the metal ion to draw electron density away from the O—H bond in a H_2O molecule in its hydration sphere, favoring the release of a H^+ ion. The smaller the cation, the more highly concentrated is the positive charge. Hence, for a given positive charge, the smaller the cation, the more acidic the solution.

The ratio of the charge on the cation to the volume of the cation is called the *charge density*.

$$\text{charge density} = \frac{\text{ionic charge}}{\text{ionic volume}}$$

The greater its charge density, the more effective a metal ion is at pulling electron density from the O—H bond and the more acidic is the hydrated cation. A highly concentrated positive charge on a small cation is better able to pull electron density from the O—H bond than is a less concentrated positive charge on a larger cation.

Thus the small (53 pm ionic radius), highly charged Al^{3+} ion produces acidic solutions, but the larger (99 pm) Na^+ cation, with a charge of just 1+, does not increase the concentration of H_3O^+. In fact, none of the group 1 cations produces appreciably acidic solutions, and only Be^{2+} of the group 2 elements is small enough to do so ($pK_a = 5.4$).

Are You Wondering?
Always ask questions about what you're learning. As you read the text, new ideas may prompt you to raise important queries. The "Are You Wondering?" boxes answer good questions that students frequently ask.

"Focus On" Boxes
You'll find one of these at the end of each chapter. They detail interesting, real-world applications of the chemistry you're learning. This will help you appreciate the enormous importance of chemistry in all aspects of your daily life.

Focus On   Coupled Reactions in Biological Systems

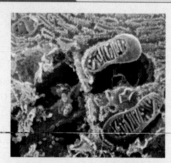

▲ **Mitochondria and endoplasmic reticulum**
A colorized scanning electron micrograph of mitochondria (blue) and rough endoplasmic reticulum (yellow) in a pancreatic cell. Mitochondria are the powerhouses of the cell. They oxidize sugars and fats, producing energy for the conversion of ADP to ATP. Rough endoplasmic reticulum is a network of folded membranes covered with protein-synthesizing ribosomes (small dots).

An important example of a coupled reaction in living organisms is the metabolism of glucose ($C_6H_{12}O_6$) that converts adenosine diphosphate (ADP) to adenosine triphosphate (ATP) in the mitochondria of cells (Fig. 20-13). The ATP is utilized in the ribosomes to produce proteins. The ATP-forming reaction

$$ADP^{3-} + HPO_4^{2-} + H^+ \longrightarrow ATP^{4-} + H_2O$$
$$\Delta G° = -9.2 \text{ kJ mol}^{-1}$$

$$\text{Adenosine}-O-\underset{\underset{O}{|}}{\overset{\overset{O^-}{|}}{P}}-O-\underset{\underset{O}{|}}{\overset{\overset{O^-}{|}}{P}}-O^-$$
$$ADP^{3-}$$

$$\text{Adenosine}-O-\underset{\underset{O}{|}}{\overset{\overset{O^-}{|}}{P}}-O-\underset{\underset{O}{|}}{\overset{\overset{O^-}{|}}{P}}-O-\underset{\underset{O}{|}}{\overset{\overset{O^-}{|}}{P}}-O^-$$
$$ATP^{4-}$$

is spontaneous under standard conditions at 37 °C, so why does the cell need to use glucose to make ATP? The answer is that cells do not operate with $[H^+] = 1$ M as required by standard conditions; in fact, the pH in a cell is about 7. When we estimate ΔG for the reaction at that pH and assume all other species are at 1.0 M (still far from actuality), we get

$$\Delta G = \Delta G° + RT \ln\left(\frac{a_{ATP}a_{H_2O}}{a_{ADP}a_P a_{H^+}}\right)$$

Note that reactions (20.15) and (20.16) are not the same, even though each has Cu(s) as a product. The purpose of coupled reactions, then, is to produce a spontaneous overall reaction by combining two other processes: one nonspontaneous and one spontaneous. Many metallurgical processes employ coupled reactions, especially those that use carbon or hydrogen as reducing agents.

To sustain life, organisms must synthesize complex molecules from simpler ones. If carried out as single-step reactions, these syntheses would generally be accompanied by increases in enthalpy, decreases in entropy, and increases in free energy—in short, they would be nonspontaneous and would not occur. In living organisms, changes in temperature and electrolysis are not viable options for dealing with nonspontaneous processes. Here, coupled reactions are crucial, as described in the Focus On feature.

Problem-Solving ▷

Worked Examples are followed by A & B Practice Exercises. Read carefully through each Worked Example. They help you understand how to solve important types of problems. To help you practice, each worked example is followed by two Practice Examples. Practice Example A is another problem very similar to the Worked Example. Practice Example B is more difficult, closer to the end-of-chapter problems and those you'll find on an exam.

Problem-Solving Notes ▷

These marginal reminders warn you of common mistakes and highlight key strategies for solving particular types of problems.

▶ Sodium phosphate is often sold under the name *trisodium phosphate* (TSP). Anyone working with TSP should wear protective gloves. Fats and greases, including those in human skin, are solubilized in strongly basic solutions.

▶ The ratio $M_b/K_b = 1.0/0.024 = 42$. This value is smaller than the minimum value of 100 that we have been using as the usual criterion.

is not difficult to calculate. It corresponds to that of $Na_3PO_4(aq)$, and PO_4^{3-} can ionize (hydrolyze) only as a base.

$$PO_4^{3-} + H_2O \rightleftharpoons HPO_4^{2-} + OH^-$$

$$K_b = K_w/K_{a_3}$$

$$= \frac{1.0 \times 10^{-14}}{4.2 \times 10^{-13}}$$

$$= 2.4 \times 10^{-2}$$

Determining the pH of a Solution Containing the Anion (A^{n-}) of a Polyprotic Acid.
Sodium phosphate, Na_3PO_4, is an ingredient of some preparations used to clean painted walls before they are repainted. What is the pH of 1.0 M $Na_3PO_4(aq)$?

Solution
In the usual fashion, we can write

$$PO_4^{3-} + H_2O \rightleftharpoons HPO_4^{2-} + OH^- \quad K_b = 2.4 \times 10^{-2}$$

initial concns:	1.0 M	—	—
changes:	$-x$ M	$+x$ M	$+x$ M
equil concns:	$(1.0 - x)$ M	x M	x M

$$K_b = \frac{[HPO_4^{2-}][OH^-]}{[PO_4^{3-}]} = \frac{x \cdot x}{1.0 - x} = 2.4 \times 10^{-2}$$

Because K_b is quite large, we should not expect the usual simplifying assumption to work here. That is, x is *not* very much smaller than 1.0. Solution of the quadratic equation $x^2 + 0.024x - 0.024 = 0$ yields $x = [OH^-] = 0.14$ M. Thus,

$$pOH = -\log[OH^-] = -\log 0.14 = +0.85$$
$$pH = 14.00 - 0.85 = 13.15$$

Practice Example A: Calculate the pH of an aqueous solution that is 1.0 M Na_2CO_3.

(*Hint:* Use data from Table 17.4 to establish K_b for CO_3^{2-}.)

Practice Example B: Calculate the pH of an aqueous solution that is 0.500 M Na_2SO_3.

(*Hint:* Use data from Table 17.4.)

It is more difficult to calculate the pH values of $NaH_2PO_4(aq)$ and $Na_2HPO_4(aq)$ than of $Na_3PO_4(aq)$. This is because with both $H_2PO_4^-$ and HPO_4^{2-}, two equilibria must be considered *simultaneously:* ionization as an acid and ionization as a base (hydrolysis). For solutions that are reasonably concentrated (say, 0.10 M or greater), the pH values prove to be *independent* of the solution concentration. Shown here (with pK_a values from Table 17.4) are general expressions, printed in blue, and their application to $H_2PO_4^-(aq)$ and $HPO_4^{2-}(aq)$:

for $H_2PO_4^-$: $pH = \frac{1}{2}(pK_{a_1} + pK_{a_2}) = \frac{1}{2}(2.15 + 7.20) = 4.68$ (18.5)

for HPO_4^{2-}: $pH = \frac{1}{2}(pK_{a_2} + pK_{a_3}) = \frac{1}{2}(7.20 + 12.38) = 9.79$ (18.6)

Integrative Example

Microwave ovens have become increasingly popular in kitchens around the world. They are also useful in the chemical laboratory, particularly in drying samples for chemical analysis. A typical microwave oven uses microwave radiation with a wavelength of 12.2 cm.

Are there any electronic transitions in the hydrogen atom that could *conceivably* produce microwave radiation of wavelength 12.2 cm?

1. *Calculate the frequency of the microwave radiation.* Microwaves are a form of electromagnetic radiation and thus travel at the speed of light, 2.998×10^8 m s^{-1}. Convert the wavelength to meters, and then use the equation $\nu = c/\lambda$.

$$\nu = \frac{2.998 \times 10^8 \text{ m s}^{-1}}{12.2 \text{ cm} \times 1 \text{ m}/100 \text{ cm}} = 2.46 \times 10^9 \text{ Hz}$$

2. *Calculate the energy associated with one photon of the microwave radiation.* This is a direct application of Planck's equation.

$$E = h\nu = 6.626 \times 10^{-34} \text{ J s} \times 2.46 \times 10^9 \text{ s}^{-1} = 1.63 \times 10^{-24} \text{ J}$$

3. *Determine if there are any electronic transitions in the hydrogen atom with an energy per photon of 1.63×10^{-24} J.* Look at Figure 9-14, the energy-level diagram for the Bohr hydrogen atom. Energy differences between the low-lying levels are of the order 10^{-19} to 10^{-20} J. These are orders of magnitude (10^4–10^5 times) greater than the energy per photon of 1.63×10^{-24} J from part **2.** Note, however, that the energy differences become progressively smaller for high-numbered orbits. As n approaches ∞, the energy differences approach zero, and some transitions between high-numbered orbits should correspond to microwave radiation (see also Exercise 101).

Integrative Examples ◁

These conclude each chapter and serve two purposes: they summarize many of the key concepts in the chapter, and they show you how to approach more complicated, integrative exercises (like those at the end of the chapter).

Media for Students

Some things are better done on a computer than on the two-dimensional surface of a page. Get the most out of this course by making full use of your Student iBook and the Companion Website for your text.

Student iBook ▶
To help you visualize chemistry in an exciting and dynamic way, the Student iBook lets you view demonstration movies, animations, and 3-D molecules; lets you run simulations; and allows you to test your understanding of the key concepts. To make it especially easy-to-use, the Student iBook is organized by chapter and section exactly like your textbook, and all the media activities are presented in the context of the Student iBook.

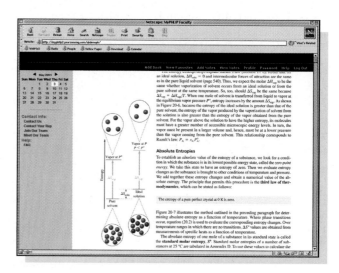

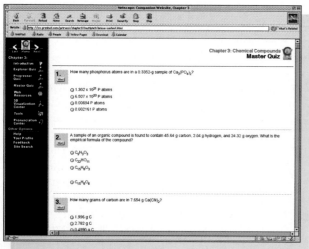

◀ **Companion Website**
(www.prenhall.com/petrucci)
This text also has an accompanying Companion Website (CW) for students. See how the chemistry you are learning applies to the real world. Solve homework problems and get instant grades and feedback on your answers. Explore websites related to the topics you are studying. This site is accessible via any web browser, was written specifically for *General Chemistry: Priciples and Applications*, and is organized by chapter. Nothing could be easier for you to use.

1

Matter—Its Properties and Measurement

Contents

The restoration of art objects, such as this painting by Raphael, is based in large measure on the physical and chemical properties of materials.

In the last few decades, the general public has become increasingly aware of chemistry because of environmental issues such as acid rain and destruction of the ozone layer. Popular accounts, however, do not usually provide much depth of understanding of the basic principles involved, although these principles are needed when applying chemical knowledge to real-world problems. Mastering the principles of chemistry requires a systematic approach to the subject. In this chapter, you will be introduced to some of the basic terminology of chemistry as well as some general methods for making chemical measurements and expressing their results. You may be familiar with some of this material from your earlier studies.

▲ Dr. Susan Solomon, a chemist and Head Project Scientist of the National Ozone Expedition to Antarctica in 1986–1987, is one of the world's leading experts in the interdisciplinary study of stratospheric ozone depletion.

▲ As a biochemist, Percy Julian (1899–1975) developed a number of pharmaceuticals based on soybean extracts. An early accomplishment was a treatment for glaucoma. Eventually, Dr. Julian founded his own company where treatments for arthritis and other conditions were developed.

1-1 The Scope of Chemistry

Chemistry is the study of matter, which includes us and everything around us. Many of our activities involve chemical reactions—changes from one chemical substance to another. Food that we cook undergoes chemical change, and after we have eaten it, our bodies carry out complex chemical reactions to extract nutrients that can be used by our bodies. The gasoline that fuels automobiles is a mixture of dozens of different chemicals. The burning of this mixture provides the energy that propels the automobile. Unfortunately, some of the substances produced in the combustion of gasoline are involved in the formation of smog. Paradoxically, although many of the environmental problems that beset modern society are of a chemical origin, the methods of controlling and correcting these problems are also largely of a chemical nature. In many ways, chemistry touches everyone.

Chemistry is sometimes called the central science because it relates to so many other fields of science and to so many areas of human endeavor. Chemists who develop new materials to improve electronic devices—such as solar cells, transistors, and fiber-optic cables—work at the interfaces of chemistry with physics and engineering. Those who develop new pharmaceuticals for use against cancer or AIDS work at the interfaces of chemistry with pharmacology and medicine. Biochemists are interested in the chemical processes that occur in living organisms. Physical chemists work with fundamental principles of physics and chemistry in an attempt to answer the basic questions that apply to all of chemistry: Why do some substances react with one another but others do not? How fast will a particular chemical reaction occur? How much useful energy can be extracted from a chemical reaction? Analytical chemists study ways to separate and identify chemical substances. Many of the techniques developed by analytical chemists are used extensively by environmental scientists. Organic chemists focus their attention on substances that contain carbon and hydrogen in combination with a few other elements. The vast majority of substances are organic chemicals. For example, living cells consist of water and organic chemicals with a small amount of various salts. Inorganic chemists focus on most of the elements other than carbon, though the fields of organic and inorganic chemistry overlap in some ways.

Although chemistry is a mature science, its landscape is dotted with unanswered questions and challenges. Modern technology calls for new materials with unusual properties, and chemists must devise methods of producing these materials. Modern medicine needs drugs to perform specific tasks in the human body, and chemists must design strategies to synthesize these drugs from relatively simple starting materials. Society requires improved methods of pollution control, substitutes for scarce materials, nonhazardous means of disposing of toxic wastes, and more efficient ways to extract energy from fuels. Chemists work in all these areas.

Progress in science depends on the way scientists do their work—asking the right questions, designing the correct experiments to supply the right answers, and formulating plausible explanations of their findings. Let's look further into this scientific method.

1-2 The Scientific Method

Science differs from other fields of study in the *method* that scientists use to acquire knowledge and the special significance of this knowledge. Scientific knowledge can be used to explain natural phenomena and, at times, to *predict* future events.

The ancient Greeks developed some powerful methods of acquiring knowledge, particularly in mathematics. The Greek approach was to start with certain basic assumptions or premises. Then, by the method known as *deduction*, certain conclusions must logically follow. For example, if $a = b$ and if $b = c$, then $a = c$. Deduction alone is not enough for obtaining scientific knowledge, however. The Greek philosopher Aristotle *assumed* four fundamental substances: air, earth, water, and fire. All other materials, he believed, were formed by combinations of these four elements. Chemists of several centuries ago (more commonly referred to as alchemists) tried, in vain, to apply the four-element idea to turn lead into gold. They failed for many reasons, one being that the four-element assumption is false.

The scientific method originated in the seventeenth century with such people as Galileo, Francis Bacon, Robert Boyle, and Isaac Newton. The key to the method is to make no initial assumptions, but rather to make careful observations of natural phenomena. When enough observations have been made so that a pattern begins to emerge, one then formulates a generalization or natural law describing the phenomenon. **Natural laws** are concise statements, often in mathematical form, of the facts of nature. The process of observations leading to a general statement or natural law is called *induction*. For example, early in the sixteenth century, the Polish astronomer Nicolas Copernicus (1473–1543), through careful study of astronomical observations, concluded that Earth revolves around the sun in a circular orbit, although the general teaching of the time, not based on scientific study, was that the sun and other heavenly bodies revolved around Earth. We can think of Copernicus's statement as a generalization or natural law. Another example of a natural law is the radioactive decay law, which dictates how long it will take for a radioactive substance to become nonradioactive.

To verify a natural law, a scientist designs a controlled situation, an *experiment*, to see if conclusions deduced from the natural law agree with experimental results. We judge the success of a natural law by its ability to summarize observations and predict new phenomena. Copernicus's work was a great success because he was able to predict future positions of the planets more accurately than his contemporaries. We should not think of a natural law as an *absolute* truth, however. Future experiments may require us to modify the law. Copernicus's ideas were refined a half-century later by Johannes Kepler, who showed that planets travel in elliptical, not circular, orbits.

A **hypothesis** is a tentative explanation of a natural law. If a hypothesis survives testing by experiments, it is often referred to as a theory. We can use this term in a broader sense, though. A **theory** is a model or way of looking at nature that can be used to explain natural laws and make further predictions about natural phenomena. When differing or conflicting theories are proposed, the one that is most successful in its predictions is generally chosen. Also, the theory that involves the smallest number of assumptions—the simplest theory—is preferred. Over time, as new evidence accumulates, most scientific theories undergo modification and some are discarded.

The **scientific method** is the combination of observation, experimentation, and the formulation of laws, hypotheses, and theories. The method is illustrated by the flow diagram in the margin. However, it is wrong to suppose that merely following a set of procedures, rather like using a cookbook, will guarantee scientific success. Occasionally scientists develop a pattern of thinking about their field, known as a *paradigm*, that is at first successful but then becomes less so. A new paradigm may be needed. For example, in psychiatry, it was long believed that all mental illness was a product of the mind, not the body. A new paradigm in psychiatry recognizes that some mental illness is caused by chemical imbalances in the body. And, finally, many discoveries (X rays, radioactivity, and penicillin, to name a few) have been made by accident. Such chance discoveries are the result of serendipity. In 1839, the American inventor Charles Goodyear was searching for a treatment of natural rubber that would make it less brittle when cold and less tacky when warm. In the course of this work, he accidentally spilled a rubber–sulfur mixture on a hot stove and found that the resulting product had exactly the properties he was seeking.

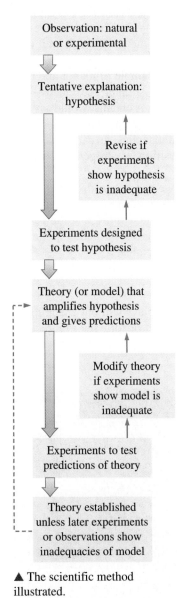

▲ The scientific method illustrated.

▲ Louis Pasteur (1822–1895). This great practitioner of the scientific method was the developer of the germ theory of disease, the sterilization of milk by pasteurization, and vaccination against rabies. He has been called the greatest physician of all time by some. He was, in fact, not a physician at all, but a chemist–by training and by profession.

So scientists (and inventors) always need to be alert to unexpected observations. Perhaps no one has been more aware of this than Louis Pasteur, who wrote, "Change favors the prepared mind."

1-3 Properties of Matter

Dictionary definitions of chemistry usually include the terms *matter*, *composition*, and *properties*, as in the statement that "chemistry is the science that deals with the composition and properties of matter." In this and the next section, we will consider some basic ideas relating to these three terms, in hopes of gaining a better understanding of what chemistry is all about.

Matter is anything that occupies space, displays a property known as *mass*, and possesses inertia. Every human being is a collection of matter. We all occupy space, and we describe our mass through a related property, weight. (Mass and weight are described in more detail in Section 1-5. Inertia is described in Appendix B.) All the objects that we see around us consist of matter. The gases of the atmosphere, even though they are invisible, are examples of matter—they occupy space and have mass. Sunlight is *not* matter; rather, it is a form of energy. We will wait until later chapters, however, to discuss the concept of energy.

Composition refers to the parts or components of a sample of matter and their relative proportions. Ordinary water is made up of two simpler substances—hydrogen and oxygen—present in certain fixed proportions. A chemist would say that the composition of water is 11.19% hydrogen and 88.81% oxygen by mass. Hydrogen peroxide, a substance used in bleaches and antiseptics, is also made up of hydrogen and oxygen, but it has a different composition. Hydrogen peroxide is 5.93% hydrogen and 94.07% oxygen by mass.

Properties are those qualities or attributes that we can use to distinguish one sample of matter from others. In some cases, we can establish properties visually. Thus, we can distinguish between the reddish brown solid, copper, and the yellow solid, sulfur, by the property of *color* (Figure 1-1). Properties of matter are generally grouped into two broad categories: physical and chemical.

Physical Properties and Physical Changes

A **physical property** is one that a sample of matter displays without changing its composition. Copper can be hammered into a thin sheet of foil (see Figure 1-1).

▶ FIGURE 1-1
Physical properties of sulfur and copper
A lump of sulfur (left) crumbles into a yellow powder when hammered. Copper (right) can be obtained as large lumps of native copper, formed into pellets, hammered into a thin foil, or drawn into a wire.

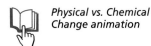
Physical vs. Chemical Change animation

▲ **FIGURE 1-2**
A chemical property of zinc and gold: reaction with hydrochloric acid
The zinc-plated (galvanized) nail reacts with hydrochloric acid, producing bubbles of hydrogen gas. The gold bracelet is unaffected by hydrochloric acid. In this photograph, the zinc plating has been consumed, exposing the underlying iron nail. The reaction of iron with hydrochloric acid imparts some color to the acid solution.

▶ The identity of an atom is established by a feature called its atomic number (see page 43). In recent years, research in nuclear physics has led to the production of several new elements. Elements with atomic numbers 110 and 111 were discovered in 1994, element 112 in 1996, and elements 114, 116, and 118 in 1999.

Solids that have this ability are said to be *malleable*. Sulfur is not malleable. If we strike a chunk of sulfur with a hammer, it crumbles into a powder. Sulfur is *brittle*. Other physical properties of copper that sulfur does not share are the ability to be drawn into a fine wire (ductility) and the ability to conduct heat and electricity.

Sometimes a sample of matter undergoes a change in its physical appearance. In such a **physical change**, some of the physical properties of the sample may change, but its composition remains *unchanged*. When liquid water freezes into solid water (ice), it certainly looks different and, in many ways, it is different. Yet, the water remains 11.19% hydrogen and 88.81% oxygen by mass.

Chemical Properties and Chemical Change

In a **chemical change**, or **chemical reaction**, one or more kinds of matter are converted to new kinds with *different* compositions. The key to identifying chemical change, then, comes in observing a *change* in composition. The burning of paper involves a chemical change. Paper is a complex material, but its principal constituents are carbon, hydrogen, and oxygen. The chief products of the combustion are two gases, one consisting of carbon and oxygen (carbon dioxide) and the other of hydrogen and oxygen (water, as steam). The ability of paper to burn is an example of a chemical property. A **chemical property** is the ability (or inability) of a sample of matter to undergo a change in composition under stated conditions.

Zinc reacts with hydrochloric acid solution to produce hydrogen gas and a water solution of zinc chloride (Figure 1-2). The ability of zinc to react with hydrochloric acid is one of zinc's distinctive chemical properties. The *inability* of gold to react with hydrochloric acid is one of gold's chemical properties. Sodium reacts not only with hydrochloric acid but also with water. In some of their physical properties, zinc, gold, and sodium are similar. For example, each is malleable and a good conductor of heat and electricity. In most of their chemical properties, though, zinc, gold, and sodium are quite different. Knowing these differences helps us to understand why zinc, which does not react with water, is used in roofing nails, for roof flashings, and in rain gutters, and sodium is not. Also, we can appreciate why gold, because of its chemical inertness, is prized for jewelry and coins: It does not tarnish or rust. In our study of chemistry, we shall see why substances differ in properties and how these differences determine the ways in which we use materials.

1-4 Classification of Matter

As we will describe more fully in later chapters, matter is made up of very tiny units called **atoms**. Presently we know of 115 different types of atoms, and *all* matter is made up of just these 115 types! These 115 different types of atoms are the basis of 115 elements. A chemical **element** is a substance made up of only a *single* type of atom. The known elements range from common substances, such as carbon, iron, and silver, to uncommon ones, such as lutetium and thulium. We can obtain about 90 of the elements from natural sources. The remainder do not occur naturally; they have been created only in laboratories. On the inside front cover you will find a complete listing of the elements and also a special tabular arrangement known as the periodic table. The periodic table is the chemist's directory of the elements. We will describe it in Chapter 2 and use it throughout most of the text.

Chemical **compounds** are substances in which atoms of two or more *different* elements are combined with one another. Scientists have identified millions of different chemical compounds. In some cases, we can isolate a molecule of a compound. A **molecule** is the smallest entity having the same proportions of the constituent atoms as does the compound as a whole. A molecule of water consists of three atoms: two hydrogen atoms joined to a single oxygen atom. A molecule of hydrogen peroxide

*Mixtures and
Compounds movie*

has two hydrogen atoms and two oxygen atoms; the two oxygen atoms are joined together and one hydrogen atom is attached to each oxygen atom. By contrast, a molecule of the blood protein gamma globulin is made up of 19,996 atoms altogether, but they are of just four types: carbon, hydrogen, oxygen, and nitrogen.

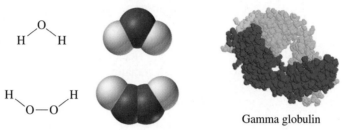

Gamma globulin

The composition and properties of an element or compound are uniform throughout a given sample and from one sample to another. Elements and compounds are called **substances**. (In the chemical sense, the term *substance* should be used only for elements and compounds.) In describing *mixtures* of substances, we use the terms **solution** or **homogeneous mixture** for mixtures that are uniform in composition and properties throughout a given sample but variable from one sample to another. A given solution of sucrose (cane sugar) in water is uniformly sweet throughout the solution, but the sweetness of another sucrose solution may be rather different if the sugar and water are present in different proportions. Ordinary air is a homogeneous mixture of several gases, principally the *elements* nitrogen and oxygen. Seawater is a solution of the *compounds* water, sodium chloride (salt), and a host of others. Gasoline is a homogeneous mixture or solution of dozens of compounds.

In **heterogeneous mixtures**—sand and water, for example—the components separate into distinct regions. Thus, the composition and physical properties vary from one part of the mixture to another. Salad dressing, a slab of concrete, and the leaf of a plant are all heterogeneous. It is usually easy to distinguish heterogeneous from homogeneous mixtures. A scheme for classifying matter into elements and compounds and homogeneous and heterogeneous mixtures is summarized in Figure 1-3.

▲ Is it homogeneous or heterogeneous? When viewed through a microscope, homogenized milk is seen to consist of globules of fat dispersed in a watery medium. Homogenized milk is a *heterogeneous* mixture.

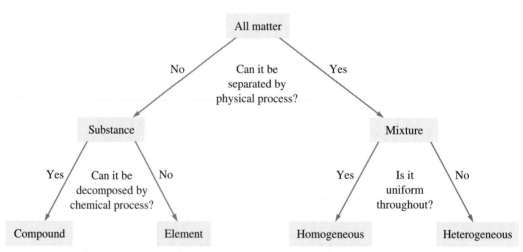

▲ **FIGURE 1-3   A classification scheme for matter**
Every sample of matter is either a single substance (an element or compound) or a mixture of substances. At the molecular level, an element consists of atoms of a single type and a compound, of two or more different types of atoms, usually joined into molecules. In a homogeneous mixture, atoms or molecules are randomly mixed at the molecular level. In heterogeneous mixtures, the components are physically separated, as in a layer of octane molecules (a constituent of gasoline) floating on a layer of water molecules.

Separating Mixtures

A mixture can be separated into its components by appropriate *physical* means. Consider again the heterogeneous mixture of sand in water. When we pour this mixture into a funnel lined with porous filter paper, the water passes through and sand is retained on the paper. This process of separating a solid from the liquid in which it is suspended is called *filtration* (Figure 1-4a). You will probably use this procedure in the laboratory. On the other hand, we cannot separate a homogeneous mixture (solution) of copper(II) sulfate in water by filtration because all components pass through the paper. We can, however, boil the solution of copper(II) sulfate and water. Pure liquid water is obtained from the vapor given off by the boiling solution. When all the water has been removed, the copper(II) sulfate remains behind. This process is called *distillation* (Figure 1-4b).

Another method of separation available to modern chemists depends on the differing abilities of compounds to adhere to the surfaces of various solid substances, such as paper and starch. The technique of *chromatography* relies on this principle. The dramatic results that can be obtained with chromatography are illustrated by the separation of ink on a filter paper (Figure 1-4c–d).

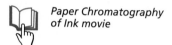
Paper Chromatography of Ink movie

Decomposing Compounds

A chemical compound retains its identity during physical changes, but it can be *decomposed* into its constituent elements by *chemical* changes. The decomposition

(a)   (b)

▶ **FIGURE 1-4**
Separating mixtures: a physical process
(a) Separation of a heterogeneous mixture by filtration: Solid copper(II) sulfate is retained on the filter paper, while liquid hexane passes through. **(b)** Separation of a homogeneous mixture by distillation: Copper(II) sulfate remains in the flask on the left as water passes to the flask on the right, by first evaporating and then condensing back to a liquid. **(c)** Separation of the components of ink using chromatography: A dark spot of black ink can be seen just above the water line as water moves up the paper. **(d)** Water has dissolved the colored components of the ink, and these components are retained in different regions on the paper according to their differing tendencies to adhere to the paper.

(c)   (d)

▲ **FIGURE 1-5**
A chemical change: decomposition of ammonium dichromate

of compounds into their constituent elements is a more difficult matter than the mere physical separation of mixtures. The extraction of iron from iron oxide ores requires a blast furnace. The preparation of pure magnesium from magnesium chloride requires electricity for its industrial production. It is generally easier to convert a compound into other compounds by a chemical reaction than it is to separate a compound into its constituent elements. For example, when heated, ammonium dichromate decomposes into the substances chromium(III) oxide, nitrogen, and water. This reaction, once used in movies to simulate a volcano, is illustrated in Figure 1-5.

States of Matter

Matter is generally found in one of three *states*—solid, liquid, or gas. In a **solid**, atoms or molecules are in close contact, sometimes in a highly organized arrangement called a *crystal*. A solid occupies a definite shape. In a **liquid**, the atoms or molecules are usually separated by somewhat greater distances than in a solid. Movement of these atoms or molecules gives a liquid its most distinctive property—the ability to flow, covering the bottom and assuming the shape of its container. In a **gas**, distances between atoms or molecules are much greater than in a liquid. A gas always expands to fill its container. Depending on conditions, a substance may exist in only one state of matter, or it may be in two or three states. Thus, as the ice in a small pond begins to melt in the spring, water is in two states: solid and liquid (actually, three states if we also consider water vapor in the air above the pond).

The three states of water are illustrated at two levels in Figure 1-6. The *macroscopic level* refers to how we perceive matter with our eyes, through the outward appearance of objects. The *microscopic level* describes matter as chemists conceive of it—in terms of atoms and molecules and their behavior. In this text, we will describe many macroscopic, observable properties of matter, but to explain these properties, we will often shift our view to the atomic or molecular level—the microscopic level.

1-5   Measurement of Matter: SI (Metric) Units

▶ Nonnumerical information is *qualitative*, such as the color blue.

Chemistry is a *quantitative* science, which means that in many cases we can measure a property of a substance and compare it with a standard having a known value of the property. We express the measurement as the product of a *number* and a *unit*. The unit indicates the standard against which the measured quantity is being compared. When we say that the length of the playing field in football is 100 yd, we mean that the field is 100 times longer than a standard of length called the yard (yd). In this section, we will introduce some basic units of measurement that are important to chemists.

The scientific system of measurement is called the *Système Internationale d'Unités* (International System of Units) and is abbreviated SI. It is a modern version of the metric system, a system based on the unit of length called a *meter* (m). The meter was originally defined as 1/10,000,000 of the distance from the Equator to the North Pole and translated into the length of a metal bar kept in Paris. Unfortunately, this length is subject to change with temperature, and it cannot be exactly reproduced. The SI system substitutes for the standard meter bar a quantity that can be reproduced anywhere: 1 meter is the distance traveled by light in a vacuum in 1/299,792,458 of a second. Length is one of the seven fundamental quantities in the SI system (see Table 1.1). All other physical quantities have units that can be derived from these seven. SI is a *decimal* system. Quantities differing from the base unit by powers of ten are noted by the use of prefixes. For example, the prefix *kilo*

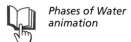

Phases of Water animation

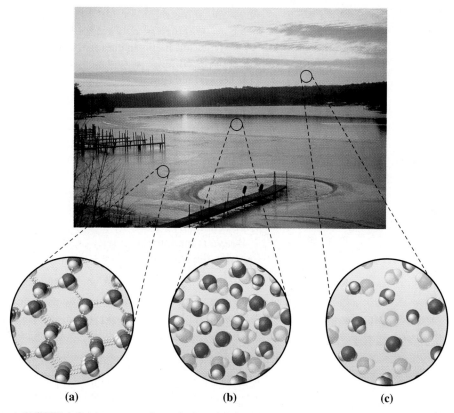

(a)   (b)   (c)

▲ **FIGURE 1-6   Macroscopic and microscopic views of matter**
The picture shows a frozen pond with the three states of water as we perceive them macro-scopically—**(a)** ice (solid water), **(b)** liquid water, and **(c)** gaseous water. The circular in-sets show how chemists conceive of these states of matter microscopically. In ice (solid water), the structure consists of rather closely packed water molecules, each molecule con-sisting of one oxygen and two hydrogen atoms. In liquid water, the units are mobile water molecules. The gaseous form of water is made up of widely separated water molecules.

TABLE 1.1 SI Base Quantities		
Physical Quantity	**Unit**	**Abbreviation**
Length	meter[a]	m
Mass	kilogram	kg
Time	second	s
Temperature	kelvin	K
Amount of substance[b]	mole	mol
Electric current[c]	ampere	A
Luminous intensity[d]	candela	cd

[a] The official spelling of this unit is "metre," but we will use the more common American spelling.

[b] The mole is introduced in Section 2-7.

[c] Electric current is described in Appendix B and in Chapter 21.

[d] Luminous intensity is not discussed in this text.

TABLE 1.2	SI Prefixes
Multiple	**Prefix**
10^{18}	exa (E)
10^{15}	peta (P)
10^{12}	tera (T)
10^{9}	giga (G)
10^{6}	mega (M)
10^{3}	kilo (k)
10^{2}	hecto (h)
10	deca (da)
10^{-1}	deci (d)
10^{-2}	centi (c)
10^{-3}	milli (m)
10^{-6}	micro $(\mu)^a$
10^{-9}	nano (n)
10^{-12}	pico (p)
10^{-15}	femto (f)
10^{-18}	atto (a)

[a] The Greek letter μ (pronounced "mew").

▶ The symbol $\propto$ means "proportional to." It can be replaced by an equality sign and a proportionality constant. In expression (1.1) the constant is the acceleration due to gravity, g. (See Appendix B.)

 SI Prefix activity

means *one thousand times* (10^3) the base unit; it is abbreviated as k. Thus 1 *kilometer* = *1000* meters, or 1 km = 1000 m. The SI prefixes are listed in Table 1.2.

Most measurements in chemistry are made in SI units. Sometimes we must convert between SI units, as when converting kilometers to meters. At other times we must convert measurements expressed in non-SI units into SI units or from SI units into non-SI units. In all of these cases we can use a *conversion factor* or a series of conversion factors in a scheme called a conversion pathway. Later in this chapter, we will apply the conversion pathway method of problem solving. The method itself is described in some detail in Appendix A.

Mass

Mass describes the quantity of matter in an object. In SI the standard of mass is 1 *kilogram* (kg), which is a fairly large unit for most applications in chemistry. More commonly we use the unit *gram* (g) (about the mass of three aspirin tablets).

Weight is the force of gravity on an object. It is directly proportional to mass, as shown in the mathematical equations.

$$W \propto m \quad \text{and} \quad W = g \cdot m \tag{1.1}$$

An object has a fixed mass (m), which is independent of where or how the mass is measured. Its weight (W), on the other hand, may vary because the acceleration due to gravity (g) varies slightly from one point on Earth to another. Thus, an object that weighs 100.0 kg in St. Petersburg, Russia, weighs only 99.6 kg in Panama (about 0.4% less). The same object would weigh only about 17 kg on the moon. Although weight varies from place to place, the *mass* of an object is the *same* in all three locations. The terms *weight* and *mass* are often used interchangeably, but only *mass* is a measure of the quantity of matter. A common laboratory device for measuring mass is called a balance. (A balance is often called, incorrectly, a scale.)

The principle used in a balance is that of counteracting the force of gravity on an unknown mass with a force of equal magnitude that can be precisely measured. In older types of balances, this counterbalancing is achieved through the force of gravity acting on *weights*, objects of precisely known mass. In the type of balance most commonly seen in laboratories today—the electronic balance—the counterbalancing force is a magnetic force produced by passing an electric current through an electromagnet. First, an initial balance condition is achieved when no object is present on the balance pan. When the object to be weighed is placed on the pan, the initial balance condition is upset. To restore the balance condition, additional electric current must be passed through the electromagnet. The magnitude of this additional current is proportional to the mass of the object being weighed and is translated into a mass reading that is displayed on the balance.

▶ An electronic balance.

Time

In daily use we measure time in seconds, minutes, hours, and years, depending on whether we are dealing with short intervals (such as the time for a 100-m race) or long ones (such as the time before the next appearance of Halley's comet in 2062). We use all these units in scientific work, although in SI the standard of time is the *second* (s). A time interval of 1 second is not easily established. At one time it was based on the length of a day, but this is not constant because the rate of Earth's rotation undergoes slight variations. Later, in 1956, the second was defined as 1/31,556,925.9747 of the length of the year 1900. With the advent of atomic clocks, a more precise definition became possible. The second is the duration of 9,192,631,770 cycles of a particular radiation emitted by atoms of the element cesium known as cesium-133.

► Electromagnetic radiation is discussed in Section 9-1.

Temperature

To establish a temperature scale, we arbitrarily set certain fixed points and temperature increments called degrees. Two commonly used fixed points are the temperature at which ice melts and the temperature at which water boils, both at standard atmospheric pressure.*

On the *Fahrenheit* temperature scale, the melting point of ice is 32 °F, the boiling point of water is 212 °F, and the interval between is divided into 180 equal parts called Fahrenheit degrees. On the *Celsius* (centigrade) scale, the melting point of ice is 0 °C, the boiling point of water is 100 °C, and the interval between is divided into 100 equal parts called Celsius degrees. Figure 1-7 compares the Fahrenheit and Celsius temperature scales.

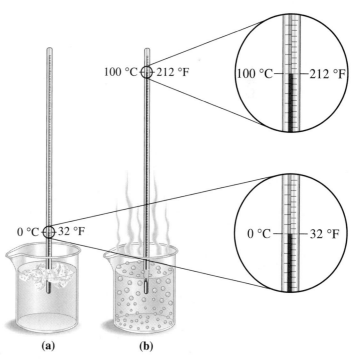

100 °C ⊕ 212 °F

100 °C — 212 °F

0 °C ⊕ 32 °F

0 °C — 32 °F

(a)   (b)

▲ FIGURE 1-7 **A comparison of temperature scales**
(a) The melting point of ice. (b) The boiling point of water.

---

*Standard atmospheric pressure is defined in Section 6-1. The effect of pressure on melting and boiling points is described in Chapter 13.

▶ We reintroduce and use Kelvin temperature in Chapter 6.

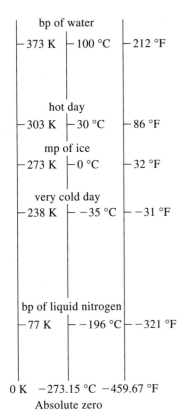

bp of water
373 K ── 100 °C ── 212 °F

hot day
303 K ── 30 °C ── 86 °F

mp of ice
273 K ── 0 °C ── 32 °F

very cold day
238 K ── −35 °C ── −31 °F

bp of liquid nitrogen
77 K ── −196 °C ── −321 °F

0 K   −273.15 °C −459.67 °F
Absolute zero

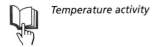

Temperature activity

▶ Answers to Practice Examples are given in Appendix F.

The SI temperature scale, called the Kelvin scale, assigns a value of zero to the lowest conceivable temperature. This zero—0 K—comes at −273.15 °C. The Kelvin scale is an absolute temperature scale; there are no negative Kelvin temperatures. The interval on the Kelvin scale, called a *kelvin*, is the same size as the Celsius degree (although the degree symbol is not used with Kelvin temperatures).

In the laboratory, temperature is most commonly measured in Celsius degrees; however, these temperatures must often be converted to the Kelvin scale (in describing the behavior of gases, for example). At other times, particularly in many engineering applications, temperatures must be converted between the Celsius and Fahrenheit scales. Temperature conversions can be made in a straightforward way by using the algebraic equations shown below.

$$\text{Kelvin from Celsius} \qquad T(\text{K}) = t(°\text{C}) + 273.15$$

$$\text{Celsius from Fahrenheit} \quad t(°\text{C}) = \frac{5}{9}\left[t(°\text{F}) - 32\right]$$

The factor of 5/9 arises because the Celsius scale uses 100 degrees between the two chosen reference points and the Fahrenheit scale uses 180 degrees: 100/180 = 5/9. The diagram in the margin illustrates the relationship between the three scales for several temperatures.

EXAMPLE 1-1

Converting Between Fahrenheit and Celsius Temperatures A recipe calls for roasting a cut of meat at 350 °F. What is this temperature on the Celsius scale?

Solution

We are given a Fahrenheit temperature and seek a Celsius temperature. We need the algebraic equation given above that expresses $t(°\text{C})$ as a function of $t(°\text{F})$.

$$t(°\text{C}) = \frac{5}{9}\left[t(°\text{F}) - 32\right] = \frac{5}{9}\left[350 - 32\right] = 177\ °\text{C}$$

If a problem requires converting Celsius to Fahrenheit temperature, we need to use the equation in this alternate form,

$$t(°\text{F}) = \frac{9}{5}t(°\text{C}) + 32$$

Practice Example A: The predicted high temperature for New Delhi, India, on a given day is 41 °C. Is this temperature higher or lower than the predicted daytime high of 103 °F in Phoenix, Arizona?

Practice Example B: A particular automobile engine coolant has antifreeze protection to a temperature of −22 °C. Will this coolant offer protection to temperatures as low as −15 °F?

Derived Units

The seven units listed in Table 1.1 are the SI units for the fundamental quantities of length, mass, time, and so on. Many measured properties are expressed as combinations of these fundamental, or base, quantities. We refer to the units of such properties as *derived units*. For example, velocity is a distance divided by the time required to travel that distance. The unit of velocity is length divided by time, such as m/s or m s^{-1}.

An important measurement that uses derived units is *volume*. Volume has the unit (length)^3, and the SI standard unit of volume is the *cubic meter* (m^3). More commonly used volume units are the *cubic centimeter* (cm^3) and the *liter* (L). One

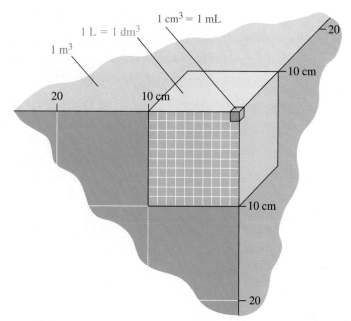

▲ **FIGURE 1-8   Some metric volume units compared**
The largest volume, shown in part, is the SI standard—1 cubic meter (m^3). A cube with a length of 10 cm (1 dm) on edge (in blue) has a volume of 1000 cm^3 (1 dm^3) and is called 1 liter (1 L). The smallest cube is 1 cm on edge (red) and has a volume of 1 cm^3 = 1 mL.

liter is defined as a volume of 1000 cm^3, which means that one *milliliter* (1 mL) is equal to 1 cm^3. The liter is also equal to one *cubic decimeter* (1 dm^3). Several volume units are depicted in Figure 1-8.

Non-SI Units

Although in the United States we are growing more accustomed to expressing distances in kilometers and volumes in liters, most units used in everyday life are still non-SI. Masses are given in pounds, room dimensions in feet, and so on. In this

 Are You Wondering ...

Why attaching the units to a number is so important?

In 1993, NASA started the Mars Surveyor program to conduct an ongoing series of missions to explore Mars. In 1995, two missions were scheduled that would be launched in late 1998 and early 1999. The missions were the Mars Climate Orbiter (MCO) and the Mars Polar Lander (MPL). The MCO was launched December 11, 1998, and the MPL, January 3, 1999.

Nine and a half months after launch, the MCO was to fire its main engine to achieve an elliptical orbit around Mars. The MCO engine start occurred on September 23, 1999, but the MCO mission was lost when the orbiter entered the Martian atmosphere on a lower-than-expected trajectory. The MCO entered the low orbit because the computer on Earth used British Engineering units, whereas the MCO computer used SI units!

This error in units brought the MCO 56 km above the surface of Mars instead of the desired 250 km. At 250 km, the MCO would have successfully entered the desired elliptic orbit, and the $168 million orbiter would probably not have been lost.

▲ **FIGURE 1-9   Some familiar non-SI and SI units compared**
The green ribbon is 1 cm wide and is wrapped around a stick that is 1 m long. The yellow
ribbon is 1 inch (in.) wide and is wrapped around a stick 1 yard (yd) long. The meterstick
is about 10% longer than the yardstick (1 in. = 2.54 cm, *exactly*). Of the two identical
beakers, the beaker on the left contains 1 kg of candy and the one on the right, 1 pound
(lb) (1 lb = 0.4536 kg = 453.6 g). The two identical volumetric flasks hold 1 L when
filled to the mark. The flask on the left and the carton behind it each contain 1 quart (qt)
of milk. The flask on the right and the bottle behind it each contain 1 L of orange soda
(1 qt = 0.9464 L).

book, we will not routinely use these non-SI units, but we will occasionally introduce them in examples and end-of-chapter exercises. In such cases, any necessary
relationships between non-SI and SI units will be given or can be found on the inside back cover. Figure 1-9 may help you to develop a frame of reference between
some non-SI and SI units.

1-6   Density and Percent Composition: Their Use in Problem Solving

Throughout this text, we will encounter new concepts about the structure and behavior of matter. One means of firming up our understanding of some of these concepts is to work problems relating concepts that we already know to those we are
trying to understand. In this section, we will introduce two quantities frequently required in problem solving: density and percent composition.

Density

Here is an old riddle: "What weighs more, a ton of bricks or a ton of feathers?" If
you answer that they weigh the same, you demonstrate a clear understanding of
the meaning of mass—a measure of a quantity of matter. Anyone who answers that
the bricks weigh more than the feathers has confused the concepts of mass and

density. Matter in a brick is more concentrated than in a feather—that is, the matter in a brick is confined to a smaller volume. Bricks are more dense than feathers. **Density** is the ratio of mass to volume.

$$\text{density } (d) = \frac{\text{mass } (m)}{\text{volume } (V)} \tag{1.2}$$

Mass and volume are both extensive properties. An **extensive property** depends on the quantity of matter observed. However, if we divide the mass of a substance by its volume, we obtain density, an intensive property. An **intensive property** is *independent* of the amount of matter observed. Thus, the density of pure water at 25 °C has a unique value, whether the sample fills a small beaker or a swimming pool. Intensive properties are especially useful in chemical studies because they can often be used to identify substances.

The SI base units of mass and volume are kilograms and cubic meters, respectively, but in practice, chemists generally express mass in grams and volume in cubic centimeters or milliliters. The most commonly encountered density unit is grams per cubic centimeter (g/cm^3) or the identical grams per milliliter (g/mL).

The mass of 1.000 L of water at 4 °C is 1.000 kg. The density of water at 4 °C is 1000 g/1000 mL, or 1.000 g/mL. At 20 °C, the density of water is 0.9982 g/mL. *Density is a function of temperature* because volume varies with temperature, whereas mass remains constant. One reason that global warming is a concern is because if the average temperature of seawater increases, the seawater will become less dense. The volume of seawater must increase and sea level will rise—all before any ice melts at the polar caps.

In addition to temperature, the state of matter affects the density of a substance. In general, solids are denser than liquids and both are denser than gases. There are significant overlaps, however. Following are the ranges of values generally observed for densities; this information should prove useful in solving problems.

- Solid densities: from about 0.2 g/cm^3 to 20 g/cm^3
- Liquid densities: from about 0.5 g/mL to 3–4 g/mL
- Gas densities: mostly in the range of a few grams per *liter*

In general, densities of liquids are known more precisely than those of solids (which may have imperfections in their microscopic structures). Also, densities of elements and compounds are known more precisely than densities of materials with variable compositions (such as wood or rubber).

An important consequence of the differing densities of solids and liquids is that liquids and solids of lower density will float on a liquid of higher density (as long as they do not dissolve in the higher-density liquid).

Density in Conversion Pathways

If we measure the mass of an object and its volume, simple division gives us its density. On the other hand, if we know the density of an object, we can use density as a conversion factor to determine the object's mass or volume. For example, a cube of osmium 1.000 cm on edge weighs 22.48 g. The density of osmium (the densest of the elements) is 22.48 g/cm^3. What would be the mass of a cube of osmium that is 1.25 in. on edge (1 in. = 2.54 cm)? To solve this problem, we begin by relating the volume of a cube to its length, that is, $V = l^3$. Then we can map out the *conversion pathway*:

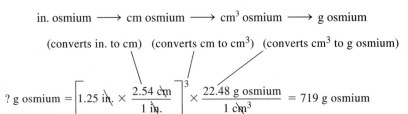

$$? \text{ g osmium} = \left[1.25 \text{ in.} \times \frac{2.54 \text{ cm}}{1 \text{ in.}}\right]^3 \times \frac{22.48 \text{ g osmium}}{1 \text{ cm}^3} = 719 \text{ g osmium}$$

KEEP IN MIND ▶
that in a conversion pathway all
units must cancel except for the
desired unit in the final result
(see Appendix A-5: *Using Conversion Factors*).

The density of mercury, the only liquid metal, is 13.5 g/mL at 25 °C. Suppose we wish to know the volume, in mL, of 1.000 kg of mercury at 25 °C. We proceed by (a) identifying the known information: 1.000 kg of mercury and d = 13.5 g/mL (at 25 °C); (b) noting what we are trying to determine—a volume in milliliters (which we designate mL mercury); and (c) looking for the relevant conversion factors. Outlining the conversion pathway will help us find these conversion factors:

$$\text{kg mercury} \longrightarrow \text{g mercury} \longrightarrow \text{mL mercury}$$

We need the factor 1000 g/kg to convert from kilograms to grams. Density provides the factor to convert from mass to volume. But in this instance, we need to use density in the *inverted* form. That is,

$$? \text{ mL mercury} = 1.000 \text{ kg} \times \frac{1000 \text{ g}}{1 \text{ kg}} \times \frac{1 \text{ mL mercury}}{13.5 \text{ g}} = 74.1 \text{ mL mercury}$$

Examples 1-2 and 1-3 further illustrate that numerical calculations involving density are generally of two types: determining density from mass and volume measurements and using density as a conversion factor to relate mass and volume.

EXAMPLE 1-2

Calculating the Density of an Object from Its Mass and Volume. The block of wood pictured in Figure 1-10 has a mass of 2.52 kg. What is the density of the wood in grams per cubic centimeter?

Solution

To determine density, we need to know a mass and the corresponding volume. We are given the mass, which we can easily convert from kilograms to grams, and to calculate the volume of the rectangular block, we can use the geometric formula in Figure 1-10. First, however, we must change the length from meters to centimeters.

$$l = 1.08 \text{ m} \times \frac{100 \text{ cm}}{1 \text{ m}} = 108 \text{ cm}$$

Now we can calculate the volume in cubic centimeters.

$$V = 108 \text{ cm} \times 5.1 \text{ cm} \times 6.2 \text{ cm} = 3400 \text{ cm}^3$$

The mass of the block, expressed in grams, is

$$m = 2.52 \text{ kg} \times \frac{1000 \text{ g}}{1 \text{ kg}} = 2520 \text{ g}$$

The density of the wood is

$$d = \frac{m}{V} = \frac{2520 \text{ g}}{3400 \text{ cm}^3} = 0.74 \text{ g/cm}^3$$

Check: Note that if we had mistakenly taken the mass as 2.52 g, the result would be $d = 7.4 \times 10^{-4} \text{ g/cm}^3$. If we had taken the length as 1.08 cm, the result would be $d = 74 \text{ g/cm}^3$. Both values are well outside the range of densities expected for a solid.

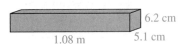

▲ **FIGURE 1-10**
**Measuring the volume of a
regularly shaped object—
Example 1-2 visualized**
The volume of the rectangular
solid is the product of its length,
width, and height:
$V = l \times w \times h.$

Density activity

Practice Example A: To determine the density of trichloroethylene, a liquid used to degrease electronic components, a flask is first weighed empty (108.6 g). Then it is filled with 125 mL of the trichloroethylene to give a total mass of 291.4 g. What is the density of trichloroethylene in grams per milliliter?

Practice Example B: A graduated cylinder contains 33.8 mL of water. A stone with a mass of 28.4 g is placed in the cylinder. The new level of water in the cylinder is 44.1 mL. What is the density of the stone?

EXAMPLE 1-3

Calculating the Mass of a Liquid from Its Volume and Density What is the mass of a 275-mL sample of ethanol (ethyl alcohol) at 20 °C? The density of ethanol at 20 °C is 0.789 g/mL.

Solution

Multiplying the volume of liquid by its density (the conversion factor) gives us our result.

$$? \text{ g ethanol} = 275 \text{ mL ethanol} \times \frac{0.789 \text{ g ethanol}}{1 \text{ mL ethanol}} = 217 \text{ g ethanol}$$

Practice Example A: What is the mass of 125 mL of a sugar solution (sucrose in water) that has a density of 1.081 g/mL at 20 °C?

Practice Example B: What is the volume, in liters, occupied by 50.0 kg ethanol at 20 °C? The density of ethanol at 20 °C is 0.789 g/mL.

Percent as a Conversion Factor

In Section 1-3, we described composition as an identifying characteristic of a sample of matter. A common way of referring to composition is through percentages. The Latin word *centum* means 100. **Percent** (*percentum*) is the number of parts of a constituent in 100 parts of the whole. To say that a seawater sample contains 3.5% sodium chloride by mass means there is 3.5 g of sodium chloride in every 100 g of the seawater. We make the statement in terms of grams because we are talking about percent *by mass*. We can express this percent by writing the following ratios.

$$\frac{3.5 \text{ g sodium chloride}}{100 \text{ g seawater}} \quad \text{and} \quad \frac{100 \text{ g seawater}}{3.5 \text{ g sodium chloride}} \tag{1.3}$$

In Example 1-4, we will use this type of ratio as a conversion factor.

EXAMPLE 1-4

Using Percent as a Conversion Factor A 75-g sample of sodium chloride (table salt) is to be produced by evaporating to dryness a quantity of seawater containing 3.5% sodium chloride by mass. How many *liters* of seawater must be taken for this purpose? Assume a density of 1.03 g/mL for seawater.

Solution

To convert from grams of sodium chloride to grams of seawater, we need the conversion factor with grams of seawater in the numerator and grams of sodium chloride in the denominator. In addition, we need to make the conversions g seawater ⟶ mL seawater ⟶ L seawater.

$$? \text{ L seawater} = 75 \text{ g sodium chloride} \times \frac{100 \text{ g seawater}}{3.5 \text{ g sodium chloride}}$$

$$\times \frac{1 \text{ mL seawater}}{1.03 \text{ g seawater}} \times \frac{1 \text{ L seawater}}{1000 \text{ mL seawater}}$$

$$= 2.1 \text{ L seawater}$$

Practice Example A: How many kilograms of ethanol are present in 25 L of a gasohol solution that is 90% gasoline–10% ethanol by mass? The density of the gasohol is 0.71 g/mL.

Practice Example B: Common rubbing alcohol is a solution of 70.0% isopropyl alcohol by mass in water. If a 25.0-mL sample of rubbing alcohol contains 15.0 g of isopropyl alcohol, what is the density of the rubbing alcohol?

Are You Wondering...

In doing a problem with percentages, when to multiply and when to divide?

A common way of dealing with a percentage is to convert it to decimal form (3.5% becomes 0.035) and then to multiply or divide by this decimal. Students sometimes can't decide which to do. Expressing percentage as a conversion factor and using it to produce a cancellation of units of physical quantities gets around this difficulty. Also, remember that

- The quantity of a *component* must always be *less* than the quantity of the whole mixture. (*Multiply by percentage.*)

- The quantity of a *mixture* must always be *greater* than the quantity of any of its components. (*Divide by percentage.*)

If, in Example 1-4, we had not been careful about the cancellation of units and had multiplied by percentage (3.5/100) instead of dividing by it (100/3.5), we would have gotten the numerical answer 2.5×10^{-3}. This would be a 2.5-mL sample of seawater, weighing about 2.5 g. Clearly, a sample of seawater that *contains* 75 g of sodium chloride must have a mass *greater than* 75 g.

KEEP IN MIND

that a numerical answer that defies common sense is probably wrong. ▶

1-7 Uncertainties in Scientific Measurements

All measurements are subject to error. To some extent, measuring instruments have built-in, or inherent, errors, called **systematic errors**. (For example, a kitchen scale might consistently yield results that are 25 g too high or a thermometer a reading that is 2° too low.) Limitations in an experimenter's skill or ability to read a scientific instrument also lead to errors and give results that may be either too high or too low. Such errors are called **random errors**.

Precision refers to the degree of reproducibility of a measured quantity—that is, the closeness of agreement when the same quantity is measured several times. The precision of a series of measurements is *high* (or good) if each of a series of measurements deviates by only a small amount from the average. Conversely, if there is wide deviation among the measurements, the precision is *poor* (or low). **Accuracy** refers to how close a measured value is to the accepted, or real, value. High-precision measurements are not always accurate—a large systematic error could be present. Still, it is likely that measurements of high precision are more accurate than those of low precision.

To illustrate these ideas, consider the mass of an object measured on two different balances. One is a relatively crude balance called a platform balance, and the other is a sophisticated analytical balance.

	Platform Balance	Analytical Balance
Three measurements	10.4, 10.2, 10.3 g	10.3107, 10.3108, 10.3106 g
Their average	10.3 g	10.3107 g
Reproducibility	±0.1 g	±0.0001 g
Precision	low or poor	high or good

1-8   Significant Figures

For the measurements on the platform balance just considered—that is, 10.4, 10.2, and 10.3 g—the result we would report is the average of the three: 10.3 g. A scientist would interpret these results to mean that the first two digits—10—are known with certainty and the last digit—3—is uncertain because it was estimated. That is, the mass is known only to the nearest 0.1 g, a fact that we could also express by writing 10.3 ± 0.1 g. To a scientist, the measurement 10.3 g is said to have *three* **significant figures**. If this mass is reported in kilograms rather than in grams, 10.3 g = 0.0103 kg, the measurement is still expressed to *three* significant figures even though more than three digits are shown. For the measurements on the analytical balance, the corresponding reported value would be 10.3107 g—a value with *six* significant figures. The number of significant figures in a measured quantity gives an indication of the capabilities of the measuring device and the precision of the measurements.

We will frequently need to determine the number of significant figures in an expressed quantity. The rules for doing this are outlined in Figure 1-11. Even with the ideas in Figure 1-11, we remain uncertain over how many significant figures to assign to the measurement 7500 m. Do we mean 7500 m, measured to the nearest meter? 10 meters? If all the zeros are significant—if the value has *four* significant figures—we can write 7500. m. That is, by writing a decimal point that is not otherwise needed, we show that all zeros preceding the decimal point are significant. This technique does not help if only one of the zeros, or if neither zero, is significant. The best approach here is to use exponential notation. (Review Appendix A if necessary.) The coefficient establishes the number of significant figures, and the power of ten locates the decimal point.

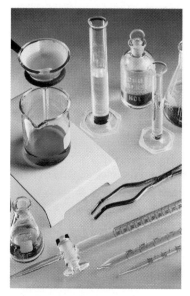

▲ **Glassware of varying degrees of precision.**
Beakers and flasks offer the *least* precise means of measuring liquid volumes. Graduated cylinders are much more precise measuring tools. In the foreground of the photo are two pipets. They provide one of the most precise methods of measuring liquid volume. Just behind the pipets is another precise measuring tool, the buret.

 Significant Figure activity

2 significant figures	3 significant figures	4 significant figures
7.5×10^3 m	7.50×10^3 m	7.500×10^3 m

▶ **FIGURE 1-11**
Determining the number of significant figures in a quantity
The quantity shown here, 0.004004500, has *seven* significant figures. All nonzero digits are significant, as are the indicated zeros.

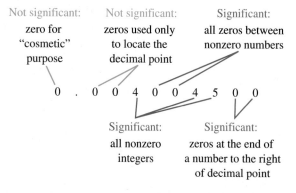

Significant Figures in Numerical Calculations

An important requirement in calculations is that precision can be neither gained nor lost in calculations involving measured quantities. There are several exact methods of determining how precisely to express the result of a calculation, but it is usually sufficient just to observe some simple rules involving significant figures.

▶ A more exact rule on multiplication/division is that the result should have about the same percent error as the least precisely known quantity. Usually the significant figure rule conforms to this requirement; occasionally, it does not (see Exercise 73).

The result of multiplication or division may contain only as many significant figures as the *least* precisely known quantity in the calculation.

In the following chain multiplication to determine the volume of a rectangular block of wood, we should round off the result to *three* significant figures. Figure 1-12 may help you to understand this.

$$14.79 \text{ cm} \times 12.11 \text{ cm} \times 5.05 \text{ cm} = 904 \text{ cm}^3$$

(4 sig. fig.)   (4 sig. fig.)   (3 sig. fig.)   (3 sig. fig.)

In adding and subtracting numbers, the applicable rule is that

The result of addition or subtraction must be expressed with the same number of digits beyond the decimal point as the quantity carrying the *smallest* number of such digits.

Consider the following sum of masses.

$$
\begin{array}{r}
15.02 \text{ g} \\
9986.0 \text{ g} \\
3.518 \text{ g} \\
\hline
10{,}004.538 \text{ g}
\end{array}
$$

▲ FIGURE 1-12   **Significant figure rule in multiplication**
In forming the product 14.79 cm × 12.11 cm × 5.05 cm, the least precisely known quantity is 5.05 cm. Shown on the calculators are the products of 14.79 and 12.11 with 5.04, 5.05, and 5.06, respectively. In the three results, only the first two digits, 90..., are identical. Variations begin in the third digit. We are certainly not justified in carrying digits beyond the third. We express the volume as 904 cm³. Usually, instead of a detailed analysis of the type done here, we can use a simpler idea: *The result of a multiplication may contain only as many significant figures as does the least precisely known quantity.*

The sum has the same uncertainty, ±0.1 g, as does the term with the *smallest* number of digits beyond the decimal point, 9986.0 g. Note that this calculation is *not* limited by significant figures. In fact, the sum has more significant figures (6) than do any of the terms in the addition.

There are two situations when a quantity appearing in a calculation may be *exact*, that is, not subject to errors in measurement. This may occur

- By definition (such as 3 ft = 1 yd, or 1 in. = 2.54 cm)

or as a result of

- Counting (such as *six* faces on a cube, or *two* hydrogen atoms in a water molecule).

Exact numbers can be considered to have an unlimited number of significant figures.

▶ Later in the text, we will need to apply ideas about significant figures to logarithms. This matter is discussed in Appendix A.

Rounding Off Numerical Results

To three significant figures, we should express 15.453 as 15.5 and 14,775 as 1.48×10^4. If we need to drop just one digit, that is, to round off a number, the simplest rule to follow is to increase the final digit by one unit if the digit dropped is 5, 6, 7, 8, or 9 and to leave the final digit unchanged if the digit dropped is 0, 1, 2, 3, or 4. To three significant figures, 15.55 rounds off to 15.6 and 15.54 rounds off to 15.5.

▶ As added practice in working with significant figures, review the calculations in Section 1-6. You will note that they conform to the significant figure rules presented here.

EXAMPLE 1-5

Applying Significant Figure Rules: Multiplication/Division   Express the result of the following calculation with the correct number of significant figures.

$$\frac{0.225 \times 0.0035}{2.16 \times 10^{-2}} = ?$$

Solution

By inspecting the three quantities, we see that the least precisely known is 0.0035, with *two* significant figures. Our result must also contain only *two* significant figures. The fact that two of the quantities are given in decimal form and one in exponential form is immaterial. Simply enter the numbers, as written, into an electronic calculator and read off the answer: 0.0364583. Express the result to *two* significant figures: 0.036. Alternatively, write the answer in exponential form: 3.6×10^{-2}.

Practice Example A: Perform the following calculation, and express the result with the appropriate number of significant figures.

$$\frac{62.356}{0.000456 \times 6.422 \times 10^3} = ?$$

Practice Example B: Perform the following calculation, and express the result with the appropriate number of significant figures.

$$\frac{8.21 \times 10^4 \times 1.3 \times 10^{-3}}{0.00236 \times 4.071 \times 10^{-2}} = ?$$

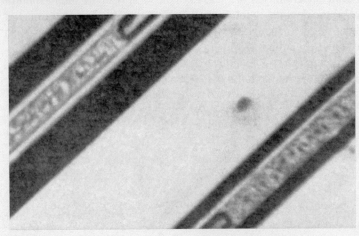

About 30 years ago, scientists were heatedly debating the reported discovery of a new form of water called polywater. The initial experiments had been done in Russia by scientists who were highly regarded as careful experimenters. These scientists reported that if a sample of water were confined in a quartz capillary tube (a very small-diameter tube of a special high-purity glass), some liquid water condensed in another part of the tube. The recondensed water had some peculiar properties, including a high boiling point (over 150 °C), a low freezing point (below −30 °C), and a density greater than that of ordinary water.

▲ The melting behavior of polywater–water mixtures. Two polywater–water mixtures at −10 °C. The sample at the left has just begun to melt. At −6 °C, both samples will have completely melted. Pure water, by contrast, does not melt until the temperature reaches 0 °C.

At first, this discovery was largely overlooked by other scientists, probably because it was published in specialized Russian-language journals. Only after the Russian scientists presented an interpretation of their results at an international meeting did other scientists begin to take note. The Russian results were repeated by many laboratories and thus met the first criterion of a valid scientific result: It must be reproducible. There

EXAMPLE 1-6

Applying Significant Figure Rules: Addition/Subtraction   Express the result of the following calculation with the correct number of significant figures.

$$(2.06 \times 10^2) + (1.32 \times 10^4) - (1.26 \times 10^3) = ?$$

Solution

If the calculation if performed with an electronic calculator, the quantities can be entered just as they are written and the answer obtained adjusted to the correct number of significant figures. We can see that this correct number is *three* by writing the three factors with the same power of ten.

$$(2.06 \times 10^2) + (1.32 \times 10^4) - (1.26 \times 10^3)$$
$$= (0.0206 \times 10^4) + (1.32 \times 10^4) - (0.126 \times 10^4)$$
$$= (0.0206 + 1.32 - 0.126) \times 10^4$$
$$= 1.2146 \times 10^4$$
$$= 1.21 \times 10^4$$

Check: The answer can have no more digits beyond the decimal point (*two*) than the factor with the smallest number of such digits (1.32×10^4).

was no dispute, then, about the facts—but their interpretation sparked active debate.

The properties of the recondensed water could be explained in one of two ways, and each explanation was consistent with the observations. One explanation was that the water contained impurities that had dissolved during the preparation of the water sample. The other was that the water was of a form in which simple water molecules (each consisting of two hydrogen atoms and one oxygen atom) aggregate into large clusters. Such "super" molecules are known as *polymers*.

In one critical experiment, some American researchers subjected recondensed water to very careful examination by a method known as spectroscopy. The method is designed to reveal the structure of a substance at the molecular level. The spectroscopic data suggested that the recondensed water was a pure substance, not a solution. Further, the data suggested a polymer-like structure for the water. Hence the name polywater was coined. The publication of these results in 1969 created a sensation, both among scientists and in the popular press.

Those who firmly believed in polywater employed the reasoning of Sherlock Holmes, a fictional but ardent follower of the scientific method: "When you have eliminated the impossible, whatever remains, however improbable, must be the truth." The trouble was that the impossible had not been eliminated. Within a year or so, new reports pointed to dissolved impurities in polywater. At about the same time, new spectroscopic data indicated that the results interpreted as due to polywater could not be due to any form of water. They were now attributed to impurities in the water. Finally, in 1973, the Russian scientists who had made the original observations reported that they could not produce polywater in capillary tubes that had been freed of all impurities. And so ended this brief but fascinating chapter in the history of science.

The polywater story may seem to reveal a failure of science, but actually it represents a great success of the scientific method. A novel idea was throughly examined by many scientists by a variety of techniques. Hypotheses were proposed, experiments were conducted, and the experimental results were discussed openly. After all of this, the scientific community reached a consensus: Polywater is an impure form of water. A crucial aspect of the scientific method was also revealed: the self-correcting nature of science. Science may get off course at times, but eventually it always gets back on the right track.

Practice Example A: Express the result of the following calculation with the appropriate number of significant figures.

$$0.236 + 128.55 - 102.1 = ?$$

Practice Example B: Perform the following calculation, and express the result with the appropriate number of significant figures.

$$\frac{(1.302 \times 10^3) + 952.7}{(1.57 \times 10^2) - 12.22} = ?$$

Are You Wondering…

At what point(s) to round off numerical results during a calculation?

What's nice about using electronic calculators is that we don't have to write down intermediate results. On occasion, however, when we take a stepwise approach to an in-text example, we will write down intermediate results. We may round off these intermediate results as well as the final result. In general, though, you should store all intermediate results in your electronic calculator without regard to significant figures. Round off to the correct number of significant figures only in the final answer.

KEEP IN MIND

that addition/subtraction is governed by one significant figure rule and multiplication/division by a different rule. ▶

Summary

Matter is classified as either a substance (element or compound) or a mixture (homogeneous or heterogeneous) and in a certain state (solid, liquid, or gas). Chemistry is the study of the structure and properties of matter and how these may be changed by physical or chemical means. Like other branches of science, chemistry makes use of the scientific method, a series of activities involving observations and experimentation that culminate in the expression of natural laws and theories to explain and predict natural phenomena.

The system of measurement used in science is called SI. Four of the fundamental, or base, quantities in SI are length, mass, time, and temperature. Other quantities described in the chapter have derived units, such as volume, expressed through the unit (length)^3. When measuring a property of matter, the precision of the measurement needs to be shown. This is done through the proper use of significant figures. Moreover, calculations must be performed so that the calculated result is stated no more precisely than warranted by the measured quantities.

The method of problem solving employed in this chapter, called the conversion factor method, is described in detail in Appendix A. Many of the conversion factors introduced are based on relationships among SI units or between SI and non-SI units. Also used as conversion factors are density and percent composition.

Integrative Example

Methanol (methyl alcohol or wood alcohol) is a potential automotive fuel, either in pure form or mixed with gasoline. Some fleet vehicles, such as municipal buses, have been modified to burn methanol-containing fuels. These fuels are also used by some race cars.

An automobile modified to use a mixture of 85.0% methanol and 15.0% gasoline by mass gets 25.5 mi/gal. The fuel has a density of 0.775 g/mL. How many kilograms of methanol does the auto consume in a trip of 808 km?

1. *Convert the trip length from kilometers to miles.* This can be a single-step conversion if a conversion factor between kilometers and miles is available. Alternatively, a lengthier, but just as acceptable, pathway of conversions can be used, such as km $\longrightarrow$ m $\longrightarrow$ cm $\longrightarrow$ in. $\longrightarrow$ ft $\longrightarrow$ mi.

$$? \text{ mi} = 808 \text{ km} \times \frac{1000 \text{ m}}{1 \text{ km}} \times \frac{100 \text{ cm}}{1 \text{ m}}$$
$$\times \frac{1 \text{ in.}}{2.54 \text{ cm}} \times \frac{1 \text{ ft}}{12 \text{ in.}} \times \frac{1 \text{ mi}}{5280 \text{ ft}}$$
$$= 502 \text{ mi}$$

2. *Determine the volume of fuel consumed, in gallons.* The fuel mileage can be expressed as the conversion factor

1 gal/25.5 mi. Then take the product of this factor and the distance from step 1.

$$? \text{ gal} = 502 \text{ mi} \times \frac{1 \text{ gal}}{25.5 \text{ mi}} = 19.7 \text{ gal}$$

3. *Convert the volume of fuel to mass of fuel.* Convert 19.7 gal to an equivalent volume in milliliters (for example, gal $\longrightarrow$ qt $\longrightarrow$ L $\longrightarrow$ mL). Then multiply by the density of the fuel.

$$? \text{ g fuel} = 19.7 \text{ gal} \times \frac{4 \text{ qt}}{1 \text{ gal}} \times \frac{0.9464 \text{ L}}{1 \text{ qt}}$$
$$\times \frac{1000 \text{ mL}}{1 \text{ L}} \times \frac{0.775 \text{ g fuel}}{1 \text{ mL}}$$
$$= 5.78 \times 10^4 \text{ g fuel}$$

4. *Determine the mass of methanol in the fuel.* Multiply the result of step 3 by the percent composition of the fuel, that is, by the conversion factor 85.0 g methanol/100.0 g fuel.

$$? \text{ kg methanol} = 5.78 \times 10^4 \text{ g fuel}$$
$$\times \frac{85.0 \text{ g methanol}}{100 \text{ g fuel}} \times \frac{1 \text{ kg methanol}}{1000 \text{ g methanol}}$$
$$= 49.1 \text{ kg methanol}$$

Key Terms   *(see Glossary for definitions of these terms)*

accuracy (1-7)
atom (1-4)
chemical change (reaction) (1-3)
chemical property (1-3)
composition (1-3)
compound (1-4)
density (1-6)
element (1-4)
extensive property (1-6)
gas (1-4)
heterogeneous mixture (1-4)

homogeneous mixture (solution) (1-4)
hypothesis (1-2)
intensive property (1-6)
liquid (1-4)
mass (1-5)
matter (1-3)
molecule (1-4)
natural law (1-2)
percent (1-6)
physical change (1-3)
physical property (1-3)

precision (1-7)
property (1-3)
random error (1-7)
scientific method (1-2)
significant figures (1-8)
solid (1-4)
substance (1-4)
systematic error (1-7)
theory (1-2)

Review Questions *(see also Appendices A-1 and A-5)*

1. In your own words, define or explain the following terms or symbols: **(a)** m^3; **(b)** % by mass; **(c)** °C; **(d)** density; **(e)** element.
2. Briefly describe each of the following ideas: **(a)** SI base units; **(b)** significant figures; **(c)** natural law.
3. Explain the important distinctions between each pair of terms: **(a)** mass and weight; **(b)** intensive and extensive property; **(c)** substance and mixture; **(d)** precision and accuracy; **(e)** hypothesis and theory.
4. Perform and following conversions.
   - **(a)** 1.55 kg = _____ g
   - **(b)** 642 g = _____ kg
   - **(c)** 2896 mm = _____ cm
   - **(d)** 0.086 cm = _____ mm
5. Perform and following conversions.
   - **(a)** 0.127 L = _____ mL
   - **(b)** 15.8 mL = _____ L
   - **(c)** 981 cm^3 = _____ L
   - **(d)** 2.65 m^3 = _____ cm^3
6. Perform the following conversions from non-SI to SI units. (Use information from Figure 1-9, as necessary.)
   - **(a)** 68.4 in. = _____ cm
   - **(b)** 94 ft = _____ m
   - **(c)** 1.42 lb = _____ g
   - **(d)** 248 lb = _____ kg
   - **(e)** 1.85 gal = _____ L
   - **(f)** 3.72 qt = _____ mL
7. Determine the number of
   - **(a)** square meters (m^2) in 1 square kilometer (km^2)
   - **(b)** square centimeters (cm^2) in 1 square meter (m^2)
   - **(c)** square meters (m^2) in 1 square mile (mi^2) (1 mi = 5280 ft, 1 ft = 12 in., 1 in. = 2.54 cm)
8. *Without doing a calculation*, explain which of the following represents the higher temperature: 204 °F or 102 °C.
9. *Without doing a detailed calculation*, explain which of the following represents the greatest *mass*: 80.0 g ethanol (d = 0.79 g/mL), 100.0 mL benzene (d = 0.87 g/mL), or 90.0 mL carbon disulfide (d = 1.26 g/mL).
10. A 2.18-L sample of butyric acid, a substance present in rancid butter, has a mass of 2088 g. What is the density of butyric acid in grams per milliliter?
11. A 385-mL sample of liquid mercury has a mass of 5.23 kg. What is the density of liquid mercury in grams per milliliter?
12. Ethylene glycol, an antifreeze, has a density of 1.11 g/mL at 20 °C.
    - **(a)** What is the mass, in grams, of 452 mL ethylene glycol?
    - **(b)** What is the mass, in kilograms, of 18.6 L ethylene glycol?
    - **(c)** What is the volume, in milliliters, occupied by 65.0 g ethylene glycol?

(d) What is the volume, in liters, occupied by 23.9 kg ethylene glycol?

13. A solution consisting of 8.50% acetone and 91.5% water by mass has a density of 0.9867 g/mL. What mass of acetone, in kilograms, is present in 7.50 L of the solution?
14. A bottle of vinegar is found to contain 5.4% acetic acid by mass. What mass of acetic acid, in grams, is contained per pound of this vinegar? (1 lb = 453.6 g.)
15. A solution contains 12.62% sucrose (cane sugar) by mass. What mass of the solution, in grams, is needed for an application that requires 1.00 kg sucrose?
16. A fertilizer contains 21% nitrogen by mass. What mass of this fertilizer, in kilograms, is required for an application requiring 775 g of nitrogen?
17. Express each number in exponential notation (see Appendix A). **(a)** 8950.; **(b)** 10,700.; **(c)** 0.0240; **(d)** 0.0047; **(e)** 938.3; **(f)** 275,482
18. Express each number in common decimal form (see Appendix A). **(a)** 3.21×10^{-2}; **(b)** 5.08×10^{-4}; **(c)** 121.9×10^{-5}; **(d)** 16.2×10^{-2}
19. How many significant figures are shown in each of the following? If this is indeterminate, tell why. **(a)** 450; **(b)** 98.6; **(c)** 0.0033; **(d)** 902.10; **(e)** 0.02173; **(f)** 7000; **(g)** 7.02; **(h)** 67,000,000
20. Express each of the following to *four* significant figures. **(a)** 3984.6; **(b)** 422.04; **(c)** 186,000; **(d)** 33,900; **(e)** 6.321×10^4; **(f)** 5.0472×10^{-4}
21. Perform the following calculations; express each answer in exponential form and with the appropriate number of significant figures.
    - **(a)** $0.406 \times 0.0023 =$
    - **(b)** $0.1357 \times 16.80 \times 0.096 =$
    - **(c)** $0.458 + 0.12 - 0.037 =$
    - **(d)** $32.18 + 0.055 - 1.652 =$
22. Perform the following calculations; express each number and the answer in exponential form and with the appropriate number of significant figures.
    - **(a)** $\dfrac{320 \times 24.9}{0.080} =$
    - **(b)** $\dfrac{432.7 \times 6.5 \times 0.002300}{62 \times 0.103} =$
    - **(c)** $\dfrac{32.44 + 4.9 - 0.304}{82.94} =$
    - **(d)** $\dfrac{8.002 + 0.3040}{13.4 - 0.066 + 1.02} =$
23. Calculate the mass of a block of iron (d = 7.86 g/cm^3) with dimensions of 52.8 cm × 6.74 cm × 3.73 cm.
24. Calculate the mass of a cylinder of stainless steel (d = 7.75 g/cm^3) with a height of 18.35 cm and radius 1.88 cm.
    (*Hint:* Refer to the inside back cover.)

Exercises (*see also Appendices A-1 and A-5*)

The Scientific Method

25. Is it possible to predict how many experiments are required to verify a natural law? Explain.

26. What are the principal reasons that one theory might be adopted over a conflicting one?

27. An important premise of science is that there exists an underlying order to nature. Einstein described this belief in the words "God is subtle, but He is not malicious." What do you think Einstein meant by this remark?

28. Explain why the common saying, "The exception proves the rule," is incompatible with the scientific method.

29. If you wish to test a theory, describe the necessary characteristics of a suitable experiment.

30. If you wish to propose a scientific theory, describe the necessary characteristics of your theory.

Properties and Classification of Matter

31. State whether each property is physical or chemical.
  (a) An iron nail is attracted to a magnet.
  (b) Charcoal lighter fluid is ignited with a match.
  (c) A bronze statue develops a green coating (patina) over time.
  (d) A block of wood floats on water.

32. State whether each property is physical or chemical.
  (a) A piece of sliced apple turns brown.
  (b) A slab of marble feels cool to the touch.
  (c) A sapphire is blue.
  (d) A clay pot fired in a kiln becomes hard.

33. Indicate whether each sample of matter listed is a substance or a mixture, and, if a mixture, whether homogeneous or heterogeneous.
  (a) a wooden beam
  (b) red ink
  (c) deionized water
  (d) freshly squeezed orange juice

34. Indicate whether each sample of matter listed is a substance or a mixture, and, if a mixture, whether homogeneous or heterogeneous.
  (a) A breath of clean fresh air
  (b) A brass doorknob
  (c) garlic salt
  (d) ice

35. What type of change—physical or chemical—is necessary to bring about the following separations? (*Hint:* Refer to a listing of the elements.)
  (a) sugar and sand
  (b) iron from iron oxide (rust)
  (c) pure water from seawater
  (d) water and sand

36. Suggest physical changes by which the following mixtures can be separated.
  (a) iron filings and wood chips
  (b) ground glass from sucrose (cane sugar)
  (c) pure water from a slurry of ice and salt
  (d) gold flakes and water

Exponential Arithmetic

37. Express each value in exponential form. Where appropriate, include units in your answer.
  (a) speed of sound (sea level): 34,000 centimeters per second
  (b) equatorial radius of Earth: six thousand three hundred seventy-eight kilometers
  (c) the distance between the two hydrogen atoms in the hydrogen molecule: seventy-four trillionths of a meter
  (d) $\dfrac{(2.2 \times 10^3) + (4.7 \times 10^2)}{5.8 \times 10^{-3}} =$

38. Express each value in exponential form. Where appropriate, include units in your answer.
  (a) solar radiation received by Earth: 173 thousand trillion watts
  (b) average human cell diameter: ten millionths of a meter
  (c) the distance between the centers of the atoms in silver metal: one hundred and forty-two trillionths of a meter
  (d) $\dfrac{5.07 \times 10^4 \times (1.8 \times 10^{-3})^2}{0.065 + (3.3 \times 10^{-2})} =$

Significant Figures

39. Indicate whether each of the following is an exact number or a measured quantity subject to uncertainty.
  (a) the number of soda cans in a case
  (b) the volume of milk in a gallon jug
  (c) the distance between Earth and the sun
  (d) the distance between the centers of the two hydrogen atoms in the hydrogen molecule

40. Indicate whether each of the following is an exact number or a measured quantity subject to uncertainty.
  (a) the number of pages in this text
  (b) the number of days in the month of January
  (c) the area of a city lot
  (d) the distance between the centers of the atoms in silver metal

41. Perform the following calculations and retain the appropriate number of significant figures in each result.

(a) $38.4 \times 10^{-3} \times 6.36 \times 10^{5} =$

(b) $\dfrac{1.45 \times 10^{2} \times 8.76 \times 10^{-4}}{(9.2 \times 10^{-3})^{2}} =$

(c) $24.6 + 18.35 - 2.98 =$

(d) $(1.646 \times 10^{3}) - (2.18 \times 10^{2})$
$+ (1.36 \times 10^{4} \times 5.17 \times 10^{-2}) =$

(e)

$$\dfrac{-7.29 \times 10^{-4} + \sqrt{(7.29 \times 10^{-4})^{2} + 4(1.00)(2.7 \times 10^{-5})}}{2 \times (1.00)}$$

(*Hint:* The significant figure rule for the extraction of a root is the same as for multiplication.)

42. Express the result of each of the following calculations in exponential form and with the appropriate number of significant figures.

(a) $4.65 \times 10^{4} \times 2.95 \times 10^{-2} \times 6.663 \times 10^{-3} \times 8.2 =$

(b) $\dfrac{1912 \times (0.0077 \times 10^{4}) \times 3.12 \times 10^{-3}}{(4.18 \times 10^{-4})^{3}} =$

(c) $(3.46 \times 10^{3}) \times 0.087 \times 15.26 \times 1.0023 =$

(d) $\dfrac{(4.505 \times 10^{-2})^{2} \times 1.080 \times 1545.9}{0.03203 \times 10^{3}} =$

(e)

$$\dfrac{-3.61 \times 10^{-4} + \sqrt{(3.61 \times 10^{-4})^{2} + 4(1.00)(1.9 \times 10^{-5})}}{2 \times (1.00)}$$

(*Hint:* The significant figure rule for the extraction of a root is the same as for multiplication.)

43. A press release describing the 1986 nonstop, round-the-world trip by the ultra-lightweight aircraft *Voyager* included the following data.

flight distance: 25,012 mi
flight time: 9 days, 3 minutes, 44 seconds
fuel capacity: nearly 9000 lb
fuel remaining at end of flight: 14 gal

To the maximum number of significant figures permitted, calculate
(a) the average speed of the aircraft, in miles per hour
(b) the fuel consumption, in miles per pound of fuel (Assume a density of 0.70 g/mL for the fuel.)

44. Use the concept of significant figures to criticize the way in which the following information was presented. "The estimated proved reserve of natural gas as of January 1, 1982, was 2,911,346 trillion cubic feet."

Units of Measurements

45. Which is the greater mass, 2172 μg or 0.00515 mg? Explain.
46. Which is the greater mass, 3257 mg or 0.00475 kg? Explain.
47. The non-SI unit, the hand (used in measuring horses), is 4 inches. What is the height, in meters, of a horse that stands 15 hands high? (1 in. = 2.54 cm.)
48. The unit *furlong* is used in horse racing. The units *chain* and *link* are used in surveying. There are exactly 8 furlongs in 1 mi, 10 chains in 1 furlong, and 100 links in 1 chain. To three significant figures, what is the length of 1 link in inches?
49. A sprinter runs the 100-yd dash in 9.3 s. At this same rate,
(a) How long would it take the sprinter to run 100.0 m?
(b) What is the sprinter's speed in meters per second?
(c) How long would it take the sprinter to run a distance of 1.45 km?
50. A non-SI unit of mass used in pharmaceutical work is the grain (gr). (15 gr = 1.0 g) An aspirin tablet contains 5.0 gr of aspirin. A 155-lb arthritic individual takes two aspirin tablets per day.
(a) What is the quantity of aspirin in two tablets, expressed in milligrams?
(b) What is the dosage rate of aspirin, expressed in milligrams of aspririn per kilogram of body mass?
(c) At the given rate of consumption of aspirin tablets, how many days would it take to consume 1.0 lb of aspirin?

51. In SI units, land area is measured in *hectares*, defined as $1 \times 10^4 \ \mathrm{m}^2$ (1 hectometer = 100 m). How many acres correspond to 1 hectare? ($1 \ \mathrm{mi}^2 = 640$ acres, 1 mi = 5280 ft, 1 ft = 12 in.)

52. In an engineering reference book, you find that the density of iron is 0.284 $\mathrm{lb/in}^3$. What is the density in $\mathrm{g/cm}^3$.

53. A typical pressure for optimal performance of automobile tires is 32 $\mathrm{lb/in}^2$. What is this pressure in grams per square centimeter and kilograms per square meter?

54. The volume of a red blood cell is about $90.0 \times 10^{-12} \ \mathrm{cm}^3$. Assuming that red blood cells are spherical, what is the diameter of a red blood cell in inches?

Temperature Scales

55. The highest and lowest temperatures on record for San Bernardino, California, are 118 and 17 °F, respectively. What are these temperatures on the Celsius scale?

56. We wish to mark off a thermometer in both Celsius and Fahrenheit temperatures. On the Celsius scale, the lowest temperature mark is at −15 °C, and the highest temperature mark is at 60 °C. What are the equivalent Fahrenheit temperatures?

57. A home economics class is given an assignment in candy making that requires a sugar mixture to be brought to a "soft ball" stage (234–240 °F). A student borrows a thermometer having a range from −10 to 110 °C from the chemistry laboratory to do this assignment. Will this thermometer serve the purpose? Explain.

58. The absolute zero of temperature is found at −273.15 °C. Should it be possible to achieve a temperature of −465 °F? Explain.

59. You decide to establish a new temperature scale on which the melting point of mercury (−38.9 °C) is 0 °M and the boiling point of mercury (356.9 °C) is 100 °M. What would be the boiling point of water in degrees M? the temperature of absolute zero in degrees M?

60. You decide to establish a new temperature scale on which the melting point of ammonia (−77.75 °C) is 0 °A and the boiling point of ammonia (−33.35 °C) is 100 °A. What would be the boiling point of water in degrees A? the temperature of absolute zero in degrees A?

Density

61. To determine the density of acetone, a 55.0-gal drum is weighed twice. The drum weighs 75.0 lb when empty and 437.5 lb when filled with acetone. What is the density of acetone, expressed in grams per milliliter?

62. To determine the volume of an irregularly shaped glass vessel, the vessel is weighed empty (121.3 g) and when filled with carbon tetrachloride (283.2 g). What is the volume capacity of the vessel, in milliliters, given that the density of carbon tetrachloride is 1.59 g/mL?

63. The following densities are given at 20 °C: water, 0.998 $\mathrm{g/cm}^3$; iron, 7.86 $\mathrm{g/cm}^3$; aluminum 2.70 $\mathrm{g/cm}^3$. Arrange the following items in terms of *increasing* mass.
  (1) a rectangular bar of iron, 81.5 cm × 2.1 cm × 1.6 cm
  (2) a sheet of aluminum foil, 12.12 m × 3.62 m × 0.003 cm
  (3) 4.051 L of water

64. The density of aluminum is 2.70 $\mathrm{g/cm}^3$. A square piece of aluminum foil, 9.0 in. on a side, is found to weigh 2.568 g. What is the thickness of this foil, in millimeters?

65. To determine the approximate mass of a small spherical shot of copper, the following experiment is performed. When 125 pieces of the shot are counted out and added to 8.4 mL of water in a graduated cylinder, the total volume becomes 8.9 mL. The density of copper is 8.92 $\mathrm{g/cm}^3$. Determine the approximate mass of a single piece of shot, assuming that all the pieces are of the same dimensions.

66. The angle iron pictured here is made of steel with a density of 7.78 $\mathrm{g/cm}^3$. What is the mass, in grams, of this object?

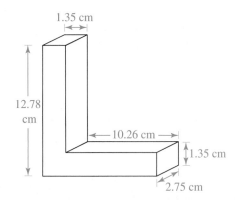

67. In normal blood, there are about 5.4×10^9 red blood cells per milliliter. The volume of a red cell is about $90.0 \times 10^{-12} \ \mathrm{cm}^3$, and the density of a red cell is 1.096 g/mL. How many liters of whole blood would be needed to collect 0.5 kg of red cells?

68. A technique once used by geologists to measure the density of a mineral is to mix two dense liquids in such proportions that the mineral grains just float. When a sample of the mixture in which the mineral calcite just floats is put in a special density bottle, the weight is 15.4448 g. When empty, the bottle weighs 12.4631 g, and when filled with water, it weighs 13.5441 g. What is the density of the calcite sample? (All measurements were carried out at 25 °C, and the density of water at 25 °C is 0.9970 g/mL).

▶ At the left, grains of the mineral *calcite* float on the surface of the liquid bromoform ($d = 2.890$ g/mL). At the right, the grains sink to the bottom of liquid chloroform ($d = 1.444$ g/mL). By mixing bromoform and chloroform in just the proportions required so that the grains barely float, the density of the calcite can be determined (see Exercise 68).

Percent Composition

69. In a class of 76 students, the results of a particular examination were 9 A's, 21 B's, 36 C's, 8 D's, 2 F's. What was the percent distribution of grades, that is, % A's, % B's, and so on?

70. A class of 84 students had a final grade distribution of 18% A's, 25% B's, 32% C's, 13% D's, 12% F's. How many students received each grade?

71. A water solution that is 28.0% sucrose by mass has a density of 1.118 g/mL. What mass of sucrose, in grams, is contained in 2.75 L of this solution?

72. A water solution containing 12.0% sodium hydroxide by mass has a density of 1.131 g/mL. What volume of this solution, in liters, must be used in an application requiring 3.50 kg of sodium hydroxide?

Integrative and Advanced Exercises

73. According to the rules on significant figures, the product 99.9×1.008 should be expressed to three significant figures—101. Yet, in this case, it would be more appropriate to express the result to *four* significant figures—100.7. Explain why.
(*Hint:* Recall the marginal note on page 20.)

74. A solution used to chlorinate a home swimming pool contains 7% chlorine by mass. An ideal chlorine level for the pool is one part per million (1 ppm). (Think of 1 ppm as being 1 g chlorine per million grams of water.) If you assume densities of 1.10 g/mL for the chlorine solution and 1.00 g/mL for the swimming pool water, what volume of the chlorine solution, in liters, is required to produce a chlorine level of 1 ppm in a 18,000-gallon swimming pool.

75. A standard kilogram mass is to be cut from a cylindrical bar of steel with a diameter of 1.50 in. The density of the steel is 7.70 g/cm³. How many inches long must the section be?

76. The volume of seawater on Earth is about 330,000,000 mi³. If seawater is 3.5% sodium chloride by mass and has a density of 1.03 g/mL, what is the approximate mass of sodium chloride, in tons, dissolved in the seawater on Earth? (1 ton = 2000 lb.)

77. The diameter of metal wire is often referred to by its American wire gauge number. A 16-gauge wire has a diameter of 0.05082 in. What length of wire, in meters, is there in a 1.00-lb spool of 16-gauge copper wire? The density of copper is 8.92 g/cm³.

78. Magnesium metal can be extracted from seawater by the Dow process (described in Section 22-2). Magnesium occurs in seawater to the extent of 1.4 g magnesium per kilogram of seawater. The annual production of magnesium in the United States is about 10^5 tons. If all this magnesium were extracted from seawater, what volume of seawater, in cubic meters, would have to be processed? (1 ton = 2000 lb. Assume a density of 1.025 g/mL for seawater.)

79. A typical rate of deposit of dust ("dustfall") from unpolluted air is 10 tons per square mile per month. (**a**) Express this dustfall in milligrams per square meter per hour. (**b**) If the dust has an average density of 2 g/cm³, how long would it take to accumulate a layer of dust 1 mm thick?

80. The volume of irrigation water is usually expressed in acre-feet. One acre-foot is a volume of water sufficient to cover 1 acre of land to a depth of 1 ft (640 acres = 1 mi²; 1 mi = 5280 ft). The principal lake in the California Water Project is Lake Oroville, whose water storage capacity is listed as 3.54×10^6 acre-feet. Express the volume of Lake Oroville in (**a**) cubic feet; (**b**) cubic meters; (**c**) gallons.

81. A Fahrenheit and a Celsius thermometer are immersed in the same medium. At what Celsius temperature will the numerical reading on the Fahrenheit thermometer be
(**a**) the same as that on the Celsius thermometer?
(**b**) twice that on the Celsius thermometer?
(**c**) one-eighth that on the Celsius thermometer?
(**d**) 300° more than that on the Celsius thermometer?

82. The accompanying illustration shows a 100.0-mL graduated cylinder half-filled with 8.0 g of diatomaceous earth, a material consisting mostly of silica and used as a filtering medium in swimming pools. How many milliliters of water are required to fill the cylinder to the 100.0-mL mark? The diatomaceous earth is insoluble in water and has a density of 2.2 g/cm³.

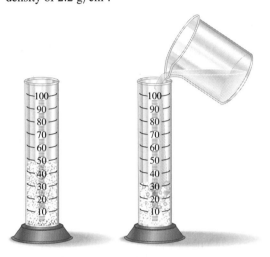

83. The simple device pictured here, a pycnometer, is used for precise density determinations. From the data presented, together with the fact that the density of water at 20 °C is 0.99821 g/mL, determine the density of methanol, in grams per milliliter.

Empty	Filled with water	Filled with methanol
25.601 g	at 20 °C: 35.552 g	at 20 °C: 33.490 g

84. A pycnometer (see Exercise 83) weighs 25.60 g empty and 35.55 g when filled with water at 20 °C. The density of water at 20 °C is 0.9982 g/mL. When 10.20 g lead is placed in the pycnometer and the pycnometer again filled with water at 20 °C, the total mass is 44.83 g. What is the density of the lead in grams per cubic centimeter?

85. The Greater Vancouver Regional District (GVRD) chlorinates the water supply of the region at the rate of 1 ppm, that is, 1 kilogram of chlorine per million kilograms of water. The chlorine is introduced in the form of sodium hypochlorite, which is 47.62% chlorine. The population of the GVRD is 1.8 million persons. If each person uses 750 L of water per day, how many kilograms of sodium hypochlorite must be added to the water supply each week to have the required chlorine level of 1 ppm?

86. A Boeing 767 due to fly from Montreal to Edmonton required refueling. Since the fuel gauge on the aircraft was not working, a mechanic, using a dipstick, determined that 7682 L of fuel were left on the plane. The plane required 22,300 kg of fuel to make the trip. In order to determine the volume of fuel required, the pilot asked for the conversion factor needed to convert a volume of fuel to a mass of fuel. The mechanic gave the factor as 1.77. Assuming that this factor was in metric units (kg/L), the pilot calculated the volume to be added as 4916 L. This volume of fuel was added and the 767 subsequently ran out the fuel, but landed safely by gliding into Gimli Airport near Winnipeg. The error arose because the factor 1.77 was in units of pounds per liter. What volume of fuel should have been added?

87. Using the microscopic visualization shown in Figure 1-6, sketch the arrangement of atoms or molecules in the following substances. The sketch can be two-dimensional and should include at least 10 particles (atoms or molecules). Atoms should be represented by a circle and different atoms distinguished by color or shading.
(a) a sample of pure oxygen gas (which consists of O_2 molecules)
(b) a sample of solid copper (which consists of copper atoms)
(c) a sample of liquid ammonia (which consists of NH_3 molecules)
(d) a heterogeneous mixture consisting of liquid water (which consists of H_2O molecules) and solid copper
(e) a homogeneous mixture of oxygen (which consists of O_2 molecules) dissolved in liquid water

88. The following equation can be used to relate the density of liquid water to Celsius temperature in the range from 0 °C to about 20 °C:

$$d(\text{g/cm}^3)$$
$$= \frac{0.99984 + (1.6945 \times 10^{-2}t) - (7.987 \times 10^{-6}t^2)}{1 + (1.6880 \times 10^{-2}t)}$$

(a) To four significant figures, determine the density of water at 10 °C.
(b) At what temperature does water have a density of 0.99860 g/cm³?
(c) In the following ways, show that the density passes through a maximum somewhere in the temperature range to which the equation applies.
  (i) by estimation
  (ii) by a graphical method
  (iii) by a method based on differential calculus

Feature Problems

89. In an attempt to determine any possible relationship between the year in which a U.S. penny was minted and its current mass, students weighed an assortment of pennies and obtained the following data.

1968	1973	1977	1980	1982	1983	1985
3.11	3.14	3.13	3.12	3.12	2.51	2.54
3.08	3.06	3.10	3.11	2.53	2.49	2.53
3.09 g	3.07 g	3.06 g	3.08 g	2.54 g	2.47 g	2.53 g

What valid conclusion(s) might they have drawn about the relationship between the masses of the pennies within a given year and from year to year?

90. In the third century BC, the Greek mathematician Archimedes is said to have discovered an important principle that is useful in density determinations. The story told is that King Hiero of Syracuse (in Sicily) asked Archimedes to verify that an ornate crown made for him by a goldsmith consisted of pure gold and not a gold–silver alloy. He had to do this, of course, without damaging the crown in any way. Describe how Archimedes did this, or if you don't know the rest of the story, rediscover Archimedes's principle and explain how it can be used to settle the question.

91. The Galileo thermometer shown in the photograph below is based on the dependence of density on temperature. The liquid in the outer cylinder and the liquid in the partially filled floating glass balls are the same except that a colored dye has been added to the liquid in the balls. Can you explain how the Galileo thermometer works?

92. The canoe gliding gracefully along the water in the photograph at the top of the next column is made of concrete, which has a density of about 2.4 g/cm³. Explain why the canoe does not sink.

93. As mentioned on page 13, the MCO was lost because of a mix-up in the units used to calculate the force needed to correct its trajectory. Ground-based computers generated the force correction file. On September 29, 1999, it was discovered that the forces reported by the ground-based computer for use in MCO navigation software were low by a factor of 4.45. The erroneous trajectory brought the MCO 56 km above the surface of Mars; the correct trajectory would have brought the MCO approximately 250 km above the surface. At 250 km, the MCO would have successfully entered the desired elliptic orbit. The data contained in the force correction file were delivered in lb-sec instead of the required SI units of newton-sec for the MCO navigation software. The newton is the SI unit of force and is described in Appendix B. The British Engineering (gravitational) system uses a pound (lb) as a unit of force and ft/s² as a unit of acceleration. In turn, the pound is defined as the pull of Earth on a unit of mass at a location where the acceleration due to gravity is 32.174 ft/s². The unit of mass in this case is the slug, which is 14.59 kg. Thus,

BE unit of force = 1 pound = 1 (slug)(ft/s²)

Use this information to confirm that:

BE unit of force = 4.45 × SI unit of force

1 pound = 4.45 newton

eMedia Exercises

94. Using the **Separation of Matter** activity *(eChapter 1.4)*, classify the following as a compound, element, homogeneous mixture, or heterogeneous mixture. For mixtures, propose a method of separating the individual components.
    (a) Tea sweetened with sugar ($C_{12}H_{22}O_{11}$)
    (b) Magnesium oxide (MgO)
    (c) Metallic brass (containing zinc and copper)
    (d) Iron

95. (a) View the **Phases of Water** illustration *(eChapter 1.5)* and describe the different physical properties of the three states of water. (b) On some winter days, ice on the roads can "disappear" when the temperature remains below the melting temperature that would cause a change to the liquid phase. Based upon the representations of the states of matter shown in the diagram, propose a mechanism by which the ice disappears.

96. Using the **Density activity** to calculate densities *(eChapter 1.6)*, determine the mass (in kg) required to produce a change in volume of 2.50 mL for each element available in the activity. From this information, formulate a mathematical statement that describes the relationship between density and the relative mass of objects or elements of the same volume.

2

Atoms and the Atomic Theory

Contents

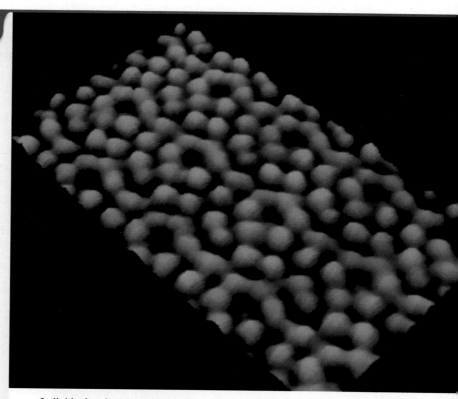

Individual carbon atoms in graphite. The hypothesis that all matter is made up of atoms has existed for over 2000 years. It is only within the last 35 years, however, that techniques have been developed that can make individual atoms visible.

Beginning more than 200 years ago, chemists developed the first theories of atomic structure. After a brief survey of these early chemical discoveries, we will describe physical evidence leading to the modern picture of a *nuclear* atom: protons and neutrons combined into a nucleus with electrons found outside this nucleus. We will also introduce the periodic table as the primary means of organizing elements into groups with similar properties. Finally, we will introduce the concept of the mole and the Avogadro constant, which are the principal tools for counting atoms and molecules and measuring amounts of substances. We will use these tools throughout the text.

2-1   Early Chemical Discoveries and the Atomic Theory

Chemistry has been practiced for a very long time, even if its practitioners were much more interested in its applications than in its underlying principles. The blast furnace for extracting iron from iron ore appeared as early as 1300 AD, and such important chemicals as sulfuric acid (oil of vitriol), nitric acid (aqua fortis), and sodium sulfate (Glauber's salt) were all well known and used several hundred years ago. Before the end of the eighteenth century, the principal gases of the atmosphere—nitrogen and oxygen—had been isolated, and natural laws had been proposed describing the physical behavior of gases. Yet chemistry cannot be said to have entered the modern age until the process of combustion was explained. In this section, we explore the direct link between the explanation of combustion and Dalton's atomic theory.

Law of Conservation of Mass

The process of combustion—burning—is so familiar that it is hard to realize what a difficult riddle it posed for early scientists. Some of the difficult-to-explain observations are described in Figure 2-1.

In 1774, Antoine Lavoisier (1743–1794) performed an experiment in which he heated a sealed glass vessel containing a sample of tin and some air. He found that the mass before heating (glass vessel + tin + air) and after heating (glass vessel + "tin calx" + remaining air) were the same. Through further experiments, he showed that the product of the reaction, tin calx (tin oxide), consisted of the original tin together with a portion of the air. Experiments like this proved to Lavoisier that oxygen from air is essential to combustion, and it also led him to formulate the **law of conservation of mass:**

▲ FIGURE 2-1
Two combustion reactions
The apparent product of the combustion of the match—the ash—weighs *less* than the match. The product of the combustion of the magnesium ribbon (the "smoke") weighs *more* than the ribbon. Actually, in each case, the total mass remains *unchanged*. To understand this, you have to know that oxygen gas enters into both combustions and that water and carbon dioxide are also products of the combustion of the match.

> The total mass of substances present after a chemical reaction is the same as the total mass of substances before the reaction.

This law is illustrated in Figure 2-2, where the reaction between silver nitrate and sodium chloride to give a white solid (silver chloride) is monitored by placing the

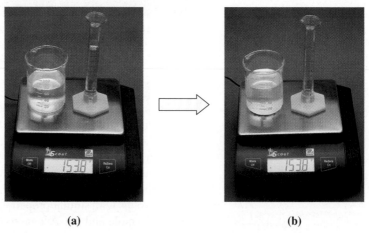

(a)   (b)

▲ FIGURE 2-2   **Mass is conserved during a chemical reaction**
(a) Before the reaction, a beaker with sodium chloride solution and a graduated cylinder with silver nitrate solution are placed on a single-pan balance, which displays their combined mass. (b) When the solutions are mixed, a chemical reaction occurs that forms silver chloride (white precipitate) and a sodium nitrate solution. Note that the total mass remains unchanged.

reactants on a single-pan balance—the total mass does not change. Stated another way, the law of conservation of mass says that matter is neither created nor destroyed in a chemical reaction.

EXAMPLE 2-1

Applying the Law of Conservation of Mass. A 0.455-g sample of magnesium is allowed to burn in 2.315 g of oxygen gas. The sole product is magnesium oxide. After the reaction, no magnesium remains and the mass of unreacted oxygen is 2.015 g. What mass of magnesium oxide is produced?

Solution

To answer this question, we need to identify the substances present before and after the reaction. The total mass is unchanged.

$$\text{mass before reaction} = 0.455 \text{ g magnesium} + 2.315 \text{ g oxygen} = 2.770 \text{ g}$$
$$\text{mass after reaction} = ? \text{ g magnesium oxide} + 2.015 \text{ g oxygen} = 2.770 \text{ g}$$
$$? \text{ g magnesium oxide} = 2.770 \text{ g} - 2.015 \text{ g} = 0.755 \text{ g}$$

Practice Example A: A 0.382-g sample of magnesium reacts with 2.652 g of nitrogen gas. The sole product is magnesium nitride. After the reaction, the mass of unreacted nitrogen is 2.505 g. What mass of magnesium nitride is produced?

Practice Example B: A 7.12-g sample of magnesium is heated with 1.80 g of bromine. All the bromine is used up, and 2.07 g of magnesium bromide is the only product. What mass of magnesium remains *unreacted*?

Law of Constant Composition

In 1799, Joseph Proust (1754–1826) reported, "One hundred pounds of copper, dissolved in sulfuric or nitric acids and precipitated by the carbonates of soda or potash, invariably gives 180 pounds of green carbonate."[*] This and similar observations became the basis of the **law of constant composition**, or the **law of definite proportions**:

> All samples of a compound have the same composition—the same proportions by mass of the constituent elements.

To see how the law of constant composition works, consider the compound water. Water is made up of two atoms of hydrogen (H) for every atom of oxygen (O), a fact that can be represented symbolically by a *chemical formula*, the familiar H_2O. The two samples described below have the same proportions of the two elements, expressed as percentages by mass. To determine the percent by mass of hydrogen, for example, simply divide the mass of hydrogen by the sample mass and multiply by 100%. For each sample, you will obtain the same result: 11.19% H.

(a)

(b)

▲ The mineral *malachite* **(a)** and the green patina on a copper roof **(b)** are both basic copper carbonate and have the same composition (the very same percent of copper, for example) as did the basic copper carbonate prepared by Proust in 1799.

Sample A	Composition	Sample B
10.000 g		27.000 g
1.119 g H	% H = 11.19	3.021 g H
8.881 g O	% O = 88.81	23.979 g O

[*]The substance Proust worked with is actually a more complex substance called *basic* copper carbonate. Proust's results were valid because, like all compounds, basic copper carbonate has a constant composition.

EXAMPLE 2-2

Using the Law of Constant Composition. A 0.100-g sample of magnesium, when combined with oxygen, yields 0.166 g of magnesium oxide. A second magnesium sample with a mass of 0.144 g is also combined with oxygen. What mass of magnesium oxide is produced from this second sample?

Solution

We can use data from the first experiment to establish the proportion of magnesium in magnesium oxide.

$$\frac{0.100 \text{ g magnesium}}{0.166 \text{ g magnesium oxide}}$$

This proportion must be the same in *all* samples. Note that for the second sample we must invert the factor before we use it, because we must convert from mass of magnesium to mass of magnesium oxide.

$$0.144 \text{ g magnesium} \times \frac{0.166 \text{ g magnesium oxide}}{0.100 \text{ g magnesium}} = 0.239 \text{ g magnesium oxide}$$

Check: Note that the conversion factor must have a value greater than 1 because the mass of magnesium oxide must be greater than the mass of magnesium. If the factor had been mistakenly inverted, the mass of magnesium oxide would have been only 0.0867 g—an impossible result.

Practice Example A: Use information provided in the example to determine the mass of magnesium contained in 0.500 g of magnesium oxide.

Practice Example B: What masses of magnesium and oxygen must be combined to make exactly 2.00 g of magnesium oxide?

▲ John Dalton (1766–1844)—developer of the atomic theory. Dalton has not been considered a particularly good experimenter, perhaps because of his color blindness (a condition sometimes called daltonism). However, he did skillfully use the data of others in formulating his atomic theory.

Dalton's Atomic Theory

From 1803 to 1808, John Dalton, an English schoolteacher, used the two fundamental laws of chemical combination just described, as the basis of an atomic theory. His theory involved three assumptions.

1. Each chemical element is composed of minute, indestructible particles called atoms. Atoms can be neither created nor destroyed during a chemical change.
2. All atoms of an element are alike in mass (weight) and other properties, but the atoms of one element are different from those of all other elements.
3. In each of their compounds, different elements combine in a simple numerical ratio: for example, one atom of A to one of B (AB), or one atom of A to two of B (AB_2).

If atoms of an element are indestructible (assumption 1), then the *same* atoms must be present after a chemical reaction as before. The total mass remains unchanged. Dalton's theory explains the law of conservation of mass. If all atoms of an element are alike in mass (assumption 2) and if atoms unite in *fixed* numerical ratios (assumption 3), the percent composition of a compound must have a unique value, regardless of the origin of the sample analyzed. Dalton's theory also explains the law of constant composition.

Like all good theories, Dalton's atomic theory led to a prediction—the **law of multiple proportions**.

Multiple Proportions animation

To illustrate, consider two oxides of carbon (an oxide is a combination of an element with oxygen). In one oxide, 1.000 g of carbon is combined with 1.333 g of oxygen, and in the other, with 2.667 g of oxygen. We see that the second oxide is richer in oxygen; in fact, it contains twice as much oxygen, 2.667 g/1.333 g = 2.00. We now know that the first oxide has the molecular formula CO and the second, CO_2 (Figure 2-3).

The characteristic relative masses of the atoms of the various elements became known as atomic weights, and throughout the nineteenth century, chemists worked at establishing reliable values of relative atomic weights. Mostly, however, chemists directed their attention to discovering new elements, synthesizing new compounds, developing techniques for analyzing materials, and, in general, building up a vast body of chemical facts. Efforts to unravel the structure of the atom became the focus of physicists, as we see in the next several sections.

KEEP IN MIND ▶

that all we know is the second oxide is twice as rich in oxygen as the first. If the first is CO, the possibilities for the second are CO_2, C_2O_4, C_3O_6, and so on. (See also Exercise 40.)

2-2 Electrons and Other Discoveries in Atomic Physics

Fortunately, we can acquire a qualitative understanding of atomic structure without having to retrace all the discoveries that preceded atomic physics. We do, however, need a few key ideas about the interrelated phenomena of electricity and magnetism, which we briefly discuss here. Electricity and magnetism were used in the experiments that led to the current theory of atomic structure.

Certain objects display a property called electric charge, which can be either positive (+) or negative (−). Positive and negative charges attract (neutralize) each other, while two positive or two negative charges repel each other. As we learn in this section, all objects of matter are made up of charged particles. An electrically *neutral* object has equal numbers of positively and negatively charged particles and carries no net charge. If the number of positive charges exceeds the number of negative charges, an object has a net positive charge. If negative charges exceed positive charges, the object has a net negative charge. We commonly observe that when we rub one substance against another, as when we comb our hair, a static electric charge builds up, implying that rubbing separates some positive and negative charges (Figure 2-4). Moreover, when we build up a positive charge somewhere, a negative charge of equal size appears somewhere else; charge is balanced.

▲ FIGURE 2-3

The molecules CO and CO_2 illustrating the law of multiple proportions

Pictured are two oxides of carbon formed by the combination of carbon and oxygen. These compounds illustrate the law of multiple proportions because the masses of oxygen in the two compounds, relative to a fixed mass of carbon, are in a ratio of small whole numbers.

(a)

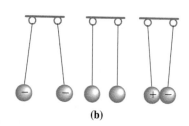

(b)

▲ FIGURE 2-4 **Forces between electrically charged objects**
(a) Charged comb. If you comb your hair on a dry day, a static charge appears on the comb and causes bits of paper to be attracted to the comb. (b) Both objects on the left carry a negative electric charge. Objects with like charge repel each other. The objects in the center lack any electric charge and exert no forces on each other. The objects on the right carry opposite charges—one positive and one negative—and attract each other.

Coulomb's Law simulation

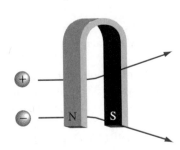

▲ **FIGURE 2-5**
Effect of a magnetic field on charged particles
When charged particles travel through a magnetic field so that their path is perpendicular to the field, they are deflected by the field. Negatively charged particles are deflected in one direction, and positively charged particles in the opposite direction. Several phenomena described in this section depend on this behavior.

Figure 2-5 shows how charged particles behave when they move through the field of a magnet. They are deflected from their straight-line path into a curved path in a plane perpendicular to the field. Think of the field or region of influence of the magnet as represented by a series of invisible "lines of force" running from the north pole to the south pole of the magnet.

The Discovery of Electrons

CRT, the abbreviation for cathode-ray tube, has become a familiar acronym. The CRT is the heart of computer monitors and TV sets. The first cathode-ray tube was made by Michael Faraday (1791–1867) about 150 years ago. In passing electricity through evacuated glass tubes, Faraday discovered **cathode rays**, a type of radiation emitted by the negative terminal, or *cathode*, that crossed the evacuated tube to the positive terminal, or *anode*. Later scientists found that cathode rays travel in straight lines and have properties that are independent of the cathode material (that is, whether it is iron, platinum, and so on). The construction of a CRT is shown in Figure 2-6. The cathode rays produced in the CRT are invisible, and they can be detected only by the light emitted by materials that they strike. These materials, called *phosphors*, are painted on the end of the CRT so that the path of the cathode rays can be revealed. (*Fluorescence* is the term used to describe the emission of light by a phosphor when it is struck by energetic radiation). Another significant observation about cathode rays is that they are deflected by electric and magnetic fields in the manner expected for negatively charged particles (Figure 2-7a, b).

In 1897, by the method outlined in Figure 2-7c, J.J. Thomson (1856–1940) established the ratio of mass (m) to electric charge (e) for cathode rays, that is, m/e. Also, Thomson concluded that cathode rays are negatively charged *fundamental particles of matter found in all atoms*. (The properties of cathode rays are *independent* of the composition of the cathode.) Cathode rays subsequently became known as **electrons**, a term first proposed by George Stoney in 1874.

Robert Millikan (1868–1953) determined the electronic charge e through a series of oil-drop experiments (1906–1914), described in Figure 2-8. The currently accepted value of the electronic charge e (to five significant figures) is -1.6022×10^{-19} C. By combining this value with an accurate value of the mass-to-charge ratio for an electron, we find that the mass of an electron is 9.1094×10^{-28} g.

The coulomb (C) is the SI unit of electric charge (see also Appendix B). ▶

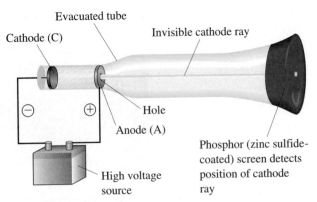

Evacuated tube

Cathode (C)

Invisible cathode ray

Hole

Anode (A)

High voltage source

Phosphor (zinc sulfide-coated) screen detects position of cathode ray

▲ **FIGURE 2-6   A cathode-ray tube**
The high-voltage source of electricity creates a negative charge on the electrode at the left (cathode) and a positive charge on the electrode at the right (anode). Cathode rays pass from the cathode (C) to the anode (A), which is perforated to allow the passage of a narrow beam of cathode rays. The rays are visible only through the green fluorescence that they produce on the zinc sulfide–coated screen at the end of the tube. They are invisible in other parts of the tube.

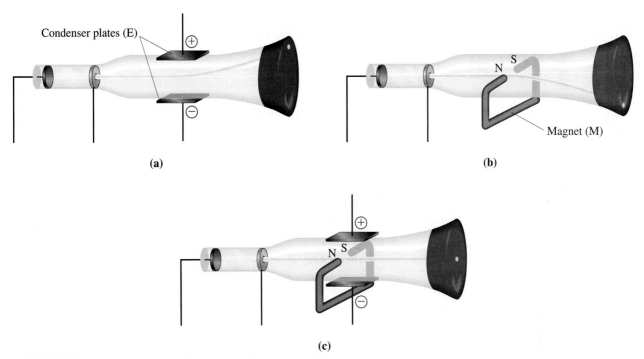

(a)

(b)

(c)

▲ FIGURE 2-7   **Cathode rays and their properties**
(a) Deflection of cathode rays in an electric field. The beam of cathode rays is deflected as it travels from left to right in the field of the electrically charged condenser plates (E). The deflection corresponds to that expected of negatively charged particles. **(b)** Deflection of cathode rays in a magnetic field. The beam of cathode rays is deflected as it travels from left to right in the field of the magnet (M). The deflection corresponds to that expected of negatively charged particles. **(c)** Determining the mass-to-charge ratio, m/e, for cathode rays. The cathode-ray beam strikes the end screen undeflected if the forces exerted on it by the electric and magnetic fields are counterbalanced. By knowing the strengths of the electric and magnetic fields, together with other data, a value of m/e can be obtained. Precise measurements yield a value of -5.6857×10^{-9} gram per coulomb. (Because cathode rays carry a negative charge, the sign of the mass-to-charge ratio is also negative.)

*Millikan Oil-Drop
Experiment animation*

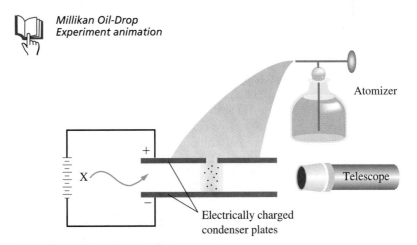

▲ FIGURE 2-8   **Millikan's oil-drop experiment**
Ions (charged atoms or molecules) are produced by energetic radiation such as X rays (X). Some of these ions become attached to oil droplets, giving them a net charge. The fall of a droplet in the electric field between the condenser plates is speeded up or slowed down, depending on the magnitude and sign of the charge on the droplet. By analyzing data from a large number of droplets, Millikan concluded that the magnitude of the charge, q, on a droplet is an *integral* multiple of the electronic charge, e. That is, $q = ne$ (where $n = 1, 2, 3,\ldots$).

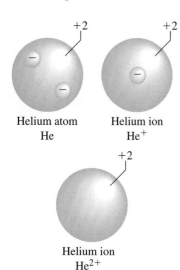

▲ **FIGURE 2-9**
The plum pudding atomic model
According to this model, a helium atom would have a +2 cloud of positive charge and two electrons (-2). If a helium atom loses one electron, it becomes charged and is called an *ion*. This ion, referred to as He$^+$, has a net charge of 1+. If the helium atom loses both electrons, the He^{2+} ion forms.

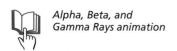

Alpha, Beta, and Gamma Rays animation

Once the electron was seen to be a fundamental particle of matter found in all atoms, atomic physicists began to speculate on how these particles were incorporated into atoms. The commonly accepted model was that proposed by J.J. Thomson. Thomson thought that the positive charge necessary to counterbalance the negative charges of electrons in a neutral atom was in the form of a nebulous cloud. Electrons, he suggested, floated in this diffuse cloud of positive charge (rather like a lump of Jello with electron "fruit" embedded in it). This model became known as the plum pudding model because of its similarity to a popular English dessert. The plum pudding model is illustrated in Figure 2-9 for a neutral atom and for atomic species, called *ions*, which carry a net charge.

X Rays and Radioactivity

Cathode-ray research had many important spin-offs. In particular, two natural phenomena of immense theoretical and practical significance were discovered in the course of other investigations.

In 1895, Wilhelm Roentgen (1845–1923) noticed that when cathode-ray tubes were operating, certain materials *outside* the tubes glowed or fluoresced. He showed that this fluorescence was caused by radiation emitted by the cathode-ray tubes. Because of the unknown nature of this radiation, Roentgen coined the term *X ray*. We now recognize the X ray as a form of high-energy electromagnetic radiation, which is discussed in Chapter 9.

Antoine Henri Becquerel (1852–1908) associated X rays with fluorescence and wondered if naturally fluorescent materials produce X rays. To test this idea, he wrapped a photographic plate with black paper, placed a coin on the paper, covered the coin with a uranium-containing fluorescent material, and exposed the entire assembly to sunlight. When he developed the film, a clear image of the coin could be seen. The fluorescent material had emitted radiation (presumably X rays) that penetrated the paper and exposed the film. On one occasion, because the sky was overcast, Becquerel placed the experimental assembly inside a desk drawer for a few days while waiting for the weather to clear. On resuming the experiment, Becquerel decided to replace the original photographic film, expecting that it may have become slightly exposed. He developed the original film, however, and found that instead of the expected feeble image, he got a very sharp one. The film had become strongly exposed. The uranium-containing material had emitted radiation continuously, even when it was not fluorescing. Becquerel had discovered **radioactivity**.

Ernest Rutherford (1871–1937) identified two types of radiation from radioactive materials, alpha (α) and beta (β). **Alpha particles** carry two fundamental units of positive charge and have essentially the same mass as helium atoms. Alpha particles are identical to He^{2+} ions. **Beta particles** are negatively charged particles produced by changes occurring within the nuclei of radioactive atoms and have the same properties as electrons. A third form of radiation that is not affected by an electric field was discovered in 1900 by Paul Villard. This radiation, called **gamma rays** (γ), is not made up of particles; it is electromagnetic radiation of extremely high penetrating power. These three forms of radioactivity are illustrated in Figure 2-10.

By the early 1900s, additional radioactive elements were discovered, principally by Marie and Pierre Curie. Rutherford and Frederick Soddy made another profound finding: The chemical properties of a radioactive element *change* as it undergoes radioactive decay. This observation suggests that radioactivity involves fundamental changes at the *subatomic* level—in radioactive decay, one element is changed into another, a process known as *transmutation*.

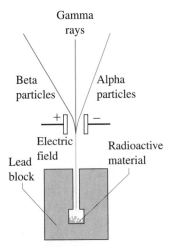

▲ FIGURE 2-10
Three types of radiation from radioactive materials
The radioactive material is enclosed in a lead block. All the radiation except that passing through the narrow opening is absorbed by the lead. When the escaping radiation is passed through an electric field, it splits into three beams. One beam is undeflected—these are gamma (γ) rays. A second beam is attracted to the negatively charged plate. These are the positively charged alpha (α) particles. The third beam, of negatively charged beta (β) particles, is deflected toward the positive plate.

▶ Perhaps because he found it tedious to sit in the dark and count spots of light on a zinc sulfide screen, Geiger was motivated to develop an automatic radiation detector. The result was the well-known Geiger counter.

Rutherford Experiment animation

2-3   The Nuclear Atom

In 1909, Rutherford, with his assistant Hans Geiger, began a line of research using α particles as probes to study the inner structure of atoms. Based on Thomson's plum pudding model, Rutherford expected that a beam of α particles would pass through thin sections of matter largely undeflected, but that some α particles would be slightly scattered or deflected as they encountered electrons. By studying these scattering patterns, he hoped to deduce something about the distribution of electrons in atoms.

The apparatus used for these studies is pictured in Figure 2-11. Alpha particles were detected by the flashes of light they produced when they struck a zinc sulfide screen mounted on the end of a telescope. When Geiger and Ernst Marsden, a student, bombarded very thin foils of gold with α particles, they observed that

- The majority of α particles penetrated the foil undeflected.
- Some α particles experienced slight deflections.
- A few (about one in every 20,000) suffered rather serious deflections as they penetrated the foil.
- A similar number did not pass through the foil at all, but bounced back in the direction from which they had come.

▶ FIGURE 2-11
The scattering of α particles by metal foil
The telescope travels in a circular track around an evacuated chamber containing the metal foil. Most α-particles pass through the metal foil undeflected, but some are deflected through large angles.

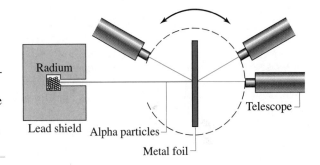

The large-angle scattering greatly puzzled Rutherford. As he commented some years later, this observation was "about as credible as if you had fired a 15-inch shell at a piece of tissue paper and it came back and hit you." By 1911, though, Rutherford had an explanation. He based his explanation on a model of the atom known as the *nuclear atom* and having these features.

1. Most of the mass and all of the positive charge of an atom are centered in a very small region called the *nucleus. The atom is mostly empty space.*
2. The magnitude of the positive charge is different for different atoms and is approximately one-half the atomic weight of the element.
3. There are as many electrons outside the nucleus as there are units of positive charge on the nucleus. The atom as a whole is electrically neutral.

Rutherford's initial expectation and his explanation of the α particle experiments are described in Figure 2-12.

Protons and Neutrons

Rutherford's nuclear atom suggested the existence of positively charged fundamental particles of matter in the nuclei of atoms. Rutherford himself discovered these particles, called **protons**, in 1919 in studies involving the scattering of α particles by nitrogen atoms in air. The protons were freed as a result of collisions

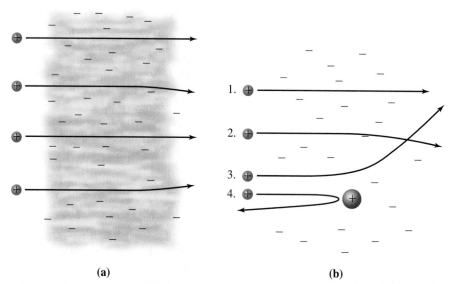

(a)   (b)

▲ FIGURE 2-12   **Explaining the results of α-particle scattering experiments**
(a) Rutherford's expectation was that small, positively charged α-particles should pass
through the nebulous, positively charged cloud of the Thomson atom largely undeflected.
Some would be slightly deflected by passing near electrons (present to neutralize the positive charge of the cloud). (b) Rutherford's explanation was based on a nuclear atom.
With an atomic model having a small, dense, positively charged nucleus and extranuclear
electrons, one would expect the four different types of paths actually observed.
1. undeflected straight-line paths exhibited by most of the α particles
2. slight deflections of α particles passing close to electrons
3. severe deflections of α particles passing close to a nucleus
4. reflections from the foil of α particles approaching a nucleus head-on

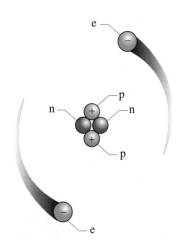

▲ FIGURE 2-13
**The nuclear atom—illustrated
by the helium atom**
In this drawing, electrons are
shown much closer to the nucleus than is the case. The actual
situation is more like this: If the
entire atom were represented by
a room, 5 m × 5 m × 5 m, the
nucleus would occupy only
about as much space as the period at the end of this sentence.

between α particles and the nuclei of nitrogen atoms. At about this same time,
Rutherford predicted the existence in the nucleus of electrically neutral fundamental
particles. In 1932, James Chadwick showed that a newly discovered penetrating
radiation consisted of beams of *neutral* particles. These particles, called **neutrons**,
originated from the nuclei of atoms. Thus, it has been only for about the past 70
years that we have had the atomic model suggested by Figure 2-13.

Fundamental Particles: A Summary

Table 2.1 presents the charges and masses of protons, neutrons, and electrons in two
ways. An electron carries an atomic unit of negative charge. A proton carries an
atomic unit of positive charge. The **atomic mass unit** (described more fully on
page 45) is defined as exactly 1/12 of the mass of the atom known as carbon-12 (carbon twelve). An atomic mass unit is abbreviated as amu and denoted by the symbol u. As we see from Table 2.1, the proton and neutron masses are just slightly

TABLE 2.1 **Properties of Three Fundamental Particles**				
	Electric Charge		**Mass**	
	SI (C)	Atomic	SI (g)	Atomic (u)[a]
Proton	$+1.602 \times 10^{-19}$	$+1$	1.673×10^{-24}	1.0073
Neutron	0	0	1.675×10^{-24}	1.0087
Electron	-1.602×10^{-19}	-1	9.109×10^{-28}	0.0005486

[a] u is the SI symbol for atomic mass unit (abbreviated as amu).

greater than 1 u. By comparison, the mass of an electron is only about 1/2000th the mass of the proton or neutron.

The number of protons in an atom is called the **atomic number**, or the **proton number**, Z. In an atom, which must be electrically neutral, the number of electrons is also equal to Z. The total number of protons and neutrons in an atom is called the **mass number**, A. The number of neutrons, the **neutron number**, is $A - Z$.

The three subatomic particles considered in this section are the only ones involved in the phenomena of interest to us in this text. You should be aware, however, that a study of matter at its most fundamental level must consider many additional subatomic particles. This is the subject of elementary particle physics.

2-4 Chemical Elements

Now that we have acquired some fundamental ideas about atomic structure, we can more thoroughly discuss the concept of chemical elements.

All atoms of a particular element have the same atomic number, Z, conversely, all atoms with the same number of protons are atoms of the same element. The 115 elements presently known include every atomic number from $Z = 1$ to 112, and $Z = 114$, 116, and 118. Each element has a name and a distinctive symbol. **Chemical symbols** are one- or two-letter abbreviations of the name (usually the English name). The first (but never the second) letter of the symbol is capitalized; for example: carbon, C; oxygen, O; neon, Ne; and silicon, Si. Some elements known since ancient times have symbols based on their Latin names, such as Fe for iron (*ferrum*) and Pb for lead (*plumbum*). The element sodium has the symbol Na, based on the Latin *natrium* for sodium carbonate. Potassium has the symbol K, based on the Latin *kalium* for potassium carbonate. The symbol for tungsten, W, is based on the German *wolfram*.

Elements beyond uranium ($Z = 92$) do not occur naturally and must be synthesized in particle accelerators (described in Chapter 26). Elements of the very highest atomic numbers have been produced only on a limited number of occasions, a few atoms at a time. Inevitably, controversies have arisen about which research team discovered a new element and, in fact, whether a discovery was made at all. However, international agreement has been reached on the first 109 elements, each of which now has an official name and symbol.

Isotopes

To represent the composition of any particular atom, we need to specify its number of protons (p), neutrons (n), and electrons (e), We can do this with the symbolism

$$\text{number p + number n} \searrow \atop \text{number p} \searrow \quad {}_Z^A\text{E} \longleftarrow \text{symbol of element} \qquad (2.1)$$

This symbolism indicates that the atom is of the element E and it has an atomic number Z and a mass number A. For example, an atom of aluminum represented as ${}_{13}^{27}\text{Al}$ has 13 protons and 14 neutrons in its nucleus and 13 electrons outside the nucleus. (Recall that an atom has the same number of electrons as protons.)

Contrary to what Dalton thought, we now know that atoms of an element do not necessarily all have the same mass. In 1912, J.J. Thomson measured the mass-to-charge ratios of positive ions formed in neon gas. He found that about 91% of the atoms had one mass and that the remaining atoms were about 10% heavier. All neon atoms have 10 protons in their nuclei, and most have 10 neutrons as well. A

Isotopes activity

▶ Because neon is the only element with $Z = 10$, the symbols ^{20}Ne, ^{21}Ne, and ^{22}Ne convey the same meaning as $^{20}_{10}$Ne, $^{21}_{10}$Ne, and $^{22}_{10}$Ne.

very few neon atoms, however, have 11 neutrons and some have 12. We can represent these three different types of neon atoms as

$$^{20}_{10}\text{Ne} \qquad ^{21}_{10}\text{Ne} \qquad ^{22}_{10}\text{Ne}$$

Atoms that have the *same* atomic number (Z) but *different* mass numbers (A) are called **isotopes**. Of all Ne atoms on Earth, 90.51% are $^{20}_{10}$Ne. The percentages of $^{21}_{10}$Ne and $^{22}_{10}$Ne are 0.27% and 9.22%, respectively. These percentages—90.51%, 0.27%, 9.22%—are the **percent natural abundances** of the three neon isotopes. Sometimes the mass numbers of isotopes are incorporated into the names of elements, such as neon-20 (read as neon twenty). Percent natural abundances are always based on *numbers*, not masses. Thus, 9051 of every 10,000 neon atoms are neon-20 atoms. Some elements, as they exist in nature, consist of just a single type of atom and therefore do not have naturally occurring isotopes.* Aluminum, for example, consists only of aluminum-27 atoms.

Ions

When an atom loses or gains electrons, the species formed is called an **ion** and carries a net charge. Because an electron is negatively charged, adding electrons to an electrically neutral atom produces a negatively charged ion. Removing electrons results in a positively charged ion. The number of protons does not change when an atom becomes an ion. For example, ^{20}Ne$^+$ and ^{22}Ne^{2+} are ions. The first one has 10 protons, 10 neutrons, and 9 electrons. The second one also has 10 protons, but 12 neutrons and 8 electrons. The charge on an ion is equal to the number of protons *minus* the number of electrons. That is

$$\text{number p} + \text{number n} \longrightarrow {}^{A}_{Z}\text{E}^{\pm?} \longleftarrow \text{number p} - \text{number e} \qquad (2.2)$$
$$\text{number p} \longrightarrow$$

Another example is the ^{16}O^{2-} ion. In this ion, there are 8 protons (atomic number 8), 8 neutrons (mass number − atomic number), and *10* electrons ($8 - 10 = -2$).

EXAMPLE 2-3

Relating the Numbers of Protons, Neutrons, and Electrons in Atoms and Ions. **(a)** Indicate the number of protons, neutrons, and electrons in $^{35}_{17}$Cl. **(b)** Write an appropriate symbol for the species consisting of 29 protons, 34 neutrons, and 27 electrons.

Solution

In using the symbolism $^{A}_{Z}$E, pay particular attention to whether the species is a neutral atom or an ion. If it is a neutral atom, number e = number p = Z (the atomic number). If the species is an ion, determine whether the number of electrons is smaller (positive ion) or larger (negative ion) than the number of protons. Whether the species is an atom or ion, the number of neutrons is equal to $A - Z$.

(a) $^{35}_{17}$Cl: $Z = 17$, $A = 35$, and no charge indicated: a neutral atom.

number p = 17   number e = 17   number n = $A - Z = 35 - 17 = 18$

---

*Nuclide is the general term used to describe an atom with a particular atomic number and mass number. Although there are several elements with only one naturally occurring nuclide, it is possible to produce additional nuclides of these elements—isotopes—by artificial means (Section 26-3). The artificial isotopes are radioactive, however. In all, the number of synthetic isotopes exceeds the number of naturally occurring ones by several fold.

(b) The element with $Z = 29$ is copper (symbol: Cu, see inside front cover). The mass number A = number p + number n = 29 + 34 = 63. Because the species has only 27 electrons, it must be an ion with a net charge = number p − number e = 29 − 27 = +2. It should be represented as $^{63}_{29}Cu^{2+}$.

Practice Example A: Write an appropriate symbol for the species with 47 protons, 61 neutrons, and 47 electrons.

Practice Example B: Determine the numbers of protons, neutrons, and electrons in an ion of sulfur-35 that carries the charge 2−.

Isotopic Masses

▶ Ordinarily we expect like-charged objects (such as protons) to repel each other. The forces holding protons and neutrons together in the nucleus are very much stronger than ordinary electrical forces (Section 26-6).

We cannot determine the mass of an individual atom just by adding up the masses of its fundamental particles. When protons and neutrons combine to form a nucleus, a very small portion of their original mass is converted to energy and released. However, we cannot predict exactly how much this so-called nuclear binding energy will be. Determining the masses of individual atoms, then, is something that must be done *by experiment*, in the following way:

We *arbitrarily* choose one atom and assign it a certain mass. By international agreement, this standard is an atom of the isotope carbon-12, which is assigned a mass of *exactly* 12 atomic mass units, that is, *12 u*. Next, we determine the masses of other atoms relative to carbon-12. To do this, we use a **mass spectrometer**. In this device, as a beam of gaseous ions passes through electric and magnetic fields, it separates into components of differing mass. The separated ions are focused on a measuring instrument, which records their presence and amounts. Figure 2-14 illustrates mass spectrometry. A typical mass spectrum is pictured in Figure 2-15.

▶ This definition also establishes that one atomic mass unit (1 u) is *exactly* 1/12 the mass of a carbon-12 atom.

 Mass Spectrometer simulation

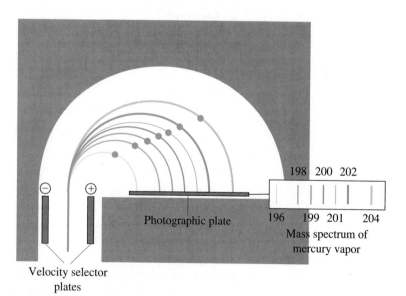

▲ FIGURE 2-14 **A mass spectrometer**
A gaseous sample is ionized by bombardment with electrons in the lower part of the apparatus (not shown). The positive ions thus formed are subjected to an electrical force by the electrically charged velocity selector plates and a magnetic force by a magnetic field perpendicular to the page. Only ions with a particular velocity pass through and are deflected into circular paths by the magnetic field. Ions with different masses strike the detector (here a photographic plate) in different regions. The more ions of a given type, the greater the response of the detector (intensity of line on the photographic plate).

▶ **FIGURE 2-15**

Mass spectrum for mercury

The response of the ion detector in Figure 2-14 (intensity of lines on photographic plate) has been converted to a scale of relative numbers of atoms. The percent natural abundances of the mercury isotopes are ^{196}Hg, 0.146%; ^{198}Hg, 10.02%; ^{199}Hg, 16.84%; ^{200}Hg, 23.13%; ^{201}Hg, 13.22%; ^{202}Hg, 29.80%; ^{204}Hg, 6.85%.

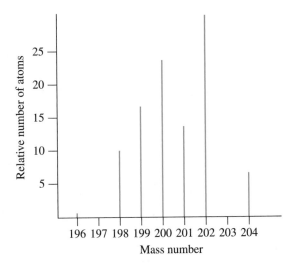

Although mass numbers are whole numbers, the *actual* masses of individual atoms (in atomic mass units, u) are never *exact* whole numbers, except for carbon-12. However, they are very close in value to the corresponding mass numbers. This means, for example, that we should expect the mass of oxygen-16 to be very nearly 16 u. In Example 2-4, we see that it is.

EXAMPLE 2-4

Establishing Isotopic Masses by Mass Spectrometry. With mass spectral data, the ratio of the mass of ^{16}O to ^{12}C is found to be 1.33291. What is the mass of an ^{16}O atom?

Solution

The ratio of the masses is $^{16}O/^{12}C$ = 1.33291. The mass of the ^{16}O atom is 1.33291 times the mass of ^{12}C.

$$\text{mass of } ^{16}O = 1.33291 \times 12 \text{ u} = 15.9949 \text{ u}$$

Practice Example A: By mass spectrometry, an atom of ^{16}O is found to have 1.06632 times the mass of ^{15}N. What is the mass of an ^{15}N atom, expressed in atomic mass units?

Practice Example B: What is the ratio of masses for $^{202}Hg/^{12}C$, if the isotopic mass for ^{202}Hg is 201.970617 u?

2-5 Atomic Masses

In a table of atomic masses, the value listed for carbon is 12.011, yet the atomic mass standard is *exactly* 12. Why the difference? The atomic mass standard is based on a sample of carbon containing *only* atoms of carbon-12, whereas *naturally occurring* carbon contains some carbon-13 atoms as well. The existence of these two isotopes causes the *observed* atomic mass to be greater than 12. The **atomic mass (weight)**[*] of an element is the average of the isotopic masses, *weighted* according to the naturally occurring abundances of the isotopes of the element. In a weighted average, we must assign greater importance—give greater weight—to the quantity that occurs more frequently. Since carbon-12 atoms are much more abundant than carbon-13, the weighted average must lie much closer to 12 than to 13. This

▶ Carbon-14, used for radiocarbon dating, is formed in the upper atmosphere. The amount of carbon-14 on Earth is too small to affect the atomic mass of carbon.

[*] Since Dalton's time, atomic masses have been called atomic weights. They still are by most chemists, yet what we are describing here is mass, not weight. Old habits die hard.

is the result that we get by applying the following general equation, where the right-hand side of the equation includes one term for each naturally occurring isotope.

KEEP IN MIND ▶
that the fractional abundance is
the percent abundance
divided by 100 percent.

$$\begin{pmatrix} \text{at. mass} \\ \text{of an} \\ \text{element} \end{pmatrix} = \begin{pmatrix} \text{fractional} & \text{mass of} \\ \text{abundance of} \times \text{isotope} \\ \text{isotope (1)} & \text{(1)} \end{pmatrix} + \begin{pmatrix} \text{fractional} & \text{mass of} \\ \text{abundance of} \times \text{isotope} \\ \text{isotope (2)} & \text{(2)} \end{pmatrix} + \dots \quad (2.3)$$

The mass spectrum of carbon shows that 98.892% of carbon atoms are carbon-12 with a mass of 12 u, *exactly*, and 1.108% are carbon-13 with a mass of 13.00335 u. In the setup below, we calculate the contribution of each isotope to the weighted average separately and then add these contributions together, as required by equation (2.3).

$$\begin{aligned} \frac{\text{contribution to}}{\text{at. mass by } ^{12}\text{C}} &= \frac{\text{fraction of carbon}}{\text{atoms that are } ^{12}\text{C}} \times \text{mass } ^{12}\text{C atom} \\ &= 0.98892 \qquad\qquad \times 12 \text{ u} \qquad\qquad = 11.867 \text{ u} \end{aligned}$$

$$\begin{aligned} \frac{\text{contribution to}}{\text{at. mass by } ^{13}\text{C}} &= \frac{\text{fraction of carbon}}{\text{atoms that are } ^{13}\text{C}} \times \text{mass } ^{13}\text{C atom} \\ &= 0.01108 \qquad\qquad \times 13.00335 \text{ u} \qquad = 0.1441 \text{ u} \end{aligned}$$

$$\begin{aligned} \text{at. mass of naturally occurring carbon} &= (\text{contribution by } ^{12}\text{C}) + (\text{contribution by } ^{13}\text{C}) \\ &= 11.867 \text{ u} \qquad\qquad + 0.1441 \text{ u} \\ &= 12.011 \text{ u} \end{aligned}$$

To determine the atomic mass of an element having three naturally occurring isotopes, such as potassium, we would have to combine three contributions to the weighted average, and so on.

The percent natural abundances of the elements remain remarkably constant from one sample of matter to another. For example, the proportions of ^{12}C and ^{13}C atoms are the same in samples of pure carbon (diamond), carbon dioxide gas, and a mineral form of calcium carbonate (calcite). We can treat all natural carbon-containing materials as if there were a single *hypothetical* type of carbon atom with a mass of 12.011 u. This means that once weighted-average atomic masses have been determined and tabulated,[*] we can simply use these values in calculations requiring atomic masses.

Sometimes a qualitative understanding of the relationship between isotopic masses, percent natural abundances, and weighted-average atomic mass is all that we need, and no calculation is necessary, as illustrated in Example 2-5. The Practice Examples accompanying Example 2-5 provide additional applications of Equation (2.3).

EXAMPLE 2-5

Understanding the Meaning of a Weighted-Average Atomic Mass. The two naturally occurring isotopes of lithium, lithium-6 and lithium-7, have masses of 6.01513 and 7.01601 u, respectively. Which of these two occurs in greater abundance?

Solution

From a table of atomic masses (inside the front cover) we see that the atomic mass of lithium is 6.941 u. Because this value—a weighted-average atomic mass—is much closer to 7.01601 than to 6.01513, lithium-7 must be the more abundant isotope.

---

[*]Atomic mass (atomic weight) values in tables are often written without units, especially if they are referred to as *relative* atomic masses. This simply means that the values listed are in relation to *exactly* 12 (rather than 12 u) for carbon-12. We will use the atomic mass unit (u) when referring to atomic masses (atomic weights). Most chemists do.

Practice Example A: The two naturally occurring isotopes of boron, boron-10 and boron-11, have masses of 10.012937 u and 11.009305 u, respectively. Which of these two occurs in greater abundance?

Practice Example B: Use the weighted-average atomic mass of 6.941 u in Equation (2.3) to determine the percent natural abundances of lithium-6 and lithium-7.

KEEP IN MIND ▶
that the sum of the fractional abundances must equal one.

Are You Wondering ...

Why some atomic masses (for example, F = 18.9984032 u) are stated so much more precisely than others?

There is *one* naturally occurring type of fluorine atom: fluorine-19. Determining the atomic mass of fluorine means establishing the mass of this type of atom as precisely as possible. On the other hand, krypton, with an atomic mass of 83.80 u, has *six* naturally occurring isotopes. Because the percent distribution of the isotopes of krypton may differ very slightly from one sample to another, the weighted-average atomic mass of krypton cannot be stated with high precision.

2-6   Introduction to the Periodic Table

Interactive Periodic Table activity

Scientists spend a lot of time organizing information into useful patterns. Before they can organize information, however, they must possess it, and it must be correct. Botanists had enough information about plants to organize their field in the eighteenth century. Because of uncertainties in atomic masses and because many elements remained undiscovered, chemists were not able to organize the elements until a century later.

We can distinguish one element from all others by its particular set of observable physical properties. For example, sodium has a low density of 0.971 g/cm³ and a low melting point of 97.81°C. No other element has this same combination of density and melting point. Potassium, though, also has a low density (0.862 g/cm³) and low melting point (63.65°C), much like sodium. Sodium and potassium further resemble each other in that both are good conductors of heat and electricity, and both react vigorously with water to liberate hydrogen gas. Gold, on the other hand, has a density (19.32 g/cm³) and melting point (1064°C) that are very much higher than those of sodium or potassium, and gold does not react with water or even with ordinary acids. It does resemble sodium and potassium in its ability to conduct heat and electricity, however. Chlorine is very different still from sodium, potassium, and gold. It is a gas under ordinary conditions, which means that the melting point of solid chlorine is far below room temperature (−101°C). Also, chlorine is a nonconductor of heat and electricity.

Even from these very limited data, we get an inkling of a useful classification scheme of the elements. If the scheme is to group together elements with similar properties, then sodium and potassium should appear in the same group. And if the classification scheme is in some way to distinguish between elements that are good conductors of heat and electricity and those that are not, chlorine should be set apart from sodium, potassium, and gold. The classification system we seek is the one shown in Figure 2-16 (and inside the front cover), known as the **periodic table** of the elements. In a later chapter, we will describe how the periodic table was formulated, and we will also learn its theoretical basis. For the present, we will consider only a few features of the table.

1 1A	2 2A	3 3B	4 4B	5 5B	6 6B	7 7B	8	9 8B	10	11 1B	12 2B	13 3A	14 4A	15 5A	16 6A	17 7A	18 8A
1 H 1.00794																	2 He 4.00260
3 Li 6.941	4 Be 9.01218											5 B 10.811	6 C 12.011	7 N 14.0067	8 O 15.9994	9 F 18.9984	10 Ne 20.1797
11 Na 22.9898	12 Mg 24.3050											13 Al 26.9815	14 Si 28.0855	15 P 30.9738	16 S 32.066	17 Cl 35.4527	18 Ar 39.948
19 K 39.0983	20 Ca 40.078	21 Sc 44.9559	22 Ti 47.88	23 V 50.9415	24 Cr 51.9961	25 Mn 54.9381	26 Fe 55.847	27 Co 58.9332	28 Ni 58.693	29 Cu 63.546	30 Zn 65.39	31 Ga 69.723	32 Ge 72.61	33 As 74.9216	34 Se 78.96	35 Br 79.904	36 Kr 83.80
37 Rb 85.4678	38 Sr 87.62	39 Y 88.9059	40 Zr 91.224	41 Nb 92.9064	42 Mo 95.94	43 Tc (98)	44 Ru 101.07	45 Rh 102.906	46 Pd 106.42	47 Ag 107.868	48 Cd 112.411	49 In 114.818	50 Sn 118.710	51 Sb 121.757	52 Te 127.60	53 I 126.904	54 Xe 131.29
55 Cs 132.905	56 Ba 137.327	57 *La 138.906	72 Hf 178.49	73 Ta 180.948	74 W 183.84	75 Re 186.207	76 Os 190.23	77 Ir 192.22	78 Pt 195.08	79 Au 196.967	80 Hg 200.59	81 Tl 204.383	82 Pb 207.2	83 Bi 208.980	84 Po (209)	85 At (210)	86 Rn (222)
87 Fr (223)	88 Ra 226.025	89 †Ac 227.028	104 Rf (261)	105 Db (262)	106 Sg (263)	107 Bh (262)	108 Hs (265)	109 Mt (266)	110 (269)	111 (272)	112 (272)	114 (287)		116 (289)		118 (293)	

	58 Ce 140.115	59 Pr 140.908	60 Nd 144.24	61 Pm (145)	62 Sm 150.36	63 Eu 151.965	64 Gd 157.25	65 Tb 158.925	66 Dy 162.50	67 Ho 164.930	68 Er 167.26	69 Tm 168.934	70 Yb 173.04	71 Lu 174.967
*Lanthanide series														
†Actinide series	90 Th 232.038	91 Pa 231.036	92 U 238.029	93 Np 237.048	94 Pu (244)	95 Am (243)	96 Cm (247)	97 Bk (247)	98 Cf (251)	99 Es (252)	100 Fm (257)	101 Md (258)	102 No (259)	103 Lr (260)

▲ **FIGURE 2-16** **Periodic table of the elements**
Atomic masses are relative to carbon-12. For certain radioactive elements, the numbers listed in parentheses are the mass numbers of the most stable isotopes. Metals are shown in orange, nonmetals in blue, and metalloids in green. The noble gases (also nonmetals) are shown in purple.

Features of the Periodic Table

In the periodic table, elements are listed according to increasing atomic number starting at the upper left and arranged in a series of horizontal rows. This arrangement places similar elements in vertical **groups**, or **families**. For example, sodium and potassium are found together in a group labeled 1 (called the alkali metals). We should expect other members of the group, such as cesium and rubidium, to have properties similar to sodium and potassium. Chlorine is found at the other end of the table in a group labeled 17. Some of the groups are given distinctive names, mostly related to an important property of the elements in the group. For example, the group 17 elements are called the halogens, a term derived from Greek, meaning "salt former".

Each element is listed by placing its symbol in the middle of a box in the table. The atomic number (Z) of the element is shown above the symbol, and the weighted-average atomic mass of the element is shown below its symbol. Some periodic tables provide other information, such as density and melting point, but the atomic number and atomic mass are generally sufficient for our needs. Elements with atomic masses in parentheses, such as plutonium, Pu (244), are produced synthetically, and the number shown is the mass number of the most stable isotope.

It is customary also to divide the elements into two broad categories known as **metals** and **nonmetals**. In Figure 2-16, colored backgrounds are used to distinguish the metals (tan) from the nonmetals (blue and pink). Except for mercury, a liquid, metals are solids at room temperature. They are generally malleable (capable of being flattened into thin sheets), ductile (capable of being drawn into fine wires), good conductors of heat and electricity, and have a lustrous or shiny appearance. Nonmetals generally have the opposite properties of metals; for example, they are poor conductors of heat and electricity. Several of the nonmetals, such as nitrogen, oxygen, and chlorine, are gases at room temperature. Some, such as silicon and sulfur, are brittle solids. One—bromine—is a liquid. We will have more to say about metals and nonmetals later in the text.

 *Categories of Elements activity*

Two other highlighted categories in Figure 2-16 are a special group of nonmetals known as the *noble gases* (pink), and a small group of elements, often called *metalloids* (green), that have some metallic and some nonmetallic properties.

The *horizontal* rows of the table are called **periods**. (The periods are numbered at the extreme left in the periodic table inside the front cover.) The first period of the table consists of just two elements, hydrogen and helium. This is followed by two periods of eight elements each, lithium through neon and sodium through argon. The fourth and fifth periods contain 18 elements each, ranging from potassium through krypton and from rubidium through xenon. The sixth period is a long one of 32 members. To fit this period to a table that is held to a maximum width of 18 members, we extract 14 members of the period and place them at the bottom of the table. This series of 14 elements follows lanthanum ($Z = 57$), and these elements are called the **lanthanides**. The seventh and final period is incomplete (some members are yet to be discovered), but it is known to be a long one. A 14-member series is also extracted from the seventh period and placed at the bottom of the table. Because the elements in this series follow actinium ($Z = 89$), they are called the **actinides**.

The labeling of the groups has been a matter of some debate among chemists. The numbering system used in the previous paragraph is the most recently adopted one. Group labels previously used in the United States consisted of a letter and a number, closely following the method adopted by Mendeleev, the discoverer of the table. As seen in Figure 2-16, the A groups 1 and 2 are separated from the remaining A groups (3 to 8) by B groups 1 through 8. The International Union of Pure and Applied Chemistry (IUPAC) recommended the simple 1–18 numbering scheme in order to avoid confusion between the American number and letter system and that used in Europe, where some of the A and B designations were switched! Currently, the IUPAC system is officially recommended by the American Chemical Society (ACS) and chemical societies in other nations. Because both numbering systems are in use, we show both in Figure 2-16 and in the periodic table inside the front cover. However, except for an occasional reminder of the earlier system, we will use the IUPAC numbering system in this text.

Useful Relationships from the Periodic Table

The periodic table helps chemists describe and predict the properties of chemical compounds and the outcomes of chemical reactions. Throughout this text, we will use it as an aid to understanding chemical concepts. One application of the table worth mentioning here is how it can be used to predict likely charges on simple ions.

Main-group elements are those in groups 1, 2, and 13–18. When main-group metal atoms in groups 1 and 2 form ions, they lose the same number of electrons as the IUPAC group number. Thus, Na atoms (group 1) lose one electron to become Na^+, and Ca atoms (group 2) lose two electrons to become Ca^{2+}. Aluminum, in group 13, loses three electrons to form Al^{3+} (here the charge is "group number minus 10"). The few other metals in groups 13 and higher form more than one possible ion, a matter that we deal with in Chapter 10.

When nonmetal atoms form ions, they gain electrons. The number of electrons gained is normally 18 minus the IUPAC group number. Thus, an O atom gains $18 - 16 = 2$ electrons to become O^{2-}, and a Cl atom, gains $18 - 17 = 1$ electron to become Cl^-. The "18 minus group number" rule suggests that an atom of Ne in group 18 gains no electrons: $18 - 18 = 0$. The very limited tendency of the noble gas atoms to form ions is one of several characteristics of this family of elements that we will consider from time to time.

The elements in groups 3–12 are the **transition elements**, and because all of them are metals, they are also called the **transition metals**. Like the main-group metals, the transition metals form positive ions, but the number of electrons lost is not related in any simple way to the group number, mostly because transition metals can form two or more ions of differing charge. We will discuss transition metal ions in more detail in several later chapters.

EXAMPLE 2-6

Describing Relationships Based on the Periodic Table. Refer to the periodic table on the inside front cover, and indicate

(a) the element that is in group 14 and the fourth period

(b) two elements with properties similar to those of molybdenum (Mo)

(c) the ion most likely formed from a strontium atom

Solution

(a) The elements in the fourth period range from K ($Z = 19$) to Kr ($Z = 36$). Those in group 14 are C, Si, Ge, Sn, and Pb. The only element that is common to both of these groupings is Ge ($Z = 32$).

(b) Molybdenum is in group 6. Two other members of this group that should resemble it are chromium (Cr) and tungsten (W).

(c) Strontium (Sr) is in group 2. It should form the ion Sr^{2+}.

Practice Example A: Write a symbol for the ion most likely formed by an atom of each of the following: Li, S, Ra, F, I, and Al.

Practice Example B: Classify each of the following elements as a main-group or transition element. Also, specify whether they are metals, metalloids, or nonmetals: Na, Re, S, I, Kr, Mg, U, Si, B, Al, As, H.

2-7 The Concept of the Mole and the Avogadro Constant

Our most frequent use of the periodic table will undoubtedly be for its listing of atomic masses. As we will learn in subsequent chapters, atomic masses are essential to determining the compositions of chemical compounds and the quantities of substances produced in chemical reactions. In this section, though, we introduce another concept that is as fundamental as atomic mass.

Starting with Dalton, chemists have recognized the importance of relative numbers of atoms, as in the statement that *two* hydrogen atoms and *one* oxygen atom combine to form *one* molecule of water. Yet we cannot physically count the atoms in a given sample in the usual sense. We must resort to some other measurement, usually mass. This means that we need a relationship between a *measured* mass of an element and some *known*, but *uncountable*, number of atoms. Consider a practical example of mass substituting for a desired number of items: If you want to nail down new floorboards on the deck of a mountain cabin, you need a certain number of nails. However, if you have some idea of how much the nails weigh you can buy them by the pound.

The SI quantity that describes an amount of substance by relating it to a number of particles of that substance is called the *mole* (abbreviated *mol*). A **mole** is an amount of substance that contains the same number of elementary entities as there are carbon-12 atoms in *exactly* 12 g of carbon-12. The "number of elementary entities (atoms, molecules, ...)" in a mole is the **Avogadro constant**, N_A.

$$N_A = 6.02214199 \times 10^{23} \text{ mol}^{-1}$$

▶ The number $6.02214199 \times 10^{23}$ is known as Avogadro's *number*. The Avogadro *constant* is the number and unit (mol^{-1}) together.

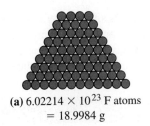

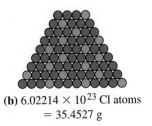

(a) 6.02214×10^{23} F atoms
$= 18.9984$ g

(b) 6.02214×10^{23} Cl atoms
$= 35.4527$ g

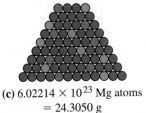

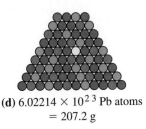

(c) 6.02214×10^{23} Mg atoms
$= 24.3050$ g

(d) 6.02214×10^{23} Pb atoms
$= 207.2$ g

▲ **FIGURE 2-17** **Distribution of isotopes in four elements**
(a) There is only one type of fluorine atom, ^{19}F (shown in red). **(b)** In chlorine, 75.77% of the atoms are ^{35}Cl (red) and the remainder are ^{37}Cl (blue). **(c)** Magnesium has one principal isotope, ^{24}Mg (red), and two minor ones, ^{25}Mg (gray) and ^{26}Mg (blue). **(d)** Lead has four naturally occurring isotopes: 1.4% ^{204}Pb (yellow), 24.1% ^{206}Pb (blue), 22.1% ^{207}Pb (gray), and 52.4% ^{208}Pb (red).

Often we will round of the value of N_A to 6.022×10^{23} mol^{-1}, or even to 6.02×10^{23} mol^{-1}. The unit mol^{-1} signifies that the entities being counted are those present in one mole.

If a substance contains atoms of a single isotope, we may write

$$1 \text{ mol } ^{12}\text{C} = 6.02214 \times 10^{23} \, {}^{12}\text{C atoms} = 12.0000 \text{ g}$$

$$1 \text{ mol } ^{16}\text{O} = 6.02214 \times 10^{23} \, {}^{16}\text{O atoms} = 15.9949 \text{ g and so on.}$$

▶ We established the atomic mass of ^{16}O relative to that of ^{12}C in Example 2-4.

Most elements are composed of mixtures of two or more isotopes. The atoms to be "counted out" to yield one mole are not all of the same mass. They must be taken in their naturally occurring proportions. Thus, in one mole of carbon, most of the atoms are carbon-12, but some are carbon-13. In one mole of oxygen, most of the atoms are oxygen-16, but some are oxygen-17 and some are oxygen-18. As a result,

$$1 \text{ mol of C} = 6.02214 \times 10^{23} \text{ C atoms} = 12.011 \text{ g}$$

$$1 \text{ mol of O} = 6.02214 \times 10^{23} \text{ O atoms} = 15.9994 \text{ g and so on.}$$

▶ We calculated the weighted-average atomic mass of carbon on page 47.

We can easily establish the mass of one mole of atoms, called the **molar mass**, M, from a table of atomic masses—for example, 6.941 g Li/mol Li. Figure 2-17 attempts to portray the distribution of isotopes of an element, and Figure 2-18 pictures one mole each of four common elements.

KEEP IN MIND ▶
that molar mass has the unit g/mol.

Thinking About Avogadro's Number

Avogadro's number $(6.022142 \times 10^{23})$ is an enormously large number and practically inconceivable in terms of ordinary experience. Suppose we were counting garden peas instead of atoms. If the typical pea had a volume of about 0.1 cm^3, the required pile of peas would cover the United States to a depth of about 6 km (4 mi). Or imagine that grains of wheat could be counted at the rate of 100 per minute. A given individual might be able to count out about 4 billion grains in a lifetime. Even so, if all the people currently on Earth were to spend their lives counting grains of wheat, they could not reach

▲ **FIGURE 2-18   One mole of an element**
The watch glasses contain one mole of copper atoms (left) and one mole of sulfur atoms (right). The beaker contains one mole of mercury atoms as liquid mercury, and the balloon contains one mole of helium atoms in the gaseous state.

Avogadro's number. In fact, if all the people who ever lived on Earth had spent their lifetimes counting grains of wheat, the total would still be far less than Avogadro's number. (And Avogadro's number of wheat grains is far more wheat than has been produced in human history.) Now consider a much more efficient counting device, a modern personal computer; it is capable of counting at a rate of about 1 billion units per second. The task of counting out Avogadro's number would still take about 20 million years!

Avogadro's number is clearly not a useful number for counting ordinary objects. On the other hand, when this inconceivably large number is used to count inconceivably small objects such as atoms and molecules, the result is quantities of materials that are easily within our grasp.

2-8   Using the Mole Concept in Calculations

Throughout the text, the mole concept will provide us with conversion factors for problem-solving situations. As we encounter each new situation, we will explore how the mole concept applies. For now, we will deal with the relationship between numbers of atoms and the mole. Consider the statement: $1 \text{ mol S} = 6.022 \times 10^{23}$ S atoms $= 32.07 \text{ g S}$. This allows us to write the conversion factors

$$\frac{1 \text{ mol S}}{6.022 \times 10^{23} \text{ S atoms}} \quad \text{and} \quad \frac{32.07 \text{ g S}}{1 \text{ mol S}}.$$

Example 2-7 is perhaps the simplest possible application of the mole concept: relating the number of atoms in a sample to the number of moles of atoms. We use both conversion factors written above in Example 2-8.

In calculations requiring the Avogadro constant, students often ask when to multiply and when to divide by N_A. One answer is to always use the constant in a way that gives you the proper cancellation of units. Another answer is to think in terms of the expected result. In calculating a number of atoms, we expect the answer to be a *very large* number, and certainly *never* smaller than 1. The number of *moles* of atoms, on the other hand, is generally a number of more modest size. It will often be smaller than 1.

EXAMPLE 2-7

Relating Number of Moles and Total Number of Atoms. A sample of iron metal is described as being 2.35 mol Fe. How many iron atoms are present in this sample?

Solution

The conversion factor we need is based on the fact that 1 mol Fe $= 6.022 \times 10^{23}$ Fe atoms.

$$? \text{ Fe atoms} = 2.35 \text{ mol Fe} \times \frac{6.022 \times 10^{23} \text{ Fe atoms}}{1 \text{ mol Fe}}$$

$$= 1.42 \times 10^{24} \text{ Fe atoms}$$

Practice Example A: How many atoms of gold are present in 5.07×10^{-3} mol Au?

Practice Example B: How many lead-206 atoms are present in 8.27×10^{-3} mol Pb? (*Hint:* What proportion of all Pb atoms are ^{206}Pb? See Figure 2-17)

KEEP IN MIND ▶
that you should always check for the proper cancellation of units when solving a problem. From this point on, we will not routinely show the cancellation of units.

EXAMPLE 2-8

Relating Number of Atoms of an Element to an Amount in Moles and a Mass in Grams.
(a) How many moles of sulfur are present in a sample containing 7.65×10^{22} S atoms?
(b) What is the mass of this sample?

Solution

(a) As in Example 2-7, we need the Avogadro constant as a conversion factor, but here the factor is *inverted* from the form used there.

$$? \text{ mol S} = 7.65 \times 10^{22} \text{ S atoms} \times \frac{1 \text{ mol S}}{6.022 \times 10^{23} \text{ S atoms}}$$

$$= 0.127 \text{ mol S}$$

(b) We can begin with the result of part (a)—0.127 mol S—and use molar mass as a conversion factor.

$$? \text{ g S} = 0.127 \text{ mol S} \times \frac{32.07 \text{ g S}}{1 \text{ mol S}} = 4.07 \text{ g S}$$

Or, we can simply combine the setups for parts (a) and (b).

$$? \text{ g S} = 7.65 \times 10^{22} \text{ S atoms} \times \frac{1 \text{ mol S}}{6.022 \times 10^{23} \text{ S atoms}} \times \frac{32.07 \text{ g S}}{1 \text{ mol S}}$$

$$= 4.07 \text{ g S}$$

Figure 2-19 shows how we can obtain this quantity of sulfur by weighing.

Practice Example A: What is the mass of 2.35×10^{24} atoms of Cu?

Practice Example B: How many He atoms are present in a 22.6-g sample of He gas?

▲ FIGURE 2-19
Measurement of 7.65×10^{22} S atoms (0.127 mol S)—Example 2-8 illustrated
The balance is set to zero (tared) when just the weighing boat is present. The sample of sulfur weighs 4.07 g.

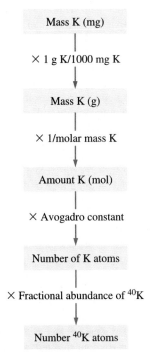

Mass K (mg)

$\times$ 1 g K/1000 mg K

Mass K (g)

$\times$ 1/molar mass K

Amount K (mol)

$\times$ Avogadro constant

Number of K atoms

$\times$ Fractional abundance of ^{40}K

Number ^{40}K atoms

▲ **FIGURE 2-20**
**Outlining a calculation—
Example 2-9 visualized.**

Are You Wondering ...

How many significant figures to carry in atomic masses or the Avogadro constant when using them in calculations?

As a general rule, to ensure the maximum precision allowable, carry at least one more significant figure in well-known physical constants than in other measured quantities. For example, in calculating the mass of 0.127 mol of sulfur, the answers 0.127 mol S $\times$ 32.066 g S/mol S = 4.07 g S or 0.127 mol S $\times$ 32.07 g S/mol S = 4.07 g S are more accurate responses than 0.127 mol S $\times$ 32.1 g S/mol S = 4.08 g S.

Example 2-9 is perhaps the most representative way in which we use the mole concept—as part of a larger problem that requires other, unrelated, conversion factors as well. One approach is to outline a conversion pathway to get from the given to the desired information. The path in Example 2-9, from mg K $\longrightarrow$ g K $\longrightarrow$ mol K $\longrightarrow$ K atoms $\longrightarrow$ ^{40}K atoms, is represented schematically in Figure 2-20. The first step requires a conversion factor between SI mass units that you should know. The second step requires molar mass as a conversion factor. You can obtain this number from a table of atomic masses. In the third step, you need Avogadro's constant, a quantity that you should remember. In the fourth step, the necessary conversion factor, the fractional abundance of ^{40}K, is related to the percent natural abundance of ^{40}K, which must be given in the statement of the problem.

▶ The radioactivity associated with milk is part of the natural background radiation to which we are all exposed, and it is not considered a matter of concern.

EXAMPLE 2-9

Combining Several Factors in a Calculation—Molar Mass, the Avogadro Constant, Percent Abundances. Potassium-40 is one of the few naturally occurring radioactive isotopes of elements of low atomic number. Its percent natural abundance among K isotopes is 0.012%. How many ^{40}K atoms do you ingest by drinking one cup of whole milk containing 371 mg K?

Solution

Following the path charted in Figure 2-20, convert the mass of K to moles of K. For this, use molar mass in the *inverse* manner to Example 2-8.

$$? \text{ mol K} = 371 \text{ mg K} \times \frac{1 \text{ g K}}{1000 \text{ mg K}} \times \frac{1 \text{ mol K}}{39.10 \text{ g K}} = 9.49 \times 10^{-3} \text{ mol K}$$

Then convert moles of K to the number of K atoms.

$$? \text{ K atoms} = 9.49 \times 10^{-3} \text{ mol K} \times \frac{6.022 \times 10^{23} \text{ K atoms}}{1 \text{ mol K}}$$

$$= 5.71 \times 10^{21} \text{ K atoms}$$

Finally, use the percent natural abundance of ^{40}K to formulate a factor to convert from the number of K atoms to the number of ^{40}K atoms.

$$? \,^{40}\text{K atoms} = 5.71 \times 10^{21} \text{ K atoms} \times \frac{0.012 \,^{40}\text{K atoms}}{100 \text{ K atoms}}$$

$$= 6.9 \times 10^{17} \,^{40}\text{K atoms}$$

▲ Iron is one of the most abundant elements in Earth's crust. Iron compounds give soils a characteristic red color, as in the Navajo Sandstone Paria Canyon Wilderness Area of Arizona.

What is the most abundant element? This seemingly simple question does not have a simple answer. If we consider the entire universe, hydrogen accounts for about 90% of all the atoms and 75% of the mass, and helium accounts for most of the rest. If we consider only the elements present on Earth, iron is probably the most abundant element. However, most of this iron is believed to be in Earth's core. The currently accessible elements are those present in Earth's atmosphere, oceans, and solid continental crust. The relative abundances of some elements and common materials containing these elements are listed in Table 2.2.

Not all the known elements exist in Earth's crust. Trace amounts of neptunium ($Z = 93$) and plutonium ($Z = 94$) are found in uranium minerals, but for practical purposes, elements with atomic numbers higher than 92 can be produced only artificially by nuclear processes. Moreover, most of the elements do not occur *free* in nature—that is, as the uncombined element. Only about 20% of them do. The remaining elements occur only in chemical combination with other elements.

We cannot assume that the ease and cost of obtaining a pure element from its natural sources are necessarily related to the relative abundance of the element. Aluminum is the most abundant of the metals in Earth's crust, but it cannot be produced as cheaply as iron. This is partly because concentrated deposits

Outlining an approach to a multistep problem, as in Figure 2-20, may also make it easier to see how to do a problem through a single setup. Thus, we can write

$$? \, ^{40}\text{K atoms} = 371 \text{ mg K} \times \frac{1 \text{ g K}}{1000 \text{ mg K}} \times \frac{1 \text{ mol K}}{39.10 \text{ g K}}$$
$$\times \frac{6.022 \times 10^{23} \text{ K atoms}}{1 \text{ mol K atoms}} \times \frac{0.012 \, ^{40}\text{K atoms}}{100 \text{ K atoms}}$$
$$= 6.9 \times 10^{17} \, ^{40}\text{K atoms}$$

Practice Example A: How many Pb atoms are present in a small piece of lead with a volume of 0.105 cm³? The density of Pb $= 11.34$ g/cm³.

Practice Example B: Rhenium-187 is a radioactive isotope that can be used to determine the age of meteorites. A 0.100-mg sample of Re contains 2.02×10^{17} atoms of ^{187}Re. What is the percent abundance of rhenium-187 in the sample?

of iron-containing compounds—iron *ores*—are more common than are ores of aluminum. Some elements whose percent abundances are quite low are nevertheless widely used because their ores are fairly common. This is the case with copper, for example, whose abundance in Earth's crust is only 0.005%. On the other hand, some elements have a fairly high abundance but no characteristic ores of their own. They are not easily obtainable, as in the case of rubidium, the twenty third most abundant element.

In later chapters, we will describe the ways in which elements are obtained from natural sources: oxygen, nitrogen, and argon from the atmosphere; hydrogen from natural gas; magnesium and bromine from seawater; sulfur from underwater deposits; sodium and chlorine from ordinary rock salt; and several metals from their ores.

TABLE 2.2 Abundances of the Elements in Earth's Crust[a]

Element	Abundance, Mass (%)	Principal Materials Containing the Element
Oxygen	49.3	Water; silica; silicates; metal oxides; the atmosphere
Silicon	25.8	Silica (sand, quartz, agate, flint); silicates (feldspar, clay, mica)
Aluminum	7.6	Silicates (clay, feldspar, mica); oxide (bauxite)
Iron	4.7	Oxide (hematite, magnetite)
Calcium	3.4	Carbonate (limestone, marble, chalk); sulfate (gypsum); fluoride (fluorite); silicates (feldspar, zeolites)
Sodium	2.7	Chloride (rock salt, ocean waters); silicates (feldspar, zeolites)
Potassium	2.4	Chloride; silicates (feldspar, mica)
Magnesium	1.9	Carbonate; chloride (seawater); sulfate (Epsom salts)
Hydrogen	0.7	Oxide (water); natural gas and petroleum; organic matter
Titanium	0.4	Oxide
Chlorine	0.2	Common salt (rock salt, ocean waters)
Phosphorus	0.1	Phosphate rock; organic matter
all others[b]	0.8	

[a] Earth's crust is taken to consist of the solid crust, terrestrial waters, and the atmosphere.
[b] This figure includes C, N, and S—all essential to life—and less abundant, although commercially important, elements, such as B, Be, Cr, Cu, F, I, Pb, Sn, and Zn.

Summary

Modern chemistry began with eighteenth-century discoveries that led to the formulation of the law of conservation of mass and the law of constant composition, followed by Dalton's atomic theory. Important nineteenth-century developments leading to an understanding of the structure of the atom occurred mostly in the field of physics, however.

Cathode-ray research led to the discovery of the electron—a fundamental particle of all matter and the basic unit of negative electric charge. The discoveries of X rays and radioactivity were a consequence of cathode-ray studies, as were the discovery of isotopes and the development of modern mass spectrometry. Studies on the scattering of α particles by thin metal foils led to the concept of the nuclear atom. A more complete description of the atomic nucleus was made possible by the later discoveries of the proton and neutron.

By assigning a mass of *exactly* 12 u to a carbon-12 atom, the masses of other atoms can be established by mass spectrometry. From the masses of the different isotopes of an element and their percent natural abundances, the weighted-average atomic mass (weight) of an element can be determined. It is these weighted averages that are listed in tables of atomic masses (weights).

The periodic table is a listing of the elements organized into horizontal rows (periods) and vertical columns (groups) and according to atomic number. Each group in the table consists of elements with similar physical and chemical properties. The periodic table can also be subdivided into broad categories of elements: metals, nonmetals, metalloids, and noble gases. The table has many uses, which are examined throughout the text. Emphasis in this chapter is on using the periodic table as a listing of atomic masses and as an aid in writing symbols for simple ions.

The Avogadro constant, $N_A = 6.02214 \times 10^{23}$ mol^{-1}, represents the number of carbon-12 atoms in *exactly* 12 g of carbon-12. More generally, it is the number of elementary entities (for example, atoms or molecules) present in one mole. The mass of one mole of atoms of an element is called its molar mass. Molar mass and the Avogadro constant are used in a variety of calculations involving the mass, amount (in moles), and number of atoms in a sample of an element.

Integrative Example

A stainless steel ball bearing has a radius of 6.35 mm and a density of 7.75 g/cm³. Iron is the principal element in steel. Carbon is a key minor element. The ball bearing consists of 0.25% carbon, by mass.

Given that the percent natural abundance of ^{13}C is 1.108%, how many ^{13}C atoms are present in the ball bearing?

1. *Determine the volume of the ball bearing in cubic centimeters.* The formula for the volume (V) of a sphere is $V = 4/3\,\pi\,r^3$. In this formula, the radius (r) must be expressed in centimeters.

$$V = \frac{4\pi}{3}\left[6.35\ \text{mm} \times \frac{1\ \text{cm}}{10\ \text{mm}}\right]^3 = 1.07\ \text{cm}^3$$

2. *Determine the mass of carbon present.* Use the density of steel and the volume of the ball bearing to calculate its mass. Then use the percent carbon in the steel to convert from mass of steel to mass of carbon.

$$?\ \text{g C} = 1.07\ \text{cm}^3 \times \frac{7.75\ \text{g steel}}{1\ \text{cm}^3\ \text{steel}} \times \frac{0.25\ \text{g C}}{100\ \text{g steel}} = 0.021\ \text{g C}$$

3. *Determine the total number of carbon atoms present.* Use the molar mass of carbon to convert the mass of carbon to the amount of carbon expressed in moles. Then use the Avogadro constant to establish the number of carbon atoms.

$$?\ \text{C atoms}$$
$$= 0.021\ \text{g C} \times \frac{1\ \text{mol C}}{12.011\ \text{g C}} \times \frac{6.02 \times 10^{23}\ \text{C atoms}}{1\ \text{mol C}}$$
$$= 1.1 \times 10^{21}\ \text{C atoms}$$

4. *Determine the number of ^{13}C atoms.* The percent natural abundance leads to the conversion factor: 1.108 ^{13}C atoms/100 C atoms. Multiply the result of step 3 by this factor.

$$?\ ^{13}\text{C atoms} = 1.1 \times 10^{21}\ \text{C atoms} \times \frac{1.108\ ^{13}\text{C atoms}}{100\ \text{C atoms}}$$
$$= 1.2 \times 10^{19}\ ^{13}\text{C atoms}$$

Key Terms

actinides (2-6)
alpha (α) particle (2-2)
atomic mass (weight) (2-5)
atomic mass unit (u) (2-3)
atomic number (proton number), Z (2-3)
Avogadro constant, N_A (2-7)
beta (β) particle (2-2)
cathode ray (2-2)
chemical symbol (2-4)
electron (2-2)
family (2-6)
gamma (γ) ray (2-2)

group (2-6)
ion (2-4)
isotope (2-4)
lanthanides (2-6)
law of conservation of mass (2-1)
law of constant composition (definite proportions) (2-1)
law of multiple proportions (2-1)
main-group elements (2-6)
mass number, A (2-3)
mass spectrometer (2-4)
metals (2-6)
molar mass, M (2-7)

mole (2-7)
neutron (2-3)
neutron number (2-3)
nonmetals (2-6)
nuclide (2-4)
percent natural abundance (2-4)
period (2-6)
periodic table (2-6)
proton (2-3)
radioactivity (2-2)
transition elements (transition metals) (2-6)

Review Questions

1. In your own words, define or explain the following terms or symbols: **(a)** $^A_Z E$; **(b)** β particle; **(c)** isotope; **(d)** ^{16}O; **(e)** molar mass.
2. Briefly describe each of the following ideas: **(a)** law of conservation of mass; **(b)** Rutherford's nuclear atom; **(c)** weighted-average atomic mass; **(d)** radioactivity.
3. Explain the important distinctions between each pair of terms: **(a)** cathode rays and X rays; **(b)** protons and neutrons; **(c)** nuclear charge and ionic charge; **(d)** period and group of the periodic table; **(e)** metal and nonmetal; **(f)** Avogadro's constant and the mole.

4. A 0.406-g sample of magnesium reacts with oxygen, producing 0.674 g of magnesium oxide as the only product. What mass of oxygen was consumed in the reaction?
5. A 1.205-g sample of potassium reacts with 6.815 g of chlorine to produce potassium chloride as the only product. After the reaction, 3.300 g of chlorine remains unreacted. What mass of potassium chloride was formed?
6. When a sample of pure sulfur is burned, there is no measurable solid residue. How can this observation conform to the law of conservation of mass?

7. If sodium and chlorine atoms combine in the proportions 1 : 1 in sodium chloride, what must be the percent by mass of sodium in the compound?
   (*Hint:* Use the tabulated atomic masses of Na and Cl.)

8. In Example 2-2, we established that the proportion of magnesium oxide is 0.100 g magnesium/0.166 g magnesium oxide.
   (a) What is the proportion of oxygen in magnesium oxide?
   (b) What is the proportion of oxygen to magnesium (that is, g O/g Mg) in magnesium oxide?
   (c) What is the percent by mass of magnesium in magnesium oxide?

9. Samples of pure carbon weighing 3.62, 5.91, and 7.07 g were burned in an excess of air. The masses of carbon dioxide obtained (the sole product in each case) were 13.26, 21.66, and 25.91 g, respectively.
   (a) Do these data establish that carbon dioxide has a fixed composition?
   (b) What is the composition of carbon dioxide, expressed in %C and %O, by mass?

10. Sulfur forms two compounds with oxygen. In the first compound, 1.000 g sulfur is combined with 0.998 g oxygen, and in the second, 1.000 g sulfur is combined with 1.497 g oxygen. Show that these results are consistent with Dalton's law of multiple proportions.

11. Phosphorus forms two compounds with chlorine. In the first compound, 1.000 g of phosphorus is combined with 3.433 g chlorine, and in the second, 2.500 g phosphorus is combined with 14.308 g chlorine. Show that these results are consistent with Dalton's law of multiple proportions.

12. When 4.15 g magnesium and 82.6 g bromine react, (1) all the magnesium is used up, (2) some bromine remains unreacted, and (3) magnesium bromide is the only product. With this information alone, is it possible to deduce the mass of magnesium bromide produced? Explain.

13. Complete the following table. What minimum amount of information is required to characterize completely an atom or ion? (Note that not all rows can be completed).

Name	Symbol	Number Protons	Number Electrons	Number Neutrons	Mass Number
Sodium	$_{11}^{23}$Na	11	11	12	23
Silicon	—	—	—	14	—
—	—	37	—	—	85
—	^{40}K	—	—	—	—
—	—	—	33	42	—
—	^{20}Ne^{2+}	—	—	—	—
—	—	—	—	—	80
—	—	—	—	126	—

14. Arrange the following species in order of increasing (a) number of electrons; (b) number of neutrons; (c) mass. $_{50}^{112}$Sn $_{18}^{40}$Ar $_{52}^{122}$Te $_{29}^{59}$Cu $_{48}^{120}$Cd $_{27}^{58}$Co $_{19}^{39}$K

15. All of these radioactive isotopes have applications in medicine. Write their symbols in the form $_Z^A$E. (a) cobalt-60; (b) phosphorus-32; (c) iodine-131; (d) sulfur-35.

16. For the nuclide $_{56}^{138}$Ba, express the percentage of the fundamental particles in the nucleus that are neutrons.

17. There are *two* principal isotopes of iridium (atomic mass = 192.22 u). One of these, ^{191}Ir, has a mass of 190.9609 u. Which of these must be the second isotope: ^{190}Ir, ^{192}Ir, ^{193}Ir? Explain.

18. An isotope with mass number 38 has two more neutrons than protons. This is an isotope of what element?

19. The following data on isotopic masses are given in a handbook. What is the ratio of each of these masses to that of $_6^{12}$C? (a) $_{17}^{35}$Cl, 34.96885 u; (b) $_{12}^{26}$Mg, 25.98259 u; (c) $_{86}^{222}$Rn, 222.0175 u.

20. The following ratios of masses were obtained with a mass spectrometer: $_9^{19}$F/$_6^{12}$C = 1.5832; $_{17}^{35}$Cl/$_9^{19}$F = 1.8406; $_{35}^{81}$Br/$_{17}^{35}$Cl = 2.3140. Determine the mass of a $_{35}^{81}$Br atom in atomic mass units.
   (*Hint:* What is the mass of a ^{12}C atom?)

21. In naturally occurring argon, 99.600% of the atoms are $_{18}^{40}$Ar with mass 39.9624 u; 0.337%, $_{18}^{36}$Ar with mass 35.96755 u; and 0.063%, $_{18}^{38}$Ar with mass 37.96272 u. Calculate the weighted-average atomic mass of naturally occurring argon.

22. Refer to the periodic table inside the front cover and identify
   (a) an element that is in both group 13 and the fifth period
   (b) one element similar to and one unlike sulfur
   (c) the alkali metal in the sixth period
   (d) the halogen element in the fifth period
   (e) an element with atomic number greater than 50 that has properties similar to the element with atomic number 18
   (f) the group number of the element E that forms an ion E^{3-}
   (g) an element M that you would expect to form the ion M^{2+}

23. Assuming that the seventh period of the periodic table has 32 members, what should be the atomic number of the noble gas following radon (Rn)? of the alkali metal following francium (Fr)?

24. What is the total number of atoms in each of the following samples? (a) 12.7 mol Ca; (b) 0.00361 mol Ne; (c) 1.8×10^{-12} mol Pu.

25. Calculate the quantities indicated.
   (a) the number of moles represented by 2.18×10^{26} Fe atoms
   (b) the mass, in grams, of 7.71 mol Kr
   (c) the mass, in mg, of a sample containing 6.15×10^{19} Au atoms
   (d) the number of atoms in 112 cm^3 of Fe ($d = 7.86$ g/cm^3)

26. *Without doing detailed calculations*, indicate which of the following quantities contains the greatest number of atoms: 6.02×10^{23} Ni atoms, 25.0 g nitrogen, 52.0 g Cr, 10.0 cm^3 Fe ($d = 7.86$ g/cm^3). Explain your reasoning.

27. How many ^{204}Pb atoms are present in a piece of lead weighing 215 mg? The percent natural abundance of ^{204}Pb is 1.4%.

28. A particular lead–cadmium alloy is 8.0% cadmium by mass. What mass of this alloy, in grams, must you weigh out to obtain a sample containing 6.50×10^{23} Cd atoms?

Exercises

Law of Conservation of Mass

29. When an iron object rusts, its mass increases. When a match burns, its mass decreases. Do these observations violate the law of conservation of mass? Explain.

30. When a strip of magnesium metal is burned in air (recall Figure 2-1), it produces a white powder that weighs more than the original metal. When a strip of magnesium is burned in a flashbulb, the bulb weighs the same before and after it is flashed. Explain the difference in these observations.

31. When a solid mixture consisting of 10.500 g calcium hydroxide and 11.125 g ammonium chloride is strongly heated, gaseous products are evolved and 14.336 g of a solid residue remains. The gases are passed into 62.316 g water, and the mass of the resulting solution is 69.605 g. Within the limits of experimental error, show that these data conform to the law of conservation of mass.

32. Within the limits of experimental error, show that the law of conservation of mass was obeyed in the following experiment: 10.00 g calcium carbonate (found in limestone) was dissolved in 100.0 mL hydrochloric acid ($d = 1.148$ g/mL). The products were 120.40 g solution (a mixture of hydrochloric acid and calcium chloride) and 2.22 L carbon dioxide gas ($d = 1.9769$ g/L).

Law of Constant Composition

33. In one experiment, 2.18 g sodium was allowed to react with 16.12 g chlorine. All the sodium was used up, and 5.54 g sodium chloride (salt) was produced. In a second experiment, 2.10 g chlorine was allowed to react with 10.00 g sodium. All the chlorine was used up, and 3.46 g sodium chloride was produced. Show that these results are consistent with the law of constant composition.

34. When 3.06 g hydrogen was allowed to react with an excess of oxygen, 27.35 g water was obtained. In a second experiment, a sample of water was decomposed by electrolysis, resulting in 1.45 g hydrogen and 11.51 g oxygen. Are these results consistent with the law of constant composition? Demonstrate why or why not.

35. In one experiment, the burning of 0.312 g sulfur produced 0.623 g sulfur dioxide as the sole product of the reaction. In a second experiment, 0.842 g sulfur dioxide was obtained. What mass of sulfur must have been burned in the second experiment?

36. In one experiment, the reaction of 1.00 g mercury and an excess of sulfur yielded 1.16 g of a sulfide of mercury as the sole product. In a second experiment, the same sulfide was produced in the reaction of 1.50 g mercury and 1.00 g sulfur.
(a) What mass of the sulfide of mercury was produced in the second experiment?
(b) What mass of which element (mercury or sulfur) remained *unreacted* in the second experiment?

Law of Multiple Proportions

37. The following data were obtained for compounds of nitrogen and hydrogen for the mass of nitrogen given:

Compound	Mass of Nitrogen (g)	Mass of Hydrogen (g)
A	0.500	0.108
B	1.000	0.0720
C	0.750	0.108

(a) Show that these data are consistent with the law of multiple proportions.
(b) If the formula of compound B is N_2H_2, what are the formulas of compounds A and C?

38. The following data were obtained for compounds of iodine and fluorine for the mass of iodine given:

Compound	Mass of Iodine (g)	Mass of Fluorine (g)
A	1.000	0.1497
B	0.500	0.2246
C	0.750	0.5614
D	1.000	1.0480

(a) Show that these data are consistent with the law of multiple proportions.
(b) If the formula for compound A is IF, what are the formulas for compounds B, C, and D?

39. There are two oxides of copper. One oxide has 20% oxygen, by mass. The second oxide has a *smaller* percent of oxygen than the first. What is the probable percent of oxygen in the second oxide?

40. The two oxides of carbon described on page 37 were CO and CO_2. Another oxide of carbon has 1.106 g of oxygen in a 2.350-g sample. In what ratio are carbon and oxygen atoms combined in molecules of this third oxide? Explain.

Fundamental Particles

41. Cite the evidence that most convincingly established that electrons are fundamental particles of all matter.

42. Why could the same methods that had been used to characterize electrons not be used to isolate and detect neutrons?

Fundamental Charges and Mass-to-Charge Ratios

43. These observations were made for a series of 5 oil drops in an experiment similar to Millikan's (see Figure 2-8). Drop 1 carried a charge of 1.28×10^{-18} C; drops 2 and 3 each carried $\frac{1}{2}$ the charge of drop 1; drop 4 carried 1/8 the charge of drop 1; drop 5 had a charge four times that of drop 1. Are these data consistent with the value of the electronic charge given in the text? Could Millikan have inferred the charge on the electron from this particular series of data? Explain.

44. In an experiment similar to that described in Exercise 43, drop 1 carried a charge of 6.41×10^{-19} C; drop 2 had $\frac{1}{2}$ the charge of drop 1; drop 3 had twice the charge of drop 1; drop 4 had a charge of 1.44×10^{-18} C; and drop 5 had $\frac{1}{3}$ the

charge of drop 4. Are these data consistent with the value of the electronic charge given in the text? Could Millikan have inferred the charge on the electron from this particular series of data? Explain.

45. Use data from Table 2.1 to verify the following statements.
(a) The mass of electrons is about 1/2000 that of H atoms.
(b) The mass-to-charge ratio (m/e) for positive ions is considerably larger than that for electrons.

46. Determine the approximate value of m/e in coulombs per gram for the ions $^{127}_{53}I^-$ and $^{32}_{16}S^{2-}$. Why are these values only approximate?

Atomic Number, Mass Number, and Isotopes

47. For the atom ^{108}Pd with mass 107.90389 u, determine
(a) the numbers of protons, neutrons, and electrons in the atom;
(b) the ratio of the mass of this atom to that of an atom of $^{12}_{6}C$.

48. For the ion $^{228}Ra^{2+}$ with a mass of 228.030 u, determine
(a) the numbers of protons, neutrons, and electrons in the atom;
(b) the ratio of the mass of this atom to that of an atom of ^{16}O (refer to Example 2-4).

49. An isotope of silver has a mass that is 6.68374 times that of oxygen-16. What is the mass in u of this isotope? (Refer to Example 2-4.)

50. The ratio of the masses of the two naturally occurring isotopes of indium is 1.0177:1. The heavier of the two iso-

topes has 7.1838 times the mass of ^{16}O. What are the masses in u of the two isotopes? (Refer to Example 2-4.)

51. Which of the following species: $^{24}Mg^{2+}$, ^{47}Cr, $^{60}Co^{3+}$, $^{35}Cl^-$, $^{120}Sn^{2+}$, ^{226}Th, ^{90}Sr
(a) has equal numbers of neutrons and protons?
(b) has protons contributing more than 50% of the mass?
(c) has 50% more neutrons than protons?

52. Given the same species as listed in Exercise 51, which
(a) has equal numbers of neutrons and electrons?
(b) has protons, neutrons, and electrons in the ratio 9:11:8?
(c) has a number of neutrons equal to the number of protons plus one-half the number of electrons?

Atomic Mass Units, Atomic Masses

53. The mass of a carbon-12 atom is taken to be *exactly* 12 u. Are there likely to be any other atoms with an *exact* integral (whole number) mass, expressed in u? Explain.

54. Which statement is probably true concerning the masses of *individual* copper atoms: *All*, *some*, or *none* has a mass of 63.546 u? Explain.

55. There are *three* naturally occurring isotopes of magnesium. Their masses and percent natural abundances are

23.985042 u, 78.99%; 24.985837 u, 10.00%; and 25.982593 u, 11.01%. Calculate the weighted-average atomic mass of magnesium.

56. There are *four* naturally occurring isotopes of chromium. Their masses and percent natural abundances are 49.9461 u, 4.35%; 51.9405 u, 83.79%; 52.9407 u, 9.50%; and 53.9389 u, 2.36%. Calculate the weighted-average atomic mass of chromium.

57. The *two* naturally occurring isotopes of silver have the following abundances: ^{107}Ag, 51.84%; ^{109}Ag, 48.16%. The mass of ^{107}Ag is 106.905092 u. What is the mass of ^{109}Ag?

58. Bromine has *two* naturally occurring isotopes. One of them, bromine-79, has a mass of 78.918336 u and a natural abundance of 50.69%. What must be the mass and percent natural abundance of the other isotope, bromine-81?

59. The three naturally occurring isotopes of potassium are ^{39}K, 38.963707 u; ^{40}K, 39.963999 u; and ^{41}K. The percent natural abundances of ^{39}K and ^{41}K are 93.2581% and 6.7302%, respectively. Determine the isotopic mass of ^{41}K.

60. What are the percent natural abundances of the two naturally occurring isotopes of boron, ^{10}B and ^{11}B? These isotopes have masses of 10.012937 u and 11.009305 u, respectively.

Mass Spectrometry

61. A mass spectrum of germanium displayed peaks at mass numbers 70, 72, 73, 74, and 76, with relative heights of 20.5, 27.4, 7.8, 36.5, and 7.8, respectively.
(a) In the manner of Figure 2-15, sketch this mass spectrum.
(b) Estimate the weighted-average atomic mass of germanium, and state why this result is only approximately correct.

62. Hydrogen and chlorine atoms react to form simple diatomic molecules in a 1:1 ratio, that is, HCl. The natural abundances of the chlorine isotopes are 75.77% ^{35}Cl and 24.23% ^{37}Cl. The natural abundances of ^{2}H and ^{3}H are 0.015% and less than 0.001%, respectively.
(a) How many different HCl molecules are possible, and what are their mass numbers (that is, the sum of the mass numbers of the H and Cl atoms)?
(b) Which is the most abundant of the possible HCl molecules? Which is the second most abundant?

The Avogadro Constant and the Mole

63. Determine
(a) the number of moles of Rb in a 167.0-g sample of rubidium metal
(b) the number of Fe atoms in 363.2 kg iron
(c) the mass of a one-trillion (1.0×10^{12})-atom sample of metallic silver
(d) the mass of one fluorine atom

64. Determine
(a) the number of Ar atoms in a 5.25-mg sample of argon
(b) the molar mass, M, of an element if the mass of a 2.80×10^{22}-atom sample of the element is 4.24 g
(c) the mass of a sample of aluminum that contains the same number of atoms as 35.55 g of zinc

65. How many Ag atoms are present in a piece of sterling silver jewelry weighing 38.7 g? Sterling silver contains 92.5% Ag by mass.

66. How many atoms are present in a 75.0-cm^3 sample of plumber's solder, an alloy containing 67% Pb and 33% Sn by mass and having a density of 9.4 g/cm^3?

67. Medical experts generally believe a level of 30 μg Pb per deciliter of blood poses a significant health risk (1 dL = 0.1 L). Express this level (a) in the unit mol Pb/L blood; (b) as the number of Pb atoms per milliliter blood.

68. During a severe air pollution episode, the concentration of lead in air was observed to be 3.01 μg Pb/m^3. How many Pb atoms would be present in a 0.500-L sample of this air (the approximate lung capacity of a human adult)?

69. How many atoms are present in a 1.00-m length of 20-gauge copper wire? A 20-gauge wire has a diameter of 0.03196 in., and the density of copper is 8.92 g/cm^3.

70. *Without doing detailed calculations,* determine which of the following samples has the greatest number of atoms:
(a) A cube of iron with a length of 10.0 cm $(d = 7.86 \text{ g/cm}^3)$; (b) 1.00 kg of hydrogen contained in a 10,000-L balloon; (c) A mound of sulfur weighing 20.0 kg; (d) A 76-lb sample of liquid mercury $(d = 13.5 \text{ g/mL})$.

Integrative and Advanced Exercises

71. A solution is prepared by dissolving 2.50 g potassium chlorate (a substance used in fireworks and flares) in 100.0 mL water at 40°C. When the solution was cooled to 20°C, its volume was found to still be 100.0 mL, but some of the potassium chlorate had crystallized (deposited from the solution as a solid). At 40°C, the density of water is 0.9922 g/mL, and at 20°C, the potassium chlorate solution had a density of 1.0085 g/mL.
(a) Estimate, to two significant figures, the mass of potassium perchlorate that crystallized.
(b) Why can't the answer in (a) be given more precisely?

72. William Prout (1815) proposed that all other atoms are built up of hydrogen atoms, suggesting that all elements should have integral atomic masses based on an atomic mass of 1 for hydrogen. This hypothesis appeared discredited by the discovery of atomic masses, such as 24.3 u for magnesium and 35.5 u for chlorine. In terms of modern knowledge, explain why Prout's hypothesis is actually quite reasonable.

73. Fluorine has a single atomic species, ^{19}F. Determine the atomic mass of ^{19}F by summing the masses of its protons, neutrons, and electrons, and compare with the value listed on the inside front cover. Explain why the agreement is poor.

74. Use 1×10^{-13} cm as the approximate diameter of the spherical nucleus of the hydrogen-1 atom, together with data from Table 2.1, to estimate the density of matter in a proton.

75. Use fundamental definitions and statements from the text to establish the fact that 6.022×10^{23} u = 1.000 g.

76. In each case, identify the element in question.
(a) The mass number of an atom is 234 and the atom has 60.0% more neutrons than protons.
(b) An ion with a 2+ charge has 10.0% more protons than electrons.
(c) An ion with a mass number of 110 and a 2+ charge has 25.0% more neutrons than electrons.

77. Identify the isotope E if its nucleus contains one more neutron than protons and the mass number is nine times larger than the charge on the ion E^{3+}.

78. Determine the only possible 2+ ion for which the following two conditions are both satisfied:
• The net ionic charge is *one-tenth* the nuclear charge
• The number of neutrons is *four* more than the number of electrons.

79. Determine the only possible isotope (E) for which the following conditions are met:
• The mass number of E is 2.50 times its atomic number.
• The atomic number of E is equal to the mass number of another isotope (Y). In turn, the isotope Y has a neutron number that is 1.33 times the atomic number of Y and equal to the neutron number of selenium-82.

80. Suppose we redefined the atomic mass scale by arbitrarily assigning to the naturally occurring *mixture* of chlorine isotopes an atomic mass of 35.00000 u.
(a) What would be the atomic masses of helium, sodium, and iodine on this new atomic mass scale?
(b) Why do these three elements have nearly integral (whole number) atomic masses based on carbon-12 but not based on naturally occurring chlorine?

81. The two naturally occurring isotopes of nitrogen have masses of 14.0031 and 15.0001 u, respectively. Determine the percentage of ^{15}N atoms in naturally occurring nitrogen.

82. The masses of the naturally occurring mercury isotopes are ^{196}Hg, 195.9658 u; ^{198}Hg, 197.9668 u; ^{199}Hg, 198.9683 u; ^{200}Hg, 199.9683 u; ^{201}Hg, 200.9703 u; ^{202}Hg, 201.9706 u; and ^{204}Hg, 203.9735 u. Use these data, together with data from Figure 2-15, to calculate the weighted-average atomic mass of mercury.

83. Silicon has one major isotope, ^{28}Si (27.97693 u, 92.21% natural abundance), and two minor ones, ^{29}Si (28.97649 u) and ^{30}Si (29.97376 u). What are the percent natural abundances of the two minor isotopes?

84. From the densities of the lines in the mass spectrum of krypton gas, the following observations were made.
(a) Somewhat more than 50% of the atoms were krypton-84.
(b) The numbers of krypton-82 and krypton-83 atoms were essentially equal.
(c) The number of krypton-86 atoms was 1.50 times greater than the number of krypton-82 atoms.
(d) The number of krypton-80 atoms was 19.6% of the number of krypton-82 atoms.
(e) The number of krypton-78 atoms was 3.0% of the number of krypton-82 atoms.
The masses of the isotopes are ^{78}Kr, 77.9204 u; ^{80}Kr, 79.9164 u; ^{82}Kr, 81.9135 u; ^{83}Kr, 82.9141 u; ^{84}Kr, 83.9115 u; ^{86}Kr, 85.9106 u. The weighted-average atomic mass of Kr is 83.80. Use these data to calculate the percent natural abundances of the krypton isotopes.

85. The two naturally occurring isotopes of chlorine are ^{35}Cl (34.9689 u, 75.77%) and ^{37}Cl (36.9658 u, 24.23%). The two naturally occurring isotopes of bromine are ^{79}Br (78.9183 u, 50.69%) and ^{81}Br (80.9163 u, 49.31%). Chlorine and bromine combine to form bromine monochloride, BrCl. Sketch a mass spectrum for BrCl with relative number of molecules plotted against molecular mass (similar to Figure 2-15).

86. Monel metal is a corrosion-resistant copper–nickel alloy used in the electronics industry. A particular alloy with a density of 8.80 g/cm^3 and containing 0.022% Si by mass is used to make a rectangular plate that is 15.0 cm long, 12.5 cm wide, 3.00 mm thick, and has a 2.50-cm diameter hole drilled through its center. How many silicon-30 atoms are found in this plate? The mass of a silicon-30 atom is 29.97376 u, and the percent natural abundance of silicon-30 is 3.10%.

87. An alloy that melts at about the boiling point of water has Bi, Pb, and Sn atoms in the ratio 10:6:5, respectively. What mass of alloy contains a total of one mole of atoms?

88. A particular silver solder (used in the electronics industry to join electrical components) is to have the *atom* ratio of 5.00 Ag/4.00 Cu/1.00 Zn. What masses of the three metals must be melted together to prepare 1.00 kg of the solder?

89. A low-melting Sn–Pb–Cd alloy called *eutectic alloy* is analyzed. The *mole* ratio of tin to lead is 2.73:1.00, and the *mass* ratio of lead to cadmium is 1.78:1.00. What is the mass percent composition of this alloy?

Feature Problems

90. The data Lavoisier obtained in the experiment described on page 34 are as follows:

Before heating: glass vessel + tin + air
= 13 onces, 2 gros, 2.50 grains
After heating: glass vessel + tin calx + remaining air
= 13 onces, 2 gros, 5.62 grains

How closely did Lavoisier's results conform to the law of conservation of mass? 1 livre = 16 onces; 1 once = 8 gros; 1 gros = 72 grains. In modern terms, 1 livre = 30.59 g.

91. Some of Millikan's oil-drop data are shown here. The measured quantities were not actual charges on oil drops but were proportional to these charges. Show that these data are consistent with the idea of a fundamental electronic charge.

Observation	Measured Quantity	Observation	Measured Quantity
1	19.66	8	53.91
2	24.60	9	59.12
3	29.62	10	63.68
4	34.47	11	68.65
5	39.38	12	78.34
6	44.42	13	83.22
7	49.41		

92. Prior to 1961, physicists used as the standard for atomic masses the isotope ^{16}O, to which they assigned a value of exactly 16. At the same time, chemists assigned a value of exactly 16 to the naturally occurring mixture of the isotopes ^{16}O, ^{17}O, and ^{18}O. Would you expect atomic masses listed in a 50-year old text to be the same, generally higher, or generally lower than in this text? Explain.

93. The German chemist Fritz Haber proposed paying off the reparations imposed against Germany after World War I by extracting gold from seawater. Given that (a) the amount of the reparations was $28.8 billion dollars, (b) the value of gold at the time was about $21.25 per troy ounce (1 troy ounce = 31.103 g), and (c) gold occurs in seawater to the extent of 4.67×10^{17} atoms per ton of seawater (1 ton = 2000 lb), how many cubic kilometers of seawater would have had to be processed to obtain the required amount of gold? Assume that the density of seawater is 1.03 g/cm^3. (Haber's scheme proved to be commercially infeasible, and the reparations were never fully paid.)

94. Mass spectrometry is one of the most versatile and powerful tools in chemical analysis because of its capacity to discriminate between atoms of different masses. When a sample containing a mixture of isotopes is introduced into a mass spectrometer, the ratio of the peaks observed reflects the ratio of the percent natural abundances of the isotopes. This ratio provides an internal standard from which the amount of a certain isotope present in a sample can be determined. This is accomplished by deliberately introducing a known quantity of a particular isotope into the sample to be analyzed. A comparison of the new isotope ratio to the first ratio allows the determination of the amount of the isotope present in the original sample.

An analysis was done on a rock sample to determine its rubidium content. The rubidium content of a portion of rock weighing 0.350 g was extracted, and to the extracted sample was added an additional 29.45 μg of ^{87}Rb. The mass spectrum of this spiked sample showed a ^{87}Rb peak that was 1.12 times as high as the peak for ^{85}Rb. Assuming that the two isotopes react identically, what is the Rb content of the rock (expressed in parts per million by mass)? The natural abundances and isotopic masses are shown in the table.

Isotope	% Natural Abundance	Atomic Mass (u)
^{87}Rb	27.83	86.909
^{85}Rb	72.17	84.912

eMedia Exercises

95. After viewing the **Multiple Proportions** animation (*eChapter 2.1*), consider a similar process being carried out with solid phosphorus. In the reaction with phosphorus, two compounds containing only oxygen and phosphorus form. The ratio of the oxygen/phosphorus mass ratios of the two compounds is 1.66. If both formula units contain 4 phosphorus atoms, determine the compositions of these two compounds. Why is the ratio of mass ratios not equal to a whole number in this case?

96. Referring to the **Rutherford Experiment** animation (*eChapter 2.3*), describe the conclusions about the nuclear atom reached from the following pieces of experimental data:
   (i) the majority of alpha particles passed through the gold foil
   (ii) a few alpha particles were reflected at large angles
   (iii) the angle of deflection of the alpha particles varied

97. Review the animation of the **Millikan Oil Drop Experiment** (*eChapter 2.2*) and provide an explanation for each of the following:

(a) Oil droplets could be localized in the field of view of the microscope objective by controlling an external voltage.
(b) The calculated charge of the electron did not depend directly upon the mass of the oil droplet.
(c) Oil droplets that became doubly charged were observed to rise.

98. Using information in the **Mass Spectrometer** simulation (*eChapter2.4*), calculate the average atomic mass of fluorine and compare your answer to the value listed in the periodic table.

99. Explore the **Interactive Periodic Table** activity (*eChapter 2.6*) to provide the answer to the following questions.
   (a) Which group has the lowest average melting point?
   (b) Can the majority of known elements be categorized as metals or nonmetals?
   (c) Which is greater, the number of main-group elements or the number of transition elements?
   (d) Which element has the highest boiling point?

3

Chemical Compounds

Contents

Blue crystals of copper(II) sulfate pentahydrate, $CuSO_4 \cdot 5H_2O$. Chemical compounds, their formulas, and names are topics discussed in this chapter.

Water, ammonia, carbon monoxide, and carbon dioxide—all familiar substances—are rather simple chemical compounds. Only slightly less familiar are sucrose (cane sugar), acetylsalicylic acid (aspirin), and ascorbic acid (vitamin C). They too are chemical compounds. In fact, the study of chemistry is mostly about chemical compounds, and in this chapter, we will consider a number of ideas about compounds.

The common feature of all compounds is that they are composed of two or more elements. The full range of compounds can be divided into a few broad categories by applying ideas from the periodic table of the elements. Compounds are represented by chemical formulas, which in turn are derived from the symbols of their constituent elements. In this chapter, you will learn how to deduce and write chemical formulas and how to use the information incorporated into chemical formulas. The chapter ends with an overview of the relationship between names and formulas—chemical nomenclature.

3-1 Types of Chemical Compounds and Their Formulas

▶ In our later study of chemical bonding, we will find that the distinction between covalent and ionic bonding is not as clear-cut as these statements imply, but this matter need not concern us at this time.

When atoms approach each other in a chemical reaction, the electrons of the atoms interact to form chemical bonds. Generally speaking, there are two fundamental kinds of chemical bonds between atoms in chemical compounds—covalent bonds and ionic bonds. Covalent bonds involve a sharing of electrons between atoms, and ionic bonds involve a transfer of electrons from one atom to another. Covalent bonds are commonly formed between two nonmetallic elements, and ionic bonds between a metal and a nonmetal. We will study types of bonding in later chapters.

Molecular Compounds

A **molecular compound** is made up of discrete units called molecules, which typically consist of a small number of nonmetal atoms held together by covalent bonds. To represent a molecular compound, we use a **chemical formula,** a symbolic representation that, at minimum, indicates

- the elements present
- the relative number of atoms of each element

In the following formula for water, the elements are denoted by their symbols. The relative numbers of atoms are indicated by *subscripts*. Where no subscript is written, the number 1 is understood.

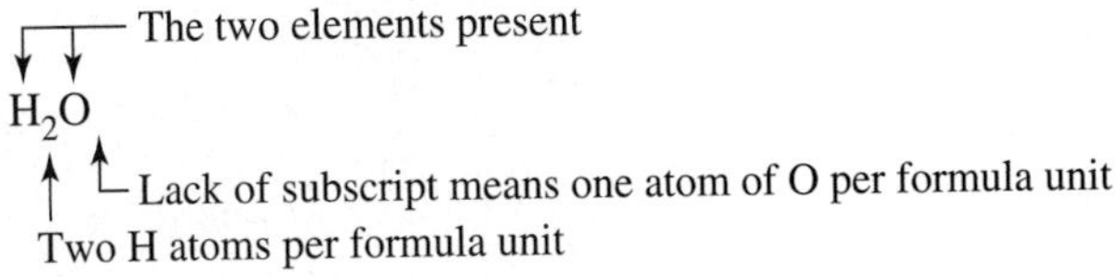

The two elements present
H_2O
Lack of subscript means one atom of O per formula unit
Two H atoms per formula unit

Another example of a chemical formula is CCl_4, which represents the compound carbon tetrachloride. The formulas H_2O and CCl_4 both represent distinct entities—*molecules*. Thus, we can refer to water and carbon tetrachloride as molecular compounds.

An **empirical formula** is the simplest formula for a compound; it shows the types of atoms present and their relative numbers. The subscripts in an empirical formula are reduced to their simplest whole-number ratio. For example, P_2O_5 is the empirical formula for a compound whose molecules have the formula P_4O_{10}. Generally, the empirical formula does not tell us a great deal about a compound. Acetic acid ($C_2H_4O_2$), formaldehyde (CH_2O, used to make certain plastics and resins), and glucose ($C_6H_{12}O_6$, blood sugar) all have the empirical formula CH_2O.

A **molecular formula** is based on an *actual* molecule of a compound. In some cases, the empirical and molecular formulas are identical, such as CH_2O for formaldehyde. In other cases, the molecular formula is a multiple of the empirical formula. A molecule of acetic acid, for example, consists of eight atoms—two C atoms, four H atoms, and two O atoms. This is twice the number of atoms in the formula unit (CH_2O). The molecular formula of acetic acid is $C_2H_4O_2$. That of glucose, on the other hand, is $C_6H_{12}O_6$.

Empirical and molecular formulas tell us the combining ratio of the atoms in the compound, but they show nothing about how the atoms are attached to each other. There are other types of formulas, however, that do convey this additional information. To illustrate, consider the carbon-oxygen-hydrogen compound acetic acid, the acid constituent that gives vinegar its sour taste. Figure 3-1 shows several representations of this compound.

Molecular model:
("ball and stick")

Empirical formula: CH_2O

Molecular formula: $C_2H_4O_2$

Structural formula:

$$H - \overset{\displaystyle H}{\underset{\displaystyle H}{\overset{|}{\underset{|}{C}}}} - \overset{\displaystyle O}{\overset{\|}{C}} - O - H$$

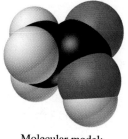

Molecular model:
("space-filling")

   Acetic Acid model

▲ FIGURE 3-1   **Several representations of the compound acetic acid**
In the molecular models, the black spheres are carbon, the red are oxygen, and the white are hydrogen. To show that one H atom in the molecule is fundamentally different from the other three, the formula of acetic acid is often written as $HC_2H_3O_2$ (see Section 5-3). To show that this H atom is bonded to an O atom, the formulas CH_3COOH and CH_3CO_2H are also used. For a few chemical compounds you may find different versions of chemical formulas in different sources.

A **structural formula** shows the order in which atoms are bonded together in a molecule and by what types of bonds. Thus, the structural formula of acetic acid tells us that three of the four H atoms are bonded to one of the C atoms, and the remaining H atom is bonded to an O atom. Both of the O atoms are bonded to one of the C atoms, and the two C atoms are bonded to each other. The covalent bonds in the structural formula are represented by lines or dashes (—). One of the bonds is represented by a double dash (=) and is called a *double* covalent bond. Differences between single and double bonds are discussed later in the text. For now, just think of a double bond as being a stronger or tighter bond than a single bond.

An alternative, less cumbersome way of showing how the atoms are connected in the acetic acid molecule is to use a *condensed structural formula*: either CH_3COOH or CH_3CO_2H. With this type of formula, the different ways in which the H atoms are attached are still apparent, and the formula can be written on a single line. The structural formula for butane, C_4H_{10}, is shown in Figure 3-2a; the condensed structural formula showing the atoms' connectivities is $CH_3(CH_2)_2CH_3$. Condensed structural formulas can also be used to show how a group of atoms is attached to another atom. Consider methylpropane, C_4H_{10}, which has the structural formula in Figure 3-2b. There is a — CH_3 group of atoms attached to the central carbon atom. This is shown in the condensed structural formula by enclosing the CH_3 in parentheses to the right of the atom to which it is attached, thus $CH_3CH(CH_3)CH_3$. Alternatively, because the central C atom is bonded to each of the other three C atoms, we can write the condensed structural formula $CH(CH_3)_3$.

In *organic* compounds, compounds containing carbon and hydrogen as their principal elements, the carbon atom *always forms four covalent bonds*. Organic compounds can be very complex, and a way of simplifying their structural formulas is to write structures without showing the C and H atoms explicitly. We do this by drawing lines to indicate chemical bonds, and wherever a line ends or meets another line, we assume that a carbon atom exists. We assume that each carbon atom has enough H atoms to satisfy the carbon atoms' need for four bonds. The male hormone *testosterone* is represented by such a structure in Figure 3-2c.

Finally, molecules occupy space and have a three-dimensional shape. Empirical and molecular formulas do not convey any information about the spatial arrangements

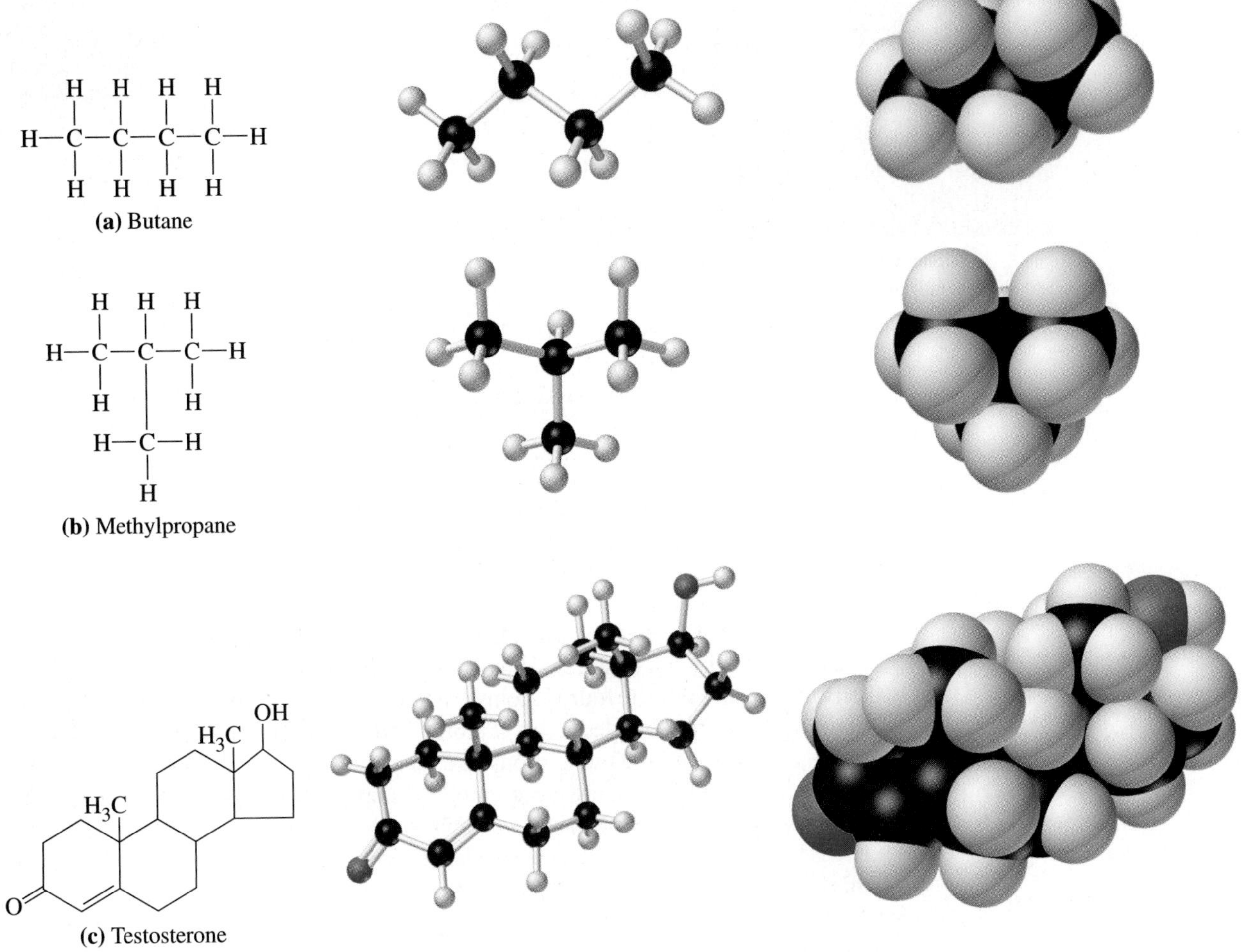

(a) Butane

(b) Methylpropane

H₃C OH

H₃C

O

(c) Testosterone

▲ FIGURE 3-2   **Visualizations of three molecules**

Butane, Methylpropane, Testosterone models

of atoms. Structural formulas can sometimes show this, but usually the only satisfactory way to represent the three-dimensional structures of molecules is by constructing models. In a *ball-and stick-model*, the centers of the bonded atoms are represented by small balls, and the bonds between atoms by sticks. Such models help us to visualize distances between the centers of atoms (bond lengths) and the geometrical shapes of molecules. A ball-and-stick model of acetic acid, constructed from a model set, is shown in Figure 3-1. The ball-and-stick model is easy to draw and interpret, but the atoms in a molecule are in contact, not held apart as implied by a ball-and-stick model.

A *space-filling model* shows that the atoms in a molecule occupy space and that they are in actual contact with one onother. Specially designed computer programs generate space-filling models such as those shown in Figures 3-1 and 3-2. A space-filling model is the most accurate representation of the size and shape of a molecule because it is constructed to scale (that is, a nanometer-size molecule is magnified to a millimeter or centimeter scale).

The acetic acid molecule is made up of three types of atoms (C, H, and O) and models of the molecule should reflect this fact. We generally adopt a color scheme to distinguish the various types of atoms when we use ball-and-stick and space-

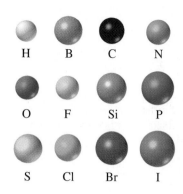

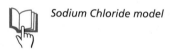

filling models. A common color scheme is shown in the margin of this page. You will notice that the colored spheres have different sizes, corresponding to the size differences between the various atoms in the periodic table.

The various depictions of molecules discussed above will be used throughout this book. In fact, visualization of the sizes and shapes of molecules and interpretation of the physical and chemical properties in terms of molecular sizes and shapes is one of the most important aspects of modern chemistry.

Ionic Compounds

Chemical combination of a metal and a nonmetal usually results in an *ionic compound*. An **ionic compound** is made up of positive and negative ions joined together by electrostatic forces of attraction (recall the attraction of oppositely charged objects pictured in Figure 2-4). The atoms of metallic elements tend to lose one or more electrons when they combine with nonmetal atoms, and the nonmetal atoms tend to gain one or more electrons. As a result of this electron transfer, the metal atom becomes a positive ion (called a **cation**), and the nonmetal atom becomes a negative ion (called an **anion**). We can usually deduce the charge on a main-group cation or anion from the group to which an element belongs (recall Section 2-6), and thus the periodic table can help us to write the formulas of ionic compounds.

In the formation of sodium chloride—ordinary table salt—each sodium atom gives up one electron to become a sodium ion, Na^+, and each chlorine atom gains one electron to become a chloride ion Cl^-. This fact conforms to the relationship between periodic table locations of the elements and the charges on their simple ions (see page 49). For sodium chloride to be electrically neutral, we require one Na^+ ion for each Cl^- ion ($+1 - 1 = 0$). Thus, the formula is NaCl.

The structure of NaCl is shown in Figure 3-3. We observe that each Na^+ ion in sodium chloride is surrounded by six Cl^- ions (and vice versa), and we cannot say that any one of these six Cl^- ions belongs exclusively to a given Na^+ ion. Yet, the ratio of Cl^- to Na^+ ions in sodium chloride is 1:1, and so we arbitrarily select a combination of one Na^+ ion and one Cl^- ion as a *formula unit*. The **formula unit** of an ionic compound is the smallest electrically neutral collection of ions. The ratio of atoms (ions) in the formula unit is the same as in the chemical formula. Because it is buried in a vast network of ions, called a crystal, a formula unit of an ionic compound does not exist as a distinct entity. Thus it is inappropriate to call a formula unit of solid sodium chloride a molecule.

The situation with magnesium chloride is similar. In magnesium chloride, found as a trace impurity in table salt, magnesium atoms lose two electrons to become magnesium ions, Mg^{2+} (Mg is in group 2). To obtain an electrically neutral formula unit, we need two Cl^- ions, each with a charge of $1-$, for every Mg^{2+} ion. The formula of magnesium chloride is $MgCl_2$.

The ions Na^+, Mg^{2+}, and Cl^- are *monatomic*, meaning that each consists of a single ionized atom. By contrast, a *polyatomic* ion is made up of two or more atoms. In the nitrate ion, NO_3^-, the subscripts signify that *three* O atoms and *one* N atom are joined into the single ion NO_3^-. Magnesium nitrate is an ionic compound made up of magnesium and nitrate ions. An electrically neutral formula unit of this compound must consist of *one* Mg^{2+} ion and *two* NO_3^- ions. The formula based on this formula unit is denoted by enclosing NO_3 in parentheses, followed by the subscript 2; thus, $Mg(NO_3)_2$. Polyatomic ions are discussed further in Section 3-6.

Sodium Chloride model

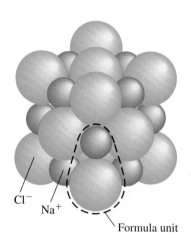

▲ **FIGURE 3-3**
Portion of an ionic crystal and a formula unit of NaCl
Solid sodium chloride consists of enormous numbers of Na^+ and Cl^- ions in a network called a crystal. The combination of one Na^+ and one Cl^- ion is the smallest collection of ions from which we can deduce the formula NaCl. It is a formula unit.

Are You Wondering ...

If a compound can be formed between different metal atoms?

In a metal, electrons of the atoms interact to form *metallic* bonds. The bonded atoms are usually of the same element, but they may also be of different elements, giving rise to *intermetallic* compounds. The metallic bond gives metals and intermetallic compounds their characteristic properties of electrical and heat conductivity. These bonds are described in Chapter 12.

3-2   The Mole Concept and Chemical Compounds

▶ The terms *formula weight* and *molecular weight* are often used in place of formula mass and molecular mass.

Once we know the chemical formula of a compound, we can determine its formula mass. **Formula mass** is the mass of a *formula unit* in atomic mass units. It is always appropriate to use the term formula mass, but for a molecular compound, the formula unit is an actual molecule and we can speak of its molecular mass. **Molecular mass** is the mass of a *molecule* in atomic mass units.

Weighted-average formula and molecular masses can be obtained just by adding up weighted-average atomic masses (those on the inside front cover). Thus, for the molecular compound water, H_2O,

$$\text{molecular mass } H_2O = 2 \text{ (atomic mass H)} + \text{(atomic mass O)}$$
$$= 2(1.008 \text{ u}) + 15.999 \text{ u}$$
$$= 18.015 \text{ u}$$

For the ionic compound magnesium chloride, $MgCl_2$,

▶ The terms *formula mass* and *molecular mass* have essentially the same meaning, although when referring to ionic compounds, such as NaCl and $MgCl_2$, *formula mass* is the term that should be used.

$$\text{formula mass } MgCl_2 = \text{atomic mass Mg} + 2 \text{ (atomic mass Cl)}$$
$$= 24.305 \text{ u} + 2 (35.453 \text{ u})$$
$$= 95.211 \text{ u}$$

and for the ionic compound magnesium nitrate, $Mg(NO_3)_2$,

$$\text{formula mass } Mg(NO_3)_2 = \text{atomic mass Mg}$$
$$+ 2 \text{ [atomic mass N} + 3 \text{ (atomic mass O)]}$$
$$= 24.305 \text{ u} + 2 [14.007 \text{ u} + 3 (15.999 \text{ u})]$$
$$= 148.313 \text{ u}$$

Mole of a Compound

KEEP IN MIND

that although *molecular* mass and *molar* mass sound similar and are related, they aren't the same. Molecular mass is the weighted-average mass of one molecule (expressed in atomic mass units, u). Molar mass is the mass of Avogadro's number of molecules (expressed in grams per mole, g/mol). The two terms have the same numerical value but different units. ▶

Recall that in Chapter 2 a mole was defined as an amount of substance having the same number of elementary entities as there are atoms in exactly 12 g of pure carbon-12. This definition carefully avoids saying that the entities to be counted are always atoms. As a result, we can apply the concept of a mole to any quantity that we can represent by a symbol or formula—atoms, ions, formula units, or molecules. Specifically, a *mole of compound* is an amount of compound containing Avogadro's number (6.02214×10^{23}) of formula units or molecules. The **molar mass** is the mass of one mole of compound—one mole of molecules of a molecular compound and one mole of formula units of an ionic compound.

The weighted-average molecular mass of H_2O is 18.015 u compared with a mass of exactly 12 u for a carbon-12 atom. If we compare samples of water molecules and carbon atoms using Avogadro's number of each, we get a mass of 18.015 g H_2O compared with exactly 12 g for carbon-12. The molar mass of H_2O is 18.015 g H_2O/mol H_2O. If we know the formula of a compound, we can equate the following terms, as illustrated for H_2O, $MgCl_2$, and $Mg(NO_3)_2$.

$$1 \text{ mol } H_2O = 18.015 \text{ g } H_2O = 6.02214 \times 10^{23} \ H_2O \text{ molecules}$$

$$1 \text{ mol } MgCl_2 = 95.211 \text{ g } MgCl_2 = 6.02214 \times 10^{23} \ MgCl_2 \text{ formula units}$$

$$1 \text{ mol } Mg(NO_3)_2 = 148.313 \text{ g } Mg(NO_3)_2$$

$$= 6.02214 \times 10^{23} \ Mg(NO_3)_2 \text{ formula units}$$

Expressions such as these provide several different types of conversion factors that can be applied in a variety of problem-solving situations. The strategy that works best for a particular problem will depend, in part, on how the necessary conversions are visualized. For instance, in Example 3-1 mass is converted to an amount in moles and then to a number of formula units. The central focus of a problem is generally the conversion of a mass in grams to an amount in moles, or vice versa. This conversion must often be preceded or followed by other conversions involving volumes, densities, percentages, and so on. As we saw in Chapter 2, one helpful tool in problem solving is to establish a conversion pathway (recall Figure 2-20).

EXAMPLE 3-1

Relating Molar Mass, the Avogadro Constant, and Formula Units of an Ionic Compound. An analytical balance can detect a mass of 0.1 mg. What is the total number of ions present in this minimally detectable quantity of $MgCl_2$?

Solution

After making the mass conversion, mg $\longrightarrow$ g, we can use the molar mass to convert from mass to number of moles of $MgCl_2$. Then, with the Avogadro constant as a conversion factor, we can convert from moles to number of formula units. The final factor we need is based on the fact that there are *three* ions (*one* Mg^{2+} and *two* Cl^-) per formula unit (fu) of $MgCl_2$.

As previously noted, it is often helpful to map out a conversion pathway that starts with the information given and proceeds through a series of conversion factors to the information sought. In Example 2-9, this was done with a diagram (Figure 2-20). However, the pathway can usually be mapped more simply. For this problem, we can begin with milligrams of $MgCl_2$ and make the following conversions:

$$mg \longrightarrow g \longrightarrow mol \longrightarrow fu \longrightarrow \text{number of ions.}$$

$$? \text{ ions} = 0.1 \text{ mg } MgCl_2 \times \frac{1 \text{ g } MgCl_2}{1000 \text{ mg } MgCl_2} \times \frac{1 \text{ mol } MgCl_2}{95 \text{ g } MgCl_2}$$

$$\times \frac{6.0 \times 10^{23} \text{ f.u. } MgCl_2}{1 \text{ mol } MgCl_2} \times \frac{3 \text{ ions}}{1 \text{ f.u. } MgCl_2}$$

$$= 2 \times 10^{18} \text{ ions}$$

Practice Example A: Zinc oxide, ZnO, is used in sunscreen preparations. What is the total number of ions present in a 1.0-g. sample of ZnO?

Practice Example B: How many grams of $MgCl_2$ would you need to obtain 5.0×10^{23} Cl^- ions?

In Example 3-2, we use additional factors before using factors based on the mole concept. Overall, the conversion pathway is $\mu L \longrightarrow mL \longrightarrow g \longrightarrow mol \longrightarrow$ number of molecules. Again, notice that we start with a value given in the problem and use conversion factors to make a path to the desired solution.

EXAMPLE 3-2

▶ Given that a normal drop of liquid is about 0.05 mL, a volume of 1.0 μL is only *0.02 of a drop.*

Combining Several Factors in a Calculation Involving Molar Mass. The volatile liquid ethyl mercaptan, C_2H_6S, is one of the most odoriferous substances known. It can be added to natural gas to make gas leaks detectable. How many C_2H_6S molecules are contained in a 1.0-μL sample? ($d = 0.84$ g/mL.)

Solution

First, we convert from *micro*liters, and then to milliliters. At this point, we bring in density as a conversion factor. The remaining conversions are from the mass of the substance to the number of moles and, finally, to the number of molecules.

$$? \text{ molecules } C_2H_6S = 1.0 \ \mu L \times \frac{1 \times 10^{-6} \text{ L}}{1 \ \mu L} \times \frac{1000 \text{ mL}}{1 \text{ L}} \times \frac{0.84 \text{ g } C_2H_6S}{1 \text{ mL}}$$

$$\times \frac{1 \text{ mol } C_2H_6S}{62.1 \text{ g } C_2H_6S} \times \frac{6.02 \times 10^{23} \text{ molecules } C_2H_6S}{1 \text{ mol } C_2H_6S}$$

$$= 8.1 \times 10^{18} \text{ molecules } C_2H_6S$$

Practice Example A: Gold has a density of 19.32 g/cm^3. A piece of gold leaf is 2.50 cm on each side and 0.100 mm thick. How many atoms of gold are in this piece of gold leaf?

Practice Example B: If 1.0 μL of liquid ethyl mercaptan, C_2H_6S ($d = 0.84$ g/mL), is allowed to evaporate and distribute itself throughout a 1500-m^3 chemistry lecture hall, will the vapor be detectable in the room? The limit of detectability is $9 \times 10^{-4} \ \mu mol/m^3$. (*Hint:* 1 $\mu mol = 1 \times 10^{-6}$ mol. How many micromoles of C_2H_6S are there in 1.0 μL?)

Mole of an Element—A Second Look

In Chapter 2, we took one mole of an element to be 6.02214×10^{23} *atoms* of the element. This is the only definition possible for elements such as iron, magnesium, sodium, and copper, in which enormous numbers of individual spherical atoms are clustered together, much like marbles in a can. But the atoms of some elements are joined together to form molecules. Bulk samples of these elements are composed of collections of molecules. The molecules of P_4 and S_8 are represented in Figure 3-4. The molecular formulas of elements that you should become familiar with are

$$H_2 \quad O_2 \quad N_2 \quad F_2 \quad Cl_2 \quad Br_2 \quad I_2 \quad P_4 \quad S_8$$

For these elements, we speak of an *atomic* mass or a *molecular* mass, and molar mass can be expressed in two ways. Hydrogen, for example, has an atomic mass of 1.008 u and a molecular mass of 2.016 u; its molar mass can be expressed as 1.008 g H/mol H or 2.016 g H_2/mol H_2.

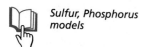
Sulfur, Phosphorus models

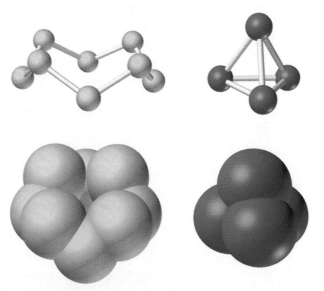

▲ **FIGURE 3-4   Molecular forms of elemental sulfur and phosphorus**
In a sample of solid sulfur, there are *eight* sulfur atoms in a sulfur molecule. In solid white phosphorus, there are *four* phosphorus atoms per molecule.

? Are You Wondering...

What to use for the molar mass when dealing with "a mole of hydrogen?"

The phrase "a mole of hydrogen" is ambiguous. One must always specify either a mole of hydrogen *atoms* or a mole of hydrogen *molecules*. It is better still to write 1 mol H or 1 mol H_2. Remember to use 1.008 g H/mol H when dealing with H atoms and 2.016 g H_2/mol H_2 when working with H_2 molecules. The distinction between hydrogen atoms and hydrogen molecules is very much like the distinction between one dozen socks and one dozen pairs of socks.

▶ The structural formula of halothane is

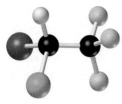

3-3   Composition of Chemical Compounds

A chemical formula conveys considerable quantitative information about a compound and its constituent elements. We have already learned how to determine the molar mass of a compound, and in this section, we consider some other types of calculations based on the chemical formula.

The colorless, volatile liquid halothane has been used as a fire extinguisher and also as an inhalation anesthetic. Both its empirical and molecular formulas are $C_2HBrClF_3$; its molecular mass is 197.38 u and its molar mass is 197.38 g/mol, as calculated below.

$$M_{C_2HBrClF_3} = 2M_C + M_H + M_{Br} + M_{Cl} + 3M_F$$
$$= (2 \times 12.01) + 1.01 + 79.90 + 35.45 + (3 \times 19.00)\,g$$
$$= 197.38\ g/mol$$

The molecular formula of $C_2HBrClF_3$ tells us that *per mole* of halothane there are two moles of C atoms, one mole each of H, Br, and Cl atoms, and three moles of F atoms. This factual statement can be turned into conversion factors to answer questions such as, "How many C atoms are present per mole of halothane?" In this case, the factor needed is 2 mol C/mol $C_2HBrClF_3$. That is,

$$? \text{ C atoms} = 1.000 \text{ mol } C_2HBrClF_3 \times \frac{2 \text{ mol C}}{1 \text{ mol } C_2HBrClF_3} \times \frac{6.022 \times 10^{23} \text{ C atoms}}{1 \text{ mol C}}$$

$$= 1.204 \times 10^{24} \text{C atoms}$$

In Example 3-3, we use another conversion factor derived from the formula for halothane. This factor is shown in blue in the setup, which includes other familiar factors to make the conversion pathway: mL $\longrightarrow$ g $\longrightarrow$ mol $C_2HBrClF_3 \longrightarrow$ mol F.

EXAMPLE 3-3

Using Relationships Derived from a Chemical Formula. How many moles of F atoms are in a 75.0-mL sample of halothane ($d = 1.871$ g/mL)?

Solution

First, convert the volume of the sample to mass; this requires density as a conversion factor. Next, convert the mass of halothane to its amount in moles; this requires the inverse of the molar mass as a conversion factor. The final conversion factor is based on the formula of halothane.

$$? \text{ mol F} = 75.0 \text{ mL } C_2HBrClF_3 \times \frac{1.871 \text{ g } C_2HBrClF_3}{1 \text{ mL } C_2HBrClF_3}$$

$$\times \frac{1 \text{ mol } C_2HBrClF_3}{197.4 \text{ g } C_2HBrClF_3} \times \frac{3 \text{ mol F}}{1 \text{ mol } C_2HBrClF_3}$$

$$= 2.13 \text{ mol F}$$

Practice Example A: How many grams of C are contained in 75.0 mL of halothane ($d = 1.871$ g/mL)?

Practice Example B: How many milliliters of halothane would contain 100.0 g Br?

Calculating Percent Composition from a Chemical Formula

When a chemist believes that he or she has synthesized a new compound, a sample is generally sent to an analytical laboratory where its percent composition is determined. This experimentally determined percent composition is then compared with the percent composition calculated from the formula of the expected compound. In this way, one can see if the compound obtained could be the one expected. Let us outline how the calculation is done.

Establish the molar mass of the compound, keeping track of the contribution of each element to the molar mass. For each element, formulate the ratio of its mass contribution to the mass of the compound as a whole. Multiply this ratio by 100% to obtain the mass percent of the element. This approach is illustrated in Example 3-4.

Mass % element =

$$\dfrac{\left[\begin{array}{c}\text{number of}\\\text{atoms of element}\\\text{per formula unit}\end{array}\right] \times \left[\begin{array}{c}\text{molar mass}\\\text{of element}\end{array}\right]}{\text{molar mass of compd}} \times 100\%$$

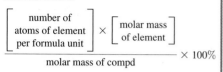

Molecular Mass and Mass Percent activity

EXAMPLE 3-4

Calculating the Mass Percent Composition of a Compound. What is the mass percent composition of halothane, $C_2HBrClF_3$?

Solution

First, determine the molar mass of $C_2HBrClF_3$. This was shown on page 73 to be 197.38 g/mol.

Then, for one mole of compound, formulate mass ratios and percents.

$$\% \, C = \frac{(2 \times 12.01) \, g}{197.38 \, g} \times 100\% = 12.17\%$$

$$\% \, H = \frac{1.01 \, g}{197.38 \, g} \times 100\% = 0.51\%$$

$$\% \, Br = \frac{79.90 \, g}{197.38 \, g} \times 100\% = 40.48\%$$

$$\% \, Cl = \frac{35.45 \, g}{197.38 \, g} \times 100\% = 17.96\%$$

$$\% \, F = \frac{(3 \times 19.00) \, g}{197.38 \, g} \times 100\% = 28.88\%$$

Practice Example A: Calculate the mass percent composition of acetic acid, the compound featured in Figure 3-1.

Practice Example B: Adenosine triphosphate (ATP) is the main energy-storage molecule in cells. Its chemical formula is $C_{10}H_{16}N_5P_3O_{13}$. What is its mass percent composition?

The percentages of the elements in a compound should add up to 100.00%, and we can use this fact in one of two ways.

1. Check the accuracy of the computations by ensuring that the percentages do total 100.00%. As applied to the results of Example 3-4,

$$12.17\% + 0.51\% + 40.48\% + 17.96\% + 28.88\% = 100.00\%$$

2. Determine the percentages of all the elements but one. Obtain that one by difference (subtraction). In Example 3-4,

$$\% \, H = 100.00\% - \% \, C - \% \, Br - \% \, Cl - \% \, F$$
$$= 100.00\% - 12.17\% - 40.48\% - 17.96\% - 28.88\%$$
$$= 0.51\%$$

Establishing Formulas from the Experimentally Determined Percent Composition of Compounds

At times, a chemist isolates a chemical compound—say, from an exotic tropical plant—and has no idea what it is. A report from an analytical laboratory on the percent composition of the compound yields data needed to determine its formula.

Percent composition establishes the relative proportions of the elements in a compound on a *mass* basis. A chemical formula requires these proportions to be on a *mole* basis, that is, in terms of *numbers* of atoms. Consider the following five-step approach to determining a formula from the experimentally determined percent composition of a compound. Let's apply it to the compound 2-deoxyribose, a sugar that is a basic constituent of DNA (deoxyribonucleic acid). The mass percent composition of 2-deoxyribose is 44.77% C, 7.52% H, and 47.71% O.

1. Although we could choose any sample size, if we take one of *exactly* 100 g, the masses of the elements are numerically equal to their percentages, that is, 44.77 g C, 7.52 g H, and 47.71 g O.

2. Convert the masses of the elements in the 100.0-g sample to amounts in moles.

$$? \text{ mol C} = 44.77 \text{ g C} \times \frac{1 \text{ mol C}}{12.011 \text{ g C}} = 3.727 \text{ mol C}$$

$$? \text{ mol H} = 7.52 \text{ g H} \times \frac{1 \text{ mol H}}{1.008 \text{ g H}} = 7.46 \text{ mol H}$$

$$? \text{ mol O} = 47.71 \text{ g O} \times \frac{1 \text{ mol O}}{15.999 \text{ g O}} = 2.982 \text{ mol O}$$

3. Write a tentative formula based on the numbers of moles just determined.

$$C_{3.727}H_{7.46}O_{2.982}$$

4. Attempt to convert the subscripts in the tentative formula to small whole numbers. This requires dividing each of the subscripts by the smallest one (2.982)

$$C_{\frac{3.727}{2.982}} H_{\frac{7.46}{2.982}} O_{\frac{2.982}{2.982}} = C_{1.25}H_{2.50}O$$

5. If the subscripts at this point differ only slightly from whole numbers, round them off to whole numbers to obtain the final formula. If one or more subscripts is not a whole number, multiply all subscripts by a small whole number that will make all subscripts integral. Here, we must multiply by 4.

$$C_{(4\times1.25)}H_{(4\times2.50)}O_{(4\times1)} = C_5H_{10}O_4$$

The formula that we get by the method just outlined, $C_5H_{10}O_4$, is the simplest possible formula—the *empirical formula*. The actual *molecular formula* may be equal to, or some multiple of, the empirical formula, such as $C_{10}H_{20}O_8$, $C_{15}H_{30}O_{12}$, $C_{20}H_{40}O_{16}$, and so on. To find the multiplying factor, we must compare the formula mass based on the empirical formula with the true molecular mass of the compound. We can establish the molecular mass from a separate experiment (by methods introduced in Chapters 6 and 14). The experimentally determined molecular mass of 2-deoxyribose is 134 u. The formula mass based on the empirical formula, $C_5H_{10}O_4$, is 134.1 u. The measured molecular mass is the same as the empirical formula mass. The molecular formula is also $C_5H_{10}O_4$.

We apply the five-step approach outlined above in Example 3-5, where we will find that the empirical formula and the molecular formula are not the same.

EXAMPLE 3-5

Determining the Empirical and Molecular Formulas of a Compound from Its Mass Percent Composition.   Dibutyl succinate is an insect repellent used against household ants and roaches. Its composition is 62.58% C, 9.63% H, and 27.79% O. Its experimentally determined molecular mass in 230 u. What are the empirical and molecular formulas of dibutyl succinate?

Solution
The first five steps are the same as those illustrated previously.

Step 1. Determine the mass of each element in a 100.0-g sample.

$$62.58 \text{ g C}, \quad 9.63 \text{ g H}, \quad 27.79 \text{ g O}$$

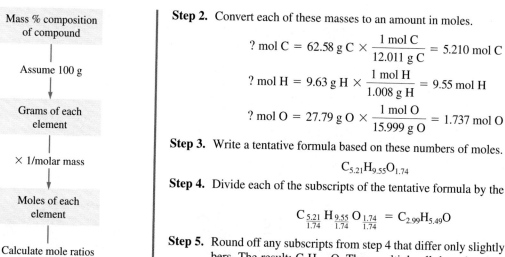

Mass % composition
of compound

↓

Assume 100 g

↓

Grams of each
element

↓

× 1/molar mass

↓

Moles of each
element

↓

Calculate mole ratios

↓

Empirical
formula

Step 2. Convert each of these masses to an amount in moles.

$$? \text{ mol C} = 62.58 \text{ g C} \times \frac{1 \text{ mol C}}{12.011 \text{ g C}} = 5.210 \text{ mol C}$$

$$? \text{ mol H} = 9.63 \text{ g H} \times \frac{1 \text{ mol H}}{1.008 \text{ g H}} = 9.55 \text{ mol H}$$

$$? \text{ mol O} = 27.79 \text{ g O} \times \frac{1 \text{ mol O}}{15.999 \text{ g O}} = 1.737 \text{ mol O}$$

Step 3. Write a tentative formula based on these numbers of moles.

$$C_{5.21}H_{9.55}O_{1.74}$$

Step 4. Divide each of the subscripts of the tentative formula by the smallest (1.74).

$$C_{\frac{5.21}{1.74}} H_{\frac{9.55}{1.74}} O_{\frac{1.74}{1.74}} = C_{2.99}H_{5.49}O$$

Step 5. Round off any subscripts from step 4 that differ only slightly from whole numbers. The result: $C_3H_{5.49}O$. Then multiply all the subscripts by a small whole number chosen to make them integral. Here, we multiply by 2:

$$2 \times 5.49 = 10.98 \approx 11.$$

$$C_{2\times3}H_{2\times5.49}O_{2\times1} = C_6H_{10.98}O_2$$

Empirical formula: $C_6H_{11}O_2$

The empirical formula mass is $[(6 \times 12.0) + (11 \times 1.0) + (2 \times 16.0)] = 115$ u. Because the experimentally determined molecular mass (230 u) is twice the empirical formula mass, we conclude

Molecular formula: $C_{12}H_{22}O_4$

Practice Example A: Diacetone glucose has a molecular mass of 260 u and a mass percent composition of 55.37% C, 7.75% H, and 36.88% O. What are the empirical and molecular formulas of this substance?

Practice Example B: Sorbitol, used as a sweetener in some "sugar-free" foods, has a molecular mass of 182 u and a mass percent composition of 39.56% C, 7.74% H, and 52.70% O. What are the empirical and molecular formulas of sorbitol?

Are You Wondering...

How much rounding off to do to get integral subscripts in an empirical formula and what factors to use to convert fractional to whole numbers?

How much rounding off is justified depends on how precisely the elemental analysis is done. As a result, there is no ironclad rule on the matter. For the examples in this text, if you carry all the significant figures allowable in a calculation, you can generally round off a subscript that is within a few hundredths of a whole number (for example, 3.98 rounds off to 4). If the deviation is more than this, you may need to adjust subscripts to integral values by multiplying by the appropriate constant. In choosing this constant, you'll find it helpful to recognize decimal equivalents of some common fractions: $0.50 = 1/2$; $0.333 = 1/3$; $0.25 = 1/4$; $0.20 = 1/5$; and so on. For example, for the subscript 1.25, we use the factor 4. That is, $1.25 = 5/4$, and $5/4 \times 4 = 5$.

Combustion Analysis

Figure 3-5 illustrates an experimental method for establishing an empirical formula for compounds that are easily burned, such as compounds containing carbon and hydrogen with oxygen, nitrogen, and a few other elements. In *combustion*

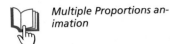

Multiple Proportions animation

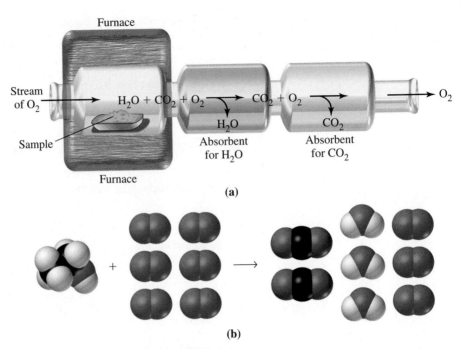

Furnace

Stream of O_2 → $H_2O + CO_2 + O_2$ → $CO_2 + O_2$ →   → O_2

H_2O

Sample

Absorbent for H_2O

CO_2

Absorbent for CO_2

Furnace

(a)

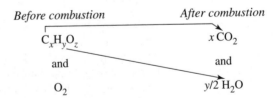

(b)

▲ **FIGURE 3-5 Apparatus for combustion analysis**
(a) Oxygen gas passes through the combustion tube containing the sample being analyzed. This portion of the apparatus is enclosed in a high-temperature furnace. Products of the combustion are absorbed as they leave the furnace—water vapor by magnesium perchlorate, and carbon dioxide gas by sodium hydroxide (producing sodium carbonate). The differences in mass of the absorbers, after and before the combustion, yield the masses of H_2O and CO_2 produced in the combustion reaction. **(b)** A molecular picture of the combustion of ethanol. Each molecule of ethanol produces two CO_2 molecules and three H_2O molecules. Combustion takes place in an excess of oxygen, so that there are oxygen molecules present at the end of the reaction. Note the conservation of mass.

analysis, a weighed sample of a compound is burned in a stream of oxygen gas. The water vapor and carbon dioxide gas produced in the combustion are absorbed by appropriate substances. The increases in mass of these absorbers correspond to the masses of water and carbon dioxide. We can think of the matter as shown below. (The subscripts x, y, and z are integers whose values we do not know initially.)

Before combustion   *After combustion*

$C_xH_yO_z$ $x\,CO_2$

and   and

O_2 $y/2\,H_2O$

After combustion, all the carbon atoms in the sample are found in the CO_2. All the H atoms are in the H_2O. Moreover, the only source of the carbon and hydrogen atoms was the sample being analyzed. Oxygen atoms in the CO_2 and H_2O could have come partly from the sample and partly from the oxygen gas consumed in the combustion. Thus, the quantity of oxygen in the sample has to be determined indirectly. These ideas are applied in Example 3-6.

EXAMPLE 3-6

Determining an Empirical Formula from Combustion Analysis Data. Vitamin C is essential for the prevention of scurvy (and large doses may be effective in preventing colds). Combustion of a 0.2000-g sample of this carbon–hydrogen–oxygen compound yields 0.2998 g CO_2 and 0.0819 g H_2O. What is the empirical formula of vitamin C?

Solution

We can base our calculation directly on the 0.2000-g sample. All we need to do is determine the number of moles of C, H, and O present. Once we have done this, we can write the empirical formula by the method presented earlier. There is a complication, however. We can easily find the numbers of moles of C and H from the quantities of CO_2 and H_2O produced. To find the number of moles of O in the sample, however, we must know the mass of O present. We can find the mass of O only by difference. This means that the masses of C and H must be determined, as well as their amounts in moles.

$$? \text{ mol C} = 0.2998 \text{ g } CO_2 \times \frac{1 \text{ mol } CO_2}{44.010 \text{ g } CO_2} \times \frac{1 \text{ mol C}}{1 \text{ mol } CO_2}$$

$$= 0.006812 \text{ mol C}$$

$$? \text{ g C} = 0.006812 \text{ mol C} \times \frac{12.011 \text{ g C}}{1 \text{ mol C}}$$

$$= 0.08182 \text{ g C}$$

$$? \text{ mol H} = 0.0819 \text{ g } H_2O \times \frac{1 \text{ mol } H_2O}{18.02 \text{ g } H_2O} \times \frac{2 \text{ mol H}}{1 \text{ mol } H_2O}$$

$$= 0.00909 \text{ mol H}$$

$$? \text{ g H} = 0.00909 \text{ mol H} \times \frac{1.008 \text{ g H}}{1 \text{ mol H}}$$

$$= 0.00916 \text{ g H}$$

Now we can find the mass of O in the vitamin C sample by difference,

$$? \text{ g O} = 0.2000 \text{ g sample} - 0.08182 \text{ g C} - 0.00916 \text{ g H} = 0.1090 \text{g O}$$

and the number of moles of O.

$$? \text{ mol O} = 0.1090 \text{ g O} \times \frac{1 \text{ mol O}}{15.999 \text{ g O}} = 0.006813 \text{ mol O}$$

As trial subscripts for the empirical formula, we can write

$$C_{0.006812}H_{0.00909}O_{0.006813}$$

Next, divide each subscript by the smallest—0.006812—to obtain

$$CH_{1.33}O$$

Finally, multiply each subscript by 3 (because $3 \times 1.33 = 3.99 \approx 4.00$).

Empirical formula of vitamin C: $C_3H_4O_3$

Practice Example A: Isobutyl propionate is the substance that provides the flavor for rum extract. Combustion of a 1.152-g sample of this carbon–hydrogen–oxygen compound yields 2.726 g CO_2 and 1.116 g H_2O. What is the empirical formula of isobutyl propionate?

Practice Example B: Complete combustion of a 1.505-g sample of thiophene, a carbon–hydrogen–sulfur compound, yields 3.149 g CO_2, 0.645 g H_2O, and 1.146 g SO_2. What is the empirical formula of thiophene? (*Hint:* All the sulfur appears as SO_2.)

▶ An alternative approach to establishing an empirical formula from combustion analysis data is to calculate the mass percent composition of a compound and then to proceed as in Example 3-5.

We have just seen how combustion reactions can be used to analyze chemical substances, but not all samples can be easily burned. Fortunately, several other types of reactions can be used for chemical analyses. Also, modern methods in chemistry rely much more on physical measurements with instruments than on chemical reactions. We will cite some of these methods later in the text.

3-4   Oxidation States: A Useful Tool in Describing Chemical Compounds

Oxidation States activity

Most basic concepts in chemistry deal with measurable properties or phenomena. In a few instances, though, a concept has been devised more for convenience than because of any fundamental significance. This is the case with **oxidation state** (oxidation number),[*] which is related to the number of electrons that an atom loses, gains, or otherwise appears to use in joining with other atoms in compounds.

Consider NaCl. In this compound a Na atom, a metal, loses one electron to a Cl atom, a nonmetal. The compound consists of the ions Na^+ and Cl^- (see Figure 3-3). Na is in a +1 oxidation state and Cl^- is in a −1 state.

In $MgCl_2$, a Mg atom loses two electrons to become Mg^{2+}, and each Cl atom gains one electron to become Cl^-. As in NaCl, the oxidation state of Cl is −1, but that of Mg is +2. If we take the *total* of the oxidation states of all the atoms (ions) in a formula unit of $MgCl_2$, we get $+2 - 1 - 1 = 0$.

In the molecule Cl_2, the two Cl atoms are identical and should have the *same* oxidation state. But if their total is to be zero, each oxidation state must itself be 0. Thus, the oxidation state of an element can vary, depending on the compounds in which it occurs. In the molecule H_2O, we *arbitrarily* assign H the oxidation state +1. Then, because the total of the oxidation states of the atoms must be zero, the oxidation state of oxygen must be −2.

From these examples you can see that we need some conventions or rules for assigning oxidation states. The following seven rules are sufficient to deal with most cases in this text, with this important qualification: *Whenever two rules appear to contradict each other (which they often will), follow the rule that appears higher on the list.* Some examples are given for each rule, and the rules are applied together in Example 3-7.

1. *The oxidation state (O.S.) of an individual atom in a free element (uncombined with other elements) is 0.*
    [*Examples:* The O.S. of an isolated Cl atom is 0; the two Cl atoms in the molecule Cl_2 both have an O.S. of 0.]

2. *The total of the oxidation states of all the atoms in*
    (a) *neutral species, such as isolated atoms, molecules, and formula units, is 0.*
    [*Examples:* The sum of the O.S. of all the atoms in CH_3OH and of all the ions in $MgCl_2$ is 0.]
    (b) *an ion is equal to the charge on the ion.*
    [*Examples:* The O.S. of Fe in Fe^{3+} is +3. The sum of the O.S. of all atoms in MnO_4^- is −1.]

3. *In their compounds, the group 1 metals have an O.S. of +1 and the group 2 metals have an O.S. of +2.*
    [*Examples:* The O.S. of K is +1 in KCl and K_2CO_3; the O.S. of Mg is +2 in $MgBr_2$ and $Mg(NO_3)_2$.]

4. *In its compounds, the O.S. of fluorine is −1.*
    [*Examples:* The O.S. of F is −1 in HF, ClF_3, and SF_6.]

---

[*]Because oxidation state refers to a number, the term *oxidation number* is often used synonymously. We will use the two terms interchangeably.

▶ The principal exceptions to rule 5 occur when H is bonded to metals, as in LiH, NaH, and CaH_2; exceptions to rule 6 occur in compounds where O atoms are bonded to one another, as in H_2O_2, and KO_2.

5. *In its compounds, hydrogen has an O.S. of +1.*
   [*Examples:* The O.S. of H is +1 in HI, H_2S, NH_3, and CH_4.]

6. *In its compounds, oxygen has an O.S. of −2.*
   [*Examples:* The O.S. of O is −2 in H_2O, CO_2 and $KMnO_4$.]

7. *In binary (two-element) compounds with metals, group 17 elements have an O.S. of −1; group 16 elements, −2; and group 15 elements, −3.*
   [*Examples:* The O.S. of Br is −1 in $MgBr_2$; the O.S. of S is −2 in Li_2S; and the O.S. of N is −3 in Li_3N.]

 Assigning Oxidation States activity

EXAMPLE 3-7

Assigning Oxidation States. What is the oxidation state of the underlined element in each of the following? (a) $\underline{P}_4$; (b) $\underline{Al}_2O_3$; (c) $\underline{Mn}O_4^-$; (d) $Na\underline{H}$; (e) $H_2\underline{O}_2$; (f) $\underline{Fe}_3O_4$.

Solution

(a) P_4: This formula represents a molecule of elemental phosphorus. For an atom of a free element, the O.S. = 0 (rule 1). The O.S. of P in P_4 is 0.

(b) Al_2O_3: The total of the oxidation states of all the atoms in this formula unit is 0 (rule 2). The O.S. of oxygen is −2 (rule 6). The total for three O atoms is −6. The total for two Al atoms is +6. The O.S. of Al is +3.

(c) MnO_4^-: This is the formula for permanganate *ion*. The total of the oxidation states of all the atoms in the ion is −1 (rule 2). The total for the four O atoms is −8. The O.S. of Mn is +7.

(d) NaH: This is a formula unit of the *ionic* compound sodium hydride. Rule 3 states that the O.S. of Na is +1. Rule 5 indicates that H should also have O.S. +1. If both atoms had O.S. +1, the total for the formula unit would be +2. This violates rule 2. *Rules 2 and 3 take precedence over rule 5.* Na has O.S. +1; the total for the formula unit is 0; and the O.S. of H is −1.

(e) H_2O_2: This is hydrogen peroxide. Rule 5, stating that H has O.S. +1, takes precedence over rule 6 (which says that oxygen has O.S. −2). The sum of the oxidation states of the two H atoms is +2 and that of the two O atoms must be −2. The O.S. of O is −1.

(f) Fe_3O_4: The total of the oxidation states of four O atoms is −8. For three Fe atoms, the total must be +8. The O.S. per Fe atom is 8/3 or $+2\frac{2}{3}$.

Practice Example A: What is the oxidation state of the underlined element in each of the following: $\underline{S}_8$; $\underline{Cr}_2O_7^{2-}$; $\underline{Cl}_2O$; $K\underline{O}_2$?

Practice Example B: What is the oxidation state of the underlined element in each of the following: $\underline{S}_2O_3^{2-}$; $\underline{Hg}_2Cl_2$; $K\underline{Mn}O_4$; $H_2\underline{C}O$?

In part (f) of Example 3-7, we got the somewhat surprising answer of $+2\frac{2}{3}$ for the oxidation state of the iron atoms in Fe_3O_4. Prior to that, we saw only integral values for oxidation states. How does this fractional value come about? Generally, it comes from the assumption that all the atoms of an element have the same oxidation state in a given compound. Usually they do, but not always. Fe_3O_4, for example, is probably better represented as $FeO \cdot Fe_2O_3$, that is, through a combination of two simpler formula units. In FeO, the O.S. of the Fe atom is +2. In Fe_2O_3, the O.S. of each of *two* Fe atoms is +3. When we *average* the oxidation states over all three Fe atoms, we get a nonintegral value: $(2 + 3 + 3)/3 = 8/3 = 2\frac{2}{3}$.

Also, we may at times need to "fragment" a formula into its constituent parts before assigning oxidation states. The ionic compound NH_4NO_3, for instance, consists of the ions NH_4^+ and NO_3^-. The oxidation state of N in NH_4^+ is −3, and in NO_3^-, +5, and we do *not* want to average them. It is far more useful to know the oxidation states of the individual N atoms than it is to deal with an average oxidation state of $+1$ for the two N atoms.

Our first use of oxidation states comes in the naming of chemical compounds in the next section.

3-5 Naming Compounds: Organic and Inorganic Compounds

Throughout this chapter, we have referred to compounds mostly by their formulas, but we do need to give them names. When we know the name of a compound, we can look up its properties in a handbook, locate a chemical on a storeroom shelf, or discuss an experiment with a colleague. Later in the text we will see cases in which different compounds have the same formula. In these instances, we will find it essential to distinguish among compounds by name. We cannot give two substances the same name, yet we do want some similarities in the names of similar substances (Figure 3-6). If all compounds were referred to by a common or trivial name, such as water (H_2O), ammonia (NH_3) or glucose ($C_6H_{12}O_6$), we would have to learn millions of unrelated names—an impossibility.

What we need is a systematic method of assigning names—a system of *nomenclature*. Several systems are used, and we will introduce each at an appropriate point in the text. Compounds formed by carbon and hydrogen or carbon and hydrogen together with oxygen, nitrogen, and a few other elements are **organic compounds**. They are generally considered in a special branch of chemistry with its own set of nomenclature rules—*organic* chemistry. Compounds that do not fit this description are **inorganic compounds**. The branch of chemistry that concerns itself with the study of these compounds is called *inorganic* chemistry. In the next section, we will consider the naming of inorganic compounds, and in Section 3-7, we will introduce the naming of organic compounds.

3-6 Names and Formulas of Inorganic Compounds

Binary Compounds of Metals and Nonmetals

Binary compounds are those formed between *two* elements. If one of the elements is a metal and the other a nonmetal, the binary compound is usually made up of ions;

▲ **FIGURE 3-6 Two oxides of lead**
These two compounds contain the same elements—lead and oxygen—but in different proportions. Their names and formulas must convey this fact: lead(IV) oxide = PbO_2 (red-brown); lead(II) oxide = PbO (yellow).

that is, it is a binary ionic compound. To name a binary compound of a metal and a nonmetal

- Write the *unmodified* name of the metal.
- Then write the name of the nonmetal, modified to end in *ide*.

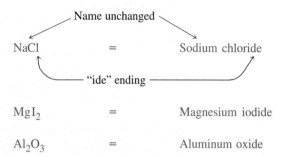

$$MgI_2 \qquad = \qquad \text{Magnesium iodide}$$

$$Al_2O_3 \qquad = \qquad \text{Aluminum oxide}$$

Inorganic Compounds activity

Ionic compounds, though made up of positive and negative ions, must be *electrically neutral.* The net, or total, charge of the ions in a formula unit must be *zero.* This means *one* Na^+ to *one* Cl^- (in NaCl); *one* Mg^{2+} to *two* I^- (in MgI_2); *two* Al^{3+} to *three* O^{2-} in Al_2O_3; and so on. Table 3.1 lists the names and symbols of simple ions formed by metals and nonmetals. You will find this list useful when writing names and formulas of binary compounds of metals and nonmetals. Notice that the main-group metals and nonmetals form ions related to their group number, as discussed in Section 2-6. The transition metals, however, may form several ions, and it is important to distinguish between them in naming their compounds. The metal iron, for example, forms *two* common ions, Fe^{2+} and Fe^{3+}. The first is called *iron(II)* ion, and the second is *iron(III)* ion. The Roman numeral immediately following

TABLE 3.1   Some Simple Ions

Name	Symbol	Name	Symbol
Positive ions (cations)			
Lithium ion	Li^+	Chromium(II) ion	Cr^{2+}
Sodium ion	Na^+	Chromium(III) ion	Cr^{3+}
Potassium ion	K^+	Iron(II) ion	Fe^{2+}
Rubidium ion	Rb^+	Iron(III) ion	Fe^{3+}
Cesium ion	Cs^+	Cobalt(II) ion	Co^{2+}
Magnesium ion	Mg^{2+}	Cobalt(III) ion	Co^{3+}
Calcium ion	Ca^{2+}	Copper(I) ion	Cu^+
Strontium ion	Sr^{2+}	Copper(II) ion	Cu^{2+}
Barium ion	Ba^{2+}	Mercury(I) ion	Hg_2^{2+}
Aluminum ion	Al^{3+}	Mercury(II) ion	Hg^{2+}
Zinc ion	Zn^{2+}	Tin(II) ion	Sn^{2+}
Silver ion	Ag^+	Lead(II) ion	Pb^{2+}
Negative ions (anions)			
Hydride ion	H^-	Iodide ion	I^-
Fluoride ion	F^-	Oxide ion	O^{2-}
Chloride ion	Cl^-	Sulfide ion	S^{2-}
Bromide ion	Br^-	Nitride ion	N^{3-}

the name of the metal indicates its oxidation state or simply the charge on the ion. Thus, $FeCl_2$ is *iron(II)* chloride, while $FeCl_3$ is *iron(III)* chloride.

An earlier system of nomenclature that is still used to some extent uses two different word endings to distinguish between two binary compounds containing the same two elements but in different proportions, such as Cu_2O and CuO. In Cu_2O, the oxidation state of copper is +1, and in CuO it is +2. Cu_2O is assigned the name cupr*ous* oxide, and CuO is cupr*ic* oxide. Similarly, $FeCl_2$ is ferr*ous* chloride, and $FeCl_3$ is ferr*ic* chloride. The idea is to use the *ous* ending for the lower oxidation state of the metal and *ic* for the higher oxidation state. The *ous*/*ic* system has several inadequacies though, and we will not use it in this text. For example, the *ous* and *ic* endings do not help in naming the four oxides of vanadium: VO, V_2O_3, VO_2, and V_2O_5. We will therefore not use it in this text.

EXAMPLE 3-8

Writing Formulas When Names of Compounds Are Given. Write formulas for the compounds barium oxide, calcium fluoride, and iron(III) sulfide.

Solution

In each case, identify the cations and their charges, based on periodic table group numbers or on oxidation states appearing as Roman numerals in names: Ba^{2+}, Ca^{2+}, and Fe^{3+}. Then identify the anions and their charges: O^{2-}, F^-, and S^{2-}. Combine the cations and anions in the relative numbers required to produce electrically *neutral* formula units.

barium oxide:	*one* Ba^{2+} and *one* O^{2-}	= BaO
calcium fluoride:	*one* Ca^{2+} and *two* F^-	= CaF_2
iron(III) sulfide:	*two* Fe^{3+} and *three* S^{2-}	= Fe_2S_3

Note that in the first case, an electrically neutral formula unit results from the combination of the charges 2+ and 2−; in the second case, 2+ and $2 \times (1-)$; and in the third case, $2 \times (3+)$ and $3 \times (2-)$.

Practice Example A: Write formulas for lithium oxide, tin(II) fluoride, and lithium nitride.

Practice Example B: Write formulas for the compounds aluminum sulfide, magnesium nitride, and vanadium(III) oxide.

EXAMPLE 3-9

Naming Compounds When Their Formulas Are Given. Write acceptable names for the compounds Na_2S, AlF_3, Cu_2O.

Solution

This task is generally easier than that of Example 3-8 because all you need to do is name the ions present. However, you must recognize that copper forms *two* different ions and that the cation in Cu_2O is Cu^+, copper(I).

Na_2S:	sodium sulfide
AlF_3:	aluminum fluoride
Cu_2O:	copper(I) oxide

Practice Example A: Write acceptable names for CsI, CaF_2, FeO, $CrCl_3$.

Practice Example B: Write acceptable names for CaH_2, $CuCl$, Ag_2S, Hg_2Cl_2.

Are You Wondering ...

Why we don't use names such as sodium(I) chloride for NaCl and magnesium(II) chloride for MgCl$_2$?

Each proposed name clearly indicates the compound in question, but as a general rule, chemists always write the *simplest name possible*. The metals of periodic table group 1 (including Na) and group 2 (including Mg) have *only one ionic form,* one oxidation state. Roman numerals designating these oxidation states are superfluous. Later in the text, we will discuss which elements can exist in several oxidation states. For now, use the information in Table 3.1 as a guide.

Binary Compounds of Two Nonmetals

If the two elements in a binary compound are both nonmetals instead of a metal and a nonmetal, the compound is a molecular compound. The method of naming these compounds is similar to that just discussed. For example.

$$HCl = \text{hydrogen chloride}$$

In both the formula and the name, we write the element with the positive oxidation state first: HCl and *not* ClH.

Some pairs of nonmetals form more than a single binary molecular compound, and we need to distinguish among them.

Generally, we indicate relative numbers of atoms through *prefixes: mono* = 1, *di* = 2, *tri* = 3, *tetra* = 4, *penta* = 5, *hexa* = 6, and so on. Thus, for the two principal oxides of sulfur we write

$$SO_2 = \text{sulfur dioxide}$$

$$SO_3 = \text{sulfur trioxide}$$

and for the following boron—bromine compound

$$B_2Br_4 = \underline{di}\text{boron }\underline{tetra}\text{bromide}$$

Additional examples are given in Table 3.2. Note that in these examples the prefix *mono* is treated in a special way. We do not use it for the first named element. Thus, NO is called nitrogen monoxide, *not mono*nitrogen *mono*xide. Finally, several substances have common or trivial names that are so well established that we almost never use their systematic names. For example,

$$H_2O = \text{water (dihydrogen monoxide)}$$

$$NH_3 = \text{ammonia (}H_3N = \text{trihydrogen mononitride)}$$

Are You Wondering ...

Why isn't MgCl$_2$ called magnesium dichloride and FeCl$_3$, iron trichloride?

The names do give a clear indication of the compounds we mean, but they do *not* conform to the conventions that chemists use. Because the magnesium ion can be only Mg^{2+}, the name *magnesium chloride* can only mean MgCl$_2$. Similarly, *iron*(III) *chloride* can refer only to a compound of Fe^{3+} and Cl$^-$, that is, FeCl$_3$. On the other hand, to name the several oxides of nitrogen, we do need prefixes (*mono, di, tri*).

TABLE 3.2 Naming Binary Molecular Compounds

Formula	Name[a]
BCl_3	boron trichloride
CCl_4	carbon tetrachloride
CO	carbon monoxide
CO_2	carbon dioxide
NO	nitrogen monoxide
NO_2	nitrogen dioxide
N_2O	dinitrogen monoxide
N_2O_3	dinitrogen trioxide
N_2O_4	dinitrogen tetroxide
N_2O_5	dinitrogen pentoxide
PCl_3	phosphorus trichloride
PCl_5	phosphorus pentachloride
SF_6	sulfur hexafluoride

[a] When the prefix ends in *a* or *o* and the element name begins with *a* or *o*, the final vowel of the prefix is dropped for ease of pronunciation. For example, carbon *mon*oxide, not carbon *mono*oxide, and dinitrogen *tetr*oxide, not dinitrogen *tetra*oxide. However, PI_3 is phosphorus *tri*iodide, not phosphorus *tri*odide.

Binary Acids

Even though we use names like hydrogen chloride for pure binary molecular compounds, we sometimes want to emphasize that their aqueous solutions are acids. Although acids will be discussed in depth later in the text, for now, so that we can recognize these substances and name them when we see their formulas, let us say that an *acid* is a substance that produces hydrogen ions (H^+)[*] when dissolved in water. *Binary acids,* then, are certain compounds of H with other nonmetal atoms. HCl, when dissolved in water, ionizes, or breaks down, into hydrogen ions (H^+) and chloride ions (Cl^-); it is an acid. NH_3 in water is *not* an acid. It shows practically no tendency to produce H^+ under any conditions. NH_3 belongs to a complementary category of substances called bases. As we shall see in Chapter 5, bases yield hydroxide ion (OH^-) in aqueous solutions.

In naming acids we use the prefix *hydro* followed by the name of the other nonmetal modified with an *ic* ending. The most important binary acids are listed below.

▶ The symbol (aq) signifies a substance in aqueous (water) solution.

$$HF(aq) = hydrofluoric \text{ acid}$$

$$HCl(aq) = hydrochloric \text{ acid}$$

$$HBr(aq) = hydrobromic \text{ acid}$$

$$HI(aq) = hydroiodic$$

$$H_2S(aq) = hydrosulfuric \text{ acid}$$

Polyatomic Ions

With the exception of $Hg_2{}^{2+}$, the ions listed in Table 3.1 are *monatomic*—each consists of a single atom. In **polyatomic ions,** two or more atoms are joined together by covalent bonds. These ions are commonly encountered, especially among the

[*] The species produced in aqueous solution is actually more complex than the simple ion H^+. The H^+ combines with an H_2O molecule to produce an ion known as the hydronium ion, H_3O^+. Chemists often use H^+ for H_3O^+ and that is what we will do until we discuss this matter more fully in Chapter 5.

TABLE 3.3 Some Common Polyatomic Ions

Name	Formula	Typical
Cation		
Ammonium ion	NH_4^+	NH_4Cl
Anions		
Acetate ion	$C_2H_3O_2^-$	$NaC_2H_3O_2$
Carbonate ion	CO_3^{2-}	Na_2CO_3
Hydrogen carbonate ion[a] (or bicarbonate ion)	HCO_3^-	$NaHCO_3$
Hypochlorite ion	ClO^-	$NaClO$
Chlorite ion	ClO_2^-	$NaClO_2$
Chlorate ion	ClO_3^-	$NaClO_3$
Perchlorate ion	ClO_4^-	$NaClO_4$
Chromate ion	CrO_4^{2-}	Na_2CrO_4
Dichromate ion	$Cr_2O_7^{2-}$	$Na_2Cr_2O_7$
Cyanide ion	CN^-	$NaCN$
Hydroxide ion	OH^-	$NaOH$
Nitrite ion	NO_2^-	$NaNO_2$
Nitrate ion	NO_3^-	$NaNO_3$
Oxalate ion	$C_2O_4^{2-}$	$Na_2C_2O_4$
Permanganate ion	MnO_4^-	$NaMnO_4$
Phosphate ion	PO_4^{3-}	Na_3PO_4
Hydrogen phosphate ion[a]	HPO_4^{2-}	Na_2HPO_4
Dihydrogen phosphate ion[a]	$H_2PO_4^-$	NaH_2PO_4
Sulfite ion	SO_3^{2-}	Na_2SO_3
Hydrogen sulfite ion[a] (or bisulfite ion)	HSO_3^-	$NaHSO_3$
Sulfate ion	SO_4^{2-}	Na_2SO_4
Hydrogen sulfate ion[a] (or bisulfate ion)	HSO_4^-	$NaHSO_4$
Thiosulfate ion	$S_2O_3^{2-}$	$Na_2S_2O_3$

[a] These anion names are sometimes written as a single word—for example, hydrogencarbonate, hydrogenphosphate, and so forth.

nonmetals. A number of polyatomic ions and compounds containing them are listed in Table 3.3. From this table, you can see that

1. Polyatomic anions are more common than polyatomic cations. The most familiar polyatomic cation is the ammonium ion, NH_4^+.
2. Very few polyatomic anions carry the *ide* ending in their names. Of those listed, only OH^- (hydroxide ion) and CN^- (cyanide ion) do. The common endings are *ite* and *ate*, and some names carry prefixes, *hypo* or *per*.
3. An element common to many polyatomic anions is *oxygen*, usually in combination with another nonmetal. Such anions are called **oxoanions.**
4. Certain nonmetals (such as Cl, N, P, and S) form a series of oxoanions containing different numbers of oxygen atoms. Their names are related to the oxidation state of the nonmetal atom to which the O atoms are bonded, ranging from *hypo* (lowest) to *per* (highest) according to the scheme

Increasing oxidation state of nonmetal ⟶

hypo__ite   __ite   __ate   per__ate

Increasing number of oxygen atoms ⟶

TABLE 3.4   Nomenclature of Some Oxoacids and Their Salts

Oxidation State	Formula of Acid[a]	Name of Acid[b]	Formula of Salt[b]	Name of Salt
Cl: +1	HClO	*Hypo*chlor*ous* acid	NaClO	Sodium *hypo*chlor*ite*
Cl: +3	HClO$_2$	Chlor*ous* acid	NaClO$_2$	Sodium chlor*ite*
Cl: +5	HClO$_3$	Chlor*ic* acid	NaClO$_3$	Sodium chlor*ate*
Cl: +7	HClO$_4$	*Per*chlor*ic* acid	NaClO$_4$	Sodium *per*chlor*ate*
N: +3	HNO$_2$	Nitr*ous* acid	NaNO$_2$	Sodium nitr*ite*
N: +5	HNO$_3$	Nitr*ic* acid	NaNO$_3$	Sodium nitr*ate*
S: +4	H$_2$SO$_3$	Sulfur*ous* acid	Na$_2$SO$_3$	Sodium sulf*ite*
S: +6	H$_2$SO$_4$	Sulfur*ic* acid	Na$_2$SO$_4$	Sodium sulf*ate*

[a] In all these acids, H atoms are bonded to O atoms, not the central nonmetal atom. Often formulas are written to reflect this fact, for instance, HOCl instead of HClO and HOClO instead of HClO$_2$.

[b] In general, the *ic* and *ate* names are assigned to compounds in which the central nonmetal atom has an oxidation state equal to the periodic group number minus 10. Halogen compounds are exceptional in that the *ic* and *ate* names are assigned to compounds in which the halogen has an oxidation state of +5 (even though the group number is 17).

5. All the common oxoanions of Cl, Br, and I carry a charge of 1−.

6. Some series of oxoanions also contain varying numbers of H atoms and are named accordingly. For example, HPO$_4^{2-}$ is the *hydrogen phosphate* ion and H$_2$PO$_4^-$, the *dihydrogen phosphate* ion.

7. The prefix *thio* signifies that a sulfur atom has been substituted for an oxygen atom. (The sulfate ion has *one* S and *four* O atoms; thiosulfate ion has *two* S and *three* O atoms.)

Oxoacids

The majority of acids are **ternary compounds.** They contain *three* different elements—hydrogen, and two other nonmetals. If one of the nonmetals is oxygen, the acid is called an **oxoacid.** Or, think of oxoacids as combinations of hydrogen ions (H$^+$) and oxoanions. The scheme for naming oxoacids is similar to that outlined for oxoanions, except that the ending *ous* is used instead of *ite* and *ic* instead of *ate*. Several oxoacids are listed in Table 3.4. Also listed are the names and formulas of compounds in which the hydrogen of the oxoacid has been replaced by a metal such as sodium. These compounds are called *salts;* we will say much more about them later in the text. Acids are molecular compounds, and salts are ionic compounds.

EXAMPLE 3-10

Applying Various Rules in Naming Compounds. Name the compounds (a) CuCl$_2$; (b) ClO$_2$; (c) HIO$_4$; (d) Ca(H$_2$PO$_4$)$_2$.

Solution

(a) In this compound, the oxidation state of Cu is +2. Because Cu can also exist in the oxidation state +1, we must clearly distinguish between the two possible chlorides. CuCl$_2$ is copper(II) chloride.

(b) Both Cl and O are nonmetals. ClO$_2$ is a binary molecular compound called chlorine dioxide.

(c) The oxidation state of I is +7. By analogy to the chlorine-containing oxoacids in Table 3.4, we should name this compound periodic acid (pronounced "purr eye oh dic" acid).

KEEP IN MIND

that ClO$_2$ is *not* chlorite ion (it carries no net charge). Furthermore, there can be no compound called chlorite. A compound cannot consist of just a single type of ion. ▶

(d) The polyatomic anion $H_2PO_4^-$ is dihydrogen phosphate ion. Two of these ions are present for every Ca^{2+} ion in the compound calcium dihydrogen phosphate.

Practice Example A: Name the compounds SF_6, HNO_2, $Ca(HCO_3)_2$, $FeSO_4$.

Practice Example B: Name the compounds NH_4NO_3, PCl_3, $HBrO$, $AgClO_4$, $Fe_2(SO_4)_3$.

EXAMPLE 3-11

Applying Various Rules in Writing Formulas. Write the formula of the compound **(a)** tetranitrogen tetrasulfide; **(b)** ammonium chromate; **(c)** bromic acid; **(d)** calcium hypochlorite.

Solution

(a) Molecules of this compound consist of *four* N atoms and *four* S atoms. The formula is N_4S_4.

(b) Two ammonium ions (NH_4^+) must be present for every chromate ion (CrO_4^{2-}). Place parentheses around NH_4^+, followed by the subscript 2. The formula is $(NH_4)_2CrO_4$. (This formula is read as "N-H-4, taken twice, C-R-O-4.")

(c) The ic acid for the oxoacids of the halogens (group 17) has the halogen in oxidation state +5. Bromic acid is $HBrO_3$ (analogous to $HClO_3$ in Table 3.4).

(d) Here there are *one* Ca^{2+} and *two* ClO^- ions in a formula unit. This leads to the formula $Ca(ClO)_2$.

Practice Example A: Write formulas for the following compounds: boron trifluoride, potassium dichromate, sulfuric acid, calcium chloride.

Practice Example B: Write formulas for the following compounds: aluminum nitrate, tetraphosphorous decoxide, chromium(III) hydroxide, iodic acid.

> **KEEP IN MIND**
> the importance of the proper placement of parentheses in chemical formulas. Without parentheses, we would have $CaClO_2$, an *incorrect* formula for both calcium hypochlorite and calcium chlorite. ▶

Some Compounds of Greater Complexity

The copper compound that Joseph Proust used to establish the law of constant composition (page 35) is referred to in different ways. If you look up Proust's compound in a handbook of minerals, you will find it listed as *malachite,* with the formula $Cu_2(OH)_2CO_3$. In a handbook that specializes in pharmaceutical applications, this same compound is listed as *basic cupric carbonate,* with the formula $CH_2Cu_2O_5$. In a chemistry handbook, it is listed as *copper (II) carbonate dihydroxide,* with the formula $CuCO_3 \cdot Cu(OH)_2$. All you need to understand at this point is that regardless of the formula you use, you should obtain the same molar mass (221.12 g/mol), the same mass percent copper (57.48% Cu), the same H-to-O mole ratio (2 mol H/5 mol O), and so on. In short, you should be able to interpret a formula, no matter how complex its appearance.

> ▶ In general, we use dots ($\cdot$) to show that a formula is a composite of two or more simpler formulas.

Some complex substances you are certain to encounter are known as hydrates. In a **hydrate,** each formula unit of the compound has associated with it a certain number of water molecules. This does not mean that the compounds are "wet", however. The water molecules are incorporated in the solid structure of the compound. The formula shown below signifies *six* H_2O molecules per formula unit of $CoCl_2$.

$$CoCl_2 \cdot 6\ H_2O$$

Using the prefix for six—*hexa*—we call this compound cobalt(II) chloride *hexa*hydrate. Its formula mass is that of $CoCl_2$ *plus* that associated with six

▲ **FIGURE 3-7**
Effect of moisture on CoCl$_2$
The piece of filter paper was soaked in a water solution of cobalt(II) chloride and then allowed to dry. When kept in dry air, the paper is blue in color (anhydrous CoCl$_2$). In humid air, the paper changes to pink (CoCl$_2$ · 6 H$_2$O).

H$_2$O: 129.8 u + (6 × 18.02 u) = 237.9 u. We can speak of the mass percent water in a hydrate. For CoCl$_2 \cdot$ 6H$_2$O this is

$$\% \ H_2O = \frac{(6 \times 18.02) \ g \ H_2O}{237.9 \ g \ CoCl_2 \cdot 6H_2O} \times 100\% = 45.45\%$$

The water present in compounds as water of hydration can generally be removed, in part or totally, by heating. When the water is totally removed, the resulting compound is said to be *anhydrous* (without water). Anhydrous compounds can be used as water absorbers, as in the use of anhydrous magnesium perchlorate in combustion analysis (recall Figure 3-5). CoCl$_2$ gains and loses water quite readily and indicates this through a color change. Anhydrous CoCl$_2$ is blue, whereas the hexahydrate is pink. This fact can be used to make a simple moisture detector (Figure 3-7).

3-7 Names and Formulas of Organic Compounds

Organic compounds abound in nature. The foods we eat are made up almost exclusively of organic compounds, including not only energy-producing fats and carbohydrates and muscle-building proteins, but also trace compounds that impart color, odor, and flavor to these foods. Almost all fuels, whether used to power automobiles, trucks, trains, or airplanes, are mixtures of organic compounds of a type called hydrocarbons. Most of the drugs produced by pharmaceutical companies are complex organic compounds, as are common plastics. The multiplicity of organic compounds is so vast that *organic chemistry* exists as a separate field of chemistry.

The great diversity of organic compounds arises from the ability of carbon atoms to combine readily with other carbon atoms and with atoms of a number of other elements. Carbon atoms join together to form a framework of chains or rings to which other atoms are attached. All organic compounds contain carbon atoms; almost all contain hydrogen atoms; and many common ones also have oxygen, nitrogen, or sulfur atoms. These possibilities allow for a limitless number of different organic compounds. Organic compounds are mostly molecular; a few are ionic.

The naming of organic compounds uses a different set of rules than inorganic compounds. There are millions of organic compounds, many comprised of highly complex molecules. Their names are equally complicated. A systematic approach to naming these compounds is crucial. The usual name, often called the common, or *trivial*, name, for a familiar sweetener is sucrose (sugar). The systematic name is α-D-glucopyranosyl-β-D-fructofuranoside. At this point, however, we only need to recognize organic compounds and use their common names, together with an occasional systematic name. We will look at the systematic nomenclature of organic compounds in more detail in Chapter 27.

Hydrocarbons

Compounds containing only carbon and hydrogen are called **hydrocarbons.** The simplest hydrocarbon contains one carbon atom and four hydrogen atoms. It is methane, CH$_4$ (Figure 3-8a). As the number of carbon atoms increases, the number of hydrogen atoms also increases in a systematic way, depending on the type of hydrocarbon. The complexity of organic chemistry arises because carbon atoms can form chains and rings, and the nature of the chemical bonds between the carbon atoms can vary. Hydrocarbons containing only single bonds are called **alkanes.** The simplest alkane is methane, followed by ethane, C$_2$H$_6$ (Figure 3-8b), and then propane, C$_3$H$_8$, (Figure 3-8c). The fourth member of the series, butane, C$_4$H$_{10}$, was

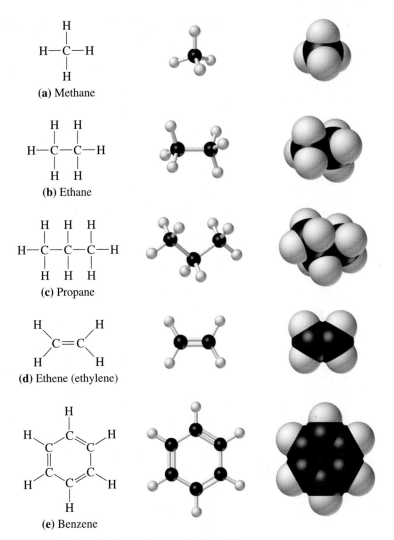

(a) Methane

(b) Ethane

(c) Propane

(d) Ethene (ethylene)

(e) Benzene

▲ FIGURE 3-8   **Visualizations of some hydrocarbons**

TABLE 3.5 Word stem (or prefix) indicating the number of carbon atoms in simple organic molecules	
Stem (or prefix)	**Number of C Atoms**
Meth	1
Eth	2
Prop	3
But	4
Pent	5
Hex	6
Hept	7
Oct	8
Non	9
Dec	10

shown in Figure 3-2a. Notice that each succeeding member of the alkane series is formed by the addition of one C atom and two H atoms to the preceding member.

The names of the alkanes are composed of two parts: a word stem indicating the number of carbon atoms (Table 3.5) and the ending (suffix) *ane* indicating that the molecule is an alk*ane*. Thus C_3H_8 is propane and C_7H_{16} is heptane.

Hydrocarbon molecules with one or more double bonds between carbon atoms are called *alkenes*. The simplest of the alkenes is ethene (Figure 3-8d); its name consists of the stem *eth* and the ending *ene*. Benzene, C_6H_6 (Figure 3-8e), is a molecule with six carbon atoms arranged in a hexagonal ring. Molecules with structures related to benzene make up a large proportion of known organic compounds.

Refer back to Figure 3-2, and you will notice that butane and methylpropane have the same molecular formula, C_4H_{10}, but different structural formulas. Butane is based on a four-carbon chain, whereas in methylpropane a —CH_3 group, called a *methyl* group, is attached to the middle carbon atom of the three-carbon propane chain. Butane and methylpropane are *isomers*. **Isomers** are molecules that have the same molecular formula but different arrangements of atoms in space. As organic molecules become more complex, the possibilities for isomerism increase very rapidly.

EXAMPLE 3-12

Recognizing Isomers. Are the following pairs of molecules isomers or not?

(a) $CH_3CH(CH_3)(CH_2)_3CH_3$ and $CH_3CH_2CH(CH_3)(CH_2)_3CH_3$

(b) 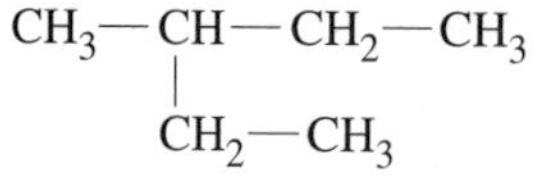 $CH_3\!-\!CH\!-\!CH_2\!-\!CH_3$ and $CH_3\!-\!CH\!-\!CH_2\!-\!CH_2\!-\!CH_3$
            $|$ $|$
           $CH_2\!-\!CH_3$ CH_3

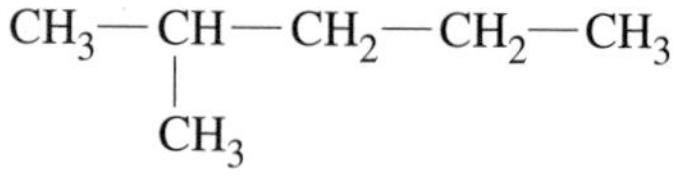

Solution

Isomers have the same molecular formula but different structures. We first check to see if the molecular formulas are the same. If the formulas are *not* the same, they signify *different* compounds. If the formulas *are* the same, they *may* signify isomers, but only if the structures are *different*.

(a) The molecular formula of the first compound is C_7H_{16}, and the second compound has the formula C_8H_{18}. The molecules are not isomers.

(b) These molecules have the same formula, C_6H_{14}, but they differ in structure. They are isomers. The difference in structure is that in the first structure, a methyl side chain is on the third carbon atom of a five-carbon chain and in the second structure it is on the second carbon atom. In making this assessment, we identify the longest continuous chain of carbon atoms and imagine that the C atoms are numbered so that the methyl group appears at the lowest numbered C atom possible.

Practice Example A: Are the following pairs of molecules isomers?

(a) $CH_3C(CH_3)_2(CH_2)_3CH_3$ and $CH_3CH(CH_3)CH(CH_3)(CH_2)_3CH_3$

(b) 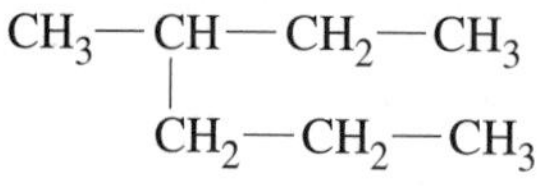 $CH_3\!-\!CH\!-\!CH_2\!-\!CH_3$ and $CH_3\!-\!CH\!-\!CH_2\!-\!CH\!-\!CH_3$
            $|$ $|$ $|$
           $CH_2\!-\!CH_2\!-\!CH_3$ CH_3 CH_3

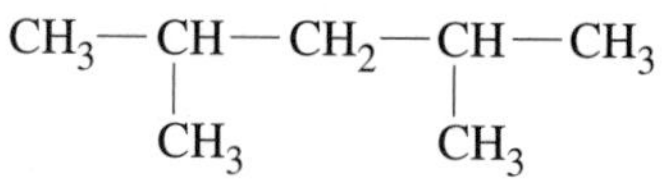

Practice Example B: Are the pairs of molecules represented by the following structural formulas isomers?

(a)

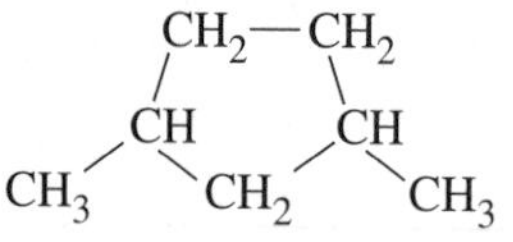

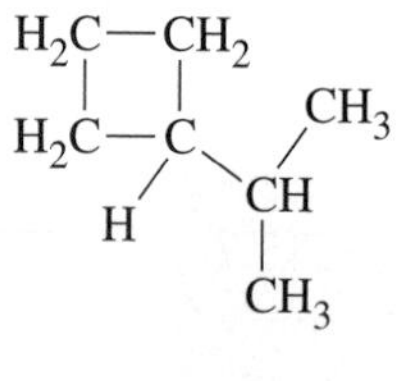

(b)

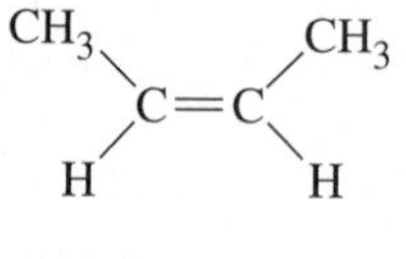

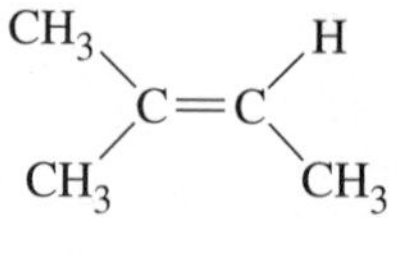

Functional Groups

Carbon chains provide the framework upon which other atoms or groups of atoms can replace one or more of the hydrogen atoms. We can illustrate this with the common molecule alcohol that occurs in beer, wine, and spirits. The molecule is *ethanol*, CH_3CH_2OH, in which one of the H atoms of ethane is replaced by an $-OH$ group (Figure 3-9a). The systematic name ethanol is derived from the name of the alkane,

 *Functional Group activity*

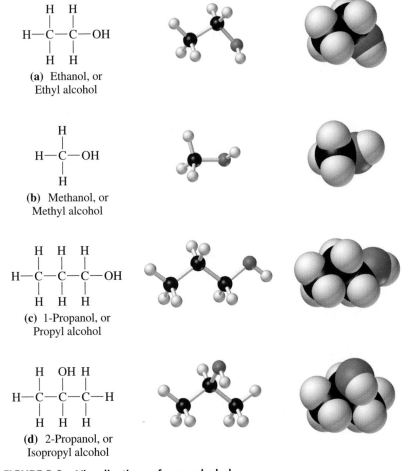

(a) Ethanol, or Ethyl alcohol

(b) Methanol, or Methyl alcohol

(c) 1-Propanol, or Propyl alcohol

(d) 2-Propanol, or Isopropyl alcohol

▲ FIGURE 3-9   **Visualizations of some alcohols**

ethane, with the final *e* replaced by the suffix *ol*. The suffix *ol* designates the presence of the —OH group in a class of organic molecules called **alcohols**.

Ethyl alcohol, the common name of ethanol, also indicates attachment of the —OH group to the ethane hydrocarbon chain. To name the alkane chain as a group, replace the final *e* by *yl*, so that ethane becomes *ethyl*; thus *ethyl alcohol* for CH_3CH_2OH. It is often the case that the common name of one compound, alcohol in this case, will provide the generic name for a complete class of compounds, that is, all alcohols contain at least one —OH group.

Another common alcohol is *methanol,* or wood alcohol, which has the formula CH_3OH (Figure 3-9b). The common name for methanol is methyl alcohol. It is interesting to note that wood alcohol is a dangerous poison, whereas the grain alcohol in beer and wine is safe to consume in moderate quantities.

The —OH group in alcohols is one of the many *functional groups* found in organic compounds. **Functional groups** are individual atoms or groupings of atoms that are attached to the carbon chains or rings of organic molecules and give the molecules their characteristic properties. Compounds with the same functional group generally have similar properties. The —OH group is called the *hydroxyl* group. At this point we will introduce only a few functional groups, and we will deal with functional groups in more detail in Chapter 27.

The presence of functional groups also increases the possibility of isomers. For example, there is only one propane molecule, C_3H_8. However, if one of the H atoms is replaced by a hydroxyl group, two possibilities exist for the point of attachment: at one of the end C atoms or at the middle C atom (Figure 3-9c,d). This leads to two isomers. The alcohol with the —OH group attached to the end carbon atom is commonly called propyl alcohol or, systematically, 1-propanol; the prefix 1- indicates that the —OH group is on the first, or end C atom. The alcohol with the —OH group attached to the middle carbon atom is commonly called isopropyl alcohol or, systematically, 2-propanol; the prefix 2- indicates that the —OH group is on the second C atom from the end.

Another important functional group is the *carboxyl* group, —COOH, which confers acidic properties on a molecule. The C atom in the carboxyl group is bound to the two O atoms in two ways. One bond is a single bond to an oxygen atom that is also attached to a H atom, and the other is a double bond to a lone O atom (Figure 3-10a). The *ionizable*, or *acidic*, hydrogen atom in a carboxyl group is the one that is attached to an O atom. Compounds containing the carboxyl group are called **carboxylic acids.** The first carboxylic acid based on alkanes is *methanoic* acid, HCOOH (Figure 3-10b). In the systematic name, the *methan* indicates *one* carbon atom and the *oic* acid indicates a carboxylic acid. The common name for methanoic acid is formic acid, deriving from the Latin word *formica*, meaning "ant." Formic acid is injected by an ant when it bites. This leads to the burning sensation that accompanies the bite.

The simplest carboxylic acid containing two carbon atoms is *ethanoic* acid, more commonly known as acetic acid. The molecular formula is CH_3COOH, and the structure is shown in Figure 3-10c. Vinegar is a solution of acetic acid in water. An additional functional group is introduced in the examples that follow—a halogen atom (F, Cl, Br, I) substituting for one or more H atoms. When present as functional groups, the halogens carry the names, *fluoro, chloro, bromo,* and *iodo*.

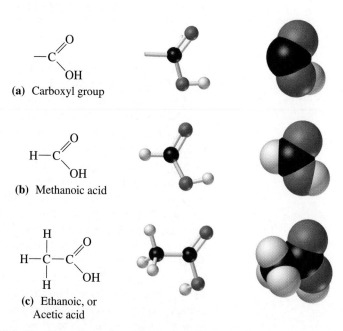

(a) Carboxyl group

(b) Methanoic acid

(c) Ethanoic, or Acetic acid

▲ FIGURE 3-10   **The carboxylic acid group and visualizations of two carboxylic acids**

EXAMPLE 3-13

Recognizing Types of Organic Compounds. What type of compound is each of the following?

(a) $CH_3CH_2CH_2CH_3$

(b) $CH_3CHClCH_2CH_3$

(c) $CH_3CH_2CO_2H$

(d) $CH_3CH_2CH(OH)CH_2CH_3$

Solution

(a) The carbon-to-carbon bonds are all single bonds in this hydrocarbon. This compound is an alkane.

(b) In its molecules, there are only single bonds, and one H atom has been replaced by a Cl atom. This compound is a chloroalkane.

(c) The presence of the carboxyl group, $—CO_2H$, in its molecules means that this compound is a carboxylic acid.

(d) The presence of the hydroxyl group, $—OH$, in its molecules means that this compound is an alcohol.

Practice Example A: What type of molecules correspond to the following formulas?

(a) $CH_3CH_2CH_3$

(b) $ClCH_2CH_2CH_3$

(c) $CH_3CH_2CH_2CO_2H$

(d) $CH_3CHCHCH_3$

Practice Example B: What type of molecules correspond to these formulas?

(a) $CH_3CH(OH)CH_3$

(b) $CH_3CH(OH)CH_2CO_2H$

(c) $CH_2ClCH_2CO_2H$

(d) $BrCHCHCH_3$

EXAMPLE 3-14

Naming Organic Compounds. Name these compounds.

(a) $CH_3CH_2CH_2CH_2CH_3$

(b) $CH_3CHFCH_2CH_3$

(c) $CH_3CH_2CO_2H$

(d) $CH_3CH_2CH(OH)CH_2CH_3$

Solution

(a) The structure is that of an alkane molecule with a five-carbon chain, so the compound is pentane.

(b) The structure is that of a fluoroalkane molecule with the F atom on the second carbon atom of a four-carbon chain. The compound is called 2-fluorobutane.

(c) The carbon chain in this structure is three carbon atoms long with the end C atom in a carboxyl group. The compound is propanoic acid.

(d) This structure is that of an alcohol with the hydroxyl group on the third carbon atom of a five-carbon chain. The compound is called 3-pentanol.

Practice Example A: Name the following compounds: (a) $CH_3CH(OH)CH_3$, (b) $ICH_2CH_2CH_3$, (c) $CH_3CH(CH_3)CH_2CO_2H$, and (d) CH_3CHCH_2.

Practice Example B: Give plausible names for the molecules that correspond to the following ball-and-stick models.

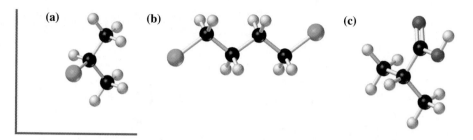

EXAMPLE 3-15

Writing Structural Formulas from the Names of Organic Compounds. Write the condensed structural formula for each of the following organic compounds: **(a)** butane, **(b)** butanoic acid, **(c)** 1-chloropentane, **(d)** 1-hexanol.

Solution

(a) The word stem *but* indicates a structure with a four-carbon chain, and the suffix *ane* indicates an alkane. There are no functional groups indicated; hence the condensed structural formula is $CH_3(CH_2)_2CH_3$.

(b) The *oic* ending indicates that the end carbon atom of the four-carbon chain is part of a carboxylic acid group. The condensed structural formula is $CH_3(CH_2)_2CO_2H$.

(c) The prefix *chloro* indicates the substitution of a chlorine atom for a H atom, and the 1-designates that it is on the first C atom of the carbon chain. The carbon chain is five C atoms long, as signified by the word stem *pent*. The condensed structural formula is $CH_3(CH_2)_3CH_2Cl$.

(d) The suffix *ol* indicates the presence of a hydroxyl group in place of a H atom, and the 1-designates that it is on the first C atom of the carbon chain. The word stem *hex* signifies that the carbon chain is six C atoms long. The condensed structural formula is $CH_3(CH_2)_4CH_2OH$.

Practice Example A: Write the condensed structural formula for each of the following organic compounds: **(a)** pentane, **(b)** ethanoic acid, **(c)** 1-iodooctane, **(d)** 1-pentanol.

Practice Example B: Write the condensed structural formula for each of the following organic compounds: **(a)** propene, **(b)** 1-heptanol, **(c)** chloroacetic acid, **(d)** hexanoic acid.

Summary

Chemical formulas, once established, provide a wealth of information about chemical compounds, serving as the basis for determining

- formula (molecular) masses
- molar masses
- percent compositions
- various relationships among the elements present

Chemical formulas can be related to experimentally determined percent compositions of compounds. For organic compounds, this often involves combustion analysis. The formulas determined from experimental data are empirical formulas—the simplest formulas that can be written. Molecular formulas can be related to empirical formulas when experimentally determined molecular masses are available.

Compounds are designated by name as well as by formula. The periodic table of the elements, the classification of elements as metals and nonmetals, and oxidation state are all used in naming and writing formulas for binary ionic and molecular compounds, polyatomic ions, oxoacids and their salts, and hydrates.

Organic compounds are based on the element carbon; they frequently exhibit isomerism. Hydrocarbons contain only carbon and hydrogen. Alkane hydrocarbon molecules contain only single bonds, and alkenes contain at least one double bond.

Functional groups confer distinctive properties on organic molecules when the groups are substituted for H atoms on carbon chains or rings. The hydroxyl group —OH is present in *alcohols*, and the carboxyl group —COOH, in *carboxylic acids*.

Integrative Example

Molecules of a dicarboxylic acid have *two* carboxyl groups (—COOH). A 2.250-g sample of a dicarboxylic acid was burned in an excess of oxygen and yielded 4.548 g CO_2 and 1.629 g H_2O. In a separate experiment, the molecular mass of the acid was found to be 174 u. From these data, what can we deduce about the structural formula of this acid?

Our approach will require several steps, which we can identify in this general way. (1) Use the combustion data to determine the percent composition of the compound (similar to Example 3-6). (2) Determine the empirical formula from the percent composition (similar to Example 3-5). (3) Obtain the molecular formula from the empirical formula and the molecular mass. (4) Determine how the C, H, and O atoms represented in the molecular formula might be assembled into a dicarboxylic acid.

1. *Determine the percent composition.* Shown below is the calculation of the % H in the compound.

$$? \text{ g H} = 1.629 \text{ g } H_2O \times \frac{1 \text{ mol } H_2O}{18.015 \text{ g } H_2O}$$

$$\times \frac{2 \text{ mol H}}{1 \text{ mol } H_2O} \times \frac{1.00794 \text{ g H}}{1 \text{ mol H}} = 0.1823 \text{ g H}$$

$$\% \text{ H} = \frac{0.1823 \text{ g H}}{2.250 \text{ g compd.}} \times 100\% = 8.102\% \text{ H}$$

The mass of C and the % C in the compound burned are calculated in a similar manner, yielding 1.241 g C and 55.16% C in the compound.

The % O in the compound is obtained as a difference, that is,

$$\% \text{ O} = 100.00\% - 55.16\% \text{ C} - 8.102\% \text{ H} = 36.74\% \text{ O}$$

2. *Obtain the empirical formula from the percent composition.* The masses of the elements in 100.0 g of the compound are

    55.16 g C   8.102 g H   36.74 g O

The numbers of moles of the elements in 100.0 g of the compound are

$$55.16 \text{ g C} \times \frac{1 \text{ mol C}}{12.01 \text{ g C}} = 4.593 \text{ mol C}$$

$$8.102 \text{ g H} \times \frac{1 \text{ mol H}}{1.00794} = 8.038 \text{ mol H}$$

$$36.74 \text{ g O} \times \frac{1 \text{ mol O}}{15.999 \text{ g O}} = 2.296 \text{ mol O}$$

The tentative formula based on the numbers of moles determined here is

$$C_{4.593}H_{8.038}O_{2.296}$$

Divide all the subscripts by 2.296 to obtain

$$C_2H_{3.50}O$$

Multiply all subscripts by two. The result is the empirical formula $C_4H_7O_2$ (87 u).

3. *Obtain the molecular formula.* The experimentally determined molecular mass is 174 u, just twice the molecular mass based on the empirical formula. The molecular formula is $C_8H_{14}O_4$.

4. *Assemble the atoms in $C_8H_{14}O_4$ into a plausibile structural formula.* The structure must contain two —COOH groups; this accounts for all four O atoms, two C atoms and two H atoms. The remainder of the structure is based on C_6H_{12}. The simplest possibility is to arrange the six —CH_2 into a six-carbon chain and attach the —COOH groups at the ends of the chain.

$$HOOC — CH_2(CH_2)_4CH_2 — COOH$$

However, there are other possibilities based on shorter chains with branches, for example

$$HOOC—CH_2—\overset{\displaystyle CH_3}{\underset{\displaystyle CH_3}{\overset{|}{\underset{|}{C}}}}—CH_2—CH_2—COOH$$

From the available data, we can go no further with our answer.

Key Terms

anion (3-1)
alcohol (3-7)
alkane (3-7)
binary compound (3-6)
carboxylic acid (3-7)
cation (3-1)
chemical formula (3-1)
empirical formula (3-1)
formula mass (3-2)

formula unit (3-1)
functional group (3-7)
hydrate (3-6)
hydrocarbon (3-7)
inorganic compound (3-5)
ionic compound (3-1)
isomer (3-7)
molar mass (3-2)
molecular compound (3-1)

molecular formula (3-1)
molecular mass (3-2)
organic compound (3-5)
oxidation state (3-4)
oxoacid (3-6)
oxoanion (3-6)
polyatomic ion (3-6)
structural formula (3-1)
ternary compound (3-6)

▲ Instrumentation for mass spectometry

Through Figures 2-14 and 2-15, we learned how the masses and relative abundances of the isotopes of an element are determined by mass spectrometry, thereby making it possible to establish weighted-average atomic masses.

If the ions analyzed by a mass spectrometer are *molecular* ions, the experimental data yield a *molecular* mass. For example, through mass spectrometry the molecular mass of vitamin C is found to be 176 u. Combining this fact with the empirical formula from Example 3-6 ($C_3H_4O_3$), we obtain the true molecular formula, $C_6H_8O_6$ (that is, 176 u = 2 × 88 u).

Actually, mass spectrometry is capable of providing much more than molecular masses. It can tell us about the *structures* of molecules. To see how this is possible, consider 2-methylpentane, which has a molecular mass of 86 u. Ioniza-

Ionization:
$$CH_3CH_2CH_2\overset{\overset{\displaystyle CH_3}{|}}{C}HCH_3 + e^- \longrightarrow$$

$$\left[CH_3CH_2CH_2\overset{\overset{\displaystyle CH_3}{|}}{C}HCH_3 \right]^+ + 2\,e^-$$

tion occurs when an energetic electron in a cathode-ray beam strikes a molecule. The molecule becomes the ion $[CH_3CH_2CH_2CH(CH_3)_2]^+$, and the released electron joins the original electron in the cathode-ray beam.

An impinging electron in the cathode-ray beam has more than enough energy to induce ionization and transfer some of its excess energy to the molecular ion. When this energy is distributed throughout the ion, some of the bonds in the ion rupture, producing fragments. Some of these fragments carry a unit of positive charge and are detected in the mass spectrum. Others are neutral species and are not detected. The fragmentation pattern outlined below includes mass-to-charge ratios (m/z) of a few of the several ions produced.

Fragmentation:

$$\left[CH_3CH_2CH_2\overset{\overset{\displaystyle CH_3}{|}}{C}H \right]^+ + CH_3$$
$$m/z = 71$$

$$\left[CH_3CH_2CH_2\overset{\overset{\displaystyle CH_3}{|}}{C}HCH_3 \right]^+ \longrightarrow CH_3CH_2CH_2 + \left[\overset{\overset{\displaystyle CH_3}{|}}{C}HCH_3 \right]^+$$
$$m/z = 86 \qquad\qquad m/z = 43$$

$$CH_3CH_2 + \left[CH_2\overset{\overset{\displaystyle CH_3}{|}}{C}HCH_3 \right]^+$$
$$m/z = 57$$

Review Questions

1. In your own words, define or explain the following terms or symbols: (a) formula unit; (b) S_8; (c) ionic compound; (d) oxoacid; (e) hydrate.

2. Briefly describe each of the following ideas or methods: (a) molecule of an element; (b) structural formula; (c) oxidation state; (d) carbon-hydrogen-oxygen determination by combustion analysis.

3. Explain the important distinctions between each pair of terms: (a) ionic and molecular compounds; (b) empirical and molecular formulas; (c) systematic and trivial, or common, name; (d) binary and ternary acid.

4. Explain each of the following terms as it applies to the element oxygen: (a) atomic mass; (b) molecular mass; (c) molar mass.

5. Calculate the total number of (a) atoms in one molecule of nitroglycerin, $C_3H_5(NO_3)_3$; (b) atoms in 0.00102 mol C_2H_6; (c) F atoms in 12.15 mol $C_2HBrClF_3$.

6. Determine the *mass*, in grams, of (a) 7.34 mol N_2O_4; (b) 3.16×10^{24} O_2 molecules; (c) 18.6 mol $CuSO_4 \cdot 5H_2O$; (d) 4.18×10^{24} molecules of $C_2H_4(OH)_2$.

7. Determine the number of moles of Br_2 in a sample consisting of (a) 8.08×10^{22} Br_2 molecules; (b) 2.17×10^{24} Br atoms; (c) 11.3 kg bromine; (d) 2.65 L liquid bromine ($d = 3.10$ g/mL).

8. The amino acid methionine, which is essential in human diets, has the molecular formula $C_5H_{11}NO_2S$. Determine (a) its molecular mass; (b) the number of moles of H atoms per mole of methionine; (c) the number of grams of C per

In the mass spectrum in Figure 3-11, we arbitrarily assign a value of 100 to the ion present in the greatest abundance ($m/z = 43$) and relate the abundances of the other ions to this base peak. The mass spectrum of 3-methylpentane, shown in Figure 3-12, differs in significant ways from that of its isomer, 2-methylpentane. Notably, the base peak with value 100 now has $m/z = 57$. This is because there are *two* possible modes of fragmentation that produce the $[CH_2CH(CH_3)_2]^+$ ion, as shown at the right.

Altogether there are *five* isomers of the alkane hydocarbon C_6H_{14}. Each has a unique mass spectrum, and the differences among them can be explained by applying a set of rules for analyzing fragmentation patterns. However, we need go no further to emphasize the main point of this discussion: Mass spectrometry enables us to establish both molecular formulas *and* structural formulas.

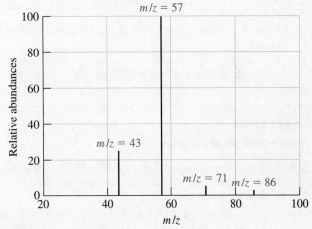

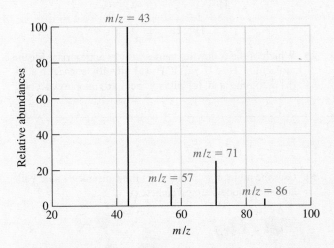

▲ FIGURE 3-11
Partial mass spectrum of 2-methylpentane
Peaks are shown only for the four molecular ions described in the text. Several additional ions are present in the complete spectrum.

▲ FIGURE 3-12
Partial spectrum of 3-methylpentane
The same four peaks are shown as in Figure 3-11.

mole of methionine; **(d)** the number of C atoms in 9.07 mol methionine.

9. Determine the mass percent H in the hydrocarbon decane, $C_{10}H_{22}$.

10. Determine the mass percent O in the mineral malachite, $Cu_2(OH)_2CO_3$.

11. Determine the mass percent H_2O in the hydrate $Cr(NO_3)_3 \cdot 9H_2O$.

12. Determine the percent, by mass, of the indicated element in each of the following.
   (a) Pb in tetraethyl lead, $Pb(C_2H_5)_4$, once extensively used as an antiknock additive in gasoline
   (b) Fe in Prussian blue, $Fe_4[Fe(CN)_6]_3$, a pigment used in paints and printing inks
   (c) Mg in chlorophyll, $C_{55}H_{72}MgN_4O_5$, the green pigment in plant cells

13. *Without doing detailed calculations,* explain which of the following has the *greatest* mass percent of sulfur: SO_2, S_2Cl_2, Na_2S, $Na_2S_2O_3$, or CH_3CH_2SH.

14. Diethylene glycol, used to deice aircraft, is a carbon–hydrogen–oxygen compound with 45.27% C and 9.50% H by mass. What is its empirical formula?

15. The food flavor enhancer monosodium glutamate (MSG) has the composition 13.6% Na, 35.5% C, 4.8% H, 8.3% N, 37.8% O, by mass. What is the empirical formula of MSG?

16. A compound of C, H, and O, known as terephthalic acid and used in the manufacture of Dacron, has a molecular mass of 166.1 u. By combustion analysis, it is found to have 57.83% C and 3.64% H, by mass. What is the molecular formula of terephtalic acid?

17. Ibuprofen is a carbon–hydrogen–oxygen compound used in painkillers. When a 2.174-g sample is burned completely, it yields 6.029 g CO_2 and 1.709 g H_2O.

(a) What is the percent composition, by mass, of ibuprofen?

(b) What is the empirical formula of ibuprofen?

18. An oxide of chromium used in chrome plating has a formula mass of 100.0 u and contains *four* atoms per formula unit. Establish the formula of this compound, *with a minimum of calculation.*

19. Two oxides of sulfur have nearly identical molecular masses. One oxide consists of 40.05% S. What are the simplest possible formulas for the two oxides?

20. Supply the missing information (name or formula) for each *ion*

(a) lead(II), __   (b) __, Co^{3+}

(c) __, Ba^{2+} (d) chromium(II), __

(e) periodate, __   (f) __, ClO_2^-

(g) gold(III), __   (h) __, HSO_3^-

(i) hydrogen carbonate, __   (j) cyanide, __

21. Name the compounds (a) KBr; (b) $SrCl_2$; (c) ClF_3; (d) N_2O_4; (e) PCl_5.

22. Name the compounds (a) KCN; (b) HClO; (c) $(NH_4)_2SO_4$; (d) KIO_3.

23. Indicate the oxidation state of the underlined element in (a) <u>Zn</u>; (b) Ba<u>S</u>; (c) <u>N</u>O_2; (d) H<u>N</u>O_2; (e) <u>V</u>O^{2+}; (f) H_2<u>P</u>O_4^-.

24. Write correct formulas for the compounds (a) magnesium bromide; (b) barium oxide; (c) mercury(II) acetate; (d) iron(III) oxalate; (e) strontium perchlorate; (f) potassium hydrogen sulfate; (g) nitrogen trichloride; (h) bromine pentafluoride.

25. Name of each of the following acids: (a) $HClO_2$; (b) H_2SO_3; (c) H_2Se; (d) HNO_2.

26. Supply the formula for each of the following acids: (a) hydroiodic acid; (b) nitric acid; (c) phosphoric acid; (d) sulfuric acid.

27. Which of the following names is most appropriate for the molecule with the structure shown below: (a) butyl alcohol, (b) 2-butanol, (c) 1-butanol, (d) ethyl ethanol?

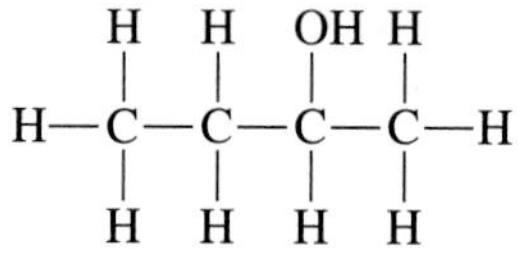

$$H-\overset{\displaystyle H}{\underset{\displaystyle H}{C}}-\overset{\displaystyle H}{\underset{\displaystyle H}{C}}-\overset{\displaystyle OH}{\underset{\displaystyle H}{C}}-\overset{\displaystyle H}{\underset{\displaystyle H}{C}}-H$$

28. Which of the following names is most appropriate for the molecule $CH_3(CH_2)_2COOH$: (a) dimethyleneacetic acid, (b) propionic acid, (c) butanoic acid, (d) oxobutylalcohol?

Exercises

Representing Molecules

29. Refer to the color scheme on page 69, and give the molecular formulas for the molecules whose ball-and -stick models are given here.

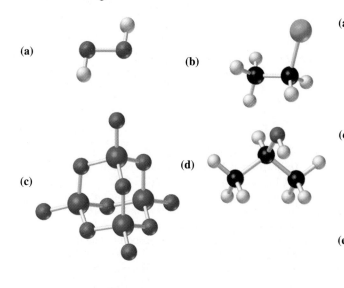

(a)

(b)

(c)

(d)

(e)

30. Give the molecular formulas for the molecules whose ball-and-stick models are given here.

(a)

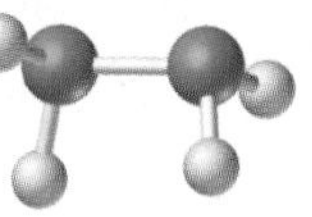

(b)

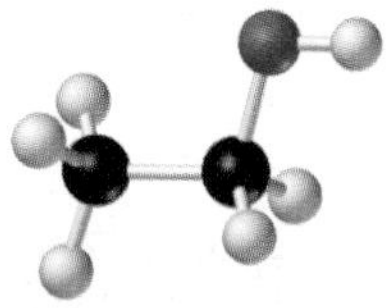

(c)

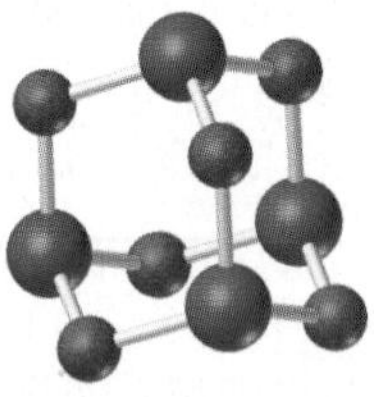

(d)

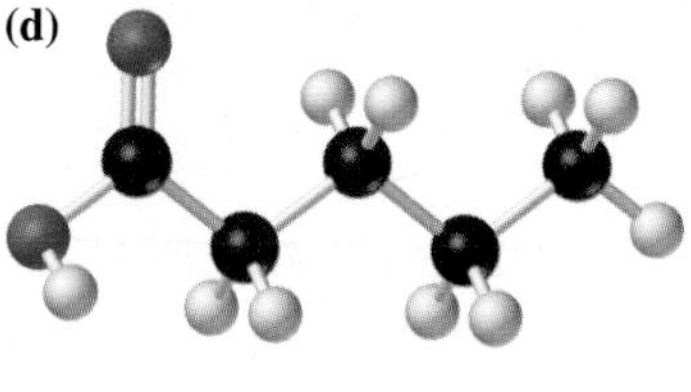

(e)

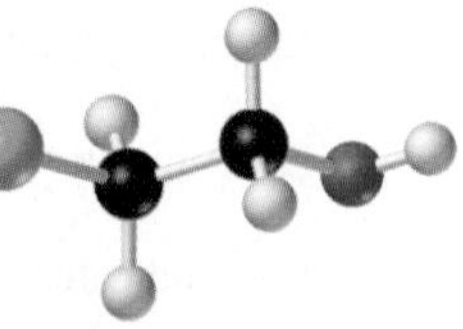

31. Give the structural formulas of the molecules shown in Exercise 29 (b), (d), and (e).

32. Give the structural formulas of the molecules shown in Exercise 30 (b), (d), and (e).

The Avogadro Constant and the Mole

33. *Without doing detailed calculations,* determine which of the following has the greatest number of S atoms: 0.12 mol of solid sulfur, S_8; 0.50 mol of gaseous S_2O; 65 g of gaseous SO_2; 75 mL of liquid thiophene, C_4H_4S ($d = 1.064$ g/mL).

34. *Without doing detailed calculation,* explain which of the following has the greatest number of N atoms: 50.0 g N_2O; 17.0 g NH_3; 150 mL of liquid pyridine, C_5H_5N ($d = 0.983$ g/mL); 1.0 mol N_2.

35. Determine the *mass,* in grams, of
 (a) 6.25×10^{-2} mol P_4;
 (b) 4.03×10^{24} molecules of stearic acid, $C_{18}H_{36}O_2$;
 (c) a quantity of the amino acid lysine, $C_6H_{14}N_2O_2$, containing 1.15 mol N atoms.

36. Determine the number of *moles* of

 (a) N_2O_4 in a 82.5-g sample
 (b) N atoms in 106 g of $Mg(NO_3)_2$
 (c) N atoms in a sample of $C_7H_5(NO_2)_3$ that has the same number of O atoms as 56.5 g $C_6H_{12}O_6$

37. In rhombic sulfur, S atoms are joined into S_8 molecules (see Figure 3-4). If the density of rhombic sulfur is 2.07 g/cm^3, for a crystal of volume 0.568 mm^3 determine: (a) the number of moles of S_8 present; (b) the total number of S atoms.

38. The hemoglobin content of blood is about 15.5 g/100 mL blood. The molar mass of hemoglobin is about 64,500 g/mol, and there are four iron (Fe) atoms in a hemoglobin molecule. Approximately how many Fe atoms are present in the 6 L of blood in a typical adult?

Chemical Formulas

39. Explain which of the following statement(s) is(are) correct concerning glucose (blood sugar), $C_6H_{12}O_6$.
 (a) The percentages, by mass, of C and O are the same as in CO.
 (b) The ratio of C to H to O atoms is the same as in dihydroxyacetone, $(CH_2OH)_2CO$.
 (c) The proportions, by mass, of C and O are equal.
 (d) The highest percentage, by mass, is that of H.

40. Explain which of the following statement(s) is (are) correct for sorbic acid, $C_6H_8O_2$, a mold and yeast inhibitor.
 (a) It has a C:H:O mass ratio of 3:4:1.
 (b) It has the same mass percent composition as the aquatic herbicide, acrolein, C_3H_4O.
 (c) It has same empirical formula as aspidinol, $C_{12}H_{16}O_4$, a drug used to kill parasitic worms.

 (d) It has four times as many H atoms as O atoms, but four times more O than H by mass.

41. For the compound halothane, $C_2HBrClF_3$, determine
 (a) the total number of atoms in one formula unit
 (b) the ratio, by number, of F atoms to C atoms
 (c) the ratio, by mass, of Br to F
 (d) the element present in the greatest mass percent
 (e) the mass of compound required to contain 1.00 g F

42. For the compound $Ge[S(CH_2)_4CH_3]_4$, determine
 (a) the total number of atoms in one formula unit
 (b) the ratio, by number, of C atoms to H atoms
 (c) the ratio, by mass, of Ge to S
 (d) the number of g S in 1 mol of the compound
 (e) the number of C atoms in 33.10 g of the compound

Percent Composition of Compounds

43. Determine the mass percent of each of the elements in the fatty acid, stearic acid, $C_{18}H_{36}O_2$.

44. Determine the mass percent of each of the elements in the antimalarial drug quinine, $C_{20}H_{24}N_2O_2$.

45. All of the following minerals are semiprecious or precious stones. Determine the mass percent of the indicated element in each one. (a) Zr in zircon, $ZrSiO_4$; (b) Be in beryl (emerald), $Be_3Al_2Si_6O_{18}$; (c) Fe in almandine (garnet), $Fe_3Al_2Si_3O_{12}$; (d) S in lazurite (lapis lazuli), $Na_4SSi_3Al_3O_{12}$.

46. Determine the mass percent of the indicated elements in the following forms of penicillin. (a) K in penicillin G

potassium, $C_{16}H_{17}KN_2O_4S$; (b) N in penicillin N, $C_{14}H_{21}N_3O_6S$; (c) S in penicillin S potassium, $C_{14}H_{18}ClKN_2O_4S_2$; (d) C in penicillin G calcium, $Ca(C_{16}H_{17}N_2O_4S)_2$.

47. *Without doing detailed calculations,* arrange the following in order of increasing % Cr, by mass, and explain your reasoning: CrO, Cr_2O_3, CrO_2, CrO_3.

48. *Without doing detailed calculations,* arrange the following in order of increasing % P, by mass, and explain your reasoning: $Ca(H_2PO_4)_2$, $(NH_4)_2HPO_4$, H_3PO_4, Na_3PO_4.

Chemical Formulas from Percent Composition

49. A compound of carbon and hydrogen consists of 93.71% C and 6.29% H, by mass. The molecular mass of the compound is found to be 128 u. What is its molecular formula?

50. Selenium, an element used in the manufacture of photoelectric cells and solar energy devices, forms two oxides. One has 28.8% O, by mass, and the other, 37.8% O. What are the formulas of these oxides? Propose acceptable names for them.

51. Determine the empirical formula of
    (a) the rodenticide (rat killer) warfarin, which consists of 74.01% C, 5.23% H, and 20.76% O, by mass.
    (b) the chemical warfare agent, mustard gas, which consists of 30.20% C, 5.07% H, 44.58% Cl, and 20.16% S, by mass.
52. Determine the empirical formula of
    (a) benzo[a]pyrene, a suspected carcinogen found in cigarette smoke, consisting of 95.21% C and 4.79% H, by mass.
    (b) hexachlorophene, used in germicidal soaps, which consists of 38.37% C, 1.49% H, 52.28% Cl, and 7.86% O by mass.
53. Indigo, the dye for blue jeans, has a percent composition, by mass, of 73.27% C, 3.84% H, 10.68% N, and the remainder as oxygen. Its molecular mass is 262.3 u. What is the molecular formula of indigo?

54. Beta-carotene provides the orange color in carrots and is important for good vision in humans. Its percent composition, by mass, is 89.49% C and 10.51% H. The molecular mass of β-carotene is 536.9 u. What is its molecular formula?
55. The compound XF_3 consist of 65% F, by mass. What is the atomic mass of X?
56. The element X forms the chloride XCl_4 containing 75.0% Cl, by mass. What is the element X?
57. Chlorophyll contains 2.72% Mg by mass. Assuming one Mg atom per chlorophyll molecule, what is the molecular mass of chlorophyll?
58. Two compounds of Cl and X are found to have molecular masses and % Cl, by mass, as follows: 137 u, 77.5% Cl; 208 u, 85.1% Cl. What is element X? What is the formula for each compound?

Combustion Analysis

59. A 0.2612-g sample of a hydrocarbon produces 0.8661 g CO_2 and 0.2216 g H_2O in combustion analysis. Its molecular mass is found to be 106 u. For this hydrocarbon, determine (a) its mass percent composition: (b) its empirical formula; (c) its molecular formula.
60. *Para*-cresol (*p*-cresol) is used as a disinfectant and in the manufacture of herbicides. A 0.3654-g sample of this carbon–hydrogen–oxygen compound yields 1.0420 g CO_2 and 0.2437 g H_2O in combustion analysis. Its molecular mass is 108.1 u. For *p*-cresol, determine (a) its mass percent composition; (b) its empirical formula; (c) its molecular formula.
61. Dimethylhydrazine is a carbon–hydrogen–nitrogen compound used in rocket fuels. When burned completely, a 0.505-g sample yields 0.741 g CO_2 and 0.605 g H_2O. The nitrogen content of a 0.486-g sample is converted to 0.226 g N_2. What is the empirical formula of dimethylhydrazine?
62. The organic solvent thiophene is a carbon–hydrogen–sulfur compound that yields CO_2, H_2O, and SO_2 on complete combustion. When subjected to combustion analysis,

a 1.086-g sample of thiophene produces 2.272 g CO_2, 0.465 g H_2O, and 0.827 g SO_2. What is the empirical formula of thiophene?
63. *Without doing detailed calculations,* explain which of the following compounds produces the *greatest* mass of CO_2 when 1.00 mol of the compound undergoes complete combustion: CH_4, C_2H_5OH, $C_{10}H_8$, C_6H_5OH.
64. *Without doing detailed calculations,* explain which of the following compounds produces the *greatest* mass of H_2O when 1.00 g of the compound undergoes complete combustion: CH_4, C_2H_5OH, $C_{10}H_8$, C_6H_5OH.
65. A 1.562-g sample of the hydrocarbon C_7H_{16} is burned in an excess of oxygen. What masses of CO_2 and H_2O should be obtained?
66. Refer to the compound ethyl mercaptan described in Example 3-2. Assuming that the complete combustion of this compound produces CO_2, H_2O, and SO_2, what masses of each of these three products would be produced in the complete combustion of 1.50 mL of ethyl mercaptan?

Oxidation States

67. Indicate the oxidation state of the underlined element in (a) $\underline{C}H_4$; (b) $\underline{S}F_4$; (c) Na_2O_2; (d) $\underline{C}_2H_3O_2^-$; (e) $\underline{Fe}O_4^{2-}$.
68. Indicate the oxidation state of S in each of the following: (a) SO_3^{2-}; (b) $S_2O_3^{2-}$; (c) $S_2O_8^{2-}$; (d) HSO_4^-; (e) $S_4O_6^{2-}$.

69. Chromium forms three principal oxides. Write appropriate formulas for these compounds in which the oxidation states of Cr are +3, +4, and +6, respectively.
70. Nitrogen forms five oxides. Write appropriate formulas for these compounds in which the oxidation states of N are +1, +2, +3, +4, and +5, respectively.

Nomenclature

71. Name the compounds: (a) SrO; (b) ZnS; (c) K_2CrO_4; (d) Cs_2SO_4; (e) Cr_2O_3; (f) $Fe_2(SO_4)_3$; (g) $Mg(HCO_3)_2$; (h) $(NH_4)_2HPO_4$; (i) $Ca(HSO_3)_2$; (j) $Cu(OH)_2$; (k) HNO_3; (l) $KClO_4$; (m) $HBrO_3$; (n) H_3PO_3.
72. Name the compounds: (a) $Ba(NO_3)_2$; (b) HNO_2; (c) CrO_2; (d) KIO_3; (e) LiCN; (f) KIO; (g) $Fe(OH)_2$; (h)

$Ca(H_2PO_4)_2$; (i) H_3PO_4; (j) $NaHSO_4$; (k) $Na_2Cr_2O_7$; (l) $NH_4C_2H_3O_2$; (m) MgC_2O_4; (n) $Na_2C_2O_4$.
73. Assign suitable names to the compounds: (a) CS_2; (b) SiF_4; (c) ClF_5; (d) N_2O_5; (e) SF_6; (f) I_2Cl_6.
74. Assign suitable names to the compounds: (a) ICl; (b) ClF_3; (c) SF_4; (d) BrF_5; (e) N_2O_4; (f) S_4N_4.

75. Write formulas for the compounds: **(a)** aluminum sulfate; **(b)** ammonium dichromate; **(c)** silicon tetrafluoride; **(d)** iron(III) oxide; **(e)** tricarbon disulfide; **(f)** cobalt(II) nitrate; **(g)** strontium nitrite; **(h)** hydrobromic acid; **(i)** iodic acid; **(j)** phosphorus dichloride trifluoride.

76. Write formulas for the compounds: **(a)** magnesium perchlorate; **(b)** lead(II) acetate; **(c)** tin(IV) oxide; **(d)** hydroiodic acid; **(e)** chlorous acid; **(f)** sodium hydrogen sulfite; **(g)** calcium dihydrogen phosphate; **(h)** aluminum phosphate; **(i)** dinitrogen tetroxide; **(j)** disulfur dichloride.

77. Write a formula for **(a)** the chloride of titanium having Ti in the O.S. +4; **(b)** the sulfate of iron having Fe in the O.S. +3; **(c)** an oxide of chlorine with Cl in the O.S. +7; **(d)** an oxoanion of sulfur in which the apparent O.S. of S is +7 and the ionic charge is 2−.

78. Write a formula for **(a)** an oxide of nitrogen with N in the O.S. +5; **(b)** an oxoacid of nitrogen with N in the O.S. +3; **(c)** an oxide of carbon in which the apparent O.S. of C is +4/3; **(d)** a sulfur-containing oxoanion in which the apparent O.S. of S is +2.5 and the ionic charge is 2−.

Hydrates

79. *Without performing detailed calculations,* indicate which of the following hydrates has the greatest % H_2O by mass: $CuSO_4 \cdot 5\ H_2O$, $Cr_2(SO_4)_3 \cdot 18\ H_2O$, $MgCl_2 \cdot 6\ H_2O$, and $LiC_2H_3O_2 \cdot 2\ H_2O$.

80. *Without performing detailed calculations,* determine the hydrate of Na_2SO_3 that contains almost exactly 50% H_2O, by mass.

81. Anhydrous $CuSO_4$ can be used to dry liquids in which it is insoluble. The $CuSO_4$ is converted to $CuSO_4 \cdot 5\ H_2O$, which can be filtered off from the liquid. What is the minimum mass of anhydrous $CuSO_4$ needed to remove 8.5 g H_2O from a tankful of gasoline?

82. Anhydrous sodium sulfate, Na_2SO_4, absorbs water vapor and is converted to the *deca*hydrate, $Na_2SO_4 \cdot 10\ H_2O$. How much would the mass of 3.50 g of anhydrous Na_2SO_4 increase if converted completely to the decahydrate?

83. A certain hydrate is found to have the composition 20.3% Cu, 8.95% Si, 36.3% F, and 34.5% H_2O by mass. What is the empirical formula of this hydrate?

84. A sample of $MgSO_4 \cdot x\ H_2O$ weighing 8.129 g is heated until all the water of hydration is driven off. The resulting anhydrous compound, $MgSO_4$, weighs 3.967 g. What is the formula of the hydrate?

Organic Compounds and Organic Nomenclature

85. Which of the following structures are isomers?

(a) CH_3—CH—CH_2—OH
         |
        CH_2—CH_3

(b) CH_3—CH—CH_2—CH_2—OH
         |
        CH_3

(c) CH_3—CH_2—CH—CH_2—OH
                  |
                 CH_3

(d) CH_3—CH—CH_2—O—CH_3
         |
        CH_3

(e) CH_3—CH—CH_2—CH—CH_3
         | |
        CH_3 OH

86. Which of the following structures are isomers?

(a) CH_3—CH—CH_2—Cl
         |
        CH_2—CH_3

(b) CH_3—CH—CH_2—CH_3
         |
        CH_2Cl

(c) CH_3—CH—$CHClCH_3$
         |
        CH_3

(d) CH_3—CH—CH_2—CH—CH_3
         | |
        CH_3 Cl

87. Write the condensed structural formulas for the following organic compounds.
(a) hexane   **(b)** methanoic acid
(c) 2-methyl-1-butanol   **(d)** chloroethane

88. Write the condensed structural formulas for the following organic compounds.
(a) octane   **(b)** heptanoic acid
(c) pentanol   **(d)** chloromethane

89. Give the name, condensed structural formula, and molecular mass of the molecule whose ball-and-stick model is shown. Refer to the color scheme on page 69.

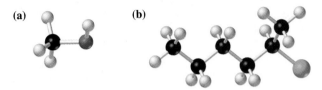

(a)   (b)

(c)   (d)

90. Give the name, condensed structural formula, and molecular mass of the molecule whose ball-and-stick model is shown. Refer to the color scheme on page 69.

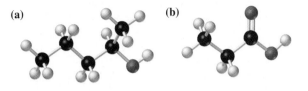

(a)   (b)

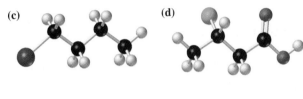

(c)   (d)

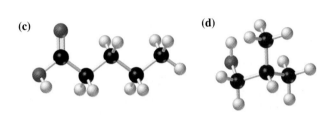

Integrative and Advanced Exercises

91. The mineral spodumene has the empirical formula $LiAlSi_2O_6$. Given that the percentage of lithium-6 atoms in naturally occuring lithium is 7.40%, how many lithium-6 atoms are present in a 518-g sample of spodumene?

92. A particular type of brass contains Cu, Sn, Pb, and Zn. A 1.1713-g sample is treated in such a way as to convert the Sn to 0.245 g SnO_2, the Pb to 0.115 g $PbSO_4$, and the Zn to 0.246 g $Zn_2P_2O_7$. What is the mass percent of each element in the sample?

93. To deposit exactly one mole of Ag from an aqueous solution containing Ag^+ requires a quantity of electricity known as one faraday (F). The electrodeposition requires that each Ag^+ ion gain one electron to become an Ag atom. Use appropriate physical constants listed on the inside back cover to obtain a precise value of the Avogadro constant, N_A.

94. Can there be an alkane hydrocarbon that yields a greater mass of H_2O than CO_2 upon its complete combustion? Explain.

95. A hydrocarbon mixture consists of 60.0% by mass of C_3H_8 and 40.0% of C_xH_y. When 10.0 g of this mixture is burned, 29.0 g CO_2 and 18.8 g H_2O are the only products. What is the formula of the unknown hydrocarbon?

96. A 0.732-g mixture of methane, CH_4, and ethane, C_2H_6, is burned, yielding 2.064 g CO_2. What is the percent composition of this mixture, **(a)** by mass; **(b)** on a mole basis?

97. A sample of the compound MSO_4 weighing 0.1131 g reacts with barium chloride and yields 0.2193 g $BaSO_4$. What must be the atomic mass of the metal M? (*Hint:* All the SO_4^{2-} from the MSO_4 appears in the $BaSO_4$.)

98. The metal M forms the sulfate $M_2(SO_4)_3$. A 0.738-g sample of this sulfate is converted to 1.511 g $BaSO_4$. What is the atomic mass of M? (*Hint:* Refer to Exercise 97.)

99. A 0.622-g sample of a metal oxide with the formula M_2O_3 is converted to 0.685 g of the sulfide, MS. What is the atomic mass of the metal M?

100. $MgCl_2$ often occurs as an impurity in table salt (NaCl) and is responsible for caking of the salt. A 0.5200-g sample of table salt is found to contain 61.10% Cl, by mass. What is the % $MgCl_2$ in the sample? Why is the precision of this calculation so poor?

101. When 2.750 g of the oxide of lead Pb_3O_4 is strongly heated, it decomposes and produces 0.0640 g of oxygen gas and 2.686 g of a second oxide of lead. What is the empirical formula of this second oxide?

102. A 1.013-g sample of $ZnSO_4 \cdot xH_2O$ is dissolved in water and the sulfate ion precipitated as $BaSO_4$. The mass of pure, dry $BaSO_4$ obtained is 0.8223 g. What is the formula of the zinc sulfate hydrate?

103. The iodide ion in a 1.552-g sample of the ionic compound MI is removed through precipitation. The precipitate is found to contain 1.186 g I. What is the element M?

104. An oxoacid with the formula $H_xE_yO_z$ has a formula mass of 178 u, has 13 atoms in its formula unit, contains 34.80% by mass, and 15.38% by number of atoms, of the element E. What is the element E, and what is the formula of this oxoacid?

105. The insecticide dieldrin contains carbon, hydrogen, oxygen, and chlorine. Upon complete combustion, a 1.510-g sample yields 2.094 g CO_2 and 0.286 g H_2O. The compound has a molecular mass of 381 u and has half as many chlorine atoms as carbon atoms. What is the molecular formula of dieldrin?

106. A thoroughly dried 1.271-g sample of Na_2SO_4 is exposed to the atmosphere and found to gain 0.387 g in mass. What is the percent, by mass, of $Na_2SO_4 \cdot 10\ H_2O$ in the resulting mixture of anhydrous Na_2SO_4 and the decahydrate?

107. The atomic mass of Bi is to be determined by converting the compound $Bi(C_6H_5)_3$ to Bi_2O_3. If 5.610 g of $Bi(C_6H_5)_3$ yields 2.969 g Bi_2O_3, what is the atomic mass of Bi?

Feature Problems

108. All-purpose fertilizers contain the essential elements nitrogen, phosphorus, and potassium. A typical fertilizer carries numbers on its label, such as "5-10-5". These numbers represent the % N, % P_2O_5, and % K_2O, respectively. The N is contained in the form of a nitrogen compound such as $(NH_4)_2SO_4$, NH_4NO_3, or $CO(NH_2)_2$ (urea). The P is generally present as a phosphate, and the K as KCl. The expressions "% P_2O_5" and "% K_2O" were devised in the nineteenth century, before the nature of chemical compounds was fully understood. To convert from % P_2O_5 to % P and from % K_2O to % K, we must use the factors 2 mol P/mol P_2O_5 and 2 mol K/mol K_2O, together with molar masses.
(a) What is the percent composition of the 5-10-5 fertilizer in % N, % P, and % K?
(b) What is the % P_2O_5 in the following compounds, both common fertilizers: **(1)** $Ca(H_2PO_4)_2$; **(2)** $(NH_4)_2HPO_4$?

109. A hydrate of copper(II) sulfate, when heated, goes through the succession of changes suggested by the photograph. In this photograph, **(a)** is the original fully hydrated copper(II) sulfate; **(b)** is the product obtained by heating the original hydrate to 140 °C; **(c)** is the product obtained by further heating to 400 °C; and **(d)** is the product obtained at 1000 °C.

   (a)   **(b)**   **(c)**   **(d)**

A 2.574-g sample of $CuSO_4 \cdot xH_2O$ was heated to 140 °C, cooled, and reweighed. The resulting solid was reheated to 400 °C, cooled, and reweighed. Finally, this solid was heated to 1000 °C, cooled, and weighed for the last time.

Original sample:	2.574 g
After heating to 140 °C	1.833 g
After re-heating to 400 °C:	1.647 g
After re-heating to 1000 °C	0.812 g

(a) Assuming that all the water of hydration is driven off at 400 °C, what is the formula of the original hydrate?
(b) What is the formula of the hydrate obtained when the original hydrate is heated to only 140 °C?
(c) The black residue obtained at 1000 °C is an oxide of copper. What is its percent composition and empirical formula?

110. Some substances that are only very slightly soluble in water will spread over the surface of water to produce a film called a *monolayer* because it is only one molecule thick. A practical use of this phenomenon is to cover ponds to reduce the loss of water by evaporation. Stearic acid forms a monolayer on water. The molecules are arranged upright and in contact with one another, rather like pencils tightly packed and standing upright in a coffee mug. The model below represents an individual stearic acid molecule in the monolayer.
(a) How many square meters of water surface would be covered by a monolayer made from 10.0 g of stearic acid? (*Hint:* What is the formula of stearic acid?)
(b) If stearic acid has a density of 0.85 g/cm^3, estimate the length (in nanometers) of a stearic acid molecule. (*Hint:* What is the thickness of the monolayer described in Part [a]?)
(c) A very dilute solution of oleic acid in liquid pentane is prepared in the following way: 1.00 mL oleic acid + 9.00mL pentane $\longrightarrow$ solution(1); 1.00 mL solution(1) + 9.00 mL pentane $\longrightarrow$ solution(2); 1.00 mL solution(2) + 9.00 mL pentane $\longrightarrow$ solution(3); 1.00 mL solution(3) + 9.00 mL pentane $\longrightarrow$ solution(4). A 0.10-mL sample of solution(4) is spread in a monolayer on water. The area covered by the monolayer is 85 cm^2. Assume the oleic acid molecules are arranged in the same way as described above for stearic acid, and that the cross-sectional area of the molecule is $4.6 \times 10^{-15}\ cm^2$. The density of oleic acid is 0.895 g/mL. Use these data to obtain an approximate value of Avogadro's number.

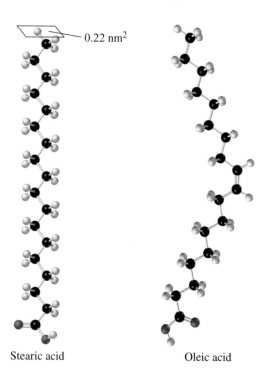

0.22 nm^2

Stearic acid   Oleic acid

eMedia Exercises

111. Observe the three-dimensional models of **acetic acid** and **sodium chloride** found in *eChapter 3.1*. By using the pop-up menu (right-click for Windows, click and hold for Macintosh), change the representation of the models to both the ball-and-stick and space-filling representations. **(a)** Describe the general differences between a molecular compound and an ionic compound in terms of the number of atoms present in the compound. **(b)** What is different about the way the two types of compounds are represented in the animations?

112. Referring to the **Molar Mass** activity *(eChapter 3.2)*, suggest the additional calculation step needed to determine the number of moles of oxygen atoms found in 32.5 grams of V_2O_5. Is this number larger or smaller than the number of moles of V_2O_5?

113. **(a)** Use the **Assigning Oxidation States** activity *(eChapter 3.4)* to help estimate the different oxidation states of manganese in the different oxides of this transition metal: Mn_2O_3, MnO_2, MnO_4^{2-}, MnO_4^-. Adjust the number of metal and oxygen atoms and their relative charges to obtain the desired net charge. **(b)** If the last two species were combined with sodium, what would be the expected compositions of the compounds?

114. Because of the need to exactly describe a structure, some molecular names become awkward to pronounce. Consider for example the eight-carbon alcohol that is known to attract mosquitoes, 1-octen-3-ol. **(a)** Use the **Functional Groups** activity *(eChapter3.7)* to deduce the structure of this molecule by observing first the structure of the alkene and then the structure of the alcohol. **(b)** What would the name of the structure be if the carbon chain contained two double bonds (at 1 and 4) and a carboxyl group on carbon 8? **(c)** Draw the structures of both of these molecules.

4

Chemical Reactions

In this vigorous reaction between iron(III) oxide and powdered aluminum metal, called the thermite reaction, the products are liquid iron and aluminum oxide. The thermite reaction is used in on-site welding of large iron objects.

We are all aware that iron rusts and natural gas (for cooking or heating) burns. These processes are chemical reactions. Chemical reactions are the central concern, not just of this chapter but of the entire science of chemistry.

In this chapter we will first learn to represent chemical reactions by chemical equations, and then we will use chemical equations to establish quantitative (numerical) relationships among the reactants and products of a reaction, a topic known as reaction stoichiometry. Because many chemical reactions occur in solution, we will also introduce a method of describing the composition of a solution, called solution molarity. Throughout the chapter, we will discuss new aspects of problem solving and more uses for the mole concept.

▲ **FIGURE 4-1**

Precipitation of silver chromate When aqueous solutions of silver nitrate and potassium chromate are mixed, the disappearance of the distinctive yellow color of chromate ion and the appearance of red-brown solid silver chromate provide physical evidence of a reaction.

▶ Sometimes products react, partially or completely, to reform the original reactants. Such reactions are called *reversible* reactions and are designated by a double arrow ($\rightleftharpoons$). In this chapter we assume that any reverse reactions are negligible and that reactions go only in the forward direction.

4-1 Chemical Reactions and Chemical Equations

A **chemical reaction** is a process in which one set of substances called **reactants** is converted to a new set of substances called **products.** In other words, a chemical reaction is the *process* by which a chemical change occurs. In many cases, though, nothing happens when substances are mixed; each retains its original composition and properties. We need evidence before we can say that a reaction has occurred. Some of the types of physical evidence to look for are shown here.

- a color change (Figure 4-1)
- formation of a solid (precipitate) within a clear solution (Figure 4-1)
- evolution of a gas (Figure 4-2a)
- evolution or absorption of heat (Figure 4-2b)

When none of these signs of a chemical reaction appears, we need *chemical* evidence to determine if a reaction has occurred. This requires a detailed chemical analysis of the reaction mixture to discover whether any new substances are present.

Just as we use symbols for elements and formulas for compounds, we have a symbolic, or shorthand, way of representing a chemical reaction—the **chemical equation.** In a chemical equation, formulas for the reactants are written on the *left* side of the equation and formulas for the products are written on the *right*. The two sides of the equation are joined by an arrow ($\longrightarrow$) or equal sign (=). We say that the reactants *yield* the products. Consider the reaction of colorless nitrogen monoxide and oxygen gases to form red-brown nitrogen dioxide gas, a reaction that occurs in the manufacture of nitric acid

$$\text{nitrogen monoxide} + \text{oxygen} \longrightarrow \text{nitrogen dioxide}$$

1. Substitute chemical *formulas* for names, to obtain the following expression.

$$NO + O_2 \longrightarrow NO_2$$

In this expression, there are *three* O atoms on the left side (*one* in the molecule NO and *two* in the molecule O_2). On the right side there are only *two* O atoms (in the molecule NO_2). Because atoms are neither created nor destroyed in a chemical reaction, this expression needs to be balanced.

Reactions with Oxygen movie

▶ **FIGURE 4-2**

Evidence of a chemical reaction
(a) Evolution of a gas: When a copper penny reacts with nitric acid, the red-brown gas NO_2 is evolved. **(b)** Evolution of heat: When iron gauze (steel wool) is ignited in an oxygen atmosphere, evolved heat and light provide physical evidence of a reaction.

(a)

(b)

2. *Balance* numbers of atoms to obtain a *chemical equation.*[*]

In this step, we place the coefficient 2 in front of the formulas NO and NO_2. This means that *two* molecules of NO are consumed and *two* molecules of NO_2 are produced for every molecule of O_2 consumed. In the balanced equation there are *two* N atoms and *four* O atoms on each side. In a **balanced equation**, for each element present, the total number of atoms of the element is the same on both sides. We see this below, both in the symbolic equation and in the molecular representation of the reaction.

$$2 NO + O_2 \longrightarrow 2 NO_2$$

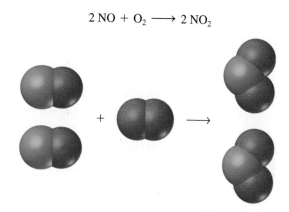

The coefficients required to balance a chemical equation are called **stoichiometric coefficients**. These coefficients are essential in relating the amounts of reactants used and products formed in a chemical reaction, through a variety of calculations.

In balancing a chemical equation, keep in mind that the equation can be balanced *only* by adjusting the *coefficients* of formulas, as necessary. In particular,

Never introduce extraneous atoms or molecules into an equation.

$$\text{(incorrect)} \quad NO + O_2 \longrightarrow NO_2 + O$$

Although the above equation is balanced, it is *incorrect*. No atomic oxygen (O) is produced in this reaction; NO_2 is the only product.

Never change formulas for the purpose of balancing an equation.

$$\text{(incorrect)} \quad NO + O_2 \longrightarrow NO_3$$

[*]An equation—whether mathematical or chemical—must have the left and right sides equal. We should not call an expression an equation until it is balanced. And the term chemical equation automatically signifies that this balance exists. Although unnecessary, the term *balanced* is commonly used when referring to a chemical equation.

Again, the equation is balanced, but it is *incorrect*. The formula for nitrogen dioxide, the sole product of the reaction, can only be NO_2. We must not change it to NO_3 just to balance an equation.

The method of equation balancing we have been describing is called *balancing by inspection*. Balancing by inspection means to adjust stoichiometric coefficients by trial and error until a balanced condition is found. Although we can generally balance the elements in any order, equation balancing need not be a hit-or-miss affair. Here are some useful strategies for balancing equations.

- If an element occurs in only one compound on each side of the equation, try balancing this element *first*.
- When one of the reactants or products exists as the *free* element, balance this element *last*.
- In some reactions, certain groups of atoms (for example, polyatomic ions) remain unchanged. In such cases, balance these groups as a unit.
- It is permissible to use fractional as well as integral numbers as coefficients. At times, an equation can be balanced most easily by using one or more fractional coefficients and then, if desired, clearing the fractions by multiplying *all* coefficients by a common multiplier.

▶ We will encounter a few situations in Chapters 7 and 20 where some fractional coefficients are actually required.

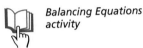

Balancing Equations activity

At this point, you should be able to write formulas for the reactants and products of a reaction and balance the equation. A third task, which we will undertake in later chapters, is to *predict* the products formed when certain reactants are brought together under the appropriate conditions. Even now, however, based on what we learned in the previous chapter, we can predict the products of a *combustion reaction*. In a plentiful supply of oxygen gas, the combustion of hydrocarbons and of carbon–hydrogen–oxygen compounds produces carbon dioxide and water as the *only* products. If the compound contains sulfur as well, sulfur dioxide will also be a product. These ideas and an equation-balancing strategy are illustrated in Examples 4-1 and 4-2.

EXAMPLE 4-1

Balancing an Equation. Ammonia and oxygen can react in several different ways, including the following. Balance the equation for this reaction.

$$NH_3 + O_2 \longrightarrow N_2 + H_2O$$

Solution

One N atom appears in NH_3 on the left, and *two* N atoms are found in N_2 on the right. We need the coefficient 2 in front of NH_3.

$$2\,NH_3 + O_2 \longrightarrow N_2 + H_2O$$

Now we see *six* H atoms on the left (in $2\,NH_3$) and only *two* on the right (in H_2O). We need the coefficient 3 in front of H_2O.

$$2\,NH_3 + O_2 \longrightarrow N_2 + 3\,H_2O$$

At this point, we have *two* O atoms on the left (in O_2) and *three* O atoms on the right (in $3\,H_2O$). We can get three O atoms on the left by placing the coefficient $3/2$ in front of O_2.

$$2\,NH_3 + 3/2\,O_2 \longrightarrow N_2 + 3\,H_2O \qquad \text{(balanced)}$$

To write an equation in which all the coefficients are integers, we can multiply them all by 2. The balanced equation follows, both symbolically and in a molecular representation.

$$4 \, NH_3 + 3 \, O_2 \longrightarrow 2 \, N_2 + 6 \, H_2O \qquad \text{(balanced)}$$

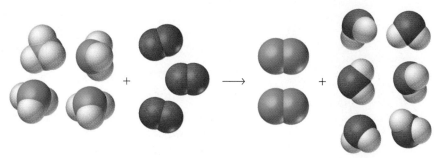

Check: From both the symbolic and molecular representations, we see that the equation is balanced, with *four* N atoms, *twelve* H atoms, and *six* O atoms on *each* side.

Practice Example A: Balance the following equations:

    (a) $H_3PO_4 + CaO \longrightarrow Ca_3(PO_4)_2 + H_2O$

    (b) $C_3H_8 + O_2 \longrightarrow CO_2 + H_2O$

Practice Example B: Balance the following equations:

    (a) $NH_3 + O_2 \longrightarrow NO_2 + H_2O$

    (b) $NO_2 + NH_3 \longrightarrow N_2 + H_2O$

Triethylene glycol

KEEP IN MIND ▶

that substituting correct formulas for the names of reactants and products is an integral part of equation writing. The nomenclature rules from Chapter 3 can be helpful here.

EXAMPLE 4-2

Writing and Balancing an Equation: The Combustion of a Carbon–Hydrogen–Oxygen Compound. Liquid triethylene glycol, $C_6H_{14}O_4$, is used as a solvent and plasticizer for vinyl and polyurethane plastics. Write a balanced chemical equation for its complete combustion.

Solution

Carbon–hydrogen–oxygen compounds, like hydrocarbons, yield carbon dioxide and water when burned in oxygen gas.

    Starting expression: $C_6H_{14}O_4 + O_2 \longrightarrow CO_2 + H_2O$

    Balance C: $C_6H_{14}O_4 + O_2 \longrightarrow 6 \, CO_2 + H_2O$

    Balance H: $C_6H_{14}O_4 + O_2 \longrightarrow 6 \, CO_2 + 7 \, H_2O$

At this point, the right side of the expression has 19 O atoms (12 in six CO_2 molecules and 7 in seven H_2O molecules). To get 19 O atoms on the left, we start with 4 in a molecule of $C_6H_{14}O_4$ and need 15 more. This requires a *fractional* coefficient of $15/2$ for O_2.

    Balance O: $C_6H_{14}O_4 + \dfrac{15}{2} O_2 \longrightarrow 6 \, CO_2 + 7 \, H_2O \qquad \text{(balanced)}$

To remove the fractional coefficient, multiply all coefficients by 2, the denominator of the fractional coefficient, 15/2.

$$2 \, C_6H_{14}O_4 + 15 \, O_2 \longrightarrow 12 \, CO_2 + 14 \, H_2O \qquad \text{(balanced)}$$

Check:

Left: $(2 \times 6) = 12 \, C;$ $(2 \times 14) = 28 \, H;$ $\left[(2 \times 4) + (15 \times 2) \right] = 38 \, O$

Right: $(12 \times 1) = 12 \, C;$ $(14 \times 2) = 28 \, H;$ $\left[(12 \times 2) + (14 \times 1) \right] = 38 \, O$

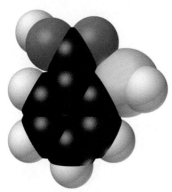

Thiosalicylic acid

▶ At the high combustion temperature, water is present as $H_2O(g)$. But when the reaction products are returned to the initial temperature, the water condenses to a liquid, $H_2O(l)$.

▶ In a *decomposition* reaction a substance is broken down into simpler substances (for instance, into its elements).

▶ In a *synthesis* reaction a new compound is formed from the reaction of two or more simpler substances, usually called the reactants or starting materials.

Practice Example A: Write a balanced equation to represent the reaction of mercury(II) sulfide and calcium oxide to produce calcium sulfide, calcium sulfate, and mercury metal.

Practice Example B: Write a balanced equation for the combustion of thiosalicylic acid, $C_7H_6O_2S$, used in the manufacture of indigo dyes.

States of Matter

Ammonia (Example 4-1) is a gas, but triethylene glycol (Example 4-2) is a liquid. Sometimes we need to represent this kind of information in a chemical equation. We can show the state of matter or physical form of reactants and products through symbols in parentheses. Four common symbols and their meanings are

(g) gas   (l) liquid   (s) solid   (aq) aqueous (water) solution

Thus, for the combustion of triethylene glycol, we can write

$$2\ C_6H_{14}O_4(l)\ +\ 15\ O_2(g)\ \longrightarrow\ 12\ CO_2(g)\ +\ 14\ H_2O(l)$$

Reaction Conditions

The equation for a chemical reaction does not provide enough information to enable you to carry out the reaction in a laboratory or chemical plant. An important aspect of modern chemical research involves working out the conditions for a reaction. We often write the reaction conditions above or below the arrow in an equation. For example, the Greek capital letter delta, Δ, means that a high temperature is required. That is, the reaction mixture must be heated, as in the *decomposition* of silver oxide.

$$2\ Ag_2O(s)\ \xrightarrow{\ \Delta\ }\ 4\ Ag(s)\ +\ O_2(g)$$

An even more explicit statement of reaction conditions is shown below for the BASF (Badische Anilin- & Soda-Fabrik) process for the *synthesis* of methanol from CO and H_2. This reaction occurs at 350 °C, under a total gas pressure that is 340 times greater than the normal pressure of the atmosphere, and on the surface of a mixture of ZnO and Cr_2O_3 acting as a catalyst. (A *catalyst* is a substance that enters into a reaction in such a way as to make the reaction go faster without being consumed in the reaction.)

$$CO(g)\ +\ 2\ H_2(g)\ \xrightarrow[\substack{340\ atm \\ ZnO,\ Cr_2O_3}]{350\ °C}\ CH_3OH(g)$$

It is important to be able to calculate how much of a particular product will be produced when certain quantities of the reactants are consumed. In the next section, we will see how to use chemical equations to set up conversion factors that we can use for these and related calculations.

4-2 Chemical Equations and Stoichiometry

In Greek, the word *stoicheion* means element. The term **stoichiometry** (stoy-key-om'-eh-tree) means, literally, to measure the elements, but from a more practical standpoint, it includes all the quantitative relationships involving atomic and formula masses, chemical formulas, and the chemical equation. We considered the quantitative meaning of chemical formulas in Chapter 3, and now we will explore some quantitative aspects of chemical equations.

The coefficients in the chemical equation

$$2\ H_2(g)\ +\ O_2(g)\ \longrightarrow\ 2\ H_2O(l) \tag{4.1}$$

mean that

$$2x \text{ molecules } H_2 + x \text{ molecules } O_2 \longrightarrow 2x \text{ molecules } H_2O.$$

Suppose that we let $x = 6.02214 \times 10^{23}$ (Avogadro's number). Then x molecules represents *1 mole*. Thus the chemical equation also means that

$$2 \text{ mol } H_2 + 1 \text{ mol } O_2 \longrightarrow 2 \text{ mol } H_2O$$

The coefficients in the chemical equation allow us to make statements, such as

- *Two* moles of H_2O are *produced* for every *two* moles of H_2 *consumed*.
- *Two* moles of H_2O are *produced* for every *one* mole of O_2 *consumed*.
- *Two* moles of H_2 are *consumed* for every *one* mole of O_2 *consumed*.

Moreover, we can turn such statements into conversion factors, called stoichiometric factors. A **stoichiometric factor** relates the amounts, on a mole basis, of any two substances involved in a chemical reaction; thus a stoichiometric factor is a mole ratio. In the examples that follow, stoichiometric factors are printed in blue.

EXAMPLE 4-3

Relating the Numbers of Moles of Reactant and Product. How many moles of H_2O are produced in reaction (4.1) by burning 2.72 mol H_2 in an excess of O_2?

Solution

The statement "an excess of O_2" means that there is more than enough O_2 available to permit the complete conversion of 2.72 mol H_2 to H_2O. The conversion factor we need from equation (4.1) is based on the fact that 2 mol H_2O is produced for every 2 mol H_2 burned. Think of this as 2 mol H_2O = 2 mol H_2.

$$? \text{ mol } H_2O = 2.72 \text{ mol } H_2 \times \frac{2 \text{ mol } H_2O}{2 \text{ mol } H_2} = 2.72 \text{ mol } H_2O$$

Practice Example A: How many moles of O_2 are produced from the decomposition of 1.76 moles of potassium chlorate: $2 KClO_3(s) \longrightarrow 2 KCl(s) + 3 O_2(g)$?

Practice Example B: How many moles of Ag are produced in the decomposition of 1.00 kg of silver(I) oxide: $2 Ag_2O(s) \longrightarrow 4 Ag(s) + O_2(g)$?

KEEP IN MIND ▶

that in using the equal sign we don't mean, literally, that 2 mol H_2O and 2 mol H_2 are identical. We mean that in this reaction, 2 mol H_2O is *equivalent* to 2 mol H_2. Sometimes this distinction is made by using the equivalent sign $\simeq$, that is, 2 mol $H_2O \simeq 2$ mol H_2.

Most reaction stoichiometry calculations are more complex than Example 4-3, or at least they may seem so. However, there is a key to every such calculation: a stoichiometric factor—a mole ratio—as a central conversion factor. Because a stoichiometric factor is always written on a mole-to-mole basis, the part of the setup preceding the conversion factor involves one or more conversions from the original quantity to moles of a substance. The part of the setup following the conversion factor involves one or more conversions from moles of a substance to the final quantity sought. Throughout the setup, units cancel in such a way that one unit remains—the one required for the final answer.

A general strategy for reaction stoichiometry calculations is suggested by the flow diagram in Figure 4-3 and is applied in Examples 4-4 and 4-5. In these examples, the object is to find the mass of one substance involved in a reaction from the given mass of another substance. This is a practical type of calculation because we usually measure quantities of substances by their masses.

In the scheme shown in Figure 4-3, the mass of the given compound is converted to moles, then the balanced chemical equation is used to obtain the appropriate

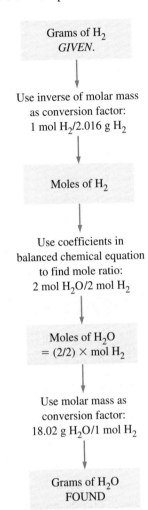

Grams of H_2
GIVEN.

↓

Use inverse of molar mass
as conversion factor:
1 mol H_2/2.016 g H_2

↓

Moles of H_2

↓

Use coefficients in
balanced chemical equation
to find mole ratio:
2 mol H_2O/2 mol H_2

↓

Moles of H_2O
= (2/2) × mol H_2

↓

Use molar mass as
conversion factor:
18.02 g H_2O/1 mol H_2

↓

Grams of H_2O
FOUND

▲ **FIGURE 4-3**
**Strategy for a stoichiometric
calculation**

$$2H_2(g) + O_2(g) \longrightarrow 2H_2O(l)$$

mole ratio (stoichiometric factor), and finally the moles (or grams) of products or other reactants are calculated. The three conversion factors required in Example 4-4 must follow the pathway

$$g\ H_2 \longrightarrow mol\ H_2 \longrightarrow mol\ H_2O \longrightarrow g\ H_2O$$

The three conversion factors required in Example 4-5 must follow the pathway

$$g\ H_2 \longrightarrow mol\ H_2 \longrightarrow mol\ O_2 \longrightarrow g\ O_2$$

EXAMPLE 4-4

Relating the Mass of a Reactant and a Product. What mass of H_2O is formed in the reaction of 4.16 g H_2 with an excess of O_2?

Solution

The general strategy for reaction stoichiometry problems, outlined earlier and illustrated in Figure 4-3, suggests these three steps.

1. Convert the quantity of H_2 from grams to moles. (Use the inverse of molar mass of H_2.)
2. From the number of moles of H_2, calculate the number of moles of H_2O formed. [Use the stoichiometric factor from equation (4.1).]
3. Convert the quantity of H_2O from moles to grams. (Use the molar mass of H_2O.)

Although we think in terms of these steps, we can just as easily write a single setup. The individual steps are noted by the small numbers above the arrows.

$$(g\ H_2 \xrightarrow{\ 1\ } mol\ H_2 \xrightarrow{\ 2\ } mol\ H_2O \xrightarrow{\ 3\ } g\ H_2O)$$

$$?\ g\ H_2O = 4.16\ g\ H_2 \times \frac{1\ mol\ H_2}{2.016\ g\ H_2} \times \frac{2\ mol\ H_2O}{2\ mol\ H_2} \times \frac{18.02\ g\ H_2O}{1\ mol\ H_2O}$$

$$= 37.2\ g\ H_2O$$

Check: Because the molar mass of H_2O is so much larger than that of H_2 and because one mole of H_2O is produced for every mole of H_2 consumed, the mass of H_2O (37.2 g) should be much greater than that of the original H_2 (4.16 g). It is.

Practice Example A: How many grams of magnesium nitride are produced by the reaction of 3.82 g Mg with an excess of N_2? $3\ Mg + N_2 \longrightarrow Mg_3N_2$

Practice Example B: How many grams of $H_2(g)$ are required to produce 1.00 kg methanol, CH_3OH, by the reaction $CO + 2\ H_2 \longrightarrow CH_3OH$?

EXAMPLE 4-5

Relating the Masses of Two Reactants to Each Other. What mass of O_2 is consumed in the complete combustion of 6.86 g H_2 in reaction (4.1)?

Solution

As in Example 4-4, we need molar masses and the appropriate stoichiometric factor from the balanced equation (4.1) to use as conversion factors. We use them in the following three-stage setup.

$$(g\ H_2 \longrightarrow mol\ H_2 \longrightarrow mol\ O_2 \longrightarrow g\ O_2)$$

$$?\ g\ O_2 = 6.86\ g\ H_2 \times \frac{1\ mol\ H_2}{2.016\ g\ H_2} \times \frac{1\ mol\ O_2}{2\ mol\ H_2} \times \frac{32.00\ g\ O_2}{1\ mol\ O_2}$$

$$= 54.4\ g\ O_2$$

Reduction of CuO movie

Check: Even though only one mole of O_2 is consumed for every two moles of H_2, because the molar mass of O_2 is so much greater than that of H_2, the mass of O_2 consumed (54.4 g) still should be significantly greater than that of H_2 (6.86 g). It is.

Practice Example A: How many grams of H_2 are consumed per gram of O_2 in reaction (4.1)?

Practice Example B: How many grams of O_2 are consumed per gram of octane, C_8H_{18}, in the combustion of octane?

(*Hint:* What is the balanced chemical equation?)

What lends great variety to stoichiometric calculations is that many other conversions may be required before and after the mol A $\longrightarrow$ mol B step at the center of the stoichiometric scheme shown in Figure 4-3. These other conversions may use such factors as volume, density, and percent composition. To complete the calculation, however, we must still use the appropriate stoichiometric factor from the chemical equation.

The reaction pictured in Figure 4-4, a simple laboratory method for preparing small volumes of hydrogen gas, provides a range of possibilities for calculations.

$$2\,Al(s) + 6\,HCl(aq) \longrightarrow 2\,AlCl_3(aq) + 3\,H_2(g) \tag{4.2}$$

Examples 4-6 and 4-7 are based on this reaction.

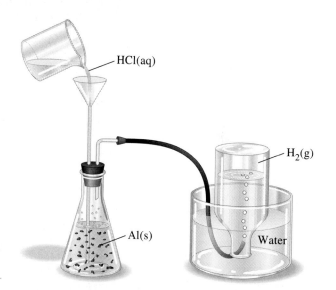

▶ **FIGURE 4-4**

The reaction 2 Al(s) + 6 HCl(aq) $\longrightarrow$ 2 AlCl₃(aq) + 3 H₂(g)
HCl(aq) is introduced to the flask on the left. The reaction occurs within the flask. The liberated $H_2(g)$ is conducted to a gas-collection apparatus, where it displaces water. Hydrogen is only very slightly soluble in water.

EXAMPLE 4-6

Additional Conversion Factors in a Stoichiometric Calculation: Volume, Density, and Percent Composition. An alloy used in aircraft structures consists of 93.7% Al and 6.3% Cu by mass. The alloy has a density of 2.85 g/cm³. A 0.691-cm³ piece of the alloy reacts with an excess of HCl(aq). If we assume that *all* the Al but *none* of the Cu reacts with HCl(aq), what is the mass of H_2 obtained?

Solution

One approach to this calculation is to follow the strategy diagrammed in Figure 4-5. Although such diagrams are helpful, we can generally make a simpler outline by using arrows to represent conversions. Each numbered arrow refers to a conversion factor that changes the unit on the left to the one on the right. That is,

$$\text{cm}^3 \text{ alloy} \xrightarrow{\ 1\ } \text{g alloy} \xrightarrow{\ 2\ } \text{g Al} \xrightarrow{\ 3\ } \text{mol Al} \xrightarrow{\ 4\ } \text{mol H}_2 \xrightarrow{\ 5\ } \text{g H}_2$$

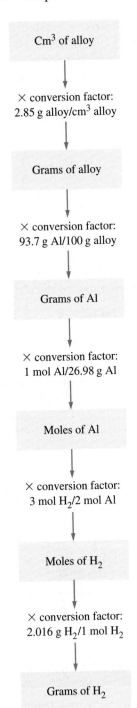

× conversion factor:
2.85 g alloy/cm³ alloy

Grams of alloy

× conversion factor:
93.7 g Al/100 g alloy

Grams of Al

× conversion factor:
1 mol Al/26.98 g Al

Moles of Al

× conversion factor:
3 mol H₂/2 mol Al

Moles of H₂

× conversion factor:
2.016 g H₂/1 mol H₂

Grams of H₂

▲ FIGURE 4-5
Strategy for the calculation in Example 4-6

$$2\ Al(s) + 6\ HCl(aq) \longrightarrow 2\ AlCl_3(aq) + 3\ H_2(g)$$

We can do the calculation in five distinct steps, reaching our final answer in step 5. Alternatively, we can do the calculation in a single setup in which five conversions are performed in sequence. This approach, which is used below, avoids the need to write down intermediate results and reduces rounding errors.

$$(cm^3\ alloy \xrightarrow{1} g\ alloy \xrightarrow{2} g\ Al \xrightarrow{3} mol\ Al$$

$$?\ g\ H_2 = 0.691\ cm^3\ alloy \times \frac{2.85\ g\ alloy}{1\ cm^3\ alloy} \times \frac{93.7\ g\ Al}{100.0\ g\ alloy} \times \frac{1\ mol\ Al}{26.98\ g\ Al}$$

$$\xrightarrow{4} mol\ H_2 \xrightarrow{5} g\ H_2)$$

$$\times \frac{3\ mol\ H_2}{2\ mol\ Al} \times \frac{2.016\ g\ H_2}{1\ mol\ H_2} = 0.207\ g\ H_2$$

Practice Example A: What volume of the aluminum–copper alloy described in Example 4-6 must be dissolved in an excess of HCl(aq) to produce 1.00 g H₂?

(*Hint:* You may think of this as the "inverse" of Example 4-6.)

Practice Example B: A fresh sample of the aluminum–copper alloy described in Example 4-6 yielded 1.31 g H₂. How many grams of *copper* were present in the sample?

EXAMPLE 4-7

Additional Conversion Factors in a Stoichiometric Calculation: Volume, Density, and Percent Composition of a Solution. A hydrochloric acid solution consists of 28.0% HCl by mass and has a density of 1.14 g/mL. What volume of this solution is required to react completely with 1.87 g Al in reaction (4.2)?

Solution

The first challenge here is to determine where to begin. Although the problem refers to 28.0% HCl and a density of 1.14 g/mL, the appropriate starting point is with the "given" information–1.87 g Al. The goal of our calculation is a solution volume—mL HCl solution.

$$g\ Al \xrightarrow{1} mol\ Al \xrightarrow{2} mol\ HCl \xrightarrow{3}$$

$$g\ HCl \xrightarrow{4} g\ HCl\ solution \xrightarrow{5} mL\ HCl\ solution$$

The conversion factors in the calculation involve (1) the molar mass of Al, (2) stoichiometric coefficients from Equation (4.2), (3) the molar mass of HCl, (4) the percent composition of the HCl solution, and (5) the density of the HCl solution. The strategy is outlined below.

$$(g\ Al \xrightarrow{1} mol\ Al \xrightarrow{2} mol\ HCl \xrightarrow{3} g\ HCl$$

$$?\ mL\ HCl\ soln = 1.87\ g\ Al \times \frac{1\ mol\ Al}{26.98\ g\ Al} \times \frac{6\ mol\ HCl}{2\ mol\ Al} \times \frac{36.46\ g\ HCl}{1\ mol\ HCl}$$

$$\xrightarrow{4} g\ HCl\ soln \xrightarrow{5} mL\ HCl\ soln)$$

$$\times \frac{100.0\ g\ HCl\ soln}{28.0\ g\ HCl} \times \frac{1\ mL\ HCl\ soln}{1.14\ g\ HCl\ soln}$$

$$= 23.8\ mL\ HCl\ soln$$

Practice Example A: How many milligrams of H₂ are produced when 1 drop (0.05 mL) of the hydrochloric acid solution described in Example 4-7 reacts with an excess of aluminum in reaction (4.2)?

Practice Example B: A particular vinegar contains 4.0% $HC_2H_3O_2$, by mass. It reacts with $NaHCO_3$ (baking soda) to produce carbon dioxide.

$$HC_2H_3O_2(aq) + NaHCO_3(s) \longrightarrow NaC_2H_3O_2(aq) + H_2O(l) + CO_2(g)$$

How many grams of CO_2 are produced by the reaction of 5.00 mL of this vinegar with an excess of $NaHCO_3$? The density of the vinegar is 1.01 g/mL.

4-3   Chemical Reactions in Solution

Most chemical reactions in the general chemistry laboratory are carried out in solutions. This is partly because mixing the reactants in solution helps to achieve the close contact between atoms, ions, or molecules necessary for a reaction to occur. To describe the stoichiometry of reactions in solutions, we can use the same ideas as for other reactions. In fact, we did this in Example 4-7. We also need a few new ideas that apply specifically to solution stoichiometry.

Components of a Solution animation

One component of a solution, called the **solvent**, determines whether the solution exists as a solid, liquid, or gas. In this discussion we will limit ourselves to *aqueous* solutions—solutions in which liquid water is the solvent. The other components of a solution, called **solutes**, are dissolved in the solvent. By writing NaCl(aq), for example, we are describing a solution in which liquid water is the solvent and NaCl is the solute. The term *aqueous* does not convey any information about the relative proportions of NaCl and H_2O in the solution. For this purpose, we prefer something other than the percent by mass composition that we have used previously.

Molarity

The concentration, or **molarity**, of a solution is defined as

$$\text{molarity } (M) = \frac{\text{amount of solute (in moles)}}{\text{volume of solution (in liters)}} \tag{4.3}$$

If 0.444 mol urea, $CO(NH_2)_2$, is dissolved in enough water to make 1.000 L of solution, the solution concentration, or molarity, is

$$\frac{0.440 \text{ mol } CO(NH_2)_2}{1 \text{ L soln}} = 0.440 \text{ M } CO(NH_2)_2$$

Alternatively, if 0.110 mol urea is present in 250.0 mL of solution, the solution is also 0.440 M.

$$\frac{0.110 \text{ mol } CO(NH_2)_2}{0.2500 \text{ L soln}} = 0.440 \text{ M } CO(NH_2)_2$$

The symbol M is often used in place of the unit mol/L and the word "molar" in place of molarity. Thus, we might describe a solution that has 0.440 mol $CO(NH_2)_2$/L as 0.440 M $CO(NH_2)_2$ or 0.440 molar $CO(NH_2)_2$.

Of course, we must relate an amount in moles to other quantities that we can measure directly. Also, we need not always work with exactly 1 liter of solution. Sometimes a much smaller volume is sufficient; sometimes we need larger volumes. In Example 4-8, we relate the mass of a liquid solute to its volume by using density as a conversion factor. Then, we use molar mass to convert from the mass of solute to an amount of solute in moles. Also, we convert the solution volume from milliliters (mL) to liters (L).

KEEP IN MIND ▶

that the volume used in molarity is that of the *solution*, not of the solvent. If we add 25.0 mL ethanol to 250.0 mL water, the solution volume will not be 250.0 mL, nor will it be if we add 25.0 mL of ethanol to 225.0 mL of water. We learn why in Chapter 14.

Acetone

EXAMPLE 4-8

Calculating Molarity from Measured Quantities. A solution is prepared by dissolving 25.0 mL ethanol, C_2H_5OH ($d = 0.789$ g/mL), in enough water to produce 250.0 mL solution. What is the molarity of ethanol in the solution?

Solution

To determine the molarity, we must first calculate how many moles of ethanol are in the solution. To calculate the number of moles of ethanol in a 25.0-mL sample, we need conversion factors based on density and molar mass.

$$? \text{ mol } C_2H_5OH = 25.0 \text{ mL } C_2H_5OH \times \frac{0.789 \text{ g } C_2H_5OH}{1 \text{ mL } C_2H_5OH} \times \frac{1 \text{ mol } C_2H_5OH}{46.07 \text{ g } C_2H_5OH}$$

$$= 0.428 \text{ mol } C_2H_5OH$$

Now, we note that 250.0 mL = 0.2500 L, and then we use the definition of molarity in expression (4.3).

$$\text{Molarity} = \frac{0.428 \text{ mol } C_2H_5OH}{0.2500 \text{ L soln}} = 1.71 \text{ M } C_2H_5OH$$

Practice Example A: A 22.3-g sample of acetone, $(CH_3)_2CO$, is dissolved in enough water to produce 1.25 L of solution. What is the molarity of acetone in this solution?

Practice Example B: 15.0 mL of acetic acid, CH_3COOH ($d = 1.048$ g/mL), is dissolved in enough water to produce 500.0 mL of solution. What is the molarity of acetic acid in the solution?

Figure 4-6 illustrates a method commonly used to prepare a solution. We weigh out a solid sample and dissolve it in sufficient water to produce a solution of *known* volume—250.0 mL in Figure 4-6. To fill a beaker to a 250-mL mark is not nearly precise enough as a volume measurement. As discussed in Section 1-7, there would be a systematic error because the beaker is not calibrated with sufficient precision. (The error in volume could be 10–20 mL or more.) To dissolve the solute in water and bring the solution level to the 250-mL mark in a graduated cylinder is also not sufficiently precise. Although the graduated cylinder is calibrated more precisely than the beaker, the error might still be 1–2 mL or more. On the other hand, when filled to the calibration mark, the volumetric flask pictured in Figure 4-6 contains 250.0 mL with only about 0.1 mL of error.

In Example 4-9, we calculate the mass of solute needed to produce the solution in Figure 4-6. The key to this calculation is to formulate a conversion factor based on molarity. Thus, when we write 0.250 M K_2CrO_4, we are in effect saying that 1 liter of the solution is equivalent to 0.250 mol K_2CrO_4. The conversion pathway required in Example 4-9 is L soln $\longrightarrow$ mol solute $\longrightarrow$ g solute.

EXAMPLE 4-9

Calculating the Mass of Solute in a Solution of Known Molarity. We want to prepare exactly 0.2500 L (250.0 mL) of an 0.250 M K_2CrO_4 solution in water. What mass of K_2CrO_4 should we use (see Figure 4-6)?

Solution

As just stated, we can use solution molarity (printed in blue) to convert between volume of solution and amount of solute. The other conversion factor we need is the molar mass of K_2CrO_4.

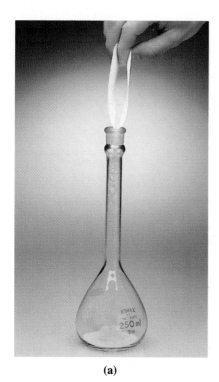
(a)

(b)

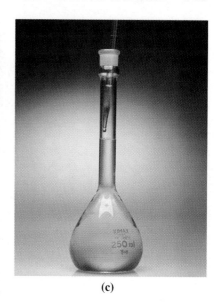
(c)

Solution Formation from a Solid animation

▲ **FIGURE 4-6   Preparation of 0.250 M K₂CrO₄—Example 4-9 illustrated**
The solution cannot be prepared just by adding 12.1 g $K_2CrO_4(s)$ to 250.0 mL water. Instead, **(a)** the weighed quantity of $K_2CrO_4(s)$ is first added to a clean, dry volumetric flask; **(b)** the $K_2CrO_4(s)$ is dissolved in less than 250 mL of water; and **(c)** the flask is filled to the 250.0-mL calibration mark by the careful addition (dropwise) of the remaining water.

$$(\text{L soln} \longrightarrow \text{mol K}_2\text{CrO}_4 \longrightarrow \text{g K}_2\text{CrO}_4)$$

$$? \text{ g K}_2\text{CrO}_4 = 0.2500 \text{ L soln} \times \frac{0.250 \text{ mol K}_2\text{CrO}_4}{1 \text{ L soln}} \times \frac{194.2 \text{ g K}_2\text{CrO}_4}{1 \text{ mol K}_2\text{CrO}_4}$$

$$= 12.1 \text{ g K}_2\text{CrO}_4$$

Practice Example A: An aqueous solution saturated with $NaNO_3$ at 25 °C is 10.8 M $NaNO_3$. How many grams of $NaNO_3$ are present in 125 mL of this solution at 25 °C?

Practice Example B: How many grams of $Na_2SO_4 \cdot 10H_2O$ are needed to prepare 355 mL of 0.445 M Na_2SO_4?

Solution Dilution

A common sight in chemistry storerooms and laboratories is rows of bottles containing solutions for use in chemical reactions. It is not practical, however, to store solutions of every possible concentration. Instead, most labs store fairly concentrated solutions, so-called *stock* solutions, which can then be used to prepare more dilute solutions by adding water. The principle involved, suggested by Figure 4-7, is that *all the solute taken from the initial, more concentrated solution appears in the final diluted solution.*

This statement and the definition of molarity are all that you need to work out dilution problems. You may, however, prefer a method based on expression (4.3).

▶ A *concentrated* solution has a relatively large amount of dissolved solute; a *dilute* solution has a small amount.

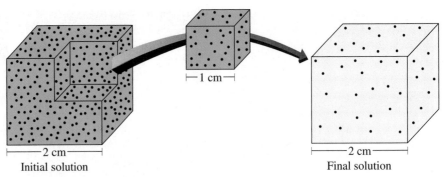

▲ **FIGURE 4-7 Visualizing the dilution of a solution**
The final solution is prepared by extracting $\frac{1}{8}$ of the initial solution—1 cm³—and diluting it with water to a volume of 8 cm³. The number of dots in the 8 cm³ of final solution, representing the number of solute particles, is the same as in the 1 cm³ of initial solution.

Suppose we refer to a solution concentration (molarity) as M, a solution volume, in liters, as V, and the amount of solute, in moles, as n. Then, expression (4.3) becomes, $M = n/V$, which we can rearrange to

$$n = M \times V$$

Solution Formation by Dilution animation

When a solution is diluted, the amount of solute *remains constant* between the initial (i) solution taken and the final (f) solution produced. That is,

$$M_i V_i = n_i = n_f = M_f V_f$$

or

▶ An alternative expression uses "conc" and "dil" as subscripts in place of "i" and "f".

$$M_i \times V_i = M_f \times V_f \qquad (4.4)$$

Figure 4-8 illustrates the laboratory procedure for preparing a solution by dilution. Example 4-10 explains the necessary calculation.

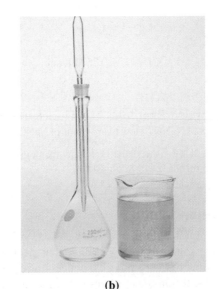

(a)   (b)   (c)

▲ **FIGURE 4-8 Preparing a solution by dilution—Example 4-10 illustrated**
(a) A pipet is used to withdraw a 10.0-mL sample of 0.250 M K_2CrO_4(aq). **(b)** The pipetful of 0.250 M K_2CrO_4 is discharged into a 250.0-mL volumetric flask. **(c)** Following this, water is added to bring the level of the solution to the calibration mark on the neck of the flask. At this point, the solution is 0.0100 M K_2CrO_4.

EXAMPLE 4-10

Preparing a Solution by Dilution. A particular analytical chemistry procedure requires 0.0100 M K_2CrO_4. What volume of 0.250 M K_2CrO_4 must be diluted with water to prepare 0.250 L of 0.0100 M K_2CrO_4?

Solution

To solve this problem, imagine reversing the arrow in Figure 4-7, that is, start with the final solution because we know the most about it. Then, determine something about the initial solution. First, calculate the amount of solute that must be present in the *final* solution.

$$? \text{ mol } K_2CrO_4 = 0.250 \text{ L soln} \times \frac{0.0100 \text{ mol } K_2CrO_4}{1 \text{ L soln}} = 0.00250 \text{ mol } K_2CrO_4$$

Because all the solute in the final, diluted solution comes from the initial, more concentrated solution, we must answer the question: What volume of 0.250 M K_2CrO_4 contains 0.00250 mol K_2CrO_4?

$$? \text{ L soln} = 0.00250 \text{ mol } K_2CrO_4 \times \frac{1 \text{ L soln}}{0.250 \text{ mol } K_2CrO_4} = 0.0100 \text{ L soln}$$

Alternatively, we can use equation (4.4). We know the volume of solution that we wish to prepare ($V_f = 250.0$ mL) and the concentrations of the final (0.0100 M) and initial (0.250 M) solutions. We must solve for the initial volume, V_i. Note that, although in deriving equation (4.4) we expressed volumes in liters, in applying it we can use any volume unit as long as we use the same unit for both V_i and V_f (milliliters in the present case). The term needed to convert volumes to liters would appear on each side of the equation and cancel out.

$$V_i = V_f \times \frac{M_f}{M_i} = 250.0 \text{ mL} \times \frac{0.0100 \text{ M}}{0.250 \text{ M}} = 10.0 \text{ mL}$$

Practice Example A: A 15.00-mL sample of 0.450 M K_2CrO_4 is diluted to 100.00 mL. What is the concentration of the new solution?

Practice Example B: When left in an open beaker for a period of time, the volume of 275 mL of 0.105 M NaCl is found to decrease to 237 mL due to the evaporation of water. What is the new concentration of the solution?

Stoichiometry of Reactions in Solution

When we measure the volume of a solution of known molarity, we obtain a fixed number of moles of solute. Therefore, because stoichiometric calculations require us to work in moles, we can use molarities and volumes instead. Example 4-11 illustrates this idea. The central conversion factor in Example 4-11 is the same as in previous examples—the appropriate stoichiometric factor. What differs from previous examples is that we need to use molarity as a conversion factor from solution volume to number of moles of reactant.

EXAMPLE 4-11

Relating the Mass of a Product to the Volume and Molarity of a Reactant Solution. A 25.00-mL pipetful of 0.250 M K_2CrO_4(aq) is added to an excess of $AgNO_3$(aq). What mass of Ag_2CrO_4(s) will precipitate from the solution?

$$K_2CrO_4(aq) + 2\,AgNO_3(aq) \longrightarrow Ag_2CrO_4(s) + 2\,KNO_3(aq)$$

Solution

(1) Determine the amount of K_2CrO_4 that reacts; (2) use a stoichiometric factor from the chemical equation to determine the amount (in moles) of Ag_2CrO_4(s) produced;

(3) convert the amount of $Ag_2CrO_4(s)$ to its mass in grams. This conversion pathway is mL K_2CrO_4 $\longrightarrow$ mol K_2CrO_4 $\longrightarrow$ mol Ag_2CrO_4 $\longrightarrow$ g Ag_2CrO_4. We can combine these steps into a single setup.

$$? \text{ g } Ag_2CrO_4 = 25.00 \text{ mL} \times \frac{1 \text{ L}}{1000 \text{ mL}} \times \frac{0.250 \text{ mol } K_2CrO_4}{1 \text{ L}}$$

$$\times \frac{1 \text{ mol } Ag_2CrO_4}{1 \text{ mol } K_2CrO_4} \times \frac{331.7 \text{ g } Ag_2CrO_4}{1 \text{ mol } Ag_2CrO_4}$$

$$= 2.07 \text{ g } Ag_2CrO_4(s)$$

Practice Example A: How many milliliters of 0.250 M K_2CrO_4 must be added to excess $AgNO_3(aq)$ to produce 1.50 g Ag_2CrO_4?

(*Hint:* Think of this as the inverse of Example 4-11.)

Practice Example B: How many milliliters of 0.150 M $AgNO_3(aq)$ are required to react completely with 175 mL of 0.0855 M $K_2CrO_4(aq)$? What mass of $Ag_2CrO_4(s)$ is formed?

We will consider a number of additional examples of stoichiometric calculations involving solutions in Chapter 5.

4-4   Determining the Limiting Reactant

When all the reactants are completely and simultaneously consumed in a chemical reaction, the reactants are said to be in **stoichiometric proportions**—in the mole ratios dictated by the coefficients in the balanced equation. This condition is sometimes required, for example, in certain chemical analyses. At other times, as in a precipitation reaction, one of the reactants is completely converted into products by using an *excess* of all the other reactants. The reactant that is completely consumed—the **limiting reactant**—determines the quantities of products formed. In the reaction described in Example 4-11, K_2CrO_4 is the limiting reactant and $AgNO_3$ is present in excess. Up to this point, we have stated the reactant(s) in excess and, by implication, the limiting reactant. In some cases, however, the limiting reactant will not be indicated explicitly; that is, you will be given quantities of two or more reactants. You must determine for yourself which is the limiting reactant, as suggested by the analogy in Figure 4-9.

▶ By an excess of a reactant we mean that more of the reactant is present than is consumed in the reaction. Some is left over.

▶ Chemicals are often referred to as *reagents*, and the limiting reactant in a reaction is sometimes called the *limiting reagent*.

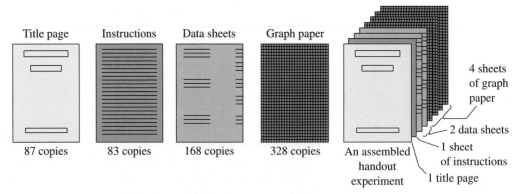

Title page   Instructions   Data sheets   Graph paper

87 copies   83 copies   168 copies   328 copies   An assembled handout experiment

4 sheets of graph paper

2 data sheets

1 sheet of instructions

1 title page

▲ **FIGURE 4-9   An analogy to determining the limiting reactant in a chemical reaction—assembling a handout experiment**
From the numbers of copies available and the instructions on how to assemble the handout, can you see that only 82 complete handouts are possible and that the graph paper is the limiting reactant?

EXAMPLE 4-12

Phosphorus trichloride

Determining the Limiting Reactant in a Reaction. Phosphorus trichloride, PCl_3, is a commercially important compound used in the manufacture of pesticides, gasoline additives, and a number of other products. It is made by the direct combination of phosphorus and chlorine.

$$P_4(s) + 6\,Cl_2(g) \longrightarrow 4\,PCl_3(l)$$

What mass of $PCl_3(l)$ forms in the reaction of 125 g P_4 with 323 g Cl_2?

Solution

One approach, outlined in Figure 4-10, is to compare the initial mole ratio of the two reactants to the ratio in which the reactants combine—6 mol Cl_2 to 1 mol P_4. If more than 6 mol Cl_2 is available per mole of P_4, chlorine is in excess and P_4 is the limiting reactant. If less than 6 mol Cl_2 is available per mole of P_4, chlorine is the limiting reactant.

$$?\ mol\ Cl_2 = 323\ g\ Cl_2 \times \frac{1\ mol\ Cl_2}{70.91\ g\ Cl_2} = 4.56\ mol\ Cl_2$$

$$?\ mol\ P_4 = 125\ g\ P_4 \times \frac{1\ mol\ P_4}{123.9\ g\ P_4} = 1.01\ mol\ P_4$$

We see rather clearly that there is *less than* 6 mol Cl_2 per mole of P_4. Chlorine is the limiting reactant. The remainder of the calculation is to determine the mass of PCl_3 formed in the reaction of 323 g Cl_2 with an excess of P_4.

$$?\ g\ PCl_3 = 323\ g\ Cl_2 \times \frac{1\ mol\ Cl_2}{70.91\ g\ Cl_2} \times \frac{4\ mol\ PCl_3}{6\ mol\ Cl_2} \times \frac{137.3\ g\ PCl_3}{1\ mol\ PCl_3} = 417\ g\ PCl_3$$

A different approach that leads to the same result is to do two separate calculations: Calculate the mass of PCl_3 produced by the reaction of 323 g Cl_2 with an excess of P_4 (*answer*: 417 g PCl_3), and then calculate the mass of PCl_3 produced by the reaction of 125 g P_4 with an excess of Cl_2 (*answer*: 554 g PCl_3). Only one answer can be correct, and it must be the *smaller* of the two.

*Limiting Reactant
animation*

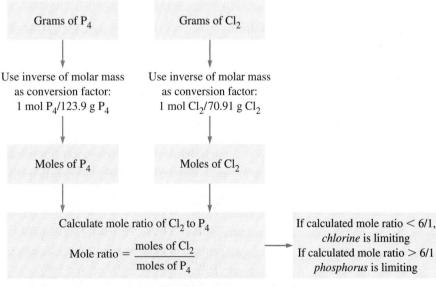

▲ **FIGURE 4-10**   **Determining the limiting reactant for the reaction**
$$P_4(s) + 6\,Cl_2(g) \longrightarrow 4\,PCl_3(l)$$

Practice Example A: If 215 g P_4 is allowed to react with 725 g Cl_2 in the reaction in Example 4-12, how many grams of PCl_3 are formed?

Practice Example B: If 1.00 kg each of PCl_3, Cl_2, and P_4O_{10} are allowed to react, how many kilograms of $POCl_3$ will be formed?

$$6\ PCl_3(l)\ +\ 6\ Cl_2(g)\ +\ P_4O_{10}(s)\ \longrightarrow\ 10\ POCl_3(l)$$

Sometimes our interest in a limiting reactant problem is in determining how much of an excess reactant remains, as well as how much product is formed. This additional calculation is illustrated in Example 4-13.

EXAMPLE 4-13

Determining the Quantity of Excess Reactant(s) Remaining After a Reaction. What mass of P_4 remains in excess following the reaction in Example 4-12?

Solution

The key to this problem is to calculate the mass of P_4 that is consumed, and we can base this calculation either on the mass of Cl_2 consumed

$$?\ g\ P_4 = 323\ g\ Cl_2 \times \frac{1\ mol\ Cl_2}{70.91\ g\ Cl_2} \times \frac{1\ mol\ P_4}{6\ mol\ Cl_2} \times \frac{123.9\ g\ P_4}{1\ mol\ P_4} = 94.1\ g\ P_4$$

or on the mass of PCl_3 produced.

$$?\ g\ P_4 = 417\ g\ PCl_3 \times \frac{1\ mol\ PCl_3}{137.3\ g\ PCl_3} \times \frac{1\ mol\ P_4}{4\ mol\ PCl_3} \times \frac{123.9\ g\ P_4}{1\ mol\ P_4} = 94.1\ g\ P_4$$

The mass of P_4 remaining after the reaction is simply the difference between what was originally present and what was consumed; that is,

$$125\ g\ P_4\ \text{initially}\ -\ 94.1\ g\ P_4\ \text{consumed} = 31\ g\ P_4\ \text{remaining}$$

Practice Example A: In Practice Example 4-12A, which reactant is in excess and what mass of that reactant remains after the reaction to produce PCl_3?

Practice Example B: If 12.2 g H_2 and 154 g O_2 are allowed to react, which gas and what mass of that gas remains after the reaction?

$$2\ H_2(g)\ +\ O_2(g)\ \longrightarrow\ 2\ H_2O(l)$$

KEEP IN MIND ▶

that in subtraction, the quantity with the smallest number of digits beyond the decimal point determines how the answer should be expressed. Because there is no digit beyond the decimal point in 125, there can be none after the decimal point in 31.

4-5 Other Practical Matters in Reaction Stoichiometry

There are a few added matters of concern regarding reaction stoichiometry in both the laboratory and the manufacturing plant. For one, the calculated outcome of a reaction may not be what is actually observed. Specifically, the amount of product may be, unavoidably, less than expected. Also, the route to producing a desired chemical may require several reactions carried out in sequence. And in some cases, two or more reactions may occur simultaneously. These are the matters we describe in this section.

Theoretical Yield, Actual Yield, and Percent Yield

The **theoretical yield** of a reaction is the calculated quantity of product that one expects from given quantities of reactants. The quantity of product that is *actually* produced is called the **actual yield**. The **percent yield** is defined as

$$\text{percent yield} = \frac{\text{actual yield}}{\text{theoretical yield}} \times 100\% \qquad (4.5)$$

In many reactions the actual yield almost exactly equals the theoretical yield, and the reactions are said to be *quantitative*. We can use such reactions in performing quantitative chemical analyses, for example. On the other hand, in some reactions the actual yield is less than the theoretical yield, and the percent yield is less than 100%. The yield may be less than 100% for many reasons. The product of a reaction rarely appears in a pure form, and in the necessary purification steps, some product may be lost through handling. This reduces the yield. In many cases the reactants may participate in reactions other than the one of central interest. These are called *side reactions*, and the unintended products are called **by-products**. To the extent that side reactions occur, the yield of the main product is reduced. Finally, if a reverse reaction occurs, some of the expected product may react to re-form the reactants, and again the yield is less than expected.

At times, the apparent yield is greater than 100%. Because one cannot get something for nothing, this situation usually indicates an error in technique. Some products are formed as a precipitate from a solvent. The product may be wet with solvent, thus giving a larger-than-expected mass for the wet product. More thorough drying of the product would give a more accurate yield determination. Another possibility is that the product is contaminated with an excess reactant or a by-product. This makes the mass of product appear larger than expected. In any case, a product must be purified before its yield is determined.

In Example 4-14, we establish the theoretical, actual, and percent yields of an important industrial process.

EXAMPLE 4-14

Determining Theoretical, Actual, and Percent Yields. Billions of pounds of urea, $CO(NH_2)_2$, are produced annually for use as a fertilizer. The reaction used is

$$2\,NH_3 + CO_2 \longrightarrow CO(NH_2)_2 + H_2O$$

The typical starting reaction mixture has a $3:1$ mole ratio of NH_3 to CO_2. If 47.7 g urea forms *per mole* of CO_2 that reacts, what is the **(a)** theoretical yield; **(b)** actual yield; and **(c)** percent yield in this reaction?

Solution

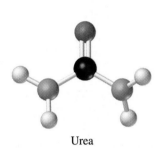

Urea

(a) The stoichiometric proportions are 2 mol NH_3 : 1 mol CO_2. Because the mole ratio of NH_3 to CO_2 used is $3:1$, NH_3 is in excess and CO_2 is the limiting reactant. Because the quantity of urea is given per mole of CO_2, we should base the calculation on 1.00 mol CO_2.

$$\text{theoretical yield} = 1.00\ \text{mol}\ CO_2 \times \frac{1\ \text{mol}\ CO(NH_2)_2}{1\ \text{mol}\ CO_2} \times \frac{60.1\ \text{g}\ CO(NH_2)_2}{1\ \text{mol}\ CO(NH_2)_2}$$

$$= 60.1\ \text{g}\ CO(NH_2)_2$$

(b) actual yield $= 47.7\ \text{g}\ CO(NH_2)_2$

(c) $\%$ yield $= \dfrac{47.7\ \text{g}\ CO(NH_2)_2}{60.1\ \text{g}\ CO(NH_2)_2} \times 100\% = 79.4\%$

Practice Example A: Formaldehyde, CH_2O, can be made from methanol by the following reaction, using a copper catalyst.

$$CH_3OH(g) \longrightarrow CH_2O(g) + H_2(g)$$

If 25.7 g $CH_2O(g)$ is produced per mole of methanol that reacts, what are **(a)** the theoretical yield, **(b)** the actual yield, and **(c)** the percent yield in this reaction?

Practice Example B: What is the percent yield if the reaction of 25.0 g P_4 with 91.5 g Cl_2 produces 104 g PCl_3? $P_4 + 6\,Cl_2 \longrightarrow 4\,PCl_3$

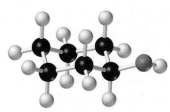

Cyclohexanol

Cyclohexene

▶ Think of this first step as an algebra problem. If 83% of a number is equal to 25, what is the number? $0.83x = 25$; $x = 25/0.83 = 30$.

If we know that a yield will be less than 100%, we need to make an adjustment in the amounts of reactants that we use to produce the desired amount of product. That is, we cannot simply use the theoretical amounts of reactants; we must use more. This point is illustrated in Example 4-15.

EXAMPLE 4-15

Adjusting the Quantities of Reactants in Accordance with the Percent Yield of a Reaction. When heated with sulfuric or phosphoric acid, cyclohexanol, $C_6H_{11}OH$, is converted to cyclohexene, C_6H_{10}.

$$C_6H_{11}OH(l) \longrightarrow C_6H_{10}(l) + H_2O(l) \tag{4.6}$$

If the percent yield of this reaction is 83%, what mass of cyclohexanol must we use to obtain 25 g cyclohexene?

Solution

First, let us answer this question: If the 25 g of C_6H_{10} represents only 83% of the theoretical yield, what is the theoretical yield? Then we can calculate the quantity of $C_6H_{11}OH$ required to produce this theoretical yield of C_6H_{10}.

Step 1. Rearrange equation (4.5), and calculate the theoretical yield.

$$\text{theoretical yield} = \frac{\text{actual yield} \times 100\%}{\text{percent yield}} = \frac{25 \text{ g} \times 100\%}{83\%} = 30 \text{ g}$$

Step 2. Calculate the quantity of $C_6H_{11}OH$ needed to produce 30 g C_6H_{10}.

$$? \text{ g } C_6H_{11}OH = 30 \text{ g } C_6H_{10} \times \frac{1 \text{ mol } C_6H_{10}}{82.1 \text{ g } C_6H_{10}} \times \frac{1 \text{ mol } C_6H_{11}OH}{1 \text{ mol } C_6H_{10}}$$
$$\times \frac{100.2 \text{ g } C_6H_{11}OH}{1 \text{ mol } C_6H_{11}OH}$$
$$= 37 \text{ g } C_6H_{11}OH$$

Practice Example A: If the percent yield for the formation of urea in Example 4-14 were 87.5%, what mass of CO_2, together with an excess of NH_3, would have to be used to obtain 50.0 g $CO(NH_2)_2$?

Practice Example B: Calculate the mass of cyclohexanol ($C_6H_{11}OH$) needed to produce 45.0 g cyclohexene (C_6H_{10}) by reaction (4.6) if the reaction has a 86.2% yield and the cyclohexanol is 92.3% pure.

Consecutive Reactions, Simultaneous Reactions, and Overall Reactions

Where possible, whether in the laboratory or the manufacturing plant, we prefer processes that yield a product through a single reaction. Often such processes give a higher yield because there is no need to remove products from one reaction mixture for further processing in subsequent reactions. But in many cases a multistep reaction is unavoidable. Reactions that are carried out one after another in sequence to yield a final product are called **consecutive reactions**. In **simultaneous reactions**, two or more substances react independently of one another in separate reactions occurring at the same time.

Example 4-16 presents an industrial process that is carried out in two reactions occurring consecutively. The key to the calculation is to use a stoichiometric factor for each reaction. Practice Example 4-16B deals with two reactions that occur simultaneously to produce a common product, hydrogen gas.

EXAMPLE 4-16

Titanium tetrachloride

Calculating the Quantity of a Substance Produced by Reactions Occurring Consecutively. Titanium dioxide, TiO_2, is the most widely used white pigment for paints, having displaced most lead-based pigments, which are environmental hazards. Before it can be used, however, naturally occurring TiO_2 must be freed of colored impurities. One process for doing this converts impure $TiO_2(s)$ to $TiCl_4(g)$, which is then converted back to pure $TiO_2(s)$. What mass of carbon is consumed in producing 1.00 kg of pure $TiO_2(s)$ in this process?

$$2\,TiO_2\,(impure) + 3\,C(s) + 4\,Cl_2(g) \longrightarrow 2\,TiCl_4(g) + CO_2(g) + 2\,CO(g)$$
$$TiCl_4(g) + O_2(g) \longrightarrow TiO_2(s) + 2\,Cl_2(g)$$

Solution

This calculation requires us to begin with the product, TiO_2, and work backwards to one of the reactants, carbon. First, we convert from moles of TiO_2 to moles of $TiCl_4$ in the *second* reaction, and then from moles of $TiCl_4$ to moles of C in the *first* reaction. Together with other conversions between grams and moles of substances, the overall conversion pathway is

$$? \text{ g C} = 1.00 \text{ kg TiO}_2 \times \frac{1000 \text{ g TiO}_2}{1 \text{ kg TiO}_2} \times \frac{1 \text{ mol TiO}_2}{79.88 \text{ g TiO}_2} \times \overset{(a)}{\frac{1 \text{ mol TiCl}_4}{1 \text{ mol TiO}_2}} \times \overset{(b)}{\frac{3 \text{ mol C}}{2 \text{ mol TiCl}_4}}$$

$$\times \frac{12.01 \text{ g C}}{1 \text{ mol C}} = 226 \text{ g C}$$

Practice Example A: Nitric acid, HNO_3, is produced from ammonia and oxygen by the following consecutive reactions.

$$4\,NH_3(g) + 5\,O_2(g) \longrightarrow 4\,NO(g) + 6\,H_2O(g)$$
$$2\,NO(g) + O_2(g) \longrightarrow 2\,NO_2(g)$$
$$3\,NO_2(g) + H_2O(l) \longrightarrow 2\,HNO_3(aq) + NO(g)$$

How many grams of nitric acid can be obtained from 1.00 kg $NH_3(g)$, if $NO(g)$ in the third reaction is *not* recycled?

Practice Example B: Magnalium alloys are widely used in aircraft construction. One magnalium alloy contains 70.0% Al and 30.0% Mg, by mass. How many grams of $H_2(g)$ are produced in the reaction of a 0.710-g sample of this alloy with excess $HCl(aq)$?

$$2\,Al(s) + 6\,HCl(aq) \longrightarrow 2\,AlCl_3(aq) + 3\,H_2(g)$$
$$Mg(s) + 2\,HCl(aq) \longrightarrow MgCl_2(aq) + H_2(g)$$

Usually, we can combine a series of chemical equations to obtain a single equation to represent the overall reaction. The equation for this overall reaction is the overall equation. At times we can use the overall equation for problem solving instead of working with the individual equations. This strategy does not work, however, if the substance of interest is not a starting material or final product but appears only in one of the intermediate reactions. Any substance that is produced in one step and consumed in another step of a multistep process is called an **intermediate**. In Example 4-16, $TiCl_4(g)$ is an intermediate.

To write an overall equation for Example 4-16, multiply the coefficients in the second equation by the factor 2, add the second to the first equation, and cancel any substances that appear on both sides of the overall equation.

$$2\,\cancel{TiO_2}\,(impure) + 3\,C(s) + 4\,\cancel{Cl_2}(g) \longrightarrow 2\,\cancel{TiCl_4}(g) + CO_2(g) + 2\,CO(g)$$

$$\underline{2\,\cancel{TiCl_4}(g) + 2\,O_2(g) \longrightarrow 2\,\cancel{TiO_2}(s) + 4\,\cancel{Cl_2}(g)\cancel{Cl_2}(g)}$$

Overall equation: $3\,C(s) + 2\,O_2(g) \longrightarrow CO_2(g) + 2\,CO(g)$

Chemicals are manufactured on a large scale and sold for use in everything from food to pharmaceuticals. Every day, tons of chemicals are made to be used directly or to serve as reactants in the production of other materials (Table 4.1). It is important that these chemicals be made in high yield and be of high purity, but this is not always easy. Industrial chemists and engineers must consider many factors when making decisions about building and operating a chemical plant.

Let us examine some of these factors using as an example the production of hydrazine, N_2H_4, by a manufacturing method known as the Raschig process. Hydrazine is used as a rocket fuel, as a starting material to make pesticides, and in water treatment to remove dissolved oxygen from water in hot-water heating systems.

TABLE 4.1   The Top Ten Chemicals Produced in the United States (1999)

Rank	Billions of Pounds	Chemical	Formula	Principal Uses and End Products
1	90.2	Sulfuric acid	H_2SO_4	Fertilizers; metallurgy; petroleum refining; manufacture of chemicals
2	866[*]	Nitrogen	N_2	Inert atmospheres for manufacture of chemicals; electronics manufacturing; petroleum recovery; metals processing; freezing foods
3	691[*]	Oxygen	O_2	Steel making and other metallurgical processes; chemical manufacture; medical uses
4	55.8	Ethylene	C_2H_4	Manufacture of plastics, antifreeze, fibers, and solvents
5	45.2	Calcium oxide (lime)	CaO	Metallurgy; pollution control and wastewater treatment; manufacture of chemicals; municipal water treatment
6	37.9	Ammonia	NH_3	Fertilizers; manufacture of plastics, fibers, and explosives
7	29.1	Propylene	C_3H_6	Plastics, fibers, and solvents
8	27.5	Phosphoric acid	H_3PO_4	Fertilizers; phosphates for the food industry
9	26.6	Chlorine	Cl_2	Manufacture of organic chemicals, plastics, pulp, and paper
10	22.8	Sodium hydroxide	$NaOH$	Manufacture of chemicals, pulp, paper, soaps, and detergents; oil refining

[*]Quantity in billions of cubic feet

This is an interesting result. It suggests that (a) we should obtain as much TiO_2 in the second reaction as we started with in the first; (b) the $Cl_2(g)$ produced in the second reaction can be recycled back into the first reaction; and (c) the only substances actually consumed in the overall reaction are $C(s)$ and $O_2(g)$. You can also see that we could not have used the overall equation in the calculation of Example 4-16; $TiO_2(s)$ does not appear in it.

Stepwise Reactions, Intermediates, and the Overall Chemical Equation

An industrial process is usually carried out in steps. The Raschig process involves three steps.

1. $Cl_2(g)$ and $NaOH(aq)$ react to produce sodium hypochlorite, $NaOCl(aq)$.

$$2\,NaOH(aq) + Cl_2(g) \longrightarrow NaOCl(aq) + NaCl(aq) + H_2O(l)$$

2. $NH_3(aq)$ reacts with $NaOCl(aq)$ to form chloramine, $NH_2Cl(aq)$.

$$NaOCl(aq) + NH_3(aq) \longrightarrow NH_2Cl(aq) + NaOH(aq)$$

3. Additional $NH_3(l)$ reacts with $NH_2Cl(aq)$ to form hydrazine, $N_2H_4(aq)$.

$$NH_3(l) + NH_2Cl(aq) + NaOH(aq) \longrightarrow N_2H_4(aq) + NaCl(aq) + H_2O(l)$$

$NaOCl(aq)$ and $NH_2Cl(aq)$ are intermediates in the process. Their presence is crucial to the overall process, but they are consumed immediately after their formation. Hydrazine is the *end product* of the process, and the overall process can be represented through the overall chemical equation.

$$2\,NaOH(aq) + Cl_2(g) + 2\,NH_3(aq) \longrightarrow N_2H_4(aq) + 2\,NaCl(aq) + 2\,H_2O \quad (4.7)$$

Notice that the intermediates do not show up as part of the overall chemical equation for this reaction.

By-Products and Side Reactions

Equation (4.7) shows two products in addition to N_2H_4 ($NaCl$ and H_2O). Substances formed along with the desired end product are called *by-products*. The by-products themselves may be commercially valuable, a feature that improves the economic success of a process.

By-products may also form as a result of *side reactions* that compete with the main reaction. For example, one intermediate, chloramine, NH_2Cl, reacts with hydrazine to produce NH_4Cl and N_2 as by-products.

$$2\,NH_2Cl(aq) + N_2H_4(aq) \longrightarrow 2\,NH_4Cl(aq) + N_2(g) \quad (4.8)$$

Side reactions lower the yield of the desired product, so it is important for industrial chemists and chemical engineers to find conditions that minimize them.

Reaction Conditions

Even if the conditions of a chemical reaction are shown in a chemical equation (above the arrow), the reasons for these conditions are probably not given. Temperature, pressure, and the presence of catalysts that help speed up the reaction are all important conditions that need to be controlled in order to get the best yield of the end product.

A modern modification of the reaction shown in equation (4.7), has the reaction being carried out in acetone, $(CH_3)_2CO$, rather than water. This change of solvent changes the mechanism of the reaction, that is, the path by which the reactant molecules become the product molecules. As a result, the side reaction, equation (4.8), is largely eliminated and the yield of hydrazine increases from 60–80% to nearly 100%. Currently most chemical plants in the United States use the modified process, but some still use the original process. This illustrates two points.

1. Industrial methods of manufacturing chemicals undergo constant change.

2. The changeover to a new process takes place over time, as new plants are built or old ones converted.

Purification

Purifying the product is an essential last step of any process. Rarely is the end product sufficiently pure for its intended uses. Hydrazine produced in reaction (4.7) is in a dilute aqueous solution, together with NH_3, $NaCl$, NH_4Cl, and traces of $NaOH$. Ammonia and water are removed by evaporating them from the solution (distillation). Substances such as $NaCl$, NH_4Cl, and $NaOH$ deposit as solids when water is evaporated from solution (crystallization). The solids are removed by filtration. Ultimately, a product that is more than 98% N_2H_4 is obtained. The NH_3 recovered in the purification process is reused in step 2 of the process. This illustrates another important principle of industrial chemistry. *Materials are recycled whenever possible.*

Summary

Chemical reactions are represented by chemical equations, and the equations must be balanced to show that the total number of atoms of each element remains unchanged in the reaction. Physical states of the reactants and products and reaction conditions can also be indicated. Calculations based on the chemical equation use conversion factors derived from the equation (stoichiometric factors). These calculations often require the use of molar masses, densities, and percent compositions as well.

The molarity of a solution is the amount of solute (in moles) per liter of solution. Molarity can be treated as a conversion factor between solution volume and amount of solute. Molarity as a conversion factor may be applied to individual solutions, in cases where solutions are mixed or diluted by adding more solvent, and to reactions occurring in solution.

Additional features sometimes arise in stoichiometric calculations: The single reactant that determines the amount of product formed—the limiting reactant—may have to be identified. Some reactions yield exactly the quantity of product calculated. They have 100% yield. Others have an actual yield less than the theoretical value. Their percent yield is less than 100%. In some cases, a final product may be produced in a sequence of reactions. Sometimes it is possible to replace a series of equations for consecutive reactions by a single overall equation.

Integrative Example

Sodium nitrite is used in the production of dyes for coloring fabrics, as a preservative in meat processing (to prevent botulism), as a bleach for fibers, and in photography. It can be prepared by passing nitrogen monoxide and oxygen gases into an aqueous solution of sodium carbonate. Carbon dioxide gas is another product of the reaction.

In one reaction with a 95.0% yield, 225 mL of 1.50 M aqueous solution of sodium carbonate, 22.1 g of nitric oxide, and a large excess of oxygen gas are allowed to react. What mass of sodium nitrite is obtained?

1. *Write a chemical equation for the reaction.* This requires substituting formulas for names of substances and balancing the formula expression.

The reactants are sodium carbonate solution [$Na_2CO_3(aq)$], nitrogen monoxide [$NO(g)$], and oxygen [$O_2(g)$]. The products are sodium nitrite solution [$NaNO_2(aq)$] and carbon dioxide [$CO_2(g)$]. The balanced equation for the reaction is

$$2\ Na_2CO_3(aq) + 4\ NO(g) + O_2(g) \longrightarrow$$
$$4\ NaNO_2(aq) + 2\ CO_2(g)$$

2. *Determine the limiting reactant.* To do this, first determine the number of moles of Na_2CO_3 in 225 mL 1.50 M Na_2CO_3 and the number of moles of NO in 22.1 g NO. (We are told that oxygen is in excess.)

$$? \text{ mol } Na_2CO_3 = 0.225 \text{ L soln} \times \frac{1.50 \text{ mol } Na_2CO_3}{\text{L soln}}$$
$$= 0.338 \text{ mol } Na_2CO_3$$

$$? \text{ mol } NO = 22.1 \text{ g } NO \times \frac{1 \text{ mol } NO}{30.01 \text{ g } NO} = 0.736 \text{ mol } NO$$

The mole ratio of NO to Na_2CO_3 is $0.736/0.338 = 2.18$. Because this ratio exceeds the $2:1$ ratio from the chemical equation, we see that NO is in excess and Na_2CO_3 is the limiting reactant.

3. *Determine the theoretical yield of the reaction.* This calculation is based on the amount of Na_2CO_3, the limiting reactant.

$$? \text{ g } Na_2CO_3 = 0.338 \text{ mol } Na_2CO_3 \times \frac{4 \text{ mol } NaNO_2}{2 \text{ mol } Na_2NCO_3}$$
$$\times \frac{69.00 \text{ g } NaNO_2}{1 \text{ mol } NaNO_2} = 46.6 \text{ g } NaNO_2$$

4. *Determine the actual yield.* Use expression (4.5), which relates percent yield (95.0%), theoretical yield (46.6 g $NaNO_2$), and the actual yield.

$$\text{actual yield} = \frac{\% \text{ yield} \times \text{theoretical yield}}{100\%}$$
$$= \frac{95.0\% \times 46.6 \text{ g } NaNO_2}{100\%} = 44.3 \text{ g } NaNO_2$$

Key Terms

actual yield (4-5)	**limiting reactant** (4-4)	**solute** (4-3)
balanced equation (4-1)	**molarity** (4-3)	**solvent** (4-3)
by-product (4-5)	**overall reaction** (4-5)	**stoichiometric coefficient** (4-1)
chemical equation (4-1)	**percent yield** (4-5)	**stoichiometric factor** (4-2)
chemical reaction (4-1)	**product** (4-1)	**stoichiometric proportions** (4-4)
consecutive reactions (4-5)	**reactant** (4-1)	**stoichiometry** (4-2)
intermediate (4-5)	**simultaneous reactions** (4-5)	**theoretical yield** (4-5)

Review Questions

1. In your own words, define or explain the following terms or symbols:

(a) $\xrightarrow{\Delta}$;   (b) (aq);
(c) stoichiometric coefficient;   (d) overall equation.

2. Briefly describe each of the following ideas or methods: (a) balancing a chemical equation; (b) preparing a solution by dilution; (c) determining the limiting reactant in a reaction.

3. Explain the important distinctions between each pair of terms: (a) chemical formula and chemical equation; (b) decomposition and synthesis reaction; (c) solute and solvent; (d) actual yield and percent yield.

4. Balance the following equations by inspection.
(a) $Na_2SO_4(s) + C(s) \longrightarrow Na_2S(s) + CO(g)$
(b) $HCl(g) + O_2(g) \longrightarrow H_2O(l) + Cl_2(g)$
(c) $PCl_5(s) + H_2O(l) \longrightarrow H_3PO_4(aq) + HCl(aq)$
(d) $PbO(s) + NH_3(g) \longrightarrow Pb(s) + N_2(g) + H_2O(l)$
(e) $Mg_3N_2(s) + H_2O(l) \longrightarrow Mg(OH)_2(s) + NH_3(g)$

5. Write balanced equations for the following reactions.
    (a) magnesium + oxygen $\longrightarrow$ magnesium oxide
    (b) nitrogen monoxide + oxygen $\longrightarrow$ nitrogen dioxide
    (c) ethane (C_2H_6) + oxygen $\longrightarrow$

    carbon dioxide + water
    (d) aqueous silver sulfate + aqueous barium iodide $\longrightarrow$

    solid barium sulfate + solid silver iodide

6. Write a balanced chemical equation to represent
    (a) the complete combustion of C_7H_{16}
    (b) the complete combustion of C_4H_9OH
    (c) the reaction (neutralization) of hydroiodic acid with aqueous sodium carbonate to produce sodium iodide, water, and carbon dioxide gas
    (d) the precipitation of iron(III) hydroxide by the mixing of aqueous solutions of iron(III) chloride and sodium hydroxide

7. On an examination, students were asked to write a balanced equation for the decomposition of potassium chlorate to yield potassium chloride and oxygen gas. Among the incorrect responses were the following three. Explain what is wrong with each expression.
    (a) $KClO_3(s) \longrightarrow KCl(s) + 3\ O(g)$
    (b) $KClO_3(s) \longrightarrow KCl(s) + O_2(g) + O(g)$
    (c) $KClO_3(s) \longrightarrow KClO(s) + O_2(g)$

8. Which of the following are correct statements for the reaction: $2\ H_2S + SO_2 \longrightarrow 3\ S + 2\ H_2O$. Explain your reasoning. **(1)** 3 mol S is produced per mole of H_2S; **(2)** 3 g S is produced for every gram of SO_2 consumed; **(3)** 1 mol H_2O is produced per mole of H_2S consumed; **(4)** two-thirds of the S produced comes from H_2S; **(5)** the total number of moles of products equals the number of moles of reactants consumed.

9. Iron metal reacts with chlorine gas as follows.

    $$2\ Fe(s) + 3\ Cl_2(g) \longrightarrow 2\ FeCl_3(s)$$

    How many moles of $FeCl_3$ are obtained when 7.26 mol Cl_2 reacts with excess Fe?

10. How many grams of O_2 are produced in the decomposition of 43.4 g $KClO_3$?

    $$2\ KClO_3(s) \xrightarrow{\Delta} 2\ KCl(s) + 3\ O_2(g)$$

11. If 0.337 mol PCl_3 is produced by the reaction

    $$6\ Cl_2 + P_4 \longrightarrow 4\ PCl_3$$

    how many grams each of Cl_2 and P_4 are consumed?

12. The following reaction of potassium superoxide, KO_2, is used in life-support systems to replace $CO_2(g)$ in expired air by $O_2(g)$.

    $$4\ KO_2(s) + 2\ CO_2(g) \longrightarrow 2\ K_2CO_3(s) + 3\ O_2(g)$$

    (a) How many moles of $O_2(g)$ are produced by the reaction of 156 g $CO_2(g)$ with excess $KO_2(s)$?
    (b) How many grams of $KO_2(s)$ are consumed per 100.0 g $CO_2(g)$ removed from expired air?
    (c) How many O_2 molecules are produced per milligram of KO_2 consumed?

13. What are the molarities of the following solutes when dissolved in water?
    (a) 2.92 mol CH_3OH in 7.16 L of solution
    (b) 7.69 mmol C_2H_5OH in 50.00 mL of solution
    (c) 25.2 g $CO(NH_2)_2$ in 275 mL of solution

    (d) 18.5 mL of $C_3H_5(OH)_3$ ($d = 1.26$ g/mL) in 375 mL of solution

14. How much
    (a) NaI, in moles, is needed to prepare 2.55×10^3 L of 0.125 M NaI?
    (b) Na_2CO_3, in grams, is needed to produce 475 mL of 0.398 M Na_2CO_3?
    (c) $CaCl_2$, in mg, is present *per milliliter* of 0.148 M $CaCl_2(aq)$?

15. *Without doing detailed calculations*, explain which of the following represents a 1.00 M KCl solution: 1.00 L of solution containing 100 g KCl; 500 mL of solution containing 74.6 g KCl; a solution containing 7.46 mg KCl/mL; 5.00 L of solution containing 373 g KCl.

16. What volume of 0.650 M $AgNO_3$, in milliliters, must be diluted with water to prepare 250.0 mL of 0.423 M $AgNO_3$?

17. Water is evaporated from 135 mL of 0.188 M K_2SO_4 solution until the volume becomes 105 mL. What is the molarity of K_2SO_4 in the solution that results?

18. Excess $NaHCO_3(s)$ is added to 415 mL of 0.275 M $Cu(NO_3)_2(aq)$. How many grams of $CuCO_3(s)$ will precipitate from solution?

    $Cu(NO_3)_2(aq) + 2\ NaHCO_3(s) \longrightarrow$
    $$CuCO_3(s) + 2\ NaNO_3(aq) + H_2O(l) + CO_2(g)$$

19. How many milliliters of 2.35 M HCl(aq) are required to dissolve a 1.75-g piece of calcium carbonate (marble)?

    $CaCO_3(s) + 2\ HCl(aq) \longrightarrow$
    $$CaCl_2(aq) + H_2O(l) + CO_2(g)$$

20. A 0.78-mol sample of calcium hydride (CaH_2) and 1.52 mol of water are allowed to react. How many moles of H_2 will be produced?

    $$CaH_2(s) + 2\ H_2O(l) \longrightarrow Ca(OH)_2(s) + 2\ H_2(g)$$

21. A 0.696-mol sample of Cu is added to 136 mL of 6.0 M $HNO_3(aq)$. Assuming the following reaction is the only one that occurs, will the Cu react completely?

    $3\ Cu(s) + 8\ HNO_3(aq) \longrightarrow$
    $$3\ Cu(NO_3)_2(aq) + 4\ H_2O(l) + 2\ NO(g)$$

22. In the reaction of 1.80 mol CCl_4 with an excess of HF, 1.55 mol CCl_2F_2 is obtained. What are the (a) theoretical, (b) actual, and (c) percent yields of this reaction?

    $$CCl_4 + 2\ HF \longrightarrow CCl_2F_2 + 2\ HCl$$

23. In the following reaction, 100.0 g $C_6H_{11}OH$ yielded 64.0 g C_6H_{10}.

    $$C_6H_{11}OH \longrightarrow C_6H_{10} + H_2O$$

    (a) What is the theoretical yield of the reaction?
    (b) What is the percent yield?
    (c) What mass of $C_6H_{11}OH$ should have been used to produce 100.0 g C_6H_{10} if the percent yield is that determined in part (b)?

24. Chalkboard chalk is made from calcium carbonate and calcium sulfate, with minor impurities such as SiO_2. Only the $CaCO_3$ reacts with dilute HCl(aq). What is the mass percent $CaCO_3$ in a piece of chalk if a 3.28-g sample yields 0.981 g $CO_2(g)$?

    (*Hint:* What mass of $CaCO_3$ produces 0.981 g CO_2?)

    $CaCO_3(s) + 2\ HCl(aq) \longrightarrow$
    $$CaCl_2(aq) + H_2O(l) + CO_2(g)$$

Exercises

Writing and Balancing Chemical Equations

25. Balance the following equations by inspection.
(a) $Cr_2O_3(s) + Al(s) \xrightarrow{\Delta} Al_2O_3(s) + Cr(l)$
(b) $CaC_2(s) + H_2O(l) \longrightarrow Ca(OH)_2(s) + C_2H_2(g)$
(c) $H_2(g) + Fe_2O_3(s) \xrightarrow{\Delta} Fe(l) + H_2O(g)$
(d) $NCl_3(g) + H_2O(l) \longrightarrow NH_3(g) + HOCl(aq)$

26. Balance the following equations by inspection.
(a) $(NH_4)_2Cr_2O_7(s) \xrightarrow{\Delta} Cr_2O_3(s) + N_2(g) + H_2O(g)$
(b) $NO_2(g) + H_2O(l) \longrightarrow HNO_3(aq) + NO(g)$
(c) $H_2S(g) + SO_2(g) \longrightarrow S(g) + H_2O(g)$
(d) $SO_2Cl_2 + HI \longrightarrow H_2S + H_2O + HCl + I_2$

27. Write balanced equations to represent the complete combustion of (a) butane, C_4H_{10}; (b) isopropyl alcohol, $CH_3CH(OH)CH_3$; (c) lactic acid, $CH_3CH(OH)COOH$.

28. Write balanced equations to represent the complete combustion of (a) propylene, C_3H_6; (b) glycerol, $CH_2(OH)CH(OH)CH_2OH$; (c) thiobenzoic acid, C_6H_5COSH.

29. Write balanced equations to represent the following reactions.
(a) the decomposition, by heating, of solid ammonium nitrate to produce dinitrogen monoxide gas (laughing gas) and water vapor
(b) the reaction of aqueous sodium carbonate with hydrochloric acid to produce water, carbon dioxide gas, and aqueous sodium chloride
(c) the reaction of methane (CH_4), ammonia, and oxygen gases to form gaseous hydrogen cyanide and water vapor

30. Write balanced equations to represent the following reactions.
(a) the reaction of sulfur dioxide gas with oxygen gas to produce sulfur trioxide gas (one reaction involved in the industrial preparation of sulfuric acid)
(b) the dissolving of limestone (calcium carbonate) in water containing dissolved carbon dioxide to produce calcium hydrogen carbonate (a reaction producing temporary hardness in groundwater)
(c) The reaction of ammonia and nitrogen monoxide to form nitrogen and water vapor

31. Write a balanced chemical equation for the reaction depicted below.

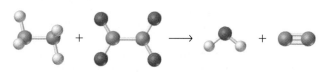

32. Write a balanced chemical equation for the reaction depicted below.

Stoichiometry of Chemical Reactions

33. A laboratory method of preparing $O_2(g)$ involves the decomposition of $KClO_3(s)$.
$$2\ KClO_3(s) \xrightarrow{\Delta} 2\ KCl(s) + 3\ O_2(g)$$
(a) How many moles of $O_2(g)$ can be produced by the decomposition of 32.8 g $KClO_3$?
(b) How many grams of $KClO_3$ must be decomposed to produce 50.0 g O_2?
(c) How many grams of KCl are formed, together with 28.3 g O_2, in the decomposition of $KClO_3$?

34. A commercial method of manufacturing hydrogen involves the reaction of iron and steam.
$$3\ Fe(s) + 4\ H_2O(g) \xrightarrow{\Delta} Fe_3O_4(s) + 4\ H_2(g)$$
(a) How many moles of H_2 can be produced from 42.7 g Fe and an excess of $H_2O(g)$ [steam]?
(b) How many grams of H_2O are consumed in the conversion of 63.5 g Fe to Fe_3O_4?
(c) If 7.36 mol H_2 is produced, how many grams of Fe_3O_4 must also be produced?

35. How many grams of Ag_2CO_3 must have been decomposed if 75.1 g Ag was obtained in the reaction

$$Ag_2CO_3(s) \xrightarrow{\Delta} Ag(s) + CO_2(g) + O_2(g) \text{ (not balanced)}$$

36. How many kilograms of HNO_3 are consumed to produce 125 kg $Ca(H_2PO_4)_2$ in the reaction

$$Ca_3(PO_4)_2 + HNO_3 \longrightarrow$$
$$Ca(H_2PO_4)_2 + Ca(NO_3)_2 \text{ (not balanced)}$$

37. Iron ore is impure Fe_2O_3. When Fe_2O_3 is heated with an excess of carbon (coke), iron metal and carbon monoxide gas are produced. From a sample of ore weighing 938 kg, 523 kg of pure iron is obtained. What is the % Fe_2O_3, by mass, in the ore sample?
(*Hint:* Write a balanced equation for the reaction.)

38. Silver oxide decomposes at temperatures in excess of 300 °C, yielding metallic silver and oxygen gas. A 3.13-g sample of *impure* silver oxide yields 0.187 g $O_2(g)$. If one assumes that $Ag_2O(s)$ is the only source of $O_2(g)$, what is the % Ag_2O, by mass, in the sample?
(*Hint:* Write a balanced equation for the reaction that occurs.)

39. A piece of aluminum foil measuring 10.25 cm × 5.50 cm × 0.601 mm is dissolved in excess HCl(aq). What mass of

$H_2(g)$ is produced? Use equation (4.2) and 2.70 g/cm^3 as the density of Al.

40. An excess of aluminum foil is allowed to react with 225 mL of an aqueous solution of HCl that has a density of 1.088 g/mL and contains 18.0% HCl by mass. What mass of $H_2(g)$ is produced?
[*Hint:* Use equation (4.2).]

41. *Without performing detailed calculations,* determine which of the following produces the maximum quantity of $O_2(g)$ *per gram of reactant.*

(a) $2 NH_4NO_3(s) \xrightarrow{\Delta} 2 N_2(g) + 4 H_2O(l) + O_2(g)$

(b) $2 Ag_2O(s) \xrightarrow{\Delta} 4 Ag(s) + O_2(g)$

(c) $2 HgO(s) \xrightarrow{\Delta} 2 Hg(l) + O_2(g)$

(d) $2 Pb(NO_3)_2(s) \xrightarrow{\Delta} 2 PbO(s) + 4 NO_2(g) + O_2(g)$

42. *Without performing detailed calculations,* determine which of the following metals yields the greatest amount of H_2 *per gram* of metal reacting with HCl(aq): Na, Mg, Al, or Zn?
[*Hint:* Write equations similar to (4.2).]

Molarity

43. What are the molarities of the following solutes?
(a) sucrose ($C_{12}H_{22}O_{11}$) if 150.0 g is dissolved per 250.0 mL of water solution
(b) urea, $CO(NH_2)_2$, if 98.3 mg of the 97.9% pure solid is dissolved in 5.00 mL of aqueous solution
(c) methanol, CH_3OH, ($d = 0.792$ g/mL) if 125.0 mL is dissolved in enough water to make 15.0 L of solution

44. What are the molarities of the following solutes?
(a) aspartic acid ($H_2C_4H_5NO_4$) if 0.405 g is dissolved in enough water to make 100.0 mL of solution
(b) acetone, C_3H_6O, ($d = 0.790$ g/mL) if 35.0 mL is dissolved in enough water to make 425 mL of solution
(c) diethyl ether, $(C_2H_5)_2O$, if 8.8 mg is dissolved in enough water to make 3.00 L of solution

45. How much
(a) glucose, $C_6H_{12}O_6$, in grams, must be dissolved in water to produce 75.0 mL of 0.350 M $C_6H_{12}O_6$?
(b) methanol, CH_3OH ($d = 0.792$ g/mL), in milliliters, must be dissolved in water to produce 2.25 L of 0.485 M CH_3OH?

46. How much
(a) ethanol, C_2H_5OH ($d = 0.789$ g/mL), in liters, must be dissolved in water to produce 200.0 L of 1.65 M C_2H_5OH?
(b) concentrated hydrochloric acid solution (36.0% HCl by mass; $d = 1.18$ g/mL), in milliliters, is required to produce 12.0 L of 0.234 M HCl?

47. Which has the higher concentration of sucrose: a 46% sucrose solution by mass, with a density of 1.21 g/mL, or 1.50 M $C_{12}H_{22}O_{11}$? Explain your reasoning.

48. Which has the greater molarity of ethanol: a white wine ($d = 0.95$ g/mL) with 11% C_2H_5OH by mass, or the solution described in Example 4-8? Explain your reasoning.

49. A 10.00-mL sample of 2.05 M KNO_3 is diluted to a volume of 250.0 mL. What is the concentration of the diluted solution?

50. A 25.0-mL sample of HCl(aq) is diluted to a volume of 500.0 mL. If the concentration of the diluted solution is found to be 0.085 M HCl, what was the concentration of the original solution?

51. Given a 0.250 M K_2CrO_4 stock solution, describe how you would prepare a solution that is 0.0125 M K_2CrO_4. That is, what combination(s) of pipet and volumetric flask would you use? Typical sizes of volumetric flasks found in a general chemistry laboratory are 100.0, 250.0, 500.0, and 1000.0 mL, and typical sizes of volumetric pipets are 1.00, 5.00, 10.00, 25.00, and 50.00 mL.

52. Given a two-liter supply of 0.496 M KCl, describe how you would use this solution to prepare 250.0 mL of 0.175 M KCl. Give sufficient details so that another student could follow your instructions.

Chemical Reactions in Solution

53. For the reaction
$2 AgNO_3(aq) + Na_2S(aq) \longrightarrow Ag_2S(s) + 2 NaNO_3(aq)$
(a) How many grams of $Na_2S(s)$ are required to react completely with 27.8 mL of 0.163 M $AgNO_3$?
(b) How many grams of $Ag_2S(s)$ are obtained from the reaction in part (a)?

54. For the reaction
$Ca(OH)_2(s) + 2 HCl(aq) \longrightarrow CaCl_2(aq) + 2 H_2O(l)$
(a) How many grams of $Ca(OH)_2$ are required to react completely with 415 mL of 0.477 M HCl?
(b) How many kilograms of $Ca(OH)_2$ are required to react with 324 L of a HCl solution that is 24.28% HCl, by mass, and has a density of 1.12 g/mL?

55. Refer to Example 4-7 and equation (4.2). For the conditions stated in Example 4-7, determine (a) the number of moles of $AlCl_3$ and (b) the molarity of the $AlCl_3(aq)$ if the solution volume is simply the 23.8 mL calculated in the example.

56. Refer to the Integrative Example. If 138 g Na_2CO_3 in 1.42 L of aqueous solution is treated with an excess of NO(g) and $O_2(g)$, what is the molarity of the $NaNO_2(aq)$ solution that results? (Assume that the reaction goes to completion.)

57. How many milliliters of 0.650 M K_2CrO_4 are needed to precipitate all the silver ion in 415 mL of 0.186 M $AgNO_3$ as $Ag_2CrO_4(s)$?
$2 AgNO_3(aq) + K_2CrO_4(aq) \longrightarrow$
$Ag_2CrO_4(s) + 2 KNO_3(aq)$

58. How many grams of Ag_2CrO_4 will precipitate if excess $K_2CrO_4(aq)$ is added to the 415 mL of 0.186 M $AgNO_3$ in Exercise 57?

59. How many grams of sodium must react with 155 mL H_2O to produce a solution that is 0.175 M NaOH? (Assume a final solution volume of 155 mL.)

$$2 \text{ Na(s)} + 2 \text{ H}_2\text{O(l)} \longrightarrow 2 \text{ NaOH(aq)} + \text{H}_2\text{(g)}$$

60. A method of adjusting the concentration of HCl(aq) is to allow the solution to react with a small quantity of Mg.

$$\text{Mg(s)} + 2 \text{ HCl(aq)} \longrightarrow \text{MgCl}_2\text{(aq)} + \text{H}_2\text{(g)}$$

How many milligrams of Mg must be added to 250.0 mL of 1.023 M HCl to reduce the solution concentration to exactly 1.000 M HCl?

61. A 0.3126-g sample of oxalic acid, $H_2C_2O_4$, requires 26.21 mL of a particular concentration of NaOH(aq) to complete the following reaction. What is the molarity of the NaOH(aq)?

$$\text{H}_2\text{C}_2\text{O}_4\text{(s)} + 2 \text{ NaOH(aq)} \longrightarrow \text{Na}_2\text{C}_2\text{O}_4\text{(aq)} + 2 \text{ H}_2\text{O(l)}$$

62. A 25.00-mL sample of HCl(aq) was added to a 0.1000-g sample of $CaCO_3$. All the $CaCO_3$ reacted, leaving some excess HCl(aq).

$$\text{CaCO}_3\text{(s)} + 2 \text{ HCl(aq)} \longrightarrow \text{CaCl}_2\text{(aq)} + \text{H}_2\text{O(l)} + \text{CO}_2\text{(g)}$$

The excess HCl(aq) required 43.82 mL of 0.01185 M $Ba(OH)_2(aq)$ to complete the following reaction.

$$2 \text{ HCl(aq)} + \text{Ba(OH)}_2\text{(aq)} \longrightarrow \text{BaCl}_2\text{(aq)} + 2 \text{ H}_2\text{O(l)}$$

What was the molarity of the original HCl(aq)?

Determining the Limiting Reactant

63. How many grams of NO(g) can be produced in the reaction of 1.00 mol $NH_3(g)$ and 1.00 mol $O_2(g)$?

$$4 \text{ NH}_3\text{(g)} + 5 \text{ O}_2\text{(g)} \xrightarrow{\Delta} 4 \text{ NO(g)} + 6 \text{ H}_2\text{O(l)}$$

64. How many grams of $H_2(g)$ are produced by the reaction of 1.84 g Al with 75.0 mL of 2.95 M HCl?
[*Hint:* Use equation (4.2).]

65. A side reaction in the manufacture of rayon from wood pulp is

$$3 \text{ CS}_2 + 6 \text{ NaOH} \longrightarrow 2 \text{ Na}_2\text{CS}_3 + \text{Na}_2\text{CO}_3 + 3 \text{ H}_2\text{O}$$

How many grams of Na_2CS_3 are produced in the reaction of 92.5 mL of liquid CS_2 ($d = 1.26$ g/mL) and 2.78 mol NaOH?

66. Lithopone is a brilliant white pigment used in water-based interior paints. It is a mixture of $BaSO_4$ and ZnS produced by the reaction

$$\text{BaS(aq)} + \text{ZnSO}_4\text{(aq)} \longrightarrow \underset{\text{lithopone}}{\text{ZnS(s)} + \text{BaSO}_4\text{(s)}}$$

How many grams of lithopone are produced in the reaction of 315 mL of 0.275 M $ZnSO_4$ and 285 mL of 0.315 M BaS?

67. Ammonia can be generated by heating together the solids NH_4Cl and $Ca(OH)_2$; $CaCl_2$ and H_2O are also formed. If a mixture containing 33.0 g each of NH_4Cl and $Ca(OH)_2$ is heated, how many grams of NH_3 will form? Which reactant remains in excess, and in what mass?
(*Hint:* Write a balanced equation for the reaction.)

68. Chlorine can be generated by heating together calcium hypochlorite and hydrochloric acid. Calcium chloride and water are also formed. If 50.0 g $Ca(OCl)_2$ and 275 mL of 6.00 M HCl are allowed to react, how many grams of chlorine gas will form? Which reactant, $Ca(OCl)_2$ or HCl, remains in excess, and in what mass?
(*Hint:* Write a balanced equation for the reaction.)

Theoretical, Actual, and Percent Yields

69. The reaction of 15.0 g C_4H_9OH, 22.4 g NaBr, and 32.7 g H_2SO_4 yields 17.1 g C_4H_9Br in the reaction

$$\text{C}_4\text{H}_9\text{OH} + \text{NaBr} + \text{H}_2\text{SO}_4 \longrightarrow$$
$$\text{C}_4\text{H}_9\text{Br} + \text{NaHSO}_4 + \text{H}_2\text{O}$$

What are **(a)** the theoretical yield, **(b)** the actual yield, and **(c)** the percent yield of this reaction?

70. Azobenzene, an intermediate in the manufacture of dyes, can be prepared from nitrobenzene by reaction with triethylene glycol in the presence of Zn and KOH. In one reaction, 0.10 L of nitrobenzene ($d = 1.20$ g/mL) and 0.30 L of triethylene glycol ($d = 1.12$ g/mL) yields 55 g azobenzene. What are **(a)** the theoretical yield, **(b)** the actual yield, and **(c)** the percent yield of this reaction?

$$\underset{\text{nitrobenzene}}{2 \text{ C}_6\text{H}_5\text{NO}_2} + \underset{\text{triethylene glycol}}{4 \text{ C}_6\text{H}_{14}\text{O}_4} \xrightarrow[\text{KOH}]{\text{Zn}}$$

$$\underset{\text{azobenzene}}{(\text{C}_6\text{H}_5\text{N})_2} + 4 \text{ C}_6\text{H}_{12}\text{O}_4 + 4 \text{ H}_2\text{O}$$

71. How many grams of commercial acetic acid (97% $C_2H_4O_2$ by mass) must be allowed to react with an excess of PCl_3 to produce 75 g of acetyl chloride (C_2H_3OCl), if the reaction has a 78.2% yield?

$$\text{C}_2\text{H}_4\text{O}_2 + \text{PCl}_3 \longrightarrow \text{C}_2\text{H}_3\text{OCl} + \text{H}_3\text{PO}_3 \text{ (not balanced)}$$

72. Suppose that each of the following reactions has a 92% yield.
(a) $\text{CH}_4 + \text{Cl}_2 \longrightarrow \text{CH}_3\text{Cl} + \text{HCl}$
(b) $\text{CH}_3\text{Cl} + \text{Cl}_2 \longrightarrow \text{CH}_2\text{Cl}_2 + \text{HCl}$
Starting with 112 g CH_4 in reaction (1) and an excess of $Cl_2(g)$, how many grams of CH_2Cl_2 are formed in reaction (2)?

73. An essentially 100% yield is necessary for a chemical reaction used to *analyze* a compound, but it is almost never expected for a reaction that is used to *synthesize* a compound. Explain this difference.

74. Suppose we carry out the precipitation of $Ag_2CrO_4(s)$ described in Example 4-11. If we obtain 2.058 g of precipitate, we might conclude that it is nearly pure $Ag_2CrO_4(s)$, but if we obtain 2.112 g, we can be quite sure that the precipitate is not pure. Explain this difference.

Consecutive Reactions, Simultaneous Reactions

75. How many grams of HCl are consumed in the reaction of 425 g of a mixture containing 35.2% $MgCO_3$ and 64.8% $Mg(OH)_2$, by mass?

$$Mg(OH)_2 + 2\,HCl \longrightarrow MgCl_2 + 2\,H_2O$$
$$MgCO_3 + 2\,HCl \longrightarrow MgCl_2 + H_2O + CO_2(g)$$

76. How many grams of CO_2 are produced in the complete combustion of 406 g of a bottled gas that consists of 72.7% propane (C_3H_8) and 27.3% butane (C_4H_{10}), by mass? (*Hint:* Write balanced equations for the combustion of the two gases.)

77. Dichlorodifluoromethane, once widely used as a refrigerant, can be prepared by the following reactions.

$$CH_4 + Cl_2 \longrightarrow CCl_4 + HCl \text{ (not balanced)}$$
$$CCl_4 + HF \longrightarrow CCl_2F_2 + HCl \text{ (not balanced)}$$

How many moles of Cl_2 must be consumed in the first reaction to produce 2.25 kg CCl_2F_2 in the second? Assume that all the CCl_4 produced in the first reaction is consumed in the second.

78. $CO_2(g)$ produced in the combustion of a sample of ethane is absorbed in $Ba(OH)_2(aq)$, producing 0.506 g $BaCO_3(s)$.

How many grams of ethane (C_2H_6) must have been burned?

$$C_2H_6(g) + O_2(g) \longrightarrow CO_2(g) + H_2O(l) \text{ (not balanced)}$$
$$CO_2(g) + Ba(OH)_2(aq) \longrightarrow BaCO_3(s) + H_2O(l)$$

79. The following process has been used to obtain iodine from oil-field brines in California.

sodium iodide + silver nitrate $\longrightarrow$
  silver iodide + sodium nitrate
silver iodide + iron $\longrightarrow$ iron(II) iodide + silver
iron(II) iodide + chlorine gas $\longrightarrow$
  iron(III) chloride + solid iodine

How many grams of silver nitrate are required in the first step for every kilogram of iodine produced in the third step?

80. Sodium bromide, used to produce silver bromide for use in photography, can be prepared as follows.

$$Fe + Br_2 \longrightarrow FeBr_2$$
$$FeBr_2 + Br_2 \longrightarrow Fe_3Br_8 \text{ (not balanced)}$$
$$Fe_3Br_8 + Na_2CO_3 \longrightarrow NaBr + CO_2 + Fe_3O_4$$
$$\text{(not balanced)}$$

How many kilograms of iron are consumed to produce 2.50×10^3 kg NaBr?

Integrative and Advanced Exercises

81. Write chemical equations to represent the following reactions.
 (a) Limestone rock (calcium carbonate) is heated (calcined) and decomposes to calcium oxide and carbon dioxide gas.
 (b) Zinc sulfide ore is heated in air (roasted) and is converted to zinc oxide and sulfur dioxide gas. (Note that oxygen gas in the air is also a reactant.)
 (c) Propane gas reacts with gaseous water to produce a mixture of carbon monoxide and hydrogen gases (called *synthesis gas* and used in the chemical industry to produce a variety of other chemicals).
 (d) Sulfur dioxide gas is passed into an aqueous solution containing sodium sulfide and sodium carbonate. The reaction products are carbon dioxide and an aqueous solution of sodium thiosulfate.

82. Write chemical equations to represent the following reactions.
 (a) (Tri)calcium phosphate is heated with silicon dioxide and carbon, producing calcium silicate ($CaSiO_3$), phosphorus (P_4), and carbon monoxide. The phosphorus and chlorine react to form phosphorus trichloride, and the phosphorus trichloride and water react to form phosphorus acid.
 (b) Copper metal reacts with gaseous oxygen, carbon dioxide, and water to form green basic copper carbonate, $Cu_2(OH)_2CO_3$ (a reaction responsible for the formation of the green patina, or coating, often seen on outdoor bronze statues).

(c) White phosphorus and oxygen gas react to form tetraphosphorus decoxide. The tetraphosphorus decoxide reacts with water to form an aqueous solution of phosphoric acid.
 (d) Calcium dihydrogen phosphate reacts with sodium hydrogen carbonate (bicarbonate), producing (tri)calcium phosphate, (di)sodium hydrogen phosphate, carbon dioxide, and water (the principal reaction occurring in the action of ordinary baking powder in the baking of cakes, bread, and biscuits).

83. $H_2(g)$ is passed over $Fe_2O_3(s)$ at 400 °C. Water vapor is formed, together with a black residue—a compound consisting of 72.3% Fe and 27.7% O. Write a balanced equation for this reaction.

84. A sulfide of iron, containing 36.5% S, by mass, is heated in $O_2(g)$, and the products are sulfur dioxide and an oxide of iron containing 27.6% O, by mass. Write a balanced chemical equation for this reaction.
 (*Hint:* The formulas of the sulfide and oxide are not necessarily those that you would predict from Table 3-1.)

85. What is the molarity of NaCl(aq) if a solution has 1.52 ppm Na? [Assume that NaCl is the only source of Na and that the solution density is 1.00 g/mL. Parts per million (ppm) can be taken to mean g Na per million grams of solution.]

86. How many milligrams $Ca(NO_3)_2$ must be present in 50.0 L of a solution containing 2.35 ppm Ca? (*Hint:* See also Exercise 85.)

87. A drop (0.05 mL) of 12.0 M HCl is spread over a sheet of thin aluminum foil. Assume that all the acid reacts with, and thus dissolves through, the foil. What will be the area, in cm^2, of the cylindrical hole produced? (density of Al = 2.70 g/cm^3; thickness of the foil = 0.10 mm.)

$$2\ Al(s) + 6\ HCl(aq) \longrightarrow 2\ AlCl_3(aq) + 3\ H_2(g)$$

88. A small piece of zinc is dissolved in 50.00 mL of 1.035 M HCl.

$$Zn(s) + 2\ HCl(aq) \longrightarrow ZnCl_2(aq) + H_2(g)$$

At the conclusion of the reaction, the concentration of the 50.00-mL sample is redetermined and found to be 0.812 M HCl. What must have been the mass of the piece of zinc that dissolved?

89. How many milliliters of 0.715 M NH_4NO_3 solution must be diluted with water to produce 1.00 L of a solution with a concentration of 2.37 mg N per mL?

90. A seawater sample has a density of 1.03 g/mL and 2.8% NaCl by mass. A saturated solution of NaCl in water is 5.45 M NaCl. How many liters of water would have to be evaporated from 1.00×10^6 L of the seawater before NaCl would begin to crystallize? (A saturated solution contains the maximum amount of dissolved solute possible.)

91. A 99.8-mL sample of a solution that is 12.0% KI by mass (d = 1.093 g/mL) is added to 96.7 mL of another solution that is 14.0% $Pb(NO_3)_2$ by mass (d = 1.134 g/mL). How many grams of PbI_2 should form?

$$Pb(NO_3)_2(aq) + 2\ KI(aq) \longrightarrow PbI_2(s) + 2\ KNO_3(aq)$$

92. $CaCO_3(s)$ reacts with HCl(aq) to form H_2O, $CaCl_2(aq)$ and $CO_2(g)$. If a 45.0-g sample of $CaCO_3(s)$ is added to 1.25 L of HCl(aq) that is 25.7% HCl by mass (d = 1.13 g/mL), what will be the molarity of HCl in the solution after the reaction is completed? (Assume that the solution volume remains constant.)

93. A 2.05-g sample of an iron–aluminum alloy (ferroaluminum) is dissolved in excess HCl(aq) to produce 0.105 g $H_2(g)$. What is the percent composition, by mass, of the ferroaluminum?

$$Fe(s) + 2\ HCl(aq) \longrightarrow FeCl_2(aq) + H_2(g)$$
$$2\ Al(s) + 6\ HCl(aq) \longrightarrow 2\ AlCl_3(aq) + 3\ H_2(g)$$

94. A 0.155-g sample of an Al–Mg alloy reacts with an excess of HCl(aq) to produce 0.0163 g H_2. What is the percent Mg in the alloy?
[*Hint:* Write equations similar to (4.2).]

95. An organic liquid is either methyl alcohol (CH_3OH), ethyl alcohol (C_2H_5OH), or a mixture of the two. A 0.220-g sample of the liquid is burned in an excess of $O_2(g)$ and yields 0.352 g $CO_2(g)$. Is the liquid a pure alcohol or a mixture of the two?

96. The manufacture of ethyl alcohol, C_2H_5OH, yields diethyl ether, $(C_2H_5)_2O$, as a by-product. The complete combustion of a 1.005-g sample of the product of this process yields 1.963 g CO_2. What must be the mass percents of C_2H_5OH and of $(C_2H_5)_2O$ in this sample?

97. Under appropriate conditions, copper sulfate, potassium chromate, and water react to form a product containing Cu^{2+}, CrO_4^{2-}, and OH^- ions. Analysis of the compound yields 48.7% Cu^{2+}, 35.6% CrO_4^{2-}, and 15.7% OH^-.
    (a) Determine the empirical formula of the compound.
    (b) Write a plausible equation for the reaction.

98. Write a chemical equation to represent the complete combustion of malonic acid, a compound with 34.62% C, 3.88% H, and 61.50% O, by mass.

99. A mixture of 4.800 g H_2 and 36.40 g O_2 reacts.
    (a) Write a balanced equation for this reaction.
    (b) What substances are present after the reaction?
    (c) Show that the total mass of substances present before and after the reaction is the same.

100. Silver nitrate is a very expensive chemical. For a particular experiment, a student needs 100.0 mL of 0.0750 M $AgNO_3(aq)$. She has available 60 mL of 0.0500 M $AgNO_3$. She decides to (1) pipet exactly 50.00 mL of the solution that she does have into a 100.0-mL flask, (2) add an appropriate mass of $AgNO_3$, and (3) dilute the resulting solution to exactly 100.0 mL. What mass of $AgNO_3$ must she use?

101. When sulfur (S_8) and chlorine are mixed in a reaction vessel, disulfur dichloride is the sole product. The starting mixture below is represented by yellow spheres for the S_8 molecules and green spheres for the chlorine molecules.

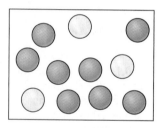

Which of the following is (are) a valid representation(s) of the contents of the reaction vessel after some disulfur dichloride (represented by red spheres) has formed?

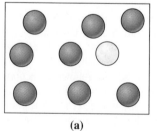

(a)

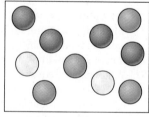

(b)

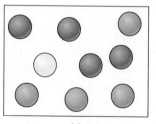

(c)

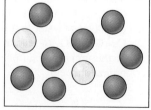

(d)

102. Refer to Figure 1-5 on page 8 and explain why, if ammonium dichromate is the only reactant and chromium(III) oxide and water are two of the products, the only possible nitrogen-containing product is nitrogen gas. How many grams of nitrogen are produced per kilogram of ammonium dichromate decomposed?

103. It is desired to produce as *large* a volume of 1.25 M urea [$CO(NH_2)_2$(aq)] as possible from these three sources: 345 mL of 1.29 M $CO(NH_2)_2$, 485 mL of 0.653 M $CO(NH_2)_2$, and 835 mL of 0.775 M $CO(NH_2)_2$. How can this be done? What is the maximum volume of this solution obtainable?

104. The mineral ilmenite, $FeTiO_3$, is an important source of titanium dioxide for use as a white pigment. In the first step in its conversion to titanium dioxide, ilmenite is treated with sulfuric acid and water to form $TiOSO_4$ and iron(II) sulfate heptahydrate. Titanium dioxide is obtained in two subsequent steps. How many kilograms of iron(II) sulfate heptahydrate are produced for every 1.00×10^3 kg of ilmenite processed?

105. Refer to Exercise 104. Iron(II) sulfate heptahydrate formed in the processing of ilmenite ore cannot be dumped into the environment. Its further treatment involves dehydration by heating to produce anhydrous iron(II) sulfate. Upon further heating, the iron(II) sulfate decomposes to iron(III) oxide, and sulfur dioxide and oxygen gases. The iron(III) oxide is used in the production of iron and steel. How many kilograms of iron(III) oxide are obtained for every 1.00×10^3 kg of iron(II) sulfate heptahydrate?

106. Refer to Exercises 104 and 105. How many kilograms of H_2SO_4(l) can be produced from the sulfur dioxide by-product formed in the processing of 1.00×10^3 kg of ilmenite ore?

107. Melamine [$C_3N_3(NH_2)_3$] is used in adhesives and resins. It is manufactured in a two-step process in which urea [$CO(NH_2)_2$] is the sole starting material, isocyanic acid [HNCO] is an intermediate, and ammonia and carbon dioxide gases are by-products.
    (a) Write a balanced equation for the *overall* reaction.
    (b) What mass of melamine will be obtained from 100.0 kg of urea if the yield of the overall reaction is 84%?

108. Acrylonitrile is used in the production of synthetic fibers, plastics, and rubber goods. It can be prepared from propylene (propene), ammonia, and oxygen in the reaction illustrated below through molecular models.

    (a) Write a balanced chemical equation for this reaction.
    (b) The actual yield of the reaction is 0.73 lb acrylonitrile per pound of propylene. What is the minimum mass of ammonia required to produce 1.00 ton (2000 lb) of acrylonitrile?

Feature Problems

109. It is often difficult to determine the concentration of a species in solution, particularly if it is a biological species that takes part in complex reaction pathways. One way to do this is through a dilution experiment with labeled molecules. Instead of molecules, however, we will use fish.

    A fisher wants to know the number of fish in a particular pond. The fisher puts an indelible mark on 100 fish and adds them to the pond. After waiting for the fish to spread throughout the pond, the fisher starts fishing. Eventually, the fisher catches 18 fish. Of these, five are marked. What is the total number of fish in the pond?

110. Lead nitrate and potassium iodide react in aqueous solution to form a yellow precipitate of lead iodide. In one series of experiments, the masses of the two reactants were varied, but the *total* mass of the two was held constant at 5.000 g. The lead iodide formed was filtered from solution, washed, dried, and weighed. Data for a series of reactions are presented below.

Experiment	Mass of Lead Nitrate, g	Mass of Lead Iodide, g
1	0.500	0.692
2	1.000	1.388
3	1.500	2.093
4	3.000	2.778
5	4.000	1.391

(a) Plot the data in a graph of mass of lead iodide versus mass of lead nitrate, and draw the appropriate curve(s) connecting the data points. What is the maximum mass of precipitate that can be obtained?
(b) Explain why the maximum mass of precipitate is obtained when the reactants are in their stoichiometric proportions. What are these stoichiometric proportions expressed (1) as a mass ratio, and (2) as a mole ratio?
(c) Show how the stoichiometric proportions determined in part (b) are related to the balanced equation for the reaction.

111. Some situations in the business world require balancing supplies (reactants) and products. For example, a business firm makes necklaces using beads of three colors: red, blue, and green. Each necklace contains 70 beads, shown in the sketch. The firm has on hand 8 boxes of red beads, 11 boxes of blue beads, and 10 boxes of green beads. Each box contains 10.0 kg of beads, and the average masses of the beads are red: 1.98 g; blue: 3.05 g; green: 1.82 g.
    (a) Write a "balanced equation" for forming a necklace from individual beads.
    (b) How many necklaces can be made with the supplies on hand?
    (c) Explain the limitations on the precision of your answer.

112. Baking soda, $NaHCO_3$, is made from soda ash, a common name for sodium carbonate. The soda ash is obtained in one of two ways: It can be manufactured in a process in which carbon dioxide, ammonia, sodium chloride, and water are the starting materials. Alternatively, it is mined as a mineral called *trona* (photo below on left).

Whether the soda ash is mined or manufactured, it is dissolved in water and carbon dioxide is bubbled through the solution. Sodium bicarbonate, $NaHCO_3$, precipitates from the solution.

Trona	Baking soda
$Na_2CO_3 \cdot NaHCO_3 \cdot 2H_2O$	$NaHCO_3$

As a chemical analyst you are presented with two samples of sodium bicarbonate, one is from the manufacturing process and the other is derived from trona. You are asked to determine which is purer. You are told that the impurity is sodium carbonate. You decide to treat the samples with hydrochloric acid to convert all the sodium carbonate and bicarbonate to sodium chloride, carbon dioxide, and water. You then precipitate silver chloride in the reaction of sodium chloride with silver nitrate.

A 6.93-g sample of baking soda derived from trona gave 11.89 g of silver chloride. A 6.78-g sample from manufactured sodium carbonate gave 11.77 g of silver chloride. Which baking soda sample is purer; that is, which has the greater mass percent $NaHCO_3$?

 # eMedia Exercises

113. View the **Reactions with Oxygen** movie *(eChapter 4-1)*. **(a)** Describe the physical evidence indicative of a chemical reaction in each case. **(b)** Why is a proportionately different amount of oxygen incorporated into the different products in the reactions presented in this movie?

114. The reaction of carbon and copper oxide is illustrated in the **Reduction of CuO** movie *(eChapter 4-2)*. **(a)** If the reaction were carried out with 124.3 grams of CuO in the presence of excess carbon, how many grams of copper would be present at the end of the experiment? **(b)** Describe two reasons why the mass of the solid poured from the test tube may be greater than the amount you calculated.

115. The steps involved in preparing an aqueous solution of known concentration are discussed in the **Solution Formation from a Solid** animation *(eChapter 4-3)*. What would be the effect on the actual concentration of the solution if:
    (a) the mass of the weighing paper was neglected?
    (b) water was added past the line indicating a volume of 250 mL?
    (c) Propose an additional source of error related to composition of the starting compound that would lead to a concentration less than 1 M.

116. After viewing the **Limiting Reactant** animation *(eChapter 4-4)*, **(a)** determine the mass of H_2 gas produced in the three different experiments. **(b)** Where a limiting reagent is involved, calculate the mass of the excess species present at the completion of the experiment.

5 Introduction to Reactions in Aqueous Solutions

Contents

When clear, colorless aqueous solutions of lead nitrate and potassium iodide are mixed, a yellow cloud of solid lead iodide is formed. Precipitation reactions are one of the three types of reactions considered in this chapter.

Precipitation, the formation of a solid when solutions are mixed, is probably the most common evidence of a chemical reaction that general chemistry students see. A practical application of precipitation is in determining the presence of certain ions in solution. If, for example, we do not know whether a bottle contains distilled water or a barium chloride solution, we can easily find out by adding a few drops of silver nitrate solution to a small sample of the liquid. If a white solid (AgCl) forms, the sample is a barium chloride solution; if nothing happens, it is water. Precipitation reactions are the first reaction type we will study in this chapter. $Mg(OH)_2(s)$ is insoluble in water but soluble in hydrochloric acid, $HCl(aq)$, as a result of an acid–base reaction. This is the reaction by which milk of magnesia neutralizes excess stomach acid.

Magnesium hydroxide is a base, and a survey of acids, bases, and acid–base reactions is another topic in this chapter. The third category of reactions presented in this chapter, oxidation–reduction reactions, can be found in all aspects of life, from reactions in organisms to processes for manufacturing chemicals, to such practical matters as bleaching fabrics, purifying water, and destroying toxic chemicals.

5-1 The Nature of Aqueous Solutions

Reactions in aqueous (water) solution are important because (1) water is inexpensive and able to dissolve a vast number of substances; (2) in aqueous solution, many substances are dissociated into ions, which can participate in chemical reactions; and (3) aqueous solutions are found everywhere, from seawater to living systems.

Let's try to form a mental image of a solution at the molecular level. The greatest number of molecules in a solution are generally solvent molecules, which are packed together rather tightly. Water is the solvent in an aqueous solution, and our mental image of water might look something like Figure 5-1a. Solute particles—molecules or ions—are present in much smaller number and are randomly distributed among the solvent molecules. Our mental image of an aqueous solution of oxygen might look something like Figure 5-1b.

We have learned that ions are individual atoms or groupings of atoms that acquire a net electric charge by losing or gaining electrons. Thus, Mg^{2+} is a positively

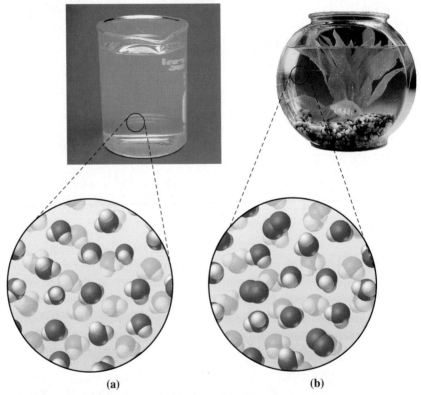

(a)   (b)

▲ FIGURE 5-1   **Molecular view of water and an aqueous solution**
(a) Water molecules are in close proximity in liquid water. (b) Oxygen molecules dissolved in water are far apart, separated by water molecules.

Electrolytes and Nonelectrolytes animation

charged ion (a cation) formed when a Mg atom loses two electrons. Cl^- is a negatively charged ion (an anion) formed when a Cl atom gains one electron. When a group of three O atoms covalently bonded to one N atom acquires an additional electron, it becomes the polyatomic nitrate anion, NO_3^-.

Unlike metallic conductors, where electrons carry the electric charge, in electrically conducting aqueous solutions it is ions that carry the electric charge. Pure water contains so few ions that it does not conduct electric current. However, some solutes dissociate into ions in water, thereby making the aqueous solution an electrical conductor; these solutes are called *electrolytes*. The manner in which the ions conduct electric current is suggested by Figure 5-2. With the apparatus pictured in Figure 5-3, we can detect the presence of ions in an aqueous solution by measuring how well the solution conducts electricity. Depending on which of the following observations we make, we can label a solute as a nonelectrolyte, strong electrolyte, or weak electrolyte.

- *The lamp fails to light up.* Conclusion: There are no ions present (or, if there are some present, their concentration is extremely low). A **nonelectrolyte** is a substance that is not ionized and does not conduct electric current (Figure 5-3a).
- *The lamp lights up brightly.* Conclusion: The concentration of ions in solution is *high*. A **strong electrolyte** is a substance that is essentially completely ionized in aqueous solution, and the solution is a good electrical conductor (Figure 5-3b).
- *The lamp lights up only dimly.* Conclusion: The concentration of ions in solution is *low*. A **weak electrolyte** is *partially* ionized in aqueous solution, and the solution is only a fair conductor of electricity (Figure 5-3c).

The following generalization is helpful when deciding whether a particular solute in an aqueous solution is most likely to be a nonelectrolyte, a strong electrolyte, or a weak electrolyte.

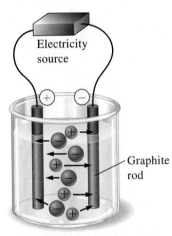

▲ **FIGURE 5-2**
Conduction of electricity through a solution
Two graphite rods called electrodes are placed in a solution. The external source of electricity pulls electrons from one graphite rod and forces them onto the other. As a result, a positive charge is created on one electrode and a negative charge on the other. In the solution, positive ions (cations) are attracted to the negative electrode, or cathode, and negative ions (anions) are attracted to the positive electrode, or anode. Thus, electric charge is carried through the solution by the migration of ions. Other important aspects of electrical conductivity, including what happens at the interface between the electrodes and the solution, are discussed in Chapter 21.

- Essentially all soluble ionic compounds and only a relatively few molecular compounds are *strong electrolytes*.
- Most molecular compounds are either *nonelectrolytes* or *weak electrolytes*.

Now let us consider how best to represent these three types of substances in chemical equations. For water-soluble $MgCl_2$, an ionic compound, we may write

$$MgCl_2(s) \xrightarrow{H_2O} Mg^{2+}(aq) + 2\,Cl^-(aq)$$

By this equation we mean that, in the presence of water, formula units of $MgCl_2$ are dissociated into the separate ions. The best representation of $MgCl_2(aq)$, then, is $Mg^{2+}(aq) + 2\,Cl^-(aq)$.[*]

For a molecular compound that is a strong electrolyte, we similarly can write

$$HCl(g) \xrightarrow{H_2O} H^+(aq) + Cl^-(aq)$$

which means that the best representation of hydrochloric acid, $HCl(aq)$, is $H^+(aq) + Cl^-(aq)$.

---

[*]To say that a strong electrolyte is completely dissociated into individual ions in aqueous solution is a good approximation, but somewhat of an oversimplification. Some of the cations and anions in solution may become associated into units called *ion pairs*. Generally, though, at the low solution concentrations we will be using, to assume complete dissociation will not seriously affect our results.

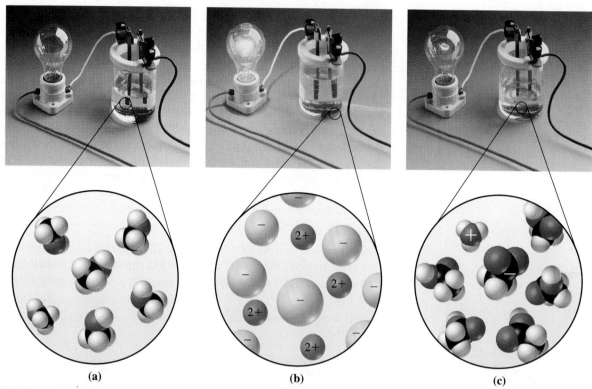

(a) (b) (c)

▲ **FIGURE 5-3** **Three types of electrolytes**

In (a), there are no ions present to speak of—only molecules. Methanol (methyl alcohol), CH_3OH, is a *nonelectrolyte* in aqueous solutions. In (b), the solute is present almost entirely as individual ions. $MgCl_2$ is a *strong electrolyte* in aqueous solutions. In (c), although most of the solute is present as molecules, a small fraction of the molecules ionize. CH_3COOH is a *weak electrolyte* in aqueous solution. The CH_3COOH molecules that ionize produce acetate ions CH_3COO^- and H^+ ions, and the H^+ ions attach themselves to water molecules to form hydronium ions, H_3O^+.

KEEP IN MIND ▶
that the solvent molecules are densely packed. In diagrams such as Figure 5-3, we will often show the solvent as a uniformly colored background and depict only the solute particles.

Hydronium ion
H_3O^+

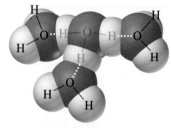

A hydrated proton
$H_9O_4^+$

▲ **The hydrated proton**
The hydronium ion, H_3O^+, interacts with other water molecules through electrostatic attractions.

The hydrogen cation H^+ is an interesting and important species that has been the subject of intensive research. The ion H^+—a bare proton—is a small particle that interacts with the water molecules surrounding it. The simple hydrogen ion, H^+, does not exist in aqueous solutions. Its actual form is as *hydronium* ion, H_3O^+, in which a H^+ ion is attached to a H_2O molecule. The hydronium ion, in turn, interacts with the water molecules surrounding it to form additional species such as $H_5O_2^+$, $H_7O_3^+$, $H_9O_4^+$ (shown in the margin) and many others. These interactions are called *hydration*, and we will represent the hydrated proton as $H^+(aq)$, a shorthand notation. When we wish to emphasize the interaction of the proton with a single water molecule, we will write $H_3O^+(aq)$. The basis for these interactions will be discussed in Chapter 13 and employed extensively in Chapter 17.

For a *weak electrolyte*, the situation is best described as a reaction that does not go to completion. Of all the molecules placed in solution, only a fraction are ionized. An equation written with a double arrow, such as

$$HC_2H_3O_2(aq) \rightleftharpoons H^+(aq) + C_2H_3O_2^-(aq)$$

indicates that a process is *reversible*. While some $HC_2H_3O_2$ (acetic acid) molecules ionize, some H^+ and $C_2H_3O_2^-$ ions in solution recombine to form new $HC_2H_3O_2$

KEEP IN MIND ▶

that there are different formulas for acetic acid. The molecular formula, $C_2H_4O_2$, makes no distinction between the types of H atoms; $HC_2H_3O_2$, indicates one ionizable H atom; and CH_3COOH and CH_3CO_2H show that the ionizable H atom is part of the carboxyl group.

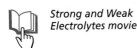

Strong and Weak Electrolytes movie

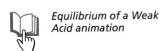

Equilibrium of a Weak Acid animation

molecules. Nevertheless, in any given acetic acid solution, the relative proportions of ionized and nonionized (molecular) acid remain fixed. The predominant species is the molecule $HC_2H_3O_2$, and the solution is best represented by $HC_2H_3O_2(aq)$, *not* $H^+(aq) + C_2H_3O_2{}^-(aq)$. Think of the matter this way: If we could tag a particular $C_2H_3O_2$ group and watch it over time, we would sometimes see it as the ion $C_2H_3O_2{}^-$, but most of the time it would be in a molecule, $HC_2H_3O_2$.

For a *nonelectrolyte*, we simply write the molecular formula. Thus, for a solution of methanol in water we would write $CH_3OH(aq)$.

With this new information about the nature of aqueous solutions, we can introduce a useful notation for solution concentrations. In a solution that is 0.0050 M $MgCl_2$, we assume that the $MgCl_2$ is completely dissociated into ions. Because there are two Cl^- ions for every Mg^{2+} ion, we say that the solution is 0.0050 M Mg^{2+} and 0.0100 M Cl^-. Better still, let us introduce a special symbol for the concentration of a species in solution—the bracket symbol [ ]. The statement $[Mg^{2+}] = 0.0050$ M means that the concentration of the species within the brackets—that is, Mg^{2+}—is 0.0050 mol/L. Thus,

in 0.0050 M MgCl$_2$: $[Mg^{2+}] = 0.0050$ M $[Cl^-] = 0.0100$ M $[MgCl_2] = 0$ M

Although we do not usually write expressions like $[MgCl_2] = 0$, we do so here to emphasize that there is essentially no undissociated $MgCl_2$ in the solution.

Example 5-1 shows how to calculate the concentrations of ions in a strong electrolyte solution.

EXAMPLE 5-1

Calculating Ion Concentrations in a Solution of a Strong Electrolyte. What are the aluminum and sulfate ion concentrations in 0.0165 M $Al_2(SO_4)_3(aq)$?

Solution

First, identify the solute as a strong electrolyte, and write an equation to represent its dissociation. The solute is an ionic compound consisting of the ions Al^{3+} and $SO_4{}^{2-}$.

$$Al_2(SO_4)_3(s) \xrightarrow{H_2O} 2\,Al^{3+}(aq) + 3\,SO_4{}^{2-}(aq)$$

Now, with stoichiometric factors from this equation, we can devise a conversion pathway to relate $[Al^{3+}]$ and $[SO_4{}^{2-}]$ to the given molarity of $Al_2(SO_4)_3$. The stoichiometric factors shown in blue in the following equations are derived from the fact that 1 mol $Al_2(SO_4)_3$ consists of 2 mol Al^{3+} and 3 mol $SO_4{}^{2-}$.

$$[Al^{3+}] = \frac{0.0165 \text{ mol } Al_2(SO_4)_3}{1 \text{ L}} \times \frac{2 \text{ mol } Al^{3+}}{1 \text{ mol } Al_2(SO_4)_3} = \frac{0.0330 \text{ mol } Al^{3+}}{1 \text{ L}}$$

$$= 0.0330 \text{ M}$$

$$[SO_4^{2-}] = \frac{0.0165 \text{ mol } Al_2(SO_4)_3}{1 \text{ L}} \times \frac{3 \text{ mol } SO_4^{2-}}{1 \text{ mol } Al_2(SO_4)_3} = \frac{0.0495 \text{ mol } SO_4^{2-}}{1 \text{ L}}$$

$$= 0.0495 \text{ M}$$

Practice Example A: The chief ions in seawater are Na^+, Mg^{2+}, and Cl^-. Seawater is approximately 0.438 M NaCl and 0.0512 M $MgCl_2$. What is the molarity of Cl^-— that is, the total $[Cl^-]$—in seawater?

Practice Example B: A water treatment plant adds fluoride ion to the water to the extent of 1.5 mg F^-/L.

    (a) What is the molarity of fluoride ion in this water?

    (b) If the fluoride ion in the water is supplied by calcium fluoride, what mass of calcium fluoride is present in 1.00×10^6 liters of this water?

▲ A qualitative test for Cl⁻ ion in tap water involves the addition of a few drops of $AgNO_3(aq)$. The formation of a precipitate of $AgCl(s)$ confirms the presence of Cl⁻.

KEEP IN MIND ▶

that, although the insoluble solid consists of ions, we don't represent ionic charges in the whole formula. That is, we write $AgI(s)$, not $Ag^+I^-(s)$.

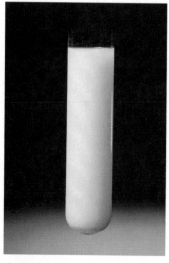

▲ **FIGURE 5-4**
A precipitate of silver iodide
When an aqueous solution of $AgNO_3$ is added to one of NaI, insoluble yellow $AgI(s)$ precipitates from solution.

5-2 Precipitation Reactions

Some metal salts, such as NaCl, are quite soluble in water, while others, such as AgCl, are not very soluble at all. In fact, so little AgCl will dissolve in water that we generally consider this compound to be *insoluble*. Precipitation reactions occur when certain cations and anions combine to produce an insoluble ionic solid called a **precipitate**. One laboratory use of precipitation reactions is in identifying the ions present in a solution, an aspect of chemical analysis we consider in a later chapter. In industry, precipitation reactions are used in the manufacture of numerous chemicals. In the extraction of magnesium metal from seawater, for instance, the first step is to precipitate Mg^{2+} as $Mg(OH)_2(s)$. Our objective in this section is to represent precipitation reactions by chemical equations and to apply some simple rules for predicting precipitation reactions.

Net Ionic Equations

The reaction of silver nitrate and sodium iodide in an aqueous water solution yields sodium nitrate in solution and a yellow precipitate of silver iodide (Figure 5-4). Applying the principles of equation writing from Chapter 4, we can write

$$AgNO_3(aq) + NaI(aq) \longrightarrow AgI(s) + NaNO_3(aq) \qquad (5.1)$$

You might note a contradiction, however, between equation (5.1) and something we learned earlier in this chapter. In their aqueous solutions, the soluble ionic compounds $AgNO_3$, NaI, and $NaNO_3$—all *strong* electrolytes—should be represented by their separate ions.

$$Ag^+(aq) + NO_3^-(aq) + Na^+(aq) + I^-(aq) \longrightarrow AgI(s) + Na^+(aq) + NO_3^-(aq) \quad (5.2)$$

We might say that equation (5.1) is the "whole formula" form of the equation, whereas equation (5.2) is the "ionic" form. Notice also that in equation (5.2) $Na^+(aq)$ and $NO_3^-(aq)$ appear on both sides of the equation. These ions are not reactants; they go through the reaction unchanged. We might call them *spectator ions*. If we eliminate the spectator ions, all that remains is the net ionic equation (5.3).

$$Ag^+(aq) + I^-(aq) \longrightarrow AgI(s) \qquad (5.3)$$

A **net ionic equation** is an equation that includes only the actual participants in a reaction, with each participant denoted by the symbol or formula that best represents it. Symbols are written for individual ions, such as $Ag^+(aq)$, and whole formulas are written for insoluble solids such as $AgI(s)$. Because net ionic equations include electrically charged species—ions—a net ionic equation must be balanced *both* for the numbers of atoms of all types and for electric charge. The same net electric charge must appear on both sides of the equation. Throughout the remainder of this chapter, we will represent most chemical reactions in aqueous solution by net ionic equations.

Predicting Precipitation Reactions

Suppose we are asked whether precipitation occurs when the following aqueous solutions are mixed.

$$AgNO_3(aq) + KBr(aq) \longrightarrow ? \qquad (5.4)$$

A good way to begin is to rewrite expression (5.4) in the ionic form.

$$Ag^+(aq) + NO_3^-(aq) + K^+(aq) + Br^-(aq) \longrightarrow ? \qquad (5.5)$$

There are only *two* possibilities. Either some cation–anion combination leads to an insoluble solid—a precipitate—or no such combination is possible, and there is no reaction at all.

To *predict* what will happen, that is, without going into the laboratory to do experiments, we need some information about which sorts of ionic compounds are water soluble and which are water insoluble. We expect the insoluble ones to form when the appropriate ions are mixed in solution. The most concise form for this information is a set of *solubility rules*. Following are some of the simpler rules.

▶ Some solubility rules.

Ionic Compound activity

Compounds that are **soluble**

- those of the alkali metals (group 1) and the ammonium ion (NH_4^+)
- nitrates, perchlorates, and acetates

Compounds that are mostly **soluble**

- chlorides, bromides, and iodides, *except* those of Pb^{2+}, Ag^+, and Hg_2^{2+}, which are *insoluble*
- sulfates *except* those of Sr^{2+}, Ba^{2+}, Pb^{2+}, and Hg_2^{2+}, which are *insoluble* ($CaSO_4$ is slightly soluble)

▶ In principle, all ionic compounds dissolve in water to some extent, though this may be very slight. For practical purposes, we consider a compound to be insoluble if the maximum amount that can dissolve is less than about 0.01 mole per liter.

Compounds that are mostly **insoluble**

- hydroxides and sulfides
  (Those of the group 1 metals and NH_4^+ are *soluble*. Sulfides of the group 2 metals are soluble. The hydroxides of Ca^{2+}, Sr^{2+}, and Ba^{2+} are slightly soluble.)
- carbonates and phosphates
  (Those of the group 1 metals and NH_4^+ are *soluble*.)

According to these rules, AgBr(s) is insoluble in water and should precipitate, whereas KNO_3(s) is soluble. Written as an ionic equation, expression (5.5) becomes

$$Ag^+(aq) + NO_3^-(aq) + K^+(aq) + Br^-(aq) \longrightarrow AgBr(s) + K^+(aq) + NO_3^-(aq)$$

KEEP IN MIND ▶

that when two ionic compounds form a solid precipitate, they do so by exchanging ions. In the formation of AgBr from KBr and $AgNO_3$, the following exchange takes place.

For the net ionic equation, we have

$$Ag^+(aq) + Br^-(aq) \longrightarrow AgBr(s) \qquad (5.6)$$

The three predictions concerning precipitation reactions made in Example 5-2 are verified in Figure 5-5.

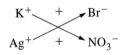

Precipitation Reactions movie

▶ **FIGURE 5-5**
Verifying the predictions made in Example 5-2
(a) When NaOH(aq) is added to $MgCl_2$(aq), a white precipitate of $Mg(OH)_2$(s) forms. **(b)** When colorless BaS(aq) is added to blue $CuSO_4$(aq), a dark precipitate forms. The precipitate is a mixture of white $BaSO_4$(s) and black CuS(s). (A slight excess of $CuSO_4$ remains in solution.) **(c)** No reaction occurs when colorless $(NH_4)_2SO_4$(aq) is added to colorless $ZnCl_2$(aq).

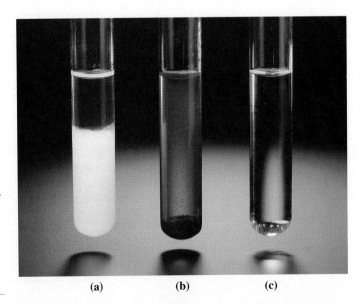

(a)   (b)   (c)

EXAMPLE 5-2

Using Solubility Rules to Predict Precipitation Reactions. Predict whether a reaction will occur in each of the following cases. If so, write a net ionic equation for the reaction.

(a) $NaOH(aq) + MgCl_2(aq) \longrightarrow$?

(b) $BaS(aq) + CuSO_4(aq) \longrightarrow$?

(c) $(NH_4)_2SO_4(aq) + ZnCl_2(aq) \longrightarrow$?

Solution

▶ As you gain experience, you should be able to go directly to a net ionic equation without first writing an ionic equation that includes spectator ions.

(a) Because all common Na compounds are water soluble, Na^+ remains in solution. The combination of Mg^{2+} and OH^- produces *insoluble* $Mg(OH)_2$. With this information, we can write

$$2\,Na^+(aq) + 2\,OH^-(aq) + Mg^{2+}(aq) + 2\,Cl^-(aq) \longrightarrow$$
$$Mg(OH)_2(s) + 2\,Na^+(aq) + 2\,Cl^-(aq)$$

With the elimination of spectator ions, we obtain

$$2\,OH^-(aq) + Mg^{2+}(aq) \longrightarrow Mg(OH)_2(s)$$

(b) From the solubility rules, we conclude that the insoluble combinations formed when Ba^{2+}, S^{2-}, Cu^{2+}, and SO_4^{2-} are found together in solution are $BaSO_4(s)$ and $CuS(s)$. The net ionic equation is

$$Ba^{2+}(aq) + S^{2-}(aq) + Cu^{2+}(aq) + SO_4^{2-}(aq) \longrightarrow BaSO_4(s) + CuS(s)$$

(c) A careful review of the solubility rules shows that all the possible ion combinations lead to water-soluble compounds.

$$2\,NH_4^+(aq) + SO_4^{2-}(aq) + Zn^{2+}(aq) + 2\,Cl^-(aq) \longrightarrow \text{no reaction}$$

KEEP IN MIND ▶
that a solution made by combining $ZnSO_4(aq)$ and $NH_4Cl(aq)$ is indistinguishable from the combination of $ZnCl_2(aq)$ and $(NH_4)_2SO_4(aq)$.

Practice Example A: Indicate whether a precipitate forms, by completing each of the following as a net ionic equation. If no reaction occurs, so state.

(a) $AlCl_3(aq) + KOH(aq) \longrightarrow$?

(b) $K_2SO_4(aq) + FeBr_3(aq) \longrightarrow$?

(c) $CaI_2(aq) + Pb(NO_3)_2(aq) \longrightarrow$?

Practice Example B: Indicate through a net ionic equation whether a precipitate forms when the following compounds in aqueous solution are mixed. If no reaction occurs, so state.

(a) sodium phosphate + aluminum chloride $\longrightarrow$?

(b) aluminum sulfate + barium chloride $\longrightarrow$?

(c) ammonium carbonate + lead nitrate $\longrightarrow$?

5-3 Acid–Base Reactions

Ideas about acids and bases (or alkalis) date back to ancient times. The word *acid* is derived from the Latin *acidus* (sour). *Alkali* (base) comes from the Arabic *al-qali*, referring to the ashes of certain plants from which alkaline substances can be extracted. The acid–base concept is a major theme in the history of chemistry. The view we describe in this section, proposed by Svante Arrhenius in 1884, is still used in many applications. Later in the text, we will explore more modern theories.

Acids

From a practical standpoint, acids can be identified by their sour taste, their ability to react with a variety of metals and carbonate minerals, and their effect on the colors of substances called *acid–base indicators* (Figure 5-6). From a chemist's

▲ **FIGURE 5-6   An acid, a base, and an acid–base indicator**
The acidic nature of lemon juice is shown by the red color of the acid–base indicator *methyl red*. The basic nature of soap is indicated by the change in color of the indicator from red to yellow.

point of view, however, an **acid** can be defined as a substance that provides hydrogen ions (H^+) in aqueous solution.

When hydrogen chloride dissolves in water, complete ionization into H^+ and Cl^- occurs (HCl is a strong electrolyte).

$$HCl(g) \xrightarrow{\;H_2O\;} H^+(aq) + Cl^-(aq)$$

An aqueous solution of nitric acid, also a strong electrolyte, can be represented as

$$H^+(aq) + NO_3^-(aq)$$

Strong acids are molecular compounds that are almost completely ionized in aqueous solution—for example HCl and HNO_3. In dilute solution, we will consider them to be completely ionized. There are actually so few common strong acids that they make only a short list (Table 5.1).

As we described on page 142, the ionization of acetic does not go to completion; it is *reversible*. Acetic acid is a *weak electrolyte*.

$$\underset{\text{acetic acid}}{HC_2H_3O_2(aq)} \rightleftharpoons H^+(aq) + C_2H_3O_2^-(aq)$$

Acids that are incompletely ionized in aqueous solution are called **weak acids**. Like acetic acid, the vast majority of acids are weak acids.

We can use the same method to calculate ion concentrations in a strong acid as we did in Example 5-1 for other strong electrolytes. Because the ionization of a weak acid does not go to completion, however, it is a more difficult matter to calculate ion concentrations in a weak acid solution. We will not take up these calculations at this time.

Bases

From a practical standpoint, we can identify bases through their bitter taste, slippery feel, and effect on the colors of acid–base indicators (Figure 5-6). The Arrhenius definition of a **base** is a substance that produces hydroxide ions (OH^-) in aqueous solution. Consider a soluble ionic hydroxide such as NaOH. In the solid state, this compound consists of Na^+ and OH^- ions. When the solid dissolves in water, the ions dissociate.

$$NaOH(s) \xrightarrow{\;H_2O\;} Na^+(aq) + OH^-(aq)$$

KEEP IN MIND ▶
that H^+(aq) actually represents a hydrated proton, that is, H_3O^+ formed by the transfer of a proton from an HCl to a H_2O molecule and the several additional H_2O molecules associated with the H_3O^+ ion.

TABLE 5.1   Common Strong Acids and Strong Bases

Acids	Bases
HCl	LiOH
HBr	NaOH
HI	KOH
$HClO_4$	RbOH
HNO_3	CsOH
H_2SO_4[a]	$Ca(OH)_2$
	$Sr(OH)_2$
	$Ba(OH)_2$

[a] H_2SO_4 ionizes in two distinct steps. It is a strong acid only in its first ionization step (see Section 17-6).

A base that dissociates completely—or very nearly so—in aqueous solution is a **strong base**. As is true of strong acids, the number of common strong bases is small (Table 5.1). They are primarily the hydroxides of group 1 and some group 2 metals.

Certain substances produce OH^- ions by *reacting* with water, not just by dissolving in it. Such substances, for example, ammonia, are also bases.

$$NH_3(aq) + H_2O(l) \rightleftharpoons NH_4^+(aq) + OH^-(aq) \qquad (5.7)$$

NH_3 is a weak electrolyte; its reaction with water does not go to completion. A base that is incompletely ionized in aqueous solution is a **weak base**. Most basic substances are weak bases.

Neutralization

Perhaps the most significant property of acids and bases is the ability of each to cancel or neutralize the properties of the other. In a **neutralization** reaction, an acid and a base react to form water and an aqueous solution of an ionic compound called a **salt**. Thus, in ionic form,

$$\underbrace{H^+(aq) + Cl^-(aq)}_{\text{(acid)}} + \underbrace{Na^+(aq) + OH^-(aq)}_{\text{(base)}} \longrightarrow \underbrace{Na^+(aq) + Cl^-(aq)}_{\text{(salt)}} + \underbrace{H_2O(l)}_{\text{(water)}}$$

By eliminating the spectator ions, we discover the essential nature of the neutralization of a strong acid by a strong base: H^+ ions from the acid and OH^- ions from the base combine to form water.

$$H^+(aq) + OH^-(aq) \longrightarrow H_2O(l)$$

In a neutralization involving the weak base $NH_3(aq)$, we can think of H^+ from an acid combining directly with NH_3 molecules to form NH_4^+. The neutralization can be represented by an ionic equation

$$\underbrace{H^+(aq) + Cl^-(aq)}_{\text{(acid)}} + \underbrace{NH_3(aq)}_{\text{(base)}} \longrightarrow \underbrace{NH_4^+(aq) + Cl^-(aq)}_{\text{(salt)}}$$

or by a net ionic equation

$$H^+(aq) + NH_3(aq) \longrightarrow NH_4^+(aq)$$

Recognizing Acids and Bases

Acids contain *ionizable* hydrogen atoms, which are generally identified by the way in which the formula of an acid is written. Ionizable H atoms are separated from other H atoms in the formula either by writing them first in the molecular formula or by indicating where they are found in the molecule. Thus, there are two ways that we can show that *one* H atom in the acetic molecule is ionizable and the other three H atoms are not.

$$\underbrace{HC_2H_3O_2 \quad \text{or} \quad CH_3COOH}_{\text{acetic acid}}$$

In contrast to acetic acid, methane has four H atoms, but they are *not* ionizable. CH_4 is *not* an acid, nor is it a base.

We expect a substance to be a base if its formula indicates a combination of OH^- ions with metal ions (for example, NaOH). To identify a weak base, we usually need a chemical equation for the ionization reaction, as in equation (5.7). The main weak base we will work with for the present is NH_3. Metal salts of carbonate (CO_3^{2-}) and hydrogen carbonate (HCO_3^-) ions are also common weak bases. Sodium hydrogen carbonate, $NaHCO_3$, for instance, is often used in the laboratory to neutralize acid spills. Note that ethanol, C_2H_5OH, is *not* a base. The OH group is not present as OH^-, either in pure ethanol or in its aqueous solutions.

KEEP IN MIND ▶
that $NH_4^+(aq)$ is formed by the transfer of a proton from a H_2O to a NH_3 molecule and NH_4^+ interacts with water in much the same way as the hydronium ion does.

Ammonium ion

▶ The formula of $NH_3(aq)$ is sometimes written as NH_4OH (ammonium hydroxide), and its ionization represented as

$$NH_4OH(aq) \rightleftharpoons NH_4^+(aq) + OH^-(aq)$$

There is no hard evidence for the existence of NH_4OH, however, and we will use only the formula $NH_3(aq)$.

Introduction to Acids and Bases animation

More Acid–Base Reactions

$Mg(OH)_2$ is a base because it contains OH^-, but this compound is quite insoluble in water. Its finely divided solid particles form a suspension in water that is the familiar milk of magnesia, used as an antacid. In this suspension, $Mg(OH)_2(s)$ does dissolve very slightly, producing some OH^- in solution. If an acid is added, H^+ from the acid combines with this OH^- to form water—neutralization occurs. More $Mg(OH)_2(s)$ dissolves to produce more OH^- in solution, which is neutralized by more H^+, and so on. In this way, the neutralization reaction results in the dissolving of otherwise insoluble $Mg(OH)_2(s)$. The net ionic equation for the reaction of $Mg(OH)_2(s)$ with a strong acid is

$$Mg(OH)_2(s) + 2\,H^+(aq) \longrightarrow Mg^{2+}(aq) + 2\,H_2O(l) \tag{5.8}$$

$Mg(OH)_2(s)$ also reacts with a weak acid such as acetic acid. In the net ionic equation, acetic acid is written in its molecular form. But remember that some H^+ and $C_2H_3O_2^-$ ions are always present in an acetic acid solution. The H^+ ions react with OH^- ions, as in reaction (5.8), followed by further ionization of $HC_2H_3O_2$, more neutralization, and so on. If enough acetic acid is present, the $Mg(OH)_2$ will dissolve completely.

$$Mg(OH)_2(s) + 2\,HC_2H_3O_2(aq) \longrightarrow Mg^{2+}(aq) + 2\,C_2H_3O_2^-(aq) + 2\,H_2O(l) \tag{5.9}$$

Calcium carbonate (present in limestone and marble) is another water-insoluble solid that is soluble in strong and weak acids. Here the solid produces a low concentration of CO_3^{2-} ions, which combine with H^+ to form the weak acid H_2CO_3. This causes more of the solid to dissolve, and so on. H_2CO_3, carbonic acid, is a very unstable substance that decomposes into H_2O and $CO_2(g)$. Thus, a gas is given off when $CaCO_3(s)$ reacts with an acid and dissolves. Calcium carbonate, like magnesium hydroxide, is used as an antacid.

$$CaCO_3(s) + 2\,H^+(aq) \longrightarrow Ca^{2+}(aq) + H_2O(l) + CO_2(g) \tag{5.10}$$

To treat reaction (5.10) as an acid–base reaction, we must think of CO_3^{2-} as a base. The definition that we have been using recognizes only OH^- as a base, but when we reconsider acids and bases (Chapter 17), our expanded definitions will identify CO_3^{2-} and many other anions as bases. Table 5.2 lists several common anions and one cation that produce gases in acid–base reactions.

▲ This marble statue has been eroded by acid rain. Marble consists primarily of $CaCO_3$. Acids react with and dissolve marble through the reaction described in equation (5.10).

TABLE 5.2	Some Common Gas-Forming Reactions
Ion	**Reaction**
HSO_3^-	$HSO_3^- + H^+ \longrightarrow SO_2(g) + H_2O(l)$
SO_3^{2-}	$SO_3^{2-} + 2\,H^+ \longrightarrow SO_2(g) + H_2O(l)$
HCO_3^-	$HCO_3^- + H^+ \longrightarrow CO_2(g) + H_2O(l)$
CO_3^{2-}	$CO_3^{2-} + 2\,H^+ \longrightarrow CO_2(g) + H_2O(l)$
S^{2-}	$S^{2-} + 2\,H^+ \longrightarrow H_2S(g)$
NH_4^+	$NH_4^+ + OH^- \longrightarrow NH_3(g) + H_2O(l)$

EXAMPLE 5-3

Writing Equation(s) for Acid–Base Reactions. Write a net ionic equation to represent the reaction of **(a)** aqueous strontium hydroxide with nitric acid; **(b)** solid aluminum hydroxide with hydrochloric acid.

Solution

In each case we begin by writing the reactants in the whole-formula form. Then we substitute ionic forms, where appropriate. Finally we complete the equation as a net ionic equation.

(a) *Whole-formula form:* $HNO_3(aq) + Sr(OH)_2(aq) \longrightarrow ?$

Ionic form:

$$2\,H^+(aq) + 2\,NO_3^-(aq) + Sr^{2+}(aq) + 2\,OH^-(aq) \longrightarrow$$
$$Sr^{2+}(aq) + 2\,NO_3^-(aq) + 2\,H_2O(l)$$

Net ionic equation: Remove the spectator ions (Sr^{2+} and NO_3^-).

$$2\,H^+(aq) + 2\,OH^-(aq) \longrightarrow 2\,H_2O(l)$$

or, more simply,

$$H^+(aq) + OH^-(aq) \longrightarrow H_2O(l)$$

(b) *Whole-formula form:* $Al(OH)_3(s) + HCl(aq) \longrightarrow ?$

Ionic form:

$$Al(OH)_3(s) + 3\,H^+(aq) + 3\,Cl^-(aq) \longrightarrow Al^{3+}(aq) + 3\,Cl^-(aq) + 3\,H_2O(l)$$

Net ionic equation: Remove the spectator ion (Cl^-).

$$Al(OH)_3(s) + 3\,H^+(aq) \longrightarrow Al^{3+}(aq) + 3\,H_2O(l)$$

Practice Example A: Write a net ionic equation to represent the reaction of aqueous ammonia with propionic acid, $HC_3H_5O_2$. What is the formula and name of the salt that results from this neutralization?

Practice Example B: Calcium carbonate is a major constituent of the hard water deposits found in teakettles and automatic coffeemakers. Vinegar, which is essentially a dilute aqueous solution of acetic acid, is commonly used to remove such deposits. Write a net ionic equation for the reaction that occurs.
[*Hint:* Recall equations (5.9) and (5.10).]

5-4   Oxidation–Reduction: Some General Principles

Iron ores are minerals with a high iron content. One of them, *hematite*, Fe_2O_3, is chemically very similar to ordinary iron rust. In simplified fashion, the reaction in which iron metal is produced from hematite in a blast furnace is described as

$$Fe_2O_3(s) + 3\,CO(g) \xrightarrow{\Delta} 2\,Fe(l) + 3\,CO_2(g) \tag{5.11}$$

In this reaction, we can think of the $CO(g)$ as taking O atoms away from Fe_2O_3 to produce $CO_2(g)$ and the free element iron. A commonly used term to describe a reaction in which a substance gains O atoms is *oxidation*, and one in which a substance loses O atoms, *reduction*. In reaction (5.11), $CO(g)$ is oxidized and $Fe_2O_3(s)$ is reduced. An oxidation and a reduction must always occur together, and such a reaction is called an **oxidation–reduction**, or **redox**, **reaction**.

Definitions of oxidation and reduction based solely on the transfer of O atoms are too restrictive. By using broader definitions, we can, for example, describe many reactions in aqueous solution as oxidation–reduction reactions, even when the reactions do not involve oxygen. We can look at oxidation–reduction reactions in two other ways, which we describe next.

▲ In the *thermite* reaction, the iron atoms of iron(III) oxide give up O atoms to Al atoms, producing Al_2O_3.

$$Fe_2O_3(s) + 2\,Al(s) \longrightarrow Al_2O_3(s) + 2\,Fe(l)$$

▶ Because it is easier to say, the term *redox* is often used instead of oxidation–reduction.

Oxidation State Changes

Suppose we rewrite equation (5.11) and indicate the oxidation states (O.S) of the elements on both sides of the equation by using the rules listed on page 80.

$$\overset{+3\ \ -2}{Fe_2O_3} + 3\ \overset{+2\ -2}{CO} \longrightarrow 2\ \overset{0}{Fe} + 3\ \overset{+4\ -2}{CO_2}$$

The O.S. of oxygen is -2 everywhere it appears in this equation. That of iron (shown in red) changes. It *decreases* from $+3$ in Fe_2O_3 to 0 in the free element, Fe. The O.S. of carbon (shown in blue) also changes. It *increases* from $+2$ in CO to $+4$ in CO_2. In terms of oxidation state changes, in an *oxidation* process, the O.S. of some element *increases*, and in a *reduction* process, the O.S. of some element *decreases*.

EXAMPLE 5-4

Identifying Oxidation–Reduction Reactions. Indicate whether each of the following is an oxidation–reduction reaction.

(a) $MnO_2(s) + 4\ H^+(aq) + 2\ Cl^-(aq) \longrightarrow Mn^{2+}(aq) + 2\ H_2O(l) + Cl_2(g)$

(b) $H_2PO_4^-(aq) + OH^-(aq) \longrightarrow HPO_4^{2-}(aq) + H_2O(l)$

Solution

In each case, indicate the oxidation states of the elements on both sides of the equation, and look for changes.

KEEP IN MIND ▶

that even though we assess oxidation state changes by element, oxidation and reduction involve the entire species in which the element is found. Thus, MnO_2 is reduced, not just Mn; and Cl^- is oxidized, not Cl.

(a) The O.S. of Mn *decreases* from $+4$ in MnO_2 to $+2$ in Mn^{2+}. MnO_2 is reduced to Mn^{2+}. The O.S. of O remains at -2 throughout the reaction, and that of H, at $+1$. The O.S. of Cl *increases* from -1 in Cl^- to 0 in Cl_2. Cl^- is oxidized to Cl_2. The reaction is an oxidation–reduction reaction.

(b) The O.S. of H is $+1$ on both sides of the equation. Oxygen remains at O.S. -2 throughout. The O.S. of phosphorus is $+5$ in $H_2PO_4^-$ and also $+5$ in HPO_4^{2-}. There are no changes in O.S. This is *not* an oxidation–reduction reaction. (It is, in fact, an acid–base reaction.)

Practice Example A: Identify whether each of the following is an oxidation–reduction reaction.

(a) $(NH_4)_2SO_4(aq) + Ba(NO_3)_2(aq) \longrightarrow BaSO_4(s) + 2\ NH_4NO_3(aq)$

(b) $2\ Pb(NO_3)_2(s) \longrightarrow 2\ PbO(s) + 4\ NO_2(g) + O_2(g)$

Practice Example B: Identify the species that is oxidized and the species that is reduced in the reaction

$$5\ VO^{2+}(aq) + MnO_4^-(aq) + H_2O \longrightarrow 5\ VO_2^+(aq) + Mn^{2+}(aq) + 2\ H^+(aq)$$

Oxidation and Reduction Half-Reactions

The reaction illustrated in Figure 5-7 is an oxidation–reduction reaction.

$$Zn(s) + Cu^{2+}(aq) \longrightarrow Zn^{2+}(aq) + Cu(s)$$

We can show this by evaluating changes in oxidation state, but there is another especially useful way to establish that it is an oxidation–reduction reaction. Think of the reaction as involving two **half-reactions** occurring at the same time. The overall reaction is the sum of the two half-reactions. We can represent the half-reactions by half-equations and the overall reaction by an overall equation.

(a) (b) (c)

Oxidation–Reduction Chemistry of Tin and Zinc movie

▲ **FIGURE 5-7** **An oxidation–reduction reaction**
(a) A zinc rod above an aqueous solution of copper(II) sulfate. **(b)** Following immersion of the Zn rod in the $CuSO_4(aq)$ for several hours, the blue color of $Cu^{2+}(aq)$ disappears and a deposit of copper forms on the rod. In the microscopic view of the reaction (bottom), Zn atoms lose electrons to the metal surface, entering the solution as Zn^{2+} ions. Cu^{2+} ions from solution pick up electrons and deposit on the metal surface as atoms of solid copper. **(c)** The pitted zinc rod (providing evidence that zinc entered into a chemical reaction) and the collected copper metal.

Oxidation:	$Zn(s) \longrightarrow Zn^{2+}(aq) + 2\,e^-$	(5.12)
Reduction:	$\underline{Cu^{2+}(aq) + 2\,e^- \longrightarrow Cu(s)}$	(5.13)
Overall:	$Zn(s) + Cu^{2+}(aq) \longrightarrow Zn^{2+}(aq) + Cu(s)$	(5.14)

In half-reaction (5.12), Zn is *oxidized*—its oxidation state *increases* from 0 to +2. This change corresponds to a *loss* of two electrons by each zinc atom. In half-reaction (5.13), Cu^{2+} is *reduced*—its oxidation state *decreases* from +2 to 0. This change corresponds to the *gain* of two electrons by each Cu^{2+} ion. To summarize,

- **Oxidation** is a process in which the O.S. of some element *increases* and in which electrons appear on the *right* side of a half-equation.

- **Reduction** is a process in which the O.S. of some element *decreases* and in which electrons appear on the *left* side of a half-equation.
- Oxidation and reduction half-reactions must always occur together, and the total number of electrons associated with the oxidation must equal the total number associated with the reduction.

EXAMPLE 5-5

Expressing an Oxidation–Reduction Reaction Through Half-Equations and an Overall Equation. Show the oxidation and reduction that occur, and write an overall ionic equation for the reaction of iron with hydrochloric acid solution to produce $H_2(g)$ and $Fe^{2+}(aq)$ (Figure 5-8).

Solution

$$\text{Oxidation:} \qquad Fe(s) \longrightarrow Fe^{2+}(aq) + 2\,e^-$$
$$\text{Reduction:} \qquad 2\,H^+(aq) + 2\,e^- \longrightarrow H_2(g)$$
$$\text{Overall:} \qquad Fe(s) + 2\,H^+(aq) \longrightarrow Fe^{2+}(aq) + H_2(g)$$

Practice Example A: Represent the reaction of aluminum with hydrochloric acid (reaction 4.2) by oxidation and reduction half-equations and an overall equation.

Practice Example B: Represent the reaction of chlorine gas with aqueous sodium bromide to produce liquid bromine and aqueous sodium chloride by oxidation and reduction half-equations and an overall equation.

Figure 5-8 and Example 5-5 suggest some fundamental questions about oxidation–reduction. For example,

- Why does Fe react with $HCl(aq)$, displacing $H_2(g)$, whereas Cu does not?
- Why does Fe form Fe^{2+} and not Fe^{3+} in this reaction?

(a) $\qquad\qquad$ (b) $\qquad\qquad$ (c)

▲ **FIGURE 5-8   Displacement of $H^+(aq)$ by iron—Example 5-5 illustrated**
(a) An iron nail is wrapped in a piece of copper screen. **(b)** The nail and screen are placed in $HCl(aq)$. Hydrogen gas is evolved as the nail reacts. **(c)** The nail reacts completely and produces $Fe^{2+}(aq)$, but the copper does not react.

TABLE 5.3 Behavior of Some Common Metals with Nonoxidizing Acids[a]

React to Produce $H_2(g)$	Do Not React
Alkali metals (group 1)[b]	Cu, Ag, Au, Hg
Alkaline earth metals (group 2)[b]	
Al, Zn, Fe, Sn, Pb	

[a] A nonoxidizing acid (for example, HCl, HBr, HI) is one in which the only possible reduction half-reaction is the reduction of H^+ to H_2. Additional possibilities for metal–acid reactions are considered in Chapter 21.

[b] With the exception of Be and Mg, all group 1 and group 2 metals also react with cold water to produce $H_2(g)$. (The metal hydroxide is the other product.)

Even now, you can probably see that the answers to these questions lie in the relative abilities of Fe and Cu atoms to give up electrons—to become oxidized. Fe gives up electrons more easily than does Cu; also, Fe is more readily oxidized to Fe^{2+} than it is to Fe^{3+}. We can give more complete answers after we develop specific criteria describing electron loss and gain in Chapter 21. For now, the information in Table 5.3 should be helpful. The table lists some common metals that do react with acids to displace $H_2(g)$ and a few that do not. As noted in the table, most of the group 1 and 2 metals are so reactive that they will react with cold water to produce $H_2(g)$ and a solution of the metal hydroxide.

5-5 Balancing Oxidation–Reduction Equations

The same principles of equation balancing apply to oxidation–reduction (redox) equations as to other equations—balance for numbers of atoms and balance for electric charge. However, it is often a little more difficult to apply these principles in redox equations. In fact, only a small proportion of redox equations can be balanced by simple inspection. We need a systematic approach, and while several methods are available, we emphasize the one described here.

The Half-Reaction (Ion-Electron) Method

Balancing Redox Equations activity

The basic steps in this method of balancing a redox equation are as follows.

- Write and balance separate half-equations for oxidation and reduction.
- Adjust coefficients in the two half-equations so that the same number of electrons appears in each half-equation.
- Add together the two half-equations (canceling out electrons) to obtain the balanced overall equation.

This method is applied in a stepwise fashion in Example 5-6.

▶ Try balancing expression (5.15) by inspection just for numbers of atoms. You will see that the electric charge does not balance.

EXAMPLE 5-6

Balancing the Equation for a Redox Reaction in Acidic Solution. The reaction described by expression (5.15) is used to determine the sulfite ion concentration present in wastewater from a papermaking plant. Write the balanced equation for this reaction in acidic solution.

$$SO_3^{2-}(aq) + MnO_4^-(aq) \longrightarrow SO_4^{2-}(aq) + Mn^{2+}(aq) \qquad (5.15)$$

Solution

Step 1. *Write skeleton half-equations based on the species undergoing oxidation and reduction.* The O.S. of sulfur increases from +4 in SO_3^{2-} to +6 in SO_4^{2-}. The O.S. of Mn

▶ In most cases, you should be able to identify the key species in skeleton half-equations without having to evaluate oxidation states.

decreases from +7 in MnO_4^- to +2 in Mn^{2+}. The skeleton half-equations are

$$SO_3^{2-}(aq) \longrightarrow SO_4^{2-}(aq)$$
$$MnO_4^-(aq) \longrightarrow Mn^{2+}(aq)$$

Step 2. *Balance each half-equation for numbers of atoms, in this order.*

- atoms other than H and O
- O atoms, by adding H_2O with the appropriate coefficient
- H atoms, by adding H^+ with the appropriate coefficient

The other atoms (S and Mn) are already balanced in the skeleton half-equations. To balance O atoms, we add one H_2O molecule to the left side of the first half-equation and four to the right side of the second.

$$SO_3^{2-}(aq) + H_2O(l) \longrightarrow SO_4^{2-}(aq)$$
$$MnO_4^-(aq) \longrightarrow Mn^{2+}(aq) + 4 H_2O(l)$$

To balance H atoms, we add two H^+ ions to the right side of the first half-equation and eight to the left side of the second.

$$SO_3^{2-}(aq) + H_2O(l) \longrightarrow SO_4^{2-}(aq) + 2 H^+(aq)$$
$$MnO_4^-(aq) + 8 H^+(aq) \longrightarrow Mn^{2+}(aq) + 4 H_2O(l)$$

Step 3. *Balance each half-equation for electric charge.* Add the number of electrons necessary to get the same electric charge on both sides of each half-equation. By doing this, you will see that the half-equation in which electrons appear on the right side is the *oxidation* half-equation. The other half-equation, with electrons on the left side, is the *reduction* half-equation.

Oxidation: $SO_3^{2-}(aq) + H_2O(l) \longrightarrow SO_4^{2-}(aq) + 2 H^+(aq) + 2 e^-$
(net charge on each side, −2)

Reduction: $MnO_4^-(aq) + 8 H^+(aq) + 5 e^- \longrightarrow Mn^{2+}(aq) + 4 H_2O(l)$
(net charge on each side, +2)

Step 4. *Obtain the overall redox equation by combining the half-equations.* Multiply through the oxidation half-equation by *5* and through the reduction half-equation by *2*. This results in 10 e^- on each side of the overall equation. These terms cancel out. *Electrons must not appear in the final equation.*

$$5 SO_3^{2-}(aq) + 5 H_2O(l) \longrightarrow 5 SO_4^{2-}(aq) + 10 H^+(aq) + \cancel{10 e^-}$$
$$2 MnO_4^-(aq) + 16 H^+(aq) + \cancel{10 e^-} \longrightarrow 2 Mn^{2+}(aq) + 8 H_2O(l)$$

$$5 SO_3^{2-}(aq) + 2 MnO_4^-(aq) + 5 H_2O(l) + 16 H^+(aq) \longrightarrow$$
$$5 SO_4^{2-}(aq) + 2 Mn^{2+}(aq) + 8 H_2O(l) + 10 H^+(aq)$$

Step 5. *Simplify.* The overall equation should not contain the same species on both sides. Subtract *five* H_2O from each side of the equation in step 4. This leaves *three* H_2O on the right. Also subtract *ten* H^+ from each side, leaving *six* on the left.

$$5 SO_3^{2-}(aq) + 2 MnO_4^-(aq) + 6 H^+(aq) \longrightarrow 5 SO_4^{2-}(aq) + 2 Mn^{2+}(aq) + 3 H_2O(l)$$

Step 6. *Verify.* Check the overall equation to ensure that it is balanced both for numbers of atoms and electric charge. For example, show that in the balanced equation from Step 5, the net charge on each side of the equation is minus six: $(5 \times 2-) + (2 \times 1-) + (6 \times 1+) = (5 \times 2-) + (2 \times 2+) = -6$.

Practice Example A: Balance the equation for this reaction in acidic solution.

$$Fe^{2+}(aq) + MnO_4^-(aq) \longrightarrow Fe^{3+}(aq) + Mn^{2+}(aq)$$

Practice Example B: Balance the equation for this reaction in acidic solution.

$$UO^{2+}(aq) + Cr_2O_7^{2-}(aq) \longrightarrow UO_2^{2+}(aq) + Cr^{3+}(aq)$$

For ready reference, the procedure used in Example 5-6 is summarized below.

Balancing Equations for Redox Reactions in Acidic Aqueous Solutions by the Half-Reaction Method: A Summary

- Write the equations for the oxidation and reduction half-reactions.
- In each half-equation:
  (1) Balance atoms of all the elements except H and O.
  (2) Balance oxygen using H_2O.
  (3) Balance hydrogen using H^+.
  (4) Balance charge, using electrons.
- If necessary, equalize the number of electrons in the oxidation and reduction half-equations by multiplying one or both half-equations by appropriate integers.
- Add the half-equations, then cancel species common to both sides of the overall equation.
- Check that numbers of atoms and charges balance.

Balancing Redox Equations in Basic Solution

To balance equations for redox reactions in *basic* solution, we generally have to add a step or two to the procedure used in Example 5-6. The problem is this: In basic solution, OH^-, not H^+, must appear in the final balanced equation. Because *both* OH^- and H_2O contain H and O atoms, at times it is hard to decide on which side of the half-equations to put each one. One simple approach is to treat the reaction *as if* it were occurring in an acidic solution, and balance it as in Example 5-6. Then, add to each side of the overall redox equation a number of OH^- ions equal to the number of H^+ ions. Where H^+ and OH^- appear on the same side of the equation, combine them to produce H_2O molecules. If H_2O now appears on both sides of the equation, subtract the same number of H_2O molecules from each side, leaving a remainder of H_2O on just one side. This method is illustrated in Example 5-7.

EXAMPLE 5-7

Balancing the Equation for a Redox Reaction in Basic Solution. Balance the equation for the reaction in which permanganate ion oxidizes cyanide ion to cyanate ion in basic solution and is itself reduced to $MnO_2(s)$.

$$MnO_4^-(aq) + CN^-(aq) \longrightarrow MnO_2(s) + OCN^-(aq) \qquad (5.16)$$

Solution

Initially, we treat the half-reactions and the overall reaction *as if* they were occurring in an acidic solution, and finally, we adjust the overall equation to a basic solution.

Step 1. *Write two skeleton half-equations and balance them for Mn, C, and N atoms.*

$$MnO_4^-(aq) \longrightarrow MnO_2(s)$$
$$CN^-(aq) \longrightarrow OCN^-(aq)$$

Note that in this case, the skeleton half-equations as initially written are balanced for Mn, C, and N atoms.

Step 2. *Balance the half-equations for O and H atoms. Add H_2O and/or H^+ as required.*

$$MnO_4^-(aq) + 4\,H^+(aq) \longrightarrow MnO_2(s) + 2\,H_2O(l)$$
$$CN^-(aq) + H_2O(l) \longrightarrow OCN^-(aq) + 2\,H^+(aq)$$

Step 3. *Balance the half-equations for electric charge by adding the appropriate numbers of electrons.*

Reduction: $MnO_4^-(aq) + 4\,H^+(aq) + 3\,e^- \longrightarrow MnO_2(s) + 2\,H_2O(l)$
Oxidation: $CN^-(aq) + H_2O(l) \longrightarrow OCN^-(aq) + 2\,H^+(aq) + 2\,e^-$

Step 4. *Combine the half-equations to obtain an overall redox equation.* Multiply the reduction half-equation by *2* and the oxidation half-equation by *3*. Make the appropriate cancellations of H^+ and H_2O.

$$2\,MnO_4^-(aq) + 8\,H^+(aq) + \cancel{6\,e^-} \longrightarrow 2\,MnO_2(s) + 4\,H_2O(l)$$
$$3\,CN^-(aq) + 3\,H_2O(l) \longrightarrow 3\,OCN^-(aq) + 6\,H^+(aq) + \cancel{6\,e^-}$$

Overall: $2\,MnO_4^-(aq) + 3\,CN^-(aq) + 2\,H^+(aq) \longrightarrow$
$$2\,MnO_2(s) + 3\,OCN^-(aq) + H_2O(l)$$

Step 5. *Change from an acidic to a basic medium by adding 2 OH^- to both sides of the overall equation; combine 2 H^+ and 2 OH^- to form 2 H_2O, and simplify.*

$2\,MnO_4^-(aq) + 3\,CN^-(aq) + 2\,H^+(aq) + 2\,OH^-(aq) \longrightarrow$
$$2\,MnO_2(s) + 3\,OCN^-(aq) + H_2O(l) + 2\,OH^-(aq)$$
$2\,MnO_4^-(aq) + 3\,CN^-(aq) + 2\,H_2O(l) \longrightarrow$
$$2\,MnO_2(s) + 3\,OCN^-(aq) + H_2O(l) + 2\,OH^-(aq)$$

Subtract one H_2O molecule from each side to obtain the overall balanced redox equation for reaction (5.16).

$$2\,MnO_4^-(aq) + 3\,CN^-(aq) + H_2O(l) \longrightarrow 2\,MnO_2(s) + 3\,OCN^-(aq) + 2\,OH^-(aq)$$

Step 6. *Verify.* Check the final overall equation to ensure that it is balanced both for number of atoms and for electric charge. For example, show that in the balanced equation from Step. 5, the net charge on each side of the equation is 5−.

Practice Example A: Balance the equation for this reaction in basic solution.

$$S(s) + OCl^-(aq) \longrightarrow SO_3^{2-}(aq) + Cl^-(aq)$$

Practice Example B: Balance the equation for this reaction in basic solution.

$$MnO_4^-(aq) + SO_3^{2-}(aq) \longrightarrow MnO_2(s) + SO_4^{2-}(aq)$$

For ready reference, the procedure used in Example 5-7 is summarized below.

Balancing Equations for Redox Reactions in Basic Aqueous Solutions by the Half-Reaction Method: A Summary

- Balance the equation *as if* the reaction were occurring in *acidic* medium, by using the method for acidic aqueous solutions.
- To both sides of the overall equation obtained, add a number of OH^- that is equal to the number of H^+ ions.
- On the side of the overall equation containing both H^+ and OH^- ions, combine them to form H_2O molecules. If H_2O molecules now appear on both sides of the overall equation, cancel the same number from each side, leaving a remainder of H_2O on just one side.
- Check that numbers of atoms and charges balance.

Disproportionation Reactions

In some oxidation–reduction reactions, called **disproportionation reactions**, the same substance is both oxidized *and* reduced. Some of these reactions have practical significance. The decomposition of hydrogen peroxide, H_2O_2, produces $O_2(g)$. It is the $O_2(g)$ that has a germicidal effect when a dilute aqueous solution of hydrogen peroxide (usually 3%) is used as an antiseptic.

$$2\,H_2O_2(aq) \longrightarrow 2\,H_2O(l) + O_2(g) \tag{5.17}$$

In reaction (5.17), the oxidation state of oxygen changes from -1 in H_2O_2 to -2 in H_2O (a reduction) and to 0 in $O_2(g)$ (an oxidation). H_2O_2 is both oxidized and reduced.

Solutions of sodium thiosulfate ($Na_2S_2O_3$) are often used in the laboratory in redox reactions. The disproportionation of $S_2O_3^{2-}$ produces sulfur as one of its products, which accounts for the fact that old stock solutions of $Na_2S_2O_3$ sometimes contain a light yellow deposit.

$$S_2O_3^{2-}(aq) + 2\,H^+(aq) \longrightarrow S(s) + SO_2(g) + H_2O(l) \tag{5.18}$$

The oxidation states of S are $+2$ in $S_2O_3^{2-}$, 0 in $S(s)$, and $+4$ in $SO_2(g)$.

The same substance appears on the left side in each half-equation for a disproportionation reaction. The balanced half-equations and overall equation for reaction (5.18) are

Oxidation:	$S_2O_3^{2-}(aq) + H_2O(l) \longrightarrow 2\,SO_2(g) + 2\,H^+(aq) + 4\,e^-$
Reduction:	$S_2O_3^{2-}(aq) + 6\,H^+(aq) + 4\,e^- \longrightarrow 2\,S(s) + 3\,H_2O(l)$

$$2\,S_2O_3^{2-}(aq) + 4\,H^+(aq) \longrightarrow 2\,S(s) + 2\,SO_2(g) + 2\,H_2O(l)$$

Overall: $S_2O_3^{2-}(aq) + 2\,H^+(aq) \longrightarrow S(s) + SO_2(g) + H_2O(l)$

 ## Are You Wondering …

Whether you can still use the half-reaction method if the reaction occurs in a medium other than an aqueous solution?

You can. All you need to do is treat the reaction *as if* it were occurring in an aqueous acidic solution. H^+ should cancel out in the overall equation. Consider, for example, the oxidation of $NH_3(g)$ to $NO(g)$, the first step in the commercial production of nitric acid.

$$NH_3(g) + O_2(g) \longrightarrow NO(g) + H_2O(g)$$

By the half-reaction method,

Oxidation:	$4\{NH_3 + H_2O \longrightarrow NO + 5\,H^+ + 5\,e^-\}$
Reduction:	$5\{O_2 + 4\,H^+ + 4\,e^- \longrightarrow 2\,H_2O\}$
Overall:	$4\,NH_3 + 5\,O_2 \longrightarrow 4\,NO + 6\,H_2O$

Some people prefer a method called the *oxidation-state change method*[*] for reactions of this type. However, we have just demonstrated that the half-reaction works just as well.

[*]In this method, changes in oxidation states are identified. That of nitrogen increases from -3 in NH_3 to $+2$ in NO, corresponding to a "loss" of 5 electrons per N atom. That of oxygen decreases from 0 in O_2 to -2 in NO and H_2O, corresponding to a "gain" of 2 electrons per O atom. The proportion of N to O atoms must be 2 N (loss of 10 e^-) to 5 O (gain of 10 e^-).

$$2\,NH_3 + \frac{5}{2}O_2 \longrightarrow 2\,NO + 3\,H_2O \quad or \quad 4\,NH_3 + 5\,O_2 \longrightarrow 4\,NO + 6\,H_2O$$

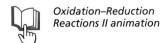

Oxidation–Reduction Reactions II animation

5-6   Oxidizing and Reducing Agents

Chemists frequently use the terms *oxidizing agent* and *reducing agent* to describe certain of the reactants in redox reactions, as in statements like "fluorine gas is a powerful oxidizing agent," or "calcium metal is a good reducing agent." Let us briefly consider the meaning of these terms.

In a redox reaction, the substance that makes it possible for some other substance to be oxidized is called the **oxidizing agent**, or **oxidant**. In doing so, the oxidizing agent is itself reduced. Similarly, the substance that causes some other substance to be reduced is called the **reducing agent**, or **reductant**. In the reaction, the reducing agent is itself oxidized. Or, stated in other ways,

An oxidizing agent (oxidant)

- contains an element whose oxidation state *decreases* in a redox reaction
- gains electrons (electrons are found on the left side of its half-equation)

A reducing agent (reductant)

- contains an element whose oxidation state *increases* in a redox reaction
- loses electrons (electrons are found on the right side of its half-equation)

In general, a substance with an element in one of its highest possible oxidation states is an oxidizing agent. If the element is in one of its lowest possible oxidation states, the substance is a reducing agent. Figure 5-9 shows the range of oxidation states of nitrogen and the species to which they correspond. We see that the oxidation state of nitrogen in dinitrogen tetroxide (N_2O_4) is nearly the maximum value attainable, and hence N_2O_4 is generally an oxidizing agent. On the other hand, the nitrogen atom in hydrazine (N_2H_4) is in nearly the lowest oxidation state, and hence hydrazine is generally a reducing agent. When these two liquid compounds are mixed, a vigorous reaction takes place.

$$N_2O_4(l) + 2\,N_2H_4(l) \longrightarrow 3\,N_2(g) + 4\,H_2O(g)$$

In this reaction, one of the nitrogen-containing molecules is the oxidizing agent and the other the reducing agent. So much heat is released during this reaction that the reaction is used in some rocket propulsion systems.

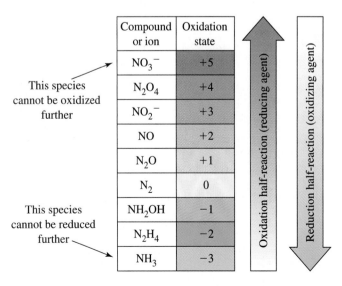

▶ **FIGURE 5-9**
Oxidation states of nitrogen: identifying oxidizing and reducing agents

Certain substances in which the oxidation state of an element is between its highest and lowest possible values may act as an oxidizing agent in some instances and a reducing agent in others. For example, in the reaction of hydrazine with hydrogen to produce ammonia, hydrazine acts as an oxidizing agent

$$N_2H_4 + H_2 \longrightarrow 2\,NH_3$$

EXAMPLE 5-8

Identifying Oxidizing and Reducing Agents. Hydrogen peroxide, H_2O_2, is a versatile chemical. Its uses include bleaching wood pulp and fabrics and substituting for chlorine in water purification. One reason for its versatility is that it can be either an oxidizing or a reducing agent. For the following reactions, identify whether hydrogen peroxide is an oxidizing or reducing agent.

(a) $H_2O_2(aq) + 2\,Fe^{2+}(aq) + 2\,H^+(aq) \longrightarrow 2\,H_2O(l) + 2\,Fe^{3+}(aq)$

(b) $5\,H_2O_2(aq) + 2\,MnO_4^-(aq) + 6\,H^+(aq) \longrightarrow$
$$8\,H_2O(l) + 2\,Mn^{2+}(aq) + 5\,O_2(g)$$

Solution

▶ When H_2O_2 acts as an oxidizing agent, it is reduced to H_2O in acidic solution or to OH^- in basic solution. When it acts as a reducing agent, it is oxidized to $O_2(g)$.

(a) Fe^{2+} is oxidized to Fe^{3+}, and because H_2O_2 makes this possible, it is an oxidizing agent. Looking at the matter another way, we see that the oxidation state of oxygen in H_2O_2 is -1. In H_2O, it is -2. Hydrogen peroxide is *reduced* and thereby acts as an oxidizing agent.

(b) MnO_4^- is reduced to Mn^{2+}, and H_2O_2 makes this possible. In this situation, hydrogen peroxide is a reducing agent. Or, the oxidation state of oxygen *increases* from -1 in H_2O_2 to 0 in O_2. Hydrogen peroxide is *oxidized* and thereby acts as a reducing agent.

Practice Example A: Is $H_2(g)$ an oxidizing or reducing agent in the following reaction? Explain.

$$2\,NO_2(g) + 7\,H_2(g) \longrightarrow 2\,NH_3(g) + 4\,H_2O(g)$$

Practice Example B: Identify the oxidizing agent and the reducing agent in the following reaction.

$$4\,Au(s) + 8\,CN^-(aq) + O_2(g) + 2\,H_2O(l) \longrightarrow 4\,[Au(CN)_2]^-(aq) + 4\,OH^-(aq)$$

Permanganate ion, MnO_4^-, is a versatile oxidizing agent that has many uses in the chemical laboratory. In the next section, we describe its use in the quantitative analysis of iron—that is, the determination of the exact (quantitative) amount of iron in an iron-containing material. *Ozone*, $O_3(g)$, a triatomic form of oxygen, is an oxidizing agent used in water purification, as in the oxidation of the organic compound phenol, C_6H_5OH.

$$C_6H_5OH(aq) + 14\,O_3(g) \longrightarrow 6\,CO_2(g) + 3\,H_2O(l) + 14\,O_2(g)$$

Thiosulfate ion, $S_2O_3^{2-}$, is an important reducing agent. One of its industrial uses is as an antichlor to destroy residual chlorine from the bleaching of fibers.

$$S_2O_3^{2-}(aq) + 4\,Cl_2(aq) + 5\,H_2O \longrightarrow 2\,HSO_4^-(aq) + 8\,H^+(aq) + 8\,Cl^-(aq)$$

Oxidizing and reducing agents also play important roles in biological systems: in photosynthesis (to store the sun's energy), metabolism (oxidizing glucose), and the transport of oxygen.

5-7   Stoichiometry of Reactions in Aqueous Solutions: Titrations

If our objective is to get the maximum yield of a product, we generally choose one of the reactants (usually the most expensive one) as the limiting reactant and use excess amounts of the other reactants. This is the case in most precipitation reactions. In some instances, as in determining the concentration of a solution, we may not be interested in the products of a reaction but only in the relationship between two reactants. Then we have to carry out the reaction in such a way that *neither* reactant is in excess. A method that has long been used for doing this is known as *titration*.

We place a solution of one reactant in a small beaker or flask. We place another reactant, also in solution, in a *buret*, a long, graduated tube equipped with a stopcock valve. We can slowly add the second solution to the first by manipulating the stopcock. **Titration** is a reaction carried out by the carefully controlled addition of one solution to another. The trick is to stop the titration at the point where both reactants have reacted completely, a condition called the **equivalence point** of the titration. In a titration we need some means of signaling when the equivalence point is reached. In modern chemical laboratories this is commonly done with an appropriate measuring instrument. Still widely used, though, is a technique in which a very small quantity of a substance added to the reaction mixture changes color at or very near the equivalence point. Such substances are called **indicators**. Figure 5-10 illustrates the neutralization of an acid by a base by the titration technique. Calculations that use titration data are much the same as those introduced in Chapter 4, as illustrated in Example 5-9.

▶ The key to a successful acid–base titration is in selecting the right indicator. We learn how to do this when we consider theoretical aspects of titration in Chapter 18.

Acid-Base Titration animation

|(a)|(b)|(c)|

▲ **FIGURE 5-10   An acid–base titration—Example 5-9 illustrated**
(a) A 5.00-mL sample of vinegar, a small quantity of water, and a few drops of phenolphthalein indicator are added to a flask. **(b)** 0.1000 M NaOH from a previously filled buret is slowly added. **(c)** As long as the acid is in excess, the solution in the flask remains colorless. When the acid has been neutralized, an additional drop of NaOH(aq) causes the solution to become slightly basic. The phenolphthalein indicator turns a light pink. The first lasting appearance of the pink color is taken to be the equivalence point of the titration.

EXAMPLE 5-9

Using Titration Data to Establish the Concentrations of Acids and Bases. Vinegar is a dilute aqueous solution of acetic acid produced by the bacterial fermentation of apple cider, wine or other carbohydrate material. The legal minimum acetic acid content of vinegar is 4% by mass. A 5.00-mL sample of a particular vinegar was titrated with 38.08 mL of 0.1000 M NaOH(aq). Does this sample exceed the minimum limit? (Vinegar has a density of about 1.01 g/mL.)

Solution

With ideas from Section 5-3, we can write a net ionic equation for the neutralization reaction.

$$HC_2H_3O_2(aq) + OH^-(aq) \longrightarrow C_2H_3O_2^-(aq) + H_2O(l)$$

Key to a titration calculation is this fact: *At the equivalence point, two reactants have combined in their stoichiometric proportions until both have been consumed and neither remains in excess.* From titration data, we calculate the amount of one reactant, and by using a stoichiometric factor, we then determine the amount of the other. In the present case, we know the volume and molarity of NaOH(aq), from which we determine the number of moles of OH^-. Then, using the factor 1 mol $HC_2H_3O_2$/1 mol OH^-, we calculate the quantity of $HC_2H_3O_2$, first in moles and then in grams. Finally, as a separate calculation, we calculate the mass percent composition $HC_2H_3O_2$ in the 5.00-mL sample.

The stepwise conversions for the first part of the calculation are

$$mL\ NaOH \longrightarrow L\ NaOH \longrightarrow mol\ NaOH \longrightarrow$$
$$mol\ OH^- \longrightarrow mol\ HC_2H_3O_2 \longrightarrow g\ HC_2H_3O_2.$$

$$? \text{ g } HC_2H_3O_2 = 38.08 \text{ mL} \times \frac{1 \text{ L}}{1000 \text{ mL}} \times \frac{0.1000 \text{ mol NaOH}}{1 \text{ L}} \times \frac{1 \text{ mol } OH^-}{1 \text{ mol NaOH}}$$

$$\times \frac{1 \text{ mol } HC_2H_3O_2}{1 \text{ mol } OH^-} \times \frac{60.05 \text{ g } HC_2H_3O_2}{1 \text{ mol } HC_2H_3O_2}$$

$$= 0.2287 \text{ g } HC_2H_3O_2$$

This mass of $HC_2H_3O_2$ is found in 5.00 mL of vinegar of density 1.01 g/mL. The percent mass of $HC_2H_3O_2$ is

$$\% HC_2H_3O_2 = \frac{0.2287 \text{ g } HC_2H_3O_2}{5.00 \text{ mL vinegar}} \times \frac{1 \text{ mL vinegar}}{1.01 \text{ g vinegar}} \times 100\%$$

$$= 4.53\% \ HC_2H_3O_2$$

The vinegar sample exceeds the legal minimum limit, but only slightly. There is also a standard for the maximum amount of acetic acid allowed in vinegar. A vinegar producer might use this titration technique to ensure that the vinegar stays between these limits. The competitors' products might also be monitored in this way.

Practice Example A: A particular solution of NaOH is supposed to be approximately 0.100 M. To determine the exact molarity of the NaOH(aq), a 0.5000-g sample of $KHC_8H_4O_4$ is dissolved in water and titrated with 24.03 mL of the NaOH(aq). What is the actual molarity of the NaOH(aq)?

$$HC_8H_4O_4^-(aq) + OH^-(aq) \longrightarrow C_8H_4O_4^{2-}(aq) + H_2O(l)$$

Practice Example B: A 0.235-g sample of a solid that is 92.5% NaOH and 7.5% $Ca(OH)_2$, by mass, requires 45.6 mL of a HCl(aq) solution for its titration. What is the molarity of the HCl(aq)?

▶ Pure acetic acid is about 20 times more concentrated than is vinegar and must be handled with care (for example, it should not be inhaled). At times, whether a material is harmless or hazardous is primarily a matter of concentration.

(a)   (b)   (c)

▲ FIGURE 5-11   **Standardizing a solution of an oxidizing agent through a redox titration—Example 5-10 illustrated**
(a) The solution contains a known amount of Fe^{2+}, and the buret is filled with the intensely colored $KMnO_4(aq)$ to be standardized. (b) As it is added to the strongly acidic solution of $Fe^{2+}(aq)$, the $KMnO_4(aq)$ is immediately decolorized as a result of reaction (5.19). (c) When all the Fe^{2+} has been oxidized to Fe^{3+}, additional $KMnO_4(aq)$ has nothing left to oxidize and the solution turns a distinctive pink. Even a fraction of a drop of the $KMnO_4(aq)$ beyond the equivalence point is sufficient to cause this pink coloration.

Suppose we need a $KMnO_4(aq)$ solution of exactly known molarity, close to 0.020 M. We cannot prepare this solution by weighing out the required amount of $KMnO_4(s)$ and dissolving it in water. The solid is *not pure*, and its actual purity (that is, % $KMnO_4$) is *not known*. On the other hand, we can obtain iron wire in essentially pure form and allow the wire to react with an acid to yield $Fe^{2+}(aq)$. $Fe^{2+}(aq)$ is oxidized to $Fe^{3+}(aq)$ by $KMnO_4(aq)$ in an acidic solution. By determining the volume of $KMnO_4(aq)$ required to oxidize a known quantity of $Fe^{2+}(aq)$, we can calculate the exact molarity of the $KMnO_4(aq)$. Example 5-10 and Figure 5-11 illustrate this procedure, which is called **standardization of a solution**.

EXAMPLE 5-10

Standardizing a Solution for Use in Redox Titrations. A piece of iron wire weighing 0.1568 g is converted to $Fe^{2+}(aq)$ and requires 26.24 mL of a $KMnO_4(aq)$ solution for its titration. What is the molarity of the $KMnO_4(aq)$?

$$5\,Fe^{2+}(aq) + MnO_4^-(aq) + 8\,H^+(aq) \longrightarrow 5\,Fe^{3+}(aq) + Mn^{2+}(aq) + 4\,H_2O(l) \quad (5.19)$$

Focus On   Water Treatment

Water is not just water. Water often contains impurities that make it unsuitable for some purposes. For instance, a chemist would not use ordinary tap water to prepare an aqueous solution of silver nitrate. The solution would acquire a milky cast from the reaction of $Ag^+(aq)$ with traces of $Cl^-(aq)$ to form $AgCl(s)$. Yet this same tap water is probably perfectly safe to drink.

◀ Aeration and chemicals are used to treat effluent wastewater from papermaking.

The way in which water is purified depends on how it is to be used or on how it has been used. Water for a certain industrial application may require a different prior treatment than does water for domestic use. Also, water that has been used in a chemical plant may require a different treatment after use than does domestic sewage. Still, these varying treatments of water share a common feature. They include reactions in aqueous solutions of the general types described in this chapter. A few specific examples follow.

Removal of Iron from Drinking Water
Certain wells deliver water with up to 25 mg of iron per liter, but the federal recommended limit in drinking water in the United States is 0.3 mg/L. Methods of removing excess iron generally involve (a) making the chlorinated water slightly basic with slaked lime, $Ca(OH)_2$, (b) oxidizing Fe^{2+} to Fe^{3+} with hypochlorite ion, OCl^-, and (c) precipitating $Fe(OH)_3(s)$ from the basic solution. The reactions can be represented by these equations.

Solution
First, determine the amount of $KMnO_4$ consumed in the titration.

$$? \text{ mol } KMnO_4 = 0.1568 \text{ g Fe} \times \frac{1 \text{ mol Fe}}{55.847 \text{ g Fe}} \times \frac{1 \text{ mol Fe}^{2+}}{1 \text{ mol Fe}}$$

$$\times \frac{1 \text{ mol MnO}_4^-}{5 \text{ mol Fe}^{2+}} \times \frac{1 \text{ mol KMnO}_4}{1 \text{ mol MnO}_4^-}$$

$$= 5.615 \times 10^{-4} \text{ mol } KMnO_4$$

The volume of solution containing the 5.615×10^{-4} mol $KMnO_4$ is 26.24 mL $= 0.02624$ L, which means that

$$\text{concn } KMnO_4(aq) = \frac{5.615 \times 10^{-4} \text{ mol } KMnO_4}{0.02624 \text{ L}} = 0.02140 \text{ M } KMnO_4(aq)$$

KEEP IN MIND ▶
that the key to a titration calculation is that the amounts of two reactants consumed in the titration are stoichiometrically equivalent—neither reactant is in excess. Here, 1 mol MnO_4^- is stoichiometrically equivalent to 5 mol Fe^{2+}.

Practice Example A: A 0.376-g sample of an iron ore is dissolved in acid, the iron reduced to $Fe^{2+}(aq)$, and then titrated with 41.25 mL of 0.02140 M $KMnO_4$. Determine the % Fe by mass in the iron ore.
[*Hint:* Use equation (5.19).]

Practice Example B: Another substance that may be used to standardize $KMnO_4(aq)$ is sodium oxalate, $Na_2C_2O_4$. If 0.2482 g $Na_2C_2O_4$ is dissolved in water and titrated with 23.68 mL $KMnO_4$, what is the molarity of the $KMnO_4(aq)$?

$$MnO_4^-(aq) + C_2O_4^{2-}(aq) + H^+(aq) \longrightarrow Mn^{2+}(aq) + H_2O(l) + CO_2(g) \quad \text{(not balanced)}$$

$$Cl_2(g) + 2\,OH^-(aq) \longrightarrow Cl^-(aq) + OCl^-(aq) + H_2O(l)$$

$$2\,Fe^{2+}(aq) + OCl^-(aq) + H_2O(l) \longrightarrow$$
$$2\,Fe^{3+}(aq) + Cl^-(aq) + 2\,OH^-(aq)$$

$$Fe^{3+}(aq) + 3\,OH^-(aq) \longrightarrow Fe(OH)_3(s)$$

While all of this is occurring, the OCl^- is also serving its primary function: destroying pathogenic microorganisms in the water.

Removal of Oxygen from Boiler Water

High-temperature boilers are used to convert water to steam in electric power plants. Dissolved oxygen in the water is highly objectionable because it promotes corrosion of the steel from which many of the boiler parts are made. Because $O_2(g)$ is a good oxidizing agent, a reducing agent such as hydrazine, N_2H_4, is needed to remove it.

$$O_2(aq) + N_2H_4(aq) \longrightarrow 2\,H_2O(l) + N_2(g)$$

Removal of Phosphates from Domestic Sewage

Eutrophication is the term used to describe a series of events leading to the rapid growth of algae, the killing of fish, and other deleterious effects in bodies of freshwater. These events are caused by an excess of nutrients, principally phosphates. The treatment of domestic sewage includes the removal of phosphates. A particularly simple method is to precipitate the phosphates with slaked lime, $Ca(OH)_2$. The phosphates may be present in a number of different forms, including hydrogen phosphate ion, HPO_4^{2-}.

$$5\,Ca^{2+}(aq) + 3\,HPO_4^{2-}(aq) + 4\,OH^-(aq) \longrightarrow$$
$$Ca_5OH(PO_4)_3(s) + 3\,H_2O(l)$$

The precipitate has the composition of the mineral hydroxyapatite.

Destruction of Cyanide Ion in Industrial Operations

Cyanide compounds are used in metal-cleaning operations, in electroplating, and in extracting gold from gold-bearing rocks in mining operations. The cyanide ion must be destroyed in waste solutions from these operations. This can be done through an oxidation–reduction reaction, such as

$$2\,CN^-(aq) + 5\,OCl^-(aq) + 2\,OH^-(aq) \longrightarrow$$
$$N_2(g) + 2\,CO_3^{2-}(aq) + 5\,Cl^-(aq) + H_2O(l)$$

The poisonous $CN^-(aq)$ is converted to innocuous $N_2(g)$ and $CO_3^{2-}(aq)$.

Later in the text, we will consider additional examples of water treatment that have important domestic and industrial uses, such as reverse osmosis and water softening. Also, we must not overlook the biological processes that play an important role in water treatment, particularly in the treatment of wastewater.

Summary

Substances in aqueous solution are nonelectrolytes, weak electrolytes, or strong electrolytes, depending on the extent to which they produce ions. Strong electrolytes are almost completely dissociated into ions, and the concentration of a solution can be expressed in terms of these ions.

Some reactions in solution involve the combination of ions to yield water-insoluble solids—precipitates. Precipitation reactions can be predicted through a few solubility rules.

Still other reactions involve the combination of H^+ and OH^- ions to form $H_2O(HOH)$. The source of H^+ is called an acid, and the source of OH^- a base. The reaction is an acid–base, or neutralization, reaction. By extending the definition of a base to certain ions other than OH^-, some reactions in which gases are evolved can also be treated as acid–base reactions.

In an oxidation–reduction (redox) reaction certain atoms undergo an increase in oxidation state, a process called oxidation. Others undergo a decrease in oxidation state—reduction. An especially useful representation of redox reactions is through separate half-equations for oxidation and reduction and an overall equation obtained by combining the two half-equations. This approach also underlies a technique for balancing redox equations.

A common technique for carrying out a reaction in solution is known as titration. Titration data can be used to establish the molarities of solutions or to provide other information about the compositions of samples being analyzed.

Key Terms

acid (5-3)
base (5-3)
disproportionation reaction (5-5)
equivalence point (5-7)
half-reaction (5-4)
indicator (5-7)
net ionic equation (5-2)
neutralization (5-3)
nonelectrolyte (5-1)

oxidation (5-4)
oxidation–reduction (redox) reaction (5-4)
oxidizing agent (oxidant) (5-6)
precipitate (5-2)
reducing agent (reductant) (5-6)
reduction (5-4)
salt (5-3)
standardization of a solution (5-7)

strong acid (5-3)
strong base (5-3)
strong electrolyte (5-1)
titration (5-7)
weak acid (5-3)
weak base (5-3)
weak electrolyte (5-1)

Integrative Example

▲ White solid sodium dithionite, $Na_2S_2O_4$, is added to a yellow solution of potassium chromate, $K_2CrO_4(aq)$ (left). A product of the reaction is gray-green chromium(III) hydroxide, $Cr(OH)_3(s)$ (right).

Sodium dithionite, $Na_2S_2O_4$, is an important reducing agent. One interesting use is the reduction of chromate ion to insoluble chromium(III) hydroxide by dithionite ion, $S_2O_4^{2-}$, in basic solution. Sulfite ion is another product. The chromate ion may be present in wastewater from a chromium-plating plant, for example.

What mass of $Na_2S_2O_4$ is consumed in a reaction with 100.0 L of wastewater having $[CrO_4^{2-}] = 0.0148$ M?

1. *Write an ionic expression representing the reaction.* The participants in the reaction are named in the opening paragraph. Use information from Chapter 3 to substitute symbols and formulas for names.

$$CrO_4^{2-}(aq) + S_2O_4^{2-}(aq) + OH^-(aq) \longrightarrow Cr(OH)_3(s) + SO_3^{2-}(aq)$$

2. *Balance the redox equation written in part 1 as if the reaction were occurring in acidic solution.* Begin by writing skeleton half-equations.

$$CrO_4^{2-} \longrightarrow Cr(OH)_3$$
$$S_2O_4^{2-} \longrightarrow SO_3^{2-}$$

Balance the half-equations for Cr, S, O, and H atoms.

$$CrO_4^{2-} + 5\,H^+ \longrightarrow Cr(OH)_3 + H_2O$$
$$S_2O_4^{2-} + 2\,H_2O \longrightarrow 2\,SO_3^{2-} + 4\,H^+$$

Balance the half-equations for charge, and label them as oxidation and reduction.

Oxidation: $S_2O_4^{2-} + 2\,H_2O \longrightarrow 2\,SO_3^{2-} + 4\,H^+ + 2\,e^-$

Reduction: $CrO_4^{2-} + 5\,H^+ + 3\,e^- \longrightarrow Cr(OH)_3 + H_2O$

Combine the half-equations into an overall equation.

$$3 \times [S_2O_4^{2-} + 2\,H_2O \longrightarrow 2\,SO_3^{2-} + 4\,H^+ + 2\,e^-]$$
$$2 \times [CrO_4^{2-} + 5\,H^+ + 3\,e^- \longrightarrow Cr(OH)_3 + H_2O]$$

$$\overline{3\,S_2O_4^{2-} + 2\,CrO_4^{2-} + 4\,H_2O \longrightarrow}$$
$$6\,SO_3^{2-} + 2\,Cr(OH)_3 + 2\,H^+$$

3. *Change the conditions to basic solution.* Add $2\,OH^-$ to each side of the above equation, and combine $2\,H^+$ and $2\,OH^-$ to form $2\,H_2O$ on the right.

$$3\,S_2O_4^{2-} + 2\,CrO_4^{2-} + 4\,H_2O + 2\,OH^- \longrightarrow$$
$$6\,SO_3^{2-} + 2\,Cr(OH)_3 + 2\,H_2O$$

Subtract $2\,H_2O$ from each side of the equation to obtain the final balanced equation.

$$3\,S_2O_4^{2-}(aq) + 2\,CrO_4^{2-}(aq) + 2\,H_2O(l) + 2\,OH^-(aq) \longrightarrow$$
$$6\,SO_3^{2-}(aq) + 2\,Cr(OH)_3(s)$$

4. *Complete the stoichiometric calculation based on the balanced chemical equation.*

$$? \text{ g } Na_2S_2O_4 = 100.0 \text{ L} \times \frac{0.0148 \text{ mol } CrO_4^{2-}}{1 \text{ L}}$$

$$\times \frac{3 \text{ mol } S_2O_4^{2-}}{2 \text{ mol } CrO_4^{2-}} \times \frac{1 \text{ mol } Na_2S_2O_4}{1 \text{ mol } S_2O_4^{2-}}$$

$$\times \frac{174.1 \text{ g } Na_2S_2O_4}{1 \text{ mol } Na_2S_2O_4} = 387 \text{ g } Na_2S_2O_4$$

Review Questions

1. In your own words, define or explain the following terms or symbols: (**a**) ⇌; (**b**) [ ]; (**c**) spectator ion; (**d**) weak acid.
2. Briefly describe each of the following ideas or methods: (**a**) half-reaction method of balancing redox equations; (**b**) disproportionation reaction; (**c**) titration; (**d**) standardization of a solution.

3. Explain the important distinctions between each pair of terms: (**a**) strong electrolyte and strong acid; (**b**) oxidizing agent and reducing agent; (**c**) precipitation reactions and neutralization reactions; (**d**) half-reaction and overall reaction.

4. From the following solutions, select the **(a)** best and **(b)** poorest electrical conductor, and explain the reason for your choices: 0.10 M NH_3; 0.10 M NaCl; 0.10 M $HC_2H_3O_2$ (acetic acid); 0.10 M C_2H_5OH (ethanol).

5. Identify each of the following substances as a strong acid, weak acid, strong base, weak base, or salt: **(a)** Na_2SO_4; **(b)** $Ba(OH)_2$; **(c)** $Ba(NO_3)_2$; **(d)** H_3PO_4; **(e)** HBr; **(f)** HNO_2; **(g)** NH_3; **(h)** NH_4I; **(i)** KOH.

6. *Without doing detailed calculations*, indicate which of the following solutions has the greatest $[SO_4^{2-}]$: 0.075 M H_2SO_4; 0.22 M $MgSO_4$; 0.15 M Na_2SO_4; 0.080 M $Al_2(SO_4)_3$; 0.20 M $CuSO_4$.

7. Determine the concentration of the ion indicated in each of the following solutions: **(a)** $[K^+]$ in 0.238 M KNO_3; **(b)** $[NO_3^-]$ in 0.167 M $Ca(NO_3)_2$; **(c)** $[Al^{3+}]$ in 0.083 M $Al_2(SO_4)_3$; **(d)** $[Na^+]$ in 0.209 M Na_3PO_4.

8. Which of the following contains the greatest amount of chloride ion? 200.0 mL of 0.35 M NaCl; 500.0 mL of 0.065 M $MgCl_2$; 1.00 L of 0.068 M HCl.

9. A solution is prepared by dissolving 0.132 g $Ba(OH)_2 \cdot 8\,H_2O$ in 275 mL of water solution. What is $[OH^-]$ in this solution?

10. A solution is 0.126 M KCl and 0.148 M $MgCl_2$. What are $[K^+]$, $[Mg^{2+}]$, and $[Cl^-]$ in this solution?

11. How many milligrams of MgI_2 must be added to 250.0 mL of 0.0876 M KI to produce a solution with $[I^-] = 0.1000$ M?

12. Which of the following compounds is (are) *insoluble* in water? Explain. $CuCl_2$; NaI; $BaSO_4$; $Zn(NO_3)_2$; $Pb(C_2H_3O_2)_2$; $Al(OH)_3$.

13. Which of the following react(s) with HCl(aq) to produce a gas? Explain. Na_2SO_4; $KHSO_3$; $Zn(OH)_2$; Ca; $CaCl_2$.

14. Complete each of the following as a *net ionic equation*, indicating whether a precipitate forms. If no reaction occurs, so state.
    (a) $Na^+ + Br^- + Pb^{2+} + 2\,NO_3^- \longrightarrow$
    (b) $Mg^{2+} + 2\,Cl^- + Cu^{2+} + SO_4^{2-} \longrightarrow$
    (c) $Fe^{3+} + 3\,NO_3^- + Na^+ + OH^- \longrightarrow$
    (d) $Ca^{2+} + 2\,I^- + 2\,Na^+ + CO_3^{2-} \longrightarrow$
    (e) $Ba^{2+} + S^{2-} + 2\,Na^+ + SO_4^{2-} \longrightarrow$
    (f) $2\,K^+ + S^{2-} + Ca^{2+} + 2\,Cl^- \longrightarrow$

15. Complete each of the following as a *net ionic equation*. If no reaction occurs, so state.
    (a) $Ba^{2+} + 2\,OH^- + HC_2H_3O_2 \longrightarrow$
    (b) $H^+ + Cl^- + HC_3H_5O_2 \longrightarrow$
    (c) $FeS(s) + H^+ + I^- \longrightarrow$
    (d) $K^+ + HCO_3^- + H^+ + NO_3^- \longrightarrow$
    (e) $Mg(s) + H^+ \longrightarrow$
    (f) $Cu(s) + H^+ \longrightarrow$

16. Which of the following solutions would you use to precipitate Mg^{2+} from an aqueous solution of $MgCl_2$? Explain your choice. **(a)** $KNO_3(aq)$; **(b)** $NH_3(aq)$; **(c)** $H_2SO_4(aq)$; **(d)** $HC_2H_3O_2(aq)$.

17. What volume of 0.0962 M NaOH is required to exactly neutralize 10.00 mL of 0.128 M HCl?

18. The exact neutralization of 10.00 mL of 0.1012 M $H_2SO_4(aq)$ requires 23.31 mL of NaOH(aq). What must be the molarity of the NaOH(aq)?
    $H_2SO_4(aq) + 2\,NaOH(aq) \longrightarrow Na_2SO_4(aq) + 2\,H_2O(l)$

19. A 23.58-mL sample of 0.1278 M KOH is added to 25.13 mL of 0.1264 M HCl. Is the resulting mixture acidic, basic, or exactly neutral? Explain.
    (*Hint:* Is there a limiting reactant?)

20. In the equation
    $?\,Fe^{2+}(aq) + O_2(g) + 4\,H^+(aq) \longrightarrow ?\,Fe^{3+}(aq) + 2\,H_2O(l)$
    the missing coefficients **(a)** are each 4; **(b)** are each 2; **(c)** can have any values as long as they are the same; **(d)** must be determined by experiment.

21. Assign oxidation states to the elements involved in the following reactions. Indicate which are redox reactions and which are not.
    (a) $MgCO_3(s) + 2\,H^+(aq) \longrightarrow$
    $$Mg^{2+}(aq) + H_2O + CO_2(g)$$
    (b) $Cl_2(aq) + 2\,Br^-(aq) \longrightarrow 2\,Cl^-(aq) + Br_2(aq)$
    (c) $Ag(s) + 2\,H^+(aq) + NO_3^-(aq) \longrightarrow$
    $$Ag^+(aq) + H_2O + NO_2(g)$$
    (d) $2\,Ag^+(aq) + CrO_4^{2-}(aq) \longrightarrow Ag_2CrO_4(s)$

22. Assign oxidation states to the elements in the following redox reactions. Indicate which species is the oxidizing and which is the reducing agent.
    (a) $2\,NO(g) + 5\,H_2(g) \longrightarrow 2\,NH_3(g) + 2\,H_2O(g)$
    (b) $3\,Cu(s) + 8\,H^+(aq) + 2\,NO_3^-(aq) \longrightarrow$
    $$3\,Cu^{2+}(aq) + 4\,H_2O(l) + 2\,NO(g)$$
    (c) $3\,Cl_2(g) + 6\,OH^-(aq) \longrightarrow$
    $$5\,Cl^-(aq) + ClO_3^-(aq) + 3\,H_2O(l)$$

23. Complete and balance the following half-equations, and indicate whether oxidation or reduction is involved.
    (a) $SO_3^{2-} \longrightarrow S_2O_3^{2-}$ (acidic solution)
    (b) $HNO_3 \longrightarrow N_2O(g)$ (acidic solution)
    (c) $I^- \longrightarrow IO_3^-$ (acidic solution)
    (d) $Al(s) \longrightarrow Al(OH)_4^-$ (basic solution)

24. Balance these equations for redox reactions in acidic solution.
    (a) $Zn(s) + NO_3^- \longrightarrow Zn^{2+} + NO(g)$
    (b) $Zn(s) + NO_3^- \longrightarrow Zn^{2+} + NH_4^+$
    (c) $Fe^{2+} + Cr_2O_7^{2-} \longrightarrow Fe^{3+} + Cr^{3+}$
    (d) $H_2O_2 + MnO_4^- \longrightarrow Mn^{2+} + O_2(g)$

25. Balance these equations for redox reactions in basic solution.
    (a) $MnO_2(s) + ClO_3^- \longrightarrow MnO_4^- + Cl^-$
    (b) $Fe(OH)_3(s) + OCl^- \longrightarrow FeO_4^{2-} + Cl^-$
    (c) $ClO_2 \longrightarrow ClO_3^- + Cl^-$

26. A 25.12-mL sample of a $KMnO_4(aq)$ solution is required to titrate 0.2879 g sodium oxalate, $Na_2C_2O_4$, in a redox reaction occurring in acidic solution.
    $C_2O_4^{2-}(aq) + MnO_4^-(aq) \longrightarrow Mn^{2+}(aq) + CO_2(g)$
    (not balanced)
    What is the molarity of the $KMnO_4(aq)$?

Exercises

Strong Electrolytes, Weak Electrolytes, and Nonelectrolytes

27. Using information from this chapter, indicate whether each of the following substances in aqueous solution is a nonelectrolyte, weak electrolyte, or strong electrolyte. **(a)** HC_6H_5O; **(b)** Li_2SO_4; **(c)** MgI_2; **(d)** $(CH_3CH_2)_2O$; **(e)** $Sr(OH)_2$.

28. $NH_3(aq)$ conducts electric current only weakly. The same is true for $HC_2H_3O_2(aq)$. When these solutions are mixed, however, the resulting solution is a good conductor. How do you explain this?

29. The sketches are molecular views of the solute in an aqueous solution. For each of the sketches, indicate whether the solute is a strong, weak, or nonelectrolyte and which of the following substances it is: sodium chloride, propionic acid, hypochlorous acid, ammonia, barium bromide, ammonium chloride, methanol.

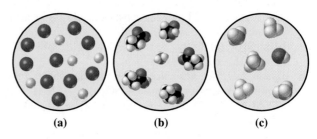

(a)   (b)   (c)

30. After identifying the three substances represented by the sketches in Exercise 29, sketch molecular views of aqueous solutions of the remaining four substances.

Ion Concentrations

31. These data are given for several cations in solution. Express them as molarities. **(a)** 35.0 mg Ca^{2+}/L; **(b)** 25.6 mg K^+/100 mL; **(c)** 0.168 mg Zn^{2+}/mL.

32. What molarity of $NaF(aq)$ corresponds to a fluoride ion content of 0.9 mg F^-/L, the federal government's recommended limit for fluoride ion in drinking water?

33. Which of the following aqueous solutions has the highest concentration of Na^+? **(a)** 0.208 M Na_2SO_4; **(b)** a solution containing 1.05 g $NaCl$/100 mL; **(c)** a solution having 14.7 mg Na^+/mL.

34. Which of the following aqueous solutions has the greatest $[H^+]$? Explain your choice. **(a)** 0.011 M $HC_2H_3O_2$; **(b)** 0.010 M HCl; **(c)** 0.010 M H_2SO_4; **(d)** 1.00 M NH_3.

35. If one assumes the volumes are additive, what is the $[Cl^-]$ in a solution obtained by mixing 225 mL of 0.625 M KCl and 615 mL of 0.385 M $MgCl_2$?

36. If one assumes the volumes are additive, what is the $[NO_3^-]$ in a solution obtained by mixing 275 mL of 0.283 M KNO_3, 328 mL of 0.421 M $Mg(NO_3)_2$, and 784 mL of H_2O?

Predicting Precipitation Reactions

37. Predict whether a reaction is likely to occur in each of the following cases. If so, write a net ionic equation.
(a) $HI(aq) + Zn(NO_3)_2(aq) \longrightarrow$
(b) $CuSO_4(aq) + Na_2CO_3(aq) \longrightarrow$
(c) $Cu(NO_3)_2(aq) + Na_3PO_4(aq) \longrightarrow$

38. Predict whether a reaction is likely to occur in each of the following cases. If so, write a net ionic equation.
(a) $AgNO_3(aq) + CuCl_2(aq) \longrightarrow$
(b) $Na_2S(aq) + FeCl_2(aq) \longrightarrow$
(c) $Na_2CO_3(aq) + AgNO_3(aq) \longrightarrow$

39. What reagent solution might you use to separate the cations in the following mixtures, that is, with one ion appearing in solution and the other in a precipitate?
(*Hint:* Refer to the solubility rules on page 145, and consider water also to be a reagent.)
(a) $BaCl_2(s)$ and $CaCl_2(s)$

(b) $MgCO_3(s)$ and $Na_2CO_3(s)$
(c) $AgNO_3(s)$ and $Cu(NO_3)_2(s)$

40. What reagent solution might you use to separate the cations in each of the following mixtures?
(*Hint:* Refer to Exercise 39.)
(a) $PbSO_4(s)$ and $Cu(NO_3)_2(s)$
(b) $Mg(OH)_2(s)$ and $BaSO_4(s)$
(c) $PbCO_3(s)$ and $CaCO_3(s)$

41. You are provided with $NaOH(aq)$, $K_2SO_4(aq)$, $Mg(NO_3)_2(aq)$, $BaCl_2(aq)$, $NaCl(aq)$, $Sr(NO_3)_2(aq)$, $Ag_2SO_4(s)$, and $BaSO_4(s)$. Write net ionic equations to show how you would use one or more of those reagents to obtain **(a)** $SrSO_4(s)$; **(b)** $Mg(OH)_2(s)$; **(c)** $KCl(aq)$.

42. Write net ionic equations to show how you would use one or more of the reagents in Exercise 41 to obtain **(a)** $BaSO_4(s)$; **(b)** $AgCl(s)$; **(c)** $KNO_3(aq)$.

Acid–Base Reactions

43. Every antacid contains one or more ingredients capable of reacting with excess stomach acid (HCl). The essential neutralization products are CO_2 and/or H_2O. Write net ionic equations to represent the neutralizing action of the following popular antacids.
(a) Alka-Seltzer (sodium bicarbonate)
(b) Tums (calcium carbonate)
(c) milk of magnesia (magnesium hydroxide)
(d) Maalox (magnesium hydroxide; aluminum hydroxide)
(e) Rolaids [$NaAl(OH)_2CO_3$]

44. Suppose you are given the following separate solids and solvents: solids, Na_2CrO_4, $BaCO_3$, $Al(OH)_3$, and $ZnSO_4$; solvents, $H_2O(l)$, HCl(aq), and H_2SO_4(aq). Your task is to prepare four solutions, each containing one of the cations (that is, one with Na^+, one with Ba^{2+}, and so on). What solvent would you use to prepare each solution? Explain.

45. In this chapter, we described an acid as a substance capable of producing H^+ and a salt as the ionic compound formed by the neutralization of an acid by a base. Write ionic equations to show that sodium hydrogen sulfate has the characteristics of both a salt and an acid (sometimes called an *acid salt*).

46. A neutralization reaction between an acid and a base is a common method of preparing useful salts. Give net ionic equations showing how the following salts could be prepared in this way: $(NH_4)_2HPO_4$; NH_4NO_3; and $(NH_4)_2SO_4$.

Oxidation–Reduction (Redox) Equations

47. Explain why the following reactions cannot occur as written.
(a) $Fe^{3+}(aq) + MnO_4^-(aq) + H^+(aq) \longrightarrow$
$$Mn^{2+}(aq) + Fe^{2+}(aq) + H_2O(l)$$
(b) $H_2O_2(aq) + Cl_2(aq) \longrightarrow$
$$ClO^-(aq) + O_2(g) + H^+(aq)$$

48. A newspaper account of an accidental spill of hydrochloric acid in an area where sodium hydroxide solution was also stored spoke of the potential hazardous release of chlorine gas if the two solutions should come into contact. Was this an accurate accounting of the hazard involved? Explain.

49. The following reactions do not occur in aqueous solutions. Balance their equations by the half-reaction method, as suggested in the Are You Wondering feature on page 158.
(a) $CH_4(g) + NO(g) \longrightarrow CO_2(g) + N_2(g) + H_2O(g)$
(b) $H_2S(g) + SO_2(g) \longrightarrow S_8(s) + H_2O(g)$
(c) $Cl_2O(g) + NH_3(g) \longrightarrow N_2(g) + NH_4Cl(s) + H_2O(l)$

50. The following reactions do not occur in aqueous solutions. Balance their equations by the half-reaction method, as suggested in the Are You Wondering feature on page 158.
(a) $CH_4(g) + NH_3(g) + O_2(g) \longrightarrow HCN(g) + H_2O(g)$
(b) $NO(g) + H_2(g) \longrightarrow NH_3(g) + H_2O(g)$
(c) $Fe(s) + H_2O(l) + O_2(g) \longrightarrow Fe(OH)_3(s)$

51. Balance the following equations for redox reactions occurring in *acidic* solution.
(a) $MnO_4^- + I^- \longrightarrow Mn^{2+} + I_2(s)$
(b) $BrO_3^- + N_2H_4 \longrightarrow Br^- + N_2$
(c) $VO_4^{3-} + Fe^{2+} \longrightarrow VO^{2+} + Fe^{3+}$
(d) $UO^{2+} + NO_3^- \longrightarrow UO_2^{2+} + NO(g)$

52. Balance the following equations for redox reactions occurring in *acidic* solution.
(a) $P_4(s) + NO_3^- \longrightarrow H_2PO_4^- + NO(g)$
(b) $S_2O_3^{2-} + MnO_4^- \longrightarrow SO_4^{2-} + Mn^{2+}$

(c) $HS^- + HSO_3^- \longrightarrow S_2O_3^{2-}$
(d) $Fe^{3+} + NH_3OH^+ \longrightarrow Fe^{2+} + N_2O(g)$

53. Balance the following equations for redox reactions occurring in *basic* solution.
(a) $CN^- + MnO_4^- \longrightarrow MnO_2(s) + CNO^-$
(b) $[Fe(CN)_6]^{3-} + N_2H_4 \longrightarrow [Fe(CN)_6]^{4-} + N_2(g)$
(c) $Fe(OH)_2(s) + O_2(g) \longrightarrow Fe(OH)_3(s)$
(d) $C_2H_5OH(aq) + MnO_4^- \longrightarrow C_2H_3O_2^- + MnO_2(s)$

54. Balance the following equations for *disproportionation* reactions occurring in *basic* solution, except where otherwise noted.
(a) $Cl_2(g) \longrightarrow Cl^- + ClO_3^-$
(b) $S_2O_4^{2-} \longrightarrow S_2O_3^{2-} + HSO_3^-$ (acidic solution)
(c) $MnO_4^{2-} \longrightarrow MnO_2(s) + MnO_4^-$
(d) $P_4(s) \longrightarrow H_2PO_2^- + PH_3(g)$

55. Balance the following equations for redox reactions occurring in *acidic solution*
(a) $S_2O_3^{2-} + Cl_2(g) \longrightarrow SO_4^{2-} + Cl^-$
(b) $Sn^{2+} + Cr_2O_7^{2-} \longrightarrow Sn^{4+} + Cr^{3+}$
Basic solution
(c) $S_8(s) \longrightarrow S^{2-} + S_2O_3^{2-}$
(d) $As_2S_3 + H_2O_2 \longrightarrow AsO_4^{3-} + SO_4^{2-}$

56. Write a balanced equation for each of the following redox reactions.
(a) the oxidation of nitrite ion to nitrate ion by permanganate ion, MnO_4^-, in acidic solution (MnO_4^- ion is reduced to Mn^{2+})
(b) the reaction of manganese(II) ion and permanganate ion in basic solution to form solid manganese dioxide
(c) the reaction of sodium metal with hydroiodic acid
(d) the reduction of vanadyl ion (VO^{2+}) to vanadic ion (V^{3+}) in acidic solution with zinc metal as the reducing agent

Oxidizing and Reducing Agents

57. What are the oxidizing and reducing agents in the following redox reactions?

(a) $5\,SO_3^{2-} + 2\,MnO_4^- + 6\,H^+ \longrightarrow$
$$5\,SO_4^{2-} + 2\,Mn^{2+} + 3\,H_2O$$

(b) $2\,NO_2(g) + 7\,H_2(g) \longrightarrow 2\,NH_3(g) + 4\,H_2O(g)$

(c) $2\,[Fe(CN)_6]^{4-} + H_2O_2 + 2\,H^+ \longrightarrow$
$$2\,[Fe(CN)_6]^{3-} + 2\,H_2O$$

58. Thiosulfate ion, $S_2O_3^{2-}$, is a reducing agent that can be oxidized to different products, depending on the strength of the oxidizing agent and other conditions. By adding H^+, H_2O, and/or OH^- as necessary, write redox equations to show the oxidation of $S_2O_3^{2-}$ to

(a) $S_4O_6^{2-}$ by I_2; iodide ion is another product.

(b) HSO_4^- by Cl_2; chloride ion is another product.

(c) SO_4^{2-} by OCl^- in basic solution; chloride ion is another product.

Neutralization and Acid–Base Titrations

59. A NaOH(aq) solution cannot be made up to an exact concentration simply by weighing out the required mass of NaOH, because the NaOH is not pure. Also, water vapor condenses on the solid as it is being weighed. The solution must be standardized by titration. For this purpose, a 25.00-mL sample of a NaOH(aq) solution requires 28.34 mL of 0.1085 M HCl. What is the molarity of the NaOH(aq)?
$$HCl(aq) + NaOH(aq) \longrightarrow NaCl(aq) + H_2O(l)$$

60. Household ammonia, used as a window cleaner and for other cleaning purposes, is $NH_3(aq)$. A 28.72-mL sample of 1.021 M HCl(aq) is required to neutralize the NH_3 present in a 5.00-mL sample. The net ionic equation for the neutralization is
$$NH_3(aq) + H^+(aq) \longrightarrow NH_4^+(aq)$$
What is the molarity of NH_3 in the sample?

61. How many milliliters of 2.155 M KOH(aq) are required to titrate 25.00 mL of 0.3057 M $HC_3H_5O_2$(aq) (propionic acid)?

62. How many milliliters of 0.0844 M $Ba(OH)_2$(aq) are required to titrate 50.00 mL of 0.0526 M HNO_3(aq)?

63. We want to determine the acetylsalicyclic acid content of a series of aspirin tablets by titration with NaOH(aq).
$$HC_9H_7O_4(aq) + OH^-(aq) \longrightarrow C_9H_7O_4^-(aq) + H_2O(l)$$
Each of the tablets is expected to contain about 0.32 g of $HC_9H_7O_4$. What molarity NaOH should we use for titration volumes of about 23 mL? (This procedure ensures good precision and allows the titration of two samples with the contents of a 50-mL buret.)

64. For use in titrations, we wish to prepare 20 L of HCl(aq) a concentration known to *four* significant figures. We use a two-step procedure. First, we prepare a solution of about 0.10 M HCl. Then, we titrate a sample of this dilute HCl(aq) with a NaOH(aq) solution of known concentration.

(a) How many milliliters of concentrated HCl(aq) ($d = 1.19$ g/mL; 38% HCl, by mass) must we dilute to 20.0 L with water to prepare 0.10 M HCl?

(b) A 25.00 mL sample of the approximately 0.10 M HCl prepared in part (a) requires 20.93 mL of 0.1186 M NaOH for its titration. What is the molarity of the HCl(aq)?

(c) Why is a titration necessary? That is, why could we not prepare a standard solution of 0.1000 M HCl simply by an appropriate dilution of the concentrated HCl(aq)?

65. A 25.00-mL sample of 0.132 M HNO_3 is mixed with 10.00 mL of 0.318 M KOH. Is the resulting solution acidic, basic, or exactly neutralized?

66. A 7.55-g sample of Na_2CO_3(s) is added to 125 mL of a vinegar that is 0.762 M $HC_2H_3O_2$. Will the resulting solution still be acidic? Explain.

67. Refer to Example 5-9. Suppose the analysis of all vinegar samples uses 5.00 mL of the vinegar and 0.1000 M NaOH for the titration. What volume of the 0.1000 M NaOH would represent the legal minimum 4.0%, by mass, acetic acid content of the vinegar? That is, calculate the volume of 0.1000 M NaOH so that if a titration requires more than this volume, the legal minimum limit is met (less than this volume, and the limit is not met).

68. The electrolyte in a lead storage battery must have a concentration between 4.8 and 5.3 M H_2SO_4 if the battery is to be most effective. A 5.00-mL sample of a battery acid requires 49.74 mL of 0.935 M NaOH for its complete reaction (neutralization). Does the concentration of the battery acid fall within the desired range?
(*Hint:* The H_2SO_4 produces two H^+ ions per formula unit.)

69. Which of the following points in a titration is represented by the molecular view shown in the sketch?

(a) 20% of the necessary titrant added in the titration of NH_4Cl(aq) with HCl(aq)

(b) 20% of the necessary titrant added in the titration of NH_3(aq) with HCl(aq)

(c) the equivalence point in the titration of NH_3(aq) with HCl(aq)

(d) 120% of the necessary titrant added in the titration of NH_3(aq) with HCl(aq)

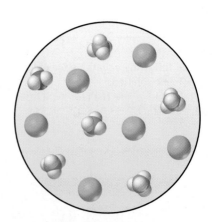

70. Using the sketch in Exercise 69 as a guide, sketch the molecular view of a solution in which
(a) HCl(aq) is titrated to the equivalence point with KOH(aq).

(b) $CH_3COOH(aq)$ is titrated halfway to the equivalence point with NaOH(aq).

Stoichiometry of Oxidation–Reduction Reactions

71. A $KMnO_4(aq)$ solution is to be standardized by titration against $As_2O_3(s)$. A 0.1078-g sample of As_2O_3 requires 22.15 mL of the $KMnO_4(aq)$ for its titration. What is the molarity of the $KMnO_4(aq)$?
$$5 As_2O_3 + 4 MnO_4^- + 9 H_2O + 12 H^+ \longrightarrow$$
$$10 H_3AsO_4 + 4 Mn^{2+}$$

72. Refer to Example 5-6. Assume that the only reducing agent present in a particular wastewater is SO_3^{2-}. If a 25.00-mL sample of this wastewater requires 31.46 mL of 0.02237 M $KMnO_4$ for its titration, what is the molarity of SO_3^{2-} in the wastewater?

73. An iron ore sample weighing 0.9132 g is dissolved in HCl(aq), and the iron is obtained as $Fe^{2+}(aq)$. This solution is then titrated with 28.72 mL of 0.05051 M $K_2Cr_2O_7(aq)$. What is the % Fe by mass in the ore sample?
$$6 Fe^{2+} + 14 H^+ + Cr_2O_7^{2-} \longrightarrow 6 Fe^{3+} + 2 Cr^{3+} + 7 H_2O$$

74. The concentration of $Mn^{2+}(aq)$ can be determined by titration with $MnO_4^-(aq)$ in basic solution.
$$Mn^{2+} + MnO_4^- \longrightarrow MnO_2(s) \quad \text{(not balanced)}$$
A 25.00-mL sample of $Mn^{2+}(aq)$ requires 37.21 mL of 0.04162 M $KMnO_4(aq)$ for its titration. What is $[Mn^{2+}]$ in the sample?

75. The titration of 50.0 mL of a saturated solution of sodium oxalate, $Na_2C_2O_4$, requires 25.8 mL of 0.02140 M $KMnO_4$ in acidic solution. What mass of $Na_2C_2O_4$, in grams, would be present in 1.00 L of this saturated solution?
$$C_2O_4^{2-} + MnO_4^- \longrightarrow Mn^{2+} + CO_2(g) \quad \text{(not balanced)}$$

76. Refer to the Integrative Example. In the treatment of 100×10^2 L of a wastewater solution that is 0.0126 M CrO_4^{2-},
(a) how many grams of $Cr(OH)_3(s)$ would precipitate?
(b) how many grams of $Na_2S_2O_4$ would be consumed?

Integrative and Advanced Exercises

77. Write net ionic equations for the reactions depicted in the photos.
(a) Sodium metal reacts with water to produce hydrogen.
(b) An excess of aqueous iron(III) chloride is added to the solution in (a).
(c) The precipitate from (b) is collected and treated with an excess of HCl(aq).

(a)   (b)   (c)

78. Following are some laboratory methods occasionally used for the preparation of small quantities of chemicals. Write a balanced equation for each.
(a) Preparation of $H_2S(g)$: HCl(aq) is heated with FeS(s).
(b) Preparation of $Cl_2(g)$: HCl(aq) is heated with $MnO_2(s)$; $MnCl_2(aq)$ and H_2O are other products.
(c) Preparation of N_2: Br_2 and NH_3 react in aqueous solution; NH_4Br is another product.

(d) Preparation of chlorous acid: An aqueous suspension of solid barium chlorite is treated with dilute $H_2SO_4(aq)$. (*Hint:* What is the other probable product in addition to chlorous acid?)

79. When concentrated $CaCl_2(aq)$ is added to $Na_2HPO_4(aq)$, a white precipitate forms that is 38.7% Ca by mass. Write a net ionic equation representing the probable reaction that occurs.

80. You have available a solution that is 0.0250 M $Ba(OH)_2$ and the following pieces of equipment: 1.00-, 5.00-, 10.00-, 25.00-, and 50.00-mL pipets and 100.0-, 250.0-, 500.0-, and 1000.0-mL volumetric flasks. Describe how you would use this equipment to produce a solution in which $[OH^-]$ is 0.0100 M.

81. Sodium hydroxide used to make standard NaOH(aq) solutions for acid–base titrations is invariably contaminated with some sodium carbonate: (a) Explain why, except in the most precise work, the presence of this sodium carbonate generally does not seriously affect the results obtained, for example, when NaOH(aq) is used to titrate HCl(aq). (b) On the other hand, show that if Na_2CO_3 comprises more than 1–2% of the solute in NaOH(aq), the titration results are affected.

82. A 110.520-g sample of mineral water is analyzed for its magnesium content. The Mg^{2+} in the sample is first precipitated as $MgNH_4PO_4$, and this precipitate is then converted to $Mg_2P_2O_7$, which is found to weigh 0.0549 g. Express the quantity of magnesium in the sample in parts per million (that is, in grams of Mg per million grams of H_2O).

83. What volume of 0.248 M $CaCl_2$ must be added to 335 mL of 0.186 M KCl to produce a solution with a concentration of 0.250 M Cl^-?

84. A *white* solid unknown consists of *two* compounds, each containing a different cation. As suggested in the illustration, the unknown is partially soluble in water. The solution is treated with NaOH(aq) and yields a *white* precipitate. The part of the original solid that is insoluble in water dissolves in HCl(aq) with the evolution of a gas. The resulting solution is then treated with $(NH_4)_2SO_4$(aq) and yields a *white* precipitate. Is it possible that any of the following cations were present in the original unknown: Mg^{2+}, Cu^{2+}, Ba^{2+}, Na^+, NH_4^+? Explain your reasoning. What compounds could be in the unknown mixture? (That is, what anions might be present?)

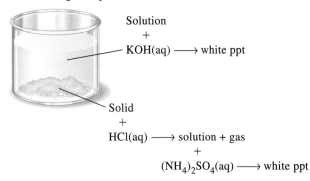

Solution
+
KOH(aq) ⟶ white ppt

Solid
+
HCl(aq) ⟶ solution + gas
+
$(NH_4)_2SO_4$(aq) ⟶ white ppt

85. Balance the following equations for reactions in acidic solution.
    (a) $IBr + BrO_3^- + H^+ \longrightarrow IO_3^- + Br^- + H_2O$
    (b) $C_2H_5NO_3 + Sn \longrightarrow NH_2OH + C_2H_5OH + Sn^{2+}$
    (c) $As_2S_3 + NO_3^- \longrightarrow H_3AsO_4 + S + NO$
    (d) $H_5IO_6 + I_2 \longrightarrow IO_3^- + H^+ + H_2O$
    (e) $S_2F_2 + H_2O \longrightarrow S_8 + H_2S_4O_6 + HF$

86. Balance the following equations for reactions in basic solution.
    (a) $Fe_2S_3 + H_2O + O_2 \longrightarrow Fe(OH)_3 + S$
    (b) $O_2^- + H_2O \longrightarrow OH^- + O_2$
    (c) $CrI_3 + H_2O_2 \longrightarrow CrO_4^{2-} + IO_4^-$
    (d) $Ag + CN^- + O_2 + OH^- \longrightarrow$
       $\qquad\qquad\qquad\qquad\qquad [Ag(CN)_2]^- + H_2O$
    (e) $B_2Cl_4 + OH^- \longrightarrow BO_2^- + Cl^- + H_2O + H_2$
    (f) $C_2H_5OH + MnO_4^- \longrightarrow C_2H_3O_2^- + MnO_2$

87. A method of producing phosphine, PH_3, from elemental phosphorus, P_4, involves heating the P_4 with H_2O. An additional product is phosphoric acid, H_3PO_4. Write a balanced equation for this reaction.

88. When iron pyrites, waste products from the mining of coal, are exposed to the environment, their sulfur content is oxidized to sulfuric acid. This creates the environmental problem known as *acid mine drainage*. Following are two of the main reactions involved. Balance these equations for the reactions.
    (a) $FeS_2(s) + O_2 + H_2O \longrightarrow Fe^{2+} + SO_4^{2-} + H^+$
    (b) $FeS_2(s) + Fe^{3+} + H_2O \longrightarrow Fe^{2+} + SO_4^{2-} + H^+$

89. A sample of battery acid is to be analyzed for its sulfuric acid content. A 1.00-mL sample weighs 1.239 g. This 1.00-mL sample is diluted to 250.0 mL, and 10.00 mL of this diluted acid requires 32.44 mL of 0.00498 M $Ba(OH)_2$ for its titration. What is the mass percent of H_2SO_4 in the battery acid? (Assume that complete ionization and neutralization of the H_2SO_4 occurs.)

90. A piece of marble (assume it is pure $CaCO_3$) reacts with 2.00 L of 2.52 M HCl. After dissolution of the marble, a 10.00-mL sample of the remaining HCl(aq) is withdrawn, added to some water, and titrated with 24.87 mL of 0.9987 M NaOH. What must have been the mass of the piece of marble? Comment on the precision of this method; that is, how many significant figures are justified in the result?

91. The reaction below can be used as a laboratory method of preparing small quantities of Cl_2(g). If a 62.6-g sample that is 98.5% $K_2Cr_2O_7$ by mass is allowed to react with 325 mL of HCl(aq) with a density of 1.15 g/mL and 30.1% HCl by mass, how many grams of Cl_2(g) are produced?

$$Cr_2O_7^{2-} + H^+ + Cl^- \longrightarrow Cr^{3+} + H_2O + Cl_2(g)$$
$$\text{(not balanced)}$$

92. Refer to Example 5-10. Suppose that the $KMnO_4$(aq) described in this example were standardized by reaction with As_2O_3 instead of iron wire. If a 0.1304-g sample that is 99.96% As_2O_3, by mass, had been used in the titration, how many milliliters of the $KMnO_4$(aq) would have been required?

$$As_2O_3 + MnO_4^- + H^+ + H_2O \longrightarrow H_3AsO_4 + Mn^{2+}$$
$$\text{(not balanced)}$$

93. A new method under development for water treatment uses chlorine dioxide rather than chlorine itself. One method of producing ClO_2 involves passing Cl_2(g) into a concentrated solution of sodium chlorite; NaCl(aq) is the other product. If the reaction has a 97% yield, what mass of ClO_2 is produced per gallon of 2.0 M $NaClO_2$(aq) treated in this way?

94. A 0.4324-g sample of a potassium hydroxide–lithium hydroxide mixture requires 28.28 mL of 0.3520 M HCl for its titration to the equivalence point. What is the mass percent lithium hydroxide in this mixture?

95. Chile saltpeter is a natural source of $NaNO_3$; it also contains $NaIO_3$. The $NaIO_3$ can be used as a source of iodine. Iodine is produced from sodium iodate in the following two-step process occurring under acidic conditions and pictured in the illustration.

$$IO_3^-(aq) + HSO_3^-(aq) \longrightarrow$$
$$\qquad\qquad I^-(aq) + SO_4^{2-}(aq) \quad \text{(not balanced)}$$

$$I^-(aq) + IO_3^-(aq) \longrightarrow I_2(s) + H_2O(l) \quad \text{(not balanced)}$$

A 5.00-L sample of a $NaIO_3$(aq) solution containing 5.80 g $NaIO_3$/L is treated with the stoichiometric quantity of

NaHSO₃ (no excess of either reactant). Then, a further quantity of the initial $NaIO_3(aq)$ is added to the reaction mixture to bring about the second reaction. How many grams of NaHSO₃ are required in the first step, and what additional volume of the starting solution must be added in the second step?

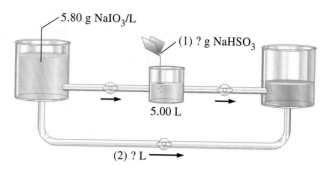

5.80 g NaIO₃/L

(1) ? g NaHSO₃

5.00 L

(2) ? L ⟶

Feature Problems

96. Sodium cyclopentadienide (NaCp), NaC_5H_5, is a common reducing agent in the chemical laboratory, but there is a problem in using it: NaCp is contaminated with tetrahydrofuran (THF), C_4H_8O, a solvent used in its preparation. The THF is present as $NaCp \cdot (THF)_x$, and it is generally necessary to know exactly how much of this $NaCp \cdot (THF)_x$ is present. This is accomplished by allowing a small amount of the $NaCp \cdot (THF)_x$ to react with water,

$$NaCp \cdot (C_4H_8O)_x + H_2O \longrightarrow$$
$$NaOH(aq) + Cp\text{---}H + x\ C_4H_8O$$

followed by titration of the NaOH(aq) with a standard acid.

From the sample data tabulated below, determine the value of x in the formula $NaCp \cdot (THF)_x$.

	Trial 1	Trial 2
Mass of $NaCp \cdot (THF)_x$	0.242 g	0.199 g
Volume of 0.1001 M HCl required to titrate NaOH(aq)	14.92 mL	11.99 mL

97. Manganese is derived from pyrolusite ore, an impure manganese dioxide. The following procedure is used to analyze a pyrolusite ore for its MnO_2 content: A 0.533-g sample is treated with 1.651 g oxalic acid ($HC_2O_4 \cdot 2\ H_2O$) in an acidic medium. Following this reaction, the excess oxalic acid is titrated with 0.1000 M KMnO₄, 30.06 mL being required. What is the % MnO_2, by mass, in the ore?

$$H_2C_2O_4 + MnO_2 + H^+ \longrightarrow Mn^{2+} + H_2O + CO_2$$
$$\text{(not balanced)}$$

$$H_2C_2O_4 + MnO_4^- + H^+ \longrightarrow Mn^{2+} + H_2O + CO_2$$
$$\text{(not balanced)}$$

98. In describing titration reactions, we noted that the equivalence point is often detected by measuring a physical property of the solution. One possibility is to measure the electrical conductance. The graph shows how the total electrical conductance of a solution varies as NaOH(aq) is added

to HCl(aq). The measured conductances are represented by the two straight red lines. The two black lines lying below the red lines divide the graph into four areas, each area representing the contribution of a particular ion to the total conductance of the solution. Thus, the conductance at the start of the titration is high, with H^+ contributing most to the solution conductance, Cl^- making a secondary contribution, and Na^+ not yet contributing anything. Until the equivalence point is reached, Na^+ replaces H^+ and the total conductance drops. At the equivalence point, where the two red lines meet, the contribution of the H^+ ion has fallen to zero and the total conductance is that of NaCl(aq). Beyond the equivalence point, the conductance again rises as excess NaOH accumulates in the solution. An important requirement to obtain straight lines in the graph is that the titrant solution [NaOH(aq)] be much more concentrated than the solution being titrated [HCl(aq)]. In this way, the total volume of solution changes little and dilution effects can be ignored.

To summarize the ideas suggested by the graph,

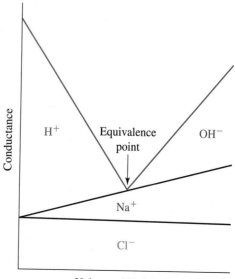

Volume of NaOH(aq)

(1) Ions differ in their intrinsic abilities to conduct electric current (H^+ and OH^- are much better conductors than are Na^+ and Cl^-).

(2) The greater the concentration of an ion, the greater is its contribution to the conductance of a solution.

(3) The conductance of the solution as a whole can be expressed as the sum of the conductances of the ions present.

Assume that the contributions toward the total electrical conductance of a solution by several different ions are roughly as follows: $H^+ = 350$; $NH_4^+ = 73$; $Ag^+ = 62$; $Na^+ = 50$; $OH^- = 198$; $Cl^- = 76$; $NO_3^- = 71$; $C_2H_3O_2^- = 41$.

Sketch the graphs you would expect for the following titrations.

(a) $NaOH(aq)$ titrated with $HNO_3(aq)$

(b) $HC_2H_3O_2(aq)$ titrated with $NaOH(aq)$

(c) $HCl(aq)$ titrated with $AgNO_3(aq)$

(d) $HCl(aq)$ titrated with $NH_3(aq)$

 eMedia Exercises

99. The **Equilibrium of a Weak Acid** animation *(eChapter 5-1)* illustrates both forward and reverse reactions taking place. It is important to keep a mental picture of this process in mind as you continue to view different reactions. Propose a method for quantitatively describing the fraction form in solution when equilibrium is reached.

100. In the **Precipitation Reactions** movie *(eChapter 5-2)*, what are the spectator ions present following the formation of each of the precipitates? Why do these species remain in solution?

101. The **Introduction to Acids and Bases** animation *(eChapter 5-3)* shows a relatively large number of species undergoing dissociation and recombination. **(a)** Using a density of water of 1.0 g/mL, calculate the number of protons produced in an aqueous solution of 0.10 M HNO_3. **(b)** According to your calculation, approximately how many ions would appear in the field of view of the animation?

102. From the **Oxidation–Reduction Chemistry of Tin and Zinc** movie *(eChapter 5-4)*, **(a)** identify the species being oxidized and reduced in the reaction. Notice that the reaction begins immediately but slows over time. **(b)** Considering the different steps to an oxidation–reduction reaction, what causes the reaction to slow down? **(c)** What could be changed about the reactants to speed up the reaction?

103. The measurement of a quantity called pH represents a convenient way of following relative concentrations of acid and base species in solution. **(a)** In the **Acid-Base Titration** animation *(eChapter 5-7)*, what causes the rapid change in pH upon the addition of the single drop of base after the equivalence point has been reached? **(b)** If the concentration of the base being added were doubled, at what approximate volume would the equivalence point be reached?

6

Gases

Contents

Hot air balloons have intrigued people from the time the simple gas laws fundamental to their operation came to be understood over 200 years ago.

A bicycle tire must not be overinflated, because it might rupture. An aerosol can should not be discarded in an incinerator. Carbon dioxide gas vaporizing from a block of dry ice sinks to the floor. A balloon filled with helium or hot air rises in air. One should not search for a gas leak with an open flame. These and many other observations concerning gases can be explained by concepts considered in this chapter. For example, the behavior of the bicycle tire and the aerosol can are based on relationships between pressure, temperature, volume, and amount of gas. An understanding of the lifting power of lighter-than-air balloons comes in large part from knowledge of gas densities and their dependence on molar mass, temperature, and pressure. Predicting how far and how fast gas molecules migrate through air requires knowing something about the phenomenon of diffusion.

To do quantitative calculations about the behavior of gases, we will use some simple gas laws and a more general expression called the ideal gas equation. To explain these laws, we will use the kinetic-molecular theory of gases. Ideas presented in this chapter reappear in new contexts in several later chapters.

Motions of a Gas animation

6-1   Properties of Gases: Gas Pressure

Several characteristics of gases are familiar to everyone. Gases expand to fill and assume the shapes of their containers. They diffuse into one another and mix in all proportions. We cannot see individual particles of a gas, although we can see the bulk gas if it is colored (Figure 6-1). Some gases, such as hydrogen and methane, are combustible; whereas others, such as helium and neon, are chemically unreactive.

Four properties determine the physical behavior of a gas: the amount of gas (in moles) and the gas volume, temperature, and pressure. If we know any three of these, we can usually calculate the value of the remaining one. To do this, we can use a mathematical equation called an equation of state, as we will see later in the chapter. To some extent we have already discussed the properties of amount, volume, and temperature, but we need to consider the idea of pressure.

▲ FIGURE 6-1   **The gaseous states of three halogens (group 17)**
The greenish yellow gas is $Cl_2(g)$; the brownish red gas is $Br_2(g)$ above a small pool of liquid bromine; the violet gas is $I_2(g)$ in contact with grayish black solid iodine. Most other common gases, such as H_2, O_2, N_2, CO, and CO_2, are colorless.

▲ FIGURE 6-2
Illustrating the pressure exerted by a solid
The two cylinders have the same mass and exert the same force on the supporting surface $(F = gm)$. The tall, thin one has a smaller area of contact, however, and exerts a greater pressure $(P = F/A)$.

The Concept of Pressure

A balloon expands when it is inflated with air, but what keeps the balloon in its distended shape? A plausible hypothesis is that molecules of a gas are in constant motion, frequently colliding with one another and with the walls of their container. In their collisions, the gas molecules exert a force on the container walls. This force keeps the balloon distended. It is not easy, however, to measure the total force exerted by a gas. Instead of focusing on this total force, we consider instead the gas pressure. **Pressure** is defined as a force per unit area, that is, a force divided by the area over which the force is distributed. Figure 6-2 illustrates the idea of pressure exerted by a solid.

In SI units, force is expressed in *newtons* (N) and area in square meters (m²). The corresponding force per unit area—the pressure—is in the unit N/m². A pressure of one newton per square meter is defined as one **pascal (Pa)**. Thus, a pressure in pascals is

$$P(\text{Pa}) = \frac{F(\text{N})}{A(\text{m}^2)} \tag{6.1}$$

A pascal is a rather small pressure unit, so the **kilopascal (kPa)** is more commonly used. The pascal honors Blaise Pascal (1623–1662), who studied pressure and its transmission through fluids—the basis of modern hydraulics.

Liquid Pressure

Because it is difficult to measure the total force exerted by gas molecules, it is also difficult to apply equation (6.1) to gases. The pressure of a gas is usually measured *indirectly*, by comparing it with a liquid pressure. Figure 6-3 illustrates the concept of liquid pressure and suggests that the pressure of a liquid depends only on the height of the liquid column and the density of the liquid. To confirm this statement, consider a liquid with density d, contained in a cylinder with cross-sectional area A, filled to a height h.

Now recall these facts. (1) Weight is a force, and weight and mass are proportional: $W = g \cdot m$. (2) The mass of a liquid is the product of its volume and density: $m = V \cdot d$. (3) The volume of a cylinder is the product of its height and cross-sectional area: $V = h \cdot A$. We use these ideas to derive equation (6.2).

$$P = \frac{F}{A} = \frac{W}{A} = \frac{g \cdot m}{A} = \frac{g \cdot V \cdot d}{A} = \frac{g \cdot h \cdot A \cdot d}{A} = g \cdot h \cdot d \tag{6.2}$$

Thus, because g is a constant, *liquid pressure is directly proportional to the liquid density and the height of the liquid column.*

Barometric Pressure

In 1643, Evangelista Torricelli constructed the device pictured in Figure 6-4 to measure the pressure exerted by the atmosphere. This device is called a **barometer**.

If we stand a glass tube, *open at both ends*, upright in a container of mercury (Figure 6-4a), the mercury levels inside and outside the tube are the same. To create the situation in Figure 6.4b, we seal one end of a long glass tube, completely fill the

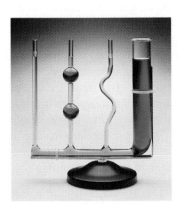

▲ FIGURE 6-3
The concept of liquid pressure
All the interconnected vessels fill to the same height. As a result, the liquid pressures are the same despite the different shapes and volumes of the containers.

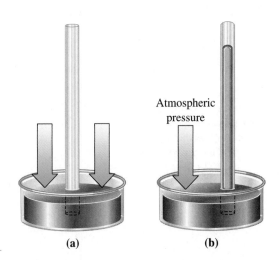

▶ FIGURE 6-4
Measurement of atmospheric pressure with a mercury barometer
Arrows represent the pressure exerted by the atmosphere. **(a)** The liquid mercury levels are equal inside and outside the open-end tube. **(b)** A column of mercury 760 mm high is maintained in the closed-end tube. The pressure at the bottom of the column of liquid mercury is equal to that exerted by the atmosphere. The region above the mercury column is devoid of air and contains only a trace of mercury vapor.

tube with Hg(l), cover the open end, and invert the tube into a container of Hg(l). Then we reopen the end under the mercury. The mercury level in the tube falls to a certain height and stays there. Something keeps the mercury at a greater height inside the tube than outside. Some tried to ascribe this phenomenon to forces *within* the tube, but Torricelli understood that these forces exist *outside* the tube.

In the open-end tube (Figure 6-4a), the atmosphere exerts the same pressure on the surface of the mercury, both inside and outside the tube. The liquid levels are equal. Inside the closed-end tube (Figure 6-4b), there is no air above the mercury (only a trace of mercury vapor). The atmosphere exerts a force on the surface of the mercury in the outside container that is transmitted through the liquid, holding up a column of mercury. The column exerts a downward pressure that depends on its height and the density of Hg(l). For a particular height, the pressure at the bottom of the mercury column and that of the atmosphere are equal. Thus, the column height is maintained.

The height of mercury in a barometer, called the **barometric pressure**, varies with atmospheric conditions and with altitude. The **standard atmosphere (atm)** is defined as the pressure exerted by a mercury column of exactly 760 mm in height when the density of mercury $= 13.5951$ g/cm³ (0 °C) and the acceleration due to gravity $g = 9.80665$ m s⁻², exactly. This statement relates two useful units of pressure, the standard atmosphere (atm) and the **millimeter of mercury (mmHg)**.

$$1 \text{ atm} = 760 \text{ mmHg} \qquad (6.3)$$

The pressure unit called a **torr** and denoted by the symbol **Torr** honors Torricelli. It is defined as exactly 1/760 of a standard atmosphere. Or, 1 atm $=$ 760 Torr. Thus, we can use the pressure units torr and millimeter of mercury interchangeably.

Mercury is a relatively rare, expensive, and poisonous liquid. Why use it rather than water in a barometer? As we will see in Example 6-1, the extreme height required for a water barometer is a distinct disadvantage.

EXAMPLE 6-1

Comparing Liquid Pressures. What is the height of a column of water that exerts the same pressure as a column of mercury 76.0 cm (760 mm) high?

Solution

We can use equation (6.2) to describe the pressure of the mercury column of known height and the pressure of the water column of unknown height. Then we can set the two pressures equal to each other.

$$\text{pressure of Hg column} = g \cdot h_{\text{Hg}} \cdot d_{\text{Hg}} = g \times 76.0 \text{ cm} \times 13.6 \text{ g/cm}^3$$
$$\text{pressure of H}_2\text{O column} = g \cdot h_{\text{H}_2\text{O}} \cdot d_{\text{H}_2\text{O}} = g \times h_{\text{H}_2\text{O}} \times 1.00 \text{ g/cm}^3$$
$$\cancel{g} \times h_{\text{H}_2\text{O}} \times 1.00 \text{ g/cm}^3 = \cancel{g} \times 76.0 \text{ cm} \times 13.6 \text{ g/cm}^3$$

$$h_{\text{H}_2\text{O}} = 76.0 \text{ cm} \times \frac{13.6 \text{ g/cm}^3}{1.00 \text{ g/cm}^3} = 1.03 \times 10^3 \text{ cm} = 10.3 \text{ m}$$

Thus, we see that the column height is inversely proportional to the liquid density: for a given liquid pressure, the lower the liquid density, the greater the height of the liquid column.

Practice Example A: A barometer is filled with diethylene glycol ($d = 1.118$ g/cm³). The liquid height is found to be 9.25 m. What is the barometric (atmospheric) pressure expressed in millimeters of mercury?

Practice Example B: A barometer is filled with triethylene glycol. The liquid height is found to be 7.39 m when the atmospheric pressure is 757 mmHg. What is the density of triethylene glycol?

▶ Whereas atmospheric pressure can be measured with a mercury barometer less than 1 meter high, a water barometer would have to be as tall as a three-story building.

When you use a drinking straw, you reduce the air pressure above the liquid inside the straw by inhaling. Atmospheric pressure outside the straw then pushes the liquid into your mouth. An old-fashioned hand suction pump for pumping water (once common in rural areas) works on the same principle. The result of Example 6-1 indicates, however, that even if all the air could be removed from inside a pipe, atmospheric pressure outside the pipe could not raise water to a height of more than about 10 m. A suction pump works only for shallow wells. To pump water from a deep well, a mechanical pump is required. The force of the mechanical pump pushing the water upward is greater than the force of the atmosphere pushing the water downward.

Manometers

Although a mercury barometer is indispensable for measuring the pressure of the atmosphere, we can rarely use it alone to measure other gas pressures. The difficulty is in placing a barometer inside the container of gas whose pressure we wish to measure. We can, however, *compare* the gas pressure and barometric pressure, using a **manometer**. Figure 6-5 illustrates the principle of an open-end manometer. As long as the gas pressure being measured and the prevailing atmospheric (barometric) pressure are equal, the heights of the mercury columns in the two arms of the manometer are equal. A difference in height of the two arms signifies a difference between the gas pressure and barometric pressure.

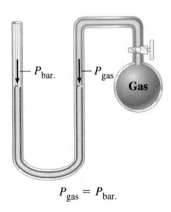

$$P_{gas} = P_{bar.}$$

(a) Gas pressure equal to barometric pressure

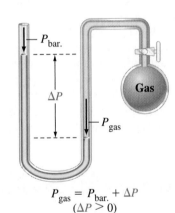

$$P_{gas} = P_{bar.} + \Delta P$$
$$(\Delta P > 0)$$

(b) Gas pressure greater than barometric pressure

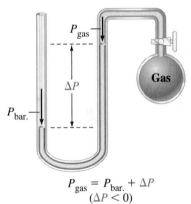

$$P_{gas} = P_{bar.} + \Delta P$$
$$(\Delta P < 0)$$

(c) Gas pressure less than barometric pressure

▲ FIGURE 6-5   **Measurement of gas pressure with an open-end manometer**
The possible relationships between barometric pressure and a gas pressure under measurement are pictured here and described in Example 6-2.

EXAMPLE 6-2

Using a Manometer to Measure Gas Pressure. When the manometer in Figure 6-5c is filled with liquid mercury ($d = 13.6$ g/cm^3), the barometric pressure is 748.2 mmHg, and the difference in mercury levels is 8.6 mmHg. What is the gas pressure P_{gas}?

Solution

Whether the situation corresponds to Figure 6-5b or c, the pressure we seek is $P_{gas} = P_{bar} + \Delta P$. The crucial difference between the two situations is that in Figure 6-5b the measured gas pressure is greater than barometric pressure, and ΔP is positive. In Figure 6-5c, P_{gas} is less than barometric pressure, and ΔP is negative. In the question we have been asked, ΔP is *negative*. (Because all pressures are expressed in millimeters of mercury, the density of mercury does not enter into the calculation.)

$$P_{gas} = P_{bar} + \Delta P = 748.2 \text{ mmHg} - 8.6 \text{ mmHg} = 739.6 \text{ mmHg}$$

Practice Example A: Suppose that the mercury level in Example 6-2 is 7.8 mm higher in the arm open to the atmosphere than in the closed arm. What would be the value of P_{gas}?

Practice Example B: Suppose P_{bar} and P_{gas} are those described in Example 6-2, but the manometer is filled with liquid glycerol ($d = 1.26 \text{ g/cm}^3$) instead of mercury. What would be the difference in the two levels of the liquid?

Units of Pressure: A Summary

Table 6.1 lists several different units used to express pressure. In this text, we will use primarily the first three units (shown in red). When we use millimeter of mercury and torr, we substitute the height of a liquid column for an actual pressure. The units shown in blue are those preferred in the SI system. To understand why their values are all based on the sequence of digits 101325, let us calculate the pressure exerted by a column of mercury that is exactly 0.76 m (760 mm) high. To do this, we use equation (6.2) as follows;

$$P = d \cdot g \cdot h = (1.35951 \times 10^4 \text{ kg m}^{-3})(9.80665 \text{ m s}^{-2})(0.760000 \text{ m})$$
$$= 1.01325 \times 10^5 \text{ kg m}^{-1} \text{ s}^{-2}$$

The unit that arises in this calculation, $\text{kg m}^{-1} \text{ s}^{-2}$, is the natural SI unit for pressure and, as mentioned earlier, is called the pascal (Pa). One standard atmosphere (1 atm) is defined, in SI units, as exactly 1.01325×10^5 Pa or 101.325 kPa. The unit *bar* is one hundred times as large as a kilopascal. The unit *millibar* is commonly used by meteorologists. Although we can generally choose freely among the pressure units in Table 6.1 when doing calculations involving gases, occasionally a situation arises that requires SI units. This is the case in Example 6-3.

TABLE 6.1 Some Common Pressure Units	
Atmosphere (atm)	
Millimeter of mercury (mmHg)	1 atm = 760 mmHg
Torr (Torr)	= 760 Torr
Newton per square meter (N/m²)	= 101,325 N/m²
Pascal (Pa)	= 101,325 Pa
Kilopascal (kPa)	= 101.325 kPa
Bar (bar)	= 1.01325 bar
Millibar (mb)	= 1013.25 mb

EXAMPLE 6-3

Obtaining an SI unit of pressure. The 1.000-kg red cylinder in Figure 6-2 has a diameter of 4.10 cm. What pressure, expressed in Torr, does this cylinder exert on the surface beneath it?

Solution

The basic expression that defines pressure is $P = F/A$. We can readily calculate the circular area of contact between the cylinder and the surface; we know the radius of the cylinder—2.05 cm (half the diameter).

$$A = \pi r^2 = 3.1416 \times \left[2.05 \text{ cm} \times \frac{1 \text{ m}}{100 \text{ cm}} \right]^2 = 1.32 \times 10^{-3} \text{ m}^2$$

The force exerted by the cylinder is its weight, that is, $F = W = m \times g$. The mass is 1.000 kg, and the value of g (the acceleration due to gravity) is 9.81 m s^{-2}. The product of these two terms is the force in newtons.

$$F = m \times g = 1.000 \text{ kg} \times 9.81 \text{ m s}^{-2} = 9.81 \text{ N}$$

When we divide this force by the area (in square meters), we have the pressure in pascals.

$$P = \frac{F}{A} = \frac{9.81 \text{ N}}{1.32 \times 10^{-3} \text{ m}^2} = 7.43 \times 10^3 \text{ Pa}$$

Then, from Table 6.1, we can establish the relationship between the units torr and pascal.

$$P = 7.43 \times 10^3 \text{ Pa} \times \frac{760 \text{ Torr}}{101,325 \text{ Pa}} = 55.7 \text{ Torr}$$

Check: It's difficult to tell at a glance whether this pressure is too high, too low, or about right, but we might reason in this way: A pressure of 55.7 Torr is also 55.7 mmHg. Then we can determine the mass of mercury in a 55.7-mm cylindrical column with 2.05 cm radius. The volume of this column is $V = \pi r^2 h = \pi (2.05 \text{ cm})^2 \times 5.57 \text{ cm} = 73.5 \text{ cm}^3$. The density of mercury is about 13.6 g/cm^3 (page 178). The mass of the mercury column is 73.5 cm^3 $\times$ 13.6 g/cm^3 = 1.00 $\times$ 10^3 g = 1.00 kg, and this is exactly the mass of the steel cylinder.

Practice Example A: The 1.000-kg green cylinder in Figure 6-2 has a diameter of 2.60 cm. What pressure, expressed in torr, does this cylinder exert on the surface beneath it?

Practice Example B: We want to increase the pressure exerted by the 1.000-kg red cylinder in Example 6-3 to 100.0 mb by placing a weight atop it. What must be the mass of this weight? Must the added weight have the same cross-sectional area as the cylinder? Explain.

6-2   The Simple Gas Laws

In this section, we consider relationships involving the pressure, volume, temperature, and amount of a gas. Specifically, we will see how one variable depends on another, as the remaining two are held fixed. Collectively, we refer to these relationships as the simple gas laws. You can use these laws in problem solving, but you will probably prefer the equation we develop in the next section—the ideal gas equation. You may find that the greatest use of the simple gas laws is in solidifying your qualitative understanding of the behavior of gases.

Boyle's Law

In 1662, working with air, Robert Boyle discovered the first of the simple gas laws, now known as **Boyle's law**.

> For a fixed amount of gas at a constant temperature, gas volume is inversely proportional to gas pressure.   (6.4)

Consider the gas in Figure 6-6. It is confined in a cylinder closed off by a freely moving "weightless" piston. The pressure of the gas depends on the total weight placed on top of the piston. This weight (a force), divided by the area of the piston,

P–V Relationships animation

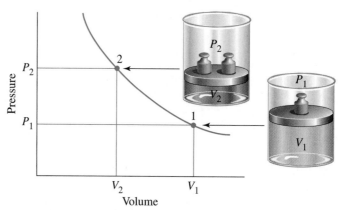

▲ **FIGURE 6-6** **Relationship between gas volume and pressure—Boyle's law**
When the temperature and amount of gas are held constant, gas volume is inversely proportional to the pressure: A doubling of the pressure causes the volume to decrease to one-half its original value.

yields the gas pressure. If the weight on the piston is doubled, the pressure *doubles* and the gas volume decreases to *one-half* its original value. If the pressure of the gas is tripled, the volume decreases to *one third*. On the other hand, if the pressure is reduced to *one half*, the volume *doubles*, and so on. Mathematically, we can express this inverse relationship between pressure and volume as

$$P \propto \frac{1}{V} \quad \text{or} \quad PV = a \text{ (a constant)} \tag{6.5}$$

When we replace the proportionality sign ($\propto$) by an equal sign and a proportionality constant, we see that the product of the pressure and volume of a fixed amount of gas at a given temperature is a constant (a). The value of a depends on the amount of gas and the temperature. The graph in Figure 6-6 is that of $PV = a$. It is called a hyperbola.

EXAMPLE 6-4

Relating Gas Volume and Pressure—Boyle's Law. The volume of the large, irregularly shaped, closed tank pictured in Figure 6-7 is determined as follows. The tank is first evacuated, and then it is connected to a 50.0-L cylinder of compressed nitrogen gas. The gas pressure in the cylinder, originally at 21.5 atm, falls to 1.55 atm after it is connected to the evacuated tank. What is the volume of the tank?

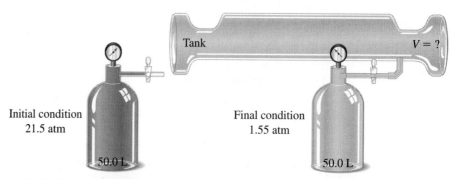

▲ **FIGURE 6-7** **An application of Boyle's law—Example 6-4 illustrated**
The final volume is the volume of the cylinder plus the volume of the tank. The amount of gas and its temperature remain constant when the cylinder is connected to the tank, but the pressure drops from 21.5 atm (the initial pressure) to 1.55 atm.

Solution

One way to solve this problem is by writing equation (6.5) twice: for the initial condition (i) and for the final condition (f).

$$P_i V_i = a = P_f V_f$$

Now solve for the final volume, V_f.

$$V_f = V_i \times \frac{P_i}{P_f} = 50.0 \text{ L} \times \frac{21.5 \text{ atm}}{1.55 \text{ atm}} = 694 \text{ L}$$

Of this volume, 50.0 L is that of the cylinder. The volume of the tank is 694 L − 50.0 L = 644 L.

Check: Reasoning qualitatively from Boyle's law, we expect a large increase in volume as the decrease in pressure allows the gas to expand into the tank. The final gas volume (694 L) should be larger than the initial volume (50.0 L).

Practice Example A: A cylinder like that pictured in Figure 6-6 contains a gas at 5.25 atm pressure. When the gas is allowed to expand to a final volume of 12.5 L, the pressure drops to 1.85 atm. What was the original volume of the gas?

Practice Example B: 1.50 L of gas at 2.25 atm pressure expands to a final volume of 8.10 L. What is the final gas pressure in millimeters of mercury?

Charles experimented with the first hydrogen-filled balloons, much like the one shown here, though smaller. He also invented most of the features of modern ballooning, including the suspended basket and the valve to release excess gas.

▶ Charles's ideas about the effect of temperature on the volume of a gas were probably influenced by his passion for hot air balloons, a popular craze of the late eighteenth century.

Charles's Law

The relationship between the volume of a gas and temperature was discovered by the French physicist Jacques Charles in 1787 and, independently, by Joseph Louis Gay-Lussac, who published it in 1802.

Figure 6-8 pictures a fixed amount of gas confined in a cylinder. The pressure is held constant at 1 atm while the temperature is varied. The volume of gas increases as the temperature is raised and decreases as the temperature is lowered. The relationship is linear (a straight line). Figure 6-8 shows the linear dependence of volume on temperature for three gases at three different initial conditions. One point in common to the three lines is their intersection with the temperature axis. Although they differ at every other temperature, the gas volumes all reach a value of zero at the same temperature. The temperature at which the volume of a *hypothetical* gas becomes zero is the absolute zero of temperature[*]: −273.15 °C on the Celsius scale or 0 K on the **absolute**, or **Kelvin**, **scale**.

$$T(\text{K}) = t(^\circ\text{C}) + 273.15 \tag{6.6}$$

Thus, from a volume of zero at 0 K, the gas volume is *directly* proportional to the temperature. Our statement of **Charles's law** then becomes

> The volume of a fixed amount of gas at constant pressure is directly proportional to the Kelvin (absolute) temperature. (6.7)

In mathematical terms

$$V \propto T \qquad \text{or} \qquad V = bT \text{ (where } b \text{ is a constant)} \tag{6.8}$$

---

[*]All gases condense to liquids or solids before the temperature approaches absolute zero. Also, when we speak of the volume of a gas, we mean the free volume among the gas molecules, not the volume of the molecules themselves. Thus, the gas we refer to here is *hypothetical*. It is a gas whose molecules have mass but no volume and that does not condense to a liquid or solid.

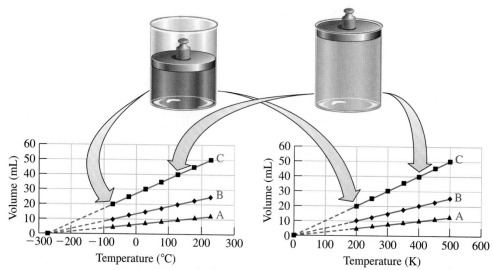

▲ **FIGURE 6-8   Gas volume as a function of temperature**
Volume is plotted against temperature on two different scales—Celsius and Kelvin. Gases
A, *B*, and *C*, are shown at three different initial conditions, all at 1 atm. Gas *C*, for example, starts at 500 K (227 °C) and 50 mL. As the temperature is lowered, the volume decreases as predicted by Charles's law. Thus, at 250 K (−23 °C), one-half the initial
temperature of 500 K, the volume has become 25 mL—one-half of the original 50 mL.
Although the relationship between volume and temperature is linear for both the Celsius
and Kelvin temperature scales, the volume is directly proportional to temperature only on
the Kelvin scale.

The value of the constant *b* depends on the amount of gas and the pressure. It does
not depend on the identity of the gas.

From either expression (6.7) or (6.8), we see that doubling the Kelvin temperature of a gas causes its volume to double. Reducing the Kelvin temperature by one-
half (say, from 300 to 150 K) causes the volume to decrease to one-half, and so on.
It is not difficult to see that the temperature used in equation (6.8) must be on the
Kelvin scale. We certainly do not expect that increasing the temperature of a gas
from 1 to 2 °C or from 1 to 2 °F (in each case, an apparent doubling) would cause
the gas volume to double.

EXAMPLE 6-5

Relating a Gas Volume and Temperature—Charles's Law.   A balloon is inflated to a volume of 2.50 L in a warm living room (24 °C). Then it is taken outside on a very cold winter day (−25 °C). Assume that the quantity of air in the balloon and its pressure both
remain constant. What will be the volume of the balloon when it is outdoors?

Solution

Similar to the method used in Example 6-4, write equation (6.8) twice: for the initial (i)
and the final (f) conditions. In doing this, though, first rearrange the equation and solve
for *b*. That is, $b = V/T$.

$$\frac{V_i}{T_i} = b = \frac{V_f}{T_f}$$

Now, solve the above equation for V_f, but do not forget to change temperatures to the Kelvin scale—that is, $T_i = 24\,°C + 273 = 297\,K$, and $T_f = -25\,°C + 273 = 248\,K$.

$$V_f = V_i \times \frac{T_f}{T_i} = 2.50\,L \times \frac{248\,K}{297\,K} = 2.09\,L$$

Check: Reasoning qualitatively from Charles's law, because the gas temperature is *lowered*, we expect the volume to *decrease*. We expect V_f to be less than V_i.

Practice Example A: A gas at 25 °C and 0.987 atm pressure is confined in a cylinder by a piston. When the cylinder is heated, the gas volume expands from 0.250 L to 1.65 L. What is the new temperature of the gas, assuming the pressure remains constant?

Practice Example B: To what temperature must 50.5 mL of N_2 gas at 33.4 °C and 1.00 atm pressure be heated to produce the same percentage change in volume as observed when 35.0 mL of O_2 gas at 1.00 atm pressure is heated from -13.5 °C to 25.0 °C?

Standard Conditions of Temperature and Pressure

Because gas properties depend on temperature and pressure, at times we find it useful to choose standard conditions of temperature and pressure. This is especially true when we compare gases to one another. The standard temperature for gases is taken to be 0 °C = 273.15 K and standard pressure, 1 atm = 760 mmHg, exactly. **Standard conditions of temperature and pressure** are usually abbreviated as **STP**.

Avogadro's Law

In 1808, Gay-Lussac reported that gases react by volumes in the ratio of small whole numbers. One proposed explanation was that equal volumes of gases at the same temperature and pressure contain equal numbers of atoms. Dalton did not agree with this proposition, however. He viewed the reaction of hydrogen and oxygen to be $H(g) + O(g) \longrightarrow HO(g)$, for which the combining volumes should have been in the ratio $1:1:1$, not the $2:1:2$ ratio that was observed.

In 1811, Amedeo Avogadro resolved this dilemma by proposing not only the "equal volumes–equal numbers" hypothesis, but that molecules of a gas may break up into half-molecules when they react. Using modern terminology, we would say that O_2 molecules split into atoms, which then combine with molecules of H_2 to form H_2O molecules. In this way, the volume of oxygen needed is only one-half that of hydrogen. Avogadro's reasoning is outlined in Figure 6-9.

"Volumi eguali di gas nelle stesse condizioni di temperatura e di pressione contengono lo stesso numero di molecole."

AMEDEO AVOGADRO
POSTE ITALIANE L.25

Amedeo Avogadro (1776–1856)—a scientist ahead of his time
Avogadro's hypothesis and its ramifications were not understood by his contemporaries, but were effectively communicated by Stanislao Cannizzaro (1826–1910) about fifty years later.

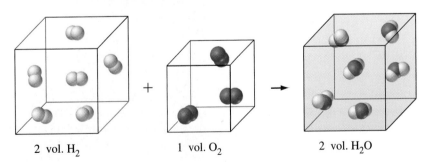

2 vol. H_2 1 vol. O_2 2 vol. H_2O

▲ **FIGURE 6-9   Formation of water—actual observation and Avogadro's hypothesis**
In the reaction $2\,H_2(g) + 1\,O_2(g) \longrightarrow 2\,H_2O(g)$, only one-half as many O_2 molecules are required as are H_2 molecules. If equal volumes of gases contain equal numbers of molecules, this means the volume of $O_2(g)$ is one-half that of $H_2(g)$. The combining ratio by volume is $2:1:2$.

Avogadro's equal volumes–equal numbers hypothesis can be stated in either of two ways.

▶ Avogadro's hypothesis and statements derived from it apply *only to gases*. There is no similar relationship for liquids or solids.

1. Equal volumes of different gases compared at the same temperature and pressure contain equal numbers of molecules.

2. Equal numbers of molecules of different gases compared at the same temperature and pressure occupy equal volumes.

A relationship that follows from **Avogadro's hypothesis**, often called **Avogadro's law**, is as follows.

At a fixed temperature and pressure, the volume of a gas is directly proportional to the amount of gas. (6.9)

If the number of moles of gas (n) is doubled, the volume doubles, and so on. A mathematical statement of this fact is

$$V \propto n \qquad \text{and} \qquad V = c \cdot n$$

At STP, the number of molecules contained in 22.4 L of a gas is 6.02×10^{23}, that is, *1 mol*.

$$1 \text{ mol gas} = 22.4 \text{ L gas (at STP)} \qquad (6.10)$$

Figure 6-10 should help you to visualize 22.4 L of a gas.

▲ **FIGURE 6-10   Molar volume of a gas visualized**
The wooden cube has the same volume as a mole of gas at STP: 22.4 L. By contrast, the volume of the basketball is 7.5 L; the soccer ball, 6.0 L; and the football, 4.4 L.

EXAMPLE 6-6

Using the Molar Volume of a Gas at STP as a Conversion Factor. What is the mass of 1.00 L of cyclopropane gas, C_3H_6 (used as an anesthetic), when measured at STP?

Solution

> If a gas is *not* at STP, relating the amount of gas and its volume is best done with the ideal gas equation. (Section 6-3).

Expression (6.10) allows us to convert directly from volume at STP to number of moles of gas. The conversion from moles of gas to mass of gas requires the molar mass.

$$? \text{ g } C_3H_6 = 1.00 \text{ L} \times \frac{1 \text{ mol } C_3H_6}{22.4 \text{ L}} \times \frac{42.08 \text{ g } C_3H_6}{1 \text{ mol } C_3H_6} = 1.88 \text{ g } C_3H_6$$

Practice Example A: Propane, C_3H_8, is stored as a liquid under pressure for use in backyard barbecue units. The liquid vaporizes to a gas when the pressure is released. A small tank of propane is opened and releases 30.0 L of gas at STP. What mass of gas was released?

Practice Example B: A 128-g piece of solid carbon dioxide (dry ice) sublimes (evaporates without first melting) into $CO_2(g)$. How many liters of gas are formed at STP?

Ideal Gas Behavior activity

6-3 Combining the Gas Laws: The Ideal Gas Equation and the General Gas Equation

Each of the three simple gas laws describes the effect on the gas volume of changes in one variable when the other two are held constant.

1. Boyle's law describes the effect of pressure $V \propto 1/P$.
2. Charles's law describes the effect of temperature $V \propto T$.
3. Avogadro's law describes the effect of the amount of gas $V \propto n$.

We can combine these three laws into a single equation—**the ideal gas equation**—that includes all four gas variables: volume, pressure, temperature, and amount of gas.

The Ideal Gas Equation

From the three simple gas laws, it seems reasonable that the volume of a gas should be *directly* proportional to the amount of gas, *directly* proportional to the Kelvin temperature, and *inversely* proportional to pressure. That is,

$$V \propto \frac{nT}{P} \quad \text{and} \quad V = \frac{RnT}{P}$$

> The ideal gas equation.

$$PV = nRT \tag{6.11}$$

Four Common Values of R

0.082057 L atm mol^{-1} K^{-1}

62.364 L Torr mol^{-1} K^{-1}

8.3145 m^3 Pa mol^{-1} K^{-1}

8.3145 J mol^{-1} K^{-1}

A gas whose behavior conforms to the ideal gas equation is called an **ideal** or **perfect gas**. Before we can apply equation (6.11), we need a value for the constant R, called the **gas constant**. The simplest way to obtain this is to substitute into equation (6.11) the molar volume of an ideal gas at STP. However, the value of R will then depend on what units we use to express the pressure and volume. With a molar volume of 22.4140 L and pressure in atmospheres, we get

$$R = \frac{PV}{nT} = \frac{1 \text{ atm} \times 22.4140 \text{ L}}{1 \text{ mol} \times 273.15 \text{ K}} = 0.082057 \frac{\text{L atm}}{\text{mol K}} = 0.082057 \text{ L atm } mol^{-1} \text{ } K^{-1}$$

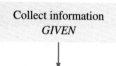

Collect information
GIVEN

↓

Use conversion factors
to convert data to
correct units,
e.g., convert g to mol,
to use R in L atm mol^{-1} K^{-1}.

↓

Identify the variable that
is to be determined.

↓

Rearrange the IDEAL
GAS EQUATION to solve
for the desired variable.
The four possibilities are

$$P = \frac{nRT}{V}, \quad V = \frac{nRT}{P}$$

$$T = \frac{PV}{nR}, \quad n = \frac{PV}{RT}$$

↓

Carry units
throughout
the calculation.

This enables you to check
your calculation.

▲ **Applying the ideal gas
equation**

KEEP IN MIND ▶
that when canceling units, a unit
such as mol^{-1} is the same as
1/mol. Thus, mol × mol^{-1} = 1,
and also K × K^{-1} = 1.

If we use the SI units of m^3 for volume and Pa for pressure, we get

$$R = \frac{PV}{nT} = \frac{101{,}325 \text{ Pa} \times 2.24140 \times 10^{-2} \text{ m}^3}{1 \text{ mol} \times 273.15 \text{ K}} = 8.3145 \text{ m}^3 \text{ Pa mol}^{-1} \text{ K}^{-1}$$

The units m^3 Pa mol^{-1} K^{-1} also have another significance. The pascal has units kg m^{-1} s^{-2}, so the units m^3 Pa become kg m^2 s^{-2}, which is the SI unit of energy—the joule. Thus R also has the value

$$R = 8.3145 \text{ J mol}^{-1} \text{ K}^{-1}.$$

We will use this value of R when we consider the energy involved in gas expansion and compression.

Four common values of the gas constant are listed in the margin on page 187; you will have a chance to use each of these values in Practice Examples accompanying Examples 6-7, 6-8, and some of the other practice examples in this chapter. However, we will use $R = 0.082057$ L atm mol^{-1} K^{-1} in the examples themselves to help you become more familiar with this particular value of R. Because we are rarely justified in carrying all the significant figures that appear in the value of R, we will generally round off its numerical value to 0.08206, or even to 0.0821.

In Examples 6-7 and 6-8, to help organize the data, we first list the known quantities and then the unknown. As you become more familiar with the ideal gas equation, you may be able to bypass this step and substitute directly into the equation. A useful check of the calculated result is to make certain the units cancel properly. This general strategy is illustrated in the diagram in the margin.

EXAMPLE 6-7

Calculating a Gas Volume with the Ideal Gas Equation. What is the volume occupied by 13.7 g Cl$_2$(g) at 45 °C and 745 mmHg?

Solution

$$P = 745 \text{ mmHg} \times \frac{1 \text{ atm}}{760 \text{ mmHg}} = \frac{745}{760} \text{ atm} = 0.980 \text{ atm}$$

$$V = ?$$

$$n = 13.7 \text{ g Cl}_2 \times \frac{1 \text{ mol Cl}_2}{70.91 \text{ g Cl}_2} = 0.193 \text{ mol Cl}_2$$

$$R = 0.08206 \text{ L atm mol}^{-1} \text{ K}^{-1}$$

$$T = 45 \,°C + 273 = 318 \text{ K}$$

Divide both sides of the ideal gas equation by P to solve for V.

$$\frac{P V}{P} = \frac{nRT}{P} \quad \text{and} \quad V = \frac{nRT}{P}$$

$$V = \frac{nRT}{P} = \frac{0.193 \text{ mol} \times 0.08206 \text{ L atm mol}^{-1} \text{ K}^{-1} \times 318 \text{ K}}{0.980 \text{ atm}} = 5.14 \text{ L}$$

Check: Note that all units cancel in the setup above, except for L. This is a unit of volume.

Practice Example A: What is the volume occupied by 20.2 g NH$_3$(g) at −25 °C and 752 mmHg?

Practice Example B: At what temperature will a 13.7-g Cl$_2$ sample exert a pressure of 745 mmHg when confined in a 7.50-L container?

EXAMPLE 6-8

Calculating a Gas Pressure with the Ideal Gas Equation. What is the pressure exerted by 1.00×10^{20} molecules of N_2 in a 305-mL flask at 175 °C?

Solution

Before we can use the ideal gas equation, we must convert from molecules to moles of a gas, from milliliters to liters, and from Celsius to Kelvin. After the ideal gas equation is rearranged to the form, $P = nRT/V$, the appropriate data are entered into the equation and the equation is solved.

$$P = ?$$

$$V = 305 \text{ mL} \times \frac{1 \text{ L}}{1000 \text{ mL}} = 0.305 \text{ L}$$

$$n = 1.00 \times 10^{20} \text{ molecules } N_2 \times \frac{1 \text{ mol } N_2}{6.022 \times 10^{23} \text{ molecules } N_2} = 0.000166 \text{ mol } N_2$$

$$R = 0.08206 \text{ L atm mol}^{-1} \text{ K}^{-1}$$

$$T = 175 \text{ °C} + 273 = 448 \text{ K}$$

$$P = \frac{nRT}{V} = \frac{0.000166 \text{ mol} \times 0.08206 \text{ L atm mol}^{-1} \text{ K}^{-1} \times 448 \text{ K}}{0.305 \text{ L}} = 0.0200 \text{ atm}$$

Check: Again, we see from the cancellation of units above that only the desired unit—a pressure unit—remains.

Practice Example A: How many moles of He(g) are in a 5.00-L storage tank filled with helium at 10.5 atm pressure at 30.0 °C?

Practice Example B: How many molecules of $N_2(g)$ remain in an ultrahigh vacuum chamber of 3.45 m³ volume when the pressure is reduced to 6.67×10^{-7} Pa at 25 °C?

The General Gas Equation

In Examples 6-7 and 6-8, we applied the ideal gas equation to a *single* set of conditions (P, V, n, and T). Sometimes a gas is described under *two* different sets of conditions. Here, we have to apply the ideal gas equation *twice*, to an initial condition and a final condition. That is,

(i) *Initial condition*	(f) *Final condition*
$P_i V_i = n_i R T_i$	$P_f V_f = n_f R T_f$
$R = \dfrac{P_i V_i}{n_i T_i}$	$R = \dfrac{P_f V_f}{n_f T_f}$

The above expressions are equal to each other because each is equal to R.

$$\frac{P_i V_i}{n_i T_i} = \frac{P_f V_f}{n_f T_f} \tag{6.12}$$

Expression (6.12) is called the **general gas equation**. We often apply it in cases in which one or two of the gas properties are held constant, and we can simplify the equation by eliminating these constants. For example, if a constant mass of gas is subject to changes in temperature, pressure, and volume we have

$$\frac{P_i V_i}{T_i} = \frac{P_f V_f}{T_f}$$

(a) Ice bath

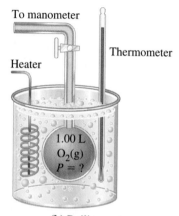

(b) Boiling water

▲ **FIGURE 6-11**

Pressure of a gas as a function of temperature—Example 6-9 visualized

The amount of gas and volume are held constant. **(a)** 1.00 L $O_2(g)$ at STP; **(b)** 1.00 L $O_2(g)$ at 100 °C.

where n_i and n_f have canceled because they are equal (constant mass). This equation is sometimes referred to as the *combined gas law*. In Example 6-9, both volume and mass are constant, and this establishes the simple relationship between gas pressure and temperature known as *Amontons's law*: The pressure of a fixed amount of gas confined to a fixed volume is directly proportional to the Kelvin temperature.

EXAMPLE 6-9

Applying the General Gas Equation. Pictured in Figure 6-11 is a 1.00-L flask of $O_2(g)$, first at STP and then at 100 °C. What is the pressure at 100 °C?

Solution

Identify the terms in the general gas equation that remain constant. Cancel out these terms and solve the remaining equation. In this case, the amount of $O_2(g)$ is constant $(n_i = n_f)$ and the volume is constant $(V_i = V_f)$.

$$\frac{P_i V_i}{n_i T_i} = \frac{P_f V_f}{n_f T_f} \quad \text{and} \quad \frac{P_i}{T_i} = \frac{P_f}{T_f} \quad \text{and} \quad P_f = P_i \times \frac{T_f}{T_i}$$

Because P_i is standard pressure = 1.00 atm.

$$P_f = 1.00 \text{ atm} \times \frac{(100 + 273) \text{ K}}{273 \text{ K}} = 1.00 \text{ atm} \times \frac{373 \text{ K}}{273 \text{ K}} = 1.37 \text{ atm}$$

Check: We can base our check on a qualitative, intuitive understanding about what happens when a gas is heated in a closed container. Its pressure increases (possibly to the extent that the container bursts). If, by error, we had used the ratio of temperatures 273 K/373 K, we would have obtained a final pressure smaller than 1.00 atm—an impossible result.

Practice Example A: A 1.00-mL sample of $N_2(g)$ at 36.2 °C and 2.14 atm is heated to 37.8 °C, and the pressure changed to 1.02 atm. What volume does the gas occupy at this final temperature and pressure?

Practice Example B: Suppose that in Figure 6-11 we wish the pressure to remain at 1.00 atm when the $O_2(g)$ is heated to 100 °C. What mass of $O_2(g)$ must we release from the flask?

Using the Gas Laws

When confronted with a problem involving gases, students sometimes wonder which gas equation to use. Gas law problems can often be thought of in more than one way. If your preference is for a brief verbal statement, you can generally proceed in this way. When a problem involves a comparison of two gases, or two states (initial and final) of a single gas, use the general gas equation (6.12) after eliminating any term (n, P, T, V) that remains constant. Otherwise, use the ideal gas equation (6.11). If you prefer a diagrammatic approach to a problem-solving strategy, you should find Figure 6-12 helpful. The key step in this diagram is to determine whether the amount of gas (in mass or moles) is given in the problem or asked for in the answer to the problem.

6-4 Applications of the Ideal Gas Equation

Although the ideal gas equation can always be used as it was presented in equation (6.11), it is useful to recast it into slightly different forms for some applications. We will consider two such applications in this section: determination of molar masses and gas densities.

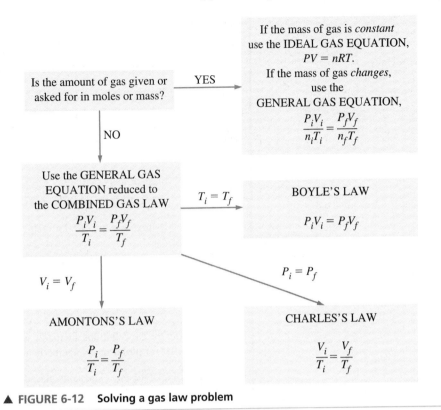

▲ FIGURE 6-12   **Solving a gas law problem**

Molar Mass Determination

If we know the volume occupied by a gas at a fixed temperature and pressure, we can solve the ideal gas equation for the amount of gas in moles. Because the number of moles of gas (n) is equal to the mass of gas (m) divided by the molar mass (M), if we know the mass and number of moles of gas, we can solve the expression $n = m/M$ for the molar mass, M. An alternative is to make the substitution $n = m/M$ directly into the ideal gas equation.

$$PV = \frac{mRT}{M} \qquad (6.13)$$

We apply equation (6.13) to a molar mass determination in Example 6-10, but note that the equation can also be used when the molar mass of a gas is known and the mass of a particular sample of the gas is sought.

EXAMPLE 6-10

Determining a Molar Mass with the Ideal Gas Equation. Propylene is an important commercial chemical (about ninth in the amount produced among manufactured chemicals). It is used in the synthesis of other organic chemicals and in plastics production (polypropylene). A glass vessel weighs 40.1305 g when clean, dry, and evacuated; it weighs 138.2410 g when filled with water at 25.0 °C (density of water = 0.9970 g/mL) and 40.2959 g when filled with propylene gas at 740.3 mmHg and 24.0 °C. What is the molar mass of propylene?

Solution

Our first task is to determine the volume of the glass vessel and hence the volume of the gas.

$$\text{mass of water to fill vessel} = 138.2410 \text{ g} - 40.1305 \text{ g} = 98.1105 \text{ g}$$

$$\text{volume of water (volume of vessel)} = 98.1105 \text{ g H}_2\text{O} \times \frac{1 \text{ mL H}_2\text{O}}{0.9970 \text{ g H}_2\text{O}}$$

$$= 98.41 \text{ mL} = 0.09841 \text{ L}$$

Now we can list values for the other variables, starting with the mass of gas, which we get by difference.

$$\text{mass of gas} = 40.2959 \text{ g} - 40.1305 \text{ g} = 0.1654 \text{ g}$$
$$\text{temperature} = 24.0\,°\text{C} + 273.15 = 297.2 \text{ K}$$
$$\text{pressure} = 740.3 \text{ mmHg} \times \frac{1 \text{ atm}}{760 \text{ mmHg}} = 0.9741 \text{ atm}$$

Finally, we substitute into a rearranged form of equation (6.13).

$$M = \frac{mRT}{PV} = \frac{0.1654 \text{ g} \times 0.08206 \text{ L atm mol}^{-1} \text{ K}^{-1} \times 297.2 \text{ K}}{0.9741 \text{ atm} \times 0.09841 \text{ L}}$$
$$= 42.08 \text{ g/mol}$$

Check: After the cancellations are made, we are left with g and mol^{-1}. The unit g mol^{-1} or g/mol is that for molar mass, again, the quantity we are seeking.

Practice Example A: The same glass vessel used in Example 6-10 is filled with an unknown gas at 772 mmHg and 22.4 °C. The vessel filled with gas weighs 40.4868 g. What is the molar mass of the gas?

Practice Example B: A 1.27-g sample of an oxide of nitrogen, believed to be either NO or N_2O, occupies a volume of 1.07 L at 25 °C and 737 mmHg. Which oxide is it?

Suppose that we want to determine the formula of an unknown hydrocarbon. With combustion analysis we can establish the mass percent composition, and from this, we can determine the empirical formula. The method of Example 6-10 gives us a molar mass (g/mol), which is numerically equal to the molecular mass (u). This is all the information we need to establish the true molecular formula of the hydrocarbon (see Exercise 90).

Gas Densities

To determine the density of a gas, we can start with the density equation, $d = m/V$. Then we can express the mass of gas as the product of the number of moles of gas and the molar mass: $m = n \times M$. This leads to

$$d = \frac{m}{V} = \frac{n \times M}{V} = \frac{n}{V} \times M$$

Now, with the ideal gas equation, we can replace n/V by its equivalent, P/RT, to obtain

$$d = \frac{m}{V} = \frac{MP}{RT} \tag{6.14}$$

KEEP IN MIND
that gas densities are typically much smaller than those of liquids and solids. Gas densities are usually expressed in grams per *liter* rather than grams per *milliliter*. ▶

The density of a gas at STP can easily be calculated by dividing its molar mass by the molar volume (22.4 L/mol). For $O_2(g)$ at STP, for example, the density is $32.0 \text{ g}/22.4 \text{ L} = 1.43$ g/L. Under other conditions of temperature and pressure, we can use equation (6.14).

EXAMPLE 6-11

Using the Ideal Gas Equation to Calculate a Gas Density. What is the density of oxygen gas (O_2) at 298 K and 0.987 atm?

Solution

We can get the terms for the right side of equation (6.14) easily. Density (m/V) is the left side of the equation.

$$d = \frac{m}{V} = \frac{MP}{RT} = \frac{32.00 \text{ g mol}^{-1} \times 0.987 \text{ atm}}{0.08206 \text{ L atm mol}^{-1} \text{ K}^{-1} \times 298 \text{ K}} = 1.29 \text{ g/L}$$

Check: The calculated density (1.29 g/L) should be somewhat less than the density of $O_2(g)$ at STP (1.43 g/L). The gas pressure is somewhat less than 1.00 atm, and the temperature is somewhat greater than 273 K. From equation (6.14), note that decreasing pressure and increasing temperature both cause density to decrease.

Practice Example A: What is the density of helium gas at 298 K and 0.987 atm? Based on your answer, explain why we can say that helium is "lighter than air."

Practice Example B: At what temperature will the density of $O_2(g)$ be 1.00 g/L if the pressure is kept at 745 mmHg?

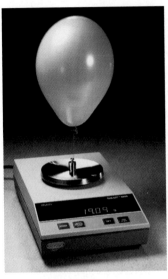

▲ **FIGURE 6-13**
The helium-filled balloon exerts a lifting force on the 20.00-g weight, so that the balloon and the weight together weigh only 19.09 g.

Gas densities differ from solid and liquid densities in two important ways.

1. Gas densities depend strongly on pressure and temperature, increasing as the gas pressure increases and decreasing as the temperature increases. Densities of liquids and solids also depend somewhat on temperature, but they depend far less on pressure.

2. The density of a gas is directly proportional to its molar mass. No simple relationship exists between density and molar mass for liquids and solids.

An important application of gas densities is in establishing conditions for lighter-than-air balloons. A gas-filled balloon will rise in the atmosphere only if the density of the gas is less than that of air. Because gas densities are directly proportional to molar masses, the lower the molar mass of the gas, the greater its lifting power. The lowest molar mass is that of hydrogen, but hydrogen is flammable and forms explosive mixtures with air. The explosion of the dirigible *Hindenburg* in 1937 spelled the end to transoceanic travel by hydrogen-filled airships. Now, airships such as the Goodyear blimps use helium, which has a molar mass only twice that of hydrogen (Figure 6-13). Hydrogen is still used for weather and other observational balloons.

Another alternative is to fill a balloon with *hot* air. Equation (6.14) indicates that the density of a gas is *inversely* proportional to temperature. Hot air is less dense than cold air. However, because the density of air decreases rapidly with altitude, there is a limit to how high a hot air balloon or any gas-filled balloon can rise.

6-5   Gases in Chemical Reactions

Reactions involving gases as reactants or products (or both) are no strangers to us. Now, though, we have a new tool to apply to reaction stoichiometry calculations: the ideal gas equation. Specifically, we can now handle information about gases in terms of volumes, temperatures, and pressures, as well as by mass and amount in moles.

In many cases, the best approach is to (a) use stoichiometric factors to relate the amount of a gas to amounts of other reactants or products, and (b) use the ideal gas equation to relate the amount of gas to volume, temperature, and pressure. In Example 6-12, we determine the number of moles of $N_2(g)$ from the stoichiometry of a reaction. Then we use the ideal gas equation to determine a gas volume.

Air Bags animation

EXAMPLE 6-12

Using the Ideal Gas Equation in Reaction Stoichiometry Calculations. The decomposition of sodium azide, NaN_3, at high temperatures produces $N_2(g)$. Together with the necessary devices to initiate the reaction and trap the sodium metal formed, this reaction is used in air-bag safety systems (see Focus On, page 210). What volume of $N_2(g)$, measured at 735 mmHg and 26 °C, is produced when 70.0 g NaN_3 is decomposed?

$$2\,NaN_3(s) \xrightarrow{\Delta} 2\,Na(l) + 3\,N_2(g)$$

Solution

$$? \text{ mol } N_2 = 70.0 \text{ g } NaN_3 \times \frac{1 \text{ mol } NaN_3}{65.01 \text{ g } NaN_3} \times \frac{3 \text{ mol } N_2}{2 \text{ mol } NaN_3} = 1.62 \text{ mol } N_2$$

$$P = 735 \text{ mmHg} \times \frac{1 \text{ atm}}{760 \text{ mmHg}} = 0.967 \text{ atm}$$

$$V = ?$$

$$n = 1.62 \text{ mol}$$

$$R = 0.08206 \text{ L atm mol}^{-1} \text{ K}^{-1}$$

$$T = 26\,°C + 273 = 299 \text{ K}$$

$$V = \frac{nRT}{P} = \frac{1.62 \text{ mol} \times 0.08206 \text{ L atm mol}^{-1} \text{ K}^{-1} \times 299 \text{ K}}{0.967 \text{ atm}} = 41.1 \text{ L}$$

Check: 70.0 g NaN_3 is slightly more than one mole ($M \approx 65$ g/mol). From this amount of NaN_3 we should expect about 1.5 mol $N_2(g)$. At STP, 1.5 mol $N_2(g)$ would occupy a volume of $1.5 \times 22.4 = 33.6$ L. The gas collected is at a *lower* pressure and a *higher* temperature than STP; it should have a volume somewhat greater than 33.6 L.

Practice Example A: How many grams of NaN_3 are needed to produce 20.0 L of $N_2(g)$ at 30.0 °C and 776 mmHg?

Practice Example B: How many grams of Na(l) are produced *per liter* of $N_2(g)$ formed in the decomposition of sodium azide if the gas is collected at 25 °C and 751 Torr?

Law of Combining Volumes

If the reactants and products involved in a stoichiometric calculation are gases, we can, at times, use a particularly simple approach. Consider this reaction.

$$2\,NO(g) + O_2(g) \longrightarrow 2\,NO_2(g)$$

$$2 \text{ mol } NO(g) + 1 \text{ mol } O_2(g) \longrightarrow 2 \text{ mol } NO_2(g)$$

Suppose the gases are compared at the same *T* and *P*. Under these conditions, one mole of gas occupies a particular volume, call it *V* liters; 2 mol gas, 2*V* liters; and so on.

$$2V \text{ L } NO(g) + V \text{ L } O_2(g) \longrightarrow 2V \text{ L } NO_2(g)$$

Now divide each coefficient by *V*.

$$2 \text{ L } NO(g) + 1 \text{ L } O_2(g) \longrightarrow 2 \text{ L } NO_2(g)$$

When we describe the chemical reaction in this way, we can write conversion factors based on the statements:

$$2 \text{ L } NO_2(g) \text{ is equivalent to } (\Leftrightarrow) \ 2 \text{ L } NO(g)$$

$$2 \text{ L } NO_2(g) \Leftrightarrow 1 \text{ L } O_2(g); \ 2 \text{ L } NO(g) \Leftrightarrow 1 \text{ L } O_2(g)$$

What we have just done is to develop, in modern terms, Gay-Lussac's **law of combining volumes**. We previewed this law on page 185 by suggesting that the volumes of gases involved in a reaction are in the ratio of small whole numbers. We apply this law in Example 6-13.

EXAMPLE 6-13

Applying the Law of Combining Volumes. Zinc blende, ZnS, is the most important zinc ore. Roasting (strong heating) of ZnS is the first step in the commercial production of zinc.

$$2 \, ZnS(s) + 3 \, O_2(g) \xrightarrow{\Delta} 2 \, ZnO(s) + 2 \, SO_2(g)$$

What volume of $SO_2(g)$ forms *per liter* of $O_2(g)$ consumed? Both gases are measured at 25 °C and 745 mmHg.

Solution

The reactant and product being compared are both gases *and both are at the same temperature and pressure*, so we can derive a ratio of combining volumes from the balanced equation (in blue) and use it as follows.

$$? \, L \, SO_2(g) = 1.00 \, L \, O_2(g) \times \frac{2 \, L \, SO_2(g)}{3 \, L \, O_2(g)} = 0.667 \, L \, SO_2(g)$$

Practice Example A: The first step in making nitric acid is to convert ammonia to nitrogen monoxide. This is done under conditions of high temperature and in the presence of a platinum catalyst. What volume of $O_2(g)$ is consumed per liter of NO(g) formed?

$$4 \, NH_3(g) + 5 \, O_2(g) \xrightarrow[850 \, °C]{Pt} 4 \, NO(g) + 6 \, H_2O(g)$$

Practice Example B: If all gases are measured at the same temperature and pressure, what volume of $NH_3(g)$ is produced when 225 L $H_2(g)$ is consumed in the reaction $N_2(g) + H_2(g) \longrightarrow NH_3(g)$ (not balanced)?

Are You Wondering ...

How to proceed if the gases involved in a stoichiometric calculation are *not* at the same temperature and pressure?

If the gases are not at *identical* temperatures and pressures, you can't use the law of combining volumes. In such cases, the best approach is to convert information about the gases to a mole basis. Then use a mole ratio from the chemical equation, as in Example 6-12. Note also that when the law of combining volumes does apply, you don't need to know the temperature and pressure values. Thus, in Example 6-13 we didn't use the conditions, 25 °C and 745 mmHg.

6-6 Mixtures of Gases

The simple gas laws, such as Boyle's and Charles's laws, were based on the behavior of air—a mixture of gases. So, the simple gas laws and the ideal gas equation apply to a *mixture* of nonreactive gases as well as to individual gases. Where possible, the simplest approach to working with gaseous mixtures is just to use for the value of n the *total* number of moles of the gaseous mixture (n_{tot}).

EXAMPLE 6-14

Applying the Ideal Gas Equation to a Mixture of Gases. What is the pressure exerted by a mixture of 1.0 g H_2 and 5.00 g He when the mixture is confined to a volume of 5.0 L at 20 °C?

Solution

$$n_{tot} = \left(1.0 \text{ g H}_2 \times \frac{1 \text{ mol H}_2}{2.02 \text{ g H}_2}\right) + \left(5.00 \text{ g He} \times \frac{1 \text{ mol He}}{4.003 \text{ g He}}\right)$$

$$= 0.50 \text{ mol H}_2 + 1.25 \text{ mol He} = 1.75 \text{ mol gas}$$

$$P = \frac{n_{tot}RT}{V} \tag{6.15}$$

$$= \frac{1.75 \text{ mol} \times 0.0821 \text{ L atm mol}^{-1} \text{ K}^{-1} \times 293 \text{ K}}{5.0 \text{ L}} = 8.4 \text{ atm}$$

Practice Example A: If we add 12.5 g Ne to the mixture of gases described in Example 6-14 and then raise the temperature to 55 °C, what will be the total gas pressure? (*Hint:* What is the new number of moles of gas? What effect does raising the temperature have on the pressure of a gas at constant volume?)

Practice Example B: 2.0 L of $O_2(g)$ and 8.0 L of $N_2(g)$, each at STP, are mixed together. The nonreactive gaseous mixture is compressed to occupy 2.0 L at 298 K. What is the pressure exerted by this mixture?

John Dalton made an important contribution to the study of gaseous mixtures. He proposed that in a mixture, each gas expands to fill the container and exerts the same pressure (called its **partial pressure**) that it would if it were alone in the container. **Dalton's law of partial pressures** states that the total pressure of a mixture of gases is the sum of the partial pressures of the components of the mixture (Figure 6-14). For a mixture of gases, A, B, ... ,

KEEP IN MIND ►
that when this expression is used, $V_A = V_B = \ldots = V_{tot}$

$$P_{tot} = P_A + P_B + \ldots \tag{6.16}$$

In a gaseous mixture of n_A moles of A, n_B moles of B, and so on, the volume each gas would individually occupy at a pressure equal to P_{tot} is

$$V_A = n_A RT/P_{tot}; V_B = n_B RT/P_{tot}; \text{ and so on.}$$

KEEP IN MIND
that when using this expression, $P_A = P_B = \ldots = P_{tot}$ ►

The total volume of the gaseous mixture is

$$V_{tot} = V_A + V_B \ldots$$

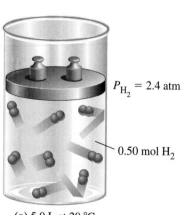

(a) 5.0 L at 20 °C

$P_{H_2} = 2.4$ atm

0.50 mol H_2

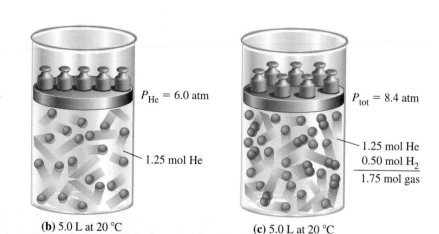

(b) 5.0 L at 20 °C

$P_{He} = 6.0$ atm

1.25 mol He

(c) 5.0 L at 20 °C

$P_{tot} = 8.4$ atm

1.25 mol He
0.50 mol H_2
1.75 mol gas

▲ **FIGURE 6-14   Dalton's law of partial pressures illustrated**
The pressure of each gas is proportional to the number of moles of gas. The total pressure is the sum of the partial pressures of the individual gases.

and the commonly used expression *percent by volume* is

$$\text{volume } \%A = \frac{V_A}{V_{tot}} \times 100\%; \text{ volume } \%B = \frac{V_B}{V_{tot}} \times 100\%; \text{ and so on.}$$

We can derive a particularly useful expression from the following ratios,

$$\frac{P_A}{P_{tot}} = \frac{n_A(RT/V_{tot})}{n_{tot}(RT/V_{tot})} = \frac{n_A}{n_{tot}} \quad \text{and} \quad \frac{V_A}{V_{tot}} = \frac{n_A(RT/P_{tot})}{n_{tot}(RT/P_{tot})} = \frac{n_A}{n_{tot}}$$

which means that

$$\frac{n_A}{n_{tot}} = \frac{P_A}{P_{tot}} = \frac{V_A}{V_{tot}} = x_A \qquad (6.17)$$

The term n_A/n_{tot} is given a special name, the mole fraction of A, x_A. The **mole fraction** of a component in a mixture is the fraction of all the molecules in the mixture contributed by that component. The sum of all the mole fractions in a mixture is 1.

As illustrated in Example 6-15, we can often think about mixtures of gases in more than one way.

EXAMPLE 6-15

Calculating the Partial Pressures in a Gaseous Mixture. What are the partial pressures of H_2 and He in the gaseous mixture described in Example 6-14?

Solution

One approach involves a direct application of Dalton's law. That is, calculate the pressure each gas would exert if it were alone in the container.

$$P_{H_2} = \frac{n_{H_2} \cdot RT}{V} = \frac{0.50 \text{ mol} \times 0.0821 \text{ L atm mol}^{-1} \text{ K}^{-1} \times 293 \text{ K}}{5.0 \text{ L}} = 2.4 \text{ atm}$$

$$P_{He} = \frac{n_{He} \cdot RT}{V} = \frac{1.25 \text{ mol} \times 0.0821 \text{ L atm mol}^{-1} \text{ K}^{-1} \times 293 \text{ K}}{5.0 \text{ L}} = 6.0 \text{ atm}$$

Expression (6.17) gives us a simpler way to answer the question because we already know the number of moles of each gas and the total pressure from Example 6-14 ($P_{tot} = 8.4$ atm).

$$P_{H_2} = \frac{n_{H_2}}{n_{tot}} \times P_{tot} = \frac{0.50}{1.75} \times 8.4 \text{ atm} = 2.4 \text{ atm}$$

$$P_{He} = \frac{n_{He}}{n_{tot}} \times P_{tot} = \frac{1.25}{1.75} \times 8.4 \text{ atm} = 6.0 \text{ atm}$$

Check: An effective way of checking an answer is to obtain the same answer when the problem is done in different ways, as was the case here.

Practice Example A: A mixture of 0.197 mol $CO_2(g)$ and 0.00278 mol $H_2O(g)$ is held at 30.0 °C and 2.50 atm. What is the partial pressure of each gas?

Practice Example B: The percent composition of air by volume is 78.08% N_2, 20.95% O_2, 0.93% Ar, and 0.036% CO_2. What are the partial pressures of these four gases in a sample of air at a barometric pressure of 748 mmHg?

▶ Some early experimenters used mercury in the pneumatic trough, so as to be able to collect gases soluble in water.

The device pictured in Figure 6-15, a *pneumatic trough*, played a crucial role in isolating gases in the early days of chemistry. The method works, of course, only for gases that are insoluble in and do not react with the liquid being displaced. But many important gases meet these criteria. For example, H_2, O_2, and N_2 are essentially insoluble in and unreactive with water.

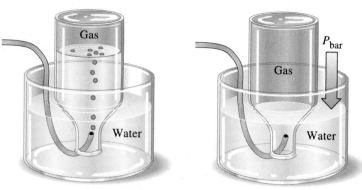

▲ FIGURE 6-15   **Collecting a gas over water**
The bottle is filled with water and its open end is held below the water level in the container. Gas from a gas-generating apparatus is directed into the bottle. As gas accumulates in the bottle, water is displaced from the bottle into the container. To make the total gas pressure in the bottle equal to barometric pressure, the position of the bottle must be adjusted so that the water levels inside and outside the bottle are the same.

A gas collected in a pneumatic trough filled with water is said to be *collected over water* and is "wet." It is a mixture of two gases—the desired gas and water vapor. The gas being collected expands to fill the container and exerts its partial pressure, P_{gas}. Water vapor, formed by the evaporation of liquid water, also fills the container and exerts a partial pressure, P_{H_2O}. The pressure of the water vapor depends only on the temperature of the water. Water vapor pressure data are readily available in tabulated form (see the table in the margin).

According to Dalton's law, the total pressure of the wet gas is the sum of the two partial pressures. The total pressure can be made equal to the prevailing pressure of the atmosphere (barometric pressure).

$$P_{tot} = P_{bar} = P_{gas} + P_{H_2O} \qquad \text{or} \qquad P_{gas} = P_{bar} - P_{H_2O}$$

Once P_{gas} has been established, it can be used in stoichiometric calculations as illustrated in Example 6-16.

▶ **Vapor Pressure of Water at Various Temperatures**

Temperature (°C)	Vapor Pressure (mmHg)
15.0	12.79
17.0	14.53
19.0	16.48
21.0	18.65
23.0	21.07
25.0	23.76
30.0	31.82
50.0	92.51

EXAMPLE 6-16

Collecting a Gas over a Liquid (Water). In the following reaction, 81.2 mL of $O_2(g)$ is collected over water at 23 °C and barometric pressure 751 mmHg. What must have been the mass of $Ag_2O(s)$ decomposed? (The vapor pressure of water at 23 °C is 21.1 mmHg.)

$$2\ Ag_2O(s) \longrightarrow 4\ Ag(s) + O_2(g)$$

Solution

First, we need to calculate the number of moles of $O_2(g)$ produced. We can do this with the ideal gas equation. The key to this calculation is to see that the gas collected is wet, that is, a *mixture* of $O_2(g)$ and water vapor.

$$P_{O_2} = P_{bar} - P_{H_2O} = 751\ \text{mmHg} - 21.1\ \text{mmHg} = 730\ \text{mmHg}$$

$$P_{O_2} = 730\ \text{mmHg} \times \frac{1\ \text{atm}}{760\ \text{mmHg}} = 0.961\ \text{atm}$$

$$V = 81.2\ \text{mL} = 0.0812\ \text{L}$$

$$n = ?$$

$$R = 0.08206\ \text{L atm mol}^{-1}\ \text{K}^{-1}$$

$$T = 23\,°C + 273 = 296\ K$$

$$n = \frac{PV}{RT} = \frac{0.961\ \text{atm} \times 0.0812\ L}{0.08206\ L\ \text{atm mol}^{-1}\ K^{-1} \times 296\ K} = 0.00321\ \text{mol}$$

From the chemical equation we obtain a factor to convert from moles of O_2 to moles of Ag_2O. The molar mass of Ag_2O provides the final factor.

$$? \text{ g } Ag_2O = 0.00321\ \text{mol } O_2 \times \frac{2\ \text{mol } Ag_2O}{1\ \text{mol } O_2} \times \frac{231.7\ \text{g } Ag_2O}{1\ \text{mol } Ag_2O}$$

$$= 1.49\ \text{g } Ag_2O$$

Practice Example A: The reaction of aluminum with hydrochloric acid produces hydrogen gas.

$$2\ Al(s) + 6\ HCl(aq) \longrightarrow 2\ AlCl_3(aq) + 3\ H_2(g)$$

If 35.5 mL of $H_2(g)$ is collected over water at 26 °C and a barometric pressure of 755 mmHg, how many moles of HCl must have been consumed? (The vapor pressure of water at 26 °C is 25.2 mmHg.)

Practice Example B: An 8.07-g sample that is 88.3% Ag_2O by mass decomposes into solid silver and $O_2(g)$. The $O_2(g)$ is collected over water at 25 °C and 749.2 mmHg barometric pressure. The vapor pressure of water at 25 °C is 23.8 mmHg. What is the volume of gas collected?

6-7 Kinetic-Molecular Theory of Gases

Let us apply some terminology that we introduced in Section 1-2 on the scientific method: The simple gas laws and the ideal gas equation are used to predict gas behavior. They are *natural laws*. To explain the gas laws, we need a *theory*. One theory developed during the mid-nineteenth century is called the **kinetic-molecular theory of gases**. It is based on the *model* illustrated in Figure 6-16 and outlined as follows.

1. A gas is composed of a very large number of extremely small particles (molecules or, in some cases, atoms) in constant, random, straight-line motion.

2. Molecules of a gas are separated by great distances. The gas is mostly empty space. (The molecules are treated as so-called point masses, as though they have mass but no volume.)

3. Molecules collide with one another and with the walls of their container. Because these collisions occur very rapidly, however, most of the time molecules are not engaged in collisions.

4. There are assumed to be no forces between molecules except very briefly during collisions. That is, each molecule acts independently of all the others and is unaffected by their presence, except during collisions.

5. Individual molecules may gain or lose energy as a result of collisions. In a collection of molecules at constant temperature, however, *the total energy remains constant*.

Because pressure is a force per unit area, the key to deriving a kinetic-molecular theory equation for pressure is in assessing the forces associated with molecular collisions. These depend on several factors.

1. The amount of *translational kinetic energy* of the molecules. Translational kinetic energy is energy possessed by objects moving through space. Like speeding bullets, gas molecules have this energy of motion. The translational kinetic

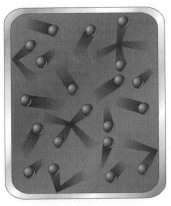

▲ **FIGURE 6-16**
Visualizing molecular motion
Molecules of a gas are in constant motion and undergo collisions with one another and with the container wall.

energy of a molecule is represented as e_k and has the value, $e_k = \dfrac{1}{2}mu^2$, where m is the mass of the molecule and u is its speed. The faster the molecules move, the greater their translational kinetic energies and the greater the forces exerted between molecules during collisions.

2. The *frequency* of molecular collisions—the number of collisions per second. The higher this frequency, the greater the total force of these collisions. Collision frequency increases with the number of molecules per unit volume and with molecular speeds.

$$\text{Collision frequency} \propto (\text{molecular speed}) \times (\text{molecules per unit volume})$$

$$\text{Collision frequency} \propto (u) \times (N/V)$$

3. When a molecule hits the wall of a vessel, momentum is transferred as the molecule reverses direction. This momentum transfer is called an *impulse*. The magnitude of the impulse is directly proportional to the mass of a molecule and its velocity.

$$\text{Impulse (momentum transfer)} \propto (\text{molecular speed}) \times (\text{mass of particle})$$

$$\text{Impulse} \propto (mu)$$

The pressure of a gas (P) is the product of the impulse and collision frequency. Thus, the complete proportionality for factors that affect pressure is

$$P \propto (mu) \times (u) \times (N/V) \propto (N/V)mu^2$$

At any instant, however, not all molecules are moving at the same speed. The pressure depends on the average of all molecules with different speeds, so we must use the *average* of the *squares* of their speeds in the expression for pressure. The average of the squares of a group of speeds is called the *mean-square speed*, $\overline{u^2}$. To better understand the concept of the average of the squares of speeds, consider five molecules with speeds 400, 450, 525, 585, and 600 m/s. We find the average of the squares of these speeds by squaring the speeds, adding the squares, and dividing by the number of particles, in this case 5.

▶ The bar over a quantity means that the quantity can have a range of values and that the *average* value is intended.

$$\overline{u^2} = \frac{(400 \text{ m/s})^2 + (450 \text{ m/s})^2 + (525 \text{ m/s})^2 + (585 \text{ m/s})^2 + (600 \text{ m/s})^2}{5}$$

$$= 2.68 \times 10^5 \text{ m}^2/\text{s}^2$$

Thus the proportionality expression for pressure becomes

$$P \propto \frac{N}{V} m\overline{u^2}$$

KEEP IN MIND ▶
that mean-*square* speed is different from *square* mean speed. The mean or average speed is $(400 + 450 + 525 + 585 + 600)/5 = 512$ m/s. The square of this average is $(512 \text{ m/s})^2 = 2.62 \times 10^5 \text{ m}^2/\text{s}^2$.

A final factor is that the direction in which every molecule moves has a component in each of the three perpendicular dimensions (x, y, and z). Our calculation of pressure must be based on motion in just one of these dimensions, leading to a factor of $1/3$, which in fact is the proportionality constant in the above expression. Thus, when all factors are properly considered, the result is the following expression for the pressure of a gas, the basic equation of the kinetic-molecular theory

$$P = \frac{1}{3}\frac{N}{V}m\overline{u^2} \tag{6.18}$$

Equation (6.18) leads to some interesting results, as we see next.

Boltzmann Distribution simulation

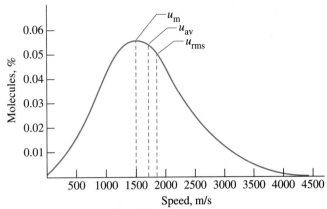

▲ **FIGURE 6-17**   **Distribution of molecular speeds—hydrogen gas at 0 °C**
The percentages of molecules with a certain speed are plotted as a function of the speed. Three different speeds are noted on the graph and discussed in the text.

Distribution of Molecular Speeds

The averaging of molecular speeds referred to in establishing equation (6.18) is required because molecules of a gas do not all have the same speed. There is a distribution of speeds, as shown in Figure 6-17. Also, three characteristic speeds are identified in Figure 6-17. More molecules have the speed u_m, the most probable, or modal, speed, than any other single speed. The average speed, u_{av} (or $\bar{u}$), is the simple average. The **root-mean-square speed**, u_{rms}, is the *square root* of the *average* of the *squares* of the speeds of all the molecules in a sample.

▶ If the postulates of the kinetic-molecular theory hold, a gas is automatically an ideal gas.

We can derive a useful equation for u_{rms} by combining equation (6.18) with the ideal gas equation. Consider 1 mol of an ideal gas. The number of molecules present is $N = N_A$ (Avogadro's number), and the ideal gas equation becomes $PV = RT$ (that is, $n = 1$ in the equation $PV = nRT$). First, replace N by N_A, and multiply both sides of equation (6.18) by V. This leads to

$$PV = \frac{1}{3} N_A m \overline{u^2} \tag{6.19}$$

Next, replace PV by RT and multiply both sides of the equation by 3.

$$3RT = N_A m \overline{u^2}$$

Now, note that the product $N_A m$ represents the mass of 1 mol of molecules, the molar mass, M.

$$3RT = M \overline{u^2}$$

Finally, solve for $\overline{u^2}$ and then, $\sqrt{\overline{u^2}}$ which is u_{rms}.

$$u_{rms} = \sqrt{\overline{u^2}} = \sqrt{\frac{3RT}{M}} \tag{6.20}$$

Equation (6.20) shows that u_{rms} of a gas is directly proportional to the square root of its Kelvin temperature and inversely proportional to the square root of its molar mass. This means that lighter gas molecules have greater speeds than heavier ones, but that all molecular speeds increase as the temperature rises (Figure 6-18).

To use equation (6.20) in calculating a root-mean-square speed, we must express the gas constant as

$$R = 8.3145 \text{ J mol}^{-1} \text{ K}^{-1}$$

Kinetic Energies in a Gas animation

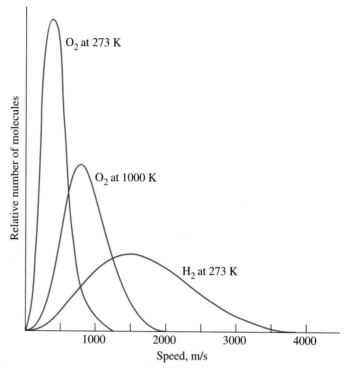

▲ **FIGURE 6-18   Distribution of molecular speeds—the effect of mass and temperature**
The relative numbers of molecules with a certain speed are plotted as a function of the speed. Note the effect of temperature on the distribution for oxygen molecules and the effect of mass—oxygen must be heated to a very high temperature to have the same distribution of speeds as does hydrogen at 273 K.

The joule has the units $(kg)(m/s)^2$, and consequently we must express molar mass in *kilograms* per mole, as we show in Example 6-17.

EXAMPLE 6-17

Calculating a Root-Mean-Square Speed. Which is the greater speed, that of a bullet fired from a high-powered M-16 rifle (2180 mi/h) or the root-mean-square speed of H_2 molecules at 25 °C?

Solution

Determine u_{rms} of H_2 with equation (6.20). In doing so, recall the points about R and M noted above.

$$u_{rms} = \sqrt{\frac{3 \times 8.3145 \text{ kg m}^2 \text{ s}^{-2} \text{ mol}^{-1} \text{ K}^{-1} \times 298 \text{ K}}{2.016 \times 10^{-3} \text{ kg mol}^{-1}}}$$
$$= \sqrt{3.69 \times 10^6 \text{ m}^2/\text{s}^2} = 1.92 \times 10^3 \text{ m/s}$$

The remainder of the problem requires us either to convert 1.92×10^3 m/s to a speed in miles per hour, or 2180 mi/h to meters per second. Then we can compare the two speeds. When we do this, we find that 1.92×10^3 m/s corresponds to 4.29×10^3 mi/h. The root-mean-square speed of H_2 molecules at 25 °C is greater than the speed of the high-powered rifle bullet.

Practice Example A: Which has the greater root-mean-square speed at 25 °C, $NH_3(g)$ or $HCl(g)$? Calculate u_{rms} for the one with the greater speed.

Practice Example B: At what temperature are u_{rms} of H_2 and the speed of the M-16 rifle bullet given in Example 6-17 the same?

Are You Wondering...

How the distribution of molecular speeds can be demonstrated experimentally?

This can be done with the apparatus shown in Figure 6-19. An oven and attached evacuated chamber are separated by a wall with a small hole in it. Gas molecules are heated in the oven, emerge through the hole, and pass through a series of slits, called *collimators*, that herd the molecules into a beam. The number of molecules in the beam is kept low so that collisions between them will not disturb the beam.

The molecular beam passes through a series of rotating disks. Each disk has a slit cut in it. The slits on successive disks are offset from each other by a certain angle. A molecule passing through the first rotating disk will pass through the second disk only if the molecular velocity is such that the molecule arrives at the disk at the exact moment that the second slit appears. Thus, for a given rotation speed, only those molecules with the appropriate velocity can pass through the entire series of disks.

The number of molecules that pass through the disks and arrive at the detector is recorded for each chosen speed of rotation. The number of molecules for each speed of rotation is then plotted against the rotation speed. From the dimensions of the apparatus, the rotation speeds of the disks can be converted to molecular speeds, and a plot similar to that in Figure 6-17 can be obtained.

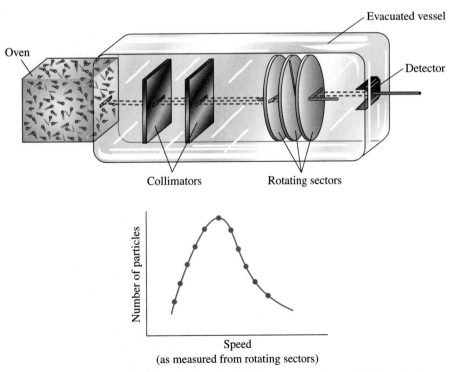

▲ **FIGURE 6-19   Distribution of molecular speeds—an experimental determination**
Only those molecules with the correct speed to pass through all rotating sectors will reach the detector, where they can be counted. By changing the rate of rotation of the sectors, the complete distribution of molecular speeds can be determined.

The Meaning of Temperature

We can gain an important insight into the meaning of temperature by starting with equation (6.19), the basic equation of the kinetic-molecular theory written for 1 mol of gas . We modify it slightly by replacing the fraction $\frac{1}{3}$ by the equivalent product $\frac{2}{3} \times \frac{1}{2}$.

$$PV = \frac{1}{3} N_A \overline{mu^2} = \frac{2}{3} N_A \left(\frac{1}{2} \overline{mu^2} \right)$$

The terms in this equation that are gathered into $\left(1/2\ \overline{mu^2} \right)$ represent the average translational kinetic energy, $\overline{e_k}$, of a collection of molecules. When we make this substitution and replace PV by RT, we obtain

$$RT = \frac{2}{3} N_A \overline{e_k}$$

Then, we can solve this equation for e_k.

$$\overline{e_k} = \frac{3}{2} \frac{R}{N_A} (T) \tag{6.21}$$

Because R and N_A are constants, equation (6.21) simply states that $\overline{e_k} = constant \times T$. This leads to an interesting new idea about temperature.

The Kelvin temperature (T) of a gas is directly proportional to the average translational kinetic energy $\left(\overline{e_k} \right)$ of its molecules. (6.22)

Also, we have a new concept of what changes in temperature mean—changes in the intensity of translational molecular motion. When heat flows from one body to another, molecules in the hotter body (higher temperature) give up some of their kinetic energy through collisions with molecules in the colder body (lower temperature). The flow of heat continues until the average translational kinetic energies of the molecules become equal, that is, until the temperatures become equalized. Finally, we have a new way of looking at the absolute zero of temperature: *It is the temperature at which translational molecular motion should cease.*

Are You Wondering...

What's the lowest temperature we can reach?

Charles's Law suggests that there is an absolute zero of temperature, 0 K, but can we attain this temperature? The answer is no, but we can come mighty close. Current attempts have resulted in temperatures as low as just a few nanokelvins! However, it is not simply a matter of putting some hot atoms in a "refrigerator" operating at 0 K. We have seen that the temperature of a sample of gas is proportional to the kinetic energy of the gas molecules; therefore to cool atoms down, we have to remove their kinetic energy. Simple cooling will not do the trick, because the refrigerator would always have to be at a lower temperature than are the atoms being cooled. The way that extremely cold atoms can be created is to remove their kinetic energy by stopping them in their tracks. This has been accomplished using a technique called laser cooling, in which a laser light is directed at a beam of atoms, hitting them head-on and dramatically slowing them down. Once the atoms are cooled, intersecting beams of six lasers are used to reduce their energies still further. The sample of cold atoms is then trapped by a magnetic field for about 1 s. In 1995, a team at the University of Colorado successfully cooled a beam of rubidium (Rb) atoms to 1.7×10^{-7} K using this procedure.

6-8 Gas Properties Relating to the Kinetic-Molecular Theory

A molecular speed of 1500 m/s corresponds to about 1 mi/s, or 3600 mi/h. From this, it might seem that a given gas molecule could travel very long distances over a very short time, but this is not quite the case. Every gas molecule undergoes frequent collisions with other gas molecules and, as a result, keeps changing direction. Gas molecules follow a tortuous path, which slows them down in getting from one point to another in a gas. Still, the net rate at which gas molecules move in a particular direction does depend on their average speeds.

Diffusion is the migration of molecules as a result of random molecular motion. The diffusion of two or more gases results in an intermingling of the molecules and, in a closed container, soon produces a homogeneous mixture. Figure 6-20 pictures a phenomenon commonly seen in a chemistry laboratory. A related phenomenon, **effusion**, is the escape of gas molecules from their container through a tiny orifice or pinhole. The effusion of a hypothetical mixture of two gases is suggested by Figure 6-21. The rate at which effusion occurs is directly proportional to molecular speeds. That is, molecules with high speeds effuse faster than molecules with low speeds. Let us consider the effusion of two different gases at the same temperature and pressure. We can first compare

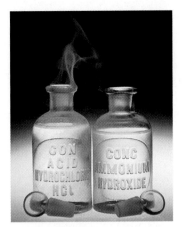

▲ **FIGURE 6-20**
Diffusion of $NH_3(g)$ and $HCl(g)$
$NH_3(g)$ escapes from $NH_3(aq)$ (but labeled ammonium hydroxide in this photograph), and $HCl(g)$ escapes from $HCl(aq)$. The gases diffuse toward each other, and where they meet, a white cloud of ammonium chloride forms:
$NH_3(g) + HCl(g) \longrightarrow NH_4Cl(s)$.
Because of their greater average speed, NH_3 molecules diffuse faster than HCl. As a result, the cloud forms close to the mouth of the $HCl(aq)$ container.

Gas Diffusion animation

Diffusion of Bromine Vapor movie

▶ **FIGURE 6-21 Diffusion and effusion**
(a) Diffusion is the passage of one substance through another. In this case, the H_2 initially diffuses farther through the N_2 because it is lighter, although eventually a complete random mixing occurs.
(b) Effusion is the passage of a substance through a pinhole or porous membrane into a vacuum. In this case, the lighter H_2 effuses faster across the empty space than does the N_2.

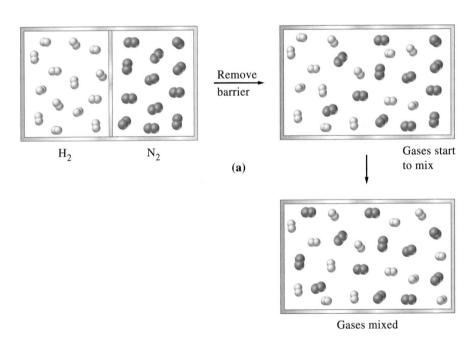

H₂   N₂
(a)

Remove barrier →

Gases start to mix

Gases mixed

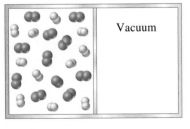

Vacuum

Open pinhole →

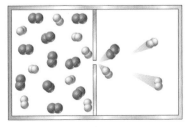

(b)

effusion rates with root-mean-square speeds and then substitute expression (6.20) for these speeds.

$$\frac{\text{rate of effusion of A}}{\text{rate of effusion of B}} = \frac{(u_{\text{rms}})_A}{(u_{\text{rms}})_B} = \sqrt{\frac{3RT/M_A}{3RT/M_B}} = \sqrt{\frac{M_B}{M_A}} \qquad (6.23)$$

Any appropriate units (for example, g/s, mol/min) can be used to express a rate of effusion in equation (6.23) because in the ratio of two rates, the units cancel. Equation (6.23) is a kinetic-theory statement of a nineteenth-century law called **Graham's law**.

> The rates of effusion of two different gases are inversely proportional to the square roots of their molar masses.   (6.24)

There are serious limitations to Graham's law that you should be aware of. It can be used to describe *effusion* only for gases at very low pressures, so that molecules escape through an orifice individually, not as a jet of gas. Also, the orifice must be tiny so that no collisions occur as molecules pass through. Graham proposed his law in 1831 to describe the diffusion of gases, but the law actually does *not* apply to diffusion. Molecules of a diffusing gas undergo collisions with each other and with the gas into which they are diffusing. Some even move in the opposite direction to the net flow. Nevertheless, diffusion does occur, and gases of low molar mass do diffuse faster than those of higher molar mass. We just cannot use Graham's law to make quantitative predictions about rates of diffusion.

When compared at the same temperature, two different gases have the same value of $\overline{e_k} = \frac{1}{2}m\overline{u^2}$. This means that molecules with a smaller mass (m) have a higher speed (u_{rms}). When effusion takes place under the restrictions described above, we can use equation (6.23) to determine which of two gases effuses *faster*, which does so in a *shorter* period of time, which travels *farther* in a given period of time, and so on. An effective way to do this is to note that in every case, a ratio of effusion rates, times, distances, and so on is equal to the *square root* of a ratio of molar masses. That is,

a ratio of   (1) molecular speeds

           (2) effusion rates

           (3) effusion times $= \sqrt{\text{ratio of two molar masses}}$ (6.25)

           (4) distances traveled by molecules

           (5) amounts of gas effused

In using equation (6.25), first reason *qualitatively* whether the ratio of properties should be greater or less than one. Then set up the ratio of molar masses accordingly. Examples 6-18 and 6-19 illustrate this line of reasoning.

EXAMPLE 6-18

Comparing Amounts of Gases Effusing Through an Orifice. 2.2×10^{-4} mol $N_2(g)$ effuses through a tiny hole in 105 s. How much $H_2(g)$ would effuse through the same orifice in 105 s?

Solution

First, let us reason qualitatively. H_2 molecules have less mass than N_2 molecules, so they should have the greater speed when the gases are compared at the same temperature. Because $H_2(g)$ effuses faster, more H_2 than N_2 molecules should effuse in a given length of time. In the following setup, we need a ratio of molar masses *greater than one*.

$$\frac{?\ \text{mol } H_2}{2.2 \times 10^{-4}\ \text{mol } N_2} = \sqrt{\frac{M_{N_2}}{M_{H_2}}} = \sqrt{\frac{28.014}{2.016}} = 3.728$$

$$?\ \text{mol } H_2 = 3.728 \times 2.2 \times 10^{-4} = 8.2 \times 10^{-4}\ \text{mol } H_2$$

Practice Example A: In Example 6-18, how much $O_2(g)$ would effuse through the same orifice in 105 s?

Practice Example B: In Example 6-18, how long would it take for 2.2×10^{-4} mol H_2 to effuse through the same orifice as the 2.2×10^{-4} mol N_2?

EXAMPLE 6-19

Relating Effusion Times and Molar Masses. A sample of $Kr(g)$ escapes through a tiny hole in 87.3 s, and an unknown gas requires 42.9 s under identical conditions. What is the molar mass of the unknown gas?

Solution

Again, start with qualitative reasoning. Because the unknown gas effuses faster, it must have a smaller molar mass than Kr. The ratio of molar masses in the following setup must be *smaller than one*. M_{unk} goes in the numerator.

$$\frac{\text{effusion time for unknown}}{\text{effusion time for Kr}} = \frac{42.9\ \text{s}}{87.3\ \text{s}} = \sqrt{\frac{M_{\text{unk}}}{M_{\text{Kr}}}} = 0.491$$

$$M_{\text{unk}} = (0.491)^2 \times M_{\text{Kr}} = (0.491)^2 \times 83.80 = 20.2\ \text{g/mol}$$

Practice Example A: Under the same conditions as in Example 6-19, another unknown gas requires 131.3 s to escape. What is the molar mass of this unknown gas?

Practice Example B: Given all the same conditions as in Example 6-19, how long would it take for a sample of ethane gas, C_2H_6, to effuse?

Applications of Diffusion

The diffusion of gases into one another has many practical applications. Natural gas and liquefied petroleum gas (LPG) are odorless; for commercial use, a small quantity of a gaseous organic sulfur compound, methyl mercaptan, CH_3SH, is added to them. The mercaptan has an odor that can be detected in parts per billion (ppb) or less. When a leak occurs, which can lead to asphyxiation or an explosion, we rely on the diffusion of this odorous compound for a warning.

During World War II, the Manhattan Project (the secret, government-run program for developing the atomic bomb) used a method called gaseous diffusion to separate the desired isotope ^{235}U from the predominant ^{238}U. The method is based on the fact that uranium hexafluoride is one of the few compounds of uranium that can be obtained as a gas at moderate temperatures. When high-pressure $UF_6(g)$ is forced through a barrier having millions of submicroscopic holes per square centimeter,

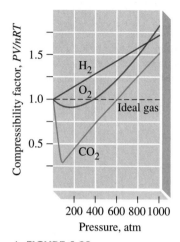

▲ **FIGURE 6-22**
The behavior of real gases—compressibility factor as a function of pressure at 0 °C
Values of the compressibility factor less than one signify that intermolecular forces of attraction are largely responsible for deviations from ideal gas behavior. Values greater than one are found when the volume of the gas molecules themselves is a significant fraction of the total gas volume.

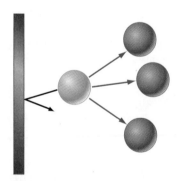

▲ **FIGURE 6-23**
Intermolecular forces of attraction
Attractive forces of the red molecules for the green molecule cause the green molecule to exert less force when it collides with the wall than if these attractions did not exist.

Nonideal Gas Behavior simulation

molecules containing the isotope ^{235}U pass through the barrier slightly faster than those containing ^{238}U, just as expected from expression (6.25). The $UF_6(g)$ contains a slightly higher ratio of ^{235}U to ^{238}U than it did previously. The gas has become enriched in ^{235}U. Carrying this process through several thousand passes yields a product highly enriched in ^{235}U.

6-9   Nonideal (Real) Gases

In introducing the ideal gas equation, we remarked that real gases obey the ideal gas law under suitable conditions. We should comment briefly on what those "suitable" conditions are and on what to do when the conditions are not suitable. A useful measure of how much a gas deviates from ideal gas behavior is found in its compressibility factor. The *compressibility factor* of a gas is the ratio PV/nRT. From the ideal gas equation $(PV = nRT)$, we see that for an ideal gas, $PV/nRT = 1$. For a real gas, the closeness of the experimentally determined ratio, PV/nRT, to 1 is a measure of how closely the behavior of the gas approaches that of an ideal gas. In Figure 6-22, the compressibility factor is plotted as a function of pressure for three different gases. Our principal conclusion from this figure is that all gases behave ideally at sufficiently low pressures, say, below 1 atm, but that deviations set in at increased pressures. At very high pressures, the compressibility factor is always greater than one.

Here is how we might explain nonideal gas behavior: Boyle's law predicts that at very high pressures, a gas volume becomes extremely small and approaches zero. This cannot be, however, because the molecules themselves occupy space and are practically incompressible. The PV product is larger than predicted for an ideal gas, and the compressibility factor is greater than one. We must also allow for the fact that intermolecular forces exist in gases. Figure 6-23 suggests that because of attractive forces between the molecules, the force of the collisions of gas molecules with the container walls is less than expected for an ideal gas. Intermolecular forces of attraction account for compressibility factors less than one. These forces become increasingly important at *low temperatures*, where translational molecular motion slows down. To summarize:

- Gases tend to behave *ideally at high temperatures and low pressures.*
- Gases tend to behave *nonideally at low temperatures and high pressures.*

The van der Waals Equation

A number of equations can be used for real gases, equations that apply over a wider range of temperatures and pressures than the ideal gas equation. Such equations are not as general as the ideal gas equation. They contain terms that have specific, but different, values for different gases. Such equations must correct for the volume associated with the molecules themselves and for intermolecular forces of attraction. In the **van der Waals equation**,

$$\left(P + \frac{n^2a}{V^2} \right)(V - nb) = nRT \qquad (6.26)$$

V is the volume of n moles of gas. The term n^2a/V^2 is related to intermolecular forces of attraction. It is added to P because the measured pressure is lower than expected (recall Figure 6-23). The value b, called the *excluded* volume per mole, is related to the volume of the gas molecules, and nb is subtracted from the measured volume to represent the *free* volume within the gas: $V - nb$. Both a and b have specific values for particular gases, values that vary somewhat with temperature and pressure. In Example 6-20 we calculate the pressure of a real gas, using the van der Waals equation. Solving the equation for either n or V is more difficult, however (see Exercise 106).

EXAMPLE 6-20

Using the van der Waals Equation to Calculate the Pressure of a Gas. Use the van der Waals equation to calculate the pressure exerted by 1.00 mol $Cl_2(g)$ confined to a volume of 2.00 L at 273 K. The value of $a = 6.49 \ L^2$ atm mol^{-2}, and that of $b = 0.0562 \ L \ mol^{-1}$.

Solution

Solve equation (6.26) for *P*.

$$P = \frac{nRT}{V - nb} - \frac{n^2 a}{V^2}$$

Then substitute the following values into the equation.

$$n = 1.00 \text{ mol}; V = 2.00 \text{ L}; T = 273 \text{ K}; R = 0.08206 \text{ L atm mol}^{-1} \text{ K}^{-1}$$

$$n^2 a = (1.00)^2 \text{ mol}^2 \times 6.49 \frac{L^2 \text{ atm}}{\text{mol}^2} = 6.49 \ L^2 \text{ atm}$$

$$nb = 1.00 \text{ mol} \times 0.0562 \text{ L mol}^{-1} = 0.0562 \text{ L}$$

$$P = \frac{1.00 \text{ mol} \times 0.08206 \text{ L atm mol}^{-1} \text{ K}^{-1} \times 273 \text{ K}}{(2.00 - 0.0562)\text{L}} - \frac{6.49 \ L^2 \text{ atm}}{(2.00)^2 \ L^2}$$

$$P = 11.5 \text{ atm} - 1.62 \text{ atm} = 9.9 \text{ atm}$$

▶ Although the deviation from ideality here is rather large, in problem-solving situations, you can generally assume that the ideal gas equation will give satisfactory results.

The pressure calculated with the ideal gas equation is 11.2 atm. By including only the *b* term in the van der Waals equation, we get a value of 11.5 atm. Including the *a* term reduces the calculated pressure by 1.62 atm. Under the conditions of this problem, intermolecular forces of attraction are the main cause of the departure from ideal behavior.

Practice Example A: Substitute $CO_2(g)$ for $Cl_2(g)$ in Example 6-20. The values of *a* and *b* are $a = 3.59 \ L^2$ atm mol^{-2} and $b = 0.0427 \ L \ mol^{-1}$. Which gas, CO_2 or Cl_2, shows the greater departure from ideal gas behavior?
(*Hint:* For which gas do you find the greater difference in calculated pressures, using first the ideal gas equation and then the van der Waals equation?)

Practice Example B: Substitute $CO(g)$ for $Cl_2(g)$ in Example 6-20. The values of *a* and *b* are $a = 1.49 \ L^2$ atm mol^{-2} and $b = 0.0399 \ L \ mol^{-1}$. Including CO_2 from Practice Example 6-20A, which of the three gases—Cl_2, CO_2, or CO—shows the greatest departure from ideal gas behavior?

Summary

A gas is described in terms of its pressure, temperature, volume, and amount. Gas pressure is most readily measured by comparing it with the pressure exerted by a liquid column, usually mercury. Atmospheric pressure is measured with a mercury barometer and other gas pressures with a manometer. Pressure can be expressed in a variety of units.

The most common simple gas laws are Boyle's law, relating gas pressure and volume; Charles's law, relating gas volume and temperature; and Avogadro's law, relating volume and amount of gas. Some important ideas that originate from the simple gas laws are the Kelvin temperature scale, the standard conditions of temperature and pressure (STP), and the molar volume of a gas at STP—22.4 L/mol.

The simple gas laws can be combined into the ideal gas equation: $PV = nRT$. This equation can be solved for any one of the variables when the others are known. It can also be applied in determining molar masses and gas densities. Still other uses of the ideal gas equation are in describing (1) the gaseous reactants or products of a chemical reaction and (2) mixtures of gases. Collecting gases over water is a common procedure involving a mixture of gases—the particular gas being isolated and water vapor.

The kinetic-molecular theory establishes one relationship involving the root-mean-square speed of molecules, temperature, and molar mass of a gas and another relationship between average molecular translational kinetic energy and Kelvin temperature. Also, the diffusion and effusion of gases can be related to their molar masses through the kinetic-molecular theory.

Real gases generally behave ideally only at high temperatures and low pressures. Other equations of state, such as the van der Waals equation, often work where the ideal gas equation fails.

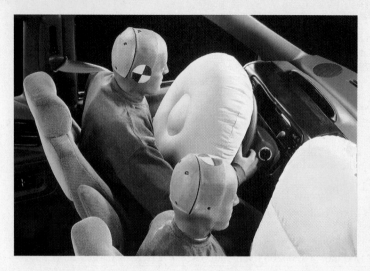

◀ Air bags inflating during a crash test.

Air bags in automobiles have saved thousands of lives. The idea behind them is simple: When a crash occurs, a plastic bag rapidly inflates with a gas, protecting the driver from hitting the dashboard or steering column. However, the development of a workable air-bag system required the combined efforts of chemists and engineers.

The air-bag system has many special requirements. The air bag must not inflate accidentally. The gas used must be nontoxic in case of leakage after inflation. The gas must be "cool," so as not to produce burn injuries. The gas must be produced very rapidly, ideally inflating the bag within 20–60 ms. Finally, the gas-producing chemicals must be easy to handle and be stable for long periods.

Integrative Example

Some fleet vehicles have been modified to burn natural gas, which is mostly methane, $CH_4(g)$. The combustion can be controlled to produce CO_2 and H_2O with a minimum of pollutants (CO and oxides of nitrogen). The ideal air–fuel ratio uses methane and oxygen (from the air) in *stoichiometric* proportions (meaning that neither is in excess).

What volume of air, measured at 22 °C and 745 mmHg, is required for the complete combustion of 1.00 L of compressed $CH_4(g)$ at 22 °C and 3.55 atm? [Air contains 20.95% $O_2(g)$, by volume.]

1. *Write a chemical equation for the complete combustion of methane.*

$$CH_4(g) + 2\,O_2(g) \longrightarrow CO_2(g) + 2\,H_2O(g)$$

2. *Determine the volume of $O_2(g)$ consumed in the combustion.* First, we can use Gay-Lussac's law of combining volumes to

determine the volume of $O_2(g)$ *at 22 °C and 3.55 atm.*

$$? \text{ L } O_2(g) = 1.00 \text{ L } CH_4(g) \times \frac{2 \text{ L } O_2(g)}{1 \text{ L } CH_4(g)} = 2.00 \text{ L } O_2(g)$$

Then, we can use Boyle's law to determine the volume of $O_2(g)$ at the temperature and pressure of the air from which the $O_2(g)$ is drawn: *22 °C and 745 mmHg pressure.*

$$V_f = V_i \times \frac{P_i}{P_f} = 2.00 \text{ L} \times \frac{3.55 \text{ atm}}{(745/760) \text{ atm}} = 7.24 \text{ L } O_2$$

3. *Determine the volume of air required.* The volume percent $O_2(g)$ in air provides the final conversion factor.

$$? \text{ L air} = 7.24 \text{ L } O_2(g) \times \frac{100.0 \text{ L air}}{20.95 \text{ L } O_2(g)} = 34.6 \text{ L air}$$

Key Terms

Avogadro's law (hypothesis) (6-2)
barometer (6-1)
barometric pressure (6-1)
Boyle's law (6-2)
Charles's law (6-2)
Dalton's law of partial pressures (6-6)
diffusion (6-8)
effusion (6-8)
gas constant, *R* (6-3)
general gas equation (6-3)

Graham's law (6-8)
ideal (perfect) gas (6-3)
ideal gas equation (6-3)
Kelvin (absolute) scale (6-2)
kilopascal (kPa) (6-1)
kinetic-molecular theory of gases (6-7)
law of combining volumes (6-5)
manometer (6-1)
millimeter of mercury (mmHg) (6-1)
mole fraction (6-6)

partial pressure (6-6)
pascal (Pa) (6-1)
pressure (6-1)
root-mean-square speed (6-7)
standard atmosphere (atm) (6-1)
standard conditions of temperature and pressure (STP) (6-2)
torr (Torr) (6-1)
van der Waals equation (6-9)

Nitrogen is the best choice for a nontoxic gas. After all, nitrogen makes up about 78% of air by volume. A good source of nitrogen is the decomposition of alkali metal azides, such as sodium azide, NaN_3.

$$2\,NaN_3(s) \xrightarrow{\Delta} 2\,Na(l) + 3\,N_2(g)$$

Sensors that detect the initial crash activate the air-bag system by electrically initiating the explosion of a small charge. This, in turn, starts the rapid burning of a pellet containing sodium azide, which releases a large volume of $N_2(g)$ to fill the bag (recall Example 6-12).

By 1980, the engineering challenges of the air-bag safety system had been resolved, but not the chemical challenges. The problems posed by the use of sodium azide were that it does not make a good pellet, its reaction does not go to completion rapidly, and one of the reaction products—sodium metal—reacts violently with water.

To solve these problems, investigators tried adding other compounds to the sodium azide. To make a good pellet-forming mixture, a lubricant, typically molybdenum disulfide (MoS_2),

was added to the sodium azide. This mixture did not burn well, however. Sulfur, a familiar constituent of gunpowder, was then added to produce smooth-burning pellets. The nitrogen gas produced was cool, and the sodium metal was converted mainly to the sulfate, though the solid residue was finely powdered and difficult to contain. Some air-bag systems on the market today use the $MoS_2 - S - NaN_3$ pellet, but the latest ones use a pellet that is still more complex.

In earlier experimental work, a mixture of sodium azide and iron(III) oxide had proved satisfactory for trapping the sodium metal by converting it to an easy-to-handle solid residue. This mixture did not burn well, however. Researchers then tried the obvious solution: Mix together all the compounds that give the gas-forming pellets their most desirable properties: sodium azide, iron(III) oxide, molybdenum disulfide, and sulfur. In scientific research, the obvious often turns up some unexpected results. In this case, however, the hoped-for final result was achieved: a rapidly burning pellet producing cool, odor-free nitrogen gas and a nonreactive solid residue that is easily trapped. At this point, the widespread use of air-bag collision systems in automobiles became a reality.

Review Questions

1. In your own words, define or explain the following terms or symbols: (a) atm; (b) STP; (c) R; (d) partial pressure; (e) u_{rms}.

2. Briefly describe each of the following ideas, phenomena, or methods: (a) absolute zero of temperature; (b) collection of a gas over water; (c) effusion of a gas; (d) law of combining volumes.

3. Explain the important distinctions between each pair of terms: (a) barometer and manometer; (b) Celsius and Kelvin temperature; (c) ideal gas equation and general gas equation; (d) ideal gas and real gas.

4. Convert each pressure to an equivalent pressure in standard atmospheres: (a) 736 mmHg; (b) 58.2 cm Hg; (c) 892 Torr; (d) 225 kPa.

5. Calculate the height of a mercury column required to produce a pressure (a) of 0.984 atm; (b) of 928 Torr; (c) equal to that of a column of water 142 ft high.

6. What is P_{gas} for the manometer reading (millimeters of mercury) in the figure when barometric pressure is 744 mmHg?

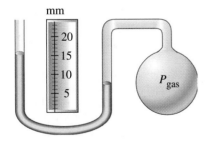

7. A sample of $O_2(g)$ has a volume of 26.7 L at 762 Torr. What is the new volume if, with the temperature and amount of gas held constant, the pressure is (a) lowered to 385 Torr; (b) increased to 3.68 atm?

8. An 886-mL sample of Ne(g) is at 752 mmHg and 26 °C. What will be the new volume if, with the pressure and amount of gas held constant, the temperature is (a) increased to 98 °C; (b) lowered to −20 °C?

9. We want to increase the volume of a fixed amount of gas from 57.3 to 165 mL while holding the pressure constant. To what temperature must we heat this gas if the initial temperature is 22 °C?

10. What is the volume at STP of a 49.6-g sample of acetylene gas, C_2H_2?

11. What volume of $Cl_2(g)$ at STP would you need to obtain a 250.0-g sample of $Cl_2(g)$?

12. *Without doing detailed calculations*, determine which of the following gases has the greatest density at STP: Cl_2, SO_3, N_2O, PF_3? Explain.

13. What is the volume, in milliliters, occupied by 89.2 g $CO_2(g)$ at 37 °C and 737 mmHg?

14. A 40.0-L cylinder contains 285 g $SO_2(g)$ at 27 °C. What is the pressure, in atmospheres, exerted by this gas?

15. A 0.418-g sample of gas has a volume of 115 mL at 66.3 °C and 743 mmHg. What is the molar mass of this gas?

16. What is the density, in grams per liter, of $CO_2(g)$ at 32.7 °C and 758 mmHg?

17. How many liters of $H_2(g)$ at STP are produced per gram of Al(s) consumed in the following reaction?

$$2\,Al(s) + 6\,HCl(aq) \longrightarrow 2\,AlCl_3(aq) + 3\,H_2(g)$$

18. A method of removing $CO_2(g)$ from a spacecraft is to allow the CO_2 to react with LiOH. How many liters of $CO_2(g)$ at 25.9 °C and 751 Torr can be removed per kilogram of LiOH consumed?

$$2 LiOH(s) + CO_2(g) \longrightarrow Li_2CO_3(s) + H_2O(l)$$

19. What is the volume, in liters, occupied by a mixture of 15.2 g Ne(g) and 34.8 g Ar(g) at 7.15 atm pressure and 26.7 °C?

20. A balloon filled with $H_2(g)$ at STP has a volume of 2.24 L. To the balloon is added 0.10 mol He(g). Then the temperature is raised to 100 °C while the pressure and amount of gas are held constant. What is the final gas volume?

21. An 89.3-mL sample of "wet" $O_2(g)$ is collected over water at 21.3 °C at a barometric pressure of 756 mmHg (vapor pressure of water at 21.3 °C = 19 mmHg). **(a)** What is the partial pressure of $O_2(g)$, in millimeters of mercury, in the sample collected? **(b)** What is the volume percent O_2 in the gas collected? **(c)** How many grams of O_2 are present in the sample?

22. Which of the following is/are true when comparing 0.50 mol $H_2(g)$ and 1.0 mol He(g) at STP? The two gases have equal average molecular kinetic energies, molecular speeds, volumes, and effusion rates.

23. A sample of $Cl_2(g)$ effuses through a tiny hole in 28.6 s. How long would it take for a sample of NO(g) to effuse under the same conditions?

24. Under which conditions is Cl_2 most likely to behave like an ideal gas? Explain. (1) 100 °C and 10.0 atm; (2) 0 °C and 0.50 atm; (3) 200 °C and 0.50 atm; (4) −100 °C and 10.0 atm.

Exercises

Pressure and Its Measurement

25. Calculate the height of a column of liquid benzene ($d = 0.879$ g/cm³), in meters, required to exert a pressure of 0.970 atm.

26. Calculate the height of a column of liquid glycerol ($d = 1.26$ g/cm³), in meters, required to exert the same pressure as 3.02 m of $CCl_4(l)$ ($d = 1.59$ g/cm³).

27. The mercury level in the open arm of an open-end manometer is 276 mm above a reference point. In the arm connected to a container of gas, the level is 49 mm above the same reference point. If barometric pressure is 749 mmHg, what is the pressure of the gas in the container? (*Hint:* Refer to the drawing with Review Question 6.)

28. A gas is collected over water when the barometric pressure is 756.2 mmHg; but the water level inside the container of gas is 4.5 cm higher than outside. What is the total pressure of the gas inside the container, in millimeters of mercury? (*Hint:* Refer to Figure 6-15.)

29. At times, a pressure is stated in units of *mass* per unit area rather than force per unit area. What is the standard atmosphere of pressure expressed in the unit kg/cm²? (*Hint:* How is a mass in kilograms related to a force?)

30. What is the standard atmosphere of pressure expressed in pounds per square inch (psi)? (*Hint:* Refer to Exercise 29.)

The Simple Gas Laws

31. A 35.8-L cylinder of Ar(g) is connected to an evacuated 1875-L tank. If the temperature is held constant and the final pressure is 721 mmHg, what must have been the original gas pressure in the cylinder, in atmospheres?

32. A sample of $N_2(g)$ occupies a volume of 42.0 mL under the existing barometric pressure. Increasing the pressure by 85 mmHg reduces the volume to 37.7 mL. What is the prevailing barometric pressure, in millimeters of mercury?

33. On page 184, the statement is made that doubling the Celsius temperature of a fixed amount of gas at a fixed pressure does *not* cause the gas volume to double. What is the *percent* increase in volume under these conditions for a temperature change from 1.00 °C to 2.00 °C? Is this the same percent increase as produced by a temperature change from 10.00 °C to 20.00 °C? Explain.

34. The photographs show the contraction of a gas-filled balloon when it is cooled by liquid nitrogen at its boiling point of −196 °C. To what fraction of its original volume will the balloon shrink when it is cooled from a room temperature of 71 °F to the boiling point of liquid nitrogen?

35. A 27.6-mL sample of $PH_3(g)$ (used in the manufacture of flame-retardant chemicals) is obtained at STP.
    (a) What is the mass of this gas, in milligrams?
    (b) How many molecules of PH_3 are present?

36. A 5.0×10^{17}-atom sample of radon gas is obtained.
    (a) What is the mass of this sample, in micrograms?
    (b) What is the volume of this sample at STP, in microliters?

37. You purchase a bag of potato chips at an ocean beach to take on a picnic in the mountains. At the picnic, you notice

that the bag has become inflated, almost to the point of bursting. Use your knowledge of gas behavior to explain this phenomenon.

38. Scuba divers know that they must not ascend quickly from deep underwater because of a condition known as the bends, discussed in Chapter 14. Another concern is that they must constantly exhale during their ascent to prevent damage to the lungs and blood vessels. Describe what would happen to the lungs of a diver who inhaled compressed air at a depth of 30 meters and held her breath while rising to the surface.

General Gas Equation

39. A sample of gas has a volume of 4.25 L at 25.6 °C and 748 mmHg. What will be the volume of this gas at 26.8 °C and 742 mmHg?

40. A 10.0-g sample of a gas has a volume of 5.25 L at 25 °C and 762 mmHg. If to this *constant* 5.25-L volume is added 2.5 g of the same gas and the temperature raised to 62 °C, what is the new gas pressure?

41. A constant-volume vessel contains 12.5 g of a gas at 21°C. If the pressure of the gas is to remain constant as the temperature is raised to 210 °C, how many grams of gas must be released?

42. A 34.0-L cylinder contains 305 g $O_2(g)$ at 22 °C. How many grams of $O_2(g)$ must be released to reduce the pressure in the cylinder to 1.15 atm if the temperature remains constant?

Ideal Gas Equation

43. A 12.8-L cylinder contains 35.8 g O_2 at 46 °C. What is the pressure of this gas, in atmospheres?

44. Kr(g) in a 18.5-L cylinder exerts a pressure of 11.2 atm at 28.2 °C. How many grams of gas are present?

45. A 72.8-L constant-volume cylinder containing 1.85 mol He is heated until the pressure reaches 3.50 atm. What is the final temperature in degrees Celsius?

46. What is the pressure, in pascals, exerted by 1242 g CO(g) when confined at −25 °C to a cylindrical tank 25.0 cm in diameter and 1.75 m high?

Determining Molar Mass

47. Refer to Example 6-10. By combustion analysis, the mass percent composition of propylene is found to be 85.63% C and 14.37% H. What is the *molecular* formula of propylene?

48. A 2.650-g sample of a gaseous compound occupies 428 mL at 24.3 °C and 742 mmHg. The compound consists of 15.5% C, 23.0% Cl, and 61.5% F, by mass. What is its molecular formula?

49. A gaseous hydrocarbon weighing 0.231 g occupies a volume of 102 mL at 23 °C and 749 mmHg. What is the molar mass of this compound? What conclusion can you draw about its molecular formula?

50. A 132.10-mL glass vessel weighs 56.1035 g when evacuated and 56.2445 g when filled with the gaseous hydrocarbon acetylene at 749.3 mmHg and 20.02 °C. What is the molar mass of acetylene? What conclusion can you draw about its molecular formula?

Gas Densities

51. A particular application calls for $N_2(g)$ with a density of 1.80 g/L at 32 °C. What must be the pressure of the $N_2(g)$ in millimeters of mercury?

52. Monochloroethylene is used to make polyvinylchloride (PVC). It has a density of 2.56 g/L at 22.8 °C and 756 mmHg. What is the molar mass of monochloroethylene?

53. In order for a gas-filled balloon to rise in air, the density of the gas in the balloon must be less than that of air.
    (a) Consider air to have a molar mass of 28.96 g/mol air, and determine the density of air at 25 °C and 1 atm, in g/L.
    (b) Show by calculation that a balloon filled with carbon dioxide at 25 °C and 1 atm could not be expected to rise in air at 25 °C.

54. Refer to Exercise 53, and determine the minimum temperature to which the balloon described in part (b) would have to be heated before it could begin to rise in air. (Ignore the mass of the balloon itself.)

55. The density of phosphorus vapor at 310 °C and 775 mmHg is 2.64 g/L. What is the molecular formula of the phosphorus under these conditions?

56. A particular gaseous hydrocarbon that is 82.7% C and 17.3% H by mass has a density of 2.33 g/L at 23 °C and 746 mmHg. What is the molecular formula of this hydrocarbon?

Gases in Chemical Reactions

57. What volume of $O_2(g)$ is consumed in the combustion of 75.6 L $C_3H_8(g)$ if both gases are measured at STP?

58. Calculate the volume of $H_2(g)$, measured at 26 °C and 751 Torr required to react with 28.5 L $CO(g)$, measured at 0 °C and 760 Torr, in the following reaction.

$$3\ CO(g) + 7\ H_2(g) \longrightarrow C_3H_8(g) + 3\ H_2O(l)$$

59. A particular coal sample contains 3.28% S by mass. When the coal is burned, the sulfur is converted to $SO_2(g)$. What volume of $SO_2(g)$, measured at 23 °C and 738 mmHg, is produced by burning 2.7×10^6 lb of this coal?

60. A 3.57-g sample of a KCl–$KClO_3$ mixture is decomposed by heating and produces 119 mL $O_2(g)$, measured at 22.4 °C and 738 mmHg? What is the mass percent of $KClO_3$ in the mixture?

$$2\ KClO_3(s) \longrightarrow 2\ KCl(s) + 3\ O_2(g)$$

61. Hydrogen peroxide, H_2O_2, is used to disinfect contact lenses.

$$2\ H_2O_2(aq) \longrightarrow 2\ H_2O(l) + O_2(g)$$

How many milliliters of $O_2(g)$ at 22 °C and 752 mmHg can be liberated from 10.0 mL of an aqueous solution containing 3.00% H_2O_2 by mass. The density of the aqueous solution of H_2O_2 is 1.01 g/mL.

62. The Haber process is the principal method for fixing nitrogen (converting N_2 to nitrogen compounds).

$$N_2(g) + 3\ H_2(g) \longrightarrow 2\ NH_3(g)$$

Assume that the reactant gases are completely converted to $NH_3(g)$ and that the gases behave ideally.
(a) What volume of $NH_3(g)$ can be produced from 313 L of $H_2(g)$ if the gases are measured at 315 °C and 5.25 atm?
(b) What volume of $NH_3(g)$, measured at 25 °C and 727 mmHg, can be produced from 313 L $H_2(g)$, measured at 315 °C and 5.25 atm?

Mixtures of Gases

63. A gas cylinder of 53.7 L volume contains $N_2(g)$ at a pressure of 28.2 atm and 26 °C. How many grams of $Ne(g)$ must we add to this same cylinder to raise the total pressure to 75.0 atm?

64. A 2.35-L container of $H_2(g)$ at 762 mmHg and 24 °C is connected to a 3.17-L container of $He(g)$ at 728 mmHg and 24 °C. After mixing, what is the *total* gas pressure, in millimeters of mercury, with the temperature remaining at 24 °C?

65. Which of the following actions would you take to establish a pressure of 2.00 atm in a 2.24-L cylinder containing 1.60 g $O_2(g)$ at 32 °F? **(a)** add 1.60 g O_2; **(b)** release 0.80 g O_2; **(c)** add 2.00 g He; **(d)** add 0.60 g He.

66. A mixture of 4.0 g $H_2(g)$ and 10.0 g $He(g)$ in a 4.3-L flask is maintained at 0 °C.
(a) What is the total pressure in the container?
(b) What is the partial pressure of each gas?

67. A 2.00-L container is filled with $Ar(g)$ at 752 mmHg and 35 °C. A 0.728-g sample of C_6H_6 vapor is then added.
(a) What is the total pressure in the container?
(b) What is the partial pressure of Ar and of C_6H_6?

68. The chemical composition of air that is exhaled (expired) is different from ordinary air. A typical analysis of expired air at 37 °C and 1.00 atm, expressed as percent *by volume*, is 74.2% N_2, 15.2% O_2, 3.8% CO_2, 5.9% H_2O, and 0.9% Ar. The composition of ordinary air is given in Practice Example 6-15B.
(a) What is the ratio of the partial pressure of $CO_2(g)$ in expired air to that in ordinary air?

(b) Is the density of expired air greater or less than that of ordinary air at the same temperature and pressure? Explain.
(c) Calculate the densities of ordinary air and expired air at 37 °C and 1.00 atm.

69. In the drawing below (on the left), 1.00 g $H_2(g)$ is maintained at 1 atm pressure in a cylinder closed off by a freely moving piston. Which of the sketches below (on the right) best represents the mixture obtained when 1.00 g $He(g)$ is added? Explain.

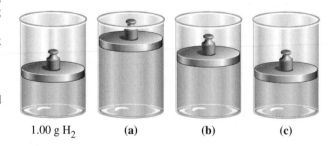

1.00 g H_2 **(a)**   **(b)**   **(c)**

70. To the drawings shown in Exercise 69, sketch in H_2 molecules and He atoms. Specifically, choose some small number of symbols to represent 1.00 g H_2. Then, sketch in the appropriate number of symbols of He atoms to account for the conditions indicated for each case: (a), (b), and (c). Make your sketches reflect the mixtures at the molecular level to the best of your ability.

Collecting Gases over Liquids

71. A 1.65-g sample of Al reacts with excess HCl, and the liberated H_2 is collected over water at 25 °C at a barometric pressure of 744 mmHg. What *total* volume of gas, in liters, is collected?

$$2\,Al(s) + 6\,HCl(aq) \longrightarrow 2\,AlCl_3(aq) + 3\,H_2(g)$$

72. A 367-mL sample of Ar(g) at 25 °C and a barometric pressure of 748 mmHg is passed through water at 25 °C. What is the volume of gas when saturated with water vapor, and again measured at 25 °C and 748 mmHg barometric pressure?

73. A sample of $O_2(g)$ is collected over water at 24 °C. The volume of gas is 1.16 L. In a subsequent experiment, it is determined that the mass of O_2 present is 1.46 g. What must have been barometric pressure at the time the gas was collected? (Vapor pressure of water = 22.4 Torr.)

74. A 1.072-g sample of He(g) is found to occupy a volume of 8.446 L when collected over hexane at 25.0 °C and 738.6 mmHg barometric pressure. Use these data to determine the vapor pressure of hexane at 25 °C.

Kinetic-Molecular Theory

75. Calculate u_{rms}, in meters per second, for $Cl_2(g)$ molecules at 30 °C.

76. The u_{rms} of H_2 molecules at 273 K is 1.84×10^3 m/s. At what temperature is u_{rms} for H_2 twice this value?

77. Refer to Example 6-17. What must be the molecular mass of a gas if its molecules are to have a root-mean-square speed at 25 °C equal to that of the M-16 rifle bullet?

78. Refer to Example 6-17. Noble gases (group 18) exist as atoms, not molecules (they are monatomic). Cite one noble gas whose u_{rms} at 25 °C is higher than the rifle bullet's and one whose u_{rms} is lower.

79. At what temperature will u_{rms} for Ne(g) be the same as u_{rms} for He at 300 K?

80. Determine u_m, $\bar{u}$, and u_{rms} for a group of ten automobiles clocked by radar at speeds of 38, 44, 45, 48, 50, 55, 55, 57, 58, and 60 mi/h, respectively.

Diffusion and Effusion of Gases

81. If 0.00484 mol $N_2O(g)$ effuses through an orifice in a certain period of time, how much $NO_2(g)$ would effuse in the same time under the same conditions?

82. A sample of $N_2(g)$ effuses through a tiny hole in 38 s. What must be the molar mass of a gas that requires 64 s to effuse under identical conditions?

83. What are the ratios of the diffusion rates for the following pairs of gases? **(a)** N_2 and O_2; **(b)** H_2O and D_2O (D = deuterium, i.e., $_1^2H$); **(c)** $^{14}CO_2$ and $^{12}CO_2$; **(d)** $^{235}UF_6$ and $^{238}UF_6$.

84. Which of the following visualizations best represents the distribution of O_2 and SO_2 molecules near an orifice some time after effusion occurs in the direction indicated by the arrows? The initial condition was one of equal numbers of O_2 molecules (•) and SO_2 molecules (•). Explain.

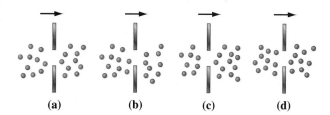

(a)   (b)   (c)   (d)

Nonideal Gases

85. Refer to Example 6-20. Recalculate the pressure of $Cl_2(g)$ using both the ideal gas equation and van der Waals equation at the following temperatures: **(a)** 100 °C; **(b)** 200 °C; **(c)** 400 °C. From the results, confirm the statement that a gas tends to be more ideal at high temperatures than at low temperatures.

86. Use both the ideal gas equation and the van der Waals equation to calculate the pressure exerted by 1.50 mol of $SO_2(g)$ when it is confined at 298 K to a volume of **(a)** 100.0 L, **(b)** 20.0 L, **(c)** 5.0 L, **(d)** 1.0 L, **(e)** 0.50 L. Under which of these conditions is the pressure calculated with the ideal gas equation within a few percent of that calculated with the van der Waals equation? For SO_2, the value of $a = 6.71$ L^2 atm mol^{-2}, and that of $b = 0.0564$ L mol^{-1}.

Integrative and Advanced Exercises

87. Explain why it is necessary to include the density of Hg(l) and the value of the acceleration due to gravity, *g*, in a precise definition of a standard atmosphere of pressure (page 178).

88. Start with the conditions at points *A*, *B*, and *C* in Figure 6-8. Use Charles's law to calculate the volume of each gas at 0, −100, −200, −250, and −270 °C; show that the volume of each gas becomes zero at −273.15 °C.

89. The "initial" sketch illustrates, both at the macroscopic and molecular levels, an initial condition: 1 mol of a gas at STP. The "final" sketch suggests a final condition. Supply the sketch with as much detail as possible to represent the final condition after each of the following changes.

(a) The pressure is changed to 250 mmHg while standard temperature is maintained.

(b) The temperature is changed to 140 K while standard pressure is maintained.

(c) The pressure is changed to 0.5 atm while the temperature is changed to 550 K.

(d) An additional 0.5 mol of gas is introduced into the cylinder, the temperature is changed to 135 °C, and the pressure is changed to 2.25 atm.

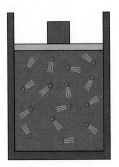 

Initial   Final

90. Complete combustion of 1.110 g of a gaseous hydrocarbon yields 3.613 g CO_2 and 1.109 g H_2O. A 0.288-g sample of the hydrocarbon occupies a volume of 131 mL at 24.8 °C and 753 mmHg. What is the molecular formula of the hydrocarbon?

91. A balloon is inflated with 2.15 ft^3 of He(g) at STP and released. What is the gas pressure in the balloon when it has expanded to a volume of 155 L? Assume a temperature of −20 °C at this altitude.

92. In the drawings at the top of the next column, the diver toy floats at the top of the liquid (left), but when the bottle is squeezed, the diver descends (right). Explain the phenomena illustrated by these drawings.

93. In the photographs below, we see two balloons connected by a stopcock. In the top, one balloon is inflated to a greater volume than the other. In the bottom, we see what happens when the stopcock is opened. Explain this observation, which seems contrary to what we would expect intuitively.

94. A 3.05-g sample of $NH_4NO_3(s)$ is introduced into an evacuated 2.18-L flask and then heated to 250 °C. What is the *total* gas pressure, in atmospheres, in the flask at 250 °C when the NH_4NO_3 has completely decomposed?

$$NH_4NO_3(s) \longrightarrow N_2O(g) + 2 H_2O(g)$$

95. A mixture of 1.00 g H_2 and 8.60 g O_2 is introduced into a 1.500-L flask at 25 °C. **(a)** What is the total gas pressure in the flask? **(b)** When the mixture is ignited, an explosive reaction occurs in which water is the only product. What is the total gas pressure when the flask is returned to 25 °C? (The vapor pressure of water at 25 °C is 23.8 mmHg.)

96. In the reaction of $CO_2(g)$ and sodium peroxide (Na_2O_2), sodium carbonate and oxygen gas are formed.

$$2\,Na_2O_2(s) + 2\,CO_2(g) \longrightarrow 2\,Na_2CO_3(s) + O_2(g)$$

This reaction is used in submarines and space vehicles to remove expired $CO_2(g)$ and to generate some of the $O_2(g)$ required for breathing. Assume the following. Volume of gases exchanged in the lungs: 4.0 L/min; CO_2 content of expired air: 3.8% CO_2 by volume. If the $CO_2(g)$ and $O_2(g)$ in the above reaction are measured at the same temperature and pressure, **(a)** how many milliliters of $O_2(g)$ are produced per minute; **(b)** at what rate is the $Na_2O_2(s)$ consumed, in grams per hour? Assume that the gases are at 25 °C and 735 mmHg.

97. What is the partial pressure of $Cl_2(g)$, in millimeters of mercury, at STP in a gaseous mixture that consists of 46.5% N_2, 12.7% Ne, and 40.8% Cl_2, *by mass?*

98. A gaseous mixture of He and O_2 has a density of 0.518 g/L at 25 °C and 721 mmHg. What is the % He, by mass, in the mixture?

99. When working with a mixture of gases, it is sometimes convenient to use an *apparent molar mass* (a weighted-average molar mass). Think in terms of replacing the mixture by a *hypothetical* single gas. What is the apparent molar mass of air, given that air is 78.08% N_2, 20.95% O_2, 0.93% Ar, and 0.036% CO_2, by volume?

100. Gas cylinder A has a volume of 48.2 L and contains $N_2(g)$ at 8.35 atm at 25 °C. Gas cylinder B, of unknown volume, contains He(g) at 9.50 atm at 25 °C. When the two cylinders are connected and the gases mixed, the pressure in each cylinder becomes 8.71 atm. What is the volume of cylinder B?

101. The accompanying sketch is that of a closed-end manometer. Describe how it is constructed and how the gas pressure is measured. Why is a measurement of P_{bar} not necessary when using this manometer? Explain why the closed-end manometer is more suitable for measuring low pressures and the open-end manometer more suitable for measuring pressures nearer atmospheric pressure.

102. Producer gas is a type of fuel gas made by passing air or steam through a bed of hot coal or coke. A typical producer gas has the following composition in percent

by volume: 8.0% CO_2, 23.2% CO, 17.7% H_2, 1.1% CH_4, and 50.0% N_2.
(a) What is the density of this gas at 23 °C and 763 mmHg, in grams per liter?
(b) What is the partial pressure of CO in this mixture at STP?
(c) What volume of air, measured at 23 °C and 741 Torr, is required for the complete combustion of 1.00×10^3 L of this producer gas, also measured at 23 °C and 741 Torr? (*Hint:* Which three of the constituent gases are combustible?)

103. A *mixture* of $H_2(g)$ and $O_2(g)$ is prepared by electrolyzing 1.32 g water, and the mixture of gases is collected over water at 30 °C when the barometric pressure is 748 mmHg. The volume of "wet" gas obtained is 2.90 L. What must be the vapor pressure of water at 30 °C?

$$2\,H_2O(l) \xrightarrow{\text{electrolysis}} 2\,H_2(g) + O_2(g)$$

104. A 0.168-L sample of $O_2(g)$ is collected over water at 26 °C and a barometric pressure of 737 mmHg. In the gas that is collected, what is the percent water vapor **(a)** by volume; **(b)** by number of molecules; **(c)** by mass? (Vapor pressure of water at 26 °C = 25.2 mmHg.)

105. We have noted that atmospheric pressure depends on altitude. Atmospheric pressure as a function of altitude can be calculated with an equation known as the barometric formula.

$$P = P_0 \times 10^{-Mgh/2.303RT}$$

In this equation, P and P_0 can be in any pressure units, for example, Torr. P_0 is the pressure at sea level, generally taken to be 1.00 atm or its equivalent. The units in the exponential term must be SI units, however. Use the barometric formula to
(a) estimate the barometric pressure at the top of Mt. Whitney in California (altitude: 14,494 ft; assume a temperature of 10 °C)
(b) show that barometric pressure decreases by one-thirtieth in value for every 900-ft increase in altitude

106. If the van der Waals equation is solved for volume, a cubic equation is obtained.
(a) Derive this equation by rearranging equation (6.26).

$$V^3 - n\left(\frac{RT + bP}{P}\right)V^2 + \left(\frac{n^2 a}{P}\right)V - \frac{n^3 ab}{P} = 0$$

(b) What is the volume, in liters, occupied by 185 g $CO_2(g)$ at a pressure of 12.5 atm and 286 K? For $CO_2(g)$, $a = 3.59\ L^2\ atm\ mol^{-2}$ and $b = 0.0427\ L\ mol^{-1}$.

107. A particular equation of state for $O_2(g)$ has the form

$$P\bar{V} = RT\left\{1 + \frac{B}{\bar{V}} + \frac{C}{\bar{V}^2}\right\}$$

where $\bar{V}$ is the molar volume, $B = -21.89\ cm^3/mol$ and $C = 1230\ cm^6/mol$.
(a) Use the equation to calculate the pressure exerted by 1 mol $O_2(g)$ confined to a volume of 500 cm^3 at 273 K.
(b) Is the result calculated in **(a)** consistent with that suggested for $O_2(g)$ by Figure 6-22? Explain.

108. A sounding balloon is a rubber bag, filled with $H_2(g)$ and carrying a set of instruments (the payload). Because this combination of bag, gas, and payload has a smaller mass than a corresponding volume of air, the balloon rises. As the balloon rises, it expands. From the following data, estimate the maximum height to which a spherical balloon can rise: mass of balloon, 1200 g; payload, 1700 g: quantity of $H_2(g)$ in balloon, 120 ft³ at STP; diameter of balloon at maximum height, 25 ft. Air pressure and temperature as function of altitude are:

Alt., km	Pressure, mb	Temp., K
0	1.0×10^3	288
5	5.4×10^2	256
10	2.7×10^2	223
20	5.5×10^1	217
30	1.2×10^1	230
40	2.9×10^0	250
50	8.1×10^{-1}	250
60	2.3×10^{-1}	256

109. A 0.156-g sample of a magnesium–aluminum alloy dissolves completely in an excess of HCl(aq). The liberated $H_2(g)$ is collected over water at 5 °C when the barometric pressure is 752 Torr. After the gas is collected, the water and gas gradually warm to the prevailing room temperature of 23 °C. The pressure of the collected gas is again equalized against the barometric pressure of 752 Torr, and its volume is found to be 202 mL. What is the percent composition of the magnesium–aluminum alloy? (Vapor pressure of water: 6.54 mmHg at 5 °C and 21.07 mmHg at 23 °C)

Feature Problems

110. Shown here is a diagram of Boyle's original apparatus. At the start of the experiment, the length of the air column (A) on the left was 30.5 cm and the heights of mercury in the arms of the tube were equal. When mercury was added to the right arm of the tube, a difference in mercury levels (B) was produced, and the entrapped air on the left was compressed into a shorter length of the tube (smaller volume). For example, in the illustration, A = 27.9 cm and B = 7.1 cm. Boyle's values of A and B, in centimeters, are listed as follows:

A:	30.5	27.9	25.4	22.9	20.3
B:	0.0	7.1	15.7	25.7	38.3
A:	17.8	15.2	12.7	10.2	7.6
B:	53.8	75.4	105.6	147.6	224.6

Barometric pressure at the time of the experiment was 739.8 mmHg. Assuming that the length of the air column (A) is proportional to the volume of air, show that these data do, in fact, conform reasonably well to Boyle's law.

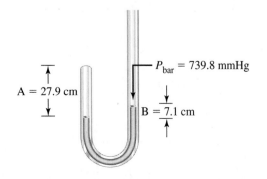

111. An alternative to Figure 6-6 is to plot P against $1/V$. The resulting graph is a straight line passing through the origin. Use Boyle's data from Feature Problem 110 to draw such a straight-line graph. What factors would affect the *slope* of this straight line? Explain.

112. In 1860, Stanislao Cannizzaro showed how Avogadro's hypothesis could be used to establish the atomic masses of elements in *gaseous* compounds. Cannizzaro took the atomic mass of hydrogen to be exactly one and assumed that hydrogen exists as H_2 molecules (molecular mass, 2). Next, he determined the volume of $H_2(g)$ at STP that has a mass of exactly 2 g. This volume is 22.4 L. Now he assumed that 22.4 L of any other gas would have the same number of molecules as in 22.4 L of $H_2(g)$. (Here is where Avogadro's hypothesis entered in.) Finally, he reasoned that the ratio of the mass of 22.4 L of any other gas to the mass of 22.4 L of $H_2(g)$ should be the same as the ratio of their molecular masses. The sketch on the next page illustrates Cannizzaro's reasoning in establishing the atomic weight of oxygen as 16. The gases in the table all contain the element X. Their molecular masses were determined by Cannizzaro's method. Use the percent composition data to deduce the atomic mass of X, the number of atoms of X in each of the gas molecules, and the identity of X.

Compound	Molecular Mass, u	Mass Percent X, %
Nitryl fluoride	65.01	49.4
Nitrosyl fluoride	49.01	32.7
Thionyl fluoride	86.07	18.6
Sulfuryl fluoride	102.07	31.4

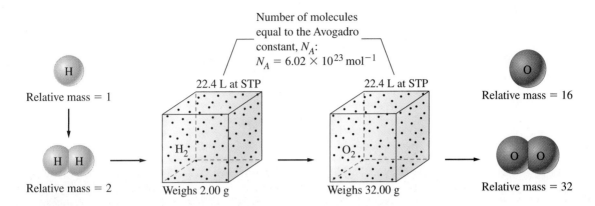

Number of molecules equal to the Avogadro constant, N_A: $N_A = 6.02 \times 10^{23}\ \text{mol}^{-1}$

H — Relative mass = 1

H H — Relative mass = 2

H_2 — 22.4 L at STP — Weighs 2.00 g

O_2 — 22.4 L at STP — Weighs 32.00 g

O — Relative mass = 16

O O — Relative mass = 32

113. The equation $d/P = M/RT$, which can be derived from equation (6.14), suggests that the ratio of the density (d) to pressure (P) of a gas at constant temperature should be a constant. The following gas density data were obtained for $O_2(g)$ at various pressures at 273.15 K.

P, mmHg:	760.00	570.00	380.00	190.00
d, g/L:	1.428962	1.071485	0.714154	0.356985

(a) Calculate values of d/P, and with a graph or by other means determine the ideal value of the term d/P for $O_2(g)$ at 273.15 K.
[*Hint:* The ideal value is that associated with a perfect (ideal) gas.]
(b) Use the value of d/P from part (a) to calculate a precise value for the atomic mass of oxygen, and compare this value with that listed on the inside front cover.

 # eMedia Exercises

114. (a) View the **Motion of a Gas** animation *(eChapter 6-1)* and list the different forces experienced by the molecules. **(b)** Would any of the forces be potentially attractive in nature? **(c)** In what conditions would attractive forces affect the pressure of the system predicted by the ideal gas law?

115. Use the **Ideal Gas Behavior** simulation *(eChapter 6-3)* to simulate the relationships between the different variables found in the ideal gas law. Which variable pairs, when plotted against one another, give rise to data that can be fit with a straight line? Propose a means to extract the value of the ideal gas constant from this activity.

116. Consider the reaction of solid sodium azide as seen in the **Air Bags** movie *(eChapter 6-5)*. **(a)** Calculate the volume of gas produced from the reaction of 30 grams of sodium azide at a pressure of 1.1 atm and a temperature of 25 °C. **(b)** If the exothermic nature of the reaction raised the temperature of the same volume of gas to 100 °C, what would be the resulting pressure within the same volume of the air bag?

117. (a) Use the **Boltzman Distribution** simulation *(eChapter 6-7)* to confirm the relationship given in equation (6.20). Do this by determining the u_{rms} of each gas at a fixed temperature and plotting this value against the square root of the molecular masses of the different gases. **(b)** What is the slope of this plot and how does this relationship confirm the relationship described in equation (6.20)? **(c)** Does the relationship hold at more than one temperature?

118. In the **Diffusion of Bromine Vapor** movie *(eChapter 6-8)* the gas phase molecules experience a resistance to their upward propagation. **(a)** What is the primary source of resistance to the upward propagation of bromine in the test tube? **(b)** Describe the difference between this resistance to propagation with that found in a situation where a gas undergoes effusion. What are the physical properties of a gas which influence both rates of diffusion and effusion?

7

Thermochemistry

Contents

▶ Thermochemistry is a subfield of a larger discipline called *thermodynamics*. We consider broader aspects of thermodynamics in Chapters 20 and 21.

A stream of boiling water transfers heat to the ice cube, causing it to melt. The transfer of heat between substances is an important aspect of thermochemistry.

Natural gas consists mostly of methane, CH_4. As we learned in Chapter 4, the complete combustion of a hydrocarbon such as methane yields carbon dioxide and water as products. More important, however, is another "product" of this reaction that we have not previously mentioned: heat. We can use this heat to produce hot water in a water heater, to heat a house, or to cook food.

Thermochemistry is the branch of chemistry concerned with heat effects accompanying chemical reactions, and much of this chapter deals with determining quantities of heat—by measurement and by calculation. Some of these calculations will allow us to establish, indirectly, a quantity of heat that would be difficult or impossible to measure directly. This type of calculation, which relies on compilations of tabulated data, will come up again in later chapters. Finally, we can use concepts introduced in this chapter to answer a host of practical questions, such as why natural gas is a better fuel than coal and why the energy value of fats is greater than that of carbohydrates and proteins.

7-1 Getting Started: Some Terminology

In this section, we introduce and define some very basic terms. Most are discussed in greater detail in later sections, and your understanding of these terms should grow as you proceed through the chapter.

We call the part of the universe chosen for study a **system**. A system can be as large as all the oceans on Earth or as small as the contents of a beaker. Most of the systems we will examine will be small. Our study will focus on the system's interactions, that is, on the transfer of *energy* (as heat and work) and *matter* between a system and its surroundings. The **surroundings** are that part of the universe outside the system in which these interactions can be detected. Figure 7-1 pictures three common systems: first, as we see them and, then, in an abstract form that chemists commonly use. The system in Figure 7-1a is an *open* system—it can freely exchange energy and matter with its surroundings. The one in Figure 7-1b is a *closed* system—it can exchange energy with its surroundings, but not matter. And the one in Figure 7-1c is an approximation of an *isolated* system—a system that does not interact with its surroundings.

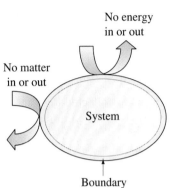

▲ **Isolated system**
No interactions across the system boundary.

 Types of Energy activity

▶ FIGURE 7-1
Systems and their surroundings.
(a) Open system. The beaker of hot coffee transfers energy to the surroundings—it loses heat as it cools. Matter is also transferred in the form of water vapor. **(b) Closed system**. The flask of hot coffee transfers energy (heat) to the surroundings as it cools. Because the flask is stoppered, no water vapor escapes and no matter is transferred. **(c) Isolated system**. Hot coffee in an insulated flask approximates an isolated system. No water vapor escapes, and for a time at least, little heat is transferred to the surroundings. (Eventually, though, the coffee in the flask cools to room temperature.)

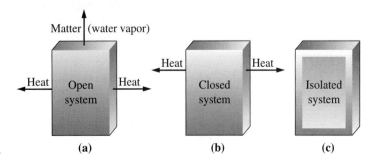

In the remainder of this section, we will say more, in a general way, about energy and its relationship to work. Like many other scientific terms, *energy* is derived from Greek. It means "work within." **Energy** is the capacity to do work. **Work** is done when a force acts through a distance. Moving objects do work when they slow down or are stopped. Thus, when one billiard ball strikes another and sets it in motion, work is done. The energy of a moving object is called **kinetic energy** (the word *kinetic* means "motion" in Greek). We can see the relationship between work and energy by comparing the units for these two quantities. The kinetic energy of an object is based on its mass (m) and velocity (u) through the first equation below; work (w) is related to force [mass (m) × acceleration (a)] and distance (d) by the second.

$$\text{kinetic energy} = \tfrac{1}{2} \times m\,(\text{kg}) \times [u(\text{m/s})]^2 \tag{7.1}$$

$$\text{work} = \text{force} \times \text{distance} = [m\,(\text{kg}) \times a\,(\text{m s}^{-2})] \times d\,(\text{m})$$

When the units for the expressions for work and energy are collected together, in both cases, the resultant unit is kg m^{2}s^{-2}. This corresponds to the SI unit of energy, called the joule (J). That is, 1 *joule* (J) = 1 kg m^{2}s^{-2}.

The bouncing ball in Figure 7-2 suggests something about the nature of energy and work. First, to lift the ball to the starting position, we have to apply a force through a distance (to overcome the force of gravity). The work we do is "stored" in the ball as energy. This stored energy has the potential to do work when released and is therefore called potential energy. **Potential energy** is energy due to condition, position, or composition; it is an energy associated with forces of attraction or repulsion between objects.

When we release the ball, it is pulled toward Earth's center by the force of gravity—it falls. Potential energy is converted to kinetic energy during this fall. The kinetic energy reaches its maximum just as the ball strikes the surface. On its rebound, the kinetic energy of the ball decreases (the ball slows down) and its potential energy increases (the ball rises). If the collision of the ball with the surface were perfectly *elastic*, like collisions between molecules in the kinetic-molecular theory, the sum of the potential and kinetic energies of the ball would remain constant. The ball would reach the same maximum height on each rebound, and the ball would bounce forever. But we know this doesn't happen—the bouncing ball soon comes to rest. All the energy originally invested in the ball as potential energy (by raising it to its initial position) eventually appears as additional kinetic energy of the atoms and molecules that make up the ball, the surface, and the surrounding air. This kinetic energy, which is associated with random molecular motion, is called **thermal energy**.

▶ **FIGURE 7-2**
Potential energy (P.E.) and kinetic energy (K.E.)
The energy of the bouncing tennis ball changes continuously from potential to kinetic energy, back to potential energy, and so on. The maximum in potential energy is at the top of each bounce, and the maximum in kinetic energy occurs at the moment of impact. The sum of P.E. and K.E. decreases with each bounce as the thermal energies of the ball and the surroundings increase. The ball soon comes to rest.

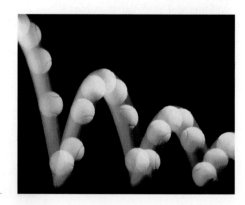

In general, thermal energy is proportional to the temperature of a system, as suggested by the kinetic theory of gases. The more vigorous the motion of the molecules in the system, the hotter the sample and the greater is its thermal energy. However, the thermal energy of a system also depends on the number of particles present, so that a small sample at a high temperature (for example, a cup of coffee at 75 °C) may have less thermal energy than a larger sample at a lower temperature (for example, a swimming pool at 30 °C). Thus, temperature and thermal energy must be carefully distinguished.

Equally important, we need to distinguish between energy changes produced by the action of forces through distances—*work*—and those involving the transfer of thermal energy—*heat*. We discuss several matters pertaining to heat in the next two sections, and to work in the section following those.

Work and Heat simulation

7-2   Heat

Heat is energy transferred between a system and its surroundings as a result of a temperature difference. Energy, as heat, passes from a warmer body (with a higher temperature) to a colder body (with a lower temperature). At the molecular level, molecules of the warmer body, through collisions, lose kinetic energy to those of the colder body. Thermal energy is transferred—heat "flows"—until the average molecular kinetic energies of the two bodies become the same, until the temperatures become equal. Heat, like work, describes energy in transit between a system and its surroundings.

Not only can heat transfer cause a change in temperature, but in some instances, it can change a state of matter. For example, when a solid is heated, the molecules, atoms, or ions of the solid move with greater vigor and eventually break free from their neighbors by overcoming the attractive forces between them. Energy is required to overcome these attractive forces. During the process of melting, the temperature remains constant as a thermal energy transfer (heat) is used to overcome the forces holding the solid together. A process occurring at a constant temperature is said to be *isothermal*. Once a solid has melted completely, any further heat flow will raise the temperature of the resulting liquid.

Although we commonly use expressions like "heat is lost," "heat is gained," "heat flows," and "the system loses heat to the surroundings," you should not take these statements to mean that a system contains heat. It does not. The energy content of a system, as we shall see in Section 7-5, is a quantity called the internal energy. Heat is simply a form in which a quantity of energy may be *transferred* across a boundary between a system and its surroundings.

It is reasonable to expect that the quantity of heat, q, required to change the temperature of a substance depends on

- how much the temperature is to be changed
- the quantity of substance
- the nature of the substance (type of atoms or molecules)

Historically, the quantity of heat required to change the temperature of one *gram* of water by one degree Celsius has been called the **calorie (cal)**. The calorie is a small unit of energy, and the unit *kilocalorie* (kcal) has also been widely used. The SI unit for heat is simply the basic SI energy unit, the joule (J).

▲ **James Joule (1818–1889)— an amateur scientist**
Joule's primary occupation was running a brewery, but he also conducted scientific research in a home laboratory. His precise measurements of quantities of heat formed the basis of the law of conservation of energy.

$$1 \text{ cal} = 4.184 \text{ J} \tag{7.2}$$

Although we will use the joule almost exclusively in this text, you should also be familiar with the calorie. It is widely encountered in older scientific literature and still used to some extent. Kilocalories are commonly used for measuring the energy content of foods.

The quantity of heat required to change the temperature of a system by one degree is called the **heat capacity** of the system. If the system is a mole of substance, we can use the term *molar heat capacity*. If it is one gram of substance, we call it the *specific heat capacity*, or more commonly, **specific heat** (sp ht).[*] The specific heat of water depends somewhat on temperature, but, over the range from 0 to 100 °C, its value is about

$$\frac{1.00 \text{ cal}}{\text{g} \,°\text{C}} = 1 \text{ cal g}^{-1}\,°\text{C}^{-1} = \frac{4.18 \text{ J}}{\text{g} \,°\text{C}} = 4.18 \text{ J g}^{-1}\,°\text{C}^{-1} \qquad (7.3)$$

In Example 7-1, our objective is to calculate a quantity of heat, based on how much of a substance we have, the specific heat of that substance, and its temperature change. In some later examples in the chapter, we will routinely do calculations of the type in Example 7-1 as part of a larger problem.

EXAMPLE 7-1

Calculating a Quantity of Heat. How much heat is required to raise the temperature of 7.35 g of water from 21.0 to 98.0 °C? (Assume the specific heat of water is 4.18 J g^{-1} °C^{-1} throughout this temperature range.)

Solution

The specific heat is the heat capacity of 1.00 g water: $\dfrac{4.18 \text{ J}}{\text{g water} \,°\text{C}}$. The heat capacity of the system (7.35 g water) is $7.35 \text{ g water} \times \dfrac{4.18 \text{ J}}{\text{g water} \,°\text{C}} = 30.7 \dfrac{\text{J}}{°\text{C}}$. The required temperature change in the system is $(98.0 - 21.0)\,°\text{C} = 77.0\,°\text{C}$. The heat required to produce this temperature change $= 30.7 \dfrac{\text{J}}{°\text{C}} \times 77.0\,°\text{C} = 2.36 \times 10^3 \text{ J}$.

Practice Example A: How much heat, in kilojoules (kJ), is required to raise the temperature of 8.00 ounces of ice water (237 g) from 4.0 to 37.0 °C (body temperature)?

Practice Example B: How much heat, in kilojoules (kJ), is required to raise the temperature of 2.50 kg Hg(l) from −20.0 to −6.0 °C? Assume a density of 13.6 g/mL and a molar heat capacity of 28.0 J mol^{-1} °C^{-1} for Hg(l).

The line of reasoning we used in Example 7-1 can be summarized in the following basic equation relating a mass of substance, a temperature change, and a quantity of heat.

$$\text{quantity of heat} = \underbrace{\text{mass of substance} \times \text{specific heat}}_{\text{heat capacity} = C} \times \text{temperature change} \quad (7.4)$$

▶ The Greek letter *delta*, Δ, indicates a *change* in some quantity.

$$q = m \times \text{specific heat} \times \Delta T = C \times \Delta T \qquad (7.5)$$

[*]The original meaning of specific heat was that of a *ratio:* the quantity of heat required to change the temperature of a mass of substance divided by the quantity of heat required to produce the same temperature change in the same mass of water. This would make specific heat dimensionless. The meaning given here is more commonly used.

In equation (7.5), the temperature change is expressed as $\Delta T = T_f - T_i$, where T_f is the final temperature and T_i is the initial temperature. When the temperature of a system increases $(T_f > T_i)$, ΔT is *positive*. A positive q signifies that heat is absorbed or *gained* by the system. When the temperature of a system decreases $(T_f < T_i)$, ΔT is *negative*. A negative q signifies that heat is evolved or *lost* by the system.

▶ The symbol > means "greater than," and < means "less than."

Another idea that enters into calculations of quantities of heat is the **law of conservation of energy:** In interactions between a system and its surroundings, the total energy remains *constant*—energy is neither created nor destroyed. Applied to the exchange of heat, this means that

$$q_{\text{system}} + q_{\text{surroundings}} = 0 \tag{7.6}$$

Thus, heat *lost* by a system is *gained* by its surroundings, and vice versa.

$$q_{\text{system}} = -q_{\text{surroundings}} \tag{7.7}$$

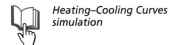
Heating–Cooling Curves simulation

Experimental Determination of Specific Heats

Let us consider how the law of conservation of energy is used in the experiment outlined in Figure 7-3. The object is to determine the specific heat of lead. The transfer of energy, as heat, from the lead to the cooler water causes the temperature of the lead to decrease and that of the water to increase, until the lead and water are at the same temperature. Either the lead or the water can be considered the system. If we consider lead to be the system, we can write $q_{\text{lead}} = q_{\text{system}}$. Furthermore, if the lead and water are maintained in a thermally insulated enclosure, we can assume that $q_{\text{water}} = q_{\text{surroundings}}$. Then, applying equation (7.7), we have

$$q_{\text{lead}} = -q_{\text{water}} \tag{7.8}$$

We complete the calculation in Example 7-2.

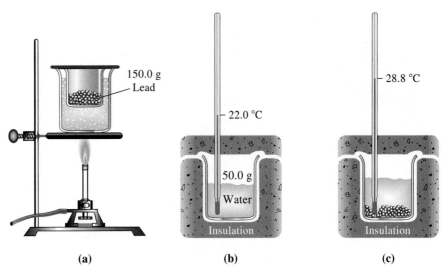

(a) (b) (c)

▲ **FIGURE 7-3   Determining the specific heat of lead—Example 7-2 illustrated**
(a) A 150.0-g sample of lead is heated to the temperature of boiling water (100.0 °C).
(b) A 50.0-g sample of water is added to a thermally insulated beaker, and its temperature is found to be 22.0 °C. **(c)** The hot lead is dumped into the cold water, and the temperature of the final lead–water mixture is 28.8 °C.

TABLE 7.1 Specific Heats of Several Solid Elements (in $J\,g^{-1}\,°C^{-1}$)	
Metals	
Lead	0.128
Copper	0.385
Iron	0.449
Aluminum	0.903
Magnesium	1.024
Nonmetals	
Selenium	0.321
Sulfur	0.706
Phosphorus	0.777
Metalloids	
Tellurium	0.202
Arsenic	0.329

KEEP IN MIND ▶

that if you know any four of the five quantities—q, m, specific heat, T_f, T_i—you can solve equation (7.5) for the remaining one.

▲ An ice cube thawing on a block of aluminum alloy (miracle thaw).

EXAMPLE 7-2

Determining a Specific Heat from Experimental Data. Use data presented in Figure 7-3 to calculate the specific heat of lead.

Solution

First, let us use equation (7.5) to calculate q_{water}.

$$q_{water} = 50.0 \text{ g water} \times \frac{4.18 \text{ J}}{\text{g water °C}} \times (28.8 - 22.0)\,°C = 1.4 \times 10^3 \text{ J}$$

From equation (7.8) we can write

$$q_{lead} = -q_{water} = -1.4 \times 10^3 \text{ J}$$

Now, from equation (7.5) again, we obtain

$$q_{lead} = 150.0 \text{ g lead} \times \text{specific heat of lead} \times (28.8 - 100.0)\,°C$$
$$= -1.4 \times 10^3 \text{ J}$$

$$\text{specific heat of lead} = \frac{-1.4 \times 10^3 \text{ J}}{150.0 \text{ g lead} \times (28.8 - 100.0)\,°C}$$

$$= \frac{-1.4 \times 10^3 \text{ J}}{150.0 \text{ g lead} \times -71.2\,°C} = 0.13 \text{ J g}^{-1}\,°C^{-1}$$

Practice Example A: When 1.00 kg lead (specific heat = 0.13 J g^{-1} °C^{-1}) at 100.0 °C is added to a quantity of water at 28.5 °C, the final temperature of the lead–water mixture is 35.2 °C. What is the mass of water present?

Practice Example B: A 100.0-g copper sample (specific heat = 0.385 J g^{-1} °C^{-1}) at 100.0 °C is added to 50.0 g water at 26.5 °C. What is the final temperature of the copper–water mixture?

Significance of Specific-Heat Values

Table 7.1 lists the specific heats of several solid elements. The relatively high specific heat of aluminum compared with other metals helps to account for its use in "miracle thaw" products designed to thaw frozen foods rapidly. The aluminum cools only slowly as it transfers heat to the frozen food, and the food thaws more quickly than when simply exposed to air.

Because of their greater complexity at the molecular level, compounds generally have more ways of storing internal energy than do the elements; they tend to have higher specific heats. Water, for example, has a specific heat that is more than 30 times greater than that of lead. We need a much larger quantity of heat to change the temperature of a sample of water than of an equal mass of a metal. To return to the food-thawing example, we expect frozen food placed in room-temperature water to thaw much faster than if just exposed to room-temperature air.

An environmental consequence of the high specific heat of water is found in the effect of large lakes on local climates. Because a lake takes much longer to heat up in summer and cool in winter than do other types of terrain, lakeside communities tend to be cooler in summer and warmer in winter than communities more distant from the lake.

7-3 Heats of Reaction and Calorimetry

In Section 7-1, we introduced the notion of *thermal energy*—kinetic energy associated with random molecular motion. Another type of energy that contributes to the internal energy of a system is **chemical energy**. This is energy associated with chemical bonds and intermolecular attractions. If we think of a chemical reaction as a process in which some chemical bonds are broken and others are formed, then, in general, we expect the chemical energy of a system to change as a result of a reaction. Furthermore, we might expect some of this energy change to appear as heat. A **heat of reaction**, q_{rxn}, is the quantity of heat exchanged between a system and its surroundings when a chemical reaction occurs within the system, at *constant temperature*. One of the most common reactions studied is the combustion reaction. This is such a common reaction that we often refer to the *heat of combustion* when describing the heat released by a combustion reaction.

If a reaction occurs in an *isolated* system, that is, one that exchanges no matter or energy with its surroundings, the reaction produces a change in the thermal energy of the system—the temperature either increases or decreases. Imagine that the previously isolated system is allowed to interact with its surroundings. The heat of reaction is the quantity of heat exchanged between the system and its surroundings as the system is restored to its initial temperature. In actual practice, we do not physically restore the system to its initial temperature. We calculate the quantity of heat that *would be* exchanged in this restoration. We do this by having a probe (thermometer) within the system to record the temperature change produced by the reaction.

At this point, let us introduce two widely used terms related to heats of reaction— exothermic and endothermic reactions. An **exothermic** reaction is one that produces a temperature increase in an isolated system or, in a nonisolated system, gives off heat to the surroundings. For an exothermic reaction, the heat of reaction is a negative quantity ($q_{rxn} < 0$). In an **endothermic** reaction, the corresponding situation is a temperature decrease in an isolated system or a gain of heat from the surroundings by a nonisolated system. In this case, the heat of reaction is a positive quantity ($q_{rxn} > 0$). Figure 7-4 pictures one exothermic and one endothermic reaction.

Heats of reaction are experimentally determined in a **calorimeter**, a device for measuring quantities of heat. We will consider two types of calorimeters in this section, and we will treat both of them as *isolated* systems.

Bomb Calorimetry

The type of calorimeter shown in Figure 7-5, called a **bomb calorimeter**, is ideally suited for measuring the heat evolved in a combustion reaction. The *system* is everything within the double-walled outer jacket of the calorimeter. This includes the bomb and its contents, the water in which the bomb is immersed, the thermometer,

(a)

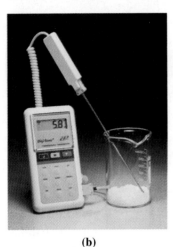

(b)

▲ **FIGURE 7-4**
Exothermic and endothermic reactions
(a) An exothermic reaction.
Slaked lime [$Ca(OH)_2$] is produced by the action of water on quicklime (CaO). The reactants are mixed at room temperature, but the temperature of the mixture rises to 40.5 °C.

$$CaO(s) + H_2O(l) \longrightarrow Ca(OH)_2(s)$$

(b) An endothermic reaction.
$Ba(OH)_2 \cdot 8\,H_2O(s)$ and $NH_4Cl(s)$ are mixed at room temperature, and the temperature falls to 5.8 °C in the reaction:

$$Ba(OH)_2 \cdot 8\,H_2O(s) + 2\,NH_4Cl(s) \longrightarrow BaCl_2 \cdot 2\,H_2O(s) + 2\,NH_3(aq) + 8\,H_2O(l)$$

▶ **FIGURE 7-5**
A bomb calorimeter assembly
An iron wire is embedded in the sample in the lower half of the bomb. The bomb is assembled and filled with $O_2(g)$ at high pressure. The assembled bomb is immersed in water in the calorimeter, and the initial temperature is measured. A short pulse of electric current heats the sample, causing it to ignite. The final temperature of the calorimeter assembly is determined after the combustion.

Because the bomb confines the reaction mixture to a fixed volume, the reaction is said to occur at *constant volume*. We explore the significance of this fact on page 235.

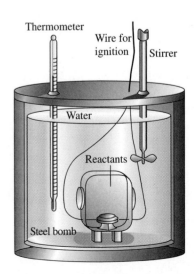

Thermometer
Wire for ignition
Stirrer
Water
Reactants
Steel bomb

KEEP IN MIND ▶

that the temperature of a reaction mixture usually changes during a reaction, so that we must return the mixture to the initial temperature (actually or hypothetically) before we assess how much heat is exchanged with the surroundings.

Bomb Calorimetry simulation

the stirrer, and so on. The system is *isolated* from its surroundings. When the combustion reaction occurs, chemical energy is converted to thermal energy, and the temperature of the system rises. The heat of reaction, as we described earlier, is the quantity of heat that the system would have to *lose* to its surroundings to be restored to its initial temperature. This quantity of heat, in turn, is just the *negative* of the thermal energy gained by the calorimeter and its contents ($q_{calorim}$).

$$q_{rxn} = -q_{calorim} \text{ (where } q_{calorim} = q_{bomb} + q_{water} \ldots) \qquad (7.9)$$

If we assemble the calorimeter in exactly the same way each time we use it—that is, use the same bomb, the same quantity of water, and so on—we can define a *heat capacity of the calorimeter*. This is the quantity of heat required to raise the temperature of the calorimeter assembly by one degree Celsius. When we multiply this heat capacity by the observed temperature change, we get $q_{calorim}$.

$$q_{calorim} = \text{heat capacity of calorim} \times \Delta T \qquad (7.10)$$

And from $q_{calorim}$, we then establish q_{rxn}, as in Example 7-3, where we determine the heat of combustion of sucrose (table sugar).

EXAMPLE 7-3

Using Bomb Calorimetry Data to Determine a Heat of Reaction. The combustion of 1.010 g sucrose, $C_{12}H_{22}O_{11}$, in a bomb calorimeter causes the temperature to rise from 24.92 to 28.33 °C. The heat capacity of the calorimeter assembly is 4.90 kJ/°C. **(a)** What is the heat of combustion of sucrose, expressed in kilojoules per mole of $C_{12}H_{22}O_{11}$? **(b)** Verify the claim of sugar producers that one teaspoon of sugar (about 4.8 g) contains only 19 Calories.

Solution

(a) First we can calculate $q_{calorim}$ with equation (7.10).

$$q_{calorim} = 4.90 \text{ kJ/°C} \times (28.33 - 24.92)°C = (4.90 \times 3.41) \text{ kJ} = 16.7 \text{ kJ}$$

Now, using equation (7.9), we get

$$q_{rxn} = -q_{calorim} = -16.7 \text{ kJ}$$

This is the heat of combustion of the 1.010-g sample.
Per gram $C_{12}H_{22}O_{11}$

$$q_{rxn} = \frac{-16.7 \text{ kJ}}{1.010 \text{ g } C_{12}H_{22}O_{11}} = -16.5 \text{ kJ/g } C_{12}H_{22}O_{11}$$

Per mole $C_{12}H_{22}O_{11}$

$$q_{rxn} = \frac{-16.5 \text{ kJ}}{\text{g } C_{12}H_{22}O_{11}} \times \frac{342.3 \text{ g } C_{12}H_{22}O_{11}}{1 \text{ mol } C_{12}H_{22}O_{11}}$$
$$= -5.65 \times 10^3 \text{ kJ/mol } C_{12}H_{22}O_{11}$$

(b) To determine the caloric content of sucrose, we can use the heat of combustion per gram of sucrose determined in part **(a)**, together with a factor to convert from kilojoules to kilocalories. (Because 1 cal = 4.184 J, 1 kcal = 4.184 kJ.)

$$? \text{ kcal} = \frac{4.8 \text{ g } C_{12}H_{22}O_{11}}{\text{tsp}} \times \frac{-16.5 \text{ kJ}}{\text{g } C_{12}H_{22}O_{11}} \times \frac{1 \text{ kcal}}{4.184 \text{ kJ}} = \frac{-19 \text{ kcal}}{\text{tsp}}$$

1 food Calorie (1 Calorie with a capital C) is actually 1000 cal, or 1 kcal. Therefore, 19 kcal = 19 Calories. The claim is justified.

Practice Example A: Vanillin is a natural constituent of vanilla. It is also manufactured for use in artificial vanilla flavoring. The combustion of 1.013 g of vanillin, $C_8H_8O_3$ in the same bomb calorimeter as in Example 7-3 causes the temperature to rise from 24.89 to 30.09 °C. What is the heat of combustion of vanillin, expressed in kilojoules per mole of $C_8H_8O_3$?

Practice Example B: The heat of combustion of benzoic acid is -26.42 kJ/g. The combustion of a 1.176-g sample of benzoic acid ($HC_7H_5O_2$) causes a temperature *increase* of 4.96 °C in a bomb calorimeter assembly. What is the heat capacity of the assembly?

▶ The heat capacity of a bomb calorimeter must be determined by experiment.

▲ **FIGURE 7-6**
A Styrofoam "coffee-cup" calorimeter
The reaction mixture is in the inner cup. The outer cup provides additional thermal insulation from the surrounding air. The cup is closed off with a cork stopper through which a thermometer and a stirrer are inserted and immersed into the reaction mixture.
  The reaction in the calorimeter occurs under the *constant pressure* of the atmosphere. We consider the difference between constant-volume and constant-pressure reactions in Section 7-6.

The "Coffee-Cup" Calorimeter

In the general chemistry laboratory you are much more likely to run into the simple calorimeter pictured in Figure 7-6 than a bomb calorimeter. We mix the reactants (generally in aqueous solution) in a Styrofoam cup and measure the temperature change. Styrofoam is a good heat insulator, so there is very little heat transfer between the cup and the surrounding air. We treat the system—the cup and its contents—as an *isolated* system.

As with the bomb calorimeter, we define the heat of reaction as the quantity of heat that would be exchanged with the surroundings in restoring the calorimeter to its initial temperature. But, again, we do not physically restore the calorimeter to its initial conditions. We simply take the heat of reaction to be the *negative* of the quantity of heat producing the temperature change in the calorimeter. That is, we use equation (7.9): $q_{rxn} = -q_{calorim}$.

In Example 7-4, we make certain assumptions to simplify the calculation, but for more precise measurements, these assumptions would not be made (see Exercise 45).

EXAMPLE 7-4

Determining a Heat of Reaction from Calorimetric Data. In the neutralization of a strong acid with a strong base, the essential reaction is the combination of $H^+(aq)$ and $OH^-(aq)$ to form water (recall page 148).

$$H^+(aq) + OH^-(aq) \longrightarrow H_2O(l)$$

Two solutions, 100.0 mL of 1.00 M HCl(aq) and 100.0 mL of 1.00 M NaOH(aq), both initially at 21.1 °C, are added to a Styrofoam cup calorimeter and allowed to react. The temperature rises to 27.8 °C. Determine the heat of the neutralization reaction, expressed per mole of H_2O formed. Is the reaction endothermic or exothermic?

In addition to assuming that the calorimeter is an isolated system, assume that all there is in the system to absorb heat is 200.0 mL of water. This assumption ignores the fact that 0.10 mol each of NaCl and H_2O are formed in the reaction, that the density of the resulting NaCl(aq) is not quite 1.00 g/mL, and that its specific heat is not quite 4.18 J g^{-1} °C^{-1}. Also, ignore the small heat capacity of the Styrofoam cup itself.

Solution

Because the reaction is a neutralization reaction, let us call the heat of reaction q_{neutr}. Now, according to equation (7.9), $q_{neutr} = -q_{calorim}$, and if we make the assumptions described above,

$$q_{calorim} = 200.0 \text{ mL} \times \frac{1.00 \text{ g}}{\text{mL}} \times \frac{4.18 \text{ J}}{\text{g °C}} \times (27.8 - 21.1)\,°C = 5.6 \times 10^3 \text{ J}$$

$$q_{neutr} = -q_{calorim} = -5.6 \times 10^3 \text{ J} = -5.6 \text{ kJ}$$

In 100.0 mL of 1.00 M HCl, the amount of H^+ is

$$? \text{ mol } H^+ = 0.1000 \text{ L} \times \frac{1.00 \text{ mol HCl}}{1 \text{ L}} \times \frac{1 \text{ mol } H^+}{1 \text{ mol HCl}} = 0.100 \text{ mol } H^+$$

Similarly, in 100.0 mL of 1.00 M NaOH there is 0.100 mol OH^-. Thus, the H^+ and the OH^- combine to form 0.100 mol H_2O. (The two are in *stoichiometric* proportions; neither is in excess.)

The amount of heat produced per mole of H_2O is

$$q_{neutr} = \frac{-5.6 \text{ kJ}}{0.100 \text{ mol } H_2O} = -56 \text{ kJ/mol } H_2O$$

Because q_{neutr} is a *negative* quantity, the neutralization reaction is *exothermic*.

Practice Example A: Two solutions, 100.0 mL of 1.00 M $AgNO_3(aq)$ and 100.0 mL of 1.00 M NaCl(aq), both initially at 22.4 °C, are added to a Styrofoam cup calorimeter and allowed to react. The temperature rises to 30.2 °C. Determine q_{rxn} per mole of AgCl(s) in the reaction.

$$Ag^+(aq) + Cl^-(aq) \longrightarrow AgCl(s)$$

Practice Example B: Two solutions, 100.0 mL of 1.020 M HCl and 50.0 mL of 1.988 M NaOH, both initially at 24.52 °C, are mixed in a Styrofoam cup calorimeter. What will be the final temperature of the mixture? Make the same assumptions, and use the heat of neutralization established in Example 7-4.

(*Hint:* The reactants are not in stoichiometric proportions. Which is the limiting reactant?)

7-4 Work

We have just learned that chemical reactions are generally accompanied by heat effects. In some reactions, **work** is also involved. That is, the system may do work on its surroundings, or vice versa. Consider the decomposition of potassium chlorate to potassium chloride and oxygen. Suppose that this decomposition is carried out in the strange vessel pictured in Figure 7-7. The walls of the container resist moving under the pressure of the expanding $O_2(g)$ except for the piston that closes off the cylindrical top of the vessel. The pressure of the $O_2(g)$ exceeds the atmospheric pressure and the piston is lifted—the system does work on the surroundings. Can you see that even if the piston were removed, work still would be done as the expanding $O_2(g)$ pushed aside other atmospheric gases? Work involved in the expansion or compression of gases is called **pressure–volume work**.

Now let's switch to a somewhat simpler situation to see how we can calculate a quantity of P–V work.

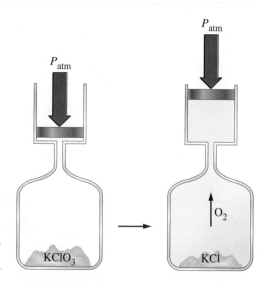

▶ **FIGURE 7-7**
Illustrating work (expansion) during a chemical reaction
The oxygen gas that is formed pushes back the weight and, in doing so, does work on the surroundings.

Work of Gas Expansion
animation

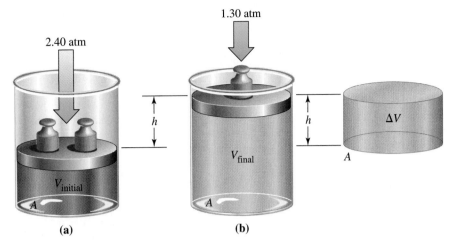

(a)   (b)

▲ **FIGURE 7-8   Pressure–volume work**
When the external pressure on the confined gas is suddenly lowered to 1.30 atm, the gas expands, pushing the piston up by the distance, h. The increase in volume of the gas (ΔV) is the product of the cross-sectional area of the cylinder (A) and the distance (h).

In Figure 7-8a, a quantity of gas is confined in a cylinder by a freely moving piston. The gas is maintained at its initial volume by an external pressure of 2.40 atm. When the external pressure is reduced to 1.30 atm, the gas expands by pushing up the piston by the distance h (Figure 7-8b). Now let us see how pressure and volume enter into calculating how much *pressure–volume work* the expanding gas does.

The "pressure" part of the pressure–volume work is the final external pressure, 1.30 atm. This is the external pressure against which the gas must work. Now we can use equation (7.1) to calculate the work done.

$$\text{Work } (w) = \text{force } (F) \times \text{distance } (h) = P \times A \times h$$

Equation (7.11) is the expression we will use to represent a quantity of pressure-volume work.

$$w = -P_{ext} \times \Delta V \tag{7.11}$$

Two significant features to note in equation (7.11) are (1) the *negative* sign and (2) the term P_{ext}. The negative sign is necessary to conform to sign conventions that we introduce in the next section. When a gas expands, ΔV is positive and w is negative, signifying that energy leaves the system as work. When a gas is compressed, ΔV is negative and w is positive, signifying that energy (as work) enters the system. The term P_{ext} is the *external* pressure—the pressure against which a system expands or the applied pressure that compresses a system. In many instances the internal pressure in a system will be essentially equal to the external pressure, in which case we will express the pressure in equation (7.11) simply as P.

If pressure is stated in atmospheres and volume in liters, the unit of work is the liter-atmosphere, L atm. The SI unit of work is the joule, J. The conversion factor between these two units of work can be obtained from the gas constant, R.

$$8.3145 \text{ m}^3 \text{ Pa mol}^{-1} \text{ K}^{-1} = 8.3145 \text{ m}^3 \text{ N m}^{-2} \text{ mol}^{-1} \text{ K}^{-1} = 8.3145 \text{ kg m}^2 \text{ s}^{-2} \text{ mol}^{-1} \text{ K}^{-1}$$

$$8.3145 \text{ J mol}^{-1} \text{ K}^{-1} = 0.082057 \text{ L atm mol}^{-1} \text{ K}^{-1}$$

and

$$\frac{8.3145 \text{ J}}{0.082057 \text{ L atm}} = 101.33 \text{ J/L atm}$$

that the ideal gas equation embodies Boyle's law: The volume of a fixed amount of gas at a fixed temperature is inversely proportional to the pressure. Thus, we could simply write that

$$V_{final} = 1.02 \text{ L} \times \frac{2.40 \text{ atm}}{1.30 \text{ atm}}$$

$$V_{final} = 1.88 \text{ L} \quad \blacktriangleright$$

Translational

Rotational

Vibrational

Electrostatic
(Intermolecular attractions)

▲ **FIGURE 7-9**
Some contributions to the internal energy of a system
The models represent water molecules, and the arrows, the types of motion they can undergo. In the intermolecular attractions between water molecules, the symbols $\delta+$ and $\delta-$ signify a separation of charge, producing centers of positive and negative charge that are smaller than ionic charges. These intermolecular attractions are discussed in Chapter 13.

EXAMPLE 7-5

Calculating Pressure–Volume Work. Suppose the gas in Figure 7-8 is 0.100 mol He at 298 K. How much work, in joules, is associated with its expansion at constant temperature?

Solution

We are given enough data to calculate the initial and final gas volumes (note that the identity of the gas does not enter into the calculations because we are assuming ideal gas behavior). Once we have these volumes, we can obtain ΔV. The external pressure term in the pressure–volume work is the *final* pressure—1.30 atm. Finally, the product $-P_{ext} \times \Delta V$ must be multiplied by a factor to convert work in L atm to work in joules.

$$V_{initial} = \frac{nRT}{P_{initial}} = \frac{0.100 \text{ mol} \times 0.0821 \text{ L atm mol}^{-1} \text{ K}^{-1} \times 298 \text{ K}}{2.40 \text{ atm}} = 1.02 \text{ L}$$

$$V_{final} = \frac{nRT}{P_{final}} = \frac{0.100 \text{ mol} \times 0.0821 \text{ L atm mol}^{-1} \text{ K}^{-1} \times 298 \text{ K}}{1.30 \text{ atm}} = 1.88 \text{ L}$$

$$\Delta V = V_{final} - V_{initial} = 1.88 \text{ L} - 1.02 \text{ L} = 0.86 \text{ L}$$

$$w = -P_{ext} \times \Delta V = -1.30 \text{ atm} \times 0.86 \text{ L} \times \frac{101 \text{ J}}{1 \text{ L atm}} = -1.1 \times 10^2 \text{ J}$$

The negative value signifies that the expanding gas does work on its surroundings.

Practice Example A: How much work, in joules, is involved when 0.225 mol N_2 at a constant temperature of 23 °C is allowed to expand 1.50 L in volume against an external pressure of 0.750 atm?

(*Hint:* How much of this information is required?)

Practice Example B: How much work, in joules, is involved in compressing 50.0 g $N_2(g)$ in a 75.0-L cylinder like that in Figure 7-8 when an external pressure of 2.50 atm is applied at a constant temperature of 20.0 °C?

Pressure–volume work is the type of work performed by explosives and by the gases formed in the combustion of gasoline in an automobile engine. We will say more about pressure–volume work in chemical reactions in Section 7-6.

7-5 The First Law of Thermodynamics

The absorption or evolution of heat and the performance of work require changes in energy of a system and its surroundings. When considering the energy of a system, we use the concept of internal energy and how heat and work are related to it.

Internal energy, U, is the total energy (both kinetic and potential) in a system, including translational kinetic energy of molecules, the energy associated with molecular rotations and vibrations, the energy stored in chemical bonds and intermolecular attractions, and the energy associated with electrons in atoms. Some of these forms of internal energy are illustrated in Figure 7-9. Internal energy also includes energy associated with the interactions of protons and neutrons in atomic nuclei, although this component is unchanged in chemical reactions. A system contains *only* internal energy. A system does not contain energy in the form of heat or work. Heat and work are the means by which a system exchanges energy with its surroundings. Heat and work exist only during a *change* in the system. The relationship between heat (q), work (w), and changes in internal energy (ΔU) is dictated by the law of conservation of energy, expressed in the form known as the **first law of thermodynamics**.

$$\Delta U = q + w \qquad (7.12)$$

If we consider that an isolated system is unable to exchange either heat or work with its surroundings, then $\Delta U_{\text{isolated system}} = 0$, and we can say

> The energy of an isolated system is constant.

This last statement is an alternative form of the first law of thermodynamics.
   In using equation (7.12) we must keep these important points in mind.

- Any energy *entering* the system carries a *positive* sign. Thus, if heat is *absorbed* by the system, $q > 0$. If work is done *on* the system, $w > 0$.
- Any energy *leaving* the system carries a *negative* sign. Thus, if heat is *given off* by the system, $q < 0$. If work is done *by* the system, $w < 0$.
- In general, the internal energy of a system changes as a result of energy entering or leaving the system as heat and/or work. If, on balance, more energy enters the system than leaves, ΔU is *positive*. If more energy leaves than enters, ΔU is *negative*.

These ideas are summarized in Figure 7-10 and illustrated in Example 7-6.

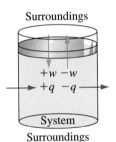

▲ FIGURE 7-10
Sign conventions used in thermodynamics
Arrows represent the direction of flow of heat (⟶) and work (⟶). The + signs refer to energy entering the system from the surroundings. The − signs signify energy leaving the system and entering the surroundings. These sign conventions are consistent with the expression $\Delta U = q + w$.

EXAMPLE 7-6

Relating ΔU, q, and w Through the First Law of Thermodynamics. A gas, while expanding (recall Figure 7-8), absorbs 25 J of heat and does 243 J of work. What is ΔU for the gas?

Solution
The key to problems of this type lies in assigning the correct signs to the quantities of heat and work. Because heat is absorbed by (enters) the system, q is *positive*. Because work done *by* the system represents energy *leaving* the system, w is *negative*. You may find it useful to represent the values of q and w, with their correct signs, within parentheses. Then complete the algebra.

$$\Delta U = q + w = (+25 \text{ J}) + (-243 \text{ J}) = 25 \text{ J} - 243 \text{ J} = -218 \text{ J}$$

Practice Example A: In compressing a gas, 355 J of work is done on the system. At the same time, 185 J of heat escapes from the system. What is ΔU for the system?

Practice Example B: If the internal energy of a system *decreases* by 125 J at the same time that the system *absorbs* 54 J of heat, does the system do work or have work done on it? How much?

Functions of State

To describe a system completely, we must indicate its temperature, pressure, and the kinds and amounts of substances present. When we have done this, we have specified the *state* of the system. Any property that has a unique value for a specified state of a system is said to be a **function of state** or a **state function**. For example, a sample of pure water at 20 °C (293.15 K) and under one standard atmosphere of pressure is in a specified state. The density of water in this state is 0.99820 g/mL. We can establish that this density is a unique value—a function of state—in this way:

Obtain three different samples of water—one purified by extensive distillation of groundwater; one synthesized by burning pure $H_2(g)$ in pure $O_2(g)$; and one prepared by driving off the water of hydration from $CuSO_4 \cdot 5\ H_2O$ and condensing the gaseous water to a liquid. The densities of the three different samples for the state that we specified will all be the same—0.99820 g/mL. Thus, the value of a function of state depends on the state of the system, and not on how that state was established.

The internal energy of a system is a function of state, although there is no simple measurement or calculation that we can use to establish its value. That is, we cannot write down a value of U for a system in the same way that we can write $d = 0.99820$ g/mL for the density of water at 20 °C. Fortunately, we don't need to know actual values of U. Consider, for example, heating 10.0 g of ice at 0 °C to a final temperature of 50 °C. The internal energy of the ice at 0 °C has one unique value, U_1, while that of the liquid water at 50 °C has another, U_2. The *difference* in internal energy between these two states also has a unique value, $\Delta U = U_2 - U_1$, and this difference *is* something that we can precisely measure. It is the quantity of energy (as heat) that must be transferred from the surroundings to the system during the change from state 1 to state 2. As a further illustration, consider the scheme outlined here and illustrated by the diagram below. Imagine that a system changes from state 1 to state 2 and then back to state 1.

$$\text{State 1}\left(U_1\right) \xrightarrow{\ \Delta U\ } \text{State 2}\left(U_2\right) \xrightarrow{\ -\Delta U\ } \text{State 1}\left(U_1\right)$$

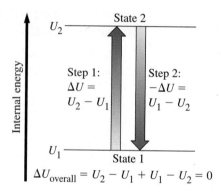

$$\Delta U_{\text{overall}} = U_2 - U_1 + U_1 - U_2 = 0$$

Because U has a unique value in each state, ΔU also has a unique value; it is $U_2 - U_1$. The change in internal energy when the system is returned from state 2 to state 1 is $-\Delta U = U_1 - U_2$. Thus, the *overall* change in internal energy is

$$\Delta U + (-\Delta U) = (U_2 - U_1) + (U_1 - U_2) = 0$$

This means that the internal energy returns to its initial value of U_1, which it must do, since it is a function of state. It is important to note here that when we reverse the direction of change, we change the sign of ΔU.

Path-Dependent Functions

Unlike internal energy and changes in internal energy, heat (q) and work (w) are *not* functions of state. Their values depend on the path followed when a system undergoes a change. We can see why this is so by considering again the process described by Figure 7-8 and Example 7-5. Think of the 0.100 mol of He at 298 K and under a pressure of 2.40 atm as *state 1*, and under a pressure of 1.30 atm as *state 2*. The change from state 1 to state 2 occurred in a single step. Suppose that in another instance, we allowed the expansion to occur through the intermediate stage pictured in Figure 7-11. That is, suppose we first reduced the external pres-

▲ **FIGURE 7-11**
An intermediate stage in the expansion of the gas in Figure 7-8
As described in the text, this intermediate stage in the expansion of the gas in Figure 7-8 helps us to establish that the work of expansion depends on the path taken.

sure on the gas from 2.40 atm to 1.80 atm (at which point, the gas volume would be 1.36 L). Then, in a second stage, we could reduce the pressure to 1.30 atm, arriving at state 2.

We calculated the amount of work done by the gas in a single-stage expansion in Example 7-5; it was $w = -1.1 \times 10^2$ J. The amount of work done in the two-stage process is the sum of two terms: the pressure–volume work for each stage of the expansion.

$$w = -1.80 \text{ atm} \times (1.36 \text{ L} - 1.02 \text{ L}) - 1.30 \text{ atm} \times (1.88 \text{ L} - 1.36 \text{ L})$$

$$= -0.61 \text{ L atm} - 0.68 \text{ L atm}$$

$$= -1.3 \text{ L atm} \times \frac{101 \text{ J}}{1 \text{ L atm}} = -1.3 \times 10^2 \text{ J}$$

The value of ΔU is the same for the single- and two-stage expansion processes because internal energy is a function of state. However, we see that slightly more work is done in the two-stage expansion. Work is not a function of state; it is path dependent. In the next section, we will stress that heat is also path dependent.

KEEP IN MIND ▶
that if w differs in the two different expansion processes, q must also differ, and in such a way that $q + w$ for the one-stage and two-stage expansions are the same. This then makes $q + w = \Delta U$ a unique quantity, as required by the first law of thermodynamics.

7-6   Heats of Reaction: ΔU and ΔH

Think of the reactants in a chemical reaction as the initial state of a system and the products as the final state.

$$\text{reactants} \quad \longrightarrow \quad \text{products}$$

$$\text{(initial state)} \qquad\qquad \text{(final state)}$$

$$U_i \qquad\qquad\qquad U_f$$

$$\Delta U = U_f - U_i$$

According to the first law of thermodynamics, we can also say that $\Delta U = q + w$. We have previously identified a heat of reaction as q_{rxn}, and so we can write

$$\Delta U = q_{rxn} + w$$

Now consider again a combustion reaction carried out in a bomb calorimeter (see Figure 7-5). The original reactants and products are confined within the bomb, and we say that the reaction occurs at *constant volume*. Because the volume is constant, $\Delta V = 0$, and no work is done. That is, $w = -P\Delta V = 0$. Denoting the heat of reaction for a constant–volume reaction as q_V, we see that $\Delta U = q_V$.

$$\Delta U = q_{rxn} + w = q_{rxn} + 0 = q_{rxn} = q_V \qquad (7.13)$$

The heat of reaction measured in a bomb calorimeter is equal to ΔU.

But we do not ordinarily carry out chemical reactions in bomb calorimeters. The metabolism of sucrose occurs under the conditions present in the human body. The combustion of methane (natural gas) in a water heater occurs in an open flame. This question then arises: How does the heat of a reaction measured in a bomb calorimeter compare with the heat of reaction if the reaction is carried out in some other way? The usual other way is in beakers, flasks, and other containers open to the atmosphere and under the *constant pressure* of the atmosphere. We live in a constant-pressure world! The neutralization reaction of Example 7-4 is typical of this more common method of conducting chemical reactions.

In many reactions carried out at constant pressure, a small amount of pressure–volume work is done as the system expands or contracts (recall Figure 7-7). In these cases, the heat of reaction, q_P, is different from q_V. We know that the change in internal energy (ΔU) for a reaction carried out between a given initial and a

▶ **FIGURE 7-12**
Two different paths leading to the same internal energy change in a system
In path (a), the volume of the system remains constant and no internal energy is converted into work—think of burning gasoline in a bomb calorimeter. In path (b), the system does work, so some of the internal energy change is used to do work—think of burning gasoline in an automobile engine to produce heat and work.

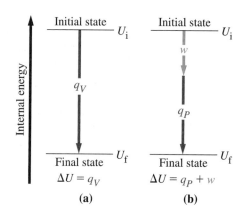

given final state has a unique value. Furthermore, we have seen that for a reaction at constant volume $\Delta U = q_V$. From Figure 7-12 and the first law of thermodynamics, we see that for the same reaction at constant pressure $\Delta U = q_P + w$, which means $\Delta U = q_V = q_P + w$. Thus, unless $w = 0$, q_V and q_P must be different. The fact that q_V and q_P for a reaction may differ, even though ΔU has the same value, underscores that U is a function of state and q and w are not.

We can use the relationship between q_V and q_P to devise another state function that represents the heat flow for a process at constant pressure. To do this, we begin by writing

$$q_V = q_P + w$$

Now, using $\Delta U = q_V$, $w = -P\Delta V$ and rearranging terms, we obtain

$$\Delta U = q_P - P\Delta V$$

$$q_P = \Delta U + P\Delta V$$

The quantities $U, P,$ and V are all state functions, so it should be possible to derive the expression $\Delta U + P\Delta V$ from yet another state function. This state function, called **enthalpy**, H, is the sum of the internal energy and the pressure–volume product of a system: $H = U + PV$. The **enthalpy change**, ΔH, for a process between initial and final states is

$$\Delta H = H_f - H_i = \left(U_f + P_f V_f\right) - \left(U_i + P_i V_i\right)$$

$$\Delta H = \left(U_f - U_i\right) + \left(P_f V_f - P_i V_i\right)$$

$$\Delta H = \Delta U + \Delta PV$$

If the process is carried out at a constant temperature and pressure $\left(P_i = P_f\right)$ and with work limited to pressure–volume work, the enthalpy change is

$$\Delta H = \Delta U + P\Delta V$$

and the heat flow for the process under these conditions is

$$\Delta H = q_P \qquad (7.14)$$

Enthalpy (ΔH) and Internal Energy (ΔU) Changes in a Chemical Reaction

We have noted that the heat of reaction at constant pressure, ΔH, and the heat of reaction at constant volume, ΔU, are related by the expression

$$\Delta U = \Delta H - P\Delta V \qquad (7.15)$$

Are You Wondering...

Why introduce the term ΔH anyway? Why not work just with ΔU, q, and w?

It's mainly a matter of convenience. Think of an analogous situation from daily life—buying gasoline at a filling station. The gasoline price posted on the pump is actually the sum of a base price and various federal, state, and perhaps local taxes. This breakdown is important to the accountants who must determine how much tax is to be paid to which agencies. To the consumer, however, it's easier to be given just the total cost per gallon. After all, this determines what he or she must pay. In thermochemistry, our chief interest is generally in heats of reaction, not pressure–volume work. And because most reactions are carried out under atmospheric pressure, it's helpful to have a function of state, enthalpy, H, whose change is exactly equal to something we can measure: q_P.

The last term in this expression is the energy associated with the change in volume of the system under a constant external pressure. To assess just how significant pressure–volume work is, let's consider the following reaction, which is also illustrated in Figure 7-13.

$$2\,CO(g) + O_2(g) \longrightarrow 2\,CO_2(g)$$

If we measure the heat of this reaction under constant–pressure conditions at a constant temperature of 298 K, we get -566.0 kJ, indicating that 566.0 kJ of energy

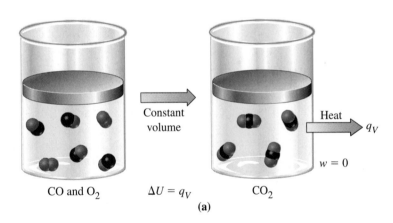

CO and O_2 $\Delta U = q_V$ CO_2

Constant volume   Heat q_V

$w = 0$

(a)

▶ **FIGURE 7-13**

Comparing heats of reaction at constant volume and constant pressure for the reaction:
$2\,CO(g) + O_2(g) \longrightarrow 2\,CO_2(g)$

(a) *No* work is performed at constant volume because the piston cannot move; $q_V = \Delta U = -563.5$ kJ **(b)** When the reaction is carried out at constant pressure, the surroundings do work on the system as the system shrinks into a smaller volume. More heat is evolved than in the constant-volume reaction; $q_P = \Delta H = -566.0$ kJ.

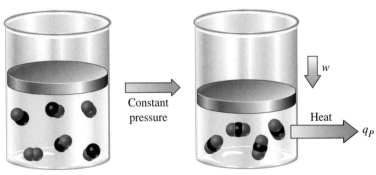

Constant pressure w Heat q_P

$\Delta U = q_P - P\Delta V$
$q_V = q_P - P\Delta V$

(b)

has left the system as heat: $\Delta H = -566.0$ kJ. To evaluate the pressure–volume work, we begin by writing

$$P\Delta V = P(V_f - V_i)$$

Then we can use the ideal gas equation to write this alternate expression.

$$P\Delta V = RT(n_f - n_i)$$

Here, n_f is the number of moles of gas in the products (*2* mol CO_2) and n_i is the number of moles of gas in the reactants (*2* mol CO + *1* mol O_2). Thus,

$$P\Delta V = 0.0083145\ \text{kJ mol}^{-1}\ \text{K}^{-1} \times 298\ \text{K} \times [2 - (2 + 1)]\ \text{mol} = -2.5\ \text{kJ}$$

The change in internal energy is

$$\Delta U = \Delta H - P\Delta V$$

$$= -566.0\ \text{kJ} - (-2.5\ \text{kJ})$$

$$= -563.5\ \text{kJ}$$

This calculation shows that the $P\Delta V$ term is quite small compared to ΔH and that ΔU and ΔH are almost the same. An additional interesting fact here is that the volume of the system decreases as a consequence of the work done by the surroundings on the system.

Now let's consider the combustion of sucrose. At a fixed temperature, the heat of combustion of sucrose turns out to be the same, whether at constant volume (q_V) or constant pressure (q_P). Only heat is transferred between the reaction mixture and the surroundings; no pressure–volume work is done. This is because the volume of a system is almost entirely determined by the volume of gas and because 12 mol $CO_2(g)$ occupies the same volume as 12 mol $O_2(g)$. There is no change in volume in the combustion of sucrose: $q_P = q_V$. Thus, we can represent the result of Example 7–3 as

Heats of Reaction simulation

$$C_{12}H_{22}O_{11}(s) + 12\ O_2(g) \longrightarrow 12\ CO_2(g) + 11\ H_2O(l) \quad \Delta H = -5.65 \times 10^3\ \text{kJ} \quad (7.16)$$

That is, 1 mol $C_{12}H_{22}O_{11}(s)$ reacts with 12 mol $O_2(g)$ to produce 12 mol $CO_2(g)$, 11 mol $H_2O(l)$, and 5.65×10^3 kJ of evolved heat.[*]

In summary, in most reactions, the heat of reaction we measure is ΔH. In some reactions, notably combustion reactions, we measure ΔU (that is, q_V). In reaction (7.16), $\Delta U = \Delta H$, but this is not always the case. Where it is not, we can obtain a value of ΔH from ΔU by the method we illustrated in our discussion of expression (7.15), but even in those cases, ΔH and ΔU will be nearly equal. In this text, we will treat all heats of reactions as ΔH values unless there is an indication to the contrary.

Example 7-7 shows how enthalpy changes can provide conversion factors for problem solving.

EXAMPLE 7-7

Stoichiometric Calculations Involving Quantities of Heat. How much heat is associated with the complete combustion of 1.00 kg of sucrose, $C_{12}H_{22}O_{11}$?

---

[*]Strictly speaking, the unit for ΔH should be kilojoules per mole, meaning per mole of reaction. "One mole of reaction" relates to the amounts of reactants and products in the equation as written. Thus, reaction (7.16) involves 1 mol $C_{12}H_{22}O_{11}(s)$, 12 mol $O_2(g)$, 12 mol $CO_2(g)$, 11 mol $H_2O(l)$, and -5.65×10^3 kJ of enthalpy change *per mol reaction*. The mol^{-1} part of the unit of ΔH is often dropped, but there are times we need to carry it to achieve the proper cancellation of units. We will find this to be the case in Chapters 20 and 21.

Solution

First, express the quantity of sucrose in moles.

$$? \text{ mol } = 1.00 \text{ kg } C_{12}H_{22}O_{11} \times \frac{1000 \text{ g } C_{12}H_{22}O_{11}}{1 \text{ kg } C_{12}H_{22}O_{11}} \times \frac{1 \text{ mol } C_{12}H_{22}O_{11}}{342.3 \text{ g } C_{12}H_{22}O_{11}}$$

$$= 2.92 \text{ mol } C_{12}H_{22}O_{11}$$

Now, formulate a conversion factor (shown in blue) based on the information in equation (7.16)—that is, -5.65×10^3 kJ of heat is associated with the combustion of 1 mol $C_{12}H_{22}O_{11}$.

$$? \text{ kJ } = 2.92 \text{ mol } C_{12}H_{22}O_{11} \times \frac{-5.65 \times 10^3 \text{ kJ}}{1 \text{ mol } C_{12}H_{22}O_{11}} = -1.65 \times 10^4 \text{ kJ}$$

The negative sign denotes that heat is given off in the combustion.

Practice Example A: What mass of sucrose must be burned to produce 1.00×10^3 kJ of heat?

[*Hint:* Again, use the enthalpy change of equation (7.16) as a conversion factor. Here, however, you must invert it.]

Practice Example B: A 25.0-mL sample of 0.1045 M HCl(aq) was neutralized by NaOH(aq). Use the result of Example 7-4 to determine the heat evolved in this neutralization.

Enthalpy Change (ΔH) Accompanying a Change in State of Matter

When a liquid is in contact with the atmosphere, energetic molecules at the surface of the liquid can overcome forces of attraction to their neighbors and pass into the gaseous, or vapor, state. We say that the liquid *vaporizes*. If the temperature of the liquid is to remain constant, the liquid must absorb heat from its surroundings to replace the energy carried off by the vaporizing molecules. The heat required to vaporize a fixed quantity of liquid is called the enthalpy (or heat) of vaporization. Usually the fixed quantity of liquid chosen is one mole, and we can call this quantity the *molar enthalpy of vaporization*. For example,

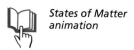
States of Matter animation

$$H_2O(l) \longrightarrow H_2O(g) \qquad \Delta H = 44.0 \text{ kJ at 298 K}$$

We described the melting of a solid in a similar fashion, on page 223. The energy requirement in this case is called the enthalpy (or heat) of fusion. For the melting of one mole of ice, we can write

$$H_2O(s) \longrightarrow H_2O(l) \qquad \Delta H = 6.01 \text{ kJ at 273.15 K}$$

We can use the data represented in these equations, together with other appropriate data, to answer questions like those posed in Example 7-8 and its accompanying Practice Examples.

EXAMPLE 7-8

Enthalpy Changes Accompanying Changes in States of Matter. Calculate ΔH for the process in which 50.0 g of water is converted from liquid at 10.0 °C to vapor at 25.0 °C?

Solution

The key to this calculation is to view the process as proceeding in two steps: raising the temperature of liquid water from 10.0 to 25.0 °C, and then completely vaporizing the liquid at 25.0 °C. The total enthalpy change is the sum of the changes in each step. For a process at constant pressure, $\Delta H = q_P$, so we need to calculate the heat absorbed in each step.

Heating water from 10.0 to 25.0 °C We determine this heat requirement in the same way as shown in Example 7-1; that is, we apply equation (7.5).

$$? \text{ kJ} = 50.0 \text{ g H}_2\text{O} \times \frac{4.18 \text{ J}}{\text{g H}_2\text{O °C}} \times (25.0 - 10.0)\,°\text{C} \times \frac{1 \text{ kJ}}{1000 \text{ J}} = 3.14 \text{ kJ}$$

Vaporizing water at 25.0 °C For this part of the calculation, we need to express the quantity of water on a mole basis, so that we can then use the molar enthalpy of vaporization at 25 °C: 44.0 kJ/mol.

$$? \text{ kJ} = 50.0 \text{ g H}_2\text{O} \times \frac{1 \text{ mol H}_2\text{O}}{18.02 \text{ g H}_2\text{O}} \times \frac{44.0 \text{ kJ}}{1 \text{ mol H}_2\text{O}} = 122 \text{ kJ}$$

Total enthalpy change

$$\Delta H = 3.14 \text{ kJ} + 122 \text{ kJ} = 125 \text{ kJ}$$

Practice Example A: What is the enthalpy change when a cube of ice, 2.00-cm on edge, is brought from -10.0 °C to a final temperature of 23.2 °C? For ice, use a density of 0.917 g/cm^3, a specific heat of 2.01 J g^{-1} °C^{-1}, and an enthalpy of fusion of 6.01 kJ/mol.

Practice Example B: What is the maximum mass of ice at -15.0 °C that can be completely converted to water vapor at 25.0 °C if the available heat for this transition is 5.00×10^3 kJ?

Standard States and Standard Enthalpy Changes

The measured enthalpy change for a reaction has a unique value *only* if the initial state (reactants) and final state (products) are precisely described. If we define a particular state as *standard* for the reactants and products, we can then say that the standard enthalpy change is the enthalpy change in a reaction in which the reactants and products are in their standard states. This so-called **standard enthalpy of reaction** is denoted with a superscript degree symbol, $\Delta H°$.

The **standard state** of a solid or liquid substance is the pure element or compound at a pressure of *1 bar* (10^5 Pa)* and at the temperature of interest. For a gas, the standard state is the pure gas behaving as an (hypothetical) ideal gas at a pressure of *1 bar* and the temperature of interest. While temperature is not part of the definition of a standard state, it still must be specified in tabulated values of $\Delta H°$, because $\Delta H°$ depends on temperature. The values given in this text are all for 298.15 K (25 °C) unless otherwise stated.

In the rest of this chapter, we will use standard enthalpy changes. We will explore the details of nonstandard conditions in Chapter 20.

Enthalpy Diagrams

The negative sign of ΔH in equation (7.16) means that the enthalpy of the products is lower than that of the reactants. This *decrease* in enthalpy appears as heat evolved to the surroundings. The combustion of sucrose is an exothermic reaction. In the reaction

$$\text{N}_2(g) + \text{O}_2(g) \longrightarrow 2 \text{ NO}(g) \quad \Delta H = +180.50 \text{ kJ} \tag{7.17}$$

the products have a *higher* enthalpy than the reactants; ΔH is positive. To produce this increase in enthalpy, heat is absorbed from the surroundings. The reaction is

---

*The International Union of Pure and Applied Chemistry (IUPAC) recommended that the standard-state pressure be changed from 1 atm to 1 bar about 20 years ago, but some data tables are still based on the 1-atm standard. Fortunately, the differences in values resulting from this change in standard-state pressure are very small, almost always small enough to be ignored.

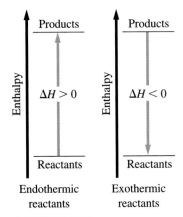

▲ FIGURE 7-14
Enthalpy diagrams
Horizontal lines represent absolute values of enthalpy. The higher a horizontal line, the greater the value of H that it represents. Vertical lines or arrows represent changes in enthalpy (ΔH). Arrows pointing up signify increases in enthalpy—endothermic reactions. Arrows pointing down signify decreases in enthalpy—exothermic reactions.

Are You Wondering...

Why ΔH° depends on temperature?

The difference in $\Delta H°$ for a reaction at two different temperatures is determined by how much heat is involved in changing the reactants and products from one temperature to the other under constant pressure. This quantity of heat can be calculated with equation (7.5): q_P = heat capacity × temperature change = $C_P \times \Delta T$. We write an expression of this type for each reactant and product, and combine these expressions to determine the difference in $\Delta H°$ between two temperatures. Usually, if the temperature interval is small, these differences are small and can be ignored. This is not the case, however, at high temperatures, as is common in geological applications.

endothermic. An **enthalpy diagram** is a diagrammatic representation of enthalpy changes in a process. Figure 7-14 shows how we can represent exothermic and endothermic reactions through such diagrams.

7-7   Indirect Determination of ΔH: Hess's Law

One of the reasons the enthalpy concept is so useful is that we can calculate large numbers of heats of reaction from a small number of measurements. The following features of enthalpy change (ΔH) make this possible.

1. **ΔH is an Extensive Property.** Consider the standard enthalpy change in the formation of NO(g) from its elements at 25 °C.

$$N_2(g) + O_2(g) \longrightarrow 2\,NO(g) \quad \Delta H = 180.50 \text{ kJ}$$

   To express the enthalpy change in terms of *one mole* of NO(g), we divide all coefficients *and the ΔH value* by *two*.

$$\frac{1}{2} N_2(g) + \frac{1}{2} O_2(g) \longrightarrow NO(g) \quad \Delta H = \frac{1}{2} \times 180.50 = 90.25 \text{ kJ}$$

   Enthalpy change is directly proportional to the amounts of substances in a system.

▶ Although we have avoided fractional coefficients previously, we need them here. We require the coefficient of NO(g) to be one.

2. **ΔH Changes Sign When a Process Is Reversed.** As we learned on page 234, if we reverse a process, the change in a function of state reverses sign. Thus, ΔH for the *decomposition* of one mole of NO(g) is $-\Delta H$ for the *formation* of one mole of NO(g).

$$NO(g) \longrightarrow \frac{1}{2} N_2(g) + \frac{1}{2} O_2(g) \quad \Delta H° = -90.25 \text{ kJ}$$

3. **Hess's Law of Constant Heat Summation.** To describe the standard enthalpy change for the formation of $NO_2(g)$ from $N_2(g)$ and $O_2(g)$,

$$\frac{1}{2} N_2(g) + O_2(g) \longrightarrow NO_2(g) \quad \Delta H = ?$$

   we can think of the reaction as proceeding in two steps: First we form NO(g) from $N_2(g)$ and $O_2(g)$, and then $NO_2(g)$ from NO(g) and $O_2(g)$. When we add the equations for these two steps, together with their individual and distinctive ΔH values, we get the overall equation and ΔH value that we are seeking.

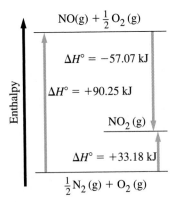

▲ FIGURE 7-15
An enthalpy diagram illustrating Hess's law
Whether the reaction occurs through a single step (blue arrow) or in two steps (black arrows), the enthalpy change is $\Delta H° = 33.18$ kJ for the overall reaction:
$$\frac{1}{2}N_2(g) + O_2(g) \longrightarrow NO_2(g).$$

KEEP IN MIND ▶
that $\Delta H°$ is an extensive property. In a chemical equation, the stoichiometric coefficients specify the amounts involved, and the unit kJ suffices for $\Delta H°$. When $\Delta H°$ is not accompanied by an equation, the amount involved must somehow be specified, such as per mole of $C_3H_8(g)$ in the expression
$\Delta H°_{comb.}$
$= -2219.9$ kJ/mol $C_3H_8(g)$.

$$\frac{1}{2}N_2(g) + \frac{1}{2}O_2(g) \longrightarrow NO(g) \qquad \Delta H = +90.25 \text{ kJ}$$

$$NO(g) + \frac{1}{2}O_2(g) \longrightarrow NO_2(g) \qquad \Delta H = -57.07 \text{ kJ}$$

$$\frac{1}{2}N_2(g) + O_2(g) \longrightarrow NO_2(g) \qquad \Delta H = +33.18 \text{ kJ}$$

Note that in summing the two equations, we canceled out NO(g), a species that would have appeared on both sides of the overall equation. Figure 7-15 illustrates what we just did through an enthalpy diagram. **Hess's law** states the principle we used.

If a process occurs in stages or steps (even if only hypothetically), the enthalpy change for the overall process is the sum of the enthalpy changes for the individual steps.

Hess's law is simply a consequence of the state function property of enthalpy. Regardless of the path we take in going from the initial state to the final state, ΔH (or $\Delta H°$ if the process is carried out under standard conditions) has the same value.
Suppose we want the standard enthalpy change for the reaction

$$3 \text{ C(graphite)} + 4 H_2(g) \longrightarrow C_3H_8(g) \quad \Delta H° = ? \qquad (7.18)$$

How should we proceed? If we try to get graphite and hydrogen to react, a slight reaction will occur, but it will not go to completion. Furthermore, the product will not be limited to propane (C_3H_8). Several other hydrocarbons will form as well. The fact is that we cannot measure $\Delta H°$ for reaction (7.18) directly. Instead, we must resort to an *indirect calculation* from $\Delta H°$ values that can be established by experiment. Here is where Hess's law is of greatest value. It permits us to calculate ΔH values that we cannot measure directly. In Example 7-9, we use the following standard heats of combustion to calculate $\Delta H°$ for reaction (7.18).

$$\Delta H° \text{ combustion:} \qquad C_3H_8(g) = -2219.9 \text{ kJ/mol } C_3H_8(g)$$

$$C(\text{graphite}) = -393.5 \text{ kJ/mol C(graphite)}$$

$$H_2(g) = -285.8 \text{ kJ/mol } H_2(g)$$

EXAMPLE 7-9

Applying Hess's Law. Use the heat of combustion data listed above to determine $\Delta H°$ for reaction (7.18): $3 \text{ C(graphite)} + 4 H_2(g) \longrightarrow C_3H_8(g) \qquad \Delta H° = ?$

Solution

To determine an enthalpy change with Hess's law, we need to combine the appropriate chemical equations. A good starting point is to write chemical equations for the given combustion reactions, based on *one mole* of the indicated reactant. Recall from Section 3-4 that the products of the combustion of carbon–hydrogen–oxygen compounds are $CO_2(g)$ and $H_2O(l)$.

(a) $C_3H_8(g) + 5 O_2(g) \longrightarrow 3 CO_2(g) + 4 H_2O(l) \qquad \Delta H° = -2219.9 \text{ kJ}$
(b) $C(\text{graphite}) + O_2(g) \longrightarrow CO_2(g) \qquad \Delta H° = -393.5 \text{ kJ}$

(c) $H_2(g) + \frac{1}{2}O_2(g) \longrightarrow H_2O(l) \qquad \Delta H° = -285.8 \text{ kJ}$

Because our objective in reaction (7.18) is to *produce* $C_3H_8(g)$, the next step is to find a reaction in which $C_3H_8(g)$ is formed—the *reverse* of reaction (a).

−(a): $3 CO_2(g) + 4 H_2O(l) \longrightarrow C_3H_8(g) + 5 O_2(g)$ $\Delta H° = -(-2219.9)$ kJ
$= +2219.9$ kJ

Now, we turn our attention to the reactants, C(graphite) and $H_2(g)$. To get the proper number of moles of each, we must multiply equation (b) by 3 and equation (c) by 4.

$3 \times$ (b): $3 C(graphite) + 3 O_2(g) \longrightarrow 3 CO_2(g)$ $\Delta H° = 3(-393.5$ kJ$)$
$= -1181$ kJ

$4 \times$ (c): $4 H_2(g) + 2 O_2(g) \longrightarrow 4 H_2O(l)$ $\Delta H° = 4(-285.8$ kJ$)$
$= -1143$ kJ

Here is the overall change we have described: 3 mol C(graphite) and 4 mol $H_2(g)$ have been consumed, and 1 mol $C_3H_8(g)$ has been produced. This is exactly what is required in Equation (7.18). We can now combine the three modified equations.

−(a): $3 CO_2(g) + 4 H_2O(l) \longrightarrow C_3H_8(g) + 5 O_2(g)$ $\Delta H° = +2219.9$ kJ

$3 \times$ (b): $3 C(graphite) + 3 O_2(g) \longrightarrow 3 CO_2(g)$ $\Delta H° = -1181$ kJ

$4 \times$ (c): $4 H_2(g) + 2 O_2(g) \longrightarrow 4 H_2O(l)$ $\Delta H° = -1143$ kJ

$3 C(graphite) + 4 H_2(g) \longrightarrow C_3H_8(g)$ $\Delta H° = -104$ kJ

Practice Example A: The standard heat of combustion of propene, $C_3H_6(g)$, is −2058 kJ/mol $C_3H_6(g)$. Use this value and other data from this example to determine $\Delta H°$ for the hydrogenation of propene to propane.

$$CH_3CH{=}CH_2(g) + H_2(g) \longrightarrow CH_3CH_2CH_3(g) \quad \Delta H° = ?$$

Practice Example B: From the data in Practice Example A and the following reaction, determine the standard enthalpy of combustion of one mole of 1-propanol, $CH_3CHOHCH_3(l)$.

$$CH_3CH{=}CH_2(g) + H_2O(l) \longrightarrow CH_3CHOHCH_3(l) \quad \Delta H° = -52.3 \text{ kJ}$$

7-8   Standard Enthalpies of Formation

In the enthalpy diagrams we have drawn, we have not written any numerical values on the enthalpy axis. This is because we cannot determine *absolute* values of enthalpy, H. However, enthalpy *is* a function of state, so *changes* in enthalpy, ΔH, have unique values. We can deal just with these changes. Nevertheless, as with many other properties, it is still useful to have a starting point, a zero value.

Consider a map-making analogy: What do we list as the height of a mountain? Do we mean by this the vertical distance between the mountaintop and the center of Earth? Between the mountaintop and the deepest trench in the ocean? No. By agreement, we mean the vertical distance between the mountaintop and mean sea level. We *arbitrarily* assign to mean sea level an elevation of zero and to all other points on Earth, an elevation relative to this zero. The elevation of Mt. Everest is +8848 m; that of Badwater, Death Valley, California, is −86 m. We do something similar with enthalpies. In effect, we relate our zero to the enthalpies of certain forms of the elements and determine the enthalpies of other substances relative to this zero.

The **standard enthalpy of formation** $(\Delta H_f°)$ of a substance is the enthalpy *change* that occurs in the formation of one mole of the substance in the standard state from the *reference* forms of the elements in their standard states. The reference forms of the elements in all but a few cases are the most stable forms of the elements at one bar and the given temperature. The superscript degree symbol denotes that the enthalpy change is a standard enthalpy change, and the subscript "f" signifies

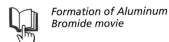

Formation of Aluminum Bromide movie

KEEP IN MIND ▶

that we use the expression "standard enthalpy of formation" even though what we are describing is actually a standard enthalpy *change*.

that the reaction is one in which a substance is formed from its elements. Because the formation of the most stable form of an element from itself is no change at all,

the standard enthalpy of formation of a pure element in its reference form is 0.

Listed here are the most stable forms of several elements at 298 K, the temperature at which thermochemical data are commonly tabulated.

$$Na(s) \quad H_2(g) \quad N_2(g) \quad O_2(g) \quad C(graphite) \quad Br_2(l)$$

The situation with carbon is an interesting one. In addition to graphite, carbon also exists naturally in the form of diamond. However, because we can measure an enthalpy difference between them, we cannot assign $\Delta H_f^\circ = 0$ to both forms.

$$C(graphite) \longrightarrow C(diamond) \qquad \Delta H^\circ = 1.9 \text{ kJ}$$

We choose as the reference form the more stable form, the one with the lower enthalpy. Thus, we assign $\Delta H_f^\circ(graphite) = 0$ and $\Delta H_f^\circ(diamond) = 1.9 \text{ kJ/mol}$. Although we can obtain bromine in either the gaseous or liquid state at 298 K, $Br_2(l)$ is the most stable form. $Br_2(g)$, if obtained at 298 K and 1 bar pressure, immediately condenses to $Br_2(l)$.

$$Br_2(l) \longrightarrow Br_2(g) \qquad \Delta H_f^\circ = 30.91 \text{ kJ}$$

The enthalpies of formation are $\Delta H_f^\circ[Br_2(l)] = 0$ and $\Delta H_f^\circ[Br_2(g)] = 30.91 \text{ kJ/mol}$.

We see one of the few cases in which the reference form is not the most stable form with the element phosphorus. Although over time it converts to solid red phosphorus, solid white phosphorus has been chosen as the reference form.

▲ Two of the distinctive physical forms of carbon, diamond and graphite (used in pencils).

▲ Liquid bromine vaporizing.

▶ Two of the distinctive physical forms of phosphorus.

$$P(s, white) \longrightarrow P(s, red) \qquad \Delta H^\circ = -17.6 \text{ kJ}$$

The standard enthalpies of formation are $\Delta H_f^\circ[P(s, white)] = 0$ and $\Delta H_f^\circ [P(s, red)] = -17.6 \text{ kJ/mol}$.

Enthalpies of formation of some common substances are presented in Table 7.2. A more extensive listing can be found in Appendix D. Figure 7-16 is a three-dimensional representation of a few ΔH_f° values. We will use standard enthalpies of formation in a variety of calculations. Often, the first thing we must do is write the chemical equation to which a ΔH_f° value applies, as in Example 7-10.

TABLE 7.2 Some Standard Enthalpies of Formation at 298 K			
Substance	$\Delta H^{\circ}_{f, 298}$, kJ/mol[a]	**Substance**	$\Delta H^{\circ}_{f, 298}$, kJ/mol[a]
$CO(g)$	−110.5	$HBr(g)$	−36.40
$CO_2(g)$	−393.5	$HI(g)$	26.48
$CH_4(g)$	−74.81	$H_2O(g)$	−241.8
$C_2H_2(g)$	226.7	$H_2O(l)$	−285.8
$C_2H_4(g)$	52.26	$H_2S(g)$	−20.63
$C_2H_6(g)$	−84.68	$NH_3(g)$	−46.11
$C_3H_8(g)$	−103.8	$NO(g)$	90.25
$C_4H_{10}(g)$	−125.6	$N_2O(g)$	82.05
$CH_3OH(l)$	−238.7	$NO_2(g)$	33.18
$C_2H_5OH(l)$	−277.7	$N_2O_4(g)$	9.16
$HF(g)$	−271.1	$SO_2(g)$	−296.8
$HCl(g)$	−92.31	$SO_3(g)$	−395.7

[a] Values are for reactions in which one mole of substance is formed. Most of the data have been rounded off to four significant figures.

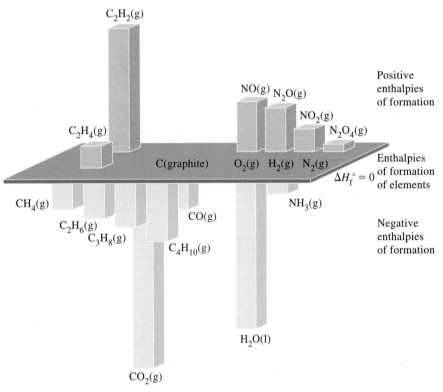

▲ **FIGURE 7-16   Some standard enthalpies of formation at 298 K**
Enthalpies of formation of elements are shown in the central plane, with $\Delta H^{\circ}_f = 0$. Substances with positive enthalpies of formation are above the plane, while those with negative enthalpies of formation are below the plane.

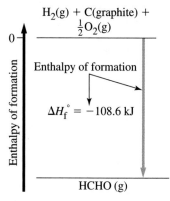

▲ FIGURE 7-17
Standard enthalpy of formation of formaldehyde, HCHO(g)
The formation of HCHO(g) from its elements in their standard states is an exothermic reaction. The heat evolved per mole of HCHO(g) formed is the standard enthalpy (heat) of formation.

EXAMPLE 7-10

Relating a Standard Enthalpy of Formation to a Chemical Equation. The enthalpy of formation of formaldehyde at 298 K is $\Delta H_f^\circ = -108.6 \, \text{kJ/mol HCHO(g)}$. Write the chemical equation to which this value applies.

Solution

The equation must be written for the formation of one mole of gaseous HCHO. The most stable forms of the elements at 298 K and 1 bar are gaseous H_2 and O_2 and solid carbon in the form of graphite (Figure 7-17). Note that we need one fractional coefficient in this equation.

$$H_2(g) + \frac{1}{2} O_2(g) + C(\text{graphite}) \longrightarrow HCHO(g) \qquad \Delta H_f^\circ = -108.6 \, \text{kJ}$$

Practice Example A: The standard enthalpy of formation for the amino acid leucine is $-637.3 \, \text{kJ/mol } C_6H_{13}O_2N(s)$. Write the chemical equation to which this value applies.

Practice Example B: How is ΔH° for the following reaction related to the standard enthalpy of formation of $NH_3(g)$ listed in Table 7.2? What is the value of ΔH°?

$$2 \, NH_3(g) \longrightarrow N_2(g) + 3 \, H_2(g) \qquad \Delta H^\circ?$$

 ## Are You Wondering ...

What is the significance of the sign of a ΔH_f° value?

A compound having a positive value of ΔH_f° is formed from its elements by an endothermic reaction. If the reaction is reversed, the compound decomposes into its elements in an exothermic reaction. We say that the compound is unstable with respect to its elements. This does not mean that the compound cannot be made, but it does suggest a tendency for the compound to enter into chemical reactions yielding products with lower enthalpies of formation.

When no other criteria are available, chemists sometimes use enthalpy charge as a rough indicator of the likelihood of a chemical reaction occurring—exothermic reactions generally being more likely to occur unassisted than endothermic ones. We'll present much better criteria later in the text.

Standard Enthalpies of Reaction

We have learned that if the reactants and products of a reaction are in their standard states, the enthalpy change is a *standard* enthalpy change, which we can denote as ΔH° or ΔH_{rxn}°. One of the primary uses of standard enthalpies of formation is in calculating standard enthalpy changes in chemical reactions. As a convenience, we will henceforth simply call them **standard enthalpies of reaction**.

Let us use Hess's law to calculate the standard enthalpy of reaction for the decomposition of sodium bicarbonate, a minor reaction that occurs when baking soda is used in baking.

$$2 \, NaHCO_3(s) \longrightarrow Na_2CO_3(s) + H_2O(l) + CO_2(g) \qquad \Delta H^\circ = ? \qquad (7.19)$$

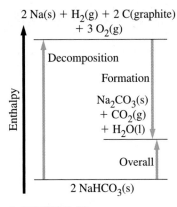

▲ **FIGURE 7-18**
Computing heats of reaction from standard enthalpies of formation
Enthalpy is a state function, hence ΔH° for the overall reaction

$2 \, NaHCO_3(s) \longrightarrow$
$Na_2CO_3(s) + CO_2(g) + H_2O(l)$

is the sum of the enthalpy changes for the two steps shown.

From Hess's law, we see that the following four equations yield equation (7.19) when added together.

(a) $2 \, NaHCO_3(s) \longrightarrow 2 \, Na(s) + H_2(g) + 2 \, C(graphite) + 3 \, O_2(g)$
$$\Delta H^\circ = -2 \times \Delta H_f^\circ[NaHCO_3(s)]$$

(b) $2 \, Na(s) + C(graphite) + \dfrac{3}{2} O_2(g) \longrightarrow Na_2CO_3(s) \quad \Delta H^\circ = \Delta H_f^\circ[Na_2CO_3(s)]$

(c) $H_2(g) + \dfrac{1}{2} O_2(g) \longrightarrow H_2O(l) \qquad\qquad\qquad \Delta H^\circ = \Delta H_f^\circ[H_2O(l)]$

(d) $C(graphite) + O_2(g) \longrightarrow CO_2(g) \qquad\qquad\qquad \Delta H^\circ = \Delta H_f^\circ[CO_2(g)]$

$2 \, NaHCO_3(s) \longrightarrow Na_2CO_3(s) + H_2O(l) + CO_2(g) \qquad\qquad \Delta H^\circ = \, ?$

Equation (a) is the *reverse* of the equation representing the formation of two moles of $[NaHCO_3(s)]$ from its elements. This means that ΔH° for reaction (a) is the *negative* of twice $\Delta H_f^\circ[NaHCO_3(s)]$. Equations (b), (c) and (d) represent the formation of *one* mole of $Na_2CO_3(s)$, $CO_2(g)$ and $H_2O(l)$, respectively. Thus, we can express the value of ΔH° for the decomposition reaction as

$$\Delta H^\circ = \Delta H_f^\circ[Na_2CO_3(s)] + \Delta H_f^\circ[H_2O(l)] + \Delta H_f^\circ[CO_2(g)]$$
$$- 2 \times \Delta H_f^\circ[NaHCO_3(s)] \quad (7.20)$$

We can use the enthalpy diagram in Figure 7-18 to visualize the Hess's law procedure and to show how the state function property of enthalpy enables us to arrive at equation (7.20). Imagine the decomposition of sodium bicarbonate taking place in two steps. In the first step, suppose a vessel contains 2 mol $NaHCO_3$, which is allowed to decompose into 2 mol $Na(s)$, 2 mol $C(graphite)$, 1 mol $H_2(g)$, and 3 mol $O_2(g)$, as in equation (a) above. In the second step, combine the 2 mol $Na(s)$, 2 mol $C(graphite)$, 1 mol $H_2(g)$, and 3 mol $O_2(g)$ to form the products according to equations (b),(c) and (d) above.

The pathway shown in Figure 7-18 *is not* how the reaction actually occurs. This does not matter, though, because enthalpy is a state function and the change of any state function is independent of the path chosen. The enthalpy change for the overall reaction is the sum of the standard enthalpy changes of the individual steps.

$$\Delta H^\circ = \Delta H_{decomposition}^\circ + \Delta H_{formation}^\circ$$

$$\Delta H_{decomposition}^\circ = -2 \times \Delta H_f^\circ[NaHCO_3(s)]$$

$$\Delta H_{formation}^\circ = \Delta H_f^\circ[Na_2CO_3(s)] + \Delta H_f^\circ[H_2O(l)] + \Delta H_f^\circ[CO_2(g)]$$

so that

$$\Delta H^\circ = \Delta H_f^\circ[Na_2CO_3(s)] + \Delta H_f^\circ[H_2O(l)] + \Delta H_f^\circ[CO_2(g)] - 2 \times \Delta H_f^\circ[NaHCO_3(s)]$$

Equation (7.20) is a specific application of the following more general relationship for a standard enthalpy of reaction.

$$\Delta H^\circ = \Sigma v_p \Delta H_f^\circ(products) - \Sigma v_r \Delta H_f^\circ(reactants) \quad (7.21)$$

The symbol Σ (Greek, sigma) means "the sum of." The terms that are added together are the products of the standard enthalpies of formation $\left(\Delta H_f^\circ\right)$ and their stoichiometric coefficients, v. One sum is required for the reaction products, and another for the initial reactants. The enthalpy change of the reaction is the sum of terms for the products *minus* the sum of terms for the reactants. Equation 7.21 avoids the manipulation of a number of chemical equations. The state function basis for equation (7.21) is shown in Figure 7-19 and is applied in Example 7-11.

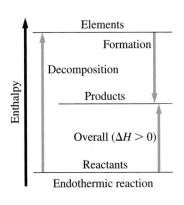

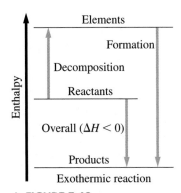

▲ **FIGURE 7-19**
Diagrammatic representation of equation (7.21)

EXAMPLE 7-11

Calculating $\Delta H°$ from Tabulated Values of $\Delta H_f^°$. Let us apply equation 7.21 to calculate the standard enthalpy of combustion of ethane, $C_2H_6(g)$, a component of natural gas.

Solution

The reaction is

$$C_2H_6(g) + \frac{7}{2}O_2(g) \longrightarrow 2\,CO_2(g) + 3\,H_2O(l)$$

The relationship we need is equation (7.21). The data we substitute into the relationship are from Table 7.2.

$$\Delta H° = \{2\text{ mol }CO_2 \times \Delta H_f^°[CO_2(g)] + 3\text{ mol }H_2O \times \Delta H_f^°[H_2O(l)]\}$$
$$- \{1\text{ mol }C_2H_6 \times \Delta H_f^°[C_2H_6(g)] + \frac{7}{2}\text{ mol }O_2 \times \Delta H_f^°[O_2(g)]\}$$
$$= 2\text{ mol }CO_2 \times (-393.5\text{ kJ/mol }CO_2) + 3\text{ mol }H_2O \times (-285.8\text{ kJ/mol }H_2O)$$
$$-1\text{ mol }C_2H_6 \times (-84.7\text{ kJ/mol }C_2H_6) - \frac{7}{2}\text{ mol }O_2 \times 0\text{ kJ/mol }O_2$$
$$= -787.0\text{ kJ} - 857.4\text{ kJ} + 84.7\text{ kJ} = -1559.7\text{ kJ}$$

Practice Example A: Use data from Table 7.2 to calculate the standard enthalpy of combustion of ethanol, $CH_3CH_2OH(l)$, at 298 K.

Practice Example B: Calculate the standard enthalpy of combustion at 298 K *per mole* of a gaseous fuel that contains C_3H_8 and C_4H_{10} in the mole fractions 0.62 and 0.38, respectively.

KEEP IN MIND ▶
that the standard enthalpy of formation of an element in its reference state is 0. Thus, we can drop the term involving $\Delta H_f^°[O_2(g)]$ at any time in the calculation.

A type of calculation as important as the one illustrated in Example 7-11 is the determination of an unknown $\Delta H_f^°$ value from a set of known $\Delta H_f^°$ values and a known standard enthalpy of reaction, $\Delta H°$. As shown in Example 7-12, the essential step is to rearrange expression (7.21) to isolate the unknown $\Delta H_f^°$ on one side of the equation. Also shown is a way of organizing the data that you may find helpful.

EXAMPLE 7-12

Calculating an unknown $\Delta H_f^°$ value. Use the data here and in Table 7.2 to calculate $\Delta H_f^°$ of benzene, $C_6H_6(l)$.

$$2\,C_6H_6(l) + 15\,O_2(g) \longrightarrow 12\,CO_2(g) + 6\,H_2O(l) \qquad \Delta H° = -6535\text{ kJ}$$

Solution

To organize the data needed in the calculation, let's begin by writing the chemical equation for the reaction, with $\Delta H_f^°$ data listed under the chemical formulas.

$$2\,C_6H_6(l) + 15\,O_2(g) \longrightarrow 12\,CO_2(g) + 6\,H_2O(l) \qquad \Delta H° = -6535\text{ kJ}$$

$\Delta H_f^°$, kJ/mol   ?   0 -393.5 -285.8

Now, we can substitute known data into expression (7.21) and rearrange the equation to obtain a lone term on the left: $\Delta H_f^°[C_6H_6(l)]$. The remainder of the problem simply involves numerical calculations.

$$\Delta H° = \{12\text{ mol }CO_2 \times (-393.5\text{ kJ/mol }CO_2) + 6\text{ mol }H_2O \times$$
$$(-285.8\text{ kJ/mol }H_2O)\} - 2\text{ mol }C_2H_6 \times \Delta H_f^°[C_6H_6(l)] = -6535\text{ kJ}$$

$$\Delta H_f^°[C_6H_6(l)] = \frac{\{-4722\text{ kJ} - 1715\text{ kJ}\} + 6535\text{ kJ}}{2\text{ mol }C_6H_6} = 49\text{ kJ/mol }C_6H_6(l)$$

Practice Example A: The overall reaction that occurs in photosynthesis in plants is

$$6\ CO_2(g)\ +\ 6\ H_2O(l)\ \longrightarrow\ C_6H_{12}O_6(s)\ +\ 6\ O_2(g)\qquad \Delta H^\circ = 2803\ kJ$$

Determine the standard enthalpy of formation of glucose, $C_6H_{12}O_6(s)$, at 298 K.

Practice Example B: A handbook lists the standard enthalpy of combustion of gaseous dimethyl ether at 298 K as $-31.70\ kJ/g\ (CH_3)_2\ O(g)$. What is the standard molar enthalpy of formation of dimethyl ether at 298 K?

Ionic Reactions in Solutions

Many chemical reactions in aqueous solution are best thought of as reactions between ions and best represented by net ionic equations. Consider the neutralization of a strong acid by a strong base. Using a somewhat more accurate enthalpy of neutralization than we obtained in Example 7-4, we can write

$$H^+(aq)\ +\ OH^-(aq)\ \longrightarrow\ H_2O(l)\qquad \Delta H^\circ = -55.8\ kJ\qquad (7.22)$$

We should also be able to calculate this enthalpy of neutralization by using enthalpy of formation data in expression (7.21). But this requires us to have enthalpy of formation data for individual ions. And there is a slight problem in getting these. We cannot create ions of a single type in a chemical reaction. We always produce cations and anions simultaneously, as in the reaction of sodium and chlorine to produce Na^+ and Cl^- in NaCl. We must choose a particular ion to which we assign an enthalpy of formation of *zero* in its aqueous solutions. We then compare the enthalpies of formation of other ions to this reference ion. The ion we arbitrarily choose for our zero is $H^+(aq)$. Now let us see how we can use expression (7.21) and data from equation (7.22) to determine the enthalpy of formation of $OH^-(aq)$.

$$\Delta H^\circ = 1\ mol\ H_2O \times \Delta H_f^\circ[H_2O(l)] - 1\ mol\ H^+ \times \Delta H_f^\circ[H^+(aq)]$$
$$- 1\ mol\ OH^- \times \Delta H_f^\circ[OH^-(aq)] = -55.8\ kJ$$

$$\Delta H_f^\circ[OH^-(aq)] = \frac{55.8\ kJ + (1\ mol\ H_2O \times \Delta H_f^\circ[H_2O(l)]) - (1\ mol\ H^+ \times \Delta H_f^\circ[H^+(aq)])}{1\ mol\ OH^-}$$

$$\Delta H_f^\circ[OH^-(aq)] = \frac{55.8\ kJ - 285.8\ kJ - 0\ kJ}{1\ mol\ OH^-} = -230.0\ kJ/mol\ OH^-$$

Table 7.3 lists data for several common ions in aqueous solution. Enthalpies of formation in solution depend on the solute concentration. These data are representative for *dilute* aqueous solutions (about 1 M), the type of solution that we normally deal with. Some of these data are used in Example 7-13.

EXAMPLE 7-13

Calculating the Enthalpy Change in an Ionic Reaction. Given that $\Delta H_f^\circ[BaSO_4(s)] = -1473\ kJ/mol$, what is the standard enthalpy change for the precipitation of barium sulfate?

Solution

First, let's write the net ionic equation for the reaction and introduce the relevant data.

$$Ba^{2+}(aq)\ +\ SO_4^{2-}(aq)\ \longrightarrow\ BaSO_4(s)\qquad \Delta H^\circ = ?$$

ΔH_f°, kJ/mol -537.6 -909.3 -1473

Then we can substitute data into equation (7.21).

$$\Delta H^\circ = 1\ mol\ BaSO_4 \times \Delta H_f^\circ[BaSO_4(s)] - 1\ mol\ Ba^{2+} \times \Delta H_f^\circ[Ba^{2+}(aq)]$$
$$- 1\ mol\ SO_4^{2-} \times \Delta H_f^\circ[SO_4^{2-}(aq)]$$

TABLE 7.3 Some Standard Enthalpies of Formation of Ions in Aqueous Solution

Ion	$\Delta H_{f, 298}^{\circ}$, kJ/mol	Ion	$\Delta H_{f, 298}^{\circ}$, kJ/mol
H^+	0	OH^-	−230.0
Li^+	−278.5	Cl^-	−167.2
Na^+	−240.1	Br^-	−121.6
K^+	−252.4	I^-	−55.19
NH_4^+	−132.5	NO_3^-	−205.0
Ag^+	+105.6	CO_3^{2-}	−677.1
Mg^{2+}	−466.9	S^{2-}	+33.05
Ca^{2+}	−542.8	SO_4^{2-}	−909.3
Ba^{2+}	−537.6	$S_2O_3^{2-}$	−648.5
Cu^{2+}	+64.77	PO_4^{3-}	−1277
Al^{3+}	−531		

$$= 1 \text{ mol } BaSO_4 \times (-1473 \text{ kJ/mol } BaSO_4) - 1 \text{ mol } Ba^{2+}$$
$$\times (-537.6 \text{ kJ/mol } Ba^{2+}) - 1 \text{ mol } SO_4^{2-} \times (-909.3 \text{ kJ/mol } SO_4^{2-})$$

$$= -1473 \text{ kJ} + 537.6 \text{ kJ} + 909.3 \text{ kJ} = -26 \text{ kJ}$$

Practice Example A: Given that $\Delta H_f^{\circ}[AgI(s)] = -61.84$ kJ/mol, what is the standard enthalpy change for the precipitation of silver iodide?

Practice Example B: The standard enthalpy change for the precipitation of $Ag_2CO_3(s)$ is −39.9 kJ per mole of $Ag_2CO_3(s)$ formed. What is $\Delta H_f^{\circ}[Ag_2CO_3(s)]$?

7-9 Fuels as Sources of Energy

One of the most important uses of thermochemical measurements and calculations is in assessing materials as energy sources. For the most part, these materials, called fuels, liberate heat through the process of combustion. We will briefly survey some common fuels, emphasizing matters that a thermochemical background helps us to understand.

Fossil Fuels

The bulk of current energy needs are met by petroleum, natural gas, and coal—so-called fossil fuels. These fuels are derived from plant and animal life of millions of years ago. The original source of the energy locked into these fuels is solar energy. In the process of *photosynthesis*, CO_2 and H_2O, in the presence of enzymes, the pigment chlorophyll, and sunlight, are converted to *carbohydrates*. These are compounds with formulas $C_m(H_2O)_n$, where m and n are integers. For example, in the sugar glucose $m = n = 6$, that is, $C_6(H_2O)_6 = C_6H_{12}O_6$. Its formation through photosynthesis is an *endothermic* process, represented as

▶ Although the formula $C_m(H_2O)_n$ suggests a "hydrate" of carbon, in carbohydrates, there are no H_2O units as there are in hydrates such as $CuSO_4 \cdot 5\, H_2O$. H and O atoms are simply found in the same numerical ratio as in H_2O.

$$6\, CO_2(g) + 6\, H_2O(l) \xrightarrow[\text{sunlight}]{\text{chlorophyll}} C_6H_{12}O_6(s) + 6\, O_2(g) \quad \Delta H^{\circ} = +2.8 \times 10^3 \text{ kJ} \quad (7.23)$$

When reaction (7.23) is reversed, as in the combustion of glucose, heat is evolved. The combustion reaction is *exothermic*.

The complex carbohydrate cellulose, with molecular masses ranging up to 500,000 u, is the principal structural material of plants. When plant life decomposes in the presence of bacteria and out of contact with air, O and H atoms are removed and the approximate carbon content of the residue increases in the progression

peat ⟶ lignite (32% C) ⟶ sub-bituminous coal (40% C) ⟶
bituminous coal (60% C) ⟶ anthracite coal (80% C)

For this process to proceed all the way to anthracite coal may take about 300 million years. Coal, then, is a combustible organic rock consisting of carbon, hydrogen, and oxygen, together with small quantities of nitrogen, sulfur, and mineral matter (ash). (One proposed formula for a "molecule" of bituminous coal is $C_{153}H_{115}N_3O_{13}S_2$.)

Petroleum and natural gas have formed in a somewhat different way. The remains of plants and animals living in ancient seas fell to the ocean floor, where they were decomposed by bacteria and covered with sand and mud. Over time, the sand and mud were converted to sandstone by the weight of overlying layers of sand and mud. The high pressures and temperatures resulting from this overlying sandstone rock formation transformed the original organic matter into petroleum and natural gas. The ages of these deposits range from about 250 million to 500 million years.

A typical natural gas consists of about 85% methane (CH_4), 10% ethane (C_2H_6), 3% propane (C_3H_8), and small quantities of other combustible and noncombustible gases. A typical petroleum consists of several hundred different hydrocarbons that range in complexity from C_1 molecules (CH_4) to C_{40} or higher (such as $C_{40}H_{82}$).

One way to compare different fuels is through their heats of combustion: In general, *the higher the heat of combustion, the better the fuel.* Table 7.4 lists approximate heats of combustion for the fossil fuels. These data show that *biomass* (living matter or materials derived from it—wood, alcohols, municipal waste) is a viable fuel, but that fossil fuels yield more energy per unit mass.

Problems Posed by Fossil Fuel Use.   There are two fundamental problems with the use of fossil fuels. First, the formation of new fossil fuels, if it is occurring at all, cannot possibly match the rate at which existing resources are being depleted. Fossil fuels are essentially *nonrenewable* energy sources. As shown in Figure 7-20, they will be used up in a relatively short time (at least compared with the span of human history). The second problem is their environmental effect. Sulfur impurities in fuels produce oxides of sulfur. The high temperatures associated with combustion cause the reaction of N_2 and O_2 in air to form oxides of nitrogen. Oxides of sulfur and nitrogen are implicated in air pollution and are important contributors to the environmental problem known as acid rain. Another inevitable product of the combustion of fossil fuels is carbon dioxide, one of the "greenhouse" gases that may lead to global warming and changes in Earth's climate.

TABLE 7.4   Approximate Heats of Combustion of Some Fuels

Fuel	Heat of Combustion kJ/g
Municipal waste	−12.7
Cellulose	−17.5
Pinewood	−21.2
Methanol	−22.7
Peat	−20.8
Bituminous coal	−28.3
Isooctane (a component of gasoline)	−47.8
Natural gas	−49.5

▶ Each of these environmental issues is discussed more fully elsewhere in the next.

▶ **FIGURE 7-20**
Estimated world production of fossil fuels
The exact shape of each curve and the predicted time when the fuel will be used up depend on the estimate used for the total quantity of recoverable fuel and the rate of production. These curves are based on no increases in current rates of consumption, which is probably an unrealistic assumption. Even a growth rate of 1% per year would cause the rate of consumption to double in 70 years and shorten the time until the remaining supplies are exhausted.

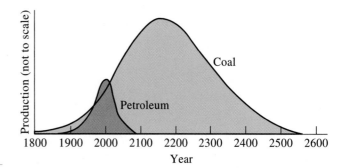

In the United States, coal reserves far exceed those of petroleum and natural gas. Despite this relative abundance, however, the use of coal has not increased significantly in recent years. In addition to the environmental effects cited above, the expense and hazards involved in the deep mining of coal are considerable. Surface mining, which is less hazardous and expensive than deep mining, is also more damaging to the environment. One promising possibility for using coal reserves is to convert coal to gaseous or liquid fuels, either in surface installations or while the coal is still underground.

Gasification of Coal. Before cheap natural gas became available in the 1940s, gas produced from coal (variously called producer gas, town gas, or city gas) was widely used in the United States. This gas was manufactured by passing steam and air through heated coal and involved reactions such as

$$C(s) + H_2O(g) \longrightarrow CO(g) + H_2(g) \qquad \Delta H° = +131.3 \text{ kJ} \qquad (7.24)$$

$$CO(g) + H_2O(g) \longrightarrow CO_2(g) + H_2(g) \qquad \Delta H° = -41.2 \text{ kJ} \qquad (7.25)$$

$$2\, C(s) + O_2(g) \longrightarrow 2\, CO(g) \qquad \Delta H° = -221.0 \text{ kJ} \qquad (7.26)$$

$$C(s) + 2\, H_2(g) \longrightarrow CH_4(g) \qquad \Delta H° = -74.8 \text{ kJ} \qquad (7.27)$$

The principal gasification reaction (7.24) is highly endothermic. The heat requirements for this reaction are met by the carefully controlled partial burning of coal (reaction 7.26).

A typical producer gas consists of about 23% CO, 18% H_2, 8% CO_2, and 1% CH_4, by volume. It also contains about 50% N_2 because air is used in its production. Because the N_2 and CO_2 are noncombustible, producer gas has only about 10–15% of the heat value of natural gas. Modern gasification processes include the following features.

1. They use $O_2(g)$ instead of air, thereby eliminating $N_2(g)$ in the product.
2. They provide for the removal of noncombustible $CO_2(g)$ and sulfur impurities. For example,

$$CaO(s) + CO_2(g) \longrightarrow CaCO_3(s)$$

$$2\, H_2S(g) + SO_2(g) \longrightarrow 3\, S(s) + 2\, H_2O(g)$$

3. They include a step (called *methanation*) to convert CO and H_2, in the presence of a catalyst, to CH_4.

$$CO(g) + 3\, H_2(g) \longrightarrow CH_4(g) + H_2O(l)$$

The product is called *substitute natural gas* (SNG), a gaseous mixture with composition and heat value similar to that of natural gas.

Liquefaction of Coal. The first step in obtaining liquid fuels from coal generally involves gasification of coal, as in reaction (7.24). This step is followed by catalytic reactions in which liquid hydrocarbons are formed.

$$n\, CO + (2n + 1)\, H_2 \longrightarrow C_nH_{2n+2} + n\, H_2O$$

In still another process, liquid methanol is formed.

$$CO(g) + 2\, H_2(g) \longrightarrow CH_3OH(l) \qquad (7.28)$$

In 1942, some 32 million gallons of aviation fuel were made from coal in Germany. In South Africa, the Sasol process for coal liquefaction has been a major source of gasoline and a variety of other petroleum products and chemicals for 50 years.

Methanol

▶ Isooctane, $(CH_3)_3CCH_2CH(CH_3)_2$, is an excellent, smooth-burning automotive fuel; it is given an octane number of 100. Heptane, $CH_3(CH_2)_5CH_3$, burns very unevenly and causes engine knock; it is given an octane number of 0. Gasolines are rated against mixtures of these two hydrocarbons: A 92-octane premium gasoline is equivalent to a mixture of 92% isooctane and 8% heptane.

Methanol, CH_3OH, can be obtained from coal by reaction (7.28). It can also be produced by thermal decomposition (pyrolysis) of wood, manure, sewage, or municipal waste. The heat of combustion of methanol is only about one-half that of a typical gasoline on a mass basis, but methanol has a high octane number—106—compared with 100 for the gasoline hydrocarbon isooctane and about 92 for premium gasoline. Methanol has been tested and used as a fuel in internal combustion engines and is cleaner burning than gasoline. Methanol can also be used for space heating, electric power generation, fuel cells, and as a reactant to make a variety of other organic compounds.

Ethanol

Ethanol, C_2H_5OH, is produced mostly from ethylene, C_2H_4, which in turn is derived from petroleum. Current interest centers on the production of ethanol by the fermentation of organic matter, a process known throughout recorded history. Ethanol production by fermentation is probably most advanced in Brazil, where sugarcane and cassava (manioc) are the plant matter (biomass) used. In the United States, ethanol is used chiefly as a 90% gasoline–10% ethanol mixture called *gasohol*. Ethanol is also used to raise the octane number of gasoline.

Hydrogen

Another fuel with great potential is hydrogen. Its most attractive features are the following:

- On a per gram basis, its heat of combustion is more than twice that of methane and about three times that of gasoline.
- The product of its combustion is H_2O, not CO and CO_2 as with gasoline.

Currently, the bulk of hydrogen used commercially is made from petroleum and natural gas. (Alternative methods of producing hydrogen, and the prospects of developing an economy based on hydrogen are discussed in the next chapter.)

Alternative Energy Sources

Combustion reactions are only one means of extracting useful energy from materials. An alternative, for example, is to carry out reactions that yield the same products as combustion reactions in electrochemical cells called *fuel cells*. The energy is released as electricity rather than as heat (see Section 21-5). Solar energy can be used directly, without recourse to photosynthesis. Nuclear processes can be used in place of chemical reactions (Chapter 26).

Alternative energy sources, including those intended for automobiles, are likely to become increasingly important in the twenty-first century.

Focus On — Fats, Carbohydrates, and Energy Storage

◀ Aerobic exercise is an excellent way to burn fat and build muscle.

Aerobics, tennis, weight lifting, and jogging are popular forms of exercise, which is important to maintaining a healthy body. Where do we get the energy to do these things? Surprisingly, mostly from fat, the body's chief energy storage system.

During exercise, fat molecules react with water (hydrolyze) to form a group of compounds called fatty acids. Through a complex series of reactions, these fatty acids are converted to carbon dioxide and water. Energy released in these reactions is used to power muscles. A typical human fatty acid is palmitic acid, $CH_3(CH_2)_{14}COOH$.

The direct combustion of $CH_3(CH_2)_{14}COOH$ in a bomb calorimeter yields the same products as its metabolism in the body, together with a large quantity of heat.

$$CH_3(CH_2)_{14}COOH + 23\ O_2(g) \longrightarrow$$
$$16\ CO_2(g) + 16\ H_2O(l) \quad \Delta H° = -9977\ kJ$$

A hydrocarbon with a similar carbon and hydrogen content, $C_{16}H_{34}$, yields a similar quantity of heat on complete combustion, $-10{,}700$ kJ. The fat stored in our bodies (our fuel) is comparable to the jet fuel in an airplane. In both cases, the fuel adds weight but can be burned to produce the large amount of energy needed for movement and life support.

Simple carbohydrates are broken down and their energy released more rapidly than fats, and this is why we consume sugars (fruit juice, candy bars...) for a quick burst of energy. The combustion of 1 mol $C_{12}H_{22}O_{11}$ (sucrose or cane sugar),

Summary

Thermochemistry is concerned with the heat and work accompanying chemical reactions. Reactions in which heat is evolved by the system and released to the surroundings are called exothermic. Those in which heat is absorbed by the system are called endothermic. Heat effects are most commonly measured through temperature changes in a calorimeter. The most common type of work in chemical reactions is that associated with the expansion and compression of gases, pressure–volume work.

Combustion reactions can be carried out in a bomb calorimeter. Because these reactions occur at constant volume, the measured heats of reaction are $q_V = \Delta U$. Most reactions are carried out in containers open to the atmosphere, and for these reactions, we use the enthalpy change (ΔH). For a reaction conducted at constant pressure and with work limited to pressure–volume work, the heat of reaction is $q_P = \Delta H$.

Thermochemical calculations are of two main types:

1. Calculating heat effects from calorimetric data—heat capacities, specific heats, masses, and temperature changes.
2. Manipulating known ΔH values to obtain an unknown value. (In some cases, these manipulations require the use of Hess's law of constant heat summation. In others, they are based on the concepts of standard states and standard enthalpies of formation of pure substances and of ions in aqueous solution.)

One of the chief applications of thermochemistry is in the study of the combustion of fuels as energy sources. Currently, the principal fuels are the fossil fuels, but potential alternative fuels are also mentioned in this chapter and discussed in more depth later in the text.

however, yields much less energy (-5640 kJ/mol) than does a fatty acid of similar carbon and hydrogen content. In the enthalpy diagram of Figure 7-21, we can think of the sucrose molecule as in a condition somewhere between a hydrocarbon and the ultimate oxidation products: CO_2 and H_2O. A fatty acid, on the other hand, is in a condition closer to that of the hydrocarbon.

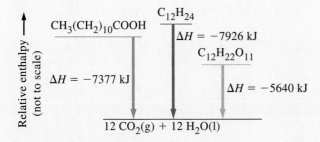

▲ **FIGURE 7-21   Energy values of a fatty acid, a carbohydrate, and a hydrocarbon**
Each of these three comparable 12-carbon compounds produces 12 mol $CO_2(g)$ and some $H_2O(l)$ as combustion products. The hydrocarbon, $C_{12}H_{24}$, releases the most energy on combustion. The fatty acid lauric acid, $CH_3(CH_2)_{10}COOH$, yields nearly as much, but the carbohydrate sucrose, $C_{12}H_{22}O_{11}$, yields considerably less.

A common energy unit for expressing the energy values of food is the food Calorie (Cal), which is actually a kilocalorie.

$$1 \text{ Cal} = 1 \text{ kcal} = 4.184 \text{ kJ}$$

As energy sources, fats yield about 9 Cal/g (38 kJ/g), whereas carbohydrates and proteins both yield about 4 Cal/g (17 kJ/g).

Consider a 65-kg (143-lb) person of average height and build. Typically this person would have about 11 kg (24 lb) of stored fats—a fuel reserve of about 4.2×10^5 kJ. On the other hand, if the body stored only carbohydrates, the stored reservoir would have to be about 25 kg. The 65-kg person would have to weigh about 80 kg (176 lb). A more dramatic picture might be that of a bird carrying its energy reserves as carbohydrates instead of fat. The excess "baggage" might make it impossible for the bird to fly.

It is important to understand some other matters about fat. First, the quantity of stored fat that an individual typically carries is greater than what is needed to meet energy requirements. Excess body fat leads to an increased risk of heart disease, diabetes, and other medical problems. Second, one does not have to consume fats in order to store body fats. All food types are broken down into small molecules during digestion, and these small molecules are reassembled into the more complex structures that the body needs. In the body, fat molecules can be synthesized from the small molecules produced from carbohydrates, for example. Current medical findings indicate that the most healthful diet provides the bulk of food calories through carbohydrates, with the recommended percent of caloric intake from fats being less than 30%.

Integrative Example

The partial burning of coal in the presence of O_2 and H_2O produces a mixture of CO(g) and $H_2(g)$ called *synthesis gas*. This gas can be used to synthesize organic compounds, or it can be burned as a fuel. A typical synthesis gas consists of 55.0% CO(g), 33.0% $H_2(g)$, and 12.0% noncombustible gases (mostly CO_2), *by volume*.

To what temperature can 1.00 kg water at 25.0 °C be heated with the heat liberated by the combustion of 1.00 L (STP) of this typical synthesis gas?

1. *Determine the number of moles of CO(g) and $H_2(g)$ in a 1.00-L sample.* First, we can use the molar volume at STP to determine the total number of moles of gas.

$$? \text{ mol gas} = 1.00 \text{ L} \times \frac{1 \text{ mol gas}}{22.4 \text{ L}} = 0.0446 \text{ mol gas}$$

Then, we can use expression (6.17) to find the number of moles of the combustible gases.

$$? \text{ mol CO} = 0.0446 \text{ mol gas} \times \frac{55.0 \text{ mol CO}}{100 \text{ mol gas}}$$

$$= 0.0245 \text{ mol CO}$$

$$? \text{ mol H}_2 = 0.0446 \text{ mol gas} \times \frac{33.0 \text{ mol H}_2}{100 \text{ mol gas}}$$

$$= 0.0147 \text{ mol H}_2$$

2. *Use tabulated data to determine the enthalpy of combustion of the 1.00 L of synthesis gas.* Let's write an equation for the combustion of CO(g), list ΔH_f° data beneath the equation, and then determine ΔH_{comb}° for the CO.

$$CO(g) + \frac{1}{2} O_2(g) \longrightarrow CO_2(g)$$

ΔH_f°, kJ/mol -110.5 0 -393.5

$\Delta H_{comb}^\circ = 1 \text{ mol CO}_2 \times (-393.5 \text{ kJ/mol CO}_2)$
$\qquad - 1 \text{ mol CO} \times (-110.5 \text{ kJ/mol CO}) = -283.0 \text{ kJ}$

Thus, the enthalpy of combustion of 0.0245 mol CO is

$$0.0245 \text{ mol CO} \times \frac{-283.0 \text{ kJ}}{1 \text{ mol CO}} = -6.93 \text{ kJ}$$

Now, repeat these calculations for the combustion of the 0.0147 mol $H_2(g)$.

$$H_2(g) + \frac{1}{2} O_2(g) \longrightarrow H_2O(l)$$

$\Delta H_f^\circ, \text{kJ/mol} \qquad 0 \qquad 0 \qquad -285.8$

$$\Delta H_{comb}^\circ = 1 \text{ mol } H_2O \times (-285.8 \text{ kJ/mol } H_2O)$$
$$= -285.8 \text{ kJ, and for } 0.0147 \text{ mol } H_2$$

$$0.0147 \text{ mol } H_2 \times \frac{-285.8 \text{ kJ}}{1 \text{ mol } H_2} = -4.20 \text{ kJ}$$

The enthalpy of combustion of 1.00 L of the synthesis gas is
$\Delta H^\circ = -6.93 \text{ kJ} - 4.20 \text{ kJ} = -11.13 \text{ kJ}$

3. *Apply* equation (7.6). The combustion reactions represent the system, and the 1.00 kg of water, the surroundings.

$$q_{water} = -q_{system} = -(-11.13 \text{ kJ}) = 11.13 \text{ kJ}$$
$$= 1.113 \times 10^4 \text{ J}$$

Also,

$$q_{water} = \text{mass water} \times \text{specific heat of water} \times \Delta T$$

$$1.113 \times 10^4 \text{ J} = 1.00 \times 10^3 \text{ g } H_2O \times \frac{4.18 \text{ J}}{\text{g } H_2O \text{ }^\circ C} \times \Delta T$$

$$\Delta T = 2.66 \text{ }^\circ C, \text{ where } \Delta T = T_f - T_i$$

$$T_f = T_i + \Delta T = 25.0 \text{ }^\circ C + 2.66 \text{ }^\circ C = 27.7 \text{ }^\circ C$$

Key Terms

bomb calorimeter (7-3)	**first law of thermodynamics** (7-5)	**potential energy** (7-1)
calorie (cal) (7-2)	**function of state (state function)** (7-5)	**pressure–volume work** (7-4)
calorimeter (7-3)	**heat** (7-2)	**specific heat** (7-2)
chemical energy (7-3)	**heat capacity** (7-2)	**standard enthalpy of formation** (7-8)
closed system (7-1)	**heat of reaction** (7-3)	**standard enthalpy of reaction (ΔH°)**
endothermic (7-3)	**Hess's law** (7-7)	(7-6, 7-8)
energy (7-1)	**internal energy (U)** (7-5)	**standard state** (7-6)
enthalpy (H) (7-6)	**isolated system** (7-1)	**surroundings** (7-1)
enthalpy change (ΔH) (7-6)	**kinetic energy** (7-1)	**system** (7-1)
enthalpy diagram (7-6)	**law of conservation of energy** (7-2)	**thermal energy** (7-1)
exothermic (7-3)	**open system** (7-1)	**work** (7-1, 7-4)

Review Questions

1. In your own words, define or explain the following terms or symbols: **(a)** ΔH; **(b)** $P\Delta V$; **(c)** ΔH_f°; **(d)** standard state; **(e)** fossil fuel.

2. Briefly describe each of the following ideas or methods: **(a)** law of conservation of energy; **(b)** bomb calorimetry; **(c)** function of state; **(d)** enthalpy diagram; **(e)** Hess's law.

3. Explain the important distinctions between each pair of terms: **(a)** system and surroundings; **(b)** heat and work; **(c)** specific heat and heat capacity; **(d)** endothermic and exothermic; **(e)** constant-volume and constant-pressure process.

4. Calculate the quantity of heat
   (a) in *kilocalories*, required to raise the temperature of 9.25 L of water from 22.0 to 29.4 °C;
   (b) in *kilojoules*, associated with a 33.5 °C decrease in temperature in a 5.85-kg aluminum bar (specific heat of aluminum = 0.903 J g^{-1} °C^{-1}).

5. Calculate the final temperature that results when
   (a) a 12.6-g sample of water at 22.9 °C absorbs 875 J of heat
   (b) a 1.59-kg sample of platinum at 78.2 °C gives off 1.05 kcal of heat (specific heat of Pt = 0.032 cal g^{-1} °C^{-1}).

6. From the information given, determine
   (a) the specific heat of toluene (C_7H_8), given that 186 J of heat is required to raise the temperature of a 15.0-g sample from 22.3 to 29.6 °C
   (b) the final temperature, when 2.75 kcal of heat is removed from a 2.25-kg sample of water initially at 23.1 °C

7. A mixture of 118 g of copper and 197 g of water is heated from 22.7 to 79.2 °C. How much heat, in kilojoules, is absorbed by the copper–water mixture? (Specific heats: water, 4.18 J g^{-1} °C^{-1}; copper, 0.385 J g^{-1} °C^{-1}.)

8. A 1.22-kg piece of iron at 126.5 °C is dropped into 981 g water at 22.1 °C. The temperature rises to 34.4 °C. Determine the specific heat of iron, in J g^{-1} °C^{-1}.

9. *Without doing detailed calculations*, decide which of the following is a plausible final temperature when 75.0 mL of water at 80.0 °C is added to 100.0 mL of water at 20.0 °C: (1) 40 °C; (2) 46 °C; (3) 50 °C; (4) 28 °C. Explain your reasoning.

10. What are the internal energy changes, ΔU, if a system **(a)** absorbs 67 J of heat and does 67 J of work?

(b) absorbs 356 J of heat and does 592 J of work?

(c) loses 38 J of heat and has 171 J of work done on it?

(d) absorbs no heat and does 416 J of work?

11. Upon complete combustion, the substances indicated evolve the given quantities of heat. Express each heat of combustion in kilojoules per mole of substance.
    (a) 0.584 g of propane, $C_3H_8(g)$, yields 29.4 kJ
    (b) 0.136 g of camphor, $C_{10}H_{16}O$, yields 1.26 kcal
    (c) 2.35 mL of acetone, $(CH_3)_2CO(l)$ ($d = 0.791$ g/mL), yields 58.3 kJ

12. A sample gives off 5228 cal when burned in a bomb calorimeter. The temperature of the calorimeter assembly increases by 4.39 °C. Calculate the heat capacity of the calorimeter, in kilojoules per degree Celsius.

13. The following substances undergo complete combustion in a bomb calorimeter. The calorimeter assembly has a heat capacity of 5.136 kJ/°C. In each case, what is the final temperature if the initial water temperature is 22.43 °C?
    (a) 0.3268 g caffeine, $C_8H_{10}O_2N_4$; heat of combustion = −1014.2 kcal/mol caffeine
    (b) 1.35 ml of methyl ethyl ketone, $C_4H_8O(l)$ ($d = 0.805$ g/mL); heat of combustion = −2444 kJ/mol methyl ethyl ketone

14. A bomb calorimetry experiment is performed with xylose, $C_5H_{10}O_5(s)$, as the combustible substance. The data obtained are
    mass of xylose burned: 1.183 g
    heat capacity of calorimeter: 4.728 kJ/°C
    initial calorimeter temperature: 23.29 °C
    final calorimeter temperature: 27.19 °C
    (a) What is the heat of combustion of xylose, in kilojoules per mole?
    (b) Write the chemical equation for the complete combustion of xylose, and represent the value of ΔH in this equation. (Assume for this reaction that $\Delta U \approx \Delta H$.)

15. A "coffee-cup" calorimeter contains 100.0 mL of 0.300 M HCl at 20.3 °C. When 1.82 g Zn(s) is added, the temperature rises to 30.5 °C. What is the heat of reaction per mol Zn? Make the same assumptions as in Example 7-4, and also that there is no heat lost to the $H_2(g)$ that escapes.
    $$Zn(s) + 2\,H^+(aq) \longrightarrow Zn^{2+}(aq) + H_2(g)$$

16. A 0.75-g sample of KCl is added to 35.0 g H_2O in a Styrofoam cup and stirred until it dissolves. The temperature of the solution drops from 24.8 to 23.6 °C.
    (a) Is the process endothermic or exothermic?
    (b) What is the heat of solution of KCl expressed in kilojoules per mole of KCl?

17. The heat of solution of potassium acetate in water is −15.3 kJ/mol $KC_2H_3O_2$. If 0.136 mol $KC_2H_3O_2$ is dissolved in 525 mL water that is initially at 25.1 °C, what will be the final solution temperature?

18. Write the balanced chemical equations that have the following as their standard enthalpy changes.
    (a) $\Delta H_f^\circ = +82.05$ kJ/mol $N_2O(g)$

(b) $\Delta H_f^\circ = -394.1$ kJ/mol $SO_2Cl_2(l)$

(c) $\Delta H_{combustion}^\circ = -1527$ kJ/mol $CH_3CH_2COOH(l)$

19. The complete combustion of butane, $C_4H_{10}(g)$, is represented by the equation
    $$C_4H_{10}(g) + \frac{13}{2}\,O_2(g) \longrightarrow 4\,CO_2(g) + 5\,H_2O(l)$$
    $$\Delta H = -2877 \text{ kJ}$$
    How much heat, in kilojoules, is evolved in the complete combustion of **(a)** 1.325 g $C_4H_{10}(g)$; **(b)** 28.4 L $C_4H_{10}(g)$ at STP; **(c)** 12.6 L $C_4H_{10}(g)$ at 23.6 °C and 738 mmHg?

20. The standard enthalpy of formation of $NH_3(g)$ is −46.11 kJ/mol NH_3. What is ΔH° for the following reaction?
    $$\frac{2}{3}\,NH_3(g) \longrightarrow \frac{1}{3}\,N_2(g) + H_2(g) \qquad \Delta H^\circ =$$

21. Use Hess's law to determine ΔH° for the reaction
    $$CO(g) + \frac{1}{2}\,O_2(g) \longrightarrow CO_2(g), \text{ given that}$$
    (1) $C(graphite) + \frac{1}{2}\,O_2(g) \longrightarrow CO(g)$
    $$\Delta H^\circ = -110.54 \text{ kJ}$$
    (2) $C(graphite) + O_2(g) \longrightarrow CO_2(g)$
    $$\Delta H^\circ = -393.51 \text{ kJ}$$

22. Use Hess's law to determine ΔH° for the reaction $C_3H_4(g) + 2\,H_2(g) \longrightarrow C_3H_8(g)$, given that
    (1) $H_2(g) + \frac{1}{2}\,O_2(g) \longrightarrow H_2O(l)$ $\quad \Delta H^\circ = -285.8$ kJ
    (2) $C_3H_4(g) + 4\,O_2(g) \longrightarrow 3\,CO_2(g) + 2\,H_2O(l)$
    $$\Delta H^\circ = -1937 \text{ kJ}$$
    (3) $C_3H_8(g) + 5\,O_2(g) \longrightarrow 3\,CO_2(g) + 4\,H_2O(l)$
    $$\Delta H^\circ = -2219.1 \text{ kJ}$$

23. Given the following information:
    $$\frac{1}{2}\,N_2(g) + \frac{3}{2}\,H_2(g) \longrightarrow NH_3(g) \qquad\qquad \Delta H_1^\circ$$
    $$NH_3(g) + \frac{5}{4}\,O_2(g) \longrightarrow NO(g) + \frac{3}{2}\,H_2O(l) \quad \Delta H_2^\circ$$
    $$H_2(g) + \frac{1}{2}\,O_2(g) \longrightarrow H_2O(l) \qquad\qquad \Delta H_3^\circ$$
    Determine ΔH° for the following reaction, expressed in terms of ΔH_1°, ΔH_2°, and ΔH_3°.
    $$N_2(g) + O_2(g) \longrightarrow 2\,NO(g) \qquad \Delta H^\circ = ?$$

24. Use standard enthalpies of formation from Table 7.2 in equation (7.21) to determine the enthalpy changes in the following reactions.
    (a) $C_3H_8(g) + H_2(g) \longrightarrow C_2H_6(g) + CH_4(g)$
    (b) $2\,H_2S(g) + 3\,O_2(g) \longrightarrow 2\,SO_2(g) + 2\,H_2O(l)$

25. Use standard enthalpies of formation from Tables 7.2 and 7.3 and equation (7.21) to determine the standard enthalpy change of the following reaction.
    $$NH_4^+(aq) + OH^-(aq) \longrightarrow H_2O(l) + NH_3(g)$$

26. Use the information given here, data from Appendix D, and equation (7.21) to calculate the standard enthalpy of formation, *per mole*, of ZnS(s).
    $$2\,ZnS(s) + 3\,O_2(g) \longrightarrow 2\,ZnO(s) + 2\,SO_2(g)$$
    $$\Delta H^\circ = -878.2 \text{ kJ}$$

Exercises

Heat Capacity (Specific Heat)

27. Refer to Example 7-2. The experiment is repeated with several different metals substituting for the lead. The masses of metal and water and the initial temperatures of the metal and water are the same as in Figure 7-3. The final temperatures are **(a)** Zn, 38.9 °C; **(b)** Pt, 28.8 °C; **(c)** Al, 52.7 °C. What is the specific heat of each metal, expressed in $J\ g^{-1}\ °C^{-1}$?

28. A 75.0-g piece of Ag metal is heated to 80.0 °C and dropped into 50.0 g of water at 23.2 °C. The final temperature of the Ag–H_2O mixture is 27.6 °C. What is the specific heat of silver?

29. A 465-g chunk of iron is removed from an oven and plunged into 375 g water in an insulated container. The temperature of the water increases from 26 to 87 °C. If the specific heat of iron is $0.449\ J\ g^{-1}\ °C^{-1}$, what must have been the original oven temperature?

30. A piece of stainless steel (specific heat $= 0.50\ J\ g^{-1}\ °C^{-1}$) is transferred from an oven at 183 °C into 125 mL of water at 23.2 °C. The water temperature rises to 51.5 °C. What is the mass of the steel? How precise is this method of mass determination? Explain.

31. A 1.00-kg sample of magnesium at 40.0 °C is added to 1.00 L of water maintained at 20.0 °C in an insulated container. What will be the final temperature of the Mg–H_2O mixture? (Specific heat of Mg $= 1.024\ J\ g^{-1}\ °C^{-1}$.)

32. Brass has a density of $8.40\ g/cm^3$ and a specific heat of $0.385\ J\ g^{-1}\ °C^{-1}$. A 15.2-$cm^3$ piece of brass at an initial temperature of 163 °C is dropped into an insulated container with 150.0 g water initially at 22.4 °C. What will be the final temperature of the brass–water mixture?

33. A 74.8-g sample of copper at 143.2 °C is added to an insulated vessel containing 165 mL of glycerol, $C_3H_8O_3(l)$ ($d = 1.26\ g/mL$), at 24.8 °C. The final temperature is 31.1 °C. The specific heat of copper is $0.385\ J\ g^{-1}\ °C^{-1}$. What is the heat capacity of glycerol in $J\ mol^{-1}\ °C^{-1}$?

34. What volume of 18.5 °C water must be added, together with a 1.23-kg piece of iron at 68.5 °C, so that the temperature of the water in the insulated container shown in the figure remains constant at 25.6 °C?

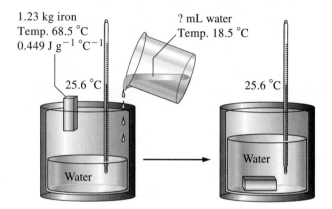

1.23 kg iron
Temp. 68.5 °C
$0.449\ J\ g^{-1}\ °C^{-1}$

? mL water
Temp. 18.5 °C

25.6 °C

25.6 °C

Water

Water

Heats of Reaction

35. How much heat, in kilojoules, is associated with the production of 283 kg of slaked lime, $Ca(OH)_2$?

$$CaO(s) + H_2O(l) \longrightarrow Ca(OH)_2(s) \quad \Delta H° = -65.2\ kJ$$

36. The standard enthalpy change in the combustion of the hydrocarbon octane is $\Delta H° = -5.48 \times 10^3\ kJ/mol\ C_8H_{18}(l)$. How much heat, in kilojoules, is liberated *per gallon* of octane burned? (Density of octane $= 0.703\ g/mL$; 1 gal $= 3.785\ L$.)

37. The combustion of methane gas, the principal constituent of natural gas, is represented by the equation

$$CH_4(g) + 2\ O_2(g) \longrightarrow CO_2(g) + 2\ H_2O(l)$$
$$\Delta H° = -890.3\ kJ$$

(a) What mass of methane, in kilograms, must be burned to liberate $2.80 \times 10^7\ kJ$ of heat?
(b) What quantity of heat, in kilojoules, is liberated in the complete combustion of $1.65 \times 10^4\ L$ of $CH_4(g)$, measured at 18.6 °C and 768 mmHg?
(c) If the quantity of heat calculated in part (b) could be transferred with 100% efficiency to water, what volume of water, in liters, could be heated from 8.8 to 60.0 °C as a result?

38. Refer to the Integrative Example. What volume of the synthesis gas, measured at STP and burned in an open flame (constant-pressure process), is required to heat 40.0 gal of water from 15.2 to 65.0 °C? (1 gal $= 3.785\ L$.)

39. The combustion of hydrogen–oxygen mixtures is used to produce very high temperatures (ca. 2500 °C) needed for certain types of welding operations. Consider the reaction to be

$$H_2(g) + \frac{1}{2}\ O_2(g) \longrightarrow H_2O(g) \quad \Delta H° = -241.8\ kJ$$

What is the quantity of heat evolved, in kilojoules, when a 180-g mixture containing equal parts of H_2 and O_2 by mass is burned?

40. Thermite mixtures are used for certain types of welding. The thermite reaction is highly exothermic.

$$Fe_2O_3(s) + 2\ Al(s) \longrightarrow Al_2O_3(s) + 2\ Fe(s)$$
$$\Delta H = -852\ kJ$$

1.00 mol Fe_2O_3 and 2.00 mol Al are mixed at room temperature (25 °C), and a reaction is initiated. The liberated heat is retained within the products, whose combined specific heat over a broad temperature range is about $0.8\ J\ g^{-1}\ °C^{-1}$. The melting point of iron is 1530 °C. Show that the quantity of heat liberated is more than sufficient to raise the temperature of the products to the melting point of iron.

41. A 0.205-g pellet of potassium hydroxide, KOH, is added to 55.9 g water in a Styrofoam coffee cup. The water temperature rises from 23.5 to 24.4 °C. [Assume that the specific heat of dilute KOH(aq) is the same as that of water.]
(a) What is the approximate heat of solution of KOH, expressed as kilojoules per mole of KOH?
(b) How could the precision of this measurement be improved *without* modifying the apparatus?

42. The heat of solution of KI(s) in water is +20.3 kJ/mol KI. If a quantity of KI is added to sufficient water at 23.5 °C in a Styrofoam cup to produce 150.0 mL of 2.50 M KI, what will be the final temperature? (Assume a density of 1.30 g/mL and a specific heat of 2.7 J g^{-1} °C^{-1} for 2.50 M KI.)

43. You are planning a lecture demonstration to illustrate an endothermic process. You desire to lower the temperature of 1400 mL water in an insulated container from 25 to 10 °C. Approximately what mass of NH$_4$Cl(s) should you dissolve in the water to achieve this result? The heat of solution of NH$_4$Cl is +14.7 kJ/mol NH$_4$Cl.

44. Care must be taken in preparing solutions of solutes that liberate heat on dissolving. The heat of solution of NaOH is -44.5 kJ/mol NaOH. To what maximum temperature may a sample of water, originally at 21 °C, be raised in the preparation of 500 mL of 7.0 M NaOH? Assume the solution has a density of 1.08 g/mL and specific heat of 4.00 J g^{-1} °C^{-1}.

45. Refer to Example 7-4. The product of the neutralization is 0.500 M NaCl. For this solution, assume a density of 1.02 g/mL and a specific heat of 4.02 J g^{-1} °C^{-1}. Also, assume a heat capacity for the Styrofoam cup of 10 J/°C, and recalculate the heat of neutralization.

46. The heat of neutralization of HCl(aq) by NaOH(aq) is -55.84 kJ/mol H$_2$O produced. If 50.00 mL of 1.05 M NaOH is added to 25.00 mL of 1.86 M HCl, with both solutions originally at 24.72 °C, what will be the final solution temperature? (Assume that no heat is lost to the surrounding air and that the solution produced in the neutralization reaction has a density of 1.02 g/mL and a specific heat of 3.98 J g^{-1} °C^{-1}.)

Enthalpy Changes and States of Matter

47. What mass of ice can be melted with the same quantity of heat as required to raise the temperature of 3.50 mol H$_2$O(l) by 50.0 °C? [$\Delta H^\circ_{fusion} = 6.01$ kJ/mol H$_2$O(s)]

48. What will be the final temperature of the water in an insulated container as the result of passing 5.00 g of steam [H$_2$O(g)] at 100.0 °C into 100.0 g of water at 25.0 °C? ($\Delta H^\circ_{vap} = 40.6$ kJ/mol H$_2$O)

49. A 125-g stainless steel ball bearing (sp ht $= 0.50$ J g^{-1} °C^{-1}) at 525 °C is dropped into 75.0 mL of water at 28.5 °C in an open Styrofoam cup. The water is brought to a boil, with the temperature rising to 100.0 °C. What mass of water vaporizes while the boiling continues? ($\Delta H^\circ_{vap} = 40.6$ kJ/mol H$_2$O)

50. If the ball bearing described in Exercise 49 is dropped onto a large block of ice at 0 °C, what mass of liquid water will form? ($\Delta H^\circ_{fusion} = 6.01$ kJ/mol H$_2$O(s))

Bomb Calorimetry

51. A 1.620-g sample of naphthalene, C$_{10}$H$_8$(s), is completely burned in a bomb calorimeter assembly and a temperature increase of 8.44 °C is noted. If the heat of combustion of naphthalene is -5156 kJ/mol C$_{10}$H$_8$, what is the heat capacity of the bomb calorimeter?

52. Salicylic acid, C$_7$H$_6$O$_3$, has been suggested as a calorimetric standard. Its heat of combustion is -3.023×10^3 kJ/mol C$_7$H$_6$O$_3$. From the following data determine the heat capacity of a bomb calorimeter assembly (that is, the bomb, water, stirrer, thermometer, wires, and so forth).

mass of salicylic acid burned:	1.201 g
initial calorimeter temperature:	23.68 °C
final calorimeter temperature:	29.82 °C

53. Refer to Example 7-3. Based on the heat of combustion of sucrose established in the example, what should be the temperature change (ΔT) produced by the combustion of 1.227 g C$_{12}$H$_{22}$O$_{11}$ in a bomb calorimeter assembly with a heat capacity of 3.87 kJ/°C?

54. A 1.397-g sample of thymol, C$_{10}$H$_{14}$O(s) (a preservative and a mold and mildew preventative), is burned in a bomb calorimeter assembly. The temperature increase is 11.23 °C, and the heat capacity of the bomb calorimeter is 4.68 kJ/°C. What is the heat of combustion of thymol, expressed in kilojoules per mole of C$_{10}$H$_{14}$O?

Pressure–Volume Work

55. Calculate the quantity of work associated with a 3.5-L expansion of a gas (ΔV) against a pressure of 748 mmHg in the units: **(a)** liter atmospheres (L atm); **(b)** joules (J); **(c)** calories (cal).

56. Calculate the quantity of work, in joules, associated with the compression of a gas from 5.62 L to 3.37 L by a constant pressure of 1.23 atm.

57. A 1.00-g sample of Ne(g) at 1 atm pressure and 27 °C is allowed to expand into an *evacuated* vessel of 2.50-L volume. Does the gas do work? Explain.

58. Compressed air in aerosol cans is used to free electronic equipment of dust. Does the air do any work as it escapes from the can?

59. In each of the following processes, is any work done when the reaction is carried out at constant pressure in a vessel open to the atmosphere? If so, is work done by the reacting system or on it? **(a)** neutralization of $Ba(OH)_2(aq)$ by $HCl(aq)$; **(b)** conversion of gaseous nitrogen dioxide to gaseous dinitrogen tetroxide; **(c)** decomposition of calcium carbonate to calcium oxide and carbon dioxide gas.

60. In each of the following processes, is any work done when the reaction is carried out at constant pressure in a vessel open to the atmosphere? If so, is work done by the reacting system or on it? **(a)** the reaction of nitrogen monoxide and oxygen gases to form gaseous nitrogen dioxide; **(b)** precipitation of magnesium hydroxide by the reaction of aqueous solutions of NaOH and $MgCl_2$; **(c)** the reaction of copper(II) sulfate and water vapor to form copper(II) sulfate pentahydrate.

First Law of Thermodynamics

61. What is the change in internal energy of a system if the system **(a)** absorbs 58 J of heat and does 58 J of work? **(b)** absorbs 125 J of heat and does 687 J of work? **(c)** evolves 280 cal of heat and has 1.25 kJ of work done on it?

62. What is the change in internal energy of a system if the *surroundings* **(a)** transfer 235 J of heat and 128 J of work to the system? **(b)** absorb 145 J of heat from the system while doing 98 J of work on the system? **(c)** exchange no heat, but receive 1.07 kJ of work from the system?

63. The internal energy of a fixed quantity of an ideal gas depends only on its temperature. If a sample of an ideal gas is allowed to expand at a constant temperature (isothermal expansion): **(a)** Does the gas do work? **(b)** Does the gas exchange heat with its surroundings? **(c)** What happens to the temperature of the gas? **(d)** What is ΔU for the gas?

64. In an *adiabatic* process, a system is thermally insulated— there is no exchange of heat between system and surroundings. For the adiabatic expansion of an ideal gas: **(a)** Does the gas do work? **(b)** Does the internal energy of the gas increase, decrease, or remain constant? **(c)** What happens to the temperature of the gas? (*Hint:* Refer also to Exercise 63.)

65. Do you think the following observation is in any way possible? An ideal gas is expanded isothermally and is observed to do twice as much work as the heat absorbed from its surroundings. Explain your answer. (*Hint:* Refer also to Exercises 63 and 64.)

66. Do you think the following observation is any way possible? A gas absorbs heat from its surroundings while being compressed. Explain your answer. (*Hint:* Refer also to Exercises 63 and 64.)

Relating ΔH and ΔU

67. Only one of the following expressions holds true for the heat of a chemical reaction, *regardless of how the reaction is carried out.* Which is the correct expression and why? **(a)** q_V; **(b)** q_P; **(c)** $\Delta U - w$; **(d)** ΔU; **(e)** ΔH.

68. Determine whether ΔH is equal to, greater than, or less than ΔU for the following reactions. Keep in mind that "greater than" means more positive or less negative and "less than" means less positive or more negative. Assume that the only significant change in volume during a reaction at constant pressure is that associated with changes in the amounts of gases.
(a) The complete combustion of one mole of the liquid, 1-butanol.
(b) The complete combustion of one mole of glucose, $C_6H_{12}O_6(s)$.
(c) The decomposition of solid ammonium nitrate to produce liquid water and gaseous dinitrogen monoxide.

69. The heat of combustion of 2-propanol at 298.15 K, determined in a bomb calorimeter, is -33.41 kJ/g. For the combustion of one mole of 2-propanol, determine **(a)** ΔU and **(b)** ΔH.

70. Write an equation to represent the combustion of thymol referred to in Exercise 54. Include in this equation the values for $\Delta U°$ and $\Delta H°$.

Hess's Law

71. For the reaction $C_2H_4(g) + Cl_2(g) \longrightarrow C_2H_4Cl_2(l)$, determine $\Delta H°$, given that

$$4\,HCl(g) + O_2(g) \longrightarrow 2\,Cl_2(g) + 2\,H_2O(l)$$
$$\Delta H° = -202.4 \text{ kJ}$$

$$2\,HCl(g) + C_2H_4(g) + \frac{1}{2}O_2(g) \longrightarrow$$
$$C_2H_4Cl_2(l) + H_2O(l) \qquad \Delta H° = -318.7 \text{ kJ}$$

72. Determine $\Delta H°$ for the reaction
$$N_2H_4(l) + 2\,H_2O_2(l) \longrightarrow N_2(g) + 4\,H_2O(l)$$
from these data.

$$N_2H_4(l) + O_2(g) \longrightarrow N_2(g) + 2\,H_2O(l)$$
$$\Delta H° = -622.2 \text{ kJ}$$

$$H_2(g) + \frac{1}{2}O_2(g) \longrightarrow H_2O(l) \qquad \Delta H° = -285.8 \text{ kJ}$$

$$H_2(g) + O_2(g) \longrightarrow H_2O_2(l) \qquad \Delta H° = -187.8 \text{ kJ}$$

73. Substitute natural gas (SNG) is a gaseous mixture containing $CH_4(g)$ that can be used as a fuel. One reaction for the production of SNG is

$$4\ CO(g) + 8\ H_2(g) \longrightarrow$$
$$3\ CH_4(g) + CO_2(g) + 2\ H_2O(l) \qquad \Delta H° = ?$$

Use appropriate data from the following listing to determine $\Delta H°$ for this SNG reaction.

$$C(\text{graphite}) + \frac{1}{2} O_2(g) \longrightarrow CO(g) \quad \Delta H° = -110.5\ \text{kJ}$$

$$CO(g) + \frac{1}{2} O_2(g) \longrightarrow CO_2(g) \qquad \Delta H° = -283.0\ \text{kJ}$$

$$H_2(g) + \frac{1}{2} O_2(g) \longrightarrow H_2O(l) \qquad \Delta H° = -285.8\ \text{kJ}$$

$$C(\text{graphite}) + 2\ H_2(g) \longrightarrow CH_4(g) \quad \Delta H° = -74.81\ \text{kJ}$$
$$CH_4(g) + 2\ O_2(g) \longrightarrow CO_2(g) + 2\ H_2O(l)$$
$$\Delta H° = -890.3\ \text{kJ}$$

74. CCl_4, an important commercial solvent, is prepared by the reaction of $Cl_2(g)$ with a carbon compound. Determine $\Delta H°$ for the reaction

$$CS_2(l) + 3\ Cl_2(g) \longrightarrow CCl_4(l) + S_2Cl_2(l)$$

Use appropriate data from the following listing.

$$CS_2(l) + 3\ O_2(g) \longrightarrow CO_2(g) + 2\ SO_2(g)$$
$$\Delta H° = -1077\ \text{kJ}$$
$$2\ S(s) + Cl_2(g) \longrightarrow S_2Cl_2(l) \qquad \Delta H° = -58.2\ \text{kJ}$$

$$C(s) + 2\ Cl_2(g) \longrightarrow CCl_4(l) \qquad\qquad \Delta H° = -135.4\ \text{kJ}$$
$$S(s) + O_2(g) \longrightarrow SO_2(g) \qquad\qquad \Delta H° = -296.8\ \text{kJ}$$
$$SO_2(g) + Cl_2(g) \longrightarrow SO_2Cl_2(l) \qquad \Delta H° = +97.3\ \text{kJ}$$
$$C(s) + O_2(g) \longrightarrow CO_2(g) \qquad\qquad \Delta H° = -393.5\ \text{kJ}$$
$$CCl_4(l) + O_2(g) \longrightarrow COCl_2(g) + Cl_2O(g)$$
$$\Delta H° = -5.2\ \text{kJ}$$

75. Use Hess's law and the following data

$$CH_4(g) + 2\ O_2(g) \longrightarrow CO_2(g) + 2\ H_2O(g)$$
$$\Delta H° = -802\ \text{kJ}$$
$$CH_4(g) + CO_2(g) \longrightarrow 2\ CO(g) + 2\ H_2(g)$$
$$\Delta H° = +247\ \text{kJ}$$
$$CH_4(g) + H_2O(g) \longrightarrow CO(g) + 3\ H_2(g)$$
$$\Delta H° = +206\ \text{kJ}$$

to determine $\Delta H°$ for the following reaction, an important source of hydrogen gas.

$$CH_4(g) + \frac{1}{2} O_2(g) \longrightarrow CO(g) + 2\ H_2(g)$$

76. The standard heats of combustion $(\Delta H°)$ per mole of 1,3-butadiene [$C_4H_6(g)$], butane [$C_4H_{10}(g)$], and $H_2(g)$ are -2540.2, -2877.6, and -285.8 kJ, respectively. Use these data to calculate the heat of hydrogenation of 1,3-butadiene to butane.

$$C_4H_6(g) + 2\ H_2(g) \longrightarrow C_4H_{10}(g) \qquad \Delta H° = ?$$

[*Hint:* Write equations for the combustion reactions. In each combustion, the products are $CO_2(g)$ and $H_2O(l)$.]

Standard Enthalpies of Formation

77. Why do the standard enthalpies of formation of some compounds have positive values, whereas others have negative values? Are there likely to be many compounds that have a standard enthalpy of formation of zero? Explain.

78. Only one of the following statements is correct. Choose the correct one, and explain what is wrong with the others. The standard enthalpy of formation of $CO_2(g)$ is **(a)** 0; **(b)** the standard enthalpy of combustion of C(graphite); **(c)** the sum of the standard enthalpies of formation of $CO(g)$ and $O_2(g)$; **(d)** the standard enthalpy of combustion of $CO(g)$.

79. Use standard enthalpies of formation from Table 7.2 to determine the enthalpy change at 25 °C for the reaction

$$2\ Cl_2(g) + 2\ H_2O(l) \longrightarrow 4\ HCl(g) + O_2(g) \quad \Delta H° = ?$$

80. Use data from Appendix D to calculate the standard enthalpy change for the following reaction at 25 °C.

$$Fe_2O_3(s) + 3\ CO(g) \longrightarrow 2\ Fe(s) + 3\ CO_2(g) \quad \Delta H° = ?$$

81. Use data from Table 7.2 to determine the standard heat of combustion of $C_2H_5OH(l)$, if reactants and products are maintained at 25 °C and 1 atm.

82. Use data from Table 7.2, together with the fact that $\Delta H° = -3509$ kJ for the complete combustion of one mole of pentane, $C_5H_{12}(l)$, to calculate $\Delta H°$ for the synthesis of 1 mol $C_5H_{12}(l)$ from $CO(g)$ and $H_2(g)$.

$$5\ CO(g) + 11\ H_2(g) \longrightarrow C_5H_{12}(l) + 5\ H_2O(l)\ \Delta H° = ?$$

(*Hint:* You can get $\Delta H_f°$ [$C_5H_{12}(l)$] from the combustion equation.)

83. Use data from Table 7.2 and $\Delta H°$ for the following reaction to determine the standard enthalpy of formation of $CCl_4(g)$ at 25 °C and 1 atm.

$$CH_4(g) + 4\ Cl_2(g) \longrightarrow CCl_4(g) + 4\ HCl(g)$$
$$\Delta H° = -397.3\ \text{kJ}$$

84. Use data from Table 7.2 and $\Delta H°$ for the following reaction to determine the standard enthalpy of formation of hexane, $C_6H_{14}(l)$, at 25 °C and 1 atm.

$$2\ C_6H_{14}(l) + 19\ O_2(g) \longrightarrow 12\ CO_2(g) + 14\ H_2O(l)$$
$$\Delta H° = -8326\ \text{kJ}$$

85. Use data from Table 7.3 and Appendix D to determine the standard enthalpy change in the reaction

$$Al^{3+}(aq) + 3\ OH^-(aq) \longrightarrow Al(OH)_3(s) \qquad \Delta H° = ?$$

86. Use data from Table 7.3 and Appendix D to determine the standard enthalpy change in the reaction

$$Mg(OH)_2(s) + 2\ NH_4^+ \longrightarrow$$
$$Mg^{2+}(aq) + 2\ H_2O(l) + 2\ NH_3(g) \qquad \Delta H° = ?$$

87. The decomposition of limestone, $CaCO_3(s)$, into quicklime, $CaO(s)$, and $CO_2(g)$ is carried out in a gas-fired kiln. Use data from Appendix D to determine how much heat is required to decompose 1.35×10^3 kg $CaCO_3(s)$. (Assume that heats of reaction are the same as at 25 °C and 1 atm.)

88. Use data from Table 7.2 to calculate the volume of butane, $C_4H_{10}(g)$, measured at 24.6 °C and 756 mmHg, that must be burned to liberate 5.00×10^4 kJ of heat.

Integrative and Advanced Exercises

89. A British thermal unit (Btu) is defined as the quantity of heat required to change the temperature of 1 lb of water by 1 °F. Assume the specific heat of water to be independent of temperature. How much heat is required to raise the temperature of the water in a 40-gal water heater from 48 to 145 °F **(a)** in Btu; **(b)** in kcal; **(c)** in kJ?

90. A 7.26-kg shot (as used in the sporting event, the shot put) is dropped from the top of a building 168 m high. What is the maximum temperature increase that could occur in the shot? Assume a specific heat of $0.47 \ J \ g^{-1} \ °C^{-1}$ for the shot. Why would the actual measured temperature increase likely be less than the calculated value?

91. An alternative approach to bomb calorimetry is to establish the heat capacity of the calorimeter, *exclusive* of the water it contains. The heat absorbed by the water and by the rest of the calorimeter must be calculated separately and then added together.

   A bomb calorimeter assembly containing 983.5 g water is calibrated by the combustion of 1.354 g anthracene. The temperature of the calorimeter rises from 24.87 to 35.63 °C. When 1.053 g citric acid is burned in the same assembly, but containing 968.6 g water, the temperature increases from 25.01 to 27.19 °C. The heat of combustion of anthracene, $C_{14}H_{10}(s)$, is $-7067 \ kJ/mol \ C_{14}H_{10}$. What is the heat of combustion of citric acid, $C_6H_8O_7$, expressed in kilojoules per mole?

92. The method of Exercise 91 is used in some bomb calorimetry experiments. A 1.148-g sample of benzoic acid is burned in an excess of $O_2(g)$ in a bomb immersed in 1181 g of water. The temperature of the water rises from 24.96 to 30.25 °C. The heat of combustion of benzoic acid is $-26.42 \ kJ/g$. In a second experiment, a 0.895-g powdered coal sample is burned in the same calorimeter assembly. The temperature of 1162 g of water rises from 24.98 to 29.81 °C. How many metric tons (1 metric ton = 1000 kg) of this coal would have to be burned to release $2.15 \times 10^9 \ kJ$ of heat?

93. A handbook lists two different values for the heat of combustion of hydrogen: 33.88 kcal/g H_2 if $H_2O(l)$ is formed, and 28.67 kcal/g H_2 if $H_2O(g)$ (steam) is formed. Explain why these two values are different, and indicate what property this difference represents. Devise a means of verifying your conclusions.

94. Determine the missing values of $\Delta H°$ in the diagram shown below.

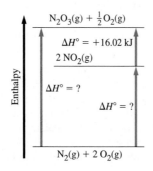

95. A particular natural gas consists, in mole percents, of 83.0% CH_4, 11.2% C_2H_6, and 5.8% C_3H_8. A 385-L sample of this gas, measured at 22.6 °C and 739 mmHg, is burned at constant pressure in an excess of oxygen gas. How much heat, in kilojoules, is evolved in the combustion reaction?

96. An overall reaction for a coal gasification process is

$$2 \ C(s) + 2 \ H_2O(g) \longrightarrow CH_4(g) + CO_2(g)$$

Show that this overall equation can be established by an appropriate combination of equations from Section 7-9.

97. Which of the following gases has the greater fuel value on a per liter (STP) basis? That is, which has the greater heat of combustion?
   (*Hint:* The only combustible gases are CH_4, C_3H_8, CO, and H_2.)

   (a) coal gas: 49.7% H_2, 29.9% CH_4, 8.2% N_2, 6.9% CO, 3.1% C_3H_8, 1.7% CO_2, and 0.5% O_2, by volume.
   (b) sewage gas: 66.0% CH_4, 30.0% CO_2, and 4.0% N_2, by volume.

98. A calorimeter that measures an exothermic heat of reaction by the quantity of ice that can be melted is called an ice calorimeter. Now consider that 0.100 L of methane gas, $CH_4(g)$, at 25.0 °C and 744 mmHg is burned at constant pressure in air. The heat liberated is captured and used to melt 9.53 g ice at 0 °C. (ΔH_{fusion} of ice = 6.01 kJ/mol)
   (a) Write an equation for the complete combustion of CH_4, and show that combustion is incomplete in this case.
   (b) Assume that CO(g) is produced in the incomplete combustion of CH_4, and represent the combustion as best you can through a single equation with small whole numbers as coefficients. $H_2O(l)$ is another product of the combustion.

99. For the reaction

$$C_2H_4(g) + 3 \ O_2(g) \longrightarrow 2 \ CO_2(g) + 2 \ H_2O(l)$$
$$\Delta H° = -1410.9 \ kJ$$

   if the H_2O were obtained as a gas rather than a liquid, **(a)** would the heat of reaction be greater (more negative) or smaller (less negative) than that indicated in the equation? **(b)** Explain your answer. **(c)** Now calculate the value of ΔH in this case.

100. Some of the butane, $C_4H_{10}(g)$, in a 200.0-L cylinder at 26.0 °C is withdrawn and burned at a constant pressure in an excess of air. As a result, the pressure of the gas in the cylinder falls from 2.35 atm to 1.10 atm. The liberated heat is used to raise the temperature of 132.5 L of water from 26.0 to 62.2 °C. Assume that the combustion products are $CO_2(g)$ and $H_2O(l)$ exclusively, and determine the efficiency of the water heater. (That is, what percent of the heat of combustion was absorbed by the water?)

101. The metabolism of glucose, $C_6H_{12}O_6$, yields $CO_2(g)$ and $H_2O(l)$ as products. Heat released in the process is converted to useful work with about 70% efficiency. Calculate the mass of glucose metabolized by a 58.0-kg person in climbing a mountain with an elevation gain of 1450 m. Assume that the work performed in the climb is about four times that required to simply lift 58.0 kg by 1450 m. ΔH_f° of $C_6H_{12}O_6(s)$ is -1273.3 kJ/mol.

102. An alkane hydrocarbon has the formula C_nH_{2n+2}. The enthalpies of formation of the alkanes decrease (become more negative) as the number of C atoms increases. Starting with butane, $C_4H_{10}(g)$, for each additional CH_2 group in the formula, the enthalpy of formation, ΔH_f°, changes by about -21 kJ/mol. Use this fact and data from Table 7.2 to estimate the heat of combustion of heptane, $C_7H_{16}(l)$.

103. Upon complete combustion, a 1.00-L sample (at STP) of a natural gas gives off 43.6 kJ of heat. If the gas is a mixture of $CH_4(g)$ and $C_2H_6(g)$, what is its percent composition, *by volume?*

104. Under the entry "H_2SO_4," a reference source lists many values for the standard enthalpy of formation. For example, for pure $H_2SO_4(l)$, $\Delta H_f^\circ = -814.0$ kJ/mol; for a solution with 1 mol H_2O per mole of H_2SO_4, -841.8; with 10 mol H_2O, -880.5; with 50 mol H_2O; -886.8; with 100 mol H_2O, -887.7; with 500 mol H_2O, -890.5; with 1000 mol H_2O, -892.3, with 10,000 mol H_2O, -900.8, and with 100,000 mol H_2O, -907.3.
(a) Explain why these values are not all the same.
(b) The value of $\Delta H_f^\circ[H_2SO_4(aq)]$ in an infinitely dilute solution is -909.3 kJ/mol. What data from this chapter can you cite to confirm this value? Explain.
(c) If 500.0 mL of 1.00 M $H_2SO_4(aq)$ is prepared from pure $H_2SO_4(l)$, what is the approximate change in temperature that should be observed? [Assume that the $H_2SO_4(l)$ and $H_2O(l)$ are at the same temperature initially and that the specific heat of the $H_2SO_4(aq)$ is about 4.2 J g^{-1} °C^{-1}.]

105. Refer to the discussion of the gasification of coal (page 252), and show that some of the heat required in the gasification reaction (equation 7.24) can be supplied by the *methanation* reaction. This fact contributes to the success of modern processes to produce *synthetic natural gas* (SNG)

106. A 1.103-g sample of a gaseous carbon–hydrogen–oxygen compound that occupies a volume of 582 mL at 765.5 Torr and 25.00 °C is burned in an excess of $O_2(g)$ in a bomb calorimeter. The products of the combustion are 2.108 g $CO_2(g)$, 1.294 g $H_2O(l)$, and enough heat to raise the temperature of the calorimeter assembly from 25.00 to 31.94 °C. The heat capacity of the calorimeter is 5.015 kJ/°C. Write an equation for the combustion reaction, and indicate $\Delta H°$ for this reaction at 25.00 °C.

107. Several factors are involved in determining the cooking times required for foods in a microwave oven. One of these factors is specific heat. Determine the approximate time required to warm 250 mL of chicken broth from 4 °C (a typical refrigerator temperature) to 50 °C in a 700-W microwave oven. Assume that the density of chicken broth is about 1 g/mL and that its specific heat is approximately 4.2 J g^{-1} °C^{-1}.

108. Pictured here is an ideal gas at 25.0 °C confined in a cylinder 12.00 cm in diameter by a piston surmounted by a steel cylinder 10.00 cm in diameter. The density of the steel is 7.75 g/cm³, atmospheric pressure is 745 Torr, and other data are indicated on the sketch. How much work is done when the steel cylinder is suddenly removed?

109. When one mole of sodium carbonate decahydrate (washing soda) is gentle warmed, 155.3 kJ of heat is absorbed, water vapor is formed, and sodium carbonate heptahydrate remains. On more vigorous heating, the heptahydrate absorbs 320.1 kJ of heat and loses more water vapor to give the monohydrate. Continued heating gives the anhydrous salt (soda ash) while 57.3 kJ of heat is absorbed. Calculate ΔH for the conversion of one mole of washing soda into soda ash. Estimate ΔU for this process. Why is the value of ΔU only an estimate?

Feature Problems

110. James Joule published his definitive work related to the first law of thermodynamics in 1850, stating that "the quantity of heat capable of increasing the temperature of one pound of water by 1 °F requires for its evolution the expenditure of a mechanical force represented by the fall of 772 lb through the space of one foot." Validate this statement by relating it to information given in this text.

111. Based on specific heat measurements, Pierre Dulong and Alexis Petit proposed in 1818 that the specific heat of an element is inversely related to its atomic weight (atomic mass). Thus, by measuring the specific heat of a new element, its atomic weight could be readily established.
(a) Use data from Table 7.1 and inside the front cover to plot a *straight-line* graph relating atomic mass and specific heat. Write the equation for this straight line.
(b) Use the measured specific heat of 0.23 J g^{-1} °C^{-1} and the equation derived in part (a) to obtain an approximate value of the atomic mass of cadmium, an element discovered in 1817.
(c) To raise the temperature of 75.0 g of a particular metal by 15 °C requires 450 J of heat. What might this metal be?

112. We can use the heat liberated by a neutralization reaction as a means of establishing the stoichiometry of the reaction. The data in the table below are for the reaction of 1.00 M NaOH with 1.00 M citric acid, $C_6H_8O_7$, in a total solution volume of 60.0 mL.

mL 1.00 M NaOH Used	mL 1.00 M Citric Acid Used	ΔT, °C
20.0	40.0	4.7
30.0	30.0	6.3
40.0	20.0	8.2
50.0	10.0	6.7
55.0	5.0	2.7

(a) Plot ΔT versus mL 1.00 M NaOH, and identify the exact stoichiometric proportions of NaOH and citric acid at the equivalence point of the neutralization reaction.
(b) Why is the temperature change in the neutralization greatest when the reactants are in their exact stoichiometric proportions? That is, why not use an excess of one of the reactants to ensure that the neutralization has gone to completion to achieve the maximum temperature increase?
(c) Rewrite the formula of citric acid to reflect more precisely its acidic properties. Then write a balanced net ionic equation for the neutralization reaction.

113. In a student experiment to confirm Hess's law, the reaction

$$NH_3 \text{ (concd aq)} + HCl(aq) \longrightarrow NH_4Cl(aq)$$

was carried out in two different ways. First, 8.00 mL of concentrated NH_3(aq) was added to 100.0 mL of 1.00 M HCl in a calorimeter. [The NH_3(aq) was slightly in excess.] The reactants were initially at 23.8 °C, and the final temperature after neutralization was 35.8 °C.

In a second experiment, outlined in part in the sketch, air was bubbled through 100.0 mL of concentrated NH_3(aq), sweeping out NH_3(g). The NH_3(g) was neutralized in 100.0 mL of 1.00 M HCl. The temperature of the concentrated NH_3(aq) fell from 19.3 to 13.2 °C. At the same time, the temperature of the 1.00 M HCl rose from 23.8 to 42.9 °C as it was neutralized by NH_3(g).

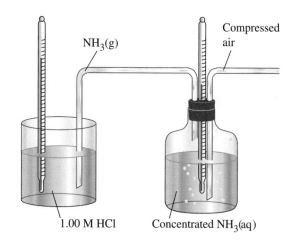

Assume that all solutions have densities of 1.00 g/mL and specific heats of 4.18 J g^{-1} °C^{-1}.
(a) Write the two equations and ΔH values for the processes occurring in the second experiment. Show that the sum of these two equations is the same as the equation for the reaction in the first experiment.
(b) Show that, within the limits of experimental error, ΔH for the overall reaction is the same in the two experiments, thereby confirming Hess's law.

114. When an ideal gas is heated, the change in internal energy is limited to increasing the average translational kinetic energy of the gas molecules. Thus, there is a simple relationship between ΔU of the gas and the change in temperature that occurs. Derive this relationship with the help of ideas about the kinetic-molecular theory of gases developed in Chapter 6. After doing so, obtain numerical values (in J mol^{-1} K^{-1}) for the following molar heat capacities.

(a) The heat capacity, C_v, for one mole of gas under constant-volume conditions.
(b) The heat capacity, C_p, for one mole of gas under constant-pressure conditions.

115. Refer to Example 7-5, dealing with the work done by 0.100 mol He at 298 K in expanding in a single step from 2.40 to 1.30 atm. Review also, the two-step expansion (2.40 atm $\longrightarrow$ 1.80 atm $\longrightarrow$ 1.30 atm) described on page 234. Now,
(a) Determine the total work that would be done if the He expanded in a series of steps, at 0.10 atm intervals, from 2.40 atm to 1.30 atm.
(b) Represent this total work on the accompanying graph, in which the quantities of work done in the one-step and two-step expansions are already shown.
(c) Show that the maximum amount of work would occur if the expansion occurred in an infinite number of steps. To do this, express each infinitesimal quantity of work as $dw = -PdV$ and use the methods of integral calculus (integration) to sum these quantities. Assume ideal behavior for the gas.
(d) Imagine reversing the process, that is, compressing the He from 1.30 to 2.40 atm.
What are the maximum and minimum amounts of work required to produce this compression? Explain.
(e) In the isothermal compression described in part (d), what is the change in internal energy assuming ideal gas behavior? What is the value of q?

(f) Using the formula for the work derived in part (c), obtain an expression for q/T. Is this new function a state function? Explain.

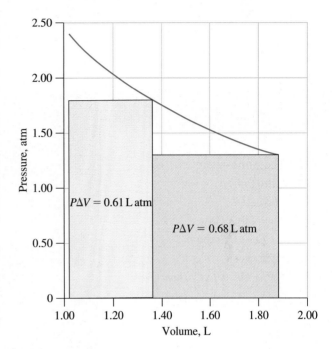

eMedia Exercises

116. From the **Types of Energy** activity *(eChapter 7-1)*, describe the differences in the motions of particles possessing predominantly high kinetic energy, predominantly high thermal energy, and predominantly high potential energy.

117. From the **Bomb Calorimetry** simulation *(eChapter 7-3)*, estimate which of the compounds, benzoic acid or sucrose, has a higher heat of combustion.
(Hint: Consider the reaction of equal masses of material.)

118. Consider the gas contained in the cylinder shown in the **Work of Gas Expansion** animation *(eChapter 7-4)*. Calculate the amount of work done by the system through the expansion of the gas if the cylinder has a radius of 10 cm and the piston is displaced upwards by 20 cm against an opposing pressure of 1.2 atm.

119. Using only the graphs generated through the States of Matter Animation *(eChapter 7-6)*, **(a)** estimate whether ethanol or ethane has a higher heat capacity. **(b)** Which one has a higher molar enthalpy of vaporization? **(c)** Which part of the graphs is used to derive this information?

120. For the reaction described in the **Formation of Aluminum Bromide** movie *(eChapter 7-8)*, **(a)** calculate the amount of heat released from the complete reaction of 27.2 g of Al in the presence of excess bromine, given that the heat of formation of $AlBr_3(s)$ is -527 kJ/mol. **(b)** What aspects of this reaction make it difficult to quantify the reaction yield of both products, heat and aluminum bromide?

8

The Atmospheric Gases and Hydrogen

Lightning is an important source of nitrates. The lighting provides the energy required for the chemical reactions involved. Molecular nitrogen is converted into oxides of nitrogen that, when they react with water, are converted to nitric acid. The nitric acid ends up as nitrates in the soil. It is estimated that about 30 million tons of HNO_3 are produced this way annually.

The first seven chapters of this book presented a number of fundamental ideas or principles—the "nuts and bolts" of chemistry. Although many of these ideas are interesting in themselves, the main reason for studying principles is to provide a basis for learning about chemical substances. The application of principles to describing and explaining chemical behavior is often called *descriptive chemistry*.

Our aim in this chapter is to convey the essence of descriptive chemistry by studying the properties and uses of a small number of elements and compounds—hydrogen and some of the elements and compounds associated with the atmosphere. Among the topics we will consider are several important environmental issues, such as smog formation, ozone destruction, and global warming. The principles we will apply are primarily those dealing with stoichiometry, gases, and thermochemistry.

8-1   The Atmosphere

When astronauts look at Earth from space, they see it surrounded by a thin blue shell—the atmosphere. Commonly, we use the word *air* to describe the substances that make up the atmosphere. Air is a mixture of nitrogen and oxygen gases, with smaller quantities of argon, carbon dioxide, and the other substances listed in Table 8.1. This blanket of air surrounding Earth protects us from harmful radiation and is a major source of a number of chemicals necessary for life.

TABLE 8.1 Composition of Dry Air (Near Sea Level)	
Component	**Volume Percent[a]**
Nitrogen (N_2)	78.084
Oxygen (O_2)	20.946
Argon (Ar)	0.934
Carbon dioxide (CO_2)	0.037
Neon (Ne)	0.001818
Helium (He)	0.000524
Methane (CH_4)	0.0002
Krypton (Kr)	0.000114
Hydrogen (H_2)	0.00005
Dinitrogen monoxide (N_2O)	0.00005
Xenon (Xe)	0.000009
Ozone (O_3)	
Sulfur dioxide (SO_2)	
Nitrogen dioxide (NO_2)	trace
Ammonia (NH_3)	
Carbon monoxide (CO)	
Iodine (I_2)	

[a] Recall the meaning of percent by volume of a gaseous mixture introduced in Section 6-6.

Structure of the Atmosphere

About 90% of the mass of the atmosphere is in the first 12 km above Earth's surface. This familiar layer of air is called the **troposphere**. It is the region in which weather phenomena occur. Temperatures in the troposphere fall continuously with increasing altitude to a minimum of about 220 K (Figure 8-1).

The region from about 12 to 55 km above Earth's surface, the **stratosphere**, is less familiar to us, although supersonic aircraft fly in its lower reaches. The temperature–altitude profile of the stratosphere is much different from that of the troposphere. At altitudes from about 12 to 25 km, the temperature is fairly constant at about 220 K; it then rises to about 280 K at 50 km. In part, this is due to exothermic reactions that occur between the atoms in the thinning atmosphere and ultraviolet radiation from the sun. These reactions produce the *ozone layer* and will be discussed in more detail later.

Beyond the stratosphere, the atmosphere is very thin, with densities rapidly falling to the range of micrograms and nanograms per liter. In the region from 55 to 80 km, the *mesophere*, the temperature falls continuously to about 180 K. Beyond this, in a region known as the *thermosphere*, or *ionosphere*, temperatures rise to about 1500 K. Here, the atmosphere consists of positive and negative ions, free electrons, neutral atoms, and molecules. The dissociation of molecules into atoms and the ionization of atoms into positive ions and free electrons require the

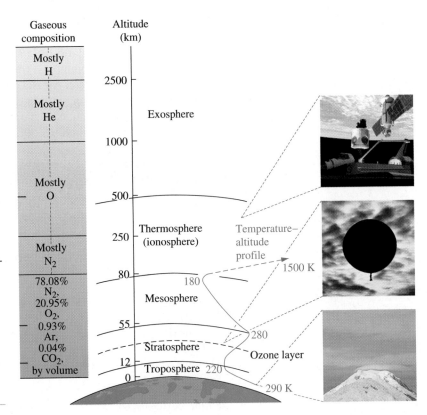

▶ **FIGURE 8-1**

Regions of the atmosphere: composition and temperature–altitude profile

The left side of the figure shows the composition of the atmosphere at different altitudes. The temperature–altitude profile in the center gives approximate temperatures for the first 80 km and indicates that the temperature rises to 1500 K in the thermosphere. The small photographs at the right picture some familiar objects and indicate the regions of the atmosphere associated with them. The values given here are only approximate. For example, the height of the troposphere varies from about 8 km at the poles to 16 km at the equator. Also, temperatures in the thermosphere vary greatly between day and night.

absorption of energy. The source of that energy is electromagnetic radiation from the sun. An interesting natural phenomenon associated with the ionosphere is the *aurora borealis*, or northern lights, pictured in Figure 8-2.

Normally, at 1500 K an iron object glows a bright red, but a cold iron object brought to the thermosphere would not glow. From the kinetic theory of gases (Section 6-7), we know that at 1500 K gaseous atoms and molecules travel at high speeds. And these energetic atoms and molecules transfer energy as heat when they collide with a cold object. Because the concentration of gaseous particles in the thermosphere is so low, however, molecular collisions are infrequent and very little heat is transferred. An object does not "heat up" the way it would in the lower atmosphere. Thus, as meteors fall toward Earth, they do not begin to glow in the high-temperature thermosphere. They do so only at lower elevations where collisions with molecules of air cause surface atoms of the meteor to vaporize, ionize, and emit light.

▶ **FIGURE 8-2**

The aurora borealis

The aurora borealis, or northern lights, observed at high northern latitudes are light emissions from atoms, ions, and molecules in the ionosphere.

▲ **Dew and frost formation**
When the temperature drops below the value at which the relative humidity is 100%, water vapor condenses as (top) *dew,* or, if the ambient temperature is below the freezing point of water, (bottom), *frost.*

Water Vapor in the Atmosphere

Table 8.1 lists the components of dry air, but air is not ordinarily dry. It contains water vapor, $H_2O(g)$, in quantities ranging from traces to as much as 4% by volume. This water vapor plays an essential role in the hydrologic (water) cycle: Ocean water evaporates; air masses move from the oceans to the land; water vapor in the atmosphere, when cooled, forms clouds; and the clouds produce rain. Water falling on land returns to the oceans through systems of lakes, rivers, and groundwater.

The amount of water vapor (n_{H_2O}) in a sample of air and the partial pressure of the water vapor (P_{H_2O}) are proportional, as we see from the ideal gas equation: $P = nRT/V$. The maximum possible partial pressure of water vapor at a given temperature is a quantity that we have previously introduced: the vapor pressure of water (see page 198). For example, at 25 °C, the vapor pressure of water is 23.8 mmHg. This is the highest partial pressure of water vapor that can be maintained at 25 °C. If a sample of air at 25 °C has a water vapor pressure higher than 23.8 mmHg, we expect some of the vapor to condense to liquid water. A common method of describing the water vapor content of air is through its relative humidity. **Relative humidity** is the ratio of the partial pressure of water vapor to the vapor pressure of water at the same temperature, expressed on a percent basis. Thus, if the partial pressure of water vapor in air at 25 °C is 12.2 mmHg, the relative humidity of the air is 51.3%.

$$\text{Relative humidity} = \frac{\text{partial pressure of water vapor}}{\text{vapor pressure of water}} \times 100\%$$

$$= \frac{12.2 \text{ mmHg}}{23.8 \text{ mmHg}} \times 100\% = 51.3\%$$

Because the vapor pressure of water increases with temperature, air having $P_{H_2O} = 12.2$ mmHg would have a relative humidity of only 38.4% at 30 °C, where the vapor pressure of water is 31.8 mmHg. Conversely, at 10 °C, where the vapor pressure of water is only 9.2 mmHg, some water vapor in the air sample would condense as liquid. This fluctuation of the relative humidity with temperature accounts for the formation of dew that is sometimes observed in the early morning hours.

The Atmosphere as a Source of Chemicals

We do not normally think of water as a commercial chemical, but it is one, in addition to being essential to life. As we indicated previously, freshwater on Earth comes mostly from seawater, by way of the atmosphere. Nitrogen and oxygen are among the most widely used manufactured chemicals. They, as well as argon and other noble gases, are obtained by the fractional distillation of liquid air. This process involves only physical changes and is described in Figure 8-3.

EXAMPLE 8-1

Using the Atmosphere as a Source of Chemicals. What volume of air must be processed to produce 5.00 L of $N_2(g)$?

Solution

From Table 8.1 we find that air contains 78.084% N_2 by volume. To find the volume of air needed, we must use this value as a conversion factor.

$$\text{volume of air} = 5.00 \text{ L } N_2 \times \frac{100.000 \text{ L air}}{78.084 \text{ L } N_2} = 6.40 \text{ L}$$

▶ **FIGURE 8-3**

The fractional distillation of liquid air—a simplified representation

Clean air is fed into a compressor and cooled by refrigeration. The cold air expands through a nozzle and is cooled still further—enough to cause it to liquefy. The liquid air is filtered to remove solid CO_2 and hydrocarbons and then distilled. Liquid air enters the top of the column where nitrogen, the most volatile component (lowest boiling point), passes off as a gas. In the middle of the column, gaseous argon is removed. Liquid oxygen, the least volatile component, collects at the bottom. The normal boiling points of nitrogen, argon, and oxygen are 77.4, 87.5, and 90.2 K, respectively.

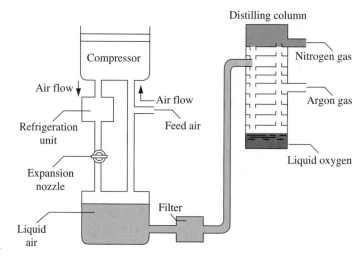

Practice Example A: What volume of air must be processed to produce 5.00 L of Ar(g)?

Practice Example B: What volume of air, measured at 1 atm and 298 K, must be processed to produce 5.00 g of $CO_2(s)$?

8-2 Nitrogen

Nitrogen occurs mainly in the atmosphere. Its abundance in Earth's solid crust is only 0.002% by mass. The only important nitrogen-containing minerals are KNO_3 (niter, or saltpeter) and $NaNO_3$ (soda niter, or Chile saltpeter), found in a few desert regions. Other natural sources of nitrogen-containing compounds are plant and animal protein and the fossilized remains of ancient plant life, such as coal.

Until about 100 years ago, sources of pure nitrogen and its compounds were quite limited. This all changed with the invention of a process for the liquefaction of air in 1895 (see Figure 8-3) and a process for converting nitrogen to ammonia in 1908. A host of nitrogen compounds can be made from ammonia. Nitrogen gas has many important uses of its own in addition to being a precursor of manufactured nitrogen compounds. Some of these uses are listed in Table 8.2.

The substance from which all nitrogen compounds are ultimately derived, $N_2(g)$, is unusually stable. One explanation of the limited reactivity of the N_2 molecule is based on its electronic structure, a topic that we will consider in Chapter 11. There, we will learn that the bond between the two N atoms in N_2 is a *triple* covalent bond, an unusually strong bond that is difficult to break. In thermochemical terms, we say

▲ Viruses for use in medical research are frozen in liquid nitrogen.

TABLE 8.2 Production and Uses of Nitrogen Gas

Production
1999: 34.91×10^6 tons
Ranks No. 2, by mass, among manufactured chemicals in U.S.
Uses
Provide a blanketing (inert) atmosphere for the production
   of chemicals and electronic components
Pressurized gas for enhanced oil recovery
Metals treatment
Refrigerant (e.g., fast freezing of foods)

that the enthalpy change associated with breaking the bonds in one mole of N_2 molecules is very high—the dissociation reaction is highly endothermic.

$$N\equiv N(g) \longrightarrow 2\,N(g) \qquad \Delta H° = +945.4 \text{ kJ}$$

Also, the enthalpies of formation of many nitrogen compounds are positive, which means that their formation reactions are endothermic. For NO(g),

$$\frac{1}{2}N_2(g) + \frac{1}{2}O_2 \longrightarrow NO(g) \qquad \Delta H° = 90.25 \text{ kJ}$$

As we noted in Chapter 7 (page 246), our general expectation is that highly endothermic reactions, such as the formation of NO(g) from its elements, do not occur to any significant extent at normal temperatures. Just imagine the situation if $\Delta H_f°$ [NO(g)] = −90.25 kJ/mol instead of +90.25 kJ/mol. The reaction of $N_2(g)$ and $O_2(g)$ to form NO(g) would likely proceed to a far greater extent. With an atmosphere depleted in $O_2(g)$ and rich in noxious NO(g), life as we know it would not be possible on Earth.

In the remainder of this section, we consider several important nitrogen compounds and their uses.

Ammonia and Related Compounds

Ammonia, NH_3, is one of the most useful chemicals we know. Each year it ranks about sixth, by mass, among manufactured chemicals in the United States. Ammonia's importance stems from the ease with which it can be converted into a wide variety of other nitrogen-containing chemicals.

It was not until 1908 that Fritz Haber worked out the conditions for the ammonia synthesis reaction and tested the reaction in a laboratory. Converting Haber's method into a manufacturing process was one of the most challenging engineering problems of its time. This work was led by Carl Bosch at Badische Anilin & Soda Fabrik (BASF) in Germany. By 1913, a plant producing 30,000 kg NH_3 per day was in operation. A modern ammonia plant has about 50 times this capacity.

The essential difficulty in the ammonia synthesis reaction is that under most conditions, the reaction does not go to completion. As soon as some NH_3 is produced, it tends to decompose back to N_2 and H_2. The reaction is *reversible* and reaches a condition of equilibrium that we describe more fully in Chapter 16. (We represent the reversible nature of the reaction by using a double arrow in the equation for the reaction.)

$$N_2(g) + 3\,H_2(g) \rightleftharpoons 2\,NH_3(g) \qquad (8.1)$$

A high yield of ammonia requires (a) a high temperature (400 °C), (b) a catalyst to speed up the reaction, and (c) a high pressure (about 200 atm). The key to achieving essentially 100% yield is continuous removal of NH_3 and recycling of the unreacted $N_2(g)$ and $H_2(g)$. The NH_3 is removed by liquefaction. The Haber–Bosch process is outlined in Figure 8-4. A critical aspect of the process is having a source of $H_2(g)$. Mostly, this is made from natural gas (see page 287).

Ammonia is the starting material in the manufacture of most other nitrogen compounds, but it has some direct uses of its own. Its most important use is as a fertilizer. The highest concentration in which nitrogen fertilizer can be applied to fields is as pure liquid NH_3, known as "anhydrous ammonia." $NH_3(aq)$ is also applied in a variety of household cleaning products, such as commercial glass cleaners. In these products, the ammonia acts as an inexpensive base to produce $OH^-(aq)$. The $OH^-(aq)$ reacts with grease and oil molecules to convert them into compounds that are more soluble in water. In addition, the aqueous ammonia solution dries quickly, leaving few streaks on glass.

▲ **Fritz Haber (1868–1934)**
Haber's perfection of the ammonia synthesis reaction, which made the manufacture of inexpensive explosives possible, was of critical importance to Germany in World War I. After the war, Haber again applied his chemical knowledge for his country's benefit by attempting, unsuccessfully, to extract gold from seawater for use in paying war reparations. Despite his past services, this Jewish scientist was driven from his academic post by the Nazi regime in 1933.

▶ **FIGURE 8-4**

Ammonia synthesis reaction—the Haber–Bosch process

The gaseous N_2–H_2 mixture is introduced into a reactor at high temperature and pressure in the presence of a catalyst. The gaseous N_2–H_2–NH_3 mixture leaves the reactor and is cooled as it passes through a condenser. Liquefied NH_3 is removed, and the remaining N_2–H_2 mixture is compressed and returned to the reactor. The yield is essentially 100%.

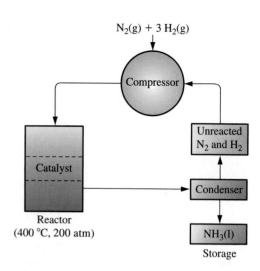

▲ Anhydrous liquid ammonia being applied directly to the soil.

Because ammonia is a base, a simple approach to producing certain nitrogen compounds is to neutralize ammonia with an appropriate acid. The acid–base reaction that forms ammonium sulfate, an important solid fertilizer, is

$$2\,NH_3(aq) + H_2SO_4(aq) \longrightarrow (NH_4)_2SO_4(aq) \tag{8.2}$$

Ammonium chloride, made by the reaction of $NH_3(aq)$ and $HCl(aq)$, is used in the manufacture of dry cell batteries, in cleaning metals, and as an agent to help solder flow smoothly when soldering metals. Ammonium nitrate, made by the reaction of $NH_3(aq)$ and $HNO_3(aq)$, is used both as a fertilizer and as an explosive. The explosive power of ammonium nitrate was not fully appreciated until a shipload of this material exploded in Texas City, Texas, in 1947, killing many people. More recently, mixtures of ammonium nitrate and fuel oil were used as explosives in the terrorist attacks on the World Trade Center in New York City in 1993 and the Murrah Federal Building in Oklahoma City in 1995. The reaction of $NH_3(aq)$ and $H_3PO_4(aq)$ yields ammonium phosphates [such as $NH_4H_2PO_4$ and $(NH_4)_2HPO_4$]. These compounds are good fertilizers because they supply two vital plant nutrients, N and P; they are also used as fire retardants.

Urea, which contains 46% nitrogen by mass, is often manufactured at ammonia synthesis plants by a simple reaction.

$$2\,NH_3 + CO_2 \longrightarrow CO(NH_2)_2 + H_2O \tag{8.3}$$
$$\text{urea}$$

Urea is an excellent fertilizer, either as a pure solid, as a solid mixed with ammonium salts, or in a very concentrated aqueous solution mixed with NH_4NO_3 or NH_3 (or both). Urea is also used as a feed supplement for cattle and in the production of polymers and pesticides. Production figures for the United States in 1999 for ammonium sulfate, ammonium nitrate, and urea were about 2.9×10^6, 8.2×10^6, and 9.3×10^6 ton, respectively.

Nitrogen Oxides

Nitrogen forms a series of oxides in which the oxidation state of N can have every value ranging from +1 to +5. Table 8.3 lists their formulas and names them according to the scheme introduced in Chapter 3. The oxides of nitrogen are not as familiar as several other nitrogen compounds, but we do encounter them in a number of situations. N_2O has anesthetic properties and finds some use in dentistry (laughing gas). NO_2 is employed in the manufacture of nitric acid. N_2O_4 is used ex-

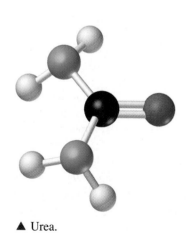

▲ Urea.

TABLE 8.3	Nomenclature of Some Common Oxides of Nitrogen	
O.S. of N	**Formula**	**Name**
+1	N_2O	Dinitrogen monoxide
+2	NO	Nitrogen monoxide
+3	N_2O_3	Dinitrogen trioxide
+4	NO_2	Nitrogen dioxide
+4	N_2O_4	Dinitrogen tetroxide
+5	N_2O_5	Dinitrogen pentoxide

tensively as an oxidizer in rocket fuels. NO is the most important oxide of nitrogen from a biological standpoint. In humans, it plays a role in maintaining blood pressure, aids in the immune response in killing foreign organisms, and is essential to the establishment of long-term memory. In 1996, scientists discovered that hemoglobin carries NO as well as O_2. The NO causes the walls of blood vessels to thin, making it easier for oxygen to be transported to the surrounding tissues. This discovery has resulted in the development of several new medications.

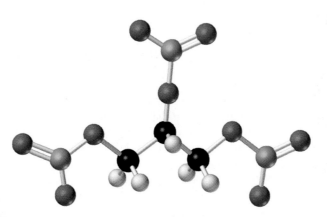

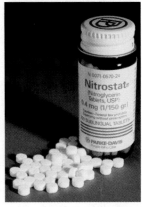

▲ Nitroglycerin is used to dilate coronary arteries in the treatment of angina. It exerts its effect by virtue of its metabolic conversion to nitric oxide.

▲ A copper penny reacting with nitric acid. The reaction is that given by Equation (8.6). The blue-green color of the solution is due to Cu^{2+}(aq), and the reddish brown color is that of nitrogen dioxide, NO_2(g).

Dinitrogen monoxide (nitrous oxide), N_2O(g), can be produced in the laboratory by an interesting disproportionation reaction, the decomposition of NH_4NO_3(s).

$$NH_4NO_3(s) \xrightarrow{200\text{--}260\,°C} N_2O(g) + 2\,H_2O(g)$$

The N atom in NH_4^+ is in the oxidation state (O.S.) −3; in NO_3^- the O.S. of N is +5. In N_2O, both N atoms are in the O.S. +1. The change in O.S. in one N atom is just matched by the change of O.S. in the other, which makes the redox equation very easy to balance.

Nitrogen monoxide (nitric oxide), NO(g), is produced commercially by the oxidation of NH_3(g) in the presence of a catalyst (the Ostwald process). This oxidation is the first step in converting NH_3 to a number of other nitrogen compounds.

$$4\,NH_3(g) + 5\,O_2(g) \xrightarrow[850\,°C]{Pt} 4\,NO(g) + 6\,H_2O(g) \qquad (8.4)$$

Another source of NO, usually unwanted, is in high-temperature combustion processes, such as those that occur in automobile engines and in electric power plants. At the same time that the fuel combines with oxygen from air to produce a high temperature, $N_2(g)$ and $O_2(g)$ in the hot air combine to a limited extent to form $NO(g)$.

$$N_2(g) + O_2(g) \xrightarrow{\Delta} 2\,NO(g) \tag{8.5}$$

We often see brown nitrogen dioxide, $NO_2(g)$, in reactions involving nitric acid. An example is the reaction of $Cu(s)$ with warm concentrated $HNO_3(aq)$.

$$Cu(s) + 4\,H^+(aq) + 2\,NO_3^-(aq) \longrightarrow Cu^{2+}(aq) + 2\,H_2O + 2\,NO_2(g) \tag{8.6}$$

Of particular interest to atmospheric chemists is the key role of $NO_2(g)$ in the formation of photochemical smog (page 275).

Nitric Acid and Nitrates

An **acid anhydride** is a nonmetallic oxide that reacts with water to produce a ternary acid as the sole product. The acid anhydride of nitric acid is N_2O_5.

$$N_2O_5(s) + H_2O(l) \longrightarrow 2\,HNO_3(aq)$$

The commercial synthesis of nitric acid does not use N_2O_5, however. It involves the following three reactions, the first of which—the Ostwald process—we described previously. $NO(g)$ from reaction (8.8) is recycled into reaction (8.7).

$$4\,NH_3(g) + 5\,O_2(g) \xrightarrow[850\,°C]{Pt} 4\,NO(g) + 6\,H_2O(g) \tag{8.4}$$

$$2\,NO(g) + O_2(g) \longrightarrow 2\,NO_2(g) \tag{8.7}$$

$$3\,NO_2(g) + H_2O(l) \longrightarrow 2\,HNO_3(aq) + NO(g) \tag{8.8}$$

Nitric acid is used in the preparation of various dyes; drugs; fertilizers (ammonium nitrate); and explosives, such as nitroglycerin, nitrocellulose, and trinitrotoluene (TNT). It is also used in metallurgy and in reprocessing spent nuclear fuels. Nitric acid is about twelfth, by mass, among the top chemicals produced in the United States.

Nitric acid is also a good oxidizing agent. For example, copper reacts with dilute $HNO_3(aq)$, producing primarily NO or, with concentrated $HNO_3(aq)$, NO_2 (reaction 8.6). With a more active metal such as Zn, the reduction product has N in one of its lower oxidation states, as in NH_4^+ in some cases. Nitrates can be made by neutralizing nitric acid with appropriate bases.

The preceding paragraph refers to several chemical reactions without explicitly supplying equations for them. If you read the descriptions of these reactions carefully and compare them with other reactions for which equations are given, however, you should be able to supply your own equations. Example 8-2 illustrates how to do this.

▲ Flash paper is used by magicians for dramatic effect. It can be made by treating paper with nitric and sulfuric acids. This process converts the cellulose fibers into nitrocellulose, which burns cleanly and rapidly.

▲ Trinitrotoluene (TNT).

EXAMPLE 8-2

Writing Chemical Equations. Write an equation to represent the synthesis of a calcium nitrate solution by the reaction of aqueous nitric acid and solid calcium hydroxide.

Solution

As we learned in Chapter 4, when a chemical reaction is described through the names of its reactants and products, we can substitute formulas for names, complete the formula expression, and then balance the equation.

$$Ca(OH)_2(s) + 2\,HNO_3(aq) \longrightarrow Ca(NO_3)_2(aq) + 2\,H_2O(l)$$

Practice Example A: Lead(IV) oxide reacts with nitric acid to produce water, oxygen gas, and an aqueous solution of lead(II) nitrate. Write a chemical equation for this reaction.

Practice Example B: Write an equation to represent the reaction of zinc with dilute nitric acid solution to form aqueous zinc(II) nitrate, ammonium nitrate, and water.

[*Hint:* This is an oxidation–reduction reaction similar to reaction (8.6). Balance it by the half-reaction method of Section 5-5.]

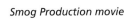

Smog Production movie

▶ Air quality in London has been greatly improved by control measures such as elimination of coal as a household fuel, introduced after a severe smog episode in 1952.

▲ **FIGURE 8-5**
Smog in Mexico City
At times, the topographical features, climatic conditions, traffic congestion, and heavy industrial pollution combine to create severe smog conditions in Mexico City.

Smog—An Environmental Issue Involving Oxides of Nitrogen

About 100 years ago, a new word entered the English language—**smog**. It referred to a condition, common in London, in which a combination of *smoke* and *fog* obscured visibility and produced health hazards (including death). These conditions are often associated with heavy industry, and this type of smog is now called *industrial* smog.

The form of air pollution more commonly thought of as smog results from the action of sunlight on the products of combustion. Chemical reactions brought about by light are called *photochemical* reactions, and the smog formed by such reactions is photochemical smog. Photochemical smog results from high-temperature combustion processes, such as those that occur in automobile engines. Because the combustion of gasoline takes place in air rather than in pure oxygen, $NO(g)$ produced by reaction (8.5) is inevitably found in the exhaust from automobiles. Other products found in the exhaust are hydrocarbons (unburned gasoline) and partially oxidized hydrocarbons. These, then, are the starting materials—the precursors—of photochemical smog.

Many substances have been identified in smoggy air, including NO, NO_2, ozone, (a form of the element oxygen with the formula O_3, discussed further in Section 8-3), and a variety of organic compounds derived from gasoline hydrocarbons. Ozone is very reactive and is largely responsible for the breathing difficulties that some people experience during smog episodes. Another noxious substance found in smog is an organic compound known as peroxyacetyl nitrate (PAN). PAN is a powerful lacrimator—that is, it causes tear formation in the eyes. Photochemical smog components cause heavy crop damages (to oranges, for example) and the deterioration of rubber goods. And, of course, the best known symbol of photochemical smog is the hazy brown air that results in reduced visibility (Figure 8-5).

Chemists who have been studying photochemical smog formation over the past several decades have determined that the precursors listed earlier are converted to the observable smog components through the action of sunlight. Because the chemical reactions involved are very complex and still not totally understood, we will give only a very brief, simplified scheme showing how photochemical smog is formed.

The precursor in smog formation is $NO(g)$, produced by reaction (8.5).

$$N_2(g) + O_2(g) \xrightarrow{\Delta} 2\ NO(g) \tag{8.5}$$

$NO(g)$ is then converted to $NO_2(g)$, which absorbs ultraviolet radiation from sunlight and decomposes.

$$NO_2 + \text{sunlight} \longrightarrow NO + O \tag{8.9}$$

This is followed by a reaction forming ozone, O_3.

$$O + O_2 \longrightarrow O_3 \tag{8.10}$$

Catalytic Destruction of Stratospheric Ozone animation

▶ We will discuss the rate of reaction (8.7) in Chapter 15.

A large buildup of ozone in photochemical smog, then, requires a plentiful source of NO_2. At one time, this source was believed to be reaction (8.7).

$$2\,NO + O_2 \longrightarrow 2\,NO_2 \qquad (8.7)$$

It is now well established, however, that reaction (8.7) occurs much too slowly to yield the required levels of NO_2 in photochemical smog. NO is rapidly converted to NO_2 when it reacts with O_3,

$$O_3 + NO \longrightarrow O_2 + NO_2 \qquad (8.11)$$

but even though this reaction accounts for the formation of NO_2, it leads to the destruction of ozone. Thus, photochemical smog formation cannot occur just through the reaction sequence: (8.5), (8.11), (8.9), and (8.10). The ozone would be consumed as quickly as it was formed, and there would be no ozone buildup at all.

We now know that organic compounds, particularly unburned hydrocarbons in automotive exhaust, provide a pathway for the conversion of NO to NO_2. The reaction sequence that follows involves some highly reactive molecular fragments known as *free radicals* and represented by formulas written with a bold dot. RH represents a hydrocarbon molecule, and R· is a fragment of a hydrocarbon molecule, a free radical. Oxygen atoms, fragments of the O_2 molecule, are also represented as free radicals, as are hydroxyl groups, fragments of the H_2O molecule.

$$RH + O\cdot \longrightarrow R\cdot + \cdot OH$$

$$RH + \cdot OH \longrightarrow R\cdot + H_2O$$

$$R\cdot + O_2 \longrightarrow RO_2\cdot$$

$$RO_2\cdot + NO \longrightarrow RO\cdot + NO_2$$

The final step in this sequence accounts for the rapid conversion of NO to NO_2 that seems essential to smog formation.

The role of NO_2 in the formation of the smog component PAN is suggested by the equation

$$\underset{\text{PAN}}{CH_3\overset{\overset{O}{\parallel}}{C}-O-O\cdot \;+\; NO_2 \longrightarrow CH_3\overset{\overset{O}{\parallel}}{C}-O-ONO_2}$$

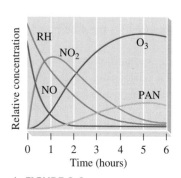

▲ **FIGURE 8-6**

Smog component profile
Data from a smog chamber show how the concentrations of smog components change with time. For example, the concentrations of hydrocarbon (RH) and nitrogen monoxide (NO) fall continuously, whereas that of nitrogen dioxide (NO_2) rises to a maximum and then drops off. The concentrations of ozone (O_3) and peroxyacetyl nitrate (PAN) build up more slowly. Any reaction scheme proposed to explain smog formation must be consistent with observations such as these. Under actual smog conditions, the pattern of concentration changes shown here repeats itself on a daily basis.

The details of smog formation have been worked out in part through the use of smog chambers. By varying experimental conditions in these chambers, scientists have been able to create polluted atmospheres very similar to smog. For example, they have found that if hydrocarbons are omitted from the starting materials in the smog chamber, no ozone is formed. The reaction scheme just proposed is consistent with this observation. Figure 8-6 gives a typical result from a smog chamber.

To control smog, automobiles are now provided with *catalytic converters*. CO and hydrocarbons are oxidized to CO_2 and H_2O in the presence of an oxidation catalyst such as platinum or palladium metal. NO must be reduced to N_2, and this requires a reduction catalyst. A dual-catalyst system uses both types of catalysts. Alternatively, the air–fuel ratio of the engine is set to produce some CO and unburned hydrocarbons. These then act as reducing agents to reduce NO to N_2.

$$2\,CO(g) + 2\,NO(g) \longrightarrow 2\,CO_2(g) + N_2(g)$$

Next, the exhaust gases are passed through an oxidation catalyst to oxidize the remaining hydrocarbons and CO to CO_2 and H_2O. Future control measures may include the use of alternative fuels, such as methanol or hydrogen, and the development of electric-powered automobiles.

8-3   Oxygen

▶ Most elements in Earth's crust, because they are in contact with a highly oxygenated atmosphere, occur combined with oxygen.

Although nitrogen is the most abundant element in the atmosphere, it occurs to a very limited extent in Earth's crust. By contrast, oxygen, also a major component of the atmosphere, is found far more extensively in compounds in Earth's crust. In fact, oxygen is the most abundant of the elements, constituting 45.5% by mass of Earth's solid crust.

Oxygen is central to a study of chemistry. It forms compounds with all the elements except the group 18 elements of low atomic numbers (He, Ne, Ar). We constantly find ourselves considering properties of oxygen and its compounds as we encounter new chemical principles. Most of the discussion of oxygen and its compounds, then, comes in later chapters. Because of its many uses and its role in atmospheric chemistry, however, it merits discussion here also.

Preparation and Uses

Even though oxygen occurs and is used mostly in combined form, oxygen gas is itself an important commercial chemical. Some of its more important uses are listed in Table 8.4. It is obtained mostly from air (page 270). With its ready commercial availability, O_2 is not commonly prepared in the laboratory.

TABLE 8.4   Production and Uses of Oxygen Gas

Production
1999: 29.33×10^6 tons
Ranking, by mass, among manufactured chemicals in the United States: No. 3
Uses
Manufacture of iron and steel
Manufacture and fabrication of other metals (cutting and welding)
Chemicals manufacture and other oxidation processes
Water treatment
Oxidizer of rocket fuels
Medicinal uses
Petroleum refining

In a submarine, in spacecraft, and in an emergency breathing apparatus, however, it is necessary to generate small quantities of oxygen from solids. The reaction of potassium *superoxide* with CO_2 works well for this purpose; it removes CO_2 while O_2 is being formed.

$$4 KO_2(s) + 2 CO_2(g) \longrightarrow 2 K_2CO_3(s) + 3 O_2(g)$$

▶ Potassium superoxide contains the superoxide ion, O_2^-, formed when an O_2 molecule gains an electron (see page 879).

Oxygen and hydrogen gases can be obtained simultaneously through the electrolysis of water. **Electrolysis** is the decomposition of a substance through the passage of electric current. This method is cost-effective for producing only small quantities of these gases, though, unless electric power is unusually cheap. (Many other materials, because of their higher value, can be economically produced using electrolysis.)

$$2 H_2O(l) \xrightarrow[\text{H}_2\text{SO}_4(aq)]{\text{electrolysis}} 2 H_2(g) + O_2(g)$$

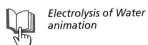
Electrolysis of Water animation

The relationship between electricity and chemical change, referred to as *electrochemistry*, is so important that we devote an entire later chapter to a discussion of this topic. Figure 8-7 illustrates the electrolysis of water and provides some background information.

▶ **FIGURE 8-7**
The electrolysis of water

The passage of electric current through a liquid involves the migration of ions. To make water an electrical conductor, an electrolyte such as H_2SO_4 must be added. The electrolysis occurs in $H_2SO_4(aq)$. H^+ ions in this acidic solution are attracted to the negative electrode (cathode). Here they gain electrons to form H atoms, and H atoms join to form molecules of $H_2(g)$. *Reduction* occurs at the cathode. SO_4^{2-} ions are attracted to the positive electrode (anode), but they undergo no change. Instead, a reaction occurs in which H_2O molecules decompose to replace the H^+ ions lost at the cathode and release $O_2(g)$. *Oxidation* occurs at the anode. The overall reaction is:

$$2\,H_2O(l) \longrightarrow 2\,H_2(g) + O_2(g)$$

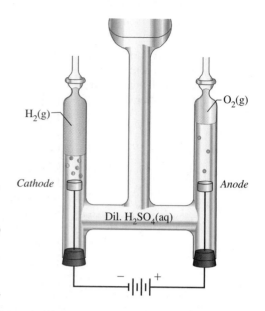

EXAMPLE 8-3

Redox Chemistry. From the description given in the caption to Figure 8-7, write equations for the half-reactions occurring at each electrode, and then an overall equation for the electrolysis of water.

Solution

The description of the electrolysis indicates the type of half-reaction occurring at each electrode and names the reactants and products. This information and the general approach to writing half-equations presented in Section 5-5 lead us to write

Reduction (at cathode): $2\,H^+(aq) + 2\,e^- \longrightarrow H_2(g)$
Oxidation (at anode): $2\,H_2O(l) \longrightarrow 4\,H^+(aq) + O_2(g) + 4\,e^-$

To obtain the overall equation, multiply the reduction half-equation by two, add it to the oxidation half-equation, and simplify.

$$\cancel{4\,H^+(aq)} + 2\,H_2O(l) + \cancel{4\,e^-} \longrightarrow 2\,H_2(g) + \cancel{4\,H^+(aq)} + O_2(g) + \cancel{4\,e^-}$$

Overall equation: $2\,H_2O(l) \longrightarrow 2\,H_2(g) + O_2(g)$

Practice Example A: The electrolysis of dilute $NaOH(aq)$ yields $H_2(g)$ at the cathode (reduction electrode) and $O_2(g)$ at the anode (oxidation electrode). Write plausible half-equations for this electrolysis.

(*Hint:* The overall equation is the same as in Example 8-3.)

Practice Example B: Small amounts of oxygen for use in spacecraft can be obtained from lithium peroxide, Li_2O_2. Write a balanced chemical equation for the reaction of lithium peroxide with carbon dioxide to form lithium carbonate and oxygen as products.

Ozone: An Allotrope of Oxygen

▶ Bonding in O_2 and O_3 will be described in Chapter 11.

Although we usually think of the formula for oxygen as O_2, there are actually two different oxygen molecules. Familiar oxygen is *dioxygen*, O_2; the other is *trioxygen*—ozone, O_3. The term used to describe the existence of two or more forms of an element that differ in their bonding and molecular structure is **allotropy**. O_2 and O_3 are *allotropes* of oxygen.

Normally, the quantity of O_3 in the atmosphere is quite limited at low altitudes, about 0.04 part per million (ppm). As we saw on page 275, however, its level increases (perhaps severalfold) in smog situations. Ozone levels exceeding 0.12 ppm are considered unhealthful.

The reaction producing $O_3(g)$ directly from $O_2(g)$ is highly endothermic and occurs only rarely in the lower atmosphere.

$$3\,O_2(g) \longrightarrow 2\,O_3(g) \qquad \Delta H° = +285\ kJ$$

This reaction does occur in high-energy environments such as electrical storms. The pungent odor you may have smelled around heavy-duty electrical equipment or xerographic office copiers was probably O_3. The chief method of producing ozone in the laboratory, in fact, is to pass an electric discharge (high-energy electrons) through $O_2(g)$. Because ozone is unstable and decomposes back to $O_2(g)$, it is always generated at the point where it is to be used.

Ozone is an excellent oxidizing agent. Its oxidizing ability is surpassed by few other substances (two are F_2 and OF_2). Its most important use is as a substitute for chlorine in purifying drinking water. Its advantages are that it does not impart a taste to the water and it does not form the potentially carcinogenic chlorination products that chlorine can. Its main disadvantage is that O_3 is unstable and quickly disappears from the water after it is treated. Thus, the water is not as well protected against bacterial contamination after leaving the waterworks as is water treated with chlorine.

Stratospheric Ozone animation

The Ozone Layer and Its Environmental Role

In the stratosphere, at altitudes between 25 and 35 km, the concentration of O_3 (expressed in molecules per cubic centimeter) is several times higher than at ground level. When expressed in proportion to the other gases present, the O_3 content of this part of the stratosphere—as much as 8 ppm—is considerably greater than at ground level (0.04 ppm). This belt of the stratosphere is called the **ozone layer**.

Stratospheric ozone plays a vital role in protecting life on Earth. First, ozone absorbs certain ultraviolet (UV) radiation that at Earth's surface causes skin cancer and eye damage in humans and is harmful to other biological organisms as well. Second, in absorbing UV radiation, O_3 molecules dissociate, evolving heat, and thus help to maintain a heat balance in the atmosphere.

The chemical reactions that produce $O_3(g)$ in the upper atmosphere are

$$O_2 + UV\ radiation \longrightarrow O + O \tag{8.12}$$

$$O_2 + O + M \longrightarrow O_3 + M \tag{8.13}$$

Equation (8.12) describes how an O_2 molecule absorbs UV and dissociates. Atomic and molecular oxygen then react to form ozone (8.13). The "third body," M [for example, $N_2(g)$], carries off excess energy; otherwise, the O_3 formed would be too energetic and simply decompose.

Equation (8.14) illustrates the first of the two important functions of ozone: absorbing UV radiation. The second function, that of releasing heat into the atmosphere, is illustrated by Equation (8.15).

▶ The UV radiations absorbed by O_2 in reaction (8.12) and by O_3 in reaction (8.14) differ in a quantity known as wavelength, as we will describe in the next chapter.

$$O_3 + UV\ radiation \longrightarrow O_2 + O \tag{8.14}$$

$$O_3 + O \longrightarrow 2\,O_2 \qquad \Delta H = -389.8\ kJ \tag{8.15}$$

Reaction (8.15) is an ozone-destroying reaction that occurs naturally. There are other natural ozone-destroying processes that have (8.15) as their overall

reaction, such as

$$NO + O_3 \longrightarrow NO_2 + O_2$$

$$NO_2 + O \longrightarrow NO + O_2$$

Overall reaction: $\quad O_3 + O \longrightarrow 2\,O_2$

What is interesting about this pair of reactions is that NO consumed in the first reaction is replenished in the second. A little NO goes a long way. Atmospheric NO is produced mainly from N_2O released by soil bacteria. Thus, natural events account for the continuous formation and destruction of stratospheric ozone and the maintenance of a steady-state concentration of about 8 ppm.

It has recently become apparent that certain gases produced by human activities are contributing to ozone depletion and threatening the integrity of the ozone layer. For example, NO produced by combustion in supersonic jets operating in the stratosphere may contribute to ozone destruction. Most worrisome, though, are gaseous chlorofluorocarbons (CFCs). These molecules have a very long lifetime in the atmosphere and eventually rise to the stratosphere in low concentrations. In the stratosphere, they can absorb ultraviolet radiation and dissociate. For example,

$$CCl_2F_2 + UV \text{ radiation} \longrightarrow CClF_2 + Cl$$

Cl atoms from this reaction can then set up an ozone-destroying cycle.

$$Cl + O_3 \longrightarrow ClO + O_2$$

$$ClO + O \longrightarrow Cl + O_2$$

Overall reaction: $\quad O_3 + O \longrightarrow 2\,O_2 \qquad (8.15)$

Currently, the most convincing evidence that depletion of stratospheric ozone is indeed occurring comes from studies in Antarctica. With the coming of spring (October) a large depletion of O_3 occurs over a period of several weeks, before more normal levels return (Figure 8-8). The chemical reactions postulated to account for

▶ Dichlorodifluoromethane, CCl_2F_2, once widely used in refrigeration and air-conditioning systems, is a typical chlorofluorocarbon. Its desirability was based, in part, on the fact that it is nontoxic and was thought to be completely inert.

 CFCs and Stratospheric Ozone animation

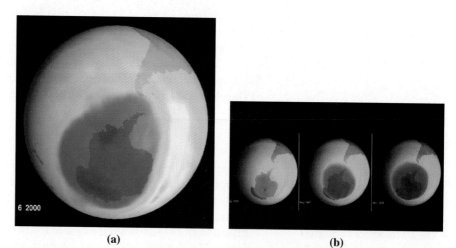

(a) (b)

▲ **FIGURE 8-8** **Ozone levels over Antarctica in the year 2000**
(a) The light blue and dark blue at the center constitute the ozone hole, with the concentration of ozone being lowest in the dark blue region. Higher concentrations of ozone are found in the yellow and green regions. **(b)** Growth of the ozone hole since it was first detected. The ozone hole now extends into populated regions of Chile. In October 2000, the citizens of Southern Chile were warned to stay indoors due to possible burns from UV radiation.

▲ Gaseous atoms in the region of an electric discharge emit light. The light emitted by neon atoms is red-orange in color. Neon lights of other colors use other noble gases or mixtures of gases.

▲ **William Ramsay (1852–1916)** This distinguished Scottish chemist received the 1904 Nobel Prize in Chemistry for his work on the noble gases.

this are much more complex than the simplified scheme outlined here. In addition, meteorological conditions play an important role. Evidence is mounting that ozone destruction also occurs in Arctic regions and perhaps worldwide. The most significant measures taken to date to rectify the problem of ozone depletion in the stratosphere are international agreements on chlorofluorocarbons that have led to significant reductions in their production and use.

8-4   The Noble Gases

In 1785, Henry Cavendish, the discoverer of hydrogen, passed electric discharges through air to form oxides of nitrogen. (A similar process occurs during lightning storms.) He then dissolved these oxides in water to form nitric acid. Even by using excess oxygen, Cavendish was unable to get all the air to react. He suggested that air contained an unreactive gas making up "not more than 1/120 of the whole." John Rayleigh and William Ramsay isolated this gas one century later (1894) and named it argon. The name argon is derived from the Greek *argos*, "lazy one," meaning "inert." Its *inability* to form chemical compounds with any of the other elements— its chemical inertness—was found to be argon's most notable feature. Because argon resembled no other known element, Ramsay placed it in a separate group of the periodic table and reasoned that there should be other members of this group.

Ramsay then began a systematic search for other inert gases. In 1895, he extracted helium from a uranium mineral. A few years later, by very carefully distilling liquid argon, he was able to extract three additional inert gases: neon, krypton, and xenon. The final member of the group of inert gases, a radioactive element called radon, was discovered in 1900. In 1962, compounds of Xe were first prepared, and so the inert gases proved not to be completely inert after all. Since that time, this group of gases has been called the *noble gases*. They are found in group 18 of the periodic table shown on the inside front cover.

Occurrence

Air contains 0.000524% He, 0.001818% Ne, and 0.934% Ar, by volume. The proportion of Kr is about 1 ppm by volume, and that of Xe, 0.05 ppm. The atmosphere is the only source of all these gases except helium. The main source of helium is certain natural gas wells in the western United States that produce natural gas containing up to 8% He by volume. It is cost-effective to extract helium from natural gas even down to levels of about 0.3%. Underground helium accumulates as a result of α-particle emission by radioactive elements in Earth's crust. Whereas the abundance of He on Earth is very limited, it is second only to hydrogen in the universe as a whole.

Are You Wondering...

Why argon is so much more abundant in the atmosphere than all the other noble gases?

Most of the noble gases have escaped from the atmosphere since Earth was formed, but Ar is an exception. The concentration of Ar remains quite high because it is constantly being formed by the radioactive decay of potassium-40, a reasonably abundant, naturally occurring radioactive isotope. Helium is also constantly produced through α-particle emissions by radioactive decay processes, but because the molar mass of He is 10 times less than that of Ar, it escapes from the atmosphere into outer space at a higher rate.

▶ Using helium–oxygen mixtures instead of air for deep-sea diving eliminates nitrogen and prevents nitrogen narcosis (rapture of the deep). Also, because helium is more rapidly and smoothly expelled from the blood than is nitrogen, helium–oxygen mixtures prevent a condition called the *bends* (page 549).

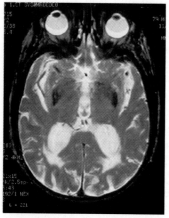

▲ An MRI image of a head.

▶ We first mentioned the emission of α, β, and γ rays on page 40.

▲ Because of the explosive nature of hydrogen, helium is now used in airships.

Properties and Uses

The lighter noble gases are commercially important, in part because they are chemically inert. The efficiency and life of electric lightbulbs are increased when they are filled with an argon–nitrogen mixture. Electric discharge through neon-filled glass or plastic tubes produces a distinctive red light (neon light). Krypton and xenon are used in lasers and in flashlamps in photography. Helium has several unique physical properties. Best known of these is the fact that it exists as a liquid at temperatures approaching 0 K. All other substances freeze to solids at temperatures well above 0 K. (The melting point of solid H_2, for example, is 14 K.) Because of their inertness, both He and Ar are used to blanket materials that need to be protected from nitrogen and oxygen in the air, as in certain types of welding, in metallurgical processes, and in the preparation of ultrapure Si and Ge and other semiconductor materials. Helium mixed with oxygen is used as a breathing mixture for deep-sea diving and in certain medical applications. Large quantities of liquid helium are used to maintain low temperatures (cryogenics). Metals essentially lose their electrical resistivity at liquid He temperatures and become *superconductors*. Powerful magnets can be made by immersing the coils of electromagnets in liquid helium. Such magnets are used in particle accelerators and in nuclear fusion research. More familiar uses of large electromagnets cooled by liquid helium are nuclear magnetic resonance (NMR) instruments in research laboratories and magnetic resonance imaging (MRI) devices in hospitals. Helium is also used to fill lighter-than-air airships (blimps).

Compounds of xenon are of special interest to research chemists because the compounds can be fairly easily made and are useful in studies of chemical bonding. We will consider some xenon compounds in Chapter 23.

An Environmental Issue Involving Radon

All atoms with an atomic number greater than 83 are radioactive. The nuclei of these atoms are unstable and emit α, β, and γ radiation, eventually breaking down to more stable elements with lower atomic numbers. Radon-222, a colorless, odorless gas is produced by the loss of α particles from radium-226, which in turn results from the radioactive decay, through several steps, of uranium-238.

In December 1984, a worker at a nuclear power plant in New Jersey registered high readings on a radiation detector during a routine safety check. But the radiation to which he had been exposed came not from within the plant but from his own home. This incident led to the recognition that a number of individuals may be exposed to high levels of radioactivity from radon. The possible harmful effects of this exposure, primarily an increased risk of lung cancer, are fairly well documented, but the topic remains one of continuing research and debate.

In some instances, the source of radon is in wastes from uranium mining or phosphate production. In most cases, it is emitted by the radioactive decay of ^{238}U present in small amounts in rocks and soils. Because radon is a gas, it readily passes through air passages in the body and is breathed in and out. The product formed when a ^{222}Rn atom gives up an α particle is the isotope polonium-218, which also emits α particles. Unlike radon, polonium is a solid. Health hazards posed by radon seem to be from ^{218}Po and other radioactive decay products becoming attached to dust particles in the air and then being breathed into the lungs.

Fortunately, indoor radon can be rather easily detected by its radioactivity. The chief method of reducing radon levels is through improved ventilation and by venting subsoil radon to keep it from concentrating within a building. In the future, minimizing indoor radon should become a conscious part of building construction.

8-5 Oxides of Carbon

The chief oxides of carbon are carbon monoxide, CO, and carbon dioxide, CO_2. There are about 370 ppm of CO_2 in air (0.037% by volume). CO occurs to a much lesser extent. Although they are only minor constituents of air, these two oxides are important in many ways, as we will see in this section.

Combustion of Carbon Compounds

Carbon dioxide is the only oxide of carbon formed when carbon or carbon-containing compounds are burned in an *excess* of air (providing an abundance of O_2). This condition exists when a fuel-lean mixture is burned in an automobile engine. Thus, for the combustion of the gasoline component octane,

$$C_8H_{18}(l) + \frac{25}{2}\,O_2(g) \longrightarrow 8\,CO_2(g) + 9\,H_2O(l) \qquad (8.16)$$

If the combustion occurs in a *limited* quantity of air, carbon monoxide is also produced. This condition prevails when a fuel-rich mixture is burned in an automobile engine. One possibility for the incomplete combustion of octane is

$$C_8H_{18}(l) + 12\,O_2(g) \longrightarrow 7\,CO_2(g) + CO(g) + 9\,H_2O(l) \qquad (8.17)$$

Heme model

CO as an air pollutant comes chiefly from the incomplete combustion of fossil fuels in automobile engines. CO is an inhalation poison because CO molecules bond to Fe atoms in hemoglobin in blood and displace the O_2 molecules that the hemoglobin normally carries (Figure 8-9).

Not only does incomplete combustion of gasoline contribute to air pollution, but it represents a loss of efficiency. A given quantity of gasoline evolves less heat if CO(g) is formed as a combustion product rather than $CO_2(g)$.

Preparation and Uses

Although carbon dioxide can be obtained directly from the atmosphere as a by-product of the liquefaction of air (see Figure 8-3), this is not an important source. Some of the principal commercial sources of CO_2 are summarized in Table 8.5.

The major use of carbon dioxide (about 50%) is as a refrigerant in the form of dry ice for freezing, preserving, and transporting food. Carbonated beverages account for about 20% of CO_2 consumption. Other important uses are in oil recovery in oil fields and in fire-extinguishing systems. Of course, the major use is not

▶ **FIGURE 8-9**
CO bound to hemoglobin
Carbon monoxide binds to the iron atoms in hemoglobin more strongly than does oxygen. Thus, toxic amounts of carbon monoxide can cause death by oxygen deprivation. The portion of the hemoglobin molecule shown here is called a *heme* group. An iron atom (yellow) is at the center of the group and is surrounded by four nitrogen atoms. In hemoglobin, an O_2 molecule projects above the plane of the iron and nitrogen atoms, but here it has been replaced by a CO molecule (black and red).

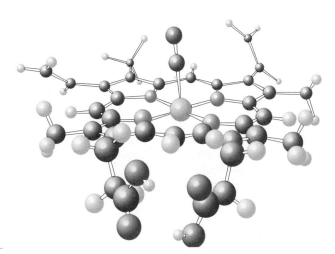

TABLE 8.5 Some Industrial Methods of Preparing CO_2	
Method	**Chemical Reaction**
Recovery from exhaust stack gases in the combustion of carbonaceous fuels, such as the combustion of coke	$C(s) + O_2(g) \longrightarrow CO_2(g)$
Recovery in ammonia plants from steam–reforming reactions used to produce hydrogen	$CH_4(g) + 2\,H_2O(g) \longrightarrow CO_2(g) + 4\,H_2(g)$
Decomposition (calcination) of limestone at about 900 °C	$CaCO_3(s) \longrightarrow CaO(s) + CO_2(g)$
Fermentation by-product in the production of ethanol	$\underset{\text{a sugar}}{C_6H_{12}O_6(aq)} \longrightarrow 2\,C_2H_5OH(aq) + 2\,CO_2(g)$

by humans but by algae and plants. Atmospheric CO_2 is the source of all the carbon-containing compounds (organic compounds) in plant life (see Focus On The Carbon Cycle, page 290).

A modern method of making carbon monoxide is the steam reforming of natural gas, represented as

$$CH_4(g) + H_2O(g) \longrightarrow CO(g) + 3\,H_2(g)$$

The term *steam* refers to gaseous water. *Reforming* refers to the restructuring of a carbon compound, such as CH_4 to CO. The reforming of natural gas (mostly CH_4) is an important source of $H_2(g)$ for use in the synthesis of NH_3 (page 271).

There are three main uses of carbon monoxide: One is in synthesizing other compounds. For example, a mixture of CO and H_2 produced by reforming methane or some other hydrocarbon and known as **synthesis gas**, can be converted to a new organic chemical product, such as methanol.

$$CO(g) + 2\,H_2(g) \longrightarrow CH_3OH(l)$$

Another use of CO is as a reducing agent. In a blast furnace, for instance, **coke**, a rather pure form of carbon produced by heating coal in the absence of air, is converted to CO. The CO then reduces iron oxide to iron.

$$Fe_2O_3(s) + 3\,CO(g) \longrightarrow 2\,Fe(l) + 3\,CO_2(g)$$

A third use of CO is as a fuel, usually mixed with CH_4, H_2, and other combustible gases (recall Section 7-9).

Carbon Dioxide and Carbonates

$CO_2(g)$ dissolves in water to produce a solution generally referred to as carbonic acid, $H_2CO_3(aq)$. H_2CO_3 is a weak acid that ionizes in two steps. When the acid is neutralized in the first step, a salt—variously called a hydrogen carbonate, an acid carbonate, or a bicarbonate—is obtained. If the neutralization is carried through the second step by neutralizing the acid carbonate, a carbonate is obtained.

$$H_2CO_3(aq) + Na^+OH^- \longrightarrow \underset{\text{sodium hydrogen carbonate}}{\underbrace{Na^+ + HCO_3^-}} + H_2O(l)$$

$$Na^+ + HCO_3^- + Na^+ + OH^- \longrightarrow \underset{\text{sodium carbonate}}{\underbrace{2\,Na^+ + CO_3^{2-}}} + H_2O(l)$$

KEEP IN MIND ▶

that H_2CO_3 does not exist as a stable molecule. $CO_2(aq)$ is often referred to as $H_2CO_3(aq)$ to emphasize its acidic properties. We will discuss this matter further in Chapter 17.

 Carbon Dioxide Behaves as an Acid in Water movie

The reactions just described can be reversed by adding an acid to a carbonate. The carbonic acid breaks down into CO_2 and H_2O, and this is a simple way of preparing $CO_2(g)$ in the laboratory.

$$Na_2CO_3(aq) + 2\,H^+(aq) \longrightarrow 2\,Na^+(aq) + H_2O(l) + CO_2(g) \qquad (8.18)$$

Group 1 metal carbonates are water soluble and occur in natural saltwater solutions called brines. Group 2 and other metal carbonates are water insoluble, and many of these are naturally occurring minerals. Calcite ($CaCO_3$) is one such mineral. Dolomite ($CaCO_3 \cdot MgCO_3$) is another. Carbonate minerals are discussed later in the text, as are other aspects of the chemistry of carbonic acid and carbonates. Aspects of carbonic acid/carbonate chemistry range from the maintenance of the proper acidity (pH) of blood to the formation of limestone caves and the formation of hard water.

▶ This type of reaction accounts for the damage done by acid rain to marble ($CaCO_3$) statues.

Global Warming—An Environmental Issue Involving Carbon Dioxide

We do not think of CO_2 as an air pollutant because it is essentially nontoxic. Its ultimate effect on the environment, however, could be very significant. A buildup of $CO_2(g)$ in the atmosphere may disturb the energy balance on Earth.

Earth's atmosphere is largely transparent to visible and UV radiation from the sun. This radiation is absorbed at Earth's surface and warms it. But some of this absorbed energy is reradiated as infrared radiation. Certain atmospheric gases, primarily CO_2 and water vapor, absorb some of this infrared radiation. Energy thus retained in the atmosphere produces a warming effect. This process, outlined in Figure 8-10, is often compared to the retention of thermal energy in a greenhouse and is called the "greenhouse effect."[*] The natural greenhouse effect is essential in maintaining the proper temperature for life on Earth. Without it, Earth would be permanently covered with ice.

From 1880 to 1980, the CO_2 content of the atmosphere increased from 275 to 339 ppm. The current content is about 370 ppm (Figure 8-11). These increases come from the burning of carbon-containing fuels (wood, coal, natural gas, gasoline, and so forth) and from the deforestation of tropical regions (plants, through photosynthesis, *consume* CO_2 from the atmosphere). The expected effect of a CO_2 buildup is an increase in Earth's average temperature, a **global warming**. Some estimates are that a doubling of the CO_2 content over that of preindustrial times could occur before the end of the present century and that this doubling could produce an average global temperature increase of from 1.5 to 4.5 °C.

Predicting the probable effects of a CO_2 buildup in the atmosphere is done largely through computer models, and it is very difficult to know all the factors that should be included in these models and the relative importance of these factors. For example, global warming could lead to the increased evaporation of water and increased cloud formation. In turn, an increased cloud cover could reduce the amount of solar radiation reaching Earth's surface and, to some extent, offset global warming.

Some of the significant possible effects of global warming are

- local temperature changes. The average number of days of daily maximum temperatures in excess of 90 °F in New York City might increase from 15 days per year at the close of the twentieth century to 48 days per year before the close of the twenty-first.

- a rise in sea level caused by the thermal expansion of seawater and increased melting of the polar ice caps. A potential increase in sea level of up to 1 m by 2100 would displace tens of millions of inhabitants in Bangladesh alone.

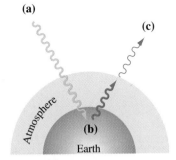

▲ FIGURE 8-10

The "greenhouse" effect
(a) Sunlight is received by Earth. Some incoming radiation is reflected back into space by the atmosphere, and some is absorbed, such as certain UV light by stratospheric ozone. Much of the radiation, however, reaches Earth's surface. **(b)** Earth's surface emits infrared radiation. **(c)** Infrared radiation leaving Earth's atmosphere is less intense than that emitted by Earth's surface. The infrared radiation absorbed by CO_2 and other greenhouse gases warms the atmosphere.

[*] Glass, like CO_2, is transparent to visible and some UV light but absorbs infrared radiation. The glass in a greenhouse, though, acts primarily to prevent the bulk flow of warm air out of the greenhouse.

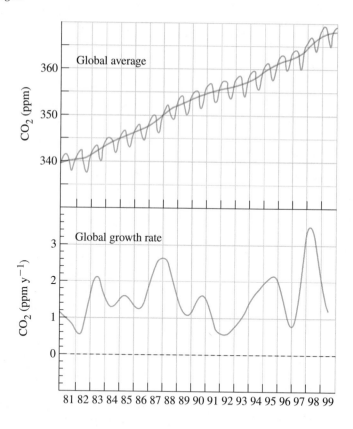

▶ **FIGURE 8-11**

Increasing carbon dioxide content of the atmosphere
Shown in the top graph is the global average atmospheric carbon dioxide level (blue line) measured by a worldwide cooperative sampling network. The peaks occur in April, and the valleys in October. They reflect different levels of photosynthetic activity at different times of the year (photosynthesis consumes CO_2). The red line represents the long-term trend. Shown in the bottom graph is the global average growth rate in atmospheric carbon dioxide. The interpretation of the year-to-year variability in the growth rate of CO_2 has no simple explanation. The variability is most likely due to changes in natural sources and sinks resulting from large-scale events such as El Niño, strong temperature or precipitation anomalies, and the occasional strong volcanic eruption. The mechanisms linking climate events and change in CO_2 growth rate are currently being debated in the literature.

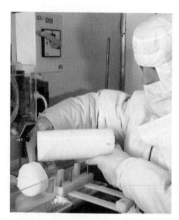

▲ An ice core from the ice sheet in Antarctica is cut into sections in a refrigerated clean room. The ice core will be analyzed to determine the amount and type of trapped gases and trace elements it contains. These data provide information regarding past changes in climate and current trends in the pollution of the atmosphere.

- the migration of plant and animal species. Vegetation now characteristic of certain areas of the globe could migrate into regions several hundred kilometers closer to the poles. The areas in which diseases such as malaria are endemic could also expand.

Although some of the current thinking involves speculation, a growing body of evidence supports the likelihood of global warming. For example, analyses of tiny air bubbles trapped in the Antarctic ice cap show a strong correlation between the atmospheric CO_2 content and temperature for the past 160,000 years—low temperatures during periods of low CO_2 levels and higher temperatures with higher levels.

CO_2 is not the only greenhouse gas. Several gases are even stronger infrared absorbers—specifically, methane (CH_4), ozone (O_3), nitrous oxide (N_2O), and chlorofluorocarbons (CFCs). Furthermore, atmospheric concentrations of some of these gases have been growing at a faster rate than that of CO_2. No strategies beyond curtailing the use of chlorofluorocarbons and fossil fuels have emerged for countering a possible global warming. Like several other major environmental issues, some aspects of global warming are not well understood, and research, debate, and action are all likely to occur simultaneously for a long time to come.

8-6 Hydrogen

Think how important hydrogen is in the study of chemistry. John Dalton based atomic masses on a value of 1 for the H atom. Humphry Davy (1810) proposed that hydrogen is the key element in acids. In the next chapter we will see that theoretical studies of the H atom provide us with our modern view of atomic structure. In Chapter 12 we will find that the H_2 molecule is the usual starting point for modern theories of molecular structure. For all of its theoretical significance, though, hydrogen is also of great practical importance, as we will emphasize in this section.

Occurrence and Preparation

Hydrogen is a very minor component of the atmosphere, about 0.5 ppm at Earth's surface. At altitudes above 2500 km, the atmosphere is mostly atomic hydrogen at extremely low pressures. In the universe as a whole, hydrogen accounts for about 90% of the atoms and 75% of the mass. On Earth, *hydrogen occurs in more compounds than does any other element.*

The free element can easily be produced, but from only a few of its compounds. Our first choice might be H_2O—the most abundant hydrogen compound. To extract hydrogen from water means reducing the oxidation state of H from +1 in H_2O to 0 in H_2. To do this, we need an appropriate *reducing* agent, such as carbon (coal or coke), carbon monoxide, or a hydrocarbon—particularly methane (natural gas). The first pair of reactions that follow are called the **water gas** reactions; they represent a way of making combustible gases—CO and H_2—from steam.

$$\text{Water gas reactions:} \quad C(s) + H_2O(g) \longrightarrow CO(g) + H_2(g) \tag{8.19}$$

$$CO(g) + H_2O(g) \longrightarrow CO_2(g) + H_2(g) \tag{8.20}$$

$$\text{Reforming of methane:} \quad CH_4(g) + H_2O(g) \longrightarrow CO(g) + 3\,H_2(g)$$

Another source of $H_2(g)$ is as a by-product in petroleum refining.

Often we use methods in the chemical laboratory that are not commercially feasible. Electrolysis of water is one useful laboratory method; another involves the reaction of active metals (recall the list in Table 5.3) in acidic solutions, such as

$$Zn(s) + 2\,H^+(aq) \longrightarrow Zn^{2+}(aq) + H_2(g)$$

Compounds of Hydrogen

Hydrogen forms binary compounds called **hydrides**, with most of the other elements. Binary hydrides are usually grouped into three broad categories: covalent, ionic, and metallic. *Covalent hydrides* are those formed between hydrogen and nonmetals. Some of these hydrides are simple molecules that can be formed by the direct union of hydrogen and the second element, such as

$$H_2(g) + Cl_2(g) \longrightarrow 2\,HCl(g)$$

$$3\,H_2(g) + N_2(g) \rightleftharpoons 2\,NH_3(g)$$

Ionic hydrides form between hydrogen and the most active metals, particularly those of groups 1 and 2. In these compounds hydrogen exists as the hydride *ion*, H^-.

$$2\,M(s) + H_2(g) \longrightarrow 2\,MH(s) \qquad\qquad M(s) + H_2(g) \longrightarrow MH_2(s)$$
$$\text{(M is any group 1 metal)} \qquad\qquad\qquad \text{(M is Ca, Sr, or Ba)}$$

Ionic hydrides react vigorously with water to produce $H_2(g)$. CaH_2, a gray solid, has been used as a portable source of $H_2(g)$ for filling weather observation balloons.

$$CaH_2(s) + 2\,H_2O(l) \longrightarrow Ca(OH)_2(s) + 2\,H_2(g) \tag{8.21}$$

Metallic hydrides are commonly formed with the transition elements—groups 3–12. A distinctive feature of these hydrides is that in many cases they are *nonstoichiometric*—the ratio of H atoms to metal atoms is variable, not fixed. This is because H atoms can enter the voids or holes among the metal atoms in a crystalline lattice and fill some but not others.

▲ **Reaction of CaH₂ with water**
The pink color of phenolphthalein indicator added to the water signals the production of $Ca(OH)_2$ in reaction (8.21).

▲ Liquid vegetable oils contain long molecules with some carbon-to-carbon double bonds. When some of these double bonds in the molecules are hydrogenated to give carbon-to-carbon single bonds, the result is conversion of the liquid to a solid "partially hydrogenated vegetable oil."

Uses of Hydrogen

Hydrogen is not listed among the top chemicals produced, because only a small percentage is ever sold to customers. Most hydrogen is produced and used on the spot. In these terms, its most important use (about 42%) is in the manufacture of NH_3 (reaction 8.1). The next most important use of H_2 (about 38%) is in petroleum refining, where it is produced in some operations and consumed in others, such as in the production of the high-octane gasoline component, isooctane, from diisobutylene.

$$CH_3-\underset{\underset{CH_3}{|}}{\overset{\overset{CH_3}{|}}{C}}-\underset{\underset{H}{|}}{C}=\underset{\underset{CH_3}{|}}{C}-CH_3 \ + \ H_2(g) \ \xrightarrow{catalyst} \ CH_3-\underset{\underset{CH_3}{|}}{\overset{\overset{CH_3}{|}}{C}}-\underset{\underset{H}{|}}{\overset{\overset{H}{|}}{C}}-\underset{\underset{CH_3}{|}}{\overset{\overset{H}{|}}{C}}-CH_3$$

diisobutylene   isooctane

In similar reactions, called **hydrogenation reactions**, hydrogen atoms, in the presence of a catalyst, can be added to double or triple bonds in other molecules. This type of reaction, for example, will convert liquid oleic acid, $C_{17}H_{33}COOH$, to solid stearic acid, $C_{17}H_{35}COOH$.

$$CH_3(CH_2)_7CH{=}CH(CH_2)_7COOH + H_2(g) \xrightarrow{Ni} CH_3(CH_2)_{16}COOH \quad (8.22)$$

oleic acid   stearic acid

Similar reactions serve as the basis for converting oils that contain carbon–carbon double bonds, such as vegetable oils, into solid or semisolid fats, such as shortening.

Another important chemical manufacturing process that uses hydrogen is the synthesis of methyl alcohol (methanol), an alternative fuel.

$$CO(g) + 2 H_2(g) \xrightarrow{catalyst} CH_3OH(g)$$

Hydrogen gas is an excellent reducing agent and in some cases is used to produce metals from their oxide ores. For example, at 850 °C

$$WO_3(s) + 3 H_2(g) \longrightarrow W(s) + 3 H_2O(g)$$

The uses of hydrogen described here, together with several others, are listed in Table 8.6.

Surface Reaction—Hydrogenation animation

TABLE 8.6   Some Uses of Hydrogen

Synthesis of:
   ammonia, NH_3
   hydrogen chloride, HCl
   methanol, CH_3OH
Hydrogenation reactions in:
   petroleum refining
   converting oils to fats
Reduction of metal oxides, such as those of iron, cobalt, nickel, copper, tungsten, molybdenum
Metal cutting and welding with atomic and oxyhydrogen torches
Rocket fuel, usually $H_2(l)$ in combination with $O_2(l)$
Fuel cells for generating electricity, in combination with $O_2(g)$

A Hydrogen Economy—An Environmental Issue Involving Hydrogen

As we contemplate the eventual decline of the world's supplies of fossil fuels, hydrogen emerges as an attractive means of storing, transporting, and using energy. For example, when an automobile engine burns hydrogen rather than gasoline, its exhaust is essentially pollution-free. The range of supersonic aircraft could be increased if they used liquid hydrogen as a fuel. A hypersonic airplane (the space plane) might also become possible. One method of using hydrogen that is already available combines H_2 and O_2 to form H_2O in an electrochemical device called a *fuel cell*. This device, used in space vehicles, converts chemical energy to electricity rather than heat. The subsequent conversion of electrical energy to mechanical energy (work) can be carried out much more efficiently than can the conversion of heat to mechanical energy. Fuel cells are described more fully in Chapter 21.

The basic problems are in finding a cheap source of hydrogen and an effective means of storing it. One possibility is to use hydrogen made by the electrolysis of seawater. This possibility requires an abundant energy source, however—perhaps nuclear fusion energy if it can be developed. Another alternative is the thermal decomposition of water. The problem here is that even at 2000 °C, water is only about 1% decomposed. What is needed is a thermochemical cycle, a series of reactions that have as their overall reaction: $2 H_2O(l) \longrightarrow 2 H_2(g) + O_2(g)$. Ideally, no single reaction in the cycle would require a very high temperature. Still another alternative being studied involves the use of solar energy to decompose water—photodecomposition.

Storage of gaseous hydrogen is difficult because of the bulk of the gas. When liquefied, hydrogen occupies a much smaller volume, but because of its very low boiling point (-253 °C), the $H_2(l)$ must be stored at very low temperatures. Also, hydrogen must be maintained out of contact with oxygen or air, with which it forms explosive mixtures. One approach may be to dissolve $H_2(g)$ in a metal or metal alloy, such as an iron–titanium alloy. The gas can be released by mild heating. In an automobile, this storage system would replace the gasoline tank. The heat required to release hydrogen from the metal would come from the engine exhaust.

If the problems described here can be solved, not only can hydrogen be used to supplant gasoline as a fuel for transportation, but it can also replace natural gas for space heating. Because H_2 is a good reducing agent, it can replace carbon (as coal or coke) in metallurgical processes, and, of course, it would be abundantly available for reaction with N_2 to produce NH_3 for the manufacture of fertilizers. The combination of all these potential uses of hydrogen could lead to a fundamental change in our way of life and give rise to what is called a **hydrogen economy**.

Summary

The atmosphere consists of several regions that differ in their temperature–altitude profiles and composition. In the portions of the atmosphere closest to Earth's surface, the major components are nitrogen, oxygen, and argon. These gases are obtained by the fractional distillation of liquid air. Important minor constituents of air are carbon dioxide, carbon monoxide, and the noble gases other than argon. All of these gases, as well as hydrogen, are the subjects of this chapter.

Some of the topics considered are the methods of obtaining nitrogen compounds—particularly NH_3 and HNO_3—the uses of these compounds, and the role of nitrogen oxides in smog formation. The discussion of oxygen centers on methods of preparing the allotropes O_2 and O_3 and their uses. The environmental threat resulting from ozone destruction in the stratosphere is also explored. The major point of interest concerning the noble gases is some of the special uses made possible by their physical properties. The greatest interest concerning radon gas is in the potential health hazard associated with its radioactivity.

The focus of the discussion of CO and CO_2 is their formation in combustion processes, their industrial preparation and uses, and the chemistry of carbonates. The environmental issue associated with CO is air pollution, and with CO_2, global warming. The discussion of hydrogen emphasizes its preparation—both commercially and in the laboratory—its uses, and some of its binary compounds, hydrides. Hydrogen is so useful that a hydrogen-based economy replacing a petroleum-based one is a future possibility.

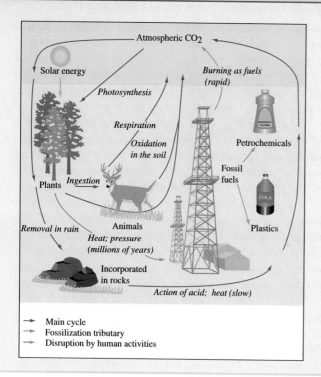

Main cycle
Fossilization tributary
Disruption by human activities

Unlike energy, which flows to Earth constantly as sunlight, the pool of elements essential to life is fixed. After these elements have served their purposes in living matter, they are recycled by natural processes. Perhaps the most familiar of these so-called nutrient cycles is the nitrogen cycle, through which elemental nitrogen from the atmosphere is converted to plant and animal protein and subsequently returned to the air as N_2 (see page 656). But other natural cycles are of interest as well. Here we briefly consider the *carbon cycle*, a portion of which is represented at the left—the major exchanges that occur between the atmosphere and the land masses on Earth.

The only source of carbon available to plants for making organic compounds, through a process known as *photosynthesis*, is atmospheric CO_2. The process is extremely complicated, and its details have been known for only a few decades (Melvin Calvin, Nobel Prize, 1961). It involves up to 100 sequential steps for the conversion of 6 mol CO_2 to 1 mol $C_6H_{12}O_6$ (glucose). The overall change is

$$6\,CO_2 + 6\,H_2O \xrightarrow[\text{sunlight}]{\text{chlorophyll}} C_6H_{12}O_6 + 6\,O_2$$

$$\Delta H = +2.8 \times 10^3\ \text{kJ}$$

Integrative Example

The oxidation of $NH_3(g)$ to $NO(g)$ in the Ostwald process (reaction 8.4) must be very carefully controlled in terms of temperature, pressure, and contact time with the catalyst. This is because the oxidation of $NH_3(g)$ can yield any one of the products $N_2(g)$, $N_2O(g)$, $NO(g)$, and $NO_2(g)$, depending on conditions. Show that oxidation of $NH_3(g)$ to $N_2(g)$ is the most exothermic (has the most negative ΔH°) of the four possible reactions.

1. *Write equations for each reaction.* The first step is to write four balanced chemical equations with $NH_3(g)$ and $O_2(g)$ as reactants and the nitrogen-containing species and $H_2O(g)$ as products.

$$4\,NH_3(g) + 3\,O_2(g) \longrightarrow 2\,N_2(g) + 6\,H_2O(g)$$

$$2\,NH_3(g) + 2\,O_2(g) \longrightarrow N_2O(g) + 3\,H_2O(g)$$

$$4\,NH_3(g) + 5\,O_2(g) \longrightarrow 4\,NO(g) + 6\,H_2O(g)$$

$$4\,NH_3(g) + 7\,O_2(g) \longrightarrow 4\,NO_2(g) + 6\,H_2O(g)$$

These equations are perfectly acceptable for most purpose, but to compare the four reactions, it is best to write each equation in terms of *one* mole of $NH_3(g)$. In each case, dividing all the stoichiometric coefficients by that of ammonia, we get the result:

$$NH_3(g) + \frac{3}{4}O_2(g) \longrightarrow \frac{1}{2}N_2(g) + \frac{3}{2}H_2O(g)$$

$$NH_3(g) + O_2(g) \longrightarrow \frac{1}{2}N_2O(g) + \frac{3}{2}H_2O(g)$$

$$NH_3(g) + \frac{5}{4}O_2(g) \longrightarrow NO(g) + \frac{3}{2}H_2O(g)$$

$$NH_3(g) + \frac{7}{4}O_2(g) \longrightarrow NO_2(g) + \frac{3}{2}H_2O(g)$$

2. *Use standard enthalpies of formation from Appendix D and expression (7.21) to obtain enthalpies of reaction.* The direct approach is to do four separate calculations using the expression $\Delta H^\circ = \Sigma \Delta H_f^\circ(\text{products}) - \Sigma \Delta H_f^\circ(\text{reactants})$, and choose the most negative result. For example, for the first reaction, in which N_2 and H_2O are produced, we have

$$\Delta H_{\text{rxn}}^\circ = \frac{1}{2}\Delta H_f^\circ[N_2(g)] + \frac{3}{2}\Delta H_f^\circ[H_2O(g)]$$

$$- \Delta H_f^\circ[NH_3(g)] - \frac{3}{4}\Delta H_f^\circ[O_2(g)]$$

The overall reaction is highly endothermic. The required energy comes from sunlight. Chlorophyll, a green pigment in plants, is crucial to the process.

Here are some of the ideas illustrated in the diagram. When animals consume plants, carbon atoms pass on to the animals. Some carbon is returned to the atmosphere as CO_2 when the animals breathe and when they expel gas (methane). Additional CO_2 returns to the atmosphere as plants and animals die and their remains are broken down by bacteria. Some carbon in decaying organic matter is converted to coal, petroleum, and natural gas. This carbon is unavailable for photosynthesis.

Not represented in the drawing is the cycle of CO_2 through the oceans of the world. Phytoplankton (small, floating green organisms) also carry on photosynthesis, converting CO_2 to organic compounds. The plankton are the base of the ocean food chain, directly and indirectly supporting all the animals in the oceans.

Huge quantities of carbon have accumulated in the form of carbonate rocks (mostly $CaCO_3$). These come from the shells of decayed mollusks in ancient seas.

Human activities now play a far more significant role in the carbon cycle than in preindustrial times. Slash-and-burn agricultural methods—particularly in rain forests—are returning CO_2 to the atmosphere more rapidly than naturally occurring fires do. The combustion of fossil fuels is replacing stored carbon by carbon dioxide to an even greater extent. We have already seen the possible consequences of this distortion of the carbon cycle: an increased level of atmospheric CO_2 and a possible future global warming (page 285). Human disruption of the natural carbon cycle has become a widely debated issue.

The enthalpies of formation of the elements are zero, so we obtain

$$\Delta H^\circ_{rxn} = \frac{3}{2} \Delta H^\circ_f[H_2O(g)] - \Delta H^\circ_f[NH_3(g)]$$

$$= 1.5 \text{ mol} \times (-241.8 \text{ kJ mol}^{-1})$$

$$- 1.0 \text{ mol} (-46.1 \text{ kJ mol}^{-1})$$

$$= -316.6 \text{ kJ}$$

For the other reactions we have, ignoring the enthalpies of formation of the elements,

$$\Delta H^\circ_{rxn} = \frac{1}{2} \Delta H^\circ_f[N_2O(g)] + \frac{3}{2} \Delta H^\circ_f[H_2O(g)]$$
$$- \Delta H^\circ_f[NH_3(g)] = -275.6 \text{ kJ}$$

$$\Delta H^\circ_{rxn} = \Delta H^\circ_f[NO(g)] + \frac{3}{2} \Delta H^\circ_f[H_2O(g)]$$
$$- \Delta H^\circ_f[NH_3(g)] = -226.4 \text{ kJ}$$

$$\Delta H^\circ_{rxn} = \Delta H^\circ_f[NO_2(g)] + \frac{3}{2} \Delta H^\circ_f[H_2O(g)]$$
$$- \Delta H^\circ_f[NH_3(g)] = -283.4 \text{ kJ}$$

Thus, we can conclude from these calculations that the most exothermic reaction is

$$NH_3(g) + \frac{3}{4} O_2(g) \longrightarrow \frac{1}{2} N_2(g) + \frac{3}{2} H_2O(g)$$

Throughout the calculations you probably have noticed that

$$\frac{3}{2} \Delta H^\circ_f[H_2O(g)] - \Delta H^\circ_f[NH_3(g)]$$

is a recurring term. This suggests that there is a simpler way to solve the problem. In each application of (7.21), the terms based on $\frac{3}{2} H_2O(g)$ and $NH_3(g)$ will be the same. The most negative ΔH° will be for the reaction in which ΔH°_f of the nitrogen-containing species is *smallest*. Refer to Appendix D, and you will see that all oxides of nitrogen have *positive* values of ΔH°_f. The smallest ΔH°_f (zero) is for the element $N_2(g)$. Thus, the most exothermic reaction is

$$NH_3(g) + \frac{3}{4} O_2(g) \longrightarrow \frac{1}{2} N_2(g) + \frac{1}{2} H_2O(g)$$

This example points up an important problem-solving strategy: Always look for terms that are constant, and eliminate them to avoid unnecessary computations. Look ahead!

Key Terms

acid anhydride (8-2)
allotropy (8-3)
coke (8-5)
electrolysis (8-3)
global warming (8-5)

hydride (8-6)
hydrogen economy (8-6)
hydrogenation reactions (8-6)
ozone layer (8-3)
relative humidity (8-1)

smog (8-2)
stratosphere (8-1)
synthesis gas (8-5)
troposphere (8-1)
water gas (8-6)

Review Questions

1. In your own words, define the following terms: **(a)** relative humidity; **(b)** noble gas; **(c)** chlorofluorocarbon; **(d)** non-stoichiometric compound.
2. Briefly describe each of the following ideas, methods, or phenomena: **(a)** fractional distillation; **(b)** electrolysis; **(c)** hydrogenation reaction; **(d)** dew and frost formation.
3. Explain the important distinctions between each pair of terms: **(a)** troposphere and stratosphere; **(b)** allotrope and isotope; **(c)** fuel-lean and fuel-rich; **(d)** metallic and ionic hydride.
4. Supply a name or formula for **(a)** O_3; **(b)** dinitrogen monoxide; **(c)** potassium superoxide; **(d)** CaH_2; **(e)** Mg_3N_2; **(f)** potassium carbonate; **(g)** ammonium dihydrogen phosphate.
5. Describe these commercial materials by chemical formulas: **(a)** water gas; **(b)** coke; **(c)** urea; **(d)** limestone; **(e)** synthesis gas.
6. Write the formula of a compound mentioned in this chapter that has the indicated element in the designated oxidation state: **(a)** N, +4; **(b)** O, $-\frac{1}{2}$ **(c)** N, +1; **(d)** H, −1; **(e)** C, +2.
7. Give a practical laboratory method that you might use to produce small quantities of **(a)** O_2; **(b)** N_2O; **(c)** H_2; **(d)** CO_2.
8. Which of the following gases is the best oxidizing agent, and why? **(a)** H_2; **(b)** NO_2; **(c)** NO; **(d)** NH_3.
9. Complete and balance these equations for the reactions of substances with water.
   (a) $LiH(s) + H_2O \longrightarrow$
   (b) $C(s) + H_2O \xrightarrow{\Delta}$
   (c) $NO_2(g) + H_2O \longrightarrow$
10. Complete and balance these equations for the reactions of substances with acids.
    (a) $Mg(s) + HCl(aq) \longrightarrow$
    (b) $NH_3(g) + HNO_3(aq) \longrightarrow$
    (c) $MgCO_3(s) + HCl(aq) \longrightarrow$
    (d) $NaHCO_3(s) + HC_2H_3O_2(aq) \longrightarrow$
11. Write an equation to represent the complete neutralization of an aqueous sulfuric acid solution by an aqueous ammonia solution.
12. Identify the oxidizing and reducing agents in reactions (8.4), (8.8), (8.16), and (8.21).

13. Among the equations displayed in the text, identify one that represents a disproportionation reaction.
14. Hydrogen gas reduces an oxide of iron with Fe in the oxidation state +3 to pure iron metal. Write a plausible equation for this reaction.
15. Write an equation similar to (8.6) in which copper reacts with nitric acid to produce NO(g).
16. The following compounds give off a gas when heated: $KClO_3$, $CaCO_3$, NH_4NO_3. Identify the gaseous product from each, and write an equation to show its production.
17. Describe the basic cause of the environmental problem known as **(a)** photochemical smog; **(b)** ozone layer depletion; **(c)** global warming.
18. What is the catalytic converter in an automotive exhaust system, and how does it work?
19. The noble gases have many characteristics in common, such as their inertness. Why, then, is helium found in certain natural gas deposits, whereas none of the other noble gases is ?
20. Describe some advantages and disadvantages of replacing petroleum products with hydrogen as a fuel for transportation.

▲ The burning of the dirigible Hindenberg.

Exercises

The Atmosphere

21. Show that the composition of air on a mole percent basis is the same as given in Table 8.1 on a volume percent basis.

22. Explain why the composition of air expressed on a mass percent basis is not the same as on a volume percent basis.

Nitrogen

23. Write balanced equations for the following important commercial reactions involving nitrogen and its compounds.
  (a) the principal artificial method of fixing atmospheric N_2
  (b) oxidation of ammonia to NO
  (c) preparation of nitric acid from NO

24. When heated, each of the following substances decomposes to the products indicated. Write balanced equations for these reactions.
  (a) $NH_4NO_3(s)$ to $N_2(g)$, $O_2(g)$, and $H_2O(g)$
  (b) $NaNO_3(s)$ to sodium nitrite and oxygen gas
  (c) $Pb(NO_3)_2(s)$ to lead(II) oxide, nitrogen dioxide, and oxygen

25. Pure liquid HNO_3 decomposes, even at low temperatures, to produce N_2O_4, O_2, and water. Write a balanced equation for this reaction.

26. Sodium nitrite can be made by passing oxygen and nitrogen monoxide gases into an aqueous solution of sodium carbonate. Write a balanced equation for this reaction.

27. A chemistry magazine listed the 1995 demand for $N_2(g)$ as 9.39×10^{11} ft^3 at STP. What is this quantity of N_2 expressed in kilograms?

28. Concentrated $HNO_3(aq)$ used in laboratories is usually 15 M HNO_3 and has a density of 1.41 g/mL. What is the percent by mass of HNO_3 in this concentrated acid?

29. In 1968, before pollution controls were introduced, over 75 billion gallons of gasoline were used in the United States as a motor fuel. Assume an emission of oxides of nitrogen of 5 grams per vehicle mile and an average mileage of 15 mi/gal of gasoline. How many kilograms of nitrogen oxides were released into the atmosphere in the United States in 1968?

30. One reaction that competes with reaction (8.4), the Ostwald process, is the reaction of gaseous ammonia and nitrogen monoxide to produce gaseous nitrogen and gaseous water. Use data from Appendix D to determine $\Delta H°$ for this reaction, per mole of ammonia consumed.

Oxygen

31. Each of the following compounds decomposes to produce $O_2(g)$ when heated. Write plausible equations for these reactions. **(a)** HgO(s); **(b)** $KClO_4(s)$.

32. $O_3(g)$ is a powerful oxidizing agent. Write equations to represent oxidation of **(a)** I^- to I_2 in acidic solution; **(b)** sulfur in the presence of moisture to sulfuric acid; **(c)** $[Fe(CN)_6]^{4-}$ to $[Fe(CN)_6]^{3-}$ in basic solution. In each case $O_3(g)$ is reduced to $O_2(g)$.

33. *Without performing detailed calculations*, determine which of the following compounds has the greatest percent oxygen by mass: dinitrogen tetroxide, aluminum oxide, tetraphosphorus hexoxide, or carbon dioxide.

34. *Without performing detailed calculations*, determine which decomposition yields the most $O_2(g)$ **(a)** per mole and **(b)** per gram of substance.
  (1) ammonium nitrate $\longrightarrow$ nitrogen + oxygen + water
  (2) hydrogen peroxide $\longrightarrow$ oxygen + water
  (3) potassium chlorate $\longrightarrow$ potassium chloride + oxygen

35. The natural abundance of O_3 in unpolluted air at ground level is about 0.04 parts per million (ppm) by volume. What is the approximate partial pressure of O_3 under these conditions, expressed in millimeters of mercury?

36. A typical concentration of O_3 in the ozone layer is 5×10^{12} O_3 molecules/cm^3. What is the partial pressure of O_3, expressed in millimeters of mercury, in that layer? Assume a temperature of 220 K.

37. Explain why the volumes of $H_2(g)$ and $O_2(g)$ in the electrolysis of water pictured in Figure 8-7 are not the same.

38. In the electrolysis of a sample of water in an apparatus similar to that of Figure 8-7, 22.83 mL of $O_2(g)$ was collected at 25.0 °C at an oxygen partial pressure of 736.7 mmHg. Determine the mass of water that was decomposed.

The Noble Gases

39. A 55-L cylinder contains Ar at 145 atm and 26 °C. What minimum volume of air at STP must have been liquefied and distilled to produce this Ar? Air contains 0.934% Ar, by volume.

40. Some sources of natural gas contain 8% He by volume. How many liters of such a natural gas must be processed at STP to produce 5.00 g of He?

41. A breathing mixture is prepared in which He is substituted for N_2. The gas is 79% He and 21% O_2, by volume.

What is the density of this mixture, in grams per liter, at 25 °C and 1 atm?

42. Concerning the breathing mixture in Exercise 41, at what pressure would the He–O_2 mixture have the same density as air at 25 °C and 1 atm?
  (*Hint:* What are the apparent molar masses of air and of the He–O_2 mixture?)

Carbon

43. Write equations for the reactions that you would expect when
(a) $C_6H_{14}(l)$ is burned in an excess of air.
(b) $CO(g)$ is heated with $PbO(s)$.
(c) $CO_2(g)$ is bubbled into $KOH(aq)$.
(d) $MgCO_3(s)$ is added to $HCl(aq)$.

44. Refer to Section 8-5, and write equations for the following reactions.
(a) The action of vinegar (aqueous acetic acid) on baking soda (sodium hydrogen carbonate)
(b) Carbon monoxide acting as a reducing agent in the reduction of zinc oxide to zinc metal
(c) The production from synthesis gas of CH_2OHCH_2OH (ethylene glycol), used as an antifreeze

45. A laboratory method for preparing CO_2 is the reaction of an acid with a carbonate. Would you expect any CO to form in this reaction? Explain.

46. Compare the environmental effects of chlorofluorocarbons and carbon dioxide with respect to ozone destruction and global warming.

47. Determine the quantity of heat evolved in the complete combustion of 1.00 gal of octane, $C_8H_{18}(l)$, in reaction (8.16). Use $\Delta H_f°[C_8H_{18}(l)] = -250.0$ kJ/mol and data for other substances from Appendix D. The density of $C_8H_{18}(l)$ is 0.703 g/mL; 1 gal = 3.785 L.

48. Use Hess's law to show that if 1.00 mol $C_8H_{18}(l)$ burns according to equation (8.17) rather than (8.16), the heat of combustion is less, by an amount equal to $\Delta H°$ for the reaction $CO(g) + \frac{1}{2} O_2(g) \longrightarrow CO_2(g)$. What is this quantity of heat? What percent of the maximum possible heat of combustion [$\Delta H°$ for reaction (8.16)] does this represent? Use data from Appendix D together with the value $\Delta H_f°[C_8H_{18}(l)] = -250.0$ kJ/mol.

Hydrogen

49. Use data from Table 7.2 (page 245) to calculate the standard enthalpies of combustion of the four alkane hydrocarbons listed there.

50. Based on the results of Exercise 49, which alkane evolves the greatest amount of heat upon combustion on (a) a *per mole* basis and (b) a *per gram* basis? Which is the most desirable alkane from the standpoint of reducing the emission of carbon dioxide to the atmosphere? Explain.

51. Write chemical equations for the following reactions.
(a) the displacement of $H_2(g)$ from $HCl(aq)$ by $Al(s)$
(b) the reforming of propane gas (C_3H_8) with steam
(c) the reduction of $MnO_2(s)$ to $Mn(s)$ with $H_2(g)$

52. Write equations to show how to prepare $H_2(g)$ from each of the following substances: (a) H_2O; (b) $HI(aq)$; (c) $Mg(s)$; (d) $CO(g)$. Use other common laboratory reactants as necessary, that is, water, acids or bases, metals, and so on.

53. $CaH_2(s)$ reacts with water to produce $Ca(OH)_2$ and $H_2(g)$. $Ca(s)$ reacts with water to produce the same products. $Na(s)$ reacts with water to form NaOH and $H_2(g)$. *Without doing detailed calculations*, determine (a) which of these reactions produces the most H_2 per liter of water used, and (b) which solid—CaH_2, Ca, or Na—produces the most $H_2(g)$ *per gram* of the solid.

54. What volume of $H_2(g)$ at 25 °C and 752 mmHg is required to hydrogenate oleic acid, $C_{17}H_{33}COOH(l)$, to produce one mole of stearic acid, $C_{17}H_{35}COOH(s)$. Assume reaction (8.22) proceeds with a 95% yield.

55. *Without doing detailed calculations*, explain in which of the following materials you would expect to find the greatest mass percent of hydrogen: seawater, the atmosphere, natural gas (CH_4), ammonia.

56. How many grams of $CaH_2(s)$ are required to generate sufficient $H_2(g)$ to fill a 235-L weather observation balloon at 722 mmHg and 19.7 °C?
$CaH_2(s) + 2 H_2O(l) \longrightarrow Ca(OH)_2(aq) + 2 H_2(g)$

Integrative and Advanced Exercises

57. Explain why, expressed as a *mass percent*, the % N_2 in air is less than the 78.084% listed in Table 8.1 whereas the % O_2 is greater than the 20.946%.

58. Recompute the data in Table 8.1 to express the percentages of atmospheric gases on a mass basis.

59. A 0.25-mL sample of $H_2O(l)$ at 20 °C ($d = 0.998$ g/mL) is allowed to vaporize into a closed 18.5-L container of dry air at 20 °C. What is the relative humidity of this sample of air? The vapor pressure of water at 20 °C = 17.5 mmHg.

60. A 0.1052-g sample of $H_2O(l)$ in an 8.050-L sample of dry air at 30.1 °C evaporates completely. To what temperature must the air be cooled to give a relative humidity of 80.0%? Vapor pressures of water: 20 °C, 17.54 mmHg; 19 °C,

16.48 mmHg; 18 °C, 15.48 mmHg; 17 °C, 14.53 mmHg; 16 °C, 13.63 mmHg; 15 °C, 12.79 mmHg.

61. Suppose that no attempt is made to separate the $H_2(g)$ and $O_2(g)$ produced by the electrolysis of water. What volume of a H_2/O_2 mixture, saturated with $H_2O(g)$ and collected at 23 °C and 755 mmHg, would be produced by electrolyzing 17.3 g water? Assume that the water vapor pressure of the dilute electrolyte solution is 20.5 mmHg.

62. One measure for reducing $CO_2(g)$ emissions is to use fuels that produce a large quantity of heat per mole of $CO_2(g)$ produced. Viewed in this way, which is the better fuel, gasoline [assume it to be octane, $C_8H_{18}(l)$] or methane, $CH_4(g)$? The enthalpy of formation of $C_8H_{18}(l)$ is -250.0 kJ/mol.

63. The photograph was taken after a few drops of a deep-purple acidic solution of $KMnO_4(aq)$ were added to $NaNO_3(aq)$ (left) and to $NaNO_2(aq)$ (right). Explain the difference in the results shown.

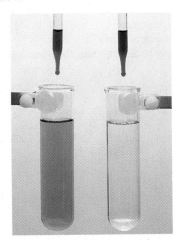

64. Carbon suboxide, a compound consisting of 52.96% C and 47.04% O by mass, is the acid anhydride of malonic acid, an acid with two ionizable H atoms per molecule and consisting of 34.62% C, 61.50% O, and 3.87% H by mass. Write an equation for the reaction of carbon suboxide with water. (*Hint:* Recall the definition of an acid anhydride on page 274.)

65. For every volume of $O_2(g)$ inhaled, a person exhales 0.82 volume of $CO_2(g)$. Oxygen-generating systems for enclosed spaces, such as spacecraft, should therefore have the capacity to consume 0.82 L of $CO_2(g)$ for every liter of $O_2(g)$

produced. Does the reaction of potassium superoxide and carbon dioxide, producing potassium carbonate and oxygen, meet this requirement? Explain.

66. Zn can reduce NO_3^- to $NH_3(g)$ in basic solution. (The following equation is *not* balanced.)

$$NO_3^- + Zn(s) + OH^- + H_2O \longrightarrow [Zn(OH)_4]^{2-} + NH_3(g)$$

The NH_3 can be neutralized with an excess of $HCl(aq)$. Then, the unreacted HCl can be titrated with NaOH. In this way a quantitative determination of NO_3^- can be achieved. A 25.00-mL sample of nitrate solution was treated with zinc in basic solution. The $NH_3(g)$ was passed into 50.00 mL of 0.1500 M HCl. The excess HCl required 32.10 mL of 0.1000 M NaOH for its titration. What was the $[NO_3^-]$ in the original sample?

67. Write equations for the consecutive reactions involved in the production of nitric acid from ammonia and an equation for the overall reaction.

68. Refer to the smog formation discussion on page 276 to explain why the time at which O_3 levels reach a maximum in the smog component profile in Figure 8-6 corresponds to the time at which NO levels reach a minimum.

69. Oxygen atoms are an important constituent of the thermosphere, a layer of the atmosphere with temperatures up to 1500 K. Calculate the average translational kinetic energy of O atoms at 1500 K.

70. One reaction for the production of adipic acid, $HOOC(CH_2)_4COOH$, used in the manufacture of nylon, involves the oxidation of cyclohexanone, $C_6H_{10}O$, in a nitric acid solution. Assume that dinitrogen monoxide is also formed, and write a balanced equation for this reaction.

Feature Problems

71. The figure shows how the emission of pollutants is related to the air–fuel ratio in an internal combustion engine.

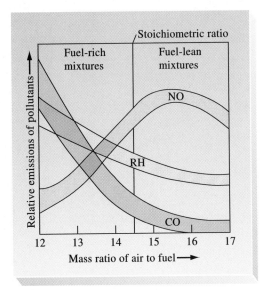

(a) Use information given in this chapter to provide a general interpretation of the figure.
(b) Establish by calculation that the stoichiometric ratio of air to fuel (on a mass basis) is about 14.5 : 1. To do so, assume that (1) octane, C_8H_{18}, is a typical gasoline molecule; (2) the maximum quantity of heat is evolved when $C_8H_{18}(l)$ and $O_2(g)$ are in their stoichiometric proportions and $CO_2(g)$ and $H_2O(l)$ are the products; (3) air used in the combustion is the naturally occurring mixture of $O_2(g)$, $N_2(g)$, and Ar(g).

72. In some research that required the careful measurement of gas densities, John Rayleigh, a physicist, found that the density of $O_2(g)$ had the same value whether the gas was obtained from air or derived from one of its compounds. The situation with $N_2(g)$ was different, however. The density of $N_2(g)$ had the same value when the $N_2(g)$ was derived from any of various compounds, but a different value when the $N_2(g)$ was extracted from air. In 1894, Rayleigh enlisted the aid of William Ramsay, a chemist, to solve this apparent mystery; in the course of their work, they discovered the noble gases.

(a) Why do you suppose that the $N_2(g)$ extracted from liquid air did not have the same density as $N_2(g)$ obtained from its compounds?

(b) Which gas do you suppose had the greater density, $N_2(g)$ extracted from air or $N_2(g)$ prepared from nitrogen compounds? Explain.

(c) The way in which Ramsay proved that nitrogen gas extracted from air was itself a mixture of gases involved allowing this nitrogen to react with magnesium metal to form magnesium nitride. Explain the significance of this experiment.

(d) Use data on the composition of air from Table 8.1 to calculate the *percent difference* in the densities at STP of Rayleigh's $N_2(g)$ extracted from air and $N_2(g)$ derived from nitrogen compounds.

73. Various thermochemical cycles are being explored as possible sources of $H_2(g)$. The object is to find a series of reactions that can be conducted at moderate temperatures (about 500 °C) and results in the decomposition of water into H_2 and O_2. Show that the following series of reactions meets this requirement.

(*Hint:* Balance the equations, multiply by the appropriate coefficients, and combine them into an overall equation.)

$$FeCl_2 + H_2O \longrightarrow Fe_3O_4 + HCl + H_2$$

$$Fe_3O_4 + HCl + Cl_2 \longrightarrow FeCl_3 + H_2O + O_2$$

$$FeCl_3 \longrightarrow FeCl_2 + Cl_2$$

 eMedia Exercises

74. The regions of our atmosphere are illustrated in *eChapter 8-1*. The space shuttle orbits the Earth at an approximate altitude of 600 km. What conditions does the vehicle encounter as it orbits and as it descends back through the atmosphere?

75. A common method of producing gaseous hydrogen and oxygen is seen in the **Electrolysis of Water** animation (*eChapter 8-3*). (a) What is the volume (mL) of water that must undergo electrolysis in order to produce 0.50 L of a mixture of hydrogen and oxygen gas under STP conditions? (b) What is the volume of oxygen produced in this procedure?

76. The process shown in the **Electrolysis of Water** animation (*eChapter 8-3*) typically is not an economically favorable process. (a) Calculate the enthalpy of the electrolysis reaction for one mole of water. (b) What is the enthalpy change associated with the formation of a total volume of 0.50 L of gas under STP conditions?

77. About 40% of hydrogen used in industrial processes stems from petroleum refining. An example of a refining step is shown in the **Surface Reaction—Hydrogenation** animation (*eChapter 8-5*). (a) What role does the metal surface play in this reaction? (b) In some experimental studies, deuterium (D_2) is used in place of hydrogen. What would be the product of a reaction of ethylene with deuterium?

9

Electrons in Atoms

Contents

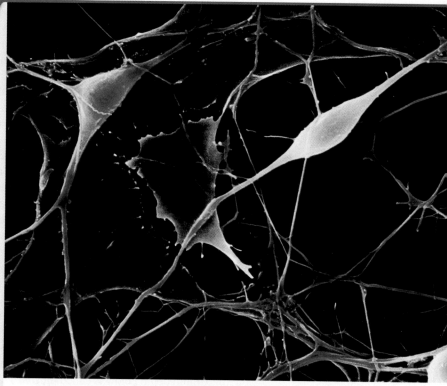

This image of three neurons (gray objects) is produced by an electron microscope that relies on the wave properties of electrons discussed in this chapter.

Some observers of the scientific scene at the close of the nineteenth century believed the time was near for closing the books on the field of physics. With the accumulated knowledge of the previous two or three centuries, all that remained was to apply this body of physics—classical physics—to fields such as chemistry and biology.

Only a few fundamental problems remained, such as explaining certain details of light emission and a phenomenon known as the photoelectric effect. But the solution to these problems, rather than marking an end, spelled the beginning of a new golden age of physics. These problems were solved through a bold new proposal—the quantum theory—a scientific breakthrough of epic proportions. In this chapter, we will see that to explain phenomena at the atomic and molecular level, classical physics is inadequate—only the quantum theory will do.

The aspect of quantum mechanics that we will emphasize is how electrons are described through features known as quantum numbers and electron orbitals. The model of atomic structure we develop here will help us to explain many of the topics discussed in the next several chapters: periodic trends in the physical and chemical properties of the elements, chemical bonding, and intermolecular forces.

9-1   Electromagnetic Radiation

Our main topic in this chapter is the electronic structures of atoms. We can learn about electrons in atoms by studying the interactions of electromagnetic radiation and matter. We will begin with background information about electromagnetic radiation and then consider connections between electromagnetic radiation and atomic structure in the following sections. It is perhaps best to approach the subject matter of this chapter at two levels. Concentrate on the basic ideas relating to atomic structure, many of which are illustrated through the in-text examples. At the same time, you can pursue some of the details that are presented in the Are You Wondering features and portions of Sections 9-6, 9-8, and 9-10.

Electromagnetic radiation is a form of energy transmission in which electric and magnetic fields are propagated as waves through empty space (a vacuum) or through a medium such as glass. A **wave** is a disturbance that transmits energy through a medium. Anyone who has sat in a small boat on a large body of water has experienced wave motion. The wave moves across the surface of the water, and the disturbance alternately lifts the boat and allows it to drop. Although water waves may be more familiar, let us use a simpler example to illustrate some important ideas and terminology about waves—a traveling wave in a rope.

Imagine tying one end of a long rope to a post and holding the other end in your hand (Figure 9-1). Imagine also that you have marked one small segment of the rope with red ink. As you move your hand up and down, you set up a wave motion in the rope. The wave travels along the rope, but the colored segment simply moves up and down. In relation to the center line (the broken line in Figure 9-1), the wave consists of *crests*, or high points, where the rope is at its greatest distance above the center line, and *troughs*, or low points, where the rope is at its greatest distance below the center line. The maximum height of the wave above the center line or the maximum depth below is called the *amplitude*. The distance between the tops of two successive crests (or the bottoms of two troughs) is called the **wavelength**, designated by the Greek letter lambda, λ.

Wavelength is one important characteristic of a wave. Another feature, **frequency**, designated by the Greek letter nu, ν, is the number of crests or troughs that pass through a given point per unit of time. Frequency has the unit, time^{-1}, usually s^{-1} (per second), meaning the number of events or cycles per second. The product of the length of a wave (λ) and the frequency (ν) shows how far the wave front travels in a unit of time. This is the *velocity* of the wave. Thus, if the wavelength in Figure 9-1 were 0.5 m and the frequency, 3 s^{-1} (meaning three complete up-and-down hand motions per second), the velocity of the wave would be 0.5 m $\times$ 3 s^{-1} = 1.5 m/s.

We cannot actually see an electromagnetic wave as we do the traveling wave in a rope, but we can try to represent it as in Figure 9-2. As the figure shows, the radiation component associated with the magnetic field lies in a plane perpendicular to that of the electric field component. An electric field is the region around an electrically charged particle. We can detect the presence of an electric field by measuring the force on an electrically charged object when it is brought into the field. A magnetic field is found in the region surrounding a magnet. According to a theory proposed by James Clerk Maxwell (1831–1879) in 1865, electromagnetic radiation—a propagation of electric and magnetic fields—is produced by an

Water waves, sound waves, and seismic waves (which produce earthquakes) are unlike electromagnetic radiation. They require a material medium for their transmission. ▶

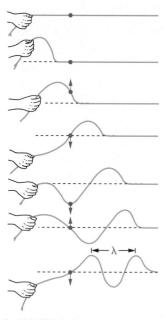

▲ **FIGURE 9-1**
The simplest wave motion—traveling wave in a rope
As a result of the up-and-down hand motion (top to bottom), waves pass along the long rope from left to right. This one-dimensional moving wave is called a traveling wave. The wavelength of the wave, λ—the distance between two successive crests—is identified.

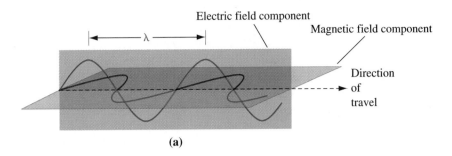

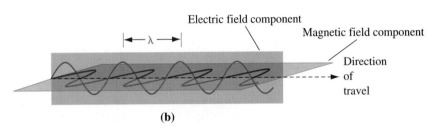

▲ **FIGURE 9-2   Electromagnetic waves**
This sketch of two different electromagnetic waves shows the propagation of mutually perpendicular oscillating electric and magnetic fields. For a given wave, the wavelengths, frequencies, and amplitudes of the electric and magnetic field components are identical. If these views are of the same instant of time, we would say that **(a)** has the *longer* wavelength and *lower* frequency, and **(b)**, the *shorter* wavelength and *higher* frequency.

accelerating electrically charged particle (a charged particle whose velocity changes). Radio waves, for example, are a form of electromagnetic radiation produced by causing oscillations (fluctuations) of the electric current in a specially designed electrical circuit. With visible light, another form of electromagnetic radiation, the accelerating charged particles are the electrons in atoms or molecules.

Electromagnetic Spectrum activity

Frequency, Wavelength, and Velocity of Electromagnetic Radiation

The SI unit for frequency, s^{-1}, is the **hertz (Hz)**, and the basic SI wavelength unit is the meter (m). Because many types of electromagnetic radiation have very short wavelengths, however, smaller units, including those listed below, are also used. The angstrom, named for the Swedish physicist Anders Ångström (1814–1874), is not an SI unit.

$$1 \text{ centimeter (cm)} = 1 \times 10^{-2} \text{ m}$$
$$1 \text{ micrometer } (\mu m) = 1 \times 10^{-6} \text{ m}$$
$$1 \text{ nanometer (nm)} = 1 \times 10^{-9} \text{ m} = 1 \times 10^{-7} \text{ cm} = 10 \text{ Å}$$
$$1 \text{ picometer (pm)} = 1 \times 10^{-12} \text{ m} = 1 \times 10^{-10} \text{ cm} = 10^{-2} \text{ Å}$$
$$1 \text{ angstrom (Å)} = 1 \times 10^{-10} \text{ m} = 1 \times 10^{-8} \text{ cm} = 100 \text{ pm}$$

▶ The speed of light is commonly rounded off to 3.00×10^8 m s^{-1}.

A distinctive feature of electromagnetic radiation is its *constant* velocity of 2.997925×10^8 m s^{-1} in a vacuum, often referred to as the *speed of light*. The speed of light is represented by the symbol c, and the relationship between this speed and the frequency and wavelength of electromagnetic radiation is

$$c = \nu \cdot \lambda \tag{9.1}$$

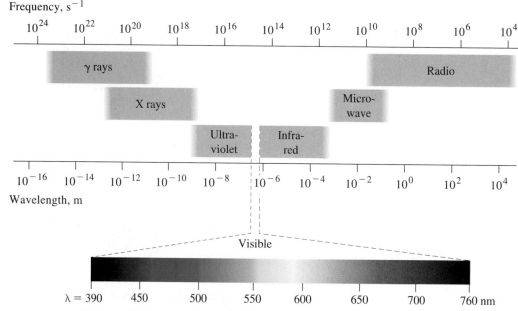

Frequency, s^{-1}

 Visible Spectrum activity

▲ **FIGURE 9-3**   **The electromagnetic spectrum**
The visible region, which extends from violet at the shortest wavelength to red at the
longest wavelength, is only a small portion of the entire spectrum. The approximate
wavelength and frequency ranges of some other forms of electromagnetic radiation are
also indicated.

Figure 9-3 indicates the wide range of possible wavelengths and frequencies for
some common types of electromagnetic radiation and illustrates this important fact:
The wavelength of electromagnetic radiation is shorter for high frequencies and
longer for low frequencies. Example 9-1 illustrates the use of equation (9.1).

EXAMPLE 9-1

Relating the Frequency and Wavelength of Electromagnetic Radiation. Most of the
light from a sodium vapor lamp has a wavelength of 589 nm. What is the frequency of
this radiation?

Solution

We can first convert the wavelength of the light from nanometers to meters and then
apply equation (9.1).

$$\lambda = 589 \text{ nm} \times \frac{1 \times 10^{-9} \text{ m}}{1 \text{ nm}} = 5.89 \times 10^{-7} \text{ m}$$
$$c = 2.998 \times 10^8 \text{ m/s}$$
$$\nu = ?$$

Rearrange equation (9.1) to the form $\nu = c/\lambda$, and solve for ν.

$$\nu = \frac{c}{\lambda} = \frac{2.998 \times 10^8 \text{ m s}^{-1}}{5.89 \times 10^{-7} \text{ m}} = 5.09 \times 10^{14} \text{ s}^{-1} = 5.09 \times 10^{14} \text{ Hz}$$

Practice Example A: The light from red LEDs (light-emitting diodes) is commonly seen in many electronic devices. A typical LED produces 690-nm light. What is
the frequency of this light?

Practice Example B: An FM radio station broadcasts on a frequency of 91.5 mega-
hertz (MHz). What is the wavelength of these radio waves in meters?

An Important Characteristic of Electromagnetic Waves

The properties of electromagnetic radiation that we will use most extensively are those just introduced—amplitude, wavelength, frequency, and velocity. There is another essential characteristic of the radiation that will underpin our discussion of atomic structure later in the chapter. We introduce that characteristic here.

If we drop two pebbles close together into a pond, we observe ripples (waves) emerging from the points of impact of the two stones. The two sets of waves intersect, and as we see in Figure 9-4a, there are places where the waves disappear and places where the waves persist, creating a crisscross pattern. Where the waves are "in step" upon meeting, their crests coincide, as do their troughs. The waves combine to produce the highest crests and deepest troughs in the water. The waves are said to be *in phase*, and the addition of the waves is called *constructive interference* (Figure 9-5a). Where the waves meet in such a way that the peak of one wave occurs at the trough of another, the waves cancel and the water is flat (Figure 9-5b). These out-of-step waves are said to be *out of phase*, and the cancellation of the waves is called *destructive interference*.

An everyday illustration of interference involving electromagnetic waves is seen in the rainbow of colors that shine from the surface of a compact disc (Figure 9-4b). White light, such as sunlight, contains all the colors of the rainbow. These colors differ in wavelength (and frequency), and when these different wavelength components are reflected off the tightly spaced grooves of the CD, they travel slightly different distances. This creates phase differences that depend on the angle at which we hold the CD to the light source. The light waves in the beam interfere with each other, and for a given angle between the incoming and reflected light, all colors cancel except one. Light waves of that color interfere constructively and reinforce one another. Thus, as we change the angle of the CD to the light source, we see different colors. The dispersion of different wavelength components of a light beam through the interference produced by reflection from a grooved surface is called **diffraction**.

Diffraction is a phenomenon that can be explained only as a property of waves. Both the physical picture and mathematics of interference and diffraction are the same for water waves and electromagnetic waves.

KEEP IN MIND ▶

that destructive interference occurs when the waves are out of phase by one-half wavelength. If the waves are out of phase by more or less than this, but also not completely in phase, then only partial destructive interference occurs.

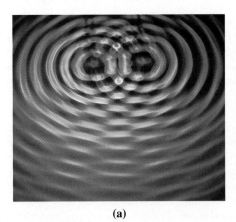
(a)

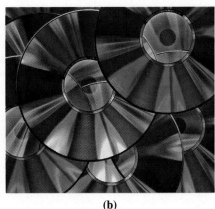
(b)

▲ FIGURE 9-4   **Examples of interference**
(a) Stones and ripples. (b) CD reflection.

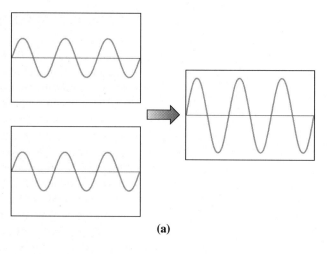

(a)

▶ **FIGURE 9-5**
Interference in two overlapping light waves
(a) In constructive interference, the troughs and crests are in step (in phase), leading to addition of the two waves. **(b)** In destructive interference, the troughs and crests are out of step (out of phase), leading to cancellation of the two waves.

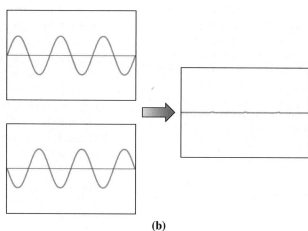

(b)

Are You Wondering...

What happens to the energy of an electromagnetic wave when interference takes place?

As we noted in Figure 9-2, the wave is made up of oscillating electric (E) and magnetic (B) fields. The magnitudes of E and B continuously oscillate, or switch, from positive to negative. These oscillating fields create an oscillating electromagnetic force. A charged particle will interact with the electromagnetic force and will oscillate back and forth with a constantly changing velocity The changing velocity gives the particle a changing kinetic energy that is proportional to the square of the velocity (that is, $e_k = \frac{1}{2} mu^2$). Consequently the energy of a wave depends not on the values of E and B alone, but on the sum of their squares, that is, on $E^2 + B^2$. The energy is also related to the *intensity* (I) of a wave, a quantity which, in turn, is related to the *square* of the amplitude of the wave. Suppose we let the amplitude of the waves be 1. Then for each wave, whether a pair of waves are in phase or out of phase, the energy is proportional to $1^2 + 1^2 = 2$. The average energy of the pair of waves is also proportional to 2 [that is, $(2 + 2)/2$]. In constructive interference, the amplitudes become 2, so that the energy is proportional to 4. In destructive interference, the amplitude is zero and the energy is zero. Note, however, that the *average* between the two situations is still 2 [that is, $(4 + 0)/2$], so that energy is conserved, as it must be.

▲ **FIGURE 9-6**
Refraction of light
Light is refracted (bent) as it passes from air into the glass prism, and again as it emerges from the prism into air. This photograph shows that red light is refracted the least and blue light the most. The blue light strikes the prism at such an angle that the beam undergoes an internal reflection before it emerges from the prism.

The Visible Spectrum

The speed of light is lower in any medium than it is in a vacuum. Also, the speed is different in different media. As a consequence, light is refracted, or bent, when it passes from one medium to another (Figure 9-6). Also, although electromagnetic waves all have the same speed in a vacuum, waves of differing wavelengths have slightly differing speeds in air and other media. Thus, when a beam of white light is passed through a transparent medium, the wavelength components are refracted differently. The light is dispersed into a band of colors, a *spectrum*. In Figure 9-7a, a beam of white light (for example, sunlight) is dispersed by a glass prism into a continuous band of colors corresponding to all the wavelength components from red to violet. This is the visible spectrum shown in Figure 9-3 and also seen in a rainbow, where the medium that disperses the sunlight is droplets of water (Figure 9-7b).

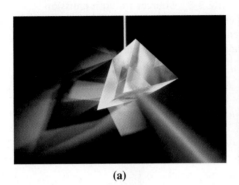

(a)   (b)

▲ **FIGURE 9-7**   **The spectrum of "white" light**
(a) Dispersion of light through a prism. Red light is refracted the least and violet light the most when "white" light is passed through a glass prism. The other colors of the visible spectrum are found between red and violet. (b) Rainbow near a waterfall. Here, water droplets are the dispersion medium.

Flame Tests for Metals movie

9-2   Atomic Spectra

The visible spectrum in Figure 9-7 is said to be continuous because the light being diffracted consists of many, many wavelength components. If the source of a spectrum produces light having only a relatively small number of wavelength components, then a *discontinuous* spectrum is observed. For example, if the light source is an electric discharge passing through a gas, only certain colors are seen in the spectrum as in Figure 9-8a. Or if the light source is a gas flame into which an ionic compound has been introduced, the flame may acquire a distinctive color indicative of the metal ion present (Figure 9-8b). In these cases, the spectra consist of only a limited number of discrete wavelength components, observed as colored lines with dark spaces between them. These *discontinuous* spectra are called **atomic**, or **line**, **spectra**.

The production of the line spectrum of helium is illustrated in Figure 9-9. The light source is a lamp containing helium gas at a low pressure. When an electric discharge is passed through the lamp, helium atoms absorb energy, which they then emit as light. The light is passed through a narrow slit and then dispersed by a prism. The colored components of the light are detected and recorded on a photographic film. Each wavelength component appears as an image of the slit—a thin line. In all, there are six lines in the spectrum of helium that can be seen with the unaided eye.

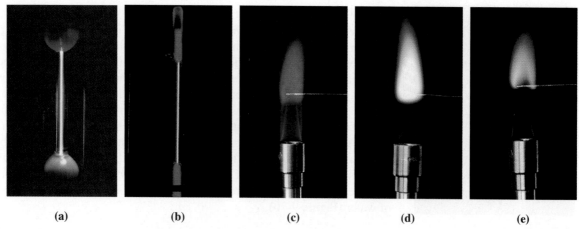

▲ FIGURE 9-8 **Sources for light emission**
Light emitted by an electric discharge through (a) hydrogen gas and (b) helium gas. Light emitted when compounds of the alkali metals are excited in gas flames: (c) lithium, (d) sodium, and (e) potassium.

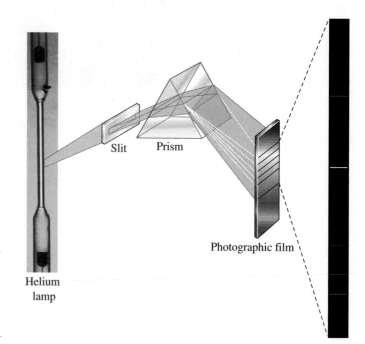

▶ FIGURE 9-9
The atomic, or line, spectrum of helium
The apparatus pictured here, in which the spectral lines are photographed, is called a *spectrograph*. If the observations are made by visual sighting alone, the device is called a *spectroscope*. If the positions and brightness of the lines are measured and recorded by other than visual or photographic means, the term generally used is *spectrometer*.

Slit   Prism

Photographic film

Helium
lamp

▶ Bunsen designed a special gas burner for his spectroscopic studies. This burner, the common laboratory Bunsen burner, produces very little background radiation to interfere with the spectral observations.

Each element has its own distinctive line spectrum—a kind of atomic finger-print. Robert Bunsen (1811–1899) and Gustav Kirchhoff (1824–1887) developed the first spectroscope and used it to identify elements. In 1860, they discovered a new element and named it cesium (Latin, *caesius*, sky blue) because of the distinctive blue lines in its spectrum. They discovered rubidium in 1861 in a similar way (Latin, *rubidius*, deepest red). Still another element characterized by its unique spectrum is helium (Greek, *helios*, the sun). Its spectrum was observed during the solar eclipse of 1868, but helium was not isolated on Earth for another 27 years.

Among the most extensively studied atomic spectra has been the hydrogen spectrum. Light from a hydrogen lamp appears to the eye as a reddish purple color (Figure 9-8a). The principal wavelength component of this light is red light of

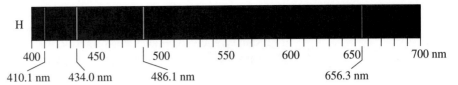

▲ FIGURE 9-10   **The Balmer series for hydrogen atoms—a line spectrum**
The four lines shown are the only ones visible to the unaided eye. Additional, closely
spaced lines lie in the ultraviolet (UV) region.

wavelength 656.3 nm. Three other lines appear in the visible spectrum of atomic
hydrogen, however: a greenish blue line at 486.1 nm, a violet line at 434.0 nm, and
another violet line at 410.1 nm. The visible atomic spectrum of hydrogen is shown
in Figure 9-10. In 1885, Johann Balmer, apparently through trial and error, deduced
a formula for the wavelengths of these spectral lines. Balmer's equation, written in
a form proposed by Johannes Rydberg, is

$$\nu = 3.2881 \times 10^{15} \, \text{s}^{-1} \left(\frac{1}{2^2} - \frac{1}{n^2} \right) \qquad (9.2)$$

In this equation, ν is the frequency of the spectral line, and n must be an *integer*
(whole number) *greater than two*. If $n = 3$ is substituted into the equation, the frequency of the red line is obtained. If $n = 4$, the frequency of the greenish blue line
is obtained, and so on.

The fact that atomic spectra consist of only limited numbers of well-defined
wavelength lines provides a great opportunity to learn about the structures of atoms.
For example, it suggests that only a limited number of energy values are available
to excited gaseous atoms. Classical (nineteenth-century) physics, however, was not
able to provide an explanation of atomic spectra. The key to this puzzle lay in a great
breakthrough of modern science—the quantum theory.

9-3   Quantum Theory

We are aware that hot objects emit light of different colors, from the dull red of an
electric-stove heating element to the bright white of a lightbulb filament. Light emitted by a hot radiating object can be dispersed by a prism to produce a *continuous*
color spectrum. As seen in Figure 9-11, the light intensity varies smoothly with
wavelength, peaking at a wavelength fixed by the source temperature. As with atomic spectra, classical physics could not provide a complete explanation of light emission by heated solids, known as *blackbody radiation*. Classical theory predicts that
the intensity of the radiation emitted would increase indefinitely, as indicated by
the dashed lines in Figure 9-11. In 1900, Max Planck (1858–1947), to explain the
fact that the intensity does not increase indefinitely, made a revolutionary proposal: *Energy, like matter, is discontinuous.* Here, then, is the essential difference between the classical physics of Planck's time and the new quantum theory that he
proposed: Classical physics places no limitations on the amount of energy a system
may possess, whereas quantum theory limits this energy to a discrete set of specific values. The *difference* between two of the allowed energies of a system also has
specific value, called a **quantum** of energy. This means that when the energy increases from one allowed value to the next, it increases by a tiny jump, or quantum.
Here is a way of thinking about a quantum of energy: It bears a similar relationship
to the total energy of a system as a single atom does to an entire sample of matter.

▲ Light emission by molten
iron.

▲ Max Planck (1858–1947)

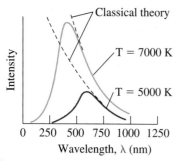

▲ FIGURE 9-11

Spectrum of radiation given off by a heated body

A red-hot object has a spectrum that peaks around 675 nm, whereas a white-hot object has a spectrum that has comparable intensities for all wavelengths in the visible region. The sun has a blackbody temperature of about 5750 K. Objects emit radiation at *all* temperatures, not just at high temperatures. For example, infrared radiation emitted by objects makes them visible in the dark through night-vision goggles.

The model Planck used for the emission of electromagnetic radiation was that of a group of atoms on the surface of the heated object oscillating together with the same frequency. Planck's assumption was that the group of atoms, the oscillator, must have an energy corresponding to the equation

$$\epsilon = nh\nu$$

where ϵ is the energy, n is a positive integer, ν is the oscillator frequency, and h is a constant that had to be determined by experiment. Using his theory and experimental data for the distribution of frequencies with temperature, Planck established the following value for the constant h. We now call it **Planck's constant**, and it has the value

$$h = 6.62607 \times 10^{-34} \text{ J s}$$

Planck's postulate can be rephrased in this more general way: The energy of a quantum of electromagnetic radiation is proportional to the frequency of the radiation— the higher the frequency, the greater the energy. This is summarized by what we now call Planck's equation.

$$E = h\nu \tag{9.3}$$

The existence of separated energy levels is a concept that is difficult to accept, because it is contrary to all ordinary experience with macroscopic physical systems. Thus it is not surprising that scientists, including Planck himself, were initially skeptical of the quantum hypothesis. It had been designed to explain radiation from heated bodies and certainly could not be accepted as a general principle until it had been tested on other applications.

Only after the quantum hypothesis was successfully applied to phenomena other than blackbody radiation did it acquire status as a great new scientific theory. The

Are You Wondering...

How Planck's ideas account for the fact that the intensity of blackbody radiation drops off at higher frequencies?

Planck was aware of the work of Ludwig Boltzmann, who, with James Maxwell, had derived an equation to account for the distribution of molecular speeds. Boltzmann had shown that the relative chance of finding a molecule with a particular speed was related to its energy by the following expression.

$$\text{relative chance} \propto e^{-\frac{\text{kinetic energy}}{k_B T}}$$

where k_B is the Boltzmann constant. You will also notice that the curve of intensity versus wavelength in Figure 9-11 bears a strong resemblance to the distribution of molecular speeds in Figure 6-17. Planck assumed that the energies of the substance oscillating to emit blackbody radiation were distributed according to the Boltzmann distribution law. That is, the relative chance of an oscillator having the energy $nh\nu$ is proportional to $e^{-\frac{nh\nu}{k_B T}}$, where n is an integer, 1, 2, 3, ... So this expression shows that the chance of an oscillator having a high frequency is lower than for oscillators having lower frequencies, since as ν increases, $e^{-\frac{nh\nu}{k_B T}}$ decreases. The assumption that the energy of the oscillators in the light-emitting source cannot have continuous values leads to excellent agreement between theory and experiment.

first of these successes came in 1905 with Albert Einstein's quantum explanation of the photoelectric effect.

The Photoelectric Effect

In 1888, Heinrich Hertz discovered that when light strikes the surface of certain metals, electrons are ejected. This phenomenon is called the **photoelectric effect** and its salient features are that

- electron emission only occurs when the frequency of the incident light exceeds a particular threshold value (ν_0). When this condition is met, it is further observed that
- the number of electrons emitted depends on the *intensity* of the incident light, but
- the kinetic energies of the emitted electrons depend on the *frequency* of the light.

These observations, especially the dependency on frequency, could not be explained by classical wave theory. However, Albert Einstein showed that they are exactly what would be expected with a particle interpretation of radiation. In 1905, Einstein proposed that electromagnetic radiation has particle-like qualities and that "particles" of light, subsequently called **photons** by G. N. Lewis, have a characteristic energy given by Planck's equation, $E = h\nu$.

▶ To escape from a photoelectric surface, an electron must do so with the energy from a single photon collision. The electron cannot accumulate the energy from several hits by photons.

In the particle model, a photon of energy $h\nu$ strikes a bound electron, which absorbs the photon energy. If the photon energy, $h\nu$, is greater than the energy binding the electron to the surface (a quantity known as the *work function*) a photoelectron is liberated. Thus, the lowest frequency light producing the photoelectric effect is the threshold frequency, and any energy in excess of the work function appears as kinetic energy in the emitted photoelectrons.

In the discussion that follows, based on the experimental setup shown in Figure 9-12, we will see how the threshold frequency and work function are evaluated. Also, we will see that the photoelectric effect provides an independent evaluation of Planck's constant, h.

In Figure 9-12, light is allowed to shine on a piece of metal in an evacuated chamber. The electrons emitted by the metal (photoelectrons) travel to the upper

Photoelectric Effect animation

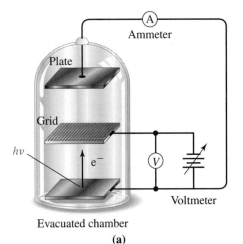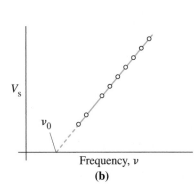

(a)   (b)

▲ **FIGURE 9-12   The photoelectric effect**
(**a**) Schematic diagram of the apparatus for photoelectric effect measurements. (**b**) Stopping voltage of photoelectrons as a function of frequency of incident radiation. The stopping voltage (V_s) is plotted against the frequency of the incident radiation. The threshold frequency, ν_0, of the metal is found by extrapolation.

plate and complete an electric circuit set up to measure the photoelectric current through an ammeter. From the magnitude of the current we can determine the rate at which photoelectrons are emitted. A second circuit is set up to measure the velocity of the photoelectrons, and hence their kinetic energy. In this circuit, a potential difference (voltage) is maintained between the photoelectric metal and an open-grid electrode placed below the upper plate. For electric current to flow, electrons must pass through the openings in the grid and onto the upper plate. The negative potential on the grid acts to slow down the approaching electrons. As the potential difference between the grid and the emitting metal is increased, a point is reached at which the photoelectrons are stopped before they reach the upper plate and the current ceases to flow. The potential difference at this point is called the *stopping voltage*, V_s. At the stopping voltage, the kinetic energy of the photoelectrons has been converted to potential energy, expressed through the following equation (in which m, u, and e are the mass, speed, and charge of an electron, respectively).

$$\frac{1}{2}mu^2 = eV_s$$

As a result of experiments of the type just described, we find that V_s is proportional to the frequency of the incident light but *independent* of the light intensity. Also, as shown in Figure 9-12, if the frequency, ν, is below the *threshold frequency*, ν_0, no photoelectric current is produced. At frequencies greater than ν_0, the empirical equation for the stopping voltage is

$$V_s = k(\nu - \nu_0)$$

The constant k is independent of the metal used, but ν_0 varies from one metal to another. Although there is no relation between V_s and the light intensity, the photoelectric current *is* proportional to the intensity (I) of the light.

The work function is a quantity of work and, hence, of energy. One way to express this quantity is as the product of Planck's constant and the threshold frequency: $E = h\nu_0$. Another way is as the product of the charge on the electron, e, and the potential, V_0, that has to be overcome in the metal: $E = eV_0$. Thus, the *threshold frequency* for the photoelectric effect is given by the expression

$$\nu_0 = \frac{eV_0}{h}$$

Since the work function (eV_0) is a characteristic of the metal used in the experiment, then ν_0 is also a characteristic of the metal, as confirmed by experiment.

When a photon of energy $h\nu$ strikes an electron, the electron overcomes the work function eV_0 and is liberated with kinetic energy $\left(\frac{1}{2}\right)mu^2$. Thus, by the law of conservation of energy, we have

$$\frac{1}{2}mu^2 + eV_0 = h\nu$$

which gives

$$eV_s = \frac{1}{2}mu^2 = h\nu - eV_0$$

which is identical to the empirically determined equation for V_s with $k = h/e$ when $h\nu_0 = eV_0$. Careful experiments showed that the constant h had the same value as determined by Planck for blackbody radiation. The additional fact that the number of photoelectrons increases with the intensity of light indicates that we should associate light intensity with the number of photons arriving at a point per unit time.

▲ Albert Einstein

Are You Wondering...

How the energy of a photon is manifested?

We begin with an important relationship between the mass and velocity of a particle given by Einstein. Let the mass of a particle when the particle and the measuring device are at rest be denoted as m_0. If we remeasure the mass when the particle moves with a velocity u, we find that its mass increases according to the equation

$$m = \frac{m_0}{\sqrt{1 - u^2/c^2}}$$

where m is the particle mass, referred to as the *relativistic mass*, and c is the speed of light. For particles moving at speeds less than 90% of the speed of light, the relativistic mass (m) is essentially the same as the rest mass (m_0).

We have seen that the kinetic energy of a particle is given by

$$E_K = \frac{1}{2} mu^2$$

However, because photons travel at the speed of light they must have *zero* rest mass (otherwise their relativistic mass m would become infinite). So where is their energy?

Although photons have zero rest mass, they do possess momentum, which is defined as the relativistic mass times the velocity of the particle. Einstein's theory of special relativity states that a particle's energy and momentum ($p = mu$, recall page 200) are related by the expression

$$E^2 = (pc)^2 + \left(m_0 c^2\right)^2$$

where m_0 is the rest mass of the particle. For photons traveling at the speed of light c, the rest mass is zero. Hence

$$E = pc = h\nu$$

$$p = \frac{h\nu}{c} = \frac{h}{\lambda}$$

Photons possess momentum, and it is this momentum that is transferred to an electron in a collision. In all collisions between photons and electrons, momentum is conserved. Thus, we see that the wave and the photon models are intimately connected. The energy of a photon is related to the frequency of the wave by Planck's equation, and the momentum of the photon is related to the wavelength of the wave by the equation just derived! When a photon collides with an electron, it transfers momentum to the electron, which accelerates to a new velocity. The energy of the photon decreases, and, as a consequence, its wavelength increases. This phenomenon, called the Compton effect, was discovered in 1923 and confirmed the particulate nature of light.

Photons of Light and Chemical Reactions

In Chapter 8 we encountered chemical reactions that are induced by light, *photochemical* reactions. In these reactions photons are "reactants" and we can indicate them in chemical equations by the symbol $h\nu$. Thus, the reactions by which ozone is produced from oxygen in the atmosphere can be represented as

$$O_2 + h\nu \longrightarrow O + O$$

$$O_2 + O + M \longrightarrow O_3 + M$$

The radiation required in the first reaction is UV radiation with wavelength less than 242.4 nm. O atoms from the first reaction then combine with O_2 to form O_3. In the second reaction, a "third body," M, such as $N_2(g)$, is needed to carry away excess energy to prevent immediate dissociation of O_3 molecules.

Photochemical reactions involving ozone are the subject of Example 9-2. There we see that the product of Planck's constant (h) and frequency (ν) yields the energy of a single photon of electromagnetic radiation in the unit joule. Invariably, this energy is only a tiny fraction of a joule. Often it is useful to deal with the much larger energy of a mole of photons (6.02214×10^{23} photons).

EXAMPLE 9-2

Using Planck's Equation to Calculate the Energy of Photons of Light. For radiation of wavelength 242.4 nm, the longest wavelength that will bring about the photodissociation of O_2, what is the energy of **(a)** one photon and **(b)** a mole of photons of this light?

Solution

(a) To use Planck's equation, we need the frequency of the radiation. This we can get from equation (9.1) after first expressing the wavelength in meters.

$$\nu = \frac{c}{\lambda} = \frac{2.998 \times 10^8 \text{ m s}^{-1}}{242.4 \times 10^{-9} \text{ m}} = 1.237 \times 10^{15} \text{ s}^{-1}$$

Planck's equation is written for *one* photon of light. We emphasize this by including the unit, photon^{-1}, in the value of h.

$$E = h\nu = 6.626 \times 10^{-34} \times \frac{\text{J s}}{\text{photon}} \times 1.237 \times 10^{15} \text{ s}^{-1}$$

$$= 8.196 \times 10^{-19} \text{ J/photon}$$

(b) Once we have the energy per photon, we can multiply it by the Avogadro constant to convert to a per-mole basis.

$$E = 8.196 \times 10^{-19} \text{ J/photon} \times 6.022 \times 10^{23} \text{ photons/mol}$$

$$= 4.936 \times 10^5 \text{ J/mol}$$

(This quantity of energy is sufficient to raise the temperature of 10.0 L of water by 11.8 °C.)

Practice Example A: The protective action of ozone in the atmosphere comes through ozone's absorption of UV radiation in the 230–290-nm wavelength range. What is the energy, in kilojoules per mole, associated with radiation in this wavelength range?

Practice Example B: Chlorophyll absorbs light at energies of 3.056×10^{-19} J/photon and 4.414×10^{-19} J/photon. To what color and frequency do these absorptions correspond?

▲ **Niels Bohr (1885–1962)**
In addition to his work on the hydrogen atom, Bohr headed the Institute of Theoretical Physics in Copenhagen, which became a mecca for theoretical physicists in the 1920s and 1930s.

9-4 The Bohr Atom

The Rutherford model of a nuclear atom (Section 2-3) does not indicate how electrons are arranged outside the nucleus of an atom. According to classical physics, stationary, negatively charged electrons would be pulled into the positively charged nucleus. This suggests that the electrons in an atom must be in motion, like the planets orbiting the sun. However, again according to classical physics, orbiting electrons should be constantly accelerating and should radiate energy. By losing energy, the electrons would be drawn ever closer to the nucleus and soon spiral into it. In 1913, Niels Bohr (1885–1962) resolved this problem by using Planck's quantum hypothesis. In an interesting blend of classical and quantum theory, Bohr postulated that for a hydrogen atom

1. The electron moves in circular orbits about the nucleus with the motion described by classical physics.

2. The electron has only a fixed set of allowed orbits, called *stationary states*. The allowed orbits are those in which certain properties of the electron have unique values. Even though classical theory would predict otherwise, *as long as an electron remains in a given orbit, its energy is constant and no energy is emitted*. The particular property of the electron having only certain allowed values, leading to only a discrete set of allowed orbits, is called the *angular momentum*. Its possible values are $nh/2\pi$, where n must be an integer. Thus $n = 1$ for the first orbit; $n = 2$ for the second orbit; and so on.

3. An electron can pass only from one allowed orbit to another. In such transitions, fixed discrete quantities of energy (quanta) are involved—either absorbed or emitted.

KEEP IN MIND ▶
that momentum (p) is the product of the mass and the velocity of a particle. If the particle undergoes a circular motion, then the particle possesses angular momentum.

The atomic model of hydrogen based on these ideas is pictured in Figure 9-13. The allowed states for the electron are numbered, $n = 1$, $n = 2$, $n = 3$, and so on. These *integral* numbers, which arise from Bohr's assumption that only certain values are allowed for the angular momentum of the electron, are called **quantum numbers**.

The Bohr theory predicts the radii of the allowed orbits in a hydrogen atom.

$$r_n = n^2 a_0, \quad \text{where } n = 1, 2, 3, \ldots \quad \text{and} \quad a_0 = 0.53 \text{ Å (53 pm)} \tag{9.4}$$

The theory also allows us to calculate the electron velocities in these orbits and, most important, the energy. When the electron is free of the nucleus, by convention, it is said to be at a *zero* of energy. When a free electron is attracted to the nucleus and confined to the orbit n, the electron energy becomes negative, with its value lowered to

 Bohr Model activity

$$E_n = \frac{-R_H}{n^2} \tag{9.5}$$

R_H is a numerical constant with a value of 2.179×10^{-18} J.

With expression (9.5), we can calculate the energies of the allowed energy states, or *energy levels*, of the hydrogen atom. These levels can be represented schematically as in Figure 9-14. This representation is called an **energy-level diagram**. Example 9-3 shows how Bohr's model can be used to predict whether certain energy levels, are possible (allowed) or impossible (not allowed).

▶ FIGURE 9-13

Bohr model of the hydrogen atom
A portion of the hydrogen atom is pictured. The nucleus is at the center, and the electron is found in one of the discrete orbits, $n = 1, 2, \ldots$ Excitation of the atom raises the electron to higher-numbered orbits, as shown through black arrows. Light is emitted when the electron falls to a lower-numbered orbit. Two transitions are shown that produce lines in the Balmer series of the hydrogen spectrum, in the approximate colors of the spectral lines.

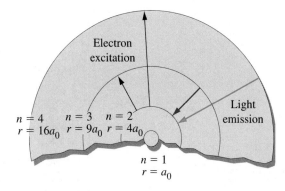

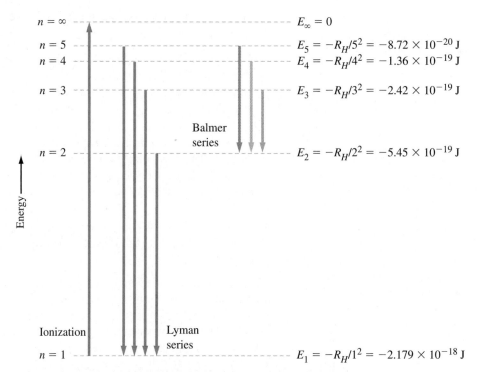

$n = \infty$ $E_{\infty} = 0$

$n = 5$ $E_5 = -R_H/5^2 = -8.72 \times 10^{-20} \text{ J}$
$n = 4$ $E_4 = -R_H/4^2 = -1.36 \times 10^{-19} \text{ J}$

$n = 3$ $E_3 = -R_H/3^2 = -2.42 \times 10^{-19} \text{ J}$

Balmer series

$n = 2$ $E_2 = -R_H/2^2 = -5.45 \times 10^{-19} \text{ J}$

Energy

Ionization

Lyman series

$n = 1$ $E_1 = -R_H/1^2 = -2.179 \times 10^{-18} \text{ J}$

▲ **FIGURE 9-14** **Energy-level diagram for the hydrogen atom**
If the electron acquires 2.179×10^{-18} J of energy, it moves to the orbit $n = \infty$; ionization of the H atom occurs (black arrow). Energy emitted when the electron falls from higher-numbered orbits to the orbit $n = 1$ is in the form of ultraviolet light, which produces a spectral series called the Lyman series (gray lines). Electron transitions to the orbit $n = 2$ yield lines in the Balmer series (recall Figure 9-10); three of the lines are shown here (in color). Transitions to $n = 3$ yield spectral lines in the infrared.

EXAMPLE 9-3

Understanding the Meaning of Quantization of Energy. Is it likely that there is an energy level for the hydrogen atom, $E_n = -1.00 \times 10^{-20}$ J?

Solution

We really do not need to do a detailed calculation to conclude that the answer is very likely *no*. We cannot expect some random value to correspond to one of the unique set of allowed energy levels. Let us rearrange equation (9.5), solve for n^2, and then for n.

$$n^2 = \frac{-R_H}{E_n}$$

$$= \frac{-2.179 \times 10^{-18} \text{ J}}{-1.00 \times 10^{-20} \text{ J}} = 2.179 \times 10^2 = 217.9$$

$$n = \sqrt{217.9} = 14.76$$

Because the value of n is not an integer, this is not an allowed energy level for the hydrogen atom.

Practice Example A: Is there an energy level for the hydrogen atom, $E_n = -2.69 \times 10^{-20}$ J?

Practice Example B: Is it likely that one of the electron orbits in the Bohr atom has a radius of 1.00 nm?

[*Hint:* Recall expression (9.4).]

▶ Think of a person on a stairway going up steps (excitation) or down steps (emission). The person must land on a step—inbetween levels are not available.

▶ Notice the resemblance of this equation to the Balmer equation (9.2). In addition to developing a theory of atomic structure to account for Rutherford's atomic model, Bohr sought a theoretical explanation of the Balmer equation.

Normally, the electron in a hydrogen atom is found in the orbit closest to the nucleus ($n = 1$). This is the lowest allowed energy, or the **ground state**. When the electron gains a quantum of energy, it moves to a higher level ($n = 2, 3, \dots$) and the atom is in an **excited state**. When the electron drops from a higher to a lower numbered orbit, a unique quantity of energy is emitted—the difference in energy between the two levels. We can use equation (9.5) to derive an expression for the *difference* in energy between two levels, where n_f is the final level and n_i is the initial one.

$$\Delta E = E_f - E_i = \frac{-R_H}{n_f^2} - \frac{-R_H}{n_i^2} = R_H\left(\frac{1}{n_i^2} - \frac{1}{n_f^2}\right) = 2.179 \times 10^{-18}\,\text{J}\left(\frac{1}{n_i^2} - \frac{1}{n_f^2}\right) \quad (9.6)$$

The energy of the photon, either absorbed or emitted, is calculated in accordance with Planck's equation,

$$\Delta E = h\nu$$

where we have used ΔE, the energy difference between the energy levels involved in the electronic transition.

In Example 9-4 we use equation (9.6) as a basis for calculating the lines in the hydrogen emission spectrum. Because the differences between energy levels are limited in number, so too are the energies of the emitted photons. Therefore, only certain wavelengths (or frequencies) are observed for the spectral lines.

EXAMPLE 9-4

Calculating the Wavelength of a Line in the Hydrogen Spectrum. Determine the wavelength of the line in the Balmer series of hydrogen corresponding to the transition from $n = 5$ to $n = 2$.

Solution

The specific data for equation (9.6) are $n_i = 5$ and $n_f = 2$.

$$\Delta E = 2.179 \times 10^{-18}\,\text{J}\left(\frac{1}{5^2} - \frac{1}{2^2}\right)$$

$$= 2.179 \times 10^{-18} \times (0.04000 - 0.25000)$$

$$= -4.576 \times 10^{-19}\,\text{J}$$

The negative sign of ΔE signifies that energy is *emitted*, just as we expect it to be. This quantity of energy is emitted as a photon hence the magnitude of ΔE, the difference between two energy levels, is equal to the energy of the photon emitted. We can obtain the photon frequency from the Planck equation: $\Delta E = E_{photon} = h\nu$, which we rearrange to the form

$$\nu = \frac{E_{photon}}{h} = \frac{4.576 \times 10^{-19}\,\text{J photon}^{-1}}{6.626 \times 10^{-34}\,\text{J s photon}^{-1}} = 6.906 \times 10^{14}\,\text{s}^{-1}$$

Finally, to calculate the wavelength of this line, we use equation (9.1).

$$\lambda = \frac{c}{\nu} = \frac{2.998 \times 10^8\,\text{m s}^{-1}}{6.906 \times 10^{14}\,\text{s}^{-1}} = 4.341 \times 10^{-7}\,\text{m} = 434.1\,\text{nm}$$

Note the good agreement between this result and data in Figure 9-10.

Practice Example A: Determine the wavelength of light *absorbed* in an electron transition from $n = 2$ to $n = 4$ in a hydrogen atom.

Practice Example B: Refer to Figure 9-14 and determine which transition produces the longest wavelength line in the Lyman series of the hydrogen spectrum. What is the wavelength of this line in nanometers and in angstroms?

The Bohr Theory and Spectroscopy

As we just saw in Example 9-4, the Bohr theory provides a model for understanding the emission spectra of atoms. Emission spectra are obtained when the individual atoms in a collection of atoms (roughly 10^{20} of them) are excited to the possible excited states of the atom. The atoms then relax to states of lower energy by emitting photons of frequency given by

$$\nu_{\text{photon}} = \frac{E_i - E_f}{h} \tag{9.7}$$

Thus, the quantization of the energy states of atoms leads to line spectra.

Earlier in this chapter, we learned how the emission spectrum of a sample can be measured by dispersing emitted light through a prism and determining the wavelengths of the individual components of that light. We can conceive of an alternate technique in which we pass electromagnetic radiation, such as white light, through a sample of atoms in their ground states and then pass the emerging light through a prism. Now we observe which frequencies of light the atoms *absorb*. This form of spectroscopy is called *absorption spectroscopy*. The two types of spectroscopy—emission and absorption—are illustrated in Figure 9-15.

For absorption of a photon to take place, the energy of the photon must exactly match the energy difference between the final and initial states, that is,

$$\nu = \frac{E_f - E_i}{h} \tag{9.8}$$

Note that the farther apart the energy levels, the shorter the wavelength of the photon needed to induce a transition.

You may have also noticed that in equation (9.7) we expressed the energy difference as $E_i - E_f$, whereas in equation (9.8) we wrote $E_f - E_i$. We did this to signify that energy is conserved during photon absorption and emission. That is, during emission $E_f = E_i - h\nu$, so that $\nu = (E_i - E_f)/h$. During photon absorption, $E_f = E_i + h\nu$, so that $\nu = (E_f - E_i)/h$.

Spectroscopic techniques have been used extensively in the study of the structure of molecules. We describe other forms of spectroscopy available to chemists elsewhere in the text.

The Bohr Theory and the Ionization Energy of Hydrogen

The Bohr model of the atom helps us to understand the formation of cations. In the special case where the energy of a photon interacting with a hydrogen atom is just enough to remove an electron from the ground state ($n = 1$), the electron is freed, the atom is ionized, and the energy of the free electron is zero.

$$h\nu_{\text{photon}} = E_i = -E_1$$

where the quantity E_i is called the *ionization energy* of the hydrogen atom. If we substitute $n_i = 1$ and $n_f = \infty$ into the Bohr expression for an electron initially in the ground state of an H atom, we have

$$h\nu_{\text{photon}} = E_i = -E_1 = \frac{R_H}{1^2} = R_H$$

We apply the ideas just developed about the ionization of atoms in Example 9-5, and we couple these ideas with another aspect of the Bohr model: The model also works for hydrogen-like species, such as the ions He^+ and Li^{2+}, which have only one electron. For these species, the nuclear charge (atomic number) appears in the energy-level expression. That is,

$$E_n = \frac{-Z^2 R_H}{n^2} \tag{9.9}$$

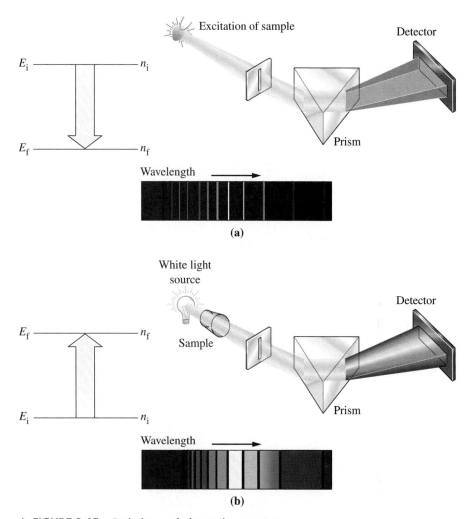

▲ FIGURE 9-15   **Emission and absorption spectroscopy**
(a) Emission spectroscopy. Bright lines are observed on a dark background of the photographic plate. (b) Absorption spectroscopy. Dark lines are observed on a bright background on the photographic plate.

Are You Wondering...

Why a photon of light is not immediately reemitted after having been absorbed?

This should certainly be possible, since for any given pair of energy levels, the exact same quantity of energy is involved, whether the photon is absorbed or emitted. It does not happen very often, however, because the excited state of an atom has a limited lifetime, and during this lifetime the atom can undergo collisions with other atoms in the sample. (Remember, even a gas at low pressures contains a lot of atoms.) During all these collisions, the energy imparted to the atom by the absorbed photon can be lost as kinetic energy, so that through a succession of such losses the atom eventually reverts to the ground state. Thus, there is a net absorption of photons.

EXAMPLE 9-5

Using the Bohr Model. Determine the kinetic energy of the electron ionized from a Li^{2+} ion in its ground state, using a photon of frequency $5.000 \times 10^{16} \ s^{-1}$.

Solution

We calculate the energy of the electron in the Li^{2+} ion, using equation 9.9.

$$E_1 = \frac{-3^2 \times 2.179 \times 10^{-18} \ J}{1^2} = -1.961 \times 10^{-17} \ J$$

The energy of a photon of frequency $5.000 \times 10^{16} \ s^{-1}$ is

$$E = h\nu = 6.626 \times 10^{-34} \times \frac{J \ s}{photon} \times 5.000 \times 10^{16} \ s^{-1} = 3.313 \times 10^{-17} \ J \ photon^{-1}$$

The ionization energy, the energy required to remove the electron, is $E_i = -E_1 = 1.961 \times 10^{-17} \ J$. The extra energy from the photon is transferred as kinetic energy to the electron. Thus, the kinetic energy of the electron is

$$\text{kinetic energy} = 3.313 \times 10^{-17} \ J - 1.961 \times 10^{-17} \ J = 1.352 \times 10^{-17} \ J$$

Practice Example A: Determine the wavelength of light *emitted* in an electron transition from $n = 5$ to $n = 3$ in a Be^{3+} ion.

Practice Example B: The frequency of the $n = 3$ to $n = 2$ transition for an unknown hydrogen-like ion occurs at a frequency 16 times that of the hydrogen atom. What is the identity of the ion?

Inadequacies of the Bohr Model

In spite of the accomplishments of the Bohr model for the hydrogen atom and hydrogen-like ions, the Bohr theory has a number of weaknesses. From an experimental point of view, the theory cannot explain the emission spectra of atoms and ions with more than one electron, despite numerous attempts to do so. In addition, the theory cannot explain the effect of magnetic fields on emission spectra. From a fundamental standpoint, the Bohr theory is an uneasy mixture of classical and nonclassical physics. Bohr understood at the time that there is no fundamental basis for the postulate of quantized angular momentum forcing an electron into a circular orbit. He made the postulate only so that his theory would agree with experiment.

Modern quantum mechanics replaced the Bohr theory in 1926. The quantization of energy and angular momentum arose out of the postulates of this new quantum theory and required no extra assumptions. However, the circular orbits of the Bohr theory do not occur in quantum mechanics. The Bohr theory gave the paradigm shift—the quantum leap—from classical physics to the new quantum physics. We must not underestimate its importance as a scientific development. The new quantum theory is based on ideas presented in the next section.

9-5 Two Ideas Leading to a New Quantum Mechanics

In the previous section, we examined some successes of the Bohr theory and pointed out its inability to deal with multielectron atoms. A decade or so after Bohr's work on hydrogen, two landmark ideas stimulated a new approach to quantum mechanics. We consider those two ideas in this section and the new quantum mechanics—wave mechanics—in the next.

Wave–Particle Duality

To explain the photoelectric effect, Einstein suggested that light has particle-like properties, which are embodied in photons. Other phenomena, however, such as the dispersion of light into a spectrum by a prism, are best understood in terms of the wave theory of light. Light, then, appears to have a *dual* nature.

In 1924, Louis de Broglie, considering the nature of light and matter, offered a startling proposition: *Small particles of matter may at times display wavelike properties*. How did de Broglie come up with such a suggestion? He was aware of Einstein's famous equation

$$E = mc^2$$

where m is the relativistic mass of the photon and c is the speed of light. He combined this equation with the Planck relationship for the energy of a photon $E = h\nu$ as follows

$$h\nu = mc^2$$

$$\frac{h\nu}{c} = mc = p$$

where p is the momentum of the photon. Using $\nu\lambda = c$, we have

$$p = \frac{h}{\lambda}$$

In order to use this equation for a material particle, such as an electron, de Broglie substituted for the momentum, p, its equivalent—the product of the mass of the particle, m, and its velocity, u. When this is done, we arrive at de Broglie's famous relationship.

$$\lambda = \frac{h}{p} = \frac{h}{mu} \tag{9.10}$$

De Broglie called the waves associated with material particles "matter waves." If matter waves exist for small particles, then beams of particles such as electrons should exhibit the characteristic properties of waves, namely diffraction (recall page 301). If the distance between the objects that the waves scatter from is about the same as the wavelength of the radiation, diffraction occurs and an interference pattern is observed. For example, X rays are highly energetic photons with an associated wavelength of about 1 Å (100 pm). X rays are scattered by the regular array of atoms in the metal aluminum, where the atoms are about 2 Å (200 pm) apart, producing the diffraction pattern shown on page 318 (top left).

In 1927, C. J. Davisson and L. H. Germer of the United States showed that a beam of slow electrons is diffracted by a crystal of nickel. In a similar experiment in that same year, G. P. Thomson of Scotland directed a beam of electrons at a thin metal foil and obtained the same diffraction pattern as with X rays of the same wavelength, as in the pattern for the diffraction of electrons by aluminum foil shown on page 318 (top right).

Thomson and Davisson shared the 1937 Nobel Prize in physics for their electron diffraction experiments. George P. Thomson was the son of J. J. Thomson, who had won the Nobel Prize in physics in 1906 for his discovery of the electron. It is interesting to note that Thomson the father showed that the electron is a particle and Thomson the son showed that the electron is a wave. Father and son together demonstrated the wave–particle duality of electrons.

▲ **Louis de Broglie**
De Broglie conceived of the wave–particle duality of small particles while working on his doctorate degree. He was awarded the Nobel Prize in physics in 1929 for this work.

KEEP IN MIND

that in equation (9.10), wavelength is in meters, mass is in kilograms, and velocity is in meters per second. Planck's constant must also be expressed in units of mass, length, and time. This requires replacing the unit, joule, by the equivalent units $kg \ m^2 \ s^{-2}$. ▶

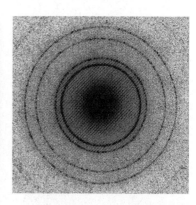

▶ (Left) Diffraction of X rays by metal foil. (Right) Diffraction of electrons by metal foil, confirming the wave-like properties of electrons.

EXAMPLE 9-6

Calculating the Wavelength Associated with a Beam of Particles. What is the wavelength associated with electrons traveling at one-tenth the speed of light?

Solution

The electron mass, expressed in kilograms, is 9.109×10^{-31} kg (recall Table 2.1). The electron velocity is $v = 0.100 \times c = 0.100 \times 3.00 \times 10^8$ m s^{-1} = 3.00×10^7 m s^{-1}. Planck's constant $h = 6.626 \times 10^{-34}$ J s = 6.626×10^{-34} kg m^2 s^{-2} s = 6.626×10^{-34} kg m^2 s^{-1}. Substituting these data into equation (9.10), we obtain

$$\lambda = \frac{6.626 \times 10^{-34} \text{ kg m}^2\text{s}^{-1}}{(9.109 \times 10^{-31} \text{ kg})(3.00 \times 10^7 \text{ m s}^{-1})}$$

$$= 2.42 \times 10^{-11} \text{ m} = 24.2 \text{ pm}$$

Practice Example A: Assuming Superman has a mass of 91 kg, what is the wavelength associated with him if he is traveling at one-fifth the speed of light?

Practice Example B: To what velocity (speed) must a beam of protons be accelerated to display a de Broglie wavelength of 10.0 pm? Obtain the proton mass from Table 2.1.

▲ **Werner Heisenberg (1901–1976)**
In addition to his enunciation of the uncertainty principle, for which he won the Nobel Prize in physics in 1932, Heisenberg also developed a mathematical description of the hydrogen atom that gave the same results as Schrödinger's equation (page 323). Heisenberg (left) is shown here dining with Niels Bohr.

The wavelength calculated in Example 9-6, 24.2 pm, is about one-half the radius of the first Bohr orbit of a hydrogen atom. It is only when wavelengths are comparable to atomic or nuclear dimensions that wave–particle duality is important. The concept has little meaning when applied to large (macroscopic) objects, such as baseballs and automobiles because their wavelengths are too small to measure. For these macroscopic objects, the laws of classical physics are quite adequate.

The Uncertainty Principle

The laws of classical physics permit us to make precise predictions. For example, we can *calculate* the exact point at which a rocket will land after it is fired. The more precisely we measure the variables that affect the rocket's trajectory (path), the more accurate our calculation (prediction) will be. In effect, there is no limit to the accuracy we can achieve. In classical physics, nothing is left to chance—physical behavior can be predicted with certainty.

During the 1920s, Niels Bohr and Werner Heisenberg considered hypothetical experiments to establish just how precisely the behavior of subatomic particles can be determined. The two variables that must be measured are the position of the particle (x) and its momentum ($p = mu$). The conclusion they reached is that there must *always* be uncertainties in measurement such that the product of the uncertainty in position, Δx, and the uncertainty in momentum, Δp, is

$$\Delta x \, \Delta p \geq \frac{h}{4\pi} \qquad (9.11)$$

▶ The uncertainty principle is not easy for most people to accept. Einstein spent a good deal of time from the middle 1920s until his death in 1955 attempting, unsuccessfully, to disprove it.

The significance of this expression, called the **Heisenberg uncertainty principle**, is that we cannot measure position and momentum with great precision simultaneously. If we design an experiment to locate the position of a particle with great precision, we cannot measure its momentum precisely, and vice versa. In simpler terms, if we know precisely where a particle is, we cannot also know precisely where it has come from or where it is going. If we know precisely how a particle is moving, we cannot also know precisely where it is. In the subatomic world, things must always be "fuzzy." Why should this be so?

Suppose we wish to study an electron in a hydrogen atom by looking at it with a microscope. What kind of microscope should this be? The smallest particle we can see with a microscope has about the same dimensions as the wavelength of the light used to illuminate it. Let us try working out the energy of electromagnetic radiation needed to see an electron and what effect that might have on the electron we are trying to observe.

The first Bohr orbit in a hydrogen atom is 53 pm, and the diameter of the atom is about 100 pm (10^{-10} m). Electrons are much smaller than the atoms in which they are found. Suppose we assume a diameter of about 10^{-14} m for an electron. Light of this wavelength would have a frequency of 3×10^{22} s^{-1} ($\nu = c/\lambda$) and an energy per photon of 2×10^{-11} J ($E = h\nu$). But from Figure 9-14, we can see that this energy is far, far in excess of the 2.179×10^{-18} J required to ionize the electron in a hydrogen atom. A photon of this energy would not help us to see the electron in the atom—it would knock the electron out of the atom! Just in trying to look at the electron, we would drastically change the situation from an electron bound in a hydrogen atom to an ionized electron free of the atom. This situation is pictured in Figure 9-16. On the other hand, in trying to see and make measurements on a golf ball rolling across a green, we can use visible light. Light with a frequency of about 5×10^{14} s^{-1} has an energy per photon of 3×10^{-19} J. This energy, even if multiplied millions of times over, is much smaller than the kinetic energy of the ball. The light would not interfere with the measurement at all. Although applicable to macroscopic objects, the uncertainty principle yields uncertainties that are insignificant.

Once we understand that the consequence of wave–particle duality is the uncertainty principle, we realize that a fundamental error of the Bohr model was to constrain an electron to a one-dimensional orbit (1-D in the sense that the electron cannot move off the orbit). We are now ready to turn our attention to a modern description of electrons in atoms.

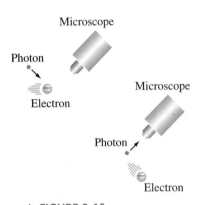

▲ **FIGURE 9-16**
The uncertainty principle
A photon of light strikes an electron and is reflected (left). In the collision, the photon transfers momentum to the electron. The reflected photon is seen through the microscope, but the electron is out of focus (right). Its exact position cannot be determined.

EXAMPLE 9-7

Calculating the Uncertainty of the Position of an Electron. A 12-*eV* electron can be shown to have a speed of 2.05×10^6 m/s. Assuming that the precision (uncertainty) of this value is 1.5%, with what precision can we simultaneously measure the position of the electron?

Solution

The electron mass, expressed in kilograms, is 9.109×10^{-31} kg (recall Table 2.1). The electron momentum is

$$p = mu = (9.11 \times 10^{-31} \text{ kg})(2.05 \times 10^6 \text{ m/s})$$
$$= 1.87 \times 10^{-24} \text{ kg m s}^{-1}$$
$$\Delta p = (0.015) \times 1.87 \times 10^{-24} \text{ kg m s}^{-1} = 2.80 \times 10^{-26} \text{ kg m s}^{-1}$$

From equation (9.11), the uncertainty in the position is

$$\Delta x = \frac{h}{4\pi\,\Delta p} = \frac{6.63 \times 10^{-34} \text{ kg m}^2 \text{ s}^{-1}}{4 \times 3.14 \times 2.80 \times 10^{-26} \text{ kg m s}^{-1}} = 1.89 \times 10^{-9} \text{ m} = 1890 \text{ pm}$$

which is about 10 atomic diameters. With this measurement of the electron's momentum, there is simply no way to pin down its position with any greater precision.

Practice Example A: Superman has a mass of 91 kg and is traveling at one-fifth the speed of light. If the speed at which Superman travels is known with a precision of 1.5% what is the uncertainty in his position?

Practice Example B: What is the uncertainty in the speed of a beam of protons whose position is known with the uncertainty of 24 nm?

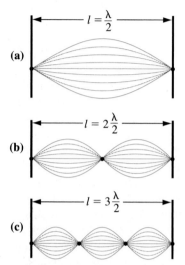

▲ FIGURE 9-17

Standing waves in a string
The string can be set into motion by plucking it. The blue boundaries outline the range of displacements at each point for each standing wave. The relationships between the wavelength, string length, and the number of nodes–points that are not displaced–are given by equation (9.12). The nodes are marked by bold dots.

9-6   Wave Mechanics

De Broglie's relationship suggests that electrons are matter waves and thus should display wavelike properties. A consequence of this wave–particle duality is the limited precision in determining an electron's position and momentum imposed by the Heisenberg uncertainty principle. How then are we to view electrons in atoms? To answer this question, we must begin by identifying two types of waves.

Standing Waves

On an ocean, the wind produces waves on the surface whose crests and troughs travel great distances. These are called *traveling waves*. In the traveling wave in Figure 9-1, every portion of a very long rope goes through an identical up-and-down motion. The wave transmits energy along the entire length of the rope. An alternative form of a wave is seen in the vibrations in a plucked guitar string, suggested by Figure 9-17. Segments of the string experience up-and-down displacements with time, and they oscillate or vibrate between the limits set by the blue curves. The important aspect of these waves is that the crests and troughs of the wave occur at fixed positions and the amplitude of the wave at the fixed ends is zero. Of special interest is the fact that the magnitudes of the oscillations differ from point to point along the wave, including certain points, called *nodes*, that undergo no displacement at all. A wave with these characteristics is called a **standing wave**.

We might say that the permitted wavelengths of standing waves are quantized. They are equal to twice the length (L) divided by a whole number (n), that is,

$$\lambda = \frac{2L}{n} \text{ where } n = 1, 2, 3, \dots \text{ and the total number of nodes} = n + 1 \quad (9.12)$$

The plucked guitar string represents a *one*-dimensional standing wave, and so does an electron in a Bohr orbit. We now understand why only an integral number of electron orbits are permitted in the Bohr theory (Figure 9-18). The fact that Bohr orbits are one-dimensional also points up a serious deficiency in the Bohr model: The matter waves of electrons in the hydrogen atom must be three-dimensional.

KEEP IN MIND ▶
that a circle is one dimensional in the sense that all points on the circumference are at the same distance from the center. Thus, only one value needs to be given to define the circle—its radius.

Standing Waves, Quantum Particles, and Wave Functions

In 1927, Erwin Schrödinger, an expert in the theory of vibrations and standing waves, suggested that an electron (or any other particle) exhibiting wavelike properties should be describable by a mathematical equation called a **wave function** and denoted by the Greek letter psi, ψ. The wave function should correspond to a standing wave within the boundary of the system being described. The simplest system for which we can write a wave function is another one-dimensional system, that of a particle confined to move in a single direction in a box. The wave function for this

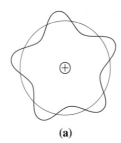 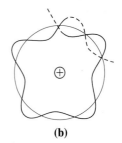

(a)   (b)

▲ **FIGURE 9-18   The electron as a matter wave**
These patterns are two-dimensional cross sections of a much more complicated three-dimensional wave. The wave pattern in **(a)**, a *standing wave*, is an acceptable representation. It has an integral number of wavelengths (five) about the nucleus; successive waves reinforce one another. The pattern in **(b)** is unacceptable. The number of wavelengths is nonintegral, and successive waves tend to cancel each other; that is, the crest in one part of the wave overlaps a trough in another part of the wave, and there is no resultant wave at all.

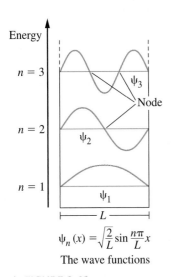

▲ **FIGURE 9-19**
The standing waves of a particle in a one-dimensional box
The first three wave functions and their energies are shown in relation to the position of the particle within the box. The wave function changes sign at the nodes.

so-called "particle in a box" looks like those of a guitar string (Figure 9-17), but now they represent the matter waves of a particle. Since the particle is constrained to be in the box, the waves also must be inside the box, as illustrated in Figure 9-19.

If the length of the box is L and the particle moves along the x direction, then the equation for the standing wave is a sine function.

$$\psi_n(x) = \sqrt{\frac{2}{L}}\sin\left(\frac{n\pi x}{L}\right) \quad n = 1, 2, 3 \dots$$

where the integer n, called a quantum number, labels the wave function.

It is reasonable that this wave function should be a sine function. A sine function undergoes the same up-and-down motion that is characteristic of waves. To illustrate, let's consider the case where $n = 2$.

When

$x = 0,$	$\sin 2\pi x/L = \sin 0 = 0,$	and $\psi_n(x) = 0$
$x = L/4,$	$\sin 2\pi(L/4)/L = \sin \pi/2 = 1,$	and $\psi_n(x) = (2/L)^{1/2}$
$x = L/2,$	$\sin 2\pi(L/2)/L = \sin \pi = 0,$	and $\psi_n(x) = 0$
$x = 3L/4$	$\sin 2\pi(3L/4)/L = \sin 3\pi/2 = -1$	and $\psi_n(x) = -(2/L)^{1/2}$
$x = L$	$\sin 2\pi(L)/L = \sin 2\pi = 0,$	and $\psi_n(x) = 0$

At one end of the box $(x = 0)$, both the sine function and the wave function are zero. At one-fourth of the length of the box $(x = L/4)$, the sine function and the wave function both reach their maximum values. At the midpoint of the box, both are again zero; the wave function has a node. At three-fourths of the box length, both functions reach their minimum values (negative quantities), and at the farther end of the box, both functions are again zero.

What sense can we make of the wave function and the quantum number? First consider the quantum number, n. What can we relate it to? The particle that we are considering is freely moving (not acted upon by any outside forces) with a kinetic energy given by the expression

$$E_k = \frac{1}{2}mu^2 = \frac{m^2u^2}{2m} = \frac{p^2}{2m}$$

Now, to associate this kinetic energy with a wave, we can use de Broglie's relationship ($\lambda = h/p$) to get

$$E_k = \frac{p^2}{2m} = \frac{h^2}{2m\lambda^2}$$

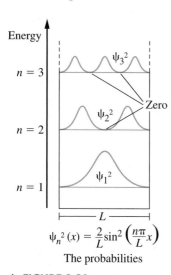

$$\psi_n^2(x) = \frac{2}{L}\sin^2\left(\frac{n\pi}{L}x\right)$$

The probabilities

▲ **FIGURE 9-20**

The probabilities of a particle in a one-dimensional box
The squares of the first three wave functions and their energies are shown in relation to the position of the particle within the box. There is no chance of finding the particle at the points where $\psi^2 = 0$.

KEEP IN MIND ▶

that the particle exhibits wave-particle duality, rendering inappropriate a question about how it gets from one side of the node to the other (an appropriate question for a *classical* particle). All we know is that it is in state $n = 2$ and in the box somewhere. When we make a measurement, we'll find the particle on one side of the node or the other!

The wavelengths of the matter wave have to fit the same standing wave conditions that we described earlier for the standing waves of a guitar string (equation 9.12). When we substitute the wavelength of the matter wave from equation (9.12) into the equation for the energy of the wave, we obtain the following.

$$E_k = \frac{h^2}{2m\lambda^2} = \frac{h^2}{2m(2\,L/n)^2} = \frac{n^2h^2}{8mL^2}$$

So we see that the standing wave condition naturally gives rise to quantization of the wave's energy, with the allowable values determined by the value of the quantum number n. Note also that as we decrease the size of the box, the kinetic energy of the particle increases, and according to the uncertainty principle, our knowledge of the momentum must decrease. A final noteworthy point is that the energy of the particle *cannot be zero*. The lowest possible energy, corresponding to $n = 1$, is called the **zero-point energy**. Because the zero-point energy is not zero, the particle cannot be at rest. This observation is consistent with the uncertainty principle because the position and momentum both must be uncertain, and there is nothing uncertain about a particle at rest.

The particle-in-a box model helps us see the origin of the quantization of energy, but how are we to interpret the wave function, ψ? What does it mean that the value of the wave function can be positive or negative? Actually, unlike the trajectory of a classical particle, the wave function of a particle has no physical significance. We need to take a different approach, one suggested by the German physicist Max Born in 1926. From the electron-as-particle standpoint, we have a special interest in the *probability* that the electron is at some particular point; from the electron-as-wave standpoint, our interest is in *electron charge density*. In a classical wave (such as visible light), the amplitude of the wave corresponds to ψ, and the intensity of the wave to ψ^2. The intensity relates to the photon density—the number of photons present in a region. For an electron wave, then, ψ^2 relates to electron charge density. Electron probability is proportional to electron charge density, and both these quantities are associated with ψ^2. Thus, in Born's interpretation of the wave function, the value of ψ^2 is the probability density of finding an electron in a small volume of space, and the total probability of finding the electron in that small volume is ψ^2 times the volume of interest.

Now let's return to a particle constrained to a one-dimensional path in a box and look at the probabilities for the wave functions. These are shown in Figure 9-20. First, notice that even where the wave function is negative, the probability density is positive, as it should be in all cases. Next, look at the probability density for the wave function corresponding to $n = 1$. The highest value of ψ^2 is at the center of the box; that is, the particle is most likely to be found there. The probability density for the state with $n = 2$ indicates that the particle is most likely to be found between the center of the box and the walls.

A final consideration of the particle-in-a-box model concerns its extension to a three-dimensional box. In this case, the particle can move in all three directions—x, y, and z—and the quantization of energy is described by the following expression.

$$E_{n_x n_y n_z} = \frac{h^2}{8m}\left[\frac{n_x^2}{L_x} + \frac{n_y^2}{L_y} + \frac{n_z^2}{L_z}\right]$$

where there is one quantum number for each dimension. Thus, a three-dimensional system needs three quantum numbers. With these particle-in-a-box ideas, we can now discuss solving the quantum mechanical problem of the hydrogen atom.

Wave Functions of the Hydrogen Atom

In 1927, Schrödinger showed that the wave functions of a quantum mechanical system can be obtained by solving a wave equation that has since become known

Are You Wondering . . .

What the Schrödinger equation looks like?

If you are familiar with differential calculus, you will recognize the equation written below as a differential equation. The solution of this equation is the wave function for a system. Specifically, this equation describes the standing wave in one dimension for our simple system of a particle in a box.

$$\frac{d^2\psi}{dx^2} = -\left(\frac{2\pi}{\lambda}\right)^2 \psi$$

Notice the form of the equation: By differentiating the wave function twice, we obtain the wave function times a constant. To obtain the Schrödinger equation, we substitute de Broglie's relationship for the wavelength of a matter wave.

$$\frac{d^2\psi}{dx^2} = -\left(\frac{2\pi}{h}p\right)^2 \psi$$

Finally, we use the relationship between momentum and kinetic energy (page 321) to obtain

$$-\frac{h^2}{8\pi^2 m}\frac{d^2\psi}{dx^2} = E_k\psi$$

This is the Schrödinger equation of a free particle moving in one dimension. If the particle is subjected to a force, $V(x)$, then we have

$$-\frac{h^2}{8\pi^2 m}\frac{d^2\psi}{dx^2} + V(x)\psi = E_{total}\psi$$

Extending this treatment to three dimensions, we obtain the Schrödinger equation for the hydrogen atom, where we understand the force $V(r)$ to be $-Ze^2/r$, the attractive force between the electron and the nucleus.

$$-\frac{h^2}{8\pi^2 m_e}\left(\frac{\partial^2\psi}{\partial x^2} + \frac{\partial^2\psi}{\partial y^2} + \frac{\partial^2\psi}{\partial z^2}\right) - \frac{Ze^2}{r}\psi = E\psi$$

This is the equation that Schrödinger obtained, and following a suggestion by Eugene Wigner, he used spherical polar coordinates to solve it rather than the Cartesian coordinates shown here. The solutions are shown in Table 9.1 on page 327.

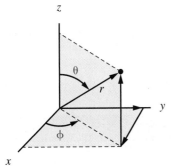

Spherical polar coordinates
$x^2 + y^2 + z^2 = r^2$
$x = r \sin\theta \cos\phi$
$y = r \sin\theta \sin\phi$
$z = r \cos\theta$

▲ **FIGURE 9-21**
The relationship between spherical polar coordinates and Cartesian coordinates
The coordinates x, y, and z are expressed in terms of the distance r and the angles θ and ϕ.

as the **Schrödinger equation**. We shall not go into the details of its solution but just describe and interpret the solution using ideas introduced in the previous discussion.

Solutions of the Schrödinger equation for the hydrogen atom give the wave functions for the electron in the hydrogen atom. These wave functions are called **orbitals** to distinguish them from the orbits of the Bohr theory. The mathematical form of these orbitals is more complex than for the particle in a box, but nonetheless we can interpret them in a straightforward way.

Wave functions are most easily analyzed in terms of the *three* variables required to define a point with respect to the nucleus. In the usual Cartesian coordinate system, these three variables are the x, y, and z dimensions. In the spherical polar coordinate system, they are r, the distance of the point from the nucleus, and the angles θ (theta) and ϕ (phi), which describe the orientation of the distance line, r, with respect to the x, y, and z axes (Figure 9-21). Either coordinate system could be used in solving the Schrödinger equation. However, whereas in the Cartesian coordinate system the orbitals would involve all three variables, x, y, and z, in the spherical polar system the orbitals can be expressed in terms of one function R that depends only on r and a second function Y that depends on θ and ϕ. That is,

$$\psi(r, \theta, \phi) = R(r)Y(\theta, \phi)$$

The function $R(r)$ is called the **radial wave function**, and the function $Y(\theta, \phi)$ is called the **angular wave function**. Each orbital has three quantum numbers to define it since the hydrogen atom is a three-dimensional system. The particular set of quantum numbers confers a particular functional form to $R(r)$ and $Y(\theta, \phi)$.

Probability densities and the spatial distribution of these densities can be derived from these functional forms. We will first discuss quantum numbers and the orbitals they define, and then the distribution of probability densities associated with the orbitals.

9-7 Quantum Numbers and Electron Orbitals

In the preceding section we stated that by specifying *three* quantum numbers in a wave function ψ we obtain an orbital. Here, we explore the combinations of quantum numbers that produce different orbitals. First, though, we need to learn more about the nature of these three quantum numbers.

Assigning Quantum Numbers

The following relationships involving the three quantum numbers arise from the solution of the Schrödinger wave equation for the hydrogen atom. In this solution the values of the quantum numbers are fixed *in the order listed*.

The first number to be fixed is the *principal* quantum number, n, which may have only a *positive, nonzero integral* value.

$$n = 1, 2, 3, 4, \ldots \tag{9.13}$$

Second is the *orbital angular momentum* quantum number, ℓ, which may be *zero* or a *positive* integer, but not larger than $n - 1$ (where n is the principal quantum number).

$$\ell = 0, 1, 2, 3, \ldots, n - 1 \tag{9.14}$$

Third is the *magnetic* quantum number, m_ℓ. It may be a negative or positive integer, including zero, and ranging from $-\ell$ to $+\ell$ (where ℓ is the orbital angular momentum quantum number).

$$m_\ell = -\ell, -\ell + 1, -\ell + 2, \ldots, 0, 1, 2, \ldots, \ell - 1, +\ell \tag{9.15}$$

▶ Chemists often refer to the three quantum numbers simply as the "n," "ℓ" and "m_ℓ" quantum numbers rather than designating them by name.

EXAMPLE 9-8

Applying Relationships Among Quantum Numbers. Can an orbital have the quantum numbers $n = 2$, $\ell = 2$, and $m_\ell = 2$?

Solution

No. The ℓ quantum number cannot be greater than $n - 1$. Thus, if $n = 2$, ℓ can be only 0 or 1. And if ℓ can be only 0 or 1, m_ℓ cannot be 2; m_ℓ must be 0 if $\ell = 0$ and may be $-1, 0$, or $+1$ if $\ell = 1$.

Practice Example A: Can an orbital have the quantum numbers $n = 3$, $\ell = 0$, and $m_\ell = 0$?

Practice Example B: For an orbital with $n = 3$ and $m_\ell = -1$, what is (are) the possible value(s) of ℓ?

Principal Shells and Subshells

All orbitals with the same value of n are in the same **principal electronic shell** or **principal level**, and all orbitals with the same n and ℓ values are in the same **subshell**, or **sublevel**.

Principal electronic shells are numbered according to the value of n. The *first* principal shell consists of orbitals with $n = 1$; the *second* principal shell, of orbitals with $n = 2$; and so on. The value of n relates to the energy and most probable distance of an electron from the nucleus. The higher the value of n, the greater the electron energy and the farther, on average, the electron is from the nucleus.

The number of subshells in a principal electronic shell is the same as the number of allowed values of the orbital angular momentum quantum number, ℓ. In the first principal shell, with $n = 1$, the only allowed value of ℓ is 0, and there is a single subshell. The second principal shell ($n = 2$), with the allowed ℓ values of 0 and 1, consists of two subshells; the third principal shell ($n = 3$) has three subshells ($\ell = 0, 1,$ and 2); and so on. Or, to put the matter in another way, because there are n possible values of the ℓ quantum number, that is, $0, 1, 2, \ldots (n - 1)$, the number of subshells in a principal shell is equal to the principal quantum number. As a result, there is one subshell in the principal shell with $n = 1$, two subshells in the principal shell with $n = 2$, and so on. The name given to a subshell, regardless of the principal shell in which it is found, depends on the value of the ℓ quantum number. The first four subshells are

s subshell	*p* subshell	*d* subshell	*f* subshell
$\ell = 0$	$\ell = 1$	$\ell = 2$	$\ell = 3$

The number of orbitals in a subshell is the same as the number of allowed values of m_ℓ for the particular value of ℓ. Recall that the allowed values of m_ℓ are 0, $\pm 1, \pm 2, \ldots \pm \ell$, and you can see that the total number of orbitals in a subshell is $2\ell + 1$. The names of the orbitals are the same as the names of the subshells in which they appear.

Quantum Number activity

s orbitals	*p* orbitals	*d* orbitals	*f* orbitals
$\ell = 0$	$\ell = 1$	$\ell = 2$	$\ell = 3$
$m_\ell = 0$	$m_\ell = 0, \pm 1$	$m_\ell = 0, \pm 1, \pm 2$	$m_\ell = 0, \pm 1, \pm 2, \pm 3$
one *s* orbital in an *s* subshell	three *p* orbitals in a *p* subshell	five *d* orbitals in a *d* subshell	seven *f* orbitals in an *f* subshell

To designate the particular principal shell in which a given subshell or orbital is found, we use a combination of a number and a letter. The symbol 1*s* designates the *s* subshell in the first principal shell, but 1*s* is also the designation for the *s* orbital in the first principal shell. The symbol 2*p* is used to designate both the *p* subshell of the second principal shell and any of the three *p* orbitals in that subshell.

The energies of the orbitals for a hydrogen atom, in joules, are given by an equation with a familiar appearance.

$$E_n = -2.178 \times 10^{-18} \left(\frac{1}{n^2} \right) \text{ J}$$

▶ In Section 9-10 and Chapter 25, we will see that orbital energies of multielectron atoms also depend on the quantum numbers ℓ and m_ℓ.

It is the same formula derived by Bohr. Orbital energies for a hydrogen atom depend only on the principal quantum number n. This means that all the subshells within a principal shell have the same energy, as do all the orbitals within a subshell. Orbitals at the same energy level are said to be **degenerate**. Figure 9-22 shows an energy-level diagram and the arrangement of shells and subshells for a hydrogen atom.

Some of the points discussed in the preceding paragraphs are illustrated in Example 9-9.

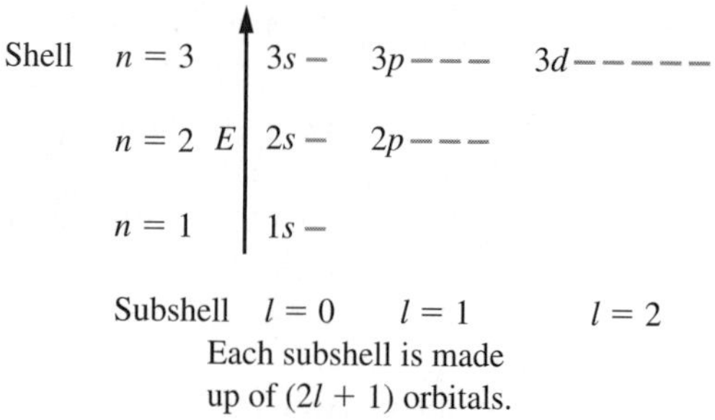

Shell $n = 3$ $3s$ — $3p$ – – – $3d$ – – – – –

$n = 2$ E $2s$ — $2p$ – – –

$n = 1$ $1s$ —

Subshell $l = 0$ $l = 1$ $l = 2$

Each subshell is made
up of $(2l + 1)$ orbitals.

▲ FIGURE 9-22   **Shells and subshells of a hydrogen atom**
The hydrogen atom orbitals are organized into shells and subshells.

EXAMPLE 9-9

Relating Orbital Designations and Quantum Numbers. Write an orbital designation corresponding to the quantum numbers $n = 4$, $\ell = 2$, $m_\ell = 0$.

Solution

The magnetic quantum number, m_ℓ, is not reflected in the orbital designation. The type of orbital is determined by the ℓ quantum number. Because $\ell = 2$, the orbital is of the d type. Because $n = 4$, the orbital designation is $4d$.

Practice Example A: Write an orbital designation corresponding to the quantum numbers $n = 3$, $\ell = 1$, and $m_\ell = 1$.

Practice Example B: Write all the combinations of quantum numbers that define hydrogen-atom orbitals with the same energy as the $3s$ orbital.

9-8 Interpreting and Representing the Orbitals of the Hydrogen Atom

Our next major undertaking will be to describe the three-dimensional probability density distributions obtained for the various orbitals in the hydrogen atom. Using the Born interpretation of wave functions (page 322), we can represent the probability densities of the orbitals of the hydrogen atom using surfaces that encompass most of the electron probability. We shall see that the probability density for each type of orbital has its own distinctive shape, and like all waves, probability densities exhibit nodes and differing phase behavior.

Throughout this discussion, remember that orbitals are wave functions, mathematical solutions of the Schrödinger wave equation. The wave function itself has no physical significance. However, the square of the wave function, ψ^2, is a quantity that we interpret through its relationship to probabilities. Probability density distributions based on ψ^2 are three-dimensional, and it is these three-dimensional regions that we mean when, as is so commonly done, we refer to the shape of an orbital.

▶ In Chapter 12, we will discover important uses of the wave function ψ itself as a basis for discussing bonding between atoms.

In studying this section, it is important for you to remember that, even though we will offer some additional quantitative information about orbitals, your primary concern should be to acquire a broad qualitative understanding. It is this qualitative understanding that you can apply in our later discussion of how orbitals enter into a description of chemical bonding.

The forms of the radial wave function $R(r)$ and the angular wave function $Y(\theta, \phi)$ for a one-electron, hydrogen-like atom are shown in Table 9.1. The first

TABLE 9.1   The Angular and Radial Wave Functions of a Hydrogen-like Atom

Angular Part $Y(\theta, \phi)$	Radial Part $R_{n,\ell}(r)$
$Y(s) = \left(\dfrac{1}{4\pi}\right)^{1/2}$	$R(1s) = 2\left(\dfrac{Z}{a_0}\right)^{3/2} e^{-\sigma/2}$
	$R(2s) = \dfrac{1}{2\sqrt{2}}\left(\dfrac{Z}{a_0}\right)^{3/2}(2 - \sigma)e^{-\sigma/2}$
	$R(3s) = \dfrac{1}{9\sqrt{3}}\left(\dfrac{Z}{a_0}\right)^{3/2}(6 - 6\sigma + \sigma^2)e^{-\sigma/2}$
$Y(p_x) = \left(\dfrac{3}{4\pi}\right)^{1/2}\sin\theta\cos\phi$	$R(2p) = \dfrac{1}{2\sqrt{6}}\left(\dfrac{Z}{a_0}\right)^{3/2}\sigma e^{-\sigma/2}$
$Y(p_y) = \left(\dfrac{3}{4\pi}\right)^{1/2}\sin\theta\sin\phi$	$R(3p) = \dfrac{1}{9\sqrt{6}}\left(\dfrac{Z}{a_0}\right)^{3/2}(4 - \sigma)\sigma e^{-\sigma/2}$
$Y(p_z) = \left(\dfrac{3}{4\pi}\right)^{1/2}\cos\theta$	
$Y(d_{z^2}) = \left(\dfrac{5}{16\pi}\right)^{1/2}(3\cos^2\theta - 1)$	$R(3d) = \dfrac{1}{9\sqrt{30}}\left(\dfrac{Z}{a_0}\right)^{3/2}\sigma^2 e^{-\sigma/2}$
$Y(d_{x^2-y^2}) = \left(\dfrac{15}{4\pi}\right)^{1/2}\sin^2\theta\cos 2\phi$	
$Y(d_{xy}) = \left(\dfrac{15}{4\pi}\right)^{1/2}\sin^2\theta\sin 2\phi$	
$Y(d_{xz}) = \left(\dfrac{15}{4\pi}\right)^{1/2}\sin\theta\cos\theta\cos\phi$	
$Y(d_{yz}) = \left(\dfrac{15}{4\pi}\right)^{1/2}\sin\theta\cos\theta\sin\phi$	
	$\sigma = \dfrac{2Zr}{na_0}$

thing to note is that the angular part of the wave function for an *s* orbital, $\left(\dfrac{1}{4}\pi\right)^{1/2}$, is always the same, regardless of the principal quantum number. Next, note that the angular parts of the *p* orbitals and *d* orbitals are also independent of the quantum number *n*. Therefore all orbitals of a given type (s, p, d, f) have the same angular behavior. Also note that the equations in Table 9.1 are in a general form where the atomic number *Z* is included. This means that the equations apply to any one-electron atom, that is, to a hydrogen atom or a hydrogen-like ion. Finally, note that the term σ appearing throughout the table is equal to $2Zr/na_0$.

To obtain the wave function for a particular state, we simply multiply the radial part by the angular part. We will now illustrate this by looking at the three major types of orbitals.

s Orbitals

To obtain a complete wave function for the hydrogen 1*s* orbital, we use $Z = 1$ and $n = 1$, and combine the angular and radial wave functions.

$$\psi(1s) = \frac{2e^{-r/a_0}}{a_0^{3/2}} \times \frac{1}{\sqrt{4\pi}} = \frac{e^{-r/a_0}}{\sqrt{(\pi a_0^3)}}$$

▶ $\psi^2(1s)$ is the probability density for a 1s electron at *one* point a distance r from the nucleus. Equally important is the probability-density distribution, which gives the total probability for *all* points at a distance r from the nucleus. In Section 9-10 we will see that this distribution is given by $4\pi r^2 \psi^2$.

The term a_0 has the same significance as in the Bohr theory; it is the first Bohr radius—53 pm. By squaring $\psi(1s)$ we obtain an expression for the probability density of finding a 1s electron at a distance r from the nucleus in a hydrogen atom.

$$\psi^2(1s) = \frac{1}{\pi}\left(\frac{1}{a_0}\right)^3 e^{-2r/a_0}$$

The top graph in Figure 9-23a shows ψ^2 for a 1s orbital as a function of distance from the nucleus along a line through the nucleus. The pattern of dots in the middle of Figure 9-23a represents the distribution of electron probability densities in a plane with the nucleus at its center. Because a very low electron charge density exists even at large distances from the nucleus, we cannot represent a 1s orbital encompassing all the electron charge. We can describe the sphere drawn at the bottom of Figure 9-23a as containing about 90% of the electron charge, or as a region in which there is a 90% probability of finding an electron. This is the way a 1s orbital is usually depicted.

Now let's look at the wave function of the 2s orbital.

$$\psi(2s) = \frac{1}{4}\left(\frac{1}{2\pi a_0^3}\right)^{1/2}\left(2 - \frac{r}{a_0}\right)e^{-r/2a_0}$$

The presence of $r/2a_0$ in the exponential term indicates that as r increases, the 2s function decreases in amplitude more slowly than the 1s function (that is, $r/2a_0$ is smaller than r/a_0). This signifies that the 2s electron tends to stay farther from the nucleus than the 1s electron.

The factor $\left(2 - \frac{r}{a_0}\right)$ in the 2s wave function controls the sign of the function.

For small values of r, r/a_0 is smaller than 2 and the wave function is positive, but for large values of r, r/a_0 is larger than 2 and the wave function is negative. At $r = 2a_0$ the preexponential factor is zero and the wave function is said to have a *radial node*. These two factors mean that the probability density sphere of a 2s orbital is bigger than that of a 1s orbital and contains a sphere of zero probability due to the radial node. To encompass 90% of the electron probability for an orbital of the 2s type, we need a larger sphere than for 1s. These features are illustrated in Figure 9-23, which compares the 1s, 2s and 3s orbitals. Note that the 3s orbital exhibits two radial nodes and is larger than both the 1s and 2s orbitals. The fact that the number of nodes increases as the energy is increased is characteristic of high-energy standing waves. We turn our attention now to the *p* orbitals.

p Orbitals

The radial part of $\psi(2p)$ for a hydrogen atom is

$$R(2p) = \frac{1}{2\sqrt{6}}\left(\frac{1}{a_0}\right)^{3/2} r e^{-r/2a_0}$$

▶ The points at which a wave function changes sign are nodes. However, even though the 2p wave function becomes zero at $r = 0$ and $r = \infty$, these points are not true nodes because the function does not change sign at these points. These points are sometimes called *trivial* nodes.

Thus, the 2p orbital has no radial nodes at finite values of r. In contrast to the s orbitals, which are non-zero at $r = 0$, the p orbitals vanish at $r = 0$. This difference will have an important consequence when we consider multielectron atoms.

In contrast to the angular function of the 2s orbital, the angular part of the 2p orbital is not a constant, but a function of θ and ϕ. This means that the probability density distribution of a p orbital is not spherically symmetric; that is, the probability density distribution does not have a spherical shape. We see this most easily in Table 9.1 in the functional form of the angular part of the $2p_z$ wave function; it is proportional to $\cos\theta$. Thus the $2p_z$ wave function has an angular maximum along the positive z axis,

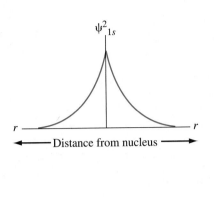

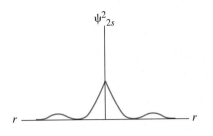

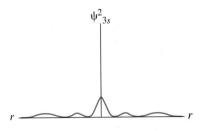

Distance from nucleus

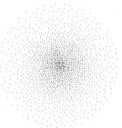

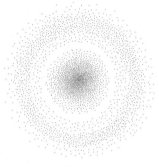

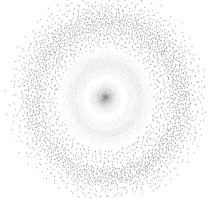

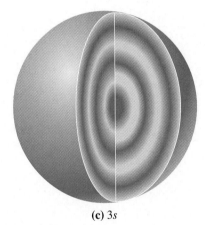

(a) 1s   **(b)** 2s   **(c)** 3s

▲ **FIGURE 9-23   Three representations of the electron probability and charge density for the 1s, 2s, and 3s orbitals**

(a) 1s orbital; **(b)** 2s orbital; **(c)** 3s orbital. *Top:* The electron probability density (ψ^2) or charge density as a function of r. *Middle:* The pattern of dots represents the electron probability or charge density in a plane with the nucleus at its center. The closer the spacing between dots, the higher is the probability of finding an electron and the greater is the electron charge density. *Bottom:* A spherical envelope containing 90% of the electron charge density or a 90% probability of finding an electron.

s Orbital model

for there $\theta = 0$ and $\cos(0) = +1$. Along the negative z axis, the p_z wave function has its most negative value, for there $\theta = \pi$ and $\cos(\pi) = -1$. The fact that the angular part has its maximum magnitude along the z axis is the reason for the designation p_z. Everywhere in the xy plane $\theta = \pi/2$ and $\cos\theta = 0$, so the xy plane is a node. Because this node arises in the angular function, it is called an *angular node.*

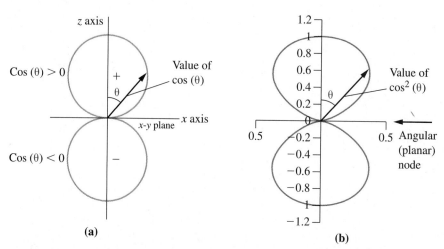

▲ **FIGURE 9-24** **Two representations of the *p* orbital angular function**
(a) A plot of $\cos \theta$ in the zx plane, representing the angular part of the $2p_z$ wave function. Note the difference in the sign of the function in the two lobes. The angular wave function has phase. (b) A plot of $\cos^2 \theta$ in the zx plane, representing the square of the wave function and proportional to the angular probability density of finding the electron.

A similar analysis of the p_x and p_y orbitals shows that they are similar to the p_z orbital, but with angular nodes in the yz and xz planes, respectively.

Figure 9-24 shows the two ways of representing the angular part of the p_z wave function. In Figure 9-24a, $\cos \theta$ is plotted as a function of θ and results in two tangential circles. In Figure 9-24b, the function $\cos^2 \theta$, which is related to the angular probability density, is plotted as a function of θ, resulting in a double teardrop shape. Both these representations are used. What is important to note in part (a) is the phase of the plot of $\cos \theta$ and in (b), the lack of phase of $\cos^2 \theta$, which is always positive. We shall see later in the text that the phase of the orbital is important in understanding chemical bonding.

The simultaneous display of both the radial and angular parts of $\psi^2(2p)$ is more difficult to achieve, but Figure 9-25 attempts to do this for the 90% probability surface of the p_x orbital. All three of the p orbitals are shown in Figure 9-26 and are seen to be directed along the three perpendicular axes of the Cartesian system. To remind ourselves of the phases of the wave functions, we often attach plus and minus signs. However, we must remember that these refer only to the phases of the original wave function, *not* to ψ^2.

d Orbitals

The *d* orbitals occur for the first time when $n = 3$. The angular function in these cases possesses two angular (or planar) nodes. Let's illustrate this with the orbital that has an angular function proportional to

$$\sin^2 \theta \cos 2\phi$$

How should we visualize this function? We can proceed by setting $\theta = \pi/2$ and plotting the function $\cos 2\phi$. The angle $\theta = \pi/2$ corresponds to xy plane, so we obtain the cross section shown in Figure 9-27. The wave function exhibits positive and negative lobes along the x and y axes. This orbital, in common with all the other d orbitals, is a function of two of the three variables (x, y, and z). It is designated

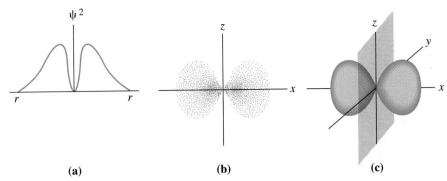

▲ **FIGURE 9-25**   **Three representations of electron probability and charge density for a 2*p* orbital**

(a) The value of ψ^2 is zero at the nucleus, rises to a maximum on either side, and then falls off with distance (r) along a line through the nucleus (i.e., along the x, y, or z axis). (b) The dots represent electron probability and charge density in a *plane* passing through the nucleus, for example, the xz plane. (c) Electron probabilities and charge densities represented in three dimensions. The greatest probability of finding an electron is within the two lobes of the dumbbell-shaped region. Note that this region is *not* spherically symmetric. Note also that the probability drops to zero in the shaded plane—the nodal plane (the yz plane).

p Orbital model

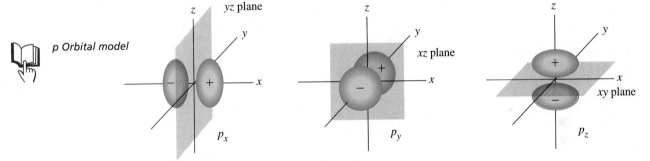

▲ **FIGURE 9-26**   **The three 2*p* orbitals**

The *p* orbitals are usually represented as directed along the perpendicular x, y, and z axes, and the symbols p_x, p_y, and p_z are often used. The p_z orbital has $m_\ell = 0$. The situation with p_x and p_y is more complex: Each of these orbitals has contributions from both $m_\ell = 1$ and $m_\ell = -1$. Our main concern is just to recognize that *p* orbitals occur in sets of three and can be represented in the orientation shown here. In higher-numbered shells, *p* orbitals have a somewhat different appearance, but we will use these general shapes for all *p* orbitals. The signs on the lobes are the phases of the original wave function and are not to be confused with electric charges.

$d_{x^2-y^2}$. The other *d* orbitals, d_{xy}, d_{xz}, d_{yz}, and d_{z^2}, are also displayed in cross section in Figure 9-27. We observe that four of them have the same basic shape except for orientation with respect to the axes and that the d_{z^2} has quite a different shape.

The 90% probability surfaces of the five *d* orbitals are shown in Figure 9-28. Two of the *d* orbitals ($d_{x^2-y^2}$ and d_{z^2}) are seen to be directed along the three perpendicular axes of the Cartesian system, and the remaining three (d_{xy}, d_{xz}, d_{yz}) are seen to point between these Cartesian axes. The *d* orbitals are important in understanding the chemistry of the transition elements.

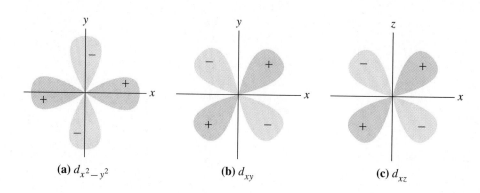

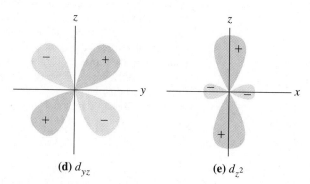

d Orbital model

▲ FIGURE 9-27 **Cross sections of the five *d* orbitals**
The two-dimensional cross sections of the angular functions of the five *d* orbitals in the planes indicated.

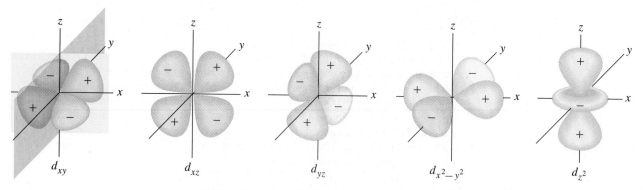

▲ FIGURE 9-28 **Representations of the five *d* orbitals**
The designations *xy*, *xz*, *yz*, and so on are related to the values of the quantum number m_ℓ, but this is a detail that we will not pursue in the text. The number of nodal surfaces for an orbital is equal to the ℓ quantum number. For *d* orbitals, there are two such surfaces. The nodal planes for the d_{xy} orbital are shown here. (The nodal surfaces for the d_{z^2} orbital are actually cone-shaped.)

9-9 Electron Spin: A Fourth Quantum Number

Wave mechanics provides three quantum numbers with which we can develop a description of electron orbitals. But we need a fourth quantum number as well. In 1925, George Uhlenbeck and Samuel Goudsmit proposed that some unexplained features of the hydrogen spectrum could be understood by assuming that an electron acts as if it spins, much as Earth spins on its axis. As suggested by Figure 9-29, there are two possibilities for **electron spin**. The electron spin quantum number, m_s, may have a value of $+\frac{1}{2}$ (also denoted by the arrow ↑) or $-\frac{1}{2}$ (denoted by the arrow ↓); the value of m_s does not depend on any of the other three quantum numbers.

What is the evidence that the phenomenon of electron spin exists? An experiment by Otto Stern and Walter Gerlach in 1920, though designed for another purpose, seems to yield this proof (Figure 9-30). Silver was vaporized in a furnace, and a beam of silver atoms was passed through a nonuniform magnetic field. The beam split in two. Here is a simplified explanation.

1. An electron, because of its spin, generates a magnetic field.
2. A pair of electrons with opposing spins has no net magnetic field.
3. In a silver atom, 23 electrons have a spin of one type and 24 of the opposite type. The direction of the net magnetic field produced depends only on the spin of the unpaired electron.
4. In a beam of a large number of silver atoms there is an equal chance that the unpaired electron will have a spin of $+\frac{1}{2}$ or $-\frac{1}{2}$. The magnetic field induced by the silver atoms interacts with the nonuniform field, and the beam of silver atoms splits into two beams.

▶ **FIGURE 9-29**
Electron spin visualized
Two possibilities for electron spin are shown with their associated magnetic fields. Two electrons with opposing spins have opposing magnetic fields that cancel, leaving no net magnetic field for the pair.

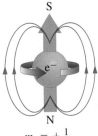

$m_s = +\frac{1}{2}$

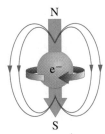

$m_s = -\frac{1}{2}$

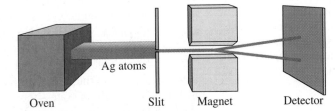

Oven   Ag atoms   Slit   Magnet   Detector

▲ **FIGURE 9-30   The Stern–Gerlach experiment**
Ag atoms vaporized in the oven are collimated into a beam by the slit, and the beam is passed through a nonuniform magnetic field. The beam splits in two. (The beam of atoms would not experience a force if the magnetic field were uniform. The field strength must be stronger in certain directions than in others.)

Electronic Structure of the H Atom: Representing the Four Quantum Numbers

Now that we have described the four quantum numbers, we are in a position to bring them together into a description of the electronic structure of the hydrogen atom. The electron in a ground-state hydrogen atom is found at the lowest energy level. This corresponds to the principal quantum number $n = 1$, and because the first principal shell consists only of an s orbital, the orbital quantum number $\ell = 0$. The only possible value of the magnetic quantum number is $m_\ell = 0$. Either spin state is possible for the electron, and we don't know which it is unless we do an experiment like that of Uhlenbeck and Goudsmit. Thus,

$$n = 1 \qquad \ell = 0 \qquad m_\ell = 0 \qquad m_s = +\frac{1}{2} \text{ or } -\frac{1}{2}$$

KEEP IN MIND ▶
that orbitals are mathematical functions and not themselves physical regions in space. However, it is customary to refer to an electron that is described by a particular orbital as being "in the orbital."

Chemists often say that the electron in the ground-state hydrogen atom is in the $1s$ orbital, or that it is a $1s$ electron, and they represent this by the notation

$$1s^1$$

where the superscript 1 indicates one electron in the $1s$ orbital. Either spin state is allowed, but we do not designate the spin state in this notation.

In the excited states of the hydrogen atom, the electron occupies orbitals with higher values of n. Thus, when excited to the level with $n = 2$, the electron can occupy either the $2s$ or one of the $2p$ orbitals; all have the same energy. Because the probability density extends farther from the nucleus in the $2s$ and $2p$ orbitals than in the $1s$ orbital, the excited-state atom is larger than is the ground-state atom. The excited states just described can be represented as

$$2s^1 \text{ or } 2p^1$$

Our task in the remaining sections of the chapter will be to extend this discussion to the electronic structures of atoms having more than one electron—*multielectron* atoms.

9-10 Multielectron Atoms

Schrödinger developed his wave equation for the hydrogen atom—an atom containing just one electron. For multielectron atoms, a new factor arises: mutual repulsion between electrons. The repulsion between the electrons means that the electrons in a multielectron atom try to stay away from each other and their motions become inextricably entangled. The approximate approach taken to solve this many-particle problem is to consider the electrons, one by one, in the environment established by the nucleus and other electrons. When this is done, the electron orbitals obtained are of the same types as those obtained for the hydrogen atom; they are called *hydrogen-like* orbitals. Compared to the hydrogen atom, the angular parts of the orbitals of a multielectron atom are unchanged, but the radial parts are different.

KEEP IN MIND ▶
that the orbital wave functions extend farther out from the nucleus as n increases. Thus, an electron in a $3s$ or $3p$ orbital has a higher probability of being farther from the nucleus than does an electron in a $1s$ orbital.

We have seen that the solution of the Schrödinger equation for a hydrogen atom gives the energies of the orbitals and that all orbitals with the same principal quantum number n are degenerate—they have the same energy. In a hydrogen atom, the orbitals $2s$ and $2p$ are degenerate, as are $3s$, $3p$, and $3d$.

In multielectron atoms, the attractive force of the nucleus for a given electron increases as the nuclear charge increases. As a result, we find that orbital energies become lower (more negative) with increasing atomic number of the atom. Also, orbital energies in multielectron atoms depend on the type of orbital; the orbitals with different values of ℓ within a principal shell are not degenerate.

Penetration and Shielding

Think about the attractive force of the atomic nucleus for one particular electron some distance from the nucleus. Electrons in orbitals closer to the nucleus *screen* or *shield* the nucleus from electrons farther away. In effect, the screening electrons reduce the effectiveness of the nucleus in attracting the particular more distant electron. They effectively reduce the nuclear charge.

The magnitude of the reduction of the nuclear charge depends on the types of orbitals the inner electrons are in and the type of orbital that the screened electron is in. We have seen that electrons in *s* orbitals have a high probability density at the nucleus, whereas *p* and *d* orbitals have zero probability densities at the nucleus. Thus electrons in *s* orbitals are more effective at screening the nucleus from outer electrons than are electrons in *p* or *d* orbitals. This ability of electrons in *s* orbitals that allows them to get close to the nucleus is called *penetration.* An electron in an orbital with good penetration is better at screening than one with low penetration.

We need to consider a different kind of probability distribution to describe the penetration to the nucleus by orbital electrons. Rather than considering the probability at a point, which we did to ascribe three-dimensional shapes to orbitals, we need to consider the probability of finding the electron anywhere in a spherical shell of radius *r* and an infinitesimal thickness. This type of probability is called a *radial probability distribution* and is found by multiplying the radial probability density, $R^2(r)$, by the factor $4\pi r^2$, the area of a sphere of radius *r*. Figure 9-31 offers a dartboard analogy that might help clarify the distinction between probability at a point and probability in a region of space.

The quantity $4\pi r^2 \times R^2(r)$ provides a different insight into the behavior of the electron. The radial probability distributions for some hydrogenic (hydrogen-like) orbitals are plotted in Figure 9-32. The radial probability density $\left[R^2(r)\right]$ for a 1*s* orbital predicts that the maximum probability for a 1*s* electron is *at* the nucleus. However, because the volume of this region is vanishingly small ($r = 0$), the radial probability distribution is zero at the nucleus. The electron in a hydrogen atom is most likely to be found 53 pm from the nucleus; this is where the radial probability

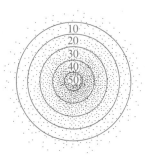

▲ **FIGURE 9-31**

Dartboard analogy to a 1s orbital

Imagine that a single dart (electron) is thrown at a dartboard 1500 times. The board contains 90% of all the holes; it is analogous to the 1*s* orbital. Where is a thrown dart most likely to hit? The number of holes per unit area is greatest in the "50" region—that is, the 50 region has the greatest probability density. The most likely score is "30," however, because the most probable area hit is in the "30" ring and not the 50 ring, which is smaller than the 30 ring. The 30 *ring* on the dartboard is analogous to a spherical *shell* of 53 pm radius within the larger sphere representing the 1*s* orbital.

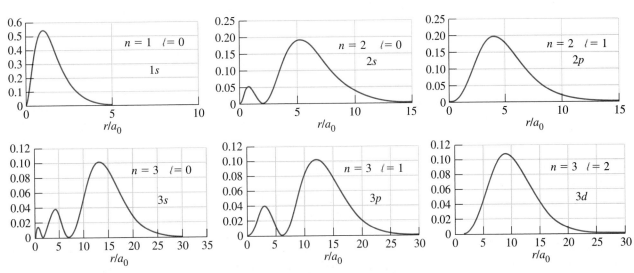

Electron Shielding activity

▲ **FIGURE 9-32 Radial probability distributions**

The value of $4\pi r^2 \ R^2(r)$ for the orbitals in the first three principal shells. Note that the smaller its orbital angular momentum quantum number, the more closely an electron approaches the nucleus. Thus, *s* electrons penetrate more and are less shielded from the nucleus than electrons in other orbitals with the same value of *n*.

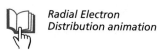

Radial Electron Distribution animation

distribution reaches a maximum. This is the same radius as the first Bohr orbit. The boundary surface within which there is a 90% probability of finding an electron (see Figure 9-23) is a much larger sphere, one with a radius of about 141 pm.

In comparing the radial probability curves for the $1s$, $2s$ and $3s$ orbitals, we find that a $1s$ electron has a greater probability of being close to the nucleus than a $2s$ electron, which in turn has a greater probability than a $3s$ electron. In comparing $2s$ and $2p$ orbitals, a $2s$ electron has a greater chance of being close to the nucleus than a $2p$ electron. The $2s$ electron exhibits greater penetration than the $2p$ electron. Electrons having a high degree of penetration effectively "block the view" of an electron in an outer orbital "looking" for the nucleus.

The nuclear charge that an electron would experience if there were no intervening electrons is Z, the atomic number. The nuclear charge that an electron actually experiences is reduced by intervening electrons to a value of Z_{eff}, called the **effective nuclear charge**. The less of the nuclear charge that an outer electron "sees" (that is, the smaller the value of Z_{eff}), the smaller is the attraction of the electron to the nucleus, and hence the *higher* is the energy of the orbital in which the electron is found.

To summarize, compared to a p electron in the same principal shell, an s electron is more penetrating and not as well screened. The s electron experiences a higher Z_{eff}, is held more tightly, and is at a lower energy than a p electron. Similarly, the p electron is at a lower energy than a d electron in the same principal shell. Thus, the energy level of a principal shell is split into separate levels for its subshells. There is no further splitting of energies within a subshell, however, because all the orbitals in the subshell have the same radial characteristics, and thereby experience the same effective nuclear charge, Z_{eff}. As a result, all three p orbitals of a principal shell have the same energy; all five d orbitals have the same energy; and so on.

In a few instances, the combined effect of the decreased spacing between successive energy levels at higher quantum numbers (due to the energy dependence on $1/n^2$) and the splitting of subshell energy levels (due to shielding and penetration) causes some energy levels to overlap. For example, because of the extra penetration of a $4s$ electron over that of a $3d$ electron, the $4s$ energy level is below the $3d$ level despite its higher principal quantum number n (Figure 9-33). In the next sections, we will see that substantial experimental support for this effect is found in the relationship between the electronic structures of atoms and their positions in the periodic table.

KEEP IN MIND ▶

that, similar to the situation in equation 9.9, the energy of an orbital $\left(E_n\right)$ is given by the proportionality $E_n \propto -\dfrac{Z_{\text{eff}}^2}{n^2}$.

9-11 Electron Configurations

The **electron configuration** of an atom is a designation of how electrons are distributed among various orbitals in principal shells and subshells. In later chapters, we will find that many of the physical and chemical properties of elements can be correlated with electron configurations. In this section, we will see how the results of wave mechanics, expressed as a set of rules, can help us to write probable electron configurations for the elements.

Rules for Assigning Electrons to Orbitals

1. **Electrons occupy orbitals in a way that minimizes the energy of the atom.** Figure 9-33 suggests the order in which electrons occupy the subshells in the principal electronic shells, first the $1s$, then $2s$, $2p$, and so on. The exact order of filling of orbitals has been established by *experiment*, principally through spectroscopy and magnetic studies, and it is this order based on experiment that we must follow in assigning electron configurations to the elements. With only a few exceptions, the order in which orbitals fill is

This order of filling corresponds roughly to the order of increasing orbital energy, but the overriding principle governing the order of filling of orbitals is that the energy of the atom as a whole be kept at a minimum. ▶

$$1s, 2s, 2p, 3s, 3p, 4s, 3d, 4p, 5s, 4d, 5p, 6s, 4f, 5d, 6p, 7s, 5f, 6d, 7p \qquad (9.16)$$

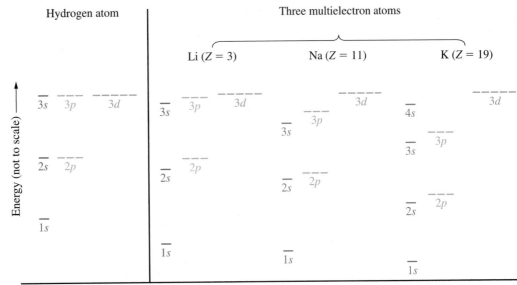

▲ **FIGURE 9-33** **Orbital energy diagram for the first three electronic shells**
Energy levels are shown for a hydrogen atom (left) and three typical multielectron atoms (right). Each multielectron atom has its own energy-level diagram. Note that for the hydrogen atom, orbital energies within a principal shell—for example, 3*s*, 3*p*, 3*d*—are alike (degenerate), but in a multielectron atom they become rather widely separated. Another feature of the diagram described in the text is the steady decrease in all orbital energies with increasing atomic number. Finally, note that the 4*s* orbital is at a lower energy than 3*d*.

Some students find the diagram pictured in Figure 9-34 a useful way to remember this order, but the best method of establishing the order of filling of orbitals is based on the periodic table, as we will see in Section 9-12.

2. **No two electrons in an atom may have all four quantum numbers alike— the Pauli exclusion principle.** In 1926, Wolfgang Pauli explained complex features of emission spectra associated with atoms in magnetic fields by proposing that no two electrons in an atom can have all four quantum numbers alike. The first three quantum numbers, n, ℓ, and m_ℓ, determine a specific orbital. Two electrons may have these three quantum numbers alike; but if they do, they must have different values of m_s, the spin quantum number. Another way to state this result is that *only two electrons may occupy the same orbital and these electrons must have opposing spins.*

Because of this limit of two electrons per orbital, the capacity of a subshell for electrons can be obtained by doubling the number of orbitals in the subshell. Thus, the *s* subshell consists of *one* orbital with a capacity of two electrons; the *p* subshell consists of *three* orbitals with a total capacity of six electrons; and so on.

3. **When orbitals of identical energy (degenerate orbitals) are available, electrons initially occupy these orbitals singly.** In line with this rule, known as **Hund's rule**, an atom tends to have as many unpaired electrons as possible. This behavior can be rationalized by saying that electrons, because they all carry the same electric charge, try to get as far apart as possible. They do this by seeking out empty orbitals of similar energy in preference to pairing up with an electron in a half-filled orbital.

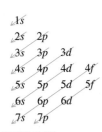

▲ **FIGURE 9-34**
The order of filling of electronic subshells
Beginning with the top line, follow the arrows, and the order obtained is the same as in expression (9.16).

Representing Electron Configurations

Before we assign electron configurations to atoms of the different elements, we need to introduce methods of representing these configurations. The electron configuration of an atom of carbon is shown in three different ways:

▶ When listed in tables, as in Appendix D, electron configurations are usually written in the condensed *spdf* notation.

spdf notation (condensed):	C	$1s^2 2s^2 2p^2$
spdf notation (expanded):	C	$1s^2 2s^2 2p_x^1 2p_y^1$
orbital diagram:	C	

1s   2s   2p

In each of these methods we assign *six* electrons because the atomic number of carbon is 6. Two of these electrons are in the $1s$ subshell, two in the $2s$, and two in the $2p$. The condensed ***spdf*** **notation** denotes only the total number of electrons in each subshell; it does not show how electrons are distributed among orbitals of equal energy. In the expanded *spdf* notation, Hund's rule is reflected in the assignment of electrons to the $2p$ subshell—two $2p$ orbitals are singly occupied and one remains empty. The **orbital diagram** breaks down each subshell into individual orbitals (drawn as boxes). This notation is similar to an energy-level diagram, except that the direction of increasing energy is from left to right instead of vertically.

Electrons in orbitals are shown as arrows. An arrow pointing up corresponds to one type of spin $\left(+\dfrac{1}{2}\right)$, and an arrow pointing down to the other $\left(-\dfrac{1}{2}\right)$. Electrons in the same orbital with opposing (opposite) spins are said to be *paired* (↑↓). The electrons in the $1s$ and $2s$ orbitals of the carbon atom are paired. Electrons in different singly occupied orbitals of the same subshell have the same or *parallel* spins (arrows pointing in the same direction). This fact is conveyed in the orbital diagram for carbon, where we write [↑][↑][ ] rather than [↑][↓][ ] for the $2p$ subshell. Both experiment and theory confirm that an electron configuration in which electrons in singly occupied orbitals have parallel spins is a better representation of the lowest energy state of an atom than any other electron configuration that we can write.

The most stable or the most energetically favorable configurations for isolated atoms, those discussed here, are called *ground-state electron configurations*. Later in the text we will briefly mention some electron configurations that are not the most stable. Atoms with such configurations are said to be in an *excited state*.

The Aufbau Process

Electron Configurations movie

To write electron configurations we will use the **aufbau process**. *Aufbau* is a German word that means "building up," and what we do is assign electron configurations to the elements in order of increasing atomic number. To proceed from one atom to the next, we add a proton and some neutrons to the nucleus and then describe the orbital into which the added electron goes.

Z = 1, H. The lowest energy state for the electron is the $1s$ orbital. The electron configuration is $1s^1$.

Z = 2, He. A second electron goes into the $1s$ orbital, and the two electrons have opposing spins, $1s^2$.

Z = 3, Li. The third electron cannot be accommodated in the $1s$ orbital (Pauli exclusion principle). It goes into the lowest energy orbital available, $2s$. The electron configuration is $1s^2 2s^1$.

Z = 4, Be. The configuration is $1s^2 2s^2$.

Z = 5, B. Now the 2p subshell begins to fill: $1s^2 2s^2 2p^1$.

Z = 6, C. A second electron goes into the 2p subshell, but into one of the remaining empty p orbitals (Hund's rule), and with a spin parallel to the first 2p electron.

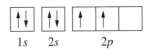

Z = 7–10, N through Ne. In this series of four elements, the filling of the 2p subshell is completed. The number of unpaired electrons reaches a maximum (3) with nitrogen and then decreases to zero with neon.

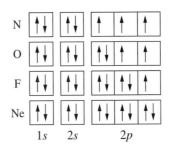

Z = 11–18, Na through Ar. This series of eight elements closely parallels what we have already seen for the eight elements from Li through Ne, except that electrons go into 3s and 3p subshells. Each element has the 1s, 2s, and 2p subshells filled. Because the configuration $1s^2 2s^2 2p^6$ is that of neon, we will call this the neon core, represent it as [Ne], and concentrate on the electrons beyond the core. Electrons that are added to the electronic shell of highest principal quantum number (the outermost or valence shell) are called **valence electrons**. The electron configuration of Na is written below in a form called a *noble-gas-core*-abbreviated electron configuration, consisting of [Ne] as the noble gas core and $3s^1$ as the configuration of the valence electron. For the other third period elements, only the valence-shell electron configurations are shown.

Na	Mg	Al	Si	P	S	Cl	Ar
[Ne]$3s^1$	$3s^2$	$3s^2 3p^1$	$3s^2 3p^2$	$3s^2 3p^3$	$3s^2 3p^4$	$3s^2 3p^5$	$3s^2 3p^6$

Z = 19 and 20, K and Ca. After argon, instead of 3d, the next subshell to fill is 4s. Using the symbol [Ar] to represent the noble gas core, $1s^2 2s^2 2p^6 3s^2 3p^6$, we get the electron configurations shown below for K and Ca.

$$\text{K: } [Ar]4s^1 \quad \text{and} \quad \text{Ca: } [Ar]4s^2$$

Z = 21–30, Sc through Zn. In this next series of elements, electrons fill the d orbitals of the third shell. The d subshell has a total capacity of ten electrons—ten elements are involved. We have two possible ways to write the electron configuration of scandium.

$$\text{(a) Sc: } [Ar]3d^1 4s^2 \quad \text{or} \quad \text{(b) Sc: } [Ar]4s^2 3d^1$$

▶ Although method (b) conforms better to the order in which orbitals fill, method (a) better represents the order in which electrons are lost on ionization, as we will see in the next chapter.

Both methods are commonly used. Method (a) groups together all the subshells of a principal shell and places last subshells of the highest principal quantum level. Method (b) lists orbitals in the apparent order in which they fill. In this text, we will use method (a).

The electron configurations of this series of ten elements are listed on page 340 in both the orbital diagram and the *spdf* notation.

		3d						4s	
Sc:	[Ar]	↑						↑↓	$[\text{Ar}]3d^14s^2$
Ti:	[Ar]	↑	↑					↑↓	$[\text{Ar}]3d^24s^2$
V:	[Ar]	↑	↑	↑				↑↓	$[\text{Ar}]3d^34s^2$
Cr:	[Ar]	↑	↑	↑	↑	↑		↑	$[\text{Ar}]3d^54s^1$
Mn:	[Ar]	↑	↑	↑	↑	↑		↑↓	$[\text{Ar}]3d^54s^2$
Fe:	[Ar]	↑↓	↑	↑	↑	↑		↑↓	$[\text{Ar}]3d^64s^2$
Co:	[Ar]	↑↓	↑↓	↑	↑	↑		↑↓	$[\text{Ar}]3d^74s^2$
Ni:	[Ar]	↑↓	↑↓	↑↓	↑	↑		↑↓	$[\text{Ar}]3d^84s^2$
Cu:	[Ar]	↑↓	↑↓	↑↓	↑↓	↑↓		↑	$[\text{Ar}]3d^{10}4s^1$
Zn:	[Ar]	↑↓	↑↓	↑↓	↑↓	↑↓		↑↓	$[\text{Ar}]3d^{10}4s^2$

The d orbitals fill in a fairly regular fashion in this series, but there are two exceptions: chromium (Cr) and copper (Cu). These exceptions are usually explained in terms of a special stability for configurations in which a $3d$ subshell is half-filled with electrons, as with Cr $\left(3d^5\right)$, or completely filled, as with Cu $\left(3d^{10}\right)$.

Z = 31–36, Ga through Kr. In this series of six elements, the $4p$ subshell is filled, ending with krypton.

$$\text{Kr: } [\text{Ar}]3d^{10}4s^24p^6$$

Z = 37–54, Rb to Xe. In this series of 18 elements, the subshells fill in the order $5s$, $4d$, and $5p$, ending with the configuration of xenon.

$$\text{Xe: } [\text{Kr}]4d^{10}5s^25p^6$$

Z = 55–86, Cs to Rn. In this series of 32 elements, with a few exceptions, the subshells fill in the order $6s$, $4f$, $5d$, $6p$. The configuration of radon is

$$\text{Rn: } [\text{Xe}]4f^{14}5d^{10}6s^26p^6$$

Electron Configuration activity

Z = 87–?, Fr to ?. Francium starts a series of elements in which the subshells that fill are $7s$, $5f$, $6d$, and presumably $7p$, although atoms in which filling of the $7p$ subshell presumably occurs have only recently been discovered and not yet characterized.

Appendix D gives a complete listing of probable electron configurations.

9-12 Electron Configurations and the Periodic Table

We have just described the aufbau process of making probable assignments of electrons to the orbitals in atoms. Although electron configurations may seem rather abstract, they actually lead us to a better understanding of the periodic table. Around 1920, Niels Bohr began to promote the connection between the periodic table and quantum theory. The chief link, he pointed out, is in electron configurations. *Elements in the same group of the table have similar electron configurations.*

Group	Element	Configuration
TABLE 9.2		Electron Configurations of Some Groups of Elements
1	H	$1s^1$
	Li	$[He]2s^1$
	Na	$[Ne]3s^1$
	K	$[Ar]4s^1$
	Rb	$[Kr]5s^1$
	Cs	$[Xe]6s^1$
	Fr	$[Rn]7s^1$
17	F	$[He]2s^22p^5$
	Cl	$[Ne]3s^23p^5$
	Br	$[Ar]3d^{10}4s^24p^5$
	I	$[Kr]4d^{10}5s^25p^5$
	At	$[Xe]4f^{14}5d^{10}6s^26p^5$
18	He	$1s^2$
	Ne	$[He]2s^22p^6$
	Ar	$[Ne]3s^23p^6$
	Kr	$[Ar]3d^{10}4s^24p^6$
	Xe	$[Kr]4d^{10}5s^25p^6$
	Rn	$[Xe]4f^{14}5d^{10}6s^26p^6$

To construct Table 9.2, we have taken three groups of elements from the periodic table and written their electron configurations. The similarity in electron configuration within each group is readily apparent. If we label the shell of highest principal quantum number—the outermost, or valence, shell—as n, then

▶ Hydrogen is found in group 1 because of its electron configuration, $1s^1$. However, it is not an alkali metal.

- The group 1 atoms (alkali metals) have a *single* outer-shell (valence) electron in an s orbital, that is, ns^1.
- The group 17 atoms (halogens) have *seven* outer-shell (valence) electrons, in the configuration ns^2np^5.
- The group 18 atoms (noble gases)—with the exception of helium, which has only two electrons—have outermost shells with *eight* electrons, in the configuration ns^2np^6.

Although it is not correct in all details, Figure 9-35 relates the aufbau process to the periodic table by dividing the table into the following four blocks of elements, according to the subshells being filled.

- ***s block.*** The s orbital of highest principal quantum number (n) fills. The s block consists of groups 1 and 2.
- ***p block.*** The p orbitals of highest quantum number (n) fill. The p block consists of groups 13, 14, 15, 16, 17, and 18.
- ***d block.*** The d orbitals of the electronic shell $n-1$ (the next to outermost) fill. The d block includes groups 3, 4, 5, 6, 7, 8, 9, 10, 11, and 12.
- ***f block.*** The f orbitals of the electronic shell $n-2$ fill. The f block elements are the lanthanides and the actinides.

Another point to notice from Table 9.2 is that the electron configuration consists of a noble gas core corresponding to the noble gas from the previous period plus the additional electrons required to satisfy the atomic number. Recognizing this and dividing the periodic table into blocks can simplify the task of assigning electron configurations. For example, strontium is in group 2, the second s block group, so that its valence-shell configuration is $5s^2$ since it is in the fifth period. The remaining

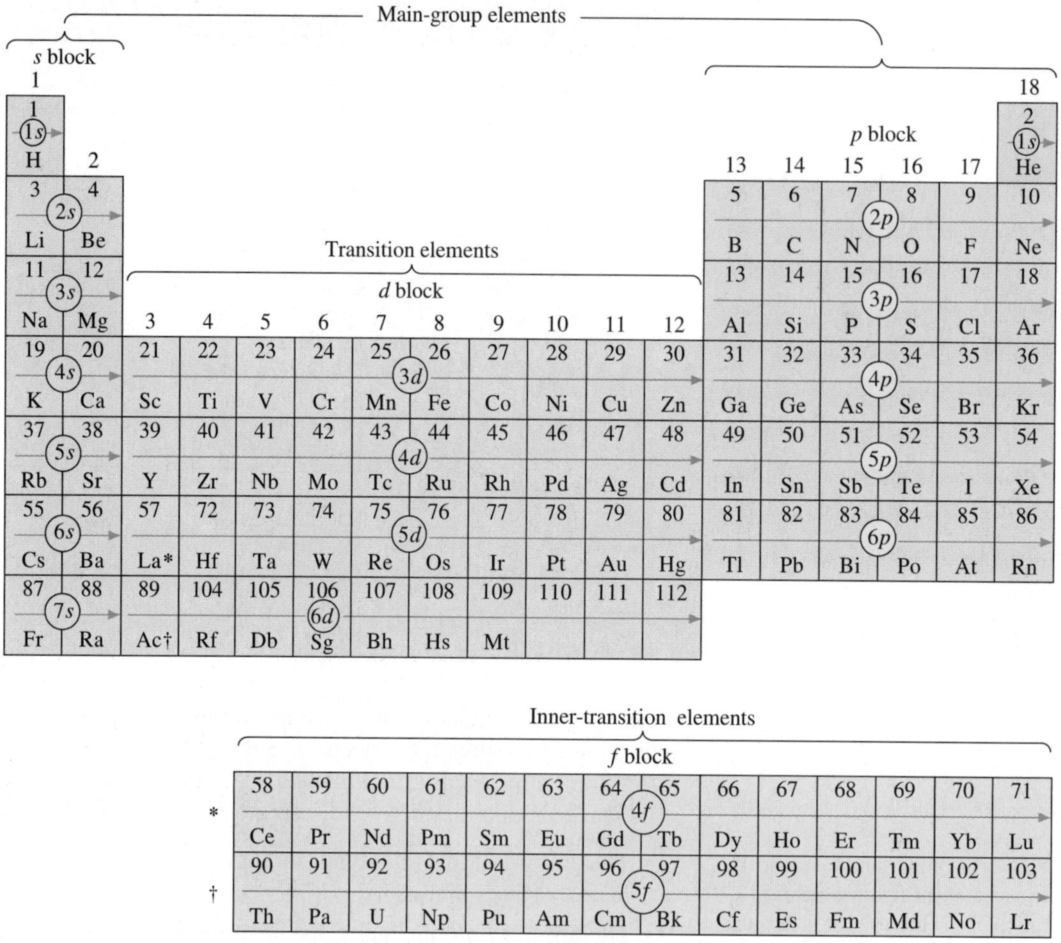

Main-group elements

s block

1																	18
1 ⓛ$1s$	2											p block					2 ⓛ$1s$
H												13	14	15	16	17	He
3	4 ②$2s$											5	6	7 ②$2p$	8	9	10
Li	Be				Transition elements							B	C	N	O	F	Ne
11	12 ③$3s$				d block							13	14	15	16	17	18
Na	Mg	3	4	5	6	7	8	9	10	11	12	Al	Si	P ③$3p$	S	Cl	Ar
19	20 ④$4s$	21	22	23	24	25 ③$3d$	26	27	28	29	30	31	32	33 ④$4p$	34	35	36
K	Ca	Sc	Ti	V	Cr	Mn	Fe	Co	Ni	Cu	Zn	Ga	Ge	As	Se	Br	Kr
37	38 ⑤$5s$	39	40	41	42	43 ④$4d$	44	45	46	47	48	49	50	51 ⑤$5p$	52	53	54
Rb	Sr	Y	Zr	Nb	Mo	Tc	Ru	Rh	Pd	Ag	Cd	In	Sn	Sb	Te	I	Xe
55	56 ⑥$6s$	57	72	73	74	75 ⑤$5d$	76	77	78	79	80	81	82	83 ⑥$6p$	84	85	86
Cs	Ba	La*	Hf	Ta	W	Re	Os	Ir	Pt	Au	Hg	Tl	Pb	Bi	Po	At	Rn
87	88 ⑦$7s$	89	104	105	106 ⑥$6d$	107	108	109	110	111	112						
Fr	Ra	Ac†	Rf	Db	Sg	Bh	Hs	Mt									

Inner-transition elements

f block

	58	59	60	61	62	63	64 ④$4f$	65	66	67	68	69	70	71
*	Ce	Pr	Nd	Pm	Sm	Eu	Gd	Tb	Dy	Ho	Er	Tm	Yb	Lu
	90	91	92	93	94	95	96 ⑤$5f$	97	98	99	100	101	102	103
†	Th	Pa	U	Np	Pu	Am	Cm	Bk	Cf	Es	Fm	Md	No	Lr

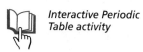

Interactive Periodic Table activity

▲ **FIGURE 9-35   Electron configurations and the periodic table**

To use this figure as a guide to the aufbau process, locate the position of an element in the table. Subshells listed ahead of this position are filled. For example, germanium $(Z = 32)$ is located in group 14 of the blue $4p$ row. The filled subshells are $1s^2$, $2s^2$, $2p^6$, $3s^2$, $3p^6$, $4s^2$, and $3d^{10}$. At $Z = 32$, a second electron has entered the $4p$ subshell. The electron configuration of Ge is $[Ar]3d^{10}4s^24p^2$.

Exceptions to the orderly filling of subshells suggested here are found among a few of the d block and some of the f block elements.

electrons are in the krypton core configuration, the noble gas in the previous period, thus the electron configuration of Sr is

$$\text{Sr: } [\text{Kr}]5s^2$$

For the p block elements in groups 13 to 18, the number of valence electrons is from 1 to 6. For example, aluminum is in period 3 and group 13, its valence-shell electron configuration is $3s^23p^1$. We use $n = 3$ since it is in the third period and we have to accommodate three electrons after the neon core, which contains 10 electrons. Thus the electron configuration of Al is

$$\text{Al: } [\text{Ne}]3s^23p^1$$

Gallium is also in group 13, but in period 4. Its valence-shell electron configuration is $4s^24p^1$. To write the electron configuration of Ga, we can start with the electron configuration of the noble gas that closes the third period, argon, and we add to it the subshells that fill in the fourth period: $4s$, $3d$, and $4p$. The $3d$ subshell must fill with 10 electrons before the $4p$ subshell begins to fill. Consequently, the electron configuration of gallium must be

$$\text{Ga: } [Ar]3d^{10}4s^24p^1$$

Thallium is in group 13 and period 6. Its valence-shell electron configuration is $6s^26p^1$. Again, we indicate the electron configuration of the noble gas that closes the fifth period as a core and add the subshells that fill in the sixth period: $6s$, $4f$, $5d$, and $6p$.

$$\text{Tl: } [Xe]4f^{14}5d^{10}6s^26p^1$$

The elements in group 13 have the common valence configuration ns^2np^1, again illustrating the repeating pattern of valence electron configurations down a group, which is the basis of the similar chemical properties of the elements within a group of the periodic table.

The transition elements correspond to the d block, and their electron configurations are established in a similar manner. To write the electron configuration of a transition element, we start with the electron configuration of the noble gas that closes the period prior to the period of the transition element and add the subshells that fill in the period of the transition element being considered. The s subshell fills immediately after the preceding noble gas; most transition metal atoms have two electrons in the s subshell of the valence shell, but some have only one. Thus, vanadium $(Z = 23)$, which has two valence electrons in the $4s$ subshell and core electrons in the configuration of the noble gas argon, must have *three* $3d$ electrons $(2 + 18 + 3 = 23)$.

$$\text{V: } [Ar]3d^34s^2$$

Chromium $(Z = 24)$, as we have seen before, has only one valence electron in the $4s$ subshell and core electrons in the argon configuration. Consequently it must have *five* $3d$ electrons $(1 + 18 + 5 = 24)$.

$$\text{Cr: } [Ar]3d^54s^1$$

Copper $(Z = 29)$ also has only one valence electron in the $4s$ subshell in addition to its argon core, so the copper atom must have *ten* $3d$ electrons $(1 + 18 + 10 = 29)$.

$$\text{Cu: } [Ar]3d^{10}4s^1$$

Chromium and copper are two exceptions to the straightforward filling of atomic subshells in the first d block row. An examination of the electron configurations of the heavier elements (Appendix D) will reveal that there are other special cases that are not easily explained—for example gadolinium has the configuration $[Xe]$ $4f^76d^16s^2$. Examples 9-10 through 9-12 provide several more illustrations of the assignment of electron configurations using the ideas presented here.

EXAMPLE 9-10

Using spdf *Notation for an Electron Configuration.*

(a) Identify the element having the electron configuration

$$1s^22s^22p^63s^23p^5$$

(b) Write out the electron configuration of arsenic.

Solution

(a) All electrons must be accounted for in an electron configuration. Add up the superscript numerals $(2 + 2 + 6 + 2 + 5)$ to obtain the atomic number 17. The element with this atomic number is chlorine.

(b) Arsenic $(Z = 33)$ is in period 4 and group 15. Its valence-shell electron configuration is $4s^24p^3$. The noble gas that closes the third period is

Ar ($Z = 18$), and the subshells that fill in the fourth period are $4s$, $3d$, and $4p$, in that order. Note that we account for 33 electrons in the configuration

$$\text{As: } [\text{Ar}]3d^{10}4s^24p^3$$

Practice Example A: Identify the element having the electron configuration $1s^22s^22p^63s^23p^63d^24s^2$.

Practice Example B: Use *spdf* notation to show the electron configuration of iodine. How many electrons does the I atom have in its $3d$ subshell? How many unpaired electrons are there in an I atom?

EXAMPLE 9-11

Representing Electron Configurations.   Write the **(a)** electron configuration of mercury and **(b)** an orbital diagram for the electron configuration of tin.

Solution

(a)   Mercury, in period 6 and group 12, is the transition element at the end of the third transition series, in which the $5d$ subshell fills $(5d^{10})$. The noble gas that closes period 5 is xenon, and the lanthanide series intervenes between xenon and mercury, in which the $4f$ subshell fills $(4f^{14})$. When we put all these facts together, we conclude that the electron configuration of mercury is

$$[\text{Xe}]4f^{14}5d^{10}6s^2$$

(b)   Tin is in period 5 and group 14. Its valence-shell electron configuration is $5s^25p^2$. The noble gas that closes the fourth period is Kr ($Z = 36$), and the subshells that fill in the fifth period are $5s$, $4d$, and $5p$. Note that all subshells are filled in the orbital diagram except for $5p$. Two of the $5p$ orbitals are occupied by single electrons with parallel spins; one $5p$ orbital remains empty.

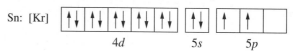

Sn: [Kr] $4d$ $5s$ $5p$

Practice Example A: Represent the electron configuration of iron with an orbital diagram.

Practice Example B: Represent the electron configuration of bismuth with an orbital diagram.

EXAMPLE 9-12

Relating Electron Configurations to the Periodic Table.   Indicate the number of **(a)** valence electrons in an atom of bromine; **(b)** $5p$ electrons in an atom of tellurium; **(c)** unpaired electrons in an atom of indium; **(d)** $3d$ and $4d$ electrons in a silver atom.

Solution

Determine the atomic number and the periodic table location of each element. Then establish the significance of the location.

(a)   Bromine ($Z = 35$) is in group 17. There are *seven* outer-shell, or valence, electrons in all atoms in this group.

(b)   Tellurium ($Z = 52$) is in period 5 and group 16. There are six outer-shell electrons, two of them, s, and the other four, p. The valence-shell electron configuration of tellurium is $5s^25p^4$; the tellurium atom has *four* $5p$ electrons.

(c)   Indium ($Z = 49$) is in period 5 and group 13. The electron configuration of its inner shells is $[\text{Kr}]4d^{10}$. All the electrons in this inner-shell configuration are paired. The valence-shell electron configuration is $5s^25p^1$. The two 5s electrons are paired, and the 5p electron is unpaired. The In atom has *one* unpaired electron.

(d) Ag $(Z = 47)$ is in period 5 and group 11. The noble gas that closes period 4 is krypton, and the 4d subshell fills $(4d^{10})$. There is one electron in the 5s orbital; thus the electron configuration of silver is

$$\text{Ag: } [\text{Kr}]4d^{10}5s^1$$

There are *ten* 3d electrons and *ten* 4d electrons in a silver atom.

Practice Example A: For an atom of Sn, indicate the number of **(a)** electronic shells that are either filled or partially filled; **(b)** 3p electrons; **(c)** 5d electrons; and **(d)** unpaired electrons.

Practice Example B: Indicate the number of **(a)** 3d electrons in Y atoms; **(b)** 4p electrons in Ge atoms; and **(c)** unpaired electrons in Au atoms.

Summary

Understanding electromagnetic radiation is important to learning about atomic structure. The dispersion of "white" light produces a continuous spectrum—a rainbow. Light produced by excited gaseous atoms yields a line spectrum when dispersed—a series of colored lines. The simplest line spectrum is that of hydrogen atoms, which can be described through the Balmer equation.

To explain phenomena at the atomic and molecular level requires thinking about energy in tiny discrete units—quanta. Einstein used the quantum theory to explain the photoelectric effect, and Bohr applied it to an atomic model that explains the observed spectrum of hydrogen.

Wave mechanics is a more sophisticated form of quantum mechanics than that used by Bohr. Two ideas contributing to wave mechanics are de Broglie's concept of wave–particle duality (matter waves) and Heisenberg's uncertainty principle. Schrödinger used these ideas to provide a new model of the hydrogen atom.

The essential feature of the Schrödinger atom is to view an electron as a matter wave described by a wave equation. Solutions to this equation are called wave functions, ψ, that are mathematical expressions containing parameters called quantum numbers. When these quantum numbers are specified, the resulting wave functions are called orbitals. Orbitals can be used to describe three-dimensional regions in atoms where there is a high probability of finding electrons. The specific orbital types are s, p, d, and f. The key parameters that distinguish among orbitals are the three quantum numbers, n, ℓ and m_ℓ. A fourth parameter required to describe an electron in an atom is the spin quantum number, m_s.

The wave mechanical model of the hydrogen atom can be modified to apply to multielectron atoms, and, through a set of three rules, electrons can be assigned to the orbitals in principal shells and subshells. These assignments—electron configurations—are made in Section 9-11 by a method known as the aufbau process. Finally, the connection between electron configurations and the periodic table is explored.

Integrative Example

Microwave ovens have become increasingly popular in kitchens around the world. They are also useful in the chemical laboratory, particularly in drying samples for chemical analysis. A typical microwave oven uses microwave radiation with a wavelength of 12.2 cm.

Are there any electronic transitions in the hydrogen atom that could *conceivably* produce microwave radiation of wavelength 12.2 cm?

1. *Calculate the frequency of the microwave radiation.* Microwaves are a form of electromagnetic radiation and thus travel at the speed of light, 2.998×10^8 m s^{-1}. Convert the wavelength to meters, and then use the equation $\nu = c/\lambda$.

$$\nu = \frac{2.998 \times 10^8 \text{ m s}^{-1}}{12.2 \text{ cm} \times 1 \text{ m}/100 \text{ cm}} = 2.46 \times 10^9 \text{ Hz}$$

2. *Calculate the energy associated with one photon of the microwave radiation.* This is a direct application of Planck's equation.

$$E = h\nu = 6.626 \times 10^{-34} \text{ J s} \times 2.46 \times 10^9 \text{ s}^{-1} = 1.63 \times 10^{-24} \text{ J}$$

3. *Determine if there are any electronic transitions in the hydrogen atom with an energy per photon of 1.63×10^{-24} J.* Look at Figure 9-14, the energy-level diagram for the Bohr hydrogen atom. Energy differences between the low-lying levels are of the order 10^{-19} to 10^{-20} J. These are orders of magnitude (10^4–10^5 times) greater than the energy per photon of 1.63×10^{-24} J from part 2. Note, however, that the energy differences become progressively smaller for high-numbered orbits. As n approaches ∞, the energy differences approach zero, and some transitions between high-numbered orbits should correspond to microwave radiation (see also Exercise 101).

◀ Helium–neon laser device for use as a bar-code scanner.

Laser devices seem to be in use everywhere—from reading bar codes in supermarket checkout stands to maintaining quality control in factory assembly lines. Lasers are used in compact disc players, laboratory instruments, and, increasingly, as a tool in surgery. The word *laser* is an acronym for *light amplification by stimulated emission of radiation*.

Consider a neon sign. An electric discharge produces high-energy electrons that strike Ne atoms and excite certain Ne electrons to a higher energy state. When an excited electron in the neon atom returns to a lower-energy orbital, a photon of light is emitted. Although we know how long Ne atoms remain in an excited state *on average*, we do not know precisely when an atom will emit a photon, nor do we know in what direction the photon will travel. Light emission in a neon sign is a spontaneous, or random, event. *Spontaneous* light emission is pictured in Figure 9-36a.

Like a neon sign, a helium–neon laser also produces red light (633 nm), but the similarity ends here. The He–Ne laser works on this principle: If a 633-nm photon interacts with an excited Ne atom *before* the atom spontaneously emits a photon, the atom will be induced, or *stimulated*, to emit its photon at the precise time of the interaction. Moreover, this second photon will be *coherent*, or *in phase*, with the first; that is, the crests and troughs of the two waves will match exactly. *Stimulated* light emission is pictured in Figure 9-36b.

A He–Ne laser consists of a tube containing a helium–neon mixture at a pressure of about 1 mmHg. One end of the tube is a totally reflecting mirror, and the other end is a mirror that allows about 1% of the light to leave. An electric discharge is used to produce excited He atoms. These atoms transfer their excitation energy to Ne atoms through collisions. The Ne atoms enter a *metastable* state, an excited state that is stable for a relatively long time before the spontaneous emission of a photon occurs. Most of the Ne atoms must remain "pumped up" to this metastable state during the operation of the laser.

Following the first spontaneous release of a photon, other excited atoms release photons of the same frequency and in phase with the stimulating photon. This is rather like lining up a row of dominoes and knocking over the first one. This is the beginning of the laser beam. As shown in Figure 9-37, the beam is *amplified* as it bounces back and forth between the two mirrors. The output is the portion of the laser beam that escapes through the partially reflecting mirror. Because the photons of light travel in phase and in the same direction, laser light can be produced at a much higher intensity than would normally be achieved with spontaneous light emission.

Shown below are the transitions that produce (1) exited He atoms, (2) Ne atoms in the metastable state, and (3) laser light. After the laser light emission, (4) Ne atoms return to the ground state by way of two additional transitions. Ground-state Ne atoms are then pumped back up to the metastable state by collisions with excited He atoms. (The symbols * and # represent electronically excited species.)

(1) $\text{He} \xrightarrow{e^-} \text{He*}$
$\quad 1s^2 \qquad\qquad 1s^1 2s^1$

(2) $\text{He*} + \text{Ne} \longrightarrow \text{He} + \text{Ne*}$ (metastable)
$\qquad [\text{He}]2s^2 2p^6 \qquad [\text{He}]2s^2 2p^5 5s^1$

(3) $\text{Ne*} \longrightarrow \text{Ne}^{\#} \quad + \quad h\nu$
$\qquad [\text{He}]2s^2 2p^5 3p^1 \qquad (633\ \text{nm})$

(4) $\text{Ne}^{\#} \longrightarrow \longrightarrow \text{Ne}$ (return to step 2)

Key Terms

angular wave function, *Y* (9-6)
atomic (line) spectra (9-2)
aufbau process (9-11)
d block (9-12)
degenerate orbitals (9-7)

diffraction (9-1)
effective nuclear charge, Z_{eff} (9-10)
electromagnetic radiation (9-1)
electron configuration (9-11)
electron spin (9-9)

energy-level diagram (9-4)
excited state (9-4)
f block (9-12)
frequency, *ν* (9-1)
ground state (9-4)

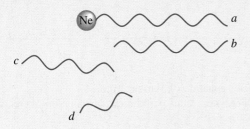

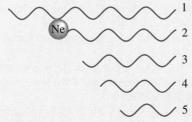

(a) Spontaneous emission

(b) Stimulated emission

◀ **FIGURE 9-36**

Spontaneous and stimulated light emission by Ne atoms
(a) The waves *a*, *b*, *c*, and *d*, although having the same frequency and wavelength, are emitted in different directions. Waves *a* and *b* are out of phase. The crests of one wave line up with the troughs of the other. The two waves cancel, and they transmit no energy at all. **(b)** Photon (1) interacts with a Ne atom in a metastable energy state and stimulates it to emit photon (2). Photon (2) stimulates another Ne atom to emit photon (3), and so on. The waves are coherent, or in phase. The crests and troughs of the waves match perfectly.

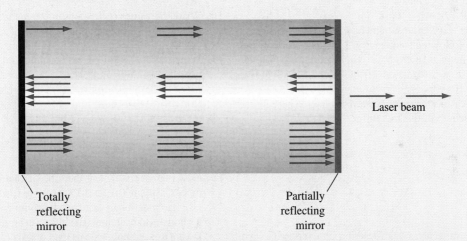

Laser beam

Totally reflecting mirror

Partially reflecting mirror

▲ **FIGURE 9-37   Operation of a He–Ne laser**
Pictured here is a hypothetical laser tube arbitrarily divided into three sections representing the same portion of the tube at different times. At the top left, a 633-nm photon (⟶) stimulates light emission from an excited Ne atom. Now there are two photons, and one of them in turn stimulates emission of a third photon. The photons are reflected by the mirror on the right, and their transit of the tube in the opposite direction is shown in the center. Three photons are amplified to five. At the bottom, the five photons are amplified to seven, and so on. A small portion of the photons is drawn off as the laser beam.

Heisenberg uncertainty principle (9-5)	**photon** (9-3)	**Schrödinger equation** (9-6)
hertz, Hz (9-1)	**Planck's constant, *h*** (9-3)	**standing wave** (9-6)
Hund's rule (9-11)	**principal electronic shell (level)** (9-7)	**subshell (sublevel)** (9-7)
orbital (9-6)	**quantum** (9-3)	**valence electrons** (9-11)
orbital diagram (9-11)	**quantum numbers** (9-4)	**wave** (9-1)
Pauli exclusion principle (9-11)	**radial wave function, R** (9-6)	**wave function, ψ** (9-6)
***p* block** (9-12)	***s* block** (9-12)	**wavelength, λ** (9-1)
photoelectric effect (9-3)	***spdf* notation** (9-1)	**zero-point energy** (9-6)

Review Questions

1. In your own words, define the following terms or symbols: **(a)** λ; **(b)** ν; **(c)** h; **(d)** ψ; **(e)** principal quantum number, n.

2. Briefly describe each of the following ideas or phenomena: **(a)** atomic (line) spectrum; **(b)** photoelectric effect; **(c)** matter wave; **(d)** Heisenberg uncertainty principle; **(e)** electron spin; **(f)** Pauli exclusion principle.

3. Explain the important distinctions between each pair of terms: **(a)** frequency and wavelength; **(b)** ultraviolet and infrared light; **(c)** continuous and discontinuous spectrum; **(d)** traveling and standing wave; **(e)** quantum number and orbital; **(f)** *spdf* notation and orbital diagram; **(g)** *s* block and *p* block; **(h)** main group and transition element.

4. Restate each of the following wavelengths in the unit indicated.
   (a) 1625 Å = ___ nm
   (b) 3880 Å = ___ μm
   (c) 7.27×10^{-3} m = ___ nm
   (d) 546 nm = ___ m
   (e) 1.12 cm = ___ nm
   (f) 2.6×10^4 Å = ___ cm

5. What are the wavelengths, in meters, associated with radiation of the following frequencies? In what portion of the electromagnetic spectrum is each radiation found? **(a)** 6.8×10^{12} s^{-1}; **(b)** 9.8×10^{15} s^{-1}; **(c)** 2.54×10^7 Hz; **(d)** 1.07×10^8 Hz.

6. A hypothetical electromagnetic wave is pictured here. What is the wavelength of this radiation?

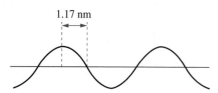

1.17 nm

7. For the electromagnetic wave described in Exercise 6, what are **(a)** the frequency, in hertz, and **(b)** the energy, in joules per photon?

8. For electromagnetic radiation transmitted through a vacuum, state whether each of the following properties is directly proportional to, inversely proportional to, or independent of the frequency: **(a)** velocity; **(b)** wavelength; **(c)** energy per mole. Explain.

9. Use the Balmer equation (9.2) to determine
   (a) the frequency, in s^{-1}, of the radiation corresponding to $n = 5$
   (b) the wavelength, in nanometers, of the line in the Balmer series corresponding to $n = 7$
   (c) the value of n corresponding to the Balmer series line at 380 nm

10. How would the Balmer equation (9.2) have to be modified to predict lines in the infrared spectrum of hydrogen? [*Hint:* Compare equations (9.2) and (9.6).]

11. Use Planck's equation (9.3) to determine
    (a) the energy, in joules per photon, of radiation of frequency 8.62×10^{15} s^{-1}
    (b) the energy, in kilojoules per mole, of radiation of frequency 1.53×10^{14} s^{-1}

12. Use Planck's equation (9.3) to determine
    (a) the frequency, in hertz, of radiation having an energy of 4.18×10^{-21} J/photon
    (b) the wavelength, in nanometers, of radiation with 215 kJ/mol of energy

13. What is ΔE for the transition of an electron from $n = 5$ to $n = 3$ in a Bohr hydrogen atom? What is the frequency of the spectral line produced?

14. If traveling at equal speeds, which of the following matter waves has the *longest* wavelength? Explain. **(a)** electron; **(b)** proton; **(c)** neutron; **(d)** α particle (He^{2+}).

15. Give a possible value for the missing quantum number in each of the following sets. **(a)** $n = 3$, $\ell = 1$, $m_\ell = ?$; **(b)** $n = 4$, $\ell = ?$, $m_\ell = -1$; **(c)** $n = ?$, $\ell = 1$, $m_\ell = +1$.

16. Write the values corresponding to n and ℓ for each of the following orbital designations. **(a)** $4s$; **(b)** $3p$; **(c)** $5f$; **(d)** $3d$.

17. Which of the following sets of quantum numbers is (are) *not* allowed? Why not?
    (a) $n = 3$, $\ell = 2$, $m_\ell = -1$
    (b) $n = 2$, $\ell = 3$, $m_\ell = -1$
    (c) $n = 3$, $\ell = 0$, $m_\ell = +1$
    (d) $n = 6$, $\ell = 2$, $m_\ell = -1$
    (e) $n = 4$, $\ell = 4$, $m_\ell = +4$
    (f) $n = 4$, $\ell = 3$, $m_\ell = -1$

18. How many orbitals can there be of each of the following types? Explain. **(a)** $2s$; **(b)** $3f$; **(c)** $4p$; **(d)** $5d$; **(e)** $5f$; **(f)** $6p$.

19. Use *spdf* notation to write out the complete electron configuration of **(a)** bromine; **(b)** sulfur; **(c)** antimony; **(d)** silicon. Do not use symbols such as [Ne] or [Ar] for noble gas cores.

20. Indicate the number of unpaired electrons in an atom of **(a)** magnesium; **(b)** thallium; **(c)** tellurium; **(d)** aluminum.

21. Refer to the periodic table on the inside front cover and identify **(a)** a main-group metal; **(b)** a main-group non-metal; **(c)** a noble gas; **(d)** a *d* block element; **(e)** an inner transition element.

22. Use the periodic table as a guide to write electron configurations for **(a)** In; **(b)** Cd; **(c)** Sb; **(d)** Au. Compare your results with the electron configurations shown in Appendix D.

23. Use Figure 9-35 as a guide to indicate the number of **(a)** $4s$ electrons in K; **(b)** $5p$ electrons in I; **(c)** $3d$ electrons in Zn; **(d)** $2p$ electrons in S; **(e)** $4f$ electrons in Pb; **(f)** $3d$ electrons in Ni.

24. The accompanying diagrams help to summarize some ideas about the electron configurations of the elements. The diagrams contain some features marked with a question mark (?) and others by leaders (straight-line pointers). Replace

each (?) by the symbol of the appropriate element or noble gas core. Use terms from the following list to attach a label to each leader. You may use a given label only once, and you will find that some labels are inappropriate.

(a) an unpaired electron (b) a filled subshell
(c) a valence shell (d) a valence electron
(e) an *f* subshell (f) a principal shell
(g) a half-filled subshell (h) a *d* orbital
(i) a *p* subshell (j) a *d* subshell
(k) an electron pair (l) an *s* subshell

1. (?): $1s^2 2s^2 2p^6 3s^2 3p^6 3d^5 4s^2$

2. Nb: [?]

3. (?): [?] 4*d*

4. (?): [?] 5*f*

Exercises

Electromagnetic Radiation

25. The magnesium spectrum has a line at 266.8 nm. Which of these statements is (are) correct concerning this radiation? Explain.
(a) It has a higher frequency than radiation with wavelength 402 nm.
(b) It is visible to the eye.
(c) It has a greater speed in a vacuum than does red light of wavelength 652 nm.
(d) Its wavelength is longer than that of X rays.

26. The most intense line in the cerium spectrum is at 418.7 nm.
(a) Determine the frequency of the radiation producing this line.
(b) In what part of the electromagnetic spectrum does this line occur?
(c) Is it visible to the eye? If so, what color is it? If not, is this line at higher or lower energy than visible light?

27. *Without doing detailed calculations*, determine which of the following wavelengths represents light of the *highest* frequency: (a) 5.9×10^{-4} cm; (b) 1.13 mm; (c) 860 Å; (d) 6.92 μm.

28. *Without doing detailed calculations*, arrange the following electromagnetic radiation sources in order of increasing frequency: (a) a red traffic light, (b) a 91.9 MHz radio transmitter, (c) light with a frequency of 3.0×10^{14} s^{-1}, (d) light with a wavelength of 485 Å.

29. How long does it take light from the sun, 93 million miles away, to reach Earth?

30. In astronomy, distances are measured in *light-years*, the distance that light travels in one year. What is the distance of one light-year expressed in kilometers?

Atomic Spectra

31. Calculate the wavelengths, in nanometers, of the first four lines of the Balmer series of the hydrogen spectrum, starting with the *longest* wavelength component.

32. A line is detected in the hydrogen spectrum at 1880 nm. Is this line in the Balmer series? Explain.

33. To what value of *n* in equation (9.2) does the line in the Balmer series at 389 nm correspond?

34. The Lyman series of the hydrogen spectrum can be represented by the equation

$$\nu = 3.2881 \times 10^{15} \text{ s}^{-1} \left(\frac{1}{1^2} - \frac{1}{n^2} \right) \quad (\text{where } n = 2, 3, \dots)$$

(a) Calculate the maximum and minimum wavelength lines, in nanometers, in this series.
(b) What value of *n* corresponds to a spectral line at 95.0 nm?
(c) Is there a line at 108.5 nm? Explain.

Quantum Theory

35. A certain radiation has a wavelength of 474 nm. What is the energy, in joules, of (a) one photon; (b) a mole of photons of this radiation?

36. What is the wavelength, in nanometers, of light with an energy content of 1799 kJ/mol? In what portion of the electromagnetic spectrum is this light?

37. *Without doing detailed calculations*, indicate which of the following electromagnetic radiations has the *greatest* energy per photon and which has the *least*. **(a)** 662 nm; **(b)** 2.1×10^{-5} cm; **(c)** 3.58 μm; **(d)** 4.1×10^{-6} m.

38. *Without doing detailed calculations*, arrange the following forms of electromagnetic radiation in *increasing* order of energy per mole of photons: **(a)** radiation with $\nu = 3.0 \times 10^{15}$ s^{-1}, **(b)** an infrared heat lamp, **(c)** radiation having $\lambda = 7000$ Å, **(d)** dental X rays.

39. In what region of the electromagnetic spectrum would you expect to find radiation having an energy per photon 100 times that associated with 988-nm radiation?

40. High-pressure sodium vapor lamps are used in street lighting. The two brightest lines in the sodium spectrum are at 589.00 and 589.59 nm. What is the *difference* in energy per photon of the radiations corresponding to these two lines?

The Photoelectric Effect

41. The lowest-frequency light that will produce the photoelectric effect is called the *threshold frequency*.
(a) The threshold frequency for indium is 9.96×10^{14} s^{-1}. What is the energy, in joules, of a photon of this radiation?
(b) Will indium display the photoelectric effect with UV light? infrared light? Explain.

42. Sir James Jeans described the photoelectric effect in this way: "It not only prohibits killing two birds with one stone, but also the killing of one bird with two stones." Comment on the appropriateness of this analogy in reference to the marginal note on page 307.

The Bohr Atom

43. Use the description of the Bohr atom given in the text to determine **(a)** the radius, in nanometers, of the sixth Bohr orbit for hydrogen; **(b)** the energy, in joules, of the electron when it is in this orbit.

44. Calculate the increase in **(a)** distance from the nucleus and **(b)** energy when an electron is excited from the first to the third Bohr orbit.

45. What are the **(a)** frequency, in s^{-1}, and **(b)** wavelength, in nanometers, of the light emitted when the electron in a hydrogen atom drops from the energy level $n = 7$ to $n = 4$? **(c)** In what portion of the electromagnetic spectrum is this light?

46. *Without doing detailed calculations*, indicate which of the following electron transitions requires the greatest amount of energy to be *absorbed* by a hydrogen atom: from **(a)** $n = 1$ to $n = 2$; **(b)** $n = 2$ to $n = 4$; **(c)** $n = 3$ to $n = 9$; **(d)** $n = 10$ to $n = 1$.

47. For the Bohr hydrogen atom determine
(a) the radius of the orbit $n = 4$
(b) whether there is an orbit having a radius of 4.00 Å
(c) the energy level corresponding to $n = 8$
(d) whether there is an energy level at -25.00×10^{-18} J

48. *Without doing detailed calculations*, indicate which of the following electron transitions in the hydrogen atom results in the emission of light of the longest wavelength. **(a)** $n = 4$ to $n = 3$; **(b)** $n = 1$ to $n = 2$; **(c)** $n = 1$ to $n = 6$; **(d)** $n = 3$ to $n = 2$.

49. What electron transition in a hydrogen atom, starting from the orbit $n = 7$, will produce infrared light of wavelength 2170 nm?

50. What electron transition in a hydrogen atom, ending in the orbit $n = 5$, will produce light of wavelength 3740 nm?

Wave–Particle Duality

51. Which must possess a greater velocity to produce matter waves of the same wavelength (such as 1 nm), protons or electrons? Explain your reasoning.

52. What must be the velocity, in meters per second, of a beam of electrons if they are to display a de Broglie wavelength of 1 μm?

53. Calculate the de Broglie wavelength, in nanometers, associated with a 145-g baseball traveling at a speed of 168 km/h. How does this wavelength compare with typical nuclear or atomic dimensions?

54. What is the wavelength, in nanometers, associated with a 1000-kg automobile traveling at a speed of 25 m/s, that is, considering the automobile to be a "matter" wave? Comment on the feasibility of an experimental measurement of this wavelength.

The Heisenberg Uncertainty Principle

55. Describe the ways in which the Bohr model of the hydrogen atom appears to violate the Heisenberg uncertainty principle.

56. Although Einstein made some early contributions to quantum theory, he was never able to accept the Heisenberg uncertainty principle. He stated, "God does not play dice with the Universe." What do you suppose Einstein meant by this remark? In reply to Einstein's remark, Niels Bohr is supposed to have said, "Albert, stop telling God what to do." What do you suppose Bohr meant by this remark?

57. A proton is accelerated to one-tenth the velocity of light, and this velocity can be measured with a precision of 1%. What is the uncertainty in the position of this proton?

58. Show that the uncertainty principle is not significant when applied to large objects such as automobiles. Assume that m is precisely known; assign a reasonable value to either the uncertainty in position or the uncertainty in velocity, and estimate a value of the other.

59. What must be the velocity of electrons if their associated wavelength is to equal the radius of the first Bohr orbit of the hydrogen atom?

60. What must be the velocity of electrons if their associated wavelength is to equal the *longest* wavelength line in the Lyman series?
(*Hint:* Refer to Figure 9-14.)

Wave Mechanics

61. A standing wave in a string 42 cm long has a *total* of six nodes (including those at the ends). What is the wavelength, in centimeters, of this standing wave?

62. What is the length of a string that has a standing wave with four nodes (including those at the ends) and $\lambda = 17$ cm.

63. In a plucked guitar string, the frequency of the *longest* wavelength standing wave is called the *fundamental frequency*. The standing wave with one interior node is called the first *overtone*, and so on. What is the wavelength of the second overtone of a 24-in. guitar string?

64. In the particle-in-a-box model, what is the wavelength of the third overtone of a box with a length of 100 pm?
(*Hint:* Refer also to Exercise 63.)

65. Describe some of the differences between the orbits of the Bohr atom and the orbitals of the wave mechanical atom. Are there any similarities?

66. The greatest probability of finding the electron in a small-volume element of the $1s$ orbital of the hydrogen atom is at the nucleus. Yet the most probable distance of the electron from the nucleus is 53 pm. How can you reconcile these two statements?

Quantum Numbers and Electron Orbitals

67. Select the correct answer and explain your reasoning. An electron having $n = 3$ and $m_\ell = 0$ **(a)** must have $m_s = +\frac{1}{2}$; **(b)** must have $\ell = 1$; **(c)** may have $\ell = 0, 1,$ or 2; **(d)** must have $\ell = 2$.

68. Write an acceptable value for each of the missing quantum numbers.
(a) $n = 3, \ell = ?, m_\ell = 2, m_s = +\frac{1}{2}$
(b) $n = ?, \ell = 2, m_\ell = -1, m_s = -\frac{1}{2}$
(c) $n = 4, \ell = 2, m_\ell = 0, m_s = ?$
(d) $n = ?, \ell = 0, m_\ell = ?, m_s = ?$

69. What type of orbital (i.e., $3s, 4p, \ldots$) is designated by these quantum numbers?
(a) $n = 5, \ell = 1, m_\ell = 0$
(b) $n = 4, \ell = 2, m_\ell = -2$
(c) $n = 2, \ell = 0, m_\ell = 0$

70. Which of the following statements is (are) correct for an electron with $n = 4$ and $m_\ell = -2$? Explain.
(a) The electron is in the fourth principal shell.

(b) The electron may be in a d orbital.
(c) The electron may be in a p orbital.
(d) The electron must have $m_s = +\frac{1}{2}$.

71. Concerning the electrons in the shells, subshells, and orbitals of an atom, how many can have
(a) $n = 3, \ell = 2, m_\ell = 0$, and $m_s = +\frac{1}{2}$?
(b) $n = 3, \ell = 2$, and $m_\ell = 0$?
(c) $n = 3$ and $\ell = 2$?
(d) $n = 3$?
(e) $n = 3, \ell = 2$, and $m_s = +\frac{1}{2}$?

72. Concerning the concept of subshells and orbitals,
(a) How many subshells are found in the $n = 4$ level?
(b) What are the names of the subshells in the $n = 3$ level?
(c) How many orbitals have the values $n = 4$ and $\ell = 3$?
(d) How many orbitals have the values $n = 4, \ell = 3$, and $m_\ell = -2$?
(e) What is the total number of orbitals in the $n = 4$ level?

The Shapes of Orbitals and Radial Probabilities

73. Calculate the finite value of r, in terms of a_0, at which the node occurs in the wave function of the $2s$ orbital of a hydrogen atom.
74. Calculate the finite value of r, in terms of a_0, at which the node occurs in the wave function of the $2s$ orbital of a Li^{2+} ion.
75. Show that the probability of finding a $2p_y$ electron in the xz plane is zero.
76. Show that the probability of finding a $3d_{xz}$ electron in the xy plane is zero.
77. Prepare a two-dimensional plot of $Y(\theta, \phi)$ for the p_x orbital in the xy plane.
78. Prepare a two-dimensional plot of $Y(\theta, \phi)$ for the p_y orbital in the xy plane.
79. Prepare a two-dimensional plot of $Y^2(\theta, \phi)$ for the p_x orbital in the xy plane.
80. Prepare a two dimensional plot of $Y^2(\theta, \phi)$ for the p_y orbital in the xy plane.
81. Using a graphical method, show that in a hydrogen atom the radius at which there is a maximum probability of finding an electron is a_0 (53 pm).
82. Use a graphical method or some other means to show that in a Li^{2+} ion, the radius at which there is a maximum probability of finding an electron is $\dfrac{a_0}{3}$ (18 pm).

Electron Configurations

83. Which of the following is the correct orbital diagram for the ground-state electron configuration of phosphorus? Explain what is wrong with each of the others.

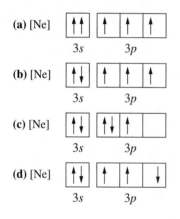

84. Which of the following is the correct orbital diagram for the ground-state electron configuration of molybdenum? Explain what is wrong with each of the others.

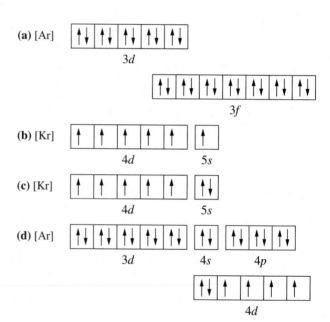

85. Use the basic rules for electron configurations to indicate the number of **(a)** unpaired electrons in an atom of P; **(b)** $3d$ electrons in an atom of Br; **(c)** $4p$ electrons in an atom of Ge; **(d)** $6s$ electrons in an atom of Ba; **(e)** $4f$ electrons in an atom of Au.
86. Use orbital diagrams to show the distribution of electrons among the orbitals in **(a)** the $4p$ subshell of Br; **(b)** the $3d$ subshell of Co^{2+}, given that the two electrons lost are $4s$; **(c)** the $5d$ subshell of Pb.

87. On the basis of the periodic table and rules for electron configurations, indicate the number of **(a)** $2p$ electrons in N; **(b)** $4s$ electrons in Rb; **(c)** $4d$ electrons in As; **(d)** $4f$ electrons in Au; **(e)** unpaired electrons in Pb; **(f)** elements in group 14 of the periodic table; **(g)** elements in the sixth period of the periodic table.

88. Based on the relationship between electron configurations and the periodic table, give the number of **(a)** outer-shell electrons in an atom of Sb; **(b)** electrons in the fourth principal electronic shell of Pt; **(c)** elements whose atoms have six outer-shell electrons; **(d)** unpaired electrons in an atom of Te; **(e)** transition elements in the sixth period.

89. The element that the recently discovered element 114 should most closely resemble is Pb.
 (a) Write the electron configuration of Pb.
 (b) Propose a plausible electron configuration for element 114.

90. Without referring to any tables or listings in the text, mark an appropriate location in the blank periodic table provided for each of the following: **(a)** the fifth-period noble gas; **(b)** a sixth-period element whose atoms have three unpaired p electrons; **(c)** a d block element having one $4s$ electron; **(d)** a p block element that is a metal.

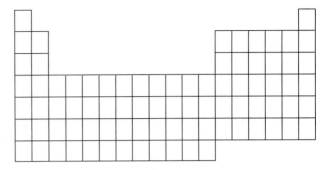

Integrative and Advanced Exercises

91. Derive the Balmer equation from equation (9.6).

92. Electromagnetic radiation can be transmitted through a vacuum or empty space. Can heat be similarly transferred? Explain.

93. The *work function* is the energy that must be supplied to cause the release of an electron from a photoelectric material. The corresponding photon frequency is the threshold frequency. The higher the energy of the incident light, the more kinetic energy the electrons have in moving away from the surface. The work function for mercury is equivalent to 435 kJ/mol photons.
 (a) Can the photoelectric effect be obtained with mercury by using visible light? Explain.
 (b) What is the kinetic energy, in joules, of the ejected electrons when light of 215 nm strikes a mercury surface?
 (c) What is the velocity, in meters per second, of the ejected electrons in (b)?

94. Infrared lamps are used in cafeterias to keep food warm. How many photons per second are produced by an infrared lamp that consumes energy at the rate of 95 W and is 14% efficient in converting this energy to infrared radiation? Assume that the radiation has a wavelength of 1525 nm.

95. In 5.0 s, a 75-watt light source emits 9.91×10^{20} photons of a monochromatic (single wavelength) radiation. What is the color of the emitted light?

96. In everyday usage, the term "quantum jump" describes a change of a very significant magnitude compared to more gradual, incremental changes; it is similar in meaning to the term "a sea change." Does quantum jump have the same meaning when applied to events at the atomic or molecular level? Explain.

97. The Pfund series of the hydrogen spectrum has as its *longest* wavelength component a line at 7400 nm. Describe the electron transitions that produce this series. That is, give a Bohr quantum number that is common to this series.

98. Between which two orbits of the Bohr hydrogen atom must an electron fall to produce light of wavelength 1876 nm?

99. Use appropriate relationships from the chapter to determine the wavelength of the line in the emission spectrum of He^+ produced by an electron transition from $n = 5$ to $n = 2$.

100. Draw an energy-level diagram that represents all the possible lines in the emission spectrum of hydrogen atoms produced by electron transitions, in one or more steps, from $n = 5$ to $n = 1$.

101. Refer to the Integrative Example. Assume that the microwave radiation could be produced by an electronic transition from the Bohr orbit $(n + 1)$ to the orbit n. What is the value of n?

102. An atom in which just one of the outer-shell electrons is excited to a very high quantum level (n) is called a "high Rydberg" atom. In some ways, all these atoms resemble a Bohr hydrogen atom with its electron in a high-numbered orbit. Explain why you might expect this to be the case.

103. If all other rules governing electron configurations were valid, what would be the electron configuration of cesium if **(a)** there were *three* possibilities for electron spin? **(b)** the quantum number ℓ could have the value n?

104. Ozone, O_3, absorbs ultraviolet radiation and dissociates into O_2 molecules and O atoms: $O_3 + h\nu \longrightarrow O_2 + O$. A 1.00-L sample of air at 22 °C and 748 mmHg contains 0.25 ppm of O_3. How much energy, in joules, must be absorbed if all the O_3 molecules in the sample of air are to dissociate? Assume that each photon absorbed causes one O_3 molecule to dissociate, and that the wavelength of the radiation is 254 nm.

105. Radio signals from *Voyager 1* in the 1970s were broadcast at a frequency of 8.4 GHz. On Earth, this radiation was received by an antenna able to detect signals as weak as

4×10^{-21} W. How many photons per second does this detection limit represent?

106. Certain metal compounds impart colors to flames–sodium compounds, yellow; lithium, red; barium, green. Flame tests can be used to detect these elements.
   (a) At a flame temperature of 800 °C, can collisions between gaseous atoms with average kinetic energies supply the energies required for the emission of visible light?
   (b) If not, how do you account for the excitation energy?

107. The angular momentum of an electron in the Bohr hydrogen atom is *mur*, where *m* is the mass of the electron, *u*, its velocity, and *r*, the radius of the Bohr orbit. The angular momentum can have only the values $nh/2\pi$, where *n* is an integer (the number of the Bohr orbit). Show that the *circumferences* of the various Bohr orbits are integral

multiples of the de Broglie wavelengths of the electron treated as a matter wave.

108. Combine ideas from Exercise 107 and other data given in the text to obtain, for an electron in the third orbit ($n = 3$) of a hydrogen atom: **(a)** its velocity; **(b)** the number of revolutions it makes about the nucleus per second.

109. Using the relationships given in Table 9.1, find the finite values of *r*, in terms of a_0, of the nodes for a 3s orbital.

110. Use a graphical method or some other means to determine the radius at which the probability of finding a 2s orbital is maximum.

111. Using the relationships given in Table 9.1, prepare a two-dimensional plot of $R(r)Y(\theta, \phi)$ for the $3p_x$ orbital in the *xy* plane; hence, sketch the 90% probability surface of a 3p orbital.

Feature Problems

112. We have noted that an emission spectrum is a kind of "atomic fingerprint." The various steels are alloys of iron and carbon, usually containing one or more other metals. Based on the principal lines of their atomic spectra, which of the metals in the table below are likely to be present in a steel sample whose hypothetical emission spectrum is pictured? Is it likely that still other metals are present in the sample? Explain.

Principal spectral lines of some period 4 transition metals, nm

V	306.64	309.31	318.40	318.54	327.11	437.92	438.47	439.00
Cr	357.87	359.35	360.53	361.56	425.44	427.48	428.97	520.45
Mn	257.61	259.37	279.48	279.83	403.08	403.31	403.45	
Fe	344.06	358.12	372.00	373.49	385.99			
Ni	341.48	344.63	345.85	346.17	349.30	351.51	352.45	361.94

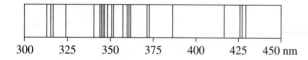

300   325   350   375   400   425   450 nm

▲ **Hypothetical emission spectrum**
In a real spectrum, the photographic images of the spectral lines would differ in depth and thickness depending on the strengths of the emissions producing them. Some of the spectral lines would not be seen because of their faintness.

113. Balmer seems to have deduced his formula for the visible spectrum of hydrogen just by manipulating numbers. A more common scientific procedure is to graph experimental data and then find a mathematical equation to describe the graph. Show that equation (9.2) describes a straight line. Indicate which variables must be plotted, and determine the numerical values of the slope and intercept of this line. Use data from Figure 9-10 to con-

firm that the four lines in the visible spectrum of hydrogen fall on the straight-line graph.

114. Follow up on the dartboard analogy of Figure 9-31 by plotting a graph of the scoring summary tabulated below. That is, plot the number of hits as a function of the scoring ring: 50, 40, ... What illustration in the text does this plot most resemble? Explain similarities and differences between the two.

Summary of scoring (1500 darts)	
200 darts score	"50"
300	"40"
400	"30"
250	"20"
200	"10"
150	off the board

115. Emission and absorption spectra of the hydrogen atom exhibit line spectra characteristic of quantized systems. In an absorption experiment, a sample of hydrogen atoms is irradiated with light with wavelengths ranging from 100 nm to 1000 nm. In an emission spectrum experiment, the hydrogen atoms are excited through an energy source that provides a range of energies from 1230 to 1240 kJ mol^{-1} to the atoms. Assume that the absorption spectrum is obtained at room temperature, when all atoms are in the ground state, and
(a) Calculate the position of the lines in the absorption spectrum.
(b) Calculate the position of the lines in the emission spectrum.

(c) Compare the line spectra observed in the two experiments. In particular, will the number of lines observed be the same?

116. Diffraction of radiation takes place when the distance between the scattering centers is comparable to the wavelength of the radiation.
(a) What velocity must helium atoms possess to be diffracted by a film of silver atoms in which the spacing is 100 pm?
(b) Electrons accelerated through a certain potential are diffracted by a thin film of gold. Would you expect a beam of protons accelerated through the same potential to be diffracted when it strikes the film of gold. If not, what would you expect to see instead?

eMedia Exercises

117. Using the **Electromagnetic Spectrum** activity *(eChapter 9-1)*, establish the mathematical relationship between the *energy* of radiation and the *wavelength* of radiation. *Hint:* You need three data points to determine whether there is a linear or inversely linear relationship.

118. After viewing the movie **Flame Tests for Metals** *(eChapter 9-2)*, **(a)** suggest approximately where dominant lines should appear (in terms of wavelength) in the emission spectrum of each metal shown. **(b)** Why are these lines not accurately described by equation (9.2)?

119. View the **Quantum Number** activity *(eChapter 9-7)*. This illustration demonstrates that the total number of populated orbitals in an energy level is equal to n^2, where n is the principal quantum number. At what value of n does

this calculation no longer provide physically applicable information? *Hint:* See the periodic table and refer to the number of known elements.

120. View the three-dimensional *s*, *p*, and *d* **Orbital** models *(eChapter 9-8)*. **(a)** Create your own two-dimensional representations of these orbitals on paper. **(b)** What information is lost in moving from three dimensions to two? Assign a set of quantum numbers (ℓ and m_ℓ) to each orbital.

121. From the **Radial Electron Distribution** animation *(eChapter 9-10)* predict the number of nodes that would be expected for krypton. From the electronic configurations of the noble gases argon and krypton, suggest what might cause a significant difference in the radial distribution curve for krypton.

10

The Periodic Table and Some Atomic Properties

Contents

Potassium, an alkali metal, reacts with water to liberate hydrogen gas, which bursts into flame. Phenolphthalein indicator in the water turns magenta, signaling the formation of hydroxide ions, another product of the reaction. The vigor of this reaction will not seem surprising, once we have explored the significance of potassium's location in group 1 of the periodic table.

By the middle of the nineteenth century, chemists had discovered a large number of elements, determined their relative atomic masses, and measured a host of their properties. Chemists had assembled what amounted to the "white pages" of a chemistry directory, but what they needed was a set of "yellow pages"—an arrangement that would group similar elements together. This tabulation would help chemists focus on similarities and differences among the known elements and predict properties of elements still undiscovered. In this chapter, we will continue to study the first successful tabulation—the periodic table of the elements. Chemists value the periodic table as a means of organizing their field, and they would continue to use it even if they had never figured out why it works. But the underlying rationale of the periodic table was discovered about 50 years after the table was proposed.

The basis of the periodic table is in the electron configurations of the elements, a topic we studied in Chapter 9. In this chapter, we will use the table as a backdrop for a discussion of some properties of the elements—atomic radii, ionization energies, and electron affinities. We will use these atomic properties in the discussion of chemical bonding in the following two chapters, and the periodic table itself will be our indispensable guide throughout much of the remainder of the text.

10-1   Classifying the Elements: The Periodic Law and the Periodic Table

In 1869, Dmitri Mendeleev and Lothar Meyer independently proposed the **periodic law:**

> When the elements are arranged in order of increasing atomic mass, certain sets of properties recur periodically.

Meyer based his periodic law on the property called atomic volume—the atomic mass of an element divided by the density of its solid form. We now call this just the molar volume.

$$\text{atomic (molar) volume (cm}^3/\text{mol)} = \text{molar mass (g/mol)} \times 1/d \text{ (cm}^3/\text{g)} \quad (10.1)$$

Meyer presented his results as a graph of atomic volume against atomic mass. Now it is customary to plot his results as molar volume against atomic number, as seen in Figure 10-1. Notice how high atomic volumes recur periodically for the alkali metals Li, Na, K, Rb, and Cs. Later, Meyer examined other physical properties of the elements and their compounds, such as hardness, compressibility, and boiling points, and found that these also vary periodically.

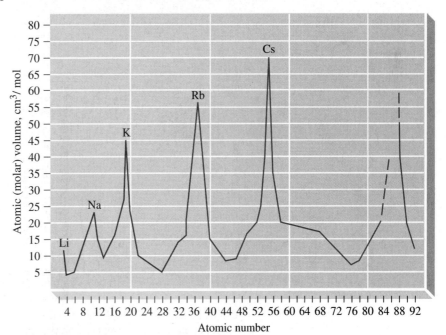

▲ **FIGURE 10-1   An illustration of the periodic law—variation of atomic volume with atomic number**

This adaptation of Meyer's 1870 graph plots atomic volumes against atomic numbers. Of course, a number of elements, such as the noble gases, were undiscovered in Meyer's time. The graph shows peaks at the alkali metals (Li, Na, K, ...). Nonmetals fall on the ascending portions of the curve and metals at the peaks, on the descending portions, and in the valleys.

Mendeleev's Periodic Table

We have already described the periodic table as a tabular arrangement of the elements that groups similar elements together. Mendeleev's work attracted more attention than Meyer's for two reasons: He left blank spaces in his table for undiscovered elements, and he corrected some atomic mass values. The blanks in his table came at atomic masses 44, 68, 72, and 100, for the elements we now know as scandium, gallium, germanium, and technetium. Two of the atomic mass values he corrected were those of indium and uranium.

In Mendeleev's table, similar elements fall in vertical groups, and the properties of the elements change gradually from top to bottom in the group. As an example, we have seen that the alkali metals (Mendeleev's group I) have high molar volumes (Figure 10-1). They also have low melting points, which decrease in the order

▶ We consider additional properties of the alkali metals in Section 10-7.

$$\text{Li (174 °C)} > \text{Na (97.8 °C)} > \text{K (63.7 °C)} > \text{Rb (38.9 °C)} > \text{Cs (28.5 °C)}$$

In their compounds, the alkali metals exhibit the oxidation state +1, forming ionic compounds such as NaCl, KBr, CsI, Li_2O, and so on.

Reihen	Gruppe I. — R^2O	Gruppe II. — RO	Gruppe III. — R^2O^3	Gruppe IV. RH^4 RO^2	Gruppe V. RH^3 R^2O^5	Gruppe VI. RH^2 RO^3	Gruppe VII. RH R^2O^7	Gruppe VIII. — RO^4
1	H = 1							
2	Li = 7	Be = 9,4	B = 11	C = 12	N = 14	O = 16	F = 19	
3	Na = 23	Mg = 24	Al = 27,3	Si = 28	P = 31	S = 32	Cl = 35,5	
4	K = 39	Ca = 40	– = 44	Ti = 48	V = 51	Cr = 52	Mn = 55	Fe = 56, Co = 59, Ni = 59, Cu = 63.
5	(Cu = 63)	Zn = 65	– = 68	– = 72	As = 75	Se = 78	Br = 80	
6	Rb = 85	Sr = 87	?Yt = 88	Zr = 90	Nb = 94	Mo = 96	– = 100	Ru = 104, Rh = 104, Pd = 106, Ag = 108
7	(Ag = 108)	Cd = 112	In = 113	Sn = 118	Sb = 122	Te = 125	J = 127	
8	Cs = 133	Ba = 137	?Di = 138	?Ce = 140	–	–	–	– – – –
9	(–)	–	–	–	–	–	–	
10	–	–	?Er = 178	?La = 180	Ta = 182	W = 184	–	Os = 195, Ir = 197, Pt = 198, Au = 199
11	(Au = 199)	Hg = 200	Tl = 204	Pb = 207	Bi = 208			
12	–	–	–	Th = 231	–	U = 240		

▲ Dmitri Mendeleev (1834–1907). Mendeleev's discovery of the periodic table came as a result of attempting to systematize properties of the elements for presentation in a chemistry textbook. His highly influential book went through eight editions in his lifetime and five more after his death.

In his periodic table Mendeleev arranged the elements into eight groups (Gruppe) and twelve rows (Reihen). The formulas are written as Mendeleev wrote them. R^2O, RO, ..., are formulas of the element oxides (such as Li_2O, MgO, ...); RH^4, RH^3, ..., are formulas of the element hydrides (such as CH_4, NH_3, ...).

▶ The term *eka* is derived from Sanskrit and means "first." That is, eka-silicon means, literally, "first comes silicon" (and then comes the unknown element).

	TABLE 10.1 Properties of Germanium: Predicted and Observed	
Property	**Predicted Eka-silicon (1871)**	**Observed Germanium (1886)**
Atomic mass	72	72.6
Density, g/cm^3	5.5	5.47
Color	dirty gray	grayish white
Density of oxide, g/cm^3	EsO_2: 4.7	GeO_2: 4.703
Boiling point of chloride	$EsCl_4$: below 100 °C	$GeCl_4$: 86 °C
Density of chloride, g/cm^3	$EsCl_4$: 1.9	$GeCl_4$: 1.887

Discovery of New Elements

Two of the elements predicted by Mendeleev were discovered shortly after the appearance of his 1871 periodic table (gallium, 1875; scandium, 1879). Table 10.1 illustrates how closely Mendeleev's predictions for eka-silicon agree with the observed properties of the element germanium, discovered in 1886. Often, new ideas in science take hold slowly, but the success of Mendeleev's predictions stimulated chemists to adopt his table fairly quickly.

A New Group for the Periodic Table

One group of elements that Mendeleev did not anticipate was the noble gases. He left no blanks for them. As we noted in Chapter 8, William Ramsay, their discoverer, proposed placing them in a separate group of the table. Because argon, the first noble gas discovered, had an atomic mass greater than that of chlorine and comparable to that of potassium, Ramsay placed the new group, which he called group 0, between the halogen elements (group VII) and the alkali metals (group I).

Atomic Number as the Basis for the Periodic Law

Mendeleev had to place certain elements out of the order of increasing atomic mass to get them into the proper groups of his periodic table. He assumed this was because of errors in atomic masses. With improved methods of determining atomic masses and with the discovery of argon (group 0, atomic mass 39.9), which was placed ahead of potassium (group I, atomic mass 39.1), it became clear that a few elements might always remain "out of order." At the time, these out-of-order placements were justified by chemical evidence. Elements were placed in the groups that their chemical behavior dictated. There was no theoretical explanation for this reordering. Matters changed in 1913 as a result of some research by H. G. J. Moseley on the X-ray spectra of the elements.

As we learned in Chapter 2, X rays are a high-frequency form of electromagnetic radiation. They are produced when a cathode-ray (electron) beam strikes the anode of a cathode-ray tube. The anode is called the target. Moseley was familiar with Bohr's atomic model, which explains X-ray emission in terms of transitions in which electrons drop into orbits close to the atomic nucleus (see Figure 9-13). Moseley reasoned that because the energies of electron orbits depend on the nuclear charge, the frequencies of emitted X rays should depend on the nuclear charges of atoms in the target. Using techniques newly developed by the father–son team of W. Henry Bragg and W. Lawrence Bragg, Moseley obtained photographic images of X-ray spectra and assigned frequencies to the spectral lines. His spectra for the elements from Ca to Zn are reproduced in Figure 10-2.

▲ **Henry G. J. Moseley (1887–1915)**
Moseley was one of a group of brilliant scientists whose careers were launched under Ernest Rutherford. He was tragically killed at Gallipoli, in Turkey, in World War I.

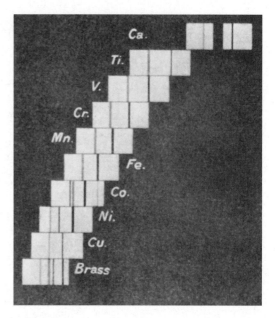

▶ FIGURE 10-2

Moseley's X-ray spectra of several elements
In this photograph from Moseley's 1913 paper, you can see two lines for each element, beginning with Ca at the top. With each successive element, the lines are displaced to the left, the direction of increasing X-ray frequency in these experiments. Where more than two lines appear, the sample had one or more other elements as an impurity. Notice, for example, that one line in the Co spectrum matches a line in the Fe spectrum, and another matches a line in the Ni spectrum. Brass, which is an alloy of copper and zinc, shows two lines for Cu and two for Zn.

Moseley was able to correlate X-ray frequencies to numbers equal to the nuclear charges and corresponding to the positions of elements in Mendeleev's periodic table. For example, aluminum, the thirteenth element in the table, was assigned an *atomic number* of 13. Moseley's equation is $\nu = A(Z - b)^2$, where ν is the X-ray frequency, Z is the atomic number, and A and b are constants. Moseley used this relationship to predict three new elements ($Z = 43, 61,$ and 75), which were discovered in 1937, 1945, and 1925, respectively. Also, he proved that in the portion of the periodic table with which he worked (from $Z = 13$ to $Z = 79$), there could be no additional new elements beyond those three. All available atomic numbers had been assigned. From the standpoint of Moseley's work, then, we should restate the periodic law.

Similar properties recur periodically when elements are arranged according to increasing atomic number.

Description of a Modern Periodic Table: The Long Form

Mendeleev's periodic table consisted of 8 groups, while most modern periodic tables are arranged in 18 groups of elements (see the inside front cover). We gave a description of the periodic table in Section 2-6. Let us briefly review that description.

The vertical groups bring together elements with similar properties. The horizontal periods of the table are arranged in order of increasing atomic number from left to right. The groups are numbered at the top, and the periods at the extreme left in the periodic table on the inside front cover. The first two groups—the *s* block—and the last six groups—the *p* block—together constitute the *main-group* elements. Because they come between the *s* block and the *p* block, the *d*-block elements are known as the *transition* elements. The *f*-block elements, sometimes called the *inner transition* elements, would extend the table to a width of 32 members if incorporated in the main body of the table. The table would generally be too wide to fit on a printed page, and so the *f*-block elements are extracted from the table and placed at the bottom. The 14 elements following lanthanum ($Z = 57$) are called the *lanthanides*, and the 14 following actinium ($Z = 89$) are called the *actinides*.

10-2 Metals and Nonmetals and Their Ions

In Section 2-6, to help us write names and formulas, we established two categories of elements: *metals* and *nonmetals*. At that time, we described *metals* and *nonmetals* in terms of physical properties: Most metals are good conductors of heat and electricity, are malleable and ductile, and have moderate to high melting points. In general, nonmetals are nonconductors of heat and electricity and are nonmalleable (brittle) solids, though a number of nonmetals are gases at room temperature.

Through the color scheme of the periodic table on the inside front cover, we see that the majority of the elements are metals (orange) and that nonmetals (blue) are confined to the right side of the table. The noble gases (purple) are treated as a special group of nonmetals. Metals and nonmetals are often separated by a stairstep diagonal line, and several elements near this line are often called metalloids (green). **Metalloids** are elements that look like metals and in some ways behave like metals, but also have some nonmetallic properties.

In the original periodic table, the positions of the elements were based on readily observable physical and chemical properties. In Chapter 9, we learned of the close correlation between electron configurations and the positions of the elements in the table. Thus, it appears that the physical and chemical properties of an element are determined largely by its electron configuration, particularly that of the *valence* (outermost) electronic shell. Adjacent members of a series of main-group elements in the same period (such as P, S, and Cl) have significantly different properties because they differ in their valence-electron configurations. Within a transition series, differences in electron configurations are mostly in inner shells, and so a transition element has some similarities to neighboring transition elements in the same period. We particularly find many similar properties for adjacent members of the same period within the f block. In fact, the strong similarities among the lanthanide elements presented a particular challenge to the nineteenth-century chemists who tried to separate and identify them.

Let's now briefly explore a few of the links between electron configurations and some other observations about the elements, starting with the noble gases.

Noble Gases

Atoms of the noble gases have the maximum number of electrons permitted in the valence shell of an atom, two in helium $(1s^2)$ and eight in the other noble gas atoms (ns^2np^6). These electron configurations are very difficult to alter and seem to confer a high degree of chemical inertness to the noble gases. It is interesting to note, then, that the s-block metals, together with Al in group 13, tend to lose enough electrons to acquire the electron configurations of the noble gases. On the other hand, nonmetals tend to gain enough electrons to achieve the same configurations.

Main-Group Metal Ions

The electron configurations of the atoms of groups 1 and 2—the most active metals—differ from those of the noble gas of the preceding period by only one and two electrons in the s orbital of a new electron shell. If a K atom is stripped of its outer-shell electron, it becomes the *positive ion* K^+, with the electron configuration [Ar]. A Ca atom acquires the [Ar] configuration following the removal of two electrons.

$$K\ ([Ar]4s^1) \longrightarrow K^+\ ([Ar]) + e^-$$

$$Ca\ ([Ar]4s^2) \longrightarrow Ca^{2+}\ ([Ar]) + 2\ e^-$$

Although metal atoms do not lose electrons spontaneously, the energy required to bring about ionization is often provided by other processes occurring at the same

Interactive Periodic Table activity

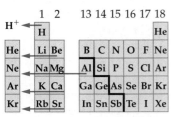
▲ Metals *tend* to lose electrons to attain noble gas electron configurations.

TABLE 10.2	Electron Configurations of Some Metal Ions[a]			
"Noble Gas"		"Pseudo-Noble Gas"[b]	"18 + 2"[c]	Other
Li^+	Be^{2+}	Ga^{3+}	In^+	Cr^{2+}, Cr^{3+}
Na^+	Mg^{2+}	Tl^{3+}	Tl^+	Mn^{2+}, Fe^{2+}
K^+	Ca^{2+}	Cu^+	Sn^{2+}	Fe^{3+}, Co^{2+}
Rb^+	Sr^{2+}	Ag^+, Au^+	Pb^{2+}	Ni^{2+}, Cu^{2+}
Cs^+	Ba^{2+}	Zn^{2+}	Sb^{3+}	
Fr^+	Ra^{2+}		Bi^{3+}	
Al^{3+}				

[a] Main-group metal ions are printed in black and transition metal ions, in blue.
[b] In the configuration labeled "pseudo-noble gas," all electrons of the outermost shell have been lost. The next-to-outermost electron shell of the atom becomes the outermost shell of the ion and contains 18 electrons, for example, Ga^{3+}: $[Ne]3s^23p^63d^{10}$.
[c] In the configuration labeled "18 + 2" all outer-shell electrons except the two s electrons are lost, producing an ion with 18 electrons in the next-to-outermost shell and 2 electrons in the outermost, for example, Sn^{2+}: $[Ar]3d^{10}4s^24p^64d^{10}5s^2$.

time (such as an attraction between positive and negative ions). Aluminum is the only p-block metal that forms an ion with a noble gas electron configuration—Al^{3+}. The electron configurations of the other p-block metal ions are summarized in Table 10.2.

Main-Group Nonmetal Ions

The atoms of groups 17 and 16—the most active nonmetals—have one and two electrons fewer than the noble gas at the end of the period. Groups 17 and 16 atoms can acquire the electron configurations of noble gas atoms by *gaining* the appropriate numbers of electrons.

$$Cl\ ([Ne]3s^23p^5) + e^- \longrightarrow Cl^-\ ([Ar])$$
$$S\ ([Ne]3s^23p^4) + 2\ e^- \longrightarrow S^{2-}\ ([Ar])$$

In most cases, a nonmetal atom will gain a single electron spontaneously, but energy is required to force it to accept additional electrons. Often other processes occurring simultaneously supply the necessary energy (such as an attraction between positive and negative ions). Nonmetal ions with a charge of 3− are rare, but some metal nitrides are described as containing the nitride ion, N^{3-}.

Transition Metal Ions

We have seen that in the aufbau process the ns subshell fills before electrons enter the $(n-1)d$ subshell (page 339), but we have also noted that the energy levels of these two subshells are nearly the same (Figure 9-33). Thus, it is not surprising that when transition metal atoms ionize, the ns subshell is emptied. For example, the electron configuration of Ti is $[Ar]3d^24s^2$, and that of Ti^{2+} is $[Ar]3d^2$. Moreover, we should expect that in some cases one or more $(n-1)d$ electrons might be lost together with the ns electrons. This happens in the formation of Ti^{4+}, which has the electron configuration $[Ar]$.

A few transition metal atoms acquire noble gas electron configurations when forming cations, as do Ti in Ti^{4+} and Sc in Sc^{3+}, but most transition metal atoms *do not* (see Table 10.2). An iron atom does not acquire a noble gas electron configuration when it loses its $4s^2$ electrons to form the ion Fe^{2+},

$$Fe\ ([Ar]3d^64s^2) \longrightarrow Fe^{2+}\ ([Ar]3d^6) + 2\ e^-$$

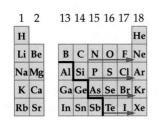

▲ Nonmetals *tend* to gain electrons to attain noble gas electron configurations.

▶ A useful mnemonic is that the electron configuration of a cation can be obtained from the electron configuration of a neutral species by removing those electrons in orbitals with the *highest* quantum number first.

Are You Wondering ...

Why hydrogen, a nonmetal, is included with the group 1 metals in the periodic table?

Although all the other elements have a definite place in the periodic table, hydrogen doesn't. Its uniqueness stems from the fact that its atoms have only one electron, in the configuration $1s^1$. This is what causes us to put hydrogen in group 1, even though we classify it as a nonmetal.[*] Because, like the halogens, hydrogen is one electron short of having a noble gas electron configuration $(1s^2)$, in some periodic tables it is placed in group 17. Hydrogen, however, doesn't resemble the halogens very much. For example, F_2 and Cl_2 are excellent oxidizing agents, but H_2 isn't. Still another alternative that one sometimes sees is hydrogen placed by itself at the top of the periodic table and near the center.

[*]Hydrogen appears to become metallic when subjected to pressures of about 2 million atm, but these are hardly ordinary laboratory conditions.

nor does it with the loss of an additional $3d$ electron to form the ion Fe^{3+}.

$$\text{Fe ([Ar]}3d^6 4s^2) \longrightarrow \text{Fe}^{3+} \text{([Ar]}3d^5) + 3\ \text{e}^-$$

The $3d$ subshell in Fe^{3+} is half-filled, a fact that helps to account for the observed ease of oxidation of iron(II) to iron(III) compounds. Electron configurations with half-filled or filled d or f subshells have a special stability, and we find that a number of transition metal ions have such configurations.

10-3   The Sizes of Atoms and Ions

In earlier chapters we discovered the importance of atomic masses in matters relating to stoichiometry. To understand certain physical and chemical properties, we need to know something about atomic sizes. In this section we describe atomic radius, the first of a group of *atomic* properties that we will examine in this chapter.

Atomic Radius

Unfortunately, atomic radius is hard to define. The probability of finding an electron decreases with increasing distance from the nucleus, but nowhere does the probability fall to zero. There is no precise outer boundary to an atom. We might describe an *effective* atomic radius as, say, the distance from the nucleus within which 90% of all the electron charge density is found, but, in fact, all that we can *measure* is the distance between the nuclei of adjacent atoms (internuclear distance). Even though it varies, depending on whether atoms are chemically bonded or merely in contact without forming a bond, we define atomic radius in terms of internuclear distance.

Because we are primarily interested in bonded atoms, we will emphasize an atomic radius based on the distance between the nuclei of two atoms joined by a chemical bond. The **covalent radius** is one-half the distance between the nuclei of two identical atoms joined by a single covalent bond. The **ionic radius** is based on the distance between the nuclei of ions joined by an ionic bond. Because the ions are not identical in size, this distance must be properly apportioned between the cation and anion. For example, starting with a radius of 140 pm for O^{2-}, the radius of Mg^{2+} can be obtained from the internuclear distance in MgO, the radius of Cl^-

Despite the SI convention, the angstrom unit, Å, is still widely used by X-ray crystallographers and others who work with atomic and molecular dimensions. ▶

Covalent radius:

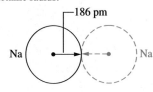

Metallic radius:

186 pm

Na — → Na

Ionic radius:

99 pm

Na⁺ — → Cl⁻

▲ **FIGURE 10-3**
Covalent, metallic, and ionic radii compared
Atomic radii are represented by the solid arrows. The covalent radius is based on the diatomic molecule $Na_2(g)$, found only in *gaseous* sodium. The metallic radius is based on adjacent atoms in solid sodium, Na(s). The value of the ionic radius of Na^+ is obtained by the comparative method described in the text.

Effective Nuclear Charge animation

from the internuclear distance in $MgCl_2$, and the radius of Na^+ from the internuclear distance in NaCl. For metals, we define a **metallic radius** as one-half the distance between the nuclei of two atoms in contact in the crystalline solid metal.

The angstrom unit, Å, has long been used for atomic dimensions (1 Å = 10^{-10} m). The angstrom, however, is not a recognized SI unit. The SI units are the nanometer (nm) and picometer (pm).

$$1 \text{ nm} = 1000 \text{ pm} = 1 \times 10^{-9} \text{ m} \qquad (10.2)$$

Figure 10-3 illustrates the definitions of covalent, ionic, and metallic radii by comparing these three radii for sodium. Figure 10-4 is a plot of atomic radius against atomic number for a large number of elements. In such a plot it is customary to use metallic radii for metals and covalent radii for nonmetals; this is what we have done. Figure 10-4 suggests certain trends in atomic radii, for example, large radii for group 1, decreasing across the periods to smaller radii for group 17. In order to interpret these trends, let us first return to a topic introduced in Chapter 9.

Screening and Penetration

In Section 9-10, we described penetration as a gauge of how close an electron gets to the nucleus. When interpreting the radial probability distributions we saw that *s* electrons, by virtue of their extra humps of probability close to the nucleus (Figure 9-32), penetrate better than *p* electrons, which in turn penetrate better than *d* electrons. Screening, or shielding, reflects how an outer electron is blocked from the nuclear charge by inner electrons. Consider the hypothetical process of building up each atom in the third period from the atom preceding it, beginning with sodium. In this process, the number of inner-shell, or core, electrons is fixed at ten in the configuration $1s^2 2s^2 2p^6$. As a first approximation, let us *assume* that the core electrons completely cancel an equivalent charge on the nucleus. In this way, the core electrons *shield*, or *screen*, the outer-shell electrons from the full attractive force of the nucleus. Let us also assume that the outer-shell electrons do not screen one another. Finally, let us redefine an **effective nuclear charge**, Z_{eff}, first introduced in Section 9-10, as the true nuclear charge minus the charge that is screened out by electrons.

$$Z_{eff} = Z - S \qquad (10.3)$$

Think of *S* as representing the number of inner electrons that appear to screen or shield an outer electron.

Based on the two assumptions just stated, in sodium ($Z = 11$) the ten core electrons would screen out 10 units of nuclear charge (that is, $S = 10$), leaving an effective nuclear charge of $11 - 10 = +1$. In magnesium ($Z = 12$), Z_{eff} would be $+2$. In aluminum, Z_{eff} would be $+3$, and so on across the period.

Actually, neither assumption we made above—full screening by inner-shell electrons and no screening by outer-shell electrons—is correct. These assumptions ignore the fact that the electrons, both inner and outer, occupy orbitals with different radial probability distributions and, consequently, different degrees of penetration. Thus, an *s* electron, with its greater penetration, will be screened by inner electrons less than will a *p* electron. Similarly a *p* electron is shielded less than a *d* electron, which has a much lower penetration. In sodium, the ten core electrons cancel only about 9 units of nuclear charge. Z_{eff} is more nearly $+2$ than $+1$. Also, outer-shell electrons do screen one another somewhat because of penetration effects. Each outer-shell electron is about one-third effective in screening the other outer-shell electrons. Thus, the Z_{eff} that each of the two outer-shell electrons in magnesium experiences is about $12 - 9 - \frac{1}{3} = 2\frac{2}{3}$ (Figure 10-5).

*Periodic Trends:
Atomic Radii
animation*

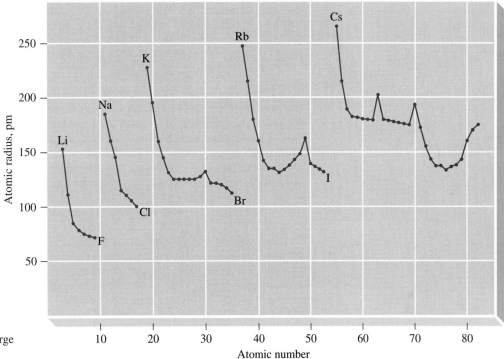

▲ FIGURE 10-4   **Atomic radii**
The values plotted are metallic radii for metals and covalent radii for nonmetals. Data for the noble gases are not included because of the difficulty of measuring covalent radii for these elements (only Kr and Xe compounds are known). (The explanations usually given for the several small peaks in the middle of some periods are beyond the scope of this discussion.)

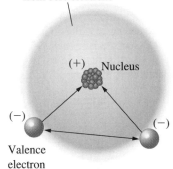

Screen of electron charge
from core electrons

Valence
electron

▲ FIGURE 10-5
**The shielding effect and
effective nuclear charge, Z_{eff}**
Two valence electrons (blue) are
attracted to the nucleus of a Mg
atom. The atom's 12+ nuclear
charge is screened by the 10 core
electrons (gray), but not perfectly.
The valence electrons also screen
each other somewhat. The result
is an *effective* nuclear charge, Z_{eff},
closer to 3+ than to 2+.

▶ The leading term of equation
(10.5) is the Bohr radius of the
nth Bohr orbit. The corresponding formula for the orbitals of
the H atom can be obtained by
setting $Z_{eff} = 1$.

The Effects of Penetration and Screening.   The wave function of multielectron atoms provides a qualitative understanding of the effects of penetration and screening. In this simplified picture, the nuclear charge is replaced by Z_{eff}, so that the orbital energy is approximated by

$$E_n = -R_H \frac{Z_{eff}^2}{n^2} \tag{10.4}$$

where Z_{eff} is the effective nuclear charge in the shell corresponding to the value of n. All other symbols have their usual meaning (see Chapter 9). Equation (10.4) has the same form as the energy of the hydrogen atom obtained as a solution to the Schrödinger equation. The multielectron atom has been reduced to a one-electron approximation, a great, but useful, oversimplification. The average size of an orbital is taken to be the average value of the distance, $\bar{r}_{nl}$, of the electron in that orbital from the nucleus.

$$\bar{r}_{nl} = \frac{n^2 a_0}{Z_{eff}} \left\{ 1 + \frac{1}{2} \left[1 - \frac{l(l+1)}{n^2} \right] \right\} \tag{10.5}$$

In equation (10.5), all the symbols have their usual meaning. Again, equation (10.5) is the equation for the hydrogen atom or hydrogen-like ions with the nuclear charge replaced by Z_{eff} to approximate multielectron effects. Equations (10.4) and (10.5) are approximate but provide for a very useful semiquantitative interpretation of atomic properties. We consider next the three most important trends among atomic radii in relation to the periodic table.

Are You Wondering...

Where estimates of the screening by electrons come from?

These estimates come from an analysis of the wave functions of multielectron atoms. An exact solution of the Schrödinger equation can be obtained for the H atom, but for multielectron atoms, only approximate solutions are possible. The principle of the calculation is to assume each electron in the atom occupies an orbital much like those of the hydrogen atom. However, the functional form of the orbital is based on another assumption: that the electron moves in an effective or average field dictated by all the other electrons. With this assumption, the complicated multielectron Schrödinger equation is converted into a set of simultaneous equations—one for each electron. Each equation contains the unknown effective field and the unknown orbital for the electron. The approach to solving such a set of equations is to guess at the functional form of the orbitals, calculate an average potential for each electron to move in, and then solve for a new set of orbitals—one for each electron. The expectation is that the new orbitals are better than the initial guess. The new orbitals are then used to calculate a new effective field for the electrons, and the whole process is repeated until the calculated orbitals do not change much. This iterative procedure, called the *self-consistent field (SCF) method*, was devised by Douglas Hartree in 1936, before the advent of computers. Currently, the wave functions of atoms and molecules are obtained by implementing the SCF procedures on computers. This has allowed the field of molecular modeling to become an important tool in modern chemical research.

The atomic orbitals obtained from SCF calculations closely resemble the atomic orbitals of the hydrogen atom in many ways. The angular dependence of the orbitals is identical, so that we can identify *s*, *p*, *d*, *f* orbitals by their characteristic shapes. The radial functions of the orbitals are different because the effective field is different from the one found in the hydrogen atom, but the principal quantum number can still be defined. Thus, each electron in a multielectron atom has associated with it the four quantum numbers n, ℓ, m_ℓ, and m_s. Estimates of screening constants are based on an analysis of the radial functions obtained from SCF calculations.

1. **Variation of Atomic Radii Within a Group of the Periodic Table.** Radial probability densities extend farther out from the nucleus as n increases, a fact seen both in Figure 9-32 and in equation (10.5). Thus, we should expect that the more electron shells in an atom occupied by electrons, the larger the atom. This idea works for the group members of lower atomic numbers, where the increase in radius from one period to the next is large (as from Li to Na to K in group 1). At higher atomic numbers, the increase in radius is smaller (as from K to Rb to Cs in group 1). In these elements of higher atomic number, outer-shell electrons are held somewhat more tightly than otherwise expected because inner-shell electrons in *d* and *f* subshells are less effective than *s* and *p* electrons in screening outer-shell electrons from the nucleus; that is, Z_{eff} is larger than expected. Nevertheless, in general, the following is true.

KEEP IN MIND

that *s* and *p* valence electrons have some probability of being near the nucleus (Figure 9-32). These electrons penetrate the inner core of electrons and experience a greater attraction to the nucleus than otherwise expected. ▶

▶ Group trends in the periodic table are largely governed by the principal quantum number, *n*.

The more electronic shells in an atom, the larger is the atom. Atomic radius increases from top to bottom through a group of elements.

2. **Variation of Atomic Radii Within a Period of the Periodic Table.** From Figure 10-4, we see that, in general, atomic radius decreases from left to right across a period. A careful look at the figure suggests that this trend does not apply to the transition elements. Let us look first at the general trend of decreasing radii and then at what is special about the transition elements.

▶ Values of Z_{eff} can be estimated using the rules set out in Feature Problem 76.

▶ Period trends in the periodic table are largely governed by the effective nuclear charge, Z_{eff}.

Across a period, the atomic number increases by one for each succeeding element. For the main-group elements, each increase in atomic number is accompanied by the addition of an electron to the valence shell. The valence-shell electrons, being in the same shell, shield each other poorly from the increasing nuclear charge. The Z_{eff} for the $2s$ electron of Li is 1.3, and that for Be is 1.9; thus, Z_{eff} increases as Z increases across the main-group portions of a period. Across a period, the principal quantum number stays constant, so that whether we use Z_{eff} or just the nuclear charge Z in equation (10.5), the result is as follows.

The atomic radius decreases from left to right through a period of elements.

3. **Variation in Atomic Radius Within a Transition Series.** With the transition elements, the situation is a little different from that described above. In Figure 10-4, it is apparent that the atomic radii of transition elements tend to be about the same across a period with a few unusual peaks. It is beyond the scope of this text to explain the exceptions; however, the general trend is not difficult to understand. In a series of transition elements, additional electrons go into an *inner* electron shell, where they participate in shielding outer-shell electrons from the nucleus. At the same time, the number of electrons in the *outer* shell tends to remain constant. Thus, the outer-shell electrons experience a roughly comparable force of attraction to the nucleus throughout a transition series. Consider Fe, Co, and Ni. Fe has 26 protons in the nucleus and 24 inner-shell electrons. In Co ($Z = 27$), there are 25 inner-shell electrons, and in Ni ($Z = 28$), there are 26. In each case, the two outer-shell electrons are under the influence of about the same net charge (about +2). That is, Z_{eff} for the $4s$ electrons of the first transition series is approximately constant. Thus, atomic radii do not change very much for this series of three elements, namely, 124 pm for Fe and 125 pm for Co and Ni.

EXAMPLE 10-1

Relating Atomic Size to Position in the Periodic Table. Refer only to the periodic table on the inside front cover, and determine which is the largest atom: Sc, Ba, or Se.

Solution

Sc and Se are both in the fourth period, and we would expect Sc to be larger than Se because atomic sizes decrease from left to right in a period. Ba is in the sixth period and so has more electronic shells than either Sc or Se. Furthermore, it lies even closer to the left side of the table (group 2) than does Sc (group 3). We can say with confidence that the Ba atom should be the largest of the three. [The actual atomic radii are Se (117 pm), Sc (161 pm), and Ba (217 pm).]

Practice Example A: Use the periodic table on the inside front cover to predict which is the smallest atom: As, I, or S.

Practice Example B: Which of the following atoms do you think is closest in size to the Na atom: Br, Ca, K, or Al? Explain your reasoning, and do not use any tabulated data from the chapter in reaching your conclusion.

Ionic Radius

When a metal atom loses one or more electrons to form a positive ion, the positive nuclear charge exceeds the negative charge of the electrons in the resulting cation. The nucleus draws the electrons in closer, and as a consequence, the following holds true.

Cations are smaller than the atoms from which they are formed.

Gain and Loss of Electrons animation

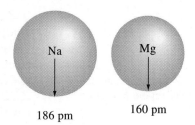

▶ **FIGURE 10-6**
A comparison of atomic and ionic sizes
Metallic radii are shown for Na and Mg and ionic radii for Na^+ and Mg^{2+}.

Figure 10-6 compares four species: the atoms Na and Mg and the ions Na^+ and Mg^{2+}. As we would expect, the Mg atom is smaller than the Na atom, and the cations are smaller than the corresponding atoms. Na^+ and Mg^{2+} are **isoelectronic**—they have equal numbers of electrons (10) in identical configurations, $1s^2 2s^2 2p^6$. Mg^{2+} is smaller than Na^+ because its nuclear charge is larger (+12, compared with +11 for Na).

> For isoelectronic cations, the more positive the ionic charge, the smaller the ionic radius.

When a nonmetal atom gains one or more electrons to form a negative ion (anion), the nuclear charge remains constant, but Z_{eff} is reduced because of the additional electron(s). The electrons are not held as tightly. Repulsions among the electrons increase. The electrons spread out more, and the size of the atom increases, as suggested in Figure 10-7.

> Anions are larger than the atoms from which they are formed. For isoelectronic anions, the more negative the charge, the larger the ionic radius.

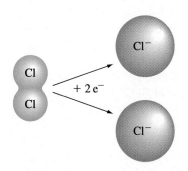

Covalent radius	Ionic radius
99 pm	181 pm

▲ **FIGURE 10-7**
Covalent and anionic radii compared
The two Cl atoms in a Cl_2 molecule gain one electron each to form two Cl^- ions.

EXAMPLE 10-2

Comparing the Sizes of Cations, Anions, and Neutral Atoms. Refer only to the periodic table on the inside front cover, and arrange the following species in order of increasing size: Ar, K^+, Cl^-, S^{2-}, and Ca^{2+}.

Solution

The key lies in recognizing that the five species are *isoelectronic*, having the electron configuration of Ar, $1s^2 2s^2 2p^6 3s^2 3p^6$. For isoelectronic cations, the higher the charge on the ion, the smaller the ion. This means that Ca^{2+} is smaller than K^+. Because K^+ has a higher nuclear charge than Ar ($Z = 19$ compared with $Z = 18$), it is smaller than Ar. And Ar is smaller than Cl^-, again because of a higher nuclear charge ($Z = 18$ compared with $Z = 17$). For isoelectronic anions, the higher the charge, the larger the ion. S^{2-} is larger than Cl^-. The order of increasing size is

$$Ca^{2+} < K^+ < Ar < Cl^- < S^{2-}$$

Practice Example A: Refer only to the periodic table on the inside front cover, and arrange the following species in order of increasing size: Ti^{2+}, V^{3+}, Ca^{2+}, Br^-, and Sr^{2+}.

Practice Example B: Refer only to the periodic table on the inside front cover, and determine which species is in the *middle* position when the following five are ranked according to size: the atoms N, Cs, and As and the ions Mg^{2+} and Br^-.

▶ We can summarize the generalizations about isoelectronic atoms and ions into a single statement: Among *isoelectronic* species, the greater the atomic number, the smaller the size.

Knowledge of atomic and ionic radii can be used to vary certain physical properties. One example concerns the strengthening of glass. Normal window glass contains Na^+ and Ca^{2+} ions. The glass is brittle and shatters easily when struck a hard blow. One way to strengthen the glass is to replace the Na^+ ions at the surface with K^+ ions. The K^+ ions are larger and fill up the surface sites, leaving less opportunity for cracking than is the case with the smaller Na^+ ions. The result is a shatter-resistant glass.

Another example is the striking result when Cr^{3+} ions replace about 1% of the Al^{3+} ions in aluminum oxide, Al_2O_3. This substitution is possible because Cr^{3+} ions are only slightly larger (by 9 pm) than Al^{3+} ions. Pure aluminum oxide is colorless, but with this small amount of chromium(III) ion, it is a beautiful red color. This impure Al_2O_3 is the gem known as a ruby. Rubies and other gemstones can be made artificially and are used as jewelry and in devices such as lasers. The color of the ruby is further discussed in Chapter 25.

Figure 10-8, arranged in the format of the periodic table, shows relative sizes of typical atoms and ions. It summarizes the generalizations described in this section.

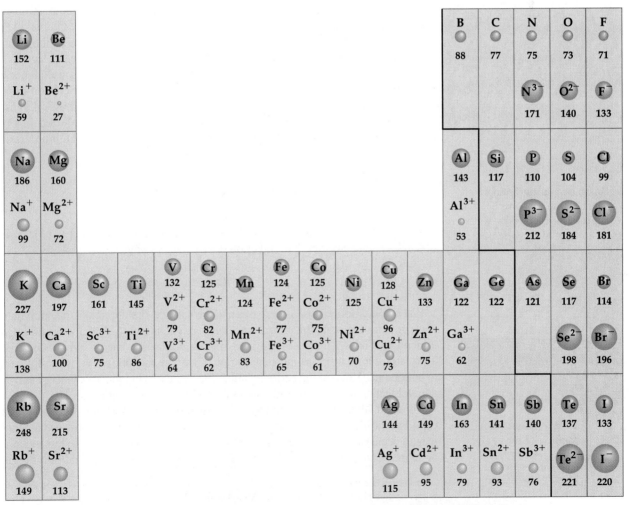

▲ **FIGURE 10-8**   **A comparison of some atomic and ionic radii**
The values given, in picometers (pm), are metallic radii for metals, single covalent radii for nonmetals, and ionic radii for the ions indicated.

10-4   Ionization Energy

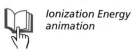

Ionization Energy animation

In discussing metals, we talked about metal atoms losing electrons and thereby altering their electron configurations. But atoms do not eject electrons spontaneously. Electrons are attracted to the positive charge on the nucleus of an atom, and energy is needed to overcome that attraction. The more easily its electrons are lost, the more metallic we consider an atom to be. The **ionization energy**, *I*, is the quantity of energy a *gaseous* atom must absorb so that an electron is stripped from it. The electron lost is the one most loosely held.

Ionization energies are measured through experiments in which gaseous atoms at low pressures are bombarded with beams of electrons (cathode rays). Here are two typical values.

$$Mg(g) \longrightarrow Mg^+(g) + e^- \qquad I_1 = 738 \text{ kJ/mol}$$

$$Mg^+(g) \longrightarrow Mg^{2+}(g) + e^- \qquad I_2 = 1451 \text{ kJ/mol}$$

The symbol I_1 stands for the *first* ionization energy—the energy required to strip one electron from a neutral gaseous atom.[*] I_2 is the *second* ionization energy—the energy to strip an electron from a gaseous ion with a charge of 1+. Further ionization energies are I_3, I_4, ... Invariably, we find that each succeeding ionization energy is larger than the preceding one. In the case of magnesium, for example, in the second ionization, the electron, once freed, has to move away from an ion with a charge of 2+ (Mg^{2+}). More energy must be invested than for a freed electron to move away from an ion with a charge of 1+ (Mg^+). This is a direct consequence of Coulomb's law, which states, in part, that the force of attraction between oppositely charged particles is directly proportional to the magnitudes of the charges.

First ionization energies (I_1) for many of the elements are plotted in Figure 10-9. In general, the farther an electron is from the nucleus, the more easily it can be extracted.

> Ionization energies decrease as atomic radii increase.

This observation reflects the effect of *n* and Z_{eff} on the ionization energy (I). Equation (10.4) suggests that the ionization energy is given by

$$I = R_H \times \frac{Z_{eff}^2}{n^2} \tag{10.6}$$

so that across a period, as Z_{eff} increases and the valence-shell principal quantum number *n* remains constant, the ionization energy should increase. And down a group, as *n* increases and Z_{eff} increases only slightly, the ionization energy should decrease. Thus, atoms lose electrons more easily (become more metallic) as we move from top to bottom in a group of the periodic table. The decreases in ionization energy and the parallel increases in atomic radii are outlined in Table 10.3 for group 1.

Table 10.4 lists ionization energies for the third-period elements. With minor exceptions, the trend in moving across a period (follow the colored stripe) is that atomic radii decrease, ionization energies increase, and the elements become less metallic, or more nonmetallic, in character. Table 10.4 lists stepwise ionization energies

▶ This generalization works well for main-group elements but less so for transition elements, where there are several exceptions.

TABLE 10.3   Atomic Radii and First Ionization Energies of the Alkali Metal (Group 1) Elements

	Atomic Radius, pm	Ionization Energy (I_1), kJ/mol
Li	152	520.2
Na	186	495.8
K	227	418.8
Rb	248	403.0
Cs	265	375.7

[*]Ionization energies are sometimes expressed in the unit electronvolt (eV). One electronvolt is the energy acquired by an electron as it falls through an electric potential difference of 1 volt. It is a very small energy unit, especially suited to describing processes involving individual atoms. When ionization is based on a *mole* of atoms, kJ/mol is the preferred unit. One eV/atom = 96.49 kJ/mol. Sometimes the term *ionization potential* is used instead of ionization energy. Further, the quantities $I_1, I_2, ...$, may be replaced by enthalpy changes, $\Delta H_1, \Delta H_2, ...$

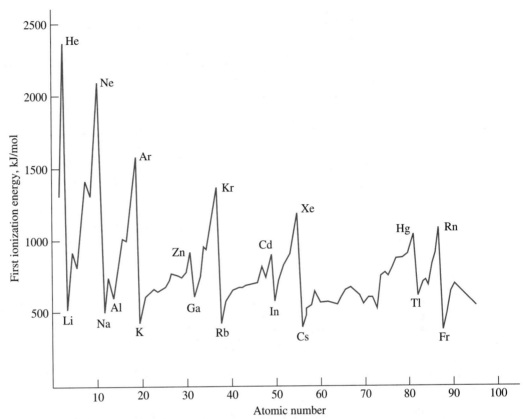

▲ **FIGURE 10-9 First ionization energies as a function of atomic number**
Because their electron configurations are so stable, more energy is required to ionize
noble gas atoms than to ionize atoms of the elements immediately preceding or following
them. The maxima on the graph come at the atomic numbers of the noble gases. The alkali metals are the most easily ionized of all groups. The minima in the graph come at
their atomic numbers.

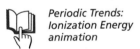

Periodic Trends:
Ionization Energy
animation

$(I_1, I_2, \dots)$. Note particularly the large breaks that occur along the zigzag diagonal
line. Consider magnesium as an example. To remove a *third* electron, as measured by
I_3, requires breaking into the especially stable noble gas subshell electron configuration
$2s^2 2p^6$. I_3 is *much* larger than I_2—so much larger that Mg^{3+} cannot be produced in ordinary chemical processes. Similarly, we do not expect to find the ions Na^{2+} or Al^{4+}.

TABLE 10.4	Ionization Energies of the Third-Period Elements (in kJ/mol)						
	Na	**Mg**	**Al**	**Si**	**P**	**S**	**Cl**	**Ar**
I_1	495.8	737.7	577.6	786.5	1012	999.6	1251.1	1520.5
I_2	4562	1451	1817	1577	1903	2251	2297	2666
I_3		7733	2745	3232	2912	3361	3822	3931
I_4			11580	4356	4957	4564	5158	5771
I_5				16090	6274	7013	6542	7238
I_6					21270	8496	9362	8781
I_7						27110	11020	12000

Are You Wondering...

If ionization energies can be used to estimate the effective nuclear charge?

One of the earliest estimates of effective nuclear charge was obtained by analyzing ionization energies in terms of equation (10.6). Thus, for example, the ionization energy of Li in its ground state is 519 kJ mol^{-1} and from equation 10.6, we have

$$I.E. = 1312.1 \frac{Z_{eff}^2}{n^2} \text{ kJ mol}^{-1}$$

so that

$$519 \text{ kJ mol}^{-1} = 1312.1 \frac{Z_{eff}^2}{2^2} \text{ kJ mol}^{-1}$$

and we get

$$Z_{eff} = 1.26$$

The Z_{eff} obtained from the ionization energy of the first excited state of Li ($1s^2 2p$), 339 kJ mol^{-1}, is 1.02. The value of Z_{eff} is much closer to unity because the inner $1s^2$ core almost perfectly screens the $2p$ electron, whereas the penetration of the $2s$ electron leads to a larger Z_{eff}.

Now let us turn to the obvious exceptions to the regular trend in I_1 values for the third-period elements and ask, Why is I_1 of Al smaller than that of Mg and I_1 of S smaller than that of P?

We expect I_1 of Al to be *larger* than for Mg. The reversal occurs because of the particular electrons lost. Mg loses a $3s$ electron, while Al loses a $3p$ electron. We expect that *more* energy is required to strip an electron from the lower energy $3s$ orbital in Mg ([↑↓]$_{3s}$[ ][ ][ ]$_{3p}$) than from a half-filled $3p$ orbital in Al ([↑↓]$_{3s}$[↑] [ ][ ]$_{3p}$). I_1 for S is slightly lower than for P for a different reason. Although the orbitals in the $3p$ subshell are degenerate, we can think of repulsion between electrons in the filled $3p$ orbital of a S atom ([↑↓][↑][↑]$_{3p}$) as making it easier to remove one of those electrons than an electron from the half-filled $3p$ subshell of a P atom ([↑][↑][↑]$_{3p}$).

KEEP IN MIND ▶
the orbital energy diagram in Figure 9-33 and the order in which electrons occupy orbitals.

EXAMPLE 10-3

Relating Ionization Energies and Atomic Radii. Refer to the periodic table on the inside front cover, and arrange the following in the expected order of increasing first ionization energy, I_1: As, Sn, Br, Sr.

Solution

If we arrange these four atoms according to decreasing radius, we will likely have arranged them according to increasing ionization energy. The largest atoms are to the left and the bottom of the periodic table. Of the four atoms, the one that best fits the large-atom category is *Sr*. The smallest atoms are to the right and toward the top of the periodic table. Although none of the four atoms is particularly close to the top of the table, *Br* is the best fit for this category because it is farthest to the right. This fixes the two extremes: Sr with the lowest ionization energy and Br with the highest. A tin atom should be larger than an arsenic atom, and thus Sn should have a lower ionization energy than As. The expected order of *increasing* ionization energies is Sr < Sn < As < Br.

Practice Example A: Refer to the periodic table on the inside front cover, and arrange the following in the expected order of increasing first ionization energy, I_1: Cl, K, Mg, S.

Practice Example B: Refer to the periodic table on the inside front cover, and determine which element is most likely in the *middle* position when the following five elements are arranged according to first ionization energy, I_1: Rb, As, Sb, Br, Sr.

10-5   Electron Affinity

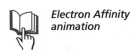

Electron Affinity animation

Ionization energy concerns the loss of electrons. **Electron affinity**, *EA*, is a measure of the energy change that occurs when a *gaseous* atom gains an electron. For example,

$$F(g) + e^- \longrightarrow F^-(g) \qquad EA = -328\,\text{kJ/mol}$$

When a F atom gains an electron, energy is given off. The process is *exothermic*, and, following the thermochemical conventions we established in Chapter 7, the electron affinity is a negative quantity.[*]

Why should a neutral fluorine atom so readily gain an electron? A plausible explanation is that when a free electron approaches a F atom from an "infinite" distance away, the electron "sees" a center of positive charge—the atomic nucleus—to which it is attracted. This attraction is offset to some extent by the repulsive effect of other electrons in the atom. But, so long as the attractive force on the additional electron exceeds the repulsive force, the electron is gained and energy is given off. In becoming F^-, a fluorine atom acquires the very stable electron configuration of the noble gas neon (Ne). That is,

$$F\left(1s^2 2s^2 2p^5\right) + e^- \longrightarrow F^-\left(1s^2 2s^2 2p^6\right)$$

Even metal atoms can form negative ions *in the gaseous state*, where they will only rarely contact one another. This is the situation for gaseous Li atoms, for example, where the added electron enters the half-filled 2s orbital.

$$Li(g) + e^- \longrightarrow Li^-(g) \qquad EA = -59.6\,\text{kJ/mol}$$
$$\left(1s^2 2s^1\right) \qquad\quad \left(1s^2 2s^2\right)$$

For some atoms, there is no tendency to gain an electron. This is the case with the noble gases, where the added electron would have to enter the empty *s* orbital of the next electronic shell; the groups 2 and 12 elements, where the electron would have to enter the *p* subshell of the valence shell; and a few other elements, such as Mn, where the electron would have to enter either the *p* subshell of the valence shell or a half-filled 3*d* subshell.

In considering the gain of a *second* electron by a nonmetal atom, we encounter *positive* electron affinities. Here the electron to be added is approaching not a neutral atom, but a negative *ion*. A strong repulsion is felt, and the energy of the system increases. Thus, for an element like oxygen, the first electron affinity is negative and the second is positive.

$$O(g) + e^- \longrightarrow O^-(g) \qquad\qquad EA_1 = -141.0\,\text{kJ/mol}$$

$$O^-(g) + e^- \longrightarrow O^{2-}(g) \qquad\qquad EA_2 = +744\,\text{kJ/mol}$$

---

[*]We have defined electron affinity to reflect the tendency for a neutral atom to gain an electron. An alternative definition refers to the energy change in the process: $X^-(g) \longrightarrow X(g) + e^-$, that is, reflecting the tendency of an anion to lose an electron. This alternative definition leads to the opposite signs for *EA* values from those written in this text. You should be prepared to see electron affinities expressed in both ways in the chemical literature.

Periodic Trends: Electron Affinity animation

1	2	13	14	15	16	17	18
H −72.8							**He** --
Li −59.6	**Be** --	**B** −26.7	**C** −153.9	**N** −7	**O** −141.0	**F** −328.0	**Ne** --
Na −52.9	**Mg** --	**Al** −42.5	**Si** −133.6	**P** −72	**S** −200.4	**Cl** −349.0	**Ar** --
K −48.4	**Ca** --	**Ga** −28.9	**Ge** −119.0	**As** −78	**Se** −195.0	**Br** −324.6	**Kr** --
Rb −46.9	**Sr** --	**In** −28.9	**Sn** −107.3	**Sb** −103.2	**Te** −190.2	**I** −295.2	**Xe** --
Cs −45.5	**Ba** --	**Tl** −19.2	**Pb** −35.1	**Bi** −91.2	**Po** −186	**At** −270	**Rn** --

▲ FIGURE 10-10   **Electron affinities of main-group elements**
Values are in kilojoules per mole for the process $X(g) + e^- \longrightarrow X^-(g)$.

The high positive value of EA_2 makes the formation of *gaseous* O^{2-} seem very unlikely. The ion O^{2-} can exist, however, in ionic compounds such as MgO(s), where formation of the ion is accompanied by other energetically favorable processes.

Some representative electron affinities are listed in Figure 10-10. Generalizations about electron affinities are more difficult to make than about ionization energies. We can see that the smaller atoms to the right of the periodic table (for example, group 17) tend to have large negative electron affinities.[*] Electron affinities tend to become less negative in progressing toward the bottom of a group, but the second period group members (for example, N, O, and F) do not conform to this trend. It is likely that for these small atoms, an incoming electron encounters strong repulsions from other electrons in the atom and is thereby not as tightly bound as we might otherwise expect.

10-6   Magnetic Properties

Another property related to the electron configurations of atoms and ions is their behavior in a magnetic field. A spinning electron is an electric charge in motion. It induces a magnetic field (recall the discussion on page 333). In a **diamagnetic** atom or ion, all electrons are paired and the individual magnetic effects cancel out. A diamagnetic species is weakly repelled by a magnetic field. A **paramagnetic** atom or ion has *unpaired* electrons, and the individual magnetic effects do not cancel out. The unpaired electrons possess a magnetic moment that causes the atom or ion to be attracted into an external magnetic field. The more unpaired electrons present, the stronger is this attraction.

KEEP IN MIND ▶
that it was the effect of the magnetic moments due to the electron's two spin quantum numbers (equal in magnitude and opposite in sign) that allowed Stern and Gerlach to detect the presence of electron spin using a magnetic field.

[*]It is somewhat awkward to speak of larger and smaller with the term *electron affinity*. A strong tendency to gain an electron, which implies a high "affinity" for an electron, as with F and Cl, is reflected through a *low* value of *EA*—a large *negative* value.

Manganese has a paramagnetism corresponding to five unpaired electrons, which is consistent with the electron configuration

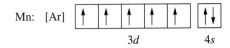

When a manganese atom loses two electrons, it becomes the ion Mn^{2+}. Mn^{2+} is paramagnetic, and the strength of its paramagnetism corresponds to five unpaired electrons.

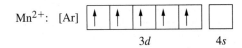

When a third electron is lost to produce Mn^{3+}, we find that the ion has a paramagnetism corresponding to four unpaired electrons. The third electron lost is one of the unpaired $3d$ electrons.

EXAMPLE 10-4

Determining the Magnetic Properties of an Atom or Ion. Which of the following would you expect to be diamagnetic, and which paramagnetic?

    (a) Na atom   **(b)** Mg atom   **(c)** Cl^- ion   **(d)** Ag atom

Solution

    (a) Paramagnetic. The Na atom has a single $3s$ electron outside the Ne core. This electron is unpaired.

    (b) Diamagnetic. The Mg atom has *two* $3s$ electrons outside the Ne core. They must be paired, as are all the other electrons.

    (c) Diamagnetic. Cl^- is isoelectronic with Ar, and Ar has all electrons paired $\left(1s^2 2s^2 2p^6 3s^2 3p^6\right)$.

    (d) Paramagnetic. We do not need to work out the exact electron configuration of Ag. Because the atom has 47 electrons—an odd number—at least one of the electrons must be unpaired (recall the Stern–Gerlach experiment, page 333).

Practice Example A: Which of the following are paramagnetic, and which are diamagnetic: Zn, Cl, K^+, O^{2-}, and Al?

Practice Example B: Which has the greater number of unpaired electrons, Cr^{2+} or Cr^{3+}? Explain.

10-7   Periodic Properties of the Elements

As we noted at the beginning of the chapter, we can use the periodic law and the periodic table to predict some of the atomic, physical, and chemical properties of elements and compounds.

Atomic Properties

In this chapter we have learned how some atomic properties—atomic radius, ionization energy, electron affinity—vary within groups and periods of elements. We

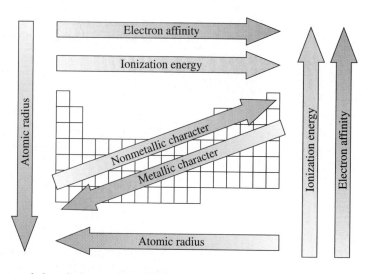

▶ **FIGURE 10-11**
Atomic properties and the periodic table—a summary
Atomic radius refers to metallic radius for metals and covalent radius for nonmetals. Ionization energies refer to first ionization energy. For electron affinity, the direction is that in which values become *more negative*. Metallic character relates generally to the ability to lose electrons, and nonmetallic character, to gain electrons.

Summary of Atomic Properties activity

summarize these trends in relation to the periodic table in Figure 10-11. Trends are generally easy to apply within a group: The atomic radius of Sr is greater than that of Mg; both elements are in group 2. Usually, there is no difficulty in applying trends within a period either: The first ionization energy of P is greater than that of Mg; both elements are in the third period. However, comparing elements that are not within the same group or period can be difficult. We have no problem seeing that the atomic radius of Sr is greater than that of P. Sr is farther down in its group of the periodic table and much farther to the left in its period than is P. Each of these directions is that of increasing atomic radius. On the other hand, we cannot easily predict whether Mg or I has the larger atomic radius. The position of Mg to the left in its period suggests that Mg should have the larger radius, but the position of I toward the bottom of its group argues for I. Despite this limitation, you should find Figure 10-11 helpful in most cases.

Variation of Physical Properties Within a Group. Table 10.5 lists some properties of three of the halogens (group 17). The table has two blank spaces for bromine. We fill in these blanks in Example 10-5 by making an assumption that works often enough to make it useful:

The value of a property often changes uniformly from the top to the bottom of a group of elements in the periodic table.

First, let us make some predictions about fluorine, the halogen not listed in Table 10.5. Its closest neighbor in group 17 is chlorine, which has a boiling point of 239 K (−34 °C); chlorine is a *gas* at room temperature (about 298 K). The other halogens are *liquid* bromine and *solid* iodine (Figure 10-12). We would expect fluorine to have a lower melting point and lower boiling point than chlorine and also to be a *gas* at room temperature. (Observed values for F_2: mp = 53 K; bp = 85 K.)

▲ **FIGURE 10-12**
Three halogen elements
Chlorine is a yellow-green gas. Bromine is a dark red liquid. Iodine is a grayish black solid.

TABLE 10.5 Some Properties of Three Halogen (Group 17) Elements

	Atomic Number	Atomic Mass, u	Molecular Form	Melting Point, K	Boiling Point, K
Cl	17	35.45	Cl_2	172	239
Br	35	79.90	Br_2	?	?
I	53	126.90	I_2	387	458

EXAMPLE 10-5

Using the Periodic Table to Estimate Physical Properties. Use data from Table 10.5 to estimate the boiling point of bromine.

Solution

The atomic number of bromine (35) is intermediate to the atomic numbers of chlorine (17) and iodine (53). Its atomic mass (79.90 u) is also about intermediate to those of chlorine and iodine. (The average of the atomic masses of Cl and I is 81.18 u.) It is reasonable to expect that the boiling point of liquid bromine might also be intermediate to the boiling points of chlorine and iodine.

$$\text{bp Br}_2 = \frac{239 \text{ K} + 458 \text{ K}}{2} = 349 \text{ K}$$

The observed boiling point is 332 K.

Practice Example A: Estimate the melting point of bromine.

Practice Example B: Estimate the boiling point of astatine, At.

▶ When evaluating trends, it is often useful to sketch a graph showing the variation of the property.

The generalization that a property varies uniformly within a group of the periodic table can work for compounds as well as for elements. Table 10.6 lists the melting points of two sets of compounds, binary carbon–halogen compounds and the *hydrogen halides*, HX (where X = F, Cl, Br, or I). We see that the melting points increase fairly uniformly with increasing molecular mass for the carbon–halogen compounds. This trend between melting point (and boiling point) and molecular mass can be explained in terms of intermolecular forces, as we shall see in Chapter 13. Based on the melting points of HCl, HBr, and HI, we should expect the melting point of HF to be about −145 °C, but the observed value is −83.6 °C. Some factor other than molecular mass must be involved here. In Chapter 13, we will find that in HF there is a special intermolecular force of attraction that is missing or unimportant in the other compounds in Table 10.6.

Variation of Physical Properties Across a Period. A few properties vary regularly across a period. The ability to conduct heat and electricity are two that do. Thus, among the third-period elements, the metals Na, Mg, and Al have good thermal and electrical conductivities. The metalloid Si is only a fair conductor of heat and electricity. The nonmetals P, S, Cl, and Ar have poor thermal and electrical conductivities.

In some cases, the trend in a property reverses direction in the period (similar to our experience above with the trend in melting points of the hydrogen halides reversing direction within a group). Consider, for example, the melting points of the third-period elements shown in a bar graph in Figure 10-13. Melting involves destruction of the orderly arrangement of atoms or molecules found in a crystalline solid. The amount of thermal energy needed for melting to occur, and hence the melting point temperature, depends on the strength of the attractive forces between atoms or molecules in the solid. For the metals Na, Mg, and Al, these forces are *metallic bonds*, which, roughly speaking, become stronger as the number of electrons available to participate in the bonding increases. Sodium, therefore, has the lowest melting point (371 K) of the third-period metals. With silicon, the forces between atoms are strong *covalent bonds* extending throughout the crystalline solid. Silicon has the highest melting point (1683 K) of the third-period elements. Phosphorus, sulfur, and chlorine exist as discrete molecules (P_4, S_8, and Cl_2). The bonds between atoms within molecules are strong, but *intermolecular forces*, the forces

TABLE 10.6 Melting Points of Two Series of Compounds

	Molecular Mass, u	Melting Point, °C
CF_4	88.0	−183.7
CCl_4	153.8	−22.9
CBr_4	331.6	90.1
CI_4	519.6	171
HF	20.0	−83.6
HCl	36.5	−114.2
HBr	80.9	−86.8
HI	127.9	−50.8

▶ Metallic bonds are described in Section 12-7; covalent bonding in substances like silicon is discussed in Section 13-7; and the topic of intermolecular forces is examined throughout Chapter 13.

Sodium and Potassium in Water movie

(a)

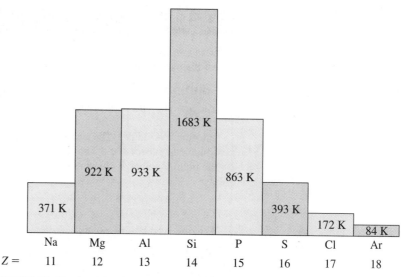

▲ **FIGURE 10-13 Melting points of the third-period elements**
The trend outlined in this bar graph is discussed in the text.

(b)

▲ **FIGURE 10-14**
The reactions of potassium and of calcium with water: a comparison
(a) Potassium, a group 1 metal, reacts so rapidly that the hydrogen evolved bursts into flame. Notice that the metal is less dense than water. **(b)** Calcium, a group 2 metal, reacts more slowly than does potassium. Also, calcium is denser than water. The pink color of the acid–base indicator phenolphthalein signals the buildup of OH^- ions.

between molecules, become progressively weaker across the period, and the melting points decrease. Argon atoms do not form molecules, and the forces between Ar atoms in solid argon are especially weak. Argon's melting point is the lowest for the entire period (84 K). The property of hardness also depends on forces between atoms and molecules in a solid. So the hardness of the solid third-period elements vary in much the same way as their melting points. Thus, on a 10-point scale in which solids are rated according to their abilities to scratch or abrade one another, sodium has a hardness of about 0.5; magnesium, 2; aluminum, 3; silicon, 7; and phosphorus and sulfur 1-2. Silicon has the greatest hardness.

Reducing Abilities of Group 1 and 2 Metals.   We learned in Chapter 5 that a reducing agent makes possible a reduction half-reaction. The reducing agent itself, by losing electrons, is oxidized. In the following reactions, M, a group 1 or 2 metal, is the reducing agent and H_2O is the substance that is reduced.

$$2\,M(s) + 2\,H_2O(l) \longrightarrow 2\,M^+(aq) + 2\,OH^-(aq) + H_2(g) \quad (M = \text{group 1 metal})$$

$$M(s) + 2\,H_2O(l) \longrightarrow M^{2+}(aq) + 2\,OH^-(aq) + H_2(g) \quad (M = \text{Ca, Sr, Ba, or Ra})$$

At a first guess, we might think that the lower the energy requirement for extracting electrons—the lower the ionization energy—the better the metal is as a reducing agent and the more vigorous its reaction with water. Potassium, for instance, has a lower ionization energy ($I_1 = 419$ kJ/mol) than does the next member of the fourth period, calcium ($I_1 = 590; I_2 = 1145$ kJ/mol). Our expectation is that potassium should react more vigorously with water than calcium. This, indeed, is the case (Figure 10-14). Mg and Be do not react with cold water as do the other alkaline earth metals. We might explain this in terms of the higher ionization energies for those two metals (Mg: $I_1 = 738, I_2 = 1451$ kJ/mol; Be: $I_1 = 900, I_2 = 1757$ kJ/mol).

Attributing these group 1 and 2 metal reactivities just to ionization energies is an oversimplification, however. As long as the differences in ionization energies are very large, we may get by with considering only this factor. Where differences in ionization energies are smaller, though, other factors must be considered when making comparisons.

▲ **FIGURE 10-15**
Reaction of sodium metal and chlorine gas
The contents of the flask glow in this exothermic reaction between Na(s) and Cl_2(g). The product is the ionic solid NaCl(s).

(a)   (b)

▲ **FIGURE 10-16**
Displacement of I^-(aq) by Cl_2(g)
(a) Cl_2(g) is bubbled through colorless, dilute I^-(aq). (b) The I_2 produced is extracted into CCl_4(l), in which it is much more soluble (purple layer).

Oxidizing Abilities of the Halogen Elements (Group 17). An oxidizing agent gains the electrons that are lost in an oxidation half-reaction. The oxidizing agent, by gaining electrons, is itself reduced. Electron affinity is the atomic property introduced in this chapter that is related to the gain of electrons. We might expect an atom with a strong tendency to gain electrons (a large *negative* electron affinity) to take electrons away from atoms with low ionization energies—metals. In these terms, it is understandable that active metals form ionic compounds with active nonmetals. If M is a group 1 metal and X a group 17 nonmetal (halogen), this exchange of an electron leads to the formation of M^+ and X^- ions. In some cases, the reaction is especially vigorous (Figure 10-15).

$$2 M + X_2 \longrightarrow 2 MX \qquad [\text{e.g., } 2 Na(s) + Cl_2(g) \longrightarrow 2 NaCl(s)]$$

Another interesting oxidation–reduction reaction involving the halogens is a *displacement* reaction. Two halogens, one in molecular form and the other in ionic form, exchange places, as in this reaction (Figure 10-16).

$$Cl_2(g) + 2 I^-(aq) \longrightarrow I_2(aq) + 2 Cl^-(aq)$$

We might think of this reaction as involving a competition between Cl and I atoms for an extra electron that only the I atoms (as I^-) have initially. The Cl atoms win out because they have a more negative electron affinity. (This is an oversimplified explanation, however, because electron affinities apply only to the behavior of isolated gaseous atoms and not to atoms in molecules or ions in solution.) By similar reasoning can you see why no reaction occurs for this combination?

$$Br_2(l) + Cl^-(aq) \longrightarrow \text{no reaction}$$

Predictions such as those in the preceding paragraph work well for the halogens Cl_2, Br_2, and I_2, but not for F_2. We cannot account for the *observed* fact that F_2 is the strongest oxidizing agent among all chemical substances, just by considering electron affinities.

Acid–Base Nature of Element Oxides. Some metal oxides, such as Li_2O, react with water to produce the metal hydroxide.

$$Li_2O(s) + H_2O(l) \longrightarrow 2 Li^+(aq) + 2 OH^-(aq)$$
a basic oxide   lithium hydroxide

These metal oxides are called *basic* oxides or *base anhydrides*. The term **anhydride** means "without water." A "base without water" becomes a base when the water is added. Thus, the base anhydride Li_2O becomes the base LiOH, and BaO becomes $Ba(OH)_2$ after reaction with water.

Some nonmetal oxides react with water to produce an acidic solution. These are *acidic* oxides or *acid anhydrides*. SO_2(g) reacts with water to produce H_2SO_3, a weak acid.

$$SO_2(g) + H_2O(l) \longrightarrow H_2SO_3(aq)$$
an acidic oxide   sulfurous acid

Now let us examine the acid–base properties of the oxides of the third-period elements. We should expect the metal oxides at the left of the period to be basic and the nonmetal oxides at the right to be acidic. But where and how does the changeover occur? Na_2O and MgO yield basic solutions in water. Cl_2O, SO_2, and P_4O_{10} produce acidic solutions. SiO_2 (quartz) does not dissolve in water. However, it does dissolve slightly in strongly basic solutions to produce silicates (similar to the carbonates formed by CO_2 in basic solutions). For this reason, we consider SiO_2 to be an acidic oxide.

Mercury is a liquid at room temperature. You are probably aware of this fact, but to chemists, it is surprising. Our expectation from the periodic law is that mercury should be a solid. Zinc and cadmium, the elements above mercury in group 12, are both solids. Their melting points are 419.6 and 320.9 °C, respectively. Based on these values and applying the predictive methods of Section 10-7, we should expect mercury to be a soft solid metal with a melting point of about 200 °C.

Suppose we base our prediction on mercury's sixth-period neighbors. To the immediate left of mercury, we find the metals Pt ($Z = 78$, mp = 1172 °C) and Au ($Z = 79$, mp = 1064 °C). To the immediate right, are two more metals: Tl ($Z = 81$, mp = 303 °C) and Pb ($Z = 82$, mp = 328 °C). From these data, we might expect a melting point for Hg ($Z = 80$) of well over 300 °C. Instead, we observe a melting point of −38.86 °C.

Aluminum, a good conductor of heat and electricity, is clearly metallic in its physical properties. Al_2O_3, however, can act as either an acidic *or* basic oxide. Oxides with this ability are called **amphoteric** (from the Greek word *amphos*, meaning "both"). Al_2O_3 is insoluble in water, but exhibits its amphoterism by reacting with both acidic and basic solutions.

$$Al_2O_3(s) + 6\ HCl(aq) \longrightarrow 2\ AlCl_3(aq) + 3\ H_2O(l)$$

base   acid

$$Al_2O_3(s) + 2\ NaOH(aq) + 3\ H_2O(l) \longrightarrow 2\ Na[Al(OH)_4](aq)$$

acid   base   sodium aluminate

The amphoterism of Al_2O_3 signifies the point at which a changeover from basic to acidic oxides occurs in the third period of elements. Figure 10-17 summarizes the acid/base properties of the oxides of the main-group elements.

1	2	13	14	15	16	17
Li	Be	B	C	N	O	F
Na	Mg	Al	Si	P	S	Cl
K	Ca	Ga	Ge	As	Se	Br
Rb	Sr	In	Sn	Sb	Te	I
Cs	Ba	Tl	Pb	Sn	Po	At

▲ FIGURE 10-17
Acidic, basic, and amphoteric oxides of the *s*- and *p*-block elements
The acidic oxides are red, the basic oxides are blue, and the amphoteric oxides are tan.

Summary

The experimental basis of the periodic table of the elements is the periodic law: Certain properties recur periodically when the elements are arranged by increasing atomic number. The theoretical basis is that the properties of an element are related to the electron configuration of its atoms, and elements in the same group of the periodic table have similar electron configurations.

The metallic and nonmetallic characters of atoms can be related to a set of atomic properties. In general, large atomic radii and low ionization energies are associated with metals; small atomic radii, high ionization energies, and large negative electron affinities are associated with nonmetals. The magnetic properties of an atom or ion stem from the presence or absence of unpaired electrons and are useful in establishing electron configurations. Several examples of predictions and comparisons of properties based on the periodic table are given in the chapter.

The reason that mercury is a liquid at room temperature is something of a mystery and not completely understood. In this limited discussion, we can indicate only the most likely explanation, and this comes from an unlikely source—Einstein's theory of relativity. According to this theory, the mass of a particle *increases* if the particle travels at speeds approaching the speed of light. The value listed in Table 2-1 for the mass of an electron is its *rest mass*. We are justified in using the rest mass as long as an electron travels at low to moderate speeds. In atoms of high atomic number, though, because of the high nuclear charge, electrons that come close to the nucleus are accelerated to high speeds and their masses increase.

In the wave–mechanical model, the relativity effect is especially pronounced for *s* electrons because these are the electrons that have a rather high probability of being found near the nucleus. With increased speed and, therefore, mass of an electron, there is a corresponding decrease in size of the electron orbital and a lowering of the orbital energy. In Hg, because of the lowering of the energy of the $6s$ orbital and the stability of the $5s^25p^65d^{10}6s^2$ electron configuration, the bonds between Hg atoms are too weak to maintain a solid structure at room temperature. Mercury becomes almost as stable as a noble gas—sort of a noble liquid.

The relativistic shrinking of *s* orbitals affects all the heavy metal atoms, but it is at a maximum with mercury. As new heavy elements are characterized, such as the recently discovered elements with $Z = 112, 114, 116$, and 118, chemists may find other significant deviations from the periodic law.

Integrative Example

Francium $(Z = 87)$ is an extremely rare, radioactive element formed when actinium $(Z = 89)$ undergoes α-particle emission. Francium occurs in natural uranium minerals, but estimates are that there is no more than 15 g of francium in the top 1 km of Earth's crust. Few of francium's properties have been measured, but some can be inferred from its position in the periodic table.

Estimate the melting point, density, and atomic (metallic) radius of francium.

1. *Locate Fr (Z = 87) in the periodic table.* The symbol Fr corresponds to the element with atomic number 87 at the bottom of group 1 in period 7.

2. *Determine the melting point of Fr.* First, refer to the melting points of the alkali metals (group 1) listed on page 358: Li, 174° C; Na, 97.8° C; K, 63.7° C; Rb, 38.9° C; Cs, 28.5° C.

We observe that both the melting points and the successive differences in melting point decrease down the group. As seen from the bar graph, the decrease from K to Rb is about 25° C, and from Rb to Cs it is about 10° C. So, following this trend of an approximate halving of the difference, we expect the difference from Cs to Fr to be about 5° C. The expected melting point of Fr is therefore about 24° C.

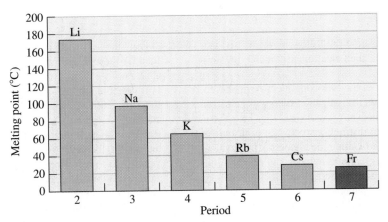

▲ Trend of melting point with period number.

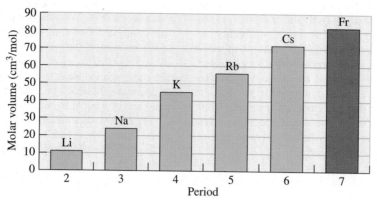

▲ Trend of molar volumes with period number.

3. *Determine the atomic volume and density of Fr.* To obtain the trend in atomic volumes for group 1, we can estimate the volumes from Figure 10-1. The estimates are: Li, 12 cm^3/mol; Na, 23 cm^3/mol; K, 45 cm^3/mol; Rb, 56 cm^3/mol; Cs, 70 cm^3/mol.

As seen from the bar graph, the molar volume increases down the group and the rate of increase is roughly constant as we approach Cs. The increase from K to Rb is about 11 cm^3/mol and from Rb to Cs, about 14 cm^3/mol. A reasonable estimate of the atomic volume of Fr might be about 82 cm^3/mol.

To estimate the density of Fr, we first need to estimate its atomic mass. We can take the molar mass of Fr ($Z = 87$) to be the average of the molar masses of Rn ($Z = 86$) and Ra

($Z = 88$): $\dfrac{(222 + 226)}{2} = 224$ g/mol. To obtain the density, we use the relationship between atomic (molar) volume, molar mass, and density (equation 10.1) to obtain

$$\text{density} = \frac{224 \text{ g/mol}}{82 \text{ cm}^3/\text{mol}} = 2.7 \text{ g/cm}^3$$

4. *Determine the atomic radius of Fr.* Using Figure 10-4, we note that the atomic radius of Cs (265 pm) is about 15 pm more than that of Rb (250 pm). We might expect the atomic radius of Fr to be 15 pm larger still, that is, 280 pm.

Key Terms

amphoteric (10-7)
anhydride (10-7)
covalent radius (10-3)
diamagnetic (10-6)
effective nuclear charge, Z_{eff} (10-3)

electron affinity, *EA* (10-5)
ionic radius (10-3)
ionization energy, *I* (10-4)
isoelectronic (10-3)
metallic radius (10-3)

metalloid (10-2)
paramagnetic (10-6)
periodic law (10-1)

Review Questions

1. In your own words, define the following terms: **(a)** isoelectronic; **(b)** valence-shell electrons; **(c)** metal; **(d)** nonmetal; **(e)** metalloid.
2. Briefly describe each of the following ideas or phenomena: **(a)** the periodic law; **(b)** ionization energy; **(c)** electron affinity; **(d)** paramagnetism.
3. Explain the important distinctions between each pair of terms: **(a)** actinide and lanthanide element; **(b)** covalent and metallic radius; **(c)** atomic number and effective nuclear charge; **(d)** ionization energy and electron affinity; **(e)** paramagnetic and diamagnetic.
4. Find three pairs of elements that are out of order in the periodic table in terms of their atomic masses. Why is it necessary to invert their order in the table?
5. For the atom $^{119}_{50}$Sn, indicate the number of **(a)** protons in the nucleus; **(b)** neutrons in the nucleus; **(c)** $4d$ electrons; **(d)** $3s$ electrons; **(e)** $5p$ electrons; **(f)** electrons in the valence shell.

6. Among the following ions, several pairs are *isoelectronic*. Identify these pairs. Fe^{2+}, Sc^{3+}, Ca^{2+}, F^-, Co^{2+}, Co^{3+}, Sr^{2+}, Cu^+, Zn^{2+}, Al^{3+}.
7. Is it possible for two different atoms to be isoelectronic? two different cations? two different anions? a cation and an anion? Explain.
8. Refer to the periodic table on the inside front cover, and indicate **(a)** the most nonmetallic element; **(b)** the transition metal with lowest atomic number; **(c)** a metalloid whose atomic number is exactly midway between those of two noble gas elements.
9. For each of the following pairs, indicate the atom that has the *larger* size: **(a)** Te or Br; **(b)** K or Ca; **(c)** Ca or Cs; **(d)** N or O; **(e)** O or P; **(f)** Al or Au.
10. Indicate the *smallest* and the *largest* species (atom or ion) in the following group: Al atom, F atom, As atom, Cs^+ ion, I^- ion, N atom.

11. Use principles established in this chapter to arrange the following atoms in order of *increasing* value of the first ionization energy: Sr, Cs, S, F, As.

12. Give the symbol of the element (a) in group 14 that has the smallest atoms; (b) in period 5 that has the largest atoms; (c) in group 17 that has the lowest first ionization energy.

13. How much energy, in joules, must be absorbed to convert to Na^+ all the atoms present in 1.00 mg of *gaseous* Na? The first ionization energy of Na is 495.8 kJ/mol.

14. Refer only to the periodic table on the inside front cover, and indicate which of the atoms Bi, S, Ba, As, and Ca (a) is most metallic; (b) is most nonmetallic; (c) has the intermediate value when the five are arranged in order of increasing first ionization energy.

15. Arrange the following elements in order of *decreasing* metallic character: Sc, Fe, Rb, Br, O, Ca, F, Te.

16. Which of the following species would you expect to be diamagnetic and which paramagnetic? (a) K^+; (b) Cr^{3+}; (c) Zn^{2+}; (d) Cd; (e) Co^{3+}; (f) Sn^{2+}; (g) Br.

17. Match each of the lettered items in the column on the left with the most appropriate numbered item(s) in the column on the right. Some of the numbered items may be used more than once and some, not at all.

(a) Tl
(b) $Z = 70$
(c) Ni
(d) $[Ar]4s^2$
(e) a metalloid
(f) a nonmetal

1. an alkaline earth metal
2. element in period 5 and group 15
3. largest atomic radius of all the elements
4. an element in period 4 and group 16
5. $3d^8$
6. one p electron in the shell of highest n
7. lowest ionization energy of all the elements
8. an f-block element

18. The boiling points of the noble gases increase down the group as follows: He, 4.2; Ne, 27.1; Ar, 87.3; Kr, 119.7; and Xe, 165 K. What would you predict to be the boiling point of radon?

Exercises

The Periodic Law

19. Use data from Figure 10-1 and equation (10.1) to estimate the density expected for the recently discovered element 114. Assume a mass number of 298.

20. Suppose that lanthanum ($Z = 57$) were a newly discovered element having a density of 6.145 g/cm³. Estimate its molar mass.

21. The following densities, in grams per cubic centimeter, are for the listed elements in their standard states at 298 K. Show that density is a periodic property of these elements:

Al, 2.699; Ar, 0.0018; As, 5.778; Br, 3.100; Ca, 1.550; Cl, 0.0032; Ga, 5.904; Ge, 5.323; Kr, 0.0037; Mg, 1.738; P, 1.823; K, 0.856; Se, 4.285; Si, 2.336; Na, 0.968; S, 2.069.

22. The following melting points are in degrees Celsius. Show that melting point is a periodic property of these elements: Al, 660; Ar, −189; Be, 1278; B, 2300; C, 3350; Cl, −101; F, −220; Li, 179; Mg, 651; Ne, −249; N, −210; O, −218; P, 590; Si, 1410; Na, 98; S, 119.

The Periodic Table

23. Mendeleev's periodic table did not preclude the possibility of a new group of elements that would fit within the existing table, as was the case with the noble gases. Moseley's work did preclude this possibility. Explain this difference.

24. Explain why the several periods in the periodic table do not all have the same number of members.

25. Assuming that the seventh period is 32 members long, what should be the atomic number of the noble gas following radon (Rn)? of the alkali metal following francium (Fr)? What would you expect their approximate atomic masses to be?

26. Concerning the incomplete seventh period of the periodic table, what should be the atomic number of the element (a) for which the filling of the 6d subshell is completed; (b) that should most closely resemble bismuth; (c) that should be a noble gas?

Atomic Radii and Ionic Radii

27. Explain why the radii of atoms do not simply increase uniformly with increasing atomic number.

28. The masses of individual atoms can be determined with great precision, yet there is considerable uncertainty about the exact size of an atom. Explain why this is the case.

29. Which is (a) the smallest atom in group 13; (b) the smallest of the following atoms: Te, In, Sr, Po, Sb? Why?

30. How would you expect the sizes of the hydrogen ion, H^+, and the hydride ion, H^-, to compare with that of the H atom and the He atom? Explain.

31. All the isoelectronic species illustrated in the text had the electron configurations of noble gases. Can two ions be isoelectronic *without* having noble gas electron configurations? Explain.

32. The following species are *isoelectronic* with the noble gas krypton. Arrange them in order of *increasing* radius, and comment on the principles involved in doing so: Rb^+, Y^{3+}, Br^-, Sr^{2+}, Se^{2-}.

33. Arrange the following in expected order of *increasing* radius: Br, Li^+, Se, I^-. Explain your answer.

34. Explain why we cannot use the generalizations presented in Figure 10-11 to answer the question, Which is larger, an Al atom or an I atom?

Ionization Energies; Electron Affinities

35. Are there any atoms for which the second ionization energy (I_2) is smaller than the first (I_1)? Explain.

36. Some electron affinities are negative quantities, and some are zero or positive. Why is this not also the case with ionization energies?

37. How much energy, in kilojoules, is required to remove all the third-shell electrons in a mole of gaseous silicon atoms?

38. What is the maximum number of Cs^+ ions that can be produced per joule of energy absorbed by a sample of gaseous Cs atoms?

39. The production of gaseous bromide ions from bromine molecules can be considered a two-step process in which the first step is

$$Br_2(g) \longrightarrow 2\,Br(g) \qquad \Delta H = +193\;kJ$$

Is the formation of $Br^-(g)$ from $Br_2(g)$ an endothermic or exothermic process?

40. Use ionization energies and electron affinities listed in the text to determine whether the following reaction is endothermic or exothermic.

$$Mg(g) + 2\,F(g) \longrightarrow Mg^{2+}(g) + 2\,F^-(g)$$

41. The Na^+ ion and the Ne atom are isoelectronic. The ease of loss of an electron by a gaseous Ne atom, I_1, has a value of 2081 kJ/mol. The ease of loss of an electron from a gaseous Na^+ ion, I_2, has a value of 4562 kJ/mol. Why are these values not the same?

42. From the data in Figure 10-10, the formation of a gaseous *anion* Li^- appears energetically favorable. That is, energy is given off when gaseous Li atoms accept electrons. Comment on the likelihood of forming a stable compound containing the Li^- ion, such as Li^+Li^- or Na^+Li^-.

Magnetic Properties

43. Unpaired electrons are found in only one of the following species. Indicate which one, and explain why: F^-, Ca^{2+}, Fe^{2+}, S^{2-}.

44. Write electron configurations consistent with the following data on numbers of unpaired electrons: Ni^{2+}, two; Cu^{2+}, one; Cr^{3+}, three.

45. Must all atoms with an odd atomic number be paramagnetic? Must all atoms with an even atomic number be diamagnetic? Explain.

46. Neither Fe^{2+} nor Fe^{3+} has $4s$ electrons in its electron configuration. How many unpaired electrons would you expect to find in each of these ions? Explain.

Predictions Based on the Periodic Table

47. Use ideas presented in this chapter to indicate **(a)** three metals that you would expect to exhibit the photoelectric effect with visible light and three that you would not; **(b)** the noble gas element that should have the highest density in the liquid state; **(c)** the approximate I_1 of fermium $(Z = 100)$; **(d)** the approximate density of solid radium $(Z = 88)$.

48. The heat of atomization for an element is the amount of energy required to convert an appropriate amount of an element in its standard state to a mole of atoms in the gaseous state. The heats of atomization, in kilojoules per mole, for three of the group 14 elements are: carbon, 717; silicon, 452; and tin, 302 kJ/mol. Predict the heat of atomization of germanium.

49. Gallium is a commercially important element (used to make gallium arsenide for the semiconductor industry). Gallium was unknown in Mendeleev's time, and he predicted properties of this element. Predict the following for gallium: **(a)** its density; **(b)** the formula and percent composition of its oxide.
[*Hint:* Use Figure 10-1, equation (10.1), and Mendeleev's periodic table (page 358).]

50. Estimate the missing boiling point in the following series of compounds.
(a) CH_4, $-164\,°C$; SiH_4, $-112\,°C$; GeH_4, $-90\,°C$; SnH_4, _____ ? $°C$.
(b) H_2O, ____ ? $°C$; H_2S, $-61\,°C$; H_2Se, $-41\,°C$; H_2Te, $-2\,°C$. Does your estimate in (b) agree with the known value?

51. For the following groups of elements, select the one that has the property noted.
(a) the largest atom: Mg, Mn, Mo, Ba, Bi, Br
(b) the lowest first ionization energy: B, Sr, Al, Br, Mg, Pb
(c) the most negative electron affinity: As, B, Cl, K, Mg, S
(d) the largest number of unpaired electrons: F, N, S^{2-}, Mg^{2+}, Sc^{3+}, Ti^{3+}

52. Match each of the lettered items on the left with an appropriate numbered item on the right. All the numbered items should be used, and some more than once.
(a) $Z = 32$
(b) $Z = 8$
(c) $Z = 53$
(d) $Z = 38$
(e) $Z = 48$
(f) $Z = 20$

1. two unpaired p electrons
2. diamagnetic
3. more negative electron affinity than elements on either side of it in the same period
4. first ionization energy lower than that of Ca but greater than that of Cs

Integrative and Advanced Exercises

53. Complete and balance the following equations. If no reaction occurs, so state.
   (a) $Rb(s) + H_2O(l) \longrightarrow$
   (b) $I_2(s) + Na^+(aq) + Br^-(aq) \longrightarrow$
   (c) $SrO(s) + H_2O(l) \longrightarrow$
   (d) $SO_3(g) + H_2O(l) \longrightarrow$

54. Write balanced equations to represent
   (a) the displacement of a halide anion from aqueous solution by liquid bromine
   (b) the reaction with water of an alkali metal with $Z > 50$
   (c) the reaction of tetraphosphorus decoxide with water
   (d) the reaction of aluminum oxide with aqueous sulfuric acid

55. Four atoms and/or ions are sketched below in accordance with their relative atomic and/or ionic radii.

Which of the following sets of species are compatible with the sketch? Explain.
   (a) C, Ca^{2+}, Cl^-, Br^-; **(b)** Sr, Cl, Br^-, Na^+; **(c)** Y, K, Ca, Na^+; **(d)** Al, Ra^{2+}, Zr^{2+}, Mg^{2+}; **(e)** Fe, Rb, Co, Cs.

56. Discuss the likelihood that element 168, should it ever be synthesized in sufficient quantity, would be a "noble liquid" at 298 K and 1 atm. Some data that may be of use are tabulated here. Might element 168 be a "noble solid" at 298 K and 1 atm? Use *spdf* notation to show the electron configuration you would expect for element 168.

Element	Atomic Mass, u	mp, K	bp, K
Argon	39.948	83.95	87.45
Helium	4.0026	—	4.25
Krypton	83.80	116.5	120.9
Neon	20.179	24.48	27.3
Radon	222	202	211.4
Xenon	131.29	161.3	166.1

57. Sketch a periodic table that would include *all* the elements in the main body of the table. How many "numbers" wide would the table be?

58. In Mendeleev's time, indium oxide, which is 82.5% In, by mass, was thought to be InO. If this were the case, in which group of Mendeleev's table (page 358) should indium be placed?

59. Instead of accepting the atomic mass of indium implied by the data in Exercise 58, Mendeleev proposed that the formula of indium oxide is In_2O_3. Show that this assumption places indium in the proper group of Mendeleev's periodic table on page 358.

60. Listed below are two atomic properties of the element germanium. Refer only to the periodic table on the inside front cover, and indicate probable values for each of the following elements, expressed as greater than, about equal to, or less than the value for Ge.

Element	Atomic Radius	First Ionization Energy
Ge	122 pm	762 kJ/mol
Al	?	?
In	?	?
Se	?	?

61. In estimating the boiling point and freezing point of bromine in Example 10-5, could we have used Celsius or Fahrenheit instead of Kelvin temperature? Explain.

62. In the formula X_2, if the two X atoms are the same halogen, the substance is a halogen element (e.g., Cl_2, Br_2). If the two X atoms are different (such as Cl and Br), we describe an *interhalogen* compound. Use data from Table 10-5 to estimate the melting points and boiling points of the *interhalogen* compounds BrCl and ICl.

63. Refer to Figure 10-8 and explain why the difference between the ionic radii of the 1− and 2− anions does not remain constant from top to bottom of the periodic table.

64. Explain why the third ionization energy of Li(g) is an easier quantity to calculate than either the first or second ionization energies. Calculate I_3 for Li, and express the result in kJ/mol.

65. Studies done in 1880 showed that a chloride of uranium had 37.34% Cl by mass and an approximate formula mass of 382 u. Other data indicated the specific heat of uranium to be $0.0276 \text{ cal g}^{-1}\,°C^{-1}$. Are these data in agreement with the atomic mass of uranium assigned by Mendeleev? (*Hint:* Refer to Feature Problem 111 of Chapter 7.)

66. Assume that atoms are hard spheres, and use the metallic radius of 186 pm for Na to estimate the volume of one Na atom and of one mole of Na atoms. How does your result compare with the atomic volume found in Figure 10-1? Why is there so much disagreement between the two values?

67. When sodium chloride is strongly heated in a flame, the flame takes on the yellow color associated with the emission spectrum of sodium atoms. The reaction that occurs in the *gaseous* state is

$$Na^+(g) + Cl^-(g) \longrightarrow Na(g) + Cl(g).$$

Calculate ΔH for this reaction.

68. Use information from Chapters 9 and 10 to calculate the *second* ionization energy for the He atom. Compare your result with the tabulated value of 5251 kJ/mol.

69. Refer only to the periodic table on the inside front cover, and arrange the following ionization energies in order of *increasing* value: I_1 for F; I_2 for Ba; I_3 for Sc; I_2 for Na; I_3 for Mg. Explain the basis of any uncertainties.

70. Refer to the footnote on page 370. Then, use values of basic physical constants and other data from the appendices to show that 1 eV/atom = 96.49 kJ/mol.

71. Plot a graph of the square roots of the ionization energies versus the nuclear charge for the series Li, Be^+, B^{2+}, C^{3+} and Na, Mg^+, Al^{2+}, Si^{3+}. Explain the observed relationship with the aid of Bohr's expression for the binding energy of an electron in a one-electron atom.

Feature Problems

72. The work functions for a number of metals are given in the table.

Metal	Work Function ($J \times 10^{19}$)
Al	6.86
Cs	3.45
Li	4.6
Mg	5.86
Na	4.40
Rb	3.46

How do the work functions vary
(a) down a group?
(b) across a period?
(c) Estimate the work function for potassium, and compare it with a published value.
(d) What periodic property is the work function most like?

73. The following are a few elements and their characteristic X-ray wavelengths:

Element	X-ray Wavelength (pm)
Mg	987
S	536
Ca	333
Cr	229
Zn	143
Rb	93

Use these data to determine the constants A and b in Moseley's relationship (page 360). Compare your value of A with the value obtained from Bohr's theory for the frequencies emitted by one-electron atoms. Suggest a reasonable interpretation of the quantity b.

74. Gaseous sodium atoms absorb quanta with the energies shown

Energy of quanta (kJ mol^{-1})	Electron Configuration
0	[Ne] $3s^1$
203	[Ne] $3p^1$
308	[Ne] $4s^1$
349	[Ne] $3d^1$
362	[Ne] $4p^1$

(a) The ionization energy of the ground state is 496 kJ mol^{-1}. Calculate the ionization energies for each of the states given in the table.
(b) Calculate Z_{eff} for each state.
(c) Calculate $\bar{r}_{nl}$ for each state.
(d) Interpret the results obtained from parts (b) and (c) in terms of penetration and screening.

75. A method for estimating electron affinities is to extrapolate Z_{eff} values for atoms and ions that contain the same number of electrons as the negative ion of interest. Using the following data.

Atom or Ion: IE(kJ mol^{-1})	Atom or Ion: IE(kJ mol^{-1})	Atom or Ion: IE(kJ mol^{-1})
Ne: 2080	F: 1681	O: 1314
Na$^+$: 4565	Ne$^+$: 3963	F$^+$: 3375
Mg^{2+}: 7732	Na^{2+}: 6912	Ne^{2+}: 6276
Al^{3+}: 11,577	Mg^{3+}: 10,548	Na^{3+}: 9540

(a) Estimate the electron affinity of F, and compare it with the experimental value.
(b) Estimate the electron affinities of O and N.
(c) Examine your results in terms of penetration and screening.

76. We have seen that the wave functions of hydrogen-like atoms contain the nuclear charge Z for hydrogen-like atoms and ions, but modified through the following equation to account for the phenomenon of shielding or screening,

$$Z_{eff} = Z - S$$

where Z_{eff} is the effective nuclear charge and S is the shielding or screening constant. In 1930, John C. Slater devised the following set of empirical rules to calculate a shielding constant for a designated electron in the orbital ns or np.
(i) Write the electron configuration of the element, and group the subshells as follows: $(1s)$, $(2s, 2p)$, $(3s, 3p)$, $(3d)$, $(4s, 4p)$, $(4d)$, $(4f)$, $(5s, 5p)$, ...
(ii) Electrons in groups to the right of the (ns, np) group contribute nothing to the shielding constant for the designated electron.
(iii) All the other electrons in the (ns, np) group shield the designated electron to the extent of 0.35 each.
(iv) All electrons in the $n - 1$ shell shield to the extent of 0.85 each.
(v) All electrons in the $n - 2$ shell, or lower, shield completely—their contributions to the shielding constant are 1.00 each.

When the designated electron being shielded is in an nd or nf group, rules (ii) and (iii) remain the same but rules (iv) and (v) are replaced by

(vi) Each electron in a group lying to the left of the nd or nf group contributes 1.00 to the shielding constant.

These rules are a simplified generalization based on the average behavior of different types of electrons. Use these rules to do the following:

(a) Calculate Z_{eff} for a valence electron of oxygen.

(b) Calculate Z_{eff} for the $4s$ electron in Cu.

(c) Calculate Z_{eff} for a $3d$ electron in Cu.

(d) Evaluate the Z_{eff} for the valence electrons in the group 1 elements (including H), and show that the ionization energies observed for this group are accounted for by using the Slater rules.

(*Hint:* Do not overlook the effect of n on the orbital energy.)

(e) Evaluate Z_{eff} for a valence electron in the elements Li through Ne, and use the results to explain the observed trend in first ionization energies for these elements.

(f) Using the radial functions given in Table 9.1 and Z_{eff} estimated with the Slater rules, compare plots of the radial probability for the $3s$, $3p$, and $3d$ orbitals for the H atom and the Na atom. What do you observe from these plots regarding the effect of shielding on radial probability distributions?

eMedia Exercises

77. For each of the classes of elements illustrated in the **Interactive Periodic Table** activity *(eChapter 10-2)*, describe the type of ions formed by the elements and the driving force behind the formation.

78. After watching the **Ionization Energy** animation *(eChapter 10-5)*, **(a)** write balanced chemical equations describing the processes associated with the first four ionization energies of aluminum. **(b)** Write the electron configuration of each atom and ion found in these equations. **(c)** What is the percent increase in ionization energy for each of these steps? **(d)** Why is I_4 of aluminum so much larger than the third ionization energy?

79. The magnitudes of the electron affinities of argon and chlorine are explained by their respective electron configurations in the **Periodic Trends: Electron Affinity** animation *(eChapter 10-6)*. Using a similar treatment, account for the irregular pattern in electron affinities of the other elements in the third row of the periodic table. For example, why would phosphorus have a less exothermic electron affinity than silicon?

80. The **Physical Properties of the Halogens** movie *(eChapter 10-7)* depicts the periodic variation of melting and boiling points with the atomic size of the elements. Referring to the **Interactive Periodic Table**, describe the trend in these physical properties for the period 2 elements and account for the variations observed across this *row* of the periodic table.

81. Of the properties described in this chapter (see animations in *eChapter 10-4 through 10-8*), which of the chemical properties exhibit trends that are dictated by common atomic factors? What are the underlying physical factors dictating each of these trends?

11 Chemical Bonding I: Basic Concepts

Contents

Computer-generated, three-dimensional molecular models of methanol (upper left), CH_3OH, and ethanol (bottom right), CH_3CH_2OH. In this chapter, we study ideas that enable us to predict the geometric shapes of molecules.

Consider all that we already know about chemical compounds. We can determine their compositions and write their formulas. We can represent the reactions of compounds by chemical equations and perform stoichiometric and thermochemical calculations based on these equations. And we can do all this without really having to consider the ultimate structure of matter—the structure of atoms and molecules. Yet the shape of a molecule—that is, the arrangement of its atoms in space— often defines its chemistry. If water had a different shape, its properties would be significantly different from those familiar to us, and life as we know it would not be possible.

In this chapter, we will describe the interactions between atoms called chemical bonds. Most of our discussion will center on one of the simplest methods of representing chemical bonding, known as the Lewis theory. We will also explore, however, another relatively simple theory, one for predicting probable molecular shapes. Throughout the chapter, we will try to relate these theories to what is known about molecular structures from experimental measurements. In Chapter 12, we will examine the subject of chemical bonding in more depth, and in Chapter 13 we will describe forces between molecules—intermolecular forces—and explore further the relationship between molecular shape and the properties of substances.

11-1   Lewis Theory: An Overview

In the period 1916–1919, two Americans, G. N. Lewis and Irving Langmuir, and a German, Walther Kossel, advanced an important proposal about chemical bonding: Something unique in the electron configurations of noble gas atoms accounts for their inertness, and atoms of other elements combine with one another to acquire electron configurations like noble gas atoms. The theory that grew out of this model has been most closely associated with G. N. Lewis and is called the **Lewis theory**. Some fundamental ideas associated with Lewis's theory are

▶ Since 1962, a number of compounds of Xe and Kr have been synthesized. As we will see in this chapter, a focus on noble gas electron configurations can still be useful, even if the idea that they confer complete inertness is invalid.

1. Electrons, especially those of the outermost (valence) electronic shell, play a fundamental role in chemical bonding.

2. In some cases, electrons are *transferred* from one atom to another. Positive and negative ions are formed and attract each other through electrostatic forces called **ionic bonds**.

▶ The term *covalent* was introduced by Irving Langmuir.

3. In other cases, one or more pairs of electrons are *shared* between atoms. A bond formed by the sharing of electrons between atoms is called a **covalent bond**.

4. Electrons are transferred or shared in such a way that each atom acquires an especially stable electron configuration. Usually this is a noble gas configuration, one with eight outer-shell electrons, or an **octet**.

Lewis Symbols and Lewis Structures

Lewis developed a special set of symbols for his theory. A **Lewis symbol** consists of a chemical symbol to represent the nucleus and *core* (inner-shell) electrons of an atom, together with dots placed around the symbol to represent the *valence* (outer-shell) electrons. Thus, the Lewis symbol for silicon, which has the electron configuration $[Ne]3s^2 3p^2$ is

·Ṣi·

▲ Gilbert Newton Lewis (1875–1946). Lewis's contribution to the study of chemical bonding is evident throughout this text. Equally important, however, was his pioneering introduction of thermodynamics into chemistry.

Electron spin had not yet been proposed when Lewis framed his theory, and so he did not show that two of the valence electrons $(3s^2)$ are paired and two $(3p^2)$ are unpaired. We will write Lewis symbols in the way Lewis did. We will place single dots on the sides of the symbol, up to a maximum of four. Then we will pair up dots until we reach an octet. Lewis symbols are commonly written for main-group elements but much less often for transition elements. Lewis symbols for several main-group elements are written in Example 11-1.

Periodic Trends: Lewis Structures activity

EXAMPLE 11-1

Writing Lewis Symbols. Write Lewis symbols for the following elements: **(a)** N, P, As, Sb, Bi; **(b)** Al, I, Se, Ar.

Solution

(a) These are the elements of group 15. Their atoms all have *five* valence electrons $(ns^2\, np^3)$. The Lewis symbols all have *five* dots.

$$\cdot \overset{\displaystyle \cdot}{\underset{\displaystyle \cdot}{N}} \cdot \qquad \cdot \overset{\displaystyle \cdot}{\underset{\displaystyle \cdot}{P}} \cdot \qquad \cdot \overset{\displaystyle \cdot}{\underset{\displaystyle \cdot}{As}} \cdot \qquad \cdot \overset{\displaystyle \cdot}{\underset{\displaystyle \cdot}{Sb}} \cdot \qquad \cdot \overset{\displaystyle \cdot}{\underset{\displaystyle \cdot}{Bi}} \cdot$$

(b) Al is in group 13; I, in group 17 ; Se, in group 16; Ar, in group 18.

$$\cdot \dot{Al} \cdot \qquad :\overset{\displaystyle \cdot}{\underset{\displaystyle \cdot}{I}} \cdot \qquad :\overset{\displaystyle \cdot}{\underset{\displaystyle \cdot}{Se}} \cdot \qquad :\overset{\displaystyle \cdot}{\underset{\displaystyle \cdot}{Ar}} :$$

Note that for main-group elements, the number of valence electrons, and hence the number of dots appearing in a Lewis symbol, is equal to the group number for the *s*-block elements and to the "group number minus 10" for the *p*-block elements.

Practice Example A: Write Lewis symbols for Mg, Ge, K, and Ne.

Practice Example B: Write Lewis symbols for Sn, Br^-, Tl^+, and S^{2-}.

(*Hint:* How are the electron configurations of the ions related to those of the atoms from which they are formed?)

A **Lewis structure** is a combination of Lewis symbols that represents either the transfer or the sharing of electrons in a chemical bond.

Ionic bonding (transfer of electrons):

$$Na\times + \cdot \overset{\displaystyle \cdot \cdot}{\underset{\displaystyle \cdot \cdot}{Cl}} : \longrightarrow [Na]^+ [\times \overset{\displaystyle \cdot \cdot}{\underset{\displaystyle \cdot \cdot}{Cl}} :]^-$$

Lewis symbols   Lewis structure   (11.1)

Covalent bonding (sharing of electrons):

$$H\times + \cdot \overset{\displaystyle \cdot \cdot}{\underset{\displaystyle \cdot \cdot}{Cl}} : \longrightarrow H\times \overset{\displaystyle \cdot \cdot}{\underset{\displaystyle \cdot \cdot}{Cl}} :$$

Lewis symbols   Lewis structure   (11.2)

In these two examples, we designated the electrons involved in bond formation differently—(×) from one atom and (·) from the other. This helps to emphasize that an electron is transferred in ionic bonding and that a pair of electrons is shared in covalent bonding. Of course, it is impossible to distinguish between electrons, and henceforth we will use only dots (·) to represent electrons in Lewis structures.

Lewis's work dealt mostly with covalent bonding, which we will emphasize throughout this chapter. However, Lewis's ideas also apply to ionic bonding, and we briefly describe this application next.

Lewis Structures for Ionic Compounds

In Section 3-2, we learned that the formula unit of an ionic compound is the simplest electrically neutral collection of cations and anions from which we can establish the chemical formula of the compound. The Lewis structure of sodium chloride (structure 11.1) represents its formula unit. In the Lewis structure of an ionic compound of main-group elements, (1) the Lewis symbol of the metal ion has no dots if all the valence electrons are lost and (2) the ionic charges of both ions are shown. These ideas are further illustrated through Example 11-2.

EXAMPLE 11-2

Writing Lewis Structures of Ionic Compounds. Write Lewis structures for the following compounds: **(a)** BaO; **(b)** $MgCl_2$; **(c)** aluminum oxide.

Solution

(a) Write the Lewis symbol, and determine how many electrons each atom must gain or lose to acquire a noble gas electron configuration. Ba loses two electrons, and O gains two.

$$Ba\cdot + \cdot\ddot{O}\!: \longrightarrow [Ba]^{2+}[\!:\!\ddot{\ddot{O}}\!:\!]^{2-}$$

Lewis structure

(b) A Cl atom can accept only *one* electron because it already has seven valence electrons. One more will give it a complete octet. On the other hand, a Mg atom must lose *two* electrons to have the electron configuration of the noble gas, neon. So two Cl atoms are required for each Mg atom.

$$Mg\cdot + \begin{array}{c} \cdot\ddot{C}l\!: \\ \\ \cdot\ddot{C}l\!: \end{array} \longrightarrow [Mg]^{2+}2[\!:\!\ddot{C}l\!:\!]^{-}$$

Lewis structure

(c) We do not need to be given the formula of aluminum oxide. It follows directly from the Lewis structure that we write. The combination of one Al atom, which loses three electrons, and one O atom, which gains two, leaves an excess of one electron lost. To match the numbers of electrons lost and gained, we need a formula unit based on *two* Al atoms and *three* O atoms.

$$\begin{array}{c} Al\cdot \quad \cdot\ddot{O}\!: \\ +\cdot\ddot{O}\!: \longrightarrow 2[Al]^{3+}3[\!:\!\ddot{\ddot{O}}\!:\!]^{2-} \\ Al\cdot \quad \cdot\ddot{O}\!: \end{array}$$

Lewis structure

Practice Example A: Write plausible Lewis structures for **(a)** Na_2S and **(b)** Mg_3N_2.

Practice Example B: Write plausible Lewis structures for **(a)** calcium iodide; **(b)** barium sulfide; **(c)** lithium oxide.

The compounds described in Example 11-2 are *binary* ionic compounds consisting of monatomic cations and monatomic anions. Commonly encountered *ternary* ionic compounds consist of monatomic and *poly*atomic ions, and bonding between atoms in the polyatomic ions is covalent. We will consider some ternary ionic compounds later in the chapter.

With the exception of ion pairs such as (Na^+Cl^-) that may be found in the *gaseous* state, formula units of ionic compounds do not exist as separate entities. Instead, each cation surrounds itself with anions, and each anion with cations. These very large numbers of ions are arranged in an orderly network called an *ionic crystal* (Figure 11-1). We will describe ionic crystal structures and energy changes accompanying the formation of ionic crystals in Chapter 13.

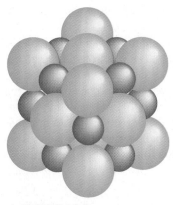

▲ **FIGURE 11-1**
Portion of an ionic crystal
This structure of alternating Na^+ and Cl^- ions extends in all directions and involves countless numbers of ions.

11-2 Covalent Bonding: An Introduction

A chlorine atom shows a tendency to gain an electron, as indicated by its electron affinity (−349 kJ/mol). From which atom, sodium or hydrogen, can the electron most readily be extracted? Neither atom gives up an electron freely, but the energy required to extract an electron from Na (I_1 = 496 kJ/mol) is much smaller than that for H (I_1 = 1312 kJ/mol). In Chapter 10 we learned that the lower its ionization energy, the more metallic an element is; sodium is much more metallic than hydrogen (recall Figure 10-11). In fact, hydrogen is considered to be a *nonmetal*. A hydrogen atom in the gaseous state does not *give up* an electron to another nonmetal atom. Bonding between a hydrogen atom and a chlorine atom involves the *sharing* of electrons. This sharing leads to a *covalent bond*.

To emphasize the sharing of electrons, let us think of the Lewis structure of HCl in this manner.

The broken circles represent the outermost electron shells of the bonded atoms. The number of dots lying on or within each circle represents the effective number of electrons in each valence shell. The H atom has two dots, as in the electron configuration of He. The Cl atom has eight dots, corresponding to the outer-shell configuration of Ar. Note that we counted the two electrons between H and Cl (:) *twice*. These two electrons are shared by the H and Cl atoms. This shared pair of electrons constitutes the covalent bond. Written below are two additional Lewis structures of simple molecules.

▶ The Lewis structures for H_2O and Cl_2O suggest that these molecules have a linear shape. They do not. Lewis theory by itself does not address the question of molecular shape (see Section 11-7).

$$\text{H·} + \text{·}\ddot{\underset{..}{\text{O}}}\text{·} + \text{·H} \longrightarrow \text{H:}\ddot{\underset{..}{\text{O}}}\text{:H} \quad \text{and} \quad \text{:}\ddot{\underset{..}{\text{C}}}\text{l·} + \text{·}\ddot{\underset{..}{\text{O}}}\text{·} + \text{·}\ddot{\underset{..}{\text{C}}}\text{l:} \longrightarrow \text{:}\ddot{\underset{..}{\text{C}}}\text{l:}\ddot{\underset{..}{\text{O}}}\text{:}\ddot{\underset{..}{\text{C}}}\text{l:}$$

water   dichlorine oxide

As was the case for Cl in HCl, the O atom in the Lewis structure of H_2O and in Cl_2O is surrounded by eight electrons (when the bond-pair electrons are double counted). In attaining these eight electrons, the O atom conforms to the **octet rule**—a requirement of eight valence-shell electrons for the atoms in a Lewis structure. Note, however, that the H atom is an exception to this rule. The H atom can accommodate only two valence-shell electrons.

H₂ Bond Formation animation

Lewis theory helps us to understand why elemental hydrogen and chlorine exist as diatomic molecules, H_2 and Cl_2. In each case, a pair of electrons is shared between the two atoms. The sharing of a single pair of electrons between bonded atoms produces a **single covalent bond**. To underscore the importance of electron pairs in the Lewis theory, we introduce the term **bond pair** for a pair of electrons in a covalent bond and **lone pair** for electron pairs that are not involved in bonding. Also, it is customary to replace some electron pairs with dashes (—), especially for bond pairs. These features are shown in the following Lewis structures.

$$\text{H·} + \text{·H} \longrightarrow \text{H:H} \quad \text{or} \quad \text{H—H} \tag{11.3}$$

Bond pair

$$\text{:}\ddot{\underset{..}{\text{C}}}\text{l·} + \text{·}\ddot{\underset{..}{\text{C}}}\text{l:} \longrightarrow \text{:}\ddot{\underset{..}{\text{C}}}\text{l:}\ddot{\underset{..}{\text{C}}}\text{l:} \quad \text{or} \quad \text{:}\ddot{\underset{..}{\text{C}}}\text{l—}\ddot{\underset{..}{\text{C}}}\text{l:} \quad \text{Lone pairs} \tag{11.4}$$

Bond pair

EXAMPLE 11-3

Writing Simple Lewis Structures. Write a Lewis structure for the ammonia molecule.

Solution

One of the requirements in writing a Lewis structure is knowledge of the formula of the molecule to be represented. The formula of ammonia is NH_3. This formula tells us how many and what type of atoms must be represented in the structure. Another requirement is knowledge of the number of valence electrons associated with the atoms present in the molecule.

These can then be represented in the Lewis symbols for those atoms, as shown here.

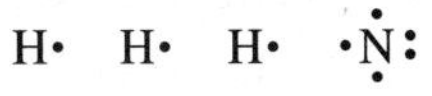

$$H\cdot \quad H\cdot \quad H\cdot \quad \cdot\ddot{N}\colon$$

Now we can assemble one N and three H atoms into a structure that gives the N atom a valence-shell octet and each of the H atoms two valence electrons (producing the electron configuration of He).

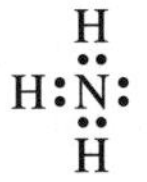

$$\begin{array}{c} H \\ \vdots \\ H\colon\!N\colon \\ \vdots \\ H \end{array}$$

Practice Example A: Write Lewis structures for Br_2, CH_4, and HOCl.

Practice Example B: Write Lewis structures for NI_3, N_2H_4, and C_2H_6.

Coordinate Covalent Bonds

The Lewis theory of bonding describes a covalent bond as the sharing of a pair of electrons, but this does not necessarily mean that each atom contributes an electron to the bond. A covalent bond in which a single atom contributes both of the electrons to a shared pair is called a **coordinate covalent bond**.

If we attempt to attach a fourth H atom to the Lewis structure of NH_3 shown in Example 11-3, we encounter a difficulty. The electron brought by the fourth H atom would raise the total number of valence electrons around the N atom to *nine*; we would no longer have an octet. The *molecule* NH_4 does not form, but the ammonium *ion*, NH_4^+, does, as suggested in Figure 11-2. That is, the lone pair of electrons on a NH_3 molecule detaches an H atom from a HCl molecule, and the electrons in the H—Cl bond remain on the Cl atom. As a result, a H^+ ion is joined to the NH_3 molecule, forming the NH_4^+ ion, and the electron pair left on the Cl atom converts it to a Cl^- ion.

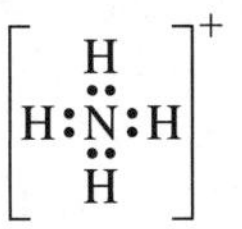

$$\left[\begin{array}{c} H \\ \vdots \\ H\colon\!N\colon\!H \\ \vdots \\ H \end{array} \right]^{+} \tag{11.5}$$

$$\begin{array}{c} H \\ \vdots \\ H\colon\!N\colon \\ \vdots \\ H \end{array} \;\curvearrowright\; H\colon\!\ddot{\underset{\cdots}{Cl}}\colon \;\longrightarrow\; \left[\begin{array}{c} H \\ \vdots \\ H\colon\!N\colon\!H \\ \vdots \\ H \end{array} \right]^{+} + \left[\colon\!\ddot{\underset{\cdots}{Cl}}\colon \right]^{-}$$

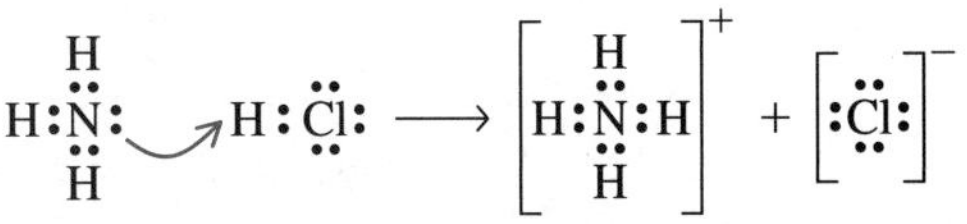

▲ FIGURE 11-2   **Formation of the ammonium ion, NH_4^+**

The H atom of HCl leaves its electron with the Cl atom and, as H^+, attaches itself to the NH_3 molecule through the lone pair electrons on the N atom. The ions NH_4^+ and Cl^- are formed.

The bond formed between the N atom of NH_3 and the H^+ ion is a *coordinate covalent bond*. It is important to note, however, that once the bond has formed we cannot say which of the four N—H bonds is the coordinate covalent bond. Thus, a coordinate covalent bond is indistinguishable from a regular covalent bond.

We find another example of coordinate covalent bonding in another familiar ion—the hydronium ion.

$$\left[H \!:\! \overset{\cdot\cdot}{\underset{\displaystyle H}{O}} \!:\! H \right]^+ \tag{11.6}$$

Multiple Covalent Bonds

In the preceding description of the Lewis model for covalent chemical bonding, we have used a single pair of electrons between two atoms to describe a single covalent bond. Often, however, more than one pair of electrons needs to be shared if an atom is to attain an octet (noble gas electron configuration). Two examples where this occurs are seen in the atmospheric molecules CO_2 and N_2.

First, let's apply the ideas about Lewis structures to CO_2. From the Lewis symbols, we see that the C atom can share a valence electron with each O atom, thus forming two carbon-to-oxygen single bonds.

$$:\!\overset{\cdot\cdot}{\underset{\cdot\cdot}{O}}\!\cdot \quad \cdot\overset{\cdot}{\underset{\cdot}{C}}\!\cdot \quad \cdot\overset{\cdot\cdot}{\underset{\cdot\cdot}{O}}\!: \quad \longrightarrow \quad :\!\overset{\cdot\cdot}{\underset{\cdot\cdot}{O}}\!:\!\overset{}{\underset{}{C}}\!:\!\overset{\cdot\cdot}{\underset{\cdot\cdot}{O}}\!:$$

But this leaves the C atom and both O atoms still shy of an octet. The problem is solved by shifting the unpaired electrons into the region of the bond, as indicated by the small arrows.

$$:\!\overset{\cdot\cdot}{\underset{\cdot\cdot}{O}}\!:\!\overset{}{\underset{}{C}}\!:\!\overset{\cdot\cdot}{\underset{\cdot\cdot}{O}}\!: \quad \longrightarrow \quad :\!\overset{\cdot\cdot}{O}\!:\!:\!C\!:\!:\!\overset{}{\underset{\cdot\cdot}{O}}\!: \quad \longrightarrow \quad :\!\overset{\cdot\cdot}{O}\!=\!C\!=\!\overset{\cdot\cdot}{O}\!: \tag{11.7}$$

In Lewis structure (11.7), the bonded atoms are seen to share *two* pairs of electrons (a total of four electrons) between them—a **double covalent bond** (=).

Now let's try our hand at writing a Lewis structure for the N_2 molecule. Our first attempt might again involve a single covalent bond and the incorrect structure shown below.

$$:\!\overset{\cdot}{\underset{\cdot}{N}}\!\cdot \; + \; \cdot\overset{\cdot}{\underset{\cdot}{N}}\!: \quad \longrightarrow \quad :\!\overset{\cdot}{\underset{\cdot}{N}}\!:\!\overset{\cdot}{\underset{\cdot}{N}}\!: \quad (\textit{Incorrect})$$

Each N atom appears to have only *six* outer-shell electrons, not the expected eight. We can correct the situation by bringing the four unpaired electrons into the region between the N atoms and using them for additional bond pairs. In all, we now show the sharing of *three* pairs of electrons between the N atoms. The bond between the N atoms in N_2 is a **triple covalent bond** (≡). Double and triple covalent bonds are known as **multiple covalent bonds**.

$$:\!\overset{\cdot}{\underset{\cdot}{N}}\!:\!\overset{\cdot}{\underset{\cdot}{N}}\!: \quad \longrightarrow \quad :\!N\!\overset{\cdot\cdot\cdot}{\underset{}{:}}\!N\!: \quad \text{or} \quad :\!N\!\equiv\!N\!: \tag{11.8}$$

The triple covalent bond in N_2 is a very strong bond that is difficult to break in a chemical reaction. The unusual strength of this bond makes $N_2(g)$ quite inert. As a result, $N_2(g)$ coexists with $O_2(g)$ in the atmosphere and forms oxides of nitrogen only in trace amounts at high temperatures. The lack of reactivity of N_2 toward O_2 is an essential condition for life on Earth. The inertness of $N_2(g)$ also makes it difficult to synthesize nitrogen compounds.

Another molecule whose Lewis structure features a multiple bond is O_2. Here we can draw a Lewis structure in which the bond is a double bond.

$$:\overset{..}{O}\cdot \ + \ \cdot\overset{..}{O}: \ \longrightarrow \ :\overset{..}{O}:\overset{..}{O}: \ \longrightarrow \ (\overset{..}{O}::\overset{..}{O}) \ \longrightarrow \ :\overset{..}{O}=\overset{..}{O}: \ _? \qquad (11.9)$$

The blue question mark suggests that there is some doubt about the validity of structure (11.9), and the source of the doubt is illustrated in Figure 11-3. The structure fails to account for the *paramagnetism* of oxygen—the O_2 molecule must have *unpaired* electrons. Unfortunately, no completely satisfactory Lewis structure is possible for O_2. In Chapter 12, however, we will describe bonding in the O_2 molecule in a way that does account for both the double bond *and* the observed paramagnetism.

We could continue applying ideas introduced in this section, but our ability to write plausible Lewis structures will be greatly aided by a couple of new ideas that we introduce in Section 11-3.

11-3   Polar Covalent Bonds

We have introduced chemical bonds as if they are of two distinctly different types: ionic bonds involving a *complete* transfer of electrons and covalent bonds involving an *equal* sharing of electron pairs. Such is not the case, however, and many chemical bonds have features of both types.

A covalent bond in which electrons are not shared equally between two atoms is called a **polar covalent bond**. In such a bond, electrons are displaced toward the more nonmetallic element. In Figure 11-4a, note the even distribution of electron charge density between the two H atoms in H_2 and between the two Cl atoms in Cl_2. In both molecules, each atom has the same electron affinity, and electrons are not displaced toward either atom. The centers of positive and negative charge coincide; each is found at a point midway between the atomic nuclei. Both the H—H and Cl—Cl bonds are *nonpolar*. In HCl, on the other hand, Cl attracts electrons more strongly than does H (Figure 11-4b). The electron charge density is greater near the Cl atom than near the H atom. The center of negative charge lies closer to the Cl nucleus than does the center of positive charge. We say that there is a separation of charge in the H—Cl bond and that the bond is *polar*. We can represent the polar bond in HCl by a Lewis structure in which the bond pair of electrons lies closer to the Cl than to the H.

$$^{\delta+}H \ :\overset{..}{\underset{..}{Cl}}:^{\delta-}$$

◀ KEEP IN MIND
that merely being able to write a plausible Lewis structure doesn't prove that it is the correct electronic structure. Proof can come only through confirming experimental evidence.

▲ **FIGURE 11-3**
Paramagnetism of oxygen
Liquid oxygen is attracted into the magnetic field of a large magnet.

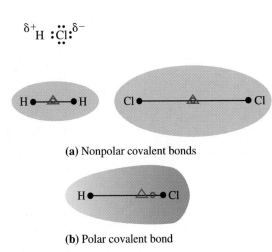

(a) Nonpolar covalent bonds

(b) Polar covalent bond

● = Atomic nucleus
△ = Center of positive charge
○ = Center of negative charge

▶ **FIGURE 11-4**
Nonpolar and polar covalent bonds
(a) In the diatomic *nonpolar* molecules H_2 and Cl_2, the center of positive charge lies midway between the atomic nuclei along a line between them. The center of negative charge comes at the same point. There is no charge separation. **(b)** In the H—Cl bond, the center of positive charge lies much closer to the Cl nucleus than to the H nucleus, but this is because the Cl nucleus has 17 units of positive charge compared with just 1 unit for H. The center of negative charge lies still closer to the Cl nucleus. This is because of the stronger attraction of Cl than of H for the electron pair in the H—Cl bond. In a *polar covalent* bond, the centers of positive and negative charge are separated.

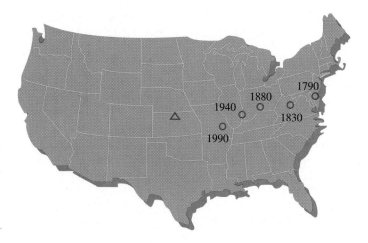

▶ **FIGURE 11-5**

An analogy to a polar covalent bond
The geographical center of the contiguous 48 states of the United States remains fixed (△), but the population center (○) is moving to the south and west. The separation between these two centers is analogous to the separation of the centers of positive and negative charge in a polar covalent bond. As the distance between the centers becomes smaller, the bond becomes less polar.

 Are You Wondering . . .

If there is an analogy you can use in thinking about polar covalent bonds?

Consider the map in Figure 11-5 that locates the geographic center of the contiguous 48 states of the United States. Here is a way to show the significance of this center: Imagine cutting out the map and then trying to balance the cut-out on the point of a pin. The geographic center is the point on the map where you would have to put the pin to achieve this balance. Also shown on the map is the center of population density in the United States at various times in its history. To locate the population center, imagine putting a pencil dot for each inhabitant at his or her geographic location on a weightless cutout of the map; then search for the new balance point.

The analogy to a polar bond is to think of the geographic center as the center of positive charge and the population center as the center of negative charge. Two hundred years ago, there was a great separation between these two centers, analogous to a highly polar bond. With time, however, the distance between these two centers has shortened, analogous to a less polar bond. If the two centers were eventually to coincide, this would be analogous to a nonpolar bond.

The δ+ signifies that the center of positive charge is displaced toward the H nucleus, and δ− signifies that the center of negative charge is displaced toward the Cl nucleus. The separation of charge is such that there is a *partial* positive charge on H and a *partial* negative charge on Cl.

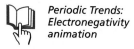

Periodic Trends: Electronegativity animation

Electronegativity

We concluded that the H—Cl bond is polar because the Cl atom has a greater affinity for electrons than does the H atom. Electron affinity is an atomic property, however, and more meaningful predictions about bond polarities are those based on a molecular property, one that relates to the ability of atoms to lose or gain electrons when they are part of a molecule rather than isolated from other atoms. **Electronegativity (EN)** describes an atom's ability to compete for electrons with other atoms to which it is bonded. As such, electronegativity is related to ionization energy (I) and electron affinity (EA). A widely used electronegativity scale, with values given in Figure 11-6, is one devised by Linus Pauling (1901–1994). Pauling's EN values range from about 0.7 to 4.0. In general, the lower its EN, the more

1																
H 2.1	2											13	14	15	16	17
Li 1.0	**Be** 1.5			below 1.0		2.0–2.4						**B** 2.0	**C** 2.5	**N** 3.0	**O** 3.5	**F** 4.0
Na 0.9	**Mg** 1.2	3	4	5	6	7	8	9	10	11	12	**Al** 1.5	**Si** 1.8	**P** 2.1	**S** 2.5	**Cl** 3.0
K 0.8	**Ca** 1.0	**Sc** 1.3	**Ti** 1.5	**V** 1.6	**Cr** 1.6	**Mn** 1.5	**Fe** 1.8	**Co** 1.8	**Ni** 1.8	**Cu** 1.9	**Zn** 1.6	**Ga** 1.6	**Ge** 1.8	**As** 2.0	**Se** 2.4	**Br** 2.8
Rb 0.8	**Sr** 1.0	**Y** 1.2	**Zr** 1.4	**Nb** 1.6	**Mo** 1.8	**Tc** 1.9	**Ru** 2.2	**Rh** 2.2	**Pd** 2.2	**Ag** 1.9	**Cd** 1.7	**In** 1.7	**Sn** 1.8	**Sb** 1.9	**Te** 2.1	**I** 2.5
Cs 0.8	**Ba** 0.9	**La*** 1.1	**Hf** 1.3	**Ta** 1.5	**W** 2.4	**Re** 1.9	**Os** 2.2	**Ir** 2.2	**Pt** 2.2	**Au** 2.4	**Hg** 1.9	**Tl** 1.8	**Pb** 1.8	**Bi** 1.9	**Po** 2.0	**At** 2.2
Fr 0.7	**Ra** 0.9	**Ac†** 1.1														

Legend: below 1.0 · 1.0–1.4 · 1.5–1.9 · 2.0–2.4 · 2.5–2.9 · 3.0–4.0

*Lanthanides: 1.1–1.3
†Actinides: 1.3–1.5

▲ **FIGURE 11-6** **Electronegativities of the elements**
As a general rule, electronegativities *decrease* from top to bottom in a group and *increase* from left to right in a period of elements. The values are from L. Pauling, *The Nature of the Chemical Bond*, 3rd edition, Cornell University, Ithaca, NY, 1960, p. 93. Values may be somewhat different when based on other electronegativity scales.

metallic an element is; and the higher the EN, the more nonmetallic it is. From Figure 11-6 we also see that electronegativity decreases from top to bottom in a group and increases from left to right in a period of the periodic table.

Electronegativity values allow us to gain an insight into the amount of polar character in a covalent bond. We do this by describing the **electronegativity difference**, **ΔEN**, as the absolute value of the difference in EN values of the bonded atoms. If ΔEN for two atoms is very small, the bond between them is essentially covalent. If ΔEN is large, the bond is essentially ionic. For intermediate values of ΔEN, the bond is described as polar covalent. A useful rough relationship between ΔEN and percent ionic character of a bond is presented in Figure 11-7.

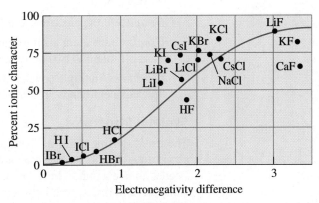

▲ **FIGURE 11-7** **Percent ionic character of a chemical bond as a function of electronegativity difference**

▶ Although Figure 11-7 suggests that the bond between two identical metal atoms should be covalent [as it is in $Li_2(g)$, for example], in *solid* metals, where bonding extends throughout a network of many, many atoms, the bonding is of a type called *metallic* (explored in the next chapter).

Large EN differences are found between the more metallic and the more non-metallic elements. Combinations of these elements are expected to produce bonds that are essentially ionic. Small EN differences are expected for two nonmetal atoms, and the bond between them should be essentially covalent. Thus, even without a compilation of EN values at hand, you should be able to predict the essential character of a bond between two atoms. Simply assess the metallic/nonmetallic characters of the bonded elements from the periodic table (recall Figure 10-11).

EXAMPLE 11-4

Assessing Electronegativity Differences and the Polarity of Bonds.

   (a) Which bond is more polar, H—Cl or H—O?

   (b) What is the percent ionic character of each of these bonds?

Solution

   (a) Look up EN values of H, Cl, and O in Figure 11-6, and compute ΔEN. $EN_H = 2.1$; $EN_{Cl} = 3.0$; $EN_O = 3.5$. For the H—Cl bond, $\Delta EN = 3.0 - 2.1 = 0.9$. For the H—O bond, $\Delta EN = 3.5 - 2.1 = 1.4$. Because its ΔEN is somewhat greater, we expect the H—O bond to be the more polar bond.

   (b) Determine the percent ionic character from Figure 11-7.

$$\text{H—Cl bond: } \Delta EN = 0.9; \qquad \approx 20\% \text{ ionic}$$
$$\text{H—O bond: } \Delta EN = 1.4; \qquad \approx 35\% \text{ ionic}$$

Practice Example A: Which of the following bonds is the most polar, that is, has the greatest ionic character: H—Br, N—H, N—O, P—Cl?

Practice Example B: Which is the most polar bond: C—S, C—P, P—O, or O—F?

11-4   Writing Lewis Structures

In this section we combine ideas introduced in the preceding three sections with a few new concepts to write a variety of Lewis structures.

Fundamental Requirements

Let us begin with a reminder of some of the essential features of Lewis structures that we have already encountered.

- *All* the valence electrons of the atoms in a Lewis structure must appear in the structure.

- *Usually*, all the electrons in a Lewis structure are paired.

- *Usually*, each atom acquires an outer shell octet of electrons. Hydrogen, however, is limited to two outer shell electrons.

- *Sometimes*, multiple covalent bonds, (double or triple bonds) are needed. Multiple covalent bonds are formed most readily by C, N, O, P, and S atoms.

Skeletal Structures

The usual starting point in writing a Lewis structure is to designate the **skeletal structure**—all the atoms in the structure arranged in the order in which they are bonded to one another. In a skeletal structure with more than two atoms, we generally need to distinguish between central and terminal atoms. A **central atom** is bonded to two or more atoms, and a **terminal atom** is bonded to just one other atom. As an example, consider ethanol, CH_3CH_2OH. Its skeletal structure is the same as the following structural formula. In this structure, the *central* atoms—both

C atoms and the O atom—are printed in red. The *terminal* atoms—all six H atoms—are printed in blue.

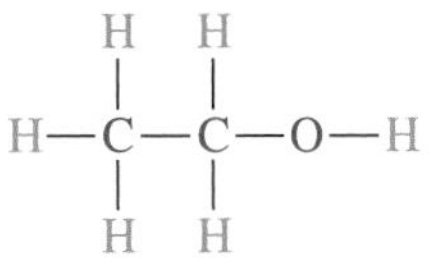

$$
\begin{array}{ccc}
& \text{H} & \text{H} \\
& | & | \\
\text{H}- & \text{C}-\text{C} & -\text{O}-\text{H} \\
& | & | \\
& \text{H} & \text{H}
\end{array}
\qquad (11.10)
$$

Here are a few additional facts about central atoms, terminal atoms, and skeletal structures.

- *Hydrogen atoms are always terminal atoms.* This is because a H atom can accommodate only two electrons in its valence shell, thereby limiting it to only one bond to another atom. (An interesting and rare exception occurs in some unusual boron–hydrogen compounds.)

- *Central atoms are generally those with the lowest electronegativity.* In the skeletal structure (11.10), the atoms of lowest electronegativity (2.1) happen to be H atoms, but as noted above, H atoms can only be terminal atoms. Next lowest in electronegativity (2.5) are the C atoms, and these are central atoms. The O atom has the highest electronegativity (3.5) but nevertheless is also a central atom. For O to be a terminal atom in structure (11.10) would require it to exchange places with a H atom, but this would make the H atom a central atom and that is not possible. The chief cases where O atoms are central atoms are in structures with a *peroxo* linkage ($-\text{O}-\text{O}-$) or a *hydroxy* group ($-\text{O}-\text{H}$). Otherwise, expect an O atom to be a *terminal* atom.

- *Carbon atoms are always central atoms.* This is a useful fact to keep in mind when writing Lewis structures of organic molecules.

- Except for the very large number of chainlike organic molecules, *molecules and polyatomic ions generally have compact, symmetrical structures.* Thus, of the two skeletal structures below, the more compact structure on the right is the one actually observed for phosphoric acid, H_3PO_4.

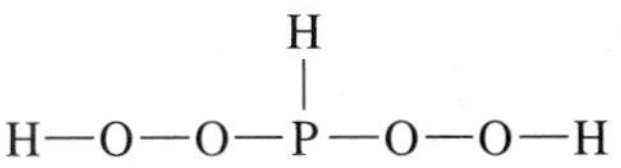

$$
\begin{array}{cc}
\text{H} & \text{O}-\text{H} \\
| & | \\
\text{H}-\text{O}-\text{O}-\text{P}-\text{O}-\text{O}-\text{H} \qquad \text{H}-\text{O}-\text{P}-\text{O}-\text{H} \\
& | \\
& \text{O} \\
(\textit{Incorrect}) & (\textit{Correct})
\end{array}
$$

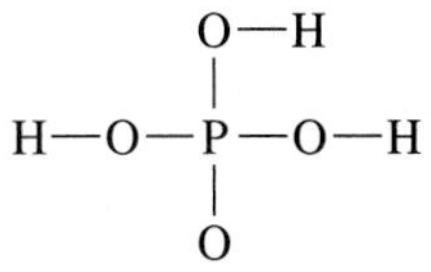

A Strategy for Writing Lewis Structures

At this point, let us incorporate a number of the ideas that we have considered so far into a specific approach to writing Lewis structures. This strategy is designed to give you a place to begin and also consecutive steps to follow to achieve a plausible Lewis structure.

1. Determine the total number of valence electrons that must appear in the structure.

   Examples: In the *molecule* CH_3CH_2OH, there are *4* valence electrons for each C atom or *8* for the two C atoms; *1* for each H atom or *6* for the six H atoms; and *6* for the lone O atom. The total number of valence electrons in the Lewis structure is

   $$8 + 6 + 6 = 20.$$

   In the *polyatomic ion* PO_4^{3-}, there are *5* valence electrons for the P atom and *6* for each O atom, or *24* for all four O atoms. To produce the charge of 3−, an additional *3* valence electrons must be brought into the structure. The total number of valence electrons in the Lewis structure is

$$5 + 24 + 3 = 32$$

In the *polyatomic ion* NH_4^+, there are *5* valence electrons for the N atom and *1* for each H atom, or *4* for all four H atoms. To account for the charge of 1+, one of the electrons must be *lost*. The total number of valence electrons is

$$5 + 4 - 1 = 8$$

2. Identify the central atoms(s) and terminal atoms.
3. Write a plausible skeletal structure. Join the atoms in the skeletal structure by *single* covalent bonds (single dashes).
4. For each bond in the skeletal structure, subtract *two* from the total number of valence electrons.
5. With the valence electrons remaining, *first* complete the octets of the terminal atoms. *Then*, to the extent possible, complete the octets of the central atoms(s). If there are just enough valence electrons to complete octets for all the atoms, the structure at this point is a satisfactory Lewis structure.
6. If one or more central atoms is left with an incomplete octet after Step 5, move lone-pair electrons from terminal atoms to form *multiple* covalent bonds to central atoms. Do this to the extent necessary to give all atoms complete octets, thereby producing a satisfactory Lewis structure.

Figure 11-8 summarizes the above procedure for writing Lewis structures.

EXAMPLE 11-5

Applying the General Strategy for Writing Lewis Structures. Write a plausible Lewis structure for *cyanogen*, C_2N_2, a poisonous gas used as a fumigant and rocket propellant.

Solution

Step 1. Determine the total number of valence electrons. Each of the two C atoms (group 14) has *four* valence electrons, and each of the two N atoms (group 15) has *five*. The total number of valence electrons is $4 + 4 + 5 + 5 = 18$.

Step 2. Identify the central atom(s) and terminal atoms. The C atoms have a lower electronegativity (2.5) than do the N atoms (3.0). C atoms are central atoms, and N atoms are terminal atoms.

Step 3. Write a plausible skeletal structure by joining atoms through *single* covalent bonds.

$$N—C—C—N$$

Step 4. Subtract *two* electrons for each bond in the skeletal structure. The three bonds in this structure account for *six* of the 18 valence electrons. This leaves *12* valence electrons to be assigned.

Step 5. Complete octets for the terminal N atoms, and to the extent possible, the central C atoms. The remaining *12* valence electrons are sufficient only to complete the octets of the N atoms.

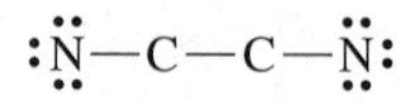

$$:\ddot{N}—C—C—\ddot{N}:$$

Step 6. Move lone pairs of electrons from the terminal N atoms to form multiple bonds to the central C atoms. Each C atom has only four electrons in its valence shell and needs four more to complete an octet. Thus, each C atom requires two additional pairs of electrons, which it acquires if we move two lone pairs from each N atom into its bond with a C atom, as shown below.

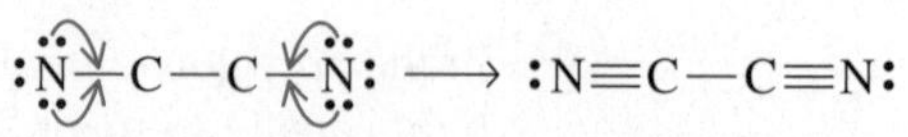

$$:\ddot{N}—C—C—\ddot{N}: \longrightarrow :N≡C—C≡N:$$

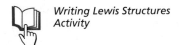

Writing Lewis Structures Activity

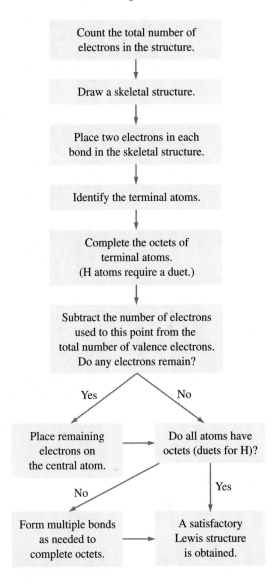

▶ **FIGURE 11-8**
Summary scheme for drawing Lewis structures

Practice Example A: Write plausible Lewis structures for (a) CS_2, (b) HCN, and (c) $COCl_2$.

Practice Example B: Write plausible Lewis structures for (a) formic acid, HCOOH, and (b) acetaldehyde, CH_3CHO.

EXAMPLE 11-6

Writing a Lewis Structure for a Polyatomic Ion. Write the Lewis structure for the *nitronium* ion, NO_2^+.

Solution

Step 1. Determine the total number of valence electrons. The N atom (group 15) has *five* valence electrons, and each of the two O atoms (group 16) has *six*. However, *one* valence electron must be removed to produce the charge of 1+.

The total number of valence electrons is *5 + 6 + 6 − 1 = 16*

Step 2. Identify the central atom(s) and terminal atoms. The N atom has a lower electronegativity (3.0) than the O atoms (3.5). N is the central atom, and the O's are the terminal atoms.

Step 3. Write a plausible skeletal structure by joining atoms through *single* covalent bonds.

$$O\!-\!N\!-\!O$$

Step 4. Subtract *two* electrons for each bond in the skeletal structure. The two bonds in this structure account for *4* of the 16 valence electrons. This leaves *12* valence electrons to be assigned.

Step 5. Complete octets for the terminal O atoms, and to the extent possible, the central N atom. The remaining *12* valence electrons are sufficient only to complete the octets of the O atoms.

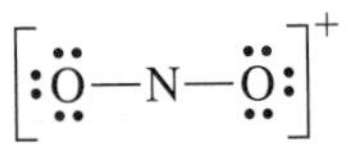

$$\left[:\ddot{\text{O}}\!-\!\text{N}\!-\!\ddot{\text{O}}:\right]^{+}$$

Step 6. Move lone pairs of electrons from the terminal O atoms to form multiple bonds to the central N atom. The N atom has only four electrons in its valence shell and needs four more to complete an octet. Thus, the N atom requires two additional pairs of electrons, which it acquires if we move one lone pair from each O atom into its bond with the N atom, as shown below.

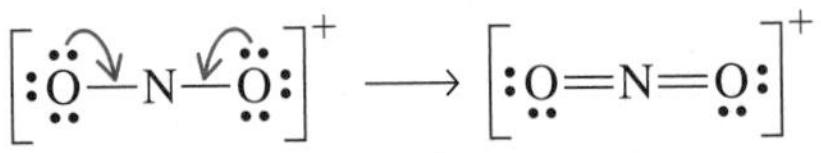

$$\left[:\ddot{\text{O}}\!-\!\text{N}\!-\!\ddot{\text{O}}:\right]^{+} \longrightarrow \left[:\text{O}\!=\!\text{N}\!=\!\text{O}:\right]^{+} \qquad (11.11)$$

Practice Example A: Write plausible Lewis structures for the following ions: **(a)** NO^{+}; **(b)** $N_2H_5^{+}$; **(c)** O^{2-}.

Practice Example B: Write plausible Lewis structures for the following ions: **(a)** BF_4^{-}; **(b)** NH_3OH^{+}; **(c)** NCO^{-}.

Formal Charges animation

Formal Charge

Instead of writing Lewis structure (11.11) for the nitronium ion in Example 11-6, we might have written the following structure.

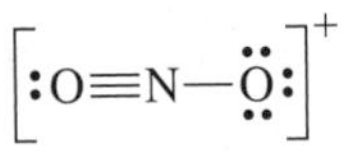

$$\left[:\text{O}\!\equiv\!\text{N}\!-\!\ddot{\text{O}}:\right]^{+} \qquad (incorrect) \quad (11.12)$$

Despite the fact that this structure satisfies the usual requirements—the correct number of valence electrons and an octet for each atom—we have marked it incorrect because it fails in one additional requirement. Have you noticed that with this strategy for writing Lewis structures that once we have determined the total number of valence electrons, we no longer need to keep track of which electrons in the structure came from which atoms? Nevertheless, after we have a plausible Lewis structure, we can go back and assess where each electron apparently came from, and in this way we can evaluate the formal charge. **Formal charges** are apparent charges on certain atoms in a Lewis structure that arise when atoms have not contributed equal numbers of electrons to the covalent bonds joining them.

> The formal charge on an atom in a Lewis structure is the number of valence electrons in the free (uncombined) atom minus the number of electrons assigned to that atom in the Lewis structure.

We assign electrons to the atoms in a Lewis structure in the following way.

- Count *lone-pair* electrons as belonging entirely to the atom on which they are found.
- Divide *bond-pair* electrons *equally* between the bonded atoms.

Assigning electrons (e⁻) in this way is equivalent to writing that

e⁻ assigned to a bonded atom in a Lewis structure

$$= \text{number of lone-pair e}^- + \frac{1}{2}\text{bond-pair e}^-$$

Because formal charge is the difference between the assignment of valence electrons to a free (uncombined) atom and to the atom in a Lewis structure, we can express formal charge (FC) as follows.

$$\text{FC} = \text{number valence e}^- \text{ in free atom} - \text{number lone-pair e}^-$$
$$- \frac{1}{2}\text{number bond-pair e}^- \qquad (11.13)$$

Now, let us assign formal charges to the atoms in structure (11.12), proceeding from left to right.

:O≡   FC = 6 valence e⁻ in O − 2 lone-pair e⁻ − $\frac{1}{2}$ (6 bond-pair e⁻) = 6 − 2 − 3 = +1

≡N—   FC = 5 valence e⁻ in N − 0 lone-pair e⁻ − $\frac{1}{2}$ (8 bond-pair e⁻) = 5 − 0 − 4 = +1

—Ö: FC = 6 valence e⁻ in O − 6 lone-pair e⁻ − $\frac{1}{2}$ (2 bond-pair e⁻) = 6 − 6 − 1 = −1

We can represent formal charges in a Lewis structure through small encircled numbers, as shown below.

$$\left[\overset{\oplus 1}{:\text{O}} \equiv \overset{\oplus 1}{\text{N}} - \overset{\ominus 1}{\ddot{\text{O}}} : \right]^+ \qquad (11.14)$$

Following are the general rules that help us determine the plausibility of a Lewis structure, based on its formal charges.

- The sum of the formal charges in a Lewis structure must equal *zero* for a neutral molecule and must equal the magnitude of the charge for a polyatomic ion. [Thus for structure (11.14), this sum is +1 + 1 − 1 = +1.]
- Where formal charges are required, they should be as small as possible.
- Negative formal charges usually appear on the most electronegative atoms; positive formal charges, on the least electronegative atoms.
- Structures having formal charges of the same sign on adjacent atoms are unlikely.

Lewis structure (11.14) conforms to the first two rules, but is not in good accordance with the third rule. Despite the fact that O is the most electronegative element in the structure, one of the O atoms has a positive formal charge. The greatest failing, though, is in the fourth rule. Both the O atom on the left and the N atom adjacent to it have positive formal charges. By contrast, the Lewis structure of NO_2^+ derived in Example 11-6 has only one formal charge, +1, on the central N atom. It conforms to the rules completely.

▶ We will see some exceptions to the idea that formal charges should be kept to a minimum in Section 11-6.

KEEP IN MIND ▶
that formal charges are not actual charges but simply "balances" in a form of "electron bookkeeping" that help us choose a probable Lewis structure.

EXAMPLE 11-7

Using the Formal Charge Concept in Writing Lewis Structures. Write the most plausible Lewis structure of nitrosyl chloride, NOCl, one of the oxidizing agents present in *aqua regia*, a mixture of concentrated nitric and hydrochloric acids capable of dissolving gold.

Solution

Although the formula is written as NOCl, we can pretty much reject the skeletal structure N—O—Cl. It places the most electronegative atom as the central atom (see also

Practice Example A). This then leaves the following as possible skeletal structures.

$$O-Cl-N \text{ and } O-N-Cl$$

Regardless of the skeletal structure we choose, the number of valence electrons (dots) that must appear in the final Lewis structure is

$$5 \text{ from N} + 6 \text{ from O} + 7 \text{ from Cl} = 18$$

When we apply the four steps listed below to the two possible skeletal structures, we obtain a total of four Lewis structures—two for each skeletal structure. This doubling occurs because in Step 4, there are two ways to complete the octets of the central atoms. The final Lewis structures obtained are labeled (a_1), (a_2), (b_1), and (b_2).

(a)		(b)
O—Cl—N	1. Assign four electrons	O—N—Cl
:Ö—Cl—N̈:	2. Assign twelve more electrons	:Ö—N—C̈l:
:Ö—C̈l—N̈:	3. Assign the last two electrons	:Ö—N̈—C̈l:
	4. Complete the octet on the central atom	

(a_1)	(a_2)	(b_1)	(b_2)
:Ö=C̈l—N̈:	:Ö—C̈l=N̈:	:Ö=N̈—C̈l:	:Ö—N̈=C̈l:

Evaluate formal charges using equation (11.13). In structure (a_1), for the N atom,

$$FC = 5 - 6 - \frac{1}{2}(2) = -2$$

for the O atom,

$$FC = 6 - 4 - \frac{1}{2}(4) = 0$$

for the Cl atom,

$$FC = 7 - 2 - \frac{1}{2}(6) = +2.$$

Proceed in a similar manner for the other three structures. Summarize the formal charges for the four structures.

	(a_1)	(a_2)	(b_1)	(b_2)
N:	−2	−1	0	0
O:	0	−1	0	−1
Cl:	+2	+2	0	+1

▶ Based on structure (b_1) ONCl is a better way to write the formula of nitrosyl chloride than NOCl.

Select the best Lewis structure in terms of formal charge rules. First, note that all four structures obey the requirement that formal charges of a neutral molecule add up to zero. In structure (a_1), the formal charges are large (+2 on Cl and −2 on N) and the negative formal charge is not on the most electronegative atom. Structure (a_2) has formal charges on all atoms, one of them large (+2 on Cl). Structure (b_1) is the ideal we seek—no formal charges. In structure (b_2), we again have formal charges. The best Lewis structure of nitrosyl chloride is

$$:Ö=N̈—C̈l:$$

Practice Example A: Write a Lewis structure for nitrosyl chloride based on the skeletal structure N—O—Cl, and show that this structure is not as plausible as the one obtained in Example 11-7.

Practice Example B: Write two Lewis structures for cyanamide, NH_2CN, an important chemical of the fertilizer and plastics industries. Use the formal charge concept to choose the more plausible structure.

11-5 Resonance

The ideas presented in the previous section allow us to write many Lewis structures, but some structures still present problems. We describe these problems in the next two sections.

As we learned in Chapter 8, although oxygen commonly occurs as diatomic molecules, O_2, it can also exist as triatomic molecules of *ozone*, O_3. Ozone is found naturally in the stratosphere and is also produced in the lower atmosphere as a constituent of smog.

When we apply the usual rules for Lewis structures for ozone, we come up with these *two* possibilities.

$$:\ddot{O}=\ddot{O}-\ddot{O}: \qquad :\ddot{O}-\ddot{O}=\ddot{O}:$$

▶ Bond lengths are discussed more fully in Section 11-8.

But there is something wrong with both structures. Each suggests that one oxygen-to-oxygen bond is single and the other is double. Yet, experimental evidence indicates that the two oxygen-to-oxygen bonds are the same; each has a length of 127.8 pm. This bond length is shorter than the O—O single-bond length of 147.5 pm in hydrogen peroxide, $H-\ddot{O}-\ddot{O}-H$ but it is longer than the double-bond length of 120.74 pm in diatomic oxygen, $:\ddot{O}=\ddot{O}:$. The bonds in ozone are intermediate between a single and a double bond. The difficulty is resolved if we say that the true Lewis structure of O_3 is *neither* of the previously proposed structures, but a composite, or *hybrid*, of the two, a fact that we can represent as

$$:\ddot{O}=\ddot{O}-\ddot{O}: \longleftrightarrow :\ddot{O}-\ddot{O}=\ddot{O}: \qquad (11.15)$$

▶ The two resonance structures in (11.15) are equivalent, that is, they contribute equally to the structure of the resonance hybrid. In many cases, there are several resonance structures that do not contribute equally. Sometimes, a single structure is the main contributor.

The situation in which two or more plausible Lewis structures can be written but the "correct" structure cannot be written at all is called **resonance**. The true structure is a *resonance hybrid* of plausible contributing structures. Acceptable contributing structures to a resonance hybrid must all have the same skeletal structure; they can differ only in how electrons are distributed within the structure. In expression (11.15), the two contributing structures are joined by a double-headed arrow. The arrow does *not* mean that the molecule has one structure part of the time and the other structure the rest of the time. It has the *same* structure *all* the time. By averaging the single bond in one structure with the double bond in the other, we might say that the oxygen-to-oxygen bonds in ozone are halfway between a single and double bond, that is, 1.5 bonds.

EXAMPLE 11-8

Representing the Lewis Structure of a Resonance Hybrid. Write the Lewis structure of the acetate ion, CH_3COO^-.

Solution

The skeletal structure has the three H atoms as terminal atoms bonded to a C atom as a central atom. The second C atom is also a central atom bonded to the first. The two O atoms are terminal atoms bonded to the second C atom.

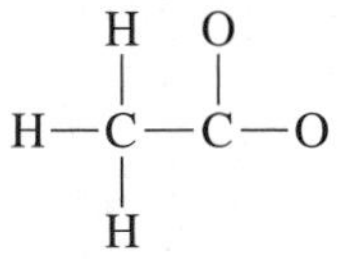

The number of valence electrons (dots) that must appear in the Lewis structure is

$$(3 \times 1) + (2 \times 4) + (2 \times 6) + 1 = 3 + 8 + 12 + 1 = 24$$

from H   from C   from O   to establish charge of 1−

Twelve of the valence electrons are used in the bonds in the skeletal structure, and the remaining twelve are distributed as lone-pair electrons on the two O atoms.

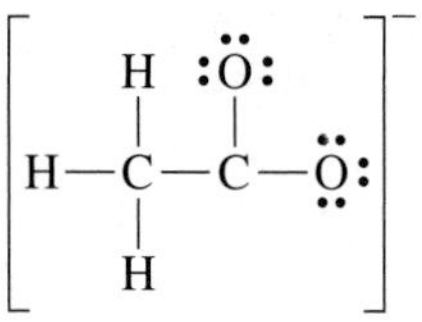

$$\begin{bmatrix} & H & :\!\ddot{O}\!: & \\ & | & | & \\ H\!-\!C\!-\!C\!-\!\ddot{O}\!: & & & \\ & | & & \\ & H & & \end{bmatrix}^{-}$$

In completing the octet of the C atom on the right, we discover that we can write two completely equivalent Lewis structures, depending on which of the two O atoms furnishes the lone pair of electrons to form a carbon-to-oxygen double bond. The true Lewis structure is a resonance hybrid of the following two contributing structures.

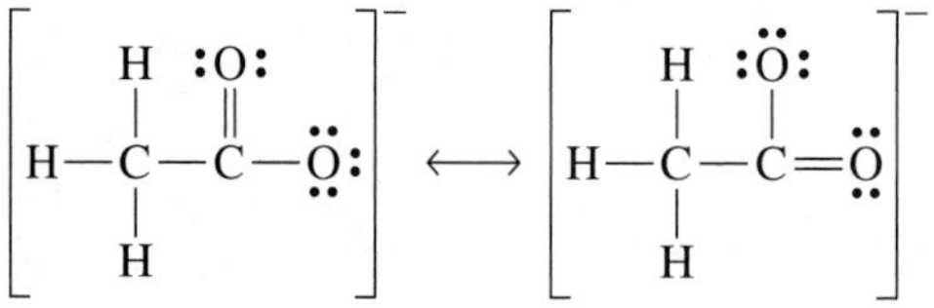

$$\begin{bmatrix} & H & :\!O\!: & \\ & | & \| & \\ H\!-\!C\!-\!C\!-\!\ddot{O}\!: & & & \\ & | & & \\ & H & & \end{bmatrix}^{-} \longleftrightarrow \begin{bmatrix} & H & :\!\ddot{O}\!: & \\ & | & | & \\ H\!-\!C\!-\!C\!=\!\ddot{O} & & & \\ & | & & \\ & H & & \end{bmatrix}^{-} \qquad (11.16)$$

Practice Example A: Draw Lewis structures to represent the resonance hybrid for the SO_2 molecule.

Practice Example B: Draw Lewis structures to represent the resonance hybrid for the nitrate ion.

11-6 Exceptions to the Octet Rule

The octet rule has been a mainstay in writing Lewis structures, and it will continue to be one. Yet at times, we must depart from the octet rule, as we will see in this section.

Odd-Electron Species

The molecule NO has 11 valence electrons, an *odd* number. If the number of valence electrons in a Lewis structure is *odd*, there must be an unpaired electron somewhere in the structure. Lewis theory deals with electron pairs and does not tell us where to put the unpaired electron; it could be on either the N or the O atom. To obtain a structure free of formal charges, however, we will put the unpaired electron on the N atom.

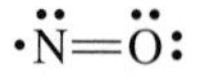

$$\cdot\ddot{N}\!=\!\ddot{O}\!:$$

The presence of unpaired electrons causes odd-electron species to be paramagnetic. NO is paramagnetic. Molecules with an *even* number of electrons are expected to have all electrons paired and to be diamagnetic. An important exception is seen in the case of O_2, which is *paramagnetic* despite having 12 valence electrons. Lewis theory does not provide a good electronic structure for O_2, but the molecular orbital theory that we will consider in the next chapter is much more successful.

▶ Experimental evidence for the paramagnetism of O_2 is shown in Figure 11-3 on page 395.

The number of stable odd-electron molecules is quite limited. More common are **free radicals**, or simply *radicals*, highly reactive molecular fragments with one or more unpaired electrons. The formulas of free radicals are usually written with a dot to emphasize the presence of an unpaired electron, such as in the *methyl* radical, $\cdot CH_3$, and the *hydroxyl* radical, $\cdot OH$. The Lewis structures of these two free radicals are

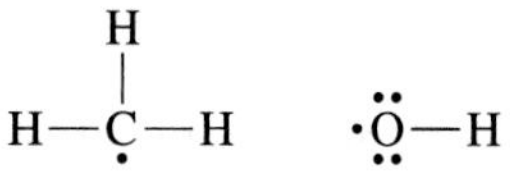

$$H\!-\!\underset{\cdot}{\overset{\displaystyle\overset{H}{|}}{C}}\!-\!H \qquad \cdot\ddot{\underset{\cdot\cdot}{O}}\!-\!H$$

Both free radicals are commonly encountered as transitory species in flames. In addition, OH is formed in the atmosphere in trace amounts as a result of photochemical reactions. Many important atmospheric reactions involve free radicals as reactants, such as in the oxidation of CO to CO_2.

$$\cdot OH + CO \longrightarrow CO_2 + \cdot H$$

Incomplete Octets

Our initial attempt to write the Lewis structure of boron trifluoride leads to a structure in which the B atom has only *six* electrons in its valence shell—an *incomplete octet*.

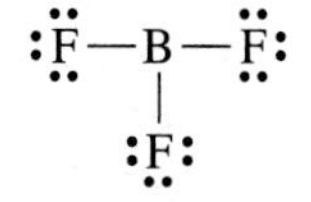

$$:\!\ddot{F}\!-\!B\!-\!\ddot{F}\!: \qquad (11.17)$$
$$\underset{\displaystyle :\!\ddot{F}\!:}{|}$$

We have learned to complete the octets of central atoms by shifting lone-pair electrons from terminal atoms to form multiple bonds. One of *three* equivalent structures with a boron-to-fluorine double bond is shown below

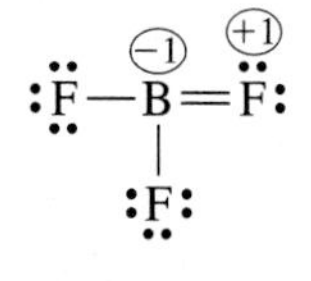

$$:\!\ddot{F}\!-\!\overset{\ominus}{B}\!=\!\overset{\oplus}{\ddot{F}}\!: \qquad (11.18)$$
$$\underset{\displaystyle :\!\ddot{F}\!:}{|}$$

▶ The primary industrial uses of BF_3 are not to produce chemicals containing the elements boron or fluorine. Rather, BF_3 is used because of properties stemming from its electronic structure. In most cases the BF_3 is recovered and recycled.

One observation in support of structure (11.18) is that the B—F bond length in BF_3 (130 pm) is less than expected for a single bond. A shorter bond suggests that more than two electrons are present, that is, that there is multiple bond character in the bond. On the other hand, the placement of formal charges in structure (11.18) breaks an important rule—negative formal charge should be found on the more electronegative atom in the bond. In this structure, the *positive* formal charge is on the most electronegative of all atoms—F.

The high electronegativity of fluorine (4.0) and the much lower one of boron (2.0) suggest an appreciable ionic character to the boron-to-fluorine bond (see Figure 11-7). This suggests the possibility of *ionic* structures such as the following.

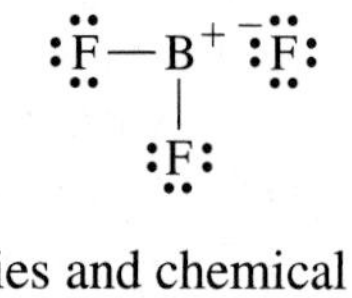

$$:\!\ddot{F}\!-\!B^+\ {}^-\!\!:\!\ddot{F}\!: \qquad (11.19)$$
$$\underset{\displaystyle :\!\ddot{F}\!:}{|}$$

In view of its molecular properties and chemical behavior, the best representation of BF_3 appears to be a resonance hybrid of structures (11.17), (11.18), and (11.19), with perhaps the most important contribution made by the structure with an incomplete octet (11.17). Whichever BF_3 structure we choose to emphasize, an important characteristic of BF_3 is its strong tendency to form a coordinate covalent bond with a species capable of donating an electron pair to the B atom. This can be seen in the formation of the BF_4^- ion.

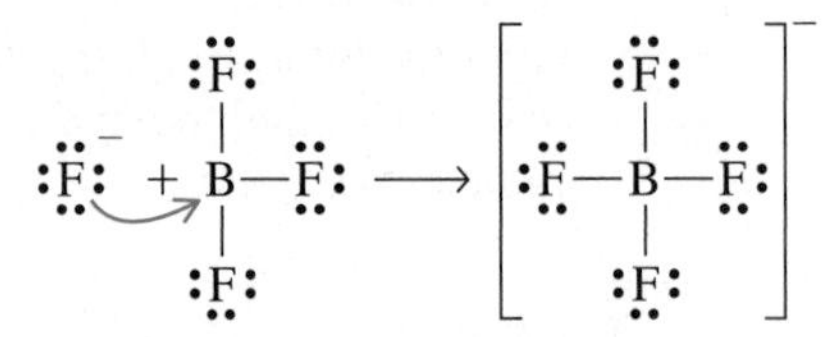

In BF_4^-, the bonds are single bonds and the bond length is 145 pm.

The number of species with incomplete octets is limited to some beryllium, boron, and aluminum compounds. Perhaps the best examples are the boron hydrides. Bonding in the boron hydrides will be discussed in Chapter 23.

Expanded Valence Shells

We have consistently tried to write Lewis structures in which all atoms except H have a complete octet, that is, in which each atom has 8 valence electrons. There are a few Lewis structures that break this rule by having 10 or even 12 valence electrons around the central atom, creating what is called an **expanded valence shell**. Describing bonding in these structures is an area of current active interest among chemists.

Molecules with expanded valence shells typically involve nonmetal atoms of the third period and beyond, that are bonded to highly electronegative atoms. For example, phosphorus forms two chlorides, PCl_3 and PCl_5. We can write a Lewis structure for PCl_3 with the octet rule. In PCl_5, with five Cl atoms bonded directly to the central P atom, the outer shell of the P atom appears to have *ten* electrons. We might say that the valence shell has expanded to 10 electrons. In the SF_6 molecule, the valence shell appears to expand to 12.

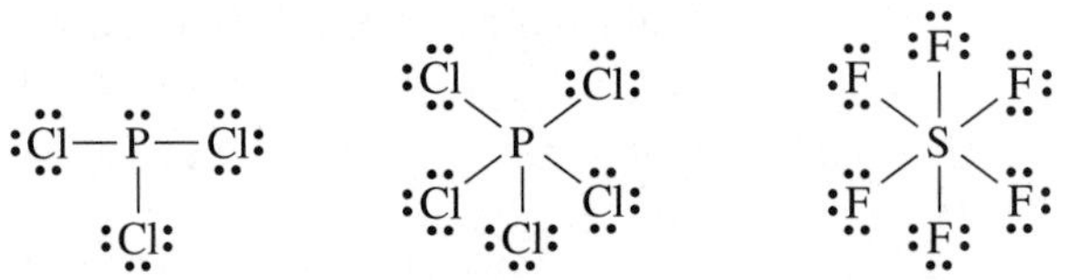

Octet   Expanded valence   Expanded valence
                shell   shell

The problem with expanded valence-shell structures, of course, is where do the "extra" electrons go? This expansion has been rationalized by assuming that after the $3s$ and $3p$ subshells fill to capacity in the central atom (eight electrons), extra electrons go into the empty $3d$ subshell. The thought is that the energy difference between the $3p$ and $3d$ levels is not very large, and so the valence-shell expansion scheme seems reasonable.

Recently, however, some chemists have used sophisticated computer modeling programs to suggest that the $3d$ subshell is not significantly involved in bonding in molecules such as PCl_5 and SF_6, but a discussion of the evidence in support of this assertion is beyond the scope of this text.[*] In any case, our need is just to recognize that to write useful Lewis structures for certain species, we may, at times, have to employ an expanded valence shell on the central atom. Expanded valence shells have also been used at times in cases where they appear to give a better Lewis structure than strict adherence to the octet rule, as suggested by the two Lewis structures for the sulfate ion that follow.

[*]L. Suidan et al., *J. Chem. Educ.*: 72, 583 (1995).

$$
\left[\begin{array}{c} \overset{\ominus}{:\ddot{O}:} \\ | \\ \ominus:\ddot{O}-\overset{+2}{S}\overset{..}{\underset{..}{O}}:\ominus \\ | \\ :\underset{..}{\overset{..}{O}}: \\ \ominus \end{array} \right]^{2-}
\qquad
\left[\begin{array}{c} :\overset{..}{O}: \\ \| \\ \ominus:\ddot{O}-S-\ddot{O}:\ominus \\ \| \\ :\overset{..}{O}: \end{array} \right]^{2-}
$$

Normal octet   Expanded valence shell

The argument for including the expanded valence-shell structure is that it reduces formal charges. Also, the experimentally determined sulfur-to-oxygen bond lengths seem to be shorter than single bonds. The expanded valence-shell structure is suggestive of this partial double-bond character, whereas the octet structure is not. A newer proposal, however, argues that the reduction in bond length can be attributed to the partial ionic character of the sulfur-to-oxygen bonds, and this partial ionic character is emphasized in the octet structure with its positive and negative formal charges.

Although unresolved questions about the expanded valence-shell concept may be unsettling, the point to keep in mind is that the unmodified octet rule works perfectly well for most of our uses of Lewis structures, and it will continue to be our mainstay for the remainder of the chapter.

11-7   The Shapes of Molecules

From the Lewis structure for water, we get the impression that the atoms are arranged in a straight line.

$$ H:\overset{..}{\underset{..}{O}}:H $$

However, the experimentally determined shape of the molecule is not linear. The molecule is *bent*, as shown in Figure 11-9. Does it really matter that the H_2O molecule is bent rather than linear? The answer is, decidedly, yes. As we will learn in Chapter 13, the bent shape of water molecules helps to account for the fact that water is a liquid rather than a gas at room temperature. In Chapter 14, we will find that it also accounts for the ability of liquid water to dissolve so many different substances.

What we seek in this section is a simple model for predicting the approximate shape of a molecule. Unfortunately, Lewis theory tells us nothing about the shapes of molecules. On the other hand, it is an excellent place to begin. The next step is to use an idea based on repulsions between valence-shell electron pairs. We will discuss this idea after defining a few terms.

Some Terminology

By molecular shape, we mean the geometric figure we get when joining the nuclei of bonded atoms by straight lines. Figure 11-9 depicts the *triatomic* (three-atom) water molecule using a ball-and-stick model. The balls represent the three atoms in the molecule, and the straight lines (sticks), the bonds between atoms. In reality, the atoms in the molecule are in close contact, but for clarity we show only the centers of the atoms. To have a complete description of the shape of a molecule, we need to know two quantities.

- **bond lengths**, the distances between the nuclei of bonded atoms
- **bond angles**, the angles between adjacent lines representing bonds

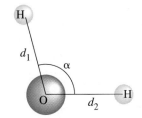

▲ **FIGURE 11-9**
Geometric shape of a molecule
To establish the shape of the triatomic H_2O molecule shown here, we need to determine the distances between the nuclei of the bonded atoms and the angle between adjacent bonds. In H_2O, the bond lengths $d_1 = d_2 = 95.8$ pm and the bond angle $\alpha = 104.45°$.

We will concentrate on bond angles in this section and bond lengths in Section 11-8.

A diatomic molecule has only one bond and no bond angle. Because the geometric shape determined by two points is a straight line, *all diatomic molecules are linear*. A triatomic molecule has two bonds and one bond angle. If the bond angle is 180°, the three atoms lie on a straight line, and the molecule is *linear*. For any other bond angle, a triatomic molecule is said to be *angular, bent*, or *V-shaped*. Some polyatomic molecules with more than three atoms have planar or even linear shapes. More commonly, however, the centers of the atoms in these molecules define a three-dimensional geometric figure.

Valence-Shell Electron-Pair Repulsion (VSEPR) Theory

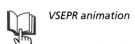
VSEPR animation

The shape of a molecule is established by experiment or by a quantum mechanical calculation confirmed by experiment. The results of these experiments and calculations are generally in good agreement with the **valence-shell electron-pair repulsion theory** (written **VSEPR** and pronounced "vesper"). In VSEPR theory, we focus on *pairs* of electrons in the *valence* electron shell of a central atom in a structure.

> Electron pairs repel each other whether they are in chemical bonds (bond pairs) or unshared (lone pairs). Electron pairs assume orientations about an atom to minimize repulsions.

This, in turn, results in particular geometric shapes for molecules.

Another aspect of VSEPR theory is that we focus not just on *pairs* of electrons but on *electron groups*. A group of electrons can be a pair, either a lone pair or a bond pair, but it can also be a single unpaired electron on an atom with an incomplete octet, as in NO. A group can also be a double or triple bond between two atoms. Thus, in the molecule $\ddot{O}{=}C{=}\ddot{O}$, the central C atom has *two* electron groups in its valence shell. Each of the double bonds with its two electron pairs is treated as *one* electron group.

Consider the methane molecule, CH_4, in which the central C atom has acquired the electron configuration of Ne by forming covalent bonds with four H atoms.

$$
\begin{array}{c}
H \\
\cdot\cdot \\
H\!:\!\overset{\cdot\cdot}{\underset{\cdot\cdot}{C}}\!:\!H \\
H
\end{array}
$$

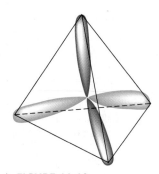

▲ **FIGURE 11-10**
Balloon analogy to valence-shell electron-pair repulsion
When two elongated balloons are twisted together, they separate into four lobes. To minimize interferences, the lobes spread out into a tetrahedral pattern. (A regular tetrahedron has four faces, each an equilateral triangle.) The lobes are analogous to valence-shell electron pairs.

What orientation will the four electron groups (bond pairs) assume? The balloon analogy of Figure 11-10 suggests that electron-group repulsions will force the groups as far apart as possible—to the corners of a tetrahedron having the C atom at its center. The VSEPR method predicts, correctly, that CH_4 is a *tetrahedral* molecule.

In NH_3 and H_2O, the central atom is also surrounded by four groups of electrons, but these molecules *do not* have a tetrahedral shape.

$$
\begin{array}{ccc}
H & & H \\
\cdot\cdot & & \cdot\cdot \\
H\!:\!\overset{\cdot\cdot}{N}\!:\!H & \text{and} & :\!\overset{\cdot\cdot}{O}\!:\!H \\
\cdot\cdot & & \cdot\cdot
\end{array}
$$

Here is the situation: VSEPR theory predicts the distribution of electron groups, and in these molecules, electron groups are arranged tetrahedrally about the central atom. The shape of a molecule, however, is determined by the location of the atomic nuclei. To avoid confusion, we will call the geometric distribution of electron groups the **electron-group geometry** and the geometric arrangement of the atomic nuclei—the actual determinant of the molecular shape—the **molecular geometry**.

In the NH_3 molecule, only *three* of the electron groups are bond pairs; the fourth is a *lone pair*. By joining the N nucleus to the H nuclei by straight lines, we outline a *pyramid* (called a trigonal pyramid). This pyramid has the N atom at the apex and the three H atoms at the base. This is not the same as a tetrahedron, which would have the N atom at the center. We say that the electron-group geometry is tetrahedral and the molecular geometry is trigonal-pyramidal.

In the H_2O molecule, two of the four electron groups are *bond* pairs and two are *lone* pairs. The molecular shape is obtained by joining the two H nuclei to the O nucleus with straight lines. For H_2O, the electron-group geometry is tetrahedral and the molecular geometry is V-shaped, or bent. The geometric shapes of CH_4, NH_3, and H_2O are shown in Figure 11-11, together with space-filling molecular models.

In the VSEPR notation used in Figure 11-11, A is the central atom, X is a terminal atom or group of atoms bonded to the central atom, and E is a lone pair of electrons. Thus, the symbol AX_2E_2 signifies that *two* atoms or groups (X) are bonded to the central atom (A). The central atom also has *two* lone pairs of electrons (E). H_2O is an example of a molecule of the AX_2E_2 type.

For tetrahedral electron-group geometry, we expect bond angles of 109.5°, known as the *tetrahedral* bond angle. In the CH_4 molecule, the measured bond angles are, in fact, 109.5°. The bond angles in NH_3 and H_2O are slightly smaller: 107° for the H—N—H bond angle and 104.5° for the H—O—H bond angle. We can explain these less-than-tetrahedral bond angles by the fact that the charge cloud of the lone-pair electrons spreads out. This forces the bond-pair electrons closer together and reduces the bond angles.

▶ VSEPR theory works best for second-period elements. The predicted bond angle of 109.5° for H_2O is close to the measured angle of 104.5°. For H_2S, however, the predicted value of 109.5° is not in good agreement with the observed 92°.

CH_4, NH_3, and H_2O models

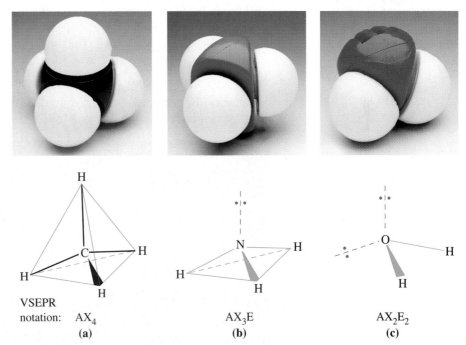

▲ **FIGURE 11-11 Molecular shapes based on tetrahedral electron-group geometry—CH_4, NH_3, and H_2O**
Molecular shapes are established by the blue lines. Lone-pair electrons are shown as red dots along broken lines originating at the central atom. **(a)** All electron groups around the central atom are bond pairs. The blue lines that outline the molecule are different from the black lines representing the carbon-to-hydrogen bonds. **(b)** The lone pair of electrons is directed to the "missing" corner of the tetrahedron. The nitrogen-to-hydrogen bonds form three of the edges of a trigonal pyramid. **(c)** The H_2O molecule is a bent molecule outlined by the two oxygen-to-hydrogen bonds.

Possibilities for Electron-Group Distributions

In general, we will encounter situations in which central atoms have two, three, four, five, or six electron groups distributed around them.

Electron-group geometries

- two electron groups: linear
- three electron groups: trigonal-planar
- four electron groups: tetrahedral
- five electron groups: trigonal-bipyramidal
- six electron groups: octahedral

Figure 11-12 extends the balloon analogy to these cases. The cases for five and six electron groups are typified by PCl_5 and SF_6, molecules with expanded valence shells.

The molecular geometry is the same as the electron-group geometry *only* when all electron groups are bond pairs. These are for the VSEPR notation AX_n (that is, AX_2, AX_3, AX_4, ...). In Table 11.1, the AX_n cases are illustrated by photographs of ball-and-stick models. If one or more electron groups are lone pairs, the molecular geometry is *different* from the electron-group geometry, although still derived from it. To understand all the cases in Table 11.1, we need two more ideas.

- *The closer together two groups of electrons are forced, the stronger the repulsion between them.* The repulsion between two electron groups is much stronger at an angle of 90° than at 120° or 180°.

- *Lone-pair electrons spread out more than do bond-pair electrons.* As a result, the repulsion of one lone pair of electrons for another lone pair is greater than, say, between two bond pairs. The order of repulsive forces, from strongest to weakest, is:

    lone pair–lone pair $>$ lone pair–bond pair $>$ bond pair–bond pair

Consider SF_4, with the VSEPR notation AX_4E. *Two* possibilities for its structure are presented in the margin, but only one is correct. The correct structure (top)

▲ **FIGURE 11-12** **Several electron-group geometries illustrated**
The electron-group geometries pictured are trigonal-planar (orange), tetrahedral (gray), trigonal-bipyramidal (pink), and octahedral (yellow). The atoms at the ends of the balloons are not shown and are not important in this model. The relationship between electron-group geometry and molecular geometry is summarized in Table 11.1.

TABLE 11.1 Molecular Geometry as a Function of Electron Group Geometry

Number of Electron Groups	Electron-Group Geometry	Number of Lone Pairs	VSEPR Notation	Molecular Geometry	Ideal Bond Angles	Example	
2	linear	0	AX_2	X—A—X (linear)	180°	$BeCl_2$	
3	trigonal-planar	0	AX_3	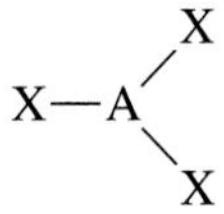 (trigonal-planar)	120°	BF_3	
	trigonal-planar	1	AX_2E	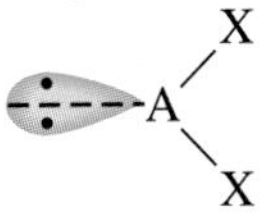 (angular)	120°	SO_2[a]	
4	tetrahedral	0	AX_4	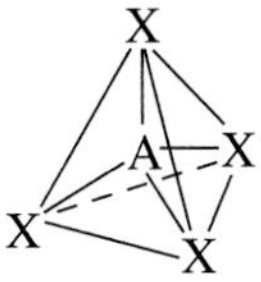 (tetrahedral)	109.5°	CH_4	
	tetrahedral	1	AX_3E	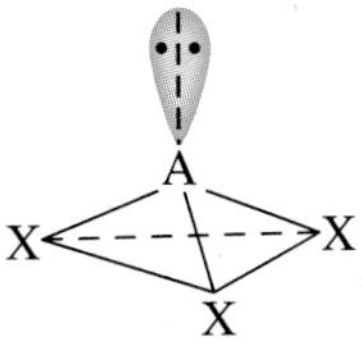 (trigonal-pyramidal)	109.5°	NH_3	
	tetrahedral	2	AX_2E_2	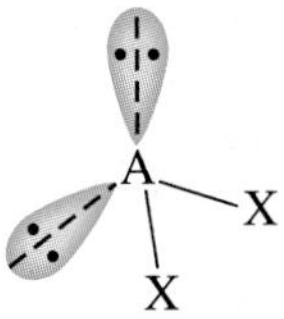 (angular)	109.5°	OH_2	
5	trigonal-bipyramidal	0	AX_5	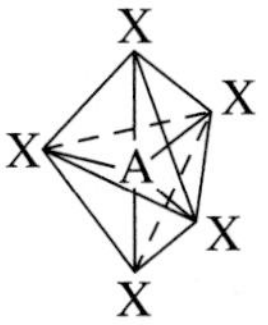 (trigonal-bipyramidal)	90°, 120°	PCl_5	

($BeCl_2$)

(BF_3)

(CH_4)

(PCl_5)

(continues)

TABLE 11.1   (Continued)

Number of Electron Groups	Electron-Group Geometry	Number of Lone Pairs	VSEPR Notation	Molecular Geometry	Ideal Bond Angles	Example
	trigonal-bipyramidal	1	AX_4E^b	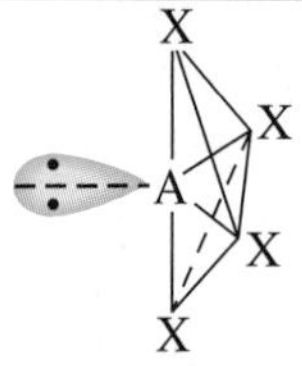 X, X, A, X, X (seesaw)	90°, 120°	SF_4
	trigonal-bipyramidal	2	AX_3E_2	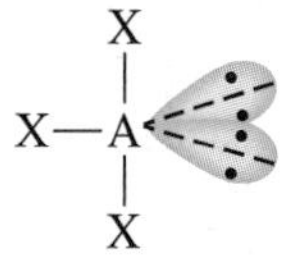 X, X—A, X (T-shaped)	90°	ClF_3
	trigonal-bipyramidal	3	AX_2E_3	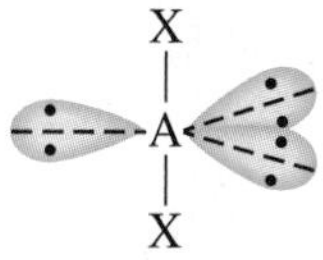 X, A, X (linear)	180°	XeF_2
6	octahedral	0	AX_6	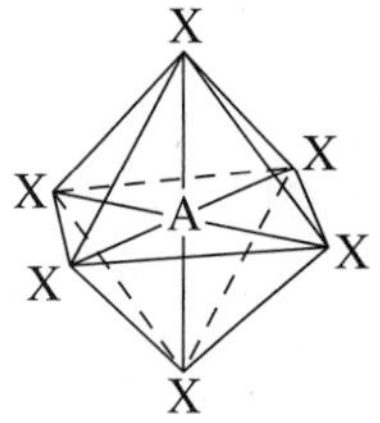 X, X, A, X, X, X (octahedral)	90°	SF_6
	octahedral	1	AX_5E	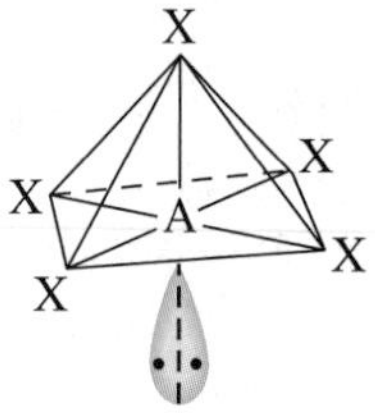 X, X, A, X, X (square-pyramidal)	90°	BrF_5
	octahedral	2	AX_4E_2	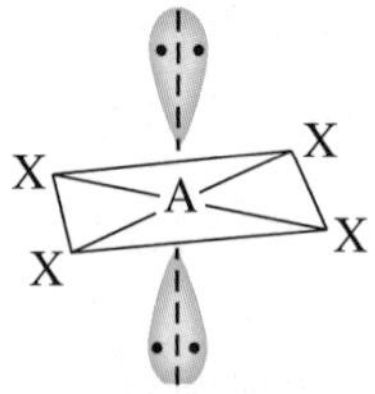 X, A, X, X (square-planar)	90°	XeF_4

(SF_6)

[a] For a discussion of the structure of SO_2, see page 416.
[b] For a discussion of the placement of the lone-pair electrons in this structure, see page 412.

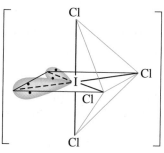

(Incorrect)

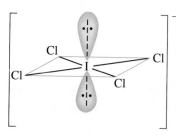

(Correct)

▲ **FIGURE 11-13**
Example 11-9 illustrated
The observed structure of ICl_4^- is the square-planar structure.

places a lone pair of electrons in the central plane of the bipyramid. As a result, *two* of the lone pair–bond pair interactions are 90°. In the incorrect structure (bottom), the lone pair of electrons is at the bottom of the bipyramid and results in *three* lone pair–bond pair interactions of 90°. This is a less favorable arrangement.

Applying VSEPR Theory

Let us use the following four-step strategy for predicting the shapes of molecules.

1. Draw a plausible Lewis structure of the species (molecule or polyatomic ion).
2. Determine the number of electron groups around the central atom, and identify them as being either *bond* electron groups or *lone* pairs of electrons.
3. Establish the electron-group geometry around the central atom—linear, trigonal-planar, tetrahedral, trigonal-bipyramidal, or octahedral.
4. Determine the molecular geometry from the positions around the central atom occupied by the other atomic nuclei—that is, from data in Table 11.1.

EXAMPLE 11-9

Using VSEPR Theory to Predict a Geometric Shape. Predict the molecular geometry of the polyatomic anion ICl_4^-.

Solution

Apply the four steps outlined above.

Step 1. Write the Lewis structure. The number of valence electrons is

From I	From Cl	To establish ionic charge of −1	
(1×7)	(4×7) +	1	= 36

To join 4 Cl atoms to the central I atom and to provide octets for all the atoms, we need 32 electrons. In order to account for all 36 valence electrons, we need to place an *additional* four electrons around the I atom as lone pairs. That is, we are forced to expand the valence shell of the I atom to accommodate all the electrons required in the Lewis structure.

KEEP IN MIND ▶
that we need to expand a valence shell only if more than eight electrons must be accommodated by the central atom in a Lewis structure. This leads to structures based on five or six electron groups. Otherwise, octet-based Lewis structures are perfectly satisfactory when applying the VSEPR theory.

$$
\left[\begin{array}{cc} \ddot{C}l & \ddot{C}l \\ & I \\ \ddot{C}l & \ddot{C}l \end{array} \right]^-
$$

Step 2. There are six electron groups around the I atom, four *bond* pairs and two *lone* pairs.
Step 3. The electron-group geometry (the orientation of six electron groups) is *octahedral*.
Step 4. The ICl_4^- anion is of the type AX_4E_2, which, according to Table 11.1, leads to a molecular geometry that is square-planar.

Figure 11-13 suggests two possibilities for distributing bond pairs and lone pairs in ICl_4^-. The square-planar structure is correct because the lone pair–lone pair interaction is kept at 180°. In the incorrect structure, this interaction is at 90°, which results in a strong repulsion.

Practice Example A: Predict the molecular geometry of nitrogen trichloride.

Practice Example B: Predict the molecular geometry of phosphoryl chloride, $POCl_3$, an important chemical in the manufacture of gasoline additives, hydraulic fluids, and fire retardants.

VSEPR models

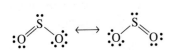

Structures with Multiple Covalent Bonds

In a multiple covalent bond, all electrons in the bond are confined to the region between the bonded atoms, and together constitute one group of electrons. Let us test this idea by predicting the molecular geometry of sulfur dioxide. S is the central atom, and the total number of valence electrons is $3 \times 6 = 18$. The Lewis structure is the resonance hybrid of the two contributing structures in the margin.

It is immaterial which structure we use. In either case, we count the electrons in the double covalent bond as one group. This bond and the sulfur-to-oxygen single bond account for *two* electron groups. The third electron group around the central S atom is a lone pair of electrons. The electron-group geometry around the central S atom is that of *three* electron groups—*trigonal-planar*. Of the three electron groups, two are bonding groups and one is a lone pair. This is the case of AX_2E (see Table 11.1). The molecular shape is *angular* or *bent*, with an expected bond angle of 120°. (The measured bond angle in SO_2 is 119°.)

EXAMPLE 11-10

Using VSEPR Theory to Predict the Shape of a Molecule with a Multiple Covalent Bond. Predict the molecular geometry of formaldehyde, H_2CO, used to make a number of polymers, such as melamine resins.

Solution

The Lewis structure in the margin has a total of 12 valence electrons and C as the central atom. If all the bonds to the carbon atom were single bonds, C would lack an octet. This deficiency is corrected by moving a lone pair of electrons from the O atom into the carbon-to-oxygen bond, making it a double bond.

We count *three* electron groups around the C atom, two groups in the carbon-to-hydrogen single bonds and the third group in the carbon-to-oxygen double bond. The electron-group geometry for three electron groups is *trigonal-planar*. Because all the electron groups are involved in bonding, the VSEPR notation for this molecule is AX_3. The molecular geometry is also trigonal-planar.

Practice Example A: Predict the shape of the COS molecule.

Practice Example B: Nitrous oxide, N_2O, is the familiar laughing gas used as an anesthetic in dentistry. Predict the shape of the N_2O molecule.

Molecules with More Than One Central Atom

Although many of the structures of interest to us have only one central atom, we can apply VSEPR theory to molecules or polyatomic anions with more than one central atom. In such cases, we have to work out the geometric distribution of terminal atoms around *each* central atom and then combine the results into a single description of the molecular shape. We use this idea in Example 11-11.

EXAMPLE 11-11

Applying VSEPR Theory to a Molecule with More Than One Central Atom. Methyl isocyanate, CH_3NCO, is used in the manufacture of insecticides, such as carbaryl (Sevin). In the CH_3NCO molecule, the three H atoms and the O atom are terminal atoms and the two C and one N atom are central atoms. Draw a sketch of this molecule.

Solution

To apply the VSEPR method, let us begin with a plausible Lewis structure. The number of valence electrons in the structure is

From C	*From N*	*From O*	*From H*

$$(2 \times 4) \quad (1 \times 5) \quad (1 \times 6) \quad (3 \times 1) = 22$$

In drawing the skeletal structure and assigning valence electrons, we first obtain a structure with incomplete octets. By shifting the indicated electrons, we can give each atom an octet.

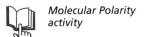

The C atom on the left has four electron groups around it—all bond pairs. The shape of this end of the molecule is *tetrahedral*. The C atom to the right, by forming two double bonds, is treated as having *two* groups of electrons around it. This distribution is *linear*. For the N atom, *three* groups of electrons are distributed in a *trigonal-planar* manner. The C—N—C bond angle should be about 120°.

Practice Example A: Draw a sketch of the methanol molecule, CH_3OH. Indicate the bond angles in this molecule.

Practice Example B: Glycine, an amino acid, has the formula H_2NCH_2COOH. Sketch the glycine molecule, and indicate the various bond angles.

Molecular Polarity activity

Molecular Shapes and Dipole Moments

Let us recall some facts that we learned about polar covalent bonds in Section 11-3. In the HCl molecule, the Cl atom is more electronegative than the H atom. Electrons are displaced toward the Cl atom. The HCl molecule is a **polar molecule**. In the following representation, we use a cross-base arrow ($\longmapsto$) that points to the atom that attracts electrons more strongly.

$$H \longmapsto Cl$$

The extent of the charge displacement in a polar covalent bond is given by the **dipole moment**, μ. The dipole moment is the product of a partial charge (δ) and distance (d).

$$\mu = \delta d \qquad (11.20)$$

If the product, δd, has a value of 3.34×10^{-30} coulomb·meter (C·m), the dipole moment, μ, has a value called 1 *debye, D*. One experimental method of determining dipole moments is based on the behavior of polar molecules in an electric field, suggested in Figure 11-14.

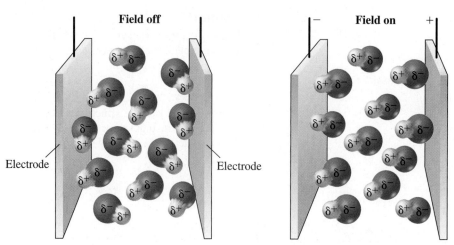

▲ FIGURE 11-14   **Polar molecules in an electric field**
The device pictured is called an electrical condenser (or capacitor). It consists of a pair of electrodes separated by a medium that does not conduct electricity but consists of polar molecules. **(a)** When the field is off, the molecules orient randomly. **(b)** When the electric field is turned on, the polar molecules orient in the field between the charged plates so that the negative ends of the molecules are toward the positive plate and vice versa.

The polarity of the H—Cl bond, as we demonstrated on page 395, involves a shift of the electron charge density toward the Cl atom, and this produces a separation of the centers of positive and negative charge. Suppose, instead, we think of an equivalent situation—the transfer of a *fraction* of the charge of an electron from the H atom to the Cl atom through the entire internuclear distance. Let us determine the magnitude of this partial charge, δ. To do this, we need the measured dipole moment, 1.03 D; the H—Cl bond length, 127.4 pm; and equation (11.20) rearranged to

$$\delta = \frac{\mu}{d} = \frac{1.03 \text{ D} \times 3.34 \times 10^{-30} \text{ C·m/D}}{127.4 \times 10^{-12} \text{ m}} = 2.70 \times 10^{-20} \text{ C}$$

This charge is about 17% of the charge on an electron (1.602×10^{-19} C) and suggests that HCl is about 17% ionic. This assessment of the percent ionic character of the H—Cl bond agrees well with the 20% we made based on electronegativity differences (recall Example 11-4).

CO₂.   Carbon dioxide molecules are *nonpolar*. To understand this observation, we need to distinguish between the displacement of electron charge density in a particular bond and in the molecule as a whole. The electronegativity difference between C and O causes a displacement of electron charge density toward the O atom in each carbon-to-oxygen bond and gives rise to a *bond dipole*. However, because the two bond dipoles are equal in magnitude and point in opposite directions, they cancel each other and lead to a *resultant* dipole moment of zero for the molecule.

$$\overset{\longleftarrow\ +}{\text{O}}\!-\!\text{C}\!-\!\overset{+\ \longrightarrow}{\text{O}} \qquad \mu = 0$$

The fact that CO₂ is nonpolar is experimental proof that CO₂ is a linear molecule. Of course, we might also predict that CO₂ is a linear molecule with the VSEPR theory, based on the Lewis structure

$$:\!\ddot{\text{O}}\!=\!\text{C}\!=\!\ddot{\text{O}}\!:$$

H₂O. Water molecules are *polar*. They have bond dipoles because of the electronegativity difference between H and O, and the bond dipoles combine to produce a resultant dipole moment of 1.84 D. The molecule cannot be linear, for this would lead to a cancelation of bond dipoles, just as with CO_2. We have predicted with the VSEPR theory that the H_2O molecule is bent, and the observation that it is a polar molecule simply confirms the prediction.

$$\overset{\uparrow}{\underset{H}{\swarrow}} \underset{104°}{\overset{O \rightarrow H}{\diagdown}}$$

CCl₄. Carbon tetrachloride molecules are *nonpolar*. Based on the electronegativity difference between Cl and C, we expect a bond dipole for the C—Cl bond. The fact that the resultant dipole moment is *zero* means that the bond dipoles must be oriented in such a way that they cancel. The tetrahedral molecular geometry of CCl_4 provides the symmetrical distribution of bond dipoles that leads to this cancelation, as shown in Figure 11-15a. Can you see that the molecule will be polar if we replace one of the Cl atoms by an atom with a different electronegativity, say H? In the molecule $CHCl_3$, there is a resultant dipole moment (Figure 11-15b).

KEEP IN MIND ▶
that the lack of a molecular dipole moment cannot distinguish between the two possible molecular geometries: tetrahedral and square-planar. To do this, other experimental evidence, such as X-ray diffraction, is required.

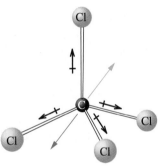

(a) CCl₄: a nonpolar molecule

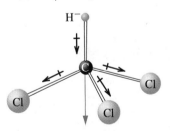

(b) CHCl₃: a polar molecule

▲ **FIGURE 11-15**
Molecular shapes and dipole moments
(a) The resultant of two of the C—Cl bond dipoles is shown as a red arrow, and that of the other two, as a blue arrow. The red and blue arrows point in opposite directions and cancel. The CCl₄ molecule is nonpolar. (b) The individual bond dipoles do combine to yield a resultant dipole moment (red arrow) of 1.04 D.

EXAMPLE 11-12

Determining the Relationship Between Geometric Shapes and the Resultant Dipole Moments of Molecules. Which of these molecules would you expect to be polar? Cl_2, ICl, BF_3, NO, SO_2.

Solution

Polar: ICl, NO, SO_2. ICl and NO are diatomic molecules with an electronegativity difference between the bonded atoms. SO_2 is a bent molecule with an electronegativity difference between the S and O atoms.

Nonpolar: Cl_2 and BF_3. Cl_2 is a diatomic molecule of identical atoms; hence no electronegativity difference. For BF_3, refer to Table 11.1. BF_3 is a symmetrical planar molecule (120° bond angles). The B—F bond dipoles cancel each other.

Practice Example A: Only one of the following molecules is polar. Which is it, and why? SF_6, H_2O_2, C_2H_4.

Practice Example B: Only one of the following molecules is *nonpolar*. Which is it, and why? Cl_3CCH_3, PCl_5, CH_2Cl_2, NH_3.

The Importance of Lone-Pair Electrons

We have seen that electronegativity differences determine whether bond dipoles exist in a molecule, and molecular shape determines whether bond dipoles cancel (nonpolar molecules) or combine to produce a resultant dipole moment (polar molecules). Thus, the ozone molecule, O_3, has no bond dipoles because all the atoms are alike. Yet, O_3 *does* have a resultant dipole moment: $\mu = 0.534$ D. There must be another factor at work here.

When writing the Lewis structure of O_3 on page 405, we saw that there are two equivalent structures contributing to a resonance hybrid. Because the central atom in these structures is surrounded by three groups of electrons, the electron-group geometry is trigonal-planar and the molecule is bent.

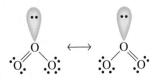

The zero bond dipoles in O_3 signify that the centers of negative and positive charge coincide along the oxygen-to-oxygen bonds. However, the lone-pair electrons on the central O atom constitute another center of negative charge that is offset from the oxygen-to-oxygen bonds. The dipole moment in the O_3 molecule is directed toward this charge center.

11-8   Bond Order and Bond Lengths

The term **bond order** describes whether a covalent bond is *single* (bond order = 1), *double* (bond order = 2), or *triple* (bond order = 3). Think of electrons as the "glue" that binds atoms together in covalent bonds. The higher the bond order—that is, the more electrons present—the more glue and the more tightly the atoms are held together.

Bond length is the distance between the centers of two atoms joined by a covalent bond. As we might expect, a double bond between atoms is shorter than a single bond, and a triple bond is shorter still. You can see this relationship clearly in Table 11.2 by comparing the three different bond lengths for the nitrogen-to-nitrogen bond. For example, the measured nitrogen-to-nitrogen bond length in N_2 is 109.8 pm—a triple bond—whereas the nitrogen-to-nitrogen bond length in hydrazine, H_2N-NH_2, is 147 pm—a single bond.

Perhaps you can now also better understand the meaning of covalent radius that we introduced in Section 10-3. The single covalent radius is one-half the distance between the centers of identical atoms joined by a *single* covalent bond. Thus, the single covalent radius of chlorine in Figure 10-8 (99 pm) is one-half the bond length given in Table 11.2, that is, $\frac{1}{2} \times 199$ pm. Furthermore, as a rough generalization,

> the length of the covalent bond between two atoms can be approximated as the sum of the covalent radii of the two atoms.

Some of these ideas about bond length are applied in Example 11-13.

TABLE 11.2   Some Average Bond Lengths[a]

Bond	Bond Length, pm	Bond	Bond Length, pm	Bond	Bond Length, pm
H—H	74.14	C—C	154	N—N	145
H—C	110	C=C	134	N=N	123
H—N	100	C≡C	120	N≡N	109.8
H—O	97	C—N	147	N—O	136
H—S	132	C=N	128	N=O	120
H—F	91.7	C≡N	116	O—O	145
H—Cl	127.4	C—O	143	O=O	121
H—Br	141.4	C=O	120	F—F	143
H—I	160.9	C—Cl	178	Cl—Cl	199
				Br—Br	228
				I—I	266

[a] Most values (C—H, N—H, C—H, ...) are averaged over a number of species containing the indicated bond and may vary by a few picometers. Where a diatomic molecule exists, the value given is the actual bond length in that molecule (H_2, N_2, HF, ...) and is known more precisely.

EXAMPLE 11-13

Estimating Bond Lengths. Provide the best estimate you can of these bond lengths: **(a)** the nitrogen-to-hydrogen bonds in NH_3; **(b)** the bromine-to-chlorine bond in BrCl.

Solution

(a) The Lewis structure of ammonia (page 393) shows the nitrogen-to-hydrogen bonds as single bonds. The value listed in Table 11.2 for the N—H bond is 100 pm, so this is the value we would predict. (The measured N—H bond length in NH_3 is 101.7 pm.)

(b) We do not find a bromine-to-chlorine bond length in Table 11.2. We need to calculate an approximate bond length with the relationship between bond length and covalent radii. BrCl contains a Br—Cl *single* bond [just imagine substituting one Br atom for one Cl atom in structure (11.4).] The length of the Br—Cl bond is one-half the Cl—Cl bond length *plus* one-half the Br—Br bond length: $\left(\frac{1}{2} \times 199 \text{ pm}\right) + \left(\frac{1}{2} \times 228 \text{ pm}\right) = 214 \text{ pm}$. (The measured bond length is 213.8 pm.)

Practice Example A: Estimate the bond lengths of the carbon-to-hydrogen bonds and the carbon-to-bromine bond in CH_3Br.

Practice Example B: In a CO_2 molecule, each O atom is bonded to the C atom with a bond length of 116.3 pm. Write a plausible Lewis structure for CO_2 that is consistent with this experimental evidence.

11-9   Bond Energies

Together with bond lengths, we can use bond energies to assess the suitability of a proposed Lewis structure. Bond energy, bond length, and bond order are interrelated properties in this sense: the higher the bond order, the shorter the bond between two atoms and the greater the bond energy.

Energy is *released* when isolated atoms join to form a covalent bond, and energy must be *absorbed* to break apart covalently bonded atoms. **Bond-dissociation energy, D,** is the quantity of energy required to break one mole of covalent bonds in a *gaseous* species. The SI units are kilojoules per mole of bonds (kJ/mol).

In the manner of Chapter 7, we can think of bond-dissociation energy as an enthalpy change or a heat of reaction. For example,

Bond breakage: $H_2(g) \longrightarrow 2\,H(g)$ $\Delta H = D(\text{H—H}) = +435.93 \text{ kJ/mol}$

Bond formation: $2\,H(g) \longrightarrow H_2(g)$ $\Delta H = -D(\text{H—H}) = -435.93 \text{ kJ/mol}$

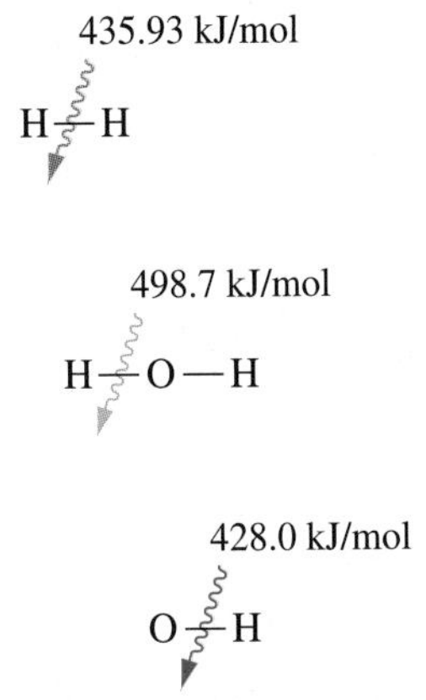

It is not hard to picture the meaning of bond energy for a *diatomic* molecule, because there is only one bond in the molecule. It is also not difficult to see that the bond-dissociation energy of a diatomic molecule can be expressed rather precisely, as is that of $H_2(g)$. With a polyatomic molecule such as H_2O, the situation is different (Figure 11-16). The energy needed to dissociate one mole of H atoms by breaking one O—H bond per H_2O molecule

$$\text{H—OH}(g) \longrightarrow \text{H}(g) + \text{OH}(g) \quad \Delta H = D(\text{H—OH}) = +498.7 \text{ kJ/mol}$$

is different from the energy required to dissociate one mole of H atoms by breaking the bonds in OH(g).

$$\text{O—H}(g) \longrightarrow \text{H}(g) + \text{O}(g) \quad \Delta H = D(\text{O—H}) = +428.0 \text{ kJ/mol}$$

The two O—H bonds in H_2O are identical; therefore, they should have identical energies. This energy, which we can call the O—H bond energy in H_2O, is the

435.93 kJ/mol

H$\rightleftharpoons$H

498.7 kJ/mol

H$\rightleftharpoons$O—H

428.0 kJ/mol

O$\rightleftharpoons$H

▲ **FIGURE 11-16**
Some bond energies compared
The same quantity of energy, 435.93 kJ/mol, is required to break all H—H bonds. In H_2O, more energy is required to break the first bond (498.7 kJ/mol) than to break the second (428.0 kJ/mol). The second bond broken is that in the OH radical. The O—H bond energy in H_2O is the average of the two values: 463.4 kJ/mol.

TABLE 11.3 Some Average Bond Energies[a]

Bond	Bond Energy, kJ/mol	Bond	Bond Energy, kJ/mol	Bond	Bond Energy, kJ/mol
H—H	436	C—C	347	N—N	163
H—C	414	C=C	611	N=N	418
H—N	389	C≡C	837	N≡N	946
H—O	464	C—N	305	N—O	222
H—S	368	C=N	615	N=O	590
H—F	565	C≡N	891	O—O	142
H—Cl	431	C—O	360	O=O	498
H—Br	364	C=O	736[b]	F—F	159
H—I	297	C—Cl	339	Cl—Cl	243
				Br—Br	193
				I—I	151

[a] Although all data are listed with about the same precision (three significant figures), some values are actually known more precisely. Specifically, the values for the diatomic molecules: H_2, HF, HCl, HBr, HI, N_2 (N≡N), O_2 (O=O), F_2, Cl_2, Br_2, and I_2 are actually bond-dissociation energies, rather that average bond energies.
[b] The value for the C=O bonds in CO_2 is 799 kJ/mol.

average of the two values listed above: 463.4 kJ/mol. The O—H bond energy in other molecules containing the OH group will be somewhat different from that in H—O—H. For example, in methanol, CH_3OH, the O—H bond-dissociation energy, which we can represent as $D(H—OCH_3)$, is 436.8 kJ/mol. The usual method of tabulating bond energies (Table 11.3) is as *averages*. An **average bond energy** is the average of bond-dissociation energies for a number of different species containing the particular bond. Understandably, average bond energies cannot be stated as precisely as specific bond-dissociation energies.

As you can see from Table 11.3, double bonds have higher bond energies than do single bonds between the same atoms, but they are *not* twice as large. Triple bonds are stronger still, but their bond energies are *not* three times as large as single bonds between the same atoms. This observation about bond order and bond energy will seem quite reasonable when we describe multiple bonds more fully in the next chapter.

Bond energies also have some interesting uses in thermochemistry. For a reaction involving *gases*, visualize the process

$$\text{gaseous reactants} \longrightarrow \text{gaseous atoms} \longrightarrow \text{gaseous products}$$

In this hypothetical process, we first break all the bonds in reactant molecules and form gaseous atoms. For this step, the enthalpy change is ΔH(bond breakage) = ΣBE (reactants), where the symbol BE stands for bond energy. Next, we allow the gaseous atoms to recombine into product molecules. In this step, bonds are formed and ΔH(bond formation) = $-\Sigma$BE (products). The enthalpy change of the reaction, then, is

$$\Delta H_{rxn} = \Delta H\text{(bond breakage)} + \Delta H \text{(bond formation)}$$

$$\approx \Sigma \text{BE (reactants)} - \Sigma \text{BE (products)}$$

(11.21)

The approximately equal sign ($\approx$) in expression (11.21) signifies that some of the bond energies used are likely to be *average* bond energies rather than true bond-dissociation energies. Also, a number of terms often cancel out because some of the

KEEP IN MIND ▶
that tabulated bond energies are for isolated molecules in the gaseous state. They do not apply to molecules in close contact in liquids and solids.

same types of bonds appear in the products as in the reactants. We can base the calculation of ΔH_{rxn} just on the *net* number and types of bonds broken and formed, as illustrated in Example 11-14.

Enthalpy of Reaction animation

EXAMPLE 11-14

Calculating an Enthalpy of Reaction from Bond Energies. The reaction of methane (CH_4) and chlorine produces a mixture of products called chloromethanes. One of these is monochloromethane, CH_3Cl, used in the preparation of silicones. Calculate ΔH for the reaction

$$CH_4(g) + Cl_2(g) \longrightarrow CH_3Cl(g) + HCl(g)$$

Solution

To assess which bonds are broken and formed, it helps to draw structural formulas (or Lewis structures), as in Figure 11-17. To apply expression (11.21) literally, we would break *four* C—H bonds and *one* Cl—Cl bond and form *three* C—H bonds, *one* C—Cl bond, and *one* H—Cl bond. The *net* change, however, is the breaking of *one* C—H bond and *one* Cl—Cl bond, followed by the formation of *one* C—Cl bond and *one* H—Cl bond.

ΔH *for net bond breakage:*	1 mol C — H bonds	+414 kJ
	1 mol Cl — Cl bonds	+243 kJ
	sum:	+657 kJ

ΔH *for net bond formation:*	1 mol C — Cl bonds	−339 kJ
	1 mol H — Cl bonds	−431 kJ
	sum:	−770 kJ

Enthalpy of reaction:	$\Delta H_{rxn} = 657 - 770 = -113$ kJ

Practice Example A: Use bond energies to estimate the enthalpy change for the reaction

$$2 H_2(g) + O_2(g) \longrightarrow 2 H_2O(g)$$

Practice Example B: Use bond energies to estimate the enthalpy of formation of $NH_3(g)$.

(*Hint:* What is the reaction having the required enthalpy change?)

There is no advantage to using bond energies over enthalpy-of-formation data. Enthalpies of formation are generally known rather precisely, whereas bond energies are only average values. But when enthalpy-of-formation data are lacking, bond energies can prove particularly useful.

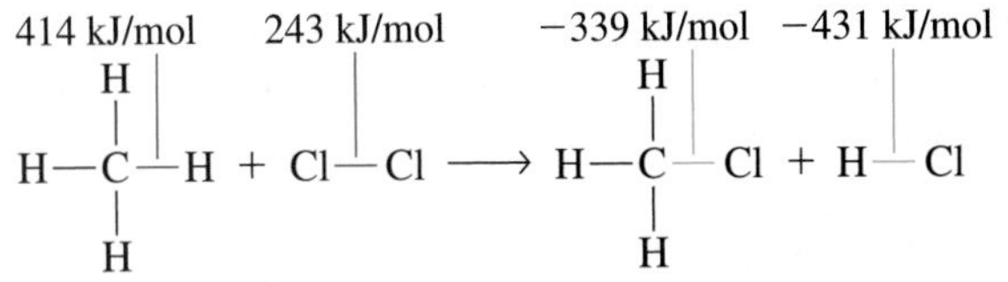

414 kJ/mol   243 kJ/mol   −339 kJ/mol   −431 kJ/mol

$$
\begin{matrix}
& H & & & & & H & & \\
& | & | & & | & & | & | & | \\
H - & C - & H & + & Cl - Cl & \longrightarrow & H - C - & Cl & + & H - Cl \\
& | & & & & & | & & \\
& H & & & & & H & &
\end{matrix}
$$

▲ **FIGURE 11-17   Net bond breakage and formation in a chemical reaction— Example 11-14 illustrated**
Bonds that are broken are shown in red, and bonds that are formed, in blue. Bonds that remain unchanged are in black. The net change is that *one* C—H and *one* Cl—Cl bond break and *one* C—Cl and *one* H—Cl bond form.

Another way to use bond energies is in predicting whether a reaction will be *endothermic* or *exothermic*. In general, if

$$\underset{\text{(reactants)}}{\text{weak bonds}} \longrightarrow \underset{\text{(products)}}{\text{strong bonds}} \qquad \Delta H < 0 \quad \textit{(exothermic)}$$

and

$$\underset{\text{(reactants)}}{\text{strong bonds}} \longrightarrow \underset{\text{(products)}}{\text{weak bonds}} \qquad \Delta H > 0 \quad \textit{(endothermic)}$$

Example 11-15 applies this idea to a reaction involving highly reactive, unstable species for which enthalpies of formation are not normally listed.

EXAMPLE 11-15

Using Bond Energies to Predict Exothermic and Endothermic Reactions. One of the steps in the formation of monochloromethane (Example 11-14) is the reaction of a gaseous chlorine *atom* (a chlorine radical) with a molecule of methane. The products are an unstable methyl radical, $\cdot CH_3$, and $HCl(g)$. Is this reaction endothermic or exothermic?

$$CH_4(g) + \cdot Cl(g) \longrightarrow \cdot CH_3(g) + HCl(g)$$

Solution

For every molecule of CH_4 that reacts, *one* C—H bond breaks, requiring 414 kJ per mole of bonds; and *one* H—Cl bond forms, releasing 431 kJ per mole of bonds. Because more energy is released in forming new bonds than is absorbed in breaking old ones, we predict that the reaction is exothermic.

Practice Example A: One naturally occurring reaction involved in the sequence of reactions leading to the destruction of ozone (page 280) is

$$NO_2(g) + O(g) \longrightarrow NO(g) + O_2(g)$$

Is this reaction endothermic or exothermic?

Practice Example B: Predict whether the following reaction should be exothermic or endothermic: $H_2O(g) + Cl_2(g) \longrightarrow \frac{1}{2} O_2(g) + 2\, HCl(g)$.

Summary

A Lewis symbol represents the valence electrons of an atom through dots placed around the chemical symbol. A Lewis structure is a combination of Lewis symbols used to represent chemical bonding. Normally, all the electrons in a Lewis structure are paired, and each atom in the structure acquires eight electrons in its valence shell (octet rule).

Most bonds have both partial ionic and partial covalent character. If the centers of positive and negative charge in a bond become separated because one member of the bond attracts electrons more strongly than the other, the bond is said to be polar. Whether a bond is polar can be predicted through the concept of electronegativity. The greater the electronegativity difference between two atoms, the more polar the bond and the more ionic the character of the bond.

To draw the Lewis structure of a covalent molecule, one needs to know the skeletal structure—that is, which is the central atom and what atoms are bonded to it. Typically, the atom with the lowest electronegativity is a central atom. At times, the concept of formal charge is useful in selecting a skeletal structure and assessing the plausibility of a Lewis structure.

Often, more than one plausible Lewis structure can be written for a species. In these cases the true structure is a resonance hybrid of two or more contributing structures. At times, a molecule may have unpaired electrons (such as NO), and in some compounds of nonmetals of the third period and beyond, the valence shell of the central atom must be expanded to 10 or 12 electrons in order to write a Lewis structure.

A powerful method for predicting molecular shapes is the valence-shell electron-pair repulsion (VSEPR) theory. The shape of a molecule (or polyatomic ion) depends on the geometric distribution of valence-shell electron groups and whether these groups are bonding electrons or lone pairs. An important use of information about the shapes of molecules is in establishing whether bonds in a molecule combine to produce a resultant dipole moment. Molecules with a resultant dipole moment are polar. Those with no resultant dipole moment are nonpolar.

Molecular properties, such as bond length and bond energy, are used to assess whether a particular Lewis structure is plausible. For example, they can be used to establish whether a covalent bond has a multiple-bond character. Also, bond energies can be used to estimate enthalpy changes for reactions involving gases.

Integrative Example

Nitryl fluoride is a reactive gas useful in rocket propellants. Its mass percent composition is 21.55% N, 49.23% O, and 29.23% F. Its density is 2.7 g/L at 20 °C and 1.00 atm pressure. Describe the nitryl fluoride molecule as completely as possible, that is, its formula, Lewis structure, molecular shape, and polarity.

1. *Determine the empirical formula.* Use the method illustrated in Example 3-5 on page 76. In 100.0 g of the compound,

$$\text{mol N} = 21.55 \text{ g N} \times \frac{1 \text{ mol N}}{14.007 \text{ g N}} = 1.539 \text{ mol N}$$

$$\text{mol O} = 49.23 \text{ g O} \times \frac{1 \text{ mol O}}{15.999 \text{ g O}} = 3.077 \text{ mol O}$$

$$\text{mol F} = 29.23 \text{ g F} \times \frac{1 \text{ mol F}}{18.998 \text{ g F}} = 1.539 \text{ mol F}$$

The empirical formula is $N_{1.539}O_{3.077}F_{1.539} = NO_2F$. The molar mass based on the empirical formula =

$$14 + 32 + 19 = 65 \text{ g/mol}$$

2. *Determine the true molecular formula.* Use the method of Example 6-10 on page 191 to establish the molar mass of the gas.

$$\text{molar mass} = \frac{mRT}{PV} = \frac{dRT}{P}$$

$$= \frac{2.7 \text{ g/L} \times 0.0821 \text{ L atm mol}^{-1} \text{ K}^{-1} \times 293 \text{ K}}{1.00 \text{ atm}}$$

$$= 65 \text{ g/mol}$$

The true molar mass and that based on the empirical formula are the same. Therefore the true molecular formula is the same as the empirical formula: NO_2F.

3. *Write a Lewis structure.* Because it has the lowest electronegativity, N should be the central atom. The other atoms are terminal atoms. There are two equivalent structures having one nitrogen-to-oxygen double bond and the remaining bonds as single bonds, that is, two contributing structures to a resonance hybrid.

4. *Apply the VSEPR theory.* The three electron groups around the central N atom have a trigonal-planar electron-group geometry. Because all the electron groups participate in bonding, the molecular geometry is also trigonal-planar, and the predicted bond angles are 120°. (The experimentally determined F—N—O bond angle is 118°.)

5. *Assess the polarity of NO$_2$F.* The molecule has a symmetrical shape, and if all the bond moments were of equal magnitude, we would expect the molecule to be nonpolar. On the other hand, because the electronegativity of F is greater than that of O, the N—F bond dipole should be greater than the N—O bond dipoles. This should lead to a small resultant dipole moment. NO_2F is a polar molecule.

Key Terms

- average bond energy (11-9)
- bond angle (11-7)
- bond-dissociation energy, D (11-9)
- bond length (11-8)
- bond order (11-8)
- bond pair (11-2)
- central atom (11-4)

- coordinate covalent bond (11-2)
- covalent bond (11-1)
- dipole moment, μ (11-7)
- double covalent bond (11-2)
- electronegativity (EN) (11-3)
- electronegativity difference (ΔEN) (11-3)

- electron-group geometry (11-7)
- expanded valence shell (11-6)
- formal charge (11-4)
- free radical (11-6)
- ionic bond (11-1)
- Lewis structure (11-1)
- Lewis symbol (11-1)

Focus On   Polymers—Macromolecular Substances

◀ In 1934, Wallace Carothers and his associates at E. I. du Pont de Nemours & Company succeeded in producing the first synthetic fiber—*nylon*. Chemistry students can now carry out a variation of this polymerization in the general chemistry laboratory.

The molecules studied in this chapter all have molecular masses ranging from 2 u (for H_2) to about 200 u. Some molecules, however, have molecular masses up to several million atomic mass units. These are macromolecules, or polymers. *Polymers* are made up of simple molecules with low molecular masses joined together into extremely large molecules. Polymers with molecular masses below about 20,000 u are called low polymers and those above 20,000 u, high polymers.

One familiar polymer is *polyethylene*. As its name implies, its basic unit, or *monomer*, is the ethylene molecule, which has the Lewis structure

$$\underset{\underset{\displaystyle H}{|}}{\overset{\overset{\displaystyle H}{|}}{C}}=\underset{\underset{\displaystyle H}{|}}{\overset{\overset{\displaystyle H}{|}}{C}}$$

We can imagine that the polymerization of ethylene begins with the "opening up" of the double bonds in ethylene molecules.

$$\cdot\underset{\underset{\displaystyle H}{|}}{\overset{\overset{\displaystyle H}{|}}{C}}-\underset{\underset{\displaystyle H}{|}}{\overset{\overset{\displaystyle H}{|}}{C}}\cdot$$

Then each C atom in the resulting molecular fragment (radical) forms an additional single covalent bond with a C atom in another molecular fragment, and so on, producing the structure shown below.

$$\cdots-\underset{\underset{\displaystyle H}{|}}{\overset{\overset{\displaystyle H}{|}}{C}}-\underset{\underset{\displaystyle H}{|}}{\overset{\overset{\displaystyle H}{|}}{C}}-\underset{\underset{\displaystyle H}{|}}{\overset{\overset{\displaystyle H}{|}}{C}}-\underset{\underset{\displaystyle H}{|}}{\overset{\overset{\displaystyle H}{|}}{C}}-\underset{\underset{\displaystyle H}{|}}{\overset{\overset{\displaystyle H}{|}}{C}}-\underset{\underset{\displaystyle H}{|}}{\overset{\overset{\displaystyle H}{|}}{C}}-\cdots$$

Lewis theory (11-1)
lone pair (11-2)
multiple covalent bond (11-2)
molecular geometry (11-7)
octet (11-1)

octet rule (11-2)
polar covalent bond (11-3)
polar molecule (11-7)
resonance (11-5)
single covalent bond (11-2)

skeletal structure (11-4)
terminal atom (11-4)
triple covalent bond (11-2)
valence-shell electron-pair repulsion (VSEPR) theory (11-7)

Review Questions

1. In your own words, define the following terms: **(a)** valence electrons; **(b)** electronegativity; **(c)** bond-dissociation energy; **(d)** double covalent bond; **(e)** coordinate covalent bond.
2. Briefly describe each of the following ideas: **(a)** formal charge; **(b)** resonance; **(c)** expanded valence shell; **(d)** bond energy.
3. Explain the important distinctions between each pair of terms: **(a)** ionic and covalent bonds; **(b)** lone-pair and bond-pair electrons; **(c)** molecular geometry and electron-group geometry; **(d)** bond dipole and resultant dipole moment; **(e)** polar molecule and nonpolar molecule.

4. Write Lewis symbols for the following atoms and ions: **(a)** H^-; **(b)** Kr; **(c)** Sn^{2+}; **(d)** K^+; **(e)** Br^-; **(f)** Ge; **(g)** N; **(h)** Ca; **(i)** Se^{2-}; **(j)** Sc^{3+}.
5. Write Lewis structures for the following ionic compounds: **(a)** calcium chloride; **(b)** barium sulfide; **(c)** lithium oxide; **(d)** sodium fluoride; **(e)** magnesium nitride.
6. Write plausible Lewis structures for the following molecules that contain only single covalent bonds: **(a)** ICl; **(b)** Br_2; **(c)** OF_2; **(d)** NI_3; **(e)** H_2Se.

426

In the following notation, the monomer unit is enclosed in square brackets and the subscript n signifies the number of monomers present in the final *macromolecule*. Typically, n might range from several hundred to several thousand.

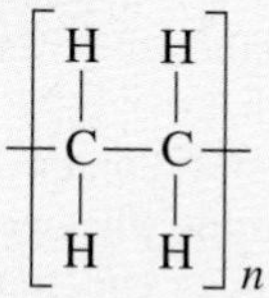

$$\left[\begin{array}{c} \text{H} \quad \text{H} \\ | \quad\;\; | \\ \text{C}-\text{C} \\ | \quad\;\; | \\ \text{H} \quad \text{H} \end{array} \right]_n$$

In Chapter 27, we will consider how the double bond in C_2H_4 is opened up and how polymer chains are propagated and terminated.

Another polymer in which monomer units join end to end is *latex*—natural rubber.

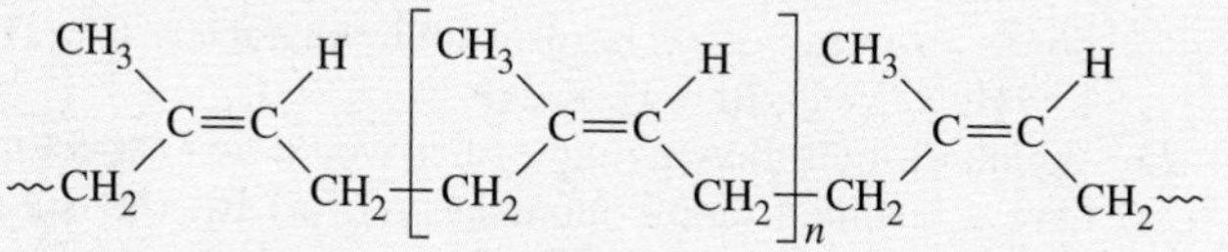

Rubber

Early rubber products were of limited use because they were sticky in hot weather and stiff in cold weather. In 1839, Charles Goodyear accidentally discovered that by heating a sulfur–rubber mixture, a product could be made that was stronger, more elastic, and more resistant to heat and cold than natural rubber. This process is now called vulcanization (after Vulcan, the

Roman god of fire). The purpose of vulcanization is to form *cross-links* between long polymer chains. An example of a cross-link through two sulfur atoms is shown below.

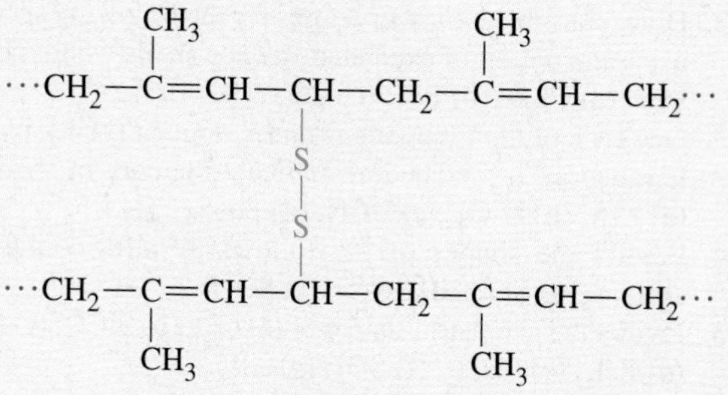

Polymers are familiar products in the modern world. Nylon, one of the first polymers developed, is like an artificial silk and is used in making clothing, ropes, and sails. The fluorine-containing polymer Teflon (polytetrafluoroethylene) is used in non-stick frying and baking pans. Polyvinyl chloride (PVC) is used in food wrap, hoses, pipes, and floor tile. In all, the polymer industry is a very large one. It has been estimated, for example, that about half of all chemists work with polymers.

In addition to latex, which comes from the rubber tree (*Hevea brasiliensis*), there are many other *natural* polymers. Cellulose, the basic structural material of plants, is a polymer of the sugar glucose ($C_6H_{12}O_6$). Proteins are high polymers with amino acids as their monomers. DNA, sometimes called the "thread of life," is also a macromolecular substance. We will learn more about these natural polymers in Chapter 28.

---

7. Each of the following molecules contains at least one multiple (double or triple) covalent bond. Give a plausible Lewis structure for **(a)** CS_2; **(b)** $(CH_3)_2CO$; **(c)** Cl_2CO; **(d)** FNO.

8. Describe what is wrong with each of the following Lewis structures.

   (a) 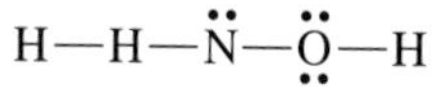 H—H—N̈—Ö—H

   (b) 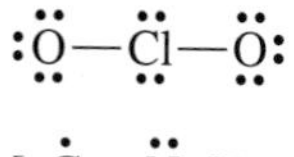 :Ö—C̈l—Ö:

   (c) [·C̈=N̈:]⁻

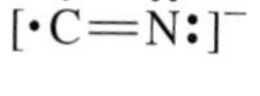

   (d) 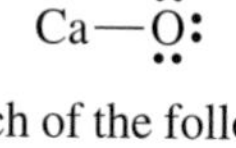 Ca—Ö:

9. Each of the following ionic compounds consists of a combination of monatomic and polyatomic ions. Represent these compounds with Lewis structures. **(a)** $Ca(OH)_2$; **(b)** NH_4Br; **(c)** $Ca(OCl)_2$.

10. Assign formal charges to each of the atoms in the following structures.

    (a) $[H-C{\equiv}C:]^-$

    (b)

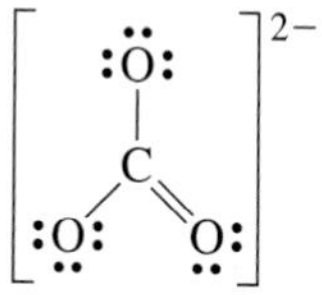

    (c) $[CH_3-CH-CH_3]^+$

    (d) 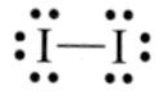 :Ï—Ï:

    (e)

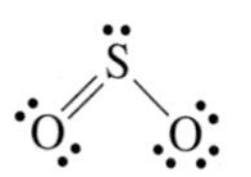

    (f)

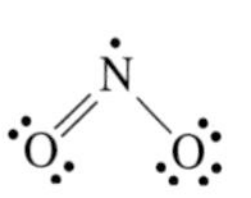

11. Which of the following species would you expect to be diamagnetic and which paramagnetic: (a) OH^-; (b) OH; (c) NO_3; (d) SO_3; (e) SO_3^{2-}; (f) HO_2?

12. Draw plausible Lewis structures for the following species, using the notion of expanded valence shells where necessary: (a) Cl_2O; (b) PF_3; (c) CO_3^{2-}; (d) BrF_5.

13. One each of the following is linear, angular (bent), planar, tetrahedral, or octahedral. Indicate the correct shape of (a) H_2S; (b) N_2O_4; (c) HCN; (d) $SbCl_6^-$; (e) BF_4^-.

14. Predict the shapes of the following sulfur-containing species: (a) SO_2; (b) SO_3^{2-}; (c) SO_4^{2-}.

15. Predict the geometric shapes of (a) CO; (b) $SiCl_4$; (c) PH_3; (d) ICl_3; (e) $SbCl_5$; (f) SO_2; (g) AlF_6^{3-}.

16. Use data from Tables 11.2 and 11.3 to determine for each bond in the following structure (a) the bond length and (b) the bond energy.

$$\begin{array}{ccc} & O & H \\ & \| & | \\ H- & C- & C-Cl \\ & & | \\ & & H \end{array}$$

17. *Without* referring to tables in the text, indicate which of the following bonds you would expect to have the greatest bond length, and give your reasons. (a) O_2; (b) N_2; (c) Br_2; (d) BrCl.

18. A reaction involved in the formation of ozone in the upper atmosphere is $O_2 \longrightarrow 2\,O$. *Without* referring to Table 11.3, indicate whether this reaction is endothermic or exothermic. Explain.

19. Use data from Table 11.3, but *without performing detailed calculations*, determine whether each of the following reactions is exothermic or endothermic.
(a) $CH_4(g) + I(g) \longrightarrow \cdot CH_3(g) + HI(g)$
(b) $H_2(g) + I_2(g) \longrightarrow 2\,HI(g)$

20. Without referring to tables or figures in the text other than the periodic table, indicate which of the following atoms, Bi, S, Ba, As, or Mg, has the intermediate value when the five are arranged in order of increasing electronegativity.

21. Use your knowledge of electronegativities, but *do not* refer to tables or figures in the text, to arrange the following bonds in terms of *increasing* ionic character: C—H; F—H; Na—Cl; Br—H; K—F.

22. Which of the following molecules would you expect to have a resultant dipole moment (μ): (a) F_2; (b) NO_2; (c) BF_3; (d) HBr; (e) H_2CCl_2; (f) SiF_4; (g) OCS? Explain.

Exercises

Lewis Theory

23. Give several examples for which the following statement proves to be incorrect. "All atoms in a Lewis structure have an octet of electrons in their valence shells."

24. Which of the following have Lewis structures that *do not* obey the octet rule: NH_3, BF_3, SF_6, SO_3, NH_4^+, SO_4^{2-}, NO_2? Explain.

25. By means of Lewis structures, represent bonding between the following pairs of elements: (a) Cs and Br; (b) H and Sb; (c) B and Cl; (d) Cs and Cl; (e) Li and O; (f) Cl and I. Your structures should show whether the bonding is essentially ionic or covalent.

26. Suggest reasons why the following do not exist as stable molecules: (a) H_3 (b) HHe; (c) He_2; (d) H_3O.

27. Only one of the following Lewis structures is correct. Select that one and indicate the errors in the others.

(a) cyanate ion $[:\ddot{O}-C{=}\ddot{N}:]^-$

(b) carbide ion $[C{\equiv}C:]^{2-}$

(c) hypochlorite ion $[:\ddot{\underset{..}{Cl}}-\ddot{\underset{..}{O}}:]^-$

(d) nitrogen(II) oxide $:\ddot{N}{=}\ddot{O}:$

28. Indicate what is wrong with each of the following Lewis structures. Replace each with a more acceptable structure.

(a) $Mg\,:\ddot{\underset{..}{O}}:$

(b) $[:\ddot{\underset{..}{Cl}}]^+[:\ddot{\underset{..}{O}}:]^{2-}[\ddot{\underset{..}{Cl}}:]^+$

(c) $[:\ddot{O}-\dot{N}{=}\ddot{O}:]^+$

(d) $[:\ddot{S}-C{=}\ddot{N}:]^-$

Ionic Bonding

29. Derive the correct formulas for the following ionic compounds by writing Lewis structures: **(a)** lithium sulfide; **(b)** sodium fluoride; **(c)** calcium iodide; **(d)** scandium chloride.

30. Under appropriate conditions, both hydrogen and nitrogen can form monatomic anions. What are the Lewis symbols for these ions? What are the Lewis structures of the compounds **(a)** lithium hydride; **(b)** calcium hydride; **(c)** magnesium nitride?

Formal Charge

31. Both oxidation state and formal charge involve conventions for assigning valence electrons to bonded atoms in compounds, but clearly they are not the same. Describe several ways in which these concepts differ.

32. Although the notion that a Lewis structure in which formal charges are zero or held to a minimum seems to apply in most instances, describe several significant situations in which this appears not to be the case.

33. What is the formal charge of the indicated atom in each of the following structures?
(a) the central O atom in O_3
(b) B in BF_4^-
(c) N in NO_3^-
(d) P in PCl_5
(e) I in ICl_4^-

34. Assign formal charges to the atoms in the following species, and then select the more likely skeletal structure.
(a) H_2NOH or H_2ONH
(b) SCS or CSS
(c) NFO or FNO
(d) $SOCl_2$ or $OSCl_2$ or OCl_2S

35. The concept of formal charge helped us to choose the more plausible of the Lewis structures for NO_2^+ given in expressions (11.11) and (11.12). Can it similarly help us to choose a single Lewis structure as most plausible for CO_2H^+? Explain.

36. Show that the idea of minimizing the formal charges in a structure is at times in conflict with the observation that compact, symmetrical structures are more commonly observed than elongated ones with many central atoms. Use ClO_4^- as an illustrative example.

Lewis Structures

37. Write acceptable Lewis structures for the following molecules: **(a)** H_2NOH; **(b)** $HOClO_2$; **(c)** HONO; **(d)** O_2SCl_2.

38. Two molecules that have the same formulas but different structures are said to be isomers. (In isomers, the same atoms are present but linked together in different ways.) Draw acceptable Lewis structures for *two* isomers of S_2F_2. [*Hint:* What is(are) the central atom(s)?]

39. The following polyatomic anions involve covalent bonds between O atoms and the central nonmetal atom. Propose an acceptable Lewis structure for each: **(a)** SO_3^{2-}; **(b)** NO_2^-; **(c)** CO_3^{2-}; **(d)** HO_2^-.

40. Represent the following ionic compounds by Lewis structures: **(a)** barium hydroxide; **(b)** sodium nitrite; **(c)** magnesium iodate; **(d)** aluminum sulfate.

41. Write a plausible Lewis structure for crotonaldehyde, $CH_3CHCHCHO$, a substance used in tear gas and insecticides.

42. Write a plausible Lewis structure for C_3O_2, a substance known as carbon suboxide.

43. Write Lewis structures for the molecules represented by the following molecular models.

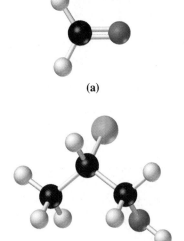

(a)

(b)

44. Write Lewis structures for the molecules represented by the following molecular models.

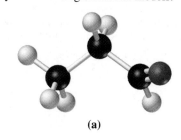

(a)

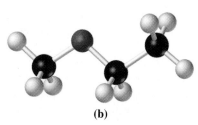

(b)

45. Write Lewis structures for the molecules represented by the following line diagram. (*Hint:* Recall page 67 and Figure 3-2.)

(a) 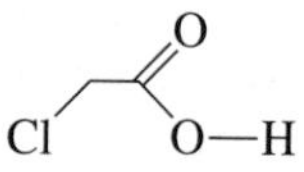 Cl $\quad$ O—H

(b) 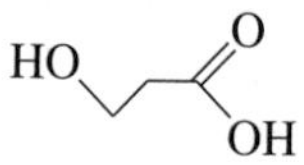 HO $\quad$ OH

46. Write Lewis structures for the molecules represented by the following line diagram. (*Hint:* Recall page 67 and Figure 3-2.)

(a) 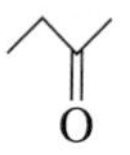 O

(b) 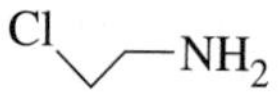 Cl —NH$_2$

Polar Covalent Bonds

47. What is the percent ionic character of each of the following bonds: **(a)** S—H; **(b)** O—Cl; **(c)** Al—O; **(d)** As—O?

48. Plot the data of Figure 11-6 as a function of atomic number. Does the property of electronegativity conform to the periodic law? Do you think it should?

Resonance

49. Through appropriate Lewis structures, show that the phenomenon of resonance is involved in the nitrite ion.

50. Which of the following species requires a resonance hybrid for its Lewis structure: **(a)** CO_2; **(b)** OCl^-; **(c)** CO_3^{2-}; **(d)** OH^-? Explain.

51. Dinitrogen oxide (nitrous oxide, or "laughing gas") is sometimes used as an anesthetic. Here are some data about the N_2O molecule: N—N bond length = 113 pm; N—O bond length = 119 pm. Use these data and other information from the chapter to comment on the plausibility of each of the following Lewis structures. Are they all valid? Which ones do you think contribute most to the resonance hybrid?

$$:N\equiv N-\ddot{\underset{..}{O}}:$$
(1)

$$:\ddot{N}=N=\ddot{O}:$$
(2)

$$:\ddot{\underset{..}{N}}-N\equiv O:$$
(3)

$$:\ddot{N}=O=\ddot{N}:$$
(4)

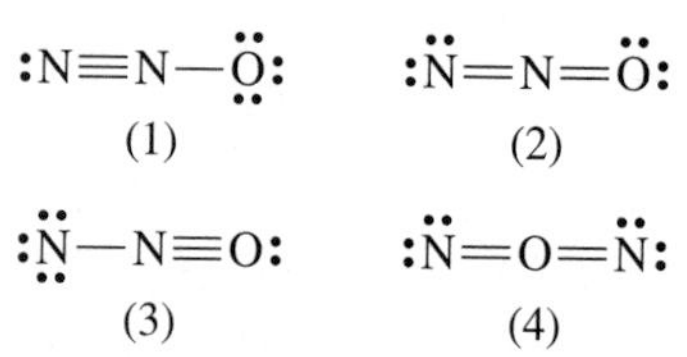

52. The Lewis structure of nitric acid, $HONO_2$, is a resonance hybrid. How important do you think the contribution of the following structure is to the resonance hybrid? Explain.

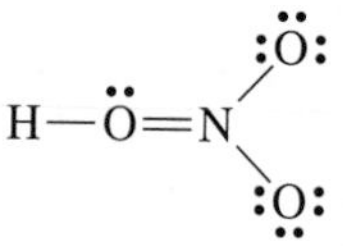

$$H-\ddot{\underset{..}{O}}=N\begin{array}{c} :\ddot{\underset{..}{O}}: \\ \\ :\underset{..}{O}: \end{array}$$

Odd-Electron Species

53. Write plausible Lewis structures for the following odd-electron species: **(a)** CH_3; **(b)** ClO_2; **(c)** NO_3.

54. Write a plausible Lewis structure for NO_2, and indicate whether the molecule is diamagnetic or paramagnetic. Two

NO_2 molecules can join together (*dimerize*) to form N_2O_4. Write a plausible Lewis structure for N_2O_4, and comment on the magnetic properties of the molecule.

Expanded Valence Shells

55. In which of the following species is it *necessary* to employ an expanded valence shell to represent the Lewis structure: PO_4^{3-}, PI_3, ICl_3, $OSCl_2$, SF_4, ClO_4^-? Explain your choices.

56. Describe the carbon-to-sulfur bond in H_2CSF_4. That is, is it most likely a single, double, or triple bond?

Molecular Shapes

57. Use VSEPR theory to predict the geometric shapes of the following molecules and ions: **(a)** N_2; **(b)** HCN; **(c)** NH_4^+; **(d)** NO_3^-; **(e)** NSF.

58. Use VSEPR theory to predict the geometric shapes of the following molecules and ions: **(a)** PCl_3; **(b)** SO_4^{2-}; **(c)** $SOCl_2$; **(d)** SO_3; **(e)** BrF_4^+.

59. One of the following ions has a *trigonal-planar* shape: SO_3^{2-}; PO_4^{3-}; PF_6^-; CO_3^{2-}. Which ion is it? Explain.

60. Two of the following have the same shape. Which two, and what is their shape? What are the shapes of the other two? NI_3, HCN, SO_3^{2-}, NO_3^-

61. Each of the following molecules contains one or more multiple covalent bonds. Draw plausible Lewis structures to represent this fact, and predict the shape of each molecule. **(a)** CO_2; **(b)** Cl_2CO; **(c)** $ClNO_2$

62. Draw a sketch of the probable geometric shape of a molecule of **(a)** N_2O_4 (O_2NNO_2); **(b)** C_2N_2 (NCCN); **(c)** C_2H_6 (H_3CCH_3); **(d)** C_2H_6O (H_3COCH_3).

63. Use the VSEPR theory to predict the shapes of the anions **(a)** ClO_4^-; **(b)** $S_2O_3^{2-}$ (i.e., SSO_3^{2-}); **(c)** PF_6^-; **(d)** I_3^-.

64. Use the VSEPR theory to predict the shape of **(a)** the molecule OSF_2; **(b)** the molecule O_2SF_2; **(c)** the ion SF_5^-; **(d)** the ion ClO_4^-; **(e)** the ion ClO_3^-.

65. The molecular shape of BF_3 (Table 11.1) is planar. If a fluoride ion is attached to the B atom of BF_3 through a coordinate covalent bond, the ion BF_4^- results. What is the shape of this ion?

66. Explain why it is not necessary to find the Lewis structure with the smallest formal charges to make a successful prediction of molecular geometry in the VSEPR theory. For example, write Lewis structures for SO_2 having different formal charges, and predict the molecular geometry based on these structures.

Shapes of Molecules with More Than One Central Atom

67. Draw a sketch of the propyne molecule, $CH_3C{\equiv}CH$. Indicate the bond angles in this molecule. What is the maximum number of atoms that can be in the same plane?

68. Draw a sketch of the propene molecule, $CH_3CH{=}CH_2$. Indicate the bond angles in this molecule What is the maximum number of atoms that can be in the same plane?

69. Lactic acid has the formula $CH_3CH(OH)COOH$. Sketch the lactic acid molecule, and indicate the various bond angles.

70. Levulinic acid has the formula $CH_3(CO)CH_2CH_2COOH$. Sketch the levulinic acid molecule, and indicate the various bond angles.

Polar Molecules

71. Predict the shapes of the following molecules, and then predict which would have resultant dipole moments: **(a)** SO_2; **(b)** NH_3; **(c)** H_2S; **(d)** C_2H_4; **(e)** SF_6; **(f)** CH_2Cl_2.

72. Which of the following molecules would you expect to be polar: **(a)** HCN; **(b)** SO_3; **(c)** CS_2; **(d)** OCS; **(e)** $SOCl_2$; **(f)** SiF_4; **(g)** POF_3? Give reasons for your conclusions.

73. The molecule H_2O_2 has a resultant dipole moment of 2.2 D. Can this molecule be linear? If not, describe a shape that might account for this dipole moment.

74. Refer to the Integrative Example. A compound related to nitryl fluoride is nitrosyl fluoride, FNO. For this molecule, indicate **(a)** a plausible Lewis structure and **(b)** the geometric shape. **(c)** Explain why the measured resultant dipole moment for FNO is larger than the value for FNO_2.

Bond Lengths

75. A relationship between bond lengths and single covalent radii of atoms is given on page 420. Use this relationship together with appropriate data from Table 11.2 to estimate these single-bond lengths: **(a)** I—Cl; **(b)** O—Cl; **(c)** C—F; **(d)** C—Br.

76. In which of the following molecules would you expect the oxygen-to-oxygen bond to be the *shortest:* **(a)** H_2O_2, **(b)** O_2, **(c)** O_3? Explain.

77. Refer to the Integrative Example. Use data from the chapter to estimate the length of the N—F bond in FNO_2.

78. Write a Lewis structure of the hydroxylamine molecule, H_2NOH. Then, with data from Table 11.2, determine all the bond lengths.

Bond Energies

79. Use data from Table 11.3 to estimate the enthalpy change (ΔH) for the following reaction.
$$C_2H_6(g) + Cl_2(g) \longrightarrow C_2H_5Cl(g) + HCl(g) \quad \Delta H = ?$$

80. One of the chemical reactions that occurs in the formation of photochemical smog is $O_3 + NO \longrightarrow NO_2 + O_2$. Estimate the enthalpy change of this reaction by using appropriate Lewis structures and data from Table 11.3.

81. Estimate the standard enthalpies of formation at 25 °C and 1 atm of **(a)** $OH(g)$; **(b)** $N_2H_4(g)$. Write Lewis structures and use data from Table 11.3, as necessary.

82. Use ΔH for the reaction in Example 11-14 and other data from Appendix D to estimate $\Delta H_f^\circ[CH_3Cl(g)]$.

83. Use bond energies from Table 11.3 to estimate the enthalpy change (ΔH) for the following reaction.
$$C_2H_2(g) + H_2(g) \longrightarrow C_2H_4(g) \quad \Delta H = ?$$

84. Equations (1) and (2) can be combined to yield the equation for the formation of $CH_4(g)$ from its elements.

(1) $C(s) \longrightarrow C(g) \quad \Delta H = 717 \text{ kJ}$
(2) $C(g) + 2 H_2(g) \longrightarrow CH_4(g) \quad \Delta H = ?$

Overall: $C(s) + 2 H_2(g) \longrightarrow CH_4(g) \quad \Delta H_f^\circ = -75 \text{ kJ/mol}$

Use the preceding data and a bond energy of 436 kJ/mol for H_2 to estimate the $C—H$ bond energy. Compare your result with the value listed in Table 11.3.

Integrative and Advanced Exercises

85. Given the bond-dissociation energies: nitrogen-to-oxygen bond in NO, 631 kJ/mol; H—H in H_2, 436 kJ/mol; N—H in NH_3, 389 kJ/mol; O—H in H_2O, 463 kJ/mol; calculate ΔH for the reaction

$$2 NO(g) + 5 H_2(g) \longrightarrow 2 NH_3(g) + 2 H_2O(g)$$

86. The following statements are not made as carefully as they might be. Criticize each one.
(a) Lewis structures with formal charges are incorrect.
(b) Triatomic molecules have a planar shape.
(c) Molecules in which there is an electronegativity difference between the bonded atoms are polar.

87. A compound consists of 47.5% S and 52.5% Cl, by mass. Write a Lewis structure based on the empirical formula of this compound, and comment on its deficiencies. Write a more plausible structure with the *same* ratio of S to Cl.

88. A 0.325-g sample of a gaseous hydrocarbon occupies a volume of 193 mL at 749 mmHg and 26.1 °C. Determine the molecular mass, and draw a plausible Lewis structure for this hydrocarbon.

89. A 1.24-g sample of a hydrocarbon, when completely burned in an excess of $O_2(g)$, yields 4.04 g CO_2 and 1.24 g H_2O. Draw a plausible Lewis structure for the hydrocarbon molecule.
(*Hint:* There is more than one possible arrangement of the C and H atoms.)

90. Draw Lewis structures for two different molecules with the formula C_3H_4. Is either of these molecules linear? Explain.

91. Sodium azide, NaN_3, is the nitrogen gas-forming substance used in automobile air-bag systems. It is an ionic compound containing the azide ion, N_3^-. In this ion, the two nitrogen-to-nitrogen bond lengths are 116 pm. Describe the resonance hybrid Lewis structure of this ion.

92. Explain what is wrong with the following statement: Because the ions ICl_2^+ and ICl_2^- differ by only two electrons and because electrons are so small compared to the size of an entire polyatomic anion, we expect these two ions to have the same geometric shape. Give a clear statement comparing the molecular geometries of the two ions.

93. Use the bond-dissociation energies of $N_2(g)$ and $O_2(g)$ in Table 11.3, together with data from Appendix D, to estimate the bond-dissociation energy of $NO(g)$.

94. Hydrogen azide, HN_3, is a liquid that explodes violently when subjected to shock. In the HN_3 molecule, one nitrogen-to-nitrogen bond length is 113 pm, and the other is 124 pm. The H—N—N bond angle is 112°. Draw Lewis structures and a sketch of the molecule consistent with these facts.

95. NO_2^- and NO_2^+ are made up of the same atoms. How do the shapes of these two ions compare? How do the nitrogen-to-oxygen bond lengths compare? How does the molecule NO_2 compare in shape and bond lengths to the two ions?

96. Of the two molecules NH_3 and NF_3, one has a dipole moment of $\mu = 0.24$ D, and the other, $\mu = 1.47$ D. Which has the larger dipole moment and why?

97. A few years ago the synthesis of a salt containing the N_5^+ ion was reported. What is the likely shape of this ion—linear, bent, zigzag, tetrahedral, seesaw, or square-planar? Explain your choice.

98. Carbon suboxide has the formula C_3O_2. The carbon-to-carbon bond lengths are 130 pm and carbon-to-oxygen, 120 pm. Propose a plausible Lewis structure to account for these bond lengths, and predict the shape of the molecule.

99. In certain polar solvents, PCl_5 undergoes an ionization reaction in which a Cl^- ion leaves one PCl_5 molecule and attaches itself to another. The products of the ionization are PCl_4^+ and PCl_6^-. Draw a sketch showing the changes in geometric shapes that occur in this ionization (i.e., give the shapes of PCl_5, PCl_4^+, and PCl_6^-).

$$2\,PCl_5 \rightleftharpoons PCl_4^+ + PCl_6^-$$

100. Estimate the enthalpy of formation of HCN using bond energies from Table 11.3, data from elsewhere in the text, and the reaction scheme outlined as follows.

(1) $\quad C(s) \qquad\qquad\longrightarrow C(g) \qquad \Delta H° = ?$

(2) $\quad C(g) + \frac{1}{2}N_2(g) + \frac{1}{2}H_2(g) \longrightarrow HCN(g)\;\; \Delta H° = ?$

Overall: $C(s) + \frac{1}{2}N_2(g) + \frac{1}{2}H_2(g) \longrightarrow HCN(g)\;\; \Delta H_f° = ?$

101. The enthalpy of formation of $H_2O_2(g)$ is -136 kJ/mol. Use this value, with other appropriate data from the text, to estimate the oxygen-to-oxygen single-bond energy. Compare your result with the value listed in Table 11.3.

102. Use the VSEPR theory to predict a probable shape of the molecule F_4SCH_2, and explain the source of any ambiguities in your prediction.

103. The enthalpy of formation of methanethiol, $CH_3SH(g)$, is -22.9 kJ/mol. Methanethiol can be synthesized by the reaction of gaseous methanol and $H_2S(g)$. Water vapor is another product. Use this information and data from elsewhere in the text to estimate the carbon-to-sulfur bond energy in methanethiol.

104. For LiBr, the dipole moment (measured in the gas phase) and the bond length (measured in the solid state) are 7.268 D and 217 pm, respectively. For NaCl, the corresponding values are 9.001 D and 236.1 pm. **(a)** Calculate the percent ionic character for each bond. **(b)** Compare these values with the expected ionic character based on differences in electronegativity (see Figure 11-7). **(c)** Account for any differences in the values obtained in these two different ways.

105. One possibility for the electron-group geometry for *seven* electron groups is pentagonal-bipyramidal, as found in the ion $[ZrF_7]^{3-}$. Write the VSEPR notation for this ion. Sketch the structure of the ion, labeling all the bond angles.

Feature Problems

106. Pauling's reasoning in establishing his original electronegativity scale went something like this: If we *assume that* the bond A—B is nonpolar, its bond energy is the average of the bond energies of A—A and B—B. The *difference* between the calculated and measured bond energies of the bond A—B is attributable to the partial ionic character of the bond and is called the *ionic resonance energy* (IRE). If the IRE is expressed in kilojoules per mole, the relationship between IRE and the electronegativity difference is $(\Delta EN)^2 = IRE/96$. To test this basis for an electronegativity scale,
(a) Use data from Table 11.3 to determine IRE for the H—Cl bond.
(b) Determine ΔEN for the H—Cl bond.
(c) Establish the approximate percent ionic character in the H—Cl bond by using the result of part (b) and Figure 11-7. Compare this result with that obtained in Example 11-4.

107. On page 419, the bond angle in the H_2O molecule is given as $104°$, and the resultant dipole moment is $\mu = 1.84$ D.
(a) By an appropriate geometric calculation, determine the value of the H—O bond dipole in H_2O.
(b) Use the same method as in part (a) to estimate the bond angle in H_2S, given that the H—S bond dipole is 0.67 D and that the resultant dipole moment is $\mu = 0.93$ D.

(c) Refer to Figure 11-15 on page 419. Given the bond moments 1.87 D for the C—Cl bond and 0.30 D for the C—H bond, together with $\mu = 1.04$ D, estimate the H—C—Cl bond angle in $CHCl_3$.

108. Alternative strategies to the one used in this chapter have been proposed for applying the VSEPR theory to molecules or ions with a single central atom. In general, these strategies do not require writing Lewis structures. In one strategy, we write that
1. total number of electron pairs = [(number of valence electrons) ± (electrons required for ionic charge)]/2
2. number of bonding electron pairs = (number of atoms) − 1
3. number of electron pairs around central atom = total number of electron pairs −3 × [number of terminal atoms (excluding H)]
4. number of lone-pair electrons = number of central atom pairs − number of bonding pairs
After evaluating items 2, 3, and 4, establish the VSEPR notation and determine the molecular shape. Use this method to predict the geometrical shapes of the following:
(a) PCl_5; **(b)** NH_3; **(c)** ClF_3; **(d)** SO_2; **(e)** ClF_4^-; **(f)** PCl_4^+. Justify each of the steps in the strategy, and explain why it yields the same results as the VSEPR method based on Lewis structures. How does the strategy deal with multiple bonds?

eMedia Exercises

109. In the **Periodic Trends: Lewis Structures** activity *(eChapter 11-1)*, the similarity between the Lewis structures of elements found within the same group is illustrated. Use this concept to describe the similarity of the Lewis structures of ionic compounds involving atoms of a similar group (i.e. NaCl, NaBr, and NaI).

110. After viewing the **H₂ Bond Formation** animation *(eChapter 11-2)*, describe the forces involved in covalent bond formation in the simplest covalent compound. Which of the forces are attractive and which are repulsive?

111. Using the principles described in the **Formal Charges** animation *(eChapter 11-4)*, generate the possible Lewis dot structures of the molecules O_3 and SO_2. What are the similarities and differences between the predicted structures of the two molecules?

112. In the **VSEPR** animation *(eChapter 11-7)* a sequential pattern in electron group geometry is observed with the addition of each pair of electrons. Although the sequential addition of electron pairs (bonding or nonbonding) is *not* a realistic mechanism by which molecules are formed, it does illustrate the key principle behind the valence-shell electron-pair repulsion theory. **(a)** Use this principle to predict the structure if the central atom of a molecule was bonded to eight identical neighboring atoms. **(b)** Can you predict an approximate bond angle?

113. Use the **Molecular Polarity activity** *(eChapter 11-7)* to predict which of the following molecules (i) possess polar covalent bonds and (ii) a net molecular dipole moment.
  (a) CF_4 **(b)** CF_2Cl_2 **(c)** CH_4
  (d) NH_3 **(e)** H_2S

12 Chemical Bonding II: Additional Aspects

Contents

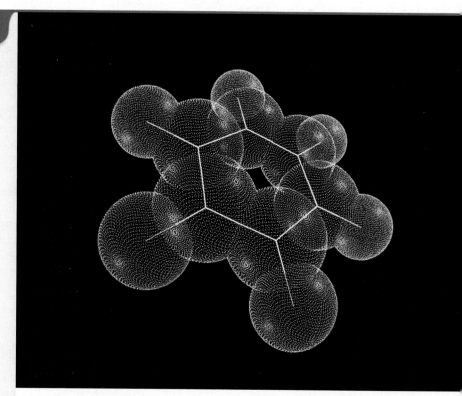

A computer graphics representation of one of the molecular orbitals of the benzene molecule showing the buildup of electron charge density between the nuclei. Another molecular orbital representation of benzene is presented on page 462.

Useful as it has been in our discussion of chemical bonding, the Lewis theory does have its shortcomings. It does not help us explain why metals conduct electricity or how a semiconductor works, for example. Although we will continue to use the Lewis theory for most purposes, we need more sophisticated approaches in some cases.

In one such method, we will work with the familiar *s*, *p*, and *d* atomic orbitals, or with mixed-orbital types called *hybrid* orbitals. In a second approach, we will create a set of orbitals that belongs to a molecule as a whole. Then we will assign electrons to these *molecular* orbitals.

Our purpose in this chapter is not to try to master theories of covalent bonding in all their details. We want simply to discover how these theories provide models that yield deeper insights into the nature of chemical bonding than do Lewis structures alone.

435

12-1 What a Bonding Theory Should Do

Imagine bringing together two atoms that are initially very far apart. Three types of interactions occur: First, the electrons are attracted to the two nuclei; second, the electrons repel each other; and third, the two nuclei repel each other. We can plot potential energy—the net energy of interaction of the atoms—as a function of the distance between the atomic nuclei. In this plot, negative energies correspond to a net attractive force between the atoms; positive energies, to a net repulsion.

Figure 12-1 shows the energy of interaction of two H atoms. This starts at zero when the atoms are very far apart. At very small internuclear distances, repulsive forces exceed attractive forces and the potential energy is positive. At intermediate distances, attractive forces predominate and the potential energy is negative. In fact, at one particular internuclear distance (74 pm) the potential energy reaches its lowest value (−436 kJ/mol). This is the condition in which the two H atoms combine into a H_2 molecule through a covalent bond. The nuclei continuously move back and forth; that is, the molecule vibrates, but the average internuclear distance remains constant. This internuclear distance corresponds to the *bond length*. The potential energy corresponds to the negative of the *bond-dissociation energy*. A theory of covalent bonding should help us understand why a given molecule has its particular set of observed properties—bond-dissociation energies, bond lengths, bond angles, and so on.

It is important to realize that there are several approaches to understanding bonding. The approach used often depends on the situation. Different methods have different strengths and weaknesses. The strength of the Lewis theory is in the ease with which we can apply it; we can write a Lewis structure rather quickly. With the VSEPR theory we can propose molecular shapes that are generally in good

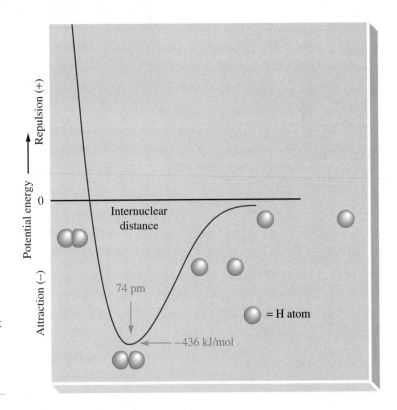

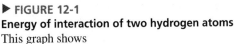

▶ **FIGURE 12-1**

Energy of interaction of two hydrogen atoms
This graph shows
• a zero of energy when two H atoms are separated by great distances
• a drop in potential energy (net attraction) as the two atoms approach each other
• a minimum in potential energy (−436 kJ/mol) at a particular internuclear distance (74 pm) corresponding to the stable molecule H_2
• an increase in potential energy as the atoms approach more closely

agreement with experimental results. However, these methods do not yield quantitative information about bond energies and bond lengths, and the Lewis theory has problems with odd-electron species and situations in which it is not possible to represent a molecule through a single structure (resonance).

12-2 Introduction to the Valence-Bond Method

▶ What we are calling "overlap" is actually an interpenetration of two orbitals.

 H₂ Bond Formation animation

Recall the region of high electron probability in a H atom that we described in Chapter 9 through the mathematical function called a $1s$ orbital (page 327). As the two H atoms pictured in Figure 12-1 approach each other, these regions begin to interpenetrate. We say that the two orbitals overlap. Furthermore, we can say that a bond is produced between the two atoms because of the high electron probability found in the region between the atomic nuclei where the $1s$ orbitals overlap. In this way, a covalent bond is formed between the two H atoms in the H_2 molecule.

A description of covalent bond formation in terms of atomic orbital overlap is called the **valence-bond method**. The creation of a covalent bond in the valence-bond method is normally based on the overlap of half-filled orbitals, but sometimes such an overlap involves a filled orbital on one atom and an empty orbital on another. The valence-bond method gives a *localized* electron model of bonding: Core electrons and lone-pair valence electrons retain the same orbital locations as in the separated atoms, and the charge density of the bonding electrons is concentrated in the region of orbital overlap.

Are You Wondering...

Why the overlap of orbitals leads to a chemical bond?

The origin of this extra stability comes from the overlap of the two orbitals in which the two atomic wave functions are *in phase*, leading to constructive interference of the wave functions between the two nuclei and hence increased electron density between the two nuclei. The increased electron density, with its negative charge, attracts the two positively charged nuclei, leading to an energy that is lower than that of the two separated atoms. Thus, the increased electron density between the nuclei produces the chemical bond. We will say more about the interaction between orbitals later in this chapter.

Figure 12-2 shows the imagined overlap of atomic orbitals involved in the formation of hydrogen-to-sulfur bonds in hydrogen sulfide. Note especially that maximum overlap between the $1s$ orbital of a H atom and a $3p$ orbital of a S atom occurs along a line joining the centers of the H and S atoms. The two half-filled $3p$ orbitals of sulfur involved in orbital overlap in H_2S are perpendicular to each other, and the valence-bond method suggests a H—S—H bond angle of 90°. This is in good agreement with the observed angle of 92°.

Because the energies of atomic orbitals differ from one type to another (recall Figure 9-33), the valence-bond method implies different bond energies for different bonds. A quantitative application of the valence-bond method would show that the $1s$–$1s$ overlap in H_2 produces a greater bond energy than the $1s$–$3p$ overlap in a H—S bond in H_2S. The Lewis structures H—O—H and H—S—H provide no information about bond energies.

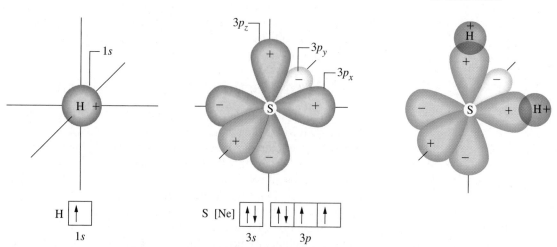

Isolated atoms

Covalent bonds

▲ **FIGURE 12-2 Bonding in H_2S represented by atomic orbital overlap**
Orbitals with a single electron are gray. Those with an electron pair are in color. For S, only $3p$ orbitals are shown. Here, and in most other places in this chapter, even when we represent the phase of the wave function (through $+$ and $-$ signs), we depict p orbitals through their probability-density plots (recall Figure 9-24b), rather than the p orbital functions themselves (recall Figure 9-24a). Bond formation occurs between orbitals that are in phase.

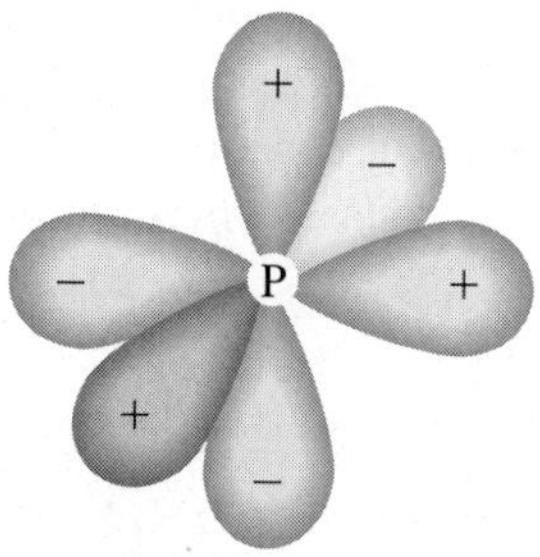

Bonding orbitals of P atom

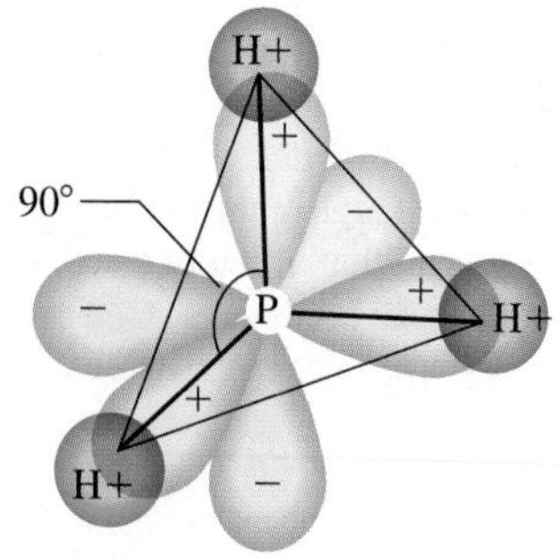

$90°$

Covalent bonds formed

▲ **FIGURE 12-3**

Bonding and structure of the PH_3 molecule—Example 12-1 illustrated

Orbitals with a single electron are gray. Those with electron pairs are in color. Only bonding orbitals are shown. The $1s$ orbitals of three H atoms overlap with the three $3p$ orbitals of the P atom. The plus and minus signs indicate the phases of the orbital.

EXAMPLE 12-1

Using the Valence-Bond Method to Describe a Molecular Structure. Describe the phosphine molecule, PH_3, by the valence-bond method.

Solution

Here is a four-step approach to applying the valence-bond method.

Step 1. Draw valence-shell orbital diagrams for the separate atoms.

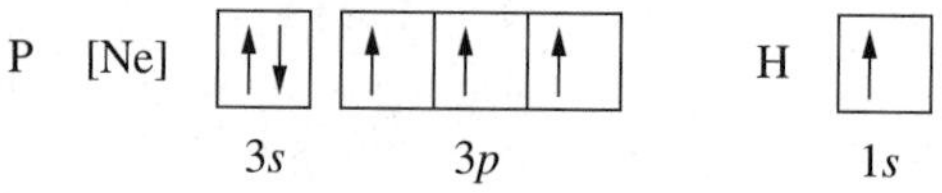

Step 2. Sketch the orbitals of the central atom (P) that are involved in the overlap. These are the half-filled $3p$ orbitals (Figure 12-3).

Step 3. Complete the structure by bringing together the bonded atoms and representing the orbital overlap.

Step 4. Describe the structure. PH_3 is a *trigonal-pyramidal* molecule. The three H atoms lie in the same plane. The P atom is situated at the top of the pyramid above the plane of the H atoms, and the three H—P—H bond angles are $90°$.

(The experimentally measured H—P—H bond angles are $93-94°$.)

Practice Example A: Use the valence-bond method to describe bonding and the expected molecular geometry in nitrogen triiodide, NI_3.

Practice Example B: Describe the molecular geometry of NH_3, first using the VSEPR method and then using the valence-bond method described above. How do your answers differ? Which method seems to be more appropriate in this case? Explain.

12-3 Hybridization of Atomic Orbitals

If we try to extend the unmodified valence-bond method of Section 12-2 to a greater number of molecules, we are quickly disappointed. In most cases, our descriptions of molecular geometry based on the simple overlap of unmodified atomic orbitals do not conform to observed measurements. For example, based on the *ground-state* electron configuration of the valence shell of carbon

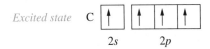

Ground state

and employing only half-filled orbitals, we expect the existence of a molecule with the formula CH_2 and a bond angle of 90°. The CH_2 molecule is stable, but is not normally observed in the laboratory. It is a highly reactive molecule observed only under specially designed circumstances.

▶ The word "stable" must be used carefully in chemistry. When we say the molecule is stable, we mean that the molecule is stable with respect to the separated atoms. The word stable should not be used without a qualifying statement if we are describing reactivity. Thus the metal sodium is stable (unreactive) in mineral oil but is unstable (highly reactive) in water.

The simplest hydrocarbon observed under normal laboratory conditions is methane, CH_4. This is a stable unreactive molecule with a molecular formula consistent with our expectation from the octet rule of the Lewis theory. To obtain this molecular formula by the valence-bond method, we need an orbital diagram for carbon in which there are *four* unpaired electrons, so that orbital overlap leads to four C—H bonds. To get such a diagram, imagine that one of the $2s$ electrons in a ground-state C atom absorbs energy and is promoted to the empty $2p$ orbital. The resulting electron configuration is that of an *excited state*, which we can represent as

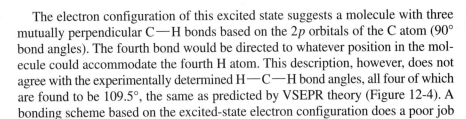

Excited state

▲ FIGURE 12-4
Ball-and-stick model of methane, CH_4
The molecule has a tetrahedral structure, and the H—C—H bond angles are 109.5°.

The electron configuration of this excited state suggests a molecule with three mutually perpendicular C—H bonds based on the $2p$ orbitals of the C atom (90° bond angles). The fourth bond would be directed to whatever position in the molecule could accommodate the fourth H atom. This description, however, does not agree with the experimentally determined H—C—H bond angles, all four of which are found to be 109.5°, the same as predicted by VSEPR theory (Figure 12-4). A bonding scheme based on the excited-state electron configuration does a poor job of explaining the bond angles in CH_4.

The problem is not with the theory but with the way we have defined the situation. We have been describing *bonded* atoms as if they have the same kinds of orbitals (that is, s, p, ...) as *isolated, nonbonded* atoms. This assumption worked rather well for H_2S and PH_3, but why should we expect these unmodified pure atomic orbitals to work equally well in all cases?

▶ The algebraic combination of wave functions is in fact a linear combination of atomic orbitals; that is, they are simply added or subtracted. The resultant linear combinations are solutions to the Schrödinger equation of the molecule.

One way that we can respond to this situation is by modifying the atomic orbitals of the bonded atoms. Recall that atomic orbitals are mathematical expressions of the electron waves in an atom. We need to algebraically combine the wave equations of the $2s$ and three $2p$ orbitals of the carbon atom to produce a new set of four identical orbitals. These new orbitals, which are directed in a tetrahedral fashion, have energies that are intermediate between those of the $2s$ and $2p$ orbitals. This mathematical process of replacing pure atomic orbitals with reformulated atomic orbitals for bonded atoms is called **hybridization**, and the new orbitals are called **hybrid orbitals**. Figure 12-5 pictures the hybridization of one s and three p orbitals into a new set of four sp^3 **hybrid orbitals**.

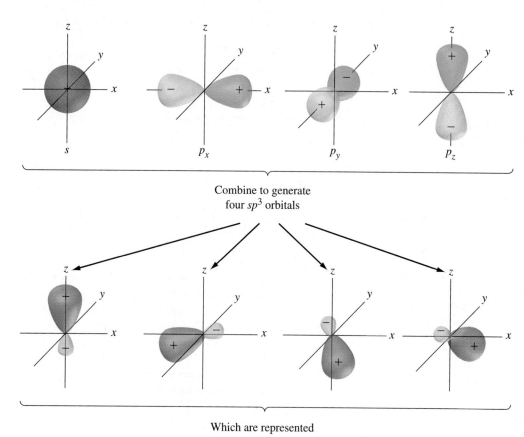

Combine to generate
four sp^3 orbitals

Which are represented
as the set

Hybridization animation

▲ **FIGURE 12-5** The ***sp*³ hybridization scheme**

In a hybridization scheme, *the number of hybrid orbitals equals the total number of atomic orbitals that are combined.* The symbols identify the numbers and kinds of orbitals involved. Thus, sp^3 signifies that *one s* and *three p* orbitals are combined. A useful representation of sp^3 hybridization of the valence-shell orbitals of carbon is

Figure 12-6 pictures sp^3 hybrid orbitals and bond formation in methane.

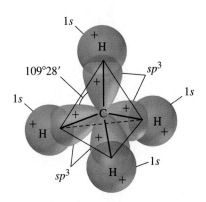

▶ FIGURE 12-6
Bonding and structure of CH₄
The four carbon orbitals are sp^3 hybrid orbitals (violet). Those of the hydrogen atoms (red) are $1s$. The structure is tetrahedral, with H—C—H bond angles of 109.5° (more precisely, 109°28′).

The objective of a hybridization scheme is an after-the-fact rationalization of the experimentally observed shape of a molecule. Hybridization is not an actual physical phenomenon. We cannot observe electron charge distributions changing from those of pure orbitals to those of hybrid orbitals. Moreover, for some covalent bonds no single hybridization scheme works well. Nevertheless, the concept of hybridization works very well for carbon-containing molecules and is therefore used a great deal in organic chemistry.

Are You Wondering...

How real are hybrid orbitals?

Are they a good description of how electrons behave, or are they just a contrivance invented to fix up a bad theory? The fact is they are just as real as the hydrogen-like orbitals we have used to describe multielectron atoms. The hydrogen-like orbitals are appropriate for describing the behavior of electrons in atoms. The hybrid orbitals are an appropriate way of describing the relative motions of electrons in molecules. Hybrid orbitals are very useful in discussions of chemical bonding in polyatomic molecules.

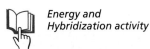

Energy and Hybridization activity

Bonding in H₂O and NH₃

Applied to H_2O and NH_3, VSEPR theory describes a tetrahedral electron-group geometry for *four* electron groups. This, in turn, requires an sp^3 hybridization scheme for the central atoms in H_2O and NH_3. This scheme suggests angles of 109.5° for the H—O—H bond in water and the H—N—H bonds in NH_3. These angles are in reasonably good agreement with the experimentally observed bond angles of 104.5° in water and 107° in NH_3. We can describe bonding in NH_3, for example, in terms of the following valence-shell orbital diagram for nitrogen.

▶ Notice in this orbital diagram how hybrid orbitals can accommodate lone-pair electrons as well as bonding electrons.

Because one of the sp^3 orbitals is occupied by a lone pair of electrons, only the three half-filled sp^3 orbitals are involved in bond formation. This suggests the trigonal-pyramidal molecular geometry depicted in Figure 12-7, just as does VSEPR theory.

Even though the sp^3 hybridization scheme seems to work quite well for H_2O and NH_3, there is both theoretical and experimental (spectroscopic) evidence that

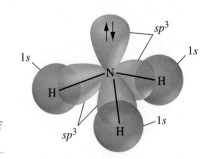

► FIGURE 12-7
sp³ Hybrid orbitals and bonding in NH₃
An *sp³* hybridization scheme yields a molecular
geometry in close agreement with what is observed experimentally. The portion of the figure
exclusive of the orbital occupied by a lone pair of
electrons is a trigonal pyramid.

favors a description based on *unhybridized p* orbitals of the central atoms. The
H—O—H and H—N—H bond angle expected for 1*s* and 2*p* atomic orbital overlaps is 90°, which does not conform to the observed bond angles. One possible explanation is that because O—H and N—H bonds have considerable ionic character,
repulsions between the positive partial charges associated with the H atoms force
the H—O—H and H—N—H bonds to "open up" to values greater than 90°. The
issue of how best to describe the bonding orbitals in H_2O and NH_3 is still unsettled
and underscores the occasional difficulty of finding a single theory that is consistent with *all* the available evidence.

sp² Hybrid Orbitals

Carbon's group 13 neighbor, boron, has *four* orbitals but only *three* electrons in its
valence shell. For most boron compounds, the appropriate hybridization scheme
combines the 2*s* and *two* 2*p* orbitals into *three sp²* **hybrid orbitals** and leaves one
p orbital unhybridized. Valence-shell orbital diagrams for this hybridization scheme
for boron are shown here, and the scheme is further outlined in Figure 12-8.

The *sp²* hybridization scheme corresponds to trigonal-planar electron-group
geometry and 120° bond angles, as in BF_3.

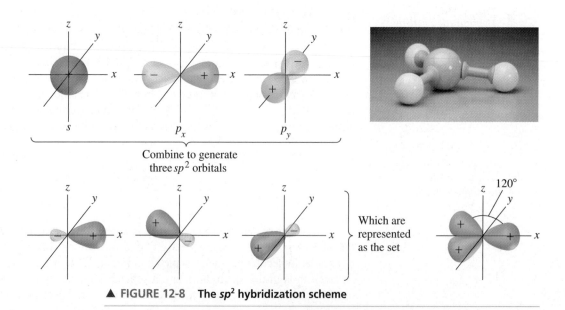

Combine to generate
three *sp²* orbitals

Which are
represented
as the set

▲ FIGURE 12-8   The *sp²* hybridization scheme

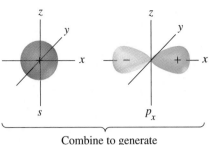

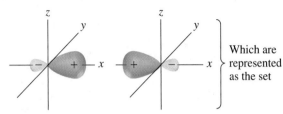

Combine to generate
two *sp* orbitals

Which are
represented
as the set

▲ **FIGURE 12-9 The *sp* hybridization scheme**

sp Hybrid Orbitals

Boron's group 2 neighbor, beryllium, has *four* orbitals and only *two* electrons in the valence shell. In the hybridization scheme that best describes certain *gaseous* beryllium compounds, the 2*s* and *one* 2*p* orbital of Be are hybridized into *two* **sp hybrid orbitals** and the remaining two 2*p* orbitals are left unhybridized. Valence-shell orbital diagrams of beryllium in this hybridization scheme are shown here, and the scheme is further outlined in Figure 12-9.

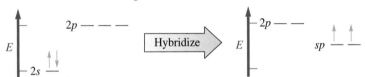

The *sp* hybridization scheme corresponds to a linear electron-group geometry and a 180° bond angle, as in $BeCl_2(g)$.

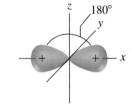

Are You Wondering ...

How the atomic orbitals mix to form a hybrid orbital?

A hybrid atomic orbital is the result of a mathematical combination (algebraic addition and subtraction) of the wave functions describing two or more atomic orbitals. When the algebraic functions that represent *s* and *p* orbitals are added, a new function is produced; this is an *sp* hybrid. When the same algebraic functions are subtracted, another new function is produced; this is a second *sp* hybrid. The hybridization process is shown in Figure 12-10, where we see the consequence of the phase of the *p* orbital when we add the *s* and *p* orbitals: The negative phase of the *p* orbital cancels part of the positive *s* orbital. This leads to the teardrop-shaped orbital pointing in the direction of the positive lobe of the *p* orbital. As shown in Figure 12-10, subtraction of the two orbitals reverses this situation. The two ways of combining an *s* and a *p* orbital generate the two equivalent *sp* hybrid orbitals, one of which has its greatest amplitude (electron density if we square it) in a direction 180° from the other. A similar procedure is used to construct the three sp^2 and the four sp^3 hybrid orbitals, although the combinations of orbitals are slightly more complicated.

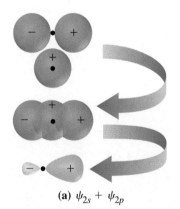

(a) $\psi_{2s} + \psi_{2p}$

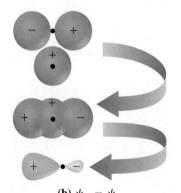

(b) $\psi_{2s} - \psi_{2p}$

▲ **FIGURE 12-10**

The formation of an *sp* hybrid orbital

Here, *p* and *s* orbitals themselves are depicted rather than their probability-density plots.

sp^3d and sp^3d^2 Hybrid Orbitals. To describe hybridization schemes to correspond to the 5- and 6-electron-group geometries of VSEPR theory, we need to go beyond the *s* and *p* subshells of the valence shell, and traditionally this has meant including *d*-orbital contributions. We can achieve the *five* half-filled orbitals of phosphorus to account for the five P—Cl bonds in PCl_5 and its trigonal-bipyramidal molecular geometry through the hybridization of the *s*, three *p*, and one *d* orbital of the valence shell into *five sp^3d* **hybrid orbitals**.

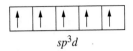

sp^3d

We can achieve the *six* half-filled orbitals of sulfur to account for the six S—F bonds in SF_6 and its octahedral molecular geometry through the hybridization of the *s*, three *p*, and two *d* orbitals of the valence shell into *six sp^3d^2* **hybrid orbitals**.

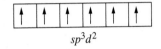

sp^3d^2

The sp^3d and sp^3d^2 hybrid orbitals and two of the molecular geometries in which they can occur are featured in Figure 12-11.

We have previously stated that hybridization is not a real phenomenon, but an after-the-fact rationalization of an experimentally determined result. Perhaps there is no better illustration of this point than the issue of the sp^3d and sp^3d^2 hybrid orbitals. In discussing the concept of the expanded valence shell in Chapter 11, we

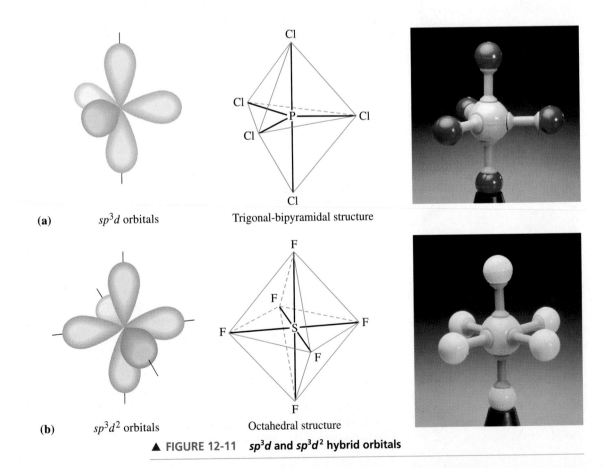

(a) sp^3d orbitals   Trigonal-bipyramidal structure

(b) sp^3d^2 orbitals   Octahedral structure

▲ **FIGURE 12-11** sp^3d and sp^3d^2 **hybrid orbitals** noted that valence-shell expansion would seem to require d electrons in bonding schemes, but recent theoretical considerations cast serious doubt on d-electron participation. The same doubt, of course, extends to the use of d orbitals in hybridization schemes.

Despite the difficulty posed by hybridization schemes involving d orbitals, the sp, sp^2, and sp^3 hybridization schemes are well established and very commonly encountered, particularly among the second-period elements.

Hybrid Orbitals and the Valence-Shell Electron-Pair Repulsion (VSEPR) Theory

In the previous section, we used either the experimental geometry or the geometry predicted by VSEPR theory to help us decide on the appropriate hybridization scheme for the central atom. The concept of hybridization arose before the formulation of the VSEPR theory as we used it in Chapter 11. In 1931, Linus Pauling introduced the concept of hybridization of orbitals to account for the known geometries of CH_4, H_2O, and NH_3. It was first suggested by N. V. Sidgwick and H. E. Powell in 1940 that molecular geometry was determined by the arrangement of electron pairs in the valence shell, and this suggestion was subsequently developed into the set of rules known as VSEPR by Ronald Gillespie and Ronald Nyholm in 1957. The advantage of VSEPR is that it has a predictive capability based on Lewis structures, whereas hybridization schemes, as described here, require a prior knowledge of the molecular geometry. So how should we proceed to describe the bonding in molecules? We can choose the likely hybridization scheme for a central atom in a structure in the valence-bond method by

- writing a plausible Lewis structure for the species of interest
- using VSEPR theory to predict the probable electron-group geometry of the central atom
- selecting the hybridization scheme corresponding to the electron-group geometry.

As suggested by Table 12.1, the hybridization scheme adopted for a central atom should be the one producing the same number of hybrid orbitals as there are valence-shell electron groups, and in the same geometric orientation. Thus, an sp^3 hybridization scheme for the central atom predicts that four hybrid orbitals are distributed in a tetrahedral fashion. This results in molecular structures that are tetrahedral, trigonal-pyramidal, or angular, depending on how many hybrid orbitals are involved in orbital overlap and how many contain lone-pair electrons, corresponding to the VSEPR notations AX_4, AX_3E, and AX_2E_2, respectively.

TABLE 12.1 Some Hybrid Orbitals and Their Geometric Orientations

Hybrid Orbitals	Geometric Orientation	Example
sp	Linear	$BeCl_2$
sp^2	Trigonal-planar	BF_3
sp^3	Tetrahedral	CH_4
sp^3d	Trigonal-bipyramidal	PCl_5
sp^3d^2	Octahedral	SF_6

The s and p orbital hybridization schemes are especially important in organic compounds, whose principal elements are C, O, and N, in addition to H. We will consider some important applications to organic chemistry in the next section.

EXAMPLE 12-2

Proposing a hybridization scheme to account for the shape of a molecule. Predict the shape of the XeF_4 molecule and a hybridization scheme consistent with this prediction.

Solution

Let's use the four-part strategy outlined here.

1. *Write a plausible Lewis structure.* The Lewis structure we write must account for *36* valence electrons—*eight* from the Xe atom and *seven* each from the *four* F atoms. To place this many electrons in the Lewis structure, we must expand the valence shell of the Xe atom to accommodate 12 electrons. The Lewis structure is

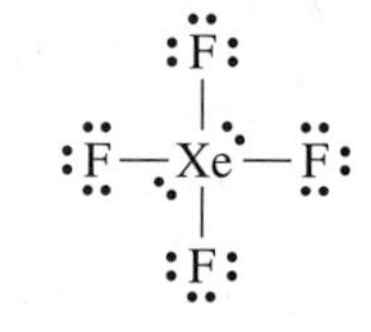

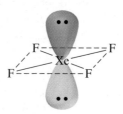

2. *Use the VSEPR theory to establish the electron-group geometry of the central atom.* From the Lewis structure we see that there are *six* electron groups around the Xe atom. Four electron groups are bond electron pairs and two are lone pairs. The electron-group geometry for *six* electron groups is *octahedral.*

3. *Describe the molecular geometry.* The VSEPR notation for XeF_4 is AX_4E_2, and the molecular geometry is *square-planar* (see Table 11.1). The four pairs of bond electrons are directed to the corners of a square, and the lone pairs of electrons are found above and below the plane of the Xe and F atoms, as shown here.

4. *Select a hybridization scheme that corresponds to the VSEPR prediction.* The only hybridization scheme consistent with an octahedral distribution of *six* electron groups is sp^3d^2. The orbital diagram for this scheme shows clearly that *four* of the eight valence electrons of the central Xe atom singly occupy four of the sp^3d^2 orbitals. The remaining *four* valence electrons of the atom occupy the remaining two sp^3d^2 orbitals as lone pairs. These are the lone pairs of electrons situated above and below the plane of the Xe and F atoms in the above sketch.

Valence shell of Xe atom $\boxed{\uparrow\downarrow\,|\,\uparrow\downarrow\,|\,\uparrow\,|\,\uparrow\,|\,\uparrow\,|\,\uparrow}$
sp^3d^2

Practice Example A: Describe the molecular geometry and propose a plausible hybridization scheme for the central atom in the ion Cl_2F^+.

Practice Example B: Describe the molecular geometry and propose a plausible hybridization scheme for the central atom in the ion BrF_4^+.

Are You Wondering...

Which method to use in rationalizing the geometric shape of a molecule, the VSEPR method (Section 11-7) or the valence-bond method?

There is no "correct" method for describing molecular structures. The only correct information is the experimental evidence from which the structure is established. Once this experimental evidence is in hand, you may find it easier to rationalize this evidence by one method or another. For H_2S, the valence-bond method, which suggests a bond angle of 90°, seems to do a better job of explaining the observed 92° bond angle than does the VSEPR theory. For the Lewis structure, $H-\ddot{S}-H$, VSEPR theory predicts a tetrahedral electron-group geometry, which in turn suggests a tetrahedral bond angle—that is, 109.5°. However, by modifying this initial VSEPR prediction to accommodate lone-pair/lone-pair and lone-pair/bond-pair repulsions (see page 412), the predicted bond angle is less than 109.5°.

VSEPR theory gives reasonably good results in the majority of cases. Unless you have specific information to suggest otherwise, describing a molecular shape with the VSEPR theory is still a good bet. It is important to remember that, in both cases, these are simply models we use to rationalize the shapes and bonding of polyatomic molecules and, as such, should be viewed with a critical eye, always keeping experimental results in sight.

KEEP IN MIND ▶

that the VSEPR method uses empirical data to give an approximate molecular geometry, whereas the valence-bond method relates to the orbitals used in bonding, based on a given geometry.

12-4   Multiple Covalent Bonds

In this section we will find that two different types of orbital overlap occur when multiple bonds are described by the valence-bond method. In our discussion we will use as specific examples the carbon-to-carbon double bond in ethylene, C_2H_4, and the carbon-to-carbon triple bond in acetylene, C_2H_2.

C_2H_4 model

Bonding in C_2H_4

Ethylene has a carbon-to-carbon double bond in its Lewis structure.

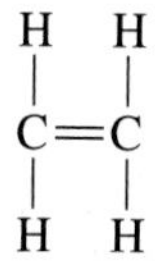

$$\begin{array}{cc} H & H \\ | & | \\ C = C \\ | & | \\ H & H \end{array}$$

Ethylene is a *planar* molecule with 120° H—C—H and H—C—C bond angles. VSEPR theory treats each C atom as being surrounded by *three* electron groups in a trigonal-planar arrangement. VSEPR theory does *not* dictate that the two —CH_2 groups be coplanar, but, as we shall see, valence-bond theory does.

The hybridization scheme that produces a set of hybrid orbitals with a trigonal-planar orientation is sp^2. The valence-shell orbital diagrams of carbon for this scheme are

The $sp^2 + p$ orbital set is pictured in Figure 12-12. One of the bonds between the carbon atoms results from the overlap of sp^2 hybrid orbitals from each atom.

*Multiple Bond
Formation activity*

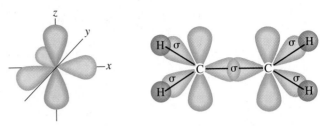

The set of orbitals $sp^2 + p$ Sigma (σ) bonds

▶ **FIGURE 12-12**
**Sigma (σ) and pi (π)
bonding in C_2H_4**
The reddish purple orbitals
are sp^2 hybrid orbitals and
the blue orbitals, $2p$. The
sp^2 hybrid orbitals overlap
along the line joining the
bonded atoms—a σ bond.
The $2p$ orbitals overlap in a
side-to-side fashion and
form a π bond.

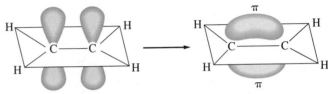

Overlap of p orbitals leading to pi (π) bond

This overlap occurs along the line joining the nuclei of the two atoms. Orbitals that overlap in this end-to-end fashion produce a **sigma bond**, designated σ **bond**. Figure 12-12 shows that a second bond between the C atoms results from the overlap of the unhybridized p orbitals. In this bond, there is a region of high electron charge density above and below the plane of the carbon and hydrogen atoms. The bond produced by this side-to-side overlap of two parallel orbitals is called a **pi bond**, designated π **bond**.

The ball-and-stick model in Figure 12-13 illustrates bonding in ethylene. This model helps to illustrate that

- the shape of a molecule is determined only by the orbitals forming σ bonds (the σ-bond framework).
- rotation about the double bond is severely restricted. In the ball-and-stick model, we could easily twist or rotate the terminal H atoms about the s bonds that join them to a C atom. To twist one —CH_2 group out of the plane of the other, however, would reduce the amount of overlap of the p orbitals and weaken the π bond. The double bond is rigid, and the C_2H_4 molecule is planar.

▲ **FIGURE 12-13**
**Ball-and-stick model of
ethylene, C_2H_4**
The H—C—H and C—C—H
bond angles are 120°. The model
also distinguishes between the
σ bond between the C atoms
(the straight plastic tube) and the
π bond extending above and
below the plane of the molecule.
The picture of the π bond suggested by the white plastic
"arches" is somewhat distorted,
but the model does convey the
idea that the π bond places a
high electron charge density
above and below the plane of the
molecule.

Additionally, regarding carbon-to-carbon multiple bonds, the σ bond involves more extensive overlap than does the π bond. As a result, a carbon-to-carbon double bond ($\sigma + \pi$) is stronger than a single bond (σ), but not twice as strong (from Table 11.3, C—C, 347 kJ/mol; C=C, 611 kJ/mol; C≡C, 837 kJ/mol).

Bonding in C_2H_2

Bonding in acetylene, C_2H_2, is similar to that in C_2H_4, but with these differences: The Lewis structure of C_2H_2 features a triple covalent bond, H—C≡C—H. The molecule is *linear*, as found by experiment and as expected from VSEPR theory. A hybridization scheme to produce hybrid orbitals in a linear orientation is sp. The valence-shell orbital diagrams representing sp hybridization are

KEEP IN MIND

that only one of the bonds in a
multiple bond is a σ bond; the
others are π bonds—one π bond
in a double bond and two in a
triple bond. ▶

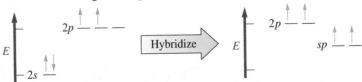

In the triple bond in C_2H_2, one of the carbon-to-carbon bonds is a σ bond and *two* are π bonds, as suggested in Figure 12-14.

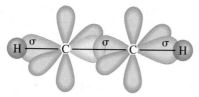

Formation of σ bonds

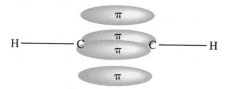

Formation of π bonds

Space-filling model

▲ **FIGURE 12-14   Sigma (σ) and pi (π) bonding in C₂H₂**
The σ-bond framework joins the atoms H—C—C—H through $1s$ orbitals of the H
atoms and sp orbitals of the C atoms. There are two π bonds. Each π bond consists of
two parallel cigar-shaped segments. The four segments shown actually merge into a hollow cylindrical shell with the carbon-to-carbon σ bond as its axis.

EXAMPLE 12-3

Proposing Hybridization Schemes Involving σ and π Bonds. Formaldehyde gas, H₂CO,
is used in the manufacture of plastics; in aqueous solution, it is the familiar biological
preservative called formalin. Describe the molecular geometry and a bonding scheme for
the H₂CO molecule.

Solution

1. *Write the Lewis structure.* C is the central atom, and H and O are terminal atoms.
    The total number of valence electrons is 12. Note that this structure requires a car-
    bon-to-oxygen double bond.

$$\begin{array}{c} H \\ | \\ H-C=\ddot{O}\!: \end{array}$$

▶ Now we see the VSEPR theory
rationale of treating a multiple
bond as a single group of electrons. The number of electron
groups around the central atom
is equal to the number of
σ bonds in the σ-bond
framework.

2. *Determine the electron-group geometry of the central C atom.* The σ-bond frame-
    work is based on three electron groups around the central C atom. VSEPR theory,
    based on the distribution of three electron groups, suggests a trigonal-planar mole-
    cule with 120° bond angles.

3. *Identify the hybridization scheme that conforms to the electron-group geometry.* A
    trigonal-planar orientation of orbitals is associated with sp^2 hybrid orbitals.

4. *Identify the orbitals of the central atom that are involved in orbital overlap.* The C
    atom is hybridized to produce the orbital set $sp^2 + p$, as in C₂H₄. Two of the sp^2
    hybrid orbitals are used to form σ bonds with the H atoms. The remaining sp^2 hy-
    brid orbital is used to form a σ bond with oxygen. The unhybridized p orbital of the
    C atom is used to form a π bond with O.

Valence shell of the C atom: ↑ ↑ ↑ | ↑
                                 sp^2 $2p$

▶ Because one of the main purposes of hybridizing orbitals is
to describe molecular geometry
(for example, bond angles), we
generally apply hybridization
schemes only to central atoms,
although hybridization of terminal atoms (except H) may also
be invoked.

5. *Sketch the bonding orbitals of the central and terminal atoms.* The bonding orbitals
    of the central C atom described above are pictured in Figure 12-15. The sp^2 hybrid
    orbitals are shown in lavender, and the pure p orbital, in blue. The H atoms have only
    $1s$ orbitals available for bonding. For oxygen, we can use a half-filled $2p$ orbital for
    end-to-end overlap in the σ bond to carbon and a half-filled $2p$ orbital to participate
    in the side-to-side overlap leading to a π bond. That is, the valence-shell orbital di-
    agram we can use for oxygen is

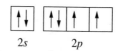

$2s$ $2p$

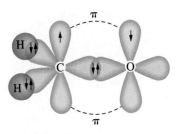

▲ **FIGURE 12-15**
Bonding and structure of the
H$_2$CO molecule—Example 12-3
illustrated
The orbital set $sp^2 + p$ is used
for the C atom, $1s$ orbitals for H,
and two half-filled $2p$ orbitals
for O. For simplicity, only bonding orbitals of the valence shells
are shown.

The bonding and structure of the formaldehyde molecule are suggested by the three-dimensional sketch in Figure 12-15.

Practice Example A: Describe a plausible bonding scheme and the molecular geometry of dimethyl ether, CH_3OCH_3.

Practice Example B: Acetic acid, the acidic component of vinegar, has the formula CH_3COOH. Describe the molecular geometry and a bonding scheme for this molecule.

Drawing three-dimensional sketches to show orbital overlaps, as in Figure 12-15, is not easy. A simpler, two-dimensional representation of the bonding scheme for formaldehyde is shown in Figure 12-16. Bonds between atoms are drawn as straight lines. They are labeled σ or π, and the orbitals that overlap are indicated.

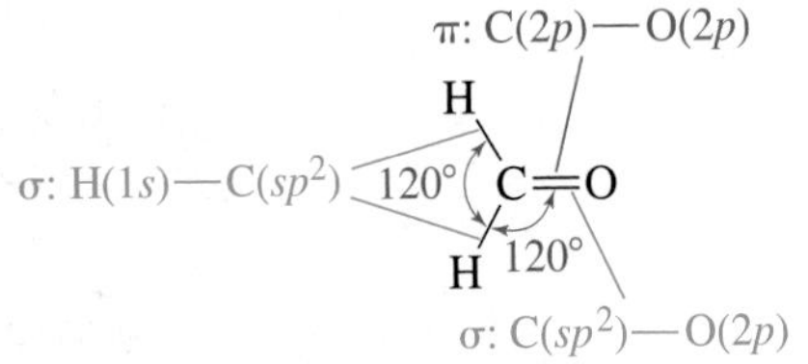

▶ **FIGURE 12-16**
Bonding in H$_2$CO—a
schematic representation

$$\pi: C(2p)\text{—}O(2p)$$

$$\sigma: H(1s)\text{—}C(sp^2) \quad 120° \quad C\text{=}O$$

$$\sigma: C(sp^2)\text{—}O(2p)$$

We have stressed a Lewis structure as the first step in describing a bonding scheme. Sometimes the starting point is a description of the species obtained *by experiment*. Example 12-4 illustrates such a case.

EXAMPLE 12-4

Using Experimental Data to Assist in Selecting a Hybridization and Bonding Scheme. Formic acid, HCOOH, is an irritating substance released by ants when they sting (Latin, *formica*, ant). A structural formula with bond angles is given here. Propose a hybridization and bonding scheme consistent with this structure.

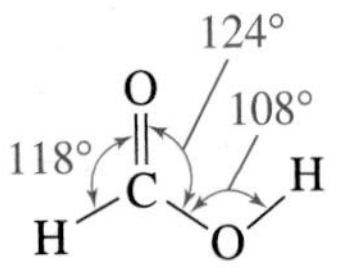

$$118° \quad \overset{124°}{O} \quad 108° \quad H$$
$$H \quad C \quad O$$

Solution

The 118° H—C—O bond angle on the left is very nearly the 120° angle for a trigonal-planar distribution of three groups of electrons. This requires an sp^2 hybridization scheme for the C atom. The 124° O—C—O bond angle is also close to the 120° expected for sp^2 hybridization. The C—O—H bond angle of 108° is close to the tetrahedral angle— 109.5°. The O atom on the right employs an sp^3 hybridization scheme. The four σ and one π bonds and the orbital overlaps producing them are indicated in Figure 12-17.

Practice Example A: Acetonitrile is an industrial solvent. Propose a hybridization and bonding scheme consistent with its structure.

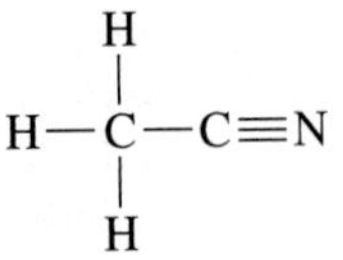

$$H$$
$$|$$
$$H\text{—}C\text{—}C\equiv N$$
$$|$$
$$H$$

Practice Example B: A reference source on molecular structures lists the following data for dinitrogen monoxide (nitrous oxide), N_2O: Bond lengths: N—N = 113 pm; N—O = 119 pm; bond angle = 180°. Show that the Lewis structure of N_2O is a resonance hybrid of two contributing structures, and describe a plausible hybridization and bonding scheme for each.

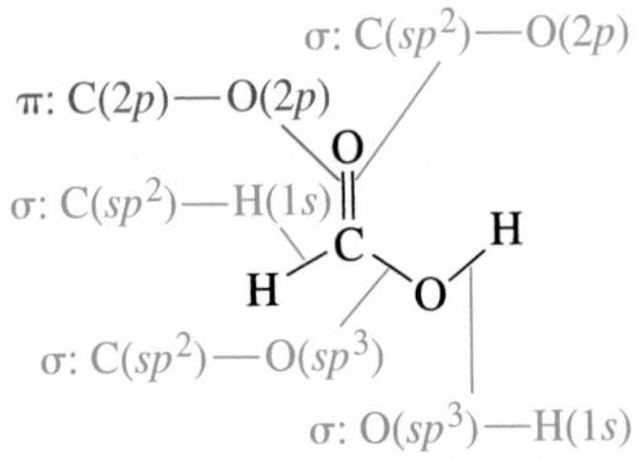

$$\sigma: C(sp^2)\text{—}O(2p)$$
$$\pi: C(2p)\text{—}O(2p)$$
$$\sigma: C(sp^2)\text{—}H(1s)$$
$$O$$
$$C \quad H$$
$$H \quad O$$
$$\sigma: C(sp^2)\text{—}O(sp^3)$$
$$\sigma: O(sp^3)\text{—}H(1s)$$

▲ **FIGURE 12-17**
Bonding and structure of
HCOOH—Example 12-4
illustrated

12-5 Molecular Orbital Theory

Lewis structures, VSEPR theory, and the valence-bond method make a potent combination for describing covalent bonding and molecular structures. They are satisfactory for most of our purposes. Sometimes, however, chemists need a greater understanding of molecular structures and properties than these methods provide. None of these methods, for instance, provides an explanation of the electronic spectra of molecules, why oxygen is paramagnetic, or why H_2^+ is a stable species. To address these questions, we need a different method of describing chemical bonding.

This method, called **molecular orbital theory**, starts with a simple picture of molecules, but it quickly becomes complex in its details. We can provide only an overview here. The theory assigns the electrons in a molecule to a series of orbitals that belong to the molecule as a whole. These are called molecular orbitals. Like atomic orbitals, molecular orbitals are mathematical functions, but we can relate them to the probability of finding electrons in certain regions of a molecule. Also like an atomic orbital, a molecular orbital can accommodate just two electrons, and the electrons must have opposing spins.

Earlier in this chapter, we described the approach of two H atoms toward one another to form a chemical bond (see Figure 12-1). What happens to the atomic orbitals as the two H atoms merge to form a chemical bond? As the atoms approach, the two 1s wave functions combine; they do this by interfering constructively or destructively. Constructive interference corresponds to adding the two mathematical functions (the positive sign puts the waves in phase). Destructive interference corresponds to subtracting the two mathematical functions (the minus sign puts the waves out of phase). These two types of combination are illustrated in Figure 12-18.

How do we interpret these two different combinations of wave functions? The constructive interference (addition) of the two wave functions leads to a greater probability of finding the electron between the nuclei. The increased electron charge density between the nuclei causes them to draw closer together, forming a chemical bond. The electron probability or electron charge density in the σ_{1s} orbital is $\left(1s_A + 1s_B\right)^2$, the square of the new function $\left(1s_A + 1s_B\right)$, where $1s_A$ and $1s_B$ are the two 1s orbitals on the two H atoms. The square is $1s_A^2 + 1s_B^2$ *plus* the extra term $2 \times 1s_A 1s_B$, which is the extra charge density between the nuclei.

The result of this constructive interference is a **bonding molecular orbital** because it places a high electron charge density between the two nuclei. A high electron charge density between atomic nuclei reduces repulsions between the positively charged nuclei and promotes a strong bond. This bonding molecular orbital, designated σ_{1s}, is at a *lower* energy than the 1s atomic orbitals

The molecular orbital formed by the subtraction of the two 1s orbitals leads to reduced electron probability between the nuclei. This produces an **antibonding molecular orbital** because it places a very low electron charge density between the two nuclei. The electron probability or electron charge density in the σ_{1s}^* orbital is

► FIGURE 12-18

Formation of bonding and antibonding orbitals

(a) The addition of two 1s orbitals in phase to form a σ_{1s} molecular orbital. This orbital produces electron density between the nuclei, leading to a chemical bond. (b) The addition of two 1s orbitals out of phase to produce a σ_{1s}^* antibonding orbital. This orbital has a nodal plane perpendicular to the internuclear axis, as do all antibonding orbitals.

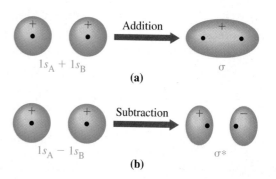

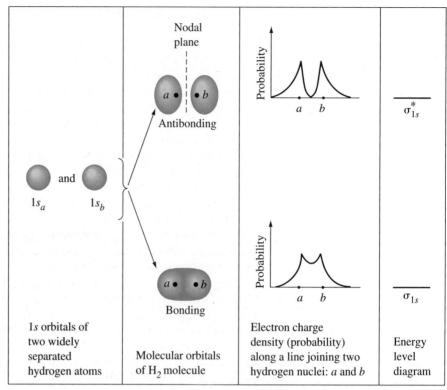

Nodal plane

Antibonding

$1s_a$ and $1s_b$

Bonding

Probability

Probability

σ_{1s}^*

σ_{1s}

| 1s orbitals of two widely separated hydrogen atoms | Molecular orbitals of H_2 molecule | Electron charge density (probability) along a line joining two hydrogen nuclei: *a* and *b* | Energy level diagram |

▲ **FIGURE 12-19   The interaction of two hydrogen atoms according to the molecular theory**
The energy of the σ_{1s} bonding molecular orbital is lower and that of the σ_{1s}^* antibonding molecular orbital is higher than the energies of the 1s atomic orbitals. Electron charge density in a bonding molecular orbital is high in the internuclear region. In an antibonding orbital, it is high in parts of the molecule away from the internuclear region.

$(1s_A - 1s_B)^2$, the square of the new function $(1s_A - 1s_B)$, where $1s_A$ and $1s_B$ are the 1s orbitals on the two H atoms. The square is $1s_A{}^2 + 1s_B{}^2$ *minus* the extra term $2 \times 1s_A 1s_B$, which is the loss of charge density between the nuclei. Notice that the $-2 \times 1s_A 1s_B$ here exactly balances the extra density $(+2 \times 1s_A 1s_B)$ in the molecular orbital formed by addition of the atomic functions.

With a low electron charge density between atomic nuclei, the nuclei are not screened from each other, strong repulsions occur, and the bond is weakened (hence the term "antibonding"). This antibonding molecular orbital, designated σ_{1s}^*, is at a *higher* energy than the 1s atomic orbitals.

The combination of two 1s orbitals of H atoms into two molecular orbitals in a H_2 molecule is summarized in Figure 12-19.

Basic Ideas Concerning Molecular Orbitals

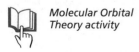

Molecular Orbital Theory activity

Here are some useful ideas about molecular orbitals and how electrons are assigned to them.

1. The number of molecular orbitals (MOs) formed is equal to the number of atomic orbitals combined.

2. Of the two MOs formed when two atomic orbitals are combined, one is a *bonding* MO at a *lower* energy than the original atomic orbitals. The other is an *antibonding* MO at a *higher* energy.

3. In ground-state configurations, electrons enter the lowest energy MOs available.

4. The maximum number of electrons in a given MO is two (Pauli exclusion principle).

5. In ground-state configurations, electrons enter MOs of identical energies *singly* before they pair up (Hund's rule).

A stable molecular species has more electrons in bonding orbitals than in antibonding orbitals. For example, if the excess of bonding over antibonding electrons is *two*, this corresponds to a *single* covalent bond in Lewis theory. In molecular orbital theory, we say that the bond order is 1. **Bond order** is one-half the difference between the number (no.) of bonding and antibonding electrons (e^-), that is,

$$\text{bond order} = \frac{\text{no. of } e^- \text{ in bonding MOs} - \text{no. of } e^- \text{ in antibonding MOs}}{2} \quad (12.1)$$

Diatomic Molecules of the First-Period Elements

Let's use the ideas just outlined to describe some molecular species of the first-period elements, H and He. Figure 12-20 helps us to do this.

Molecular Orbitals activity

▶ **FIGURE 12-20**
Molecular orbital diagrams for the diatomic molecules and ions of the first-period elements
The $1s$ energy levels of the isolated atoms are shown to the left and right of each diagram. The line segments in the middle represent the molecular orbital energy levels—lower than the $1s$ levels for σ_{1s} and higher for σ_{1s}^*.

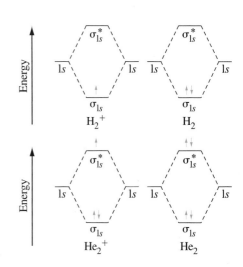

H_2^+.   This species has a single electron. It enters the σ_{1s} orbital, a bonding molecular orbital. With equation (12.1) we see that the bond order is $(1 - 0)/2 = \frac{1}{2}$. This is equivalent to a one-electron, or *half*, bond, a bond type that is not easily described by the Lewis theory.

H_2.   This molecule has two electrons, both in the σ_{1s} orbital. The bond order is $(2 - 0)/2 = 1$. With Lewis theory and the valence-bond method, we describe the bond in H_2 as single covalent.

He_2^+.   This ion has three electrons. Two electrons are in the σ_{1s} orbital, and one is in the σ_{1s}^* orbital. This species exists as a stable ion with a bond order of $(2 - 1)/2 = \frac{1}{2}$.

He_2.   Two electrons are in the σ_{1s} orbital, and two are in the σ_{1s}^*. The bond order is $(2 - 2)/2 = 0$. No bond is produced. He_2 is not a stable species.

EXAMPLE 12-5

Relating Bond Energy and Bond Order. The bond energy of H_2 is 436 kJ/mol. Estimate the bond energies of H_2^+ and He_2^+.

Solution

The bond order in H_2 is 1, equivalent to a single bond. In both H_2^+ and He_2^+, the bond order is $\frac{1}{2}$. We should expect the bonds in these two species to be only about one-half as strong as in H_2—about 220 kJ/mol. (Actual values: H_2^+, 255 kJ/mol; He_2^+, 251 kJ/mol.)

Practice Example A: The bond energy of Li_2 is 106 kJ/mol. Estimate the bond energy of Li_2^+.

Practice Example B: Do you think the ion H_2^- is stable? Explain.

Molecular Orbitals of the Second-Period Elements

For diatomic molecules and ions of H and He, we had to combine only $1s$ orbitals. In the second period, the situation is more interesting because we must work with both $2s$ and $2p$ orbitals. This results in *eight* molecular orbitals. Let's see how this comes about.

The molecular orbitals formed by combining $2s$ atomic orbitals are similar to those from $1s$ atomic orbitals, except they are at a higher energy. The situation for combining $2p$ atomic orbitals, however, is different. Two possible ways for $2p$ atomic orbitals to combine into molecular orbitals are shown in Figure 12-21: end-to-end and side-to-side. The best overlap for p orbitals is along a straight line (that is, end-to-

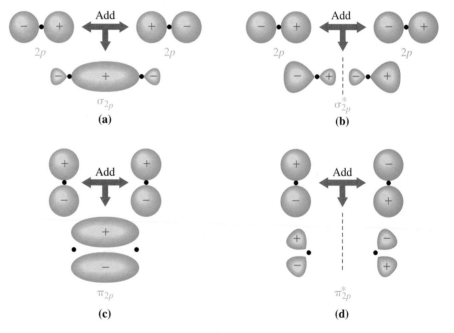

▶ In Figure 12-21, the actual p orbitals are used rather than their probability-density plots because our emphasis is on the phases of the orbitals that are being combined.

▲ **FIGURE 12-21 Formation of bonding and antibonding orbitals from $2p$ orbitals**
(a) The addition of two $2p$ orbitals in phase along the internuclear axis to form a σ_{2p} molecular orbital. This orbital produces electron density between the nuclei, leading to a chemical bond. (b) The addition of two $2p$ orbitals out of phase to produce a σ_{2p}^* antibonding orbital. This orbital has a nodal plane perpendicular to the internuclear axis, as do all antibonding orbitals. (c) The addition of two $2p$ orbitals in phase perpendicular to the internuclear axis to form a π_{2p} molecular orbital. This orbital produces electron density between the nuclei, contributing to a multiple chemical bond. (d) The addition of two $2p$ orbitals out of phase to produce a π_{2p}^* antibonding orbital. This orbital has a nodal plane perpendicular to the internuclear axis, as do all antibonding orbitals.

end). This combination produces σ-type molecular orbitals: σ_{2p} and σ_{2p}^{*}. In forming the bonding and antibonding combinations along the internuclear axis, we must take into account the phase of the $2p$ orbitals. We set up the atomic orbitals as shown in Figure 12-21(a), with the positive lobe of each function pointing to the internuclear region. Then, since the wave functions are in phase, the addition of the two wave functions leads to an increase of electron density in the internuclear region and produces a σ_{2p} orbital. When the two atomic orbitals are set up as shown in Figure 12-21(b), with lobes of opposite phase pointing into the internuclear region, a nodal plane midway between the nuclei is formed, leading to an antibonding σ_{2p}^{*} orbital.

Only one pair of p orbitals can combine in an end-to-end fashion. The other two pairs must combine in a parallel or side-to-side fashion to produce π-type molecular orbitals: π_{2p} and π_{2p}^{*}. The two possible ways for the side-to-side combination of a pair of $2p$ orbitals are shown in Figure 12-21(c and d). The π_{2p} bonding orbital (Figure 12-21c) is formed by adding the p orbital on one nucleus to a p orbital on the other nucleus, in such a way that the positive and negative lobes of one orbital are in phase with the positive and negative lobes of the other p orbital on the other nucleus. This produces additional electron density between the nuclei, but in a much less direct way than in the σ orbital because the additional electron density is not found along the internuclear axis. Typically, the π bond is weaker than the σ bond. The π_{2p}^{*} antibonding orbital is formed by subtracting the two p orbitals perpendicular to the internuclear axes, as shown in Figure 12-21(d). Now, in addition to the nodal plane that contains the nuclei, a node is formed between the nuclei, and this is a characteristic of antibonding character. There are actually *four* π-type molecular orbitals (two bonding and two antibonding) because there are *two* pairs of $2p$ atomic orbitals arranged in a parallel fashion. In Figure 12-22, we

KEEP IN MIND ▶
that subtracting two wave functions that are in phase is equivalent to *adding* the same functions when they are out of phase.

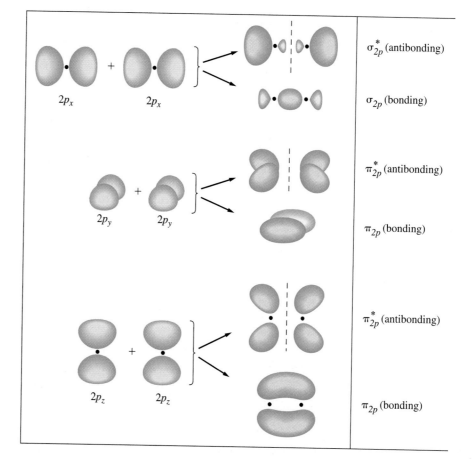

▶ FIGURE 12-22
Combining 2p atomic orbitals
These diagrams suggest the electron charge distributions for several orbitals. They are not exact in all details. Nodal planes for antibonding orbitals are represented by broken lines.

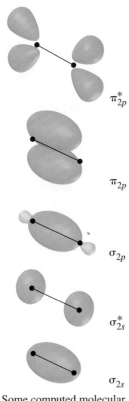

π_{2p}^*

π_{2p}

σ_{2p}

σ_{2s}^*

σ_{2s}

Some computed molecular
orbitals of F_2

▶ A *homonuclear* diatomic molecule (X_2) is one in which both atoms are of the same kind. In a *hetero*nuclear diatomic molecule (XY) the two atoms are different.

KEEP IN MIND ▶

that Hund's rule applies to molecules as well as atoms.

have depicted approximate probability distributions based on the orbital interactions described in Figure 12-21. Again, we see that bonding molecular orbitals places a high electron charge density between the atomic nuclei. In antibonding molecular orbitals, there are nodal planes between the nuclei, where the electron charge density falls to zero.

The energy-level diagram for the molecular orbitals formed from atomic orbitals of the second principal electronic shell is related to the atomic orbital energy levels. For example, molecular orbitals formed from $2s$ orbitals are at a lower energy than those formed from $2p$ orbitals—the same relationship as between the $2s$ and $2p$ atomic orbitals. Another expectation is that σ-type bonding orbitals should have lower energies than π-type because end-to-end overlap of $2p$ orbitals should be more extensive than side-to-side overlap, resulting in a lower energy. This ordering is shown in Figure 12-23a. In constructing this energy-level diagram, we have made the assumption that s orbitals mix only with s orbitals and p orbitals mix only with p orbitals. However, if we use this assumption for some diatomic molecules, we will make predictions that do not match experimental results.

We need to take into account the fact that both the $2s$ and $2p$ orbitals form molecular orbitals (σ_{2s} and σ_{2p}) that produce electron density in the same region between the nuclei. These two σ orbitals are of such similar energy and shape that they themselves mix to form modified σ orbitals. The modified σ orbitals each contain a fraction of the original σ_{2s} and σ_{2p}. The modified σ_{2s} (with some σ_{2p} mixed in) goes down in energy, and the modified σ_{2p} (with some σ_{2s} mixed in) goes up in energy, producing a different ordering of energy levels. The important aspect of this mixing is that the modified σ_{2p} is pushed up in energy above the π_{2p} orbitals (Figure 12-23b).

For the molecular orbitals in O_2 and F_2, the situation is as expected because the energy difference between the $2s$ and $2p$ orbitals is large and little s and p mixing takes place; that is, the σ_{2s} and σ_{2p} orbitals are *not* modified as described above. For other diatomic molecules of the second-period elements (for example, C_2 and N_2), the π_{2p} orbitals are at a lower energy than σ_{2p} because the energy difference between the $2s$ and $2p$ orbitals is smaller and $2s$-$2p$ orbital interactions affect the way in which atomic orbitals combine. This leads to the modified σ_{2s} and σ_{2p} orbitals described above.

Here is how we assign electrons to the molecular orbitals of the diatomic molecules of the second-period elements: We start with the σ_{1s} and σ_{1s}^* orbitals filled. Then we add electrons, in order of increasing energy, to the available molecular orbitals of the second principal shell. Figure 12-24 on page 458 shows the electron assignments for the homonuclear diatomic molecules of the second-period elements. Some molecular properties are also listed in the figure.

Just as we might arrange the valence-shell atomic orbitals of an atom, we can arrange the second-shell molecular orbitals of a diatomic molecule in the order of increasing energy. Then we can assign electrons to these orbitals, thereby obtaining a *molecular orbital diagram*. If we assign the eight valence electrons of the molecule C_2 to the diagram in Figure 12-23a, we obtain

σ_{2s}	σ_{2s}^*	σ_{2p}	π_{2p}	π_{2p}^*	σ_{2p}^*
↑↓	↑↓	↑↓	↑ ↑		

Experiment shows that the C_2 molecule is diamagnetic, not paramagnetic, and this configuration is *incorrect*. So here we see the importance of the modified energy-level diagram in Figure 12-23b. Assignment of eight electrons to the

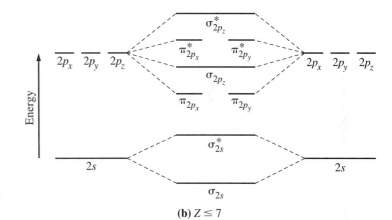

(a) $Z \geq 8$

▶ FIGURE 12-23

The two possible molecular orbital energy-level schemes for diatomic molecules of the second-period elements

(a) The expected ordering when σ_{2p} lies below the π_{2p}. This is the ordering for elements with $Z \geq 8$.
(b) The modified ordering due to s and p orbital mixing when σ_{2p} lies above the π_{2p}. This is the ordering for elements with $Z \leq 7$.

(b) $Z \leq 7$

following molecular orbital diagram is consistent with the observation that C_2 is diamagnetic.

$$\sigma_{2s} \quad \sigma_{2s}^* \quad \pi_{2p} \quad \sigma_{2p} \quad \pi_{2p}^* \quad \sigma_{2p}^*$$

This modified energy-level diagram is used for homonuclear diatomic molecules involving elements with atomic numbers from three through seven.

A Special Look at O_2

Molecular orbital theory helps us understand some of the previously unexplained features of the O_2 molecule. Each O atom brings six valence electrons to the diatomic molecule, O_2. In Figure 12-24, we see that when we assign 12 valence electrons to molecular orbitals, the molecule has *two unpaired electrons*. This explains the *paramagnetism* of O_2. The number of valence electrons in bonding orbitals is eight; the number in antibonding orbitals is four; and the bond order is two. This corresponds to a double covalent bond in O_2. This is summarized in the following molecular orbital diagram.

▲ **Paramagnetism of oxygen**
Liquid oxygen is attracted into the magnetic field of a large magnet.

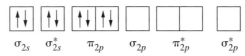

$$\sigma_{2s} \quad \sigma_{2s}^* \quad \sigma_{2p} \quad \pi_{2p} \quad \pi_{2p}^* \quad \sigma_{2p}^*$$

(12.2)

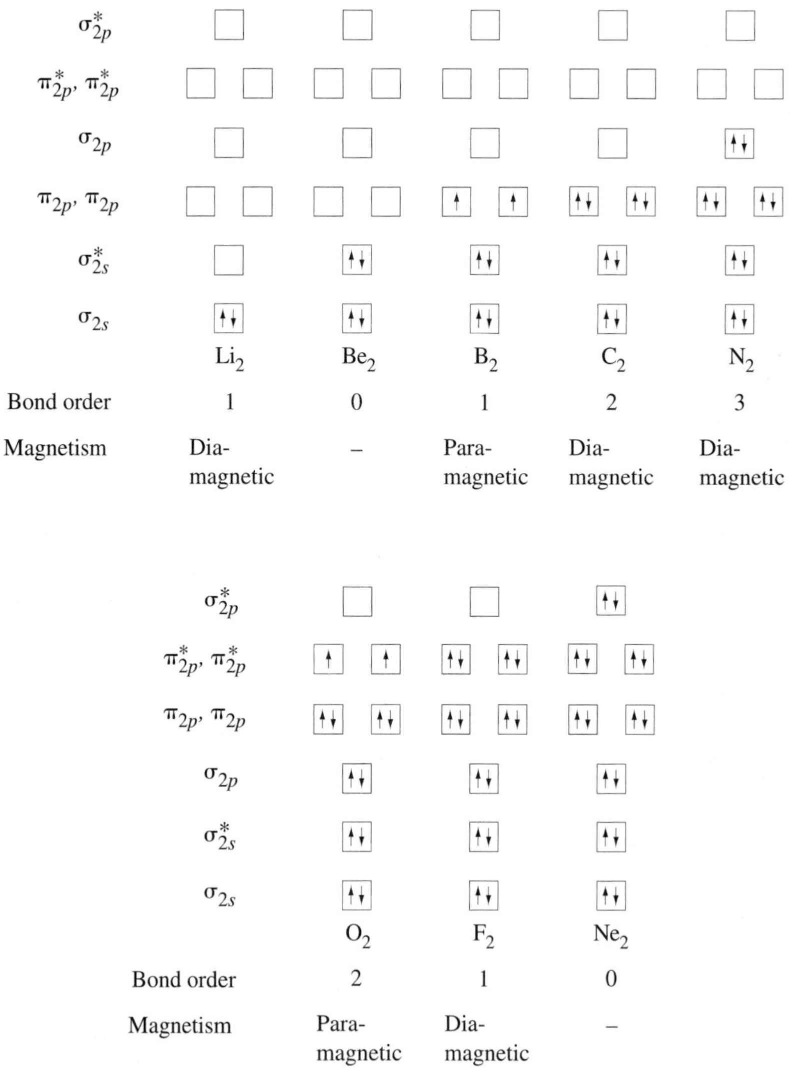

σ_{2p}^*

π_{2p}^*, π_{2p}^*

σ_{2p}

π_{2p}, π_{2p}

σ_{2s}^*

σ_{2s}

	Li$_2$	Be$_2$	B$_2$	C$_2$	N$_2$
Bond order	1	0	1	2	3
Magnetism	Dia-magnetic	–	Para-magnetic	Dia-magnetic	Dia-magnetic

σ_{2p}^*

π_{2p}^*, π_{2p}^*

π_{2p}, π_{2p}

σ_{2p}

σ_{2s}^*

σ_{2s}

	O$_2$	F$_2$	Ne$_2$
Bond order	2	1	0
Magnetism	Para-magnetic	Dia-magnetic	–

▲ FIGURE 12-24 **Molecular orbital diagrams for the homonuclear diatomic molecules of the second-period elements**

In all cases, the σ_{1s} and σ_{1s}^* molecular orbitals are filled but not shown.

EXAMPLE 12-6

Writing a Molecular Orbital Diagram and Determining Bond Order. Represent bonding in O$_2^+$ with a molecular orbital diagram, and determine the bond order in this ion.

Solution

The ion O$_2^+$ has 11 valence electrons. Assign these to the available molecular orbitals in accordance with the ideas stated on page 452. [Or simply remove one of the electrons from a π_{2p}^* orbital in expression (12.2).] In the following diagram, we find an excess of *five* bonding electrons over antibonding ones. The bond order is 2.5.

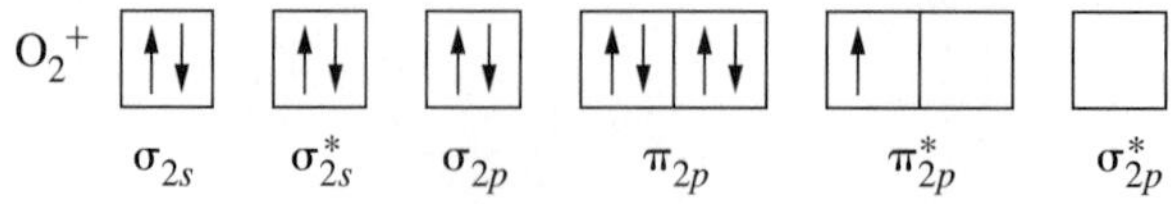

O$_2^+$

σ_{2s} σ_{2s}^* σ_{2p} π_{2p} π_{2p}^* σ_{2p}^*

Practice Example A: Refer to Figure 12-24. Write a molecular orbital diagram and determine the bond order of **(a)** N_2^+; **(b)** Ne_2^+; **(c)** C_2^{2-}.

Practice Example B: The bond lengths for O_2^+, O_2, O_2^-, and O_2^{2-} are 112, 121, 128, and 149 pm, respectively. Are these bond lengths consistent with the bond order determined from the molecular orbital diagram? Explain.

A Look at Heteronuclear Diatomic Molecules

The ideas that we have developed for homonuclear diatomic species can be extended, with some care, to give us an idea of the bonding in heteronuclear diatomic species. When we drew a diagram to illustrate the mixing of two 1s orbitals to form the σ bond in H_2, we understood that the two 1s orbitals have the same energy and mix fifty-fifty. This is not the case with heteronuclear diatomic species because, as we know from ionization energies, the orbitals of different atoms have different energies. To illustrate the differences, let us consider the construction of the σ bonding and antibonding orbitals from the 2s orbitals on C and O atoms in the bonding for CO. The bonding combination is

$$\sigma = c_O\, 2s_O + c_C\, 2s_C$$

and the antibonding combination is

$$\sigma^* = c_O^*\, 2s_O - c_C^*\, 2s_C$$

where the coefficients c_O and c_C reflect the percentage mixing of the orbitals. In the homonuclear case, the two atoms are the same and the coefficients are equal because there is an equal probability of finding the electrons in the orbital associated with either nucleus. If the nuclei are different, we expect a greater probability of finding the electrons in the orbital associated with the more electronegative element. As a result, in our example we expect c_O to be greater than c_C in the bonding orbital. Thus, the bonding molecular orbital has a greater contribution from the oxygen 2s orbital than from the carbon 2s orbital, meaning that the σ_{2s} orbitals looks more like an oxygen 2s orbital. In the antibonding orbital, the situation is reversed, with c_O^* less than c_C^* and the antibonding orbital more closely resembling a carbon 2s orbital.

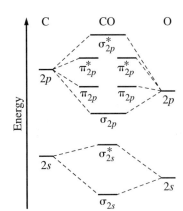

The energy of the bonding orbital is closer to that of the more electronegative element, whereas the energy of the antibonding orbital is closer to the energy of the less electronegative element, as shown in the diagram in the margin. Thus the molecular orbital diagram shown in Figure 12-23 has to be modified to take this into account. However, we can still use the molecular orbital diagrams in Figure 12-24 to discuss bonding in heteronuclear diatomic species with this restriction: The two atoms must not be too far apart in atomic number, so that the order of energy levels is not too different from that found for homonuclear diatomic species. The question arises as to which order of orbitals to use, the expected one or the modified one in which the σ_{2p} is above the π_{2p} in energy. If one of the elements is oxygen or fluorine, the appropriate energy-level diagram is that shown in Figure 12-23a, that is, the unmodified one. The reason for this is that the energy difference between the 2s and 2p orbitals of oxygen and the 2s and 2p orbitals of any other element preceding oxygen in the periodic table is too great for significant mixing to take place.

Let's apply these ideas to carbon monoxide, which has ten valence electrons. Applying the aufbau principle to the molecular orbital diagram for O_2 in Figure 12-24, we obtain this configuration.

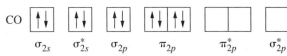

Thus CO has a bond order of three, in accord with experimental data. Notice that CO has a greater bond energy than NO, which has a bond order of 2.5, as predicted by the molecular orbital diagram

NO $\boxed{\uparrow\downarrow}$ $\boxed{\uparrow\downarrow}$ $\boxed{\uparrow\downarrow}$ $\boxed{\uparrow\downarrow\;\uparrow\downarrow}$ $\boxed{\uparrow\;\;}$ $\boxed{\;\;}$

$\qquad\quad\sigma_{2s}\qquad\sigma_{2s}^{*}\qquad\sigma_{2p}\qquad\pi_{2p}\qquad\;\pi_{2p}^{*}\qquad\sigma_{2p}^{*}$

We see that NO is also predicted to be paramagnetic, which we know to be the case from experimental data.

EXAMPLE 12-7

Writing Molecular Orbital Diagrams for Heteronuclear Diatomic Species. Write the molecular orbital diagram for the cyanide ion, CN^{-}, and determine the bond order in this ion. Compare your results with the predicted bond order from a Lewis structure.

Solution

The number of valence electrons to be assigned in the molecular orbital diagram is 10 $(4 + 5 + 1)$. Notice that here we have to use the modified molecular orbital energy-level diagram, as both elements precede oxygen in the periodic table. That is, use the MO diagram for either C_2 or N_2, and introduce 10 electrons.

CN^{-} $\boxed{\uparrow\downarrow}$ $\boxed{\uparrow\downarrow}$ $\boxed{\uparrow\downarrow\;\uparrow\downarrow}$ $\boxed{\uparrow\downarrow}$ $\boxed{\;\;}$ $\boxed{\;\;}$

$\qquad\quad\sigma_{2s}\qquad\sigma_{2s}^{*}\qquad\pi_{2p}\qquad\;\;\sigma_{2p}\qquad\pi_{2p}^{*}\qquad\sigma_{2p}^{*}$

The bond order is the difference between the number of bonding electrons and the number of antibonding electrons, divided by two, which is $(2 + 4 + 2 - 2)/2 = 3$. The bond is a triple bond. The Lewis structure also predicts a triple bond, that is, $[:C\equiv N:]^{-}$.

Practice Example A: Write a molecular orbital diagram for CN^{+}, and determine the bond order.

Practice Example B: Write a molecular orbital diagram for BN, and determine the bond order.

12-6 Delocalized Electrons: Bonding in the Benzene Molecule

▶ The term "aromatic" relates to the fragrant aromas associated with some (but by no means all) of these compounds.

In Section 12-4 we discussed localized π bonds, such as those in ethylene, C_2H_4. Some molecules have a network of π bonds, such as benzene, C_6H_6, and substances related to it—aromatic compounds. In this section we will consider bonding in benzene, with the bonding theories we have studied so far; the conclusions we reach will help us understand other cases of bonding as well.

Bonding in Benzene

In 1865, Friedrich Kekulé advanced the first good proposal for the structure of benzene. He suggested that the C_6H_6 molecule consists of a flat, hexagonal ring of six carbon atoms joined by alternating single and double covalent bonds. Each C atom is joined to two other C atoms and to one H atom. To explain the fact that the carbon-to-carbon bonds are all alike, Kekulé suggested that the single and double bonds continually oscillate from one position to the other. Today, we say that the two possible Kekulé structures are actually contributing structures to a resonance hybrid. This view is suggested by Figure 12-25.

(a)   (b)

(c)

▲ **FIGURE 12-25** **Resonance in the benzene molecule and the Kekulé structures**
(a) Lewis structure for C_6H_6, showing alternate carbon-to-carbon single and double
bonds. **(b)** Two equivalent Kekulé structures for benzene. A carbon atom is at each corner
of the hexagonal structure, and a hydrogen atom is bonded to each carbon atom. (The
symbols for carbon and hydrogen, as well as the C—H bonds, are customarily omitted in
these structures.) **(c)** A space-filling model.

C_6H_6 model

▲ Dame Kathleen Londsdale
first determined the X-ray crystal structure of benzene. Her experiment demonstrated that the
benzene molecule is flat, as predicted by theorists.

We can gain a more thorough understanding of bonding in the benzene molecule
by combining the valence-bond and molecular orbital methods. We can construct
a σ-bond framework for the observed *planar* structure with 120° bond angles by
using sp^2 hybridization at each carbon atom. End-to-end overlap of the sp^2 orbitals
produces σ bonds. Side-to-side overlap of $2p$ orbitals yields three π bonds. Figure 12-26 gives a valence-bond theory interpretation of bonding in C_6H_6.

We do not need to think in terms of an oscillation between two structures
(Kekulé) or of a resonance hybrid for benzene. The π bonds are not localized between specific carbon atoms but are spread out around the six-membered ring. To
represent this *delocalized* π bonding, the symbol for benzene is often written as a
hexagon with a circle inscribed within. (Figure 12-26c).

We can best understand delocalized π bonds through molecular orbital theory.
Six $2p$ atomic orbitals of the C atoms combine to form six molecular orbitals of the
π type. Three of these π-type molecular orbitals are bonding, and three are antibonding. The three bonding orbitals fill with six electrons (one $2p$ electron from each
C atom), and the three antibonding orbitals remain empty. The bond order associated with the six electrons in π-bonding molecular orbitals is $(6 - 0)/2 = 3$. The
three bonds are distributed among the six C atoms, which amounts to $3/6$, or a half-
bond, between each pair of C atoms. Add to this the σ bonds in the σ-bond framework, and we have a bond order of 1.5 for each carbon-to-carbon bond. This is
exactly what we also get by averaging the two Kekulé structures of Figure 12-25.
The π molecular orbital diagram for benzene is shown in Figure 12-27.

The three bonding π molecular orbitals in C_6H_6 describe the distribution of π
electron charge in the molecule. We can think of this in terms of two doughnut-
shaped regions: one above and one below the plane of the C and H atoms. Because
they are spread out among all six C atoms instead of being concentrated between

Bonding in Benzene activity

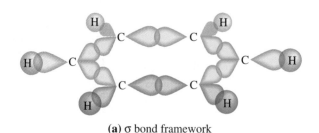

(a) σ bond framework

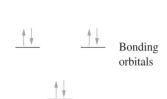

—— —— Antibonding orbitals

- - - - - - - - - - - - -

↑↓   ↑↓   Bonding orbitals

↑↓

▲ **FIGURE 12-27**

π Molecular orbital diagram for C_6H_6

Of the six π molecular orbitals, three are bonding orbitals and each of these is filled with an electron pair. The three anti-bonding molecular orbitals at a higher energy remain empty.

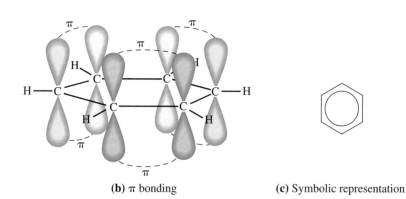

(b) π bonding   **(c)** Symbolic representation

▲ **FIGURE 12-26   Bonding in benzene, C_6H_6, by the valence-bond method**

(a) Carbon atoms use sp^2 and p orbitals. Each carbon atom forms three σ bonds, two with neighboring C atoms in the hexagonal ring and a third with a H atom. **(b)** The overlap in side-to-side fashion of $2p$ orbitals produces three π bonds. Thus, there are three double bonds (σ + π) between carbon atoms in the ring. **(c)** Because the three π bonds are delocalized around the benzene ring, the molecule is often represented by a hexagon with an inscribed circle.

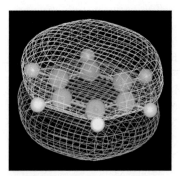

▲ **FIGURE 12-28**

Molecular orbital representation of π bonding in benzene

A computer-generated model of the benzene molecule. The planar σ-bond framework is clearly visible. The π orbitals above and below the C and H plane are highlighted.

pairs of C atoms, these molecular orbitals are called **delocalized molecular orbitals**. Figure 12-28 pictures these delocalized molecular orbitals in the benzene molecule.

Other Structures with Delocalized Molecular Orbitals

By using delocalized bonding schemes we can avoid writing two or more contributing structures to a resonance hybrid, as is so often required in the Lewis theory. Consider the ozone molecule, O_3, that we used to introduce the concept of resonance in Section 11-5. In place of the resonance hybrid based on these contributing structures,

$$\ddot{\text{:}}\overset{..}{\underset{}{\text{O}}}\overset{\overset{..}{\text{O}}}{}\text{O:} \longleftrightarrow \text{:}\overset{..}{\underset{}{\text{O}}}\overset{\overset{..}{\text{O}}}{}\ddot{\text{O:}}$$

we can write the single structure shown in Figure 12-29. Here are the ideas that lead to Figure 12-29.

1. With VSEPR theory, we predict a trigonal-planar electron-group geometry (the measured bond angle is 117°). The hybridization scheme chosen for the central O atom is sp^2, and, although we normally do not need to invoke hybridization

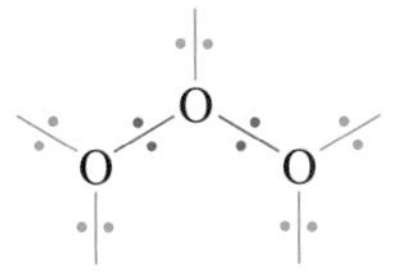

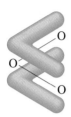

(a) σ bond framework

(b) Delocalized π molecular
orbital

▲ FIGURE 12-29
**Structure of the ozone
molecule, O_3**
(a) The σ-bond framework and
the assignment of bond-pair (:)
and lone-pair (:) electrons to sp^2
hybrid orbitals are discussed in
items 1 and 2. **(b)** The π molecular orbitals and assignments of
electrons to them are discussed
in items 3 and 4.

for terminal atoms, this case is simplified if we assume sp^2 hybridization for
the terminal O atoms as well. Thus, each O atom uses the orbital set $sp^2 + p$.

2. Of the 18 valence electrons in O_3, assign 14 to the sp^2 hybrid orbitals of the
   σ-bond framework. Four of these are bonding electrons (red) and ten are lone-
   pair electrons (blue).

3. The three unhybridized $2p$ orbitals combine to form *three* molecular orbitals
   of the π type (Figure 12-30). One of these orbitals is a bonding molecular orbital, and the second is antibonding. The third is a type we have not described
   before—a *nonbonding molecular orbital*. A nonbonding molecular orbital has
   the same energy as the atomic orbitals from which it is formed, and it neither
   adds to nor detracts from bond formation.

4. The remaining four valence electrons are assigned to the π molecular orbitals.
   Two go into the bonding orbital and two into the nonbonding orbital. The antibonding orbital remains empty.

5. The bond order associated with the π molecular orbitals is $(2 - 0)/2 = 1$.
   This π bond is distributed between the two O—O bonds and amounts to one-
   half of a π bond for each.

The points listed here lead to a total bond order of 1.5 for the O—O bonds in O_3.
This is equivalent to averaging the two Lewis structures. The O—O bond length
suggested by this method was described in Section 11-5.

EXAMPLE 12-8

Representing Delocalized Molecular Orbitals With Atomic Orbital Diagrams. Represent
π molecular orbital formation, bonding and orbital occupancy in the nitrate ion, NO_3^-.

Solution

First, we should recognize this fact: For this type of exercise, we need not concern ourselves with resonance; we can use any one of the following resonance structures.

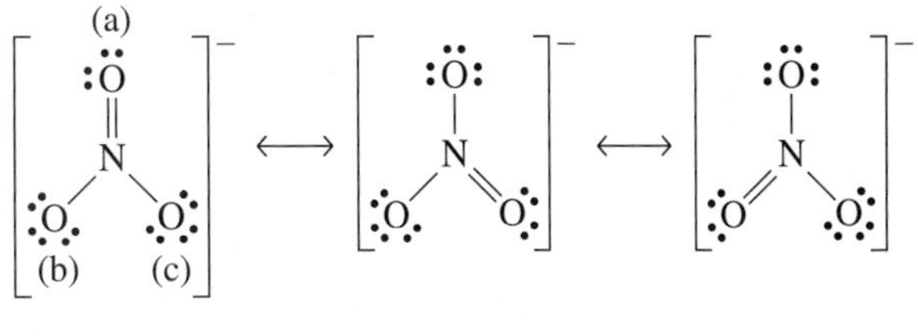

Let us use the first structure shown, in which we have labeled the oxygen atoms (a), (b),
and (c). Each atom in the structure has an electron-group geometry corresponding to
three electron groups (an electron pair in each N-to-O single bond and two electron pairs
in the N-to-O double bond). We must use sp^2 hybrid orbitals to form σ bonds and $2p$ orbitals to form π bonds. In assigning valence-shell electrons to the atoms, place one unpaired electron in an sp^2 for every σ bond that the atom forms and one unpaired electron
for every π bond the atom forms. Then place lone-pair electrons in the orbitals that remain available. Because of the existence of formal charges in the structure, the number
of valence electrons for each atom must be adjusted for the formal charge [for example,
O atom (b) has a formal charge of −1 and must show seven electrons in its valence-shell
orbitals, not the customary six].

Once you have assigned all the valence electrons to orbitals, show that all the unpaired electrons in sp^2 hybrid orbitals participate in σ-bond formation. Now combine the
p orbitals to form the appropriate number of π molecular orbitals (four). Assign the p electrons to these molecular orbitals in the usual fashion.

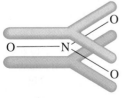

Delocalized π molecular
orbitals of NO_3^-

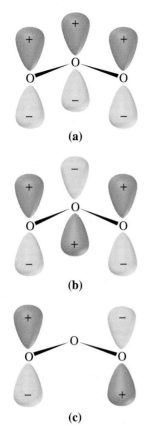

▲ FIGURE 12-30
Pi bonding orbitals of the ozone molecule O_3
(a) The π bonding molecular orbital with all the $2p$ orbitals in phase. **(b)** The π antibonding molecular orbital with the three $2p$ orbitals out of phase. **(c)** The π nonbonding molecular orbital with a single node and with zero contribution of the wave function from the central atom. This orbital is called nonbonding because there is neither a node nor a region of increased electron density between the central atom and its neighbors.

▶ A dull metal surface usually signifies that the surface is coated with a compound of the metal (for example, an oxide, sulfide, or carbonate).

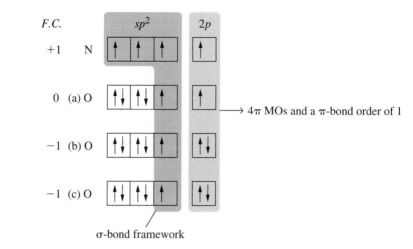

σ-bond framework

Practice Example A: Represent chemical bonding in the molecule SO_3 by using a combination of localized and delocalized orbitals.

Practice Example B: Represent chemical bonding in the ion NO_2^- by using a combination of localized and delocalized orbitals.

12-7 Bonding in Metals

In nonmetal atoms, the valence shells generally have more electrons than they do orbitals. To illustrate, an F atom has *four* valence-shell orbitals $(2s, 2p_x, 2p_y, 2p_z)$ and *seven* valence-shell electrons. Whether fluorine exists as a solid, liquid, or gas, F atoms join in pairs to form F_2 molecules. One pair of electrons is shared in the F—F bond, and the other electron pairs are lone pairs, as seen in the Lewis structure $:\overset{..}{F}—\overset{..}{F}:$. By contrast, the metal atom Li has the same four valence-shell orbitals as F but only *one* valence-shell electron $(2s^1)$. This may account for the formation of the *gaseous* molecule Li:Li, but in the solid metal, each Li atom is somehow bonded to *eight* neighbors. The challenge to a bonding theory for metals is to explain how so much bonding can occur with so few electrons. Also, the theory should account for certain properties that metals display to a far greater extent than nonmetals, such as a lustrous appearance, an ability to conduct electricity, and an ease of deformation (metals are easily flattened into sheets and drawn into wires).

The Electron Sea Model

An oversimplified theory, which can explain some of the properties of metals just cited, pictures a solid metal as a network of positive ions immersed in a "sea of electrons." In lithium, for instance, the ions are Li^+ and one electron per atom is contributed to the sea of electrons. Electrons in the sea are *free* (not attached to any particular ion), and they are mobile. Thus, if electrons from an external source enter a metal wire at one end, free electrons pass through the wire and leave the other end at the same rate. In this way, electrical conductivity is explained.

Free electrons (those in the electron sea) are not limited in their ability to absorb photons of visible light as are electrons bound to an atom. Thus, metals absorb visible light; they are opaque. Electrons at the surface of a metal are able to reradiate, at the same frequency, light that strikes the surface, which explains the lustrous appearance of metals. The ease of deformation of metals can be explained in this way: If one layer of metal ions is forced across another, perhaps by hammering, no bonds are broken, the internal structure of the metal remains essentially unchanged, and the sea of electrons rapidly adjusts to the new situation (Figure 12-31).

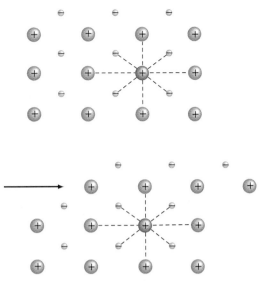

▲ **FIGURE 12-31   The electron sea model of metals**
A network of positive ions is immersed in a "sea of electrons," derived from the valence shells of the metal atoms and belonging to the crystal as a whole. One particular ion (red), its nearest neighboring ions (brown), and nearby electrons in the electron sea (blue) are emphasized.

At the bottom of the figure, a force is applied (from left to right). The highlighted cation is unaffected; its immediate environment is unchanged. The electron sea model explains the ease of deformation of metals.

Band Theory

The electron sea model is a simple qualitative description of the metallic state, but for most purposes, the theory of metallic bonding used is a form of molecular orbital theory called **band theory**.

Recall the formation of molecular orbitals and the bonding between two Li atoms (see Figure 12-24). Each Li atom contributes one $2s$ orbital to the production of two molecular orbitals: σ_{2s} and σ_{2s}^*. The electrons originally described as the $2s^1$ electrons of the Li atoms enter and half-fill these molecular orbitals. That is, they fill the σ_{2s} orbital and leave the σ_{2s}^* empty. If we extend this combination of Li atoms to a third Li atom, three molecular orbitals are formed, containing a total of three electrons. Again, the set of molecular orbitals is half-filled. We can extend this process to an enormously large number (N) of atoms—the total number of atoms in a crystal of Li. Here is the result we get: a set of N molecular orbitals with an extremely small energy separation between each pair of successive levels. This collection of very closely spaced molecular orbital energy levels is called an *energy band* (Figure 12-32).

In the band just described, there are N electrons (a $2s$ electron from each Li atom) occupying, in pairs, $N/2$ molecular orbitals of lowest energy. These are the electrons responsible for bonding the Li atoms together. They are valence electrons, and the band in which they are found is called a *valence band*. Because the energy differences between the occupied and unoccupied levels in the valence band are so small, however, electrons can be easily excited from the highest filled levels to the unfilled levels that lie immediately above them in energy. This excitation, which has the effect of producing mobile electrons, can be accomplished by applying a small electric potential difference across the crystal. This is how the band

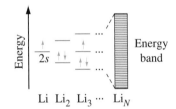

▲ **FIGURE 12-32**
Formation of an energy band in lithium metal
As more and more Li atoms are added to the growing "molecule," Li_2, Li_3, ..., additional energy levels are added and the spacing between levels becomes increasingly smaller. In an entire crystal of N atoms, the energy levels merge into a band of N closely spaced levels. The lowest $N/2$ levels are filled with electrons, and the upper $N/2$ levels are empty.

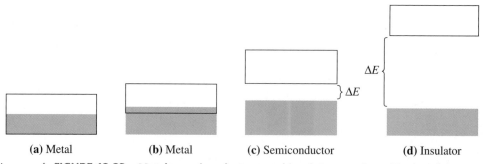

(a) Metal   (b) Metal   (c) Semiconductor   (d) Insulator

Development of Bond Structure animation

▲ **FIGURE 12-33   Metals, semiconductors, and insulators as viewed by band theory**
(a) In some metals, the valence band (blue) is only partially filled (for example, the half-filled 3s band in Na). The valence band also serves as a conduction band (outlined in black). (b) In other metals, the valence band is full, but a conduction band overlaps it (for example, the empty 2p band of Be overlaps the full 2s valence band). (c) In a semiconductor, the valence band is full and the conduction band is empty. The energy gap (ΔE) between the two is small enough, however, that some electrons make the transition between the two just by acquiring thermal energy. (d) In an insulator, the valence band is filled with electrons and a large energy gap (ΔE) separates the valence band from the conduction band. Few electrons can make the transition between bands, and the insulator does not conduct electricity.

theory explains the ability of metals to conduct electricity. The essential feature for electrical conductivity, then, is *an energy band that is only partly filled with electrons*. Such an energy band is called a *conduction band*. In lithium, the 2s band is both a valence band and a conduction band (Figure 12-33a).

Let's extend our discussion to N atoms of beryllium, which has the electron configuration $1s^2 2s^2$. We expect the band formed from 2s atomic orbitals to be filled—N molecular orbitals and $2N$ electrons. But how can we reconcile this with the fact that Be is a good electric conductor? At the same time that 2s orbitals are being combined into a 2s band, 2p orbitals combine to form an *empty* 2p band. The lowest levels of the 2p band are at a *lower* energy than the highest levels of the 2s band. The bands overlap (Figure 12-33b). As a consequence, empty molecular orbitals are available to the valence electrons in beryllium.

In an *electrical insulator*, like diamond or silica (SiO_2), not only is the valence band filled, but there is a large energy gap between the valence band and the conduction band (Figure 12-33d). Very few electrons are able to make the transition between the two.

Semiconductors

Much of modern electronics depends on the use of semiconductor materials. Light-emitting diodes (LEDs), transistors, and solar cells are among the familiar electronic components using semiconductors. Semiconductors such as cadmium yellow (CdS) and vermilion (HgS) are brilliantly colored, and artists use them in paints.

What determines the electronic properties of a semiconductor is the energy gap (band gap) between the valence band and the conduction band (Figure 12-33c). In some materials, such as CdS, this gap is of a fixed size. These materials are called *intrinsic semiconductors*. When white light interacts with the semiconductor, electrons are excited (promoted) to the conduction band. CdS absorbs violet light and some blue light, but other frequencies contain less energy than is needed to excite an electron above the energy gap. The frequencies that are not absorbed are reflected, and the color we see is yellow. Some semiconductors, such as GaAs and

PbS, have a sufficiently small band gap that all frequencies of visible light are absorbed. There is no reflected visible light, and the materials have a black color.

In a semiconductor such as silicon or germanium, the filled valence band and empty conduction band are separated by only a small energy gap. Electrons in the valence band may acquire enough thermal energy to jump to a level in the conduction band. The greater the thermal energy, the more electrons can make the transition. In this way, band theory explains the observation that the electrical conductivity of semiconductors increases with temperature.

In many semiconductors, called *extrinsic semiconductors*, the size of the band gap is controlled by carefully adding impurities—a process called *doping*. Let's consider what doping does to one of the most common semiconductors, silicon.

When silicon is doped with phosphorus, the energy level of the P atoms lies just below the conduction band of the silicon, as shown in Figure 12-34. Each P atom uses *four* of its five valence electrons to form bonds to four neighboring Si atoms, and thermal energy alone is enough to cause the "extra" valence electron to be promoted to the conduction band, leaving behind an immobile positive P^+ ion. The P atoms are called *donor* atoms, and electrical conductivity in this type of semiconductor involves primarily the movement through the conduction band of electrons coming from the donor atoms. This type of semiconductor is called an *n-type*, where *n* refers to negative—the type of electric charge carried by electrons.

When silicon is doped with aluminum, the energy level of the Al atoms, called *acceptor* atoms, lies just above the valence band of the silicon (Figure 12-34). Because an Al atom has only *three* valence electrons, it forms regular electron-pair bonds with three neighboring Si atoms but only a one-electron bond with a fourth Si atom. An electron is easily promoted from the valence band to an Al atom in the acceptor level, however, forming an immobile negative Al^- ion. When this occurs, a *positive hole* is created in the valence band. Because electrical conductivity in this type of semiconductor consists primarily of the migration of positive holes, it is called a *p-type* semiconductor.

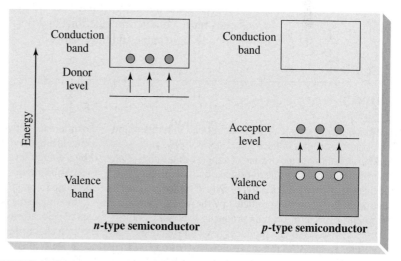

▲ FIGURE 12-34   *p-* and *n*-type semiconductors
In a semiconductor with donor atoms (for example, P in Si), the donor level lies just beneath the conduction band. Electrons (●) are easily promoted into the conduction band. The semiconductor is of the *n*-type. In a semiconductor with acceptor atoms (for example, Al in Si), the acceptor level lies just above the valence band. Electrons (●) are easily promoted to the acceptor level, leaving positive holes (○) in the valence band. The semiconductor is of the *p*-type.

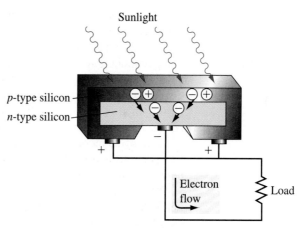

▲ **FIGURE 12-35**   **A photovoltaic (solar) cell using silicon-based semiconductors**

Figure 12-35 suggests how semiconductors are used in photovoltaic (solar) cells. A thin layer of a *p*-type semiconductor is in contact with an *n*-type semiconductor in a region called the *junction*. Normally, migration of electrons and positive holes across the junction is very limited because such a migration would lead to a separation of charge: Positive holes crossing the junction from the *p*-type semiconductor would have to move away from immobile Al^- ions, and electrons crossing from the *n*-type semiconductor would have to move away from immobile P^+ ions.

Now imagine that the *p*-type semiconductor is struck by a beam of light. Electrons in the valence band can absorb energy and be promoted to the conduction band, creating positive holes in the valence band. Conduction electrons, unlike positive holes, can easily cross the junction into the *n*-type semiconductor. This sets up a flow of electrons, an electric current. Electrons can be carried by wires through an external load (lights, electric motors, and so forth) and eventually returned to the *p*-type semiconductor, where they fill positive holes. Further light absorption creates more conduction electrons and positive holes, and the process continues as long as light shines on the solar cell.

Summary

Valence-bond theory considers a covalent bond in terms of the overlap of atomic orbitals of the bonded atoms. Some molecules can be described through the overlap of simple orbitals, but often hybrid orbitals are needed. The hybridization scheme chosen is the one that produces an orientation of hybrid orbitals to match the electron-group geometry predicted by the VSEPR theory.

End-to-end overlap of orbitals produces σ bonds. Side-to-side overlap of two *p* orbitals produces a π bond. Single covalent bonds are σ bonds. A double bond consists of one σ bond and one π bond. A triple bond consists of one σ bond and two π bonds. The geometric shape of a species is determined by its σ-bond framework, not its π bonds.

In molecular orbital theory, electrons are assigned to molecular orbitals. The numbers and kinds of molecular orbitals are related to the atomic orbitals used to generate them. Electron charge density between atoms is high in bonding molecular orbitals and very low in antibonding orbitals. Bond order is one-half

the difference between the numbers of electrons in bonding molecular orbitals and in antibonding molecular orbitals. Molecular orbital energy-level diagrams and an aufbau process can be used to describe the electronic structure of a molecule, similar to what was done for atomic electron configurations in Chapter 9.

Bonding in the benzene molecule, C_6H_6, is based on the concept of delocalized molecular orbitals. These are regions of high electron charge density that extend over several atoms in a molecule. Delocalized molecular orbitals also provide an alternative to the concept of resonance in other molecules and ions.

Finally, molecular orbital theory, in the form called band theory, can be applied to metals, semiconductors, and insulators. Band theory provides explanations of the thermal and electric conductivity, the ease of deformation, and the characteristic luster of metals. It also explains the colors of semiconductors and the fact that their electric conductivities increase with temperature.

Integrative Example

Hydrogen azide, HN_3, and its salts (metal azides) are unstable substances used in detonators for high explosives. Sodium azide, NaN_3, is used in air-bag safety systems in automobiles (see page 210). A reference source lists the following data for HN_3. (The subscripts a, b, and c distinguish the three N atoms from one another.) Bond lengths: N_a—N_b = 124 pm; N_b—N_c = 113 pm. Bond angles: H—N_a—N_b = 112.7°; N_a—N_b—N_c = 180°.

Write two contributing structures to the resonance hybrid for HN_3, and describe a plausible hybridization and bonding scheme for each structure.

1. *Compare bond lengths with values from Table 11.2.* From Table 11.2 the average bond lengths for N-to-N bonds are 145 pm for a single bond, 123 pm for a double bond, and 110 pm for a triple bond. Thus it is likely that the N_a—N_b bond (124 pm) has considerable a double bond character and the N_b—N_a bond (113 pm), considerable triple bond character.

2. *Draw plausible Lewis structures.* The HN_3 molecules has a total of 16 valence electrons in 8 electron pairs. The plausible Lewis structures have N_a and N_b atoms as central atoms, the N_c and H atoms as terminal atoms, and bonds reflecting the observed bond lengths (part 1).

(I) $H—\ddot{N}_a{=}N_b{=}\ddot{N}_c$ (II) $H—\overset{..}{\underset{..}{N}}_a—N_b{\equiv}\ddot{N}_c$

3. *Determine hybridization schemes for the central atoms.* According to VSEPR theory, both in structures (I) and (II) the electron group geometry around N_b is linear. This corresponds to sp hybridization. In structure (I), the electron group geometry around N_a is trigonal planar, corresponding to sp^2 hybridization; in structure (II), the electron group geometry around N_a is tetrahedral, corresponding to sp^3 hybridization. These hybridization schemes, the orbital overlaps, and the geometric structures of the two resonance structures are indicated below.

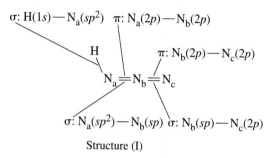

Structure (I)

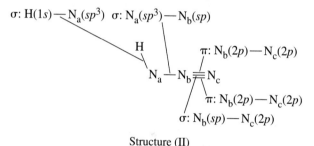

Structure (II)

Key Terms

antibonding molecular orbital (12-5)
band theory (12-7)
bond order (12-5)
bonding molecular orbital (12-5)
delocalized molecular orbital (12-6)
hybrid orbital (12-3)

hybridization (12-3)
molecular orbital theory (12-5)
pi (π) bond (12-4)
sigma (σ) bond (12-4)
sp **hybrid orbital** (12-3)
sp^2 **hybrid orbital** (12-3)

sp^3 **hybrid orbital** (12-3)
sp^{3}d **hybrid orbital** (12-3)
sp^{3}d^2 **hybrid orbital** (12-3)
valence-bond method (12-2)

Review Questions

1. In your own words, define the following terms or symbols: **(a)** sp^2; **(b)** σ_{2p}^*; **(c)** bond order; **(d)** π bond.

2. Briefly describe each of the following ideas: **(a)** hybridization of atomic orbitals; **(b)** σ-bond framework; **(c)** Kekulé structures of benzene, C_6H_6; **(d)** band theory of metallic bonding.

3. Explain the important distinctions between the terms in each of the following pairs: **(a)** σ and π bonds; **(b)** local-ized and delocalized electrons; **(c)** bonding and antibonding molecular orbitals; **(d)** metal and semiconductor.

4. The measured bond angle in H_2Se is 91°. Which orbitals of the H and Se atoms are most likely involved in orbital overlap in H—Se bonds? Explain.

5. In which of the following would you expect to find sp^2 hybridization of the central atom? Explain. CO_3^{2-}, SO_2, CCl_4, CO, NO_2^-.

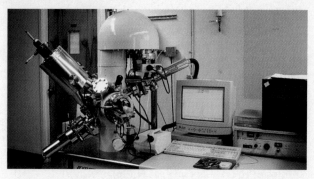

▲ Photoelectron or XPS spectrometer

We described the photoelectric effect in Chapter 9 when considering the particle-like nature of photons. Suppose, instead of shining light on the surface of a metal, we pass photons through a gaseous sample of molecules such as CO or N_2. What is likely to happen? If the energy of the photons is greater than the energy of occupied orbitals in the molecules, electrons will be ejected from these orbitals with a kinetic energy that we can measure. We can make this measurement in a manner similar to that described in Chapter 9.

A typical experimental setup is shown in Figure 12-36. The ejected electrons are passed through an electrostatic analyzer whose field strength can be changed so that electrons are directed to a detector and give a signal. We can use the strength of the field needed to obtain a signal to calculate the speed of the ejected electrons. Then we can calculate the kinetic energy of the ejected electrons, and following that, the energy of the orbital from which the electrons were ejected.

$$E_{orbital} = h\nu - E_{kinetic}$$

Typically, we use photons emitted by excited He^+ ions; they have an energy of 2104 kJ mol^{-1}. For higher energies, we can use photons emitted from excited states of other atoms, or we can use X rays. To understand why we need photons of differing energies, consider the photoelectron spectrum of neon atoms obtained with 11,900 kJ mol^{-1} radiation (from excited Mg), shown in Figure 12-37. The photoelectron spectrum consists of two peaks, one at 2080 kJ mol^{-1} and the other at 4680 kJ mol^{-1}, corresponding to the $2p$ and $2s$ orbitals, respectively. The $1s$ orbital energy is 75,000 kJ mol^{-1}, and, consequently, electrons cannot be ejected from this orbital by the radiation employed. X rays are needed to eject $1s$ electrons. The photoelectron spectrum shown in Figure 12-37 confirms the orbital structure of neon.

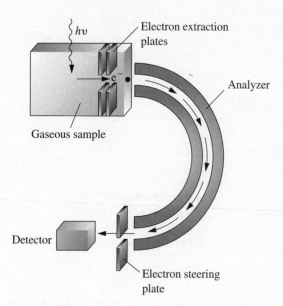

▲ FIGURE 12-36
Schematic of a photoelectron spectrometer

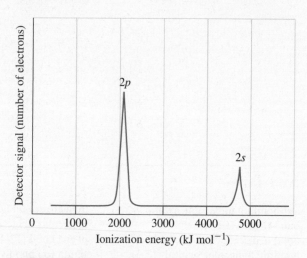

▲ FIGURE 12-37
The photoelectron spectrum of Ne
The valence electron orbital structure of Ne. The inner $1s$ electrons, sometimes called core electrons, exhibit a peak to the right of the $2s$ peak if radiation of sufficient energy is employed. The binding energy of the $1s$ electrons in Ne is at 75,000 kJ mol^{-1}.

In a similar way, we can use photoelectron spectroscopy to provide confirmation of the orbital structure of molecules. The photoelectron spectra of N_2 and CO obtained using 2104 kJ mol^{-1} radiation are shown schematically in Figure 12-38. The first thing that we notice is that there are many more lines than in the neon spectrum. This is because the molecules undergo vibrations. The vibrations of a molecule are quantized, just are its electronic motions. The excitation of quantized molecular vibrations leads to a further loss of energy by the ionizing photons, and this, in turn, leads to a range of kinetic energies corresponding to a given ionization. The spacing between these additional peaks corresponds to the energy needed to excite vibrations. The presence of these vibrational peaks is strongly indicative of whether the orbitals from which the electrons are ejected are σ (strong) bonding orbitals or π (moderate-strength) bonding orbitals or lone pairs of electrons (no bonding) in σ-type orbitals. As an ap-

proximation, the more vibrational peaks observed, the stronger the bonding orbital. In the photoelectron spectrum of N_2, the first ionization produces very few vibrational lines, corresponding to ionization from a lone pair in a σ-type orbital. The second ionization has a much richer structure, suggesting, as expected, that ionization is from a π bonding orbital. The third ionization exhibits some structure and corresponds to ionization from the σ-bonding orbital.

Thus, photoelectron spectroscopy confirms the orbital structure of molecules, but it also points out inadequacies in the simple analysis of photoelectron spectra based on an orbital picture. For example, the similarity of the photoelectron spectra of N_2 and CO suggests that ionization from the σ and π orbitals occurs in the same order in the two molecules. Reference to their orbital diagrams suggests that the reverse order should be observed. However, this is the nature of science. Better theories often are needed to explain all aspects of our experiments.

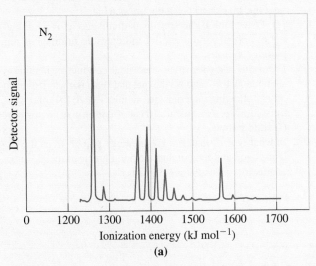

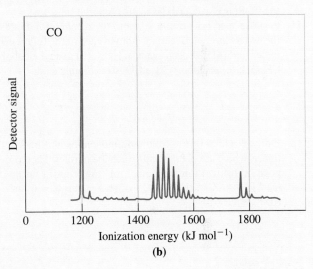

▲ FIGURE 12-38   The photoelectron spectrum of N_2 and CO
The valence electron orbital structure of (a) N_2 and (b) CO. The extra peaks compared to the photoelectron spectrum of Ne correspond to vibrational excitations in the molecules, which are also quantized.

6. Only one of the following statements is correct regarding bonding in carbon–hydrogen–oxygen compounds. Identify the correct statement, and indicate the errors in the other statements. (a) All oxygen-to-hydrogen bonds are π bonds; (b) all carbon-to-hydrogen bonds are σ bonds; (c) all carbon-to-carbon bonds consist of a σ bond and a π bond; (d) all carbon-to-carbon bonds are π bonds.

7. In the manner of Example 12-1, describe the probable structure and bonding in (a) HI; (b) BrCl; (c) H_2Se; (d) OCl_2.

8. Explain why the molecular structure of BF_3 cannot be adequately described through overlaps involving pure s and p orbitals.

9. In the manner of Figure 12-15, indicate the structures of the following molecules in terms of the overlap of simple atomic orbitals and hybrid orbitals: (a) CH_2Cl_2; (b) OCN^-; (c) BF_3.

10. Match each of the following species with one of these hybridization schemes: sp, sp^2, sp^3, sp^3d, sp^3d^2. (a) PF_6^-; (b) COS; (c) $SiCl_4$; (d) NO_3^-; (e) AsF_5.

11. Propose a hybridization scheme to account for bonds formed by the central carbon atom in each of the following molecules: (a) hydrogen cyanide, HCN; (b) methyl alcohol, CH_3OH; (c) acetone, $(CH_3)_2CO$ (d) carbamic acid,

H_2NCOH

12. Indicate which of the following molecules and ions are linear, which are planar, and which are neither. Then propose hybridization schemes for the central atoms. (a) $Cl_2C{=}CCl_2$; (b) $N{\equiv}C{-}C{\equiv}N$; (c) $F_3C{-}C{\equiv}N$; (d) $[S{-}C{\equiv}N]^-$.

13. Write Lewis structures for the following molecules, and then label each σ and π bond. (a) HCN; (b) C_2N_2; (c) $CH_3CHCHCCl_3$; (d) HONO.

14. Represent bonding in the carbon dioxide molecule, CO_2, by (a) a Lewis structure and (b) the valence-bond method, identifying σ and π bonds, the necessary hybridization scheme, and orbital overlap.

15. What is the total number of (a) σ bonds and (b) π bonds in the molecule CH_3NCO?

16. For the following pairs of molecular orbitals, indicate the one you expect to have the lower energy, and state the reason for your choice. (a) σ_{1s} or σ_{1s}^*; (b) σ_{2s} or σ_{2p}; (c) σ_{1s}^* or σ_{2s}; (d) σ_{2p} or σ_{2p}^*.

17. Which of these diatomic molecules do you think has the greater bond energy, Li_2 or C_2? Explain.

18. For each of the species C_2^+, O_2^-, F_2^+, and NO^+
   (a) Write the molecular orbital diagram (as in Example 12-6).
   (b) Determine the bond order, and state whether you expect the species to be stable or unstable.
   (c) Determine if the species is diamagnetic or paramagnetic; and if paramagnetic; indicate the number of unpaired electrons.

19. From this list of terms—electrical conductor, insulator, semiconductor—choose the one that best characterizes each of the following materials: (a) stainless steel; (b) solid sodium chloride; (c) sulfur; (d) germanium; (e) seawater; (f) solid iodine.

20. In what type of material is the energy gap between the valence band and the conduction band greatest: metal, semiconductor, or insulator? Explain.

Exercises

Valence-Bond Method

21. Indicate several ways in which the valence-bond method is superior to Lewis structures in describing covalent bonds.

22. Explain why it is necessary to hybridize atomic orbitals when applying the valence-bond method—that is, why are there so few molecules that can be described through the overlap of pure atomic orbitals only?

23. Describe the molecular geometry of H_2O suggested by each of the following methods: (a) Lewis theory; (b) valence-bond method using simple atomic orbitals; (c) VSEPR theory; (d) valence-bond method using hybridized atomic orbitals.

24. Describe the molecular geometry of CCl_4 suggested by each of the following methods: (a) Lewis theory; (b) valence-bond method using simple atomic orbitals; (c) VSEPR theory; (d) valence-bond method using hybridized atomic orbitals.

25. For each of the following species, identify the central atom(s) and propose a hybridization scheme for the atom(s): (a) CO_2; (b) $HONO_2$; (c) ClO_3^-; (d) BF_4^-.

26. Propose a plausible Lewis structure, geometric structure, and hybridization scheme for the NSF molecule.

27. Describe a hybridization scheme for the central Cl atom in the molecule ClF_3 that is consistent with the geometric shape pictured in Table 11.1. Which orbitals of the Cl atom are involved in overlaps, and which are occupied by lone-pair electrons?

28. Describe a hybridization scheme for the central S atom in the molecule SF_4 that is consistent with the geometric shape pictured in Table 11.1. Which orbitals of the S atom are involved in overlaps, and which are occupied by lone-pair electrons?

29. Use the method of Figure 12-16 to represent bonding in each of the following molecules: (a) CCl_4; (b) ONCl; (c) HONO; (d) $COCl_2$.

30. Use the method of Figure 12-16 to represent bonding in each of the following ions: **(a)** NO_2^-; **(b)** I_3^-; **(c)** $C_2O_4^{2-}$; **(d)** HCO_3^-.

31. The molecular model below represents citric acid, an acidic component of citrus juices. Represent bonding in the citric acid molecule, using the method of Figure 12-16 to indicate hybridization schemes and orbital overlaps.

32. Malic acid is a common organic acid found in unripe apples and other fruit. With the help of the molecular model shown below, represent bonding in the malic acid molecule, using the method of Figure 12-16 to indicate hybridization schemes and orbital overlaps.

33. Shown below are ball-and-stick models. Describe hybridization and orbital-overlap schemes consistent with these structures.

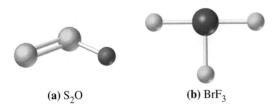

(a) S_2O **(b)** BrF_3

34. Shown below are ball-and-stick models. Describe hybridization and orbital-overlap schemes consistent with these structures.

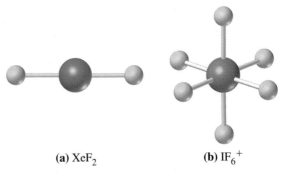

(a) XeF_2 **(b)** IF_6^+

35. Propose a bonding scheme that is consistent with the structure for propynal.
(*Hint:* Consult Table 11.2 to assess the multiple-bond character in some of the bonds.)

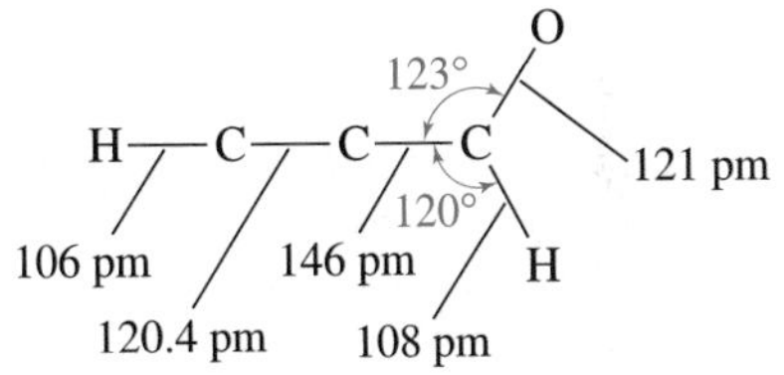

36. The structure of the molecule allene, CH_2CCH_2, is shown here. Propose hybridization schemes for the C atoms in this molecule.

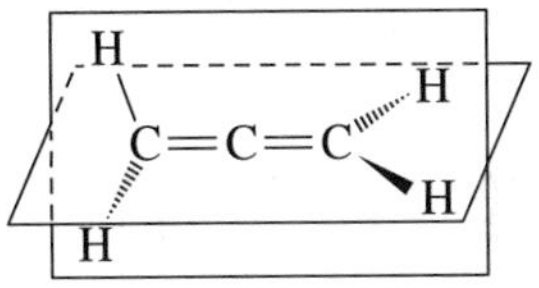

Molecular Orbital Theory

37. Explain the essential difference in the ways that the valence-bond method and molecular orbital theory describe a covalent bond.

38. Describe the bond order of diatomic carbon, C_2, with Lewis theory and molecular orbital theory, and explain why the results are different.

39. $N_2(g)$ has an exceptionally high bond energy. Would you expect either N_2^- or N_2^{2-} to be a stable diatomic species in the gaseous state? Explain.

40. The paramagnetism of gaseous B_2 has been established. Explain how this observation confirms that the π_{2p} orbitals are at a lower energy than the σ_{2p} orbital for B_2.

41. In our discussion of bonding, we have not encountered a bond order higher than triple. Use the energy-level diagrams of Figure 12-24 to show why this is to be expected.

42. Is it correct to say that when a diatomic molecule loses an electron, the bond energy always decreases, that is, that the bond is always weakened? Explain.

43. Assume that the energy-level diagrams of Figure 12-24 are applicable, and write plausible molecular orbital diagrams for the following *heteronuclear* diatomic species: **(a)** NO; **(b)** NO^+; **(c)** CO; **(d)** CN; **(e)** CN^-; **(f)** CN^+; **(g)** BN.

44. We have used the term "isoelectronic" to refer to atoms with identical electron configurations. In molecular orbital theory, this term can be applied to molecular species as well. Which of the species in Exercise 43 are isoelectronic?

45. Assume that the energy-level diagrams of Figure 12-24 apply to the diatomic ions NO^+ and N_2^+.
    (a) Predict the bond order of each.
    (b) Which of these ions is paramagnetic? diamagnetic?
    (c) Which of these ions do you think has the greater bond length? Explain.

46. Assume that the energy-level diagrams of Figure 12-24 apply to the diatomic ions CO^+ and CN^-.
    (a) Predict the bond order of each.
    (b) Which of these ions is paramagnetic? diamagnetic?
    (c) Which of these ions do you think has the greater bond length? Explain.

Delocalized Molecular Orbitals

47. Explain why the concept of delocalized molecular orbitals is essential to an understanding of bonding in the benzene molecule, C_6H_6.

48. Explain how it is possible to avoid the concept of resonance by using molecular orbital theory.

49. In which of the following molecules would you expect to find delocalized molecular orbitals: **(a)** C_2H_4; **(b)** SO_2; **(c)** H_2CO? Explain.

50. In which of the following ions would you expect to find delocalized molecular orbitals: **(a)** HCO_2^-; **(b)** CO_3^{2-}; **(c)** CH_3^+? Explain.

Metallic Bonding

51. Which of the following factors are especially important in determining whether a substance has metallic properties: **(a)** atomic number; **(b)** atomic mass; **(c)** number of valence electrons; **(d)** number of vacant atomic orbitals; **(e)** total number of electronic shells in the atom? Explain.

52. Based on the ground-state electron configurations of the atoms, how would you expect the melting points and hardnesses of sodium, iron, and zinc to compare? Explain.

53. How many energy levels are present in the $3s$ conduction band of a single crystal of sodium weighing 26.8 mg? How many electrons are present in this band?

54. Magnesium is an excellent electrical conductor even though it has a full $3s$ subshell with the electron configuration: $[Ne]3s^2$. Use band theory to explain why magnesium conducts electricity.

Semiconductors

55. Which of the following substances when added in trace amounts to silicon would produce a *p*-type semiconductor: **(a)** sulfur, **(b)** arsenic, **(c)** lead, **(d)** boron, **(e)** gallium arsenide, **(f)** gallium? Explain.

56. Which of the following substances when added in trace amounts to germanium would produce an *n*-type semiconductor: **(a)** sulfur, **(b)** aluminum, **(c)** tin, **(d)** cadmium sulfide, **(e)** arsenic, **(f)** gallium arsenide? Explain.

57. The effect of temperature change on the electrical conductivity of ultrapure silicon is quite different than it is on silicon containing a minute trace of arsenic. Why is this so?

58. Explain why the electrical conductivity of a semiconductor is significantly increased if trace amounts of either donor or acceptor atoms are present, but is unchanged if both are present in equal number.

59. The energy gap, ΔE for silicon is 110 kJ/mol. What is the minimum wavelength light that can promote an electron from the valence band to the conduction band in silicon? In what region of the electromagnetic spectrum is this light?

60. Explain why the solar cell in Figure 12-35 operates over a broad range of wavelengths rather than at a single wavelength (as is so often the case when quantum effects are involved)?

Integrative and Advanced Exercises

61. The Lewis structure of N_2 indicates that the nitrogen-to-nitrogen bond is a triple covalent bond. Other evidence suggests that the σ bond in this molecule involves the overlap of *sp* hybrid orbitals.
    (a) Draw orbital diagrams for the N atoms to describe bonding in N_2.
    (b) Can this bonding be described by either sp^2 or sp^3 hybridization of the N atoms? Can bonding in N_2 be described in terms of unhybridized orbitals? Explain.

62. Show that both the valence-bond method and molecular orbital theory provide an explanation for the existence of the covalent molecule Na_2 in the gaseous state. Would you predict Na_2 by the Lewis theory?

63. A group of spectroscopists believe that they have detected one of the following molecules: NeF, NeF^+, or NeF^-. Assume that the energy-level diagrams of Figure 12-24 apply, and describe bonding in these molecules. Indicate which of these species you would expect the spectroscopists to have observed.

64. Lewis theory is satisfactory to explain bonding in the ionic compound K_2O. It does not readily explain formation of the ionic compounds potassium superoxide, KO_2, and potassium peroxide, K_2O_2, however.

(a) Show that molecular orbital theory can provide this explanation.

(b) Write Lewis structures consistent with this explanation.

65. The compound potassium sesquoxide has the empirical formula K_2O_3. Show that this compound can be described by an appropriate combination of potassium, peroxide, and superoxide ions. Write a Lewis structure for a formula unit of the compound.
(*Hint:* Refer also to Exercise 64.)

66. Draw a Lewis structure for the urea molecule $CO(NH_2)_2$, and predict its geometric shape with the VSEPR theory. Then revise your assessment of this molecule, given the fact that all the atoms lie in the same plane and all the bond angles are 120°. Propose a hybridization and bonding scheme consistent with these experimental observations.

67. Methyl nitrate, CH_3NO_3, is used as a rocket propellant. The skeletal structure of the molecule is CH_3ONO_2. The N and three O atoms all lie in the same plane, but the CH_3 group is not in the same plane as the NO_3 group. The bond angle C—O—N is 105°, and the bond angle O—N—O is 125°. One nitrogen-to-oxygen bond length is 136 pm, and the other two are 126 pm.
(a) Draw a sketch of the molecule, showing its geometric shape.
(b) Label all the bonds in the molecule as σ or π, and indicate the probable orbital overlaps involved.
(c) Explain why all three nitrogen-to-oxygen bond lengths are not the same.

68. Fluorine nitrate, $FONO_2$, is an oxidizing agent used as a rocket propellant. A reference source lists the following data for FO_aNO_2. (The subscript "a" shows that this O atom is different from the other two.)
Bond lengths: N—O = 129 pm; N—O_a = 139 pm; O_a—F = 142 pm
Bond angles: O—N—O = 125°; F—O_a—N = 105°
NO_aF plane is perpendicular to the O_2NO_a plane.
Use these data to construct a Lewis structure(s), a three-dimensional sketch of the molecule, and a plausible bonding scheme showing hybridization and orbital overlaps.

69. Draw a Lewis structure(s) for the nitrite ion, NO_2^-. Then propose a bonding scheme to describe the σ and π bonding in this ion. What conclusion can you reach about the number and types of π molecular orbitals in this ion? Explain.

70. Think of the reaction shown here as involving the transfer of a fluoride ion from ClF_3 to AsF_5 to form the ions ClF_2^+ and AsF_6^-. As a result, the hybridization scheme of each central atom must change. For each reactant molecule and product ion, indicate (a) its geometric structure and (b) the hybridization scheme for its central atom.

$$ClF_3 + AsF_5 \longrightarrow (ClF_2^+)(AsF_6^-)$$

71. In the gaseous state, HNO_3 molecules have two nitrogen-to-oxygen bond distances of 121 pm and one of 140 pm. Draw a plausible Lewis structure(s) to represent this fact, and propose a bonding scheme in the manner of Figure 12-16.

72. He_2 does not exist as a stable molecule, but there is evidence that such a molecule can be formed between elec-

tronically excited He atoms. Write a molecular orbital diagram to account for this.

73. The molecule formamide, $HCONH_2$, has the following approximate bond angles: H—C—O, 123°; H—C—N, 113°; N—C—O, 124°; C—N—H, 119°; H—N—H, 119°. The C—N bond length is 138 pm. Two Lewis structures can be written for this molecule, with the true structure being a resonance hybrid of the two. Propose a hybridization and bonding scheme for each structure.

74. Pyridine, C_5H_5N, is used in the synthesis of vitamins and drugs. The molecule can be thought of in terms of replacing one CH unit in benzene with a N atom. Draw orbital diagrams to show the orbitals of the C and N atoms involved in the σ and π bonding in pyridine. How many bonding and antibonding π-type molecular orbitals are present? How many delocalized electrons are present?

75. One of the characteristics of antibonding molecular orbitals is the presence of a nodal plane. Which of the bonding molecular orbitals considered in this chapter have nodal planes? Explain how a molecular orbital can have a nodal plane and still be a bonding molecular orbital.

76. The ion F_2Cl^- is linear, but the ion F_2Cl^+ is bent. Describe hybridization schemes for the central Cl atom consistent with this difference in structure.

77. Melamine is a carbon–hydrogen–nitrogen compound used in the manufacture of adhesives, protective coatings, and textile finishing (such as in wrinkle-free, wash-and-wear fabrics). Its mass percent composition is 28.57% C, 4.80% H, and 66.64% N. The melamine molecule features a six-member ring with alternating carbon and nitrogen atoms. Half the nitrogen atoms and all the H atoms are outside the ring. For melamine, (a) write a plausible Lewis structure, (b) describe bonding in the molecule by the valence-bond method, and (c) describe bonding in the ring system through molecular orbital theory.

78. Ethyl cyanoacetate, a chemical used in the synthesis of dyes and pharmaceuticals, has the mass percent composition: 53.09% C, 6.24% H, 12.39% N, and 28.29% O. In the manner of Figure 12-16, show a bonding scheme for this substance. The scheme should designate orbital overlaps, σ and π bonds, and expected bond angles.

79. A certain monomer used in the production of polymers has one nitrogen atom and the mass composition: 67.90% C, 5.70% H, and 26.40% N. Sketch the probable geometric structure of this molecule, labeling all the expected bond lengths and bond angles.

80. Dimethylglyoxime (DMG) is a carbon–hydrogen–nitrogen–oxygen compound with a molecular mass of 116.12 u. In a combustion analysis, a 2.464-g sample of DMG yields 3.735 g CO_2 and 1.530 g H_2O. In a separate experiment, the nitrogen in a 1.868-g sample of DMG is converted to $NH_3(g)$, and the NH_3 is neutralized by passing it into 50.00 mL of 0.3600 M $H_2SO_4(aq)$. After neutralization of the NH_3, the excess $H_2SO_4(aq)$ requires 18.63 mL of 0.2050 M NaOH(aq) for its neutralization. Using these data, determine for dimethylglyoxime (a) the most plausible Lewis structure and, (b) in the manner of Figure 12-16, a plausible bonding scheme.

81. A solar cell that is 15% efficient in converting solar to electric energy produces an energy flow of $1.00 \ kW/m^2$ when exposed to full sunlight. If the cell has an area of $40.0 \ cm^2$,
(a) What is the power output of the cell, in watts?
(b) If the power calculated in (a) is produced at 0.45 V, how much current does the cell deliver?

82. Toluene-2,4-diisocyanate is used in the manufacture of polyurethane foam. Its structural formula is shown below. Describe the hybridization scheme for the atoms marked with an asterisk, and indicate the values of the bond angles marked α and β.

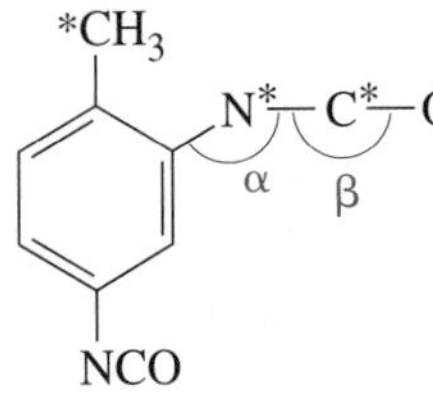

83. The anion $I_4{}^{2-}$ is linear, and the anion $I_5{}^-$ is V-shaped, with a 95° angle between the two arms of the V. For the central atoms in these ions, propose hybridization schemes that are consistent with these observations.

Feature Problems

84. Resonance energy is the difference in energy between a real molecule—a resonance hybrid—and its most important contributing structure. To determine the resonance energy for benzene, we can determine an energy change for benzene and the corresponding change for one of the Kekulé structures. The resonance energy is the difference between these two quantities.
(a) Use data from Appendix D to determine the enthalpy of hydrogenation of liquid benzene to liquid cyclohexane.
(b) Use data from Appendix D to determine the enthalpy of hydrogenation of liquid cyclohexene to liquid cyclohexane.

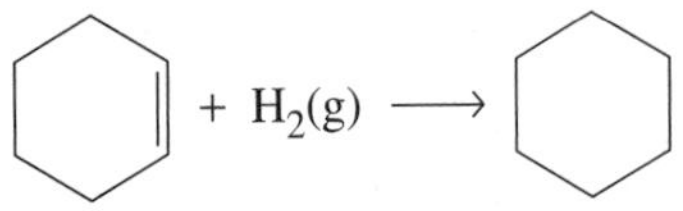

For the enthalpy of formation of liquid cyclohexene, use $\Delta H_f^\circ = -38.5 \ kJ/mol$.
(c) Assume that the enthalpy of hydrogenation of 1, 3, 5-cyclohexatriene is three times as great as that of cyclohexene, and calculate the resonance energy of benzene.
(d) Another way to assess resonance energy is through bond energies. Use bond energies from Table 11.3 (page 422) to determine the total enthalpy change required to break all the bonds in a Kekulé structure of benzene. Next, determine the enthalpy change for the dissociation of $C_6H_6(g)$ into its gaseous atoms by using data from Table 11.3 and Appendix D. Then calculate the resonance energy of benzene.

85. The 60-cycle alternating electric current (AC) commonly used in households changes direction 120 times per second. That is, in a one-second time period a terminal at an electric outlet is positive 60 times and negative 60 times. In direct electric current (DC), the flow between terminals is in one direction only. A *rectifier* is a device that converts alternating to direct current. One type of rectifier is the *p–n* junction rectifier. It is commonly incorporated in adapters required to operate electronic devices from ordinary house current. In the operation of this rectifier, a *p*-type semiconductor and an *n*-type semiconductor are in contact along a boundary or junction. Each semiconductor is connected to one of the terminals in an AC electrical outlet.

Describe how this rectifier works. That is, show that when the semiconductors are connected to the terminals in an AC outlet, half the time a large flow of charge occurs and half the time essentially no charge flows across the *p–n* junction.

86. Furan, C_4H_4O, is a substance derivable from oat hulls, corn cobs, and other cellulosic waste. It is a starting material for the synthesis of other chemicals used as pharmaceuticals and herbicides. The furan molecule is planar and the C and O atoms are bonded into a five-membered pentagonal ring. The H atoms are attached to the C atoms. The chemical behavior of the molecule suggests that it is a resonance hybrid of several contributing structures. These structures show that the double bond character is associated with the entire ring in the form of a π electron cloud.
(a) Draw Lewis structures for the several contributing structures to the resonance hybrid mentioned above.
(b) Draw orbital diagrams to show the orbitals that are involved in the σ and π bonding in furan.
(*Hint:* You need use only one of the contributing structures, such as the one with no formal charges.)
(c) How many π electrons are there in the furan molecule? Show that this number of π electrons is the same, regardless of the contributing structure you use for this assessment.

87. As discussed in the Are You Wondering feature on page 443, the *sp* hybrid orbitals are algebraic combinations of the *s* and *p* orbitals. The required combinations of 2s and 2p orbitals are

$$\psi_1(sp) = \frac{1}{\sqrt{2}} \left[\psi(2s) + \psi(2p_z) \right]$$

$$\psi_2(sp) = \frac{1}{\sqrt{2}} \left[\psi(2s) - \psi(2p_z) \right]$$

(a) By combining appropriate angular functions in Table 9.1, construct a polar plot in the manner of Figure 9-24 for each of the above functions in the *xz* plane. In a polar plot, the value of r/a_0 is set at a fixed value (for example, 1). Describe the shapes and phases of the different portions of the hybrid orbitals, and compare them with those shown in Figure 12-10.

(b) Convince yourself that the combinations employing the $2p_x$ or $2p_y$ orbital also give similar hybrid orbitals but pointing in different directions.

(c) The combinations for the sp^2 hybrids in the *xy* plane are

$$\psi_1(sp^2) = \frac{1}{\sqrt{3}}\psi(2s) + \frac{\sqrt{2}}{\sqrt{3}}\psi(2p_x)$$

$$\psi_2(sp^2) = \frac{1}{\sqrt{3}}\psi(2s) - \frac{1}{\sqrt{6}}\psi(2p_x) + \frac{1}{\sqrt{2}}\psi(2p_y)$$

$$\psi_3(sp^2) = \frac{1}{\sqrt{3}}\psi(2s) - \frac{1}{\sqrt{6}}\psi(2p_x) - \frac{1}{\sqrt{2}}\psi(2p_y)$$

By constructing polar plots (in the *xy* plane), show that these functions correspond to the sp^2 hybrids depicted in Figure 12-8. (Inclusion of radial functions is more complicated, requiring a spread sheet or graphing calculator.)

 eMedia Exercises

88. The **H₂ Bond Formation** animation *(eChapter 12-2)* illustrates the formation of the simplest covalent bond in terms of bond energy and bond length. Describe the same bond formation in the language of the valence-bond method (atomic orbitals, orbital overlap, and electron density).

89. In the **Hybridization** animation *(eChapter 12-3)*, the hybrid orbitals involved in bonding of NH_3 are identified as sp^3 orbitals. **(a)** Identify different molecules that would involve *sp* and sp^2 hybrid orbitals. **(b)** How many total electron pairs are associated with the central atom in each case?

90. **(a)** Construct an energy level diagram for each of the geometries shown in the **Hybridization** animation *(eChapter 12-3)*. In each case, indicate the energy level and occupation of the atomic orbitals on the central atom before

and after hybridization. **(b)** What is the general relationship between the energy of hybridized orbitals and the energy of the atomic orbitals from which they originated?

91. In the **Multiple Bond Formation** activity of acetylene *(eChapter 12-4)* we see that bonding occurs between both hybridized and unhybridized orbitals. Using acetylene as a model, predict the orbital that is occupied by the lone pairs of the diatomic molecule, nitrogen.

92. For each of the atomic and molecular orbitals represented in the **Molecular Orbitals** activity *(eChapter 12-5)*, **(a)** construct electron charge density diagrams (a plot of density versus internuclear distance). **(b)** What is the primary difference between the diagrams describing the σ_{1s} and σ_{1s}^* molecular orbitals?

13 Liquids, Solids, and Intermolecular Forces

Contents

In this scene from Antarctica, water exists in all three states of matter—solid in the iceberg, liquid in the sea, and gas in the atmosphere.

▶ Two of the many natural phenomena described in this chapter include the more ordered structure of the solid compared to the liquid state and the variation of density with the state of matter.

478

I n our study of gases, we intentionally sought conditions in which the forces between molecules—intermolecular forces—were negligible. This approach allowed us to describe gases with the ideal gas equation and to explain their behavior with the kinetic–molecular theory of gases.

To describe the other states of matter—liquids and solids—we must seek situations in which the intermolecular forces are significant. We then consider some interesting properties of liquids and solids related to the strengths of these forces. Learning about these properties is a worthwhile goal in itself, for they have some important applications. By studying intermolecular forces, however, we also can understand *why* some liquids and solids have the properties they do, and we can even *predict* some properties. We will not formulate any general equations of state for liquids and solids; our treatment will be less mathematical and more qualitative than was the case for gases.

13-1 Intermolecular Forces and Some Properties of Liquids

▶ Solids, liquids, and gases were compared at the macroscopic and microscopic levels in Figure 1-5.

In our study of gases, we noted that at high pressures and low temperatures intermolecular forces cause gas behavior to depart from ideality. When these forces are sufficiently strong, a gas condenses to a liquid. That is, the intermolecular forces keep the molecules in close enough proximity that the molecules are confined to a definite volume, as expected for the liquid state.

Intermolecular forces are important in establishing the surface tension and viscosity of a liquid, two properties described in this section. Other properties of liquids closely associated with intermolecular forces will be described in Section 13-2.

Surface Tension

The observation of a needle floating on water, as pictured in Figure 13-1, is puzzling. Steel is much denser than water and should not float. Something must overcome the force of gravity on the needle, allowing it to remain suspended on the surface of the water. The floating needle is possible because of a special quality associated with the surface of a liquid.

Figure 13-2 suggests an important difference in the forces between molecules within the bulk of a liquid and at the surface: Interior molecules have more neighbors to which they are attracted by intermolecular forces than do surface molecules. This increased attraction by its neighboring molecules places an interior molecule in a lower energy state than a surface molecule. As a consequence, as many molecules as possible try to get into the bulk of a liquid, and as few as possible remain at the surface. Thus, there is a tendency for liquids to maintain a minimum surface area. This is most clearly seen in the formation of spherical drops in a rain shower. A sphere has a smaller ratio of surface area to volume than does any other geometric structure. To increase the surface area of a liquid requires that molecules be moved from the interior to the surface of a liquid, and this requires that work be done. The steel needle of Figure 13-1 remains suspended on the surface

▲ FIGURE 13-1
An effect of surface tension illustrated
Despite being denser than water, the needle is supported on the surface of the water. The property of surface tension accounts for this unexpected behavior.

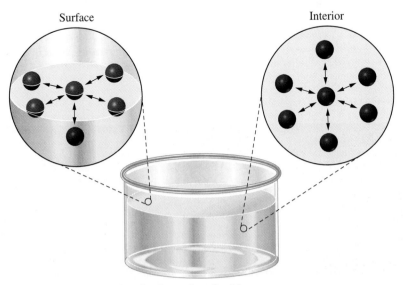

▲ FIGURE 13-2   **Intermolecular forces in a liquid**
Molecules at the surface are attracted only by other surface molecules and by molecules below the surface. Molecules in the interior experience forces from neighboring molecules in all directions.

of the water because the work that would be done by the force of gravity acting on the needle is not as great as the energy required to spread the surface of the water over the top of the needle.

Surface tension is the energy, or work, required to increase the surface area of a liquid. Surface tension is often represented by the Greek letter gamma (γ) and has the units of energy per unit area, typically joules per square meter (J/m^2). As the temperature—and hence the intensity of molecular motion—increases, intermolecular forces become less effective. Less work is required to extend the surface of a liquid, meaning that surface tension *decreases* with *increased* temperature.

When a drop of liquid spreads into a film across a surface, we say that the liquid *wets* the surface. Whether a drop of liquid wets a surface or retains its spherical shape and stands on the surface depends on the strengths of two types of intermolecular forces. **Cohesive forces** are the intermolecular forces between *like* molecules, and **adhesive forces** are the forces between *unlike* molecules. If cohesive forces are strong compared with adhesive forces, a drop maintains its shape. If adhesive forces are strong enough, on the other hand, the energy requirement for spreading the drop into a film is met through the work done by the collapsing drop.

Water wets many surfaces, such as glass and certain fabrics. This characteristic is essential to its use as a cleaning agent. If glass is coated with a film of oil or grease, water no longer wets the surface and water droplets stand on the glass, as shown in Figure 13-3. When we clean glassware in the laboratory, we have done a good job if water forms a uniform thin film on the glass. When we wax a car, we have done a good job if water uniformly beads up all along the surface.

Adding a detergent to water has two effects: The detergent solution dissolves grease to expose a clean surface, and the detergent lowers the surface tension of water. Lowering the surface tension means lowering the energy required to spread drops into a film. Substances that reduce the surface tension of water and allow it to spread more easily are known as *wetting agents*. They are used in applications ranging from dish washing to industrial processes.

Figure 13-4 illustrates another familiar observation. If the liquid in the glass tube is water, the water is drawn slightly up the walls of the tube by adhesive forces between water and glass. The interface between the water and the air above it, called a *meniscus*, is concave, or curved in. With liquid mercury, the meniscus is convex, or curved out. Cohesive forces in mercury, consisting of metallic bonds between Hg atoms, are strong; mercury does not wet glass. The effect of meniscus

▶ The surface tension of water at 20 °C, for example, is $7.28 \times 10^{-2} J/m^2$, and that of mercury is $47.2 \times 10^{-2} J/m^2$.

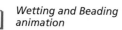
Wetting and Beading animation

▲ FIGURE 13-3   **Wetting of a surface**
Water spreads into a thin film on a clean glass surface (left). If the glass is coated with oil or grease, the adhesive forces between the water and oil are not strong enough to spread the water. Drops of water stand on the surface (right).

▲ FIGURE 13-4
Meniscus formation
Water wets glass (left). The meniscus is concave—the bottom of the meniscus is below the level of the water–glass contact line. Mercury does not wet glass. The meniscus is convex—the top of the meniscus is above the mercury–glass contact line.

▶ **FIGURE 13-5**
Capillary action
A thin film of water spreads up the inside walls of the capillary because of strong adhesive forces between water and glass (water wets glass). The pressure below the meniscus falls slightly. Atmospheric pressure then pushes a column of water up the tube to eliminate the pressure difference. The *smaller* the diameter of the capillary, the *higher* the liquid rises. Because its magnitude is also directly proportional to surface tension, capillary rise provides a simple experimental method of determining surface tension, described in Exercise 120.

formation is greatly magnified in tubes of small diameter, called *capillary* tubes. In the *capillary action* shown in Figure 13-5, the water level inside the capillary tubes is noticeably higher than outside. The soaking action of a sponge depends on the rise of water into capillaries of a fibrous material such as cellulose. The penetration of water into soils also depends in part on capillary action. On the other hand, mercury, with its strong cohesive forces and weaker adhesive forces, does not show a capillary rise. Rather, mercury in a glass capillary tube will have a lower level than the mercury outside the capillary.

Viscosity

▲ **FIGURE 13-6**
Measuring viscosity
By measuring the velocity of a ball dropping through a liquid a measure of the liquid viscosity can be obtained.

Another property at least partly related to intermolecular forces is **viscosity**—a liquid's resistance to flow. The stronger the intermolecular forces of attraction, the greater the viscosity. When a liquid flows, one portion of the liquid moves with respect to neighboring portions. Cohesive forces within the liquid create an "internal friction," which reduces the rate of flow. The effect is weak in liquids of low viscosity, such as ethyl alcohol and water. They flow easily. Liquids like honey and heavy motor oil flow much more sluggishly. We say that they are *viscous*. One method of measuring viscosity is to time the fall of a steel ball through a certain depth of liquid (Figure 13-6). The greater the viscosity of the liquid, the longer it takes for the ball to fall. Because intermolecular forces of attraction can be offset by higher molecular kinetic energies, viscosity generally *decreases* with increased temperature.

13-2 Vaporization of Liquids: Vapor Pressure

In our study of the kinetic–molecular theory (Section 6-7), we saw that the speeds and kinetic energies of molecules vary over a wide range at any given temperature (Figure 6-17). Then in Chapter 7, we learned that molecules having kinetic energies sufficiently above the average value are able to overcome intermolecular forces of attraction and escape from the surface of the liquid into the gaseous state. This passage of molecules from the surface of a liquid into the gaseous, or vapor, state is called **vaporization**. The term **evaporation** is also used. As we might expect, vaporization occurs more readily with

▲ Honey is a viscous liquid.

* *increased* temperature (More molecules have sufficient kinetic energy to overcome intermolecular forces of attraction in the liquid.)
* *increased* surface area of the liquid (A greater proportion of the liquid molecules are at the surface.)

• *decreased* strength of intermolecular forces (The kinetic energy needed to overcome intermolecular forces of attraction is less, and more molecules have enough energy to escape.)

Enthalpy of Vaporization

Because the molecules lost through evaporation are much more energetic than average, the average kinetic energy of the remaining molecules decreases. The temperature of the evaporating liquid falls. This accounts for the cooling sensation you feel when a volatile liquid such as ethyl alcohol evaporates on your skin.

Suppose we allow a liquid to vaporize but wish to keep its temperature constant. How can we do this? We must replace the excess kinetic energy carried away by the vaporizing molecules by adding heat to the liquid. The *enthalpy of vaporization* is the quantity of heat that must be absorbed if a certain quantity of liquid is vaporized at a constant temperature. Stated in another way,

$$\Delta H_{vaporization} = H_{vapor} - H_{liquid}$$

Because vaporization is an *endothermic* process, $\Delta H_{vaporization}$, or ΔH_{vap}, as it is more commonly denoted, is always positive. We will generally express enthalpies of vaporization in terms of one mole of liquid vaporized, as seen in Table 13.1.

KEEP IN MIND ▶
that absolute enthalpies such as H_{vapor} and H_{liquid} cannot be measured. However, because enthalpy is a function of state, the difference between the absolute enthalpies has a unique value, and it *can* be measured.

TABLE 13.1 Some Enthalpies of Vaporization at 298 K[a]	
Liquid	ΔH_{vap}, **kJ/mol**
Diethyl ether, $(C_2H_5)_2O$	29.1
Methyl alcohol, CH_3OH	38.0
Ethyl alcohol, CH_3CH_2OH	42.6
Water, H_2O	44.0

[a] ΔH_{vap} values are somewhat temperature-dependent (see Exercise 96).

The conversion of a gas or vapor to a liquid is called **condensation**. From a thermochemical standpoint, condensation is the reverse of vaporization.

$$\Delta H_{condensation} = H_{liquid} - H_{vapor} = -\Delta H_{vap}$$

Because it is opposite in sign but equal in magnitude to ΔH_{vap}, $\Delta H_{condensation}$ is always negative. Condensation is an *exothermic* process. This explains why burns produced by a given mass of steam (vaporized water) are very much more severe than burns produced by the same mass of hot water. Hot water burns only by releasing heat as it cools. Steam releases a large quantity of heat when it condenses to liquid water, followed by the further release of heat as the hot water cools.

Vapor Pressure

We know that water left in an open beaker will completely evaporate. A different condition results if the beaker is placed in a closed container. As shown in Figure 13-7, in a container with both liquid and vapor present, vaporization and condensation occur simultaneously. If sufficient liquid is present, eventually a condition is reached in which the amount of vapor remains constant. This condition is one of *dynamic equilibrium*. Dynamic equilibrium always implies that two opposing processes are occurring simultaneously and at equal rates. As a result, there is no net change with time once equilibrium has been established. A symbolic representation of the liquid–vapor equilibrium is shown below.

$$\text{liquid} \xrightleftharpoons[\text{condensation}]{\text{vaporization}} \text{vapor}$$

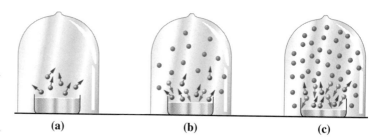

- Molecules in vapor state
- Molecules undergoing vaporization
- Molecules undergoing condensation

▶ **FIGURE 13-7**

Establishing liquid–vapor equilibrium

(a) A liquid is allowed to evaporate into a closed container. Initially, only vaporization occurs. **(b)** Condensation begins. The rate at which molecules evaporate is greater than the rate at which they condense, and the number of molecules in the vapor state continues to increase. **(c)** The rate of condensation is equal to the rate of vaporization. The number of vapor molecules remains constant over time, as does the pressure exerted by this vapor.

▶ Gasoline is a mixture of volatile hydrocarbons and is an important precursor of smog (Section 8-2), whether it is vaporized from oil refineries, filling-station operations, automobile gas tanks, or power lawn mowers.

The pressure exerted by a vapor in dynamic equilibrium with its liquid is called the **vapor pressure**. Liquids with high vapor pressures at room temperature are said to be *volatile*, and those with very low vapor pressures are *nonvolatile*. Whether a liquid is volatile or not is determined primarily by the strengths of intermolecular forces—the weaker these forces, the more volatile the liquid (the higher its vapor pressure). Diethyl ether and acetone are volatile liquids; at 25 °C their vapor pressures are 534 and 231 mmHg, respectively. Water at ordinary temperatures is a moderately volatile liquid; at 25 °C, its vapor pressure is 23.8 mmHg. Mercury is essentially a nonvolatile liquid; at 25 °C, its vapor pressure is 0.0018 mmHg.

As an excellent first approximation, the vapor pressure of a liquid depends only on the particular liquid and its temperature. Vapor pressure depends neither on the amount of liquid or vapor, as long as some of each is present at equilibrium. These statements are illustrated in Figure 13-8. A graph of vapor pressure as a function of temperature is known as a **vapor pressure curve**. Vapor pressure curves always have the appearance of those in Figure 13-9: *vapor pressure increases with temperature*. Vapor pressures of water at different temperatures are presented in Table 13.2.

When a pan of water is put on the stove to boil, small bubbles are usually observed as the water begins to warm. These are bubbles of dissolved air being expelled. Once the water boils, however, all the dissolved air is expelled and the bubbles consist only of water vapor. ▶

Boiling and the Boiling Point

When a liquid is heated in a container *open to the atmosphere*, there is a particular temperature at which vaporization occurs throughout the liquid rather than simply at the surface. Vapor bubbles form within the bulk of the liquid, rise to the surface, and escape. The pressure exerted by escaping molecules equals that exerted by molecules of the atmosphere, and **boiling** is said to occur. During boiling, energy absorbed as heat is used only to convert molecules of liquid to vapor.

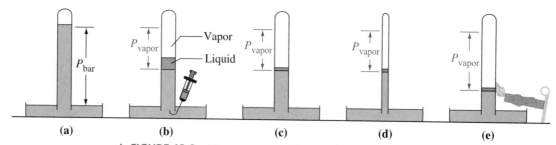

▲ **FIGURE 13-8** **Vapor pressure illustrated**

(a) A mercury barometer. **(b)** The pressure exerted by the vapor in equilibrium with a liquid injected to the top of the mercury column depresses the mercury level. **(c)** Compared to (b), the vapor pressure is independent of the volume of liquid injected. **(d)** Compared to (c), the vapor pressure is independent of the volume of vapor present. **(e)** Vapor pressure increases with an increase in temperature.

Equilibrium Vapor Pressure simulation

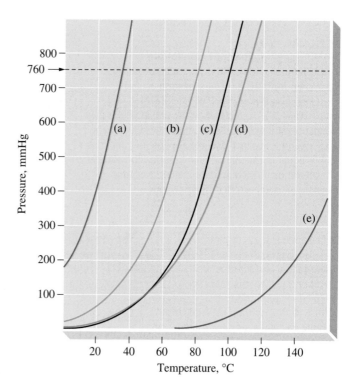

▶ **FIGURE 13-9**

Vapor pressure curves of several liquids
(a) Diethyl ether, $C_4H_{10}O$; **(b)** benzene, C_6H_6; **(c)** water, H_2O; **(d)** toluene, C_7H_8; **(e)** aniline, C_6H_7N. The normal boiling points are the temperatures at the intersection of the dashed line at $P = 760$ mmHg with the vapor pressure curves.

The temperature remains constant until all the liquid has boiled away, as is dramatically illustrated in Figure 13-10. The temperature at which the vapor pressure of a liquid is equal to standard atmospheric pressure (1 atm = 760 mmHg) is the **normal boiling point**. In other words, the normal boiling point is the boiling point of a liquid at 1 atm pressure. The normal boiling points of several liquids can be determined from the intersection of the dashed line in Figure 13-9 with the vapor pressure curves for the liquids.

Figure 13-9 also helps us see that the boiling point of a liquid varies significantly with barometric pressure. Shift the dashed line shown at $P = 760$ mmHg to higher or lower pressures, and the new points of intersection with the vapor pressure curves come at different temperatures. Barometric pressures below 1 atm are commonly encountered at high altitudes. At an altitude of 1609 m (that of Denver,

TABLE 13.2	Vapor Pressure of Water at Various Temperatures				
Temperature, °C	Pressure, mmHg	Temperature, °C	Pressure, mmHg	Temperature, °C	Pressure, mmHg
0.0	4.6	29.0	30.0	93.0	588.6
10.0	9.2	30.0	31.8	94.0	610.9
20.0	17.5	40.0	55.3	95.0	633.9
21.0	18.7	50.0	92.5	96.0	657.6
22.0	19.8	60.0	149.4	97.0	682.1
23.0	21.1	70.0	233.7	98.0	707.3
24.0	22.4	80.0	355.1	99.0	733.2
25.0	23.8	90.0	525.8	100.0	760.0
26.0	25.2	91.0	546.0	110.0	1074.6
27.0	26.7	92.0	567.0	120.0	1489.1
28.0	28.3				

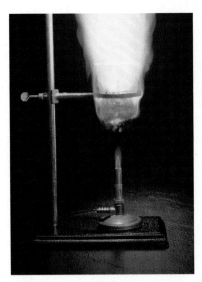

▲ **FIGURE 13-10** **Boiling water in a paper cup**
An empty paper cup heated over a Bunsen burner quickly bursts into flame. If a paper cup is filled with water, it can be heated for an extended time as the water boils. This is possible for three reasons: (1) Because of the high heat capacity of water, heat from the burner goes primarily into heating the water, not the cup. (2) As the water boils, large quantities of heat (ΔH_{vap}) are required to convert the liquid to its vapor. (3) The temperature of the cup does not rise above the boiling point of water as long as liquid water remains. The boiling point of 99.9 instead of 100.0 °C suggests that the prevailing barometric pressure was slightly below 1 atm.

▶ A more extreme case is that on the summit of Mt. Everest, where a climber would barely be able to heat a cup of tea to 70 °C.

▲ **A liquid boils at low pressure**
Water boils when its vapor pressure equals the pressure on its surface. Bubbles form throughout the liquid.

Colorado), barometric pressure is about 630 mmHg. The boiling point of water at this pressure is 95 °C (203 °F). It takes longer to cook foods under conditions of lower boiling-point temperatures. A 3-minute boiled egg takes longer than 3 minutes to cook. We can counteract the effect of high altitudes by using a pressure cooker. In a pressure cooker, the cooking water is maintained under higher-than-atmospheric pressure and its boiling temperature increases, for example, to about 120 °C at 2 atm pressure.

The Critical Point

In describing boiling, we made an important qualification: Boiling occurs "in a container open to the atmosphere." If a liquid is heated in a *sealed* container, boiling does not occur. Instead, the temperature and vapor pressure rise continuously. Pressures many times atmospheric pressure may be attained. If we seal just the right quantity of liquid in a glass tube and heat it, as in Figure 13-11, we observe these phenomena.

- The density of the liquid decreases, that of the vapor increases, and eventually the two densities become equal.
- The surface tension of the liquid approaches zero. The interface between the liquid and vapor becomes less distinct and eventually disappears.

The **critical point** is the point at which these conditions are reached and the liquid and vapor become indistinguishable. The temperature at the critical point is the critical temperature, T_c, and the pressure is the critical pressure, P_c. The critical point is the highest point on a vapor pressure curve and represents the highest temperature at which the liquid can exist. Several critical temperatures and pressures are listed in Table 13.3.

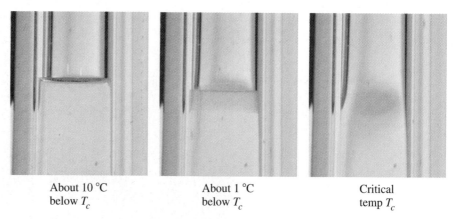

| About 10 °C below T_c | About 1 °C below T_c | Critical temp T_c |

▲ FIGURE 13-11 **Attainment of the critical point**
The meniscus separating a liquid from its vapor disappears at the critical point. The liquid and vapor become indistinguishable.

TABLE 13.3 Some Critical Temperatures, T_c, and Critical Pressures, P_c		
Substance	T_c, **K**	P_c, **atm**
"Permanent" gases[a]		
H_2	33.3	12.8
N_2	126.2	33.5
O_2	154.8	50.1
CH_4	191.1	45.8
"Nonpermanent" gases[b]		
CO_2	304.2	72.9
HCl	324.6	82.1
NH_3	405.7	112.5
SO_2	431.0	77.7
H_2O	647.3	218.3

[a] "Permanent" gases cannot be liquefied at 25 °C (298 K).
[b] "Nonpermanent" gases can be liquefied at 25 °C.

▶ Although the term "gas" can be used exclusively, sometimes the term "vapor" is used for the gaseous state at temperatures *below* T_c and gas at temperatures *above* T_c.

A gas can be liquefied only at temperatures *below* its critical temperature, T_c. If room temperature is *below* T_c, this liquefaction can be accomplished just by applying sufficient pressure. If room temperature is *above* T_c, however, added pressure *and* a lowering of temperature to a value below T_c are required. We comment further on the liquefaction of gases on page 494.

Measuring Vapor Pressure Data

Figure 13-8 suggests one method of determining vapor pressure—inject a small sample of the target liquid at the top of a mercury barometer, and measure the depression of the mercury level. The method does not give very precise results, however, and it is not useful for measuring vapor pressures that are either very low or quite high. Better results are obtained in methods in which the pressure above a liquid is continuously varied and measured, and the liquid–vapor equilibrium temperature is recorded. In short, the boiling point of the liquid changes in accordance with the change in the pressure above the liquid, and we thereby trace out the vapor pressure curve of the liquid. The pressure measurements are made with either a closed-end or open-end manometer (page 179). A method that is useful for determining very low vapor pressures is based on the rate of effusion of a gas through a tiny orifice. In this method, equations from the kinetic–molecular theory (Section 6-7) are applied. Example 13-1 illustrates a method (called the transpiration method) in which an inert gas is saturated with the vapor under study. Then the ideal gas equation is used to calculate the vapor pressure.

EXAMPLE 13-1

Using the Ideal Gas Equation to Calculate a Vapor Pressure. A sample of 113 L of helium gas at 1360 °C and prevailing barometric pressure is passed through molten silver at the same temperature. The gas becomes saturated with silver vapor, and the liquid silver loses 0.120 g in mass. What is the vapor pressure of liquid silver at 1360 °C?

Solution

Let's assume that after the gas has become saturated with silver vapor, its volume remains at 113 L. This assumption will be valid if the vapor pressure of the silver is quite low compared to the barometric pressure. According to Dalton's law of partial pressures (page 196), we can deal with the silver vapor as if it were a single gas occupying a volume of 113 L. The data required in the ideal gas equation are listed below.

$$P = ? \qquad\qquad V = 113 \text{ L}$$
$$R = 0.08206 \text{ L atm mol}^{-1} \text{ K}^{-1} \quad T = 1360 + 273.15 = 1633 \text{ K}$$
$$n = 0.120 \text{ g Ag} \times \frac{1 \text{ mol Ag}}{107.9 \text{ g Ag}} = 0.00111 \text{ mol Ag}$$
$$P = \frac{nRT}{V}$$
$$P = \frac{0.00111 \text{ mol} \times 0.08206 \text{ L atm mol}^{-1} \text{ K}^{-1} \times 1633 \text{ K}}{113 \text{ L}}$$
$$= 1.32 \times 10^{-3} \text{ atm (1.00 Torr)}$$

Practice Example A: Equilibrium is established between liquid hexane, C_6H_{14}, and its vapor at 25.0 °C. A sample of the vapor is found to have a density of 0.701 g/L. Calculate the vapor pressure of hexane at 25.0 °C, expressed in Torr.

Practice Example B: With the help of Figure 13-9, estimate the density of the vapor in equilibrium with liquid diethyl ether at 20.0 °C.

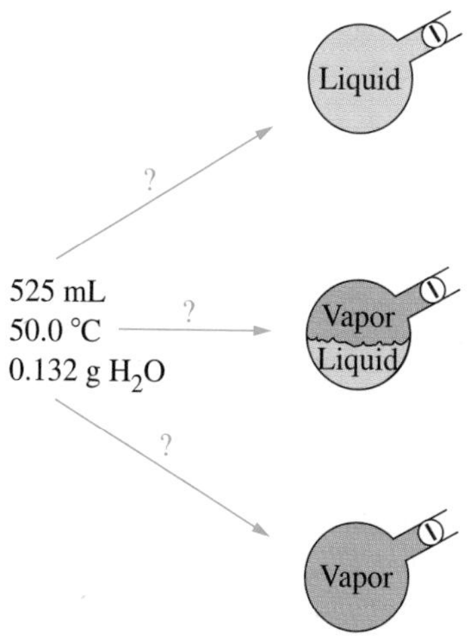

Using Vapor Pressure Data

One use of vapor pressure data is in calculations dealing with the collection of gases over liquids, particularly water (Section 6-6). Another use, illustrated in Example 13-2, is in predicting whether a substance exists solely as a gas (vapor) or as a liquid and vapor in equilibrium.

▲ FIGURE 13-12
**Predicting states of matter—
Example 13-2 illustrated**
For the conditions given on the left, which of the final conditions pictured on the right will result?

525 mL
50.0 °C
0.132 g H_2O

EXAMPLE 13-2

Making Predictions with Vapor Pressure Data. As a result of a chemical reaction, 0.132 g H_2O is produced and maintained at a temperature of 50.0 °C in a closed flask of 525-mL volume. Will the water be present as liquid only, as vapor only, or as liquid and vapor in equilibrium (Figure 13-12)?

Solution

Let's consider each of the three possibilities in the order that they are given.

Liquid Only

With a density of about 1 g/mL, a 0.132-g sample of H_2O has a volume of only about 0.13 mL. There is no way that the sample could completely fill a 525-mL flask. The condition of liquid only is *impossible*.

Vapor Only

The portion of the flask that is not occupied by liquid water must be filled with something (it cannot remain a vacuum). That something is water vapor. The question is, Will the sample vaporize completely, leaving no liquid? Let's use the ideal gas equation to calculate the pressure that would be exerted if the entire 0.132 g H_2O were present in the gaseous state.

$$P = \frac{nRT}{V}$$

$$= \frac{0.132 \text{ g } H_2O \times \dfrac{1 \text{ mol } H_2O}{18.02 \text{ g } H_2O} \times 0.08206 \text{ L atm mol}^{-1} \text{ K}^{-1} \times 323.2 \text{ K}}{0.525 \text{ L}}$$

$$= 0.370 \text{ atm} \times \frac{760 \text{ mmHg}}{1 \text{ atm}} = 281 \text{ mmHg}$$

Now compare this calculated pressure with the vapor pressure of water at 50 °C (Table 13.2). The calculated pressure—281 mmHg—greatly exceeds the vapor pressure—92.5 mmHg. Water formed in the reaction as $H_2O(g)$ condenses to $H_2O(l)$ when the gas pressure reaches 92.5 mmHg, for this is the pressure at which the liquid and vapor are in equilibrium at 50.0 °C. The condition of vapor only is *impossible*.

Liquid and Vapor

This is the only possibility for the final condition in the flask. Liquid water and water vapor coexist in equilibrium at 50.0 °C and 92.5 mmHg.

Practice Example A: If the reaction described in this example resulted in H_2O produced and maintained at 80.0 °C, would the water be present as vapor only or as liquid and vapor in equilibrium? Explain.

Practice Example B: For the situation described in Example 13-2, what mass of water is present as liquid and what mass as vapor?

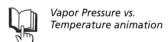

Vapor Pressure vs. Temperature animation

An Equation for Expressing Vapor Pressure Data

If you seek vapor pressure data on a liquid in a handbook or in data tables, you are unlikely to find graphs like Figure 13-9. Also, with the exception of a few liquids, such as water and mercury, you are unlikely to find data tables like Table 13.2. What you will find, instead, are mathematical equations relating vapor pressures and temperatures. Such an equation can summarize in one line data that might otherwise take a full page. A particularly common form of vapor pressure equation is the one shown here, which expresses the natural logarithm (ln) of vapor pressure as a function of the reciprocal of the Kelvin temperature $(1/T)$. The relationship is that of a straight line, and the straight-line plots for the liquids featured in Figure 13-9 are drawn in Figure 13-13.

$$\underbrace{\ln P}\; = \;\underbrace{-A}\;\underbrace{\left(\frac{1}{T}\right)}\;\underbrace{+B} \tag{13.1}$$

Equation of straight line: $y \;\;=\;\; m \;\times\; x \;+\; b$

To use equation (13.1), we need to have values for the two constants, A and B. The constant A is related to the enthalpy of vaporization of the liquid: $A = \Delta H_{vap}/R$,

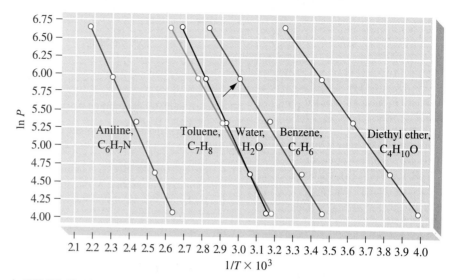

▲ **FIGURE 13-13**   **Vapor pressure data plotted as ln *P* versus 1/*T***

Pressures are in millimeters of mercury, and temperatures are in kelvins. Data from Figure 13-9 have been recalculated and replotted as in the following example: For benzene at 60 °C, the vapor pressure is 400 mmHg; $\ln P = \ln 400 = 5.99$. $T = 60 + 273 = 333\ K$; $1/T = 1/333 = 0.00300 = 3.00 \times 10^{-3}$; $1/T \times 10^3 = 3.00 \times 10^{-3} \times 10^3 = 3.00$. The point corresponding to $(3.00, 5.99)$ is marked by the arrow ($\longrightarrow$).

▶ Refer to Appendix A-4 to see how the constant *B* is eliminated to give equation (13.2).

KEEP IN MIND ▶

that the heat of vaporization in this equation cannot be ΔH_{vap}°, since in general the pressure is not 1 bar.

where ΔH_{vap} is expressed in the unit J/mol and the value used for *R* is 8.3145 J $mol^{-1}\ K^{-1}$. It is customary to eliminate *B* by rewriting (13.1) in a form called the Clausius–Clapeyron equation. We apply equation (13.2) in Example 13-3.

$$\ln \frac{P_2}{P_1} = \frac{\Delta H_{vap}}{R}\left(\frac{1}{T_1} - \frac{1}{T_2}\right) \tag{13.2}$$

EXAMPLE 13-3

Applying the Clausius–Clapeyron Equation. Calculate the vapor pressure of water at 35.0 °C with data from Tables 13.1 and 13.2.

Solution

Let's designate the unknown vapor pressure as P_1 at the temperature T_1. That is,

$$P_1 = ? \qquad T_1 = (35.0 + 273.15)\ K = 308.2\ K$$

Then, let's choose for P_2 and T_2 known data for a temperature close to 35.0 °C, for example, 40.0 °C.

$$P_2 = 55.3\ mmHg \qquad T_2 = (40.0 + 273.15)\ K = 313.2\ K$$

For ΔH_{vap}, let's assume that the value given in Table 13.1 applies throughout the temperature range from 30.0 °C to 40.0 °C.

$$\Delta H_{vap} = 44.0\ kJ/mol \times \frac{1000\ J}{1\ kJ} = 44.0 \times 10^3\ J/mol$$

Now we can substitute these values into equation (13.2) to obtain

$$\ln \frac{55.3\ mmHg}{P_1} = \frac{44.0 \times 10^3\ J\ mol^{-1}}{8.3145\ J\ mol^{-1}\ K^{-1}}\left(\frac{1}{308.2} - \frac{1}{313.2}\right)K^{-1}$$

$$= 5.29 \times 10^3\,(0.003245 - 0.003193) = 0.28$$

Next, determine that $e^{0.28} = 1.32$ (see Appendix A). Thus,

$$\frac{55.3 \text{ mmHg}}{P_1} = e^{0.28} = 1.32$$

$$P_1 = 55.3 \text{ mmHg}/1.32 = 41.9 \text{ mmHg}$$

(The experimentally determined vapor pressure of water at 35.0 °C is 42.175 mmHg.)

Practice Example A: A handbook lists the vapor pressure of methyl alcohol as 100 mmHg at 21.2 °C. What is its vapor pressure at 25.0 °C?

Practice Example B: A handbook lists the normal boiling point of isooctane, a gasoline component, as 99.2 °C and its enthalpy of vaporization (ΔH_{vap}) as 35.76 kJ/mol C_8H_{18}. Calculate the vapor pressure of isooctane at 25 °C.

13-3 Some Properties of Solids

We mentioned some properties of solids (for example, malleability, ductility) at the beginning of this text, and we will continue to consider additional properties. For now, we will comment on some properties that allow us to think of solids in relation to the other states of matter—liquids and gases.

Melting, Melting Point, and Heat of Fusion

As a crystalline solid is heated, its atoms, ions, or molecules vibrate more vigorously. Eventually a temperature is reached at which these vibrations disrupt the ordered crystalline structure. The atoms, ions, or molecules can slip past one another, and the solid loses its definite shape and is converted to a liquid. This process is called **melting**, or fusion, and the temperature at which it occurs is the **melting point**. The reverse process, the conversion of a liquid to a solid, is called **freezing**, or solidification, and the temperature at which it occurs is the **freezing point**. The melting point of a solid and the freezing point of its liquid are identical. At this temperature, solid and liquid coexist in equilibrium.

If we add heat uniformly to a solid–liquid mixture at equilibrium, the temperature remains constant while the solid melts. Only when all the solid has melted does the temperature begin to rise. Conversely, if we remove heat uniformly from a solid–liquid mixture at equilibrium, the liquid freezes at a constant temperature. The quantity of heat required to melt a solid is the *enthalpy of fusion*, ΔH_{fus}. Some typical enthalpies of fusion, expressed in kilojoules per mole, are listed in Table 13.4. Perhaps the most familiar example of a melting (and freezing) point is that of water, 0 °C. This is the temperature at which liquid and solid water, in contact with air and under standard atmospheric pressure, are in equilibrium. The enthalpy of fusion of water is 6.01 kJ/mol, which we can express as

$$H_2O(s) \longrightarrow H_2O(l) \qquad \Delta H_{fus} = +6.01 \text{ kJ/mol} \qquad (13.3)$$

Here is an easy way to determine the freezing point of a liquid. Allow the liquid to cool, and measure the liquid temperature as it falls with time. When freezing begins, the temperature *remains constant* until all the liquid has frozen. Then the temperature is again free to fall as the solid cools. If we plot temperatures against time, we get a graph known as a *cooling curve*. Figure 13-14 is a cooling curve for water. We can also run this process backwards, that is, by starting with the solid and adding heat. Now the temperature remains constant while melting occurs. This temperature–time plot is called a *heating curve*. Generally speaking,

KEEP IN MIND ▶

that here P_1 must be *smaller than* P_2 because $T_1 < T_2$. Thus, regardless of how we write equation (13.2)—different formulations are possible—or choose (T_1, P_1) and (T_2, P_2), we are guided by the fact that vapor pressure always increases with temperature.

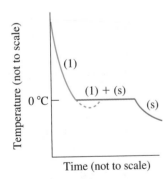

▲ **FIGURE 13-14**
Cooling curve for water
The broken-line portion represents the condition of supercooling that occasionally occurs.
(l) = liquid; (s) = solid.

TABLE 13.4 Some Enthalpies of Fusion		
Substance	Melting Point, °C	ΔH_{fus}, kJ/mol
Mercury, Hg	−38.9	2.30
Sodium, Na	97.8	2.60
Methyl alcohol, CH_3OH	−97.7	3.21
Ethyl alcohol, CH_3CH_2OH	−114	5.01
Water, H_2O	0.0	6.01
Benzoic acid, C_6H_5COOH	122.4	18.08
Naphthalene, $C_{10}H_8$	80.2	18.98

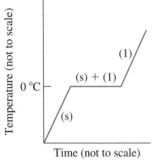

▲ **FIGURE 13-15**
Heating curve for water
This curve traces the changes that occur as ice is heated from below the melting point to produce liquid water somewhat above the melting point.

the appearance of the heating curve is that of a cooling curve that has been flipped from left to right. A heating curve for water is sketched in Figure 13-15.

Often, an experimentally determined cooling curve does not look quite like the solid-line plot in Figure 13-14. The temperature may drop below the freezing point without any solid appearing. This condition is known as *supercooling*. For a crystalline solid to start forming from a liquid at the freezing point, the liquid must contain some small particles (for example, suspended dust particles) on which crystals can form. If a liquid contains a very limited number of particles on which crystals can grow, it may supercool for a time before freezing. When a supercooled liquid does begin to freeze, however, the temperature rises back to the normal freezing point while freezing is completed. We can always recognize supercooling through a slight dip in a cooling curve just before the straight-line portion.

Sublimation

Like liquids, solids can also give off vapors, although because of the stronger intermolecular forces present, solids are generally not as volatile as liquids at a given temperature. The direct passage of molecules from the solid to the vapor state is called **sublimation**. The reverse process, the passage of molecules from the vapor to the solid state, is called **deposition**. When sublimation and deposition occur at equal rates, a dynamic equilibrium exists between a solid and its vapor. The vapor exerts a characteristic pressure called the *sublimation pressure*. A plot of sublimation pressure as a function of temperature is called a *sublimation curve*. The *enthalpy of sublimation* is the quantity of heat needed to convert a solid to vapor. At the melting point, sublimation (solid ⟶ vapor) is equivalent to melting (solid ⟶ liquid) followed by vaporization (liquid ⟶ vapor). This suggests the following relationship between ΔH_{fus}, ΔH_{vap}, and ΔH_{sub} at the melting point.

$$\Delta H_{sub} = \Delta H_{fus} + \Delta H_{vap} \tag{13.4}$$

▲ **FIGURE 13-16**
Sublimation of iodine
Even at temperatures well below its melting point of 114 °C, solid iodine exhibits an appreciable sublimation pressure. Here, purple iodine vapor is produced at about 70 °C. Deposition of the vapor to solid iodine occurs on the colder walls of the flask.

The value of ΔH_{sub} obtained with equation (13.4) can replace the enthalpy of vaporization in the Clausius–Clapeyron equation [equation (13.2)], enabling us to calculate sublimation pressures as a function of temperature.

Two familiar solids with significant sublimation pressures are ice and dry ice (solid carbon dioxide). If you live in a cold climate, you are aware that snow may disappear from the ground even though the temperature may fail to rise above 0 °C. Under these conditions, the snow does not melt; it sublimes. The sublimation pressure of ice at 0 °C is 4.58 mmHg. That is, the solid ice has a vapor pressure of 4.58 mmHg at 0 °C. If the air is not already saturated with water vapor, the ice will sublime. The sublimation and deposition of iodine are pictured in Figure 13-16.

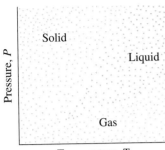

▲ **FIGURE 13-17**
Temperatures, pressures, and states of matter
The outline of a phase diagram is suggested by the distribution of dots. (See also Figures 13-18 and 13-19.)

13-4 Phase Diagrams

Imagine constructing a pressure–temperature graph in which each point on the graph represents a condition under which a substance might be found. At low temperatures and high pressures, such as the green points in Figure 13-17, we expect the atoms, ions, or molecules of a substance to be in a close orderly arrangement—a solid. At high temperatures and low pressures—the tan points in Figure 13-17—we expect the gaseous state; and at intermediate temperatures and pressures, we expect a liquid (blue points in Figure 13-17).

Figure 13-17 suggests a **phase diagram**, a graphical representation of the conditions of temperature and pressure at which solids, liquids, and gases (vapors) exist, either as single phases or states of matter or as two or more phases in equilibrium with one another. The different regions of the diagram correspond to single phases, or states, of matter. Straight or curved lines where single-phase regions adjoin represent two phases in equilibrium.

Iodine

One of the simplest phase diagrams is that of iodine shown in Figure 13-18. The curve *OC* is the vapor pressure curve of liquid iodine, and *C* is the critical point. *OB* is the sublimation curve of solid iodine. The nearly vertical line *OD* represents the effect of pressure on the melting point of iodine; it is called the *fusion curve*. The point *O* has a special significance. It defines the *unique* temperature and pressure at which the *three* states of matter, solid, liquid, and gas, coexist in equilibrium. It is called a **triple point**. For iodine, the triple point is at 113.6 °C and 91.6 mmHg. The normal melting point (113.6 °C) and the boiling point (184.4 °C) are the temperatures at which a line at *P* = 1 atm intersects the fusion and vapor pressure curves, respectively. Melting is essentially unaffected by pressure in the

 Phase Diagram activity

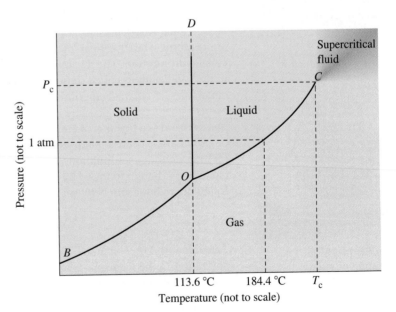

▲ **FIGURE 13-18** **Phase diagram for iodine**
Note that the melting-point and triple-point temperatures for iodine are essentially the same. Generally, large pressure increases are required to produce even small changes in solid–liquid equilibrium temperatures. The pressure and temperature axes on a phase diagram are generally not drawn to scale so that the significant features of the diagram can be more readily emphasized.

limited range from 91.6 mmHg to 1 atm, and the normal melting point and the triple point are at almost the same temperature.

The extreme range of temperatures and pressures required for the entire phase diagram preclude plotting it to scale. This is why the axes are labeled "not to scale."

 Are You Wondering...

If the sublimation curve can ever be just an extension of the vapor pressure curve?

Although the sublimation curve for iodine in Figure 13-18 would look almost like a continuation of the vapor pressure curve if the data were plotted to scale, there is a discontinuity at the triple point O. Moreover, this must *always* be the case. If these two curves were continuous, then the lines representing the variation of $\ln P$ with $1/T$ (Figure 13-13) would have the same slope, but this is not possible. The value of ΔH_{vap} determines the slope of the vapor pressure line (recall equation 13.1), whereas ΔH_{sub} determines the slope of the sublimation line. However, these two enthalpy changes can never be the same, because $\Delta H_{sub} = \Delta H_{vap} + \Delta H_{fus}$.

Carbon Dioxide

The case of carbon dioxide, shown in Figure 13-19, differs from that of iodine in one important respect—the pressure at the triple point O is greater than 1 atm. A line at $P = 1$ atm intersects the *sublimation curve*, not the vapor pressure curve. If solid CO_2 is heated in an open container, it sublimes away at a constant temperature of $-78.5\,°C$. *It does not melt* at atmospheric pressure (and so is called "dry ice"). Because it maintains a low temperature and does not produce a liquid by melting, dry ice is widely used in freezing and preserving foods.

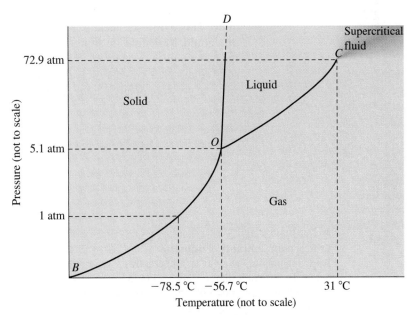

▲ **FIGURE 13-19   Phase diagram for carbon dioxide**
Several aspects of this diagram are described in the text. An additional feature not shown here is the curvature of the fusion curve *OD* to the right at very high pressures, ultimately reaching temperatures above the critical temperature.

We can obtain liquid CO_2 at pressures above 5.1 atm and encounter it most frequently in CO_2 fire extinguishers. All three states of matter are involved in the action of these fire extinguishers. When the liquid CO_2 is released, most of it quickly vaporizes. The heat required for this vaporization is extracted from the remaining $CO_2(l)$, which has its temperature lowered to the point that it freezes and falls as a $CO_2(s)$ "snow." In turn, the $CO_2(s)$ quickly sublimes to $CO_2(g)$. All of this helps to quench a fire by (1) displacing the air around the fire with a "blanket" of $CO_2(g)$ and (2) cooling the area somewhat.

Supercritical Fluids

Because the liquid and gaseous states become identical and indistinguishable at the critical point, it is difficult to know what to call the state of matter at temperatures and pressures above the critical point. For example, this state of matter has the high density of a liquid and the low viscosity of a gas. The term that is now commonly used is "*supercritical fluid*" (SCF). Above the critical temperature, no amount of pressure can liquefy a supercritical fluid. Consider the generic phase diagram in Figure 13-20. The path of dots starting with a vapor below the critical isotherm takes us to a low-density gas above the isotherm. When the pressure is greatly increased, we produce a supercritical fluid of much greater density. If, while the pressure exceeds the critical pressure, P_c, the temperature is reduced below the critical isotherm, we obtain a liquid. Even with further reduction of pressure, the sample remains in the liquid phase. In following the path described, we have gone from a gas to a liquid without observing a liquid–gas interface. The only way to observe the liquid–vapor interface is to cross the phase boundary below the critical isotherm. Note that in the present case we could observe the liquid–vapor interface by lowering the pressure on the liquid to a point on the vapor pressure curve.

Although we do not ordinarily think of liquids or solids as being soluble in gases, volatile ones are. The mole fraction solubility is simply the ratio of the vapor pressure (or the sublimation pressure) to the total gas pressure. And liquids and solids become much more soluble in a gas above its critical pressure and temperature. This is mostly because the density of the SCF is high and approaches that of a liquid. Molecules in supercritical fluids, being in much closer proximity than in ordinary gases, can exert strong attractive forces on the molecules of a liquid or solid solute. SCFs display solvent properties similar to ordinary liquid solvents. To vary the pressure of an SCF means to vary its density and also its solvent properties.

▶ **FIGURE 13-20**
Critical point and critical isotherm
Applying pressure to a gas at temperatures below the critical isotherm causes a liquid to form with the appearance of a meniscus, a discontinuous phase change. Applying pressure above the critical isotherm simply increases the density of the supercritical fluid. In a path traced by the small arrows, gas changes to liquid without exhibiting a discontinuous phase transition.

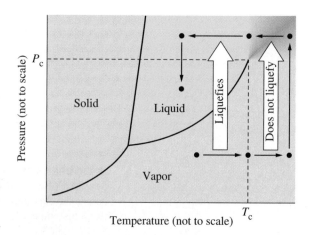

▲ Decaffeinated coffee
"Naturally" decaffeinated coffee is made through a process that uses supercritical fluid CO_2 as a solvent to dissolve the caffeine in green coffee beans. Afterwards the beans are roasted and sold to consumers.

▶ An increase in pressure to 125 atm lowers the freezing point of water by only about 1 °C.

Thus, a given SCF, such as carbon dioxide, can be made to behave like many different solvents.

Until recently, the principal method of decaffeinating coffee was to extract the caffeine with a solvent such as methylene chloride (CH_2Cl_2). This solvent is objectionable because it is hazardous in the workplace and difficult to remove completely from the coffee. Now, supercritical fluid CO_2 is used. In one process, green coffee beans are brought into contact with CO_2 at about 90 °C and 160–220 atm. The caffeine content of the coffee is reduced from its normal 1–3% to about 0.02%. When the temperature and pressure of the CO_2 are reduced, the caffeine precipitates. Then the CO_2 is recycled.

Water

The phase diagram of water (Figure 13-21) presents several new features. One is the fact that the fusion curve *OD* has a *negative* slope; that is, it slopes toward the pressure axis. The melting point of ice *decreases* with an increase in pressure, and this is rather unusual behavior for a solid (bismuth and antimony also behave in this way). However, because large changes in pressure are required to produce even small decreases in the melting point, we do not commonly observe this melting behavior of ice. One example that has been given comes from ice-skating. Presumably, the pressure of the skate blades melts the ice, and the skater skims along on a thin lubricating film of liquid water. This explanation is unlikely, however, because the pressure of the blades doesn't produce a significant lowering of the melting point and certainly cannot explain the ability to skate on ice at temperatures much below the freezing point. (Recent experimental evidence suggests that molecules in a very thin surface layer on ice are mobile in the same way as in liquid water, and this mobility persists even at very low temperatures.)

Another feature illustrated in the phase diagram of water is **polymorphism**, the existence of a solid substance in more than one form. Ordinary ice, called ice I, exists under ordinary pressures. The other forms exist only at high pressures. Polymorphism is more the rule than the exception among solids. Where it occurs, a phase diagram has triple points in addition to the usual solid–liquid–vapor triple point. For example, ice I, ice III, and liquid H_2O are in equilibrium at −22.0 °C and 2045 atm (point *D* in Figure 13-21). Note that the fusion curves for the forms of ice other than ice I have *positive* slopes. Thus, the triple point with ice VI, ice VII, and liquid water is at 81.6 °C and 21,700 atm.

▶ FIGURE 13-21
Phase diagram for water
Point *O*, the triple point, is at 0.0098 °C and 4.58 mmHg. (The normal melting point is at exactly 0 °C and 760 mmHg.) The critical point, *C*, is at 374.1 °C and 218.2 atm. At point *D* the temperature is −22.0 °C and the pressure is 2045 atm. The negative slope of the fusion curve, *OD*, (greatly exaggerated here) and the significance of the broken straight lines are discussed in the text.

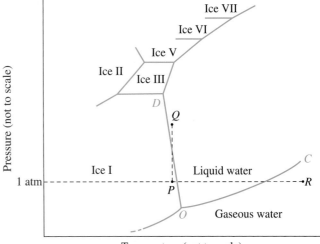

Changes of State animation

Phases and Phase Transitions

What's the difference between a phase and a state of matter? These terms tend to be used synonymously, but there is a small distinction between them. As we have already noted, there are just *three* states of matter: solid, liquid, and gas. A *phase* is any sample of matter with definite composition and uniform properties that is distinguishable from other phases with which it is in contact. Thus, we can describe liquid water in equilibrium with its vapor as a two-phase mixture. The liquid is one phase and the gas, or vapor, is the other. In this case, the phases (liquid and gas) are the same as the states of matter present (liquid and gas).

We can describe the equilibrium mixture at the triple point *D* in Figure 13-21 as a *three-phase* mixture, even though only *two* states of matter are present (solid and liquid). Two of the phases are in the solid state—the polymorphic forms ice I and ice III. For mixtures of two or more components, different phases may exist in the liquid state as well as in the solid state. Because the pressure–temperature diagrams we have been describing can accommodate all the phases in a system, we call them *phase* diagrams. We call the crossing of a two-phase curve in a phase diagram a *phase transition.*

Listed below are six common names assigned to phase transitions.

melting (s $\longrightarrow$ l)   freezing (l $\longrightarrow$ s)

vaporization (l $\longrightarrow$ g)   condensation (g $\longrightarrow$ l)

sublimation (s $\longrightarrow$ g)   deposition (g $\longrightarrow$ s)

Following are two useful generalizations about the changes that occur when crossing a two-phase equilibrium curve in a phase diagram.

- From lower to higher temperatures along a *constant-pressure* line (an isobar), enthalpy *increases*. (Heat is absorbed.)
- From lower to higher pressures along a *constant-temperature* line (an isotherm), volume *decreases*. (The phase at the higher pressure has the higher density.)

The latter of these two generalizations helps us to understand why a fusion curve generally has a positive slope. Typical behavior is for a solid to have a greater density than the corresponding liquid. Example 13-4 illustrates how we can use a phase diagram to describe the phase transitions that a substance can undergo.

EXAMPLE 13-4

Interpreting a Phase Diagram. A sample of ice is maintained at 1 atm and at a temperature represented by point *P* in Figure 13-21. Describe what happens when **(a)** the temperature is raised, at constant pressure, to point *R*, and **(b)** the pressure is raised, at constant temperature, to point *Q*. The sketches in Figure 13-22 suggest the conditions at points *P*, *Q*, and *R*.

Solution

(a) When the temperature reaches a point on the fusion curve *OD* (0 °C), ice begins to melt. The temperature remains constant as ice is converted to liquid. When melting is complete, the temperature again increases. No vapor appears in the cylinder until the temperature reaches 100 °C, at which point the vapor pressure is 1 atm. When all the liquid has vaporized, the temperature is again free to rise to a final value of *R*.

(b) Because solids are not very compressible, very little change occurs until the pressure reaches the point of intersection of the constant-temperature line *PQ* with the fusion curve *OD*. Here melting begins. A significant *decrease* in vol-

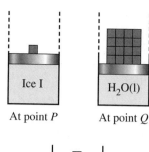

▲ FIGURE 13-22

Example 13-4 illustrated
A sample of pure water is confined in a cylinder by a freely moving piston surmounted by weights to establish the confining pressure. Sketched here are conditions at the points labeled *P*, *Q*, and *R* in Figure 13-21. The transition from point *P* to *Q* is accomplished by changing the pressure at constant temperature (isothermal). The transition from point *P* to *R* is accomplished by changing the temperature at constant pressure (isobaric).

KEEP IN MIND ▶

that a bond dipole results from the separation of centers of positive and negative charge in a covalent bond, that a resultant dipole moment, μ, is a summation of bond dipoles taking into account their magnitudes and directions, and that a *polar* molecule (also called a *dipole*) is one that has a permanent resultant dipole moment.

ume occurs (about 10%) as ice is converted to liquid water. After melting, additional pressure produces very little change because liquids are not very compressible.

Practice Example A: With as much detail as possible, describe the phase changes that would occur if a sample of water represented by the point *R* in Figure 13-21 were brought first to the point *P* and then to the point *Q*.

Practice Example B: Draw a sketch showing the condition prevailing along the line *PR* when 1.00 mol of water has been brought to the point where exactly one-half of it has vaporized. Compare this to the condition at point *R* in Figure 13-22, assuming that this is also based on 1.00 mol of water. For example, is the volume of the system the same as that in Figure 13-22? If not, is it larger or smaller, and by how much? Assume that the temperature at point *R* is the same as the critical temperature of water and that water vapor behaves as an ideal gas.

13-5   Van der Waals Forces

Because helium forms no stable chemical bonds, we might expect it to remain a gas right down to 0 K. Although helium remains gaseous to very low temperatures, it does condense to a liquid at 4 K and freeze to a solid (at 25 atm pressure) at 1 K. These data suggest that intermolecular forces, even though very weak, must exist among He atoms. If the temperature is sufficiently low, these forces overcome thermal agitation and cause helium to condense. In this section, we will examine the types of intermolecular forces known collectively as **van der Waals forces**. The intermolecular forces contributing to the term n^2a/V^2 in the van der Waals equation for nonideal gases (equation 6.26) are of this type.

Instantaneous and Induced Dipoles

In describing electronic structures, we speak of electron charge density or the probability that an electron is in a certain region at a given time. One probability is that at some particular instant—purely by chance—electrons are concentrated in one region of an atom or molecule. This displacement of electrons causes a normally nonpolar species to become momentarily polar. An *instantaneous dipole* is formed. That is, the molecule has an instantaneous dipole moment. After this, electrons in a neighboring atom or molecule may be displaced to also produce a dipole. This is a process of induction (Figure 13-23), and the newly formed dipole is called an *induced dipole*.

Taken together, these two events lead to an intermolecular force of attraction (Figure 13-24). We can call this an instantaneous dipole–induced dipole attraction, but the names more commonly used are **dispersion force** and **London force**. (Fritz London offered a theoretical explanation of these forces in 1928.)

Polarizability is the term used to describe the tendency for charge separation to occur in a molecule. The greater this tendency, the more polarizable the molecule is said to be. Polarizability increases with increased numbers of electrons, and the number of electrons increases with increased molecular mass. Also, in large molecules, some electrons, being farther from atomic nuclei, are less firmly held. These electrons are more easily displaced, and the polarizability of the molecule increases. Because dispersion forces become stronger as polarizability increases, molecules attract each other more strongly, with the result that melting points and boiling points of covalent substances generally increase with increasing molecular mass. For instance, helium, with a molecular (atomic) mass of 4 u, has a boiling point of 4 K, whereas radon (atomic mass, 222 u) has a boiling point of 211 K. The melting points and boiling points of the halogens increase in a similar way in the series F_2, Cl_2, Br_2, I_2 (recall Table 10.5).

▲ **FIGURE 13-23**
The phenomenon of induction
The attraction of a balloon to a surface is a commonplace example of induction. The balloon is charged by rubbing, and the charged balloon induces an opposite charge on the surface. (See also Appendix B).

The strength of dispersion forces also depends on *molecular shape*. Electrons in elongated molecules are more easily displaced than are those in small, compact, symmetrical molecules; the elongated molecules are more polarizable. Two substances with identical numbers and kinds of atoms but different molecular shapes (*isomers*) may have different properties. This idea is illustrated in Figure 13-25.

Dipole–Dipole Interactions

In a *polar* substance, the molecules have permanent dipole moments, so the molecules tend to line up with the positive end of one dipole directed toward the negative ends of neighboring dipoles (Figure 13-26). This additional partial ordering of molecules can cause a substance to persist as a solid or liquid at temperatures higher than otherwise expected. Consider N_2, O_2, and NO. There are no electronegativity differences in N_2 and O_2, and both substances are nonpolar. In NO, on the other hand, there is an electronegativity difference and the molecule has a slight dipole

N_2	NO	O_2
$\mu = 0$ (nonpolar)	$\mu = 0.153$ D (polar)	$\mu = 0$ (nonpolar)
mol. mass 28 u	mol. mass 30 u	mol. mass 32 u
bp 77.34 K	bp 121.39 K	bp 90.19 K

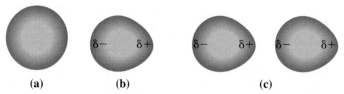

      (a)   (b)   (c)

▲ **FIGURE 13-24   Instantaneous and induced dipoles**
(a) *Normal condition*. A nonpolar molecule has a symmetrical charge distribution. **(b)** *Instantaneous condition*. A displacement of the electronic charge produces an instantaneous dipole with a charge separation represented as $\delta+$ and $\delta-$. **(c)** *Induced dipole*. The instantaneous dipole on the left induces a charge separation in the molecule on the right. The result is a dipole–dipole attraction.

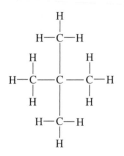

(a) Neopentane
bp = 9.5 °C

(b) Pentane
bp = 36.1 °C

▲ **FIGURE 13-25   Molecular shapes and polarizability**
The elongated pentane molecule is more easily polarized than is the compact neopentane molecule. Intermolecular forces are stronger in pentane than they are in neopentane. As a result, pentane boils at a higher temperature than neopentane.

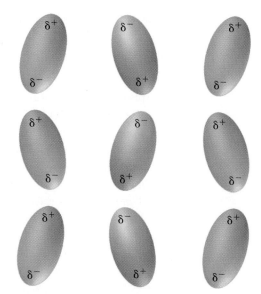

▲ FIGURE 13-26 **Dipole–dipole interactions**
Dipoles tend to arrange themselves with the positive end of one dipole pointed toward the negative end of a neighboring dipole. Ordinarily, thermal motion upsets this orderly array. Nevertheless, this tendency for dipoles to align themselves can affect physical properties, such as the melting points of solids and the boiling points of liquids.

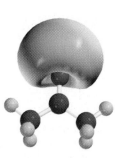

▲ **Butane and acetone**
The lower diagrams are electro-static potential diagrams for bu-tane and acetone. The red color indicates regions of high nega-tive electrostatic potential.

moment. Considering only dispersion forces, we would expect the boiling point of NO(l) to be intermediate to those of $N_2(l)$ and $O_2(l)$, but in the comparison on page 498, we see that it is not. NO(l) has the highest boiling point of the three.

EXAMPLE 13-5

Comparing Physical Properties of Polar and Nonpolar Substances. Which would you expect to have the higher boiling point, the hydrocarbon fuel butane, C_4H_{10}, or the or-ganic solvent acetone, $(CH_3)_2CO$?

Solution

Ordinarily, our first clue comes in a comparison of molecular masses. However, because the two substances have the same molecular mass (58 u), we have to look elsewhere for a factor on which to base our prediction.

  Next, let's see whether either molecule is polar. The electronegativity difference be-tween C and H is so small that we generally expect hydrocarbons to be nonpolar. The ace-tone molecule contains an O atom, and we do expect a strong carbon-to-oxygen bond dipole. At times, we need to sketch the structure of a molecule to see whether symmet-rical features cause bond dipoles to cancel. However, we do not need to do so for the ace-tone molecule because only the C=O bond has a large bond dipole, and this cannot be offset by other bond dipoles. Acetone must have a resultant dipole moment and is there-fore a *polar* molecule. Given two substances with the same molecular mass, one polar and one nonpolar, we expect the polar substance—acetone—to have the higher boiling point. (The measured boiling points are butane, −0.5 °C; acetone, 56.2 °C.)

Practice Example A: Which of the following substances would you expect to have the highest boiling point: C_3H_8, CO_2, CH_3CN? Explain.

Practice Example B: Arrange the following in the expected order of increasing boiling point: C_8H_{18}, $CH_3CH_2CH_2CH_3$, $(CH_3)_3CH$, C_6H_5CHO, SO_3 (octane, butane, isobutane, benzaldehyde, and sulfur trioxide, respectively).

Summary of van der Waals Forces

When assessing the importance of van der Waals forces, consider the following statements.

- *Dispersion (London) forces exist between all molecules.* They involve displacements of all the electrons in molecules, and they increase in strength with increasing molecular mass. The forces also depend on molecular shapes.

- *Forces associated with permanent dipoles* involve displacements of electron pairs in bonds rather than in molecules as a whole. These forces are found only in substances with resultant dipole moments (polar molecules). Their existence *adds* to the effect of dispersion forces also present.

- *When comparing substances of roughly comparable molecular masses,* dipole forces can produce significant differences in properties such as melting point, boiling point, and enthalpy of vaporization.

- *When comparing substances of widely different molecular masses,* dispersion forces are usually more significant than dipole forces.

Let's see how these statements relate to the data in Table 13.5, which includes a rough breakdown of van der Waals forces into dispersion forces and those due to dipoles. HCl and F_2 have comparable molecular masses, but because HCl is polar, it has a significantly larger ΔH_{vap} and a higher boiling point. Within the series HCl, HBr, and HI, molecular mass increases sharply and ΔH_{vap} and boiling points increase in the order HCl < HBr < HI. The more polar nature of HCl and HBr relative to HI is not sufficient to reverse the trends produced by the increasing molecular masses—dispersion forces are the predominant intermolecular forces.

TABLE 13.5   Intermolecular Forces and Properties of Selected Substances

	Molecular Mass, u	Dipole Moment, D	van der Waals Forces		ΔH_{vap}, kJ/mol	Boiling Point, K
			% Dispersion	% Dipole		
F_2	38.00	0	100	0	6.86	85.01
HCl	36.46	1.08	81.4	18.6	16.15	188.11
HBr	80.92	0.82	94.5	5.5	17.61	206.43
HI	127.91	0.44	99.5	0.5	19.77	237.80

EXAMPLE 13-6

Relating Intermolecular Forces and Physical Properties. Arrange the following substances in the order in which you would expect their boiling points to increase: CCl_4, Cl_2, ClNO, N_2.

Solution

Three of the substances are nonpolar. For these, the strengths of dispersion forces, and hence the boiling points, should increase with increasing molecular mass, that is, $N_2 < Cl_2 < CCl_4$. ClNO has a molecular mass (65.5 u) comparable to that of Cl_2 (70.9 u), but the ClNO molecule is polar (bond angle $\approx 120°$). This suggests stronger intermolecular forces and a higher boiling point for ClNO than for Cl_2. We should not expect the boiling point of ClNO to be higher than that of CCl_4, however, because of the large difference in their molecular masses (65.5 u compared with 154 u). The expected order is $N_2 < Cl_2 < ClNO < CCl_4$. (The observed boiling points are 77.3, 239.1, 266.7, and 349.9 K, respectively.)

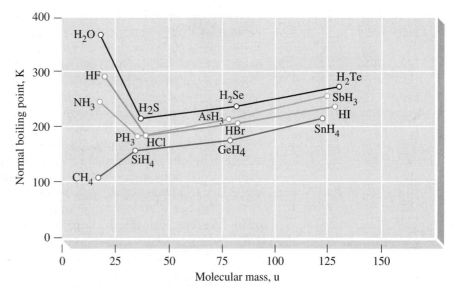

▲ **FIGURE 13-27   Comparison of boiling points of some hydrides of the elements of groups 14, 15, 16, and 17**
The values for NH_3, H_2O, and HF are unusually high compared with those of other members of their groups.

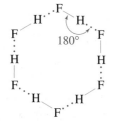

▲ **FIGURE 13-28**
Hydrogen bonding in gaseous hydrogen fluoride
In gaseous hydrogen fluoride, many of the HF molecules are associated into cyclic $(HF)_6$ structures of the type pictured here. Each H atom is bonded to one F atom by a single covalent bond (—) and to another F atom through a hydrogen bond (···).

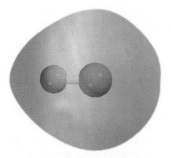

▲ **Electron density distribution in HF**

Practice Example A: Arrange the following in the expected order of increasing boiling point: Ne, He, Cl_2, $(CH_3)_2CO$, O_2, O_3.

Practice Example B: Following are some values of ΔH_{vap} for several liquids at their normal boiling points: H_2, 0.92 kJ/mol; CH_4, 8.16 kJ/mol; C_6H_6, 31.0 kJ/mol; CH_3NO_2, 34.0 kJ/mol. Explain the differences among these values.

13-6   Hydrogen Bonding

Figure 13-27, in which the boiling points of a series of similar compounds are plotted as a function of molecular mass, demonstrates some features that we can't explain by the types of intermolecular forces considered to this point. The hydrogen compounds (hydrides) of the group 14 elements display normal behavior; that is, the boiling points increase regularly as the molecular mass increases. But there are three striking exceptions in groups 15, 16, and 17. The boiling points of NH_3, H_2O and HF are as high or higher than that of any other hydride in their group—not lowest, as we might expect. A special type of intermolecular force causes this exceptional behavior, as we see for hydrogen fluoride in Figure 13-28. Following are the main points established in the figure.

1. The alignment of HF dipoles places a H atom between two F atoms. Because of the very small size of the H atom, the dipoles come close together and produce strong *dipole–dipole* attractions.

2. Although a H atom is covalently bonded to one F atom, it is also weakly bonded to the F atom of a nearby HF molecule. This occurs through a lone pair of electrons on the F atom. Each H atom acts as a bridge between two F atoms.

3. The bond angle between two F atoms bridged by a H atom (that is, the angle F—H···F) is about 180°.

*Hydrogen Bonding
activity*

The type of intermolecular force we have been describing is called a hydrogen bond. A **hydrogen bond** is formed when a H atom bonded to one highly electronegative atom is simultaneously attracted to a highly electronegative atom in a neighboring molecule. In hydrogen-bond formation, the highly electronegative atom to which a H atom is covalently bonded attracts electron density away from the H nucleus, a proton, which is then attracted to a lone pair of electrons on a highly electronegative atom of a neighboring molecule.

Hydrogen bonds are possible only with certain hydrogen-containing compounds because all atoms other than H have inner-shell electrons to shield their nuclei from attraction by lone-pair electrons of nearby atoms. Only F, O, and N easily meet the requirements for hydrogen-bond formation. Weak hydrogen bonding is occasionally encountered between a H atom of one molecule and a Cl or S atom in a neighboring molecule. Compared to other intermolecular forces, hydrogen bonds are relatively strong; their energies are of the order of 15 to 40 kJ/mol. By contrast, single covalent bonds are much stronger still—greater than 150 kJ/mol. (See Table 11.3 for further comparisons.)

Hydrogen Bonding in Water

Ordinary water is certainly the most common substance in which hydrogen bonding occurs. Figure 13-29 shows how one water molecule is held to four neighbors in a tetrahedral arrangement by hydrogen bonds. In ice, hydrogen bonds hold the water molecules in a rigid but rather open structure. As ice melts, only a fraction of the hydrogen bonds are broken. One indication of this is the relatively low heat of fusion of ice (6.01 kJ/mol). It is much less than we would expect if all the hydrogen bonds were to break during melting.

The open structure of ice shown in Figure 13-29 gives ice a low density. When ice melts, some of the hydrogen bonds are broken. This allows the water molecules to be more compactly arranged, accounting for the increase in density when ice

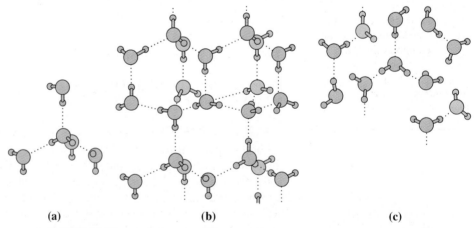

(a)   (b)   (c)

▲ FIGURE 13-29   **Hydrogen bonding in water**
(a) Each water molecule is linked to four others through hydrogen bonds. The arrangement is tetrahedral. Each H atom is situated along a line joining two O atoms, but closer to one O atom (100 pm) than to the other (180 pm). (b) The crystal structure of ice. H atoms lie between pairs of O atoms, again closer to one O atom than to the other. Molecules behind the plane of the page are shaded light blue. O atoms are arranged in bent hexagonal rings arranged in layers. This characteristic pattern is similar to the hexagonal shapes of snowflakes. (c) In the liquid, water molecules have hydrogen bonds to only some of their neighbors. This allows the water molecules to pack more densely in the liquid than in the solid.

Ice model

▲ **FIGURE 13-30**
Solid and liquid densities compared
The sight of ice cubes floating on liquid water (left) is a familiar one. Ice is less dense than liquid water. The more common situation, however, is that of paraffin wax (right). Solid paraffin is denser than the liquid and sinks to the bottom of the beaker.

▲ **FIGURE 13-31**
An acetic acid dimer
Electron density contour showing hydrogen bonding.

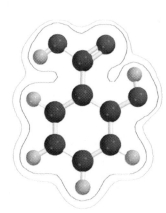

▲ Electron density contour showing intramolecular hydrogen bonding in salicylic acid.

melts. That is, the number of H_2O molecules per unit volume is greater in the liquid than in the solid.

As liquid water is heated above the melting point, hydrogen bonds continue to break. The molecules become even more closely packed, and the density of the liquid water continues to increase. Liquid water attains its maximum density at 3.98 °C. Above this temperature, the water behaves in a "normal" fashion: Its density decreases as temperature increases. The unusual freezing-point behavior of water explains why a freshwater lake freezes from the top down. When the water temperature falls below 4 °C, the denser water sinks to the bottom of the lake and the colder surface water freezes. The ice over the top of the lake then tends to insulate the water below from further heat loss. This allows fish to survive the winter in a lake that has been frozen over. Without hydrogen bonding, all lakes would freeze from the bottom up; and fish, small bottom-feeding animals, and aquatic plants would not survive the winter. The density relationship between liquid water and ice is compared with the more common liquid–solid density relationship in Figure 13-30.

Other Properties Affected by Hydrogen Bonding

Water is one example of a substance whose properties are affected by hydrogen bonding. There are numerous others. In acetic acid, CH_3COOH, pairs of molecules tend to join together into *dimers* (double molecules), both in the liquid and vapor states (Figure 13-31). Not all the hydrogen bonds are disrupted when liquid acetic acid vaporizes, and as a result, the heat of vaporization is abnormally low.

Certain trends in viscosity can also be explained by hydrogen bonding. In alcohols, the H atom in a —OH group in one molecule can form a hydrogen bond to the O atom in a neighboring alcohol molecule. An alcohol molecule with two —OH groups (a *diol*) has more possibilities for hydrogen-bond formation than a comparable alcohol with a single —OH group. Having stronger intermolecular forces, we expect the diol to flow more slowly, that is, to have a greater viscosity, than the simple alcohol. When still more —OH groups are present (*polyols*), we expect a further increase in viscosity. These comparisons are illustrated by the three common alcohols below. (The unit cP is a centipoise. The SI unit of viscosity is $1 \text{ N} \cdot \text{s} \cdot \text{m}^{-2} = 10 \text{ P.}$)

| Ethyl alcohol (ethanol) at 20 °C: 1.20 cP | Ethylene glycol (1,2-ethanediol) at 20 °C: 19.9 cP | Glycerol (1,2,3-propanetriol) at 20 °C: 1490 cP |

*Inter*molecular and *Intra*molecular Hydrogen Bonding

All the examples of hydrogen bonding presented to this point have involved an intermolecular force *between* two molecules, and we call this an *inter*molecular hydrogen bond. We find another possibility in molecules with a H atom covalently bonded to one highly electronegative atom (for example, O or N) and with another highly electronegative atom nearby in the same molecule. This type of hydrogen bonding *within* the same molecule is called *intra*molecular hydrogen bonding. As we see from the molecular model of salicylic acid in the margin, an intermolecular

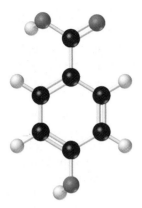

▲ *para*-Hydroxybenzoic acid (no intramolecular hydrogen bonding).

hydrogen bond (represented by a dotted line) joins the —OH group to the doubly bonded oxygen atom of the —COOH group on the same molecule. To underscore the importance of molecular geometry in establishing the conditions necessary for intramolecular hydrogen bonding, we need only turn to an isomer of salicylic acid called *para*-hydroxybenzoic acid. In this molecule, the H atom of the —COOH group is too close to the doubly bonded O atom of the same group to form a hydrogen bond, and the H atom of the —OH group on the opposite side of the molecule is too far away. Intramolecular hydrogen bonding does not occur in this situation.

Hydrogen Bonding in Living Matter

Some chemical reactions in living matter involve complex structures such as proteins and DNA, and in these reactions certain bonds must be easily broken and reformed. Hydrogen bonding is the only type of bonding with energies of just the right magnitude to allow this, as we will discover in Chapter 28. We will also find that both intra- and intermolecular hydrogen bonding is involved in these complex structures.

Hydrogen bonding seems to provide an answer to the puzzle of how some trees are able to grow to great heights. In Chapter 6 (page 179), we learned that atmospheric pressure is capable of pushing a column of water to a maximum height of only about 10 meters. Other factors must be involved in transporting water to the tops of redwood trees up to 100 m tall. In the next chapter, we will learn about osmotic pressure and its ability to force water through membranes, but hydrogen bonding seems also to be a factor in transporting water in trees. Thin columns of water (in xylem) extend from the roots to the leaves in the very tops of trees. In these columns, the water molecules are hydrogen bonded to one another, with each water molecule acting like a link in a cohesive chain. When one water molecule evaporates from a leaf, another molecule in the chain moves to take its place and all the other molecules are pulled up the chain. Ultimately, a new water molecule joins the chain in the root system.

▲ **Sequoia trees**
The mystery as to how these trees can bring water to leaves that are hundreds of feet up may be explained by hydrogen bonding.

13-7 Chemical Bonds as Intermolecular Forces

In most covalent substances, intermolecular forces are quite weak compared with the bonds between atoms within molecules. This is why covalent substances of low molecular mass (those with weak dispersion forces) are generally gaseous at room temperature. Others, usually of somewhat higher molecular masses (with stronger dispersion forces), are liquids. Still others are solids with moderately low melting points.

In a few substances, known as *network covalent* solids, covalent bonds extend throughout a crystalline solid. In these cases, the entire crystal is held together by strong forces. Consider, for example, two of the allotropic forms in which pure carbon occurs—diamond and graphite.

Diamond. Figure 13-32 shows one way that carbon atoms can bond one to another in a very extensive array or crystal. The two-dimensional Lewis structure (Figure 13-32a) is useful only in suggesting that the bonding scheme involves ever increasing numbers of C atoms leading to a giant molecule. It does not give us any insight into the three-dimensional structure of the molecule. For this, we need the portion of the crystal shown in Figure 13-32b. Each atom is bonded to four others. Atoms 1, 2, and 3 lie in a plane, with atom 4 above the plane. Atoms 1, 2, 3, and 5 define a tetrahedron with atom 4 at its center. When viewed from a particular direction, a nonplanar hexagonal arrangement of carbon atoms (gray) is also seen.

Diamond, Graphite, and C$_{60}$ models

KEEP IN MIND ▶
that four bonds directed from a central atom to the corners of a tetrahedron correspond to the sp^3 hybridization scheme.

▶ Another silicon-containing network covalent solid is ordinary silica—silicon dioxide, SiO_2.

▶ **FIGURE 13-32**
The diamond structure
(a) A portion of the Lewis structure.
(b) Crystal structure. Each carbon atom is bonded to four others in a tetrahedral fashion. The segment of the entire crystal shown here is called a unit cell.

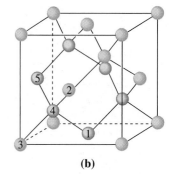

(a)   **(b)**

If silicon atoms are substituted for half the carbon atoms in this structure, the resulting structure is that of silicon carbide (carborundum). Both diamond and silicon carbide are extremely hard, and this accounts for their extensive use as abrasives. In fact, diamond is the hardest substance known. To scratch or break diamond or silicon carbide crystals, covalent bonds must be broken. These two materials are also nonconductors of electricity and do not melt or sublime except at very high temperatures. SiC sublimes at 2700 °C, and diamond melts above 3500 °C.

Graphite. Carbon atoms can bond together in a different way to produce a solid with properties very different from those of diamond. This bonding involves the orbital set $sp^2 + p$. The three sp^2 orbitals are directed in a plane at angles of 120°. The p orbital is perpendicular to the plane, directed above and below it. These orbitals are the same ones used for carbon atoms in benzene, C_6H_6 (described in Figure 12-26). This type of bonding produces the crystal structure shown in Figure 13-33. Each carbon atom forms strong covalent bonds with three neighboring carbon atoms in the same plane, which gives rise to layers of carbon atoms in a hexagonal arrangement. The p electrons of the carbon atoms are *delocalized*; they are not restricted to the region between two C atoms, but are shared among many atoms. Bonding within layers is strong, but between layers it is much weaker. We can see this through bond distances. The C—C bond distance within a layer is 142 pm (compared with 139 pm in benzene); between layers, it is 335 pm.

Its unique crystal structure gives graphite some distinctive properties. Because bonding between layers is weak, the layers can glide over one another rather easily. As a result, graphite is a good lubricant, either in dry form or in an oil suspension.[*] If we apply a mild pressure to a piece of graphite, layers of the graphite flake off; this is what happens when we use a graphite pencil. Also, because the p electrons are delocalized, they migrate through the planes of carbon atoms when an electric field is applied; graphite conducts electricity. An important use of graphite is as electrodes in batteries and in industrial electrolysis processes. Diamond is not an electrical conductor because all of its valence electrons are localized or permanently fixed into single covalent bonds.

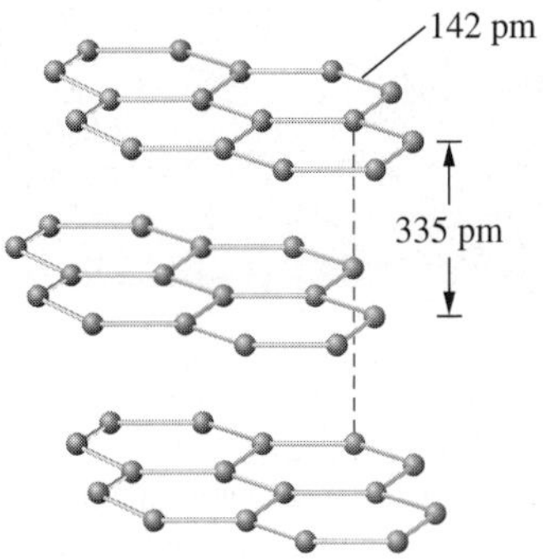

Other Allotropes of Carbon

In 1985, the first of what is now known to be an extensive series of allotropes of carbon was discovered. In experiments designed to mimic conditions found near giant red stars, a number of carbon-containing molecules were discovered and characterized through mass spectroscopy. The strongest peak in the mass spectrum came at 720 u, corresponding to the molecule C_{60}. For a time, proposing a plausible

─142 pm

335 pm

▲ **FIGURE 13-33**
The graphite structure

─────────────

[*]The lubricating properties of graphite also appear to depend on the presence of oxygen molecules between the layers of carbon atoms. When graphite is strongly heated in a vacuum, it becomes a much poorer lubricant.

▲ **Graphite conducts electricity**

Delocalized electrons in graphite allow the conduction of electricity. In this photo, pencil "lead," a mixture of graphite and clay, is used as electrodes to complete the circuit. The beaker contains a solution of ions that carry the current between the pencil electrodes.

structure for this molecule proved to be a challenge. Neither diamond- nor graphite-type structures could account for a molecule with 60 carbon atoms, because "dangling" bonds would remain at the edges of the structures. The structure that was finally proposed, and confirmed by X-ray crystallography, is that of a *truncated icosohedron*—a three-dimensional figure composed of 12 pentagonal and 20 hexagonal faces, with a carbon atom at each of its 60 vertices (Figure 13-34). This figure resembles a soccer ball and also certain geodesic domes. In fact, the resemblance to the geodesic dome led first to the proposed name "buckminsterfullerenes," then simply, *fullerenes*, and finally the colloquial expression "buckyballs. " (The geodesic dome is an architectural form pioneered by R. Buckminster Fuller.) Since 1985, many other fullerenes have been discovered, including C_{70}, C_{74}, and C_{82}. Fullerenes can also form compounds, some by attaching atoms or groups of atoms to their surfaces, others by encasing an atom inside the fullerene structure. To date, several thousand fullerene compounds have been prepared.

Research on fullerenes has led to the discovery of a related type of carbon allotrope—*nanotubes*. We can think of the structure of a nanotube in this way: Imagine a two-dimensional array of hexagonal rings of carbon atoms, called a graphene sheet. An analogous macroscopic structure is a sheet of chicken wire. Now imagine rolling the graphene sheet into a cylinder (something that a section cut from a roll of chicken wire seems to do so naturally). Finally, cap off each end of the cylindrical graphene sheet with a fullerene (Figure 13-35). The diameters of these tubes are of the order of a few nanometers (hence the name *nano*tube). Their lengths can vary from a few nanometers to a micrometer or more. Nanotubes possess unusual electronic and mechanical properties that offer the promise of some applications in the macroscopic world and probably many more in the submicroscopic world of *nanotechnology*. For example, nanotubes might one day be used to channel molecules into the interior of cells.

Interionic Forces

When predicting properties of an ionic solid, we often face this question: How difficult is it to break up an ionic crystal and separate its ions? This question is addressed by the *lattice energy* of a crystal. **Lattice energy** is the energy given off

(a)

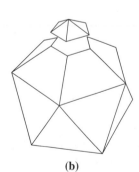
(b)

(c)

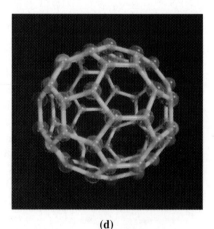
(d)

▲ **FIGURE 13-34** Fullerenes

(a) An icosahedron, a figure formed by 20 equilateral triangles. Five triangles meet at each of the 12 vertices. (b) Truncating or cutting off a vertex reveals a new pentagonal face. (c) The truncated icosahedron. Twelve pentagons have replaced the original 12 vertices, and the 20 equilateral triangles have been replaced by 20 hexagons. (d) The C_{60} molecule.

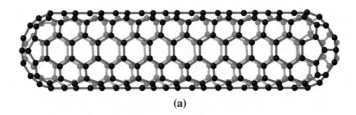

(a)

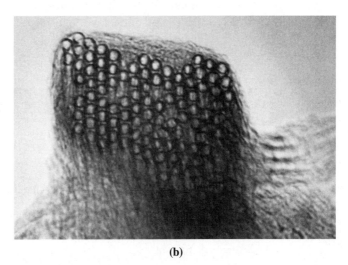

(b)

▲ FIGURE 13-35   **Nanotubes**
(a) Ball-and-stick model of a small nanotube. **(b)** A bundle of single-wall nanotubes.

when separated *gaseous* ions, positive and negative, come together to form *one mole* of a *solid* ionic compound. Lattice energies can be useful in making predictions about the melting points and water solubilities of ionic compounds. We will examine how to calculate lattice energies in Section 13-9. At times, however, we need to make only *qualitative* comparisons of interionic forces, and the following generalization works quite well.

The attractive force between a pair of oppositely charged ions increases with increased charge on the ions and with decreased ionic sizes.

This idea is illustrated in Figure 13-36.

▶ FIGURE 13-36
Interionic forces of attraction
Because of the higher charges on the ions and the closer proximity of their centers, the interionic attractive force between Mg^{2+} and O^{2-} is about seven times as great as between Na^+ and Cl^-.

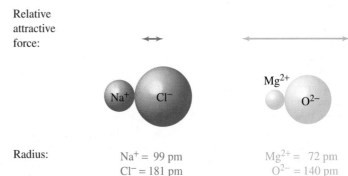

Relative attractive force:

	Na^+ Cl^-	Mg^{2+} O^{2-}
Radius:	$Na^+ = 99$ pm	$Mg^{2+} = 72$ pm
	$Cl^- = 181$ pm	$O^{2-} = 140$ pm
Radius sum = distance between center of ions:	280 pm	212 pm

▶ We consider lattice energy in more detail in Section 13-9.

For most ionic compounds, lattice energies are great enough that ions do not readily detach themselves from the crystal and pass into the gaseous state. Ionic solids do not sublime at ordinary temperatures. We can melt ionic solids by supplying enough thermal energy to disrupt the crystalline lattice. In general, the higher the lattice energy of an ionic compound, the higher is its melting point.

The energy required to break up an ionic crystal when it dissolves results from the interaction of ions in the crystal with molecules of the solvent. The extent to which an ionic solid dissolves in a solvent, however, depends only in part on the lattice energy of the ionic solid. As a rough rule, though, the lower the lattice energy, the greater the quantity of an ionic solid that can be dissolved in a given quantity of solvent.

EXAMPLE 13-7

Predicting Physical Properties of Ionic Compounds. Which has the higher melting point, KI or CaO?

Solution

Ca^{2+} and O^{2-} are more highly charged than K^+ and I^-. Also, Ca^{2+} is smaller than K^+, and O^{2-} is smaller than I^-. We certainly expect the interionic forces in crystalline CaO to be much larger than in KI. CaO should have the higher melting point. (The observed melting points are 677 °C for KI and 2590 °C for CaO.)

Practice Example A: Cite one ionic compound that you would expect to have a lower melting point than KI and one with a higher melting point than CaO.

Practice Example B: Which would you expect to have the greater solubility in water, NaI or $MgCl_2$? Explain.

13-8   Crystal Structures

Crystals—whether as ice, rock salt, quartz, or gemstones—have aroused interest from earliest times. Yet, only in relatively recent times have we come to a fundamental understanding of the crystalline state. This understanding started with the invention of the optical microscope and was greatly expanded following the discovery of X rays. The key idea, now supported by countless experiments, is that the regularity we observe in crystals at the macroscopic level is due to an underlying regular pattern in the arrangement of atoms, ions, or molecules.

Crystal Lattices

You can probably think of a number of situations in which you have had to deal with repeating patterns in one or two dimensions. These might include projects like sewing a decorative border on a piece of material, wallpapering a room, or creating a design with floor tiles.

To describe the structures of crystals, however, we have to work with *three*-dimensional patterns. We do this through three sets of parallel planes called a *lattice*. We have chosen a special case for Figure 13-37: The planes are equidistant and mutually perpendicular (intersecting at 90° angles). This is called a *cubic* lattice. We can use it to describe a few crystals. For others, the appropriate lattice may involve planes that are not equidistant or that intersect at angles other than 90°. In all, there are seven possibilities for crystal lattices, but we will emphasize only the cubic lattice.

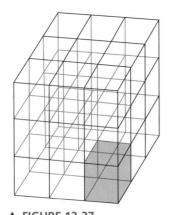

▲ FIGURE 13-37
The cubic space lattice
One parallelepiped formed by the intersection of mutually perpendicular planes is shaded in green. It is a cube. An endless lattice can be generated by simple displacements of the green cube in the three perpendicular directions (that is, left and right, up and down, and forward and backward).

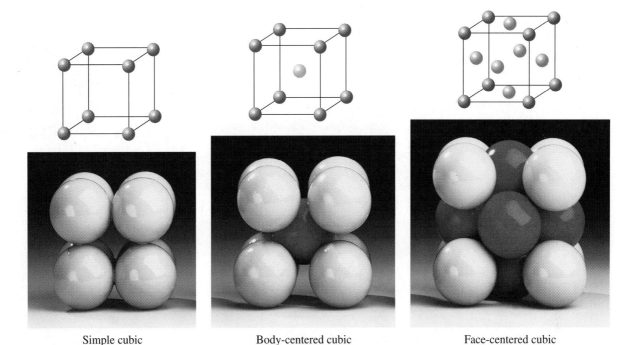

Simple cubic

Body-centered cubic

Face-centered cubic

Unit Cells activity

▲ FIGURE 13-38 **Unit cells in the cubic crystal system**
In the line-and-ball drawings in the top row, only the centers of spheres (atoms) are shown at their respective positions in the unit cells. The space-filling models in the bottom row show contacts between spheres (atoms). In the simple cubic cell, spheres come into contact along each edge. In the body-centered cubic (bcc) cell, contact of the spheres is along the cube diagonal. In the face-centered cubic (fcc) cell, contact is along the diagonal of each face. The spheres shown here are identical atoms; color is used only for emphasis.

Lattice planes intersect to produce three-dimensional figures having six faces arranged in three sets of parallel planes. These figures are called *parallelepipeds*. In Figure 13-37, these parallelepipeds are cubes. A parallelepiped that can be used to generate the entire lattice by simple straight-line displacements is called a **unit cell**. Where possible, we arrange the three-dimensional space lattice so that the centers of the structural particles of the crystal (atoms, ions, or molecules) are situated at lattice points. If a unit cell has structural particles only at its corners, it is called a *primitive* unit cell or simple cubic cell, the simplest unit cell that we can consider. But sometimes we find unit cells that have more structural particles. In the **body-centered cubic (bcc)** structure, a structural particle of the crystal is found at the center of the cube as well as at each corner. In a **face-centered cubic (fcc)** structure, there is a structural particle at the center of each face as well as at each corner. These unit cells are shown in Figure 13-38.

Closest Packed Structures

Unlike boxes, which can be stacked to fill all space, when spheres are stacked together, there must always be some unfilled space. In some arrangements, however, the spheres come into as close contact as possible, and the holes or voids are kept to a minimum. These are known as closest packed structures and are the basis of a number of crystal structures.

To analyze the closest packed structures in Figure 13-39a, let us imagine one layer of spheres, layer A (red), in which each sphere is in contact with six others arranged in a hexagonal fashion around it. Among the spheres, we see open spaces

▲ A closest packed pyramid of cannonballs. Oranges at the fruit stand are often packed in hexagonal closest packed pyramids so that they will not slip.

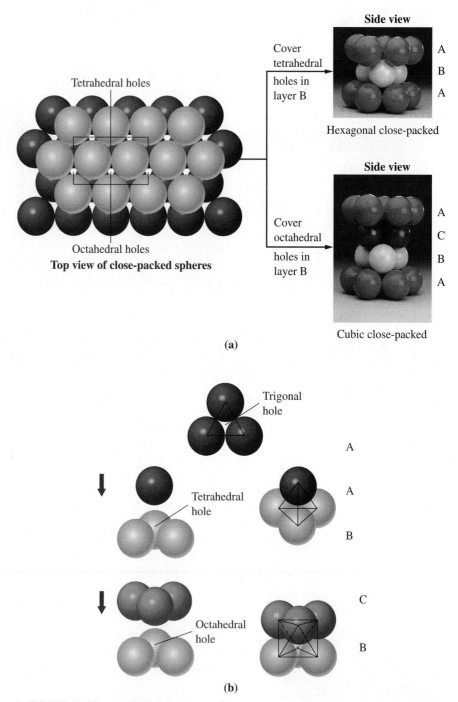

▲ FIGURE 13-39 **Closest packed structures**
(a) Spheres in layer A are red. Those in layer B are yellow, and in layer C, blue. (b) The holes in closest packed structures. The trigonal hole is formed by three spheres in one of the layers. The tetrahedral hole is formed when a sphere in the upper layer sits in the dimple of the lower layer. The octahedral hole is formed between two groups of three spheres in two layers.

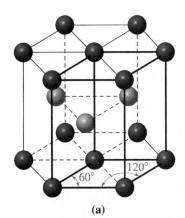

▲ **FIGURE 13-40   A face-centered cubic unit cell for the cubic closest packing of spheres**

The 14 spheres on the left are extracted from a larger array of spheres in a cubic closest packed structure. The two middle layers each have six atoms; the top and bottom layers, one. Rotation of the group of 14 spheres reveals the fcc unit cell (right).

(a)

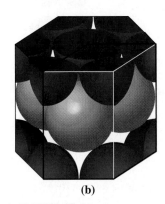

(b)

▲ **FIGURE 13-41**
The hexagonal closest packed (hcp) crystal structure
(a) A unit cell is highlighted in heavy black. The atoms that are part of that cell are in solid color. Note that the unit cell is a parallelepiped, but not a cube. Three adjoining unit cells are also shown. The highlighted unit cell and broken-line regions together show the layering (ABA) described in Figure 13-39.
(b) The hexagonal prism showing parts of the shared spheres at the corners and the single sphere at the center of the unit cell.

or voids. The space between three spheres forming a triangle is called a *trigonal hole* (Figure 13-39). Once the first sphere is placed in the next layer, layer B (yellow), the entire pattern for that layer is fixed. Again, there are holes or voids in layer B, but the holes are of two different types. *Tetrahedral* holes fall directly over spheres in layer A (Figure 13-39). *Octahedral* holes fall directly over holes in layer A (Figure 13-39).

There are two possibilities for C, the third layer. In one arrangement, called **hexagonal closest packed (hcp)**, all the tetrahedral holes are covered. Layer C is identical to layer A, and the structure begins to repeat itself. In the other arrangement, called **cubic closest packed**, all the octahedral holes are covered. The spheres in layer C (blue) are out of line with those in layer A. Only when the fourth layer is added does the structure begin to repeat itself.

Study Figure 13-40 and you will see that the cubic closest packed structure has a face-centered cubic unit cell. The unit cell of the hexagonal closest packed structure is shown in Figure 13-41. In both the hcp and the fcc structures, voids account for only 25.96% of the total volume. Another arrangement in which the packing of spheres is not quite so close has a body-centered cubic unit cell. In this structure, voids account for 31.98% of the total volume. The best examples of crystal structures based on the closest packing of spheres are found among the metals. Some examples are listed in Table 13.6.

TABLE 13.6   Some Features of Close-Packed Structures in Metals

	Coordination Number	Number of Atoms per Unit Cell	Examples
Hexagonal closest packed (hcp)	12	2	Cd, Mg, Ti, Zn
Face-centered cubic (fcc)	12	4	Al, Cu, Pb, Ag
Body-centered cubic (bcc)	8	2	Fe, K, Na, W

Coordination Number and Number of Atoms per Unit Cell

In close-packed structures of atoms, each atom is in contact with several others. For example, can you see in Figure 13-38 that the center atom in the bcc unit cell is in contact with each corner atom? The number of atoms with which a given atom is in contact is called its *coordination number*. For the bcc structure, this is 8. For the fcc and hcp structures, the coordination number is 12. The easiest way to see this is from the layering of spheres described in Figure 13-39. Each sphere is in contact with *six* others in the same layer, *three* in the layer above, and *three* in the layer below.

Although it takes nine atoms to draw the bcc unit cell, it is wrong to conclude that the unit cell consists of nine atoms. As shown in Figure 13-42a, only the center atom belongs *entirely* to the bcc unit cell. The other atoms are shared with other unit cells. The corner atoms are shared among eight adjoining unit cells. Only one-eighth of each corner atom should be thought of as belonging entirely to a given unit cell (Figure 13-42b). Thus, the eight corner atoms collectively contribute the equivalent of

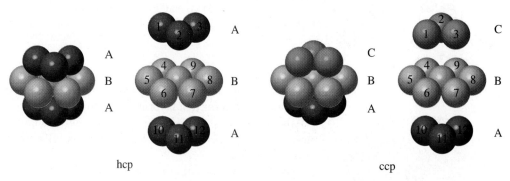

hcp   ccp

▲ **Illustrating the coordination number for the hcp and ccp structures**

Are You Wondering…

How to calculate the volume of the voids in a structure?

To illustrate this, consider the bcc structure. The ratio of the occupied volume to the unit cell volume is

$$f_v = \frac{\text{volume of spheres in unit cell}}{\text{volume of unit cell}}$$

If the radius of the atom is r, the volume of a sphere is $(4/3)\pi r^3$, and by the construction shown in Figure 13-45, the cube edge is $r\,\dfrac{4}{\sqrt{3}}$. So that we have

$$f_v = \frac{2 \times \dfrac{4}{3}\pi r^3}{\left(r\,\dfrac{4}{\sqrt{3}}\right)^3} = 0.6802$$

since there are two complete spheres in the unit cell. Thus, 68.02% of the unit cell is occupied and 31.98% of the unit cell is empty. Note also that this is the same regardless of the radius of the sphere.

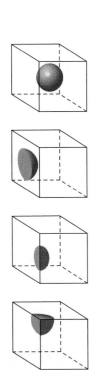

▲ **How spheres are shared between or among unit cells**

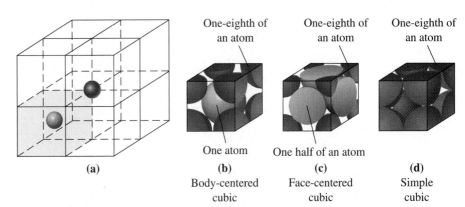

One-eighth of an atom

One-eighth of an atom

One-eighth of an atom

One atom

(a)

One half of an atom

(b)
Body-centered cubic

(c)
Face-centered cubic

(d)
Simple cubic

▲ **FIGURE 13-42   Apportioning atoms among unit cells**
(a) Eight unit cells are outlined. Attention is directed to the blue unit cell. For clarity, only the centers of two atoms are pictured. The atom in the center of the blue cell belongs entirely to that cell. The corner atom is seen to be shared by all eight unit cells. **(b)** The shared spheres of a body-centered unit cell. **(c)** The shared spheres of a face-centered unit cell. **(d)** The shared spheres of a simple cubic unit cell. The spheres shown here are identical atoms; color is used only for emphasis.

one atom to the unit cell. The total number of atoms in a bcc unit cell, then, is *two* [that is, $1 + (8 \times \frac{1}{8})$]. For the hcp unit cell of Figure 13-41, we also get *two* atoms per unit cell if we use the correct counting procedure. The corner atoms account for $\frac{1}{8} \times 8 = 1$ atom, and the central atom belongs entirely to the unit cell. In the fcc unit cell, the corner atoms account for $\frac{1}{8} \times 8 = 1$ atom, and those in the center of the faces for $\frac{1}{2} \times 6 = 3$ atoms. The fcc unit cell contains *four* atoms (Figure 13-42c). The simple cubic unit cell contains only one atom per unit cell (Figure 13-42d).

X-Ray Diffraction

We can see macroscopic objects using visible light and our eyes. To "see" how atoms, ions, or molecules are arranged in a crystal, we need light of much shorter wavelength. When a beam of X rays encounters atoms, X rays interact with electrons in the atoms and the original beam is scattered in all directions. The pattern of this scattered radiation is related to the distribution of electronic charge in the atoms and/or molecules. Our eyes and brains cannot perceive X rays, however. We must make the scattered X rays produce a visible pattern, as on a photographic film. Then we have to infer the microscopic structure of the substance from this visible pattern. How successful we are in making inferences depends on the amount of the scattered radiation we recover, that is, on how much "information" we gather. The power of the X-ray diffraction method has been greatly increased by the use of high-speed computers to process vast amounts of X-ray data.

Figure 13-43 suggests a method of scattering X rays from a crystal. X-ray data can be explained by a geometric analysis proposed by W. H. Bragg and W. L. Bragg in 1912 and illustrated in Figure 13-44. The figure shows two rays in a monochromatic (single-wavelength) X-ray beam, labeled *a* and *b*. Wave *a* is diffracted or scattered by one plane of atoms or ions in a crystal and wave *b* from the next plane below. Wave *b* travels a greater distance than wave *a*. The additional distance is $2d \sin \theta$. The intensity of the scattered radiation will be greatest if waves *a* and *b* reinforce each other, that is, if their crests and troughs line up. To satisfy this requirement, the additional distance traveled by wave *b* must be an integral multiple of the wavelength of the X rays.

$$n\lambda = 2d \sin \theta \qquad (13.5)$$

▶ The X-ray diffraction method was originated by Max von Laue (Nobel Prize, 1914), but carried further by the Braggs. William Lawrence Bragg was only 25 years old when he and his father, William Henry Bragg, won the Nobel Prize in 1915.

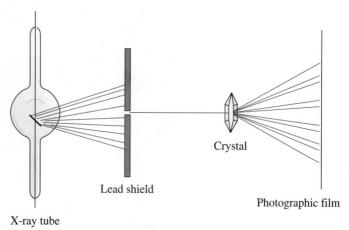

▲ FIGURE 13-43 **Diffraction of X rays by a crystal**

From the measured angle θ, which yields the maximum intensity for the scattered X rays, and the X-ray wavelength (λ), we can calculate the spacing (d) between atomic planes. With different orientations of the crystal, we can determine atomic spacings and electron densities for different directions through the crystal, in short, the crystal structure.

Once a crystal structure is known, certain other properties can be determined by calculation. In Example 13-8 we calculate a metallic radius, and in Example 13-9 we estimate the density of a crystalline solid. For both of these calculations, we need to sketch, or in some way visualize, a unit cell of the crystal. In particular, we need to see which atoms are in direct contact.

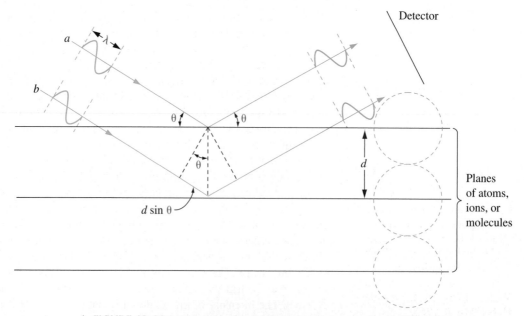

▲ FIGURE 13-44 **Determination of crystal structure by X-ray diffraction**
The two triangles outlined by dotted lines are identical. The hypotenuse of each triangle is equal to the interatomic distance, d. The side opposite the angle θ thus has a length of $d \sin \theta$. Wave b travels farther than wave a by the distance $2d \sin \theta$.

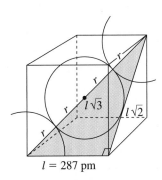

$l = 287$ pm

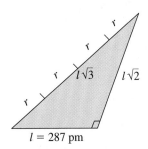

$l = 287$ pm

▲ **FIGURE 13-45**
Determination of the atomic radius of iron—Example 13-8 illustrated
The right triangle must conform to the Pythagorean formula $a^2 + b^2 = c^2$. That is, with l as an edge of the cube,

$$(l)^2 + (l\sqrt{2})^2 = (l\sqrt{3})^2, \text{ or}$$
$$(l)^2 + 2(l^2) = 3(l^2).$$

KEEP IN MIND ▶
that the type of unit cell adopted by a metal can be deduced by using the experimental density and experimentally determined cell edge length.

EXAMPLE 13-8

Using X-ray Data to Determine an Atomic Radius. At room temperature, iron crystallizes in a bcc structure. By X-ray diffraction, the edge of the cubic cell corresponding to Figure 13-45 is found to be 287 pm. What is the radius of an iron atom?

Solution

Nine atoms are associated with a bcc unit cell. One atom is located at each of the eight corners of the cube and one at the center. The three atoms along a cube diagonal are in contact. The length of the cube diagonal (the distance from the farthest upper-right corner to the nearest lower-left corner) is four times the atomic radius. Also shown in Figure 13-45 is the fact that the diagonal of a cube is equal to $\sqrt{3} \times l$. The length of an edge, l, is what is given.

$$4r = l\sqrt{3} \qquad r = \frac{\sqrt{3} \times 287 \text{ pm}}{4} = \frac{1.732 \times 287 \text{ pm}}{4} = 124 \text{ pm}$$

Practice Example A: Potassium crystallizes in the bcc structure. What is the length of the unit cell in this structure? Use the metallic radius of potassium given in Figure 10-8.

Practice Example B: Aluminum crystallizes in an fcc structure. Given that the atomic radius of Al is 143.1 pm, what is the volume of a unit cell?

EXAMPLE 13-9

Relating Density to Crystal Structure Data. Use data from Example 13-8, together with the molar mass of Fe and the Avogadro constant, to calculate the density of iron.

Solution

In Example 13-8, we saw that the length of a unit cell is $l = 287$ pm $= 287 \times 10^{-12}$ m $= 2.87 \times 10^{-8}$ cm. The volume of the unit cell is $V = l^3 = (2.87 \times 10^{-8})^3$ cm$^3 = 2.36 \times 10^{-23}$ cm^3.

From Table 13.6, we find that there are *two* Fe atoms per bcc unit cell. We need the mass of these two atoms, and the key to getting this is a conversion factor based on the fact that 1 mol Fe = 6.022×10^{23} Fe atoms = 55.85 g Fe.

$$\text{Mass} = 2 \text{ Fe atoms} \times \frac{55.85 \text{ g Fe}}{6.022 \times 10^{23} \text{ Fe atoms}} = 1.855 \times 10^{-22} \text{ g Fe}$$

Density is the ratio of mass to volume.

$$\text{density of Fe} = \frac{m}{V} = \frac{1.855 \times 10^{-22} \text{ g Fe}}{2.36 \times 10^{-23} \text{ cm}^3} = 7.86 \text{ g Fe/cm}^3$$

Practice Example A: Use the result of Practice Example 13-8A, together with the molar mass of K and the Avogadro constant, to calculate the density of potassium.

Practice Example B: Use the result of Practice Example 13-8B, together with the molar mass of Al and its density (2.6984 g/cm^3), to evaluate the Avogadro constant, N_A.

(*Hint:* From the volume of a unit cell and the density of Al, you can determine the mass of a unit cell. Knowing the number of Al atoms in the fcc unit cell, you can determine the mass per Al atom.)

Ionic Crystal Structures

If we try to apply the packing-of-spheres model to an ionic crystal, we run into two complications: (1) Some of the ions are positively charged and some are negatively charged, and (2) the cations and anions are of different sizes. What we can expect, however, is that oppositely charged ions will come into close proximity.

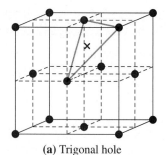

(a) Trigonal hole

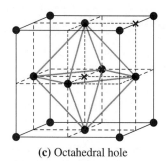

(b) Tetrahedral hole

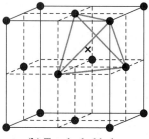

(c) Octahedral hole

▲ **FIGURE 13-46**

Holes in a face-centered cubic unit cell

(a) The trigonal hole is formed by two face-centered spheres and one corner sphere. **(b)** The tetrahedral hole is formed by three face-centered spheres and one corner sphere. **(c)** The octahedral hole is formed by all six face-centered spheres in the cube.

Generally we think of them as being in contact. Like-charged ions, because of mutual repulsions, are not in direct contact. We can think of some ionic crystals in this way: a fairly closely packed arrangement of ions of one type with holes or voids filled by ions of the opposite charge. The relative sizes of cations and anions are important in establishing a particular packing arrangement.

A common arrangement adopted by binary ionic solids is the cubic closest packed arrangement. Very often one of the ions, usually the anion, can be viewed as adopting the cubic closest packed structure and the cation fits in one of the holes between the close-packed spheres. The three types of holes of the cubic closest packed structure—trigonal, tetrahedral, and octahedral—are shown in Figure 13-46. The size of the holes is related to the radius, R, of the anions used to form the structure. Figure 13-47 shows a cross section through an octahedral hole. The radius of the cation, r, that can just fit into the hole can be found using the Pythagorean formula, as follows.

$$(2R)^2 + (2R)^2 = (2R + 2r)^2$$

$$2\sqrt{2}R = 2R + 2r$$

$$(2\sqrt{2} - 2)R = 2r$$

$$(\sqrt{2} - 1)R = r$$

$$r = 0.414R$$

Similar calculations can be used for tetrahedral and trigonal holes, for which $r = 0.225\ R$ and $r = 0.155\ R$, respectively. These calculations show that in the cubic closest packed structure, the octahedral hole is bigger than the tetrahedral hole.

Another arrangement adopted by binary ionic solids is the simple cubic arrangement. The simple cubic arrangement is not a closest packed structure and has larger holes than the cubic closest packed arrangement. The simple cubic structure has a cubic hole at the center of the unit cell. The size of the cubic hole is $r = 0.732\ R$; of the cubic unit cells considered here, this is the largest hole.

Which hole does the cation occupy in a close-packed array of anions? The cation occupies a hole that maximizes the attractions between the cation and anion and minimizes the repulsions between the anions. This can be accomplished by accommodating cations into holes that are slightly smaller than the actual size of the ion. This pushes the anions of the closest packed array slightly apart, reducing repulsions; while the anion and cation are in contact, maximizing attractions. Therefore, if a cation is to occupy a tetrahedral hole, the ion should be bigger than the tetrahedral hole but smaller than the octahedral hole; that is,

$$0.225\ R_{\text{anion}} < r_{\text{cation}} < 0.414\ R_{\text{anion}}$$

or in terms of the *radius ratio* of the cation (r) to the anion (R)

$$0.225 < (r_{\text{cation}}/R_{\text{anion}}) < 0.414$$

Similarly, if a cation is to occupy an octahedral hole, the radii will be governed by the radius ratio inequality

$$0.414 < (r_{\text{cation}}/R_{\text{anion}}) < 0.732$$

where the upper limit of the inequality corresponds to the hole in a simple cubic lattice. When the cation is too large, that is, bigger than $0.732\ R$, we expect the structure that the anions adopt to be that of a simple cubic structure so the cation can be accommodated in the cubic hole of the lattice.

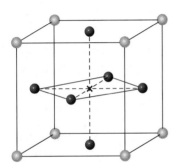

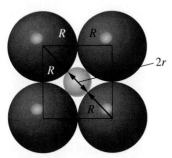

▲ **FIGURE 13-47**
Cross section of an octahedral hole

To summarize, if

$$0.225 < \left(r_{\text{cation}}/R_{\text{anion}}\right) < 0.414$$ tetrahedral hole of fcc array of anions occupied by the cation

$$0.414 < \left(r_{\text{cation}}/R_{\text{anion}}\right) < 0.732$$ octahedral hole of fcc array of anions occupied by the cation

$$0.732 < \left(r_{\text{cation}}/R_{\text{anion}}\right)$$ cubic hole of simple cubic array of anions occupied by the cation

The criteria given here provide a useful way of rationalizing the structures of binary ionic solids. However, as with all simplified models, we must be aware of the limitations of the model. In developing the criteria given, we have assumed that there are no interactions other than coulombic attractions between the ions. The criteria will fail if this is not the case. Nonetheless we will find the criteria to be very useful.

In defining a unit cell of an ionic crystal we must choose a unit cell that

- by translation in three dimensions, generates the entire crystal
- is consistent with the formula of the compound
- indicates the coordination numbers of the ions

Unit cells of crystalline NaCl and CsCl are pictured in Figures 13-48 and 13-49. We can investigate these structures for their consistency with the formula of the compound and the type of hole the cation occupies.

= Cl⁻

= Na⁺

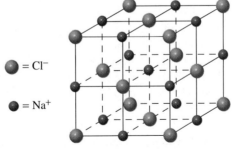

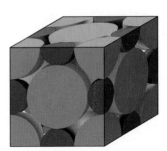

▲ **FIGURE 13-48   The sodium chloride unit cell**
For clarity, only the centers of the ions are shown. Oppositely charged ions are actually in contact. We can think of this structure as an fcc lattice of Cl^- ions, with Na^+ ions filling the octahedral holes.

= Cl⁻

= Cs⁺

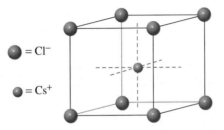

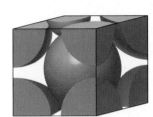

▲ **FIGURE 13-49   The cesium chloride unit cell**
The Cs^+ ion is in the center of the cube, with Cl^- ions at the corners. In reality, each Cl^- is in contact with the Cs^+ ion. An alternative unit cell has Cl^- at the center and Cs^+ at the corners.

Are You Wondering...

Why we don't select one of the smaller cubes in Figure 13-48 as the unit cell of NaCl?

There are eight smaller cubes in the unit cell, each of which has a Na^+ ion at four of the corners and a Cl^- ion at the other four. Such a cube is consistent with the formula NaCl but will not generate the entire lattice by simple displacements. Move any one of these cubes to the adjoining cube in any direction, and a Na^+ ion takes the position where one should find a Cl^- ion.

The radius ratio for NaCl is

$$\frac{r_{Na^+}}{R_{Cl^-}} = \frac{99 \text{ pm}}{181 \text{ pm}} = 0.55$$

KEEP IN MIND ▶
that a cubic closest packed array of spheres produces a face-centered cubic unit cell.

We expect Na^+ ions to occupy the octahedral holes of the cubic closest packed arrays of Cl^- ions. The sodium chloride unit cell is shown in Figure 13-48. To establish the formula of the compound, we must apportion the 27 ions in Figure 13-48 among the unit cell and its neighboring unit cells in the following way: Each Cl^- ion in a corner position is shared by *eight* unit cells, and each Cl^- in the center of a face is shared by *two* unit cells. This leads to a total number of Cl^- ions in the unit cell of $\left(8 \times \frac{1}{8}\right) + \left(6 \times \frac{1}{2}\right) = 1 + 3 = 4$. There are 12 Na^+ ions along the edges of the unit cell, and each edge is shared by *four* unit cells. The Na^+ ion in the very center of the unit cell belongs entirely to that cell. Thus, the total number of Na^+ ions in a unit cell is $\left(12 \times \frac{1}{4}\right) + (1 \times 1) = 3 + 1 = 4$. The unit cell has the equivalent of 4 Na^+ and 4 Cl^- ions. The ratio of Na^+ to Cl^- is $4:4 = 1:1$, corresponding to the formula NaCl.

To establish the coordination number in an ionic crystal, count the number of nearest neighbor ions of opposite charge to any given ion in the crystal. In NaCl, each Na^+ is surrounded by *six* Cl^- ions. The coordination numbers of both Na^+ and Cl^- are *six*. By contrast, the coordination numbers of Cs^+ and Cl^- in Figure 13-49 are *eight*. The difference in the structure of CsCl from that of NaCl can be accounted for in terms of the radius ratio for this compound.

$$\frac{r_{Cs^+}}{R_{Cl^-}} = \frac{169 \text{ pm}}{181 \text{ pm}} = 0.934$$

We expect from the radius ratio inequalities above that the Cs^+ ion will occupy a cubic hole in a simple cubic lattice of Cl^- ions. This unit cell is in accord with the one-to-one ratio of Cs^+ to Cl^- ions since there is one Cs^+ in the center of the unit cell and $8 \times \left(\frac{1}{8}\right) Cl^-$ ions at the corners.

EXAMPLE 13-10

Relating Ionic Radii and the Dimensions of a Unit Cell of an Ionic Crystal. The ionic radii of Na^+ and Cl^- in NaCl are 99 and 181 pm, respectively. What is the length of the unit cell of NaCl?

Solution

Again, the key to solving this problem lies in understanding geometric relationships in the unit cell. Along each edge of the unit cell (see Figure 13-48), two Cl^- ions are in contact with one Na^+. The edge length is equal to the radius of one Cl^- plus the diameter of Na^+, plus the radius of another Cl^-. That is,

$$\text{Length} = (r_{Cl^-}) + (r_{Na^+}) + (r_{Na^+}) + (r_{Cl^-})$$
$$= 2(r_{Na^+}) + 2(r_{Cl^-})$$
$$= (2 \times 99) + (2 \times 181) = 560 \text{ pm}$$

Practice Example A: The ionic radius of Cs^+ is 167 pm. Use Figure 13-49 and information in Examples 13-8 and 13-10 to determine the length of the unit cell of CsCl.

Practice Example B: Use the length of the unit cell of NaCl obtained in Example 13-10, together with the molar mass of NaCl and the Avogadro constant, to estimate the density of NaCl.

(*Hint:* What are the mass and volume of a unit cell?)

Ionic compounds of the type $M^{2+}X^{2-}$ (for example, MgO, BaS, CaO) may form crystals of the NaCl type. However, if the cation is small enough, as in the case of Zn^{2+}, it can occupy the tetrahedral holes. The radius ratio for ZnS is 0.35, so to satisfy the stoichiometry only half of the tetrahedral holes (there are 8 of them) are occupied, to correspond to the four S^{2-} forming the face-centered cubic array (Figure 13-50). For substances with the formulas MX_2 or M_2X, the crystal structures are more complex. Because the cations and anions occur in unequal numbers, the crystals have *two* coordination numbers, one for the cation and another for the anion.

In CaF_2 (the fluorite structure), there are twice as many fluoride ions as calcium ions. The coordination number of Ca^{2+} is *eight*, and that of F^- is *four*. This is easiest to see by looking at the Ca^{2+} ion in the middle of a face. There are four F^- ions within the unit cell that are nearest neighbors. In addition, the four F^- ions in

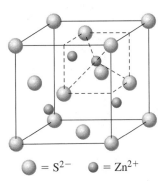

= S^{2-} = Zn^{2+}

(a) Unit cell of ZnS, the zinc blend structure

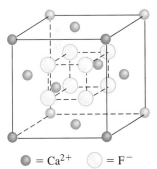

= Ca^{2+} = F^-

(b) Unit cell of CaF_2, the fluorite structure

Fluorite, Rutile, and Zinc Blend models

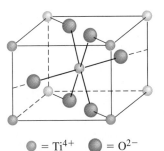

= Ti^{4+} = O^{2-}

(c) Unit cell of TiO_2, the rutile structure

▶ **FIGURE 13-50**
Some unit cells of greater complexity

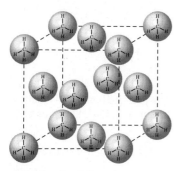

▲ **A sketch of the unit cell of methane**

The unit cell of methane is a face-centered cubic array of CH_4 molecules. In the diagram, the spheres containing the methane molecules are meant to emphasize the fcc array of molecules.

the next unit cell (the one that shares the face-centered Ca^{2+} ion) are also nearest neighbors. This gives a coordination number of eight for the Ca^{2+}. The F^- ions each have one corner Ca^{2+} ion and three face-centered Ca^{2+} ions as nearest neighbors, giving a coordination number of four. In TiO_2 (the rutile structure), Ti^{4+} has a coordination number of *six* and O^{2-}, *three*. In this structure, two of the O^{2-} ions are within the interior of the cell, two are in the top face, and two in the bottom face of the cell. Ti^{4+} ions are at the corners and the center of the cell.

Types of Crystalline Solids: A Summary

In this section, we have emphasized crystal structures composed of metal atoms and those composed of ions. But the structural particles of crystalline solids can be atoms, ions, or molecules. To picture a crystal structure having molecules as its structural units, consider solid methane, CH_4. The crystal structure is fcc, which means that the unit cell has a CH_4 molecule at each corner and at the center of each face. Each CH_4 molecule occupies a volume equivalent to a sphere with a radius of 228 pm. Substances that adopt fcc structures are relatively uncommon. Most complex molecules adopt less symmetric unit cells, but a further discussion of this matter is beyond the scope of this text.

The intermolecular forces operating among the structural units of a crystal may be metallic bonds, interionic attractions, van der Waals forces, hydrogen bonds, or covalent bonds. Table 13.7 lists the basic types of crystalline solids, the intermolecular forces within them, some of their characteristic properties, and examples of each.

TABLE 13.7 Characteristics of Crystalline Solids

Type	Structural Particles	Intermolecular Forces	Typical Properties	Examples
Metallic	Cations and delocalized electrons	Metallic bonds	Hardness varies from soft to very hard; melting point varies from low to very high; lustrous; ductile; malleable; very good conductors of heat and electricity	Na, Mg, Al, Fe, Sn, Cu, Ag, W
Ionic	Cations and anions	Electrostatic attractions	Hard; moderate to very high melting points; nonconductors as solids, but good electric conductors as liquids; many are soluble in polar solvents like water.	NaCl, MgO, $NaNO_3$
Network covalent	Atoms	Covalent bonds	Most are very hard and either sublime or melt at very high temperatures; most are nonconductors of electricity	C (diamond), C (graphite), SiC, AlN, SiO_2
Molecular *Nonpolar*	Atoms or nonpolar molecules	Dispersion forces	Soft; extremely low to moderate melting points (depending on molar mass); sublime in some cases; soluble in some nonpolar solvents	He, Ar, H_2, CO_2, CCl_4, CH_4, I_2
Polar	Polar molecules	Dispersion forces and dipole–dipole attractions	Low to moderate melting points; soluble in some polar and some nonpolar solvents	$(CH_3)_2O$, $CHCl_3$, HCl
Hydrogen– Bonded	Molecules with H bonded to N, O, or F	Hydrogen bonds	Low to moderate melting points; soluble in some hydrogen-bonded solvents and some polar solvents	H_2O, NH_3

13-9 Energy Changes in the Formation of Ionic Crystals

The concept of lattice energy, which we introduced qualitatively in Section 13-7, is most useful when stated in quantitative terms. It is difficult, however, to calculate a lattice energy directly. The problem is that oppositely charged ions attract one another and like-charged ions repel one another, and these interactions must be considered at the same time. More commonly, lattice energy is determined *indirectly* through an application of Hess's law known as the Born–Fajans–Haber cycle, named after its originators Max Born, Kasimir Fajans, and Fritz Haber. The crux of the method is to design a sequence of steps in which enthalpy changes are known for all the steps but one—the step in which a crystal lattice is formed from gaseous ions.

Figure 13-51 illustrates the method for NaCl through five steps.

1. Sublime one mole of solid Na.
2. Dissociate 0.5 mole of $Cl_2(g)$ into one mole of $Cl(g)$.
3. Ionize one mole of $Na(g)$ to $Na^+(g)$.
4. Convert one mole of $Cl(g)$ to $Cl^-(g)$.
5. Allow the $Na^+(g)$ and $Cl^-(g)$ to form one mole of $NaCl(s)$.

The overall change in these five steps is the same as the reaction in which NaCl(s) is formed from its elements in their standard states—that is, $\Delta H_{overall} = \Delta H_f^\circ[\text{NaCl(s)}]$. From Appendix D, we see that $\Delta H_f^\circ[\text{NaCl(s)}] = -411$ kJ/mol, so in the following setup, the lattice energy of NaCl is the only unknown.

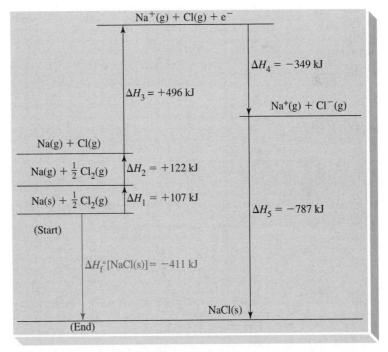

▲ **FIGURE 13-51   Enthalpy diagram for the formation of an ionic crystal**
Shown here is a five-step sequence for the formation of one mole of NaCl(s) from its elements in their standard states. The sum of the five enthalpy changes gives $\Delta H_f^\circ[\text{NaCl(s)}]$. The equivalent one-step reaction for the formation of NaCl(s) directly from Na(s) and $Cl_2(g)$ is shown in color. (The vertical arrows representing ΔH values are not to scale.)

1. $$Na(s) \longrightarrow Na(g)$$ $$\Delta H_1 = \Delta H_{\text{sublimation}} = +107 \text{ kJ}$$

2. $$\frac{1}{2}Cl_2(g) \longrightarrow Cl(g)$$ $$\Delta H_2 = \frac{1}{2}Cl—Cl \text{ bond energy} = +122 \text{ kJ}$$

3. $$Na(g) \longrightarrow Na^+(g) + e^-$$ $$\Delta H_3 = \text{1st ioniz. energy} = +496 \text{ kJ}$$

4. $$Cl(g) + e^- \longrightarrow Cl^-(g)$$ $$\Delta H_4 = \text{electron affinity of Cl} = -349 \text{ kJ}$$

5. $$Na^+(g) + Cl^-(g) \longrightarrow NaCl(s)$$ $$\Delta H_5 = \text{lattice energy of NaCl} = ?$$

overall: $$Na(s) + \frac{1}{2}Cl_2(g) \longrightarrow NaCl(s)$$

$$\Delta H_{\text{overall}} = -411 \text{ kJ} = \Delta H_1 + \Delta H_2 + \Delta H_3 + \Delta H_4 + \Delta H_5$$

$$-411 \text{ kJ} = 107 \text{ kJ} + 122 \text{ kJ} + 496 \text{ kJ} - 349 \text{ kJ} + \Delta H_5$$

$$\Delta H_5 = \text{lattice energy} = (-411 - 107 - 122 - 496 + 349) \text{ kJ} = -787 \text{ kJ}$$

▶A commonly used convention defines lattice energy in terms of the break-up of a crystal rather that its formation. By this convention all lattice energies are positive quantities and $\Delta H_5 = -$ (lattice energy).

One way to use the concept of lattice energy is in making predictions about the possibility of synthesizing ionic compounds. In Example 13-11, we predict the likelihood of obtaining the compound MgCl(s) by evaluating its enthalpy of formation.

EXAMPLE 13-11

Relating Enthalpy of Formation, Lattice Energy, and Other Energy Quantities. With the following data, calculate ΔH_f° per mol MgCl(s): Enthalpy of sublimation of 1 mol Mg(s): +146 kJ; enthalpy of dissociation of $\frac{1}{2}$ mol $Cl_2(g)$: +122 kJ; first ionization energy of 1 mol Mg(g): +738 kJ; electron affinity of 1 mol Cl(g): −349 kJ; lattice energy of 1 mol MgCl(s): −676 kJ.

Solution

In this case the lattice energy (ΔH_5) is known, and the unknown is $\Delta H_{\text{overall}}$, which is the enthalpy of formation of MgCl(s).

$$Mg(s) \longrightarrow Mg(g)$$ $$\Delta H_1 = +146 \text{ kJ}$$

$$\frac{1}{2}Cl_2(g) \longrightarrow Cl(g)$$ $$\Delta H_2 = +122 \text{ kJ}$$

$$Mg(g) \longrightarrow Mg^+(g) + e^-$$ $$\Delta H_3 = +738 \text{ kJ}$$

$$Cl(g) + e^- \longrightarrow Cl^-(g)$$ $$\Delta H_4 = -349 \text{ kJ}$$

$$Mg^+(g) + Cl^-(g) \longrightarrow MgCl(s)$$ $$\Delta H_5 = -676 \text{ kJ}$$

overall: $$Mg(s) + \frac{1}{2}Cl_2(g) \longrightarrow MgCl(s)$$

$$\Delta H_{\text{overall}} = \Delta H_f^\circ[\text{MgCl(s)}] = \Delta H_1 + \Delta H_2 + \Delta H_3 + \Delta H_4 + \Delta H_5$$

$$= 146 \text{ kJ} + 122 \text{ kJ} + 738 \text{ kJ} - 349 \text{ kJ} - 676 \text{ kJ} = -19 \text{ kJ}$$

Practice Example A: The enthalpy of sublimation of cesium is 78.2 kJ/mol, and $\Delta H_f^\circ[\text{CsCl(s)}] = -442.8$ kJ/mol. Use these values, together with other data from the text, to calculate the lattice energy of CsCl(s).

Practice Example B: Given the following data, together with data included in Example 13-11, calculate ΔH_f° per mol $CaCl_2(s)$: Enthalpy of sublimation of Ca(s), +178.2 kJ/mol; first ionization energy of Ca(g), +590 kJ/mol; second ionization energy of Ca(g), +1145 kJ/mol; lattice energy of $CaCl_2(s)$, −2223 kJ/mol.

Example 13-11 suggests that we can obtain MgCl(s) as a stable compound—it has a slightly negative enthalpy of formation. Why have we been writing $MgCl_2$ all this time instead of MgCl? You might think that because MgCl has a Mg-to-Cl ratio of 1 : 1 and $MgCl_2$ has a ratio of 1 : 2, MgCl should form if Mg(s) reacts with a limited amount of $Cl_2(g)$. But this is not the case. No matter how limited the amount of $Cl_2(g)$ available, the only compound that forms is $MgCl_2$. To understand this, repeat the calculation of Example 13-11 for the formation of $MgCl_2(s)$, and you will obtain an enthalpy of formation that is very much more negative than that for MgCl(s) (see Exercise 89). Even though the energy requirement to produce Mg^{2+} is larger than to produce Mg^+, the lattice energy is very much greater for $MgCl_2(s)$ than for MgCl(s). This is because the *doubly* charged Mg^{2+} ions exert a much stronger force on Cl^- ions than do *singly* charged Mg^+ ions. The reaction between Mg and Cl atoms does not stop at MgCl, but continues on to the more stable $MgCl_2$.

Is $NaCl_2$ a stable compound? Here, the answer is no. The additional lattice energy associated with $NaCl_2$ over NaCl is not nearly enough to compensate for the very high second ionization energy of sodium (see Exercise 115).

Summary

Two liquid properties related to intermolecular forces are surface tension and viscosity. Familiar phenomena such as drop shape, meniscus formation, and capillary action depend on surface tension. Vapor pressure—the pressure exerted by a vapor in equilibrium with a liquid—is a measure of the volatility of a liquid. It too is related to the strength of intermolecular forces. Properties of a solid affected by intermolecular forces are its sublimation (vapor) pressure and its melting point.

A phase diagram is a graphical plot of conditions under which solids, liquids, and gases (vapors) exist as single phases or in equilibrium with one another. Especially significant points on a phase diagram are the triple point, melting point, boiling point, and critical point. The critical point is the condition of temperature and pressure at which a liquid and its vapor become indistinguishable.

The most common intermolecular forces of attraction are those between instantaneous and induced dipoles (dispersion forces). In polar substances, there are also dipole–dipole forces. In some substances containing H and certain highly electronegative elements, hydrogen bonding is most significant. In network covalent solids, chemical bonds extend throughout a crystalline structure, as they do in ionic crystals. For these substances, the chemical bonds are themselves intermolecular forces.

Some crystal structures can be described in terms of the close packing of spheres. With ionic crystals, we must also note that the ions are not all of the same size or charge. An important idea for crystals of all types is the unit cell. Its dimensions can be used in calculating atomic radii and densities. Lattice energies of ionic crystals can be related to certain atomic and thermodynamic properties.

Integrative Example

Use data for hydrazine, N_2H_4, from the following table to calculate a value for the vapor pressure of hydrazine when it is placed in a closed container in an ice-water bath.

Property	Value
Freezing point	2.0 °C
Boiling point	113.5 °C
Critical temperature	380 °C
Critical pressure	145.4 atm
Enthalpy of fusion	12.66 kJ/mol
Heat capacity of liquid	98.84 J mol^{-1} °C^{-1}
Density of liquid at 25.0 °C	1.0036 g/mL
Vapor pressure at 25.0 °C	14.4 torr

Let's first outline the calculation to see which of these data are actually required. The temperature of the ice-water bath will be 0 °C. Because its freezing point is 2.0 °C, the hydrazine will exist primarily as a solid, $N_2H_4(s)$. We need to calculate the sublimation pressure at 0 °C. The Clausius–Clapeyron equation expresses vapor pressure (or sublimation pressure) as a function of temperature; but to use it, we need to know the vapor pressure (or sublimation pressure) at one temperature and also the enthalpy of vaporization (or sublimation).

Neither the enthalpy of vaporization nor the enthalpy of sublimation is given, but the enthalpy of fusion is. If we can find a value of ΔH_{vap}, we have $\Delta H_{sub} = \Delta H_{vap} + \Delta H_{fus}$. We can get ΔH_{vap} from the Clausius–Clapeyron equation if we have two values for the vapor pressure and the corresponding temperatures. We are given the vapor pressure at 25 °C and at 113.5 °C

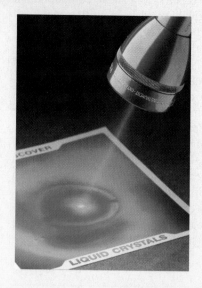

◀ The flashlight provides enough radiant energy to slightly raise the temperature of this liquid crystalline material. As their temperature changes, the liquid crystals undergo a change in color. This effect can be used to make liquid crystal thermometers.

discovered as laboratory curiosities about 100 years ago, are forms of matter that are intermediate between the liquid and solid state. Liquid crystals have the fluid properties of liquids and the optical properties of solids.

Liquid crystals are observed most commonly in organic compounds that have cylindrically shaped (rodlike) molecules with masses of 200 to 500 u and lengths four to eight times their diameters. Potentially, this represents about 0.5% of all organic compounds.

In the *nematic* (meaning threadlike) form of the liquid crystalline state, the rodlike molecules are arranged in a parallel fashion. They are free to move in all directions, but they can rotate only on their long axes. (Imagine the ways in which you might move a particular pencil in a box of loosely packed pencils.) In the *smectic* (meaning greaselike) form, rodlike molecules are arranged in layers, with the long axes of the molecules perpendicular to the planes of the layers. The molecular motions possible here are a translation within, but not between, layers and a rotation about the long axis. The *cholesteric* form is related to the smectic form, but the orientation in each layer is different from that in the layer above and below. These three forms of liquid crystals are illustrated in Figure 13-52.

In a cholesteric structure, each particular orientation is repeated over a span of several layers. The distance between planes with molecules in the same orientation is a distinctive feature of a cholesteric liquid crystal. Certain properties of light

The acronym LCD needs little introduction. Most people know that this term stands for liquid crystal display and that such displays are found in many places—calculators, watches and clocks, thermometers, and many more. After a study of this chapter, though, "liquid crystal" sounds like a contradiction. Liquids are one state of matter, and crystals are something we associate with another state of matter—solids. Liquid crystals,

(the normal boiling point, where the vapor pressure is 760 Torr). The temperature at which we can obtain a value for the sublimation pressure is 2.0 °C, the freezing point (triple point). At this temperature, the sublimation pressure of the solid is equal to the vapor pressure of the liquid. To obtain the vapor pressure of $N_2H_4(l)$ at 2.0 °C, we can use the Clausius–Clapeyron equation. Thus, we can proceed in the following four steps.

1. *Obtain ΔH_{vap}.* In the setup below, $T_2 = 113.5\ °C = 386.7\ K$, and $P_2 = 760$ Torr; $T_1 = 25.0\ °C = 298.2\ K$, and $P_1 = 14.4$ Torr

$$\ln \frac{760\ \text{Torr}}{14.4\ \text{Torr}} = \frac{\Delta H_{vap}}{8.3145\ \text{J mol}^{-1}\ \text{K}^{-1}} \times \left(\frac{1}{298.2\ K} - \frac{1}{386.7\ K} \right)$$
$$= 3.967$$

$$\Delta H_{vap} = \frac{8.3145\ \text{J mol}^{-1}\ \text{K}^{-1} \times 3.967}{(0.003353 - 0.002586)\ \text{K}^{-1}}$$
$$= 4.30 \times 10^4\ \text{J mol}^{-1}\ (43.0\ \text{kJ/mol})$$

2. *Calculate vapor pressure of $N_2H_4(l)$ at 2.0 °C.* Let this be $P_2 = ?$ at $T_2 = 2.0\ °C = 275.2\ K$. Use $P_1 = 14.4$ Torr at $T_1 = 298.2\ K$.

$$\ln \frac{P_2}{14.4} = \frac{4.30 \times 10^4\ \text{J mol}^{-1}}{8.3145\ \text{J mol}^{-1}\ \text{K}^{-1}} \times \left(\frac{1}{298.2\ K} - \frac{1}{275.2\ K} \right)$$
$$= 5.17 \times 10^3 \times (-2.80 \times 10^{-4}) = -1.45$$
$$P_2/14.4 = e^{-1.45} = 0.235 \qquad P_2 = 3.4\ \text{Torr}$$

3. *Determine the enthalpy of sublimation of $N_2H_4(s)$.* From the expression $\Delta H_{sub} = \Delta H_{vap} + \Delta H_{fus}$, we obtain

$$\Delta H_{sub} = 43.0\ \text{kJ/mol} + 12.66\ \text{kJ/mol} = 55.7\ \text{kJ/mol}$$
$$= 5.57 \times 10^4\ \text{J mol}^{-1}$$

4. *Calculate the sublimation pressure of $N_2H_4(s)$ at 0 °C.* Use the Clausius–Clapeyron equation with the known enthalpy of sublimation and sublimation pressure at 2.0 °C and an unknown pressure P_2 at $T_2 = 0\ °C = 273.2\ K$.

$$\ln \frac{P_2}{3.4} = \frac{5.57 \times 10^4\ \text{J mol}^{-1}}{8.3145\ \text{J mol}^{-1}\ \text{K}^{-1}} \times \left(\frac{1}{275.2\ K} - \frac{1}{273.2\ K} \right)$$
$$= 6.70 \times 10^3 \times (-2.66 \times 10^{-5}) = -0.178$$
$$P_2/3.4 = e^{-0.178} = 0.837 \qquad P_2 = 2.8\ \text{Torr}$$

reflected by a liquid crystal depend on this characteristic distance. For example, because this distance is very temperature sensitive, the reflected light changes color with changing temperature. This phenomenon is the basis of liquid crystal temperature-sensing devices, which can detect temperature changes as small as 0.01 °C.

The orientation of molecules in a thin film of nematic liquid crystals is easily altered by pressure and by an electric field. The altered orientation affects the optical properties of the film, such as causing the film to become opaque. Suppose electrodes are arranged in certain patterns (say, in the shape of numbers). When an electric field is imposed through these electrodes onto a thin film of liquid crystals, the patterns of the electrodes become visible. This principle is used in liquid crystal display devices.

Liquid crystals occur widely in living matter. Cell membranes and certain tissues have structures that can be described as liquid crystalline. Hardening of the arteries is caused by the deposition of liquid crystalline compounds of cholesterol. Liquid crystalline properties have also been identified in various synthetic polymers, such as Du Pont Kevlar fiber.

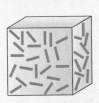

(a) Orientation of molecules in liquid

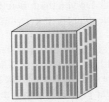

Smectic liquid crystal

Nematic liquid crystal

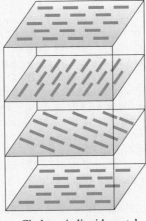

Cholesteric liquid crystal

(b) Orientation of molecules in liquid crystals

◀ FIGURE 13-52
The liquid crystalline state

Key Terms

adhesive force (13-1)
body-centered cubic (bcc) (13-8)
boiling (13-2)
cohesive force (13-1)
condensation (13-2)
critical point (13-2)
cubic closest packed (13-8)
deposition (13-3)
dispersion (London) forces (13-5)
face-centered cubic (fcc) (13-8)

freezing (13-3)
freezing point (13-3)
hexagonal closest packed (13-8)
hydrogen bond (13-6)
lattice energy (13-7)
melting (13-3)
melting point (13-3)
normal boiling point (13-2)
phase diagram (13-4)
polarizability (13-5)

polymorphism (13-4)
sublimation (13-3)
surface tension (13-1)
triple point (13-4)
unit cell (13-8)
van der Waals forces (13-5)
vaporization (evaporation) (13-2)
vapor pressure (13-2)
vapor pressure curve (13-2)
viscosity (13-1)

Review Questions

1. In your own words, define or explain the following terms or symbols: (a) ΔH_{vap}; (b) T_c; (c) instantaneous dipole; (d) coordination number; (e) unit cell.

2. Briefly describe each of the following phenomena or methods: (a) capillary action; (b) polymorphism; (c) sublimation; (d) supercooling; (e) determining the freezing point of a liquid from a cooling curve.

3. Explain the important distinctions between each pair of terms: (a) adhesive and cohesive forces; (b) vaporization and condensation; (c) triple point and critical point;

(d) face-centered and body-centered cubic unit cell; (e) tetrahedral and octahedral hole.

4. Three types of intermolecular forces considered in this chapter were instantaneous dipole–induced dipole, dipole–dipole, and hydrogen bonds. Describe the essential nature of each of these forces.

5. Based on the discussion of phase diagrams, must every substance have (a) a normal melting point; (b) a normal boiling point; (c) a critical point? Explain.

6. Which of the following quantities (expressed in kilojoules per mole) would you expect to be largest for a substance: (a) heat capacity of the liquid; (b) enthalpy of fusion; (c) enthalpy of vaporization; (d) enthalpy of sublimation? Explain.

7. Which of the following factors affect the vapor pressure of a liquid? Explain.
   (a) intermolecular forces in the liquid
   (b) volume of liquid in the liquid–vapor equilibrium
   (c) volume of vapor in the liquid–vapor equilibrium
   (d) the size of the container holding the liquid–vapor equilibrium mixture
   (e) temperature of the liquid

8. At its normal boiling point, the enthalpy of vaporization of chloroform, $CHCl_3$, is 247 J/g.
   (a) How many grams of $CHCl_3$ can be vaporized with 6.62 kJ of heat?
   (b) What is ΔH_{vap} of $CHCl_3$ expressed in kilojoules per mole?
   (c) How much heat, in kilojoules, is evolved when 19.6 g $CHCl_3(g)$ condenses?

9. From Figure 13-9, estimate (a) the vapor pressure of C_6H_7N at 100 °C; (b) the normal boiling point of C_7H_8.

10. Use data in Figure 13-13 to estimate (a) the normal boiling point of aniline; (b) the vapor pressure of diethyl ether at 25 °C.

11. Equilibrium is established between $Br_2(l)$ and $Br_2(g)$ at 25.0 °C. A 250.0-mL sample of the vapor weighs 0.486 g. What is the vapor pressure of bromine at 25.0 °C, in millimeters of mercury.

12. Cyclohexanol has a vapor pressure of 10.0 mmHg at 56.0 °C and 100.0 mmHg at 103.7 °C. Calculate its enthalpy of vaporization, ΔH_{vap}.

13. The vapor pressure of methyl alcohol is 40.0 mmHg at 5.0 °C. Estimate its normal boiling point. $\Delta H_{vap} = 38.0$ kJ/mol.

14. How much heat, in kilojoules, is required to melt a cube of ice that measures 5.08 cm on an edge. The density of ice is 0.92 g/cm³, and the enthalpy of fusion is 6.01 kJ/mol.

15. At its normal melting point, ΔH_{fus} of Cu is 13.05 kJ/mol.
   (a) How much heat, in kilojoules, is evolved when a 3.78-kg sample of molten Cu freezes?
   (b) How much heat, in kilojoules, must be absorbed to melt a bar of copper that is 75 cm × 15 cm × 12 cm? (Assume $d = 8.92$ g/cm³ for Cu.)

16. An 80.0-g piece of dry ice, $CO_2(s)$, is placed in a 0.500-L container, and the container is sealed. If this container is held at 25 °C, what state(s) of matter must be present? (*Hint:* Refer to Table 13.3 and Figure 13-19.)

17. In each of the following pairs, which would you expect to have the higher boiling point? (a) C_7H_{16} or $C_{10}H_{22}$; (b) C_3H_8 or $(CH_3)_2O$; (c) CH_3CH_2SH or CH_3CH_2OH.

18. One of the substances is out of order in the following list based on increasing boiling point. Identify it, and put it in its proper place: N_2, O_3, F_2, Ar, Cl_2. Explain your reasoning.

19. Arrange the following substances in the expected order of increasing melting point: KI, Ne, K_2SO_4, C_3H_8, CH_3CH_2OH, MgO, $CH_2OHCHOHCH_2OH$.

20. Hydrazine, $H_2N—NH_2$, has a normal boiling point of 113.5 °C. Do you think hydrogen bonding is an important intermolecular force in hydrazine? Explain.

21. Place each of the following substances in the appropriate category in Table 13.7, and state your reason for each placement. (a) Si; (b) $SiCl_4$; (c) $CaCl_2$; (d) Ag; (e) SO_2.

22. Only one of the following statements is true about the unit cell of an ionic crystal. Which is the correct statement, and what is wrong with the others? The unit cell (a) is the same as the formula unit; (b) is any portion of the ionic crystal with a cubic shape; (c) shares some of its ions with other unit cells; (d) always contains the same number of cations and anions.

23. In the manner illustrated in the text for NaCl, show that the formula of CsCl is consistent with the unit cell in Figure 13-49.

24. The fcc unit cell is a cube with atoms at each of the corners and in the center of each face, as shown here. Copper has the fcc crystal structure. Assume an atomic radius of 128 pm for a Cu atom.

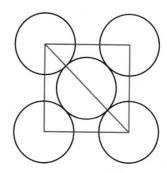

   (a) What is the length of the unit cell of Cu?
   (b) What is the volume of the unit cell?
   (c) How many atoms belong to the unit cell?
   (d) What percentage of the volume of the unit cell is occupied?
   (e) What is the mass of a unit cell of copper?
   (f) Calculate the density of copper.

Exercises

Surface Tension; Viscosity

25. Silicone oils are used in water repellents for treating tents, hiking boots, and similar items. Explain how they function.

26. Surface tension, viscosity, and vapor pressure are all related in some way to intermolecular forces. Why do surface tension and viscosity decrease with temperature, whereas vapor pressure increases with temperature?

27. Is there any scientific basis to the colloquial expression, "slower than molasses in January"? Explain.

28. A television commercial claims that a product makes water "wetter." Can there be any basis to this claim? Explain

Vaporization

29. When a liquid evaporated from an open container, the liquid temperature was observed to remain roughly constant. When the same liquid evaporated from a thermally insulated container (a vacuum bottle or Dewar flask), its temperature was observed to drop. How would you account for this difference?

30. Explain why vaporization occurs only at the surface of a liquid until the boiling point temperature is reached. That is, why does vapor not form throughout the liquid at all temperatures?

31. The enthalpy of vaporization of benzene, $C_6H_6(l)$, is 33.9 kJ/mol at 298 K. How many liters of $C_6H_6(g)$, measured at 298 K and 95.1 mmHg, are formed when 1.54 kJ of heat is absorbed by $C_6H_6(l)$ at a constant temperature of 298 K?

32. A vapor volume of 1.17 L forms when a sample of liquid acetonitrile, CH_3CN, absorbs 1.00 kJ of heat at its normal boiling point (81.6 °C and 1 atm). What is ΔH_{vap} in kilojoules per mole of CH_3CN?

33. Use data from the Integrative Example (page 523) to determine how much heat is required to convert 25.00 mL of liquid hydrazine at 25.0 °C to hydrazine vapor at its normal boiling point.

34. How much heat is required to raise the temperature of 215 g $CH_3OH(l)$ from 20.0 to 30.0 °C and then vaporize it at 30.0 °C? Use data from Table 13.1 and a molar heat capacity of $CH_3OH(l)$ of 81.1 J mol^{-1} K^{-1}.

35. How many liters of $CH_4(g)$, measured at 23.4 °C and 768 mmHg, must be burned to provide the heat needed to vaporize 3.78 L of water at 100 °C? $\Delta H_{combustion} = -8.90 \times 10^2$ kJ/mol CH_4. For $H_2O(l)$ at 100 °C, $d = 0.958$ g/cm^3 and $\Delta H_{vap} = 40.7$ kJ/mol.

36. A 50.0-g piece of iron at 152 °C is dropped into 20.0 g $H_2O(l)$ at 89 °C in an open, thermally insulated container. How much water would you expect to vaporize, assuming no water splashes out? The specific heats of iron and water are 0.45 and 4.21 J g^{-1} °C^{-1}, respectively, and $\Delta H_{vap} = 40.7$ kJ/mol H_2O.
(*Hint:* What is the final temperature?)

Vapor Pressure and Boiling Point

37. A double boiler is used when a careful control of temperature is required in cooking. Water is boiled in an outside container to produce steam, and the steam condenses on the outside walls of an inner container in which cooking occurs. (A related laboratory device is called a steam bath.)
(a) How is heat energy conveyed to the food to be cooked?
(b) What is the maximum temperature that can be reached in the inside container?

38. One popular demonstration in chemistry labs is performed by boiling a small quantity of water in a metal can (such as a used soda can), picking up the can with tongs and quickly submerging it upside down in cold water. The can collapses with a loud and satisfying pop. Give an explanation of this crushing of the can. (Note: If you try this demonstration, do not heat the can over an open flame.)

39. Pressure cookers achieve a high cooking temperature to speed the cooking process by heating a small amount of water under a constant pressure. If the pressure is set at 2 atm, what is the boiling point of the water? Use information from Table 13.2.

40. Use data from Table 13.2 to estimate **(a)** the boiling point of water in Santa Fe, New Mexico, if the prevailing atmospheric pressure is 640 mmHg; **(b)** the prevailing atmospheric pressure at Lake Arrowhead, California, if the observed boiling point of water is 94 °C.

41. A 25.0-L volume of He(g) at 30.0 °C is passed through 6.220 g of liquid aniline ($C_6H_5NH_2$) at 30.0 °C. The liquid remaining after the experiment weighs 6.108 g. Assume that the He(g) becomes saturated with aniline vapor and that the total gas volume and temperature remain constant. What is the vapor pressure of aniline at 30.0 °C?

42. A 7.53-L sample of $N_2(g)$ at 742 mmHg and 45.0 °C is bubbled through $CCl_4(l)$ at 45.0 °C. Assuming the gas becomes saturated with $CCl_4(g)$, what is the volume of the resulting gaseous mixture, if the total pressure remains at 742 mmHg and the temperature remains at 45 °C? The vapor pressure of CCl_4 at 45 °C is 261 mmHg.

43. Some vapor pressure data for Freon-12, CCl_2F_2, once a common refrigerant, are −12.2 °C, 2.0 atm; 16.1 °C, 5.0 atm; 42.4 °C, 10.0 atm; 74.0 °C, 20.0 atm. Also, bp = −29.8 °C, $T_c = 111.5$ °C, $P_c = 39.6$ atm. Use these data to plot the vapor pressure curve of Freon-12. What approximate pressure would be required in the compressor of a refrigeration system to convert Freon-12 vapor to liquid at 25 °C?

44. A 10.0-g sample of liquid water is sealed in a 1515-mL flask and allowed to come to equilibrium with its vapor at 27 °C. What is the mass of $H_2O(g)$ present when equilibrium is established? Use vapor pressure data from Table 13.2.

The Clausius–Clapeyron Equation

45. The normal boiling point of acetone, an important laboratory and industrial solvent, is 56.2 °C and its ΔH_{vap} is 25.5 kJ/mol. At what temperature does acetone have a vapor pressure of 375 mmHg?

46. The vapor pressure of trichloromethane (chloroform) is 40.0 Torr at −7.1 °C. Its enthalpy of vaporization is 29.2 kJ mol^{-1}. Calculate its normal boiling point.

47. Benzaldehyde, C_6H_5CHO, has a normal boiling point of 179.0 °C and a critical point at 422 °C and 45.9 atm. Estimate its vapor pressure at 100.0 °C.

48. With reference to Figure 13-13, which is the more volatile liquid, benzene or toluene? At approximately what temperature does the less volatile liquid have the same vapor pressure as the more volatile one at 65 °C?

Critical Point

49. Which substances listed in Table 13.3 can exist as liquids at room temperature (about 20 °C)? Explain.

50. Can SO_2 be maintained as a liquid under a pressure of 100 atm at 0 °C? Can liquid methane be obtained under the same conditions?

States of Matter and Phase Diagrams

51. Describe what happens to the following samples in situations like those pictured in Figure 13-22. Be as specific as you can about the temperatures and pressures at which changes occur.
(a) A sample of water is heated from −20 to 200 °C at a constant pressure of 600 Torr.
(b) The pressure on a sample of iodine is increased from 90 mmHg to 100 atm at a constant temperature of 114 °C.
(c) A sample of carbon dioxide at 35 °C is cooled to −100 °C at a constant pressure of 50 atm.
(*Hint:* Refer also to Table 13.3.)

52. Shown here is a portion of the phase diagram for phosphorus.
(a) Indicate the phases present in the regions labeled (?).
(b) A sample of solid red phosphorus cannot be melted by heating in a container open to the atmosphere. Explain why this is so.
(c) Trace the phase changes that occur when the pressure on a sample is reduced from point *A* to *B*, at constant temperature.

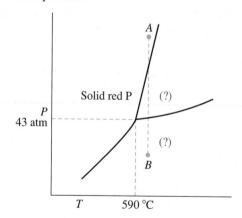

53. A 0.240-g sample of $H_2O(l)$ is sealed into an evacuated 3.20-L flask. What is the pressure of the vapor in the flask if the temperature is (a) 30.0 °C; (b) 50.0 °C; (c) 70.0 °C?

54. A 2.50-g sample of $H_2O(l)$ is sealed in a 5.00-L flask at 120.0 °C.
(a) Show that the sample exists completely as vapor.
(b) Estimate the temperature to which the flask must be cooled before liquid water condenses.

55. Use appropriate phase diagrams and data from Table 13.3 to determine whether any of the following is likely to occur naturally at or near Earth's surface anywhere on Earth: (a) $CO_2(s)$; (b) $CH_4(l)$; (c)$SO_2(g)$; (d) $I_2(l)$; (e) $O_2(l)$. Explain.

56. Trace the phase changes that occur as a sample of $H_2O(g)$, originally at 1.00 mmHg and −0.10 °C, is compressed at constant temperature until the pressure reaches 100 atm.

57. To an insulated container with 100.0 g $H_2O(l)$ at 20.0 °C, 175 g steam at 100.0 °C and 1.65 kg of ice at 0.0 °C are added .
(a) What mass of ice remains unmelted after equilibrium is established?
(b) What *additional* mass of steam should be introduced into the insulated container to just melt all of the ice?

58. A 54-cm^3 ice cube at −25.0 °C is added to a thermally insulated container with 400.0 mL $H_2O(l)$ at 32.0 °C. What will be the final temperature and what state(s) of matter will be present? Specific heats: $H_2O(s)$, 2.01 J g^{-1} °C^{-1}; $H_2O(l)$, 4.18 J g^{-1} °C^{-1}. Densities: $H_2O(s)$, 0.917 g/cm^3; $H_2O(l)$, 0.998 g/cm^3. Also, ΔH_{fus} of ice = 6.01 kJ/mol.

59. You decide to cool a can of soda pop quickly in the freezer compartment of a refrigerator. When you take out the can, the soda pop is still liquid; but when you open the can, the soda pop immediately freezes. Explain why this happens.

60. Why is the triple point of water (ice–liquid–vapor) a better fixed point for establishing a thermometric scale than either the melting point of ice or the boiling point of water?

Intermolecular Forces

61. For each of the following substances describe the importance of dispersion (London) forces, dipole–dipole interactions, and hydrogen bonding: (a) HCl; (b) Br_2; (c) ICl; (d) HF; (e) CH_4.

62. When another atom or group of atoms is substituted for one of the hydrogen atoms in benzene, C_6H_6, the boiling point changes. Explain the order of the following boiling points: C_6H_6, 80 °C; C_6H_5Cl, 132 °C; C_6H_5Br, 156 °C; C_6H_5OH, 182 °C.

63. Arrange the liquids represented through the following molecular models in the expected order of increasing viscosity at 25 °C.

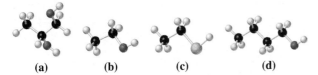

    (a)   **(b)**   **(c)**   **(d)**

64. Arrange the liquids represented by the following molecular models in the expected order of increasing normal boiling point.

    (a)   **(b)**   **(c)**   **(d)**

65. One of the following substances is a liquid at room temperature, whereas the others are gaseous: CH_3OH; C_3H_8; N_2; N_2O. Which do you think is the liquid? Explain.

66. In which of the following compounds do you think that intramolecular hydrogen bonding is an important factor: **(a)** $CH_3CH_2CH_2CH_3$; **(b)** $HOOCCH_2CH_2CH_2CH_2COOH$; **(c)** CH_3COOH; **(d)** *ortho*-phthalic acid? Explain.

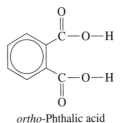

ortho-Phthalic acid

Network Covalent Solids

67. Based on data presented in the text, would you expect diamond or graphite to have the greater density? Explain.

68. Diamond is often used as a cutting medium in glass cutters. What property of diamond makes this possible? Could graphite function as well?

69. Silicon carbide, SiC, crystallizes in a form similar to diamond, whereas boron nitride, BN, crystallizes in a form similar to graphite.

(a) Sketch the SiC structure as in Figure 13-32.
(b) Propose a bonding scheme for BN.

70. Are the fullerenes network covalent solids? What makes them different from diamond and graphite? It has been shown that carbon can form chains in which every other carbon atom is bonded to the next carbon atom by a triple bond. Is this allotrope of carbon a network covalent solid? Explain.

Ionic Bonding and Properties

71. The melting points of NaF, NaCl, NaBr, and NaI are 988, 801, 755, and 651 °C, respectively. Are these data consistent with ideas developed in Section 13-7? Explain.

72. Use Coulomb's law (see Appendix B) to verify the conclusion concerning the relative strengths of the attractive forces in the ion pairs Na^+Cl^- and $Mg^{2+}O^{2-}$ presented in Figure 13-36.

Crystal Structures

73. Explain why there are *two* arrangements for the closest packing of spheres rather than a single one.

74. Argon, copper, sodium chloride, and carbon dioxide all crystallize in the fcc structure. How can this be when their physical properties are so different?

75. For the two-dimensional lattice shown here,

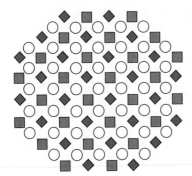

(a) Identify a unit cell.
(b) How many of each of the following elements are in the unit cell: ◆, ■, and ○?
(c) Indicate some simpler units than the unit cell, and explain why they cannot function as a unit cell.

76. As we saw in Section 13-8, the stacking of spheres always leaves open space or voids. Consider the corresponding situation in two dimensions: Squares can be arranged to cover all the area, but circles cannot. For the arrangement of circles pictured here, what percentage of the area remains uncovered?

77. Tungsten has a body-centered cubic crystal structure. Using a metallic radius of 139 pm for the W atom, calculate the density of tungsten.

78. Magnesium crystallizes in the hcp arrangement shown in Figure 13-41. The dimensions of the unit cell are height, 520 pm; length on an edge, 320 pm. Calculate the density of Mg(s), and compare with the measured value of 1.738 g/cm³.

79. Polonium (Po) is the largest member of group 16 and is the only element known to take on the simple cubic crystal system. The distance between nearest neighbor Po atoms in this structure is 335 pm.
    (a) What is the diameter of a Po atom?
    (b) What is the density of Po metal?
    (c) At what angle (in degrees) to the parallel faces of the Po unit cells would first-order diffraction be observed when using X rays of wavelength 1.785×10^{-10} m?

80. Germanium has a cubic unit cell with a side edge of 565 pm. The density of germanium is 5.36 g/cm³. What is the crystal system adopted by germanium?

Ionic Crystal Structures

81. Show that the unit cells for CaF_2 and TiO_2 in Figure 13-50 are consistent with their formulas.

82. Using methods similar to Examples 13-9 and 13-10, calculate the density of CsCl. Use 169 pm as the radius of Cs^+.

83. The crystal structure of magnesium oxide, MgO, is of the NaCl type (Figure 13-48). Use this fact, together with ionic radii from Figure 10-8, to establish the following.
    (a) the coordination numbers of Mg^{2+} and O^{2-}
    (b) the number of formula units in the unit cell
    (c) the length and volume of a unit cell
    (d) the density of MgO

84. Potassium chloride has the same crystal structure as NaCl. Careful measurement of the internuclear distance between K^+ and Cl^- ions gave a value of 314.54 pm. The density of KCl is 1.9893 g/cm³. Use these data to evaluate the Avogadro constant, N_A.

85. Use data from Figure 10-8 to predict the type of cubic unit cell adopted by (a) CaO; (b) CuCl; (c) LiO_2 (the radius of the O_2^- ion is 128 pm).

86. Use data from Figure 10-8 to predict the type of cubic unit cell adopted by (a) BaO(radius of Ba^{2+} is 135 pm); (b) CuI; (c) LiS_2 (the radius of the S_2^- ion is 198 pm).

Lattice Energy

87. *Without doing calculations*, indicate how you would expect the lattice energies of LiCl(s), KCl(s), RbCl(s), and CsCl(s) to compare with the value of −787 kJ/mol determined for NaCl(s) on page 522.
    (*Hint:* Assume that the enthalpies of sublimation of the alkali metals are comparable in value. What atomic properties from Chapter 10 should you compare?)

88. Determine the lattice energy of KF(s) from the following data: $\Delta H_f°[KF(s)] = -567.3$ kJ/mol; enthalpy of sublimation of K(s), 89.24 kJ/mol; enthalpy of dissociation of $F_2(g)$, 159 kJ/mol F_2; I_1 for K(g), 418.9 kJ/mol; EA for F(g), −328 kJ/mol.

89. Refer to Example 13-11. Together with data given there, use the data here to calculate $\Delta H_f°$ for 1 mol $MgCl_2(s)$. Explain why you would expect $MgCl_2$ to be a much more stable compound than MgCl. Second ionization energy of Mg, $I_2 = 1451$ kJ/mol; lattice energy of $MgCl_2(s) = -2526$ kJ/mol $MgCl_2$.

90. In ionic compounds with certain metals, hydrogen exists as the hydride ion, H^-. Determine the electron affinity of hydrogen [that is, ΔH for the process: $H(g) + e^- \longrightarrow H^-(g)$]. To do so, use (1) data from Section 13-9; (2) the bond energy of $H_2(g)$ from Table 11.3; (3) −812 kJ/mol NaH for the lattice energy of NaH(s); and (4) −57 kJ/mol NaH for the enthalpy of formation of NaH(s).

Integrative and Advanced Exercises

91. When a wax candle is burned, the fuel consists of *gaseous* hydrocarbons appearing at the end of the candle wick. Describe the phase changes and processes by which the solid wax is ultimately consumed.

92. Is it possible to obtain a sample of ice from liquid water without ever putting the water in a freezer or other enclosure at a temperature below 0 °C? If so, how might this be done?

93. The normal boiling point of isooctane (a gasoline component with a high octane rating) is 99.2 °C, and its ΔH_{vap} is 35.76 kJ/mol. Because isooctane and water have nearly identical boiling points, will they have nearly equal vapor pressures at room temperature? If not, which would you expect to be more volatile? Explain.

94. A supplier of cylinder gases warns customers to determine how much gas remains in a cylinder by weighing the cylinder and comparing this mass to the original mass of the full cylinder. In particular, the customer is told not to try to estimate the mass of gas available from the measured gas pressure. Explain the basis of this warning. Are there cases where a measurement of the gas pressure *can* be used as a measure of the remaining available gas? If so, what are they?

95. Use the following data and data from Appendix D to determine the quantity of heat needed to convert 15.0 g of solid mercury at −50.0 °C to mercury vapor at 25 °C. Specific heats: Hg(s), 24.3 J mol⁻¹ K⁻¹; Hg(l), 28.0 J mol⁻¹ K⁻¹. Melting point of Hg(s), −38.87 °C. Heat of fusion, 2.33 kJ mol⁻¹.

96. To vaporize 1.000 g water at 20 °C requires 2447 J of heat. At 100 °C, 10.00 kJ of heat will convert 4.430 g $H_2O(l)$ to $H_2O(g)$. Do these observations conform to your expectations? Explain.

97. Estimate how much heat is absorbed when 1.00 g of Instant Car Kooler vaporizes. Comment on the effectiveness of this spray in cooling the interior of a car. Assume the spray is 10% $C_2H_5OH(aq)$ by mass, the temperature is 55 °C, the heat capacity of air is 29 J mol^{-1} K^{-1}, and use ΔH_{vap} data from Table 13.1.

98. Because solid *p*-dichlorobenzene, $C_6H_4Cl_2$, sublimes rather easily, it has been used as a moth repellent. From the data given below, estimate the sublimation pressure of $C_6H_4Cl_2(s)$ at 25 °C. For $C_6H_4Cl_2$: mp = 53.1 °C; vapor pressure of $C_6H_4Cl_2(l)$ at 54.8 °C is 10.0 mmHg: ΔH_{fus} = 17.88 kJ/mol; ΔH_{vap} = 72.22 kJ/mol.

99. What is the difference in the boiling point temperature of water at normal atmospheric pressure (1 atm) and at a pressure of 1 bar?

100. One handbook lists the sublimation pressure of *solid* benzene as a function of *Kelvin* temperature, T, as $\log P$ (mmHg) = $9.846 - 2309/T$. Another handbook lists the vapor pressure of *liquid* benzene as a function of *Celsius* temperature, t, as $\log P$ (mmHg) = $6.90565 - 1211.033/(220.790 + t)$. Use these equations to estimate the normal melting point of benzene, and compare your result with the listed value of 5.5 °C.

101. By the method used to graph Figure 13-13, plot $\ln P$ versus $1/T$ for liquid white phosphorus, and estimate **(a)** its normal boiling point and **(b)** its enthalpy of vaporization, ΔH_{vap}, in kJ/mol. Vapor–pressure data: 76.6 °C, 1 mmHg; 128.0 °C, 10; 166.7 °C, 40; 197.3 °C, 100; 251.0 °C, 400.

102. Assume that a skater has a mass of 80 kg and that his skates make contact with 2.5 cm^2 of ice. **(a)** Calculate the pressure in atm exerted by the skates on the ice. **(b)** If the melting point of ice decreases by 1.0 °C for every 125 atm of pressure, what would be the melting point of the ice under the skates?

103. Estimate the boiling point of water in Leadville, Colorado, elevation 3170 m. To do this, use the barometric formula relating pressure and altitude: $P = P_0 \times 10^{-Mgh/2.303RT}$ (where P = pressure in atm; P_0 = 1 atm; g = acceleration due to gravity; molar mass of air, M = 0.02896 kg/ mol; R = 8.3145 J mol^{-1} K^{-1}; and T is the Kelvin temperature). Assume the air temperature is 10.0 °C and that ΔH_{vap} = 41 kJ/mol H_2O.

104. Inspection of the straight-line graphs in Figure 13-13 suggests that the graphs for benzene and water intersect at some point that falls off the page. At this point, the two liquids have the same vapor pressure. Estimate the temperature and the vapor pressure at this point by a calculation based on data obtainable from the graphs.

105. A cylinder containing 151 lb Cl_2 has an inside diameter of 10 in. and a height of 45 in. The gas pressure is 100 psi (1 atm = 14.7 psi) at 20 °C. Cl_2 melts at −103 °C, boils at −35 °C, and has its critical point at 144 °C and 76 atm. In what state(s) of matter does the Cl_2 exist in the cylinder?

106. In acetic acid vapor, some molecules exist as monomers and some as dimers (see Figure 13-31). If the density of the vapor at 350 K and 1 atm is 3.23 g/L, what percent of the molecules must exist as dimers? Would you expect this percent to increase or decrease with temperature?

107. A 685-mL sample of Hg(l) at 20 °C is added to a large quantity of liquid N_2 kept at its boiling point in a thermally insulated container. What mass of $N_2(l)$ is vaporized as the Hg is brought to the temperature of the liquid N_2? For the specific heat of Hg(l) from 20 to −39 °C use 0.138 J g^{-1} °C^{-1}, and for Hg(s) from −39 to −196 °C, 0.126 J g^{-1} °C^{-1}. The density of Hg(l) is 13.6 g/mL, its melting point is −39 °C, and its enthalpy of fusion is 2.30 kJ/mol. The boiling point of $N_2(l)$ is −196 °C, and its ΔH_{vap} is 5.58 kJ/mol.

108. Sketched here are two hypothetical phase diagrams for a substance, but neither of these diagrams is possible. Indicate what is wrong with each of them.

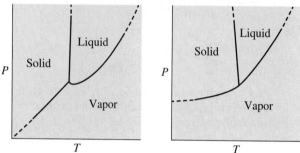

109. The crystal structure of lithium sulfide (Li_2S), is pictured here. The length of the unit cell is 5.88×10^2 pm. For this structure, determine

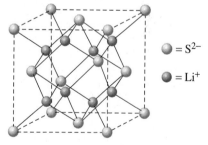

(a) the coordination numbers of Li$^+$ and S^{2-}
(b) the number of formula units in the unit cell
(c) the density of Li_2S

110. Refer to Figures 13-44 and 13-48. Suppose that the two planes of ions pictured in Figure 13-44 correspond to the top and middle planes of ions in the NaCl unit cell in Figure 13-48. If the X rays used have a wavelength of 154.1 pm, at what angle θ would the diffracted beam have its greatest intensity?
[*Hint:* Use $n = 1$ in equation (13.5).]

111. Use the analyses of a bcc structure on page 512 and the fcc structure in Exercise 24 to determine the percent voids in the packing-of-spheres arrangement found in the fcc crystal structure.

112. One way to describe ionic crystal structures is in terms of cations filling voids among closely packed anions. Show that in order for cations to fill the tetrahedral voids in a close packed arrangement of anions, the radius ratio of cation, r_c, to anion, r_a, must fall between the following limits: $0.225 < r_c/r_a < 0.414$.

113. Use the unit cell of diamond in Figure 13-32b (page 505) and a carbon-to-carbon bond length of 154.45 pm, together with other relevant data from the text, to calculate the density of diamond.

114. The enthalpy of formation of $NaI(s)$ is -288 kJ/mol. Use this value, together with other data in the text, to calculate the lattice energy of $NaI(s)$.
    [*Hint:* Use data from Appendix D also.]

115. Show that the formation of $NaCl_2(s)$ is very unfavorable, that is, $\Delta H_f^\circ[NaCl_2(s)]$ is a large *positive* quantity. To do this, use data from Section 13-9 and assume that the lattice energy for $NaCl_2$ would be about the same as that of $MgCl_2$, -2.5×10^3 kJ/mol.

116. We cannot measure the second electron affinity of oxygen directly.

$$O^-(g) + e^- \longrightarrow O^{2-}(g) \qquad EA_2 = ?$$

The O^{2-} ion can exist in the solid state, however, where the high energy requirement for its formation is offset by the large lattice energies of ionic oxides.

(a) Show that EA_2 can be calculated from the enthalpy of formation and lattice energy of $MgO(s)$, enthalpy of sublimation of $Mg(s)$, ionization energies of Mg, bond energy of O_2, and EA_1 for $O(g)$.
(b) The lattice energy of MgO is -3925 kJ/mol. Combine this with other values in the text to estimate EA_2 for oxygen.

117. A crystalline solid contains three types of ions, Na^+, O^{2-} and Cl^-. The solid is made up of cubic unit cells that have O^{2-} ions at each corner, Na^+ ions at the center of each face, and Cl^- ions at the center of the cells. What is the chemical formula of the compound? What are the coordination numbers for the O^{2-} and Cl^- ions? If the length of one edge of the unit cell is a, what is the shortest distance from the center of a Na^+ ion to the center of an O^{2-} ion? Similarly, what is the shortest distance from the center of a Cl^- ion to the center of an O^{2-} ion?

118. A certain mineral has a cubic unit cell with calcium at each corner, oxygen at the center of each face, and titanium at its body center. What is the formula of the mineral? An alternate way of drawing the unit cell has calcium at the center of each cubic unit cell. What are the positions of titanium and oxygen in such a representation of the unit cell? How many oxygens surround a particular titanium atom in either representation?

119. Calculate the radius ratio (r_+/r_-) for CaF_2. Suggest an alternative structure to that shown in Figure 13-50 that better conforms to the radius ratio you compute.

Feature Problems

120. In a capillary rise experiment, the height (h) to which a liquid rises depends on the density (d) and surface tension (γ) of the liquid and the radius of the capillary (r). The equation relating these quantities and the acceleration due to gravity (g) is $h = 2\gamma/dgr$. The sketch provides data obtained with ethanol. What is the surface tension of ethanol?

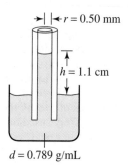

→| |← $r = 0.50$ mm

$h = 1.1$ cm

$d = 0.789$ g/mL

121. We have learned that the enthalpy of vaporization of a liquid is generally a function of temperature. If we wish to take this temperature variation into account, we cannot use the Clausius–Clapeyron equation in the form given in the text [that is, equation (13.2)]. Instead, we must go back to the differential equation upon which the Clausius–Clapeyron equation is based and reintegrate it into a new expression. Our starting point is the following equation describing the rate of change of vapor pressure with temperature in terms of the enthalpy of vaporization, the difference in molar volumes of the vapor and liquid, and the temperature.

$$dP/dT = \Delta H_{vap}/T(V_g - V_l)$$

Because in most cases the volume of one mole of vapor, V_g, greatly exceeds the molar volume of liquid, V_l, we can treat the V_l term as if it were zero. Also, unless the vapor pressure is unusually high, we can treat the vapor as if it were an ideal gas; that is, for one mole of vapor,

$PV = RT$. Make the appropriate substitutions into the above expression, and separate the P and dP terms from the T and dT terms. The appropriate substitution for ΔH_{vap} means expressing it as a function of temperature. Finally, integrate the two sides of the equation between the limits P_1 and P_2 on one side and T_1 and T_2 on the other.

(a) Derive an equation for the vapor pressure of $C_2H_4(l)$ as a function of temperature, if $\Delta H_{vap} = 15{,}971 + 14.55\,T - 0.160\,T^2$ (in J/mol).

(b) Use the equation derived in (a), together with the fact that the vapor pressure of $C_2H_4(l)$ at 120 K is 10.16 Torr, to determine the normal boiling point of ethylene.

122. All solids contain defects or imperfections of structure or composition. Defects are important because they influence properties, such as mechanical strength. Two common types of defects are a missing ion in an otherwise perfect lattice, and the slipping of an ion from its normal site to a hole in the lattice. The holes discussed in this chapter are often called *interstitial sites*, since the holes are in fact interstices in the array of spheres. The two types of defects described here are called *point defects* because they occur within specific sites. In the 1930s, two solid-state physicists, W. Schottky and J. Fraenkel, studied the two types of point defects: A Schottky defect corresponds to a missing ion in a lattice, while a Fraenkel defect corresponds to an ion that is displaced into an interstitial site.

(a) An example of a Schottky defect is the absence of a Na^+ ion in the NaCl structure. The absence of a Na^+ ion means that a Cl^- ion must also be absent to preserve electrical neutrality. If one NaCl unit is missing per unit cell, does the overall stoichiometry change, and what is the change in density?

(b) An example of a Fraenkel defect is the movement of a Ag^+ ion to a tetrahedral interstitial site from its normal octahedral site in AgCl, which has a structure like NaCl. Does the overall stoichiometry of the compound change, and do you expect the density to change?

(c) Titanium monoxide (TiO) has a sodium chloride–like structure. X-ray diffraction data show that the edge length of the unit cell is 418 pm. The density of the crystal is 4.92 g/cm³. Do the data indicate the presence of vacancies? If so what type of vacancies?

 eMedia Exercises

123. In the **Vapor Pressure vs. Temperature** animation *(eChapter 13-2)*, the equation relating vapor pressure and temperature is graphed. If actual data of pressure and temperature were given, describe how one would experimentally determine the constants A and B for different molecules.

124. Mixtures of liquids can at times be heated in order to separate the mixture into its individual components. **(a)** Making use of the **Equilibrium Vapor Pressure** simulation *(eChapter 13-2)*, suggest the role of vapor pressure in this process. **(b)** Which two liquids in this simulation would be the easiest to separate? **(c)** Which two liquids would be most difficult to separate? The separation of liquids by fractional distillation is discussed in Chapter 14.

125. The phase diagrams for several compounds are shown in the **Phase Diagram** activity *(eChapter 13-4)*, but are not shown with linear axes. **(a)** Using several key points, recreate these diagrams with linear pressure and temperature axes. **(b)** Which has the greatest effect on the rate of exchange between liquid and gaseous H_2O at a temperature above the triple point: a one percent change in the pressure or a one percent change in the temperature?

126. View the **Graphite, Diamond and C_{60}** models of carbon *(eChapter 13-7)*. Rotate the structure to view the directional nature of bonding in each form. Mechanical properties such as hardness can be related to the atomic scale bonding geometry. From these representations, suggest an order of increasing hardness for the three forms of carbon and justify your ordering with a discussion of local bonding in the solid.

127. View the animations of the simple cubic and NaCl structures in the **Unit Cells** activity *(eChapter 13-8)*. **(a)** Assuming that all the atoms in these structures have the same radius, which structure has the largest unit cell dimension? **(b)** Why aren't they the same?

14 Solutions and Their Physical Properties

Contents

The dissolving of a cube of sugar (sucrose) is seen here as swirls of a higher density sucrose solution falling through the lower density water.

Residents of cold climate regions know they must add antifreeze to the water in the cooling system of an automobile in the winter. The antifreeze–water mixture has a much lower freezing point than does pure water. In this chapter we will learn why.

To restore body fluids to a dehydrated individual by intravenous injection, pure water cannot be used. A solution with just the right value of a physical property known as osmotic pressure is necessary, and this requires a solution of a particular concentration. Again, in this chapter we will learn why.

Altogether, we will explore several solution properties whose values depend on solution concentration. Our emphasis will be on describing solution phenomena and their applications and explaining these phenomena at the molecular level.

14-1 Types of Solutions: Some Terminology

Dissolution of NaCl in Water

In Chapters 1 and 4 we learned that a solution is a *homogeneous mixture*. It is *homogeneous* because its composition and properties are uniform, and it is a *mixture* because it contains two or more substances in proportions that can be varied. The **solvent** is the component that is present in the greatest quantity or that determines the state of matter in which a solution exists. Other solution components, called **solutes**, are said to be dissolved in the solvent. A *concentrated* solution has a relatively large quantity of dissolved solute(s), and a *dilute* solution has only a small quantity. Consider solutions containing sucrose (cane sugar) as one of the solutes in the solvent water: Pancake syrup is a concentrated solution, whereas a sweetened cup of coffee is much more dilute.

Although liquid solutions are most common, solutions can exist in gaseous and solid states as well. For instance, the U.S. five-cent coin, the nickel, is a solid solution of 75% Cu and 25% Ni. Solid solutions with a metal as the solvent are also called **alloys**.[*] Table 14.1 lists a few common solutions.

TABLE 14.1	Some Common Solutions
Solution	**Components**
Gaseous solutions	
Air	N_2, O_2, and several others
Natural gas	CH_4, C_2H_6, and several others
Liquid solutions	
Seawater	H_2O, NaCl, and many others
Vinegar	H_2O, $HC_2H_3O_2$ (acetic acid)
Soda pop	H_2O, CO_2, $C_{12}H_{22}O_{11}$ (sucrose), and several others
Solid solutions	
Yellow brass	Cu, Zn
Palladium–hydrogen	Pd, H_2

14-2 Solution Concentration

In Chapters 4 and 5 we learned that to describe a solution fully we must know its *concentration*—a measure of the quantity of solute in a given quantity of solvent (or solution). The concentration unit we stressed in those chapters was molarity. In this section, we describe several methods of expressing concentration, each of which serves a different purpose.

Mass Percent, Volume Percent, and Mass/Volume Percent

If we dissolve 5.00 g NaCl in 95.0 g H_2O, we get 100.0 g of a solution that is 5.00% NaCl, by *mass*. Mass percent is widely used in industrial chemistry. Thus, we might read that the action of 78% H_2SO_4(aq) on phosphate rock [$3Ca_3(PO_4)_2 \cdot CaF_2$] produces 46% H_3PO_4(aq).

Because liquid volumes are so easily measured, some solutions are prepared on a *volume* percent basis. For example, a handbook lists a freezing point of $-15.6\ ^\circ\text{C}$ for a methyl alcohol–water antifreeze solution that is 25.0% CH_3OH, by volume. This solution is prepared by dissolving 25.0 mL CH_3OH(l) for every 100.0 mL of aqueous solution.

---

[*]The term *alloy* can also apply to certain heterogeneous mixtures, such as the common two-phase solid mixture of lead and tin known as *solder*, or to intermetallic compounds, such as the silver–tin compound Ag_3Sn that is mixed with mercury in *dental amalgam*.

Another possibility is to express the *mass* of solute and *volume* of solution. An aqueous solution with 0.9 g NaCl in 100.0 mL of solution is said to be 0.9% NaCl (*mass/volume*). Mass/volume percent is extensively used in the medical and pharmaceutical fields.

Parts per Million, Parts per Billion, and Parts per Trillion

In solutions where the mass or volume percent of a component is very low, we often switch to other units to describe solution concentration. For example, 1 mg solute/L solution amounts to only 0.001 g/L. A solution that is this dilute will have the same density as water, approximately 1 g/mL; therefore, the solution concentration is 0.001 g solute/1000 g solution, which is the same as 1 g solute/1,000,000 g solution. We can describe the solute concentration more succinctly as 1 *part per million* (**ppm**). For a solution with only 1μg solute/L solution, the situation is 1×10^{-6} g solute/1000 g solution, or 1.0 g solute/1×10^9 g solution. Here, the solute concentration is 1 *part per billion* (**ppb**). If the solute concentration is only 1 ng solute/L solution. the concentration is 1 *part per trillion* (**ppt**).

Because these terms are widely used in environmental reporting, they may be more familiar than other units that chemists use. For example, a consumer in California might read in an annual water quality report from the municipal water department that the maximum contaminant level allowed for nitrate ion is 45 ppm and for carbon tetrachloride, 0.5 ppb.

KEEP IN MIND ▶
that 1 ppm = 1 mg/L,
1 ppb = 1 μg/L, and
1 ppt = 1 ng/L.

Mole Fraction and Mole Percent

To relate certain physical properties (such as vapor pressure) to solution concentration, we need a unit in which all solution components are expressed on a mole basis. We can do this with the mole fraction. The **mole fraction** of component i, designated x_i, is the fraction of all the molecules in a solution that are of type i. The mole fraction of component j is x_j, and so on. The mole fraction of a solution component is defined as

$$x_i = \frac{\text{amount of component } i \text{ (in moles)}}{\text{total amount of all solution components (in moles)}}$$

The sum of the mole fractions of all the solution components is 1.

$$x_i + x_j + x_k + \ldots = 1$$

The **mole percent** of a solution component is the percent of all the molecules in solution that are of a given type. Mole percents are mole fractions multiplied by 100%.

Molarity

In Chapters 4 and 5, we introduced molarity to provide a conversion factor relating the amount of solute and the volume of solution. We used it in various stoichiometric calculations. As we learned at that time,

$$\text{molarity(M)} = \frac{\text{amount of solute (in moles)}}{\text{volume of solution (in liters)}}$$

Molality

Suppose we prepare a solution at 20 °C by using a volumetric flask calibrated at 20 °C. Then suppose we warm this solution to 25 °C. As the temperature increases from 20 to 25 °C, the amount of solute remains constant, but the solution volume increases slightly (by about 0.1%). The number of moles of solute per liter—the molarity—*decreases* slightly (by about 0.1%). This temperature dependence of molarity can be a problem in experiments demanding a high precision. That is, the solution might be used at a temperature different from the one at which it was prepared, and so its molarity is not exactly the one written on the label. A concentration unit that is *independent* of temperature, and also proportional to mole fraction in dilute solutions, is **molality (*m*)**—the number of moles of solute per kilogram of *solvent* (not of solution). A solution in which 1.00 mol of urea, $CO(NH_2)_2$, is dissolved in 1.00 kg of water is described as a 1.00 molal solution and designated as 1.00 *m* $CO(NH_2)_2$. Molality is defined as

$$\text{molality } (m) = \frac{\text{amount of solute (in moles)}}{\text{mass of solvent (in kilograms)}}$$

Illustrative Examples

The concentration of a solution is expressed in several different ways in Example 14-1. The calculation in Example 14-2 is perhaps more typical: A concentration is converted from one unit (molarity) to another (mole fraction).

EXAMPLE 14-1

Expressing a Solution Concentration in Various Units. An ethanol–water solution is prepared by dissolving 10.00 mL of ethanol, C_2H_5OH ($d = 0.789$ g/mL), in a sufficient volume of water to produce 100.0 mL of a solution with a density of 0.982 g/mL (Figure 14-1). What is the concentration of ethanol in this solution, expressed as **(a)** volume percent; **(b)** mass percent; **(c)** mass/volume percent; **(d)** mole fraction; **(e)** mole percent; **(f)** molarity; **(g)** molality?

Solution

(a) Volume percent ethanol

$$\text{volume percent ethanol} = \frac{10.00 \text{ mL ethanol}}{100.0 \text{ mL solution}} \times 100\% = 10.00\%$$

(b) Mass percent ethanol

$$\text{mass ethanol} = 10.00 \text{ mL ethanol} \times \frac{0.789 \text{ g ethanol}}{1.00 \text{ mL ethanol}}$$

$$= 7.89 \text{ g ethanol}$$

$$\text{mass soln} = 100.0 \text{ mL soln} \times \frac{0.982 \text{ g soln}}{1.0 \text{ mL solution}} = 98.2 \text{ g soln}$$

$$\text{mass percent ethanol} = \frac{7.89 \text{ g ethanol}}{98.2 \text{ g solution}} \times 100\% = 8.03\%$$

(c) Mass/volume percent ethanol

$$\text{mass/volume percent ethanol} = \frac{7.89 \text{ g ethanol}}{100.0 \text{ mL solution}} \times 100\% = 7.89\%$$

(d) Mole fraction of ethanol

Convert the mass of ethanol from part (b) to an amount in moles.

$$? \text{ mol } C_2H_5OH = 7.89 \text{ g } C_2H_5OH \times \frac{1 \text{ mol } C_2H_5OH}{46.07 \text{ g } C_2H_5OH}$$

$$= 0.171 \text{ mol } C_2H_5OH$$

Determine the mass of water present in 100.0 mL of solution.

$$98.2 \text{ g soln} - 7.89 \text{ g ethanol} = 90.3 \text{ g water}$$

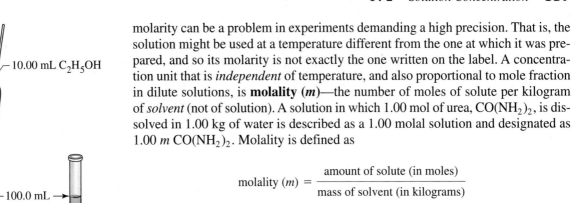

— 10.00 mL C_2H_5OH

← 100.0 mL →

Water

Ethanol–water solution: $d = 0.982$ g/mL

▲ **FIGURE 14-1**
Preparation of an ethanol–water solution— Example 14-1 illustrated
A 10.00-mL sample of C_2H_5OH is added to some water in the volumetric flask. The solution is mixed, and more water is added to bring the total volume to 100.0 mL.

Convert the mass of water to the number of moles present.

$$? \text{ mol } H_2O = 90.3 \text{ g } H_2O \times \frac{1 \text{ mol } H_2O}{18.02 \text{ g } H_2O} = 5.01 \text{ mol } H_2O$$

$$x_{C_2H_5OH} = \frac{0.171 \text{ mol } C_2H_5OH}{0.171 \text{ g } C_2H_5OH + 5.01 \text{ g } H_2O} = \frac{0.171}{5.18} = 0.0330$$

(e) Mole percent ethanol

$$\text{mole percent } C_2H_5OH = x_{C_2H_5OH} \times 100\% = 0.0330 \times 100\% = 3.30\%$$

(f) Molarity of ethanol

Divide the number of moles of ethanol from part (d) by the solution volume, 100.0 mL = 0.1000 L.

$$\text{molarity} = \frac{0.171 \text{ mol } C_2H_5OH}{0.1000 \text{ L soln}} = 1.71 \text{ M } C_2H_5OH$$

(g) Molality of ethanol

First, convert the mass of water present in 100.0 mL of solution [from part (d)] to the unit kg.

$$? \text{ kg } H_2O = 90.3 \text{ g } H_2O \times \frac{1 \text{ kg } H_2O}{1000 \text{ g } H_2O} = 0.0903 \text{ kg } H_2O$$

Use this result and the number of moles of C_2H_5OH from part (d) to establish the molality.

$$\text{molality} = \frac{0.171 \text{ mol } C_2H_5OH}{0.0903 \text{ kg } H_2O} = 1.89 \text{ } m \text{ } C_2H_5OH$$

Practice Example A: A solution that is 20.0% ethanol, by volume, is found to have a density of 0.977 g/mL. Use this fact, together with data from Example 14-1, to determine the mass percent ethanol in the solution.

Practice Example B: A 11.3-mL sample of CH_3OH ($d = 0.793$ g/mL) is dissolved in enough water to produce 75.0 mL of a solution with a density of 0.980 g/mL. What is the solution concentration expressed as **(a)** mole fraction H_2O; **(b)** molarity of CH_3OH; **(c)** molality of CH_3OH?

EXAMPLE 14-2

Converting Molarity to Mole Fraction. Laboratory ammonia is 14.8 M NH_3(aq) with a density of 0.8980 g/mL. What is x_{NH_3} in this solution?

Solution

No volume of solution is stated, suggesting that our calculation can be based on any fixed volume of our choice. A convenient volume to work with is one liter. We need to determine the number of moles of NH_3 and of H_2O in one liter of the solution.

$$\text{moles of } NH_3 = 1.00 \text{ L} \times \frac{14.8 \text{ mol } NH_3}{1 \text{ L}} = 14.8 \text{ mol } NH_3$$

To get the number of moles of H_2O, we can proceed as follows.

$$\text{mass of soln} = 1000.0 \text{ mL soln} \times \frac{0.8980 \text{ g soln}}{1.0 \text{ mL solution}} = 898.0 \text{ g soln}$$

$$\text{mass of } NH_3 = 14.8 \text{ mol } NH_3 \times \frac{17.03 \text{ g } NH_3}{1 \text{ mol } NH_3} = 252 \text{ g } NH_3$$

$$\text{mass of } H_2O = 898.0 \text{ g soln} - 252 \text{ g } NH_3 = 646 \text{ g } H_2O$$

$$\text{moles of } H_2O = 646 \text{ g } H_2O \times \frac{1 \text{ mol } H_2O}{18.02 \text{ g } H_2O} = 35.8 \text{ mol } H_2O$$

$$x_{NH_3} = \frac{14.8 \text{ mol } NH_3}{14.8 \text{ mol } NH_3 + 35.8 \text{ mol } H_2O} = 0.292$$

Practice Example A: A 16.00% aqueous solution of glycerol, $C_3H_5(OH)_3$, by mass, has a density of 1.037 g/mL. What is the mole fraction of $C_3H_5(OH)_3$ in this solution?

Practice Example B: A 10.00% aqueous solution of sucrose, $C_{12}H_{22}O_{11}$, by mass, has a density of 1.040 g/mL. What is **(a)** the molarity; **(b)** the molality; and **(c)** the mole fraction of $C_{12}H_{22}O_{11}$ in this solution?

14-3   Intermolecular Forces and the Solution Process

If there is even a little water in the fuel tank of an automobile, the engine will misfire. This problem would not occur if water were soluble in gasoline. Why does water not form solutions with gasoline? We can often understand a process if we analyze the process's energy requirements; this approach can help us to explain why some substances mix to form solutions and others do not. In this section, we focus on the behavior of molecules in solution, specifically on intermolecular forces and their contribution to the energy required for the dissolution process.

Enthalpy of Solution

In the formation of some solutions, heat is given off to the surroundings; in other cases, heat is absorbed. An enthalpy of solution, ΔH_{soln}, can be rather easily measured—for example, in the coffee-cup calorimeter of Figure 7-6—but why should some solution processes be exothermic, whereas others are endothermic?

Let's think in terms of a three-step approach to ΔH_{soln}. First, solvent molecules must be separated from one another to make room for the solute molecules. Some energy is required to overcome the forces of attraction between solvent molecules. As a result, this step should be an endothermic one: $\Delta H > 0$. Second, the solute molecules must be separated from one another. This step, too, will take energy and should be endothermic. Finally, we can imagine that we allow the separated solvent and solute molecules to be attracted to one another. These attractions will bring the molecules closer together and energy should be released. This is an exothermic step: $\Delta H < 0$. The enthalpy of solution is the sum of the three enthalpy changes just described, and depending on their relative values, ΔH_{soln} is either positive (endothermic) or negative (exothermic). This three-step process is summarized by equation (14.1) and Figure 14-2

(a)	pure solvent	$\longrightarrow$ separated solvent molecules	$\Delta H_a > 0$
(b)	pure solute	$\longrightarrow$ separated solute molecules	$\Delta H_b > 0$
(c)	separated solvent and solute molecules	$\longrightarrow$ solution	$\Delta H_c < 0$

Overall: pure solvent+pure solute $\longrightarrow$ solution

$$\Delta H_{soln} = \Delta H_a + \Delta H_b + \Delta H_c \tag{14.1}$$

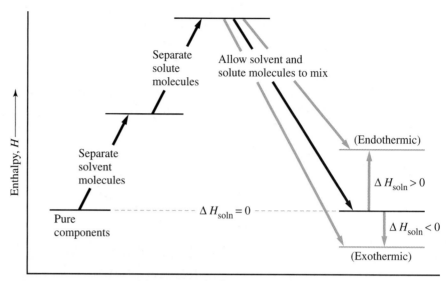

▲ FIGURE 14-2 **Enthalpy diagram for solution formation**
The solution process is endothermic (blue arrow), exothermic (red arrow), or has $\Delta H_{soln} = 0$ (black arrow), depending on the magnitude of the enthalpy change in the mixing step.

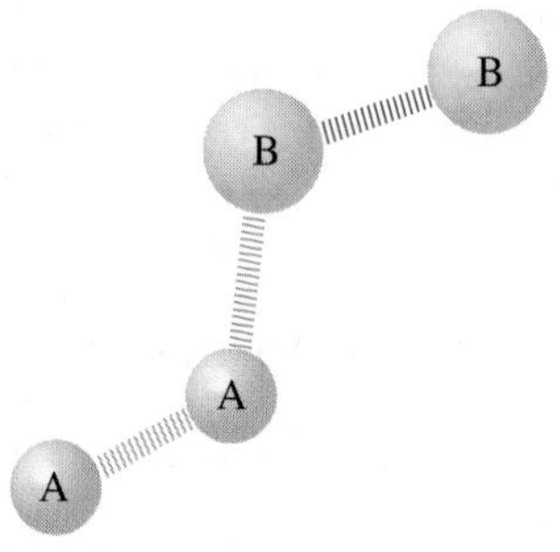

Intermolecular Forces in Mixtures

We see from equation (14.1) that the magnitude and sign of ΔH_{soln} depends on the values of the three terms ΔH_a, ΔH_b, and ΔH_c. These, in turn, depend on the strengths of the *three* kinds of intermolecular forces of attraction represented in Figure 14-3. Four possibilities for the relative strengths of these intermolecular forces are described in the discussion that follows.

▲ FIGURE 14-3
Representing intermolecular forces in a solution
The intermolecular forces of attraction represented here by springs are between: (1) solvent molecules A (yellow), (2) solute molecules B (red), and (3) solvent A and solute B (orange).

1. If the intermolecular forces of attraction shown in Figure 14-3 are of the same type and of equal strength, the solute and solvent molecules mix randomly. A homogeneous mixture or solution results. Because properties of solutions of this type can generally be predicted from the properties of the pure components, they are called **ideal solutions**. There is no overall enthalpy change in forming an ideal solution from its components, and $\Delta H_{soln} = 0$. This means that ΔH_c in equation (14.1) is equal in magnitude and opposite in sign to the sum of ΔH_a and ΔH_b. Many mixtures of liquid hydrocarbons fit this description, or very nearly so (Figure 14-4).

2. If forces of attraction between unlike molecules *exceed* those between like molecules, a solution also forms. The properties of such solutions generally cannot be predicted, however, and they are called *nonideal solutions*. Interac-

▲ FIGURE 14-4
Two components of a nearly ideal solution
Think of the —CH_3 group in toluene **(b)** as a small "bump" on the planar benzene ring **(a)**. Substances with similar molecular structures have similar intermolecular forces of attraction.

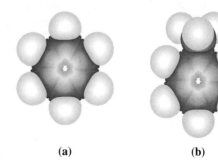

(a)   **(b)**

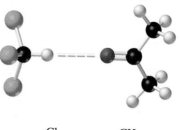

▶ FIGURE 14-5

Intermolecular force between unlike molecules leading to a nonideal solution

Hydrogen bonding between $CHCl_3$ (chloroform) and $(CH_3)_2CO$ (acetone) molecules produces forces of attraction between unlike molecules that exceed those between like molecules.

$$Cl-\underset{\underset{Cl}{|}}{\overset{\overset{Cl}{|}}{C}}-H\cdots O=\underset{\underset{CH_3}{|}}{\overset{\overset{CH_3}{|}}{C}}$$

tions between solute and solvent molecules (ΔH_c) release more heat than the heat absorbed to separate the solvent and solute molecules ($\Delta H_a + \Delta H_b$). The solution process is exothermic ($\Delta H_{soln} < 0$). Solutions of acetone and chloroform fit this type. As suggested by Figure 14-5, weak hydrogen bonding occurs between the two kinds of molecules, but the conditions for hydrogen bonding are not met in either of the pure liquids alone.[*]

3. If forces of attraction between solute and solvent molecules are *somewhat weaker* than between molecules of the same kind, complete mixing may still occur, but the solution formed is *nonideal*. The solution has a higher enthalpy than the pure components, and the solution process is endothermic. This type of behavior is observed in mixtures of carbon disulfide (CS_2), a nonpolar liquid, and acetone, a polar liquid. In these mixtures, the acetone molecules are attracted to other acetone molecules by dipole–dipole interactions and hence show a preference for other acetone molecules as neighbors. A possible explanation of how a solution process can be endothermic and still occur is found on page 544.

4. Finally, if forces of attraction between unlike molecules are *much* weaker than those between like molecules, the components remain segregated in a *heterogeneous mixture*. Dissolution does not occur to any significant extent. In a mixture of water and octane (a constituent of gasoline), strong hydrogen bonds hold water molecules together in clusters. The nonpolar octane molecules cannot exert a strong attractive force on the water molecules, and the two liquids do not mix. Thus, we now have an answer to the question posed at the beginning of this section of why water does not dissolve in gasoline or vice versa.

As an oversimplified summary of the four cases described in the preceding paragraphs, we can say that "like dissolves like." That is, substances with similar molecular structures are likely to exhibit similar intermolecular forces of attraction and to be soluble in one another. Substances with dissimilar structures are likely not to form solutions. Of course, in many cases, parts of the structures may be similar and parts may be dissimilar. Then it is a matter of trying to establish which are the more important parts, a matter we explore in Example 14-3.

[*]In most cases, H atoms bonded to C atoms cannot participate in hydrogen bonding. In a molecule like $CHCl_3$, however, the three Cl atoms have a strong electron-withdrawing effect on electrons in the C—H bond ($\mu = 1.01$ D). The H atom is then attracted to a lone pair of electrons on the O atoms of $(CH_3)_2CO$ (but not to Cl atoms in other $CHCl_3$ molecules).

? Are You Wondering...

What sort of intermolecular forces exist in a mixture of carbon disulfide and acetone?

Carbon disulfide is a nonpolar molecule, so in the pure substance the only intermolecular forces are weak London dispersion forces; carbon disulfide is a volatile liquid. Acetone is a polar molecule, and in the pure substance dipole–dipole forces are strong. Acetone is somewhat less volatile than carbon disulfide. In a solution of acetone in carbon disulfide, the dipoles of acetone molecules polarize carbon disulfide molecules, giving rise to *dipole-induced dipole* interactions.

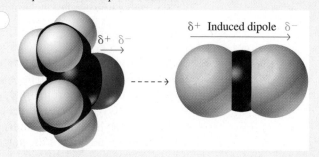

The dipole-induced dipole forces between acetone and carbon disulfide molecules are weaker than the dipole-dipole interactions among acetone molecules, causing the acetone molecules to be relatively less stable in their solutions with carbon disulfide than they are in pure acetone. As a result, acetone-carbon disulfide mixtures are nonideal solutions.

EXAMPLE 14-3

Using Intermolecular Forces to Predict Solution Formation. Predict whether or not a solution will form in each of the following mixtures and whether the solution is likely to be ideal. **(a)** ethyl alcohol, CH_3CH_2OH, and water (HOH); **(b)** the hydrocarbons hexane, $CH_3(CH_2)_4CH_3$, and octane, $CH_3(CH_2)_6CH_3$; **(c)** octyl alcohol, $CH_3(CH_2)_6CH_2OH$, and water (HOH).

Solution

KEEP IN MIND ▶
that ideal or nearly ideal solutions are not too common. They require the solvent and solute(s) to be quite similar in structure.

(a) If we think of water as H—OH, ethyl alcohol is similar to water. (Just substitute the group CH_3CH_2— for one of the H atoms in water.) Both molecules meet the requirements of hydrogen bonding as an important intermolecular force. The strengths of the hydrogen bonds between like molecules and between unlike molecules are likely to differ, however. We should expect ethyl alcohol and water to form *nonideal* solutions.

(b) In hexane, the carbon chain is six atoms long, and in octane it is eight. Both substances are virtually nonpolar, and intermolecular attractive forces (of the dispersion type) should be quite similar both in the pure liquids and in the solution. We expect a solution to form and for it to be nearly *ideal*.

▶ Butyl alcohol, $CH_3CH_2CH_2CH_2OH$, has a limited solubility in water (9 grams per 100 grams of water). The aqueous solubilities of alcohols fall off fairly rapidly as the hydrocarbon chain length increases beyond four.

(c) At first sight, this case may seem similar to (a), with the substitution of a hydrocarbon group for a H atom in H—OH. Here, however, the carbon chain is *eight* members long. This long carbon chain is much more important than the terminal —OH group in establishing the physical properties of octyl alcohol. Viewed from this perspective, octyl alcohol and water are quite *dissimilar*. We do not expect a solution to form.

Practice Example A: Which of the following organic compounds do you think is most readily soluble in water? Explain.

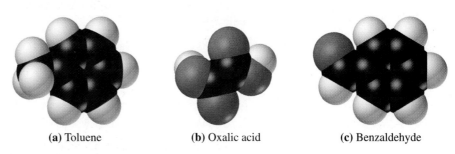

(a) Toluene   (b) Oxalic acid   (c) Benzaldehyde

Practice Example B: In which solvent is solid iodine likely to be more soluble, water or carbon tetrachloride? Explain.

Formation of Ionic Solutions

To assess the energy requirements for the formation of aqueous solutions of ionic compounds, we turn to the process pictured in Figure 14-6. Water dipoles are shown clustered around ions at the surface of a crystal. The negative ends of water dipoles are pointed toward the positive ions, and the positive ends of water dipoles toward negative ions. If these ion–dipole forces of attraction are strong enough to overcome the interionic forces of attraction in the crystal, dissolving will occur. Moreover, these ion–dipole forces also persist in the solution. An ion surrounded by a cluster of water molecules is said to be *hydrated*. Energy is *released* when ions become hydrated. The greater the hydration energy compared with the energy needed to separate ions from the ionic crystal, the more likely that the ionic solid will dissolve in water.

We can again use a *hypothetical* three-step process to describe the dissolution of an ionic solid. The energy requirement to dissociate a mole of an ionic solid into

Dissolution of NaCl in Water

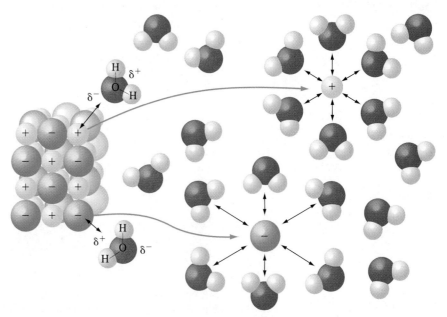

▲ **FIGURE 14-6   An ionic crystal dissolving in water**
Clustering of water dipoles around the surface of the ionic crystal and the formation of hydrated ions in solution are the key factors in the dissolution process.

▶ We discussed lattice energy in Section 13-9.

separated gaseous ions, an endothermic process, is the *negative* of the lattice energy. Energy is released in the next two steps—hydration of the gaseous cations and anions. The enthalpy of solution is the sum of these three ΔH values, described below for NaCl.

$$NaCl(s) \longrightarrow Na^+(g) + Cl^-(g) \qquad \Delta H_1 = (-\text{lattice energy of NaCl}) > 0$$

$$Na^+(g) \xrightarrow{H_2O} Na^+(aq) \qquad \Delta H_2 = (\text{hydration energy of } Na^+) < 0$$

$$Cl^-(g) \xrightarrow{H_2O} Cl^-(aq) \qquad \Delta H_3 = (\text{hydration energy of } Cl^-) < 0$$

$$NaCl(s) \xrightarrow{H_2O} Na^+(aq) + Cl^-(aq) \quad \Delta H_{soln} = \Delta H_1 + \Delta H_2 + \Delta H_3 \approx +5 \text{ kJ/mol}$$

The dissolution of sodium chloride in water is *endothermic*, and this is also the case for the vast majority (about 95%) of soluble ionic compounds. Why does NaCl dissolve in water if the process is endothermic? We might think that an endothermic process will not occur because of the increase in enthalpy. Since NaCl does actually dissolve in water, there must be a factor that we overlooked or are unaware of. In fact, we need to consider *two* factors to determine if a process will occur spontaneously. Enthalpy change is only one of them. The other factor, which we will introduce in Section 20-2 and call *entropy*, concerns the natural tendency for microscopic particles—atoms, ions, or molecules—to spread themselves out in the space available to them. The dispersed condition of the microscopic particles in NaCl(aq) compared with pure NaCl(s) and $H_2O(l)$ offsets the +5 kJ/mol increase in enthalpy in the solution process. In summary, if the hypothetical three-step process for solution formation is *exothermic*, we expect dissolution to occur; but we also expect a solution to form for an endothermic solution process, as long as ΔH_{soln} is not too large.

14-4   Solution Formation and Equilibrium

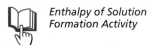
Enthalpy of Solution Formation Activity

In the previous section, we described what happens at the molecular (microscopic) level when solutions form. In this section, we will describe solution formation in terms of phenomena that we can actually observe, that is, a "macroscopic" view.

Figure 14-7 suggests what happens when a solid solute and liquid solvent are mixed. At first, only dissolution occurs, but soon the reverse process of crystallization becomes increasingly important; and some dissolved atoms, ions, or mol-

▲ **FIGURE 14-7   Formation of a saturated solution**
The lengths of the arrows represent the rate of dissolution (↑) and the rate of crystallization (↓). **(a)** When solute is first placed in the solvent, only dissolution occurs. **(b)** After a time, the rate of crystallization becomes significant. **(c)** The solution becomes saturated when the rates of dissolution and crystallization become equal.

ecules return to the undissolved state. When dissolution and crystallization occur at the same rate, the solution is in a state of dynamic equilibrium. The quantity of dissolved solute remains constant with time, and the solution is said to be a **saturated solution**. The concentration of the saturated solution is called the **solubility** of the solute in the given solvent. Solubility varies with temperature, and a solubility–temperature graph is called a *solubility curve*. Some typical solubility curves are shown in Figure 14-8.

If, in preparing a solution, we start with less solute than would be present in the saturated solution, the solute completely dissolves, and the solution is an **unsaturated solution**. On the other hand, suppose we prepare a saturated solution at one temperature and then change the temperature to a value at which the solubility is lower (this generally means a lower temperature). Usually, the excess solute crystallizes from solution, but occasionally all the solute may remain in solution. In these cases, because the quantity of solute is greater than in a saturated solution, the solution is said to be a **supersaturated solution**. A supersaturated solution is unstable, and if a few crystals of solute are added to serve as particles on which crystallization can occur, the excess solute crystallizes. Figure 14-8 shows how unsaturated and supersaturated solutions can be represented with a solubility curve.

▶ In some solutions, the solute and solvent are miscible (dissolve) in all proportions. The solution never becomes saturated. Ethyl alcohol–water solutions are an example.

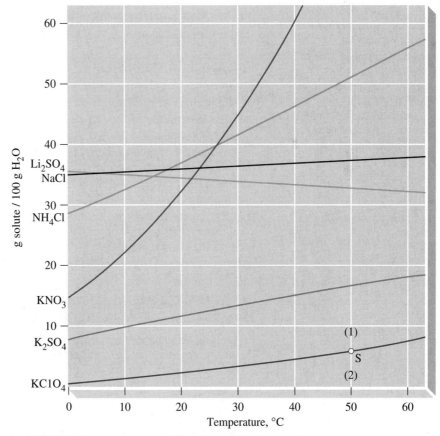

▲ **FIGURE 14-8   Aqueous solubility of several salts as a function of temperature**
Solubilities can be expressed in many ways: molarities, mass percent, or, as in this figure, grams of solute per 100 g H_2O. For each solubility curve (as shown here for $KClO_4$), points on the curve (S) represent saturated solutions. Regions above the curve (1) correspond to supersaturated solutions and below the curve (2), to unsaturated solutions.

Solubility as a Function of Temperature

As a general observation, the solubilities of ionic substances (about 95% of them) *increase* with increasing temperature. Exceptions to this generalization tend to be found among compounds containing the anions SO_3^{2-}, SO_4^{2-}, AsO_4^{3-}, and PO_4^{3-}.

In Chapter 16, we will learn to predict how an equilibrium condition changes with such variables as temperature and pressure by using an idea known as *Le Châtelier's principle*. One statement of the principle is that heat added to a system at equilibrium stimulates the heat-absorbing, or endothermic, reaction. This suggests that when $\Delta H_{soln} > 0$, raising the temperature stimulates dissolving and *increases* the solubility of the solute. Conversely, if $\Delta H_{soln} < 0$ (exothermic), the solubility *decreases* with increasing temperature. In this case, crystallization—being endothermic—is favored over dissolving.

We must be careful in applying the relationship we just described. The particular value of ΔH_{soln} that establishes whether solubility increases or decreases with increased temperature is that associated with dissolving a small quantity of solute in a solution that is already very nearly saturated. In some cases, this heat effect is altogether different from what we observe by adding a solute to the pure solvent. For example, when we dissolve NaOH in water, we observe a sharp increase in temperature—an *exothermic* process. This fact suggests that the solubility of NaOH in water should decrease as the temperature is raised. What we observe, though, is that the solubility of NaOH in water *increases* with increased temperature. This is because when a small quantity of NaOH is added to a solution that is already nearly saturated, heat is *absorbed*, not evolved.[*]

▲ FIGURE 14-9
Recrystallization of KNO₃
Colorless crystals of KNO₃ separated from an aqueous solution of KNO₃ and CuSO₄ (an impurity). The pale blue color of the solution is produced by Cu²⁺, which remains in solution.

Fractional Crystallization

Compounds synthesized in chemical reactions are generally impure, but the fact that the solubilities of most solids increase with increased temperature provides the basis for one simple method of purification. Usually, the impure solid consists of a high proportion of the desired compound and lesser proportions of the impurities. Suppose both the compound and its impurities are soluble in a particular solvent and that we prepare a concentrated solution at a high temperature. Then we let the concentrated solution cool. At lower temperatures, the solution becomes saturated in the desired compound. The excess compound crystallizes from solution. The impurities remain in solution because the temperature is still too high for these to crystallize.[†] This method of purifying a solid, called **fractional crystallization**, or **recrystallization**, is pictured in Figure 14-9. Example 14-4 illustrates how solubility curves can be used to predict the outcome of a fractional crystallization.

EXAMPLE 14-4

Applying Solubility Data in Fractional Crystallization. A solution is prepared by dissolving 95 g NH₄Cl in 200.0 g H₂O at 60 °C. **(a)** What mass of NH₄Cl will recrystallize when the solution is cooled to 20 °C? **(b)** How might we improve the yield of NH₄Cl?

---

[*]The solid in equilibrium with saturated NaOH(aq) over a range of temperatures around 25 °C is NaOH·H₂O(s). It is actually the temperature dependence of the solubility of this hydrate that we have been discussing.

[†]This is the usual behavior, but at times, one or more impurities may form a solid solution with the compound being recrystallized. In these cases simple recrystallization does not work as a method of purification.

Solution

(a) From Figure 14-8 we see that the solubility of NH_4Cl at 20 °C is 37 g NH_4Cl/100 g H_2O. The quantity of NH_4Cl in the saturated solution at 20 °C is

$$200.0 \text{ g } H_2O \times \frac{37 \text{ g } NH_4Cl}{100 \text{ g } H_2O} = 74 \text{ g } NH_4Cl$$

The mass of NH_4Cl recrystallized is $95 - 74 = 21$ g NH_4Cl.

(b) The yield of NH_4Cl in (a) is rather poor—21 g out of 95 g, or 22%. We can do better: (1) The solution at 60 °C, although concentrated, is not saturated. A saturated solution at 60 °C has 55 g NH_4Cl/100 g H_2O. Thus, the 95 g NH_4Cl requires less than 200.0 g H_2O to make a saturated solution. At 20 °C, a smaller quantity of saturated solution would contain less NH_4Cl than in (a), and the yield of recrystallized NH_4Cl would be greater. (2) Instead of cooling the solution to 20 °C, we might cool it to 0 °C. Here the solubility of NH_4Cl is less than at 20 °C, and more solid would recrystallize. (3) Still another possibility is to start with a solution at a temperature higher than 60 °C, say closer to 100 °C. The mass of water needed for the saturated solution would be less than at 60 °C. Note that options (1) and (3) both require changing the conditions by using a different amount of water from that originally specified.

Practice Example A: Calculate the quantity of NH_4Cl that would be obtained if suggestions (1) and (2) in part (b) were followed.

(*Hint:* Use data from Figure 14-8. What mass of water is needed to produce a saturated solution containing 95 g NH_4Cl at 60 °C?)

Practice Example B: Use Figure 14-8 to examine the solubility curves for the three potassium salts: $KClO_4$, K_2SO_4, and KNO_3. If saturated solutions of these salts at 40 °C are cooled to 20 °C, rank the salts in order of highest percent yield for the recrystallization.

KEEP IN MIND ▶
that fractional crystallization works best when (a) the quantities of impurities are small and (b) the solubility curve of the desired solute rises steeply with temperature.

14-5   Solubilities of Gases

Why does a freshly opened can of soda pop fizz, and why does the soda go flat after a time? To answer questions like these requires an understanding of the solubilities of gases. As we see in this section, the effect of temperature is generally different from what we find with solid solutes, and the pressure of a gas strongly affects its solubility.

Effect of Temperature

We cannot make an all-inclusive generalization about the effect of temperature on the solubilities of gases in solvents. It is certainly true, though, that the solubilities of most gases in water *decrease* with an increase in temperature. This is true of $N_2(g)$ and $O_2(g)$—the major components of air—and of air itself (Figure 14-10). This fact helps to explain why many types of fish can survive only in cold water. There is not enough dissolved air (oxygen) in warm water to sustain them.

For solutions of gases in organic solvents, the situation is often the reverse of that just described; that is, gases may become more soluble at higher temperatures. The solubility behavior of the noble gases in water is more complex. The solubility of each gas decreases with an increase in temperature, reaching a minimum at a certain temperature; then the solubility trend reverses direction, with the gas becoming more soluble with an increase in temperature. For example, for helium at 1 atm pressure, this minimum solubility in water comes at 35 °C.

▲ **FIGURE 14-10**
Effect of temperature on the solubilities of gases
Dissolved air is released as water is heated, even at temperatures well below the boiling point.

▲ The unopened bottle of soda water is under a high pressure of $CO_2(g)$. When a similar bottle is opened, the pressure quickly drops and some of the $CO_2(g)$ is released from solution (bubbles).

Effect of Pressure

Pressure affects the solubility of a gas in a liquid much more than does temperature. The English chemist William Henry (1775–1836) found that *the solubility of a gas increases with increasing pressure.* A mathematical statement of **Henry's law** is

$$C = k \cdot P_{gas} \qquad (14.2)$$

In this equation, C represents the solubility of a gas in a particular solvent at a fixed temperature, P_{gas} is the partial pressure of the gas above the solution, and k is a proportionality constant. To evaluate the proportionality constant k, we need to have one measurement of the solubility of the gas at a known pressure and temperature. For example, the aqueous solubility of $N_2(g)$ at 0 °C and 1.00 atm is 23.54 mL N_2 per liter. The Henry's law constant, k, is

$$k = \frac{C}{P_{gas}} = \frac{23.54 \text{ mL } N_2/L}{1.00 \text{ atm}}$$

Suppose we wish to increase the solubility of the $N_2(g)$ to a value of 100.0 mL N_2 per liter. Equation (14.2) suggests that to do so, we must increase the pressure of $N_2(g)$ above the solution. That is,

$$P_{N_2} = \frac{C}{k} = \frac{100.0 \text{ mL } N_2/L}{(23.54 \text{ mL } N_2/L)/1.00 \text{ atm}} = 4.25 \text{ atm}$$

At times, we are required to change the units used to express a gas solubility at the same time that the pressure is changed. This variation is illustrated in Example 14-5.

EXAMPLE 14-5

Using Henry's Law. At 0 °C and an O_2 pressure of 1.00 atm, the aqueous solubility of $O_2(g)$ is 48.9 mL O_2 per liter. What is the molarity of O_2 in a saturated water solution when the O_2 is under its normal partial pressure in air, 0.2095 atm?

Solution

Think of this as a two-part problem. (1) Determine the molarity of the saturated O_2 solution at 0 °C and 1 atm. (2) Use Henry's law in the manner outlined above.

Determine the molarity of O_2 at 0 °C when $P_{O_2} = 1$ atm.

Henry's Law animation

$$\text{molarity} = \frac{0.0489 \text{ L } O_2 \times \dfrac{1 \text{ mol } O_2}{22.4 \text{ L } O_2 \text{ (STP)}}}{1 \text{ L soln}} = 2.18 \times 10^{-3} \text{ M } O_2$$

Evaluate the Henry's law constant.

$$k = \frac{C}{P_{gas}} = \frac{2.18 \times 10^{-3} \text{ M } O_2}{1.00 \text{ atm}}$$

Apply Henry's law.

$$C = k \times P_{gas} = \frac{2.18 \times 10^{-3} \text{ M } O_2}{1.00 \text{ atm}} \times 0.2095 \text{ atm} = 4.57 \times 10^{-4} \text{ M } O_2$$

Practice Example A: Use data from Example 14-5 to determine the partial pressure of O_2 above its saturated aqueous solution at 0 °C. The solution is $8.23 \times 10^{-4} \text{ M } O_2$.

Practice Example B: A handbook lists the solubility of carbon monoxide in water at 0 °C and 1 atm pressure as 0.0354 mL CO per milliliter of H_2O. What pressure of CO(g) must be maintained above the solution to obtain 0.0100 M CO?

▲ To avoid the painful and dangerous condition of the bends, divers must not surface too quickly from great depths (see page 549).

We can rationalize Henry's law as follows: In a saturated solution, the rate of evaporation of gas molecules from solution and the rate of condensation of gas molecules into the solution are equal. Both of these rates depend on the number of molecules per unit volume. As the number of molecules per unit volume increases in the gaseous state (through an increase in the gas pressure), the number of molecules per unit volume must also increase in the solution (through an increase in concentration). Figure 14-11 illustrates this rationalization.

We see a practical application of Henry's law in carbonated beverages. The dissolved gas is carbon dioxide, and the higher the gas pressure maintained above the soda pop, the more CO_2 that dissolves. When we open a can of pop, we release some gas. As the gas pressure above the solution drops, dissolved CO_2 is expelled, usually fast enough to cause fizzing. In sparkling wines, the dissolved CO_2 is also under pressure, but rather than being added artificially as in soda pop, the CO_2 is produced by a fermentation process within the bottle.

Deep-sea diving provides us with still another example of Henry's law. Divers must carry a supply of air to breathe while underwater. If they are to stay submerged for any period of time, they must breathe compressed air. High-pressure air, however, is much more soluble in the blood and other body fluids than is air at normal pressures. When a diver returns to the surface, excess dissolved $N_2(g)$ is released as tiny bubbles from body fluids. When the ascent to the surface is made too quickly, N_2 diffuses out of the blood too quickly, causing severe pain in the limbs and joints, probably by interfering with the nervous system. This dangerous condition, known as "the bends," can be avoided if the diver ascends very slowly or spends time in a decompression chamber. Another effective method is to substitute a helium–oxygen mixture for compressed air. Helium is less soluble in blood than is nitrogen.

Henry's law (equation 14.2) fails for gases at high pressures; it also fails if the gas ionizes in water or reacts with water. For example, at 20 °C and with $P_{HCl} = 1$ atm, a saturated solution of HCl(aq) is about 20 M. But to prepare 10 M HCl, we do not need to maintain $P_{HCl} = 0.5$ atm above the solution, nor is $P_{HCl} = 0.05$ atm above 1 M HCl. We cannot even detect HCl(g) above 1 M HCl by its odor. The reason we cannot is that HCl ionizes in aqueous solutions, and in dilute solutions there are almost no molecules of HCl.

$$HCl(g) \xrightarrow{\text{H}_2\text{O}} H^+(aq) + Cl^-(aq)$$

We expect Henry's law to apply only to equilibrium between molecules of a gas and the same *molecules* in solution.

14-6 Vapor Pressures of Solutions

Separating compounds from one another is a task that chemists commonly face. If the compounds are volatile liquids, this separation often can be achieved by *distillation*. To understand how distillation works, we need to know something about the vapor pressures of solutions. An understanding of the vapor pressures of solutions will also enable us to deal with other important solution properties, such as boiling points, freezing points, and osmotic pressures.

To simplify the discussion that follows, we will consider only solutions with two components, solvent A and solute B. In the 1880s, the French chemist F. M. Raoult found that a dissolved solute *lowers* the vapor pressure of the solvent. **Raoult's law** states that the partial pressure exerted by solvent vapor above an ideal solution, P_A, is the product of the mole fraction of solvent in the solution, x_A, and the vapor pressure of the pure solvent at the given temperature, P_A°.

$$P_A = x_A P_A^\circ \tag{14.3}$$

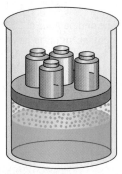

▲ FIGURE 14-11
Effect of pressure on the solubility of a gas
The concentration of dissolved gas (suggested by the depth of color) is proportional to the pressure of the gas above the solution (suggested by the density of the dots).

Equation (14.3) relates to Raoult's observation that a dissolved solute lowers the vapor pressure of the solvent because, if $x_A + x_B = 1.00$, x_A must be less than 1.00, and P_A must be smaller than P_A°. Strictly speaking, Raoult's law applies only to ideal solutions, and to all volatile components of the solutions. However, even in nonideal solutions, the law often works reasonably well for the *solvent* in *dilute* solutions, for example, solutions in which $x_{solv} > 0.98$. A more detailed discussion of Raoult's law requires the notion of entropy, which was briefly mentioned on page 544. Rather than attempt the explanation now, however, we will wait until Section 20-3, after we have said more about entropy.

EXAMPLE 14-6

Predicting Vapor Pressures of Ideal Solutions. The vapor pressures of pure benzene and toluene at 25 °C are 95.1 and 28.4 mmHg, respectively. A solution is prepared in which the mole fractions of benzene and toluene are both 0.500. What are the partial pressures of the benzene and toluene above this solution? What is the total vapor pressure?

Solution

We saw in Figure 14-4 that benzene–toluene solutions should be ideal. We expect Raoult's law to apply to both solution components.

$$P_{benz} = x_{benz} P_{benz}^\circ = 0.500 \times 95.1 \text{ mmHg} = 47.6 \text{ mmHg}$$
$$P_{tol} = x_{tol} P_{benz}^\circ = 0.500 \times 28.4 \text{ mmHg} = 14.2 \text{ mmHg}$$
$$P_{total} = P_{benz} + P_{tol} = 47.6 \text{ mmHg} + 14.2 \text{ mmHg} = 61.8 \text{ mmHg}$$

▶ Because the mole fractions of the two components are both 0.500, the total vapor pressure is just the average of the vapor pressures of the two components. This is a very special case, however.

Practice Example A: The vapor pressure of pure hexane and pentane at 25 °C are 149.1 mmHg and 508.5 mmHg, respectively. If a hexane–pentane solution has a mole fraction of hexane of 0.750, what are the vapor pressures of hexane and pentane above the solution? What is the total vapor pressure?

Practice Example B: Calculate the vapor pressures of benzene, C_6H_6, and toluene, C_7H_8, and the total pressure at 25 °C above a solution with equal *masses* of the two liquids. Use the vapor pressure data given in Example 14-6.

EXAMPLE 14-7

Calculating the Composition of Vapor in Equilibrium with a Liquid Solution. What is the composition of the vapor in equilibrium with the benzene–toluene solution of Example 14-6?

Solution

The ratio of each partial pressure to the total pressure is the mole fraction of that component in the vapor. (This is another application of equation 6.17.) The mole-fraction composition of the vapor is

$$x_{benz} = \frac{P_{benz}}{P_{total}} = \frac{47.6 \text{ mmHg}}{61.8 \text{ mmHg}} = 0.770$$

$$x_{tol} = \frac{P_{tol}}{P_{total}} = \frac{14.2 \text{ mmHg}}{61.8 \text{ mmHg}} = 0.230$$

Practice Example A: What is the composition of the vapor in equilibrium with the hexane–pentane solution described in Practice Example 14-6A?

Practice Example B: What is the composition of the vapor in equilibrium with the benzene–toluene solution described in Practice Example 14-6B?

Liquid–Vapor Equilibrium: Ideal Solutions

The results of Examples 14-6 and 14-7, together with similar data for other benzene–toluene solutions, are plotted in Figure 14-12. This figure consists of four lines—three straight and one curved—spanning the entire concentration range.

The red line shows how the vapor pressure of benzene varies with the solution composition. Because benzene in benzene–toluene solutions obeys Raoult's law, the red line has the equation $P_{benz} = x_{benz}P°_{benz}$. The blue line shows how the vapor pressure of toluene varies with solution composition and indicates that toluene also obeys Raoult's law. The dashed black line shows how the *total* vapor pressure varies with the solution composition. Can you see that each pressure on this black line is the sum of the pressures on the two straight lines that lie below it? Point 3 represents the total vapor pressure (point 1 + point 2) of a benzene–toluene solution in which $x_{benz} = 0.500$ (recall Example 14-6).

As we calculated in Example 14-7, the vapor in equilibrium with a solution in which $x_{benz} = 0.500$ is richer still in benzene. The vapor has $x_{benz} = 0.770$ (point 4). The line joining points 3 and 4 is called a *tie line*. Imagine establishing a series of tie lines throughout the composition range. The vapor ends of these tie lines can be joined to form the green curve in Figure 14-12. From the relative placement of the liquid and vapor curves, we see that for ideal solutions of two components, *the vapor phase is richer in the more volatile component than is the liquid phase*.

Vapor Pressure of Solutions activity

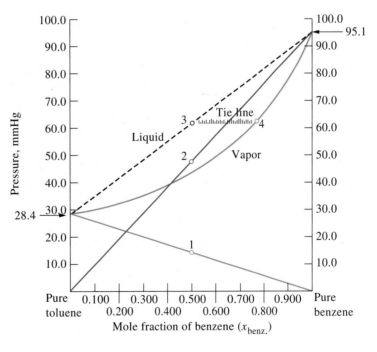

Vapor pressure of benzene
Vapor pressure of toluene
Total vapor pressure (and liquid composition)
Vapor composition

▲ **FIGURE 14-12** **Liquid–vapor equilibrium for benzene–toluene mixtures at 25 °C**
In this diagram, partial pressures and the total pressure of the vapor are plotted as a function of the solution and vapor compositions.

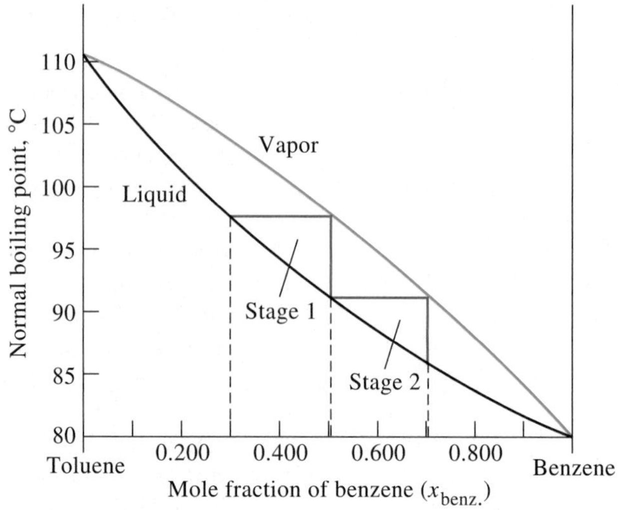

▶ **FIGURE 14-13**
Liquid–vapor equilibrium for benzene–toluene mixtures at 1 atm
In this diagram, the normal boiling points of solutions are plotted as a function of solution and vapor compositions.

Fractional Distillation

Let's look at liquid–vapor equilibrium in benzene–toluene mixtures in a somewhat different way. Instead of plotting vapor pressures as a function of the solution and vapor compositions, let's plot normal boiling temperature—the temperature at which the *total* vapor pressure of the solution is 1 atm. The resulting graph is shown in Figure 14-13. This graph is useful in explaining **fractional distillation**, a procedure for separating volatile liquids from one another.

Notice that the graph starts at a high temperature—110.6 °C, the boiling point of toluene—and ends at a lower temperature—80.0 °C, the boiling point of benzene. This is the reverse of the situation in Figure 14-12. Also, the vapor curve lies above the liquid curve in Figure 14-13, not below, as is the case in Figure 14-12.

Figure 14-13 indicates that a benzene–toluene solution with $x_{benz} = 0.30$ boils at a temperature of 98.6 °C and is in equilibrium with a vapor in which $x_{benz} = 0.51$. Imagine extracting some of this vapor and cooling it to the point where it condenses to a liquid. The new liquid will have $x_{benz} = 0.51$ and represents the conclusion of stage 1 in Figure 14-13. Now imagine repeating the process, that is, vaporizing a solution with $x_{benz} = 0.51$ and condensing the vapor. The new liquid at the end of stage 2 has $x_{benz} = 0.71$. By repeating the cycle, the vapor becomes progressively richer in benzene. As pictured in Figure 14-14, boiling solutions in equilibrium with vapor can be spread out over a long column, called a *fractionating column*, in which the equilibrium temperatures range from lowest at the top of the column to highest at the bottom. The most volatile component in the solution emerges from the top of the column as a vapor that is condensed to a liquid and removed. The least volatile component concentrates in the pot at the bottom of the column. Fractional distillation of a solution of many volatile components, such as petroleum, can be carried out in such a way that the components are withdrawn from the top of the column and condensed, one by one.

Liquid–Vapor Equilibrium: Nonideal Solutions

We cannot construct a liquid–vapor equilibrium diagram for nonideal solutions in the simple manner illustrated in Figure 14-12. For example, vapor pressures in acetone–chloroform solutions are *lower* than we would predict for ideal solutions and boiling temperatures are correspondingly *higher*. In acetone–carbon disulfide solutions, on the other hand, vapor pressures are higher than predicted and boiling temperatures are correspondingly lower. In Figure 14-5, we saw that forces of

KEEP IN MIND ▶

that the placement of the two curves in liquid–vapor equilibrium diagrams is such that the vapor is richer in the more volatile component than is the liquid. The more volatile component is the one with the higher vapor pressure or lower boiling point.

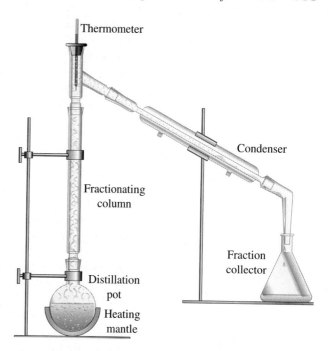

▶ **FIGURE 14-14**
Fractional distillation
The fractionating column is packed with glass beads or stainless steel turnings. Initially, as vapor rises from the pot and encounters these cooler objects, it condenses to a liquid. As the beads or turnings heat up, the liquid–vapor equilibrium front moves progressively up the column. Soon, liquid–vapor equilibrium occurs throughout the column, but with the equilibrium temperature changing continuously from the hottest regions at the bottom of the column to the coolest at the top. The vapor emerging from the top of the column is condensed to a liquid in the water-cooled condenser. The first fraction collected contains the most volatile component (lowest boiling point). Later fractions are less volatile liquids. The least volatile (highest boiling point) components remain as a residue in the distillation pot.

▲ **Fractional distillation is used in many industrial processes**

attraction between unlike molecules are greater than between like molecules in acetone–chloroform mixtures. It is not unreasonable to expect the components in such solutions to show a reduced tendency to vaporize and to have lower-than-predicted vapor pressures. With acetone–carbon disulfide solutions, the situation is the reverse: Forces of attraction between unlike molecules are weaker than between like molecules. This leads to greater tendencies for vaporization and higher vapor pressures than predicted by Raoult's law.

If the departures from ideal solution behavior are sufficiently great, some solutions may have vapor pressures that pass through either a maximum or a minimum in vapor-pressure-composition graphs. Correspondingly, their boiling points pass through either a minimum or maximum in boiling-point-composition graphs. The solutions corresponding to these maxima or minima boil at a constant temperature and produce a vapor having the *same* composition as the liquid. These solutions are called **azeotropes**. The boiling-point diagram of a minimum boiling-point azeotrope is illustrated in Figure 14-15.

▶ **FIGURE 14-15**
A minimum boiling point azeotrope
A solution of propanol in water having 71.69% $CH_3CH_2CH_2OH$ by mass—an azeotrope—has a lower boiling point than any other solution of these two components. In fractional distillation, solutions having *less than* 71.69% of the alcohol yield the azeotrope and water as ultimate products. Solutions with *more* than 71.69% of the alcohol yield the azeotrope and propanol. In each case, the azeotrope is drawn off through the condenser (Figure 14-14), and the other component remains in the pot.

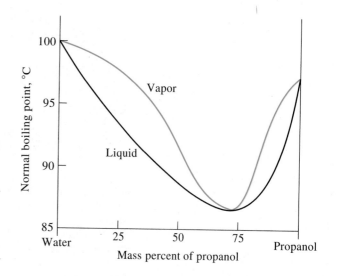

One of the most familiar azeotropes consists of 96.0% ethanol (C_2H_5OH) and 4.0% water, by mass, and has a boiling point of 78.174 °C. Pure ethanol has a boiling point of 78.3 °C. Dilute ethanol–water solutions can be distilled to produce the azeotrope, but the remaining water cannot be removed by ordinary distillation. As a result, most ethanol used in the laboratory or in industry is only 96.0% C_2H_5OH. To obtain absolute, or 100%, C_2H_5OH requires special measures.

14-7 Osmotic Pressure

Up to this point in our discussion, we have emphasized solutions containing a volatile solvent and volatile solute. Another common type of solution is one with a volatile solvent, such as water, but a *nonvolatile* solute(s), such as glucose, sucrose, or urea. Raoult's law still applies to the solvent in such solutions—the vapor pressure of the solvent is lowered.

Figure 14-16a pictures two aqueous solutions of a nonvolatile solute within the same enclosure. They are labeled A and B. The curved arrow indicates that water vaporizes from A and condenses into B. What is the driving force behind this? It must be that the vapor pressure of H_2O above A is greater than that above B. Solution A is more dilute; it has a higher mole fraction of H_2O. How long will this transfer of water continue? Solution A becomes more concentrated as it loses water, and solution B becomes more dilute as it gains water. When the mole fraction of H_2O is the same in each solution, the net transfer of H_2O stops.

A related phenomenon occurs when $CaCl_2 \cdot 6H_2O(s)$ is exposed to air (Figure 14-16b). Water vapor from the air condenses on the solid, and the solid begins to dissolve, a phenomenon known as *deliquescence*. In order for a solid to deliquesce, the partial pressure of water vapor in the air must be greater than the vapor pressure of water above a saturated aqueous solution of the solid. This requirement is often met for certain solids under conditions of an appropriate relative humidity. The deliquescence of $CaCl_2 \cdot 6H_2O$ occurs when the relative humidity exceeds 32%. (Relative humidity was described in Section 8-1.)

Like the case just described, Figure 14-17 also pictures the flow of solvent molecules. Here, however, the flow is not through the vapor phase. An aqueous sucrose (sugar) solution in a long glass tube is separated from pure water by a

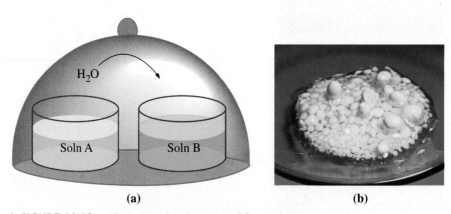

(a)	(b)

▲ FIGURE 14-16 **Observing the direction of flow of water vapor**
(a) Water passes, as vapor, from the more dilute solution (higher mole fraction of H_2O) to the more concentrated solution. (b) Water vapor in air condenses onto solid calcium chloride hexahydrate, $CaCl_2 \cdot 6H_2O$. The liquid water dissolves some of the solid. The eventual result could be an unsaturated solution.

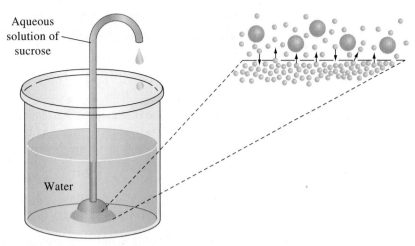

▲ **FIGURE 14-17** **An illustration of osmosis**
Water molecules pass through pores in the membrane and create a pressure within the funnel that causes the solution to rise and overflow. After a time, the solution inside the funnel becomes more dilute and the pure water outside becomes a sucrose solution. Liquid flow stops when the compositions of the solutions separated by the membrane have become nearly equal. Can you see the similarity between this phenomenon and that pictured in Figure 14-16a? The only difference is that here water remains in the liquid phase throughout.

Semipermeable Membrane animation

▶ Semipermeable membranes are materials containing submicroscopic holes, such as pig's bladder, parchment, or cellophane. The holes permit the passage of solvent molecules but not those of the solute.

▶ The adjustment required to apply equation (14.4) to electrolyte solutions is discussed in Section 14-9.

semipermeable membrane (permeable to water only). Water molecules can pass through the membrane in either direction, and they do. But because the concentration of water molecules is greater in the pure water than in the solution, there is a net flow from the pure water into the solution. This net flow, called **osmosis**, causes the solution to rise in the tube. The more concentrated the sucrose solution, the higher the solution level rises.

Applying a pressure to the sucrose solution slows down the net flow of water across the membrane into the solution. With a sufficiently high pressure, the net influx of water can be stopped altogether. The necessary pressure to stop osmotic flow is called the **osmotic pressure** of the solution. For a 20% sucrose solution, this pressure is about 15 atm. The magnitude of osmotic pressure depends only on the number of solute particles per unit volume of solution. It does not depend on the identity of the solute. Properties of this sort, whose values depend only on the concentration of solute particles in solution and *not* on what the solute is, are called **colligative properties**. The following equation works quite well for calculating osmotic pressures of *dilute* solutions of nonelectrolytes. The osmotic pressure is represented by the symbol π; R is the gas constant (0.08206 L atm mol^{-1} K^{-1}); and T is the Kelvin temperature. The term n represents the amount of solute (in moles), and V is the volume (in liters) of solution. Notice that this equation is similar to the equation for the ideal gas law. In this case, however, it is convenient to rearrange terms to yield equation (14.4). The ratio, n/V, then, is the *molarity* of the solution, represented by the symbol M.

$$\pi V = nRT$$

$$\pi = \frac{n}{V} RT = \text{M} \times RT \tag{14.4}$$

EXAMPLE 14-8

Calculating Osmotic Pressure. What is the osmotic pressure at 25 °C of an aqueous solution that is 0.0010 M $C_{12}H_{22}O_{11}$ (sucrose)?

Solution

We just need to substitute the data into equation (14.4).

$$\pi = \frac{0.0010 \text{ mol} \times 0.08206 \text{ L atm mol}^{-1} \text{ K}^{-1} \times 298 \text{ K}}{1 \text{ L}}$$

$$\pi = 0.024 \text{ atm (18 mmHg)}$$

Practice Example A: What is the osmotic pressure at 25 °C of an aqueous solution that contains 1.50 g $C_{12}H_{22}O_{11}$ in 125 mL of solution?

Practice Example B: What mass of urea $[CO(NH_2)_2]$ would you dissolve in 225 mL of solution to obtain an osmotic pressure of 0.015 atm at 25 °C?

The pressure difference of 18 mmHg that we calculated in Example 14-8 is easy to measure. (It corresponds to a solution height of about 25 cm.) This means that we can easily use the measurement of osmotic pressure for determining molar masses when we are dealing with very dilute solutions or solutes with high molar masses (or both). Example 14-9 shows how osmotic pressure measurements can be used to determine molar mass.

EXAMPLE 14-9

Establishing a Molar Mass from a Measurement of Osmotic Pressure. A 50.00-mL sample of an aqueous solution contains 1.08 g of human serum albumin, a blood plasma protein. The solution has an osmotic pressure of 5.85 mmHg at 298 K. What is the molar mass of the albumin?

Solution

First we need to express the osmotic pressure in atmospheres.

$$\pi = 5.85 \text{ mmHg} \times \frac{1 \text{ atm}}{760 \text{ mmHg}} = 7.70 \times 10^{-3} \text{ atm}$$

Now, we can modify equation (14.4) slightly [that is, with the number of moles of solute (n) represented by the mass of solute (m) divided by the molar mass (M)], and solve it for M.

$$\pi = \frac{(m/M) RT}{V} \quad \text{and} \quad M = \frac{mRT}{\pi V}$$

$$M = \frac{1.08 \text{ g} \times 0.08206 \text{ L atm mol}^{-1} \text{ K}^{-1} \times 298 \text{ K}}{7.70 \times 10^{-3} \text{ atm} \times 0.0500 \text{ L}} = 6.86 \times 10^4 \text{ g/mol}$$

Practice Example A: Creatinine is a by-product of nitrogen metabolism and can be used to provide an indication of renal function. A 4.04-g sample of creatinine is dissolved in enough water to make 100.0 mL of solution. The osmotic pressure of the solution is 8.73 mmHg at 298 K. What is the molar mass of creatinine?

Practice Example B: What would be the osmotic pressure of a solution containing 2.12 g of human serum albumin in 75.00 mL of water at 37.0 °C? Use the molar mass determined in Example 14-9.

▲ A normal red blood cell (top) and a red blood cell from a hypertonic solution (center).

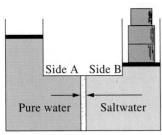

▲ **FIGURE 14-18**
Desalination of saltwater by reverse osmosis
The membrane is permeable to water but not to ions. The normal flow of water is from side A to side B. If we exert a pressure on side B that exceeds the osmotic pressure of the saltwater, a net flow of water occurs in the *reverse* direction—from the saltwater to the pure water. The lengths of the arrows suggest the magnitudes of the flow of water molecules in each direction.

Practical Applications

Some of the best examples of osmosis are those associated with living organisms. For instance, consider red blood cells. If we place red blood cells in pure water, the cells expand and eventually burst as a result of water that enters through osmosis. The osmotic pressure associated with the fluid inside the cell is equivalent to that of 0.92% (mass/volume) NaCl(aq). Thus, if we place a cell in a sodium chloride (saline) solution of this concentration, there is no net flow of water through the cell membrane, and the cell remains stable. Such a solution is said to be *isotonic*. If we place cells in a solution with a concentration greater than 0.92% NaCl, water flows out of the cells and the cells shrink. The solution is *hypertonic*. If the NaCl concentration is less than 0.92%, the solution is *hypotonic*, and water flows into the cells. Fluids that are intravenously injected into patients to combat dehydration or to supply nutrients must be adjusted so that they are isotonic with blood. The osmotic pressure of the fluids must be the same as that of 0.92% (mass/volume) NaCl.

One recent application of osmosis goes to the very definition of osmotic pressure. Suppose in the device shown in Figure 14-18 we apply a pressure to the right side (side B) that is less than the osmotic pressure of the saltwater. The net flow of water molecules through the membrane will be from side A to side B. Ordinary osmosis occurs. If we apply a pressure greater than the osmotic pressure to side B, we can cause a net flow of water in the *reverse* direction, from the saltwater into the pure water. This is the condition known as **reverse osmosis**. Reverse osmosis can be used in the *desalination* of seawater to supply drinking water for emergency situations or as an actual source of municipal water. Another application of reverse osmosis is the removal of dissolved materials from industrial or municipal wastewater before it is discharged into the environment.

14-8 Freezing-Point Depression and Boiling-Point Elevation of Nonelectrolyte Solutions

In Section 14-6 we examined the lowering of the vapor pressure of a solvent produced by a dissolved solute. Vapor pressure lowering is not so frequently measured as are certain properties directly related to it. To aid in this discussion, let's refer to Figure 14-19. The blue curves represent the vapor pressure, fusion, and sublimation curves in the phase diagram for a pure solvent. The red curves represent the vapor pressure and fusion curves of the solvent in a solution. The sublimation curve for the solid solvent that freezes from the solution is shown in purple. Two assumptions are implicit in Figure 14-19. One is that the solute is nonvolatile, and the other is that the solid that freezes from a solution is pure solvent. For many mixtures, these requirements are easily met.[*]

The vapor pressure curve of the solution (red) intersects the sublimation curve at a lower temperature than is the case for the pure solvent. The solid–liquid fusion curve, because it originates at the intersection of the sublimation and vapor pressure curves, is also displaced to lower temperatures. Now recall how we establish normal melting points and boiling points in a phase diagram. They are the temperatures at which a line at $P = 1$ atm intersects the fusion and vapor pressure curves, respectively. We have highlighted four points of intersection in Figure 14-19—the freezing points and the boiling points of the pure solvent and of the solvent in a solution. The freezing point of the solvent in solution is *depressed*, and the boiling point is *elevated*.

The extent to which the freezing point is lowered or the boiling point raised is proportional to the mole fraction of solute (just as is vapor pressure lowering). In *dilute* solutions, the solute mole fraction is proportional to its molality, and so we can write

[*] Actually, the equation for freezing-point depression (14.5) applies even if the solute is volatile.

▲ A small reverse osmosis unit used to desalinize seawater.

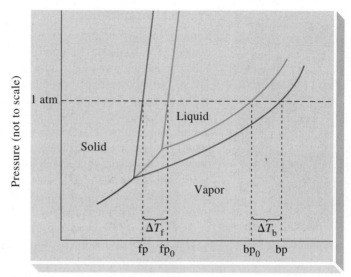

▲ **FIGURE 14-19** **Vapor pressure lowering by a nonvolatile solute**
The normal freezing point and normal boiling point of the pure solvent are fp_0 and bp_0, respectively. The corresponding points for the solution are fp and bp. The freezing-point depression, ΔT_f, and the boiling-point elevation, ΔT_b, are indicated. Because the solute is assumed to be insoluble in the solid solvent, the sublimation curve of the solvent is unaffected by the presence of solute in the liquid solution phase. That is, the sublimation curve is the same for the two phase diagrams.

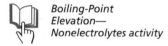

Boiling-Point Elevation— Nonelectrolytes activity

$$\Delta T_f = -K_f \times m \tag{14.5}$$

$$\Delta T_b = K_b \times m \tag{14.6}$$

KEEP IN MIND ▶

that freezing-point depression (ΔT_f) is defined as $T - T_f$, where T is the freezing point of the solution and T_f is the freezing point of the pure solvent, and similarly the boiling-point elevation (ΔT_b) is defined as $T - T_b$ where T_b is the boiling point of the pure solvent. So the need for the negative sign in equation 14.5 is evident.

In these equations, ΔT_f and ΔT_b are the freezing-point depression and boiling-point elevation, respectively; m is the solute molality; and K_f and K_b are proportionality constants. The value of K_f depends on the melting point, enthalpy of fusion, and molar mass of the solvent. The value of K_b depends on the boiling point, enthalpy of vaporization, and molar mass of the solvent. The units of K_f and K_b are $°C\ m^{-1}$, and you can think of their values as representing the freezing-point depression and boiling-point elevation for a 1 m solution. In practice, though, equations (14.5) and (14.6) often fail for solutions as concentrated as 1 m. Table 14.2 lists some typical values of K_f and K_b.

TABLE 14.2 Freezing-Point Depression and Boiling-Point Elevation Constants		
Solvent	K_f	K_b
Acetic acid	3.90	3.07
Benzene	5.12	2.53
Nitrobenzene	8.1	5.24
Phenol	7.27	3.56
Water	1.86	0.512

Values correspond to freezing-point depressions and boiling-point elevations, in degrees Celsius, due to 1 mol of solute particles dissolved in 1 kg of solvent in an ideal solution. Units: $°C$ kg solvent (mol solute)$^{-1}$ or $°C\ m^{-1}$.

Historically, chemists have used the group of colligative properties—vapor pressure lowering, freezing-point depression, boiling-point elevation, and osmotic pressure—for molecular mass determinations. In Example 14-9, we showed how this could be accomplished with osmotic pressure. Example 14-10 shows how freezing-point depression can be used to determine a molar mass and, with other information, a molecular formula. To help you understand how this is done, we present a three-step procedure in the form of answers to three separate questions. In other cases, you should be prepared to work out your own stepwise procedure.

EXAMPLE 14-10

Establishing a Molecular Formula with Freezing-Point Data. Nicotine, extracted from tobacco leaves, is a liquid completely miscible with water at temperatures below 60 °C. **(a)** What is the *molality* of nicotine in an aqueous solution that starts to freeze at -0.450 °C? **(b)** If this solution is obtained by dissolving 1.921 g of nicotine in 48.92 g H_2O, what must be the molar mass of nicotine? **(c)** Combustion analysis shows nicotine to consist of 74.03% C, 8.70% H, and 17.27% N, by mass. What is the molecular formula of nicotine?

Solution

(a) We can establish the molality of the nicotine by using equation (14.5) with the value of K_f for water listed in Table 14.2. Note that $T_f = -0.450$ °C, and that $\Delta T_f = -0.450$ °C $- 0.000$ °C $= -0.450$ °C.

$$\text{molality} = \frac{\Delta T_f}{-K_f} = \frac{-0.450 \text{ °C}}{-1.86 \text{ °C m}^{-1}} = 0.242 \ m$$

(b) Here we can use the definition of molality, but with a known molality $(0.242 \ m)$ and an unknown molar mass of solute (M). The number of moles of solute is simply $1.921 \text{ g}/M$.

$$\text{molality} = \frac{1.921 \text{ g } M^{-1}}{0.04892 \text{ kg water}} = \frac{0.242 \text{ mol}}{\text{kg water}}$$

$$M = \frac{1.921 \text{ g}}{(0.04892 \times 0.242)\text{ mol}} = 162 \text{ g/mol}$$

(c) To establish the empirical formula of nicotine, we need to use the method of Example 3-5. This calculation is left as an exercise for you to do. The result you should obtain is C_5H_7N. The formula mass based on this empirical formula is 81 u. The molecular mass obtained from the molar mass in part (b) is exactly twice this value—162 u. The molecular formula is twice C_5H_7N, or $C_{10}H_{14}N_2$.

Practice Example A: Vitamin B_2, riboflavin, is soluble in water. If 0.833 g of riboflavin is dissolved in 18.1 g H_2O, the resulting solution has a freezing point of -0.227 °C. **(a)** What is the molality of the solution? **(b)** What is the molar mass of riboflavin? **(c)** What is the molecular formula of riboflavin if combustion analysis shows it to consist of 54.25% C, 5.36% H, 25.51% O, and 14.89% N?

Practice Example B: An aqueous solution that is 0.205 m urea $[CO(NH_2)_2]$ is found to boil at 100.025 °C. Is the prevailing barometric pressure above or below 760.0 mmHg?

[*Hint:* At what temperature should the solution begin to boil if atmospheric pressure is 760.0 mmHg?]

▶ The adjustment required to apply these equations to electrolyte solutions is discussed in Section 14-9.

Molar mass determination by freezing-point depression or boiling-point elevation has its limitations. First, equations (14.5) and (14.6) apply only to dilute solutions of nonelectrolytes, usually much less than 1 *m*. This requires the use of special thermometers, so that temperatures can be measured very precisely, say to ±0.001 °C. Because boiling points depend on barometric pressure, precise measurements require that pressure be held constant. As a consequence, boiling-point elevation is not much used. The precision of the freezing-point depression method can be improved by using a solvent with a larger K_f value than that of water. For example, for cyclohexane $K_f = 20.0 \ °C \ m^{-1}$ and for camphor $K_f = 37.7 \ °C \ m^{-1}$.

Practical Applications

The typical automobile antifreeze is ethylene glycol, $C_2H_4(OH)_2$. It is a good idea to leave the ethylene glycol–water mixture in the cooling system at all times to provide all-weather protection. In summer, the ethylene glycol helps by raising the boiling point of water and preventing cooling system boilover.

▲ Water sprayed on citrus fruit releases its heat of fusion as it freezes into a layer of ice that acts as a thermal insulator. For a time, the temperature remains at 0 °C. The juice of the fruit, having a freezing point below 0 °C, is protected from freezing.

▲ A typical aircraft deicer is propylene glycol, $CH_3CHOHCH_2OH$, diluted with water and applied as a hot, high-pressure spray.

▲ Lowering the freezing point of water on roads.

Citrus growers faced with an impending freeze know they must take preventive measures only if the temperature drops below 0 °C by several degrees. The juice in the fruit has enough dissolved solutes to lower the freezing point by a degree or two. The growers also know they must protect lemons sooner than oranges because lemons have a lower concentration of dissolved solutes (sugars) than do oranges.

Salts such as NaCl can be used to prepare a *slush bath*, a mixture used to cool or freeze something. One example is the mixture of ice and NaCl(s) used to freeze ice cream in a home ice cream maker. Because the slush bath is at a temperature well below 0 °C, it is easy to freeze the sugar and milk mixture that makes up the ice cream. NaCl is also useful for deicing roads. It is effective in melting ice at temperatures as low as −21 °C (−6 °F). This is the lowest freezing point of a NaCl(aq) solution.

14-9   Solutions of Electrolytes

Our discussion of the electrical conductivities of solutions in Section 5-1 retraced some of the work done by the Swedish chemist Svante Arrhenius for his doctoral dissertation (1883). Prevailing opinion at the time was that ions form only with the passage of electric current. Arrhenius, however, reached the conclusion that ions exist in a solid substance and become dissociated from each other when the solid dissolves in water. Such is the case with NaCl, for example. In other cases, as with

Are You Wondering...

If there is a molecular interpretation of the depression of the freezing point of a solvent by a solute?

When the solid and liquid phases of a pure substance coexist at a particular temperature, two processes occur. First, liquid molecules that collide with the solid sometimes are captured and added to the solid phase. At the same time, molecules on the surface of the solid sometimes become detached and enter the liquid phase. There is a state of dynamic equilibrium in which, at any given time, the number of molecules leaving the surface of the solid matches the number of molecules that enter the solid from the liquid phase. There is no net change even though individual molecules continue to move back and forth between the phases.

Now, imagine adding a solute to the liquid coexisting with its solid phase. In the solution formed above the solid phase, solute molecules replace some of the solvent molecules, and as a consequence, a given volume of solution contains a smaller number of solvent molecules than does the same volume of pure solvent. The dynamic equilibrium between liquid and solid solvent that existed in the pure solvent is disrupted since fewer solvent molecules in the liquid can get to the surface of the solid in a given time. The rate at which molecules leave the liquid phase is reduced and is no longer equal to the rate at which molecules leave the pure solid phase. However, cooling the solution restores the dynamic equilibrium because it simultaneously reduces the number of molecules that have sufficient energy to break away from the surface of the solid and increases the number of molecules in the liquid phase with sufficiently low kinetic energy to be captured by the solid. The lowered equilibrium temperature corresponds to the freezing-point depression.

Although this kinetic-molecular interpretation of freezing-point depression is an appealing one, the phenomenon is better described by the thermodynamic concept of entropy, which we will introduce in Chapter 20.

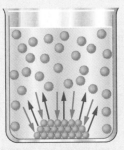  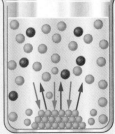

| Equilibrium at freezing point of solvent | Equilibrium disrupted by solute added to solvent | Equilibrium restored at a lower temperature |

▲ Molecular view of freezing-point depression. The addition of solute to the liquid solvent does not change the rate of escape of molecules from the solid phase, but it does decrease the rate at which solvent molecules can enter the solid phase. The dynamic equilibrium between solid and liquid solvent is disrupted, and can be re-established only at a lower temperature.

▲ **Svante Arrhenius (1859–1927)**
At the time Arrhenius was awarded the Nobel Prize in chemistry (1903), his results were described thus: "Chemists would not recognize them as chemistry; nor physicists as physics. They have in fact built a bridge between the two." The field of physical chemistry had its origins in Arrhenius's work.

▶ Later in this section, we explain why the experimentally determined i for 0.0100 m NaCl is 1.94 instead of 2.

 Boiling-Point Elevation—Electrolytes activity

▶ Because the value of i for $MgCl_2$ is not exactly 3, we are not justified in carrying more than one or two significant figures in our answer.

HCl, ions do not exist in the substance but are formed when it dissolves in water. In any case, electricity is not required to produce ions.

Although Arrhenius developed his theory of electrolytic dissociation to explain the electrical conductivities of solutions, he was able to apply it more widely. One of his first successes came in explaining certain anomalous values of colligative properties described by the Dutch chemist Jacobus van't Hoff (1852–1911).

Anomalous Colligative Properties

Certain solutes produce a greater effect on colligative properties than expected. For example, consider a 0.0100 m aqueous solution. The predicted freezing-point depression of this solution is

$$\Delta T_f = -K_f \times m = -1.86 \,°C \, m^{-1} \times 0.0100 \, m = -0.0186 \,°C$$

We expect the solution to have a freezing point of $-0.0186 \,°C$. If the 0.0100 m solution is 0.0100 m urea, the measured freezing point is just about $-0.0186 \,°C$. If the solution is 0.0100 m NaCl, however, the measured freezing point is about $-0.0361 \,°C$.

Van't Hoff defined the factor i as the ratio of the measured value of a colligative property to the expected value if the solute is a nonelectrolyte. For 0.0100 m NaCl,

$$i = \frac{\text{measured } \Delta T_f}{\text{expected } \Delta T_f} = \frac{-0.0361 \,°C}{-1.86 \,°C \, m^{-1} \times 0.0100 \, m} = 1.94$$

Arrhenius's theory of electrolytic dissociation allows us to explain different values of the van't Hoff factor i for different solutes. For solutes such as urea, glycerol, and sucrose (all nonelectrolytes), $i = 1$. For a strong electrolyte such as NaCl, which produces *two* moles of ions in solution per mole of solute dissolved, we would expect the effect on freezing-point depression to be twice as great as for a nonelectrolyte. We would expect that $i = 2$. Similarly, for $MgCl_2$, our expectation would be that $i = 3$. For the weak acid $HC_2H_3O_2$ (acetic acid), which is only slightly ionized in aqueous solution, we expect i to be slightly larger than 1 but not nearly equal to 2.

This discussion suggests that equations (14.4), (14.5), and (14.6) should all be rewritten in the form

$$\pi = i \times M \times RT$$

$$\Delta T_f = -i \times K_f \times m$$

$$\Delta T_b = i \times K_b \times m$$

If these equations are used for nonelectrolytes, simply substitute $i = 1$. For strong electrolytes, predict a value of i as suggested in Example 14-11.

EXAMPLE 14-11

Predicting Colligative Properties for Electrolyte Solutions. Predict the freezing point of aqueous 0.00145 m $MgCl_2$.

Solution

First determine the value of i for $MgCl_2$. We can do this by writing an equation to represent the dissociation of $MgCl_2(s)$.

$$MgCl_2(s) \xrightarrow{H_2O} Mg^{2+}(aq) + 2 \, Cl^-(aq)$$

Because *three* moles of ions are obtained per mole of formula units dissolved, we expect the value $i = 3$.

Now use the expression

$$\Delta T_f = -i \times K_f \times m$$
$$= -3 \times 1.86 \,°C\, m^{-1} \times 0.00145 \, m$$
$$= -0.0081 \,°C$$

The predicted freezing point is 0.0081 °C below the normal freezing point of water. The predicted freezing point is -0.0081 °C.

Practice Example A: What is the expected osmotic pressure of a 0.0530 M MgCl$_2$ solution at 25 °C?

Practice Example B: You wish to prepare an aqueous solution that has a freezing point of -0.100 °C. How many milliliters of 12.0 M HCl would you use to prepare 250.0 mL of such a solution?

(*Hint:* Note that in a *dilute* aqueous solution, molality and molarity are essentially numerically equal.)

Interionic Attractions

Despite its initial successes, there were apparent deficiencies in Arrhenius's theory. The electrical conductivities of concentrated solutions of strong electrolytes are not as great as expected, and values of the van't Hoff factor i depend on the solution concentrations, as shown in Table 14.3. For strong electrolytes that exist completely in ionic form in aqueous solutions, we would expect $i = 2$ for NaCl, $i = 3$ for MgCl$_2$, and so on, regardless of the solution concentration.

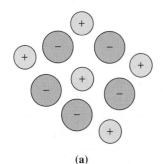

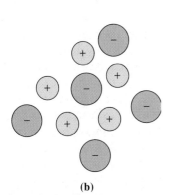

(a)

(b)

▲ **FIGURE 14-20**
Interionic attractions in aqueous solution
(a) A positive ion in aqueous solution is surrounded by a shell of negative ions. **(b)** A negative ion attracts positive ions to its immediate surroundings.

TABLE 14.3	Variation of the van't Hoff Factor, *i*, with Solution Molality					
	Molality, *m*					
Solute	**1.0**	**0.10**	**0.010**	**0.0010**	**...**	**Inf dil***
NaCl	1.81	1.87	1.94	1.97	...	2
MgSO$_4$	1.09	1.21	1.53	1.82	...	2
Pb(NO$_3$)$_2$	1.31	2.13	2.63	2.89	...	3

*The limiting values: $i = 2, 2$, and 3 are reached when the solution is infinitely dilute. Note that a solute whose ions are singly charged (for example, NaCl) approaches its limiting value more quickly than does a solute whose ions carry higher charges. Interionic attractions are greater in solutes with more highly charged ions.

These difficulties can be resolved with a theory of electrolyte solutions proposed by Peter Debye and Erich Hückel in 1923. This theory continues to view strong electrolytes as existing only in ionic form in aqueous solution, but the ions in solution *do not behave independently of one another*. Instead, each cation is surrounded by a cluster of ions in which anions predominate, and each anion is surrounded by a cluster in which cations predominate. In short, each ion is enveloped by an *ionic atmosphere* with a net charge opposite its own (Figure 14-20).

In an electric field, the mobility of each ion is reduced because of the attraction or drag exerted by its ionic atmosphere. Similarly, the magnitudes of colligative properties are reduced. This explains why, for example, the value of i for 0.010 m NaCl is 1.94 rather than 2.00. What we can say is that each type of ion in an aqueous

Are You Wondering...

How much you need to know about activities?

First, it is worth noting that in sufficiently dilute solutions, activity coefficients approximate one: Activities and stoichiometric concentrations are the same. In doing calculations, we'll work only with stoichiometric concentrations because a quantitative treatment of activity coefficients is beyond the scope of this text. Mainly, you should be aware of the distinction between activity and stoichiometric concentration because it helps to explain a number of chemical phenomena, as we shall see again in a few other instances in the text.

solution has two "concentrations." One is called the *stoichiometric concentration* and is based on the amount of solute dissolved. The other is the "effective" concentration, called the *activity*, which takes into account interionic attractions. Stoichiometric calculations of the type presented in Chapters 4 and 5 can be made with great accuracy using stoichiometric concentrations. However, no calculations involving solution properties are 100% accurate if stoichiometric concentrations are used. Activities are needed instead. The activity of a solution is related to its stoichiometric concentration through a factor called an *activity coefficient*. We will discuss the importance of activities in more detail in Chapter 20.

14-10 Colloidal Mixtures

Everyone knows that in a mixture of sand and water, the sand (silica, SiO_2) quickly settles to the bottom of the mixture. Yet mixtures can be prepared containing up to 40% by mass of SiO_2, and the silica may remain dispersed in the aqueous medium for many years. In these mixtures, the silica is *not* present as ions or molecules. Rather, much larger particles of silica are present, though they are still submicroscopic in size. The mixture is said to be a **colloid**. To be classified *colloidal*, a material must have one or more of its dimensions (length, width, or thickness) in the approximate range of 1–1000 nm. If all the dimensions are smaller than 1 nm, the particles are of molecular size. If all the dimensions exceed 1000 nm, the particles are of ordinary, or macroscopic, size (even if they are visible only under a microscope). One method of determining whether a mixture is a true solution or a colloid is illustrated in Figure 14-21. When light passes through a true solution, an observer viewing from a direction perpendicular to the light beam sees no light. In a colloidal dispersion, light is scattered in many directions and is readily seen. This effect, first studied by John Tyndall in 1869, is known as the *Tyndall effect*. A common example is the scattering of light by dust particles in a flashlight beam.

The particles in colloidal silica have a spherical shape. Some colloidal particles are rod-shaped, and some, like gamma globulin in human blood plasma, have a disclike shape. Thin films, like an oil slick on water, are colloidal. And some colloids, such as cellulose fibers, are randomly coiled filaments.

What keeps the SiO_2 particles suspended in colloidal silica? The most important factor is that the surfaces of the particles *adsorb*, or attach to themselves, ions from the solution, and they preferentially adsorb one type of ion over others. In the case of SiO_2, the preferred ions that are adsorbed are OH^-. As a result, the particles acquire a net negative charge. Having like charges, the particles repel one another. These mutual repulsions overcome the force of gravity, and the particles remain suspended indefinitely. Figure 14-22 represents the surface of a colloidal silica particle.

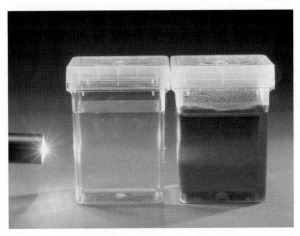

▲ **FIGURE 14-21 The Tyndall effect**
The flashlight beam is not visible as it passes through a true solution (left), but it is readily seen as it passes through the colloidal dispersion of Fe_2O_3 (right).

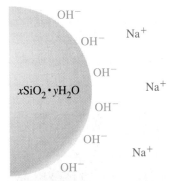

▲ **FIGURE 14-22**
Surface of SiO₂ particle in colloidal silica
The points made in this simplified drawing are: (1) The SiO_2 particles are hydrated; (2) OH^- ions are preferentially adsorbed on the surface; (3) in the immediate vicinity of the particle, negative ions outnumber positive ions and the particle carries a net negative charge. Not illustrated here are the facts that (1) some of the negative charge comes from silicate anions, for example, SiO_3^{2-}; (2) as a whole, the solution in which these particles are found is electrically neutral.

Although electric charge can be important in stabilizing a colloid, a high concentration of ions can also bring about the *coagulation*, or precipitation, of a colloid (Figure 14-23). The ions responsible for the coagulation are those carrying a charge opposite that on the colloidal particles themselves. *Dialysis*, a process similar to osmosis, can be used to remove excess ions from a colloidal mixture. As suggested by Figure 14-24, molecules of solvent and molecules or ions of solute pass through a semipermeable membrane, but the much larger colloidal particles do not. In some cases, the process is more effective when carried out in an electric field. In *electrodialysis*, ions are attracted out of a colloidal mixture to an electrode carrying the opposite charge. A human kidney dialyzes blood, a colloidal mixture, to remove excess electrolytes produced by metabolic processes. Certain diseases causes the kidneys to lose this ability, but a dialysis machine, external to the body, can function for the kidneys.

Table 14.4 lists some common colloids, and as so aptly put by Wilder Bancroft, an American pioneer in the field of colloid chemistry, "... colloid chemistry is essential to anyone who really wishes to understand ... oils, greases, soaps, ... glue, starch, adhesives, ... paints, varnishes, lacquers, ... cream, butter, cheese, ... cooking, washing, dyeing, ... colloid chemistry is the chemistry of life."

▶ **FIGURE 14-23**
Coagulation of colloidal iron oxide
On the left is red colloidal hydrous Fe_2O_3, obtained by adding $FeCl_3(aq)$ to boiling water. When a few drops of $Al_2(SO_4)_3(aq)$ are added, the suspended particles rapidly coagulate into a precipitate of $Fe_2O_3(s)$ (right).

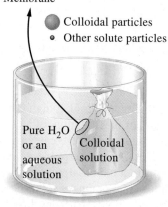

Membrane

⬤ Colloidal particles
◦ Other solute particles

Pure H₂O
or an
aqueous
solution

Colloidal
solution

▲ **FIGURE 14-24**
The principle of dialysis
Water molecules, other solute molecules, and dissolved ions are all free to pass through the pores of the membrane (for example, a film of cellophane) in either direction. The direction of net flow of these species depends on their relative concentrations on either side of the membrane. Colloidal particles, however, cannot pass through the pores of the membrane.

TABLE 14.4	Some Common Types of Colloids		
Dispersed phase	Dispersion medium	Type	Examples
Solid	Liquid	Sol	Clay sols[a], colloidal silica, colloidal gold
Liquid	Liquid	Emulsion	Oil in water, milk, mayonnaise
Gas	Liquid	Foam	Soap and detergent suds, whipped cream, meringues
Solid	Gas	Aerosol[b]	Smoke, dust-laden air[c]
Liquid	Gas	Aerosol[b]	Fog, mist (as in aerosol products)
Solid	Solid	Solid sol	Ruby glass, certain natural and synthetic gems, blue rock salt, black diamond
Liquid	Solid	Solid emulsion	Opal, pearl
Gas	Solid	Solid foam	Pumice, lava, volcanic ash

[a] In water purification, it is sometimes necessary to precipitate clay particles or other suspended colloidal materials. This is often done by treating the water with an appropriate electrolyte. Clay sols are suspected of adsorbing organic substances, such as pesticides, and distributing them in the environment.

[b] Smogs are complex materials that are at least partly colloidal. The suspended particles are both solid and liquid. Other constituents of smog are molecular, for example, sulfur oxides, nitrogen oxides, and ozone.

[c] The bluish haze of tobacco smoke and the brilliant sunsets in desert regions are both attributable to the scattering of light by colloidal particles suspended in air.

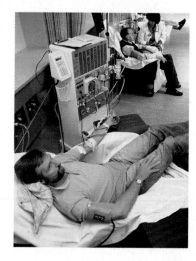

▶ **Hemodialysis**
An artificial kidney machine cleaning the blood of patients with impaired kidney function.

Summary

To describe the composition of a solution, we must indicate how much solute and solvent (or solution) are present. Percentage composition units have practical importance, but the units with more scientific value are molarity, molality, and mole fraction. At times, it is necessary to convert among these various units.

To predict whether two substances will mix to form a solution, we must compare intermolecular forces involving like and unlike molecules. This approach also allows us to distinguish between ideal and nonideal solutions. Generally, a solvent has a limited ability to dissolve a solute to produce a saturated solution. Solute solubilities as a function of temperature are represented by solubility curves. These curves can be used when planning the fractional crystallization of a solute. The solubilities of gases depend on pressure as well as temperature, and many familiar phenomena are related to gas solubilities. A simple mathematical relationship, called Henry's law, can be used to relate the concentration of a gas in solution to its pressure above the solution.

The physical properties studied in this chapter—vapor pressure lowering, osmotic pressure, freezing-point depression, and boiling-point elevation—have many practical applications. The values of these colligative properties depend only on the concentration of solute particles and not on the identity of the solute. In dealing with the physical properties of electrolyte solutions, however, we must work with the total concentration of ions in solution. Simple equations can be used to calculate the several colligative properties. For electrolyte solutions these equations are modified through the van't Hoff factor i.

Colloidal mixtures represent an intermediate state between true solutions and heterogeneous mixtures and are encountered in a broad range of contexts, from biological fluids in living organisms to pollutants in large air masses.

Integrative Example

Four aqueous solutions of acetone, CH_3COCH_3, are prepared at different concentrations: **(a)** 0.100% CH_3COCH_3 by mass; **(b)** 0.100 M CH_3COCH_3; **(c)** 0.100 m CH_3COCH_3; and **(d)** $x_{acetone} = 0.100$. Estimate the highest partial pressure of water at 25 °C to be found in the equilibrium vapor above these solutions. Also estimate the lowest freezing point to be found among these solutions.

At the outset, we can say that the solution having the highest mole fraction of water (the most dilute solution) will have the highest partial pressure of water in its equilibrium vapor. Also, the solution with the lowest mole fraction of water (the most concentrated solution) will have the lowest freezing point. First, we need to identify these two solutions, and for this we need a simple way to compare concentrations, which are given here in disparate units. Let's look at the solutions one at a time. Keep in mind that 1 L of water contains about 1000 g H_2O/18.02 g H_2O $mol^{-1} \approx 55.49$ mol H_2O.

Solution (a): The concentration 0.100% CH_3COCH_3 by mass is equivalent to 1.00 g CH_3COCH_3/1000 g solution, which in turn is about 1.00 g CH_3COCH_3/L solution. (The density of the solution will be very nearly the same as that of water.) A 1.00-g sample of CH_3COCH_3 is the same as 1.00 g/58.1 g $mol^{-1} \ll 0.10$ mol CH_3COCH_3.

Solution (b): The 0.100 M CH_3COCH_3 has 0.100 mol CH_3COCH_3 per liter. Solution (b) is more concentrated than (a), so (b) cannot be the solution with the highest water vapor pressure.

Solution (c): The 0.100 m CH_3COCH_3 has 0.100 mol CH_3COCH_3 per kg H_2O. Because water is the preponderant component in the solution, the density of the solution will be very nearly that of water, and the 0.100 m $CH_3COCH_3 \approx 0.100$ M

CH_3COCH_3. So, as with (b), solution (c) cannot be the solution with the highest water vapor pressure.

Solution (d): In the solution $x_{acetone} = 0.100$, acetone molecules comprise one-tenth of all the molecules, or there is one acetone molecule for every nine water molecules. In a liter of this solution there would be about $55.5/9 \approx 6$ mol CH_3COCH_3. This is clearly the most concentrated of the solutions.

To summarize, solution (a) has the highest water vapor pressure and solution (d) has the lowest freezing point.

Our next task is to apply Raoult's law to solution (a), but for this we need the mole fraction of water in solution (a). One kilogram of this solution contains 1.00 g/58.1 g $mol^{-1} = 0.0172$ mol CH_3COCH_3 and 999 g/18.02 g $mol^{-1} = 55.44$ mol H_2O. The mole fraction of the water is $x_{water} = 55.44/(55.44 + 0.0172) \cong 1.00$. The mole fraction is only very slightly less than 1.00, so the water vapor pressure above this solution will be only very slightly less than the vapor pressure of pure water at 25 °C—23.8 mmHg.

To apply equation (14.5) for freezing-point depression to solution (d), we must first express its concentration in molality. In a solution that has 1.00 mol CH_3COCH_3 for every 9.00 mol H_2O, there is 1.00 mol of the solute for every $(9.00 \times 18.02) = 162$ g H_2O. The molality of the solution is 1.00 mol CH_3COCH_3/0.162 kg H_2O = 6.17 m CH_3COCH_3. The approximate freezing-point depression for this solution is

$$\Delta T_f = -K_f \times m = -1.86\,°C\,m^{-1} \times 6.17\,m \approx -11.5\,°C$$

and the approximate freezing point of the solution is −11.5 °C. This result is only approximate because we have used equation (14.5) for a much more concentrated solution than the equation is intended.

Key Terms

alloy (14-1)
azeotrope (14-6)
chromatography (Focus On)
colligative properties (14-7)
colloid (14-10)
fractional crystallization (recrystallization) (14-4)
fractional distillation (14-6)
Henry's law (14-5)

ideal solution (14-3)
molality (*m*) (14-2)
mole fraction (14-2)
mole percent (14-2)
osmosis (14-7)
osmotic pressure (14-7)
ppb (14-2)
ppm (14-2)
ppt (14-2)

Raoult's law (14-6)
reverse osmosis (14-7)
saturated solution (14-4)
solubility (14-4)
solute (14-1)
solvent (14-1)
supersaturated solution (14-4)
unsaturated solution (14-4)

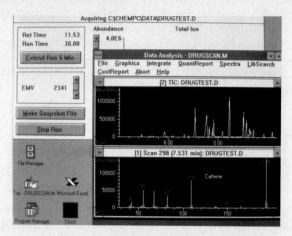

▲ A chromatograph obtained in a laboratory drug test.

Chromatography is a technique for separating the components of a mixture present in one phase, called the *mobile phase*, as it moves relative to another phase, called the *stationary phase*. As originally conceived by the Russian botanist Mikhail Tsvett in 1903, the method is based on the ability of a column of finely powdered solid material, such as Al_2O_3, (the stationary phase) to *adsorb* substances from a solution (the mobile phase) allowed to trickle through it. In general, the attractive forces at the solid surface differ for different species in the solution. The substances that the solid adsorbs most strongly move through the stationary phase much more slowly than do those that are not so strongly adsorbed. This means that although the various solution components start out together, they soon become separated when in contact with the stationary phase. Tsvett used this method to separate plant pigments into colored components. His column developed bands of color, and he named this separation technique **chromatography**. (Coincidentally, in Greek, *chromatography* means "written in color," and in Russian, *tsvett* means color.)

You can demonstrate chromatography in a very simple way by replicating the experiment pictured in Figure 1-4c-d (page 7). Use a strip of chemistry filter paper (or a paper filter for an automatic coffee maker) as the stationary phase. Water is the mobile phase. Start the demonstration by placing a drop of black ink in the center of the paper strip just above the water level. Water is drawn up the paper by capillary action, dissolving the colored components of the ink as it passes through the ink spot. The components then become separated, based on differences in the strengths of the intermolecular forces between them and the cellulose in the paper.

Tsvett's method remained largely unknown until it was reintroduced by biochemists in the 1930s. This first form of chromatography is known as *adsorption* chromatography. In 1942, Archer Martin and Richard Synge used a liquid adsorbed on a solid as the stationary phase, and another liquid immiscible with the stationary phase as the mobile phase. With this technique, called *liquid–liquid partition chromatography*, the components being separated distribute themselves between the two liquid phases. In 1952, Martin and A. T. James substituted a gas for the liquid mobile phase and thus introduced *gas–liquid partition chromatography*. The effectiveness of liquid–liquid partition chromatography was greatly increased in the 1970s through the use of smaller particles for the solid support and pumps to force the mobile phase through the stationary phase under high pressure. This method is known as *high performance liquid chromatography* (HPLC). More recently, methods have been developed that use supercritical fluids as the mobile phases—*supercritical fluid chromatography* (SFC). With the various chromatographic techniques now available, it is possible to separate components in solid, liquid, or gaseous mixtures of organic, inorganic, or biological compounds and to detect components in the range of parts per billion or even parts per trillion. The effectiveness of chromatography is even greater when the chromatograph is combined with another instrument. For example, in the combination of gas chromatography and mass spectrometry (GC–MS), the effluent from the gas chromatograph column is passed into a mass spectrometer for analysis. Figure 14-25 is a *chromatogram*—the data output—of a gas chromatography experiment.

Review Questions

1. In your own words, define or explain the following terms or symbols: **(a)** x_B; **(b)** P_A°; **(c)** K_f; **(d)** i; **(e)** activity.
2. Briefly describe each of the following ideas or phenomena: **(a)** Henry's law; **(b)** freezing-point depression; **(c)** recrystallization; **(d)** hydrated ion; **(e)** deliquescence.
3. Explain the important distinctions between each pair of terms: **(a)** molality and molarity; **(b)** ideal and nonideal solution; **(c)** unsaturated and supersaturated solution; **(d)** fractional crystallization and fractional distillation; **(e)** osmosis and reverse osmosis.
4. A saturated aqueous solution of NaBr at 20 °C contains 116 g NaBr/100 g H_2O. Express this composition in the more conventional percent by mass, that is, as grams of NaBr per 100 grams of solution.
5. An aqueous solution with density 0.988 g/mL at 20 °C is prepared by dissolving 12.8 mL $CH_3CH_2CH_2OH$

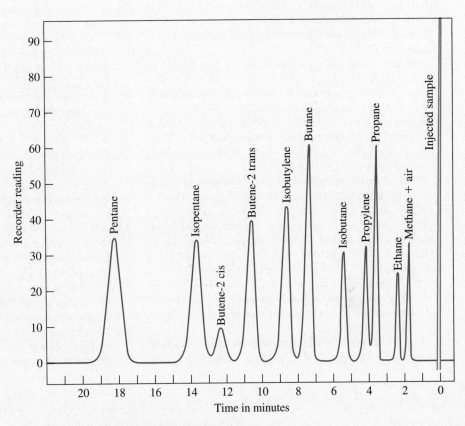

FIGURE 14-25   A typical gas chromatogram
The times plotted here are called *retention times*. A retention time for a component in a mixture is the time that elapses between injection of the sample into the chromatographic column and the appearance of that component at the detector. In the hydrocarbon mixture represented here, we see that retention times tend to increase with increasing molecular mass of the components. We might expect this to be the case because intermolecular forces of the dispersion type also increase with increasing molecular mass.

($d = 0.803$ g/mL) in enough water to produce 75.0 mL of solution. What is the percent $CH_3CH_2CH_2OH$ expressed as **(a)** percent by volume; **(b)** percent by mass; **(c)** percent (mass/volume)?

6. A certain brine has 3.87% NaCl by mass. A 75.0-mL sample weighs 76.9 g. How many liters of this solution should be evaporated to dryness to obtain 725 kg NaCl?

7. You are asked to prepare 125.0 mL of 0.0321 M $AgNO_3$. How many grams would you need of a sample known to be 99.81% $AgNO_3$ by mass?

8. What is the molality of *para*-dichlorobenzene in a solution prepared by dissolving 2.65 g $C_6H_4Cl_2$ in 50.0 mL of benzene ($d = 0.879$ g/mL)?

9. An aqueous solution is 6.00% methanol (CH_3OH) by mass, with $d = 0.988$ g/mL. What is the molarity of CH_3OH in this solution?

10. A solution is prepared by mixing 1.28 mol C_7H_{16}, 2.92 mol C_8H_{18}, and 2.64 mol C_9H_{20}. What are the **(a)** mole fraction and **(b)** mole percent of each component of the solution?

11. A solution ($d = 1.159$ g/mL) is 62.0% glycerol, $C_3H_8O_3$, and 38.0% H_2O, by mass. Determine (a) the molarity of $C_3H_8O_3$ (with H_2O as the solvent); (b) the molarity of H_2O (with $C_3H_8O_3$ as the solvent); (c) the molality of H_2O in $C_3H_8O_3$; (d) the mole fraction of $C_3H_8O_3$; (e) the mole percent of H_2O.

12. A solution prepared by dissolving 1.12 mol NH_4Cl in 150.0 g H_2O is brought to a temperature of 30 °C. Use Figure 14-8 to determine whether the solution is unsaturated or whether excess solute will crystallize.

13. Which of the following mixtures in a single liquid phase is most likely to be an ideal solution: (a) $NaCl–H_2O$; (b) $C_2H_5OH(l)–C_6H_6(l)$; (c) $C_7H_{16}(l)–H_2O(l)$; (d) $C_7H_{16}(l)–C_8H_{18}(l)$? Explain.

14. Which of the following do you expect to be most water soluble and why? $C_{10}H_8(s)$, $NH_2OH(s)$, $C_6H_6(l)$, $CaCO_3(s)$.

15. Which one of the following do you expect to be moderately soluble both in water and in benzene [$C_6H_6(l)$], and why? (a) butyl alcohol, C_4H_9OH; (b) naphthalene, $C_{10}H_8$; (c) hexane, C_6H_{14}; (d) $NaCl(s)$.

16. Under an $O_2(g)$ pressure of 1.00 atm, 28.31 mL of $O_2(g)$ at 25 °C dissolves in 1.00 L H_2O at 25 °C. What will be the molarity of O_2 in the saturated solution at 25 °C when the O_2 pressure is 3.86 atm? (Assume that the solution volume remains at 1.00 L.)

17. What are the partial and total vapor pressures of a solution obtained by mixing 35.8 g benzene, C_6H_6, and 56.7 g

toluene, C_7H_8, at 25 °C? Vapor pressures at 25 °C: $C_6H_6 = 95.1$ mmHg; $C_7H_8 = 28.4$ mmHg.

18. Determine the composition of the vapor above the benzene–toluene solution described in Exercise 17.

19. *Without doing detailed calculations*, determine which of the following aqueous solutions probably has the *lowest* freezing point. (a) 0.010 *m* $MgSO_4$; (b) 0.011 *m* NaCl; (c) 0.050 *m* C_2H_5OH (ethanol); (d) 0.010 *m* MgI_2.

20. Use your knowledge of strong, weak, and nonelectrolytes to arrange the following 0.0010 *m* aqueous solutions in the probable order of *decreasing* freezing point: C_2H_5OH, NaCl, $MgBr_2$, $HC_2H_3O_2$, and $Al_2(SO_4)_3$.

21. NaCl(aq) isotonic with blood is 0.92% NaCl (mass/volume). For this solution, what is (a) $[Na^+]$; (b) the total molarity of ions; (c) the osmotic pressure at 37 °C; (d) the approximate freezing point? (Assume that the solution has a density of 1.005 g/mL.)

22. A 0.72-g sample of polyvinyl chloride (PVC) is dissolved in 250.0 mL of a suitable solvent at 25 °C. The solution has an osmotic pressure of 1.67 mmHg. What is the molar mass of the PVC?

23. 1.10 g of an unknown compound reduces the freezing point of 75.22 g benzene from 5.53 to 4.92 °C. What is the molar mass of the compound?

24. The freezing point of a 0.010 *m* aqueous solution of a nonvolatile solute is -0.072 °C. What would you expect the normal boiling point of this same solution to be?

Exercises

Homogeneous and Heterogeneous Mixtures

25. Substances that dissolve in water do not generally dissolve in benzene. Some substances are moderately soluble in both solvents, however. One of the following is such a substance. Which do you think it is and why?

(a) *para*-Dichlorobenzene
(a moth repellent)

(b) Salicyl alcohol
(a local anesthetic)

(c) Diphenyl
(a heat transfer agent)

(d) Hydroxyacetic acid
(used in textile dyeing)

26. Some vitamins are water soluble and some are fat soluble. (Fats are substances whose molecules have long hydrocarbon chains.) The structural formulas of two vitamins are

shown here. One is water soluble and one is fat soluble. Identify each, and explain your reasoning.

Vitamin C

Vitamin E

27. Two of the substances listed here are highly soluble in water; two are only slightly soluble in water; and two are insoluble in water. Indicate the situation you expect for each one.
    (a) iodoform, CHI_3

    (b) benzoic acid,

    (c) formic acid, $H-C-OH$
    (d) butyl alcohol, $CH_3CH_2CH_2CH_2OH$

    (e) chlorobenzene,

    (f) propylene glycol, $CH_3CHOHCH_2OH$

28. Benzoic acid, C_6H_5COOH, is much more soluble in NaOH(aq) than it is in pure water. Can you suggest a reason for this? The structural formula for benzoic acid is given in Exercise 27(b).
29. In light of the factors outlined on page 544, which of the following ionic fluorides would you expect to be most water soluble on a moles-per-liter basis: MgF_2, NaF, KF, CaF_2? Explain your reasoning.
30. Explain the observation that all metal nitrates are water soluble, whereas many metal sulfides are not water soluble. Among metal sulfides, which would you expect to be most soluble?

Percent Concentration

31. According to Example 14-1, the mass percent ethanol in a particular aqueous solution is less than the volume percent in the same solution. Explain why this is also true for *all* aqueous solutions of ethanol. Would it be true of all ethanol solutions, regardless of the other component? Explain.
32. Is either mass percent or volume percent independent of temperature? Explain your answer.
33. A certain vinegar is 6.02% acetic acid ($HC_2H_3O_2$) by mass. How many grams of $HC_2H_3O_2$ are contained in a 355-mL bottle of vinegar. Assume a density of 1.01 g/mL.

34. 6.00 M sulfuric acid, H_2SO_4(aq), has a density of 1.338 g/mL. What is the percent by mass of sulfuric acid in this solution?
35. The sulfate ion level in a municipal water supply is given as 46.1 ppm. What is $[SO_4^{2-}]$ in this water?
36. A water sample is found to have 9.4 ppb of chloroform, $CHCl_3$. How many grams of $CHCl_3$ would be found in a glassful (250 mL) of this water?

Molarity

37. How many milliliters of the ethanol–water solution described in Example 14-1 should be diluted with water to produce 825 mL of 0.235 M C_2H_5OH?

38. A 30.00%-by-mass solution of nitric acid, HNO_3, in water has a density of 1.18 g/cm³ at 20 °C. What is the molarity of HNO_3 in this solution?

Molality

39. How many grams of iodine, I_2, must be dissolved in 725 mL of carbon disulfide, CS_2 ($d = 1.261$ g/mL), to produce a 0.236 *m* solution?
40. How many grams of water would you add to 1.00 kg of 1.38 *m* CH_3OH(aq) to reduce the molality to 1.00 *m* CH_3OH?
41. An aqueous solution is 34.0% H_3PO_4 by mass and has a density of 1.209 g/mL. What are the molarity and molality of this solution?

42. A 10.00%-by-mass solution of ethanol, C_2H_5OH, in water has a density of 0.9831 g/mL at 15 °C and 0.9804 g/mL at 25 °C. Calculate the molality of the ethanol–water solution at these two temperatures. Does the molality differ at the two temperatures (that is, at 15 and 25 °C)? Would you expect the molarities to differ? Explain.

Mole Fraction, Mole Percent

43. Calculate the mole fraction of solute in the following aqueous solutions: (a) 21.7% C_2H_5OH, by mass; (b) 0.684 *m* $CO(NH_2)_2$ (urea).
44. Calculate the mole fraction of the solute in the following aqueous solutions: (a) 0.112 M $C_6H_{12}O_6$ ($d = 1.006$ g/mL); (b) 3.20% ethanol, by volume ($d = 0.993$ g/mL; pure C_2H_5OH, $d = 0.789$ g/mL).

45. Refer to Example 14-1. How many grams of C_2H_5OH must be added to 100.0 mL of the solution described in part (d) to increase the mole fraction of C_2H_5OH to 0.0525?
46. How many milliliters of glycerol, $C_3H_8O_3$ ($d = 1.26$ g/mL), must be added per kilogram of water to produce a solution with 4.85 mol % $C_3H_8O_3$?

Solubility Equilibrium

47. Refer to Figure 14-8, and determine the molality of NH_4Cl in a saturated aqueous solution at 40 °C.
48. Refer to Figure 14-8, and estimate the temperature at which a saturated aqueous solution of $KClO_4$ is 0.200 *m*.
49. A solution of 20.0 g $KClO_4$ in 500.0 g of water is brought to a temperature of 40 °C.
    (a) Refer to Figure 14-8, and determine whether the solution is unsaturated or supersaturated at 40 °C.
    (b) Approximately what mass of $KClO_4$, in grams, must be added to saturate the solution (if originally unsaturated), or what mass of $KClO_4$ can be crystallized (if originally supersaturated)?
50. One way to recrystallize a solute from a solution is to change the temperature. Another way is to evaporate solvent from the solution. A 335-g sample of a saturated solution of $KNO_3(s)$ in water is prepared at 25.0 °C. If 55 g H_2O is evaporated from the solution at the same time as the temperature is reduced from 25.0 to 0.0 °C, what mass of $KNO_3(s)$ will recrystallize? (Refer to Figure 14-8.)

Solubility of Gases

51. Most natural gas consists of about 90% methane, CH_4. Assume that the solubility of natural gas at 20 °C and 1 atm gas pressure is about the same as that of CH_4, 0.02 g/kg water. If a sample of natural gas under a pressure of 20 atm is kept in contact with 1.00×10^3 kg of water, what mass of natural gas will dissolve?
52. At 1.00 atm, the solubility of O_2 in water is 2.18×10^{-3} M at 0 °C and 1.26×10^{-3} M at 25 °C. What volume of $O_2(g)$, measured at 25 °C and 1.00 atm, is expelled when 515 mL of water saturated with O_2 is heated from 0 to 25 °C?
53. The aqueous solubility at 20 °C of Ar at 1.00 atm is equivalent to 33.7 mL Ar(g), measured at STP, per liter of water. What is the molarity of Ar in water at 20 °C that is saturated with air at 1.00 atm? Air contains 0.934% Ar, by volume. Assume that the volume of water does not change when it becomes saturated with air.
54. The aqueous solubility of CO_2 at 20 °C and 1.00 atm is equivalent to 87.8 mL $CO_2(g)$, measured at STP, per 100 mL of water. What is the molarity of CO_2 in water that is at 20 °C and saturated with air at 1.00 atm? The volume percent of CO_2 in air is 0.0360%. Assume that the volume of the water does not change when it becomes saturated with air.
55. Henry's law can be stated this way: The mass of a gas dissolved by a given quantity of solvent at a fixed temperature is directly proportional to the pressure of the gas. Show how this statement is related to equation (14.2).
56. Still another statement of Henry's law is this: At a fixed temperature, a given quantity of liquid dissolves the same volume of gas at all pressures. What is the connection between this statement and the one given in Exercise 55? Under what conditions is this second statement not valid?

Raoult's Law and Liquid–Vapor Equilibrium

57. Calculate the vapor pressure at 25 °C of a solution containing 165 g of the *nonvolatile* solute, glucose, $C_6H_{12}O_6$, in 685 g H_2O. The vapor pressure of water at 25 °C is 23.8 mmHg.
58. Calculate the vapor pressure at 20 °C of a saturated solution of the *nonvolatile* solute, urea, $CO(NH_2)_2$, in methanol, CH_3OH. The solubility is 17 g urea/100 mL methanol. The density of methanol is 0.792 g/mL, and its vapor pressure at 20 °C is 95.7 mmHg.
59. Styrene, used in the manufacture of polystyrene plastics, is made by the extraction of hydrogen atoms from ethylbenzene. The product obtained contains about 38% styrene (C_8H_8) and 62% ethylbenzene (C_8H_{10}), by mass. The mixture is separated by fractional distillation at 90 °C. Determine the composition of the vapor in equilibrium with this 38%–62% mixture at 90 °C given the vapor pressures of the two components: ethylbenzene, 182 mmHg; styrene, 134 mmHg.
60. Calculate $x_{C_6H_6}$ in a benzene–toluene liquid solution that is in equilibrium at 25 °C with a vapor phase that contains 62.0 mol % C_6H_6. (Use data from Exercise 17.)
61. A benzene–toluene solution with $x_{benz} = 0.300$ has a normal boiling point of 98.6 °C. The vapor pressure of pure toluene at 98.6 °C is 533 mmHg. What must be the vapor pressure of pure benzene at 98.6 °C? (Assume ideal solution behavior.)
62. The two NaCl(aq) solutions pictured are at the same temperature.

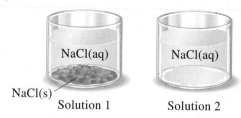

NaCl(s)

Solution 1   Solution 2

(a) Above which solution is the vapor pressure of water, P_{H_2O}, greater? Explain.
(b) Above one of these solutions, the vapor pressure of water, P_{H_2O}, remains *constant*, even as water evaporates from solution. Which solution is this? Explain.
(c) Which of these solutions has the higher boiling point? Explain.

Osmotic Pressure

63. When the stems of cut flowers are held in concentrated NaCl(aq), the flowers wilt. In a similar solution a fresh cucumber shrivels up (becomes pickled). Explain the basis of these phenomena.

64. Some fish live in saltwater environments and some in freshwater, but in either environment they need water to survive. Saltwater fish drink water, but freshwater fish do not. Explain this difference between the two types of fish.

65. In what volume of water must 1 mol of a nonelectrolyte be dissolved if the solution is to have an osmotic pressure of 1 atm at 273 K? Which of the gas laws does this result resemble?

66. The molecular mass of hemoglobin is 6.86×10^4 u. What mass of hemoglobin must be present per 100.0 mL of a solution to exert an osmotic pressure of 7.25 mmHg at 25 °C?

67. A 0.50-g sample of polyisobutylene (a polymer used in synthetic rubber) in 100.0 mL of benzene solution has an osmotic pressure at 25 °C that supports a 5.1-mm column of solution ($d = 0.88$ g/mL). What is the molar mass of the polyisobutylene? (For Hg, $d = 13.6$ g/mL.)

68. Use the concentration of an isotonic saline solution, 0.92% NaCl (mass/volume), to determine the osmotic pressure of blood at body temperature, 37.0 °C.

(*Hint:* Assume that NaCl is completely dissociated in aqueous solutions.)

69. What approximate pressure is required in the reverse osmosis depicted in Figure 14-18 if the saltwater contains 2.5% NaCl, by mass.
(*Hint:* Assume that NaCl is completely dissociated in aqueous solutions. Also, assume a temperature of 25 °C.)

70. The two solutions pictured here are separated by a semipermeable membrane that permits only the passage of water molecules. In what direction will a net flow of water occur; that is, from left to right or right to left?

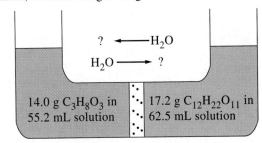

Freezing-Point Depression and Boiling-Point Elevation

71. Adding 1.00 g of benzene, C_6H_6, to 80.00 g cyclohexane, C_6H_{12}, lowers the freezing point of the cyclohexane from 6.5 to 3.3 °C.
(a) What is the value of K_f for cyclohexane?
(b) Which is the better solvent for molar mass determinations by freezing-point depression, benzene or cyclohexane? Explain.

72. The boiling point of water at 749.2 mmHg is 99.60 °C. What mass percent sucrose ($C_{12}H_{22}O_{11}$) should be present in an aqueous sucrose solution to raise the boiling point to 100.00 °C at this pressure?

73. A compound is 42.9% C, 2.4% H, 16.7% N, and 38.1% O, by mass. Addition of 6.45 g of this compound to 50.0 mL benzene, C_6H_6 ($d = 0.879$ g/mL), lowers the freezing point from 5.53 to 1.37 °C. What is the molecular formula of this compound?

74. Nicotinamide is a water-soluble vitamin important in metabolism. A deficiency in this vitamin results in the debilitating condition known as pellagra. Nicotinamide is 59.0% C, 5.0% H, 22.9% N, 13.1% O, by mass. Addition of 3.88 g of nicotinamide to 30.0 mL nitrobenzene, $C_6H_5NO_2$ ($d = 1.204$ g/mL), lowers the freezing point from 5.7 to −1.4 °C. What is the molecular formula of this compound?

75. Thiophene (fp, −38.3; bp, 84.4 °C) is a sulfur-containing hydrocarbon sometimes used as a solvent in place of benzene. Combustion of a 2.348-g sample of thiophene produces 4.913 g CO_2, 1.005 g H_2O, and 1.788 g SO_2. When

a 0.867-g sample of thiophene is dissolved in 44.56 g of benzene (C_6H_6), the freezing point is lowered by 1.183 °C. What is the molecular formula of thiophene?

76. Coniferin is a glycoside (a derivative of a sugar) found in conifers such as fir trees. When a 1.205-g sample of coniferin is subjected to combustion analysis (recall Section 3-3), the products are 0.698 g H_2O and 2.479 g CO_2. A 2.216-g sample is dissolved in 48.68 g H_2O, and the normal boiling point of this solution is found to be 100.068 °C. What is the molecular formula of coniferin?

77. Cooks often add some salt to water before boiling it. Some people say this helps the cooking process by raising the boiling point of the water. Others say not enough salt is usually added to make any noticeable difference. Approximately how many grams of NaCl must be added to a liter of water at 1 atm pressure to raise the boiling point by 2 °C? Is this a typical amount of salt that you might add to cooking water?

78. An important test for the purity of an organic compound is to measure its melting point. Usually, if the compound is not pure, it begins to melt at a *lower* temperature than the pure compound.
(a) Why is this the case, rather than the melting point being higher in some cases and lower in others?
(b) Are there any conditions under which the melting point of the impure compound may be *higher* than that of the pure compound? Explain.

Strong Electrolytes, Weak Electrolytes, and Nonelectrolytes

79. Predict the approximate freezing points of 0.10 m solutions of the following solutes dissolved in water: **(a)** $CO(NH_2)_2$ (urea); **(b)** NH_4NO_3; **(c)** HCl; **(d)** $CaCl_2$; **(e)** $MgSO_4$; **(f)** C_2H_5OH (ethanol); **(g)** $HC_2H_3O_2$ (acetic acid).

80. Calculate the van't Hoff factors of the following weak electrolyte solutions.
(a) 0.050 m $HCHO_2$, which begins to freeze at $-0.0986\,°C$
(b) 0.100 M HNO_2, which has a hydrogen ion (and nitrite ion) concentration of 6.91×10^{-3} M

81. $NH_3(aq)$ conducts electric current only weakly. The same is true for acetic acid, $HC_2H_3O_2(aq)$. When these solutions are mixed, however, the resulting solution conducts electric current very well. Propose an explanation.

82. An isotonic solution is described as 0.92% NaCl (mass/volume). Would this also be the required concentration for isotonic solutions of other salts, such as KCl, $MgCl_2$, or $MgSO_4$? Explain

Integrative and Advanced Exercises

83. A typical root beer contains 0.13% of a 75% H_3PO_4 solution by mass. How many milligrams of phosphorus are contained in a 12-oz can of this root beer? Assume a solution density of 1.00 g/mL; also, 1 oz $= 29.6$ mL.

84. Water and phenol are only partially miscible at temperatures below 66.8 °C. In a mixture prepared at 29.6 °C from 50.0 g water and 50.0 g phenol, 32.8 g of a phase consisting of 92.50% water and 7.50% phenol by mass is obtained. This is a saturated solution of phenol in water. What is the mass percent of water in the second phase—a saturated solution of water in phenol?

85. An aqueous solution has 109.2 g KOH/L solution. The solution density is 1.09 g/mL. Your task is to use 100.0 mL of this solution to prepare 0.250 m KOH. What mass of which component, KOH or H_2O, would you add to the 100.0-mL of solution?

86. The term "proof," still used to describe the ethanol content of alcoholic beverages, originated in seventeenth-century England. A sample of whiskey was poured on gunpowder and set afire. If the gunpowder ignited after the whiskey had burned off, this "proved" that the whiskey had not been watered down. The minimum ethanol content for a positive test was about 50%, by volume. The 50% ethanol solution became known as "100 proof." Thus, an 80-proof whiskey would be 40% CH_3CH_2OH by volume. Listed in the table are some data for several aqueous solutions of ethanol. *With a minimum amount of calculation,* determine which of the solutions are more than 100 proof. Assume that the density of pure ethanol is 0.79 g/mL.

Molarity of Ethanol (M)	Density of Solution (g/mL)
4.00	0.970
5.00	0.963
6.00	0.955
7.00	0.947
8.00	0.936
9.00	0.926
10.00	0.913

87. A solid mixture consists of 85.0% KNO_3 and 15.0% K_2SO_4, by mass. A 60.0-g sample of this solid is added to 130.0 g of water at 60 °C. Refer to Figure 14-8.
(a) Will all the solid dissolve at 60 °C?
(b) If the resulting solution is cooled to 0 °C, what mass of KNO_3 should crystallize?
(c) Will K_2SO_4 also crystallize at 0 °C?

88. Suppose you have available 2.50 L of a solution ($d = 0.9767$ g/mL) that is 13.8% ethanol (C_2H_5OH), by mass. From this solution you would like to make the *maximum* quantity of ethanol–water antifreeze solution that will offer protection to -2.0 °C. Would you add more ethanol or more water to the solution? What mass of liquid would you add?

89. Hydrogen chloride is a colorless gas, yet when a bottle of concentrated hydrochloric acid [HCl(conc aq)] is opened, mistlike fumes are often seen to escape from the bottle. How do you account for this?

90. Use the following information to confirm that the triple point temperature of water is about 0.0098 °C.
(a) The slope of the fusion curve of water in the region of the normal melting point of ice (Figure 13-21, page 495) is -0.00750 °C/atm.
(b) The solubility of air in water at 1 atm pressure and 0 °C is 0.02918 mL of air (measured at STP) per mL of water.

91. Stearic acid ($C_{18}H_{36}O_2$) and palmitic acid ($C_{16}H_{32}O_2$) are common fatty acids. Commercial grades of stearic acid usually contain palmitic acid as well. A 1.115-g sample of a commercial-grade stearic acid is dissolved in 50.00 mL benzene ($d = 0.879$ g/mL). The freezing point of the solution is found to be 5.072 °C. The freezing point of pure benzene is 5.533 °C, and K_f for benzene is 5.12 °C m^{-1}. What is the mass percent of palmitic acid in the stearic acid sample?

92. At a constant temperature of 25.00 °C, a current of dry air was passed through pure water and then through a drying tube, D_1, followed by passage through 1.00 m sucrose and another drying tube, D_2. After the experiment, D_1 had gained 11.7458 g in mass and D_2, 11.5057 g. Given that the vapor pressure of water is 23.76 mmHg at 25.00 °C, **(a)** what was the vapor pressure lowering in the 1.00 m sucrose, and **(b)** what was the expected lowering?

93. Nitrobenzene, $C_6H_5NO_2$, and benzene, C_6H_6, are completely miscible in each other. Other properties of the two liquids are *nitrobenzene*: fp = 5.7 °C, K_f = 8.1 °C m^{-1}; *benzene*: fp = 5.5 °C, K_f = 5.12 °C m^{-1}. It is possible to prepare *two* different solutions with these two liquids having a freezing point of 0.0 °C. What are the compositions of these two solutions, expressed as mass percent nitrobenzene?

94. Refer to Figure 14-16a. Initially, solution A contains 0.515 g urea, $CO(NH_2)_2$, dissolved in 92.5 g H_2O; solution B contains 2.50 g sucrose, $C_{12}H_{22}O_{11}$, dissolved in 85.0 g H_2O. What are the compositions of the two solutions when equilibrium is reached; that is, when the two have the same vapor pressure?

95. In Figure 14-17, why does the net transfer of water stop when the two solutions are of *nearly* equal concentrations rather than of *exactly* equal concentrations?

96. Shown below is a typical cooling curve for an aqueous solution. Why is there no straight-line portion comparable to that seen in the cooling curve for pure water in Figure 13-14?

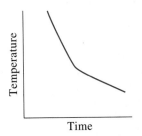

97. Suppose that 1.00 mg of gold is obtained in a colloidal dispersion in which the gold particles are spherical, with a radius of 1.00×10^2 nm. (The density of gold is 19.3 g/cm³.)
(a) What is the total surface area of the particles?
(b) What is the surface area of a single cube of gold weighing 1.00 mg?

98. At 20 °C, liquid benzene has a density of 0.879 g/cm³; liquid toluene, 0.867 g/cm³. Assume ideal solutions.
(a) Calculate the densities of solutions containing 20, 40, 60, and 80 volume percent benzene.
(b) Plot a graph of density versus volume percent composition.
(c) Write an equation that relates the density (d) to the volume percent benzene (V) in benzene–toluene solutions at 20 °C.

99. The two compounds whose structures are depicted here are isomers. When derived from petroleum, they always occur mixed together. *meta*-Xylene is used in aviation fuels and in the manufacture of dyes and insecticides. The principal use of *para*-xylene is in the manufacture of polyester resins and fibers (for example, Dacron). Comment on the effectiveness of fractional distillation as a method of separating these two xylenes. What other method(s) might be used to separate them?

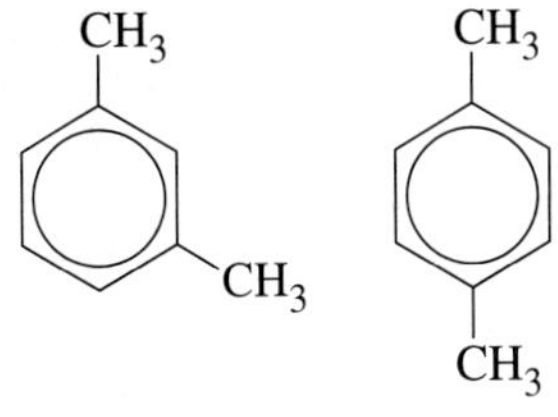

meta-Xylene
fp, −47.9 °C
bp, 139.1 °C
d, 0.864 g/mL

para-Xylene
fp, 13.3 °C
bp, 138.4 °C
d, 0.861 g/mL

100. Instructions on a container of Prestone® (ethylene glycol; fp, −12.6 °C, bp, 197.3 °C) give the following volumes of Prestone® to be used in protecting a 12-qt cooling system against freeze-up at different temperatures (the remaining liquid is water): 10 °F, 3 qt; 0 °F, 4 qt; −15 °F, 5 qt; −34 °F, 6 qt. Since the freezing point of the coolant is successively lowered by using more Prestone®, why not use even more than 6 qt of Prestone® (and proportionately less water) to ensure the maximum protection against freezing?

101. Demonstrate that:
(a) For a *dilute aqueous* solution, the numerical value of the molality is essentially equal to that of the molarity.
(b) In a *dilute* solution, the solute mole fraction is proportional to the molality.
(c) In a *dilute aqueous* solution, the solute mole fraction is proportional to the molarity.

102. At 25 °C and under an $O_2(g)$ pressure of 1 atm, the solubility of $O_2(g)$ in water is 28.31 mL/1.00 L H_2O. Under an $N_2(g)$ pressure of 1 atm, the solubility of $N_2(g)$ at 25 °C is 14.34 mL/1.00 L H_2O. The composition of the atmosphere is 78.08% N_2 and 20.95% O_2, by volume. What is the composition of air dissolved in water at 25 °C, expressed as volume percents of N_2 and O_2?

103. We noted in Figure 14-13 that the liquid and vapor curves taken together outline a lens-shaped region when the normal boiling points of benzene–toluene solutions are plotted as a function of mole fraction of benzene. That is, unlike Figure 14-12, the liquid curve is not a straight line. Use data from Figure 14-13, and show by calculation that this should be the case.

104. A saturated solution prepared at 70 °C contains 32.0 g $CuSO_4$ per 100.0 g solution. A 335-g sample of this solution is then cooled to 0 °C and $CuSO_4 \cdot 5H_2O$ crystallizes out. If the concentration of a saturated solution at 0 °C is 12.5 g $CuSO_4/100$ g soln, what mass of $CuSO_4 \cdot 5H_2O$ would be obtained?
(*Hint:* Note that the solution composition is stated in terms of $CuSO_4$ and that the solid that crystallizes is the hydrate $CuSO_4 \cdot 5H_2O$.)

Feature Problems

105. Cinnamaldehyde is the chief constituent of cinnamon oil, which is obtained from the twigs and leaves of cinnamon trees grown in tropical regions. Cinnamon oil is used in the manufacture of food flavorings, perfumes, and cosmetics. The normal boiling point of cinnamaldehyde, C_9H_8O is 246.0 °C, but at this temperature it begins to decompose. As a result, cinnamaldehyde cannot be easily purified by ordinary distillation. A method that can be used instead is *steam distillation*. A heterogeneous mixture of cinnamaldehyde and water is heated until the sum of the vapor pressures of the two liquids is equal to barometric pressure. At this point, the temperature remains constant as the liquids vaporize. The mixed vapor condenses to produce two immiscible liquids; one liquid is essentially pure water and the other, pure cinnamaldehyde. The following vapor pressures of cinnamaldehyde are given: 1 mmHg at 76.1 °C; 5 mmHg at 105.8 °C; and 10 mmHg at 120.0 °C. Vapor pressures of water are given in Table 13.2.
(a) What is the approximate temperature at which the steam distillation occurs?
(b) The proportions of the two liquids condensed from the vapor is *independent* of the composition of the boiling mixture, as long as both liquids are present in the boiling mixture. Explain why this is so.
(c) Which of the two liquids, water or cinnamaldehyde, condenses in the greater quantity, by mass? Explain.

106. The phase diagram shown is for mixtures of HCl and H_2O at a pressure of 1 atm. The red curve represents the normal boiling points of solutions of HCl(aq) of various mole fractions. The blue curve represents the compositions of the vapors in equilibrium with boiling solutions.

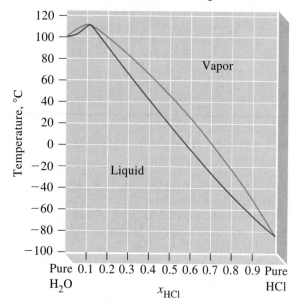

(a) As a solution containing $x_{HCl} = 0.50$ begins to boil, will the vapor have a mole fraction of HCl equal to, less than, or greater than 0.50? Explain.

(b) In the boiling of a pure liquid, there is no change in composition. However, as a solution of HCl(aq) boils in an open container, the composition changes. Explain why this is so.
(c) One particular solution (the azeotrope) is an exception to the observation stated in part (b); that is, its composition remains *unchanged* during boiling. What are the approximate composition and boiling point of this solution?
(d) A 5.00-mL sample of the azeotrope ($d = 1.099$ g/mL) requires 30.32 mL of 1.006 M NaOH for its titration in an acid–base reaction. Use these data to determine a more precise value of the composition of the azeotrope, expressed as the mole fraction of HCl.

107. The laboratory device pictured is called a *desiccator*. It can be used to maintain a constant relative humidity within an enclosure. The material(s) used to control the relative humidity are placed in the bottom compartment, and the substance being subjected to a controlled relative humidity is placed on the platform in the container.

(a) If the material in the bottom compartment is a saturated solution of NaCl(aq) in contact with NaCl(s), what will be the approximate relative humidity in the container at a temperature of 20 °C? Obtain the solubility of NaCl from Figure 14-8 and vapor pressure data for water from Table 13.2; use the definition of Raoult's law from page 549 and that of relative humidity from page 269; assume that the NaCl is completely dissociated into its ions.
(b) If the material placed on the platform is dry $CaCl_2 \cdot 6H_2O(s)$, will the solid deliquesce? Explain.
(*Hint:* Recall the discussion on page 554.)
(c) To maintain $CaCl_2 \cdot 6H_2O(s)$ in the dry state in a desiccator, should the substance in the saturated solution in the bottom compartment be one with a high or a low water solubility? Explain.

108. Every year, oral rehydration therapy (ORT)—the feeding of an electrolyte solution—saves the lives of countless children worldwide who become severely dehydrated as a result of diarrhea. One requirement of the solution used is that it be *isotonic* with human blood.

(a) One definition of an isotonic solution given in the text is that it have the same osmotic pressure as 0.92% NaCl(aq) (mass/volume). Another definition is that the solution have a freezing point of −0.52 °C. Show that these two definitions are in reasonably close agreement, given that we are using solution concentrations rather than activities.

(b) Use the freezing-point definition from part (a) to show that an ORT solution containing 3.5 g NaCl, 1.5 g KCl, 2.9 g $Na_3C_6H_5O_7$ (sodium citrate), and 20.0 g $C_6H_{12}O_6$ (glucose) per liter meets the requirement of being isotonic.
(*Hint:* Which of the solutes are nonelectrolytes and which are strong electrolytes?)

 eMedia Exercises

109. In the **Dissolution of NaCl in Water** animation *(eChapter 14-1)*, **(a)** what types of intermolecular forces compete in the solvation of this ionic compound? **(b)** How would the solution process be influenced by a change in the polarity of the solvent? **(c)** What insight does this observation give you into the saying, "like dissolves like"?

110. In the **Enthalpy of Solution Formation** activity *(eChapter 14-3)*, the dissolution of five different compounds in water is illustrated. The dissolution of some compounds show a temperature rise, while the dissolution of others show a temperature fall. **(a)** Account for the different behaviors with a discussion of the intermolecular forces at play in the process. **(b)** What trend can you discern by comparing this series of compounds?

111. The diffusion of water through a semipermeable membrane is illustrated in the **Osmosis** animation *(eChapter 14-7)*. **(a)** In terms of molecular motion, how would this behavior be influenced by an increase in the temperature of the system? **(b)** What is the quantitative relationship between temperature and the osmotic pressure of a solution?

112. In the **Boiling Point Elevation** activity for nonelectrolyte solutions *(eChapter 14-8)*, the addition of similar masses of different compounds alters the boiling point by different amounts. How is this observation related to the nature of colligative properties?

113. In the **Boiling Point Elevation** activity for electrolyte solutions *(eChapter 14-9)*, **(a)** which of the compounds illustrated gives rise to the greatest change in boiling point? **(b)** Why does this one have that characteristic?

15 Chemical Kinetics

Contents

The decomposition of hydrogen peroxide, H_2O_2, to H_2O and O_2 is a highly exothermic reaction that is strongly catalyzed by platinum metal. Both the rate of this decomposition and the nature of catalysis are explored in this chapter.

Rocket fuel is designed to give a rapid release of gaseous products and energy to give a rocket maximum thrust. Milk is stored in a refrigerator to slow down the chemical reactions that cause it to spoil. Current strategies to reduce the rate of deterioration of the ozone layer seek to deprive the ozone-consuming reaction cycle of key intermediates that come from chlorofluorocarbons (CFCs). These examples illustrate the importance of the rates of chemical reactions. Moreover, how fast a reaction occurs depends on the reaction mechanism—the step-by-step molecular pathway leading from reactants to products. Thus, *chemical kinetics* concerns how rates of chemical reactions are measured, how they can be predicted, and how reaction-rate data are used to deduce probable reaction mechanisms.

We will begin the chapter by describing what we mean by rate of reaction and presenting some ideas about measuring rates of reaction. We will follow this by introducing mathematical equations, called rate laws, that relate the rates of reactions to the concentrations of the reactants. Finally, with this information as background, we will turn to one of our central purposes: relating rate laws to plausible reaction mechanisms.

15-1 The Rate of a Chemical Reaction

Rate, or speed, refers to something that happens in a unit of time. A car traveling at 60 mph, for example, covers a distance of 60 miles in one hour. For chemical reactions, the rate of reaction describes how fast the concentration of a reactant or product changes with time.

To illustrate, let's consider the reaction that begins immediately after the ions Fe^{3+} and Sn^{2+} are simultaneously introduced into an aqueous solution.

$$2\,Fe^{3+}(aq) + Sn^{2+}(aq) \longrightarrow 2\,Fe^{2+}(aq) + Sn^{4+}(aq) \qquad (15.1)$$

Suppose that 38.5 s after the reaction starts, $[Fe^{2+}]$ is found to be 0.0010 M. During the period of time, $\Delta t = 38.5$ s, the *change* in concentration of Fe^{2+}, which we can designate as $\Delta[Fe^{2+}]$, is $\Delta[Fe^{2+}] = 0.0010\ M - 0 = 0.0010\ M$. The *average* rate at which Fe^{2+} is formed in this interval is the change in concentration of Fe^{2+} divided by the change in time.

$$\text{rate of formation of } Fe^{2+} = \frac{\Delta[Fe^{2+}]}{\Delta t} = \frac{0.0010\ M}{38.5\ s} = 2.6 \times 10^{-5}\ M\ s^{-1}$$

How has the concentration of Sn^{4+} changed during the 38.5 s we were monitoring the Fe^{2+}? Can you see that in 38.5 s, $\Delta[Sn^{4+}]$ will be $0.00050\ M - 0 = 0.00050\ M$? Because only *one* Sn^{4+} ion is produced for every *two* Fe^{2+} ions, the buildup of $[Sn^{4+}]$ will be only one-half that of $[Fe^{2+}]$. Consequently the rate of formation of Sn^{4+} is 1.3×10^{-5} mole per liter per second.

$$\text{rate of formation of } Sn^{4+} = 1.3 \times 10^{-5}\ M\ s^{-1}$$

We can also follow the course of the reaction by monitoring the concentrations of the starting reactants. Thus, the amount of Fe^{3+} consumed is the same as the amount of Fe^{2+} produced. The *change* in concentration of Fe^{3+} is $\Delta[Fe^{3+}] = -0.0010\ M$. The average rate at which Fe^{3+} disappears in the reaction is given by the expression

$$\text{rate of disappearance of } Fe^{3+} = \frac{\Delta[Fe^{3+}]}{\Delta t} = \frac{-0.0010\ M}{38.5\ s} = -2.6 \times 10^{-5}\ M\ s^{-1}$$

The rate of disappearance is a *negative* quantity because concentration decreases with time—the concentration at the end of a time period is less than it was at the start of the period. In the same way that we related the rate of formation of Sn^{4+} to that of Fe^{2+}, we can relate the rate of disappearance of Sn^{2+} to that of Fe^{3+}. That is, the rate of disappearance of Sn^{2+} is half that of Fe^{3+}, giving

$$\text{rate of disappearance of } Sn^{2+} = -1.3 \times 10^{-5}\ M\ s^{-1}$$

When we refer to the rate of reaction (15.1), which of the four quantities described here should we use? To avoid confusion in this matter, the International Union of Pure and Applied Chemistry (IUPAC) recommends that we use a *general* rate of reaction, which, for the hypothetical reaction represented by the balanced equation,

$$a\mathrm{A} + b\mathrm{B} \longrightarrow g\mathrm{G} + h\mathrm{H}$$

▶ Recall that the symbol [ ] means "the molarity of." Also, Δ means "the change in," that is, the final value minus the initial value.

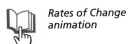

Rates of Change animation

is

$$\text{rate of reaction} = -\frac{1}{a}\frac{\Delta[A]}{\Delta t} = -\frac{1}{b}\frac{\Delta[B]}{\Delta t} = \frac{1}{g}\frac{\Delta[G]}{\Delta t} = \frac{1}{h}\frac{\Delta[H]}{\Delta t} \qquad (15.2)$$

In this expression, we take the *negative* values of rates of disappearance, positive values of rates of formation, and *divide* all rates by the appropriate stoichiometric coefficients from the balanced equation. The result is a single, positive-valued quantity that we call the **rate of reaction**. Thus, for reaction (15.1) we can write

$$\text{rate of reaction} = -\frac{1}{2}\frac{\Delta[Fe^{3+}]}{\Delta t} = -\frac{\Delta[Sn^{2+}]}{\Delta t}$$

$$= \frac{1}{2}\frac{\Delta[Fe^{2+}]}{\Delta t} = \frac{\Delta[Sn^{4+}]}{\Delta t} = 1.3 \times 10^{-5}\,\text{M s}^{-1}$$

EXAMPLE 15-1

Expressing the Rate of a Reaction. Suppose that at some point in the reaction

$$\text{A} + 3\,\text{B} \longrightarrow 2\,\text{C} + 2\,\text{D}$$

[B] = 0.9986 M, and that 13.20 min later [B] = 0.9746 M. What is the average rate of reaction during this time period, expressed in M s^{-1}?

Solution

The rate of disappearance of B is the *change* in molarity, $\Delta[B]$, divided by the time interval, Δt, over which this change occurs $\Delta[B] = 0.9746\,\text{M} - 0.9986\,\text{M} = -0.0240\,\text{M}$; $\Delta t = 13.20$ min, and

$$\text{rate of reaction} = -\frac{1}{3}\frac{\Delta[B]}{\Delta t} = -\frac{1}{3} \times \frac{-0.0240\,\text{M}}{13.20\,\text{min}} = 6.06 \times 10^{-4}\,\text{M min}^{-1}$$

To express the rate of reaction in moles per liter per second, we must convert from min^{-1} to s^{-1}. We can do this with the conversion factor 1 min/60 s.

$$\text{rate of reaction} = 6.06 \times 10^{-4}\,\text{M min}^{-1} \times \frac{1\,\text{min}}{60\,\text{s}} = 1.01 \times 10^{-5}\,\text{M s}^{-1}$$

Alternatively, we could have converted 13.20 min to 792 s and used $\Delta t = 792$ s in evaluating the rate of reaction.

Practice Example A: At some point in the reaction $2\,\text{A} + \text{B} \longrightarrow \text{C} + \text{D}$, [A] = 0.3629 M. At a time 8.25 min later [A] = 0.3187 M. What is the average rate of reaction during this time interval, expressed in M s^{-1}?

Practice Example B: In the reaction $2\,\text{A} \longrightarrow 3\,\text{B}$, [A] drops from 0.5684 M to 0.5522 M in 2.50 min. What is the average rate of formation of B during this time interval, expressed in M s^{-1}?

15-2 Measuring Reaction Rates

To determine a rate of reaction, we need to measure changes in concentration over time. A change in time can be measured with a stopwatch or other timing device, but how do we measure concentration changes during a chemical reaction? Also, why have we used the term *average* in referring to a rate of reaction? These are two of the questions we will answer in this section.

Following a Chemical Reaction

A 3% aqueous solution of hydrogen peroxide is a common antiseptic. Its antiseptic action results from the release of $O_2(g)$ as the H_2O_2 decomposes. $O_2(g)$ escapes from the $H_2O_2(aq)$, and ultimately the reaction goes to completion.[*]

$$H_2O_2(aq) \longrightarrow H_2O(l) + \frac{1}{2} O_2(g) \qquad (15.3)$$

We can follow the progress of the reaction by focusing either on the formation of $O_2(g)$ or on the disappearance of H_2O_2. For example, we can

- Measure the volumes of $O_2(g)$ produced (Figure 15-1) at different times and relate these volumes to decreases in concentration of H_2O_2.
- Remove small samples of the reaction mixture from time to time, and analyze these samples for their H_2O_2 content. One way to do this is by titration with $KMnO_4$ in acidic solution. The net ionic equation for this oxidation–reduction reaction is

$$2\,MnO_4^- + 5\,H_2O_2 + 6\,H^+ \longrightarrow 2\,Mn^{2+} + 8\,H_2O + 5\,O_2(g) \qquad (15.4)$$

Table 15.1 lists typical data for the decomposition of H_2O_2, and Figure 15-2 displays comparable data graphically.

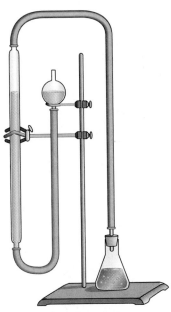

▲ **FIGURE 15–1**
Experimental setup for determining the rate of decomposition of H_2O_2
Oxygen gas given off by the reaction mixture is trapped, and its volume is measured in the gas buret. The amount of H_2O_2 consumed and the remaining concentration of H_2O_2 can be calculated from the measured volume of $O_2(g)$.

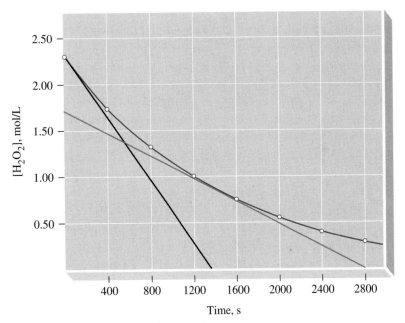

▲ **FIGURE 15–2**   **Graphical representation of kinetic data for the reaction:**

$$H_2O_2(aq) \longrightarrow H_2O(l) + \frac{1}{2} O_2(g)$$

This is the usual form in which concentration–time data are plotted. Reaction rates are determined from the slopes of the tangent lines. The blue line has a slope of -1.70 M/ 2800 s $= -6.1 \times 10^{-4}$ M s^{-1}. The slope of the black line and its relation to the initial rate of reaction are described in Example 15-2.

TABLE 15.1
Decomposition of H_2O_2

Time, s	$[H_2O_2]$, M
0	2.32
200	2.01
400	1.72
600	1.49
1200	0.98
1800	0.62
3000	0.25

[*] Although reaction (15.3) goes to completion, it does so very slowly. Generally, a catalyst is used to speed up the reaction. We describe the catalysis in this reaction in Section 15-11. When $H_2O_2(aq)$ is applied to an open wound, the peroxisomal enzyme catalase in blood catalyzes its decomposition.

Reaction Rates simulation

Rate of Reaction Expressed as Concentration Change over Time

We have extracted some data from Figure 15-2 and listed them in Table 15.2 (in blue). Column III lists the molarities of H_2O_2 at the times shown in column I. Column II lists the *arbitrary* time interval we have chosen between data points—400 s. Column IV reports the concentration changes that occur for each 400-s interval. The rates of reaction, expressed as the negative of the rate of disappearance of H_2O_2, are shown in column V. The data show that the reaction rate is not constant—the lower the remaining concentration of H_2O_2, the more slowly the reaction proceeds.

TABLE 15.2		Decomposition of H_2O_2—Derived Rate Data		
I	**II**	**III**	**IV**	**V Reaction Rate**
Time, s	Δt, s	$[H_2O_2]$, M	$\Delta[H_2O_2]$, M	$-\Delta[H_2O_2]/\Delta t$, $M\ s^{-1}$
0		2.32		
	400		−0.60	15.0×10^{-4}
400		1.72		
	400		−0.42	10.5×10^{-4}
800		1.30		
	400		−0.32	8.0×10^{-4}
1200		0.98		
	400		−0.25	6.3×10^{-4}
1600		0.73		
	400		−0.19	4.8×10^{-4}
2000		0.54		
	400		−0.15	3.8×10^{-4}
2400		0.39		
	400		−0.11	2.8×10^{-4}
2800		0.28		

Rate of Reaction Expressed as the Slope of a Tangent Line

When we express the rate of reaction as $-\Delta[H_2O_2]/\Delta t$, we simply get an *average* value for the time interval Δt. For example, in the interval from 1200 to 1600 s, the rate averages $6.3 \times 10^{-4}\ M\ s^{-1}$ (fourth entry in column V of Table 15.2). We can think of this as the reaction rate at about the middle of the interval—1400 s. We could just as well have chosen $\Delta t = 200$ s in Table 15.2. In this case, to obtain the rate of reaction at $t = 1400$ s, we would use concentration data at $t = 1300$ s and $t = 1500$ s. The rate of reaction would be slightly less than $6.3 \times 10^{-4}\ M\ s^{-1}$. The smaller the time interval we choose, the closer we will come to the actual rate at $t = 1400$ s. As Δt approaches 0 s, the rate of reaction approaches the *negative of the slope of the tangent line* to the curve of Figure 15-2. The rate of reaction determined from the slope of a tangent line to a concentration–time curve is the **instantaneous rate of reaction** at the point where the tangent line touches the curve. Our best estimate of the rate of reaction of H_2O_2 at 1400 s, then, is from the slope of the blue tangent line in Figure 15-2: $-(-6.1 \times 10^{-4}\ M\ s^{-1}) = 6.1 \times 10^{-4}\ M\ s^{-1}$.

One's instantaneous speed can result in a speeding ticket even if the average speed never exceeds the speed limit. ▶

To better understand the difference between average and instantaneous reaction rates, think of taking a 108-mi highway trip in 2.00 h. The *average* speed is 54.0 mph. The *instantaneous* speed is the speedometer reading at any instant.

Are You Wondering...

If there is a way to designate through symbols whether a rate of reaction is an average rate or an instantaneous rate?

If you are familiar with differential calculus, you probably know the answer. If we write the rate of reaction (15.3) as

$$\lim_{\Delta t \longrightarrow 0} \frac{-\Delta[H_2O_2]}{\Delta t}$$

we can replace the delta quantities by the differentials $d[H_2O_2]$ and dt, leading to the expression

$$\frac{-d[H_2O_2]}{dt}$$

Thus, the delta notation (not taken to the limit of $\Delta t \longrightarrow 0$) signifies an *average* rate and the differential notation, an *instantaneous* rate.

Initial Rate of Reaction

Sometimes we simply want to find the rate of reaction when the reactants are first brought together—the **initial rate of reaction**. One way to get this is from the tangent line to the concentration–time curve at $t = 0$. An alternative way is to measure the concentration of the chosen reactant as soon as possible after mixing, in this way obtaining Δ[reactant] for a very short time interval (Δt) at essentially $t = 0$. These two approaches give the same result if we limit ourselves to the time interval in which the tangent line and the concentration–time curves practically coincide. In Figure 15-2, this condition occurs for about the first 200 s.

EXAMPLE 15-2

Determining and Using an Initial Rate of Reaction. From the data in Table 15.1 and Figure 15-2 for the decomposition of H_2O_2, **(a)** determine the initial rate of reaction, and **(b)** $[H_2O_2]_t$ at $t = 100$ s.

Solution

(a) To determine the initial rate of reaction from the slope of the tangent line, we use the intersections of the black tangent line with the axes: $t = 0$, $[H_2O_2] = 2.32$ M; $t = 1360$ s, $[H_2O_2] = 0$.

$$\text{initial rate of reaction} = -(\text{slope of tangent line}) = \frac{-(0 - 2.32)\,\text{M}}{(1360 - 0)\,\text{s}}$$

$$= 1.71 \times 10^{-3}\,\text{M s}^{-1}$$

An alternative method is to use data from Table 15.1: $[H_2O_2] = 2.32$ M at $t = 0$ and $[H_2O_2] = 2.01$ M at $t = 200$ s.

$$\text{initial rate} = \frac{-\Delta[H_2O_2]}{\Delta t} = \frac{-(2.01 - 2.32)\,\text{M}}{200\,\text{s}}$$

$$= 1.6 \times 10^{-3}\,\text{M s}^{-1}$$

▶ Another reason for favoring a graphical method is that it tends to minimize the effect of errors that may be found in individual data points.

The agreement between the two methods is fairly good, although it might be better if the time interval were less than 200 s. Of the two results, the one based on the tangent line is presumably more precise because it is expressed with more significant figures. On the other hand, the reliability of the tangent line depends on how carefully the tangent line is constructed.

(b) We assume that the rate determined in (a) remains essentially constant for at least 100 s. Because

$$\text{rate of reaction} = \frac{-\Delta[H_2O_2]}{\Delta t}$$

then

$$1.71 \times 10^{-3} \text{ M s}^{-1} = \frac{-\Delta[H_2O_2]}{100 \text{ s}}$$

$$-(1.71 \times 10^{-3} \text{ M s}^{-1})(100 \text{ s}) = \Delta[H_2O_2] = [H_2O_2]_t - [H_2O_2]_0$$

$$-1.71 \times 10^{-1} \text{ M} = [H_2O_2]_t - 2.32 \text{ M}$$

$$[H_2O_2]_t = 2.32 \text{ M} - 0.17 \text{ M} = 2.15 \text{ M}$$

Practice Example A: For reaction (15.3), determine **(a)** the *instantaneous* rate of reaction at 2400 s and **(b)** $[H_2O_2]$ at 2450 s.

(*Hint:* Assume that the instantaneous rate of reaction at 2400 s holds constant for the next 50 s.)

Practice Example B: Use data *only* from Table 15.2 to determine $[H_2O_2]$ at $t = 100$ s. Compare this value to the one calculated in Example 15-2b. Explain the reason for the difference.

15-3   Effect of Concentration on Reaction Rates: The Rate Law

One of the goals in a chemical kinetics study is to derive an equation that can be used to predict the relationship between the rate of reaction and the concentrations of reactants. Such an experimentally determined equation is called a **rate law**, or **rate equation**.

Consider the hypothetical reaction

$$a\text{A} + b\text{B} \cdots \longrightarrow g\text{G} + h\text{H} \cdots \tag{15.5}$$

where $a, b, \ldots$ stand for coefficients in the balanced equation. We can often express the rate of such a reaction as[*]

$$\text{rate of reaction} = k[\text{A}]^m[\text{B}]^n \cdots \tag{15.6}$$

The terms [A], [B], ... represent reactant molarities. The exponents, $m, n, \ldots$ are generally small, positive whole numbers, although in some cases they may be zero, fractional, and/or negative. They are generally *not* related to the stoichiometric coefficients $a, b, \ldots$. That is, often $m \neq a$, $n \neq b$, and so on.

The term *order* is used in two ways in describing a rate of reaction: (1) If $m = 1$, we say that the reaction is *first order in A*. If $n = 2$, the reaction is *second order in B*, and so on. (2) The *overall* **order of reaction** is the sum of all the exponents: $m + n + \cdots$. The proportionality constant k relates the rate of reaction to reactant concentrations and is called the **rate constant** of the reaction. Its value depends on the specific reaction, the presence of a catalyst (if any), and the

---

[*]We assume that reaction (15.5) goes to completion. If it is reversible, the rate equation is more complex than (15.6). Even for reversible reactions, though, equation (15.6) applies to the *initial* rate of reaction because in the early stages of the reaction there are not enough products for a significant reverse reaction to occur.

temperature. *The larger the value of k, the faster a reaction goes.* The units of k depend on the order of the reaction (that is, on the values of the exponents $m, n, \ldots$).

With the rate law for a reaction, we can

- calculate rates of reaction for known concentrations of reactants
- derive an equation that expresses a reactant concentration as a function of time

But how do we establish the rate law? We need to use *experimental* data of the type described in Section 15-2. The method we describe next works especially well.

Method of Initial Rates

As its name implies, this method requires us to work with *initial* rates of reaction. As an example, let's look at a specific reaction: that between mercury(II) chloride and oxalate ion.

$$2\,HgCl_2(aq) + C_2O_4^{2-}(aq) \longrightarrow 2\,Cl^-(aq) + 2\,CO_2(g) + Hg_2Cl_2(s) \qquad (15.7)$$

The tentative rate law that we can write for this reaction is

$$\text{rate of reaction} = k[HgCl_2]^m[C_2O_4^{2-}]^n \qquad (15.8)$$

We can follow the reaction by measuring the quantity of $Hg_2Cl_2(s)$ formed as a function of time. Some representative data are given in Table 15.3, which we can assume are based on either the rate of formation of Hg_2Cl_2 or the rate of disappearance of $C_2O_4^{2-}$. In Example 15-3, we will use some of these data to illustrate the method of initial rates.

TABLE 15.3   Kinetic Data for the Reaction:
$2\,HgCl_2 + C_2O_4^{2-} \longrightarrow 2\,Cl^- + 2\,CO_2 + Hg_2Cl_2$

Experiment	$[HgCl_2]$, M	$[C_2O_4^{2-}]$, M	Initial rate, M min^{-1}
1	$[HgCl_2]_1 = 0.105$	$[C_2O_4^{2-}]_1 = 0.15$	1.8×10^{-5}
2	$[HgCl_2]_2 = 0.105$	$[C_2O_4^{2-}]_2 = 0.30$	7.1×10^{-5}
3	$[HgCl_2]_3 = 0.052$	$[C_2O_4^{2-}]_3 = 0.30$	3.5×10^{-5}

EXAMPLE 15-3

Establishing the Order of a Reaction by the Method of Initial Rates. Use data from Table 15.3 to establish the order of reaction (15.7) with respect to $HgCl_2$ and $C_2O_4^{2-}$ and also the overall order of the reaction.

Solution

We need to determine the values of m and n in equation (15.8). In comparing Experiment 2 with Experiment 3, note that $[HgCl_2]$ is essentially doubled ($0.105\ M \approx 2 \times 0.052\ M$) while $[C_2O_4^{2-}]$ is held constant (at $0.30\ M$). Note also that $R_2 = 2 \times R_3$ ($7.1 \times 10^{-5} \approx 2 \times 3.5 \times 10^{-5}$). Rather than use the actual concentrations and rates in the following rate equation, let's work with their symbolic equivalents.

$$R_2 = k \times [HgCl_2]_2^m \times [C_2O_4^{2-}]_2^n = k \times (2 \times [HgCl_2]_3)^m \times [C_2O_4^{2-}]_3^n$$

$$R_3 = k \times [HgCl_2]_3^m \times [C_2O_4^{2-}]_3^n$$

$$\frac{R_2}{R_3} = \frac{2 \times \cancel{R_3}}{\cancel{R_3}} = 2 = \frac{\cancel{k} \times 2^m \times \cancel{[HgCl_2]_3^m} \times \cancel{[C_2O_4^{2-}]_3^n}}{\cancel{k} \times \cancel{[HgCl_2]_3^m} \times \cancel{[C_2O_4^{2-}]_3^n}} = 2^m$$

In order that $2m = 2$, $m = 1$.

To determine the value of n, we can form the ratio R_2/R_1. Now, $[C_2O_4^{2-}]$ is doubled and $[HgCl_2]$ is held constant. This time, let's use actual concentrations instead of symbolic equivalents. Also, we now have the value $m = 1$.

$$R_2 = k \times [HgCl_2]_2^1 \times [C_2O_4^{2-}]_2^n = k \times (0.105)^1 \times (2 \times 0.15)^n$$
$$R_1 = k \times [HgCl_2]_1^1 \times [C_2O_4^{2-}]_1^n = k \times (0.105)^1 \times (0.15)^n$$
$$\frac{R_2}{R_1} = \frac{7.1 \times 10^{-5}}{1.8 \times 10^{-5}} \approx 4 = \frac{k \times (0.105)^1 \times 2^n \times (0.15)^n}{k \times (0.105)^1 \times (0.15)^n} = 2^n$$

In order that $2^n = 4$, $n = 2$.

In summary, the reaction is *first* order in $HgCl_2$ ($m = 1$), *second* order in $C_2O_4^{2-}$ ($n = 2$), and *third* order overall ($m + n = 1 + 2 = 3$).

Practice Example A: The decomposition of N_2O_5 is given by the following equation:

$$2 N_2O_5 \longrightarrow 4 NO_2 + O_2$$

At an initial $[N_2O_5] = 3.15$ M, the initial rate of reaction $= 5.45 \times 10^{-5}$ M s^{-1}, and when $[N_2O_5] = 0.78$ M, the initial rate of reaction $= 1.35 \times 10^{-5}$ M s^{-1}. Determine the order of this decomposition reaction.

Practice Example B: Consider a hypothetical Experiment 4 in Table 15.3, in which the initial conditions are $[HgCl_2]_4 = 0.025$ M and $[C_2O_4^{2-}]_4 = 0.045$ M. Predict the initial rate of reaction.

(*Hint:* Work with a ratio of two initial rates, but now with known values of m and n.)

We made an important observation in Example 15-3: If a reaction is *first order* in one of the reactants, doubling the initial concentration of that reactant causes the initial rate of reaction to double. Following is the general effect of *doubling* the initial concentration of a particular reactant (with other reactant concentrations held constant).

- *Zero* order in the reactant—there is *no effect* on the initial rate of reaction.
- *First* order in the reactant—the initial rate of reaction *doubles*.
- *Second* order in the reactant—the initial rate of reaction *quadruples*.
- *Third* order in the reactant—the initial rate of reaction *increases eightfold*.

Another consequence of the form of the rate law for a reaction is that it establishes the units of the rate constant, k. That is, if on the left side of the rate law the rate of reaction has the units M (time)$^{-1}$, on the right side, the units of k must provide for the cancellations that also lead to M (time)$^{-1}$. Thus, for the rate law established in Example 15-3,

rate law: rate of reaction $= k \times [HgCl_2] \times [C_2O_4^{2-}]^2$

units: M min^{-1} M^{-2} min^{-1} M   M^2

Once we have the exponents in a rate equation, we can determine the value of the rate constant, k. To do this, all we need is the rate of reaction corresponding to known initial concentrations of reactants, as illustrated in Example 15-4.

EXAMPLE 15-4

Using the Rate Law. Use the results of Example 15-3 and data from Table 15.3 to establish the value of k in the rate law (15.8).

Solution

We can use data from any one of the three experiments of Table 15.3, together with the values $m = 1$ and $n = 2$. First, we solve equation (15.8) for k.

$$k = \frac{R_1}{[HgCl_2][C_2O_4^{2-}]^2} = \frac{1.8 \times 10^{-5} \text{ M min}^{-1}}{0.105 \text{ M} \times (0.15)^2 \text{ M}^2}$$

$$= 7.6 \times 10^{-3} \text{ M}^{-2} \text{ min}^{-1}$$

Practice Example A: A reaction has the rate law: rate = $k[A]^2[B]$. When $[A] = 1.12$ M and $[B] = 0.87$ M, the rate of reaction = 4.78×10^{-2} M s^{-1}. What is the value of the rate constant, k?

Practice Example B: What is the rate of reaction (15.7) at the point where $[HgCl_2] = 0.050$ M and $[C_2O_4^{2-}] = 0.025$ M?

KEEP IN MIND ▶
that if the rate data in Table 15.3 were based on the disappearance of $HgCl_2$ instead of $C_2O_4^{2-}$, R_1 in this setup would be twice as great. Then k for the general rate of reaction would have to be based on $-\dfrac{1}{2} \times$ (rate of disappearance of $HgCl_2$).

The dependence of k on temperature is discussed in Section 15-9. ▶

? Are You Wondering ...

Just what is the difference between the rate of a reaction and the rate *constant* of a reaction?

Many students have difficulty with this distinction. Remember, a *rate of reaction* tells how the amount of a reactant or product changes with time and is usually expressed as moles per liter per time. A rate of reaction can be established through an expression of the type $-\Delta[A]/\Delta t$, from a tangent to a concentration–time curve, and by calculation from a rate law. In most cases, the rate of a reaction strongly depends on reactant concentrations.

The *rate constant* of a reaction (k) relates the rate of a reaction to reactant concentrations. Generally, it is not itself a rate of reaction, but we can use it to calculate rates of reaction. Once the value of k at a given temperature has been established, *this value stays fixed*. Whereas the units of rates of reaction do not depend on the order of a reaction, those of k do, as we shall see in the next three sections.

15-4   Zero-Order Reactions

An overall **zero-order reaction** has a rate law in which the sum of the exponents, $m + n \cdots$ is equal to 0. As an example, let's take a reaction in which a single reactant A decomposes to products.

$$A \longrightarrow \text{products}$$

If the reaction is zero order, the rate law is

$$\text{rate of reaction} = k[A]^0 = k = \text{constant} \qquad (15.9)$$

Other features of this zero-order reaction are

- The concentration–time graph is a straight line with a *negative* slope (Figure 15-3).
- The rate of reaction, which is equal to k and remains constant throughout the reaction, is the *negative* of the slope of this line.
- The units of k are the same as the units of the rate of a reaction: mol L^{-1} (time)$^{-1}$, for example, mol L^{-1} s^{-1} or M s^{-1}.

Equation (15.9) is the *rate law* for a zero-order reaction. Another useful equation, called an **integrated rate law**, expresses the concentration of a reactant as a

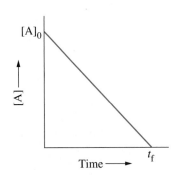

▲ FIGURE 15–3
A zero-order reaction:
A ⟶ products
The initial concentration of the reactant A is $[A]_0$, that is, $[A] = [A]_0$ at $t = 0$. $[A]$ decreases at a constant rate until the reaction stops. This occurs at the time, t_f, where $[A] = 0$. The slope of the line is $\big(0 - [A]_0\big)/\big(t_f - 0\big) = -[A]_0/t_f$. The rate constant is the *negative* of the slope: $k = -\text{slope} = [A]_0/t_f$.

▶ Perhaps the most important examples of zero-order reactions are found in the action of enzymes. Enzyme-catalyzed reactions are discussed in Section 15-11.

function of time. We can establish this equation rather easily from the graph in Figure 15-3. Let's start with the general equation for a straight line

$$y = mx + b$$

and substitute $y = [A]_t$ (the concentration of A at some time t); $x = t$ (time); $b = [A]_0$ (the initial concentration of A at time $t = 0$); and $m = -k$ (m, the slope of the straight line, is obtained as indicated in Figure 15-3).

$$[A]_t = -kt + [A]_0 \qquad (15.10)$$

Are You Wondering...

If the term "integrated" rate law has anything to do with integral calculus?

Not surprisingly, it does. The rate of reaction in a rate law such as (15.9) is an *instantaneous* rate, which we learned in the "Are You Wondering" on page 583, can be represented through differentials. When we substitute $-d[A]/dt$ for the rate of reaction in the rate law for a zero-order reaction, we get this equation, $-d[A]/dt = k$. We can separate the differentials to obtain $d[A] = -kdt$. At this point, we can apply the calculus procedure of integration to obtain, successively, the following expressions. The final one is the integrated rate law for a zero-order reaction (15.10).

$$\int_{[A]_0}^{[A]_t} d[A] = -k \int_0^t dt, \qquad [A]_t - [A]_0 = -kt, \qquad [A]_t = -kt + [A]_0 \quad (15.10)$$

15-5   First-Order Reactions

An overall **first-order reaction** has a rate law in which the sum of the exponents, $m + n \cdots$ is equal to 1. A particularly common type of first-order reaction, and the only type we will consider, is one in which a single reactant decomposes into products. Reaction (15.3), the decomposition of H_2O_2 that we described in Section 15-2, is a first-order reaction.

$$H_2O_2(aq) \longrightarrow H_2O(l) + \frac{1}{2} O_2(g)$$

The rate of reaction depends on the concentration of H_2O_2 raised to the *first* power, that is,

$$\text{rate of reaction} = k[H_2O_2] \qquad (15.11)$$

It is easy to establish that reaction (15.3) is first order by the method of initial rates, but there are also other ways of recognizing a first-order reaction.

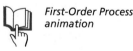

First-Order Process animation

An Integrated Rate Law for a First-Order Reaction

Let's begin our discussion of first-order reactions as we did zero-order reactions, by examining a hypothetical reaction

$$A \longrightarrow \text{products}$$

for which the rate law is

$$\text{rate of reaction} = k[A] \qquad (15.12)$$

We can obtain the *integrated* rate law for this first-order reaction by applying the calculus technique of *integration* to equation (15.12). The result of this derivation (shown in the Are You Wondering feature on page 590) is

KEEP IN MIND

that the scatter of experimental data on a straight-line graph often makes a value of k calculated from equation (15.13) less reliable than a value obtained from the slope of the straight line. ▶

$$\ln \frac{[A]_t}{[A]_0} = -kt \quad \text{or} \quad \ln[A]_t = -kt + \ln[A]_0 \qquad (15.13)$$

$[A]_t$ is the concentration of A at time t, $[A]_0$ is its concentration at $t = 0$, and k is the rate constant. Because the logarithms of numbers have no units (are dimensionless), the product $-k \times t$ must also be without units. This means that the unit of k in a first-order reaction is (time)$^{-1}$, such as s^{-1} or min^{-1}. Equation (15.13) is that of a straight line.

$$\underbrace{\ln[A]_t}_{y} = \underbrace{(-k)t}_{m \cdot x} + \underbrace{\ln[A]_0}_{b}$$

Equation of straight line

An easy test for a first-order reaction is to plot the natural logarithm of a reactant concentration versus time and see if the graph is linear. The data from Table 15.1 are plotted in Figure 15-4, and the rate constant k is derived from the slope of the line: $k = -\text{slope} = -(-7.30 \times 10^{-4} \text{ s}^{-1}) = 7.30 \times 10^{-4} \text{ s}^{-1}$. An alternative, nongraphical approach, illustrated in Practice Example 15-5B, is to substitute data points into equation (15.13) and solve for k.

EXAMPLE 15-5

Using the Integrated Rate Law for a First-Order Reaction. H_2O_2(aq), initially at a concentration of 2.32 M, is allowed to decompose. What will [H_2O_2] be at $t = 1200$ s? Use $k = 7.30 \times 10^{-4} \text{ s}^{-1}$ for this first-order decomposition.

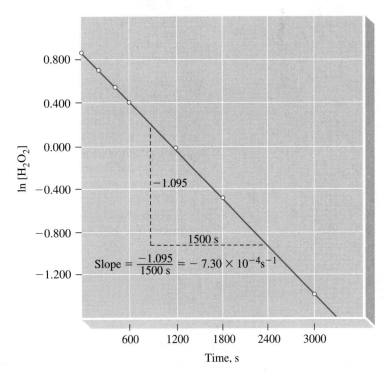

▶ **FIGURE 15–4**

Test for a first-order reaction: decomposition of H_2O_2(aq)

Here we plot ln [H_2O_2] versus t. The data are based on Table 15.1 and are listed below. The slope of the line is used in the text.

t, s	[H_2O_2], M	ln[H_2O_2]
0	2.32	0.842
200	2.01	0.698
400	1.72	0.542
600	1.49	0.399
1200	0.98	−0.020
1800	0.62	−0.48
3000	0.25	−1.39

> ### Are You Wondering ...
>
> **How to obtain the integrated rate law for a first-order reaction?**
>
> In differential form, the rate law for the reaction A $\longrightarrow$ products is $d[A]/dt = -k[A]$
> Separation of the differentials leads to the expression $d[A]/[A] = -kdt$
> Integration of this expression between the limits $[A]_0$ at time $t = 0$ and $[A]_t$ at time t
> is indicated through the expression
>
> $$\int_{[A]_0}^{[A]_t} \frac{d[A]}{[A]} = -k \int_0^t dt, \quad \text{yielding} \ \ln \frac{[A]_t}{[A]_0} = -kt$$
>
> The result of the integration is the integrated rate law.
>
> $$\ln \frac{[A]_t}{[A]_0} = -kt \tag{15.13}$$

Solution

We have values for three of the four terms in equation (15.13)

$$k = 7.30 \times 10^{-4} \ \text{s}^{-1} \qquad t = 1200 \ \text{s}$$
$$[H_2O_2]_0 = 2.32 \ \text{M} \qquad [H_2O_2]_t = ?$$

which we substitute into the expression

$$\begin{aligned}
\ln[H_2O_2]_t &= -kt + \ln[H_2O_2]_0 \\
&= -(7.30 \times 10^{-4} \ \text{s}^{-1} \times 1200 \ \text{s}) + \ln 2.32 \\
&= \qquad -0.876 \qquad\quad + 0.842 = -0.034 \\
[H_2O_2]_t &= e^{-0.034} = 0.967 \ \text{M}
\end{aligned}$$

▶ To find the number whose natural logarithm is −0.034, raise e to the −0.034 power.

This calculated value agrees well with the experimentally determined value of 0.98 M.

Practice Example A: The reaction A $\longrightarrow$ 2 B + C is first order. If the initial $[A] = 2.80$ M and $k = 3.02 \times 10^{-3} \ \text{s}^{-1}$, what is the value of $[A]$ after 325 s?

Practice Example B: Use data tabulated in Figure 15-4, together with equation (15.13), to show that the decomposition of H_2O_2 is a first-order reaction.

(*Hint:* Use a pair of data points for $[H_2O_2]_0$ and $[H_2O_2]_t$, and their corresponding times to solve for k. Repeat this calculation using other sets of data. How should the results compare?)

Although until now we have used only molar concentrations in kinetics equations, we can sometimes work directly with the masses of reactants. Another possibility is to work with a fraction of reactant consumed, as is done in the concept of half-life.

The **half-life** of a reaction is the time required for one-half of a reactant to be consumed. It is the time during which the amount of reactant or its concentration decreases to one-half of its initial value. That is, at $t = t_{1/2}$, $[A]_t = \frac{1}{2}[A]_0$. At this time, equation (15.13) takes the form

$$\ln \frac{[A]_t}{[A]_0} = \ln \frac{\frac{1}{2}[A]_0}{[A]_0} = \ln \frac{1}{2} = -\ln 2 = -k \times t_{1/2}$$

$$t_{1/2} = \frac{\ln 2}{k} = \frac{0.693}{k} \tag{15.14}$$

For the decomposition of $H_2O_2(aq)$, the reaction described in Section 15-2, we conclude that the half-life is

$$t_{1/2} = \frac{0.693}{7.30 \times 10^{-4}\ \text{s}^{-1}} = 9.49 \times 10^2\ \text{s} = 949\ \text{s}$$

Equation (15.14) indicates that *the half-life is constant for a first-order reaction.* Thus, regardless of the value of $[A]_0$ at the time we begin to follow a reaction, at $t = t_{1/2}$, $[A] = \frac{1}{2}[A]_0$. After *two* half-lives, that is, at $t = 2 \times t_{1/2}$, $[A] = \frac{1}{2} \times \frac{1}{2}[A]_0 = \frac{1}{4}[A]_0$. At $t = 3 \times t_{1/2}$, $[A] = \frac{1}{8}[A]_0$, and so on.

The constancy of the half-life and its independence of the initial concentration can be used as a test for a first-order reaction. Try it with the simple concentration–time graph of Figure 15-2. That is, starting with $[H_2O_2] = 2.32$ M at $t = 0$, at what time is $[H_2O_2] \approx 1.16$ M, ≈ 0.58 M, ≈ 0.29 M? Starting with $[H_2O_2] = 1.50$ M at $t = 600$ s, at what time is $[H_2O_2] = 0.75$ M?

As illustrated in Example 15-6, a first-order reaction can also be described in terms of the percent of a reactant consumed or remaining.

EXAMPLE 15-6

Expressing Fraction (or Percent) of Reactant Consumed in a First-Order Reaction. Use a value of $k = 7.30 \times 10^{-4}\ \text{s}^{-1}$ for the first-order decomposition of $H_2O_2(aq)$ to determine the percent H_2O_2 that has decomposed in the first 500.0 s after the reaction begins.

Solution

The ratio $[H_2O_2]_t/[H_2O_2]_0$ represents the fractional part of the initial amount of H_2O_2 that remains *unreacted* at time t. Our problem is to evaluate this ratio at $t = 500.0$ s.

$$\ln \frac{[H_2O_2]_t}{[H_2O_2]_0} = -kt = -7.30 \times 10^{-4}\ \text{s}^{-1} \times 500.0\ \text{s} = -0.365$$

$$\frac{[H_2O_2]_t}{[H_2O_2]_0} = e^{-0.365} = 0.694 \quad \text{and} \quad [H_2O_2]_t = 0.694[H_2O_2]_0$$

The fractional part of the H_2O_2 remaining is 0.694, or 69.4%. The percent of H_2O_2 that has decomposed is $100.0\% - 69.4\% = 30.6\%$.

Practice Example A: Consider the first-order reaction A $\longrightarrow$ products, with $k = 2.95 \times 10^{-3}\ \text{s}^{-1}$. What percent of A remains after 150 s?

Practice Example B: At what time after the start of the reaction is a sample of $H_2O_2(aq)$ two-thirds decomposed? $k = 7.30 \times 10^{-4}\ \text{s}^{-1}$?

Reactions Involving Gases

For gaseous reactions, rates are often measured in terms of gas pressures. For the hypothetical reaction, A(g) $\longrightarrow$ products, the initial partial pressure, $(P_A)_0$, and the partial pressure at some time t, $(P_A)_t$, are related through the expression

$$\ln \frac{(P_A)_t}{(P_A)_0} = -kt \tag{15.15}$$

To see how this equation is derived, start with the ideal gas equation written for reactant A: $P_A V = n_A RT$. Note that the ratio n_A/V is the same as $[A]$. So, $[A]_0 = (P_A)_0/RT$ and $[A]_t = (P_A)_t/RT$. Substitute these terms into equation (15.13), and note that the RT terms cancel in the numerator and denominator and leave the simple ratio $(P_A)_t/(P_A)_0$.

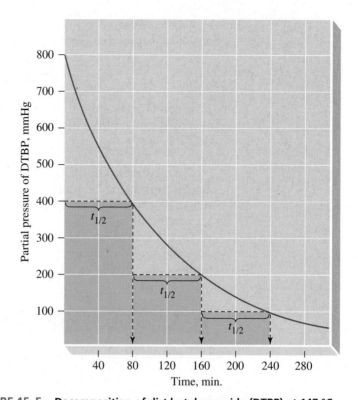

▲ **FIGURE 15–5 Decomposition of di-*t*-butyl peroxide (DTBP) at 147 °C**
The decomposition reaction is described through equation (15.16). In this graph of the partial pressure of DTBP as a function of time, three successive half-life periods of 80 min each are indicated. This constancy of the half-life is proof that the reaction is first order.

Di-*t*-butyl peroxide (DTBP) is used as a catalyst in the manufacture of polymers. In the gaseous state, DTBP decomposes into acetone and ethane by a first-order reaction.

$$C_8H_{18}O_2(g) \longrightarrow 2\ C_3H_6O(g) + C_2H_6(g) \qquad (15.16)$$

$$\text{DTBP} \qquad\qquad \text{acetone} \qquad \text{ethane}$$

Partial pressures of DTBP are plotted as a function of time in Figure 15-5, and the half-life of the reaction is indicated.

As shown in the Are You Wondering feature, the partial pressure of the reactant DTBP can be obtained from two experimentally determined quantities: the initial pressure P_0, and the total pressure P_{total}.

EXAMPLE 15-7

Applying First-Order Kinetics to a Reaction Involving Gases. Reaction (15.16) is started with pure DTBP at 147 °C and 800.0 mmHg pressure in a flask of constant volume. **(a)** What is the value of the rate constant k? **(b)** At what time will the partial pressure of DTBP be 50.0 mmHg?

Solution

(a) From Figure 15-5, we see that $t_{1/2} = 8.0 \times 10^1$ min. For a first-order reaction, $t_{1/2} = 0.693/k$, or

$$k = 0.693/t_{1/2} = 0.693/8.0 \times 10^1 \text{ min} = 8.7 \times 10^{-3} \text{ min}^{-1}$$

Are You Wondering...

How the partial pressures can be obtained experimentally in a reaction involving gases?

In the study of a reaction like the decomposition of DTBP, the *total* pressure is typically measured as a function of time. At any time t, the total pressure is

$$(P_{total})_t = (P_{DTBP})_t + (P_{acetone})_t + (P_{ethane})_t$$

If the initial pressure of DTBP is P_0 then using the stoichiometry of the balanced chemical equation, the partial pressure of DTBP is $P_{DTBP} = (P_0 - P_{ethane})$ since for every mole of DTBP that decomposes, a mole of ethane is produced. The partial pressure of acetone is $P_{acetone} = 2\,P_{ethane}$, again using the stoichiometry of the reaction. The total pressure is then given by

$$P_{total} = (P_0 - P_{ethane}) + 2\,P_{ethane} + P_{ethane} = P_0 + 2\,P_{ethane}$$

and

$$P_{ethane} = (P_{total} - P_0)/2$$

So that

$$P_{DTBP} = (P_0 - P_{ethane}) = P_0 - (P_{total} - P_0)/2 = (3\,P_0 - P_{total})/2$$

▶ Note that the total pressure in reaction (15.16) increases from P_0 at the start of the reaction to $3\,P_0$ when P_{DTBP} has fallen to zero.

(b) A DTBP partial pressure of 50.0 mmHg is $\frac{1}{16}$ of the starting pressure of 800.0 mmHg, that is, $P_{DTBP} = \left(\frac{1}{2}\right)^4 \times 800.0 = 50.0$ mmHg. The reaction must go through *four* half-lives; $t = 4 \times t_{1/2} = 4 \times 8.0 \times 10^1$ min $= 3.2 \times 10^2$ min.

Practice Example A: Start with DTBP at a pressure of 800.0 mmHg at 147 °C. What will be the pressure of DTBP at $t = 125$ min, if $t_{1/2} = 8.0 \times 10^1$ min?

[*Hint:* Because 125 min is not an exact multiple of the half-life, you must use equation (15.15). Can you see that the answer is between 200 and 400 mmHg?]

Practice Example B: Use data from Table 15.4 to determine **(a)** the partial pressure of ethylene oxide, and **(b)** the total gas pressure after 30.0 h in a reaction vessel at 415 °C if the initial partial pressure of $(CH_2)_2O(g)$ is 782 mmHg.

TABLE 15.4 Some Typical First-Order Processes

Process	Half-Life, $t_{1/2}$	Rate Constant k, s^{-1}
Radioactive decay of $^{238}_{92}U$	4.51×10^9 years	4.87×10^{-18}
Radioactive decay of $^{14}_{6}C$	5.73×10^3 years	3.83×10^{-12}
Radioactive decay of $^{32}_{15}P$	14.3 days	5.61×10^{-7}
$C_{12}H_{22}O_{11}(aq) + H_2O(l) \xrightarrow{15\,°C} C_6H_{12}O_6(aq) + C_6H_{12}O_6(aq)$ sucrose $\qquad\qquad$ glucose $\qquad$ fructose	8.4 h	2.3×10^{-5}
$(CH_2)_2O(g) \xrightarrow{415\,°C} CH_4(g) + CO(g)$ ethylene oxide	56.3 min	2.05×10^{-4}
$2\,N_2O_5 \xrightarrow[45\,°C]{in\ CCl_4} 2\,N_2O_4 + O_2(g)$	18.6 min	6.21×10^{-4}
$HC_2H_3O_2(aq) \longrightarrow H^+(aq) + C_2H_3O_2^-(aq)$	8.9×10^{-7} s	7.8×10^5

▶ We will explore radioactive decay in some detail in Chapter 26.

Examples of First-Order Reactions

One of the most familiar examples of a first-order process is radioactive decay. For example, the radioactive isotope iodine-131, used in treating thyroid disorders, has a half-life of 8.04 days. Whatever number of iodine-131 atoms we have in a sample at this moment, we will have half that number in 8.04 days; one-quarter of that number in $8.04 + 8.04 = 16.08$ days; and so on. The rate constant for the decay is $k = 0.693/t_{1/2}$, and in equation (15.13) we can use numbers of atoms, that is, N_t for $[A]_t$ and N_0 for $[A]_0$. Table 15.4 lists several examples of first-order processes. Note the great range of values of $t_{1/2}$ and k. The processes range from very slow to ultrafast.

15-6   Second-Order Reactions

An overall **second-order reaction** has a rate law with the sum of the exponents, $m + n \cdots$, equal to 2. As with zero- and first-order reactions, we limit our discussion to reactions involving the decomposition of a single reactant

$$A \longrightarrow \text{products}$$

that follow the rate law

$$\text{rate of reaction} = k[A]^2 \tag{15.17}$$

Again, our primary interest will be in the *integrated* rate law that is derived from the rate law. This proves to be the equation of a straight-line graph.

$$\frac{1}{[A]_t} = kt + \frac{1}{[A]_0} \tag{15.18}$$

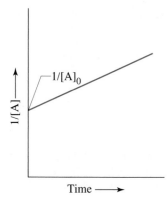

Figure 15-6 is a plot of $1/[A]_t$ against time. The slope of the line is k, and the intercept is $1/[A]_0$. From the graph, we can see that the units of k must be the *reciprocal* of concentration divided by time: $M^{-1}/(\text{time})$ or $M^{-1}(\text{time})^{-1}$—for example, $M^{-1} s^{-1}$ or $M^{-1} \min^{-1}$. We can reach this same conclusion by determining the units of k that produce the required units for the rate of a reaction, that is, moles per liter per time. From equation (15.17):

$$\textit{rate law:}\quad \text{rate of reaction} = k \times [A]^2$$
$$\textit{units:}\qquad\qquad M\,\text{time}^{-1}\quad M^{-1}\,\text{time}^{-1}\,M^2$$

▲ **FIGURE 15–6**
A straight-line plot for the second-order reaction
A ⟶ products
The reciprocal of the concentration, $1/[A]$, is plotted against time. As the reaction proceeds, $[A]$ decreases and $1/[A]$ increases in a linear fashion. The slope of the line is the rate constant k.

For the half-life of the second-order reaction A ⟶ products, we can substitute $t = t_{1/2}$ and $[A] = \frac{1}{2}[A]_0$ into equation (15.18).

$$\frac{1}{[A]_0/2} = kt_{1/2} + \frac{1}{[A]_0}; \qquad \frac{2}{[A]_0} = kt_{1/2} + \frac{1}{[A]_0} \quad \text{and} \quad t_{1/2} = \frac{1}{k[A]_0} \tag{15.19}$$

From equation (15.19), we see that the half-life depends on both the rate constant *and* the initial concentration $[A]_0$. The *half-life is not constant*. Its value depends on the concentration of reactant at the start of each half-life interval. Because the starting concentration is always one-half that of the previous half-life, each successive half-life is twice as long as the one before it.

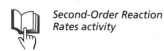

Second-Order Reaction Rates activity

Pseudo-First-Order Reactions

At times, it's possible to simplify the kinetic study of complex reactions by getting them to behave like reactions of a lower order. Then their rate laws become easier to work with. Consider the hydrolysis of ethyl acetate, which is second order overall.

$$\underset{\text{ethyl acetate}}{CH_3COOC_2H_5} + H_2O \longrightarrow \underset{\text{acetic acid}}{CH_3COOH} + \underset{\text{ethanol}}{C_2H_5OH}$$

Are You Wondering...

How we obtain the integrated rate law for the second-order reaction A $\longrightarrow$ products?

In differential form the rate law for the reaction A $\longrightarrow$ products, is $d[A]/dt = -k[A]^2$
Separation of the differentials leads to the expression $d[A]/[A]^2 = -kdt$
Integration of this expression between the limits $[A]_0$ at time $t = 0$ and $[A]_t$ at time t
is indicated through the expression

$$\int_{[A]_0}^{[A]_t} \frac{d[A]}{[A]^2} = -\int_0^t kdt$$

The result of the integration is the integrated rate law.

$$-\frac{1}{[A]_t} + \frac{1}{[A]_0} = -kt \quad \text{or} \quad \frac{1}{[A]_t} = kt + \frac{1}{[A]_0} \tag{15.18}$$

Suppose we follow the hydrolysis of 1 L of aqueous 0.01 M ethyl acetate to completion. $[CH_3COOC_2H_5]$ decreases from 0.01 M to essentially zero. This means that 0.01 mol $CH_3COOC_2H_5$ is consumed, and along with it, 0.01 mol H_2O. Now consider what happens to the molarity of the H_2O. Initially, the solution contains about 1000 g H_2O, or about 55.5 mol H_2O. When the reaction is completed, there is still 55.5 mol H_2O (that is, $55.5 - 0.01 \approx 55.5$). The molarity of the water remains essentially constant throughout the reaction—55.5 M. The rate of reaction does not appear to depend on $[H_2O]$. So, the reaction appears to be *zero* order in H_2O, *first* order in $CH_3COOC_2H_5$, and *first* order overall. A second-order reaction that is made to behave like a first-order reaction by holding one reactant concentration constant is called a *pseudo*-first-order reaction. We can treat the reaction with the methods of first-order reaction kinetics. Other reactions of higher order can be made to behave like reactions of lower order under certain conditions. Thus, a third-order reaction might be converted to pseudo-second order or even to pseudo-first order.

▶ The rate constant obtained from a study using an excess of one reactant, sometimes called Ostwald's isolation method after the kineticist who invented it, is a pseudo-rate constant and is concentration-dependent.

15-7 Reaction Kinetics: A Summary

Let's pause briefly to review what we have learned about rates of reaction, rate constants, and reaction orders. Although a problem often can be solved in several different ways, these approaches are generally most direct.

1. To calculate a rate of reaction when the rate law is known, use this expression:
   rate of reaction $= k[A]^m[B]^n \ldots$.

2. To determine a rate of reaction when the rate law is not given, use
   - the slope of an appropriate tangent line to the graph of [A] versus t
   - the expression $-\Delta[A]/\Delta t$, with a short time interval Δt

3. To determine the order of a reaction, use one of the following methods.
   - Use the method of initial rates if the experimental data are given in the form of reaction rates at different initial concentrations.
   - Find the graph of rate data that yields a straight line (Table 15.5).
   - Test for the constancy of the half-life (good only for first-order).
   - Substitute rate data into integrated rate laws to find the one that gives a constant value of k.

TABLE 15.5 Reaction Kinetics: A Summary For the Hypothetical Reaction A ⟶ products

Order	Rate law	Integrated rate equation	Straight line	$k =$	Units of k	Half-life
0	Rate $= k$	$[A]_t = -kt + [A]_0$	$[A]$ v. time	$-$slope	mol L^{-1} s^{-1}	$[A]_0/2k$
1	Rate $= k[A]$	$\ln[A]_t = -kt + \ln[A]_0$	$\ln [A]$ v. time	$-$slope	s^{-1}	$0.693/k$
2	Rate $= k[A]^2$	$\dfrac{1}{[A]_t} = kt + \dfrac{1}{[A]_0}$	$\dfrac{1}{[A]}$ v. time	slope	L mol^{-1} s^{-1}	$\dfrac{1}{k[A]_0}$

4. To find the rate constant k for a reaction, use one of the following methods.
- Obtain k from the slope of a straight-line graph.
- Substitute concentration-time data into the appropriate integrated rate law.
- Obtain k from the half-life of the reaction (good only for a first-order reaction).

5. To relate reactant concentrations and times, use the appropriate integrated rate law after first determining k.

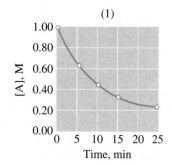

(1)

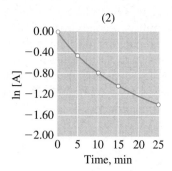

(2)

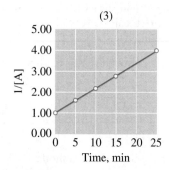

(3)

▲ **FIGURE 15–7**
Testing for the order of a reaction—Example 15-8 illustrated
The straight-line plot is obtained for 1/[A] versus t, graph (3). The reaction is second order.

EXAMPLE 15-8

Graphing Data to Determine the Order of a Reaction. The data listed in Table 15.6 were obtained for the decomposition reaction, A ⟶ products. **(a)** Establish the order of the reaction. **(b)** What is the rate constant, k? **(c)** What is the half-life, $t_{1/2}$, if $[A]_0 = 1.00$ M?

TABLE 15.6 Kinetic Data for Example 15-8

Time, min	[A], M	ln [A]	1/[A]
0	1.00	0.00	1.00
5	0.63	−0.46	1.6
10	0.46	−0.78	2.2
15	0.36	−1.02	2.8
25	0.25	−1.39	4.0

Solution

(a) Plot the following three graphs.
1. [A] versus time (If a straight line, reaction is zero order.)
2. ln[A] versus time (If a straight line, reaction is first order.)
3. 1/[A] versus time (If a straight line, reaction is second order.)

These graphs are plotted in Figure 15-7. The reaction is second order.

(b) The slope of graph 3 in Figure 15-7 is

$$k = \frac{(4.00 - 1.00)\,\text{L/mol}}{25\,\text{min}} = 0.12\ \text{M}^{-1}\,\text{min}^{-1}$$

(c) According to equation (15.19),

$$t_{1/2} = \frac{1}{k[A]_0} = \frac{1}{0.12\ \text{M}^{-1}\,\text{min}^{-1} \times 1.00\ \text{M}} = 8.3\ \text{min}$$

Are You Wondering...

If you can determine the order of a reaction just from a single graph of [A] versus time?

You can for reactions of the type A $\longrightarrow$ products and orders limited to zero, first, and second. If the [A] versus t graph is a straight line, the reaction is *zero order*. If the graph is not linear, apply the half-life test, as in Figure 15-5. If $t_{1/2}$ is constant, the reaction is *first order*. If the half-life is not constant, the reaction is *second order*. (For a second-order reaction, you'll find that the half-life *doubles* for every successive half-life period.)

Practice Example A: In the decomposition reaction B $\longrightarrow$ products, the following data are obtained: $t = 0$ s, [B] $= 0.88$ M; $t = 25$ s, 0.74 M; $t = 50$ s, 0.62 M; $t = 75$ s, 0.52 M; $t = 100$ s, 0.44 M; $t = 150$ s, 0.31 M; $t = 200$ s, 0.22 M; $t = 250$ s, 0.16 M. What are the order of this reaction and its rate constant k?

(*Hint:* You should be able to answer this question without graphing the data.)

Practice Example B: The following data are obtained for the reaction A $\longrightarrow$ products: $t = 0$ min, [A] $= 0.250$ M; $t = 4.22$ min, [A] $= 0.210$ M; $t = 6.60$ min, [A] $= 0.188$ M; $t = 10.61$ min, [A] $= 0.150$ M; $t = 14.48$ min, [A] $= 0.114$ M; $t = 18.00$ min, [A] $= 0.083$ M. What are the order of this reaction and its rate constant, k?

15-8 Theoretical Models for Chemical Kinetics

We can describe practical aspects of reaction kinetics—rate laws, rate constants—without considering the behavior of individual molecules, but to acquire insight into the processes involved, we must turn to the molecular level. For example, we can show by experiment that the decomposition of H_2O_2 is first order, but we may wonder *why* it is. In the remainder of the chapter, we will consider theoretical aspects of chemical kinetics that help us to find such answers.

Collision Theory

Although we will not attempt to do so, with the kinetic–molecular theory (Section 6-7) it is possible to calculate the number of molecular collisions per unit time—the **collision frequency**. In a typical reaction involving gases, the collision frequency is of the order of 10^{30} collisions per second. If each collision yielded product molecules, the rate of reaction would be about 10^6 M s^{-1}, an extremely rapid rate. The typical gas-phase reaction would go essentially to completion in a fraction of a second. Gas-phase reactions generally proceed at a much slower rate, perhaps on the order of 10^{-4} M s^{-1}. This must mean that, generally, *only a fraction of the collisions among gaseous molecules lead to chemical reaction.* This is a reasonable conclusion; we should not expect every collision to result in a reaction.

For a reaction to occur following a collision between molecules, there must be a redistribution of energy that puts enough energy into certain key bonds to break them. We would not expect two slow-moving molecules to bring enough kinetic energy into their collision to permit bond breakage. We would expect two fast-moving molecules to do so, however, or perhaps one extremely fast molecule colliding with a slow-moving one. The **activation energy** of a reaction is the minimum energy above the average kinetic energy that molecules must bring to their collisions for a chemical reaction to occur.

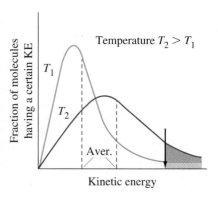

▶ **FIGURE 15–8**
Distribution of molecular kinetic energies
At both temperatures, the fraction of all molecules having kinetic energies in excess of the value marked by the heavy black arrow is small. (Note the shaded areas on the right.) At the higher temperature T_2 (red), however, this fraction is considerably larger than at the lower temperature T_1 (blue).

The kinetic–molecular theory can be used to establish the fraction of all the molecules in a mixture that possess certain kinetic energies. The results of this calculation are depicted in Figure 15-8. On this graph, a hypothetical energy is noted and the fraction of all molecules having energies in excess of this value is identified. Let's assume that these are the molecules whose molecular collisions are most likely to lead to chemical reaction. The rate of a reaction, then, depends on the product of the collision frequency *and* the fraction of these "activated" molecules—in other words, on how often molecules with sufficient kinetic energy to react are likely to collide with each other. Because the fraction of high-energy molecules is generally so small, the rate of reaction is usually much smaller than the collision frequency. Moreover, the *higher* the activation energy of a reaction, the *smaller* is the fraction of energetic collisions and the slower the reaction.

Another factor that can strongly affect the rate of a reaction is the *orientation* of molecules at the time of their collision. In a reaction in which two hydrogen atoms combine to form hydrogen molecules no bonds are broken and a H—H bond forms.

$$H\cdot + \cdot H \longrightarrow H_2$$

The H atoms are spherically symmetrical, and all approaches of one H atom to another prior to collision are equivalent. Orientation is *not* a factor, and the reaction occurs about as rapidly as the atoms collide. On the other hand, orientation of the colliding molecules is a crucial matter in the reaction of N_2O and NO, represented here in an equation highlighting chemical bonds.

$$N\equiv N-O + N=O \longrightarrow N\equiv N + O-N\overset{\displaystyle O}{\diagup\!\!\!\!\!\parallel} \qquad (15.20)$$

The fundamental changes that occur during a successful collision is that the N—O bond in N_2O breaks and a new O—N bond is established to the NO molecule. As a result of the collision, the molecules N_2 and NO_2 are formed. As suggested by Figure 15-9, a favorable collision requires the N atom of the NO molecule to strike the O atom of N_2O during a collision. Other orientations, such as the N atom of NO striking the terminal N atom of N_2O, do not produce a reaction. The number of unfavorable collisions in the reaction mixture exceeds the number of favorable ones.

Transition State Theory

In a theory proposed by Henry Eyring (1901–1981) and others, special emphasis is placed on a hypothetical species believed to exist in a transitory state that lies between the reactants and the products. We call this state the **transition state**, and the hypothetical species, the **activated complex**. The activated complex, formed through

Molecular Collisions activity

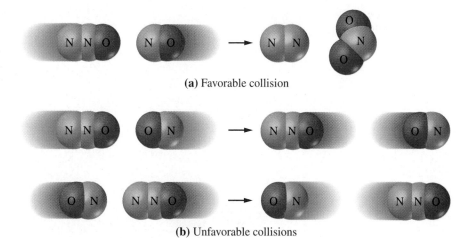

(a) Favorable collision

(b) Unfavorable collisions

▲ **FIGURE 15–9   Molecular collisions and chemical reactions**
(a) A favorable collision between N_2O and NO molecules, resulting in the products N_2 and NO_2. **(b)** Two unfavorable collisions between N_2O and NO molecules; no reaction follows the collisions.

collisions, either dissociates back into the original reactants or forms product molecules. We can represent an activated complex for reaction (15.20) in this way.

$$N\equiv N—O + N=O \rightleftharpoons N\equiv N\cdots O\cdots N\diagup\!\!\!{}^O \longrightarrow N\equiv N + O—N\diagup\!\!\!{}^O$$
reactants   activated complex   products

In the reactants, there is no bond between the O atom of N_2O and the N atom of NO. In the activated complex, the O atom has been partially removed from the N_2O molecule and partially joined to the NO molecule, as indicated by the *partial bonds* ($\cdots$). The formation of the activated complex is a reversible process. Once formed, some molecules of the activated complex may dissociate back into the reactants, but others may dissociate into the product molecules, where the partial bond of the O atom in N_2O has been completely severed and the partial bond between the O atom and NO has become a complete bond.

Figure 15-10 suggests a graphical way of looking at activation energy, called the reaction profile. In a **reaction profile**, energies are plotted on the vertical axis against a quantity called "progress of reaction" on the horizontal axis. Think of the progress of reaction as representing the extent of the reaction. That is, the reaction starts with reactants on the left, progresses through a transition state, and ends with products on the right.

The difference in energies between the reactants and products is ΔH for the reaction. Reaction (15.20) is an exothermic reaction with $\Delta H = -139$ kJ. The difference in energy between the activated complex and the reactants, 209 kJ, is the *activation energy* of the reaction. Thus, a large energy barrier separates the reactants from the products, and only very energetic molecules can pass over this barrier. Figure 15-11 suggests an analogy to activation energy and the reaction profile.

Figure 15-10 describes both the forward reaction and its reverse—the reaction of N_2 and NO_2 to form N_2O and NO. The activation energy for the reverse reaction is 348 kJ; this reverse reaction is highly endothermic. Figure 15-10 also illustrates two useful ideas. (1) The enthalpy change of a reaction is equal to the difference in activation energies of the forward and reverse reactions. (2) For an *endothermic*

KEEP IN MIND ▶
that the difference in potential energy between reactants and product is ΔE of a reaction. For this reaction $\Delta E = \Delta H$ because the number of product gas molecules is equal to the number of reactant gas molecules. As we learned in Chapter 7, even when this is not the case, the differences between ΔE and ΔH are usually quite small.

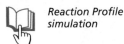
Reaction Profile simulation

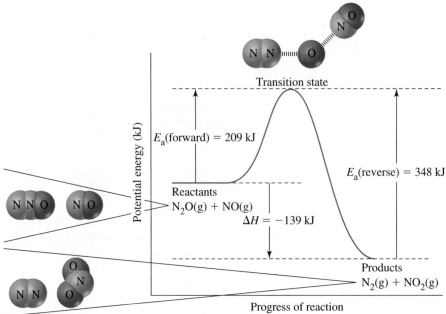

▲ **FIGURE 15–10** **A reaction profile for the reaction**
$$N_2O(g) + NO(g) \longrightarrow N_2(g) + NO_2(g)$$
This simplified reaction profile traces energy changes during the course of reaction (15.20). The reactant and product molecules are depicted by the molecular models, as is the activated complex.

▲ **FIGURE 15–11**
An analogy to reaction profile and activation energy
A hike (red path) is taken from the valley on the left (reactants) over the ridge to the valley on the right (products). The ridge above the starting point corresponds to the transition state. It is probably the height of this ridge (activation energy) more than anything else that determines how many people are willing to take the hike, regardless of the fact that it is all downhill on the other side.

▶ This equation indicates that a rate constant *increases* as the temperature *increases* and as the activation energy *decreases*.

reaction, the activation energy must be equal to or greater than the enthalpy of reaction (and usually it is greater).

Attempts at purely theoretical predictions of rate constants have not been very successful. The principal value of reaction-rate theories is to help us explain experimentally observed reaction-rate data. For example, in the next section we will see how the concept of activation energy enters into a discussion of the effect of temperature on reaction rates.

15-9 The Effect of Temperature on Reaction Rates

From practical experience, we expect chemical reactions to go faster at higher temperatures. To speed up the biochemical reactions involved in cooking, we raise the temperature, and to slow down other reactions, we lower the temperature—as in refrigerating milk to prevent it from souring.

In 1889, Svante Arrhenius demonstrated that the rate constants of many chemical reactions vary with temperature in accordance with the expression

$$k = Ae^{-E_a/RT}$$

By taking the natural logarithm of both sides of this equation, we obtain the following expression.

$$\ln k = \frac{-E_a}{RT} + \ln A \tag{15.21}$$

A graph of $\ln k$ versus $1/T$ is a straight line, and we can use equation (15.21) for a graphical determination of the activation energy of a reaction, as shown in Figure 15-12. We can also derive a useful variation of the equation by writing it twice—

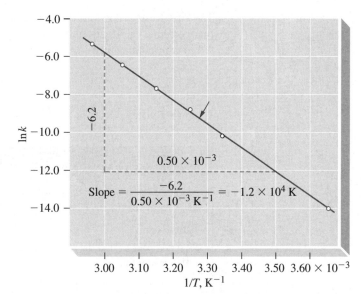

▲ FIGURE 15–12   **Temperature dependence of the rate constant k for the reaction**

$$N_2O_5 \text{ (in } CCl_4) \longrightarrow N_2O_4 \text{ (in } CCl_4) + \frac{1}{2}O_2(g)$$

Data are plotted as follows, for the representative point in black.

$$t = 25\,°C = 298\,K$$
$$1/T = 1/298 = 0.00336 = 3.36 \times 10^{-3}\,K^{-1}$$
$$k = 3.46 \times 10^{-5}\,s^{-1}; \ln k = \ln 3.46 \times 10^{-5} = -10.272$$

To evaluate E_a,

$$\text{slope of line} = -E_a/R = -1.2 \times 10^4\,K$$
$$E_a = 8.3145\,J\,mol^{-1}\,K^{-1} \times 1.2 \times 10^4\,K$$
$$= 1.0 \times 10^5\,J/mol = 1.0 \times 10^2\,kJ/mol$$

(A more precise plot yields a value of $E_a = 106$ kJ/mol. The arrow points to data referred to in Example 15-9.)

▶ This is the same technique used for the Clausius–Clapeyron equation on page 489 and illustrated in Appendix A-4.

each time for a different value of k and the corresponding temperature—and then eliminating the constant $\ln A$. The result, also called the Arrhenius equation, is

$$\ln \frac{k_2}{k_1} = \frac{E_a}{R}\left(\frac{1}{T_1} - \frac{1}{T_2}\right) \tag{15.22}$$

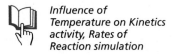

Influence of Temperature on Kinetics activity, *Rates of Reaction simulation*

In equation (15.22)* T_2 and T_1 are two Kelvin temperatures; k_2 and k_1 are the rate constants at these temperatures; and E_a is the activation energy in joules per mole. R is the gas constant expressed as $8.3145\,J\,mol^{-1}\,K^{-1}$.

Equation (15.22) can be rewritten in several ways.

$$\ln \frac{k_2}{k_1} = -\frac{E_a}{R}\left(\frac{1}{T_2} - \frac{1}{T_1}\right) = -\frac{E_a}{R}\left(\frac{T_1 - T_2}{T_2 T_1}\right) = \frac{E_a}{R}\left(\frac{T_2 - T_1}{T_2 T_1}\right)$$

*To see that this expression is dimensionally correct, note that on the right side the units in the numerator are $J\,mol^{-1}$ (for E_a) and K^{-1} (for the quantity in parentheses) and that in the denominator the units are $J\,mol^{-1}\,K^{-1}$ (for R). Cancellation yields a dimensionless quantity on the right, to match the dimensionless logarithmic term on the left.

The first form brings out the minus sign and reverses the T_1 and T_2, and the next two combine the fractions within the parentheses. Also, the equation can be written in exponential form.

$$\frac{k_2}{k_1} = e^{-\frac{E_a}{R}\left(\frac{1}{T_2} - \frac{1}{T_1}\right)}$$

Which equation is used is a matter of calculational convenience.

EXAMPLE 15-9

Applying the Arrhenius Equation. Use data from Figure 15-12 to determine the temperature at which $t_{1/2}$ for the first-order decomposition of N_2O_5 in CCl_4 is 2.00 h.

Solution

First we need to find the rate constant k corresponding to a 2.00-h half-life. For a *first-order reaction,*

$$k = \frac{\ln 2}{t_{1/2}} = \frac{0.693}{2.00 \text{ h}} = \frac{0.693}{7200 \text{ s}} = 9.63 \times 10^{-5} \text{ s}^{-1}$$

Now we can proceed in one of two ways.

Graphical Method. We want the temperature at which $k = 9.63 \times 10^{-5}$ and $\ln k = \ln 9.63 \times 10^{-5} = -9.248$. We have marked this point on Figure 15-12 with an arrow. Corresponding to $\ln k = -9.248$, $1/T = 3.28 \times 10^{-3} \text{ K}^{-1}$.

$$T = 1/(3.28 \times 10^{-3}) \text{ K} = 305 \text{ K} = 32 \text{ °C}$$

Using Equation (15.22). Take T_2 to be the temperature at which $k = k_2 = 9.63 \times 10^{-5} \text{ s}^{-1}$. T_1 is some other temperature at which a value of k is known. Suppose we take $T_1 = 298 \text{ K}$ and $k_1 = 3.46 \times 10^{-5} \text{ s}^{-1}$, a point referred to in the caption of Figure 15-12. The activation energy is $106 \text{ kJ/mol} = 1.06 \times 10^5 \text{ J/mol}$ (the more precise value given in Figure 15-12). Now we can solve equation (15.22) for T_2. (For simplicity, we have omitted units below, but the temperature is obtained in kelvins.)

$$\ln \frac{k_2}{k_1} = \frac{E_a}{R}\left(\frac{1}{T_1} - \frac{1}{T_2}\right)$$

$$\ln \frac{9.63 \times 10^{-5}}{3.46 \times 10^{-5}} = \frac{1.06 \times 10^5}{8.3145}\left(\frac{1}{298} - \frac{1}{T_2}\right)$$

$$1.024 = 1.27 \times 10^4 \left(0.00336 - \frac{1}{T_2}\right) = 42.7 - \frac{1.27 \times 10^4}{T_2}$$

$$\frac{1.27 \times 10^4}{T_2} = 42.7 - 1.024 = 41.7$$

$$T_2 = \frac{(1.27 \times 10^4)}{41.7} = 305 \text{ K}$$

Practice Example A: What is the half-life of the first-order decomposition of N_2O_5 at 75.0 °C? Use data from Example 15-9.

Practice Example B: At what temperature will it take 1.50 h for two-thirds of a sample of N_2O_5 in CCl_4 to decompose in Example 15-9?

▲ Both the rate of chirping of tree crickets and the flashing of fireflies roughly double for a 10 °C temperature rise. This corresponds to an activation energy of about 50 kJ/mol and suggests that the physiological processes governing these phenomena involve chemical reactions.

Arrhenius established equation (15.22) by fitting experimental data into his equation. This was before the collision theory of chemical reactions had been developed, but his equation is consistent with the collision theory. In the preceding section, we discussed the importance of (1) the frequency of molecular collisions, (2) the fraction of collisions energetic enough to produce a reaction, and (3) the need for favorable orientations during collisions. Let's represent the collision frequency by the symbol Z_0. From kinetic–molecular theory, the fraction of sufficiently energetic collisions proves to be $e^{-E_a/RT}$. The probability of favorable orientations of colliding molecules is p. In collision theory, the rate constant of a reaction can be expressed as the product of these three terms. If we replace the product $Z_0 \times p$ by the term A, collision theory yields a result identical to Arrhenius's experimentally determined equation.

$$k = Z_0 \cdot p \cdot e^{-E_a/RT} = Ae^{-E_a/RT}$$

15-10 Reaction Mechanisms

In the discussion of photochemical smog in Section 8-2, we indicated a key role played by $NO_2(g)$, but it is unlikely that very much of this gas is formed in the atmosphere by the direct reaction

$$2\,NO(g) + O_2(g) \longrightarrow 2\,NO_2(g) \tag{15.23}$$

For this reaction to occur in a single step in the manner suggested by equation (15.23), *three* molecules would have to collide simultaneously, or very nearly so. A three-molecule collision is an unlikely event. The reaction appears to follow a different mechanism or pathway. One of the main purposes in determining rate laws of chemical reactions is to relate them to probable reaction mechanisms.

A **reaction mechanism** is a step-by-step detailed description of a chemical reaction. Each step in a mechanism is called an elementary process. An **elementary process** is any molecular event that significantly alters a molecule's energy or geometry or produces a new molecule. Two requirements of a plausible reaction mechanism are that it must

- be consistent with the stoichiometry of the overall reaction
- account for the *experimentally determined* rate law

In this section, we will first explore the nature of elementary processes and then apply these processes to two simple types of reaction mechanisms.

Elementary Processes

The characteristics of elementary processes are as follows:

1. Elementary processes are either **unimolecular**—a process in which a single molecule dissociates—or **bimolecular**—a process involving the collision of two molecules. A *termolecular* process, which would involve the simultaneous collision of three molecules, is relatively rare as an elementary process.

2. The exponents of the concentration terms in the rate law for an *elementary* process are the *same* as the stoichiometric coefficients in the balanced equation for the process. (Note that this is unlike the case of the overall rate law, for which the exponents are *not* necessarily related to the stoichiometric coefficients in the overall equation.)

3. Elementary processes are reversible, and some may reach a condition of equilibrium in which the rates of the forward and reverse processes are equal.

4. Certain species are produced in one elementary process and consumed in another. In a proposed reaction mechanism, such intermediates must not appear in either the overall chemical equation or the overall rate law.

5. One elementary process may occur much more slowly than all the others, and in some cases may determine the rate of the overall reaction. Such a process is called the **rate-determining step** (Figure 15-13).

Biomolecular Reaction

▲ FIGURE 15–13
The San Ysidro/Tijuana border station: an analogy to a rate-determining step
Note the tie-up of traffic on the Mexican side of the border (top) and the relatively few cars on the United States side (bottom). This station is a bottleneck and hence the rate-determining part of the trip by car from Tijuana, Mexico, to San Diego, California, two cities located only about a dozen miles apart.

Keep these characteristics in mind as we will apply them in our analysis of different mechanisms below.

A Mechanism with a Slow Step Followed by a Fast Step

The reaction between gaseous iodine monochloride and gaseous hydrogen produces iodine and hydrogen chloride as gaseous products.

$$H_2(g) + 2\ ICl(g) \longrightarrow I_2(g) + 2\ HCl(g)$$

The experimentally determined rate law for this reaction is

$$\text{rate of reaction} = k[H_2][ICl]$$

Let's *postulate* the following two-step mechanism.

(1) Slow: $H_2 + ICl \longrightarrow HI + HCl$

(2) Fast: $HI + ICl \longrightarrow I_2 + HCl$

Overall: $H_2 + 2\ ICl \longrightarrow I_2 + 2\ HCl$

First, it is important to note that the sum of two steps yields the *overall* reaction. Each step in the above mechanism is bimolecular. Because each step is an elementary step, we can write

$$\text{rate(1)} = k_1[H_2][ICl] \quad \text{and} \quad \text{rate(2)} = k_2[HI][ICl]$$

Now, let's further postulate that step (1) occurs *slowly* but step (2) occurs *rapidly*. This suggests that HI is consumed in the second step just as fast as it is formed in the first. The first step is the rate-determining step, and the rate of the overall reaction is governed just by the rate at which HI is formed in this first step, that is, by rate (1). This explains why the observed rate law for the net reaction is rate of reaction $= k[H_2][ICl]$. The proposed mechanism gives a rate law that is in agreement with experiment, as it should if we have made a reasonable proposal.

The species HI is called a **reaction intermediate**, and it does not occur in the experimental rate law. In this case, the intermediate species is a well-known stable molecule. Often, when postulating mechanisms, we have to invoke less well known and less stable species; and in these instances, we have to rely on the chemical reasonableness of the basic assumptions. The presence of a reaction intermediate leads to a slightly more complicated reaction profile. The reaction profile for the two steps in the proposed mechanism is shown in Figure 15-14. We see that there are two transition states and one reaction intermediate. The activation energy for the first step is greater than that for the second step, reflecting the fact that the first step in the reaction mechanism is the slowest. It is important to understand the difference between a transition state (activated complex) and a reaction intermediate. The transition state represents the highest energy structure involved in a reaction (or step in a mechanism). While transition states exist only momentarily and can never be isolated, reaction intermediates can sometimes be isolated. Transition states have *partially formed bonds*, whereas reaction intermediates have *fully formed bonds*.

A Mechanism with a Fast Reversible First Step Followed by a Slow Step

The rate law for the reaction of $NO(g)$ and $O_2(g)$

$$2\ NO(g) + O_2(g) \longrightarrow 2\ NO_2(g) \tag{15.23}$$

is found to be

$$\text{rate of reaction} = k[NO]^2[O_2] \tag{15.24}$$

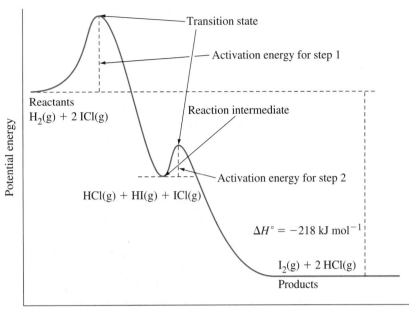

▲ **FIGURE 15–14   A reaction profile for a two-step mechanism**

Even though it is consistent with this rate law, we have already noted that the one-step *termolecular* mechanism suggested by equation (15.23) is highly improbable. Let's explore instead the following mechanism.

$$\textit{Fast:} \qquad 2\,NO(g) \underset{k_2}{\overset{k_1}{\rightleftharpoons}} N_2O_2(g) \qquad\qquad (15.25)$$

$$\textit{Slow:} \quad \underline{N_2O_2(g) + O_2(g) \xrightarrow{k_3} 2\,NO_2(g)} \qquad (15.26)$$

$$\textit{Overall:} \quad 2\,NO(g) + O_2(g) \longrightarrow 2\,NO_2(g) \qquad (15.23)$$

In this mechanism, there is a rapid equilibrium as the first step, but some of the N_2O_2 is slowly drawn off and consumed in the second, slow step. The rate law for the slow, or *rate-determining*, step (15.26) is

$$\text{rate of reaction} = k_3[N_2O_2][O_2] \qquad (15.27)$$

Because N_2O_2 is an *intermediate*, however, we must eliminate it from the rate law. We are told that the first step of the mechanism consists of a fast reversible reaction, so we can assume that this step progresses rapidly to equilibrium. If this is the case, the forward and reverse rates of reaction in the first step become equal and we write

$$\text{rate of forward reaction} = \text{rate of reverse reaction}$$

$$k_1[NO]^2 = k_2[N_2O_2]$$

Now, let us arrange this equation into an expression having a ratio of rate constants on one side and a ratio of concentration terms on the other. Also, we can replace the ratio of rate constants by a single constant, which we will represent as K_1.

$$K_1 = \frac{k_1}{k_2} = \frac{[N_2O_2]}{[NO]^2}$$

Equilibrium constant expressions are of fundamental importance throughout chemistry. Here, we see their significance in chemical kinetics. In Chapter 16, we describe the experimental basis of equilibrium constant expressions and their application to the stoichiometry of reversible reactions. In Chapter 20, we explore their thermodynamic basis. ▶

The above expression is known as an equilibrium constant expression; the numerical constant, K_1, is an equilibrium constant. Next, we rearrange the expression to solve for the term $[N_2O_2]$.

$$[N_2O_2] = K_1[NO]^2$$

Then, substituting this into equation (15.27) we obtain the experimentally observed rate law.

$$\text{rate of reaction} = k_3[N_2O_2][O_2] = k_3K_1[NO]^2[O_2]$$

The experimentally observed rate constant, k, is related to the other constants in the proposed mechanism, as follows:

$$k = k_3K_1 = \frac{k_1}{k_2} \times k_3$$

The type of mechanism described here with a *rapid pre-equilibrium* is a very common mechanism, and is to be expected when the overall stoichiometry suggests an unlikely termolecular collision.

We have just shown that the proposed mechanism is consistent with (a) the reaction stoichiometry and (b) the experimentally determined rate law. Whether this mechanism is the actual reaction path, we cannot say, however. All that we can say is that it is *plausible*; it has not been ruled out by kinetics.

EXAMPLE 15-10

Testing a Reaction Mechanism. An alternative mechanism of the reaction $2\,NO(g) + O_2(g) \longrightarrow 2\,NO_2(g)$ follows. Show that this mechanism is consistent with the rate law (15.24).

$$\textit{Fast:} \qquad NO(g) + O_2(g) \underset{k_2}{\overset{k_1}{\rightleftharpoons}} NO_3(g)$$

$$\textit{Slow:} \qquad \underline{NO_3(g) + NO(g) \overset{k_3}{\longrightarrow} 2\,NO_2(g)}$$

$$\textit{Overall:} \quad 2\,NO(g) + O_2(g) \longrightarrow 2\,NO_2(g)$$

Solution

The rate equation for the rate-determining step is

$$\text{rate of reaction} = k_3[NO_3][NO]$$

To eliminate $[NO_3]$, we assume that the pre-equilibrium is rapidly established and we write

$$\text{rate of forward reaction} = \text{rate of reverse reaction}$$

$$k_1[NO][O_2] = k_2[NO_3]$$

Rearranging this to give the expression for an equilibrium constant (K_2) in terms of the rate constants k_1 and k_2, we obtain

$$K_2 = \frac{k_1}{k_2} = \frac{[NO_3]}{[NO][O_2]}$$

Then we can rearrange this expression to solve for $[NO_3]$.

$$[NO_3] = K_2[NO][O_2]$$

Finally, we substitute this value of $[NO_3]$ into the rate equation for the rate-determining step: rate $= k_3[NO_3][NO]$

$$\text{rate of reaction} = k_3K_2[NO]^2[O_2] = k_3 \times \frac{k_1}{k_2}[NO]^2[O_2] = k[NO]^2[O_2] \quad (15.24)$$

We note that this mechanism also produces the experimental rate law. To decide between the two mechanisms, we would have to perform additional experiments.

Practice Example A: In a proposed two-step mechanism for the reaction $CO + NO_2 \longrightarrow CO_2 + NO$, the second, fast step is $NO_3 + CO \longrightarrow NO_2 + CO_2$. What must be the *slow* step? What would you expect the rate law of the reaction to be? Explain.

Practice Example B: Show that the proposed mechanism for the reaction $2\ NO_2(g) + F_2(g) \longrightarrow 2\ NO_2F(g)$ is plausible. The rate of reaction is rate $= k[NO_2][F_2]$.

Fast:	$NO_2(g) + F_2(g) \rightleftharpoons NO_2F_2(g)$
Slow:	$NO_2F_2(g) \longrightarrow NO_2F(g) + F(g)$
Fast:	$F(g) + NO_2(g) \longrightarrow NO_2F(g)$

The Steady-State Approximation

The reaction mechanisms that we have considered so far have had one particular rate-determining step, and the rate law of the reaction could be deduced from the rate of this step after the relationships for the concentrations of any intermediates had been established. In complex multistep reaction mechanisms, however, more than one step may control the rate of a reaction.

To illustrate, let's re-consider the first mechanism presented for the reaction of nitric oxide with oxygen, but this time we will make no assumptions about the relative rates of the steps in the mechanism. The proposed mechanism is

$$NO + NO \xrightarrow{k_1} N_2O_2$$
$$N_2O_2 \xrightarrow{k_2} NO + NO$$
$$N_2O_2 + O_2 \xrightarrow{k_3} 2\,NO_2$$

where, for clarity, we have written the first reversible reaction as two forward steps.

We choose one of the steps of the mechanism that provides a convenient relationship to the observed rate of reaction. In this case, this is the third step, as it involves the disappearance of O_2. Therefore, the rate of the reaction from this mechanism is

$$\text{rate of reaction} = k_3[N_2O_2][O_2] \tag{15.27}$$

▶ **KEEP IN MIND**
that if the rate of change of the concentration of a substance is zero then the concentration of that substance is constant.

As before, we must eliminate the intermediate N_2O_2 from this rate law. We can do this by assuming that the $[N_2O_2]$ reaches a *steady-state condition*, in which N_2O_2 is produced and consumed at equal rates. That is, $[N_2O_2]$ remains constant throughout most of the reaction. We can use the steady-state assumption to express $[N_2O_2]$ in terms of $[NO]$.

$$\Delta[N_2O_2]/\Delta t = \text{rate of formation of } N_2O_2 + \text{rate of disappearance of } N_2O_2 = 0$$

$$\text{rate of formation of } N_2O_2 = -(\text{rate of disappearance of } N_2O_2)$$

The rate of disappearance of N_2O_2 is made up of two parts—the reverse step of equation (15.25) and the forward step of (15.26)—so we write

$$\text{rate of disappearance of } N_2O_2 = -\left(k_3[N_2O_2][O_2] + k_2[N_2O_2]\right)$$

where we have added the two rates for the steps depleting the concentration of N_2O_2; the minus sign signifies a decrease in concentration. Now, as dictated by the steady-state assumption, we equate the negative of the rate of disappearance of N_2O_2 with the rate of appearance of N_2O_2, which is $k_1[NO]^2$.

$$k_1[NO]^2 = k_2[N_2O_2] + k_3[N_2O_2][O_2] = [N_2O_2]\left(k_2 + k_3[O_2]\right)$$

Rearranging to solve for $[N_2O_2]$, we have

$$[N_2O_2] = \frac{k_1[NO]^2}{\left(k_2 + k_3[O_2]\right)}$$

We now substitute this into equation (15.27) to obtain

$$\text{rate} = k_3[O_2][N_2O_2] = k_3[O_2]\left(\frac{k_1[NO]^2}{(k_2 + k_3[O_2])}\right)$$

or

$$\text{rate} = \frac{k_1 k_3[O_2][NO]^2}{(k_2 + k_3[O_2])}$$

So, this is the rate law for the proposed mechanism based on our steady-state analysis. This rate law is more complicated than the observed rate law. What happened? In carrying out the steady-state calculation, we did not make any assumptions about the relative rates of the three steps in the mechanism. If we now make the assumption that the rate of disappearance of N_2O_2 in the second step of the proposed mechanism is greater than the rate of disappearance of N_2O_2 in the third step of the proposed mechanism, then

$$k_2[N_2O_2] > k_3[N_2O_2][O_2]$$

which means

$$k_2 > k_3[O_2]$$

and

$$k_2 + k_3[O_2] \cong k_2$$

so that

$$[N_2O_2] = \frac{k_1[NO]^2}{k_2}$$

If we substitute this value of $[N_2O_2]$ into the rate equation (15.27) and replace $k_1 k_3 / k_2$ by k, we obtain for the overall reaction

$$\text{rate of reaction} = \frac{k_1 k_3}{k_2}[NO]^2[O_2] = k[NO]^2[O_2] \qquad (15.24)$$

The result of a steady-state analysis of any mechanism in which no rate-determining step can be identified will often be a complicated rate law. The use of this type of rate law is illustrated in the section on enzyme catalysis later in this chapter.

15-11   Catalysis

A reaction can generally be made to go faster by increasing the temperature. Another way to speed up a reaction is to use a catalyst. A **catalyst** provides an alternative reaction pathway of lower activation energy. The catalyst participates in a chemical reaction without undergoing permanent change. As a result, the formula of a catalyst does not appear in the overall chemical equation. (Its formula is generally placed over the reaction arrow.)

The success of a chemical process often hinges on finding the right catalyst, as in the manufacture of nitric acid. By conducting the oxidation of $NH_3(g)$ very quickly (less than 1 ms) in the presence of a Pt–Rh catalyst, $NO(g)$ can be obtained as a product instead of $N_2(g)$. The formation of $HNO_3(aq)$ from $NO(g)$ then follows easily (see page 274).

In this section, the two basic types of catalysis—homogeneous and heterogeneous—are described first. This is followed by discussions of the catalyzed decomposition of $H_2O_2(aq)$ and the biological catalysts called enzymes.

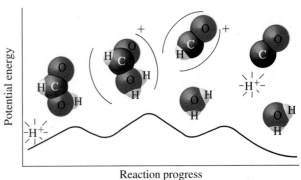

(a) Uncatalyzed reaction   (b) Catalyzed reaction

▲ **FIGURE 15–15   An example of homogeneous catalysis**
The activation energy is lowered in the presence of H^+, a catalyst for the decomposition of HCOOH.

Homogeneous Catalysis

Figure 15-15 shows reaction profiles for the decomposition of formic acid (HCOOH). In the uncatalyzed reaction, a H atom must be transferred from one part of the formic acid molecule to another, shown by the arrow. Then a C—O bond breaks. Because the energy requirement for this atom transfer is high, the activation energy is high and the reaction is slow.

In the acid-catalyzed decomposition of formic acid, a hydrogen ion from solution attaches itself to the O atom that is singly bonded to the C atom to form $(HCOOH_2)^+$. The C—O bond breaks, and a H atom attached to a carbon atom in the intermediate species $(HCO)^+$ is released to the solution as H^+.

$$H-\overset{\overset{O}{\|}}{C}-O-H \;+\; H^+ \;\longrightarrow\; \left(H-\overset{\overset{O}{\|}}{C}-\overset{\overset{H}{|}}{O}-H\right)^{\!+} \;\longrightarrow\; \left(H-\overset{\overset{O}{\|}}{C}\right)^{\!+} \;+\; H_2O$$

$$\downarrow$$

$$H^+ \;+\; C\!\equiv\!O$$

This catalyzed reaction pathway does not require a H atom to be transferred within the formic acid molecule. It has a lower activation energy than does the uncatalyzed reaction and proceeds at a faster rate. Because the reactants and products of this reaction are all present throughout the solution or homogeneous mixture, this type of catalysis is called *homogeneous* catalysis.

Heterogeneous Catalysis

Many reactions can be catalyzed by allowing them to occur on an appropriate solid surface. Essential reaction intermediates are found on the surface. This type of catalysis is called *heterogeneous* catalysis because the catalyst is present in a different phase of matter than are the reactants and products. Catalytic activity is associated with many transition elements and their compounds. The precise mechanism of heterogeneous catalysis is not totally understood, but in many cases the availability of electrons in *d* orbitals in surface atoms may play a role.

A key feature of heterogeneous catalysis is that reactants from a gaseous or solution phase are adsorbed, or attached, to the surface of the catalyst. Not all surface atoms are equally effective for catalysis; those that are effective are called **active sites**. Basically, heterogeneous catalysis involves (1) *adsorption* of reactants; (2) diffusion of reactants along the surface; (3) reaction at an active site to form adsorbed product; and (4) *desorption* of the product.

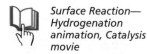
Surface Reaction—Hydrogenation animation, Catalysis movie

In Chapter 8, we described the oxidation of CO to CO$_2$ and reduction of NO to N$_2$ in automotive exhaust gases as a smog-control measure (page 276). Figure 15-16 shows how this reaction is thought to occur on the surface of rhodium metal in a catalytic converter. In general, the reaction profile for a surface-catalyzed reaction resembles that shown in Figure 15-17.

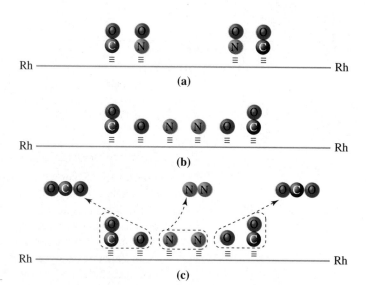

▶ **FIGURE 15–16**
Heterogeneous catalysis in the reaction

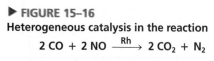

$$2\ CO + 2\ NO \xrightarrow{\ Rh\ } 2\ CO_2 + N_2$$

(a) Molecules of CO and NO are adsorbed on the rhodium surface. **(b)** The adsorbed NO molecules dissociate into adsorbed N and O atoms. **(c)** Adsorbed CO molecules and O atoms combine to form CO$_2$ molecules, which desorb into the gaseous state. Two N atoms combine and are desorbed as a N$_2$ molecule.

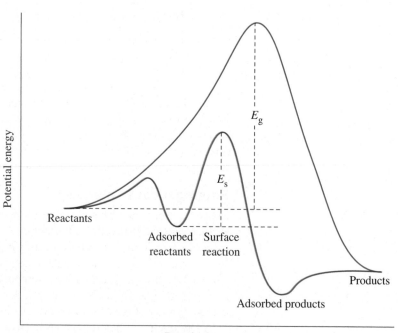

▲ **FIGURE 15–17** **Reaction profile for a surface-catalyzed reaction**
In the reaction profile (blue) for the surface-catalyzed reaction, the activation energy for the reaction step, E_s, is considerably less than in the reaction profile (red) for the uncatalyzed gas-phase reaction, E_g.

The Catalyzed Decomposition of Hydrogen Peroxide

As we have previously noted (footnote on page 581), the decomposition of $H_2O_2(aq)$ is a slow reaction and generally must be catalyzed. Iodide ion is a good catalyst that seems to function by means of the following two-step mechanism.

Slow: $H_2O_2 + I^- \longrightarrow OI^- + H_2O$

Fast: $H_2O_2 + OI^- \longrightarrow H_2O + I^- + O_2(g)$

Overall: $2\ H_2O_2 \longrightarrow 2\ H_2O + O_2(g)$

As required for a catalyzed reaction, the formula of the catalyst does not appear in the overall equation. Neither does the intermediate species OI^-. The rate of reaction of H_2O_2 is determined by the rate of the slow first step.

$$\text{rate of reaction of } H_2O_2 = k[H_2O_2][I^-] \qquad (15.28)$$

Because I^- is constantly regenerated, its concentration is constant throughout a given reaction. If we replace the product of the constant terms $k[I^-]$ by a new constant k', we can rewrite the rate law as

$$\text{rate of reaction of } H_2O_2 = k'[H_2O_2] \qquad (15.29)$$

Equation (15.28) indicates that the rate of decomposition of $H_2O_2(aq)$ is affected by the initial concentration of I^-. For each initial concentration of I^-, we get a different rate constant, k', in equation (15.29).

We have just described the homogeneous catalysis of the decomposition of hydrogen peroxide. The decomposition can also be catalyzed by heterogeneous catalysis (see photograph on page 578).

Enzymes as Catalysts

Unlike platinum, which catalyzes a wide variety of reactions, the catalytic action of high molar mass proteins known as **enzymes** is very specific. For example, in the digestion of milk, lactose, a more complex sugar, breaks down into two simpler ones, glucose and galactose. This occurs in the presence of the enzyme *lactase*.

$$\text{lactose} \xrightarrow{\text{lactase}} \text{glucose} + \text{galactose}$$
"milk sugar"

Biochemists describe enzyme activity with the "lock-and-key" model (Figure 15-18). The reacting substance, the **substrate** (S), attaches itself to the enzyme

Many people lose the ability to produce lactase when they become adults. In them, lactose passes through the small intestine into the colon, where it ferments and may cause severe gastric disturbances. ▶

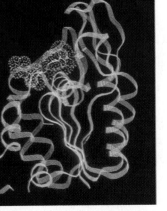

▲ A computer graphics representation of the enzyme phosphoglycerate kinase (carbon backbone shown as blue ribbon). A molecule of ATP, the substrate, is shown in green.

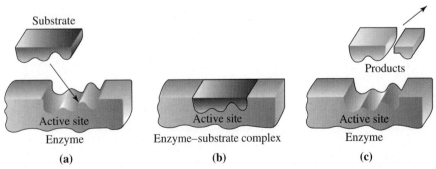

▲ **FIGURE 15–18** **Lock-and-key model of enzyme action**
(a) The substrate attaches itself to an active site on an enzyme molecule. **(b)** Reaction occurs. **(c)** Product molecules detach themselves from the site, freeing the enzyme molecule to attach another molecule of substrate. The substrate and enzyme must have complementary structures to produce a complex, hence the term *lock-and-key*.

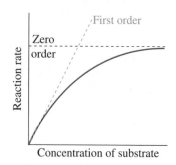

▲ FIGURE 15–19
Effect of substrate concentration on the rate of an enzyme reaction

(E) at a particular point called an active site to form the complex (ES). The complex decomposes to form products (P) and regenerate the enzyme.

$$E + S \underset{k_{-1}}{\overset{k_1}{\rightleftharpoons}} ES$$

$$ES \xrightarrow{k_2} E + P$$

Most human enzyme reactions proceed fastest at about 37 °C (body temperature). If the temperature is raised much higher than that, the structure of the enzyme changes, the active sites become distorted, and the catalytic activity is lost.

Determining the rates of enzyme reactions is an important part of enzyme studies. Figure 15-19, a plot of reaction rate against substrate concentration, illustrates what we generally observe. Along the rising portion of the graph, the rate of reaction is proportional to the substrate concentration, [S]. The reaction is first order: rate of reaction = k[S]. At high substrate concentrations, the rate is independent of [S]. The reaction follows a zero-order rate equation: rate of reaction = k.

This behavior can be understood in terms of the three-step mechanism given above. The rate at which the product appears, which is often called the velocity (V) of the reaction by biochemists, is given by

$$\text{rate of production of } P = V = k_2[ES]$$

To proceed further, we need an expression for the enzyme–substrate complex, and we can get this by applying the steady-state approximation to the concentration of the enzyme–substrate complex.

$$\text{rate of formation of ES} = \text{rate of destruction of ES}$$

$$k_1[E][S] = (k_{-1} + k_2)[ES] \tag{15.30}$$

We can solve this equation for [ES], but the solution contains the concentration of free enzyme E, which is unknown. However, we do know the *total* concentration of enzyme in an experiment, $[E_0]$. Then, by the condition known as *material balance*, we have

$$[E_0] = [E] + [ES]$$

Solving this equation for [E] and substituting the result into equation (15.30), we get

$$k_1[S]([E_0] - [ES]) = (k_{-1} + k_2)[ES]$$

$$[ES] = \frac{k_1[E_0][S]}{(k_{-1} + k_2) + k_1[S]}$$

for the concentration of the enzyme–substrate complex. Substituting this value into the rate of reaction, we get

$$V = \frac{k_2 k_1[E_0][S]}{(k_{-1} + k_2) + k_1[S]}$$

This equation can be put into a more convenient form by dividing the numerator and denominator by k_1 and replacing a ratio of rate constants by the single constant K_M.

$$K_M = \frac{k_{-1} + k_2}{k_1}$$

▶ The reaction mechanism outlined here was proposed by Michaelis and Menten in 1913, accounting for the subscript M in the constant K_M.

Our final result is

$$V = \frac{k_2[E_0][S]}{K_M + [S]} \tag{15.31}$$

Now, let's test whether the reaction velocity V, given by equation (15.31), depends on the substrate concentration in the manner suggested by Figure 15–19. At sufficiently low concentrations of S, we have the inequality.

$$K_M \gg [S]$$

We can ignore [S] with respect to K_M in the denominator and obtain the following for the reaction velocity.

$$V = \frac{k_2}{K_M}[E_0][S]$$

Because the total enzyme concentration is constant, the rate law is first order with respect to substrate, as observed experimentally.

The other limiting case is that in which the velocity of the reaction becomes independent of substrate concentration. At sufficiently high concentrations of substrate,

$$[S] \gg K_M$$

and

$$V = k_2[E_0]$$

Here, for a chosen concentration of enzyme, the reaction velocity is constant, and is the maximum reaction velocity attainable for the particular enzyme. This velocity corresponds to the experimentally observed plateau at high substrate concentrations in a plot such as Figure 15–19. Thus, we have satisfactory agreement between the predictions of the postulated mechanism and the experimental results. As is typical of the scientific method, postulated mechanisms are continually tested by subsequent experiment and modified when necessary.

Are You Wondering...

Why the reaction order of an enzyme reaction changes from first to zero order, depending on substrate concentration?

This situation is much like what you observe at a bank of telephones in an airport terminal. The telephones are like active sites, and the callers are like substrate molecules. When the terminal is relatively empty (low [S]), the rate at which calls are made depends on the number of people present. When the terminal is crowded, with callers waiting in line, calls are made at essentially a constant rate, which depends on the number of telephones, not the number of patrons.

Summary

The rate of a reaction is related to the rate of change of concentration of a reactant or product. Rates can be calculated through an expression of the type $\Delta[A]/\Delta t$ or determined from the slope of a tangent line to a concentration–time curve. The relationship between the rate of a reaction and the concentrations of the reactants is called a rate law and has the form: rate of reaction = $k[A]^m[B]^n \ldots$.

The order of a reaction refers to the exponents m, n, … in the rate law. In this chapter, main consideration is given to zero-, first-, and second-order reactions, mostly of the type $A \longrightarrow$ products. The order of a reaction can be determined by measuring the initial rates of reaction for different initial concentrations of reactants; by plotting an appropriate function of concentration against time; or by doing a series of calculations of the rate constant k, using the appropriate integrated rate law.

The theoretical basis of chemical kinetics is that reactions involve molecular collisions and that only collisions in which the molecules have sufficient energy and a proper orientation result in reaction. A reaction profile charts the energy changes during the course of reaction, rising from the energy of the reactants to a maximum for the transition state, and then falling back to that of the products. The difference between the energy of the

◀ The main thrusters that provide the energy to put the space shuttle into orbit are powered by the controlled combustion of hydrogen and oxygen.

At times, only a fine line separates combustion reactions and explosions. The typical explosion is a self-sustaining combustion reaction that proceeds at an ever-increasing rate and cannot be stopped until the reactants are consumed. An understanding of the mechanisms of combustion and explosion reactions can help us to achieve the one and avoid the other.

A dramatic contrast can be seen in the reaction

$$2\,H_2 + O_2 \longrightarrow 2\,H_2O$$

When it is carried out in a controlled fashion, this reaction provides the energy and thrust needed to propel the space shuttle into orbit.

One type of explosion, called a *thermal* explosion, occurs because the exothermic heat of a reaction cannot be conducted away from the reaction mixture fast enough. The temperature of the reaction mixture rises, and with the increase in temperature, the rate constant increases unchecked. Another cause of explosive reactions is found in the *branched-chain* mechanism, outlined here in simplified form for the hydrogen-oxygen reaction.

activated complex in the transition state and the energy of the reactants is the activation energy for the reaction. The difference in the energy states of the products and reactants is the enthalpy change for the reaction. The rates of chemical reactions can be increased by raising the temperature or by using a catalyst. A catalyst changes the reaction pathway to one with a lower activation energy. For reactions in living systems, these catalysts are called enzymes.

To propose a mechanism for a chemical reaction, a series of elementary processes is postulated. Rate laws are written for these elementary processes and combined into a rate law for the overall reaction. To be *plausible*, the reaction mechanism must be consistent with the stoichiometry of the overall reaction and its experimentally determined rate law.

Integrative Example

Peroxyacetyl nitrate (PAN) is an air pollutant produced in photochemical smog by the reaction of hydrocarbons, oxides of nitrogen, and sunlight. PAN is unstable and dissociates into peroxyacetyl radicals and $NO_2(g)$. Its presence in polluted air is like a reservoir for NO_2 storage.

$$\underset{\text{PAN}}{CH_3\overset{\overset{\displaystyle O}{\|}}{C}OONO_2} \longrightarrow \underset{\substack{\text{peroxyacetyl}\\\text{radical}}}{CH_3\overset{\overset{\displaystyle O}{\|}}{C}OO} + NO_2$$

The first-order decomposition of PAN has a half-life of 35 h at 0 °C and 30.0 min at 25 °C. At what temperature will a sample of air containing 5.0×10^{14} PAN molecules per liter decompose at the rate of 1.0×10^{12} PAN molecules per liter per minute?

1. *Determine the values of k at 0 °C and 25 °C.* For a first-order reaction, $k = 0.693/t$. Thus, at 0 °C, $k = (0.693/35\ \text{h}) \times 1\ \text{h}/60\ \text{min} = 3.3 \times 10^{-4}\ \text{min}^{-1}$; and at 25 °C, $k = 0.693/30.0\ \text{min} = 2.31 \times 10^{-2}\ \text{min}^{-1}$.

2. *Determine the activation energy for the reaction.* We can substitute the two k values and their corresponding Kelvin temperatures into the Arrhenius equation (15.22), which we solve for the activation energy, E_a.

$$\ln \frac{2.31 \times 10^{-2}\ \text{min}^{-1}}{3.3 \times 10^{-4}\ \text{min}^{-1}}$$

$$= \frac{E_a}{8.3145\ \text{J mol}^{-1}\ \text{K}^{-1}} \times \left(\frac{1}{273\ \text{K}} - \frac{1}{298\ \text{K}}\right)$$

$$= \frac{E_a}{8.3145\ \text{J mol}^{-1}}(0.00366 - 0.00336) = 4.25$$

$$E_a = \frac{8.3145\ \text{J mol}^{-1} \times 4.25}{0.00030} = 1.2 \times 10^5\ \text{J mol}^{-1}$$

Initiation:

(1) $H_2 + O_2 \longrightarrow HO_2\cdot + H\cdot$

Propagation:

(2) $HO_2\cdot + H_2 \longrightarrow HO\cdot + H_2O$

(3) $HO\cdot + H_2 \longrightarrow H\cdot + H_2O$

Branching:

(4) $H\cdot + O_2 \longrightarrow O\cdot + HO\cdot$

(5) $O\cdot + H_2 \longrightarrow HO\cdot + H\cdot$

Termination:

(6) $O\cdot + O\cdot + M \longrightarrow O_2 + M^*$
            etc.

As we have learned previously (page 407), a *free radical* is a highly reactive molecular fragment with one or more unpaired electrons. Free radicals are produced in reactions such as step (1) above. In steps (2) and (3), for each free radical consumed, a molecule of the product H_2O is formed, together with another free radical. These steps continue the "chain" of reaction and are called *propagation* steps.

In steps (4) and (5), a free radical and a molecule combine to form *two* new free radicals—the chain branches. As more and more branches are produced in the chain of reaction, more and more free radicals are produced, and the reaction quickly becomes explosive.

One method of preventing a combustion reaction from becoming explosive is to keep the concentration of one of the reactants very low (*lean*). The rate of reaction is slowed down because of this low concentration of the lean reactant. Another method is to introduce into the reaction mixture a species M that can remove energy from free radicals and cause them to combine, as in the *termination* step (6). Energy from the excited species M* is dissipated without the further formation of free radicals. Also, a species M can be chosen that reacts with free radicals; this again reduces their number. Still another way to avoid explosions is to keep the gaseous mixture at a low pressure. Then, many of the free radicals formed are able to migrate to the walls of the container without undergoing further reaction. At the walls, the free radicals lose energy and combine in reactions similar to step (6). At higher pressures, however, the free radicals are more likely to collide with molecules of the reactants before they reach the container walls. Here again, an explosion rather than a steady combustion is the likely result.

---

3. *Determine k at the unknown temperature.* Because the reaction is first order, the rate law is rate $= k$[PAN]. Also, because the quantities mol/L and molecules/L are proportional to one another, we can work with the concentration of PAN in molecules per liter. For the first-order reaction, rate of reaction $= k$[PAN], which we can rearrange as follows.

$$k = \frac{\text{rate of reaction}}{[\text{PAN}]} = \frac{1.0 \times 10^{12} \text{ molecules L}^{-1} \text{ min}^{-1}}{5.0 \times 10^{14} \text{ molecules L}^{-1}}$$

$$= 2.0 \times 10^{-3} \text{ min}^{-1}$$

4. *Determine the unknown temperature.* We can use the value $k = 2.0 \times 10^{-3}$ min^{-1} as k_2. For k_1, we can use the value of k at either 0 °C or 25 °C. T_2 is the unknown temperature, and T_1 is the temperature corresponding to the value we use for k_1. We know E_a from part 2. Finally, we can solve the Arrhenius equation (15.22) for the unknown temperature. The answer is 283 K.

Key Terms

activated complex (15–8)
activation energy (15–8)
active sites (15–11)
bimolecular (15–10)
catalyst (15–11)
collision frequency (15–8)
elementary process (15–10)
enzyme (15–11)
first-order reaction (15–5)

half-life (15–5)
initial rate of reaction (15–2)
instantaneous rate of reaction (15–2)
integrated rate law (15–4)
order of reaction (15–3)
rate constant, *k* (15–3)
rate-determining step (15–10)
rate law (rate equation) (15–3)
rate of reaction (15–1)

reaction intermediate (15–10)
reaction mechanism (15–10)
reaction profile (15–8)
second-order reaction (15–6)
substrate (15–11)
transition state (15–8)
unimolecular (15–10)
zero-order reaction (15–4)

Review Questions

1. In your own words, define or explain the following terms or symbols: **(a)** $[A]_0$; **(b)** k; **(c)** $t_{1/2}$; **(d)** zero-order reaction; **(e)** catalyst.

2. Briefly describe each of the following ideas, phenomena, or methods: **(a)** the method of initial rates; **(b)** activated complex; **(c)** reaction mechanism; **(d)** heterogeneous catalysis; **(e)** rate-determining step.

3. Explain the important distinctions between each pair of terms: **(a)** first-order and second-order reactions; **(b)** rate law and integrated rate law; **(c)** activation energy and enthalpy of reaction; **(d)** elementary process and overall reaction; **(e)** enzyme and substrate.

4. In the reaction A $\longrightarrow$ products, the initial concentration of A is 0.2643 M; 35 min later it is 0.1832 M. What is the initial rate of reaction expressed in **(a)** M min^{-1}; **(b)** M s^{-1}?

5. In the reaction 2 A + B $\longrightarrow$ C + 3 D, reactant A is found to react at the rate of 6.2×10^{-4} M s^{-1}. **(a)** What is the rate of reaction at this point? **(b)** What is the rate of disappearance of B? **(c)** What is the rate of formation of D?

6. From Figure 15-2 estimate the rate of reaction at **(a)** $t = 800$ s; **(b)** the time at which $[H_2O_2] = 0.50$ M.

7. A first-order reaction, A $\longrightarrow$ products, has a half-life of 75 s. Which of the following statements correctly applies to the reaction, and what is wrong with the other three? **(a)** The reaction goes to completion in 150 s; **(b)** the quantity of A remaining after 150 s is half of what remains after 75 s; **(c)** the same quantity of A is consumed for every 75 s of the reaction; **(d)** one-quarter of the original quantity of A is consumed in the first 37.5 s of the reaction.

8. The reaction A + B $\longrightarrow$ C + D is second order in A and zero order in B. The value of k is 0.0103 M^{-1} min^{-1}. What is the rate of this reaction when $[A] = 0.116$ M and $[B] = 3.83$ M?

9. A first-order reaction, A $\longrightarrow$ products, has a half-life of 19.8 min. What is the rate of the reaction when $[A] = 0.632$ M?

10. A reaction is 50% complete in 30.0 min. How long after its start will the reaction be 75% complete if it is **(a)** first order; **(b)** zero order?

11. Substance A decomposes by a first-order reaction. Starting initially with $[A] = 2.00$ M, after 126 min $[A] = 0.250$ M. For this reaction, what is **(a)** $t_{1/2}$; **(b)** k?

12. The reaction A $\longrightarrow$ products is first order in A.
(a) If 1.60 g A is allowed to decompose for 38 min, the mass of A remaining undecomposed is found to be 0.40 g. What is the half-life, $t_{1/2}$, of this reaction?
(b) Starting with 1.60 g A, what is the mass of A remaining undecomposed after 1.00 h?

13. In the first-order reaction A $\longrightarrow$ products, $[A] = 0.816$ M initially and 0.632 M after 16.0 min.
(a) What is the value of the rate constant, k?
(b) What is the half-life of this reaction?
(c) At what time will $[A] = 0.235$ M?
(d) What will $[A]$ be after 2.5 h?

14. The initial rate of the reaction A + B $\longrightarrow$ C + D is determined for different initial conditions, with the results listed in the following table.

Expt	[A], M	[B], M	Initial rate, M s^{-1}
1	0.185	0.133	3.35×10^{-4}
2	0.185	0.266	1.35×10^{-3}
3	0.370	0.133	6.75×10^{-4}
4	0.370	0.266	2.70×10^{-3}

(a) What is the order of reaction with respect to A and to B?
(b) What is the overall reaction order?
(c) What is the value of the rate constant, k?

15. A kinetic study of the reaction A $\longrightarrow$ products yields the data: $t = 0$ s, $[A] = 2.00$ M; 500 s, 1.00 M; 1500 s, 0.50 M; 3500 s, 0.25 M. *Without performing detailed calculations*, determine the order of this reaction and indicate your method of reasoning.

16. The reaction A + 2 B $\longrightarrow$ C + 2 D has $\Delta H = +25$ kJ. Which of the following is true concerning the activation energy of the reaction: $E_a = $ **(a)** -25 kJ; **(b)** $+25$ kJ; **(c)** less than $+25$ kJ; **(d)** more than $+25$ kJ? Explain.

17. The rate constant for the reaction $H_2(g) + I_2(g) \longrightarrow 2\, HI(g)$ has been determined at the following temperatures: 599 K, $k = 5.4 \times 10^{-4}$ M^{-1} s^{-1}; 683 K, $k = 2.8 \times 10^{-2}$ M^{-1} s^{-1}.
(a) Calculate the activation energy for the reaction.
(b) At what temperature will the rate constant have the value $k = 5.0 \times 10^{-3}$ M^{-1} s^{-1}?

Three different sets of data of [A] versus time are given in the following table for the reaction A $\longrightarrow$ products. (*Hint:* There are several ways of arriving at answers for each of the following six questions.)

Data for Exercises 18–23					
I		**II**		**III**	
Time, s	[A], M	Time, s	[A], M	Time, s	[A], M
0	1.00	0	1.00	0	1.00
25	0.78	25	0.75	25	0.80
50	0.61	50	0.50	50	0.67
75	0.47	75	0.25	75	0.57
100	0.37	100	0.00	100	0.50
150	0.22			150	0.40
200	0.14			200	0.33
250	0.08			250	0.29

18. Which of these sets of data corresponds to a **(a)** zero-order, **(b)** first-order, **(c)** second-order reaction?
19. What is the value of the rate constant k of the zero-order reaction?
20. What is the approximate half-life of the first-order reaction?
21. What is the approximate initial rate of the second-order reaction?
22. What is the approximate rate of reaction at $t = 75$ s for the **(a)** zero-order, **(b)** first-order, **(c)** second-order reaction?
23. What is the approximate concentration of A remaining after 110 s in the **(a)** zero-order, **(b)** first-order, **(c)** second-order reaction?

24. For the reaction A + 2 B $\longrightarrow$ C + D, the rate law is rate of reaction = $k[A][B]$.
    (a) Show that the following mechanism is consistent with the stoichiometry of the overall reaction and with the rate law.
    $$A + B \longrightarrow I \quad \text{(slow)}$$
    $$I + B \longrightarrow C + D \quad \text{(fast)}$$
    (b) Show that the following mechanism is consistent with the stoichiometry of the overall reaction, but *not* with the rate law.
    $$2 B \underset{k_2}{\overset{k_1}{\rightleftharpoons}} B_2 \quad \text{(fast)}$$
    $$A + B_2 \xrightarrow{k_3} C + D \text{ (slow)}$$

Exercises

Rates of Reactions

25. In the reaction A $\longrightarrow$ products, [A] is found to be 0.485 M at $t = 71.5$ s and 0.474 M at $t = 82.4$ s. What is the average rate of the reaction during this time interval?
26. In the reaction A $\longrightarrow$ products, at $t = 0$, [A] = 0.1565 M. After 1.00 min, [A] = 0.1498 M, and after 2.00 min, [A] = 0.1433 M.
    (a) Calculate the average rate of the reaction during the first minute and during the second minute.
    (b) Why are these two rates not equal?
27. In the reaction A $\longrightarrow$ products, 4.40 min after the reaction is started, [A] = 0.588 M. The rate of reaction at this point is found to be rate = $-\Delta[A]/\Delta t = 2.2 \times 10^{-2}$ M min^{-1}. Assume that this rate remains constant for a short period of time.
    (a) What is [A] 5.00 min after the reaction is started?
    (b) At what time after the reaction is started will [A] = 0.565 M?
28. Refer to Experiment 2 of Table 15.3 and to reaction (15.7) and rate law (15.8). Exactly 1.00 h after the reaction is started, what are **(a)** [HgCl$_2$] and **(b)** [C$_2$O$_4^{2-}$] in the mixture?
29. For the reaction A + 2 B $\longrightarrow$ 2 C, the rate of reaction is 1.76×10^{-5} M s^{-1} at the time when [A] = 0.3580 M.
    (a) What is the rate of formation of C?
    (b) What will [A] be 1.00 min later?

    (c) Assume the rate remains at 1.76×10^{-5} M s^{-1}. How long would it take for [A] to change from 0.3580 to 0.3500 M?
30. If the rate of reaction (15.3) is 5.7×10^{-4} M s^{-1}, what is the rate of production of O$_2$(g) from 1.00 L of the H$_2$O$_2$(aq), expressed as **(a)** mol O$_2$ s^{-1}; **(b)** mol O$_2$ min^{-1}; **(c)** mL O$_2$(STP) min^{-1}?
31. In the reaction A(g) $\longrightarrow$ 2 B(g) + C(g), the *total* pressure increases while the *partial* pressure of A(g) decreases. If the initial pressure of A(g) in a vessel of constant volume is 1.000×10^3 mmHg,
    (a) What will be the total pressure when the reaction has gone to completion?
    (b) What will be the total gas pressure when the partial pressure of A(g) has fallen to 8.00×10^2 mmHg?
32. At 65 °C, the half-life for the first-order decomposition of N$_2$O$_5$(g) is 2.38 min.
    $$N_2O_5(g) \longrightarrow 2\ NO_2(g) + \frac{1}{2}\ O_2(g)$$
    If 1.00 g of N$_2$O$_5$ is introduced into an evacuated 15-L flask at 65 °C,
    (a) What is the initial partial pressure, in mmHg, of N$_2$O$_5$(g)?
    (b) What is the partial pressure, in mmHg, of N$_2$O$_5$(g) after 2.38 min?
    (c) What is the total gas pressure, in mmHg, after 2.38 min?

Method of Initial Rates

33. For the reaction A + B $\longrightarrow$ C + D, the following initial rates of reaction were found. What is the rate law for this reaction?

Expt	[A], M	[B], M	Initial Rate, M min^{-1}
1	0.50	1.50	4.2×10^{-3}
2	1.50	1.50	1.3×10^{-2}
3	3.00	3.00	5.2×10^{-2}

34. The following rates of reaction were obtained in three experiments with the reaction 2 NO(g) + Cl$_2$(g) $\longrightarrow$ 2 NOCl(g).

Expt	Initial [NO], M	Initial [Cl$_2$], M	Initial rate of reaction, M s^{-1}
1	0.0125	0.0255	2.27×10^{-5}
2	0.0125	0.0510	4.55×10^{-5}
3	0.0250	0.0255	9.08×10^{-5}

What is the rate law for this reaction?

35. The following data are obtained for the initial rates of reaction in the reaction A + 2 B + C ⟶ 2 D + E.

Expt	[A], M	[B], M	[C], M	Initial Rate
1	1.40	1.40	1.00	R_1
2	0.70	1.40	1.00	$R_2 = \frac{1}{2} \times R_1$
3	0.70	0.70	1.00	$R_3 = \frac{1}{4} \times R_2$
4	1.40	1.40	0.50	$R_4 = 16 \times R_3$
5	0.70	0.70	0.50	$R_5 = ?$

(a) What are the reaction orders with respect to A, B, and C?
(b) What is the value of R_5 in terms of R_1?

36. In the thermal decomposition of acetaldehyde,
$$CH_3CHO(g) \longrightarrow CH_4(g) + CO(g)$$
the rate of reaction is found to increase by a factor of about 2.8 when the initial concentration of acetaldehyde is doubled. What is the order of this reaction?

First-Order Reactions

37. One of the following statements is true and the other is false regarding the first-order reaction 2 A ⟶ B + C. Identify the true statement and the false one, and explain your reasoning.
(a) The rate of the reaction decreases as more and more of B and C form.
(b) The time required for one-half of substance A to react is directly proportional to the quantity of A present initially.

38. One of the following statements is true and the other is false regarding the first-order reaction 2 A ⟶ B + C. Identify the true statement and the false one, and explain your reasoning.
(a) A graph of [A] versus time is a straight line.
(b) The rate of the reaction is one-half the rate of disappearance of A.

39. The first-order reaction A ⟶ products has $t_{1/2} = 180$ s.
(a) What percent of a sample of A remains *unreacted* 900 s after a reaction has been started?
(b) What is the rate of reaction when [A] = 0.50 M?

40. The reaction A ⟶ products is first order in A. Initially, [A] = 0.800 M; and after 54 min,[A] = 0.100 M.
(a) At what time is [A] = 0.025 M?
(b) What is the rate of reaction when [A] = 0.025 M?

41. In the first-order reaction A ⟶ products, it is found that 99% of the original amount of reactant A decomposes in 137 min. What is the half-life, $t_{1/2}$, of this decomposition reaction?

42. The half-life of the radioactive isotope phosphorus-32 is 14.3 days. How long does it take for a sample of phosphorus-32 to lose 99% of its radioactivity?

43. Acetoacetic acid, CH_3COCH_2COOH, a reagent used in organic synthesis, decomposes in acidic solution, producing acetone and $CO_2(g)$.
$$CH_3COCH_2COOH(aq) \longrightarrow CH_3COCH_3(aq) + CO_2(g)$$
This first-order decomposition has a half-life of 144 min.

(a) How long will it take for a sample of acetoacetic acid to be 65% decomposed?
(b) How many liters of $CO_2(g)$, measured at 24.5 °C and 748 Torr, are produced as a 10.0-g sample of CH_3COCH_2COOH decomposes for 575 min? (Neglect the aqueous solubility of $CO_2(g)$.)

44. The following first-order reaction occurs in $CCl_4(l)$ at 45 °C: $N_2O_5 \longrightarrow N_2O_4 + \frac{1}{2} O_2(g)$. The rate constant is $k = 6.2 \times 10^{-4}$ s^{-1}. An 80.0-g sample of N_2O_5 in $CCl_4(l)$ is allowed to decompose at 45 °C.
(a) How long does it take for the quantity of N_2O_5 to be reduced to 2.5 g?
(b) How many liters of O_2, measured at 745 mmHg and 45 °C, are produced up to this point?

45. For the reaction A ⟶ products, the following data give [A] as a function of time: $t = 0$ s, [A] = 0.600 M; 100 s, 0.497 M; 200 s, 0.413 M; 300 s, 0.344 M; 400 s, 0.285 M; 600 s, 0.198 M; 1000 s, 0.094 M.
(a) Show that the reaction is first order.
(b) What is the value of the rate constant, k?
(c) What is [A] at $t = 750$ s?

46. The decomposition of dimethyl ether at 504 °C is
$$(CH_3)_2O(g) \longrightarrow CH_4(g) + H_2(g) + CO(g)$$
The following data are partial pressures of dimethyl ether (DME) as a function of time: $t = 0$ s, $P_{DME} = 312$ mmHg; 390 s, 264 mmHg; 777 s, 224 mmHg; 1195 s, 187 mmHg; 3155 s, 78.5 mmHg.
(a) Show that the reaction is first order.
(b) What is the value of the rate constant, k?
(c) What is the total gas pressure at 390 s?
(d) What is the total gas pressure when the reaction has gone to completion?
(e) What is the total gas pressure at $t = 1000$ s?

Reactions of Various Orders

47. The decomposition of HI(g) at 700 K is followed for 400 s, yielding the following data: at $t = 0$, [HI] = 1.00 M; at $t = 100$ s, [HI] = 0.90 M; at $t = 200$ s, [HI] = 0.81 M; at $t = 300$ s, [HI] = 0.74 M; at $t = 400$ s, [HI] = 0.68 M. What are the reaction order and the rate constant for the reaction: $HI(g) \longrightarrow \frac{1}{2} H_2(g) + \frac{1}{2} I_2(g)$? Write the rate law for the reaction at 700 K.

48. For the disproportionation of *p*-toluenesulfinic acid,
$$3 \, ArSO_2H \longrightarrow ArSO_2SAr + ArSO_3H + H_2O$$
[where Ar = p-$CH_3C_6H_4$—], the following data were obtained: $t = 0$ min, [ArSO$_2$H] = 0.100 M; 15 min, 0.0863 M; 30 min, 0.0752 M; 45 min, 0.0640 M; 60 min, 0.0568 M; 120 min, 0.0387 M; 180 min, 0.0297 M; 300 min, 0.0196 M.
(a) Show that this reaction is second order.
(b) What is the value of the rate constant, k?
(c) At what time would [ArSO$_2$H] = 0.0500 M?
(d) At what time would [ArSO$_2$H] = 0.0250 M?
(e) At what time would [ArSO$_2$H] = 0.0350 M?

49. For the reaction A $\longrightarrow$ products, the following data were obtained: $t = 0$ s, [A] = 0.715 M; 22 s, 0.605 M; 74 s, 0.345 M; 132 s, 0.055 M. **(a)** What is the order of this reaction? **(b)** What is the half-life of the reaction?

50. The following data were obtained for the dimerization of 1,3-butadiene, $2 \, C_4H_6(g) \longrightarrow C_8H_{12}(g)$, at 600 K: $t = 0$ min, [C$_4$H$_6$] = 0.0169 M; 12.18 min, 0.0144 M; 24.55 min, 0.0124 M; 42.50 min, 0.0103 M; 68.05 min, 0.00845 M.
(a) What is the order of this reaction?
(b) What is the value of the rate constant, k?
(c) At what time would [C$_4$H$_6$] = 0.00423 M?
(d) At what time would [C$_4$H$_6$] = 0.0050 M?

51. For the reaction A $\longrightarrow$ products, the following data are obtained.

First Experiment

[A] = 1.512 M	$t = 0$ min
[A] = 1.490 M	$t = 1.0$ min
[A] = 1.469 M	$t = 2.0$ min

Second Experiment

[A] = 3.024 M	$t = 0$ min
[A] = 2.935 M	$t = 1.0$ min
[A] = 2.852 M	$t = 2.0$ min

(a) Determine the initial rate of reaction (that is, $-\Delta[A]/\Delta t$) in each of the two experiments.
(b) Determine the order of the reaction.

52. For the reaction A $\longrightarrow$ 2 B + C, the following data are obtained for [A] as a function of time: $t = 0$ min, [A] = 0.80 M; 8 min, 0.60 M; 24 min, 0.35 M; 40 min, 0.20 M.
(a) By suitable means, establish the order of the reaction.
(b) What is the value of the rate constant, k?
(c) Calculate the rate of formation of B at $t = 30$ min.

53. In three different experiments, the following results were obtained for the reaction A $\longrightarrow$ products: [A]$_0$ = 1.00 M, $t_{1/2} = 50$ min; [A]$_0$ = 2.00 M, $t_{1/2} = 25$ min; [A]$_0$ = 0.50 M, $t_{1/2} = 100$ min. Write the rate equation for this reaction, and indicate the value of k.

54. Ammonia decomposes on the surface of a hot tungsten wire. Following are the half-lives that were obtained at 1100 °C for different initial concentrations of NH$_3$: [NH$_3$]$_0$ = 0.0031 M, $t_{1/2} = 7.6$ min; 0.0015 M, 3.7 min; 0.00068 M, 1.7 min. For this decomposition reaction, what is **(a)** the order of the reaction; **(b)** the rate constant k?

55. The half-lives of both zero-order and second-order reactions depend on the initial concentration, as well as on the rate constant. In one case, the half-life gets longer as the initial concentration increases, and in the other it gets shorter. Which is which, and why isn't the situation the same for both?

56. Consider three hypothetical reactions, A $\longrightarrow$ products, all having the same numerical value of the rate constant k. One of the reactions is zero order, one is first order, and one is second order. What must be the initial concentration [A]$_0$ if **(a)** the zero- and first-order; **(b)** the zero- and second-order; **(c)** the first- and second-order reactions are to have the same half-life?

Collision Theory; Activation Energy

57. Explain why
(a) A reaction rate cannot be calculated from the collision frequency alone.
(b) The rate of a chemical reaction may increase dramatically with temperature, whereas the collision frequency increases much more slowly.
(c) The addition of a catalyst to a reaction mixture can have such a pronounced effect on the rate of a reaction, even if the temperature is held constant.

58. If even a tiny spark is introduced into a mixture of $H_2(g)$ and $O_2(g)$, a highly exothermic explosive reaction occurs. Without the spark, the mixture remains unreacted indefinitely.
(a) Explain this difference in behavior.
(b) Why is the nature of the reaction independent of the size of the spark?

59. For the reversible reaction A + B $\rightleftharpoons$ C + D, the enthalpy change of the forward reaction is +21 kJ/mol. The activation energy of the forward reaction is 84 kJ/mol.

(a) What is the activation energy of the reverse reaction?
(b) In the manner of Figure 15-10, sketch the reaction profile of this reaction.

60. By an appropriate sketch, indicate why there is some relationship between the enthalpy change and the activation energy for an endothermic reaction but not for an exothermic reaction.

61. By inspection of the reaction profile for the reaction A to D given below, answer the following questions.

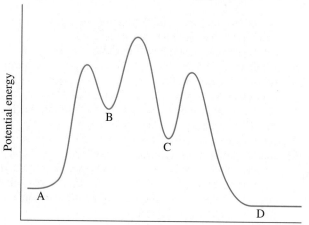

Reaction progress

(a) How many intermediates are there in the reaction?
(b) How many transition states are there?
(c) Which is the fastest step in the reaction?
(d) What is the reactant in the rate-determining step?

(e) Is the first step of the reaction exothermic or endothermic?
(f) Is the overall reaction exothermic or endothermic?

62. By inspection of the reaction profile for the reaction A to D given below answer the following questions.

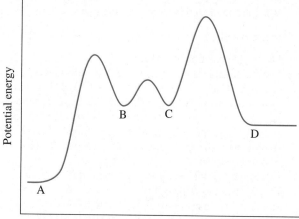

Reaction progress

(a) How many intermediates are there in the reaction?
(b) How many transition states are there?
(c) Which is the fastest step in the reaction?
(d) What is the reactant in the rate-determining step?
(e) Is the first step of the reaction exothermic or endothermic?
(f) Is the overall reaction exothermic or endothermic?

Effect of Temperature on Rates of Reaction

63. A treatise on atmospheric chemistry lists the following rate constants for the decomposition of PAN described in the Integrative Example on page 614: 0 °C, $k = 5.6 \times 10^{-6}\,\text{s}^{-1}$; 10 °C, $k = 3.2 \times 10^{-5}\,\text{s}^{-1}$; 20 °C, $k = 1.6 \times 10^{-4}\,\text{s}^{-1}$; 30 °C, $k = 7.6 \times 10^{-4}\,\text{s}^{-1}$.
(a) Construct a graph of $\ln k$ versus $1/T$.
(b) What is the activation energy, E_a, of the reaction?
(c) Calculate the half-life of the decomposition reaction at 40 °C?

64. The reaction $C_2H_5I + OH^- \longrightarrow C_2H_5OH + I^-$ was studied in an ethanol (C_2H_5OH) solution, and the following rate constants were obtained: 15.83 °C, $k = 5.03 \times 10^{-5}$; 32.02 °C, 3.68×10^{-4}; 59.75 °C, 6.71×10^{-3}; 90.61 °C, $0.119\,\text{M}^{-1}\,\text{s}^{-1}$.
(a) Determine E_a for this reaction by a graphical method.
(b) Determine E_a by the use of equation (15.22).
(c) Calculate the value of the rate constant k at 100.0 °C.

65. The first-order reaction A $\longrightarrow$ products has a half-life, $t_{1/2}$, of 46.2 min at 25 °C and 2.6 min at 102 °C.
(a) Calculate the activation energy of this reaction.
(b) At what temperature would the half-life be 10.0 min?

66. For the first-order reaction

$$N_2O_5(g) \longrightarrow 2\,NO_2(g) + \frac{1}{2}\,O_2(g)$$

$t_{1/2} = 22.5$ h at 20 °C and 1.5 h at 40 °C.
(a) Calculate the activation energy of this reaction.
(b) If the Arrhenius constant A $= 2.05 \times 10^{13}\,\text{s}^{-1}$, determine the value of k at 30 °C.

67. A commonly stated rule of thumb is that reaction rates double for a temperature increase of about 10 °C. (This rule is very often wrong.)
(a) What must be the approximate activation energy for this statement to be true for reactions at about room temperature?
(b) Would you expect this rule of thumb to apply at room temperature for the reaction profiled in Figure 15-10? Explain.

68. Concerning the rule of thumb stated in Exercise 67, estimate how much faster cooking occurs in a pressure cooker with the vapor pressure of water at 2.00 atm instead of in water under normal boiling conditions.
(*Hint:* Refer to Table 13.2.)

Catalysis

69. The following statements about catalysis are not stated as carefully as they might be. What slight modifications would you make in them?
  (a) A catalyst is a substance that speeds up a chemical reaction but does not take part in the reaction.
  (b) The function of a catalyst is to lower the activation energy for a chemical reaction.

70. The following substrate concentration [S] versus time data were obtained during an enzyme-catalyzed reaction: $t = 0$ min, [S] = 1.00 M; 20 min, 0.90 M; 60 min, 0.70 M; 100 min, 0.50 M; 160 min, 0.20 M. What is the order of this reaction with respect to S in the concentration range studied?

71. What are the similarities and differences between the catalytic activity of platinum metal and of an enzyme?

72. Certain gas-phase reactions on a heterogeneous catalyst are first order at low gas pressures and zero order at high pressures. Can you suggest a reason for this?

73. The graph shows the effect of enzyme concentration on the rate of an enzyme reaction. What reaction conditions are necessary to account for this graph?

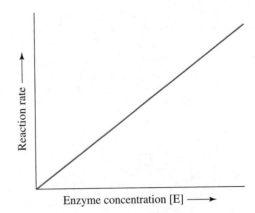

74. The graph shows the effect of temperature on enzyme activity. Explain why the graph has the general shape shown. For human enzymes, at what temperature would you expect the maximum in the curve to appear?

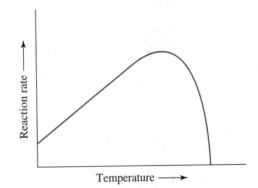

Reaction Mechanisms

75. We have used the terms *order of a reaction* and *molecularity of an elementary process* (that is, unimolecular, bimolecular). What is the relationship, if any, between these two terms?

76. According to collision theory, chemical reactions occur through molecular collisions. A unimolecular elementary process in a reaction mechanism involves dissociation of a *single* molecule. How can these two ideas be compatible? Explain.

77. The reaction $2 NO + 2 H_2 \longrightarrow N_2 + 2 H_2O$ is second order in [NO] and first order in [H$_2$]. A *three-step* mechanism has been proposed. The first, fast step is the elementary process given as equation (15.25) on page 605. The third step, also fast, is $N_2O + H_2 \longrightarrow N_2 + H_2O$. Propose an entire three-step mechanism, and show that it conforms to the experimentally determined reaction order.

78. The mechanism proposed for the reaction of $H_2(g)$ and $I_2(g)$ to form HI(g) consists of a fast reversible first step involving $I_2(g)$ and $I(g)$, followed by a slow step. Propose a two-step mechanism for the reaction $H_2(g) + I_2(g) \longrightarrow 2 HI(g)$, which is known to be first order in H_2 and first order in I_2.

79. The reaction $2 NO + Cl_2 \longrightarrow 2 NOCl$ has the rate law: rate of reaction $= k[NO]^2[Cl_2]$. Propose a two-step mechanism for this reaction consisting of a fast reversible first step, followed by a slow step.

80. A simplified rate law for the reaction $2 O_3(g) \longrightarrow 3 O_2(g)$ is

$$\text{rate} = k \frac{[O_3]^2}{[O_2]}$$

  For this reaction, propose a two-step mechanism that consists of a fast, reversible first step, followed by a slow second step.

Integrative and Advanced Exercises

81. $[A]_t$ as a function of time for the reaction A $\longrightarrow$ products is plotted in the graph shown here. Use data from this graph to determine **(a)** the order of the reaction; **(b)** the rate constant, k; **(c)** the rate of the reaction at $t = 3.5$ min, using the results of parts (a) and (b); **(d)** the rate of the reaction at $t = 5.0$ min, from the slope of the tangent line; **(e)** the initial rate of the reaction.

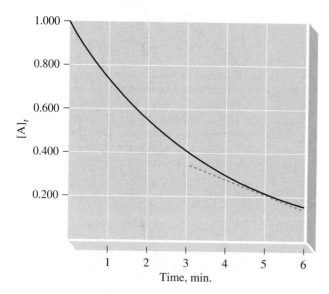

Time, min.

82. Exactly 300 s after decomposition of H_2O_2(aq) begins (reaction 15.3), a 5.00-mL sample is removed and immediately titrated with 37.1 mL of 0.1000 M KMnO$_4$. What is $[H_2O_2]$ at this 300-s point in the reaction?

83. Use the method of Exercise 82 to determine the volume of 0.1000 M KMnO$_4$ required to titrate 5.00-mL samples of H_2O_2(aq) for each of the entries in Table 15.1. Plot these volumes of KMnO$_4$(aq) as a function of time, and show that from this graph you can get the same rate of reaction at 1400 s as that obtained in Figure 15-2.

84. The initial rate of reaction (15.3) is found to be 1.7×10^{-3} M s^{-1}. Assume that this rate holds for 2 minutes. Start with 175 mL of 1.55 M H_2O_2(aq) at $t = 0$. How many milliliters of O_2(g), measured at 24 °C and 757 mmHg, are released from solution in the first minute of the reaction?

85. We have seen that the unit of k depends on the overall order of a reaction. Derive a general expression for the units of k for a reaction of any overall order, based on the order of the reaction (o) and the units of concentration (M) and time (s).

86. At room temperature (20 °C), milk turns sour in about 64 hours. In a refrigerator at 3 °C, milk can be stored three times as long before it sours. **(a)** Estimate the activation energy of the reaction that causes the souring of milk. **(b)** How long should it take milk to sour at 40 °C?

87. Hydroxide ion is involved in the mechanism of the following reaction but is not consumed in the overall reaction.

$$OCl^- + I^- \xrightarrow{\ OH^-\ } OI^- + Cl^-$$

(a) From the data given, determine the order of the reaction with respect to OCl$^-$, I$^-$, and OH$^-$.
(b) What is the overall reaction order?
(c) Write the rate equation, and determine the value of the rate constant, k.

[OCl$^-$], M	[I$^-$], M	[OH$^-$], M	Rate formation OI$^-$, M s^{-1}
0.0040	0.0020	1.00	4.8×10^{-4}
0.0020	0.0040	1.00	5.0×10^{-4}
0.0020	0.0020	1.00	2.4×10^{-4}
0.0020	0.0020	0.50	4.6×10^{-4}
0.0020	0.0020	0.25	9.4×10^{-4}

88. The half-life for the first-order decomposition of nitramide, NH_2NO_2(aq) $\longrightarrow$ N_2O(g) + H_2O(l), is 123 min at 15 °C. If 165 mL of a 0.105 M NH_2NO_2 solution is allowed to decompose, how long must the reaction proceed to yield 50.0 mL of N_2O(g) collected over water at 15 °C and a barometric pressure of 756 mmHg? (The vapor pressure of water at 15 °C is 12.8 mmHg.)

89. The decomposition of ethylene oxide at 690 K is monitored by measuring the *total* gas pressure as a function of time. The data obtained are $t = 10$ min, $P_{tot} = 139.14$ mmHg; 20 min, 151.67 mmHg; 40 min, 172.65 mmHg; 60 min, 189.15 mmHg; 100 min, 212.34 mmHg; 200 min, 238.66 mmHg; ∞, 249.88 mmHg. What is the order of the reaction $(CH_2)_2O$(g) $\longrightarrow$ CH_4(g) + CO(g)?

90. Refer to Example 15-7. For the decomposition of di-*t*-butyl peroxide (DTBP), determine the time at which the *total* gas pressure is 2100 mmHg.

91. The following data are for the reaction 2 A + B $\longrightarrow$ products. Establish the order of this reaction with respect to A and to B.

Expt 1, [B] = 1.00 M		Expt 2, [B] = 0.50 M	
Time, min	[A], M	Time, min	[A], M
0	1.000×10^{-3}	0	1.000×10^{-3}
1	0.951×10^{-3}	1	0.975×10^{-3}
5	0.779×10^{-3}	5	0.883×10^{-3}
10	0.607×10^{-3}	10	0.779×10^{-3}
20	0.368×10^{-3}	20	0.607×10^{-3}

92. Show that the following mechanism is consistent with the rate law established for the iodide–hypochlorite reaction in Exercise 87.

(Fast) $\quad OCl^- + H_2O \underset{k_2}{\overset{k_1}{\rightleftharpoons}} HOCl + OH^-$

(Slow) $\quad I^- + HOCl \xrightarrow{k_3} HOI + Cl^-$

(Fast) $\quad HOI + OH^- \underset{k_5}{\overset{k_4}{\rightleftharpoons}} H_2O + OI^-$

93. In the hydrogenation of a compound containing a carbon-to-carbon triple bond, two products are possible, as in the reaction

$$CH_3—C{\equiv}C—CH_3 + H_2 \longrightarrow$$

The amount of each product can be controlled by using an appropriate catalyst. The Lindlar catalyst is a *heterogeneous* catalyst that produces only one of these products. Which product, I or II, do you think is produced and why? Draw a sketch of how the reaction might occur.

94. Derive a plausible mechanism for the following reaction in aqueous solution, $Hg_2^{2+} + Tl^{3+} \longrightarrow 2\ Hg^{2+} + Tl^+$, for which the observed rate law is: rate $= k[Hg_2^{2+}][Tl^{3+}] / [Hg^{2+}]$.

95. The following three-step mechanism has been proposed for the reaction of chlorine and chloroform.

(1) $Cl_2(g) \underset{k_{-1}}{\overset{k_1}{\rightleftharpoons}} 2\ Cl(g)$

(2) $Cl(g) + CHCl_3(g) \xrightarrow{k_2} HCl(g) + CCl_3(g)$

(3) $CCl_3(g) + Cl(g) \xrightarrow{k_3} CCl_4(g)$

The numerical values of the rate constants for these steps are $k_1 = 4.8 \times 10^3$; $k_{-1} = 3.6 \times 10^3$; $k_2 = 1.3 \times 10^{-2}$; $k_3 = 2.7 \times 10^2$. Derive the rate law and the magnitude of k for the overall reaction.

96. For the reaction $A \longrightarrow$ products, derive the integrated rate law and an expression for the half-life if the reaction is third order.

97. The reaction $A + B \longrightarrow$ products is first order in A, first order in B, and second order overall. Consider that the starting concentrations of the reactants are $[A]_0$ and $[B]_0$, and that x represents the decrease in these concentrations at the time t. That is, $[A]_t = [A]_0 - x$ and $[B]_t = [B]_0 - x$. Show that the integrated rate law for this reaction can be expressed as shown below.

$$\ln \frac{[A]_0 \times [B]_t}{[B]_0 \times [A]_t} = ([B]_0 - [A]_0) \times kt$$

98. The rate of the reaction

$$2\ CO(g) \longrightarrow CO_2(g) + C(s)$$

was studied by injecting CO(g) into a reaction vessel and measuring the total pressure at constant volume.

P_{total} (Torr)	Time (s)
250	0
238	398
224	1002
210	1801

What is the rate constant of this reaction?

99. The kinetics of the decomposition of phosphine at 950 K

$$4\ PH_3(g) \longrightarrow P_4(g) + 6\ H_2(g)$$

was studied by injecting $PH_3(g)$ into a reaction vessel and measuring the total pressure at constant volume.

P_{total} (Torr)	Time (s)
100	0
150	40
167	80
172	120

What is the rate constant of this reaction?

100. The rate of an enzyme-catalyzed reaction can be slowed down by the presence of an inhibitor (I) that reacts with the enzyme in a rapid equilibrium process.

$$E + I \rightleftharpoons EI$$

By adding this step to the mechanism for enzyme catalysis on page 612, determine the effect of adding the concentration $[I_0]$ on the rate of an enzyme-catalyzed reaction.

101. By taking the reciprocal of both sides of equation 15.31, obtain an expression for $1/V$. Using the resulting equation, suggest a strategy for determining the Michaelis–Menten constant K_M and the value of k_2.

102. The following mechanism can be used to account for the change in apparent order of unimolecular reactions, such as the conversion of cyclopropane (A) into propene (P), where A* is an energetic form of cyclopropane that can either react or return to unreacted cyclopropane.

$$A + A \underset{k_{-1}}{\overset{k_1}{\rightleftharpoons}} A^* + A$$

$$A^* \xrightarrow{k_2} P$$

Show that at low pressures of cyclopropane the rate law is second order in A and at high pressures it is first order in A.

103. You wish to test the following proposed mechanism for the oxidation of HBr.

$$HBr + O_2 \xrightarrow{k_1} HOOBr$$

$$HOOBr + HBr \xrightarrow{k_2} 2\ HOBr$$

$$HOBr + HBr \xrightarrow{k_3} H_2O + Br_2$$

You find that the rate is first order with respect to HBr and with respect to O_2. You cannot detect HOBr among the products.
(a) If the proposed mechanism is correct, which must be the rate-determining step?

(b) Can you prove the mechanism from these observations?
(c) Can you disprove the mechanism from these observations?

Feature Problems

104. Benzenediazonium chloride decomposes by a first-order reaction in water, yielding $N_2(g)$ as a product.

$$C_6H_5N_2Cl \longrightarrow C_6H_5Cl + N_2(g)$$

The reaction can be followed by measuring the volume of $N_2(g)$ as a function of time. The data in the table were obtained for the decomposition of a 0.071 M solution at 50 °C, where $t = \infty$ corresponds to the completed reaction.

Time, min	$N_2(g)$, mL	Time, min	$N_2(g)$, mL
0	0	18	41.3
3	10.8	21	44.3
6	19.3	24	46.5
9	26.3	27	48.4
12	32.4	30	50.4
15	37.3	∞	58.3

(a) Convert the information given here into a table with one column for time and the other for $[C_6H_5N_2Cl]$.
(b) Construct a table similar to Table 15.2, in which the time interval is $\Delta t = 3$ min.
(c) Plot graphs similar to Figure 15-2, showing both the formation of $N_2(g)$ and the disappearance of $C_6H_5N_2Cl$ as a function of time.
(d) From the graph of part (c), determine the rate of reaction at $t = 21$ min, and compare your result with the reported value of 1.1×10^{-3} M min^{-1}.
(e) Determine the initial rate of reaction.
(f) Write the rate law for the first-order decomposition of $C_6H_5N_2Cl$, and estimate a value of k based on the rate determined in parts (d) and (e).
(g) Determine $t_{1/2}$ for the reaction, by estimation from the graph of the rate data and by calculation.
(h) At what time would the decomposition of the sample be three-fourths complete?
(i) Plot $\ln[C_6H_5N_2Cl]$ versus time, and show that the reaction is indeed first order.
(j) Determine k from the slope of the graph of part (i).
105. Our object is to study the kinetics of the reaction between peroxodisulfate and iodide ions.
(a) $S_2O_8^{2-}(aq) + 3\,I^-(aq) \longrightarrow 2\,SO_4^{2-}(aq) + I_3^-(aq)$
The I_3^- formed in reaction (a) is actually a complex of iodine, I_2, and iodide ion, I^-. Thiosulfate ion, $S_2O_3^{2-}$, also present in the reaction mixture, reacts with I_3^- just as fast as it is formed.

(b) $2\,S_2O_3^{2-}(aq) + I_3^-(aq) \longrightarrow S_4O_6^{2-} + 3\,I^-(aq)$
When all of the thiosulfate ion present initially has been consumed by reaction (b), a third reaction occurs between $I_3^-(aq)$ and starch, which is also present in the reaction mixture.
(c) $I_3^-(aq) + \text{starch} \longrightarrow \text{blue complex}$
The rate of reaction (a) is inversely related to the time required for the blue color of the starch–iodine complex to appear. That is, the faster reaction (a) proceeds, the more quickly the thiosulfate ion is consumed in reaction (b), and the sooner the blue color appears in reaction (c). One of the photographs shows the initial colorless solution and an electronic timer set at $t = 0$; the other photograph shows the very first appearance of the blue complex (after 49.89 s). Tables I and II list some actual student data obtained in this study.

Table I
Reaction conditions at 24 °C: 25.0 mL of the $(NH_4)_2S_2O_8(aq)$ listed, 25.0 mL of the KI(aq) listed, 10.0 mL of 0.010 M $Na_2S_2O_3(aq)$, and 5.0 mL starch solution are mixed. The time is that of the first appearance of the starch–iodine complex.

Experiment	Initial Concentrations, M		Time, s
	$(NH_4)_2S_2O_8$	KI	
1	0.20	0.20	21
2	0.10	0.20	42
3	0.050	0.20	81
4	0.20	0.10	42
5	0.20	0.050	79

Table II

Reaction conditions: those listed in Table I for Experiment 4, but at the temperatures listed.

Experiment	Temperature, °C	Time, s
6	3	189
7	13	88
8	24	42
9	33	21

(a) Use the data in Table I to establish the order of reaction (a) with respect to $S_2O_8^{2-}$ and to I^-. What is the overall reaction order?
(*Hint:* How are the times required for the blue complex to appear related to the actual rates of reaction?)
(b) Calculate the initial rate of reaction in Experiment 1, expressed in M s^{-1}.
[*Hint:* You must take into account the dilution that occurs when the various solutions are mixed, as well as the reaction stoichiometry indicated by equations (a), (b), and (c).]

(c) Calculate the value of the rate constant, k, based on Experiments 1 and 2.
(d) Calculate the rate constant, k, for the four different temperatures in Table II.
(e) Determine the activation energy, E_a, of the peroxodisulfate–iodide ion reaction.
(f) The following mechanism has been proposed for reaction (a). The first step is slow, and the others are fast.

$$I^- + S_2O_8^{2-} \longrightarrow IS_2O_8^{3-}$$

$$IS_2O_8^{3-} \longrightarrow 2\ SO_4^{2-} + I^+$$

$$I^+ + I^- \longrightarrow I_2$$

$$I_2 + I^- \longrightarrow I_3^-$$

Show that this mechanism is consistent with both the stoichiometry and the rate law of reaction (a). Explain why it is reasonable to expect the first step in the mechanism to be slower than the others.

 # eMedia Exercises

106. Different approaches to describing a reaction rate are illustrated in the **Reaction Rates** simulation in *eChapter 15-2*. **(a)** At which time is the reaction rate greatest? What is the physical reason behind this observation? **(b)** Are short- or long-time intervals more useful in approximating an instantaneous reaction rate?

107. In the animation describing a **First-Order Process** (*eChapter 15-5*), the half-life of the reaction is illustrated. **(a)** How does this relationship between time and concentration differ from that of zero- and second-order processes? **(b)** What is the ratio between the half-life and quarter-life (the time required for the concentration to fall to 75% of its initial value) for a first-order process?.

108. In the **Plotting Rate Data** activity found in *eChapter 15-6*, **(a)** determine the order of each reaction by plotting the data in the forms used to describe different orders of reaction. **(b)** Does varying the initial concentration of the reactant vary the result of this graphing exercise? Why or why not?

109. The influence of reaction conditions and energetics is simulated in the **Rates of Reactions** simulation found in *eChapter 15-8*. **(a)** What set of adjustable conditions in this simulation favor the greatest rate of reaction? What set of conditions favor the lowest rate of reaction? **(b)** What is the general relationship between ΔH_{rxn} and reaction rate? Under what conditions will endothermic reactions proceed at an appreciable rate?

110. The reaction of OH^- and CH_3Cl is illustrated in the **Bimolecular Reaction** animation (*eChapter 15-10*). **(a)** Identify the transition state or activated complex in this reaction. **(b)** Describe the hybridization of the central carbon atom as a reactant, activated complex, and product. What does this tell you about the stability of the transition state?

111. The reaction of hydrogen and ethylene over an alloy surface is illustrated in the **Surface Reaction—Hydrogenation** animation (*eChapter 15-11*). **(a)** Write out a balanced chemical equation describing this reaction. Where does the metal surface appear in this equation? **(b)** What is the primary role that the metal surface plays in the reaction? Why would this process proceed more slowly in the absence of the metal catalyst?

16

Principles of Chemical Equilibrium

Contents

Ammonia is an important industrial chemical. The reversible chemical reaction between nitrogen and hydrogen to give ammonia is unfavorable for the efficient production of ammonia. The production of ammonia can be improved by using high temperatures and high pressures in ammonia production plants such as the one depicted here. The condition of equilibrium in such reactions is the subject of this chapter.

Until now, we have stressed reactions that go to completion and the concepts of stoichiometry that allow us to calculate the outcomes of such reactions. We have made occasional references to situations involving both a forward and a reverse reaction—reversible reactions—but in this chapter, we will look at them in a detailed and systematic way.

Our emphasis will be on the equilibrium condition reached when forward and reverse reactions proceed at the same rate. Our main tool in dealing with equilibrium will be the equilibrium constant. We will begin with some key relationships involving equilibrium constants; then we will make qualitative predictions about the condition of equilibrium; and finally we will do various equilibrium calculations. As we will dis-

Dynamic Equilibrium animation

cover throughout the remainder of the text, the equilibrium condition plays a role in numerous natural phenomena and affects the methods used to produce many important industrial chemicals.

16-1   Dynamic Equilibrium

Let's begin by describing three simple physical phenomena that give a feeling of what happens in a system at **equilibrium**—two opposing processes take place at equal rates.

1. When a liquid vaporizes within a closed container, after a time, vapor molecules condense to the liquid state at the same rate at which liquid molecules vaporize. Even though molecules continue to pass back and forth between liquid and vapor (a *dynamic* process), the pressure exerted by the vapor remains constant with time. *The vapor pressure of a liquid is a property associated with an equilibrium condition.*

2. When a solute dissolves in a solvent, the system reaches a point at which the rate of dissolution is just matched by the rate at which dissolved solute crystallizes—that is, the solution is saturated. Even though solute particles continue to pass back and forth between the saturated solution and the undissolved solute, the concentration of dissolved solute remains constant. *The solubility of a solute is a property associated with an equilibrium condition.*

3. When an aqueous solution of iodine, I_2, is shaken with pure carbon tetrachloride, $CCl_4(l)$, I_2 molecules move into the CCl_4 layer. As the concentration of I_2 builds up in the CCl_4, the rate of return of I_2 to the water layer becomes significant. When I_2 molecules pass between the two liquids at equal rates—a condition of dynamic equilibrium—the concentration of I_2 in each layer remains constant. At this point, the concentration of I_2 in the CCl_4 is about 85 times greater than in the H_2O (Figure 16-1). The ratio of concentrations of a solute in two immiscible solvents is called the distribution coefficient. *The distribution coefficient of a solute between two immiscible solvents is a property associated with an equilibrium condition.*

(a)   (b)

▲ **FIGURE 16-1**
Dynamic equilibrium in a physical process
(a) A yellow-brown saturated solution of I_2 in water (top layer) is brought into contact with colorless $CCl_4(l)$ (bottom layer). **(b)** I_2 molecules distribute themselves between the H_2O and CCl_4. When equilibrium is reached, $[I_2]$ in the CCl_4 (violet, bottom layer) is about 85 times greater than in the water (colorless, top layer).

The properties in the three situations just described—vapor pressure, solubility, distribution coefficient—are all examples of a general quantity known as an *equilibrium constant*, the subject of the next section.

16-2   The Equilibrium Constant Expression

Methanol (methyl alcohol) is synthesized from a carbon monoxide–hydrogen mixture called synthesis gas. This reaction is likely to become increasingly important as methanol and its mixtures with gasoline find greater use as motor fuels. Methanol has a high octane rating, and its combustion produces much less air pollution than does gasoline.

Methanol synthesis is a *reversible* reaction, which means that at the same time $CH_3OH(g)$ is being formed,

$$CO(g) + 2 H_2(g) \longrightarrow CH_3OH(g) \qquad (16.1)$$

it decomposes in the reverse reaction

$$CH_3OH(g) \longrightarrow CO(g) + 2 H_2(g) \qquad (16.2)$$

Initially, only the forward reaction (16.1) occurs, but as soon as some CH_3OH forms, the reverse reaction (16.2) begins. With passing time, the forward reaction

▲ Methanol is actively being considered as an alternative to gasoline as a fuel.

Are You Wondering...

How we know that an equilibrium is dynamic—that forward and reverse reactions continue even after equilibrium is reached?

Suppose we have an equilibrium mixture of AgI(s) and its saturated aqueous solution.

$$AgI(s) \rightleftharpoons AgI(satd\ aq)$$

Now let's add to this mixture some saturated solution of AgI made from AgI containing radioactive iodine-131 as iodide ion, as illustrated in Figure 16-2. If both the forward and reverse processes stopped at equilibrium, radioactivity would be confined to the solution. What we find, though, is that radioactivity shows up in the solid in contact with the saturated solution. Over time, the radioactive "hot" spots distribute themselves throughout the solution and undissolved solid. The only way this can happen is if the dissolving of the solid solute and its crystallization from the saturated solution continue indefinitely. The equilibrium condition is *dynamic*.

Chemical Equilibrium animation

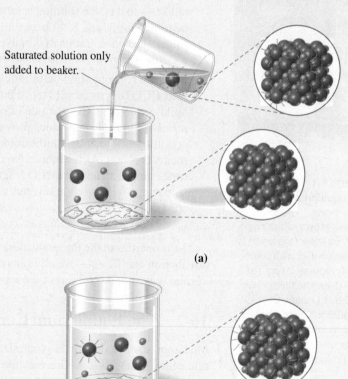

Saturated solution only added to beaker.

(a)

(b)

▲ **FIGURE 16-2   Dynamic equilibrium illustrated**
(a) A saturated solution of radioactive AgI is added to a saturated solution of AgI.
(b) The radioactive iodide ions distribute themselves throughout the solution and the solid AgI, showing that the equilibrium is dynamic.

TABLE 16.1   Three Approaches to Equilibrium in the Reaction[a]
$CO(g) + 2 H_2(g) \rightleftharpoons CH_3OH(g)$

	CO(g)	H₂(g)	CH₃OH(g)
Experiment 1			
Initial amounts, mol	1.000	1.000	0.000
Equilibrium amounts, mol	0.911	0.822	0.0892
Equilibrium concentrations, mol/L	0.0911	0.0822	0.00892
Experiment 2			
Initial amounts, mol	0.000	0.000	1.000
Equilibrium amounts, mol	0.753	1.506	0.247
Equilibrium concentrations, mol/L	0.0753	0.151	0.0247
Experiment 3			
Initial amounts, mol	1.000	1.000	1.000
Equilibrium amounts, mol	1.380	1.760	0.620
Equilibrium concentrations, mol/L	0.138	0.176	0.0620

The concentrations printed in blue are used in the calculations in Table 16.2.
[a] Reaction carried out in a 10.0-L flask at 483 K.

slows because of the decreasing concentrations of CO and H_2 and the reverse reaction speeds up as more CH_3OH accumulates. Eventually, the forward and reverse reactions proceed at equal rates, and the reaction mixture reaches a condition of dynamic equilibrium, which we can represent with a double arrow $\rightleftharpoons$.

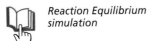
Reaction Equilibrium simulation

$$CO(g) + 2 H_2(g) \rightleftharpoons CH_3OH(g) \tag{16.3}$$

One consequence of the equilibrium condition is that the amounts of the reactants and products remain constant with time. These equilibrium amounts, however, depend on the quantities of reactants and products present initially. For example, Table 16.1 lists data for three hypothetical experiments. All three experiments are conducted in a 10.0-L flask at 483 K. In the first experiment, only CO and H_2 are present initially; in the second, only CH_3OH; and in the third, CO, H_2, and CH_3OH. The data from Table 16.1 are plotted in Figure 16-3, and from these graphs we see that

- in no case is any reacting species completely consumed
- the equilibrium amounts of reactants and products in these three cases appear to have nothing in common

Although it is not obvious from a cursory inspection of the data, a particular ratio involving equilibrium *concentrations* of product and reactants has a *constant* value, independent of how the equilibrium is reached. This ratio, which is central to the study of chemical equilibrium, can be derived theoretically, but it can also be established by trial and error. Three reasonable attempts at formulating the desired ratio for reaction (16.3) are outlined in Table 16.2, and the ratio that works is identified.

For the methanol synthesis reaction, the ratio of equilibrium concentrations shown in the following equation has a constant value of 14.5 at 483 K.

$$K_c = \frac{[CH_3OH]}{[CO][H_2]^2} = 14.5 \tag{16.4}$$

This ratio is called the **equilibrium constant expression**, and its numerical value is the **equilibrium constant**, denoted by the symbol K_c. The subscript "c" indicates that concentrations (expressed as molarities) are used.

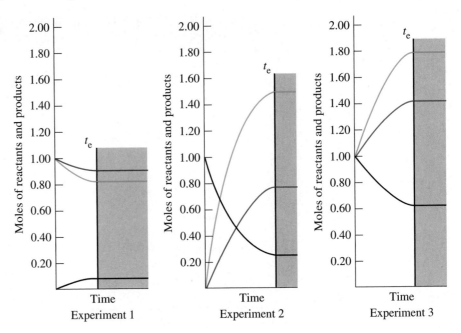

t_e = time for equilibrium to be reached
— mol CO
— mol H_2
— mol CH_3OH

▲ **FIGURE 16-3 Three approaches to equilibrium in the reaction**
$$CO(g) + 2\,H_2(g) \rightleftharpoons CH_3OH(g)$$
The initial and equilibrium amounts for each of these three cases are listed in Table 16.1.
t_e = time for equilibrium to be reached.

TABLE 16.2 Three Attempts to Find a Constant Ratio of Equilibrium Concentrations in the Reaction $CO + 2\,H_2 \rightleftharpoons CH_3OH$

Expt	Trial 1: $\dfrac{[CH_3OH]}{[CO][H_2]}$	Trial 2: $\dfrac{[CH_3OH]}{[CO](2 \times [H_2])}$	Trial 3: $\dfrac{[CH_3OH]}{[CO][H_2]^2}$
1	$\dfrac{0.00892}{0.0911 \times 0.0822} = 1.19$	$\dfrac{0.00892}{0.0911 \times (2 \times 0.0822)} = 0.596$	$\dfrac{0.00892}{0.0911 \times (0.0822)^2} = 14.5$
2	$\dfrac{0.0247}{0.0753 \times 0.151} = 2.17$	$\dfrac{0.0247}{0.0753 \times (2 \times 0.151)} = 1.09$	$\dfrac{0.0247}{0.0753 \times (0.151)^2} = 14.4$
3	$\dfrac{0.0620}{0.138 \times 0.176} = 2.55$	$\dfrac{0.0620}{0.138 \times (2 \times 0.176)} = 1.28$	$\dfrac{0.0620}{0.138 \times (0.176)^2} = 14.5$

Equilibrium concentration data are from Table 16.1. In Trial 1, the equilibrium concentration of CH_3OH is placed in the numerator and the product of the equilibrium concentrations, $[CO][H_2]$, in the denominator. In Trial 2, each concentration is multiplied by its stoichiometric coefficient. In Trial 3, each concentration is raised to a power equal to its stoichiometric coefficient. Trial 3 has essentially the same value for each experiment. This value is the equilibrium constant K_c.

What makes the equilibrium constant expression useful is that it allows us to calculate equilibrium concentrations of reactants and products. We will concentrate on this application in Section 16-7, but as a preview, consider Example 16-1.

KEEP IN MIND ▶
that any concentrations substituted into an equilibrium constant expression, or obtained from it, must be *equilibrium* concentrations.

EXAMPLE 16-1

Relating Equilibrium Concentrations of Reactants and Products. These equilibrium concentrations are measured in reaction (16.3) at 483 K: [CO] = 1.03 M and $[CH_3OH]$ = 1.56 M. What is the equilibrium concentration of H_2?

Solution

Substitute the known equilibrium concentrations into the equilibrium constant expression (16.4). Then solve for the unknown concentration, that is, $[H_2]$.

$$K_c = \frac{[CH_3OH]}{[CO][H_2]^2} = \frac{1.56}{1.03[H_2]^2} = 14.5$$

$$[H_2]^2 = \frac{1.56}{1.03 \times 14.5} = 0.104 \qquad [H_2] = \sqrt{0.104} = 0.322 \text{ M}$$

Practice Example A: In another experiment, equal concentrations of CH_3OH and CO are found at equilibrium in reaction (16.3). What must be the equilibrium concentration of H_2?

Practice Example B: At a certain temperature, $K_c = 1.8 \times 10^4$ for the reaction $N_2(g) + 3\,H_2(g) \rightleftharpoons 2\,NH_3(g)$. If the equilibrium concentrations of N_2 and NH_3 are 0.015 M and 2.00 M, respectively, what is the equilibrium concentration of H_2?

A General Expression for K_c

The equilibrium constant expression (16.4) for the methanol synthesis reaction is just a specific example of a more general case. For the hypothetical, generalized reaction

$$aA + bB \cdots \rightleftharpoons gG + hH \cdots$$

The equilibrium constant expression has the form

$$K_c = \frac{[G]^g[H]^h \cdots}{[A]^a[B]^b \cdots} \tag{16.5}$$

The *numerator* of an equilibrium constant expression is the product of the concentrations of the species on the *right* side of the equation ([G], [H], ...), and each concentration is raised to a power given by the stoichiometric coefficient ($g, h, \ldots$). The *denominator* is the product of the concentrations of the species on the *left* side of the equation ([A], [B], ...), and again, each concentration term is raised to a power given by the stoichiometric coefficient ($a, b, \ldots$).

The numerical value of an equilibrium constant, K_c, depends on the particular reaction and on the temperature. We will explore the significance of these numerical values in Section 16-4.

The Thermodynamic Equilibrium Constant, K_{eq}: A Preview

As we will learn in Chapter 20, the most fundamental description of equilibrium is provided by thermodynamics through the *thermodynamic equilibrium constant*, K_{eq}. The terms used in the thermodynamic equilibrium constant are dimensionless quantities known as *activities*. Activities are defined in such a way that, for solutes in aqueous solutions, numerical values of molarities (mol L^{-1}) can generally be substituted into K_c expressions. In the next section, when we describe equilibrium constant expressions in terms of gas pressures, we will substitute numerical values of pressures in atmospheres (atm) for the activities of gases. We will assign activities of 1 to pure solids and pure liquids. The main point to understand now is why we do not write a unit to accompany the numerical value of an equilibrium constant:

Are You Wondering...

If there is a relationship between the equilibrium constant and rate constants?

Given the requirement that the rates of the forward and reverse reactions become equal at equilibrium, it seems that there should be a relationship. The rate laws for the forward and reverse reactions include numerical constants (rate constants) and concentration terms raised to powers. Moreover, if we set the rate equations equal to one another at equilibrium, we should be able to derive an expression involving a ratio of concentration terms (raised to powers) and a ratio of constants. A major difficulty, however, is that the exponents of the concentration terms in the equilibrium constant expression must be the same as the coefficients in the balanced equation, whereas the exponents in rate laws are generally *not* the same as these coefficients. The trick to discovering the relationship between rate constants and an equilibrium constant is to work with the detailed mechanism for the reaction in the manner outlined in Exercise 91. Still, it is generally easier to obtain K_c directly from measurements on equilibrium conditions than to attempt a calculation based on rate constants. Moreover, in Chapters 20 and 21 we will learn of much more direct measurements and calculations leading to values of equilibrium constants.

We are anticipating the thermodynamic form of the equilibrium constant based on activities, and activities are dimensionless quantities.

16-3 Relationships Involving Equilibrium Constants

Before assessing an equilibrium situation, it may be necessary to make some preliminary calculations or decisions to get the appropriate equilibrium constant expression. This section presents some useful ideas in working with equilibrium constants.

Relationship of K_c to the Balanced Chemical Equation

We must always make certain that the expression for K_c matches the corresponding balanced equation. In doing so, the following hold true.

- When we *reverse* an equation, we *invert* the value of K_c.
- When we *multiply* the coefficients in a balanced equation by a common factor (2, 3, ...), we raise the equilibrium constant to the *corresponding power* (2, 3, ...).
- When we *divide* the coefficients in a balanced equation by a common factor (2, 3, ...), we take the *corresponding root* of the equilibrium constant (square root, cube root, ...).

Suppose that in discussing the synthesis of CH_3OH from CO and H_2 we had written the reverse of equation (16.3)—that is,

$$CH_3OH(g) \rightleftharpoons CO(g) + 2 H_2(g) \quad K'_c = ?$$

Now, according to the generalized equilibrium constant expression (16.5), we should write

$$K'_c = \frac{[CO][H_2]^2}{[CH_3OH]} = \frac{1}{\frac{[CH_3OH]}{[CO][H_2]^2}} = \frac{1}{K_c} = \frac{1}{14.5} = 0.0690$$

Are You Wondering...

How using activities makes the equilibrium constant dimensionless?

As we have indicated, equilibrium constants are made dimensionless when activities are used. For example, a_X, the activity of the component X in a solution, is the dimensionless ratio $[X]/c^0$, where c^0 corresponds to the concentration in a chosen reference state. With the reference state for solutions taken as 1 mol L^{-1}, the net effect is that the units cancel when concentrations are measured in mol L^{-1}. To illustrate, consider the reaction

$$a\text{A} + b\text{B} \cdots \rightleftharpoons g\text{G} + h\text{H} \cdots$$

The equilibrium constant expression in terms of activities has the form

$$K = \frac{(a_G)^g (a_H)^h \cdots}{(a_A)^a (a_B)^b \cdots}$$

Where, for example,

$$a_G = \frac{[\text{G}]}{c^0}$$

so that K becomes

$$K = \frac{\left(\dfrac{[\text{G}]}{c^0}\right)^g \left(\dfrac{[\text{H}]}{c^0}\right)^h \cdots}{\left(\dfrac{[\text{A}]}{c^0}\right)^a \left(\dfrac{[\text{B}]}{c^0}\right)^b \cdots}$$

With all the concentrations expressed in mol/L^{-1}, we have, using the value $c^0 = 1$ mol/L^{-1} for the reference state

$$K = \frac{\left(\dfrac{[\text{G}]\,\text{mol L}^{-1}}{1 \text{ mol L}^{-1}}\right)^g \left(\dfrac{[\text{H}]\,\text{mol L}^{-1}}{1 \text{ mol L}^{-1}}\right)^h \cdots}{\left(\dfrac{[\text{A}]\,\text{mol L}^{-1}}{1 \text{ mol L}^{-1}}\right)^a \left(\dfrac{[\text{B}]\,\text{mol L}^{-1}}{1 \text{ mol L}^{-1}}\right)^b \cdots}$$

Canceling the units mol L^{-1} gives

$$K = \frac{[\text{G}]^g [\text{H}]^h \cdots}{[\text{A}]^a [\text{B}]^b \cdots} = K_c$$

Thus, we have recovered the general expression for the equilibrium constant, except that the value of K_c is now seen to be dimensionless.

In the preceding expression, the terms printed in blue are the equilibrium constant expression and K_c value originally written as expression (16.4). We see that $K_c' = 1/K_c$.

Suppose that for a certain application we want an equation based on synthesizing *two* moles of $CH_3OH(g)$.

$$2\,CO(g) + 4\,H_2(g) \rightleftharpoons 2\,CH_3OH(g) \qquad K_c'' = ?$$

Here, $K_c'' = (K_c)^2$. That is,

$$K_c'' = \frac{[CH_3OH]^2}{[CO]^2 [H_2]^4} = \left(\frac{[CH_3OH]}{[CO][H_2]^2}\right)^2 = (K_c)^2 = (14.5)^2 = 2.10 \times 10^2$$

EXAMPLE 16-2

Relating K_c to the Balanced Chemical Equation. The following K_c value is given at 298 K for the synthesis of $NH_3(g)$ from its elements.

$$N_2(g) + 3\,H_2(g) \rightleftharpoons 2\,NH_3(g) \qquad K_c = 3.6 \times 10^8$$

What is the value of K_c at 298 K for the following reaction?

$$NH_3 \rightleftharpoons \frac{1}{2}N_2(g) + \frac{3}{2}H_2(g) \qquad K_c = ?$$

Solution

First, let's reverse the given equation. This will put $NH_3(g)$ on the left side of the equation, where we need it. The equilibrium constant K_c' becomes $1/(3.6 \times 10^8)$.

$$2\,NH_3(g) \rightleftharpoons N_2(g) + 3\,H_2(g) \qquad K_c' = 1/(3.6 \times 10^8) = 2.8 \times 10^{-9}$$

Then, to base the equation on 1 mol $NH_3(g)$, let's divide all coefficients by 2. This requires that we take the square root of K_c'.

$$NH_3(g) \rightleftharpoons \frac{1}{2}N_2(g) + \frac{3}{2}H_2(g) \qquad K_c = \sqrt{2.8 \times 10^{-9}} = 5.3 \times 10^{-5}$$

Practice Example A: Use data from Example 16-2 to determine the value of K_c at 298 K for the reaction

$$\frac{1}{3}N_2(g) + H_2(g) \rightleftharpoons \frac{2}{3}NH_3(g)$$

Practice Example B: For the reaction $NO(g) + \frac{1}{2}O_2(g) \rightleftharpoons NO_2(g)$ at 184 °C, $K_c = 7.5 \times 10^2$. What is the value of K_c at 184 °C for the reaction $2\,NO_2(g) \rightleftharpoons 2\,NO(g) + O_2(g)$?

Combining Equilibrium Constant Expressions

In Section 7-7, through Hess's law, we showed how to combine a series of equations into a single overall equation. To obtain the enthalpy change of the overall reaction, we added together the enthalpy changes of the individual reactions. We can use a similar procedure when working with equilibrium constants, but with this important difference:

> When individual equations are combined (that is, added), their equilibrium constants are *multiplied* to obtain the equilibrium constant for the overall reaction.

Suppose we seek the equilibrium constant for the reaction

$$N_2O(g) + \frac{1}{2}O_2(g) \rightleftharpoons 2\,NO(g) \qquad K_c = ? \tag{16.6}$$

and know the K_c values of these two equilibria.

$$N_2(g) + \frac{1}{2}O_2(g) \rightleftharpoons N_2O(g) \qquad K_c = 2.7 \times 10^{-18} \tag{16.7}$$

$$N_2(g) + O_2(g) \rightleftharpoons 2\,NO(g) \qquad K_c = 4.7 \times 10^{-31} \tag{16.8}$$

We can get equation (16.6) by reversing equation (16.7) and adding it to (16.8). When we do this, we must also take the *reciprocal* of the K_c value of equation (16.7).

(a) $\qquad\qquad\qquad N_2O(g) \rightleftharpoons N_2(g) + \dfrac{1}{2} O_2(g) \quad K_c(a) = 1/(2.7 \times 10^{-18})$

$\qquad\qquad\qquad\qquad\qquad\qquad\qquad\qquad\qquad\qquad = 3.7 \times 10^{17}$

(b) $\qquad\qquad N_2(g) + O_2(g) \rightleftharpoons 2\,NO(g) \qquad\qquad K_c(b) = 4.7 \times 10^{-31}$

Overall: $\quad N_2O(g) + \dfrac{1}{2} O_2(g) \rightleftharpoons 2\,NO(g) \qquad\qquad K_c(\text{overall}) = ? \qquad\qquad$ (16.6)

According to the general expression (16.5),

$$K_c(\text{overall}) = \frac{[NO]^2}{[N_2O][O_2]^{1/2}} = \underbrace{\frac{[N_2][O_2]^{1/2}}{[N_2O]}}_{K_c(a)} \times \underbrace{\frac{[NO]^2}{[N_2][O_2]}}_{K_c(b)} = K_c(a) \times K_c(b)$$

$$= 3.7 \times 10^{17} \times 4.7 \times 10^{-31} = 1.7 \times 10^{-13}$$

Equilibria Involving Gases: The Equilibrium Constant, K_p

Mixtures of gases are as much solutions as are mixtures with a liquid solvent. Thus, concentrations in a gaseous mixture can be expressed on a mole-per-liter basis, and this is what we do when we write and use K_c expressions. Yet in our preview of the thermodynamic equilibrium constant expression on page 631, we indicated a need to describe gases through their partial pressures in atmospheres. We do this when we write an equilibrium constant expression, K_p, based on the *partial pressures* of gaseous reactants and products.

To derive this alternative form of an equilibrium constant, let's consider the formation of $SO_3(g)$, a key step in the manufacture of sulfuric acid.

$$2\,SO_2(g) + O_2(g) \rightleftharpoons 2\,SO_3(g) \tag{16.9}$$

$$K_c = \frac{[SO_3]^2}{[SO_2]^2[O_2]} = 2.8 \times 10^2 \text{ (at 1000 K)}$$

Now, let's use the ideal gas law, $PV = nRT$, to write

$$[SO_3] = \frac{n_{SO_3}}{V} = \frac{P_{SO_3}}{RT} \qquad [SO_2] = \frac{n_{SO_2}}{V} = \frac{P_{SO_2}}{RT} \qquad [O_2] = \frac{n_{O_2}}{V} = \frac{P_{O_2}}{RT}$$

Next, substitute these terms for the concentration terms in K_c.

$$K_c = \frac{\left(P_{SO_3}/RT\right)^2}{\left(P_{SO_2}/RT\right)^2\left(P_{O_2}/RT\right)} = \frac{\left(P_{SO_3}\right)^2}{\left(P_{SO_2}\right)^2\left(P_{O_2}\right)} \times RT \tag{16.10}$$

The ratio of equilibrium partial pressures, expressed in atmospheres and shown in blue, is the equilibrium constant, K_p. The relationship between K_p and K_c for reaction (16.9) is

$$K_c = K_p \times RT \qquad \text{or} \qquad K_p = \frac{K_c}{RT} = K_c(RT)^{-1} \tag{16.11}$$

If we carried out a similar derivation for the general reaction,

$$aA(g) + bB(g) + \cdots \rightleftharpoons gG(g) + hH(g) + \cdots$$

this is the result we would get.

$$K_p = K_c(RT)^{\Delta n_{\text{gas}}} \tag{16.12}$$

KEEP IN MIND ▶
that $K_p = K_c$ only if $\Delta n_{gas} = 0$.
That is, $K_p = K_c(RT)^0 = K_c$ because any number raised to the
zero power equals one.

where Δn_{gas} is the difference in the stoichiometric coefficients of *gaseous* products and reactants; that is, $\Delta n_{gas} = (g + h + \cdots) - (a + b + \cdots)$. In reaction (16.9), $\Delta n_{gas} = 2 - (2 + 1) = -1$, just as we found in equation (16.11).

Although we do not use units for the equilibrium constants, we do need to use the correct dimensionless values of concentrations or pressures for terms within equilibrium constant expressions. These are *molarity* for K_c expressions and pressures in *atmospheres* for K_p expressions. These choices then require that we use a value of $R = 0.08206$ L atm mol^{-1} K^{-1} in equation (16.12).

EXAMPLE 16-3

Relating K_c and K_p. Complete the calculation of K_p for reaction (16.9) from the data given.

Solution

$$K_p = K_c(RT)^{-1} = 2.8 \times 10^2(0.08206 \times 1000)^{-1} = \frac{2.8 \times 10^2}{0.08206 \times 1000}$$

$$= 3.4$$

Practice Example A: For the reaction $2\,NH_3(g) \rightleftharpoons N_2(g) + 3\,H_2(g)$ at 298 K, $K_c = 2.8 \times 10^{-9}$. What is the value of K_p for this reaction?

Practice Example B: At 1065 °C, for the reaction $2\,H_2S(g) \rightleftharpoons 2\,H_2(g) + S_2(g)$, $K_p = 1.2 \times 10^{-2}$. What is the value of K_c for the reaction $H_2(g) + \frac{1}{2}S_2(g) \rightleftharpoons H_2S(g)$ at 1065 °C?

Equilibria Involving Pure Liquids and Solids

Up to this point in the chapter, all our examples have involved gaseous reactions. In subsequent chapters, we will emphasize equilibria in aqueous solutions. Gas-phase reactions and reactions in aqueous solution are *homogeneous* reactions: They occur within a single phase. Let's extend our coverage now to include reactions involving one or more condensed phases—solids and liquids—in contact with a gas or solution phase. These are called *heterogeneous* reactions. One of the most important ideas about heterogeneous reactions is that

Equilibrium constant expressions do not *contain concentration terms for solid or liquid phases of a single component (that is, pure solids and liquids).*

▶ Still another way to think about solids and liquids is through their densities. Density, the mass per unit volume of a substance, can be expressed in moles per liter by converting the unit volume from milliliter to liter and dividing the mass in grams by the molar mass. The resultant molar density (mol/L) is a concentration term and, at a fixed temperature, is a constant that would be incorporated in the K_c value.

We can think about this statement in either of two ways: (1) An equilibrium constant expression includes terms only for reactants and products whose concentrations and/or partial pressures can *change* during a chemical reaction. The concentration of the single component within a pure solid or liquid phase *cannot change*. (2) Alternatively, if we use the idea of activities, the activities of pure liquids and solids are set equal to 1; thus the effect on the numerical value of the thermodynamic equilibrium constant is the same as not including terms for pure solids and liquids at all.

The water-gas reaction, used to make combustible gases from coal (see Section 8-6), has reacting species in both gaseous and solid phases.

$$C(s) + H_2O(g) \rightleftharpoons CO(g) + H_2(g)$$

Although solid carbon must be present for the reaction to occur, the equilibrium constant expression contains terms only for the species in the homogeneous gas phase: H_2O, CO, and H_2.

$$K_c = \frac{[CO][H_2]}{[H_2O]}$$

The decomposition of calcium carbonate (limestone) is also a heterogeneous reaction. The equilibrium constant expression contains just a single term.

$$CaCO_3(s) \rightleftharpoons CaO(s) + CO_2(g) \qquad K_c = [CO_2] \qquad (16.13)$$

We can write K_p for reaction (16.13) by using equation (16.12), with $\Delta n_{gas} = 1$.

$$K_p = P_{CO_2} \quad \text{and} \quad K_p = K_c(RT) \qquad (16.14)$$

Equation (16.14) indicates that the equilibrium pressure of $CO_2(g)$ in contact with $CaCO_3(s)$ and $CaO(s)$ is a constant equal to K_p. Its value is *independent* of the quantities of $CaCO_3$ and CaO (as long as both solids are present). Figure 16-4 offers a conceptualization of this decomposition reaction.

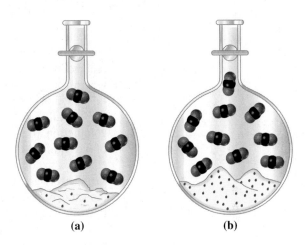

▶ **FIGURE 16-4**
Equilibrium in the reaction
$CaCO_3(s) \rightleftharpoons CaO(s) + CO_2(g)$
(a) Decomposition of $CaCO_3(s)$ upon heating in a closed vessel yields a few granules of $CaO(s)$, together with $CO_2(g)$, which soon exerts its equilibrium partial pressure. **(b)** Introduction of additional $CaCO_3(s)$ and/or more $CaO(s)$ has no effect on the partial pressure of the $CO_2(g)$, which remains the same as in (a).

(a)   (b)

One of our examples in Section 16-1 was liquid–vapor equilibrium. This is a physical equilibrium because no chemical reactions are involved. Consider the liquid–vapor equilibrium for water.

$$H_2O(l) \rightleftharpoons H_2O(g)$$

$$K_c = [H_2O(g)] \qquad K_p = P_{H_2O} \qquad K_p = K_c(RT)$$

So, equilibrium vapor pressures such as P_{H_2O} are just values of K_p. As we have seen before, these values do *not* depend on the quantities of liquid or vapor at equilibrium, as long as some of each is present.

EXAMPLE 16-4

Writing Equilibrium Constant Expressions for Reactions Involving Pure Solids or Liquids. At equilibrium in the following reaction at 60 °C, the partial pressures of the gases are found to be $P_{HI} = 3.65 \times 10^{-3}$ atm and $P_{H_2S} = 9.96 \times 10^{-1}$ atm. What is the value of K_p for the reaction?

$$H_2S(g) + I_2(s) \rightleftharpoons 2\,HI(g) + S(s) \qquad K_p = ?$$

Solution

Recall that terms for pure solids do *not* appear in an equilibrium constant expression. The value of K_p is

$$K_p = \frac{(P_{HI})^2}{(P_{H_2S})} = \frac{(3.65 \times 10^{-3})^2}{9.96 \times 10^{-1}} = 1.34 \times 10^{-5}$$

Practice Example A: Teeth are made principally from the mineral hydroxyapatite, $Ca_5(PO_4)_3OH$, which can be dissolved in acidic solution such as that produced by bacteria in the mouth. The reaction is $Ca_5(PO_4)_3OH(s) + 4\ H^+(aq) \rightleftharpoons 5\ Ca^{2+}(aq) + 3\ HPO_4^{2-}(aq) + H_2O(l)$. Write the equilibrium constant expression K_c for this reaction.

Practice Example B: The steam–iron process is used to generate $H_2(g)$, mostly for use in hydrogenating oils. Iron metal and steam [$H_2O(g)$] react to produce $Fe_3O_4(s)$ and $H_2(g)$. Write expressions for K_c and K_p for this reversible reaction. How are the values of K_c and K_p related to each other? Explain.

16-4 The Significance of the Magnitude of an Equilibrium Constant

In principle, every chemical reaction has an equilibrium constant, but often the constants are not used. Why is this so? Table 16.3 lists equilibrium constants for several reactions mentioned in this chapter or previously in the text. The first of these reactions is the synthesis of H_2O from its elements. We have always assumed that this reaction goes to completion, that is, that the reverse reaction is negligible and the overall reaction proceeds only in the forward direction. If a reaction goes to completion, one (or more) of the reactants is used up. A term in the *denominator* of the equilibrium constant expression approaches *zero* and makes the value of the equilibrium constant very large. A very large numerical value of K_c or K_p signifies that the forward reaction, as written, *goes to completion* or very nearly so. Because the value of K_p for the water synthesis reaction is 1.4×10^{83}, we are entirely justified in saying that the reaction goes to completion at 298 K.

If the equilibrium constant is so large, why is a mixture of hydrogen and oxygen gases stable at room temperature? The value of the equilibrium constant relates to thermodynamic stability: $H_2O(l)$ is much more thermodynamically stable than a mixture of $H_2(g)$ and $O_2(g)$ because it lies at a lower energy state. As noted in Chapter 15, however, the rate of a chemical reaction is strongly governed by the activation energy, E_a. Because E_a is very high for the synthesis of $H_2O(l)$ from $H_2(g)$ and $O_2(g)$, the rate of reaction is inconsequential at 298 K. To get the reaction to occur at a measurable rate, we must either raise the temperature or use a catalyst. A chemist would say that the synthesis of $H_2O(l)$ at 298 K is a *kinetically* controlled reaction (as opposed to *thermodynamically* controlled).

 Equilibrium Constant activity

TABLE 16.3 Equilibrium Constants of Some Common Reactions	
Reaction	**Equilibrium constant, K_p**
$2\ H_2(g) + O_2(g) \rightleftharpoons 2\ H_2O(l)$	1.4×10^{83} at 298 K
$CaCO_3(s) \rightleftharpoons CaO(s) + CO_2(g)$	1.9×10^{-23} at 298 K
	1.0 at about 1200 K
$2\ SO_2(g) + O_2(g) \rightleftharpoons 2\ SO_3(g)$	3.4 at 1000 K
$C(s) + H_2O(g) \rightleftharpoons CO(g) + H_2(g)$	1.6×10^{-21} at 298 K
	10.0 at about 1100 K

From Table 16.3, we see that K_p for the decomposition of $CaCO_3(s)$ (limestone) is very small at 298 K (only 1.9×10^{-23}). To account for a very small numerical value of an equilibrium constant, the *numerator* must be very small (approaching zero). A very small numerical value of K_c or K_p signifies that the forward reaction, as written, *does not occur to any significant extent*. Although limestone does not decompose at ordinary temperatures, the partial pressure of $CO_2(g)$ in equilibrium with $CaCO_3(s)$ and $CaO(s)$ increases with temperature. It becomes 1 atm at about 1200 K. An important application of this decomposition reaction is in the commercial production of quicklime (CaO). The conversion of $SO_2(g)$ and $O_2(g)$ to $SO_3(g)$ at 1000 K has an equilibrium constant such that we expect significant amounts of both reactants and products to be present at equilibrium (see Table 16.3). Both the forward and reverse reactions are important. A similar situation exists for the reaction of C(s) and $H_2O(g)$ at 1100 K, but not at 298 K where the forward reaction does not occur to any significant extent $\left(K_p = 1.6 \times 10^{-21}\right)$. In conclusion,

A reaction is most likely to reach a state of equilibrium in which significant quantities of both reactants and products are present if the numerical value of K_c or K_p is *neither very large nor very small*, roughly in the range of about 10^{-10} to 10^{10}.

Thus, we see that equilibrium calculations are not required for all reactions. At times, we can use simple stoichiometric calculations to determine the outcome of a reaction, and in some cases there may be no reaction at all.

16-5 The Reaction Quotient, *Q*: Predicting the Direction of Net Change

Let's return briefly to the set of three experiments that we discussed in Section 16-2, involving the reaction

$$CO(g) + 2\,H_2(g) \rightleftharpoons CH_3OH(g) \qquad K_c = 14.5$$

In Experiment 1, we start with just the reactants CO and H_2. An overall, or net, change has to occur in which some CH_3OH forms. Only in this way can an equilibrium condition be reached in which all reacting species are present. We say that a net change occurs in the *forward* direction (*to the right*).

In Experiment 2 ,we start with just the product, CH_3OH. Here, some of the CH_3OH must decompose back to CO and H_2 before equilibrium can be established. We say that a net change occurs in the *reverse* direction (*to the left*).

In Experiment 3, all the reacting species are present initially—CO, H_2, and CH_3OH. In this system, it is not obvious in what direction a net change occurs to establish equilibrium.

The ability to predict the direction of net change in establishing equilibrium is important to us for two reasons.

- At times we do not need detailed equilibrium calculations. We may need only a qualitative description of the changes that occur in establishing equilibrium from a given set of initial conditions.

- In some equilibrium calculations, it is helpful to determine the direction of net change as a first step.

For any set of *initial* concentrations in a reaction mixture, we can set up a ratio of concentrations having the same form as the equilibrium constant expression. This ratio is called the **reaction quotient** and is designated Q_c. For a hypothetical generalized reaction, the reaction quotient is

$$Q_c = \frac{[G]_{\text{init}}^g [H]_{\text{init}}^h \cdots}{[A]_{\text{init}}^a [B]_{\text{init}}^b \cdots} \qquad (16.15)$$

If $Q_c = K_c$, *a reaction is at equilibrium*, but our primary interest in the relationship between Q_c and K_c is for a reaction mixture that is *not* at equilibrium. To see what this relationship is, let's turn again to the experiments in Table 16.1.

In Experiment 1, the initial concentrations of CO and H_2 are $1.000 \text{ mol}/10.0 \text{ L} = 0.100 \text{ M}$. Initially there is *no* CH_3OH. The value of Q_c is

$$Q_c = \frac{[CH_3OH]_{\text{init}}}{[CO]_{\text{init}}[H_2]_{\text{init}}^2} = \frac{0}{(0.100)(0.100)^2} = 0 \qquad (16.16)$$

We know that a net reaction occurs *to the right*, producing some CH_3OH. As it does, the numerator in expression (16.16) increases, the denominator decreases, the value of Q_c *increases*, and eventually $Q_c = K_c$.

If $Q_c < K_c$, a net change occurs from left to right (the direction of the forward reaction).

In Experiment 2, the initial concentration of CH_3OH is $1.000 \text{ mol}/10.0 \text{ L} = 0.100 \text{ M}$. Initially, there is *no* CO or H_2. The value of Q_c is

$$Q_c = \frac{[CH_3OH]_{\text{init}}}{[CO]_{\text{init}}[H_2]_{\text{init}}^2} = \frac{0.100}{0 \times 0} = \infty \qquad (16.17)$$

We know that a net reaction occurs *to the left*, producing some CO and H_2. As it does, the numerator in expression (16.17) decreases, the denominator increases, the value of Q_c *decreases*, and eventually $Q_c = K_c$.

If $Q_c > K_c$, a net change occurs from right to left (the direction of the reverse reaction).

Now let us turn to a case for which we really do need a criterion for the direction of net change. In Experiment 3, the initial concentrations of all three species are $1.000 \text{ mol}/10.0 \text{ L} = 0.100 \text{ M}$. The value of Q_c is

$$Q_c = \frac{[CH_3OH]_{\text{init}}}{[CO]_{\text{init}}[H_2]_{\text{init}}^2} = \frac{0.100}{(0.100)(0.100)^2} = 100$$

Because $Q_c > K_c$ (100 compared with 14.5), a net change occurs in the *reverse direction*. Note that you can verify this conclusion from Figure 16-3. The amounts of CO and H_2 at equilibrium are greater than they were initially, and the amount of CH_3OH is less.

The criteria for predicting the direction of a net chemical change in a reversible reaction are summarized in Figure 16-5 and applied in Example 16-5.

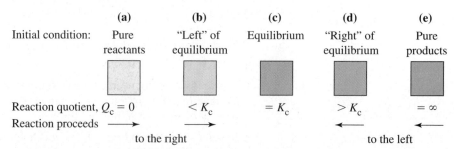

	(a)	(b)	(c)	(d)	(e)
Initial condition:	Pure reactants	"Left" of equilibrium	Equilibrium	"Right" of equilibrium	Pure products
Reaction quotient,	$Q_c = 0$	$< K_c$	$= K_c$	$> K_c$	$= \infty$
Reaction proceeds	→	→		←	←
		to the right			to the left

▲ **FIGURE 16-5** **Predicting the direction of net change in a reversible reaction**
Five possibilities for the relationship of initial and equilibrium conditions are shown. From Table 16.1 and Figure 16-3, Experiment 1 corresponds to initial condition (a); Experiment 2 to condition (e); and Experiment 3 to (d). The situation in Example 16-5 also corresponds to condition (d).

EXAMPLE 16-5

Predicting the Direction of a Net Chemical Change in Establishing Equilibrium. To increase the yield of $H_2(g)$ in the water-gas reaction—the reaction of $C(s)$ and $H_2O(g)$ to form $CO(g)$ and $H_2(g)$—a follow-up reaction called the "water-gas shift reaction" is generally used. In this reaction, some of the $CO(g)$ of the water gas is replaced by $H_2(g)$.

$$CO(g) + H_2O(g) \rightleftharpoons CO_2(g) + H_2(g)$$

$K_c = 1.00$ at about 1100 K. The following amounts of substances are brought together and allowed to react at this temperature: 1.00 mol CO, 1.00 mol H_2O, 2.00 mol CO_2, and 2.00 mol H_2. Compared with their initial amounts, which of the substances will be present in a greater amount and which in a lesser amount when equilibrium is established?

Solution

Our task is to determine the direction of net change. This means evaluating Q_c. To substitute concentrations into the expression for Q_c, we can assume an arbitrary volume V. Its value is immaterial because the volume cancels out in this case.

$$Q_c = \frac{[CO_2][H_2]}{[CO][H_2O]} = \frac{(2.00/V)(2.00/V)}{(1.00/V)(1.00/V)} = 4.00$$

Because $Q_c > K_c$ (that is, $4.00 > 1.00$), a net change occurs to the *left*. When equilibrium is established, the amounts of CO and H_2O will be *greater* than the initial quantities and the amounts of CO_2 and H_2 will be *less*.

Practice Example A: For the reaction $PCl_5(g) \rightleftharpoons PCl_3(g) + Cl_2(g)$, $K_c = 0.0454$ at 261 °C. If a vessel is filled with these gases such that the initial concentrations are $[PCl_3] = 0.50$ M, $[Cl_2] = 0.20$ M, and $[PCl_5] = 4.50$ M, in which direction will a net change occur?

Practice Example B: In Example 16-5, equal masses of CO, H_2O, CO_2, and H_2 are mixed at a temperature of about 1100 K. When equilibrium is established, which substance(s) will show an increase in quantity and which will show a decrease compared with the initial quantities?

KEEP IN MIND ▶

that volume terms cancel in a reaction quotient or equilibrium constant expression whenever the sum of the exponents in the numerator equals that in the denominator. This can simplify problem solving in some instances.

16-6 Altering Equilibrium Conditions: Le Châtelier's Principle

At times, we want only to make qualitative statements about a reversible reaction: the direction of a net change, whether the amount of a substance will have increased or decreased when equilibrium is reached, and so on. Also, we may not have the data needed for a quantitative calculation. In these cases, we can use a statement attributed to the French chemist Henri Le Châtelier (1884). **Le Châtelier's principle** is hard to state unambiguously, but its essential meaning is stated here.

When an equilibrium system is subjected to a change in temperature, pressure, or concentration of a reacting species, the system responds by attaining a new equilibrium that *partially* offsets the impact of the change.

As we will see in the examples that follow, it is generally not difficult to predict the outcome of changing one or more variables in a system at equilibrium.

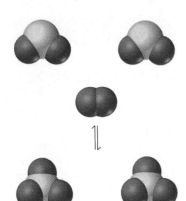

▲ **2 SO₂(g) + O₂(g) ⇌
2 SO₃(g)**

*Le Châtelier's Principle
simulation*

Effect of Changing the Amounts of Reacting Species on Equilibrium

Let's return to the reaction

$$2\,SO_2(g) + O_2(g) \rightleftharpoons 2\,SO_3(g) \qquad K_c = 2.8 \times 10^2 \text{ at } 1000\text{ K} \qquad (16.9)$$

Suppose we start with certain equilibrium amounts of SO_2, O_2, and SO_3, as suggested by Figure 16-6a. Now let's create a disturbance in the equilibrium mixture by forcing an additional 1.00 mol SO_3 into the 10.0-L flask (Figure 16-6b). How will the amounts of the reacting species change to reestablish equilibrium?

According to Le Châtelier's principle, if the system is to partially offset an action that increases the equilibrium concentration of one of the reacting species, it must do so by favoring the reaction in which that species is consumed. This is the *reverse* reaction—conversion of some of the added SO_3 to SO_2 and O_2. In the new equilibrium, there will be greater amounts of all the substances than in the original equilibrium, although, of course, the additional amount of SO_3 will be less than the 1.00 mol that was added.

Another way to look at the matter is to evaluate the reaction quotient immediately after adding the SO_3.

Original equilibrium   *Following disturbance*

$$Q_c = \frac{[SO_3]^2}{[SO_2]^2[O_2]} = K_c \qquad Q_c = \frac{[SO_3]^2}{[SO_2]^2[O_2]} > K_c$$

Adding any quantity of SO_3 to a constant-volume equilibrium mixture makes Q_c larger than K_c. A net change occurs in the direction that reduces $[SO_3]$, that is, to the left, or in the *reverse* direction. Notice that reaction in the reverse direction increases $[SO_2]$ and $[O_2]$, further decreasing the value of Q_c.

*Le Châtelier's Principle
movie*

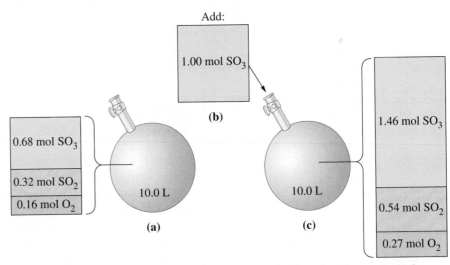

▲ **FIGURE 16-6** **Changing equilibrium conditions by changing the amount of a reactant**

$$2\,SO_2(g) + O_2(g) \rightleftharpoons 2\,SO_3(g),\ K_c = 2.8 \times 10^2 \text{ at } 1000\text{ K}$$

(a) The original equilibrium condition. **(b)** Disturbance caused by adding 1.00 mol SO_3. **(c)** The new equilibrium condition. The amount of SO_3 in the new equilibrium mixture, 1.46 mol, is greater than the original 0.68 mol but it is not as great as immediately after the addition of 1.00 mol SO_3. The effect of adding SO_3 to an equilibrium mixture is *partially* offset when equilibrium is restored.

EXAMPLE 16-6

Applying Le Châtelier's Principle: Effect of Adding More of a Reactant to an Equilibrium Mixture. Predict the effect of adding more $H_2(g)$ to a constant-volume equilibrium mixture of N_2, H_2, and NH_3.

$$N_2(g) + 3\,H_2(g) \rightleftharpoons 2\,NH_3(g)$$

Solution

The action of increasing $[H_2]$ stimulates the forward reaction and a shift of the equilibrium condition to the right. However, only a portion of the added H_2 is consumed in this reaction. When equilibrium is reestablished, there will be more H_2 than was present originally. The amount of NH_3 will also be greater, but the amount of N_2 will be *smaller*. Some of the original N_2 must be consumed in converting some of the added H_2 to NH_3.

Practice Example A: Given the reaction $2\,CO(g) + O_2(g) \rightleftharpoons 2\,CO_2(g)$, what is the effect of adding $O_2(g)$ to a constant-volume equilibrium mixture?

Practice Example B: Calcination of limestone (decomposition by heating), $CaCO_3(s) \rightleftharpoons CaO(s) + CO_2(g)$, is the commercial source of quicklime, $CaO(s)$. After this equilibrium has been established in a constant-temperature, constant-volume container, what is the effect on the equilibrium amounts of materials caused by *adding* some **(a)** $CaO(s)$; **(b)** $CO_2(g)$; **(c)** $CaCO_3(s)$?

Effect of Changes in Pressure or Volume on Equilibrium

There are three ways we can change the pressure of a constant-temperature equilibrium mixture.

1. **Add or remove a gaseous reactant or product.** The effect of these actions on the equilibrium condition is simply that due to adding or removing a reaction component, as described previously.

2. **Add an inert gas to the constant-volume reaction mixture.** This has the effect of increasing the *total* pressure, but the partial pressures of the reacting species are all unchanged. An inert gas added to a constant-volume equilibrium mixture has no effect on the equilibrium condition.

3. **Change the pressure by changing the volume of the system.** Decreasing the volume of the system increases the pressure, and increasing the system volume decreases the pressure. Thus, the effect of this type of pressure change is simply that of a volume change.

Let's explore the third situation first. Consider, again, the formation of $SO_3(g)$ from $SO_2(g)$ and $O_2(g)$.

$$2\,SO_2(g) + O_2(g) \rightleftharpoons 2\,SO_3(g) \qquad K_c = 2.8 \times 10^2 \text{ at 1000 K} \qquad (16.9)$$

The equilibrium mixture in Figure 16-7a has its volume reduced to one-tenth of its original value by increasing the external pressure. To see how the equilibrium amounts of the gases change, let's first rearrange the equilibrium constant expression to the form

$$K_c = \frac{[SO_3]^2}{[SO_2]^2[O_2]} = \frac{(n_{SO_3}/V)^2}{(n_{SO_2}/V)^2(n_{O_2}/V)} = \frac{(n_{SO_3})^2}{(n_{SO_2})^2(n_{O_2})} \times V = 2.8 \times 10^2 \quad (16.18)$$

From equation (16.18), we see that if V is *reduced* by a factor of 10, the ratio

$$\frac{(n_{SO_3})^2}{(n_{SO_2})^2(n_{O_2})}$$

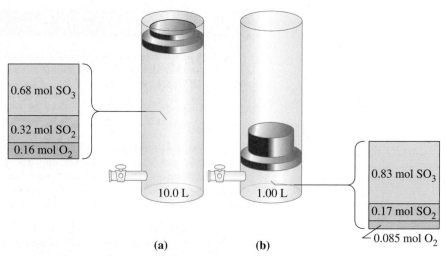

(a)   (b)

▲ **FIGURE 16-7**   **Effect of pressure change on equilibrium in the reaction**
$$2\ SO_2(g)\ +\ O_2(g)\ \rightleftharpoons\ 2\ SO_3(g)$$
An increase in external pressure causes a decrease in the reaction volume and a shift in equilibrium "to the right." (See Exercise 73 for a calculation of the new equilibrium amounts.)

must *increase* by a factor of 10. In this way, the value of K_c is restored, as it must be to restore equilibrium. There is only one way in which the ratio of moles of gases will increase in value: The number of moles of SO_3 must increase, and the numbers of moles of SO_2 and O_2 must decrease. The equilibrium shifts in the direction producing more SO_3—to the right.

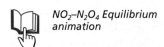

NO₂–N₂O₄ Equilibrium animation

Notice that three moles of *gas* on the left produce two moles of *gas* on the right in reaction (16.9). When compared at the same temperature and pressure, two moles of $SO_3(g)$ occupies a smaller volume than does a mixture of two moles of $SO_2(g)$ and one mole of $O_2(g)$. Given this fact and the observation from equation (16.18) that a decrease in volume favors the production of additional SO_3, we can formulate a statement that is especially easy to apply.

> When the volume of an equilibrium mixture of gases is *reduced*, a net change occurs in the direction that produces *fewer moles of gas*. When the volume is *increased*, a net change occurs in the direction that produces more *moles of gas*.

Figure 16-7 suggests a way of *decreasing* the volume of gaseous mixture at equilibrium—by increasing the external pressure. One way to *increase* the volume is to lower the external pressure. Another way is to transfer the equilibrium mixture from its original container to one of larger volume. A third method is to add an inert gas at *constant pressure*; the volume of the mixture must increase to make room for the added gas. The effect on the equilibrium, however, is the same for all three methods: Equilibrium shifts in the direction of the reaction producing the greater number of moles of gas.

KEEP IN MIND ▶
that an inert gas has no effect on an equilibrium condition if the gas is added to a system maintained at constant volume, but it can have an effect if added at constant pressure.

Equilibria between condensed phases are not affected much by changes in external pressure because solids and liquids are not easily compressible. Also, we cannot assess whether the forward or reverse reaction is favored by these changes by examining only the chemical equation.

EXAMPLE 16-7

Applying Le Châtelier's Principle: The Effect of Changing Volume. An equilibrium mixture of $N_2(g)$, $H_2(g)$, and $NH_3(g)$ is transferred from a 1.50-L flask to a 5.00-L flask. In which direction does a net change occur to restore equilibrium?

$$N_2(g) + 3 H_2(g) \rightleftharpoons 2 NH_3(g)$$

Solution

When the gaseous mixture is transferred to the larger flask, the partial pressure of each gas and the total pressure drop. Whether we think in terms of a decrease in pressure or an increase in volume, we reach the same conclusion. Equilibrium shifts in such a way as to produce a larger number of moles of gas. Some of the NH_3 originally present decomposes back to N_2 and H_2. A net change occurs in the direction of the reverse reaction—to the left—in restoring equilibrium.

Practice Example A: The reaction $N_2O_4(g) \rightleftharpoons 2 NO_2(g)$ is at equilibrium in a 3.00-L cylinder. What would be the effect on the concentrations of $N_2O_4(g)$ and $NO_2(g)$ if the pressure were doubled (cylinder volume decreased to 1.50 L)?

Practice Example B: How is the equilibrium amount of $H_2(g)$ produced in the water-gas shift reaction affected by changing the total gas pressure or the system volume? Explain.

$$CO(g) + H_2O(g) \rightleftharpoons CO_2(g) + H_2(g)$$

Effect of Temperature on Equilibrium

We can think of changing the temperature of an equilibrium mixture in terms of adding heat (raising the temperature) or removing heat (lowering the temperature). According to Le Châtelier's principle, adding heat favors the reaction in which heat is absorbed (*endothermic* reaction). Removing heat favors the reaction in which heat is evolved (*exothermic* reaction). Stated in terms of changing temperature,

> *Raising the temperature* of an equilibrium mixture shifts the equilibrium condition in the direction of the *endothermic* reaction. *Lowering the temperature* causes a shift in the direction of the *exothermic* reaction.

The principal effect of temperature on equilibrium is in changing the value of the equilibrium constant. In Chapter 20, we will learn how to calculate equilibrium constants as a function of temperature. For now, we will limit ourselves to making qualitative predictions.

EXAMPLE 16-8

Applying Le Châtelier's Principle: Effect of Temperature on Equilibrium. Consider the reaction

$$2 SO_2(g) + O_2(g) \rightleftharpoons 2 SO_3(g) \qquad \Delta H° = -197.8 \text{ kJ}$$

Will the amount of $SO_3(g)$ formed from given amounts of $SO_2(g)$ and $O_2(g)$ be greater at high or low temperatures?

Solution

Raising the temperature favors the endothermic reaction, the *reverse* reaction. Lowering the temperature favors the forward (exothermic) reaction. Therefore, an equilibrium mixture would have a higher concentration of SO_3 at lower temperatures. The conversion of SO_2 to SO_3 is favored at *low* temperatures.

Practice Example A: The reaction $N_2O_4(g) \rightleftharpoons 2 NO_2(g)$ has $\Delta H° = +57.2$ kJ. Will the amount of $NO_2(g)$ formed from $N_2O_4(g)$ be greater at high or low temperatures?

Practice Example B: The enthalpy of formation of NH_3 is $\Delta H_f°[NH_3(g)] = -46.11$ kJ/mol NH_3. Will the concentration of NH_3 in an equilibrium mixture with its elements be greater at -100 or at 300 °C? Explain.

Effect of a Catalyst on Equilibrium

Adding a catalyst to a reaction mixture speeds up *both* the forward and reverse reactions. Equilibrium is achieved more rapidly, but the equilibrium amounts are *unchanged* by the catalyst. Consider again reaction (16.9)

$$2 SO_2(g) + O_2(g) \rightleftharpoons 2 SO_3(g) \qquad K_c = 2.8 \times 10^2 \text{ at } 1000 \text{ K}$$

▲ **Sulfuric acid is produced from SO_3**
$$SO_3(g) + H_2O(l) \rightleftharpoons H_2SO_4(aq)$$
The catalyst used to speed up the conversion of SO_2 to SO_3 in the commercial production of sulfuric acid is $V_2O_5(s)$.

For a given set of reaction conditions, the equilibrium amounts of SO_2, O_2, and SO_3 have fixed values. This is true whether the reaction is carried out by a slow homogeneous reaction, catalyzed in the gas phase, or conducted as a heterogeneous reaction on the surface of a catalyst. Stated another way, the presence of a catalyst does not change the numerical value of the equilibrium constant.

We now have two thoughts about a catalyst to reconcile: one from the preceding chapter and one from this discussion.

- A catalyst changes the mechanism of a reaction to one with a lower activation energy.
- A catalyst has no effect on the condition of equilibrium in a reversible reaction.

Taken together, these two statements must mean that an equilibrium condition is *independent* of the reaction mechanism. Thus, even though we have described equilibrium in terms of opposing reactions occurring at equal rates, we do not have to concern ourselves with the kinetics of chemical reactions to work with the equilibrium concept. This observation is still another indication that the equilibrium constant is a thermodynamic quantity, as we shall describe more fully in Chapter 20.

16-7 Equilibrium Calculations: Some Illustrative Examples

We are now ready to tackle the problem of describing, in quantitative terms, the condition of equilibrium in a reversible reaction. Part of the approach we use may seem unfamiliar at first—it has an algebraic look to it. But as you adjust to this "new look," do not lose sight of the fact that we continue to use some familiar and important ideas—molar masses, molarities, and stoichiometric factors from the balanced equation, for example.

The five numerical examples that follow apply the general equilibrium principles described earlier in the chapter. The first four involve gases, and the fifth deals with equilibrium in an aqueous solution. (The study of equilibria in aqueous solutions is the principal topic of the next three chapters.) Each example includes a brief section labeled "comments," which is printed on a colored background. Think of the comments as the basic methodology of equilibrium calculations. You may want to refer back to these comments from time to time while you are studying later chapters.

Example 16-9 is relatively straightforward. It demonstrates how to determine the equilibrium constant of a reaction when the equilibrium concentrations of the reactants and products are known.

EXAMPLE 16-9

Nitrogen Dioxide and Dinitrogen Tetroxide movie

Determining a Value of K_c from the Equilibrium Quantities of Substances. Dinitrogen tetroxide, $N_2O_4(l)$, is an important component of rocket fuels—for example, as an oxidizer of liquid hydrazine in the Titan rocket. At 25 °C, N_2O_4 is a colorless gas that partially dissociates into NO_2, a red-brown gas. The color of an equilibrium mixture of these two gases depends on their relative proportions, which in turn depends on the temperature (Figure 16-8).

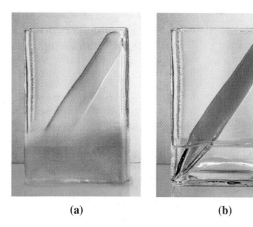

(a)   (b)

▲ FIGURE 16-8 The equilibrium $N_2O_4(g) \rightleftharpoons 2\ NO_2(g)$
(a) At dry ice temperatures, N_2O_4 exists as a solid. The gas in equilibrium with the solid is mostly colorless N_2O_4, with only a trace of brown NO_2. (b) When warmed to room temperature and above, the N_2O_4 melts and vaporizes. The proportion of $NO_2(g)$ at equilibrium increases over that at low temperatures, and the equilibrium mixture of $N_2O_4(g)$ and $NO_2(g)$ has a red-brown color.

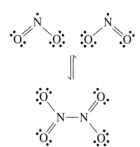

▲ The Lewis structures of N_2O_4 and $NO_2(g)$
Nitrogen dioxide is a free radical that combines exothermically to form dinitrogen tetroxide.

Equilibrium is established in the reaction $N_2O_4(g) \rightleftharpoons 2\ NO_2(g)$ at 25 °C. The quantities of the two gases present in a 3.00-L vessel are 7.64 g N_2O_4 and 1.56 g NO_2. What is the value of K_c for this reaction?

Solution

For both N_2O_4 and NO_2, we need the conversions g $\longrightarrow$ mol $\longrightarrow$ mol/L.

$$[N_2O_4] = \frac{7.64 \text{ g } N_2O_4 \times \dfrac{1 \text{ mol } N_2O_4}{92.01 \text{ g } N_2O_4}}{3.00 \text{ L}} = 0.0277 \text{ M}$$

$$[NO_2] = \frac{1.56 \text{ g } NO_2 \times \dfrac{1 \text{ mol } NO_2}{46.01 \text{ g } NO_2}}{3.00 \text{ L}} = 0.0113 \text{ M}$$

Now we can write the K_c expression and substitute in the equilibrium concentrations.

$$K_c = \frac{[NO_2]^2}{[N_2O_4]} = \frac{(0.0113)^2}{(0.0277)} = 4.61 \times 10^{-3}$$

Practice Example A: Equilibrium is established in a 3.00-L flask at 1405 K for the reaction $2 H_2S(g) \rightleftharpoons 2 H_2(g) + S_2(g)$. At equilibrium, there is 0.11 mol $S_2(g)$, 0.22 mol $H_2(g)$, and 2.78 mol $H_2S(g)$. What is the value of K_c for this reaction?

Practice Example B: Equilibrium is established at 25 °C in the reaction $N_2O_4(g) \rightleftharpoons 2 NO_2(g)$, $K_c = 4.61 \times 10^{-3}$. If $[NO_2] = 0.0236$ M in a 2.26-L flask, how many grams of N_2O_4 are also present?

Comments

1. The quantities required in an equilibrium constant expression, K_c, are equilibrium *concentrations* in *moles per liter*, not simply equilibrium amounts in moles or masses in grams. You will find it helpful to organize all the equilibrium data and carefully label each item.

Example 16-10 is somewhat more involved than Example 16-9. We are still interested in determining the equilibrium constant for a reaction, but we do not have the same sort of information as in Example 16-9. We are given the *initial* concentrations of all the reactants and products, but the equilibrium concentration of only one substance. In this case, we need a little algebra and some careful bookkeeping. We will introduce a tabular system for keeping track of the changing concentrations of the reactants and products that some call an **ICE table**. It notes the **i**nitial, **c**hange in, and **e**quilibrium concentration of each species. It is a helpful device that we will use throughout the next three chapters.

EXAMPLE 16-10

Determining a Value of K_p from Initial and Equilibrium Amounts of Substances: Relating K_c and K_p. The equilibrium condition for $SO_2(g)$, $O_2(g)$, and $SO_3(g)$ is important in sulfuric acid production. When a 0.0200-mol sample of SO_3 is introduced into an evacuated 1.52-L vessel at 900 K, 0.0142 mol SO_3 is found to be present at equilibrium. What is the value of K_p for the dissociation of $SO_3(g)$ at 900 K?

$$2 SO_3(g) \rightleftharpoons 2 SO_2(g) + O_2(g) \qquad K_p = ?$$

Solution

Let's first determine K_c and then convert to K_p by using equation (16.12). In the following ICE table of data, the key item is the *change* in amount of SO_3: In progressing from 0.0200 mol SO_3 to 0.0142 mol SO_3, 0.0058 mol SO_3 is dissociated. We use the *negative* sign (−0.0058 mol) to indicate that this amount of SO_3 is consumed in establishing equilibrium. In the row labeled "changes," we must relate the changes in amounts of SO_2 and O_2 to the change in amount of SO_3. For this, we use the stoichiometric coefficients from the balanced equation: 2, 2, and 1. That is, *two* moles of SO_2 and *one* mole of O_2 are produced for every *two* moles of SO_3 that dissociate.

The reaction:	$2 SO_3(g)$	$\rightleftharpoons$	$2 SO_2(g)$	+	$O_2(g)$
initial amounts:	0.0200 mol		0.00 mol		0.00 mol
changes:	−0.0058 mol		+0.0058 mol		+0.0029 mol
equil amounts:	0.0142 mol		0.0058 mol		0.0029 mol
equil concns:	$[SO_3] = 0.0142\ mol/1.52\ L$		$[SO_2] = 0.0058\ mol/1.52\ L$		$[O_2] = 0.0029\ mol/1.52\ L$
	$= 9.34 \times 10^{-3}\ M$		$= 3.8 \times 10^{-3}\ M$		$= 1.9 \times 10^{-3}\ M$

$$K_c = \frac{[SO_2]^2[O_2]}{[SO_3]^2} = \frac{(3.8 \times 10^{-3})^2(1.9 \times 10^{-3})}{(9.34 \times 10^{-3})^2} = 3.1 \times 10^{-4}$$
$$K_p = K_c(RT)^{\Delta n_{gas}} = 3.1 \times 10^{-4}(0.0821 \times 900)^{(2+1)-2}$$
$$= 3.1 \times 10^{-4}(0.0821 \times 900)^1 = 2.3 \times 10^{-2}$$

Practice Example A: A 5.00-L evacuated flask is filled with 1.86 mol NOBr. At equilibrium at 25 °C, there is 0.082 mol of Br_2 present. Determine K_c and K_p for the reaction $2 NOBr(g) \rightleftharpoons 2 NO(g) + Br_2(g)$.

Practice Example B: 0.100 mol SO_2 and 0.100 mol O_2 are introduced into an evacuated 1.52-L flask at 900 K. When equilibrium is reached, the amount of SO_3 found is 0.0916 mol. Use these data to determine K_p for the reaction $2 SO_3(g) \rightleftharpoons 2 SO_2(g) + O_2(g)$.

Comments

2. The chemical equation for a reversible reaction serves *both* to establish the form of the equilibrium constant expression *and* to provide the conversion factors (stoichiometric factors) to relate the equilibrium quantity of one species to equilibrium quantities of the others.

3. For equilibria involving gases, you can use either K_c or K_p. In general, if the data given involve amounts of substances and volumes, it is easier to work with K_c. If data are given as partial pressures, then work with K_p. Whether working with K_c or K_p or the relationship between them, you must always base these expressions on the given chemical equation, not on equations you may have used in other situations.

The methods used in Examples 16-9 and 16-10 are summarized in Figure 16-9. Example 16-11 demonstrates that we can often determine several pieces of useful information about an equilibrium system from just the equilibrium constant and the reaction equation.

EXAMPLE 16-11

Determining Equilibrium Partial and Total Pressures from a Value of K_p. Ammonium hydrogen sulfide, $NH_4HS(s)$, used as a photographic developer, is unstable and dissociates at room temperature.

$$NH_4HS(s) \rightleftharpoons NH_3(g) + H_2S(g) \qquad K_p(atm) = 0.108 \text{ at } 25 \text{ °C}$$

A sample of $NH_4HS(s)$ is introduced into an evacuated flask at 25 °C. What is the total gas pressure at equilibrium?

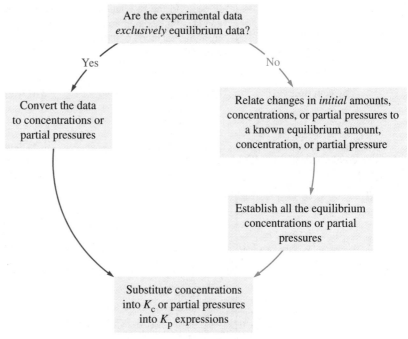

▲ FIGURE 16-9   **Determining K_c or K_p from experimental data**

Solution

K_p for this reaction is just the product of the equilibrium partial pressures of $NH_3(g)$ and $H_2S(g)$, each stated in atmospheres. (There is no term for NH_4HS because it is a solid.) Moreover, because these gases are produced in equimolar amounts, $P_{NH_3} = P_{H_2S}$. Let us first find P_{NH_3} and then P_{tot}.

$$K_p = (P_{NH_3})(P_{H_2S}) = (P_{NH_3})(P_{NH_3}) = (P_{NH_3})^2 = 0.108$$

$$P_{NH_3} = \sqrt{0.108} = 0.329 \text{ atm} \qquad P_{H_2S} = P_{NH_3} = 0.329 \text{ atm}$$

$$P_{tot} = P_{NH_3} + P_{H_2S} = 0.329 \text{ atm} + 0.329 \text{ atm} = 0.658 \text{ atm}$$

Practice Example A: Sodium hydrogen carbonate (baking soda) decomposes at elevated temperatures and is one of the sources of $CO_2(g)$ when this compound is used in baking.

$$2\,NaHCO_3(s) \rightleftharpoons Na_2CO_3(s) + H_2O(g) + CO_2(g) \qquad K_p(atm) = 0.231 \text{ at } 100\,°C$$

What is the partial pressure of $CO_2(g)$ when this equilibrium is established starting with $NaHCO_3(s)$?

Practice Example B: If enough additional $NH_3(g)$ is added to the flask in Example 16-11 to raise its partial pressure to 0.500 atm at equilibrium, what will be the *total* gas pressure when equilibrium is reestablished?

Comments

4. When using K_p expressions, look for relationships among partial pressures of the reactants. If you need to relate the total pressure to the partial pressures of the reactants, you should be able to do this with some equations presented in Chapter 6 (for example, equations 6.15, 6.16, and 6.17).

Example 16-12 brings back the ICE format that we introduced in Example 16-10, but with a twist. This time we *know* the value of the equilibrium constant and an *initial* amount of the reactant, but we have *no* information about the equilibrium amount of the reactant or the product. That means that we do not know how much the initial value will change. We show this by using an "*x*" in that part of the table. The setup will be quite algebraic; in fact, we must use the quadratic formula to obtain a solution.

EXAMPLE 16-12

Calculating Equilibrium Concentrations from Initial Conditions.   A 0.0240-mol sample of $N_2O_4(g)$ is allowed to come to equilibrium with $NO_2(g)$ in a 0.372-L flask at 25 °C. Calculate the amount of N_2O_4 present at equilibrium (Figure 16-10).

$$N_2O_4(g) \rightleftharpoons 2\, NO_2(g) \qquad K_c = 4.61 \times 10^{-3} \text{ at 25 °C}$$

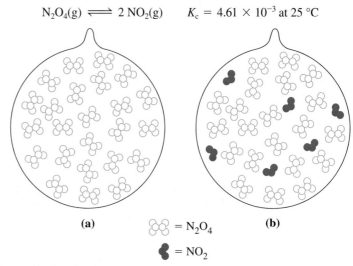

(a)

$\bigcirc\!\!\bigcirc = N_2O_4$ **(b)**

$\blacklozenge = NO_2$

▲ **FIGURE 16-10   Equilibrium in the reaction**
$$N_2O_4(g) \rightleftharpoons 2\, NO_2(g)$$
at 25 °C—Example 16-12 illustrated
Each "molecule" illustrated represents 0.001 mol. **(a)** Initially, the bulb contains 0.024 mol N_2O_4, represented by 24 "molecules." **(b)** At equilibrium, some molecules of N_2O_4 have dissociated to NO_2. The 21 "molecules" of N_2O_4 and 6 of NO_2 correspond to 0.021 mol N_2O_4 and 0.006 mol NO_2 at equilibrium.

Solution

We need to determine the amount of N_2O_4 that dissociates to establish equilibrium. For the first time, we introduce an algebraic unknown, *x*. Suppose we let $x =$ the number of moles of N_2O_4 that dissociate. In the following ICE table, we enter the value $-x$ into the row labeled "changes." The amount of NO_2 produced is $+2x$ because the stoichiometric coefficient of NO_2 is 2 and that of N_2O_4 is *1*.

The reaction:	$N_2O_4(g)$	$\rightleftharpoons$	$2\, NO_2(g)$
initial amounts:	0.0240 mol		0.00 mol
changes:	$-x$ mol		$+2x$ mol
equil amounts:	$(0.0240 - x)$ mol		$2x$ mol
equil concns:	$[N_2O_4] = (0.0240 - x \text{ mol})/0.372 \text{ L}$		$[NO_2] = 2x \text{ mol}/0.372 \text{ L}$

$$K_c = \frac{[NO_2]^2}{[N_2O_4]} = \frac{\left(\dfrac{2x}{0.372}\right)^2}{\left(\dfrac{0.0240 - x}{0.372}\right)} = \frac{4x^2}{0.372(0.0240 - x)} = 4.61 \times 10^{-3}$$

$$4x^2 = 4.12 \times 10^{-5} - \left(1.71 \times 10^{-3}\right)x$$

$$x^2 + \left(4.28 \times 10^{-4}\right)x - 1.03 \times 10^{-5} = 0$$

▶ For a discussion of quadratic equations, see Appendix A.

▶ The symbol ± signifies that there are two possible roots. In this problem, x must be a *positive* quantity smaller than 0.0240.

$$x = \frac{-4.28 \times 10^{-4} \pm \sqrt{(4.28 \times 10^{-4})^2 + 4 \times 1.03 \times 10^{-5}}}{2}$$

$$= \frac{-4.28 \times 10^{-4} \pm \sqrt{(1.83 \times 10^{-7}) + 4.12 \times 10^{-5}}}{2}$$

$$x = \frac{-4.28 \times 10^{-4} \pm \sqrt{4.14 \times 10^{-5}}}{2}$$

$$= \frac{-4.28 \times 10^{-4} \pm 6.43 \times 10^{-3}}{2}$$

$$= \frac{-4.28 \times 10^{-4} + 6.43 \times 10^{-3}}{2} = \frac{6.00 \times 10^{-3}}{2}$$

$$= 3.00 \times 10^{-3} \text{ mol } N_2O_4$$

The amount of N_2O_4 at equilibrium is $(0.0240 - x) = (0.0240 - 0.0030) = 0.0210$ mol N_2O_4

Practice Example A: If 0.150 mol $H_2(g)$ and 0.200 mol $I_2(g)$ are introduced into a 15.0-L flask at 445 °C and allowed to come to equilibrium, how many moles of HI(g) will be present?

$$H_2(g) + I_2(g) \rightleftharpoons 2\,HI(g) \qquad K_c = 50.2 \text{ at } 445\,°C$$

Practice Example B: Suppose the equilibrium mixture of Example 16-12 is transferred to a 10.0-L flask. **(a)** Will the equilibrium amount of N_2O_4 increase or decrease? Explain. **(b)** Calculate the number of moles of N_2O_4 in the new equilibrium condition.

Comments

5. When you need to introduce an algebraic unknown, x, into an equilibrium calculation, follow these steps.

- Introduce x into the ICE setup in the row labeled "changes."
- Decide which change to label as x, that is, the amount of a reactant consumed or of a product formed. Usually, we base this on the species that has the smallest stoichiometric coefficient in the balanced chemical equation.
- Use stoichiometric factors to relate the other changes to x (that is, $2x, 3x, \ldots$).
- Consider that equilibrium amounts = initial amounts + "changes." (If you have assigned the correct signs to the changes, equilibrium amounts will also be correct.)
- After substitutions have been made into the equilibrium constant expression, the equation will often be a quadratic equation in x, which you can solve by the quadratic formula. Occasionally you may encounter a higher-degree equation. Appendix A–3 outlines a straightforward method of dealing with these.

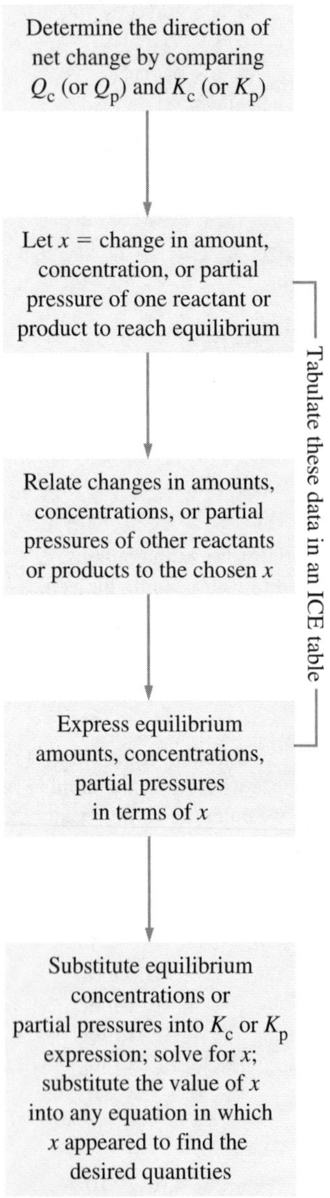

Determine the direction of net change by comparing Q_c (or Q_p) and K_c (or K_p)

↓

Let x = change in amount, concentration, or partial pressure of one reactant or product to reach equilibrium

↓

Relate changes in amounts, concentrations, or partial pressures of other reactants or products to the chosen x

↓

Express equilibrium amounts, concentrations, partial pressures in terms of x

} Tabulate these data in an ICE table

↓

Substitute equilibrium concentrations or partial pressures into K_c or K_p expression; solve for x; substitute the value of x into any equation in which x appeared to find the desired quantities

▲ **FIGURE 16-11**
Determining equilibrium concentrations and partial pressures

Our final example is similar to the previous one, but with this slight complication: Initially, we don't know whether a net change occurs to the right or to the left to establish equilibrium. We can find out, though, by using the reaction quotient, Q_c, and proceeding in the manner suggested in Figure 16-11. Also, because the reactants and products are in solution, we can work exclusively with concentrations in formulating the K_c expression.

EXAMPLE 16-13

Using the Reaction Quotient, Q_c, in an Equilibrium Calculation. Solid silver is added to a solution with these initial concentrations: $[Ag^+] = 0.200$ M, $[Fe^{2+}] = 0.100$ M, and $[Fe^{3+}] = 0.300$ M. The following reversible reaction occurs.

$$Ag^+(aq) + Fe^{2+}(aq) \rightleftharpoons Ag(s) + Fe^{3+}(aq) \qquad K_c = 2.98$$

What are the ion concentrations when equilibrium is established?

Solution

Because all reactants and products are present initially, we need to use the reaction quotient Q_c to determine the direction in which a net change occurs.

$$Q_c = \frac{[Fe^{3+}]}{[Ag^+][Fe^{2+}]} = \frac{0.300}{(0.200)(0.100)} = 15.0$$

Because Q_c (15.0) is larger than K_c (2.98), a net change must occur in the direction of the reverse reaction, *to the left*. Let's define x as the change in molarity of Fe^{3+}. Because the net change occurs *to the left*, we designate the changes for the species on the left side of the equation as positive and those on the right side as negative. The relevant data are tabulated as follows.

The reaction:	$Ag^+(aq)$	+	$Fe^{2+}(aq)$	$\rightleftharpoons$ $Ag(s)$ +	$Fe^{3+}(aq)$
Initial concns:	0.0200 M		0.100 M		0.300 M
changes:	$+x$ M		$+x$ M		$-x$ M
equil concns:	$(0.0200 + x)$ M		$(0.100 + x)$ M		$(0.300 - x)$ M

$$K_c = \frac{[Fe^{3+}]}{[Ag^+][Fe^{2+}]} = \frac{(0.300 - x)}{(0.200 + x)(0.100 + x)} = 2.98$$

This equation, which is solved in Appendix A–3, is a quadratic equation for which the acceptable root is $x = 0.11$. To obtain the equilibrium concentrations, we substitute this value of x into the terms shown in the table of data.

$$[Ag^+]_{equil} = 0.200 + 0.11 = 0.31 \text{ M}$$
$$[Fe^{2+}]_{equil} = 0.100 + 0.11 = 0.21 \text{ M}$$
$$[Fe^{3+}]_{equil} = 0.300 - 0.11 = 0.19 \text{ M}$$

Checking the Results: If we have done the calculation correctly, we should obtain a value very close to that given for K_c when we substitute the *calculated* equilibrium concentrations into the reaction quotient, Q_c. We do.

$$Q_c = \frac{[Fe^{3+}]}{[Ag^+][Fe^{2+}]} = \frac{(0.19)}{(0.31)(0.21)} = 2.9 \quad (K_c = 2.98)$$

Practice Example A: Excess $Ag(s)$ is added to 1.20 M $Fe^{3+}(aq)$. Given that

$$Ag^+(aq) + Fe^{2+}(aq) \rightleftharpoons Ag(s) + Fe^{3+}(aq) \qquad K_c = 2.98$$

what are the equilibrium concentrations of the species in solution?

Practice Example B: A solution is prepared with $[V^{3+}] = [Cr^{2+}] = 0.0100$ M and $[V^{2+}] = [Cr^{3+}] = 0.150$ M. The following reaction occurs.

$$V^{3+}(aq) + Cr^{2+}(aq) \rightleftharpoons V^{2+}(aq) + Cr^{3+}(aq) \qquad K_c = 7.2 \times 10^2$$

What are the ion concentrations when equilibrium is established?

(*Hint:* The algebra can be greatly simplified by extracting the square root of both sides of an equation at the appropriate point.)

KEEP IN MIND ▶
that if one or more of the substances appearing in Q_c (or Q_p) is not present initially, a net change must occur to produce some of the substance(s). You need to compare Q_c (or Q_p) with K_c (or K_p) only if *all* the substances appearing in these expressions are present initially.

Comments

6. It is sometimes helpful to compare the reaction quotient, Q_c (or Q_p), to the equilibrium constant, K_c (or K_p), to determine the direction of the net change.

7. In many equilibrium calculations—often those in aqueous solutions—you can work with molarities directly, without having to work with moles of reactants and solution volumes.

8. Where possible, check your calculation, for instance, by substituting *calculated* equilibrium concentrations into the reaction quotient, Q_c (or Q_p), to see if its numerical value is close to that of K_c (or K_p).

The Nitrogen Cycle and the Synthesis of Nitrogen Compounds

▲ Use of liquid ammonia as a fertilizer, by direct injection into the soil.

Nitrogen is an element essential both in living things and in industry. The atmosphere contains a vast pool of elemental nitrogen, but most organisms cannot use nitrogen in that form. Instead, they require certain nitrogen-containing compounds. The conversion of elemental atmospheric nitrogen to nitrogen compounds is called *nitrogen fixation,* and in nature, this process is carried out only by certain types of bacteria. Nitrogen fixation is part of the nitrogen cycle, which is the path of nitrogen through the environment and a variety of living organisms and back into the environment. A simplified version of the nitrogen cycle is depicted in Figure 16-12, which shows that elemental nitrogen from the atmosphere is fixed, becomes part of plants, animals, and other organisms, and is then returned to the atmosphere. The numbers in parentheses refer to the various parts of the cycle described below.

A few leguminous plants, such as beans, peas, and alfalfa, have bacteria residing on their roots that convert elemental nitrogen from the air into compounds used to make plant protein (1). Plants also convert nitrates in the soil to proteins (5). Animals obtain the nitrogen that they use to make their own proteins by feeding on plants and/or other animals (2). The decay of plant and animal proteins produces ammonia (3). Through a series of bacterial actions, ammonia is converted to nitrites and nitrates (4). Denitrifying bacteria decompose nitrites and nitrates, returning N_2O and N_2 to the atmosphere, thus completing the cycle (6). Some atmospheric nitrogen is converted to nitrates during electrical storms (1a).

Reversible chemical reactions play an important role in the nitrogen cycle. One example is the reaction of $N_2(g)$ and $O_2(g)$ to form $NO(g)$.

$$N_2(g) + O_2(g) \rightleftharpoons 2\,NO(g)$$

$$K_p = 4.7 \times 10^{-31} \text{ at } 298 \text{ K}$$

$$= 1.3 \times 10^{-4} \text{ at } 1800 \text{ K}$$

The reaction does not occur to any measurable extent at 298 K, but at 1800 K the situation is a little different. An equilibrium

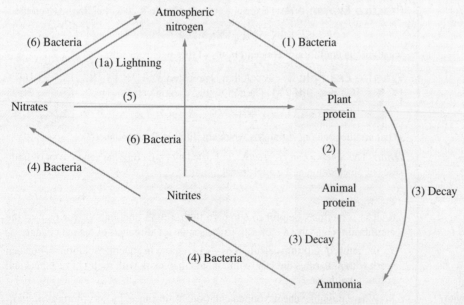

FIGURE 16-12
The nitrogen cycle

▲ During electrical storms, $N_2(g)$ and $O_2(g)$ combine to produce small quantities of $NO(g)$ in a reversible chemical reaction.

mixture of $N_2(g)$ and $O_2(g)$ at 1800 K contains about 1 or 2% $NO(g)$. $NO(g)$ is introduced into the nitrogen cycle through this high-temperature reaction occurring *naturally* in lightning flashes during thunderstorms and *artificially* in high-temperature combustion processes, as in internal combustion engines. Further reactions account for the formation of $HNO_3(aq)$ in rainstorms and the introduction of nitrates into the soil.

$$2 NO(g) + O_2(g) \longrightarrow 2 NO_2(g)$$

$$3 NO_2(g) + H_2O(l) \longrightarrow 2 HNO_3(aq) + NO(g)$$

Ammonia is an important industrial chemical. Its synthesis is achieved by the following reaction.

$$N_2(g) + 3 H_2(g) \rightleftharpoons 2 NH_3(g)$$

$$\Delta H° = -92.22 \text{ kJ} \qquad K_p = 6.2 \times 10^5 \text{ at 298 K}$$

Since ammonia can be converted by soil bacteria to nitrites and nitrates that can be used by plants, one of the principal uses of NH_3 is as a fertilizer that is injected directly into the soil. NH_3 is also used in the production of other nitrogen compounds, such as urea, hydrazine, ammonium sulfate, ammonium nitrate, ammonium dihydrogen phosphate, and ammonium hydrogen phosphate. Several of these compounds are used as fertilizers, and others are used in the manufacture of explosives, pharmaceuticals, and plastics. So much nitrogen is now being fixed artificially that fixed nitrogen is accumulating in the environment somewhat faster than it is being returned to the atmosphere. This causes environmental problems such as the buildup of nitrates in groundwater.

As described in Section 8-2 (page 271), the commercial synthesis of ammonia uses the Haber–Bosch process, but let's consider this process again from the standpoint of chemical equilibrium and kinetics. In the synthesis reaction, two moles of gaseous product are formed for every four moles of gaseous reactants. Carrying out the reaction at high pressure favors the production of NH_3. Because the forward reaction is exothermic,

the equilibrium yield of NH_3 is greatest at low temperatures. Thus, the optimum conditions for the equilibrium production of NH_3 are *high pressures* and *low temperatures*. However, these "optimum" conditions do not take into account the rate of reaction. Although the equilibrium production of NH_3 is favored at low temperatures, equilibrium is achieved so slowly that the synthesis is not feasible at these temperatures. One way to speed up the reaction is to raise the temperature, even though doing so decreases the equilibrium concentration of NH_3. Another way is to use a catalyst. The usual industrial operating conditions are a temperature of about 550 °C, pressures ranging from 150 to 350 atm, and a catalyst—usually iron in the presence of Al_2O_3, MgO, CaO, and K_2O. Figure 16-13 suggests the dramatic difference between the theoretical optimum conditions and the actual operating conditions.

Another essential feature of the Haber–Bosch method is to remove NH_3 as it forms. This is done by liquefying the $NH_3(g)$ (recall Figure 8-4). In fact, the mixture need not be brought to equilibrium at all, and nearly 100% conversion of N_2 and H_2 to NH_3 is achieved.

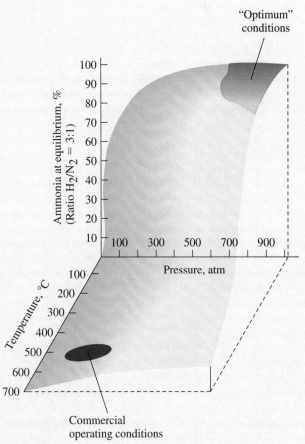

FIGURE 16-13
Equilibrium conversion of $N_2(g)$ and $H_2(g)$ to $NH_3(g)$ as a function of temperature and pressure

Summary

A reversible reaction reaches a point at which the rates of the forward and reverse reactions become equal. This condition of dynamic equilibrium is described through an equilibrium constant expression. In one case—K_c—the equilibrium condition is expressed through concentrations. In another—K_p—partial pressures of gases are used. The form of an equilibrium constant expression is established from the balanced chemical equation. Its numerical value is determined by experiment. Other ideas in the chapter deal with the effect of the following situations on K_c or K_p.

- the presence of pure solids or liquids
- reversing a chemical equation
- multiplying a chemical equation by a factor
- combining several equations into an overall chemical equation

Le Châtelier's principle is used to make qualitative predictions of the effects of different variables on an equilibrium condition. This principle describes how an equilibrium condition is modified, or "shifts," when an equilibrium is disturbed by adding or removing reactants or by changing the reaction volume, external pressure, or temperature. Catalysts, by speeding up both the forward and reverse reactions equally, have no effect on an equilibrium condition.

Another qualitative prediction concerns the direction of net change leading to equilibrium. This prediction can be made with a ratio of initial concentrations or partial pressures, formed in the same manner as K_c or K_p and called the reaction quotient, Q_c or Q_p.

For quantitative equilibrium calculations, a few basic principles and algebraic techniques are required. These are illustrated in the concluding section of the chapter.

Integrative Example

In the manufacture of ammonia, the chief source of hydrogen gas is this reaction for the reforming of methane at high temperatures.

$$CH_4(g) + 2 H_2O(g) \rightleftharpoons CO_2(g) + 4 H_2(g) \quad (16.19)$$

The following data are also given.

(a) $CO(g) + H_2O(g) \rightleftharpoons CO_2(g) + H_2(g)$
$$\Delta H° = -40 \text{ kJ}; K_c = 1.4 \text{ at } 1000 \text{ K}$$

(b) $CO(g) + 3 H_2(g) \rightleftharpoons H_2O(g) + CH_4(g)$
$$\Delta H° = -230 \text{ kJ}; K_c = 190 \text{ at } 1000 \text{ K}$$

At 1000 K, 1.00 mol each of CH_4 and H_2O are allowed to come to equilibrium in a 10.0-L vessel. Calculate the number of moles of H_2 present at equilibrium. Would the yield of H_2 increase if we raised the temperature above 1000 K?

First, let's assemble the data we need to solve this problem. Because we are given amounts of substances and a reaction vol-

ume, we should be able to work with a K_c expression. However, because we are not given the K_c value for the reaction of interest, we will have to derive this value from other data. We can do this by combining the two equations for which we are given data.

(a) $\qquad CO(g) + H_2O(g) \rightleftharpoons CO_2(g) + H_2(g)$
$$\Delta H = -40 \text{ kJ} \quad K_c = 1.4$$

(b) $\qquad CH_4(g) + H_2O(g) \rightleftharpoons CO(g) + 3 H_2(g)$
$$\Delta H = 230 \text{ kJ} \quad K_c = 1/190$$

Overall: $\quad CH_4(g) + 2 H_2O(g) \rightleftharpoons CO_2(g) + 4 H_2(g)$
$$\Delta H = 190 \text{ kJ} \quad K_c = 1.4/190 = 7.4 \times 10^{-3}$$

Next we can set up an ICE table in which we let x represent the number of moles of CH_4 consumed in reaching equilibrium.

The reaction:	$CH_4(g)$	+	$2 H_2O(g)$	$\rightleftharpoons$	$CO_2(g)$	+	$4 H_2(g)$
initial amounts:	1.00 mol		1.00 mol		0.00 mol		0.00 mol
changes:	$-x$ mol		$-2x$ mol		x mol		$4x$ mol
equil amounts:	$(1.00 - x)$ mol		$(1.00 - 2x)$ mol		x mol		$4x$ mol
equil concns, M:	$(1.00 - x)/10.0$		$(1.00 - 2x)/10.0$		$x/10.0$		$4x/10.0$

Now we can set up K_c and make substitutions into the expression.

$$K_c = \frac{[CO_2][H_2]^4}{[CH_4][H_2O]^2}$$

$$= \frac{(x/10.0)(4x/10.0)^4}{[(1.00 - x)/10.0][(1.00 - 2x)/10.0]^2}$$

$$= \frac{x(4x)^4}{100(1.00 - x)(1.00 - 2x)^2} = 7.4 \times 10^{-3}$$

The above equation reduces to
$$256x^5 = 0.74 \left[(1.00 - x)(1.00 - 2x)^2\right]$$

and then to
$$256x^5 - 0.74\left[(1.00 - x)(1.00 - 2x)^2\right] = 0$$

The above equation looks impossibly difficult to solve, but it's not. It can be solved rather simply by the method of successive approximations. This is done in Appendix A–3. The result is $x = 0.23$. The number of moles of H_2 at equilibrium is $4x = 0.92$.

Because the reaction is endothermic ($\Delta H = 190$ kJ), the forward reaction is favored at higher temperatures. The equilibrium yield of H_2 will increase if the temperature is raised above 1000 K.

Key Terms

equilibrium (16-1)	**ICE table** (16-7)	**Le Châtelier's principle** (16-6)
equilibrium constant (16-2)	K_c (16-2)	**reaction quotient,** Q_c (16-5)
equilibrium constant expression (16-2)	K_p (16-3)	

Review Questions

1. In your own words, define or explain the following terms or symbols: (a) K_p; (b) Q_c; (c) Δn_{gas}.
2. Briefly describe each of the following ideas or phenomena: (a) dynamic equilibrium; (b) direction of a net chemical change; (c) Le Châtelier's principle; (d) effect of a catalyst on equilibrium.
3. Explain the important distinctions between each pair of terms: (a) reaction that goes to completion and reversible reaction; (b) K_c and K_p; (c) reaction quotient (Q) and equilibrium constant expression (K); (d) homogeneous and heterogeneous reaction.
4. Write equilibrium constant expressions, K_c, for the reactions
   (a) $2\,NO(g) + O_2(g) \rightleftharpoons 2\,NO_2(g)$
   (b) $Zn(s) + 2\,Ag^+(aq) \rightleftharpoons Zn^{2+}(aq) + 2\,Ag(s)$
   (c) $Mg(OH)_2(s) + CO_3^{2-}(aq) \rightleftharpoons$
   $$MgCO_3(s) + 2\,OH^-(aq)$$
5. Write equilibrium constant expressions, K_p, for the reactions
   (a) $CS_2(g) + 4\,H_2(g) \rightleftharpoons CH_4(g) + 2\,H_2S(g)$
   (b) $Ag_2O(s) \rightleftharpoons 2\,Ag(s) + \frac{1}{2}O_2(g)$
   (c) $2\,NaHCO_3(s) \rightleftharpoons Na_2CO_3(s) + CO_2(g) + H_2O(g)$
6. Write an equilibrium constant expression, K_c, for the formation of 1 mol of each of the following *gaseous* compounds from its *gaseous* elements: (a) N_2O; (b) HBr; (c) NH_3; (d) ClF_3; (e) NOF_2.
7. From these values of K_c,
   $$CO(g) + H_2O(g) \rightleftharpoons CO_2(g) + H_2(g)$$
   $$K_c = 23.2 \text{ at } 600 \text{ K}$$
   $$SO_2(g) + \frac{1}{2}O_2(g) \rightleftharpoons SO_3(g)$$
   $$K_c = 56 \text{ at } 900 \text{ K}$$
   $$2\,H_2S(g) \rightleftharpoons 2\,H_2(g) + S_2(g)$$
   $$K_c = 2.3 \times 10^{-4} \text{ at } 1405 \text{ K}$$
   determine values of K_c for the following reactions at the temperatures given.
   (a) $CO_2(g) + H_2(g) \rightleftharpoons CO(g) + H_2O(g)$
   (b) $2\,SO_2(g) + O_2(g) \rightleftharpoons 2\,SO_3(g)$
   (c) $H_2(g) + \frac{1}{2}S_2(g) \rightleftharpoons H_2S(g)$

8. Determine K_c for the reaction
   $$\frac{1}{2}N_2(g) + \frac{1}{2}O_2(g) + \frac{1}{2}Br_2(g) \rightleftharpoons NOBr(g)$$
   from the following information (at 298 K).
   $$2\,NO(g) \rightleftharpoons N_2(g) + O_2(g)$$
   $$K_c = 2.1 \times 10^{30}$$
   $$NO(g) + \frac{1}{2}Br_2(g) \rightleftharpoons NOBr(g) \qquad K_c = 1.4$$
9. In the reversible reaction $H_2(g) + I_2(g) \rightleftharpoons 2\,HI(g)$, an initial mixture contains 2 mol H_2 and 1 mol I_2. Which of the following is the amount of HI expected at equilibrium? Explain. (a) 1 mol; (b) 2 mol; (c) more than 2 but less than 4 mol; (d) less than 2 mol.
10. Equilibrium is established in the reversible reaction $2\,A + B \rightleftharpoons 2\,C$. The equilibrium concentrations are $[A] = 0.55$ M, $[B] = 0.33$ M, $[C] = 0.43$ M. What is the value of K_c for this reaction?
11. A 0.0040-mol sample of $S_2(g)$ is allowed to dissociate in a 0.500-L flask at 1000 K. When equilibrium is established, 2.0×10^{-11} mol S(g) is present. What is the value of K_c for the reaction $S_2(g) \rightleftharpoons 2\,S(g)$?
12. An equilibrium mixture of SO_2, SO_3, and O_2 gases is maintained in a 2.05-L flask at a temperature at which $K_c = 35.5$ for the reaction
    $$2\,SO_2(g) + O_2(g) \rightleftharpoons 2\,SO_3(g)$$
    (a) If the numbers of moles of SO_2 and SO_3 in the flask are equal, how many moles of O_2 are present?
    (b) If the number of moles of SO_3 in the flask is twice the number of moles of SO_2, how many moles of O_2 are present?
13. Determine the numerical values of K_p for reactions (a), (b), and (c) in Review Question 7.
14. For the reaction $2\,NO_2(g) \rightleftharpoons 2\,NO(g) + O_2(g)$, $K_c = 1.8 \times 10^{-6}$ at 184 °C. What is the value of K_p for this reaction at 184 °C?
    $$NO(g) + \frac{1}{2}O_2(g) \rightleftharpoons NO_2(g)$$

15. An equilibrium mixture at 1000 K contains 0.276 mol H_2, 0.276 mol CO_2, 0.224 mol CO, and 0.224 mol H_2O.
$$CO_2(g) + H_2(g) \rightleftharpoons CO(g) + H_2O(g)$$
    (a) Show that for this reaction, K_c is independent of the reaction volume, V.
    (b) Determine the value of K_c and of K_p.

16. The two common chlorides of phosphorus, PCl_3 and PCl_5, both important in the production of other phosphorus compounds, coexist in equilibrium through the reaction
$$PCl_3(g) + Cl_2(g) \rightleftharpoons PCl_5(g)$$
    At 250 °C, an equilibrium mixture in a 2.50-L flask contains 0.105 g PCl_5, 0.220 g PCl_3, and 2.12 g Cl_2. What are the values of (a) K_c and (b) K_p for this reaction at 250 °C?

17. When 1.00 mol $I_2(g)$ is introduced into an evacuated 1.00-L flask at 1200 °C, 5% of the I_2 molecules dissociate into iodine atoms. For the reaction $I_2(g) \rightleftharpoons 2\,I(g)$ at 1200 °C, what are the values of (a) K_c and (b) K_p?

18. 0.455 mol SO_2, 0.183 mol O_2, and 0.568 mol SO_3 are introduced simultaneously into a 1.90-L vessel at 1000 K.
    (a) Is this mixture at equilibrium?
    (b) If not, in which direction must a net change occur?
$$2\,SO_2(g) + O_2(g) \rightleftharpoons 2\,SO_3(g) \quad K_c = 2.8 \times 10^2$$
    at 1000 K

19. What effect does increasing the volume of the system have on the equilibrium condition in each of the following reactions?
    (a) $C(s) + H_2O(g) \rightleftharpoons CO(g) + H_2(g)$
    (b) $Ca(OH)_2(s) + CO_2(g) \rightleftharpoons CaCO_3(s) + H_2O(g)$
    (c) $4\,NH_3(g) + 5\,O_2(g) \rightleftharpoons 4\,NO(g) + 6\,H_2O(g)$

20. For which of the following reactions would you expect the extent of the forward reaction to increase with increasing temperatures? Explain.

    (a) $NO(g) \rightleftharpoons \frac{1}{2} N_2(g) + \frac{1}{2} O_2(g) \quad \Delta H° = -90.2$ kJ
    (b) $SO_3(g) \rightleftharpoons SO_2(g) + \frac{1}{2} O_2(g) \quad \Delta H° = +98.9$ kJ
    (c) $N_2H_4(g) \rightleftharpoons N_2(g) + 2\,H_2(g) \quad \Delta H° = -95.4$ kJ
    (d) $COCl_2(g) \rightleftharpoons CO(g) + Cl_2(g) \quad \Delta H° = +108.3$ kJ

21. The Deacon process for producing chlorine gas from hydrogen chloride is used in situations where HCl is available as a by-product from other chemical processes.
$$4\,HCl(g) + O_2(g) \rightleftharpoons 2\,H_2O(g) + 2\,Cl_2(g)$$
$$\Delta H° = -114 \text{ kJ}$$
    A mixture of HCl, O_2, H_2O, and Cl_2 is brought to equilibrium at 400 °C. What is the effect on the equilibrium amount of $Cl_2(g)$ if
    (a) Additional $O_2(g)$ is added to the mixture at constant volume?
    (b) HCl(g) is removed from the mixture at constant volume?
    (c) The mixture is transferred to a vessel of twice the volume?
    (d) A catalyst is added to the reaction mixture?
    (e) The temperature is raised to 500 °C?

22. A mixture consisting of 0.150 mol H_2 and 0.150 mol I_2 is brought to equilibrium at 445 °C in a 3.25-L flask. What are the equilibrium amounts of H_2, I_2, and HI?
$$H_2(g) + I_2(g) \rightleftharpoons 2\,HI(g) \quad K_c = 50.2 \text{ at } 445 \text{ °C}$$

23. A 0.150-mol sample of HI is brought to equilibrium at 445 °C in a 3.25-L flask. How many moles of I_2 will be present?
$$H_2(g) + I_2(g) \rightleftharpoons 2\,HI(g) \quad K_c = 50.2 \text{ at } 445 \text{ °C}$$

24. A sample of $NH_4HS(s)$ is introduced into a 2.58-L flask containing 0.100 mol NH_3. What will be the total gas pressure when equilibrium is established at 25 °C?
$$NH_4HS(s) \rightleftharpoons NH_3(g) + H_2S(g)$$
$$K_p(\text{atm}) = 0.108 \text{ at } 25 \text{ °C}$$

Exercises

Writing Equilibrium Constant Expressions

25. Based on these descriptions, write a balanced equation and the corresponding K_c expression for each reversible reaction.
    (a) Carbonyl fluoride, $COF_2(g)$, decomposes into gaseous carbon dioxide and gaseous carbon tetrafluoride.
    (b) Copper metal displaces silver(I) ion from aqueous solution, producing silver metal and an aqueous solution of copper(II) ion.
    (c) Peroxodisulfate ion, $S_2O_8^{2-}$, oxidizes iron(II) ion to iron(III) ion in aqueous solution and is itself reduced to sulfate ion.

26. Based on these descriptions, write a balanced equation and the corresponding K_p expression for each reversible reaction.
    (a) Oxygen gas oxidizes gaseous ammonia to gaseous nitrogen and water vapor.
    (b) Hydrogen gas reduces gaseous nitrogen dioxide to gaseous ammonia and water vapor.

    (c) Nitrogen gas reacts with the solids sodium carbonate and carbon to produce solid sodium cyanide and carbon monoxide gas.

27. Determine values of K_c from the K_p values given.
    (a) $SO_2Cl_2(g) \rightleftharpoons SO_2(g) + Cl_2(g)$
$$K_p = 2.9 \times 10^{-2} \text{ at } 303 \text{ K}$$
    (b) $2\,NO(g) + O_2(g) \rightleftharpoons 2\,NO_2(g)$
$$K_p = 1.48 \times 10^4 \text{ at } 184 \text{ °C}$$
    (c) $Sb_2S_3(s) + 3\,H_2(g) \rightleftharpoons 2\,Sb(s) + 3\,H_2S(g)$
$$K_p = 0.429 \text{ at } 713 \text{ K}$$

28. Determine the values of K_p from the K_c values given.
    (a) $N_2O_4(g) \rightleftharpoons 2\,NO_2(g)$
$$K_c = 4.61 \times 10^{-3} \text{ at } 25 \text{ °C}$$
    (b) $2\,CH_4(g) \rightleftharpoons C_2H_2(g) + 3\,H_2(g)$
$$K_c = 0.154 \text{ at } 2000 \text{ K}$$
    (c) $2\,H_2S(g) + CH_4(g) \rightleftharpoons 4\,H_2(g) + CS_2(g)$
$$K_c = 5.27 \times 10^{-8} \text{ at } 973 \text{ K}$$

29. The vapor pressure of water at 25 °C is 23.8 mmHg. Write K_p for the vaporization of water, with pressures in atmospheres. What is the value of K_c for the vaporization process?

30. If $K_c = 5.12 \times 10^{-3}$ for the equilibrium established between liquid benzene and its vapor at 25 °C, what is the vapor pressure of C_6H_6 at 25 °C, expressed in millimeters of mercury?

31. Given the equilibrium constant values

$$N_2(g) + \frac{1}{2} O_2(g) \rightleftharpoons N_2O(g) \qquad K_c = 2.7 \times 10^{-18}$$

$$N_2O_4(g) \rightleftharpoons 2\,NO_2(g) \qquad K_c = 4.6 \times 10^{-3}$$

$$\frac{1}{2} N_2(g) + O_2(g) \rightleftharpoons NO_2(g) \qquad K_c = 4.1 \times 10^{-9}$$

Determine a value of K_c for the reaction

$$2\,N_2O(g) + 3\,O_2(g) \rightleftharpoons 2\,N_2O_4(g)$$

32. Use the following data at 1200 K to estimate a value of K_p for the reaction $2\,H_2(g) + O_2(g) \rightleftharpoons 2\,H_2O(g)$.

$$C(graphite) + CO_2(g) \rightleftharpoons 2\,CO(g)$$
$$K_c = 0.64$$

$$CO_2(g) + H_2(g) \rightleftharpoons CO(g) + H_2O(g)$$
$$K_c = 1.4$$

$$C(graphite) + \frac{1}{2} O_2(g) \rightleftharpoons CO(g)$$
$$K_c = 1 \times 10^8$$

Experimental Determination of Equilibrium Constants

33. 1.00×10^{-3} mol PCl_5 is introduced into a 250.0-mL flask, and equilibrium is established at 284 °C: $PCl_5(g) \rightleftharpoons PCl_3(g) + Cl_2(g)$. The quantity of $Cl_2(g)$ present at equilibrium is found to be 9.65×10^{-4} mol. What is the value of K_c for the dissociation reaction at 284 °C?

34. A mixture of 1.00 g H_2 and 1.06 g H_2S in a 0.500-L flask comes to equilibrium at 1670 K: $2\,H_2(g) + S_2(g) \rightleftharpoons 2\,H_2S(g)$. The equilibrium amount of $S_2(g)$ found is 8.00×10^{-6} mol. Determine the value of K_p at 1670 K.

Equilibrium Relationships

35. Equilibrium is established at 1000 K, where $K_c = 281$ for the reaction $2\,SO_2(g) + O_2(g) \rightleftharpoons 2\,SO_3(g)$. The equilibrium amount of $O_2(g)$ in a 0.185-L flask is 0.00247 mol. What is the ratio of $[SO_2]$ to $[SO_3]$ in this equilibrium mixture?

36. For the dissociation of $I_2(g)$ at about 1200 °C, $I_2(g) \rightleftharpoons 2\,I(g)$, $K_c = 1.1 \times 10^{-2}$. What volume flask should we use if we want 0.37 mol I to be present for every 1.00 mol I_2 at equilibrium?

37. In the Ostwald process for oxidizing ammonia, a variety of products is possible—N_2, N_2O, NO, and NO_2—depending on the conditions. One possibility is

$$NH_3(g) + \frac{5}{4} O_2(g) \rightleftharpoons NO(g) + \frac{3}{2} H_2O(g)$$

$$K_p(atm) = 2.11 \times 10^{19} \text{ at } 700 \text{ K}$$

For the decomposition of NO_2 at 700 K,

$$NO_2(g) \rightleftharpoons NO(g) + \frac{1}{2} O_2(g) \qquad K_p(atm) = 0.524$$

(a) Write a chemical equation for the oxidation of $NH_3(g)$ to $NO_2(g)$.
(b) Determine K_p for the chemical equation you have written.

38. At 2000 K, $K_c = 0.154$ for the reaction $2\,CH_4(g) \rightleftharpoons C_2H_2(g) + 3\,H_2(g)$. If a 1.00-L equilibrium mixture at 2000 K contains 0.10 mol each of $CH_4(g)$ and $H_2(g)$,
(a) What is the mole fraction of $C_2H_2(g)$ present?
(b) Is the conversion of $CH_4(g)$ to $C_2H_2(g)$ favored at high or low pressures?
(c) If the equilibrium mixture at 2000 K is transferred from a 1.00-L flask to a 2.00-L flask, will the number of moles of $C_2H_2(g)$ increase, decrease, or remain unchanged?

Direction and Extent of Chemical Change

39. Can a mixture of 2.2 mol O_2, 3.6 mol SO_2, and 1.8 mol SO_3 be maintained indefinitely in a 7.2-L flask at a temperature at which $K_c = 100$ in this reaction? Explain.
$$2\,SO_2(g) + O_2(g) \rightleftharpoons 2\,SO_3(g)$$

40. Is a mixture of 0.0205 mol $NO_2(g)$ and 0.750 mol $N_2O_4(g)$ in a 5.25-L flask at 25 °C at equilibrium? If not, in which direction will the reaction proceed—toward products or reactants?
$$N_2O_4(g) \rightleftharpoons 2\,NO_2(g) \qquad K_c = 4.61 \times 10^{-3} \text{ at } 25 \text{ °C}$$

41. Starting with 0.280 mol $SbCl_3$ and 0.160 mol Cl_2, how many moles of $SbCl_5$, $SbCl_3$, and Cl_2 are present when equilibrium is established at 248 °C in a 2.50-L flask?
$$SbCl_5(g) \rightleftharpoons SbCl_3(g) + Cl_2(g)$$
$$K_c = 2.5 \times 10^{-2} \text{ at } 248 \text{ °C}$$

42. Starting with 0.3500 mol CO(g) and 0.05500 mol $COCl_2(g)$ in a 3.050-L flask at 668 K, how many moles of $Cl_2(g)$ will be present at equilibrium?
$$CO(g) + Cl_2(g) \rightleftharpoons COCl_2(g)$$
$$K_c = 1.2 \times 10^3 \text{ at } 668 \text{ K}$$

43. 1.00 g *each* of CO, H_2O, and H_2 are sealed in a 1.41-L vessel and brought to equilibrium at 600 K. How many grams of CO_2 will be present in the equilibrium mixture?

$$CO(g) + H_2O(g) \rightleftharpoons CO_2(g) + H_2(g) \qquad K_c = 23.2$$

44. Equilibrium is established in a 2.50-L flask at 250 °C for the reaction

$$PCl_5(g) \rightleftharpoons PCl_3(g) + Cl_2(g) \qquad K_c = 3.8 \times 10^{-2}$$

How many moles of PCl_5, PCl_3, and Cl_2 are present at equilibrium, if

(a) 0.550 mol each of PCl_5 and PCl_3 are initially introduced into the flask?

(b) 0.610 mol PCl_5 alone is introduced into the flask?

45. For the following reaction, $K_c = 2.00$ at 1000 °C.

$$2\,COF_2(g) \rightleftharpoons CO_2(g) + CF_4(g)$$

If a 5.00-L mixture contains 0.145 mol COF_2, 0.262 mol CO_2, and 0.074 mol CF_4 at a temperature of 1000 °C,

(a) Will the mixture be at equilibrium?

(b) If the gases are not at equilibrium, in what direction will a net change occur?

(c) How many moles of each gas will be present at equilibrium?

46. In the following reaction, $K_c = 4.0$.

$$C_2H_5OH + CH_3COOH \rightleftharpoons CH_3COOC_2H_5 + H_2O$$

A reaction is allowed to occur in a mixture of 17.2 g C_2H_5OH, 23.8 g CH_3COOH, 48.6 g $CH_3COOC_2H_5$, and 71.2 g H_2O.

(a) In what direction will a net change occur?

(b) How many grams of each substance will be present at equilibrium?

47. A gaseous mixture containing 0.125 mol *each* of $H_2(g)$ and $I_2(g)$ is introduced into a 6.14-L flask at 445 °C, and equilibrium is established. What is the mole percent HI in the equilibrium mixture?

$$H_2(g) + I_2(g) \rightleftharpoons 2\,HI(g) \qquad K_c = 50.2$$

48. The N_2O_4–NO_2 equilibrium mixture in the flask on the left in the figure is allowed to expand into the evacuated flask on the right. What is the composition of the gaseous mixture when equilibrium is reestablished in the system consisting of the two flasks?

$$N_2O_4(g) \rightleftharpoons 2\,NO_2(g) \qquad K_c = 4.61 \times 10^{-3} \text{ at } 25 \text{ °C}$$

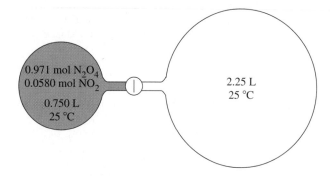

0.971 mol N_2O_4
0.0580 mol NO_2
0.750 L
25 °C

2.25 L
25 °C

49. Formamide, used in the manufacture of pharmaceuticals, dyes, and agricultural chemicals, decomposes at high temperatures.

$$HCONH_2(g) \rightleftharpoons NH_3(g) + CO(g)$$
$$K_c = 4.84 \text{ at } 400 \text{ K}$$

If 0.186 mol $HCONH_2(g)$ dissociates in a 2.16-L flask at 400 K, what will be the *total* pressure at equilibrium?

50. A mixture of 1.00 mol $NaHCO_3(s)$ and 1.00 mol $Na_2CO_3(s)$ is introduced into a 2.50-L flask in which the partial pressure of $CO_2(g)$ is 2.10 atm and that of $H_2O(g)$ is 715 mmHg. When equilibrium is established at 100 °C, will the partial pressures of $CO_2(g)$ and $H_2O(g)$ be greater or less than their initial partial pressures? Explain.

$$2\,NaHCO_3(s) \rightleftharpoons Na_2CO_3(s) + CO_2(g) + H_2O(g)$$
$$K_p(\text{atm}) = 0.23 \text{ at } 100 \text{ °C}$$

51. Cadmium metal is added to 0.350 L of an aqueous solution in which $[Cr^{3+}] = 1.00$ M. What are the concentrations of the different ionic species at equilibrium? What is the minimum mass of cadmium metal required to establish this equilibrium?

$$2\,Cr^{3+}(aq) + Cd(s) \rightleftharpoons 2\,Cr^{2+}(aq) + Cd^{2+}(aq)$$
$$K_c = 0.288$$

52. Lead metal is added to 0.100 M $Cr^{3+}(aq)$. What are $[Pb^{2+}]$, $[Cr^{2+}]$, and $[Cr^{3+}]$ when equilibrium is established in the reaction?

$$Pb(s) + 2\,Cr^{3+}(aq) \rightleftharpoons Pb^{2+}(aq) + 2\,Cr^{2+}(aq)$$
$$K_c = 3.2 \times 10^{-10}$$

53. The sketch at the left below represents an initial nonequilibrium mixture of $SO_2(g)$, $Cl_2(g)$ and $SO_2Cl_2(g)$ at a temperature at which $K_c = 4.0$ in the reversible reaction

$$SO_2(g) + Cl_2(g) \rightleftharpoons SO_2Cl_2(g)$$

Assume that the concentrations of the species in the reaction are in the same proportions as the numbers of molecules represented, and determine which of the three lettered sketches best represents an equilibrium mixture of this reaction. Explain your reasoning

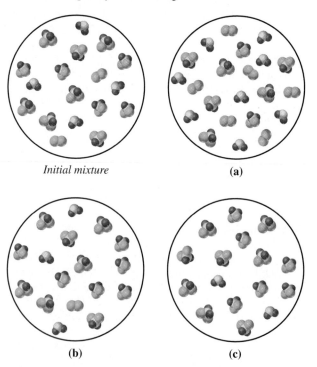

Initial mixture

(a)

(b)

(c)

54. The sketch at the upper left represents an initial nonequilibrium mixture of NO(g), Br_2(g), and NOBr(g) at a temperature at which $K_c = 3.0$ in the reversible reaction
$$2 \, NO(g) + Br_2(g) \rightleftharpoons 2 \, NOBr(g)$$
Assume that the concentrations of the species in the reaction are in the same proportions as the numbers of molecules represented, and determine which of the three lettered sketches best represents an equilibrium mixture of this reaction. Explain your reasoning.

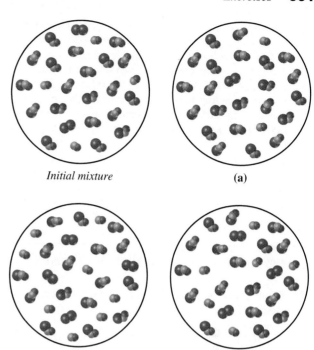

Initial mixture   **(a)**

(b)   **(c)**

Partial Pressure Equilibrium Constant, K_p

55. The following reaction is used in some self-contained breathing devices as a source of O_2(g).
$$4 \, KO_2(s) + 2 \, CO_2(g) \rightleftharpoons 2 \, K_2CO_3(s) + 3 \, O_2(g)$$
$$K_p = 28.5 \text{ at } 25 \, °C$$
Suppose that a sample of CO_2(g) is added to an evacuated flask containing KO_2(s) and equilibrium is established. If the equilibrium partial pressure of CO_2(g) is found to be 0.0721 atm, what are the equilibrium partial pressure of O_2(g) and the total gas pressure?

56. Concerning the reaction in Exercise 55, if KO_2(s) and K_2CO_3(s) are maintained in contact with air at 1.00 atm and 25 °C, in which direction will a net change occur to establish equilibrium, to the left or to the right? Explain.

[*Hint:* Recall equation (6.17) and the composition of air (Table 8.1.)]

57. 1.00 mol *each* of CO and Cl_2 are introduced into an evacuated 1.75-L flask, and the following equilibrium is established at 668 K.
$$CO(g) + Cl_2(g) \rightleftharpoons COCl_2(g) \quad K_p = 22.5$$
For this equilibrium, calculate **(a)** the partial pressure of $COCl_2$(g); **(b)** the total gas pressure.

58. Refer to Example 16-4. H_2S(g) at 747.6 mmHg pressure and a 1.85-g sample of I_2(s) are introduced into a 725-mL flask at 60 °C. What will be the total pressure in the flask at equilibrium?
$$H_2S(g) + I_2(s) \rightleftharpoons 2 \, HI(g) + S(s)$$
$$K_p(atm) = 1.34 \times 10^{-5} \text{ at } 60 \, °C$$

Le Châtelier's Principle

59. Continuous removal of one of the products of a chemical reaction has the effect of causing the reaction to go to completion. Explain this fact in terms of Le Châtelier's principle.

60. We can represent the freezing of H_2O(l) at 0 °C as H_2O (l, $d = 1.00$ g/cm³) $\rightleftharpoons$ H_2O(s, $d = 0.92$ g/cm³). Explain why increasing the pressure on ice causes it to melt. Is this the behavior you expect for solids in general? Explain.

61. Explain how each of the following affect the amount of H_2 present in an equilibrium mixture in the reaction
$$3 \, Fe(s) + 4 \, H_2O(g) \rightleftharpoons Fe_3O_4(s) + 4 \, H_2(g)$$
$$\Delta H° = -150 \text{ kJ}$$
(a) Raising the temperature of the mixture; **(b)** introducing more H_2O(g); **(c)** doubling the volume of the container holding the mixture; **(d)** adding an appropriate catalyst.

62. In the gas phase, iodine reacts with cyclopentene (C_5H_8) by a free radical mechanism to produce cyclopentadiene (C_5H_6) and hydrogen iodide. Explain how each of the following affects the amount of HI(g) present in the equilibrium mixture in the reaction
$$I_2(g) + C_5H_8(g) \rightleftharpoons C_5H_6(g) + 2 \, HI(g)$$
$$\Delta H° = 92.5 \text{ kJ}$$
(a) Raising the temperature of the mixture; **(b)** introducing more C_5H_6(g); **(c)** doubling the volume of the container holding the mixture; **(d)** adding an appropriate catalyst; **(e)** adding an inert gas such as He to a constant-volume reaction mixture.

63. The reaction $N_2(g) + O_2(g) \rightleftharpoons 2 \, NO(g)$, $\Delta H° = +181$ kJ, occurs in high-temperature combustion processes carried out in air. Oxides of nitrogen produced from the nitrogen and oxygen in air are intimately involved in the production of photochemical smog. What effect does increasing the temperature have on (a) the equilibrium production of $NO(g)$; (b) the rate of this reaction?

64. Use data from Appendix D to determine whether the forward reaction is favored by high temperatures or low temperatures.
   (a) $PCl_3(g) + Cl_2(g) \rightleftharpoons PCl_5(g)$
   (b) $SO_2(g) + 2 H_2S(g) \rightleftharpoons 2 H_2O(g) + 3 S(s)$
   (c) $2 N_2(g) + 3 O_2(g) + 4 HCl(g) \rightleftharpoons$
   $$4 NOCl(g) + 2 H_2O(g)$$

65. If the volume of an equilibrium mixture of $N_2(g)$, $H_2(g)$, and $NH_3(g)$ is reduced by doubling the pressure, will P_{N_2}

have increased, decreased, or remained the same when equilibrium is reestablished? Explain.
$$N_2(g) + 3 H_2(g) \rightleftharpoons 2 NH_3(g)$$

66. For the reaction
$$A(s) \rightleftharpoons B(s) + 2 C(g) + \frac{1}{2} D(g) \qquad \Delta H^\circ = 0$$
   (a) Will K_p increase, decrease, or remain constant with temperature? Explain.
   (b) If a *constant-volume* mixture at equilibrium at 298 K is heated to 400 K and equilibrium reestablished, will the number of moles of $D(g)$ increase, decrease, or remain constant? Explain.

Integrative and Advanced Exercises

67. Explain why the percent of molecules that dissociate into atoms in reactions of the type $I_2(g) \rightleftharpoons 2 I(g)$ *always* increases with an increase in temperature.

68. A 1.100-L flask at 25 °C and 1.00 atm pressure contains $CO_2(g)$ in contact with 100.0 mL of a saturated aqueous solution in which $[CO_2(aq)] = 3.29 \times 10^{-2}$ M.
   (a) What is the value of K_c at 25 °C for the equilibrium $CO_2(g) \rightleftharpoons CO_2(aq)$?
   (b) If 0.01000 mol of radioactive $^{14}CO_2$ is added to the flask, how many moles of the $^{14}CO_2$ will be found in the gas phase and in the aqueous solution when equilibrium is reestablished?
   (*Hint:* The radioactive $^{14}CO_2$ distributes itself between the two phases in exactly the same manner as the nonradioactive $^{12}CO_2$.)

69. Refer to Example 16-13. Suppose that 0.100 L of the equilibrium mixture is diluted to 0.250 L with water. What will be the new concentrations when equilibrium is reestablished?

70. In the equilibrium described in Example 16-12, the percent dissociation of N_2O_4 can be expressed as
$$\frac{3.00 \times 10^{-3} \text{ mol } N_2O_4}{0.0240 \text{ mol } N_2O_4 \text{ initially}} \times 100\% = 12.5\%$$

What must be the total pressure of the gaseous mixture if $N_2O_4(g)$ is to be 10.0% dissociated at 298 K?
$$N_2O_4 \rightleftharpoons 2 NO_2(g) \qquad K_p(\text{atm}) = 0.113 \text{ at 298 K}$$

71. Starting with $SO_3(g)$ at 1.00 atm, what will be the total pressure when equilibrium is reached in the following reaction at 700 K?
$$2 SO_3(g) \rightleftharpoons 2 SO_2(g) + O_2(g) \qquad K_p = 1.6 \times 10^{-5}$$

72. A sample of air with a mole ratio of N_2 to O_2 of 79:21 is heated to 2500 K. When equilibrium is established in a closed container with air initially at 1.00 atm, the mole percent of NO is found to be 1.8%. Calculate K_p for the reaction.
$$N_2(g) + O_2(g) \rightleftharpoons 2 NO(g)$$

73. Derive, by calculation, the equilibrium amounts of SO_2, O_2, and SO_3 listed in (a) Figure 16-6(c); (b) Figure 16-7(b).

74. The decomposition of salicylic acid to phenol and carbon dioxide was carried out at 200.0 °C, a temperature at which the reactant and products are all gaseous. A 0.300-g sample of salicylic acid was introduced into a 50.0-mL reaction vessel, and equilibrium was established. The equilibrium mixture was rapidly cooled to condense salicylic acid and phenol as solids; the $CO_2(g)$ was collected over mercury and its volume was measured at 20.0 °C and 730 mmHg. In two identical experiments, the volumes of $CO_2(g)$ obtained were 48.2 and 48.5 mL, respectively. Calculate K_p for this reaction.

75. One of the key reactions in the gasification of coal is the methanation reaction, in which methane is produced from synthesis gas—a mixture of CO and H_2.
$$CO(g) + 3 H_2(g) \rightleftharpoons CH_4(g) + H_2O(g)$$
$$\Delta H = -230 \text{ kJ}; \qquad K_c = 190 \text{ at 1000 K}$$
   (a) Is the equilibrium conversion of synthesis gas to methane favored at higher or lower temperatures? Higher or lower pressures?
   (b) Assume you have 4.00 mol of synthesis gas with a 3:1 mol ratio of $H_2(g)$ to $CO(g)$ in a 15.0-L flask. What will be the mole fraction of $CH_4(g)$ at equilibrium at 1000 K?

76. A sample of pure $PCl_5(g)$ is introduced into an evacuated flask and allowed to dissociate.
$$PCl_5(g) \rightleftharpoons PCl_3(g) + Cl_2(g)$$
   If the fraction of PCl_5 molecules that dissociate is denoted by α, and if the total gas pressure is P, show that
$$K_p = \frac{\alpha^2 P}{1 - \alpha^2}$$

77. Nitrogen dioxide obtained as a cylinder gas is always a mixture of $NO_2(g)$ and $N_2O_4(g)$. A 5.00-g sample obtained from such a cylinder is sealed in a 0.500-L flask at 298 K. What is the mole fraction of NO_2 in this mixture?

$$N_2O_4(g) \rightleftharpoons 2\ NO_2(g) \qquad K_c = 4.61 \times 10^{-3}$$

78. What is the apparent molar mass of the gaseous mixture that results when $COCl_2(g)$ is allowed to dissociate at 395 °C and a total pressure of 3.00 atm?

$$COCl_2(g) \rightleftharpoons CO(g) + Cl_2(g)$$
$$K_p = 4.44 \times 10^{-2} \text{ at } 395\ °C$$

Think of the apparent molar mass as the molar mass of a hypothetical single gas that is equivalent to the gaseous mixture.

79. Show that in terms of mole fractions of gases and *total* gas pressure the equilibrium constant expression for

$$N_2(g) + 3\ H_2(g) \rightleftharpoons 2\ NH_3(g)$$

is

$$K_p = \frac{(x_{NH_3})^2}{(x_{N_2})(x_{H_2})^3} \times \frac{1}{(P_{tot})^2}$$

80. For the synthesis of ammonia at 500 K, $N_2(g) + 3\ H_2(g) \rightleftharpoons 2\ NH_3(g)$, $K_p = 9.06 \times 10^{-2}$. Assume that N_2 and H_2 are mixed in the mole ratio $1:3$ and that the total pressure is maintained at 1.00 atm. What is the mole percent NH_3 at equilibrium?
(*Hint:* Use the equation from Exercise 79 .)

81. A mixture of $H_2S(g)$ and $CH_4(g)$ in the mole ratio $2:1$ was brought to equilibrium at 700 °C and a total pressure of 1 atm. The *equilibrium* mixture was analyzed for the amount of H_2S, and 9.54×10^{-3} mol H_2S was found. The CS_2 present at equilibrium was converted successively to H_2SO_4 and then to $BaSO_4$: 1.42×10^{-3} mol $BaSO_4$ was obtained. Use these data to determine K_p at 700 °C for the reaction

$$2\ H_2S(g) + CH_4(g) \rightleftharpoons CS_2(g) + 4\ H_2(g)$$
$$K_p \text{ at } 700\ °C = ?$$

82. A solution is prepared having these initial concentrations: $[Fe^{3+}] = [Hg_2^{2+}] = 0.5000$ M; $[Fe^{2+}] = [Hg^{2+}] = 0.03000$ M. The following reaction occurs among the ions at 25 °C.

$$2\ Fe^{3+}(aq) + Hg_2^{2+}(aq) \rightleftharpoons 2\ Fe^{2+}(aq) + 2\ Hg^{2+}(aq)$$
$$K_c = 9.14 \times 10^{-6}$$

What will be the ion concentrations at equilibrium?

83. Refer to the Integrative Example. A gaseous mixture is prepared containing 0.100 mol each of $CH_4(g)$, $H_2O(g)$, $CO_2(g)$, and $H_2(g)$ in a 5.00-L flask. Then the mixture is allowed to come to equilibrium at 1000 K in reaction (16.19). What will be the equilibrium amount, in moles, of each gas?

84. Concerning the reaction in Exercise 38 and the situation described in part (c) of that exercise, will the mole fraction of $C_2H_2(g)$ increase, decrease, or remain unchanged when equilibrium is reestablished? Explain.

85. The formation of nitrosyl chloride is given by the following equation: $2\ NO(g) + Cl_2(g) \rightleftharpoons 2\ NOCl(g)$; $K_c = 4.6 \times 10^4$ at 298 K. In a 1.50-L flask, there are 4.125 mol of NOCl and 0.1125 mol of Cl_2 present at equilibrium (298 K).
(a) Determine the partial pressure of NO at equilibrium.
(b) What is the total pressure of the system at equilibrium?

86. The method of extracting liquid ammonia from equilibrium mixtures in the ammonia synthesis is suggested by Figure 8-4. At 500 K, a 10.0-L equilibrium mixture contains 0.424 mol N_2, 1.272 mol H_2, and 1.152 mol NH_3. The mixture is quickly chilled to a temperature at which the NH_3 liquefies, and the $NH_3(l)$ is completely removed. The 10.0-L gaseous mixture is then returned to 500 K, and equilibrium is reestablished. How many moles of $NH_3(g)$ will be present in the new equilibrium mixture?

$$N_2(g) + 3\ H_2(g) \rightleftharpoons 2\ NH_3 \qquad K_c = 152 \text{ at } 500\ K$$

87. Recall the formation of methanol from synthesis gas, the reversible reaction with which we began our discussion of the equilibrium constant expression on page 627.

$$CO(g) + 2\ H_2(g) \rightleftharpoons CH_3OH(g) \qquad K_c = 14.5 \text{ at } 483\ K$$

A particular synthesis gas consisting of 35.0 mole percent $CO(g)$ and 65.0 mole percent $H_2(g)$ at a total pressure of 100.0 atm at 483 K is allowed to come to equilibrium. Determine the partial pressure of $CH_3OH(g)$ in the equilibrium mixture.

Feature Problems

88. A classic experiment in equilibrium studies dating from 1862 involved the reaction in solution of ethanol (C_2H_5OH) and acetic acid (CH_3COOH) to produce ethyl acetate and water.

$$C_2H_5OH + CH_3COOH \rightleftharpoons CH_3COOC_2H_5 + H_2O$$

The reaction can be followed by analyzing the equilibrium mixture for its acetic acid content.

$$2\ CH_3COOH(aq) + Ba(OH)_2(aq) \longrightarrow$$
$$Ba(CH_3COO)_2(aq) + 2\ H_2O(l)$$

In one experiment, a mixture of 1.000 mol acetic acid and 0.5000 mol ethanol is brought to equilibrium. A sample containing exactly one-hundredth of the equilibrium mixture requires 28.85 mL 0.1000 M $Ba(OH)_2$ for its titration. Calculate the equilibrium constant, K_c, for the ethanol–acetic acid reaction based on this experiment.

89. The decomposition of HI(g) is represented by the equation

$$2\ HI(g) \rightleftharpoons H_2(g) + I_2(g)$$

HI(g) is introduced into five identical 400-cm³ glass bulbs, and the five bulbs are maintained at 623 K. Each bulb is opened after a period of time and analyzed for I_2 by titration with 0.0150 M $Na_2S_2O_3(aq)$.

$$I_2(aq) + 2\ Na_2S_2O_3(aq) \longrightarrow Na_2S_4O_6(aq) + 2\ NaI(aq)$$

Data for this experiment are provided in the table below. What is the value of K_c at 623 K?

Bulb Number	Initial Mass of HI(g), g	Time Bulb Opened (h)	Volume 0.0150 M $Na_2S_2O_3$ Required for Titration (in mL)
1	0.300	2	20.96
2	0.320	4	27.90
3	0.315	12	32.31
4	0.406	20	41.50
5	0.280	40	28.68

90. In one of Fritz Haber's experiments to establish the conditions required for the ammonia synthesis reaction, pure $NH_3(g)$ was passed over an iron catalyst at 901 °C and 30.0 atm. The gas leaving the reactor was bubbled through 20.00 mL of a HCl(aq) solution. In this way, the $NH_3(g)$ present was removed by reaction with HCl. The remaining gas occupied a volume of 1.82 L at STP. The 20.00 mL of HCl(aq) through which the gas had been bubbled required 15.42 mL of 0.0523 M KOH for its titration. Another 20.00-mL sample of the same HCl(aq) through which no gas had been bubbled required 18.72 mL of 0.0523 M KOH for its titration. Use these data to obtain a value of K_p at 901 °C for the reaction $N_2(g) + 3 H_2(g) \rightleftharpoons 2 NH_3(g)$.

91. The following is an approach to establishing a relationship between the equilibrium constant and rate constants mentioned in the Are You Wondering feature on page 632.
- Work with the detailed mechanism for the reaction.
- Use the principle of microscopic reversibility, the idea that every step in a reaction mechanism is reversible. (In the presentation of elementary reactions in Chapter 15, we treated some reaction steps as reversible and others as going to completion. However, as noted in Table 16.3, every reaction has an equilibrium constant even though a reaction is generally considered to go to completion if its equilibrium constant is very large.)
- Use the idea that when equilibrium is attained in an overall reaction, it is also attained in each step of its mechanism. Moreover, we can write an equilibrium constant expression for each step in the mechanism, similar to what we did with the steady-state assumption in describing reaction mechanisms.
- Combine the K_c expressions for the elementary steps into a K_c expression for the overall reaction. The numerical value of the overall K_c can thereby be expressed as a ratio of rate constants, k.

Use this approach to establish the equilibrium constant expression for the overall reaction,

$$H_2(g) + I_2(g) \rightleftharpoons 2 HI(g)$$

The mechanism of the reaction appears to be the following:

Fast: $\qquad\qquad I_2(g) \rightleftharpoons 2 I(g)$

Slow: $2 I(g) + H_2(g) \rightleftharpoons 2 HI(g)$

 eMedia Exercises

92. After viewing the **Dynamic Equilibrium** animation of equilibrium through a physical change *(eChapter 16-1)*, describe the influence of temperature on rates of change.

93. After viewing the **Chemical Equilibrium** animation *(eChapter 16-2)*, estimate the value of the equilibrium constant of the reaction depicted. What information did you use to make this estimate?

94. (a) Describe how the stoichiometry of the reaction in the **Chemical Equilibrium** animation *(eChapter 16-2)* influences the values of concentrations at equilibrium. **(b)** If the initial concentrations of both reactants are doubled in this exercise, does this change the value of the equilibrium constant? **(c)** Does this change, if any, affect the concentration of the product at equilibrium?

95. Using equal initial concentrations of reactant and product in the **Equilibrium Constant** activity *(eChapter 16-4)*, describe how the value of the equilibrium constant is related to the concentration of the product once equilibrium has been reached.

96. (a) What would be the change in the results seen in the $NO_2 - N_2O_4$ **Equilibrium** animation *(eChapter 16-6)* if the pressure were *decreased* by a factor of two instead of increased? **(b)** Given that this forward reaction is exothermic, what would be the effect on the ratio of the NO_2 and N_2O_4 concentrations by increasing the temperature of the system?

17 Acids and Bases

Contents

Fruit juices contain a variety of acids. Orange juice, for example, is an excellent source of citric acid, $H_3C_6H_5O_7$. Citric acid is a polyprotic acid, a type of acid discussed in Section 17-6.

The general public is familiar with the concepts of acids and bases. The environmental problem of acid rain is a popular topic in newspapers and magazines, and television commercials mention pH in relation to such products as deodorants, shampoos, and antacids.

Chemists have been classifying substances as acids and bases for a long time. Antoine Lavoisier thought that the common element in all acids was oxygen, a fact conveyed by its name. Oxygen means "acid former" in Greek. In 1810, Humphry Davy showed that hydrogen instead is the element that acids have in common. In 1884, Svante Arrhenius developed the theory of acids and bases that we introduced in Chapter 5. There, we emphasized the stoichiometry of acid–base reactions.

Some of the topics we will study in this chapter are modern acid–base theories, factors affecting the strengths of acids and bases, the pH scale, and the calculation of ion concentrations in solutions of weak acids and bases. At the end of the chapter, we will bring some of these ideas together in a discussion of acid rain.

17-1   The Arrhenius Theory: A Brief Review

Some aspects of the behavior of acids and bases can be explained adequately with the theory developed by Arrhenius as part of his studies of electrolytic dissociation (Section 14-9). Arrhenius proposed that in an aqueous solution, a strong electrolyte exists only in the form of ions, whereas a weak electrolyte exists partly as ions and partly as molecules. When the acid HCl dissolves in water, the HCl molecules ionize completely, yielding hydrogen ions, H^+, as one of the products.

$$HCl(g) \xrightarrow{\text{H}_2\text{O}} H^+(aq) + Cl^-(aq)$$

When the base NaOH dissolves in water, the Na^+ and OH^- ions in the solid become dissociated from one another through the action of H_2O molecules (recall Figure 14-6).

$$NaOH(s) \xrightarrow{\text{H}_2\text{O}} Na^+(aq) + OH^-(aq)$$

We can represent the neutralization reaction of HCl and NaOH with the ionic equation

$$\underset{\text{an acid}}{H^+(aq) + Cl^-(aq)} + \underset{\text{a base}}{Na^+(aq) + OH^-(aq)} \longrightarrow \underset{\text{a salt}}{Na^+(aq) + Cl^-(aq)} + \underset{\text{water}}{H_2O(l)}$$

or, perhaps better still, with the net ionic equation

$$\underset{\text{an acid}}{H^+(aq)} + \underset{\text{a base}}{OH^-(aq)} \longrightarrow \underset{\text{water}}{H_2O(l)} \tag{17.1}$$

Equation (17.1) illustrates an essential idea of the Arrhenius theory: *A neutralization reaction involves the combination of hydrogen ions and hydroxide ions to form water.*

Despite its early successes and continued usefulness, the Arrhenius theory does have limitations. One of the most glaring is in its treatment of the weak base ammonia, NH_3. The Arrhenius theory suggests that all bases contain OH^-. Where is the OH^- in NH_3? To get around this difficulty, chemists began to think of aqueous solutions of NH_3 as containing the compound ammonium hydroxide, NH_4OH, which as a weak base is partially ionized into NH_4^+ and OH^- ions:

$$NH_3(g) + H_2O(l) \longrightarrow NH_4OH(aq)$$

$$NH_4OH(aq) \rightleftharpoons NH_4^+(aq) + OH^-(aq)$$

The problem with this formulation is that there is no compelling evidence that NH_4OH exists in aqueous solutions. We should always question a hypothesis or theory that postulates the existence of hypothetical substances. As we shall see in Section 17-2, the essential failure of the Arrhenius theory is in not recognizing the key role of the *solvent* in the ionization of a solute.

▲ A holdover of the Arrhenius theory. Although there is no compelling evidence that NH_4OH molecules exist in $NH_3(aq)$, solutions are common-

KEEP IN MIND
that a "proton donor" is a donor of H^+ ions. That is, a hydrogen atom consists of one proton and one electron, and the hydrogen ion, H^+, is simply a proton. ▶

17-2   Brønsted–Lowry Theory of Acids and Bases

In 1923, J. N. Brønsted in Denmark and T. M. Lowry in Great Britain independently proposed a new acid–base theory. According to their theory, an acid is a **proton donor** and a base is a **proton acceptor**. To describe the behavior of ammonia as a base, which we found difficult to do with the Arrhenius theory, we can write

$$NH_3 + H_2O \longrightarrow NH_4^+ + OH^- \qquad (17.2)$$
$$\text{base} \qquad \text{acid}$$

In reaction (17.2), H_2O acts as an *acid*. It gives up a proton, H^+, which is taken away by NH_3, a *base*. As a result of this transfer, the polyatomic ions NH_4^+ and OH^- are formed—the same ions produced by the ionization of the hypothetical NH_4OH of the Arrhenius theory. Because NH_3 is a *weak* base, we need also to consider the *reverse* of reaction (17.2). In the reverse reaction, NH_4^+ is an *acid* and OH^- is a *base*.

$$NH_4^+ + OH^- \longrightarrow NH_3 + H_2O \qquad (17.3)$$
$$\text{acid} \qquad \text{base}$$

The conventional way to represent a reversible reaction is to use the double arrow notation. In addition, we can place acid and base labels under all four species in the reaction.

$$NH_3 + H_2O \rightleftharpoons NH_4^+ + OH^- \qquad (17.4)$$
$$\text{base(1) \quad acid(2)} \qquad \text{acid(1) \quad base(2)}$$

KEEP IN MIND ▶

that in designating conjugate pairs, it does not matter which conjugate pair we call (1) and which we call (2). Nor does it matter in what order the acid and base are written on each side of the equation.

We have labeled the NH_3/NH_4^+ combination as "(1)" and H_2O/OH^- as "(2)." Each combination is called a *conjugate pair*. An NH_3 molecule acts as a base by accepting a proton, and an NH_4^+ ion is the **conjugate acid** of NH_3. Similarly, in reaction (17.4) H_2O is an acid and OH^- is its **conjugate base**. Figure 17-1 illustrates the proton transfer involved in the forward and reverse reactions of (17.4).

On the basis of what we learned in Chapter 16, we might write as the equilibrium constant expression for reaction (17.4)

$$K_c = \frac{[NH_4^+][OH^-]}{[NH_3][H_2O]}$$

In aqueous ammonia, however, H_2O molecules are present in such overwhelming numbers compared with NH_3 molecules and NH_4^+ and OH^- ions that water, the solvent, is essentially a pure liquid with an activity of one (recall page 631). For this reason, the equilibrium constant expression we want does not include the term $[H_2O]$.

$$K_b = \frac{[NH_4^+][OH^-]}{[NH_3]} = 1.8 \times 10^{-5}$$

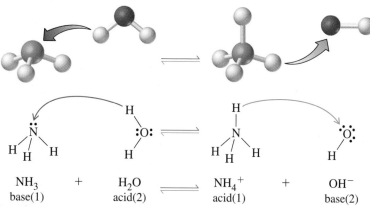

$$NH_3 \qquad + \qquad H_2O \qquad \rightleftharpoons \qquad NH_4^+ \qquad + \qquad OH^-$$
$$\text{base(1)} \qquad\qquad \text{acid(2)} \qquad\qquad \text{acid(1)} \qquad\qquad \text{base(2)}$$

▲ **FIGURE 17-1   Brønsted–Lowry acid–base reaction: weak base**
The arrows represent the proton transfer in reaction (17.4). The red arrows represent the forward reaction; the blue arrows, the reverse reaction. Because NH_4^+ is a stronger acid than H_2O and OH^- is a stronger base than NH_3, the reverse reaction proceeds to a greater extent than does the forward reaction. Hence, NH_3 is only slightly ionized.

The equilibrium constant K_b is called the **base ionization constant**. We can express the ionization of acetic acid as

$$HC_2H_3O_2 + H_2O \rightleftharpoons C_2H_3O_2^- + H_3O^+$$

$$\text{acid(1)} \qquad \text{base(2)} \qquad \text{base(1)} \qquad \text{acid(2)}$$

Here, acetate ion, $C_2H_3O_2^-$, is the conjugate base of the acid $HC_2H_3O_2$. This time, H_2O acts as a base. Its conjugate acid is **hydronium ion, H_3O^+**. In Chapter 5, we discussed the formation of the hydronium ion when an acid dissociates. Because the H^+ ion is very small, the positive charge of this ion is concentrated in a small region; the ion has a high positive charge density. We should expect H^+ ions (protons) to seek out centers of negative charge with which to bond. When a H^+ ion attaches to a lone pair of electrons in an O atom in H_2O, the resulting hydronium ion forms hydrogen bonds with several water molecules (Figure 17-2). Figure 17-3 illustrates the proton transfer involved in the forward and reverse reactions of the ionization of acetic acid.

Using the same approach as for $NH_3(aq)$, we can describe the ionization of acetic acid in the following way.

$$K_a = \frac{[C_2H_3O_2^-][H_3O^+]}{[HC_2H_3O_2]} = 1.8 \times 10^{-5}$$

The equilibrium constant K_a is called the **acid ionization constant**. (The fact that K_a of acetic acid and K_b of ammonia have the same value is just a coincidence.)

We can represent the ionization of HCl in the same way that we did for acetic acid. In this case, however, because K_a is so large (about 10^6), we can treat the ionization of HCl as a reaction that goes to completion. We denote this by writing the ionization equation with a single arrow.

$$HCl + H_2O \longrightarrow Cl^- + H_3O^+ \qquad (17.5)$$

▶ The Lewis structure of the hydronium ion is

$$\left[H - \overset{\displaystyle ..}{\underset{\displaystyle |}{O}} - H \atop H \right]^+$$

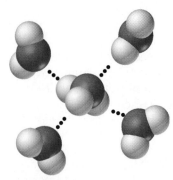

▲ **FIGURE 17-2**

A hydrated hydronium ion
This species, $H_{11}O_5^+$, consists of a central H_3O^+ ion hydrogen-bonded to four H_2O molecules.

Hydronium Ion model

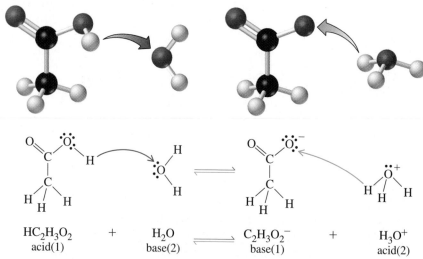

$$\begin{array}{ccccccc}
HC_2H_3O_2 & + & H_2O & \rightleftharpoons & C_2H_3O_2^- & + & H_3O^+ \\
\text{acid(1)} & & \text{base(2)} & & \text{base(1)} & & \text{acid(2)}
\end{array}$$

▲ **FIGURE 17-3 Brønsted–Lowry acid–base reaction: weak acid**
The arrows represent the proton transfer in the ionization of acetic acid. The red arrows represent the forward reaction; the blue arrows, the reverse reaction. Because H_3O^+ is a stronger acid than $HC_2H_3O_2$ and $C_2H_3CO^-$ is a stronger base than H_2O, the reverse reaction proceeds to a greater extent than does the forward reaction. Hence, $HC_2H_3O_2$ is only slightly ionized.

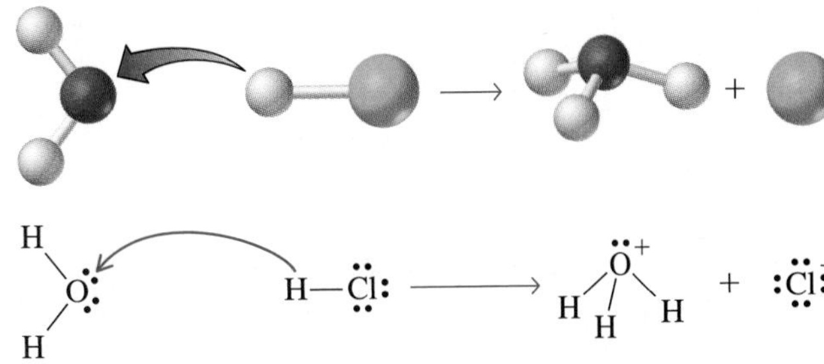

▲ FIGURE 17-4 Brønsted–Lowry acid–base reaction: strong acid
The red arrows represent the proton transfer in the ionization of hydrochloric acid. Because H_3O^+ is a weaker acid than HCl and Cl^- is a much weaker base than H_2O, the forward reaction proceeds almost to completion. Hence, HCl is essentially completely ionized.

Figure 17-4 illustrates the proton transfer involved in the complete ionization of hydrochloric acid.

In Example 17-1, we identify acids and bases in some typical acid–base reactions. In working through this example, notice the following additional features: (1) Any species that is an acid according to the Arrhenius theory is also an acid according to the Brønsted–Lowry theory; the same is true of bases. (2) Certain species, even though they do not contain the OH group, produce OH^- in aqueous solution—for example, OCl^-. As such, they are Brønsted–Lowry bases. (3) The Brønsted–Lowry theory accounts for substances that can act either as an acid or a base; they are said to be **amphiprotic**. The Arrhenius theory does not account for the behavior of amphiprotic substances.

▶ The term *amphiprotic* is similar to the term *amphoteric*, which indicates the ability of a substance to behave as both an acid and a base. Amphiprotic conveys the notion of proton transfer embodied in the Brønsted–Lowry theory of acids and bases.

EXAMPLE 17-1

Identifying Brønsted–Lowry Acids and Bases and Their Conjugates. For each of the following, identify the acids and bases in both the forward and reverse reactions in the manner shown in equation (17.4).

(a) $HClO_2 + H_2O \rightleftharpoons ClO_2^- + H_3O^+$

(b) $OCl^- + H_2O \rightleftharpoons HOCl + OH^-$

(c) $NH_3 + H_2PO_4^- \rightleftharpoons NH_4^+ + HPO_4^{2-}$

(d) $HCl + H_2PO_4^- \rightleftharpoons Cl^- + H_3PO_4$

Solution

Consider $HClO_2$ in reaction (a). It gives up a proton, H^+, to become ClO_2^-. Therefore, $HClO_2$ is an acid, and ClO_2^- is its conjugate base. Now consider H_2O. It takes the proton from $HClO_2$ and becomes H_3O^+. Thus, H_2O is a base, and H_3O^+ is its conjugate acid. In reaction (b), OCl^- is a base and gains a proton from water. OH^- produced in this reaction is the conjugate base of H_2O. Taken together, reactions (a) and (b) show that H_2O is amphiprotic. Reactions (c) and (d) illustrate the same for $H_2PO_4^-$.

(a) $\quad HClO_2 \; + \; H_2O \rightleftharpoons ClO_2^- \; + \; H_3O^+$
$\qquad$ acid(1) $\quad$ base(2) $\qquad$ base(1) $\quad$ acid(2)

(b) $\quad OCl^- \; + \; H_2O \rightleftharpoons HOCl \; + \; OH^-$
$\qquad$ base(1) $\quad$ acid(2) $\qquad$ acid(1) $\quad$ baase(2)

(c) $NH_3 + H_2PO_4^- \rightleftharpoons NH_4^+ + HPO_4^{2-}$

$\quad$ base(1) $\quad$ acid(2) $\quad\quad$ acid(1) $\quad$ base(2)

(d) $HCl + H_2PO_4^- \rightleftharpoons Cl^- + H_3PO_4$

$\quad$ acid(1) $\quad$ base(2) $\quad\quad$ base(1) $\quad$ acid(2)

Practice Example A: For each of the following reactions, identify the acids and bases in both the forward and reverse directions.

(a) $HF + H_2O \rightleftharpoons F^- + H_3O^+$

(b) $HSO_4^- + NH_3 \rightleftharpoons SO_4^{2-} + NH_4^+$

(c) $C_2H_3O_2^- + HCl \rightleftharpoons HC_2H_3O_2 + Cl^-$

Practice Example B: Of the following species, one is acidic, one is basic, and one is amphiprotic in their reactions with water: HNO_2, PO_4^{3-}, HCO_3^-. Write the *four* equations needed to represent these facts.

The ionization of HCl in aqueous solution (reaction 17.5) goes to completion because HCl is a strong acid; it readily gives up protons to H_2O. At the same time, Cl^- ion, the conjugate base of HCl, has very little tendency to take a proton from H_3O^+; Cl^- is a very weak base. This observation suggests the generalization that follows.

> In an acid–base reaction, the favored direction of the reaction is from the stronger to the weaker member of a conjugate acid–base pair.

With this generalization, we can predict that the neutralization of HCl by OH^- should go to completion.

$$HCl + OH^- \longrightarrow Cl^- + H_2O$$

$\quad$ acid(1) $\quad$ base(2) $\quad\quad$ base(1) $\quad$ acid(2)

$\quad$ strong $\quad$ strong $\quad\quad$ weak $\quad\quad$ weak

And we would predict that the following reaction should occur almost exclusively in the *reverse* direction.

$$H_2O + I^- \longleftarrow OH^- + HI$$

$\quad$ acid(1) $\quad$ base(2) $\quad\quad$ base(1) $\quad$ acid(2)

$\quad$ weak $\quad\quad$ weak $\quad\quad\quad$ strong $\quad$ strong

► The acid–base strengths listed in Table 17.1 are the result of experiments carried out by many chemists.

To be able to apply the generalization more broadly, though, we need a tabulation of acid and base strengths, such as that in Table 17.1. The strongest acids are at the top of the column on the left, and the strongest bases are at the bottom of the column on the right. It is important to note that *the stronger an acid, the weaker its conjugate base*.

Both HCl and $HClO_4$ are strong acids because H_2O is a sufficiently strong base to take protons from either acid in a reaction that goes to completion. Because both HCl and $HClO_4$ react to completion with water, yielding H_3O^+ (the strongest acid possible in water), the solvent water is said to have a *leveling effect* on these two acids. That is, we cannot say which is the stronger acid of the two in water. How can we say, then, as we indicate in Table 17.1, that $HClO_4$ is a stronger acid than HCl?

TABLE 17.1   Relative Strengths of Some Common Brønsted–Lowry Acid and Bases

Acid		Conjugate Base	
Perchloric acid	$HClO_4$	Perchlorate ion	ClO_4^-
Hydroiodic acid	HI	Iodide ion	I^-
Hydrobromic acid	HBr	Bromide ion	Br^-
Hydrochloric acid	HCl	Chloride ion	Cl^-
Sulfuric acid	H_2SO_4	Hydrogen sulfate ion	HSO_4^-
Nitric acid	HNO_3	Nitrate ion	NO_3^-
Hydronium ion[a]	H_3O^+	Water[a]	H_2O
Hydrogen sulfate ion	HSO_4^-	Sulfate ion	SO_4^{2-}
Nitrous acid	HNO_2	Nitrite ion	NO_2^-
Acetic acid	$HC_2H_3O_2$	Acetate ion	$C_2H_3O_2^-$
Carbonic acid	H_2CO_3	Hydrogen carbonate ion	HCO_3^-
Ammonium ion	NH_4^+	Ammonia	NH_3
Hydrogen carbonate ion	HCO_3^-	Carbonate ion	CO_3^{2-}
Water	H_2O	Hydroxide ion	OH^-
Methanol	CH_3OH	Methoxide ion	CH_3O^-
Ammonia	NH_3	Amide ion	NH_2^-

Increasing acid strength (↑)   *Increasing base strength* (↓)

[a] The hydronium ion–water combination refers to the ease with which a proton is passed from one water molecule to another; that is, $H_3O^+ + H_2O \rightleftharpoons H_2O + H_3O^+$

To determine whether $HClO_4$ or HCl is the stronger acid, we need to use a solvent that is a weaker base than water—a solvent that will take protons from the stronger of the two acids more readily than from the weaker one. In the solvent $(C_2H_5)_2O$, diethyl ether, $HClO_4$ is completely ionized, but HCl is only partially ionized. Thus, $HClO_4$ is a stronger acid than is HCl.

$$HClO_4 + C_2H_5 - \overset{..}{\underset{..}{O}} - C_2H_5 \longrightarrow ClO_4^- + [C_2H_5 - \overset{\overset{H}{|}}{\underset{..}{O}} - C_2H_5]^+$$

$$HCl + C_2H_5 - \overset{..}{\underset{..}{O}} - C_2H_5 \rightleftharpoons Cl^- + [C_2H_5 - \overset{\overset{H}{|}}{\underset{..}{O}} - C_2H_5]^+$$

17-3   The Self-Ionization of Water and the pH Scale

Even when it is pure, water contains a very low concentration of ions that can be detected in precise electrical conductivity measurements. The ions form as a result of the amphiprotic nature of water; some water molecules donate protons and others accept protons. In the **self-ionization** (or *autoionization*) of water, for each H_2O molecule that acts as an acid, another H_2O molecule acts as a base, and hydronium (H_3O^+) and hydroxide (OH^-) ions are formed. The reaction is reversible, and in the reverse reaction, H_3O^+ releases a proton to OH^-. The reverse reaction is far more significant than the forward reaction. *Equilibrium is displaced far to the left*. In reaction (17.6), acid(1) and base(2) are *much* stronger than are acid(2) and base(1).

Self-Ionization of Water animation

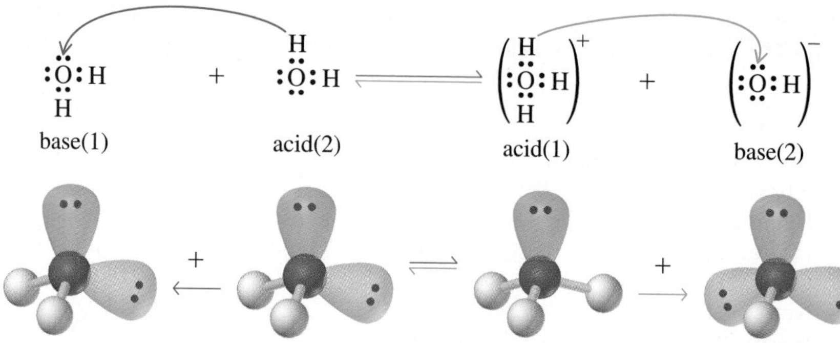

$$\underset{\text{base(1)}}{\text{:O:H}} \quad + \quad \underset{\text{acid(2)}}{\text{:O:H}} \quad \rightleftharpoons \quad \underset{\text{acid(1)}}{\left(\text{:O:H}\right)^{+}} \quad + \quad \underset{\text{base(2)}}{\left(\text{:O:H}\right)^{-}} \quad (17.6)$$

Again, we follow the approach we used in writing equilibrium constants for the ionization of NH_3 and $HC_2H_3O_2$, namely we assume an activity of one for H_2O molecules and replace activities of other species by their molarities. For the self-ionization of water

$$H_2O + H_2O \rightleftharpoons H_3O^+ + OH^-$$

we can write

$$K = [H_3O^+][OH^-]$$

Equation (17.6) indicates that $[H_3O^+]$ and $[OH^-]$ are equal in pure water. There are several experimental methods of determining these concentrations. All lead to this result.

At 25 °C in pure water: $\quad [H_3O^+] = [OH^-] = 1.0 \times 10^{-7}\,M$

The equilibrium constant for the self-ionization of water is called the **ion product of water**. It is symbolized as K_w. At 25 °C,

$$K_w = [H_3O^+][OH^-] = 1.0 \times 10^{-14} \quad (17.7)$$

Since K_w is an equilibrium constant, the product of the concentrations of the hydronium and hydroxide ions must always equal 10^{-14}. If we increase the concentration of H_3O^+ by adding an acid, then the concentration of OH^- must decrease to maintain the value of K_w. If we increase the concentration of OH^- by adding a base, then the concentration of H_3O^+ must decrease. Equation (17.7) connects the concentrations of H_3O^+ and OH^- and applies to *all* aqueous solutions, not just to pure water, as we shall see shortly.

pH and pOH

Because their product in an aqueous solution is only 1.0×10^{-14}, we expect $[H_3O^+]$ and $[OH^-]$ also to be small. Typically, they are less than 1 M—often very much less. Exponential notation is useful in these situations; for example, $[H_3O^+] = 2.2 \times 10^{-13}\,M$. But we now want to consider an even more convenient way to describe hydronium and hydroxide ion concentrations.

In 1909, the Danish biochemist Søren Sørensen proposed the term **pH** to refer to the "potential of hydrogen ion." He defined pH as the *negative of the logarithm of* $[H^+]$. Restated in terms of $[H_3O^+]^*$,

$$pH = -\log[H_3O^+] \quad (17.8)$$

KEEP IN MIND ▶

that this definition of pH is one of the few scientific expressions that uses logarithms to the base 10 (log) rather than natural logarithms (ln).

---

*Strictly speaking, we should use the *activity* of H_3O^+, $a_{H_3O^+}$, a dimensionless quantity. But we will not use activities here, just as we did not use them in Chapter 16. We will substitute the numerical value of the molarity of H_3O^+ for its activity and recognize that some pH calculations may be only approximations.

Thus, in a solution that is 0.0025 M HCl,

$$[H_3O^+] = 2.5 \times 10^{-3} \text{ M} \quad \text{and} \quad \text{pH} = -\log(2.5 \times 10^{-3}) = 2.60$$

To determine the $[H_3O^+]$ that corresponds to a particular pH value, we do an inverse calculation. In a solution with pH = 4.50,

$$\log[H_3O^+] = -4.50 \quad \text{and} \quad [H_3O^+] = 10^{-4.50} = 3.2 \times 10^{-5} \text{ M}$$

We can also define the quantity **pOH**:

$$\text{pOH} = -\log[OH^-] \quad (17.9)$$

And we can derive still another useful expression by taking the *negative logarithm* of the K_w expression (written for 25 °C) and introducing the symbol pK_w.

$$K_w = [H_3O^+][OH^-] = 1.0 \times 10^{-14}$$
$$-\log K_w = -(\log[H_3O^+][OH^-]) = -\log(1.0 \times 10^{-14})$$
$$\text{p}K_w = -(\log[H_3O^+] + \log[OH^-]) = -(-14.00)$$
$$= -\log[H_3O^+] - \log[OH^-] = 14.00$$

$$\text{p}K_w = \text{pH} + \text{pOH} = 14.00 \quad (17.10)$$

An aqueous solution with $[H_3O^+] = [OH^-]$ is said to be *neutral*. In pure water at 25 °C, $[H_3O^+] = [OH^-] = 1.0 \times 10^{-7}$ M and pH = 7.00. Thus at 25 °C, all aqueous solutions with pH = 7.00 are neutral. If the pH is less than 7.00, the solution is *acidic*; if the pH is greater than 7.00, the solution is *basic*, or alkaline. Equation 17.10 is a restatement of the interrelationship between $[H_3O^+]$, $[OH^-]$, and K_w in terms of pH, pOH, and pK_w. If we know the value of either $[H_3O^+]$ or $[OH^-]$, we can calculate the value of the other (Figure 17-5).

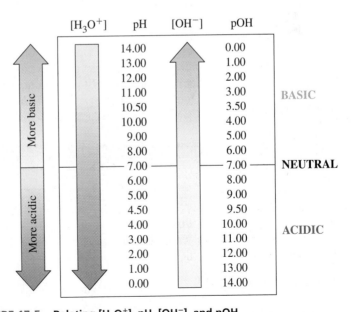

▲ FIGURE 17-5   Relating [H₃O⁺], pH, [OH⁻], and pOH
In aqueous solutions, the sum of the pH and pOH values always gives pK_w = 14 because of the self-ionization equilibrium of water.

▶ The determination of logarithms and inverse logarithms (antilogarithms) is discussed in Appendix A. Significant figure rules for logarithms are also presented there.

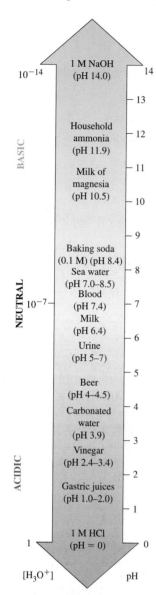

▲ FIGURE 17-6

The pH scale and pH values of some common materials

The scale shown here ranges from pH 0 to pH 14. Slightly negative pH values, perhaps to about −1 (corresponding to $[H_3O^+] \approx 10$ M), are possible. Also possible are pH values up to about 15 (corresponding to $[OH^-] \approx 10$ M). For practical purposes, however, the pH scale is useful only in the range $2 < pH < 12$, because the molarities of H_3O^+ and OH^- in concentrated acids and bases may differ significantly from their true activities.

The pH values of a number of materials are depicted in Figure 17-6. These values and the many examples in this chapter and the next should familiarize you with the pH concept. Later, we will consider two methods for measuring pH: by means of acid–base indicators (Section 18-3) and electrical measurements (Section 21-4).

EXAMPLE 17-2

Relating $[H_3O^+]$, $[OH^-]$, pH, and pOH. In a laboratory experiment, students measured the pH of samples of rainwater and household ammonia. Determine **(a)** $[H_3O^+]$ in the rainwater, with pH measured at 4.35; **(b)** $[OH^-]$ in the ammonia, with pH measured at 11.28.

Solution

(a) By definition, $pH = -\log[H_3O^+]$, or

$$\log[H_3O^+] = -pH = -4.35$$
$$[H_3O^+] = 10^{-4.35} = 4.5 \times 10^{-5} \text{ M}$$

(b) First, determine pOH with equation (17.10).

$$pOH = 14.00 - pH = 14.00 - 11.28 = 2.72$$

Now, use the definition $pOH = -\log[OH^-]$.

$$\log[OH^-] = -pOH = -2.72$$
$$[OH^-] = 10^{-2.72} = 1.9 \times 10^{-3} \text{ M}$$

Practice Example A: Students found a sample of yogurt to have a pH of 2.85. What are $[H^+]$ and $[OH^-]$ in the yogurt?

Practice Example B: The pH of a solution of HCl in water is found to be 2.50. What volume of water would you add to 1.00 L of this solution to raise the pH to 3.10?

17-4 Strong Acids and Strong Bases

As equation (17.5) indicated, the ionization of HCl in dilute aqueous solutions

$$HCl + H_2O \longrightarrow Cl^- + H_3O^+$$

goes essentially to completion.* In contrast, equation (17.6) suggested that the self-ionization of water occurs only to a very slight extent. As a result, we conclude that in calculating $[H_3O^+]$ in an aqueous solution of a strong acid, the strong acid is the only significant source of H_3O^+. The contribution due to the self-ionization of water can generally be ignored *unless the solution is extremely dilute.*

EXAMPLE 17-3

Calculating Ion Concentrations in an Aqueous Solution of a Strong Acid. Calculate $[H_3O^+]$, $[Cl^-]$, and $[OH^-]$ in 0.015 M HCl(aq).

Solution

We can assume that HCl is completely ionized and is the sole source of H_3O^+ in solution. Therefore,

$$[H_3O^+] = 0.015 \text{ M}$$

---

*In very concentrated aqueous solutions, HCl does not exist exclusively as the separated ions H_3O^+ and Cl^-. One indication of this is that we can smell HCl in the vapor above such solutions.

Furthermore, because one Cl^- ion is produced for every H_3O^+ ion,

$$[Cl^-] = [H_3O^+] = 0.015 \text{ M}$$

To calculate $[OH^-]$, we must use the following facts.

1. All the OH^- is derived from the self-ionization of water, per reaction (17.6).

2. $[OH^-]$ and $[H_3O^+]$ must have values consistent with K_w for water.

$$K_w = [H_3O^+][OH^-] = 1.0 \times 10^{-14}$$
$$(0.015)[OH^-] = 1.0 \times 10^{-14}$$
$$[OH^-] = \frac{1.0 \times 10^{-14}}{1.5 \times 10^{-2}} = 0.67 \times 10^{-12} = 6.7 \times 10^{-13} \text{ M}$$

Practice Example A: A 0.0025 M solution of HI(aq) has $[H_3O^+] = 0.0025$ M. Calculate the $[I^-]$, $[OH^-]$, and pH of the solution.

Practice Example B: If 535 mL of *gaseous* HCl, at 26.5 °C and 747 mmHg, is dissolved in enough water to prepare 625 mL of solution, what is the pH of this solution?

Aqueous Acids and Bases animation

The common strong bases are ionic hydroxides. When these bases dissolve in water, H_2O molecules completely dissociate the cations and anions (OH^-) of the base from each other. The self-ionization of water, because it occurs to so very limited an extent, is an inconsequential source of OH^-. This means that in calculating $[OH^-]$ in an aqueous solution of a strong base, the strong base is the only significant source of OH^- *unless the solution is extremely dilute.*

As we noted in Chapter 5, the number of common strong acids and strong bases is quite small. Try to memorize the listing in Table 17.2.

TABLE 17.2
The Common Strong Acids and Strong Bases

Acids	Bases
HCl	LiOH
HBr	NaOH
HI	KOH
$HClO_4$	RbOH
HNO_3	CsOH
$H_2SO_4^a$	$Mg(OH)_2$
	$Ca(OH)_2$
	$Sr(OH)_2$
	$Ba(OH)_2$

$^a H_2SO_4$ ionizes in two distinct steps. It is a strong acid only in its first ionization (see page 687).

► A common error is to assume that you have calculated the pH when in fact you have found the pOH. Thus, the pH cannot be 1.36. The solution must be *basic*; it must have pH > 7. Use qualitative reasoning to avoid simple errors.

EXAMPLE 17-4

Calculating the pH of an Aqueous Solution of a Strong Base. Calcium hydroxide (slaked lime), $Ca(OH)_2$, is the cheapest strong base available. It is generally used for industrial operations in which a high concentration of OH^- is not required. $Ca(OH)_2(s)$ is soluble in water only to the extent of 0.16 g $Ca(OH)_2$/100.0 mL solution at 25 °C. What is the pH of saturated $Ca(OH)_2(aq)$ at 25 °C?

Solution

First, we need to express the solubility of $Ca(OH)_2$ on a molar basis.

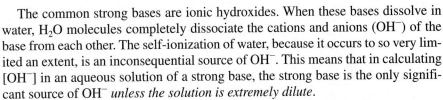

$$\text{molarity} = \frac{0.16 \text{ g } Ca(OH)_2 \times \dfrac{1 \text{ mol } Ca(OH)_2}{74.1 \text{ g } Ca(OH)_2}}{0.1000 \text{ L}} = 0.022 \text{ M } Ca(OH)_2$$

Next, we can relate the molarity of OH^- to the molarity of $Ca(OH)_2$.

$$[OH^-] = \frac{0.022 \text{ mol } Ca(OH)_2}{1 \text{ L}} \times \frac{2 \text{ mol } OH^-}{1 \text{ mol } Ca(OH)_2} = 0.044 \text{ M } OH^-$$

Now we can calculate the pOH and, from it, the pH.

$$pOH = -\log[OH^-] = -\log 0.044 = 1.36$$
$$pH = 14.00 - pOH = 14.00 - 1.36 = 12.64$$

Practice Example A: Milk of magnesia is a saturated solution of $Mg(OH)_2$. Its solubility is 9.63 mg $Mg(OH)_2$/100.0 mL solution at 20 °C. What is the pH of saturated $Mg(OH)_2$ at 20 °C?

Practice Example B: Calculate the pH of an aqueous solution that is 3.00% KOH, by mass, and has a density of 1.0242 g/mL.

Are You Wondering...

How to calculate $[H_3O^+]$ in an extremely dilute solution of a strong acid?

The method of Example 17-3 won't work for calculating the pH of a solution as dilute as 1.0×10^{-8} M HCl. We would write $[H_3O^+] = 1.0 \times 10^{-8}$ M, and pH = 8.00. But how can a solution of a strong acid, no matter how dilute, have a pH greater than 7? The difficulty is that at this extreme dilution, we must consider two sources of H_3O^+. The sources of H_3O^+ and the ion concentrations from both sources are indicated as follows:

$$H_2O \; + \; H_2O \; \rightleftharpoons \; H_3O^+ \; + \; OH^-$$

Molarity: x x

$$HCl \; + \; H_2O \; \longrightarrow \; H_3O^+ \; + \; Cl^-$$

Molarity: 1.0×10^{-8} 1.0×10^{-8}

To satisfy the K_w expression for water in this solution, we use equation (17.7) to get

$$[H_3O^+][OH^-] = x(x + 1.0 \times 10^{-8}) = 1.0 \times 10^{-14}$$

This expression rearranges to the quadratic form

$$x^2 + (1.0 \times 10^8 x) - (1.0 \times 10^{-14}) = 0$$

The solution to this equation is $x = 9.5 \times 10^{-8}$ M. Therefore, we combine $[H_3O^+]$ from both sources to get $[H_3O^+] = (9.5 \times 10^{-8}) + (1.0 \times 10^{-8}) = 1.05 \times 10^{-7}$ M, and pH = 6.98.

From this result, we conclude that the pH is slightly less than 7, as expected for a very dilute acid, and that the self-ionization of water contributes nearly ten times as much hydronium ion to the solution as does the strong acid.

17-5   Weak Acids and Weak Bases

Figure 17-7 illustrates two ways of showing that ionization has occurred in an aqueous solution of an acid: One is by the color of an acid–base indicator; the other, the response of a pH meter. The pink color of the solution in Figure 17-7a tells us the pH of 0.10 M HCl is *less than 1.2*. The pH meter registers a value of 1.0, just what we expect for a strong acid solution with $[H_3O^+] = 0.10$ M. The yellow color of the solution in Figure 17-7b indicates that the pH of 0.10 M $HC_2H_3O_2$ (acetic acid) is *2.8 or greater*. The pH meter registers 2.8.

So we see that two acids can have the same molarity but different pH values. The acid's molarity simply indicates what was put into the solution, but $[H_3O^+]$ and pH depend on what *happens* in the solution. In both solutions, some self-ionization of water occurs, but this reaction is negligible. Ionization of HCl, a strong acid, can be assumed to go to completion, as indicated in equation (17.5). As we previously noted, ionization of $HC_2H_3O_2$, a weak acid, is a reversible reaction that reaches a condition of equilibrium.*

$$HC_2H_3O_2 + H_2O \rightleftharpoons H_3O^+ + C_2H_3O_2^- \qquad (17.11)$$

▶ Most laboratory pH meters can be read to the nearest 0.01 unit. Some pH meters for research work can be read to 0.001 unit, but unless unusual precautions are taken, the reading of the meter may not correspond to the true pH.

 Equilibrium Constant activity

---

*We have been writing ionization equations in the form: acid(1) + base(2) $\rightleftharpoons$ base(1) + acid(2). Here we have written acid(1) + base(2) $\rightleftharpoons$ acid(2) + base(1) to highlight H_3O^+, the species that is usually the subject of a calculation.

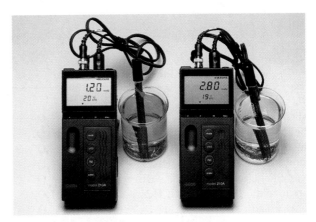

▲ **FIGURE 17-7   Strong and weak acids compared**
The color of thymol blue indicator, which is present in both solutions, depends on the pH of the solution.

$$pH < 1.2 < pH < 2.8 < pH$$

red   orange   yellow

The principle of the pH meter is discussed in Section 21-4. **(left)** 0.10 M HCl has pH $\approx$ 1. **(right)** 0.10 M $HC_2H_3O_2$ has pH $\approx$ 2.8.

The equilibrium constant expression for reaction (17.11) is

$$K_a = \frac{[H_3O^+][C_2H_3O_2^-]}{[HC_2H_3O_2]} = 1.8 \times 10^{-5} \tag{17.12}$$

Just as pH is a convenient shorthand designation related to $[H_3O^+]$, pK is related to an equilibrium constant. That is, **pK = $-\log K$**. Thus, for acetic acid,

$$pK_a = -\log K_a = -\log(1.8 \times 10^{-5}) = -(-4.74) = 4.74$$

As with other equilibrium constants, the larger the value of K_a (or K_b for a base), the farther the equilibrium condition lies in the direction of the forward reaction. And the more extensive the ionization, the greater are the concentrations of the ions produced. Ionization constants must be determined *by experiment*.

A few ionization constants for weak acids and weak bases are listed in Table 17.3. A more extensive listing is given in Appendix D.

Lactic acid

▲ Lactic acid,
$CH_3CH(OH)COOH$.

Glycine

▲ Glycine, NH_2CH_2COOH.

Identifying Weak Acids and Bases

A large number of weak acids have the same structural feature as acetic acid: that is, a —COOH group as part of the molecule. This *carboxyl group* is a common feature of many organic acids, including such biologically important acids as lactic acid and all the amino acids. We will use a number of carboxylic acids as examples in this and later chapters.

In general, to distinguish a weak acid from a strong acid, you need recall only that the half dozen strong acids listed in Table 17.2 are the most common strong acids. Unless you are informed to the contrary, assume that any acid not listed in Table 17.2 is a weak acid.

At first glance, weak bases seem more difficult to identify than weak acids: There is no distinctive element such as H written first in the formula. Yet, if you study the bases in Table 17.3, you will see that all but one of them (pyridine) can be viewed as an ammonia molecule in which some other group (—C_6H_5, —C_2H_5, —OH, or

TABLE 17.3   Ionization Constants of Some Weak Acids and Weak Bases in Water at 25 °C

	Ionization Equilibrium	Ionization Constant K	pK
Acid		$K_a =$	p$K_a =$
Iodic acid	$HIO_3 + H_2O \rightleftharpoons H_3O^+ + IO_3^-$	1.6×10^{-1}	0.80
Chlorous acid	$HClO_2 + H_2O \rightleftharpoons H_3O^+ + ClO_2^-$	1.1×10^{-2}	1.96
Chloroacetic acid	$HC_2H_2ClO_2 + H_2O \rightleftharpoons H_3O^+ + C_2H_2ClO_2^-$	1.4×10^{-3}	2.85
Nitrous acid	$HNO_2 + H_2O \rightleftharpoons H_3O^+ + NO_2^-$	7.2×10^{-4}	3.14
Hydrofluoric acid	$HF + H_2O \rightleftharpoons H_3O^+ + F^-$	6.6×10^{-4}	3.18
Formic acid	$HCHO_2 + H_2O \rightleftharpoons H_3O^+ + CHO_2^-$	1.8×10^{-4}	3.74
Benzoic acid	$HC_7H_5O_2 + H_2O \rightleftharpoons H_3O^+ + C_7H_5O_2^-$	6.3×10^{-5}	4.20
Hydrazoic acid	$HN_3 + H_2O \rightleftharpoons H_3O^+ + N_3^-$	1.9×10^{-5}	4.72
Acetic acid	$HC_2H_3O_2 + H_2O \rightleftharpoons H_3O^+ + C_2H_3O_2^-$	1.8×10^{-5}	4.74
Hypochlorous acid	$HOCl + H_2O \rightleftharpoons H_3O^+ + OCl^-$	2.9×10^{-8}	7.54
Hydrocyanic acid	$HCN + H_2O \rightleftharpoons H_3O^+ + CN^-$	6.2×10^{-10}	9.21
Phenol	$HOC_6H_5 + H_2O \rightleftharpoons H_3O^+ + C_6H_5O^-$	1.0×10^{-10}	10.00
Hydrogen peroxide	$H_2O_2 + H_2O \rightleftharpoons H_3O^+ + HO_2^-$	1.8×10^{-12}	11.74
Base		$K_b =$	p$K_b =$
Diethylamine	$(C_2H_5)_2NH + H_2O \rightleftharpoons (C_2H_5)_2NH_2^+ + OH^-$	6.9×10^{-4}	3.16
Ethylamine	$C_2H_5NH_2 + H_2O \rightleftharpoons C_2H_5NH_3^+ + OH^-$	4.3×10^{-4}	3.37
Ammonia	$NH_3 + H_2O \rightleftharpoons NH_4^+ + OH^-$	1.8×10^{-5}	4.74
Hydroxylamine	$HONH_2 + H_2O \rightleftharpoons HONH_3^+ + OH^-$	9.1×10^{-9}	8.04
Pyridine	$C_5H_5N + H_2O \rightleftharpoons C_5H_5NH^+ + OH^-$	1.5×10^{-9}	8.82
Aniline	$C_6H_5NH_2 + H_2O \rightleftharpoons C_6H_5NH_3^+ + OH^-$	7.4×10^{-10}	9.13

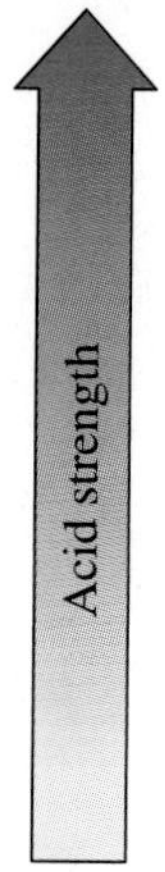

Acid strength ↑

Base strength ↓

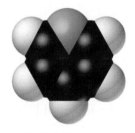

—CH_3) has been substituted for one of the H atoms. The substitution of a methyl group, —CH_3, for a H atom is suggested in the following structural formulas.

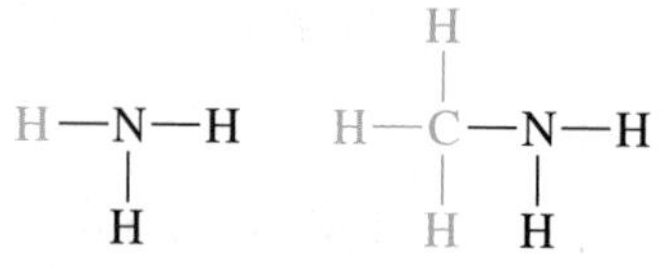

Ammonia   Methylamine

We can represent the ionization of methylamine in this way.

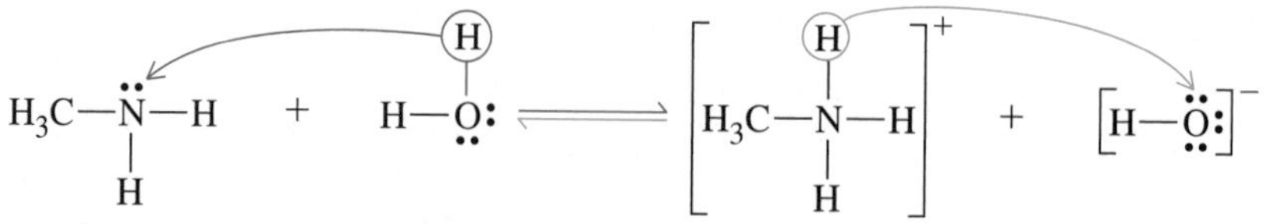

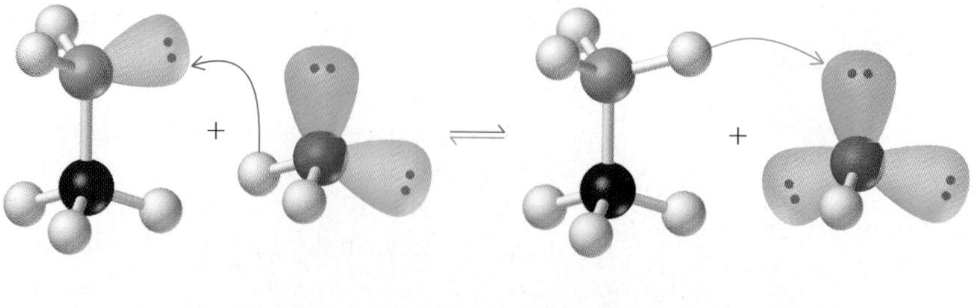

H_3C—N—H + H—O: $\rightleftharpoons$ $[H_3C$—N—H$]^+$ + $[H$—O:$]^-$

base   acid   acid   base

The ionization constant expression is

$$K_b = \frac{[CH_3NH_3^+][OH^-]}{[CH_3NH_2]} = 4.2 \times 10^{-4} \qquad (17.13)$$

Not all weak bases contain N. Yet, so many of them do that the similarity to NH_3 outlined here is well worth remembering. These weak bases derived from ammonia are known as *amines*.

Illustrative Examples

For some students, solution equilibrium calculations are among the most challenging in general chemistry. At times, the difficulty is in sorting out what is relevant to a given problem. The number of types of calculations seems very large, although in fact it is quite limited. The key to solving solution equilibrium problems is to be able to imagine what is going on. Here are some questions to ask yourself.

- Which are the principal species in solution?
- What are the chemical reactions that produce them?
- Can some reactions (for example, the self-ionization of water) be ignored?
- Can you make any assumptions that allow you to simplify the equilibrium calculations?
- What is a reasonable answer to the problem? For instance, should the final solution be acidic (pH < 7) or basic (pH > 7)?

In short, first think through a problem *qualitatively*. At times, you may not even have to do a calculation. Next, organize the relevant data in a clear, logical manner. This action alone can get you on the right track. We recommend that you write the balanced chemical equation and use it as the basis for an ICE table, as discussed in Chapter 16 (page 648). In this way, many problems that at first appear new to you will take on a familiar pattern. Look for other helpful hints as you proceed through this chapter and the following two chapters.

EXAMPLE 17-5

Determining a Value of K_a from the pH of a Solution of a Weak Acid. Butyric acid, $HC_4H_7O_2$, is used to make compounds employed in artificial flavorings and syrups. A 0.250 M aqueous solution of $HC_4H_7O_2$ is found to have a pH of 2.72. Determine K_a for butyric acid.

$$HC_4H_7O_2 + H_2O \rightleftharpoons H_3O^+ + C_4H_7O_2^- \qquad K_a = ?$$

Solution

For $HC_4H_7O_2$, K_a is likely to be much larger than K_w. Therefore, we can assume that self-ionization of water is unimportant and that ionization of the butyric acid is the only source of H_3O^+. Let's treat the situation as if $HC_4H_7O_2$ first dissolves in molecular form and then the molecules ionize until equilibrium is reached. We will represent the concentrations of H_3O^+ and $C_4H_7O_2^-$ at equilibrium as x M.

$$\mathbf{HC_4H_7O_2 + H_2O \rightleftharpoons H_3O^+ + C_4H_7O_2^-}$$

	$HC_4H_7O_2$		H_3O^+	$C_4H_7O_2^-$
initial concns:	0.250 M		—	—
changes:	$-x$ M		$+x$ M	$+x$ M
equil concns:	$(0.250 - x)$ M		x M	x M

But x is *not* unknown. It is the $[H_3O^+]$ in solution, which we can determine from the pH.

$$\log[H_3O^+] = -pH = -2.72$$
$$[H_3O^+] = 10^{-2.72} = 1.9 \times 10^{-3} = x$$

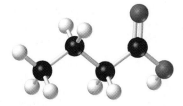

▲ Butyric acid, $CH_3CH_2CH_2COOH$.

Now we can solve the following expression for K_a, substituting in the known value for x.

$$K_a = \frac{[H_3O^+][C_4H_7O_2^-]}{[HC_4H_7O_2]} = \frac{x \cdot x}{0.250 - x}$$

$$= \frac{(1.9 \times 10^{-3})(1.9 \times 10^{-3})}{0.250 - (1.9 \times 10^{-3})} = 1.5 \times 10^{-5}$$

Notice that our original assumption was correct: K_a is much larger than K_w.

Practice Example A: Hypochlorous acid, HOCl, is used in water treatment and as a disinfectant in swimming pools. A 0.150 M solution of HOCl has a pH of 4.18. Determine K_a for hypochlorous acid.

Practice Example B: The much-abused drug cocaine is an alkaloid. Alkaloids are noted for their bitter taste, an indication of their basic properties. Cocaine, $C_{17}H_{21}O_4N$, is soluble in water to the extent of 0.17 g/100 mL solution, and a saturated solution has a pH = 10.08. What is the value of K_b for cocaine?

$$C_{17}H_{21}O_4N + H_2O \rightleftharpoons C_{17}H_{21}O_4NH^+ + OH^- \qquad K_b = ?$$

▶ For a weak base, $[OH^-]$ and the molarity of a cation appear in the numerator of the K_b expression.

Example 17-6 presents a common problem involving weak acids and weak bases: calculating the pH of a solution of known molarity. The calculation invariably involves a quadratic equation, but very often we can make a simplifying assumption leading to a shortcut that saves both time and effort.

EXAMPLE 17-6

Calculating the pH of a Weak Acid Solution. Show by calculation that the pH of 0.100 M $HC_2H_3O_2$ should be about the value shown on the pH meter in Figure 17-7; that is, pH ≈ 2.8.

Solution

Here, we know that K_a is much larger than K_w. Let's again treat the situation as if $HC_2H_3O_2$ first dissolves in molecular form and then ionizes until equilibrium is reached. In this case, the quantity x is an unknown that we must obtain by an algebraic solution.

$$HC_2H_3O_2 + H_2O \rightleftharpoons H_3O^+ + C_2H_3O_2^-$$

initial concns:	0.100 M	—	—
changes:	$-x$ M	$+x$ M	$+x$ M
equil concns:	$(0.100 - x)$ M	x M	x M

$$K_a = \frac{[H_3O^+][C_2H_3O_2^-]}{[HC_2H_3O_2]} = \frac{x \cdot x}{0.100 - x} = 1.8 \times 10^{-5}$$

We could solve this equation by using the quadratic formula, but let's instead make a simplifying assumption that is often valid. Assume that x is very small compared with 0.100. That is, assume that $(0.100 - x) \approx 0.100$.

$$x^2 = 0.100 \times 1.8 \times 10^{-5} = 1.8 \times 10^{-6}$$
$$x = [H_3O^+] = \sqrt{1.8 \times 10^{-6}} = 1.3 \times 10^{-3} \text{ M}$$

Now, we must check our assumption: $0.100 - 0.0013 = 0.099 \approx 0.100$. Our assumption is good to about 1 part per 100 (1%) and is valid for a calculation involving two or three significant figures. Finally,

$$pH = -\log[H_3O^+] = -\log(1.3 \times 10^{-3}) = -(-2.89) = 2.89$$

Practice Example A: Substituting halogen atoms for hydrogen atoms bound to carbon increases the strength of carboxylic acids. Show that the pH of 0.100 M $HC_2H_2FO_2$,

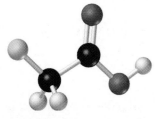

▲ Fluoroacetic acid, CH_2FCOOH.

▲ Acetylsalicylic acid, $C_6H_4(OOCCH_3)COOH$.

fluoroacetic acid, is lower than that calculated in Example 17-6 for 0.100 M $HC_2H_3O_2$.

$$HC_2H_2FO_2 + H_2O \rightleftharpoons H_3O^+ + C_2H_2FO_2^- \qquad K_a = 2.6 \times 10^{-3}$$

Practice Example B: Acetylsalicylic acid, $HC_9H_7O_4$, is the active component in aspirin. This acid is the cause of the stomach upset some people get when taking aspirin. Two extra-strength aspirin tablets, each containing 500 mg of acetylsalicylic acid, are dissolved in 325 mL of water. What is the pH of this solution?

$$HC_9H_7O_4 + H_2O \rightleftharpoons H_3O^+ + C_9H_7O_4^- \qquad K_a = 3.3 \times 10^{-4}$$

EXAMPLE 17-7

Dealing with the Failure of a Simplifying Assumption. What is the pH of a solution that is 0.00250 M $CH_3NH_2(aq)$? For methylamine, $K_b = 4.2 \times 10^{-4}$.

Solution

We can begin by describing the ionization equilibrium in the customary format.

$$CH_3NH_2 + H_2O \rightleftharpoons CH_3NH_3^+ + OH^-$$

initial concns:	0.00250 M	—	—
changes:	$-x$ M	$+x$ M	$+x$ M
equil concns:	$(0.00250 - x)$ M	x M	x M

$$K_b = \frac{[CH_3NH_3^+][OH^-]}{[CH_3NH_2]} = \frac{x \cdot x}{0.00250 - x} = 4.2 \times 10^{-4}$$

Now let's assume that x is very much less than 0.00250 and that $0.00250 - x \approx 0.00250$.

$$\frac{x^2}{0.00250} = 4.2 \times 10^{-4} \qquad x^2 = 1.1 \times 10^{-6} \qquad [OH^-] = x = 1.0 \times 10^{-3} \text{ M}$$

The value of x is nearly half as large as 0.00250—too large to ignore.

$$\frac{0.0010}{0.00250} \times 100\% = 40\%$$

Our assumption failed, so we must use an exact method. This means using the *quadratic* formula.

$$\frac{x^2}{0.00250 - x} = 4.2 \times 10^{-4}$$

$$x^2 + \left(4.2 \times 10^{-4}x\right) - \left(1.1 \times 10^{-6}\right) = 0$$

$$x = \frac{\left(-4.2 \times 10^{-4}\right) \pm \sqrt{\left(4.2 \times 10^{-4}\right)^2 + 4 \times 1.1 \times 10^{-6}}}{2}$$

$$x = [OH^-] = \frac{\left(-4.2 \times 10^{-4}\right) \pm \left(2.1 \times 10^{-3}\right)}{2} = 8.4 \times 10^{-4} \text{ M}$$

$$pOH = -\log[OH^-] = -\log\left(8.4 \times 10^{-4}\right) = 3.08$$
$$pH = 14.00 - pOH = 14.00 - 3.08 = 10.92$$

Practice Example A: What is the pH of 0.015 M $HC_2H_2FO_2(aq)$?

$$HC_2H_2FO_2 + H_2O \rightleftharpoons H_3O^+ + C_2H_2FO_2^- \qquad K_a = 2.6 \times 10^{-3}$$

Practice Example B: Piperidine is a base found in small amounts in black pepper. What is the pH of 315 mL of an aqueous solution containing 114 mg piperidine?

$$C_5H_{11}N + H_2O \rightleftharpoons C_5H_{11}NH^+ + OH^- \qquad K_b = 1.6 \times 10^{-3}$$

KEEP IN MIND

that the relationship pH + pOH = pK can be used because the equilibria

$$CH_3NH_2 + H_2O \rightleftharpoons$$
$$CH_3NH_3^+ + OH^-$$
$$H_2O + H_2O \rightleftharpoons H_3O^+ + OH^-$$

are satisfied simultaneously. That is, the same value of $[OH^-]$ goes into the equilibrium expressions for both reactions. ▶

▲ Piperidine, $C_5H_{10}NH$.

Are You Wondering...

How to calculate the pH of a very dilute solution of a weak acid?

Think of this as a companion question to the one posed in the Are You Wondering feature on page 676, except here the acid in question is weak (represented as HA) rather than strong. The initial approach is similar: Write two equations representing the sources of H_3O^+, and indicate the concentrations of various species in the solution.

$$H_2O + H_2O \rightleftharpoons H_3O^+ + OH^-$$

Molarity: $\qquad\qquad\qquad\qquad\quad x \qquad x$

$$HA + H_2O \rightleftharpoons H_3O^+ + A^-$$

Molarity: $\qquad M - y \qquad\qquad y \qquad y$

Our ultimate objective is to determine $[H_3O^+]$ in the solution, and this is $x + y$. From $[H_3O^+]$, we can easily get the pH.

Our principal task is to solve a pair of equations simultaneously for x and y. The two equations are

$$K_w = [H_3O^+][OH^-] = (x + y) \times x = 1.0 \times 10^{-14}$$

$$K_a = \frac{[H_3O^+][A^-]}{[HA]} = \frac{(x + y) \times y}{(M - y)}$$

Note the following three points in these equations. (1) There can be only one value of $[H_3O^+]$ in the solution, and it is this value $(x + y)$ that appears in each equation. (2) The stoichiometric concentration of the acid is M (its molarity), and its *equilibrium* concentration is $[HA] = M - y$. The numerical value of M depends on the particular case considered. (3) Similarly, the numerical value of K_a depends on the particular case.

When we solve the K_a expression for x, we obtain

$$x = \frac{K_a(M - y)}{y} - y \quad \text{and} \quad x + y = \frac{K_a(M - y)}{y}$$

And when we substitute these values of x and $x + y$ into the K_w equation, we get

$$K_w = \frac{K_a(M - y)}{y} \times \left(\frac{K_a(M - y)}{y} - y \right) = 1.0 \times 10^{-14}$$

The value of y satisfying this equation is not difficult to obtain by the method of successive approximations, illustrated in Appendix A.

The solution of this problem is left for you to do (see Integrative and Advanced Exercise 93), but you will find that for $1.0 \times 10^{-5} \, M \, HCN \left(K_a = 6.2 \times 10^{-10} \right)$, $y \cong 4.8 \times 10^{-8}$, $(x + y) \cong 1.3 \times 10^{-7}$, and pH $\cong 6.90$. This certainly seems like a reasonable pH for a very dilute solution of a weak acid in water—just below the neutral pH of 7.00.

Finally, we can use the results of the previous discussion to establish a criterion for ignoring the self-ionization of water in calculations. When we do this, we are assuming that $y \gg x$, so that $y \approx x + y$, and $yx \approx (x + y)x = [H_3O^+][OH^-] = K_w$. Also, if $y \gg x$, then $y^2 \gg K_w$. For what values of y can we say that $y \gg x$? Let's take the maximum value of x consistent with ignoring the self-ionization of water to be 1/100 of y which, as shown below, means that $[H_3O^+]$ must be greater than 10^{-6} M.

$$y^2 > y \times x = y \times \frac{y}{100} = K_w = 1 \times 10^{-14}$$

$$y^2 > 1 \times 10^{-12} \quad \text{and} \quad y > 1 \times 10^{-6}$$

In acid–base calculations, we should first use the simplifying assumptions (ignore K_w and assume that [HA] can be replaced by the molarity of the acid). Then we should check our answer to see if $[H_3O^+]$ is greater than 1×10^{-6} M and the criterion presented on page 683 is also met.

More on Simplifying Assumptions

The usual simplifying assumption is that of treating a weak acid or weak base as though it remains essentially nonionized (so that $M - x \approx M$). In general, this assumption will work if the molarity of the weak acid, M_A, or that of the weak base, M_B, exceeds the value of K_a or K_b by a factor of at least 100. That is,

$$\frac{M_A \ (\text{or } M_B)}{K_a \ (\text{or } K_b)} > 100$$

In any case, it is important to test the validity of any assumption that you make. If the assumption is good to within a few percent (say, less than 5%), then it is generally valid. In Example 17-6, the simplifying assumption was good to about 1%, but in Example 17-7 it was off by 40%.

Percent Ionization

We can describe the extent of ionization of a weak acid or weak base by determining the degree of ionization or percent ionization. It is convenient to introduce the generic symbol HA for any weak acid, and A^- as the conjugate base of the acid HA.

For the ionization $HA + H_2O \rightleftharpoons H_3O^+ + A^-$, the degree of ionization is the fraction of the acid molecules that ionize. Thus, if in 1.00 M HA, ionization produces $[H_3O^+] = [A^-] = 0.05$ M, the degree of ionization = 0.05 M/1.00 M = 0.05. **Percent ionization** gives the proportion of ionized molecules on a percentage basis.

$$\text{percent ionization} = \frac{\text{molarity of } H_3O^+ \text{ derived from HA}}{\text{initial molarity of HA}} \times 100\% \qquad (17.14)$$

A weak acid with a degree of ionization of 0.05 has a percent ionization of 5%.

Figure 17-8 compares percent ionization and solution molarity for a weak acid and a strong acid. Example 17-8 shows by calculation that the percent ionization of a weak acid or a weak base *increases* as the solution becomes *more dilute*, a fact that we can also demonstrate by a simple analysis of the ionization reaction

$$HA + H_2O \rightleftharpoons H_3O^+ + A^-$$

At equilibrium, n_{HA} moles of the acid HA, $n_{H_3O^+}$ moles of H_3O^+, and n_{A^-} moles of A^- are present in a volume of V liters. The K_a expression is

$$K_a = \frac{[H_3O^+][A^-]}{[HA]} = \frac{(n_{H_3O^+}/V)(n_{A^-}/V)}{n_{HA}/V} = \frac{(n_{H_3O^+})(n_{A^-})}{n_{HA}} \times \frac{1}{V}$$

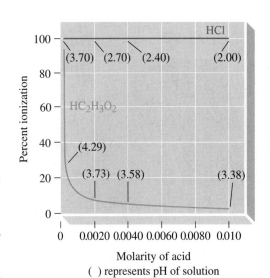

▶ **FIGURE 17-8**

Percent ionization of an acid as a function of concentration

Over the concentration range shown, HCl, a strong acid, is essentially 100% ionized. The percent ionization of $HC_2H_3O_2$, a weak acid, increases from about 4% in 0.010 M to essentially 100% when the solution is extremely dilute. Note that in extremely dilute solutions, the two acids would have virtually the same pH.

When we dilute the solution, V increases, $1/V$ decreases, and the ratio $(n_{H_3O^+})(n_{A^-})/n_{HA}$ must increase to maintain the constant value of K_a. In turn, $n_{H_3O^+}$ and n_{A^-} must increase and n_{HA} must decrease, signifying an increase in the percent ionization.

EXAMPLE 17-8

Determining Percent Ionization as a Function of Weak Acid Concentration. What is the percent ionization of acetic acid in 1.0 M, 0.10 M, and 0.010 M $HC_2H_3O_2$?

Solution

Let us use the ICE format to describe 1.0 M $HC_2H_3O_2$:

$$HC_2H_3O_2 + H_2O \rightleftharpoons H_3O^+ + C_2H_3O_2^-$$

initial concns:	1.0 M	—	—
changes:	$-x$ M	$+x$ M	$+x$ M
equil concns:	$(1.0 - x)$ M	x M	x M

We need to calculate $x = [H_3O^+] = [C_2H_3O_2^-]$. In doing so, let's make the usual assumption: $1.0 - x \approx 1.0$.

$$K_a = \frac{[H_3O^+][C_2H_3O_2^-]}{[HC_2H_3O_2]} = \frac{x \cdot x}{1.0 - x} = \frac{x^2}{1.0} = 1.8 \times 10^{-5}$$

$$x = [H_3O^+] = [C_2H_3O_2^-] = \sqrt{1.8 \times 10^{-5}} = 4.2 \times 10^{-3} \text{ M}$$

The percent ionization of 1.0 M $HC_2H_3O_2$ is

$$\% \text{ ionization} = \frac{[H_3O^+]}{[HC_2H_3O_2]} \times 100\% = \frac{4.2 \times 10^{-3} \text{ M}}{1.0 \text{ M}} \times 100\% = 0.42\%$$

The calculations for 0.10 M $HC_2H_3O_2$ and 0.010 M $HC_2H_3O_2$ are very similar and yield the results

0.10 M $HC_2H_3O_2$ is 1.3% ionized; 0.010 M $HC_2H_3O_2$ is 4.2% ionized.

Practice Example A: What is the percent ionization of hydrofluoric acid in 0.20 M HF and in 0.020 M HF?

Practice Example B: An 0.0284 M aqueous solution of lactic acid, a substance that accumulates in the blood and muscles during physical activity, is found to be 6.7% ionized. Determine K_a for lactic acid.

$$HC_3H_5O_3 + H_2O \rightleftharpoons H_3O^+ + C_3H_5O_3^- \qquad K_a = ?$$

17-6 Polyprotic Acids

All the acids listed in Table 17.3 are weak *monoprotic acids*, meaning that their molecules have only one ionizable H atom, even though several of these acids contain more than one H atom. But some acids have more than one ionizable H atom per molecule. These are **polyprotic acids**. Table 17.4 lists ionization constants for several polyprotic acids. Additional listings can be found in Appendix D. We will focus on phosphoric acid, H_3PO_4.

▲ Phosphoric acid, H_3PO_4.

TABLE 17.4	Ionization Constants of Some Polyprotic Acids		
Acid	**Ionization Equilibria**	**Ionization Constants, K**	**pK**
Hydrosulfuric[a]	$H_2S + H_2O \rightleftharpoons H_3O^+ + HS^-$	$K_{a_1} = 1.0 \times 10^{-7}$	$pK_{a_1} = 7.00$
	$HS^- + H_2O \rightleftharpoons H_3O^+ + S^{2-}$	$K_{a_2} = 1 \times 10^{-19}$	$pK_{a_2} = 19.0$
Carbonic[b]	$H_2CO_3 + H_2O \rightleftharpoons H_3O^+ + HCO_3^-$	$K_{a_1} = 4.4 \times 10^{-7}$	$pK_{a_1} = 6.36$
	$HCO_3^- + H_2O \rightleftharpoons H_3O^+ + CO_3^{2-}$	$K_{a_2} = 4.7 \times 10^{-11}$	$pK_{a_2} = 10.33$
Phosphoric	$H_3PO_4 + H_2O \rightleftharpoons H_3O^+ + H_2PO_4^-$	$K_{a_1} = 7.1 \times 10^{-3}$	$pK_{a_1} = 2.15$
	$H_2PO_4^- + H_2O \rightleftharpoons H_3O^+ + HPO_4^{2-}$	$K_{a_2} = 6.3 \times 10^{-8}$	$pK_{a_2} = 7.20$
	$HPO_4^{2-} + H_2O \rightleftharpoons H_3O^+ + PO_4^{3-}$	$K_{a_3} = 4.2 \times 10^{-13}$	$pK_{a_3} = 12.38$
Sulfurous[c]	$H_2SO_3 + H_2O \rightleftharpoons H_3O^+ + HSO_3^-$	$K_{a_1} = 1.3 \times 10^{-2}$	$pK_{a_1} = 1.89$
	$HSO_3^- + H_2O \rightleftharpoons H_3O^+ + SO_3^{2-}$	$K_{a_2} = 6.2 \times 10^{-8}$	$pK_{a_2} = 7.21$
Sulfuric[d]	$H_2SO_4 + H_2O \rightleftharpoons H_3O^+ + HSO_4^-$	$K_{a_1} = $ very large	$pK_{a_1} < 0$
	$HSO_4^- + H_2O \rightleftharpoons H_3O^+ + SO_4^{2-}$	$K_{a_2} = 1.1 \times 10^{-2}$	$pK_{a_2} = 1.96$

Acid strength

[a] The value for K_{a_2} of H_2S most commonly found in older literature is about 1×10^{-14}, but current evidence suggests that the value is considerably smaller.

[b] H_2CO_3 cannot be isolated. It is in equilibrium with H_2O and dissolved CO_2. The value given for K_{a_1} is actually for the reaction

$$CO_2(aq) + 2\,H_2O \rightleftharpoons H_3O^+ + HCO_3^-$$

Generally, aqueous solutions of CO_2 are treated *as if* the $CO_2(aq)$ were first converted to H_2CO_3, followed by ionization of the H_2CO_3.

[c] H_2SO_3 is a hypothetical, nonisolable species. The value listed for K_{a_1} is actually for the reaction

$$SO_2(aq) + 2\,H_2O \rightleftharpoons H_3O^+ + HSO_3^-$$

[d] H_2SO_4 is completely ionized in the first step.

Phosphoric Acid

Phosphoric acid, H_3PO_4, ranks second only to sulfuric acid among the important commercial acids. Annual production of H_3PO_4 in the United States typically exceeds 10 million tons. Its principal use is in the manufacture of phosphate fertilizers. In addition, various sodium, potassium, and calcium phosphates are used in the food industry.

The H_3PO_4 molecule has *three* ionizable H atoms; it is a *triprotic acid*. It ionizes in three steps. For each step, we can write an ionization equation and an acid ionization constant expression with a distinctive value of K_a.

(1) $H_3PO_4 + H_2O \rightleftharpoons H_3O^+ + H_2PO_4^-$ $K_{a_1} = \dfrac{[H_3O^+][H_2PO_4^-]}{[H_3PO_4]} = 7.1 \times 10^{-3}$

(2) $H_2PO_4^- + H_2O \rightleftharpoons H_3O^+ + HPO_4^{2-}$ $K_{a_2} = \dfrac{[H_3O^+][HPO_4^{2-}]}{[H_2PO_4^-]} = 6.3 \times 10^{-8}$

(3) $HPO_4^{2-} + H_2O \rightleftharpoons H_3O^+ + PO_4^{3-}$ $K_{a_3} = \dfrac{[H_3O^+][PO_4^{3-}]}{[HPO_4^{2-}]} = 4.2 \times 10^{-13}$

There is a ready explanation for the relative magnitudes of the ionization constants—that is, for the fact that $K_{a_1} > K_{a_2} > K_{a_3}$. When ionization occurs in step (1), a proton (H^+) moves away from an ion with a 1− charge ($H_2PO_4^-$). In step (2), the proton moves away from an ion with a 2− charge (HPO_4^{2-}), a more difficult separation. As a result, the ionization constant in the second step is smaller than that in the first. Ionization is more difficult still in step (3).

We can make three key statements about the ionization of phosphoric acid, as illustrated in Example 17-9.

1. K_{a_1} is so much larger than K_{a_2} and K_{a_3} that essentially all the H_3O^+ is produced in the first ionization step.

2. So little of the $H_2PO_4^-$ forming in the first ionization step ionizes any further that we can assume $[H_2PO_4^-] = [H_3O^+]$ in the solution.

3. $[HPO_4^{2-}] \approx K_{a_2}$, regardless of the molarity of the acid.[*]

Although statement (1) seems essential if statements (2) and (3) are to be valid, it is not as crucial as might first appear. Even for polyprotic acids whose K_a values do not differ greatly between successive ionizations, $[H_3O^+]$ is often still determined almost exclusively by the K_{a_1} expression, and statements (2) and (3) remain valid. As long as the polyprotic acid is weak in its first ionization step, the concentration of the anion produced in this step will be so much less than the molarity of the acid that additional $[H_3O^+]$ produced in the second ionization remains negligible.

EXAMPLE 17-9

Calculating Ion Concentrations in a Polyprotic Acid Solution. For a 3.0 M H_3PO_4 solution, calculate **(a)** $[H_3O^+]$; **(b)** $[H_2PO_4^-]$; **(c)** $[HPO_4^{2-}]$; and **(d)** $[PO_4^{3-}]$.

Solution

(a) Because K_{a_1} is so much larger than K_{a_2}, let's assume that all the H_3O^+ forms in the *first* ionization step. This is equivalent to thinking of H_3PO_4 as though it were a monoprotic acid, ionizing only in the first step.

$$H_3PO_4 + H_2O \rightleftharpoons H_3O^+ + H_2PO_4^-$$

initial concns:	3.0 M	—	—
changes:	$-x$ M	$+x$ M	$+x$ M
after first ionization:	$(3.0 - x)$ M	x M	x M

Following the usual assumption that x is much smaller than 3.0 and that $3.0 - x \approx 3.0$, we obtain

$$K_{a_1} = \frac{[H_3O^+][H_2PO_4^-]}{[H_3PO_4]} = \frac{x \cdot x}{(3.0 - x)} = \frac{x^2}{3.0} = 7.1 \times 10^{-3}$$

$$x^2 = 0.021 \qquad x = [H_3O^+] = 0.14 \text{ M}$$

▶ In the assumption $3.0 - x \approx 3.0$, $x = 0.14$, which is 4.7% of 3.0. This is about the maximum error that can be tolerated for an acceptable assumption.

(b) From part (a), $x = [H_2PO_4^-] = [H_3O^+] = 0.14$ M.

(c) To determine $[H_3O^+]$ and $[H_2PO_4^-]$, we assumed that the second ionization is insignificant. Here we must consider the second ionization, no matter how slight; otherwise we would have no source of the ion HPO_4^{2-}. We can represent the second ionization, as shown in the following table. *Note especially how the results of the first ionization enter in.* We start with a solution in which $[H_2PO_4^-] = [H_3O^+] = 0.14$ M.

$$H_2PO_4^- + H_2O \rightleftharpoons H_3O^+ + HPO_4^{2-}$$

from first ionization:	0.14 M	0.14 M	—
changes:	$-y$ M	$+y$ M	$+y$ M
after second ionization:	$(0.14 - y)$ M	$(0.14 + y)$ M	y M

---

[*]If we assume that $[H_2PO_4^-] = [H_3O^+]$, the second ionization expression reduces to

$$\frac{[H_3O^+][HPO_4^{2-}]}{[H_2PO_4^-]} = K_{a_2}$$

If we assume that y is much smaller than 0.14, then $(0.14 + y) \approx (0.14 - y) \approx 0.14$. We then get

$$K_{a_2} = \frac{[H_3O^+][HPO_4^{2-}]}{[H_2PO_4^-]} = \frac{(0.14 + y)(y)}{(0.14 - y)} = \frac{(0.14)(y)}{(0.14)} = 6.3 \times 10^{-8}$$

$$y = [HPO_4^{2-}] = 6.3 \times 10^{-8} \text{ M}$$

Note that the assumption is valid.

(d) The PO_4^{3-} ion forms only in the third ionization step. When we write this acid ionization constant expression, we see that we have already calculated the ion concentrations other than $[PO_4^{3-}]$. We can simply solve the K_{a_3} expression for $[PO_4^{3-}]$.

$$K_{a_3} = \frac{[H_3O^+][PO_4^{3-}]}{[HPO_4^{2-}]} = \frac{0.14 \times [PO_4^{3-}]}{6.3 \times 10^{-8}} = 4.2 \times 10^{-13}$$

$$[PO_4^{3-}] = \frac{4.2 \times 10^{-13} \times 6.3 \times 10^{-8}}{0.14} = 1.9 \times 10^{-19} \text{ M}$$

Practice Example A: Malonic acid, $HOOCCH_2COOH$, is a diprotic acid used in the manufacture of barbiturates.

$$HOOCCH_2COOH + H_2O \rightleftharpoons H_3O^+ + HOOCCH_2COO^- \qquad K_{a_1} = 1.4 \times 10^{-3}$$

$$HOOCCH_2COO^- + H_2O \rightleftharpoons H_3O^+ + {}^-OOCCH_2COO^- \qquad K_{a_2} = 2.0 \times 10^{-6}$$

Calculate $[H_3O^+]$, $[HOOCCH_2COO^-]$, and $[{}^-OOCCH_2COO^-]$ in a 1.00 M solution of malonic acid.

Practice Example B: Oxalic acid, found in the leaves of rhubarb and other plants, is a diprotic acid.

$$H_2C_2O_4 + H_2O \rightleftharpoons H_3O^+ + HC_2O_4^- \qquad K_{a_1} = ?$$

$$HC_2O_4^- + H_2O \rightleftharpoons H_3O^+ + C_2O_4^{2-} \qquad K_{a_2} = ?$$

An aqueous solution that is 1.05 M $H_2C_2O_4$ has pH = 0.67. The free oxalate ion concentration in this solution is $[C_2O_4^{2-}] = 5.3 \times 10^{-5}$ M. Determine K_{a_1} and K_{a_2} for oxalic acid.

The fact that the concentration of anion produced in the second ionization step of a polyprotic acid is equal to K_{a_2} can be used to get this result directly. ▶

A Somewhat Different Case: H_2SO_4

Sulfuric acid differs from most polyprotic acids in this important respect: It is a *strong* acid in its first ionization and a *weak* acid in its second. Ionization is complete in the first step, which means that in most $H_2SO_4(aq)$ solutions, $[H_2SO_4] \approx 0$ M. Thus, if a solution is 0.50 M H_2SO_4, we can treat it as though it were 0.50 M H_3O^+ and 0.50 M HSO_4^- initially. Then we can determine the extent to which ionization of HSO_4^- produces additional H_3O^+ and SO_4^{2-}.

▲ Sulfuric acid, H_2SO_4.

EXAMPLE 17-10

Calculating Ion Concentrations in Sulfuric Acid Solutions: Strong Acid Ionization Followed by Weak Acid Ionization.   Calculate $[H_3O^+]$, $[HSO_4^-]$, and $[SO_4^{2-}]$ in 0.50 M H_2SO_4.

Solution

Let us use an approach similar to that of Example 17-9.

$$H_2SO_4 + H_2O \longrightarrow H_3O^+ + HSO_4^-$$

	$H_2SO_4 + H_2O$	H_3O^+	HSO_4^-
initial concn:	0.50 M	—	—
changes:	−0.50 M	+0.50 M	+0.50 M
after first ionization:	≈ 0	0.50 M	0.50 M

$$HSO_4^- + H_2O \rightleftharpoons H_3O^+ + SO_4^{2-}$$

from first ionization:	0.50 M	0.50 M	—
changes:	$-x$ M	$+x$ M	$+x$ M
after second ionization:	$(0.50 - x)$ M	$(0.50 + x)$ M	x M

We need to deal only with the ionization constant expression for K_{a_2}. If we assume that x is much smaller than 0.50, then $(0.50 + x) \approx (0.50 - x) \approx 0.50$ and

$$K_{a_2} = \frac{[H_3O^+][SO_4^{2-}]}{[HSO_4^-]} = \frac{(0.50 + x) \cdot x}{(0.50 - x)} = \frac{0.50 \cdot x}{0.50} = 1.1 \times 10^{-2}$$

Our results, then, are

$$[H_3O^+] = 0.50 + x = 0.51 \text{ M}; \qquad [HSO_4^-] = 0.50 - x = 0.49 \text{ M}$$
$$[SO_4^{2-}] = x = K_{a_2} = 0.011 \text{ M}$$

Practice Example A: Calculate $[H_3O^+]$, $[HSO_4^-]$, and $[SO_4^{2-}]$ in 0.20 M H_2SO_4.

Practice Example B: Calculate $[H_3O^+]$, $[HSO_4^-]$, and $[SO_4^{2-}]$ in 0.020 M H_2SO_4.

(*Hint:* Is the assumption that $[HSO_4^-] = [H_3O^+]$ valid?)

A General Approach to Solution Equilibrium Calculations

Suppose we were required to determine the stoichiometric molarity of $H_2SO_4(aq)$ that is required to produce a solution with pH = 2.15. We could start by determining the hydronium ion concentration in the solution: $[H_3O^+] = 10^{-pH} = 10^{-2.15} = 7.1 \times 10^{-3}$ M. Then, what would we do? We could not follow an approach similar to that in Example 17-6 because H_2SO_4 *is not* a monoprotic acid. Instead, we would have to use a method like the methods outlined in the Are You Wondering features on pages 676 and 682. However, an additional alternative is worth considering—one that has the appeal of being very straightforward in getting us on the right track for all kinds of solution equilibrium calculations. The method has the following format.

1. Identify the species present in any significant amount in the solution (excluding H_2O molecules). Consider the concentrations of these species as unknowns.
2. Write equations that include these species. The number of equations involving these species should match the number of unknowns. The equations are of three types.
   (a) equilibrium constant expressions
   (b) material balance equations
   (c) an electroneutrality condition
3. Solve the system of equations for the unknowns.

Let's apply this approach to the $H_2SO_4(aq)$ solution mentioned in the first sentence of this section.

Possible Species:

$$H_2SO_4, H_3O^+, HSO_4^-, SO_4^{2-}, OH^-$$

We can eliminate H_2SO_4 because its ionization goes to completion in the first step. We can also eliminate OH^- because $[OH^-]$ is exceedingly small in an *acidic* solution that has a pH = 2.15.

Unknowns:

$$[H_3O^+], [HSO_4^-], [SO_4^{2-}], \text{ and M [the molarity of the } H_2SO_4(aq)]$$

We can eliminate $[H_3O^+]$ because we know its value from the outset from pH = 2.15: $[H_3O^+]$ = 0.0071 M. Thus, we are left with three unknowns, and we need three equations.

Equations:

(a) The K_a expression for the ionization $HSO_4^- + H_2O \rightleftharpoons H_3O^+ + SO_4^{2-}$ is

$$K_{a_2} = \frac{[H_3O^+][SO_4^{2-}]}{[HSO_4^-]} = 1.1 \times 10^{-2}$$

(b) The following material balance equation accounts for the fact that the sum of the concentrations of the sulfur-containing species must equal the stoichiometric molarity of the $H_2SO_4(aq)$.

$$[HSO_4^-] + [SO_4^{2-}] = x \text{ M}$$

(c) The electroneutrality condition simply verifies that the solution carries no net charge. The sum of the positive charges must equal the sum of the negative charges. We can sum these charges on a mol/liter basis. For example, because there is one positive charge for every H_3O^+ ion, the number of moles per liter of positive charge is the same as the number of moles per liter of H_3O^+, 0.0071 M. We multiply $[SO_4^{2-}]$ by *two* because each SO_4^{2-} ion carries two units of negative charge.

$$[H_3O^+] = [HSO_4^-] + (2 \times [SO_4^{2-}]) = 0.0071$$

Solving the Equations:

Solve equation (c) for $[HSO_4^-]$: $[HSO_4^-]$ = 0.0071 − $2[SO_4^{2-}]$. Substitute this result, together with $[H_3O^+]$ = 0.0071, into equation (a) and obtain the expression

$$\frac{0.0071 \times [SO_4^{2-}]}{0.0071 - 2[SO_4^{2-}]} = 1.1 \times 10^2$$

Solve this equation to find that $[SO_4^{2-}]$ = 0.0027 M. Then substitute this result into equation (c) to obtain $[HSO_4^-]$ = 0.0017 M. Finally, according to equation (b), $[HSO_4^-] + [SO_4^{2-}]$ = 0.0044 M. The required molarity is 0.0044 M H_2SO_4.

KEEP IN MIND ▶
that although this method gives you a quick way to set up a solution equilibrium calculation, more information may be needed for you to arrive at an answer without undue effort. For example, answers to the general questions posed on page 679 may point the way to simplifying the algebraic solution.

Check:

There are usually ways to check the reasonableness of an answer obtained by this method. In this case, we can easily determine the possible pH range for 0.0044 M H_2SO_4. If the acid ionized only in the first step, $[H_3O^+]$ = 0.0044 M (pH = 2.36); if the second ionization step also went to completion, $[H_3O^+]$ = 0.0088 M (pH = 2.06). The observed pH, 2.15, falls squarely in this range.

The alternative method outlined here is ideal for computerized calculation. Moreover, because the additional manipulations required to convert stoichiometric concentrations to activities can be incorporated into the calculations, the solutions obtained are generally both more accurate and more readily obtained than are those derived by traditional methods.

17-7   Ions as Acids and Bases

In our discussion to this point, we have emphasized the behavior of neutral molecules as acids (for example, HCl, $HC_2H_3O_2$, H_3PO_4) or as bases (for example, NH_3, CH_3NH_2). We have also seen, however, that ions can act as acids or bases. For instance, in the second and subsequent ionization steps of a polyprotic acid, an anion acts as an acid.

$$H_2PO_4^- + H_2O \rightleftharpoons H_3O^+ + HPO_4^{2-} \qquad K = K_{a_2} = 6.3 \times 10^{-8}$$

Now let's think about how each of the following can be described as an acid–base reaction.

$$NH_4^+ + H_2O \rightleftharpoons NH_3 + H_3O^+ \tag{17.15}$$
$$\text{acid(1)} \quad \text{base(2)} \quad\quad \text{base(1)} \quad \text{acid(2)}$$

$$C_2H_3O_2^- + H_2O \rightleftharpoons HC_2H_3O_2 + OH^- \tag{17.16}$$
$$\text{base(1)} \quad\; \text{acid(2)} \quad\quad \text{acid(1)} \quad\;\; \text{base(2)}$$

In reaction (17.15), NH_4^+ is an *acid*, giving up a proton to water, a *base*. We describe equilibrium in this reaction by means of the *acid ionization constant* of the ammonium ion, NH_4^+.

$$K_a = \frac{[NH_3][H_3O^+]}{[NH_4^+]} = ? \tag{17.17}$$

Two of the concentration terms in equation (17.17)—$[NH_3]$ and $[NH_4^+]$—are the same as in the K_b expression for NH_3, the conjugate base of NH_4^+. It seems that K_a for NH_4^+ and K_b for NH_3 should bear some relationship to each other, and they do. The easiest way to see this is to multiply both the numerator and denominator of (17.17) by $[OH^-]$. The product $[H_3O^+] \times [OH^-]$ is the ion product of water, K_w, shown in red. The other terms, shown in blue, represent the *inverse* of K_b for NH_3. The value obtained, 5.6×10^{-10}, is the missing value of K_a in expression (17.17).

$$K_a = \frac{[NH_3][H_3O^+][OH^-]}{[NH_4^+][OH^-]} = \frac{K_w}{K_b} = \frac{1.0 \times 10^{-14}}{1.8 \times 10^{-5}} = 5.6 \times 10^{-10}$$

The result we just obtained is an important consequence of the Brønsted–Lowry theory.

The product of the ionization constants of an acid and its conjugate base equals the ion product of water.

$$K_a \text{ (acid)} \times K_b \text{ (its conjugate base)} = K_w$$
$$K_b \text{ (base)} \times K_a \text{ (its conjugate acid)} = K_w \tag{17.18}$$

In reaction (17.16), $C_2H_3O_2^-$ acts as a *base* by taking a proton from water, an *acid*. Here, equilibrium is described by means of the *base ionization constant* of the acetate ion, $C_2H_3O_2^-$. With expression (17.18) we can evaluate K_b.

$$K_b = \frac{[HC_2H_3O_2][OH^-]}{[C_2H_3O_2^-]} = \frac{K_w}{K_a(HC_2H_3O_2)} = \frac{1.0 \times 10^{-14}}{1.8 \times 10^{-5}} = 5.6 \times 10^{-10}$$

In many tabulations of ionization constants, only K_a values are listed, whether for neutral molecules or for ions. We can use equation (17.18) to obtain the values of their conjugates.

Hydrolysis

We have learned that in pure water at 25 °C, $[H_3O^+] = [OH^-] = 1.0 \times 10^{-7}$ M and pH = 7.00. *Pure water is pH neutral.* When NaCl dissolves in water at 25 °C, complete dissociation into Na^+ and Cl^- ions occurs, and the pH of the solution remains 7.00. We can represent this fact with the equation

$$Na^+ + Cl^- + H_2O \longrightarrow \text{no reaction}$$

▲ **FIGURE 17-9   Ions as acids and bases**
Each of these 1 M solutions contains bromthymol blue indicator, which has the following colors:

pH < 7	pH = 7	pH > 7
yellow	green	blue

(left) $NH_4Cl(aq)$ is acidic. **(center)** $NaCl(aq)$ is neutral. **(right)** $NaC_2H_3O_2(aq)$ is basic.

As shown in Figure 17-9, when NH_4Cl is added to water, the pH falls below 7. This means that $[H_3O^+] > [OH^-]$ in the solution. A reaction producing H_3O^+ must occur.

$$Cl^- + H_2O \longrightarrow \text{no reaction}$$

$$NH_4^+ + H_2O \rightleftharpoons NH_3 + H_3O^+$$

The reaction between NH_4^+ and H_2O is fundamentally no different from other acid–base reactions. A reaction between an ion and water, however, is often called a **hydrolysis** reaction. We say that ammonium ion *hydrolyzes* (and chloride ion does not).

When sodium acetate is dissolved in water, the pH rises above 7 (see Figure 17-9c). This means that $[OH^-] > [H_3O^+]$ in the solution. Here, acetate ion hydrolyzes.

$$Na^+ + H_2O \longrightarrow \text{no reaction}$$

$$C_2H_3O_2^- + H_2O \rightleftharpoons HC_2H_3O_2 + OH^-$$

The pH of Salt Solutions

We are now in a position to make both qualitative predictions and quantitative calculations concerning the pH values of aqueous solutions of salts. Whichever of these tasks is called for, note that hydrolysis takes place only if a chemical reaction producing a weak acid or weak base occurs. The following generalizations are useful.

- Salts of strong bases and strong acids (for example, NaCl) *do not hydrolyze:* for the solution, pH = 7.
- Salts of strong bases and *weak* acids (for example, $NaC_2H_3O_2$) *hydrolyze:* pH > 7. (The *anion* acts as a *base*.)
- Salts of *weak* bases and strong acids (for example, NH_4Cl) *hydrolyze:* pH < 7. (The *cation* acts as an *acid*.)
- Salts of *weak* bases and *weak* acids (for example, $NH_4C_2H_3O_2$) *hydrolyze.* (The cations are acids, and the anions are bases. Whether the solution is acidic or basic, however, depends on the relative values of K_a and K_b for the ions.)

EXAMPLE 17-11

Making Qualitative Predictions About Hydrolysis Reactions. Predict whether each of the following solutions is acidic, basic, or pH neutral: **(a)** NaOCl(aq); **(b)** KCl(aq); **(c)** NH_4NO_3(aq).

Solution

(a) The ions present are Na^+, which does not hydrolyze, and OCl^-, which does. OCl^- is the conjugate base of HOCl and forms a basic solution.

$$OCl^- + H_2O \rightleftharpoons HOCl + OH^-$$

(b) Neither K^+ nor Cl^- hydrolyzes. KCl(aq) is neutral—that is, pH = 7.

(c) NH_4^+ hydrolyzes, but NO_3^- does not (HNO_3 is a strong acid).

$$NH_4^+ + H_2O \rightleftharpoons NH_3 + H_3O^+$$

In this case, $[H_3O^+] > [OH^-]$, and the solution is acidic.

Practice Example A: Predict whether each of the following 1.0 M solutions is acidic, basic, or pH neutral: **(a)** $CH_3NH_3^+NO_3^-$ (aq); **(b)** NaI(aq); **(c)** $NaNO_2$(aq).

Practice Example B: An aqueous solution containing $H_2PO_4^-$ has a pH of about 4.7. Write equations for *two* reactions of $H_2PO_4^-$ with water, and explain which reaction occurs to the greater extent.

EXAMPLE 17-12

Evaluating Ionization Constants for Hydrolysis Reactions. Both sodium nitrite, $NaNO_2$, and sodium benzoate, $NaC_7H_5O_2$, are used as food preservatives. If solutions of these two salts have the same molarity, which solution will have the *higher* pH?

Solution

Each of these substances is the salt of a strong base (NaOH) and a *weak acid*. The anions should ionize as *bases*, making their solutions somewhat basic.

$$NO_2^- + H_2O \rightleftharpoons HNO_2 + OH^- \qquad K_b(NO_2^-) = \ ?$$
$$C_7H_5O_2^- + H_2O \rightleftharpoons HC_7H_5O_2 + OH^- \qquad K_b(C_7H_5O_2^-) = \ ?$$

Our task is to determine the K_b values, neither of which is listed in a table in this chapter. Table 17.3 does list K_a for the conjugate acids, however. We can use equation (17.18) to write

$$K_b \text{ of } NO_2^- = \frac{K_w}{K_a(HNO_2)} = \frac{1.0 \times 10^{-14}}{7.2 \times 10^{-4}} = 1.4 \times 10^{-11}$$

$$K_b \text{ of } C_7H_5O_2^- = \frac{K_w}{K_a(HC_7H_5O_2)} = \frac{1.0 \times 10^{-14}}{6.3 \times 10^{-5}} = 1.6 \times 10^{-10}$$

The K_b of $C_7H_5O_2^-$ is about 10 times larger than the K_b of NO_2^-. We expect the benzoate ion to hydrolyze to a somewhat greater extent than the nitrite ion and to give a solution with a higher $[OH^-]$. A sodium benzoate solution should be more basic and have a higher pH than a sodium nitrite solution of the same concentration.

Practice Example A: The organic bases cocaine ($pK_b = 8.41$) and codeine ($pK_b = 7.95$) react with hydrochloric acid to form salts (similar to the formation of NH_4Cl by the reaction of NH_3 and HCl). If solutions of the following salts have the same molarity, which solution would have the higher pH: cocaine hydrochloride, $C_{17}H_{21}O_4NH^+Cl^-$, or codeine hydrochloride, $C_{18}H_{21}ClO_3NH^+Cl^-$?

Practice Example B: Predict whether the solution NH_4CN(aq) is acidic, basic, or neutral; and explain the basis of your prediction.

EXAMPLE 17-13

▶ Solutions containing cyanide ion must be handled with extreme caution. They should be handled only in a fume hood by an operator wearing protective clothing.

Calculating the pH of a Solution in Which Hydrolysis Occurs. Sodium cyanide, NaCN, is extremely poisonous, but it has very useful applications in gold and silver metallurgy and in the electroplating of metals. Aqueous solutions of cyanides are especially hazardous if they become acidified, because toxic hydrogen cyanide gas, HCN(g), is released. Are NaCN(aq) solutions normally acidic, basic, or pH neutral? What is the pH of 0.50 M NaCN(aq)?

Solution

Na^+ does not hydrolyze, but as represented below, CN^- does hydrolyze, producing a basic solution. In the tabulation of the concentrations of the species involved in the hydrolysis reaction, we let $[OH^-] = x$.

$$CN^- + H_2O \rightleftharpoons HCN + OH^-$$

initial concns:	0.50 M	—	—
changes:	$-x$ M	$+x$ M	$+x$ M
equil concns:	$(0.50 - x)$ M	x M	x M

To obtain a value of K_b, we must use equation (17.18).

$$K_b = \frac{K_w}{K_a(HCN)} = \frac{1.0 \times 10^{-14}}{6.2 \times 10^{-10}} = 1.6 \times 10^{-5}$$

Now we can return to the tabulated data.

$$K_b = \frac{[HCN][OH^-]}{[CN^-]} = \frac{x \cdot x}{0.50 - x} = \frac{x^2}{0.50 - x} = 1.6 \times 10^{-5}$$

▶ The symbol ≪ means "much less than."

Next let's make a familiar assumption: $x \ll 0.50$ and $0.50 - x \approx 0.50$.

▶ The value of x: 2.8×10^{-3} is less than 1% of 0.500 M, so our assumption is valid.

$$x^2 = 0.50 \times 1.6 \times 10^{-5} = 0.80 \times 10^{-5} = 8.0 \times 10^{-6}$$
$$x = [OH^-] = (8.0 \times 10^{-6})^{1/2} = 2.8 \times 10^{-3}$$
$$pOH = -\log[OH^-] = -\log(2.8 \times 10^{-3}) = 2.55$$
$$pH = 14.00 - pOH = 14.00 - 2.55 = 11.45$$

Practice Example A: Sodium fluoride, NaF, is found in some toothpaste formulations as an anticavity agent. What is the pH of 0.10 M NaF(aq)?

Practice Example B: The pH of an aqueous solution of NaCN is 10.38. What is $[CN^-]$ in this solution?

17-8   Molecular Structure and Acid–Base Behavior

We have now dealt with a number of aspects of acid–base chemistry, both qualitatively and quantitatively. Yet some very fundamental questions still remain to be answered, such as these: Why is HCl a strong acid, whereas HF is a weak acid? Why is acetic acid (CH_3COOH) a stronger acid than ethanol (CH_3CH_2OH) but a weaker acid than chloroacetic acid ($ClCH_2COOH$)?

These questions involve relative acid strengths. In this section, we will examine the relationship between molecular structure and the strengths of acids and bases.

Strengths of Binary Acids

Because the behavior of acids requires the loss of a proton through bond breakage, we expect acid strength to be related to bond strength. To relate the acidities of the HX binary acids to the mere breaking of the H—X bond is an oversimplification, however. For one thing, bond energies are based on the dissociation of gaseous species, and here we are dealing with the ionization of species in solution. Nevertheless, it does seem reasonable that the stronger the H—X bond, the *weaker* the acid should be, and that strong bonds are characterized by short bond lengths and high bond-dissociation energies. For the binary acids of group 17, bond lengths *decrease* and bond-dissociation energies *increase* in the following order:

	HI		**HBr**		**HCl**		**HF**
bond length:	160.9	>	141.4	>	127.4	>	91.7 pm
bond dissociation energy:	297	<	368	<	431	<	569 kJ/mol

and acid strengths *decrease* in the order

HI HBr HCl   HF

$$K_a = \underbrace{10^9 > 10^8 > 1.3 \times 10^6}_{\text{strong}} > \underbrace{6.6 \times 10^{-4}}_{\text{weak}}$$

That HF is a weaker acid than the other hydrogen halides is expected, but that it should be so much weaker has always seemed an anomaly. Explanations of this behavior center on the tendency for hydrogen bonding in HF (recall Figures 13-27). For example, in HF(aq), ion pairs are held together by strong hydrogen bonds, which keeps the concentration of free H_3O^+ from being as large as otherwise expected.

$$HF + H_2O \longrightarrow (^-F \dots H_3O^+) \rightleftharpoons H_3O^+ + F^-$$

ion pair

When we compare the strengths of binary acids within a *period*, bond polarity proves to be the dominant factor. The greater the electronegativity difference (ΔEN) in the bond H—X, the more polar is the bond. We might expect the loss of H^+ to a base to occur more readily from a polar bond, where partial ionic charges already exist, than from a nonpolar bond.

CH_4 NH_3 H_2O HF

ΔEN: 0.4 < 0.9 < 1.4 < 1.9

CH_4 and NH_3 do not have acidic properties in water. The acidity of H_2O is very limited ($K_w = 1.0 \times 10^{-14}$), but HF is an acid of moderate strength ($K_a = 6.6 \times 10^{-4}$).

Strengths of Oxoacids

▶ The term oxoacid was defined in Chapter 3.

To describe the relative strengths of oxoacids, we must focus on the attraction of electrons from the O—H bond toward the central atom. The following factors promote this electron withdrawal from O—H bonds: (1) a high electronegativity (EN) of the central atom and (2) a large number of terminal O atoms in the acid molecule.

Neither HOCl nor HOBr has a terminal O atom. The major difference between the two acids is one of electronegativity—Cl is slightly more electronegative than Br. As expected, HOCl is more acidic than HOBr.

H—O̤—C̤l̤:   H—O̤—B̤r̤:
$EN_{Cl} = 3.0$ $EN_{Br} = 2.8$
$K_a = 2.9 \times 10^{-8}$ $K_a = 2.1 \times 10^{-9}$

To compare the acid strengths of H_2SO_4 and H_2SO_3, we must look beyond the central atom, which is S in each acid.

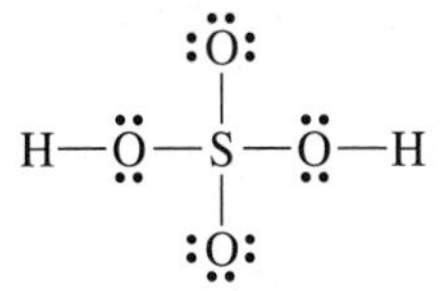

$$K_{a_1} \approx 10^3 \qquad\qquad K_{a_1} = 1.3 \times 10^{-2}$$

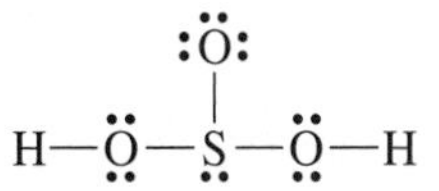

Our expectation is that a highly electronegative terminal O atom tends to withdraw electrons from the O—H bonds, weakening the bonds and increasing the acidity of the molecule. Because H_2SO_4 has *two* terminal O atoms to only one in H_2SO_3, we should expect the electron-withdrawing effect to be greater in H_2SO_4. As a result, H_2SO_4 should be a stronger acid than H_2SO_3.

Strengths of Organic Acids

We conclude our discussion of the relationship between molecular structure and acid strength with a brief consideration of some organic compounds. Consider first the case of acetic acid and ethanol. Both have an O—H group bonded to a carbon atom, but acetic acid is a much stronger acid than ethanol.

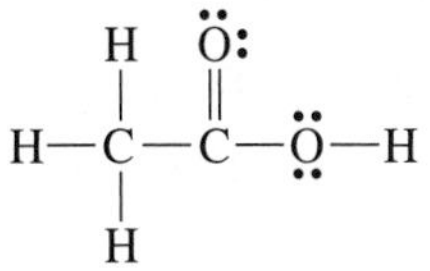

Acetic acid
$K_a = 1.8 \times 10^{-5}$

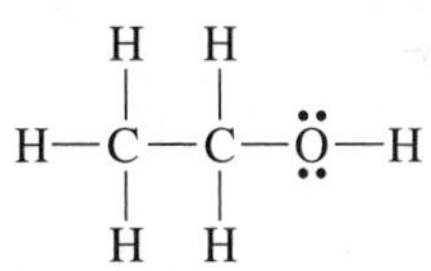

Ethanol
$K_a = 1.3 \times 10^{-16}$

To rationalize the large difference in the acidities of these two compounds, we can say that the highly electronegative terminal O atom in acetic acid withdraws electrons from the O—H bond. The bond is weakened, and a proton (H^+) is more readily taken from a molecule of the acid by a base. Acetic acid is more acidic than ethanol. A more satisfactory explanation focuses on the anions formed in the ionization.

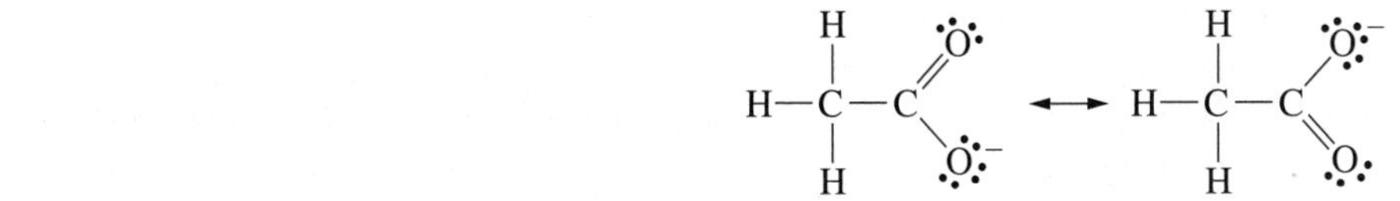

Acetate ion

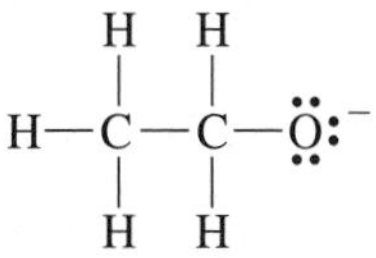

Ethoxide ion

We can write two plausible structures of the acetate ion. These structures suggest that each carbon-to-oxygen bond is a "$\frac{3}{2}$" bond and that each O atom carries "$\frac{1}{2}$" unit of negative charge. In short, the excess unit of negative charge in CH_3COO^- is spread out. This arrangement reduces the ability of either O atom to attach a proton and makes acetate ion only a weak Brønsted–Lowry base. In ethoxide ion, conversely, the unit of negative charge is localized on the single O atom. Ethoxide ion is a much stronger base than is acetate ion. The stronger the conjugate base, the weaker the corresponding acid.

The length of the carbon chain in a carboxylic acid has little effect on the acid strength, as we see in a comparison of acetic acid and octanoic acid.

$$CH_3COOH \qquad\qquad CH_3(CH_2)_6COOH$$

acetic acid: $K_a = 1.8 \times 10^{-5}$ octanoic acid: $K_a = 1.3 \times 10^{-5}$

Yet, other atoms or groups of atoms substituted onto the carbon chain may strongly affect acid strength. If we substitute a Cl atom for one of the H atoms that is bonded to carbon in acetic acid, the result is chloroacetic acid.

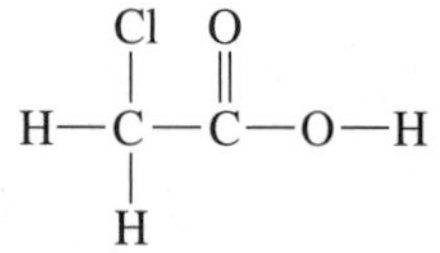

$$\underset{\begin{array}{c}|\\H\end{array}}{\overset{\begin{array}{c}Cl\\|\end{array}}{H-C}}-\overset{\overset{\displaystyle O}{\|}}{C}-O-H$$

Chloroacetic acid
$K_a = 1.4 \times 10^{-3}$

The highly electronegative Cl atom helps draw electrons away from the O—H bond. The O—H bond is weakened, the proton is lost more readily, and the acid is a stronger acid than acetic acid. This effect falls off rapidly as the distance increases between the substituted atom or group and the O—H bond in an organic acid.

Example 17-14 illustrates some of the factors affecting acid strength that are discussed in this section.

EXAMPLE 17-14

Identifying Factors That Affect the Strengths of Acids. Explain which member of each of the following pairs is the stronger acid.

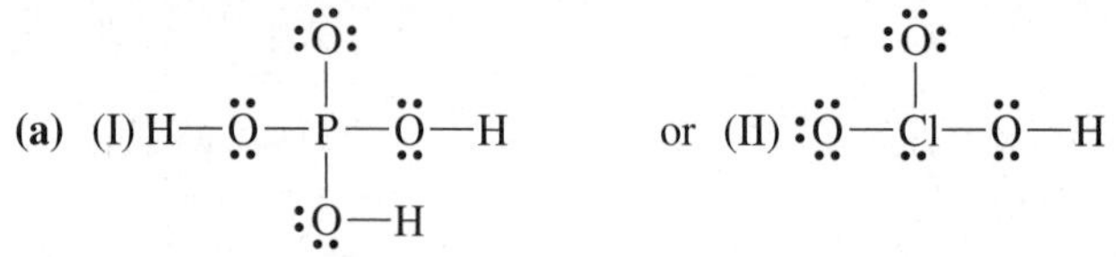

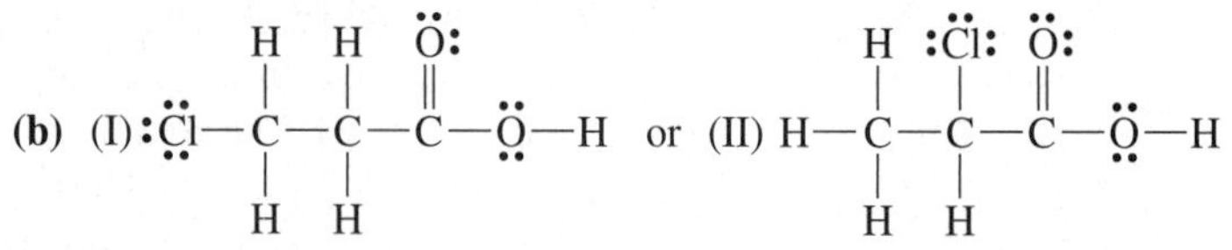

(a) (I) H—Ö—P—Ö—H   or (II) :Ö—Cl—Ö—H
         :Ö—H

(b) (I) :Cl—C—C—C—Ö—H   or (II) H—C—C—C—Ö—H

Solution

(a) Phosphoric acid, H_3PO_4, has four O atoms to the three in $HClO_3$, but it is the number of *terminal* O atoms that we must consider, not just the total number of O atoms in the molecule. $HClO_3$ has *two* terminal O atoms and H_3PO_4 has *one*. Also, the Cl atom (EN = 3.0) is considerably more electronegative than the P atom (EN = 2.1). These facts point to chloric acid (II) as being the stronger of the two acids. ($K_a \approx 5 \times 10^2$ for $HClO_3$ and $K_{a_1} = 7.1 \times 10^{-3}$ for H_3PO_4.)

(b) The Cl atom withdraws electrons more strongly when it is directly adjacent to the carboxyl group. Compound (II) (2-chloropropanoic acid, $K_a = 1.4 \times 10^{-3}$) is a stronger acid than compound (I) (3-chloropropanoic acid, $K_a = 1.0 \times 10^{-4}$).

Practice Example A: Explain which is the stronger acid, HNO_3 or $HClO_4$; CH_2FCOOH or $CH_2BrCOOH$.

(*Hint:* Draw plausible Lewis structures.)

Practice Example B: Explain which is the stronger acid, H_3PO_4 or H_2SO_3; CCl_3CH_2COOH or CCl_2FCH_2COOH.

(*Hint:* Draw plausible Lewis structures.)

Strengths of Amines as Bases

The fundamental factor affecting the strength of an amine as a base concerns the ability of the lone pair of electrons on the N atom to bind a proton taken from an acid. When an atom or group of atoms more electronegative than H replaces one of the H atoms of NH_3, the electronegative group withdraws electron density from the N atom. The lone-pair electrons cannot bind a proton as strongly, and the base is weaker. Thus, bromamine, in which the electronegative Br atom is attached to the amine group (NH_2), is a *weaker* base than ammonia.

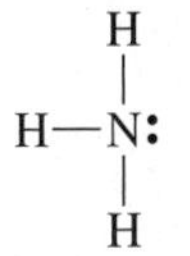

Ammonia, $pK_b = 4.74$ 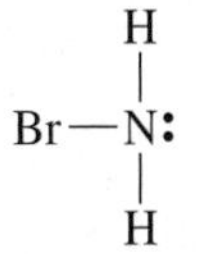   Bromamine, NH_2Br, $pK_b = 7.61$

Hydrocarbon chains have little electron-withdrawing ability. When they are attached to the amine group, the pK_b values are comparable to (actually, somewhat lower than) those of ammonia.

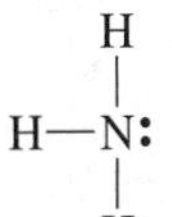

Ammonia
$pK_b = 4.74$

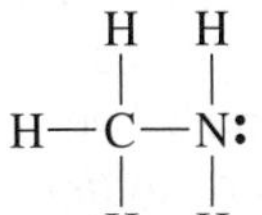

Methylamine
CH_3NH_2, $pK_b = 3.38$

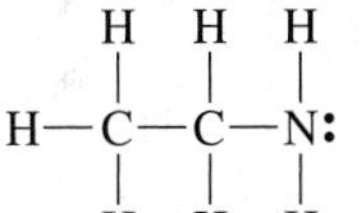

Ethylamine
$CH_3CH_2NH_2$, $pK_b = 3.37$

An additional electron-withdrawing effect is seen in amines that are based on the benzene ring or related structures. Such amines are called *aromatic* amines. Cyclohexylamine, $C_6H_{11}NH_2$, is based on cyclohexane, C_6H_{12}, which is a six-carbon alkane that forms a ring rather than a chain. Aniline, $C_6H_5NH_2$, is based on benzene, C_6H_6, which is a six-carbon ring molecule with unsaturation in the carbon-to-carbon bonds. The electrons associated with this unsaturation are said to be *delocalized*. As suggested by the following structures, to some extent even the lone-pair electrons of the NH_2 group participate in the "spreading out" of delocalized electrons. (The curved arrows suggest the progressive movement of electrons around the ring.)

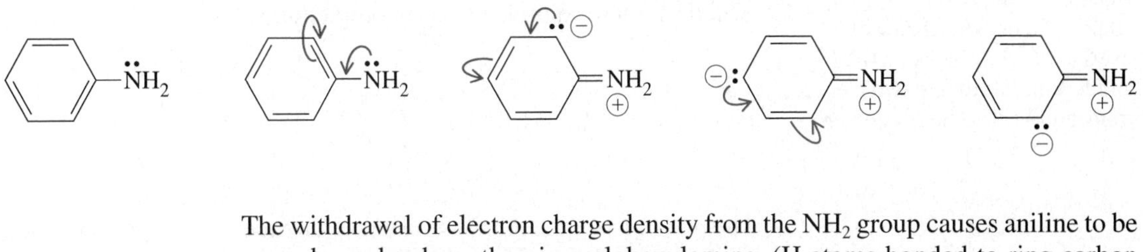

The withdrawal of electron charge density from the NH_2 group causes aniline to be a much weaker base than is cyclohexylamine. (H atoms bonded to ring carbon atoms are not shown in the following structures.)

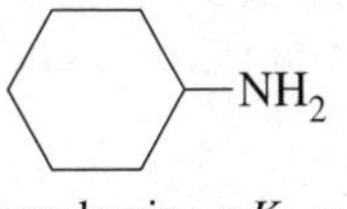

Cyclohexylamine, $pK_b = 3.36$ 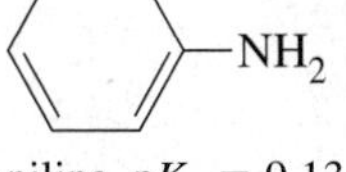   Aniline, $pK_b = 9.13$

Replacement of a ring-bound H atom in aniline with an atom or group that has a high electronegativity causes even more electron density to be drawn away from the NH_2 group, further reducing the base strength. Also, the closer this ring substituent is to the NH_2 group, the greater is the effect.

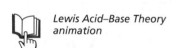

para-Chloroaniline, $pK_b = 10.01$ *ortho*-Chloroaniline, $pK_b = 11.36$

17-9   Lewis Acids and Bases

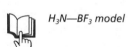

In the previous section, we presented ideas about the molecular structures of acids and bases. In 1923, G. N. Lewis proposed an acid–base theory closely related to bonding and structure. The Lewis acid–base theory is not limited to reactions involving H^+ and OH^-: It extends acid–base concepts to reactions in gases and in solids. It is especially important in describing certain reactions between organic molecules.

A **Lewis acid** is a species (an atom, ion, or molecule) that is an electron-pair *acceptor*, and a **Lewis base** is a species that is an electron-pair *donor*. A reaction between a Lewis acid (A) and a Lewis base (B:) results in the formation of a covalent bond between them. The product of a Lewis acid–base reaction is called an **adduct** (or *addition compound*). The reaction can be represented as

$$B: + A \longrightarrow B:A$$

where B:A is the adduct. The formation of a covalent chemical bond by one species donating a pair of electrons to another is called *coordination*, and the bond joining the Lewis acid and Lewis base is called a *coordinate covalent bond* (see page 393). In general, to identify Lewis *acids*, we should look for species with vacant orbitals that can accommodate electron pairs; to identify Lewis *bases*, we look for species having lone-pair electrons available for sharing.

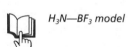

By these definitions, OH^-, a Brønsted–Lowry base, is also a Lewis base because lone-pair electrons are present on the O atom. So too is NH_3 a Lewis base. HCl, conversely, is not a Lewis acid: It is not an electron-pair acceptor. We can think of HCl as producing H^+, however, and H^+ is a Lewis acid. H^+ forms a coordinate covalent bond with an available electron pair.

We might expect substances with an incomplete valence shell to be Lewis acids. When a coordinate covalent bond is formed with a Lewis base, the octet is completed. A good example of octet completion is the reaction of BF_3 and NH_3.

Bonding in the H_3N—BF_3 adduct can be described by the overlap of sp^3 orbitals on the N and B atoms, with the two electrons donated by the N atom. ▶

The reaction of lime (CaO) with sulfur dioxide is an important reaction for reducing SO_2 emissions from coal-fired power plants. This reaction between a solid and a gas underscores that Lewis acid–base reactions can occur in all states of matter. The small curved red arrow in reaction (17.19) suggests that an electron pair in the Lewis structure is rearranged.

$$Ca^{2+} :\overset{..}{\underset{..}{O}}:^{2-} + \quad S: \longrightarrow Ca^{2+} \left[:\overset{..}{\underset{..}{O}} - S: \right]^{2-} \qquad (17.19)$$

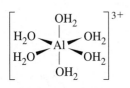

▲ **FIGURE 17-10**
The Lewis structure of [Al(H$_2$O)$_6$]$^{3+}$ and a ball-and-stick representation

An important application of the Lewis acid–base theory involves the formation of *complex ions*. Complex ions are polyatomic ions that contain a central metal ion to which other ions or small molecules are attached. *Hydrated metal ions* form in aqueous solution because the water acts as a Lewis base and the metal ion as a Lewis acid. The water molecules attach themselves to the metal ion by means of coordinate covalent bonds. Thus, for example, when anhydrous AlCl$_3$ is added to water, the resultant solution becomes hot because of the heat evolved in the formation of the hydrated metal ion [Al(H$_2$O)$_6$]$^{3+}$(aq) (Figure 17-10).

The interaction between the metal ion and the water molecules is so strong that when the salt is crystallized from the solution, the water molecules crystallize along with the metal ion, forming the hydrated metal salt AlCl$_3$ · 6 H$_2$O. In aqueous solution, the hydrated metal ions can act as Brønsted acids. For instance, the hydrolysis of hydrated Al^{3+} is given by

$$[\text{Al}(\text{H}_2\text{O})_6]^{3+} + \text{H}_2\text{O} \rightleftharpoons [\text{Al}(\text{OH})(\text{H}_2\text{O})_5]^{2+} + \text{H}_3\text{O}^+$$

The OH bond in a water molecule in the hydrated metal ion becomes weakened. The reason is that in forming the coordinate covalent bond with the O atom of the water, the metal ion causes electron density to be drawn toward it; hence, electron density is drawn away from the OH bond. As a consequence, the coordinated H$_2$O molecule can donate a H$^+$ to a solvent H$_2$O molecule (Figure 17-11). The H$_2$O molecule that has ionized is converted to OH$^-$, which remains attached to the Al^{3+}; the charge on the complex ion is reduced from 3+ to 2+. The extent of ionization of [Al(H$_2$O)$_6$]$^{3+}$, measured by its K_a value and as pictured in Figure 17-12, is essentially the same as that of acetic acid ($K_a = 1.74 \times 10^{-5}$). Many other metal ions, especially the transition metal ions, hydrolyze. We will discuss these and other hydrated metal ions acting as acids in later chapters.

 ## Are You Wondering...

Why Na$^+$(aq) does not act as an acid in aqueous solution?

Whether an aqueous solution of a metal ion is acidic depends on two principal factors. The first is the amount of charge on the cation; the second is the size of the ion. The greater the charge on the cation, the greater is the ability of the metal ion to draw electron density away from the O—H bond in a H$_2$O molecule in its hydration sphere, favoring the release of a H$^+$ ion. The smaller the cation, the more highly concentrated is the positive charge. Hence, for a given positive charge, the smaller the cation, the more acidic the solution.

The ratio of the charge on the cation to the volume of the cation is called the *charge density.*

$$\text{charge density} = \frac{\text{ionic charge}}{\text{ionic volume}}$$

The greater its charge density, the more effective a metal ion is at pulling electron density from the O—H bond and the more acidic is the hydrated cation. A highly concentrated positive charge on a small cation is better able to pull electron density from the O—H bond than is a less concentrated positive charge on a larger cation.

Thus the small (53 pm ionic radius), highly charged Al^{3+} ion produces acidic solutions, but the larger (99 pm) Na$^+$ cation, with a charge of just 1+, does not increase the concentration of H$_3$O$^+$. In fact, none of the group 1 cations produces appreciably acidic solutions, and only Be^{2+} of the group 2 elements is small enough to do so (pK_a = 5.4).

▲ On the left is a marble statue of George Washington as it appeared in 1935. On the right is the same statue 59 years later. The effects of acid rain are clearly evident.

Pure water has pH = 7, but not rainwater—it is acidic. The natural pH of rainwater should be about 5.6. In recent years, however, the pH of rainfall in northeastern United States has averaged about 4.5. Some rainfall has been found to be even more acidic than that; in one episode on the West Coast in 1982, the pH of fog was 1.7. Let us briefly consider both the natural sources of acidity in rainwater and how human activities have contributed to acid rain.

$CO_2(g)$ is an acidic oxide. Although H_2CO_3 cannot be isolated, it is customary to consider that CO_2 reacts with water to form carbonic acid. The acidity of rainwater arises mostly through the first ionization step of this acid.

$$CO_2(g) + H_2O \longrightarrow H_2CO_3$$
$$H_2CO_3 + H_2O \rightleftharpoons H_3O^+ + HCO_3^-$$
$$K_{a_1} = 4.4 \times 10^{-7}$$

Carbon Dioxide Acts as an Acid in Water movie

▲ FIGURE 17-12
Acidic properties of hydrated metal ions
The yellow color of bromthymol blue indicator in $Al_2(SO_4)_3(aq)$ denotes that the solution is acidic. The pH meter gives a more precise indication of the pH.

$$Al(H_2O)_6^{3+} + H_2O \rightleftharpoons [Al(OH)(H_2O)_5]^{2+} + H_3O^+$$

▲ FIGURE 17-11 **Hydrolysis of $[Al(H_2O)_6]^{3+}$ to produce H_3O^+**
An uncoordinated water molecule removes a proton from a coordinated water molecule.

Complex ions can also form between transition metal ions and other Lewis bases, such as NH_3. For instance, Zn^{2+} combines with NH_3 to form the complex ion $[Zn(NH_3)_4]^{2+}$. The central Zn^{2+} ion accepts electrons from the Lewis base NH_3 to form coordinate covalent bonds; it is a Lewis acid. We will discuss the application of Lewis acid–base theory to complex ions in Chapter 25.

EXAMPLE 17-15

Identifying Lewis Acids and Bases. According to the Lewis theory, each of the following is an acid–base reaction. Which species is the acid and which is the base?

(a) $BF_3 + F^- \longrightarrow BF_4^-$

(b) $OH^-(aq) + CO_2(aq) \longrightarrow HCO_3^-(aq)$

Rainwater in equilibrium with atmospheric $CO_2(g)$ at its normal partial pressure of 0.0037 atm has a pH of about 5.6.

$NO(g)$ is formed in lightning storms through the high-temperature reaction of $N_2(g)$ and $O_2(g)$ in air. This reaction is followed by the oxidation of $NO(g)$ to $NO_2(g)$ in air. Nitric acid forms by this reaction.

$$3 NO_2(g) + H_2O(l) \longrightarrow 2 HNO_3(aq) + NO(g)$$

To some extent, then, naturally produced nitric acid contributes to the acidity of rain. More important, though, is $HNO_3(aq)$ derived from artificial sources of $NO(g)$—primarily, high-temperature combustion processes in air, such as those that occur in automobile engines and power plants. In southern California, for example, where the oxides of nitrogen are seriously implicated in the formation of photochemical smog, HNO_3 is a major contributor to acid rain. On average, nitric acid accounts for about one-quarter of the acidity of acid rain.

$SO_2(g)$ is normally only a trace constituent of the atmosphere, being produced mostly by biological decay and volcanic activity. Artificial sources of $SO_2(g)$ are substantial, however, and create higher than normal concentrations in many areas. The chief source of SO_2 is the combustion of sulfur-containing coal (up to 8% S) in electric power plants. Another source is the extraction of certain metals from their sulfide ores. In the atmosphere, $SO_2(g)$ can be oxidized by any of several routes to $SO_3(g)$, and SO_3 is the acid anhydride of sulfuric acid. Sulfuric acid accounts for well over one-half of the acidity of acid rain.

Some effects of acid rain are clearly visible in urban areas, particularly in the degradation of statues, monuments, and buildings made of sandstone, limestone, or marble. In India, acidic air pollutants from nearby industrial plants threaten to destroy the Taj Mahal, the famous marble mausoleum that is more than 350 years old. The most serious ecological effects, though, are on forests and on freshwater lakes and streams. Fish and other aquatic life may be killed in acidic lakes. To maintain healthy forests, the soil must furnish certain nutrients while it immobilizes other species, such as Al^{3+}, that interfere with the nutritional cycle of trees. Acidic rainwater typically leaches important nutrients from the soil and at the same time frees up or mobilizes undesirable soil species.

Measures to reduce $SO_2(g)$ and $NO(g)$ emissions include the use of low-sulfur fuels, the control of combustion temperatures to reduce $NO(g)$ emissions, and the trapping of offending flue gases by various means.

Solution

(a) In BF_3, the B atom has a vacant orbital and an incomplete octet. The fluoride ion has an outer-shell octet of electrons. BF_3 is the electron-pair acceptor—the acid. F^- is the electron-pair donor—the base.

(b) We have already identified OH^- as a Lewis base, so we might suspect that it is the base and that $CO_2(aq)$ is the Lewis acid. The following Lewis structures show this to be so. As in reaction (17.19), a rearrangement of an electron pair at one of the double bonds is also required, as indicated by the red arrow.

Practice Example A: Identify the Lewis acids and bases in these reactions.

(a) $BF_3 + NH_3 \longrightarrow F_3BNH_3$

(b) $Cr^{3+} + 6 H_2O \longrightarrow [Cr(H_2O)_6]^{3+}$

Practice Example B: Identify the Lewis acids and bases in these reactions.

(a) $Al(OH)_3 + OH^- \longrightarrow [Al(OH)_4]^-$

(b) $SnCl_4 + 2 Cl^- \longrightarrow [SnCl_6]^{2-}$

(*Hint:* Draw plausible Lewis structures for the reactants and products.)

Summary

The Brønsted–Lowry theory describes an acid as a proton donor and a base as a proton acceptor. In an acid–base reaction, a base takes a proton (H^+) from an acid. In general, acid–base reactions are reversible, but equilibrium is displaced in the direction from the *stronger* acids and bases to their *weaker* conjugates.

In pure water and in aqueous solutions, self-ionization occurs to a very slight extent, producing H_3O^+ and OH^-, as described by the equilibrium constant K_w. The designations pH and pOH are often used to describe the concentrations of H_3O^+ and OH^- in aqueous solutions. In aqueous solutions, strong acids ionize completely to produce H_3O^+, and strong bases dissociate completely to produce OH^-. With weak acids and weak bases, ionization is reversible and must be described by means of the ionization constants, K_a and K_b. In polyprotic acids, different constants K_{a_1}, K_{a_2}, ... apply to each ionization step. In reactions between ions and water—hydrolysis reactions—the ions react as weak acids or weak bases. Calculations involving ionization equilibria are in many ways similar to those introduced in Chapter 16, although some additional considerations are necessary for polyprotic acids.

Molecular composition and structure are the keys to determining whether a substance is acidic, basic, or amphiprotic. In addition, molecular structure affects whether an acid or base is strong or weak. In assessing acid strength, for example, we must consider factors that affect the strength of the bond that must be broken to release H^+. In assessing base strength, factors that affect the ability of lone-pair electrons to bind a proton are of primary concern.

The Lewis acid–base theory views an acid as an electron-pair acceptor and a base as an electron-pair donor. Its greatest use is in situations that cannot be described by means of proton transfers, for example, in reactions involving gases and solids and in reactions between organic compounds (considered in Chapter 27).

Integrative Example

Sometimes the questions that prove to be most challenging do not require extensive calculations, but only a good insight into the concepts involved. The following question requires the application of many of the ideas discussed in this chapter.

Several aqueous solutions are prepared. *Without consulting any tables in the text*, arrange these ten solutions in order of *increasing* pH: 1.0 M NaBr, 0.05 M $HC_2H_3O_2$ (CH_3COOH), 0.05 M NH_3, 0.02 M $KC_2H_3O_2$, 0.05 M $Ba(OH)_2$, 0.05 M H_2SO_4, 0.10 M HI, 0.06 M NaOH, 0.05 M NH_4Cl, and 0.05 M $HC_2H_2ClO_2$ ($CH_2ClCOOH$).

1. *Establish the ordering that is sought.* Low pH corresponds to strongly acidic solutions. Arranging the solutions by increasing pH means starting with the most acidic and proceeding to the most basic.

2. *Indicate whether each solution is acidic, basic, or neutral.* Our assignments should be as follows:

acidic: 0.05 M $HC_2H_3O_2$, 0.05 M H_2SO_4, 0.10 M HI, 0.05 M NH_4Cl, 0.05 M $HC_2H_2ClO_2$

neutral: 1.0 M NaBr

basic: 0.05 M NH_3, 0.02 M $KC_2H_3O_2$, 0.05 M $Ba(OH)_2$, 0.06 M NaOH

3. *Sort out the solutions in each category. Which is most acidic, next most acidic, and so on?*

acidic: For the strong acid, 0.10 M HI, $[H_3O^+] = 0.10$ M. In 0.05 M H_2SO_4, 0.05 M $< [H_3O^+] < 0.10$ M because H_2SO_4 is a strong acid in the first ionization and weak in the second. $[H_3O^+]$ in 0.05 M $HC_2H_2ClO_2$ is greater than in 0.05 M $HC_2H_3O_2$ because the Cl atom has an electron-withdrawing effect on the O—H bond, but in both weak acids $[H_3O^+] < 0.05$ M. The 0.05 M NH_4Cl is acidic because the NH_4^+ ion hydrolyzes, but only weakly so. Therefore, the relative pH values for these five acids are

0.10 M HI $<$ 0.05 M H_2SO_4 $<$ 0.05 M $HC_2H_2ClO_2$
$<$ 0.05 M $HC_2H_3O_2$ $<$ 0.05 M NH_4Cl

neutral: In 1.0 M NaBr, neither Na^+ or Br^- hydrolyzes.

basic: The 0.02 M $KC_2H_3O_2$ is weakly basic because the $C_2H_3O_2^-$ ion hydrolyzes. The 0.05 M NH_3 is a weak base, but more basic than the acetate ion. In 0.06 M NaOH, $[OH^-] = 0.06$ M, but in 0.05 M $Ba(OH)_2$, $[OH^-] = 2 \times 0.05 = 0.10$ M. Therefore, the relative pH values of the four basic solutions are

0.02 M $KC_2H_3O_2$ $<$ 0.05 M NH_3 $<$
0.06 M NaOH $<$ 0.05 M $Ba(OH)_2$

4. *Combine the three sets of results from Step 3 into a final order of increasing pH.*

Answer: 0.10 M HI $<$ 0.05 M H_2SO_4 $<$ 0.05 M $< HC_2H_2ClO_2$ $<$ 0.05 M $< HC_2H_3O_2$ $<$ 0.05 M $< NH_4Cl$ $<$ 1.0 M NaBr $<$ 0.05 M $KC_2H_3O_2$ $<$ 0.05 M NH_3 $<$ 0.06 M NaOH $<$ 0.05 M $Ba(OH)_2$.

Key Terms

acid ionization constant, K_a (17-2)
adduct (17-9)
amphiprotic (17-2)
base ionization constant, K_b (17-2)
conjugate acid (17-2)
conjugate base (17-2)
hydrolysis (17-7)

hydronium ion, H_3O^+ (17-2)
ion product of water, K_w (17-3)
Lewis acid (17-9)
Lewis base (17-9)
percent ionization (17-5)
pH (17-3)
pK (17-5)

pOH (17-3)
polyprotic acid (17-6)
proton acceptor (17-2)
proton donor (17-2)
self-ionization (17-3)

Review Questions

1. In your own words, define or explain the following terms or symbols: (a) K_w; (b) pH; (c) pK_a; (d) hydrolysis; (e) Lewis acid.
2. Briefly describe each of the following ideas or phenomena: (a) conjugate base; (b) percent ionization of an acid or base; (c) self-ionization; (d) amphiprotic behavior.
3. Explain the important distinctions between each pair of terms: (a) Brønsted–Lowry acid and base; (b) $[H_3O^+]$ and pH; (c) K_a for NH_4^+ and K_b for NH_3; (d) leveling effect and electron-withdrawing effect.
4. According to the Brønsted–Lowry theory, label each of the following as an acid or a base. (a) HNO_2; (b) OCl^-; (c) NH_2^-; (d) NH_4^+; (e) $CH_3NH_3^+$
5. Write the formula of the conjugate base in the reaction of each acid with water. (a) HIO_3; (b) C_6H_5COOH; (c) HPO_4^{2-}; (d) $C_2H_5NH_3^+$
6. Calculate $[H_3O^+]$ and $[OH^-]$ for each solution: (a) 0.00165 M HNO_3; (b) 0.0087 M KOH; (c) 0.00213 M $Sr(OH)_2$; (d) 5.8×10^{-4} M HI.
7. What is the pH of each of the following solutions? (a) 0.0045 M HCl; (b) 6.14×10^{-4} M HNO_3; (c) 0.00683 M NaOH; (d) 4.8×10^{-3} M $Ba(OH)_2$
8. *Without doing detailed calculations*, which value would you expect for $[H_3O^+]$ in 0.10 M H_2SO_4: 0.10 M, 0.05 M, 0.11 M, 0.20 M? Explain.
9. What is the pH of the solution obtained by mixing 24.80 mL of 0.248 M HNO_3 and 15.40 mL of 0.394 M KOH? (*Hint:* What reaction occurs? Is the final solution acidic, basic, or neutral?)
10. Only one of the following expressions is correct concerning 0.10 M CH_3NH_2(aq): (1) $[H_3O^+]$ = 0.10 M; (2) $[OH^-]$ = 0.10 M; (3) pH < 7; (4) pH < 13. Choose the correct one, and explain your choice.
11. A 625-mL sample of an aqueous solution containing 0.275 mol propionic acid, $HC_3H_5O_2$, has $[H_3O^+]$ = 0.00239 M. What is the value of K_a for propionic acid?
$$HC_3H_5O_2 + H_2O \rightleftharpoons H_3O^+ + C_3H_5O_2^- \quad K_a = ?$$
12. For the ionization of phenylacetic acid,
$$HC_8H_7O_2 + H_2O \rightleftharpoons H_3O^+ + C_8H_7O_2^- \quad K_a = 4.9 \times 10^{-5}$$

(a) What is $[C_8H_7O_2^-]$ in 0.186 M $HC_8H_7O_2$?
(b) What is the pH of 0.121 M $HC_8H_7O_2$?
13. What mass of benzoic acid, $HC_7H_5O_2$, would you dissolve in 350.0 mL of water to produce a solution with a pH = 2.85?
$$HC_7H_5O_2 + H_2O \rightleftharpoons H_3O^+ + C_7H_5O_2^- \quad K_a = 6.3 \times 10^{-5}$$
14. What must be the molarity of an aqueous solution of trimethylamine, $(CH_3)_3N$, if it has a pH = 11.12?
$$(CH_3)_3N + H_2O \rightleftharpoons (CH_3)_3NH^+ + OH^- \quad K_b = 6.3 \times 10^{-5}$$
15. For 0.045 M H_2CO_3, a weak diprotic acid, calculate (a) $[H_3O^+]$, (b) $[HCO_3^-]$, and (c) $[CO_3^{2-}]$. Use data from Table 17.4 as necessary.
16. Complete the following equations in those instances in which a reaction (hydrolysis) will occur. If no reaction occurs, so state.
(a) NH_4^+(aq) + NO_3^-(aq) + $H_2O \longrightarrow$
(b) Na^+(aq) + NO_2^-(aq) + $H_2O \longrightarrow$
(c) K^+(aq) + $C_7H_5O_2^-$(aq) + $H_2O \longrightarrow$
(d) K^+(aq) + Cl^-(aq) + Na^+(aq) + I^-(aq) + $H_2O \longrightarrow$
(e) $C_6H_5NH_3^+$(aq) + Cl^-(aq) + $H_2O \longrightarrow$
17. Calculate the pH of an aqueous solution that is 1.68 M NH_4Cl.
$$NH_4^+ + H_2O \rightleftharpoons H_3O^+ + NH_3 \quad K_a = 5.6 \times 10^{-10}$$
18. From data in Table 17.3, determine (a) K_a for $C_5H_5NH^+$; (b) K_b for CHO_2^-; (c) K_b for $C_6H_5O^-$.
19. Determine the pH of 2.05 M $NaC_2H_2ClO_2$. (Use data from Table 17.3, as necessary.)
20. Explain why trichloroacetic acid, CCl_3COOH, is a stronger acid than acetic acid, CH_3COOH.
21. Which is the stronger acid of each of the following pairs of acids? Explain your reasoning. (a) HBr or HI; (b) HOClO or HOBr; (c) $I_3CCH_2CH_2COOH$ or $CH_3CH_2CCl_2COOH$.
22. Indicate whether each of the following is a Lewis acid or base. (a) OH^-; (b) $(C_2H_5)_3B$; (c) CH_3NH_2
23. Each of the following is a Lewis acid–base reaction. Which reactant is the acid, and which is the base? Explain.
(a) $SO_3 + H_2O \longrightarrow H_2SO_4$
(b) $Zn(OH)_2$(s) + 2 OH^-(aq) $\longrightarrow [Zn(OH)_4]^{2-}$(aq)
24. Indicate which of the following 0.10 M aqueous solutions has the highest pH and which has the lowest. Explain your reasoning. (a) NH_4Cl, (b) NH_3, (c) $NaC_2H_3O_2$, (d) KCl.

Exercises

Brønsted–Lowry Theory of Acids and Bases

25. For each of the following, identify the acids and bases involved in both the forward and reverse directions.
    (a) $HOBr + H_2O \rightleftharpoons OBr^- + H_3O^+$
    (b) $HSO_4^- + H_2O \rightleftharpoons SO_4^{2-} + H_3O^+$
    (c) $HS^- + H_2O \rightleftharpoons H_2S + OH^-$
    (d) $C_6H_5NH_3^+ + OH^- \rightleftharpoons C_6H_5NH_2 + H_2O$

26. Which of the following species are *amphiprotic* in aqueous solution? For each one that is, write two representative equations: one showing it acting as an acid, and the other as a base. OH^-, NH_4^+, H_2O, HS^-, NO_2^-, HCO_3^-, HBr

27. With which of the following bases will the ionization of acetic acid, $HC_2H_3O_2$, proceed furthest toward completion (to the right): (1) H_2O; (2) NH_3; (3) Cl^-; (4) NO_3^-? Explain your answer.

28. In a manner similar to equation (17.6), represent the self-ionization of the following liquid solvents: (a) NH_3; (b) HF; (c) CH_3OH; (d) $HC_2H_3O_2$; (e) H_2SO_4.

29. With the aid of Table 17.1, predict the direction (forward or reverse) favored in each of the following acid–base reactions.
    (a) $NH_4^+ + OH^- \rightleftharpoons H_2O + NH_3$
    (b) $HSO_4^- + NO_3^- \rightleftharpoons HNO_3 + SO_4^{2-}$
    (c) $CH_3OH + C_2H_3O_2^- \rightleftharpoons HC_2H_3O_2 + CH_3O^-$

30. With the aid of Table 17.1, predict the direction (forward or reverse) favored in each of the following acid–base reactions.
    (a) $HC_2H_3O_2 + CO_3^{2-} \rightleftharpoons HCO_3^- + C_2H_3O_2^-$
    (b) $HNO_2 + ClO_4^- \rightleftharpoons HClO_4 + NO_2^-$
    (c) $H_2CO_3 + CO_3^{2-} \rightleftharpoons HCO_3^- + HCO_3^-$

Strong Acids, Strong Bases, and pH

31. Calculate $[H_3O^+]$ and pH in saturated $Ba(OH)_2(aq)$, which contains 3.9 g $Ba(OH)_2 \cdot 8\,H_2O$ per 100 mL of solution.

32. A saturated aqueous solution of $Ca(OH)_2$ has a pH of 12.35. What is the solubility of $Ca(OH)_2$, expressed in milligrams per 100 mL of solution?

33. What is $[H_3O^+]$ in a solution obtained by dissolving 205 mL $HCl(g)$, measured at 23 °C and 751 mmHg, in 4.25 L of aqueous solution?

34. What is the pH of the solution obtained when 125 mL of 0.606 M NaOH is diluted to 15.0 L with water?

35. How many milliliters of concentrated $HCl(aq)$ (36.0% HCl by mass, $d = 1.18$ g/mL) are required to produce 12.5 L of a solution with pH = 2.10?

36. How many milliliters of a 15.0%, by mass, solution of $KOH(aq)$ ($d = 1.14$ g/mL) are required to produce 25.0 L of a solution with pH = 11.55?

37. What volume of 6.15 M HCl(aq) is required to exactly neutralize 1.25 L of 0.265 M $NH_3(aq)$?
    $$NH_3(aq) + H_3O^+(aq) \longrightarrow NH_4^+(aq) + H_2O$$

38. A 28.2-L volume of $HCl(g)$, measured at 742 mmHg and 25.0 °C, is dissolved in water. What volume of $NH_3(g)$, measured at 762 mmHg and 21.0 °C, must be absorbed by the same solution to neutralize the HCl?
    (*Hint:* Does the volume of solution make any difference?)

39. 50.00 mL of 0.0155 M HI(aq) is mixed with 75.00 mL of 0.0106 M KOH(aq). What is the pH of the final solution?

40. 25.00 mL of a $HNO_3(aq)$ solution with a pH of 2.12 is mixed with 25.00 mL of a KOH(aq) solution with a pH of 12.65. What is the pH of the final solution?

Weak Acids, Weak Bases, and pH (*Use data from Table 17.3 as necessary.*)

41. What are the $[H_3O^+]$ and pH of 0.143 M HNO_2?

42. What are the $[H_3O^+]$ and pH of 0.085 M $C_2H_5NH_2$?

43. Fluoroacetic acid occurs in gifblaar, one of the most poisonous of all plants. A 0.318 M solution of the acid is found to have a pH = 1.56. Calculate K_a of fluoroacetic acid.
    $$CH_2FCOOH(aq) + H_2O \rightleftharpoons$$
    $$H_3O^+(aq) + CH_2FCOO^-(aq) \quad K_a = ?$$

44. Caproic acid, $HC_6H_{11}O_2$, found in small amounts in coconut and palm oils, is used in making artificial flavors. A saturated aqueous solution of the acid contains 11 g/L and has pH = 2.94. Calculate K_a for the acid.
    $$HC_6H_{11}O_2 + H_2O \rightleftharpoons H_3O^+ + C_6H_{11}O_2^- \quad K_a = ?$$

45. What are $[H_3O^+]$, $[OH^-]$, pH, and pOH of 0.55 M $HClO_2$?

46. What are $[H_3O^+]$, $[OH^-]$, pH, and pOH of 0.386 M CH_3NH_2?

47. The solubility of 1-naphthylamine, $C_{10}H_7NH_2$, a substance used in the manufacture of dyes, is given in a handbook as 1 g per 590 g H_2O. What is the approximate pH of a saturated aqueous solution of 1-naphthylamine?

$$C_{10}H_7NH_2 + H_2O \rightleftharpoons C_{10}H_7NH_3^+ + OH^-$$
$$pK_a = 3.92$$

48. A saturated aqueous solution of *o*-nitrophenol, $HOC_6H_4NO_2$, has pH = 4.53. What is the solubility of *o*-nitrophenol in water, in grams per liter?
    $$HOC_6H_4NO_2 + H_2O \rightleftharpoons H_3O^+ + {}^-OC_6H_4NO_2$$
    $$pK_a = 7.23$$

49. A particular vinegar is found to contain 5.7% acetic acid, $HC_2H_3O_2$, by mass. What mass of this vinegar should be diluted with water to produce 0.750 L of a solution with pH = 4.52?

50. A particular household ammonia solution ($d = 0.97$ g/mL) is 6.8% NH_3 by mass. How many milliliters of this solution should be diluted with water to produce 625 mL of a solution with pH = 11.55?

51. A 275-mL sample of vapor in equilibrium with 1-propylamine at 25.0 °C is removed and dissolved in 0.500 L H_2O. For 1-propylamine, $pK_b = 3.43$ and v.p. = 316 Torr.

(a) What should be the pH of the aqueous solution?
(b) How many mg of NaOH dissolved in 0.500 L of water give the same pH?

52. One handbook lists a value of 9.5 for pK_b of quinoline, C_9H_7N, a weak base used as a preservative for anatomical specimens and to make dyes. Another handbook lists the solubility of quinoline in water at 25 °C as 0.6 g/100 mL. Use this information to calculate the pH of a saturated solution of quinoline in water.

53. The sketch on the far left represents the $[H_3O^+]$ present in an acetic acid solution of molarity M. If the molarity of the solution is doubled, which of the sketches on the right best represents the resulting solution?

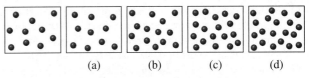

(a)   (b)   (c)   (d)

54. The sketch on the far left represents the $[OH^-]$ present in an ammonia solution of molarity M. If the solution is diluted to half its original molarity, which of the sketches on the right best represents the resulting solution?

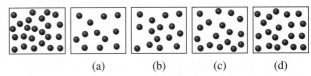

(a)   (b)   (c)   (d)

Percent Ionization

55. What is the (a) degree of ionization and (b) percent ionization of propionic acid in a solution that is 0.45 M $HC_3H_5O_2$?
$$HC_3H_5O_2 + H_2O \rightleftharpoons H_3O^+ + C_3H_5O_2^-$$
$$pK_a = 4.89$$

56. What is the (a) degree of ionization and (b) percent ionization of ethylamine, $C_2H_5NH_2$, in an 0.85 M aqueous solution?

57. What must be the molarity of an aqueous solution of NH_3 if it is 4.2% ionized?

58. What must be the molarity of an acetic acid solution if it has the same percent ionization as 0.100 M $HC_3H_5O_2$ (propionic acid, $K_a = 1.3 \times 10^{-5}$)?

59. Continuing the dilutions described in Example 17-8, should we expect the percent ionization to be 13% in 0.0010 M $HC_2H_3O_2$ and 42% in 0.00010 M $HC_2H_3O_2$? Explain.

60. What is the (a) degree of ionization and (b) percent ionization of trichloroacetic acid in a 0.035 M $HC_2Cl_3O_2$ solution?
$$HC_2Cl_3O_2 + H_2O \rightleftharpoons H_3O^+ + C_2Cl_3O_2^-$$
$$pK_a = 0.52$$

Polyprotic Acids *(Use data from Table 17.4 as necessary.)*

61. Explain why $[PO_4^{3-}]$ in 1.00 M H_3PO_4 is *not* simply $\frac{1}{3}[H_3O^+]$, but much, much less than $\frac{1}{3}[H_3O^+]$.

62. Cola drinks have a phosphoric acid content that is described as "from 0.057 to 0.084% of 75% phosphoric acid, by mass." Estimate the pH range of cola drinks corresponding to this range of H_3PO_4 content.

63. Determine $[H_3O^+]$, $[HS^-]$, and $[S^{2-}]$ for the following $H_2S(aq)$ solutions: (a) 0.075 M H_2S; (b) 0.0050 M H_2S; (c) 1.0×10^{-5} M H_2S.

64. Calculate $[H_3O^+]$, $[HSO_4^-]$, and $[SO_4^{2-}]$ in (a) 0.75 M H_2SO_4; (b) 0.075 M H_2SO_4; (c) 7.5×10^{-4} M H_2SO_4. (*Hint:* Check any assumptions that you make.)

65. Adipic acid, $HOOC(CH_2)_4COOH$, is among the top 50 manufactured chemicals in the United States (nearly 1 million tons annually). Its chief use is in the manufacture of nylon. It is a *diprotic* acid having $K_{a_1} = 3.9 \times 10^{-5}$ and $K_{a_2} = 3.9 \times 10^{-6}$. A saturated solution of adipic acid is about 0.10 M $HOOC(CH_2)_4COOH$. Calculate the concentration of each ionic species in this solution.

66. The antimalarial drug quinine, $C_{20}H_{24}O_2N_2$, is a *diprotic base* with a water solubility of 1.00 g/1900 mL of solution.
(a) Write equations for the ionization equilibria corresponding to $pK_{b_1} = 6.0$ and $pK_{b_2} = 9.8$.
(b) What is the pH of saturated aqueous quinine?

Ions as Acids and Bases (Hydrolysis)

67. Predict whether a solution of each of the following salts is acidic, basic, or pH neutral: (a) KCl; (b) KF; (c) NaNO$_3$; (d) Ca(OCl)$_2$; (e) NH$_4$NO$_2$.

68. Arrange the following 0.010 M solutions in order of *increasing* pH: NH$_3$(aq), HNO$_3$(aq), NaNO$_2$(aq), HC$_2$H$_3$O$_2$(aq), NaOH(aq), NH$_4$C$_2$H$_3$O$_2$(aq), NH$_4$ClO$_4$(aq).

69. What is the pH of an aqueous solution that is 0.089 M NaOCl?

70. What is the pH of an aqueous solution that is 0.123 M NH$_4$Cl?

71. Sorbic acid, $HC_6H_7O_2$ ($pK_a = 4.77$), is widely used in the food industry as a preservative. For example, its potassium salt (potassium sorbate) is added to cheese to inhibit the formation of mold. What is the pH of 0.37 M $KC_6H_7O_2$(aq)?

72. Pyridine, C_5H_5N ($pK_b = 8.82$), forms a salt, pyridinium chloride, as a result of a reaction with HCl. Write an ionic equation to represent the hydrolysis of the pyridinium ion, and calculate the pH of 0.0482 M $C_5H_5NH^+Cl^-$(aq).

73. For each of the following ions, write two equations—one showing its ionization as an acid and the other as a base: **(a)** HSO_3^-; **(b)** HS^-; **(c)** HPO_4^-. Then use data from Table 17.4 to predict whether each ion makes the solution acidic or basic.

74. Suppose you wanted to produce an aqueous solution of pH = 8.65 by dissolving one of the following salts in water. Which salt would you use, and at what molarity? **(a)** NH_4Cl; **(b)** $KHSO_4$; **(c)** KNO_2; **(d)** $NaNO_3$.

Molecular Structure and Acid–Base Behavior

75. Predict which is the stronger acid: **(a)** $HClO_2$ or $HClO_3$; **(b)** H_2CO_3 or HNO_2; **(c)** H_2SiO_3 or H_3PO_4. Explain.

76. Indicate which of the following is the *weakest* acid, and give reasons for your choice: HBr; $CH_2ClCOOH$; CH_3CH_2COOH; CH_2FCH_2COOH; Cl_3COOH.

77. From the following bases, select the one with the *smallest* K_b and the one with the *largest* K_b, and give reasons for your choices.

(a) [benzene ring]—NH_2 with Cl substituent (b) H_3C—[benzene ring]—NH_2

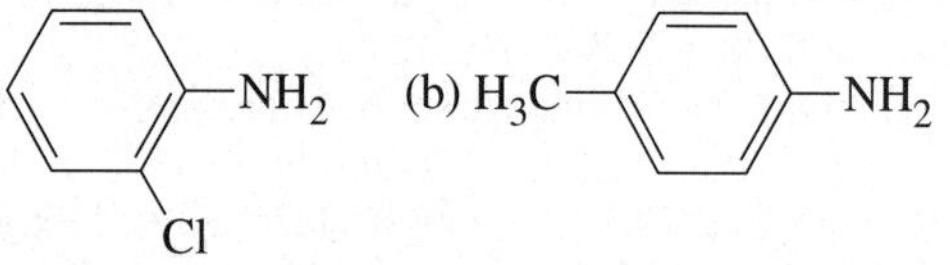

(c) $CH_3CH_2CH_2NH_2$ (d) $N{\equiv}CCH_2NH_2$

78. From the space-filling models shown on the right, write the formula of the species that is the most acidic and the one that is most basic, and give reasons for your choices.

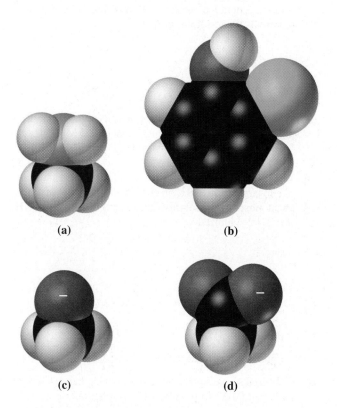

(a) (b)

(c) (d)

Lewis Theory of Acids and Bases

79. The three following reactions are acid–base reactions according to the Lewis theory. Draw Lewis structures, and identify the Lewis acid and Lewis base in each reaction.
   (a) $B(OH)_3 + OH^- \longrightarrow [B(OH)_4]^-$
   (b) $N_2H_4 + H_3O^+ \longrightarrow N_2H_5^+ + H_2O$
   (c) $(C_2H_5)_2O + BF_3 \longrightarrow (C_2H_5)_2OBF_3$

80. CO_2(g) can be removed from confined quarters (such as a spacecraft) by allowing it to react with an alkali metal hydroxide. Show that this is a Lewis acid–base reaction. For example,
$$CO_2(g) + LiOH(s) \longrightarrow LiHCO_3(s)$$

81. The molecular solid I_2(s) is only slightly soluble in water but will dissolve to a much greater extent in an aqueous solution of KI, because the I_3^- anion forms. Write an equation for the formation of the I_3^- anion, and indicate the Lewis acid and Lewis base.

82. The following very strong acids are formed by the reactions indicated:

$$HF + SbF_5 \longrightarrow HSbF_6$$
   (called "super acid," hexafluoroantimonic acid)
$$HF + BF_3 \longrightarrow HBF_4 \qquad \text{(tetrafluoroboric acid)}$$

   (a) Identify the Lewis acids and bases.
   (b) To which atom is the H atom bonded in each acid?

83. Use Lewis structures to diagram the following reaction in the manner of reaction 17.19.
$$H_2O + SO_2 \longrightarrow H_2SO_3$$
Identify the Lewis acid and Lewis base.

84. Use Lewis structures to diagram the following reaction in the manner of reaction 17.19.
$$2 NH_3 + Ag^+ \longrightarrow [Ag(NH_3)_2]^+$$
Identify the Lewis acid and Lewis base.

Integrative and Advanced Exercises

85. The Brønsted–Lowry theory can be applied to acid–base reactions in *nonaqueous* solvents. In nonaqueous solvents, the relative strengths of acids and bases can differ from what they are in aqueous solutions. Indicate whether each of the following would be an acid, a base, or amphiprotic in pure liquid acetic acid, $HC_2H_3O_2$, as a solvent. **(a)** $C_2H_3O_2^-$; **(b)** H_2O; **(c)** $HC_2H_3O_2$; **(d)** $HClO_4$. (*Hint:* Refer to Table 17.1.)

86. The pH of saturated $Sr(OH)_2(aq)$ is found to be 13.12. A 10.0-mL sample of saturated $Sr(OH)_2(aq)$ is diluted to 250.0 mL in a volumetric flask. A 10.0-mL sample of the diluted $Sr(OH)_2(aq)$ is transferred to a beaker, and some water is added. The resulting solution requires 25.1 mL of a HCl solution for its titration. What is the molarity of this HCl solution?

87. Several approximate pH values are marked on the following pH scale.

Some of the following solutions can be matched to one of the approximate pH values marked on the scale; others cannot. For solutions that can be matched to a pH value, identify each solution and its pH value. Identify the solutions that cannot be matched, and give reasons why matches are not possible. **(a)** $0.010\ M\ H_2SO_4$; **(b)** $1.0\ M\ NH_4Cl$; **(c)** $0.050\ M\ KI$; **(d)** $0.0020\ M\ CH_3NH_2$; **(e)** $1.0\ M\ NaOCl$; **(f)** $0.10\ M\ C_6H_5OH$; **(g)** $0.10\ M\ HOCl$; **(h)** $0.050\ M\ HC_2H_2ClO_2$; **(i)** $0.050\ M\ HCHO_2$

88. Show that when $[H_3O^+]$ is reduced to half its original value, the pH of a solution increases by 0.30 unit, *regardless of the initial pH*. Is it also true that when any solution is diluted to half its original concentration, the pH increases by 0.30 unit? Explain.

89. Explain why $[H_3O^+]$ in a strong acid solution *doubles* as the total acid concentration doubles, whereas in a weak acid solution, $[H_3O^+]$ increases only by about a factor of $\sqrt{2}$.

90. Use data from Appendix D to determine whether the ion product of water, K_w, increases, decreases, of remains unchanged with increasing temperature.

91. From the observation that 0.0500 M vinylacetic acid has a freezing point of $-0.096\ °C$, determine K_a for this acid.

$$HC_4H_5O_2 + H_2O \rightleftharpoons H_3O^+ + C_4H_5O_2^- \qquad K_a = ?$$

92. You are asked to prepare a 100.0-mL sample of a solution with a pH of 5.50 by dissolving the appropriate amount of a solute in water with pH = 7.00. Which of these solutes would you use, and in what quantity? Explain your choice. **(a)** $15\ M\ NH_3(aq)$; **(b)** $12\ M\ HCl(aq)$; **(c)** $NH_4Cl(s)$; **(d)** glacial (pure) acetic acid, $HC_2H_3O_2$.

93. Determine the pH of **(a)** $1.0 \times 10^{-5}\ M$ HCN and **(b)** $1.0 \times 10^{-5}\ M\ C_6H_5NH_2$ (aniline).

94. The solubility of $CO_2(g)$ in H_2O at 25 °C and under a $CO_2(g)$ pressure of 1 atm is 1.45 g CO_2/L. Air contains 0.037% CO_2, by volume. Use this information, together with data from Table 17.4, to show that rainwater saturated with CO_2 has a pH $\approx$ 5.6 (referred to in the Focus On feature on page 700 as the normal pH for rainwater). [*Hint:* Recall Henry's law. What is the partial pressure of $CO_2(g)$ in air?]

95. It is possible to write simple equations to relate pH, p*K*, and molarities (M) of various solutions. Three such equations are shown here.

Weak acid: $\quad pH = \dfrac{1}{2} pK_a - \dfrac{1}{2} \log M$

Weak base: $\quad pH = 14.00 - \dfrac{1}{2} pK_b + \dfrac{1}{2} \log M$

Salt of weak acid (pK_a) *and strong base:* $\quad pH = 14.00 - \dfrac{1}{2} pK_w + \dfrac{1}{2} pK_a + \dfrac{1}{2} \log M$

(a) Derive these three equations, and point out the assumptions involved in the derivations.
(b) Use these equations to determine the pH of 0.10 M $HC_2H_3O_2(aq)$, 0.10 M $NH_3(aq)$, and 0.10 M $NaC_2H_3O_2$. Verify that the equations give correct results by determining these pH values in the usual way.

96. A handbook lists the following formula for the percent ionization of a weak acid.

$$\% \text{ ionized} = \frac{100}{1 + 10^{(pK - pH)}}$$

(a) Derive this equation. What assumptions must you make in this derivation?
(b) Use the equation to determine the percent ionization of a formic acid solution, HCOOH(aq), with a pH of 2.50.
(c) A 0.150 M solution of propionic acid has a pH of 2.85. What is K_a for propionic acid?

$$HC_3H_5O_2 + H_2O \rightleftharpoons H_3O^+ + C_3H_5O_2^- \qquad K_a = ?$$

97. Often the following generalization applies to oxoacids with the formula $EO_m(OH)_n$ (where E is the central atom): If $m = 0$, $K_a \approx 10^{-7}$; if $m = 1$ $K_a \approx 10^{-2}$; if $m = 2$, K_a is large; and if $m = 3$, K_a is very large.
(a) Show that this generalization works well for the oxoacids of chlorine: HOCl, pK_a = 7.52; HOClO, pK_a = 1.92; $HOClO_2$, pK_a = -3; $HOClO_3$; pK_a = -8.
(b) Estimate the value of K_{a_1} for H_3AsO_4.
(c) Write a Lewis structure for hypophosphorous acid, H_3PO_2, for which pK_a = 1.1.

98. Oxalic acid, HOOCCOOH, a weak diprotic acid, has pK_{a_1} = 1.25 and pK_{a_2} = 3.81. A related diprotic acid, suberic acid, $HOOC(CH_2)_8COOH$ has pK_{a_1} = 4.21 and pK_{a_2} = 5.40. Offer a plausible reason as to why the *difference* between pK_{a_1} and pK_{a_2} is so much greater for oxalic acid than for suberic acid.

99. Here is a way to test the validity of the statement made on page 686 in conjunction with the three key ideas governing the ionization of polyprotic acids. Determine the pH of 0.100 M succinic acid in two ways: first by assuming that H_3O^+ is produced only in the first ionization step, and then by allowing for the possibility that some H_3O^+ is also produced in the second ionization step. Compare the results, and discuss the significance of your finding.

$$H_2C_4H_4O_4 + H_2O \rightleftharpoons H_3O^+ + HC_4H_4O_4^-$$
$$K_{a_1} = 6.2 \times 10^{-5}$$

$$HC_4H_4O_4^- + H_2O \rightleftharpoons H_3O^+ + C_4H_4O_4^{2-}$$
$$K_{a_2} = 2.3 \times 10^{-6}$$

100. What mass of acetic acid, $HC_2H_3O_2$, must be dissolved per liter of aqueous solution if the solution is to have the same freezing point as 0.150 M $HC_2H_2ClO_2$ (chloroacetic acid)?

101. What is the pH of a solution that is 0.68 M H_2SO_4 and 1.5 M $HCHO_2$ (formic acid)?

102. An aqueous solution of two weak acids has a stoichiometric molarity, M, in each acid. If one acid has a K_a value twice as large as the other, show that the pH of the solution is given by the equation $pH = -\frac{1}{2} \log 3M\, K_a$. Assume that the criteria for the simplifying assumption on page 683 are met.

103. Use the concept of hybrid orbitals to describe the bonding in the strong acids given in Exercise 82.

104. Phosphorous acid is listed in Appendix D as a *di*protic acid. Propose a Lewis structure for phosphorous acid that is consistent with this fact.

Feature Problems

105. Maleic acid is a carbon–hydrogen–oxygen compound used in dyeing and finishing fabrics and as a preservative of oils and fats. In a combustion analysis, a 1.054-g sample of maleic acid yields 1.599 g CO_2 and 0.327 g H_2O. In a freezing-point depression experiment, a 0.615-g sample of maleic acid dissolved in 25.10 g of glacial acetic acid, $CH_3COOH(l)$ (which has the freezing-point depression constant $K_f = 3.90\ °C\ m^{-1}$ and in which maleic acid does not ionize), lowers the freezing point by 0.82 °C. In a titration experiment, a 0.4250-g sample of maleic acid is dissolved in water and requires 34.03 mL of 0.2152 M KOH for its complete neutralization. The pH of a 0.215-g sample of maleic acid dissolved in 50.00 mL of aqueous solution is found to be 1.80.
(a) Determine the empirical and molecular formulas of maleic acid.
[*Hint:* Which experiment(s) provide the necessary data?]
(b) Use the results of part (a) and the titration data to rewrite the molecular formula to reflect the number of ionizable H atoms in the molecule.
(c) Given that the ionizable H atom(s) is(are) associated with the carboxyl group(s), write the plausible condensed structural formula of maleic acid.
(d) Determine the ionization constant(s) of maleic acid. If the data supplied are insufficient, indicate what additional data would be needed.
(e) Calculate the expected pH of a 0.0500 M aqueous solution of maleic acid. Indicate any assumptions required in this calculation.

106. In Example 17-7, rather than use the quadratic formula to solve the quadratic equation, we could have proceeded in the following way. Substitute the value yielded by our failed assumption—$x = 0.0010$—into the *denominator* of the quadratic equation; that is, use $(0.00250 - 0.0010)$ as the value of $[CH_3NH_2]$ and solve for a new value of x. Use this second value of x to reevaluate $[CH_3NH_2]$: $[CH_3NH_2] = (0.00250 - $ second value of x). Solve the simple quadratic equation for a third value of x, and so on. After three or four trials, you will find that the value of x no longer changes. This is the answer you are seeking.
(a) Complete the calculation of the pH of 0.00250 M CH_3NH_2 by this method, and show that the result is the same as that obtained by using the quadratic formula.
(b) Use this method to determine the pH of 0.500 M $HClO_2$.

107. Apply the general method for solution equilibrium calculations outlined on page 688 to determine the pH values of the following solutions. In applying the method, look for valid assumptions that may simplify the numerical calculations.
(a) A solution that is 0.315 M $HC_2H_3O_2$ and 0.250 M $HCHO_2$
(b) A solution that contains 1.55 g CH_3NH_2 and 12.5 g NH_3 in 375 mL
(c) 1.0 M $NH_4CN(aq)$

eMedia Exercises

108. In the **Self-Ionization of Water** animation, the ionization process is shown with an unrealistically high concentration of the products *(eChapter 17-3).* **(a)** According to the known value of K_w at 25 °C, estimate how many ions would be present in the field of view of the animation? **(b)** Estimate what the value of K_w would be if, in fact, the animation as shown were accurately depicting the concentrations of ions produced by self-ionization.

109. **(a)** In the **Aqueous Acids and Bases** animation *(eChapter 17-4),* write a balanced chemical equation for each of the reactions depicted. **(b)** In which reaction(s) is dynamic equilibrium an accurate description of the final state? **(c)** Which species are best described by the Arrhenius theory? **(d)** Which are best described by Brønsted-Lowry theory? **(e)** Which can be described by both theories?

110. **(a)** Using the **Equilibrium Constant** activity *(eChapter 17-5),* estimate the ranges of equilibrium constants that are typical of weak acids. **(b)** At roughly what value of K_a would the acid be more accurately described as a strong acid? **(c)** What is the corresponding percent ionization at this K_a value?

111. Several **Organic Acids and Bases** models are shown in *eChapter 17-8.* After viewing the electron density distributions of each pair of acids and bases, explain the different acidic and basic strengths of these compounds.

112. In the **Lewis Acid–Base Theory** animation *(eChapter 17-9),* three different species are identified as Lewis bases. **(a)** Identify these bases and the molecular trait that distinguishes them as Lewis bases. **(b)** How does this description differ from the other acid and base characterizations given earlier in the chapter? **(c)** Would any of these species also be classified as Arrhenius or Brønsted-Lowry bases?

18

Additional Aspects of Acid–Base Equilibria

Contents

HCl(aq) is slowly added to an aqueous solution containing the base NH_3 and the indicator methyl red. The indicator color changes from yellow to red as the pH changes from 6.2 to 4.4. The equivalence point of the neutralization is reached when the solution turns orange. The selection of indicators for acid–base titrations is one of the topics considered in this chapter.

In our study of acid rain (page 700), we learned that a very small amount of atmospheric $CO_2(g)$ dissolves in rainwater. Yet this amount is sufficient to lower the pH of rainwater by nearly 2 units. And when acid-forming air pollutants, such as SO_2, SO_3, and NO_2, also dissolve in rainwater, it becomes even more acidic. A chemist would say that water has no "buffer capacity." Water is unable to resist a change in pH when acids or bases are dissolved in it.

One of the main topics of this chapter is buffer solutions—solutions that *are* able to resist a change in pH when acids or bases are added to them. We will consider how such solutions are prepared, how they maintain a nearly constant pH, and ways in which they are used. At the end of the chapter, we will consider perhaps the most important buffer of all to humans: the buffer that maintains the constant pH of blood.

A second topic that we shall explore is acid–base titrations. Here, our aim will be to calculate how pH changes during a titration. We can use this information to select an appropriate indicator for a titration and to determine, in general, which acid–base titrations work well and which do not. For the most part, we will find the calculations in this chapter to be extensions of those in Chapter 17.

18-1 The Common-Ion Effect in Acid–Base Equilibria

The questions answered in Chapter 17 were mostly of the type, "What is the pH of 0.10 M $HC_2H_3O_2$, of 0.10 M NH_3, of 0.10 M H_3PO_4, of 0.10 M NH_4Cl?" In each of these cases, we think of dissolving a *single* substance in aqueous solution and determining the concentrations of the species present at equilibrium. In most situations in this chapter, a solution of a weak acid or weak base initially contains a second source of one of the ions produced in the ionization of the acid or base. We say that the added ions are *common* to the weak acid or weak base. As we shall see, the presence of a common ion can have some important consequences.

Solutions of Weak Acids and Strong Acids

Consider a solution that is at the same time 0.100 M $HC_2H_3O_2$ and 0.100 M HCl. We can write separate equations for the ionizations of the acids, one weak and the other strong.

$$HC_2H_3O_2 \quad + \quad H_2O \quad \rightleftharpoons \quad H_3O^+ \quad + \quad C_2H_3O_2^- \qquad K_a = 1.8 \times 10^{-5}$$
$$(0.100 - x)\ M \qquad\qquad\qquad\qquad x\ M \qquad\quad x\ M$$

$$HCl \quad + \quad H_2O \quad \longrightarrow \quad H_3O^+ \quad + \quad Cl^-$$
$$\qquad\qquad\qquad\qquad\qquad 0.100\ M \qquad 0.100\ M$$

Of course, there can be only a single concentration of H_3O^+ in the solution, and this must be $[H_3O^+] = (0.100 + x)$ M. Because H_3O^+ is formed in both ionization processes, we say that it is a *common ion*. The weak acid–strong acid mixture described here is pictured in Figure 18-1. Although we might think that the pH would be lower than 1.0, this figure indicates that such is not the case.

In Example 18-1, we calculate the concentrations of the species present in this mixture of a weak acid and a strong acid. Then we will comment on the significance of the result.

▲ **FIGURE 18-1**
A weak acid–strong acid mixture
The solution pictured is 0.100 M $HC_2H_3O_2$ and 0.100 M HCl. The reading on the pH meter (1.0) indicates that essentially all the H_3O^+ comes from the strong acid HCl. The red color of the solution is that of thymol blue indicator. Compare this photo with Figure 17–7, in which the separate acids 0.100 M $HC_2H_3O_2$ and 0.100 M HCl are shown.

EXAMPLE 18-1

Demonstrating the Common-Ion Effect: A Solution of a Weak Acid and a Strong Acid.
(a) Determine $[H_3O^+]$ and $[C_2H_3O_2^-]$ in 0.100 M $HC_2H_3O_2$. **(b)** Then determine these same quantities in a solution that is 0.100 M in both $HC_2H_3O_2$ and HCl.

Solution

(a) We did this calculation in Example 17–6 (page 680). We found that in 0.100 M $HC_2H_3O_2$, $[H_3O^+] = [C_2H_3O_2^-] = 1.3 \times 10^{-3}$ M.

(b) Instead of writing two separate ionization equations, as we did just prior to this example, let's write only the ionization equation for $HC_2H_3O_2$ and enter information about the common ion, H_3O^+, in the following format.

$$\mathbf{HC_2H_3O_2} \quad + \quad \mathbf{H_2O} \quad \rightleftharpoons \quad \mathbf{H_3O^+} \quad + \quad \mathbf{C_2H_3O_2^-}$$

	$HC_2H_3O_2$	H_3O^+	$C_2H_3O_2^-$
initial concns:			
weak acid:	0.100 M	—	—
strong acid:	—	0.100 M	—
changes:	$-x$ M	$+x$ M	$+x$ M
equil concns:	$(0.100 - x)$ M	$(0.100 + x)$ M	x M

KEEP IN MIND ▶
that by the criterion
$M_a > 100 K_a$ (see page 683), we
expect this assumption to be
valid. Actually, the assumption is
even better than in Example 17–6 because the ionization
of the weak acid is suppressed
by the presence of the common
ion, H_3O^+.

As is customary, we begin with the assumption that x is very small compared with 0.100. Thus, $0.100 - x \approx 0.100 + x \approx 0.100$.

$$K_a = \frac{[H_3O^+][C_2H_3O_2^-]}{[HC_2H_3O_2]} = \frac{(0.100 + x)(x)}{0.100 - x} = \frac{0.100\,(x)}{0.100} = 1.8 \times 10^{-5}$$

$$x = [C_2H_3O_2^-] = 1.8 \times 10^{-5}\,M \qquad 0.100 + x = [H_3O^+] = 0.100\,M$$

Practice Example A: Determine $[H_3O^+]$ and $[HF]$ in 0.500 M HF. Then determine these concentrations in a solution that is 0.100 M HCl and 0.500 M HF.

Practice Example B: How many drops of 12 M HCl would you add to 1.00 L of 0.100 M $HC_2H_3O_2$ to make $[C_2H_3O_2^-] = 1.0 \times 10^{-4}\,M$? Assume that 1 drop = 0.050 mL and that the volume of solution remains 1.00 L after the 12 M HCl is diluted.

(*Hint:* What must be the $[H_3O^+]$ in the solution?)

Now we see the consequence of adding a strong acid (HCl) to a weak acid ($HC_2H_3O_2$): The concentration of the anion $[C_2H_3O_2^-]$ is greatly reduced. Between parts (a) and (b) of Example 18-1, $[C_2H_3O_2^-]$ is lowered from 1.3×10^{-3} M to 1.8×10^{-5} M—almost a 100-fold decrease. Another way to state this result is through Le Châtelier's principle (see Chapter 16, page 641). Increasing the concentration of one of the *products* of a reaction—the common ion—shifts the equilibrium condition in the *reverse* direction. The **common-ion effect** is the suppression of the ionization of a weak electrolyte caused by adding more of an ion that is a product of this ionization. The common-ion effect of H_3O^+ on the ionization of acetic acid is suggested as follows.

Common Ion Effect animation

When a strong acid supplies the common ion H_3O^+,
the equilibrium shifts to form more $HC_2H_3O_2$.

Added H_3O^+

$$HC_2H_3O_2 + H_2O \rightleftharpoons H_3O^+ + C_2H_3O_2^- \qquad K_a = 1.8 \times 10^{-5}$$

Equilibrium shifts to form
more $HC_2H_3O_2$

The ionization of a weak base such as NH_3 is suppressed when a strong base such as NaOH is added. Here, OH^- is the common ion, and its increased concentration shifts the equilibrium to the left.

When a strong base supplies the common ion OH^-,
the equilibrium shifts to form more NH_3.

Added OH^-

$$NH_3 + H_2O \rightleftharpoons NH_4^+ + OH^- \qquad K_b = 1.8 \times 10^{-5}$$

Equilibrium shifts to
form more NH_3

Solutions of Weak Acids and Their Salts

The salt of a weak acid is a strong electrolyte—its ions become completely dissociated from one another in aqueous solution. One of the ions, the *anion*, is an ion common to the ionization equilibrium of the weak acid. The presence of this common ion suppresses the ionization of the weak acid. For example, we can represent the effect of acetate salts on the acetic acid equilibrium as

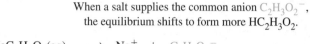

When a salt supplies the common anion $C_2H_3O_2^-$, the equilibrium shifts to form more $HC_2H_3O_2$.

$$NaC_2H_3O_2(aq) \longrightarrow Na^+ + C_2H_3O_2^-$$

Added $C_2H_3O_2^-$

$$HC_2H_3O_2 + H_2O \rightleftharpoons H_3O^+ + C_2H_3O_2^- \qquad K_a = 1.8 \times 10^{-5}$$

Equilibrium shifts to form more $HC_2H_3O_2$

(a)   **(b)**

▲ **FIGURE 18-2**
A mixture of a weak acid and its salt
Bromphenol blue indicator is present in both solutions. Its color dependence on pH is

pH < 3.0 < pH < 4.6 < pH

yellow   green   blue-violet

(a) 0.100 M $HC_2H_3O_2$ has a calculated pH of 2.89, but **(b)** if the solution is also 0.100 M in $NaC_2H_3O_2$, the calculated pH is 4.74. (The readability of the pH meters used here is 0.1 unit, and their accuracy is probably somewhat less than that. The discrepancy between 4.74 and the 4.9 value shown here is a result.)

The common-ion effect of acetate ion on the ionization of acetic acid is depicted in Figure 18-2 and demonstrated in Example 18-2. In solving common-ion problems such as Example 18-2, assume that ionization of the weak acid (or base) does not begin until both the weak acid (or base) and its salt have been placed in solution. Then consider that ionization occurs until equilibrium is reached.

EXAMPLE 18-2

Demonstrating the Common-Ion Effect: A Solution of a Weak Acid and a Salt of That Weak Acid. Calculate $[H_3O^+]$ and $[C_2H_3O_2^-]$ in a solution that is 0.100 M in both $HC_2H_3O_2$ and $NaC_2H_3O_2$.

Solution

The setup shown here is very similar to that in Example 18-1(b), except that $NaC_2H_3O_2$ is the source of the common ion.

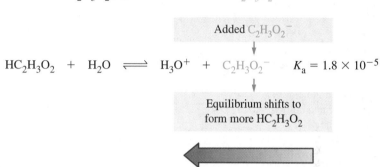

$$HC_2H_3O_2 + H_2O \rightleftharpoons H_3O^+ + C_2H_3O_2^-$$

initial concns:			
weak acid:	0.100 M	—	—
salt:	—	—	0.100 M
changes:	$-x$ M	$+x$ M	$+x$ M
equil concns:	$(0.100 - x)$ M	x M	$(0.100 + x)$ M

Because the salt suppresses the ionization of $HC_2H_3O_2$, we expect $[H_3O^+] = x$ to be very small and $0.100 - x \approx 0.100 + x \approx 0.100$. This proves to be a valid assumption.

$$K_a = \frac{[H_3O^+][C_2H_3O_2^-]}{[HC_2H_3O_2]} = \frac{(x)(0.100 + x)}{0.100 - x} = \frac{(x)\,0.100}{0.100} = 1.8 \times 10^{-5}$$

$$x = [H_3O^+] = 1.8 \times 10^{-5}\,M \qquad 0.100 + x = [C_2H_3O_2^-] = 0.100\,M$$

As in Example 18-1, ionization of $HC_2H_3O_2$ is reduced about 100-fold. This time the large decrease in concentration is that of H_3O^+.

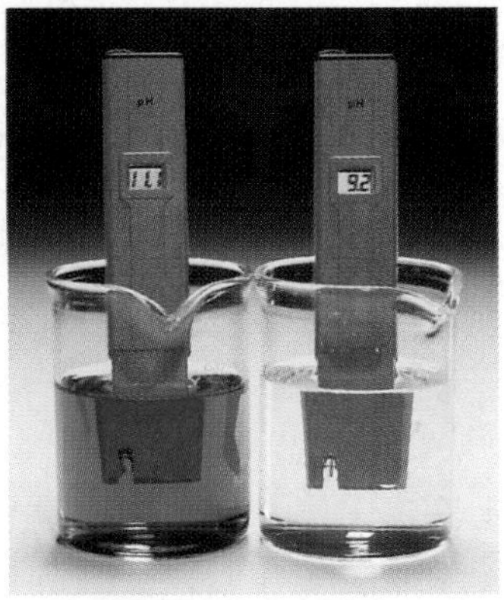

(a)   **(b)**

▲ **FIGURE 18-3**

A mixture of a weak base and its salt

Thymolphthalein indicator is blue if pH > 10 and colorless if pH < 10. **(a)** The pH of 0.100 M NH_3 is above 10 (calculated value: 11.11). **(b)** If the solution is also 0.100 M NH_4Cl, the pH drops below 10 (calculated value: 9.26). The ionization of NH_3 is suppressed in the presence of added NH_4^+. [OH^-] decreases, [H_3O^+] increases, and the pH is lowered.

Practice Example A: Calculate [H_3O^+] and [CHO_2^-] in a solution that is 0.100 M $HCHO_2$ and 0.150 M $NaCHO_2$.

Practice Example B: What mass of $NaC_2H_3O_2$ should be added to 1.00 L of 0.100 M $HC_2H_3O_2$ to produce a solution with pH = 5.00? Assume that the volume remains 1.00 L.

Solutions of Weak Bases and Their Salts

The common-ion effect of a salt of a weak base is similar to the weak acid–anion situation just described. The suppression of the ionization of NH_3 by the common *cation*, NH_4^+, is pictured in Figure 18-3 and represented as follows.

When a salt supplies the common cation NH_4^+, the equilibrium shifts to form more NH_3.

$$NH_4Cl(aq) \longrightarrow NH_4^+ + Cl^-$$

Added NH_4^+

$$NH_3 + H_2O \rightleftharpoons NH_4^+ + OH^- \quad K_b = 1.8 \times 10^{-5}$$

Equilibrium shifts to form more NH_3

18-2   Buffer Solutions

Figure 18-4 illustrates a statement made in the chapter introduction: Pure water has no buffer capacity. There are some water solutions, however, called **buffer** (or **buffered**) **solutions**, whose pH values change only very slightly on the addition of small amounts of either an acid or a base.

What buffer solutions require are two components; one component is able to neutralize acids, and the other is able to neutralize bases. But the two components must not neutralize each other. This requirement rules out mixtures of a strong acid and a strong base. Instead, common buffer solutions are described as either of these combinations.

- a weak acid and its conjugate base

- a weak base and its conjugate acid

To show that such mixtures function as buffer solutions, let's consider a solution that has the equilibrium concentrations [$HC_2H_3O_2$] = [$C_2H_3O_2^-$]. As summarized in expression (18.1), in this solution [H_3O^+] = K_a = 1.8×10^{-5} M.

$$K_a = \frac{[H_3O^+][C_2H_3O_2^-]}{[HC_2H_3O_2]} = 1.8 \times 10^{-5}$$

$$[H_3O^+] = K_a \times \frac{[HC_2H_3O_2]}{[C_2H_3O_2^-]} = 1.8 \times 10^{-5} \text{ M} \qquad (18.1)$$

As a result, pH = $-\log[H_3O^+]$ = $-\log K_a$ = $-\log 1.8 \times 10^{-5}$ = 4.74.

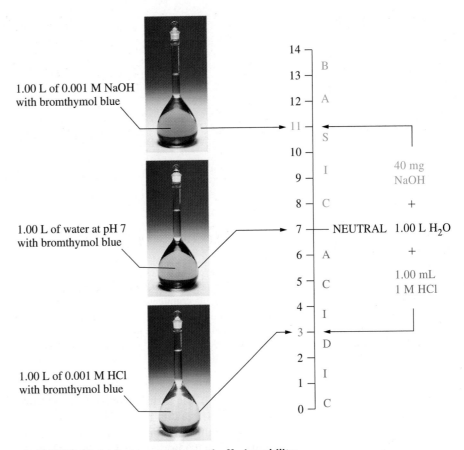

▲ FIGURE 18-4 **Pure water has no buffering ability**

Bromthymol blue indicator is blue at pH > 7, green at pH = 7, and yellow at pH < 7.
Pure water has pH = 7.0. The addition of 0.001 mol H_3O^+ (1.00 mL of 1 M HCl) to
1.00 L water produces $[H_3O^+]$ = 0.001 M and pH = 3.0. The addition of 0.001 mol OH^-
(40 mg NaOH) to 1.00 L of water produces $[OH^-]$ = 0.001 M and pH = 11.0.

Now, imagine adding a *small* amount of a strong acid to this buffer solution. A
reaction occurs in which a *small* amount of the base $C_2H_3O_2^-$ is converted to its conjugate acid $HC_2H_3O_2$.

$$C_2H_3O_2^- + H_3O^+ \longrightarrow HC_2H_3O_2 + H_2O$$

After the neutralization of the added H_3O^+, we find that in expression (18.1)
$[HC_2H_3O_2]$ has increased *slightly* and $[C_2H_3O_2^-]$ has decreased *slightly*. The ratio
$[HC_2H_3O_2]/[C_2H_3O_2^-]$ is only *slightly* greater than 1, and $[H_3O^+]$ has barely
changed. The buffer solution has resisted a change in pH following the addition of
a small amount of acid; the pH remains close to the original 4.74.

Next, imagine adding a *small* amount of a strong base to the original buffer solution with $[HC_2H_3O_2]$ = $[C_2H_3O_2^-]$. A reaction occurs in which a *small* amount
of the weak acid $HC_2H_3O_2$ is converted to its conjugate base $C_2H_3O_2^-$.

$$HC_2H_3O_2 + OH^- \longrightarrow C_2H_3O_2^- + H_2O$$

Here we find that $[C_2H_3O_2^-]$ has increased *slightly*, and $[HC_2H_3O_2]$ has decreased
slightly. The ratio $[HC_2H_3O_2]/[C_2H_3O_2^-]$ is only *slightly* smaller than 1, and again
$[H_3O^+]$ has barely changed. The buffer solution has resisted a change in pH following the addition of a small amount of base; again, the pH remains close to the

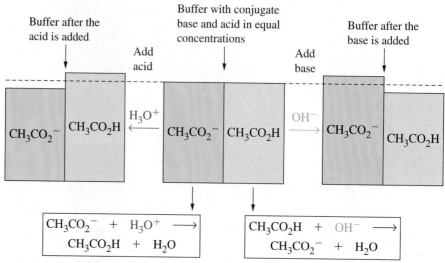

▲ FIGURE 18-5   **How a buffer works**
Acetate ion, the conjugate base of acetic acid, acts as a proton sink when strong acid is added. In this way, the ratio [conjugate base]/[acid] is kept approximately constant, so there is a minimal change in pH. Similarly, acetic acid acts as a proton donor when strong base is added, keeping the ratio [conjugate base]/[acid] approximately constant and minimizing the change in pH.

original 4.74. The variation of the concentration of the weak acid and its conjugate base are illustrated in Figure 18-5.

Later in this section, we will be more specific about what constitutes *small* additions of an acid or a base and *slight* changes in the concentrations of the buffer components and pH. Also we will discover that an acetic acid–sodium acetate buffer is good only for maintaining a nearly constant pH in a range of about 2 pH units centered on a pH $= pK_a = 4.74$. To prepare a buffer solution that maintains a nearly constant pH outside this range, we must use different buffer components, as suggested in Example 18-3.

EXAMPLE 18-3

Predicting Whether a Solution Is a Buffer Solution.   Show that an NH_3–NH_4Cl solution is a buffer solution. Over what pH range would you expect it to function?

Solution

To show that a solution has buffer properties, first identify a component in the solution that neutralizes acids and a component that neutralizes bases. In this case, these components are NH_3 and NH_4^+, respectively.

$$NH_3 + H_3O^+ \longrightarrow NH_4^+ + H_2O$$
$$NH_4^+ + OH^- \longrightarrow NH_3 + H_2O$$

In *all* aqueous solutions containing NH_3 and NH_4^+, we know that

$$NH_3 + H_2O \rightleftharpoons NH_4^+ + OH^- \quad \text{and}$$

$$K_b = \frac{[NH_4^+][OH^-]}{[NH_3]} = 1.8 \times 10^{-5}$$

If a solution has approximately equal concentrations of NH_4^+ and NH_3, then $[OH^-] \approx 1.8 \times 10^{-5}$ M; pOH ≈ 4.74; and pH ≈ 9.26. Ammonia–ammonium chloride solutions are *basic* buffer solutions that function over the approximate pH range of 8 to 10.

Practice Example A: Describe how a mixture of a strong acid (such as HCl) and the salt of a weak acid (such as $NaC_2H_3O_2$) can be a buffer solution.

(*Hint:* What is the reaction that produces $HC_2H_3O_2$? What proportions of the HCl and $NaC_2H_3O_2$ are needed to produce a buffer?)

Practice Example B: Describe how a mixture of NH_3 and HCl can result in a buffer solution.

Buffer Solutions activity

We commonly need to calculate the pH of a buffer solution. At a minimum, such a calculation requires use of the ionization constant expression for a weak acid or weak base. Aspects of solution stoichiometry may also be required.

In Example 18-4, we first determine the stoichiometric concentrations of the buffer components. Then we perform an equilibrium calculation in the same fashion as in Examples 18-1 and 18-2.

EXAMPLE 18-4

Calculating the pH of a Buffer Solution.   What is the pH of a buffer solution prepared by dissolving 25.5 g $NaC_2H_3O_2$ in a sufficient volume of 0.550 M $HC_2H_3O_2$ to make 500.0 mL of the buffer?

Solution

First, we must determine the molarity of $C_2H_3O_2^-$ corresponding to 25.5 g $NaC_2H_3O_2$ in 500.0 mL of solution.

$$\text{amount of } C_2H_3O_2^- = 25.5 \text{ g } NaC_2H_3O_2 \times \frac{1 \text{ mol } NaC_2H_3O_2}{82.04 \text{ g } NaC_2H_3O_2}$$

$$\times \frac{1 \text{ mol } C_2H_3O_2^-}{1 \text{ mol } NaC_2H_3O_2}$$

$$= 0.311 \text{ mol } C_2H_3O_2^-$$

$$[C_2H_3O_2^-] = \frac{0.311 \text{ mol } C_2H_3O_2^-}{0.500 \text{ L}} = 0.622 \text{ M } C_2H_3O_2^-$$

Equilibrium Calculation:

$$HC_2H_3O_2 \ + \ H_2O \ \rightleftharpoons \ H_3O^+ \ + \ C_2H_3O_2^-$$

initial concns:				
weak acid:	0.550 M		—	—
salt:	—		—	0.622 M
changes:	$-x$ M		$+x$ M	$+x$ M
equil concns:	$(0.550 - x)$ M		$+x$ M	$(0.622 + x)$ M

In our customary fashion, let's assume that x is very small, so $0.550 - x \approx 0.550$ and $0.622 + x \approx 0.622$. We will find this assumption to be valid.

$$K_a = \frac{[H_3O^+][C_2H_3O_2^-]}{[HC_2H_3O_2]} = \frac{(x)(0.622)}{0.550} = 1.8 \times 10^{-5}$$

$$x = [H_3O^+] = \frac{0.550}{0.622} \times 1.8 \times 10^{-5} = 1.6 \times 10^{-5}$$

$$\text{pH} = -\log[H_3O^+] = -\log(1.6 \times 10^{-5}) = 4.80$$

Check: We have seen that pH = pK_a = 4.74 when acetic acid and acetate ion are present in equal concentrations. Here the concentration of the conjugate *base* (acetate ion) is greater than that of the acetic acid. The solution should be somewhat more basic (less acidic) than pH = 4.74. A pH of 4.80 is a reasonable answer.

Practice Example A: What is the pH of a buffer solution prepared by dissolving 23.1 g $NaCHO_2$ in a sufficient volume of 0.432 M $HCHO_2$ to make 500.0 mL of the buffer?

Practice Example B: A handbook states that to prepare 100.0 mL of a particular buffer solution, mix 63.0 mL of 0.200 M $HC_2H_3O_2$ with 37.0 mL of 0.200 M $NaC_2H_3O_2$. What is the pH of this buffer?

An important point worth noting in Example 18-4 is that if a solution is to be an effective buffer, the assumptions $(M - x) \approx M$ and $(M + x) \approx M$ will always be valid. That is, the *equilibrium* concentrations of the buffer components will be very nearly the same as their *stoichiometric* concentrations. As a result, in Example 18-4 we could have gone directly from the stoichiometric concentrations of the buffer components to the expression

$$K_a = \frac{[H_3O^+][C_2H_3O_2^-]}{[HC_2H_3O_2]} = \frac{[H_3O^+](0.622)}{0.550} = 1.8 \times 10^{-5}$$

without setting up the ICE table. We can formalize this procedure through the special equation that we introduce next.

An Equation for Buffer Solutions: The Henderson–Hasselbalch Equation

Although we can continue to use the format demonstrated in Example 18-4 for buffer calculations, it is often useful to describe a buffer solution by means of an equation known as the **Henderson–Hasselbalch equation**. Biochemists and molecular biologists commonly use this equation. To derive this variation of the ionization constant expression, let's consider a mixture of a hypothetical weak acid, HA (such as $HC_2H_3O_2$), and its salt, NaA (such as $NaC_2H_3O_2$). We start with the familiar expressions

$$HA + H_2O \rightleftharpoons H_3O^+ + A^-$$

$$K_a = \frac{[H_3O^+][A^-]}{[HA]}$$

and rearrange the right side of the K_a expression to obtain

$$K_a = [H_3O^+] \times \frac{[A^-]}{[HA]}$$

Next, we take the *negative logarithm* of each side of this equation.

$$-\log K_a = -\log[H_3O^+] - \log \frac{[A^-]}{[HA]}$$

Now, recalling that $pH = -\log[H_3O^+]$ and that $pK_a = -\log K_a$, we get

$$pK_a = pH - \log \frac{[A^-]}{[HA]}$$

Then we solve for pH by rearranging the equation.

$$pH = pK_a + \log \frac{[A^-]}{[HA]}$$

By recognizing that A^- is the conjugate base of the weak acid HA, we can write the more general equation (18.2), the Henderson–Hasselbalch equation.

$$pH = pK_a + \log \frac{[\text{conjugate base}]}{[\text{acid}]} \qquad (18.2)$$

To apply this equation to an acetic acid–sodium acetate buffer, we use pK_a for $HC_2H_3O_2$ and these concentrations: $[HC_2H_3O_2]$ for [acid] and $[C_2H_3O_2^-]$ for [conjugate base]. To apply it to an ammonia–ammonium chloride buffer, we use pK_a for NH_4^+ and these concentrations: $[NH_4^+]$ for [acid] and $[NH_3]$ for [conjugate base].

Equation (18.2) is useful only when we can substitute *stoichiometric* or initial concentrations for equilibrium concentrations to give

$$pH = pK_a + \log \frac{[\text{conjugate base}]_{\text{initial}}}{[\text{acid}]_{\text{initial}}}$$

KEEP IN MIND ▶

that the Henderson–Hasselbalch equation is very useful, but should probably not be committed to memory; it is easy to get the conjugate base and acid terms inverted. It is most important to understand the principles that lead to this equation, thereby avoiding the pitfalls of using the equation incorrectly or when it is not valid.

thus avoiding the need to set up an ICE table. This constraint places important limitations on the equation's validity, however. Later, we will see that there are also conditions that must be met if a mixture is to be an effective buffer solution. Although the following rules may be overly restrictive in some cases, a reasonable approach to the twin concerns of effective buffer action and the validity of equation (18.2) is to ensure that

1. the ratio [conjugate base]/[acid] is within the limits

$$0.10 < \frac{[\text{conjugate base}]}{[\text{acid}]} < 10 \tag{18.3}$$

2. the molarity of each buffer component exceeds the value of K_a by a factor of at least 100

Viewed another way, equation (18.2) works only for cases in which the assumption $M - x \approx M$ is valid. If a quadratic equation is required to solve the equilibrium constant expression, equation (18.2) will likely fail.

Preparing Buffer Solutions

Select a weak acid with a pK_a close to the desired pH.

↓

Calculate the necessary ratio $\dfrac{[\text{conjugate base}]}{[\text{acid}]}$ to give the desired pH.

↓

Calculate the necessary concentrations of conjugate base and acid.

Suppose we need a buffer solution with pH = 5.09. Equation (18.2) suggests two alternatives. One is to find a weak acid, HA, that has $pK_a = 5.09$ and prepare a solution with equal molarities of the acid and its salt.

$$pH = pK_a + \log \frac{[A^-]}{[HA]} = 5.09 + \log 1 = 5.09$$

Although this alternative is simple in concept, generally, it is not practical. We are not likely to find a readily available, water-soluble weak acid with exactly $pK_a = 5.09$. The second alternative, summarized in the margin, is to use a cheap common weak acid such as acetic acid, $HC_2H_3O_2$ ($pK_a = 4.74$), and establish an appropriate ratio of $[C_2H_3O_2^-]/[HC_2H_3O_2]$ to obtain a pH of 5.09. Example 18-5 demonstrates this second alternative.

EXAMPLE 18-5

Preparing a Buffer Solution of a Desired pH. What mass of $NaC_2H_3O_2$ must be dissolved in 0.300 L of 0.25 M $HC_2H_3O_2$ to produce a solution with pH = 5.09? (Assume that the solution volume remains constant at 0.300 L.)

Solution

Equilibrium among the buffer components is expressed by the equation

$$HC_2H_3O_2 + H_2O \rightleftharpoons H_3O^+ + C_2H_3O_2^- \qquad K_a = 1.8 \times 10^{-5}$$

and by the ionization constant expression for acetic acid.

$$K_a = \frac{[H_3O^+][C_2H_3O_2^-]}{[HC_2H_3O_2]} = 1.8 \times 10^{-5}$$

Each of the three concentration terms appearing in a K_a expression should be an equilibrium concentration. The $[H_3O^+]$ corresponding to a pH of 5.09 is the equilibrium concentration. For $[HC_2H_3O_2]$, we will assume that the equilibrium concentration is equal to the stoichiometric or initial concentration. The value of $[C_2H_3O_2^-]$ that we calculate with the K_a expression is the equilibrium concentration, and we will assume that it is also the same as the stoichiometric concentration. Thus, we assume that neither the ionization of $HC_2H_3O_2$ to form $C_2H_3O_2^-$ nor the hydrolysis of $C_2H_3O_2^-$ to form $HC_2H_3O_2$ produces much of a difference between the stoichiometric (initial) and equilibrium concentrations of the buffer components. These assumptions work well if the conditions stated in expression (18.3) are met. The relevant concentration terms, then, are

$$[H_3O^+] = 10^{-pH} = 10^{-5.09} = 8.1 \times 10^{-6} \text{ M}$$
$$[HC_2H_3O_2] = 0.25 \text{ M}$$
$$[C_2H_3O_2^-] = ?$$

The required acetate ion concentration in the buffer solution is

$$[C_2H_3O_2^-] = K_a \times \frac{[HC_2H_3O_2]}{[H_3O^+]} = 1.8 \times 10^{-5} \times \frac{0.25}{8.1 \times 10^{-6}} = 0.56 \text{ M}$$

We complete the calculation of the mass of sodium acetate with some familiar ideas of solution stoichiometry.

$$\text{mass} = 0.300 \text{ L} \times \frac{0.56 \text{ mol } C_2H_3O_2^-}{1 \text{ L}} \times \frac{1 \text{ mol } NaC_2H_3O_2}{1 \text{ mol } C_2H_3O_2^-}$$
$$\times \frac{82.0 \text{ g } NaC_2H_3O_2}{1 \text{ mol } NaC_2H_3O_2} = 14 \text{ g } NaC_2H_3O_2$$

Practice Example A: How many grams of $(NH_4)_2SO_4$ must be dissolved in 0.500 L of 0.35 M NH_3 to produce a solution with pH = 9.00? (Assume that the solution volume remains at 0.500 L.)

Practice Example B: In Practice Example 18-3A, we established that an appropriate mixture of a strong acid and the salt of a weak acid is a buffer solution. Show that a solution made by adding 33.05 g $NaC_2H_3O_2 \cdot 3H_2O(s)$ to 300 mL of 0.250 M HCl should have pH $\approx$ 5.1.

In Example 18-5, we achieved the desired ratio of $[C_2H_3O_2^-]/[HC_2H_3O_2]$ by adding 14 g of sodium acetate to the previously prepared 0.25 M $HC_2H_3O_2$ solution. This is a common method of obtaining a buffer solution. Other methods are sometimes useful as well. Sufficient NaOH(aq) could be added to $HC_2H_3O_2(aq)$ to neutralize the acid partially, *producing* $C_2H_3O_2^-$ as a product. Or enough $NaC_2H_3O_2(s)$ could be added to HCl(aq) to convert all the HCl to $HC_2H_3O_2$ and leave some $C_2H_3O_2^-$ in excess. As we saw in Chapter 17, amines are weak bases, so an aqueous mixture of an amine and its conjugate acid is a buffer solution. Buffer solutions based on amines can be prepared in ways analogous to those based on weak acids. The methods available for making buffer solutions are summarized in Figure 18-6.

Calculating pH Changes in Buffer Solutions

To calculate how the pH of a buffer solution changes when small amounts of a strong acid or base are added, we must first use *stoichiometric* principles to establish how much of one buffer component is consumed and how much of the other component is produced. Then we can use the new concentrations of weak acid (or weak base) and its salt to calculate the pH of the buffer solution. Essentially, we solve this problem in two steps. First, we assume that the neutralization reaction proceeds *to completion* and determine new stoichiometric concentrations. Second, we

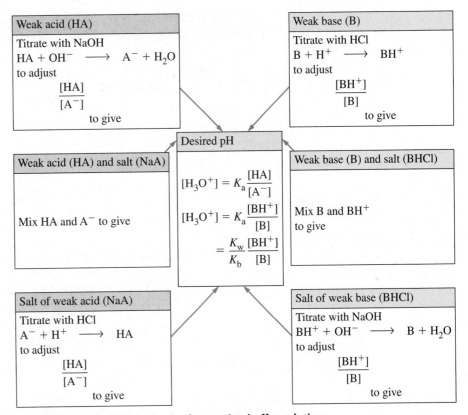

▲ **FIGURE 18-6 Six methods of preparing buffer solutions**
Depending on the pH range required and the type of experiment the buffer is to be used
for, either a weak acid or a weak base can be used to prepare a buffer solution.

use the new stoichiometric concentrations in the equilibrium constant expression to
solve for $[H_3O^+]$ and determine the pH. This method is applied in Example 18-6 and
illustrated in Figure 18-7.

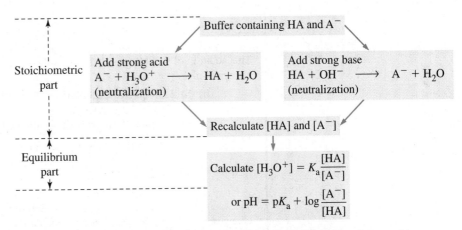

▲ **FIGURE 18-7 The calculation of the new pH of a buffer after strong acid
or base is added**
The stoichiometric and equilibrium parts of the calculation are indicated. This scheme
can also be applied to the conjugate acid–base pair BH^+/B, where B is a base.

EXAMPLE 18-6

Calculating pH Changes in a Buffer Solution. What are the effects on the pH of adding **(a)** 0.0060 mol HCl and **(b)** 0.0060 mol NaOH to 0.300 L of a buffer solution that is 0.250 M $HC_2H_3O_2$ and 0.560 M $NaC_2H_3O_2$?

Solution

To judge the effect of adding either (a) acid or (b) base on the pH of the buffer, the value we must keep in mind is the pH of the original buffer. We can obtain this value by using equation (18.2).

$$pH = pK_a + \log \frac{[C_2H_3O_2^-]}{[HC_2H_3O_2]}$$

$$= 4.74 + \log \frac{0.560}{0.250} = 4.74 + 0.35 = 5.09$$

KEEP IN MIND ▶
that $[HC_2H_3O_2] = 0.250$ M and $[C_2H_3O_2^-] = 0.560$ M are *stoichiometric* concentrations. The corresponding *equilibrium* concentrations are $0.250 - x$ and $0.560 + x$, where $x = [H_3O^+]$. We can substitute stoichiometric for equilibrium concentrations because x is very small.

(a) Stoichiometric Calculation: Let's convert all concentrations to amounts in moles, and assume that the neutralization goes to completion. Essentially, this is a limiting reactant calculation, but perhaps simpler than many of those in Chapter 4. In neutralizing the added H_3O^+, 0.0060 mol $C_2H_3O_2^-$ is converted to 0.0060 mol $HC_2H_3O_2$.

	$C_2H_3O_2^-$	$+$	H_3O^+	$\longrightarrow$	$HC_2H_3O_2$	$+$	H_2O
original buffer:	$\underline{0.300\ L \times 0.560\ M}$				$\underline{0.300\ L \times 0.250\ M}$		
	0.168 mol				0.0750 mol		
add:			0.0060 mol				
changes:	−0.0060 mol		−0.0060 mol		+0.0060 mol		
final buffer:							
amount:	0.162 mol		(?)		0.0810 mol		
concns:	$\underline{0.162\ mol/0.300\ L}$		(?)		$\underline{0.0810\ mol/0.300\ L}$		
	0.540 M				0.270 M		

Equilibrium Calculation: We can redetermine the pH with equation (18.2), using the new equilibrium concentrations.

$$pH = pK_a + \log \frac{[C_2H_3O_2^-]}{[HC_2H_3O_2]}$$

$$= 4.74 + \log \frac{0.540}{0.270} = 4.74 + 0.30 = 5.04$$

This addition of 0.0060 mol HCl *lowers* the pH from 5.09 to 5.04; this is only a small change in pH.

(b) Stoichiometric Calculation: In neutralizing the added OH^-, 0.0060 mol $HC_2H_3O_2$ is converted to 0.0060 mol $C_2H_3O_2^-$. The calculation of the new initial concentrations is shown on the last line of the following table.

	$HC_2H_3O_2$	$+$	OH^-	$\longrightarrow$	$C_2H_3O_2^-$	$+$	H_2O
original buffer:	$\underline{0.300\ L \times 0.250\ M}$				$\underline{0.300\ L \times 0.560\ M}$		
	0.0750 mol				0.168 mol		
add:			0.0060 mol				
changes:	−0.0060 mol		−0.0060 mol		+0.0060 mol		
final buffer:							
amount:	0.0690 mol		(?)		0.174 mol		
concns:	$\underline{0.0690\ mol/0.300\ L}$		(?)		$\underline{0.174\ mol/0.300\ L}$		
	0.230 M				0.580 M		

Equilibrium Calculation: This is the same type of calculation as in part (a), but with slightly different concentrations.

$$pH = 4.74 + \log \frac{0.580}{0.230} = 4.74 + 0.40 = 5.14$$

The addition of 0.0060 mol OH^- *raises* the pH from 5.09 to 5.14—another small change.

Check: The most important criteria to confirm in a buffer calculation are that the magnitude of the change in pH is small and that the change occurs in the correct direction: *lowering* of the pH by an acid and *raising* of the pH by a base. The results are indeed reasonable.

Practice Example A: A 1.00-L volume of buffer is made with concentrations of 0.350 M $NaCHO_2$ (sodium formate) and 0.550 M $HCHO_2$ (formic acid). **(a)** What is the initial pH? **(b)** What is the pH after the addition of 0.0050 mol HCl(aq)? (Assume that the volume remains 1.00 L.) **(c)** What would be the pH after the addition of 0.0050 mol NaOH to the original buffer?

Practice Example B: How many milliliters of 6.0 M HNO_3 would you add to 300.0 mL of the buffer solution of Example 18-6 to change the pH from 5.09 to 5.03?

Perhaps you have already noticed a way to simplify the calculation in Example 18-6. Because the buffer components are always present in the same solution of volume V, we can substitute numbers of moles directly into equation (18.2) without regard for the particular value of V. Thus, in Example 18-6(b),

$$pH = 4.74 + \log \frac{[C_2H_3O_2^-]}{[HC_2H_3O_2]} = 4.74 + \log \frac{0.174 \; \cancel{mol}/\cancel{V}}{0.0690 \; \cancel{mol}/\cancel{V}} = 4.74 + 0.40 = 5.14$$

This expression is also consistent with the observation that, on dilution, buffer solutions resist pH changes. Diluting a buffer solution means increasing its volume V by adding water. This action produces the same change in the numerator and the denominator of the ratio [conjugate base]/[acid]. The ratio itself remains unchanged, as does the pH.

Buffer Capacity and Buffer Range

It is not difficult to see that if we add more than 0.0750 mol OH^- to the buffer solution described in Example 18-6, the 0.0750 mol $HC_2H_3O_2$ will be completely converted to 0.0750 mol $C_2H_3O_2^-$. An excess of OH^- will remain, and the solution will become rather strongly basic.

Buffer capacity refers to the amount of acid or base that a buffer can neutralize before its pH changes appreciably. In general, the maximum buffer capacity exists when the concentrations of a weak acid and its conjugate base are kept *large* and *approximately equal to each other*. The **buffer range** is the pH range over which a buffer effectively neutralizes added acids and bases and maintains a fairly constant pH. As equation (18.2) suggests,

$$pH = pK_a + \log \frac{[\text{conjugate base}]}{[\text{acid}]}$$

when the ratio [conjugate base]/[acid] = 1, pH = pK_a. When the ratio falls to 0.10, the pH *decreases* by 1 pH unit from pK_a because log 0.10 = -1. If the ratio increases to a value of 10, the pH *increases* by 1 unit because log 10 = 1. For practical purposes, this range of 2 pH units is the maximum range to which a buffer solution should be exposed. For acetic acid–sodium acetate buffers, the effective range is about pH 3.7–5.7; for ammonia–ammonium chloride buffers, it is about pH 8.3–10.3.

KEEP IN MIND ▶
that as a rule of thumb, the amounts of the buffer components should be at least ten times greater than the amount of acid or base to be neutralized.

▶ The buffer range for the ammonia–ammonium chloride solution is based on the pK_a of NH_4^+, 9.26.

▲ A master brewer inspecting wort temperature and pH in the making of beer.

Applications of Buffer Solutions

An important example of a buffered system is that found in blood, which must be maintained at a pH of 7.4 in humans. We will consider the buffering of blood in the Focus On feature at the end of this chapter. But buffers have other important applications, too.

Protein studies often must be performed in buffered media because the structures of protein molecules, including the magnitude and kind of electric charges they carry, depend on the pH (see Section 28–4). The typical enzyme is a protein capable of catalyzing a biochemical reaction, so enzyme activity is closely linked to protein structure and hence to pH. Most enzymes in the body have their maximum activity between pH 6 and pH 8. Studying enzyme activity in the laboratory usually means working with media buffered in this pH range.

The control of pH is often important in industrial processes. For example, in the mashing of barley malt, the first step of making beer, the pH of the solution must be maintained at 5.0 to 5.2, so that the protease and peptidase enzymes can hydrolyze the proteins from the barley. The inventor of the pH scale, Søren Sørensen, was a research scientist in a brewery.

We will consider the importance of buffer solutions in solubility/precipitation processes in Chapter 19.

18-3 Acid–Base Indicators

An **acid–base indicator** is a substance whose color depends on the pH of the solution to which it is added. Several of the photographs in this and the preceding chapter have shown acid–base indicators in use. The indicator chosen depended on just how acidic or basic the solution was. In this section, we will consider how an acid–base indicator works and how an appropriate indicator is selected for a pH measurement.

Acid–base indicators exist in two forms: (1) a weak acid, represented symbolically as HIn and having one color, and (2) its conjugate base, represented as In$^-$ and having a different color. When just a small amount of indicator is added to a solution, the indicator does not affect the pH of the solution. Instead, the ionization equilibrium of the indicator is itself affected by the prevailing [H$_3$O$^+$] in solution.

$$\underset{\text{acid color}}{\text{HIn}} + \text{H}_2\text{O} \rightleftharpoons \text{H}_3\text{O}^+ + \underset{\text{base color}}{\text{In}^-}$$

From Le Châtelier's principle, we see that *increasing* [H$_3$O$^+$] in a solution displaces the equilibrium to the left, increasing the proportion of HIn and hence the acid color. *Decreasing* [H$_3$O$^+$] in a solution displaces the equilibrium to the right, increasing the proportion of In$^-$ and hence the base color. The color of the solution depends on the relative proportions of the acid and base. The pH of the solution can be related to these relative proportions and to the pK_a of the indicator by means of an equation similar to equation (18.2).

$$\text{pH} = \text{p}K_{\text{HIn}} + \log \frac{[\text{In}^-]}{[\text{HIn}^-]} \qquad (18.4)$$

▶ The acid "color" of a few indicators is colorless.

In general, if 90% or more of an indicator is in the form HIn, the solution will take on the acid color. If 90% or more is in the form In$^-$, the solution takes on the base (or anion) color. If the concentrations of HIn and In$^-$ are about equal, the indicator is in the process of changing from one form to the other and has an intermediate color. The complete change in color occurs over a range of about *2 pH units*, with pH = pK_{HIn} at about the middle of the range. The colors and pH ranges

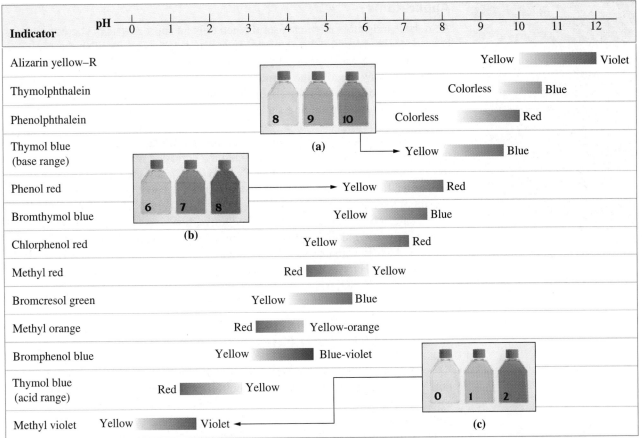

▲ **FIGURE 18-8   pH and color changes for some common acid–base indicators**
The indicators pictured and the pH values at which they change color are (a) thymol blue
(pH 8–10); (b) phenol red (pH 6–8); and (c) methyl violet (pH 0–2).

 Natural Indicators movie

TABLE 18.1 pH and the Colors of Acid–Base Indicators		
Acid Color	**Intermediate Color**	**Base Color**
$[In^-]/[HIn] < 0.10$	$[In^-]/[HIn] \approx 1$	$[In^-]/[HIn] > 10$
$pH < pK_{HIn} + \log 0.10$	$pH \approx pK_{HIn} + \log 1$	$pH > pK_{HIn} + \log 10$
$pH < pK_{HIn} - 1$	$pH \approx pK_{HIn}$	$pH > pK_{HIn} + 1$

of several acid–base indicators are shown in Figure 18-8. A summary of these ideas
is presented in Table 18.1, and an example of their use is given below.
Bromthymol blue, $pK_{HIn} = 7.1$

pH < 6.1 (yellow)   pH ≈ 7.1 (green)   pH > 8.1 (blue)

An acid–base indicator is usually prepared as a solution (in water, ethanol, or
some other solvent). In acid–base titrations, a few drops of the indicator solution
are added to the solution being titrated. In other applications, porous paper is im-
pregnated with an indicator solution and dried. When this paper is moistened with
the solution being tested, it acquires a color determined by the pH of the solution.
This paper is usually called *pH test paper*.

▲ Testing swimming pool water for its chlorine content and pH.

Applications

Acid–base indicators are most useful when only an approximate pH determination is needed. For example, they are used in soil-testing kits to establish the approximate pH of soils. Soils are usually acidic in regions of high rainfall and heavy vegetation, and they are alkaline in more arid regions. The pH can vary considerably with local conditions, however. If a soil is found to be too acidic for a certain crop, its pH can be raised by adding slaked lime [$Ca(OH)_2$]. To reduce the pH of a soil, organic matter might be added.

In swimming pools, chlorinating agents are most effective at a pH of about 7.4. At this pH, the growth of algae is avoided, and the corrosion of pool plumbing is minimized. Phenol red (see Figure 18-8) is a common indicator used in testing swimming pool water. If chlorination is carried out with $Cl_2(g)$, the pool water becomes acidic as a result of the reaction of Cl_2 with H_2O: $Cl_2 + 2\,H_2O \longrightarrow H_3O^+ + Cl^- + HOCl$. In this case, a basic substance such as sodium carbonate is used to raise the pH. Another widely used chlorinating agent is sodium hypochlorite, $NaOCl(aq)$, made by the reaction of $Cl_2(g)$ with excess $NaOH(aq)$: $Cl_2 + 2\,OH^- \longrightarrow Cl^- + OCl^- + H_2O$. The excess NaOH raises the pH of the pool water. The pH is adjusted by adding an acid such as HCl or H_2SO_4.

18-4   Neutralization Reactions and Titration Curves

As we learned in our discussion of the stoichiometry of titration reactions (Section 5–7), the **equivalence point** of a neutralization reaction is the point in the reaction at which both acid and base have been consumed. In other words, it is the point at which *neither* acid nor base is in excess.

In a titration, one of the solutions to be neutralized—say, the acid—is placed in a flask or beaker, together with a few drops of an acid–base indicator. The other solution (the base) used in a titration is added from a buret and is called the **titrant**. The titrant is added to the acid, first rapidly and then drop by drop, up to the equivalence point (recall Figure 5–10). The equivalence point is located by noting the color change of the acid–base indicator. The point in a titration at which the indicator changes color is called the **end point** of the indicator. The end point must match the equivalence point of the neutralization. That is, if the indicator's end point is near the equivalence point of the neutralization, the color change marked by that end point will signal the attainment of the equivalence point. We can achieve this match by choosing an indicator whose color change occurs over a pH range that includes the pH of the equivalence point.

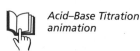

Acid–Base Titration animation

A graph of pH versus volume of titrant (the solution in the buret) is called a **titration curve**. Titration curves are most easily constructed by measuring the pH during a titration with a pH meter and plotting the data with a recorder. In this section we will emphasize calculating the pH at various points in a titration. These calculations will serve as a review of aspects of acid–base equilibria considered earlier in this chapter and in the preceding chapter.

The Millimole

In a typical titration, the volume of solution delivered from a buret is less than 50 mL (usually about 20–25 mL). The molarity of the solution used for the titration is generally less than 1 M. The typical amount of OH^- (or H_3O^+) delivered from the buret during a titration is only a few thousandths of a mole—for example, 5.00×10^{-3} mol. In calculations it is often easier to work with millimoles instead of moles. The symbol **mmol** stands for a **millimole**, which is one thousandth of a mole, or 10^{-3} mol.

Recall from Chapter 4 that molarity is defined as the number of moles per liter. We can use an alternative definition of molarity by converting from moles to millimoles and from liters to milliliters.

$$M = \frac{mol}{L} = \frac{mol/1000}{L/1000} = \frac{mmol}{mL}$$

Thus, the expression from Chapter 4 that the amount of solute is the product of molarity and solution volume (page 120) can be based either on $mol/L \times L = mol$ or on $mmol/mL \times mL = mmol$.

Titration of a Strong Acid with a Strong Base

Suppose we place 25.00 mL of 0.100 M HCl (a *strong* acid) in a small flask or beaker and then add 0.100 M NaOH (a *strong* base) from a buret. We can calculate the pH of the accumulated solution at different points in the titration and plot these pH values against the volume of NaOH added. From this titration curve we can establish the pH at the equivalence point and identify an appropriate indicator for the titration. Some typical calculations are outlined in Example 18-7.

EXAMPLE 18-7

Calculating Points on a Titration Curve: Strong Acid Titrated with a Strong Base. What is the pH at each of the following points in the titration of 25.00 mL of 0.100 M HCl with 0.100 M NaOH?

   (a) before the addition of any NaOH (*initial pH*)
   (b) after the addition of 24.00 mL 0.100 M NaOH (*before equiv point*)
   (c) after the addition of 25.00 mL 0.100 M NaOH (*at equiv point*)
   (d) after the addition of 26.00 mL 0.100 M NaOH (*beyond equiv point*)

Solution

First, let's write the titration equation in the ionic and net ionic form.

Ionic form: $H_3O^+(aq) + Cl^-(aq) + Na^+(aq) + OH^-(aq) \longrightarrow$
$$Na^+(aq) + Cl^-(aq) + 2\,H_2O(l)$$
Net ionic form: $H_3O^+(aq) + OH^-(aq) \longrightarrow 2\,H_2O(l)$

   (a) Before any NaOH is added, we are dealing with 0.100 M HCl. This solution has $[H_3O^+] = 0.100$ M and pH $= 1.00$.

   (b) The number of millimoles of H_3O^+ to be titrated is

$$25.00 \text{ mL} \times \frac{0.100 \text{ mmol } H_3O^+}{1 \text{ mL}} = 2.50 \text{ mmol } H_3O^+$$

The number of millimoles of OH^- present in 24.00 mL of 0.100 M NaOH is

$$24.00 \text{ mL} \times \frac{0.100 \text{ mmol } OH^-}{1 \text{ mL}} = 2.40 \text{ mmol } OH^-$$

Now we can represent the net ionic equation of the neutralization reaction in a familiar format.

	H_3O^+	$+$	OH^-	$\longrightarrow$	$2\,H_2O$
initially present:	2.50 mmol		—		
add:			2.40 mmol		
changes:	−2.40 mmol		−2.40 mmol		
after reaction:	0.10 mmol		≈ 0		

The remaining 0.10 mmol of H_3O^+ is present in 49.00 mL of solution (25.00 mL original acid + 24.00 mL added base).

$$[H_3O^+] = \frac{0.10 \text{ mmol } H_3O^+}{49.00 \text{ mL}} = 2.0 \times 10^{-3} \text{ M}$$
$$pH = -\log[H_3O^+] = -\log(2.0 \times 10^{-3}) = 2.70$$

(c) The equivalence point is the point at which the HCl is completely neutralized and no excess NaOH is present. As seen in the ionic form of the equation for the neutralization reaction, the solution at the equivalence point is simply NaCl(aq). And, as we learned in Section 17–7, because neither Na^+ nor Cl^- hydrolyzes in water, pH = 7.00.

(d) To determine the pH of the solution beyond the equivalence point, we can return to the format in (b), except that now OH^- is in excess. The amount of OH^- added is 26.00 mL × 0.100 mmol/L = 2.60 mmol.

	H_3O^+	+	OH^-	$\longrightarrow$	$2 H_2O$
initially present:	2.50 mmol		—		
add:			2.60 mmol		
changes:	−2.50 mmol		−2.50 mmol		
after reaction:	≈0		0.10 mmol		

The excess 0.10 mmol of NaOH is present in 51.00 mL of solution (25.00 mL original acid + 26.00 mL added base). The concentration of OH^- in this solution is

$$[OH^-] = \frac{0.10 \text{ mmol } OH^-}{51.00 \text{ mL}} = 2.0 \times 10^{-3} \text{ M}$$
$$pOH = -\log(2.0 \times 10^{-3}) = 2.70 \qquad pH = 14.00 - 2.70 = 11.30$$

Practice Example A: For the titration of 25.00 mL of 0.150 M HCl with 0.250 M NaOH, calculate (a) the initial pH; (b) the pH when neutralization is 50.0% complete; (c) the pH when neutralization is 100.0% complete; and (d) the pH when 1.00 mL of NaOH is added beyond the equivalence point.

Practice Example B: For the titration of 50.00 mL of 0.00812 M $Ba(OH)_2$ with 0.0250 M HCl, calculate (a) the initial pH; (b) the pH when neutralization is 50.0% complete; (c) the pH when neutralization is 100.0% complete.

Figure 18-9 presents pH–volume data and the titration curve for the HCl–NaOH titration. From this figure, we can establish these principal features of the titration curve for the titration of a *strong acid with a strong base.*

• The pH has a low value at the beginning of the titration.
• The pH changes slowly until just before the equivalence point.
• At the equivalence point, the pH rises very sharply, perhaps by 6 units for an addition of only 0.10 mL (2 drops) of base.
• Beyond the equivalence point, the pH again rises only slowly.
• Any acid–base indicator whose color changes in the pH range from about 4 to 10 is suitable for this titration.

In the titration of a strong base with a strong acid, we can obtain a titration curve essentially identical to Figure 18-9 by plotting pOH against volume of titrant (the strong acid). Also, we can make a similar set of statements as those listed above, except that we would substitute pOH for pH. Alternatively, if we plot pH against

Titration Data	
mL NaOH(aq)	**pH**
0.00	1.00
10.00	1.37
20.00	1.95
22.00	2.19
24.00	2.70
25.00	7.00
26.00	11.30
28.00	11.75
30.00	11.96
40.00	12.36
50.00	12.52

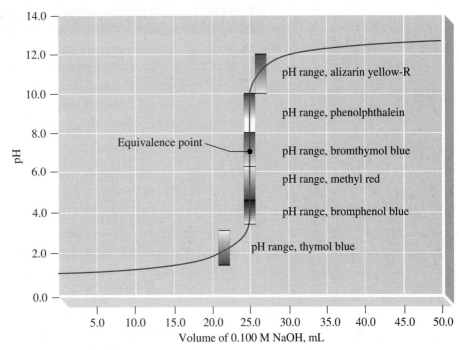

▲ **FIGURE 18-9** **Titration curve for the titration of a strong acid with a strong base—25.00 mL of 0.100 M HCl with 0.100 M NaOH**
All indicators whose color ranges fall along the steep portion of the titration curve are suitable for this titration. Thymol blue changes color too soon; alizarin yellow-R, too late.

Titration simulation

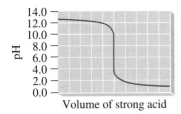

▲ **FIGURE 18-10**
Titration curve for the titration of a strong base with a strong acid

volume of titrant (the strong acid), the titration curve looks like Figure 18-9 flipped over from top to bottom, as shown in Figure 18-10.

Titration of a Weak Acid with a Strong Base

Several important differences exist between the titration of a weak acid with a strong base and a strong acid with a strong base, but one feature is *unchanged* when we compare the two titrations.

> For equal volumes of acid solutions of the same molarity, the volume of base required to titrate to the equivalence point is independent of the strength of the acid.

We can think of the neutralization of a weak acid such as $HC_2H_3O_2$ as involving the direct transfer of protons from $HC_2H_3O_2$ molecules to OH^- ions. In the neutralization of a strong acid, the protons are transferred from H_3O^+ ions. In either case, the acid and base react in a 1 : 1 mole ratio.

$$HC_2H_3O_2 + OH^- \longrightarrow H_2O + C_2H_3O_2^-$$

$$H_3O^+ + OH^- \longrightarrow H_2O + H_2O$$

When dealing with the titration of a weak acid with a strong base, we will divide the calculation into a stoichiometric part and an equilibrium part to take into account the partial ionization of the weak acid. The calculation strategy is analogous to that adopted when we considered the addition of a strong base to a buffer solution. In Example 18-8 and in Figure 18-11 (page 732), we consider the titration of 25.00 mL of 0.100 M $HC_2H_3O_2$ with 0.100 M NaOH.

EXAMPLE 18-8

Calculating Points on a Titration Curve: Weak Acid Titrated with a Strong Base. What is the pH at each of the following points in the titration of 25.00 mL of 0.100 M $HC_2H_3O_2$ with 0.100 M NaOH?

(a) before the addition of any NaOH (*initial pH*)
(b) after the addition of 10.00 mL 0.100 M NaOH (*before equiv point*)
(c) after the addition of 12.50 mL 0.100 M NaOH (*half-neutralization*)
(d) after the addition of 25.00 mL 0.100 M NaOH (*equiv point*)
(e) after the addition of 26.00 mL 0.100 M NaOH (*beyond equiv point*)

Solution

(a) The initial $[H_3O^+]$ is obtained by the calculation in Example 17–6 (page 680). $pH = -\log(1.3 \times 10^{-3}) = 2.89$.

(b) The number of millimoles of $HC_2H_3O_2$ to be neutralized is

$$25.00 \text{ mL} \times \frac{0.100 \text{ mmol } HC_2H_3O_2}{1 \text{ mL}} = 2.50 \text{ mmol } HC_2H_3O_2$$

At this point in the titration, the number of millimoles of OH^- added is

$$10.00 \text{ mL} \times \frac{0.100 \text{ mmol } OH^-}{1 \text{ mL}} = 1.00 \text{ mmol } OH^-$$

The total solution volume = 25.00 mL original acid + 10.00 mL added base = 35.00 mL. We enter this information at appropriate points into the following setup.

Stoichiometric Calculation:

	$HC_2H_3O_2$	+	OH^-	$\longrightarrow$	$C_2H_3O_2^-$	H_2O
initially present:	2.50 mmol		—		—	
add:			1.00 mmol			
changes:	−1.00 mmol		−1.00 mmol		+1.00 mmol	
after reaction:						
mmol:	1.50 mmol				1.00 mmol	
concns:	1.50 mmol/35.00 mL		≈ 0		1.00 mmol/35.00 mL	
	0.0429 M				0.0286 M	

Equilibrium Calculation: The most direct approach is to recognize that the acetic acid–sodium acetate solution is a buffer solution whose pH can be calculated with the Henderson–Hasselbalch equation. We are justified in using that equation for two reasons: (1) The ratio: $[C_2H_3O_2^-]/[HC_2H_3O_2] = 0.0286/0.0429 = 0.667$ (satisfying the requirement on page 719 that it be between 0.10 and 10), and (2) $[C_2H_3O_2^-]$ and $[HC_2H_3O_2]$ exceed K_a (1.8×10^{-5}) by the factors 1.6×10^3 and 2.4×10^3, respectively (satisfying the requirement on page 683 that the factor exceed 100). So

$$pH = pK_a + \log \frac{[A^-]}{[HA]} = 4.74 + \log \frac{0.0286}{0.0429} = 4.74 - 0.18 = 4.56$$

Simpler still would be to substitute the numbers of millimoles of $C_2H_3O_2^-$ and $HC_2H_3O_2$ directly into the Henderson–Hasselbalch equation, without converting to molarities. That is,

$$pH = pK_a + \log \frac{[A^-]}{[HA]} = 4.74 + \log \frac{1.00 \text{ mmol}/V}{1.50 \text{ mmol}/V} = 4.74 - 0.18 = 4.56$$

KEEP IN MIND ▶
that the Henderson–Hasselbalch equation (equation 18.2) might be inaccurate in the very early stages of the titration and near the equivalence point. In the first instance, $[C_2H_3O_2^-]/[HC_2H_3O_2] < 0.10$; in the second instance, $[C_2H_3O_2^-]/[HC_2H_3O_2] > 10$.

(c) When we have added 12.50 mL of 0.100 M NaOH, we have added $12.50 \times 0.100 = 1.25$ mmol OH^-. As the following setup shows, this is enough base to neutralize exactly *half* of the acid.

$$HC_2H_3O_2 \; + \; OH^- \; \longrightarrow \; C_2H_3O_2^- \; + \; H_2O$$

initially present:	2.50 mmol	—	—	
add:		1.25 mmol		
changes:	−1.25 mmol	−1.25 mmol	+1.25 mmol	
after reaction:	1.25 mmol		1.25 mmol	

Again, applying the Henderson–Hasselbalch equation, we get

$$pH = pK_a + \log \frac{[C_2H_3O_2^-]}{[HC_2H_3O_2]} = 4.74 + \log \frac{1.25 \; mmol/V}{1.25 \; mmol/V} = 4.74 + \log 1 = 4.74$$

(d) At the equivalence point, neutralization is complete and 2.50 mmol $NaC_2H_3O_2$ has been produced in 50.00 mL of solution, leading to 0.0500 M $NaC_2H_3O_2$. The question becomes, "What is the pH of 0.0500 M $NaC_2H_3O_2$?" To answer this question, we must recognize that $C_2H_3O_2^-$ hydrolyzes (and Na^+ does not). The hydrolysis reaction and value of K_b are

$$C_2H_3O_2^- + H_2O \rightleftharpoons HC_2H_3O_2 + OH^-$$

$$K_b = \frac{K_w}{K_a} = \frac{1.0 \times 10^{-14}}{1.8 \times 10^{-5}} = 5.6 \times 10^{-10}$$

With a format similar to that used in the hydrolysis calculation of Example 17–13 (page 693), we obtain the following expression, where $x = [OH^-]$ and $x \ll 0.0500$.

$$K_b = \frac{[HC_2H_3O_2]}{[C_2H_3O_2^-]} = \frac{x \cdot x}{0.0500 - x} = 5.6 \times 10^{-10}$$

$$x^2 = 2.8 \times 10^{-11} \qquad x = [OH^-] = 5.3 \times 10^{-6} \; M$$
$$pOH = -\log(5.3 \times 10^{-6}) = 5.28$$
$$pH = 14.00 - pOH = 14.00 - 5.28 = 8.72$$

(e) The amount of OH^- added is 26.00 mL $\times$ 0.100 mmol/mL = 2.60 mmol. The volume of solution is 25.00 mL acid + 26.00 mL base = 51.00 mL. The 2.60 mmol OH^- neutralizes the 2.50 mmol of available acid, and 0.10 mmol OH^- remains in *excess*. Beyond the equivalence point, the pH of the solution is determined by the excess *strong* base.

$$[OH^-] = \frac{0.10 \; mmol \; OH^-}{51.00 \; mL} = 2.0 \times 10^{-3} \; M$$

$$pOH = -\log(2.0 \times 10^{-3}) = 2.70 \qquad pH = 14.00 - 2.70 = 11.30$$

Practice Example A: A 20.00-mL sample of 0.150 M HF solution is titrated with 0.250 M NaOH. Calculate **(a)** the initial pH and the pH when neutralization is **(b)** 25.0%, **(c)** 50.0%, **(d)** 100.0% complete.

(*Hint:* What is the initial amount of HF, and what amounts remain un-neutralized at the points in question?)

Practice Example B: For the titration of 50.00 mL of 0.106 M NH_3 with 0.225 M HCl, calculate **(a)** the initial pH and the pH when neutralization is **(b)** 25.0% complete; **(c)** 50.0% complete; **(d)** 100.0% complete.

Here are the principal features of the titration curve for a weak acid titrated with a strong base (Figure 18-11).

Titration Data	
mL NaOH(aq)	pH
0.00	2.89
5.00	4.14
10.00	4.57
12.50	4.74
15.00	4.92
20.00	5.35
24.00	6.12
25.00	8.72
26.00	11.30
30.00	11.96
40.00	12.36
50.00	12.52

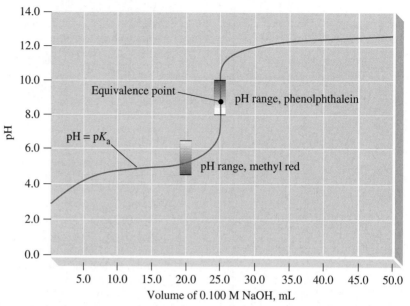

▲ **FIGURE 18-11** **Titration curve for the titration of a weak acid with a strong base—25.00 mL of 0.100 M $HC_2H_3O_2$ with 0.100 M NaOH**
Phenolphthalein is a suitable indicator for this titration, but methyl red is not. When exactly half of the acid is neutralized, $[HC_2H_3O_2] = [C_2H_3O_2^-]$ and pH = pK_a = 4.74.

1. The initial pH is higher (less acidic) than in the titration of a strong acid. (The weak acid is only partially ionized).

2. There is an initial rather sharp increase in pH at the start of the titration. (The anion produced by the neutralization of the weak acid is a common ion that reduces the extent of ionization of the acid.)

3. Over a long section of the curve preceding the equivalence point, the pH changes only gradually. (Solutions corresponding to this portion of the curve are buffer solutions.)

4. Because [HA] = [A⁻], at the point of half-neutralization, pH = pK_a.

5. At the equivalence point, pH > 7. (The conjugate base of a weak acid hydrolyzes, producing OH⁻.)

6. Beyond the equivalence point, the titration curve is identical to that of a strong acid with a strong base. (In this portion of the titration, the pH is established entirely by the concentration of unreacted OH⁻.)

7. The steep portion of the titration curve at the equivalence point occurs over a relatively short pH range (from about pH 7 to pH 10).

8. The selection of indicators available for the titration is more limited than in a strong acid–strong base titration. (An indicator whose color change occurs below pH 7 cannot be used.)

As illustrated in Example 18-8 and suggested by Figure 18-12, the necessary calculations for a weak acid–strong base titration curve are of four distinct types, depending on the portion of the titration curve being described. One type of titration that generally cannot be performed successfully is that of a weak acid with a weak base (or vice versa). The equivalence point cannot be located precisely because the change in pH with volume of titrant is too gradual.

▶ FIGURE 18-12
Constructing the titration curve for a weak acid with a strong base
The calculations needed to plot this graph, illustrated in Example 18-8, can be divided into *four* types.

1. pH of a pure weak acid (initial pH)
2. pH of a buffer solution of a weak acid and its salt (over a broad range before the equivalence point)
3. pH of a salt solution undergoing hydrolysis (equivalence point)
4. pH of a solution of a strong base (over a broad range beyond the equivalence point)

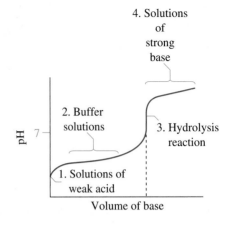

Titration of a Weak Polyprotic Acid

▶ This stepwise neutralization is observed only if successive ionization constants (K_{a_1}, K_{a_2}, …) differ significantly in magnitude (for example, by a factor of 10^3 or more). Otherwise, the second neutralization step begins before the first step is completed, and so on.

The most striking evidence that a polyprotic acid ionizes in distinct steps comes by way of its titration curve. For a polyprotic acid, we expect to see a separate equivalence point for each acidic hydrogen. Thus, we expect to see three equivalence points when H_3PO_4 is titrated with NaOH(aq). In the neutralization of phosphoric acid by sodium hydroxide, essentially all the H_3PO_4 molecules are first converted to its salt, NaH_2PO_4. Then all the NaH_2PO_4 is converted to Na_2HPO_4; and finally the Na_2HPO_4 is converted to Na_3PO_4.

The titration of 10.0 mL of 0.100 M H_3PO_4 with 0.100 M NaOH is pictured in Figure 18-13. Notice that the first two equivalence points come at equal intervals on

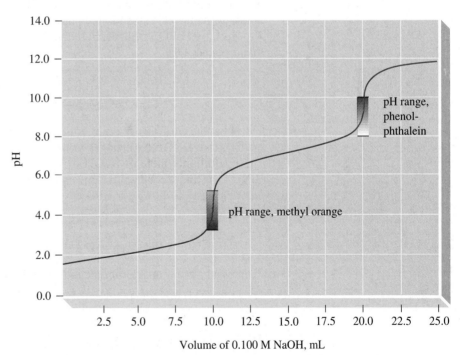

▲ FIGURE 18-13   **Titration of a weak polyprotic acid—10.0 mL of 0.100 M H_3PO_4 with 0.100 M NaOH**
A 10.0-mL volume of 0.100 M NaOH is required to reach the first equivalence point. The additional volume of 0.100 M NaOH required to reach the second equivalence point is also 10.0 mL.

Are You Wondering ...

How to calculate the pH values at various points on the titration curve in Figure 18-13?

The pH of the initial 0.100 M H_3PO_4 can be calculated with the K_{a_1} expression of H_3PO_4 (recall Example 17–9). For the slowly rising portion of the curve before the first equivalence point, the predominant species are H_3PO_4 and $H_2PO_4^-$, acting as a buffer solution. For this portion of the curve, we can calculate the pH by using equation (18.2). Thus, halfway to the first equivalence point, where $[H_3PO_4] = [H_2PO_4^-]$, pH = pK_{a_1}. Between the first and second equivalence points, the predominant species are $H_2PO_4^-$ and HPO_4^{2-}, and calculations can be based on the K_{a_2} expression. At the midway point for this section of the curve, pH = pK_{a_2}. Beyond the second equivalence point, the predominant species are HPO_4^{2-} and PO_4^{3-}. Although pH calculations can be made in this region, titrations are generally not carried this far (for reasons stated in the text). The calculations required at the equivalence points are described in Section 18-5.

the volume axis, at 10.0 mL and at 20.0 mL. Although we expect a third equivalence point at 30.0 mL, it is not realized in this titration. The pH of the strongly hydrolyzed Na_3PO_4 solution at the third equivalence point—approaching pH 13—is higher than can be reached by adding 0.100 M NaOH to water. $Na_3PO_4(aq)$ is nearly as basic as the NaOH(aq) used in the titration (as we shall see in Section 18-5).

Let's focus on a few details of this titration. For each mole of H_3PO_4, 1 mol NaOH is required to reach the first equivalence point. At this first equivalence point, the solution is essentially $NaH_2PO_4(aq)$. This is an *acidic* solution because $K_{a_2} > K_b$ for $H_2PO_4^-$: the reaction that produces H_3O^+ predominates over the one that produces OH^-.

$$H_2PO_4^- + H_2O \rightleftharpoons H_3O^+ + HPO_4^{2-} \qquad K_{a_2} = 6.3 \times 10^{-8}$$

$$H_2PO_4^- + H_2O \rightleftharpoons H_3PO_4 + OH^- \qquad K_b = 1.4 \times 10^{-12}$$

The pH at the equivalence point falls within the pH range over which the color of methyl orange indicator changes from red to orange.

An additional mole of NaOH is required to convert 1 mol $H_2PO_4^-$ to 1 mol HPO_4^{2-}. At this second equivalence point in the titration of H_3PO_4, the solution is *basic* because $K_b > K_{a_3}$ for HPO_4^{2-}.

$$HPO_4^{2-} + H_2O \rightleftharpoons H_2PO_4^- + OH^- \qquad K_b = 1.6 \times 10^{-7}$$

$$HPO_4^{2-} + H_2O \rightleftharpoons H_3O^+ + PO_4^{3-} \qquad K_{a_3} = 4.2 \times 10^{-13}$$

Phenolphthalein is an appropriate indicator for this equivalence point; the color of this indicator changes from colorless to light pink.

18-5 Solutions of Salts of Polyprotic Acids

In discussing the neutralization of phosphoric acid by a strong base, we found that the first equivalence point should come in a somewhat acidic solution and the second in a mildly basic solution. We reasoned that the third equivalence point could be reached only in a strongly basic solution. The pH at this third equivalence point is not difficult to calculate. It corresponds to that of $Na_3PO_4(aq)$, and PO_4^{3-} can ionize (hydrolyze) only as a base.

$$PO_4^{3-} + H_2O \rightleftharpoons HPO_4^{2-} + OH^- \qquad K_b = K_w/K_{a_3}$$

$$= \frac{1.0 \times 10^{-14}}{4.2 \times 10^{-13}}$$

$$= 2.4 \times 10^{-2}$$

EXAMPLE 18-9

▶ Sodium phosphate is often sold under the name *tri*sodium *phosphate* (TSP). Anyone working with TSP should wear protective gloves. Fats and greases, including those in human skin, are solubilized in strongly basic solutions.

Determining the pH of a Solution Containing the Anion (A^{n-}) of a Polyprotic Acid. Sodium phosphate, Na_3PO_4, is an ingredient of some preparations used to clean painted walls before they are repainted. What is the pH of 1.0 M $Na_3PO_4(aq)$?

Solution
In the usual fashion, we can write

$$PO_4^{3-} + H_2O \rightleftharpoons HPO_4^{2-} + OH^- \quad K_b = 2.4 \times 10^{-2}$$

initial concns:	1.0 M	—	—
changes:	$-x$ M	$+x$ M	$+x$ M
equil concns:	$(1.0 - x)$ M	x M	x M

$$K_b = \frac{[HPO_4^{2-}][OH^-]}{[PO_4^{3-}]} = \frac{x \cdot x}{1.0 - x} = 2.4 \times 10^{-2}$$

▶ The ratio $M_b/K_b = 1.0/0.024 = 42$. This value is smaller than the minimum value of 100 that we have been using as the usual criterion.

Because K_b is quite large, we should not expect the usual simplifying assumption to work here. That is, x is *not* very much smaller than 1.0. Solution of the quadratic equation $x^2 + 0.024x - 0.024 = 0$ yields $x = [OH^-] = 0.14$ M. Thus,

$$pOH = -\log[OH^-] = -\log 0.14 = +0.85$$
$$pH = 14.00 - 0.85 = 13.15$$

Practice Example A: Calculate the pH of an aqueous solution that is 1.0 M Na_2CO_3.

(*Hint:* Use data from Table 17.4 to establish K_b for CO_3^{2-}.)

Practice Example B: Calculate the pH of an aqueous solution that is 0.500 M Na_2SO_3.

(*Hint:* Use data from Table 17.4.)

It is more difficult to calculate the pH values of $NaH_2PO_4(aq)$ and $Na_2HPO_4(aq)$ than of $Na_3PO_4(aq)$. This is because with both $H_2PO_4^-$ and HPO_4^{2-}, two equilibria must be considered *simultaneously:* ionization as an acid and ionization as a base (hydrolysis). For solutions that are reasonably concentrated (say, 0.10 M or greater), the pH values prove to be *independent* of the solution concentration. Shown here (with pK_a values from Table 17.4) are general expressions, printed in blue, and their application to $H_2PO_4^-(aq)$ and $HPO_4^{2-}(aq)$:

for $H_2PO_4^-$: $\quad pH = \frac{1}{2}(pK_{a_1} + pK_{a_2}) = \frac{1}{2}(2.15 + 7.20) = 4.68 \qquad (18.5)$

for HPO_4^{2-}: $\quad pH = \frac{1}{2}(pK_{a_2} + pK_{a_3}) = \frac{1}{2}(7.20 + 12.38) = 9.79 \qquad (18.6)$

Are You Wondering...

How to derive equations (18.5) and (18.6)?

Here is good place to use the general problem-solving method introduced in Section 17–6 (page 688). Consider a solution of NaH_2PO_4 of molarity M. The principal concentrations that we must account for are $[Na^+]$, $[H_3O^+]$, $[H_3PO_4]$, $[H_2PO_4^-]$, $[HPO_4^{2-}]$, and $[OH^-]$. Of these, two have very simple values: $[Na^+] = M$, and $[OH^-] = K_w/[H_3O^+]$. Additionally, we can write the following equations.

(1) *Acid Ionization:* $H_2PO_4^- + H_2O \rightleftharpoons H_3O^+ + HPO_4^{2-}$

$$K_{a_2} = \frac{[H_3O^+][HPO_4^{2-}]}{[H_2PO_4^-]}$$

(2) *Hydrolysis:* $H_2PO_4^- + H_2O \rightleftharpoons H_3PO_4 + OH^-$

$$K_b = K_w/K_{a_1} = \frac{[H_3PO_4][OH^-]}{[H_2PO_4^-]}$$

(3) *Material Balance:* The total concentration of the phosphorus-containing species is the stoichiometric molarity, M.
$$[H_3PO_4] + [H_2PO_4^-] + [HPO_4^{2-}] = M$$

(4) *Electroneutrality Condition:*
$[H_3O^+] + [Na^+] = [H_2PO_4^-] + (2 \times [HPO_4^{2-}])$. However, since $[Na^+] = M$, we can also write $[H_3O^+] = [H_2PO_4^-] + (2 \times [HPO_4^{2-}]) - M$

Solving the set of equations: Begin by substituting equation (3) into equation (4).
$$[H_3O^+] = [H_2PO_4^-] + (2 \times [HPO_4^{2-}]) - [H_3PO_4] - [H_2PO_4^-] - [HPO_4^{2-}]$$
$$= [HPO_4^{2-}] - [H_3PO_4]$$

To continue from here, rearrange equation (1) to obtain $[HPO_4^{2-}]$ in terms of $[H_3O^+]$, $[H_2PO_4^-]$, and K_{a_2}; then rearrange equation (2) to obtain $[H_3PO_4]$ in terms of $[H_3O^+]$, $[H_2PO_4^-]$, and K_{a_1}. Next, substitute the results into the expression, $[H_3O^+] = [HPO_4^{2-}] - [H_3PO_4]$. At this point, you will have an equation in terms of $[H_3O^+]$, $[H_2PO_4^-]$, K_{a_1}, and K_{a_2}. Assume that $[H_2PO_4^-] \approx M$, and you will get an equation from which you can derive equation (18.5). The remainder of this derivation and the derivation of equation (18.6) are left for you to do (see Exercise 89).

18-6 Acid–Base Equilibrium Calculations: A Summary

In this and the preceding chapter, we have considered a variety of acid–base equilibrium calculations. When you are faced with a new problem-solving situation, you might find it helpful to relate the new problem to a type that you have encountered before. It is best not to rely exclusively on "labeling" a problem, however. Some problems might not fit a recognizable category. Instead, keep in mind some principles that apply regardless of the particular problem, as suggested by these questions.

1. **Which species are potentially present in solution, and how large are their concentrations likely to be?**

In a solution containing similar amounts of HCl and $HC_2H_3O_2$, the only significant *ionic* species are H_3O^+ and Cl^-. HCl is a completely ionized strong acid, and in the presence of a strong acid, the weak acid $HC_2H_3O_2$ is only very slightly ionized because of the common-ion effect. In a mixture containing similar amounts of two *weak* acids of similar strengths, such as $HC_2H_3O_2$ and HNO_2, each acid partially ionizes. All of these concentrations would be significant: $[HC_2H_3O_2]$, $[C_2H_3O_2^-]$, $[HNO_2]$, $[NO_2^-]$, and $[H_3O^+]$. In a solution containing phosphoric acid or a phosphate salt (or both), H_3PO_4, $H_2PO_4^-$, HPO_4^{2-}, PO_4^{3-}, OH^-, H_3O^+, and possibly other cations might be present. If the solution is simply $H_3PO_4(aq)$, however, the only species present in significant concentrations are those associated with the first ionization: H_3PO_4, H_3O^+, and $H_2PO_4^-$. If the solution is instead described as $Na_3PO_4(aq)$, the significant species are Na^+, PO_4^{3-}, and the ions that are associated with the hydrolysis of PO_4^{3-}, that is, HPO_4^{2-} and OH^-.

2. **Are reactions possible among any of the solution components; if so, what is their stoichiometry?**

Suppose that you are asked to calculate $[OH^-]$ in a solution made to be 0.10 M NaOH and 0.20 M NH_4Cl. Before you answer that $[OH^-] = 0.10$ M, consider whether a solution can be *simultaneously* 0.10 M in OH^- and 0.20 M in NH_4^+. It cannot; any solution containing both NH_4^+ and OH^- must also contain NH_3. The OH^- and NH_4^+ react in a 1 : 1 mole ratio until OH^- is almost totally consumed:

$$NH_4^+ + OH^- \longrightarrow NH_3 + H_2O$$

You are now dealing with the buffer solution 0.10 M NH_3–0.10 M NH_4^+.

3. **Which equilibrium equations apply to the particular situation? Which are the most significant?**

One equation that applies to all acids and bases in aqueous solutions is $K_w = [H_3O^+][OH^-] = 1.0 \times 10^{-14}$. In many calculations, however, this equation is not significant compared with others. One situation in which it *is* significant is in calculating $[OH^-]$ in an *acidic* solution or $[H_3O^+]$ in a *basic* solution. After all, an acid does not produce OH^-, and a base does not produce H_3O^+. Another situation in which K_w is likely to be significant is in a solution with a pH near 7.

Often you will find that the ionization equilibrium with the largest K value is the most significant, but this will not always be the case. The amounts of the various species in solution are also an important consideration. When one drop of 1.00 M H_3PO_4 $\left(K_{a_1} = 7.1 \times 10^{-3}\right)$ is added to 1.00 L of 0.100 M $HC_2H_3O_2$ $\left(K_a = 1.8 \times 10^{-5}\right)$, the acetic acid ionization is most important in establishing the pH of the solution. The solution contains far more acetic acid than it does phosphoric acid.

▲ Extreme hyperventilation is needed to compensate for the very low partial pressures of O_2 encountered by high-altitude mountain climbers. This hyperventilation leads to the condition called alkalosis. The blood pH of climbers reaching the summit of Mt. Everest (8848 m = 29,028 ft) without supplemental oxygen can rise to 7.7 or 7.8.

One characteristic of blood that we rarely think about is its pH. Yet, maintaining the proper pH in blood and in intracellular fluids is crucial not only to human health but to life itself, primarily because the functioning of enzymes is sharply pH-dependent. The normal pH of blood is 7.4. Severe illness or death can result from sustained variations of just a few tenths of a pH unit from this normal value.

Among the conditions that can lead to *acidosis*, a condition in which the pH of blood decreases below normal, are heart failure, kidney failure, diabetes, persistent diarrhea, and a long-term high-protein diet. Prolonged, extensive exercise also can produce a temporary acidosis. *Alkalosis*, a condition of an increased pH of blood, may occur as a result of severe vomiting, hyperventilation, or exposure to high altitudes (altitude sickness).

Blood as a buffered solution

Human blood has a high buffer capacity. The addition of 0.01 mol HCl to 1 L of blood lowers the pH from 7.4 only to 7.2. The same amount of HCl added to a saline (NaCl) solution isotonic with blood lowers the pH from 7.0 to 2.0. The saline solution has no buffer capacity.

Summary

The ionization of a weak acid, HA, is suppressed by the presence in solution of a common ion—either H_3O^+ (from a strong acid) or A^- (from a salt of the weak acid). The weak acid–conjugate base combination is a buffer solution, as is the combination weak base–conjugate acid. A buffer solution maintains a nearly constant pH upon dilution or in the presence of small added amounts of acids or bases.

The pH at which a buffer solution functions is determined by the pK_a value and the molarities of the two buffer components. The buffer exhibits its greatest capacity to neutralize added acids and bases when the concentrations of the buffer components are equal. The effective buffer range is about 1 pH unit on either side of the pK_a.

An acid–base indicator is a weak acid that has one color when present as the nonionized acid, HIn, and a different color when present as the conjugate base, the anion In^-. The observed indicator color depends on the pH of the solution.

An acid–base titration curve is a graph of pH versus volume of titrating agent (titrant) added. At the equivalence point of the titration, the solution contains only the salt formed in the neutralization. Whether the solution at the equivalence point is acidic, basic, or neutral depends on whether ions of the salt can ionize (hydrolyze) as acids or bases. Generally, a titration curve shows a sharp change in pH at the equivalence point. In the titration of a weak acid (or weak base) with a strong base (or strong acid), $pH = pK_a$ at the point of half-neutralization. In the titration of a polyprotic acid, one steep portion of the curve is generally observed for each ionizable H atom. The appropriate acid–base indicator for a titration is an indicator whose pK_{HIn} is close to the pH at the equivalence point.

Integrative Example

The structural formula on the right is of *para*-hydroxybenzoic acid, a weak diprotic acid used as a food preservative. Titration of 25.00 mL of a dilute aqueous solution of this acid requires 16.24 mL of 0.0200 M NaOH to reach the first equivalence point. The measured pH after the addition of 8.12 mL of the base is 4.57; after 16.24 mL, the pH is 7.02. **(a)** What are the values of pK_{a_1} and pK_{a_2} of *para*-hydroxybenzoic acid? **(b)** What are the pH values at the two equivalence points in the titration?

$$H-O-\!\!\!\bigcirc\!\!\!-\overset{\displaystyle O}{\underset{\displaystyle \|}{C}}-O-H$$

Several factors are involved in the control of blood pH. A particularly important one is the ratio of dissolved HCO_3^- (hydrogen carbonate ion) to H_2CO_3 (carbonic acid). Even though $CO_2(g)$ is only partially converted to H_2CO_3 when it dissolves in water, we generally treat the solution as if this conversion were complete. Moreover, although H_2CO_3 is a weak *di*protic acid, we deal only with the first ionization step in the carbonic acid–hydrogen carbonate buffer system: H_2CO_3 is the weak acid, and HCO_3^- is the conjugate base.

$$CO_2(g) + H_2O \longrightarrow H_2CO_3(aq)$$

$$H_2CO_3 + H_2O \rightleftharpoons H_3O^+ + HCO_3^-$$

$$K_{a_1} = 4.4 \times 10^{-7}$$

Carbon dioxide enters the bloodstream from tissues as the by-product of metabolic reactions. In the lungs, $CO_2(g)$ is exchanged for $O_2(g)$, which the blood transports throughout the body.

Using equation (18.2), a value of $pK_a = -\log(4.4 \times 10^{-7}) = 6.4$, and a normal blood pH of 7.4, we can write

$$pH = 7.4 = 6.4 + 1.0$$

$$= pK_{a_1} + \log \frac{[HCO_3^-]}{[H_2CO_3]}$$

$$= 6.4 + \log\left(\frac{10}{1}\right)$$

The large ratio of $[HCO_3^-]$ to $[H_2CO_3]$ (10 : 1) seems to place this buffer somewhat outside the range of its maximum buffer capacity. (Recall the discussion of buffer capacity on page 723.) The situation is rather complex, but some of the factors involved are listed below.

1. The need to neutralize excess acid (lactic acid produced by exercise) is generally greater than the need to neutralize excess base. The high proportion of HCO_3^- helps in this regard.

2. If additional H_2CO_3 is needed to neutralize excess alkalinity, $CO_2(g)$ in the lungs can be reabsorbed to build up the H_2CO_3 content of the blood.

3. Other components, such as some plasma proteins and the phosphate buffer system, $H_2PO_4^- - HPO_4^{2-}$, contribute to maintaining the pH of blood at 7.4.

(a) *Determining pK_{a_1} and pK_{a_2}.* The only data required for this determination are the pH values given. To picture a titration curve for this weak acid, imagine something like the curve shown for the titration of H_3PO_4 in Figure 18-13. If a volume of 16.24 mL is required to reach the first equivalence point, when 8.12 mL of base has been added, the acid is half-neutralized in its first ionization step. At this point, $pH = pK_{a_1}$. So, $pK_{a_1} = 4.57$.

At the first equivalence point, the pH is that of an aqueous solution of HOC_6H_4COONa: 7.02. Now we can use equation 18.5. That is, the pH of an aqueous solution of the ion $HOC_6H_6COO^-$ is given by the expression

$$pH = \tfrac{1}{2}\left(pK_{a_1} + pK_{a_2}\right) = \tfrac{1}{2}\left(4.57 + pK_{a_2}\right) = 7.02$$

$$pK_{a_2} = (2 \times 7.02) - 4.57 = 9.47$$

Thus, $pK_{a_1} = 4.57$ and $pK_{a_2} = 9.47$.

(b) *Determining the pH values at the equivalence points.* The pH at the first equivalence point is given; it is 7.02.

Determining the pH at the second equivalence point involves additional calculations. We begin by noting that, at the second equivalence point, the solution is one of $NaOC_6H_4COONa$. The pH of the solution is established by the hydrolysis of $^-OC_6H_4COO^-$.

$$^-OC_6H_4COO^- + H_2O \rightleftharpoons HOC_6H_4COO^- + OH^-$$

$$K_b = K_w/K_{a_2}$$

To evaluate K_b, let's first obtain K_{a_2} from pK_{a_2}.

$$pK_{a_2} = -\log K_{a_2} = 9.47 \quad \text{and} \quad K_{a_2} = 10^{-9.47} = 3.4 \times 10^{-10}$$

$$K_b = K_w/K_{a_2} = 1.0 \times 10^{-14}/3.4 \times 10^{-10} = 2.9 \times 10^{-5}$$

We can get the pH of this solution by first calculating $[OH^-]$ and pOH. However, to do this, we still need one more piece of data—the molarity of the $NaOC_6H_4COONa(aq)$. We can get this from data for titration to the first equivalence point.

$$? \text{ mmol } OH^- = 16.24 \text{ mL} \times 0.0200 \text{ mmol}$$
$$OH^-/\text{mL} = 0.325 \text{ mmol } OH^-$$

$$? \text{ mmol } HOC_6H_4COOH = 0.325 \text{ mmol } OH^- \times 1 \text{ mmol}$$
$$HOC_6H_4COOH/\text{mmol } OH^- = 0.325 \text{ mmol } HOC_6H_4COOH$$

The amount of $^-OC_6H_4COO^-$ at the second equivalence point is the same as the amount of acid at the start of the titration.

$$0.325 \text{ mmol } ^-OC_6H_4COO^-$$

The volume of solution at the second equivalence point is $25.00 \text{ mL} + 16.24 \text{ mL} + 16.24 \text{ mL} = 57.48 \text{ mL}$. Thus, $[^-OC_6H_4COO^-] = 0.325 \text{ mmol } ^-OC_6H_4COO^-/57.48 \text{ mL} = 5.65 \times 10^{-3} \text{ M}$

Now we can return to the hydrolysis equation and the expression for K_b, using the method of Example 17–13.

$$^-OC_6H_4COO^- + H_2O \rightleftharpoons HOC_6H_4COO^- + OH^-$$

initial concns:	5.65×10^{-3} M	—	—
changes:	$-x$ M	$+x$ M	$+x$ M
equil concns:	$(5.65 \times 10^{-3} - x)$ M	x M	x M

$$K_b = \frac{x \cdot x}{(5.65 \times 10^{-3} - x)} = 2.9 \times 10^{-5}$$

The solution to this quadratic equation is $x = [OH^-] = 3.9 \times 10^{-4}$, corresponding to pOH = 3.41 and pH = 10.59.

Key Terms

acid–base indicator (18-3)
buffer capacity (18-2)
buffer range (18-2)
buffer solution (18-2)
common-ion effect (18-1)

end point (18-4)
equivalence point (18-4)
Henderson–Hasselbalch
  equation (18-2)
millimole (mmol) (18-4)

titrant (18-4)
titration curve (18-4)

Review Questions

1. In your own words, define or explain the following terms or symbols: (a) mmol; (b) HIn; (c) equivalence point of a titration; (d) titration curve.

2. Briefly describe each of the following ideas, phenomena, or methods: (a) the common-ion effect; (b) the use of a buffer solution to maintain a constant pH; (c) the determination of pK_a of a weak acid from a titration curve; (d) the measurement of pH with an acid–base indicator.

3. Explain the important distinctions between each pair of terms: (a) buffer capacity and buffer range; (b) hydrolysis and neutralization; (c) first and second equivalence point in the titration of a weak diprotic acid; (d) equivalence point of a titration and end point of an indicator.

4. For a solution that is 0.275 M $HC_3H_5O_2$ (propionic acid, $K_a = 1.3 \times 10^{-5}$) and 0.0892 M HI, calculate (a) $[H_3O^+]$; (b) $[OH^-]$; (c) $[C_3H_5O_2^-]$; (d) $[I^-]$.

5. For a solution that is 0.164 M NH_3 and 0.102 M NH_4Cl, calculate (a) $[OH^-]$; (b) $[NH_4^+]$; (c) $[Cl^-]$; (d) $[H_3O^+]$.

6. Write equations to show how each of the following buffer solutions reacts with a small added amount of a strong acid or a strong base: (a) $HCHO_2$–$KCHO_2$; (b) $C_6H_5NH_2$–$C_6H_5NH_3^+Cl^-$; (c) KH_2PO_4–Na_2HPO_4.

7. Calculate the pH of a buffer that is (a) 0.012 M $HC_7H_5O_2$ $(K_a = 6.3 \times 10^{-5})$ and 0.033 M $NaC_7H_5O_2$; (b) 0.408 M NH_3 and 0.153 M NH_4Cl.

8. What concentration of formate ion, $[CHO_2^-]$, should be present in 0.366 M $HCHO_2$ to produce a buffer solution with pH = 4.06?
$$HCHO_2 + H_2O \rightleftharpoons H_3O^+ + CHO_2^-$$
$$K_a = 1.8 \times 10^{-4}$$

9. What concentration of ammonia, $[NH_3]$, should be present in a solution with $[NH_4^+] = 0.732$ M to produce a buffer solution with pH = 9.12? For NH_3, $K_b = 1.8 \times 10^{-5}$.

10. *Without performing detailed calculations*, determine which of the following will raise the pH of 1.00 L of 0.50 M HCl to the greatest extent: 0.40 mol NaOH; 0.50 mol $HC_2H_3O_2$; 0.60 mol $NaC_2H_3O_2$; 0.70 mol NaCl. Explain.

11. Lactic acid, $HC_3H_5O_3$, is found in sour milk. A solution containing 1.00 g $NaC_3H_5O_3$ in 100.0 mL of 0.0500 M $HC_3H_5O_3$ has pH = 4.11. What is K_a of lactic acid?

12. A $HCHO_2$–$NaCHO_2$ buffer solution is to be prepared. For $HCHO_2$, $K_a = 1.8 \times 10^{-4}$.
(a) How many grams of $NaCHO_2$ must be dissolved in 0.250 L of 0.465 M $HCHO_2$ to produce a pH of 3.82?
(b) If one small pellet of NaOH (0.20 g) is added to the 0.250 L of buffer solution in part (a), what will be the new pH?

13. A handbook lists the following data:

Indicator	K_{HIn}	Color Change Acid $\longrightarrow$ Anion	
Bromphenol blue	1.4×10^{-4}	yellow	$\longrightarrow$ blue
Bromcresol green	2.1×10^{-5}	yellow	$\longrightarrow$ blue
Bromthymol blue	7.9×10^{-8}	yellow	$\longrightarrow$ blue
2,4-Dinitrophenol	1.3×10^{-4}	colorless	$\longrightarrow$ yellow
Chlorphenol red	1.0×10^{-6}	yellow	$\longrightarrow$ red
Thymolphthalein	1.0×10^{-10}	colorless	$\longrightarrow$ blue

(a) Which of these indicators change color in acidic solution, which in basic solution, and which near the neutral point?
(b) What is the approximate pH of a solution if bromcresol green indicator assumes a green color? if chlorphenol red assumes an orange color?

14. With reference to the indicators listed in Exercise 13, what would be the color of each combination?
(a) 2,4-dinitrophenol in 0.100 M HCl(aq)
(b) chlorphenol red in 1.00 M NaCl(aq)
(c) thymolphthalein in 1.00 M NH_3(aq)
(d) bromcresol green in seawater (recall Figure 17–6)

15. What volume of 0.146 M KOH is needed for the complete neutralization of (a) 25.00 mL of 0.212 M HI; (b) 20.00 mL of 0.0942 M H_2SO_4?

16. Sketch the titration curves (pH versus volume of titrant) that you would expect to obtain in the following titrations. Select a suitable indicator for each titration from Figure 18-8.
(a) NaOH(aq) titrated with HNO_3(aq)
(b) NH_3(aq) titrated with HCl(aq)
(c) $HC_2H_3O_2$(aq) titrated with KOH(aq)
(d) NaH_2PO_4 titrated with KOH(aq)

17. Calculate the pH at the points in the titration of 25.00 mL of 0.160 M HCl when (a) 10.00 mL and (b) 15.00 mL of 0.242 M KOH have been added.

18. Calculate the pH at the points in the titration of 20.00 mL of 0.275 M KOH when (a) 15.00 mL and (b) 20.00 mL of 0.350 M HCl have been added.

19. Calculate the pH at the points in the titration of 25.00 mL of 0.132 M HNO_2 when (a) 10.00 mL and (b) 20.00 mL of 0.116 M NaOH have been added. For HNO_2, $K_a = 7.2 \times 10^{-4}$.
$$HNO_2 + OH^- \longrightarrow H_2O + NO_2^-$$

20. Calculate the pH at the points in the titration of 20.00 mL of 0.318 M NH_3 when **(a)** 10.00 mL and **(b)** 15.00 mL of 0.475 M HCl have been added. For NH_3, $K_b = 1.8 \times 10^{-5}$.
$$NH_3(aq) + HCl(aq) \longrightarrow NH_4^+(aq) + Cl^-(aq)$$
21. A 25.00-mL sample of 0.0100 M $HC_7H_5O_2$ $(K_a = 6.3 \times 10^{-5})$ is titrated with 0.0100 M $Ba(OH)_2$. Calculate the pH **(a)** of the initial acid solution; **(b)** after the addition of 6.25 mL of 0.0100 M $Ba(OH)_2$; **(c)** at the equivalence point; **(d)** after the addition of a total of 15.00 mL of 0.0100 M $Ba(OH)_2$.
22. *Without performing detailed calculations*, determine which of the following 0.10 M aqueous solutions is the most acidic: Na_2S; $NaHSO_4$; $NaHCO_3$; Na_2HPO_4. Explain your choice.

Exercises

The Common-Ion Effect

(Use data from Table 17.3 as necessary.)

23. Calculate the *change* in pH that results from adding **(a)** 0.100 mol $NaNO_2$ to 1.00 L of 0.100 M $HNO_2(aq)$; **(b)** 0.100 mol $NaNO_3$ to 1.00 L of 0.100 M $HNO_3(aq)$. Why are the changes not the same? Explain.
24. In Example 17–8, we calculated the percent ionization of $HC_2H_3O_2$ in **(a)** 1.0 M; **(b)** 0.10 M; and **(c)** 0.010 M $HC_2H_3O_2$ solutions. Recalculate those percent ionizations if each solution also contains 0.10 M $NaC_2H_3O_2$. Explain why the results are different from those of Example 17–8.
25. Calculate $[H_3O^+]$ in a solution that is **(a)** 0.035 M HCl and 0.075 M HOCl; **(b)** 0.100 M $NaNO_2$ and 0.0550 M HNO_2; **(c)** 0.0525 M HCl and 0.0768 M $NaC_2H_3O_2$.
26. Calculate $[OH^-]$ in a solution that is **(a)** 0.0062 M $Ba(OH)_2$ and 0.0105 M $BaCl_2$; **(b)** 0.315 M $(NH_4)_2SO_4$ and 0.486 M NH_3; **(c)** 0.196 M NaOH and 0.264 M NH_4Cl.

Buffer Solutions

(Use data from Tables 17.3 and 17.4 as necessary.)

27. Indicate which of the following aqueous solutions are buffer solutions, and explain your reasoning.
(*Hint:* Consider any reactions that might occur between solution components.)
   (a) 0.100 M NaCl
   (b) 0.100 M NaCl–0.100 M NH_4Cl
   (c) 0.100 M CH_3NH_2–0.150 M $CH_3NH_3^+Cl^-$
   (d) 0.100 M HCl–0.050 M $NaNO_2$
   (e) 0.100 M HCl–0.200 M $NaC_2H_3O_2$
   (f) 0.100 M $HC_2H_3O_2$–0.125 M $NaC_3H_5O_2$
28. The $H_2PO_4^-$–HPO_4^{2-} combination plays a role in maintaining the pH of blood.
   (a) Write equations to show how a solution containing these ions functions as a buffer.
   (b) Verify that this buffer is most effective at pH 7.2.
   (c) Calculate the pH of a buffer solution in which $[H_2PO_4^-] = 0.050$ M and $[HPO_4^{2-}] = 0.150$ M.
   (*Hint:* Focus on the second step of the phosphoric acid ionization.)
29. What is the pH of a solution obtained by adding 1.15 mg of aniline hydrochloride ($C_6H_5NH_3^+Cl^-$) to 3.18 L of 0.105 M aniline ($C_6H_5NH_2$)?
   (*Hint:* Check any assumptions that you make.)
30. What is the pH of a solution prepared by dissolving 8.50 g of aniline hydrochloride ($C_6H_5NH_3^+Cl^-$) in 750 mL of 0.215 M aniline ($C_6H_5NH_2$)? Would this solution be an effective buffer? Explain.
31. You wish to prepare a buffer solution with pH = 9.45.
   (a) How many grams of $(NH_4)_2SO_4$ would you add to 425 mL of 0.258 M NH_3 to do this? Assume that the solution's volume remains constant.
   (b) Which buffer component, and how much (in grams), would you add to 0.100 L of the buffer in part (a) to change its pH to 9.30? Assume that the solution's volume remains constant.
32. You prepare a buffer solution by dissolving 2.00 g each of benzoic acid, $HC_7H_5O_2$, and sodium benzoate, $NaC_7H_5O_2$, in 750.0 mL of water.
   (a) What is the pH of this buffer? Assume that the solution's volume is 750.0 mL.
   (b) Which buffer component, and how much (in grams), would you add to the 750.0 mL of buffer solution to change its pH to 4.00?
33. If 0.55 mL of 12 M HCl is added to 0.100 L of the buffer solution in Exercise 31a, what will be the pH of the resulting solution?
34. If 0.35 mL of 15 M NH_3 is added to 0.750 L of the buffer solution in Exercise 32a, what will be the pH of the resulting solution?
35. You are asked to prepare a buffer solution with pH 3.50, and available to you are the following solutions, all 0.100 M: $HCHO_2$, $HC_2H_3O_2$, H_3PO_4, $NaCHO_2$, $NaC_2H_3O_2$, and NaH_2PO_4. Describe how you would prepare this buffer solution.
   (*Hint:* What volumes of which solutions would you use?)

36. You are asked to reduce the pH of the 0.300 L of buffer solution in Example 18-5 from 5.09 to 5.00. How many milliliters, and which of these solutions, would you use: 0.100 M NaCl, 0.150 M HCl, 0.100 M $NaC_2H_3O_2$, 0.125 M NaOH? Explain your reasoning.

37. Given 1.00 L of a solution that is 0.100 M $HC_3H_5O_2$ and 0.100 M $KC_3H_5O_2$,
(a) Over what pH range will this solution be an effective buffer?
(b) What is the buffer capacity of the solution? That is, how many millimoles of strong acid or strong base can be added to the solution before any significant change in pH occurs?

38. Given 125 mL of a solution that is 0.0500 M CH_3NH_2 and 0.0500 M $CH_3NH_3^+Cl^-$,
(a) Over what pH range will this solution be an effective buffer?
(b) What is the buffer capacity of the solution? That is, how many millimoles of strong acid or strong base can be added to the solution before any significant change in pH occurs?

39. A solution of volume 75.0 mL contains 15.5 mmol $HCHO_2$ and 8.50 mmol $NaCHO_2$.
(a) What is the pH of this solution?
(b) If 0.25 mmol $Ba(OH)_2$ is added to the solution, what will be the pH?
(c) If 1.05 mL of 12 M HCl is added to the original solution, what will be the pH?

40. A solution of volume 0.500 L contains 1.68 g NH_3 and 4.05 g $(NH_4)_2SO_4$.
(a) What is the pH of this solution?

(b) If 0.88 g NaOH is added to the solution, what will be the pH?
(c) How many milliliters of 12 M HCl must be added to 0.500 L of the original solution to change its pH to 9.00?

41. A handbook lists various procedures for preparing buffer solutions. To obtain a pH = 9.00, the handbook says to mix 36.00 mL of 0.200 M NH_3 with 64.00 mL of 0.200 M NH_4Cl.
(a) Show by calculation that the pH of this solution is 9.00.
(b) Would you expect the pH of this solution to remain at pH = 9.00 if the 100.00 mL of buffer solution were diluted to 1.00 L? To 1000 L? Explain.
(c) What will be the pH of the original 100.00 mL of buffer solution if 0.20 mL of 1.00 M HCl is added to it?
(d) What is the maximum volume of 1.00 M HCl that can be added to 100.00 mL of the original buffer solution so that the pH does not drop below 8.90?

42. An acetic acid–sodium acetate buffer can be prepared by the reaction
$$C_2H_3O_2^- + H_3O^+ \longrightarrow HC_2H_3O_2 + H_2O$$
$$\text{(from } NaC_2H_3O_2) \quad \text{(from HCl)}$$
(a) If 12.0 g $NaC_2H_3O_2$ is added to 0.300 L of 0.200 M HCl, what is the pH of the resulting solution?
(b) If 1.00 g $Ba(OH)_2$ is added to the solution in part (a), what is the new pH?
(c) What is the maximum mass of $Ba(OH)_2$ that can be neutralized by the buffer solution of part (a)?
(d) What is the pH of the solution in part (a) following the addition of 5.50 g $Ba(OH)_2$?

Acid–Base Indicators

(Use data from Tables 17.3 and 17.4 as necessary.)

43. In the use of acid–base indicators,
(a) Why is it generally sufficient to use a *single* indicator in an acid–base titration but often necessary to use *several* indicators to establish the approximate pH of a solution?
(b) Why must the quantity of indicator used in a titration be kept as small as possible?

44. The indicator methyl red has a $pK_{HIn} = 4.95$. It changes from red to yellow over the pH range from 4.4 to 6.2.
(a) If the indicator is placed in a buffer solution of pH = 4.55, what percent of the indicator will be present in the acid form, HIn, and what percent will be present in the base or anion form, In^-?
(b) Which form of the indicator has the "stronger" (that is, more visible) color—the acid (red) form or base (yellow) form? Explain.

45. Phenol red indicator changes from yellow to red in the pH range from 6.6 to 8.0. *Without making detailed calculations*, state what color the indicator will assume in each of the following solutions: (a) 0.10 M KOH; (b) 0.10 M $HC_2H_3O_2$; (c) 0.10 M NH_4NO_3; (d) 0.10 M HBr; (e) 0.10 M NaCN; (f) 0.10 M $HC_2H_3O_2$–0.10 M $NaC_2H_3O_2$.

46. Thymol blue indicator has *two* pH ranges. It changes color from red to yellow in the pH range from 1.2 to 2.8, and from yellow to blue in the pH range from 8.0 to 9.6. What is the color of the indicator in each of the following situations?
(a) The indicator is placed in 350.0 mL of 0.205 M HCl.
(b) To the solution in part (a) is added 250.0 mL of 0.500 M $NaNO_2$.
(c) To the solution in part (b) is added 150.0 mL of 0.100 M NaOH.
(d) To the solution in part (c) is added 5.00 g $Ba(OH)_2$.

47. In the titration of 10.00 mL of 0.04050 M HCl with 0.01120 M $Ba(OH)_2$ in the presence of the indicator 2,4-dinitrophenol, the solution changes from colorless to yellow when 17.90 mL of the base has been added. What is the approximate value of pK_{HIn} for 2,4-dinitrophenol? Is this a good indicator for the titration?

48. Solution (a) is 100.0 mL of 0.100 M HCl and solution (b) is 150.0 mL of 0.100 M $NaC_2H_3O_2$. A few drops of thymol blue indicator are added to each solution. What is the color of each solution? What is the color of the solution obtained when these two solutions are mixed?

Neutralization Reactions

49. A 25.00-mL sample of H_3PO_4(aq) requires 31.15 mL of 0.2420 M KOH for titration to the second equivalence point. What is the molarity of the H_3PO_4(aq)?

50. A 20.00-mL sample of H_3PO_4(aq) requires 18.67 mL of 0.1885 M NaOH for titration from the first to the second equivalence point. What is the molarity of the H_3PO_4(aq)?

51. Two aqueous solutions are mixed: 50.0 mL of 0.0150 M H_2SO_4 and 50.0 mL of 0.0385 M NaOH. What is the pH of the resulting solution?

52. Two solutions are mixed: 100.0 mL of HCl(aq) with pH 2.50 and 100.0 mL of NaOH(aq) with pH 11.00. What is the pH of the resulting solution?

Titration Curves

53. Explain why the volume of 0.100 M NaOH required to reach the equivalence point in the titration of 25.00 mL of 0.100 M HA is the same regardless of whether HA is a strong or a weak acid, yet the pH at the equivalence point is not the same.

54. Explain whether the equivalence point of each of the following titrations should be below, above, or at pH 7:
(a) $NaHCO_3$(aq) titrated with NaOH(aq); (b) HCl(aq) titrated with NH_3(aq); (c) KOH(aq) titrated with HI(aq).

55. Sketch the titration curves of the following mixtures. Indicate the initial pH and the pH corresponding to the equivalence point. Indicate the volume of titrant required to reach the equivalence point, and select a suitable indicator from Figure 18-8.
(a) 25.0 mL of 0.100 M KOH with 0.200 M HI
(b) 10.0 mL of 1.00 M NH_3 with 0.250 M HCl

56. Determine the following characteristics of the titration curve for 20.0 mL of 0.275 M NH_3(aq) titrated with 0.325 M HI(aq).
(a) the initial pH
(b) the volume of 0.325 M HI(aq) at the equivalence point
(c) the pH at the half-neutralization point
(d) the pH at the equivalence point

57. In the titration of 20.00 mL of 0.175 M NaOH, calculate the number of milliliters of 0.200 M HCl that must be added to reach a pH of (a) 12.55, (b) 10.80, (c) 4.25.
(*Hint:* Solve an algebraic equation in which the number of milliliters is x. Which reactant is in excess at each pH?)

58. In the titration of 25.00 mL of 0.100 M $HC_2H_3O_2$, calculate the number of milliliters of 0.200 M NaOH that must be added to reach a pH of (a) 3.85, (b) 5.25, (c) 11.10.
(*Hint:* Solve an algebraic equation in which the number of milliliters is x. Which reactant is in excess at each pH?)

59. Sketch a titration curve (pH versus mL of titrant) for each of the following three hypothetical weak acids when titrated with 0.100 M NaOH. Select suitable indicators for the titrations from Figure 18-8.
(*Hint:* Select a few key points at which to estimate the pH of the solution.)
(a) 10.00 mL of 0.100 M HX; $K_a = 7.0 \times 10^{-3}$
(b) 10.00 mL of 0.100 M HY; $K_a = 3.0 \times 10^{-4}$
(c) 10.00 mL of 0.100 M HZ; $K_a = 2.0 \times 10^{-8}$

60. Sketch a titration curve (pH versus mL of titrant) for each of the following hypothetical weak bases when titrated with 0.100 M HCl. (Think of these bases as involving the substitution of organic groups, R, for one of the H atoms of NH_3.) Select suitable indicators for the titrations from Figure 18-8.
(*Hint:* Select a few key points at which to estimate the pH of the solution.)
(a) 10.00 mL of 0.100 M RNH_2; $K_b = 1 \times 10^{-3}$
(b) 10.00 mL of 0.100 M $R'NH_2$; $K_b = 3 \times 10^{-6}$
(c) 10.00 mL of 0.100 M $R''NH_2$; $K_b = 7 \times 10^{-8}$

61. For the titration of 25.00 mL of 0.100 M NaOH with 0.100 M HCl, calculate the pOH at a few representative points in the titration, sketch the titration curve of pOH versus volume of titrant, and show that it has exactly the same form as Figure 18-9. Then, using this curve and the simplest method possible, sketch the titration curve of pH versus volume of titrant.

62. For the titration of 25.00 mL 0.100 M NH_3 with 0.100 M HCl, calculate the pOH at a few representative points in the titration, sketch the titration curve of pOH versus volume of titrant, and show that it has exactly the same form as Figure 18-11. Then, using this curve and the simplest method possible, sketch the titration curve of pH versus volume of titrant.

Salts of Polyprotic Acids

(Use data from Table 17.4 or Appendix D as necessary.)

63. Is a solution that is 0.10 M Na_2S(aq) likely to be acidic, basic, or pH neutral? Explain.

64. Is a solution of sodium dihydrogen citrate, NaH_2Cit, likely to be acidic, basic, or neutral? Explain. Citric acid, H_3Cit, is $H_3C_6H_5O_7$.

65. Sodium phosphate, Na_3PO_4, is made commercially by first neutralizing phosphoric acid with sodium carbonate to obtain Na_2HPO_4. The Na_2HPO_4 is further neutralized to Na_3PO_4 with NaOH.
(a) Write net ionic equations for these reactions.
(b) Na_2CO_3 is a much cheaper base than is NaOH. Why do you suppose that NaOH must be used as well as Na_2CO_3 to produce Na_3PO_4?

66. Both sodium hydrogen carbonate (sodium bicarbonate) and sodium hydroxide can be used to neutralize acid spills. What is the pH of 1.00 M $NaHCO_3(aq)$ and of 1.00 M $NaOH(aq)$? On a per-liter basis, do these two solutions have an equal capacity to neutralize acids? Explain. On a per-gram basis, do the two solids, $NaHCO_3(s)$ and $NaOH(s)$, have an equal capacity to neutralize acids? Explain. Why do you suppose that $NaHCO_3$ is often preferred to NaOH in neutralizing acid spills?

67. The pH of a solution of 19.5 g of malonic acid in 0.250 L is 1.47. The pH of a 0.300 M solution of sodium hydrogen malonate is 4.26. What are the values of K_{a_1} and K_{a_2} for malonic acid?

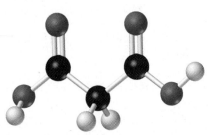

Malonic acid

68. The ionization constants of *ortho*-phthalic acid are $K_{a_1} = 1.1 \times 10^{-3}$ and $K_{a_2} = 3.9 \times 10^{-6}$.

(1) $H_2C_8H_4O_4 + H_2O \rightleftharpoons H_3O^+ + HC_8H_4O_4^-$
(2) $HC_8H_4O_4^- + H_2O \rightleftharpoons H_3O^+ + C_8H_4O_4^{2-}$

What are the pH values of the following aqueous solutions: **(a)** 0.350 M potassium hydrogen *ortho*-phthalate; **(b)** a solution containing 36.35 g potassium *ortho*-phthalate per liter?

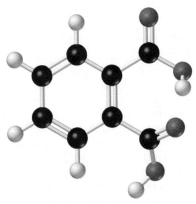

ortho-Phthalic acid

General Acid–Base Equilibria

69. What stoichiometric concentration of the indicated substance is required to obtain an aqueous solution with the pH value shown: **(a)** $Ba(OH)_2$ for pH = 11.88; **(b)** $HC_2H_3O_2$ in 0.294 M $NaC_2H_3O_2$ for pH = 4.52?

70. What stoichiometric concentration of the indicated substance is required to obtain an aqueous solution with the pH value shown: **(a)** aniline, $C_6H_5NH_2$, for pH = 8.95; **(b)** NH_4Cl for pH = 5.12?

71. Using appropriate equilibrium constants but *without doing detailed calculations*, determine whether a solution can be simultaneously
(a) 0.10 M NH_3 and 0.10 M NH_4Cl, with pH = 6.07
(b) 0.10 M $NaC_2H_3O_2$ and 0.058 M HI
(c) 0.10 M KNO_2 and 0.25 M KNO_3
(d) 0.050 M $Ba(OH)_2$ and 0.65 M NH_4Cl

(e) 0.018 M $HC_7H_5O_2$ and 0.018 M $NaC_7H_5O_2$, with pH = 4.20
(f) 0.68 M KCl, 0.42 M KNO_3, 1.2 M NaCl, and 0.55 M $NaC_2H_3O_2$, with pH = 6.4

72. This single equilibrium equation applies to different phenomena described in this or the preceding chapter.

$$HC_2H_3O_2 + H_2O \rightleftharpoons H_3O^+ + C_2H_3O_2^-$$

Of these four phenomena, ionization of pure acid, common-ion effect, buffer solution, and hydrolysis, indicate which occurs if
(a) $[H_3O^+]$ and $[HC_2H_3O_2]$ are high, but $[C_2H_3O_2^-]$ is very low.
(b) $[C_2H_3O_2^-]$ is high, but $[HC_2H_3O_2]$ and $[H_3O^+]$ are very low.
(c) $[HC_2H_3O_2]$ is high, but $[H_3O^+]$ and $[C_2H_3O_2^-]$ are low.
(d) $[HC_2H_3O_2]$ and $[C_2H_3O_2^-]$ are high, but $[H_3O^+]$ is low.

Integrative and Advanced Exercises

73. Sodium hydrogen sulfate, $NaHSO_4$, is an acidic salt with a number of uses, such as metal pickling (removal of surface deposits). $NaHSO_4$ is made by the reaction of H_2SO_4 with NaCl. To determine the percent NaCl impurity in $NaHSO_4$, a 1.016-g sample is titrated with NaOH(aq); 36.56 mL of 0.225 M NaOH is required.
(a) Write the net ionic equation for the neutralization reaction.
(b) What is the percent NaCl in the sample titrated?
(c) Select a suitable indicator(s) from Figure 18-8.

74. You are given 250.0 mL of 0.100 M $HC_3H_5O_2$ (propionic acid, $K_a = 1.35 \times 10^{-5}$). You wish to adjust its pH by adding an appropriate solution. What volume would you add of **(a)** 1.00 M HCl to lower the pH to 1.00; **(b)** 1.00 M $NaC_3H_5O_3$ to raise the pH to 4.00; **(c)** water to raise the pH by 0.15 unit?

75. Even though the carbonic acid–hydrogen carbonate buffer system is crucial to the maintenance of the pH of blood, it has no practical use as a laboratory buffer solution. Can you think of a reason(s) for this?
(*Hint:* Refer to data in Exercise 94 of Chapter 17.)

76. Thymol blue in its acid range is not a suitable indicator for the titration of HCl by NaOH. Suppose that a student uses thymol blue by mistake in the titration of Figure 18-9 and that the indicator end point is taken to be pH = 2.0.
(a) Would there be a sharp color change, produced by the addition of a single drop of NaOH(aq)?
(b) Approximately what percent of the HCl remains unneutralized at pH = 2.0?

77. Rather than calculate the pH for different volumes of titrant, a titration curve can be established by calculating the volume of titrant required to reach certain pH values. Determine the volumes of 0.100 M NaOH required to reach the following pH values in the titration of 20.00 mL of 0.150 M HCl: pH = **(a)** 2.00; **(b)** 3.50; **(c)** 5.00; **(d)** 10.50; **(e)** 12.00. Then plot the titration curve.

78. Use the method of Exercise 77 to determine the volume of titrant required to reach the indicated pH values in the following titrations.
(a) 25.00 mL of 0.250 M NaOH titrated with 0.300 M HCl; pH = 13.00, 12.00, 10.00, 4.00, 3.00
(b) 50.00 mL of 0.0100 M benzoic acid ($HC_7H_5O_2$) titrated with 0.0500 M KOH: pH = 4.50, 5.50, 11.50 ($K_a = 6.3 \times 10^{-5}$)

79. A buffer solution can be prepared by starting with a weak acid, HA, and converting some of the weak acid to its salt (for example, NaA) by titration with a strong base. The *fraction* of the original acid that is converted to the salt is designated f.
(a) Derive an equation similar to equation (18.2) but expressed in terms of f rather than concentrations.
(b) What is the pH at the point in the titration of phenol, HOC_6H_5, at which $f = 0.27$ (pK_a of phenol = 10.00)?

80. You are asked to prepare a KH_2PO_4–Na_2HPO_4 solution that has the same pH as human blood, 7.40.
(a) What should be the ratio of concentrations $[HPO_4^{2-}]/[H_2PO_4^-]$ in this solution?
(b) Suppose you have to prepare 1.00 L of the solution described in part **(a)** and that this solution must be isotonic with blood (with the same osmotic pressure as blood). What mass of KH_2PO_4 and of $Na_2HPO_4 \cdot 12H_2O$ would you use? (*Hint:* Refer to the definition of isotonic on page 557. Recall that a solution of NaCl with 9.2 g NaCl/L solution is isotonic with blood, and assume that NaCl is completely ionized in aqueous solution.)

81. You are asked to bring the pH of 0.500 L of 0.500 M NH_4Cl(aq) to 7.00. How many drops (1 drop = 0.05 mL) of which of the following solutions would you use: 10.0 M HCl or 10.0 M NH_3?

82. Because an acid–base indicator is a weak acid, it can be titrated with a strong base. Suppose you titrate 25.00 mL of a 0.0100 M solution of the indicator p-nitrophenol, $HC_6H_4NO_3$, with 0.0200 M NaOH. The pK_a of p-nitrophenol is 7.15, and it changes from colorless to yellow in the pH range from 5.6 to 7.6.
(a) Sketch the titration curve for this titration.
(b) Show the pH range over which p-nitrophenol changes color.

(c) Explain why p-nitrophenol cannot serve as its own indicator in this titration.

83. The neutralization of NaOH by HCl is represented in equation (1), and the neutralization of NH_3 by HCl in equation (2).
(1) $OH^- + H_3O^+ \rightleftharpoons 2 H_2O \qquad K = ?$
(2) $NH_3 + H_3O^+ \rightleftharpoons NH_4^+ + H_2O \qquad K = ?$
(a) Determine the equilibrium constant K for each reaction.
(b) Explain why each neutralization reaction can be considered to go to completion.

84. The titration of a weak acid by a weak base is not a satisfactory procedure because the pH does not increase sharply at the equivalence point. Demonstrate this fact by sketching a titration curve for the neutralization of 10.00 mL of 0.100 M $HC_2H_3O_2$ with 0.100 M NH_3.

85. At times, a salt of a weak base can be titrated by a strong base. Use appropriate data from the text to sketch a titration curve for the titration of 10.00 mL of 0.0500 M $C_6H_5NH_3^+Cl^-$ with 0.100 M NaOH.

86. Sulfuric acid is a diprotic acid, strong in the first ionization step and weak in the second ($K_{a_2} = 1.1 \times 10^{-2}$). By using appropriate calculations, determine whether it is feasible to titrate 10.00 mL of 0.100 M H_2SO_4 to two distinct equivalence points with 0.100 M NaOH.

87. Carbonic acid is a weak diprotic acid (H_2CO_3) with $K_{a_1} = 4.43 \times 10^{-7}$ and $K_{a_2} = 4.73 \times 10^{-11}$. The equivalence points for the titration come at approximately pH 4 and 9. Suitable indicators for use in titrating carbonic acid or carbonate solutions are methyl orange and phenolphthalein.
(a) Sketch the titration curve that would be obtained in titrating a sample of $NaHCO_3$(aq) with 1.00 M HCl.
(b) Sketch the titration curve for Na_2CO_3(aq) with 1.00 M HCl.
(c) What volume of 0.100 M HCl is required for the complete neutralization of 1.00 g $NaHCO_3$(s)?
(d) What volume of 0.100 M HCl is required for the complete neutralization of 1.00 g Na_2CO_3(s)?
(e) A sample of NaOH contains a small amount of Na_2CO_3. For titration to the phenolphthalein end point, 0.1000 g of this sample requires 23.98 mL of 0.1000 M HCl. An additional 0.78 mL is required to reach the methyl orange end point. What is the percent Na_2CO_3, by mass, in the sample?

88. Piperazine is a diprotic weak base used as a corrosion inhibitor and an insecticide. Its ionization is described by the following equations.

$HN(C_4H_8)NH + H_2O \rightleftharpoons [HN(C_4H_8)NH_2]^+ + OH^-$
$[HN(C_4H_8)NH_2]^+ + H_2O \rightleftharpoons$
$\qquad\qquad\qquad\qquad [H_2N(C_4H_8)NH_2]^{2+} + OH^-$

The piperazine used commercially is a hexahydrate, $C_4H_{10}N_2 \cdot 6H_2O$. A 1.00-g sample of this hexahydrate is dissolved in 100.0 mL of water and titrated with 0.500 M HCl. Sketch a titration curve for this titration, indicating **(a)** the initial pH; **(b)** the pH at the half-neutralization point of the first neutralization; **(c)** the volume of HCl(aq) required to reach the first equivalence point; **(d)** the pH at the first equivalence point; **(e)** the pH when the second step of the neutralization is half completed; **(e)** the pH at the half-neutralization point of the second neutralization; **(f)** the volume of HCl(aq) required to reach the second equivalence point; **(g)** the pH at the second equivalence point.

89. Complete the derivation of equation (18.5) outlined in the Are You Wondering feature on page 736. Then derive equation (18.6).

90. Explain why equation (18.5) fails when applied to dilute solutions—for example, when you calculate the pH of 0.010 M NaH_2PO_4.
    (*Hint:* Refer also to Exercise 89.)

91. A solution is prepared that is 0.150 M $HC_2H_3O_2$ and 0.250 M $NaCHO_2$.
    (a) Show that this is a buffer solution.
    (b) Calculate the pH of this buffer solution.
    (c) What is the final pH if 1.00 L of 0.100 M HCl is added to 1.00 L of this buffer solution?

92. A series of titrations of lactic acid, $CH_3CH(OH)COOH$ ($pK_a = 3.86$) is planned. About 1.00 mmol of the acid will be titrated with NaOH(aq) to a final volume of about 100 mL at the equivalence point. **(a)** Which acid–base indicator from Figure 18-8 would you select for the titration? To assist in locating the equivalence point in the titration, a buffer solution is to be prepared having the same pH as that at the equivalence point. A few drops of the indicator in this buffer will produce the color to be matched in the titrations. **(b)** Which of the following combinations would be suitable for the buffer solutions: $HC_2H_3O_2$–$C_2H_3O_2^-$, $H_2PO_4^-$–HPO_4^{2-}, or NH_4^+–NH_3? **(c)** What ratio of conjugate base to acid is required in the buffer?

93. Hydrogen peroxide, H_2O_2, is a somewhat stronger acid than water. Its ionization is represented by the equation

$$H_2O_2 + H_2O \rightleftharpoons H_3O^+ + HO_2^-$$

In 1912, the following experiments were performed to obtain an approximate value of pK_a for this ionization at 0 °C. A sample of H_2O_2 was shaken together with a mixture of water and 1-pentanol. The mixture settled into two layers. At equilibrium, the hydrogen peroxide had distributed itself between the two layers such that the water layer contained 6.78 times as much H_2O_2 as the 1-pentanol layer. In a second experiment, a sample of H_2O_2 was shaken together with 0.250 M NaOH(aq) and 1-pentanol. At equilibrium, the concentration of H_2O_2 was 0.00357 M in the 1-pentanol

layer and 0.259 M in the aqueous layer. In a third experiment, a sample of H_2O_2 was brought to equilibrium with a mixture of 1-pentanol and 0.125 M NaOH(aq); the concentrations of the hydrogen peroxide were 0.00198 M in the 1-pentanol and 0.123 M in the aqueous layer. For water at 0 °C, $pK_w = 14.94$. Find an approximate value of pK_a for H_2O_2 at 0 °C.
    (*Hint:* The hydrogen peroxide concentration in the aqueous layers is the total concentration of H_2O_2 and HO_2^-. Assume that the 1-pentanol solutions contain no ionic species.)

94. Sodium ammonium hydrogen phosphate, $NaNH_4HPO_4$, is a salt in which one of the ionizable H atoms of H_3PO_4 is replaced by Na^+, another is replaced by NH_4^+, and the third remains in the anion HPO_4^{2-}. Calculate the pH of 0.100 M $NaNH_4HPO_4$(aq).
    (*Hint:* You can use the general method introduced on page 688 of Chapter 17. First, identify all the species that could be present and the equilibria involving these species. Then identify the two equilibrium expressions that will predominate, and eliminate all the species whose concentrations are likely to be negligible. At that point, only a few algebraic manipulations are required.)

95. Consider a solution containing two weak monoprotic acids with dissociation constants K_{HA} and K_{HB}. Find the charge balance equation for this system, and use it to derive an expression that gives the concentration of H_3O^+ as a function of the concentrations of HA and HB and the various constants.

96. Calculate the pH of a solution that is 0.050 M acetic acid and 0.010 M phenylacetic acid.

97. A very common buffer agent used in the study of biochemical processes is the weak base TRIS, $(HOCH_2)_3CNH_2$, which has a pK_b of 5.91 at 25 °C. A student is given a sample of the hydrochloride of TRIS, together with standard solutions of 10 M NaOH and HCl.
    (a) Using TRIS, how might the student prepare 1 L of a buffer of pH = 7.79?
    (b) In one experiment, 30 mmol of protons are released into 500 mL of the buffer prepared in part **(a)**. Is the capacity of the buffer sufficient? What is the resulting pH?
    (c) Another student accidentally adds 20 mL of 10 M HCl to 500 mL of the buffer solution prepared in part **(a)**. Is the buffer ruined? If so, how could the buffer be regenerated?

Feature Problems

98. The graph that follows, which is related to a titration curve, shows the fraction (f) of the stoichiometric amount of acetic acid present as nonionized $HC_2H_3O_2$ and as acetate ion, $C_2H_3O_2^-$, as a function of the pH of the solution containing these species.
    (a) Explain the significance of the point at which the two curves cross. What are the fractions and the pH at that point?

    (b) Sketch a comparable set of curves for carbonic acid, H_2CO_3.
    (*Hint:* How many carbonate-containing species should appear in the graph? How many points of intersection should there be? at what pH values?)
    (c) Sketch a comparable set of curves for phosphoric acid, H_3PO_4.

(Hint: How many phosphate-containing species should appear in the graph? How many points of intersection should there be? at what pH values?)

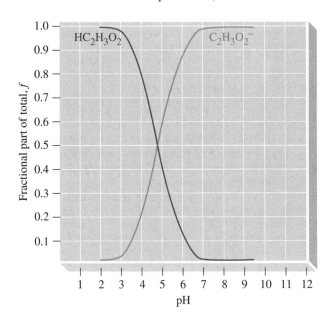

99. In some cases, the titration curve for a mixture of *two* acids has the same appearance as that for a single acid; in other cases it does not.

(a) Sketch the titration curve (pH versus volume of titrant) for the titration with 0.200 M NaOH of 25.00 mL of a solution that is 0.100 M in HCl and 0.100 M in HNO_3. Does this curve differ in any way from what would be obtained in the titration of 25.00 mL of 0.200 M HCl with 0.200 M NaOH? Explain.

(b) The titration curve shown was obtained when 10.00 mL of a solution containing both HCl and H_3PO_4 was titrated with 0.216 M NaOH. From this curve, determine the stoichiometric molarities of both the HCl and the H_3PO_4.

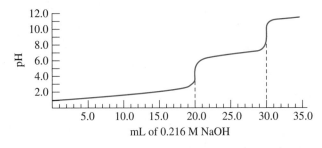

(c) A 10.00-mL solution that is 0.0400 M H_3PO_4 and 0.0150 M NaH_2PO_4 is titrated with 0.0200 M NaOH. Sketch the titration curve.

100. Amino acids contain both an acidic carboxylic acid group (—COOH) and a basic amino group (—NH_2). The amino group can be *protonated* (that is, it has an extra proton attached) in a strongly acidic solution. This produces a diprotic acid, of the form H_2A^+, as exemplified by the protonated amino acid alanine.

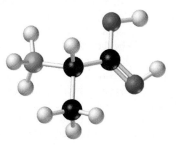

Protonated alanine

The protonated amino acid has two ionizable protons that can be titrated with OH^-.

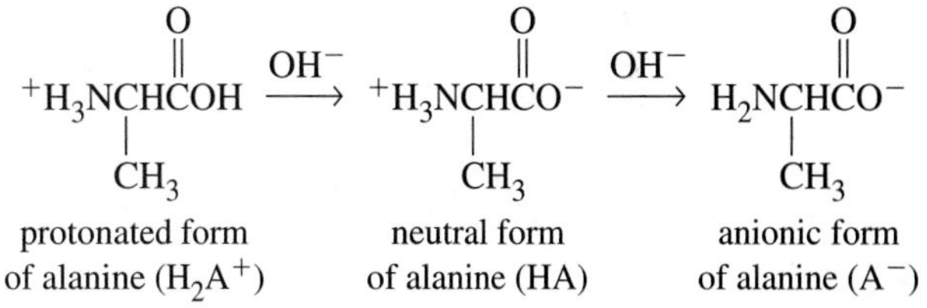

$$\underset{\substack{\text{protonated form} \\ \text{of alanine } (H_2A^+)}}{\overset{O}{\underset{CH_3}{\overset{\parallel}{^+H_3NCHCOH}}}} \xrightarrow{OH^-} \underset{\substack{\text{neutral form} \\ \text{of alanine (HA)}}}{\overset{O}{\underset{CH_3}{\overset{\parallel}{^+H_3NCHCO^-}}}} \xrightarrow{OH^-} \underset{\substack{\text{anionic form} \\ \text{of alanine } (A^-)}}{\overset{O}{\underset{CH_3}{\overset{\parallel}{H_2NCHCO^-}}}}$$

For the —COOH group, $pK_{a_1} = 2.34$; for the —NH_3^+ group, $pK_{a_2} = 9.69$. Consider the titration of a 0.500 M solution of alanine hydrochloride with 0.500 M NaOH solution. What is the pH of **(a)** the 0.500 M alanine hydrochloride; **(b)** the solution at the first half-neutralization point; **(c)** the solution at the first equivalence point?

The dominant form of alanine present at the first equivalence point is electrically neutral, despite the positive charge and negative charge it possesses. The point at which the neutral form is produced is called the *isoelectric point*. Confirm that the pH at the isoelectric point is

$$pH = \tfrac{1}{2} \left(pK_{a_1} + K_{a_2} \right)$$

What is the pH of the solution **(d)** halfway between the first and second equivalence points? **(e)** at the second equivalence point?

(f) Calculate the pH values of the solutions when the following volumes of the 0.500 M NaOH have been added to 50 mL of the 0.500 M alanine hydrochloride solution: 10.0 mL, 20.0 mL, 30.0 mL, 40.0 mL, 50.0 mL, 60.0 mL, 70.0 mL, 80.0 mL, 90.0 mL, 100.0 mL, and 110.0 mL.

(g) Sketch the titration curve for the 0.500 M solution of alanine hydrochloride, and label significant points on the curve.

eMedia Exercises

101. In the **Common Ion Effect** animation *(eChapter 18-1),* the influence of adding a second salt to solution is shown. **(a)** What are the requirements on the second ionic compound if it is to have an effect on the existing equilibrium? **(b)** What would be the influence of removing one of the ions present in the initial solution?

102. By exploring the **Buffer Solutions activity** *(eChapter 18-2),* **(a)** determine the buffer capacity (in terms of moles of an added component) of an acetic acid/sodium acetate buffer containing 0.10 M of each component. **(b)** What is the effect on the buffer capacity if the initial concentrations of the buffer components are increased?

103. The color changes of two natural acid-base indicators are shown in the **Natural Indicators** movie *(eChapter 18-3).* Estimate the pH range over which these indicators would be useful in quantitative titrations.

104. The titration of an acidic solution with a base is shown in the **Acid–Base Titration** animation *(eChapter 18-4).* **(a)** In stoichiometric terms, describe why there is a change in slope of the titration curve as the reaction progresses. **(b)** How many regions of the curve with roughly constant slope are present? **(c)** Why are the slopes dramatically different between these regions?

19

Solubility and Complex-Ion Equilibria

Contents

The White Cliffs of Dover, England, are made of chalk, a soft form of limestone, $CaCO_3$. These beautiful cliffs were formed about 65 million to 100 million years ago. They have persisted all this time as a result of the low aqueous solubility of $CaCO_3$. Equilibria between sparingly soluble solutes and their ions in solution, expressed through the solubility product constant expression, K_{sp}, are discussed in this chapter.

The dissolution and precipitation of limestone ($CaCO_3$) underlie a variety of natural phenomena, such as the formation of limestone caverns. Whether a solution containing Ca^{2+} and CO_3^{2-} ions undergoes precipitation depends on the concentrations of these ions. In turn, the CO_3^{2-} ion concentration depends on the pH of the solution. To develop a better understanding of the conditions under which $CaCO_3$ dissolves or precipitates, we must consider equilibrium relationships between Ca^{2+} and CO_3^{2-}, and between CO_3^{2-}, H_3O^+, and HCO_3^-. This requirement suggests a need to combine ideas about acid–base equilibria from Chapters 17 and 18 with ideas about the new types of equilibria to be introduced in this chapter.

Silver chloride is a familiar precipitate in the general chemistry laboratory, yet it does *not* precipitate from a solution that has a moderate to high concentration of $NH_3(aq)$. The silver ion and ammonia combine instead to form a species, called a *complex ion*, that remains in solution. Complex-ion formation and equilibria involving complex ions are additional topics discussed in this chapter.

19-1 The Solubility Product Constant, K_{sp}

Gypsum, $CaSO_4 \cdot 2H_2O$, is an important calcium mineral. It is slightly soluble in water, and groundwater that comes into contact with gypsum often contains some dissolved calcium sulfate. This water cannot be used for certain applications, such as in evaporative cooling systems in power plants, because the calcium sulfate might precipitate from the water and block pipes. The equilibrium between $Ca^{2+}(aq)$ and $SO_4^{2-}(aq)$ and undissolved $CaSO_4(s)$ can be represented as

$$CaSO_4(s) \rightleftharpoons Ca^{2+}(aq) + SO_4^{2-}(aq)$$

We can write the equilibrium constant expression for this equilibrium as we learned to do in Chapter 16 (page 636)—that is, by including concentration terms for ions that are in solution, but not for the pure solid solute. Also, it is customary to represent the equilibrium constant by a special symbol, K_{sp}.

$$K_{sp} = [Ca^{2+}][SO_4^{2-}] = 9.1 \times 10^{-6} \text{ (at 25 °C)} \tag{19.1}$$

The **solubility product constant**, K_{sp}, is the equilibrium constant for the equilibrium established between a solid solute and its ions in a saturated solution. Table 19.1 lists several K_{sp} values and the solubility equilibria to which they apply.

TABLE 19.1 Several Solubility Product Constants at 25 °C*

Solute	Solubility Equilibrium	K_{sp}
Aluminum hydroxide	$Al(OH)_3(s) \rightleftharpoons Al^{3+}(aq) + 3\,OH^-(aq)$	1.3×10^{-33}
Barium carbonate	$BaCO_3(s) \rightleftharpoons Ba^{2+}(aq) + CO_3^{2-}(aq)$	5.1×10^{-9}
Barium sulfate	$BaSO_4(s) \rightleftharpoons Ba^{2+}(aq) + SO_4^{2-}(aq)$	1.1×10^{-10}
Calcium carbonate	$CaCO_3(s) \rightleftharpoons Ca^{2+}(aq) + CO_3^{2-}(aq)$	2.8×10^{-9}
Calcium fluoride	$CaF_2(s) \rightleftharpoons Ca^{2+}(aq) + 2\,F^-(aq)$	5.3×10^{-9}
Calcium sulfate	$CaSO_4(s) \rightleftharpoons Ca^{2+}(aq) + SO_4^{2-}(aq)$	9.1×10^{-6}
Chromium(III) hydroxide	$Cr(OH)_3(s) \rightleftharpoons Cr^{3+}(aq) + 3\,OH^-(aq)$	6.3×10^{-31}
Iron(III) hydroxide	$Fe(OH)_3(s) \rightleftharpoons Fe^{3+}(aq) + 3\,OH^-(aq)$	4×10^{-38}
Lead(II) chloride	$PbCl_2(s) \rightleftharpoons Pb^{2+}(aq) + 2\,Cl^-(aq)$	1.6×10^{-5}
Lead(II) chromate	$PbCrO_4(s) \rightleftharpoons Pb^{2+}(aq) + CrO_4^{2-}(aq)$	2.8×10^{-13}
Lead(II) iodide	$PbI_2(s) \rightleftharpoons Pb^{2+}(aq) + 2\,I^-(aq)$	7.1×10^{-9}
Magnesium carbonate	$MgCO_3(s) \rightleftharpoons Mg^{2+}(aq) + CO_3^{2-}(aq)$	3.5×10^{-8}
Magnesium fluoride	$MgF_2(s) \rightleftharpoons Mg^{2+}(aq) + 2\,F^-(aq)$	3.7×10^{-8}
Magnesium hydroxide	$Mg(OH)_2(s) \rightleftharpoons Mg^{2+}(aq) + 2\,OH^-(aq)$	1.8×10^{-11}
Magnesium phosphate	$Mg_3(PO_4)_2(s) \rightleftharpoons 3\,Mg^{2+}(aq) + 2\,PO_4^{3-}(aq)$	1×10^{-25}
Mercury(I) chloride	$Hg_2Cl_2(s) \rightleftharpoons Hg_2^{2+}(aq) + 2\,Cl^-(aq)$	1.3×10^{-18}
Silver bromide	$AgBr(s) \rightleftharpoons Ag^+(aq) + Br^-(aq)$	5.0×10^{-13}
Silver carbonate	$Ag_2CO_3(s) \rightleftharpoons 2\,Ag^+(aq) + CO_3^{2-}(aq)$	8.5×10^{-12}
Silver chloride	$AgCl(s) \rightleftharpoons Ag^+(aq) + Cl^-(aq)$	1.8×10^{-10}
Silver chromate	$Ag_2CrO_4(s) \rightleftharpoons 2\,Ag^+(aq) + CrO_4^{2-}(aq)$	1.1×10^{-12}
Silver iodide	$AgI(s) \rightleftharpoons Ag^+(aq) + I^-(aq)$	8.5×10^{-17}
Strontium carbonate	$SrCO_3(s) \rightleftharpoons Sr^{2+}(aq) + CO_3^{2-}(aq)$	1.1×10^{-10}
Strontium sulfate	$SrSO_4(s) \rightleftharpoons Sr^{2+}(aq) + SO_4^{2-}(aq)$	3.2×10^{-7}

*A more extensive listing of K_{sp} values is given in Appendix D.

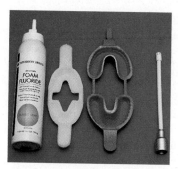

▲ A fluoride treatment kit for dental care.

EXAMPLE 19-1

Writing Solubility Product Constant Expressions for Slightly Soluble Solutes. Write the solubility product constant expression for the solubility equilibrium of

(a) Calcium fluoride, CaF_2 (one of the products formed when a fluoride treatment is applied to teeth)

(b) Copper arsenate, $Cu_3(AsO_4)_2$ (used as an insecticide and fungicide)

Solution

The K_{sp} expression is formulated for the ionic species appearing in the equation for the solubility equilibrium. This equation is written for one mole of the slightly soluble solute. That is, the coefficient "1" is understood for the slightly soluble solute. The coefficients for the ions in solution are whatever is needed to balance the equation. The coefficients then establish the powers to which the ion concentrations are raised in the K_{sp} expression.

(a) $\qquad CaF_2(s) \rightleftharpoons Ca^{2+}(aq) + 2\,F^-(aq) \qquad K_{sp} = [Ca^{2+}][F^-]^2$

(b) $Cu_3(AsO_4)_2(s) \rightleftharpoons 3\,Cu^{2+}(aq) + 2\,AsO_4^{3-}(aq) \quad K_{sp} = [Cu^{2+}]^3[AsO_4^{3-}]^2$

Practice Example A: Write the solubility product constant expression for (a) $MgCO_3$ (one of the components of dolomite, a form of limestone), and (b) Ag_3PO_4 (used in photographic emulsions).

Practice Example B: A handbook lists $K_{sp} = 1 \times 10^{-7}$ for calcium hydrogen phosphate, a substance used in dentifrices and as an animal feed supplement. Write (a) the equation for the solubility equilibrium and (b) the solubility product constant expression for this slightly soluble solute.

19-2   The Relationship Between Solubility and K_{sp}

Is there a relationship between the solubility product constant, K_{sp}, of a solute and the solute's *molar solubility*—its molarity in a saturated aqueous solution? As shown in Examples 19-2 and 19-3, there is a definite relationship between them. As we will point out in Section 19-4, calculations involving K_{sp} are generally more subject to error than are those involving other equilibrium constants, but the results are suitable for many purposes. In Example 19-2, we start with an experimentally determined solubility and obtain a value of K_{sp}.

EXAMPLE 19-2

Calculating K_{sp} of a Slightly Soluble Solute from Its Solubility. A handbook lists the aqueous solubility of $CaSO_4$ at 25 °C as 0.20 g $CaSO_4$/100 mL. What is the K_{sp} of $CaSO_4$ at 25 °C?

$$CaSO_4(s) \rightleftharpoons Ca^{2+}(aq) + SO_4^{2-}(aq) \qquad K_{sp} = ?$$

Solution

We need to construct a conversion pathway that begins with finding $[Ca^{2+}]$ and $[SO_4^{2-}]$, which we can then substitute into the K_{sp} expression.

$$\text{g } CaSO_4/100 \text{ mL} \longrightarrow \text{mol } CaSO_4/\text{L} \longrightarrow [Ca^{2+}] \text{ and } [SO_4^{2-}] \longrightarrow K_{sp}$$

In the first step, we obtain the molar solubility of $CaSO_4$. (Note that in this setup we have replaced 100 mL by 0.100 L.)

$$\text{mol } CaSO_4/\text{L satd soln} = \frac{0.20 \text{ g } CaSO_4}{0.100 \text{ L soln}} \times \frac{1 \text{ mol } CaSO_4}{136 \text{ g } CaSO_4}$$

$$= 0.015 \text{ M } CaSO_4$$

The key factors in the next step are shown in blue. They indicate that one mole of Ca^{2+} and one mole of SO_4^{2-} appear in solution for each mole of $CaSO_4$ that dissolves.

$$[Ca^{2+}] = \frac{0.015 \text{ mol } CaSO_4}{1 \text{ L}} \times \frac{1 \text{ mol } Ca^{2+}}{1 \text{ mol } CaSO_4} = 0.015 \text{ M}$$

$$[SO_4^{2-}] = \frac{0.015 \text{ mol } CaSO_4}{1 \text{ L}} \times \frac{1 \text{ mol } SO_4^{2-}}{1 \text{ mol } CaSO_4} = 0.015 \text{ M}$$

Finally, we substitute these ion concentrations into the solubility product expression.

$$K_{sp} = [Ca^{2+}][SO_4^{2-}] = (0.015)(0.015) = 2.3 \times 10^{-4}$$

▶ The rather large discrepancy between this result and the K_{sp} value in equation (19.1) is explained on page 757.

Practice Example A: A handbook lists the aqueous solubility of AgOCN as 7 mg/100 mL at 20 °C. What is the K_{sp} of AgOCN at 20 °C?

Practice Example B: A handbook lists the aqueous solubility of lithium phosphate at 18 °C as 0.034 g Li_3PO_4/100 mL soln. What is the K_{sp} of Li_3PO_4 at 18 °C?

The "inverse" of Example 19-2 is the calculation of the solubility of a solute from its K_{sp} value. When we do this, as in Example 19-3, the result is always a molar solubility—a molarity. If we need the solubility in units other than moles per liter, as in Practice Example 19-3B, additional conversions are required.

EXAMPLE 19-3

Calculating the Solubility of a Slightly Soluble Solute from Its K_{sp} Value. Lead iodide, PbI_2, is a dense, golden yellow, "insoluble" solid used in bronzing and in ornamental work requiring a golden color (such as mosaic gold). Calculate the molar solubility of lead iodide in water at 25 °C, given that its $K_{sp} = 7.1 \times 10^{-9}$.

Solution

The solubility equilibrium equation

$$PbI_2(s) \rightleftharpoons Pb^{2+}(aq) + 2\, I^-(aq)$$

shows that for each mole of PbI_2 that dissolves, *one* mole of Pb^{2+} and *two* moles of I^- appear in solution. If we let s represent the number of moles of PbI_2 dissolved per liter of saturated solution, we have,

$$[Pb^{2+}] = s \quad \text{and} \quad [I^-] = 2s$$

These concentrations must also satisfy the K_{sp} expression.

$$K_{sp} = [Pb^{2+}][I^-]^2 = (s)(2s)^2 = 7.1 \times 10^{-9}$$
$$4s^3 = 7.1 \times 10^{-9}$$
$$s^3 = 1.8 \times 10^{-9}$$
$$s = (1.8 \times 10^{-9})^{1/3} = 1.2 \times 10^{-3}$$
$$= \text{molar solubility of } PbI_2 = 1.2 \times 10^{-3} \text{ M}$$

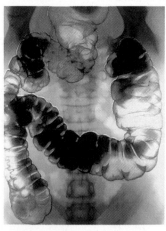

▲ Barium sulfate, $BaSO_4$, is a good absorber of X rays. As a component of a "barium milk shake," $BaSO_4$ coats the intestinal tract so that this soft tissue will show up when X-rayed. Even though Ba^{2+}(aq) is poisonous, $BaSO_4$(s) is harmless because its aqueous solubility is very low.

Practice Example A: The K_{sp} of $Fe(OH)_3$ at 25 °C is 4×10^{-38}. What is the molar solubility of $Fe(OH)_3$ in H_2O at 25 °C?

Practice Example B: How many milligrams of $BaSO_4$ are dissolved in a 225-mL sample of saturated $BaSO_4$(aq)? $K_{sp} = 1.1 \times 10^{-10}$.

Are You Wondering...

When comparing molar solubilities, is a solute with a larger value of K_{sp} always more soluble than one with a smaller value?

If the solutes being compared are of the same type (all of the type MX, MX_2, M_2X, ...), their molar solubilities will be related in the same way as their K_{sp} values. That is, the solute with the largest K_{sp} value will have the greatest molar solubility. Thus, AgCl ($K_{sp} = 1.8 \times 10^{-10}$) is more soluble than AgBr ($K_{sp} = 5.0 \times 10^{-13}$). For these particular solutes, the molar solubility is $s = \sqrt{K_{sp}}$.

If the solutes are *not* of the same type, you'll have to calculate, or at least estimate, each molar solubility and compare the results. Thus, even though its solubility product constant is smaller, Ag_2CrO_4 ($K_{sp} = 1.1 \times 10^{-12}$) is more soluble than AgCl ($K_{sp} = 1.8 \times 10^{-10}$). For Ag_2CrO_4, the molar solubility is $s = (K_{sp}/4)^{1/3} = 6.5 \times 10^{-5}$ M, whereas for AgCl, it is $s = \sqrt{K_{sp}} = 1.3 \times 10^{-5}$ M.

19-3 The Common-Ion Effect in Solubility Equilibria

In Examples 19-2 and 19-3, the ions in the saturated solutions came from a *single* source, the pure solid solute. Suppose that to the saturated solution of PbI_2 in Example 19-3, we add some I^-—a *common ion*—from a source such as KI(aq). The situation is similar to our first encounter with the common-ion effect in Chapter 18.

According to Le Châtelier's principle, an equilibrium mixture responds to a forced increase in the concentration of one of its reactants by shifting in the direction in which that reactant is consumed. In the lead iodide solubility equilibrium, if some of the common ion, I^-, is added, the reverse reaction is favored, leading to a new equilibrium.

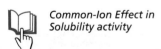
Common-Ion Effect in Solubility activity

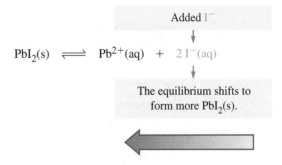

The addition of the common ion shifts the equilibrium of a slightly soluble ionic compound toward the undissolved compound, causing more to precipitate. Thus, the solubility of the compound is reduced.

> The solubility of a slightly soluble ionic compound is lowered in the presence of a second solute that furnishes a common ion.

The common-ion effect is illustrated in Figure 19-1, and it is applied quantitatively in Example 19-4.

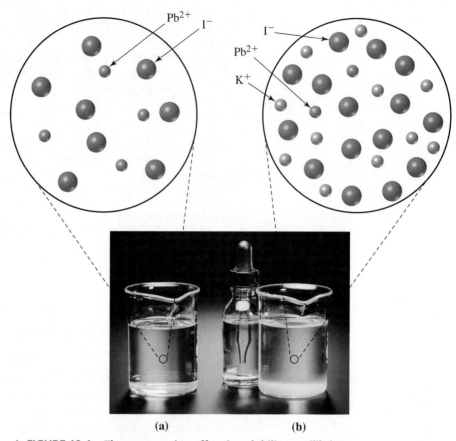

*Common-Ion Effect
animation*

(a)   **(b)**

▲ FIGURE 19-1   **The common-ion effect in solubility equilibrium**
(a) A clear saturated solution of lead(II) iodide from which excess undissolved solute has
been filtered off. **(b)** When a small volume of a concentrated solution of KI (containing
the common ion, I^-) is added, a small quantity of $PbI_2(s)$ precipitates. A common ion reduces the solubility of a sparingly soluble solute.

EXAMPLE 19-4

Calculating the Solubility of a Slightly Soluble Solute in the Presence of a Common Ion.
What is the molar solubility of PbI_2 in 0.10 M KI(aq)?

Solution

To solve this problem, let's set up the familiar ICE table with s instead of x to represent
changes in concentrations. Think of producing a saturated solution of PbI_2, but instead
of using pure water as the solvent, we will use 0.10 M KI(aq). Thus, we begin with
$[I^-] = 0.10$ M. Now let s represent the amount of PbI_2, in moles, that dissolves to produce 1 L of saturated solution. The additional concentrations appearing in this solution
are s mol Pb^{2+}/L and $2s$ mol I^-/L.

$$PbI_2(s) \rightleftharpoons Pb^{2+}(aq) \quad + \quad 2\,I^-(aq)$$

initial concns, M:		0.10
from PbI_2, M:	s	$2s$
equil concns, M:	s	$(0.10 + 2s)$

The usual K_{sp} relationship must be satisfied.

$$K_{sp} = [Pb^{2+}][I^-]^2 = (s)(0.10 + 2s)^2 = 7.1 \times 10^{-9}$$

To simplify the solution to this equation, let's assume that s is much smaller than 0.10 M, so that $0.10 + 2s \approx 0.10$.

$$s(0.10)^2 = 7.1 \times 10^{-9}$$

$$s = \frac{7.1 \times 10^{-9}}{(0.10)^2} = 7.1 \times 10^{-7} \text{ M}$$

Our assumption is well justified: 7.1×10^{-7} is much smaller than 0.10, and

$$s = \text{molar solubility of PbI}_2 = 7.1 \times 10^{-7} \text{ M}$$

Practice Example A: What is the molar solubility of PbI_2 in 0.10 M $Pb(NO_3)_2(aq)$?

(*Hint:* To which ion concentration should the solubility be related?)

Practice Example B: What is the molar solubility of $Fe(OH)_3$ in a buffered solution with pH = 8.20?

The solubility of PbI_2 in the presence of 0.10 M I^-, as calculated in Example 19-4, is about 2000 times less than its value in pure water (Example 19-3). If you work out Practice Example 19-4A, you will see that the effect of added Pb^{2+} in reducing the solubility of PbI_2 is not as striking as that of I^-, but it is significant nevertheless.

A typical error in problems such as Example 19-4 is to double the common-ion concentration—that is, to write $[I^-] = (2 \times 0.10)$ M instead of $[I^-] = 0.10$ M. While it is true that in *any* aqueous solution of PbI_2, the $[I^-]$ derived from PbI_2 is *twice* the molarity of the PbI_2, the $[I^-]$ that comes from a soluble strong electrolyte is determined only by the molarity of the strong electrolyte. Thus, $[I^-]$ in 0.10 M KI(aq) is 0.10 M. In 0.10 M KI(aq) that is also saturated with PbI_2, the *total* $[I^-] = (0.10 + 2s)$ M. In short, no relationship exists between the stoichiometry of the dissolution of PbI_2, which requires a factor of 2 in establishing $[I^-]$, and that of KI, which does not.

19-4   Limitations of the K_{sp} Concept

We have repeatedly used the term *slightly soluble* in describing the solutes for which we have written K_{sp} expressions. You might wonder if we can write K_{sp} expressions for moderately or highly soluble ionic compounds, such as NaCl, KNO_3, and NaOH. The answer is yes, but the K_{sp} must be based on ion *activities* rather than on concentrations. In ionic solutions of moderate to high concentrations, activities and concentrations are generally far from equal (recall Section 14-9). If we cannot use molarities in place of activities, much of the simplicity of the solubility product concept is lost. Thus, K_{sp} values are usually limited to slightly soluble (essentially insoluble) solutes, and ion molarities are used in place of activities. In addition, the K_{sp} concept has several other limitations, which are discussed in the following sections.

The Diverse ("Uncommon") Ion Effect: The Salt Effect

We have explored the effect of common ions on a solubility equilibrium, but what effect do ions different from those involved in the equilibrium have on solute solubilities? The effect of "uncommon" ions, or *diverse ions*, is not as striking as the common-ion effect. Moreover, uncommon ions tend to increase rather than decrease solubility. As the total ionic concentration of a solution increases, interionic attractions become more important. Activities (effective concentrations) become smaller than the stoichiometric or measured concentrations. For the ions involved in the solution process, this means that higher concentrations must appear in solution before equilibrium is established—*the solubility increases*. Figure 19-2 compares the effects of common ions and uncommon ions.

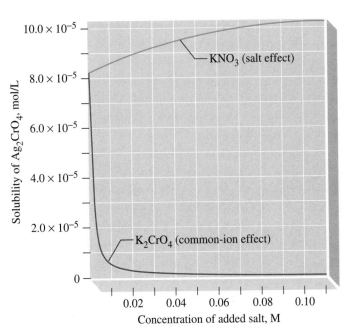

▶ **FIGURE 19-2**
Comparison of the common-ion effect and the salt effect on the molar solubility of Ag_2CrO_4
The presence of CrO_4^{2-} ions, derived from $K_2CrO_4(aq)$, reduces the solubility of Ag_2CrO_4 by a factor of about 35 over the concentration range shown (from 0 to 0.10 M added salt). Over the same concentration range, the solubility of Ag_2CrO_4 is increased by the presence of the uncommon, or diverse, ions from KNO_3, but only by about 25%.

The diverse ion effect is more commonly called the **salt effect**. As a result of the salt effect, the numerical value of a K_{sp} based on molarities will vary depending on the ionic atmosphere. Most tabulated values of K_{sp} are based on activities rather than on molarities, thus avoiding the problem of the salt effect.

Incomplete Dissociation of Solute into Ions

In performing calculations involving K_{sp} values and solubilities, we have assumed that all the dissolved solute appears in solution as separated cations and anions. However, this assumption is often not valid. The solute might not be 100% ionic, and some of the solute might enter solution in molecular form. Alternatively, some ions in solution might join together into **ion pairs**. An ion pair is two oppositely charged ions that are held together by the electrostatic attraction between the ions. For example, in a saturated solution of magnesium fluoride, although most of the solute exists as Mg^{2+} and F^- ions, some exists as the ion pair MgF^+.

To the extent that a solution contains cations and anions of a solute as ion pairs, the concentrations of the dissociated ions are reduced from the stoichiometric expectations. Thus, although the measured solubility of MgF_2 is about 4×10^{-3} M, we cannot safely assume that $[Mg^{2+}] = 4 \times 10^{-3}$ M and that $[F^-] = 8 \times 10^{-3}$ M, because some of the Mg^{2+} and F^- ions are involved in ion pairing. This means that additional solute must be present for the product of ion concentrations to equal K_{sp} of the solute, making the true solubility of the solute greater than otherwise expected on the basis of K_{sp}.

The degree of ion-pair formation increases as mutual electrostatic attraction of the anion and cation increases. Hence, ion-pair formation is increasingly likely when the cations or anions in solution carry multiple charges (for example, Mg^{2+} or SO_4^{2-}).

Simultaneous Equilibria

The reversible reaction between a solid solute and its ions in aqueous solution is never the sole process occurring. At the very least, the self-ionization of water also occurs, although we can generally ignore it. Other equilibrium processes that may occur include reactions between solute ions and other solution species. Two possi-

Ion pair

▲ An ion pair, MgF^+, in a magnesium fluoride aqueous solution.

bilities are acid–base reactions (Section 19-7) and complex-ion formation (Section 19-8). Calculations based on the K_{sp} expression may be in error if we fail to take into account other equilibrium processes that occur simultaneously with solution equilibrium.

Assessing the Limitations of K_{sp}

Let's assess the importance of the effects discussed in this section, some of which apply to $CaSO_4$. Recall that in Example 19-2 we calculated K_{sp} for $CaSO_4$ on the basis of its measured solubility. Our result was $K_{sp} = 2.3 \times 10^{-4}$. This value is about 25 times larger than the value listed in Table 19.1, which is $K_{sp} = 9.1 \times 10^{-6}$.

These conflicting results for $CaSO_4$ are understandable. The K_{sp} value listed in Table 19.1 is based on ion activities, whereas the K_{sp} value calculated from the experimentally determined solubility is based on ion concentrations, assuming complete dissociation of the solute into ions and no ion-pair formation. We will continue to substitute molarities for activities of ions, and the case of $CaSO_4$ simply suggests that some of our results, although of the appropriate general magnitude (that is, within a factor of 10 or 100), may not be highly accurate. These order-of-magnitude results, however, still allow us to make some correct predictions and to apply the K_{sp} concept in useful ways.

19-5   Criteria for Precipitation and Its Completeness

Silver iodide is a light-sensitive compound used in photographic film and also in cloud seeding to produce rain. Its solubility equilibrium and K_{sp} are represented as

$$AgI(s) \rightleftharpoons Ag^+(aq) + I^-(aq)$$

$$K_{sp} = [Ag^+][I^-] = 8.5 \times 10^{-17}$$

Suppose we mix solutions of $AgNO_3(aq)$ and $KI(aq)$ to obtain a mixed solution that has $[Ag^+] = 0.010$ M and $[I^-] = 0.015$ M. Is this solution unsaturated, saturated, or supersaturated?

Recall the reaction quotient, Q, that we introduced in Chapter 16. It has the same form as an equilibrium constant expression but uses initial concentrations rather than equilibrium concentrations. Initially,

$$Q_{sp} = [Ag^+]_{init} \times [I^-]_{init} = (0.010)(0.015) = 1.5 \times 10^{-4} > K_{sp}$$

The fact that $Q_{sp} > K_{sp}$ indicates that the concentrations of Ag^+ and I^- are already higher than they would be in a saturated solution and that a net change should occur to the left. The solution is *supersaturated*. As is generally the case with supersaturated solutions, excess AgI should precipitate from solution. If we had found $Q_{sp} < K_{sp}$, the solution would have been *unsaturated*. No precipitate would form from such a solution.

When applied to solubility equilibria, Q_{sp} is generally called the **ion product** because its form is that of the product of ion concentrations raised to appropriate powers. The criteria for determining whether ions in a solution will combine to form a precipitate require us to compare the ion product with K_{sp}.

- Precipitation *should occur* if $Q_{sp} > K_{sp}$.
- Precipitation *cannot occur* if $Q_{sp} < K_{sp}$.
- A solution is just saturated if $Q_{sp} = K_{sp}$.

These criteria are illustrated in Figure 19-3 and Example 19-5. The example emphasizes the important point that *any possible dilutions must be considered before the criteria for precipitation are applied*.

(a)

(b)

▲ FIGURE 19-3
Applying the criteria for precipitation from solution—Example 19-5 illustrated
(a) When three drops of 0.20 M KI are first added to 100.0 mL of 0.010 M $Pb(NO_3)_2$, a precipitate forms because K_{sp} is exceeded in the immediate vicinity of the drops. (b) When the KI becomes uniformly mixed in the $Pb(NO_3)_2(aq)$, K_{sp} is no longer exceeded and the precipitate redissolves. The criteria for precipitation must be applied *after* dilution has occurred.

Precipitation Reactions Movie

EXAMPLE 19-5

Applying the Criteria for Precipitation of a Slightly Soluble Solute. Three drops of 0.20 M KI are added to 100.0 mL of 0.010 M $Pb(NO_3)_2$. Will a precipitate of lead iodide form? (Assume 1 drop = 0.05 mL.)

$$PbI_2(s) \rightleftharpoons Pb^{2+}(aq) + 2\,I^-(aq) \qquad K_{sp} = 7.1 \times 10^{-9}$$

Solution

We need to compare the product $[Pb^{2+}][I^-]^2$ formulated for the initial concentrations with the K_{sp} for PbI_2. For $[Pb^{2+}]$ we can simply use 0.010 M. For $[I^-]$, however, we must consider the great reduction in concentration that occurs when the three drops of 0.20 M KI are diluted to 100.0 mL.

Dilution Calculation

KEEP IN MIND ▶
that any possible dilutions must be considered *before* the criteria for precipitation are applied.

$$\text{amount } I^- = 3 \text{ drops} \times \frac{0.05 \text{ mL}}{1 \text{ drop}} \times \frac{1 \text{ L}}{1000 \text{ mL}} \times \frac{0.20 \text{ mol KI}}{\text{L}}$$

$$\times \frac{1 \text{ mol } I^-}{1 \text{ mol KI}} = 3 \times 10^{-5} \text{ mol } I^-$$

$$[I^-] = \frac{3 \times 10^{-5} \text{ mol } I^-}{0.1000 \text{ L}} = 3 \times 10^{-4} \text{ M}$$

Applying Precipitation Criteria

$$Q_{sp} = [Pb^{2+}][I^-]^2 = (0.010)(3 \times 10^{-4})^2 = 9 \times 10^{-10}$$

Because a Q_{sp} of 9×10^{-10} is *smaller than* a K_{sp} of 7.1×10^{-9}, we conclude that $PbI_2(s)$ should *not* precipitate.

Practice Example A: Three drops of 0.20 M KI are added to 100.0 mL of a 0.010 M solution of $AgNO_3$. Will a precipitate of silver iodide form?

Practice Example B: We saw in Example 19-5 that a 3-drop volume of 0.20 M KI is insufficient to cause precipitation in 100.0 mL of 0.010 M $Pb(NO_3)_2$. What minimum number of drops would be required to produce the first precipitate?

Precipitation of a solute is considered to be complete only if the amount remaining in solution is very small. An arbitrary rule of thumb is that precipitation is complete if 99.9% or more of a particular ion has precipitated, leaving less than 0.1% of the ion in solution. In Example 19-6, we will calculate the concentration of Mg^{2+} that remains in a solution from which $Mg(OH)_2(s)$ has precipitated. We will compare this remaining $[Mg^{2+}]$ to the initial $[Mg^{2+}]$ to determine the completeness of the precipitation.

EXAMPLE 19-6

▶ One method of maintaining a constant pH during a precipitation is to carry out the precipitation from a buffer solution.

Assessing the Completeness of a Precipitation Reaction. The first step in a commercial process in which magnesium is obtained from seawater involves precipitating Mg^{2+} as $Mg(OH)_2(s)$. The magnesium ion concentration in seawater is about 0.059 M. If a seawater sample is treated so that its $[OH^-]$ is maintained at 2.0×10^{-3} M, **(a)** what will be $[Mg^{2+}]$ remaining in solution when precipitation stops ($K_{sp} = 1.8 \times 10^{-11}$)? **(b)** Can we say that precipitation of $Mg(OH)_2(s)$ is complete under these conditions?

Solution

(a) There is no question that precipitation will occur, because the ion product
$$Q_{sp} = [Mg^{2+}][OH^-]^2 = (0.059)(2.0 \times 10^{-3})^2 = 2.4 \times 10^{-7} \text{ exceeds } K_{sp}.$$
Precipitation of $Mg(OH)_2(s)$ will continue as long as the ion product exceeds K_{sp}, but will stop when that product is equal to K_{sp}. At the point at

which the ion product equals K_{sp}, whatever $[Mg^{2+}]$ is in solution remains in solution.

$$[Mg^{2+}][OH^-]^2 = [Mg^{2+}](2.0 \times 10^{-3})^2 = 1.8 \times 10^{-11} = K_{sp}$$

$$[Mg^{2+}]_{remaining} = \frac{1.8 \times 10^{-11}}{(2.0 \times 10^{-3})^2} = 4.5 \times 10^{-6} \text{ M}$$

(b) $[Mg^{2+}]$ in seawater is reduced from 0.059 M to 4.5×10^{-6} M as a result of the precipitation reaction. Expressed as a percentage,

$$\% \; [Mg^{2+}]_{remaining} = \frac{4.5 \times 10^{-6} \text{ M}}{0.059 \text{ M}} \times 100\% = 0.0076\%$$

Because less than 0.1% of the Mg^{2+} remains, we conclude that precipitation is essentially complete.

Practice Example A: A typical Ca^{2+} concentration in seawater is 0.010 M. Will the precipitation of $Ca(OH)_2$ be complete from a seawater sample in which $[OH^-]$ is maintained at 0.040 M?

Practice Example B: What $[OH^-]$ should be maintained in a solution if, after precipitation of Mg^{2+} as $Mg(OH)_2(s)$, the remaining Mg^{2+} is to be at a level of 1 μg Mg^{2+}/L?

Are You Wondering...

What conditions favor completeness of precipitation?

The key factors in determining whether the target ion is essentially completely removed from solution in a precipitation are (1) the value of K_{sp}, (2) the initial concentration of the target ion, and (3) the concentration of the common ion. In general, completeness of precipitation is favored by

- *a very small value of K_{sp}.* (The concentration of the target ion remaining in solution will be very small.)

- *a high initial concentration of the target ion.* (The concentration of the target ion remaining in solution will be only a very small fraction of the initial value.)

- *a concentration of common ion much larger than that of the target ion.* (The common-ion concentration will remain nearly constant during the precipitation.)

19-6 Fractional Precipitation

If a large excess of $AgNO_3(s)$ is added to a solution containing the ions CrO_4^{2-} and Br^-, a mixed precipitate of $Ag_2CrO_4(s)$ and $AgBr(s)$ is obtained. There is a method of adding $AgNO_3$, however, that will cause $AgBr(s)$ to precipitate but leave CrO_4^{2-} in solution. That method is fractional precipitation.

Fractional precipitation is a technique in which two or more ions in solution, each capable of being precipitated by the same reagent, are separated by the proper use of that reagent: *One ion is precipitated, while the other(s) remains in solution.* The primary condition for a successful fractional precipitation is that there be a significant difference in the solubilities of the substances being separated. (Usually this means a significant difference in their K_{sp} values.) The key to the technique is the slow addition of a concentrated solution of the precipitating reagent to the solution from which precipitation is to occur, as from a buret (Figure 19-4).

▶ Fractional precipitation is also called *selective precipitation.*

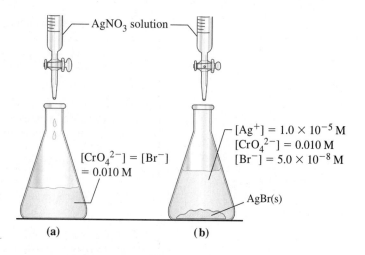

▶ FIGURE 19-4
Fractional precipitation—Example 19-7 illustrated
(a) $AgNO_3(aq)$ is slowly added to a solution that is
0.010 M in Br^- and 0.010 M in CrO_4^{2-}. **(b)** Essentially
all the Br^- has precipitated as pale yellow AgBr(s), with
$[Br^-]$ in solution = 5.0×10^{-8} M. Red-brown
$Ag_2CrO_4(s)$ is just about to precipitate.

Example 19-7 considers the separation of $CrO_4^{2-}(aq)$ and $Br^-(aq)$ through the use
of $Ag^+(aq)$.

EXAMPLE 19-7

Separating Ions by Fractional Precipitation. $AgNO_3(aq)$ is slowly added to a solution
that has $[CrO_4^{2-}] = 0.010$ M and $[Br^-] = 0.010$ M.

(a) Show that AgBr(s) should precipitate before $Ag_2CrO_4(s)$ does.

(b) When $Ag_2CrO_4(s)$ begins to precipitate, what is $[Br^-]$ remaining in solution?

(c) Is complete separation of $Br^-(aq)$ and $CrO_4^{2-}(aq)$ by fractional precipitation feasible?

Solution

The necessary data are the solubility equilibrium equations and the K_{sp} values of
Ag_2CrO_4 and AgBr.

$$Ag_2CrO_4(s) \rightleftharpoons 2\,Ag^+(aq) + CrO_4^{2-}(aq) \qquad K_{sp} = 1.1 \times 10^{-12}$$
$$AgBr(s) \rightleftharpoons Ag^+(aq) + Br^-(aq) \qquad K_{sp} = 5.0 \times 10^{-13}$$

(a) The required values of $[Ag^+]$ for precipitation to start are

AgBr ppt: $Q_{sp} = [Ag^+][Br^-] = [Ag^+](0.010) = 5.0 \times 10^{-13} = K_{sp}$
$[Ag^+] = 5.0 \times 10^{-11}$ M

Ag_2CrO_4 ppt: $Q_{sp} = [Ag^+]^2[CrO_4^{2-}] = [Ag^+]^2(0.010)$
$= 1.1 \times 10^{-12} = K_{sp}$
$[Ag^+]^2 = 1.1 \times 10^{-10}$ and $[Ag^+] = 1.0 \times 10^{-5}$ M

Because the $[Ag^+]$ required to start the precipitation of AgBr(s) is much less
than that for $Ag_2CrO_4(s)$, AgBr(s) precipitates first. As long as AgBr(s) is
forming, the silver ion concentration can only slowly approach the value
required for the precipitation of $Ag_2CrO_4(s)$.

(b) As AgBr(s) precipitates, $[Br^-]$ gradually decreases, and this permits $[Ag^+]$
to increase. When $[Ag^+]$ reaches 1.0×10^{-5} M, precipitation of $Ag_2CrO_4(s)$
begins. To determine $[Br^-]$ at the point at which $[Ag^+] = 1.0 \times 10^{-5}$ M,
we use K_{sp} for AgBr and solve for $[Br^-]$.

$$K_{sp} = [Ag^+][Br^-] = (1.0 \times 10^{-5})[Br^-] = 5.0 \times 10^{-13}$$

$$[Br^-] = \frac{5.0 \times 10^{-13}}{1.0 \times 10^{-5}} = 5.0 \times 10^{-8} \text{ M}$$

The following labels appear with the figure:
- AgNO₃ solution
- $[CrO_4^{2-}] = [Br^-] = 0.010$ M
- $[Ag^+] = 1.0 \times 10^{-5}$ M
- $[CrO_4^{2-}] = 0.010$ M
- $[Br^-] = 5.0 \times 10^{-8}$ M
- AgBr(s)
- (a)
- (b)

(c) Before $Ag_2CrO_4(s)$ begins to precipitate, $[Br^-]$ will have been reduced from 1.0×10^{-2} M to 5.0×10^{-8} M. Essentially, all the Br^- will have precipitated from solution as $AgBr(s)$, whereas the CrO_4^{2-} remains in solution. Fractional precipitation is feasible for separating mixtures of Br^- and CrO_4^{2-}.

Practice Example A: $AgNO_3(aq)$ is slowly added to a solution with $[Cl^-] = 0.115$ M and $[Br^-] = 0.264$ M. What percent of the Br^- remains unprecipitated at the point at which $AgCl(s)$ begins to precipitate?

$$AgCl: \quad K_{sp} = 1.8 \times 10^{-10} \qquad AgBr: \quad K_{sp} = 5.0 \times 10^{-13}$$

Practice Example B: A solution has $[Ba^{2+}] = [Sr^{2+}] = 0.10$ M. Use data from Appendix D to choose the best precipitating agent to separate these two ions. What is the concentration of the first ion to precipitate when the second ion begins to precipitate?

Are You Wondering ...

How the titration illustrated in Figure 19-4 and Example 19-7 might be carried out?

That is, how can we stop the titration just as $Ag_2CrO_4(s)$ begins to precipitate? One possibility, suggested in Figure 19-4, is to look for a color change in the precipitate from pale yellow (AgBr) to red-brown (Ag_2CrO_4). A more effective method is to follow $[Ag^+]$ during the titration. $[Ag^+]$ increases very rapidly between the points at which AgBr has finished precipitating and Ag_2CrO_4 is about to begin. We will discuss an electrometric method of determining very low ion concentrations in Chapter 21.

19-7 Solubility and pH

▲ Milk of magnesia, an aqueous suspension of $Mg(OH)_2(s)$.

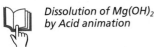

Dissolution of $Mg(OH)_2$ by Acid animation

The pH of a solution can affect the solubility of a salt to a large degree. That is especially true when the anion of the salt is the conjugate base of a weak acid or the base OH^- itself. An interesting example is the highly insoluble $Mg(OH)_2(s)$, which, when suspended in water, is the popular antacid known as milk of magnesia. Hydroxide ions from the dissolved magnesium hydroxide react with hydronium ions (in stomach acid) to form water.

$$Mg(OH)_2(s) \rightleftharpoons Mg^{2+}(aq) + 2\,OH^-(aq) \tag{19.2}$$

$$OH^-(aq) + H_3O^+(aq) \longrightarrow 2\,H_2O(l) \tag{19.3}$$

According to Le Châtelier's principle, we expect reaction (19.2) to be displaced to the right—that is, additional $Mg(OH)_2$ dissolves to replace OH^- ions drawn off by the neutralization reaction (19.3). We can obtain the overall net ionic equation by doubling equation (19.3) and adding it to equation (19.2). At the same time, we can apply the method of combining equilibrium constants that we learned in Chapter 16 (page 634). The result we obtain is

$$Mg(OH)_2(s) \rightleftharpoons Mg^{2+}(aq) + 2\,\cancel{OH^-(aq)} \qquad K_{sp} = 1.8 \times 10^{-11}$$

$$2\,\cancel{OH^-(aq)} + 2\,H_3O^+(aq) \rightleftharpoons 4\,H_2O(l) \qquad K' = 1/K_w^2 = 1.0 \times 10^{28}$$

$$Mg(OH)_2(s) + 2\,H_3O^+ \rightleftharpoons Mg^{2+}(aq) + 4\,H_2O(l) \tag{19.4}$$

$$K = K_{sp}/K_w^2 = 1.8 \times 10^{-11} \times 1.0 \times 10^{28} = 1.8 \times 10^{17}$$

The large value of K for reaction (19.4) indicates that the reaction goes essentially to completion and that $Mg(OH)_2$ is highly soluble in acidic solutions.

Other slightly soluble solutes having basic anions (such as $ZnCO_3$, MgF_2, and CaC_2O_4) also become more soluble in acidic solutions. For these solutes, we can write overall net ionic equations for solubility equilibria and corresponding K values based on K_{sp} for the solutes and K_a for the conjugate acids of the anions.

Although $Mg(OH)_2$ is soluble in acidic solution, in moderately or strongly basic solutions it is not. In Example 19-8, we calculate $[OH^-]$ in a solution of the weak base NH_3 and then use the criteria for precipitation to see if $Mg(OH)_2(s)$ will precipitate. In Example 19-9, we determine how to adjust $[OH^-]$ to prevent precipitation of $Mg(OH)_2(s)$. This adjustment is made by adding NH_4^+ to the $NH_3(aq)$, thereby converting it to a *buffer* solution.

EXAMPLE 19-8

Determining Whether a Precipitate Will Form in a Solution in Which There Is Also an Ionization Equilibrium. Should $Mg(OH)_2(s)$ precipitate from a solution that is 0.010 M $MgCl_2$ and also 0.10 M NH_3?

Solution

The key here is in understanding that $[OH^-]$ is established by the ionization of $NH_3(aq)$.

$$NH_3(aq) + H_2O(l) \rightleftharpoons NH_4^+(aq) + OH^-(aq) \qquad K_b = 1.8 \times 10^{-5}$$

If we set this up in the usual way, the equilibrium values are $x = [NH_4^+] = [OH^-]$ and $[NH_3] = (0.10 - x) \approx 0.10$. Then we obtain

$$K_b = \frac{[NH_4^+][OH^-]}{[NH_3]} = \frac{x \cdot x}{0.10} = 1.8 \times 10^{-5}$$

$$x^2 = 1.8 \times 10^{-6} \qquad x = [OH^-] = 1.3 \times 10^{-3} \text{ M}$$

Now we can rephrase the original question: Should $Mg(OH)_2(s)$ precipitate from a solution in which $[Mg^{2+}] = 1.0 \times 10^{-2}$ M and $[OH^-] = 1.3 \times 10^{-3}$ M? We must compare the ion product, Q_{sp}, with K_{sp}.

$$Q_{sp} = [Mg^{2+}][OH^-]^2 = (1.0 \times 10^{-2})(1.3 \times 10^{-3})^2$$
$$= 1.7 \times 10^{-8} > K_{sp} = 1.8 \times 10^{-11}$$

Precipitation should occur.

Practice Example A: Should $Mg(OH)_2(s)$ precipitate from a solution that is 0.010 M $MgCl_2(aq)$ and also 0.10 M $NaC_2H_3O_2$? $K_{sp} = [Mg(OH)_2] = 1.8 \times 10^{-11}$; $K_a(HC_2H_3O_2) = 1.8 \times 10^{-5}$.

(*Hint:* What equilibrium expression establishes $[OH^-]$ in the solution?)

Practice Example B: Will a precipitate of $Fe(OH)_3$ form from a solution that is 0.013 M Fe^{3+} in a buffer solution that is 0.150 M $HC_2H_3O_2$ and 0.250 M $NaC_2H_3O_2$?

EXAMPLE 19-9

Controlling an Ion Concentration, Either to Cause Precipitation or to Prevent It. What $[NH_4^+]$ must be maintained to prevent precipitation of $Mg(OH)_2(s)$ from a solution that is 0.010 M $MgCl_2$ and 0.10 M NH_3?

Solution

The maximum value of the ion product, Q_{sp}, before precipitation occurs is 1.8×10^{-11}, the value of K_{sp} for $Mg(OH)_2$. This fact allows us to determine the maximum concentration of OH^- that can be tolerated.

$$[Mg^{2+}][OH^-]^2 = (1.0 \times 10^{-2})[OH^-]^2 = 1.8 \times 10^{-11}$$
$$[OH^-]^2 = 1.8 \times 10^{-9}$$
$$[OH^-] = 4.2 \times 10^{-5} \, M$$

Next let's determine what $[NH_4^+]$ must be present in 0.10 M NH_3 to maintain $[OH^-] = 4.2 \times 10^{-5} \, M$.

$$NH_3(aq) + H_2O(l) \rightleftharpoons NH_4^+(aq) + OH^-(aq) \qquad K_b = 1.8 \times 10^{-5}$$

$$K_b = \frac{[NH_4^+][OH^-]}{[NH_3]} = \frac{[NH_4^+](4.2 \times 10^{-5})}{0.10} = 1.8 \times 10^{-5}$$

$$[NH_4^+] = \frac{0.10 \times 1.8 \times 10^{-5}}{4.2 \times 10^{-5}} = 0.043 \, M$$

To keep the $[OH^-]$ at $4.2 \times 10^{-5} \, M$ *or less*, and thus prevent the precipitation of $Mg(OH)_2(s)$, $[NH_4^+]$ should be maintained at 0.043 M or *greater*.

Practice Example A: What minimum $[NH_4^+]$ must be present to prevent precipitation of $Mn(OH)_2(s)$ from a solution that is 0.0050 M $MnCl_2$ and 0.025 M NH_3? For $Mn(OH)_2$, $K_{sp} = 1.9 \times 10^{-13}$.

Practice Example B: What is the molar solubility of $Mg(OH)_2(s)$ in a solution that is 0.250 M NH_3 and 0.100 M NH_4Cl?

(*Hint:* Use equation (19.4).]

19-8   Equilibria Involving Complex Ions

As shown in Figure 19-5, when moderately concentrated $NH_3(aq)$ is added to a saturated solution of silver chloride in contact with undissolved $AgCl(s)$, the solid dissolves. The key to this dissolving action is that Ag^+ ions from $AgCl$ combine with NH_3 molecules to form the ions $[Ag(NH_3)_2]^+$, which, together with Cl^- ions, remain in solution as the soluble compound $Ag(NH_3)_2Cl$.

$$AgCl(s) + 2 \, NH_3(aq) \longrightarrow [Ag(NH_3)_2]^+(aq) + Cl^-(aq) \qquad (19.5)$$

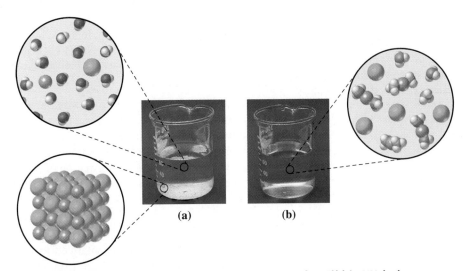

▲ **FIGURE 19-5   Complex-ion formation: dissolution of AgCl(s) in NH₃(aq)**
(a) A saturated solution of silver chloride in contact with excess AgCl(s). **(b)** When $NH_3(aq)$ is added, the excess AgCl(s) dissolves through the formation of the complex ion $[Ag(NH_3)_2]^+$.

A coordination compound

Complex cation   Anions

$[Co(NH_3)_6]^{3+}$ $3\ Cl^-$

Central
ion   Ligands

▲ **FIGURE 19-6**
Reprecipitating AgCl(s)
The reagent being added to the
solution containing $[Ag(NH_3)_2]^+$
and Cl^- is $HNO_3(aq)$. H_3O^+ from
the acid reacts with $NH_3(aq)$ to
form $NH_4^+(aq)$. As a result,
equilibrium between
$[Ag(NH_3)_2]^+$, Ag^+, and NH_3 is
upset. The complex ion is destroyed, $[Ag^+]$ quickly rises to
the point at which K_{sp} for AgCl
is exceeded, and a precipitate
forms.

Here $[Ag(NH_3)_2]^+$ is called a complex ion, and $Ag(NH_3)_2Cl$ is called a coordination compound. A **complex ion** is a polyatomic cation or anion composed of a central metal ion to which other groups (molecules or ions) called *ligands* are bonded.
Coordination compounds are substances containing complex ions. It helps to
think of reaction (19.5) as involving two simultaneous equilibria.

$$AgCl(s) \rightleftharpoons Ag^+(aq) + Cl^-(aq) \qquad (19.6)$$

$$Ag^+(aq) + 2\ NH_3(aq) \rightleftharpoons [Ag(NH_3)_2]^+(aq) \qquad (19.7)$$

The equilibrium in reaction (19.7) is shifted far to the right—$[Ag(NH_3)_2]^+$ is a stable complex ion. The equilibrium concentration of $Ag^+(aq)$ in (19.7) is kept so low
that the ion product $[Ag^+][Cl^-]$ fails to exceed K_{sp} and AgCl remains in solution.
Let's first apply additional qualitative reasoning of this sort in Example 19-10. Then
we can turn to some of the quantitative calculations that are also possible.

EXAMPLE 19-10

Predicting Reactions Involving Complex Ions. Predict what will happen if nitric acid is
added to a solution of $[Ag(NH_3)_2]Cl$ in $NH_3(aq)$.

Solution
Nitric acid neutralizes free NH_3 in the solution. Because $HNO_3(aq)$ is a strong acid, we
represent it as being completely ionized and write only the net ionic equation.

$$H_3O^+(aq) + NH_3(aq) \longrightarrow NH_4^+(aq) + H_2O(l)$$

To replace free NH_3 lost in this neutralization, equilibrium in reaction (19.7) shifts to the
left. As a result, $[Ag^+]$ increases. When $[Ag^+]$ increases to the point at which the ion
product $[Ag^+][Cl^-]$ exceeds K_{sp}, AgCl(s) precipitates (Figure 19-6).

Practice Example A: Copper(II) ion forms both an insoluble hydroxide and the
complex ion $[Cu(NH_3)_4]^{2+}$. Write equations to represent the expected reaction when
(a) $CuSO_4(aq)$ and $NaOH(aq)$ are mixed; **(b)** an excess of $NH_3(aq)$ is added to the product of part (a); and **(c)** an excess of $HNO_3(aq)$ is added to the product of part (b).

Practice Example B: Zinc(II) ion forms both an insoluble hydroxide and the complex ions $[Zn(OH)_4]^{2-}$ and $[Zn(NH_3)_4]^{2+}$. Write four equations to represent the reactions
of **(a)** $NH_3(aq)$ with $ZnSO_4(aq)$, followed by; **(b)** enough $HNO_3(aq)$ to make the product of part (a) acidic; **(c)** enough $NaOH(aq)$ to make the product of part (b) slightly basic;
(d) enough $NaOH(aq)$ to make the product of part (c) strongly basic.

To describe the ionization of a weak acid, we use the ionization constant K_a.
For a solubility equilibrium, we use the solubility product constant K_{sp}. The equilibrium constant that is used to deal with a complex-ion equilibrium is called the
formation constant. The **formation constant**, K_f, of a complex ion is the equilibrium constant describing the formation of a complex ion from a central ion and its
attached groups. For reaction (19.7) this equilibrium constant expression is

$$K_f = \frac{[[Ag(NH_3)_2]^+]}{[Ag^+][NH_3]^2} = 1.6 \times 10^7$$

Table 19.2 lists some representative formation constants, K_f.
One feature that distinguishes K_f from most other equilibrium constants that we
have been considering is the fact that K_f values are usually *large* numbers. This
fact can affect the way we approach certain calculations. We sometimes find that
when we work with large K values, it is convenient to solve a problem in two steps.

TABLE 19.2 Formation Constants for Some Complex Ions[a]

Complex Ion	Equilibrium Reaction[b]	K_f
$[Co(NH_3)_6]^{3+}$	$Co^{3+} + 6\,NH_3 \rightleftharpoons [Co(NH_3)_6]^{3+}$	4.5×10^{33}
$[Cu(NH_3)_4]^{2+}$	$Cu^{2+} + 4\,NH_3 \rightleftharpoons [Cu(NH_3)_4]^{2+}$	1.1×10^{13}
$[Fe(CN)_6]^{4-}$	$Fe^{2+} + 6\,CN^- \rightleftharpoons [Fe(CN)_6]^{4-}$	1×10^{37}
$[Fe(CN)_6]^{3-}$	$Fe^{3+} + 6\,CN^- \rightleftharpoons [Fe(CN)_6]^{3-}$	1×10^{42}
$[PbCl_3]^-$	$Pb^{2+} + 3\,Cl^- \rightleftharpoons [PbCl_3]^-$	2.4×10^1
$[Ag(NH_3)_2]^+$	$Ag^+ + 2\,NH_3 \rightleftharpoons [Ag(NH_3)_2]^+$	1.6×10^7
$[Ag(CN)_2]^-$	$Ag^+ + 2\,CN^- \rightleftharpoons [Ag(CN)_2]^-$	5.6×10^{18}
$[Ag(S_2O_3)_2]^{3-}$	$Ag^+ + 2\,S_2O_3^{2-} \rightleftharpoons [Ag(S_2O_3)_2]^{3-}$	1.7×10^{13}
$[Zn(NH_3)_4]^{2+}$	$Zn^{2+} + 4\,NH_3 \rightleftharpoons [Zn(NH_3)_4]^{2+}$	4.1×10^8
$[Zn(CN)_4]^{2-}$	$Zn^{2+} + 4\,CN^- \rightleftharpoons [Zn(CN)_4]^{2-}$	1×10^{18}
$[Zn(OH)_4]^{2-}$	$Zn^{2+} + 4\,OH^- \rightleftharpoons [Zn(OH)_4]^{2-}$	4.6×10^{17}

[a] A more extensive tabulation is given in Appendix D.
[b] Tabulated here are *overall* formation reactions and the corresponding *overall* formation constants. In Section 25-7, we describe the formation of complex ions in a *stepwise* fashion and introduce formation constants for individual steps.

First, assume that the forward reaction goes to completion; second, assume that a small change occurs in the reverse direction to establish the equilibrium. This approach is demonstrated in Example 19-11.

EXAMPLE 19-11

Determining Whether a Precipitate Will Form in a Solution Containing Complex Ions. A 0.10-mol sample of $AgNO_3$ is dissolved in 1.00 L of 1.00 M NH_3. If 0.010 mol NaCl is added to this solution, will AgCl(s) precipitate?

Solution

Let's begin by assuming that, because the value of K_f for $[Ag(NH_3)_2]^+$ is very large, the following reaction initially goes to completion, with the results shown.

$$Ag^+(aq) + 2\,NH_3(aq) \longrightarrow [Ag(NH_3)_2]^+(aq)$$

initial concns, M:	0.10	1.00	
changes, M:	−0.10	−0.20	+0.10
after reaction, M:	≈0	0.80	0.10

However, the concentration of uncomplexed silver ion, though very small, is not zero. To determine the value of $[Ag^+]$, let's start with $[[Ag(NH_3)_2]^+]$ and $[NH_3]$ in solution and establish $[Ag^+]$ at equilibrium.

$$Ag^+ + 2\,NH_3 \rightleftharpoons [Ag(NH_3)_2]^+$$

initial concns, M:		0.80	0.10
changes, M:	+x	+2x	−x
equil concns, M:	x	0.80 + 2x	0.10 − x

When substituting into the following expression, we make the assumption that $x \ll 0.10$, which we will find to be the case.

$$\frac{[[Ag(NH_3)_2]^+]}{[Ag^+][NH_3]^2} = \frac{0.10 - x}{x(0.80 + 2x)^2} \approx \frac{0.10}{x(0.80)^2} = 1.6 \times 10^7$$

$$x = [Ag^+] = \frac{0.10}{(1.6 \times 10^7)(0.80)^2} = 9.8 \times 10^{-9}\ M$$

Finally, we must compare $Q_{sp} = [Ag^+][Cl^-]$ with K_{sp} for AgCl (that is, 1.8×10^{-10}). $[Ag^+]$ is the value of x that we just calculated. Because the solution contains 0.010 mol NaCl/L, $[Cl^-] = 0.010$ M $= 1.0 \times 10^{-2}$ M, and

$$Q_{sp} = (9.8 \times 10^{-9})(1.0 \times 10^{-2}) = 9.8 \times 10^{-11} < 1.8 \times 10^{-10}$$

AgCl will not precipitate.

Practice Example A: Will AgCl(s) precipitate from 1.50 L of a solution that is 0.100 M $AgNO_3$ and 0.225 M NH_3 if 1.00 mL of 3.50 M NaCl is added?

(*Hint:* What are $[Ag^+]$ and $[Cl^-]$ immediately after the addition of the 1.00 mL of 3.50 M NaCl? Take into account the dilution of the NaCl(aq), but assume the total volume remains at 1.50 L.)

Practice Example B: A solution is prepared that is 0.100 M in $Pb(NO_3)_2$ and 0.250 M in the ethylenediaminetetraacetate anion, $EDTA^{4-}$. Together, Pb^{2+} and $EDTA^{4-}$ form the complex ion $[PbEDTA]^{2-}$. If the solution is also made 0.10 M in I^-, will PbI_2(s) precipitate? For PbI_2, $K_{sp} = 7.1 \times 10^{-9}$; for $[PbEDTA]^{2-}$, $K_f = 2 \times 10^{18}$.

Just as some precipitation reactions can be controlled by using a buffer solution (Example 19-9), precipitation from a solution of complex ions can be controlled by fixing the concentration of the complexing agent. Such a case is demonstrated for AgCl in Example 19-12.

EXAMPLE 19-12

Controlling a Concentration to Cause or Prevent Precipitation from a Solution of Complex Ions. What is the *minimum* concentration of NH_3 required to prevent AgCl(s) from precipitating from 1.00 L of a solution containing 0.10 mol $AgNO_3$ and 0.010 mol NaCl?

Solution

The $[Cl^-]$ that must be maintained in solution is 1.0×10^{-2} M. If no precipitation is to occur, $[Ag^+][Cl^-] \le K_{sp}$.

$$[Ag^+](1.0 \times 10^{-2}) \le K_{sp} = 1.8 \times 10^{-10} \qquad [Ag^+] \le 1.8 \times 10^{-8} \text{ M}$$

Thus, the maximum concentration of *uncomplexed* Ag^+ permitted in solution is 1.8×10^{-8} M. This means that essentially all the Ag^+ (0.10 mol/L) must be tied up (complexed) in the complex ion $[Ag(NH_3)_2]^+$. We need to solve the following expression for $[NH_3]$.

$$K_f = \frac{[[Ag(NH_3)_2]^+]}{[Ag^+][NH_3]^2} = \frac{1.0 \times 10^{-1}}{1.8 \times 10^{-8}[NH_3]^2} = 1.6 \times 10^7$$

$$[NH_3]^2 = \frac{1.0 \times 10^{-1}}{1.8 \times 10^{-8} \times 1.6 \times 10^7} = 0.35 \qquad [NH_3] = 0.59 \text{ M}$$

The concentration just calculated is that of *free, uncomplexed* NH_3. Considering as well the 0.20 mol NH_3/L complexed in the 0.10 M $[Ag(NH_3)_2]^+$, we see that the total concentration of NH_3(aq) required is

$$[NH_3]_{tot} = 0.59 \text{ M} + 0.20 \text{ M} = 0.79 \text{ M}$$

Practice Example A: What $[NH_3]_{total}$ is necessary to keep AgCl from precipitating from a solution that is 0.13 M $AgNO_3$ and 0.0075 M NaCl?

Practice Example B: What minimum concentration of thiosulfate ion, $S_2O_3^{2-}$, should be present in 0.10 M $AgNO_3$(aq) so that AgCl(s) does not precipitate when the solution is also made 0.010 M in Cl^-? For AgCl, $K_{sp} = 1.8 \times 10^{-10}$; for $[Ag(S_2O_3)_2]^{3-}$, $K_f = 1.7 \times 10^{13}$.

On page 763, we described qualitatively how the solubility of AgCl increases in the presence of $NH_3(aq)$. In Example 19-13, we show how to calculate the actual solubility of AgCl in $NH_3(aq)$.

Determining the Solubility of a Solute When Complex Ions Form.   What is the molar solubility of AgCl in 0.100 M $NH_3(aq)$?

Solution

As we have already seen, equation (19.5) describes the solubility equilibrium.

$$AgCl(s) + 2\,NH_3(aq) \rightleftharpoons [Ag(NH_3)_2]^+(aq) + Cl^-(aq) \qquad (19.5)$$

Let's base our calculation on the equilibrium constant K for reaction (19.5). There are two ways to obtain this value. By one method, we write equation (19.5) as the sum of equations (19.6) and (19.7) on page 764. Then we obtain its K value as the product of a K_{sp} and a K_f.

▶ This method is similar to the method we used to obtain K for reaction (19.4) on page 761.

$$AgCl(s) \rightleftharpoons Ag^+(aq) + Cl^-(aq) \qquad K_{sp} = 1.8 \times 10^{-10}$$

$$\underline{Ag^+(aq) + 2\,NH_3(aq) \rightleftharpoons [Ag(NH_3)_2]^+(aq) \qquad K_f = 1.6 \times 10^7}$$

$$AgCl(s) + 2\,NH_3(aq) \rightleftharpoons [Ag(NH_3)_2]^+(aq) + Cl^-(aq)$$

$$K = K_{sp} \times K_f = 1.8 \times 10^{-10} \times 1.6 \times 10^7 = 2.9 \times 10^{-3}$$

By a second method, we first write the equilibrium constant expression for reaction (19.5) and then multiply numerator and denominator by $[Ag^+]$.

▶ This method is similar to the method we used to obtain K_a for NH_4^+ from K_b for NH_3 and the ion product of water, K_w, on page 690.

$$K = \frac{[Ag(NH_3)_2^+][Cl^-]}{[NH_3]^2} = \frac{[Ag(NH_3)_2^+][Cl^-][Ag^+]}{[NH_3]^2[Ag^+]} = K_f \times K_{sp} = 2.9 \times 10^{-3}$$

The expression printed in red is K_f for $[Ag(NH_3)_2]^+$; the one in blue is K_{sp} for AgCl. The K value for reaction (19.5) is the product of the two.

According to equation (19.5), if s mol AgCl/L dissolves (the molar solubility), the expected concentrations of $[Ag(NH_3)_2]^+$ and Cl^- are also equal to s.

	AgCl(s)	+ 2 NH_3(aq)	$\rightleftharpoons$	$[Ag(NH_3)_2]^+$(aq)	+	Cl^-(aq)
initial concns, M:		0.100				
changes, M:		$-2s$		$+s$		$+s$
equil concns, M:		$(0.100 - 2s)$		s		s

$$K = \frac{[[Ag(NH_3)_2]^+][Cl^-]}{[NH_3]^2} = \frac{s \cdot s}{(0.100 - 2s)^2} = \left(\frac{s}{0.100 - 2s}\right)^2 = 2.9 \times 10^{-3}$$

We can solve this equation by taking the square root of both sides.

$$\frac{s}{0.100 - 2s} = \sqrt{2.9 \times 10^{-3}} = 5.4 \times 10^{-2}$$

$$s = 5.4 \times 10^{-3} - 0.11s$$

$$1.11s = 5.4 \times 10^{-3}$$

$$s = 4.9 \times 10^{-3}$$

KEEP IN MIND

that the molar solubility is actually the *total* concentration of silver in solution: $[Ag^+] + [Ag(NH_3)_2]^+$. Only when K_f is large and the concentration of complexing agent sufficiently high can we ignore the concentration of uncomplexed metal ion, as was the case here. ▶

The molar solubility of AgCl(s) in 0.100 M $NH_3(aq)$ is 4.9×10^{-3} M.

Note: The usual simplifying assumption, that is, $(0.100 - 2s) \approx 0.100$, would not have worked well in this calculation. The value of s obtained would have been 5.4×10^{-3} M, and $0.100 - (2 \times 0.0054) \neq 0.100$. Fortunately, the quadratic equation was easily solved by the square root extraction. If that had not been possible, we could have used either the quadratic formula or a method of successive approximations.

Practice Example A: What is the molar solubility of $Fe(OH)_3$ in a solution containing 0.100 M $C_2O_4^{2-}$? For $[Fe(C_2O_4)_3]^{3-}$, $K_f = 2 \times 10^{20}$.

Practice Example B: *Without doing detailed calculations,* show that the order of *decreasing* solubility in 0.100 M $NH_3(aq)$ should be $AgCl > AgBr > AgI$.

19-9 Qualitative Cation Analysis

Qualitative Analysis simulation

In *qualitative* analysis, we determine what substances are present in a mixture but *not* their quantities. An analysis that aims at identifying the cations present in a mixture is called **qualitative cation analysis**. Qualitative cation analysis is not as important a method as it once was, because now most qualitative and quantitative analyses are done with instruments. Its current value is in the wealth of examples it provides of precipitation (and dissolution) equilibria, acid–base equilibria, and oxidation–reduction reactions. Also, in the general chemistry laboratory it offers the challenge of unraveling a mystery—solving a qualitative analysis "unknown."

In the scheme in Figure 19-7, about 25 common cations are divided into five groups, depending on differing solubilities of their compounds. The first cations separated, Pb^{2+}, Hg_2^{2+}, and Ag^+, are those with insoluble *chlorides*. The reagent used

▶ In the Hg_2^{2+} ion, a covalent bond links a Hg atom to a Hg^{2+} ion.

Ionic Compounds activity

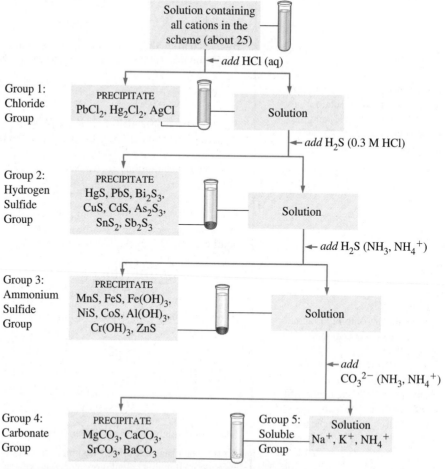

▲ FIGURE 19-7 **Outline of a qualitative cation analysis**
Various aspects of this scheme are described in the text. A sample containing all 25 cations can be separated into five groups by the indicated reagents.

is HCl(aq). All other cations remain in solution because their chlorides are soluble. After the chloride group precipitate is removed, the remaining solution is treated with H_2S in an acidic medium. Under these conditions, a group of sulfides precipitates. They are known as the hydrogen sulfide group. Next, the solution containing the remaining cations is treated with H_2S in a buffer mixture of ammonia and ammonium ion, yielding a mixture of insoluble hydroxides and sulfides.

This group is called the ammonium sulfide group. The sulfide of aluminum(III) and chromium(III) are unstable, reacting with water to form the hydroxides.

Treatment of the ammonium sulfide group filtrate with CO_3^{2-} yields the fourth group precipitate, which consists of the carbonates of Mg^{2+}, Ca^{2+}, Sr^{2+}, and Ba^{2+}. It is called the carbonate group because the aqueous carbonate anion is the key reagent. At the end of this series of precipitations, the resulting solution contains only Na^+, K^+, and NH_4^+, all of whose common salts are water soluble.

In this section, we will discuss the chemistry of the chloride and sulfide groups. The chemistry of metal carbonates is discussed in Chapter 22.

Cation Group 1: The Chloride Group

If a precipitate forms when a solution is treated with HCl(aq), one or more of these cations must be present: Pb^{2+}, Hg_2^{2+}, Ag^+. To establish the presence or absence of each of these three cations, the chloride group precipitate is filtered off and subjected to further testing.

 Are You Wondering . . .

How to test for the presence of Na^+, K^+, and NH_4^+ cations?

Tests by precipitation are difficult because the salts of these cations exhibit near-universal solubility. Both Na^+ and K^+ ions are most easily detected with a flame test. When a solution containing sodium ions is brought into contact with a flame, the characteristic orange-yellow color of the emission spectrum of sodium ions is observed. For potassium ions, a pale violet color results. To detect the presence of NH_4^+, we use the fact that the ammonium ion is the conjugate acid of the weak volatile base ammonia. Heating some of the original solution (not the final solution, which contains NH_4^+ ions added in the fractional precipitation scheme) with excess strong base will liberate ammonia.

$$NH_4^+(aq) + OH^- \longrightarrow NH_3(g) + H_2O(l)$$

The NH_3 is detected by its characteristic odor and its effect on the color of an acid–base indicator such as litmus.

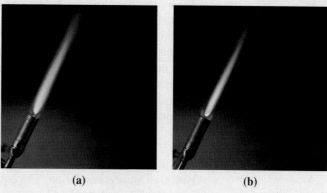

(a)   (b)

▲ Flame colors of **(a)** sodium and **(b)** potassium.

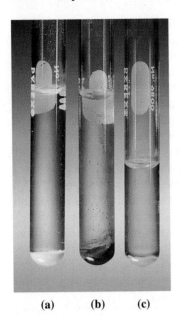

(a)   (b)   (c)

▲ **FIGURE 19-8**
Chloride group precipitates
(a) Group precipitate: a mixture
of $PbCl_2$ (white), Hg_2Cl_2 (white),
and AgCl (white). **(b)** Test for
Hg_2^{2+}: a mixture of Hg (black)
and $HgNH_2Cl$ (white). **(c)** Test
for Pb^{2+}: a yellow precipitate of
$PbCrO_4$(s).

▶ H_2S(g) has a familiar rotten
egg odor, especially noticeable
in volcanic areas and near sulfur
hot springs. It can cause nausea
and headaches at levels of
10 ppm in air, and it can produce
paralysis or death at levels of
100 ppm. The gas is detectable
by its odor at levels of 1 ppm, although exposure to the gas soon
deadens the sense of smell.

Of the three chlorides in the group precipitate, $PbCl_2$(s) is the most soluble; its
K_{sp} is much larger than those of AgCl and Hg_2Cl_2. When the precipitate is washed
with hot water, a sufficient quantity of $PbCl_2$ dissolves to permit a test for Pb^{2+} in
the solution. In this test, a lead compound less soluble than $PbCl_2$, such as lead
chromate, precipitates (Figure 19-8).

$$Pb^{2+}(aq) + CrO_4^{2-}(aq) \longrightarrow PbCrO_4(s)$$

The portion of the chloride group precipitate that is insoluble in hot water is
then treated with NH_3(aq). Two things happen. One is that any AgCl(s) present dissolves and forms the complex ion $[Ag(NH_3)_2^+]$, as described by equation (19.5).

$$AgCl(s) + 2 NH_3(aq) \longrightarrow [Ag(NH_3)_2]^+(aq) + Cl^-(aq)$$

At the same time, any Hg_2Cl_2(s) present undergoes an oxidation–reduction reaction.
One of the products of the reaction is finely divided black mercury. A black color
is common for finely divided metals.

$$Hg_2Cl_2(s) + 2 NH_3(aq) \longrightarrow \underbrace{Hg(l) + HgNH_2Cl(s)}_{\text{dark gray}} + NH_4^+(aq) + Cl^-(aq)$$

The appearance of a dark gray mixture of black mercury and white $HgNH_2Cl$ [mer-
cury(II) amidochloride] is the qualitative analysis confirmation of mercury(I) ion
(see Figure 19-8).

When the solution from reaction (19.5) is acidified with HNO_3(aq), any silver
ion present precipitates as AgCl(s). This is the reaction we predicted in Example 19-10 and pictured in Figure 19-6.

Cation Groups 2 and 3: Equilibria Involving Hydrogen Sulfide

Figure 19-7 suggests that aqueous hydrogen sulfide (hydrosulfuric acid) is the key
reagent in the analysis of cation groups 2 and 3. In aqueous solution, H_2S is a *weak
diprotic acid.*

$$H_2S(aq) + H_2O(l) \rightleftharpoons H_3O^+(aq) + HS^-(aq) \qquad K_{a_1} = 1.0 \times 10^{-7}$$

$$HS^-(aq) + H_2O(l) \rightleftharpoons H_3O^+(aq) + S^{2-}(aq) \qquad K_{a_2} = 1 \times 10^{-19}$$

The extremely small value of K_{a_2} suggests that sulfide ion is a very strong base,
as seen in the following hydrolysis reaction and K_b value.

$$S^{2-} + H_2O \rightleftharpoons HS^- + OH^- \qquad K_b = K_w/K_{a_2}$$
$$= (1.0 \times 10^{-14})/(1 \times 10^{-19}) = 1 \times 10^5$$

The hydrolysis of S^{2-} should go nearly to completion, which means that very little
S^{2-} can exist in an aqueous solution and that sulfide ion is probably not the precipitating agent for sulfides.

One way to handle the precipitation and dissolution of sulfide precipitates is to re-
strict our discussion to acidic solutions. Then we can write an equilibrium constant
expression in which we eliminate concentration terms for HS^- and S^{2-}. This approach
is reasonable because most sulfide separations are carried out in acidic solution.

Consider (1) the solubility equilibrium equation for PbS written to reflect hydrolysis of S^{2-}, (2) an equation written for the reverse of the first ionization of H_2S,
and (3) an equation that is the reverse of the self-ionization of water. We can combine these three equations into an overall equation that shows the dissolving of
PbS(s) in an acidic solution. The equilibrium constant for this overall equation is
generally referred to as K_{spa}.*

---

*See R. J. Myers, *J. Chem. Educ.* **63**, 687 (1986).

$$(1) \quad PbS(s) + H_2O(l) \rightleftharpoons Pb^{2+}(aq) + HS^-(aq) + OH^-(aq) \quad K_{sp} = 3 \times 10^{-28}$$

$$(2) \; H_3O^+(aq) + HS^-(aq) \rightleftharpoons H_2S(aq) + H_2O(l) \qquad\qquad 1/K_{a_1} = 1/1.0 \times 10^{-7}$$

$$(3) \, H_3O^+(aq) + OH^-(aq) \rightleftharpoons H_2O(l) + H_2O(l) \qquad\qquad 1/K_w = 1/1.0 \times 10^{-14}$$

overall: $PbS(s) + 2\, H_3O^+(aq) \rightleftharpoons Pb^{2+}(aq) + H_2S(aq) + 2\, H_2O(l) \qquad K_{spa} = ?$

$$K_{spa} = \frac{K_{sp}}{K_{a_1} \times K_w} = \frac{3 \times 10^{-28}}{(1.0 \times 10^{-7})(1.0 \times 10^{-14})} = 3 \times 10^{-7}$$

Example 19-14 illustrates the use of K_{spa} in the type of calculation needed to sort the sulfides into two qualitative analysis groups. Pb^{2+} is in qualitative analysis cation group 2, and Fe^{2+} is in group 3. The conditions cited in the example are those generally used.

> PbCl$_2$ is sufficiently soluble that enough Pb^{2+} ions from cation group 1 remain in solution to precipitate again as PbS(s) in cation group 2. ▶

KEEP IN MIND ▶
that the value of K_{spa} of FeS can be derived from K_{sp} of FeS (6×10^{-19}) by the method outlined for PbS.

EXAMPLE 19-14

Separating Metal Ions by Selective Precipitation of Metal Sulfides. Show that PbS(s) should precipitate but FeS(s) should not precipitate from a solution that is 0.010 M in Pb^{2+}, 0.010 M in Fe^{2+}, saturated in H_2S (0.10 M H_2S), and maintained with $[H_3O^+] = 0.30$ M. For PbS, $K_{spa} = 3 \times 10^{-7}$; for FeS, $K_{spa} = 6 \times 10^2$.

Solution

We must determine whether, for the stated conditions, equilibrium is displaced in the forward or reverse direction in reactions of the type

$$MS(s) + 2\, H_3O^+(aq) \rightleftharpoons M^{2+}(aq) + H_2S(aq) + 2\, H_2O(l)$$
$$(19.8)$$

where M represents either Pb or Fe. In each case, we can compare the Q_{spa} expression to the appropriate K_{spa} value for reaction (19.8). If $Q_{spa} > K_{spa}$, a net change should occur to the *left*. Hence MS(s) should precipitate. If $Q_{spa} < K_{spa}$, a net change should occur to the *right*. Hence, some of the metal sulfide could actually be dissolved in the solution, and no precipitation should occur.

$$Q_{spa} = \frac{[M^{2+}][H_2S]}{[H_3O^+]^2} = \frac{0.010 \times 0.10}{(0.30)^2} = 1.1 \times 10^{-2}$$

For PbS: Q_{spa} of $1.1 \times 10^{-2} > K_{spa}$ of 3×10^{-7}. We expect PbS(s) to precipitate.

For FeS: Q_{spa} of $1.1 \times 10^{-2} < K_{spa}$ of 6×10^2. FeS(s) should not precipitate.

Practice Example A: Show that $Ag_2S(s)$ $\left(K_{spa} = 6 \times 10^{-30}\right)$ should precipitate and that FeS(s) $\left(K_{spa} = 6 \times 10^2\right)$ should not precipitate from a solution that is 0.010 M Ag^+ and 0.020 M Fe^{2+}, but otherwise under the same conditions as in Example 19-14.

Practice Example B: What is the minimum pH of a solution that is 0.015 M Fe^{2+} and saturated in H_2S (0.10 M) from which FeS(s) $\left(K_{spa} = 6 \times 10^2\right)$ can be precipitated?

Dissolving Metal Sulfides

In the qualitative cation analysis, it is necessary to both precipitate and redissolve sulfides. Here, we look at several methods of dissolving metal sulfides. One way to increase the solubility of any sulfide is to allow it to react with an acid, as equation (19.8) suggests. According to Le Châtelier's principle, the solubility increases as the solution is made more acidic—equilibrium is shifted to the right. As a result, some water-insoluble sulfides, such as FeS, are readily soluble in strongly acidic solutions. Others, such as PbS and HgS, cannot be dissolved in acidic solutions because their K_{sp} values are too low. In these cases, $[H_3O^+]$ cannot be made high enough to force reaction (19.8) very far to the right.

Although most of Earth's minerals have formed by cooling from molten material called magma, a few have been produced by precipitation processes. For example, warm seawater is nearly saturated in Ca^{2+} and HCO_3^- ions and can yield the mineral *calcite* ($CaCO_3$) by precipitation.

$$Ca^{2+}(aq) + 2\ HCO_3^-(aq) \longrightarrow$$
$$CaCO_3(s) + H_2O(l) + CO_2(g)$$

Chemical precipitation, however, is not the principal source of calcite deposits in the ocean. Most calcite and related carbonate deposits originate from the biological precipitation of minerals to form the shells of certain marine organisms. These organisms can extract Ca^{2+} and HCO_3^- ions from seawater and concentrate them in solution in specialized cells. The organisms then secrete crystals of $CaCO_3$ to build the shells within which they live.

Some organisms secrete harder materials than carbonates and use them to form bones and teeth. Human bones and tooth enamel, for example, are made principally of the mineral *hydroxyapatite*, $Ca_5(PO_4)_3OH$. Hydroxyapatite is somewhat

▲ A fossilized chambered nautilus, a marine animal that forms its shell by precipitating calcium carbonate.

Another way to promote the dissolving of metal sulfides is to use an *oxidizing* acid such as $HNO_3(aq)$. In this case, sulfide ion is oxidized to elemental sulfur and the free metal ion appears in solution, as in the dissolving of $CuS(s)$.

$$3\ CuS(s) + 8\ H^+(aq) + 2\ NO_3^-(aq) \longrightarrow$$
$$3\ Cu^{2+}(aq) + 3\ S(s) + 2\ NO(g) + 4\ H_2O(l) \qquad (19.9)$$

To render the $Cu^{2+}(aq)$ more visible (Figure 19-9), it is converted to the deeply colored complex ion $[Cu(NH_3)_4]^{2+}(aq)$ in a reaction in which NH_3 molecules replace H_2O molecules in a complex ion.

▲ FIGURE 19-9   Complex-ion formation: a test for $Cu^{2+}(aq)$

Dilute $CuSO_4(aq)$ (left) derives its pale blue color from the complex ions $[Cu(H_2O)_4]^{2+}$. When $NH_3(aq)$ is added (here labeled "conc ammonium hydroxide"), the color changes to a deep violet, signaling the presence of $[Cu(NH_3)_4]^{2+}$ (right). The deep violet color is detectable at much lower concentrations than the pale blue; the formation of $[Cu(NH_3)_4]^{2+}$ is a sensitive test for the presence of Cu^{2+}.

soluble in acidic solutions, and this property is a matter of concern in good dental health. Bacteria in the mouth produce acids from food trapped between teeth. These acids cause tooth enamel to dissolve through reactions such as

$$Ca_5(PO_4)_3OH(s) + 4 H_3O^+(aq) \longrightarrow$$
$$5 Ca^{2+}(aq) + 3 HPO_4^{2-}(aq) + 5 H_2O(l)$$

If it proceeds unchecked, dissolution of tooth enamel, referred to as *demineralization*, results in the formation of cavities. Fortunately, the removal of the bacteria and food particles from teeth by brushing and flossing is one effective strategy for combating tooth decay.

Another strategy is to alter the composition of the tooth enamel to make it more resistant to attack by acids. *Fluorapatite*, $Ca_5(PO_4)_3F$, is less soluble in acids than is hydroxyapatite because F^- is a weaker base than OH^-. Brushing the teeth with a fluoride-containing toothpaste sets up a competitive equilibrium that results in the conversion of some hydroxyapatite to fluorapatite, a process called *remineralization*.

$$Ca_5(PO_4)_3OH(s) + F^-(aq) \rightleftharpoons$$
$$Ca_5(PO_4)_3F(s) + OH^-(aq) \qquad (19.10)$$

In addition to its lower solubility in acids, fluorapatite forms crystals that pack more densely than do crystals of hydroxyapatite, making the fluorapatite structure less porous and more resistant to penetration by acidic solutions.

Another source of fluoride ions for promoting the remineralization reaction (19.10) is drinking water. Some groundwater is naturally fluoridated, but many communities fluoridate drinking water by adding a soluble fluoride, enough to raise the F^- concentration in the water to about 1 ppm.

The same replacement of OH^- by F^- as described for fighting tooth decay occurs in the fossilization of bones and in natural deposits of hydroxyapatite. Fluoride ions in groundwater promote reaction (19.10). As a result, phosphate rock, used in the manufacture of phosphorus and phosphorus-containing fertilizers, is a mixture of $Ca_5(PO_4)_3OH$ and $Ca_5(PO_4)_3F$. As a further consequence, industrial operations that use phosphate rock as a source of phosphorus invariably generate large quantities of CaF_2 as an unwanted by-product. Ancient bones recovered by archeologists are composed almost entirely of fluorapatite. The fossilized (remineralized) bones retain the form of the original bones, including nicks, cuts, and breaks, and thus preserve a record of the life of the original animal.

$$[Cu(H_2O)_4]^{2+}(aq) + 4 NH_3(aq) \longrightarrow [Cu(NH_3)_4]^{2+}(aq) + 4 H_2O(l)$$

pale blue   deep blue

A few metal sulfides dissolve in a basic solution with a high concentration of HS^-, just as acidic oxides dissolve in solutions with a high concentration of OH^-. This property is used to advantage in separating the eight sulfides of cation group 2, the hydrogen sulfide group, into two subgroups. The subgroup consisting of HgS, PbS, CuS, CdS, and Bi_2S_3 remains unchanged after treatment with an alkaline solution with an excess of HS^-, but As_2S_3, SnS_2, and Sb_2S_3 dissolve.

Summary

Equilibrium between a slightly soluble ionic compound and its ions in solution is expressed by means of the solubility product constant, K_{sp}. Although K_{sp} and molar solubility are not equal, they are related to each other in a way that makes it possible to calculate one when the other is known. The solubility of a slightly soluble solute is greatly reduced in a solution containing an ion in common with the solubility equilibrium—a common ion.

A comparison of the ion product (reaction quotient) Q_{sp} with K_{sp} provides a criterion for precipitation: If $Q_{sp} > K_{sp}$, precipitation will occur; if $Q_{sp} < K_{sp}$, the solution will remain unsaturated. A comparison of K_{sp} values is a factor in determining the feasibility of fractional precipitation, a process in which one type of ion is removed by precipitation while others remain in solution. At times, because of acid–base equilibria that occur simultaneously, the selective precipitation or dissolving of ionic solutes depends on the pH.

The formation of a complex ion is an equilibrium process with an equilibrium constant called the formation constant, K_f. In general, if the formation constant of a complex ion is large, the concentration of uncomplexed metal ion in equilibrium with the complex ion is very small. Complex-ion formation can render certain insoluble materials quite soluble in appropriate aqueous solutions, such as AgCl in $NH_3(aq)$.

Precipitation, acid–base, oxidation–reduction, and complex-ion formation reactions are all used extensively in qualitative cation analysis. Such an analysis can provide a rapid means of determining the presence or absence of certain cations in an unknown material.

Integrative Example

Lime (quicklime), CaO, is obtained from the high-temperature decomposition of limestone ($CaCO_3$). Quicklime is the cheapest source of basic substances, but it is water insoluble. It does react with water, however, producing $Ca(OH)_2$ (slaked lime). Unfortunately, $Ca(OH)_2(s)$ has limited solubility, so it cannot be used to prepare aqueous solutions of high pH.

$$Ca(OH)_2(s) \rightleftharpoons Ca^{2+}(aq) + 2\,OH^-(aq) \quad K_{sp} = 5.5 \times 10^{-6}$$

When $Ca(OH)_2(s)$ reacts with a *soluble* carbonate, such as $Na_2CO_3(aq)$, however, the solution produced has a much higher pH. Equilibrium is displaced to the right in reaction (19.11) because $CaCO_3$ is much less soluble than $Ca(OH)_2$.

$$Ca(OH)_2(s) + CO_3^{2-}(aq) \rightleftharpoons$$
$$CaCO_3(s) + 2\,OH^-(aq) \quad (19.11)$$

Assume an initial $[CO_3^{2-}] = 1.0$ M in reaction (19.11), and show that the equilibrium pH should indeed be higher than that in saturated $Ca(OH)_2(aq)$.

1. *Calculate the pH of saturated $Ca(OH)_2(aq)$.* Write the K_{sp} expression for $Ca(OH)_2$. Let s be the molar solubility, so $[OH^-] = 2s$.

$$K_{sp} = [Ca^{2+}][OH^-]^2 = (s)(2s)^2 = 4s^3 = 5.5 \times 10^{-6}$$
$$s = \left(5.5 \times 10^{-6}/4\right)^{1/3} = 0.011 \text{ M}$$
$$[OH^-] = 2s = 0.022 \text{ M} \quad pOH = -\log[OH^-] = -\log 0.022$$
$$pOH = 1.66 \quad pH = 14.00 - pOH = 14.00 - 1.66 = 12.34$$

2. *Determine the value of K_c for reaction (19.11).* Combine solubility equilibrium equations for $Ca(OH)_2$ and $CaCO_3$ to obtain (19.11) as the overall net ionic equation. Combine K_{sp} values of $Ca(OH)_2$ and $CaCO_3$ in the appropriate way to obtain the overall K_c.

$$Ca(OH)_2(s) \rightleftharpoons Ca^{2+}(aq) + 2\,OH^-(aq)$$
$$K_{sp} = 5.5 \times 10^{-6}$$
$$Ca^{2+}(aq) + CO_3^{2-}(aq) \rightleftharpoons CaCO_3(s)$$
$$K = 1/K_{sp} = 1/(2.8 \times 10^{-9})$$
$$Ca(OH)_2(s) + CO_3^{2-}(aq) \rightleftharpoons CaCO_3(s) + 2\,OH^-(aq)$$
$$K = 5.5 \times 10^{-6}/2.8 \times 10^{-9} = 2.0 \times 10^3$$

3. *Calculate the equilibrium pH in reaction (19.11).* Start with $[CO_3^{2-}] = 1.0$ M, and proceed in this familiar fashion.

$$\textbf{Ca(OH)}_2\textbf{(s)} + \textbf{CO}_3^{2-}\textbf{(aq)} \rightleftharpoons \textbf{CaCO}_3\textbf{(s)} + \textbf{2 OH}^-\textbf{(aq)}$$

initial concns, M:	1.0		≈ 0
changes, M:	$-x$		$+2x$
equil concns, M:	$1.0 - x$		$2x$

$$K = \frac{[OH^-]^2}{[CO_3^{2-}]} = \frac{(2x)^2}{(1.0 - x)} = 2.0 \times 10^3$$

The usual simplifying assumption that x is small compared to 1.0 definitely will *not* work here. The value of K is *large*, and equilibrium will be displaced far to the right. The value of x is likely to be much closer to 1.0 than to zero. We must solve the quadratic equation $4x^2 + (2.0 \times 10^3)x - (2.0 \times 10^3) = 0$. The solution to this equation is $x = 0.998 \approx 1$. The hydroxide ion concentration is $[OH^-] = 2x \approx 2$ M. The pOH of the solution is $-\log 2 = -0.3$, and $pH = 14.00 - pOH = 14.00 - (-0.3) = 14.3$, higher than the 12.34 that we obtained for the pH of saturated $Ca(OH)_2(aq)$. Reaction (19.11) produces a solution with nearly 100 times as much OH^- as is found in a saturated solution of $Ca(OH)_2$.

Key Terms

complex ion (19-8)
coordination compound (19-8)
formation constant, K_f (19-8)

fractional precipitation (19-6)
ion pairs (19-4)
ion product, Q_{sp} (19-5)

qualitative cation analysis (19-9)
salt effect (19-4)
solubility product constant, K_{sp} (19-1)

Review Questions

1. In your own words, define the following terms or symbols: **(a)** K_{sp}; **(b)** K_f; **(c)** Q_{sp}; **(d)** complex ion.
2. Briefly describe each of the following ideas, methods, or phenomena: **(a)** common-ion effect in solubility equilibrium; **(b)** fractional precipitation; **(c)** ion-pair formation; **(d)** qualitative cation analysis.
3. Explain the important distinction between each pair of terms: **(a)** solubility and solubility product constant; **(b)** common-ion effect and salt effect; **(c)** ion pair and ion product.
4. Write K_{sp} expressions for the following equilibria. For example, for the reaction $AgCl(s) \rightleftharpoons Ag^+(aq) + Cl^-(aq)$, $K_{sp} = [Ag^+][Cl^-]$.
   (a) $Ag_2SO_4(s) \rightleftharpoons 2\,Ag^+(aq) + SO_4^{2-}(aq)$
   (b) $Ra(IO_3)_2(s) \rightleftharpoons Ra^{2+}(aq) + 2\,IO_3^-(aq)$
   (c) $Ni_3(PO_4)_2(s) \rightleftharpoons 3\,Ni^{2+}(aq) + 2\,PO_4^{3-}(aq)$
   (d) $PuO_2CO_3(s) \rightleftharpoons PuO_2^{2+}(aq) + CO_3^{2-}(aq)$

5. Write solubility equilibrium equations that are described by the following K_{sp} expressions. For example, $K_{sp} = [Ag^+][Cl^-]$ represents $AgCl(s) \rightleftharpoons Ag^+(aq) + Cl^-(aq)$.
   (a) $K_{sp} = [Fe^{3+}][OH^-]^3$
   (b) $K_{sp} = [BiO^+][OH^-]$
   (c) $K_{sp} = [Hg_2^{2+}][I^-]^2$
   (d) $K_{sp} = [Pb^{2+}]^3[AsO_4^{3-}]^2$
6. The following K_{sp} values are found in a handbook. Write the solubility product expression to which each one applies. For example, $K_{sp}(AgCl) = [Ag^+][Cl^-] = 1.8 \times 10^{-10}$.
   (a) $K_{sp}(CrF_3) = 6.6 \times 10^{-11}$
   (b) $K_{sp}[Au_2(C_2O_4)_3] = 1 \times 10^{-10}$
   (c) $K_{sp}[Cd_3(PO_4)_2] = 2.1 \times 10^{-33}$
   (d) $K_{sp}(SrF_2) = 2.5 \times 10^{-9}$
7. Calculate the aqueous solubility, in moles per liter, of each of the following.

(a) $BaCrO_4$, $K_{sp} = 1.2 \times 10^{-10}$
(b) $PbBr_2$, $K_{sp} = 4.0 \times 10^{-5}$
(c) CeF_3, $K_{sp} = 8 \times 10^{-16}$
(d) $Mg_3(AsO_4)_2$, $K_{sp} = 2.1 \times 10^{-20}$

8. The following aqueous solubility data, expressed as molarities, are from a handbook: (a) $CsMnO_4$, 3.8×10^{-3} M; (b) $Pb(ClO_2)_2$, 2.8×10^{-3} M; (c) $In(IO_3)_3$, 1.0×10^{-3} M. What are the K_{sp} values of these solutes?

9. Pure water is saturated with slightly soluble PbI_2. Which of the following is a correct statement concerning the lead ion concentration in the solution, and what is wrong with the others? (a) $[Pb^{2+}] = [I^-]$; (b) $[Pb^{2+}] = K_{sp}$ of PbI_2; (c) $[Pb^{2+}] = \sqrt{K_{sp}}$ of PbI_2; (d) $[Pb^{2+}] = 0.5[I^-]$.

10. Calculate the molar solubility of $Mg(OH)_2$ $\left(K_{sp} = 1.8 \times 10^{-11}\right)$ in (a) pure water; (b) 0.0862 M $MgCl_2$; (c) 0.0355 M KOH(aq).

11. How would you expect the presence of each of the following solutes to affect the molar solubility of $CaCO_3$ in water: (a) Na_2CO_3; (b) HCl; (c) $NaHSO_4$? Explain.

12. Predict whether a precipitate would form in a solution with the following ion concentrations:
(a) $[Mg^{2+}] = 0.0037$ M, $[CO_3^{2-}] = 0.0068$ M
$$K_{sp} \text{ of } MgCO_3 = 3.5 \times 10^{-8}$$
(b) $[Ag^+] = 0.018$ M, $[SO_4^{2-}] = 0.0062$ M
$$K_{sp} \text{ of } Ag_2SO_4 = 1.4 \times 10^{-5}$$
(c) $[Cr^{3+}] = 0.038$ M, pH $= 3.20$
$$K_{sp} \text{ of } Cr(OH)_3 = 6.3 \times 10^{-31}$$

13. A solution that is 0.103 M $CaCl_2$ is also made to be 0.750 M K_2SO_4. What percentage of the Ca^{2+} remains *unprecipitated*? Will the precipitation go essentially to completion? Assume that $[SO_4^{2-}]$ remains constant at 0.750 M. K_{sp} of $CaSO_4 = 9.1 \times 10^{-6}$.

14. KI(aq) is slowly added to a solution with $[Pb^{2+}] = [Ag^+] = 0.10$ M. For PbI_2, $K_{sp} = 7.1 \times 10^{-9}$; for AgI, $K_{sp} = 8.5 \times 10^{-17}$.
(a) Which precipitate should form first, PbI_2 or AgI?
(b) What $[I^-]$ is required for the *second* cation to begin to precipitate?
(c) What concentration of the first cation to precipitate remains in solution at the point at which the second cation begins to precipitate?

(d) Can Pb^{2+}(aq) and Ag^+(aq) be effectively separated by fractional precipitation of their iodides?

15. In which of the following solutions would you expect $Mg(OH)_2$(s) to be most soluble: NaOH(aq), Na_2CO_3(aq), or NH_4Cl(aq)? Explain your choice.

16. Complete and balance the following equations. If no reaction occurs, so state.
(a) Ag^+(aq) $+ NO_3^-$(aq) $+ Na^+$(aq) $+ Br^-$(aq) $\longrightarrow$
(b) Cu^{2+}(aq) $+ NO_3^-$(aq) $+ H_3O^+$(aq) $+ Cl^-$(aq) $\longrightarrow$
(c) Fe^{2+}(aq) $+ H_2S$ (aq, in 0.3 M HCl) $\longrightarrow$
(d) $Cu(OH)_2$(s) $+ NH_3$(aq) $\longrightarrow$
(e) Fe^{3+}(aq) $+ NH_3$(aq) $+ H_2O$(l) $\longrightarrow$
(f) Ag_2SO_4(s) $+ NH_3$(aq) $\longrightarrow$
(g) $CaSO_3$(s) $+ H_3O^+$(aq) $\longrightarrow$

17. $Cu(OH)_2$(s) is insoluble in water but reacts to dissolve in each of the following: HCl(aq), NH_3(aq), and HNO_3(aq). Write net ionic equations for each of these reactions.

18. Both Cu^{2+} and Ag^+ are present in the same aqueous solution. Which of the following reagents would work best in separating these ions, precipitating one and leaving the other in solution: $(NH_4)_2CO_3$(aq), HCl(aq), or NaOH(aq)? Explain your choice.

19. Both Mg^{2+} and Cu^{2+} are present in the same aqueous solution. Which of the following reagents would work best in separating these ions, precipitating one and leaving the other in solution: NaOH(aq), HCl(aq), or NH_3(aq)? Explain your choice.

20. Should $Al(OH)_3$(s) precipitate from a buffer solution that is 0.45 M $HC_2H_3O_2$ and 0.35 M $NaC_2H_3O_2$ and also 0.275 M in Al^{3+}(aq)? For $Al(OH)_3$, $K_{sp} = 1.3 \times 10^{-33}$; for $HC_2H_3O_2$, $K_a = 1.8 \times 10^{-5}$.

21. Should AgI(s) precipitate from a solution with $[[Ag(CN)_2]^-] = 0.012$ M, $[CN^-] = 1.05$ M, and $[I^-] = 2.0$ M? For AgI, $K_{sp} = 8.5 \times 10^{-17}$; for $[Ag(CN)_2]^-$, $K_f = 5.6 \times 10^{18}$.

22. A solution is 1.6 M in $[Ag(NH_3)_2]^+$ and 1.25 M in *free* NH_3. What is the maximum $[Cl^-]$ that can be maintained in the solution without forming a precipitate of AgCl(s)? For AgCl, $K_{sp} = 1.8 \times 10^{-10}$; for $[Ag(NH_3)_2]^+$, $K_f = 1.6 \times 10^7$.

Exercises *(Use data from Chapters 17 and 19 and Appendix D, as needed.)*

K_{sp} and Solubility

23. Arrange the following solutes in order of increasing molar solubility in water: AgCN, $AgIO_3$, AgI, $AgNO_2$, Ag_2SO_4. Explain your reasoning.

24. Which of the following saturated aqueous solutions would have the highest $[Mg^{2+}]$: (a) $MgCO_3$; (b) MgF_2; (c) $Mg_3(PO_4)_2$? Explain.

25. Fluoridated drinking water contains about 1 part per million (ppm) of F^-. Is CaF_2 sufficiently soluble in water to be used as the source of fluoride ion for the fluoridation of drinking water? Explain.
(*Hint:* Think of 1 ppm as signifying 1 g F^- per 10^6 g solution.)

26. In the qualitative cation analysis procedure, Bi^{3+} is detected by the appearance of a white precipitate of bismuthyl hydroxide, BiOOH(s):
$$BiOOH(s) \rightleftharpoons BiO^+(aq) + OH^-(aq) \quad K_{sp} = 4 \times 10^{-10}$$
Calculate the pH of a saturated aqueous solution of BiOOH.

27. A solution is saturated with magnesium palmitate $[Mg(C_{16}H_{31}O_2)_2$, a component of bathtub ring] at 50 °C. How many milligrams of magnesium palmitate will precipitate from 965 mL of this solution when it is cooled to 25 °C? For $Mg(C_{16}H_{31}O_2)_2$, $K_{sp} = 4.8 \times 10^{-12}$ at 50 °C and 3.3×10^{-12} at 25 °C.

28. A 725-mL sample of a saturated aqueous solution of calcium oxalate, CaC_2O_4, at 95 °C is cooled to 13 °C. How many milligrams of calcium oxalate will precipitate? For CaC_2O_4, $K_{sp} = 1.2 \times 10^{-8}$ at 95 °C and 2.7×10^{-9} at 13 °C.

29. A 25.00-mL sample of a clear *saturated* solution of PbI_2 requires 13.3 mL of a certain $AgNO_3(aq)$ for its titration. What is the molarity of this $AgNO_3(aq)$?

$$I^-(\text{satd } PbI_2) + Ag^+(\text{from } AgNO_3) \longrightarrow AgI(s)$$

30. A 250-mL sample of saturated $CaC_2O_4(aq)$ requires 4.8 mL of 0.00134 M $KMnO_4(aq)$ for its titration in an acidic solution. What is the value of K_{sp} for CaC_2O_4 obtained with these data? In the titration reaction, $C_2O_4^{2-}$ is oxidized to CO_2 and MnO_4^- is reduced to Mn^{2+}.

31. To precipitate as $Ag_2S(s)$, all the Ag^+ present in 338 mL of a saturated solution of $AgBrO_3$ requires 30.4 mL of $H_2S(g)$ measured at 23 °C and 748 mmHg. What is K_{sp} for $AgBrO_3$?

32. Excess $Ca(OH)_2(s)$ is shaken with water to produce a saturated solution. A 50.00-mL sample of the clear saturated solution is withdrawn and requires 10.7 mL of 0.1032 M HCl for its titration. What is K_{sp} for $Ca(OH)_2$?

The Common-Ion Effect

33. Describe the effects of the salts KI and $AgNO_3$ on the solubility of AgI in water.

34. Describe the effect of the salt KNO_3 on the solubility of AgI in water, and explain why it is different from the effects noted in Exercise 33.

35. A 0.150 M Na_2SO_4 solution that is saturated with Ag_2SO_4 has $[Ag^+] = 9.7 \times 10^{-3}$ M. What is the value of K_{sp} for Ag_2SO_4 obtained with these data?

36. If 100.0 mL of 0.0025 M $Na_2SO_4(aq)$ is saturated with $CaSO_4$, how many grams of $CaSO_4$ would be present in the solution?

(*Hint:* Does the usual simplifying assumption hold?)

37. What $[Pb^{2+}]$ should be maintained in $Pb(NO_3)_2(aq)$ to produce a solubility of 1.5×10^{-4} mol PbI_2/L when $PbI_2(s)$ is added?

38. What $[I^-]$ should be maintained in KI(aq) to produce a solubility of 1.5×10^{-5} mol PbI_2/L when $PbI_2(s)$ is added?

39. Can the solubility of Ag_2CrO_4 be lowered to 5.0×10^{-8} mol Ag_2CrO_4/L by using CrO_4^{2-} as the common ion? by using Ag^+? Explain.

40. A handbook lists the K_{sp} values 1.1×10^{-10} for $BaSO_4$ and 5.1×10^{-9} for $BaCO_3$. When 0.50 M $Na_2CO_3(aq)$ is added to saturated $BaSO_4(aq)$, a precipitate of $BaCO_3(s)$ forms. How do you account for this fact, given that $BaCO_3$ has a larger K_{sp} than does $BaSO_4$?

41. A particular water sample that is saturated in CaF_2 has a Ca^{2+} content of 115 ppm (that is, 115 g Ca^{2+} per 10^6 g of water sample). What is the F^- ion content of the water in ppm?

42. Assume that, to be visible to the unaided eye, a precipitate must weigh more than 1 mg. If you add 1.0 mL of 1.0 M NaCl(aq) to 100.0 mL of a clear saturated aqueous AgCl solution, will you be able to see AgCl(s) precipitated as a result of the common-ion effect? Explain.

Criteria for Precipitation from Solution

43. Should precipitation of $MgF_2(s)$ occur if a 22.5-mg sample of $MgCl_2 \cdot 6H_2O$ is added to 325 mL of 0.035 M KF?

44. Should $PbCl_2(s)$ precipitate when 155 mL of 0.016 M KCl(aq) is added to 245 mL of 0.175 M $Pb(NO_3)_2(aq)$?

45. What is the minimum pH at which $Cd(OH)_2(s)$ will precipitate from a solution that is 0.0055 M in $Cd^{2+}(aq)$?

46. What is the minimum pH at which $Cr(OH)_3(s)$ will precipitate from a solution that is 0.086 M in $Cr^{3+}(aq)$?

47. Should precipitation occur in the following cases?
(a) 0.10 mg NaCl is added to 1.0 L of 0.10 M $AgNO_3(aq)$.
(b) One drop (0.05 mL) of 0.10 M KBr is added to 250 mL of a saturated solution of AgCl.
(c) One drop (0.05 mL) of 0.0150 M NaOH(aq) is added to 3.0 L of a solution with 2.0 mg Mg^{2+} per liter.

48. The electrolysis of $MgCl_2(aq)$ can be represented as
$$Mg^{2+}(aq) + 2\,Cl^-(aq) + 2\,H_2O(l) \longrightarrow$$
$$Mg^{2+}(aq) + 2\,OH^-(aq) + H_2(g) + Cl_2(g)$$

The electrolysis of a 315-mL sample of 0.185 M $MgCl_2$ is continued until 0.652 L $H_2(g)$ at 22 °C and 752 mmHg has been collected. Will $Mg(OH)_2(s)$ precipitate when electrolysis is carried to this point?

(*Hint:* Notice that $[Mg^{2+}]$ remains constant throughout the electrolysis, but $[OH^-]$ *increases*.)

49. Determine whether 1.50 g $H_2C_2O_4$ (oxalic acid: $K_{a_1} = 5.2 \times 10^{-2}$, $K_{a_2} = 5.4 \times 10^{-5}$) can be dissolved in 0.200 L of 0.150 M $CaCl_2$ without the formation of $CaC_2O_4(s)$ ($K_{sp} = 1.3 \times 10^{-9}$).

50. If 100.0 mL of a clear saturated solution of Ag_2SO_4 is added to 250.0 mL of a clear saturated solution of $PbCrO_4$, will any precipitate form?

(*Hint:* Take into account the dilutions that occur. What are the possible precipitates?)

Completeness of Precipitation

51. When 200.0 mL of 0.350 M K_2CrO_4(aq) is added to 200.0 mL of 0.0100 M $AgNO_3$(aq), what percentage of the Ag^+ is left *unprecipitated*?

52. What percentage of the original Ag^+ remains in solution if 175 mL 0.0208 M $AgNO_3$ is added to 250 mL 0.0380 M K_2CrO_4?

53. If a constant $[Cl^-] = 0.100$ M is maintained in a solution in which the initial $[Pb^{2+}] = 0.065$ M, what percentage of

the Pb^{2+} will remain in solution after $PbCl_2$(s) precipitates? What $[Cl^-]$ should be maintained to ensure that only 1.0% of the Pb^{2+} remains unprecipitated?

54. The ancient Romans added calcium sulfate to wine to clarify it and to remove dissolved lead. What is the maximum $[Pb^{2+}]$ that might be present in wine to which calcium sulfate has been added?

Fractional Precipitation

55. For the fractional precipitation described in Example 19-7, explain why it was not necessary to specify the concentration of the $AgNO_3$(aq) that was used.

56. Which one of the following solutions can be used to separate the cations in an aqueous solution in which $[Ba^{2+}] = [Ca^{2+}] = 0.050$ M: 0.10 M NaCl(aq), 0.05 M Na_2SO_4(aq), 0.001 M NaOH(aq), or 0.50 M Na_2CO_3(aq)? Explain why.

57. A solution is 0.010 M in both CrO_4^{2-} and SO_4^{2-}. To this solution, 0.50 M $Pb(NO_3)_2$(aq) is slowly added.
(a) What should be the first anion to precipitate from solution?

(b) What is $[Pb^{2+}]$ at the point at which the second anion begins to precipitate?
(c) Are the two anions effectively separated by this fractional precipitation?

58. $AgNO_3$(aq) is slowly added to a solution that is 0.250 M NaCl and also 0.0022 M KBr.
(a) Which anion should precipitate first, Cl^- or Br^-?
(b) What is $[Ag^+]$ at the point at which the second anion begins to precipitate?
(c) Can the Cl^- and Br^- be separated effectively by this fractional precipitation?

Solubility and pH

59. Which of the following solids is (are) likely to be more soluble in an acidic solution than in pure water: KCl, $MgCO_3$, FeS, $Ca(OH)_2$, or C_6H_5COOH? Explain.

60. Which of the following solids is (are) likely to be more soluble in a basic solution than in pure water: $BaSO_4$, $H_2C_2O_4$, $Fe(OH)_3$, $NaNO_3$, or MnS? Explain.

61. The solubility of $Mg(OH)_2$ in a particular buffer solution is 0.65 g/L. What must be the pH of the buffer solution?

62. To 0.350 L of 0.150 M NH_3 is added 0.150 L of 0.100 M $MgCl_2$. How many grams of $(NH_4)_2SO_4$ should be present to prevent precipitation of $Mg(OH)_2$(s)?

63. For the equilibrium
$$Al(OH)_3(s) \rightleftharpoons Al^{3+}(aq) + 3\,OH^-(aq)$$
$$K_{sp} = 1.3 \times 10^{-33}$$

(a) What is the *minimum pH* at which $Al(OH)_3$(s) will precipitate from a solution that is 0.075 M in Al^{3+}?
(b) A solution has $[Al^{3+}] = 0.075$ M and $[HC_2H_3O_2] = 1.00$ M. What is the maximum quantity of $NaC_2H_3O_2$ that can be added to 250.0 mL of this solution before precipitation of $Al(OH)_3$(s) begins?

64. Should the following precipitates form under the given conditions?
(a) PbI_2(s) from a solution that is 1.05×10^{-3} M HI, 1.05×10^{-3} M NaI, and 1.1×10^{-3} M $Pb(NO_3)_2$
(b) $Mg(OH)_2$(s), from 2.50 L of 0.0150 M $Mg(NO_3)_2$ to which is added 1 drop (0.05 mL) of 6.00 M NH_3
(c) $Al(OH)_3$(s), from a solution that is 0.010 M in Al^{3+}, 0.010 M $HC_2H_3O_2$, and 0.010 M $NaC_2H_3O_2$

Complex-Ion Equilibria

65. $PbCl_2$(s) is considerably more soluble in HCl(aq) than in pure water, but its solubility in HNO_3(aq) is not much different from what it is in water. Explain this difference in behavior.

66. Which of the following would be most effective, and which would be least effective, in reducing the concentration of $[Zn(NH_3)_4]^{2+}$ in a solution containing that complex ion: HCl, NH_3, or NH_4Cl? Explain your choices.

67. In a solution that is 0.0500 M in $[Cu(CN)_4]^{3-}$ and 0.80 M in free CN^-, the concentration of Cu^+ is 6.1×10^{-32} M. Calculate K_f of $[Cu(CN)_4]^{3-}$.
$$Cu^+(aq) + 4\,CN^-(aq) \rightleftharpoons [Cu(CN)_4]^{3-}(aq) \quad K_f = ?$$

68. Calculate $[Cu^{2+}]$ in a 0.10 M $CuSO_4$(aq) solution that is also 6.0 M in free NH_3.
$$Cu^{2+}(aq) + 4\,NH_3(aq) \rightleftharpoons [Cu(NH_3)_4]^{2+}(aq)$$
$$K_f = 1.1 \times 10^{13}$$

69. Can the following ion concentrations be maintained in the same solution without a precipitate forming: $[[Ag(S_2O_3)_2]^{3-}] = 0.048$ M, $[S_2O_3^{2-}] = 0.76$ M, and $[I^-] = 2.0$ M?

70. A solution is 0.10 M in *free* NH_3, 0.10 M in NH_4Cl, and 0.015 M in $[Cu(NH_3)_4]^{2+}$. Should $Cu(OH)_2(s)$ precipitate from this solution? K_{sp} of $Cu(OH)_2$ is 2.2×10^{-20}.

71. A 0.10-mol sample of $AgNO_3(s)$ is dissolved in 1.00 L of 1.00 M NH_3. How many grams of KI can be dissolved in this solution without a precipitate of $AgI(s)$ forming?

72. A solution is prepared that has $[NH_3] = 1.00$ M and $[Cl^-] = 0.100$ M. How many grams of $AgNO_3$ can be dissolved in 1.00 L of this solution without a precipitate of $AgCl(s)$ forming?

Precipitation and Solubilities of Metal Sulfides

73. Can Fe^{2+} and Mn^{2+} be separated by precipitating $FeS(s)$ and not $MnS(s)$? Assume $[Fe^{2+}] = [Mn^{2+}] = [H_2S] = 0.10$ M. Choose a $[H_3O^+]$ that ensures maximum precipitation of $FeS(s)$ but not $MnS(s)$. Will the separation be complete? For FeS, $K_{spa} = 6 \times 10^2$; for MnS, $K_{spa} = 3 \times 10^7$.

74. A solution is 0.05 M in Cu^{2+}, in Hg^{2+}, and in Mn^{2+}. Which sulfides will precipitate if the solution is made to be 0.10 M $H_2S(aq)$ and 0.010 M HCl(aq)? For CuS, $K_{spa} = 6 \times 10^{-16}$; for HgS, $K_{spa} = 2 \times 10^{-32}$; for MnS, $K_{spa} = 3 \times 10^7$.

75. A buffer solution is 0.25 M $HC_2H_3O_2$–0.15 M $NaC_2H_3O_2$, saturated in H_2S (0.10 M), and it has $[Mn^{2+}] = 0.15$ M.
    (a) Show that MnS will *not* precipitate from this solution (for MnS, $K_{spa} = 3 \times 10^7$).

(b) Which buffer component would you increase in concentration, and to what minimum value, to ensure that precipitation of MnS(s) begins? Assume that the concentration of the other buffer component is held constant. [*Hint:* Recall equation (19.8).]

76. The following expressions pertain to the precipitation or dissolving of metal sulfides. Use information about the qualitative cation analysis scheme to predict whether a reaction proceeds to a significant extent in the forward direction and what the products are in each case.
    (a) $Cu^{2+}(aq) + H_2S(satd\ aq) \longrightarrow$
    (b) $Mg^{2+}(aq) + H_2S(satd\ aq) \xrightarrow{0.3\ M\ HCl}$
    (c) $PbS(s) + HCl\ (0.3\ M) \longrightarrow$
    (d) $ZnS(s) + HNO_3(aq) \longrightarrow$

Qualitative Cation Analysis

77. Suppose you did a group 1 qualitative cation analysis and treated the chloride precipitate with $NH_3(aq)$ without first treating it with hot water. What might you observe, and what valid conclusions could you reach about cations present, cations absent, and cations in doubt?

78. Show that in qualitative cation analysis group 1, if you obtain 1.00 mL of saturated $PbCl_2(aq)$ at 25 °C, sufficient Pb^{2+} should be present to produce a precipitate of $PbCrO_4(s)$. Assume that you use 1 drop (0.05 mL) of 1.0 M K_2CrO_4 for the test.

79. The addition of HCl(aq) to a solution containing several different cations produces a white precipitate. The filtrate is removed and treated with $H_2S(aq)$ in 0.3 M HCl. No pre-

cipitate forms. Which of the following conclusions is (are) valid? Explain.
    (a) Ag^+ or Hg_2^{2+} (or both) is probably present.
    (b) Mg^{2+} is probably not present.
    (c) Pb^{2+} is probably not present.
    (d) Fe^{2+} is probably not present.

80. Write net ionic equations for the following qualitative cation analysis procedures.
    (a) precipitation of $PbCl_2(s)$ from a solution containing Pb^{2+}
    (b) dissolution of $Zn(OH)_2(s)$ in a solution of NaOH(aq)
    (c) dissolution of $Fe(OH)_3(s)$ in HCl(aq)
    (d) precipitation of CuS(s) from an acidic solution of Cu^{2+} and H_2S

Integrative and Advanced Exercises (*Use data from Chapters 17 and 19 and Appendix D as needed.*)

81. A particular water sample has 131 ppm of $CaSO_4$ (131 g $CaSO_4$ per 10^6 g water). If this water is boiled in a teakettle, approximately what fraction of the water must be evaporated before $CaSO_4(s)$ begins to precipitate? Assume that the solubility of $CaSO_4(s)$ does not change much in the temperature range 0 to 100 °C.

82. A handbook lists the solubility of $CaHPO_4$ as 0.32 g $CaHPO_4 \cdot 2H_2O/L$ and lists $K_{sp} = 1 \times 10^{-7}$.
$$CaHPO_4(s) \rightleftharpoons Ca^{2+}(aq) + HPO_4^{2-}(aq)$$
$$K_{sp} = 1 \times 10^{-7}$$

(a) Are these data consistent? (That is, are the molar solubilities the same when derived in two different ways?)
(b) If there is a discrepancy, how do you account for it?

83. A 50.0-mL sample of 0.0152 M Na_2SO_4(aq) is added to 50.0 mL of 0.0125 M $Ca(NO_3)_2$(aq). What percentage of the Ca^{2+} remains unprecipitated?

84. What percentage of the Ba^{2+} in solution is precipitated as $BaCO_3$(s) if equal volumes of 0.0020 M Na_2CO_3(aq) and 0.0010 M $BaCl_2$(aq) are mixed?

85. Determine the molar solubility of lead azide, $Pb(N_3)_2$, in a buffer solution with pH = 3.00, given that

$$Pb(N_3)_2(s) \rightleftharpoons Pb^{2+}(aq) + 2 N_3^-(aq)$$
$$K_{sp} = 2.5 \times 10^{-9}$$
$$HN_3(aq) + H_2O(l) \rightleftharpoons H_3O^+(aq) + N_3^-(aq)$$
$$K_a = 1.9 \times 10^{-5}$$

86. Calculate the molar solubility of $Mg(OH)_2$ in 1.00 M NH_4Cl(aq).

87. The chief compound in marble is $CaCO_3$. Marble has been widely used for statues and ornamental work on buildings, including the Taj Mahal shown below (see also, page 701). However, marble is readily attacked by acids. Determine the solubility of marble (that is, $[Ca^{2+}]$ in a saturated solution) in **(a)** normal rainwater of pH = 5.6; **(b)** acid rainwater of pH = 4.20. Assume that the overall reaction that occurs is
$$CaCO_3(s) + H_3O^+(aq) \rightleftharpoons$$
$$Ca^{2+}(aq) + HCO_3^-(aq) + H_2O(l).$$

88. What is the solubility of MnS, in grams per liter, in a buffer solution that is 0.100 M $HC_2H_3O_2$–0.500 M $NaC_2H_3O_2$? For MnS, $K_{spa} = 3 \times 10^7$.

89. Write net ionic equations to represent each of the following observations.
(a) When concentrated $CaCl_2$(aq) is added to Na_2HPO_4(aq), a white precipitate forms that is 38.7% Ca by mass.
(b) When a piece of dry ice [CO_2(s)] is placed in a clear dilute solution of limewater [$Ca(OH)_2$(aq)], bubbles of gas evolve. At first, a white precipitate forms, but then it redissolves.

90. Concerning the reactions described in Exercise 89(b),
(a) Will the same observations be made if $Ca(OH)_2$(aq) is replaced by $CaCl_2$(aq)? Explain.
(b) Show that the white precipitate will redissolve if the $Ca(OH)_2$(aq) is about 0.005 M, but not if the solution is saturated.

91. Reaction (19.11), described in the Integrative Example, is called a *carbonate transposition*. In such a reaction, anions of a slightly soluble compound (for example, hydroxides and sulfates) are obtained in a sufficient concentration in aqueous solution that they can be identified by qualitative analysis tests. Suppose that 3 M Na_2CO_3 is used and that an anion concentration of 0.050 M is sufficient for its detection. Predict whether carbonate transposition will be effective for detecting **(a)** SO_4^{2-} from $BaSO_4$(s); **(b)** Cl^- from AgCl(s); **(c)** F^- from MgF_2(s).

92. For the titration in Example 19-7, verify the assertion in the Are You Wondering feature on page 761 that $[Ag^+]$ increases very rapidly between the point at which AgBr has finished precipitating and Ag_2CrO_4 is about to begin.

93. Aluminum compounds are soluble in acidic solution, where aluminum(III) exists as the complex ion $[Al(H_2O)_6]^{3+}$, which we generally represent simply as Al^{3+}(aq). They are also soluble in basic solutions, where the aluminum(III) is present as the complex ion $[Al(OH)_4]^-$. At certain intermediate pH values, the concentration of aluminum(III) that can exist in solution is at a minimum. Thus, a plot of the total concentration of aluminum(III) in solution as a function of pH yields a U-shaped curve. Demonstrate that this is the case with a few calculations, including that of the approximate pH at the minimum point on the curve.

94. The solubility of AgCN(s) in 0.200 M NH_3(aq) is 8.8×10^{-6} mol/L. Calculate K_{sp} for AgCN.

95. The solubility of $CdCO_3$(s) in 1.00 M KI(aq) is 1.2×10^{-3} mol/L. Given that K_{sp} of $CdCO_3$ is 5.2×10^{-12}, what is K_f for $[CdI_4]^{2-}$?

96. Use K_{sp} for $PbCl_2$ and K_f for $[PbCl_3]^-$ to determine the molar solubility of $PbCl_2$ in 0.10 M HCl(aq).
(*Hint:* What is the total concentration of lead species in solution?)

97. A mixture of $PbSO_4$(s) and PbS_2O_3(s) is shaken with pure water until a saturated solution is formed. Both solids remain in excess. What is $[Pb^{2+}]$ in the saturated solution? For $PbSO_4$, $K_{sp} = 1.6 \times 10^{-8}$; for PbS_2O_3, $K_{sp} = 4.0 \times 10^{-7}$.

98. Use the method of Exercise 97 to determine $[Pb^{2+}]$ in a saturated solution in contact with a mixture of $PbCl_2$(s) and $PbBr_2$(s).

99. A 2.50-g sample of Ag_2SO_4(s) is added to a beaker containing 0.150 L of 0.025 M $BaCl_2$.
(a) Write an equation for any reaction that occurs.
(b) Describe the final contents of the beaker—that is, the masses of any precipitates present and the concentrations of the ions in solution.

Feature Problems

100. In an experiment to measure K_{sp} of $CaSO_4$ [D. Masterman, *J. Chem. Educ.* **64**, 409 (1987)], a saturated solution of $CaSO_4(aq)$ is poured into the ion-exchange column pictured (and described in Chapter 22). As the solution passes through the column, Ca^{2+} is retained by the ion-exchange medium and H_3O^+ is released; two H_3O^+ ions appear in the effluent solution for every Ca^{2+} ion. As the drawing suggests, a 25.00-mL sample is added to the column, and the effluent is collected and diluted to 100.0 mL in a volumetric flask. A 10.00-mL portion of the diluted solution requires 8.25 mL of 0.0105 M NaOH for its titration. Use these data to obtain a value of K_{sp} for $CaSO_4$.

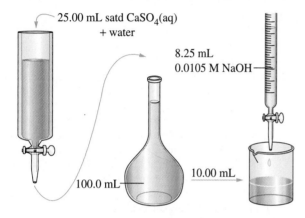

25.00 mL satd $CaSO_4(aq)$
+ water

8.25 mL
0.0105 M NaOH

100.0 mL

10.00 mL

101. In the *Mohr titration*, $Cl^-(aq)$ is titrated with $AgNO_3(aq)$ in solutions that are at about pH = 7. Thus, it is suitable for determining the chloride ion content of drinking water. The indicator used in the titration is $K_2CrO_4(aq)$. A red-brown precipitate of $Ag_2CrO_4(s)$ forms after all the Cl^- has precipitated. The titration reaction is $Ag^+(aq) + Cl^-(aq) \longrightarrow AgCl(s)$. At the equivalence point of the titration, the titration mixture consists of $AgCl(s)$ and a solution having neither Ag^+ nor Cl^- in excess. Also, no $Ag_2CrO_4(s)$ is present, but it should form immediately after the equivalence point.
(a) How many mL of 0.01000 M $AgNO_3(aq)$ would be required to titrate 100.0 mL of a municipal water sample having 29.5 mg Cl^-/L?
(b) What is $[Ag^+]$ at the equivalence point of the Mohr titration?

(c) What should $[CrO_4^{2-}]$ in the titration mixture be to meet the requirement of no precipitation of $Ag_2CrO_4(s)$ until immediately after the equivalence point?
(d) Describe the expected effect on the results of the titration if $[CrO_4^{2-}]$ were (1) greater than that calculated in part (c) or (2) less than that calculated?
(e) Do you think the Mohr titration would work if the reactants were exchanged—that is, with $Cl^-(aq)$ as the titrant and $Ag^+(aq)$ in the sample being analyzed? Explain.

102. The accompanying drawing suggests a series of manipulations starting with saturated $Mg(OH)_2(aq)$. Calculate $[Mg^{2+}](aq)$ at each of the lettered stages.
(a) 0.500 L of saturated $Mg(OH)_2(aq)$ is in contact with $Mg(OH)_2(s)$.
(b) 0.500 L of H_2O is added to the 0.500 L of solution in part (a), and the solution is vigorously stirred. Undissolved $Mg(OH)_2(s)$ remains.
(c) 100.0 mL of the clear solution in part (b) is removed and added to 0.500 L of 0.100 M HCl(aq).
(d) 25.00 mL of the clear solution in part (b) is removed and added to 250.0 mL of 0.065 M $MgCl_2(aq)$.
(e) 50.00 mL of the clear solution in part (b) is removed and added to 150.0 mL of 0.150 M KOH(aq).

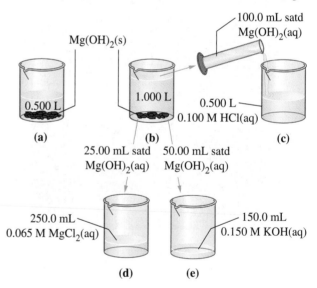

$Mg(OH)_2(s)$

0.500 L

(a)

1.000 L

(b)

100.0 mL satd $Mg(OH)_2(aq)$

0.500 L
0.100 M HCl(aq)

(c)

25.00 mL satd $Mg(OH)_2(aq)$

50.00 mL satd $Mg(OH)_2(aq)$

250.0 mL
0.065 M $MgCl_2(aq)$

(d)

150.0 mL
0.150 M KOH(aq)

(e)

eMedia Exercises

103. In the **Common-Ion Effect** animation *(eChapter 19-3)*, the concentration of the solvated ions is affected by the addition of a second ionic salt. Identify other salts that would also alter the equilibrium concentrations and how they would act to do so.

104. The formation of AgI is depicted in the **Precipitation Reactions** movie *(eChapter 19-5)*. **(a)** How do you recognize this as a precipitate? **(b)** From what you observe in the movie, what do you learn about the relative concentrations of the Ag^+ and I^- ions in the initial solution?

105. **(a)** Write an ionic equilibrium constant expression for the net ionic reaction seen in the **Dissolution of $Mg(OH)_2$ by Acid** animation *(eChapter 19-7)*. **(b)** From this relationship, describe in qualitative terms the pH dependence of the solubility of $Mg(OH)_2$.

106. Use the tests provided in the **Qualitative Analysis** simulation *(eChapter 19-9)* to determine the identity of Unknown Compound #2.

107. Use the data represented in the **Ionic Compounds** activity *(eChapter 19-9)* to write a specific set of precipitation rules for the aluminum cation.

20

Spontaneous Change: Entropy and Free Energy

Contents

Thermodynamics originated in the early nineteenth century to address the problem of improving the efficiencies of steam engines, but the laws of thermodynamics are much more widely useful throughout the field of chemistry and into biology as well, as we discover in this chapter.

In both Chapters 8 and 16, we noted that the reaction of nitrogen and oxygen gases, which does not occur appreciably in the forward direction at room temperature, produces significant equilibrium amounts of NO(g) at *high* temperatures.

$$N_2(g) + O_2(g) \rightleftharpoons 2\,NO(g)$$

Another reaction involving oxides of nitrogen is the conversion of NO(g) to $NO_2(g)$:

$$2\,NO(g) + O_2(g) \rightleftharpoons 2\,NO_2(g)$$

This reaction, unlike the first, yields its greatest equilibrium amounts of $NO_2(g)$ at *low* temperatures.

What is there about these two reactions that causes the forward reaction of the one to be favored at high temperatures but that of the other to be favored at low temperatures? Our primary objective in this chapter is to develop concepts to help us answer questions like this. This chapter, taken together with ideas from Chapter 7, shows the great power of thermodynamics to provide explanations of many chemical phenomena.

20-1 Spontaneity: The Meaning of Spontaneous Change

Most of us have played with spring-wound toys, whether a toy automobile, top, or music box. In every case, once the wound-up toy is released, it keeps running until the stored energy in the spring has been released; then the toy stops. The toy never rewinds itself. Human intervention is necessary (winding by hand). The running down of a wound-up spring is an example of a *spontaneous* process. The rewinding of the spring is a *nonspontaneous* process. Let's explore the scientific meaning of these two terms.

A **spontaneous process** is a process that occurs in a system left to itself; once started, no action from outside the system (external action) is necessary to make the process continue. Conversely, a **nonspontaneous process** will not occur *unless* some external action is continuously applied. Consider the rusting of an iron pipe exposed to the atmosphere. Although the process occurs only slowly, it does so continuously. As a result, the amount of iron decreases and the amount of rust increases until a final state of equilibrium is reached in which essentially all the iron has been converted to iron(III) oxide. We say that the reaction

$$4 \text{ Fe(s)} + 3 \text{ O}_2\text{(g)} \longrightarrow 2 \text{ Fe}_2\text{O}_3\text{(s)}$$

is *spontaneous*. Now consider the reverse situation: the extraction of pure iron from iron(III) oxide. We should not say that the process is impossible, but it is certainly *nonspontaneous*. In fact, this nonspontaneous reverse process is involved in the manufacture of iron from iron ore.

We will consider specific quantitative criteria for spontaneous change later in the chapter, but even now we can identify some spontaneous processes intuitively. For example, in the neutralization of NaOH(aq) with HCl(aq), the net change that occurs is

$$\text{H}_3\text{O}^+\text{(aq)} + \text{OH}^-\text{(aq)} \longrightarrow 2 \text{ H}_2\text{O(l)}$$

There is very little tendency for the reverse reaction (self-ionization) to occur, so the neutralization reaction is a spontaneous reaction. The melting of ice, however, is spontaneous at temperatures above 0 °C but nonspontaneous below 0 °C.

From our discussion of spontaneity to this point, we can reach these conclusions.

- If a process is spontaneous, the reverse process is nonspontaneous.
- Both spontaneous and nonspontaneous processes are possible, but only spontaneous processes will occur *without intervention*. Nonspontaneous processes require the system to be acted on by an external agent.

We would like to do more, however. We want to be able to predict whether the forward or the reverse direction is the direction of spontaneous change in a process, so we need a criterion for spontaneous change. To begin, let's look to mechanical systems for a clue. A ball rolls downhill, and water flows to a lower level. A common feature of these processes is that *potential energy decreases*.

For chemical systems, the property analogous to the potential energy of a mechanical system is the internal energy (U) or the closely related property enthalpy (H).

▲ When the spring of this toy monkey unwinds, the toy remains stationary. It cannot spontaneously rewind the spring.

▶ Spontaneous: "proceeding from natural feeling or native tendency without external constraint ...; developing without apparent external influence, force, cause, or treatment." *Merriam-Webster's Collegiate Dictionary*, on-line, 2000.

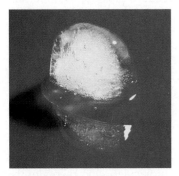

▲ The melting of an ice cube occurs spontaneously at temperatures above 0 °C.

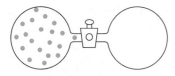

(a) Initial condition

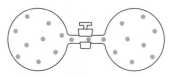

(b) After expansion into vacuum

▲ **FIGURE 20-1**
Expansion of an ideal gas into a vacuum
(a) Initially, an ideal gas is confined to the bulb on the left at 1.00 atm pressure. **(b)** When the stopcock is opened, the gas expands into the identical bulb on the right. The final condition is one in which the gas is distributed between the two bulbs at a pressure of 0.50 atm.

KEEP IN MIND ▶
that U and H are related as follows: $H = U + PV$, and $\Delta H = \Delta U + \Delta(PV)$. For a fixed amount of an ideal gas at a constant temperature, PV = constant, $\Delta(PV) = 0$, and $\Delta H = \Delta U$.

Mixing of Gases animation

In the 1870s, P. Berthelot and J. Thomsen proposed that the direction of spontaneous change is the direction in which the enthalpy of a system decreases. In a system in which enthalpy decreases, heat is given off by the system to the surroundings. Bertholet and Thomsen concluded that exothermic reactions should be spontaneous. In fact, many exothermic processes are spontaneous, but some are not. Also, some endothermic reactions *are* spontaneous. Thus, we cannot predict whether a process is spontaneous from its enthalpy change alone. Here are three examples of spontaneous, *endothermic* processes.

- the melting of ice at room temperature
- the evaporation of liquid diethyl ether from an open beaker
- the dissolving of ammonium nitrate in water

We will have to look to thermodynamic functions other than enthalpy change (ΔH) as criteria for spontaneous change.

20-2 The Concept of Entropy

To continue our search for criteria for spontaneous change, let's turn to Figure 20-1, which depicts two identical glass bulbs joined by a stopcock. Initially, the bulb on the left contains an ideal gas at 1.00 atm pressure and the bulb on the right is evacuated. When the valve is opened, the gas immediately expands into the evacuated bulb. After this expansion, the molecules are dispersed throughout the apparatus, with essentially equal numbers of molecules in both bulbs and a pressure of 0.50 atm. What causes this spontaneous expansion of the gas at a constant temperature?

One of the characteristics of an ideal gas is that its internal energy (U) does not depend on the gas pressure, but only on the temperature. Therefore, in this expansion $\Delta U = 0$. Also, the enthalpy change is zero: $\Delta H = 0$. This means that the expansion is not caused by the system dropping to a lower energy state. A convenient mental image to "explain" the expansion is that the gas molecules tend to spread out into the larger volume available to them at the reduced pressure. A more fundamental description of the underlying cause is that, for the same total energy, more translational energy levels among which the gas molecules can be distributed are available in the expanded volume. The tendency is for the energy of the system to spread out over a larger number of energy levels.

A similar situation—the mixing of ideal gases—is depicted in Figure 20-2. In this case, the two bulbs initially are filled with different ideal gases at 1.00 atm. When the stopcock is opened, the gases mix. The resulting change is essentially that of the expansion of the ideal gas pictured in Figure 20-1, but twice over. That is, each gas expands into the new volume available to it, without regard for the other gas (recall Dalton's law of partial pressures, page 196). Again, each expanded gas has more translational energy levels available to its molecules—the energy of the system has spread out. And again, the internal energy and enthalpy of the system are not changed by the expansion.

Entropy

The thermodynamic property related to the way in which the energy of a system is distributed among the available microscopic energy levels in called **entropy**.

> The greater the number of configurations of the microscopic particles (atoms, ions, molecules) among the energy levels in a particular state of a system, the greater the entropy of the system.

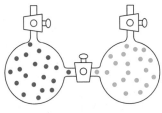

(a) Before mixing

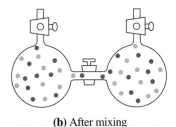

(b) After mixing

• Gas A   • Gas B

▲ **FIGURE 20-2**
The mixing of ideal gases
The total volume of the system
and the total gas pressure remain
fixed. The net change is that
(a) before mixing, each gas is
confined to half of the total volume (a single bulb) at a pressure
of 1.00 atm, and **(b)** after mixing,
each gas has expanded into the
total volume (both bulbs) and exerts a partial pressure of 0.50 atm.

▲ **A bust marking Ludwig
Boltzmann's tomb in Vienna**
Boltzmann's famous equation is
inscribed on the tomb. At the
time of Boltzmann's death, the
term "log" was used for both natural logarithms and logarithms to
the base ten; the symbol "ln" had
not yet been adopted.

Entropy is denoted by the symbol **S**. Like internal energy and enthalpy, entropy is
a function of state (see page 233). It has a unique value for a system whose temperature, pressure, and composition are specified. The **entropy change**, ΔS, is the
difference in entropy between two states of a system, and it also has a unique value.

In the gas expansion of Figure 20-1, the entropy of the gas *increases* and $\Delta S > 0$.
In the mixing of gases as carried out in Figure 20-2, entropy also increases, a fact
that we can represent symbolically.

$$A(g) + B(g) \longrightarrow \text{mixture of A(g) and B(g)}$$

$$\Delta S = S_{\text{mix of gases}} - [S_{A(g)} + S_{B(g)}] > 0$$

Because both of these expansions occur spontaneously and neither is accompanied
by a change in internal energy or enthalpy, we begin to suspect that *increases in entropy underlie spontaneous processes*. This is a proposition that we will have to
examine more closely later, but let's accept it tentatively for now.

The Boltzmann Equation for Entropy

The connection between macroscopic changes, such as the mixing of gases, and the
microscopic nature of matter was enunciated by Ludwig Boltzmann. The conceptual breakthrough that Boltzmann made was to associate the number of energy levels in the system with the number of ways of arranging the particles (atoms, ions,
or molecules) in these energy levels. The microscopic energy levels are also called
states, and the particular way a number of particles are distributed among these
states is called a *microstate*. The more states a given number of particles can occupy,
the more microstates the system has. The more microstates that exist, the greater
the entropy. Boltzmann derived the relationship

$$S = k \ln W$$

where S is the entropy, k is the Boltzmann constant, and W is the number of microstates. We can think of the Boltzmann constant as the gas constant per molecule;
that is, $k = R/N_A$. (Although we didn't specifically introduce k in the discussion
of kinetic–molecular theory, R/N_A appears in equation 6.21.) The number of microstates, W, is the number of ways that the atoms or molecules can be positioned
in the states available and still give rise to the same total energy. Each permissible
arrangement of the particles constitutes one of the microstates, so W is the total
number of microstates that correspond to the same energy.

How can we use Boltzmann's equation to think about the distribution of microscopic particles among the energy levels of a system? Let's consider again the
particle-in-a-box model for a matter wave (page 321). Specifically, we can use the
equation $E_W = n^2 h^2 / 8mL^2$ to calculate a few energy levels for a matter wave in a one-
dimensional box. Representative energy levels are shown in the energy-level diagrams in Figure 20-3a for a particle in boxes of lengths L, $2L$, and $3L$. We see the
following relationship between the length of the box and the number of levels: L,
three levels; $2L$, *six* levels; and $3L$, *nine* levels. As the boundaries of the box are expanded, the number of available energy levels increases and the separation between
levels decreases. Extending this model to three-dimensional space and large numbers of gas molecules, we find less crowding of molecules into a limited number of
energy levels when the pressure of the gas drops and the molecules expand into a larger volume. Thus, there is a greater spreading of the energy, and the entropy increases.

We can use the same particle-in-a-box model to understand the effect of raising
the temperature of a substance on the entropy of the system. We will consider a
gas, but our conclusions are equally valid for a liquid and for a solid. At low temperatures, at which molecules have a low energy, the gas molecules can occupy

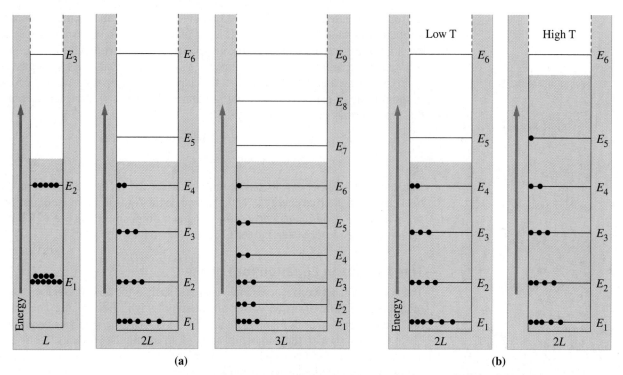

▲ FIGURE 20-3 Energy levels for a particle in a one-dimensional box
(a) The energy levels of a particle in a box become more numerous and closer together as the length of the box increases. The range of thermally accessible levels is indicated by the tinted band. The solid circles signify a system consisting of 15 particles. Each drawing represents a single microstate of the system. Can you see that as the box length increases there are many more microstates available to the particles? As the number of possible microstates for a given total energy increases, so does the entropy. **(b)** More energy levels become accessible in a box of fixed length as the temperature is raised. Because the average energy of the particles also increases, the internal energy and entropy both increase as the temperature is raised.

Lattice Entropy simulation

only a few of the energy levels; the value of W is small, and the entropy is low. As the temperature is raised, the energy of the molecules increases and the molecules have access to a larger number of energy levels. Thus the number of accessible microstates (W) increases and the entropy rises (Figure 20-3b).

In summary, we can describe the state of a thermodynamic system in two ways: the macroscopic description, in terms of state functions P, V, and T; and the microscopic description, requiring a knowledge of the position and velocity of every particle (atom or molecule) in the system. Boltzmann's equation provides the connection between the two.

Entropy Change

An entropy change is based on two measurable quantities: heat (q) and temperature (T). Both of these factors affect the availability of energy levels to the microscopic particles of a system. The following equation relates these factors to an entropy change,

$$\Delta S = \frac{q_{rev}}{T} \qquad (20.1)$$

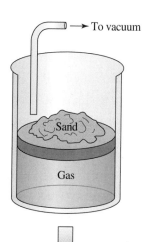

▲ **FIGURE 20-4**
**Pressure–volume work
conducted nearly reversibly**
Sand is removed one grain at a
time, and the gas slowly expands. If at any time one grain of
sand is added instead of removed, the piston reverses direction and the gas is compressed.
This process is not strictly reversible because the grains have
more than an infinitesimal mass.
Note that the expansion of a gas
in Figure 7-8 (page 231) is *not*
reversible. There, the confining
pressure is reduced in one step
rather than in an infinite number
of stages.

where T is the Kelvin temperature. Notice that ΔS is directly proportional to the quantity of heat because the more energy added to a system (as heat), the greater the number of energy levels available to the microscopic particles. Raising the temperature also increases the availability of energy levels, but for a given quantity of heat the proportional increase in number of energy levels is greatest at low temperatures. That is why ΔS is inversely proportional to the Kelvin temperature.

Equation (20.1) appears simple, but it is not. If S is to be a function of state, ΔS for a system must be independent of the path by which heat is lost or gained. Conversely, because the value of q ordinarily depends on the path chosen (recall page 235), equation (20.1) holds only for a carefully defined path. The path must be of a type called *reversible*, for which $q = q_{rev}$. A **reversible process** is one that can be made to reverse its direction when just an infinitesimal change in the opposite direction is made in some system property (Figure 20-4). Because q has the unit J and $1/T$ has the unit K^{-1}, the unit of entropy change, ΔS, is J/K, or J K^{-1}.

 ## Are You Wondering...

**If there is a natural way to incorporate the notion of infini-
tesimal changes in deriving equation (20.1)?**

The infinitesimal change in entropy, dS, that accompanies an infinitesimal reversible heat flow dq_{rev} is $dS = dq_{rev}/T$. Now imagine the change in a system from state 1 to state 2 is carried out in a series of such infinitesimal reversible steps. Summation of all these infinitesimal quantities through the calculus technique of integration yields ΔS.

$$\Delta S = \int \frac{dq_{rev}}{T}$$

If the change of state is isothermal (carried out at constant temperature) we can write

$$\Delta S = \int \frac{dq_{rev}}{T} = \frac{1}{T} \int dq_{rev} = \frac{q_{rev}}{T}$$

and we have recovered the definition of entropy change in (20.1).

Starting with equivalent expressions for dq_{rev}, ΔS can be related to other system properties. For the isothermal, reversible expansion of an ideal gas (as depicted in Figure 20-4), $dq_{rev} = -dw_{rev}$, leading to equation (20.9) on page 800, which describes ΔS in terms of gas volumes. In Exercise 96, the substitution $dq_{rev} = C_p dT$ leads to ΔS accompanying a reversible change in temperature.

In some instances, it is difficult to construct mental pictures to assess how the entropy of a system changes during a process. However, in many cases, an increase or decrease in the accessibility of energy levels for the microscopic particles of a system parallels an increase or decrease in the *number* of microscopic particles and the *space* available to them. As a consequence, we can often make qualitative predictions about entropy change by focusing on these two factors. Let's test this idea by considering again the three spontaneous, endothermic processes listed at the conclusion of Section 20-1 and illustrated in Figure 20-5.

In the melting of ice, a crystalline solid is replaced by a less structured liquid. Molecules that were relatively fixed in position in the solid, being limited to vibrational motion, are now free to move about a bit. The molecules have gained some translational and rotational motion. The number of accessible microscopic energy levels has increased, and so has the entropy.

In the vaporization process, a liquid is replaced by an even less structured gas. Molecules in the gaseous state, because they can move within a large free volume, have many more accessible energy levels than do those in the liquid state. Energy can be

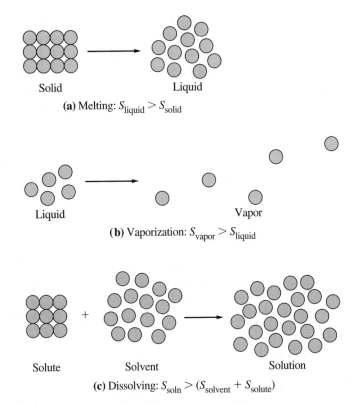

(a) Melting: $S_{liquid} > S_{solid}$

(b) Vaporization: $S_{vapor} > S_{liquid}$

(c) Dissolving: $S_{soln} > (S_{solvent} + S_{solute})$

▲ **FIGURE 20-5   Three processes in which entropy increases**
Each of the processes pictured—**(a)** the melting of a solid, **(b)** the evaporation of a liquid, and **(c)** the dissolving of a solute—results in an increase in entropy. For part (c), the generalization works best for nonelectrolyte solutions, in which ion–dipole forces do not exist.

spread over a much greater number of microscopic energy levels in the gas than in the liquid. The entropy of the gaseous state is much higher than that of the liquid state.

In the dissolving of ammonium nitrate in water, for example, a crystalline solid and a pure liquid are replaced by a mixture of ions and water molecules in the liquid (solution) state. This situation is somewhat more involved than the first two because some decrease in entropy is associated with the clustering of water molecules around the ions due to ion–dipole forces. The increase in entropy that accompanies the destruction of the solid's crystalline lattice predominates, however, and for the overall dissolution process, $\Delta S > 0$. In each of the three spontaneous, endothermic processes discussed here, the increase in entropy $(\Delta S > 0)$ outweighs the fact that heat must be absorbed $(\Delta H > 0)$, and each process is spontaneous.

In summary, we generally expect that entropy *increases* in each of these situations.

- Pure liquids or liquid solutions are formed from solids.

- Gases are formed from either solids or liquids.

- The number of molecules of gas increases as a result of a chemical reaction.

- The temperature of a substance increases. (Increased temperature means an increased number of accessible energy levels for the increased molecular motion, whether it be vibrational motion of atoms or ions in a solid, or translational and rotational motion of molecules in a liquid or gas.)

We apply these generalizations in Example 20-1.

EXAMPLE 20-1

Making Qualitative Predictions of Entropy Changes in Chemical Processes. Predict whether each of the following processes involves an increase or decrease in entropy or whether the outcome is uncertain.

(a) The decomposition of ammonium nitrate (a fertilizer and a highly explosive compound): $2 NH_4NO_3(s) \longrightarrow 2 N_2(g) + 4 H_2O(g) + O_2(g)$

(b) The conversion of SO_2 to SO_3 (a key step in the manufacture of sulfuric acid): $2 SO_2(g) + O_2(g) \longrightarrow 2 SO_3(g)$

(c) The extraction of sucrose from cane sugar juice: $C_{12}H_{22}O_{11}(aq) \longrightarrow C_{12}H_{22}O_{11}(s)$

(d) The "water gas shift" reaction (involved in the gasification of coal): $CO(g) + H_2O(g) \longrightarrow CO_2(g) + H_2(g)$

Solution

(a) Here, a solid yields a large quantity of gas. Entropy increases.

(b) Three moles of gaseous reactants produce two moles of gaseous products. The loss of one mole of gas indicates a loss of volume available to a smaller number of gas molecules. This loss reduces the number of possible configurations for the molecules in the system and the number of accessible microscopic energy levels. Entropy decreases.

(c) The sucrose molecules are reduced in mobility and in the number of forms in which their energy can be stored when they leave the solution and arrange themselves into a crystalline state. Entropy decreases.

(d) The entropies of the four gases are likely to be different because their molecular structures are different. The number of moles of gases is the same on both sides of the equation, however, so the entropy change is likely to be small if the temperature is constant. On the basis of just the generalizations listed on page 788, we cannot determine whether entropy increases or decreases.

Practice Example A: Predict whether entropy increases or decreases in each of the following reactions. **(a)** The Claus process for removing H_2S from natural gas: $2 H_2S(g) + SO_2(g) \longrightarrow 3 S(s) + 2 H_2O(g)$; **(b)** the decomposition of mercury(II) oxide: $2 HgO(s) \longrightarrow 2 Hg(l) + O_2(g)$.

Practice Example B: Predict whether entropy increases or decreases or whether the outcome is uncertain in each of the following reactions. **(a)** $Zn(s) + Ag_2O(s) \longrightarrow ZnO(s) + 2 Ag(s)$; **(b)** the chlor-alkali process, $2 Cl^-(aq) + 2 H_2O(l) \xrightarrow{\text{electrolysis}} 2 OH^-(aq) + H_2(g) + Cl_2(g)$.

20-3   Evaluating Entropy and Entropy Changes

On page 787, we remarked on the difficulty of calculating an entropy change with equation (20.1), and we then shifted our emphasis to making qualitative predictions about entropy changes. In a few instances, though, a simple direct calculation of ΔS is possible, as we will see in this section. Also, we will find that unlike the case with internal energy and enthalpy, it is possible to determine *absolute* entropy values.

Phase Transitions

In the equilibrium between two phases, the exchange of heat can be carried out reversibly, and the quantity of heat proves to be equal to the enthalpy change for the transition, ΔH_{tr}. In these cases, we can write equation (20.1) as

$$\Delta S_{tr} = \frac{\Delta H_{tr}}{T_{tr}} \qquad (20.2)$$

Rather than use the general symbol "tr" to represent a transition, we can be more specific about just which phases are involved, such as "fus" for the melting of a solid and "vap" for the vaporization of a liquid. If the transitions involve standard-state conditions (1 bar $\cong$ 1 atm pressure), we also use the degree sign ($\circ$). Thus for the melting (fusion) of ice at its normal melting point,

$$H_2O(s, 1\ atm) \rightleftharpoons H_2O(l, 1\ atm) \qquad \Delta H_{fus}^\circ = 6.02\ kJ\ at\ 273.15\ K$$

the standard entropy change is

$$\Delta S_{fus}^\circ = \frac{\Delta H_{fus}^\circ}{T_{mp}} = \frac{6.02\ kJ\ mol^{-1}}{273.15\ K} = 2.20 \times 10^{-2}\ kJ\ mol^{-1}\ K^{-1}$$

$$= 22.0\ J\ mol^{-1}\ K^{-1}$$

Entropy changes depend on the quantities of substances involved and are usually expressed on a per-mole basis.

KEEP IN MIND ▶

that the normal melting point and normal boiling point are determined at 1 atm pressure. The difference between 1 atm and the standard state pressure of 1 bar is so small that we can usually ignore it.

EXAMPLE 20-2

Determining the Entropy Change for a Phase Transition. What is the standard molar entropy of vaporization of water at 373 K, given that the standard molar enthalpy of vaporization is 40.7 kJ mol^{-1}?

Solution

Although we do not specifically need a chemical equation, writing one helps us see the process for which we seek the value of ΔS_{vap}°.

$$H_2O(l, 1\ atm) \rightleftharpoons H_2O(g, 1\ atm) \qquad \Delta H_{vap}^\circ = 40.7\ kJ/mol\ H_2O$$

$$\Delta S_{vap}^\circ = ?$$

$$\Delta S_{vap}^\circ = \frac{\Delta H_{vap}^\circ}{T_{bp}} = \frac{40.7\ kJ\ mol^{-1}}{373\ K} = 0.109\ kJ\ mol^{-1}\ K^{-1}$$

$$= 109\ J\ mol^{-1}\ K$$

Practice Example A: What is the standard molar entropy of vaporization, ΔS_{vap}°, for CCl_2F_2, a chlorofluorocarbon that once was heavily used in refrigeration systems? Its normal boiling point is $-29.79\ °C$, and $\Delta H_{vap}^\circ = 20.2\ kJ\ mol^{-1}$.

Practice Example B: The entropy change for the transition from solid rhombic sulfur to solid monoclinic sulfur at $95.5\ °C$ is $\Delta S_{tr}^\circ = 1.09\ J\ mol^{-1}\ K^{-1}$. What is the standard molar enthalpy change, ΔH_{tr}°, for this transition?

A useful generalization known as **Trouton's rule** states that for many liquids at their normal boiling points, the standard molar *entropy of vaporization* has a value of about 87 J mol^{-1} K^{-1}.

$$\Delta S_{vap}^\circ = \frac{\Delta H_{vap}^\circ}{T_{bp}} \approx 87\ J\ mol^{-1}\ K^{-1} \qquad (20.3)$$

For instance, the values of ΔS_{vap}° for benzene (C_6H_6) and octane (C_8H_{18}) are 87.1 and 86.2 J mol^{-1} K^{-1}, respectively. If the increased accessibility of microscopic energy levels produced in transferring one mole of molecules from liquid to vapor at

1 atm pressure is roughly comparable for different liquids, then we should expect similar values of ΔS°_{vap}.

Instances in which Trouton's rule fails are also understandable. In water and in ethanol, for example, hydrogen bonding among molecules produces a lower entropy than would otherwise be expected in the liquid state. Consequently, the entropy increase in the vaporization process is greater than normal, and $\Delta S^\circ_{vap} > 87$ J mol^{-1} K^{-1}.

The entropy concept helps explain Raoult's law (Section 14-6). Recall that for an ideal solution, $\Delta H_{soln} = 0$ and intermolecular forces of attraction are the same as in the pure liquid solvent (page 540). Thus, we expect the molar ΔH_{vap} to be the same whether vaporization of solvent occurs from an ideal solution or from the pure solvent at the same temperature. So, too, should ΔS_{vap} be the same because $\Delta S_{vap} = \Delta H_{vap}/T$. When one mole of solvent is transferred from liquid to vapor at the equilibrium vapor pressure P°, entropy increases by the amount ΔS_{vap}. As shown in Figure 20-6, because the entropy of the ideal solution is greater than that of the pure solvent, the entropy of the vapor produced by the vaporization of solvent from the solution is also greater than the entropy of the vapor obtained from the pure solvent. For the vapor above the solution to have the higher entropy, its molecules must have a greater number of accessible microscopic energy levels. In turn, the vapor must be present in a larger volume and, hence, must be at a lower pressure than the vapor coming from the pure solvent. This relationship corresponds to Raoult's law: $P_A = x_A P^\circ_A$.

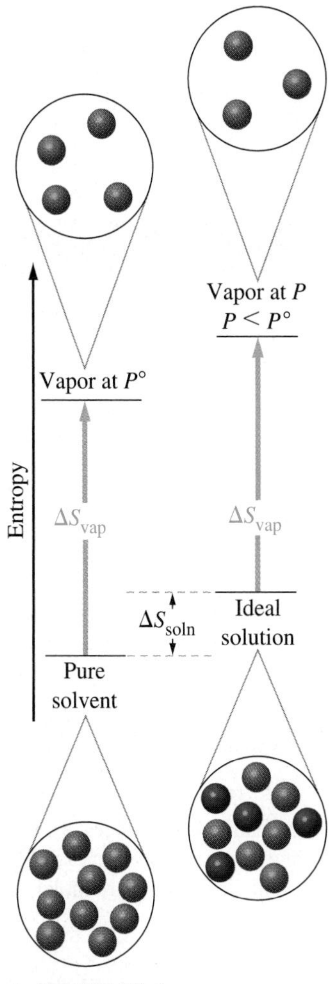

Absolute Entropies

To establish an *absolute* value of the entropy of a substance, we look for a condition in which the substance is in its lowest possible energy state, called the *zero-point energy*. We take this state to have an entropy of zero. Then we evaluate entropy changes as the substance is brought to other conditions of temperature and pressure. We add together these entropy changes and obtain a numerical value of the absolute entropy. The principle that permits this procedure is the **third law of thermodynamics**, which can be stated as follows:

> The entropy of a pure perfect crystal at 0 K is zero.

Figure 20-7 illustrates the method outlined in the preceding paragraph for determining absolute entropy as a function of temperature. Where phase transitions occur, equation (20.2) is used to evaluate the corresponding entropy changes. Over temperature ranges in which there are no transitions, ΔS° values are obtained from measurements of specific heats as a function of temperature.

The absolute entropy of one mole of a substance in its standard state is called the **standard molar entropy**, S°. Standard molar entropies of a number of substances at 25 °C are tabulated in Appendix D. To use these values to calculate the entropy change of a reaction, we use an equation with a familiar form (recall equation 7.21).

$$\Delta S^\circ = \left[\sum \nu_p S^\circ(\text{products}) - \sum \nu_r S^\circ(\text{reactants}) \right] \tag{20.4}$$

The symbol $\sum$ means "the sum of," and the terms added are the products of the standard molar entropies and the corresponding stoichiometric coefficients, ν. Example 20-3 shows how to use this equation.

▶ We found the value of ΔS°_{vap} for water at 373 K to be 109 J mol^{-1} K^{-1} in Example 20-2.

▲ **FIGURE 20-6**
An entropy-based rationale of Raoult's law
If ΔS_{vap} has the same value for vaporization from the pure solvent and from an ideal solution, the equilibrium vapor pressure is lower above the solution:
$P < P^\circ$.

Temperature Dependence of Entropy simulation

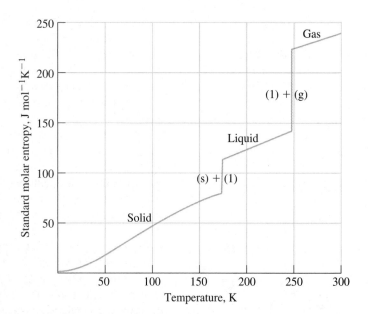

▶ **FIGURE 20-7**

Molar entropy as a function of temperature
The standard molar entropy of methyl chloride, $CHCl_3$, is plotted at various temperatures from 0 to 300 K, with the phases noted. The vertical segment between the solid and liquid phases corresponds to ΔS_{fus}; the other vertical segment, to ΔS_{vap}. By the third law of thermodynamics, an entropy of *zero* is expected at 0 K. Experimental methods cannot be carried to that temperature, however, so an extrapolation is required.

EXAMPLE 20-3

Calculating Entropy Changes from Standard Molar Entropies. Use data from Appendix D to calculate the standard molar entropy change for the conversion of nitrogen monoxide to nitrogen dioxide (a step in the manufacture of nitric acid).

$$2\,NO(g) + O_2(g) \longrightarrow 2\,NO_2(g) \qquad \Delta S^{\circ}_{298\,K} = ?$$

Solution

Equation (20.4) takes the form

$$\Delta S^{\circ} = 2S^{\circ}_{NO_2(g)} - 2S^{\circ}_{NO(g)} - S^{\circ}_{O_2(g)}$$
$$= (2 \times 240.1) - (2 \times 210.8) - 205.1 = -146.5\,J\,K^{-1}$$

As a useful check on this calculation, we can apply some qualitative reasoning: Because three moles of gaseous reactants produce only two moles of gaseous products, we should expect the entropy to decrease; that is, ΔS° should be negative.

Practice Example A: Use data from Appendix D to calculate the standard molar entropy change for the synthesis of ammonia from its elements.

$$N_2(g) + 3\,H_2(g) \longrightarrow 2\,NH_3(g) \qquad \Delta S^{\circ}_{298\,K} = ?$$

Practice Example B: N_2O_3 is an unstable oxide that readily decomposes. The decomposition of 1.00 mol of N_2O_3 to nitrogen monoxide and nitrogen dioxide at 25 °C is accompanied by the entropy change $\Delta S^{\circ} = 138.5\,J\,K^{-1}$. What is the standard molar entropy of $N_2O_3(g)$ at 25 °C?

Formation of Aluminum Bromide movie

▶ In general, at low temperatures translational energies are most important in establishing the entropy of gaseous molecules. As the temperature increases, first rotational energies become important, and finally, at still higher temperatures, vibrational modes of motion start to contribute to the entropy.

In Example 20-3, we used the standard molar entropies of $NO_2(g)$ and $NO(g)$. We might wonder why the value for $NO_2(g)$, 240.0 J mol^{-1} K^{-1}, is greater than that of $NO(g)$, 210.7 J mol^{-1} K^{-1}. We have learned that entropy increases when a substance absorbs heat (recall that $\Delta S = q_{rev}/T$) and that some of this heat goes simply into raising the average translational kinetic energies of molecules. But there are other ways for heat energy to be used. One possibility, pictured in Figure 20-8, is that the vibrational energies of molecules can be increased. In the *diatomic* molecule $NO(g)$, only one type of vibration is possible; in the *triatomic* molecule $NO_2(g)$, *three* types are possible. Because there are more possible ways of distributing en-

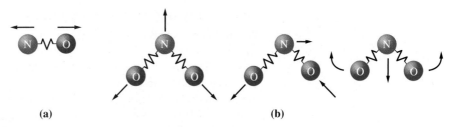

(a)   **(b)**

▲ **FIGURE 20-8**   **Vibrational energy and entropy**
The movement of atoms is suggested by the arrows. **(a)** The NO molecule has only one type of vibrational motion, whereas **(b)** the NO_2 molecule has three. This difference helps account for the fact that the molar entropy of $NO_2(g)$ is greater than that of $NO(g)$.

Methane, CH_4
$S° = 186.3 \text{ J mol}^{-1} \text{ K}^{-1}$

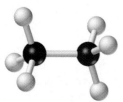

Ethane, C_2H_6
$S° = 229.6 \text{ J mol}^{-1} \text{ K}^{-1}$

Propane, C_3H_8
$S° = 270.3 \text{ J mol}^{-1} \text{ K}^{-1}$

ergy among NO_2 molecules than among NO molecules, $NO_2(g)$ has a higher molar entropy than does NO(g) at the same temperature. We should add the following statement to our generalizations about entropy on page 788.

- In general, the more complex their molecules (that is, the greater the number of atoms present), the greater the molar entropies of substances.

20-4 Criteria for Spontaneous Change: The Second Law of Thermodynamics

In Section 20-2, we came to the tentative conclusion that processes in which the entropy of a system increases should be spontaneous (and that processes in which entropy decreases should be nonspontaneous). But now we must face the vexing difficulties that this statement can present. For example, how do we explain the spontaneous freezing of water at -10 °C? Because crystalline ice has a lower molar entropy than does liquid water, the freezing of water is a process for which entropy *decreases*. The way out of this dilemma is to recognize that we must always simultaneously consider *two* entropy changes—the entropy change of the system itself and the entropy change of the surroundings. Our criteria for spontaneous change must be based on the sum of the two, called the entropy change of the "universe":

$$\Delta S_{\text{total}} = \Delta S_{\text{universe}} = \Delta S_{\text{system}} + \Delta S_{\text{surroundings}} \tag{20.5}$$

Although it is beyond the scope of this discussion to verify the following expression, it provides the basic criterion for spontaneous change. In a spontaneous change,

$$\Delta S_{\text{univ}} = \Delta S_{\text{sys}} + \Delta S_{\text{sur}} > 0 \tag{20.6}$$

Equation (20.6) is one way of stating the **second law of thermodynamics**. Another way is through the following statement.

All spontaneous processes produce an increase in the entropy of the universe.

According to expression (20.6), if a process produces positive entropy changes in both the system and its surroundings, the process is surely spontaneous. And if both these entropy changes are negative, the process is just as surely nonspontaneous. The freezing of water produces a negative entropy change in the system, but in the surroundings, which absorb heat, the entropy change is positive. As long as the temperature is below 0 °C, the entropy of the surroundings increases more than the entropy of the system decreases. Because the *total* entropy change is positive, the freezing of water below 0 °C is indeed spontaneous.

Free Energy and Free Energy Change

We could use expression (20.6) as our basic criterion for spontaneity (spontaneous change), but we would find it very difficult to apply. To evaluate a total entropy change (ΔS_{univ}), we always have to evaluate ΔS for the surroundings. At best, this process is tedious, and in many cases it is not even possible, because we cannot figure out all the interactions between a system and its surroundings. Surely it would be preferable to have a criterion that we could apply *to the system itself*, without having to worry about changes that occur in the surroundings.

To develop this new criterion, let's explore a hypothetical process conducted at constant temperature and pressure and with work limited to pressure–volume work. This process is accompanied by a heat effect, q_p, which, as we saw in Section 7-6, is equal to ΔH for the system (ΔH_{sys}). The heat effect experienced by the surroundings is the *negative* of that for the system: $q_{surr} = -q_p = -\Delta H_{sys}$. Furthermore, if the hypothetical surroundings are large enough, the path by which heat enters or leaves the surroundings can be made *reversible*. That is, the quantity of heat can be made to produce only an infinitesimal change in the temperature of the surroundings. In this case, according to equation (20.2), the entropy change in the surroundings is $\Delta S_{surr} = -\Delta H_{sys}/T$.[*] Now we substitute this value of ΔS_{surr} into equation (20.5). We next multiply by T to obtain

$$T\,\Delta S_{univ} = T\,\Delta S_{sys} - \Delta H_{sys} = -(\Delta H_{sys} - T\,\Delta S_{sys})$$

and then multiply by -1 (change signs).

$$-T\,\Delta S_{univ} = \Delta H_{sys} - T\,\Delta S_{sys} \qquad (20.7)$$

This is the significance of equation (20.7). The right side of this equation has terms involving *only the system*. On the left side appears the term ΔS_{univ}, which embodies the criterion for spontaneous change, that for a spontaneous process, $\Delta S_{univ} > 0$.

Equation (20.7) is generally cast in a somewhat different form, for which we must introduce a new thermodynamic function, called the Gibbs **free energy**, G. The Gibbs free energy for a system is defined by the equation

$$G = H - TS$$

The **free energy change**, ΔG, for a process at constant T is

$$\Delta G = \Delta H - T\,\Delta S \qquad (20.8)$$

In equation (20.8), all the terms apply to measurements *on the system*. All reference to the surroundings has been eliminated. Also, when we compare equations (20.7) and (20.8), we get

$$\Delta G = -T\,\Delta S_{univ}$$

Now, by noting that ΔG is *negative* when ΔS_{univ} is *positive*, we have our final criterion for spontaneous change, based on properties of only the system itself.

For a process occurring at constant T and P, these statements hold true.

- If $\Delta G < 0$ (*negative*), the process is *spontaneous*.
- If $\Delta G > 0$ (*positive*), the process is *nonspontaneous*.
- If $\Delta G = 0$ (*zero*), the process is *at equilibrium*.

▲ **J. Willard Gibbs (1839–1903)—a great "unknown" scientist of the United States**

Gibbs, a Yale University professor of mathematical physics, lived most of his career without recognition, partly because his work was abstract and partly because his important publications were in little-read journals. Yet, today, Gibbs's ideas serve as the basis of most of chemical thermodynamics.

---

[*]We cannot similarly substitute $\Delta H_{sys}/T$ for ΔS_{sys}. A process that occurs spontaneously is generally far removed from an equilibrium condition and is therefore *irreversible*. We cannot substitute q for an irreversible process into equation (20.2).

▶ Sometimes the term "*TΔS*" is referred to as "organizational energy," because Δ*S* is related to the way the energy of a system is distributed among the available energy levels.

We can see that free energy is indeed an energy term by evaluating the units in equation (20.8). Δ*H* has the unit joules (J), and the product *T* Δ*S* has the units K × J/K = J. Δ*G* is the difference in two quantities with units of energy.

Applying the Free Energy Criteria for Spontaneous Change

Later, we will look at quantitative applications of equation (20.8), but for now let's use the equation to make some qualitative predictions. Altogether there are four possibilities for Δ*G* on the basis of the signs of Δ*H* and Δ*S*. These possibilities are outlined in Table 20.1 and demonstrated in Example 20-4.

TABLE 20.1	Criteria for Spontaneous Change: Δ*G* = Δ*H* − *T* Δ*S*				
Case	Δ*H*	Δ*S*	Δ*G*	Result	Example
1	−	+	−	spontaneous at all temp	$2 N_2O(g) \longrightarrow 2 N_2(g) + O_2(g)$
2	−	−	−	spontaneous at low temp	$H_2O(l) \longrightarrow H_2O(s)$
			+	nonspontaneous at high temp	
3	+	+	+	nonspontaneous at low temp	$2 NH_3(g) \longrightarrow N_2(g) + 3 H_2(g)$
			−	spontaneous at high temp	
4	+	−	+	nonspontaneous at all temp	$3 O_2(g) \longrightarrow 2 O_3(g)$

 Gibbs Free Energy simulation

If Δ*H* is *negative* and Δ*S* is *positive*, the expression Δ*G* = Δ*H* − *T* Δ*S* is negative at all temperatures. The process is spontaneous at all temperatures. This corresponds to the situation we noted previously in which both Δ*S*$_{sys}$ and Δ*S*$_{surr}$ are positive and Δ*S*$_{univ}$ is also positive.

Unquestionably, if a process is accompanied by an *increase* in enthalpy (heat is absorbed) and a *decrease* in entropy, Δ*G* is positive at all temperatures and the process is nonspontaneous. This corresponds to a situation in which both Δ*S*$_{sys}$ and Δ*S*$_{surr}$ are negative and Δ*S*$_{univ}$ is also negative.

▶ For cases 2 and 3, there is a particular temperature at which a process switches from being spontaneous to being nonspontaneous. In Section 20-6 we will show how to determine such a temperature.

The questionable cases are those in which the entropy and enthalpy factors work in opposition—that is, with Δ*H* and Δ*S* *both* negative or *both* positive. In these cases, whether a reaction is spontaneous or not (that is, whether Δ*G* is negative or positive) depends on temperature. In general, if a reaction has negative values for both Δ*H* and Δ*S*, it is spontaneous at *lower* temperatures, whereas if Δ*H* and Δ*S* are both positive, the reaction is spontaneous at *higher* temperatures.

EXAMPLE 20-4

Using Enthalpy and Entropy Changes to Predict the Direction of Spontaneous Change. Under what temperature conditions would the following reactions occur spontaneously?

(a) $2 NH_4NO_3(s) \longrightarrow 2 N_2(g) + 4 H_2O(g) + O_2(g)$ Δ*H*° = −236.0 kJ

(b) $I_2(g) \longrightarrow 2 I(g)$

Solution

▶ The high activation energy of the decomposition reaction accounts for the fact that $NH_4NO_3(s)$ exists at all.

(a) The reaction is exothermic, and in Example 20-1(a) we concluded that Δ*S* > 0 because large quantities of gases are produced. With Δ*H* < 0 and Δ*S* > 0, this reaction should be spontaneous at all temperatures (case 1 of Table 20.1).

(b) Because one mole of gaseous reactant produces two moles of gaseous product, we expect entropy to increase. But what is the sign of ΔH? We could calculate ΔH from enthalpy of formation data, but there is no need to. In the reaction, covalent bonds in $I_2(g)$ are broken and no new bonds are formed. Because energy is absorbed to break bonds, ΔH must be positive. With $\Delta H > 0$ and $\Delta S > 0$, case 3 in Table 20.1 applies. ΔH is larger than $T \Delta S$ at low temperatures, and the reaction is nonspontaneous. At high temperatures, the $T \Delta S$ term becomes larger than ΔH, ΔG becomes negative, and the reaction is spontaneous.

Practice Example A: Which of the four cases in Table 20.1 would apply to each of the following reactions: **(a)** $N_2(g) + 3 H_2(g) \longrightarrow 2 NH_3(g)$, $\Delta H° = -92.22$ kJ; **(b)** $2 C(graphite) + 2 H_2(g) \longrightarrow C_2H_4(g)$, $\Delta H° = 52.26$ kJ?

Practice Example B: Under what temperature conditions would the following reactions occur spontaneously? **(a)** The decomposition of calcium carbonate into calcium oxide and carbon dioxide. **(b)** The "roasting" of zinc sulfide in oxygen to form zinc oxide and sulfur dioxide. This exothermic reaction releases 439.1 kJ for every mole of zinc sulfide that reacts.

Example 20-4(b) helps us understand why there is an upper temperature limit for the stabilities of chemical compounds. No matter how positive the value of ΔH for dissociation of a molecule into its atoms, the term $T \Delta S$ will eventually exceed ΔH in magnitude as the temperature increases. Known temperatures range from near absolute zero to the interior temperatures of stars (about 3×10^7 K). Molecules exist only at limited temperatures (up to about 1×10^4 K or about 0.03% of this total temperature range).

The method of Example 20-4 is adequate for making predictions about the sign of ΔG, but we will also want to use equation (20.8) to calculate numerical values. We will do that in Section 20-5.

▶ A related observation is that only a small fraction of the mass of the universe is in molecular form.

Are You Wondering...

What the term *free energy* signifies?

We might think that the quantity of energy available to do work in the surroundings as a result of a chemical process is $-\Delta H$. This would be the same as the quantity of heat that an exothermic reaction releases to the surroundings. (In thinking along those lines, we would say that an endothermic reaction is incapable of doing work.) However, that quantity of heat must be adjusted for the heat requirement in producing the entropy change in the system $(q_{rev} = T \Delta S)$. If an exothermic reaction is accompanied by an *increase* in entropy, the amount of energy available to do work in the surroundings is *greater* than $-\Delta H$. If entropy *decreases* in the exothermic reaction, the amount of energy available to do work is *less* than $-\Delta H$. But notice that in either case, this amount of energy is equal to $-\Delta G$. Thus, the amount of work that we are free to extract from a chemical process is $-\Delta G$, so the Gibb's function G is called the *free energy function*. Notice also that this interpretation of free energy allows for the possibility of work being done in an endothermic process if $T \Delta S$ exceeds ΔH. In Chapter 21, we will see how the free energy change of a reaction can be converted to electrical work. In any case, do not think of free energy as being "free" energy. Costs are always involved in tapping an energy source.

20-5 Standard Free Energy Change, $\Delta G°$

Because free energy is related to enthalpy ($G = H - TS$), we cannot establish absolute values of G, any more than we can for H. We must work with free energy changes, ΔG. We will find a special use for the **standard free energy change**, **$\Delta G°$**, corresponding to reactants and products in their standard states. The standard state conventions were introduced and applied to enthalpy change in Chapter 7.

The **standard free energy of formation**, **$\Delta G_f°$**, is the free energy change for a reaction in which a substance in its standard state is formed from its elements in their reference forms in their standard states. And, as was the case when we established enthalpies of formation in Section 7-8, this definition leads to values of *zero* for the free energies of formation of the elements in their reference forms at a pressure of 1 bar. Other free energies of formation are related to this condition of zero and are generally tabulated per mole of substance (see Appendix D).

Some additional relationships involving free energy changes are similar to those presented for enthalpy in Section 7-7: (1) ΔG changes sign when a process is reversed; and (2) ΔG for an overall process can be obtained by summing the ΔG values for the individual steps. The two expressions that follow are useful in calculating $\Delta G°$ values, depending on the data available. We can use the first expression at any temperature for which $\Delta H°$ and $\Delta S°$ values are known. We can use the second expression only at temperatures at which $\Delta G_f°$ values are known. The only temperature at which tabulated data are commonly given is 298.15 K. The first expression is applied in Example 20-5 and Practice Example 20-5A, and the second expression in Practice Example 20-5B.

KEEP IN MIND ▶

that a pressure of 1 bar is very nearly the same as 1 atm. The difference in these two pressures on the values of properties is generally so small that we can use the two pressure units almost interchangeably.

$$\Delta G° = \Delta H° - T\Delta S°$$

$$\Delta G° = \left[\sum \nu_p \Delta G_f°(\text{products}) - \sum \nu_r \Delta G_f°(\text{reactants}) \right]$$

EXAMPLE 20-5

Calculating $\Delta G°$ for a Reaction. Determine $\Delta G°$ at 298.15 K for the reaction

$$2\,NO(g) + O_2(g) \longrightarrow 2\,NO_2(g) \quad \text{(at 298.15 K)} \qquad \Delta H° = -114.1 \text{ kJ}$$
$$\Delta S° = -146.5 \text{ J K}^{-1}$$

Solution

Because we have values of $\Delta H°$ and $\Delta S°$, the most direct method of calculating $\Delta G°$ is to use the expression $\Delta G° = \Delta H° - T\,\Delta S°$. In doing so, we must first convert all the data to a common energy unit (for instance, kJ).

$$\Delta G° = -114.1 \text{ kJ} - (298.15 \text{ K} \times -0.1465 \text{ kJ K}^{-1})$$
$$= -114.1 \text{ kJ} + 43.68 \text{ kJ}$$
$$= -70.4 \text{ kJ}$$

Practice Example A: Determine $\Delta G°$ at 298.15 K for the reaction $4\,Fe(s) + 3\,O_2(g) \longrightarrow 2\,Fe_2O_3(s)$. $\Delta H° = -1648$ kJ and $\Delta S° = -549.3$ J K^{-1}.

Practice Example B: Determine $\Delta G°$ for the reaction in Example 20-5 by using data from Appendix D. Compare the two results.

20-6 Free Energy Change and Equilibrium

We have seen that $\Delta G < 0$ for spontaneous processes and that $\Delta G > 0$ for non-spontaneous processes. If $\Delta G = 0$, the forward and reverse processes show an equal tendency to occur, and the system is at *equilibrium*. At this point, even an infinitesimal change in one of the system variables (such as temperature or pressure) will cause a net change to occur. But if a system at equilibrium is left undisturbed, no net change occurs with time.

Let's consider the hypothetical process outlined in Figure 20-9. If we start at the left-hand side of the figure, we see that ΔH exceeds $T\,\Delta S$ and that ΔG is positive; the process is *nonspontaneous*. The magnitude of ΔG decreases with increasing temperature. At the right-hand side of the figure, $T\,\Delta S$ exceeds ΔH and ΔG is negative; the process is *spontaneous*. At the temperature at which the two lines intersect, $\Delta G = 0$ and the system is at equilibrium.

For the vaporization of water, *with both liquid and vapor in their standard states* (which means that $\Delta G = \Delta G°$), the intersection of the two lines in Figure 20-9 is at $T = 373.15$ K (100.00 °C). That is, for the vaporization of water at 1 atm,

$$H_2O(l,\ 1\ atm) \rightleftharpoons H_2O(g,\ 1\ atm) \qquad \Delta G° = 0 \text{ at } 373.15 \text{ K}$$

▶ Here, we have used a standard-state pressure of 1 atm. See Feature Problem 98 for an appraisal of this matter using a standard-state pressure of 1 bar.

At 25 °C, the $\Delta H°$ line lies above the $T\,\Delta S$ line in Figure 20-9. This means that $\Delta G° > 0$.

$$H_2O(l,\ 1\ atm) \longrightarrow H_2O(g,\ 1\ atm) \qquad \Delta G° = +8.590 \text{ kJ at } 298.15 \text{ K}$$

The positive value of $\Delta G°$ does not mean that vaporization of water will not occur. From common experience, we know that water spontaneously evaporates at room temperature. What the positive value means is that liquid water will not spontaneously produce $H_2O(g)$ at 1 atm pressure at 25 °C. Instead, $H_2O(g)$ is produced with a vapor pressure that is less than 1 atm pressure. The equilibrium vapor pressure of water at 25 °C is 23.76 mmHg = 0.03126 atm; that is,

▶ The liquid–vapor equilibrium represented here is out of contact with the atmosphere. In the presence of the atmosphere, the pressure on the liquid would be barometric pressure, whereas that of the vapor would remain essentially unchanged at 0.03126 atm.

$$H_2O(l,\ 0.03126\ atm) \rightleftharpoons H_2O(g,\ 0.03126\ atm) \qquad \Delta G = 0$$

Figure 20-10 offers a schematic summary of these ideas concerning the transition between liquid and gaseous water at 25 °C.

▶ **FIGURE 20-9 Free energy change as a function of temperature**
The value of ΔG is given by the distance between the two lines; that is, $\Delta G = \Delta H - T\,\Delta S$. In this illustration, both ΔH and ΔS have positive values at all temperatures. If ΔH is greater than the $T\,\Delta S$ value, $\Delta G > 0$ and the reaction is *nonspontaneous*. If ΔH is less than $T\,\Delta S$, $\Delta G < 0$ and the reaction is *spontaneous*. Equilibrium, $\Delta G = 0$, occurs at the temperature at which the two lines intersect. An assumption made here is that ΔH and ΔS are essentially independent of temperature.

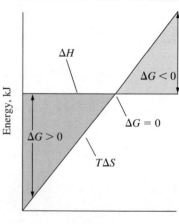

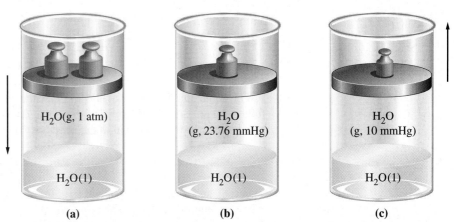

▲ FIGURE 20-10   **Liquid–vapor equilibrium and the direction of spontaneous change**
(a) For the vaporization of water at 298.15 K and 1 atm, $H_2O(l, 1\ atm) \longrightarrow H_2O(g, 1\ atm)$, $\Delta G° = 8.590$ kJ. The direction of spontaneous change is the *condensation* of $H_2O(g)$. (b) At 298.15 K and 23.76 mmHg, the liquid and vapor are in equilibrium and $\Delta G° = 0$. (c) At 298.15 K and 10 mmHg, the vaporization of $H_2O(l)$ occurs spontaneously: $H_2O(l, 10\ mmHg) \longrightarrow H_2O(g, 10\ mmHg)$, and $\Delta G < 0$.

Relationship of $\Delta G°$ to ΔG for Nonstandard Conditions

If you think about the situation just described for the vaporization of water, there is not much value in describing equilibrium in a process in terms of its $\Delta G°$ value. At only one temperature are the reactants in their standard states in equilibrium with products in their standard states; that is, at only one temperature does $\Delta G° = 0$. We want to be able to describe equilibrium for a variety of conditions, typically *nonstandard* conditions. Many reactions, such as processes occurring under physiological conditions, take place under nonstandard conditions. How, under such circumstances, can a biochemist decide which processes are spontaneous? For this we need to work with ΔG, not $\Delta G°$.

To obtain the relationship between ΔG and $\Delta G°$, we will consider a reaction between ideal gas molecules, assuming this to be the case in the reaction between nitrogen and hydrogen that produces ammonia.

$$2\ N_2(g) + 3\ H_2(g) \rightleftharpoons 2\ NH_3(g)$$

The expressions for ΔG and $\Delta G°$ are $\Delta G = \Delta H - T\Delta S$ and $\Delta G° = \Delta H° - T\Delta S°$, respectively. First consider how the enthalpy terms ΔH and $\Delta H°$ are related for an ideal gas. The enthalpy of an ideal gas, as we have seen, is a function of temperature only; it is independent of pressure. Thus, under any mixing conditions for an ideal gas, we have $\Delta H = \Delta H°$. We can write

$$\Delta G = \Delta H° - T\Delta S \qquad \textit{(ideal gas)}$$

We now need to obtain a relationship between ΔS and $\Delta S°$. To do so, let's consider the isothermal expansion of an ideal gas, for which $q = -w$ and $\Delta U = 0$. If the expansion occurs reversibly (recall Figure 20-4), the work of expansion for one mole of an ideal gas is given by an equation derived in Feature Problem 115 in Chapter 7.

$$w = -RT \ln \frac{V_f}{V_i} \qquad \textit{(reversible, isothermal)}$$

The reversible, isothermal heat of expansion is

$$q_{rev} = -w = RT \ln \frac{V_f}{V_i}$$

From equation (20.1), we obtain the entropy change for the isothermal expansion of one mole of an ideal gas.

$$\Delta S = \frac{q_{rev}}{T} = R \ln \frac{V_f}{V_i} \qquad (20.9)$$

Using equation (20.9), we can now evaluate the entropy of an ideal gas under any conditions of pressure. From the ideal gas equation, we know that the volume of an ideal gas is inversely proportional to the pressure, so we can recast equation (20.9) as

$$\Delta S = S_f - S_i = R \ln \frac{V_f}{V_i} = R \ln \frac{P_i}{P_f} = -R \ln \frac{P_f}{P_i}$$

where P_i and P_f are the initial and final pressures, respectively. If we set $P_i = 1$ bar and designate P_i as $P°$ and S_i as $S°$, we obtain for the entropy at any pressure P

$$S = S° - R \ln \frac{P}{P°} = S° - R \ln \frac{P}{1} = S° - R \ln P \qquad (20.10)$$

Are You Wondering ...

If there is a microscopic approach to obtaining equation (20.9)?

To do this, we use the ideas of Ludwig Boltzmann. Consider an ideal gas at an initial volume V_i and allow the gas to expand isothermally to a final volume V_f. Using the Boltzmann equation, we find that for the change in entropy,

$$\Delta S = S_f - S_i = k \ln W_f - k \ln W_i$$

$$\Delta S = k \ln \frac{W_f}{W_i}$$

where k is the Boltzmann constant, S_i and S_f are the initial and final entropies, respectively, and W_i and W_f are the number of microstates for the initial and final macroscopic states of the gas, respectively. We must now obtain a value for the ratio W_f/W_i. To do that, suppose that there is only a single gas molecule in a container. The number of microstates available to this single molecule should be proportional to the number of positions where the molecule can be and, hence, to the volume of the vessel. That is also true for each molecule in a system of N_A particles—Avogadro's number of particles. The number of microstates available to the whole system is

$$W_{total} = W_{particle\ 1} \times W_{particle\ 2} \times W_{particle\ 3} \times \ldots$$

Because the number of microstates for each particle is proportional to the volume V of the container, the number of microstates for N_A (Avogadro's number) ideal gas molecules is

$$W \propto V^{N_A}$$

Thus, the ratio of the microstates for isothermal expansion is

$$\frac{W_f}{W_i} = \left(\frac{V_f}{V_i} \right)^{N_A}$$

We can now calculate ΔS as follows:

$$\Delta S = k \ln \frac{W_f}{W_i} = k \ln \left(\frac{V_f}{V_i} \right)^{N_A} = N_A k \ln \frac{V_f}{V_i} = R \ln \frac{V_f}{V_i}$$

where R is the ideal gas constant. This equation, which gives the entropy change for the expansion of one mole of gas is simply equation (20.9).

Now, let's return to the ammonia synthesis reaction and calculate the entropy change for that reaction. We begin by applying equation (20.10) to each of the three gases.

$$S_{NH_3} = S^\circ_{NH_3} - R \ln P_{NH_3} \qquad S_{N_2} = S^\circ_{N_2} - R \ln P_{N_2} \qquad S_{H_2} = S^\circ_{H_2} - R \ln P_{H_2}$$

Then we substitute the above values into the equation $\Delta S = 2S_{NH_3} - S_{N_2} - 3S_{H_2}$ to obtain

$$\Delta S = 2S^\circ_{NH_3} - 2R \ln P_{NH_3} - S^\circ_{N_2} + R \ln P_{N_2} - 3S^\circ_{H_2} + 3R \ln P_{H_2}$$

By rearranging the terms, we get

$$\Delta S = 2S^\circ_{NH_3} - S^\circ_{N_2} - 3S^\circ_{H_2} - 2R \ln P_{NH_3} + R \ln P_{N_2} + 3R \ln P_{H_2}$$

and, since the first three terms on the right-hand side of the above equation represent ΔS°, we have

$$\Delta S = \Delta S^\circ - 2R \ln P_{NH_3} + R \ln P_{N_2} + 3R \ln P_{H_2}$$

$$\Delta S = \Delta S^\circ - R \ln P^2_{NH_3} + R \ln P_{N_2} + R \ln P^3_{H_2}$$

$$\Delta S = \Delta S^\circ + R \ln \frac{P_{N_2} P^3_{H_2}}{P^2_{NH_3}}$$

Finally, we can write the equation for ΔG by substituting the expression for ΔS into the equation

$$\Delta G = \Delta H^\circ - T \, \Delta S \qquad\qquad \textit{(ideal gas)}$$

This leads to

$$\Delta G = \Delta H^\circ - T \, \Delta S^\circ - RT \ln \frac{P_{N_2} P^3_{H_2}}{P^2_{NH_3}}$$

$$\Delta G = \underbrace{\Delta H^\circ - T \, \Delta S^\circ} + RT \ln \frac{P^2_{NH_3}}{P_{N_2} P^3_{H_2}}$$

$$\Delta G = \Delta G^\circ + RT \ln \frac{P^2_{NH_3}}{P_{N_2} P^3_{H_2}}$$

To simplify, we designate the quotient in the logarithmic term as the *reaction quotient Q* (recall page 639).

$$\Delta G = \Delta G^\circ + RT \ln Q \qquad\qquad (20.11)$$

Equation (20.11) is the relationship between ΔG and ΔG° that we have been seeking in this section, and we see that the key term in the equation is the reaction quotient formulated for the actual, nonstandard conditions. We can use equation (20.11) to decide on the spontaneity of a reaction under any conditions of composition, provided that the temperature and pressure at which we observe the reaction are constant. We turn now to describing how the standard Gibbs free energy change is related to the equilibrium constant.

Relationship of ΔG° to the Equilibrium Constant K_{eq}

We encounter an interesting situation when we apply equation (20.11) to a reaction at equilibrium. We have learned that at equilibrium $\Delta G = 0$, and in Chapter 16 we saw that if a system is at equilibrium, $Q = K_c$ or $Q = K_P$, or, more generally, $Q = K_{eq}$. So, we can write that, *at equilibrium*,

$$\Delta G = \Delta G^\circ + RT \ln K_{eq} = 0$$

which means that

$$\Delta G° = -RT \ln K_{eq} \qquad (20.12)$$

If we have a value of $\Delta G°$ at a given temperature, we can use equation (20.12) to calculate an equilibrium constant K_{eq}. This means that the tabulation of thermodynamic data in Appendix D can serve as a direct source of countless equilibrium constant values at 298.15 K.

We need to say a few words about the units required in equation (20.12). Because logarithms can be taken of dimensionless numbers only, K_{eq} has no units; neither does $\ln K_{eq}$. The right-hand side of equation (20.12) has the unit of "RT": J mol^{-1} K^{-1} × K = J mol^{-1}. $\Delta G°$, on the left-hand side of the equation, must have the same unit: J mol^{-1}. The "mol^{-1}" part of this unit means "per mole of reaction." One mole of reaction is simply the reaction based on the stoichiometric coefficients chosen for the balanced equation. When a $\Delta G°$ value is accompanied by a chemical equation, the "mol^{-1}" portion of the unit is often dropped, but there are a few times when we need it, as for the proper cancellation of units in equation (20.12).

Criteria for Spontaneous Change: Our Search Concluded

The graphs plotted in Figure 20-11 represent the culmination of our quest for criteria for spontaneous change. Unfortunately, to construct these plots in all their detail is beyond the scope of this text. However, we can rationalize their general shape

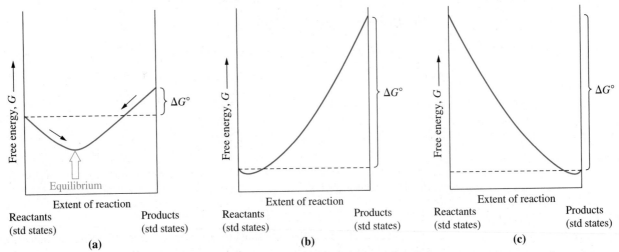

(a) (b) (c)

▲ FIGURE 20-11 Free energy change, equilibrium, and the direction of spontaneous change

Free energy is plotted against the extent of reaction for a hypothetical reaction. $\Delta G°$ is the difference between the standard molar free energies of formation of products and reactants. The equilibrium point lies somewhere between pure reactants and pure products. **(a)** $\Delta G°$ is small, so the equilibrium mixture lies about midway between the two extremes of pure products or reactants in their standard states. The effect of nonstandard conditions can be deduced from the slope of the curve. Mixtures with $Q > K_{eq}$ are to the right of the equilibrium point, and undergo spontaneous change in the direction of lower free energy, eventually coming to equilibrium. Similarly, mixtures with $Q < K_{eq}$ are to the left of the equilibrium point and spontaneously yield more products before reaching equilibrium. **(b)** $\Delta G°$ is large and positive, so the equilibrium point lies close to the extreme of pure reactants in their standard states. Consequently, very little reaction takes place before equilibrium is reached. **(c)** $\Delta G°$ is large and negative, so the equilibrium point lies close to the extreme of pure products in their standard states; the reaction goes essentially to completion.

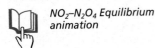

NO$_2$–N$_2$O$_4$ Equilibrium animation

on the basis of two ideas: (1) Every chemical reaction consists of both a forward and a reverse reaction, even if one of these occurs only to a very slight extent. (2) The direction of spontaneous change in both the forward and reverse reactions is the direction in which free energy decreases ($\Delta G < 0$). As a consequence, free energy reaches a minimum at some point between the left-hand and right-hand sides of the graph. This minimum is the equilibrium point in the reaction.

Now consider the vertical distance between the two end points of the graph; this distance represents $\Delta G°$ of the reaction. If, as in Figure 20-11a, $\Delta G°$ of a reaction is small, either positive or negative, the equilibrium condition is one in which significant amounts of both reactants and products will be found. If $\Delta G°$ is a *large, positive* quantity, as in Figure 20-11b, the equilibrium point lies far to the left—that is, very close to the reactants side. We can say that the reaction occurs hardly at all. If $\Delta G°$ is a *large, negative* quantity, as in Figure 20-11c, the equilibrium point lies far to the right—that is, very close to the products side. We can say that the reaction goes essentially to completion. Table 20.2 summarizes the conclusions of this discussion, gives approximate magnitudes to the terms *small* and *large*, and relates $\Delta G°$ values to values of K_{eq}.

TABLE 20.2   Significance of the Magnitude of $\Delta G°$ (at 298 K)

$\Delta G°$	K_{eq}	Significance
+200 kJ/mol	9.1×10^{-36}	No reaction
+100	3.0×10^{-18}	
+50	1.7×10^{-9}	
+10	1.8×10^{-2}	
+1.0	6.7×10^{-1}	Equilibrium calculation is necessary
0	1.0	
−1.0	1.5	
−10	5.6×10^{1}	
−50	5.8×10^{8}	
−100	3.3×10^{17}	Reaction goes to completion
−200	1.1×10^{35}	

$\Delta G°$ and ΔG: Predicting the Direction of Chemical Change

We have considered both $\Delta G°$ and ΔG in relation to the spontaneity of chemical reactions, and this is a good time to summarize some ideas about them.

$\Delta G < 0$ signifies that a reaction or process is spontaneous in the forward direction (to the right) for the stated conditions.

$\Delta G° < 0$ signifies that the forward reaction is spontaneous when reactants and products are in their standard states. It further signifies that $K_{eq} > 1$, whatever the initial concentrations or pressures of reactants and products.

$\Delta G = 0$ signifies that the reaction is at equilibrium under the stated conditions.

$\Delta G° = 0$ signifies that the reaction is at equilibrium when reactants and products are in their standard states. It further signifies that $K_{eq} = 1$, which can occur only at a particular temperature.

$\Delta G > 0$ signifies that the reaction or process is nonspontaneous in the forward direction under the stated conditions.

$\Delta G° > 0$ signifies that the forward reaction is nonspontaneous when reactants and products are in their standard states. It further signifies that $K_{eq} < 1$, whatever the initial concentrations or pressures of reactants and products.

$\Delta G = \Delta G°$ only when all reactants and products are in their standard states. Otherwise, $\Delta G = \Delta G° + RT \ln Q$.

The Thermodynamic Equilibrium Constant: Activities

When we derived the equation $\Delta G = \Delta G° + RT \ln Q$, we used the relationship given in equation (20.10),

$$S = S° - R \ln \frac{P}{P°} = S° - R \ln \frac{P}{1}$$

where for a gas, we defined a standard state of 1 bar, the reference for values of entropy. The ratio $P/P°$ is dimensionless, which is essential for a term appearing in a logarithm. At the time, we were considering a gas-phase reaction, but we must be able to discuss reactions in solutions also, so we need a more general approach. For this, we need the concept of *activity*. We write

$$S = S° - R \ln a$$

where a is the activity, defined as

$$a = \frac{\text{the effective concentration of a substance in the system}}{\text{the effective concentration of that substance in a standard reference state}}$$

In a gas-phase reaction, we take the effective concentration to be the pressure in bars and the reference state to be 1 bar. In this way, the activity is seen to be a dimensionless quantity.

In extending this approach to solutions, we define the reference state as a 1 M solution, so the activity of a substance is the numerical value of its molarity. Thus, the activity of protons in a 0.1 M solution of HCl in water is

$$a_{H^+} = \frac{0.1 \text{ M}}{1 \text{ M}} = 0.1$$

Another situation that we have encountered is that of a heterogeneous equilibrium, such as

$$CaCO_3(s) \rightleftharpoons CaO(s) + CO_2(g)$$

How do we handle solid substances? In this case, we choose the pure solids as the reference states, but the effective concentrations of the $CaCO_3(s)$ and $CaO(s)$ in the system are also those of the pure solids. Consequently, the activity of a solid is unity. This conclusion agrees with our observation in Figure 16-3 that the addition of either $CaO(s)$ or $CaCO_3(s)$ to an equilibrium mixture of $CaCO_3(s)$, $CO_2(g)$, and $CaO(s)$ has no effect on the pressure of $CO_2(g)$. The activities of $CaO(s)$ and $CaCO_3(s)$ are constant (unity).

In summary, we can make these statements.

- *For pure solids and liquids:* The activity $a = 1$. The reference state is the pure solid or liquid.

- *For gases:* With ideal gas behavior assumed, the activity is replaced by the numerical value of the gas pressure in bars. The reference state is the gas at 1 bar at the temperature of interest. Thus, the activity of a gas at 0.50 bar pressure is

▶ We introduced activities in Section 14-9 as "effective concentrations" designed to give the best agreement between the compositions and physical properties of solutions.

KEEP IN MIND ▶
that the ideal gas equation $PV = nRT$ can be rewritten as $(n/V) = [\text{concentration}] = (P/RT)$ so that pressure is an effective concentration.

KEEP IN MIND ▶
that the precise definition of pH is

$$pH = -\log a_{H^+}$$
$$= -\log \left(\frac{[H^+] M}{1 \text{ M}} \right)$$
$$= -\log [H^+]$$

By using this definition of activity, we remove the units and can employ the log term in a correct manner.

    $a = (0.50\ \text{bar})/(1\ \text{bar}) = 0.50$. (Recall also, that 1 bar of pressure is almost identical to 1 atm.)

- *For solutes in aqueous solution:* With ideal solution behavior assumed (for example, no interionic attractions), the activity is replaced by the numerical value of the molarity. The reference state is a 1 M solution. Thus, the activity of the solute in a 0.25 M solution is $a = (0.25\ \text{M})/(1\ \text{M}) = 0.25$.

When we write an equilibrium expression in terms of activities, we call the value of K_{eq} the **thermodynamic equilibrium constant**. The thermodynamic equilibrium constant is dimensionless and thus appropriate for use in equation (20.12).

Thermodynamic equilibrium constants, K_{eq}, are sometimes identical to K_c and K_p values, as in parts (a) and (b) of Example 20-6. In other instances, such as part (c) of Example 20-6, this is not the case. In working through Example 20-6, keep in mind that in this text our sole reason for writing thermodynamic equilibrium constants is to get the proper value to use in equation (20.12). Note that we must also write the reaction quotient Q in the same manner as K_{eq} when we use equation (20.11), as we will demonstrate in Example 20-7.

EXAMPLE 20-6

Writing Thermodynamic Equilibrium Constant Expressions. For the following reversible reactions, write thermodynamic equilibrium constant expressions, making appropriate substitutions for activities. Then equate K_{eq} to K_c or K_p, where this can be done.

(a) The water gas reaction

$$C(s) + H_2O(g) \rightleftharpoons CO(g) + H_2(g)$$

(b) Formation of a saturated aqueous solution of lead iodide, a very slightly soluble solute

$$PbI_2(s) \rightleftharpoons Pb^{2+}(aq) + 2\,I^-(aq)$$

(c) Oxidation of sulfide ion by oxygen gas (used in removing sulfides from wastewater, as in pulp and paper mills)

$$O_2(g) + 2\,S^{2-}(aq) + 2\,H_2O(l) \rightleftharpoons 4\,OH^-(aq) + 2\,S(s)$$

▲ A paper plant in Rumford, Maine.

Solution

In each case, once we have made the appropriate substitutions for activities, if all terms are molarities, the thermodynamic equilibrium constant is the same as K_c. If all terms are partial pressures, $K_{eq} = K_p$. If both molarities *and* partial pressures appear in the expression, however, the equilibrium constant expression can be designated only as K_{eq}.

(a) The activity of solid carbon is 1. Partial pressures are substituted for the activities of the gases.

$$K_{eq} = \frac{a_{CO(g)} a_{H_2(g)}}{a_{C(s)} a_{H_2O(g)}} = \frac{(P_{CO})(P_{H_2})}{(P_{H_2O})} = \frac{(P_{CO})(P_{H_2})}{(P_{H_2O})} = K_p$$

(b) The activity of solid lead(II) iodide is 1. Molarities are substituted for activities of the ions in aqueous solution.

$$K_{eq} = \frac{a_{Pb^{2+}(aq)} a^2_{I^-(aq)}}{a_{PbI_2(s)}} = [Pb^{2+}][I^-]^2 = K_c = K_{sp}$$

(c) The activity of both the solid sulfur and the liquid water is 1. Molarities are substituted for the activities of $OH^-(aq)$ and $S^{2-}(aq)$. The partial pressure of $O_2(g)$ is substituted for its activity. Thus, the resulting K_{eq} is neither a K_c nor a K_p.

$$K_{eq} = \frac{a^4_{OH^-(aq)} a^2_{S(s)}}{a_{O_2(g)} a^2_{S^{2-}(aq)} a^2_{H_2O(l)}} = \frac{[OH^-]^4 \cdot (1)^2}{P_{O_2} \cdot [S^{2-}]^2 \cdot (1)^2} = \frac{[OH^-]^4}{P_{O_2} \cdot [S^{2-}]^2}$$

Practice Example A: Write thermodynamic equilibrium constant expressions for each of the following reactions. Relate these to K_c or K_p where appropriate.
(a) $Si(s) + 2 Cl_2(g) \rightleftharpoons SiCl_4(g)$;
(b) $Cl_2(g) + H_2O(l) \rightleftharpoons HOCl(aq) + H^+(aq) + Cl^-(aq)$.

Practice Example B: Write a thermodynamic equilibrium constant expression to represent the reaction of solid lead(II) sulfide with aqueous nitric acid to produce solid sulfur, a solution of lead(II) nitrate, and nitrogen monoxide gas. Base the expression on the balanced net ionic equation for the reaction.

EXAMPLE 20-7

Assessing Spontaneity for Nonstandard Conditions. For the decomposition of 2-propanol to form propanone (acetone) and hydrogen,

$$(CH_3)_2CHOH(g) \rightleftharpoons (CH_3)_2CO(g) + H_2(g)$$

the equilibrium constant is 0.444 at 452 K. Is this reaction spontaneous under standard conditions? Will the reaction be spontaneous when the partial pressures of 2-propanol, propanone, and hydrogen are 1.0, 0.1 and 0.1 bar, respectively?

Solution

In each case, we first must obtain the value of $\Delta G°$, which we can get from equation (20.12).

$$\Delta G° = -RT \ln K_{eq} = -8.3145 \text{ J mol}^{-1} \text{K}^{-1} \times 452 \text{ K} \times \ln(0.444) = 3.05 \times 10^3 \text{ J mol}^{-1}$$

This result enables us to state categorically that the reaction will not proceed spontaneously if all reactants and products are in their standard states—that is, with the partial pressures of reactants and products at 1 bar.

To determine whether the reaction is spontaneous under the nonstandard-state conditions given, we must calculate ΔG. We first write Q in terms of activities and then substitute the partial pressures of the gases for the activities of the gases.

$$Q = \frac{a_{(CH_3)_2CO} a_{H_2}}{a_{(CH_3)_2CHOH}} = \frac{P_{(CH_3)_2CO} P_{H_2}}{P_{(CH_3)_2CHOH}}$$

We then use this expression in equation (20.11).

$$\Delta G = \Delta G^\circ + RT \ln \frac{P_{(CH_3)_2CO}P_{H_2}}{P_{(CH_3)_2CHOH}}$$

$$= 3.05 \times 10^3 \text{ J mol}^{-1} + \left(8.3145 \text{ J mol}^{-1} \text{ K}^{-1} \times 452 \text{ K} \times \ln \frac{0.1 \times 0.1}{1} \right)$$

$$= -1.43 \times 10^4 \text{ J mol}^{-1}$$

The value of ΔG is negative, so we can conclude that this reaction should proceed spontaneously. Remember, however, that thermodynamics says nothing about the rate of the reaction, only that the reaction will proceed in its own good time!

Practice Example A: Use the data in Appendix D to decide whether the following reaction is spontaneous under standard conditions at 298.15 K.

$$N_2O_4(g) \longrightarrow 2 NO_2(g)$$

Practice Example B: If a gaseous mixture of N_2O_4 and NO_2, both at a pressure of 0.5 bar, is introduced into a previously evacuated vessel, which of the two gases will spontaneously convert into the other at 298.15 K?

We have now acquired all the tools with which to perform one of the most practical calculations of chemical thermodynamics: *determining the equilibrium constant for a reaction from tabulated data*. In Example 20-8, where we demonstrate this application, we use thermodynamic properties of ions in aqueous solution as well as of compounds. An important idea to note about the thermodynamic properties of ions is that they are relative to $H^+(aq)$, which, by convention, is assigned values of *zero* for ΔH_f°, ΔG_f°, *and* S°. This means that entropies listed for ions are not absolute entropies, as they are for compounds. Negative values of S° simply denote an entropy less than that of $H^+(aq)$.

EXAMPLE 20-8

Calculating the Equilibrium Constant of a Reaction from the Standard Free Energy Change. Determine the equilibrium constant at 298.15 K for the dissolution of magnesium hydroxide in an acidic solution.

$$Mg(OH)_2(s) + 2 H^+(aq) \rightleftharpoons Mg^{2+}(aq) + 2 H_2O(l)$$

Solution

The key to solving this problem is to find a value of ΔG° and then to use the expression $\Delta G^\circ = -RT \ln K_{eq}$. We can obtain ΔG° from standard free energies of formation listed in Appendix D. Note that because its value is zero, the term $\Delta G_f^\circ[H^+(aq)]$ is not included.

$$\Delta G^\circ = 2\Delta G_f^\circ[H_2O(l)] + \Delta G_f^\circ[Mg^{2+}(aq)] - \Delta G_f^\circ[Mg(OH)_2(s)]$$

$$= 2(-237.1 \text{ kJ mol}^{-1}) + (-454.8 \text{ kJ mol}^{-1}) - (-833.5 \text{ kJ mol}^{-1})$$

$$= -95.5 \text{ kJ mol}^{-1}$$

Now solve for $\ln K_{eq}$ and K_{eq}.

$$\Delta G^\circ = -RT \ln K_{eq} = -95.5 \text{ kJ mol}^{-1} = -95.5 \times 10^3 \text{ J mol}^{-1}$$

$$\ln K_{eq} = \frac{-\Delta G^\circ}{RT} = \frac{-(-95.5 \times 10^3 \text{ J mol}^{-1})}{8.3145 \text{ J mol}^{-1} \text{ K}^{-1} \times 298.15 \text{ K}} = 38.5$$

$$K_{eq} = e^{38.5} = 5 \times 10^{16}$$

The value of K_{eq} obtained here is the thermodynamic equilibrium constant. According to the conventions we have established, the activity of both $Mg(OH)_2(s)$ and $H_2O(l)$ is 1, and molarities can be substituted for the activities of the ions.

$$K_{eq} = \frac{a_{Mg^{2+}(aq)}a^2_{H_2O(l)}}{a_{Mg(OH)_2}a^2_{H^+(aq)}} = \frac{[Mg^{2+}]}{[H^+]^2} = K_c = 5 \times 10^{16}$$

Practice Example A: Determine the equilibrium constant at 298.15 K for $AgI(s) \rightleftharpoons Ag^+(aq) + I^-(aq)$. Compare your answer to the K_{sp} for AgI in Appendix D.

Practice Example B: At 298.15 K, should manganese dioxide react to an appreciable extent with 1 M HCl(aq), producing manganese(II) ion in solution and chlorine gas?

When a question requires the use of thermodynamic properties, it is a good idea to think qualitatively about the problem before diving into calculations. The dissolving of $Mg(OH)_2(s)$ in acidic solution considered in Example 20-8 is an acid–base reaction. It is an example we used in Chapter 19 to illustrate the effect of pH on solubility. We also mentioned it in Chapter 5 as the basis for using milk of magnesia as an antacid. We should certainly expect the reaction to be spontaneous. This means that the value of K_{eq} should be large, which we found to be the case. If we had made an error in sign in our calculation (an easy error to make when using the expression $\Delta G° = -RT \ln K_{eq}$), we would have obtained $K_{eq} = 2 \times 10^{-17}$. But we would have seen immediately that this is the wrong answer. This value suggests a reaction in which the concentration of products is extremely low at equilibrium.

The data listed in Appendix D are for 25 °C. Thus, values of $\Delta G°$ and K_{eq} obtained with these data are also at 25 °C. Most chemical reactions are carried out at temperatures other than 25 °C, however. In Section 20-7, we will learn how to calculate values of equilibrium constants at various temperatures.

20-7 $\Delta G°$ and K_{eq} as Functions of Temperature

In Chapter 16, we used Le Châtelier's principle to make qualitative predictions of the effect of temperature on an equilibrium condition. We can now describe a *quantitative* relationship between the equilibrium constant and temperature. In the method illustrated in Example 20-9, we assume that $\Delta H°$ is practically independent of temperature. Although absolute entropies depend on temperature, we assume that the entropy *change* $\Delta S°$ for a reaction is also independent of temperature. Yet, the term "$T \Delta S°$" is strongly temperature-dependent because of the temperature factor T. As a result, $\Delta G°$, which is equal to $\Delta H° - T \Delta S°$, is also dependent on temperature.

EXAMPLE 20-9

Determining the Relationship Between an Equilibrium Constant and Temperature by Using Equations for Free Energy Change. At what temperature will the equilibrium constant for the formation of NOCl(g) be $K_{eq} = K_p = 1.00 \times 10^3$? Data for this reaction at 25 °C are

$$2 NO(g) + Cl_2(g) \rightleftharpoons 2 NOCl(g)$$
$$\Delta G° = -40.9 \text{ kJ mol}^{-1} \qquad \Delta H° = -77.1 \text{ kJ mol}^{-1}$$
$$\Delta S° = -121.3 \text{ J mol}^{-1} \text{ K}^{-1}$$

Solution

To determine an unknown temperature from a known equilibrium constant, we need an equation in which both of these terms appear. The required equation is $\Delta G° = -RT \ln K_{eq}$. However, to solve for the unknown temperature, we need the value of $\Delta G°$ at that temperature. We know the value of $\Delta G°$ at 25 °C (-40.9 kJ mol^{-1}), but we also know that this value will be different at other temperatures. We can assume, however, that the values of $\Delta H°$ and $\Delta S°$ will not change much with temperature. This means that we can obtain a value of $\Delta G°$ from the equation $\Delta G° = \Delta H° - T\Delta S°$, where T is the *unknown* temperature and the values of $\Delta H°$ and $\Delta S°$ are those at 25 °C. Now we have two equations that we can set equal to each other. That is,

$$\Delta G° = \Delta H° - T\Delta S° = -RT \ln K_{eq}$$

We can gather the terms with T on the right,

$$\Delta H° = T\Delta S° - RT \ln K_{eq} = T(\Delta S° - R \ln K_{eq})$$

and solve for T.

$$T = \frac{\Delta H°}{\Delta S° - R \ln K_{eq}}$$

Now let's substitute values for $\Delta H°$, $\Delta S°$, R, and $\ln K_{eq}$.

$$T = \frac{-77.1 \times 10^3 \text{ J mol}^{-1}}{-121.3 \text{ J mol}^{-1} \text{ K}^{-1} - \left[8.3145 \text{ J mol}^{-1} \text{ K}^{-1} \times \ln(1.00 \times 10^3)\right]}$$

$$= \frac{-77.1 \times 10^3 \text{ J mol}^{-1}}{-121.3 \text{ J mol}^{-1}\text{K}^{-1} - (8.3145 \times 6.908) \text{ J mol}^{-1} \text{ K}^{-1}}$$

$$= \frac{-77.1 \times 10^3 \text{ J mol}^{-1}}{-178.7 \text{ J mol}^{-1} \text{ K}^{-1}} = 431 \text{ K}$$

Although we have shown three significant figures in our answer, we should probably round the final result to just two significant figures. The assumption we made about the constancy of $\Delta H°$ and $\Delta S°$ is probably no more valid than that.

Practice Example A: At what temperature will the formation of $NO_2(g)$ from $NO(g)$ and $O_2(g)$ have $K_p = 1.50 \times 10^2$? For the reaction $2 NO(g) + O_2(g) \rightleftharpoons 2 NO_2(g)$ at 25 °C, $\Delta H° = -114.1$ kJ mol^{-1} and $\Delta S° = -146.5$ J mol^{-1} K^{-1}.

Practice Example B: For the reaction $2 NO(g) + Cl_2(g) \rightleftharpoons 2 NOCl(g)$, what is the value of K_{eq} at **(a)** 25 °C; **(b)** 75 °C? Use data from Example 20-9.

[*Hint:* The solution to part (a) can be done somewhat more simply than that for (b).]

An alternative to the method outlined in Example 20-9 is to relate the equilibrium constant and temperature directly, without specific reference to a free energy change. To do this, we start with the same two expressions as in Example 20-9,

$$-RT \ln K_{eq} = \Delta G° = \Delta H° - T\Delta S°$$

and divide by $-RT$.

$$\ln K_{eq} = \frac{-\Delta H°}{RT} + \frac{\Delta S°}{R} \tag{20.13}$$

If we assume that $\Delta H°$ and $\Delta S°$ are constant, equation (20.13) describes a straight line with a slope of $-\Delta H°/R$ and a y-intercept of $\Delta S°/R$. Table 20.3 lists equilibrium constants as a function of the reciprocal of Kelvin temperature for the reaction of $SO_2(g)$ and $O_2(g)$ that forms $SO_3(g)$. The $\ln K_p$ and $1/T$ data from Table 20.3 are plotted in Figure 20-12 and yield the expected straight line.

Now we can follow the procedure used in Appendix A-4 to derive the Clausius–Clapeyron equation. We can write equation (20.13) twice, for two different temperatures and with the corresponding equilibrium constants. Then, if we subtract one equation from the other, we obtain the result shown here,

$$\ln \frac{K_2}{K_1} = \frac{\Delta H°}{R}\left(\frac{1}{T_1} - \frac{1}{T_2}\right) \tag{20.14}$$

KEEP IN MIND

that the Clausius–Clapeyron equation (13.2) is just a special case of equation (20.14) in which the equilibrium constants are equilibrium vapor pressures and $\Delta H° = \Delta H_{vap}$. ▶

T, K	1/T, K⁻¹	K_p	$\ln K_p$
	TABLE 20.3 Equilibrium Constants, K_p, for the Reaction 2 SO₂(g) + O₂(g) ⇌ 2 SO₃(g) at Several Temperatures		
800	12.5×10^{-4}	9.1×10^2	6.81
850	11.8×10^{-4}	1.7×10^2	5.14
900	11.1×10^{-4}	4.2×10^1	3.74
950	10.5×10^{-4}	1.0×10^1	2.30
1000	10.0×10^{-4}	3.2×10^0	1.16
1050	9.52×10^{-4}	1.0×10^0	0.00
1100	9.09×10^{-4}	3.9×10^{-1}	−0.94
1170	8.55×10^{-4}	1.2×10^{-1}	−2.12

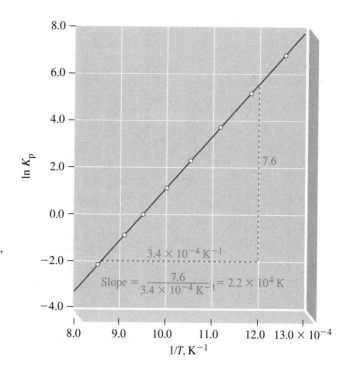

▶ **FIGURE 20-12**
Temperature dependence of the equilibrium constant K_p for the reaction

$$2 \text{ SO}_2(g) + \text{O}_2(g) \rightleftharpoons 2 \text{ SO}_3(g)$$

This graph can be used to establish the enthalpy of reaction, $\Delta H°$ (see equation 20.13).

$$\text{slope} = -\Delta H°/R = 2.2 \times 10^4 \text{ K}$$

$$\Delta H° = -8.3145 \text{ J mol}^{-1} \text{ K}^{-1} \times 2.2 \times 10^4 \text{ K}$$

$$= -1.8 \times 10^5 \text{ J mol}^{-1}$$

$$= -1.8 \times 10^2 \text{ kJ mol}^{-1}$$

where T_2 and T_1 are two Kelvin temperatures; K_2 and K_1 are the equilibrium constants at those temperatures; $\Delta H°$ is the enthalpy of reaction, expressed in J mol⁻¹; and R is the gas constant, expressed as 8.3145 J mol⁻¹ K⁻¹. Jacobus van't Hoff (1852–1911) derived equation (20.14), which is often referred to as the *van't Hoff equation.*

EXAMPLE 20-10

Relating Equilibrium Constants and Temperature Through the van't Hoff Equation. Use data from Table 20.3 and Figure 20-12 to estimate the temperature at which $K_p = 1.0 \times 10^6$ for the reaction

$$2 \text{ SO}_2(g) + \text{O}_2(g) \rightleftharpoons 2 \text{ SO}_3(g)$$

Solution

Select one known temperature and equilibrium constant from Table 20.3 and the enthalpy change of the reaction, $\Delta H°$, from Figure 20-12. The data we substitute into equation (20.14) are $T_1 = ?$, $K_1 = 1.0 \times 10^6$; $T_2 = 800$ K, $K_2 = 9.1 \times 10^2$; and $\Delta H° = -1.8 \times 10^5$ J mol^{-1}.

$$\ln \frac{K_2}{K_1} = \frac{\Delta H°}{R} \left(\frac{1}{T_1} - \frac{1}{T_2} \right)$$

$$\ln \frac{9.1 \times 10^2}{1.0 \times 10^6} = \frac{-1.8 \times 10^5}{8.3145} \left(\frac{1}{T_1} - \frac{1}{800} \right)$$

$$-7.00 = -2.2 \times 10^4 \left(\frac{1}{T_1} - \frac{1}{800} \right)$$

$$\frac{-7.00}{-2.2 \times 10^4} + \frac{1}{800} = \frac{1}{T_1}$$

$$\frac{1}{T_1} = \left(3.2 \times 10^{-4} \right) + \left(1.25 \times 10^{-3} \right) = 1.57 \times 10^{-3}$$

$$T_1 = \frac{1}{1.57 \times 10^{-3}} = 6.37 \times 10^2 \text{ K}$$

▶ For simplicity, we have dropped the units in this setup, but you should be able to show that units cancel properly.

Practice Example A: Estimate the temperature at which $K_p = 5.8 \times 10^{-2}$ for the reaction in Example 20-10. Use data from Table 20.3 and Figure 20-12.

Practice Example B: What is the value of K_p for the reaction $2 SO_2(g) + O_2(g) \rightleftharpoons 2 SO_3(g)$ at 235 °C? Use data from Table 20.3, Figure 20-12, and the van't Hoff equation (20.14).

20-8 Coupled Reactions

We have seen two ways to obtain product from a nonspontaneous reaction: (1) Change the reaction conditions to ones that make the reaction spontaneous (mostly by changing the temperature), and (2) carry out the reaction by electrolysis. But there is a third way also. Combine a pair of reactions, one with a positive ΔG and one with a negative ΔG, to obtain a spontaneous overall reaction. Such paired reactions are called **coupled reactions**. Let's consider the example of extracting a metal from its oxide.

When copper(I) oxide is heated at 673 K, no copper metal is obtained. The decomposition of Cu_2O to form products in their standard states (for instance, $P_{O_2} = 1.00$ bar) is nonspontaneous at 673 K.

$$Cu_2O(s) \xrightarrow{\Delta} 2 Cu(s) + \frac{1}{2} O_2(g) \qquad \Delta G°_{673 \text{ K}} = +125 \text{ kJ} \qquad (20.15)$$

Suppose that we couple this nonspontaneous decomposition reaction with the partial oxidation of carbon to carbon monoxide—a spontaneous reaction. The overall reaction (20.16), because it has a negative value of $\Delta G°$, is spontaneous when reactants and products are in their standard states.

$$Cu_2O(s) \longrightarrow 2 Cu(s) + \frac{1}{2} O_2(g) \qquad \Delta G°_{673 \text{ K}} = +125 \text{ kJ}$$

$$C(s) + \frac{1}{2} O_2(g) \longrightarrow CO(g) \qquad \Delta G°_{673 \text{ K}} = -175 \text{ kJ}$$

$$Cu_2O(s) + C(s) \longrightarrow 2 Cu(s) + CO(g) \qquad \Delta G°_{673 \text{ K}} = -50 \text{ kJ} \qquad (20.16)$$

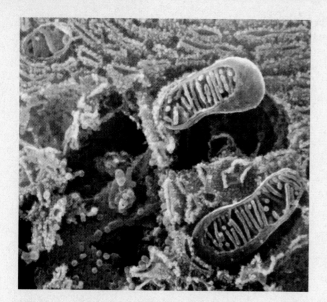

▲ Mitochondria and endoplasmic reticulum
A colorized scanning electron micrograph of mitochondria (blue) and rough endoplasmic reticulum (yellow) in a pancreatic cell. Mitochondria are the powerhouses of the cell. They oxidize sugars and fats, producing energy for the conversion of ADP to ATP. Rough endoplasmic reticulum is a network of folded membranes covered with protein-synthesizing ribosomes (small dots).

An important example of a coupled reaction in living organisms is the metabolism of glucose ($C_6H_{12}O_6$) that converts adenosine diphosphate (ADP) to adenosine triphosphate (ATP)[*] in the mitochondria of cells (Fig. 20-13). The ATP is utilized in the ribosomes to produce proteins. The ATP-forming reaction

$$ADP^{3-} + HPO_4^{2-} + H^+ \longrightarrow ATP^{4-} + H_2O$$
$$\Delta G° = -9.2 \text{ kJ mol}^{-1}$$

$$\text{Adenosine} - O - \overset{\overset{\displaystyle O^-}{|}}{\underset{\underset{\displaystyle O}{|}}{P}} - O - \overset{\overset{\displaystyle O^-}{|}}{\underset{\underset{\displaystyle O}{|}}{P}} - O^-$$

$$ADP^{3-}$$

$$\text{Adenosine} - O - \overset{\overset{\displaystyle O^-}{|}}{\underset{\underset{\displaystyle O}{|}}{P}} - O - \overset{\overset{\displaystyle O^-}{|}}{\underset{\underset{\displaystyle O}{|}}{P}} - O - \overset{\overset{\displaystyle O^-}{|}}{\underset{\underset{\displaystyle O}{|}}{P}} - O^-$$

$$ATP^{4-}$$

is spontaneous under standard conditions at 37 °C, so why does the cell need to use glucose to make ATP? The answer is that cells do not operate with $[H^+] = 1$ M as required by standard conditions; in fact, the pH in a cell is about 7. When we estimate ΔG for the reaction at that pH and assume all other species are at 1.0 M (still far from actuality), we get

$$\Delta G = \Delta G° + RT \ln\left(\frac{a_{ATP}\,a_{H_2O}}{a_{ADP}\,a_{P_i}\,a_{H^+}}\right)$$

Note that reactions (20.15) and (20.16) are not the same, even though each has Cu(s) as a product. The purpose of coupled reactions, then, is to produce a spontaneous overall reaction by combining two other processes: one nonspontaneous and one spontaneous. Many metallurgical processes employ coupled reactions, especially those that use carbon or hydrogen as reducing agents.

To sustain life, organisms must synthesize complex molecules from simpler ones. If carried out as single-step reactions, these syntheses would generally be accompanied by increases in enthalpy, decreases in entropy, and increases in free energy—in short, they would be nonspontaneous and would not occur. In living organisms, changes in temperature and electrolysis are not viable options for dealing with nonspontaneous processes. Here, coupled reactions are crucial, as described in the Focus On feature.

---

[*] The structural formulas for ADP^{3-} and ATP^{4-} are in accord with the discussion of expanded valence shells on page 408. Often, these structures are written with one P-to-O double bond per phosphate unit.

where P_i is a common abbreviation for HPO_4^{2-}. We now use $a_{ATP} = a_{ADP} = a_{P_i} = a_{H_2O} = 1$ and $a_{H^+} = 10^{-7}$ to obtain

$$\Delta G = -9.2 \text{ kJ mol}^{-1} + 41.6 \text{ kJ mol}^{-1} = +32.4 \text{ kJ mol}^{-1}$$

In biochemistry, this value of ΔG is called the *biological standard state*, designated $\Delta G^{\circ\prime}$. Thus, we can write

$$\Delta G = \Delta G^{\circ\prime} + RT \ln \left(\frac{a_{ATP} a_{H_2O}}{a_{ADP} a_{P_i} ([H^+]/10^{-7})} \right)$$

Note that the proton concentration is divided by 10^{-7} to give the activity because we have revised our definition of standard state.

Under standard biochemical conditions, the conversion of ADP to ATP is *not spontaneous*. Consequently, in cells this conversion must be coupled with another reaction to provide the necessary energy. The reaction, in the absence of oxygen (the anaerobic conditions inside a cell), is *glycolysis*—a metabolic process that converts glucose to the lactate anion. This reaction has a biochemical standard Gibbs energy change of -218 kJ mol^{-1}. In the cell, the coupling of this conversion in-

volves ten chemical reactions that are enzyme-catalyzed and form 2 moles of ATP. The overall process can be represented by two steps.

glucose $\longrightarrow$ 2 lactate$^-$ + 2 H$^+$ $\qquad \Delta G^{\circ\prime} = -218 \text{ kJ mol}^{-1}$

2 ADP^{3-} + 2 HPO$_4^{2-}$ + 2 H$^+$ $\longrightarrow$ 2 ATP^{4-} + 2 H$_2$O
$$\Delta G^{\circ\prime} = 2 \times 32.4 = 64.8 \text{ kJ mol}^{-1}$$

glucose + 2 ADP^{3-} + 2 HPO$_4^{2-}$ $\longrightarrow$
$$\text{2 lactate}^- + \text{2 ATP}^{4-} + \text{2 H}_2\text{O}$$
$$\Delta G^{\circ\prime} = -153 \text{ kJ mol}^{-1}$$

The overall reaction is therefore spontaneous. The coupling of the partial oxidation of glucose by way of glycolysis is spontaneous and provides a source of ATP. The ATP is then used in the formation of proteins. That is, the reaction

ATP^{4-} + H$_2$O $\longrightarrow$ ADP^{3-} + HPO$_4^{2-}$ + H$^+$
$$\Delta G^{\circ} = -32.4 \text{ kJ mol}^{-1}$$

is coupled with a nonspontaneous reaction to produce a protein in a ribosome as suggested by Figure 20-13.

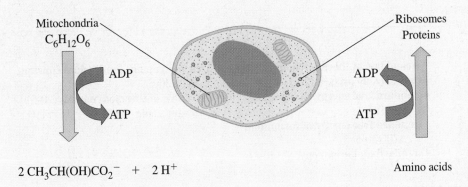

◄ FIGURE 20-13
Schematic of coupled reactions within a cell

Summary

Every spontaneous change produces an increase in the entropy of the universe. The entropy change of a system, ΔS, can be determined mathematically if the path of the reaction is reversible. Then, the entropy change equals the quantity of heat exchanged reversibly with the surroundings divided by the Kelvin temperature. That is, $\Delta S = q_{rev}/T$. For a phase transition at equilibrium, this expression becomes $\Delta S_{tr} = \Delta H_{tr}/T_{tr}$. Absolute values can be assigned to entropies of substances. This is made possible by the third law of thermodynamics, which takes as the zero of entropy that of a pure perfect crystal at 0 K.

A criterion for spontaneous change based on only the system itself is that the Gibbs free energy, G, decreases; that is, $\Delta G < 0$. At equilibrium, $\Delta G = 0$. Free energy change is related to enthalpy change and entropy change by means of the expression $\Delta G = \Delta H - T \Delta S$.

The standard free energy change, ΔG°, is based on the conversion of reactants in their standard states to products in their standard states. Tabulated free energy data are usually standard molar free energies of formation, ΔG_f°, and usually at 298.15 K (25 °C). The free energy change for a process can be determined for any chosen con-

ditions of composition by using $\Delta G = \Delta G^{\circ} + RT \ln Q$. The sign of ΔG, not of ΔG°, determines the spontaneity of a reaction, except under standard conditions, when $\Delta G = \Delta G^{\circ}$.

The relationship between the standard free energy change and the equilibrium constant for a reaction is $\Delta G^{\circ} = -RT \ln K_{eq}$. The constant, K_{eq}, is called a thermodynamic equilibrium constant. It is based on the activities of reactants and products, but these activities can be related to solution molarities and gas partial pressures by means of a few simple conventions.

Nonspontaneous processes can be made spontaneous by coupling them with spontaneous reactions and by taking advantage of the state function property of G. Coupled reactions occur in metallurgical processes and in biochemical transformations.

By starting with the relationship between standard free energy change and the equilibrium constant, the van't Hoff equation—relating the equilibrium constant and temperature—can be written. With this equation, we can use tabulated data at 25 °C to determine equilibrium constants not just at 25 °C but at other temperatures as well.

Integrative Example

The synthesis of methanol is of great importance because methanol can be used directly as a motor fuel, mixed with gasoline for fuel use, or converted to other organic compounds. The synthesis reaction, carried out at about 500 K, is

$$CO(g) + 2 H_2(g) \rightleftharpoons CH_3OH(g)$$

What is the value of K_p at 500 K?

1. *Determine $\Delta H°$ and $\Delta G°$ at 298 K.* We need $\Delta G°$ so that we can calculate K_p from it, and we need $\Delta H°$ to use in the van't Hoff equation (20.14). We must establish these values from tabulated data, and the tabulations are invariably at 298.15 K. Data from Appendix D are written beneath the equation that follows.

	$CO(g) + 2 H_2(g) \rightleftharpoons CH_3OH(g)$		
$\Delta H_f°$, kJ mol^{-1}	-110.5	0	-200.7
$\Delta G_f°$, kJ mol^{-1}	-137.2	0	-162.0

$$\Delta H° = 1 \text{ mol } CH_3OH \times (-200.7 \text{ kJ/mol } CH_3OH)$$
$$- 1 \text{ mol } CO \times (-110.5 \text{ kJ/mol } CO) = -90.2 \text{ kJ}$$

$$\Delta G° = 1 \text{ mol } CH_3OH \times (-162.0 \text{ kJ/mol } CH_3OH)$$
$$- 1 \text{ mol } CO \times (-137.2 \text{ kJ/mol } CO) = -24.8 \text{ kJ}$$

2. *Calculate K_p at 298 K.* For this calculation, we use $\Delta G°$ at 298 K, written as -24.8×10^3 J mol^{-1}, in the expression $\Delta G° = -RT \ln K_p$.

$$\ln K_p = -\Delta G°/RT = \frac{-(-24.8 \times 10^3 \text{ J mol}^{-1})}{8.3145 \text{ J mol}^{-1} \text{ K}^{-1} \times 298 \text{ K}} = 10.0$$

$$K_p = e^{10.0} = 2.2 \times 10^4$$

3. *Calculate K_p at 500 K.* Use the van't Hoff equation with $K_p = 2.2 \times 10^4$ at 298 K and $\Delta H° = -90.2 \times 10^3$ J mol^{-1}. Solve for K_p at 500 K.

$$\ln \frac{K_p}{2.2 \times 10^4} = \frac{-90.2 \times 10^3 \text{ J mol}^{-1}}{8.3145 \text{ J mol}^{-1} \text{ K}^{-1}} \left(\frac{1}{298 \text{ K}} - \frac{1}{500 \text{ K}} \right)$$
$$= -14.7$$

$$\frac{K_p}{2.2 \times 10^4} = e^{-14.7} = 4 \times 10^{-7} \qquad K_p = 9 \times 10^{-3}$$

Key Terms

coupled reactions (20-8)
entropy, *S* (20-2)
entropy change, ΔS (20-2)
free energy, *G* (20-4)
free energy change, ΔG (20-4)
nonspontaneous process (20-1)
reversible process (20-2)

second law of thermodynamics (20-4)
spontaneous process (20-1)
standard free energy change, $\Delta G°$ (20-5)
standard free energy of formation, $\Delta G_f°$ (20-5)
standard molar entropy, $S°$ (20-3)

thermodynamic equilibrium constant, K_{eq} (20-6)
third law of thermodynamics (20-3)
Trouton's rule (20-3)

Review Questions

1. In your own words, define the following symbols: **(a)** ΔS_{univ}; **(b)** $\Delta G_f°$; **(c)** K_{eq}.
2. Briefly describe each of the following ideas, methods, or phenomena: **(a)** absolute molar entropy; **(b)** reversible process; **(c)** Trouton's rule; **(d)** evaluation of an equilibrium constant from tabulated thermodynamic data.
3. Explain the important distinctions between each of the following pairs: **(a)** spontaneous and nonspontaneous processes; **(b)** the second and third laws of thermodynamics; **(c)** ΔG and $\Delta G°$.
4. Indicate whether the entropy of the system would increase or decrease in each of the following reactions. If you cannot be certain simply by inspecting the equation, explain why.
   (a) $CCl_4(l) \longrightarrow CCl_4(g)$
   (b) $CuSO_4 \cdot 3 H_2O(s) + 2 H_2O(g) \longrightarrow CuSO_4 \cdot 5 H_2O(s)$
   (c) $SO_3(g) + H_2(g) \longrightarrow SO_2(g) + H_2O(g)$
   (d) $H_2S(g) + O_2(g) \longrightarrow H_2O(g) + SO_2(g)$
   (not balanced)
5. Which substance in each of the following pairs would have the greater entropy? Explain.

   (a) at 75 °C and 1 atm: 1 mol $H_2O(l)$ *or* 1 mol $H_2O(g)$
   (b) at 5 °C and 1 atm: 50.0 g Fe(s) *or* 0.80 mol Fe(s)
   (c) 1 mol $Br_2(l$, 1 atm, 8 °C) *or* 1 mol $Br_2(s$, 1 atm, -8 °C)
   (d) 0.312 mol $SO_2(g$, 0.110 atm, 32.5 °C) *or* 0.284 mol $O_2(g$, 15.0 atm, 22.3 °C)
6. From the data given, indicate which of the four cases in Table 20.1 applies for each of the following reactions.
   (a) $H_2(g) \longrightarrow 2 H(g)$ $\quad\quad\quad\quad \Delta H° = +436.0$ kJ
   (b) $2 SO_2(g) + O_2(g) \longrightarrow 2 SO_3(g)$ $\;\; \Delta H° = -197.8$ kJ
   (c) $N_2H_4(g) \longrightarrow N_2(g) + 2 H_2(g)$ $\quad \Delta H° = -95.40$ kJ
   (d) $N_2(g) + 3 Cl_2(g) \longrightarrow 2 NCl_3(l)$ $\;\; \Delta H° = +230$ kJ
7. Which of the following changes in thermodynamic property would you expect to find for the reaction $Br_2(g) \longrightarrow 2 Br(g)$ *at all temperatures:* **(a)** $\Delta H < 0$; **(b)** $\Delta S > 0$; **(c)** $\Delta G < 0$; **(d)** $\Delta S < 0$? Explain.
8. If a reaction can be carried out only by electrolysis, which of the following changes in thermodynamic property *must* apply: **(a)** $\Delta H > 0$; **(b)** $\Delta S > 0$; **(c)** $\Delta G = \Delta H$; **(d)** $\Delta G > 0$? Explain.
9. If $\Delta G° = 0$ for a reaction, which of the following statements must also be true: **(a)** $\Delta H° = 0$; **(b)** $\Delta S° = 0$; **(c)** $K_{eq} = 0$; **(d)** $K_{eq} = 1$? Explain.

10. From the data given in the following table, determine $\Delta S°$ for the reaction $NH_3(g) + HCl(g) \longrightarrow NH_4Cl(s)$. All data are at 298 K.

	$\Delta H_f°$	$\Delta G_f°$
$NH_3(g)$	-46.11 kJ mol^{-1}	-16.48 kJ mol^{-1}
$HCl(g)$	-92.31	-95.30
$NH_4Cl(s)$	-314.4	-202.9

11. Use data from Appendix D to determine values of $\Delta G°$ for the following reactions at 25 °C.
    (a) $C_2H_2(g) + 2 H_2(g) \longrightarrow C_2H_6(g)$
    (b) $2 SO_3(g) \longrightarrow 2 SO_2(g) + O_2(g)$
    (c) $Fe_3O_4(s) + 4 H_2(g) \longrightarrow 3 Fe(s) + 4 H_2O(g)$
    (d) $2 Al(s) + 6 H^+(aq) \longrightarrow 2 Al^{3+}(aq) + 3 H_2(g)$

12. If a graph similar to Figure 20-9 were drawn for the process $I_2(s) \longrightarrow I_2(l)$ at 1 atm,
    (a) At what temperature would the two lines intersect? (*Hint:* Refer also to Figure 13-18.)
    (b) What would be the value of $\Delta G°$ at this temperature? Explain.

13. From the following data, determine $\Delta S°$ (in J mol^{-1} K^{-1}) for each transition.
    (a) the boiling of HCl(l) at -85.05 °C and 1 atm; $\Delta H_{vap}° = 3.86$ kcal mol^{-1}
    (b) the melting of Na(s) at 97.82 °C; $\Delta H_{fus}° = 27.05$ cal/g

14. Write thermodynamic equilibrium constant expressions for the following reactions. Do any of these expressions correspond to K_c or K_p?
    (a) $2 NO(g) + O_2(g) \rightleftharpoons 2 NO_2(g)$
    (b) $MgSO_3(s) \rightleftharpoons MgO(s) + SO_2(g)$
    (c) $HC_2H_3O_2(aq) + H_2O(l) \rightleftharpoons H_3O^+(aq) + C_2H_3O_2^-(aq)$
    (d) $2 NaHCO_3(s) \rightleftharpoons Na_2CO_3(s) + H_2O(g) + CO_2(g)$
    (e) $MnO_2(s) + 4 H^+(aq) + 2 Cl^-(aq) \rightleftharpoons Mn^{2+}(aq) + 2 H_2O(l) + Cl_2(g)$

15. For the reaction $Cl_2(g) \rightleftharpoons 2 Cl(g)$, $K_p = 2.45 \times 10^{-7}$ at 1000 K. What is $\Delta G°$ for this reaction at 1000 K?

16. Use data from Appendix D to determine K_p at 298 K for the reaction $N_2O(g) + \frac{1}{2} O_2(g) \rightleftharpoons 2 NO(g)$.

17. Use data from Appendix D to determine values at 298 K of $\Delta G°$ and K_{eq} for the following reactions. (*Note:* The equations are not balanced.)
    (a) $HCl(g) + O_2(g) \rightleftharpoons H_2O(g) + Cl_2(g)$
    (b) $Fe_2O_3(g) + H_2(g) \rightleftharpoons Fe_3O_4(s) + H_2O(g)$
    (c) $Ag^+(aq) + SO_4^{2-}(aq) \rightleftharpoons Ag_2SO_4(s)$

18. In Example 20-1, we were unable to conclude by inspection whether $\Delta S°$ for the reaction $CO(g) + H_2O(g) \longrightarrow CO_2(g) + H_2(g)$ should be positive or negative. Use data from Appendix D to obtain $\Delta S°$ at 298 K.

19. *Without performing calculations*, indicate whether any of the following reactions would occur to a significant extent at 298 K. (*Hint:* What is $\Delta G°$ for each reaction?)
    (a) Conversion of dioxygen to ozone:
    $$3 O_2(g) \longrightarrow 2 O_3(g)$$
    (b) Dissociation of N_2O_4 to NO_2:
    $$N_2O_4(g) \longrightarrow 2 NO_2(g)$$
    (c) Formation of BrCl:
    $$Br_2(l) + Cl_2(g) \longrightarrow 2 BrCl(g)$$

20. What must be the temperature if the following reaction has $\Delta G° = -45.5$ kJ, $\Delta H° = -24.8$ kJ, and $\Delta S° = 15.2$ J K^{-1}?
    $$Fe_2O_3(s) + 3 CO(g) \longrightarrow 2 Fe(s) + 3 CO_2(g)$$

21. Use data from Appendix D to establish for the reaction $2 N_2O_4(g) + O_2(g) \longrightarrow 2 N_2O_5(g)$:
    (a) $\Delta G°$ at 298 K for the reaction as written
    (b) K_p at 298 K

22. Use data from Appendix D to establish at 298 K for the reaction
    $$2 NaHCO_3(s) \longrightarrow Na_2CO_3(s) + H_2O(l) + CO_2(g):$$
    (a) $\Delta S°$; (b) $\Delta H°$; (c) $\Delta G°$; (d) K_{eq}.

23. A possible reaction for converting methanol to ethanol is
    $CO(g) + 2 H_2(g) + CH_3OH(g) \longrightarrow$
    $$C_2H_5OH(g) + H_2O(g)$$
    (a) Use data from Appendix D to calculate $\Delta H°$, $\Delta S°$, and $\Delta G°$ for this reaction at 25 °C.
    (b) Is this reaction thermodynamically favored at high or low temperatures? At high or low pressures? Explain.
    (c) Estimate K_p for the reaction at 750 K.

24. Estimate K_p at 100 °C for the reaction $2 SO_2(g) + O_2(g) \rightleftharpoons 2 SO_3(g)$. Use data from Table 20.3 and Figure 20-12.

Exercises

Spontaneous Change and Entropy

25. Indicate whether each of the following changes represents an increase or decrease in entropy in a system, and explain your reasoning: (a) the freezing of ethanol; (b) the sublimation of dry ice; (c) the burning of a rocket fuel.

26. Arrange the entropy changes of the following processes, all at 25 °C, in the expected order of increasing ΔS, and explain your reasoning: (a) $H_2O(l, 1 \text{ atm}) \longrightarrow H_2O(g, 1 \text{ atm})$;
    (b) $CO_2(s, 1 \text{ atm}) \longrightarrow CO_2(g, 10 \text{ mmHg})$;
    (c) $H_2O(l, 1 \text{ atm}) \longrightarrow H_2O(g, 10 \text{ mmHg})$.

27. Use ideas from this chapter to explain this famous remark attributed to Rudolf Clausius (1865): "Die Energie der Welt ist konstant; die Entropie der Welt strebt einem Maximum zu." ("The energy of the world is constant; the entropy of the world increases toward a maximum.")

28. Comment on the difficulties of solving environmental pollution problems from the standpoint of entropy changes associated with the formation of pollutants and with their removal from the environment.

29. For each of the following reactions, indicate whether ΔS for the reaction should be positive or negative. If it is not possible to determine the sign of ΔS from the information given, indicate why.
    (a) $CaO(s) + H_2O(l) \longrightarrow Ca(OH)_2(s)$
    (b) $2 HgO(s) \longrightarrow 2 Hg(l) + O_2(g)$
    (c) $2 NaCl(l) \longrightarrow 2 Na(l) + Cl_2(g)$
    (d) $Fe_2O_3(s) + 3 CO(g) \longrightarrow 2 Fe(s) + 3 CO_2(g)$
    (e) $Si(s) + 2 Cl_2(g) \longrightarrow SiCl_4(g)$

30. By analogy to ΔH_f° and ΔG_f°, how would you define entropy of formation? Which would have the largest entropy of formation: $CH_4(g)$, $CH_3CH_2OH(l)$, or $CS_2(l)$? First make a qualitative prediction; then test your prediction with data from Appendix D.

Phase Transitions

31. In Example 20-2, we dealt with ΔH_{vap}° and ΔS_{vap}° for water at 100 °C.
    (a) Use data from Appendix D to determine values for these two quantities at 25 °C.
    (b) From your knowledge of the structure of liquid water, explain the differences in ΔH_{vap}° values and in ΔS_{vap}° values between 25 °C and 100 °C.

32. Pentane is one of the most volatile of the hydrocarbons in gasoline. At 298.15 K, the following enthalpies of formation are given for pentane: $\Delta H_f^{\circ}[C_5H_{12}(l)] = -173.5$ kJ mol^{-1}; $\Delta H_f^{\circ}[C_5H_{12}(g)] = -146.9$ kJ mol^{-1}.
    (a) Estimate the normal boiling point of pentane.
    (b) Estimate ΔG° for the vaporization of pentane at 298 K.
    (c) Comment on the significance of the sign of ΔG° at 298 K.

33. Which of the following substances would obey Trouton's rule most closely: HF, $C_6H_5CH_3$ (toluene), or CH_3OH (methanol)? Explain your reasoning.

34. Estimate the normal boiling point of bromine, Br_2, in the following way: Determine ΔH_{vap}° for Br_2 with data from Appendix D. Assume that ΔH_{vap}° remains constant and that Trouton's rule is obeyed.

35. In what temperature range can the following equilibrium be established? Explain.
$$H_2O(l, 0.50 \text{ atm}) \rightleftharpoons H_2O(g, 0.50 \text{ atm})$$

36. Refer to Figures 13-19 and 20-9. Which has the lowest free energy at 1 atm and −60 °C: solid, liquid, or gaseous carbon dioxide? Explain.

Free Energy and Spontaneous Change

37. Indicate which of the four cases in Table 20.1 applies to each of the following reactions. If you are unable to decide from only the information given, state why.
    (a) $PCl_3(g) + Cl_2(g) \longrightarrow PCl_5(g)$ $\Delta H^{\circ} = -87.9$ kJ
    (b) $CO_2(g) + H_2(g) \longrightarrow CO(g) + H_2O(g)$
    $\Delta H^{\circ} = +41.2$ kJ
    (c) $NH_4CO_2NH_2(s) \longrightarrow 2 NH_3(g) + CO_2(g)$
    $\Delta H^{\circ} = +159.2$ kJ

38. Indicate which of the four cases in Table 20.1 applies to each of the following reactions. If you are unable to decide from only the information given, state why.
    (a) $H_2O(g) + \frac{1}{2} O_2(g) \longrightarrow H_2O_2(g)$ $\Delta H^{\circ} = +105.5$ kJ

    (b) $C_6H_6(l) + \frac{15}{2} O_2(g) \longrightarrow 6 CO_2(g) + 3 H_2O(g)$

    $\Delta H^{\circ} = -3135$ kJ

    (c) $NO(g) + \frac{1}{2} Cl_2(g) \longrightarrow NOCl(g) \Delta H^{\circ} = -38.54$ kJ

39. For the mixing of ideal gases (see Figure 20-2), explain whether a positive, negative, or zero value is expected for ΔH, ΔS, and ΔG.

40. What values of ΔH, ΔS, and ΔG would you expect for the formation of an ideal solution of liquid components? (Is each value positive, negative, or zero?)

41. Explain why:
    (a) Some exothermic reactions do not occur spontaneously.
    (b) Some reactions in which the entropy of the system increases do not occur spontaneously.

42. Explain why you would expect a reaction of the type $AB(g) \longrightarrow A(g) + B(g)$ always to be spontaneous at *high* rather than at low temperatures.

Standard Free Energy Change

43. At 298 K, for the reaction $2 PCl_3(g) + O_2(g) \longrightarrow 2 POCl_3(l)$, $\Delta H^{\circ} = -620.2$ kJ and the standard molar entropies are $PCl_3(g)$, 311.8 J K^{-1}; $O_2(g)$, 205.1 J K^{-1}; and $POCl_3(l)$, 222.4 J K^{-1}. Determine (a) ΔG° at 298 K and (b) whether the reaction proceeds spontaneously in the forward or the reverse direction when reactants and products are in their standard states.

44. At 298 K, for the reaction $2 H^+(aq) + 2 Br^-(aq) + 2 NO_2(g) \longrightarrow Br_2(l) + 2 HNO_2(aq)$, $\Delta H^{\circ} = -61.6$ kJ and the standard molar entropies are $H^+(aq)$, 0 J K^{-1}; $Br^-(aq)$, 82.4 J K^{-1}; $NO_2(g)$, 240.1 J K^{-1}; $Br_2(l)$, 152.2 J K^{-1}; $HNO_2(aq)$, 135.6 J K^{-1}. Determine (a) ΔG° at 298 K and (b) whether the reaction proceeds spontaneously in the forward or the reverse direction when reactants and products are in their standard states.

45. The following standard free energy changes are given for 25 °C.
    (1) $N_2(g) + 3 H_2(g) \longrightarrow 2 NH_3(g)$ $\Delta G^{\circ} = -33.0$ kJ

(2) $4 NH_3(g) + 5 O_2(g) \longrightarrow 4 NO(g) + 6 H_2O(l)$
$$\Delta G° = -1010.5 \text{ kJ}$$
(3) $N_2(g) + O_2(g) \longrightarrow 2 NO(g) \quad \Delta G° = +173.1 \text{ kJ}$
(4) $N_2(g) + 2 O_2(g) \longrightarrow 2 NO_2(g) \quad \Delta G° = +102.6 \text{ kJ}$
(5) $2 N_2(g) + O_2(g) \longrightarrow 2 N_2O(g) \quad \Delta G° = +208.4 \text{ kJ}$
Combine the preceding equations, as necessary, to obtain $\Delta G°$ values for each of the following reactions.

(a) $N_2O(g) + \dfrac{3}{2} O_2(g) \longrightarrow 2 NO_2(g) \qquad \Delta G° = ?$

(b) $2 H_2(g) + O_2(g) \longrightarrow 2 H_2O(l) \qquad \Delta G° = ?$
(c) $2 NH_3(g) + 2 O_2(g) \longrightarrow N_2O(g) + 3 H_2O(l)$
$$\Delta G° = ?$$

Of reactions (a), (b), and (c), which would tend to go to completion at 25 °C, and which would reach an equilibrium condition with significant amounts of all reactants and products present?

46. The following standard free energy changes are given for 25 °C.
(1) $SO_2(g) + 3 CO(g) \longrightarrow COS(g) + 2 CO_2(g)$
$$\Delta G° = -246.4 \text{ kJ}$$
(2) $CS_2(g) + H_2O(g) \longrightarrow COS(g) + H_2S(g)$
$$\Delta G° = -41.5 \text{ kJ}$$
(3) $CO(g) + H_2S(g) \longrightarrow COS(g) + H_2(g)$
$$\Delta G° = +1.4 \text{ kJ}$$
(4) $CO(g) + H_2O(g) \longrightarrow CO_2(g) + H_2(g)$
$$\Delta G° = -28.6 \text{ kJ}$$

Combine the preceding equations, as necessary, to obtain $\Delta G°$ values for the following reactions.
(a) $COS(g) + 2 H_2O(g) \longrightarrow$
$$SO_2(g) + CO(g) + 2 H_2(g) \qquad \Delta G° = ?$$
(b) $COS(g) + 3 H_2O(g) \longrightarrow$
$$SO_2(g) + CO_2(g) + 3 H_2(g) \qquad \Delta G° = ?$$

(c) $COS(g) + H_2O(g) \longrightarrow CO_2(g) + H_2S(g)$
$$\Delta G° = ?$$
Of reactions (a), (b), and (c), which is spontaneous in the forward direction when reactants and products are present in their standard states?

47. Write an equation for the combustion of one mole of benzene, $C_6H_6(l)$, and use data from Appendix D to determine $\Delta G°$ at 298 K if the products of the combustion are **(a)** $CO_2(g)$ and $H_2O(l)$, and **(b)** $CO_2(g)$ and $H_2O(g)$. Describe how you might determine the *difference* between the values obtained in (a) and (b) without having either to write the combustion equation or to determine $\Delta G°$ values for the combustion reactions.

48. Use molar entropies from Appendix D, together with the following data, to estimate the bond-dissociation energy of the F_2 molecule.
$$F_2(g) \longrightarrow 2 F(g) \qquad \Delta G° = 123.9 \text{ kJ}$$
Compare your result with the value listed in Table 11.3.

49. Assess the feasibility of the reaction
$$N_2H_4(g) + 2 OF_2(g) \rightleftharpoons N_2F_4(g) + 2 H_2O(g)$$
by determining each of the following quantities for this reaction at 25 °C.
(a) $\Delta S°$ (The standard molar entropy of $N_2F_4(g)$ is 301.2 J K^{-1}.)
(b) $\Delta H°$ (Use data from Table 11.3 and F—O and N—F bond energies of 222 and 301 kJ mol^{-1}, respectively.)
(c) $\Delta G°$
Is the reaction feasible? If so, is it favored at high or low temperatures?

50. Solid ammonium nitrate can decompose to dinitrogen oxide gas and liquid water. What is $\Delta G°$ at 298 K? Is the decomposition reaction favored at temperatures above or below 298 K?

The Thermodynamic Equilibrium Constant

51. For one of the following reactions, $K_c = K_p = K_{eq}$. Identify that one. For the other two, what is the relationship between K_c, K_p, and K_{eq}? Explain.
(a) $2 SO_2(g) + O_2(g) \rightleftharpoons 2 SO_3(g)$

(b) $HI(g) \rightleftharpoons \dfrac{1}{2} H_2(g) + \dfrac{1}{2} I_2(g)$

(c) $NH_4HCO_3(s) \rightleftharpoons NH_3(g) + CO_2(g) + H_2O(l)$

52. $H_2(g)$ can be prepared by passing steam over hot iron:
$3 Fe(s) + 4 H_2O(g) \rightleftharpoons Fe_3O_4(s) + 4 H_2(g)$.
(a) Write an expression for the thermodynamic equilibrium constant for this reaction.
(b) Explain why the partial pressure of $H_2(g)$ is independent of the amounts of $Fe(s)$ and $Fe_3O_4(s)$ present.
(c) Can we conclude that the production of $H_2(g)$ from $H_2O(g)$ could be accomplished regardless of the proportions of $Fe(s)$ and $Fe_3O_4(s)$ present? Explain.

Relationships Involving ΔG, $\Delta G°$, Q and K_{eq}

53. Use thermodynamic data at 298 K to decide in which direction the reaction
$$2 SO_2(g) + O_2(g) \rightleftharpoons 2 SO_3(g)$$
is spontaneous when the partial pressures of SO_2, O_2, and SO_3 are 1.0×10^{-4}, 0.20, and 0.10 atm, respectively.

54. Use thermodynamic data at 298 K to decide in which direction the reaction
$$H_2(g) + Cl_2(g) \rightleftharpoons 2 HCl(g)$$
is spontaneous when the partial pressures of H_2, Cl_2, and HCl are all 0.5 atm.

55. The standard free energy change for the reaction
$$CH_3CO_2H(aq) + H_2O(l) \rightleftharpoons$$
$$CH_3CO_2^-(aq) + H_3O^+(aq)$$
is 27.07 kJ mol^{-1} at 298 K. Use this thermodynamic quantity to decide in which direction the reaction is spontaneous when the concentrations of $CH_3CO_2H(aq)$, $CH_3CO_2^-(aq)$, and $H_3O^+(aq)$ are 0.10 M, 1.0×10^{-3} M, and 1.0×10^{-3} M, respectively.

56. The standard free energy change for the reaction
$$NH_3(aq) + H_2O(l) \rightleftharpoons NH_4^+(aq) + OH^-(aq)$$

is 29.05 kJ mol^{-1} at 298 K. Use this thermodynamic quantity to decide in which direction the reaction is spontaneous when the concentrations of $NH_3(aq)$, $NH_4^+(aq)$, and $OH^-(aq)$ are 0.10 M, 1.0×10^{-3} M, and 1.0×10^{-3} M, respectively.

57. For the reaction $2 NO(g) + O_2(g) \longrightarrow 2 NO_2(g)$, all but one of the following equations is correct. Which is *incorrect*, and why? **(a)** $K_{eq} = K_p$; **(b)** $\Delta S° = (\Delta G° - \Delta H°)/T$; **(c)** $K_p = e^{-\Delta G°/RT}$; **(d)** $\Delta G = \Delta G° + RT \ln Q$.

58. Why is $\Delta G°$ such an important property of a chemical reaction, even though we generally carry out the reaction under *nonstandard* conditions?

59. At 1000 K an equilibrium mixture in the reaction $CO_2(g) + H_2(g) \rightleftharpoons CO(g) + H_2O(g)$ contains 0.276 mol H_2, 0.276 mol CO_2, 0.224 mol CO, and 0.224 mol H_2O.
    (a) What is K_p at 1000 K?
    (b) Calculate $\Delta G°$ at 1000 K. In which direction will a spontaneous reaction occur if the following were brought together at 1000 K: 0.0750 mol CO_2, 0.095 mol H_2, 0.0340 mol CO, and 0.0650 mol H_2O?

60. For the reaction $2 SO_2(g) + O_2(g) \rightleftharpoons 2 SO_3(g)$, $K_c = 2.8 \times 10^2$ at 1000 K.
    (a) What is $\Delta G°$ at 1000 K?
    (*Hint:* What is K_p?)
    (b) If 0.40 mol SO_2, 0.18 mol O_2, and 0.72 mol SO_3 are mixed in a 2.50-L flask at 1000 K, in what direction will a net reaction occur?

61. For the following equilibrium reactions, calculate $\Delta G°$ at the indicated temperature.
    (*Hint:* How is each equilibrium constant related to a thermodynamic equilibrium constant, K_{eq}?)
    (a) $H_2(g) + I_2(g) \rightleftharpoons 2 HI(g)$ $K_c = 50.2$ at 445 °C
    (b) $N_2O(g) + \frac{1}{2} O_2(g) \rightleftharpoons 2 NO(g)$
    $$K_c = 1.7 \times 10^{-13} \text{ at } 25 °C$$
    (c) $N_2O_4(g) \rightleftharpoons 2 NO_2(g)$
    $$K_c = 4.61 \times 10^{-3} \text{ at } 25 °C$$
    (d) $2 Fe^{3+}(aq) + Hg_2^{2+}(aq) \rightleftharpoons$
    $$2 Fe^{2+}(aq) + 2 Hg^{2+}(aq)$$
    $$K_c = 9.14 \times 10^{-6} \text{ at } 25 °C$$

62. Two equations can be written for the dissolution of $Mg(OH)_2(s)$ in acidic solution.

$$Mg(OH)_2(s) + 2 H^+(aq) \rightleftharpoons Mg^{2+}(aq) + 2 H_2O(l)$$
$$\Delta G° = -95.5 \text{ kJ mol}^{-1}$$
$$\frac{1}{2} Mg(OH)_2(s) + H^+(aq) \rightleftharpoons \frac{1}{2} Mg^{2+}(aq) + H_2O(l)$$
$$\Delta G° = -47.8 \text{ kJ mol}^{-1}$$

(a) Explain why these two equations have different $\Delta G°$ values.
(b) Will K_{eq} for these two equations be the same or different? Explain.
(c) Will the solubilities of $Mg(OH)_2(s)$ in a buffer solution at pH $= 8.5$ depend on which of the two equations is used as the basis of the calculation? Explain.

63. At 298 K, $\Delta G_f°[CO(g)] = -137.2$ kJ/mol and $K_p = 6.5 \times 10^{11}$ for the reaction $CO(g) + Cl_2(g) \rightleftharpoons COCl_2(g)$. Use these data to determine $\Delta G_f°[COCl_2(g)]$, and compare your result with the value in Appendix D.

64. Use data from Appendix D to determine values of K_{sp} for the following sparingly soluble solutes: **(a)** AgBr; **(b)** $CaSO_4$; **(c)** $Fe(OH)_3$.
    (*Hint:* Begin by writing solubility equilibrium expressions.)

65. To establish the law of conservation of mass, Lavoisier carefully studied the decomposition of mercury(II) oxide:
    $$HgO(s) \longrightarrow Hg(l) + \frac{1}{2} O_2(g).$$
    At 25 °C, $\Delta H° = +90.83$ kJ and $\Delta G° = +58.54$ kJ.
    (a) Show that the partial pressure of $O_2(g)$ in equilibrium with HgO(s) and Hg(l) at 25 °C is extremely low.
    (b) What conditions do you suppose Lavoisier used to obtain significant quantities of oxygen?

66. Currently, CO_2 is being studied as a source of carbon atoms for synthesizing organic compounds. One possible reaction involves the conversion of CO_2 to methanol, CH_3OH.
    $$CO_2(g) + 3 H_2(g) \longrightarrow CH_3OH(g) + H_2O(g)$$
    With the aid of data from Appendix D, determine
    (a) if this reaction proceeds to any significant extent at 25 °C.
    (b) if the production of $CH_3OH(g)$ is favored by raising or lowering the temperature from 25 °C.
    (c) K_p for this reaction at 500 K.
    (d) the partial pressure of $CH_3OH(g)$ at equilibrium if $CO_2(g)$ and $H_2(g)$, each initially at a partial pressure of 1 atm, react at 500 K.

$\Delta G°$ and K_{eq} as Functions of Temperature

67. The synthesis of ammonia by the Haber process occurs by the reaction $N_2(g) + 3 H_2 \rightleftharpoons 2 NH_3(g)$ at 400 °C. Using data from Appendix D and assuming that $\Delta H°$ and $\Delta S°$ are essentially unchanged in the temperature interval from 25 to 400 °C, estimate K_p at 400 °C.

68. Use data from Appendix D to determine **(a)** $\Delta H°$, $\Delta S°$, and $\Delta G°$ at 298 K and **(b)** K_p at 875 K for the water gas shift reaction, used commercially to produce $H_2(g)$: $CO(g) + H_2O(g) \rightleftharpoons CO_2(g) + H_2(g)$.

(*Hint:* Assume that $\Delta H°$ and $\Delta S°$ are essentially unchanged in this temperature interval.)

69. In Example 20-10, we used the van't Hoff equation to determine the temperature at which $K_p = 1.0 \times 10^6$ for the reaction $2 SO_2(g) + O_2(g) \rightleftharpoons 2 SO_3(g)$. Obtain another estimate of this temperature with data from Appendix D and equations (20.8) and (20.12). Compare your result with that obtained in Example 20-10.

70. The following equilibrium constants have been determined for the reaction $H_2(g) + I_2(g) \rightleftharpoons 2\,HI(g)$: $K_p = 50.0$ at 448 °C and 66.9 at 350 °C. Use these data to estimate $\Delta H°$ for the reaction.

71. For the reaction $N_2O_4(g) \rightleftharpoons 2\,NO_2(g)$, $\Delta H° = +57.2$ kJ mol^{-1} and $K_p = 0.113$ at 298 K.
(a) What is K_p at 0 °C?
(b) At what temperature will $K_p = 1.00$?

72. Use data from Appendix D and the van't Hoff equation (20.14) to estimate a value of K_p at 100 °C for the reaction $2\,NO(g) + O_2(g) \rightleftharpoons 2\,NO_2(g)$.
(*Hint:* First determine K_p at 25 °C. What is $\Delta H°$ for the reaction?)

73. For the reaction

$$CO(g) + 3\,H_2(g) \rightleftharpoons CH_4(g) + H_2O(g),$$
$$K_p = 2.15 \times 10^{11} \text{ at } 200 \text{ °C}$$
$$K_p = 4.56 \times 10^{8} \text{ at } 260 \text{ °C}$$

determine $\Delta H°$ by using the van't Hoff equation (20.14) and by using tabulated data in Appendix D. Compare the two results, and comment on how good the assumption is that $\Delta H°$ is essentially independent of temperature in this case.

74. Sodium carbonate, an important chemical used in the production of glass, is made from sodium hydrogen carbonate by the reaction

$$2\,NaHCO_3(s) \rightleftharpoons Na_2CO_3(s) + CO_2(g) + H_2O(g)$$

Data for the temperature variation of K_p for this reaction are $K_p = 1.66 \times 10^{-5}$ at 30 °C; 3.90×10^{-4} at 50 °C; 6.27×10^{-3} at 70 °C; and 2.31×10^{-1} at 100 °C.
(a) Plot a graph similar to Figure 20-12, and determine $\Delta H°$ for the reaction.
(b) Calculate the temperature at which the total gas pressure above a mixture of $NaHCO_3(s)$ and $Na_2CO_3(s)$ is 2.00 atm.

Coupled Reactions

75. Titanium is obtained by the reduction of $TiCl_4(l)$, which in turn is produced from the mineral rutile (TiO_2).
(a) With data from Appendix D, determine $\Delta G°$ at 298 K for this reaction.

$$TiO_2(s) + 2\,Cl_2(g) \longrightarrow TiCl_4(l) + O_2(g)$$

(b) Show that the conversion of $TiO_2(s)$ to $TiCl_4(l)$, with reactants and products in their standard states, is spontaneous at 298 K if the reaction in (a) is coupled with the reaction

$$2\,CO(g) + O_2(g) \longrightarrow 2\,CO_2(g)$$

76. Following are some standard free energies of formation, $\Delta G_f°$, per mole of metal oxide at 1000 K: NiO, -115 kJ; MnO, -280 kJ; TiO_2, -630 kJ. The standard free energy of formation of CO at 1000 K is -250 kJ per mol CO. Use the method of coupled reactions (page 811) to determine which of these metal oxides can be reduced to the metal by a spontaneous reaction with carbon at 1000 K and with all reactants and products in their standard states.

Integrative and Advanced Exercises

77. Use data from Appendix D to estimate **(a)** the normal boiling point of mercury and **(b)** the vapor pressure of mercury at 25 °C.

78. Dinitrogen pentoxide, N_2O_5, is a solid with a high vapor pressure. Its vapor pressure at 7.5 °C is 100 mmHg, and the solid sublimes at a pressure of 1.00 atm at 32.4 °C. What is the standard free energy change for the process $N_2O_5(s) \longrightarrow N_2O_5(g)$ at 25 °C?

79. Consider the vaporization of water: $H_2O(l) \longrightarrow H_2O(g)$ at 100 °C, with $H_2O(l)$ in its standard state, but with the partial pressure of $H_2O(g)$ at 2.0 atm. Which of the following statements are true concerning this vaporization at 100 °C: **(a)** $\Delta G° = 0$, **(b)** $\Delta G = 0$, **(c)** $\Delta G° > 0$, **(d)** $\Delta G > 0$? Explain.

80. At 298 K, 1.00 mol $BrCl(g)$ is introduced into a 10.0-L vessel, and equilibrium is established in the reaction

$$BrCl(g) \rightleftharpoons \frac{1}{2} Br_2(g) + \frac{1}{2} Cl_2(g). \text{ Calculate the amounts}$$

of each of the three gases present when equilibrium is established.

(*Hint:* Use data from Appendix D as necessary.)

81. Use data from Appendix D and other information from this chapter to estimate the temperature at which the dissociation of $I_2(g)$ becomes appreciable [for example, with $I_2(g)$ 50% dissociated into I(g) at 1 atm total pressure].

82. The following table shows the enthalpies and free energies of formation of three metal oxides at 25 °C.

	$\Delta H_f°$, kJ mol^{-1}	$\Delta G_f°$, kJ mol^{-1}
PbO(red)	−219.0	−188.9
Ag$_2$O	−31.05	−11.20
ZnO	−348.3	−318.3

(a) Which of these oxides can be most readily decomposed to the free metal and $O_2(g)$?
(b) For the oxide that is most easily decomposed, to what temperature must it be heated to produce $O_2(g)$ at 1.00 atm pressure?

83. The following data are given for the two solid forms of HgI_2 at 298 K.

	ΔH_f°, kJ mol^{-1}	ΔG_f°, kJ mol^{-1}	S°, J mol^{-1} K^{-1}
HgI_2 (red)	-105.4	-101.7	180.
HgI_2 (yellow)	-102.9	(?)	(?)

Estimate values for the two missing entries. To do this, assume that for the transition HgI_2 (red) $\longrightarrow$ HgI_2 (yellow), the values of ΔH° and ΔS° at 25 °C have the same values that they do at the equilibrium temperature of 127 °C.

84. Oxides of nitrogen are produced in high-temperature combustion processes. The essential reaction is $N_2(g) + O_2(g) \rightleftharpoons 2\ NO(g)$. At what approximate temperature will an *equimolar* mixture of $N_2(g)$ and $O_2(g)$ be 1.0% converted to $NO(g)$?
(*Hint:* Use data from Appendix D as necessary.)

85. Use the following data, as appropriate, to estimate the molarity of a saturated aqueous solution of $Sr(IO_3)_2$.

	ΔH_f°, kJ mol^{-1}	ΔG_f°, kJ mol^{-1}	S°, J mol^{-1} K^{-1}
$Sr(IO_3)(s)$	-1019.2	-855.1	234
$Sr^{2+}(aq)$	-545.8	-599.5	-32.6
$IO_3^-(aq)$	-221.3	-128.0	118.4

86. A plausible reaction for the production of ethylene glycol (used as an antifreeze) is

$$2\ CO(g) + 3\ H_2(g) \longrightarrow CH_2OHCH_2OH(l)$$

The following thermodynamic properties of CH_2OHCH_2OH at 25 °C are given: $\Delta H_f^\circ = -454.8$ kJ mol^{-1} and $\Delta G_f^\circ = -323.1$ kJ mol^{-1}. Use these data, together with values from Appendix D, to obtain a value of S°, the standard molar entropy of $CH_2OHCH_2OH(l)$, at 25 °C.

87. Use the following data

	ΔH_f°, kJ mol^{-1}	ΔG_f°, kJ mol^{-1}	S°, J mol^{-1} K^{-1}
$CuSO_4 \cdot 3\ H_2O(s)$	-1684.3	-1400.0	221.3
$CuSO_4 \cdot H_2O(s)$	-1085.8	-918.1	146.0

together with other data from the text to determine the temperature at which the equilibrium pressure of water vapor above the two solids in the following reaction is 75 Torr.

$$CuSO_4 \cdot 3\ H_2O(s) \rightleftharpoons CuSO_4 \cdot H_2O(s) + 2\ H_2O(g)$$

88. For the dissociation of $CaCO_3(s)$ at 25 °C, $CaCO_3(s) \rightleftharpoons CaO(s) + CO_2(g)$ $\Delta G^\circ = +131$ kJ mol^{-1}

A sample of pure $CaCO_3(s)$ is placed in a flask and connected to an ultrahigh vacuum system capable of reducing the pressure to 10^{-9} mmHg.
(a) Would $CO_2(g)$ produced by the decomposition of $CaCO_3(s)$ at 25 °C be detectable in the vacuum system at 25 °C?
(b) What additional information do you need to determine P_{CO_2} as a function of temperature?
(c) With necessary data from Appendix D, determine the minimum temperature to which $CaCO_3(s)$ would have to be heated for $CO_2(g)$ to become detectable in the vacuum system.

89. Introduced into a 1.50-L flask is 0.100 mol of $PCl_5(g)$; the flask is held at a temperature of 227 °C until equilibrium is established. What is the total pressure of the gases in the flask at this point?

$$PCl_5(g) \rightleftharpoons PCl_3(g) + Cl_2(g)$$

(*Hint:* Use data from Appendix D and appropriate relationships from this chapter.)

90. From the data given in Exercise 74, estimate a value of ΔS° at 298 K for the reaction

$$2\ NaHCO_3(s) \longrightarrow Na_2CO_3(s) + H_2O(g) + CO_2(g)$$

91. The normal boiling point of cyclohexane, C_6H_{12}, is 80.7 °C. Estimate the temperature at which the vapor pressure of cyclohexane is 100.0 mmHg.

92. The term *thermodynamic stability* refers to the sign of ΔG_f°. If ΔG_f° is negative, the compound is stable with respect to decomposition into its elements. Use the data in Appendix D to determine whether $Ag_2O(s)$ is thermodynamically stable **(a)** at 25 °C and **(b)** at 200 °C.

93. At 0 °C, ice has a density of 0.917 g mL^{-1} and an absolute entropy of 37.95 J mol^{-1} K^{-1}. At this temperature, liquid water has a density of 1.000 g mL^{-1} and an absolute entropy of 59.94 J mol^{-1} K^{-1}. The pressure corresponding to these values is 1 bar. Calculate ΔG, ΔG°, ΔS°, ΔH° for the melting of two moles of ice at its normal melting point.

94. The decomposition of the poisonous gas phosgene is represented by the equation $COCl_2(g) \rightleftharpoons CO(g) + Cl_2(g)$. Values of K_p for this reaction are $K_p = 6.7 \times 10^{-9}$ at 99.8 °C and $K_p = 4.44 \times 10^{-2}$ at 395 °C. At what temperature is $COCl_2$ 15% dissociated when the total gas pressure is maintained at 1.00 atm?

95. Use data from Appendix D to estimate the aqueous solubility, in milligrams per liter, of $AgBr(s)$ at 100 °C.

96. The standard molar entropy of solid hydrazine at its melting point of 1.53 °C is 67.15 J mol^{-1} K^{-1}. The enthalpy of fusion is 12.66 kJ mol^{-1}. For $N_2H_4(l)$ in the interval from 1.53 °C to 298.15 K, the molar heat capacity at constant pressure is given by the expression $C_p = 97.78 + 0.0586 (T - 280)$. Determine the standard molar entropy of $N_2H_4(l)$ at 298.15 K.
(*Hint:* The heat absorbed to produce an infinitesimal change in the temperature of a substance is $dq_{rev} = C_p dT$.)

97. Use the following data to estimate the standard molar entropy of gaseous benzene at 298.15 K; that is, $S^\circ[C_6H_6(g, 1\ atm)]$.

For $C_6H_6(s, 1\ atm)$ at its melting point of 5.53 °C, $S°$ is 128.82 J mol^{-1} K^{-1}. The enthalpy of fusion is 9.866 kJ mol^{-1}. From the melting point to 298.15 K, the average heat capacity of liquid benzene is 134.0 J K^{-1} mol^{-1}. The enthalpy of vaporization of $C_6H_6(l)$ at 298.15 K is 33.85 kJ mol^{-1}, and in the vaporization, $C_6H_6(g)$ is produced at a pressure of 95.13 Torr. Imagine that this vapor could be compressed to 1 atm pressure without condensing and while behaving as an ideal gas. Calculate $S°[C_6H_6(g, 1\ atm)]$.
[*Hint:* Refer to Exercise 96, and note the following: For infinitesimal quantities, $dS = dq/dT$; for the compression of an ideal gas, $dq = -dw$; and for pressure–volume work, $dw = -PdV$.]

Feature Problems

98. A tabulation of more precise thermodynamic data than are presented in Appendix D lists the following values for $H_2O(l)$ and $H_2O(g)$ at 298.15 K, at a standard-state pressure of 1 bar.

	$\Delta H_f°$, kJ mol^{-1}	$\Delta G_f°$, kJ mol^{-1}	$S°$, J mol^{-1} K^{-1}
$H_2O(l)$	−285.830	−237.129	69.91
$H_2O(g)$	−241.818	−228.572	188.825

(a) Use these data to determine, in two different ways, $\Delta G°$ at 298.15 K for the vaporization: $H_2O(l, 1\ bar) \rightleftharpoons H_2O(g, 1\ bar)$. The value you obtain will differ slightly from that on page 798, because here, the standard state pressure is 1 bar, and there, it is 1 atm.
(b) Use the result of part (a) to obtain the value of K_{eq} for this vaporization and, hence, the vapor pressure of water at 298.15 K.
(c) The vapor pressure in part (b) is in the unit bar. Convert the pressure to millimeters of mercury.
(d) Start with the value $\Delta G° = 8.590$ kJ, given on page 798, and calculate the vapor pressure of water at 298.15 K in a fashion similar to that in parts (b) and (c). In this way, demonstrate that the results obtained in a thermodynamic calculation do not depend on the convention we choose for the standard-state pressure, as long as we use standard-state thermodynamic data consistent with that choice.

99. The graphs show how $\Delta G°$ varies with temperature for three different oxidation reactions: the oxidations of C(graphite), Zn, and Mg to CO, ZnO, and MgO, respectively. Graphs such as these can be used to show the temperatures at which carbon is an effective reducing agent to reduce metal oxides to the free metals. As a result, such graphs are important to metallurgists. Use these graphs to answer the following questions.
(a) Why can Mg be used to reduce ZnO to Zn at all temperatures, but Zn cannot be used to reduce MgO to Mg at any temperature?
(b) Why can C be used to reduce ZnO to Zn at some temperatures but not at others? What are those temperatures?
(c) Is it possible to produce Zn from ZnO by its direct decomposition, without requiring a coupled reaction? If so, at what approximate temperatures might this occur?

(d) Is it possible to decompose CO to C and O_2 in a spontaneous reaction? Explain.
(e) Add to the set of graphs straight lines representing the reactions

$$C(\text{graphite}) + O_2(g) \longrightarrow CO_2(g)$$

$$2\ CO(g) + O_2(g) \longrightarrow 2\ CO_2(g)$$

given that the three lines representing the formation of oxides of carbon intersect at about 800 °C.
(*Hint:* At what other temperature can you relate $\Delta G°$ and temperature?)
The slopes of the three lines described above differ sharply. Explain why this so—that is, explain the slope of each line in terms of principles governing free energy change.
(f) The graphs for the formation of oxides of other metals are similar to the ones shown for Zn and Mg; that is, they all have positive slopes. Explain why carbon is such a good reducing agent for the reduction of metal oxides.

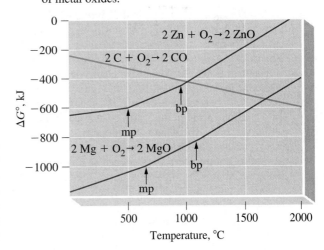

▲ $\Delta G°$ for three reactions as a function of temperature. The reactions are indicated by the equations written above the graphs. The points noted by arrows are the melting points and boiling points of zinc and magnesium.

100. In a heat engine, heat (q_h) is absorbed by a working substance (such as water) at a high temperature (T_h). Part of this heat is converted to work (w), and the rest (q_l) is released to the surroundings at the lower temperature (T_l).

The *efficiency* of a heat engine is the ratio w/q_h. The second law of thermodynamics establishes the following equation for the maximum efficiency of a heat engine, expressed on a percentage basis.

$$\text{efficiency} = \frac{w}{q_h} \times 100\% = \frac{T_h - T_l}{T_h} \times 100\%$$

In a particular electric power plant, the steam leaving a steam turbine is condensed to liquid water at 41 °C (T_l) and the water is returned to the boiler to be regenerated as steam. If the system operates at 36% efficiency,

(a) What is the minimum temperature of the steam [$H_2O(g)$] used in the plant?

(b) Why is the actual steam temperature probably higher than that calculated in part (a)?

(c) Assume that at T_h the $H_2O(g)$ is in equilibrium with $H_2O(l)$. Estimate the steam pressure at the temperature calculated in part (a).

(d) Is it possible to devise a heat engine with greater than 100% efficiency? with 100% efficiency? Explain.

101. The free energy available from the complete combustion of 1 mol of glucose to carbon dioxide and water is

$$C_6H_{12}O_6(aq) + 6\,O_2(g) \rightleftharpoons 6\,CO_2(g) + 6\,H_2O(l)$$
$$\Delta G° = -2870\ \text{kJ mol}^{-1}$$

(a) Under biological standard conditions, compute the maximum number of moles of ATP that could form from ADP and phosphate if all the energy of combustion of 1 mol of glucose could be utilized.

(b) The actual number of moles of ATP formed by a cell under aerobic conditions (that is, in the presence of oxygen) is about 38. Calculate the efficiency of energy conversion of the cell.

(c) Consider these typical physiological conditions.

$$P_{CO_2} = 0.050\ \text{bar}; P_{O_2} = 0.132\ \text{bar};$$
$$[\text{glucose}] = 1.0\ \text{mg/mL}; \text{pH} = 7.0;$$
$$[\text{ATP}] = [\text{ADP}] = [P_i] = 0.00010\ \text{M}.$$

Calculate ΔG for the conversion of 1 mol ADP to ATP and ΔG for the oxidation of 1 mol glucose under these conditions.

(d) Calculate the efficiency of energy conversion for the cell under the conditions given in part (c). Compare this efficiency with that of a diesel engine that attains 78% of the theoretical efficiency operating with T_h = 1923 K and T_l = 873 K. Suggest a reason for your result.

(Hint: See Feature Problem 100.]

 eMedia Exercises

102. Observe the motion of atoms in the **Mixing of Gases** animation *(eChapter 20-2)*. Describe why entropy increases in this process. If the choice of each atom's location in either of the two halves of the system (boxes) is considered a microstate, W, calculate the change in entropy between the initial (separated) and final (mixed) states.

103. **(a)** Describe how temperature causes a change in entropy on the atomic scale in the **Lattice Entropy** simulation *(eChapter 20-2)*. **(b)** What is the property of the crystal that represents the degree of its order (or disorder)?

104. **(a)** In the **Temperature Dependence of Entropy** simulation *(eChapter 20-3)*, what point of the curve corresponds to the greatest relative increase in entropy? **(b)** How does this relate to the values of enthalpy changes observed at this point, compared to elsewhere on the curve?

105. **(a)** Write out a balanced chemical equation for the reaction illustrated in the **Formation of Aluminum Bromide** movie *(eChapter 20-3)*. **(b)** What information is needed to calculate the entropy change associated with this reaction? **(c)** Using the tabulated thermodynamic values in Appendix D, calculate the entropy change that would occur if 25.0 grams of aluminum reacted with an excess of liquid bromine at 298 K.

106. **(a)** In the $NO_2 - N_2O_4$ **Equilibrium** animation *(eChapter 20-6)*, under which condition(s) shown will the free energy of the system be greater than the standard free energy? **(b)** Which set of conditions maximizes the free energy of the system?

21 Electrochemistry

Contents

A transit bus fitted with hydrogen-oxygen fuel cells. The use of fuel cells could dramatically reduce urban air pollution. The conversion of chemical energy into electrical energy is one of the main subjects of this chapter.

A conventional gasoline-powered automobile is only about 25% efficient in converting chemical energy into kinetic energy (energy of motion). An electric-powered auto is about three times as efficient. Unfortunately, when automotive technology was first being developed, devices for converting chemical energy to electrical energy did not perform at their intrinsic efficiencies. This fact, together with the availability of high-quality gasoline at a low cost, resulted in the preeminence of the internal combustion automobile. Now, with concern about long-term energy supplies and environmental pollution, there is a renewed interest in electric-powered buses and automobiles.

In this chapter, we will see how chemical reactions can be used to produce electricity and how electricity can be used to cause chemical

823

reactions. The practical applications of electrochemistry are countless, ranging from batteries and fuel cells as electric power sources, to the manufacture of key chemicals, the refining of metals, and the methods of controlling corrosion. Also important, however, are the theoretical implications. Because electricity involves a flow of electric charge, a study of the relationship between chemistry and electricity gives us additional insight into reactions in which electrons are transferred—*oxidation–reduction reactions*.

21-1 Electrode Potentials and Their Measurement

The criteria for spontaneous change developed in Chapter 20 apply to reactions of all types—precipitation, acid–base, and oxidation–reduction (redox). We can devise an additional useful criterion for redox reactions, however.

Figure 21-1 shows that a redox reaction occurs between $Cu(s)$ and $Ag^+(aq)$ but not between $Cu(s)$ and $Zn^{2+}(aq)$. Specifically, we see that silver ions are reduced to silver atoms on a copper surface, whereas zinc ions are *not* reduced to zinc atoms on a copper surface. We can say that Ag^+ is more readily reduced than is Zn^{2+}. In this section, we will introduce the *electrode potential*, a property related to these reduction tendencies.

▶ The term *electrode* is sometimes used for the entire half-cell assembly.

When used in electrochemical studies, a strip of metal, M, is called an **electrode**. An electrode immersed in a solution containing ions of the same metal, M^{n+}, is called a **half-cell**. Two kinds of interactions are possible between metal atoms on the electrode and metal ions in solution (Figure 21-2):

1. A metal ion M^{n+} from solution may collide with the electrode, gain n electrons from it, and be converted to a metal atom M. *The ion is reduced.*

2. A metal atom M on the surface may lose n electrons to the electrode and enter the solution as the ion M^{n+}. The *metal atom is oxidized.*

KEEP IN MIND

that although $M^{n+}(aq)$ and ne^- appear together on the right-hand side of this expression, only the ion M^{n+} enters the solution. The electrons remain on the electrode, $M(s)$. Free electrons are never found in an aqueous solution. ▶

An equilibrium is quickly established between the metal and the solution, which we can represent as

$$M(s) \underset{\text{reduction}}{\overset{\text{oxidation}}{\rightleftharpoons}} M^{n+}(aq) + ne^- \qquad (21.1)$$

 Formation of Silver Crystals movie

(a)   (b)

▶ **FIGURE 21-1**
Behavior of $Ag^+(aq)$ and $Zn^{2+}(aq)$ in the presence of copper
(a) Silver ions are displaced from colorless $AgNO_3(aq)$ as a deposit of silver metal, and copper enters the solution as blue $Cu^{2+}(aq)$.

$$Cu(s) + 2\,Ag^+(aq) \longrightarrow Cu^{2+}(aq) + 2\,Ag(s)$$

(b) $Cu(s)$ *does not* displace colorless Zn^{2+} from $Zn(NO_3)_2(aq)$.

$$Cu(s) + Zn^{2+}(aq) \longrightarrow \text{no reaction}$$

Oxidation

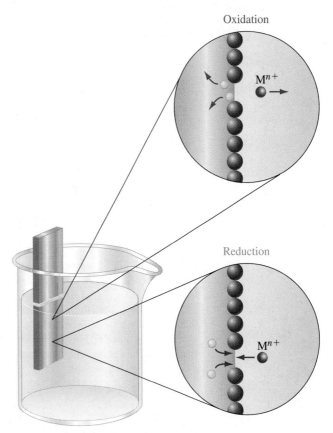

M^{n+}

Reduction

M^{n+}

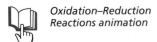

Oxidation–Reduction Reactions animation

▲ **FIGURE 21-2   An electrochemical half-cell**
The half-cell consists of a metal electrode, M, partially immersed in an aqueous solution of its ions, M^{n+}. (The anions required to maintain electrical neutrality in the solution are not shown.) The situation illustrated here is limited to metals that do not react with water.

However, any changes produced at the electrode or in the solution as a consequence of this equilibrium are too slight to measure. Instead, our measurements must be based on a combination of *two different* half-cells. Specifically, we must measure the tendency for electrons to flow from the electrode of one half-cell to the electrode of the other. Electrodes are classified according to whether oxidation or reduction takes place there. If oxidation takes place, the electrode is called the **anode**. If reduction takes place, the electrode is called the **cathode**.

Figure 21-3 depicts a combination of two half-cells, one with a Cu electrode in contact with $Cu^{2+}(aq)$, and the other with Ag and $Ag^+(aq)$. The two electrodes are joined by wires to an electric meter—here, a *voltmeter*. To complete the electric circuit, the two solutions must also be connected electrically. However, because charge is carried through solutions by the migration of *ions*, a wire cannot be used for this connection. The solutions must either be in direct contact through a porous barrier or joined by a third solution in a U-tube called a **salt bridge**. The properly connected combination of two half-cells is called an **electrochemical cell**.

Now, let's consider the changes that occur in the electrochemical cell in Figure 21-3. As the arrows suggest, Cu atoms release electrons at the anode and enter the $Cu(NO_3)_2(aq)$ as Cu^{2+} ions. Electrons lost by the Cu atoms pass through the wires and the voltmeter to the cathode, where they are gained by Ag^+ ions from the $AgNO_3(aq)$, producing a deposit of metallic silver. Simultaneously, anions (NO_3^-) from the salt bridge migrate into the copper half-cell and neutralize the positive

▶ Anions migrate toward the anode, and cations toward the cathode.

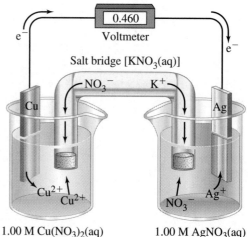

▲ **FIGURE 21-3** **Measurement of the electromotive force of an electrochemical cell**
An electrochemical cell consists of two half-cells with electrodes joined by a wire and solutions joined by a salt bridge. (The ends of the salt bridge are plugged with a porous material that allows ions to migrate but prevents the bulk flow of liquid.) Electrons flow
from the Cu electrode, where oxidation occurs (anode), to the Ag electrode, where reduction occurs (cathode). For precise measurements, the amount of electric current drawn
from the cell must be kept very small by means of either a specially designed voltmeter or
a device called a potentiometer.

charge of the excess Cu^{2+} ions; cations (K^+) migrate into the silver half-cell and neutralize the negative charge of the excess NO_3^- ions. The overall reaction that occurs
as the electrochemical cell spontaneously produces electric current is

Oxidation:	$Cu(s) \longrightarrow Cu^{2+}(aq) + 2\,e^-$
Reduction:	$2\,\{Ag^+(aq) + e^- \longrightarrow Ag(s)\}$

Overall: $\quad Cu(s) + 2\,Ag^+(aq) \longrightarrow Cu^{2+}(aq) + 2\,Ag(s) \qquad (21.2)$

KEEP IN MIND ▶
that the overall reaction occurring in the electrochemical cell
is identical to what happens in
the direct addition of Cu(s) to
$Ag^+(aq)$ pictured in
Figure 21-1a.

The reading on the voltmeter (0.460 V) is significant. It is the **cell voltage**, or the
potential difference between the two half-cells. The unit of cell voltage, **volt (V)**,
is the energy per unit charge. Thus, a potential difference of one volt signifies an
energy of one joule for every coulomb of charge passing through an electric circuit:
$1\ V = 1\ J/C$. We can think of a voltage or potential difference as the driving force
for electrons; the greater the voltage, the greater the driving force. The flow of
water from a higher to a lower level is analogous to this situation. The greater the
difference in water levels, the greater the force behind the flow of water. Cell voltage is also called **electromotive force (emf)**, or **cell potential**, and represented by
the symbol E_{cell}.

Now let's return to the question raised by Figure 21-1: Why does copper *not*
displace Zn^{2+} from solution? If we set up an electrochemical cell consisting of a
$Zn(s)/Zn^{2+}(aq)$ half-cell and a $Cu^{2+}(aq)/Cu(s)$ half-cell, we find that electrons flow
from the Zn to the Cu. The spontaneous reaction in the electrochemical cell in
Figure 21-4 is

▶ Formulations such as Zn/Zn^{2+}
and Cu^{2+}/Cu are called *couples*
and are often used as abbreviations for half-cells.

Oxidation:	$Zn(s) \longrightarrow Zn^{2+}(aq) + 2\,e^-$
Reduction:	$Cu^{2+}(aq) + 2\,e^- \longrightarrow Cu(s)$

Overall: $\quad Zn(s) + Cu^{2+}(aq) \longrightarrow Zn^{2+}(aq) + Cu(s) \qquad (21.3)$

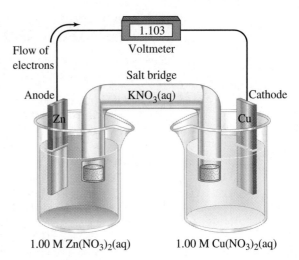

▲ **FIGURE 21-4   The reaction Zn(s) + Cu^{2+}(aq) ⟶ Zn^{2+}(aq) + Cu(s) occurring in an electrochemical cell**

Because reaction (21.3) is a spontaneous reaction, the displacement of Zn^{2+}(aq) by Cu(s)—the *reverse* of reaction (21.3)—does *not* occur spontaneously. This is the observation we made in Figure 21-1. In Section 21-3, we will discuss how to predict the direction of spontaneous change for oxidation–reduction reactions.

Cell Diagrams and Terminology

Drawing sketches of electrochemical cells, as in Figures 21-3 and 21-4, is helpful, but more often a simpler representation is used. A **cell diagram** shows the components of an electrochemical cell in a symbolic way. We will follow these generally accepted conventions in writing cell diagrams.

- The anode, the electrode at which *oxidation* occurs, is placed at the *left* side of the diagram.
- The cathode, the electrode at which *reduction* occurs, is placed at the *right* side of the diagram.
- A boundary between different phases (for example, an electrode and a solution) is represented by a *single* vertical line (|).
- The boundary between half-cell compartments, commonly a salt bridge, is represented by a *double* vertical line (‖). Species in aqueous solution are placed on either side of the double vertical line. Different species within the same solution are separated from each other by a comma.

The cell diagram corresponding to both Figure 21-4 and reaction (21.3) is customarily written as

anode ⟶ Zn(s)|Zn^{2+}(aq) ‖ Cu^{2+}(aq)|Cu(s) ⟵ cathode E_{cell} = 1.103 V   (21.4)

      half-cell   salt   half-cell   cell voltage
    (oxidation)   bridge (reduction)

The electrochemical cells of Figures 21-3 and 21-4 produce electricity as a result of spontaneous chemical reactions; as such, they are called **voltaic**, or **galvanic**, **cells**. In Section 21-7 we will consider *electrolytic cells*—electrochemical cells in which electricity is used to accomplish a nonspontaneous chemical change.

▶ Several memory devices have been proposed for the oxidation/anode and reduction/cathode relationships. Perhaps the simplest is that in the oxidation/anode relationship, both terms begin with a vowel: *o*/*a*; in the reduction/cathode relationship, both begin with a consonant: *r*/*c*.

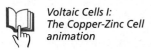

Voltaic Cells I: The Copper-Zinc Cell animation

EXAMPLE 21-1

Representing a Redox Reaction by Means of a Cell Diagram. Aluminum metal displaces zinc(II) ion from aqueous solution.

(a) Write oxidation and reduction half-equations and an overall equation for this redox reaction.

(b) Write a cell diagram for a voltaic cell in which this reaction occurs.

Solution

(a) The term *displaces* means that aluminum goes into solution as $Al^{3+}(aq)$, forcing $Zn^{2+}(aq)$ out of solution as zinc metal. Al is oxidized to Al^{3+}, and Zn^{2+} is reduced to Zn. In writing the overall equation, we must adjust coefficients so that equal numbers of electrons are involved in oxidation and in reduction. (This is the half-reaction method of balancing redox equations that we studied in Section 5-5.)

$$
\begin{aligned}
\textit{Oxidation:} &\qquad 2\,\{Al(s) \longrightarrow Al^{3+}(aq) + 3\,e^-\} \\
\textit{Reduction:} &\qquad \underline{3\,\{Zn^{2+}(aq) + 2\,e^- \longrightarrow Zn(s)\}} \\
\textit{Overall:} &\qquad 2\,Al(s) + 3\,Zn^{2+}(aq) \longrightarrow 2\,Al^{3+}(aq) + 3\,Zn(s)
\end{aligned}
$$

(b) $Al(s)$ is oxidized to $Al^{3+}(aq)$ in the anode half-cell (left), and $Zn^{2+}(aq)$ is reduced to $Zn(s)$ in the cathode half-cell (right).

$$Al(s)\,|\,Al^{3+}(aq)\,||\,Zn^{2+}(aq)\,|\,Zn(s)$$

Practice Example A: Write the overall equation for the redox reaction that occurs in the voltaic cell. $Sc(s)\,|\,Sc^{3+}(aq)\,||\,Ag^+(aq)\,|\,Ag(s)$.

Practice Example B: Draw a voltaic cell in which silver ion is displaced from solution by aluminum metal. Label the cathode, the anode, and other features of the cell. Show the direction of flow of electrons. Also, indicate the direction of flow of cations and anions from a $KNO_3(aq)$ salt bridge. Write an equation for the half-reaction occurring at each electrode, write a balanced equation for the overall cell reaction, and write a cell diagram.

Are You Wondering ...

How to ensure that you place the half-cells in the proper order in a cell diagram?

If you guess that a particular electrode is the anode in a voltaic cell, you are simply deciding to connect that electrode to the negative terminal of the voltmeter (and the other electrode to the positive terminal). If you have guessed correctly, you will get a positive reading on the voltmeter. If you get a negative reading on the voltmeter, this signifies that the electrons flow in the direction opposite that of your expectation: What you thought was the anode is actually the cathode, and vice versa. Some voltmeters will register only a positive voltage. In this case, you won't get any meter reading until you switch the connections to the terminals. You can write the cell diagram with the "misplaced" half-cells and a *negative* cell voltage, or you can switch the half-cells and write a *positive* cell voltage.

21-2 Standard Electrode Potentials

Cell voltages—potential *differences* between electrodes—are among the most precise scientific measurements possible. Potentials of individual electrodes, however, cannot be precisely established. If we could make such measurements, we could obtain cell voltages just by subtracting one electrode potential from another. We can achieve the same result, however, by *arbitrarily* choosing a particular half-cell to which we assign an electrode potential of *zero*. We can then compare other half-cells to this reference. The commonly accepted reference is the standard hydrogen electrode.

▶ This method is comparable to establishing standard enthalpies or free energies of formation on the basis of an arbitrary zero value.

The **standard hydrogen electrode (SHE)** is depicted in Figure 21-5. The SHE involves equilibrium established on the surface of an inert metal (such as platinum) between H_3O^+ ions from a solution in which they are at unit activity (that is, $a_{H_3O^+} = 1$) and H_2 molecules from the gaseous state at a pressure of 1 bar. The equilibrium reaction produces a particular potential on the metal surface, but this potential is arbitrarily taken to be *zero*.

$$2\,H^+(a = 1) + 2\,e^- \underset{\text{on Pt}}{\overset{}{\rightleftharpoons}} H_2(g,\,1\text{ bar}) \qquad E^\circ = 0 \text{ volt (V)} \qquad (21.5)$$

The diagram for this half-cell is

$$Pt\,|\,H_2(g,\,1\text{ bar})\,|\,H^+(a = 1)$$

The two vertical lines signify that three phases are present: solid platinum, gaseous hydrogen, and aqueous hydrogen ion. For simplicity, we will usually write H^+ for H_3O^+, assume that unit activity $(a = 1)$ exists at roughly $[H^+] = 1$ M, and replace a pressure of 1 bar by 1 atm.

Standard Reduction Potentials animation

By international agreement, a **standard electrode potential**, E°, measures the tendency for a *reduction* process to occur at an electrode. In all cases, the ionic species are present in aqueous solution at unit activity (approximately 1 M), and gases are at 1 bar pressure (approximately 1 atm). Where no metallic substance is indicated, the potential is established on an inert metallic electrode, such as platinum.

To emphasize that E° refers to a reduction, we will write a reduction couple as a subscript to E°, as shown in half-reaction (21.6). The substance being reduced is written on the left of the (slash) sign, and the chief reduction product on the right.

$$Cu^{2+}(1\text{ M}) + 2\,e^- \longrightarrow Cu(s) \qquad E^\circ_{Cu^{2+}/Cu} = ? \qquad (21.6)$$

To determine the value of E° for a standard electrode such as that to which half-reaction (21.6) applies, we compare it with a standard hydrogen electrode (SHE). In this comparison, the SHE is always taken as the electrode on the *left* of the cell

▶ FIGURE 21-5
The standard hydrogen electrode (SHE)
Because hydrogen is a gas at room temperature, electrodes cannot be constructed from it. The standard hydrogen electrode consists of a piece of platinum dipped into a solution containing 1 M H^+(aq) with a stream of hydrogen passing over its surface. The platinum does not react but provides a surface for the reduction of $H_3O(aq)^+$ to $H_2(g)$ and the reverse oxidation half-reaction.

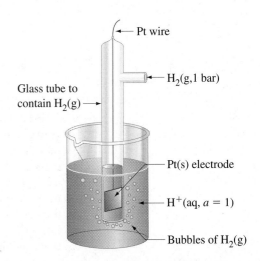

diagram—the anode—and the compared electrode is the electrode on the *right*—the cathode. In the following voltaic cell, the measured potential difference is 0.340 V, with electrons flowing from the H_2 to the Cu electrode.

$$\text{Pt} \mid H_2(g, 1 \text{ atm}) \mid H^+(1 \text{ M}) \parallel Cu^{2+}(1 \text{ M}) \mid Cu(s) \qquad E_{cell}^\circ = 0.340 \text{ V} \qquad (21.7)$$

$$\underset{\text{anode}}{} \underset{\text{cathode}}{}$$

A **standard cell potential**, E_{cell}°, is the potential difference, or voltage, of a cell formed from two *standard* electrodes. The *difference* is always taken in the following way:

$$E_{cell}^\circ = E^\circ(\text{right}) - E^\circ(\text{left})$$

$$\underset{\text{(cathode)}}{} \underset{\text{(anode)}}{}$$

Applied to the cell diagram (21.7), we get

$$E_{cell}^\circ = E_{Cu^{2+}/Cu}^\circ - E_{H^+/H_2}^\circ = 0.340 \text{ V}$$

$$= E_{Cu^{2+}/Cu}^\circ - 0 \text{ V} = 0.340 \text{ V}$$

$$E_{Cu^{2+}/Cu}^\circ = 0.340 \text{ V}$$

Thus, for the standard *reduction* half-reaction, we can write

$$Cu^{2+}(1 \text{ M}) + 2 e^- \longrightarrow Cu(s) \qquad E_{Cu^{2+}/Cu}^\circ = +0.340 \text{ V} \qquad (21.8)$$

We can represent the overall reaction occurring in the voltaic cell diagrammed in (21.7) as

$$H_2(g, 1 \text{ atm}) + Cu^{2+}(1 \text{ M}) \longrightarrow 2 H^+(1 \text{ M}) + Cu(s) \qquad E_{cell}^\circ = 0.340 \text{ V} \quad (21.9)$$

Cell reaction (21.9) indicates that $Cu^{2+}(1 \text{ M})$ is more easily reduced than is $H^+(1 \text{ M})$.

Suppose we replace the standard copper electrode in cell diagram (21.7) by a standard zinc electrode and measure the potential difference between the standard hydrogen and zinc electrodes by using the same voltmeter connections as in (21.7). In this case, we will find the voltage to be -0.763 V. The negative sign indicates that electrons flow in the direction *opposite* that in (21.7)—that is, *from* the zinc electrode *to* the hydrogen electrode. Here $H^+(1 \text{ M})$ is more easily reduced than is $Zn^{2+}(1 \text{ M})$. These findings are represented in the following cell diagram, in which the zinc electrode appears on the *right*.

$$\text{Pt} \mid H_2(g, 1 \text{ atm}) \mid H^+(1 \text{ M}) \parallel Zn^{2+}(1 \text{ M}) \mid Zn(s) \qquad E_{cell}^\circ = -0.763 \text{ V} \qquad (21.10)$$

To obtain the standard electrode potential for the Zn^{2+}/Zn couple, we can write

$$E_{cell}^\circ = E^\circ(\text{right}) - E^\circ(\text{left})$$

$$= E_{Zn^{2+}/Zn}^\circ - 0 \text{ V} = -0.763 \text{ V}$$

$$E_{Zn^{2+}/Zn}^\circ = -0.763 \text{ V}$$

Voltaic Cells II:
The Zinc-Hydrogen Cell
animation

Thus, for the standard *reduction* half-reaction, we have

$$Zn^{2+}(1 \text{ M}) + 2 e^- \longrightarrow Zn(s) \qquad E_{Zn^{2+}/Zn}^\circ = -0.763 \text{ V} \qquad (21.11)$$

In summary, the potential of the standard hydrogen electrode is set at exactly 0 V. Any electrode at which a reduction half-reaction shows a *greater* tendency to occur than does the reduction of $H^+(1 \text{ M})$ to $H_2(g, 1 \text{ atm})$ has a *positive* value for its standard electrode potential, E°. Any electrode at which a reduction half-reaction shows a *lesser* tendency to occur than does the reduction of $H^+(1 \text{ M})$ to $H_2(g, 1 \text{ atm})$ has a *negative* value for its standard reduction potential, E°. Comparisons of the standard copper and zinc electrodes to the standard hydrogen electrode are illustrated in Figure 21-6. Table 21.1 on page 832 lists some common reduction half-reactions and their standard electrode potentials at 25 °C.

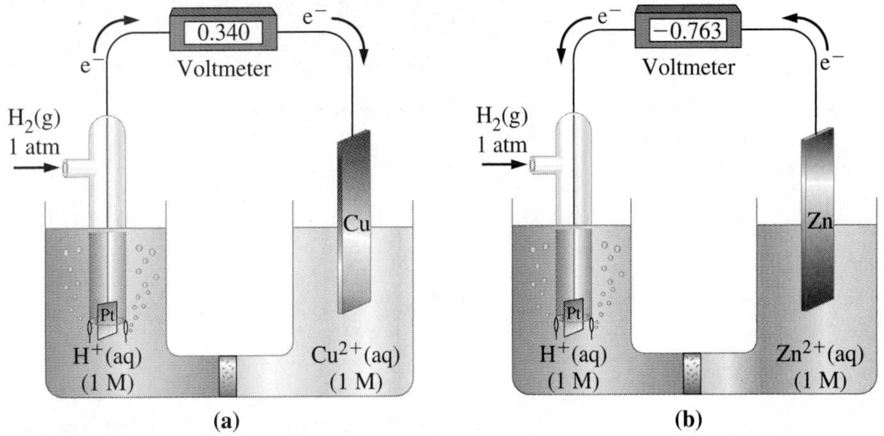

▲ FIGURE 21-6   **Measuring standard electrode potentials**
(a) A standard hydrogen electrode is the anode, and copper is the cathode. Contact between the half-cells occurs through a porous plate that prevents bulk flow of the solutions
while allowing ions to pass. (b) This cell has the same connections as that in part (a), but
with zinc substituting for copper. However, the electron flow is opposite that in (a), as
noted by the *negative* voltage. (Zinc is the anode.)

We will use standard reduction potentials throughout this chapter for many purposes. Our first objective will be to calculate standard cell potentials for redox reactions—E_{cell}° values—from standard electrode potentials for half-cell reactions—
E° values. The procedure we use is illustrated here for reaction (21.3) and cell diagram (21.4). Note that the first three equations are alternative ways of stating the
same thing; we will generally not write all of them.

▶ Reaction (21.3) is shown on
page 826 and diagram (21.4) on
page 827.

$$E_{cell}^{\circ} = E^{\circ}(\text{right}) - E^{\circ}(\text{left})$$

$$= E^{\circ}(\text{cathode}) - E^{\circ}(\text{anode})$$

$$= E^{\circ}(\text{reduction half-cell}) - E^{\circ}(\text{oxidation half-cell})$$

$$= E_{Cu^{2+}/Cu}^{\circ} - E_{Zn^{2+}/Zn}^{\circ}$$

$$= 0.340 \text{ V} - (-0.763 \text{ V}) = 1.103 \text{ V}$$

In Example 21-2, we predict E_{cell}° for a new battery system. In Example 21-3,
we use one known electrode potential and a measured E_{cell}° value to determine an
unknown E°.

EXAMPLE 21-2

Combining E° Values into E_{cell}° for a Reaction.   A new battery system currently under study
for possible use in electric vehicles is the zinc–chlorine battery. The overall reaction producing
electricity in this cell is $Zn(s) + Cl_2(g) \longrightarrow ZnCl_2(aq)$. What is E_{cell}° of this voltaic cell?

Solution

Reaction (21.12) shows that $Zn(s)$ is oxidized and that $Cl_2(g)$ is reduced. It is the overall reaction resulting from the two half-reactions indicated below. The E° values required
to establish E_{cell}° for the overall reaction are from Table 21.1.

Reduction:	$Cl_2(g) + 2 e^- \longrightarrow 2 Cl^-(aq)$
Oxidation:	$Zn(s) \longrightarrow Zn^{2+}(aq) + 2 e^-$

Overall:	$Zn(s) + Cl_2(g) \longrightarrow Zn^{2+}(aq) + 2 Cl^-(aq)$	(21.12)

▶ A more extensive listing of reduction half-reactions and their potentials is given in Appendix D.

TABLE 21.1 Some Selected Standard Electrode (Reduction) Potentials at 25 °C

Reduction Half-Reaction	$E°$, V
Acidic solution	
$F_2(g) + 2\,e^- \longrightarrow 2\,F^-(aq)$	+2.866
$O_3(g) + 2\,H^+(aq) + 2\,e^- \longrightarrow O_2(g) + H_2O(l)$	+2.075
$S_2O_8^{2-}(aq) + 2\,e^- \longrightarrow 2\,SO_4^{2-}(aq)$	+2.01
$H_2O_2(aq) + 2\,H^+(aq) + 2\,e^- \longrightarrow 2\,H_2O(l)$	+1.763
$MnO_4^-(aq) + 8\,H^+(aq) + 5\,e^- \longrightarrow Mn^{2+}(aq) + 4\,H_2O(l)$	+1.51
$PbO_2(s) + 4\,H^+(aq) + 2\,e^- \longrightarrow Pb^{2+}(aq) + 2\,H_2O(l)$	+1.455
$Cl_2(g) + 2\,e^- \longrightarrow 2\,Cl^-(aq)$	+1.358
$Cr_2O_7^{2-}(aq) + 14\,H^+(aq) + 6\,e^- \longrightarrow 2\,Cr^{3+}(aq) + 7\,H_2O(l)$	+1.33
$MnO_2(s) + 4\,H^+(aq) + 2\,e^- \longrightarrow Mn^{2+}(aq) + 2\,H_2O(l)$	+1.23
$O_2(g) + 4\,H^+(aq) + 4\,e^- \longrightarrow 2\,H_2O(l)$	+1.229
$2\,IO_3^-(aq) + 12\,H^+(aq) + 10\,e^- \longrightarrow I_2(s) + 6\,H_2O(l)$	+1.20
$Br_2(l) + 2\,e^- \longrightarrow 2\,Br^-(aq)$	+1.065
$NO_3^-(aq) + 4\,H^+(aq) + 3\,e^- \longrightarrow NO(g) + 2\,H_2O(l)$	+0.956
$Ag^+(aq) + e^- \longrightarrow Ag(s)$	+0.800
$Fe^{3+}(aq) + e^- \longrightarrow Fe^{2+}(aq)$	+0.771
$O_2(g) + 2\,H^+(aq) + 2\,e^- \longrightarrow H_2O_2(aq)$	+0.695
$I_2(s) + 2\,e^- \longrightarrow 2\,I^-(aq)$	+0.535
$Cu^{2+}(aq) + 2\,e^- \longrightarrow Cu(s)$	+0.340
$SO_4^{2-}(aq) + 4\,H^+(aq) + 2\,e^- \longrightarrow 2\,H_2O(l) + SO_2(g)$	+0.17
$Sn^{4+}(aq) + 2\,e^- \longrightarrow Sn^{2+}(aq)$	+0.154
$S(s) + 2\,H^+(aq) + 2\,e^- \longrightarrow H_2S(g)$	+0.14
$2\,H^+(aq) + 2\,e^- \longrightarrow H_2(g)$	0
$Pb^{2+}(aq) + 2\,e^- \longrightarrow Pb(s)$	−0.125
$Sn^{2+}(aq) + 2\,e^- \longrightarrow Sn(s)$	−0.137
$Fe^{2+}(aq) + 2\,e^- \longrightarrow Fe(s)$	−0.440
$Zn^{2+}(aq) + 2\,e^- \longrightarrow Zn(s)$	−0.763
$Al^{3+}(aq) + 3\,e^- \longrightarrow Al(s)$	−1.676
$Mg^{2+}(aq) + 2\,e^- \longrightarrow Mg(s)$	−2.356
$Na^+(aq) + e^- \longrightarrow Na(s)$	−2.713
$Ca^{2+}(aq) + 2\,e^- \longrightarrow Ca(s)$	−2.84
$K^+(aq) + e^- \longrightarrow K(s)$	−2.924
$Li^+(aq) + e^- \longrightarrow Li(s)$	−3.040
Basic solution	
$O_3(g) + H_2O(l) + 2\,e^- \longrightarrow O_2(g) + 2\,OH^-(aq)$	+1.246
$OCl^-(aq) + H_2O(l) + 2\,e^- \longrightarrow Cl^-(aq) + 2\,OH^-(aq)$	+0.890
$O_2(g) + 2\,H_2O(l) + 4\,e^- \longrightarrow 4\,OH^-(aq)$	+0.401
$2\,H_2O(l) + 2\,e^- \longrightarrow H_2(g) + 2\,OH^-(aq)$	−0.828

▶ The placement of *oxidizing agents* is as follows: strongest oxidizing agents (F_2, O_3, ...), *left* sides, *top* of the list; weakest oxidizing agents (Li^+, K^+, ...), *left* sides, *bottom* of list. The placement of *reducing agents* is as follows: strongest reducing agents (Li, K, ...), *right* sides, *bottom* of list; weakest reducing agents (F^-, O_2, ...), *right* sides, *top* of list.

$$E°_{cell} = E°(\text{reduction half-cell}) - E°(\text{oxidation half-cell})$$

$$= 1.358\text{ V} - (-0.763\text{ V}) = 2.121\text{ V}$$

Practice Example A: What is $E°_{cell}$ for the reaction in which $Cl_2(g)$ oxidizes $Fe^{2+}(aq)$ to $Fe^{3+}(aq)$?

$$2\,Fe^{2+}(aq) + Cl_2(g) \longrightarrow 2\,Fe^{3+}(aq) + 2\,Cl^-(aq) \qquad E°_{cell} = ?$$

Practice Example B: Use data from Table 21.1 to determine $E°_{cell}$ for the redox reaction in which $Fe^{2+}(aq)$ is oxidized to $Fe^{3+}(aq)$ by $MnO_4^-(aq)$ in acidic solution.

EXAMPLE 21-3

Determining an Unknown $E°$ from an $E°_{cell}$ Measurement. Cadmium is found in small quantities wherever zinc is found. Unlike zinc, which in trace amounts is an essential element, cadmium is an environmental poison. To determine cadmium ion concentrations by electrical measurements, we need the standard reduction potential for the Cd^{2+}/Cd electrode. The voltage of the following voltaic cell is measured.

$$Cd(s)\,|\,Cd^{2+}(1\text{ M})\,\|\,Cu^{2+}(1\text{ M})\,|\,Cu(s) \qquad E°_{cell} = 0.743\text{ V}$$

What is the standard reduction potential for the Cd^{2+}/Cd electrode?

Solution

We know one half-cell potential and $E°_{cell}$ for the overall redox reaction. We can solve for the unknown standard electrode potential, $E°_{Cd^{2+}/Cd}$.

$$E°_{cell} = E°(\text{right}) - E°(\text{left})$$
$$0.743\text{ V} = E°_{Cu^{2+}/Cu} - E°_{Cd^{2+}/Cd}$$
$$= 0.340\text{ V} - E°_{Cd^{2+}/Cd}$$
$$E°_{Cd^{2+}/Cd} = 0.340\text{ V} - 0.743\text{ V} = -0.403\text{ V}$$

Practice Example A: In acidic solution, dichromate ion oxidizes oxalic acid, $H_2C_2O_4(aq)$, to $CO_2(g)$ in a reaction with $E°_{cell} = 1.81$ V.

$$Cr_2O_7^{2-}(aq) + 3\,H_2C_2O_4(aq) + 8\,H^+(aq) \longrightarrow 2\,Cr^{3+}(aq) + 7\,H_2O + 6\,CO_2(g)$$

Use the value of $E°_{cell}$ for this reaction, together with appropriate data from Table 21.1, to determine $E°$ for the $CO_2(g)/H_2C_2O_4(aq)$ electrode.

Practice Example B: In an acidic solution, $O_2(g)$ oxidizes $Cr^{2+}(aq)$ to $Cr^{3+}(aq)$. The $O_2(g)$ is reduced to $H_2O(l)$. $E°_{cell}$ for the reaction is 1.653 V. What is the standard electrode potential for the couple Cr^{3+}/Cr^{2+}?

21-3 E_{cell}, ΔG, and K_{eq}

When a reaction occurs in a voltaic cell, the cell does work—electrical work. Think of this as the work of moving electric charges. The total work done is the product of three terms: (a) E_{cell}; (b) n, the number of moles of electrons transferred between the electrodes; and (c) the electric charge per mole of electrons, called the **Faraday constant (F)**. The Faraday constant is equal to 96,485 coulombs per mole of electrons (96,485 C/mol e⁻). Because the product volt × coulomb = joule, the unit of w_{elec} is joules (J).

$$w_{elec} = nFE_{cell} \tag{21.13}$$

Expression (21.13) applies only if the cell operates reversibly.[*] The Are You Wondering feature on page 796 described the amount of available energy (work) that can be derived from a process as equal to $-\Delta G$. Thus,

$$\Delta G = -nFE_{cell} \tag{21.14}$$

▲ **Michael Faraday (1791–1867)**
Faraday, an assistant to Humphry Davy and often called "Davy's greatest discovery," made many contributions to both physics and chemistry, including systematic studies of electrolysis.

[*]The meaning of a reversible process was illustrated by Figure 20-4 on page 787. The reversible operation of a voltaic cell requires that electric current be drawn from the cell only very very slowly.

In the special case in which the reactants and products are in their standard states,

$$\Delta G° = -nFE°_{cell} \qquad (21.15)$$

Our primary interest is not in calculating quantities of work but in using expression (21.15) as a means of evaluating free energy changes from measured cell potentials, as illustrated in Example 21-4.

EXAMPLE 21-4

Determining a Free Energy Change from a Cell Potential. Use $E°$ data to determine $\Delta G°$ for the reaction

$$Zn(s) + Cl_2(g, 1\ atm) \longrightarrow ZnCl_2(aq, 1\ M)$$

Solution

This reaction is cell reaction (21.12) occurring in the voltaic cell described in Example 21-2. In this type of problem, we generally need to separate the overall equation into two half-equations. Then we can determine the value of $E°_{cell}$ and the number of moles of electrons (n) involved in the cell reaction. Refer to Example 21-2 to see that $E°_{cell} = 2.121\ V$ and $n = 2\ mol\ e^-$. Now we can use equation (21.15).

$$\Delta G° = -nFE°_{cell} = -\left(2\ mol\ e^- \times \frac{96,485\ C}{1\ mol\ e^-} \times 2.121\ V\right)$$

$$= -4.093 \times 10^5\ J = -409.3\ kJ$$

Practice Example A: Use electrode potential data to determine $\Delta G°$ for the reaction

$$2\ Al(s) + 3\ Br_2(l) \longrightarrow 2\ Al^{3+}(aq, 1\ M) + 6\ Br^-(aq, 1\ M) \qquad \Delta G° = ?$$

Practice Example B: The hydrogen–oxygen fuel cell is a voltaic cell with a cell reaction of $2\ H_2(g) + O_2(g) \longrightarrow 2\ H_2O(l)$. Calculate $E°_{cell}$ for this reaction.

[*Hint:* Use thermodynamic data from Appendix D (Table D-2).]

Combining Reduction Half-Reactions

Not only can we use equation (21.15) to determine $\Delta G°$ from $E°_{cell}$, as in Example 21-4, but we can also reverse the calculation and determine an $E°_{cell}$ value from $\Delta G°$. Moreover, we can apply equation (21.15) to half-cell reactions and half-cell potentials—that is, to standard electrode potentials, $E°$. That is what we must do, for example, to determine $E°$ for the half-reaction

$$Fe^{3+}(aq) + 3\ e^- \longrightarrow Fe(s)$$

Both in Table 21.1 and in Appendix D, the only entries we find that deal with Fe(s) and its ions are

$$Fe^{2+}(aq) + 2\ e^- \longrightarrow Fe(s), E° = -0.440\ V \qquad and \qquad Fe^{3+}(aq) + e^- \longrightarrow Fe^{2+}(aq), E° = 0.771\ V$$

The half-equation we are seeking is simply the sum of these two half-equations, but the $E°$ value we are seeking is *not* the sum of $-0.440\ V$ and $0.771\ V$. What we *can* add together, though, are the $\Delta G°$ values for the two known half-reactions.

$$Fe^{2+}(aq) + 2\ e^- \longrightarrow Fe(s); \qquad \Delta G° = -2 \times F \times (-0.440\ V)$$

$$Fe^{3+}(aq) + e^- \longrightarrow Fe^{2+}(aq); \qquad \Delta G° = -1 \times F \times (0.771\ V)$$

$$Fe^{3+}(aq) + 3\ e^- \longrightarrow Fe(s); \qquad \Delta G° = (0.880F)\ V - (0.771F)\ V = (0.109F)\ V$$

Are You Wondering...

How the procedure for combining two $E°$ values to obtain an unknown E_{cell}° relates to combining two $E°$ values to obtain an unknown $E°$?

We have just seen to obtain an unknown $E°$ from two known values of $E°$ by working through the expression $\Delta G = -nFE°$. As shown below for a hypothetical displacement reaction, we can similarly calculate an unknown E_{cell}° through the expression $\Delta G = -nFE_{cell}^{\circ}$. (Note that for the oxidation half-reaction, ΔG_{ox}° is simply the negative of the value for the reverse half-reaction, ΔG_{red}°.)

Reduction: $M^{n+}(aq) + n\,e^- \longrightarrow M(s)$ $\qquad\qquad\qquad \Delta G_{red}^{\circ} = -nFE_{M^{n+}/M}^{\circ}$

Oxidation: $N(s) \longrightarrow N^{n+}(aq) + n\,e^-$
$$\Delta G_{ox}^{\circ} = -(\Delta G_{red}^{\circ}) = -(-nFE_{N^{n+}/N}^{\circ}) = nFE_{N^{n+}/N}^{\circ}$$

Overall: $M^{n+}(aq) + N(s) \longrightarrow M(s) + N^{n+}(aq)$

$$\Delta G^{\circ} = \Delta G_{red}^{\circ} + \Delta G_{ox}^{\circ} = -nFE_{cell}^{\circ} = -nFE_{M^{n+}/N}^{\circ} + nFE_{N^{n+}/N}^{\circ}$$

Dividing through the above equation by the term $-nF$, we obtain E_{cell}° as the familiar difference in two electrode potentials.

$$E_{cell}^{\circ} = E_{M^{n+}/M}^{\circ} - E_{N^{n+}/N}^{\circ}$$

The reason that we have been able to skip this calculation based on ΔG° values and proceed straight to the expression

$$E_{cell}^{\circ} = E°\,(\text{reduction}) - E°\,(\text{oxidation})$$

is that the term $-nF$ will always cancel out. That is, n, the number of electrons, must have the same value for the oxidation and reduction half-reactions and the overall reaction. By contrast, when obtaining an unknown $E°$ from the known $E°$ values, the value for n will not be the same in all three places where it appears, and so we do have to work through the ΔG° expressions.

Now, to get $E_{Fe^{3+}/Fe}^{\circ}$, we can again use equation (21.15) and solve for $E_{Fe^{3+}/Fe}^{\circ}$.

$$\Delta G^{\circ} = -nFE_{Fe^{3+}/Fe}^{\circ} = -3FE_{Fe^{3+}/Fe}^{\circ} = (0.109F)\ \text{V}$$

$$E_{Fe^{3+}/Fe}^{\circ} = (-0.109F/3F)\ \text{V} = -0.0363\ \text{V}$$

Spontaneous Change in Oxidation–Reduction Reactions

Our main criterion for spontaneous change is that $\Delta G < 0$. According to equation (21.14), however, redox reactions have the property that, if $\Delta G < 0$, then $E_{cell} > 0$. That is, E_{cell} must be *positive* if ΔG is to be negative. Predicting the direction of spontaneous change in a redox reaction is a relatively simple matter when we use the following ideas:

- If E_{cell} is *positive*, a reaction occurs spontaneously in the *forward* direction for the stated conditions. If E_{cell} is *negative*, the reaction occurs spontaneously in the *reverse* direction for the stated conditions. If $E_{cell} = 0$, the reaction is at equilibrium for the stated conditions.

- If a cell reaction is *reversed*, E_{cell} changes sign.

In the special case in which reactants and products are in their standard states, we work with ΔG° and E_{cell}° values, as illustrated in Examples 21-5 and 21-6.

EXAMPLE 21-5

Applying the Criterion for Spontaneous Change in a Redox Reaction. Will aluminum metal displace Cu^{2+} ion from aqueous solution? That is, will a spontaneous reaction occur in the forward direction for the following reaction?

$$2\,Al(s) + 3\,Cu^{2+}(1\,M) \longrightarrow 3\,Cu(s) + 2\,Al^{3+}(1\,M)$$

Solution

The cell diagram corresponding to the reaction is $Al(s)\,|\,Al^{3+}(aq)\,\|\,Cu^{2+}(aq)\,|\,Cu(s)$, and E°_{cell} is

$$
\begin{aligned}
E^{\circ}_{cell} &= E^{\circ}(\text{cathode}) - E^{\circ}(\text{anode}) \\
&= E^{\circ}_{Cu^{2+}/Cu} - E^{\circ}_{Al^{3+}/Al} \\
&= 0.340\,V - (-1.676\,V) = 2.016\,V
\end{aligned}
$$

Because E°_{cell} is positive, the direction of spontaneous change is that of the forward reaction. Al(s) will displace Cu^{2+} from aqueous solution under standard conditions.

Practice Example A: Name one metal ion that Cu(s) will displace from aqueous solution, and determine E°_{cell} for the reaction.

Practice Example B: When sodium metal is added to seawater, which has $[Mg^{2+}] = 0.0512\,M$, no magnesium metal is obtained. According to E° values, should this displacement reaction occur? What reaction does occur?

KEEP IN MIND ▶
that both E° and E°_{cell} are *intensive* properties. They do not depend on the quantities of materials involved, which means that their values are not affected by the way we write the cell reaction. We could just as well have written

$$Al(s) + \frac{3}{2}\,Cu^{2+}(1\,M) \longrightarrow$$
$$\frac{3}{2}\,Cu(s) + Al^{3+}(1\,M)$$

or $\frac{2}{3}\,Al(s) + Cu^{2+}(1\,M) \longrightarrow$
$$Cu(s) + \frac{2}{3}\,Al^{3+}(1\,M).$$

Even though we used electrode potentials and a cell voltage to predict a spontaneous reaction in Example 21-5, we do not have to carry out the reaction in a voltaic cell. This is an important point to keep in mind. Thus, Cu^{2+} is displaced from aqueous solution simply by adding aluminum metal, as shown in Figure 21-7. Another point, illustrated by Example 21-6, is that we can often give qualitative answers to questions concerning redox reactions without going through a complete calculation of E°_{cell}.

EXAMPLE 21-6

Making Qualitative Predictions with Electrode Potential Data. Peroxodisulfate salts, such as $Na_2S_2O_8$, are oxidizing agents used in bleaching. Dichromates such as $K_2Cr_2O_7$ have been used as laboratory oxidizing agents. Which is the better oxidizing agent in acidic solution under standard conditions, $S_2O_8^{2-}$ or $Cr_2O_7^{2-}$?

Solution

In a redox reaction, the oxidizing agent is reduced; the greater the tendency for this reduction to occur, the better the oxidizing agent. The reduction tendency, in turn, is measured by the E° value. Because the E° value for the reduction of $S_2O_8^{2-}(aq)$ to $SO_4^{2-}(aq)$ (2.01 V) is larger than that for the reduction of $Cr_2O_7^{2-}(aq)$ to $Cr^{3+}(aq)$ (1.33 V), $S_2O_8^{2-}(aq)$ should be the better oxidizing agent.

Practice Example A: An inexpensive way to produce peroxodisulfates would be to pass $O_2(g)$ through an acidic solution containing sulfate ion. Is this method feasible under standard conditions?

(*Hint:* What would be the reduction half-reaction?)

Practice Example B: Consider the following observations: (1) Aqueous solutions of Sn^{2+} are difficult to maintain because atmospheric oxygen easily oxidizes Sn^{2+} to Sn^{4+}. (2) One way to preserve the $Sn^{2+}(aq)$ solutions is to add some metallic tin. *Without doing detailed calculations*, explain these two statements by using E° data.

▲ FIGURE 21-7
Reaction of Al(s) and Cu^{2+}(aq)
Notice the holes in the foil where Al(s) has dissolved. Notice also the dark deposit of Cu(s) at the bottom of the beaker.

The Behavior of Metals Toward Acids

In discussing redox reactions in Chapter 5, we noted that most metals react with an acid such as HCl but that a few do not. We are now in a position to explain this observation. When a metal, M, reacts with an acid such as HCl, the metal is oxidized to the metal ion, such as M^{2+}. The reduction involves H^+ being reduced to $H_2(g)$. We can express these ideas as

Oxidation: $\qquad\qquad M(s) \longrightarrow M^{2+}(aq) + 2\,e^-$

Reduction: $\quad \dfrac{2\,H^+(aq) + 2\,e^- \longrightarrow H_2(g)}{}$

Overall: $\qquad M(s) + 2\,H^+(aq) \longrightarrow M^{2+}(aq) + H_2(g)$

$$E^\circ_{cell} = E^\circ_{H^+/H_2} - E^\circ_{M^{2+}/M} = 0\,V - E^\circ_{M^{2+}/M} = -E^\circ_{M^{2+}/M}$$

Metals whose standard electrode potentials are *negative* yield *positive* values of E°_{cell} in the above expression. These are the metals that should displace $H_2(g)$ from acidic solutions. Thus, all the metals listed *below* hydrogen in Table 21.1 (Pb through Li) should react with acids.

In acids such as HCl, HBr, and HI, the oxidizing agent is H^+ (that is, H_3O^+). Certain metals that will not react with HCl will react with an acid if there is present an *anion* that is a better oxidizing agent than H^+. Nitrate ion is a good oxidizing agent in acidic solution, and silver metal, which does not react with HCl(aq), readily reacts with nitric acid, $HNO_3(aq)$.

$$3\,Ag(s) + NO_3^-(aq) + 4\,H^+(aq) \longrightarrow 3\,Ag^+(aq) + NO(g) + 2\,H_2O \qquad E^\circ_{cell} = 0.156\,V$$

The Relationship Between E°_{cell} and K_{eq}

We related ΔG° and E°_{cell} through equation (21.15). In Chapter 20, we related ΔG° and K_{eq} (through equation 20.12). The three quantities are thus related in this way.

$$\Delta G^\circ = -RT \ln K_{eq} = -nFE^\circ_{cell}$$

and therefore,

$$E^\circ_{cell} = \frac{RT}{nF} \ln K_{eq} \qquad (21.16)$$

In equation (21.16), R has a value of 8.3145 J mol^{-1} K^{-1} and n represents the number of moles of electrons involved in the reaction. Then, if we specify a temperature of 25 °C = 298.15 K (the temperature at which electrode potentials are generally determined), we can replace the combined terms "RT/F" in equation (21.16) by a single constant. This constant has the value 0.025693 J/C = 0.025693 V.

$$E^\circ_{cell} = \frac{RT}{nF} \ln K_{eq} = \frac{8.3145\,J\,mol^{-1}\,K^{-1} \times 298.15\,K}{n \times 96485\,C\,mol^{-1}} \ln K_{eq}$$

$$E^\circ_{cell} = \frac{0.025693\,V}{n} \ln K_{eq} \qquad (21.17)$$

The relationship between E°_{cell} and K_{eq} is illustrated in Example 21-7. Also, Figure 21-8 summarizes several important relationships from thermodynamics, equilibrium, and electrochemistry.

EXAMPLE 21-7

Relating K_{eq} to E°_{cell} for a Redox Reaction. What is the value of the equilibrium constant K_{eq} for the reaction between copper metal and iron(III) ions in aqueous solution at 25 °C?

$$Cu(s) + 2\,Fe^{3+}(aq) \longrightarrow Cu^{2+}(aq) + 2\,Fe^{2+}(aq) \qquad K_{eq} = ?$$

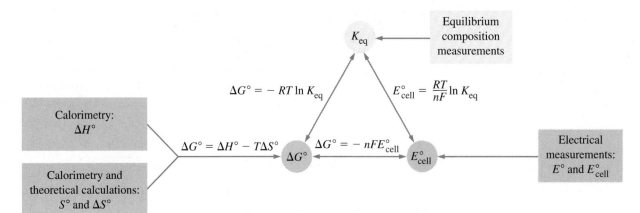

▲ **FIGURE 21-8** **A summary of important thermodynamic, equilibrium, and electro-chemical relationships**

Solution

First, use data from Table 21.1 to determine $E°_{cell}$.

$$E°_{cell} = E°(\text{reduction half-cell}) - E°(\text{oxidation half-cell})$$
$$= E°_{Fe^{3+}/Fe^{2+}} - E°_{Cu^{2+}/Cu}$$
$$= 0.771 \text{ V} - 0.340 \text{ V} = 0.431 \text{ V}$$

The number of moles of electrons for the cell reaction is 2.

$$E°_{cell} = 0.431 \text{ V} = \frac{0.02569 \text{ V}}{2} \ln K_{eq} \qquad \ln K_{eq} = \frac{2 \times 0.431 \text{ V}}{0.02569 \text{ V}} = 33.6$$
$$K_{eq} = e^{33.6} = 4 \times 10^{14}$$

Practice Example A: Should the displacement of Cu^{2+} from aqueous solution by Al(s) go to completion?

(*Hint:* Base your assessment on the value of K_{eq} for the displacement reaction. We determined $E°_{cell}$ for this reaction in Example 21-5.)

Practice Example B: Should the reaction of Sn(s) and Pb^{2+}(aq) go to completion? Explain.

21-4 E_{cell} as a Function of Concentrations

When we combine standard electrode potentials, we obtain a standard $E°_{cell}$, such as $E°_{cell} = 1.103$ V for the voltaic cell of Figure 21-4. For the following cell reaction at *nonstandard* conditions, however, the measured E_{cell} is not 1.103 V.

$$\text{Zn(s)} + Cu^{2+}(2.0 \text{ M}) \longrightarrow Zn^{2+}(0.10 \text{ M}) + \text{Cu(s)} \qquad E_{cell} = 1.142 \text{ V}$$

Experimental measurements of cell potentials are often made for nonstandard conditions; these measurements have great significance, especially for performing chemical analyses.

From Le Châtelier's principle, we might predict that *increasing* the concentration of a reactant (Cu^{2+}) while *decreasing* the concentration of a product (Zn^{2+}) should favor the forward reaction. Zn(s) should displace Cu^{2+}(aq) even more readily than for standard state conditions and $E_{cell} > 1.103$ V. E_{cell} is found to vary linearly with log ($[Zn^{2+}]/[Cu^{2+}]$), as illustrated in Figure 21-9.

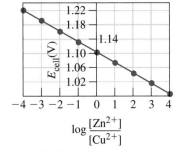

▲ **FIGURE 21-9**

Variation of E_{cell} with ion concentrations

The cell reaction is
Zn(s) + Cu^{2+}(aq) $\longrightarrow$
Zn^{2+}(aq) + Cu(s)
and has $E°_{cell} = 1.103$ V.

It is not difficult to establish the relationship between the cell potential, E_{cell}, and the concentrations of reactants and products. From Chapter 20, we can write equation (20.11).

$$\Delta G = \Delta G^\circ + RT \ln Q$$

For ΔG and ΔG°, we can substitute $-nFE_{cell}$ and $-nFE_{cell}^\circ$, respectively.

$$-nFE_{cell} = -nFE_{cell}^\circ + RT \ln Q$$

Dividing through by $-nF$, we obtain

$$E_{cell} = E_{cell}^\circ - \frac{RT}{nF} \ln Q$$

▲ **Walther Nernst (1864–1941)**
Nernst was only 25 years old when he formulated his equation relating cell voltages and concentrations. He is also credited with proposing the solubility product concept in the same year. In 1906, he announced his "heat theorem," which we now know as the third law of thermodynamics.

This equation was first proposed by Walther Nernst in 1889. It is commonly used by analytical chemists in the form in which Nernst expressed it. We can obtain the **Nernst equation** by switching from natural to common logarithms ($\ln Q = 2.3026 \log Q$).

$$E_{cell} = E_{cell}^\circ - \frac{2.3026\, RT}{nF} \log Q$$

By specifying a temperature of 298.15 K and replacing RT/F by 0.025693 V, as we did in developing equation (21.17), we find that the term $2.3026\, RT/F = 2.3026 \times 0.025693$ V $= 0.059161$ V, usually rounded off to 0.0592 V. The final form of the Nernst equation is

$$E_{cell} = E_{cell}^\circ - \frac{0.0592\ \text{V}}{n} \log Q \qquad (21.18)$$

In the Nernst equation, we make the usual substitutions into Q: $a = 1$ for the activities of pure solids and liquids, partial pressures (atm) for the activities of gases, and molarities for the activities of solution components. Example 21-8 demonstrates that the Nernst equation allows us to calculate E_{cell} for any chosen concentrations; that is, we are not restricted to standard conditions.

EXAMPLE 21-8

Applying the Nernst Equation for Determining E_{cell}. What is the value of E_{cell} for the voltaic cell pictured in Figure 21-10 and diagrammed as follows?

$$\text{Pt}\,|\,\text{Fe}^{2+}(0.10\ \text{M}),\ \text{Fe}^{3+}(0.20\ \text{M})\,\|\,\text{Ag}^{+}(1.0\ \text{M})\,|\,\text{Ag(s)} \qquad E_{cell} = ?$$

Solution

Two steps are required when using the Nernst equation. First, to determine E_{cell}°, use data from Table 21.1 to write

$$
\begin{aligned}
E_{cell}^\circ &= E^\circ(\text{cathode}) - E^\circ(\text{anode}) \\
&= E_{\text{Ag}^+/\text{Ag}}^\circ - E_{\text{Fe}^{3+}/\text{Fe}^{2+}}^\circ \qquad (21.19)\\
&= 0.800\ \text{V} - 0.771\ \text{V} = 0.029\ \text{V}
\end{aligned}
$$

Now, to determine E_{cell} for the reaction

$$\text{Fe}^{2+}(0.10\ \text{M}) + \text{Ag}^{+}(1.0\ \text{M}) \longrightarrow \text{Fe}^{3+}(0.20\ \text{M}) + \text{Ag(s)} \qquad E_{cell} = ? \qquad (21.20)$$

substitute appropriate values into the Nernst equation (21.18), starting with $E_{cell}^\circ = 0.029$ V and $n = 1$,

$$E_{cell} = 0.029\ \text{V} - \frac{0.0592\ \text{V}}{1} \log \frac{[\text{Fe}^{3+}]}{[\text{Fe}^{2+}][\text{Ag}^+]}$$

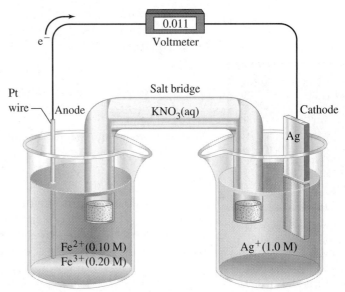

▲ FIGURE 21-10   A voltaic cell with nonstandard conditions—
Example 21-8 illustrated

and for concentrations $[Fe^{2+}]$ = 0.10 M; $[Fe^{3+}]$ = 0.20 M; $[Ag^+]$ = 1.0 M.

$$E_{cell} = 0.029 \text{ V} - 0.0592 \text{ V} \log \frac{0.20}{0.10 \times 1.0}$$

$$= 0.029 \text{ V} - 0.0592 \text{ V} \times \log 2 = 0.029 \text{ V} - 0.018 \text{ V}$$

$$= 0.011 \text{ V}$$

Practice Example A: Calculate E_{cell} for the following voltaic cell.

$$Al(s) \,|\, Al^{3+}(0.36 \text{ M}) \,\|\, Sn^{4+}(0.086 \text{ M}), Sn^{2+}(0.54 \text{ M}) \,|\, Pt$$

Practice Example B: Calculate E_{cell} for the following voltaic cell.

$$Pt(s) \,|\, Cl_2(1 \text{ atm}) \,|\, Cl^-(1.0 \text{ M}) \,\|\, Pb^{2+}(0.050 \text{ M}), H^+(0.10 \text{ M}) \,|\, PbO_2(s)$$

In Section 21-3, we developed a criterion for spontaneous change ($E_{cell} > 0$), but we used the criterion only with $E°$ data from Table 21.1. Qualitative conclusions reached with $E°_{cell}$ values often hold over a broad range of nonstandard conditions as well. However, when $E°_{cell}$ is within a few hundredths of a volt of zero, it is sometimes necessary to determine E_{cell} for nonstandard conditions in order to apply the criterion for spontaneity of redox reactions, as illustrated in Example 21-9.

EXAMPLE 21-9

Predicting Spontaneous Reactions for Nonstandard Conditions. Will the cell reaction proceed spontaneously as written for the following cell?

$$Sn(s) \,|\, Sn^{2+}(0.50 \text{ M}) \,\|\, Pb^{2+}(0.0010 \text{ M}) \,|\, Pb(s)$$

Solution

To determine $E°_{cell}$ from $E°$ data, we write

$$E°_{cell} = E°(\text{cathode}) - E°(\text{anode})$$

$$= E°_{Pb^{2+}/Pb} - E°_{Sn^{2+}/Sn}$$

$$= -0.125 \text{ V} - (-0.137) \text{ V} = 0.012 \text{ V}$$

Then, to determine E_{cell} for the reaction

$$Sn(s) + Pb^{2+}(0.0010 \text{ M}) \longrightarrow Pb(s) + Sn^{2+}(0.50 \text{ M}) \qquad E_{cell} = ?$$

we substitute the following data into the Nernst equation: $E_{cell}^{\circ} = 0.012$ V, $n = 2$, $[Sn^{2+}] = 0.50$ M, and $[Pb^{2+}] = 0.0010$ M.

$$E_{cell} = 0.012 \text{ V} - \frac{0.0592 \text{ V}}{2} \log \frac{[Sn^{2+}]}{[Pb^{2+}]}$$

$$= 0.012 \text{ V} - 0.0296 \text{ V} \log \frac{0.50}{0.0010} = 0.012 \text{ V} - 0.0296 \text{ V} \log 500$$

$$= 0.012 \text{ V} - 0.080 \text{ V} = -0.068 \text{ V}$$

Because $E_{cell} < 0$, we conclude that the reaction is nonspontaneous as written.

Practice Example A: Will the cell reaction proceed spontaneously as written for the following cell?

$$Cu(s) \mid Cu^{2+}(0.15 \text{ M}) \parallel Fe^{3+}(0.35 \text{ M}), Fe^{2+}(0.25 \text{ M}) \mid Pt(s)$$

Practice Example B: For what ratio of $[Sn^{2+}]/[Pb^{2+}]$ will the cell reaction in Example 21-9 not be spontaneous in either direction?

Concentration Cells

The voltaic cell in Figure 21-11 consists of two hydrogen electrodes. One is a standard hydrogen electrode (SHE), and the other is a hydrogen electrode immersed in a solution of unknown $[H^+]$, less than 1 M. The cell diagram is

$$Pt \mid H_2(g, 1 \text{ atm}) \mid H^+(x \text{ M}) \parallel H^+(1 \text{ M}) \mid H_2(g, 1 \text{ atm}) \mid Pt$$

The reaction occurring in this cell is

Reduction:	$2 \text{ H}^+(1 \text{ M}) + \cancel{2 e^-} \longrightarrow \cancel{H_2(g, 1 \text{ atm})}$
Oxidation:	$\cancel{H_2(g, 1 \text{ atm})} \longrightarrow 2 \text{ H}^+(x \text{ M}) + \cancel{2 e^-}$
Overall:	$2 \text{ H}^+(1 \text{ M}) \longrightarrow 2 \text{ H}^+(x \text{ M})$

(21.21)

$$E_{cell}^{\circ} = E_{H^+/H_2}^{\circ} - E_{H^+/H_2}^{\circ} = 0 \text{ V}$$

The voltaic cell in Figure 21-11 is called a concentration cell. A **concentration cell** consists of two half-cells with *identical electrodes* but different ion concentrations. Because the electrodes are identical, the standard electrode potentials are numerically

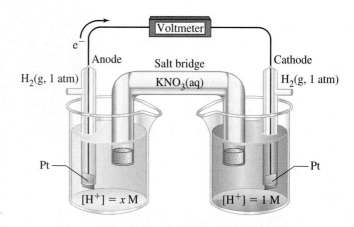

▶ **FIGURE 21-11**

A concentration cell

The cell consists of two hydrogen electrodes. The electrode on the right is a SHE. Oxidation occurs at the anode on the left, where $[H^+]$ is less than 1 M. The reading on the voltmeter is directly proportional to the pH of the solution in the anode compartment.

equal and subtracting one from the other leads to the value $E^\circ_{cell} = 0$. However, because the ion concentrations differ, there is a potential difference between the two half-cells. The spontaneous change in a concentration cell always occurs in the direction in which the concentrated solution becomes more dilute, and the dilute solution becomes more concentrated. The final result is the same as if the solutions were simply mixed. In a concentration cell, however, we use the natural tendency for entropy to increase in a mixing process as a means of producing electricity.

The Nernst equation for reaction (21.21) takes the form

$$E_{cell} = E^\circ_{cell} - \frac{0.0592 \text{ V}}{2} \log \frac{x^2}{1^2}$$

which simplifies to

$$E_{cell} = 0 - \frac{0.0592 \text{ V}}{2} \times 2 \log \frac{x}{1} = -0.0592 \text{ V} \log x$$

Because x is $[H^+]$ in the unknown solution and $-\log x = -\log[H^+] = pH$, the final result is

$$E_{cell} = (0.0592 \text{ pH}) \text{ V} \tag{21.22}$$

where the pH is that of the unknown solution. If an unknown solution has a pH of 3.50, for example, the measured cell voltage in Figure 21-11 will be $E_{cell} = (0.0592 \times 3.50) \text{ V} = 0.207 \text{ V}$.

Constructing and using a hydrogen electrode is difficult. The Pt metal surface must be specially prepared and maintained, gas pressure must be controlled, and the electrode cannot be used in the presence of strong oxidizing or reducing agents. A better approach is to replace the SHE in Figure 21-11 by a different reference electrode and the other hydrogen electrode by a *glass electrode*. A glass electrode is a thin glass membrane enclosing HCl(aq) and a silver wire coated with AgCl(s). When the glass electrode is dipped into a solution, H^+ ions are exchanged across the membrane. The potential established on the silver wire depends on $[H^+]$ in the solution being tested. Glass electrodes are most commonly encountered in laboratory pH meters (Figure 21-12).

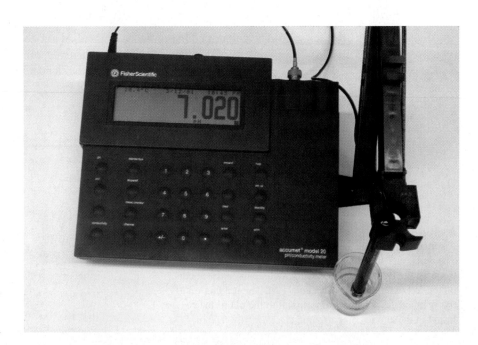

▶ **FIGURE 21-12**
A glass electrode and pH meter
The potential of the glass electrode depends on the hydrogen ion concentration of the solution being tested. The difference in potential between the glass electrode and a reference electrode is converted to a pH reading by the meter.

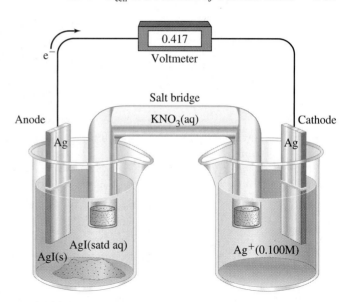

▶ FIGURE 21-13
A concentration cell for determining K_{sp} of AgI
The silver electrode in the anode compartment is in contact with a saturated solution of AgI. In the cathode compartment, $[Ag^+] = 0.100$ M.

Measurement of K_{sp}

The difference in concentration of ions in the two half-cells of a concentration cell accounts for the observed E_{cell}. It also provides a basis for determining K_{sp} values for sparingly soluble ionic compounds. Consider the following concentration cell.

$$Ag(s)\,|\,Ag^+(\text{satd AgI})\,\|\,Ag^+(0.100\text{ M})\,|\,Ag \qquad E_{cell} = 0.417\text{ V}$$

At the anode, a silver electrode is placed in a saturated aqueous solution of silver iodide. At the cathode, a second silver electrode is placed in a solution with $[Ag^+] = 0.100$ M. The two half-cells are connected by a salt bridge, and the measured cell voltage is 0.417 V (Figure 21-13). The cell reaction occurring in this *concentration* cell is

Reduction: $Ag^+(0.100\text{ M}) + e^- \longrightarrow Ag(s)$

Oxidation: $\underline{\qquad\qquad Ag(s) \longrightarrow Ag^+(\text{satd AgI})\qquad}$

Overall: $Ag^+(0.100\text{ M}) \longrightarrow Ag^+(\text{satd AgI}) \qquad\qquad (21.23)$

We complete the calculation of K_{sp} of silver iodide in Example 21-10.

EXAMPLE 21-10

Using a Voltaic Cell to Determine K_{sp} of a Slightly Soluble Solute. With the data given for reaction (21.23), calculate K_{sp} for AgI.

$$AgI(s) \rightleftharpoons Ag^+(aq) + I^-(aq) \qquad K_{sp} = ?$$

Solution

Let's represent $[Ag^+]$ in saturated silver iodide solution as x. Then we can apply the Nernst equation to the reaction (21.23). (To simplify the appearance of the equations that follow, we have dropped the unit V, which would otherwise appear in several places.)

$$E_{cell} = E^\circ_{cell} - \frac{0.0592}{n} \log \frac{[Ag^+]_{\text{sat'd AgI}}}{[Ag^+]_{0.100\text{ M soln}}}$$

$$= E^\circ_{cell} - \frac{0.0592}{1} \log \frac{x}{0.100}$$

$$0.417 = 0 - 0.0592(\log x - \log 0.100)$$

Divide both sides of the equation by 0.0592.

$$\frac{0.417}{0.0592} = -\log x + \log 0.100$$

$$\log x = \log 0.100 - \frac{0.417}{0.0592} = -1.00 - 7.04 = -8.04$$

$$[Ag^+] = 10^{-8.04} = 9.1 \times 10^{-9} \text{ M}$$

Because in saturated AgI the concentrations of Ag^+ and I^- are equal,

$$K_{sp} = [Ag^+][I^-] = (9.1 \times 10^{-9})(9.1 \times 10^{-9}) = 8.3 \times 10^{-17}$$

Practice Example A: K_{sp} for AgCl = 1.8×10^{-10}. What would be the measured E_{cell} for the voltaic cell in Example 21-10 if the contents of the anode half-cell were saturated AgCl(aq) and AgCl(s)?

Practice Example B: Calculate the K_{sp} for PbI_2 given the following concentration cell information.

$$Pb(s)\,|\,Pb^{2+}(\text{satd } PbI_2)\,||\,Pb^{2+}(0.100 \text{ M})\,|\,Pb(s) \qquad E_{cell} = 0.0567 \text{ V}$$

21-5 Batteries: Producing Electricity Through Chemical Reactions

▶ Batteries are vitally important to modern society. Annual production in developed nations has been estimated at over 10 batteries per person per year.

A **battery** is a device that stores chemical energy for later release as electricity. Some batteries consist of a single voltaic cell with two electrodes and the appropriate electrolyte(s); an example is a flashlight cell. Other batteries consist of two or more voltaic cells joined in series fashion—plus to minus—to increase the total voltage; an example is an automobile battery. In this section, we will consider three types of cells and batteries.

- **Primary batteries** (or *primary cells*). The cell reaction is not reversible. When the reactants have been mostly converted to products, no more electricity is produced and the battery is dead.

- **Secondary batteries** (or *secondary cells*). The cell reaction can be reversed by passing electricity through the battery (charging). Such a battery can be used through several hundred or more cycles of discharging followed by charging.

- **Flow batteries** and **fuel cells**. Materials (reactants, products, and electrolytes) pass through the battery, which is simply a converter of chemical energy to electric energy.

The Leclanché (Dry) Cell

The most common form of voltaic cell is the *Leclanché cell*, invented by the French chemist Georges Leclanché (1839–1882) in the 1860s. Popularly called a *dry cell* because no free liquid is present, or *flashlight battery*, the Leclanché cell is diagrammed in Figure 21-14. In this cell, oxidation occurs at a zinc anode and reduction at an inert carbon (graphite) cathode. The electrolyte is a moist paste of MnO_2, $ZnCl_2$, NH_4Cl, and carbon black (soot). The maximum cell voltage is 1.55 V. The anode (oxidation) half-reaction is simple.

$$\text{Oxidation:} \quad Zn(s) \longrightarrow Zn^{2+}(aq) + 2\,e^-$$

The reduction is more complex. Essentially, it involves the reduction of MnO_2 to compounds having Mn in a +3 oxidation state, for example,

$$\text{Reduction:} \quad 2\,MnO_2(s) + H_2O(l) + 2\,e^- \longrightarrow Mn_2O_3(s) + 2\,OH^-(aq)$$

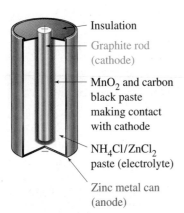

Insulation

Graphite rod (cathode)

MnO_2 and carbon black paste making contact with cathode

$NH_4Cl/ZnCl_2$ paste (electrolyte)

Zinc metal can (anode)

▲ **FIGURE 21-14**

The Leclanché (dry) cell

The chief components of the cell are a graphite (carbon) rod serving as the cathode, a zinc container serving as the anode, and an electrolyte.

▲ $E°_{cell}$ **is an intensive property**
The voltage of a dry cell battery does not depend on the size of the battery—all of those pictured here are 1.5 V batteries.

An acid–base reaction occurs between NH_4^+ (from NH_4Cl) and OH^-.

$$NH_4^+(aq) + OH^-(aq) \longrightarrow NH_3(g) + H_2O(l)$$

A buildup of $NH_3(g)$ cannot be permitted to occur around the cathode because it would disrupt the current by adhering to the cathode. That buildup is prevented by a reaction between Zn^{2+} and $NH_3(g)$ to form the complex ion $[Zn(NH_3)_2]^{2+}$, which crystallizes as the chloride salt.

$$Zn^{2+}(aq) + 2\,NH_3(g) + 2\,Cl^-(aq) \longrightarrow [Zn(NH_3)_2]Cl_2(s)$$

The Leclanché cell is a *primary* cell; it cannot be recharged. This cell is cheap to make, but it has some drawbacks. When current is drawn rapidly from the cell, products such as NH_3 build up on the electrodes, causing the voltage to drop. Also, because the electrolyte medium is acidic, zinc metal slowly dissolves.

A superior form of the Leclanché cell is the *alkaline cell*, which uses NaOH or KOH in place of NH_4Cl as the electrolyte. The reduction half-reaction is the same as that shown above, but the oxidation half-reaction involves the formation of $Zn(OH)_2(s)$, which we can think of as occurring in two-steps.

$$Zn(s) \longrightarrow Zn^{2+}(aq) + 2\,e^-$$

$$Zn^{2+}(aq) + 2\,OH^-(aq) \longrightarrow Zn(OH)_2(s)$$

$$\overline{Zn(s) + 2\,OH^-(aq) \longrightarrow Zn(OH)_2(s) + 2\,e^-}$$

The advantages of the alkaline battery are that zinc does not dissolve as readily in a basic (alkaline) medium as in an acidic medium and the battery does a better job of maintaining its voltage as current is drawn from it.

The Lead–Acid (Storage) Battery

The most common *secondary* battery is the *lead–acid battery* or *storage battery*, used in automobiles since about 1915 (Figure 21-15). A storage battery is capable of repeated use because it uses chemical reactions that are reversible. That is, the discharged energy can be restored by supplying electric current to recharge the cell.

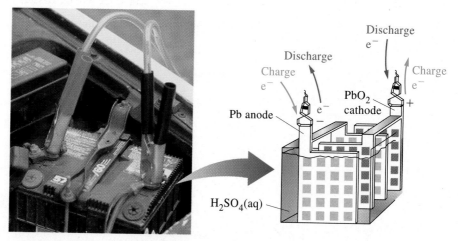

▲ **FIGURE 21-15 A lead–acid (storage) cell**
The composition of the electrodes is described in the text. The cell reaction that occurs as the cell is discharged is given in equation (21.24). The voltage of the cell is 2.02 V. In this figure, two anode plates are connected in parallel fashion, as are two cathode plates.

The reactants in a lead–acid battery are spongy lead packed into a lead grid at the anode, red-brown lead(IV) oxide packed into a lead grid at the cathode, and dilute sulfuric acid with about 35% H_2SO_4, by mass. In this strongly acidic medium, the ionization of H_2SO_4 does not go to completion. Both HSO_4^-(aq) and SO_4^{2-}(aq) are present, but HSO_4^- predominates. The half-reactions and overall reaction are

Reduction: $\quad PbO_2(s) + 3 H^+(aq) + HSO_4^-(aq) + 2 e^- \longrightarrow PbSO_4(s) + 2 H_2O(l)$

Oxidation: $\quad\quad\quad\quad\quad\quad\quad Pb(s) + HSO_4^-(aq) \longrightarrow PbSO_4(s) + H^+(aq) + 2 e^-$

Overall: $\quad PbO_2(s) + Pb(s) + 2 H^+(aq) + 2 HSO_4^-(aq) \longrightarrow 2 PbSO_4(s) + 2 H_2O(l)$ $\quad$ (21.24)

$$E_{cell} = E_{PbO_2/PbSO_4} - E_{PbSO_4/Pb} = 1.74 \text{ V} - (-0.28 \text{ V}) = 2.02 \text{ V}$$

▶ You can think of the half-reactions as occurring in two steps: (1) oxidation of Pb(s) to Pb^{2+}(aq) and reduction of PbO_2(s) to Pb^{2+}(aq), followed by (2) precipitation of $PbSO_4$(s) at each electrode.

When an automobile engine is started, the battery is at first discharged. Once the car is in motion, an alternator powered by the engine constantly recharges the battery. At times, the plates of the battery become coated with $PbSO_4$(s) and the electrolyte becomes sufficiently diluted with water that the battery must be recharged by connecting it to an external electric source. This forces the reverse of reaction (21.24), a nonspontaneous reaction.

$$2 PbSO_4(s) + 2 H_2O(l) \longrightarrow Pb(s) + PbO_2(s) + 2 H^+(aq) + 2 HSO_4^-(aq)$$
$$E_{cell} = -2.02 \text{ V}$$

To prevent the anode and cathode from coming into contact with each other, causing a *short circuit*, sheets of an insulating material are used to separate alternating anode and cathode plates. A group of anodes is connected together electrically, as is a group of cathodes. This parallel connection increases the electrode area in contact with the electrolyte solution and increases the current-delivering capacity of the cell. Cells are then joined in a series fashion, positive to negative, to produce a battery. The typical 12 V battery consists of six cells, each cell with a potential of about 2 V.

The Silver–Zinc Cell: A Button Battery

The cell diagram of a *silver–zinc cell* (Figure 21-16) is

$$Zn(s), ZnO(s) | KOH(satd) | Ag_2O(s), Ag(s)$$

The half-reactions on discharging are

Reduction: $\quad Ag_2O(s) + H_2O(l) + 2 e^- \longrightarrow 2 Ag(s) + 2 OH^-(aq)$

Oxidation: $\quad\quad\quad Zn(s) + 2 OH^-(aq) \longrightarrow ZnO(s) + H_2O(l) + 2 e^-$

Overall: $\quad\quad\quad\quad Zn(s) + Ag_2O(s) \longrightarrow ZnO(s) + 2 Ag(s)$ $\quad$ (21.25)

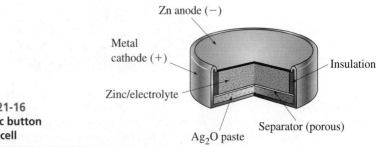

▶ **FIGURE 21-16**
A silver–zinc button (miniature) cell

Because no solution species is involved in the cell reaction, the quantity of electrolyte is very small and the electrodes can be maintained very close together. The cell voltage is 1.8 V, and its storage capacity is six times greater than that of a lead–acid battery of the same size. These characteristics make batteries such as the silver–zinc cell useful in button batteries. These miniature batteries are used in watches, hearing aids, and cameras. In addition, silver–zinc batteries fulfill the requirements of spacecraft, satellites, missiles, rockets, space launch vehicles, torpedoes, underwater vehicles, and life-support systems. On the Mars *Pathfinder* mission, the rover and the cruise system were powered by solar cells. The energy storage requirements of the lander were met by modified silver–zinc batteries with about three times the storage capacity of the standard nickel–cadmium rechargeable battery.

▲ A rechargeable nickel–cadmium cell, or nicad battery.

The Nickel–Cadmium Cell: A Rechargeable Battery

The *nickel–cadmium cell* (or *nicad battery*) is commonly used in cordless electric devices, such as electric shavers and handheld calculators. The anode in this battery is cadmium metal, and the cathode is the Ni(III) compound NiO(OH) supported on nickel metal. The half-cell reactions for a nickel–cadmium battery during discharge are

Reduction: $2\,NiO(OH)(s) + 2\,H_2O(l) + 2\,e^- \longrightarrow 2\,Ni(OH)_2(s) + 2\,OH^-(aq)$

Oxidation: $Cd(s) + 2\,OH^-(aq) \longrightarrow Cd(OH)_2(s) + 2\,e^-$

Overall: $Cd(s) + 2\,NiO(OH)(s) + 2\,H_2O(l) \longrightarrow 2\,Ni(OH)_2(s) + Cd(OH)_2(s)$

This battery gives a fairly constant voltage of 1.4 V. When recharged by connecting the battery to an external voltage source, the reactions above are reversed. Nickel–cadmium batteries can be recharged many times because the solid products adhere to the surface of the electrodes.

Fuel Cells

The two types of cells considered in the remainder of this section fall into the third category mentioned on page 844; they are flow batteries.

For most of the twentieth century, scientists explored the possibility of converting the chemical energy of fuels directly to electricity. The essential process in a fuel cell is *fuel + oxygen $\longrightarrow$ oxidation products.* The first fuel cells were based on the reaction of hydrogen and oxygen. Figure 21-17 represents such a fuel cell. The overall change is that $H_2(g)$ and $O_2(g)$ in an alkaline medium produce $H_2O(l)$.

Reduction: $O_2(g) + 2\,H_2O(l) + 4\,e^- \longrightarrow 4\,OH^-(aq)$

Oxidation: $2\,\{H_2(g) + 2\,OH^-(aq) \longrightarrow 2\,H_2O(l) + 2\,e^-\}$

Overall: $2\,H_2(g) + O_2(g) \longrightarrow 2\,H_2O(l)$ (21.26)

$$E^\circ_{cell} = E^\circ_{O_2/OH^-} - E^\circ_{H_2O/H_2} = 0.401\,V - (-0.828\,V) = 1.229\,V$$

The theoretical maximum energy available as electric energy in any electrochemical cell is the free energy change for the cell reaction, ΔG°. The maximum energy release when a fuel is burned is the enthalpy change, ΔH°. One of the measures used to evaluate a fuel cell is the *efficiency value*, $\varepsilon = \Delta G^\circ / \Delta H^\circ$. For the hydrogen–oxygen fuel cell, $\varepsilon = -474.4\,kJ/-571.6\,kJ = 0.83$.

The day is fast approaching when fuel cells based on the direct oxidation of common fuels will become a reality. For example, the half-reaction and cell reaction for a fuel cell using methane (natural gas) are

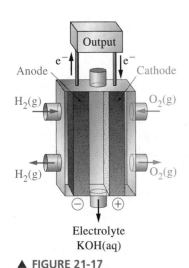

▲ **FIGURE 21-17**
A hydrogen–oxygen fuel cell
A key requirement in fuel cells is porous electrodes that allow for easy access of the gaseous reactants to the electrolyte. The electrodes chosen should also catalyze the electrode reactions.

▲ This Toyota prototype is a fuel-cell-powered electric car producing hydrogen from gasoline.

Reduction: $\quad 2\,\{O_2(g) + 4\,H^+ + 4\,e^- \longrightarrow 2\,H_2O(l)\}$

Oxidation: $\quad\quad\quad CH_4(g) + 2\,H_2O(l) \longrightarrow CO_2(g) + 8\,H^+ + 8\,e^-$

Overall: $\quad\quad\quad\quad CH_4(g) + 2\,O_2(g) \longrightarrow CO_2(g) + 2\,H_2O(l)$

$$\Delta H^\circ = -890\text{ kJ} \quad\quad \Delta G^\circ = -818\text{ kJ} \quad\quad \varepsilon = 0.92 \quad\quad\quad (21.27)$$

Although methane fuel cells are still in the research stage, an automobile engine is currently under development in which (a) a liquid hydrocarbon is vaporized; (b) the vaporized fuel is partially oxidized to $CO(g)$; (c) steam, in the presence of a catalyst, converts the $CO(g)$ to $CO_2(g)$ and $H_2(g)$; and (d) $H_2(g)$ and air are fed through a fuel cell, producing electric energy.

We should refer to a fuel cell as an *energy converter* rather than as a battery. As long as fuel and $O_2(g)$ are available, the cell will produce electricity. It does not have the limited capacity of a primary battery or the storage capacity of a secondary battery. Fuel cells based on reaction (21.26) have had their most notable successes as energy sources in space vehicles. (Water produced in the cell reaction is also a valuable product of the fuel cell.)

Air Batteries

In a fuel cell, $O_2(g)$ is the oxidizing agent that oxidizes a fuel such as $H_2(g)$ or $CH_4(g)$. Another kind of flow battery, because it uses $O_2(g)$ from air, is known as an *air battery*. The substance that is oxidized in an air battery is typically a metal.

One heavily studied battery system is the aluminum–air battery. In this battery, oxidation occurs at an aluminum anode and reduction at a carbon–air cathode. The electrolyte circulated through the battery is $NaOH(aq)$. Because it is in the presence of a high concentration of OH^-, Al^{3+} produced at the anode forms the complex ion $[Al(OH)_4]^-$. The operation of the battery is suggested by Figure 21-18. The half-reactions and the overall cell reaction are

Reduction: $\quad\quad 3\,\{O_2(g) + 2\,H_2O(l) + 4\,e^- \longrightarrow 4\,OH^-(aq)\}$

Oxidation: $\quad\quad\quad\quad 4\,\{Al(s) + 4\,OH^-(aq) \longrightarrow [Al(OH)_4]^-(aq) + 3\,e^-\}$

Overall: $\quad 4\,Al(s) + 3\,O_2(g) + 6\,H_2O(l) + 4\,OH^-(aq) \longrightarrow 4\,[Al(OH)_4]^-(aq) \quad\quad (21.28)$

The battery is kept charged by feeding chunks of Al and water into it. A typical air battery can power an automobile several hundred miles before refueling is necessary. The electrolyte is circulated outside the battery, where $Al(OH)_3(s)$ is precipitated from the $[Al(OH)_4]^-(aq)$. This $Al(OH)_3(s)$ is collected and can then be converted back to aluminum metal at an aluminum manufacturing facility.

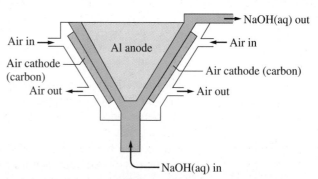

▲ **FIGURE 21-18   An aluminum–air battery, simplified**

21-6   Corrosion: Unwanted Voltaic Cells

The reactions occurring in voltaic cells (batteries) are important sources of electricity. Similar reactions underlie corrosion processes, where they are unwanted. First, we will consider the electrochemical basis of corrosion, and then we will see how electrochemical principles can be applied to control corrosion.

Figure 21-19a demonstrates the basic processes in the corrosion of an iron nail. The nail is embedded in a gel of agar in water. Incorporated in the gel are the acid–base indicator phenolphthalein and the substance $K_3[Fe(CN)_6]$ (potassium ferricyanide). Here is what we see within hours of starting the experiment: At the head and tip of the nail, a deep blue precipitate forms. Along the body of the nail, the agar gel turns pink. The blue precipitate, Turnbull's blue, establishes the presence of iron(II). The pink color is that of phenolphthalein in basic solution. From these observations, we write two simple half-equations.

Reduction: $O_2(g) + 2\,H_2O(l) + 4\,e^- \longrightarrow 4\,OH^-(aq)$

Oxidation: $2\,Fe(s) \longrightarrow 2\,Fe^{2+}(aq) + 4\,e^-$

The potential difference for these two half reactions is

$$E^\circ_{cell} = E^\circ_{O_2/OH^-} - E^\circ_{Fe^{2+}/Fe} = 0.401\,V - (-0.440\,V) = 0.841\,V$$

indicating that the corrosion process should be spontaneous when reactants and products are in their standard states. Typically, the corrosion medium has $[OH^-] \ll 1\,M$, the reduction half-reaction is even more favorable, and E_{cell} is even greater than 0.841 V. Corrosion is especially significant in acidic solutions, in which the reduction half-reaction is

$$O_2(g) + 4\,H^+(aq) + 4\,e^- \longrightarrow 2\,H_2O(l) \qquad E^\circ_{O_2/H_2O} = 1.229\,V$$

In the corroding nail of Figure 21-19a, oxidation occurs at the head and tip. Electrons given up in the oxidation move along the nail and are used to reduce dissolved O_2. The reduction product, OH^-, is detected by the phenolphthalein. In the bent nail in Figure 21-19b, oxidation occurs at *three* points: the head and tip and also the bend. The nail is preferentially oxidized at these points because the strained metal is more active (more anodic) than the unstrained metal. This situation is similar to the preferential rusting of a dented automobile fender.

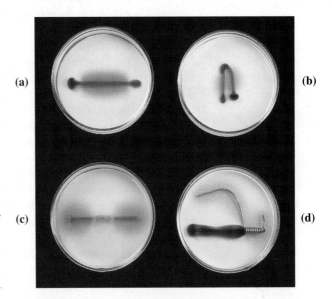

▶ **FIGURE 21-19**

Demonstration of corrosion and methods of corrosion protection

The pink color results from the indicator phenolphthalein in the presence of base; the dark blue color results from the formation of Turnbull's blue $KFe[Fe(CN)_6]$. Corrosion (oxidation) of the nail occurs at strained regions: **(a)** the head and tip and **(b)** a bend in the nail. **(c)** Contact with zinc protects the nail from corrosion. Zinc is oxidized instead of the iron (forming the faint white precipitate of zinc ferricyanide). **(d)** Copper does not protect the nail from corrosion. Electrons lost in the oxidation half-reaction distribute themselves along the copper wire, as seen by the pink color that extends the full length of the wire.

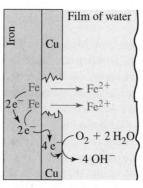

(a) Copper-plated iron   **(b)** Galvanized iron

▲ **FIGURE 21-20   Protection of iron against electrolytic corrosion**

In the *anodic* reaction, the metal that is more easily oxidized loses electrons to produce metal ions. In part (a), this is iron; in part (b), it is zinc. In the *cathodic* reaction, oxygen gas, which is dissolved in a thin film of water on the metal, is reduced to OH^-. Rusting of iron occurs in (a), but it does not in (b). When iron corrodes, Fe^{2+} and OH^- ions from the half-reactions initiate these further reactions.

$$Fe^{2+} + 2\,OH^- \longrightarrow Fe(OH)_2(s)$$

$$4\,Fe(OH)_2(s) + O_2 + 2\,H_2O \longrightarrow 4\,Fe(OH)_3(s)$$

$$2\,Fe(OH)_3(s) \longrightarrow Fe_2O_3 \cdot H_2O(s) + 2\,H_2O(l)$$
$$\text{rust}$$

▲ **Magnesium sacrificial anodes**
The small cylindrical bars of magnesium attached to the steel ship provide cathodic protection against corrosion.

Some metals, such as aluminum, form corrosion products that adhere tightly to the underlying metal and protect it from further corrosion. Iron oxide (rust), however, flakes off and constantly exposes fresh surface. This difference in corrosion behavior explains why cans made of iron deteriorate rapidly in the environment, whereas aluminum cans have an almost unlimited lifetime. The simplest method of protecting a metal from corrosion is to cover it with paint or some other protective coating impervious to water, an important reactant and solvent in corrosion processes.

Another method of protecting an iron surface is to plate it with a thin layer of a second metal. Iron can be plated with copper by electroplating or with tin by dipping the iron into molten tin. In either case, the underlying metal is protected only as long as the coating remains intact. If the coating is cracked, as when a "tin" can is dented, the underlying iron is exposed and corrodes. Iron, being more active than copper and tin, undergoes oxidation; the reduction half-reaction occurs on the plating (Figures 21-19d and 21-20a).

When iron is coated with zinc (galvanized iron), the situation is different. Zinc is more active than iron. If a break occurs in the zinc plating, the iron is still protected. Zinc is oxidized instead of the iron, and corrosion products protect the zinc from further corrosion (Figures 21-19c and 21-20b).

Still another method is used to protect large iron and steel objects in contact with water or moist soils—ships, storage tanks, pipelines, plumbing systems. This method involves connecting a chunk of magnesium, aluminum, zinc, or some other active metal to the object, either directly or through a wire. Oxidation occurs at the active metal, which slowly dissolves. The iron surface acquires electrons from the oxidation of the active metal; the iron acts as a cathode and supports a *reduction* half-reaction. As long as some of the active metal remains, the iron is protected. This type of protection is called **cathodic protection**, and the active metal is called, appropriately, a *sacrificial anode*. Millions of pounds of magnesium are used annually in the United States in sacrificial anodes.

21-7 Electrolysis: Causing Nonspontaneous Reactions to Occur

▶ **KEEP IN MIND**

that voltaic (galvanic) and electrolytic cells are the two categories subsumed under the more general term: electrochemical cell.

Until now, we have emphasized voltaic (galvanic) cells, electrochemical cells in which chemical change is used to produce electricity. Another type of electrochemical cell—the **electrolytic cell**—uses electricity to produce a *nonspontaneous* reaction. The process in which a nonspontaneous reaction is driven by the application of electric energy is called **electrolysis**.

Let's explore the relationship between voltaic and electrolytic cells by returning briefly to the cell shown in Figure 21-4. When the cell functions spontaneously, electrons flow from the zinc to the copper and the overall chemical change in the voltaic cell is

$$Zn(s) + Cu^{2+}(aq) \longrightarrow Zn^{2+}(aq) + Cu(s) \qquad E^\circ_{cell} = 1.103 \text{ V}$$

Now suppose we connect the same cell to an external electric source of voltage greater than 1.103 V, as shown in Figure 21-21. That is, the connection is made so that electrons are forced into the zinc electrode (now the cathode) and removed from the copper electrode (now the anode). The overall reaction in this case is the *reverse* of the voltaic cell reaction, and E°_{cell} is *negative*.

Reduction: $\qquad Zn^{2+}(aq) + 2 e^- \longrightarrow Zn(s)$

Oxidation: $\qquad\qquad\qquad Cu(s) \longrightarrow Cu^{2+}(aq) + 2 e^-$

Overall: $\qquad Cu(s) + Zn^{2+}(aq) \longrightarrow Cu^{2+}(aq) + Zn(s)$

$$E^\circ_{cell} = E^\circ_{Zn^{2+}/Zn} - E^\circ_{Cu^{2+}/Cu} = -0.763 \text{ V} - 0.340 \text{ V} = -1.103 \text{ V}$$

Thus, by reversing the direction of the electron flow, we change the voltaic cell into an electrolytic cell.

Predicting Electrolysis Reactions

For the cell in Figure 21-21 to function as an electrolytic cell with reactants and products in their standard states, we noted that the external voltage had to exceed 1.103 V. We can make similar calculations for other electrolyses. What actually

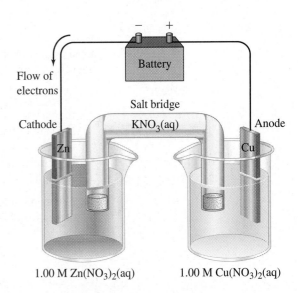

▶ **FIGURE 21-21**
An electrolytic cell
The direction of electron flow is the reverse of that in the voltaic cell of Figure 21-4, and so is the cell reaction. Now the zinc electrode is the *cathode* and the copper electrode, the *anode*. The battery must have a voltage in excess of 1.103 V in order to force electrons to flow in the reverse (nonspontaneous) direction.

Are You Wondering ...

Why the anode is (+) in an electrolytic cell but (−) in a voltaic cell?

Assigning the terms *anode* and *cathode* is not based on the electrode charges; it is based on the half-reactions at the electrode surfaces. Specifically,

- *Oxidation* always occurs at the *anode* of an electrochemical cell. Because of the buildup of electrons freed in the oxidation half-reaction, the anode of a *voltaic* cell is (−). Because electrons are withdrawn from it, the anode in an *electrolytic* cell is (+). For either type of cell, the anode is the electrode from which electrons *exit* the cell.

- *Reduction* always occurs at the cathode of an electrochemical cell. Because of the *removal* of electrons by the reduction half-reaction, the cathode of a *voltaic* cell is (+). Because of the electrons forced onto it, the cathode of an *electrolytic* cell is (−). For either type of cell, the cathode is the electrode at which electrons *enter* the cell.

The following table summarizes the relationship between a voltaic cell and an electrolytic cell.

	Voltaic Cell			**Electrolytic Cell**	
Oxidation	$A \longrightarrow A^+ + e^-$	Anode (negative)	Oxidation	$B \longrightarrow B^+ + e^-$	Anode (positive)
Reduction	$B^+ + e^- \longrightarrow B$	Cathode (positive)	Reduction	$A^+ + e^- \longrightarrow A$	Cathode (negative)
Overall	$A + B^+ \longrightarrow A^+ + B$ $\Delta G < 0$		Overall	$A^+ + B \longrightarrow A + B^+$ $\Delta G > 0$	
	Spontaneous redox reaction releases energy			Nonspontaneous redox reaction absorbs energy to drive it	
	The system (the cell) does work on the surroundings			The surroundings (the source of energy) do work on the system	

Note that the sign of each electrode in an electrolytic cell is the same as the sign of the battery electrode to which it is attached.

happens, however, does not always correspond to these calculations. We must consider four complicating factors:

1. A voltage significantly in excess of the calculated value, an **overpotential**, may be necessary to cause a particular electrode reaction to occur. Overpotentials are needed to overcome interactions at the electrode surface and are particularly common when gases are involved. For example, the overpotential for the discharge of $H_2(g)$ at a mercury cathode is approximately 1.5 V; the overpotential on a platinum cathode is practically zero.

2. Competing electrode reactions may occur. In the electrolysis of *molten* sodium chloride with inert electrodes, only one oxidation and one reduction are possible.

Reduction: $\quad 2 Na^+ + 2 e^- \longrightarrow 2 Na(l)$

Oxidation: $\qquad\qquad 2 Cl^- \longrightarrow Cl_2(g) + 2 e^-$

In the electrolysis of *aqueous* sodium chloride with inert electrodes, we must consider *two* possible reduction half-reactions and *two* oxidation half-reactions.

Reduction: $2 Na^+(aq) + 2 e^- \longrightarrow Na(s)$ $E^\circ_{Na^+/Na} = -2.71 \text{ V}$ (21.29)

$2 H_2O(l) + 2 e^- \longrightarrow H_2(g) + 2 OH^-(aq)$ $E^\circ_{H_2O/H_2} = (-0.83 \text{ V})$ (21.30)

Oxidation: $2 Cl^-(aq) \longrightarrow Cl_2(g) + 2 e^-$ $-E^\circ_{Cl_2/Cl^-} = -(1.36 \text{ V})$ (21.31)

$2 H_2O(l) \longrightarrow O_2(g) + 4 H^+(aq) + 4 e^-$ $-E^\circ_{O_2/H_2O} = -(1.23 \text{ V})$ (21.32)

▶ We have written a *minus* sign in front of the electrode potentials in (21.31) and (21.32) as a way of emphasizing the *oxidation* rather than the reduction tendency.

Electrolysis of Water animation

We can eliminate (21.29) as a possible reduction half-reaction: Unless the overpotential for $H_2(g)$ is unusually high, the reduction of Na^+ is far more difficult to accomplish than that of H_2O. This leaves two possibilities for the cell reaction.

(21.30) + (21.31):

Reduction: $2 H_2O(l) + 2 e^- \longrightarrow H_2(g) + 2 OH^-(aq)$

Oxidation: $2 Cl^-(aq) \longrightarrow Cl_2(g) + 2 e^-$

Overall: $2 Cl^-(aq) + 2 H_2O(l) \longrightarrow Cl_2(g) + H_2(g) + 2 OH^-(aq)$ (21.33)

$$E^\circ_{cell} = E^\circ_{H_2O/H_2} - E^\circ_{Cl_2/Cl^-} = -0.83 \text{ V} - (1.36 \text{ V}) = -2.19 \text{ V}$$

(21.30) + (21.32):

Reduction: $2 \{2 H_2O(l) + 2 e^- \longrightarrow H_2(g) + 2 OH^-(aq)\}$

Oxidation: $2 H_2O(l) \longrightarrow O_2(g) + 4 H^+(aq) + 4 e^-$

Overall: $2 H_2O(l) \longrightarrow 2 H_2(g) + O_2(g)$ (21.34)

$$E^\circ_{cell} = E^\circ_{H_2O/H_2} - E^\circ_{O_2/H_2O} = -0.83 \text{ V} - (1.23 \text{ V}) = -2.06 \text{ V}$$

In the electrolysis of NaCl(aq), we should expect $H_2(g)$ to be the product at the *cathode*. Because cell reactions (21.33) and (21.34) have E°_{cell} values that are so similar, we might expect a mixture of $Cl_2(g)$ and $O_2(g)$ at the *anode*. Actually, because of the high overpotential of $O_2(g)$ compared to $Cl_2(g)$, cell reaction (21.33) predominates; $Cl_2(g)$ is essentially the only product at the anode.

3. The reactants very often are in *nonstandard* states. In the industrial electrolysis of NaCl(aq), $[Cl^-] \approx 5.5 \text{ M}$, not the unit activity ($[Cl^-] \approx 1 \text{ M}$) implied in half-reaction (21.31); therefore $E_{Cl_2/Cl^-} = 1.31 \text{ V}$ (*not* 1.36 V). Also, the pH in the anode half-cell is adjusted to 4, not the unit activity ($[H_3O^+] \approx 1 \text{ M}$) implied in half-reaction (21.32); hence $E_{O_2/H_2O} = 0.99 \text{ V}$ (*not* 1.23 V). The net effect of these nonstandard conditions is to favor the production of O_2 at the

▲ The electrolysis of water into $H_2(g)$ and $O_2(g)$, shown by bubbles at the electrodes.

Are You Wondering...

What happens when no electrode reactions seem feasible in an aqueous solution?

Electrolysis requires that an electrolyte be present to carry current through the aqueous solution. If ions of the electrolyte are less easily oxidized and reduced than is water, $H_2O(l)$ will react at each electrode. The combination of half-reactions (21.30) and (21.32) occurs in the electrolysis of water: $2 H_2O(l) \longrightarrow 2 H_2(g) + O_2(g)$.

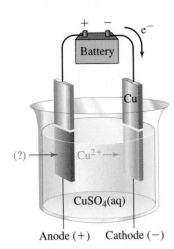

Anode (+)   Cathode (−)

▲ **FIGURE 21-22**
Predicting electrode reactions in electrolysis—Example 21-11 illustrated
Electrons are forced onto the copper cathode by the external source (battery). Cu^{2+} ions are attracted to the cathode and are reduced to Cu(s). The oxidation half-reaction depends on the metal used for the anode.

anode. In practice, however, the $Cl_2(g)$ obtained contains less than 1% $O_2(g)$, indicating the overpowering effect of the high overpotential of $O_2(g)$. Not surprisingly, the proportion of $O_2(g)$ increases significantly in the electrolysis of very dilute NaCl(aq).

4. The nature of the electrodes matters. An *inert* electrode, such as platinum, provides a surface on which an electrolysis half-reaction occurs, but the reactants themselves must come from the electrolyte solution. An *active* electrode is one that can itself participate in the oxidation or reduction half-reaction. The distinction between inert and active electrodes is explored in Figure 21-22 and Example 21-11.

EXAMPLE 21-11

Predicting Electrode Half-Reactions and Overall Reactions in Electrolysis. Refer to Figure 21-22. Predict the electrode reactions and the overall reaction when the anode is made of **(a)** copper and **(b)** platinum.

Solution

In both cases, the reduction of Cu^{2+}(aq) is the only likely reduction process.

$$Reduction: \quad Cu^{2+}(aq) + 2\,e^- \longrightarrow Cu(s) \qquad E^\circ_{Cu^{2+}/Cu} = 0.340 \text{ V}$$

(a) At the anode, Cu(s) can be oxidized to Cu^{2+}(aq), as represented by

$$Oxidation: \quad Cu(s) \longrightarrow Cu^{2+}(aq) + 2\,e^-$$

If we add the oxidation and reduction half-equations, Cu^{2+}(aq) cancels out. The electrolysis reaction is simply

$$Cu(s)[anode] \longrightarrow Cu(s)[cathode] \qquad (21.35)$$

$$E^\circ_{cell} = E^\circ_{Cu^{2+}/Cu} - E^\circ_{Cu^{2+}/Cu} = 0.340 \text{ V} - 0.340 \text{ V} = 0$$

Only a very small voltage to overcome the resistance in the electric circuit is required for this electrolysis. For every Cu atom that enters the solution at the anode, an *active* electrode, one Cu^{2+} ion deposits as a Cu atom at the cathode. Copper is transferred from the anode to the cathode through the solution as Cu^{2+}, and the concentration of $CuSO_4$(aq) remains *unchanged*.

(b) A platinum electrode is *inert*. It is not easily oxidized. The oxidation of SO_4^{2-} to $S_2O_8^{2-}$ also is not feasible ($-E^\circ_{S_2O_8^{2-}/SO_4^{2-}} = -2.01$ V). The oxidation that occurs most readily is that of H_2O, shown in reaction (21.32).

$$Oxidation: \quad 2\,H_2O(l) \longrightarrow O_2(g) + 4\,H^+(aq) + 4\,e^-$$

$$-E^\circ_{O_2/H_2O} = -1.23 \text{ V}$$

The electrolysis reaction and its E°_{cell} are

$$2\,Cu^{2+}(aq) + 2\,H_2O(l) \longrightarrow 2\,Cu(s) + 4\,H^+(aq) + O_2(g) \qquad (21.36)$$

$$E^\circ_{cell} = E^\circ(\text{reduction half-cell}) - E^\circ(\text{oxidation half-cell})$$
$$= E^\circ_{Cu^{2+}/Cu} - E^\circ_{O_2/H_2O}$$
$$= 0.340 \text{ V} - 1.23 \text{ V} = -0.89 \text{ V}$$

Practice Example A: Use data from Table 21.1 to predict the probable products when Pt electrodes are used in the electrolysis of KI(aq).

Practice Example B: In the electrolysis of $AgNO_3$(aq), what are the expected electrolysis products if the anode is silver metal and the cathode is platinum?

Quantitative Aspects of Electrolysis

We have seen how to calculate the theoretical voltage required for electrolysis. Equally important are calculations of the quantities of reactants consumed and products formed in an electrolysis. For these calculations, we will continue to use stoichiometric factors from the chemical equation, but another factor enters in as well: the quantity of electric charge associated with one mole of electrons. This factor is provided by the Faraday constant, which we can write as

$$1 \text{ mol e}^- = 96,485 \text{ C}$$

Generally, we do not measure electric charge directly; instead, we measure electric current. One *ampere* (A) of electric current represents the passage of 1 coulomb of charge per second (C/s). The product of current and time yields the total quantity of charge transferred.

$$\text{charge (C)} = \text{current (C/s)} \times \text{time (s)}$$

To determine the number of moles of electrons involved in an electrolysis reaction, we can write

$$\text{number of mol e}^- = \text{current}\left(\frac{\text{C}}{\text{s}}\right) \times \text{time (s)} \times \frac{1 \text{ mol e}^-}{96,485 \text{ C}}$$

As illustrated in Example 21-12, to determine the mass of a product in an electrolysis reaction, we can follow this conversion pathway.

$$\text{C/s} \longrightarrow \text{C} \longrightarrow \text{mol e}^- \longrightarrow \text{mol product} \longrightarrow \text{g product}$$

EXAMPLE 21-12

Calculating Quantities Associated with Electrolysis Reactions. The electrodeposition of copper can be used to determine the copper content of a sample. The sample is dissolved to produce Cu^{2+}(aq), which is electrolyzed. At the cathode, the reduction half-reaction is Cu^{2+}(aq) $+ 2 \text{ e}^- \longrightarrow$ Cu(s). What mass of copper can be deposited in 1.00 hour by a current of 1.62 A?

Solution

We can think of this as a two-part calculation. First we can determine the number of moles of electrons involved in the electrolysis in the manner outlined earlier.

$$1.00 \text{ h} \times \frac{60 \text{ min}}{1 \text{ h}} \times \frac{60 \text{ s}}{1 \text{ min}} \times \frac{1.62 \text{ C}}{1 \text{ s}} \times \frac{1 \text{ mol e}^-}{96,485 \text{ C}} = 0.0604 \text{ mol e}^-$$

Then we can calculate the mass of Cu(s) produced at the cathode by this number of moles of electrons. The key factor is printed in blue.

$$\text{mass of Cu} = 0.0604 \text{ mol e}^- \times \frac{1 \text{ mol Cu}}{2 \text{ mol e}^-} \times \frac{63.5 \text{ g Cu}}{1 \text{ mol Cu}} = 1.92 \text{ g Cu}$$

Practice Example A: If 12.3 g of Cu is deposited at the cathode of an electrolytic cell after 5.50 h, what was the current used?

Practice Example B: For how long would the electrolysis in Example 21-12 have to be carried out, using Pt electrodes and a current of 2.13 A, to produce 2.62 L O_2(g) at 26.2 °C and 738 mmHg pressure at the anode?

21-8 Industrial Electrolysis Processes

Modern industry could not function in its present form without electrolysis reactions. A number of the elements are produced almost exclusively by electrolysis—for example, aluminum, magnesium, chlorine, and fluorine. Among chemical compounds produced industrially by electrolysis are NaOH, $K_2Cr_2O_7$, $KMnO_4$, $Na_2S_2O_8$, and a number of organic compounds.

Electrorefining

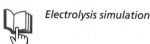

Electrolysis simulation

The *electrorefining* of metals involves the deposition of pure metal at a cathode, from a solution containing the metal ion. Copper produced by the smelting of copper ores is of sufficient purity for some uses, such as plumbing, but not for applications in which high electrical conductivity is required. For these applications, the copper must be more than 99.5% pure. The electrolysis reaction (21.35) on page 854 is used to obtain such high-purity copper.

A chunk of impure copper is taken as the anode, and a thin sheet of pure copper as the cathode. During the electrolysis, Cu^{2+} produced at the anode migrates through an aqueous sulfuric acid–copper sulfate solution to the cathode, where it is reduced to Cu(s). The pure copper cathode increases in size as the impure chunk of copper is consumed. As noted in Example 21-11a, the electrolysis is carried out at a low voltage—from 0.15 to 0.30 V. Under these conditions, Ag, Au, and Pt impurities are not oxidized at the anode, and they drop to the bottom of the tank as a sludge called *anode mud*. Sn, Bi, and Sb are oxidized, but they precipitate as oxides or hydroxides; Pb is oxidized but precipitates as $PbSO_4$(s). As, Fe, Ni, Co, and Zn are oxidized but form water-soluble species. Recovery of Ag, Au, and Pt from the anode mud helps offset the cost of the electrolysis.

▶ The refining of copper by electrolysis.

Electroplating

In *electroplating*, one metal is plated onto another, often less expensive, metal by electrolysis. This procedure is done for decorative purposes or to protect the underlying metal from corrosion. Silver-plated flatware, for example, consists of a thin coating of metallic silver on an underlying base of iron. In electroplating, the item to be plated is the cathode in an electrolytic cell. The electrolyte contains ions of the metal to be plated, which are attracted to the cathode, where they are reduced to metal atoms.

▲ A rack of metal parts being lifted from the electrolyte solution after electroplating.

In copper plating, the electrolyte is usually copper sulfate. In silver plating, it is commonly $K[Ag(CN)_2](aq)$. The concentration of free silver ion in a solution of the complex ion $[Ag(CN)_2]^-(aq)$ is very low, and electroplating under these conditions promotes a strongly adherent microcrystalline deposit of the metal. Chromium plating is useful for its resistance to corrosion as well as its appearance. Steel can be chromium-plated from an aqueous solution of CrO_3 and H_2SO_4. The plating obtained, however, is thin and porous and tends to develop cracks. In practice, the steel is first plated with a thin coat of copper or nickel, and then the chromium plating is applied. Chromium plating or cadmium plating is used to weatherproof machine parts. Metal plating can even be applied to some plastics. The plastic must first be made electrically conductive—for example, by coating it with graphite powder. Copper plating of plastics has been used to improve the quality of some microelectronic circuit boards. Electroplating is even used, quite literally, to make money. The U.S. penny is no longer copper throughout. A zinc plug is electroplated with a thin coat of copper, and the copper-plated plug is stamped to create a penny.

Electrosynthesis

Electrosynthesis is a method of producing substances through electrolysis reactions. It is useful for certain syntheses in which reaction conditions must be carefully controlled. Manganese dioxide occurs naturally as the mineral *pyrolusite*, but the small crystal size and lattice imperfections make this material inadequate for certain modern applications, such as alkaline batteries. The electrosynthesis of MnO_2 is carried out in a solution of $MnSO_4$ in $H_2SO_4(aq)$. Pure $MnO_2(s)$ is formed by the oxidation of Mn^{2+} at an inert anode (such as graphite).

Oxidation: $Mn^{2+}(aq) + 2 H_2O(l) \longrightarrow MnO_2(s) + 4 H^+(aq) + 2 e^-$

The reaction at the cathode is the reduction of H^+ to $H_2(g)$, and the overall electrolysis reaction is

$$Mn^{2+}(aq) + 2 H_2O(l) \longrightarrow MnO_2(s) + 2 H^+(aq) + H_2(g)$$

An example of electrosynthesis in organic chemistry is the reduction of acrylonitrile, $CH_2{=}CH{-}C{\equiv}N$, to adiponitrile, $N{\equiv}C(CH_2)_4C{\equiv}N$, at a lead cathode (chosen because of the high overpotential of H_2 on lead). Oxygen is released at the anode.

Reduction: $2 CH_2{=}CH{-}C{\equiv}N + 2 H_2O + 2 e^- \longrightarrow N{\equiv}C(CH_2)_4C{\equiv}N + 2 OH^-$

The commercial importance of this electrolysis is that adiponitrile can be readily converted to two other compounds: hexamethylenediamine, $H_2NCH_2(CH_2)_4CH_2NH_2$, and adipic acid, $HOOCCH_2(CH_2)_2CH_2COOH$. These two compounds are the monomers used to make the polymer *Nylon-66* (page 1107).

The Chlor–Alkali Process

On page 853 we described the electrolysis of NaCl(aq) through the reduction half-reaction (21.30) and the oxidation half-reaction (21.31).

$$2 Cl^-(aq) + 2 H_2O(l) \longrightarrow 2 OH^-(aq) + H_2(g) + Cl_2(g) \qquad E^\circ_{cell} = -2.19 \text{ V}$$

When conducted on an industrial scale, this electrolysis is called the *chlor–alkali process*, named after the two principal products: *chlor*ine and the *alkali* NaOH(aq). The chlor–alkali process is one of the most important of all electrolytic processes.

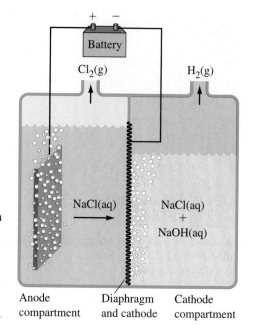

▶ **FIGURE 21-23**
A diaphragm chlor–alkali cell
The anode may be made of graphite or, in more modern technology, specially treated titanium metal. The diaphragm and cathode are generally fabricated as a composite unit consisting of asbestos or an asbestos–polymer mixture deposited on a steel wire mesh.

In the *diaphragm cell* depicted in Figure 21-23, $Cl_2(g)$ is produced in the anode compartment, and $H_2(g)$ and NaOH(aq) in the cathode compartment. If $Cl_2(g)$ comes in contact with NaOH(aq), the Cl_2 disproportionates into $ClO^-(aq)$, $ClO_3^-(aq)$, and $Cl^-(aq)$. The purpose of the diaphragm is to prevent this contact. The NaCl(aq) in the anode compartment is kept at a slightly higher level than that in the cathode compartment. This disparity creates a gradual flow of NaCl(aq) between the compartments and reduces the backflow of NaOH(aq) into the anode compartment. The solution in the cathode compartment, about 10–12% NaOH(aq) and 14–16% NaCl(aq), is concentrated and purified by evaporating some water and crystallizing the NaCl(s). The final product is 50% NaOH(aq), with up to 1% NaCl(aq).

The theoretical voltage required for this electrolysis is 2.19 V. However, as a result of the internal resistance of the cell and overpotentials at the electrodes, a voltage of about 3.5 V is used. The current is kept very high, typically about 1×10^5 A.

The NaOH(aq) produced in a diaphragm cell is not pure enough for certain uses such as rayon manufacture. A higher purity is achieved if electrolysis is carried out in a mercury cell, illustrated in Figure 21-24. This cell takes advantage of the high overpotential for the reduction of $H_2O(l)$ to $H_2(g)$ and $OH^-(aq)$ at a mercury cathode. The reduction that occurs instead is that of $Na^+(aq)$ to Na, which dissolves in Hg(l) to form an amalgam (Na–Hg alloy) with about 0.5% Na by mass.

$$2\,Na^+(aq) + 2\,Cl^-(aq) \longrightarrow 2\,Na(in\ Hg) + Cl_2(g) \qquad E^\circ_{cell} = -3.20\ V$$

When the Na amalgam is removed from the cell and treated with water, NaOH(aq) forms,

$$2\,Na(in\ Hg) + 2\,H_2O(l) \longrightarrow 2\,Na^+(aq) + 2\,OH^-(aq) + H_2(g) + Hg(l)$$

and the liquid mercury is recycled back to the electrolytic cell.

Although the mercury cell has the advantage of producing concentrated high-purity NaOH(aq), it has some disadvantages. The mercury cell requires a higher voltage (about 4.5 V) than does the diaphragm cell (3.5 V) and consumes more electrical energy, about 3400 kWh/ton Cl_2 in a mercury cell, compared with 2500 in a

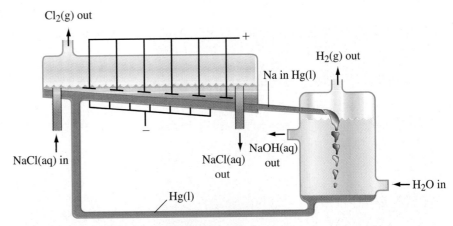

▲ **FIGURE 21-24   The mercury-cell chlor–alkali process**
The cathode is a layer of Hg(l) that flows along the bottom of the tank. Anodes, at which $Cl_2(g)$ forms, are situated in the NaCl(aq) just above the Hg(l). Sodium formed at the cathode dissolves in the Hg(l), and the sodium amalgam is decomposed with water to produce NaOH(aq) and $H_2(g)$. The regenerated Hg(l) is recycled.

diaphragm cell. Another serious drawback is the need to control mercury effluents to the environment. Mercury losses, which at one time were as high as 200 g Hg per ton Cl_2, have been reduced to about 0.25 g Hg per ton of Cl_2 in older plants and half this amount in new plants.

The ideal chlor–alkali process is one that is energy-efficient and does not use mercury. A type of cell offering these advantages is the *membrane cell*. In such a cell, the porous diaphragm of Figure 21-23 is replaced by a cation-exchange membrane, normally made of a fluorocarbon polymer. The membrane permits hydrated cations (Na^+ and H_3O^+) to pass between the anode and cathode compartments but severely restricts the backflow of Cl^- and OH^- ions.

Summary

An electrochemical cell consists of two half-cells in which electrodes are joined by a wire and the solutions are in contact, as through a salt bridge. The cell reaction involves oxidation at one electrode (anode) and reduction at the other electrode (cathode). A voltaic cell produces electricity from a spontaneous oxidation–reduction reaction. The difference in electric potential between the two electrodes is the voltage of the cell.

The reduction occurring at a standard hydrogen electrode (SHE), $2 H^+(1 M) + 2 e^- \longrightarrow H_2(g, 1\ atm)$, is arbitrarily assigned a potential of zero. A half-cell reaction with a *positive* reduction potential ($E°$) occurs more readily than does reduction of H^+ ions at the SHE. A *negative* standard reduction potential signifies a lesser tendency to undergo reduction. The voltage of a voltaic cell is the *difference* between $E°$ of the cathode and $E°$ of the anode; that is, $E°_{cell} = E°(cathode) - E°(anode)$. If $E_{cell} > 0$, the cell reaction is spontaneous in the forward direction for the stated conditions; if $E_{cell} < 0$, the forward reaction is *nonspontaneous*.

Cell voltages based on standard electrode potentials are $E°_{cell}$ values. Important relationships exist between $E°_{cell}$ and $\Delta G°$ and between $E°_{cell}$ and K_{eq}. In the Nernst equation, E_{cell} for nonstandard conditions is related to $E°_{cell}$ and the reaction quotient Q. Important applications of voltaic cells are found in various battery systems and in the phenomenon of corrosion and its control.

In an electrolytic cell, an external source of electricity forces electrons to flow in a direction opposite that in which they would flow spontaneously. As a result, electrolysis causes a normally nonspontaneous reaction to occur. $E°$ values are used to establish the theoretical voltage requirements, and the Faraday constant is used in calculating the amounts of reactants and products involved in an electrolysis. Electrolysis has many industrial applications, including electroplating and the commercial production of numerous substances, several of which are described in the chapter.

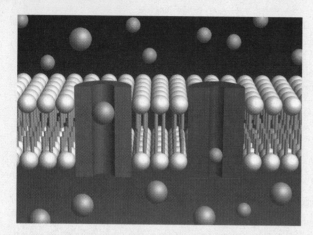

▲ A computer-generated model of a portion of a cell membrane. Na⁺ and K⁺ traverse the membrane through channels. The K⁺ channel is blue, and the Na⁺ channel is red.

Ordinary electrochemical cells consist of metallic conductors immersed in electrolyte solutions. In living systems, there are no metallic conductors. So what is the source of biological electric currents, such as those generated in muscle contraction and neuron activity? The answer is membrane potentials.

When a membrane separates two electrolyte solutions of different concentrations, in general there will be an electrical potential difference across the membrane, but various situations are possible. One possibility is that the membrane is impermeable to the ions and the solvent. In this case, inserting an electrode on each side of the membrane, joining the electrodes with a wire, and joining the solutions with a salt bridge would create a simple concentration cell. The cell potential would be given by the Nernst equation.

Figure 21-25 depicts solutions of KCl separated by a membrane. If the membrane is permeable to both K⁺ and Cl⁻ ions, their concentrations eventually become equal and there is no potential difference (Figure 21-25a). If, however, the membrane is impermeable to solvent* and permeable only to potassium ions, a net flow of K⁺ ions crosses the membrane from the more concentrated side (left-hand side of membrane in Figure 21-25b) toward the less concentrated side. As a result, the left side has a negative potential (excess anions) with respect to

*This restriction is made so that there will be no complications due to osmotic pressure.

the right. The effect of this potential difference is to prevent more K⁺ions from crossing the membrane. An equilibrium is established when the electric potential difference exactly balances the tendency of the concentrations to become equal. This potential difference is known as the *Nernst potential*, $\Delta\Phi$.

Membrane

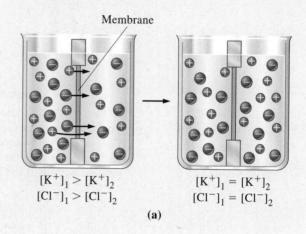

$$[K^+]_1 > [K^+]_2$$
$$[Cl^-]_1 > [Cl^-]_2$$

$$[K^+]_1 = [K^+]_2$$
$$[Cl^-]_1 = [Cl^-]_2$$

(a)

Net negative charge   Net positive charge

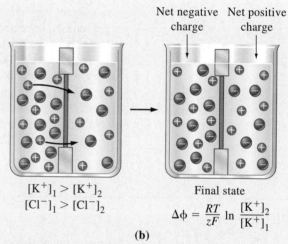

$$[K^+]_1 > [K^+]_2$$
$$[Cl^-]_1 > [Cl^-]_2$$

Final state

$$\Delta\phi = \frac{RT}{zF} \ln \frac{[K^+]_2}{[K^+]_1}$$

(b)

FIGURE 21-25

Generation of membrane potential
(a) Solutions of KCl separated by a membrane that is permeable to both K⁺ and Cl⁻. **(b)** Solutions of KCl separated by a membrane that is permeable only to K⁺. The change in [K⁺] is greatly exaggerated for the final state. Actually, [K⁺]₂ is only slightly greater than [K⁺]₁.

The Gibbs energy difference ΔG_e arising from the potential difference $\Delta \Phi$ is

$$\Delta G_e = zF\Delta \Phi$$

where z is the charge on the permeable ions ($z = 1$ for K^+) and F is the Faraday constant. In Figure 21-25b, the potential is higher on the right-hand side because more K^+ ions are there than on the left-hand side. Viewed in terms of the K^+ ions on the right side, $\Delta G_e > 0$. Also, an additional Gibbs energy difference arises from the concentration difference between the compartments.

$$\Delta G_c = RT \ln \frac{[K^+]_1}{[K^+]_2}$$

At equilibrium, the net ΔG across the membrane is zero. Therefore,

$$\Delta G = \Delta G_e + \Delta G_c = zF\Delta \Phi + RT \ln \frac{[K^+]_1}{[K^+]_2} = 0$$

Thus, the Nernst potential is given by

$$\Delta \Phi = \frac{RT}{zF} \ln \frac{[K^+]_2}{[K^+]_1}$$

Mammalian muscle cells are freely permeable to K^+ ions but much less permeable to Na^+ and Cl^- ions. Typical concentrations of K^+ ions are $[K^+] = 155$ mM inside the cells and 4.00 mM outside. The Nernst potential at 310 K (37 °C) is

$$\Delta \Phi = \frac{8.3145 \text{ J K}^{-1} \text{ mol}^{-1} \times 310 \text{ K}}{96,500 \text{ C mol}^{-1}} \ln \frac{4}{155}$$

$$= -0.0977 \text{ V} = -97.7 \text{ mV}$$

The contents of each cell in a multicellular organism are separated from other cells by a semipermeable membrane. The potential is negative inside the cell and positive outside. In actual measurements, such potentials are more like -85 mV because there is a certain amount of diffusion of Na^+ and Cl^- ions, and there is also a biological pumping mechanism that does not allow the potential difference to stray too far from a desired value. The potential difference measured is that between the interior and the exterior of the cell and is called the *transmembrane potential*.

Measurements of transmembrane potentials of various cells range from -30 to -100 mV, with the cell interior at a lower potential than the exterior. A typical value for a nerve cell (squid) undergoing no activity—a resting cell—is -70 mV. Nerve cells are permeable to all three ions: K^+, Na^+, and Cl^-. The observed concentrations of these ions are given in Table 21.2, together with the potentials calculated from the Nernst potential equation and the experimental and expected ratios of concentrations.

At a resting potential of -70 mV, Cl^- ion is nearly in equilibrium, but the Na^+ and K^+ are not. From the ratios in the table, we see that Na^+ ions spontaneously flow into the cell and K^+ ions spontaneously flow out. The observed steady-state concentrations of the Na^+ and K^+ are maintained by the active transport processes of the cell, which use some of the cell's metabolic energy to pump the Na^+ out of the cell and K^+ into the cell continuously. The cells maintain the Na^+ and K^+ in a state of nonequilibrium. At equilibrium, nothing happens. A cell that is at equilibrium is dead!

During a nerve impulse, protein activity in the membrane allows Na^+ to enter the cell and K^+ to leave, leading to a change in membrane potential toward $+60$ mV. The cell is said to be *depolarized*. After the peak of 60 mV is reached, the cell returns to its resting potential by pumping the Na^+ ions out and the K^+ ions in. The mechanisms by which cell membranes change their permeability to ions are an area of active research.

TABLE 21.2 Ion Concentrations and Membrane Potentials for a Typical Nerve Cell

Ion	External Concentration (mM)	Internal Concentration (mM)	Estimated Membrane Potential (mV)	Ratio of $[ion]_{ext}/[ion]_{int}$ Observed	Ratio of $[ion]_{ext}/[ion]_{int}$ Expected for -70 mV
Na^+	460	49	+57	9:1	1:15
K^+	10	410	-95	1:41	1:15
Cl^-	540	40	-67	14:1	15:1

Integrative Example

Two electrochemical cells are connected as shown.

Cell A

$$Zn(s)|Zn^{2+}(0.85\ M)|\ |Cu^{2+}(1.10\ M)|Cu(s)$$

$$Zn(s)|Zn^{2+}(1.05\ M)|\ |Cu^{2+}(0.75\ M)|Cu(s)$$

Cell B

1. Do electrons flow in the direction of the red arrows or the blue arrows?

2. What are the ion concentrations in the half-cells at the point at which current ceases to flow?

1. The two cells differ only in their ion concentrations. Thus, they have the same E°_{cell} value but different E_{cell} values. Consider how each cell would function alone as a voltaic cell. Zinc is the anode, copper is the cathode, and the cell reaction is

$$Zn(s) + Cu^{2+}(aq) \longrightarrow Zn^{2+}(aq) + Cu(s)$$

The E_{cell} values are given by the Nernst equation.

$$E_{cell} = E^\circ_{cell} - (0.0592/2) \log[Zn^{2+}]/[Cu^2] \quad (21.37)$$

Note that for Cell A, $[Zn^{2+}]/[Cu^{2+}] = 0.85\ M/1.10\ M < 1$, $\log[Zn^{2+}][Cu^{2+}] < 0$, and $E_{cell} > E^\circ_{cell}$. For Cell B, $[Zn^{2+}]/[Cu^{2+}] = 1.05\ M/0.75\ M > 1$, $\log[Zn^{2+}][Cu^{2+}] > 0$, and $E_{cell} < E^\circ_{cell}$. The voltage of Cell A is greater than that of Cell B.

In the connection of the two cells shown in the diagram, Cell

A is a *voltaic cell* and Cell B is an *electrolytic cell*. There is an emf from Cell B that resists that of Cell A, but on balance, the electron flow is in the direction of the red arrows.

2. As electrons flow between the two cells, the overall reaction in Cell A causes $[Zn^{2+}]$ to increase, $[Cu^{2+}]$ to decrease, and E_{cell} to decrease. The overall reaction in Cell B causes $[Zn^{2+}]$ to decrease, $[Cu^{2+}]$ to increase, and the back emf to increase. When the back emf from Cell B equals E_{cell} of Cell A, electrons cease to flow. At this point, the cell diagrams for Cell A and Cell B where x represents the *changes* in concentrations that occur to reach this balanced condition, are

Cell A: $Zn(s)|Zn^{2+}(0.85 + x)M||Cu^{2+}(1.10 - x)M|Cu(s)$

Cell B: $Zn(s)|Zn^{2+}(1.05 - x)M||Cu^{2+}(0.75 + x)M|Cu(s)$

Now we can use equation (21.37) to obtain E_{cell} for each cell, set the two expressions equal to one another, and cancel terms such as E°_{cell}, 0.0592/2, and the log function. The expression that remains is

$$(0.85 + x)/(1.10 - x) = (1.05 - x)/(0.75 + x)$$

which yields the value $x = 0.14$ when solved. When electrons no longer flow, the ion concentrations are as follows: for Cell A, $[Zn^{2+}] = 0.99\ M$; $[Cu^{2+}] = 0.96\ M$ for Cell B, $[Zn^{2+}] = 0.91\ M$; $[Cu^{2+}] = 0.89\ M$.

Key Terms

anode (21-1)
battery (21-5)
cathode (21-1)
cathodic protection (21-6)
cell diagram (21-1)
cell potential (cell voltage), E_{cell} (21-1)
concentration cell (21-4)
electrochemical cell (21-1)
electrode (21-1)

electrolysis (21-7)
electrolytic cell (21-7)
electromotive force (emf) (21-1)
Faraday constant, *F* (21-3)
flow battery (21-5)
fuel cell (21-5)
half-cell (21-1)
Nernst equation (21-4)
overpotential (21-7)

primary battery (21-5)
salt bridge (21-1)
secondary battery (21-5)
standard cell potential, E°_{cell} (21-2)
standard electrode potential, E° (21-2)
standard hydrogen electrode
  (SHE) (21-2)
volt (V) (21-1)
voltaic (galvanic) cell (21-1)

Review Questions

1. In your own words, define the following symbols or terms: **(a)** E°; **(b)** F; **(c)** anode; **(d)** cathode.

2. Briefly describe each of the following ideas, methods, or devices: **(a)** salt bridge; **(b)** standard hydrogen electrode (SHE); **(c)** cathodic protection; **(d)** fuel cell.

3. Explain the important distinctions between each pair of terms: **(a)** half-reaction and overall cell reaction; **(b)** voltaic cell and electrolytic cell; **(c)** primary battery and secondary battery; **(d)** E_{cell} and E°_{cell}.

4. Only one of the following statements is correct concerning a voltaic cell. Identify the correct statement, and tell what is wrong with the other statements. **(a)** Electrons move from the cathode to the anode; **(b)** electrons move through the salt bridge; **(c)** electrons leave the cell from either the

cathode or the anode, depending on what electrodes are used; **(d)** reduction occurs at the cathode.

5. Write equations for the oxidation and reduction half-reactions and for the overall redox reaction that describe

(a) the displacement of $Cu^{2+}(aq)$ by $Fe(s)$
(b) the oxidation of Br^- to $Br_2(aq)$ by $Cl_2(aq)$
(c) the reduction of $Fe^{3+}(aq)$ to $Fe^{2+}(aq)$ by $Al(s)$
(d) the oxidation of $Cl^-(aq)$ to $ClO_3^-(aq)$ by MnO_4^- in acidic solution (MnO_4^- is reduced to Mn^{2+})
(e) the oxidation of $S^{2-}(aq)$ to $SO_4^{2-}(aq)$ by $O_2(g)$ in basic solution

6. The sketch is of a voltaic cell consisting of a standard zinc electrode and a standard electrode of a second metal, M. For the reduction half-reaction $Zn^{2+}(1\ M) + 2\ e^- \longrightarrow$

Zn(s), $E° = -0.763$ V. For each of the given metals M and measured values of $E°_{cell}$, determine $E°$ for the reduction half-reaction

$$M^{2+}(1 \text{ M}) + 2 \text{ e}^- \longrightarrow M(s) \qquad E° = ?$$

(a) Mn, $E°_{cell} = 0.417$ V
(b) Po, $E°_{cell} = -1.13$ V
(c) Ti, $E°_{cell} = 0.87$ V
(d) V, $E°_{cell} = 0.37$ V

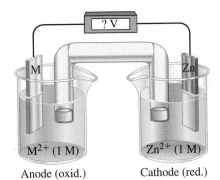

Anode (oxid.)   Cathode (red.)

7. Write cell reactions for the electrochemical cells diagrammed here, and use data from Table 21.1 to calculate $E°_{cell}$ for each reaction.
(a) $Al(s)|Al^{3+}(aq)||Sn^{2+}(aq)|Sn(s)$
(b) $Pt(s)|Fe^{2+}(aq), Fe^{3+}(aq)||Ag^+(aq)|Ag(s)$

8. From the data listed, together with data from Table 21.1, determine the quantity indicated.
(a) $E°_{cell}$ for the cell
$$Pt(s)|Cl_2(g)|Cl^-(aq)||Pb^{2+}(aq), H^+(aq)|PbO_2(s)$$
(b) $E°$ for the couple Sc^{3+}/Sc, given that
$$Mg(s)|Mg^{2+}(aq)||Sc^{3+}(aq)|Sc(s) \qquad E°_{cell} = 0.33 \text{ V}$$
(c) $E°$ for the couple Cu^{2+}/Cu^+, given that
$$Pt(s)|Cu^+(aq), Cu^{2+}(aq)||Ag^+(aq)|Ag(s)$$
$$E°_{cell} = 0.641 \text{ V}$$

9. Assume that all reactants and products are in their standard states, and use data from Table 21.1 to predict whether a spontaneous reaction will occur in the forward direction in each case.
(a) $Sn(s) + Pb^{2+}(aq) \longrightarrow Sn^{2+}(aq) + Pb(s)$
(b) $Cu^{2+}(aq) + 2 I^-(aq) \longrightarrow Cu(s) + I_2(s)$
(c) $4 NO_3^-(aq) + 4 H^+(aq) \longrightarrow$
$$3 O_2(g) + 4 NO(g) + 2 H_2O(l)$$
(d) $O_3(g) + Cl^-(aq) \longrightarrow$
$$OCl^-(aq) + O_2(g) \text{ (basic solution)}$$

10. For the reduction half-reaction $Hg_2^{2+}(aq) + 2 \text{ e}^- \longrightarrow$ $2 Hg(l)$, $E° = 0.797$ V. Will Hg(l) react with and dissolve in HCl(aq)? in HNO$_3$(aq)? Explain.

11. Use data from Table 21.1 to predict whether, to any significant extent,
(a) Mg(s) will displace Pb^{2+} from aqueous solution.
(b) Tin will react with and dissolve in 1 M HCl.
(c) SO_4^{2-} will oxidize Sn^{2+} to Sn^{4+} in acidic solution.
(d) MnO_4^-(aq) will oxidize H_2O_2(aq) to O_2(g) in acidic solution.
(e) I_2 will displace Br^-(aq) to produce Br_2(l).

12. Consider the reaction $Co(s) + Ni^{2+}(aq) \longrightarrow$ $Co^{2+}(aq) + Ni(s)$, with $E°_{cell} = 0.02$ V. If Co(s) is added to a solution with $[Ni^{2+}] = 1$ M, should the reaction go to completion? Explain.

13. Dichromate ion $(Cr_2O_7^{2-})$ in acidic solution is a good oxidizing agent. Which of the following oxidations can be accomplished with dichromate ion in acidic solution? Explain.
(a) Sn^{2+}(aq) to Sn^{4+}(aq)
(b) I_2(s) to IO_3^-(aq)
(c) Mn^{2+}(aq) to MnO_4^-(aq)

14. Determine the values of $\Delta G°$ for the following reactions carried out in voltaic cells.
(a) $2 Al(s) + 3 Cu^{2+}(aq) \longrightarrow 2 Al^{3+}(aq) + 3 Cu(s)$
(b) $O_2(g) + 4 I^-(aq) + 4 H^+(aq) \longrightarrow$
$$2 H_2O(l) + 2 I_2(s)$$
(c) $Cr_2O_7^{2-}(aq) + 14 H^+(aq) + 6 Ag(s) \longrightarrow$
$$2 Cr^{3+}(aq) + 6 Ag^+(aq) + 7 H_2O(l)$$

15. Write the equilibrium constant expression for each of the following reactions, and determine the value of K_{eq} at 25 °C. Use data from Table 21.1.
(a) $Sn^{4+}(aq) + 2 Ag(s) \longrightarrow Sn^{2+}(aq) + 2 Ag^+(aq)$
(b) $MnO_2(s) + 4 H^+(aq) + 2 Cl^-(aq) \longrightarrow$
$$Mn^{2+}(aq) + 2 H_2O(l) + Cl_2(g)$$
(c) $2 OCl^-(aq) \longrightarrow 2 Cl^-(aq) + O_2(g) \text{ (basic solution)}$

16. For each of the following combinations of electrodes (A and B) and solutions, indicate
- the overall cell reaction
- the direction in which electrons flow spontaneously (from A to B or from B to A)
- the magnitude of the voltage read on the voltmeter, V

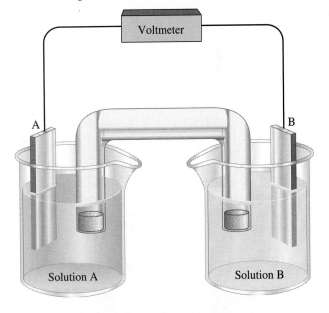

	A Solution A		B Solution B
(a)	Cu	1.0 M Cu^{2+}	Fe 1.0 M Fe^{2+}
(b)	Pt	1.0 M Sn^{2+}/1.0 M Sn^{4+}	Ag 1.0 M Ag^+
(c)	Zn	0.10 M Zn^{2+}	Fe 1.0×10^{-3} M Fe^{2+}

17. A voltaic cell represented by the following cell diagram has $E_{cell} = 1.250$ V. What must be $[Ag^+]$ in the cell?
$$Zn(s)\,|\,Zn^{2+}(1.00\ M)\,\|\,Ag^+(x\ M)\,|\,Ag(s)$$

18. For the cell pictured in Figure 21-11, what is E_{cell} if the unknown solution in the half-cell on the left **(a)** has pH = 5.25; **(b)** is 0.0103 M HCl; **(c)** is 0.158 M $HC_2H_3O_2$ ($K_a = 1.8 \times 10^{-5}$)?

19. Use data from Table 21.1, as necessary, to predict the probable products when Pt electrodes are used in the electrolysis of **(a)** $CuCl_2(aq)$; **(b)** $Na_2SO_4(aq)$; **(c)** $BaCl_2(l)$; **(d)** KOH(aq).

20. The commercial production of magnesium involves the electrolysis of molten $MgCl_2$. Why not use the electrolysis of $MgCl_2(aq)$ instead?

21. How many grams of metal are deposited at the cathode by the passage of 2.15 A of current for 75 min in the electrolysis of an aqueous solution containing **(a)** Zn^{2+}; **(b)** Al^{3+}; **(c)** Ag^+; **(d)** Ni^{2+}?

22. A quantity of electric charge brings about the deposition of 3.28 g Cu at a cathode during the electrolysis of a solution containing $Cu^{2+}(aq)$. What volume of $H_2(g)$, measured at 28.2 °C and 763 mmHg, would be produced by this same quantity of electric charge in the reduction of $H^+(aq)$ at a cathode?

Exercises   *(Use data from Table 21.1 and Appendix D as necessary.)*

Standard Electrode Potentials

23. From the observations listed, estimate the value of $E°$ for the half-reaction $M^{2+}(aq) + 2\,e^- \longrightarrow M(s)$.
(a) The metal M reacts with $HNO_3(aq)$ but not with HCl(aq); M displaces $Ag^+(aq)$ but not $Cu^{2+}(aq)$.
(b) The metal M reacts with HCl(aq), producing $H_2(g)$, but displaces neither $Zn^{2+}(aq)$ nor $Fe^{2+}(aq)$.

24. You must estimate $E°$ for the half-reaction $In^{3+}(aq) + 3\,e^- \longrightarrow In(s)$. You have no electrical equipment, but you do have all of the metals listed in Table 21.1 and aqueous solutions of their ions, as well as In(s) and $In^{3+}(aq)$. Describe the experiments you would perform and the accuracy you would expect in your result.

25. $E_{cell}° = 0.201$ V for the reaction
$$3\,Pt + 12\,Cl^- + 2\,NO_3^- + 8\,H^+ \longrightarrow$$
$$3\,[PtCl_4]^{2-} + 2\,NO(g) + 4\,H_2O(l)$$
What is $E°$ for the reduction of $[PtCl_4]^{2-}$ to Pt in acidic solution?

26. Given that $E_{cell}° = 3.20$ V for the reaction
$$2\,Na(in\ Hg) + Cl_2(g) \longrightarrow 2\,Na^+(aq) + 2\,Cl^-(aq)$$
what is $E°$ for the reduction $2\,Na^+(aq) + 2\,e^- \longrightarrow 2\,Na(in\ Hg)$?

27. Given that $E_{cell}°$ for the aluminum–air battery is 2.71 V, what is $E°$ for the reduction half-reaction $[Al(OH)_4]^- + 3\,e^- \longrightarrow Al(s) + 4\,OH^-(aq)$? [*Hint:* Refer to cell reaction (21.28).]

28. The theoretical $E_{cell}°$ for the methane–oxygen fuel cell is 1.06 V. What is $E°$ for the reduction half-reaction $CO_2(g) + 8\,H^+(aq) + 8\,e^- \longrightarrow CH_4(g) + 2\,H_2O(l)$? [*Hint:* Refer to cell reaction (21.27).]

Predicting Oxidation–Reduction Reactions

29. According to standard electrode potentials, both sodium and aluminum should displace zinc from an aqueous solution of Zn^{2+}. What is observed, however, is that aluminum does displace Zn^{2+} and sodium does not. Explain this difference in behavior.

30. In the following photograph, copper and zinc strips are joined and immersed in 1 M HCl. What reaction occurs? Why are gas bubbles evident on both metal strips?

31. Predict whether the following metals will react with the acid indicated. If a reaction does occur, write the net ionic equation for the reaction. Assume that reactants and products are in their standard states. **(a)** Ag in $HNO_3(aq)$; **(b)** Zn in HI(aq); **(c)** Au in HNO_3 (for the couple Au^{3+}/Au, $E° = 1.52$ V).

32. Predict whether, to any significant extent,
(a) Fe(s) will displace $Zn^{2+}(aq)$.
(b) MnO_4^- will oxidize $Cl^-(aq)$ to $Cl_2(g)$ in acidic solution.
(c) Ag(s) will react with 1 M HCl(aq).
(d) $O_2(g)$ will oxidize $Cl^-(aq)$ to $Cl_2(g)$ in acidic solution.

Voltaic Cells

33. In each of the following examples, sketch a voltaic cell that uses the given reaction. Label the anode and cathode; indicate the direction of electron flow; write a balanced equation for the cell reaction; and calculate E_{cell}°.
  (a) $Cu(s) + Fe^{3+}(aq) \longrightarrow Cu^{2+}(aq) + Fe^{2+}(aq)$
  (b) $Pb^{2+}(aq)$ is displaced from solution by $Al(s)$.
  (c) $Cl_2(g) + H_2O(l) \longrightarrow Cl^-(aq) + O_2(g) + H^+(aq)$
  (d) $Zn(s) + H^+ + NO_3^- \longrightarrow Zn^{2+} + H_2O(l) + NO(g)$

34. Diagram the cell, and calculate the value of E_{cell}° for a voltaic cell in which
  (a) $Cl_2(g)$ is reduced to $Cl^-(aq)$ and $Fe(s)$ is oxidized to $Fe^{2+}(aq)$.
  (b) $Ag^+(aq)$ is displaced from solution by $Zn(s)$.
  (c) The cell reaction is $2\,Cu^+(aq) \longrightarrow Cu^{2+}(aq) + Cu(s)$.
  (d) $MgBr_2(aq)$ is produced from $Mg(s)$ and $Br_2(l)$.

ΔG°, ΔE_{cell}°, and K_{eq}

35. For the reaction
$$5\,H_2O_2(aq) + 2\,Mn^{2+}(aq) \longrightarrow$$
$$2\,MnO_4^-(aq) + 6\,H^+(aq) + 2\,H_2O(l)$$
use data from Table 21.1 to determine **(a)** E_{cell}°; **(b)** ΔG°; **(c)** K_{eq}; **(d)** whether the reaction goes substantially to completion when the reactants and products are initially in their standard states.

36. For the reaction that occurs in the voltaic cell
$$Fe(s)\,|\,Fe^{2+}(aq)\,\|\,Cr^{3+}(aq),\,Cr^{2+}(aq)\,|\,Pt(s)$$
use data from Appendix D to determine **(a)** the equation for the cell reaction; **(b)** E_{cell}°; **(c)** ΔG°; **(d)** K_{eq}; **(e)** whether the reaction goes substantially to completion when the reactants and products are initially in their standard states.

37. For the reaction $2\,Cu^+(aq) + Sn^{4+}(aq) \longrightarrow$
$$2\,Cu^{2+}(aq) + Sn^{2+}(aq),\,E_{cell}^{\circ} = -0.0050\,V,$$
  (a) Can a solution be prepared at 298 K that is 0.500 M in each of the four ions?

(b) If not, in what direction will a reaction occur?

38. For the reaction $2\,V^{3+}(aq) + Ni(s) \longrightarrow$
$$2\,V^{2+}(aq) + Ni^{2+}(aq),\,E_{cell}^{\circ} = 0.0020\,V,$$
  (a) Can a solution be prepared at 298 K that has $[V^{2+}] = 0.600\,M$ and $[V^{3+}] = [Ni^{2+}] = 0.675\,M$?
  (b) If not, in what direction will a reaction occur?

39. Use thermodynamic data from Appendix D to calculate a theoretical voltage of the silver–zinc button cell described on page 846.

40. The theoretical voltage of the aluminum–air battery is $E_{cell}^{\circ} = 2.71\,V$. Use data from Appendix D and equation (21.28) to determine ΔG_f° for $[Al(OH)_4]^-$.

41. By the method illustrated on page 834, determine $E_{IrO_2/Ir}^{\circ}$, given that $E_{Ir^{3+}/Ir}^{\circ} = 1.156\,V$ and $E_{IrO_2/Ir^{3+}}^{\circ} = 0.223\,V$.

42. In a manner similar to that used on page 834, determine $E_{MoO_2/Mo^{3+}}^{\circ}$, given that $E_{H_2MoO_4/MoO_2}^{\circ} = 0.646\,V$ and $E_{H_2MoO_4/Mo^{3+}}^{\circ} = 0.428\,V$.

Concentration Dependence of E_{cell}—The Nernst Equation

43. Use the Nernst equation and Table 21.1 to calculate E_{cell} for each of the following cells.
  (a) $Al(s)\,|\,Al^{3+}(0.18\,M)\,\|\,Fe^{2+}(0.85\,M)\,|\,Fe(s)$
  (b) $Ag(s)\,|\,Ag^+(0.34\,M)\,\|\,Cl^-(0.098\,M)\,|\,Cl_2(g,\,0.55\,atm)\,|$ $Pt(s)$

44. Use the Nernst equation and data from Appendix D to calculate E_{cell} for each of the following cells.
  (a) $Mn(s)\,|\,Mn^{2+}(0.40\,M)\,\|\,Cr^{3+}(0.35\,M),\,Cr^{2+}(0.25\,M)\,|\,Pt(s)$
  (b) $Mg(s)\,|\,Mg^{2+}(0.016\,M)\,\|\,[Al(OH)_4]^-(0.25\,M),$
$$OH^-(0.042\,M)\,|\,Al(s)$$

45. Consider the reduction half-reactions listed in Appendix D, and give plausible explanations for the following observations:
  (a) For some half-reactions, E depends on pH; for others, it does not.
  (b) Whenever H^+ appears in a half-equation, that ion is on the *left* side.
  (c) Whenever OH^- appears in a half-equation, that ion is on the *right* side.

46. Write an equation to represent the oxidation of $Cl^-(aq)$ to $Cl_2(g)$ by $PbO_2(s)$ in an acidic solution. Will this reaction occur spontaneously in the forward direction if all other

reactants and products are in their standard states and
  (a) $[H^+] = 6.0\,M$; **(b)** $[H^+] = 1.2\,M$; **(c)** pH = 4.25? Explain.

47. If $[Zn^{2+}]$ is maintained at 1.0 M,
  (a) What is the minimum $[Cu^{2+}]$ for which reaction (21.3) is spontaneous in the forward direction?
  (b) Should the displacement of $Cu^{2+}(aq)$ by $Zn(s)$ go to completion? Explain.

48. Can the displacement of $Pb(s)$ from 1.0 M $Pb(NO_3)_2$ be carried to completion by tin metal? Explain.

49. A concentration cell is constructed of two hydrogen electrodes: one immersed in a solution with $[H^+] = 1.0\,M$ and the other in 0.65 M KOH.
  (a) Determine E_{cell} for the reaction that occurs.
  (b) Compare this value of E_{cell} with E° for the reduction of H_2O to $H_2(g)$ in basic solution, and explain the relationship between them.

50. If the 0.65 M KOH of Exercise 49 is replaced by 0.65 M NH_3,
  (a) Will E_{cell} be higher or lower than in the cell with 0.65 M KOH?
  (b) What will be the value of E_{cell}?

51. A voltaic cell is constructed as follows:
$$Ag(s)\,|\,Ag^+(satd\ Ag_2CrO_4)\,||\,Ag^+(0.125\ M)\,|\,Ag(s)$$
What is the value of E_{cell}? For Ag_2CrO_4, $K_{sp} = 1.1 \times 10^{-12}$.

52. A voltaic cell, with $E_{cell} = 0.180\ V$, is constructed as follows:
$$Ag(s)\,|\,Ag^+(satd\ Ag_3PO_4)\,||\,Ag^+(0.140\ M)\,|\,Ag(s)$$
What is the K_{sp} of Ag_3PO_4?

53. For the voltaic cell
$$Sn(s)\,|\,Sn^{2+}(0.075\ M)\,||\,Pb^{2+}(0.600\ M)\,|\,Pb(s)$$
(a) What is E_{cell} initially?
(b) If the cell is allowed to operate spontaneously, will E_{cell} increase, decrease, or remain constant with time? Explain.
(c) What will be E_{cell} when $[Pb^{2+}]$ has fallen to 0.500 M?
(d) What will be $[Sn^{2+}]$ at the point at which $E_{cell} = 0.020\ V$?
(e) What are the ion concentrations when $E_{cell} = 0$?

54. For the voltaic cell
$$Ag(s)\,|\,Ag^+(0.015\ M)\,||\,Fe^{3+}(0.055\ M),$$
$$Fe^{2+}(0.045\ M)\,|\,Pt(s)$$
(a) What is E_{cell} initially?
(b) As the cell operates, will E_{cell} increase, decrease, or remain constant with time? Explain.
(c) What will be E_{cell} when $[Ag^+]$ has increased to 0.020 M?
(d) What will be $[Ag^+]$ when $E_{cell} = 0.010\ V$?
(e) What are the ion concentrations when $E_{cell} = 0$?

55. Show that the oxidation of $Cl^-(aq)$ to $Cl_2(g)$ by $Cr_2O_7^{2-}(aq)$ in acidic solution, with reactants and products in their standard states, does not occur spontaneously. Explain why it is still possible to use this method to produce $Cl_2(g)$ in the laboratory? What experimental conditions would you use?

56. Refer to the Integrative Example on page 862. Calculate the net voltage of the current entering electrolytic cell B when that cell is first connected to the voltaic cell A.

Batteries and Fuel Cells

57. The iron–chromium redox battery makes use of the reaction
$$Cr^{2+}(aq) + Fe^{3+}(aq) \longrightarrow Cr^{3+}(aq) + Fe^{2+}(aq)$$
occurring at a chromium anode and an iron cathode.
(a) Write a cell diagram for this battery.
(b) Calculate the theoretical voltage of the battery.

58. Refer to the discussion of the Leclanché cell (page 844).
(a) Combine the several equations written for the operation of the Leclanché cell into a single overall equation.
(b) Given that the voltage of the Leclanché cell is 1.55 V, estimate the electrode potentials, E, for each of the half-reactions. Why are your values only estimates?

59. What is the theoretical standard cell voltage, E_{cell}°, of each of the following voltaic cells: **(a)** the hydrogen–oxygen fuel cell described by equation (21.26); **(b)** the zinc–air battery; **(c)** a magnesium–iodine battery?

60. For the alkaline Leclanché cell (page 845),
(a) Write the overall cell reaction.
(b) Determine E_{cell}° for that cell reaction.

61. One of the advantages of the aluminum–air battery over the iron–air and zinc–air batteries is the greater quantity of charge transferred per unit mass of metal consumed. Show that this is indeed the case. Assume that zinc and iron are oxidized to oxidation state +2 in air batteries.

62. Describe how you might construct batteries with each of the following voltages: **(a)** 0.10 V; **(b)** 2.5 V; **(c)** 10.0 V. Be as specific as you can about the electrodes and solution concentrations you would use, and indicate whether the battery would consist of a single cell or two or more cells connected in series.

Electrochemical Mechanism of Corrosion

63. Refer to Figure 21-19, and describe in words or with a sketch what you would expect to happen in each of the following cases.
(a) Several turns of copper wire are wrapped around the head and tip of an iron nail.
(b) A deep scratch is filed at the center of an iron nail.
(c) A galvanized nail is substituted for the iron nail.

64. When an iron pipe is partly submerged in water, the iron dissolves more readily below the waterline than at the waterline. Explain this observation by relating it to the description of corrosion given in Figure 21-20.

65. Natural gas transmission pipes are sometimes protected against corrosion by maintaining a small potential difference between the pipe and an inert electrode buried in the ground. Describe how the method works.

66. In the construction of the Statue of Liberty, a framework of iron ribs was covered with thin sheets of copper less than 2.5 mm thick. A layer of asbestos separated the copper skin and iron framework. Over time, the asbestos wore away and the iron ribs corroded. Some of the ribs lost more than half their mass in the 100 years before the statue was restored. At the same time, the copper skin lost only about 4% of its thickness. Use electrochemical principles to explain these observations.

Electrolysis Reactions

67. Which of the following reactions occur spontaneously, and which can be brought about only through electrolysis, assuming that all reactants and products are in their standard states? For those requiring electrolysis, what is the *minimum* voltage required?

(a) $2 H_2O(l) \longrightarrow 2 H_2(g) + O_2(g)$ [in 1 M H^+(aq)]
(b) $Zn(s) + Fe^{2+}(aq) \longrightarrow Zn^{2+}(aq) + Fe(s)$
(c) $2 Fe^{2+}(aq) + I_2(s) \longrightarrow 2 Fe^{3+}(aq) + 2 I^-(aq)$
(d) $Cu(s) + Sn^{4+}(aq) \longrightarrow Cu^{2+}(aq) + Sn^{2+}(aq)$

68. An aqueous solution of K_2SO_4 is electrolyzed by means of Pt electrodes.

(a) Which of the following gases should form at the *anode*: O_2, H_2, SO_2, SO_3? Explain.
(b) What product should form at the *cathode*? Explain.
(c) What is the *minimum* voltage required? Why is the actual voltage needed likely to be higher than this value?

69. If a lead storage battery is charged at too high a voltage, gases are produced at each electrode. (It is possible to recharge a lead storage battery only because of the high overpotential for gas formation on the electrodes.)

(a) What are these gases?
(b) Write a cell reaction to describe their formation.

70. A dilute aqueous solution of Na_2SO_4 is electrolyzed between Pt electrodes for 3.75 h with a current of 2.83 A. What volume of gas, saturated with water vapor at 25 °C and at a total pressure of 742 mmHg, would be collected at the *anode*? Use data from Table 13.2, as required.

71. Calculate the quantity indicated for each of the following electrolyses.

(a) the mass of Zn deposited at the cathode in 42.5 min when 1.87 A of current is passed through an aqueous solution of Zn^{2+}
(b) the time required to produce 2.79 g I_2 at the anode if a current of 1.75 A is passed through KI(aq)

72. Calculate the quantity indicated for each of the following electrolyses.

(a) $[Cu^{2+}]$ *remaining* in 425 mL of a solution that was originally 0.366 M $CuSO_4$, after passage of 2.68 A for 282 s and the deposition of Cu at the cathode

(b) the time required to reduce $[Ag^+]$ in 255 mL of $AgNO_3$(aq) from 0.196 to 0.175 M by electrolyzing the solution between Pt electrodes with a current of 1.84 A

73. In a silver *coulometer*, Ag^+(aq) is reduced to Ag(s) at a Pt cathode. If 1.206 g Ag is deposited in 1412 s by a certain quantity of electricity, (a) how much electric charge (in C) must have passed, and (b) what was the magnitude (in A) of the electric current?

74. Electrolysis is carried out for 2.00 h in the following cell. The platinum *cathode*, which has a mass of 25.0782 g, weighs 25.8639 g after the electrolysis. The platinum *anode* weighs the same before and after the electrolysis.

(a) Write plausible equations for the half-reactions that occur at the two electrodes.
(b) What must have been the magnitude of the current used in the electrolysis (assuming a constant current throughout)?
(c) A gas is collected at the anode. What is this gas, and what volume should it occupy if (when dry) it is measured at 23 °C and 755 mmHg pressure?

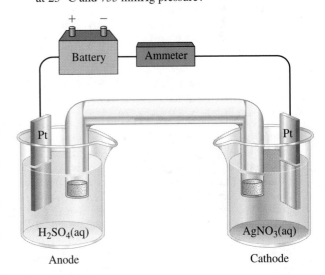

Integrative and Advanced Exercises

75. Two voltaic cells are assembled in which the following reactions occur.

$$V^{2+} + VO^{2+} + 2 H^+ \longrightarrow 2 V^{3+} + H_2O$$
$$E^\circ_{cell} = 0.616 \text{ V}$$

$$V^{3+} + Ag^+ + H_2O \longrightarrow VO^{2+} + 2 H^+ + Ag(s)$$
$$E^\circ_{cell} = 0.439 \text{ V}$$

Use these data and other values from Table 21.1 to calculate E° for the half-reaction $V^{3+} + e^- \longrightarrow V^{2+}$.

76. Suppose that a fully charged lead–acid battery contains 1.50 L of 5.00 M H_2SO_4. What will be the concentration of H_2SO_4 in the battery after 2.50 A of current is drawn from the battery for 6.0 h?

77. The energy consumption in electrolysis depends on the product of the charge and the voltage [volt × coulomb = $V \cdot C = J$ (joules)]. Determine the theoretical energy consumption per 1000 kg Cl_2 produced in a diaphragm chlor–alkali cell (page 857) that operates at 3.45 V. Express this energy in (a) kJ; (b) kilowatt-hours, kWh.

78. For the half-reaction $Cr^{3+} + e^- \longrightarrow Cr^{2+}$, $E^\circ = -0.424$ V. If excess Fe(s) is added to a solution in which $[Cr^{3+}] = 1.00$ M, what will be $[Fe^{2+}]$ when equilibrium is reached at 298 K?

$$Fe(s) + 2 Cr^{3+} \rightleftharpoons Fe^{2+} + 2 Cr^{2+}$$

79. A voltaic cell is constructed based on the following reaction and initial concentrations:

$Fe^{2+}(0.0050 \text{ M}) + Ag^+(2.0 \text{ M}) \longrightarrow$
$\qquad\qquad Fe^{3+}(0.0050 \text{ M}) + Ag(s)$

Calculate $[Fe^{2+}]$ when the cell reaction reaches equilibrium.

80. It is desired to construct a voltaic cell with $E_{cell} = 0.0860 \text{ V}$. What $[Cl^-]$ must be present in the cathode half-cell to achieve this result?

$$Ag(s)\,|\,Ag^+(\text{satd AgI})\,\|\,Ag^+(\text{satd AgCl}, x \text{ M } Cl^-)\,|\,Ag(s)$$

81. Describe a laboratory experiment that you could perform to evaluate the Faraday constant, F, and then show how you could use this value to determine the Avogadro constant, N_A.

82. The hydrazine fuel cell is based on the reaction

$$N_2H_4(aq) + O_2(g) \longrightarrow N_2(g) + 2\,H_2O(l)$$

The theoretical E°_{cell} of this fuel cell is 1.559 V. Use this information and data from Appendix D to calculate a value of ΔG°_f for $[N_2H_4(aq)]$.

83. It is sometimes possible to separate two metal ions through electrolysis. One ion is reduced to the free metal at the cathode, and the other remains in solution. In which of these cases would you expect complete or nearly complete separation: **(a)** Cu^{2+} and K^+; **(b)** Cu^{2+} and Ag^+; **(c)** Pb^{2+} and Sn^{2+}? Explain.

84. Show that for some fuel cells the efficiency value, $\varepsilon = \Delta G^\circ/\Delta H^\circ$, can have a value greater than 1.00. Can you identify one such reaction?
(*Hint:* Use data from Appendix D.)

85. In one type of Breathalyzer (alcohol meter), the quantity of ethanol in a sample is related to the amount of electric current produced by an ethanol–oxygen fuel cell. Use data from Table 21.1 and Appendix D to determine **(a)** E°_{cell} and **(b)** E° for the reduction of $CO_2(g)$ to $CH_3CH_2OH(g)$.

86. You prepare 1.00 L of a buffer solution that is 1.00 M NaH_2PO_4 and 1.00 M Na_2HPO_4. The solution is divided in half between the two compartments of an electrolytic cell. Both electrodes used are Pt. Assume that the only electrolysis is that of water. If 1.25 A of current is passed for 212 min, what will be the pH in each cell compartment at the end of the electrolysis?

87. Assume that the volume of each solution in Figure 21-21 is 100.0 mL. The cell is operated as an electrolytic cell, using a current of 0.500 A. Electrolysis is stopped after 10.00 h, and the cell is allowed to function as a voltaic cell. What is E_{cell} at this point?

88. A common reference electrode consists of a silver wire coated with AgCl(s) and immersed in 1 M KCl.

$$AgCl(s) + e^- \longrightarrow Ag(s) + Cl^-(1 \text{ M}) \quad E^\circ = 0.2223 \text{ V}$$

(a) What is E°_{cell} when this electrode is a cathode in combination with a standard zinc electrode as an anode?
(b) Cite several reasons why this electrode should be easier to use than a standard hydrogen electrode.
(c) By comparing the potential of this silver–silver chloride electrode with that of the silver–silver ion electrode, determine K_{sp} for AgCl.

89. The electrodes in the following electrochemical cell are connected to a voltmeter as shown. The half-cell on the right contains a standard silver–silver chloride electrode (see Exercise 88). The half-cell on the left contains a silver electrode immersed in 100.0 mL of 1.00×10^{-3} M $AgNO_3(aq)$. A porous plug through which ions can migrate separates the half-cells.
(a) What is the initial reading on the voltmeter?
(b) What is the voltmeter reading after 10.00 mL of 0.0100 M K_2CrO_4 has been added to the half-cell on the left and the mixture has been stirred thoroughly?
(c) What is the voltmeter reading after 10.00 mL of 10.0 M NH_3 has been added to the half-cell described in part (b) and the mixture has been stirred thoroughly?

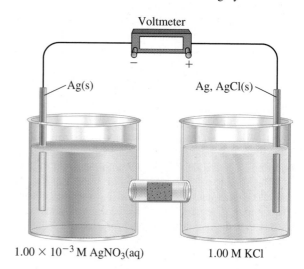

Voltmeter

Ag(s)   Ag, AgCl(s)

1.00×10^{-3} M $AgNO_3(aq)$ 1.00 M KCl

90. An important source of Ag is recovery as a by-product in the metallurgy of lead. The percentage of Ag in lead was determined as follows. A 1.050-g sample was dissolved in nitric acid to produce $Pb^{2+}(aq)$ and $Ag^+(aq)$. The solution was then diluted to 500.0 mL with water, a Ag electrode was immersed in the solution, and the potential difference between this electrode and a SHE was found to be 0.503 V. What was the percent Ag by mass in the lead metal?

91. A test for completeness of electrodeposition of Cu from a solution of $Cu^{2+}(aq)$ is to add $NH_3(aq)$. A blue color signifies the formation of the complex ion $[Cu(NH_3)_4]^{2+}$ ($K_f = 1.1 \times 10^{13}$). Let 250.0 mL of 0.1000 M $CuSO_4(aq)$ be electrolyzed with a 3.512 A current for 1368 s. At this time, add a sufficient quantity of $NH_3(aq)$ to complex any remaining Cu^{2+} and to maintain a free $[NH_3] = 0.10$ M. If $[Cu(NH_3)_4]^{2+}$ is detectable at concentrations as low as 1×10^{-5} M, should the blue color appear?

92. A solution is prepared by saturating 100.0 mL of 1.00 M $NH_3(aq)$ with AgBr. A silver electrode is immersed in this solution, which is joined by a salt bridge to a standard hydrogen electrode. What will be the measured E_{cell}? Is the standard hydrogen electrode the anode or the cathode?

93. The electrolysis of $Na_2SO_4(aq)$ is conducted in two separate half-cells joined by a salt bridge, as suggested by the cell diagram $Pt\,|\,Na_2SO_4(aq)\,\|\,Na_2SO_4(aq)\,|\,Pt$.

(a) In one experiment, the solution in the anode compartment becomes more acidic and that in the cathode compartment, more basic during the electrolysis. When the electrolysis is discontinued and the two solutions are mixed, the resulting solution has pH = 7. Write half-equations and the overall electrolysis equation.

(b) In a second experiment, a 10.00-mL sample of an unknown concentration of $H_2SO_4(aq)$ and a few drops of phenolphthalein indicator are added to the $Na_2SO_4(aq)$ in the cathode compartment. Electrolysis is carried out with a current of 21.5 mA (milliamperes) for 683 s, at which point, the solution in the cathode compartment acquires a lasting pink color. What is the molarity of the unknown $H_2SO_4(aq)$?

94. A Ni anode and an Fe cathode are placed in a solution with $[Ni^{2+}] = 1.0$ M and then connected to a battery. The Fe cathode has the shape shown. How long must electrolysis be continued with a current of 1.50 A to build a 0.050-mm-thick deposit of nickel on the iron? (Density of nickel = 8.90 g/cm^3.)

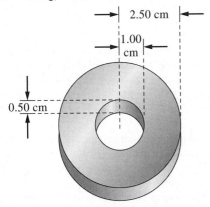

95. To each of the half-cells shown in Figure 21-21 was added a 100.0-mL sample of solution with an ion concentration of 1.000 M. The cell was operated as an electrolytic cell, with copper as the anode and zinc as the cathode. A current of 0.500 A was used. Assume that the only electrode reactions occurring were those involving Cu/Cu^{2+} and Zn/Zn^{2+}. Elec-

trolysis was stopped after 10.00 h, and the cell was allowed to function as a voltaic cell. What was E_{cell} at that point?

96. Silver tarnish is mainly Ag_2S:

$$Ag_2S(s) + 2\,e^- \longrightarrow 2\,Ag(s) + S^{2-}(aq),$$
$$E° = -0.691 \text{ V}.$$

A tarnished silver spoon is placed in contact with a commercially available metallic product in a glass baking dish. Boiling water, to which some $NaHCO_3$ has been added, is poured into the dish, and the product and spoon are completely covered. Within a short time, the removal of tarnish from the spoon begins.

(a) What metal or metals are in the product?
(b) What is the probable reaction that occurs?
(c) What do you suppose is the function of the $NaHCO_3$?
(d) An advertisement for the product appears to make two claims: (1) No chemicals are involved, and (2) the product will never need to be replaced. How valid are these claims? Explain.

97. Your task is to determine $E°$ for the reduction of $CO_2(g)$ to $C_3H_8(g)$ in two different ways and to explain why each gives the same result. **(a)** Consider a fuel cell in which the cell reaction corresponds to the complete combustion of propane gas. Write the half-cell reactions and the overall reaction. Determine $\Delta G°$ and $E°_{cell}$ for the reaction, then obtain $E°_{CO_2/C_3H_8}$. **(b)** Without considering the oxidation that occurs simultaneously, obtain $E°_{CO_2/C_3H_8}$ directly from tabulated thermodynamic data for the reduction half-reaction.

Feature Problems

98. Consider the following electrochemical cell:

$$Pt(s)\,|\,H_2(g,\,1\text{ atm})\,|\,H^+(1\text{ M})\,\|\,Ag^+(x\text{ M})\,|\,Ag(s)$$

(a) What is $E°_{cell}$—that is, the cell potential when $[Ag^+] = 1$ M?

(b) Use the Nernst equation to write an equation for E_{cell} when $[Ag^+] = x$.
(c) Now imagine titrating 50.0 mL of 0.0100 M $AgNO_3$ in the cathode half-cell compartment with 0.0100 M KI. The titration reaction is

$$Ag^+(aq) + I^-(aq) \longrightarrow AgI(s)$$

Calculate $[Ag^+]$ and then E_{cell} after addition of the following volumes of 0.0100 M KI: (i) 0.0 mL;

(ii) 20.0 mL; (iii) 49.0 mL; (iv) 50.0 mL; (v) 51.0 mL; (vi) 60.0 mL.
(d) Use the results of part (c) to sketch the titration curve of E_{cell} versus volume of titrant.

99. Ultimately, $\Delta G°_f$ values must be based on experimental results; in many cases, these experimental results are themselves obtained from $E°$ values. Early in the twentieth century, G. N. Lewis conceived of an experimental approach for obtaining reduction potentials of the alkali metals. This approach involved using a solvent with which the alkali metals do not react. Ethylamine was the solvent chosen. In the following cell diagram, Na(amalg, 0.206%) represents a solution of 0.206% Na in liquid mercury.
(1) $Na(s)\,|\,Na^+(\text{in ethylamine})\,|\,Na(\text{amalg},\,0.206\%)$ $E_{cell} = 0.8453$ V

Although Na(s) reacts violently with water to produce $H_2(g)$, at least for a short time, a sodium amalgam electrode does not react with water. This makes it possible to determine E_{cell} for the following voltaic cell.

(2) Na(amalg, 0.206%) | Na^+(1 M) || H^+(1 M) | H_2(g, 1 atm) $E_{cell} = 1.8673$ V

(a) Write equations for the cell reactions that occur in the voltaic cells (1) and (2).

(b) Use equation (21.14) to establish ΔG for the cell reactions written in part (a).

(c) Write the overall equation obtained by combining the equations of part (a), and establish $\Delta G°$ for this overall reaction.

(d) Use the $\Delta G°$ value from part (c) to obtain $E_{cell}°$ for the overall reaction. From this result, obtain $E°_{Na^+/Na}$. Compare your result with the value listed in Appendix D.

100. The following sketch is called an electrode potential diagram. Such diagrams summarize electrode potential data more efficiently than do listings such as that in Appendix D. In this diagram for bromine and its ions in basic solution,

$$BrO_4^- \xrightarrow{\ 1.025\ V\ } BrO_3^-$$

signifies

$$BrO_4^- + H_2O + 2\,e^- \longrightarrow BrO_3^- + 2\,OH^-,$$
$$E°_{BrO_4^-/BrO_3^-} = 1.025\ V$$

Similarly,

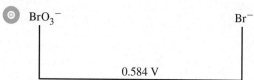

signifies

$$BrO_3^- + 3\,H_2O + 6\,e^- \longrightarrow Br^- + 6\,OH^-,$$
$$E°_{BrO_3^-/Br^-} = 0.584\ V.$$

With reference to Appendix D and to the method of determining $E°$ values outlined on page 834, supply the missing data in the following diagram.

Basic solution ([OH^-] = 1 M):

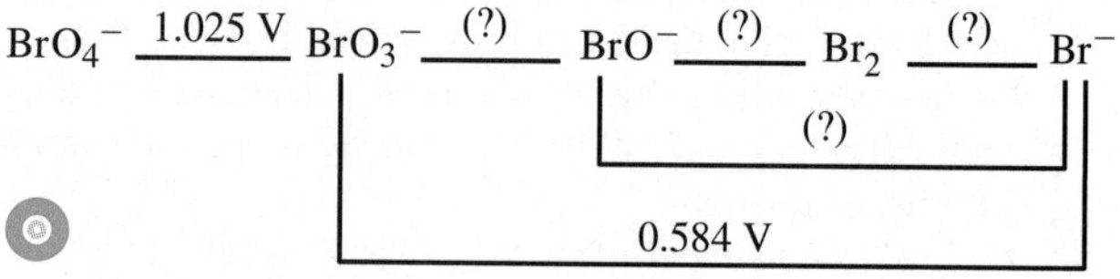

$$BrO_4^- \xrightarrow{\ 1.025\ V\ } BrO_3^- \xrightarrow{\ (?)\ } BrO^- \xrightarrow{\ (?)\ } Br_2 \xrightarrow{\ (?)\ } Br^-$$

(with $(?)$ spanning BrO^- to Br^-, and 0.584 V spanning BrO_3^- to Br^-)

101. Only a tiny fraction of the diffusible ions move across a cell membrane in establishing a Nernst potential (page 860), so there is no detectable concentration change. Consider a typical cell with a volume of 10^{-8} cm^3, a surface area (A) of 10^{-6} cm^2, and a membrane thickness (l) of 10^{-6} cm. Suppose that [K^+] = 155 mM inside the cell and [K^+] = 4 mM outside the cell and that the observed Nernst potential across the cell wall is 0.085 V. The membrane acts as a charge-storing device called a *capacitor*, with a *capacitance*, C, given by

$$C = \frac{\varepsilon_0 \varepsilon A}{l}$$

where ε_0 is the *dielectric constant* of a vacuum and the product $\varepsilon_0 \varepsilon$ is the dielectric constant of the membrane, having a typical value of $3 \times 8.854 \times 10^{-12}$ C^2 N^{-1} m^{-2} for a biological membrane. The SI unit of capacitance is the *farad*, 1 F = 1 coulomb per volt = 1 C $\times$ V^{-1} = 1C^2 N^{-1} m^{-1}.

(a) Determine the capacitance of the membrane for the typical cell described.

(b) What is the net charge required to maintain the observed membrane potential?

(c) How many K^+ ions must flow through the cell membrane to produce the membrane potential?

(d) How many K^+ ions are in the typical cell?

(e) Show that the fraction of the intracellular K^+ ions transferred through the cell membrane to produce the membrane potential is so small that it does not change [K^+] within the cell.

 **eMedia Exercises**

102. The oxidation of zinc metal by two different species is shown in the **Oxidation-Reduction Reactions (Part 2)** animation *(eChapter 21-1)*. **(a)** Describe the changes in physical state of the metals in the two reactions. **(b)** If these electrochemical processes were carried out in the laboratory, what would be the macroscopic, visible signs that a reaction is occurring?

103. In the **Voltaic Cells I: The Copper-Zinc Cell** animation *(eChapter 21-1)*, two examples of a spontaneous electrochemical reaction are shown. **(a)** Describe the factors that can impede the progress of each reaction. Are they similar for the two examples? **(b)** What do these reactions tell you about the oxidation of copper by an aqueous solution of zinc ions?

104. If the two metal half-cells in the **Standard Reduction Potentials** animation *(eChapter 21-2)* were combined, **(a)** what spontaneous reaction would occur? Write the balanced chemical equation for this reaction. **(b)** Determine the standard cell potential for this spontaneous reaction.

105. Calculate the equilibrium constant of the reaction described in eMedia Exercise 21.104(a). Is the magnitude of the equilibrium constant consistent with your understanding of the spontaneity of the electrochemical reaction? Explain.

106. Calculate the cell potential of the reaction described in eMedia Exercise 21.104(a) under conditions where $[Zn^{2+}] = 0.20$ M and $[Ag^+] = 0.10$ M. Do these conditions favor the oxidation or reduction of zinc?

107. For the reaction seen in the **Electrolysis of Water** animation *(eChapter 21-7)*, calculate the current required to produce 55 ml of oxygen gas at STP in 15 minutes?

22 Main-Group Elements I: Metals

Contents

A Fourth of July fireworks display over New York City.

Some of the dramatic colors seen in fireworks displays are the flame colors of some of the groups 1 and 2 metals. These colors, as we will see, are related to the electronic structures of those metal atoms.

Aluminum production requires prodigious quantities of electricity in a process that is not much more than one hundred years old. Conversely, lead can be obtained by chemical reduction, a method used since ancient times. Concepts of oxidation–reduction and electrochemistry help us understand which elements can be obtained from their compounds by chemical reactions and which require electrolysis. Principles of acid–base chemistry and solution equilibria help explain many common natural phenomena, such as the hardness of water. These principles also underlie methods of water softening.

TABLE 22.1 Group 1 Elements: Abundances		
	ppm*	Rank
Li	18	35
Na	22,700	7
K	18,400	8
Rb	78	23
Cs	2.6	46
Fr	Trace	—

*Grams per 1000 kg of solid crust.

KEEP IN MIND ▶

that hydrogen is often placed in group 1 of the periodic table, but this element is not an alkali metal. Francium is an alkali metal, but this highly radioactive metal is so rare that few of its properties have been measured.

Flame Tests for Metals movie

▲ The mineral spodumene, $LiAl(SiO_3)_2$.

▲ **The cutting of metallic sodium**

The sodium, an active metal, is covered with a thick oxide coating.

This chapter and the remaining chapters offer many opportunities to relate new information to principles presented earlier in the text. We will focus on the links between descriptive facts and the principles of chemistry—here, regarding a number of important and interesting metals.

22-1   Group 1: The Alkali Metals

As Table 22.1 indicates, the group 1 elements, the **alkali metals**, are relatively abundant. Some of their compounds have been known and used since prehistoric times. Yet these elements remained undiscovered until about 200 years ago. The compounds of the alkali metals are difficult to decompose by ordinary chemical means, so discovery of the elements had to await new scientific developments. Sodium (1807) and potassium (1807) were discovered through electrolysis. Cesium (1860) and rubidium (1861) were identified as new elements through their emission spectra. Francium (1939) was isolated in the radioactive decay products of actinium.

Because most alkali metal compounds are water soluble, a number of Li, Na, and K compounds, such as the chlorides, carbonates and sulfates, can be obtained from natural brines. A few alkali metal compounds, such as NaCl, KCl, and Na_2CO_3, can be mined as solid deposits. Sodium chloride is also obtained from seawater. An important source of lithium is the mineral *spodumene*, $LiAl(SiO_3)_2$. Rubidium and cesium are obtained as by-products in the processing of lithium ores.

Physical Properties of the Alkali Metals

By any measure we choose, the group 1 elements are the most active metals. Table 22.2 lists several of their properties, and a few of these properties are discussed next.

Flame Colors   The energy differences between the valence-shell *s* and *p* orbitals of the group 1 metals match those of certain wavelengths of visible light. As a result, when heated in a flame, group 1 compounds produce characteristic flame colors (recall Figure 9-8). For example, when NaCl is vaporized in a flame, ion pairs are converted to gaseous atoms. Na(g) atoms are excited to higher energies, and light with a wavelength of 589 nm (yellow) is emitted as the excited atoms (Na*) revert to their ground-state electron configurations.

$$Na^+Cl^-(g) \longrightarrow Na(g) + Cl(g)$$

$$Na(g) \longrightarrow Na^*(g)$$

$$[Ne]3s^1 \qquad [Ne]3p^1$$

$$Na^*(g) \longrightarrow Na(g) + h\upsilon \quad (589 \text{ nm; yellow})$$

Alkali metal compounds are used in pyrotechnic displays—fireworks.

Properties Influenced by Atomic Sizes   Atoms of the group 1 elements are the largest in their respective periods, and the atomic radii tend to increase from the top to the bottom within this group, as we described in Chapter 10. These large atoms make for a relatively low mass per unit volume—that is, low density. The lighter of the alkali metals (Li, Na, and K) will float on water. The large atomic sizes, together with the fact that each of these atoms has only one valence electron, makes for rather weak metallic bonding. This property, in turn, leads to soft metals with low melting points. A bar of sodium has the consistency of a frozen stick of butter and is easily cut with a knife.

Electrode Potentials   A good indicator of the extreme metallic character of the group 1 elements is their electrode potentials, which are large negative quantities.

TABLE 22.2 Some Properties of the Group 1 (Alkali) Metals					
	Li	**Na**	**K**	**Rb**	**Cs**
Atomic number	3	11	19	37	55
Valence-shell electron configuration	$2s^1$	$3s^1$	$4s^1$	$5s^1$	$6s^1$
Atomic (metallic) radius, pm	152	186	227	248	265
Ionic (M^+) radius, pm	59	99	138	149	170
Electronegativity	1.0	0.9	0.8	0.8	0.8
First ionization energy, kJ mol^{-1}	520.2	495.8	418.8	403.0	375.7
Electrode potential $E°$, V[a]	−3.040	−2.713	−2.924	−2.924	−2.923
Melting point, °C	180.54	97.81	63.65	39.05	28.4
Boiling point, °C	1347	883.0	773.9	687.9	678.5
Density, g/cm^3 at 20 °C	0.534	0.971	0.862	1.532	1.873
Hardness[b]	0.6	0.4	0.5	0.3	0.2
Electrical conductivity[c]	17.1	33.2	22.0	12.4	7.76
Flame color	Carmine	Yellow	Violet	Bluish red	Blue
Principal visible emission lines, nm	610,671	589	405,767	780,795	456,459

[a] For the reduction $M^+(aq) + e^- \longrightarrow M(s)$.
[b] Hardness measures the ability of substances to scratch, abrade, or indent one another. On the Mohs scale, ten minerals are ranked by hardness, ranging from that of talc (0) to diamond (10). Other values: wax (0 °C), 0.2; asphalt, 1–2; fingernail, 2.5; copper, 2.5–3; iron, 4–5; chromium, 9. Each substance can scratch only other substances with hardness values lower than its own.
[c] On a scale relative to silver as 100.

Sodium and Potassium in Water movie

The ions $M^+(aq)$ are very difficult to reduce to the metals $M(s)$, and in turn the metals are very easily oxidized to $M^+(aq)$. All the alkali metals easily displace $H_2(g)$ from water.

$$2\,M(s) + 2\,H_2O(l) \longrightarrow 2\,M^+(aq) + 2\,OH^-(aq) + H_2(g) \qquad (22.1)$$

$$E°_{cell} = E°_{H_2O/H_2} - E°_{M^+/M}$$

$$= -0.828\,V - E°_{M^+/M}$$

$$E°_{cell} = 2.212\,V\ (\text{for Li}) \qquad 1.885\,V\ (\text{for Na}) \qquad 2.096\,V\ (\text{for K})$$
$$2.096\,V\ (\text{for Rb}) \qquad 2.095\,V\ (\text{for Cs})$$

Production and Uses of the Alkali Metals

Lithium and sodium are produced from their molten chlorides by electrolysis. The electrolysis of $NaCl(l)$, for example, is carried out at about 600 °C.

$$2\,NaCl(l) \xrightarrow{\text{electrolysis}} 2\,Na(l) + Cl_2(g) \qquad (22.2)$$

The melting point of NaCl is 801 °C, too high a temperature to carry out this electrolysis economically. Adding $CaCl_2$ to the mixture reduces the melting point. [Calcium metal, also produced in the electrolysis, precipitates out from the Na(l) as the liquid metal is cooled. The final product is 99.95% Na.]

Potassium metal is produced by the reduction of molten KCl by liquid sodium.

$$KCl(l) + Na(l) \xrightarrow{850\,°C} NaCl(l) + K(g) \qquad (22.3)$$

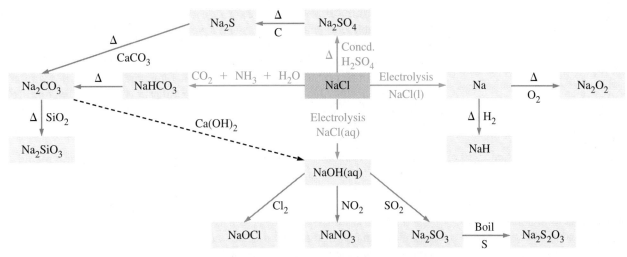

▲ **FIGURE 22-2** **Preparation of sodium compounds**
This is a simple and common method of summarizing the reactions of a compound. Most of these reactions are described in this section. A number of these compounds may be prepared by alternative methods. The conversion of Na_2CO_3 to NaOH (dashed arrow) is no longer of commercial importance.

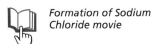

Formation of Sodium Chloride movie

Halides   All alkali metals react vigorously, sometimes explosively, with halogens to produce ionic halides, the most important of which are NaCl and KCl. Sodium chloride—salt—is the primary sodium compound. In fact, it is the most used of all minerals for the production of chemicals. It is not listed among the top chemicals, however, because it is considered a raw material, not a manufactured chemical. Annual use of NaCl in the United States amounts to about 50 million tons. Salt is used to preserve meat and fish, control ice on roads, and regenerate water softeners. In the chemical industry, NaCl is a source of many chemicals, including sodium metal, chlorine gas, hydrochloric acid, and sodium hydroxide.

Potassium chloride is obtained from naturally occurring brines (concentrated solutions of salts). It is most extensively used in plant fertilizers because potassium is a major essential element for plant growth. KCl is also used as a raw material in the manufacture of KOH, KNO_3, and other industrially important potassium compounds.

▲ Sea salt (sodium chloride) stacks that have been harvested by evaporation of seawater.

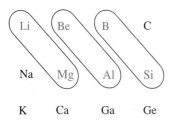

▲ FIGURE 22-3
Diagonal relationships
The two elements in each encircled pair exhibit many similar properties.

▶ The term *ash* signifies a product obtained by heating or burning. The earliest alkaline materials (notably, potassium carbonate, *potash*) were extracted from the ashes of burned plants. The first commercial source of *soda ash* was the Leblanc process, featured in Example 22-1.

Carbonates and Sulfates Except for Li_2CO_3, all the alkali metal carbonates are thermally stable compounds. In this and several other ways, lithium and its compounds differ from other members of group 1. The ability of lithium to form a nitride (Li_3N), the low aqueous solubility of its carbonate, and the instability of the carbonate with respect to the oxide at high temperature are all properties similar to those of magnesium and its compounds. This similarity is called a **diagonal relationship**. Such relationships also exist between Be and Al and between B and Si (Figure 22-3). The Li–Mg similarity probably results from the roughly comparable sizes of the Li and Mg atoms and of the Li^+ and Mg^{2+} ions.

Lithium carbonate is used in the treatment of individuals who are manic-depressive. Daily dosages of 1–2 g Li_2CO_3 maintain a level of Li^+ in the blood of about 1 millimole per liter. This treatment apparently influences the balance of Na^+ and K^+, that of Mg^{2+} and Ca^{2+}, or both, across cell membranes.

Sodium carbonate (soda ash) is used primarily in the manufacture of glass. The Na_2CO_3 produced in the United States now comes mostly from natural sources, such as the mineral *trona*, $Na_2CO_3 \cdot NaHCO_3 \cdot nH_2O$, found in dry lakes in California and in immense deposits in western Wyoming. In the past, sodium carbonate was manufactured mostly from NaCl, $CaCO_3$, and NH_3, using a process introduced by the Belgian chemist Ernest Solvay in 1863.

The great success of the *Solvay process* over the synthetic method outlined in Example 22-1 lies in the efficient use of certain raw materials through recycling. An outline of the process is shown in Figure 22-4. The key step involves the reaction of $NH_3(g)$ and $CO_2(g)$ in saturated NaCl(aq). Of the possible ionic compounds that could precipitate from such a mixture (NaCl, NH_4Cl, $NaHCO_3$, and NH_4HCO_3), the least soluble is *sodium hydrogen carbonate* (sodium bicarbonate).

$$Na^+ + Cl^- + NH_3 + CO_2 + H_2O \longrightarrow NaHCO_3(s) + NH_4 + Cl^- \quad (22.4)$$

The sodium bicarbonate can be isolated and sold or converted to sodium carbonate by heating.

$$2\,NaHCO_3(s) \xrightarrow{\Delta} Na_2CO_3(s) + H_2O(g) + CO_2(g) \quad (22.5)$$

One noteworthy feature of the Solvay process is that it involves only simple precipitation and acid–base reactions. Another feature is that materials produced in one step are efficiently recycled into a subsequent step. Recycling makes good economic sense: A process that recycles materials minimizes the use of raw materials (an expense in purchasing) and cuts down on the production of by-products (an expense in disposal). Thus, when limestone ($CaCO_3$) is heated to produce the reactant CO_2, the other reaction product, CaO, is also used. It is converted to $Ca(OH)_2$, which is then used to convert NH_4Cl (another reaction by-product) to $NH_3(g)$. The $NH_3(g)$ is recycled into the production of ammoniated brine.

The Solvay process has only one ultimate by-product—$CaCl_2$—for which the demand is very limited. In the past, some $CaCl_2$ was used for deicing roads in the winter (page 560) and for dust control on dirt roads in the summer (by means of the deliquescence of $CaCl_2 \cdot 6\,H_2O$, page 554). The bulk of the $CaCl_2$, however, was dumped into local lakes and streams (notably, Onondaga Lake, near Solvay, New York), resulting in a local environmental tragedy. Environmental regulations no longer permit such dumping. Partly due to these regulations, but mainly for economic reasons, natural sources of sodium carbonate have supplanted the Solvay process in the United States. The process is still widely used elsewhere in the world, however.

▲ The mineral *trona*, from Green River, Wyoming, is currently the principal source of Na_2CO_3 in the United States.

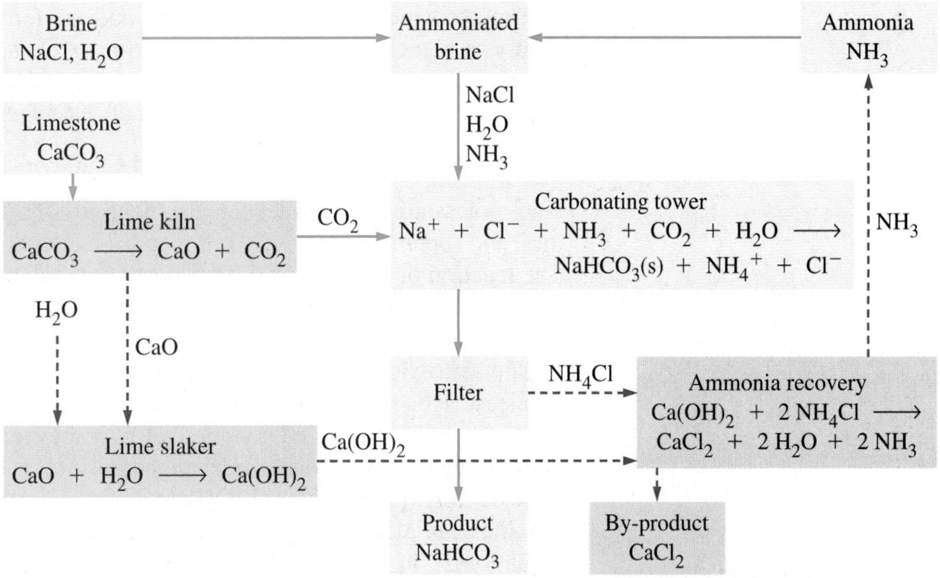

▲ **FIGURE 22-4   The Solvay process for the manufacture of NaHCO₃**
The main reaction sequence is traced by solid arrows. Recycling reactions are shown by dashed arrows.

Sodium sulfate is obtained partly from natural sources, partly from neutralization reactions, and partly through a process discovered by Johann Rudolf Glauber in 1625.

$$H_2SO_4(\text{concd aq}) + NaCl(s) \xrightarrow{\Delta} NaHSO_4(s) + HCl(g)$$

$$NaHSO_4(s) + NaCl(s) \xrightarrow{\Delta} Na_2SO_4(s) + HCl(g) \tag{22.6}$$

▶ Often, equations (22.6) are replaced by the overall equation $H_2SO_4(\text{concd aq}) + 2\,NaCl(s) \longrightarrow Na_2SO_4(s) + 2\,HCl(g)$.

The strategy behind reactions (22.6) is the production of a *volatile* acid (HCl) by heating one of its salts (NaCl) with a *nonvolatile* acid (H_2SO_4). Several other acids can be produced by similar reactions. The major use of Na_2SO_4 is in the paper industry. For instance, in the kraft process for papermaking, undesirable lignin is removed from wood by digesting the wood in an alkaline solution of Na_2S. Na_2S is produced by the reduction of Na_2SO_4 with carbon.

$$Na_2SO_4(s) + 4\,C(s) \xrightarrow{\Delta} Na_2S(s) + 4\,CO(g)$$

About 100 lb of Na_2SO_4 is required for every ton of paper produced.

Oxides and Hydroxides   The alkali metals react rapidly with oxygen to produce several different ionic oxides. Under appropriate conditions—generally by limiting the supply of oxygen—the oxide, M_2O, can be prepared for each of the alkali metals. Li reacts with excess oxygen to give Li_2O and a small amount of a *peroxide*, M_2O_2—in this case, lithium peroxide, Li_2O_2. Na reacts with excess oxygen to give mostly the peroxide Na_2O_2 and a small amount of Na_2O. K, Rb, and Cs react to form the *superoxides*, MO_2.

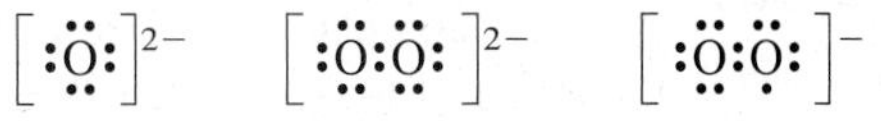

$$\left[:\overset{..}{\underset{..}{O}}:\right]^{2-} \qquad \left[:\overset{..}{\underset{..}{O}}:\overset{..}{\underset{..}{O}}:\right]^{2-} \qquad \left[:\overset{..}{\underset{.}{O}}:\overset{..}{\underset{..}{O}}:\right]^{-}$$

Oxide ion   Peroxide ion   Superoxide ion

The peroxides contain the O_2^{2-} ion and are fairly stable. Sodium peroxide is used as a bleaching agent and a powerful oxidant. Li_2O_2 and Na_2O_2 are used in emergency breathing devices in submarines and spacecraft because these compounds react with carbon dioxide to produce oxygen.

$$2\,M_2O_2(s) + 2\,CO_2(g) \longrightarrow 2\,M_2CO_3(s) + O_2(g) \qquad (M = Li, Na) \qquad (22.7)$$

KO_2, potassium superoxide, can also be used for this purpose. The oxides, peroxides, and superoxides of the alkali metals react with water to form basic solutions. The reaction of the O^{2-} ion with water is an acid–base reaction that produces hydroxide ions. The peroxide and superoxide ions react with water in oxidation–reduction reactions that produce hydroxide ions and $O_2(g)$.

The hydroxides of the group 1 metals are strong bases because they dissociate to release hydroxide ions in aqueous solution. As we learned in Section 21-8, sodium hydroxide is produced commercially by the electrolysis of NaCl(aq). $Na^+(aq)$ goes through the electrolysis unchanged; $Cl^-(aq)$ is oxidized to $Cl_2(g)$; and H_2O is reduced to $H_2(g)$. KOH and LiOH are made in a similar fashion. Alkali metal hydroxides can also be prepared by the reaction of the group 1 metals with water (equation 22.1). Alkali hydroxides are important in the manufacture of soaps and detergents, described next.

Alkali Metal Detergents and Soaps A **detergent** is a cleansing agent used primarily because it can emulsify oils. Although the term *detergent* includes common soaps, it is used primarily to describe certain synthetic products, such as sodium lauryl sulfate, whose manufacture involves these conversions.

$$CH_3(CH_2)_{10}CH_2OH \longrightarrow CH_3(CH_2)_{10}CH_2OSO_3H \longrightarrow CH_3(CH_2)_{10}CH_2OSO_3^-Na^+$$

lauryl alcohol   lauryl hydrogen sulfate   sodium lauryl sulfate

A **soap** is a specific kind of detergent that is the salt of a metal hydroxide and a fatty acid. An example is the sodium soap of palmitic acid, which we can represent as the product of the reaction of palmitic acid and NaOH.

$$\underset{\text{palmitic acid}}{CH_3(CH_2)_{14}-\overset{\overset{\displaystyle O}{\|}}{C}-O-H} + Na^+ + OH^- \longrightarrow \underset{\substack{\text{sodium palmitate}\\\text{(a soap)}}}{CH_3(CH_2)_{14}-\overset{\overset{\displaystyle O}{\|}}{C}-O^- \; Na^+} + H_2O \qquad (22.8)$$

TABLE 22.3 Group 2 Elements: Abundances		
	ppm[*]	**Rank**
Be	2	51
Mg	27,640	6
Ca	46,600	5
Sr	384	15
Ba	390	14
Ra	Trace	—

[*]Grams per 1000 kg of solid crust.

Sodium Palmitate, Lithium Stearate models

▲ An emerald crystal, which is based on the mineral *beryl*, embedded in a calcite matrix.

Figure 22-5 illustrates the detergent action of a typical soap: sodium palmitate. Sodium soaps are the familiar hard (bar) soaps. Potassium soaps have low melting points and are soft soaps. Lithium stearate, a soap not seen in ordinary household use, thickens oils into greases. These greases have excellent water-repellent and lubricating properties at both high and low temperatures. The greases remain in contact with moving metal parts under conditions in which oil by itself would run off.

22-2 Group 2: The Alkaline Earth Metals

As a group, the elements of group 2 are as common as those of group 1. Table 22.3 shows that calcium and magnesium are particularly abundant. Even beryllium, the least abundant member of group 2, is accessible because it occurs in deposits of the mineral *beryl*, $3BeO \cdot Al_2O_3 \cdot 6SiO_2$. The principal forms in which the other group 2 elements are found are as carbonates, sulfates, and silicates. Radium, like its group 1 neighbor, francium, is a radioactive element found only in trace amounts.

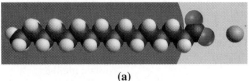

(a)

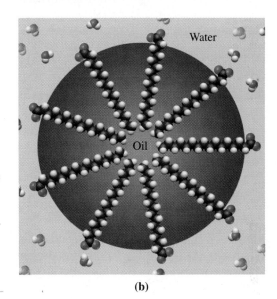

Water

Oil

(b)

▶ **FIGURE 22-5**
**Structure of soap molecules and
their cleaning action**
(a) The sodium palmitate molecule
has a long nonpolar portion buried
in a droplet of oil and a polar head
projecting into an aqueous medium.
(b) Electrostatic attractions between these polar heads and water
molecules cause the droplet to be
emulsified, or solubilized.

Radium is more interesting for its radioactive properties than for its chemical similarities to the other group 2 elements.

Even though the group 2 metal oxides and hydroxides are only very slightly soluble in water, they are basic, or *alkaline*. At one time, insoluble substances that do
not decompose on heating were called "earths." This term is the basis of the group
2 name: **alkaline earth metals**.

From a chemical standpoint (for example, in their abilities to react with water
and acids and to form ionic compounds), the heavier group 2 metals—Ca, Sr, Ba,
and Ra—are nearly as active as the group 1 metals. In terms of certain physical
properties (for example, density, hardness, and melting point), all the group 2 elements are more typically metallic than the group 1 elements, as we can see by comparing Tables 22.2 and 22.4.

Table 22.4 shows that beryllium is out of step with the other group 2 elements
in some of its physical properties. For instance, it has a higher melting point and
is much harder than the others. Its chemical properties also differ significantly.
For example,

- Be is quite unreactive toward air and water.
- BeO does not react with water, whereas the other MO oxides form $M(OH)_2$.
- Be and BeO dissolve in strongly basic solutions to form the ion BeO_2^{2-}.
- $BeCl_2$ and BeF_2 in the molten state are poor conductors of electricity; they are
  covalent substances.

We can best relate the chemical behavior of beryllium to the small size and high
ionization energy of its atoms. Beryllium shows only a limited tendency to form the
ion Be^{2+}; in fact, its ability to form covalent bonds predominates in the compounds
that beryllium forms. For example, BeO is an *acidic* oxide, even though *nonmetal*
oxides are more commonly acidic than are most metal oxides. This behavior is a

TABLE 22.4 Some Properties of the Group 2 (Alkaline Earth) Metals

	Be	Mg	Ca	Sr	Ba
Atomic number	4	12	20	38	56
Atomic (metallic) radius, pm	111	160	197	215	222
Ionic (M^{2+}) radius, pm	27	72	100	113	136
Electronegativity	1.5	1.2	1.0	1.0	0.9
First ionization energy, kJ mol^{-1}	899.4	737.7	589.7	549.5	502.8
Electrode potential $E°$, V^a	−1.85	−2.356	−2.84	−2.89	−2.92
Melting point, °C	1278	648.8	839	769	729
Boiling point, °C	2970^b	1090	1483.6	1383.9	1637
Density, g/cm³ at 20 °C	1.85	1.74	1.55	2.54	3.60
Hardness^c	~ 5	2.0	1.5	1.8	~2
Electrical conductivity^c	39.7	35.6	40.6	6.90	3.20
Flame color	None	None	Orange–red	Scarlet	Green

^{a} For the reduction $M^{2+}(aq) + 2\,e^- \longrightarrow M(s)$.
^{b} Boiling point at 5 mmHg pressure.
^{c} See footnotes of Table 22.2.

consequence of beryllium's small ionic size and relatively high ionic charge of 2+. In covalent compounds, Be atoms appear to use hybrid orbitals—*sp* orbitals in $BeCl_2(g)$ and sp^3 orbitals in $BeCl_2(s)$ (Figure 22-6).

Production and Uses of the Alkaline Earth Metals

The preferred method of producing the group 2 metals (except Mg) is by reducing their salts with other active metals. Beryl is the natural source of beryllium compounds. This mineral is processed to produce BeF_2, which is then reduced with Mg. Beryllium metal is used as an alloying agent when low density is a primary requirement. Because Be can withstand metal fatigue, an alloy of copper with about 2% Be is used in springs, clips, and electrical contacts. The Be atom does not readily absorb X rays or neutrons, so beryllium is used to make windows for X-ray tubes and for various components in nuclear reactors. Beryllium and its compounds are limited in their use, however, because they are toxic. In addition, they are suspected of being carcinogens even at a level as low as 0.002 ppm in air.

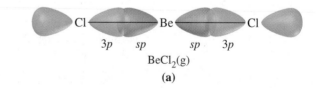

$BeCl_2(g)$
(a)

▶ **FIGURE 22-6**
Covalent bonds in BeCl₂
(a) In gaseous $BeCl_2$, discrete molecules exist with the bonding scheme shown. **(b)** In solid $BeCl_2$, two Cl atoms are bonded to one Be atom by normal covalent bonds. Two other Cl atoms are bonded by coordinate covalent bonds, using lone-pair electrons of the Cl atoms (bonds shown as arrows). Once formed, these two types of bonds are indistinguishable. $BeCl_2$ units are linked into long, chainlike polymeric molecules as $(BeCl_2)_n$.

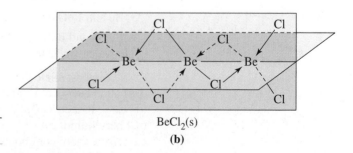

$BeCl_2(s)$
(b)

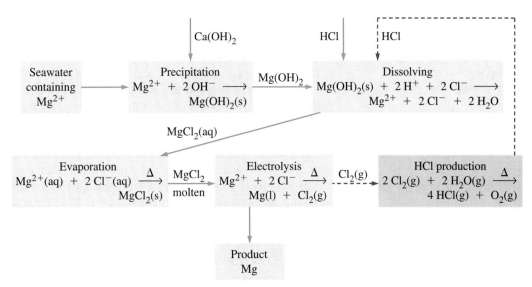

▲ **FIGURE 22-7   The Dow process for the production of Mg**
The main reaction sequence is traced by solid arrows. The recycling of $Cl_2(g)$ is shown by dashed arrows.

Calcium, strontium, and barium are obtained by the reduction of their oxides with aluminum; Ca and Sr also can be obtained by electrolysis of their molten chlorides. Calcium metal is used primarily as a reducing agent to prepare, from their oxides or fluorides, other metals such as U, Pu, and most of the lanthanides. Strontium and barium have limited use in alloys, but some of their compounds (discussed later) are quite important. Some salts of Sr and Ba provide vivid colors for pyrotechnic displays.

Magnesium metal is obtained by electrolysis of the molten chloride in the *Dow process*. The Dow process is outlined in Figure 22-7, and the electrolysis of $MgCl_2(l)$ is pictured in Figure 22-8. Like the Solvay process for making $NaHCO_3$, the Dow process takes advantage of simple chemistry and recycling.

▶ **FIGURE 22-8**
The electrolysis of molten $MgCl_2$
The electrolyte is a mixture of molten NaCl, $CaCl_2$, and $MgCl_2$. This mixture has a lower melting point and higher electrical conductivity than $MgCl_2$ alone, but if the voltage is carefully controlled, only Mg^{2+} is reduced in the electrolysis.

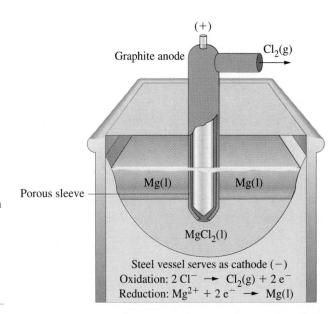

▶ In Example 19-6, we demonstrated that this precipitation goes to completion.

The source of magnesium is seawater or natural brines. The abundance of Mg^{2+} in seawater is about 1350 mg/L. The first step in the Dow process is the precipitation of $Mg(OH)_2(s)$ with slaked lime [$Ca(OH)_2$] as the source of OH^-. Slaked lime is formed by the reaction of quicklime (CaO) with water. The precipitated $Mg(OH)_2(s)$ is washed, filtered, and dissolved in HCl(aq). The resulting concentrated $MgCl_2(aq)$ is dried by evaporation, melted, and electrolyzed, yielding pure Mg metal and $Cl_2(g)$. The $Cl_2(g)$ is converted to HCl, which is recycled.

Magnesium has a lower density than that of any other metal used for structural purposes. Lightweight objects such as aircraft parts are manufactured from magnesium alloyed with aluminum and other metals. Magnesium is a good reducing agent and is used in a number of metallurgical processes, such as the production of beryllium, as mentioned earlier. The ease with which magnesium is oxidized also underlies its use in sacrificial anodes for corrosion protection (page 850). Magnesium's most spectacular use may be in fireworks.

Group 2 Compounds

The properties of group 2 compounds differ from those of group 1 compounds. In some cases, this difference is attributable to the smaller ionic size and the larger ionic charge of group 2 cations. For example, the lattice energy of $Mg(OH)_2$ is about $-3000 \ kJ \ mol^{-1}$, compared with about $-900 \ kJ \ mol^{-1}$ for NaOH. This difference in lattice energy helps account for the ready solubility of NaOH in water, to about 20 M NaOH, and the sparing solubility of $Mg(OH)_2$, with $K_{sp} = 1.8 \times 10^{-11}$. The heavier group 2 hydroxides are somewhat more soluble than $Mg(OH)_2$. Other alkaline earth compounds that are only slightly soluble include the carbonates, fluorides, and oxides. Another common characteristic of alkaline earth compounds is the formation of *hydrates*. Typical hydrates are $MX_2 \cdot 6H_2O$ (where M = Mg, Ca, or Sr, and X = Cl or Br). As noted earlier, beryllium compounds are rather different from the other group 2 compounds.

Halides The group 2 metals react directly with the halogens to form halides. Except for those of beryllium, these halides are essentially ionic.

$$M + X_2 \longrightarrow MX_2 \qquad (22.9)$$

(M = a group 2 metal, and X = F, Cl, Br, or I)

The halides have varied uses. As an example, $MgCl_2$, in addition to its use in the preparation of magnesium metal, is used in fireproofing wood, in special cements, in ceramics, in treating fabrics, and as a refrigeration brine.

Ionic Compound Solubility activity

Carbonates and Sulfates The group 2 carbonates are insoluble in water, as are the sulfates of Ca, Sr, and Ba. Because of this insolubility, these compounds are the most important minerals of the group 2 metals. The most familiar is $CaCO_3$, the principal component of the rock *limestone*. If a limestone contains more than 5% $MgCO_3$, it is usually called dolomitic limestone, or *dolomite*. Some clay, sand, or quartz may also be present in limestone. The primary use of limestone (about 70%) is as a building stone. Among other applications, limestone is used in the manufacture of quicklime and slaked lime, as an ingredient in glass, and as a flux in metallurgical processes. A *flux* is a material that combines with impurities and removes them as a free-flowing liquid—a *slag*—during production of a metal (page 962).

Portland cement, another important product of limestone, is a complex mixture of calcium silicates and aluminates. It is produced in long rotary kilns in which mixtures of limestone, clay, and sand are heated to progressively higher temperatures as they slowly move down the inclined kiln. First, moisture and, then, chemically bound water are driven off. This process is followed by decomposition (calcination) of the limestone to CaO(s) and $CO_2(g)$. Finally, the CaO combines with

▲ The decomposition (calcination) of limestone is carried out in a long rotary kiln, whether for the production of quicklime, CaO, or for the manufacture of Portland cement.

silica (SiO_2) and alumina (Al_2O_3) from the sand and clay to form silicates and aluminates. Pure cement does not have much strength. When mixed with sand, gravel, and water, however, it sets into the familiar rocklike mass called *concrete*. Portland cement is an especially valuable material for building bridge piers and other underwater structures because it hardens even when underwater—it is a *hydraulic* cement.

Pure, white $CaCO_3$ is used in a wide variety of products. For example, it is used in papermaking to impart brightness, opacity, smoothness, and good ink-absorbing qualities to paper. It is particularly suited to newer papermaking processes that produce acid-free (alkaline) paper with an expected shelf life of 300 years or more. $CaCO_3$ is used as a filler in plastics, rubber, floor tiles, putties, and adhesives as well as in foods and cosmetics. It is also used as an antacid and as a dietary supplement for the prevention of osteoporosis, a condition in which the bones become porous and brittle and break easily.

Three steps are required to obtain pure $CaCO_3$ from limestone: (1) thermal decomposition of limestone (called **calcination**), (2) reaction of CaO with water (slaking), and (3) conversion of an aqueous suspension of $Ca(OH)_2(s)$ to precipitated $CaCO_3$ (carbonation).

Calcination: $$CaCO_3(s) \xrightarrow{\Delta} CaO(s) + CO_2(g) \qquad (22.10)$$

Slaking: $$CaO(s) + H_2O(l) \longrightarrow Ca(OH)_2(s) \qquad (22.11)$$

Carbonation: $$Ca(OH)_2(s) + CO_2(g) \xrightarrow{aq} CaCO_3(s) + H_2O(l) \qquad (22.12)$$

Limestone ($CaCO_3$) is also responsible for the beautiful natural formations found in limestone caves. Natural rainwater is slightly acidic due to dissolved $CO_2(g)$ and is essentially a solution of carbonic acid, H_2CO_3.

$$CO_2 + 2\,H_2O \rightleftharpoons H_3O^+ + HCO_3^- \qquad K_{a_1} = 4.4 \times 10^{-7} \quad (22.13)$$

$$HCO_3^- + H_2O \rightleftharpoons H_3O^+ + CO_3^{2-} \qquad K_{a_2} = 4.7 \times 10^{-11} \quad (22.14)$$

Although the carbonates are not very soluble in water, they are bases; thus, they dissolve readily in acidic solutions. As mildly acidic groundwater seeps through limestone beds, insoluble $CaCO_3$ is converted to soluble $Ca(HCO_3)_2$.

$$CaCO_3(s) + H_2O + CO_2 \rightleftharpoons Ca(HCO_3)_2(aq) \qquad K_{eq} = 2.6 \times 10^{-5} \quad (22.15)$$

▲ **FIGURE 22-9**
Stalactites and stalagmites in Carlsbad Caverns, New Mexico

Over time, this dissolving action can produce a large cavity in the limestone bed—a limestone cave. Reaction (22.15) is reversible, however, and evaporation of the solution causes a loss of both water and CO_2 and conversion of $Ca(HCO_3)_2(aq)$ back to $CaCO_3(s)$. This process occurs very slowly. Yet over a period of many years, as $Ca(HCO_3)_2(aq)$ drips from the ceiling of a cave, $CaCO_3(s)$ remains as icicle-like deposits called **stalactites**. Some of the dripping solution may hit the floor of the cave before decomposition occurs; in this way, limestone deposits build up from the floor in formations called **stalagmites**. Eventually, some stalactites and stalagmites grow together into limestone columns (Figure 22-9).

A feature of the group 2 metal carbonates (MCO_3) that we have already noted for $CaCO_3(s)$ is their decomposition at high temperatures.

$$MCO_3(s) \xrightarrow{\Delta} MO(s) + CO_2(g) \qquad (22.16)$$

This is the chief method of preparing the group 2 metal oxides. The alkali metal carbonates, except Li_2CO_3, do not share this thermal instability. Thus, in the Solvay process (page 878), heating $NaHCO_3$ produces sodium carbonate, Na_2CO_3, but no further decomposition occurs.

▲ Plaster of Paris castings and the molds used to form them.

Another important calcium-containing mineral is *gypsum*, $CaSO_4 \cdot 2H_2O$. In the United States, about 50 million tons of gypsum is consumed annually. About half of this is converted to the *hemihydrate* ($\frac{1}{2}$-hydrate), **plaster of Paris**.

$$CaSO_4 \cdot 2H_2O(s) \xrightarrow{\Delta} CaSO_4 \cdot \tfrac{1}{2} H_2O(s) + \tfrac{3}{2} H_2O(g) \qquad (22.17)$$

When mixed with water, plaster of Paris reverts to gypsum. Because it expands as it sets, a mixture of plaster of Paris and water is useful in making castings where sharp details of an object must be retained. Plaster of Paris is extensively used in jewelry making and in dental work. The most important application, though, is in producing gypsum wallboard (drywall), which has all but supplanted other interior wall coverings in the construction industry.

Barium sulfate has had important applications in medical imaging because barium is opaque to X rays. Although barium ion is toxic, the very insoluble compound $BaSO_4$ is safe to use to coat the stomach or upper gastrointestinal tract with a "barium milkshake" and the lower tract with a "barium enema."

Oxides and Hydroxides All the oxides and hydroxides of the group 2 metals, except those of beryllium, are bases. Although it is not very soluble in water, calcium hydroxide is the cheapest commercial strong base. It is used in a variety of applications, such as the Solvay and Dow processes.

The term *lime* is familiar, but you may not be aware that it can refer to several compounds. CaO, called **quicklime**, is produced by the calcination of limestone (reaction 22.10). This reaction is reversible at room temperature; the reverse reaction occurs almost exclusively. So in the calcination of $CaCO_3$, a high temperature must be used and the $CO_2(g)$ produced must be continuously exhausted from the furnace (kiln) in which the reaction occurs. Among the many uses of quicklime are water treatment and the removal of $SO_2(g)$ from the smokestack gases in electric power plants.

$Ca(OH)_2$, called **slaked lime**, is formed by the action of water on CaO (reaction 22.11). A mixture of slaked lime, sand, and water is the familiar mortar used in bricklaying. First, excess water is absorbed by the bricks and then lost by evaporation. In the final setting of the mortar, $CO_2(g)$ from the air reacts with $Ca(OH)_2(s)$ and converts it back to $CaCO_3(s)$. The final form of the mortar is a complex mixture of hydrated calcium carbonate and silicate (from the sand).

$$Ca(OH)_2(s) + CO_2(g) \longrightarrow CaCO_3(s) + H_2O(g) \qquad (22.18)$$

This reaction is general for all group 2 hydroxides. Art conservators have used this reaction to preserve art objects. For example, an aqueous solution of $Ba(NO_3)_2$ is sprayed onto cracking frescos (paintings embedded in plaster). When the solution has had time to fill small cracks and spaces, an aqueous ammonia solution is applied to the surface of the fresco. The ammonia raises the pH of the solution, resulting in formation of $Ba(OH)_2$. As excess water evaporates, carbon dioxide from the air reacts with the barium hydroxide. Insoluble barium carbonate is produced, binding the cracking fresco together and strengthening it without affecting the delicate colors.

22-3 Ions in Natural Waters: Hard Water

Rainwater is not chemically pure. It contains dissolved atmospheric gases and, once it reaches the ground, dissolves some of the components of soil and rocks. It may pick up anywhere from a few to perhaps 1000 ppm of dissolved substances. If the

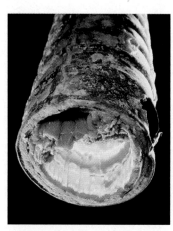

▲ Thick deposits of lime scale on the interior of a household pipe.

water contains ions capable of yielding significant quantities of a precipitate, we say that the water is *hard*.

There are two types of hard water: temporary hard water and permanent hard water. **Temporary hard water** contains bicarbonate ion, HCO_3^-. When water containing HCO_3^-(aq) is heated, the bicarbonate ion rapidly decomposes to give CO_3^{2-}, CO_2, and water. The CO_3^{2-} reacts with multivalent cations in the water, forming a mixed precipitate of $CaCO_3$, $MgCO_3$, and rust called *boiler scale*. The principal reaction that occurs is the reverse of reaction (22.15).

The formation of boiler scale can be a very serious problem in steam-driven electric power plants and in boilers producing steam for manufacturing processes. Formation of boiler scale lowers the efficiency of water heaters and can eventually cause a boiler to overheat or even to explode. Closer to home, we are familiar with this scale as a buildup observed on the inside of containers used for boiling water. Boiler scale can be removed by adding vinegar (acetic acid) to the container and heating.

Water with temporary hardness can be softened at a water treatment plant by adding slaked lime [$Ca(OH)_2$] and filtering off the precipitated metal carbonates. The OH^- reacts with bicarbonate ion to produce water and carbonate ion. The carbonate ion reacts with M^{2+} ion, such as Ca^{2+}, to precipitate a metal carbonate.

$$HCO_3^-(aq) + OH^-(aq) \longrightarrow H_2O(l) + CO_3^{2-}(aq)$$

$$CO_3^{2-}(aq) + M^{2+}(aq) \longrightarrow MCO_3(s)$$

Permanent hard water contains significant concentrations of anions other than HCO_3^-, such as SO_4^{2-}. Adding Na_2CO_3 (washing soda) softens permanent hard water by precipitating cations such as Ca^{2+} and Mg^{2+} as carbonates, leaving an aqueous solution with Na^+(aq). One annoying effect of hard water is encountered in the bath or shower. Water containing Ca^{2+} or Mg^{2+} ions forms a precipitate with soap. This familiar bathtub ring is a mixture of insoluble calcium and magnesium soaps, such as magnesium palmitate [see equation (22.8) for the structure of sodium palmitate, a soluble soap]. Formation of these precipitates also makes it difficult for other soaps or shampoos to foam up.

▲ A sodium soap is added to soft (distilled) water (*left*) and hard water (*right*). The cloudiness in the beaker on the right is caused by the formation of an insoluble calcium soap by the Ca^{2+} ions in the hard water.

Ion Exchange   One of the best ways to soften water is through **ion exchange**, a process in which the undesirable ions in hard water, typically Ca^{2+}, Mg^{2+}, and Fe^{3+}, are exchanged for ions that are not objectionable, such as Na^+. Ion exchange occurs when the hard water is passed through a column (or bed) containing an ion-exchange material. This material can be either a natural porous sodium aluminosilicate polymer called a *zeolite* or a synthetic resinous material. These polymeric materials ionize to produce two types of ions: *fixed ions*, which remain attached to the polymer surface; and free, or mobile, *counterions*. The counterions are the ones that exchange places with the undesirable ions when a sample of hard water is passed through the resin or zeolite.

Figure 22-10 depicts a resin with negatively charged fixed ions R^- and positively charged counterions. Initially, the counterions in the resin bed are Na^+. When hard water is passed through the bed, the more highly charged Ca^{2+}, Mg^{2+}, and Fe^{3+} displace Na^+ as counterions. Concentrated NaCl(aq) is used to regenerate ion-exchange resins. When present in high concentration, Na^+ can displace multivalent cations and restore an ion-exchange resin to its original condition. The ion-exchange material has an indefinite lifetime.

The only material consumed in water softening by ion exchange is the NaCl required for regenerating the ion-exchange medium. This method of water softening does have the disadvantage that the treated water has a high concentration of Na^+ and would not be appropriate for drinking by a person on a low-sodium diet.

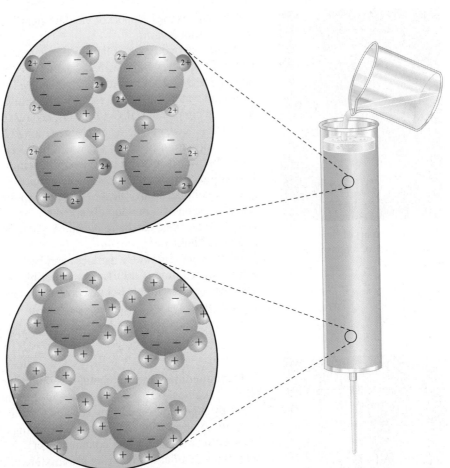

▲ **FIGURE 22-10   Ion exchange**
The resin shown here is a cation-exchange resin. Multivalent cations (green, Fe^{2+}; yellow, Ca^{2+}) replace Na^+(orange) at the top of the resin column. By the time the water has reached the bottom of the column, all the multivalent ions have been removed and only Na^+ ions remain as counterions. The exchange can be represented as $2\,NaR + M^{2+} \rightleftharpoons MR_2 + 2\,Na^+$. The reaction occurs in the forward direction during water softening. As expected from Le Châtelier's principle, in the presence of concentrated NaCl(aq), the reverse reaction is favored and the resin is recharged.

Ion exchange can be used to make **deionized water**, the familiar "DI" water found in chemistry laboratories. Deionized water is used in laboratories because ions present in ordinary tap water may interfere with chemical reactions (by forming precipitates or catalyzing reactions, for example). Deionized water is made by first replacing the cations in a water sample with H^+. Then the water is passed through a second ion-exchange column, which exchanges OH^- for all the *anions* present. The H^+ and OH^- combine to form H_2O, resulting in water that is essentially free of all ions.

EXAMPLE 22-2

Determining the Hardness of Water by Ion Exchange. A 25.00-mL sample of hard water is passed through an ion-exchange column in the acid form, HR. The water coming off the column requires 7.59 mL of 0.0133 M NaOH for its titration. What is the hardness of the water, expressed as parts per million of Ca^{2+}?

Solution

First, we assume that Ca^{2+} is the only cation present and that the ion-exchange reaction is

$$Ca^{2+} + 2\,HR \longrightarrow Ca_2R + 2\,H^+$$

Now we can use the titration data to determine the number of millimoles of H^+ that have been replaced by Ca^{2+}. The neutralization reaction is simply $H^+ + OH^- \longrightarrow H_2O$.

? mmol $H^+ =$

$$7.59\ \text{mL} \times \frac{0.0133\ \text{mmol NaOH}}{1\ \text{mL}} \times \frac{1\ \text{mmol OH}^-}{1\ \text{mmol NaOH}} \times \frac{1\ \text{mmol H}^+}{1\ \text{mmol OH}^-} = 0.101\ \text{mmol H}^+$$

Next, we can determine the mass of Ca^{2+} present in the 25.00 mL of water. In this conversion, we use the fact that *two* H^+ ions are released into the water for every Ca^{2+} ion retained on the ion-exchange resin.

$$?\ \text{mg Ca}^{2+} = 0.101\ \text{mmol H}^+ \times \frac{1\ \text{mmol Ca}^{2+}}{2\ \text{mmol H}^+} \times \frac{40.08\ \text{mg Ca}^{2+}}{1\ \text{mmol Ca}^{2+}} = 2.02\ \text{mg Ca}^{2+}$$

The mass of 25.00 mL of the hard water is 25.0 g, or 25×10^3 mg. The number of milligrams of Ca^{2+} present in 1.00×10^6 mg H_2O is the concentration of Ca^{2+} in parts per million.

$$\text{ppm Ca}^{2+} = 1.00 \times 10^6\ \text{mg H}_2\text{O} \times \frac{2.02\ \text{mg Ca}^{2+}}{25,000\ \text{mg H}_2\text{O}} = 80.8$$

Practice Example A: What was the pH of the 25.00-mL sample of water in Example 22-2 as it came off the ion-exchange column and *before* it was titrated with NaOH(aq)?

Practice Example B: A sample of water whose hardness is 185 ppm Ca^{2+} is passed through a cation-exchange resin, and the Ca^{2+} is replaced by Na^+. What is $[Na^+]$ in that water sample?

22-4 Group 13 Metals: Aluminum, Gallium, Indium, and Thallium

In their appearance and physical properties and in most of their chemical behavior, aluminum, gallium, indium, and thallium are metallic. Boron, the first element in group 13, however, is a nonmetal and will be discussed in Chapter 23. Some properties of the group 13 metals are listed in Table 22.5.

TABLE 22.5 Some Properties of the Group 13 Metals

	Al	Ga	In	Tl
Atomic number	13	31	49	81
Atomic (metallic) radius, pm	143	122	163	170
Ionic (M^{3+}) radius, pm	53	62	79	88
Electronegativity	1.5	1.6	1.7	1.8
First ionization energy, kJ mol^{-1}	577.6	578.8	558.3	589.3
Electrode potential $E°$, V[a]	−1.676	−0.56	−0.34	+0.72
Melting point, °C	660.37	29.78	156.17	303.55
Boiling point, °C	2467	2403	2080	1457
Density, g/cm^3 at 20 °C	2.698	5.907	7.310	11.85
Hardness[b]	2.75	1.5	1.2	1.25
Electrical conductivity[b]	59.7	9.1	19.0	8.82

[a] For the reduction $M^{3+}(aq) + 3\,e^- \longrightarrow M(s)$.
[b] See footnotes of Table 22.2.

Properties and Uses of the Group 13 Metals

The most important metal of the group is aluminum. The third most abundant element, aluminum comprises 8.3% by mass of Earth's solid crust. Its main use is in lightweight alloys. More than 5 million tons of aluminum are produced per year, on average, in the United States. Aluminum, like most of the other main-group metals, is an active metal. Because it is easily oxidized to the 3+ ion, aluminum is an excellent reducing agent—for example, reacting with acids to reduce $H^+(aq)$ to $H_2(g)$.

$$2\,Al(s) + 6\,H^+(aq) \longrightarrow 2\,Al^{3+}(aq) + 3\,H_2(g) \qquad (22.19)$$

Aluminum is unusual in that it also reacts with basic solutions. This behavior is due to the acid properties of $Al(OH)_3$, a topic discussed further on page 894.

▶ Certain drain cleaners are a mixture of NaOH and Al(s). When they are added to water, reaction (22.20) occurs. The evolved $H_2(g)$ helps unplug a stopped-up drain. The heat of reaction melts fats and grease, and the NaOH(aq) solubilizes them.

$$2\,Al(s) + 2\,OH^-(aq) + 6\,H_2O(l) \longrightarrow 2[Al(OH)_4]^-(aq) + 3\,H_2(g) \quad (22.20)$$

Air or other oxidants easily oxidize powdered aluminum in highly exothermic reactions that are used in some rocket fuels and explosives.

$$2\,Al(s) + \tfrac{3}{2}\,O_2(g) \longrightarrow Al_2O_3(s) \qquad \Delta H = -1676\ \text{kJ} \qquad (22.21)$$

Aluminum is such a good reducing agent that it will extract oxygen from metal oxides, producing aluminum oxide while liberating the other metal in its free state. This reaction, known as the **thermite reaction**, is used in the on-site welding of large metal objects (recall page 107).

$$Fe_2O_3(s) + 2\,Al(s) \longrightarrow Al_2O_3(s) + 2\,Fe(l) \qquad (22.22)$$

Thermite movie

Gallium metal is becoming increasingly important in the electronics industry. It is used to make gallium arsenide (GaAs), a compound that can convert light directly into electricity (photoconduction). This semiconducting material is also used in light-emitting diodes (LEDs; see Focus On, page 898) and in solid-state devices such as transistors.

Indium is a soft silvery metal used to make low-melting alloys. Like GaAs, InAs also finds use in low-temperature transistors and as a photoconductor in optical devices.

Thallium and its compounds are extremely toxic; as a result, they have few uses in industry. One possible use, however, is in high-temperature superconductors. For example, a thallium-based ceramic with the approximate formula $Tl_2Ba_2Ca_2Cu_3O_{8+x}$ exhibits superconductivity at temperatures as high as 125 K.

▶ As discussed on page 978, a superconducting material loses its electrical resistance below a certain temperature. Metals typically become superconducting only a few degrees above 0 K.

? Are You Wondering . . .

If aluminum reacts with and dissolves in both acidic and basic solutions, why does it not also dissolve in pH-neutral water?

Metallic aluminum reacts rapidly with the oxygen in air to give a thin, tough, water-insoluble coating of Al_2O_3. This oxide layer protects the metal beneath it from further reaction. In either acidic or basic solution, but not in pH-neutral water, the Al_2O_3 layer reacts and dissolves.

$$Al_2O_3(s) + 6\,H^+(aq) \longrightarrow 2\,Al^{3+}(aq) + 3\,H_2O(l)$$

$$Al_2O_3(s) + 2\,OH^-(aq) + 3\,H_2O(l) \longrightarrow 2\,[Al(OH)_4]^-(aq)$$

Once the Al_2O_3 layer has been removed, the underlying metal displays its true reactivity: Aluminum is a sufficiently active metal to displace $H_2(g)$ from pure water. We could never use aluminum metal in aircraft and building construction were it not for the protection of the Al_2O_3 surface coating.

Oxidation States

Aluminum, at the top of the group of four metals in group 13, occurs almost exclusively in the +3 oxidation state in its compounds. Gallium also favors the +3 oxidation state. Indium compounds can be found with +3 and +1 oxidation states, though the +3 is more common. In thallium, this preference is reversed. For example, thallium forms the oxide Tl_2O, the hydroxide $TlOH$, and the carbonate Tl_2CO_3. These compounds are ionic and in some respects resemble group 1 compounds. Thus, $TlOH$ is both very soluble and a strong base in aqueous solution. The higher stability of the +1 over the +3 oxidation state of thallium is often described as the **inert pair effect**. Thallium has the electron configuration $[Xe]4f^{14}5d^{10}6s^26p^1$. In forming the Tl^+ ion, a Tl atom loses the $6p$ electron and retains two electrons in its $6s$ subshell. It is this pair of electrons—$6s^2$—that is called the inert pair. The electron configuration $(n-1)s^2(n-1)p^6(n-1)d^{10}ns^2$ is commonly encountered in ions of the post-transition elements. One explanation of the inert pair effect is that the small bond energies and lattice energies associated with the large atoms and ions at the bottom of a group are not sufficiently great to offset the ionization energies of the ns^2 electrons.

Production of Aluminum

▶ Stimulated by a remark by one of his professors, Charles Martin Hall invented the electrolytic process for the production of aluminum at the age of 23, eight months after his graduation from Oberlin College. Paul Héroult, a student of Le Châtelier, and also at the age of 23, invented the identical process in the same year.

When an aluminum cap was placed atop the Washington Monument in 1884, aluminum was still a semiprecious metal. It cost several dollars per pound to produce and was used mostly in jewelry and artwork. But just two years later, all this changed. In 1886, Charles Martin Hall in the United States and Paul Héroult in France independently discovered an economically feasible method of producing aluminum from Al_2O_3 by electrolysis.

The manufacture of Al involves several interesting principles. The chief ore, *bauxite*, contains Fe_2O_3 as an impurity that must be removed. The principle used in the separation is that Al_2O_3 is an *amphoteric* oxide and dissolves in NaOH(aq), whereas the iron oxide is a *basic* oxide and does not (Figure 22-11).

| (a) | (b) | (c) |

▲ **FIGURE 22-11   Purifying bauxite**
(**a**) When an excess of OH^-(aq) is added to a solution containing Al^{3+}(aq) and Fe^{3+}(aq), the Fe^{3+} precipitates as $Fe(OH)_3$(s) and the $Al(OH)_3$(s) first formed redissolves to produce $[Al(OH)_4]^-$(aq). (**b**) The $Fe(OH)_3$(s) is filtered off, and the $[Al(OH)_4]^-$(aq) is made slightly acidic through the action of CO_2, here added as dry ice. (**c**) The precipitated $Al(OH)_3$(s) collects at the bottom of a clear, colorless solution.

$$Al_2O_3(s) + 2\,OH^-(aq) + 3\,H_2O(l) \longrightarrow 2\,[Al(OH)_4]^-(aq)$$

When the solution containing $[Al(OH)_4]^-$ is slightly acidified, $Al(OH)_3(s)$ precipitates. Pure Al_2O_3 is obtained by heating the $Al(OH)_3$.

$$[Al(OH)_4]^-(aq) + H_3O^+(aq) \longrightarrow Al(OH)_3(s) + 2\,H_2O$$
$$2\,Al(OH)_3(s) \xrightarrow{\Delta} Al_2O_3(s) + 3\,H_2O(g)$$

Al_2O_3 has a very high melting point (2020 °C) and produces a liquid that is a poor electrical conductor. Thus, its electrolysis is not feasible without a better conducting solvent. That was the crux of the discovery by Hall and Héroult. They found, independently, that up to 15% Al_2O_3, by mass, can be dissolved in the molten mineral *cryolite*, Na_3AlF_6, at about 1000 °C. The liquid is a good electrical conductor, so the electrolysis of Al_2O_3 is accomplished in molten cryolite. A typical electrolysis cell, pictured in Figure 22-12, produces aluminum of 99.6–99.8% purity.

The electrode reactions are not known with certainty, but the overall electrolysis reaction is

Oxidation:	$3\,\{C(s) + 2\,O^{2-} \longrightarrow CO_2(g) + 4\,e^-\}$
Reduction:	$4\,\{Al^{3+} + 3\,e^- \longrightarrow Al(l)\}$

Overall: $\quad 3\,C(s) + 4\,Al^{3+} + 6\,O^{2-} \longrightarrow 4\,Al(l) + 3\,CO_2(g) \qquad$ (22.23)

The energy consumed to produce aluminum by electrolysis is very high, about 15,000 kWh/ton Al [compared, for example, with 3000 kWh per ton of Cl_2 in the electrolysis of NaCl(aq)]. This means that aluminum production facilities are generally located near low-cost power sources, typically hydroelectric power. The energy required to recycle Al is only about 5% of that to produce the metal from bauxite, and currently about 45% of the Al produced in the United States is by the recycling of scrap aluminum.

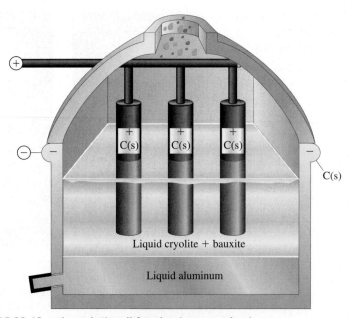

▲ **FIGURE 22-12** **Electrolysis cell for aluminum production**
The cathode is a carbon lining in a steel tank. The anodes are also made of carbon. Liquid aluminum is denser than the electrolyte medium and collects at the bottom of the tank.

Are You Wondering...

Why so much energy is consumed in the electrolytic production of aluminum?

Any process that must be carried out at a high temperature requires large amounts of energy for heating. In the electrolytic production of Al, the electrolysis bath must be kept at about 1000 °C, which is done by means of electric heating. Two other factors, however, are also involved in the large energy consumption. First, to produce one mole of Al, *three* moles of electrons must be transferred: $Al^{3+} + 3\,e^- \longrightarrow Al(l)$. Additionally, the molar mass of Al is relatively low, 27 g mol^{-1}. The electric current equivalent to the passage of one mole of electrons produces only 9 g Al. In contrast, one mole of electrons produces 12 g Mg, 20 g Ca, or 108 g Ag. Yet the same factors that make Al production a significant energy consumer make Al an outstanding energy producer when it is used in a battery. (Recall the aluminum–air battery described on page 848.)

Aluminum Halides

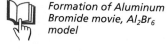

Formation of Aluminum Bromide movie, Al_2Br_6 model

Aluminum fluoride, AlF_3, has considerable ionic character. It has a high melting point (1040 °C) and when molten is a conductor of electric current. In contrast, the other aluminum halides exist as *molecular* species with the formula Al_2X_6 (for which X = Cl, Br, or I). We can think of this molecule as comprising two AlX_3 units. When two identical units combine, the resulting molecule is called a **dimer**. The dimeric structure of Al_2Cl_6 consists of two Cl atoms bonded exclusively to each Al atom and two Cl atoms bridging the two metal atoms (Figure 22-13). Bonding in this molecule can be described by sp^3 hybridization of the two Al atoms. Each bridging Cl atom appears to bond to two Al atoms in two ways. The bond to one Al atom is a conventional covalent bond: Each atom contributes one electron to the bond. The bond to a second Al atom is a coordinate covalent bond, where the chlorine atom provides the pair of electrons for the bond, noted by arrows in Figure 22-13.

The aluminum halides are very reactive Lewis acids. They easily accept a pair of electrons, forming an acid–base compound called an **adduct**. Adduct formation results in the formation of a covalent bond between the Lewis acid and Lewis base. In the following reaction, $AlCl_3$ is the Lewis acid and diethyl ether, $(C_2H_5)_2O$, is the Lewis base.

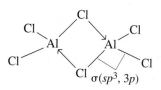

Aluminum halides are used in a number of organic reactions in which, by means of adduct formation, they act as catalysts.

▶ **FIGURE 22-13**
Bonding in Al_2Cl_6
Two Cl atoms bridge the $AlCl_3$ units and produce the dimer Al_2Cl_6. Electrons donated by these Cl atoms to Al atoms are indicated by arrows.

Bonding scheme

Space-filling model

The hydride ion, H^-, can be thought of as similar to a halide ion, and the AlH_4^- ion as an adduct of AlH_3 and H^-. Lithium aluminum hydride, $LiAlH_4$, is an important reducing agent in organic chemistry.

Another important halide *complex* of aluminum is *cryolite*, Na_3AlF_6. Cryolite exists as $3\,Na^+ + AlF_6^{3-}$. In the molten state, it is the solvent and electrolyte in the Hall–Héroult process for producing aluminum metal. For aluminum manufacturing, natural cryolite has largely been displaced by cryolite synthesized in lead-clad vessels by the reaction

$$6\,HF + Al(OH)_3 + 3\,NaOH \longrightarrow Na_3AlF_6 + 6\,H_2O \qquad (22.24)$$

Aluminum Oxide and Hydroxide

Aluminum oxide has several common names. It is often referred to as *alumina*; when in crystalline form, it is called *corundum*. Corundum, when pure, is also known as the gemstone *white sapphire*. Certain other gemstones consist of corundum with small amounts of transition metal ions as impurities: Cr^{3+} in ruby and Fe^{3+} and Ti^{4+} in blue sapphire, for example. Artificial gemstones are made by fusing corundum with carefully controlled amounts of other oxides.

▶ The origin of colors in gemstones is discussed on page 1016.

The bonding and crystal structure of alumina account for its physical properties. The small Al^{3+} ion and small O^{2-} ion make a strong ionic bond. The crystal has a cubic closest packed structure of O^{2-} ions, with Al^{3+} ions filling the octahedral holes. Because of this structure, alumina is a very hard material and is often used as an abrasive. It is also resistant to heat (mp 2020 °C) and is used in linings for high-temperature furnaces and as catalyst supports in industrial chemical processes. Aluminum oxide is relatively unreactive except at very high temperatures. Its stability at high temperatures classifies it as a *refractory* material.

As mentioned previously, aluminum is protected against reaction with water in the pH range 4.5–8.5 by a thin, impervious coating of Al_2O_3. This coating can be purposely thickened to enhance the corrosion resistance of the metal by a process known as *anodizing*. An aluminum object is used for the anode and a graphite electrode is used for the cathode in an electrolyte bath of $H_2SO_4(aq)$. The half-reaction occurring at the anode during electrolysis is

$$2\,Al(s) + 3\,H_2O(l) \longrightarrow Al_2O_3(s) + 6\,H^+(aq) + 6\,e^-$$

▲ Drinking cups made of anodized aluminum.

Al_2O_3 coatings of varying porosity and thickness can be obtained. Also, the oxide can be made to absorb pigments or other additives. Anodized aluminum is commonly used in architectural components of buildings, such as bronze or black window frames.

Aluminum hydroxide is *amphoteric*. It reacts with acids in a manner typical of metal hydroxides.

$$Al(OH)_3(s) + 3\,H_3O^+(aq) \longrightarrow [Al(H_2O)_6]^{3+}(aq) \qquad (22.25)$$

Also, it reacts with bases in a reaction best represented as the formation of a hydroxo complex ion.

$$Al(OH)_3(s) + OH^-(aq) \longrightarrow [Al(OH)_4]^-(aq) \qquad (22.26)$$

Aluminum Sulfate and Alums

Aluminum sulfate is the most important aluminum compound used commercially. It is prepared by the reaction of hot concentrated $H_2SO_4(aq)$ on $Al_2O_3(s)$. The product that crystallizes from solution is $Al_2(SO_4)_3 \cdot 18H_2O$. Over 1 million tons of aluminum sulfate are produced annually in the United States, about half of which is

▲ Alum crystals.

used in water purification. In this application, the pH of the water is adjusted so that $Al(OH)_3(s)$ precipitates when aluminum sulfate is added. As the $Al(OH)_3(s)$ settles, it removes suspended solids in the water. Another important use is in the sizing of paper. *Sizing* refers to incorporating materials such as waxes, glues, or synthetic resins into paper to make the paper more water resistant. $Al(OH)_3$ precipitated from $Al_2(SO_4)_3(aq)$ helps deposit the sizing agent in the paper.

When an aqueous solution of equimolar amounts of $Al_2(SO_4)_3$ and K_2SO_4 is allowed to crystallize, the crystals obtained are of potassium aluminum sulfate, $KAl(SO_4)_2 \cdot 12H_2O$. This is just one of a large class of double salts called alums. **Alums** have the formula $M(I)M(III)(SO_4)_2 \cdot 12H_2O$, where M(I) is a unipositive cation (other than Li^+) and M(III) is a tripositive cation—Al^{3+}, Ga^{3+}, In^{3+}, Ti^{3+}, V^{3+}, Cr^{3+}, Mn^{3+}, Fe^{3+}, Co^{3+}, Re^{3+}, or Ir^{3+}. The actual ions present in the alums are $[M(H_2O)_6]^+$, $[M(H_2O)_6]^{3+}$, and SO_4^{2-}. The most common alums have M(I) = K^+, Na^+, or NH_4^+ and M(III) = Al^{3+}. Li^+ does not form alums because the ion is too small to meet the crystal structure requirements. Sodium aluminum sulfate is the leavening acid in baking powders, and potassium aluminum sulfate is used in dyeing. The fabric to be dyed is dipped into a solution of the alum and heated in steam. Hydrolysis of $[Al(H_2O)_6]^{3+}$ deposits $Al(OH)_3$ into the fibers of the material, and the dye is adsorbed on the $Al(OH)_3$.

▶ A disadvantage of the use of aluminum sulfate in sizing paper is that its acidic character contributes to deterioration of the paper. In contrast, calcium carbonate maintains an alkaline medium in paper (page 885).

▶ In the industrial world, *alum* usually refers to simple aluminum sulfate; terms such as *potash* (potassium) *alum* and *ammonium alum* designate the double salts.

22-5 Group 14 Metals: Tin and Lead

The properties of the group 14 elements vary dramatically within the group. Tin and lead, at the bottom of the group, have mainly metallic properties. Germanium, sometimes referred to as a *metalloid*, exhibits semiconductor behavior. Silicon, although also exhibiting semiconductor properties, is mostly nonmetallic in its chemical behavior. Carbon, the first member of group 14, is a nonmetal. We will discuss carbon and silicon in Chapter 23.

The data in Table 22.6 suggest that tin and lead are rather similar to each other. Both are soft and malleable and melt at low temperatures. The ionization energies and standard electrode potentials of the two metals are also about the same. This means that their tendencies to be oxidized to the +2 oxidation state are comparable.

The fact that both tin and lead can exist in two oxidation states, +2 and +4, is an example of the inert pair effect mentioned in Section 22-4. In the +2 oxidation state, the inert pair ns^2 is not involved in bond formation, whereas in the +4 oxidation state, the pair does participate. Tin displays a stronger tendency to exist in the +4 oxidation state than does lead. That tendency is consistent with the trend observed in group 13, in which the lower oxidation state is favored farther down a group.

Another difference between tin and lead is that tin exists in two common crystalline forms (α and β), whereas lead has but a single solid form. The α (gray), or nonmetallic, form of tin is stable below 13 °C; the β (white), or metallic, form of

TABLE 22.6	Some Properties of Tin and Lead (of Group 14)	
	Sn	**Pb**
Atomic number	50	82
Atomic (metallic) radius, pm	141	175
Ionic (M^{2+}) radius, pm	93	118
First ionization energy, kJ mol^{-1}	709	716
Electrode potential $E°$, V		
$\quad$ [M^{2+}(aq) + 2 e$^-$ $\longrightarrow$ M(s)]	-0.137	-0.125
$\quad$ [M^{4+}(aq) + 2 e$^-$ $\longrightarrow$ M^{2+}(aq)]	$+0.154$	$+1.5$
Melting point, °C	232	327
Boiling point, °C	2623	1751
Density, g/cm^3 at 20 °C	5.77 (α, gray)	11.34
	7.29 (β, white)	
Hardness[a]	1.6	1.5
Electrical conductivity[a]	14.4	7.68

[a] See footnotes of Table 22.2.

tin is stable above 13 °C. Ordinarily, when a sample of β tin is cooled, it must be kept below 13 °C for a long time before the transition to α tin occurs. Once it does begin, however, the transformation takes place rather rapidly and with dramatic results. Because α tin is less dense than the β variety, the tin expands and crumbles to a powder. This transformation leads to the disintegration of objects made of tin. It has been a particular problem in churches in colder climates because some organ pipes are made of tin or tin alloys. The transformation is known in northern Europe as the *tin disease*, *tin pest*, or *tin plague*. It added to the troubles of Napoleon's army in the siege of Moscow because the soldiers' buttons were made of tin. As the cold winter wore on, the buttons disintegrated.

The chief tin ore is tin(IV) oxide, SnO_2, known as *cassiterite*. After initial purification, the tin(IV) oxide is reduced with carbon (coke) to produce tin metal.

$$SnO_2(s) + C(s) \xrightarrow{\Delta} Sn(l) + CO_2(g) \qquad (22.27)$$

Nearly 50% of the tin metal produced is used in tinplate, especially in plating iron for use in cans for storing foods. The next most important use (about 25% of the total produced) is in the manufacture of **solders**—low-melting alloys used to join wires or pieces of metal. Other important alloys of tin are *bronze* (90% Cu, 10% Sn) and *pewter* (85% Sn, 7% Cu, 6% Bi, 2% Sb). Alloys of Sn and Pb are used to make organ pipes.

Lead is found chiefly as lead(II) sulfide, PbS, an ore known as *galena*. The lead(II) sulfide is first converted to lead(II) oxide by strongly heating it in air, a process called **roasting**. The oxide is then reduced with coke to produce the metal.

$$2\,PbS(s) + 3\,O_2(g) \xrightarrow{\Delta} 2\,PbO(s) + 2\,SO_2(g) \qquad (22.28)$$

$$2\,PbO(s) + C(s) \xrightarrow{\Delta} 2\,Pb(l) + CO_2(g) \qquad (22.29)$$

Over half the lead produced is used in lead–acid (storage) batteries. Other uses include the manufacture of solder and other alloys, ammunition, and radiation shields (to protect against X rays).

Oxides

Tin forms two primary oxides, SnO and SnO_2. By heating SnO in air, it can be converted to SnO_2. One use of SnO_2 is as a jewelry abrasive.

Lead forms a number of oxides, and the chemistry of some of these oxides is not completely understood. The best known oxides of lead are yellow PbO, *litharge*, red-brown lead dioxide, PbO_2, and a mixed-valence oxide known as *red lead*, Pb_3O_4. Lead oxides are used in the manufacture of lead–acid (storage) batteries, glass, ceramic glazes, cements (PbO), metal-protecting paints (Pb_3O_4), and matches (PbO_2). Other lead compounds are generally made from the oxides.

Because lead tends to be in the +2 oxidation state, Pb(IV) compounds tend to undergo reduction to compounds of Pb(II) and are therefore good oxidizing agents. A case in point is PbO_2. In Chapter 21, we noted its use at the cathode in lead–acid storage cells. The reduction of $PbO_2(s)$ can be represented by the half-equation

$$PbO_2(s) + 4\,H^+(aq) + 2\,e^- \longrightarrow Pb^{2+}(aq) + 2\,H_2O(l) \qquad E° = +1.455\text{ V}$$

$PbO_2(s)$ is a better oxidizing agent than $Cl_2(g)$ and nearly as good as $MnO_4^-(aq)$. For example, $PbO_2(s)$ can oxidize HCl(aq) to $Cl_2(g)$.

$$PbO_2(s) + 4\,HCl(aq) \longrightarrow PbCl_2(aq) + 2\,H_2O(l) + Cl_2(g) \qquad E°_{cell} = 0.097\text{ V}$$

Halides

Both chlorides of tin—$SnCl_2$ and $SnCl_4$—have important uses. Tin(II) chloride, $SnCl_2$, is a good reducing agent and is used in the quantitative analysis of iron ores to reduce Fe(III) to Fe(II) in aqueous solution. Tin(IV) chloride, $SnCl_4$, is formed by the direct reaction of tin and $Cl_2(g)$; it is the form in which tin is recovered from scrap tinplate. Tin(II) fluoride, SnF_2 (stannous fluoride), has an important use as an anticavity additive to toothpaste.

Other Compounds

One of the few soluble lead compounds is lead(II) nitrate, $Pb(NO_3)_2$. It is formed in the reaction of PbO_2 with nitric acid.

$$2\,PbO_2(s) + 4\,HNO_3(aq) \longrightarrow 2\,Pb(NO_3)_2(aq) + 2\,H_2O(l) + O_2(g)$$

Addition of a soluble chromate salt to $Pb(NO_3)_2(aq)$ produces the pigment lead(II) chromate (*chrome yellow*), $PbCrO_4$. Another lead-based pigment used in ceramic glazes and once extensively used in the manufacture of paint is *white lead*, a basic lead carbonate, $2PbCO_3\cdot Pb(OH)_2$.

Lead Poisoning

Beginning with the ancient Romans and continuing to fairly recent times, lead has been used in plumbing systems, including those designed to transport water. Exposure to lead has also occurred through cooking and eating utensils and pottery glazes. In colonial times, lead poisoning was clearly diagnosed as the cause of "dry bellyache" suffered by some North Carolinians who consumed rum made in New England. The distilling equipment used in the manufacture of the rum had components made of lead.

Mild forms of lead poisoning produce nervousness and mental depression. More severe cases can lead to permanent nerve, brain, and kidney damage. Lead interferes with the biochemical reactions that produce the iron-containing heme group in hemoglobin. As little as 10–15 μg Pb/dL in blood seems to produce physiological effects, especially in small children. The phaseout out of leaded gasoline has resulted

▶ Another mixed-valence oxide that we have encountered on several occasions in this text is Fe_3O_4 (see, for example, page 80).

◀ Solar cells employing gallium arsenide (gold and red) placed on a silicon-based solar cell. The cells containing gallium arsenide are more efficient and can be made smaller than a comparable silicon cell.

Until the 1980s, the principal application of gallium metal was in high-temperature thermometers. Now an unlikely compound—gallium arsenide, (GaAs)—has become one of the most versatile high-technology materials of our time. Among the properties of gallium arsenide is the ability to convert electric energy into light. A crystal of GaAs can serve as the light-generating component of a light-emitting diode (LED), a device employed in indicator lights on stereo equipment and calculator displays. GaAs can also be fabricated into diode lasers, very small lasers used in compact disc systems and for transmitting infrared light through fiber-optic cables.

Recall the discussion of the band theory and semiconductors on pages 465–468. Gallium arsenide is an intrinsic semiconductor. This compound, with eight valence electrons per formula unit (three from Ga and five from As), has a filled valence band. If sufficient electric

in a dramatic drop in average lead blood levels (Figure 22-14). The principal sources of lead contamination now seem to be lead-based painted surfaces found in old buildings and soldered joints in plumbing systems. Lead has been eliminated from modern plumbing solder, which is now a mixture of 95% Sn and 5% Sb. Because lead is toxic, its disposal is closely monitored. Recycling provides about three-quarters of the current lead metal production.

▶ **FIGURE 22-14**

Lead in gasoline and in blood

The level of lead in the blood of a representative human population showed a dramatic decline that paralleled the decline in the use of lead additives in gasoline in the 1970s. (Data source: E. P. A. Office of Policy Analysis, 1984.)

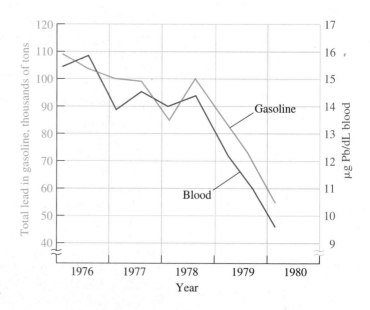

energy is supplied, however, electrons are promoted to the conduction band. When the excited electrons return to the ground state (valence band), they emit photons of light with a wavelength proportional to the energy associated with the band gap (ΔE in Figure 12-33).

What is special about GaAs semiconductors is that the band gap can be "tuned" by adding controlled amounts of another semiconductor—gallium phosphide. GaAs and GaP form solid solutions whose compositions can be varied in all proportions. GaP has a band-gap energy corresponding to that of green light (540 nm), whereas the band-gap energy in GaAs corresponds to infrared light (890 nm).* A series of solid solutions of GaAs and GaP can be represented by the general formula GaP_xAs_{1-x}. The band gap increases with increased amounts of phosphorus, so the emitted light increases in energy. Ga–P–As semiconductors whose compositions fall between those of GaP and GaAs emit light of wavelength between 540 and 890 nm. For example, the most common LED, with the formula $GaP_{0.40}As_{0.60}$, emits 660 nm red light.

The effect of the composition of Ga–P–As semiconductors on the color of the emitted light is due in part to electronegativity differences (ΔEN) and in part to differences in atomic

* G. Lisensky, R. Penn, M. Geselbracht, and A. Ellis, *J. Chem. Educ.* **69**, 151 (1992).

sizes. In GaP, ΔEN is slightly greater, and the length of the unit cell in the crystal lattice is smaller than in GaAs. These factors lead to more ionic character in the bonds and a larger band-gap energy in GaP than in GaAs. Thus the emitted light from GaP should be of a shorter wavelength than that from GaAs.

An interesting effect is noted when the temperature of an LED is changed. When an LED with the composition $GaP_{0.40}As_{0.60}$ is cooled to the boiling point of liquid nitrogen (77 K), the color shifts from that of the 660 nm red light observed at room temperature to a 639 nm *orange* light. This effect is attributed to a slight contraction of the crystal lattice of the semiconductor, which brings the atoms into closer contact and increases the band-gap energy.

These new semiconductor materials require new methods for their synthesis. Instead of precipitation from solution, the technique used is *chemical vapor deposition* (CVD). A chemical reaction is carried out between trimethylgallium, $Ga(CH_3)_3(g)$, and highly toxic arsine, $AsH_3(g)$; GaAs is deposited from the vapor phase as a thin film. If an appropriate amount of phosphine, $PH_3(g)$, is included in the reaction mixture, the film produced is GaP_xAs_{1-x}.

The application of gallium arsenide described here and its use in high-speed computers have changed the demand for GaAs from practically none in the early 1980s to a current worldwide market measured in the hundreds of millions of dollars.

Summary

The alkali (group 1) metals are the most active of the metals, as indicated by their low ionization energies and large negative electrode potentials. Most of the metals are prepared by the electrolysis of their molten salts. Electrolysis of NaCl(aq) produces NaOH(aq), from which many other sodium compounds can then be prepared. Na_2CO_3 can be produced from NaCl, NH_3, and $CaCO_3$ by the Solvay process.

The alkaline earth (group 2) metals are also very active. Some are prepared by the electrolysis of a molten salt, and some by chemical reduction. Among the most important of the alkaline earth compounds are the carbonates, especially $CaCO_3$. Reversible reactions involving CO_3^{2-}, HCO_3^-, $CO_2(g)$, and H_2O account for the formation of limestone caves and temporary hard water. Water may also have non-carbonate, or permanent,

hardness. Water can be softened either through chemical reactions or by ion exchange.

The principal metal of group 13 is aluminum, whose large-scale use is made possible by an effective method of production. The amphoterism of Al_2O_3 is the basis for separating Al_2O_3 from impurities, mostly Fe_2O_3. Electrolysis is carried out in molten Na_3AlF_6 with Al_2O_3 as a solute. Gallium has gained importance in the electronics industry because of the desirable semiconductor properties of gallium arsenide (GaAs). The use of thallium in fabricating high-temperature superconductors is also a possibility.

Tin and lead in group 14 have some similarities: They are soft metals with low melting points. They also have some differences, including the fact that tin acquires the oxidation state +4 rather easily, whereas the +2 oxidation state is favored by lead.

Integrative Example

Without performing detailed calculations, demonstrate that reaction (22.15) correctly describes the dissolving action of rainwater on limestone. (For $CaCO_3$, $K_{sp} = 2.8 \times 10^{-9}$.)

1. *Write equations for the various equilibria involved to show that reaction (22.15) is directly related to them.* The relevant equations are the solubility product expression for $CaCO_3$ and equations (22.13) and (22.14). That is,

$$CaCO_3(s) \rightleftharpoons Ca^{2+}(aq) + CO_3^{2-}(aq)$$
$$K_{sp} = 2.8 \times 10^{-9}$$

$$CO_2 + 2\,H_2O \rightleftharpoons H_3O^+ + HCO_3^-$$
$$K_{a_1} = 4.4 \times 10^{-7} \qquad (22.13)$$

$$HCO_3^- + H_2O \rightleftharpoons H_3O^+ + CO_3^{2-}$$
$$K_{a_2} = 4.7 \times 10^{-11} \qquad (22.14)$$

899

Equation (22.15) is related to the three equations above in this way: K_{sp} expression + equation (22.13) − equation (22.14).

$$CaCO_3(s) + H_2O + CO_2 \rightleftharpoons Ca^{2+}(aq) + 2\,HCO_3^-(aq)$$
$$K_{eq} = (K_{sp} \times K_{a_1})/K_{a_2} = 2.6 \times 10^{-5} \qquad (22.15)$$

With equation (22.15), we could calculate an actual solubility for $CaCO_3$ for certain stated conditions, but we seek only a qualitative assessment.

2. *Qualitative Assessment.* Consider which is the greater source of CO_3^{2-} ions in solution: a saturated aqueous solution of $CaCO_3$ or an aqueous carbonic acid solution.

[CO_3^{2-}] in saturated $CaCO_3$(aq): In this solution, $[Ca^{2+}] = [CO_3^{2-}]$;

$$K_{sp} = [Ca^{2+}][CO_3^{2-}] = [CO_3^{2-}]^2 = 2.8 \times 10^{-9}.$$
$$[CO_3^{2-}] = (2.8 \times 10^{-9})^{\frac{1}{2}} = 5.3 \times 10^{-5}\ M.$$

[CO_3^{2-}] in a carbonic acid solution:
H_2CO_3 is a weak *diprotic* acid with $K_{a_1} \ll K_{a_2}$.
Thus, $[H_3O^+] = [HCO_3^-]$, and $[CO_3^{2-}] = K_{a_2} = 4.7 \times 10^{-11}\ M$.
Carbonate ion from a saturated $CaCO_3$ solution acts as a common ion in the carbonic acid equilibrium and displaces reaction (22.14) *to the left*, converting CO_3^{2-} to *more HCO_3^-* while consuming H_3O^+. Removal of H_3O^+ in reaction (22.14) stimulates reaction (22.13) to shift *to the right*, producing more H_3O^+ and, simultaneously, *more HCO_3^-*. The overall effect is that H_2O, CO_2, and CO_3^{2-} (from $CaCO_3$) are consumed and HCO_3^- is produced, just as suggested by equation (22.15).

Key Terms

adduct (22-4)
alkali metal (22-1)
alkaline earth metal (22-2)
alum (22-4)
calcination (22-2)
deionized water (22-3)
detergent (22-1)
diagonal relationship (22-1)

dimer (22-4)
inert pair effect (22-4)
ion exchange (22-3)
permanent hard water (22-3)
plaster of Paris (22-2)
quicklime (22-2)
roasting (22-5)
slaked lime (22-2)

soap (22-1)
solders (22-5)
stalactites (22-2)
stalagmites (22-2)
temporary hard water (22-3)
thermite reaction (22-4)

Review Questions

1. In your own words, define the following terms: **(a)** dimer; **(b)** adduct; **(c)** calcination; **(d)** amphoteric oxide.
2. Briefly describe each of the following ideas, methods, or phenomena: **(a)** diagonal relationship; **(b)** preparation of deionized water by ion exchange; **(c)** thermite reaction; **(d)** inert pair effect.
3. Explain the important distinction between each pair of terms: **(a)** peroxide and superoxide; **(b)** quicklime and slaked lime; **(c)** temporary and permanent hard water; **(d)** soap and detergent.
4. Provide an acceptable name or formula for each of the following: **(a)** PbO; **(b)** SnF_2; **(c)** $CaSO_4 \cdot \frac{1}{2} H_2O$; **(d)** lithium nitride; **(e)** $Ca(OH)_2$; **(f)** potassium superoxide; **(g)** magnesium hydrogen carbonate.
5. Complete and balance the following. Write the simplest equation possible. If no reaction occurs, so state.

   (a) $Li_2CO_3(s) \xrightarrow{\Delta}$
   (b) $CaCO_3(s) + HCl(aq) \longrightarrow$
   (c) $Al(s) + NaOH(aq) \longrightarrow$
   (d) $BaO(s) + H_2O(l) \longrightarrow$
   (e) $Na_2O_2(s) + CO_2(g) \longrightarrow$

6. Assuming that water, common reagents (acids, bases, salts), and simple laboratory equipment are available, give a practical method to prepare **(a)** $MgCl_2$ from $MgCO_3(s)$; **(b)** $NaAl(OH)_4$ from $Na(s)$ and $Al(s)$; **(c)** Na_2SO_4 from $NaCl(s)$.

7. Write the simplest chemical equation to represent the reaction of **(a)** K_2CO_3(aq) and $Ba(OH)_2$(aq); **(b)** $Mg(HCO_3)_2$(aq) on heating; **(c)** tin(II) oxide when heated with carbon; **(d)** CaF_2(s) and H_2SO_4(concd aq); **(e)** $NaHCO_3$(s) and HCl(aq); **(f)** PbO_2(s) and HBr(aq).
8. Write an equation to represent the reaction of gypsum, $CaSO_4 \cdot 2H_2O$, with ammonium carbonate to produce ammonium sulfate (a fertilizer), calcium carbonate, and water.
9. All but one of the following can be used to soften temporary hard water: NH_3, Na_2CO_3, NH_4Cl, $Ca(OH)_2$. Which one cannot? Why not?
10. Write chemical equations to represent the most probable outcome in each of the following. If no reaction is likely to occur, so state.

    (a) $SrCO_3(s) \xrightarrow{\Delta}$
    (b) $Al_2O_3(s) \xrightarrow{\Delta}$
    (c) $Li_2CO_3(s) \xrightarrow{\Delta}$

11. A chemical dictionary gives the following descriptions of the production of some compounds. Write plausible chemical equations based on these descriptions.
    (a) lead carbonate: adding a solution of sodium bicarbonate to a solution of lead nitrate
    (b) lithium carbonate: reaction of lithium oxide with ammonium carbonate solution

Exercises **901**

(c) hydrogen peroxide: by the action of dilute sulfuric acid on barium peroxide
(d) lead(IV) oxide: action of an alkaline solution of calcium hypochlorite on lead(II) oxide
12. Of the metals K, Sr, Al, Na, which one does not react with cold water? Explain why not.

13. All the following substances react with water: (a) Ca and CaH_2; (b) Na and Na_2O_2; (c) K and KO_2. Which pair yields the same gaseous products?
14. Name the chemical compound(s) you would expect to be the *primary* constituent(s) of (a) stalactites; (b) gypsum; (c) bathtub ring; (d) "barium milkshake"; (e) blue sapphires.

Exercises
Group 1 (Alkali) Metals

15. Use information from the chapter to write chemical equations to represent each of the following:
(a) reaction of cesium metal with chlorine gas
(b) formation of sodium peroxide (Na_2O_2)
(c) thermal decomposition of lithium carbonate
(d) reduction of sodium sulfate to sodium sulfide
(e) combustion of potassium to form potassium superoxide
16. Use information from the chapter to write chemical equations to represent each of the following:
(a) reaction of rubidium metal with water
(b) thermal decomposition of aqueous $KHCO_3$
(c) combustion of lithium metal in oxygen gas
(d) action of concentrated aqueous H_2SO_4 on KCl(s)
(e) reaction of lithium hydride with water
17. Describe a simple test for determining whether a pure white solid is either LiCl or KCl.
18. Describe two methods for determining the identity of an unknown compound that is either Li_2CO_3 or K_2CO_3.
19. Write net ionic equations for the reactions of the normal oxide, peroxide, and superoxide ions with water that are referred to on page 880.
20. The first electrolytic process to produce sodium metal used molten NaOH as the electrolyte. Write probable half-equations and an overall equation for this electrolysis.

21. A 0.872-L NaCl(aq) solution is electrolyzed for 2.50 min with a current of 0.810 A.
(a) Calculate the pH of the solution after electrolysis.
(b) Why doesn't the result depend on the initial concentration of the NaCl(aq)?
22. A lithium battery used in a cardiac pacemaker has a voltage of 3.0 V and a capacity of 0.50 A h (ampere hour). Assume that 5.0 μW of power is needed to regulate the heartbeat. (*Hint:* See Appendix B.)
(a) How long will the implanted battery last?
(b) How many grams of lithium must be present in the battery for the lifetime calculated in part (a)?
23. An analysis of a Solvay plant shows that for every 1.00 ton of NaCl consumed, 1.03 tons of $NaHCO_3$ are obtained. The quantity of NH_3 consumed in the overall process is 1.5 lb.
(a) What is the percent efficiency of this process for converting NaCl to $NaHCO_3$?
(b) Why is so little NH_3 required?
24. Consider the reaction $Ca(OH)_2(s) + Na_2SO_4(aq) \rightleftharpoons CaSO_4(s) + 2\,NaOH(aq)$.
(a) Write a net ionic equation for this reaction.
(b) Will the reaction go essentially to completion?
(c) What will be $[SO_4^{2-}]$ and $[OH^-]$ *at equilibrium* if a slurry of $Ca(OH)_2(s)$ is mixed with 1.00 M $Na_2SO_4(aq)$?

Group 2 (Alkaline Earth) Metals

25. In the manner used to construct Figure 22-2, complete the diagram outlined. Specifically, indicate the reactants (and conditions) you would use to produce the indicated substances from $Ca(OH)_2$.

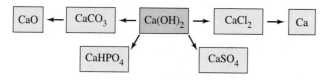

26. Replace the calcium-containing substances shown in the diagram accompanying Exercise 25 by their magnesium-containing equivalents. Then describe the reactants (and conditions) you would use to produce the indicated substances from $MgSO_4$.
27. In the Dow process (Figure 22-7), the starting material is Mg^{2+} in seawater and the final product is *Mg* metal. This

process seems to violate the principle of conservation of charge. Does it? Explain.
28. Which has the (a) higher melting point, MgO or BaO; (b) greater solubility in water, MgF_2 or $MgCl_2$? Explain.
29. Write chemical equations to represent the following:
(a) reduction of BeF_2 to Be metal with Mg as a reducing agent
(b) reaction of barium metal with Br_2(l)
(c) reduction of uranium(IV) oxide to uranium metal with calcium as the reducing agent
(d) calcination of dolomite, a mixed calcium magnesium carbonate ($MgCO_3 \cdot CaCO_3$)
(e) complete neutralization of phosphoric acid with quicklime
30. Write chemical equations for the reactions you would expect to occur when
(a) $Mg(HCO_3)_2(s)$ is heated to a high temperature.
(b) $BaCl_2$(l) is electrolyzed.

(c) Sr(s) is added to cold dilute HBr(aq).
(d) $Ca(OH)_2(aq)$ is added to $H_2SO_4(aq)$.
(e) $CaSO_4 \cdot 2H_2O(s)$ is heated.

31. *Without performing detailed calculations*, indicate whether equilibrium is displaced either far to the left or far to the right for each of the following reactions. Use data from Appendix D as necessary.
(a) $BaSO_4(s) + CO_3^{2-}(aq) \rightleftharpoons BaCO_3(s) + SO_4^{2-}(aq)$
(b) $Mg_3(PO_4)_2(s) + 3\ CO_3^{2-}(aq) \rightleftharpoons$
$3\ MgCO_3(s) + 2\ PO_4^{3-}(aq)$
(c) $Ca(OH)_2(s) + 2\ F^-(aq) \rightleftharpoons CaF_2(s) + 2\ OH^-(aq)$

32. *Without performing detailed calculations*, indicate why you would expect each of the following reactions to occur to a significant extent as written. Use data from Appendix D as necessary.
(a) $BaCO_3(s) + 2\ CH_3CO_2H(aq) \longrightarrow$
$Ba^{2+}(aq) + 2\ CH_3CO_2^-(aq) + H_2O(l) + CO_2(g)$
(b) $Ca(OH)_2(s) + 2\ NH_4^+(aq) \longrightarrow$
$Ca^{2+}(aq) + 2\ NH_3(aq) + 2\ H_2O(l)$
(c) $BaF_2(s) + 2\ H_3O^+(aq) \longrightarrow$
$Ba^{2+}(aq) + 2\ HF(aq) + 2\ H_2O(l)$

Hard Water

33. Write an equation to show the softening of temporary hard water with quicklime.

34. Write an equation to show the softening of temporary hard water with ammonia.

35. A certain hard water contains 185.0 ppm HCO_3^-, and all the cations associated with the HCO_3^- are Ca^{2+}.
(a) What is $[Ca^{2+}]$, expressed in parts per million?
(b) What is $[Ca^{2+}]$, expressed in milligrams per liter?
(c) How many grams of $Ca(OH)_2$ are required to soften 1.00×10^6 g of this water?

36. Concerning the hard water described in Exercise 35,
(a) How many grams of $CaCO_3$ would be precipitated in softening 1.00×10^6 g of this water with $Ca(OH)_2$?
(b) Show that in the $CaCO_3$, half the Ca^{2+} is derived from the $Ca(OH)_2$ used in the water softening and half from the water itself.

37. A sample of hard water is passed through a cation-exchange resin, in which cations are exchanged for H_3O^+. The water coming off the resin has a pH $= 2.37$. Assume that the cations in the water are Ca^{2+}, and determine the hardness expressed as ppm Ca^{2+}.

38. The hardness of a sample of water is due only to $CaSO_4$. When this water is passed through an anion-exchange resin, SO_4^{2-} ions are replaced by OH^-. A 25.00-mL sample of the water so treated requires 22.42 mL of 1.00×10^{-3} M H_2SO_4 for its titration. What is the hardness of the water, expressed in parts per million of $CaSO_4$? Assume that the density of the water is 1.00 g/mL.

39. What mass of bathtub ring will form if 32.1 L of water having 82.6 ppm Ca^{2+} is treated with an excess of the soap potassium stearate, $CH_3(CH_2)_{16}COO^-K^+$?

40. One of the components of bathtub ring is magnesium palmitate, $Mg(CH_3(CH_2)_{14}CO_2)_2$. A saturated solution of magnesium palmitate is found to contain 80 ppm $Mg^{2+}(aq)$. What is the K_{sp} of magnesium palmitate?

A Group 13 Metal: Aluminum

41. Write chemical equations to represent the
(a) reaction of Al(s) with HCl(aq)
(b) reaction of solid aluminum with NaOH(aq)
(c) oxidation of Al(s) to $Al^{3+}(aq)$ by an aqueous solution of sulfuric acid (the reduction product is $SO_2(g)$)

42. Write plausible equations for the
(a) reaction of Al(s) with $Br_2(l)$
(b) production of Cr from $Cr_2O_3(s)$ by the thermite reaction, with Al as the reducing agent
(c) separation of Fe_2O_3 impurity from bauxite ore

43. In some foam-type fire extinguishers, the reactants are $Al_2(SO_4)_3(aq)$ and $NaHCO_3(aq)$. When the extinguisher is activated, these reactants mix, producing $Al(OH)_3(s)$ and $CO_2(g)$. The $Al(OH)_3$–CO_2 foam extinguishes the fire. Write a net ionic equation to represent this reaction.

44. Some baking powders contain the solids $NaHCO_3$ and $NaAl(SO_4)_2$. When water is added to this mixture of compounds, $CO_2(g)$ and $Al(OH)_3(s)$ are two of the products.

Write plausible net ionic equations for the formation of these two products.

45. The maximum resistance to corrosion of aluminum is between pH 4.5 and 8.5. Explain how this observation is consistent with other facts about the behavior of aluminum presented in this text.

46. Describe a series of *simple* chemical reactions that you could use to determine whether a particular metal sample is "aluminum 2S" (99.2% Al) or "magnalium" (70% Al, 30% Mg). You are permitted to destroy the metal sample in the testing.

47. In the purification of bauxite ore, a preliminary step in the production of aluminum, $[Al(OH)_4]^-(aq)$ can be converted to $Al(OH)_3(s)$ by passing $CO_2(g)$ through the solution. Write an equation for the reaction that occurs. Could HCl(aq) be used instead of $CO_2(g)$? Explain.

48. In 1825, Hans Oersted produced aluminum chloride by passing chlorine over a heated mixture of carbon and aluminum oxide. In 1827, Friedrich Wöhler obtained aluminum by heating aluminum chloride with potassium. Write plausible equations for these reactions.

49. A description for preparing potassium aluminum alum calls for dissolving aluminum foil in KOH(aq). The solution obtained is treated with H_2SO_4(aq), and the alum is crystal-lized from the resulting solution. Write plausible equations for these reactions.

50. Handbooks and lists of chemicals do not contain entries under the formulas $Al(HCO_3)_3$ and $Al_2(CO_3)_2$. Explain why these compounds do not exist.

Some Group 14 Metals: Tin and Lead

51. Write plausible chemical equations for the **(a)** dissolving of lead(II) oxide in nitric acid; **(b)** heating of $SnCO_3$(s); **(c)** reduction of lead(II) oxide by carbon; **(d)** reduction of Fe^{3+}(aq) to Fe^{2+}(aq) by Sn^{2+}(aq); **(e)** formation of lead(II) sulfate during high-temperature roasting of lead(II) sulfide.

52. Write plausible chemical equations for preparing each compound from the indicated starting material: **(a)** $SnCl_2$ from SnO; **(b)** $SnCl_4$ from Sn; **(c)** $PbCrO_4$ from PbO_2. What reagents (acids, bases, salts) and equipment commonly available in the laboratory are needed for each reaction?

53. Lead dioxide, PbO_2, is a good oxidizing agent. Use appropriate data from Appendix D to determine whether PbO_2(s) in a solution with $[H_3O^+] = 1$ M is a sufficiently good oxidizing agent to carry the following oxidations to the point at which the concentration of the species being oxidized decreases to one-thousandth of its initial value.
(a) Fe^{2+}(1 M) to Fe^{3+}
(b) SO_4^{2-}(1 M) to $S_2O_8^{2-}$
(c) Mn^{2+}(1×10^{-4} M) to MnO_4^-

54. Tin(II) ion, Sn^{2+}, is a good reducing agent. Use data from Appendix D to determine whether Sn^{2+} is a sufficiently good reducing agent to reduce **(a)** I_2 to I^-; **(b)** Fe^{2+} to Fe(s); **(c)** Cu^{2+} to Cu(s); **(d)** Fe^{3+}(aq) to Fe^{2+}(aq). Assume that all reactants and products are in their standard states.

Integrative and Advanced Exercises

55. The melting point of NaCl(s) is 801 °C, much higher than that of NaOH (322 °C). More energy is consumed to melt and maintain molten NaCl than NaOH. Yet the preferred commercial process for the production of sodium is electrolysis of NaCl(l) rather than NaOH(l). Give a reason or reasons for this discrepancy.

56. When a 0.200-g sample of Mg is heated in air, 0.315 g of product is obtained. Assume that all the Mg appears in the product.
(a) If the product were pure MgO, what mass should have been obtained?
(b) Show that the 0.315 g product could be a mixture of MgO and Mg_3N_2.
(c) What is the mass percent of MgO in the MgO–Mg_3N_2 mixed product?

57. Comment on the feasibility of using a reaction similar to (22.3) to produce **(a)** lithium metal from LiCl; **(b)** cesium metal from CsCl, with Na(l) as the reducing agent in each case.
(*Hint:* Consider data from Table 22.2.)

58. Concerning the thermite reaction,
(a) Use data from Appendix D to calculate $\Delta H°$ at 298 K for the reaction.

$$2\,Al(s) + Fe_2O_3(s) \longrightarrow 2\,Fe(s) + Al_2O_3(s)$$

(b) Write an equation for the reaction when MnO_2(s) is substituted for Fe_2O_3(s), and calculate $\Delta H°$ for this reaction.

(c) Show that if MgO were substituted for Fe_2O_3, the reaction would be *endothermic*.

59. Use data from Appendix D (Table D-2) to calculate a value of $E°$ for the reduction of Li^+(aq) to Li(s), and compare your result with the value listed in Table 22.2.

60. The electrolysis of 0.250 L of 0.220 M $MgCl_2$ is conducted until 104 mL of gas (a mixture of H_2 and water vapor) is collected at 23 °C and 748 mmHg. Will $Mg(OH)_2$(s) precipitate if electrolysis is carried to this point? (Use 21 mmHg as the vapor pressure of the solution.)

61. A particular water sample contains 56.9 ppm SO_4^{2-} and 176 ppm HCO_3^-, with Ca^{2+} as the only cation.
(a) How many parts per million of Ca^{2+} does the water contain?
(b) How many grams of CaO are consumed in removing HCO_3^- from 602 kg of the water?
(c) Show that the Ca^{2+} remaining in the water after the treatment described in part **(b)** can be removed by adding Na_2CO_3.
(d) How many grams of Na_2CO_3 are required for the precipitation referred to in part **(c)**?

62. An Al production cell of the type pictured in Figure 22-12 operates at a current of 1.00×10^5 A and a voltage of 4.5 V. The cell is 38% efficient in using electrical energy to produce chemical change. (The rest of the electrical energy is dissipated as thermal energy in the cell.)
(a) What mass of Al can be produced by this cell in 8.00 h?
(b) If the electrical energy required to power this cell is produced by burning coal (85% C; heat of combustion of

$C = 32.8 \text{ kJ/g}$) in a power plant with 35% efficiency, what mass of coal must be burned to produce the mass of Al determined in part (a)?

63. Use data from Appendix D (Table D-2) to estimate the minimum voltage required to electrolyze Al_2O_3 in the Hall–Héroult process, reaction (22.23). Use ΔG_f° $[Al_2O_3(l)] = -1520 \text{ kJ mol}^{-1}$. Show that the oxidation of the graphite anode to $CO_2(g)$ permits the electrolysis to occur at a lower voltage than if the electrolysis reaction were $Al_2O_3(l) \longrightarrow 2\,Al(l) + \frac{3}{2}\,O_2(g)$.

64. At 20 °C, a saturated aqueous solution of $Pb(NO_3)_2$ maintains a relative humidity of 97%. What must be the composition of this solution, expressed as g $Pb(NO_3)_2/100.0$ g H_2O?

65. Use information from this chapter and elsewhere in this text to explain why neither the compound $PbBr_4$ nor the compound PbI_4 exists.

66. To prevent the air oxidation of aqueous solutions of Sn^{2+} to Sn^{4+}, metallic tin is sometimes kept in contact with the $Sn^{2+}(aq)$. Suggest how this contact helps prevent the oxidation.

67. The dissolution of $MgCO_3(s)$ in $NH_4^+(aq)$ can be represented as

$$MgCO_3(s) + NH_4^+(aq) \rightleftharpoons$$
$$Mg^{2+}(aq) + HCO_3^-(aq) + NH_3(aq)$$

Calculate the molar solubility of $MgCO_3$ in each of the following solutions: (a) 1.00 M $NH_4Cl(aq)$; (b) a buffer that is 1.00 M NH_3 and 1.00 M NH_4Cl; (c) a buffer that is 0.100 M NH_3 and 1.00 M NH_4Cl.

68. Show that, in principle, $Na_2CO_3(aq)$ can be converted almost completely to $NaOH(aq)$ by the reaction

$$Ca(OH)_2(s) + Na_2CO_3(aq) \longrightarrow$$
$$CaCO_3(s) + 2\,NaOH(aq)$$

69. Assume that the packing of spherical atoms in crystalline metals is the same for Li, Na, and K, and explain why Na has a higher density than *both* Li and K. (*Hint:* Use data from Table 22.2.)

70. Would you expect the lattice energy of $MgS(s)$ to be less than, greater than, or about the same as that of $MgO(s)$? Use appropriate data from various locations in this text to obtain values of the two lattice energies. Use a value of 456 kJ for the process $S^-(g) + e^- \longrightarrow S^{2-}(g)$.

Feature Problems

71. In Chapter 21, we examined the relationship of electrode potentials to thermodynamic data. In fact, electrode potentials can be calculated from tabulated thermodynamic data (many of which, in turn, were established from electrochemical measurements). To demonstrate this fact and to pursue further the discussion in the Are You Wondering feature on page 875, combine the three steps for the oxidation of Li(s) with a corresponding set of three steps for the reduction of $H^+(1 \text{ M})$ to $H_2(g)$. Obtain ΔH° for the overall reaction.
(a) Neglect entropy changes that occur (that is, assume that $\Delta G^{\circ} \approx \Delta H^{\circ}$), and estimate the value of $E^{\circ}_{Li^+/Li}$.
(b) Combine the calculated ΔH° value with a value of ΔS° to obtain another estimate of $E^{\circ}_{Li^+/Li}$.
(*Hint:* Hydration energies for $Li^+(g)$ and $H^+(g)$ to form 1 M solutions are -506 and -1079 kJ/mol, respectively. Also, use data from various locations in this text.)

72. Mono Lake, in eastern California, is a rather unusual salt lake. The lake has no outlets; water leaves only by evaporation. The rate of evaporation is great enough that the lake level would be lowered by four feet per year were it not for freshwater entering through underwater springs and streams originating in the nearby Sierra Nevada mountains. The principal salts in the lake are the chlorides, bicarbonates, and sulfates of sodium. An approximate "recipe" for simulating the lake water is to dissolve 18 tablespoons of sodium bicarbonate, 10 tablespoons of sodium chloride, and 8 teaspoons of Epsom salt (magnesium sulfate heptahydrate) in 1 gallon of water (although the lake water actually contains only trace amounts of magnesium ion). Assume that

1 tablespoon of any of the salts weighs about 10 g. (1 tablespoon = 3 teaspoons.)
(a) Expressed as grams of salt per liter, what is the approximate salinity of Mono Lake? How does this salinity compare with seawater, which is approximately 0.438 M NaCl and 0.0512 M $MgCl_2$?
(b) Estimate an approximate pH of Mono Lake water. How does your estimate compare to the observed pH of about 9.8? Actually, the recipe for the lake water also calls for a pinch of borax. How would its presence affect the pH? [Borax is a sodium salt, $Na_2B_4O_7 \cdot 10H_2O$, related to the weak monoprotic boric acid (pK_a = 9.25), which will be described in Section 23-6.]
(c) Mono Lake has some unusual limestone formations called *tufa*. They form at the site of underwater springs and grow only underwater, although some project above the water, having formed at a time when the lake level was higher. Explain how the tufa form.
[*Hint:* What chemical reaction(s) is(are) involved?]

 **eMedia Exercises**

73. The alkali metals are described as being the most chemically reactive elements. The **Sodium and Potassium in Water** movie *(eChapter 22-1)* shows examples of vigorous reactions. Based upon the principles described in this section, as well as those of atomic properties and thermodynamics, account for the differences in the observed reactivity between sodium and potassium.

74. The three-dimensional molecular models of two alkali soaps are shown in *eChapter 22-1*. Describe the structural differences of these molecules and account for their differing macroscopic properties and uses.

75. Use the **Ionic Compound Solubility** activity *(eChapter 22-2)* to compare the solubilities of group 1 and 2 metal compounds. **(a)** What general observations can you make about the differences in the solubilities of the two different groups of metals? **(b)** What atomic properties are related to these differences?

76. The **Thermite** movie *(eChapter 22-4)* illustrates the energetics involved in the oxidation of metallic aluminum. What do your observations of the thermite reaction indicate about the requirements for the production of metallic aluminum from ore?

77. The **Formation of Aluminum Bromide** movie *(eChapter 22-4)* depicts the physical state of this compound. From this depiction and the molecular structure shown in the **Al_2Br_6** model, predict differences in the physical properties of aluminum bromide and other ionic compounds involving aluminum (Al_2O_3, $Al(OH)_3$, AlF_3).

23

Main-Group Elements II: Nonmetals

Contents

Bromine, a nonmetal, is a reactant in the synthesis of flame retardants for use in plastics.

We have already discussed a number of aspects of the chemistry of the nonmetals. In Chapter 8, we learned that the principal gases of the atmosphere (N_2, O_2, and Ar) are nonmetals and that some of the significant trace components in the atmosphere (for example, the oxides of carbon, nitrogen, and sulfur) are compounds of the nonmetals. Also, in several instances, we have encountered chemical reactions of the nonmetals that are of natural or commercial importance, as in the natural and artificial fixation of nitrogen, the depletion of stratospheric ozone, and the formation of acid rain. In this chapter, our focus will be entirely on the nonmetals, and we will describe their behavior in a more systematic way.

After a brief survey of a special group of nonmetals (the noble gases), we will first study the most active group of nonmetals, the halogens. Then we will proceed through the periodic table from right to left, moving toward the elements with increasing metallic character. This approach will be a counterpoint to that of Chapter 22, where we progressed from the most metallic elements (group 1) toward the elements having more nonmetallic character. Among the fundamental concepts we will emphasize in the discussion are atomic, physical, and thermodynamic properties; bonding and structure; acid–base chemistry; and oxidation states, redox reactions, and electrochemistry.

23-1 Group 18: The Noble Gases

The elements of group 18 were isolated at the end of the nineteenth century and are now referred to as the *noble gases*. Their production and uses were discussed in Section 8-4. Initially, the noble gases were found to be chemically inert. This apparent inertness helped provide a theoretical framework for the Lewis theory of bonding.

In the 1930s, Linus Pauling did theoretical calculations suggesting that xenon should form oxide and fluoride compounds, but attempts to make them failed at that time. In 1962, N. Bartlett and D. H. Lohmann discovered that O_2 and PtF_6 would join in a $1:1$ mole ratio to form the compound O_2PtF_6. Properties of this compound suggest it to be ionic: $[O_2]^+[PtF_6]^-$. The energy required to extract an electron from O_2 is 1177 kJ mol^{-1}, almost identical to the first ionization energy of Xe, 1170 kJ mol^{-1}. The size of the Xe atom is also roughly the same as that of the dioxygen molecule. Thus, it was reasoned that the compound $XePtF_6$ might exist. Bartlett and Lohmann were able to prepare a yellow crystalline solid with this composition.[*]

Soon thereafter, chemists around the world synthesized several additional noble gas compounds. In general, the conditions necessary to form noble gas compounds are as Pauling predicted:

- a readily ionizable (therefore, high atomic number) noble gas atom
- highly electronegative atoms (such as F or O) to bond to it

Xenon compounds have been synthesized that have Xe in four possible oxidation states.

	+2	+4	+6	+8
Examples:	XeF_2	XeF_4, $XeOF_2$	XeF_6, XeO_3	XeO_4, H_4XeO_6

XeF₂ and XeF₄ models

Because it is difficult to oxidize Xe to these positive oxidation states, we should expect Xe compounds to be easily reduced, making them very strong oxidizing agents. For example, in aqueous acidic solution

$$XeF_2(aq) + 2\,H^+(aq) + 2\,e^- \longrightarrow Xe(g) + 2\,HF(aq) \qquad E^\circ = +2.64\ \text{V}$$

The significance of this large E° value is that XeF_2 is not very stable in aqueous solution; XeF_2 oxidizes the water, producing $O_2(g)$.

Reduction: $\quad 2\,\{XeF_2(aq) + 2\,H^+(aq) + 2\,e^- \longrightarrow Xe(g) + 2\,HF(aq)\}$

Oxidation: $\qquad\qquad\qquad 2\,H_2O(l) \longrightarrow 4\,H^+(aq) + O_2(g) + 4\,e^-$

Overall: $\qquad\quad 2\,XeF_2(aq) + 2\,H_2O(l) \longrightarrow 2\,Xe(g) + 4\,HF(aq) + O_2(g)$

$$E^\circ_{cell} = E^\circ(\text{reduction}) - E^\circ(\text{oxidation})$$

$$= E^\circ_{XeF_2/Xe} - E^\circ_{O_2/H_2O}$$

$$= 2.64\ \text{V} - (1.229\ \text{V}) = 1.41\ \text{V}$$

[*]It has since been established that this solid is more complicated than first thought. It has the formula $Xe(PtF_6)_n$, where n is between 1 and 2.

▲ Crystals of xenon tetrafluoride under magnification.

Xenon reacts directly only with F_2. When the gases are heated together at 400 °C in a sealed nickel vessel, the products depend on the Xe/F_2 mole ratio.

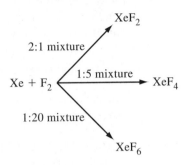

The fluorides XeF_2, XeF_4, and XeF_6 are colorless crystalline solids and are stable if kept from contact with water. The oxides of Xe are obtained from the fluorides. XeO_3 is an explosive white solid, and XeO_4 is an explosive colorless gas. The shapes of the three fluorides of xenon can be interpreted by using VSEPR theory, the results of which are summarized in Figure 23-1. The structure of XeF_6 is depicted with the possible distortion of an octahedron because of the presence of the lone pair of electrons. The prediction of the nonoctahedral shape for XeF_6 was an important success for VSEPR theory.

▶ For XeF_6, we require seven orbitals to accommodate the seven pairs of electrons; hence sp^3d^3 hybrid orbitals are used as a basis for a description of the bonding according to the valence bond model.

To conform to the shapes indicated by VSEPR theory (or experiment when the structure is known), bonding theory based on hybridization of orbitals, the valence bond method, requires the hybrid orbitals sp^3d for XeF_2, sp^3d^2 for XeF_4, and sp^3d^3 for XeF_6. However, in light of the observed bond lengths and bond energies in these fluorides and the high energy (estimated to be 1000 kJ mol^{-1}) required to promote an electron from a $5p$ to a $5d$ orbital, there is doubt as to whether d orbitals are involved in the bond formation. A molecular orbital description can be constructed that does not involve the participation of the xenon $5d$ orbitals, but in its simplest form this description cannot explain the nonoctahedral shape of XF_6. The situation described here reinforces the caveat that approximate theories of chemical bonding must be viewed critically.

23-2 Group 17: The Halogens

The halogens exist as diatomic molecules, symbolized by X_2, where X is a generic symbol for a halogen atom. That these elements occur as nonpolar diatomic molecules accounts for their relatively low melting and boiling points (Table 23.1). As expected, melting and boiling points increase from the smallest and lightest member of the group, fluorine, to the largest and heaviest, iodine.

Conversely, chemical reactivity toward other elements and compounds progresses in the *opposite* order, with fluorine being the most reactive and iodine the least reactive. Fluorine has the highest electronegativity of all the elements and a small atomic radius. Ionic bonds between fluoride and metal ions are strong, as are most covalent bonds between fluorine and other nonmetal atoms.* This tendency to form strong bonds with other atoms makes F_2 a reactive substance.

---

*The weaker-than-expected fluorine-to-fluorine bond in F_2 (159 kJ mol^{-1}) is probably the result of repulsions between the fluorine nuclei and between lone-pair $2p$ electrons. Both repulsions are rather strong because the atomic radius of fluorine is small. The weakness of the fluorine-to-fluorine bond contributes to the reactivity of F_2.

TABLE 23.1   Group 17 Elements: The Halogens

	Fluorine (F)	Chlorine (Cl)	Bromine (Br)	Iodine (I)
Physical form at room temperature	Pale yellow gas	Yellow-green gas	Dark red liquid	Violet-black solid
Melting point, °C	−220	−101	−7.2	114
Boiling point, °C	−188	−35	58.8	184
Electron configuration	$[He]2s^22p^5$	$[Ne]3s^23p^5$	$[Ar]3d^{10}4s^24p^5$	$[Kr]4d^{10}5s^25p^5$
Covalent radius, pm	71	99	114	133
Ionic (X^-) radius, pm	133	181	196	220
First ionization energy, kJ mol^{-1}	1681	1251	1140	1008
Electron affinity, kJ mol^{-1}	−328.0	−349.0	−324.6	−295.2
Electronegativity	4.0	3.0	2.8	2.5
Standard electrode potential, V ($X_2 + 2\,e^- \longrightarrow 2\,X^-$)	2.866	1.358	1.065	0.535

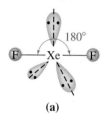

(a)

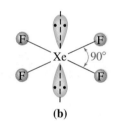

(b)

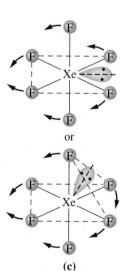

(c)

▲ **FIGURE 23-1**
Molecular shapes of the fluorides of xenon as predicted by VSEPR theory
(a) XeF_2. (b) XeF_4. (c) XeF_6.

Iodine atoms, because they are larger in size and have lower electronegativity, form weaker ionic and covalent bonds than do the other halogens. I_2 is less reactive than the other halogens. We expect astatine ($Z = 85$) also to be a halogen. Because it is radioactive and has only short-lived isotopes, only about 0.05 µg astatine has ever been prepared. Nevertheless, its chemical behavior appears to be similar to that of iodine.

Much of the reaction chemistry of the halogens involves oxidation–reduction reactions in aqueous solutions. For these reactions, standard electrode potentials are the best guides to the reactivity of the halogens. Among the properties of the halogens listed in Table 23.1 are potentials for the half-reaction

$$X_2 + 2\,e^- \longrightarrow 2\,X^-(aq)$$

By this measure, fluorine is clearly the most reactive element of the group ($E° = 2.866$ V). Of all the elements, it shows the greatest tendency to gain electrons and is therefore the most easily reduced. Given this fact, it is not surprising that fluorine occurs naturally only in combination with other elements, and only as the fluoride ion, F^-. Although both chlorine and bromine can exist in a variety of positive oxidation states, they are found in their naturally occurring compounds only as chloride and bromide ions. There are, however, naturally occurring compounds in which iodine is in a positive oxidation state (such as the iodate ion, IO_3^-, in $NaIO_3$). In the case of iodine, the tendency for I_2 to be reduced to I^- is not particularly great ($E° = 0.535$ V).

Electrode Potential Diagrams

When we summarize the reduction tendencies of main-group metals and their ions, generally one or, at most, a few $E°$ values tell the story, and these values are easily incorporated into tables such as that in Appendix D. However, the oxidation–reduction chemistry of some of the nonmetals is much richer and involves a larger number of relevant $E°$ values. In these cases, *electrode potential diagrams* are particularly useful for summarizing $E°$ data. Partial diagrams for chlorine are shown in Figure 23-2. In these diagrams, a number written above a line segment is the $E°$ value for reduction of the species on the left (higher oxidation state) to the one on the right (lower oxidation state). For a reduction involving species not joined by a line segment, we generally can calculate the appropriate value of $E°$ by the method illustrated in Example 23-1.

Acidic solution ([H⁺] = 1 M):

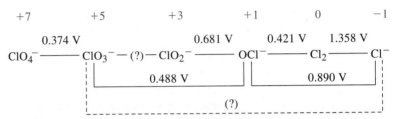

Basic solution ([OH⁻] = 1 M):

▲ **FIGURE 23-2** **Standard electrode potential diagrams for chlorine**
The numbers in color are the oxidation states of the Cl atom. $E°$ values are written above the line segments for the reduction of the species on the left to the species on the right. All reactants and products are at unit activity. Because of the basic properties of the ClO_2^- and OCl^- ions, the weak acids $HClO_2$ and $HOCl$ form in acidic solution.

EXAMPLE 23-1

Using an Electrode Potential Diagram to Determine $E°$ for a Half-Reaction. Determine $E°$ for the reduction of ClO_3^- to ClO_2^- in a basic solution [marked (?) in Figure 23-2].

Solution

We are seeking $E°$ for the reduction half-reaction $ClO_3^- + H_2O + 2\ e^- \longrightarrow ClO_2^- + 2\ OH^-$. Two half-reactions and their corresponding $E°$ values from Figure 23-2 are shown next, along with the $\Delta G°$ values (that is, $-nFE°$).

$$ClO_3^- + 2\ H_2O + 4\ e^- \longrightarrow OCl^- + 4\ OH^-$$
$$E° = 0.488\ V; \qquad \Delta G° = -4F \times 0.488\ V$$
$$ClO_2^- + H_2O + 2\ e^- \longrightarrow OCl^- + 2\ OH^-$$
$$E° = 0.681\ V; \qquad \Delta G° = -2F \times 0.681\ V$$

Recall that we cannot simply add or subtract $E°$ values when we combine two or more half-reactions to obtain yet another *half-reaction*, but we can add $\Delta G°$ values (page 834). To get the desired half-equation, we must write the reverse of the second of the two-half equations above and change the sign of its $\Delta G°$.

$$ClO_3^- + 2\ H_2O + 4\ e^- \longrightarrow OCl^- + 4\ OH^- \qquad \Delta G° = -4F \times 0.488\ V$$
$$OCl^- + 2\ OH^- \longrightarrow ClO_2^- + H_2O + 2\ e^- \qquad \Delta G° = +2F \times 0.681\ V$$

———————————————————————————————————————

$$ClO_3^- + H_2O + 2\ e^- \longrightarrow ClO_2^- + 2\ OH^-$$
$$\Delta G° = F(2 \times 0.681 - 4 \times 0.488)\ V$$

For the desired half-reaction, we can also write that $\Delta G° = -nFE°$, leading to the expression

$$\Delta G° = -2FE° = F\left[(2 \times 0.681) - (4 \times 0.488)\right]\ V$$
$$E° = \frac{F\left[(2 \times 0.681) - (4 \times 0.488)\right]\ V}{-2F} = 0.295\ V$$

KEEP IN MIND ▶
that when combining an oxidation and a reduction half-equation, electrons must cancel in the overall redox equation. However, when combining only reduction half-equations, electrons must *not* cancel out. The resultant half-equation must still be balanced for numbers of atoms and for charge.

Practice Example A: Determine the missing $E°$ value for the dashed line that joins ClO_3^- and Cl^- in basic solutions in Figure 23-2.

Practice Example B: Determine the missing $E°$ value for the dashed line that joins ClO_3^- and Cl_2 in acidic solutions in Figure 23-2.

Production and Uses of the Halogens

Although the existence of *fluorine* had been known since early in the nineteenth century, no one was able to devise a chemical reaction to extract the free element from its compounds. Finally, in 1886, H. Moissan succeeded in preparing $F_2(g)$ with an electrolysis reaction. Moissan's method, which is still the only important commercial method for fluorine extraction, involves the electrolysis of HF dissolved in molten KHF_2,

$$2\,HF \xrightarrow[KF \cdot 2HF(l)]{electrolysis} H_2(g) + F_2(g) \tag{23.1}$$

consisting of the following electrode reactions:

Anode: $\quad\quad 2\,F^- \longrightarrow F_2(g) + 2\,e^-$

Cathode: $\quad 2\,H^+ + 2\,e^- \longrightarrow H_2(g)$

The hydrogen difluoride ion in KHF_2 features a strong hydrogen bond, with a H^+ midway between two F^- ions: $[F\text{—}H\text{—}F]^-$. ▶

Moissan also developed the electric furnace and was honored for both these achievements with the Nobel Prize in chemistry in 1906. Nevertheless, the challenge of producing fluorine by means of a chemical reaction remained. In 1986, one century after Moissan isolated fluorine, the chemical synthesis of fluorine was announced (see Exercise 22).

Although *chlorine* can be prepared by chemical reactions, electrolysis of $NaCl(aq)$ is the usual industrial method, as we have mentioned previously (see page 857). The electrolysis reaction is

$$2\,Cl^-(aq) + 2\,H_2O \xrightarrow{electrolysis} 2\,OH^-(aq) + H_2(g) + Cl_2(g) \tag{23.2}$$

Bromine can be extracted from seawater, where it occurs in concentrations of about 70 ppm as Br^-, or from inland brine solutions. The seawater or brine solution is adjusted to pH 3.5 and treated with $Cl_2(g)$, which oxidizes Br^- to Br_2 in this displacement reaction.

$$Cl_2(g) + 2\,Br^-(aq) \longrightarrow Br_2(l) + 2\,Cl^-(aq) \quad\quad E°_{cell} = 0.293\,V \tag{23.3}$$

▲ **Salt formations in the Dead Sea**
The high concentrations of salts in the Dead Sea make it a good source of bromine and a number of other chemicals.

The liberated Br_2 is swept from seawater with a current of air or from brine with steam. A dilute bromine vapor forms and can be concentrated by various methods.

Certain marine plants, such as seaweed, absorb and concentrate I^- selectively in the presence of Cl^- and Br^-. *Iodine* is obtained in small quantities from such plants. In the United States, I_2 is obtained from inland brines by a process similar to that for the production of Br_2. Another abundant natural source of iodine is $NaIO_3$, found in large deposits in Chile. Because the oxidation state of iodine must be reduced from $+5$ in IO_3^- to 0 in I_2, the conversion of IO_3^- to I_2 requires the use of a reducing agent. Aqueous sodium hydrogen sulfite (bisulfite) is used as the reducing agent in the first part of a two-step procedure, followed by the reaction of I^- with additional IO_3^- to produce I_2.

$$IO_3^-(aq) + 3\,HSO_3^-(aq) \longrightarrow I^-(aq) + 3\,SO_4^{2-}(aq) + 3\,H^+(aq) \tag{23.4}$$

$$5\,I^-(aq) + IO_3^-(aq) + 6\,H^+(aq) \longrightarrow 3\,I_2(s) + 3\,H_2O(l) \tag{23.5}$$

▲ **A test for Br^-(aq)**
Reaction (23.3) is used as a qualitative test in the laboratory. The liberated Br_2 is extracted into an organic solvent (here, $CHCl_3$).

The halogen elements form a variety of useful compounds, and the elements themselves are largely used to produce these compounds. All of the halogens are used to make halogenated organic compounds. For example, elemental fluorine is used to produce compounds such as polytetrafluoroethylene, the plastic Teflon. Fluorine has been used to make chlorofluorocarbons (CFCs) as refrigerants, but international treaties have banned the production of CFCs in most countries to minimize further

TABLE 23.2	Some Important Inorganic Compounds of Fluorine
Compound	**Uses**
Na_3AlF_6	Manufacture of aluminum
BF_3	Catalyst
CaF_2	Optical components, manufacture of HF, metallurgical flux
ClF_3	Fluorinating agent, reprocessing nuclear fuels
HF	Manufacture of F_2, AlF_3, Na_3AlF_6, and fluorocarbons
LiF	Ceramics manufacture, welding, and soldering
NaF	Fluoridating water, dental prophylaxis, insecticide
SF_6	Insulating gas for high-voltage electrical equipment
SnF_2	Manufacture of toothpaste
UF_6	Manufacture of uranium fuel for nuclear reactors

Physical Properties of Halogens movie

damage to the stratospheric ozone layer from these compounds (see page 280). Now fluorine is used to make hydrochlorofluorocarbons (HCFCs), which are more environmentally benign alternatives to CFCs. Fluorinated organic compounds tend to be chemically inert, and it is this inertness that makes them useful as components in harsh chemical environments. Fluorine is a key element in a variety of useful inorganic compounds. Some of these compounds and their uses are listed in Table 23.2.

With an annual production of over 13 million tons, elemental chlorine ranks about eighth in quantity among manufactured chemicals in the United States. It has three main commercial uses: (1) production of chlorinated organic compounds (about 70%), chiefly ethylene dichloride, CH_2ClCH_2Cl, and vinyl chloride, $CH_2{=}CHCl$ (the monomer of polyvinyl chloride, PVC); (2) as a bleach in the paper and textile industries and for the treatment of swimming pools, municipal water, and sewage (about 20%); and (3) production of dozens of chlorine-containing inorganic chemicals (about 10%).

▶ The major industrial source of HCl is the chlorination of organic compounds; for example,
$CH_4 + Cl_2 \longrightarrow CH_3Cl + HCl$.

Bromine is used to make brominated organic compounds. Some of these are used as fire retardants and pesticides. Others are used extensively as dyes and pharmaceuticals. An important inorganic bromine compound is AgBr, the primary light-sensitive agent used in photographic film.

Iodine is of much less commercial importance than chlorine. Iodine and its compounds, however, do have applications as catalysts, antiseptics, and germicides and in the preparation of pharmaceuticals and photographic emulsions (as AgI).

Hydrogen Halides

We have encountered the hydrogen halides from time to time throughout this text. In aqueous solution, they are called the *hydrohalic acids*. Except for HF, hydrohalic acids are strong acids in water. The unexpected weak acid nature of HF(aq) is discussed on page 694.

One well-known property of HF is its ability to etch (and ultimately to dissolve) glass. The reaction is similar to one between HF and silica, SiO_2.

$$SiO_2(s) + 4\,HF(aq) \longrightarrow 2\,H_2O(l) + SiF_4(g) \qquad (23.6)$$

As a result of reaction (23.6), HF must be stored in special containers coated with a lining of Teflon or polyethylene.

Hydrogen fluoride is commonly produced by a method discussed in Section 22-1. When a halide salt (such as fluorite, CaF_2) is heated with a *nonvolatile* acid, such as concentrated $H_2SO_4(aq)$, a sulfate salt and the volatile hydrogen halide are produced.

▲ Glass etched with hydrofluoric acid.

$$CaF_2(s) + H_2SO_4(\text{concd aq}) \xrightarrow{\Delta} CaSO_4(s) + 2\,HF(g) \qquad (23.7)$$

This method also works for preparing HCl(g) but not for HBr(g) or HI(g). That is because concentrated H_2SO_4(aq) is a sufficiently strong oxidizing agent to oxidize Br^- to Br_2 and I^- to I_2.

$$2\,NaBr(s) + 2\,H_2SO_4(\text{concd aq}) \xrightarrow{\Delta} Na_2SO_4(s) + 2\,H_2O(l) + Br_2(g) + SO_2(g) \quad (23.8)$$

We can get around this difficulty by using a *nonoxidizing* nonvolatile acid, such as phosphoric acid. Also, all the hydrogen halides can be formed by the direct combination of the elements.

$$H_2(g) + X_2(g) \longrightarrow 2\,HX(g) \quad (23.9)$$

The reaction of $H_2(g)$ and $F_2(g)$ is very fast, however, occurring with explosive violence under some conditions. With $H_2(g)$ and $Cl_2(g)$, the reaction also proceeds rapidly (explosively) in the presence of light (photochemically initiated), although some HCl is made this way commercially. With Br_2 and I_2, the reaction occurs more slowly and a catalyst is required.

From the data in Table 23.3, we see that the standard free energies of formation of HF(g), HCl(g), and HBr(g) are large and negative, suggesting that for them reaction (23.9) goes to completion. For HI(g), conversely, ΔG_f° is small and positive. This suggests that even at room temperature HI(g) should dissociate to some extent into its elements. Because the dissociation of HI(g) has a high activation energy, however, the reaction occurs only very slowly in the absence of a catalyst. As a result, HI(g) is stable at room temperature.

TABLE 23.3 Free Energy of Formation of Hydrogen Halides at 298 K

	ΔG_f°, kJ mol^{-1}
HF(g)	−273.2
HCl(g)	−95.30
HBr(g)	−53.45
HI(g)	+1.70

EXAMPLE 23-2

Determining K_p for a Dissociation Reaction from Free Energies of Formation. What is the value of K_p for the dissociation of HI(g) into its elements at 298 K?

Solution

Dissociation of HI(g) into its elements is the reverse of the formation reaction, and ΔG° for the dissociation is the *negative* of ΔG_f° for HI(g) listed in Table 23.3.

$$HI(g) \longrightarrow \tfrac{1}{2}\,H_2(g) + \tfrac{1}{2}\,I_2(s) \qquad \Delta G^\circ = -1.70\ \text{kJ} \quad (23.10)$$

Now we can use the relationship $\Delta G^\circ = -RT \ln K_{eq}$.

$$\ln K_p = \frac{-\Delta G^\circ}{RT} = \frac{-(-1.70 \times 10^3\ \text{J mol}^{-1})}{8.3145\ \text{J mol}^{-1}\ \text{K}^{-1} \times 298\ \text{K}} = 0.686$$

$$K_p = e^{0.686} = 1.99$$

Practice Example A: What is the value of K_p for the dissociation of HF(g) into its elements at 298 K?

Practice Example B: Use data from Table 23.3 to determine K_p and the percent dissociation of HCl(g) into its elements at 298 K.

Oxoacids and Oxoanions of the Halogens

Fluorine, the most electronegative element, adopts the −1 oxidation state in its compounds. The other halogens, when bonded to a more electronegative element such as oxygen, can have any one of several positive oxidation states: +1, +3, +5, or +7. This variability of oxidation states, which we have already illustrated through electrode potential diagrams for chlorine (Figure 23-2), is emphasized again by the oxoacids listed in Table 23.4. Chlorine forms a complete set of oxoacids in all these oxidation states, but bromine and iodine do not. Only a few of the oxoacids can be isolated in pure form ($HClO_4$, HIO_3, HIO_4, H_5IO_6); the rest are stable only in aqueous solution.

TABLE 23.4	Oxoacids of the Halogens[a]		
Oxidation State of Halogen	Chlorine	Bromine	Iodine
+1	HOCl	HOBr	HOI
+3	$HClO_2$	—	—
+5	$HClO_3$	$HBrO_3$	HIO_3
+7	$HClO_4$	$HBrO_4$	HIO_4; H_5IO_6

[a] In all these acids, H atoms are bonded to O atoms—as shown for HOCl, HOBr, and HOI—not to the central halogen atom. More accurate representations of the other acids would be HOClO (instead of $HClO_2$), HOClO$_2$ (instead of $HClO_3$), and so on.

Hypochlorite, OCl^-
(linear)

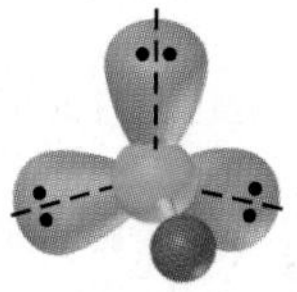

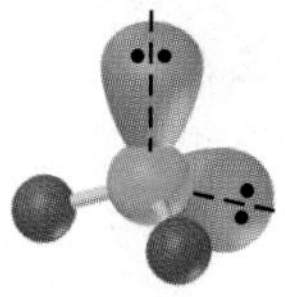

Chlorite, ClO_2^-
(angular)

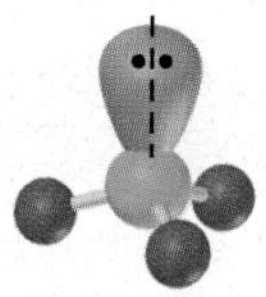

▶ Oxoanions of chlorine

Chlorate, ClO_3^-
(trigonal pyramidal)

Perchlorate, ClO_4^-
(tetrahedral)

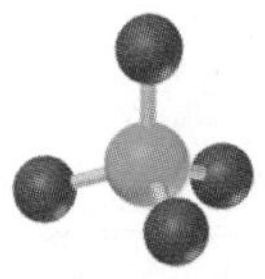

An easily prepared oxidizing agent for use in the laboratory is an aqueous solution of chlorine ("chlorine water"). The solution is not just one of Cl_2, however. The $Cl_2(aq)$ disproportionates into HOCl(aq) and HCl(aq). Although the disproportionation of Cl_2 in water is nonspontaneous when all reactants and products are in their standard states, the reaction does occur to a limited extent in solutions that are not strongly acidic, as the $E°$ and $\Delta G°$ values indicate.

Reduction: $Cl_2(g) + 2\,e^- \longrightarrow 2\,Cl^-(aq)$

Oxidation: $Cl_2(g) + 2\,H_2O(l) \longrightarrow 2\,HOCl(aq) + 2\,H^+(aq) + 2\,e^-$

Overall: $Cl_2(g) + H_2O(l) \longrightarrow HOCl(aq) + H^+(aq) + Cl^-(aq)$ (23.11)

$$E°_{cell} = E°_{Cl_2/Cl^-} - E°_{HOCl/Cl_2} = 1.358\,V - 1.611\,V = -0.253\,V$$

$$\Delta G° = -nFE° = 48.8\,kJ$$

In contrast, the disproportionation is spontaneous for standard-state conditions in basic solution.

Reduction: $Cl_2(g) + 2\,e^- \longrightarrow 2\,Cl^-(aq)$

Oxidation: $Cl_2(g) + 4\,OH^-(aq) \longrightarrow 2\,OCl^-(aq) + 2\,H_2O(l) + 2\,e^-$

Overall: $Cl_2(g) + 2\,OH^-(aq) \longrightarrow OCl^-(aq) + Cl^-(aq) + H_2O(l)$ (23.12)

$$E°_{cell} = E°_{Cl_2/Cl^-} - E°_{OCl^-/Cl_2} = 1.358\,V - 0.421\,V = 0.937\,V$$

$$\Delta G° = -nFE° = -181\,kJ$$

KEEP IN MIND
that E_{Cl_2/Cl^-} is independent of pH, but both E_{HOCl/Cl_2} and E_{OCl^-/Cl_2} are pH dependent. They become smaller as the pH increases, increasing E_{cell} and making ΔG less positive and eventually negative. In terms of Le Châtelier's principle, the smaller $[H^+]$ (greater $[OH^-]$), the more the disproportionation reaction is favored. ▶

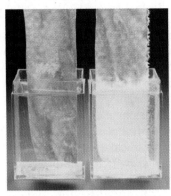

▲ Bleaching with hypochlorite ion
Both strips of cloth are heavily stained with tomato sauce. Pure water (left) has little ability to remove the stain. NaOCl (right) rapidly bleaches (oxidizes) the colored components of the sauce to colorless products.

▶ Solid household bleach normally contains Ca(OCl)Cl, which is obtained by the reaction of Cl_2 and $Ca(OH)_2$:
$Ca(OH)_2 + Cl_2 \longrightarrow$
$\qquad Ca(OCl)Cl + H_2O.$

A particularly destructive explosion of ammonium perchlorate occurred at a rocket fuel plant in Henderson, Nevada, in 1988. ▶

Are You Wondering...

If there are oxoacids of fluorine?

The hypothetical oxoacid of fluorine, HFO_2 (HOFO), would have a *positive* formal charge on the central F atom.

$$\overset{\ominus}{:\ddot{O}} - \overset{\oplus}{\ddot{F}} - \ddot{O} - H$$

This is not something that we would expect of fluorine, the most electronegative of all the elements. Hypothetical oxoacids with more O atoms would have the F atom with even higher positive formal charges. Fluorine does form HOF, an oxoacid with no formal charges $H:\ddot{O}—\ddot{F}:$, but HOF exists only in the solid and liquid states. In water, HOF decomposes to HF, H_2O_2, and $O_2(g)$.

HOCl(aq) is an effective germicide used, for example, in water purification and the treatment of swimming pools. Aqueous solutions of *hypochlorite* salts, notably NaOCl(aq), are used as common household bleaches.

Chlorine dioxide is an important bleach for paper and fibers. Its reduction with peroxide ion in aqueous solution produces *chlorite* salts.

$$2\,ClO_2(g) + O_2{}^{2-}(aq) \longrightarrow 2\,ClO_2{}^-(aq) + O_2(g) \qquad (23.13)$$

Sodium chlorite is used as a bleaching agent for textiles.

Chlorate salts form when $Cl_2(g)$ disproportionates in hot alkaline solutions. (Hypochlorites form in cold alkaline solutions; recall reaction 23.12.)

$$3\,Cl_2(g) + 6\,OH^-(aq) \longrightarrow 5\,Cl^-(aq) + ClO_3{}^-(aq) + 3\,H_2O(l) \qquad (23.14)$$

Chlorates are good oxidizing agents. Also, solid chlorates produce oxygen gas when they decompose, which makes them useful in matches and fireworks. A simple laboratory method of producing $O_2(g)$ involves heating $KClO_3(s)$ in the presence of $MnO_2(s)$, a catalyst.

$$2\,KClO_3(s) \xrightarrow[MnO_2]{\Delta} 2\,KCl(s) + 3\,O_2(g) \qquad (23.15)$$

A similar reaction is used as a source of emergency oxygen in aircraft and submarines.

Perchlorate salts are prepared mainly by electrolyzing chlorate solutions. Oxidation of $ClO_3{}^-$ occurs at a Pt anode through the half-reaction

$$ClO_3{}^-(aq) + H_2O(l) \longrightarrow ClO_4{}^-(aq) + 2\,H^+(aq) + 2\,e^- \quad -E° = -1.189\ V \qquad (23.16)$$

An interesting laboratory use of perchlorates is in aqueous solution studies in which complex-ion formation is to be avoided. $ClO_4{}^-$ has one of the lowest tendencies of any anion to act as a ligand in complex-ion formation in water. Compared with the other oxoacid salts, perchlorates are relatively stable. For example, they do not disproportionate, because no oxidation state higher than +7 is available to chlorine. At elevated temperatures or in the presence of a readily oxidizable compound, however, perchlorate salts may react explosively, so caution is advised when using them. Mixtures of ammonium perchlorate and powdered aluminum are utilized as the propellant in some solid rockets, such as those used on the space shuttle. Ammonium perchlorate is especially dangerous to handle, because an explosive reaction may occur when the oxidizing agent $ClO_4{}^-$ acts on the reducing agent $NH_4{}^+$.

Interhalogen Compounds

Not only does each halogen element form diatomic molecules X_2, but also each can form the molecules XY. In such *interhalogen compounds*, the X and Y represent two different halogens, as in ClF, BrCl, or IBr. Other interhalogen compounds consist of polyatomic molecules, such as XY_3, XY_5, or XY_7. Some of the more common ones are listed in Table 23.5. The molecular structures of the interhalogen compounds feature the large, less electronegative halogen as the central atom and the smaller halogen atoms as terminal atoms. Molecular shapes of the interhalogen compounds agree quite well with VSEPR theory predictions. Representative structures are shown in Figure 23-3, including one type that we have not seen previously: In IF_7, *seven* electron pairs are distributed around the central I atom in the form of a pentagonal-bipyramid.

Most interhalogen compounds are very reactive. ClF_3 and BrF_3, for example, react with explosive violence with water, organic materials, and some inorganic materials. These two fluorides are used to fluorinate compounds, as in the preparation of UF_6 for the separation of uranium isotopes by gaseous diffusion. ICl is used as an iodination reagent in organic chemistry.

Polyhalide Ions

Sodium Triiodide movie

The triiodide ion, I_3^-, is one of a group of species, called **polyhalide ions**, that are produced by the reaction of a halide ion with a halogen molecule. In this reaction, the halide ion acts as a *Lewis base* (an electron-pair donor) and the halogen molecule as a *Lewis acid* (an electron-pair acceptor).

$$\ddot{I}-\ddot{I}: + :\ddot{I}:^- \longrightarrow \left[:\ddot{I}-\ddot{I}-\ddot{I}:\right]^- \qquad (23.17)$$

TABLE 23.5	Some Interhalogen Compounds		
XY	**XY_3**	**XY_5**	**XY_7**
$ClF(g)^a$	$ClF_3(g)$	$ClF_5(g)$	
$BrF(g)$	$BrF_3(l)$	$BrF_5(l)$	
$BrCl(g)$			
$ICl(s)$	$ICl_3(s)$	$IF_5(l)$	$IF_7(g)$
$IBr(s)$			

^{a}The states of matter are given for 25 °C and 1 atm.

XY

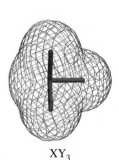
XY_3

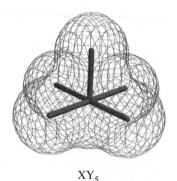
XY_5

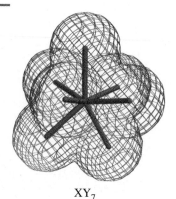
XY_7

▲ **FIGURE 23-3 Structures of some interhalogen compounds**
The thick black lines represent the molecular framework, and the boundary of the wire frame represents the electron density distribution obtained from quantum mechanical calculations.

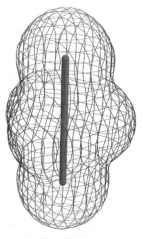

The structure of the I_3^- ion is shown in Figure 23-4. The familiar iodine solutions used as antiseptics typically contain triiodide ion, and triiodide ion solutions are widely used in analytical chemistry.

23-3 Group 16: The Oxygen Family

Oxygen and sulfur are clearly nonmetallic in their behavior, but some metallic properties begin to appear toward the bottom of group 16 with tellurium and polonium, as suggested in the periodic table on the inside front cover. On the basis of electron configurations alone, we expect oxygen and sulfur to be similar. Both elements form ionic compounds with active metals, and both form similar covalent compounds, such as H_2S and H_2O, CS_2 and CO_2, SCl_2 and Cl_2O.

Even so, there are important differences between oxygen and sulfur compounds. For example, H_2O has a very high boiling point (100 °C) for a compound of such low molecular mass (18 u), whereas the boiling point of H_2S (molecular mass, 34 u) is more normal (−61 °C). This difference in behavior can be explained in terms of the extensive hydrogen bonding that occurs in H_2O but not in H_2S (see page 502). A number of properties of sulfur and oxygen are compared in Table 23.6. In general, the differences can be attributed to these characteristics of the oxygen atom: (1) small size, (2) high electronegativity, and (3) the inability to employ an expanded valence shell in Lewis structures.

As indicated in Table 23.6, the principal oxidation states of oxygen are −2, −1, and 0. Sulfur, conversely, can exhibit all oxidation states from −2 to +6, including several "mixed" oxidation states, such as +2.5 in the tetrathionate ion, $S_4O_6^{2-}$.

▲ **FIGURE 23-4**
Structure of the I_3^- ion
The black lines represent the molecular framework of I_3^-, and the boundary of the wire frame represents the electron density distribution.

 Reactions with Oxygen movie

TABLE 23.6 Some Comparisons of Oxygen and Sulfur

Oxygen	Sulfur
• $O_2(g)$ at 298 K and 1 atm	• $S_8(s)$ at 298 K and 1 atm
• Two allotropes: $O_2(g)$ and $O_3(g)$	• Two solid crystalline forms and many different molecular species in liquid and gaseous states
• Principal oxidation states: −2, −1, 0 ($-\frac{1}{2}$ in O_2^-)	• Possible oxidation states: all values from −2 to +6
• $O_2(g)$ and $O_3(g)$ are very good oxidizing agents	• $S_8(s)$ is a poor oxidizing agent
• Forms, with metals, oxides that are mostly ionic in character	• Forms ionic sulfides with the most active metals, but many metal sulfides have partial covalent character
• O^{2-} completely hydrolyzes in water, producing OH^-	• S^{2-} strongly hydrolyzes in water to HS^- (and OH^-)
• O is not often the central atom in a structure and can never have more than four atoms bonded to it; more commonly it has two (as in H_2O) or three (as in H_3O^+)	• S is the central atom in many structures; can easily accommodate up to six atoms around itself (e.g., SO_3, SO_4^{2-}, SF_6)
• Can form only two-atom and three-atom chains, as in H_2O_2 and O_3; compounds with O—O bonds decompose readily	• Can form molecules with up to six S atoms per chain in compounds such as H_2S_n, Na_2S_n, $H_2S_nO_6$
• Forms the oxide CO_2, which reacts with NaOH(aq) to produce $Na_2CO_3(aq)$	• Forms the sulfide CS_2, which reacts with NaOH(aq), producing $Na_2CS_3(aq)$ and $Na_2CO_3(aq)$
• Forms, with hydrogen, the compound H_2O, which is a liquid at 298 K and 1 atm is extensively hydrogen-bonded has a large dipole moment is an excellent solvent for ionic solids forms hydrates and aqua complexes is oxidized with difficulty	• Forms, with hydrogen, the compound H_2S, which is a (poisonous) gas at 298 K and 1 atm is not hydrogen-bonded has a small dipole moment is a poor solvent forms no complexes is easily oxidized

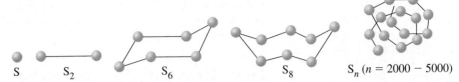

▲ **FIGURE 23-5** **Some molecular forms of sulfur**

Allotropy and Polymorphism of Sulfur

S_8 model

Sulfur has more allotropes than any other element (Figure 23-5). The most common allotrope in the solid state is the S_8 ring, although an additional half-dozen cyclic structures are known, with up to 20 S atoms per ring. In sulfur vapor, S, S_2, S_4, S_6, and S_8 can all exist under the appropriate conditions. Long-chain molecules of sulfur atoms are found in liquid sulfur. *Rhombic sulfur* (S_α), the stable solid at room temperature, is made up of cyclic S_8 molecules. At 95.5 °C, it converts to *monoclinic sulfur* (S_β), which is also made up of S_8 molecules but has a different crystal structure from S_α. At 119 °C, S_β melts, yielding *liquid sulfur* (S_λ), a straw-colored, mobile liquid comprised mostly of S_8 molecules but with other cyclic molecules containing from 6 to 20 atoms. At 160 °C, the cyclic molecules open up and recombine into long spiral-chain molecules, producing *liquid sulfur* (S_μ), a dark, viscous liquid. The chain length and viscosity reach a maximum at about 180 °C. At higher temperatures, the chains break up and the viscosity decreases. At 445 °C, the liquid boils, producing *sulfur vapor*. S_8 molecules predominate in the vapor at the boiling point but break down into smaller molecules at higher temperatures. *Plastic sulfur* forms if liquid S_μ is poured into cold water. Plastic sulfur consists of long, spiral-chain molecules and has rubberlike properties. On standing, it becomes brittle and reverts to rhombic sulfur.

To summarize,

$$S_\alpha \xrightarrow{95.5\,°C} S_\beta \xrightarrow{119} S_\lambda \xrightarrow{160} S_\mu \xrightarrow{445} S_8(g) \xrightarrow{} S_6 \xrightarrow{1000} S_4 \xrightarrow{2000} S_2 \xrightarrow{} S$$

Some of these forms of sulfur are shown in Figure 23-6.

(a)

(b)

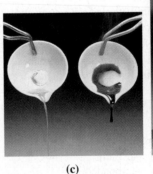
(c)

(d)

▲ **FIGURE 23-6** **Several macroscopic forms of sulfur**
(a) Rhombic sulfur. (b) Monoclinic sulfur. (c) On the left, monoclinic sulfur has just melted to form an orange liquid. On the right, after continued heating, the liquid becomes red and more viscous. (d) Liquid sulfur is poured into water to produce plastic sulfur.

Because some of these transitions, especially those in the solid state, are sluggish, additional phenomena are occasionally seen. For example, if rhombic sulfur is heated rapidly, it may melt at 113 °C and fail to convert to monoclinic sulfur. However, monoclinic sulfur may then freeze from this liquid, only to melt again at 119 °C.

Production and Uses of the Group 16 Elements

Oxygen is the most abundant element in Earth's crust, making up 45.5% by mass. It is also the most abundant element in seawater, accounting for nearly 90% of the mass. In the atmosphere, it is second only to nitrogen in abundance, accounting for 23.15% by mass and 21.04% by volume. Although it is obtained to a limited extent by the decomposition of oxygen-containing compounds and the electrolysis of water, the principal commercial source of oxygen is the fractional distillation of liquid air (Section 8-1). Its main uses were outlined in Table 8.4.

Sulfur is the sixteenth most abundant element in Earth's crust, accounting for 0.0384% by mass. Sulfur occurs as elemental sulfur, as mineral sulfides and sulfates, as $H_2S(g)$ in natural gas, and as organosulfur compounds in oil and coal. Extensive deposits of elemental sulfur are found in Texas and Louisiana, some of them in offshore sites. This sulfur is mined in an unusual way known as the **Frasch process** (Figure 23-7). Superheated water (at about 160 °C and 16 atm) is forced down the outermost of three concentric pipes into an underground bed of sulfur-containing rock. The sulfur melts and forms a liquid pool. Compressed air (at 20–25 atm) is pumped down the innermost pipe and forces the liquid sulfur–water mixture up the remaining pipe.

Although the Frasch process was once the principal source of elemental sulfur, that is no longer the case. This change has been brought about by the need to control sulfur emissions from industrial operations. H_2S is a common impurity in oil and natural gas. After being removed from the fuel, H_2S is reduced to elemental sulfur in a two-step process. A stream of H_2S gas is split into two parts. One part (about one-third of the stream) is burned to convert H_2S to SO_2. The streams are rejoined in a catalytic converter at 200–300 °C, where the following reaction occurs:

$$2\ H_2S(g) + SO_2(g) \longrightarrow 3\ S(g) + 2\ H_2O(g) \qquad (23.18)$$

▲ These vast formations of solid sulfur were formed by the solidification of liquid sulfur obtained by the Frasch process.

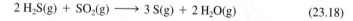
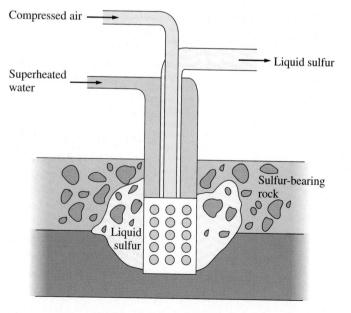

Compressed air

Liquid sulfur

Superheated water

Sulfur-bearing rock

Liquid sulfur

▶ FIGURE 23-7
The Frasch process
Sulfur is melted by using superheated water, and liquid sulfur is forced to the surface.

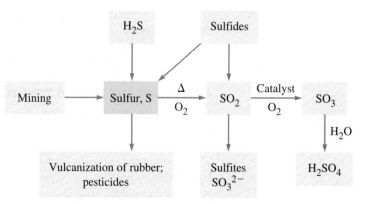

▲ **FIGURE 23-8** Sources and uses of sulfur and its oxides

About 90% of all the sulfur produced is burned to form $SO_2(g)$ (Figure 23-8), and in turn, most $SO_2(g)$ is converted to sulfuric acid, H_2SO_4. Elemental sulfur does have a few uses of its own, however. One of these is in vulcanizing rubber (page 427); another is as a fungicide used for dusting grapevines.

Selenium and *tellurium* have properties similar to those of sulfur, but they are more metallic. For example, sulfur is an electrical insulator, whereas selenium and tellurium are semiconductors. Selenium and tellurium are obtained mostly as by-products of metallurgical processes, such as in the anode mud deposited in the electrolytic refining of copper (page 856). Although there is not much use for tellurium compounds, selenium is used in the manufacture of rectifiers (devices used to convert alternating to direct electric current). Both Se and Te are employed in the preparation of alloys, and their compounds are used as additives to control the color of glass.

Selenium also displays the property of *photoconductivity*: The electrical conductivity of selenium increases in the presence of light. This property is used, for example, in photocells in cameras. In some modern photocopying machines, the light-sensitive element is a thin film of Se deposited on aluminum. The light and dark areas of the image being copied are converted into a distribution of charge on the light-sensitive element. A dry black powder (toner) coats the charged portions of the light-sensitive element, and this image is transferred to a sheet of paper. Next, the dry powder is fused to the paper. In the final step, the electrostatic charge on the light-sensitive element is neutralized to prepare it for the next cycle.

Polonium is a very rare, radioactive metal. Because it is extremely low in abundance, it has not found much practical use. Polonium was the first new radioactive element isolated from uranium ore by Marie and Pierre Curie in 1898. Madame Curie named it after her native Poland.

▲ A selenium-coated light-sensitive element from a photocopier.

Oxygen Compounds

Oxygen is so central to the study of chemistry that we constantly refer to its physical and chemical properties in developing a framework of chemical principles. For instance, our discussion of stoichiometry began with combustion reactions—reactions of substances with $O_2(g)$ to form products such as $CO_2(g)$, $H_2O(l)$, and $SO_2(g)$. Combustion reactions also figured prominently in the presentation of thermochemistry. Many of the molecules and polyatomic anions described in the chapters on chemical bonding were oxygen-containing species. Water was a primary subject in the discussion of liquids, solids, and intermolecular forces as well as in the study of

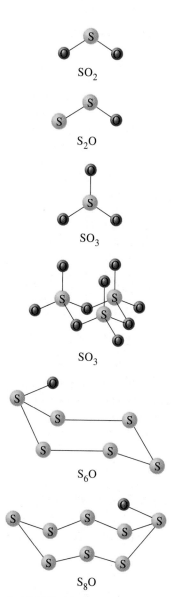

SO$_2$

S$_2$O

SO$_3$

SO$_3$

S$_6$O

S$_8$O

▲ **FIGURE 23-10**
Structures of some sulfur oxides
To conform to the observed structures of SO$_2$, S$_2$O, and SO$_3$, the hybridization scheme proposed for the central S atom is sp^2. S$_2$O has a similar structure to SO$_2$ but with a S atom substituted for one O atom. The SO$_3$ exists in equilibrium with the trimer (SO$_3$)$_3$, in which the OSO angle is approximately tetrahedral and the hybridization of the S should be sp^3. The oxides S$_6$O and S$_8$O illustrate sulfur's ability to form ring compounds.

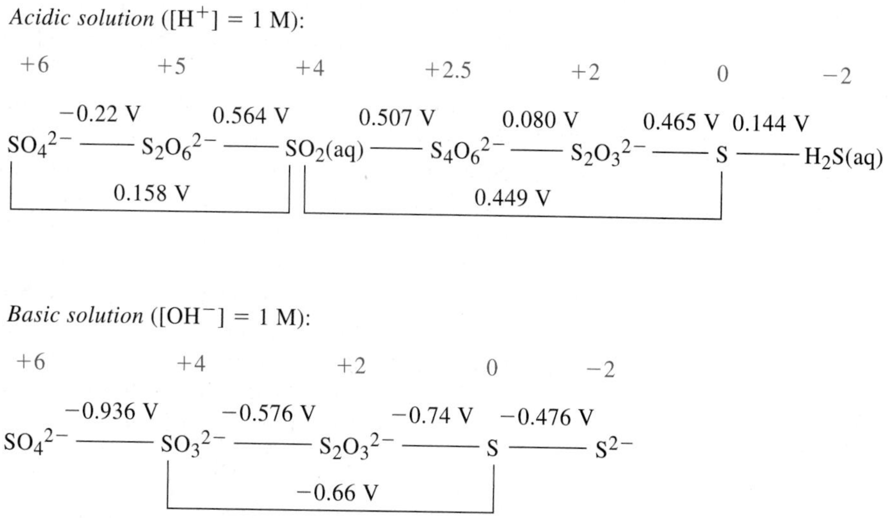

Acidic solution ([H$^+$] = 1 M):

+6   +5   +4   +2.5   +2   0   −2

$$SO_4^{2-} \xrightarrow{-0.22\ V} S_2O_6^{2-} \xrightarrow{0.564\ V} SO_2(aq) \xrightarrow{0.507\ V} S_4O_6^{2-} \xrightarrow{0.080\ V} S_2O_3^{2-} \xrightarrow{0.465\ V} S \xrightarrow{0.144\ V} H_2S(aq)$$

0.158 V

0.449 V

Basic solution ([OH$^-$] = 1 M):

+6   +4   +2   0   −2

$$SO_4^{2-} \xrightarrow{-0.936\ V} SO_3^{2-} \xrightarrow{-0.576\ V} S_2O_3^{2-} \xrightarrow{-0.74\ V} S \xrightarrow{-0.476\ V} S^{2-}$$

−0.66 V

▲ **FIGURE 23-9**   **Electrode potential diagrams for sulfur**

acid–base and other solution equilibria. The preparation and uses of oxygen and ozone were presented in Chapter 8; the dual roles of hydrogen peroxide as an oxidizing agent and a reducing agent were described in Section 5-6; and the kinetics of the decomposition of H$_2$O$_2$ was examined in detail in Chapter 15. The acidic, basic, and amphoteric properties of element oxides were outlined in Chapter 10.

A systematic study of oxygen compounds is generally done in conjunction with the study of the other elements. Thus, the oxides of nitrogen were considered in the discussion of nitrogen chemistry in Chapter 8; the oxides of carbon were also considered there. The survey of the chemistry of the active metals in Chapter 22 provided an opportunity to describe normal oxides, peroxides, and superoxides; and the important oxides of sulfur, phosphorus, silicon, and boron are discussed in this chapter.

Oxides, Oxoacids, and Oxoanions of Sulfur

As in the corresponding discussion for the halogens (Section 23-2), oxidation–reduction chemistry is a primary concern here. To assist in this discussion, we provide electrode potential diagrams for some important sulfur-containing species in Figure 23-9. More than a dozen oxides of sulfur have been reported, but only *sulfur dioxide*, SO$_2$, and *sulfur trioxide*, SO$_3$, are commonly encountered. Typical structures are shown in Figure 23-10. The main commercial methods of producing SO$_2$(g) are the direct combustion of sulfur and the "roasting" of metal sulfides.

$$S(s) + O_2(g) \xrightarrow{\Delta} SO_2(g) \tag{23.19}$$

$$2\ ZnS(s) + 3\ O_2(g) \xrightarrow{\Delta} 2\ ZnO(s) + 2\ SO_2(g) \tag{23.20}$$

The main use of SO$_2$ is in the synthesis of SO$_3$ to make sulfuric acid, H$_2$SO$_4$. In the modern process, the **contact process**, SO$_2$(g) is formed by reaction (23.19) or (23.20). Then sulfur trioxide is produced by oxidizing SO$_2$(g) in an exothermic, reversible reaction.

$$2\ SO_2(g) + O_2(g) \rightleftharpoons 2\ SO_3(g) \tag{23.21}$$

Reaction (23.21) is the key step in the process, but it occurs very slowly unless catalyzed. The principal catalyst is V$_2$O$_5$ mixed with alkali metal sulfates. The catalysis involves adsorption of the $SO_2(g)$ and $O_2(g)$ on the catalyst, followed by reaction at active sites and desorption of SO_3 (recall Figure 15-16).

SO_3 reacts with water to form H_2SO_4, but the direct reaction of $SO_3(g)$ and water produces a fine mist of $H_2SO_4(aq)$ droplets with unreacted $SO_3(g)$ trapped inside the droplets. This misting would result in a great loss of product and a tremendous pollution problem. To avoid these outcomes, $SO_3(g)$ is instead bubbled through 98% H_2SO_4 in towers packed with a ceramic material. The $SO_3(g)$ readily dissolves in the sulfuric acid and reacts with the small amount of water present to increase the concentration of the sulfuric acid. The result is a form of sulfuric acid sometimes called *oleum* but more commonly called *fuming* sulfuric acid. In a sense, the product is greater than 100% H_2SO_4. Sufficient water is added to the circulating acid in the tower to maintain the required concentration. Later, sulfuric acid of the strength desired is produced by dilution with water. If we use the formula $H_2S_2O_7$ (disulfuric acid) as an example of a particular oleum, the reactions are

$$SO_3(g) + H_2SO_4(l) \longrightarrow H_2S_2O_7(l) \qquad (23.22)$$

$$H_2S_2O_7(l) + H_2O(l) \longrightarrow 2\ H_2SO_4(l) \qquad (23.23)$$

$$H_2SO_4(l) \xrightarrow{H_2O} H_2SO_4(aq) \qquad (23.24)$$

Dilute sulfuric acid, $H_2SO_4(aq)$, enters into all the common reactions of a strong acid, such as neutralizing bases. It reacts with metals to produce $H_2(g)$ and dissolves carbonates to liberate $CO_2(g)$.

Concentrated sulfuric acid has some distinctive properties. It has a very strong affinity for water, strong enough that it will even remove H and O atoms (in the proportion H_2O) from some compounds. In the reaction of concentrated sulfuric acid with a carbohydrate like sucrose, all the H and O atoms are removed and a residue of pure carbon is left.

$$C_{12}H_{22}O_{11}(s) \xrightarrow{H_2SO_4(concd)} 12\ C(s) + 11\ H_2O(l) \qquad (23.25)$$

The concentrated acid is a moderately good oxidizing agent and is able, for example, to react with copper.

$$Cu(s) + 2\ H_2SO_4(concd) \longrightarrow Cu^{2+}(aq) + SO_4^{2-}(aq) + 2\ H_2O(l) + SO_2(g) \quad (23.26)$$

Sulfuric Acid as an Oxidizing Agent movie

(a)   (b)

▲ **(a)** Concentrated sulfuric acid is added to cane sugar. **(b)** Carbon is produced in the reaction.

▶ The current top-ranking industrial chemical, with production exceeding 55 million tons annually, is ethylene, C_2H_4, used mostly in the manufacture of plastics.

For a very long time, sulfuric acid ranked first among manufactured chemicals. It now ranks second, with annual production in the United States of about 45 million tons. Sulfuric acid continues to have many uses, but the bulk of H_2SO_4 is used in the manufacture of fertilizers. It is also used in many other ways, including various metallurgical processes, the refining of oil, the manufacture of the white pigment titanium dioxide, and in storage batteries for automobiles and emergency power supplies. We might say that sulfuric acid was the workhorse of the "old economy" but that it has a lesser role in the "new economy."

When $SO_2(g)$ reacts with water, it produces $H_2SO_3(aq)$, but this acid, *sulfurous acid*, has never been isolated in pure form. Salts of sulfurous acid, *sulfites*, are good reducing agents and are easily oxidized by $O_2(g)$. For example,

$$O_2(g) + 2\,SO_3^{2-}(aq) \longrightarrow 2\,SO_4^{2-}(aq) \tag{23.27}$$

But they can also act as oxidizing agents, as in this reaction with H_2S.

$$2\,H_2S(g) + 2\,H^+(aq) + SO_3^{2-}(aq) \longrightarrow 3\,H_2O(l) + 3\,S(s) \tag{23.28}$$

Both H_2SO_3 and H_2SO_4 are *diprotic* acids. They ionize in two steps and produce two types of salts, one in each ionization step. The term **acid salt** is sometimes used for salts such as $NaHSO_3$ and $NaHSO_4$ because their anions undergo a further *acid* ionization. H_2SO_3 is a weak acid in both ionization steps, whereas H_2SO_4 is strong in the first step and somewhat weak in the second. If a solution of H_2SO_4 is sufficiently dilute, however (less than about 0.001 M), we can treat the acid as if both ionization steps go to completion.

▶ Actually, $NaHSO_3$ cannot be isolated as a solid. When we attempt to crystallize this salt from an aqueous solution containing HSO_3^-, the reaction $2\,HSO_3^- \rightleftharpoons S_2O_5^{2-} + H_2O$ occurs. The product obtained is sodium *metabisulfite*, $Na_2S_2O_5$.

Sulfate and *sulfite* salts have a number of important uses. Calcium sulfate dihydrate (gypsum) is used to make the hemihydrate (plaster of Paris) for the building industry (page 886). Aluminum sulfate is used in water treatment and in sizing paper (page 895). Copper(II) sulfate is employed as a fungicide and algicide and in electroplating. The chief application of sulfites is in the pulp and paper industry. Sulfites solubilize *lignin*, a polymeric substance that coats the cellulose fibers in wood. This treatment frees the fibers for processing into wood pulp and then paper. Sulfites are also used as reducing agents, such as in photography and as scavengers of $O_2(aq)$ in treating boiler water (reaction 23.27).

In addition to sulfite and sulfate ions, another important sulfur–oxygen ion is *thiosulfate ion*, $S_2O_3^{2-}$. The prefix *thio* signifies that a S atom replaces an O atom in a compound. Thus, the thiosulfate ion can be viewed as a sulfate ion, SO_4^{2-}, in which a S atom replaces one of the O atoms. The formal oxidation state of S in $S_2O_3^{2-}$ is +2, but as Figure 23-11 indicates, the two S atoms are not equivalent: The central S atom is in the oxidation state +6, and the terminal S atom, −2. The structures of several other thio anions are also shown in Figure 23-11.

Thiosulfates can be prepared by boiling elemental sulfur in an alkaline solution of sodium sulfite. The sulfur is oxidized and the sulfite ion is reduced, both to thiosulfate ion.

$$SO_3^{2-}(aq) + S(s) \longrightarrow S_2O_3^{2-}(aq) \tag{23.29}$$

Thiosulfate solutions are important in photographic processing (see page 1012). They are also common analytical reagents, often used in conjunction with iodine. For example, in one method of analysis for copper, an excess of iodide ion is added to $Cu^{2+}(aq)$, producing $CuI(s)$ and triiodide ion, I_3^- (recall Figure 23-4).

$$2\,Cu^{2+}(aq) + 5\,I^-(aq) \longrightarrow 2\,CuI(s) + I_3^-(aq) \tag{23.30}$$

The excess triiodide ion is then titrated with a standard solution of $Na_2S_2O_3(aq)$, forming I^- and $S_4O_6^{2-}$, the *tetrathionate* ion.

$$I_3^-(aq) + 2\,S_2O_3^{2-}(aq) \longrightarrow 3\,I^-(aq) + S_4O_6^{2-}(aq) \tag{23.31}$$

SO_3^{2-}, sulfite

SO_4^{2-}, sulfate

$S_2O_3^{2-}$, thiosulfate

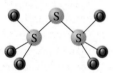

$S_2O_6^{2-}$, dithionate

$S_3O_6^{2-}$, trithionate

$S_4O_6^{2-}$, tetrathionate

▲ **FIGURE 23-11**
Structures of some oxoanions of sulfur

An Environmental Issue: SO₂ Emissions

Industrial smog consists primarily of particles (ash and smoke), $SO_2(g)$, and H_2SO_4 mist. A variety of industrial operations produce significant quantities of $SO_2(g)$. The main contributors to atmospheric releases of $SO_2(g)$, however, are power plants burning coal or high-sulfur fuel oils. SO_2 can oxidize to SO_3, especially when the reaction is catalyzed on the surfaces of airborne particles or through reaction with NO_2.

$$SO_2(g) + NO_2(g) \longrightarrow SO_3(g) + NO(g) \qquad (23.32)$$

In turn, SO_3 can react with water vapor in the atmosphere to produce H_2SO_4 mist, a component of acid rain. Also, the reaction of H_2SO_4 with airborne NH_3 produces particles of $(NH_4)_2SO_4$. The details of the effect of low concentrations of SO_2 and H_2SO_4 on the body are not well understood, but it is clear that these substances are respiratory irritants. Levels above 0.10 ppm are considered potentially harmful.

The control of industrial smog and acid rain hinges on the removal of sulfur from fuels and the control of $SO_2(g)$ emissions. Dozens of processes have been proposed for removing SO_2 from smokestack gases, one of which is illustrated in Figure 23-12.

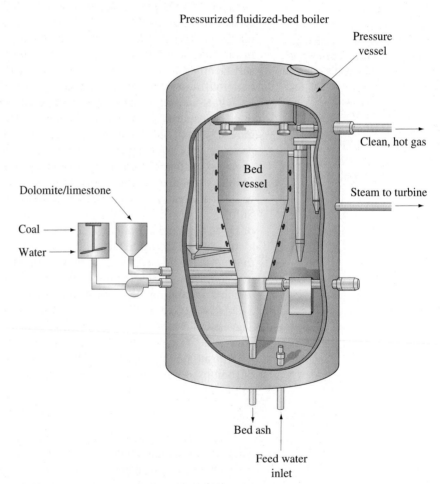

Pressurized fluidized-bed boiler

▲ **FIGURE 23-12   Fluidized-bed combustion**
Powdered coal, limestone, and air are introduced into a combustion chamber, where water circulating through coils is converted to steam. Combustion is carried out at a relatively low temperature (760–860 °C), which minimizes the production of $NO(g)$ from $N_2(g)$ and $O_2(g)$. At the same time, $SO_2(g)$ from sulfur in the coal reacts with $CaO(s)$ from decomposition of the limestone, forming $CaSO_3(s)$ in a Lewis acid–base reaction.

Halides of Oxygen and Sulfur

Both oxygen and sulfur form a number of interesting compounds with the halogens. Oxygen, for example, forms the fluorides OF_2 and O_2F_2, which have structures similar to those of water and hydrogen peroxide but which are much more reactive. Sulfur also forms compounds with the halogens; the analogous compounds S_2F_2 and SF_2 are known, as are the compounds SF_4 and SF_6. The reactivities of SF_4 and SF_6 are quite different. SF_6 is a colorless, odorless inert (unreactive) gas, whereas SF_4 is a powerful fluorinating agent.

$$3\ SF_4 + 4\ BCl_3 \longrightarrow 4\ BF_3 + 3\ SCl_2 + 3\ Cl_2$$

Sulfur and chlorine also form the compounds S_2Cl_2 and SCl_4, but the best-known di-halide is SCl_2. It is a foul-smelling, red liquid (melting point, $-122\ °C$; boiling point, $59\ °C$) that has been used in the production of the notorious poisonous mustard gas.

$$SCl_2 + 2\ CH_2CH_2 \longrightarrow S(CH_2CH_2Cl)_2$$

Mustard gas is not a gas, but a volatile liquid (melting point, $13\ °C$; boiling point, $235\ °C$). During World War I it was sprayed as a mist that stayed close to the ground and was blown by wind onto the enemy. Exposure to mustard gas causes blistering of the skin, internal and external bleeding, blindness, and, after 4 or 5 weeks, death.

23-4 Group 15: The Nitrogen Family

The chemistry of the group 15 elements is an extensive subject, especially that of the first two members of the series—nitrogen and phosphorus. We will discuss the special significance of these two elements to living matter later in the text, but even here you should get a sense of the richness of their chemistry. For example, nitrogen atoms can exist in many oxidation states (Figure 23-13).

As we move farther to the left in the periodic table, we should expect the appearance of metallic behavior to become more significant among the heavier elements in the group. We begin by examining this feature.

Assessment of Metallic–Nonmetallic Character in Group 15

All the elements in group 15 have the valence-shell electron configuration ns^2np^3. This configuration suggests nonmetallic behavior and doesn't give us a clue as to any metallic character that might exist. Table 23.7, however, indicates the usual decrease of ionization energy with increasing atomic number. These values, taken together with physical properties from the table, do suggest the order of metallic character within the group. Nitrogen and phosphorus are nonmetallic, arsenic and

Acidic solution ($[H^+] = 1\ M$):

+5	+4	+3	+2	+1	0	−1	−2	−3
	0.803 V	1.065 V	0.996 V	1.591 V	1.766 V	−1.87 V	1.42 V	1.275 V
NO_3^-	N_2O_4	HNO_2	NO	N_2O	N_2	NH_3OH^+	$N_2H_5^+$	NH_4^+

Basic solution ($[OH^-] = 1\ M$):

+5	+4	+3	+2	+1	0	−1	−2	−3
	−0.86 V	0.867 V	−0.46 V	0.76 V	0.94 V	−3.04 V	0.73 V	0.10 V
NO_3^-	N_2O_4	NO_2^-	NO	N_2O	N_2	NH_2OH	N_2H_4	NH_3

▲ **FIGURE 23-13** **Electrode potential diagrams for nitrogen**

TABLE 23.7	Selected Properties of Group 15 Elements					
Element	Covalent Radius, pm	Electronegativity	First Ionization Energy, kJ mol^{-1}	Common Physical Form(s)	Density of Solid, g/cm^3	Comparative Electrical Conductivity[a]
N	75	3.0	1402	Gas	1.03 (−252 °C)	—
P	110	2.1	1012	Waxlike white solid;	1.82	—
				Red solid	2.20	10^{-17}
As	121	2.0	947	Yellow solid;	2.03	—
				Gray solid with metallic luster	5.78	6.1
Sb	140	1.9	834	Yellow solid;	5.3	—
				Silvery white metallic solid	6.69	4.0
Bi	155	1.9	703	Pinkish white metallic solid	9.75	1.5

[a] These values are relative to an assigned value of 100 for silver.

antimony are metalloids, and bismuth is metallic. The first ionization energy of bismuth is actually somewhat less than that of magnesium, and its third ionization energy (2466 kJ mol^{-1}) is less than the third ionization energy of aluminum (2745 kJ mol^{-1}). The electronegativities indicate a high degree of nonmetallic character for nitrogen and less so for the remaining members of the group.

Three of the elements—phosphorus, arsenic, and antimony—exhibit allotropy. The common forms of phosphorus at room temperature, both nonmetallic, are white and red phosphorus. For arsenic and antimony, the more stable allotropic forms are the metallic ones. These forms have high densities, moderate thermal conductivities, and limited abilities to conduct electricity. Bismuth is a metal despite its low electrical conductivity, which, nevertheless, is better than that of manganese and almost as good as that of mercury. The nonmetals and metals in group 15 are also distinguishable by their oxides. The oxides of nitrogen and phosphorus (for example, N_2O_3 and P_4O_6) are acidic when they react with water. As with the nonmetals in groups 16 and 17, this is typical for nonmetal oxides. Arsenic(III) oxide and antimony(III) oxide are amphoteric, whereas bismuth(III) oxide acts only as a base, a property typical of metal oxides.

Allotropy of Phosphorus

White phosphorus is a white, waxy, phosphorescent solid that can be cut with a knife. (A phosphorescent material glows in the dark.) It is a nonconductor of electricity, can ignite spontaneously in air (hence it is stored under water), and is insoluble in water but soluble in some nonpolar solvents, such as CS_2. The solid has P_4 molecules as its basic structural units (Figure 23-14a). The P_4 molecule is tetrahedral, with a P atom at each corner. The phosphorus-to-phosphorus bonds in P_4 appear to involve the overlap of $3p$ orbitals almost exclusively. Such overlap normally produces 90° bond angles, but in P_4 the P—P—P bond angles are 60°. The bonds are strained, and as might be expected, species with strained bonds are reactive.

When white P is heated to about 300 °C, out of contact with air, it transforms to *red phosphorus*. What appears to happen is that one P—P bond per P_4 molecule

▲ The glow of white phosphorus gave the element its name—*phos*, light, and *phorus*, bringing. The solid has a relatively high vapor pressure, and the glow results from the slow reaction between phosphorus vapor and oxygen in air.

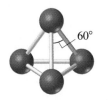

(a) White phosphorus

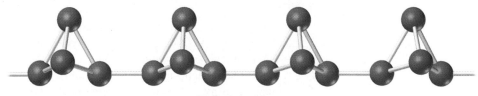

(b) Red phosphorus

▲ **FIGURE 23-14 Two forms of phosphorus**
(a) Structure of white phosphorus: the P_4 molecule. **(b)** Structure of red phosphorus

breaks, and the resulting fragments join together into long chains (Figure 23-14b). Red P is less reactive than white P. Because they have a different atomic arrangement in their basic structural units, red and white phosphorus are allotropic forms of phosphorus rather than just different solid phases. The triple point of red phosphorus is 590 °C and 43 atm. Thus, red phosphorus sublimes without melting (at about 420 °C).

Despite the fact that white P is the form obtained by condensing $P_4(g)$ and that the conversion of white P to red P is a very slow process at ambient temperatures, red P is actually the more thermodynamically stable of the two forms at 298.15 K. Nevertheless, white P is assigned values of 0 for ΔH_f° and ΔG_f°, and for red P, these values are negative.

Production and Uses of the Group 15 Elements

Nitrogen is obtained by the fractional distillation of liquefied air (Section 8-1). Its chief uses were described in Table 8.2.

Although phosphorus is the eleventh most abundant element and makes up about 0.11% of Earth's crust by mass, it was not discovered until 1669. It was originally isolated from putrefied urine—an effective if not particularly pleasant source. Today the principal source of phosphorus compounds is phosphate rock, a class of minerals known as *apatites*, such as fluorapatite [$Ca_5(PO_4)_3F$ or $3Ca_3(PO_4)_2 \cdot CaF_2$]. Elemental phosphorus is prepared by heating phosphate rock, silica (SiO_2), and coke (C) in an electric furnace. The overall change that occurs is

$$2\,Ca_3(PO_4)_2(s) + 10\,C(s) + 6\,SiO_2(s) \xrightarrow{\Delta} 6\,CaSiO_3(l) + 10\,CO(g) + P_4(g) \quad (23.33)$$

The $P_4(g)$ is condensed, collected, and stored under water as white phosphorus.

Although compounds of phosphorus are vitally important to living organisms (DNA and phosphates in bones and teeth, for example), the element itself is not widely used. Almost all the elemental phosphorus produced is reoxidized to give P_4O_{10} for the manufacture of high-purity phosphoric acid. The rest is used to make organophosphorus compounds and phosphorus sulfides (P_4S_3) in match heads.

Arsenic is obtained by heating arsenic-containing metal sulfides. For example, FeAsS yields FeS and As(g). The As(g) deposits as As(s), which can be used to make other compounds. Some arsenic is also obtained by the reduction of arsenic(III) oxide with CO(g). Antimony is obtained mainly from its sulfide ores. Bismuth is obtained as a by-product of the refining of other metals.

Both As and Sb are used in making alloys of other metals. For example, the addition of As and Sb to Pb produces an alloy that has desirable properties for use as electrodes in lead–acid batteries. Arsenic and antimony are used to produce semiconductor materials, such as GaAs, GaSb, and InSb, in electronic devices. (See page 898 for a discussion of GaAs photocells.)

Nitrides

Nitrogen forms binary compounds with most other elements, and these compounds can be grouped into four categories. In ionic (saltlike) nitrides, the nitrogen is present as the N^{3-} ion. These compounds form with lithium and the group 2 metals. Thus, when magnesium is burned in air (recall Figure 2-1), a small quantity of magnesium nitride forms, together with the principal product, magnesium oxide.

$$3\,Mg(s) + N_2(g) \xrightarrow{\Delta} Mg_3N_2(s) \qquad (23.34)$$

The nitride ion is a very strong base. In aqueous solution, it accepts protons from water molecules to form ammonia molecules and hydroxide ions.

$$N^{3-}(aq) + 3\,H_2O(l) \longrightarrow NH_3(aq) + 3\,OH^-(aq)$$

In the reaction of magnesium nitride with water, magnesium and hydroxide ions combine to form insoluble $Mg(OH)_2$, and the ammonia is released as a gas, easily detectable by its odor.

$$Mg_3N_2(s) + 6\,H_2O(l) \longrightarrow 3\,Mg(OH)_2(s) + 2\,NH_3(g)$$

When nitrogen combines with other typical nonmetals, it does so by forming covalent bonds, yielding covalent nitrides. Bonding in these nitrides can be described in terms of the general principles presented in Chapters 11 and 12. Some binary covalent nitrides are $(CN)_2$, P_3N_5, As_4N_4, S_2N_2, and S_4N_4. When nitrogen combines with elements of group 13, producing compounds of the form MN (where M = B, Al, Ga, In, or Tl), the products are isoelectronic with graphite and diamond and resemble them structurally. A fourth type of binary nitrides are the metallic nitrides with formulas such as MN, M_3N, and M_4N. These are *interstitial* compounds, in which N atoms occupy some or all of the interstices (voids) in the structure of the metal. They are hard, chemically inert, high-melting-point solids with important uses as refractory materials and catalysts. Typical examples are TiN, VN, and UN (with melting points 2950 °C, 2050 °C, and 2800 °C, respectively).

Hydrides of Nitrogen

We have said a great deal in this text about the principal hydride of nitrogen: ammonia, NH_3. Here we describe some lesser-known hydrides. If a H atom in NH_3 is replaced by the group $-NH_2$, the resulting molecule is H_2N-NH_2 or N_2H_4, *hydrazine* ($pK_{b_1} = 6.07$; $pK_{b_2} = 15.05$). Replacement of a H atom in NH_3 by $-OH$ produces NH_2OH, *hydroxylamine* ($pK_b = 8.04$). Hydrazine and hydroxylamine are weak bases; because it has two N atoms, N_2H_4 ionizes in two steps. Hydrazine and hydroxylamine form salts analogous to ammonium salts, such as $N_2H_5^+NO_3^-$, $N_2H_6^{2+}SO_4^{2-}$, and $NH_3OH^+Cl^-$. As expected, these salts hydrolyze in water to yield acidic solutions.

Hydrazine and some of its derivatives burn in air with the evolution of large quantities of heat; they are used as rocket fuels. For the combustion of hydrazine,

$$N_2H_4(l) + O_2(g) \longrightarrow N_2(g) + 2\,H_2O(l) \qquad \Delta H° = -622.2\,kJ \qquad (23.35)$$

Reaction (23.35) can also be used to remove dissolved $O_2(g)$ from boiler water. Hydrazine is particularly valued for this purpose because no salts (ionic compounds) form that would be objectionable in the water. The industrial preparation of hydrazine was described in Chapter 4 (page 128).

Both hydrazine and hydroxylamine can act as either oxidizing or reducing agents (usually the latter), depending on the pH and the substances with which they react. The oxidation of hydrazine in acidic solution by nitrite ion produces *hydrazoic acid*, $HN_3(aq)$.

$$N_2H_5^+(aq) + NO_2^-(aq) \longrightarrow HN_3(aq) + 2\,H_2O(l) \qquad (23.36)$$

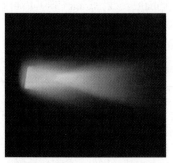

▲ The lifting thrusters of the space shuttle *Columbia*, shown here in an in-flight test, use methylhydrazine, CH_3NHNH_2, as a fuel.

Pure HN_3 is a colorless liquid that boils at 37 °C. It is very unstable and will detonate when subjected to shock. In aqueous solution, HN_3 is a weak acid; its salts are called *azides*. Azides resemble chlorides in some properties (for instance, AgN_3 is insoluble in water), but they are unstable. Some azides (such as lead azide) are used to make detonators. The release of $N_2(g)$ by the decomposition of sodium azide, NaN_3, is the basis of the air-bag safety system in automobiles (page 210).

$$:\ddot{N}\!=\!N\!=\!\ddot{N} \rightleftharpoons :N\!\equiv\!N\!-\!\ddot{N}$$
$$\qquad\qquad\qquad\quad | \qquad\qquad\qquad |$$
$$\qquad\qquad\qquad\; H \qquad\qquad\qquad\; H$$

Oxides of Nitrogen

One of the interesting features of the oxides of nitrogen is that their free energies of formation are all *positive* quantities. This feature suggests that the oxides are thermodynamically unstable, for example,

$$2\,N_2O(g) \longrightarrow 2\,N_2(g) + O_2(g) \qquad \Delta G° = -208 \text{ kJ (at 298 K)} \qquad (23.37)$$

Actually, $N_2O(g)$ is quite stable at room temperature because the activation energy for its decomposition is very high—about 250 kJ mol^{-1}. At higher temperatures (about 600 °C), its rate of decomposition becomes appreciable. Reaction (23.37) accounts for the ability of N_2O to support combustion because the $O_2(g)$ necessary for the combustion is produced by the decomposition of N_2O. Overall reactions of the following sort occur.

$$H_2(g) + N_2O(g) \longrightarrow H_2O(l) + N_2(g)$$

$$Cu(s) + N_2O(g) \longrightarrow CuO(s) + N_2(g)$$

All the oxides of nitrogen are gases at 25 °C, except N_2O_5, which is a solid with a sublimation pressure of 1 atm at 32.5 °C. It is impossible to obtain either brown $NO_2(g)$ or its colorless dimer, $N_2O_4(g)$, as a pure gas at temperatures between about -10 °C and 140 °C because of the equilibrium between them (recall Figure 16-8). At lower temperatures, N_2O_4 can be obtained as a pure solid, and above 140 °C, the gas-phase equilibrium strongly favors $NO_2(g)$. In the solid state, N_2O_3 is pale blue; in the liquid state, it is bright blue.

All nitrates decompose on heating, but only NH_4NO_3 yields $N_2O(g)$. Nitrates of active metals, such as $NaNO_3$, yield the corresponding nitrite and $O_2(g)$. Nitrates of less active metals, such as $Pb(NO_3)_2$, yield the metal oxide, $NO(g)$ and $O_2(g)$. These methods of preparing oxides of nitrogen are outlined in Table 23.8.

▲ The combustion of copper gauze in $N_2O(g)$.

Nitrogen Dioxide and Dinitrogen Tetroxide movie

TABLE 23.8   Preparation of Oxides of Nitrogen

Oxide	A Method of Preparation
N_2O	$NH_4NO_3(s) \xrightarrow{\Delta} N_2O(g) + 2\,H_2O(g)$
NO	$3\,Cu(s) + 8\,H^+(aq) + 2\,NO_3^-(aq) \longrightarrow 3\,Cu^{2+}(aq) + 2\,NO(g) + 4\,H_2O(l)$
N_2O_3	$2\,NO(g) + N_2O_4(g) \xrightarrow{-20\,°C} 2\,N_2O_3(l)$
NO_2	$2\,Pb(NO_3)_2(s) \xrightarrow{\Delta} 2\,PbO(s) + 4\,NO_2(g) + O_2(g)$
	$2\,NO(g) + O_2(g) \rightleftharpoons 2\,NO_2(g) \qquad K_p = 1.6 \times 10^{12}$ (at 298 K)
N_2O_4	$2\,NO_2(g) \rightleftharpoons N_2O_4(g) \qquad K_p = 8.84$ (at 298 K)
N_2O_5	$4\,HNO_3(l) + P_4O_{10}(s) \xrightarrow{-10\,°C} 4\,HPO_3(s) + 2\,N_2O_5(s)$

As mentioned in our assessment of the metallic/nonmetallic character of the group 15 elements, the oxides of nitrogen are acidic. N_2O_3 and N_2O_5 are the acid anhydrides of nitrous acid and nitric acid, respectively. NO_2 produces both HNO_3 and NO when it reacts with water (recall Equation 8.8). But N_2O is not an acid anhydride in the usual sense; it is related to *hyponitrous* acid, $H_2N_2O_2$ (HON=NOH), which yields N_2O and H_2O on decomposition.

$$H_2N_2O_2 \longrightarrow N_2O + H_2O$$

Phosphorus Compounds

The most important compound of phosphorus and hydrogen is *phosphine*, PH_3. This compound is analogous to ammonia, for example, by acting as a base and forming phosphonium (PH_4^+) compounds. Unlike ammonia, PH_3 is thermally unstable. Phosphine is produced by the disproportionation of P_4 in aqueous base.

▶ Phosphine is extremely poisonous and has been used as a fumigant against rodents and insects.

$$P_4(s) + 3\ OH^-(aq) + 3\ H_2O(l) \longrightarrow 3\ H_2PO_2^-(aq) + PH_3(g)$$

Another phosphorus(III) molecule is *phosphorus trichloride*, PCl_3. One typical reaction of PCl_3 is its hydrolysis, which produces hydrochloric and phosphorous acids.

$$PCl_3(l) + 6\ H_2O(l) \longrightarrow H_3PO_3(aq) + 3\ H_3O^+(aq) + 3\ Cl^-(aq)$$

PCl_3 is the most important phosphorus halide; a variety of phosphorus(III) compounds are made from it. It is produced by the direct action of $Cl_2(g)$ on elemental phosphorus. Although you may never see PCl_3, chemicals made from it are everywhere—soaps and detergents, plastics and synthetic rubber, nylon, motor oils, and insecticides and herbicides. A variety of organic groups can replace one or more chlorine atoms of PCl_3 to give a family of phosphinelike compounds. These are good Lewis bases and can act as ligands in complex-ion formation.

The halide PCl_5 is obtained by the reaction of Cl_2 with PCl_3 in tetrachloromethane (CCl_4). In the gas phase, PCl_5 exists as discrete trigonal bipyramidal molecules. In the solid state, it exists as $[PCl_4]^+[PCl_6]^-$, in which the ions are tetrahedral and octahedral, respectively.

Oxides and Oxoacids of Phosphorus

(a) P_4O_6

The simplest formulas we can write for the oxides that have phosphorus in the oxidation states +3 and +5 are P_2O_3 and P_2O_5, respectively. The corresponding names are "phosphorus trioxide" and "phosphorus pentoxide." P_2O_3 and P_2O_5 are only empirical formulas, however. The true molecular formulas of the oxides are double those just written—that is, P_4O_6 and P_4O_{10}.

The structure of each oxide molecule is based on the P_4 tetrahedron and so must have *four* P atoms, not two. As shown in Figure 23-15a, in P_4O_6 one O atom bridges each pair of P atoms in the P_4 tetrahedron, which means that there are *six* O atoms per P_4 tetrahedron. In P_4O_{10}, in addition to the six bridging O atoms, one O atom is bonded to each corner P atom (Figure 23-15b). This means that there are a total of *ten* O atoms per P_4 tetrahedron.

The reaction of P_4 with a limited quantity of $O_2(g)$ produces P_4O_6. If an excess of $O_2(g)$ is used, P_4O_{10} is obtained. Both oxides react with water to form oxoacids—both are *acid anhydrides*.

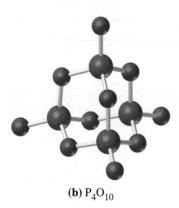

(b) P_4O_{10}

● = P ● = O

▲ **FIGURE 23-15**
Molecular structures of P_4O_6 and P_4O_{10}

$$P_4O_6(l) + 6\ H_2O(l) \longrightarrow 4\ H_3PO_3(aq)$$
$$\text{phosphorous acid}$$

$$P_4O_{10}(s) + 6\ H_2O(l) \longrightarrow 4\ H_3PO_4(l)$$
$$\text{phosphoric acid} \qquad (23.38)$$

Phosphoric acid ranks about seventh among the chemicals manufactured in the United States, with an annual production of over 13 million tons. It is used mainly to make fertilizers, but also it is used to treat metals to make them more corrosion-resistant. Phosphoric acid has many uses in the food industry: It is used to make baking powders and instant cereals, in cheese making, in curing hams, and in soft drinks to impart tartness.

If P_4O_{10} and H_2O are combined in a 1:6 mole ratio in reaction (23.38), the liquid product should be pure H_3PO_4—that is, 100% H_3PO_4, a compound called *orthophosphoric* acid. An analysis of the liquid, however, shows it to be only about 87.3% H_3PO_4. The "missing" phosphorus is still present in the liquid, but as $H_4P_2O_7$, a compound called either *diphosphoric acid* or *pyrophosphoric acid*. A molecule of diphosphoric acid forms when a molecule of H_2O is eliminated from between two molecules of orthophosphoric acid (Figure 23-16). If a third molecule of orthophosphoric acid joins in by the elimination of another H_2O molecule, the product is $H_5P_3O_{10}$, triphosphoric acid, and so on. As a class, the chainlike phosphoric acid structures are called *polyphosphoric* acids, and their salts are called *polyphosphates*. Two especially important derivatives of the polyphosphoric acids present in living organisms are the substances known as ADP and ATP. The "A" portion of the acronym stands for adenosine, a combination of an organic base called adenine and a five-carbon sugar called ribose. If this adenosine combination is linked to a diphosphate ion, the product is ADP: adenosine diphosphate. Addition of a phosphate ion to ADP yields ATP: adenosine triphosphate. These polyphosphates are described in more detail in Chapter 28.

Most phosphoric acid is made by the action of sulfuric acid on phosphate rock.

$$3Ca_3(PO_4)_2 \cdot CaF_2(s) + 10\,H_2SO_4(\text{concd aq}) + 20\,H_2O(l) \longrightarrow$$
fluorapatite

$$6\,H_3PO_4(aq) + 10\,CaSO_4 \cdot 2H_2O(s) + 2\,HF(aq) \qquad (23.39)$$
gypsum

The HF is converted to insoluble Na_2SiF_6, and the gypsum is filtered off along with other insoluble impurities. The phosphoric acid is concentrated by evaporation. Phosphoric acid obtained by this "wet process" contains a variety of metal ions as impurities and is dark green or brown. Nevertheless, it is satisfactory for the manufacture of fertilizers and for metallurgical operations.

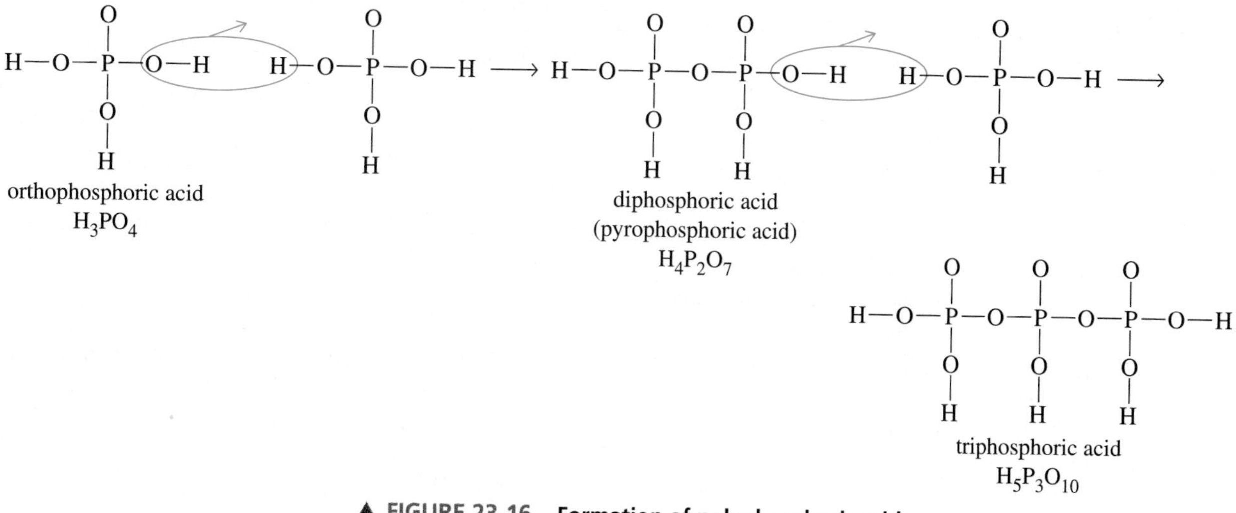

▲ **FIGURE 23-16** **Formation of polyphosphoric acids**
Removal of H_2O molecules results in P—O—P bridges.

If H_3PO_4 from reaction (23.39) is used in place of H_2SO_4 to treat phosphate rock, the principal product is calcium dihydrogen phosphate. This compound, a fertilizer containing 20 to 21% P, is marketed under the name *triple superphosphate*.

$$3Ca_3(PO_4)_2 \cdot CaF_2(s) + 14\ H_3PO_4(concd\ aq) + 10\ H_2O(l) \longrightarrow$$
$$10\ Ca(H_2PO_4)_2 \cdot H_2O(s) + 2\ HF(aq) \qquad (23.40)$$
$$\text{triple superphosphate}$$

An Environmental Issue Concerning Phosphorus

Phosphates are widely used as fertilizers because phosphorus is an essential nutrient for plant growth. Heavy fertilizer use may lead to phosphate pollution of lakes, ponds, and streams, causing an explosion of plant growth, particularly algae. The algae deplete the oxygen content of the water, eventually killing fish. This type of change, occurring in freshwater bodies as a result of their enrichment by nutrients, is called **eutrophication**. It is a natural process that occurs over geological time periods, but it can be greatly accelerated by human activities.

Natural sources of plant nutrients include animal wastes, decomposition of dead organic matter, and natural nitrogen fixation. Human sources include industrial wastes and municipal sewage plant effluents, in addition to fertilizer runoff. One way to reduce phosphate discharges into the environment is to remove them in sewage treatment plants. In the processing of sewage, polyphosphates are degraded to orthophosphates by bacterial action. The orthophosphates can then be precipitated, either as iron(III) phosphates, aluminum phosphates, or as calcium phosphate or hydroxyapatite $[Ca_5(OH)(PO_4)_3]$. The precipitating agents are generally aluminum sulfate, iron(III) chloride, or calcium hydroxide (slaked lime). In a fully equipped modern sewage treatment plant, up to 98% of the phosphates in sewage can be removed.

▲ The natural eutrophication of a lake is greatly accelerated by phosphates in wastewater and the agricultural runoff of fertilizers.

23-5 Group 14 Nonmetals: Carbon and Silicon

The differences between carbon and silicon, as outlined in Table 23.9, are perhaps the most striking between any second- and third-period elements within a group in the periodic table. As suggested by the approximate bond energies, strong C—C

TABLE 23.9   Some Comparisons of Carbon and Silicon

Carbon	Silicon
• Two principal allotropes: graphite and diamond	• One stable, diamond-type crystalline modification
• Forms two stable *gaseous* oxides, CO and CO_2, and several less stable ones, such as C_3O_2	• Forms only one *solid* oxide (SiO_2) that is stable at room temperature; a second oxide (SiO) is stable only in the temperature range 1180–2480 °C
• Insoluble in alkaline media	• Reacts in alkaline media, forming $H_2(g)$ and $SiO_4^{4-}(aq)$
• Principal oxoanion is CO_3^{2-}, which has a trigonal-planar shape	• Principal oxoanion is SiO_4^{2-}, which has a tetrahedral shape
• Strong tendency for catenation, with straight and branched chains and rings containing up to hundreds of C atoms[a]	• Less tendency for catenation, with silicon atom chains limited to about six Si atoms[a]
• Readily forms multiple bonds through use of the orbital sets $sp^2 + p$ and $sp + p^2$	• Multiple bond formation much less common than with carbon
• Approximate single-bond energies, kJ mol^{-1}: C—C, 347 C—H, 414 C—O, 360	• Approximate single-bond energies, kJ mol^{-1}: Si—Si, 226 Si—H, 318 Si—O, 464

[a]Catenation is the joining together of like atoms into chains.

and C—H bonds account for the central role of carbon-atom chains and rings in establishing the chemical behavior of carbon. A study of these chains and rings and their attached atoms is the focus of organic chemistry (Chapter 27) and biochemistry (Chapter 28). The weaker Si—Si and Si—H bonds imply a less important "organic chemistry" of silicon, and the strength of the Si—O bond accounts for the predominance of the silicates and related compounds among silicon compounds.

Production and Uses of Carbon

Graphite is widely distributed in Earth's crust, some of it in deposits rich enough for commercial exploitation. The bulk of industrial graphite, however, is synthesized from carbon-containing materials, such as coke. The key requirement is to heat the high-carbon-content material to a temperature of about 3000 °C in an electric furnace. In this process, the carbon atoms fuse into larger and larger ring systems, leading ultimately to the graphite structure.

Graphite has excellent lubricating properties, even when dry. That is because the planes of carbon atoms are held together by relatively weak forces and can easily slip past one another (recall Figure 13-33). This property is handy in pencil "lead," actually a thin rod made from a mixture of graphite and clay that glides easily on paper. Graphite's principal use is based on its ability to conduct electric current; it is used for electrodes in batteries and industrial electrolysis. Graphite's use in foundry molds, furnaces, and other high-temperature environments is based on its ability to withstand high temperatures.

A newly developed use of graphite is in the manufacture of strong, lightweight composites consisting of graphite fibers and various plastics. These composites are used in products ranging from tennis rackets to lightweight aircraft. When carbon-based fibers, such as rayon, are carefully heated to a very high temperature, all volatile matter is driven off, leaving a carbon residue with the graphite structure.

► Fabric for use in composite materials woven from carbon fibers.

As indicated in the phase diagram for carbon in Figure 23-17, graphite is the more stable form of carbon, not only at room temperature and pressure, but at temperatures up to 3000 °C and pressures of 10^4 atm and higher. Diamond is the more stable form of carbon at very high pressures. Diamonds can be synthesized from graphite by heating the graphite to temperatures of 1000–2000 °C and subjecting it to pressures of 10^5 atm or more. Usually the graphite is mixed with a metal such as iron. The metal melts, and the graphite is converted to diamond within the liquid metal. Diamonds can then be picked out of the solidified metal.

According to Figure 23-17, we might expect diamond to revert to graphite at room temperature and pressure. Fortunately for the jewelry industry and for those who treasure diamonds as gems, many phase changes that require a rearrangement in bond type and crystal structure occur extremely slowly. That is very much the case with the diamond–graphite transition.

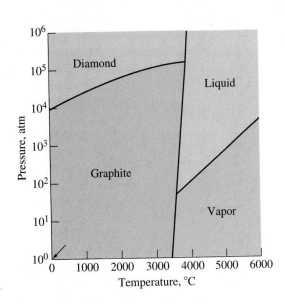

▶ **FIGURE 23-17**
Simplified phase diagram for carbon
So that the vast range of pressures required (from 1 to 1,000,000 atm) can be indicated, pressure is plotted on a logarithmic scale. The arrow marks the point 1 atm, 25 °C.

▶ Because of their expected future importance, diamond films were named "Molecule of the Year" by the journal *Science* in 1990. In 1991, the fullerenes earned that honor.

▲ **Synthetic diamonds**

Allotropes of Carbon models

Most diamonds that are used as gemstones are natural diamonds. For industrial purposes, inferior specimens of natural diamonds or, increasingly, synthetic diamonds are used. The industrial use depends on two key properties. Diamonds are used as abrasives because they are extremely hard (10 on the Mohs hardness scale). No harder substance is known. Diamonds also have a high thermal conductivity (they dissipate heat quickly), so they are used in drill bits for cutting steel and other hard materials. The rapid dissipation of heat makes the drilling process faster and increases the lifetime of the bit. A recent development has been the creation of diamond films, which can be deposited directly onto metals.

Carbon can be obtained in several other forms of mixed crystalline or amorphous structure. Incomplete combustion of natural gas, such as in an improperly adjusted Bunsen burner in the laboratory, produces a smoky flame. This smoke can be deposited as finely divided soot called **carbon black**. Carbon black is used as filler in rubber tires (several kilograms per tire), as a pigment in printing inks, and as the transfer material in carbon paper, typewriter ribbons, laser printers, and photocopying machines. Recently, new allotropes of carbon have been isolated from the decomposition of graphite. These allotropes, known as fullerenes and nanotubes, were presented on pages 506 and 507. Recall that the molecule C_{60} is remarkably stable and has a shape resembling a soccer ball (12 pentagonal faces, 20 hexagonal faces, and 60 vertices). There are a number of other fullerenes in addition to C_{60}, including C_{70}, C_{74}, and C_{82}. Fullerenes are typically produced by laser decomposition of graphite under a helium atmosphere. Because nitrogen and oxygen interfere with the process of forming fullerenes, soot (by the combustion of hydrocarbons in air) does not contain fullerenes.

Two other noteworthy carbon-containing materials are coke and charcoal. When coal is heated in the absence of air, volatile substances are driven off, leaving a high-carbon residue called *coke*. This same type of destructive distillation of wood produces *charcoal*. Currently, coke is the principal metallurgical reducing agent. It is used in blast furnaces, for instance, to reduce iron oxide to iron metal.

Some Inorganic Compounds of Carbon

In Chapters 27 and 28, we will consider the most extensive area of carbon chemistry: organic chemistry and biochemistry. In Chapter 8 and elsewhere, we discussed the oxides and carbonates, probably the most important inorganic compounds of carbon. Here, we will briefly consider a few additional inorganic carbon compounds.

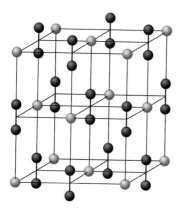

= C_2^{2-} ions

● = Ca^{2+} ions

▲ **Structure of calcium carbide**

Carbon combines with metals to form *carbides*. In many cases, the carbon atoms occupy the holes or voids in metal structures, forming *interstitial carbides*. With active metals, the carbides are ionic. *Calcium carbide* forms in the high-temperature reaction of lime and coke:

$$CaO(s) + 3\ C(s) \xrightarrow{2000\ °C} CaC_2(s) + CO(g) \tag{23.41}$$

Calcium carbide is important because acetylene ($HC{\equiv}CH$) can easily be made from it. In turn, acetylene can be used to synthesize many chemical compounds.

$$CaC_2(s) + 2\ H_2O(l) \longrightarrow Ca(OH)_2(s) + C_2H_2(g) \tag{23.42}$$

The solid CaC_2 can be regarded as a face-centered cubic array of Ca^{2+} ions with the C_2^{2-} ions in the octahedral holes.

Carbon disulfide, CS_2, can be synthesized by the reaction of methane and sulfur vapor in the presence of a catalyst.

$$CH_4(g) + 4\ S(g) \longrightarrow CS_2(l) + 2\ H_2S(g) \tag{23.43}$$

Carbon disulfide is a highly flammable, volatile liquid that acts as a solvent for sulfur, phosphorus, bromine, iodine, fats, and oils. Its uses as a solvent are decreasing, however, because it is poisonous. Other important uses are in the manufacture of rayon and cellophane.

Carbon tetrachloride, CCl_4, can be prepared by the direct chlorination of methane.

$$CH_4(g) + 4\ Cl_2(g) \longrightarrow CCl_4(l) + 4\ HCl(g) \tag{23.44}$$

Although CCl_4 has been extensively used as a solvent, dry-cleaning agent, and fire extinguisher, these uses have been steadily declining. CCl_4 causes liver and kidney damage and is a known carcinogen.

Certain groupings of atoms, several containing C atoms, have some of the characteristics of a halogen atom. They are called *pseudohalogens* and include

—CN (cyanide)　　　—OCN (cyanate)　　　—SCN (thiocyanate)　　　—N_3 (azide)

The *cyanide* ion, CN^-, is similar to the halide ions, X^-, in that it forms an insoluble salt, $AgCN$, and an acid HCN. Hydrocyanic acid is a very weak acid, unlike HCl. Another important difference is that CN^- is poisonous. Despite its extreme toxicity, HCN, a liquid that boils at about room temperature, has important uses in the manufacture of plastics. The combination of two cyanide groups produces *cyanogen*, $(CN)_2$. This gas resembles chlorine gas in undergoing a disproportionation reaction in basic solution.

$$(CN)_2 + 2\ OH^-(aq) \longrightarrow CN^-(aq) + OCN^-(aq) + H_2O(l) \tag{23.45}$$

Cyanogen is used in organic synthesis, as a fumigant, and as a rocket propellant.

Production and Uses of Silicon

Elemental silicon is produced when quartz or sand (SiO_2) is reduced by reaction with coke in an electric arc furnace.

$$SiO_2 + 2\ C \xrightarrow{\Delta} Si + 2\ CO(g) \tag{23.46}$$

Very high purity Si for solar cells can be made by reducing Na_2SiF_6 with metallic Na. Recall that Na_2SiF_6 is a by-product of the formation of phosphate fertilizers (page 931). High-purity silicon is required in the manufacture of transistors and other semiconductor devices.

Oxides of Silicon

The chemistry of the oxides of carbon (CO and CO_2) was discussed in Chapter 8. Silicon forms only one stable oxide, having the empirical formula SiO_2, **silica**. In silica, each Si atom is bonded to *four* O atoms and each O atom to *two* Si atoms.

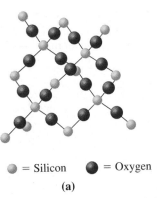

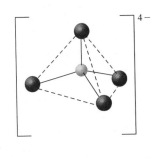

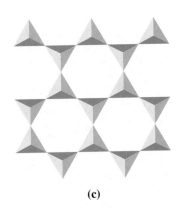

● = Silicon   ● = Oxygen

(a)

(b)

(c)

▲ FIGURE 23-18   **Structures of silica and silicates**
(a) A three-dimensional network of bonds in silica, SiO_2. (b) The silicate anion, SiO_4^{4-}, commonly found in silicate materials. The Si atom is in the center of the tetrahedron and is surrounded by four O atoms. (c) A depiction of the structure of a mica, using tetrahedra to represent the SiO_4 units. The cations K^+ and Al^{3+} are also present but are not shown. This is the usual way that materials scientists represent silicates and similar materials.

The structure is that of a network covalent solid, as suggested by Figure 23-18a. This structure is reminiscent of the diamond structure, and silica has certain properties that resemble those of diamond. For example, quartz, a form of silica, is fairly hard (with a Mohs hardness of 7, compared with 10 for diamond), has a high melting point (about 1700 °C), and is a nonconductor of electricity. Silica is the basic raw material of the glass and ceramics industries.

The central feature of all *silicates* is the SiO_4^{4-} tetrahedron depicted in Figure 23-18b. These tetrahedra may be arranged in a wide variety of ways.

- *Simple SiO₄ tetrahedra.* Typical minerals in which the anions are simple SiO_4^{4-} tetrahedra are *thorite* ($ThSiO_4$) and *zircon* ($ZrSiO_4$).

Are You Wondering...

Why silica (SiO_2) does not exist as simple molecules as CO_2 does?

Because both silicon and carbon are in group 14 of the periodic table and have four valence electrons, we might expect them to form oxides with similar properties. In CO_2, the side-to-side overlap of $2p$ orbitals of the C and O atoms is extensive (see page 448). The carbon-to-oxygen double bond in CO_2 proves to be stronger (799 kJ mol^{-1}) than two single bonds (2 × 360 kJ mol^{-1}). This results in the familiar Lewis structure of CO_2.

$$\ddot{O}{=}C{=}\ddot{O}$$

Silicon, being in the third period, would have to use $3p$ orbitals to form double bonds with oxygen. The side-to-side overlap of these orbitals with the $2p$ orbitals of oxygen is quite limited. In terms of energy, a stronger bonding arrangement results if the Si atoms form *four* single bonds with O atoms (bond energy: 4 × 464 kJ mol^{-1} = 1856 kJ total) rather than *two* double bonds (bond energy: 2 × 640 kJ mol^{-1} = 1280 kJ total). Because each O atom must simultaneously bond to two Si atoms, the result is a network of —Si—O—Si— bonds (see Figure 23-18).

On page 917 we contrasted O and S, the second-period and third-period members of group 16. Here is another example of how the second-period member of a group (here carbon) differs from the higher period members.

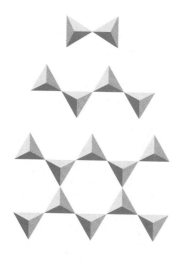

- *Two SiO_4 tetrahedra joined end-to-end.* The silicon atoms in the two tetrahedra share an O atom between them in the anion $Si_2O_7^{6-}$, found in the mineral *thortveitite* ($Sc_2Si_2O_7$).

- *SiO_4 tetrahedra joined into long chains.* Each Si atom shares an O atom with the Si atom in an adjacent tetrahedron on either side. An example is *spodumene*, the principal source of lithium and lithium compounds. Its empirical formula is $LiAl(SiO_3)_2$.

- *SiO_4 tetrahedra joined into a double chain.* Half the Si atoms share three of their four O atoms with Si atoms in adjacent tetrahedra, and half share only two. In *chrysotile asbestos*, double chains are held together by cations (chiefly Mg^{2+}); this mineral has a fibrous appearance. The empirical formula is $Mg_3(Si_2O_5)(OH)_4$.

- *SiO_4 tetrahedra are bonded together in two-dimensional sheets.* Each Si atom shares O atoms with the Si atoms in three adjacent tetrahedra. This structure is depicted in Figure 23-18c. In *muscovite*, with the empirical formula, $KAl_2(AlSi_3O_{10})(OH)_2$, the counterions that bond sheets together in layers are mostly K^+ and Al^{3+}. Bonding within the sheets is stronger than that between sheets, so mica flakes easily.

- *SiO_4 tetrahedra are bonded together in three-dimensional structures.* Three-dimensional arrays of tetrahedra result when each Si atom shares all four O atoms with the Si atoms in four adjoining tetrahedra (as in Figure 23-18a). This is the most common arrangement, occurring in silica (quartz) and in the majority of silicate minerals.

▶ H_4SiO_4 is called *ortho*silicic acid. When a combination of one H atom and one OH group (equivalent to a molecule of H_2O) is eliminated from the *ortho* acid, the resulting acid, H_2SiO_3, is called *meta*silicic acid.

SiO_2 is a weakly acidic oxide and slowly dissolves in strong bases. It forms a series of silicates, such as Na_2SiO_4 (sodium orthosilicate) and Na_2SiO_3 (sodium metasilicate). These compounds are somewhat soluble in water and, as a result, are sometimes referred to as "water glass."

Silicate anions are bases; when acidified, they produce silicic acids, which are unstable and decompose to silica. The silica obtained, however, is not a crystalline solid or powder. Depending on the acidity of the solution, the silica is obtained as a colloidal dispersion, a gelatinous precipitate, or a solidlike gel in which all the water is entrapped. These hydrated silicates are polymers of silica formed by the elimination of water molecules between neighboring molecules of silicic acid. The process begins with the reaction

$$SiO_4^{4-}(aq) + 4\,H^+(aq) \longrightarrow Si(OH)_4$$

It is followed by

where —H and —OH combine to form water (HOH), and Si—O—Si bridges are produced.

Ceramics and Glass Hydrated silicate polymers are important in the ceramics industry. A colloidal dispersion of particles in a liquid is called a *sol*. The sol can be poured into a mold and, following removal of some of the liquid, is converted to a *gel*. The gel is then processed into the final ceramic product. This sol–gel process can produce exceptionally lightweight ceramic materials.

Uses of these advanced ceramics fall into two general categories: (1) electrical, magnetic, or optical applications (as in the manufacture of integrated circuit

▲ Ceramic components of an automobile engine.

components) and (2) applications that take advantage of the ceramic's mechanical and structural properties at high temperatures. These latter properties have been explored in developing ceramic components for gas turbines and automotive engines. Quite possibly the automobile engine of the future will be a ceramic engine—lightweight and more fuel-efficient because higher operating temperatures are possible. Some engines already have several ceramic components. There is some truth to the vision of this ceramic future as the new stone age.

If sodium and calcium carbonates are mixed with sand and fused at about 1500 °C, the result is a liquid mixture of sodium and calcium silicates. When cooled, the liquid becomes more viscous and eventually becomes a solid that is transparent to light; this solid is called a **glass**. Crystalline solids have a long-range order, whereas glasses are *amorphous* solids in which order is found over relatively short distances only. Think of the structural units in glass (silicate anions) as being in a jumbled rather than a regular arrangement. A glass and a crystalline solid also differ in their melting behavior. A glass softens and melts over a broad temperature range, whereas a crystalline solid has a definite, sharp melting point. Different types of glass and methods of making them are described in the Focus On feature at the end of this chapter.

Silicon Compounds Involving Carbon

Several silicon–hydrogen compounds are known, but because Si—Si single bonds are not particularly strong, the chain length in these compounds, called *silanes*, is limited to six.

$$
\begin{array}{cccc}
\text{H} & \text{H} \quad \text{H} & \text{H} \quad \text{H} \quad \text{H} & \\
| & | \quad \ \ | & | \quad \ \ | \quad \ \ | & \\
\text{H}-\text{Si}-\text{H} & \text{H}-\text{Si}-\text{Si}-\text{H} & \text{H}-\text{Si}-\text{Si}-\text{Si}-\text{H} \ \cdots & \text{S}_6\text{H}_{14} \\
| & | \quad \ \ | & | \quad \ \ | \quad \ \ | & \\
\text{H} & \text{H} \quad \text{H} & \text{H} \quad \text{H} \quad \text{H} & \\
\text{Monosilane} & \text{Disilane} & \text{Trisilane} & \text{Hexasilane}
\end{array}
$$

Other atoms or groups of atoms can be substituted for H atoms in silanes to produce compounds called *organosilanes*. Typical is the direct reaction of Si and methyl chloride, CH_3Cl.

$$2\ CH_3Cl + Si \longrightarrow (CH_3)_2SiCl_2$$

The reaction of $(CH_3)_2SiCl_2$, dichlorodimethylsilane, with water produces an interesting compound, *dimethylsilanol*, $(CH_3)_2Si(OH)_2$. Dimethylsilanol undergoes a polymerization reaction in which H_2O molecules are eliminated from among large numbers of silanol molecules. The result of this polymerization is a material consisting of molecules with long silicon–oxygen chains: **silicones**.

$$
\underset{\underset{\text{CH}_3}{|}}{\overset{\overset{\text{CH}_3}{|}}{\text{HO}-\text{Si}-\text{O}-\text{H}}} + \underset{\underset{\text{CH}_3}{|}}{\overset{\overset{\text{CH}_3}{|}}{\text{HO}-\text{Si}-\text{OH}}} \xrightarrow{-\text{H}_2\text{O}} \underset{\underset{\text{CH}_3}{|}}{\overset{\overset{\text{CH}_3}{|}}{\text{HO}-\text{Si}-\text{O}}} \left[\underset{\underset{\text{CH}_3}{|}}{\overset{\overset{\text{CH}_3}{|}}{\text{Si}-\text{O}}} \right]_n \underset{\underset{\text{CH}_3}{|}}{\overset{\overset{\text{CH}_3}{|}}{\text{Si}-\text{OH}}}
$$

a silicone

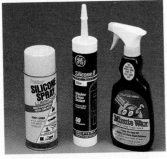

▲ Some common applications of silicones.

Silicones are important polymers because they are versatile. They can be obtained either as oils or as rubberlike materials. Silicone oils are not volatile and do not decompose when heated. Also, they can be cooled to low temperatures without solidifying or becoming viscous. Silicone oils are excellent high-temperature lubricants. In contrast, hydrocarbon oils break down at high temperatures, become very viscous, and then solidify at low temperatures. Silicone rubbers retain their elasticity at low temperatures and are chemically resistant and thermally stable. This makes them useful in caulking around windows, for instance.

23-6   The Group 13 Nonmetal: Boron

The only group 13 element that is almost exclusively nonmetallic in its physical and chemical properties is boron. The remaining members of group 13—Al, Ga, In, and Tl—are metals and were discussed in Chapter 22. Many boron compounds lack an octet about the central boron atom, which makes the compounds *electron-deficient*. This deficiency also makes them strong Lewis acids. The electron deficiency of some boron compounds leads to bonding of a type that we have not previously encountered. This type of bonding occurs in the boron hydrides.

Boron Hydrides

The molecule BH_3 (borane) may exist as a reaction intermediate, but it has not been isolated as a stable compound. The B atom in BH_3 lacks a complete octet by having only six electrons surrounding it. The simplest boron hydride that has been isolated is **diborane**, B_2H_6, but this molecule defies simple description. In the following structural formula, what holds the two borane units together?

$$
\begin{array}{ccc}
H & & H \\
| & & | \\
H-B & ? & B-H \\
| & & | \\
H & & H
\end{array}
$$

To explain the structure and bonding in B_2H_6, we need new ideas in bonding theory, particularly molecular orbital theory. The problem is this: The B_2H_6 molecule has only *12* valence electrons (three each from the two B atoms and one each from the six H atoms). The minimum number of valence-shell atomic orbitals required to make a Lewis structure for B_2H_6 resembling that of C_2H_6, however, is *14* (four each from the two B atoms and one each from the six H atoms).

The currently accepted structure of diborane is illustrated in Figure 23-19. The two B atoms and four of the H atoms lie in the same plane (a plane perpendicular to the plane of the page). The orbitals used by the B atoms to bond these particular four H atoms can be viewed as sp^3. Eight electrons are involved in these four bonds. Four electrons are left to bond the two remaining H atoms to the two B atoms and also to bond the B atoms together. This is accomplished if each of the two H atoms simultaneously bonds to *both* B atoms.

Atom "bridges" are actually fairly common, although we have not had much occasion to deal with them before (see the discussion of Al_2Cl_6, page 893). The B—H—B bridges are unusual, however, in having only *two* electrons shared among *three* atoms. For this reason, these bonds are referred to as *three-center* two-electron bonds.

We can rationalize the bonding in these three-center bonds with molecular orbital theory. *Six* atomic orbitals—an sp^3 and p orbital from each B atom and an s orbital from each H atom—are combined into *six* molecular orbitals in these two bridge bonds. Of these six molecular orbitals, *two* are bonding orbitals, and these are the orbitals into which the four electrons are placed. The concept of bridge bonds can be extended to B—B—B bonds to describe the structure of higher boranes, such as B_5H_9 (Figure 23-20).

Boron hydrides are widely used in reactions for synthesizing organic compounds. They continue to provide new and exciting developments in chemistry.

Other Boron Compounds of Interest

Boron compounds are widely distributed in Earth's crust, but concentrated ores are found in only a few locations—in Italy, Russia, Tibet, Turkey, and the desert regions of California. Typical of these ores is the hydrated borate *borax*, $Na_2B_4O_7 \cdot 10H_2O$. Figure 23-21 suggests how borax can be converted to a variety of boron compounds.

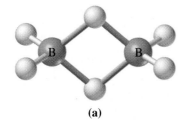

(a)

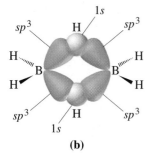

(b)

▲ **FIGURE 23-19**
Structure of diborane, B_2H_6
(a) The molecular structure.
(b) Bonding.

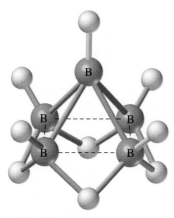

▲ **FIGURE 23-20**
Structure of pentaborane, B_5H_9
The boron atoms are joined by multicenter B—B—B bonds. Five of the H atoms are bonded to one B atom each. The other four H atoms bridge pairs of B atoms.

Stained-glass windows of the Spiral Church in Dallas, Texas.

Glass and the art of glassmaking have been known for centuries. Beautiful stained-glass windows can be seen in medieval and modern churches; ancient glass containers for perfume and oil are displayed in many museums. Today, glass is indispensable in almost every facet of life.

Soda–lime glass is the oldest form of glass. The starting material in its manufacture is a mixture of sodium carbonate (*soda* ash), calcium carbonate (which decomposes to form quick*lime* when heated), and silicon dioxide. The mixture can be fused at a relatively low temperature (1300 °C) compared with the melting point of pure silica (1710 °C), and it is easy to form into the shapes needed. The effect of the sodium ions is to break up the crystalline lattice of the SiO_2; the calcium ions render the glass insoluble in water, so it can be used for items such as drinking glasses and windows. At the high temperatures employed, chemical reactions occur that produce a mixture of sodium and calcium silicates as the ultimate glass product.

$$Na_2B_4O_7 \cdot 10H_2O$$

$$\downarrow H_2SO_4$$

$$B(OH)_3$$

$$\downarrow \Delta$$

$$BF_3 \xleftarrow[CaF_2, H_2SO_4]{\Delta} B_2O_3 \xrightarrow[C, Cl_2]{\Delta} BCl_3$$

$$\Delta \downarrow Mg \qquad\qquad \downarrow LiAlH_4$$

$$B \qquad H_2 \qquad B_2H_6$$

▶ **FIGURE 23-21**
Preparation of some boron compounds
$Na_2B_4O_7 \cdot 10H_2O$ (borax) is converted to $B(OH)_3$ by reaction with H_2SO_4. When heated strongly, $B(OH)_3$ is converted to B_2O_3. A variety of boron-containing compounds and boron itself can be prepared from B_2O_3.

$$\left[\begin{array}{c} \text{HO} \diagdown \\ \text{HO} \diagup \end{array} B \begin{array}{c} O-O \\ O-O \end{array} B \begin{array}{c} \diagup OH \\ \diagdown OH \end{array} \right]^{2-}$$

▲ Perborate ion

One useful compound that can be obtained by crystallization from a solution of borax and hydrogen peroxide is *sodium perborate*, $NaBO_3 \cdot 4H_2O$. This formula is deceptively simple; a more precise formula is $Na_2[B_2(O_2)_2(OH)_4] \cdot 6H_2O$. Sodium perborate is the bleach alternative used in many color-safe bleaches. The key to the bleaching action is the presence of the two peroxo groups (—O—O—) bridging the boron atoms in the $[B_2(O_2)_2(OH)_4]^{2-}$ ion.

One of the key compounds from which other boron compounds can be synthesized is *boric acid*, $B(OH)_3$. The weakly acidic nature of boric acid comes about in a rather unusual way: The electron-deficient $B(OH)_3$ accepts a OH^- ion from the self-ionization of water, forming the complex ion $[B(OH)_4]^-$. Thus, the source of the H_3O^+ in $B(OH)_3(aq)$ is the water itself. This ionization scheme, together with the fact that $B(OH)_3$ is a *mono*protic, not *tri*protic, acid suggests that the best formula of boric acid is $B(OH)_3$, not H_3BO_3.

$$B(OH)_3(aq) + 2\,H_2O(l) \rightleftharpoons H_3O^+(aq) + B(OH)_4^-(aq) \qquad K_a = 5.6 \times 10^{-10}$$

Borate salts, as expected of the salts of a weak acid, produce basic solutions by hydrolysis, which accounts for their use in cleaning agents. Boric acid also acts as

Glass containing even small amounts of Fe_2O_3 has a distinctive green color (bottle glass). Glass can be made colorless by incorporating MnO_2 into the glassmaking process. The MnO_2 oxidizes green $FeSiO_3$ to yellow $Fe_2(SiO_3)_3$ and is itself reduced to Mn_2O_3, which imparts a violet color. The yellow and violet are complementary colors, so the glass appears colorless. Where desired, color can be imparted by means of appropriate additives, such as CoO for cobalt blue glass. To produce an opaque glass, additives such as calcium phosphate are used. In Bohemian crystal, most of the Na^+ is replaced by K^+; a glass with exceptional transparency and brightness can be made by incorporating lead oxides (as in Waterford crystal).

A problem with soda–lime glass is its high coefficient of thermal expansion—its dimensions change significantly with temperature. The glass cannot withstand thermal shock. This limitation posed a particular problem for the lanterns used in the early days of railroads. In the rain, the hot glass in these lanterns would easily shatter. Adding B_2O_3 to the glass solved the problem. A *borosilicate* glass has a low coefficient of thermal expansion and is thus resistant to thermal shock. This is the glass commonly known by its trade name, Pyrex. It is widely used in chemical laboratories and for cookware in the home.

Most glass has small bubbles or impurities in it that decrease its ability to transmit light without scattering—a phenomenon observed in the distorted images produced by the thick bottoms of drinking glasses. In modern *fiber-optic* cables, sound waves are converted to electrical impulses, which are transmitted as laser light beams. The light must be transmitted over long distances without distortion or loss of signal. For this purpose, a special glass made of pure silica is required. The key to making this glass is in purifying silica, which can be done by a series of chemical reactions. First, impure quartz or sand is reduced to silicon by using coke as a reducing agent. The silicon is then allowed to react with $Cl_2(g)$ to form $SiCl_4(g)$.

$$SiO_2(s) + 2\,C(s) \longrightarrow Si(s) + 2\,CO(g)$$

$$Si(s) + 2\,Cl_2(g) \longrightarrow SiCl_4(g)$$

Finally, the $SiCl_4$ is burned in a methane–oxygen flame. SiO_2 deposits as a fine ash, and chlorocarbon compounds escape as gaseous products. The SiO_2, with impurity levels reduced to parts per billion, can then be melted and drawn into the fine filaments required in fiber-optic cable. Tens of millions of miles of fiber-optic cable are currently produced annually in the United States.

an insecticide, particularly to kill roaches, and as an antiseptic in eyewash solutions. Boron compounds are used in products as varied as adhesives, cement, disinfectants, fertilizers, fire retardants, glass, herbicides, metallurgical fluxes, and textile bleaches and dyes.

EXAMPLE 23-3

Writing Chemical Equations from a Summary Diagram of Reaction Chemistry. Using Figure 23-21, write chemical equations for the successive conversions of borax to (a) boric acid, (b) B_2O_3, and (c) impure boron metal.

Solution

Figure 23-21 lists the key substances involved in each reaction. We must identify other plausible reactants and products.

(a) Conversion of a salt to the corresponding acid is an acid–base reaction and does not involve changes in oxidation states. Balancing should be possible by inspection. The additional substances required in the equation are Na_2SO_4 and H_2O.

$$Na_2B_4O_7 \cdot 10H_2O + H_2SO_4 \longrightarrow 4\,B(OH)_3 + Na_2SO_4 + 5\,H_2O$$

(b) Conversion of a hydroxo compound to an oxide requires that H_2O be driven off.

$$2\,B(OH)_3 \xrightarrow{\Delta} B_2O_3 + 3\,H_2O$$

(c) The direct reduction of B_2O_3 with Mg produces impure boron.

$$3\,Mg + B_2O_3 \xrightarrow{\Delta} 2\,B + 3\,MgO$$

Practice Example A: Using Figure 23-21, write chemical equations for the sequence of reactions by which borax is converted to diborane.

Practice Example B: Using Figure 23-8 and reactions in the text, write chemical equations for the sequence of reactions by which sulfuric acid is obtained from ZnS.

Summary

Principles developed in Chapters 1–21 are applied in this chapter to several aspects of the descriptive inorganic chemistry of the nonmetals. Several differences between the first and higher members of a group of the periodic table are encountered with nonmetals. Examples cited are the failure of fluorine to form stable oxoacids and the numerous differences between O and S and between C and Si. Also, by considering several criteria, the progression of properties from nonmetallic to metallic within group 15 is established. Additional insights into chemical bonding are provided through a study of the boron hydrides.

Electrode potential diagrams are a way to summarize the oxidation–reduction chemistry of an element. These diagrams are especially useful for elements that can exist in many oxidation states, such as Cl, S, and N. Often we can obtain electrode po-

tentials for reduction processes not specifically represented in a diagram by means of calculations based on the relationship $\Delta G° = -nFE°$.

Oxoacids and oxoanions are studied in terms of the methods required to prepare them, their acid–base properties, their strengths as oxidizing or reducing agents, and their structures. With phosphorus, the elimination of H_2O from simpler acid molecules produces diphosphoric, triphosphoric, and other polyphosphoric acids and their salts.

Throughout the chapter, practical uses of the nonmetals and their compounds are described. In many cases, a particular use for a substance is a consequence of some special property that it possesses.

Integrative Example

For use in analytical chemistry, sodium thiosulfate solutions must be carefully prepared. In particular, the solutions must be kept from becoming acidic. In strongly acidic solutions, thiosulfate ion disproportionates into $SO_2(g)$ and $S_8(s)$.

▲ **Decomposition of thiosulfate ion**
When an aqueous solution of $Na_2S_2O_3$ is acidified, the sulfur is in the colloidal state when first formed (right).

Show that the disproportionation of $S_2O_3{}^{2-}(aq)$ is spontaneous for standard state conditions in acidic solution but not in basic solution.

1. *Write half-equations and an overall equation for the disproportionation reaction in acidic solution.* Base the overall equation on the description of the reaction given above.

Reduction:
$$4\,S_2O_3{}^{2-}(aq) + 24\,H^+(aq) + 16\,e^- \longrightarrow S_8(s) + 12\,H_2O(l)$$

Oxidation:
$$4\,\{S_2O_3{}^{2-}(aq) + H_2O(l) \longrightarrow 2\,SO_2(g) + 2\,H^+(aq) + 4\,e^-\}$$

Overall:
$$8\,S_2O_3{}^{2-}(aq) + 16\,H^+(aq) \longrightarrow S_8(s) + 8\,SO_2(g) + 8\,H_2O(l)$$

2. *Determine $E°_{cell}$ for the reaction described in part 1.* To do this, use data from Figure 23-9. That figure gives an $E°$ value for the reduction half-reaction (0.465 V) but no value for the oxidation. To obtain this missing $E°$, we need the method of Example 23-1. That is,

$$4\,SO_2(g) + 4\,H^+(aq) + 6\,e^- \longrightarrow S_4O_6{}^{2-}(aq) + 2\,H_2O(l)$$
$$\Delta G° = -6FE° = -6F \times 0.507\ \text{V}$$

$$S_4O_6{}^{2-}(aq) + 2\,e^- \longrightarrow 2\,S_2O_3{}^{2-}(aq)$$
$$\Delta G° = -2FE° = -2F \times 0.080\ \text{V}$$

$$4\,SO_2(g) + 4\,H^+(aq) + 8\,e^- \longrightarrow 2\,S_2O_3{}^{2-}(aq) + 2\,H_2O(l)$$
$$\Delta G° = -F\,[(6 \times 0.507) + (2 \times 0.080)]\ \text{V}$$

$$\Delta G° = -8FE° = -F\,(3.202)\ \text{V}$$

$$E° = (3.202/8)\ \text{V} = 0.400\ \text{V}$$

Now we can calculate $E°_{cell}$.

$$E°_{cell} = E°(\text{reduction}) - E°(\text{oxidation}) =$$
$$0.465\ \text{V} - 0.400\ \text{V} = 0.065\ \text{V}$$

The disproportionation is spontaneous for standard-state conditions in acidic solution.

3. *Assess the situation in basic solution.* Although we could use the Nernst equation to do a quantitative calculation, the simplest approach is to note that $E°_{cell}$ for acidic solution, though positive, is rather small in magnitude (+0.065 V). Increasing $[OH^-]$, as would be the case in making the solution basic, means decreasing $[H^+]$. In fact, $[OH^-] = 1$ M corresponds to $[H^+] = 1 \times 10^{-14}$ M. Because the overall equation in part 1 has $H^+(aq)$ on the *left* side of the equation, a decrease in $[H^+]$ favors the *reverse* reaction (by Le Châtelier's principle). At some point before the solution becomes basic, the forward reaction is no longer spontaneous (see Exercise 83).

Key Terms

acid salt (23-3)
carbon black (23-5)
contact process (23-3)
diborane (23-6)

eutrophication (23-4)
Frasch process (23-3)
glass (23-5)
polyhalide ions (23-2)

silica (23-5)
silicones (23-5)

Review Questions

1. In your own words, define the following terms: (a) polyhalide ion; (b) polyphosphate; (c) allotropy; (d) disproportionation.
2. Briefly describe each of the following terms: (a) Frasch process; (b) contact process; (c) eutrophication; (d) three-center bond.
3. Explain the important distinctions between each pair of terms: (a) acid salt and acid anhydride; (b) azide and nitride; (c) silane and silicone; (d) sol and gel.
4. Provide an acceptable name or formula for each of the following.
   (a) $KBrO_3$; (b) I_3^-; (c) sodium hypochlorite; (d) NaH_2PO_4; (e) $Pb(N_3)_2$; (f) barium thiosulfate
5. Complete and balance equations for the following reactions. If no reaction occurs, so state.
   (a) $CaCl_2(s) + H_2SO_4(\text{concd aq}) \xrightarrow{\Delta}$
   (b) $I_2(s) + Cl^-(aq) \longrightarrow$
   (c) $NH_3(aq) + HClO_4(aq) \longrightarrow$
6. Write a chemical equation to represent the reaction of (a) $Cl_2(g)$ with cold $NaOH(aq)$; (b) $NaI(s)$ with hot $H_2SO_4(\text{concd aq})$; (c) $Cl_2(g)$ with $Br^-(aq)$.
7. Write a plausible equation to represent (a) the reaction of $KI(s)$ with the nonoxidizing acid $H_3PO_4(l)$; (b) the reduction of Na_2SiF_6 to pure Si, using Na as the reducing agent.
8. Write a plausible equation(s) to represent (a) the reduction of As_4O_6 to As, using $CO(g)$ as the reducing agent; (b) the production of the fertilizer components monoammonium phosphate (MAP) and diammonium phosphate (DAP) by the reaction of H_3PO_4 and NH_3.
9. A reference work describing the production of phosphorus states that for every 8.00 tons of phosphate rock used, 1.00 ton of elemental phosphorus is obtained. The phosphorus content of the phosphate rock is given as 31% P_4O_{10}. What is the percent yield of phosphorus in this reaction?
10. If Br^- and I^- occur together in an aqueous solution, I^- can be oxidized to IO_3^- with an excess of $Cl_2(aq)$. Simultane-

ously, Br^- is oxidized to Br_2, which is extracted with $CS_2(l)$. Write chemical equations for the reactions that occur.
11. Suppose that the sulfur present in seawater as SO_4^{2-} (2650 mg/L) could be recovered as elemental sulfur. If this sulfur were then converted to H_2SO_4, how many cubic kilometers of seawater would have to be processed to yield the average U.S. annual consumption of about 45 million tons of H_2SO_4?
12. Use information from this chapter and elsewhere in the text to answer the following questions. Explain your reasoning in each case.
   (a) When $Cl_2(g)$ is added to an aqueous solution containing I^-, which is more likely to be produced, $O_2(g)$ or I_2?
   (b) When added to an acidic solution containing NH_4^+, is H_2O_2 more likely to be oxidized to $O_2(g)$ or reduced to H_2O?
13. Complete and balance each of the following equations. If no reaction occurs, so state.
   (a) $KI(s) + H_3PO_4(\text{concd aq}) \xrightarrow{\Delta}$
   (b) $K_2O(s) + H_2O(l) \longrightarrow$
   (c) $I_2(s) + KI(aq) \longrightarrow$
   (d) $HSO_3^-(aq) + MnO_4^-(aq) + H^+(aq) \longrightarrow$
14. Name each of the following compounds: (a) Na_4XeO_6; (b) BaO_2; (c) $Hg(SCN)_2$, (d) Ba_3N_2; (e) $Ag_2S_2O_3$.
15. Supply a formula for each of the following: (a) phosphorus dichloride trifluoride; (b) potassium cyanate; (c) iron(III) orthophosphate; (d) barium azide; (e) magnesium pyrophosphate.
16. Although relatively rare, all of the following compounds exist. Based on what you know about related compounds (such as from the periodic table), propose a plausible name or formula for each compound: (a) silver astatide; (b) Na_4XeO_6; (c) magnesium polonide; (d) H_2TeO_3; (e) potassium thioselenate; (f) $KAtO_4$.

Exercises

Noble Gas Compounds

17. Use VSEPR theory to predict the probable geometric structures of (a) XeO_3; (b) XeO_4; (c) XeF_5^+.
18. Use VSEPR theory to predict the probable geometric structures of the molecules (a) O_2XeF_2; (b) O_3XeF_2; (c) $OXeF_4$.
19. Write a chemical equation for the hydrolysis of XeF_4, which yields XeO_3, Xe, O_2, and HF as products.
20. Write a chemical equation for the hydrolysis in alkaline solution of XeF_6, which yields XeO_6^{4-}, Xe, O_2, F^-, and H_2O as products.

The Halogens

21. Freshly prepared solutions containing iodide ion are colorless, but over time they usually turn yellow. Describe a plausible chemical reaction (or reactions) to account for this observation.

22. Fluorine can be prepared by the reaction of hexafluoromanganate(IV) ion, MnF_6^{2-}, with antimony pentafluoride to produce manganese(IV) fluoride and SbF_6^-, followed by the disproportionation of manganese(IV) fluoride to manganese(III) fluoride and $F_2(g)$. Write chemical equations for these two reactions.

23. Make a general prediction about which of the halogen elements, F_2, Cl_2, Br_2, or I_2, displaces other halogens from a solution of halide ions. Which of the halogens is able to displace $O_2(g)$ from water? Which is able to displace $H_2(g)$ from water?

24. The following properties of astatine have been measured or estimated: (a) covalent radius; (b) ionic radius (At^-); (c) first ionization energy; (d) electron affinity; (e) electronegativity; (f) standard reduction potential. Based on periodic relationships and data in Table 23.1, what values would you expect for these properties?

25. The abundance of F^- in seawater is 1 g F^- per ton of seawater. Suppose that a commercially feasible method could be found to extract fluorine from seawater.

(a) What mass of F_2 could be obtained from 1 km^3 of seawater ($d = 1.03$ g/cm^3)?
(b) Would the process resemble that for extracting bromine from seawater? Explain.

26. Fluorine is produced chiefly from fluorite, CaF_2. Fluorine can also be obtained as a by-product of the production of phosphate fertilizers, derived from phosphate rock $[3Ca_3(PO_4)_2 \cdot CaF_2]$. What is the maximum mass of fluorine that could be extracted as a by-product from 1.00×10^3 kg of phosphate rock?

27. Show by calculation whether the disproportionation of chlorine gas to chlorate and chloride ions will occur under standard-state conditions in an acidic solution.

28. Show by calculation whether the reaction
$2 HOCl(aq) \longrightarrow HClO_2(aq) + H^+(aq) + Cl^-(aq)$ will go essentially to completion as written for standard-state conditions.

29. Predict the geometric structures of (a) BrF_3; (b) IF_5; (c) Cl_3IF^-.

30. Which of the following species has a linear structure: ClF_2^+, $IBrF^-$, OCl_2, ClF_3, or SF_4? Do any two of these species have the same structure?

Oxygen

31. Hydrogen peroxide is a somewhat stronger acid than water. For the ionization
$H_2O_2(aq) + H_2O(l) \longrightarrow H_3O^+(aq) + HO_2^-(aq)$
$pK_a = 11.75$. Calculate the expected pH of a typical antiseptic solution that is 3.0% H_2O_2 by mass.

32. In water, O^{2-} is a strong base. If 50.0 mg of Li_2O is dissolved in 750.0 mL of aqueous solution, what should be the pH of the solution?

33. The conversion of $O_2(g)$ to $O_3(g)$ can be accomplished in an electric discharge, $3 O_2(g) \longrightarrow 2 O_3(g)$. Use a bond dissociation energy of 498 kJ mol^{-1} for $O_2(g)$ and data from Appendix D to calculate an average oxygen-to-oxygen bond energy in $O_3(g)$.

34. Estimate the average bond energy in $O_3(g)$ from the structure on page 405 and data in Table 11.3. Compare this result with that of Exercise 33.

35. Use Lewis structures and other information to explain the observation that
(a) H_2S is a gas at room temperature, whereas H_2O is a liquid.
(b) O_3 is diamagnetic.

36. Use Lewis structures and other information to explain the observation that
(a) The oxygen-to-oxygen bond lengths in O_2, O_3, and H_2O_2 are 121, 128, and 148 pm, respectively.
(b) The oxygen-to-oxygen bond length of O_2 is 121 pm and for O_2^+ is 112 pm. Why is the bond length for O_2^+ so much shorter than for O_2?

37. Which of the following reactions are likely to go to completion or very nearly so?
(a) $H_2O_2(aq) + 2 I^-(aq) + 2 H^+(aq) \longrightarrow$
$$I_2(s) + 2 H_2O(l)$$
(b) $O_2(g) + 2 H_2O(l) + 4 Cl^-(aq) \longrightarrow$
$$2 Cl_2(g) + 4 OH^-(aq)$$
(c) $O_3(g) + Pb^{2+}(aq) + H_2O(l) \longrightarrow$
$$PbO_2(s) + 2 H^+(aq) + O_2(g)$$
(d) $HO_2^-(aq) + 2 Br^-(aq) + H_2O(l) \longrightarrow$
$$3 OH^-(aq) + Br_2(l)$$

38. Each of the following compounds produces $O_2(g)$ when strongly heated: (a) $HgO(s)$; (b) $KClO_4(s)$; (c) $Hg(NO_3)_2(s)$; (d) $H_2O_2(aq)$. Write a plausible equation for the reaction that occurs in each instance.

Sulfur

39. Give an appropriate name to each of the following compounds: **(a)** ZnS; **(b)** $KHSO_3$; **(c)** $K_2S_2O_3$; **(d)** SF_4.

40. Give an appropriate formula for each of the following compounds: **(a)** calcium sulfate dihydrate; **(b)** hydrosulfuric acid; **(c)** sodium hydrogen sulfate; **(d)** disulfuric acid.

41. Give a specific example of a chemical equation that illustrates the
(a) reaction of a metal sulfide with $HCl(aq)$
(b) action of a *nonoxidizing* acid on a metal sulfite
(c) oxidation of $SO_2(aq)$ to $SO_4^{2-}(aq)$ by $MnO_2(s)$ in acidic solution
(d) disproportionation of $S_2O_3^{2-}$ in acidic solution

42. Show how you would use elemental sulfur, chlorine gas, metallic sodium, water, and air to produce aqueous solutions containing **(a)** Na_2SO_3; **(b)** Na_2SO_4; **(c)** $Na_2S_2O_3$. (*Hint:* You will have to use information from other chapters as well as this one.)

43. Describe a chemical test you could use to determine whether a white solid is Na_2SO_4 or $Na_2S_2O_3$. Explain the basis of this test by writing a chemical equation or equations.

44. Explain why sulfur can occur naturally as sulfates, but not as sulfites.

45. Salts like $NaHSO_4$ are called *acid* salts because their anions undergo further ionization. What should be the pH of 250 mL of water solution containing 12.5 g $NaHSO_4$? (*Hint:* Use data from Chapter 17, as necessary.)

46. What mass of Na_2SO_3 must have been present in a sample that required 26.50 mL of 0.0510 M $KMnO_4$ for its oxidation to Na_2SO_4 in an acidic solution? MnO_4^- is reduced to Mn^{2+}.

47. A 1.100-g sample of copper ore is dissolved, and the $Cu^{2+}(aq)$ is treated with excess KI. The liberated I_3^- requires 12.12 mL of 0.1000 M $Na_2S_2O_3$ for its titration. What is the mass percent copper in the ore?

48. A 25.0-L sample of a natural gas, measured at 25 °C and 740.0 Torr, is bubbled through $Pb^{2+}(aq)$, yielding 0.535 g of $PbS(s)$. What mass of sulfur can be recovered per cubic meter of this natural gas?

Nitrogen Family

49. Use information from this chapter and previous chapters to write chemical equations to represent the following:
(a) equilibrium between nitrogen dioxide and dinitrogen tetroxide in the gaseous state
(b) the reduction of nitrous acid by $N_2H_5^+$ forming hydrazoic acid, followed by the reduction of additional nitrous acid by the hydrazoic acid, yielding nitrogen and dinitrogen monoxide
(c) the neutralization of $H_3PO_4(aq)$ to the second equivalence point by $NH_3(aq)$

50. Use information from this chapter and previous chapters to write plausible chemical equations to represent the following:
(a) the reaction of silver metal with $HNO_3(aq)$
(b) the complete combustion of the rocket fuel, unsymmetrical dimethylhydrazine, $(CH_3)_2NNH_2$
(c) the preparation of sodium triphosphate by heating a mixture of sodium dihydrogen phosphate and sodium hydrogen phosphate.

51. Draw plausible Lewis structures for
(a) dimethylhydrazine, $(CH_3)_2NNH_2$
(b) nitryl chloride, $ClNO_2$
(c) phosphorous acid, a *diprotic* acid with the empirical formula H_3PO_3

52. Both nitramide and hyponitrous acid have the formula $H_2N_2O_2$. Hyponitrous acid is a weak diprotic acid; nitramide contains the amide group ($-NH_2$). Draw plausible Lewis structures for these two substances.

53. Supply an appropriate name for each of the following: **(a)** HPO_4^{2-}; **(b)** $Ca_2P_2O_7$; **(c)** $H_6P_4O_{13}$.

54. Write an appropriate formula for each of the following: **(a)** hydroxylamine; **(b)** calcium hydrogen phosphate; **(c)** lithium nitride.

55. Use Figure 23.13 to establish $E°$ for the reduction of N_2O_4 to NO in an acidic solution.

56. Use Figure 23.13 to establish $E°$ for the reduction of NO_3^- to NO_2^- in a basic solution.

Carbon and Silicon

57. Comment on the accuracy of a jeweler's advertising that "diamonds last forever." In what sense is the statement true, and in what ways is it false?

58. A temporary fix for a "sticky" lock is to scrape a pencil point across the notches on the key and to work the key in and out of the lock a few times. What is the basis of this fix?

59. Write a chemical equation to represent
(a) the reduction of silica to elemental silicon by aluminum
(b) the preparation of potassium metasilicate by the high-temperature fusion of silica and potassium carbonate
(c) the reaction of Al_4C_3 with water to produce methane

60. Write a chemical equation to represent
(a) the reaction when potassium cyanide solution is added to silver nitrate solution
(b) the combustion of Si_3H_8 in an excess of oxygen
(c) the reaction of dinitrogen with calcium carbide to give calcium cyanamide (CaNCN)

61. Describe what is meant by the terms *silane* and *silanol*. What is their role in the preparation of silicones?

62. Describe and explain the similarities and differences between the reaction of a silicate with an acid and that of a carbonate with an acid.

63. Methane and sulfur vapor react to form carbon disulfide and hydrogen sulfide. Carbon disulfide reacts with $Cl_2(g)$ to form carbon tetrachloride and S_2Cl_2. Further reaction of carbon disulfide and S_2Cl_2 produces additional carbon tetrachloride and sulfur. Write a series of equations for the reactions described here.

64. In a manner similar to that outlined on page 938,
    (a) Write equations to represent the reaction of $(CH_3)_3SiCl$ with water, followed by the elimination of H_2O from the resulting silanol molecules.

    (b) Does a silicone polymer form?
    (c) What would be the corresponding product obtained from CH_3SiCl_3?

65. Show that the empirical formula given for muscovite is consistent with the expected oxidation states of the elements present.

66. Show that the empirical formula given for crysotile asbestos is consistent with the expected oxidation states of the elements present.

Boron

67. The molecule tetraborane has the formula B_4H_{10}.
    (a) Show that this is an electron-deficient molecule.
    (b) How many bridge bonds must occur in the molecule?
    (c) Show that butane, C_4H_{10}, is not electron-deficient.

68. Write Lewis structures for the following species, both of which involve coordinate covalent bonding.
    (a) tetrafluoroborate ion, BF_4^-, used in metal cleaning and in electroplating baths

    (b) boron trifluoride ethylamine, used in curing epoxy resins (ethylamine is $C_2H_5NH_2$)

69. Write chemical equations to represent
    (a) the preparation of boron from BBr_3
    (b) the formation of BF_3 from B_2O_3
    (c) the combustion of boron in hot $N_2O(g)$

70. Assign oxidation states to all the atoms in a perborate ion based on the structure on page 940.

Integrative and Advanced Exercises

71. Despite the fact that it has the higher molecular mass, XeO_4 exists as a gas at 298 K, whereas XeO_3 is a solid. Give a plausible explanation of this observation.

72. The text mentions that ammonium perchlorate is an explosion hazard. Assuming that NH_4ClO_4 is the sole reactant in the explosion, write a plausible equation(s) to represent the reaction that occurs.

73. The following bond energies are given for 298 K: O_2, 498; N_2, 946; F_2, 159; Cl_2, 243; ClF, 251; OF (in OF_2), 213; ClO (in Cl_2O), 205; and NF (in NF_3), 280 kJ mol^{-1}, respectively. Calculate ΔH_f° at 298 K for 1 mol of (a) ClF(g); (b) OF_2(g); (c) Cl_2O(g); (d) NF_3(g).

74. The standard electrode potential of fluorine cannot be measured directly because F_2 reacts with water, displacing O_2. Using thermodynamic data from Appendix D, obtain a value of E_{F_2/F^-}.

75. Polonium is the only element known to crystallize in the simple cubic form. In this structure, the interatomic distance between a Po atom and each of its six nearest neighbors is 335 pm. Use this description of the crystal structure to estimate the density of polonium.

76. Refer to Figure 12-24 to arrange the following species in the expected order of increasing (a) bond length and (b) bond strength (energy): O_2, O_2^+, O_2^-, O_2^{2-}. State the basis of your expectation.

77. One reaction of a chlorofluorocarbon implicated in the destruction of stratospheric ozone is $CFCl_3 + h\nu \longrightarrow CFCl_2 + Cl$.
    (a) What is the energy of the photons ($h\nu$) required to bring about this reaction, expressed in kilojoules per mole?
    (b) What is the frequency and wavelength of the light necessary to produce the reaction? In what portion of the electromagnetic spectrum is this light found?

78. The composition of a phosphate mineral can be expressed as % P, % P_4O_{10}, or % BPL [bone phosphate of lime, $Ca_3(PO_4)_2$].
    (a) Show that % P $= 0.436 \times$ (% P_4O_{10}) and % BPL $= 2.185 \times$ (% P_4O_{10}).
    (b) What is the significance of a % BPL greater than 100?
    (c) What is the % BPL of a typical phosphate rock?

79. Estimate the percent dissociation of $Cl_2(g)$ into $Cl(g)$ at 1 atm total pressure and 1000 K. Use data from Appendix D and equations found elsewhere in this text, as necessary.

80. *Peroxonitrous acid* is an unstable intermediate formed in the oxidation of HNO_2 by H_2O_2. It has the same formula as *nitric acid*, HNO_3. Show how you would expect peroxonitrous and nitric acids to differ in structure.

81. The structure of $N(SiH_3)_3$ involves a planar arrangement of N and Si atoms, whereas that of the related compound $N(CH_3)_3$ has a pyramidal arrangement of N and C atoms. Propose bonding schemes for these molecules that are consistent with this observation.

82. In the extraction of bromine from seawater (reaction 23.3), seawater is first brought to a pH of 3.5 and then treated with $Cl_2(g)$. In practice, the pH of the seawater is adjusted with H_2SO_4, and the mass of chlorine used is 15% in excess of the theoretical. Assuming a seawater sample with an initial pH of 7.0, a density of 1.03 g/cm^3, and a bromine content of 70 ppm by mass, what masses of H_2SO_4 and Cl_2 would be used in the extraction of bromine from 1.00×10^3 L of seawater?

83. Refer to the Integrative Example on page 942. Assume that the disproportionation of $S_2O_3^{2-}$ is no longer spontaneous when the partial pressure of $SO_2(g)$ above a solution with $[S_2O_3^{2-}] = 1$ M has dropped to 1×10^{-6} atm. Show that this condition is reached while the solution is still acidic.

84. The bond energies of Cl_2 and F_2 are 243 and 155 kJ mol^{-1}, respectively. Use these data to explain why XeF_2 is a much more stable compound than $XeCl_2$.
(*Hint:* Recall that Xe exists as a monatomic gas.)

85. Write plausible half-equations and a balanced oxidation–reduction equation for the disproportionation of XeF_4 to Xe and XeO_3 in aqueous acidic solution. Xe and XeO_3 are produced in a $2:1$ mol ratio, and $O_2(g)$ is also produced.

86. A handbook gives the value $E° = 0.174$ V for the reduction half-reaction $S + 2 H^+ + 2 e^- \longrightarrow H_2S(g)$. In Figure 23.9, the value given for the segment S ——— $H_2S(aq)$ is 0.144 V. Why are these $E°$ values different? Can both be correct?

87. The solubility of $Cl_2(g)$ in water is 6.4 g/L at 25 °C. Some of this chlorine is present as Cl_2, and some is found as HOCl or Cl^-. For the hydrolysis reaction

$$Cl_2(aq) + 2 H_2O(l) \longrightarrow HOCl(aq) + H_3O^+(aq) + Cl^-(aq)$$
$$K_c = 4.4 \times 10^{-4}$$

For a saturated solution of Cl_2 in water, calculate $[Cl_2]$, $[HOCl]$, $[H_3O^+]$, and $[Cl^-]$.

88. Not shown in Figure 23.13 are electrode potential data involving hydrazoic acid. Given that $E° = -3.09$ V for the reduction of HN_3 to N_2 in acidic solution, what is $E°$ for the reduction of HN_3 to NH_4^+ in acidic solution?

Feature Problems

89. The decomposition of aqueous hydrogen peroxide is catalyzed by $Fe^{3+}(aq)$. A proposed mechanism for this catalysis involves two reactions. In the first reaction, Fe^{3+} is reduced by H_2O_2. In the second, the iron is oxidized back to its original form, while hydrogen peroxide is reduced. Write an equation for the overall reaction, and show that the overall reaction is indeed spontaneous. What are the minimum and maximum values of $E°$ for a catalyst to function in this way? Which of the following should be able to catalyze the decomposition of hydrogen peroxide by the mechanism outlined here: **(a)** Cu^{2+}, **(b)** Br_2, **(c)** Al^{3+}, **(d)** Au^{3+}? In the reaction between iodic acid and hydrogen peroxide in the presence of starch indicator, the color of the reaction mixture oscillates between deep blue and colorless. What is the basis of these changes in color? Will this oscillation of color continue indefinitely? Explain.
[*Hint:* Refer to Feature Problem 105 in Chapter 15, particularly to equation (c).]

90. Both in this chapter and in Chapter 21, we have stressed the relationship between $E°$ values and thermodynamic properties. We can use this relationship to add some missing features to an electrode potential diagram. For example, note that $ClO_2(g)$, which has Cl in the oxidation state +4, is not included in Figure 23-2. Using data from Figure 23-2 and Appendix D, add $ClO_2(g)$ to the electrode potential diagram for acidic solutions, and indicate the $E°$ values that link $ClO_2(g)$ to $ClO_3^-(aq)$ and to $HClO_2(aq)$.

91. Figure 16-1 (page 627) shows that I_2 is considerably more soluble in $CCl_4(l)$ than it is in $H_2O(l)$. The concentration of I_2 in its saturated aqueous solution is 1.33×10^{-3} M, and the equilibrium achieved when I_2 distributes itself between H_2O and CCl_4 is

$$I_2(aq) \rightleftharpoons I_2(CCl_4) \qquad K_c = 85.5$$

(a) A 10.0-mL sample of saturated $I_2(aq)$ is shaken with 10.0 mL CCl_4. After equilibrium is established, the two liquid layers are separated. How many milligrams of I_2 will be in the aqueous layer?
(b) If the 10.0 mL of aqueous layer from part (a) is extracted with a second 10.0-mL portion of CCl_4, how

many milligrams of I_2 will remain in the aqueous layer when equilibrium is reestablished?
(c) If the 10.0-mL sample of saturated $I_2(aq)$ in part (a) had originally been extracted with 20.0 mL CCl_4, would the mass of I_2 remaining in the aqueous layer have been less than, equal to, or greater than that in part (b)? Explain.

92. The so-called pyroanions, $X_2O_7^{n-}$, form a series of structurally similar molecules for the elements Si, P, and S.
(a) Draw the Lewis structures of these anions, and predict the geometry of the anions. What is the maximum number of atoms that can lie in a plane?
(b) Each pyroanion in part (a) corresponds to a pyroacid, $X_2O_7H_n$. Compare each pyroacid to the acid containing only one atom of the element in its maximum oxidation state. From this comparison, suggest a strategy for the preparation of these pyroacids.
(c) What is the chlorine analogue of the pyroanions? For which acid is this species the anhydride?

93. A description of bonding in XeF_2 based on the valence bond model requires the $5d$ orbitals of Xe. A more satisfactory description uses a molecular orbital approach involving three-center bonds. Assume that bonding involves the $5p_z$ orbital of Xe and the $2p_z$ orbitals of the two F atoms. These three atomic orbitals combine to give three molecular orbitals: one bonding, one nonbonding, and one antibonding. Recall that for bonding to occur, atomic orbitals with the same phase must overlap to form bonding molecular orbitals (see Chapter 12).
(a) Construct diagrams similar to Figure 12-30 to indicate the overlap of the three atomic orbitals in forming the three molecular orbitals.
Assume that the order of energy of the molecular orbitals is bonding MO < nonbonding MO < antibonding MO.
(b) Sketch a molecular orbital energy-level diagram, and assign the appropriate number of electrons from fluorine and xenon to the molecular orbitals. What is the bond order?
(c) With the aid of VSEPR theory, show that this molecular orbital description based on three-center bonds works well for XeF_4 but not for XeF_6.

94. The sketch is a portion of the phase diagram for the element sulfur. The transition between solid orthorhombic (S_α) and solid monoclinic (S_β) sulfur, in the presence of sulfur vapor, is at 95.3 °C. The triple point involving monoclinic sulfur, liquid sulfur, and sulfur vapor is at 119 °C.
   (a) How would you modify the phase diagram to represent the melting of orthorhombic sulfur that is sometimes observed at 113 °C?
   (*Hint:* What would the phase diagram look like if the monoclinic sulfur did not form?)
   (b) Account for the observation that if a sample of rhombic sulfur is melted at 113 °C and then heated, the liquid sulfur freezes at 119 °C upon cooling.

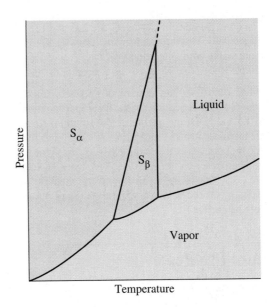

eMedia Exercises

95. Observe the three-dimensional structures of the **XeF$_2$** and **XeF$_4$** models *(eChapter 23-1)*. (a) Using the principles of VSPER theory, suggest a plausible structure for the molecule in this series, formed through the reaction with additional fluorines. (b) Will similar molecules form with other halogens? Why or why not?

96. The **Physical Properties of Halogens** movie *(eChapter 23-2)* illustrates the properties of three halogens. (a) What evidence of a periodic trend do you find in this illustration? What is the basis of this trend? (b) Based upon these observations, predict the physical state of fluorine.

97. In the **Reactions with Oxygen** movie *(eChapter 23-3)*, the oxidizing power of oxygen is illustrated. (a) Describe the change in oxidation state for each of the elements undergoing a reaction. (b) In which cases is more than one oxide compound possible? (c) What are the oxidation states of the elements in the other possible compounds?

98. Two common oxides of nitrogen are seen in the **Nitrogen Dioxide and Dinitrogen Tetroxide** movie *(eChapter 23-4)*. These oxides exist in equilibrium. (a) From the procedure carried out in the movie, predict whether the conversion from NO_2 to N_2O_4 is an endothermic or exothermic reaction. (b) Does your prediction agree with thermodynamic data for this reaction?

99. (a) View the **Allotropes of Carbon** models *(eChapter 23-5)* and describe the hybridization of carbon in each form. (b) Suggest reasons for the differences in the mechanical properties of the different forms of carbon. (c) Which of the structures of carbon is also characteristic of other elements in Group 14?

24

The Transition Elements

Contents

Whiskers of rutile, TiO_2, in quartz (left) and titanium ore (right), a source of rutile. Titanium metal, obtained from rutile, is used in industry because it has low density and high strength. Pure TiO_2 is a bright white pigment used in paints and specialty papers.

There are more transition elements—members of the d and f blocks—than main-group elements. Although some of them are rare and of limited use, others play crucial roles in many aspects of modern life. All the transition elements are metals; among them are both the chief structural metal, iron (Fe), and important alloying metals in the manufacture of steel (V, Cr, Mn, Co, Ni, Mo, W). The best electrical conductors (Ag, Cu) are transition metals. The compounds of several transition metals (Ti, Fe, Cr) are the primary constituents of paint pigments. Compounds of silver (Ag) provide the essential material for photography. Specialized materials for modern applications, such as color television screens, use compounds of the f-block elements (lanthanide oxides). Nine of the transition metals are essential elements for living organisms.

949

The chemistry of the *d*-block and *f*-block elements has both theoretical and practical significance. These elements and their compounds provide insight into fundamental aspects of bonding, magnetism, and reaction chemistry.

24-1 General Properties

The high melting points, good electrical conductivity, and moderate-to-extreme hardness of the transition elements result from the ready availability of electrons and orbitals for metallic bonding (see Section 12-7). Some similarities are found among the transition elements, but each element also has some unique properties that make it and its compounds useful in particular ways. Table 24.1 lists properties of the fourth-period transition elements—the first transition series.

TABLE 24.1 Selected Properties of Elements of the First Transition Series

	Sc	Ti	V	Cr	Mn	Fe	Co	Ni	Cu	Zn
Atomic number	21	22	23	24	25	26	27	28	29	30
Electron config.[a]	$3d^14s^2$	$3d^24s^2$	$3d^34s^2$	$3d^54s^1$	$3d^54s^2$	$3d^64s^2$	$3d^74s^2$	$3d^84s^2$	$3d^{10}4s^1$	$3d^{10}4s^2$
Metallic radius, pm	161	145	132	125	124	124	125	125	128	133
Ioniz energy, kJ/mol										
First	631	658	650	653	717	759	758	737	745	906
Second	1235	1310	1414	1592	1509	1561	1646	1753	1958	1733
Third	2389	2653	2828	2987	3248	2957	3232	3393	3554	3833
$E°$, V[b]	−2.03	−1.63	−1.13	−0.90	−1.18	−0.440	−0.277	−0.257	+0.340	−0.763
Common positive oxidation states[c]	3	2, 3, 4	2, 3, 4, 5	2, 3, 6	2, 3, 4, 7	2, 3, 6	2, 3	2, 3	1, 2	2
mp, °C	1397	1672	1710	1900	1244	1530	1495	1455	1083	420
Density, g/cm³	3.00	4.50	6.11	7.14	7.43	7.87	8.90	8.91	8.95	7.14
Hardness[d]	—	—	—	9.0	5.0	4.5	—	—	2.8	2.5
Electrical conductivity[e]	3	4	6	12	1	16	25	23	93	27

[a] Each atom has an argon inner-core configuration.
[b] For the reduction process, $M^{2+}(aq) + 2\,e^- \longrightarrow M(s)$ [except for scandium, where the ion is $Sc^{3+}(aq)$].
[c] The most important oxidation states are printed in red.
[d] Hardness values are on the Mohs scale (see Table 22.2).
[e] Electrical conductivity compared with an arbitrarily assigned value of 100 for silver.

Atomic (Metallic) Radii

In Table 24.1, with the exception of Sc and Ti, we find little variation among the atomic radii across the first transition series. The chief difference in atomic structure between successive elements involves one unit of positive charge on the nucleus and one electron in an orbital of an *inner* electron shell. This is not a major difference and does not cause much of a change in atomic radius, especially in the middle of the series.

When an element in the first transition series is compared with elements of the second and third series within the same group, important differences appear. Consider the members of group 6—Cr, Mo, and W. As we might expect, the atomic radius of Mo is larger than that of Cr; but contrary to our expectation, the atomic radius of W is the same as that of Mo, not larger. In the aufbau process, 18 electrons are added in progressing from Cr to Mo, and all of them enter *s*, *p*, and *d* subshells. Between Mo and W, however, 32 electrons must be added, and 14 of them enter the 4*f* subshell. Electrons in an *f* subshell are not very effective in screening outer-shell electrons from the nucleus. As a result, the outer-shell elec-

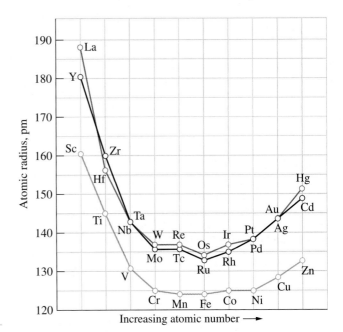

▶ FIGURE 24-1
Atomic radii of the *d*-block elements
The radii of the fourth-period transition elements (blue) are smaller than those of the corresponding group members in the succeeding periods. This trend is not seen between the fifth-period (black) and sixth-period (red) members, illustrating the contraction in atomic radii associated with the lanthanide series.

trons are held more tightly by the nucleus than we would otherwise expect. Atomic radii do not increase. In fact, in the series of elements in which the $4f$ subshell is filled, atomic radii decrease somewhat. This phenomenon occurs in the lanthanide series ($Z = 58$ to 71) and is called the **lanthanide contraction**. The lanthanide contraction is made more apparent in the graphs in Figure 24-1.

Electron Configurations and Oxidation States

The elements of the first transition series have electron configurations with the following characteristics:

- an inner core of electrons in the argon configuration
- two electrons in the $4s$ orbital for eight members and one $4s$ electron for the remaining two (Cr and Cu)
- a number of $3d$ electrons, ranging from one in Sc to ten in Cu and Zn

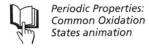

Periodic Properties: Common Oxidation States animation

As we have seen for some of the main-group elements in Chapters 22 and 23, an element may display several oxidation states. Often, however, one particular oxidation state is the most common for an element. Ti atoms, with the electron configuration $[Ar]3d^24s^2$, tend to use all four electrons beyond the argon core in compound formation and display the oxidation state +4. It is also possible, however, for Ti atoms to use fewer electrons, as through the loss of the $4s^2$ electrons to form the ion Ti^{2+}. With Ti, then, we note two features: (1) several possible oxidation states, as shown in Figure 24-2, and (2) a maximum oxidation state corresponding to the group number, 4. These two features continue with V, Cr, and Mn, for which the maximum oxidation states are +5, +6, and +7, respectively. A shift in behavior occurs in groups 8–12, however. Thus, although Fe, Co, and Ni can all exist in more than one oxidation state, they do not display the wide variety found in the earlier members of the first transition series. Nor do they exhibit a maximum oxidation state corresponding to their group number. As we cross the first transition series, the nuclear charge, number of d electrons, and energy requirement for the successive ionization of d electrons increase. Involvement of a large number of d electrons in bond formation becomes increasingly unfavorable energetically, and

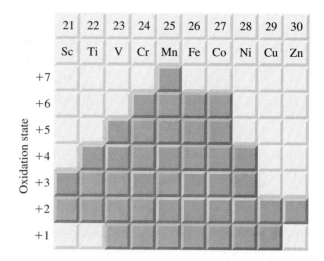

▶ **FIGURE 24-2**
Positive oxidation states of the elements of the first transition series
Common oxidation states are shown in red and less common ones in gray. Some oxidation states are rather rare, and a zero or negative oxidation state is occasionally found, chiefly in transition metal complexes. For example, the oxidation state of Cr is -2 in $Na_2[Cr(CO)_5]$, -1 in $Na_2[Cr_2(CO)_{10}]$, and 0 in $Cr(CO)_6$.

only the lower oxidation states are commonly encountered for these later elements of the first transition series.

Although the transition elements display variety in their oxidation states, the possible oxidation states differ in the ease with which they can be attained and in their stabilities. The stability of an oxidation state for a given transition metal depends on a number of factors—other atoms to which the transition metal atom is bonded; whether the compound is in solid form or in solution; and the pH of the solution. For example, $TiCl_2$ is a well-characterized compound as a solid, but the Ti^{2+} ion is oxidized to Ti^{3+} by dissolved oxygen in aqueous solutions and even by the water itself. Conversely, $Co^{3+}(aq)$ readily oxidizes water to $O_2(g)$ and is itself reduced to $Co^{2+}(aq)$. $Co^{3+}(aq)$ can be stabilized, however, in certain complex ions. Generally speaking, higher oxidation states in transition metals are stabilized when oxide or fluoride ions are bound to the metal. In Figure 24-2 and elsewhere, the term *common* oxidation state refers to an oxidation state often found in aqueous solution.

Another feature of the transition metals is the progressively increasing stability of higher oxidation states as we descend a group of the periodic table, the reverse of the trend often seen for main-group elements. Consider Cr, Mo, and W in group 6, all of which can exhibit oxidation states ranging from $+6$ to -2. The number of compounds with Cr in the $+6$ oxidation state is rather limited, whereas those of Mo and W abound. Chromium is encountered in the oxidation states $+5$ and $+4$ mostly in unstable intermediates, whereas Mo and W exhibit a rich chemistry in these states. The most stable oxidation state of chromium is $+3$. Although it is a strong reducing agent, $Cr^{2+}(aq)$ nevertheless is readily obtainable, whereas Mo and W are not obtainable as the simple $+2$ cation. This trend favoring lower oxidation states for the first group member and higher oxidation states for the later members is also found in other groups of transition metals. For instance, although Fe does not exhibit an oxidation state corresponding to the group number, Os does form the stable oxide OsO_4 with Os in the $+8$ oxidation state.

Ionization Energies and Electrode Potentials

Ionization energies are fairly constant across the first transition series. Values of the first ionization energies are about the same as for the group 2 metals. Standard electrode potentials gradually increase in value across the series. With the exception of the oxidation of Cu to Cu^{2+}, however, all these elements are more readily oxidized than hydrogen. This means that the metals displace $H_2(g)$ from $H^+(aq)$. Additional comments on electrode potentials, some supported by electrode potential diagrams, are found throughout the chapter.

Ionic and Covalent Compounds

We tend to think of metals as forming ionic compounds with nonmetals. This is certainly the case with group 1 and most group 2 metal compounds. However, we have seen that some metal compounds have significant covalent character; $BeCl_2$, and $AlCl_3$ (Al_2Cl_6), for example, are molecular compounds. Transition metal compounds display both ionic and covalent character. In general, compounds with the transition metal in lower oxidation states are essentially ionic, and those in higher oxidation states have covalent character. As an example, MnO is a green ionic solid with a melting point of 1785 °C, whereas Mn_2O_7 is a dark red, oily, molecular liquid that boils at room temperature and is highly explosive. Another feature of ionic compounds of the transition metals is that the metal atoms often occur in polyatomic cations or anions rather than as the simple, monatomic ion. Some common examples are VO_2^+, MnO_4^-, and $Cr_2O_7^{2-}$.

Catalytic Activity

*Surface Reactions—
Hydrogenation
animation*

An unusual ability to adsorb gaseous species makes some transition metals, such as Ni and Pt, good heterogeneous catalysts. The possibility of multiple oxidation states seems to account for the ability of some transition metal ions to serve as catalysts in certain oxidation–reduction reactions. In still other types of catalysis, complex-ion formation may play an important role. As we have seen in a limited way in Chapter 19 and will explore more fully in Chapter 25, complex-ion formation is a particularly distinctive feature of transition metal chemistry.

Catalysis movie

Catalysis is an essential aspect of about 90% of all chemical manufacturing processes, and the transition metals are often the key elements in the catalysts used. To name a few—Ni is used in the hydrogenation of oils (page 288); Pt, Pd, and Rh are used in catalytic converters in automobiles (pages 276 and 610); Fe_3O_4 is the main component of the catalyst used in the synthesis of ammonia (page 657); and V_2O_5 is used in the conversion of $SO_2(g)$ to $SO_3(g)$ in the manufacture of sulfuric acid (page 965).

Color and Magnetism

As we will explain more fully in Section 25-6, electronic transitions that occur within partially filled *d* subshells impart color to solid transition metal compounds and their solutions. Absence of these transitions, in turn, accounts for the fact that so many main-group metal compounds are colorless.

Because most transition elements have partially filled *d* subshells, many transition metals and their compounds are paramagnetic—that is, they have unpaired electrons. This description certainly fits Fe, Co, and Ni, but these three metals are unique among the elements in displaying a special magnetic property: the ability to be made into permanent magnets, a property known as **ferromagnetism**. A key feature of ferromagnetism is that in the solid state, the metal atoms are thought to be grouped together into small regions—called *domains*—containing rather large numbers of atoms. Instead of the individual magnetic moments of the atoms within a domain being randomly oriented, all the magnetic moments are directed in the same way. In an unmagnetized piece of iron, the domains are oriented in several directions and their magnetic effects cancel. When the metal is placed in a magnetic field, however, the domains line up and a strong resultant magnetic effect is produced. This alignment of domains may actually involve the growth of domains with

▲ Color-enhanced image of magnetic domains in a ferromagnetic garnet film.

Magnetic field absent

In presence of magnetic field

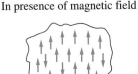
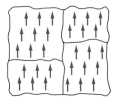

Paramagnetism

► **FIGURE 24-3**

Ferromagnetism and paramagnetism compared
In a paramagnetic material, the effect of a magnetic field is to align the magnetic moments of the individual atoms. In a ferromagnetic material, the magnetic moments are aligned within domains even in the absence of a magnetic field, but the direction of the alignment varies from one domain to another. The effect of the magnetic field is to change the orientation of these varied alignments into a single direction—the direction of the magnetic field.

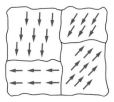

Ferromagnetism

favorable orientations at the expense of those with unfavorable orientations (rather like a recrystallization of the material). The ordering of domains can persist when the object is removed from the magnetic field, and thus permanent magnetism results. Paramagnetism and ferromagnetism are compared in Figure 24-3.

The key to ferromagnetism is in two basic factors: (1) The atoms involved have unpaired electrons (a property possessed by many atoms), and (2) interatomic distances are of just the right magnitude to make possible the ordering of atoms into domains. If atoms are too large, interactions among them are too weak to produce this ordering. With small atoms, the tendency is for atoms to pair and their magnetic moments to cancel. This critical factor of atomic size is just met in Fe, Co, and Ni. It is possible, however, to prepare *alloys* of other metals in which this condition is also met. Some examples are Al–Cu–Mn, Ag–Al–Mn, and Bi–Mn.

Comparison of Transition and Main-Group Elements

With the main-group elements, the *s* and *p* orbitals of the outermost electron shell are the most important in determining the nature of the chemical bonding that occurs. Participation in bonding by *d* orbitals is essentially nonexistent for second-period elements and for group 1 and 2 metals. With the transition elements, *d* orbitals are as important as *s* and *p* orbitals. Most of the observed behavioral differences between the transition and main-group elements—multiple versus single oxidation states, complex-ion formation, color, magnetic properties, and catalytic activity—can be traced to the orbitals that are most involved in bond formation.

24-2 Principles of Extractive Metallurgy

Many of the transition elements have important uses related to their metallic properties—iron for its structural strength and copper for its excellent electrical conductivity, for example. Unlike the more chemically reactive metals of groups 1 and 2 and aluminum in group 3, which are produced mainly by modern methods of electrolysis, the transition metals are obtained by procedures developed over many centuries.

The term *metallurgy* describes the general study of metals. **Extractive metallurgy** describes the winning of metals from their ores. There is no single method

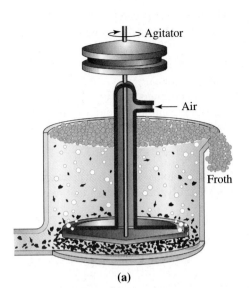

(a) (b)

▲ **FIGURE 24-4   Concentration of an ore by flotation**
(a) Powdered ore is suspended in water in a large vat, together with suitable additives, and the mixture is agitated with air. Particles of ore become attached to air bubbles, rise to the top of the vat, and are collected in the overflow froth. Particles of undesired waste rock (gangue) fall to the bottom. (b) The froth formed in the flotation process.

of extractive metallurgy, but a few basic operations generally apply. Let us illustrate them with the extractive metallurgy of zinc.

Concentration   In mining operations, the desired mineral from which a metal is to be extracted often constitutes only a small percent (or occasionally just a fraction of a percent) of the material mined. It is necessary to separate the desired ore from waste rock before proceeding with other metallurgical operations. One useful method, *flotation*, is described in Figure 24-4.

Roasting   An ore is roasted (heated to a high temperature) to convert a metal compound to its oxide, which can then be reduced. For zinc, the commercially important ores are $ZnCO_3$ (smithsonite) and ZnS (sphalerite). $ZnCO_3(s)$, like the carbonates of the group 2 metals, decomposes to $ZnO(s)$ and $CO_2(g)$ when it is strongly heated. When strongly heated in air, $ZnS(s)$ reacts with $O_2(g)$, producing $ZnO(s)$ and $SO_2(g)$. In modern smelting operations, $SO_2(g)$ is converted to sulfuric acid rather than being vented to the atmosphere.

$$ZnCO_3(s) \xrightarrow{\Delta} ZnO(s) + CO_2(g) \tag{24.1}$$

$$2\,ZnS(s) + 3\,O_2(g) \xrightarrow{\Delta} 2\,ZnO(s) + 2\,SO_2(g) \tag{24.2}$$

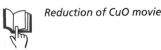
Reduction of CuO movie

Reduction   Because it is inexpensive and easy to handle, carbon, in the form of coke or powdered coal, is used as the reducing agent whenever possible. Several reactions occur simultaneously in which both $C(s)$ and $CO(g)$ act as reducing agents. The reduction of ZnO is carried out at about 1100 °C, a temperature above the boiling point of zinc. The zinc is obtained as a vapor and condensed to the liquid.

$$ZnO(s) + C(s) \xrightarrow{\Delta} Zn(g) + CO(g) \tag{24.3}$$

$$ZnO(s) + CO(g) \xrightarrow{\Delta} Zn(g) + CO_2(g) \tag{24.4}$$

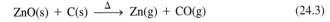

Refining The metal produced by chemical reduction is usually not pure enough for its intended uses. Impurities must be removed; that is, the metal must be refined. The refining process chosen depends on the nature of the impurities. The impurities in zinc are mostly Cd and Pb, which can be removed by the fractional distillation of liquid zinc.

Most of the zinc produced worldwide, however, is refined electrolytically, usually in a process that combines reduction and refining. ZnO from the roasting step is dissolved in $H_2SO_4(aq)$. This is represented by the ionic equation

$$ZnO(s) + 2\,H^+(aq) + SO_4^{2-}(aq) \longrightarrow Zn^{2+}(aq) + SO_4^{2-}(aq) + H_2O(l) \quad (24.5)$$

Powdered Zn is added to the solution to displace less active metals, such as Cd. Then the solution is electrolyzed. The electrode reactions are

Cathode: $\qquad\qquad\qquad\qquad Zn^{2+}(aq) + 2\,e^- \longrightarrow Zn(s)$

Anode: $\qquad\qquad\qquad\qquad H_2O \longrightarrow \frac{1}{2}\,O_2(g) + 2\,H^+(aq) + 2\,e^-$

Unchanged: $\qquad\qquad\qquad \underline{SO_4^{2-} \longrightarrow SO_4^{2-}(aq)}$

Overall: $\quad Zn^{2+}(aq) + SO_4^{2-}(aq) + H_2O \longrightarrow$
$$Zn(s) + 2\,H^+(aq) + SO_4^{2-}(aq) + \tfrac{1}{2}\,O_2(g) \quad (24.6)$$

Note that in the overall electrolysis reaction, Zn^{2+} is reduced to pure metallic zinc and sulfuric acid is regenerated. The acid is recycled in reaction (24.5).

EXAMPLE 24-1

Writing Chemical Equations for Metallurgical Processes. Write chemical equations to represent the **(a)** roasting of galena, PbS; **(b)** reduction of $Cu_2O(s)$ with charcoal as a reducing agent; **(c)** deposition of pure silver from an aqueous solution of Ag^+.

Solution

(a) We expect this process to be essentially the same as reaction (24.2).
$$2\,PbS(s) + 3\,O_2(g) \xrightarrow{\Delta} 2\,PbO(s) + 2\,SO_2(g)$$

(b) The simplest possible equation we can write is
$$Cu_2O(s) + C(s) \xrightarrow{\Delta} 2\,Cu(l) + CO(g)$$

(c) This process involves a reduction half-reaction. The accompanying oxidation half-reaction is not specified. Nor is it specified whether this is an electrolysis process or whether silver is displaced by a more active metal. In either case, the reduction half-reaction is
$$Ag^+(aq) + e^- \longrightarrow Ag(s)$$

Practice Example A: Write plausible chemical equations to represent the **(a)** roasting of Cu_2S; **(b)** reduction of WO_3 with $H_2(g)$; **(c)** thermal decomposition of HgO to its elements.

Practice Example B: Write chemical equations to represent the **(a)** reduction of Cr_2O_3 to chromium with silicon as the reducing agent; **(b)** conversion of $Co(OH)_3(s)$ to $Co_2O_3(s)$ by roasting; **(c)** production of pure $MnO_2(s)$ from $MnSO_4(aq)$ at the anode in an electrolysis cell.

[*Hint:* A few simple products are not specifically mentioned; propose plausible ones.]

Zone Refining In discussing freezing-point depression (Section 14-8), we assumed that a solute is *soluble* in a liquid solvent and *insoluble* in the solid solvent

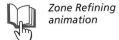

Zone Refining animation

▲ **FIGURE 24-5   Zone refining**
As a heating coil moves up the rod of material, melting occurs. Impurities concentrate in the molten zone. The portion of the rod below the molten zone is purer than that in or above the zone. With each successive passage of a heating coil, the rod becomes purer.

that freezes from solution. This behavior suggests a particularly simple way to purify a solid: Melt the solid and then refreeze a portion of it. Impurities remain in the liquid phase, and the solid that freezes is pure. In practice, the method is not quite so simple because the solid that freezes is wet with unfrozen liquid and thereby retains some impurities. Also, one or more solutes (impurities) might be slightly soluble in the solid solvent. In any case, the impurities do distribute themselves between the solid and liquid, concentrating in the liquid phase. If the solid that freezes from a liquid is remelted and the molten material refrozen, the solid obtained in the second freezing is purer than that in the first. Repeating the melting and refreezing procedure hundreds of times produces a very pure solid product.

The purification procedure we have just described implies that the melting and refreezing is done in batches, but in practice it is done *continuously*. In the method known as **zone refining**, a cylindrical rod of material is alternately melted and refrozen, as a series of heating coils passes along the rod (Figure 24-5). Impurities concentrate in the molten zones, and the portions of the rod behind these zones are somewhat purer than the portions in front of the zones. Eventually, impurities are swept to the end of the rod, which is cut off. The principle of zone refining is shown graphically in Figure 24-6. This process is capable of producing materials in which the impurity levels are as low as 10 parts per billion (ppb), a common requirement of substances used in semiconductors.

Thermodynamics of Extractive Metallurgy

It is interesting to think of the reduction of zinc oxide by carbon, as shown in reaction (24.3), as a competition between zinc and carbon for O atoms—zinc has them initially in ZnO, and carbon acquires them in forming CO. To establish the conditions under which carbon will reduce zinc oxide to zinc, we start by comparing

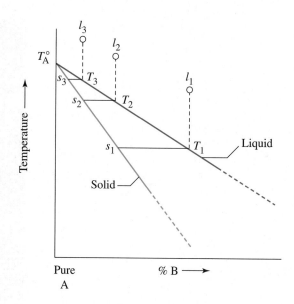

▶ **FIGURE 24-6**
The principle of zone refining
The red line shows the freezing points of solutions of impurity B in substance A. The blue line gives the composition of the solid that freezes from these solutions. In some cases, the blue line is nearly coincident with the temperature axis.

When a solution of composition l_1 is cooled to temperature T_1, it freezes to produce a solid of composition s_1. If a small quantity of this solid is removed from the solution and melted, it produces a new liquid, l_2. The freezing point of l_2 is T_2, and the composition of the solid freezing from the solution is s_2. When removed from the solution, a small portion of this solid produces liquid l_3, and so on. With each melting/freezing cycle, the melting point increases and the point representing the composition of the solid moves closer to pure A. In zone refining the melting/freezing cycles are conducted continuously, not in batches as described here.

the relative tendencies for zinc and carbon to undergo oxidation. We can assess these tendencies through free energy changes.

$$\text{(a)} \quad 2\,C(s) + O_2(g) \longrightarrow 2\,CO(g) \qquad \Delta G^\circ_{(a)}$$

$$\text{(b)} \quad 2\,Zn(g) + O_2(g) \longrightarrow 2\,ZnO(s) \qquad \Delta G^\circ_{(b)}$$

To determine if the reduction of zinc oxide to zinc by carbon is a spontaneous reaction, we need the value of ΔG° for the overall reaction, represented below by reversing equation (b) and adding it to equation (a).

(a)	$2\,C(s) + O_2(g) \longrightarrow 2\,CO(g)$	$\Delta G^\circ_{(a)}$
$-$ (b)	$2\,ZnO(s) \longrightarrow 2\,Zn(s) + O_2(g)$	$-\Delta G^\circ_{(b)}$
Overall:	$2\,ZnO(s) + 2\,C(s) \longrightarrow 2\,Zn(s) + 2\,CO(g)$	$\Delta G^\circ = \Delta G^\circ_{(a)} - \Delta G^\circ_{(b)}$

We still need some numerical data to complete our assessment. In Figure 24-7, the blue line gives $\Delta G^\circ_{(a)}$ as a function of temperature, and the top red line gives $\Delta G^\circ_{(b)}$.

Figure 24-7 shows that at low temperatures, $\Delta G^\circ_{(b)}$ is much more negative than $\Delta G^\circ_{(a)}$. This makes ΔG° for the overall reaction positive and the reaction *nonspontaneous*. At high temperatures, the situation is reversed: $\Delta G^\circ_{(a)}$ is more negative than $\Delta G^\circ_{(b)}$, and the overall reaction is *spontaneous*. The switchover from

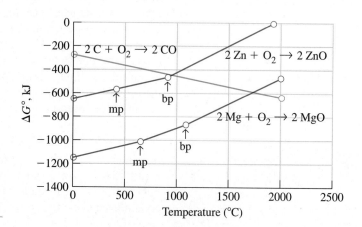

▶ **FIGURE 24-7**
ΔG° as a function of temperature for some reactions of extractive metallurgy
The points on the lines marked by arrows indicate the melting point and boiling points of Zn and Mg. At these points, the states of matter in which the metal exists change from (s) to (l) and from (l) to (g), respectively.

Are You Wondering...

Why the slope of the blue line in Figure 24-7 is negative, whereas the other slopes are positive?

Recall from Chapter 20 that $\Delta G° = \Delta H° - T \Delta S°$ and that $\Delta H°$ does not change appreciably with temperature. Thus, the temperature variation of $\Delta G°$ is determined primarily by the $-T \Delta S°$ term. We expect $\Delta S°$ to be negative for the reaction $2 \, \text{Zn(s)} + \text{O}_2\text{(g)} \longrightarrow 2 \, \text{ZnO(s)}$, because a mole of gas is lost. The term $-T\Delta S°$ is *positive*, and $\Delta G°$ *increases* with temperature. Conversely, in the reaction $2 \, \text{C(s)} + \text{O}_2\text{(g)} \longrightarrow 2 \, \text{CO(g)}$, an additional mole of gas forms, resulting in a positive value of $\Delta S°$. As a consequence, $-T\Delta S°$ is *negative* and $\Delta G°$ *decreases* with temperature.

nonspontaneous to spontaneous occurs at the point of intersection of the blue line and the red line—about 950 °C. There, $\Delta G°$ for the overall reaction is *zero*.

When we make a similar assessment for the reduction $2 \, \text{MgO(s)} + 2 \, \text{C(s)} \longrightarrow 2 \, \text{Mg(g)} + 2 \, \text{CO(g)}$, we conclude that the reaction does not become spontaneous until a temperature in excess of 1700 °C is reached. This is an exceedingly high temperature at which to carry out a chemical reaction and is not used in the metallurgy of magnesium.

Alternative Methods in Extractive Metallurgy   Some common variations of the methods previously discussed are worth mentioning. First, many ores contain several metals, and it is not always necessary to separate them. For example, a major use of vanadium, chromium, and manganese is in making alloys with iron. Obtaining each metal by itself is not commercially important. Thus, the principal chromium ore *chromite*, $\text{Fe(CrO}_2)_2$, can be reduced to give an alloy of Fe and Cr called *ferrochrome*. Ferrochrome may be added directly to iron, together with other metals, to produce one type of steel. Vanadium and manganese can be isolated as the oxides, V_2O_5 and MnO_2, respectively. When iron-containing compounds are added to these oxides and the mixtures are reduced, ferrovanadium and ferromanganese alloys form.

Extensive production of titanium was an important development of the latter half of the twentieth century, spurred first by the needs of the military and then by the aircraft industry. Steel is unsuitable as the structural metal for aircraft because it has a high density ($d = 7.8 \, \text{g/cm}^3$). Aluminum has the advantage of a low density ($d = 2.70 \, \text{g/cm}^3$), but it loses strength at high temperatures. For certain aircraft components, titanium is a good alternative to aluminum and steel because its density is moderately low ($d = 4.50 \, \text{g/cm}^3$) and it does not lose strength at high temperatures.

Titanium metal cannot be produced by reduction of TiO_2 with carbon, however, because the metal and carbon react to form titanium carbides. Also, at high temperatures the metal reacts with air to form TiO_2 and TiN. The metallurgy of titanium, then, must be conducted out of contact with air and with an active metal rather than with carbon as a reducing agent.

The first step in the production of Ti is the conversion of *rutile* ore (TiO_2) to TiCl_4 by reaction with carbon and $\text{Cl}_2\text{(g)}$.

$$\text{TiO}_2\text{(s)} + 2 \, \text{C(s)} + 2 \, \text{Cl}_2\text{(g)} \xrightarrow{800 \, °\text{C}} \text{TiCl}_4\text{(g)} + 2 \, \text{CO(g)} \qquad (24.7)$$

The purified TiCl_4 is next reduced to Ti with a good reducing agent. The *Kroll process* uses Mg.

$$\text{TiCl}_4\text{(g)} + 2 \, \text{Mg(l)} \xrightarrow{1000 \, °\text{C}} \text{Ti(s)} + 2 \, \text{MgCl}_2\text{(l)} \qquad (24.8)$$

The $\text{MgCl}_2\text{(l)}$ is removed and electrolyzed to produce Cl_2 and Mg, which are recycled in reactions (24.7) and (24.8), respectively. The Ti is obtained as a sintered

▲ Vacuum-distilled metallic titanium sponge produced by the Kroll process.

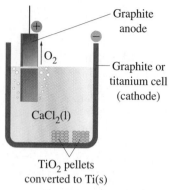

TiO$_2$ pellets
converted to Ti(s)

▲ Electrolytic production of
Ti(s) from TiO$_2$(s).

▲ Slag formed during the
smelting of copper ore.

(fused) mass called *titanium sponge*. This sponge must be subjected to further treatment and alloying with other metals before it can be used.

The Kroll process is slow; it takes a week to produce a few tons of Ti. It is also demanding in health and safety terms because it requires high-temperature vacuum distillation to remove the Mg and MgCl$_2$ from the titanium. Recently, an electrolytic process has been suggested for the production of Ti from rutile. Porous pellets of TiO$_2$ are placed at the cathode of an electrolytic cell containing molten calcium chloride. The pellets dissolve in the electrolyte and oxide ions (O^{2-}) are discharged as oxygen at a graphite anode. The Ti(IV) is reduced at the cathode, which is the vessel containing the electrolytic cell and made of either graphite or titanium. The titanium metal is obtained as sponge. This method, devised by Derek Fray, George Chen, and Tom Farthing in the United Kingdom, is being developed as a commercial process for the production of Ti(s) at a substantially lower cost than the Kroll process.

Metallurgy of Copper The extraction of copper from its ores (generally sulfides) is rather complicated, chiefly because the copper ores usually contain iron sulfides. The general scheme of extractive metallurgy just discussed produces copper contaminated with iron. For some metals, such as V, Cr, and Mn, contamination with iron is not a problem because the metals are mostly used in the manufacture of steel. Copper, however, is prized commercially for the properties of the pure metal. To avoid contamination with iron, several changes to the usual metallurgical methods are necessary.

Concentration of copper is done by flotation, and roasting converts iron sulfides to iron oxides. The copper remains as the sulfide if the temperature is kept below 800 °C. Smelting of the roasted ore in a furnace at 1400 °C causes the material to melt and separate into two layers. The bottom layer, called *copper matte*, consists chiefly of the molten sulfides of copper and iron. The top layer is a silicate slag formed by the reaction of oxides of Fe, Ca, and Al with SiO$_2$ (which typically is present in the ore or can be added). For example,

$$\text{FeO(s)} + \text{SiO}_2\text{(s)} \xrightarrow{\Delta} \text{FeSiO}_3\text{(l)} \tag{24.9}$$

A process called *conversion* is carried out in another furnace, where air is blown through the molten copper matte. First, the remaining iron sulfide is converted to the oxide, followed by formation of slag [FeSiO$_3$(l)]. The slag is poured off, and air is again blown through the furnace. The following reactions occur and yield a product that is about 98–99% Cu.

$$2\,\text{Cu}_2\text{S} + 3\,\text{O}_2\text{(g)} \xrightarrow{\Delta} 2\,\text{Cu}_2\text{O} + 2\,\text{SO}_2\text{(g)} \tag{24.10}$$

$$2\,\text{Cu}_2\text{O} + \text{Cu}_2\text{S} \xrightarrow{\Delta} 6\,\text{Cu(l)} + \text{SO}_2\text{(g)} \tag{24.11}$$

The product of reaction (24.11) is called *blister copper* because frozen bubbles of SO$_2$(g) are present. Blister copper can be used where high purity is not required (as in plumbing).

Refining blister copper to obtain high-purity copper is done electrolytically by the method outlined on page 856. High-purity copper is essential in electrical applications.

Pyrometallurgical Processes The metallurgical method based on roasting an ore, followed by reduction of the oxide to the metal, is called **pyrometallurgy**, the prefix *pyro* suggesting that high temperatures are involved. Some of the characteristics of pyrometallurgy are as follows:

- large quantities of waste materials produced in concentrating low-grade ores
- high energy consumption to maintain high temperatures necessary for roasting and reduction of ores
- gaseous emissions that must be controlled, such as SO$_2$(g) in roasting

Many of the metallurgical processes described earlier fall into this category.

Hydrometallurgical Processes   In **hydrometallurgy**, the materials handled are water and aqueous solutions at moderate temperatures rather than dry solids at high temperatures. Generally, three steps are involved in hydrometallurgy.

1. *Leaching:* Metal ions are extracted (leached) from the ore by a liquid. Leaching agents include water, acids, bases, and salt solutions. Oxidation–reduction reactions may also be involved.

2. *Purification and concentration:* Impurities are separated, and the solution produced by leaching may be made more concentrated. Methods include the adsorption of impurities on the surface of activated charcoal, ion exchange, and the evaporation of water.

3. *Precipitation:* The desired metal ions are either precipitated in an ionic solid or reduced to the free metal, often electrolytically.

Hydrometallurgy has long been used in obtaining silver and gold from natural sources. A typical gold ore currently being processed in the United States has only about 10 g Au per ton of ore. The leaching step in gold processing is known as *cyanidation*. In this process, $O_2(g)$ oxidizes the free metal to Au^+, which complexes with CN^-.

$$4\,Au(s) + 8\,CN^-(aq) + O_2(g) + 2\,H_2O(l) \longrightarrow 4\,[Au(CN)_2]^-(aq) + 4\,OH^-(aq) \tag{24.12}$$

The $[Au(CN)_2]^-(aq)$ is then filtered and concentrated. This is followed by displacement of $Au(s)$ from solution by an active metal such as zinc.

$$2\,[Au(CN)_2]^-(aq) + Zn(s) \longrightarrow 2\,Au(s) + [Zn(CN)_4]^{2-}(aq) \tag{24.13}$$

In one hydrometallurgical process for zinc, a zinc sulfide ore is leached with a sulfuric acid solution at 150 °C and an oxygen pressure of about 7 atm. The overall reaction is

$$ZnS(s) + H_2SO_4(aq) + \tfrac{1}{2}\,O_2(g) \longrightarrow ZnSO_4(aq) + S(s) + H_2O(l) \tag{24.14}$$

▲ Waste solution from the leaching operation at a gold-mining facility in the Mojave Desert of California.

In this process, there is no $SO_2(g)$ emission. Also, mercury impurities in the ZnS ore are retained in the leaching solution rather than being emitted with $SO_2(g)$ as in the traditional roasting process. Following the leaching process, $ZnSO_4(aq)$ is electrolyzed to produce pure Zn at the cathode, and $H_2SO_4(aq)$ is regenerated; see reaction (24.6). The $H_2SO_4(aq)$ is recycled into the leaching operation.

24-3 Metallurgy of Iron and Steel

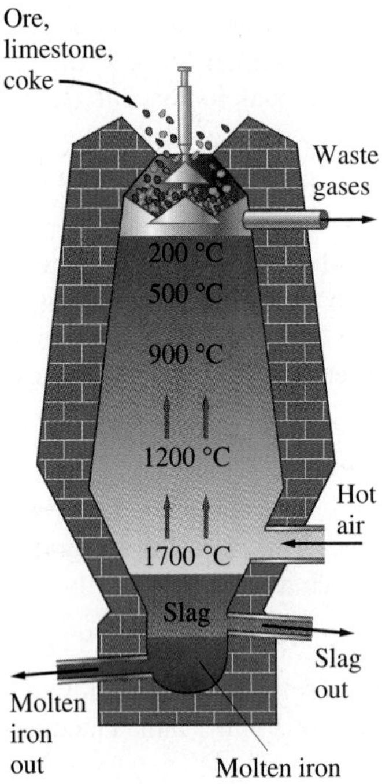

Iron is the most widely used metal from Earth's crust, and for this reason, we use this section to explore the metallurgy of iron and its principal alloy—steel—somewhat more fully. A type of steel called *wootz steel* was first produced in India about 3000 years ago; that same steel became famous in ancient times as Damascus steel, prized for making swords because of its suppleness and ability to hold a cutting edge. Many technological advances have been made since ancient times. These include introduction of the blast furnace around AD 1300, the Bessemer converter in 1856, the open-hearth furnace in the 1860s, and the basic oxygen furnace in the 1950s. A true understanding of the iron- and steelmaking processes has developed only within the past few decades, however. This understanding is based on concepts of thermodynamics, equilibrium, and kinetics.

Pig Iron

The reactions that occur in a blast furnace are complex. A highly simplified representation of the reduction of iron ore to impure iron is

$$Fe_2O_3(s) + 3\,CO(g) \longrightarrow 2\,Fe(l) + 3\,CO_2(g) \qquad (24.15)$$

A more complete description of the blast furnace reactions, including the removal of impurities as slag, is given in Table 24.2. Approximate temperatures are given for these reactions so that you can key them to regions of the blast furnace pictured in Figure 24-8.

▲ FIGURE 24-8
Typical blast furnace
Iron ore, coke, and limestone are added at the top of the furnace, and hot air is introduced through the bottom. Maximum temperatures are attained near the bottom of the furnace where molten iron and slag are drained off. The principal reactions occurring in the blast furnace are outlined in Table 24.2.

TABLE 24.2   Some Blast Furnace Reactions

Formation of gaseous reducing agents $CO(g)$ and $H_2(g)$:
  $C + H_2O \longrightarrow CO + H_2\ (>600\ °C)$
  $C + CO_2 \longrightarrow 2\,CO\ (1700\ °C)$
  $2\,C + O_2 \longrightarrow 2\,CO\ (1700\ °C)$
Reduction of iron oxide:
  $3\,CO + Fe_2O_3 \longrightarrow 2\,Fe + 3\,CO_2\ (900\ °C)$
  $3\,H_2 + Fe_2O_3 \longrightarrow 2\,Fe + 3\,H_2O\ (900\ °C)$
Slag formation to remove impurities from ore:
  $CaCO_3 \longrightarrow CaO + CO_2\ (800–900\ °C)$
  $CaO + SiO_2 \longrightarrow CaSiO_3(l)\ (1200\ °C)$
  $6\,CaO + P_4O_{10} \longrightarrow 2\,Ca_3(PO_4)_2(l)\ (1200\ °C)$
Impurity formation in the iron:
  $MnO + C \longrightarrow Mn + CO\ (1400\ °C)$
  $SiO_2 + 2\,C \longrightarrow Si + 2\,CO\ (1400\ °C)$
  $P_4O_{10} + 10\,C \longrightarrow 4\,P + 10\,CO\ (1400\ °C)$

▶ In the United States, slightly over half of all iron and steel production comes from recycled iron and steel.

The blast furnace charge—that is, the solid reactants—consists of iron ore, coke, a slag-forming flux, and perhaps some scrap iron. The exact proportions depend on the composition of the iron ore and its impurities. The common ores of iron are the oxides and carbonate: hematite (Fe_2O_3), magnetite (Fe_3O_4), limonite

$(2Fe_2O_3 \cdot 3H_2O)$, and siderite $(FeCO_3)$. The purpose of the flux is to maintain the proper ratio of acidic oxides (SiO_2, Al_2O_3, and P_4O_{10}) to basic oxides (CaO, MgO, and MnO) to obtain an easily liquefied silicate, aluminate, or phosphate slag. Because acidic oxides predominate in most ores, the flux generally employed is limestone, $CaCO_3$, or dolomite, $CaCO_3 \cdot MgCO_3$.

The iron obtained from a blast furnace is called **pig iron.** It contains about 95% Fe, 3–4% C, and varying quantities of other impurities. *Cast iron* can be obtained by pouring pig iron directly into molds of the desired shape. Cast iron is very hard and brittle and is used only where it is not subjected to mechanical or thermal shock, such as in engine blocks, brake drums, and transmission housings in automobiles.

Steel

The three fundamental changes that must be made to convert pig iron to **steel** are

1. reduction of the carbon content from 3–4% in pig iron to 0–1.5% in steel

2. removal, through slag formation, of Si, Mn, and P (each present in pig iron to the extent of 1% or so), together with other minor impurities

3. addition of alloying elements (such as Cr, Ni, Mn, V, Mo, and W) to give the steel its desired end properties

The most important method of steelmaking today is the **basic oxygen process**. Oxygen gas at about 10 atm pressure and a stream of powdered limestone are fed through a water-cooled tube (called a *lance*) and discharged above the molten pig iron (Figure 24-9). The reactions that occur (Table 24.3) accomplish the first two objectives. A typical reaction time is 22 minutes. The reaction vessel is tilted to pour off the liquid slag floating on top of the iron, and then the desired alloying elements are added.

> ▶ Thermal shock occurs when an object undergoes a rapid change of temperature. Engine blocks get quite hot, but because they cool slowly, they are not usually subject to thermal shock.

> ▶ Steel made with 18% Cr and 8% Ni resists corrosion and is commonly known as *stainless steel.*

▲ Pouring pig iron.

TABLE 24.3 Some Reactions Occurring in Steelmaking Processes

$$2\,C + O_2 \longrightarrow 2\,CO$$
$$2\,FeO + Si \longrightarrow 2\,Fe + SiO_2$$
$$FeO + Mn \longrightarrow Fe + MnO$$
$$FeO + SiO_2 \longrightarrow \underset{\text{slag}}{FeSiO_3}$$

$$MnO + SiO_2 \longrightarrow \underset{\text{slag}}{MnSiO_3}$$

$$4\,P + 5\,O_2 \longrightarrow P_4O_{10}$$
$$6\,CaO + P_4O_{10} \longrightarrow \underset{\text{slag}}{2\,Ca_3(PO_4)_2}$$

Steelmaking has been undergoing rapid technological changes. It is now possible to make iron and steel directly from iron ore in a single-step, continuous process at temperatures below the melting point of any of the materials used in the process. In the *direct reduction of iron* (DRI), CO(g) and H_2(g), obtained in the reaction of steam with natural gas, are used as reducing agents. The economic viability of the DRI process depends on an abundant supply of natural gas. Currently, only a small percentage of the world's iron production is by direct reduction, but this is a fast-growing component of the iron and steel industry, particularly in the Middle East and South America.

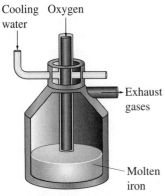

Cooling water   Oxygen

Exhaust gases

Molten iron

▲ **FIGURE 24-9**
A basic oxygen furnace

Periodic Trends:
First-Row Transition
Metals activity

24-4 First-Row Transition Metal Elements: Scandium to Manganese

The properties and uses of the first-row transition metals span a wide range, strikingly illustrating periodic behavior despite the small variation in some of the atomic properties listed in Table 24.1. The preparation, uses, and reactions of the compounds of these metals illustrate concepts we have previously discussed, including the variability of oxidation states.

Scandium

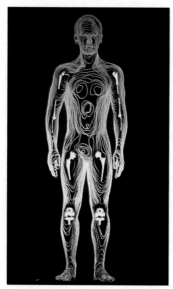

▲ A computer-generated representation of titanium joint implants at the shoulders, elbows, hips, and knees.

Scandium is a rather obscure metal, though not especially rare. Scandium constitutes about 0.0025% of Earth's crust, which makes it more abundant than many better-known metals, including lead, uranium, molybdenum, tungsten, antimony, silver, mercury, and gold. Its principal mineral form is *thortveitite*, $Sc_2Si_2O_7$. Most scandium is obtained from uranium ores, however, in which it occurs only to the extent of about 0.01% Sc by mass. The commercial uses of scandium are limited, and its production is measured in gram or kilogram quantities, not in tons. One application is in high-intensity lamps. The pure metal is usually prepared by the electrolysis of a fused mixture of $ScCl_3$ with other chlorides.

Because of its noble-gas electron configuration, the Sc^{3+} ion lacks some of the characteristic properties of transition metal ions. For instance, the ion is colorless and diamagnetic, as are most of its salts. In its chemical behavior, Sc^{3+} most closely resembles Al^{3+}, as in the hydrolysis of $[Sc(H_2O)_6]^{3+}(aq)$ to yield acidic solutions and in the formation of an amphoteric gelatinous hydroxide, $Sc(OH)_3$.

Titanium

Titanium is the ninth most abundant element, constituting 0.6% of Earth's solid crust. The metal is greatly valued for its low density, high structural strength, and corrosion resistance. The first two properties account for its extensive use in the aircraft industry, and the third for its uses in the chemical industry: in pipes, component parts of pumps, and reaction vessels. Titanium is also used in dental and other bone implants. The metal provides a strong support and bone bonds directly to a titanium implant, making it a part of the body.

Several compounds of titanium are of particular commercial importance. *Titanium tetrachloride*, $TiCl_4$, is the starting material for preparing other Ti compounds and plays a central role in the metallurgy of titanium. $TiCl_4$ is also used to formulate catalysts for the production of plastics. The usual method of preparing $TiCl_4$ involves the reaction of naturally occurring rutile (TiO_2) with carbon and $Cl_2(g)$ [reaction (24.7)].

$TiCl_4$ is a colorless liquid (mp -24 °C; bp 136 °C). In the $+4$ oxidation state, all the valence-shell electrons of Ti atoms are employed in bond formation. In this oxidation state, Ti bears a strong resemblance to the group 14 elements, with some properties and a molecular shape (tetrahedral) similar to those of CCl_4 and $SiCl_4$. The hydrolysis of $TiCl_4$, when carried out in moist air, is the basis for a type of smoke grenade in which $TiO_2(s)$ is the smoke.

$$TiCl_4(l) + 2\,H_2O(l) \longrightarrow TiO_2(s) + 4\,HCl(g)$$

$SiCl_4$ also fumes in moist air in a similar reaction.

▲ White $TiO_2(s)$, mixed with other components to produce the desired color, is the leading pigment used in paints.

Titanium dioxide, TiO_2, is bright white, opaque, inert, and nontoxic. Because of these properties and its relative low cost, it is now the most widely used white pigment for paints. In this application, TiO_2 has displaced toxic basic lead carbonate—so-called white lead. TiO_2 is also used as a paper whitener and in glass, ceramics, floor coverings, and cosmetics.

To produce pure TiO_2 for these and other uses, a gaseous mixture of $TiCl_4$ and O_2 is passed through a silica tube at about 700 °C.

$$TiCl_4(g) + O_2(g) \xrightarrow{\Delta} TiO_2(s) + 2\,Cl_2(g)$$

Vanadium

Vanadium is a fairly abundant element (0.02% of Earth's crust) found in several dozen ores. Its principal ores are rather complex, such as *vanadinite*, $3Pb_3(VO_4)_2 \cdot PbCl_2$. The metallurgy of vanadium is not simple, but vanadium of high purity (99.99%) is obtainable. For most of its applications, though, vanadium is prepared as an iron–vanadium alloy, *ferrovanadium*, containing from 35 to 95% V. About 80% of the vanadium produced is for the manufacture of steel. Vanadium-containing steels are used in applications requiring strength and toughness, such as in springs and high-speed machine tools.

The most important compound of vanadium is the pentoxide, V_2O_5, used mainly as a catalyst, as in the conversion of $SO_2(g)$ to $SO_3(g)$ in the contact method for the manufacture of sulfuric acid. The activity of V_2O_5 as an oxidation catalyst may be linked to its reversible loss of oxygen occurring from 700 to 1100 °C.

In its compounds, vanadium can exist in a variety of oxidation states. In each of these oxidation states, vanadium forms an oxide or ion. The ions display distinctive colors in aqueous solution (Figure 24-10). The acid–base properties of vanadium oxides are in accord with factors established earlier in the text: If the central metal atom is in a low oxidation state, the oxide acts as a base; in higher oxidation states for the central atom, acidic properties become important. Vanadium oxides with V in the +2 and +3 oxidation states are basic, whereas those in the +4 and +5 oxidation states are amphoteric.

Most compounds with vanadium in its highest oxidation state (+5) are good oxidizing agents. In its +2 oxidation state, vanadium (as V^{2+}) is a good reducing agent. The oxidation–reduction relationships between the ionic species pictured in Figure 24-10 are summarized in Table 24.4.

▲ **FIGURE 24-10**
Some vanadium species in solution
The yellow solution has vanadium in the +5 oxidation state, as VO_2^+. In the blue solution, the oxidation state is +4, in VO^{2+}. The green solution contains V^{3+}, and the violet solution contains V^{2+}.

TABLE 24.4	Oxidation States of Vanadium Species in Acidic Solution	
OS Change	**Reduction Half-Reaction**	**$E°$**
+5 ⟶ +4:	$VO_2^+(aq) + 2\,H^+(aq) + e^- \longrightarrow VO^{2+}(aq) + H_2O$ (yellow) .... (blue)	1.000 V
+4 ⟶ +3:	$VO^{2+}(aq) + 2\,H^+(aq) + e^- \longrightarrow V^{3+}(aq) + H_2O$ .... (green)	0.337 V
+3 ⟶ +2:	$V^{3+}(aq) + e^- \longrightarrow V^{2+}(aq)$ .... (violet)	−0.255 V
+2 ⟶ 0:	$V^{2+}(aq) + 2\,e^- \longrightarrow V(s)$	−1.13 V

EXAMPLE 24-2

Using Electrode Potential Data to Predict an Oxidation–Reduction Reaction. Can $MnO_4^-(aq)$ be used to oxidize $VO^{2+}(aq)$ to $VO_2^+(aq)$ for standard-state conditions in an acidic solution? If so, write a balanced equation for the redox reaction.

Solution

We start by writing two half-equations, one for the reduction of MnO_4^- to Mn^{2+}; the other, for the oxidation of VO^{2+} to VO_2^+. Both are in acidic solution. We find one $E°$ value in Table 24.4 and the other in Table 21.1. We then combine the $E°$ values into $E°_{cell}$.

Reduction:	$MnO_4^- + 8\,H^+ + 5\,e^- \longrightarrow Mn^{2+} + 4\,H_2O$
Oxidation:	$5(VO^{2+} + H_2O \longrightarrow VO_2^+ + 2\,H^+ + e^-)$

Overall: $5\,VO^{2+}(aq) + MnO_4^-(aq) + H_2O(l) \longrightarrow$
$$5\,VO_2^+(aq) + Mn^{2+}(aq) + 2\,H^+(aq)$$

$$E^\circ_{cell} = E^\circ_{MnO_4^-/Mn^{2+}} - E^\circ_{VO_2^+/VO^{2+}} = 1.51\,V - 1.000\,V = 0.51\,V$$

Because E°_{cell} is positive, we predict that MnO_4^- should oxidize VO^{2+} to VO_2^+ for standard-state conditions in acidic solution.

Practice Example A: Use data from Tables 21.1 and 24.4 to determine whether nitric acid can be used to oxidize $V^{3+}(aq)$ to $VO^{2+}(aq)$ for standard-state conditions. If so, write a balanced equation for the reaction.

Practice Example B: Select a *reducing* agent from Table 21.1 that can be used to reduce VO^{2+} to V^{2+} for standard-state conditions in acidic solution. Consider that the reduction occurs in two stages: $VO^{2+} \longrightarrow V^{3+} \longrightarrow V^{2+}$, but note that the V^{2+} must not be reduced to $V(s)$.

Chromium

▶ The word *chromium* is derived from the Greek *chroma*, meaning color—an apt name, given the range of colors found in chromium compounds.

Although it is found only to the extent of 122 parts per million (0.0122%) in Earth's crust, chromium is one of the most important industrial metals. The production of ferrochrome from *chromite*, $Fe(CrO_2)_2$, was discussed in Section 24-2. Chromium metal is hard and maintains a bright surface through the protective action of an invisible oxide coating. Because it is resistant to corrosion, chromium is extensively used in plating other metals.

Steel is chrome-plated from an aqueous solution containing CrO_3 and H_2SO_4. The plating obtained is thin and porous. It tends to develop cracks unless the steel is first plated with copper or nickel, which provides the true protective coating. Then chromium is plated over this layer for decorative purposes. The efficiency of chrome-plating is limited by the fact that reduction of Cr(VI) to Cr(0) produces only $\frac{1}{6}$ mol Cr per mole of electrons. In other words, large quantities of electric energy are required for chrome-plating relative to other types of metal plating.

Chromium, like vanadium, has a variety of oxidation states in aqueous solution, each having a different color.

▶ The colors may also depend on other species present in solution. For example, if [Cl$^-$] is high, $[Cr(H_2O)_6]^{3+}$ is converted to $[CrCl_2(H_2O)_4]^+$ and the violet color changes to green.

OS + 2:	$[Cr(H_2O)_6]^{2+}$, blue	
OS + 3:	(acidic) $[Cr(H_2O)_6]^{3+}$, violet	(basic) $[Cr(OH)_4]^-$, green
OS + 6:	(acidic) $Cr_2O_7^{2-}$, orange	(basic) CrO_4^{2-}, yellow

▶ Chrome-plating shown on the engine of a motorcycle.

The oxides and hydroxides of chromium conform to the general principles of acid–base behavior: CrO is basic, Cr_2O_3 is amphoteric, and CrO_3 is acidic.

Pure chromium reacts with dilute $HCl(aq)$ or $H_2SO_4(aq)$ to produce $Cr^{2+}(aq)$. Nitric acid and other oxidizing agents alter the surface of the metal (perhaps by formation of an oxide coating). They render the metal resistant to further attack—it becomes *passive*. A better source of chromium compounds than the pure metal is the alkali metal chromates, which contain Cr(VI) and can be obtained directly from chromite ore by reactions such as

$$4\, Fe(CrO_2)_2 + 8\, Na_2CO_3 + 7\, O_2(g) \xrightarrow{\Delta} 2\, Fe_2O_3 + 8\, Na_2CrO_4 + 8\, CO_2(g) \quad (24.16)$$

The *sodium chromate*, Na_2CrO_4, produced by this reaction is the source of many industrially important chromium compounds.

The Cr(VI) oxidation state is also observed in the red oxide, CrO_3. As expected, this oxide dissolves in water to produce a strongly acidic solution. The product of the reaction is not the expected chromic acid, H_2CrO_4, however, which has never been isolated in the pure state. Instead, the observed reaction is

$$2\, CrO_3(s) + H_2O(l) \longrightarrow 2\, H^+(aq) + Cr_2O_7^{2-}(aq) \quad (24.17)$$

It is possible to crystallize a *dichromate* salt from a water solution of CrO_3. If the solution is made basic, the color turns from orange to yellow. From basic solutions, only *chromate* salts can be crystallized. Thus, whether a solution contains Cr(VI) as $Cr_2O_7^{2-}$ or CrO_4^{2-} or a mixture of the two depends on the pH. The relevant equations follow.

$$2\, CrO_4^{2-}(aq) + 2\, H^+(aq) \rightleftharpoons Cr_2O_7^{2-}(aq) + H_2O(l) \quad (24.18)$$

$$K_c = \frac{[Cr_2O_7^{2-}]}{[CrO_4^{2-}]^2[H^+]^2} = 3.2 \times 10^{14} \quad (24.19)$$

Le Châtelier's principle predicts that the forward reaction of (24.18) is favored in acidic solutions and that the predominant Cr(VI) species is $Cr_2O_7^{2-}$. In basic solution, H^+ ions are removed and the reverse reaction is favored, forming CrO_4^{2-} as the principal species. Careful control of the pH is necessary when $Cr_2O_7^{2-}$ is used as an oxidizing agent or CrO_4^{2-} as a precipitating agent. In addition, equation (24.19) can be used to calculate the relative amounts of the two ions as a function of $[H^+]$.

Chromate ion in basic solution can be used to precipitate metal chromates such as $BaCrO_4(s)$ and $PbCrO_4(s)$. It is not a good oxidizing agent, however; it is not readily reduced.

$$CrO_4^{2-}(aq) + 4\, H_2O(l) + 3\, e^- \longrightarrow [Cr(OH)_4]^-(aq) + 4\, OH^-(aq) \qquad E° = -0.13\ V$$

Dichromates are poor precipitating agents but excellent oxidizing agents, which are used in a variety of industrial processes. In the chrome leather tanning process, for example, animal hides are immersed in $Na_2Cr_2O_7(aq)$, which is then reduced by $SO_2(g)$ to soluble basic chromic sulfate, $Cr(OH)SO_4$. Collagen, a protein in hides, reacts to form an insoluble chromium complex. The hides become *leather*, a tough, pliable material resistant to biological attack.

Dichromates are easily reduced to Cr_2O_3. In the case of ammonium dichromate, simply heating the compound produces Cr_2O_3 in a dramatic reaction (Figure 24-11).

Chromium(II) compounds can be prepared by the reduction of Cr(III) compounds with zinc in acidic solution or electrolytically at a lead cathode. The most distinctive feature of Cr(II) compounds is their reducing power.

$$Cr^{3+}(aq) + e^- \longrightarrow Cr^{2+}(aq) \qquad E° = -0.424\ V$$

Chromate ion (CrO_4^{2-})

Dichromate ion ($Cr_2O_7^{2-}$)

▲ **FIGURE 24-11 Decomposition of $(NH_4)_2Cr_2O_7$**
Ammonium dichromate (left) contains both an oxidizing agent, $Cr_2O_7^{2-}$, and a reducing agent, NH_4^+. The products of the reaction between these two ions are $Cr_2O_3(s)$ (right), $N_2(g)$, and $H_2O(g)$. Considerable heat and light are also evolved (center).

▶ **FIGURE 24-12**
Relationship between Cr^{2+} and Cr^{3+}
The solution on the left, containing blue $Cr^{2+}(aq)$, is prepared by dissolving chromium metal in $HCl(aq)$. Within minutes, the $Cr^{2+}(aq)$ is oxidized to green $Cr^{3+}(aq)$ by atmospheric oxygen (right). The green color is that of the complex ion $[CrCl_2(H_2O)_4]^+(aq)$.

That is, the oxidation of $Cr^{2+}(aq)$ occurs readily. In fact, Cr(II) solutions can be used to purge gases of trace amounts of $O_2(g)$, through the following reaction, illustrated in Figure 24-12.

$$4\,Cr^{2+}(aq) + O_2(g) + 4\,H^+(aq) \longrightarrow 4\,Cr^{3+}(aq) + 2\,H_2O(l) \qquad E^\circ_{cell} = +1.653\ V$$

Pure Cr can be obtained in small amounts by reducing Cr_2O_3 with Al in a reaction similar to the thermite reaction.

$$Cr_2O_3(s) + 2\,Al(s) \longrightarrow Al_2O_3(s) + 2\,Cr(l) \qquad (24.20)$$

Manganese

Manganese is a fairly abundant element, constituting about 1% of Earth's crust. Its principal ore is *pyrolusite*, MnO_2. Like V and Cr, Mn is most important in steel production, generally as the iron–manganese alloy, *ferromanganese*. Ferromanganese can be obtained by the reduction of a mixture of pyrolusite and hematite iron ores with carbon.

$$MnO_2 + Fe_2O_3 + 5\,C \xrightarrow{\Delta} Mn + 2\,Fe + 5\,CO$$
ferromanganese

Mn participates in the purification of iron by reacting with sulfur and oxygen and removing them through slag formation. In addition, Mn increases the hardness of steel. Steel containing high proportions of Mn is extremely tough and wear-resistant in such applications as railroad rails, bulldozers, and road scrapers.

The electron configuration of Mn is $[Ar]3d^5 4s^2$. By employing first the two $4s$ electrons and then, consecutively, up to all five of its unpaired $3d$ electrons, manganese exhibits all oxidation states from +2 to +7. The most important reactions of manganese compounds are oxidation–reduction reactions. Standard electrode potential diagrams are given in Figure 24-13. These diagrams help explain the following observations:

- $Mn^{3+}(aq)$ is unstable; that is, its disproportionation is spontaneous.

$$2\,Mn^{3+}(aq) + 2\,H_2O(l) \longrightarrow Mn^{2+}(aq) + MnO_2(s) + 4\,H^+(aq)$$
$$E^\circ_{cell} = 0.54\ V \qquad (24.21)$$

- Manganate ion, MnO_4^{2-} is also unstable in acidic solution; its disproportionation is spontaneous.

$$3\,MnO_4^{2-}(aq) + 4\,H^+(aq) \longrightarrow MnO_2(s) + 2\,MnO_4^- + 2\,H_2O(l)$$
$$E^\circ_{cell} = 1.70\ V \qquad (24.22)$$

- If $[OH^-]$ is kept sufficiently high, the following reaction can be reversed; thus, manganate ion can be maintained as a stable species in a strongly basic medium.

$$3\,MnO_4^{2-}(aq) + 2\,H_2O(l) \longrightarrow MnO_2(s) + 2\,MnO_4^- + 4\,OH^-(aq)$$
$$E^\circ_{cell} = 0.04\ V \qquad (24.23)$$

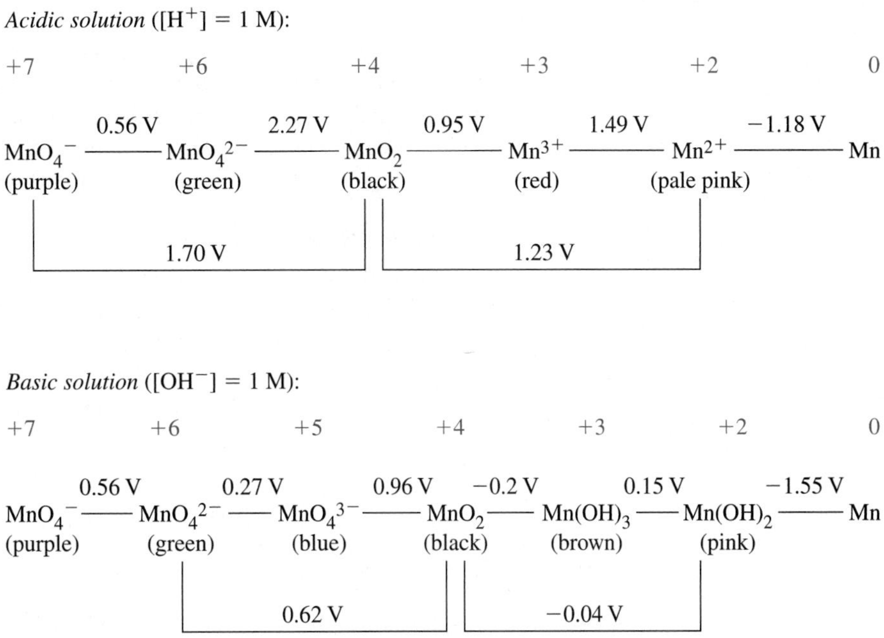

Acidic solution ($[H^+] = 1\ M$):

+7	+6	+4	+3	+2	0

$$
\begin{array}{ccccccccccc}
 & 0.56\ V & & 2.27\ V & & 0.95\ V & & 1.49\ V & & -1.18\ V & \\
MnO_4^- & \text{———} & MnO_4^{2-} & \text{———} & MnO_2 & \text{———} & Mn^{3+} & \text{———} & Mn^{2+} & \text{———} & Mn \\
\text{(purple)} & & \text{(green)} & & \text{(black)} & & \text{(red)} & & \text{(pale pink)} & &
\end{array}
$$

1.70 V   1.23 V

Basic solution ($[OH^-] = 1\ M$):

+7	+6	+5	+4	+3	+2	0

$$
\begin{array}{ccccccccccccc}
 & 0.56\ V & & 0.27\ V & & 0.96\ V & & -0.2\ V & & 0.15\ V & & -1.55\ V & \\
MnO_4^- & \text{———} & MnO_4^{2-} & \text{———} & MnO_4^{3-} & \text{———} & MnO_2 & \text{———} & Mn(OH)_3 & \text{———} & Mn(OH)_2 & \text{———} & Mn \\
\text{(purple)} & & \text{(green)} & & \text{(blue)} & & \text{(black)} & & \text{(brown)} & & \text{(pink)} & &
\end{array}
$$

0.62 V   −0.04 V

▲ **FIGURE 24-13**   **Electrode potential diagrams for manganese**

Manganese dioxide is used in dry cell batteries, in glass and ceramic glazes, and as a catalyst; it is also the principal source of manganese compounds. When MnO_2 is heated in the presence of an alkali and an oxidizing agent, a manganate salt is produced.

$$3\,MnO_2 + 6\,KOH + KClO_3 \xrightarrow{\Delta} 3\,K_2MnO_4 + KCl + 3\,H_2O(g)$$

K_2MnO_4 is extracted from the fused mass with water and can then be oxidized to $KMnO_4$, *potassium permanganate* (with Cl_2 as an oxidizing agent, for instance). Potassium permanganate, $KMnO_4$, is an important laboratory oxidizing agent. For chemical analyses, it is generally used in acidic solutions, in which it is reduced to $Mn^{2+}(aq)$. In the analysis of iron by MnO_4^-, a sample of Fe^{2+} is prepared by dissolving iron in an acid and reducing any Fe^{3+} back to Fe^{2+}. Then the sample is titrated with $MnO_4^-(aq)$.

$$5\,Fe^{2+}(aq) + MnO_4^-(aq) + 8\,H^+(aq) \longrightarrow 5\,Fe^{3+}(aq) + Mn^{2+}(aq) + 4\,H_2O \qquad (24.24)$$

$Mn^{2+}(aq)$ has a barely discernible pale pink color. $MnO_4^-(aq)$ is an intense purple color. At the end point of the titration reaction (24.24), the solution acquires a lasting light purple color with just one drop of excess $MnO_4^-(aq)$ (recall Figure 5-11). $MnO_4^-(aq)$ is less satisfactory for titrations in alkaline solutions because the insoluble reduction product, brown $MnO_2(s)$, obscures the end point.

24-5 The Iron Triad: Iron, Cobalt, and Nickel

The transition elements iron, cobalt, and nickel comprise the *iron triad*. Iron, with an annual worldwide production of over 500 million tons, is the most important metal in modern civilization. It is widely distributed in Earth's crust at an abundance of 4.7%. The major commercial use of iron is to make steel (see Section 24-3).

Cobalt is among the rarer elements. It comprises only about 0.0020% of Earth's crust, but it occurs in sufficiently concentrated deposits (ores) so that its annual production runs into the millions of pounds. Cobalt is used primarily in alloys with other metals. Like iron, cobalt is ferromagnetic. One alloy of cobalt, Co_5Sm, makes a particularly strong and lightweight permanent magnet. Because of the strength of its magnetic field, magnets of this alloy are used in the manufacture of miniature electronic devices.

Nickel ranks twenty-fourth in abundance among the elements in Earth's crust. Its ores are mainly the sulfides, oxides, silicates, and arsenides. Particularly large deposits are found in Canada. Of the 300 million pounds of nickel consumed annually in the United States, about 80% goes to the production of alloys. Another 15% is used in electroplating, and the remainder for miscellaneous purposes (for example, as catalysts).

▲ Cobalt–samarium magnets are used in high-efficiency motors.

Oxidation States

Variability of oxidation state is seen in the iron triad, even if not to the same degree as with vanadium, chromium, and manganese. The +2 oxidation state is commonly encountered in all three metals.

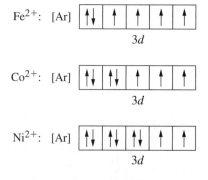

For cobalt and nickel, the +2 oxidation state is the most stable, but for iron, the most stable is +3.

$$\text{Fe}^{3+}: \quad [\text{Ar}] \quad \boxed{\uparrow \;\; \uparrow \;\; \uparrow \;\; \uparrow \;\; \uparrow}$$
$$3d$$

The electron configuration of Fe^{3+} has a half-filled d subshell with all the d electrons unpaired. This type of electron configuration has a special stability. As we can see from the orbital diagrams of Co^{2+} and Ni^{2+}, neither would have a half-filled $3d$ subshell following the loss of one additional electron. Co^{3+} and Ni^{3+} are not readily formed. To illustrate the contrast between Fe^{3+}, on the one hand, and Co^{3+} and Ni^{3+}, on the other, consider the following observations.

- $Fe^{2+}(aq)$ is spontaneously oxidized to $Fe^{3+}(aq)$ by $O_2(g)$ at 1 atm in a solution with $[H^+] = 1$ M.

$$4\,Fe^{2+}(aq) + O_2(g) + 4\,H^+(aq) \longrightarrow 4\,Fe^{3+}(aq) + 2\,H_2O(l)$$
$$E^{\circ}_{cell} = 0.44\,V \qquad (24.25)$$

  This reaction is still spontaneous with lower O_2 partial pressures and less acidic solutions.

- For the half-reaction $Co^{3+}(aq) + e^- \longrightarrow Co^{2+}(aq)$, $E^{\circ} = 1.82$ V. The reduction of $Co^{3+}(aq)$ to $Co^{2+}(aq)$ occurs readily; conversely, the oxidation of $Co^{2+}(aq)$ to Co^{3+} occurs only with difficulty. As will be noted in Chapter 25, however, the oxidation state +3 can be attained when Co^{3+} is the central metal ion in very stable complex ions.

- Nickel(III) compounds are used in batteries. For example, in the nickel–cadmium (Nicad) cell, the cathode half-reaction is $NiO(OH) + H^+ + e^- \longrightarrow Ni(OH)_2$. It is the ease of this reduction, when combined with the oxidation half-reaction, $Cd(s) + 2\,OH^-(aq) \longrightarrow Cd(OH)_2(s) + 2\,e^-$, that produces a cell with a voltage of about 1.5 V.

Some Reactions of the Iron Triad Elements

The reactions of the iron triad elements are many and varied. The metals are more active than hydrogen and liberate $H_2(g)$ from an acidic solution. Hydrated colored ions are characteristic: Co^{2+} and Ni^{2+} are red and green, respectively. In aqueous solution, Fe^{2+} is pale green and fully hydrated Fe^{3+} is purple. Generally, however, solutions of $Fe^{2+}(aq)$ are yellow to brown, but this color is probably due to the presence of species formed in the hydrolysis of $Fe^{3+}(aq)$. Like the hydrolysis of $Al^{3+}(aq)$, described on page 699, that of $Fe^{3+}(aq)$ produces an acidic solution.

$$[Fe(H_2O)_6]^{3+}(aq) + H_2O(l) \rightleftharpoons [FeOH(H_2O)_5]^{2+}(aq) + H_3O^+(aq)$$
$$K_a = 8.9 \times 10^{-4} \qquad (24.26)$$

Some reactions that can be used to identify and distinguish between $Fe^{2+}(aq)$ and $Fe^{3+}(aq)$ are summarized in Table 24.5.

▶ The blue precipitate that establishes the presence of $Fe^{2+}(aq)$ from the corroding nails in Figure 21-19 is Turnbull's blue.

TABLE 24.5	**Some Qualitative Tests for $Fe^{2+}(aq)$ and $Fe^{3+}(aq)$**	
Reagent	**$Fe^{2+}(aq)$**	**$Fe^{3+}(aq)$**
$NaOH(aq)$	Green precipitate	Red-brown precipitate
$K_4[Fe(CN)_6]$	White precipitate, turning blue rapidly	Prussian blue precipitate
$K_3[Fe(CN)_6]$	Turnbull's blue precipitate	Red-brown (no precipitate)
KSCN	No color	Deep red

▶ In Chapter 25, we will learn to write the systematic names for ferrocyanide and ferricyanide ions; they are hexacyanoferrate(II) and hexacyanoferrate(III), respectively.

An interesting set of reactions involves the complex ions $[Fe(CN)_6]^{4-}$ and $[Fe(CN)_6]^{3-}$. These ions are commonly called *ferrocyanide* and *ferricyanide*, respectively. $Fe^{3+}(aq)$ yields a dark blue precipitate called *Prussian blue* when treated with potassium ferrocyanide, $K_4[Fe(CN)_6](aq)$, whereas $Fe^{2+}(aq)$ yields a similar blue precipitate, called *Turnbull's blue*, when treated with potassium ferricyanide, $K_3[Fe(CN)_6](aq)$. Together, these two and similar related precipitates are known commercially as *iron blue*. Iron blue is used as a pigment for paints, printing inks, laundry bluing, art colors, cosmetics (eye shadow), and blueprinting. An additional sensitive test for $Fe^{3+}(aq)$ is the formation of a blood-red complex ion with thiocyanate ion, $SCN^-(aq)$.

$$[Fe(H_2O)_6]^{3+} + SCN^-(aq) \longrightarrow [FeSCN(H_2O)_5]^{2+} + H_2O(l)$$

With only a few exceptions, the transition metals form compounds with carbon monoxide (CO), called **metal carbonyls**. In the simple metal carbonyls listed in Table 24.6,

- Each CO molecule contributes an electron pair to an empty orbital of the metal atom.
- All electrons are paired (most metal carbonyls are diamagnetic).
- The metal atom acquires the electron configuration of the noble gas Kr.

TABLE 24.6 Three Metal Carbonyls			
	Number of e⁻		
	From Metal	**From CO**	**Total**
$Cr(CO)_6$	24	12	36
$Fe(CO)_5$	26	10	36
$Ni(CO)_4$	28	8	36

The structures of the simple carbonyls in Figure 24-14 are those that we would predict from VSEPR theory (based on a number of electron pairs equal to the number of CO molecules).

Metal carbonyls are produced in several ways. Nickel combines with CO(g) at ordinary temperatures and pressures in a reversible reaction.

$$Ni(s) + 4\,CO(g) \rightleftharpoons Ni(CO)_4(l)$$

▶ In the metallurgy of nickel (the Mond process), CO(g) is passed over a mixture of metal oxides. Nickel is carried off as $Ni(CO)_4(g)$, while the other oxides are reduced to the metals.

With iron, it is necessary to use higher temperatures (200 °C) and CO pressures (100 atm).

$$Fe(s) + 5\,CO(g) \rightleftharpoons Fe(CO)_5(g)$$

In other cases, the carbonyl is obtained by reducing a metal compound in the presence of CO(g).

Carbon monoxide poisoning results from a reaction similar to carbonyl formation. CO molecules coordinate with Fe atoms in hemoglobin in the blood, displacing the O_2 molecules normally carried by hemoglobin. The metal carbonyls themselves are also very poisonous.

24-6 Group 11: Copper, Silver, and Gold

Throughout the ages, Cu, Ag, and Au have been the preferred metals for coins because they are so durable and resistant to corrosion. The data in Table 24.7 help us

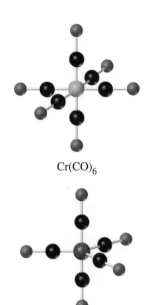

Cr(CO)₆

Fe(CO)₅

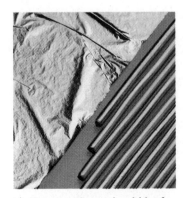

Ni(CO)₄

▲ **FIGURE 24-14**
Structures of some simple carbonyls

TABLE 24.7 Some Properties of Copper, Silver, and Gold

	Cu	Ag	Au
Electron configuration	[Ar]$3d^{10}4s^1$	[Kr]$4d^{10}5s^1$	[Xe]$4f^{14}5d^{10}6s^1$
Metallic radius, pm	128	144	144
First ioniz energy, kJ/mol	745	731	890
Electrode potential, V			
$M^+(aq) + e^- \longrightarrow M(s)$	+0.520	+0.800	+1.83
$M^{2+}(aq) + 2\,e^- \longrightarrow M(s)$	+0.340	+1.39	—
$M^{3+}(aq) + 3\,e^- \longrightarrow M(s)$	—	—	+1.52
Oxidation states[a]	+1, +2	+1, +2	+1, +3

[a] The most common oxidation states are shown in red.

understand why this is so. The metal ions are easy to reduce to the free metals, which means that the metals are difficult to oxidize.

In Mendeleev's periodic table, the alkali metals (group 1) and the coinage metals (group 11) appear together as group I. The only similarity between the two subgroups, however, is that both have a single *s* electron in the valence shells of their atoms. More significant are the differences between the group 1 and group 11 metals. For example, the first ionization energies for the group 11 metals are much larger than for the group 1 metals, and the standard electrode potentials, $E°$, are positive for the group 11 metals and negative for the group 1 metals.

Like the other transition elements that precede them in the periodic table, the group 11 metals are able to use *d* electrons in chemical bonding. Thus they can exist in different oxidation states, exhibit paramagnetism and color in some of their compounds, and form complex ions. They also possess to a high degree some of the distinctive physical properties of metals—malleability, ductility, and excellent electrical and thermal conductivity.

Copper, silver, and gold—the coinage metals—are used in jewelry making and the decorative arts. Gold, for instance, is extraordinarily malleable and can be pounded into thin translucent sheets known as gold leaf. The coinage metals are valued by the electronics industry for their ability to conduct electricity. Silver has the highest electrical conductivity of any pure element, but both copper and gold are more often used as electrical conductors because copper is inexpensive and gold does not readily corrode. The most important use of gold is as the monetary reserve of nations throughout the world.

Generally, the coinage metals are resistant to air oxidation, although silver will tarnish through reactions with sulfur compounds in air to produce black Ag_2S. In moist air, copper corrodes to produce green basic copper carbonate. This is the green color associated with copper roofing and gutters and bronze statues. (Bronze is an alloy of Cu and Sn.) Fortunately, this corrosion product forms a tough adherent coating that protects the underlying metal. The corrosion reaction is complex but may be summarized as

$$2\,Cu + H_2O + CO_2 + O_2 \longrightarrow Cu_2(OH)_2CO_3 \qquad (24.27)$$
$$\text{basic copper carbonate}$$

The group 11 metals do not react with HCl(aq), but both Cu and Ag react with concentrated H_2SO_4(aq) or HNO_3(aq). The metals are oxidized to Cu^{2+} and Ag^+, respectively, and the reduction products are SO_2(g) in H_2SO_4 and either NO(g) or NO_2(g) in HNO_3(aq).

Au does not react with either acid, but it will react with "royal water"—*aqua regia* (1 part HNO_3 and 3 parts HCl). The HNO_3(aq) oxidizes the metal and Cl⁻ from

▲ **Copper wire and gold leaf**

the HCl(aq) promotes the formation of the stable complex ion $[AuCl_4]^-$.

$$Au(s) + 4 H^+(aq) + NO_3^-(aq) + 4 Cl^-(aq) \longrightarrow$$
$$[AuCl_4]^-(aq) + 2 H_2O(l) + NO(g) \quad (24.28)$$

Trace amounts of Cu are essential to life, but larger quantities are toxic, especially to bacteria, algae, and fungi. Among the many copper compounds used as pesticides are the basic acetate, carbonate, chloride, hydroxide, and sulfate. Commercially, the most important copper compound is $CuSO_4 \cdot 5H_2O$. In addition to its agricultural applications, $CuSO_4$ is employed in batteries and electroplating, in the preparation of other copper salts, and in a variety of industrial processes.

Silver nitrate is the principal silver compound of commerce and is also an important laboratory reagent for the precipitation of anions, most of which form insoluble silver salts. These precipitation reactions can be used for the quantitative determination of anions, either gravimetrically (by weighing precipitates) or volumetrically (by titration). Most other Ag compounds are derived from $AgNO_3$. Ag compounds are used in electroplating, in the manufacture of batteries, in medicinal chemistry, as catalysts, and in cloud seeding (AgI). Their most important use by far (mostly as silver halides) is in photography (see Section 25-10).

Gold compounds are used in electroplating, photography, medicinal chemistry (as anti-inflammatory agents for severe rheumatoid arthritis, for example), and the manufacture of special glasses and ceramics (such as ruby glass).

▲ Gold plating on a space antenna.

24-7 Group 12: Zinc, Cadmium, and Mercury

The properties of the group 12 elements are consistent with elements having full subshells, $(n - 1)d^{10}ns^2$; some of those properties are summarized in Table 24.8. The low melting and boiling points of the group 12 metals can probably be attributed to the fact that, with only the ns^2 electrons participating, metallic bonding is weak. Mercury is the only metal that exists as a liquid at room temperature and below (although liquid gallium can easily be supercooled to room temperature). Mercury differs from Zn and Cd in a number of ways in addition to physical appearance.

- Mercury has little tendency to combine with oxygen. Its oxide, HgO, is thermally unstable.
- Very few mercury compounds are water soluble, and most are not hydrated.
- Many mercury compounds are covalent. Except for HgF_2, mercury halides are only slightly ionized in aqueous solution.

TABLE 24.8 Some Properties of the Group 12 Metals

	Zn	Cd	Hg
Density, g/cm^3	7.14	8.64	13.59 (liquid)
Melting point, °C	419.6	320.9	−38.87
Boiling point, °C	907	765	357
Electron configuration	$[Ar]3d^{10}4s^2$	$[Kr]4d^{10}5s^2$	$[Xe]4f^{14}5d^{10}6s^2$
Atomic radius, pm	133	149	160
Ionization energy, kJ/mol			
first	906	867	1006
second	1733	1631	1809
Principal oxidation state(s)	+2	+2	+1, +2
Electrode potential $E°$, V			
$[M^{2+}(aq) + 2 e^- \longrightarrow M]$	−0.763	−0.403	+0.854
$[M_2^{2+}(aq) + 2 e^- \longrightarrow 2 M]$	—	—	+0.796

- Mercury(I) forms a common diatomic ion with a metal–metal covalent bond, Hg_2^{2+}.
- Mercury will not displace $H_2(g)$ from $H^+(aq)$.

Some of these differences exhibited by mercury can probably be attributed to the *relativistic effect* discussed in Chapter 10 (page 381). Because the speed of the $6s$ electrons reaches a significant fraction of the speed of light, as they approach the high positive charge of the mercury nucleus, their masses increase (as predicted by Einstein's theory of relativity) and the $6s$ orbital shrinks in size. The closer approach of the $6s$ electrons to the nucleus subjects them to a greater attractive force than that experienced by the ns^2 electrons in Zn and Cd. As a result, for example, the first ionization energy of Hg is greater than that of Zn or Cd.

Uses of the Group 12 Metals

▲ A brass seagoing chronometer made by John Harrison in the eighteenth century.

▶ Iron is one of the few metals that does not form an amalgam. Mercury is generally stored and shipped in iron containers.

About one-third of the zinc produced is used in coating iron to give it corrosion protection (Section 21-6). The product is called *galvanized iron*. Large quantities of Zn are consumed in the manufacture of alloys. For example, about 20% of the production of Zn is used in *brass*, a copper alloy having 20–45% Zn and small quantities of Sn, Pb, and Fe. Brass is a good electrical conductor and is corrosion-resistant. Zinc is also employed in the manufacture of dry cell batteries, in printing (lithography), in the construction industry (roofing materials), and as sacrificial anodes in corrosion protection (Section 21-6).

Although poisonous, cadmium is substituted for zinc as a shiny and protective plating on iron in special applications. It is used in bearing alloys, in low-melting solders, in aluminum solders, and as an additive to impart strength to copper. Another application, based on its neutron-absorbing capacity, is in control rods and shielding for nuclear reactors.

The principal uses of mercury take advantage of its metallic and liquid properties and its high density. It is used in thermometers, barometers, gas-pressure regulators, and electrical relays and switches, and as electrodes, as in the chlor–alkali process (Section 21-8). Mercury vapor is used in fluorescent tubes and street lamps. Mercury alloys, called **amalgams**, can be made with most metals, and some of these amalgams are of commercial importance. A silver dental filling is an amalgam of mercury with an alloy containing about 70% Ag, 26% Sn, 3% Cu, and 1% Zn.

Compounds of the Group 12 Elements

Table 24.9 lists a few important compounds of the group 12 metals and some of their uses. Some of the most interesting compounds are the semiconductors ZnO, CdS, and HgS, which are also the artist's pigments zinc white, cadmium yellow, and vermilion (red), respectively. Like all semiconductor materials, these compounds

TABLE 24.9   Some Important Compounds of the Group 12 Metals

Compound	Uses
ZnO	Reinforcing agent in rubber; pigment; cosmetics; dietary supplement; photoconductors in copying machines
ZnS	Phosphors in X-ray and television screens; pigment; luminous paints
$ZnSO_4$	Rayon manufacture; animal feeds; wood preservative
CdO	Electroplating; batteries; catalyst
CdS	Solar cells; photoconductor in xerography; phosphors; pigment
$CdSO_4$	Electroplating; standard voltaic cells (Weston cell)
HgO	Polishing compounds; dry cells; antifouling paints; fungicide; pigment
$HgCl_2$	Manufacture of Hg compounds; disinfectant; fungicide; insecticide; wood preservative
Hg_2Cl_2	Electrodes; pharmaceuticals; fungicide

have an electronic structure consisting of a valence band and a conduction band (see Section 12-7). When light interacts with these compounds, electrons from the valence band may absorb photons and be excited into the conduction band. The energy of the light absorbed must equal or exceed the energy difference between the bands, called the *band gap*. The characteristic colors of these materials depend on the widths of the band gaps, as described on page 466.

Mercury and Cadmium Poisoning

Accumulations of mercury in the body affect the nervous system and cause brain damage. One form of chronic mercury poisoning, "hatter's disease," afflicted the Mad Hatter in *Alice's Adventures in Wonderland*. The disease was fairly common in the nineteenth century. Mercury compounds were used to convert fur to felt for making hats. Many hat makers of the time worked in hot, cramped spaces and used these compounds without special precautions. The hatters inadvertently ingested or inhaled the toxic mercury compounds while they worked.

One proposed mechanism of mercury poisoning, based on the fact that Hg has a high affinity for sulfur, involves interference with the functioning of sulfur-containing enzymes. Organic mercury compounds are generally more poisonous than inorganic ones and much more toxic than the element itself. An insidious aspect of mercury poisoning is that certain microorganisms have the ability to convert mercury(II) compounds to methylmercury (CH_3Hg^+) compounds, which then concentrate in the food chains of fish and other aquatic life. An early discovery of the environmental hazard of mercury was in Japan in the 1950s. Dozens of cases of mercury poisoning, including over 40 deaths, occurred among residents of the shores of Minamata Bay. Local seafood with up to 20 ppm of mercury was a major component of the victims' diet. The source of contamination was traced to a chemical plant discharging mercury waste into the bay.

In the free state, mercury is most poisonous as a vapor. Levels of mercury that exceed 0.05 mg Hg/m^3 air are considered unsafe. Although we think of mercury as having a low vapor pressure, the concentration of Hg in its saturated vapor far exceeds this limit, and mercury vapor levels sometimes exceed safe limits where mercury is used—as in chlor–alkali plants, thermometer factories, and smelters.

Although zinc is an essential element in trace amounts, cadmium, which so closely resembles zinc, is a poison. One effect of cadmium poisoning is an extremely painful skeletal disorder known as "itai-itai kyo" (Japanese for "ouch-ouch" disease). This disorder was discovered in an area of Japan where effluents from a zinc mine became mixed with irrigation water used in rice fields. Cadmium poisoning was discovered in people who ate the rice. Cadmium poisoning can also cause liver damage, kidney failure, and pulmonary disease. The mechanism of cadmium poisoning may involve substitution in certain enzymes of Cd (a poison) for Zn (an essential element). Concern over cadmium poisoning has increased with an awareness that some cadmium is almost always found in zinc and zinc compounds, materials that have many commercial applications.

24-8   Lanthanides

The elements from cerium ($Z = 58$) through lutetium ($Z = 71$) are inner transition elements; their electron configurations feature the filling of $4f$ orbitals. These elements, together with lanthanum ($Z = 57$), which closely resembles them, are variously called the *lanthanide, lanthanoid,* or *rare earth elements*. The rare earth elements are "rare" only relative to the alkaline earth metals (group 2). Otherwise, they are not particularly rare. Ce, Nd, and La, for example, are more abundant than lead, and Tm is about as abundant as iodine. The lanthanides occur primarily as oxides, and mineral deposits containing them are found in various locations. Large deposits near the California–Nevada border are being developed to provide oxides of the lanthanides for use as phosphors in color monitors and television sets. Another use is in magnets, such as cobalt–samarium (Co_5Sm) magnets used in high-efficiency motors.

Because the differences in electron configuration among the lanthanides are mainly in $4f$ orbitals, and because $4f$ electrons play a minor role in chemical bonding, strong similarities are found among these elements. For example, $E°$ values for the reduction process $M^{3+}(aq) + 3\,e^- \longrightarrow M(s)$ do not show much variation. All fall between -2.38 V (La) and -1.99 V (Eu). The differences in properties that do exist among the lanthanides arise mostly from the lanthanide contraction discussed in Section 24-1. This contraction is best illustrated in the radii of the ions M^{3+}. These radii decrease regularly by about 1 to 2 pm for each unit increase in atomic number, from a radius of 106 pm for La^{3+} to 85 pm for Lu^{3+}.

The lanthanides, which we can represent by the general symbol Ln, are reactive metals that liberate $H_2(g)$ from hot water and from dilute acids by undergoing oxidation to $Ln^{3+}(aq)$. The lanthanides combine with $O_2(g)$, sulfur, the halogens, $N_2(g)$, $H_2(g)$, and carbon in much the same way as expected for metals about as active as the alkaline earths. The pure metals can be prepared by electrolytic reduction of Ln^{3+} in a molten salt.

The most common oxidation state for the lanthanides is $+3$. About half the lanthanides can also be obtained in the oxidation state $+2$; the other half, $+4$. The special stability associated with an electron configuration involving half-filled f orbitals (f^7) may account for some of the observed oxidation states, but the reason for the predominance of the $+3$ oxidation state is less clear. Most of the lanthanide ions are paramagnetic and colored in aqueous solution.

The lanthanide elements are extremely difficult to extract from their natural sources and to separate from one another. All the methods for doing so are based on this principle: Species that are strongly dissimilar can often be completely separated in a one-step process, such as separating $Ag^+(aq)$ and $Cu^{2+}(aq)$ by adding $Cl^-(aq)$—AgCl is insoluble. Species that are very similar can at best be fractionated in a one-step process. That is, the ratio of the concentration of one species to that of another can be altered slightly. To achieve a complete separation may mean repeating the same basic step hundreds or even thousands of times. To achieve the separation of the lanthanides, the methods of fractional crystallization, fractional precipitation, solvent extraction, and ion exchange were brought to their highest level of performance.

Summary

More than half the elements are transition elements. The transition elements are metals, and most are more reactive than hydrogen. Transition metals tend to exist in several different oxidation states in their compounds, and they readily form complex ions (discussed in Chapter 25). Many of the transition metals and their compounds are paramagnetic, and certain of the metals (Fe, Co, and Ni) and their alloys are also ferromagnetic.

Within a group of d-block elements, the members of the second and third transition series resemble one another more than they do the group member in the first transition series. This is a consequence of the phenomenon known as the lanthanide contraction occurring in the sixth period.

The possibility of a variety of oxidation states means that oxidation–reduction reactions are commonly encountered with transition metal compounds. Two common types of oxidizing agents are the dichromates and permanganates. In aqueous solution, dichromate ion is in equilibrium with chromate ion. Although the chromate ion is not a particularly good oxidizing agent, it is a good precipitating agent for a number of metal ions. Most of the oxides and hydroxides of the transition metals are basic if the metal is in one of its lower oxidation states. In higher oxidation states, some of the transition metal oxides and hydroxy compounds are amphoteric, and in the highest oxidation states, a few are acidic (as is CrO_3, for example).

The transition metals are among the most widely used metals. Some, such as titanium and iron, display good structural strength. Others, such as copper and silver, are excellent conductors, and some metals, such as gold, are highly malleable. General and specialized methods of extractive metallurgy are presented in the chapter, with a closer look at the metallurgy of iron and steel, zinc, copper, and titanium.

Magnetically levitated trains, magnetic resonance imaging (MRI) for medical diagnoses, and particle accelerators used in high-energy physics all require high magnetic fields generated by superconducting electromagnets. Superconductors offer no resistance to an electric current, so electricity is conducted with no loss of energy.

If cooled to near absolute zero, all metals become superconducting. Several metals and alloys superconduct even at marginally higher temperatures of 10–15 K. To maintain a superconductor at these extremely low temperatures requires liquid helium (bp, 4 K) as a coolant.

In the mid-1980s, materials made of lanthanum, strontium, copper, and oxygen were found to become superconducting at 30 K. This was a much higher temperature for superconductivity than had been previously achieved. More surprising, the new materials were not metals but *ceramics!* In short order, other types of ceramic superconductors were discovered.

One of these new types was particularly easy to make. When a stoichiometric mixture of yttrium oxide (Y_2O_3), barium carbonate ($BaCO_3$), and copper(II) oxide (CuO) is heated in a stream of $O_2(g)$, a ceramic

▲ The small magnet induces an electric current in the superconductor below. Associated with this current is another magnetic field that opposes the field of the small magnet, causing it to be repelled. The magnet remains suspended above the superconductor as long as the superconducting current is present, and the current persists as long as the temperature of the superconductor is maintained at the boiling point of liquid nitrogen (77 K).

1-2-3 Superconductor model

Integrative Example

Although a number of slightly soluble copper(I) compounds (such as CuCN) can exist in contact with water, it is not possible to prepare a solution with a high concentration of Cu^+ ion.

Show that $Cu^+(aq)$ disproportionates to $Cu^{2+}(aq)$ and $Cu(s)$, and explain why a high $[Cu^+]$ cannot be maintained in aqueous solution.

1. *Describe the disproportionation reaction through half-equations and an overall equation.*

Reduction:	$Cu^+(aq) + e^- \longrightarrow Cu(s)$
Oxidation:	$Cu^+(aq) \longrightarrow Cu^{2+}(aq) + e^-$
Overall:	$2\,Cu^+(aq) \longrightarrow Cu^{2+}(aq) + Cu(s)$

2. *Obtain E°_{cell} for the disproportionation reaction.* We can find E° values for the couples $Cu^+/Cu(s)$ and Cu^{2+}/Cu^+ in Appendix D. Thus,

$$E^\circ_{cell} = E^\circ(\text{reduction}) - E^\circ(\text{oxidation}) =$$
$$E^\circ_{Cu^+/Cu(s)} - E^\circ_{Cu^{2+}/Cu^+} = 0.520\ V\ -\ 0.159\ V = 0.361\ V$$

3. *Obtain a value of K_c for the disproportionation reaction.* We need equation (21.17): $E^\circ_{cell} = (0.0257\ V/n) \times \ln K_c$, where $n = 1$.

$$\ln K_c = \frac{n \times E^\circ_{cell}}{0.0257\ V} = \frac{1 \times 0.361\ V}{0.0257\ V} = 14.0$$

$$K_c = e^{14.0} = 1.2 \times 10^6$$

4. *Interpret the significance of the K_c value.* For the disproportionation reaction,

$$K_c = \frac{[Cu^{2+}]}{[Cu^+]^2} = 1.2 \times 10^6$$

and $[Cu^{2+}] = 1.2 \times 10^6 \times [Cu^+]^2$

Thus, to maintain $[Cu^+] = 1\ M$ in solution, $[Cu^{2+}]$ would have to be more than $1 \times 10^6\ M$—a clear impossibility. As a practical matter, we could not maintain $[Cu^+]$ at much more than 0.002 M, for even this would require that $[Cu^{2+}] \approx 5\ M$.

Key Terms

amalgam (24-7)	**hydrometallurgy** (24-2)	**pyrometallurgy** (24-2)
basic oxygen process (24-3)	**metal carbonyls** (24-5)	**steel** (24-3)
extractive metallurgy (24-2)	**lanthanide contraction** (24-1)	**zone refining** (24-2)
ferromagnetism (24-1)	**pig iron** (24-3)	

is produced with the approximate formula $YBa_2Cu_3O_x$ (where x is slightly less than 7). This so-called YBCO ceramic becomes superconducting at the remarkably high temperature of 92 K. Although a temperature of 92 K is still quite low, it is far above the boiling point of helium. In fact, it is above the boiling point of nitrogen (77 K). Thus, inexpensive liquid nitrogen can be used as the coolant.

Many variations of the basic YBCO formula are possible. Almost any lanthanide element can be substituted for yttrium, and combinations of group 2 elements can be substituted for barium. All these variations yield materials that are superconducting at relatively high temperatures, but of the group, the yttrium compound is superconducting at the highest temperature.

The record high temperature for superconductivity set by the YBCO ceramics was soon eclipsed by another group of ceramics containing bismuth and copper, such as $Bi_2Sr_2CaCu_2O_8$. One of these is superconducting at 110 K, but this record was also short-lived. A ceramic containing thallium and copper, with the approximate formula $TlBa_2Ca_3Cu_4O_y$ (where y is slightly larger than 10), was found to become superconducting at 125 K. Now, the search continues for materials that might become superconducting at room temperature (about 293 K).

Structurally, ceramic superconductors have a feature in common. Copper and oxygen atoms are bonded together in planar sheets. In YBCO superconductors, the Cu–O planes are widely separated. In bismuth superconductors, the Cu–O planes occur in "sandwiches" consisting of two closely spaced sheets separated by a layer of group 2 ions. These sandwiches are separated from one another by several layers of bismuth oxide. In the thallium superconductors, the Cu–O planes are stacked in groups of three—triple-decker sandwiches.

The current theory of superconductivity, developed in the 1950s, explains the superconducting behavior of metals at very low temperatures but not the higher-temperature superconductivity of ceramics. It seems that the electrons in all known superconductors move through the material in pairs—a sort of buddy system that allows the electrons to move without resistance. The mechanism by which electron pairs form in high-temperature superconductors, however, is clearly different from that in low-temperature superconductors. Lack of a suitable theory complicates the search for higher-temperature superconductors. When the mechanism for high-temperature superconductors is better understood, new breakthroughs might be easier to accomplish. Perhaps a room-temperature superconducting material will be possible.

Despite this less-than-complete understanding of high-temperature superconductors, engineers are already building devices that use the new materials. Wires have been made that are superconducting at liquid nitrogen temperatures, and new devices for precise magnetic field measurements using ceramic superconductors are now being produced. Ultimately, ceramic superconductors may find application in low-cost, energy-efficient electric power transmission.

Review Questions

1. In your own words, define the following terms: **(a)** domain; **(b)** flotation; **(c)** leaching; **(d)** amalgam.
2. Briefly describe each of the following ideas, phenomena, or methods: **(a)** lanthanide contraction; **(b)** zone refining; **(c)** basic oxygen process; **(d)** slag formation.
3. Explain the important distinctions between each pair of terms: **(a)** ferromagnetism and paramagnetism; **(b)** roasting and reduction; **(c)** hydrometallurgy and pyrometallurgy; **(d)** chromate and dichromate.
4. Provide an acceptable name for each of the following substances: **(a)** $Sc(OH)_3$; **(b)** Cu_2O; **(c)** $TiCl_4$; **(d)** V_2O_5; **(e)** K_2CrO_4; **(f)** K_2MnO_4.
5. Provide an acceptable formula for each of the following substances.
    (a) chromium(VI) oxide
    (b) iron(II) silicate
    (c) barium dichromate
    (d) copper(I) cyanide
    (e) cobalt(II) chloride hexahydrate
6. Describe the chemical composition of the material called **(a)** pig iron; **(b)** ferromanganese; **(c)** chromite ore; **(d)** brass; **(e)** aqua regia; **(f)** blister copper; **(g)** stainless steel.
7. Complete and balance the following equations. If no reaction occurs, so state.
    (a) $TiCl_4(g) + Na(l) \longrightarrow$
    (b) $Cr_2O_3(s) + Al(s) \longrightarrow$
    (c) $Ag(s) + HCl(aq) \longrightarrow$
    (d) $K_2Cr_2O_7(aq) + KOH(aq) \longrightarrow$
    (e) $MnO_2(s) + C(s) \longrightarrow$
8. Balance the following oxidation–reduction equations.
    (a) $Fe_2S_3(s) + H_2O + O_2(g) \longrightarrow Fe(OH)_3(s) + S(s)$
    (b) $Mn^{2+}(aq) + S_2O_8^{2-}(aq) + H_2O \longrightarrow$
        $MnO_4^-(aq) + SO_4^{2-}(aq) + H^+(aq)$
    (c) $Ag(s) + CN^-(aq) + O_2(g) + H_2O \longrightarrow$
        $[Ag(CN)_2]^-(aq) + OH^-(aq)$
9. By means of a chemical equation, give an example to represent the reaction of **(a)** a transition metal with a nonoxidizing acid; **(b)** a transition metal oxide with NaOH(aq); **(c)** an inner transition metal with HCl(aq).
10. By means of orbital diagrams, write electron configurations for the following transition element atom and ions: **(a)** Ti; **(b)** V^{3+}; **(c)** Cr^{2+}; **(d)** Mn^{4+}; **(e)** Mn^{2+}; **(f)** Fe^{3+}
11. Arrange the following species according to the number of unpaired electrons they contain, starting with the one that has the greatest number: Fe, Sc^{3+}, Ti^{2+}, Mn^{4+}, Cr, Cu^{2+}.
12. Transition elements generally are expected to have which of these properties? Explain.
    (a) low melting points
    (b) high ionization energies
    (c) colored ions in solution
    (d) positive standard electrode (reduction) potentials
    (e) paramagnetic compounds

13. Which of the following ions is diamagnetic: Cr^{2+}, Fe^{3+}, Cu^{2+}, Sc^{3+}?

14. Which of the following elements is *not* expected to display an oxidation state of +6 in any of its compounds: Ti, Cr, Mn, or W? Explain.

15. Which of the following ions in aqueous solution is the best oxidizing agent: Na^+, Zn^{2+}, Ag^+, Cu^{2+}? Explain.

16. Why is +3 the most stable oxidation state for Fe, whereas it is +2 for Co and Ni?

Exercises

Properties of the Transition Elements

17. Describe how the transition elements compare with main-group metals (such as group 2) with respect to oxidation states, formation of complexes, colors of compounds, and magnetic properties.

18. With only minor irregularities, the melting points of the first series of transition metals rise from that of Sc to that of Cr and then fall to that of Zn. Give a plausible explanation for this phenomenon based on atomic structure.

19. Why do the atomic radii vary so much more for two main-group elements that differ by one unit in atomic number than they do for two transitions elements that differ by one unit?

20. The metallic radii of Ni, Pd, and Pt are 125, 138, and 139 pm, respectively. Why is the difference in radius between Pt and Pd so much less than between Pd and Ni?

21. Which of the first transition series elements exhibits the greatest number of different oxidation states in its compounds? Explain.

22. Why is the number of common oxidation states for the elements at the beginning and those at the end of the first transition series less than for elements in the middle of the series?

23. As a group, the lanthanides are more reactive metals than are those in the first transition series. How do you account for this difference?

24. The maximum difference in standard reduction potential, $E^\circ_{M^{2+}/M(s)}$, among members of the first transition series is about 2.4 V. For the lanthanides, the maximum difference in $E^\circ_{M^{3+}/M(s)}$ is only about 0.4 V. How do you account for this fact?

Reactions of Transition Metals and Their Compounds

25. Write balanced chemical equations for the following reactions described in the chapter.
 (a) the reaction of $Sc(OH)_3(s)$ with HCl(aq)
 (b) oxidation of Fe^{2+}(aq) by MnO_4^-(aq) in basic solution to give Fe^{3+}(aq) and MnO_2(s)
 (c) the reaction of TiO_2(s) with molten KOH to form K_2TiO_3
 (d) oxidation of Cu(s) to Cu^{2+}(aq) with H_2SO_4(concd aq) to form SO_2(g)

26. Write balanced equations for the following reactions described in the chapter.
 (a) Sc(l) is produced by the electrolysis of Sc_2O_3 dissolved in Na_3ScF_6(l).
 (b) Cr(s) reacts with HCl(aq) to produce a blue solution containing Cr^{2+}(aq).
 (c) Cr^{2+}(aq) is readily oxidized by O_2(g) to Cr^{3+}(aq).
 (d) Ag(s) reacts with concentrated HNO_3(aq), and NO_2(g) is evolved.

27. Suggest a series of reactions, using common chemicals, by which each of the following syntheses can be performed.
 (a) $Fe(OH)_3$(s) from FeS(s)
 (b) $BaCrO_4$(s) from $BaCO_3$(s) and $K_2Cr_2O_7$(aq)

28. Suggest a series of reactions, using common chemicals, by which each of the following syntheses can be performed.
 (a) $Cu(OH)_2$(s) from CuO(s)
 (b) $CrCl_3$(aq) from $(NH_4)_2Cr_2O_7$(s)

Extractive Metallurgy

29. One of the simplest metals to extract from its ores is mercury. Mercury vapor is produced by roasting cinnabar ore (HgS) in air. Alternatives to this simple roasting, designed to reduce or eliminate SO_2 emissions, is to roast the ore in the presence of a second substance. For example, when cinnabar is roasted with quicklime, the products are mercury vapor and calcium sulfide and calcium sulfate. Write equations for the two reactions described here.

30. According to Figure 24-7, ΔG° decreases with temperature for the reaction $2 C(s) + O_2(g) \longrightarrow 2 CO(g)$. How would you expect ΔG° to vary with temperature for the following reactions?
 (a) $C(s) + O_2(g) \longrightarrow CO_2(g)$

 (b) $2 CO(g) + O_2(g) \longrightarrow 2 CO_2(g)$

31. Calcium will reduce MgO(s) to Mg at all temperatures from 0 to 2000 °C. Use this fact, together with the melting point (839 °C) and boiling point (1484 °C) of calcium, to sketch a plausible graph of ΔG° as a function of temperature for the reaction $2 Ca + O_2(g) \longrightarrow 2 CaO(s)$.

32. One method of obtaining chromium metal from chromite ore is described as follows. Following reaction (24.16), sodium chromate is reduced to chromium(III) oxide by carbon. Then the chromium(III) oxide is reduced to chromium metal by silicon. Write plausible equations to describe these two reactions.

Oxidation–Reduction

33. Write plausible half-equations to represent each of the following in acidic solution.
  (a) VO^{2+}(aq) as an oxidizing agent
  (b) Cr^{2+}(aq) as a reducing agent

34. Write plausible half-equations to represent each of the following in basic solution.
  (a) oxidation of $Fe(OH)_3$(s) to FeO_4^{2-}
  (b) reduction of $[Ag(CN)_2]^-$ to silver metal

35. Use electrode potential data from this chapter or Appendix D to predict whether each of the following reactions will occur to any significant extent under standard-state conditions.
  (a) $2\ VO_2^+ + 6\ Br^- + 8\ H^+ \longrightarrow$
  $$2\ V^{2+} + 3\ Br_2(l) + 4\ H_2O$$
  (b) $VO_2^+ + Fe^{2+} + 2\ H^+ \longrightarrow VO^{2+} + Fe^{3+} + H_2O$
  (c) MnO_2(s) $+ H_2O_2 + 2\ H^+ \longrightarrow$
  $$Mn^{2+} + 2\ H_2O + O_2(g)$$

36. You are given these three reducing agents: Zn(s), Sn^{2+}(aq), and I^-(aq). Use data from Appendix D to determine which of them can, under standard-state conditions in acidic solution, reduce
  (a) $Cr_2O_7^{2-}$(aq) to Cr^{3+}(aq)
  (b) Cr^{3+}(aq) to Cr^{2+}(aq)
  (c) SO_4^{2-}(aq) to SO_2(g)

37. Refer to Example 24-2. Select a reducing agent (from Table 21.1 or Appendix D) that will reduce VO^{2+} to V^{3+} and no further in acidic solution.

38. The electrode potential diagram for manganese in acidic solutions in Figure 24-13 does not include a value of $E°$ for the reduction of MnO_4^- to Mn^{2+}. Use other data in the figure to establish this $E°$, and compare your result with the value found in Table 21.1.

39. Use data from the text to construct a standard electrode potential diagram relating the following chromium species in acidic solution.

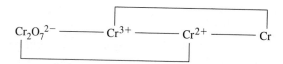

40. Use data from the text to construct a standard electrode potential diagram relating the following vanadium species in acidic solution.

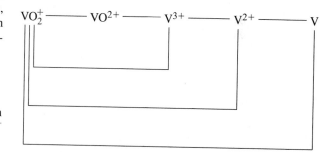

Chromium and Chromium Compounds

41. When a soluble lead compound is added to a solution containing primarily *orange* dichromate ion, *yellow* lead chromate precipitates. Describe the equilibria involved.

42. When *yellow* $BaCrO_4$ is dissolved in HCl(aq), a *green* solution is obtained. Write a chemical equation to account for the color change.

43. When Zn(s) is added to $K_2Cr_2O_7$ dissolved in HCl(aq), the color of the solution changes from orange to green, then to blue, and, over a period of time, back to green. Write equations for this series of reactions.

44. If CO_2(g) under pressure is passed into Na_2CrO_4(aq), $Na_2Cr_2O_7$(aq) is formed. What is the function of the CO_2(g)? Write a plausible equation for the net reaction.

45. Use equation (24.19) to determine $[Cr_2O_7^{2-}]$ in a solution that has $[CrO_4^{2-}] = 0.20$ M and a pH of **(a)** 6.62 and **(b)** 8.85.

46. If a solution is prepared by dissolving 1.505 g Na_2CrO_4 in 345 mL of a buffer solution with pH = 7.55, what will be $[CrO_4^{2-}]$ and $[Cr_2O_7^{2-}]$?

47. How many grams of chromium would be deposited on an object in a chrome-plating bath (see page 966) after 1.00 h at a current of 3.4 A?

48. How long would an electric current of 3.5 A have to pass through a chrome-plating bath (see page 966) to produce a chromium deposit 0.0010 mm thick on an object with a surface area of 0.375 m^2? (The density of Cr is 7.14 g/cm^3.)

49. Why is it reasonable to expect the chemistry of dichromate ion to involve mainly oxidation–reduction reactions and that of chromate ion to involve mainly precipitation reactions?

50. What products are obtained when Mg^{2+}(aq) and Cr^{3+}(aq) are each treated with a limited amount of NaOH(aq)? With an excess of NaOH(aq)? Why are the results different in these two cases?

The Iron Triad

51. Will reaction (24.25) still be spontaneous in the forward direction in a solution containing equal concentrations of Fe^{2+} and Fe^{3+}, a pH of 3.25, and under an $O_2(g)$ partial pressure of 0.20 atm?

52. Based on the description of the nickel–cadmium cell on page 971, and with appropriate data from Appendix D, estimate $E°$ for the reduction of $NiO(OH)$ to $Ni(OH)_2$.

53. Write a net ionic equation to represent the precipitation of Prussian blue, described on page 972.

54. The reaction to form Turnbull's blue (page 972) appears to occur in two stages. First, $Fe^{2+}(aq)$ is oxidized to $Fe^{3+}(aq)$ and ferricyanide ion is reduced to ferrocyanide ion. Then, the $Fe^{3+}(aq)$ and ferrocyanide ion combine. Write equations for these reactions.

Group 11 Metals

55. Write plausible equations for the following reactions occurring in the hydrometallurgy of the coinage metals.
   (a) Copper is precipitated from a solution of copper(II) sulfate by treatment with $H_2(g)$.
   (b) Gold is precipitated from a solution of Au^+ by adding iron(II) sulfate.
   (c) Copper(II) chloride solution is reduced to copper(I) chloride when treated with $SO_2(g)$ in acidic solution.

56. In the metallurgical extraction of silver and gold, an alloy of the two metals is often obtained. The alloy can be separated into Ag and Au either with concentrated HNO_3 or boiling concentrated H_2SO_4, in a process called *parting*. Write chemical equations to show how these separations work.

57. Use the result of the Integrative Example to determine whether a solution can be prepared with $[Cu^+]$ equal to **(a)** 0.20 M; **(b)** 1.0×10^{-10} M.

58. Show that the corrosion reaction in which Cu is converted to its basic carbonate (reaction 24.27) can be thought of in terms of a combination of oxidation–reduction, acid–base, and precipitation reactions.

Group 12 Metals

59. Use data from Table 24.8 to determine $E°$ for the reduction of Hg^{2+} to Hg_2^{2+} in aqueous solution.

60. At 400 °C, $\Delta G° = -25$ kJ for the reaction $2\,Hg(l) + O_2(g) \longrightarrow 2\,HgO(s)$. If a sample of $HgO(s)$ is heated to 400 °C, what will be the equilibrium partial pressure of $O_2(g)$?

61. Use Figure 24-7 to estimate for the reaction $ZnO(s) + C(s) \rightleftharpoons Zn(l) + CO(g)$, at about 800 °C, **(a)** a value of K_p and **(b)** the equilibrium pressure of $CO(g)$.

62. The vapor pressure of $Hg(l)$ as a function of temperature is $\log P(\text{mmHg}) = (-0.05223a/T) + b$, where $a = 61{,}960$ and $b = 8.118$; T is the Kelvin temperature. Show that at 25 °C, the concentration of $Hg(g)$ in equilibrium with $Hg(l)$ greatly exceeds the maximum permissible level of 0.05 mg Hg/m^3 air.

63. In ZnO, the band gap between the valence and conduction bands is 290 kJ/mol, and in CdS it is 250 kJ/mol. Show that CdS absorbs some visible light but ZnO does not. Explain the observed colors: ZnO is white and CdS is yellow.

64. CdS is yellow, HgS is red, and CdSe is black. Which of these materials has the largest band gap? the smallest? How does the band gap relate to the observed color?

Integrative and Advanced Exercises

65. Although Au reacts with and dissolves in aqua regia (3 parts HCl + 1 part HNO_3), Ag does not dissolve. What is (are) the likely reason(s) for this difference?

66. The text mentions that scandium metal is obtained from its molten chloride by electrolysis and that titanium is obtained from its chloride by reduction with magnesium. Why are these metals not obtained by the reduction of their oxides with carbon (coke), as are metals such as zinc and iron?

67. The text notes that in small quantities, zinc is an essential element (though it is toxic in higher concentrations). Tin is considered to be a toxic metal. Can you think of reasons why, for food storage, tinplate instead of galvanized iron is used in cans?

68. In an atmosphere polluted with industrial smog, Cu corrodes to a basic sulfate, $Cu_2(OH)_2SO_4$. Propose a series of chemical reactions to describe this corrosion.

69. What formulas would you expect for the metal carbonyls of **(a)** molybdenum, **(b)** osmium; **(c)** rhenium? Note that the simple carbonyls shown in Figure 24-14 have one metal atom per molecule. Some metal carbonyls are *binuclear*; that is, they have two metal atoms bonded together in the carbonyl structure. Also, **(d)** explain why iron and nickel carbonyls are liquids at room temperature, whereas that of cobalt is a solid, and **(e)** describe the probable nature of the bonding in the compound $Na[V(CO)_6]$.

70. For the straight-line graphs in Figure 24-7, explain why (a) breaks occur at the melting points and boiling points of the metals; (b) the slopes of the lines become more positive at these breaks; (c) the break at the boiling point is sharper than at the melting point.

71. Attempts to make CuI_2 by the reaction of $Cu^{2+}(aq)$ and $I^-(aq)$ produce $CuI(s)$ and $I_3^-(aq)$ instead. Without performing detailed calculations, show why this reaction should occur.

$$2\,Cu^{2+}(aq) + 5\,I^-(aq) \longrightarrow 2\,CuI(s) + I_3^-(aq)$$

72. Without performing detailed calculations, show that significant disproportionation of AuCl occurs if you attempt to make a saturated aqueous solution. Use data from Table 24.7 and $K_{sp}\,(AuCl) = 2.0 \times 10^{-13}$.

73. In acidic solution, silver(II) oxide first dissolves to produce $Ag^{2+}(aq)$. This is followed by the oxidation of H_2O to $O_2(g)$ and the reduction of Ag^{2+} to Ag^+.
    (a) Write equations for the dissolution and oxidation–reduction reactions.
    (b) Show that the oxidation–reduction reaction is indeed spontaneous.

74. Equation (24.18), which represents the chromate–dichromate equilibrium, is actually the sum of two equilibrium expressions. The first is an acid–base reaction, $H^+ + CrO_4^{2-} \rightleftharpoons HCrO_4^-$. The second reaction involves elimination of a water molecule between two $HCrO_4^-$ ions (a dehydration reaction), $2\,HCrO_4^- \rightleftharpoons Cr_2O_7^{2-} + H_2O$. If the ionization constant, K_a, for $HCrO_4^-$ is 3.2×10^{-7}, what is the value of K for the dehydration reaction?

75. Show that under the following conditions, $Ba^{2+}(aq)$ can be separated from $Sr^{2+}(aq)$ and $Ca^{2+}(aq)$ by precipitating $BaCrO_4(s)$, with the other ions remaining in solution: $[Ba^{2+}] = [Sr^{2+}] = [Ca^{2+}] = 0.10\,M$; $[HC_2H_3O_2] = [C_2H_3O_2^-] = 1.0\,M; [Cr_2O_7^{2-}] = 0.0010\,M$. $K_{sp}\,(BaCrO_4) = 1.2 \times 10^{-10}$ and $K_{sp}\,(SrCrO_4) = 2.2 \times 10^{-5}$. Also, use data from this and previous chapters, as necessary.

76. A 0.589-g sample of pyrolusite ore (impure MnO_2) is treated with 1.651 g of oxalic acid ($H_2C_2O_4 \cdot 2H_2O$) in an acidic medium (reaction 1). Following this, the excess oxalic acid is titrated with 30.06 mL of 0.1000 M $KMnO_4$ (reaction 2). What is the mass percent of MnO_2 in the pyrolusite? The following equations are neither complete nor balanced.
    (1) $H_2C_2O_4(aq) + MnO_2(s) \longrightarrow Mn^{2+}(aq) + CO_2(g)$
    (2) $H_2C_2O_4(aq) + MnO_4^-(aq) \longrightarrow Mn^{2+}(aq) + CO_2(g)$

77. Both $Cr_2O_7^{2-}(aq)$ and $MnO_4^-(aq)$ can be used to titrate $Fe^{2+}(aq)$ to $Fe^{3+}(aq)$. Suppose you have available as titrants two solutions: 0.1000 M $Cr_2O_7^{2-}(aq)$ and 0.1000 M $MnO_4^-(aq)$.

    (a) For which solution would the greater volume of titrant be required for the titration of a particular sample of $Fe^{2+}(aq)$? Explain.
    (b) How many mL of 0.1000 M $MnO_4^-(aq)$ would be required for a titration if the same titration requires 24.50 mL of 0.1000 M $Cr_2O_7^{2-}(aq)$?

78. The only important compounds of Ag(II) are AgF_2 and AgO. Why would you expect these two compounds to be stable, but not other silver(II)compounds such as $AgCl_2$, $AgBr_2$, and AgS.

79. A certain steel is to be analyzed for Cr and Mn. By suitable treatment, the Cr in the steel is oxidized to $Cr_2O_7^{2-}(aq)$ and the Mn to $MnO_4^-(aq)$. A 10.000-g sample of steel is used to produce 250.0 mL of a solution containing $Cr_2O_7^{2-}(aq)$ and $MnO_4^-(aq)$. A 10.00-mL portion of this solution is added to $BaCl_2(aq)$, and by proper adjustment of the pH, the chromium is completely precipitated as $BaCrO_4(s)$; 0.549 g is obtained. A second 10.00-mL portion of the solution requires exactly 15.95 mL of 0.0750 M $Fe^{2+}(aq)$ for its titration in acidic solution. Calculate the % Cr and % Mn in the steel sample.
    [*Hint:* In the titration $MnO_4^-(aq)$ is reduced to $Mn^{2+}(aq)$ and $Cr_2O_7^{2-}(aq)$ is reduced to $Cr^{3+}(aq)$; the $Fe^{2+}(aq)$ is oxidized to $Fe^{3+}(aq)$.]

80. Nickel can be determined as nickel dimethylglyoximate, a brilliant scarlet precipitate that has the composition 20.31%Ni, 33.26% C, 4.88% H, 22.15% O, and 19.39% N. A 15.020-g steel sample is dissolved in concentrated HCl(aq). The solution obtained is suitably treated to remove interfering ions, to establish the proper pH, and to obtain a final solution volume of 250.0 mL. A 10.00-mL sample of this solution is then treated with dimethylglyoxime. The mass of purified, dry nickel dimethylglyoximate obtained is 0.104 g.
    (a) What is the empirical formula of nickel dimethylglyoximate?
    (b) What is the mass percent nickel in the steel sample?

81. A solution is believed to contain one or more of the following ions: Cr^{3+}, Zn^{2+}, Fe^{3+}, Ni^{2+}. When the solution is treated with excess NaOH(aq), a precipitate forms. The solution in contact with the precipitate is colorless. The precipitate is dissolved in HCl(aq), and the resulting solution is treated with $NH_3(aq)$. No precipitation occurs. Based solely on these observations, what conclusions can you draw about the ions present in the original solution? That is, which ion or ions are likely present, which are most likely not present, and about which can we not be certain?
    (*Hint:* Refer to Appendix D for solubility product and complex-ion formation data.)

Feature Problems

82. As a continuation of Feature Problem 99 of Chapter 20 and the discussion on page 958, consider the three graphs of $\Delta G°$ as a function of temperature shown in the following figure.

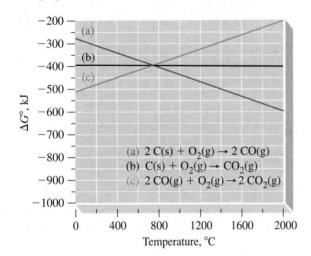

(a) $2\,C(s) + O_2(g) \rightarrow 2\,CO(g)$
(b) $C(s) + O_2(g) \rightarrow CO_2(g)$
(c) $2\,CO(g) + O_2(g) \rightarrow 2\,CO_2(g)$

(a) Explain the shapes of the three graphs. Specifically, why is one line essentially parallel to the temperature axis, why does one have a positive slope, and why does one have a negative slope?
(b) Table 24.2 lists as an additional blast furnace reaction, $C(s) + CO_2(g) \longrightarrow 2\,CO(g)$. Determine how $\Delta G°$ for this reaction is related to the three reactions shown in the figure, and plot $\Delta G°$ for this reaction as a function of temperature. If an equilibrium is established in this reaction at 1000 °C and the partial pressure of $CO_2(g)$ is 0.25 atm, what should be the equilibrium partial pressure of $CO(g)$?

83. Several transition metal ions are found in cation group 3 of the qualitative analysis scheme outlined in Figure 19-7. At one point in the separation and testing of this group, a solution containing Fe^{3+}, Co^{2+}, Ni^{2+}, Al^{3+}, Cr^{3+}, and Zn^{2+} is treated with an excess of NaOH(aq), together with $H_2O_2(aq)$. (1) The excess NaOH(aq) causes *three* of the cations to precipitate as hydroxides and *three* to form hydroxo complex ions. (2) In the presence of $H_2O_2(aq)$, the cation in one of the insoluble hydroxides is oxidized from the +2 to the +3 oxidation state, and one of the hydroxo complex ions is also oxidized. (3) The three insoluble hydroxides are found as a dark precipitate. (4) The solution above the precipitate has a yellow color. (5) The dark precipitate from (3) reacts with HCl(aq), and all the cations return to solution; one of the cations is reduced from the +3 to the +2 oxidation state. (6) The solution from (5) is treated with $6\,M\,NH_3(aq)$, and a precipitate containing one of the cations forms.
(a) Write equations for the reactions referred to in item (1).
(b) Write an equation for the most likely reaction in which a hydroxide precipitate is oxidized in item (2).
(c) What is the ion responsible for the yellow color of the solution in item (4)? Write an equation for its formation.
(d) Write equations for the dissolution of the precipitate and the reduction of the cation in item (5).
(e) Write an equation for the precipitate formation in item (6).
(*Hint:* You may need solubility product and complex-ion formation data from Appendix D, together with descriptive information from this chapter and from elsewhere in the text.)

eMedia Exercises

84. Many transition metals exhibit catalytic properties. The activity of a transition metal as a catalyst is illustrated in the **Surface Reaction—Hydrogenation** animation *(eChapter 24-1)*. **(a)** What role does the metal play in the depicted reaction? **(b)** In what sense is the metal acting as a catalyst? **(c)** Why might transition metals exhibit greater catalytic activity than other elements?

85. A method of purifying metals is shown in the **Zone Refinement** animation *(eChapter 24-2)*. **(a)** What type of impurities could be removed by this process? **(b)** Would this process be useful in the separation of metals making up an alloy? Why or why not?

86. In the **Periodic Trends: First-Row Transition Metals** activity *(eChapter 24-4)*, which of the properties can be closely correlated to changes in the electronic configuration? Explain.

25 Complex Ions and Coordination Compounds

Contents

Turquoise is a mineral of copper, $CuAl_6(PO_4)_4(OH)_8 \cdot 4H_2O$. The distinctive color of this gemstone and many others is a consequence of the nature of metal–ligand bonding in complex ions, a central topic of this chapter.

In Chapter 24, we discussed several situations involving a succession of color changes, and we attributed them to changes in oxidation state. The color changes discussed in this chapter, for the most part, are not caused by oxidation–reduction reactions. Instead, changes in color are observed with changes in the groups (ligands) bound to a metal center, even though the oxidation state of the metal remains unchanged. To explain this observation, we need to explore more fully the nature of complex ions and coordination compounds, a subject that was briefly introduced in Chapter 19. One topic we will consider is the geometric structures of complex ions. In doing so, we will discover new possibilities for isomerism: the existence of compounds having identical compositions but different structures and properties. We will also examine the nature of the bonding between ligands and the metal centers to which they are attached. It is through an understanding of bonding in complex ions that we can gain some insight into the origin of their colors.

25-1 Werner's Theory of Coordination Compounds: An Overview

Prussian blue (page 972), accidentally discovered early in the eighteenth century, was perhaps the first known coordination compound—the type we explore in this chapter. However, nearly a century passed before the uniqueness of these compounds came to be appreciated. In 1798, B. M. Tassaert obtained yellow crystals of a compound having the formula $CoCl_3 \cdot 6NH_3$ from a mixture of $CoCl_3$ and NH_3(aq). What seemed unusual was that both $CoCl_3$ and NH_3 are stable compounds capable of independent existence, yet they combine to form still another stable compound. Such compounds made up of two simpler compounds came to be called **coordination compounds**.

In 1851, another coordination compound of $CoCl_3$ and NH_3 was discovered. This one had the formula $CoCl_3 \cdot 5NH_3$ and formed purple crystals. The two compounds are shown in Figure 25-1. Their formulas follow.

$CoCl_3 \cdot 6NH_3$ $CoCl_3 \cdot 5NH_3$

(yellow)   (purple)

(a)   (b)

The mystery of coordination compounds deepened as more were discovered and studied. For example, when treated with $AgNO_3$(aq), compound (a) formed *three* moles of AgCl(s), as expected, but compound (b) formed only *two* moles of AgCl(s).

Inorganic coordination chemistry was a hot field of research in the last half of the nineteenth century, and all the pieces started to fall into place with the work of the Swiss chemist Alfred Werner. Werner's theory of coordination compounds explained the reactions of compounds (a) and (b) with $AgNO_3$(aq) by considering that, in aqueous solutions, these two compounds ionize in the following way:

(a) $[Co(NH_3)_6]Cl_3(s) \xrightarrow{H_2O} [Co(NH_3)_6]^{3+}(aq) + 3\ Cl^-(aq)$

(b) $[CoCl(NH_3)_5]Cl_2(s) \xrightarrow{H_2O} [CoCl(NH_3)_5]^{2+}(aq) + 2\ Cl^-(aq)$

Thus, compound (a) produces the three moles of Cl^- per mole of compound necessary to precipitate three moles of AgCl(s), while compound (b) produces only two moles of Cl^-. Werner's proposal of this ionization scheme was based on extensive studies of the electrical conductivity of coordination compounds. Compound (a) is a better conductor than compound (b), consistent with producing four ions per formula unit compared to three for compound (b). $CoCl_3 \cdot 4NH_3$ is a still

▲ **Alfred Werner (1866–1919)**
Werner's success in explaining coordination compounds came in large part through his application of new ideas: the theory of electrolytic dissociation and principles of structural chemistry.

▲ FIGURE 25-1   **Two coordination compounds**
The compound on the left is $[Co(NH_3)_6]Cl_3$. The compound on the right is $[CoCl(NH_3)_5]Cl_2$.

poorer conductor, corresponding to the formula $[CoCl_2(NH_3)_4]Cl$. $CoCl_3 \cdot 3NH_3$ is a *nonelectrolyte*, corresponding to the formula $[CoCl_3(NH_3)_3]$.

The heart of Werner's theory, proposed in 1893, was that certain metal atoms, primarily those of transition metals, have two types of valence or bonding capacity. One, the *primary* valence, is based on the number of electrons the atom loses in forming the metal ion. A *secondary* valence is responsible for the bonding of other groups, called **ligands**, to the central metal ion.

In modern usage, the term **complex** describes any species involving coordination of ligands to a metal center. The metal center can be an atom or an ion, and the complex can be a cation, an anion, or a neutral molecule. In a chemical formula, a complex—a metal center and attached ligands—is set off by square brackets, [ ]. Compounds that are complexes or contain complex ions are known as **coordination compounds**.

$[Co(NH_3)_6]^{3+}$	$[CoCl_4(NH_3)_2]^-$	$[CoCl_3(NH_3)_3]$	$K_4[Fe(CN)_6]$
complex cation	complex anion	neutral complex	coordination compound

The **coordination number** of a complex is the number of points around the metal center at which bonds to ligands can form. Coordination numbers ranging from 2 to 12 have been observed, although 6 is by far the most common number, followed by 4. Coordination number 2 is limited mostly to complexes of Cu(I), Ag(I), and Au(I). Coordination numbers greater than 6 are not often found in members of the first transition series but are more common in those of the second and third series. Stable complexes with coordination numbers 3 and 5 are rare. The coordination number observed in a complex depends on a number of factors, such as the ratio of the radius of the central metal atom or ion to the radii of the attached ligands.

Coordination numbers of some common ions are listed in Table 25.1. The four most commonly observed geometric shapes of complex ions are shown in Figure 25-2. One practical use of the coordination number is to assist in writing and interpreting formulas of complexes, as illustrated in Example 25-1.

TABLE 25.1 Some Common Coordination Numbers of Metal Ions

Cu^+	2, 4		
Ag^+	2		
Au^+	2, 4	Al^{3+}	4, 6
		Sc^{3+}	6
		Cr^{3+}	6
Fe^{2+}	6	Fe^{3+}	6
Co^{2+}	4, 6	Co^{3+}	6
Ni^{2+}	4, 6	Au^{3+}	4
Cu^{2+}	4, 6	Pt^{4+}	6
Zn^{2+}	4		
Pt^{2+}	4		

Coordination Number activity

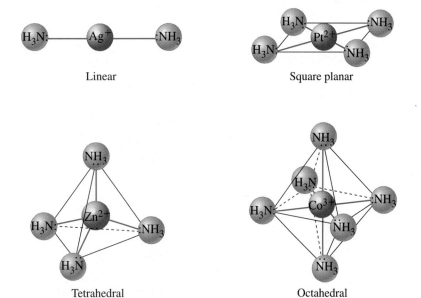

Linear

Square planar

Tetrahedral

Octahedral

▲ **FIGURE 25-2 Structures of some complex ions**

Attachment of the NH_3 molecules occurs through the lone-pair electrons on the N atoms.

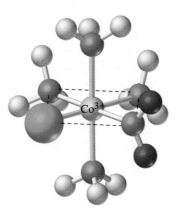

▲ The complex ion
[CoCl(NO$_2$)(NH$_3$)$_4$]$^+$

EXAMPLE 25-1

Relating the Formula of a Complex to the Coordination Number and Oxidation State of the Central Metal. What are the coordination number and oxidation state of Co in the complex ion [CoCl(NO$_2$)(NH$_3$)$_4$]$^+$?

Solution

The complex ion has as ligands *one* Cl$^-$ ion, *one* NO$_2^-$ ion, and *four* NH$_3$ molecules. The coordination number is 6. Of these six ligands, *two* carry a charge of 1$-$ each (the Cl$^-$ and NO$_2^-$ ions) and four are *neutral* (the NH$_3$ molecules). The total contribution of the anions to the net charge on the complex ion is 2$-$. Because the net charge on the complex ion is 1+, the oxidation state of the central cobalt ion is +3. Diagrammatically, we can write

Oxidation state = x Charge of 1$-$ on Cl$^-$

Charge of 1$-$ on NO$_2^-$ } Total negative charge: 2$-$

[CoCl(NO$_2$)(NH$_3$)$_4$]$^+$

Coordination number = 6   Net charge on complex ion

$$x - 2 = +1$$
$$x = +3$$

Practice Example A: What are the coordination number and oxidation state of nickel in the ion [Ni(CN)$_4$I]$^{3-}$?

Practice Example B: Write the formula of a complex with cyanide ion ligands, an iron ion with an oxidation state of +3, and a coordination number of 6.

25-2   Ligands

A common feature shared by the ligands in coordination complexes is the ability to donate electron pairs to central metal atoms or ions. Ligands are *Lewis bases*. In accepting electron pairs, central metal atoms or ions act as *Lewis acids*. A ligand that uses one pair of electrons to form one point of attachment to the central metal atom or ion is called a **monodentate** ligand. Some examples of monodentate ligands are monatomic anions such as the halide ions, polyatomic anions such as hydroxide ion, simple molecules such as ammonia (called *ammine* when it is a ligand), and more complex molecules such as methylamine, CH$_3$NH$_2$ (Table 25.2).

Ligand models

TABLE 25.2   Some Common Monodentate Ligands

Formula	Name as Ligand	Formula	Name as Ligand	Formula	Name as Ligand
Neutral molecules		**Anions**		**Anions**	
H$_2$O	Aqua	F$^-$	Fluoro	SO$_4^{2-}$	Sulfato
NH$_3$	Ammine	Cl$^-$	Chloro	S$_2$O$_3^{2-}$	Thiosulfato
CO	Carbonyl	Br$^-$	Bromo	NO$_2^-$	Nitrito-*N*-[a]
NO	Nitrosyl	I$^-$	Iodo	ONO$^-$	Nitrito-*O*-[a]
CH$_3$NH$_2$	Methylamine	O^{2-}	Oxo	SCN$^-$	Thiocyanato-*S*-[b]
C$_5$H$_5$N	Pyridine	OH$^-$	Hydroxo	NCS$^-$	Thiocyanato-*N*-[b]
		CN$^-$	Cyano		

[a] If the nitrite ion is attached through the N atom (—NO$_2$), the designation *nitrito-N-* is used; if attached through an O atom (—ONO), *nitrito-O-*.
[b] If the thiocyanate ion is attached through the S atom (—SCN), the name *thiocyanato -S-* is used; if attachment is through the N atom (—NCS), *thiocyanato-N-*.

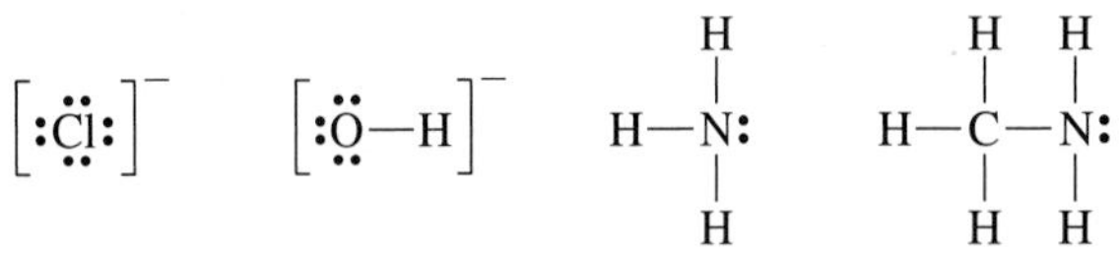

$$\left[:\ddot{\underset{\cdot\cdot}{Cl}}:\right]^{-} \qquad \left[:\ddot{\underset{\cdot\cdot}{O}}-H\right]^{-} \qquad H-\overset{\displaystyle H}{\underset{\displaystyle H}{N}}: \qquad H-\overset{\displaystyle H}{\underset{\displaystyle H}{C}}-\overset{\displaystyle H}{\underset{\displaystyle H}{N}}:$$

Ligand name: Chloro   Hydroxo   Ammine   Methylamine

KEEP IN MIND ▶

that the lone pairs of electrons of a polydentate ligand must be far enough apart to attach to the metal center at two or more points; the donated pairs of electrons must be on different atoms. Thus, a Cl⁻ ion, despite its *four* lone pairs of electrons, is always a *mono*dentate ligand.

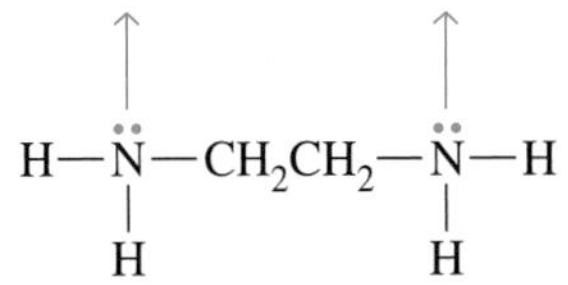

Some ligands are capable of donating more than a single electron pair from *different* atoms in the ligand and to *different* sites in the geometric structure of a complex. These are called **polydentate** ligands. The molecule *ethylenediamine (en)* can donate two electron pairs, one from each N atom. Since en attaches to the metal center at two points, it is called a **bidentate** ligand.

$$H-\overset{\displaystyle\uparrow}{\underset{\displaystyle H}{\overset{\cdot\cdot}{N}}}-CH_2CH_2-\overset{\displaystyle\uparrow}{\underset{\displaystyle H}{\overset{\cdot\cdot}{N}}}-H$$

Three common polydentate ligands are shown in Table 25.3.

TABLE 25.3   Some Common Polydentate Ligands (Chelating Agents)

Abbrevation	Name	Formula
en	Ethylenediamine	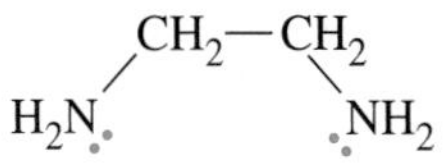 $\underset{\displaystyle H_2\overset{\cdot}{\underset{\cdot}{N}}}{}\overset{\displaystyle CH_2-CH_2}{}\underset{\displaystyle \overset{\cdot}{\underset{\cdot}{N}}H_2}{}$
ox²⁻	Oxalato	$\underset{\displaystyle ^{-}\overset{\cdot\cdot}{\underset{\cdot\cdot}{O}} \qquad \overset{\cdot\cdot}{\underset{\cdot\cdot}{O}}^{-}}{\overset{\displaystyle O \qquad\quad O}{\overset{\displaystyle \parallel \qquad\quad \parallel}{C-C}}}$
EDTA⁴⁻	Ethylenediaminetetraacetato	

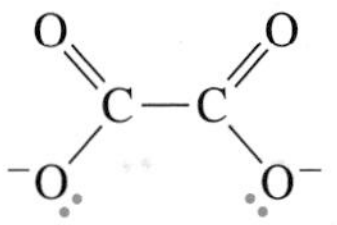

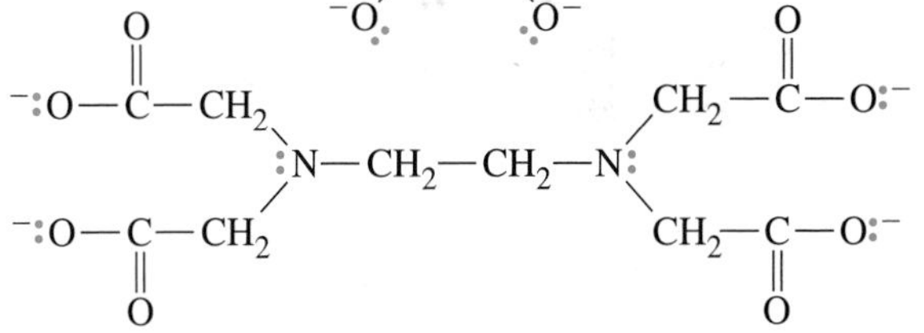

$$^{-}:O-\overset{\displaystyle O}{\overset{\parallel}{C}}-CH_2 \qquad\qquad\qquad CH_2-\overset{\displaystyle O}{\overset{\parallel}{C}}-O:^{-}$$
$$:\underset{}{N}-CH_2-CH_2-\underset{}{N}:$$
$$^{-}:O-\overset{}{\underset{\displaystyle O}{\underset{\parallel}{C}}}-CH_2 \qquad\qquad\qquad CH_2-\overset{}{\underset{\displaystyle O}{\underset{\parallel}{C}}}-O:^{-}$$

ᵃ Oxalic acid is a diprotic acid denoted H_2ox. It is ox²⁻ that binds as a bidentate ligand.
ᵇ Ethylenediaminetetraacetic acid, a tetraprotic acid, is denoted H_4EDTA.

Figure 25-3 represents the attachment of two ethylenediamine (en) ligands to a Pt^{2+} ion. Here is how we can establish that each ligand is attached to two positions in the coordination sphere around the Pt^{2+} ion.

• Because Pt^{2+} ion exhibits a coordination number of 4 with monodentate ligands, and because $[Pt(en)_2]^{2+}$ is unable to attach additional ligands such as NH_3, H_2O, or Cl^-, we conclude that each en group must be attached at two points.

• The en ligands in the complex ion exhibit no further basic properties. They cannot accept protons from water to produce OH^-, as they would if they had an available lone pair of electrons. Both $-NH_2$ groups of each en molecule must be tied up in the complex ion.

Note the two five-member rings (pentagons) outlined in Figure 25-3b. They consist of Pt, N, and C atoms. When the bonding of a polydentate ligand to a metal ion

Are You Wondering ...

How such strange terms as ligand, monodentate, and chelate got into the vocabulary of chemistry?

Ligand comes from the Latin word *ligare*, which means to bind. It is quite appropriate to describe groups that are bound to a metal center as ligands. Dentate is also derived from a Latin word, *dens*, meaning tooth. Figuratively speaking, a monodentate ligand has one tooth; a bidentate ligand has two teeth; and a polydentate ligand has several. A ligand attaches itself to the metal center in accordance with the number of "teeth" it possesses. This is an easily remembered and colorful metaphor. Chelate is derived from the Greek word *chela*, which means a crab's claw. The way in which a chelating agent attaches itself to a metal ion resembles a crab's claw—another colorful metaphor.

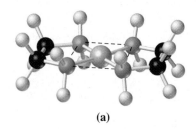

(a)

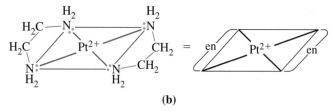

(b)

▲ FIGURE 25-3   **Three representations of the chelate [Pt(en)₂]²⁺**
(a) Overall structure. (b) The ligands attach at adjacent corners along an edge of the square. They do *not* bridge the square by attaching to opposite corners. Bonds are shown in red, and the square-planar shape is indicated by the black parallelogram.

produces a ring (normally with five or six members), we refer to the complex as a **chelate** (pronounced KEY-late). The polydentate ligand is called a **chelating agent**, and the process of chelate formation is called *chelation*.

25-3   Nomenclature

The system of naming complexes originated with Werner, but it has been modified several times over the years. Even today, usage varies somewhat. Our approach will be to consider a few rules that permit us to relate names and formulas of simple complexes. We will not consider any of the complicated cases for which writing names and formulas is more challenging.

1. *In names and formulas of coordination compounds, cations come first, followed by anions.* This is the same order as used in simple ionic compounds like sodium chloride, NaCl.

2. *Anions as ligands are named by using the ending* o. Normally, *ide* endings change to *o*, *ite* to *ito*, and *ate* to *ato* (see Table 25.2).

3. *Neutral molecules as ligands generally carry the unmodified name.* For example, the name ethylenediamine is used both for the free molecule and for the molecule as a ligand. Aqua, ammine, carbonyl, and nitrosyl are important exceptions (see Table 25.2).

4. *The number of ligands of a given type is denoted by a prefix.* The usual prefixes are *mono* = 1, *di* = 2, *tri* = 3, *tetra* = 4, *penta* = 5, and *hexa* = 6. As in many other cases, the prefix *mono* is often not used. If the ligand name is a composite name that itself contains a numerical prefix, such as ethylene*di*amine, place parentheses around the name and precede it with *bis* = 2, *tris* = 3, *tetrakis* = 4, and so on. Thus, dichloro signifies *two* Cl^- ions as ligands, and *penta*aqua signifies *five* H_2O molecules. To indicate the presence of *two* ethylenediamine (en) ligands, we write *bis*(ethylenediamine).

5. *When we name a complex, ligands are named first, in alphabetical order, followed by the name of the metal center. The oxidation state of the metal center is denoted by a Roman numeral. If the complex is an anion, the ending* ate *is attached to the name of the metal.* Prefixes (*di, tri, bis, tris,* ...) are ignored in establishing the alphabetical order. Thus, $[CrCl_2(H_2O)_4]^+$ is called tetraaquadichlorochromium(III) ion; $[CoCl_2(en)_2]^+$ is dichlorobis(ethylenediamine)cobalt(III) ion; and $[Cr(OH)_4]^-$ is tetrahydroxochromate(III) ion. For complex anions of a few of the metals, the English name is replaced by the Latin name given in Table 25.4. Thus, $[CuCl_4]^{2-}$ is the tetrachlorocuprate(II) ion.

6. *When we write the formula of a complex, the chemical symbol of the metal center is written first, followed by the formulas of anions and then neutral molecules.* If there are two or more different anions or neutral molecules as ligands, they are written in alphabetical order according to the first chemical symbols of their formulas. Thus, in the formula of the tetraaminechloronitrocobalt(III) ion, Cl^- precedes NO_2^-, and both are placed ahead of the neutral NH_3 molecules: $[CoCl(NO_2)(NH_3)_4]^+$.

▶ Occasionally, the metal center will be in the oxidation state 0, as in $[W(CO)_6]$—hexacarbonyltungsten(0).

TABLE 25.4   Names for Some Metals in Complex Anions

Iron	⟶ **Ferrate**
Copper	⟶ **Cuprate**
Tin	⟶ **Stannate**
Silver	⟶ **Argentate**
Lead	⟶ **Plumbate**
Gold	⟶ **Aurate**

EXAMPLE 25-2

Relating Names and Formulas of Complexes. **(a)** What is the name of the complex $[CoCl_3(NH_3)_3]$? **(b)** What is the formula of the compound pentaaquachlorochromium(III) chloride? **(c)** What is the name of the compound $K_3[Fe(CN)_6]$?

Solution

(a) $[CoCl_3(NH_3)_3]$ consists of *three* ammonia molecules and *three* chloride ions attached to a central Co^{3+} ion; it is electrically neutral. The name of this neutral complex is triamminetrichlorocobalt(III).

(b) The central metal ion is Cr^{3+}. There are five H_2O molecules and one Cl^- ion as ligands. The complex ion carries a net charge of 2+. Two Cl^- ions are required to neutralize the charge on this complex cation. The formula of the coordination compound is $[CrCl(H_2O)_5]Cl_2$.

(c) This compound consists of K^+ cations and complex anions having the formula $[Fe(CN)_6]^{3-}$. Each cyanide ion carries a charge of 1−, so the oxidation state of the iron must be +3. The Latin-based name "ferrate" is used because the complex ion is an anion. The name of the anion is hexacyanoferrate(III) ion. The coordination compound is potassium hexacyanoferrate(III).

Practice Example A: What is the formula of the compound potassium hexachloroplatinate(IV)?

Practice Example B: What is the name of the compound $[Co(SCN)(NH_3)_5]Cl_2$?

Although most complexes are named in the manner just outlined, some common, or trivial, names are still in use. Two such trivial names are ferrocyanide for $[Fe(CN)_6]^{4-}$ and ferricyanide for $[Fe(CN)_6]^{3-}$. These common names suggest the oxidation state of the central metal ions through the *o* and *i* designations (*o* for the ferrous ion, Fe^{2+}, in $[Fe(CN)_6]^{4-}$ and *i* for the ferric ion, Fe^{3+}, in $[Fe(CN)_6]^{3-}$). These trivial names do not indicate that the metal ions have a coordination number of 6, however. The systematic names—hexacyanoferrate(II) and hexacyanoferrate(III)—are more informative.

25-4 Isomerism

Isomerism animation

As we have noted (page 90), **isomers** are substances that have the same formulas but differ in their structures and properties. Several kinds of isomerism are found among complex ions and coordination compounds. These can be lumped into two broad categories: **Structural isomers** differ in basic structure or bond type—what ligands are bonded to the metal center and through which atoms. **Stereoisomers** have the same number and types of ligands and the same mode of attachment, but they differ in the way in which the ligands occupy the space around the metal center. Of the following five examples, the first three are types of structural isomerism and the remaining two are types of stereoisomerism.

Ionization Isomerism

The two coordination compounds whose formulas are shown here have the same central ion (Cr^{3+}), and five of the six ligands (NH_3 molecules) are the same. The compounds differ in that one has SO_4^{2-} ion as the sixth ligand, with a Cl^- ion to neutralize the charge of the complex ion, whereas the other has Cl^- as the sixth ligand and SO_4^{2-} to neutralize the charge of the complex ion.

$[CrSO_4(NH_3)_5]Cl$	$[CrCl(NH_3)_5]SO_4$
pentaamminesulfatochromium(III) chloride	pentaamminechlorochromium(III) sulfate
(a)	(b)

Coordination Isomerism

A situation somewhat similar to that just described can arise when a coordination compound is composed of both complex cations and complex anions. The ligands can be distributed differently between the two complex ions, as are NH_3 and CN^- in these two compounds.

$[Co(NH_3)_6][Cr(CN)_6]$	$[Cr(NH_3)_6][Co(CN)_6]$
hexaamminecobalt(III) hexacyanochromate(III)	hexaaminechromium(III) hexacyanocobaltate(III)
(a)	(b)

Linkage Isomerism

Some ligands may attach to the central metal ion of a complex ion in different ways. For example, the nitrite ion, a monodentate ligand, has electron pairs available for coordination both on the N and O atoms.

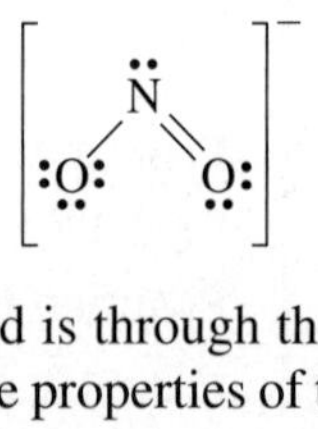

$$\left[\ddot{\underset{\displaystyle :\ddot{O} \diagdown \quad \diagup \ddot{O}:}{N}} \right]^{-}$$

Whether attachment of this ligand is through the N or an O atom, the formula of the complex ion is unaffected. The properties of the complex ion, however, may be

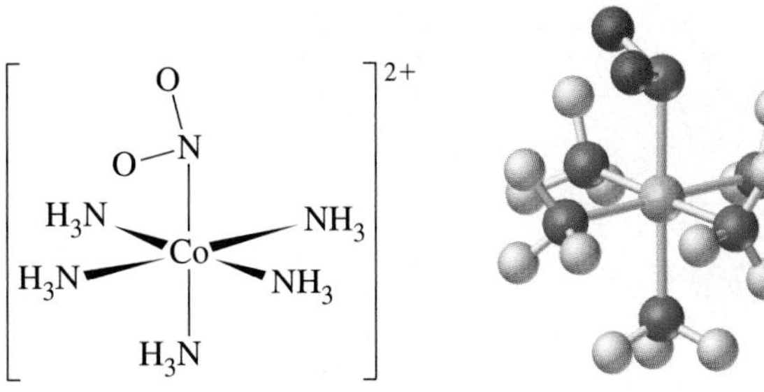

(a) $[Co(NO_2)(NH_3)_5]^{2+}$

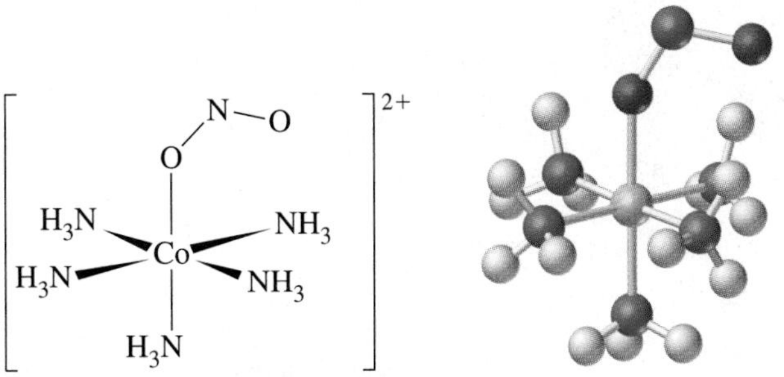

(b) $[Co(ONO)(NH_3)_5]^{2+}$

▲ **FIGURE 25-4** **Linkage isomerism illustrated**
(a) Pentaamminenitrito-*N*-cobalt(III) cation. **(b)** Pentaamminenitrito-*O*-cobalt(III) cation.

affected. When attachment occurs through the N atom, the ligand can be referred to as nitro or, more properly, nitrito-*N*- when naming the complex. Coordination through an O atom can be referred to as nitrito or, more properly, nitrito-*O*- when naming the complex.

$[Co(NO_2)(NH_3)_5]^{2+}$ $[Co(ONO)(NH_3)_5]^{2+}$

pentaamminenitrito-*N*-cobalt(III) ion   pentaamminenitrito-*O*-cobalt(III) ion

(a)   (b)

The structures of these compounds are illustrated in Figure 25-4.

Geometric Isomerism

If we substitute a single Cl^- ion for an NH_3 molecule in the square-planar complex ion $[Pt(NH_3)_4]^{2+}$ in Figure 25-2, it does not matter at which corner of the square we make this substitution. As shown in Figure 25-5a, all four possibilities are alike. If we substitute a *second* Cl^-, we now have two distinct possibilities (Figure 25-5b). The two Cl^- ions can either be along the same edge of the square (**cis**) or on opposite corners, across from each other (**trans**). To distinguish clearly between these two possibilities, we must either draw a structure or refer to the appropriate name. The formula alone will not distinguish between them. (Note that this complex is a neutral species, not an ion.)

▶ The term *cis* means "on this side" in Latin, and *trans* means "across." Domestic airline flights in the United States are cisatlantic, whereas flights to Europe are transatlantic.

$[PtCl_2(NH_3)_2]$

cis-diamminedichloroplatinum(II)

or

trans-diamminedichloroplatinum(II)

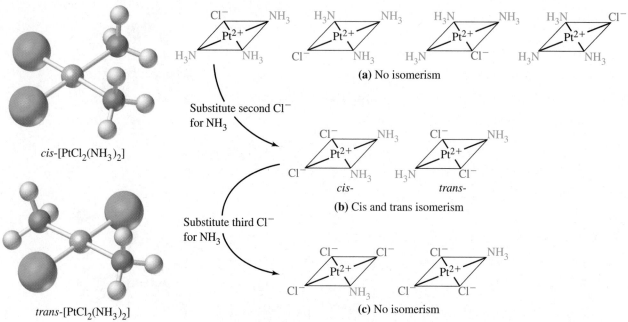

cis-[PtCl$_2$(NH$_3$)$_2$]

trans-[PtCl$_2$(NH$_3$)$_2$]

Substitute second Cl$^-$ for NH$_3$

Substitute third Cl$^-$ for NH$_3$

(a) No isomerism

cis-   *trans-*

(b) Cis and trans isomerism

(c) No isomerism

▲ **FIGURE 25-5   Geometric isomerism illustrated**
For the square-planar complexes shown here, isomerism exists only when two Cl$^-$ ions have replaced NH$_3$ molecules.

Interestingly, when we substitute a *third* Cl$^-$ ion, isomerism disappears (Figure 25-5c). There is only one complex ion with the formula [PtCl$_3$(NH$_3$)]$^-$.

With an octahedral complex, the situation is a bit more complicated. Take the complex ion [Co(NH$_3$)$_6$]$^{3+}$ of Figure 25-2 as an example. If we substitute one Cl$^-$ for an NH$_3$, we get a single structure. With *two* Cl$^-$ ions substituted for NH$_3$ molecules, we obtain cis and trans isomers. The cis isomer has two Cl$^-$ ions along the same edge of the octahedron (Figure 25-6a). The trans isomer has two Cl$^-$ ions on

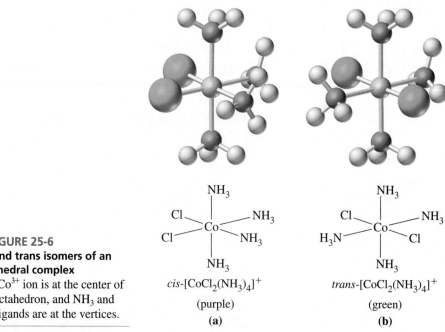

▶ **FIGURE 25-6**
Cis and trans isomers of an octahedral complex
The Co^{3+} ion is at the center of the octahedron, and NH$_3$ and Cl$^-$ ligands are at the vertices.

cis-[CoCl$_2$(NH$_3$)$_4$]$^+$
(purple)
(a)

trans-[CoCl$_2$(NH$_3$)$_4$]$^+$
(green)
(b)

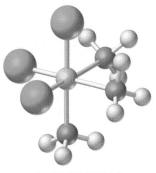

fac-[CoCl$_3$(NH$_3$)$_3$]

mer-[CoCl$_3$(NH$_3$)$_3$]

Mirror

(a)

Mirror

(b)

▲ **FIGURE 25-7**
Superimposable and nonsuperimposable objects—a cubical, open-top box
(a) You can place the box into its mirror image (hypothetically) in several ways. **(b)** In any way that you place the box into its mirror image, the stickers will not appear in the same position. The box and its mirror image are nonsuperimposable.

opposite corners, that is, at opposite ends of a line drawn through the central metal ion (Figure 25-6b). One difference between the two is that the cis isomer has a purple color and the trans has a bright green color.

Let us refer to Figure 25-6a to see what happens when we substitute a *third* Cl$^-$ for an NH$_3$. If we make this third substitution at either the top or bottom of the structure, the result is that three Cl$^-$ ions appear on the same face of the octahedron. We call this a *fac* (facial) isomer. If we make the third substitution at either of the other two positions, the result is three Cl$^-$ ions around a perimeter or meridian of the octahedron. We call this a *mer* (meridional) isomer. If we substitute a *fourth* Cl$^-$, we again get cis and trans isomers.

EXAMPLE 25-3

Identifying Geometric Isomers. Sketch structures of all the possible isomers of [CoCl(ox)(NH$_3$)$_3$].

Solution

The Co^{3+} ion exhibits a coordination number of 6. The structure is octahedral. Recall that ox (oxalate ion) is a bidentate ligand carrying a double-negative charge (see Table 25.3). Also, as we saw in Figure 25-3, a bidentate ligand must be attached in cis positions, not trans. Once the ox ligand is placed, any position is available to the Cl$^-$. This leaves two possibilities for the three NH$_3$ molecules. They can be situated (1) on the same face of the octahedron (fac isomer) or (2) around a perimeter of the octahedron (mer isomer).

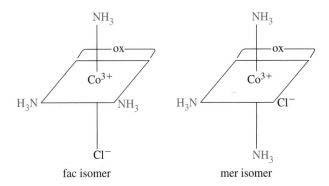

fac isomer   mer isomer

Practice Example A: Sketch the geometric isomers of [CoCl$_2$(ox)(NH$_3$)$_2$]$^-$.

Practice Example B: Sketch the geometric isomers of [MoCl$_2$(C$_5$H$_5$N)$_2$(CO)$_2$]$^+$.

Optical Isomerism

To understand optical isomerism, we need to understand the relationship between an object and its mirror image. Features on the right side of the object appear on the left side of its image in a mirror, and vice versa. Certain objects can be rotated in such a way as to be *superimposable* on their mirror images, but other objects are *nonsuperimposable* on their mirror images. An unmarked tennis ball is superimposable on its mirror image, but a left hand is nonsuperimposable on its mirror image (a right hand).

Consider the open-top cubical cardboard box pictured in Figure 25-7a. There are a number of hypothetical ways in which the box can be superimposed on its mirror image. Now imagine that a distinctive sticker is placed at a corner of one side of the box (Figure 25-7b). In this case, there is no way that the box and its mirror

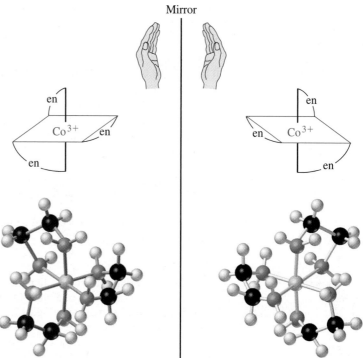

Chirality animation

▲ **FIGURE 25-8 Optical isomers**
The two structures are nonsuperimposable mirror images. Like a right hand and a left hand, one structure cannot be superimposed onto the other.

*d-Alanine, lower case
ell-Alanine models*

▶ Here is a process based on chirality: choosing a matched pair of gloves from a bin of gloves of identical sizes, 90 made for the right hand and 10 for the left hand.

▶ The prefixes *dextro* and *levo* are derived from the Latin words *dexter*, right, and *levo*, left.

image can be superimposed; they are clearly different. This is equivalent to saying that there is no way that a formfitting left glove can be worn on a right hand (turning it inside out is not allowed).

The two structures of $[Co(en)_3]^{3+}$ depicted in Figure 25-8 are related to each other as are an object and its image in a mirror. Furthermore, the two structures are *nonsuperimposable*, like a left and a right hand. The two structures represent two different complex ions; they are isomers.

Structures that are nonsuperimposable mirror images of each other are called **enantiomers** and are said to be **chiral** (pronounced KYE-rull). (Structures that are superimposable are *achiral*.) Whereas other types of isomers may differ significantly in their physical and chemical properties, enantiomers have identical properties except in a few specialized situations. These exceptions involve phenomena that are directly linked to chirality, or handedness, at the molecular level. An example is *optical activity*, pictured in Figure 25-9.

Interactions between a beam of polarized light and the electrons in an enantiomer cause the plane of the polarized light to rotate. One enantiomer rotates the plane of polarized light to the right (clockwise) and is said to be *dextrorotatory* (designated + or *d*). The other enantiomer rotates the plane of polarized light to the same extent, but to the left (counterclockwise). It is said to be *levorotatory* (− or *l*). Because they can rotate the plane of polarized light, the enantiomers are said to be *optically active* and are referred to as **optical isomers**.

When an optically active complex is synthesized, a mixture of the two optical isomers (enantiomers) is obtained, such as the two $[Co(en)_3]^{3+}$ isomers in Figure 25-8. The optical rotation of one isomer just cancels that of the other. The mixture, called a *racemic mixture*, produces no net rotation of the plane of polarized

Are You Wondering...

How to tell if its mirror image is superimposable on a molecule?

The complex ion $[Cr(NH_3)_2(H_2O)_2Br_2]^+$ has five isomers; one is shown below with its mirror image. To test if the mirror image is superimposable on the original, imagine rotating the mirror image about the vertical axis (H_2O—Cr—NH_3) by 180° so that the two Br^- ligands are in the same position as in the original molecule. We see that the NH_3 and H_2O ligands that are in the same plane as the two Br^- ligands are in reversed positions when compared with the original molecule. Thus the molecule and its mirror image are not superimposable. The molecule is potentially optically active.

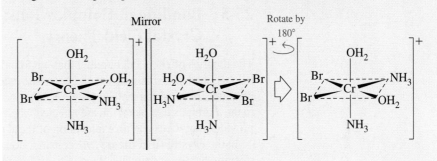

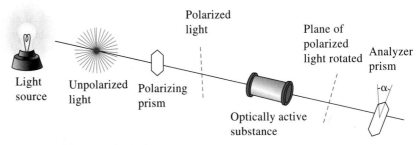

▲ **FIGURE 25-9   Optical activity**
Light from an ordinary source consists of electromagnetic waves vibrating in all planes; it is unpolarized. This light is passed through a polarizer, a material that screens out all waves except those vibrating in a particular plane. The plane of polarization of transmitted polarized light is then changed by passage through an optically active substance. The angle through which the plane of polarization has been rotated is determined by rotating an analyzer (a second polarizer) to the extent that all the polarized light is absorbed.

Optical Activity animation

▲ **FIGURE 25-10**
Hypothetical structures for $[CoCl_2(NH_3)_4]^+$
If the complex ion had this hexagonal structure, there should be *three* distinct isomers, but only two isomers exist—cis and trans.

light. Separating the *d* and *l* isomers of a racemic mixture is called *resolution*. This separation can sometimes be achieved through chemical reactions controlled by chirality, with the two enantiomers behaving differently. Many phenomena of the living state, such as the effectiveness of a drug, the activity of an enzyme, and the ability of a microorganism to promote a reaction, involve chirality. We will return to a discussion of chirality in Chapters 27 and 28.

Isomerism and Werner's Theory

The study of isomerism played a crucial role in the development of Werner's theory of coordination chemistry. Werner proposed that complexes with coordination number 6 have an octahedral structure, but other possibilities were also proposed. For example, Figure 25-10 shows a hypothetical hexagonal structure for

$[CoCl_2(NH_3)_4]^+$, similar to that of benzene. However, this hexagonal structure would require the existence of *three* isomers, but a third isomer was never found; the only two isomers are those pictured in Figure 25-6. Additional direct evidence came with the discovery of optical isomerism in the tris(ethylenediamine)cobalt(III) ion (Figure 25-8). Neither the hexagonal structure nor alternative structures can account for this isomerism—but, as we have seen, the octahedral structure does. Werner even succeeded in preparing an optically active octahedral complex with only inorganic ligands, to overcome objections that the optical activity of tris(ethylenediamine)cobalt(III) ion owed its optical activity to its carbon atoms and not to its geometric structure.

25-5 Bonding in Complex Ions: Crystal Field Theory

The theories of chemical bonding that we found so useful in earlier chapters do not help us much in explaining the characteristic colors and magnetic properties of complex ions. In transition metal ions, we need to focus our attention on how the electrons in the *d* orbitals of a metal ion are affected when they are in a complex. A theory that provides that focus and an explanation of these properties is crystal field theory.

In the **crystal field theory**, we consider bonding in a complex ion to be an electrostatic attraction between the positively charged nucleus of the central metal ion and electrons in the ligands. Repulsions occur between the ligand electrons and electrons in the central ion. In particular, the crystal field theory focuses on the repulsions between ligand electrons and *d* electrons of the central ion.

▶ Modifications of the simple crystal field theory that take into account such factors as the partial covalency of the metal–ligand bond are called *ligand field theory*. This term is often used to signify both the purely electrostatic crystal field theory and its modifications.

First, a reminder about the *d* orbitals introduced in Figure 9-28: All five of the orbitals are alike in energy when in an isolated atom or ion, but they are *unlike* in their spatial orientations. One of them, d_{z^2}, is directed along the *z* axis, and another, $d_{x^2-y^2}$, has lobes along the *x* and *y* axes. The remaining three have lobes extending into regions between the perpendicular *x*, *y*, and *z* axes. In the presence of ligands, because repulsions exist between ligand electrons and *d* electrons, the *d*-orbital energy levels of the central metal ion are raised. As we will soon see, however, they are not all raised to the same extent.

Figure 25-11 depicts six anions (ligands) approaching a central metal ion along the *x*, *y*, and *z* axes. This direction of approach leads to an octahedral complex.

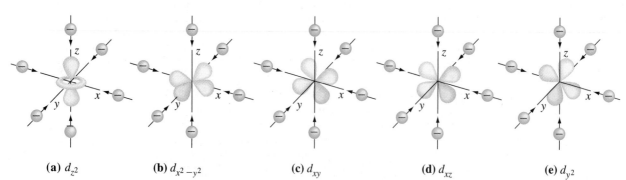

(a) d_{z^2} **(b)** $d_{x^2-y^2}$ **(c)** d_{xy} **(d)** d_{xz} **(e)** d_{y^2}

▲ **FIGURE 25-11   Approach of six anions to a metal ion to form a complex ion with octahedral structure**

The ligands (anions, in this case) approach the central metal ion along the *x*, *y*, and *z* axes. Maximum repulsion occurs with the d_{z^2} and $d_{x^2-y^2}$ orbitals, and their energies are raised. Repulsions with the other *d* orbitals are not as great. A difference in energy results between the two sets of *d* orbitals.

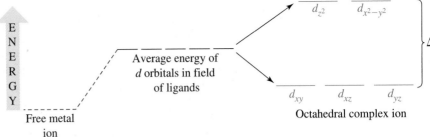

▲ **FIGURE 25-12   Splitting of *d* energy levels in the formation of an octahedral complex ion**
The *d*-orbital energy levels of the free central ion are raised in the presence of ligands to the average level shown, but the five levels are split into two groups.

Repulsions between ligand electrons and *d*-orbital electrons are strengthened in the direct, head-to-head approach of ligands to the d_{z^2} (Figure 25-11a) and $d_{x^2-y^2}$ (Figure 25-11b) orbitals. These two orbitals have their energy raised with respect to an average *d*-orbital energy for a central metal ion in the field of the ligands. For the other three orbitals (d_{xy}, d_{xz}, and d_{yz}, Figure 25-11c–e), ligands approach between the lobes of the orbitals and there is a gain in stability over the head-to-head approach; these orbital energies are lowered with respect to the average *d*-orbital energy. The difference in energy between the two groups of *d* orbitals is called *crystal field splitting* and is represented by the symbol Δ (Figure 25-12).

The removal of the degeneracy of the *d* orbitals by the crystal field has important consequences for the electron configurations of transition metal ions having between 4 and 7 *d* electrons. Consider the transition metal ion Cr^{2+} with a d^4 configuration. If we assign the four *d* electrons to the orbitals of lowest energy first, the first three electrons go into the d_{xy}, d_{xz}, and d_{yz} orbitals according to Hund's rule (page 337)—but what about the fourth electron? The aufbau process (page 338) suggests that the electron should pair up with one of the three electrons already in the d_{xy}, d_{xz}, and d_{yz} orbitals.

$$\underline{\quad} \quad \underline{\quad}$$
$$d_{x^2-y^2} \qquad d_{z^2}$$

$$\underline{\uparrow\downarrow} \quad \underline{\uparrow} \quad \underline{\uparrow}$$
$$d_{xy} \qquad d_{xz} \qquad d_{yz}$$

Placing the fourth electron in the lower level confers extra stability (lower energy) on the complex, but some of this stability is offset because it requires energy, called the **pairing energy** (P), to force an electron into an orbital that is already occupied by an electron. Alternatively, we could assign the electron to either the $d_{x^2-y^2}$ or d_{z^2} orbital, avoiding the pairing energy.

$$\underline{\uparrow} \quad \underline{\quad}$$
$$d_{x^2-y^2} \qquad d_{z^2}$$

$$\underline{\uparrow} \quad \underline{\uparrow} \quad \underline{\uparrow}$$
$$d_{xy} \qquad d_{xz} \qquad d_{yz}$$

Placing the fourth electron in the upper level requires energy and offsets the extra stability acquired by placing the first three electrons in the lower level. To pair or not to pair, that is the question.

Whether the fourth electron enters the lowest level and becomes paired or, instead, enters the upper level with the same spin as the first three electrons depends on the magnitude of Δ. If Δ is greater than the pairing energy, P, greater stability is obtained if the fourth electron is paired with one in the lower level. If Δ is less than the pairing energy, greater stability is obtained by keeping the electrons unpaired. Thus, for octahedral chromium(II) complexes, there are two possibilities for the number of unpaired electrons. In one case, there are four unpaired electrons when $\Delta < P$; this situation corresponds to the maximum number of unpaired electrons and is referred to as **high spin**. Ligands such as H_2O and F^- produce only a small crystal field splitting, leading to high-spin complexes; such ligands are said to be *weak-field ligands*. As an example, $[Cr(H_2O)_6]^{2+}$ is a weak-field complex. In the other case, there are two unpaired electrons when $\Delta > P$; this corresponds to the minimum number of unpaired electrons and is referred to as **low spin**. Ligands such as NH_3 and CN^- produce large crystal field splitting, leading to low-spin complexes; such ligands are said to be *strong-field ligands*. $[Cr(CN)_6]^{4-}$ is a strong-field complex.

Different ligands can be arranged in order of their abilities to produce a splitting of the d energy levels. This arrangement is known as the **spectrochemical series**.

> These two possibilities exist for complex ions because the crystal field splitting and pairing energies are small and of comparable value. In considering the electron configurations in atoms, the spacing between energy levels is much greater than the pairing energy. ▶

> ▶ The spectrochemical series.

Strong field

(large Δ)

$$CN^- > \underline{N}O_2^- > en > py \approx NH_3 > EDTA^{4-} > SC\underline{N}^- > H_2O >$$
$$ON\underline{O}^- > ox^{2-} > OH^- > F^- > \underline{S}CN^- > Cl^- > Br^- > I^-$$

(small Δ)

Underlining indicates the donor atom.   *Weak field*

To summarize, consider these two complexes Co(III), $[CoF_6]^{3-}$ and $[Co(NH_3)_6]^{3+}$. The F^- ion is a weak-field ligand, whereas NH_3 is a strong-field ligand. Because the crystal field splitting for NH_3 is greater than the pairing energy for Co^{3+}, we have the following situations.

> *Crystal Field Splitting activity*

If we call the energy separation for the octahedral complex Δ_o, the comparative value for the tetrahedral complex is $0.44\ \Delta_o$ and for the square-planar complex, $1.74\ \Delta_o$. The splitting between the d_{xy} and $d_{x^2-y^2}$ orbitals in a square-planar complex is Δ_o, just as in an octahedral complex, because these orbitals are equally affected by ligand repulsions in both complexes. ▶

So far, we have considered just octahedral complexes. In the formation of complex ions of other geometric structures, ligands approach from different directions and produce different patterns of splitting of the d energy level. Figure 25-13 shows the pattern for tetrahedral complexes, and Figure 25-14 shows the pattern for square-planar complexes. When we compare hypothetical complexes of different structures having the same combinations of ligands, metal ions, and metal–ligand distances, we find the greatest energy separation of the d levels for the square-planar complex and the smallest energy separation for the tetrahedral complex.

Are You Wondering...

If the two groups of orbitals split equally with respect to the average energy of the d orbitals?

The answer is no. The reason is that the total energy must be constant. If we consider a d^{10} ion such as Zn^{2+} in an octahedral field, the destabilization due to the four electrons in the $d_{x^2-y^2}$ and d_{z^2} orbitals must be offset by the stabilization gained by the six electrons in the d_{xy}, d_{xz}, and d_{yz} orbitals. This requires that

$$(6 \times \text{the energy of } d_{xy}, d_{xz}, d_{yz} \text{ orbitals}) + (4 \times \text{the energy of } d_{x^2-y^2}, d_{z^2}) = 0.$$

That is, the energy gained equals the energy lost. Also, the energy difference between the orbitals is

$$(\text{the energy of } d_{x^2-y^2}, d_{z^2}) - (\text{the energy of } d_{xy}, d_{xz}, d_{yz} \text{ orbitals}) = \Delta$$

To satisfy these two relationships, it is necessary that

$$\text{the energy of } d_{xy}, d_{xz}, d_{yz} \text{ orbitals} = -0.4\,\Delta$$

$$\text{the energy of } d_{x^2-y^2}, d_{z^2} = 0.6\,\Delta$$

The splitting is as shown.

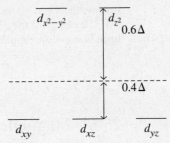

The splitting of the two groups of orbitals is not equal with respect to the average energy of the d orbitals.

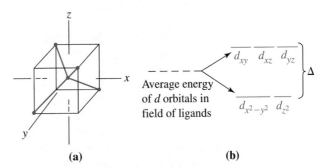

(a)   **(b)**

▲ **FIGURE 25-13   Crystal field splitting in a tetrahedral complex ion**
(a) The positions of attachment of ligands to a metal ion leading to the formation of a tetrahedral complex ion. **(b)** Interference with the d orbitals directed along the x, y, and z axes is not as great as with those that lie between the axes (see Figure 25-11). As a result, the pattern of crystal field splitting is reversed from that of an octahedral complex.

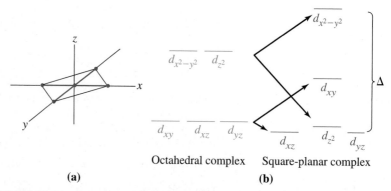

▲ FIGURE 25-14   Comparison of crystal field splitting in a square-planar and an octahedral complex

(a) The positions of attachment of ligands to a metal ion, leading to the formation of a square-planar complex. **(b)** Splitting of the d energy level in a square-planar complex can be related to that of the octahedral complex. There are no ligands along the z axis in a square-planar complex, so we expect the repulsion between ligands and d_{z^2} electrons to be much less than in an octahedral complex. The d_{z^2} energy level is lowered considerably from that in an octahedral complex. Similarly, the energy levels of the d_{xz} and d_{yz} orbitals are lowered slightly because the electrons in these orbitals are concentrated in planes perpendicular to that of the square-planar complex. The energy of the $d_{x^2-y^2}$ orbital is raised because the x and y axes represent the direction of approach of four ligands to the central ion. The energy of the d_{xy} orbital is also raised because this orbital lies in the plane of the ligands in the square-planar complex.

25-6   Magnetic Properties of Coordination Compounds and Crystal Field Theory

The paramagnetism of the dioxygen molecule was dramatically illustrated in Chapter 11 (page 395) by the interaction of liquid oxygen with the pole tips of a strong magnet. The origin of the paramagnetism is the existence of unpaired electrons in the molecule. Transition metal coordination compounds exhibit varying degrees of paramagnetism and can also be diamagnetic. A paramagnetic substance is pulled into, and a diamagnetic substance is pushed out of, a magnetic field. A straightforward way to measure magnetic properties is to weigh a substance "in" and "out" of a magnetic field, as illustrated in Figure 25-15. The mass of the substance is the same whatever magnetic property the substance possesses. However, if the substance is diamagnetic, it is slightly repelled by a magnetic field and *weighs* less within the field. If the substance is paramagnetic, it *weighs* more within the field.

The degree to which a substance weighs more in the magnetic field depends on the number of unpaired electrons. In the previous section, we saw that a high-spin d^n complex has more unpaired electrons than a low-spin d^n complex. Thus, measuring the change in weight of the complex in a magnetic field allows us to determine whether a complex is high or low spin. The magnetic properties of a complex depend on the magnitude of the crystal field splitting. Strong-field ligands tend to form low-spin, weakly paramagnetic, or even diamagnetic complexes. Weak-field ligands tend to form high-spin, strongly paramagnetic complexes. The results of measuring the magnetic properties of coordination compounds can therefore be interpreted from crystal field theory, as demonstrated in Examples 25-4 and 25-5.

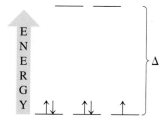

(a) No magnetic field   **(b)** Magnetic field turned on

▲ **FIGURE 25-15**   **Paramagnetism illustrated**
(a) A sample is weighed in the absence of a magnetic field. **(b)** When the field is turned on, the balanced condition is upset. The sample gains weight because it is now subjected to two attractive forces: the force of gravity *and* the force of interaction of the external magnetic field and the unpaired electrons.

EXAMPLE 25-4

Crystal Field Theory of Magnetic Properties activity

Using the Spectrochemical Series to Predict Magnetic Properties. How many unpaired electrons would you expect to find in the octahedral complex $[Fe(CN)_6]^{3-}$?

Solution

The Fe atom has the electron configuration $[Ar]3d^64s^2$. The Fe^{3+} ion has the configuration $[Ar]3d^5$. CN^- is a strong-field ligand. Because of the large energy separation in the d levels of the metal ion produced by this ligand, we expect all the electrons to be in the lowest energy level. There should be only one unpaired electron.

Practice Example A: How many unpaired electrons would you expect to find in the octahedral complex $[MnF_6]^{2-}$?

Practice Example B: How many unpaired electrons would you expect to find in the tetrahedral complex $[CoCl_4]^{2-}$? Would you expect more, fewer, or the same number of unpaired electrons as in the octahedral complex $[Co(H_2O)_6]^{2+}$?

EXAMPLE 25-5

Using the Crystal Field Theory to Predict the Structure of a Complex from Its Magnetic Properties. The complex ion $[Ni(CN)_4]^{2-}$ is diamagnetic. Use ideas from the crystal field theory to speculate on its probable structure.

Solution

We can eliminate an octahedral structure because the coordination number is 4 (not 6). Our choice is between tetrahedral and square-planar.

The electron configuration of Ni is $[Ar]3d^8 4s^2$, and that of Ni(II) is $[Ar]3d^8$. Because the complex ion is diamagnetic, all $3d$ electrons must be paired. Let us see how we would distribute these $3d$ electrons if the structure were tetrahedral (recall Figure 25-13). We would place four electrons (all paired) into the two lowest d levels. We would then distribute the remaining four electrons among the three higher-level d orbitals. Two of the electrons would be unpaired, and the complex ion would be paramagnetic.

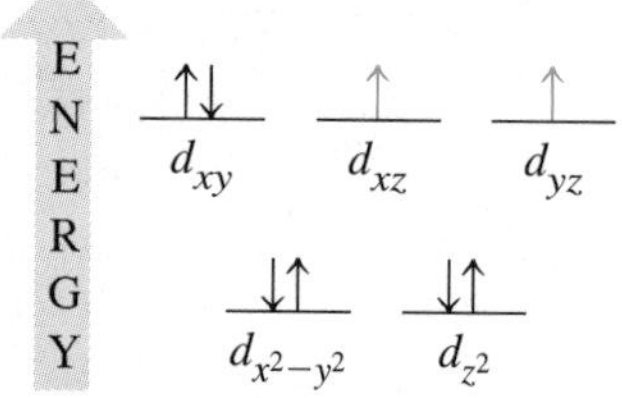

Because the tetrahedral structure would be paramagnetic, we can conclude that the structure of the diamagnetic $[Ni(CN)_4]^{2-}$ ion must be square-planar. Let's also demonstrate that this is a reasonable conclusion based on the d-orbital energy-level diagram for a square-planar complex (recall Figure 25-14). First, we fill the three lowest-energy orbitals with electrons (six), and then we assume that the energy separation between the d_{xy} and $d_{x^2-y^2}$ orbitals is large enough that the final two electrons remain paired in the d_{xy} orbital. This corresponds to a diamagnetic complex ion.

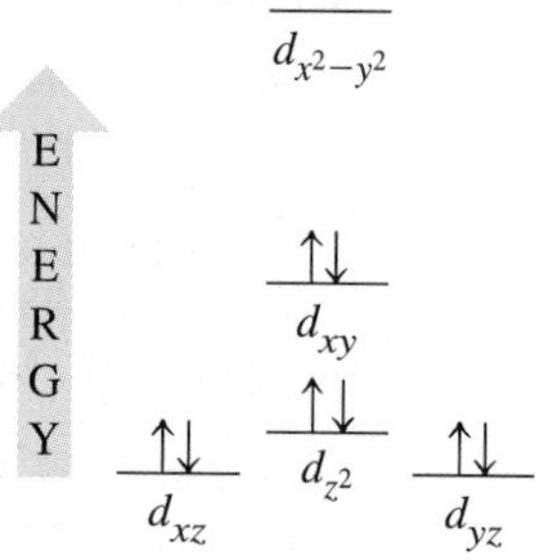

Practice Example A: The complex ion $[Co(CN)_4]^{2-}$ is paramagnetic with three unpaired electrons. Use ideas from the crystal field theory to speculate on its probable structure.

Practice Example B: Would you expect $[Cu(NH_3)_4]^{2-}$ to be diamagnetic or paramagnetic? Can you use this information about the magnetic properties of $[Cu(NH_3)_4]^{2+}$ to help you determine whether the structure of $[Cu(NH_3)_4]^{2+}$ is tetrahedral or square-planar? Explain.

25-7   Color and the Colors of Complexes

To help us understand the nature of color, let's consider the two situations for mixing colors shown in Figure 25-16. Figure 25-16a represents *additive* mixing. It is the type of color mixing that occurs when colored spotlights are superimposed. Figure 25-16b, conversely, represents *subtractive* mixing. It is the type of color mixing that occurs when paint pigments or colored solutions are mixed.

► Additive color mixing is used in color monitors and color television screens. Subtractive color mixing is used for all the photographs and art in this text.

For the additive mixing of light beams in Figure 25-16a, we define the **primary colors** as any three colors that, when combined, yield white light (W). Our choice in the figure is *red* (R), *green* (G), and *blue* (B), and we can represent their sum as R + G + B = W. The **secondary colors** are those that are produced by combining two primary colors. Figure 25-16a indicates that the secondary colors are *yellow* (Y = R + G), *cyan* (C = G + B), and *magenta* (M = B + R).

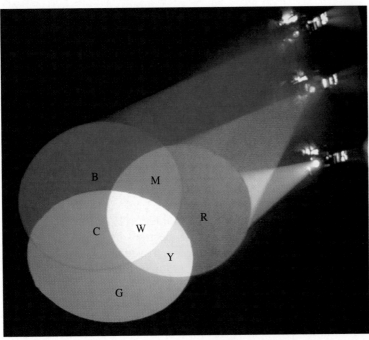

(a) Additive color mixing

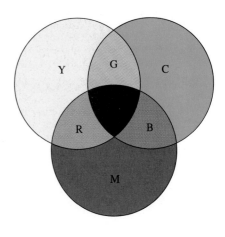

(b) Subtractive color mixing

▲ **FIGURE 25-16   The mixing of colors**
(a) The additive mixing of three beams of colored light: red (R), green (G), and blue (B). The secondary colors—yellow (Y), cyan (C), and magenta (M)—are produced in regions where two of the beams overlap. The overlap of all three beams produces white light (W). **(b)** The subtractive mixing of three pigments with the primary colors magenta (M), yellow (Y), and cyan (C). Here, the secondary colors—red (R), green (G), and blue (B)—form when two of the primary colors are mixed. A mixture of all three primary colors produces a very dark brown to black color.

In four-color printing, such as in this book and in color printers in computer systems, the basic inks used are magenta, yellow, cyan, and black. In photographs and drawings in this book, very small dots of these four basic colors, printed singly and in various combinations, produce the colored images you see. (With a magnifying glass, you can see individual dots of color.)

Each secondary color is a **complementary color** of one of the primary colors. Again, from Figure 25-16a, cyan (C) is the complementary color of red (R); magenta (M), of green (G); and yellow (Y), of blue (B). Figure 25-16a shows that when a primary color and its complementary color are mixed, the result is white light. This must be the case: Because cyan, for example, is itself the combination of two of the primary colors—green and blue (C = G + B)—the combination of cyan and red (C + R) is the same as the combination of the three primary colors: G + B + R = W.

In subtractive color mixing, some of the wavelength components of white light are removed by absorption, and the reflected light (for example, from a painted surface or colored fabric) or transmitted light (as seen through glass or a solution) is deficient in some wavelength components. The reflected or transmitted light is colored. In subtractive color mixing (Figure 25-16b), we can again define primary, secondary, and complementary colors. In this case, let's choose *magenta* (M), *yellow* (Y), and *cyan* (C) as the primary colors and *green* (G), *blue* (B), and *red* (R) as the secondary colors. If a material absorbs all three primary colors, there is essentially no light left to be reflected or transmitted; the material appears black, or nearly so. If a material absorbs one color, primary or secondary, the reflected or transmitted light is the complementary color. Thus, a magenta sweater has that color because the dye it contains strongly absorbs green light and reflects magenta, the complement of green. A solution of red food dye has that color because the dye absorbs cyan light and transmits red.

Now let's direct our attention specifically to colored solutions. Colored solutions contain species that can absorb photons of visible light and use the energy of those photons to promote electrons in the species to higher energy levels. The energies of the photons must just match the energy differences through which the electrons are to be promoted. Because the energies of photons are related to the frequencies (and wavelengths) of light (recall Planck's equation, $E = h\nu$), only certain wavelength components are absorbed as white light passes through the solution. The emerging light, because it is lacking some wavelength components, is no longer white; it is colored.

Ions having (1) a noble-gas electron configuration, (2) an outer shell of 18 electrons, or (3) the "18 + 2" configuration (18 electrons in the $n - 1$ shell and two in the n, or outermost, shell) do not have electron transitions in the energy range corresponding to visible light. White light passes through these solutions without being absorbed; these ions are colorless in solution. Examples are the alkali and alkaline earth metal ions, the halide ions, Zn^{2+}, Al^{3+}, and Bi^{3+}.

Crystal field splitting of the d energy levels produces the energy difference, Δ, that accounts for the colors of complex ions. Promotion of an electron from a lower to a higher d level results from the absorption of the appropriate components of white light; the transmitted light is colored. Subtracting one color from white light leaves the complementary color. A solution containing $[Cu(H_2O)_4]^{2+}$ absorbs most strongly in the yellow region of the spectrum (about 580 nm). The wavelength components of the light transmitted combine to produce the color *blue*. Thus, aqueous solutions of copper(II) compounds usually have a characteristic blue color. In the presence of high concentrations of Cl^-, copper(II) forms the complex ion $[CuCl_4]^{2-}$. This species absorbs strongly in the blue region of the spectrum. The transmitted light, and hence the color of the solution is *yellow*. Figure 25-17 suggests light absorption by these solutions. The colors of some complex ions of chromium are given in Table 25.5. The colors of six related coordination compounds of cobalt(III) are shown in Figure 25-18.

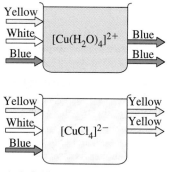

▲ **FIGURE 25-17**

Light absorption and transmission

$[Cu(H_2O)_4]^{2+}$ absorbs in the yellow region of the spectrum and transmits blue light. $[CuCl_4]^{2-}$ absorbs in the blue region of the spectrum and transmits yellow light.

TABLE 25.5   Some Coordination Compounds of Cr^{3+} and Their Colors

Isomer	Color
$[Cr(H_2O)_6]Cl_3$	Violet
$[CrCl(H_2O)_5]Cl_2$	Blue-green
$[Cr(NH_3)_6]Cl_3$	Yellow
$[CrCl(NH_3)_5]Cl_2$	Purple

EXAMPLE 25-6

Relating the Colors of Complexes to the Spectrochemical Series. Table 25.5 lists the color of $[Cr(H_2O)_6]Cl_3$ as violet, whereas that of $[Cr(NH_3)_6]Cl_3$ is yellow. Explain this difference in color.

Solution

Here are the basic facts that we must use: Both chromium(III) complexes in Table 25.5 are octahedral, and the electron configuration of Cr^{3+} is $[Ar]3d^3$. From these facts, we can

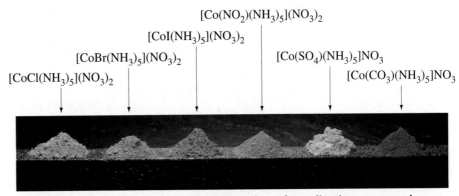

$[Co(NO_2)(NH_3)_5](NO_3)_2$

$[CoI(NH_3)_5](NO_3)_2$

$[CoBr(NH_3)_5](NO_3)_2$

$[Co(SO_4)(NH_3)_5]NO_3$

$[CoCl(NH_3)_5](NO_3)_2$

$[Co(CO_3)(NH_3)_5]NO_3$

▲ **FIGURE 25-18** **Effect of ligands on the colors of coordination compounds**
These compounds all consist of a six-coordinate cobalt complex ion in combination with
nitrate ions. In each case, the complex ion has five NH_3 molecules and one other group as
ligands.

construct the energy-level diagram shown here. The three unpaired electrons go into the
three lower-energy d orbitals. When a photon of light is absorbed, an electron is promoted from the lower to an upper energy level. The quantity of energy required for this
promotion depends on the energy level separation, Δ.

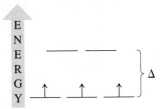

According to the spectrochemical series, NH_3 produces a greater splitting of the d energy level than does H_2O. We should expect $[Cr(NH_3)_6]^{3+}$ to absorb light of a *shorter* wavelength (higher energy) than does $[Cr(H_2O)_6]^{3+}$. Thus, $[Cr(NH_3)_6]^{3+}$ absorbs in the violet
region of the spectrum, and the transmitted light is *yellow*. $[Cr(H_2O)_6]^{3+}$ absorbs in the
yellow region of the spectrum, and the transmitted light is *violet*.

Practice Example A: The color of $[Co(H_2O)_6]^{2+}$ is pink, whereas that of tetrahedral $[CoCl_4]^{2-}$ is blue. Explain this difference in color.

Practice Example B: One of the following solids is yellow, and the other is
green: $Fe(NO_3)_2 \cdot 6H_2O$; $K_4[Fe(CN)_6] \cdot 3H_2O$. Indicate which is which, and explain
your reasoning.

25-8 Aspects of Complex-Ion Equilibria

In Chapter 19, we learned that complex-ion formation can have a great effect on the
solubilities of substances, as in the ability of NH_3(aq) to dissolve rather large quantities of AgCl(s). To calculate the solubility of AgCl(s) in NH_3(aq), we had to use
the formation constant, K_f, of $[Ag(NH_3)_2]^+$. Equations 25.1 and 25.2 illustrate how
we dealt with formation constants in Chapter 19, with $[Zn(NH_3)_4]^{2+}$ as an example.

$$Zn^{2+}(aq) + 4\,NH_3(aq) \rightleftharpoons [Zn(NH_3)_4]^{2+}(aq) \qquad (25.1)$$

$$K_f = \frac{[[Zn(NH_3)_4]^{2+}]}{[Zn^{2+}][NH_3]^4} = 4.1 \times 10^8 \qquad (25.2)$$

In fact, cations in aqueous solution exist mostly in *hydrated* form. That is, $Zn^{2+}(aq)$ is actually $[Zn(H_2O)_4]^{2+}$. As a result, when NH_3 molecules bond to Zn^{2+} and form an ammine complex ion, they do not enter an empty coordination sphere. They must displace H_2O molecules. This displacement occurs in a stepwise fashion. The reaction

$$[Zn(H_2O)_4]^{2+} + NH_3 \rightleftharpoons [Zn(H_2O)_3NH_3]^{2+} + H_2O \tag{25.3}$$

for which

$$K_1 = \frac{[[Zn(H_2O)_3NH_3]^{2+}]}{[[Zn(H_2O)_4]^{2+}][NH_3]} = 3.9 \times 10^2 \tag{25.4}$$

is followed by

$$[Zn(H_2O)_3NH_3]^{2+} + NH_3 \rightleftharpoons [Zn(H_2O)_2(NH_3)_2]^{2+} + H_2O \tag{25.5}$$

for which

$$K_2 = \frac{[[Zn(H_2O)_2(NH_3)_2]^{2+}]}{[[Zn(H_2O)_3NH_3]^{2+}][NH_3]} = 2.1 \times 10^2 \tag{25.6}$$

and so on.

The value of K_1 in equation (25.4) is often designated as β_1 and called the formation constant for the complex ion $[Zn(H_2O)_3NH_3]^{2+}$. The formation of $[Zn(H_2O)_2(NH_3)_2]^{2+}$ is represented by the *sum* of equations (25.3) and (25.5),

$$[Zn(H_2O)_4]^{2+} + 2\,NH_3 \rightleftharpoons [Zn(H_2O)_2(NH_3)_2]^{2+} + 2\,H_2O \tag{25.7}$$

The formation constant β_2, in turn, is given by the *product* of equations (25.4) and (25.6).

$$\beta_2 = \frac{[[Zn(H_2O)_2(NH_3)_2]^{2+}]}{[[Zn(H_2O)_4]^{2+}][NH_3]^2} = K_1 \times K_2 = 8.2 \times 10^4 \tag{25.8}$$

For the next ion in the series, $[Zn(H_2O)(NH_3)_3]^{2+}$, $\beta_3 = K_1 \times K_2 \times K_3$. For the final member, $[Zn(NH_3)_4]^{2+}$, $\beta_4 = K_1 \times K_2 \times K_3 \times K_4$, and it is this product of terms that we called the formation constant in Section 19-8 and listed as K_f in Table 19.2. Additional stepwise formation constants are presented in Table 25.6.

KEEP IN MIND

that the product relationship between equilibrium constants is equivalent to a summation of $\Delta G°$ values. Hence, when an overall $\Delta G°$ is used to evaluate K_n, $\Delta G°$ (overall)
$$= \Delta G_1° + \Delta G_2° + \Delta G_3° + \dots$$
$$= -RT \ln K_1 - RT \ln K_2$$
$$\qquad - RT \ln K_3 - \dots$$
$$= -RT \ln(K_1 \times K_2 \times K_3 \dots)$$
$$= -RT \ln \beta_n = -RT \ln K_f. \blacktriangleright$$

TABLE 25.6 Stepwise and Overall Formation (Stability) Constants for Several Complex Ions

Metal Ion	Ligand	K_1	K_2	K_3	K_4	K_5	K_6	β_n (or K_f)[a]
Ag^+	NH_3	2.0×10^3	7.9×10^3					1.6×10^7
Zn^{2+}	NH_3	3.9×10^2	2.1×10^2	1.0×10^2	5.0×10^1			4.1×10^8
Cu^{2+}	NH_3	1.9×10^4	3.9×10^3	1.0×10^3	1.5×10^2			1.1×10^{13}
Ni^{2+}	NH_3	6.3×10^2	1.7×10^2	5.4×10^1	1.5×10^1	5.6	1.1	5.3×10^8
Cu^{2+}	en	5.2×10^{10}	2.0×10^9					1.0×10^{20}
Ni^{2+}	en	3.3×10^7	1.9×10^6	1.8×10^4				1.1×10^{18}
Ni^{2+}	EDTA	4.2×10^{18}						4.2×10^{18}

In many tabulations in the chemical literature, formation constant data are presented as logarithms: that is, $\log K_1$, $\log K_2$, ..., and $\log \beta_n$.

[a] The β_n listed is for the number of steps shown: e.g., for $[Ag(NH_3)_2]^+$, $\beta_2 = K_f = K_1 \times K_2$; for $[Ni(en)_3]^{2+}$, $\beta_3 = K_f = K_1 \times K_2 \times K_3$; and for $[Ni(EDTA)]^{2-}$, $\beta_1 = K_f = K_1$.

The large numerical value of K_1 for reaction (25.3) indicates that Zn^{2+} has a greater affinity for NH_3 (a stronger Lewis base) than it does for H_2O. Displacement of ligand H_2O molecules by NH_3 occurs even if the number of NH_3 molecules present in aqueous solution is much smaller than the number of H_2O molecules, as in dilute $NH_3(aq)$. The fact that the successive K values decrease regularly in the displacement process, at least for displacements involving neutral molecules as ligands, is due in part to statistical factors: An NH_3 molecule has a better chance of replacing a H_2O molecule in $[Zn(H_2O)_4]^{2+}$, in which each coordination position is occupied by H_2O, than in $[Zn(H_2O)_3NH_3]^{2+}$, in which one of the positions is already occupied by NH_3. Also, once the degree of substitution of NH_3 for H_2O has become large, the chances of H_2O molecules replacing NH_3 molecules in a reverse reaction improve; again, this tends to reduce the value of K. If irregularities arise in the succession of K values, it is often because of a change in structure of the complex ion at some point in the series of displacement reactions.

If the ligand in a substitution process is polydentate, it displaces as many H_2O molecules as there are points of attachment. Thus, ethylenediamine (en) displaces H_2O molecules in $[Ni(H_2O)_6]^{2+}$ two at a time, in three steps. The first step is

$$[Ni(H_2O)_6]^{2+} + en \rightleftharpoons [Ni(en)(H_2O)_4]^{2+} + 2\,H_2O \qquad K_1(\beta_1) = 3.3 \times 10^7 \qquad (25.9)$$

Note from Table 25.6 that the complex ions with polydentate ligands have much larger formation constants than do those with monodentate ligands. For example, K_f (that is, β_3) for $[Ni(en)_3]^{2+}$ is 1.1×10^{18}, whereas K_f (that is, β_6) for $[Ni(NH_3)_6]^{2+}$ is 5.3×10^8. The additional stability of chelates over complexes with monodentate ligands is known as the **chelation (or chelate) effect**. We can attribute this effect partly to the increase in entropy associated with chelation. In the displacement of H_2O by NH_3, the entropy change is small [two particles on each side of an equation such as (25.3)]. An ethylenediamine molecule, conversely, displaces *two* H_2O molecules [two particles on the left and three on the right of equation (25.9)]. The larger, positive value of $\Delta S°$ for the displacement by ethylenediamine means a more negative $\Delta G°$ and a larger K.

25-9   Acid–Base Reactions of Complex Ions

We have described complex-ion formation in terms of Lewis acids and bases. Complex ions may also exhibit acid–base properties in the Brønsted–Lowry sense; that is, they may act as proton donors or acceptors. Figure 25-19 represents the ionization

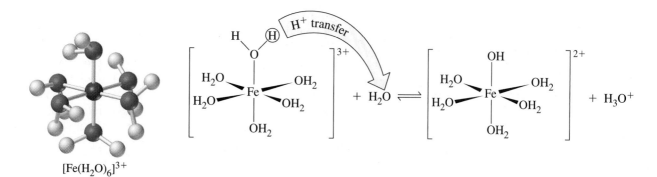

$[Fe(H_2O)_6]^{3+}$

▲ **FIGURE 25-19**   **Ionization of $[Fe(H_2O)_6]^{3+}$**

of $[Fe(H_2O)_6]^{3+}$ as an acid. A proton from a *ligand* water molecule in hexaaqua-iron(III) ion is transferred to a *solvent* water molecule. The H_2O ligand is converted to OH^-.

$$[Fe(H_2O)_6]^{3+} + H_2O \rightleftharpoons [FeOH(H_2O)_5]^{2+} + H_3O^+ \qquad K_{a_1} = 9 \times 10^{-4} \qquad (25.10)$$

The second ionization step is

$$[FeOH(H_2O)_5]^{2+} + H_2O \rightleftharpoons [Fe(OH)_2(H_2O)_4]^+ + H_3O^+ \qquad K_{a_2} = 5 \times 10^{-4} \quad (25.11)$$

From these K_a values, we see that Fe^{3+}(aq) is fairly acidic (compared, for example, with acetic acid, with $K_a = 1.8 \times 10^{-5}$). To suppress ionization (hydrolysis) of $[Fe(H_2O)_6]^{3+}$, we need to maintain a low pH by the addition of acids such as HNO_3 or $HClO_4$. The ion $[Fe(H_2O)_6]^{3+}$ is violet in color, but aqueous solutions of Fe^{3+}(aq) are generally yellow because of the presence of hydroxo complex ions.

The ions Cr^{3+} and Al^{3+} behave in a manner similar to Fe^{3+} except that, with them, hydroxo complex-ion formation can continue until complex anions are produced. $Cr(OH)_3$ and $Al(OH)_3$, as we previously noted, are soluble in alkaline as well as acidic solutions; they are amphoteric.

Regarding the acid strengths of aqua complex ions, a critical factor is the charge-to-radius ratio of the central metal ion. Thus, the small, highly charged Fe^{3+} attracts electrons away from an O—H bond in a ligand water molecule more strongly than does Fe^{2+}. Hence, $[Fe(H_2O)_6]^{3+}$ is a stronger acid ($K_{a_1} = 9 \times 10^{-4}$) than is $[Fe(H_2O)_6]^{2+}$ ($K_{a_1} = 1 \times 10^{-7}$).

25-10   Some Kinetic Considerations

When we add NH_3(aq) to an aqueous solution containing Cu^{2+}, we see a change in color from pale blue to very deep blue. The reaction involves NH_3 molecules displacing H_2O molecules as ligands.

$$[Cu(H_2O)_4]^{2+} + 4 NH_3 \longrightarrow [Cu(NH_3)_4]^{2+} + 4 H_2O$$
(pale blue)   (very deep blue)

This reaction occurs very rapidly—as rapidly as the two reactants can be brought together. The addition of HCl(aq) to an aqueous solution of Cu^{2+} produces an immediate color change from pale blue to green, or even yellow if the HCl(aq) is sufficiently concentrated.

$$Cu(H_2O)_4]^{2+} + 4 Cl^- \longrightarrow [CuCl_4]^{2-} + 4 H_2O$$
(pale blue)   (yellow)

▶ The terms *labile* and *inert* are not related to the thermodynamic stabilities of complex ions or to the equilibrium constants for ligand-substitution reactions. The terms are *kinetic* terms, referring to the rates at which ligands are exchanged.

Complex ions in which ligands can be interchanged rapidly are said to be **labile**. $[Cu(H_2O)_4]^{2+}$, $[Cu(NH_3)_4]^{2+}$, and $[CuCl_4]^{2-}$ are all labile (Figure 25-20).

In freshly prepared $CrCl_3$(aq), the ion *trans*-$[CrCl_2(H_2O)_4]^+$ produces a green color, but the color gradually turns to violet (Figure 25-21). This color change results from the very slow exchange of H_2O for Cl^- ligands. A complex ion that exchanges ligands slowly is said to be *nonlabile*, or **inert**. In general, complex ions of the first transition series, except for those of Cr(III) and Co(III), are kinetically labile. Those of the second and third transition series are generally kinetically inert. Whether a complex ion is labile or inert affects the ease with which it can be studied. The inert ones are easiest to obtain and characterize, which may explain why so many of the early studies of complex ions were based on Cr(III) and Co(III).

▲ FIGURE 25-20 **Labile complex ions**
The exchange of ligands in the coordination sphere of Cu^{2+} occurs very rapidly. The solution at the extreme left is formed by dissolving $CuSO_4$ in concentrated HCl(aq). Its yellow color is due to $[CuCl_4]^{2-}$. When a small amount of water is added, the mixture of $[Cu(H_2O)_4]^{2+}$ and $[CuCl_4]^{2-}$ ions produces a yellow-green color. When $CuSO_4$ is dissolved in water, a light blue solution of $[Cu(H_2O)_4]^{2+}$ forms. NH_3 molecules readily displace H_2O molecules as ligands and produce deep blue $[Cu(NH_3)_4]^{2+}$ (extreme right).

25-11 Applications of Coordination Chemistry

The applications of coordination chemistry are numerous and varied. They range from analytical chemistry to biochemistry. The several brief examples in this section give some idea of this diversity.

Hydrates

When a compound is crystallized from an aqueous solution of its ions, the crystals obtained are often hydrated. A hydrate is a substance that has a fixed number of water molecules associated with each formula unit. In some cases, the water molecules are ligands bonded directly to a metal ion. The coordination compound $[Co(H_2O)_6](ClO_4)_2$ may be represented as the hexahydrate, $Co(ClO_4)_2 \cdot 6H_2O$. In the hydrate $CuSO_4 \cdot 5H_2O$, four H_2O molecules are associated with copper in the complex ion $[Cu(H_2O)_4]^{2+}$, and the fifth with the SO_4^{2-} anion by hydrogen bonding. Another possibility for hydrate formation is that the water molecules may be incorporated into definite positions in the solid crystal but not associated with any particular cations or anions, as in $BaCl_2 \cdot 2H_2O$. This is called *lattice water*. Finally, part of the water may be coordinated to an ion and part of it may be lattice water. That appears to be the case with alums, such as $KAl(SO_4)_2 \cdot 12H_2O$.

Stabilization of Oxidation States

The standard electrode potential for the reduction of Co(III) to Co(II) is

$$Co^{3+}(aq) + e^- \longrightarrow Co^{2+}(aq) \qquad E^\circ = +1.82 \text{ V}$$

This large positive value suggests that $Co^{3+}(aq)$ is a strong oxidizing agent, strong enough to oxidize water to $O_2(g)$.

$$4\,Co^{3+}(aq) + 2\,H_2O(l) \longrightarrow 4\,Co^{2+}(aq) + 4\,H^+ + O_2(g) \qquad E^\circ_{cell} = +0.59 \text{ V} \quad (25.12)$$

Yet one of the complex ions featured in this chapter has been $[Co(NH_3)_6]^{3+}$. This ion is stable in water solution, even though it contains cobalt in the +3 oxidation

▲ FIGURE 25-21
Inert complex ions
The green solid $CrCl_3 \cdot 6H_2O$ produces the green aqueous solution on the left. The color is due to *trans*-$[CrCl_2(H_2O)_4]^+$. A slow exchange of H_2O for Cl^- ligands leads to a violet solution of $[Cr(H_2O)_6]^{3+}$ in one or two days (right).

state. Reaction (25.12) will not occur if the concentration of Co^{3+} is sufficiently low and $[Co^{3+}]$ is kept very low because of the great stability of the complex ion.

$$Co^{3+}(aq) + 6\,NH_3(aq) \rightleftharpoons [Co(NH_3)_6]^{3+}(aq) \qquad K_f = 4.5 \times 10^{33}$$

In fact, the concentration of free Co^{3+} is so low that for the half-reaction

$$[Co(NH_3)_6]^{3+} + e^- \longrightarrow [Co(NH_3)_6]^{2+}$$

$E°$ is only $+0.10$ V. As a consequence, not only is $[Co(NH_3)_6]^{3+}$ stable but also $[Co(NH_3)_6]^{2+}$ is rather easily oxidized to the Co(III) complex.

The ability of strong electron-pair donors (strong Lewis bases) to stabilize high oxidation states in the way that NH_3 does in Co(III) complexes and O^{2-} in Mn(VII) complexes (such as MnO_4^-) affords a means of attaining certain oxidation states that might otherwise be difficult or impossible to attain.

Photography: Fixing a Photographic Film

A black-and-white photographic film is an emulsion of a finely divided silver halide (typically AgBr) coated on a strip of polymer, such as a modified cellulose. In the *exposure* step, the film is exposed to light and some of the tiny granules of AgBr(s) absorb photons. The photons promote the oxidation of Br^- to Br and the reduction of Ag^+ to Ag. The Ag and Br atoms remain in the crystalline lattice of AgBr(s) as "defects," in numbers that depend on the intensity of the light absorbed: The brighter the light, the more Ag atoms. Because the actual number of Ag atoms produced in the exposure is not large, the silver is invisible to the eye. The pattern of distribution of the Ag atoms, however, creates a *latent* image of the object photographed. To obtain a visible image, the film is developed.

In the *developing* step, the exposed film is placed in a solution of a mild reducing agent such as hydroquinone, $C_6H_4(OH)_2$. An oxidation–reduction reaction occurs in which Ag^+ ions are reduced to Ag and the hydroquinone is oxidized. The action of the developer is such that reduction of Ag^+ to Ag occurs just in those granules of AgBr(s) that contain Ag atoms of the latent image. As a result, the number of Ag atoms in the film is greatly increased, and the latent image becomes visible. Bright regions of the photographed object appear as dark regions in the photographic image. At this stage, the film is a photographic *negative*. This negative cannot be exposed to light, however, because reduction of Ag^+ to Ag could still occur in the previously unexposed granules of AgBr(s). The negative must be fixed.

The *fixing* step requires that the black metallic silver of the negative remain on the film and the unexposed AgBr(s) be removed. A common fixer is an aqueous solution of sodium thiosulfate (also known as sodium hyposulfite or hypo). Because the complex ion $[Ag(S_2O_3)_2]^{3-}$ has a large formation constant, the following reaction is driven to completion—the AgBr(s) dissolves.

$$AgBr(s) + 2\,S_2O_3^{2-}(aq) \longrightarrow [Ag(S_2O_3)_2]^{3-}(aq) + Br^-(aq) \qquad (25.13)$$

Once the negative has been fixed, it is used to produce a *positive* image, the final photograph. This is done by projecting light through the negative onto a piece of photographic paper. Regions of the negative that are dark transmit little light to the photographic paper and will appear light when the photographic paper is subsequently developed and fixed. Conversely, light areas of the negative will appear dark in the final print. In this way, the areas of light and dark in the final print are the same as in the photographed object.

▲ FIGURE 25-22 **Qualitative tests for Co²⁺ and Fe³⁺**
(a) $[Co(SCN)_4]^{2-}$ complex ion. **(b)** $[FeSCN(H_2O)_5]^{2+}$ complex ion. **(c)** Mixture of $[FeF_6]^{3-}$ and $[Co(SCN)_4]^{2-}$.

Qualitative Analysis

In discussing qualitative cation analysis in Section 19-9 (page 768), we showed how the group 1 precipitate—$AgCl(s)$, $PbCl_2(s)$, or $Hg_2Cl_2(s)$—is separated by taking advantage of the stable complex ion formed by $Ag^+(aq)$ and $NH_3(aq)$.

$$AgCl(s) + 2\,NH_3(aq) \longrightarrow [Ag(NH_3)_2]^+(aq) + Cl^-(aq)$$

The qualitative analysis scheme abounds in other examples of complex-ion formation. For example, at a point in the procedure for cation group 3, a test is needed for Co^{2+}. In the presence of SCN^- ion, Co^{2+} forms a blue thiocyanato complex ion, $[Co(SCN)_4]^{2-}$ (Figure 25-22a). A problem develops, however, if even a trace amount of Fe^{3+} is present in the solution. Fe^{3+} reacts with SCN^- to produce $[FeSCN(H_2O)_5]^{2+}$, a strongly colored, blood-red complex ion (Figure 25-22b). Fortunately, this complication can be resolved by treating a solution containing both Co^{2+} and Fe^{3+} with an excess of F^-. The Fe^{3+} is converted to the extremely stable, pale yellow $[FeF_6]^{3-}$. The complex ion $[CoF_4]^{2-}$, being much less stable than $[Co(SCN)_4]^{2-}$, does not form. As a result, the $[Co(SCN)_4]^{2-}(aq)$ can be detected via the blue-green solution color (Figure 25-22c).

Sequestering Metal Ions

Metal ions can act as catalysts in promoting undesirable chemical reactions in a manufacturing process, or they may alter the properties of the material being manufactured. Thus, for many industrial purposes, it is imperative to remove mineral impurities from water. Often these impurities, such as Cu^{2+}, are present only in trace amounts, and precipitation of metal ions is feasible only if K_{sp} for the precipitate is very small. An alternative is to treat the water with a chelating agent. This reduces the free cation concentrations to the point at which the cations can no longer enter into objectionable reactions. The cations are said to be *sequestered*. Among the chelating agents widely employed are the salts of *ethylenediaminetetraacetic acid* (H_4EDTA), usually as the sodium salt.

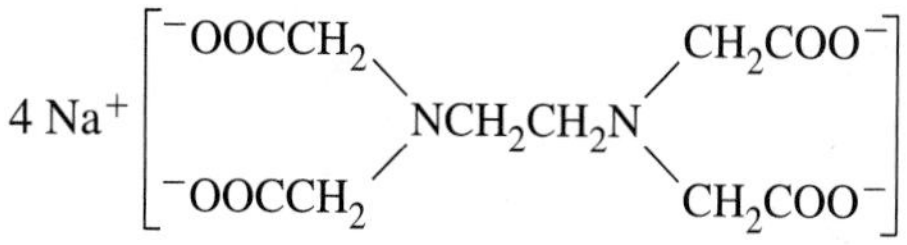

$$4\,Na^+ \begin{bmatrix} {}^-OOCCH_2 & & CH_2COO^- \\ & NCH_2CH_2N & \\ {}^-OOCCH_2 & & CH_2COO^- \end{bmatrix}$$

Mirror

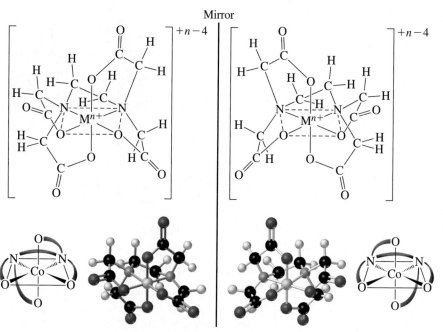

MEDTA model

▲ **FIGURE 25-23   Structure of a metal–EDTA complex**
The central metal ion M^{n+} (gray) can be Ca^{2+}, Mg^{2+}, Fe^{2+}, Fe^{3+}, and so on. The ligand is $EDTA^{4-}$, and the net charge on the complex is $+n-4$. Structural diagrams and ball-and-stick models are shown for the two optical isomers of an $[MEDTA]^{n-4}$ complex.

A representative complex ion formed by a metal ion, M^{n+}, with the hexadentate $EDTA^{4-}$ anion is depicted in Figure 25-23. The high stability of such complexes can be attributed to the presence of five, five-member chelate rings.

In the presence of $EDTA^{4-}(aq)$, Ca^{2+}, Mg^{2+}, and Fe^{3+} in hard water are unable to form boiler scale or to precipitate as insoluble soaps. The cations are sequestered in the complex ions: $[Ca(EDTA)]^{2-}$, $[Mg(EDTA)]^{2-}$, and $[Fe(EDTA)]^-$, having K_f values of 4×10^{10}, 4.0×10^8, and 1.7×10^{24}, respectively.

Chelation with EDTA can be used in treating some cases of metal poisoning. If a person with lead poisoning is fed $[Ca(EDTA)]^{2-}$, the following exchange occurs because $[Pb(EDTA)]^{2-}$ ($K_f = 2 \times 10^{18}$) is even more stable than $[Ca(EDTA)]^{2-}$ ($K_f = 4 \times 10^{10}$).

$$Pb^{2+} + [Ca(EDTA)]^{2-} \longrightarrow [Pb(EDTA)]^{2-} + Ca^{2+}$$

The body excretes the lead complex, and the Ca^{2+} remains as a nutrient. A similar method can be used to rid the body of radioactive isotopes, as in the treatment of plutonium poisoning.

Some plant fertilizers contain EDTA chelates of metals such as Cu^{2+} as a soluble form of the metal ion for the plant to use. Metal ions can catalyze reactions that cause mayonnaise and salad dressings to spoil, and the addition of EDTA reduces the concentration of metal ions by chelation.

▲ Some products containing EDTA.

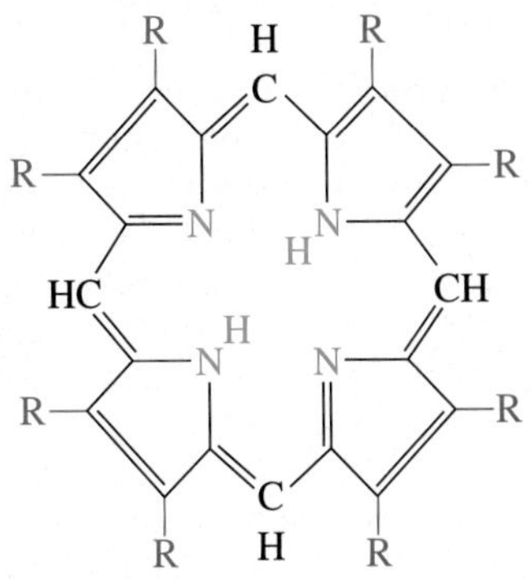

▲ FIGURE 25-24
The porphyrin structure

Biological Applications: Porphyrins

The structure in Figure 25-24 is commonly found in both plant and animal matter. If the eight R groups are all H atoms, the molecule is called *porphin*. The central N atoms can give up their H atoms, and a metal atom can coordinate simultaneously with all four N atoms. The porphin is a tetradentate ligand for the central metal, and the metal–porphin complex is called a *porphyrin*. Specific porphyrins differ in their central metals and in the R groups on the porphin rings.

In *photosynthesis*, carbon dioxide and water, in the presence of inorganic salts, a catalytic agent called *chlorophyll*, and sunlight, combine to form carbohydrates.

$$n\, CO_2 + n\, H_2O \xrightarrow[\text{chlorophyll}]{\text{sunlight}} (CH_2O)_n + n\, O_2 \qquad (25.14)$$

$$\text{carbohydrates}$$

Carbohydrates are the main structural materials of plants. Chlorophyll is a green pigment that absorbs sunlight and directs the storage of this energy into the chemical bonds of the carbohydrates. The structure of one type of chlorophyll is shown in Figure 25-25; it is a porphyrin. The central metal ion is Mg^{2+}.

Green is the complementary color of magenta—a purplish red—so we should expect chlorophyll to absorb light in the red region of the spectrum (about 670–680 nm). This suggests that green plants should grow more readily in red light than in light of other colors, and some experimental evidence indicates that this is the case. For example, the maximum rate of formation of $O_2(g)$ by reaction (25.14) occurs with red light.

In Chapter 28, we will consider another porphyrin structure that is essential to life—hemoglobin.

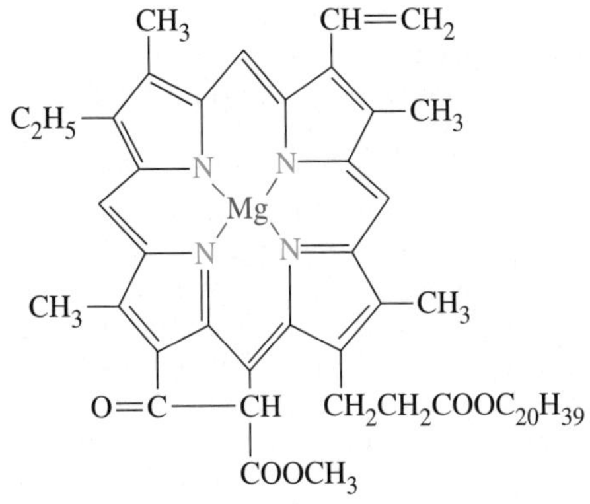

▲ FIGURE 25-25   Structure of chlorophyll *a*

▲ Emeralds in a matrix

Many of the beautifully colored stones that we consider precious or semiprecious are chemically impure. It is the impurities—transition metal ions—that are responsible for the brilliant colors. Consider ruby and emerald. Both of these gems contain small amounts of Cr^{3+} ion as the color-causing impurity. Rubies, however, have a deep red color, whereas emeralds have a distinctive blue-green color. Let us see why.

In a ruby crystal, the principal constituent is aluminum oxide (Al_2O_3), or *corundum*, with a structural pattern in which each Al^{3+} ion is surrounded by six O^{2-} ions. Although the radius of the Cr^{3+} ion is somewhat larger (62 pm) than that of Al^{3+} (53 pm), these values match well enough that Cr^{3+} is able to replace Al^{3+} in the crystal lattice. In ruby, about 1% of the Al^{3+} ions are replaced by Cr^{3+} ions.

Because Al^{3+} has no valence electrons and no readily available electron transitions for its core electrons, a pure crystal of corundum is colorless. The color of ruby is the result of electron transitions between *d* orbitals of Cr^{3+} ions. Cr^{3+} has the ground-state electron configuration: $[Ar]3d^3$. The six O^{2-} ions create an octahedral crystal field about the Cr^{3+} ion and split the energy levels of the *d* orbitals into two groups (recall Figure 25-12). In the ground state, the three valence electrons occupy the three lower-energy *d* orbitals and the two higher-energy *d* orbitals are vacant. Two excited states of the Cr^{3+} ion can be reached by absorption of two different frequencies (colors) of visible light. One transition, a low-energy transition, requires absorption of a yellow-green light, and the other, a high-energy transition, the absorption of violet light. As a consequence of these absorptions, the light that is transmitted by a ruby is *red*, with a light purple tint.

The basic structure of an emerald is that of a crystal of *beryl*, $3BeO \cdot Al_2O_3 \cdot 6SiO_2$. Here, Cr^{3+} replaces Al^{3+} in the same type

---

Summary

Many metal atoms or ions, particularly those of the transition elements, have the ability to form bonds with electron-pair donors (ligands). Polydentate ligands are able to attach simultaneously to two or more positions at the metal center. This multiple attachment produces complexes with five- or six-member rings of atoms—chelates.

To name complexes, we need rules that deal with such matters as denoting the number and kinds of ligands; the oxidation state of the metal center; and whether the complex is neutral, a cation, or an anion. The positions at which ligands are attached to the metal center are not always equivalent. In geometric isomerism, different structures with different properties result, depending on where this attachment occurs. Optical isomers differ by being nonsuperimposable mirror images of each other.

A bonding theory for complex ions that is useful in explaining the magnetic properties and characteristic colors of complex ions is the crystal field theory. This theory emphasizes the splitting of the *d* energy level of the central metal ion as a result of repulsions between *d*-orbital electrons of the central ion and electrons of the ligands. A prediction of the magnitude of *d*-level splitting produced by a ligand can be made via the spectrochemical series.

The formation of a complex ion can be viewed as a stepwise equilibrium process in which other ligands displace H_2O molecules from aqua complex ions. The stepwise constants can be combined into an overall formation constant for the complex ion, K_f. The ability of ligand water molecules to ionize causes some aqua complexes to exhibit acidic properties and helps explain amphoterism. Also important in determining properties of a complex ion is the rate at which the ion exchanges ligands between its coordination sphere and the solution. Exchange is rapid in a labile complex and slow in an inert complex.

Complex-ion formation can be used to stabilize certain oxidation states, such as Co(III). Other applications include dissolving precipitates, such as AgCl by $NH_3(aq)$ in the qualitative analysis scheme and AgBr by $Na_2S_2O_3(aq)$ in the photographic process, and sequestering ions by chelation, as with EDTA.

of octahedral field as in ruby but with dramatically different results. The presence of BeO and SiO_2 in the beryl crystal weakens the crystal field. The effect of the weaker field is a smaller energy separation (Δ) between the two groups of d orbitals in Cr^{3+} than is found in ruby. The energy per photon required to produce electron transitions between the d orbitals is less for Cr^{3+} in emerald than in ruby. The low-energy transition shifts from yellow-green to yellow-red, and the higher-energy transition is also lowered slightly in energy. As a result, when light passes through an emerald, most of the red and violet colors are absorbed and the blue and green are transmitted; the emerald takes on a blue-green color.

For rubies and emeralds to exhibit their colors continuously, there must be a mechanism by which electrons in Cr^{3+} can revert from their excited states back to the ground state. This oc-

curs in two stages. First, electrons drop from the higher-energy states to a lower, intermediate state somewhat above the ground state. The energies of these transitions are in the infrared region of the spectrum; the emissions are invisible and appear as heat, which stimulates the vibrations of the ions in the crystal. The final transition to the ground state produces a red light. This type of light emission, in which a lower energy light is emitted following the absorption of higher energy light, is called *fluorescence*. Interestingly, both ruby and emerald display a red fluorescence emission. In ruby, this red emission adds to the red color of the transmitted light, and in emerald, it is of just the right wavelength to enhance the character of the green color. (When rubies and emeralds are observed in ultraviolet, or "black," light, *both* give off a red light.) The transitions referred to in this discussion are suggested in the diagram.

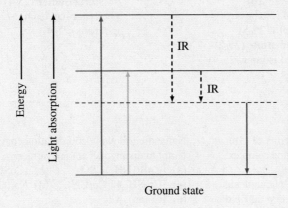

◀ **Light absorption and emission in a ruby**
The absorption of violet and yellow-green light from an incident white light beam causes the transmitted light to be red in color. In addition, the ruby emits a red light by fluorescence.

Integrative Example

Absorbance is a measure of the proportion of monochromatic (single-color) light that is absorbed as the light passes through a solution. An *absorption spectrum* is a graph of absorbance as a function of wavelength. High absorbances correspond to large proportions of the light entering a solution being absorbed. Low absorbances signify that large proportions of the light are transmitted. The absorption spectrum of $[Ti(H_2O)_6]^{3+}(aq)$ is shown in Figure 25-26a.

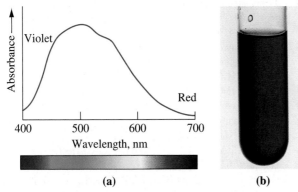

▲ **FIGURE 25-26   The color of $[Ti(H_2O)_6]^{3+}(aq)$**
(a) The absorption spectrum of $[Ti(H_2O)_6]^{3+}(aq)$. (b) A solution containing $[Ti(H_2O)_6]^{3+}(aq)$.

1. *Describe the color of light that $[Ti(H_2O)_6]^{3+}(aq)$ absorbs most strongly and the color of the solution.* The highest absorbances in the spectrum come at about 500 nm. The electromagnetic spectrum in Figure 9-3 indicates that the absorbed light should be dark green. Figure 25-16b indicates that, in subtractive color mixing, the complementary color of green is magenta. Thus, the color of the transmitted light (and hence the observed color of the solution) is a blend of red and blue.

2. *Describe the electron transition responsible for the absorption peak, and determine the energy associated with this absorption.* The electron configuration of Ti^{3+}, the central ion in the complex ion, is $[Ar]3d^1$. In the ground state, the $3d$ electron is in one of the three degenerate lower levels in the d-orbital splitting diagram for an octahedral complex (Figure 25-12). The quantity of energy we seek is that corresponding to electromagnetic radiation at the peak of the absorption spectrum: 500 nm. First, we can establish the frequency of this light, from $c = \nu \times \lambda$.

$$\nu = \frac{c}{\lambda} = \frac{2.998 \times 10^8 \text{ m s}^{-1}}{500 \times 10^{-9} \text{ m}} = 6.00 \times 10^{14} \text{ s}^{-1}$$

1017

Then, we use Planck's equation to determine E:

$E = h\nu = (6.626 \times 10^{-34} \text{ J s}) \times (6.00 \times 10^{14} \text{ s}^{-1}) =$
$$3.98 \times 10^{-19} \text{ J}$$

This is the energy per photon. If we want the energy on a per-mole basis, we can write

$E = (3.98 \times 10^{-19} \text{ J photon}^{-1}) \times$
$$(6.022 \times 10^{23} \text{ photon mol}^{-1}) \times \frac{1 \text{ kJ}}{1000 \text{ J}} = 240 \text{ kJ mol}^{-1}$$

Key Terms

bidentate (25-2)
chelate (25-2)
chelating agent (25-2)
chelation (chelate) effect (25-8)
chiral (25-4)
cis (25-4)
complementary color (25-7)
complex (25-1)
coordination compound (25-1)
coordination number (25-1)

crystal field theory (25-5)
enantiomer (25-4)
high spin (25-5)
inert (25-10)
isomer (25-4)
labile (25-10)
ligand (25-1)
low spin (25-5)
monodentate (25-2)
optical isomer (25-4)

pairing energy (25-5)
polydentate (25-2)
primary color (25-7)
secondary color (25-7)
spectrochemical series (25-5)
stereoisomer (25-4)
structural isomer (25-4)
trans (25-4)

Review Questions

1. In your own words, describe the following terms or symbols: (a) coordination number; (b) Δ; (c) ammine complex; (d) enantiomer.

2. Briefly describe each of the following ideas, phenomena, or methods: (a) spectrochemical series; (b) crystal field theory; (c) optical isomer; (d) structural isomerism.

3. Explain the important distinction between each pair: (a) coordination number and oxidation number; (b) monodentate and polydentate; (c) cis and trans isomers; (d) dextrorotatory and levorotatory; (e) low-spin and high-spin complexes.

4. Write the formula and name of
   (a) a complex ion having Cr^{3+} as the central ion and two NH_3 molecules and four Cl^- ions as ligands
   (b) a complex ion of iron(III) having a coordination number of 6 and CN^- as ligands
   (c) a coordination compound comprising two types of complex ions: one a complex of Cr(III) with ethylenediamine (en), having a coordination number of 6; the other, a complex of Ni(II) with CN^-, having a coordination number of 4

5. What are the coordination number and the oxidation state of the central metal ion in each of the following complexes? Name each complex.
   (a) $[Co(NH_3)_6]^{2+}$
   (b) $[AlF_6]^{3-}$
   (c) $[Cu(CN)_4]^{2-}$
   (d) $[CrBr_2(NH_3)_4]^+$
   (e) $[Co(ox)_3]^{4-}$
   (f) $[Ag(S_2O_3)_2]^{3-}$

6. Name the following complex ions. (Do not attempt to distinguish among isomers.)
   (a) $[AgI_2]^-$
   (b) $[Al(OH)(H_2O)_5]^{2+}$
   (c) $[Zn(CN)_4]^{2-}$
   (d) $[Pt(en)_2]^{2+}$
   (e) $[CoCl(NO_2)(NH_3)_4]^+$

7. Name the following coordination compounds. (Do not attempt to distinguish among isomers.)
   (a) $[CoBr(NH_3)_5]SO_4$
   (b) $[CoSO_4(NH_3)_5]Br$
   (c) $[Cr(NH_3)_6][Co(CN)_6]$
   (d) $Na_3[Co(NO_2)_6]$
   (e) $[Co(en)_3]Cl_3$

8. Write appropriate formulas for the following species. (Do not attempt to distinguish among isomers.)
   (a) dicyanoargentate(I) ion
   (b) triamminenitrito-N-platinum(II) ion
   (c) aquachlorobis(ethylenediamine)cobalt(III) ion
   (d) potassium hexacyanochromate(II)

9. Draw Lewis structures for the following monodentate ligands: (a) H_2O; (b) CH_3NH_2; (c) ONO^- (nitrito-O-); (d) SCN^- (thiocyanato-S-).

10. Write formulas for the following hydrates.
    (a) manganese(II) sulfate hexahydrate
    (b) potassium hexacyanochromate(II) trihydrate

11. Draw structures to represent these four complex ions: (a) $[PtCl_4]^{2-}$; (b) $[FeCl_4(en)]^-$; (c) *cis*-$[FeCl_2(ox)(en)]^-$; (d) *trans*-$[CrCl(OH)(NH_3)_4]^+$.

12. How many different structures are possible for each of the following complex ions? Draw each structure.
    (a) $[Co(H_2O)(NH_3)_5]^{3+}$
    (b) $[Co(H_2O)_2(NH_3)_4]^{3+}$
    (c) $[Co(H_2O)_3(NH_3)_3]^{3+}$
    (d) $[Co(H_2O)_4(NH_3)_2]^{3+}$

13. Indicate what type of isomerism may be found in each of the following cases. If no isomerism is possible, so indicate.
    (a) $[Zn(NH_3)_4][CuCl_4]$
    (b) $[Fe(CN)_5SCN]^{4-}$
    (c) $[NiCl(NH_3)_5]^+$
    (d) $[PtBrCl_2(py)]^-$
    (e) $[Cr(OH)_3(NH_3)_3]^-$

14. Of the complex ions $[Co(H_2O)_6]^{3+}$ and $[Co(en)_3]^{3+}$, one has a yellow color in aqueous solution; the other, blue. Match each ion with its expected color, and state your reason for doing so.

Exercises

Nomenclature

15. Supply acceptable names for the following. (Do not attempt to distinguish among isomers.)
   (a) $[Co(OH)(H_2O)_4(NH_3)]^{2+}$
   (b) $[Co(ONO)_3(NH_3)_3]$
   (c) $[Pt(H_2O)_4][PtCl_6]$
   (d) $[Fe(ox)_2(H_2O)_2]^-$
   (e) $Ag_2[HgI_4]$

16. Write appropriate formulas for the following. (Do not attempt to distinguish among isomers.)
   (a) potassium hexacyanoferrate(III)
   (b) bis(ethylenediamine)copper(II) ion
   (c) pentaaquahydroxoaluminum(III) chloride
   (d) amminechlorobis(ethylenediamine)chromium(III) sulfate
   (e) tris(ethylenediamine)iron(III) hexacyanoferrate(II)

Bonding and Structure in Complex Ions

17. Draw a plausible structure to represent
   (a) $[PtCl_4]^{2-}$
   (b) *fac*-$[Co(H_2O)_3(NH_3)_3]^{2+}$
   (c) $[CrCl(H_2O)_5]^{2+}$

18. Draw plausible structures of the following chelate complexes.
   (a) $[Pt(ox)_2]^{2-}$
   (b) $[Cr(ox)_3]^{3-}$
   (c) $[Fe(EDTA)]^{2-}$

19. Draw plausible structures corresponding to each of the following names.
   (a) pentamminesulfatochromium(III) ion
   (b) trioxalatocobaltate(III) ion
   (c) triamminedichloronitrito-*O*-cobalt(III)

20. Draw plausible structures corresponding to each of the following names.
   (a) pentamminenitrito-*N*-cobalt(III) ion
   (b) ethylenediaminedithiocyanato-*S*-copper(II)
   (c) hexaaquanickel(II) ion

Isomerism

21. Which of these general structures for a complex ion would you expect to exhibit cis and trans isomerism? Explain.
   (a) tetrahedral
   (b) square-planar
   (c) linear

22. Which of these octahedral complexes would you expect to exhibit *geometric* isomerism? Explain.
   (a) $[CrOH(NH_3)_5]^{2+}$
   (b) $[CrCl_2(H_2O)(NH_3)_3]^+$
   (c) $[CrCl_2(en)_2]^+$
   (d) $[CrCl_4(en)]^-$
   (e) $[Cr(en)_3]^{3+}$

23. If A, B, C, and D are four different ligands,
   (a) How many geometric isomers will be found for square-planar $[PtABCD]^{2+}$?
   (b) Will tetrahedral $[ZnABCD]^{2+}$ display optical isomerism?

24. Write the names and formulas of three coordination isomers of $[Co(en)_3][Cr(ox)_3]$.

25. Draw a structure for *cis*-dichlorobis(ethylenediamine)cobalt(III) ion. Is this ion chiral? Is the trans isomer chiral? Explain.

26. The structures of four complex ions are given. Each has Co^{3+} as the central ion. The ligands are H_2O, NH_3, and oxalate ion, $C_2O_4^{2-}$. Determine which, *if any*, of these complex ions are isomers (geometric or optical); which, *if any*, are identical (that is, have identical structures); and which, *if any*, are distinctly different.

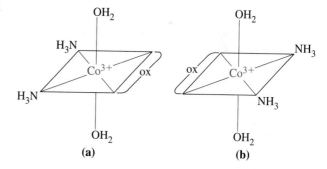
(a)   (b)

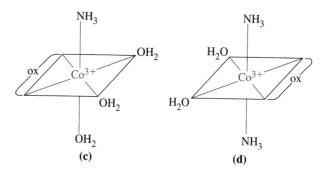
(c)   (d)

Crystal Field Theory

27. Describe how the crystal field theory makes possible an explanation of the fact that so many transition metal compounds are colored.

28. Cyano complexes of transition metal ions (such as Fe^{2+} and Cu^{2+}) are often yellow, whereas aqua complexes are often green or blue. Explain the basis for this difference in color.

29. If the ion Co^{2+} is linked with strong-field ligands to produce an octahedral complex, the complex has *one* unpaired electron. If Co^{2+} is linked with weak-field ligands, the complex has *three* unpaired electrons. How do you account for this difference?

30. In contrast to the case of Co^{2+} considered in Exercise 29, no matter what ligand is linked to Ni^{2+} to form an octahedral complex, the complex always has *two* unpaired electrons. Explain this fact.

31. Predict
  (a) which of the complex ions, $[MoCl_6]^{3-}$ and $[Co(en)_3]^{3+}$, is diamagnetic and which is paramagnetic
  (b) the number of unpaired electrons expected for the tetrahedral complex ion $[CoCl_4]^{2-}$.

32. Predict
  (a) whether the square-planar complex ion $[Cu(py)_4]^{2+}$ is diamagnetic or paramagnetic
  (b) whether octahedral $[Mn(CN)_6]^{3-}$ or tetrahedral $[FeCl_4]^-$ has the greater number of unpaired electrons.

33. In Example 25-5, we chose between a tetrahedral and a square-planar structure for $[Ni(CN)_4]^{2-}$ based on magnetic properties. Could we similarly use magnetic properties to establish whether the ammine complex of Ni(II) is octahedral $[Ni(NH_3)_6]^{2+}$ or tetrahedral $[Ni(NH_3)_4]^{2+}$? Explain.

34. In both $[Fe(H_2O)_6]^{2+}$ and $[Fe(CN)_6]^{4-}$ ions, the iron is present as Fe(II); however, $[Fe(H_2O)_6]^{2+}$ is paramagnetic, whereas $[Fe(CN)_6]^{4-}$ is diamagnetic. Explain this difference.

Complex-Ion Equilibria

35. Write equations to represent the following observations.
  (a) A mixture of $Mg(OH)_2(s)$ and $Zn(OH)_2(s)$ is treated with $NH_3(aq)$. The $Zn(OH)_2$ dissolves, but the $Mg(OH)_2(s)$ is left behind.
  (b) When $NaOH(aq)$ is added to $CuSO_4(aq)$, a pale blue precipitate forms. If $NH_3(aq)$ is added, the precipitate redissolves, producing a solution with an intense deep blue color. If this deep blue solution is made acidic with $HNO_3(aq)$, the color is converted back to pale blue.

36. Write equations to represent the following observations.
  (a) A quantity of $CuCl_2(s)$ is dissolved in concentrated $HCl(aq)$ and produces a yellow solution. The solution is diluted to twice its volume with water and assumes a green color. On dilution to ten times its original volume, the solution becomes pale blue.
  (b) When chromium metal is dissolved in $HCl(aq)$, a blue solution is produced that quickly turns green. Later the green solution becomes blue-green and then violet.

37. Which of the following complex ions would you expect to have the largest overall K_f, and why? $[Co(NH_3)_6]^{3+}$, $[Co(en)_3]^{3+}$, $[Co(H_2O)_6]^{3+}$, $[Co(H_2O)_4(en)]^{3+}$.

38. Use data from Table 25.6 to determine values of **(a)** β_4 for the formation of $[Zn(NH_3)_4]^{2+}$; **(b)** β_4 for the formation of $[Ni(H_2O)_2(NH_3)_4]^{2+}$.

39. Write a series of equations to show the stepwise displacement of H_2O ligands in $[Fe(H_2O)_6]^{3+}$ by ethylenediamine, for which $\log K_1 = 4.34$, $\log K_2 = 3.31$, and $\log K_3 = 2.05$. What is the overall formation constant, $\beta_3 = K_f$, for $[Fe(en)_3]^{3+}$?

40. A tabulation of formation constant data lists the following $\log K$ values for the formation of $[CuCl_4]^{2-}$: $\log K_1 = 2.80$, $\log K_2 = 1.60$, $\log K_3 = 0.49$, and $\log K_4 = 0.73$. What is the overall formation constant $\beta_4 = K_f$ for $[CuCl_4]^{2-}$?

41. Explain the following observations in terms of complex-ion formation.
  (a) $Al(OH)_3(s)$ is soluble in $NaOH(aq)$ but insoluble in $NH_3(aq)$.
  (b) $ZnCO_3(s)$ is soluble in $NH_3(aq)$, but ZnS(s) is not.
  (c) The molar solubility of AgCl in pure water is about 1×10^{-5} M; in 0.04 M NaCl(aq), it is about 2×10^{-6} M; but in 1 M NaCl(aq), it is about 8×10^{-5} M.

42. Explain the following observations in terms of complex-ion formation.
  (a) $CoCl_3$ is unstable in aqueous solution, being reduced to $CoCl_2$ and liberating $O_2(g)$. Yet, $[Co(NH_3)_6]Cl_3$ can be easily maintained in aqueous solution.
  (b) AgI is insoluble in water and in dilute $NH_3(aq)$, but AgI will dissolve in an aqueous solution of sodium thiosulfate.

Acid–Base Properties

43. Which of the following would you expect to react as a Brønsted–Lowry acid: $[Cu(NH_3)_4]^{2+}$, $[FeCl_4]^-$, $[Al(H_2O)_6]^{3+}$, or $[Zn(OH)_4]^{2-}$? Why?

44. Write simple chemical equations to show how the complex ion $[CrOH(H_2O)_5]^{2+}$ acts as **(a)** an acid; **(b)** a base.

Applications

45. From data in Chapter 19,
  (a) Derive an equilibrium constant for reaction (25.13), and explain why this reaction (the fixing of photographic film) is expected to go essentially to completion.
  (b) Explain why $NH_3(aq)$ cannot be used in the fixing of photographic film.

46. Show that the oxidation of $[Co(NH_3)_6]^{2+}$ to $[Co(NH_3)_6]^{3+}$ referred to on page 1012 should occur spontaneously in alkaline solution with H_2O_2 as an oxidizing agent.

Integrative and Advanced Exercises

47. From each of the following names, you should be able to deduce the formula of the complex ion or coordination compound intended. Yet, these are not the best systematic names that can be written. Replace each name with one that is more acceptable: **(a)** cupric tetraammine ion; **(b)** dichlorotetraammine cobaltic chloride; **(c)** platinic(IV) hexachloride ion; **(d)** disodium copper tetrachloride; **(e)** dipotassium antimony(III) pentachloride.

48. Magnus's green salt has the empirical formula $PtCl_2 \cdot 2NH_3$. It is a coordination compound consisting of both complex cations and complex anions. Write the probable formula of this coordination compound according to Werner's theory, and assign it a systematic name.

49. How many isomers are there of the complex ion $[CoCl_2(en)(NH_3)_2]^+$? Sketch their structures.

50. Explain the following observations through a series of equations. The green solid $CrCl_3 \cdot 6H_2O$ dissolves in water to form a green solution. The solution slowly turns blue-green; after a day or two, the solution is violet. When the violet solution evaporates to dryness, a green solid remains.

51. The cis and trans isomers of $[CoCl_2(en)_2]^+$ can be distinguished via a displacement reaction with oxalate ion. What difference in reactivity toward oxalate ion would you expect between the cis and trans isomers? Explain.

52. Write half-equations and an overall equation to represent the oxidation of tetraammineplatinum(II) ion to *trans*-tetraamminedichloroplatinate(IV) ion by Cl_2. Then make sketches of the two complex ions.

53. We learned in Chapter 17 that for polyprotic acids, ionization constants for successive ionization steps decrease rapidly. That is, $K_{a_1} \gg K_{a_2} \gg K_{a_3}$. The ionization constants for the first two steps in the ionization of $[Fe(H_2O)_6]^{3+}$ (reactions 25.10 and 25.11) are more nearly equal in magnitude. Why does this multistep ionization seem not to follow the pattern for polyprotic acids?

54. Following are the names of five coordination compounds containing complexes with platinum(II) as the central metal ion and ammonia molecules and/or chloride ions as ligands: (a) potassium amminetrichloroplatinate(II), (b) diamminedichloroplatinum(II), (c) triamminechloroplatinum(II) chloride, (d) tetraammineplatinum(II) chloride, (e) potassium tetrachloroplatinate(II). Make a rough

sketch of the expected graph when electric conductivity is plotted as a function of the chlorine content of the compounds.
[*Hint:* Your graph should be based on five points, but no quantitative data are given.]

55. For a solution that is 0.100 M in $[Fe(H_2O)_6]^{3+}$,
  (a) Assuming that ionization of the aqua complex ion proceeds only through the first step, equation (25.10), calculate the pH of the solution.
  (b) Calculate $[[FeOH(H_2O)_5]^{2+}]$ if the solution is also 0.100 M $HClO_4$. (ClO_4^- does not complex with Fe^{3+}.)
  (c) Can the pH of the solution be maintained so that $[[FeOH(H_2O)_5]^{2+}]$ does not exceed 1×10^{-6} M? Explain.

56. A solution that is 0.010 M in Pb^{2+} is also made to be 0.20 M in a salt of EDTA (that is, having a concentration of the $EDTA^{4-}$ ion of 0.20 M). If this solution is now made 0.10 M in H_2S and 0.10 M in H_3O^+, will $PbS(s)$ precipitate?

57. Without performing detailed calculations, show why you would expect the concentrations of the various ammine–aqua complex ions to be negligible compared with that of $[Cu(NH_3)_4]^{2+}$ in a solution having a total Cu(II) concentration of 0.10 M and a total concentration of NH_3 of 1.0 M. Under what conditions would the concentrations of these ammine–aqua complex ions (such as $[Cu(H_2O)_3NH_3]^{2+}$) become more significant relative to the concentration of $[Cu(NH_3)_4]^{2+}$? Explain.

58. Verify the statement on page 1014 that neither Ca^{2+} nor Mg^{2+} found in natural waters is likely to precipitate from the water on the addition of other reagents if the ions are complexed with EDTA. Assume reasonable values for the total metal ion concentration and that of *free* EDTA, such as 0.10 M each.

59. Estimate the total $[Cl^-]$ required in a solution that is initially 0.10 M $CuSO_4$ to produce a visible yellow color.

$$[Cu(H_2O)_4]^{2+} + 4\,Cl^- \rightleftharpoons [CuCl_4]^{2-} + 4\,H_2O$$
(blue)   (yellow)

$$K_f = 4.2 \times 10^5$$

Assume that 99% conversion of $[Cu(H_2O)_4]^{2+}$ to $[CuCl_4]^{2-}$ is sufficient for this to happen, and ignore the presence of any mixed aqua-chloro complex ions.

60. Refer to the stability of $[Co(NH_3)_6]^{3+}(aq)$ on page 1011, and
   (a) Verify that $E°_{cell}$ for reaction (25.12) is $+0.59$ V.
   (b) Calculate $[Co^{3+}]$ in a solution that has a total concentration of cobalt of 1.0 M and $[NH_3] = 0.10$ M.
   (c) Show that for the value of $[Co^{3+}]$ calculated in part (b), reaction (25.12) will not occur.
   [*Hint:* Assume a low, but reasonable, concentration of Co^{2+} (say, 1×10^{-4} M) and a partial pressure of $O_2(g)$ of 0.2 atm.]

61. A Cu electrode is immersed in a solution that is 1.00 M NH_3 and 1.00 M in $[Cu(NH_3)_4]^{2+}$. If a standard hydrogen electrode is the cathode, E_{cell} is $+0.08$ V. What is the value obtained by this method for the formation constant, K_f, of $[Cu(NH_3)_4]^{2+}$?

62. The following concentration cell is constructed.

$$Ag \mid Ag^+(0.10 \text{ M } [Ag(CN)_2]^-, 0.10 \text{ M CN}^-) \parallel Ag^+(0.10 \text{ M}) \mid Ag$$

   If K_f for $[Ag(CN)_2]^-$ is 5.6×10^{18}, what value would you expect for E_{cell}?
   [*Hint:* Recall that the anode is on the left.]

63. The compound $CoCl_3 \cdot 2H_2O \cdot 4NH_3$ may be one of the hydrate isomers $[Co(NH_3)_4(H_2O)Cl_2]Cl \cdot H_2O$ or $[Co(NH_3)_4(H_2O)_2Cl]Cl_2$. A 0.10 M aqueous solution of the compound is found to have a freezing point of -0.56 °C. Determine the correct formula of the compound. The freezing-point depression constant for water is 1.86 mol kg^{-1} deg, and for aqueous solutions, molarity and molality can be taken as approximately equal.

64. A compound is analyzed and found to contain 46.2% Pt, 33.6% Cl, 16.6% N, and 3.6% H. The freezing point of a 0.1 M aqueous solution of the compound is -0.74 °C. What is the structural formula of the compound? What possible isomeric forms are there for this compound?

65. The compound $Co(en)_2(NO_2)_2Cl$ has been prepared in a number of isomeric forms. One form undergoes no reaction with either $AgNO_3$ or en and is optically inactive. A second form reacts with $AgNO_3$ to form a white precipitate, does not react readily with en, and is optically inactive. A third form is optically active and reacts both with en and $AgNO_3$. Assuming a coordination number of 6 for the cobalt ion, identify each of the three isomeric forms by name, and sketch each of the structures.

66. The formation constants of the ions hexamminenickel(II), tris(ethylenediamine)nickel(II) and pentaethylenehexaminenickel(II) are $\beta_6 = 3.2 \times 10^8$, $\beta_3 = 1.6 \times 10^{18}$, and $\beta_1 = 2.0 \times 10^{19}$, respectively. Make reasonable assumptions concerning the enthalpy changes, and show that the formation constants illustrate the chelate effect. Pentaethylenehexamine (penten) is the hexadentate ligand shown below.

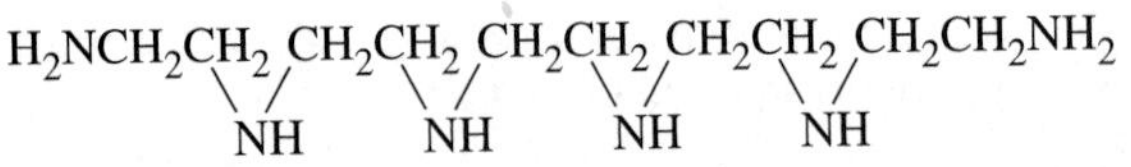

$$H_2NCH_2CH_2 \; CH_2CH_2 \; CH_2CH_2 \; CH_2CH_2 \; CH_2CH_2NH_2$$
$$\quad\quad \backslash / \quad\quad \backslash / \quad\quad \backslash / \quad\quad \backslash /$$
$$\quad\quad\quad NH \quad\quad NH \quad\quad NH \quad\quad NH$$

67. Acetyl acetone undergoes an isomerization to form a type of alcohol called an *enol*.

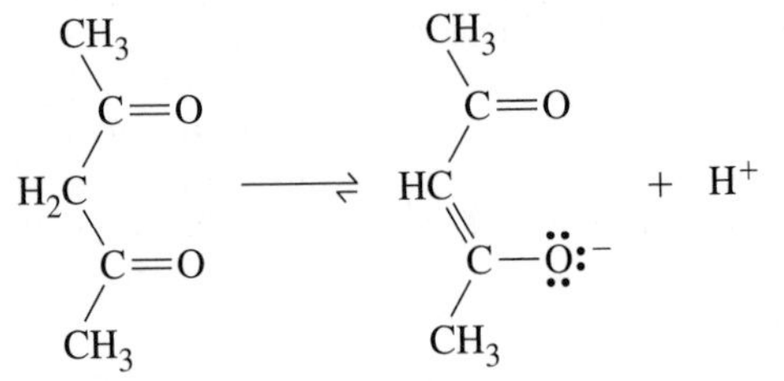

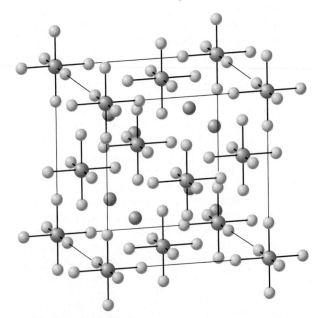

The enol, abbreviated acacH, can act as a bidentate ligand as the anion acac⁻. Which of the following compounds are optically active: $Co(acac)_3$; *trans*-$[Co(acac)_2(H_2O)_2]Cl_2$; *cis*-$[Co(acac)_2(H_2O)_2]Cl_2$?

68. We have seen that complex formation can stabilize oxidation states. An important illustration of this fact is the oxidation of water in acidic solutions by $Co^{3+}(aq)$ but not by $[Co(en)_3]^{3+}$. Use the following data.

$$[Co(H_2O)_6]^{3+} + e^- \longrightarrow [Co(H_2O)_6]^{2+}$$
$$E° = 1.82 \text{ V}$$

$$[Co(H_2O)_6]^{2+} + 3 \text{ en} \longrightarrow [Co(en)_3]^{2+} + 6 H_2O(l)$$
$$\log \beta_3 = 12.18$$

$$[Co(H_2O)_6]^{3+} + 3 \text{ en} \longrightarrow [Co(en)_3]^{3+} + 6 H_2O(l)$$
$$\log \beta_3 = 47.30$$

Calculate $E°$ for the reaction

$$[Co(en)_3]^{3+} + e^- \longrightarrow [Co(en)_3]^{2+}$$

Show that $[Co(en)_3]^{3+}$ is stable in water but $Co^{3+}(aq)$ is not.

69. The amino acid glycine ($NH_2CH_2CO_2H$, denoted Hgly) binds the anion as a bidentate ligand. Draw and name all possible isomers of $[Co(gly)_3]$. How many isomers are possible for the compound $[Co(gly)_2Cl(NH_3)]$?

70. The structure of $K_2[PtCl_6]$ in the solid state is shown here. Identify the type of cubic unit cell, and describe the structure in terms of the holes occupied by the various ions.

Feature Problems

71. A structure that Werner examined as a possible alternative to the octahedron is the trigonal prism.

(a) Does this structure predict the correct number of isomers for the complex ion $[CoCl_2(NH_3)_4]^+$? If not, why not?
(b) Does this structure account for optical isomerism in $[Co(en)_3]^{3+}$? Explain.

72. Werner demonstrated that octahedral complexes can exhibit optical isomerism, and to Werner's satisfaction, this confirmed the octahedral arrangement of ligands. However, skeptics of his theory said that because the ligands contained carbon atoms, he could not rule out carbon as the source of the optical activity. Werner devised and prepared the following compound in which the OH^- groups act as bridging groups.

$$\left[Co \left(\begin{array}{c} H \\ O \\ \diagdown \diagup \\ \diagup \diagdown \\ O \\ H \end{array} \right) Co(NH_3)_4 \right]_3^{6+}$$

Werner resolved this compound into its optical isomers, confirming his theory and confounding his critics. What are the oxidation states of the Co ions? If the complex is low spin, what is the number of unpaired electrons in the molecule? Draw the structures of the two optical isomers.

73. The crystal field model discussed in the text describes how the degeneracy of the d orbitals is removed by an octahedral field of ligands. We have seen that the d_{xy}, d_{xz}, and d_{yz} orbitals are stabilized (lower energy) with respect to the average energy of the d orbitals and that the $d_{x^2-y^2}$ and d_{z^2} orbitals are destabilized. As described in the Are You Wondering fea-

ture on page 1001, the stabilization is $-0.4\ \Delta$ and destabilization is 0.6Δ. The crystal field stabilization energy (CFSE) can be defined as CFSE = [(number of electrons in the d_{xy}, d_{xz}, and d_{yz} orbitals) $\times$ $(-0.4\ \Delta)$] + [(number of electrons in the $d_{x^2-y^2}$ and d_{z^2} orbitals) $\times$ $(0.6\ \Delta)$]. The following table contains the enthalpy of hydration for the reaction

$$M^{2+}(g) + 6\ H_2O(l) \longrightarrow [M(H_2O)_6]^{2+}$$

Dipositive Metal Ion	Hydration Energy (kJ mol^{-1})
Ca	−2468
Sc	−2673
Ti	−2750
V	−2814
Cr	−2799
Mn	−2743
Fe	−2843
Co	−2904
Ni	−2986
Cu	−2989
Zn	−2936

(a) Plot the hydration energies as a function of the atomic number of the metals shown.
(b) Assuming that all the hexaaqua complexes are high spin, which ions have zero CFSE?
(c) If lines are drawn between those ions with CFSE = 0, a line of negative slope is obtained. Can you explain this finding?
(d) The ions that do not have CFSE = 0 have heats of hydration that are more negative than the lines drawn in part (c). What is the explanation for this?
(e) Estimate the value of Δ for the Fe(II) ion in an octahedral field of water molecules.
(f) What wavelength of light would the $[Fe(H_2O)_6]^{2+}$ ion absorb?

 eMedia Exercises

74. For each of the **Ligands** models presented in *eChapter 25-2*, **(a)** draw the Lewis structures and identify the points of interaction for each ligand. **(b)** Describe the bonding interaction between these ligands and metal atoms in terms of Lewis acid-base theory.

75. **(a)** Classify ionization, coordination, and linkage isomerism in terms of the categories described in the **Isomerism** animation *(eChapter 25-4)*. **(b)** In which categories would isomers be expected to exhibit the greatest chemical differences?

76. After viewing the **Chirality** animation *(eChapter 25-4)*, describe why your two hands are not physically superimposible. Repeat this exercise for the coordination compound $[Co(en)_3]^{3+}$.

77. Suggest how the measurement procedures described in the **Optical Activity** animation *(eChapter 25-4)* could be used to follow the synthesis of enantiomers. How would this approach differ from other methods of characterization?

26 Nuclear Chemistry

Contents

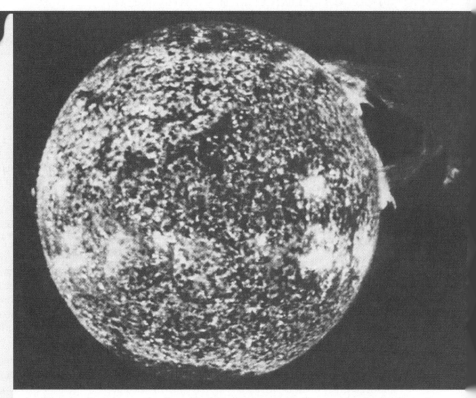

The sun, featuring a solar flare. Nuclear reactions in the stars are the source of the various chemical elements. Energy released in nuclear reactions on the sun—solar energy—is what makes life on Earth possible.

The origin of all the elements is the stars, including our sun. Nuclear fusion in stars creates heavier elements from lighter ones. The heaviest elements, those with atomic numbers greater than 83, have unstable nuclei—that is, they are *radioactive*.

Certain isotopes of the lighter elements also have this property of radioactivity. The isotope carbon-14, for example, has chemical and physical properties that are essentially identical to those of the much more abundant isotopes carbon-12 and carbon-13. Carbon-14, however, is radioactive, and it is this property that is used in the technique known as radiocarbon dating.

In this chapter, we will consider a variety of phenomena that originate within the nuclei of atoms. Collectively, we refer to these phenomena as *nuclear chemistry*. While stars are the natural source of all the elements, we will discuss how new heavy elements and radioactive

▲ Alpha particles leave a trail of liquid droplets, artificially colored green in this photograph, as they pass through a supersaturated vapor in a detector known as a cloud chamber. The chamber also contains He(g), and the trail of one α particle (yellow) is striking the nucleus of a He atom. Following the collision, the α particle and the He atom (red) move apart along lines at about a 90° angle.

isotopes of existing lighter elements can be made artificially. Another aspect of nuclear chemistry that we will discuss is the effects of ionizing radiation on matter. These effects can have both positive and negative outcomes and are a subject of society's continuing nuclear debate.

26-1   The Phenomenon of Radioactivity

The term *radioactivity* was proposed by Marie Curie to describe the emission of ionizing radiation by some of the heavier elements. Ionizing radiation, as the name implies, interacts with matter to produce ions. This means that the radiation is sufficiently energetic to break chemical bonds. Some ionizing radiation is particulate (consisting of particles), and some is nonparticulate. We introduced α, β, and γ radiation in Section 2-2. Let's describe them again in more detail, together with two other nuclear processes.

Alpha Particles

Alpha (α) particles are the nuclei of helium-4 atoms, ${}_2^4\text{He}^{2+}$, that have been accelerated artificially or created by ejection from the nucleus of a larger atom. We can think of α-particle emission as a process in which a bundle of two protons and two neutrons is emitted by a radioactive nucleus, resulting in a lighter nucleus. Alpha particles produce large numbers of ions via their collisions and near collisions with atoms as they travel through matter, but their penetrating power is low. (Generally, a sheet of paper can stop them.) Because they have a positive charge, α particles are deflected by electric and magnetic fields (recall Figure 2-10).

We can represent the production of α particles by means of a nuclear equation. A **nuclear equation** is written to conform to two rules.

1. The sum of mass numbers must be the same on both sides.
2. The sum of atomic numbers must be the same on both sides.

In equation (26.1), the alpha particle is represented as ${}_2^4\text{He}$.

$$\text{}_{92}^{238}\text{U} \longrightarrow \text{}_{90}^{234}\text{Th} + \text{}_2^4\text{He} \tag{26.1}$$

Mass numbers total 238, and atomic numbers total 92. The loss of an α particle results in a *decrease* of 2 in the atomic number and 4 in the mass number of the nucleus.

Beta Particles

Beta (β^-) particles are deflected by electric and magnetic fields in the *opposite* direction from α particles. They are less massive than α particles, so they are deflected more strongly than α particles (recall Figure 2-10). Beta ($-$) particles are electrons, but they are electrons that originate from the nuclei of atoms in nuclear decay processes. Electrons that surround the nucleus are given the familiar symbol, e^-.

The simplest decay process producing a β^- particle is the decay of a free neutron, which is unstable outside the nucleus of an atom.

$$\text{}_0^1\text{n} \longrightarrow \text{}_1^1\text{p} + \text{}_{-1}^0\beta + \nu \tag{26.2}$$

A β^- particle does not have an atomic number, but its 1$-$ charge is equivalent to an atomic number of -1. In nuclear equations, the β^- particle is represented as ${}_{-1}^0\beta$. Also, a β^- particle is small enough, compared to protons and neutrons, that its mass can be ignored in most calculations. Equation 26.2 introduces the symbol ν to

KEEP IN MIND ▶

that in a *nuclear* equation we represent only the nuclei of atoms, not the atoms as a whole. Although, we do not keep track of electrons, electric charge is conserved by the requirement that the sums of the atomic numbers be the same on the two sides of the equation.

Separation of Alpha, Beta, and Gamma Rays animation

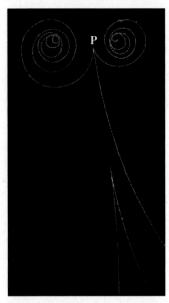

▲ **A colorized cloud chamber photograph**
Point P marks an atomic nucleus that interacts with a γ-ray photon (not visible), producing a β^- particle and a positron (spiral green and red tracks, respectively). The photon also dislodges an orbital electron (vertical green track).

represent an entity called a *neutrino*. This particle was first postulated in the 1930s as necessary for the conservation of certain properties during the β^--decay process. Because they interact so weakly with matter, neutrinos were not detected until the 1950s. Even today, little is known of their properties, including their rest mass. (Rest mass is discussed on page 309).

For a typical β^--decay process, as represented by equation (26.3), we can think of a neutron *within* the nucleus of an atom spontaneously converting to a proton and an electron. This proton remains in the nucleus, whereas the electron is emitted as a β^- particle. Because of the extra proton, the atomic number *increases* by one unit; the mass number is unchanged. The elusive neutrino is generally not included in the nuclear equation.

$$^{234}_{90}\text{Th} \longrightarrow {}^{234}_{91}\text{Pa} + {}^{0}_{-1}\beta \tag{26.3}$$

In a similar manner, in some decay processes a *proton* within the nucleus is converted to a neutron, and a β^+ particle and a neutrino* are emitted.

$$^{1}_{1}\text{p} \longrightarrow {}^{1}_{0}\text{n} + {}^{0}_{+1}\beta + \nu \tag{26.4}$$

The β^+ particle, also called a **positron**, has properties similar to the β^- particle, except that it carries a *positive* charge. (See the photograph in the margin.) This particle is also known as a *positive electron* and is designated ${}^{0}_{+1}\beta$ in nuclear equations. Positron emission is commonly encountered with artificially produced radioactive nuclei of the lighter elements. For example,

$$^{30}_{15}\text{P} \longrightarrow {}^{30}_{14}\text{Si} + {}^{0}_{+1}\beta \tag{26.5}$$

Electron Capture

Another process that achieves the same effect as positron emission is **electron capture (EC)**. In this case, an electron from an inner electron shell (usually the shell $n = 1$) is absorbed by the nucleus, where it converts a proton to a neutron. When an electron from a higher quantum level drops to the energy level vacated by the captured electron, X radiation is emitted. For example,

$$^{202}_{81}\text{Ti} + {}^{0}_{-1}\text{e} \longrightarrow {}^{202}_{80}\text{Hg} \quad \text{(followed by X radiation)} \tag{26.6}$$

Gamma Rays

Some radioactive decay processes that yield α or β^- particles leave the nucleus in an excited state. The nucleus then loses energy in the form of electromagnetic radiation—a **(gamma) γ ray**. Gamma rays are a highly penetrating form of radiation. They are *undeflected* by electric and magnetic fields (recall Figure 2-10). In the radioactive decay of $^{234}_{92}\text{U}$, 77% of the nuclei emit α particles having an energy of 4.18 MeV. The remaining 23% of the $^{234}_{92}\text{U}$ nuclei produce α particles with energies of 4.13 MeV. In the latter case, the $^{230}_{90}\text{Th}$ nuclei are left with an excess energy of 0.05 MeV. This energy is released as γ rays. If the unstable excited Th nucleus is denoted as $^{230}_{90}\text{Th}^{\ddagger}$, we can write

$$^{234}_{92}\text{U} \longrightarrow {}^{230}_{90}\text{Th}^{\ddagger} + {}^{4}_{2}\text{He} \tag{26.7}$$

$$^{230}_{90}\text{Th}^{\ddagger} \longrightarrow {}^{230}_{90}\text{Th} + \gamma \tag{26.8}$$

This γ-emission process is represented diagrammatically in Figure 26-1.

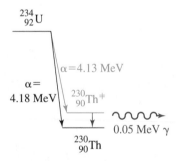

▲ **FIGURE 26-1**
Production of γ rays
The transition of a $^{230}_{90}\text{Th}$ nucleus between the two energy states shown results in the emission of 0.05 MeV of energy in the form of γ rays.

An electronvolt (eV) is the energy acquired by an electron when it falls through an electric potential difference of 1 volt:

$$1 \text{ eV} = 1.6022 \times 10^{-19} \text{ J}$$

$$1 \text{ MeV} = 1 \times 10^6 \text{ eV}$$

*There appear to be two related entities: the neutrino and antineutrino. Neutrinos accompany positron emission and electron capture; antineutrinos are associated with β^- emission.

How an α particle gets out of a nucleus?

The answer has to do with quantum theory and the nature of the forces involved. Consider the potential energy diagram shown below. The blue line represents the potential energy, where we imagine an α particle as a separate particle within a nucleus such as $^{238}_{92}U$. Region A represents the potential energy of the α particle when it is held within the nucleus by the forces inside the uranium nucleus. Region C represents the potential energy of the α particle when it is free of the nucleus. The potential energy along the downward curving portion of the blue line represents the Coulomb (electrostatic) repulsion between the positively charged α particle and the nucleus remaining after the α particle has escaped $\left(^{234}_{90}Th\right)$.

To get to region C, the α particle must get past the barrier in region B. The potential energy just beyond A, the radius of the nucleus, is greater than the energy of the α particle. (We know this from the measured energy of the α particle.) The α particle could not escape the nucleus if it were governed by classical physics because this would require an input of energy equal to the height of the barrier. Radioactive nuclei decay spontaneously, however, without an input of energy. How can the α particle get from region A to region C?

It passes through the barrier in a process known as *tunneling*. Classically, to go from A to C, the α particle would violate the principle of the conservation of energy. The α particle, however, possesses wavelike properties, as seen through the wave function at the bottom of the figure. Quantum mechanics predicts a finite probability of finding the α particle in a classically forbidden region. The wave function for the α particle trails off in the barrier region (B) and then reaches the outside, where it appears as a wave with much smaller amplitude. There is a finite probability $\left(\psi^2\right)$ of finding the α particle outside the nucleus: The α particle has tunneled through the barrier.

Moreover, an alternative form of the uncertainty principle tells us that energy conservation can be violated by an amount ΔE for the length of time Δt given by

$$\Delta E \times \Delta t = \frac{h}{4\pi}$$

That is, the wave–particle duality of quantum theory allows the conservation of energy to be violated for brief periods—long enough for an α particle to tunnel through the barrier. ΔE corresponds to the energy difference between the barrier height and the α particle's energy and Δt, to the time to pass through the barrier. The higher and wider the potential energy barrier, the less time the particle has to escape and the less likely it will do so. Thus, the height and width of the barrier control the rate of decay.

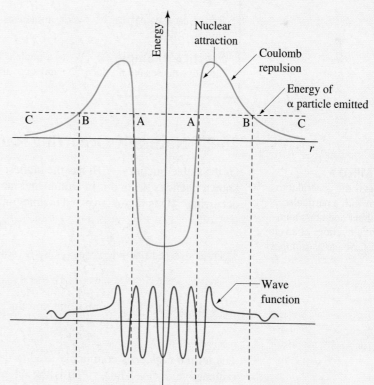

▶ Tunneling of an α particle out of the nucleus.

EXAMPLE 26-1

Writing Nuclear Equations for Radioactive Decay Processes. Write nuclear equations to represent **(a)** α-particle emission by ^{222}Rn and **(b)** radioactive decay of bismuth-215 to polonium-215.

Solution

(a) We can identify two of the species involved in this process simply from the information given. The $^{222}_{86}$Rn nucleus ejects an α particle, $^4_2$He, as shown in the following incomplete nuclear equation.

$$^{222}_{86}\text{Rn} \longrightarrow ? + {}^4_2\text{He}$$

Because the ejected α particle contains two protons, the unknown product must contain two fewer protons than $^{222}_{86}$Rn: $Z_{unk} = 86 - 2 = 84$. This atomic number identifies the element as polonium, $_{84}$Po. The mass number (A) of the product can be obtained by subtracting the mass number of the α particle from that of the radon isotope: $A = 222 - 4 = 218$. The completed nuclear equation is

$$^{222}_{86}\text{Rn} \longrightarrow {}^{218}_{84}\text{Po} + {}^4_2\text{He}$$

(b) Bismuth has the atomic number 83, and polonium, 84. We can approach this problem as we did part (a).

$$^{215}_{83}\text{Bi} \longrightarrow {}^{215}_{84}\text{Po} + ?$$

There is no change in mass number, so the particle has a zero mass number. Its atomic number is $Z = 83 - 84 = -1$. Only a $_{-1}^{0}\beta$ particle fits these parameters: Beta (−) decay is the only type of emission leading to an increase of one unit in atomic number without a change in the mass number.

$$^{215}_{83}\text{Bi} \longrightarrow {}^{215}_{84}\text{Po} + {}^{0}_{-1}\beta$$

Practice Example A: Write a nuclear equation to represent β^--particle emission by $^{241}_{94}$Pu.

Practice Example B: Write a nuclear equation to represent the decay of a radioactive nucleus to produce ^{58}Ni and a positron.

KEEP IN MIND ▶
that *nuclide* is the general term for an atom with a particular atomic number and mass number. Different nuclides of an element are referred to as *isotopes*.

26-2   Naturally Occurring Radioactive Isotopes

Of the stable nuclides, $^{209}_{83}$Bi has the highest atomic number and mass number. All known nuclides beyond it in atomic and mass numbers are radioactive. Naturally occurring $^{238}_{92}$U is radioactive and disintegrates by the loss of α particles.

$$^{238}_{92}\text{U} \longrightarrow {}^{234}_{90}\text{Th} + {}^4_2\text{He}$$

$^{234}_{90}$Th is also radioactive; it decays by β^- emission.

$$^{234}_{90}\text{Th} \longrightarrow {}^{234}_{91}\text{Pa} + {}^{0}_{-1}\beta$$

$^{234}_{91}$Pa also decays by β^- emission to produce $^{234}_{92}$U, which is also radioactive.

$$^{234}_{91}\text{Pa} \longrightarrow {}^{234}_{92}\text{U} + {}^{0}_{-1}\beta$$

The term *daughter* is commonly used to describe the new nuclide produced in a radioactive decay. Thus, ^{234}Th is the daughter of ^{238}U, ^{234}Pa is the daughter of ^{234}Th, and so on.

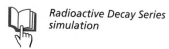
Radioactive Decay Series simulation

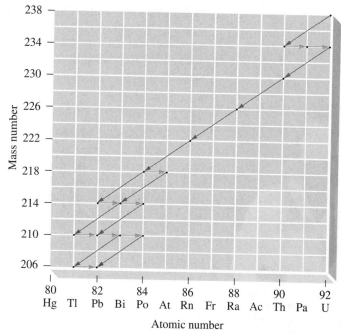

▲ **FIGURE 26-2   The natural radioactive decay series for $^{238}_{92}U$ (uranium series)**
The long arrows pointing down and to the left correspond to α-particle emissions. The short horizontal arrows represent β^- emissions. Other natural decay series originate with $^{232}_{90}Th$ (thorium series) and $^{235}_{92}U$ (actinium series).

▲ **Marie Sklodowska Curie (1867–1934)**
Marie Curie shared the 1903 Nobel Prize in physics for studies on radiation phenomena. In 1911, she won the Nobel Prize in chemistry for her discovery of polonium and radium.

The chain of radioactive decay that begins with $^{238}_{92}U$ continues through a number of steps of α and β^- emission until it eventually terminates with a stable isotope of lead—$^{206}_{82}Pb$. The entire scheme is outlined in Figure 26-2. All naturally occurring radioactive nuclides of high atomic number belong to one of three **radioactive decay series:** the *uranium* series just described, the *thorium* series, or the *actinium* series. (The actinium series actually begins with uranium-235, which was once called actino-uranium.)

Even though some of the daughters in natural radioactive decay schemes have very short half-lives, all are present because they are constantly forming as well as decaying. It is likely that only about one gram of radium-226 was present in several tons of uranium ore processed by Marie Curie in her discovery of radium in 1898. Nevertheless, she was successful in isolating it. The ore also contained only a fraction of a milligram of polonium, which she was able to detect but not isolate.

Radioactive decay schemes can be used to determine the ages of rocks and thereby the age of Earth (see Section 26-5). The appearance of certain radioactive substances in the environment can also be explained through radioactive decay series. The environmental hazards of ^{222}Rn were described in Section 8-4. The nuclides ^{210}Po and ^{210}Pb have been detected in cigarette smoke. These radioactive isotopes are derived from ^{238}U, found in trace amounts in the phosphate fertilizers used in tobacco fields. These α-emitting isotopes have been implicated in the link between cigarette smoking and cancer and heart disease.

Radioactivity, which is so common among isotopes of high atomic number, is a relatively rare phenomenon among the naturally occurring lighter isotopes. Even so, ^{40}K is a radioactive isotope, as are ^{50}V and ^{138}La. ^{40}K decays by β^- emission and by electron capture.

$$^{40}_{19}K \longrightarrow {}^{40}_{20}Ca + {}^{0}_{-1}\beta \qquad \text{and} \qquad {}^{40}_{19}K + {}^{0}_{-1}e \longrightarrow {}^{40}_{18}Ar$$

At the time Earth was formed, ^{40}K was much more abundant than it is now. It is believed that the high argon content of the atmosphere (0.934%, by volume, and almost all of it as ^{40}Ar) has been derived from the radioactive decay of ^{40}K. Aside from ^{40}K and ^{14}C (produced by cosmic radiation), the most important radioactive isotopes of the lighter elements are those that are *artificially* produced.

26-3 Nuclear Reactions and Artificially Induced Radioactivity

Ernest Rutherford discovered that atoms of one element can be transformed into atoms of another element. He did this in 1919 by bombarding $^{14}_{7}$N nuclei with α particles, producing $^{17}_{8}$O and protons. In this way, he was able to obtain protons outside atomic nuclei. The process can be represented as

$$^{14}_{7}\text{N} + ^{4}_{2}\text{He} \longrightarrow ^{17}_{8}\text{O} + ^{1}_{1}\text{H} \qquad (26.9)$$

In reaction (26.9), instead of a nucleus disintegrating spontaneously, it must be struck by another small particle to induce a nuclear reaction. $^{17}_{8}$O is a naturally occurring *nonradioactive* isotope of oxygen (0.037% natural abundance). The situation with $^{30}_{15}$P, which can also be produced by a nuclear reaction, is somewhat different.

In 1934, when bombarding aluminum with α particles, Irène Joliot-Curie and her husband, Frédéric Joliot, observed the emission of two types of particles: neutrons and positrons. The Joliots observed that when bombardment by α particles was stopped, the emission of neutrons also stopped; the emission of positrons continued, however. Their conclusion was that the nuclear bombardment produces $^{30}_{15}$P, which undergoes radioactive decay by the emission of positrons.

$$^{27}_{13}\text{Al} + ^{4}_{2}\text{He} \longrightarrow ^{30}_{15}\text{P} + ^{1}_{0}\text{n}$$
$$^{30}_{15}\text{P} \longrightarrow ^{30}_{14}\text{Si} + ^{0}_{+1}\beta$$

The first radioactive nuclide obtained by artificial means was $^{30}_{15}$P. Now, over 1000 other radioactive nuclides have been produced, and their number considerably exceeds the number of nonradioactive ones (about 280).

▲ **Irène Joliot-Curie (1897–1956)**
Irène Joliot-Curie and her husband, Frédéric Joliot, shared the 1935 Nobel Prize in chemistry for the artificial production of radioactive nuclides.

EXAMPLE 26-2

Writing Equations for Nuclear Bombardment Reactions. Write a nuclear equation for the production of ^{56}Mn by bombardment of ^{59}Co with neutrons.

Solution

First, we must realize that a particle is produced along with ^{56}Mn. To find the mass number (A) of the unknown particle, we must subtract the mass number of the Mn atom from that of the Co atom plus the neutron that initiates the reaction. Thus, for the unknown particle, $A = 59 + 1 - 56 = 4$. Subtracting the atomic number of Mn from that of Co provides the atomic number of the unknown particle: $Z = 27 - 25 = 2$. The unknown particle must have $A = 4$ and $Z = 2$; it is an α particle.

$$^{59}_{27}\text{Co} + ^{1}_{0}\text{n} \longrightarrow ^{56}_{25}\text{Mn} + ^{4}_{2}\text{He}$$

Practice Example A: Write a nuclear equation for the production of ^{147}Eu by bombardment of ^{139}La with ^{12}C.

Practice Example B: Write a nuclear equation for the production of ^{124}I by bombardment of ^{121}Sb with α particles. Also, write an equation for the subsequent decay of ^{124}I by positron emission.

26-4 Transuranium Elements

Until 1940, the only known elements were those that occur naturally. In 1940, bombardment of $^{238}_{92}\text{U}$ atoms with neutrons produced the first synthetic element. First, the unstable nucleus $^{239}_{92}\text{U}$ forms. This nucleus then undergoes β^- decay, yielding the element neptunium, with $Z = 93$.

$$^{238}_{92}\text{U} + ^{1}_{0}\text{n} \longrightarrow {}^{239}_{92}\text{U} + \gamma$$

$$^{239}_{92}\text{U} \longrightarrow {}^{239}_{93}\text{Np} + {}^{0}_{-1}\beta$$

Bombardment by neutrons is an effective way to produce nuclear reactions because these heavy uncharged particles are not repelled as they approach a nucleus.

Since 1940, all the elements from $Z = 93$ to 112, as well as elements 114, 116, and 118, have been synthesized. Many of the new elements of high atomic number have been formed by bombarding transuranium atoms with the nuclei of lighter elements. For example, an isotope of the element $Z = 105$ can be produced by bombarding atoms of $^{249}_{98}\text{Cf}$ with $^{15}_{7}\text{N}$ nuclei.

$$^{249}_{98}\text{Cf} + ^{15}_{7}\text{N} \longrightarrow {}^{260}_{105}\text{Db} + 4\,{}^{1}_{0}\text{n} \tag{26.10}$$

To bring about nuclear reactions such as (26.10) requires bombarding atomic nuclei with energetic particles. Such energetic particles can be obtained in an accelerator. A type of accelerator known as a cyclotron is illustrated in Figure 26-3.

A *charged-particle accelerator*, as the name implies, can produce only beams of charged particles (such as $^{1}_{1}\text{H}^+$) as projectiles. In many cases, neutrons are most effective as projectiles for nuclear bombardment. The neutrons required can be generated through a nuclear reaction produced by a charged-particle beam. In the following reaction, $^{2}_{1}\text{H}$ represents a beam of deuterons (actually, $^{2}_{1}\text{H}^+$) from an accelerator.

$$^{9}_{4}\text{Be} + ^{2}_{1}\text{H} \longrightarrow {}^{10}_{5}\text{B} + {}^{1}_{0}\text{n}$$

Another important source of neutrons for nuclear reactions is a nuclear reactor (as we will see in Section 26-8).

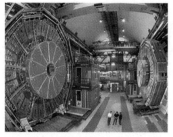

▲ The Large Electron–Positron (LEP) collider, a charged-particle accelerator at the CERN laboratories in Switzerland.

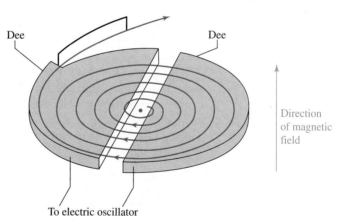

Dee   Dee

Direction of magnetic field

To electric oscillator

▲ **FIGURE 26-3   The cyclotron**

This type of accelerator consists of two hollow, flat, semicircular boxes, called *dees*, that are kept electrically charged. The entire assembly is maintained within a magnetic field. The particles to be accelerated, in the form of positive ions, are produced at the center of the opening between the dees. They are then attracted into the negatively charged dee and forced into a circular path by the magnetic field. When the particles leave the dee and enter the gap, the electric charges on the dees are reversed, so that the particles are attracted into the opposite dee. The particles are accelerated as they pass the gap and travel a wider circular path in the new dee. This process is repeated many times until the particles are brought to the required energy.

26-5   Rate of Radioactive Decay

In time, we can expect every atomic nucleus of a radioactive nuclide to disintegrate, but it is impossible to predict when any one nucleus will do so. Although we cannot make predictions for a particular atom, we can use statistical methods to make predictions for a collection of atoms. Based on experimental observations, a **radioactive decay law** has been established.

First-Order Processes animation

> *The rate of disintegration of a radioactive material—called the* activity, *A, or the* decay rate*—is directly proportional to the number of atoms present.*

In mathematical terms,

$$\text{rate of decay} \propto N \qquad \text{and} \qquad \text{rate of decay} = A = \lambda N \qquad (26.11)$$

The activity is expressed in atoms per unit time, such as atoms per second. N is the number of atoms in the sample being observed; λ is the **decay constant**, which has units of time^{-1}. Consider the case of a 1,000,000-atom sample disintegrating at the rate of 100 atoms per second. In such a case, $N = 1.0 \times 10^6$ and

$$\lambda = A/N = 100 \text{ atom s}^{-1}/1.0 \times 10^6 \text{ atoms} = 1.0 \times 10^{-4}\text{ s}^{-1}$$

Radioactive decay is a *first-order* process. To relate it to the first-order kinetics that we studied in Chapter 15, think of the activity as corresponding to a rate of reaction; the number of atoms, to the concentration of a reactant; and the decay constant, λ, to a rate constant, k. We can carry this correspondence further by writing an integrated radioactive decay law and a relationship between the decay constant and the **half-life** of the process—the length of time required for half of a radioactive sample to disappear.

$$\ln\left(\frac{N_t}{N_0}\right) = -\lambda t \qquad (26.12)$$

$$t_{1/2} = \frac{0.693}{\lambda} \qquad (26.13)$$

In these equations, N_0 represents the number of atoms at some initial time $(t = 0)$; N_t is the number of atoms at some later time, t; λ is the decay constant; and $t_{1/2}$ is the half-life.

Recall from Chapter 15 (page 590) that the half-life of a first-order process is a *constant*. Thus, if half the atoms of a radioactive sample disintegrate in 2.5 min, the number of atoms remaining will be reduced to one-fourth the original number in 5.0 min, one-eighth in 7.5 min, and so on. The shorter the half-life, the larger the value of λ and the faster the decay process. Half-lives of radioactive nuclides range from extremely short to very long, as suggested by the representative data in Table 26.1.

TABLE 26.1 Some Representative Half-Lives					
Nuclide	**Half-Life[a]**	**Nuclide**	**Half-Life[a]**	**Nuclide**	**Half-Life[a]**
$^{3}_{1}\text{H}$	12.26 y	$^{40}_{19}\text{K}$	1.25×10^9 y	$^{214}_{84}\text{Po}$	1.64×10^{-4} s
$^{14}_{6}\text{C}$	5730 y	$^{80}_{35}\text{Br}$	17.6 min	$^{222}_{86}\text{Rn}$	3.823 d
$^{13}_{8}\text{O}$	8.7×10^{-3} s	$^{90}_{38}\text{Sr}$	27.7 y	$^{226}_{88}\text{Ra}$	1.60×10^3 y
$^{28}_{12}\text{Mg}$	21 h	$^{131}_{53}\text{I}$	8.040 d	$^{234}_{90}\text{Th}$	24.1 d
$^{32}_{15}\text{P}$	14.3 d	$^{137}_{55}\text{Cs}$	30.23 y	$^{238}_{92}\text{U}$	4.51×10^9 y
$^{35}_{16}\text{S}$	88 d				

[a] s, second; min, minute; h, hour; d, day; y, year.

EXAMPLE 26-3

Using the Half-Life Concept and the Radioactive Decay Law to Describe the Rate of Radioactive Decay. The phosphorus isotope listed in Table 26.1, ^{32}P, is used in biochemical studies to determine the pathways of phosphorus atoms in living organisms. Its presence is detected through its emission of β^- particles. **(a)** What is the decay constant for ^{32}P, expressed in the unit s^{-1}? **(b)** What is the activity of a 1.00-mg sample of ^{32}P (that is, how many ^{32}P atoms disintegrate per second)? **(c)** Approximately what mass of the original 1.00-mg sample of ^{32}P will remain after 57 days? **(d)** What will be the rate of radioactive decay after 57 days?

Solution

(a) We can determine λ from $t_{1/2}$ with equation (26.13). The first result we get has the unit d^{-1}. We must convert this unit to h^{-1}, min^{-1}, and s^{-1}.

$$\lambda = \frac{0.693}{14.3\ d} \times \frac{1\ d}{24\ h} \times \frac{1\ h}{60\ min} \times \frac{1\ min}{60\ s} = 5.61 \times 10^{-7}\ s^{-1}$$

(b) First, let us find the number of atoms in 1.00 mg of ^{32}P. Then, we can multiply this number by the decay constant to get the activity or decay rate.

$$no.\ ^{32}P\ atoms = 0.00100\ g \times \frac{1\ mol\ ^{32}P}{32.0\ g} \times \frac{6.022 \times 10^{23}\ ^{32}P\ atoms}{1\ mol\ ^{32}P}$$

$$= 1.88 \times 10^{19}\ ^{32}P\ atoms$$
$$activity = \lambda N = 5.61 \times 10^{-7}\ s^{-1} \times 1.88 \times 10^{19}\ atoms$$
$$= 1.05 \times 10^{13}\ atoms/s$$

(c) A period of 57 days is $57/14.3 = 4.0$ half-lives. As shown in Figure 26-4, the quantity of radioactive material decreases by one-half for every half-life. The quantity remaining is $\left(\frac{1}{2}\right)^4$ of the original quantity.

$$?\ mg\ ^{32}P = 1.00\ mg \times \left(\frac{1}{2}\right)^4 = 1.00\ mg \times \frac{1}{16} = 0.063\ mg\ ^{32}P$$

Radioactive Decay Series simulation

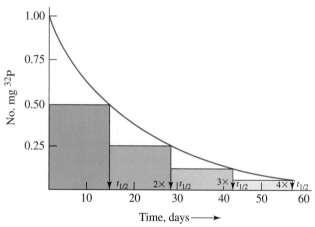

▲ **FIGURE 26-4   Radioactive decay of a hypothetical ^{32}P sample—Example 26-3 illustrated**

(d) The activity is directly proportional to the number of radioactive atoms remaining (activity $= \lambda N$), and the number of atoms is directly proportional to the mass of ^{32}P. When the mass of ^{32}P has dropped to one-sixteenth its original mass, the number of ^{32}P atoms also falls to one-sixteenth the original number, and the rate of decay is one-sixteenth the original activity.

$$rate\ of\ decay = \frac{1}{16} \times 1.05 \times 10^{13}\ atoms/s = 6.56 \times 10^{11}\ atom/s$$

Are You Wondering …

Whether, like first-order chemical reactions, radioactive decay is temperature-dependent?

We learned in Chapter 15 that an important determinant of the rate of a chemical reaction is the height of the energy barrier between the reactants and products (the activation energy). The higher the temperature, the more molecules that can surmount the barrier as a result of collisions and the faster the reaction proceeds. Although there is also an energy barrier that confines nuclear particles to the nucleus, molecular collisions do not invest any energy in nuclear particles. Moreover, in radioactive decay, nuclear particles do not escape the nucleus by surmounting an energy barrier—they tunnel through it. Thus, the rates of radioactive decay processes are independent of temperature.

Practice Example A: ^{131}I is a β^- emitter used as a tracer for radioimmunoassays in biological systems. Use information in Table 26.1 to determine **(a)** the decay constant in s^{-1}; **(b)** the activity of a 2.05-mg sample of ^{131}I; **(c)** the percentage of ^{131}I remaining after 16 days; and **(d)** the rate of β^- emission after 16 days.

Practice Example B: ^{223}Ra has a half-life of 11.4 days. How long would it take for the activity associated with a sample of ^{223}Ra to decrease to 1.0% of its current value?

Radiocarbon Dating

In the upper atmosphere, $^{14}_{6}\text{C}$ forms at a constant rate by the bombardment of $^{14}_{7}\text{N}$ with neutrons.

$$^{14}_{7}\text{N} + ^{1}_{0}\text{n} \longrightarrow ^{14}_{6}\text{C} + ^{1}_{1}\text{H}$$

The neutrons are produced by cosmic rays. $^{14}_{6}\text{C}$ disintegrates by β^- emission.

Carbon-containing compounds in living organisms are in equilibrium with ^{14}C in the atmosphere. That is, these organisms replace ^{14}C atoms that have undergone radioactive decay with "fresh" ^{14}C atoms through interactions with their environment. The ^{14}C isotope is radioactive and has a half-life of 5730 years. The activity associated with ^{14}C that is in equilibrium with its environment is about 15 disintegrations per minute (dis/min) per gram of carbon. When an organism dies (for instance, when a tree is cut down), this equilibrium is destroyed and the disintegration rate falls off because the dead organism no longer absorbs new ^{14}C. From the measured disintegration rate at some later time, we can estimate the age (that is, the elapsed time since the ^{14}C equilibrium was disrupted).

▶ The remains of a man frozen in a glacier in the Austrian Alps have been dated by the radiocarbon method as 5300 years old.

EXAMPLE 26-4

Applying the Integrated Rate Law for Radioactive Decay: Radiocarbon Dating. A wooden object found in an Indian burial mound is subjected to radiocarbon dating. The activity associated with its ^{14}C content is 10 dis min^{-1} g^{-1}. What is the age of the object? In other words, how much time has elapsed since the tree was cut down?

Solution

In this example, three equations, (26.11), (26.12), and (26.13), are required. Equation (26.13) is used to determine the decay constant.

$$\lambda = 0.693/5730 \text{ y} = 1.21 \times 10^{-4} \text{ y}^{-1}$$

Next, we use equation (26.11) to represent the actual number of atoms: N at $t = 0$ (the time when the ^{14}C equilibrium was destroyed) and N_t at time t (the present time). The activity just before the ^{14}C equilibrium was destroyed was 15 dis min^{-1} g^{-1}; at the time of the measurement, it is 10 dis min^{-1} g^{-1}. The corresponding numbers of atoms are equal to these activities divided by λ.

$$N_0 = A_0/\lambda = 15/\lambda \qquad \text{and} \qquad N_t = A_t/\lambda = 10/\lambda$$

Finally, we substitute into equation (26.12).

$$\ln \frac{N_t}{N_0} = \ln \frac{10/\lambda}{15/\log l} = \ln \frac{10}{15} = -(1.21 \times 10^{-4} \text{ y}^{-1})t$$

$$-0.41 = -(1.21 \times 10^{-4} \text{ y}^{-1})t$$

$$t = \frac{0.41}{1.21 \times 10^{-4} \text{ y}^{-1}} = 3.4 \times 10^{3} \text{ y}$$

Practice Example A: What is the age of a mummy, given a ^{14}C activity of 8.5 dis min^{-1} g^{-1}?

Practice Example B: What should be the current activity, in dis min^{-1} g^{-1}, of a wooden object believed to be 1100 years old?

▲ A lunar rock that has been radiometrically dated to be about 4.6 billion years old.

The Age of Earth

The natural radioactive decay scheme of Figure 26-2 suggests the eventual fate awaiting all the $^{238}_{92}U$ found in nature—conversion to lead. Naturally occurring uranium minerals always have associated with them some nonradioactive lead formed by radioactive decay. From the mass ratio of $^{206}_{82}Pb$ to $^{238}_{92}U$ in such a mineral, it is possible to estimate the age of the rock containing the mineral. By the age of the rock, we mean the time elapsed since molten magma solidified to form the rock. One assumption of this method is that the initial radioactive nuclide, the final stable nuclides, and all the products of a decay series remain in the rock. Another assumption is that any lead present in the rock initially consisted of the several isotopes of lead in their present, naturally occurring abundances.

The half-life of $^{238}_{92}U$ is 4.5×10^{9} years. According to the natural decay scheme of Figure 26-2, the basic change that occurs as atoms of $^{238}_{92}U$ and its daughters pass through the entire sequence of steps is

$$^{238}_{92}U \longrightarrow \, ^{206}_{82}Pb + 8\,^{4}_{2}He + 6\,^{0}_{-1}\beta$$

The decay sequence for $^{238}U \longrightarrow \, ^{206}Pb$ has 14 steps. The first step, however, has a much longer half-life than any of the other steps in the series and can thus be thought of as the rate-determining step, with the subsequent steps being "fast." As discussed in Chapter 15, we can ignore the effect on the overall rate of fast steps that occur after the slow, rate-determining step. Thus, the half-life for ^{238}U is essentially equal to the time it takes to convert half the initial ^{238}U to the ^{206}Pb isotope. Discounting the mass associated with the β^- particles, we can see that for every 238 g of uranium that undergoes complete decay, 206 g of lead and 32 g of helium are produced.

Suppose that in a hypothetical rock containing no lead initially, 1.000 g of $^{238}_{92}$U had disintegrated through one half-life, 4.51×10^9 years. At the end of that time, 0.500 g $^{238}_{92}$U would have disintegrated and another 0.500 g would remain. The quantity of $^{206}_{82}$Pb present in the rock would be

$$0.500 \text{ g } ^{238}_{92}\text{U} \times \frac{206 \text{ g } ^{206}_{82}\text{Pb}}{238 \text{ g } ^{238}_{92}\text{U}}$$

The ratio of lead-206 to uranium-238 in the rock would be

$$^{206}_{82}\text{Pb}/^{238}_{92}\text{U} = 0.433/0.500 = 0.866$$

If the $^{206}_{82}$Pb/$^{238}_{92}$U mass ratio is less than 0.866, the age of the rock is less than one half-life of $^{238}_{92}$U. A higher ratio indicates a greater age for the rock. The best estimates of the age of the oldest rocks, and presumably of Earth itself, are about 4.5×10^9 years. These estimates are based on the $^{206}_{82}$Pb/$^{238}_{92}$U ratio and on ratios for other pairs of isotopes from natural radioactive decay series.

26-6   Energetics of Nuclear Reactions

To describe the energy change accompanying a nuclear reaction, we must use the mass–energy equivalence derived by Albert Einstein.

KEEP IN MIND ▶
that the mass in equation (26.14)
is the rest mass of the particle, as
discussed on page 309.

$$E = mc^2 \tag{26.14}$$

An energy change in a process is always accompanied by a mass change, and the constant that relates them is the square of the speed of light. In chemical reactions, energy changes are so small that the equivalent mass changes are undetectable (though real nevertheless). In fact, we base the balancing of equations and stoichiometric calculations on the principle that mass is conserved (unchanged) in a chemical reaction. In nuclear reactions, energies are orders of magnitude greater than in chemical reactions. Perceptible changes in mass do occur.

If we know the exact masses of atoms, we can calculate the energy of a nuclear reaction with equation (26.14). The term m is the net change in mass, in kilograms, and c is expressed in meters per second. The resulting energy is in joules. Another common unit for expressing nuclear energy is the megaelectronvolt (MeV). (Recall from page 1026 that $1 \text{ eV} = 1.6022 \times 10^{-19}$ J.)

$$1 \text{ MeV} = 1.6022 \times 10^{-13} \text{ J} \tag{26.15}$$

Equation (26.15) is a conversion factor between megaelectronvolts and joules. A conversion factor between atomic mass units (u) and joules (J) is also helpful. We can establish this relationship by determining the energy associated with a mass of 1 u. Let's base our calculation on carbon-12 and note that 1 u is exactly $\frac{1}{12}$ of the mass of a carbon-12 atom. We can calculate the mass, in grams, corresponding to 1 u as follows:

$$1 \text{ u} \times \frac{1 \, ^{12}\text{C atom}}{12 \text{ u}} \times \frac{1 \text{ mol } ^{12}\text{C}}{6.0221 \times 10^{23} \text{ atoms } ^{12}\text{C}} \times \frac{12 \text{ g}}{1 \text{ mol } ^{12}\text{C}} = 1.6606 \times 10^{-24} \text{ g}$$

Converting this value of m to kilograms and using it in equation (26.14) gives us

$$E = 1 \text{ u} \times \frac{1.6606 \times 10^{-24} \text{ g}}{\text{u}} \times \frac{1 \text{ kg}}{1000 \text{ g}} \times (2.9979 \times 10^8)^2 \frac{\text{m}^2}{\text{s}^2}$$

$$= 1.4924 \times 10^{-10} \text{ J}$$

Thus, the energy equivalent of 1 u is

$$1 \text{ atomic mass unit (u)} = 1.4924 \times 10^{-10} \text{ J} \qquad (26.16)$$

Finally, to express this energy in MeV,

$$1 \text{ atomic mass unit (u)} = 1.4924 \times 10^{-10} \text{ J} \times \frac{1 \text{ MeV}}{1.6022 \times 10^{-13} \text{ J}} = 931.5 \text{ MeV} \qquad (26.17)$$

We use these conversion factors in Example 26-5, together with the principle that the total mass/energy of the products of a nuclear reaction is equal to the total mass/energy of the reactants. The masses required in calculations based on nuclear reactions are *nuclear* masses. The relationship of a nuclear mass to a nuclidic (atomic) mass is

$$\text{nuclear mass} = \text{nuclidic (atomic) mass} - \text{mass of extranuclear electrons}$$

EXAMPLE 26-5

Calculating the Energy of a Nuclear Reaction with the Mass–Energy Relationship. What is the energy, in joules and in megaelectronvolts, associated with the α decay of ^{238}U?

$$^{238}_{92}\text{U} \longrightarrow \,^{234}_{90}\text{Th} + \,^{4}_{2}\text{He}$$

The nuclidic (atomic) masses in atomic mass units (u) are

$$^{238}_{92}\text{U} = 238.0508 \text{ u} \qquad ^{234}_{90}\text{Th} = 234.0437 \text{ u} \qquad ^{4}_{2}\text{He} = 4.0026 \text{ u}$$

Solution

The net change in mass that accompanies the decay of a single nucleus of ^{238}U is shown below. Note that the masses of the extranuclear electrons don't enter into the calculation of the net change in mass.

change in mass = nuclear mass of $^{234}_{90}$Th + nuclear mass of $^{4}_{2}$He − nuclear mass of $^{238}_{92}$U

$$= [234.0437 \text{ u} - (90 \times \text{mass e}^-)] + [4.0026 \text{ u} - (2 \times \text{mass e}^-)]$$
$$- [238.0508 \text{ u} - (92 \times \text{mass e}^-)]$$
$$= 234.0437 \text{ u} + 4.0026 \text{ u} - 238.0508 \text{ u} - 92 \times \text{mass e}^- + 92 \times \text{mass e}^-$$
$$= -0.0045 \text{ u}$$

We can use this loss of mass and conversion factors (26.16) and (26.17) to write

$$E = -0.0045 \text{ u} \times [(1.49 \times 10^{-10} \text{ J})/\text{u}] = -6.7 \times 10^{-13} \text{ J}$$

or

$$E = -0.0045 \text{ u} \times (931.5 \text{ MeV}/\text{u}) = -4.2 \times \text{MeV}$$

The negative sign denotes that energy is lost in the nuclear reaction. This is the kinetic energy of the departing α particle.

Practice Example A: What is the energy associated with the α decay of ^{146}Sm (145.913053 u) to ^{142}Nd (141.907719 u)? Use 4.002603 u as the mass of ^{4}He.

Practice Example B: The decay of ^{222}Rn by α-particle emission is accompanied by a loss of 5.590 MeV of energy. What quantity of mass, in atomic mass units (u), is converted to energy in this process?

KEEP IN MIND ▶

that you can use either nuclear or nuclidic (atomic) masses in calculations based on a nuclear equation. The change in mass will be the same in either case because the masses of the electrons will cancel out.

Figure 26-5 suggests formation of the nucleus of a $^{4}_{2}$He atom from two protons and two neutrons. In this process, there is a *mass defect* of 0.0305 u. That is, the experimentally determined mass of a $^{4}_{2}$He nucleus is 0.0305 u *less* than the combined mass of two protons and two neutrons. This "lost" mass is liberated as energy. With expression (26.17), we can show that 0.0305 u of mass is equivalent to an energy of

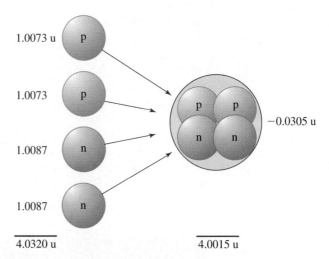

▲ **FIGURE 26-5** Nuclear binding energy in $^4_2$He
The mass of a helium nucleus ($^4_2$He) is 0.0305 u (atomic mass unit) less than the combined masses of two protons and two neutrons. The energy equivalent to this loss of mass (called the mass defect) is the nuclear energy that binds the nuclear particles together.

28.4 MeV. Because this is the energy released in forming a $^4_2$He nucleus, we can call it the **nuclear binding energy**. Viewed another way, a $^4_2$He nucleus would have to absorb 28.4 MeV to cause its protons and neutrons to become separated. If we consider the binding energy to be apportioned equally among the two protons and two neutrons in $^4_2$He, we obtain a binding energy of 7.10 MeV per nucleon. Similar calculations for other nuclei yield the data needed to plot the graph shown in Figure 26-6.

Figure 26-6 indicates that the maximum binding energy per nucleon is found in a nucleus with a mass number of approximately 60. This finding leads to two interesting conclusions: (1) If small nuclei are combined into a heavier one (up to about $A = 60$), the binding energy per nucleon increases and a certain quantity of mass must be converted to energy. The nuclear reaction is highly exothermic. This

▶ *Nucleons* are nuclear particles: protons and neutrons.

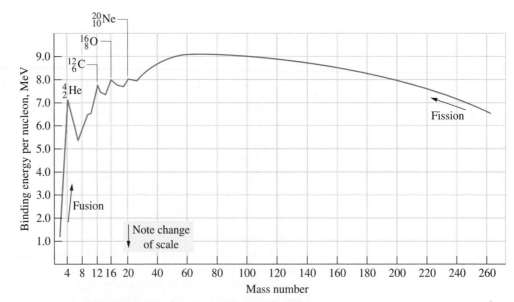

▲ **FIGURE 26-6** Average binding energy per nucleon as a function of atomic number

fusion process serves as the basis of the hydrogen bomb. (2) For nuclei with mass numbers above 60, the addition of extra nucleons to the nucleus would require the expenditure of energy (since the binding energy per nucleon decreases). However, the *disintegration* of heavier nuclei into lighter ones is accompanied by the release of energy. This nuclear *fission* process serves as the basis of the atomic bomb and conventional nuclear power reactors. Before considering nuclear fission and fusion, let's see what insights Figure 26-6 provides into the question of nuclear stability.

26-7   Nuclear Stability

A number of basic questions have probably occurred to you as we have been describing nuclear decay processes: Why do some radioactive nuclei decay by α emission, some by β^- emission, and so on? Why do the lighter elements have so few naturally occurring radioactive nuclides, whereas all those of the heavier elements seem to be radioactive?

Our first clue to answers for such questions comes from Figure 26-6, in which several nuclides are specifically noted. These nuclides have higher binding energies per nucleon than those of their neighbors. Their nuclei are especially stable. This observation is consistent with a theory of nuclear structure known as the *shell theory*. In the formation of a nucleus, protons and neutrons are believed to occupy a series of nuclear shells. This process is analogous to building up the electronic structure of an atom by the successive addition of electrons to electronic shells. Just as the aufbau process periodically produces electron configurations of exceptional stability, so do certain nuclei acquire a special stability as nuclear shells are closed. This condition of special stability of an atomic nucleus occurs for certain numbers of protons or neutrons known as **magic numbers** (Table 26.2).

Another observation concerning nuclei is that among stable nuclei, the number of protons and the number of neutrons is most commonly *even*. There are fewer stable nuclei with odd numbers of protons and of neutrons. The relationship between numbers of protons (Z), numbers of neutrons (N), and the stability of isotopes is summarized in Table 26.3. Note particularly that stable atoms with the combination Z odd–N odd are very rare. This combination is found only in the nuclides ^2_1H, ^6_3Li, $^{10}_5B$, and $^{14}_7N$. Still another observation is that elements of *odd* atomic number generally have only one or two stable isotopes, whereas those of *even* atomic number have several. Thus, F ($Z = 9$) and I ($Z = 53$) each have only one stable nuclide, and Cl ($Z = 17$) and Cu ($Z = 29$) each have two. Yet O ($Z = 8$) has three, Ca ($Z = 20$) has six, and Sn ($Z = 50$) has ten.

Neutrons are thought to provide a nuclear force to bind protons and neutrons together into a stable unit. Without neutrons, the electrostatic forces of repulsion between positively charged protons would cause the nucleus to fly apart. For the elements of lower atomic numbers (up to about $Z = 20$), the required number of neutrons for a stable nucleus is about equal to the number of protons, for example, ^4_2He, $^{12}_6C$, $^{16}_8O$, $^{28}_{14}Si$, $^{40}_{20}Ca$. For higher atomic numbers, because of increasing repulsive forces between protons, larger numbers of neutrons are required and the neutron–proton (n/p) ratio increases. For bismuth, the ratio is about $1.5:1$. Above atomic number 83, no matter how many neutrons are present, the nucleus is unstable. Thus, all isotopes of the known elements with $Z > 83$ are radioactive. Figure 26-7 indicates roughly the range of n/p ratios as a function of atomic number for stable atoms.

Using the ideas outlined here, nuclear scientists have predicted the possible existence of atoms of high atomic number that should have very long half-lives, a belt of stability in Figure 26-7. After a search of many years, such atoms have been created. In 1999, the bombardment of a plutonium-242 target with calcium-48 ions

TABLE 26.2   Magic Numbers for Nuclear Stability

Number of Protons	Number of Neutrons
2	2
8	8
20	20
28	28
50	50
82	82
114	126
	184

TABLE 26.3   Distribution of Naturally Occurring Stable Nuclides

Combination	Number of Nuclides
Z even–N even	163
Z even–N odd	55
Z odd–N even	50
Z odd–N odd	4

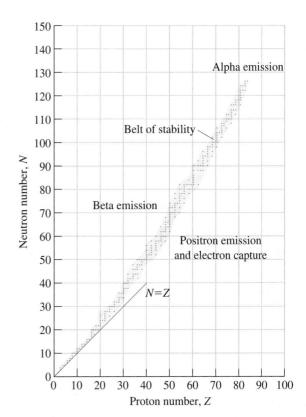

▶ **FIGURE 26-7**

Neutron-to-proton ratio and the stable nuclides up to Z = 83

The points within the belt of stability identify the stable nuclides. Some radioactive nuclides are also in this belt, but most lie outside, and the mode of their radioactive decay is indicated. The stable nuclides of low atomic numbers lie on or near the line $N = Z$; they have a neutron-to-proton ratio of one, or nearly so. At higher atomic numbers, the neutron-to-proton ratios increase to about 1.5.

produced the as-yet unnamed isotopes $^{287}114$ and $^{289}114$, with half-lives of 5 s and 30 s, respectively. Although these half-lives may appear short, they are practically an eternity compared to the half-lives of many other superheavy atoms, which are in the microsecond range.

EXAMPLE 26-6

Predicting Which Nuclei Are Radioactive. Which of the following nuclides would you expect to be stable, and which radioactive? **(a)** ^{82}As; **(b)** ^{118}Sn; **(c)** ^{214}Po.

Solution

(a) Arsenic-82 has $Z = 33$ and $N = 49$. This is an odd–odd combination that is found in only four of the lighter elements. ^{82}As is radioactive. (Note also that this nuclide is outside the belt of stability in Figure 26-7.)

(b) Tin has an atomic number of 50—a magic number. The neutron number is 68 in the nuclide ^{118}Sn. This is an even–even combination, and we should expect the nucleus to be stable. Moreover, Figure 26-7 shows that this nuclide is within the belt of stability. ^{118}Sn is a stable nuclide.

(c) ^{214}Po has an atomic number of 84. All known atoms with $Z > 83$ are radioactive. ^{214}Po is radioactive.

Practice Example A: Which of the following nuclides would you expect to be stable, and which radioactive? **(a)** ^{88}Sr; **(b)** ^{118}Cs; **(c)** ^{30}S.

Practice Example B: Write plausible nuclear equations to represent the radioactive decay of the fluorine isotopes ^{17}F and ^{22}F.

26-8   Nuclear Fission

▶ Ida Tacke Nodack, codiscoverer of the element rhenium, was the first person to suggest that Fermi's proposed experiments had produced fission. Her explanation was not generally accepted, however, until several years later.

In 1934, Enrico Fermi proposed that transuranium elements might be produced by bombarding uranium with neutrons. He reasoned that the successive loss of β^- particles would cause the atomic number to increase, perhaps to as high as 96. When such experiments were carried out, it was found that, in fact, the product did emit β^- particles. But in 1938, Otto Hahn, Lise Meitner, and Fritz Strassman found by chemical analysis that the products did not correspond to elements with $Z > 92$. Neither were they the neighboring elements of uranium—Ra, Ac, Th, and Pa. Instead, the products were radioisotopes of much lighter elements, such as Sr and Ba. Neutron bombardment of uranium nuclei causes certain of them to undergo **fission** into smaller fragments, as suggested by Figure 26-8.

The energy equivalent of the mass destroyed in a fission event is somewhat variable, but the average energy is approximately 3.20×10^{-11} J (200 MeV).

$$^{235}_{92}U + {}^{1}_{0}n \longrightarrow {}^{236}_{92}U \longrightarrow \text{fission fragments} + \text{neutrons} + 3.20 \times 10^{-11} \text{ J}$$

An energy of 3.20×10^{-11} J may seem small, but this energy is for the fission of a *single* $^{235}_{92}U$ nucleus. What if 1.00 g $^{235}_{92}U$ were to undergo fission?

$$? \text{ kJ} = 1.00 \text{ g } ^{235}U \times \frac{1 \text{ mol } ^{235}U}{235 \text{ g } ^{235}U} \times \frac{6.022 \times 10^{23} \text{ atoms } ^{235}U}{1 \text{ mol } ^{235}U} \times \frac{3.20 \times 10^{-11} \text{ J}}{1 \text{ atom } ^{235}U}$$

$$= 8.20 \times 10^{10} \text{ J} = 8.20 \times 10^{7} \text{ kJ}$$

This is an enormous quantity of energy! To release that quantity of energy would require the complete combustion of nearly 3 tons of coal.

Nuclear Reactors

In the fission of $^{235}_{92}U$, on average, 2.5 neutrons are released per fission event. These neutrons, on average, produce two or more fission events. The neutrons produced by the second round of fission produce another four or five events, and so on. The result is a *chain reaction*. If the reaction is uncontrolled, the released energy causes an explosion; this is the basis of the atomic bomb. Fission leading to an uncontrolled explosion occurs only if the quantity of ^{235}U exceeds the critical mass. The *critical mass* is the quantity of ^{235}U sufficiently large to retain enough neutrons to sustain a chain reaction. Quantities smaller than this are subcritical; neutrons escape at too great a rate to produce a chain reaction.

In a nuclear reactor, the release of fission energy is controlled. One common design, called the *pressurized water reactor* (PWR), is pictured in Figure 26-9. In the core of the reactor, rods of uranium-rich fuel are suspended in water maintained under a pressure of 70 to 150 atm. The water serves a dual purpose. First, it slows down the neutrons from the fission process so that they possess only normal thermal energy. These thermal neutrons are better able to induce fission than highly energetic ones. In this capacity, the water acts as a **moderator**. Water also functions

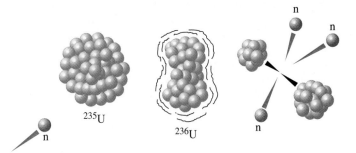

▶ **FIGURE 26-8**
Nuclear fission of $^{235}_{92}U$ with thermal neutrons
A neutron possessing ordinary thermal energy strikes a $^{235}_{92}U$ nucleus. First the unstable nucleus $^{236}_{92}U$ is produced; this then breaks up into a light fragment, a heavy fragment, and several neutrons. Various nuclear fragments are possible, but the most probable mass numbers are 97 for the light fragment and 137 for the heavy one.

▶ **The core of a nuclear reactor** The characteristic blue glow in the water surrounding the core is called *Cerenkov radiation*. It results when charged particles pass through a transparent medium faster than does light in the same medium. The charged particles are produced by nuclear fission. The radiation is analogous to the shock wave produced in a sonic boom.

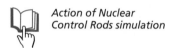

Action of Nuclear Control Rods simulation

as a heat transfer medium. Fission energy maintains the water at a high temperature (about 300 °C). The high-temperature water is brought in contact with colder water in a heat exchanger. The colder water is converted to steam, which drives a turbine, which in turn drives an electric generator. A final component of the nuclear reactor is a set of **control rods**, usually cadmium metal, whose function is to absorb neutrons. When the rods are lowered into the reactor, the fission process is slowed down. When the rods are raised, the density of neutrons and the rate of fission increase.

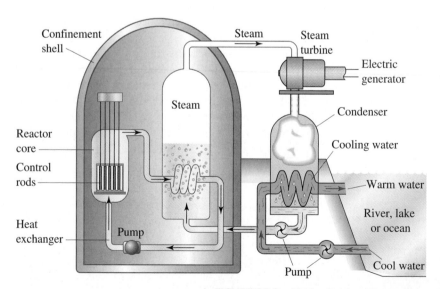

▲ **FIGURE 26-9   Pressurized water nuclear reactor**
(**a**) Schematic of a nuclear power plant. (**b**) A nuclear power plant in New York State.

(a)

(b)

(c)

▲ **(a)** The Three Mile Island nuclear reactor, site of a small nuclear accident in 1979. **(b)** A mutated pine tree in front of the Chernobyl nuclear reactor, site of a major nuclear accident in 1986. **(c)** A nuclear power plant in France that has operated for decades without mishap.

Are You Wondering...

About nuclear reactor safety?

There have been two notorious accidents at nuclear power stations. The first occurred at the Three Mile Island (TMI) generating plant near Middleton, Pennsylvania, in 1979. The TMI reactor is of the light-water type, in which water is used as the moderator and coolant. In this accident, some coolant (also the moderator) was lost; the chain reaction in the reactor stopped because there were too few slow neutrons. However, radioactive decay of the fission fragments continued, causing the fuel rods to get very hot. A partial meltdown resulted and, in turn, caused a fracture in one of the reactors. The fracture permitted the venting of a small amount of radioactive steam into the atmosphere. The reactor is now sealed, but electronic robots have discovered substantial damage to the fuel rods.

The second incident occurred at Chernobyl, in the Ukraine, in 1986. Graphite was used as the moderator there. When the coolant was turned off because of human error, the chain reaction went out of control. A tremendous rise in temperature followed, leading to a meltdown. During the meltdown, the graphite moderator surrounding the rods burned and radioactive smoke spewed out of the reactor. Radioactive materials were dispersed over much of Europe, Canada, and the United States. Although only a few dozen people were killed in the Chernobyl accident, many more will eventually die of cancer because of the related radiation. The type of accident observed at Chernobyl cannot happen in a light-water reactor, in which the coolant is the moderator.

Other nations have used nuclear power without experiencing accidents. In France and Japan, two-thirds of the electrical energy is produced by nuclear power stations. The safety record of nuclear reactors is particularly high, but the need for the storage of the nuclear waste produced is a vexing problem. Ways of dealing with these problems are discussed in the Focus On feature on page 1050.

Breeder Reactors

All that is required to initiate the fission of $^{235}_{92}U$ are neutrons of ordinary thermal energies. Conversely, nuclei of $^{238}_{92}U$, the abundant nuclide of uranium (99.28%), undergo the following reactions only when struck by energetic neutrons.

$$^{238}_{92}U + ^{1}_{0}n \longrightarrow ^{239}_{92}U$$

$$^{239}_{92}U \longrightarrow ^{239}_{93}Np + ^{0}_{-1}\beta$$

$$^{239}_{93}Np \longrightarrow ^{239}_{94}Pu + ^{0}_{-1}\beta$$

A fissionable nuclide such as $^{235}_{92}U$ is called *fissile*; $^{239}_{94}Pu$ is also fissile. A nuclide such as $^{238}_{92}U$, which can be converted into a fissile nuclide, is said to be *fertile*. In a *breeder nuclear reactor*, a small quantity of fissile nuclide provides the neutrons that convert a large quantity of a fertile nuclide into a fissile one. (The newly formed fissile nuclide then participates in a self-sustaining chain reaction.)

An obvious advantage of the breeder reactor is that the amount of uranium fuel available immediately jumps by a factor of about 100. This is the ratio of naturally occurring $^{238}_{92}U$ to $^{235}_{92}U$. But the potential advantage is even greater than this. Breeder reactors might use as nuclear fuels materials that have even very low uranium contents, such as shale deposits with about 0.006% U by mass.

There are, however, important disadvantages to breeder reactors. This is especially true of the type known as the *liquid-metal-cooled fast breeder reactor (LMFBR)*. Systems must be designed to handle a liquid metal, such as sodium, which becomes highly radioactive in the reactor. Also, both the rates of heat and neutron production are greater in the LMFBR than in the PWR, so materials deteriorate more rapidly. Perhaps the greatest unsolved problems are those of handling radioactive wastes and reprocessing plutonium fuel. Plutonium is one of the most toxic substances known. It can cause lung cancer when inhaled in even microgram (10^{-6} g) amounts. Furthermore, because plutonium has a long half-life (24,000 y), any accident involving it could leave an affected area almost permanently contaminated.

26-9   Nuclear Fusion

The **fusion** of atomic nuclei is the process that produces energy in the sun. An uncontrolled fusion reaction is the basis of the hydrogen bomb. A controlled fusion reaction could provide an almost unlimited source of energy. The nuclear reaction that holds the most immediate promise is the deuterium–tritium reaction.

$$_{1}^{2}\text{H} + {}_{1}^{3}\text{H} \longrightarrow {}_{2}^{4}\text{He} + {}_{0}^{1}\text{n}$$

The difficulties in developing a fusion energy source are probably without parallel in the history of technology. In fact, the feasibility of a controlled fusion reaction has yet to be fully demonstrated. There are a number of problems. To permit their fusion, the nuclei of deuterium and tritium must be forced into close proximity. Because atomic nuclei repel one another, this close approach requires the nuclei to have very high thermal energies. At the temperatures necessary to initiate a fusion reaction, gases are completely ionized into a mixture of atomic nuclei and electrons known as a *plasma*. Still higher plasma temperatures—over 40,000,000 K—are required to initiate a self-sustaining reaction (one that releases more energy than is required to get it started). A method must be devised to confine the plasma out of contact with other materials. The plasma loses thermal energy to any material it strikes. Also, a plasma must be at a sufficiently high density for a sufficient time to permit the fusion reaction to occur. The two methods receiving greatest attention are confinement in a magnetic field and heating of a frozen deuterium–tritium pellet with laser beams. Another series of technical problems involves the handling of liquid lithium, which is the anticipated heat-transfer medium and tritium ($_{1}^{3}\text{H}$) source.

$$\underset{\text{(fast)}}{{}_{3}^{7}\text{Li} + {}_{1}^{0}\text{n}} \longrightarrow {}_{2}^{4}\text{He} + \underset{\text{(slow)}}{{}_{1}^{3}\text{H} + {}_{0}^{1}\text{n}}$$

Finally, for the magnetic containment method, the magnetic field must be produced by superconducting magnets, which currently are very expensive to operate.

▶ In the hydrogen bomb, these high temperatures are attained by exploding an atomic (fission) bomb, which triggers the fusion reaction.

▶ The plasma chamber of a fusion reactor of the magnetic confinement type (called a *tokamak*). The chamber walls are lined with carbon-fiber composite tiles to protect against the high-temperature plasma.

The advantages of fusion over fission could be enormous. Since deuterium constitutes about one in every 6500 H atoms, the oceans of the world can supply an almost limitless amount of nuclear fuel. It is estimated that there is sufficient lithium on Earth to provide a source of tritium for about 1 million years.

26-10   Effect of Radiation on Matter

Although there are substantial differences in the way in which α particles, β particles, and γ rays interact with matter, they share an important feature: They dislodge electrons from atoms and molecules to produce ions. The ionizing power of radiation may be described in terms of the number of ion pairs formed per centimeter of path through a material. An *ion pair* consists of an ionized electron and the resulting positive ion. Alpha particles have the greatest ionizing power, followed by β particles and then γ rays. The ionized electrons produced directly by the collisions of particles of radiation with atoms are called *primary* electrons. These electrons may themselves possess sufficient energies to cause *secondary* ionizations.

Not all interactions between radiation and matter produce ion pairs. In some cases, electrons are simply raised to higher atomic or molecular energy levels. The return of these electrons to their normal states is then accompanied by radiation—X rays, ultraviolet light, or visible light, depending on the energies involved.

Some of the possibilities described here are depicted in Figure 26-10.

Radiation Detectors

The interactions of radiation with matter can serve as bases for the detection of radiation and the measurement of its intensity. One of the simplest methods is that used by Henri Becquerel in his discovery of radioactivity—the exposure of a photographic film, as in film badge radiation detectors. The effect of α and β particles, and γ rays on a photographic emulsion is similar to that of X rays.

One type of detector used to study high-energy radiation such as γ rays is the *bubble chamber*. In this device, a liquid, usually hydrogen, is kept just at its boiling point. As ion pairs are produced by the transit of an ionizing ray, bubbles of vapor

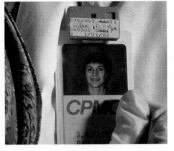

▲ A film badge used for detecting radiation.

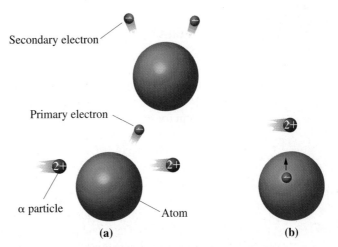

(a)   (b)

▲ FIGURE 26-10   **Some interactions of radiation with matter**
(a) The production of primary and secondary electrons by collisions. (b) The excitation of an atom by the passage of an α particle. An electron is raised to a higher energy level within the atom. The excited atom reverts to its normal state by emitting radiation.

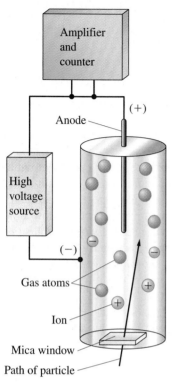

▲ **FIGURE 26-11**
A Geiger–Müller counter
Radiation enters the G–M tube through the mica window. The radiation ionizes some of the gas (usually argon) in the tube. A pulse of electric current passes through the electric circuit and is counted.

▲ A geologist using a Geiger–Müller counter to check rocks for their radioactivity.

form around the ions. The tracks of bubbles can be photographed and analyzed, and different types of radiation produce different tracks. Charged particles, for example, can be detected by their deflection in a magnetic field.

The most common device for detecting and measuring ionizing radiation is the *Geiger–Müller (G–M) counter*, depicted in Figure 26-11. The G–M counter consists of a cylindrical cathode with a wire anode running along its axis. The anode and cathode are sealed in a gas-filled glass tube. Ionizing radiation passing through the tube produces primary ions, and this is followed by secondary ionization. The positive ions are attracted to the cathode and the electrons to the anode, leading to a pulse of electric current. The tube is quickly recharged in preparation of the next ionizing event. The pulses of electric current are counted.

A detector widely used in biological studies is the *scintillation counter*. It is especially useful in detecting radiation that is not energetic enough to cause ionization. The radiation excites certain atoms in the detecting medium; when these atoms revert to their ground state, they emit pulses of light that can be counted. The light emission is similar to that produced when phosphors in a television screen are struck by cathode rays.

Effect of Ionizing Radiation on Living Matter

All life exists against a background of naturally occurring ionizing radiation—cosmic rays, ultraviolet light, and emanations from radioactive elements such as uranium in rocks. The level of this radiation varies from point to point on Earth, being greater, for instance, at higher elevations. Only in recent times have humans been able to create situations in which organisms might be exposed to radiation at levels significantly higher than natural background radiation.

The interactions of radiation with living matter are the same as with other forms of matter—ionization, excitation, and dissociation of molecules. There is no question of the effect of large doses of ionizing radiation on organisms—the organisms are killed. But even slight exposures to ionizing radiation can cause changes in cell chromosomes. Thus it is believed that, even at low dosage rates, ionizing radiation can result in birth defects, leukemia, bone cancer, and other forms of cancer. The nagging question that has eluded any definitive answers is how great an increase in the incidence of birth defects and cancers might be caused by certain levels of radiation.

Radiation Dosage

One unit long used to describe exposure to radiation is the rad. One **rad** (*r*adiation *a*bsorbed *d*ose) corresponds to the absorption of 1×10^{-2} J of energy per kilogram of matter. The effect of a dose of one rad on living matter is variable, however, and a better unit is one that takes this variability into account. The **rem** (*r*adiation *e*quivalent for *m*an) is the rad multiplied by the *relative biological effectiveness* (*Q*). The factor *Q* takes into account that equal doses of radiation of different types may have differing effects. Table 26.4 summarizes several radiation units.

It is thought that a dose of 1000 rem absorbed in a short time interval would kill 100% of the population. A short-term dose of 450 rem would probably result in death within 30 days of about 50% of the population. A single dose of 1 rem delivered to 1 million people would probably produce about 100 cases of cancer within 20–30 years. The total body radiation received by most of the world's population from normal background sources is about 0.13 rem [130 millirem (mrem)] per year. The dose delivered in a chest X-ray examination is about 20 mrem.

Some of the foregoing statements about radiation dosages and their anticipated effects are based on (1) medical histories of the survivors of the Hiroshima and

TABLE 26.4 Radiation Units[a]	
Unit	**Definition**
Radioactive decay:	
Becquerel, Bq	s^{-1} (disintegrations per second)
Curie, Ci	An amount of radioactive material decaying at the same rate as 1 g of radium (3.70×10^{10} dis/s)
	1 Ci = 3.70×10^{10} Bq
Adsorbed dose:	
Gray, Gy	One gray of radiation deposits one joule of energy per kilogram of matter
Rad	1 rad = 0.01 Gy
Equivalent dose:	
Sievert, Sv	1 Sv = 100 rem
Rem	1 rem = 1 rad $\times Q$
	The quality factor, Q, is about 1 for X rays, γ rays, and β^- particles; 3 for slow neutrons; 10 for protons and fast neutrons; and 20 for α particles

[a] SI units are shown in blue. Sources of α radiation are relatively harmless when external to the body and extremely hazardous when taken internally, as in the lungs or stomach. Other forms of radiation (X rays, γ rays), because they are highly penetrating, are hazardous even when external to the body.

Nagasaki atomic blasts, (2) the incidence of leukemia and other cancers in children whose mothers received diagnostic radiation during pregnancy, and (3) the occurrence of lung cancers among uranium miners in the United States. What does all of this tell us about a safe level of radiation exposure? One approach has been to extrapolate from these high doses to the lower doses affecting the general population. This has led the U.S. National Council on Radiation Protection and Measurements to recommend that the dosage for the general population be limited to 0.17 rem (170 mrem) per year from all sources above background level. Experts disagree, however, on how the data observed for high dosages should be extrapolated to low doses. Some experts believe that the 0.17 rem/y figure is too high. If they are right, an additional dosage of 0.17 rem/year above normal background levels might cause statistically significant increases in the incidence of birth defects and cancers.

26-11 Applications of Radioisotopes

We have described both the destructive capacity of nuclear reactions and the potential of these reactions to provide new sources of energy. Less heralded but equally important are a variety of practical applications of radioactivity. We close this chapter with a brief survey of some uses of radioisotopes.

Cancer Therapy

Ionizing radiation in low doses can induce cancers, but this same radiation, particularly γ rays, is also used in the treatment of cancer. Although ionizing radiation tends to destroy all cells, cancerous cells are more easily destroyed than normal ones. Thus, a carefully directed beam of γ rays or high-energy X rays of the appropriate dosage may be used to arrest the growth of cancerous cells. Also coming into use for some forms of cancer is radiation therapy that employs beams of protons or neutrons.

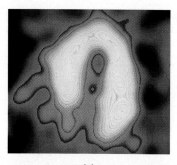

(a)

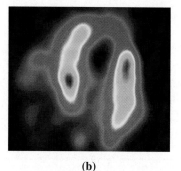

(b)

▲ **Nuclear medicine provides methods for diagnosing life-threatening conditions**
In these thallium-201 scans of a heart, γ rays released in the radioactive decay of ^{201}Tl are detected and used to provide an image of the blood flow to the heart wall (heart muscle).
(a) Normal heart wall; **(b)** heart with deficient blood flow.

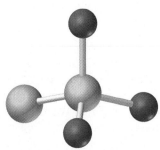

▲ **FIGURE 26-12**
Structure of thiosulfate ion, $S_2O_3^{2-}$
The central S atom is in the oxidation state +6. The terminal S atom is in the oxidation state −2.

Radioactive Tracers

The small mass differences between the isotopes of an element may produce small differences in physical properties. The differing rates of diffusion of $^{238}UF_6(g)$ and $^{235}UF_6(g)$ result from such a mass difference (page 207). Generally speaking, though, the physical and chemical properties of the various isotopes of an element are practically identical. However, if one of the isotopes is radioactive, its actions can be easily followed with radiation detectors. This is the principle behind the use of *radioactive tracers*, or tagged atoms. For example, if a small quantity of radioactive ^{32}P (as a phosphate) is added to a nutrient solution that is fed to plants, the uptake of all the phosphorus atoms—radioactive and nonradioactive alike—can be followed by charting the regions of the plant that become radioactive. Similarly, the fate of iodine (as iodide ion) in the body can be determined by having a person drink a solution of dissolved iodides containing a small quantity of a radioactive iodide as a tracer. Abnormalities in the thyroid gland can be detected in this way. Because I^- concentrates in the thyroid, people can protect themselves by ingesting nonradioactive iodides before being exposed to a radioactive iodide. The thyroid becomes saturated with the nonradioactive iodide and rejects the radioactive iodide. Iodide tablets were distributed to the general population in some regions of Europe before the arrival of fallout from the Chernobyl nuclear accident in 1986.

Industrial applications of tracers are also numerous. The fate of a catalyst in a chemical plant can be followed by incorporating a radioactive tracer in the catalyst, such as ^{192}Ir in a Pt–Ir catalyst. By monitoring the activity of the ^{192}Ir, one can determine the rate at which the catalyst is being carried away and to which parts of the plant it is being carried.

Structures and Mechanisms

Often the mechanism of a chemical reaction or the structure of a species can be inferred from experiments using radioisotopes as tracers. Consider the following experimental proof that the two S atoms in the thiosulfate ion, $S_2O_3^{2-}$, are not equivalent.

$S_2O_3^{2-}$ is prepared from radioactive sulfur (^{35}S) and sulfite ion containing the nonradioactive isotope ^{32}S.

$$^{35}S + {}^{32}SO_3^{2-} \longrightarrow {}^{35}S\,{}^{32}SO_3^{2-} \tag{26.18}$$

When the thiosulfate ion is decomposed by acidification, all the radioactivity appears in the precipitated sulfur and none in the $SO_2(g)$. The bonds of the ^{35}S atoms must be different from those of the ^{32}S atoms (Figure 26-12).

$$^{35}S\,{}^{32}SO_3^{2-} + 2\,H^+ \longrightarrow H_2O + {}^{32}SO_2(g) + {}^{35}S(s) \tag{26.19}$$

In reaction (26.20), nonradioactive KIO_4 is added to a solution containing iodide ion labeled with the radioisotope ^{128}I. All the radioactivity appears in the I_2 and none in the IO_3^-. This proves that all the IO_3^- is produced by reduction of IO_4^- and none by oxidation of I^-.

$$IO_4^- + 2\,{}^{128}I + H_2O \longrightarrow {}^{128}I_2 + IO_3^- + 2\,OH^- \tag{26.20}$$

Analytical Chemistry

The usual procedure of analyzing a substance by precipitation involves filtering, washing, drying, and weighing a pure precipitate. An alternative is to incorporate a radioactive isotope in the precipitating reagent. By measuring the activity of the precipitate and comparing it with that of the original solution, we can calculate the amount of precipitate without having to purify, dry, and weigh it.

Another method of importance in analytical chemistry is *neutron activation analysis*. In this procedure, the sample to be analyzed, normally nonradioactive, is bombarded with neutrons; the element of interest is converted to a radioisotope. The conversion of a stable isotope (X) to an unstable isotope (X*) by neutron capture can be represented as

$$^{A}X + {}^{1}_{0}n \longrightarrow {}^{(A+1)}X* \longrightarrow {}^{(A+1)}X + \gamma$$

▲ Neutron activation was used to establish that this painting was not painted by Rembrandt, but by an artist in the school of Rembrandt.

In this example, the excited nucleus that is formed decays with the emission of a γ ray with a distinctive energy. Neutron-activated nuclei may decay by other modes, however, such as β decay. The activity of the radioisotope formed is measured. This measurement, together with such factors as the rate of neutron bombardment, the half-life of the radioisotope, and the efficiency of the radiation detector, can be used to calculate the quantity of the element in the sample. The method is especially attractive because (1) trace quantities of elements can be determined (sometimes in parts per billion or less); (2) a sample can be tested without destroying it; and (3) the sample can be in any state of matter, including biological materials. Neutron activation analysis has been used, for instance, to determine the authenticity of old paintings. Old masters formulated their own paints. Differences between formulations are easily detected through the trace elements they contain.

Radiation Processing

Radiation processing describes industrial applications of ionizing radiation—γ rays from ^{60}Co or electron beams from electron accelerators. The ionizing radiation is used in the production of certain materials or to modify their properties. Its most extensive current use is in breaking, re-forming, and cross-linking polymer chains to affect the physical and mechanical properties of plastics used in foamed products, electrical insulation, and packaging materials. Ionizing radiation is used to sterilize medical supplies such as sutures, syringes, and hospital garb. In sewage treatment plants, radiation processing has been used to decrease the settling time of sewage sludge and to kill pathogens. Radiation is now being used in some instances in the preservation of foods as an alternative to canning, freeze-drying, or refrigeration. In radiation processing, the irradiated material is *not* rendered radioactive, although the ionizing radiation may produce some chemical changes.

▲ The mushrooms on the right have been preserved by irradiation.

▲ Pouring molten nuclear waste glass to demonstrate methods of immobilizing nuclear waste.

A distinct advantage of nuclear power generation over the use of fossil fuels is that it does not produce oxides of sulfur and nitro-gen as air pollutants. Nor does nuclear power generation produce $CO_2(g)$, so it does not contribute to the potential problem of global warming (see Section 7-9). However, nuclear power generation has its own waste problem. The disposal of nuclear wastes is one of the most difficult environmental problems.

Radioactive waste can be divided into two broad categories. Low-level waste consists of items used in handling radioactive materials (such as gloves and other protective clothing) and waste solutions from medical and laboratory use. The radioactive elements in these items have relatively short half-lives (30 years maximum) and low levels of activity. They can be stored in cement casks and buried. After about 300 years, these materials will no longer be radioactive.

The greater challenge is in handling spent fuel rods from nuclear reactors and other high-level radioactive wastes. In addition to significant quantities of uranium and plutonium, spent fuel rods contain a number of daughter isotopes with half-lives of hundreds of years. Plutonium, with a half-life of 24,000 years, is so toxic that any accidental release would contaminate an area for a very long time.

Summary

Radioactivity refers to the ejection of particles (α, β^-, β^+), the capture of electrons from an inner shell, or the emission of electromagnetic radiation (γ) by unstable nuclei. With the exception of γ-ray emission, radioactive decay leads to the transformation (transmutation) of one element into another.

All nuclides with $Z > 83$ are radioactive. Although a few occur naturally, most radioactive nuclides of lower atomic number are produced artificially by bombarding appropriate target nuclei with energetic particles. The basic rule in writing equations for nuclear reactions is that the sum of the atomic numbers and the sum of the mass numbers must be the same on both sides of the equation.

The rate of radioactive decay—the activity of a sample—is directly proportional to the number of atoms. Calculations of decay rates can be based on equations similar to those for first-order chemical kinetics. Measurements of decay rates of radioactive nuclides have a number of practical applications, ranging from determining the ages of rocks to the dating of certain carbon-containing objects (radiocarbon dating).

The quantity of energy released in the formation of a nucleus from protons and neutrons can be plotted as a function of mass number, yielding a distinctive graph. From this graph, we can establish that fission of heavy nuclei and fusion of lighter nuclei yield large quantities of energy. Fission is the basis of nuclear reactors, and fusion is the energy-producing process of the stars.

The stability of a nucleus depends on several factors, including whether the number of protons and neutrons is odd or even and whether either is a "magic number." Also of special importance is the neutron-to-proton ratio in the nucleus and whether this ratio lies within the belt of stable nuclides (Figure 26-7). Nuclides outside the belt of stability are radioactive.

One of the principal effects of the interaction of radiation with matter is the production of ions. This phenomenon can be used to detect radiation, and it is also the basis of radiation damage to living matter. Several methods have been developed to measure radiation doses and to predict the biological effects of these doses, but much uncertainty remains. Despite the hazards associated with radioactivity, radioactive nuclides have beneficial uses in cancer therapy, in basic studies of chemical structures and mechanisms, in analytical chemistry, and in the chemical industry.

In the United States, high-level wastes are currently stored underwater at temporary sites at power plants and elsewhere while awaiting final storage at a permanent site. An appropriate site would have to be geologically stable for tens of thousands of years. One or two such sites are now under careful study, but agreement on a final site is still a strongly debated issue.

France, which has an extensive nuclear power program including one breeder reactor, has chosen a different route for handling nuclear reactor waste: *reprocessing*. The first step in nuclear fuel reprocessing is the removal of uranium and plutonium from spent fuel rods. These elements are formed into fuel pellets for loading new fuel rods. Low-level waste is dealt with in the manner already described. What remains at this point are isotopes with long half-lives (up to 100 years). These are stabilized by incorporating them into a borosilicate (Pyrex-like) glass. The boron in the glass is a good neutron absorber. The radioactive glass is placed in sealed containers, and the containers are stored in specially designed silos. The silos are intended to contain the waste for at least 1000 years, and consideration is also being given to storage in caves known to have been geologically stable since the last ice age.

Reprocessing uranium and plutonium is a hazardous operation that must be done with extreme care, mostly through the use of remote-control equipment. An attendant problem with reprocessing is that the plutonium recovered from spent fuel rods is of weapons grade. Strict safeguards are required to ensure that none of this material is diverted for weapons production.

Thus, while nuclear energy is clean in that it does not contribute to problems such as acid rain and global warming, it does present its own set of very difficult problems. In planning a future energy strategy, the issues are as much social and political as they are scientific. How much risk are we willing to accept in return for the perceived benefits of a given course of action? Should our efforts be directed to developing a new generation of safe nuclear reactors to reduce the effects of global warming associated with fossil fuels? Should greater emphasis be given to energy conservation and the more efficient use of energy than to increasing our reliance on nuclear energy? Should we largely forgo nuclear fission as an energy source while awaiting practical development of less risky fusion power, a technology that has not yet been demonstrated? Is an expanded development of solar energy and other unconventional energy sources a way around the dilemmas posed by fossil fuels and nuclear power? These are all questions that must be seriously debated as we make decisions affecting the energy needs and resources of our society.

Integrative Example

On April 26, 1986, an explosion at the nuclear power plant at Chernobyl, Ukraine, released the greatest quantity of radioactive material ever associated with an industrial accident. (See the discussion and photo on page 1043.) One of the radioisotopes in this emission was ^{131}I, a β^- emitter with a half-life of 8.04 days.

Assume that the total quantity of ^{131}I released was 250 g, and determine the number of *curies* associated with this ^{131}I 30.0 days after the accident.

1. *Calculate the mass of ^{131}I remaining after one month.* In equation (26.12), use $\lambda = 0.693/8.04$ d and $t = 30.0$ d. Because the mass of ^{131}I is directly proportional to the number of ^{131}I atoms, if we take N_0 to be 250 g, the value of N_t will be the mass we seek.

$$\ln \frac{N_t}{N_0} = \ln \frac{N_t}{250 \text{ g}} = \frac{-0.693}{8.04 \text{ d}} \times 30.0 \text{ d} = -2.59$$

$$N_t = 250 \text{ g} \times e^{-2.59} = 250 \text{ g} \times 0.0750 = 18.8 \text{ g}$$

2. *Determine the number of atoms in 18.8 g ^{131}I.* Use the molar mass, 131 g ^{131}I/mol ^{131}I, and the Avogadro constant in the following calculation.

$$N_t = 18.8 \text{ g }^{131}\text{I} \times \frac{1 \text{ mol }^{131}\text{I}}{131 \text{ g }^{131}\text{I}} \times \frac{6.022 \times 10^{23} \, {}^{131}\text{I atoms}}{1 \text{ mol }^{131}\text{I}}$$

$$= 8.64 \times 10^{22} \, {}^{131}\text{I atoms}$$

3. *Determine the decay constant in s^{-1}.* Use the method outlined in Example 26-3a to obtain λ from $t_{1/2}$.

$$\lambda = 0.693/t_{1/2} = \frac{0.693}{8.04 \text{ d}} \times \frac{1 \text{ d}}{24 \text{ h}} \times \frac{1 \text{ h}}{60 \text{ min}} \times \frac{1 \text{ min}}{60 \text{ s}}$$

$$= 9.98 \times 10^{-7} \text{ s}^{-1}$$

4. *Determine the decay rate (activity) of the ^{131}I after 30.0 days.* Use the results of parts 2 and 3 in equation (26.11).

$$A = \lambda N = 9.98 \times 10^{-7} \text{ s}^{-1} \times 8.64 \times 10^{22} \text{ atoms}$$

$$= 8.62 \times 10^{16} \text{ dis s}^{-1}$$

5. *Express the activity of the remaining ^{131}I in curies.* Use the definition of a curie in Table 26.4 to convert from dis s^{-1} to curies.

$$A = 8.62 \times 10^{16} \text{ dis s}^{-1} \times \frac{1 \text{ Ci}}{3.70 \times 10^{10} \text{ dis s}^{-1}}$$

$$= 2.33 \times 10^6 \text{ Ci}$$

Key Terms

α (alpha) particle (26-1)
β⁻ (beta) particle (26-1)
control rod (26-8)
decay constant (26-5)
electron capture (EC) (26-1)
fission (26-8)

fusion (26-9)
γ (gamma) ray (26-1)
half-life (26-5)
magic number (26-7)
moderator (26-8)
nuclear binding energy (26-6)

nuclear equation (26-1)
positron (β⁺) (26-1)
rad (26-10)
radioactive decay law (26-5)
radioactive decay series (26-2)
rem (26-10)

Review Questions

1. In your own words, define the following symbols: (a) α; (b) β^- (c) β^+; (d) γ; (e) $t_{1/2}$.

2. Briefly describe each of the following ideas, phenomena, or methods: (a) radioactive decay series; (b) charged-particle accelerator; (c) neutron-to-proton ratio; (d) mass–energy relationship; (e) background radiation.

3. Explain the important distinctions between each pair of terms: (a) electron and positron; (b) half-life and decay constant; (c) mass defect and nuclear binding energy; (d) nuclear fission and nuclear fusion; (e) primary and secondary ionization.

4. Which of the following—α particles, β particles, or γ rays—generally has the greatest (a) penetrating power through matter; (b) ionizing power in matter; (c) deflection in a magnetic field?

5. Supply the missing information in each of the following nuclear equations representing a radioactive decay process.
   (a) $^{160}_{?}\text{W} \longrightarrow \, ^{?}_{?}\text{Hf} + ?$
   (b) $^{38}_{?}\text{Cl} \longrightarrow \, ^{?}_{?}\text{Ar} + ?$
   (c) $^{214}_{?}? \longrightarrow \, ^{?}_{?}\text{Po} + \, ^{0}_{-1}\beta$
   (d) $^{32}_{17}\text{Cl} \longrightarrow \, ^{?}_{16}? + ?$

6. Complete the following nuclear equations.
   (a) $^{23}_{11}\text{Na} + ? \longrightarrow \, ^{24}_{11}\text{Na} + \, ^{1}_{1}\text{H}$
   (b) $^{59}_{27}\text{Co} + \, ^{1}_{0}\text{n} \longrightarrow \, ^{56}_{25}\text{Mn} + ?$
   (c) $? + \, ^{2}_{1}\text{H} \longrightarrow \, ^{240}_{94}\text{Pu} + \, ^{0}_{-1}\beta$
   (d) $^{246}_{96}\text{Cm} + ? \longrightarrow \, ^{254}_{102}\text{No} + 5\, ^{1}_{0}\text{n}$
   (e) $^{238}_{92}\text{U} + ? \longrightarrow \, ^{246}_{99}\text{Es} + 6\, ^{1}_{0}\text{n}$

7. Write nuclear equations to represent
   (a) the decay of ^{214}Ra by α-particle emission
   (b) the decay of ^{205}At by positron emission
   (c) the decay of ^{212}Fr by electron capture
   (d) the reaction of two deuterium nuclei (deuterons) to produce a nucleus of $^{3}_{2}\text{He}$
   (e) the production of $^{243}_{97}\text{Bk}$ by the α-particle bombardment of $^{241}_{95}\text{Am}$

8. For the radioactive nuclides in Table 26.1,
   (a) Which one has the largest value of the decay constant, λ?
   (b) Which one displays a 75% reduction in radioactivity from its current value in approximately one month?
   (c) Which ones lose more than 99% of their radioactivity in one month?

9. Two radioisotopes are to be compared. Isotope A requires 18.0 hours for its decay rate to fall to $\frac{1}{16}$ its initial value. Isotope B has a half-life that is 2.5 times that of A. How long does it take for the decay rate of isotope B to decrease to $\frac{1}{32}$ of its initial value?

10. A sample of radioactive $^{35}_{16}\text{S}$ disintegrates at a rate of 1.00×10^3 atoms/min. The half-life of $^{35}_{16}\text{S}$ is 87.9 d. How long will it take for the activity of this sample to decrease to the point of producing (a) 253; (b) 104; and (c) 52 dis/min?

11. With appropriate equations in the text, determine
   (a) the energy in joules corresponding to the destruction of 6.02×10^{-23} g of matter
   (b) the energy in megaelectronvolts that would be released if one α particle were completely destroyed

12. The measured mass of the nucleus of an atom of silver-107 is 106.879289 u. For this atom, determine the binding energy per nucleon in megaelectronvolts.

13. Which two of the following nuclides do not occur naturally: (a) ^2H; (b) ^{32}S; (c) ^{80}Br; (d) ^{132}Cs; (e) ^{184}W?

14. Explain why
   (a) Radioactive nuclides with intermediate half-lives are generally more hazardous than those with extremely short or extremely long half-lives.
   (b) Some radioactive substances are hazardous from a distance, whereas others must be taken internally to constitute a hazard.
   (c) Argon is the most abundant noble gas in the atmosphere.
   (d) Francium is such a rare element (less than about 30 g is present in Earth's crust at any one time) and cannot be extracted from minerals containing other metals of group 1.
   (e) Such extremely high temperatures are required to develop a self-sustaining thermonuclear (fusion) process as an energy source.

Exercises

Radioactive Processes

15. What nucleus is obtained in each process?
    (a) $^{234}_{94}$Pu decays by α emission.
    (b) $^{248}_{97}$Bk decays by β^- emission.
    (c) $^{196}_{82}$Pb goes through two successive EC processes.
16. What nucleus is obtained in each process?
    (a) $^{214}_{82}$Pb decays through two successive β^- emissions.
    (b) $^{226}_{88}$Ra decays through three successive α emissions.
    (c) $^{69}_{33}$As decays by β^+ emission.
17. Based on a favorable n/p ratio for the product nucleus, write the most plausible equation for the decay of $^{14}_{6}$C.
18. Write a plausible equation for the decay of tritium, $^{3}_{1}$H, the radioactive isotope of hydrogen.

Radioactive Decay Series

19. The natural decay series starting with the radionuclide $^{232}_{90}$Th follows the sequence represented here. Construct a graph of this series, similar to Figure 26-2.

$$^{232}_{90}\text{Th}-\alpha-\beta-\beta-\alpha-\alpha-\alpha\overset{\alpha-\beta}{\underset{\beta-\alpha}{\times}}\overset{\alpha-\beta}{\underset{\beta-\alpha}{\searrow}}{}^{208}_{82}\text{Pb}$$

20. The natural decay series starting with the radionuclide $^{235}_{92}$U follows the sequence represented here. Construct a graph of this series, similar to Figure 26-2.

$$^{235}_{92}\text{U}-\alpha-\beta-\alpha\overset{\alpha-\beta}{\underset{\beta-\alpha}{\diamondsuit}}-\alpha-\alpha-\alpha-\beta\overset{\alpha-\beta}{\underset{\beta-\alpha}{\searrow}}{}^{207}_{82}\text{Pb}$$

21. The uranium series described in Figure 26-2 is also known as the "$4n + 2$" series because the mass number of each nuclide in the series can be expressed by the equation $A = 4n + 2$, where n is an integer. Show that this equation is indeed applicable to the uranium series.
22. By the description in Exercise 21, the thorium series can be called the "$4n$" series and the actinium series the "$4n + 3$" series. A "$4n + 1$" series has also been established, with $^{241}_{94}$Pu as the parent nuclide. To which series does each of the following belong: (a) $^{214}_{83}$Bi; (b) $^{216}_{84}$Po; (c) $^{215}_{85}$At; (d) $^{235}_{92}$U?

Nuclear Reactions

23. Write equations for the following nuclear reactions.
    (a) bombardment of ^{7}Li with protons to produce ^{8}Be and γ rays
    (b) bombardment of ^{9}Be with $^{2}_{1}$H to produce ^{10}B
    (c) bombardment of ^{14}N with neutrons to produce ^{14}C
24. Write equations for the following nuclear reactions.
    (a) bombardment of ^{238}U with α particles to produce ^{239}Pu
    (b) bombardment of tritium ($^{3}_{1}$H) with $^{2}_{1}$H to produce ^{4}He
    (c) bombardment of ^{33}S with neutrons to produce ^{33}P
25. Write nuclear equations to represent the formation of an isotope of element 111 with a mass number of 272 by the bombardment of bismuth-209 by nickel-64 nuclei, followed by a succession of five α-particle emissions.
26. Write nuclear equations to represent the formation of an isotope of element 118 with a mass number of 293 by the bombardment of lead-208 by krypton-86 nuclei, followed by a chain of α-particle emissions to the element seaborgium.

Rate of Radioactive Decay

27. The disintegration rate for a sample containing $^{60}_{27}$Co as the only radioactive nuclide is 6740 dis/h. The half-life of $^{60}_{27}$Co is 5.2 years. Estimate the number of atoms of $^{60}_{27}$Co in the sample.
28. How many years must the radioactive sample of Exercise 27 be maintained before the disintegration rate falls to 101 dis/min?
29. A sample containing $^{224}_{88}$Ra, which decays by α-particle emission, disintegrates at the following rate, expressed as disintegrations per minute or counts per minute (cpm): $t = 0$, 1000 cpm; $t = 1$ h, 992 cpm; $t = 10$ h, 924 cpm; $t = 100$ h, 452 cpm; $t = 250$ h, 138 cpm. What is the half-life of this nuclide?
30. Iodine-129 is a product of nuclear fission, whether from an atomic bomb or a nuclear power plant. It is a β^- emitter with a half-life of 1.7×10^7 years. How many disintegrations per second would occur in a sample containing 1.00 mg ^{129}I?
31. Suppose that a sample containing ^{32}P has an activity 1000 times the detectable limit. How long would an experiment have to be run with this sample before the radioactivity could no longer be detected?
32. What mass of carbon-14 must be present in a sample to have an activity of 1.00 mCi?

Age Determinations with Radioisotopes

33. A wooden object is claimed to have been found in an Egyptian pyramid and is offered for sale to an art museum. Radiocarbon dating of the object reveals a disintegration rate of 10.0 dis min^{-1} g^{-1}. Do you think the object is authentic? Explain.

34. The lowest level of ^{14}C activity that seems possible for experimental detection is 0.03 dis min^{-1} g^{-1}. What is the maximum age of an object that can be determined by the carbon-14 method?

35. What should be the mass ratio $^{208}Pb/^{232}Th$ in a meteorite that is approximately 2.7×10^9 years old? The half-life of ^{232}Th is 1.39×10^{10} years.
(*Hint:* One ^{208}Pb atom is the final decay product of one ^{232}Th atom.)

36. Concerning the decay of ^{232}Th described in Exercise 35, a certain rock has a $^{208}Pb/^{232}Th$ mass ratio of 0.25/1.00. Estimate the age of the rock.

Energetics of Nuclear Reactions

37. Use the electron mass from Table 2.1 and the measured mass of the nuclide $^{19}_{9}F$, 18.998403 u, to determine the binding energy per nucleon (in megaelectronvolts) of this atom.

38. Use the electron mass from Table 2.1 and the measured mass of the nuclide $^{56}_{26}Fe$, 55.934939 u, to determine the binding energy per nucleon (in megaelectronvolts) of this atom.

39. Calculate the energy, in megaelectronvolts, released in the nuclear reaction

$$^{10}_{5}B + {}^{4}_{2}He \longrightarrow {}^{13}_{6}C + {}^{1}_{1}H$$

The nuclidic masses are $^{10}_{5}B$ = 10.01294 u; $^{4}_{2}He$ = 4.00260 u; $^{13}_{6}C$ = 13.00335 u; $^{1}_{1}H$ = 1.00783 u.

40. You are given the following nuclidic masses: $^{6}_{3}Li$ = 6.01513 u; $^{4}_{2}He$ = 4.00260 u; $^{3}_{1}H$ = 3.01604 u; $^{1}_{0}n$ = 1.008665 u. How much energy, in megaelectronvolts, is released in the following nuclear reaction?

$$^{6}_{3}Li + {}^{1}_{0}n \longrightarrow {}^{4}_{2}He + {}^{3}_{1}H$$

41. Calculate the number of neutrons that could be created with 6.75×10^6 MeV of energy.

42. When β^+ and β^- particles collide, they annihilate each other, producing two γ rays that move away from each other along a straight line. What is the approximate energies of these two γ rays, in MeV?

Nuclear Stability

43. Which member of the following pairs of nuclides would you expect to be most abundant in natural sources: (a) $^{20}_{10}Ne$ or $^{22}_{10}Ne$; (b) $^{17}_{8}O$ or $^{18}_{8}O$; (c) $^{6}_{3}Li$ or $^{7}_{3}Li$? Explain your reasoning.

44. Which member of the following pairs of nuclides would you expect to be most abundant in natural sources: (a) $^{40}_{20}Ca$ or $^{42}_{20}Ca$; (b) $^{31}_{15}P$ or $^{32}_{15}P$; (c) $^{63}_{30}Zn$ or $^{64}_{30}Zn$? Explain your reasoning.

45. One member each of the following pairs of radioisotopes decays by β^- emission, and the other by positron (β^+) emission: (a) $^{29}_{15}P$ and $^{33}_{15}P$; (b) $^{120}_{53}I$ and $^{134}_{53}I$. Which is which? Explain your reasoning.

46. Each of the following isotopes is radioactive: (a) $^{28}_{15}P$; (b) $^{45}_{19}K$; (c) $^{73}_{30}Zn$. Which would you expect to decay by β^+ emission?

47. Some nuclides are said to be doubly magic. What do you suppose this term means? Postulate some nuclides that might be doubly magic, and locate them in Figure 26-7.

48. Both β^- and β^+ emissions are observed for artificially produced radioisotopes of low atomic numbers, but only β^- emission is observed with naturally occurring radioisotopes of high atomic number. Why do you suppose this is so?

Fission and Fusion

49. Refer to the Integrative Example. In contrast to the Chernobyl accident, the 1979 nuclear accident at Three Mile Island released only 170 curies of ^{131}I. How many milligrams of ^{131}I does this represent?

50. Explain why more energy is released in a fusion process than in a fission process?

Effect of Radiation on Matter

51. Explain why the rem is more satisfactory than the rad as a unit for measuring radiation dosage.
52. Discuss briefly the basic difficulties in establishing the physiological effects of low-level radiation.
53. ^{90}Sr is both a product of radioactive fallout and a radioactive waste in a nuclear reactor. This radioisotope is a β^- emitter with a half-life of 27.7 years. Suggest reasons why ^{90}Sr is such a potentially hazardous substance.
54. ^{222}Rn is an α-particle emitter with a half-life of 3.82 days. Is it hazardous to be near a flask containing this isotope? Under what conditions might ^{222}Rn be hazardous?

Applications of Radioisotopes

55. Describe how you might use radioactive materials to find a leak in the $H_2(g)$ supply line in an ammonia synthesis plant.
56. Explain why neutron activation analysis is so useful in identifying trace elements in a sample, in contrast to ordinary methods of quantitative analysis, such as precipitation or titration.
57. A small quantity of NaCl containing radioactive $^{24}_{11}$Na is added to an aqueous solution of $NaNO_3$. The solution is cooled, and $NaNO_3$ is crystallized from the solution. Would you expect the $NaNO_3(s)$ to be radioactive? Explain.
58. The following reactions are carried out with HCl(aq) containing some tritium ($^3_1$H) as a tracer. Would you expect any of the tritium radioactivity to appear in the $NH_3(g)$? In the H_2O? Explain.

$$NH_3(aq) + HCl(aq) \longrightarrow NH_4Cl(aq)$$
$$NH_4Cl(aq) + NaOH(aq) \longrightarrow$$
$$NaCl(aq) + H_2O(l) + NH_3(g)$$

Integrative and Advanced Exercises

59. In some cases, the most abundant isotope of an element can be established by rounding off the atomic mass to the nearest whole number, as in ^{39}K, ^{85}Rb, and ^{88}Sr. But in other cases, the isotope corresponding to the rounded-off atomic mass does not even occur naturally, as in ^{64}Cu. Explain the basis of this observation.
60. The overall change in the radioactive decay of $^{238}_{92}$U to $^{206}_{82}$Pb is the emission of eight α particles. Show that if this loss of eight α particles were not also accompanied by six β^- emissions, the product nucleus would still be radioactive.
61. Use data from the text to determine how many metric tons (1 metric ton = 1000 kg) of bituminous coal (85% C) would have to be burned to release as much energy as is produced by the fission of 1.00 kg $^{235}_{92}$U.
62. One method of dating rocks is based on their ^{87}Sr/^{87}Rb ratio. ^{87}Rb is a β^- emitter with a half-life of 5×10^{11} years. A certain rock has a mass ratio ^{87}Sr/^{87}Rb of 0.004/1.00. What is the age of the rock?
63. How many millicuries of radioactivity are associated with a sample containing 5.10 mg ^{229}Th, which has a half-life of 7340 years?
64. What mass of ^{90}Sr, with a half-life of 27.7 years, is required to produce 1.00 millicurie of radioactivity?
65. Refer to the Integrative Example. Another radioisotope produced in the Chernobyl accident was ^{137}Cs. If a 1.00-mg sample of ^{137}Cs is equivalent to 89.8 millicuries, what must be the half-life (in years) of ^{137}Cs?
66. The percent natural abundance of ^{40}K is 0.0117%. The radioactive decay of ^{40}K atoms occurs 89% by β^- emission; the rest is by electron capture and β^+ emission. The half-life of ^{40}K is 1.25×10^9 years. Calculate the number of β^- particles produced per second by the ^{40}K present in a 1.00-g sample of the mineral *microcline*, $KAlSi_3O_8$.
67. The carbon-14 dating method is based on the assumption that the rate of production of ^{14}C by cosmic ray bombardment has remained constant for thousands of years and that the ratio of ^{14}C to ^{12}C has also remained constant. Can you think of any effects of human activities that could invalidate this assumption in the future?
68. Calculate the minimum kinetic energy (in megaelectronvolts) that α particles must possess to produce the nuclear reaction

$$^4_2He + ^{14}_7N \longrightarrow ^{17}_8O + ^1_1H$$

The nuclidic masses are ^4_2He = 4.00260 u; $^{14}_7N$ = 14.00307 u; ^1_1H = 1.00783 u; $^{17}_8O$ = 16.99913 u.
69. Hydrogen gas is spiked with tritium to the extent of 5.00% by mass. What is the activity in curies of a 4.65-L sample of this gas at 25.0 °C and 1.05 atm pressure?
(*Hint:* Use data from the chapter and elsewhere in the text, as necessary.)
70. ^{40}K undergoes radioactive decay by electron capture to ^{40}Ar and by β^- emission to ^{40}Ca. The fraction of the decay that occurs by electron capture is 0.110. The half-life of ^{40}K is 1.25×10^9 years. Assuming that a rock in which ^{40}K has undergone decay retains all the ^{40}Ar produced, what would be the ^{40}Ar/^{40}K mass ratio in a rock that is 1.5×10^9 years old?

71. A certain shale deposit containing 0.006% U by mass is used as a potential fuel in a breeder reactor. Assuming a density of 2.5 g/cm^3, how much energy could be released from 1.00×10^3 cm^3 of this material? Assume a fission energy of 3.20×10^{-11} J per fission event (that is, per U atom).

72. An ester forms from a carboxylic acid and an alcohol.

$$RCO_2H + HOR' \longrightarrow RCO_2R' + H_2O$$

This reaction is superficially similar to the reaction of an acid with a base such as sodium hydroxide. The mechanism of the reaction can be followed by using the tracer ^{18}O. This isotope is non-radioactive, but other physical measurements can be used to detect its presence. When the *esterification* reaction is carried out with the alcohol containing oxygen-18 atoms, no oxygen-18 beyond its naturally occurring abundance is found in the water produced. How does this result affect the perception that this reaction is like an acid–base reaction?

73. The conversion of CO_2 into carbohydrates by plants via photosynthesis can be represented by the reaction

$$6\ CO_2(g) + 6\ H_2O \xrightarrow{\text{light}} C_6H_{12}O_6 + 6\ O_2(g)$$

To study the mechanism of photosynthesis, algae were grown in water containing ^{18}O, that is, $H_2^{18}O$. The oxygen evolved contained oxygen-18 in the same ratio to the other oxygen isotopes as the water in which the reaction was carried out. In another experiment, algae were grown in water containing only ^{16}O, but with oxygen-18 present in the CO_2. The oxygen evolved in this experiment contained no oxygen-18. What conclusion can you draw about the mechanism of photosynthesis from these experiments?

74. Assume that when Earth formed, uranium-238 and uranium-235 were equally abundant. Their current percent natural abundances are 99.28% uranium-238 and 0.72% uranium-235. Given half-lives of 4.5×10^9 years for uranium-238 and 7.1×10^8 years for uranium-235, determine the age of Earth corresponding to this assumption.

Feature Problems

75. The *packing fraction* of a nuclide is related to the fraction of the total mass of a nuclide that is converted to nuclear binding energy. It is defined as the fraction $(M - A)/A$, where M is the actual nuclidic mass and A is the mass number. Use data from a handbook (such as the *Handbook of Chemistry and Physics*, published by the CRC Press) to determine the packing fractions of some representative nuclides. Plot a graph of packing fraction versus mass number, and compare it with Figure 26-6. Explain the relationship between the two.

76. For medical uses, radon-222 formed in the radioactive decay of radium-226 is allowed to collect over the radium metal for a period of time. Then, the gas is withdrawn and sealed into a glass vial. Following this, the radium is allowed to disintegrate for another period of time, when a new sample of radon-222 can be withdrawn. The procedure can be continued indefinitely. The process is somewhat complicated by the fact that radon-222 itself undergoes radioactive decay to polonium-218, and so on. The half-lives of radium-226 and radon-222 are 1.60×10^3 years and 3.82 days, respectively.

(a) Beginning with pure radium-226, the number of radon-222 atoms present starts at zero, increases for a time, and then falls off again. Explain this behavior. That is, because the half-life of radon-222 is so much shorter than that of radium-226, why doesn't the radon-222 simply decay as fast as it is produced, without ever building up to a maximum concentration?

(b) Write an expression for the rate of change (dD/dt) in the number of atoms (D) of the radon-222 daughter in terms of the number of radium-226 atoms present initially (P_0) and the decay constants of the parent (λ_p) and daughter (λ_d).

(c) Integration of the expression obtained in part (b) yields the following expression for the number of atoms of the radon-222 daughter (D) present at a time t.

$$D = \frac{P_0\lambda_P(e^{-\lambda_p \times t} - e^{-\lambda_d \times t})}{\lambda_d - \lambda_P}$$

Starting with 1.00 g of pure radium-226, approximately how long will it take for the amount of radon-222 to reach its maximum value: one day, one week, one year, one century, or one millenium?

77. Most nuclear power plants use zirconium in the fuel rods because zirconium maintains its structural integrity under exposure to the radiation in the nuclear reactor. The nuclear accidents at both Three Mile Island and Chernobyl involved the evolution of hydrogen gas from the reduction of water. The half-cell reduction potential for Zr is

$$ZrO_2(s) + 4\ H_3O^+(aq) + 4e^- \longrightarrow Zr(s) + 6\ H_2O(l)$$
$$E° = -1.43\ V$$

(a) Can Zr reduce water under standard-state conditions?

(b) Calculate the equilibrium constant for the reduction of water by zirconium.

(c) Is the reaction spontaneous if the pH = 7 and Zr, ZrO_2, and water are in their standard states?

(d) Was the reduction of water by Zr the culprit in the reactor accidents mentioned above?

78. Radioactive decay and mass spectrometry are often used to date rocks after they have cooled from a magma. ^{87}Rb has a half-life of 4.8×10^{10} years and follows the radioactive decay

$$^{87}Rb \longrightarrow {}^{87}Sr + \beta^-$$

A rock was dated by assaying the product of this decay. The mass spectrum of a homogenized sample of rock showed the $^{87}Sr/^{86}Sr$ ratio to be 2.25. Assume that the original $^{87}Sr/^{86}Sr$ ratio was 0.700 when the rock cooled. Chemical analysis of the rock gave 15.5 ppm Sr and 265.4 ppm Rb, using the average atomic masses from a periodic table. The other isotope ratios were $^{86}Sr/^{88}Sr = 0.119$ and $^{84}Sr/^{88}Sr = 0.007$. The isotopic ratio for $^{87}Rb/^{85}Rb$ is 0.330. The isotopic masses are as follows:

Isotope	Atomic Mass/u
^{87}Rb	86.909
^{85}Rb	84.912
^{88}Sr	87.906
^{86}Sr	85.909
^{84}Sr	83.913
^{87}Sr	86.909

Calculate the following:
(a) the average atomic mass of Sr in the rock
(b) the original concentration of Rb in the rock in ppm
(c) the percentage of rubidium-87 decayed in the rock
(d) the time since the rock cooled

eMedia Exercises

79. Several types of radioactive emission are compared in the **Separation of Alpha, Beta, and Gamma Rays** animation *(eChapter 26-1)*. **(a)** What are the ratio of the masses of the different particles described in the movie? What effect would a difference in mass have on a particle's interaction with other matter? **(b)** Positron emission is described in the text. Where would this particle impinge on the fluorescent screen in this animation?

80. The **Radioactive Decay Series** simulation *(eChapter 26-2)* illustrates the natural decay of $^{238}_{92}U$. What is the total number and type of radioactive species produced by this series of reactions? What is the final product of the series?

81. The rate of radioactive decay is described in the **First-Order Processes** animation *(eChapter 26-5)*. On the atomic scale, describe why a radioactive decay process would be expected to exhibit first-order reaction kinetics as opposed to zero- or second-order kinetics.

82. Observe the operation of a nuclear reactor in the **Action of Nuclear Control Rods** simulation *(eChapter 26-9)*. In this application, what measures are taken to control the nuclear reaction? What are potential points of failure or points of release of radioactive material in this design of reactor?

27

Organic Chemistry

Contents

Vanillin, an aldehyde, gives vanilla beans their characteristic odor. Vanillin can be extracted from vanilla beans or synthesized from waste lignin obtained from the wood pulp industry. Synthetic vanillin is the basis of artificial vanilla flavoring. Vanillin has the structure

To early nineteenth century chemists, organic chemistry meant the study of compounds obtainable only from living matter, which was thought to have the "vital force" needed to make these compounds. In 1828, Friedrich Wöhler set out to synthesize ammonium cyanate, NH_4OCN, as in the reaction

$$AgOCN(s) + NH_4Cl(aq) \xrightarrow{\Delta} AgCl(s) + NH_4OCN(aq)$$

The white crystalline solid he obtained from the solution had none of the properties of ammonium cyanate, even though it had the same composition. The compound was not NH_4OCN but $(NH_2)_2CO$—*urea*, an organic compound. As Wöhler excitedly reported to J. J. Berzelius, "I must tell you that I can make urea without the use of kidneys, either man or dog. Ammonium cyanate is urea."

Since that time, chemists have synthesized millions of organic compounds, and today organic compounds number about 98% of all known chemical substances. In this chapter, we build on the introduction to organic compounds in Chapter 3 by exploring some of the principal types of organic compounds. In the next chapter, we will study the connection between organic compounds and living matter that once seemed so mysterious.

27-1   Organic Compounds and Structures: An Overview

As we learned in Chapter 3, organic compounds contain carbon and hydrogen atoms or carbon and hydrogen in combination with a few other types of atoms, such as oxygen, nitrogen, and sulfur. Carbon is singled out for special study because the ability of C atoms to form strong covalent bonds with one another allows them to join together into straight chains, branched chains, and rings. The nearly infinite number of possible bonding arrangements of C atoms accounts for the vast number and variety of organic compounds.

The simplest organic compounds are those of carbon and hydrogen—**hydrocarbons**—and the simplest hydrocarbon is methane, CH_4, the chief constituent of natural gas.

From VSEPR theory we expect the electron group geometry around the central C atom in CH_4 to be tetrahedral (Figure 27-1a). The four H atoms are equivalent: They are equidistant from the C atom and attached to it by covalent bonds of equal strength. The angle between any two C—H bonds is 109°28′. Figure 27-1a calls attention to a problem: How can we draw the three-dimensional structure of organic molecules in two dimensions? In a common method, often known as the **dashed-wedged line notation**, the following convention is adopted.

1. Ordinary lines are used to show bonds that lie in the plane of the paper.
2. Solid wedge lines are used to show bonds that stick out toward the viewer, that is, in front of the plane of the paper.
3. Dashed lines are used to show bonds directed away from the viewer, that is, behind the plane of the paper.

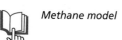
Methane model

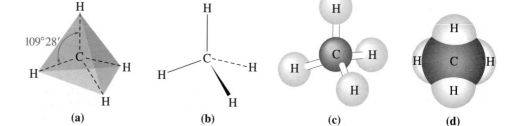

(a)   (b)   (c)   (d)

▲ **FIGURE 27-1   Representation of the methane molecule**
(a) Tetrahedral structure showing bond angle. (b) Dashed-wedged line structure convention used to suggest a three-dimensional structure through a structural formula. The solid lines represent bonds in the plane of the page. The dashed lines project *away* from the viewer (into the page), and the heavy wedge projects *toward* the viewer (out of the page). (c) Ball-and-stick model. (d) Space-filling model.

The dashed-wedged line notation is illustrated in Figure 27-1b. Figure 27-1 also shows other commonly employed depictions of methane. As shown in Figure 12-6 (page 441), the chemical bonding in methane can be described by four σ bonds formed by the overlap of four hydrogen $1s$ orbitals with four equivalent sp^3 hybrid orbitals on the carbon atom.

KEEP IN MIND
that the hybridization of the C atoms in ethane and propane is sp^3, as in methane. The tetrahedral geometry at the C atoms in alkanes means that the propane chain is not linear. ▶

The removal of one H atom from a CH_4 molecule leaves the $—CH_3$ group. Now imagine forming a covalent bond between two $—CH_3$ groups. The resulting molecule is *ethane*, C_2H_6. (Figure 27-2). By increasing the number of C atoms in the chain, we can obtain still more hydrocarbons. The three-carbon molecule propane, C_3H_8, is pictured in Figure 27-3.

Skeletal Isomerism

There are *two* ways of assembling a hydrocarbon molecule with four C and ten H atoms. There are *two* different compounds with the formula C_4H_{10}. One is called butane and the other, isobutane (recall Figure 3-2 on page 67).

As we have previously learned, compounds that have the same molecular formula but different structural formulas are called *isomers*. In the case of butane, the isomers differ in their structural skeletons, or carbon-atom frameworks—one is a straight chain and the other, a branched chain. This type of isomerism is called **skeletal isomerism**.

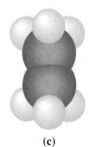

(a)

(b)

(c)

▲ **FIGURE 27-2**
The ethane molecule, C_2H_6
(a) Structural formula.
(b) Dashed-wedged line structure. **(c)** Space-filling model.

Ethane, Propane models

EXAMPLE 27-1

Identifying Isomers. Write structural formulas for all the possible isomers with the molecular formula C_5H_{12}.

Solution

A general strategy is to first write the longest chain of C atoms from left to right. This gives the straight-chain structure labeled (1). To finish this structure, we add an appropriate number of H atoms (12, in this case) to give each C atom four bonds.

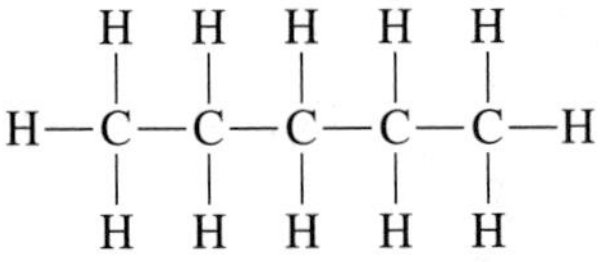

$$H—\overset{H}{\underset{H}{C}}—\overset{H}{\underset{H}{C}}—\overset{H}{\underset{H}{C}}—\overset{H}{\underset{H}{C}}—\overset{H}{\underset{H}{C}}—H$$

(1)

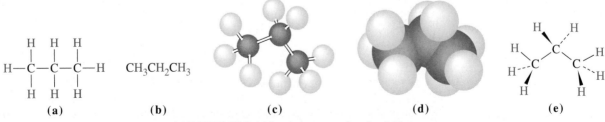

(a) $CH_3CH_2CH_3$ **(b)** **(c)** **(d)** **(e)**

▲ **FIGURE 27-3   The propane molecule, C_3H_8**
(a) Structural formula. **(b)** Condensed structural formula. **(c)** Ball-and-stick model.
(d) Space-filling model. **(e)** Dashed-wedged line notation.

Now, let's look for isomers with four C atoms in the longest chain and one C atom as a branch (five C atoms in all). There is only *one* possibility. Notice that if structure (2) is flopped from left to right, the identical structure (2′) is obtained.

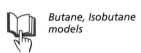

Again, we complete the structure of this isomer by adding H atoms.

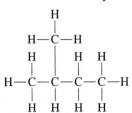

Finally, let's consider a three-carbon chain with two one-carbon branches. Again, there is only *one* possibility.

The number of isomers with the formula C_5H_{12} is three.

Practice Example A: Write condensed structural formulas for the five possible isomers with the formula C_6H_{14}.

Practice Example B: Write condensed structural formulas for the nine possible isomers with the formula C_7H_{16}.

In Chapter 3, we mentioned a way of greatly simplifying the writing of organic structures that we might consider again here. We draw lines to represent chemical bonds, and wherever a line ends or meets another line, there is a C atom. We then assume that enough H atoms are bonded to the C atoms to satisfy the need for each C atom to form *four* bonds. Such structural formulas are called *line-angle*, or *stick*, structures. Notice that in the line-angle structures written below for the three isomers in Example 27-1, we arrange the lines in a zigzag fashion, just as observed in the three-dimensional structures of these molecules.

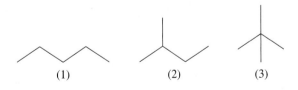

Nomenclature

Early organic chemists often assigned names related to the origin or properties of new compounds. Some of these names are still in common use. Citric acid is found in citrus fruit; uric acid is present in urine; formic acid is found in ants (from the Latin word for ant, *formica*); and morphine induces sleep (from *Morpheus*, the ancient Greek god of sleep). As thousands upon thousands of new compounds were synthesized, it became apparent that a system of common names was unworkable. Following several interim systems, one recommended by the International Union of Pure and Applied Chemistry (IUPAC) was adopted.

Butane, Isobutane models

▲ A Swiss stamp commemorating the 100th anniversary of an international congress in Geneva, at which a systematic nomenclature of organic compounds was adopted.

Nomenclature activity

In this introduction to nomenclature, we will consider only hydrocarbons with all carbon-to-carbon bonds as single bonds. These are known as **saturated hydrocarbons**, or alkanes. We have no trouble naming the first few: CH_4, methane; C_2H_6, ethane; C_3H_8, propane. We encounter our first difficulty with C_4H_{10}, which has two isomers. This problem is resolved by assigning the name *butane* to the straight-chain isomer, $CH_3CH_2CH_2CH_3$, and *isobutane* to the branched-chain isomer, $CH_3CH(CH_3)_2$. This method is inadequate for the C_5H_{12} alkanes, for which there are three structural isomers (Example 27-1), and it is even less satisfactory for alkanes with greater numbers of C atoms. We avoid unambiguous names, however, with the following rules.

▶ Do not try to memorize these rules at the outset. Refer to them as you proceed through the examples, and they will become part of your vocabulary of organic chemistry.

1. Select the *longest* continuous carbon chain in the molecule, and use the hydrocarbon name of this chain as the base name. Except for the common names methane, ethane, propane, and butane, standard Greek prefixes relate the name to the number of C atoms in the chain, as in *pent*ane (C_5), *hex*ane (C_6), *hept*ane (C_7), *oct*ane (C_8), ...

2. Consider every branch of the main chain to be a substituent derived from another hydrocarbon. For each of these substituents, change the ending of its name from *ane* to *yl*. That is, the alkane substituent becomes an **alkyl** group (Table 27.1).

3. Number the C atoms of the continuous base chain so that the substituents appear *at the lowest numbers* possible.

4. Name each substituent according to its chemical identity and the number of the C atom to which it is attached. For identical substituents use *di*, *tri*, *tetra*, and so on, and write the appropriate carbon number for each substituent.

5. Separate numbers from one another by commas, and from letters by hyphens.

6. List the substituents *alphabetically* by name.

KEEP IN MIND ▶
that in an alphabetical order, we do not consider the prefixes *di*, *tri*, ... or symbols such as *s* and *t*. Thus, butyl, even though *t*-butyl, precedes methyl in the name 4-*t*-butyl-2-methylheptane.

TABLE 27.1 Some Common Alkyl Groups	
Name	**Structural Formula**
Methyl	$-CH_3$
Ethyl	$-CH_2CH_3$
Propyl[a]	$-CH_2CH_2CH_3$
Isopropyl	$CH_3\overset{\mid}{C}HCH_3$
Butyl[a]	$-CH_2CH_2CH_2CH_3$
Isobutyl	$-CH_2\overset{\overset{\textstyle CH_3}{\mid}}{C}HCH_3$
s-Butyl[b]	$CH_3\overset{\mid}{C}HCH_2CH_3$
t-Butyl[c]	$CH_3\overset{\overset{\textstyle CH_3}{\mid}}{\underset{\mid}{C}}CH_3$

[a] In the past, the prefix *normal* or *n*- was used for a straight-chain alkyl group, such as *n*-propyl or *n*-butyl.
[b] *s* = secondary.
[c] *t* = tertiary.

EXAMPLE 27-2

Naming an Alkane Hydrocarbon. Give an appropriate IUPAC name for the following compound, an important constituent of gasoline.

$$\underset{1}{CH_3}-\underset{2}{\overset{\overset{\displaystyle CH_3}{|}}{\underset{\underset{\displaystyle CH_3}{|}}{C}}}-\underset{3}{CH_2}-\underset{4}{\overset{\overset{\displaystyle CH_3}{|}}{CH}}-\underset{5}{CH_3}$$

Solution

The C atoms are numbered in red, and the side-chain substituents to be named are shown in blue. The longest chain of C atoms is five, and the carbons are numbered so that the one with two substituent groups is number 2, instead of number 4. Each substituent is a methyl group, $-CH_3$. Two methyl groups are on the second C atom, and one methyl group is on the fourth C atom. The correct name is

$$2,2,4\text{-trimethylpentane}$$

If we had numbered the C atoms from right to left, we would have obtained the name 2,4,4-trimethylpentane. This is *not* an acceptable name, however, because it does not use the *smallest* numbers possible.

Practice Example A: Give an appropriate IUPAC name for the hydrocarbon

$$CH_3CH_2\overset{\overset{\displaystyle CH_3}{|}}{CH}CH_2CH_2\overset{\overset{\displaystyle CH_3}{|}}{\underset{\underset{\displaystyle CH_3}{|}}{C}}CH_2CH_2CH_3$$

Practice Example B: Give an appropriate IUPAC name for the hydrocarbon $CH_3CH_2CH(CH_3)CH_2CH_2CH(CH_3)CH_2CH_3$

EXAMPLE 27-3

Writing the Formula to Correspond to the Name of an Alkane Hydrocarbon. Write a condensed structural formula for 4-*t*-butyl-2-methylheptane.

Solution

Because the compound is a heptane, the longest chain of C atoms is seven.

$$C-C-C-C-C-C-C$$

Starting on the left, we attach a methyl group to the second C atom.

$$C-\overset{\overset{\displaystyle CH_3}{|}}{C}-C-C-C-C-C$$

Next, we attach a *t*-butyl group to the fourth C atom.

$$C-\overset{\overset{\displaystyle CH_3}{|}}{C}-C-\overset{\overset{\displaystyle CH_3-\overset{\overset{\displaystyle CH_3}{|}}{C}-CH_3}{|}}{C}-C-C-C$$

Finally, we add the remaining hydrogen atoms to give each C atom four bonds.

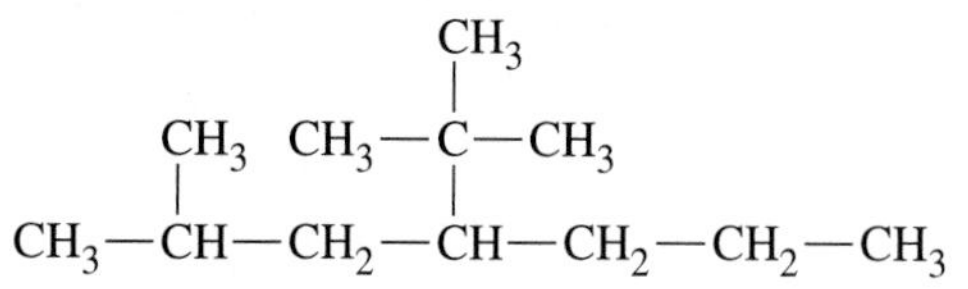

$$CH_3-CH-CH_2-CH-CH_2-CH_2-CH_3$$

(with substituents: CH_3 on second carbon, and CH_3-C-CH_3 with CH_3 above on the fourth carbon)

Practice Example A: Write a condensed structural formula for 3-ethyl-2, 6-dimethylheptane.

Practice Example B: Write a condensed structural formula for 3-isopropyl-2-methylpentane.

Positional Isomerism

A variety of atoms or groups of atoms can be substituents on carbon chains, for example, Br. The three monobromopentanes possess the same carbon skeleton. Because they differ in the position of the bromine atom on the carbon chain, these structural isomers are also called **positional isomers**.

$CH_3CH_2CH_2CH_2CH_2Br$ $CH_3CH_2CH_2CHCH_3$ $CH_3CH_2CHCH_2CH_3$
                                                    | |
                                                   Br   Br

1-Bromopentane   2-Bromopentane   3-Bromopentane

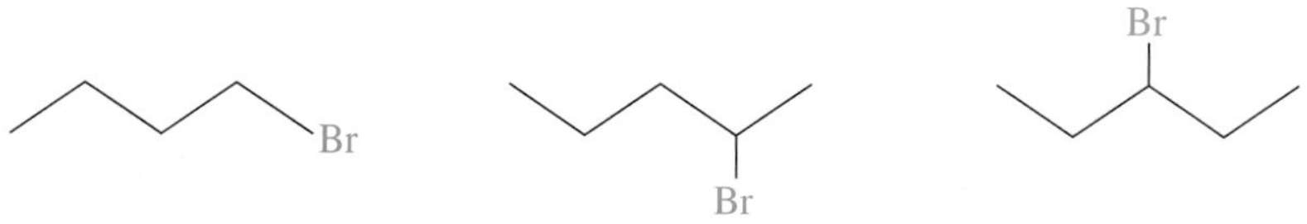

Functional Groups

Organic compounds typically contain elements in addition to carbon and hydrogen. These elements occur as distinctive groupings of one or several atoms. In some cases, these groupings of atoms are substituted for H atoms in hydrocarbon chains or rings. In other cases, they may build from the C atom itself. For example, the carbonyl group consists of a C atom in the skeletal structure to which an O atom is attached by a double bond $\left(\begin{array}{c} \diagdown \\ \diagup \end{array} C{=}O \right)$. These distinctive groupings of atoms are called

functional groups, and the remainder of the molecule is sometimes referred to by the symbol R. Usually, the R refers to an alkyl group. The physical and chemical properties of organic molecules generally depend on the particular functional groups present. The remainder of the molecule (R) often has little effect on these properties.

A convenient way to study organic chemistry, then, is to consider the properties associated with specific functional groups. We have already encountered a few functional groups in earlier chapters, such as the —OH group in alcohols and the —COOH group in carboxylic acids. Table 27.2 lists the major types of organic compounds, with their distinctive functional groups shown in blue. By combining information from Tables 27.1 and 27.2, for example, you should be able to identify $(CH_3)_2CHOCH_2CH_2CH_3$ as isopropyl propyl ether.

TABLE 27.2   Some Classes of Organic Compounds and Their Functional Groups

Class	General Structural Formula[a]	Example	Name of Example
Alkane	R—H	$CH_3CH_2CH_2CH_2CH_2CH_3$	Hexane
Alkene	$\overset{\diagdown}{\diagup}C{=}C\overset{\diagup}{\diagdown}$	$CH_2{=}CHCH_2CH_2CH_3$	1-Pentene
Alkyne	—C≡C—	$CH_3C{\equiv}CCH_2CH_2CH_2CH_2CH_3$	2-Octyne
Alcohol	R—OH	$CH_3CH_2CH_2CH_2OH$	1-Butanol
Alkyl halide	R—X[b]	$CH_3CH_2CH_2CH_2CH_2CH_2Br$	1-Bromohexane
Ether	R—O—R	$CH_3{-}O{-}CH_2CH_2CH_3$	1-Methoxypropane (methyl propyl ether)[c]
Amine	R—NH$_2$	$CH_3CH_2CH_2{-}NH_2$	1-Aminopropane (propylamine)[c]
Aldehyde	$R{-}\overset{\overset{\displaystyle O}{\|}}{C}{-}H$	$CH_3CH_2CH_2\overset{\overset{\displaystyle O}{\|}}{C}{-}H$	Butanal (butyraldehyde)[c]
Ketone	$R{-}\overset{\overset{\displaystyle O}{\|}}{C}{-}R$	$CH_3CH_2\overset{\overset{\displaystyle O}{\|}}{C}CH_2CH_2CH_3$	3-Hexanone (ethyl propyl ketone)[c]
Carboxylic acid	$R{-}\overset{\overset{\displaystyle O}{\|}}{C}{-}OH$	$CH_3CH_2CH_2\overset{\overset{\displaystyle O}{\|}}{C}{-}OH$	Butanoic acid (butyric acid)[c]
Ester	$R{-}\overset{\overset{\displaystyle O}{\|}}{C}{-}OR$	$CH_3CH_2CH_2\overset{\overset{\displaystyle O}{\|}}{C}{-}OCH_3$	Methyl butanoate (methyl butyrate)[c]
Amide	$R{-}\overset{\overset{\displaystyle O}{\|}}{C}{-}NH_2$	$CH_3CH_2CH_2\overset{\overset{\displaystyle O}{\|}}{C}{-}NH_2$	Butanamide (butyramide)[c]
Arene	Ar—H[d]	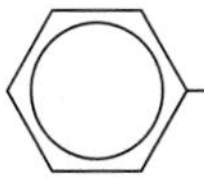 ⬡—CH_2CH_3	Ethylbenzene
Aryl halide	Ar—X[b]	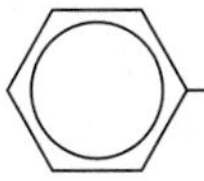 ⬡—Br	Bromobenzene
Phenol	Ar—OH	Cl—⬡—OH 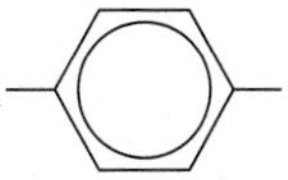	4-Chlorophenol (p-chlorophenol)[c]

[a] The functional group is shown in red. R stands for an alkyl group.
[b] X stands for a halogen atom—F, Cl, Br, or I.
[c] Common name.
[d] Ar— stands for an aromatic (*aryl*) group such as the benzene ring.

27-2   Alkanes

In this section we will explore some properties of the alkanes. The essential characteristic of **alkane** hydrocarbon molecules is that they have only single covalent bonds. The bonds in these compounds are said to be *saturated*.

Boiling Points of Organic Molecules activity

The alkanes range in complexity from methane, CH_4, to molecules containing 50 C atoms or more (found in petroleum). Most have the formula C_nH_{2n+2}, and each alkane differs from the preceding one in a sequence by a $—CH_2—$, or *methylene* group. Substances whose molecules differ only by a constant unit such as $—CH_2—$ are said to form a **homologous series**. Members of such a series usually have closely related chemical and physical properties. The data in Table 27.3 indicate that boiling points are related to molecular masses and shapes in the ways discussed in Section 13-5. For example, the straight-chain C_4, C_5, and C_6 alkane molecules are more easily polarized than their branched-chain isomers. Intermolecular attractions between the straight-chain molecules are strongest, and they have the highest boiling points. Isomers with more compact structures have lower boiling points (recall Figure 13-25).

TABLE 27.3	Boiling Points of Some Isomeric Alkanes				
Formula	Isomer	Boiling Point, °C	Formula	Isomer	Boiling Point, °C
C_4H_{10}	Butane	−0.5	C_6H_{14}	Hexane	68.7
	Methylpropane	−11.7		3-Methylpentane	63.3
C_5H_{12}	Pentane	36.1		2-Methylpentane	60.3
	2-Methylbutane	27.9		2,3-Dimethylbutane	58.0
	2,2-Dimethylpropane	9.5		2,2-Dimethylbutane	49.7

Conformations

With ball-and-stick models, we can visualize an important type of motion in alkane molecules—rotation of groups with respect to one another about the σ bond connecting them. Figure 27-4 suggests two of the many possible orientations of the two $—CH_3$ groups in the ethane molecule. In one of these configurations, one

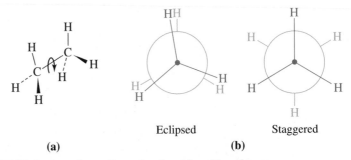

Eclipsed   Staggered
(a)   (b)

▲ **FIGURE 27-4   Rotations about the C—C bond in ethane**
(a) Side view of the molecule showing rotation about the C—C bond. **(b)** Conformations of C_2H_6 in a representation known as a Newman projection. This is based on a head-on view of a ball-and-stick model of ethane. The small solid circle (red) represents the C atom in front, and the red H atoms are those bonded to this C atom. The large open circle (blue) represents the rear C atom, and the blue H atoms are those bonded to it.

set of C—H bonds is directly behind the other when the molecule is viewed head-on. This is called the *eclipsed* conformation. The distance between the H atoms on the adjacent C atoms is at a minimum, leading to the condition of maximum repulsion between the H atoms. In another configuration, called the *staggered* conformation, the H atoms are located a maximum distance apart. Although we expect the staggered conformation to be more stable than the eclipsed, at room temperature, ethane molecules have sufficient thermal energy so that the —CH$_3$ groups can freely rotate about the C—C bond, making all conformations accessible to the molecule. At lower temperatures, however, ethane does occur mostly in the staggered conformation. Similar situations are encountered in the higher alkanes.

Ring Structures

Alkanes in chain structures have the formula C$_n$H$_{2n+2}$ and are called **aliphatic**. Alkanes can also exist in ring, or cyclic, structures; such alkanes are called **alicyclic**. Think of these rings as having formed by the joining together of the two ends of an aliphatic chain after the elimination of a H atom from each end. Simple alicyclic compounds have the formula C$_n$H$_{2n}$.

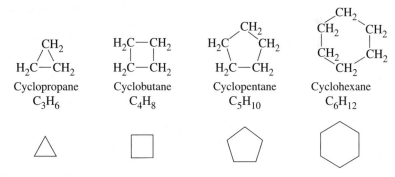

Cyclopropane
C$_3$H$_6$

Cyclobutane
C$_4$H$_8$

Cyclopentane
C$_5$H$_{10}$

Cyclohexane
C$_6$H$_{12}$

The line-angle representations of the four alicyclic compounds above are an equilateral triangle, a square, a pentagon, and a hexagon, respectively. By the nomenclature rules on page 1062, we would name the following alicyclic compound 1,1,3-trimethylcyclopentane.

The bond angles in cyclopropane are 60°, compared with the normal 109.5°; the bonds are highly strained (Figure 27-5). The inherent instability of cyclopropane is reflected in the value of ΔG_f°, which is 104 kJ mol^{-1}; this means that cyclopropane is unstable with respect to its elements. In contrast, ΔG_f° for propane is -23 kJ mol^{-1}, and it is stable with respect to the elements carbon and hydrogen. As a result, numerous reactions occur in which the ring breaks open to yield a chain molecule—propane or a propane derivative. Cyclopropane is more reactive than the other alkanes and, in several ways, its reactivity resembles that of alkenes (discussed in Section 27-3).

The cyclobutane molecule buckles slightly, so the four C atoms are not all in the same plane. In the cyclopentane molecule, the most stable arrangement is for one of the C atoms to be buckled out of the plane of the other four.

For cyclohexane, there are a number of arrangements, or *conformations*, of the molecule. Two of these are shown as ball-and-stick models in Figure 27-6. These are the *boat* and the *chair* **conformations**. The cyclohexane molecule flips between these and other conformations fairly easily at room temperature.

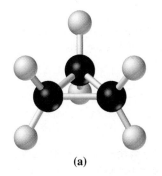

(a)

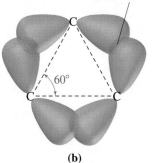

No "head-on" overlap of atomic orbitals

60°

(b)

▲ FIGURE 27-5
Ring strain in cyclopropane
(a) Ball-and-stick model.
(b) Valence-bond picture of cyclopropane using sp^3 hybrids orbitals, which leads to poor overlap of the orbitals and hence weak bonds.

Cyclobutane model

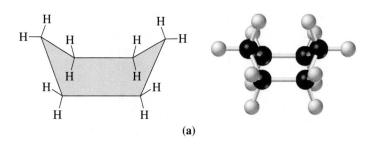

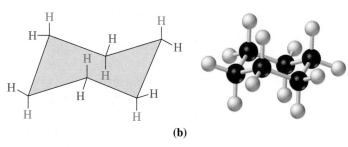

▲ FIGURE 27-6 **Two important conformations of cyclohexane**
(a) Boat form. (b) Chair form. The H atoms extending laterally from the ring—equatorial H atoms—are shown in red. The H atoms projecting above and below the ring—axial H atoms—are in blue.

Preparation of Alkanes

The chief source of alkanes is petroleum, as we describe at the end of this section, but several laboratory methods are also available for their preparation. In the presence of a catalyst, unsaturated hydrocarbons, whether containing double or triple bonds, may be converted to alkanes by the addition of H atoms to the multiple bond systems (equation 27.1). In another type of reaction, halogenated hydrocarbons react with alkali metals to produce alkanes of double the carbon content (equation 27.2). Finally, alkali metal salts of carboxylic acids may be fused with alkali metal hydroxides. Sodium carbonate and an alkane with one carbon less than the metal carboxylate are formed (equation 27.3).

$$CH_2{=}CH_2 + H_2 \xrightarrow{\text{Pt or Pd}} CH_3{-}CH_3 \qquad (27.1)$$

$$2\ CH_3CH_2Br + 2\ Na \xrightarrow{\text{heat/pressure}} 2\ NaBr + CH_3CH_2CH_2CH_3 \qquad (27.2)$$

$$CH_3{-}\overset{\overset{\displaystyle O}{\|}}{C}{-}ONa + NaOH \xrightarrow{\Delta} Na_2CO_3 + CH_4 \qquad (27.3)$$

Reactions of the Alkanes

Saturated hydrocarbons have little affinity for most chemical reactants. They are nonpolar substances that are insoluble in water and unreactive toward acids, bases, or oxidizing agents. Halogens react only slowly with alkanes at room temperature, but at higher temperatures, particularly in the presence of light, halogenation occurs. In such a reaction, called a **substitution reaction**, a halogen atom *substitutes* for a hydrogen atom. This substitution occurs by a *chain reaction*, written as follows for the chlorination of methane. (Only the electrons involved in bond breakage or formation are shown.)

$$Initiation: \qquad Cl\!:\!Cl \xrightarrow[\text{light}]{\text{heat or}} 2Cl\!\cdot$$

$$Propagation: \qquad H_3C\!:\!H + Cl\!\cdot \longrightarrow H_3C\!\cdot + H\!:\!Cl$$

$$H_3C\!\cdot + Cl\!:\!Cl \longrightarrow H_3C\!:\!Cl + Cl\!\cdot$$

$$Termination: \qquad Cl\!\cdot + Cl\!\cdot \longrightarrow Cl\!:\!Cl$$

$$H_3C\!\cdot + Cl\!\cdot \longrightarrow H_3C\!:\!Cl$$

$$H_3C\!\cdot + H_3C\!\cdot \longrightarrow H_3C\!:\!CH_3$$

▶ The reaction of H_2 with Cl_2 to form HCl, discussed on page 913, proceeds by a similar mechanism.

The reaction is initiated when some Cl_2 molecules absorb sufficient energy to dissociate into Cl atoms (represented above as $Cl\cdot$). Cl atoms collide with CH_4 molecules to produce methyl *free radicals* ($H_3C\cdot$), which combine with Cl_2 molecules to form the product, CH_3Cl. When any or all of the last three reactions proceed to the extent of consuming the free radicals present, the reaction stops. The overall equation for the formation of chloromethane is

$$CH_4 + Cl_2 \xrightarrow[\text{light}]{\text{heat or}} CH_3Cl + HCl \qquad (27.4)$$

This free-radical chain reaction usually yields a mixture of products, not just chloromethane. Polyhalogenation can also occur to form CH_2Cl_2, dichloromethane (methylene chloride, a solvent and paint remover); $CHCl_3$, trichloromethane (chloroform, a solvent and fumigant); and CCl_4, tetrachloromethane (carbon tetrachloride, a solvent).

The most commonly encountered reaction of alkanes is with oxygen. Alkanes burn; the *oxidation* of hydrocarbons underlies their important use as fuels. For example, octane reacts with oxygen as follows.

$$C_8H_{18}(l) + \frac{25}{2} O_2(g) \longrightarrow 8\,CO_2(g) + 9\,H_2O(l) \qquad \Delta H° = -5.48 \times 10^3 \text{ kJ} \qquad (27.5)$$

Alkanes from Petroleum

The lower molecular mass alkanes, methane and ethane, are found principally in natural gas. Propane and butane are found dissolved in petroleum; they can be extracted as gases and sold as liquefied petroleum gas (LPG). Higher alkanes are obtained by the fractional distillation of petroleum, a complex mixture of at least 500 compounds. The main petroleum fractions are listed in Table 27.4. (The liquid–vapor equilibrium principles that underlie fractional distillation were discussed on page 552.)

TABLE 27.4 Principal Petroleum Fractions

Boiling Range, °C	Composition	Fraction	Uses
Below 0	C_1–C_4	Gas	Gaseous fuel
0–50	C_5–C_7	Petroleum ether	Solvents
50–100	C_6–C_8	Ligroin	Solvents
70–150	C_6–C_9	Gasoline	Motor fuel
150–300	C_{10}–C_{16}	Kerosene	Jet fuel, diesel oil
Over 300	C_{16}–C_{18}	Gas–oil	Diesel oil, cracking stock
—	C_{18}–C_{20}	Wax–oil	Lubricating oil, mineral oil, cracking stock
—	C_{21}–C_{40}	Paraffin wax	Candles, wax paper
—	above C_{40}	Residuum	Roofing tar, road materials, waterproofing

▲ A catalytic cracking unit (cat cracker) at a petroleum refinery.

▶ Not only do the processes of cracking, re-forming, and alkylation produce a higher grade of gasoline, but they also increase the yield of gasoline obtained from crude oil.

Not all the gasoline components referred to in Table 27.4 are equally desirable in fuels. Some of them burn more smoothly than others. (Explosive burning results in engine knocking.) The octane hydrocarbon, *2,2,4-trimethylpentane*, has excellent engine performance, and it is given an octane rating of 100. *Heptane* has poor engine performance; its octane rating is 0. These two hydrocarbons serve as a basis for establishing the quality of automotive fuels. A gasoline that gives the same performance as a mixture of 87% 2,2,4-trimethylpentane and 13% heptane is assigned an octane number of 87, for example. In general, branched-chain hydrocarbons have higher octane numbers (burn more smoothly) than their straight-chain counterparts.

2,2,4-Trimethylpentane
Octane rating: 100

n-Heptane
Octane rating: 0

Gasoline obtained by fractional distillation of petroleum has an octane number of 50–55 and is not acceptable for use in automobiles, which require fuels with octane numbers near 90. Extensive modifications of the gasoline fraction are required. In *thermal cracking*, large hydrocarbon molecules are broken down into molecules in the gasoline range, and the presence of special catalysts promotes the production of branched-chain hydrocarbons. For example, the molecule $C_{15}H_{32}$ might be broken down into C_8H_{18} and C_7H_{14}. The process of *re-forming*, or isomerization, converts straight-chain to branched-chain hydrocarbons and other types of hydrocarbons having higher octane numbers. In thermal and catalytic cracking, some low-molecular mass hydrocarbons are rejoined into higher molecular mass hydrocarbons by a process known as *alkylation*.

The octane rating of gasoline can be further improved by adding antiknock compounds to prevent premature combustion. At one time, the preferred additive was tetraethyllead, $(C_2H_5)_4Pb$. Lead additives have now been phased out of gasoline in the United States, and substitutes such as the oxygenated hydrocarbons methanol and ethanol are used instead.

27-3 Alkenes and Alkynes

A straight- or branched-chain alkane has the maximum number of H atoms possible for its number of C atoms. In other classes of hydrocarbons, compounds with the same number of C atoms but fewer H atoms, the C atoms must either join into rings, form carbon-to-carbon multiple bonds, or both. This is essential so that the C atoms can have valence-shell octets. Hydrocarbons whose molecules contain some double or triple bonds between C atoms are said to be **unsaturated**. If there is *one double bond* in the molecules, the hydrocarbons are the simple **alkenes**, or **olefins**; they have the general formula C_nH_{2n}. Simple **alkynes** have *one triple bond* in their molecules and have the general formula C_nH_{2n-2}.

In the examples that follow, systematic names are shown in blue. The names given in parentheses are also commonly used.

$CH_2{=}CH_2$
Ethene
(ethylene)

$CH_3CH_2CH{=}CH_2$
1-Butene

$CH_3CH_2CH_2C{=}CH_2$
$|$
CH_2CH_3
2-Ethyl-1-pentene

$$HC{\equiv}CH \qquad CH_3CH_2C{\equiv}CH \qquad CH_3CHC{\equiv}CCH_3$$

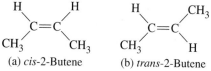

Ethyne   1-Butyne   4-Methyl-2-pentyne
(acetylene)   (ethylacetylene)   (isopropylmethylacetylene)

Following are the modifications of the rules on page 1062 required to name alkenes and alkynes.

1. Select as the base chain the longest chain containing the multiple bond.
2. Number the C atoms of the chain to place the multiple bond at the lowest possible number.
3. Use the ending *ene* for alkenes and *yne* for alkynes.

Thus, in the name 2-ethyl-1-pentene, the longest chain *containing the multiple bond* is a five-carbon chain (*pent-*). This is *not* a substituted hexane, and it is *not* a hexene. The double bond makes the molecule an alkene, and its location between the first and second carbon makes it a 1-pentene. The ethyl group is attached to the second C atom, and so the alkene is named 2-ethyl-1-pentene.

The alkenes are similar to the alkanes in physical properties. At room temperature, those containing two to four C atoms are gases; those with five to 18 are liquids; those with more than 18 are solids. In general, alkynes have higher boiling points than their alkane and alkene counterparts.

Geometric Isomerism

The molecules 2-butene, $CH_3CH{=}CHCH_3$, and 1-butene, $CH_2{=}CHCH_2CH_3$, differ in the position of the double bond and are *positional* isomers. But another kind of isomerism is possible in 2-butene, represented by these two structures.

cis-2-Butene,
trans-2-Butene models

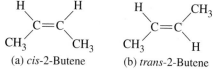

(a) *cis*-2-Butene   (b) *trans*-2-Butene

As we learned in Section 12-4, a double bond between C atoms consists of the overlap of sp^2 hybrid orbitals to form a σ bond *and* the sideways overlap of *p* orbitals to form a π bond. Rotation about this double bond is severely restricted. Molecule (a) cannot be converted into molecule (b) simply by twisting one end of the molecule through 180°, so the two molecules are distinctly different (Figure 27-7). To distinguish between these two molecules by name, we call (a) *cis*-2-butene (*cis*, Latin, on the same side), and we call (b) *trans*-2-butene (*trans*, Latin, across). This type of isomerism is called **geometric isomerism**.

In Chapter 25, we noted that geometric isomerism is only one type of a more general kind of isomerism known as **stereoisomerism** (from the Greek word *stereos*, meaning solid, or three-dimensional, in nature). The number and types of atoms and bonds in stereoisomers are the same, but certain atoms are oriented differently in space. Another type of stereoisomerism that we introduced in Chapter 25 is *optical isomerism*. As we will see in the next chapter, stereoisomerism plays an important role in the unique chemical reactions found in living organisms.

Preparation and Uses

The general laboratory preparation of alkenes uses an **elimination reaction**, a reaction in which atoms are removed from adjacent positions on a carbon chain. A small molecule is produced, and an additional bond is formed between the C atoms.

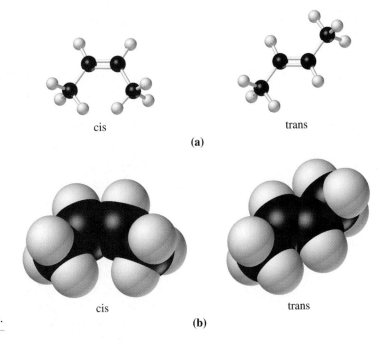

▶ FIGURE 27-7
Geometric isomerism in 2-butene
(a) Ball-and-stick models. **(b)** Space-filling models.

For example, H_2O is eliminated in the following reaction.

$$CH_3-\underset{\underset{HO}{|}}{\overset{\overset{H}{|}}{C}}-\underset{\underset{H}{|}}{\overset{\overset{H}{|}}{C}}-H \xrightarrow[H_2SO_4]{\Delta} CH_3CH{=}CH_2 + H_2O \qquad (27.6)$$

The principal alkene of the chemical industry is ethylene (ethene), the highest volume organic chemical produced in the United States. Its chief use is in the manufacture of polymers, although it is also used to manufacture other organic chemicals. Reaction (27.6) is relatively unimportant in the commercial production of ethylene, which is obtained mainly by thermal cracking of other hydrocarbons.

The simplest alkyne is acetylene, which can be prepared from coal, water, and limestone in a three-step process.

$$CaCO_3 \xrightarrow{\Delta} CaO + CO_2$$

$$CaO + 3\,C \xrightarrow[2000\,°C]{electric\ furnace} \underset{\substack{calcium\ acetylide\\(calcium\ carbide)}}{CaC_2} + CO$$

$$CaC_2 + 2\,H_2O \longrightarrow HC{\equiv}CH + Ca(OH)_2 \qquad (27.7)$$

Most other alkynes are prepared from acetylene by taking advantage of the acidity of the C—H bond. In the presence of a very strong base, such as sodium amide, the acetylene donates a proton to the amide ions to form ammonia and the salt sodium acetylide. The acetylide can then react with an alkyl halide.

$$H-C{\equiv}C-H + Na^+NH_2^- \longrightarrow H-C{\equiv}C^-Na^+ + NH_3$$

$$H-C{\equiv}C^-Na^+ + CH_3Br \longrightarrow H-C{\equiv}C-CH_3 + Na^+Br^- \qquad (27.8)$$

By continuing this reaction, the triple bond can be positioned as desired in the chain, as in the synthesis of 2-pentyne.

$$H-C{\equiv}CH-CH_3 + NaNH_2 \longrightarrow Na^+\ ^-C{\equiv}C-CH_3 + NH_3$$

$$Na^+\ ^-C{\equiv}C-CH_3 + CH_3CH_2Br \longrightarrow CH_3CH_2C{\equiv}CCH_3 + Na^+Br^- \qquad (27.9)$$

▲ Cutting steel with an oxy-acetylene torch.

At one time, acetylene was one of the most important organic raw materials in the chemical industry, but it is no longer among the top 50 industrial chemicals. At present, the chief use of acetylene is in the manufacture of other chemicals for polymer production, such as vinyl chloride, $H_2C{=}CHCl$, which is polymerized to polyvinyl chloride (PVC). Acetylene produces high-temperature flames when burned in excess oxygen, and this is the basis of oxyacetylene torches used for cutting and welding metals.

$$HC{\equiv}CH(g) + \frac{5}{2}O_2(g) \longrightarrow 2\,CO_2(g) + H_2O(l) \qquad \Delta H = -1300\ kJ$$

The large negative enthalpy of combustion of acetylene is due to acetylene's large *positive* enthalpy of formation: $\Delta H_f^\circ[C_2H_2(g)] = +226.7\ kJ/mol$.

Addition Reactions

Unlike alkanes, which react by *substitution*, alkenes react by *addition*. In an **addition reaction** of an alkene, two new groups become bonded to the C atoms at the site of the double bond, and the carbon-to-carbon double bond is converted to a single bond.

$$CH_2{=}CH_2 + Br_2 \longrightarrow \underset{\underset{Br}{|}}{CH_2}{-}\underset{\underset{Br}{|}}{CH_2} \qquad (27.10)$$

Testing for Unsaturated Hydrocarbons with Bromine movie

In this reaction, the two Br atoms are identical, as are the two $-CH_2$ groups. Br_2 and $CH_2{=}CH_2$ are known as *symmetrical* reactants. HBr and $CH_3CH{=}CH_2$ are *unsymmetrical* reactants. When at least one reactant is symmetrical, it is a fairly straightforward matter to predict the reaction product. We have a problem, however, in predicting the products when both reactants are unsymmetrical. Let's consider, for example, the reaction of *unsymmetrical* HBr with *unsymmetrical* propene. Which of the two likely products should we expect?

$$CH_3CH{=}CH_2 + H{-}Br \longrightarrow \underset{\underset{Br}{|}\ \underset{H}{|}}{CH_3CH{-}CH_2} \quad or \quad \underset{\underset{H}{|}\ \underset{Br}{|}}{CH_3CH{-}CH_2}$$

Because HBr is a polar molecule with its positive end at the H atom and its negative end at the Br atom, 2-bromopropane is the sole product of this reaction. This result is consistent with an empirical rule proposed by Vladimir Markovnikov in 1871 (modified slightly to include reactants other than HBr).

▶ A quick way to remember this rule is "them that has, gets" (referring to H atoms).

When an unsymmetrical *reactant (HX, HOH, $HOSO_3H$) is added to an* unsymmetrical *alkene or alkyne, the more positive fragment (usually H) adds to the carbon atom that has the greatest number of attached H atoms.*

Notice how the H and OH add to the double bond in accordance with Markovnikov's rule. H adds to the C atom with the greater number of attached H atoms—that is, to $-CH_2$. ▶

The addition of H_2O to a double bond is the reverse of the reaction in which a double bond is formed by the elimination of H_2O (reaction 27.6). In this reversible reaction, the addition reaction is favored in dilute acid, and the elimination reaction is favored in *concentrated* $H_2SO_4(aq)$.

$$\underset{CH_3}{\overset{CH_3}{>}}C{=}C\underset{H}{\overset{H}{<}} + HOH \xrightarrow{10\%\ H_2SO_4} CH_3{-}\underset{\underset{OH}{|}}{\overset{\overset{CH_3}{|}}{C}}{-}\underset{\underset{H}{|}}{\overset{\overset{H}{|}}{C}}{-}H \qquad (27.11)$$

t-butyl alcohol

The addition reactions of alkynes also follow Markovnikov's rule, for example,

$$CH_3C{\equiv}CH + HCl \longrightarrow CH_3\overset{\displaystyle Cl}{\underset{}{C}}{=}CH_2 + HCl \longrightarrow CH_3\overset{\displaystyle Cl}{\underset{\displaystyle Cl}{C}}{-}CH_3$$

The reaction stops at the addition of one mole of HCl unless an excess of the reagent is added. This is because the rate of addition of the hydrogen halide to the alkyne is greater than the rate of addition of the hydrogen halide to the halo-substituted alkene that is formed in the first addition reaction.

Surface Reaction—Hydrogenation animation

Hydrogen adds to both alkenes and alkynes in the presence of a metal catalyst such as palladium or platinum. This process is called *hydrogenation*. The hydrogenation of 1-butyne is as follows.

$$CH_3CH_2C{\equiv}CH + H_2 \xrightarrow{\text{Pt}} CH_3CH_2CH{=}CH_2 + H_2 \xrightarrow{\text{Pt}} CH_3CH_2CH_2CH_3$$

Since hydrogen rapidly adds to the double bond of alkenes, the reaction cannot be stopped at the alkene stage. If the alkene is the desired product, a different heterogeneous catalyst known as Lindlar's catalyst, is employed. Using this catalyst, the addition of hydrogen to an alkyne in which the triple bond is not at the terminus of the chain leads to the formation of a cis alkene (see Chapter 15, Exercise 93).

▶ Lindlar's catalyst is a heterogeneous catalyst prepared by precipitating palladium on calcium carbonate and treating this mixture with lead acetate and quinoline.

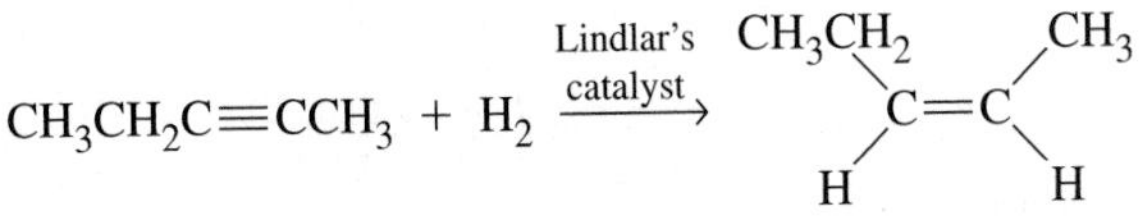

$$CH_3CH_2C{\equiv}CCH_3 + H_2 \xrightarrow[\text{catalyst}]{\text{Lindlar's}} \begin{array}{c} CH_3CH_2 \\ \diagdown \\ C{=}C \\ \diagup \quad \diagdown \\ H \qquad H \end{array} \begin{array}{c} CH_3 \\ \diagup \\ \\ \end{array}$$

2-pentyne   *cis*-2-pentene

In contrast, using sodium in liquid ammonia as a hydrogenation agent leads to the formation of the trans alkene.

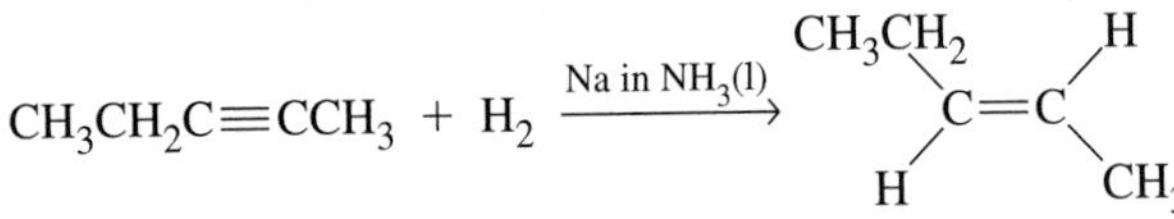

$$CH_3CH_2C{\equiv}CCH_3 + H_2 \xrightarrow{\text{Na in NH}_3\text{(l)}} \begin{array}{c} CH_3CH_2 \\ \diagdown \\ C{=}C \\ \diagup \quad \diagdown \\ H \qquad CH_3 \end{array} \begin{array}{c} H \\ \diagup \\ \\ \end{array}$$

2-pentyne   *trans*-2-pentene

The difference arises because of the different mechanisms involved. These two methods have important synthetic applications.

27-4 Aromatic Hydrocarbons

Aromatic hydrocarbons have ring structures with unsaturation (multiple-bond character) in the carbon-to-carbon bonds in the rings. Most aromatic hydrocarbons are based on the molecule benzene, C_6H_6. In Section 12-6, we discussed bonding in the benzene molecule in some detail, and we showed that there are several ways of representing the molecule, including

As in the case of alicyclic compounds, neither the C atoms at the vertices nor the H atoms bonded to them are shown in these structures. ▶

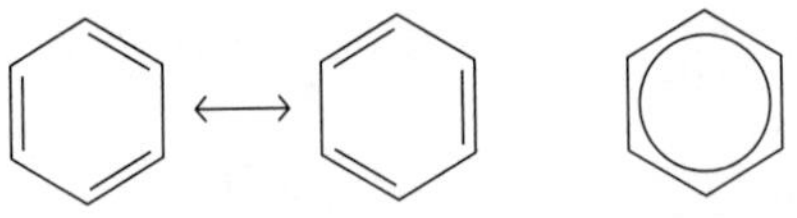

Kekulé structures   Simplified
molecular orbital
representation

▲ August Kekulé (1829–1896), who proposed the hexagonal ring structure for benzene in 1865. His representation of this molecule is still widely used.

Other examples of aromatic hydrocarbons include

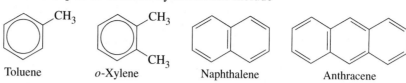

Toluene o-Xylene Naphthalene Anthracene

Toluene and *o*-xylene are *substituted benzenes*, and naphthalene and anthracene feature *fused benzene rings*. When rings are fused together, the resultant structure has two C and four H atoms fewer than the starting structures. Thus, the formula for naphthalene is $C_6H_6 + C_6H_6 - 2C - 4H = C_{10}H_8$; for anthracene: $C_{10}H_8 + C_6H_6 - 2C - 4H = C_{14}H_{10}$. In this text, we use the inscribed circle for the simple benzene ring and alternating single and double bonds for fused rings. For the fused ring systems, the bond arrangement represents one of the possible resonance structures for the molecule.

Characteristics of Aromatic Hydrocarbons

Aromatic hydrocarbons are highly flammable and should always be handled with care. Prolonged inhalation of benzene vapor results in a decreased production of both red blood cells and white blood cells, which can be fatal. Also, benzene is a carcinogen. Benzene and some other toxic aromatic compounds have been isolated in the tar formed by burning cigarettes, in polluted air, and as a decomposition product of grease in the charcoal grilling of meat.

A close examination of the structures of aromatic molecules shows that they all share two common features.

- They are *planar* (flat), *cyclic* molecules.
- They have a *conjugated* bonding system—a bonding scheme among the ring atoms that consists of alternating single and double bonds. The system must extend throughout the ring, and the π electron clouds associated with the double bonds must involve $(4n + 2)$ electrons, where $n = 1, 2, \ldots$.

Thus, the benzene molecule has six electrons in the π electron clouds: $(4 \times 1) + 2 = 6$. The naphthalene molecule has ten: $(4 \times 2) + 2 = 10$. And the anthracene molecule has 14: $(4 \times 3) + 2 = 14$. Neither of the two molecules depicted below is aromatic; both are aliphatic.

$$CH_2{=}CH{-}CH{=}CH{-}CH{=}CH_2$$

1,3,5-Hexatriene   1,3-Cyclopentadiene

The 1,3,5-hexatriene molecule has six π electrons in its conjugated bonding system, but it is not cyclic. The 1,3-cyclopentadiene molecule is cyclic but has only four π electrons in a conjugated bonding system that does not extend completely around the ring.

Benzene and its homologues are similar to other hydrocarbons in being insoluble in water but soluble in organic solvents. The boiling points of the aromatic hydrocarbons are slightly higher than those of the alkanes of similar carbon content. For example, hexane, C_6H_{14}, boils at 69 °C, whereas benzene boils at 80 °C. This can be explained by the planar structure and delocalized electron charge density of benzene, which increases the attractive forces between molecules. The symmetrical structure of benzene permits closer packing of molecules in the crystalline state and results in a higher melting point than for hexane. Benzene melts at 5.5 °C and hexane melts at −95 °C.

When one of the six equivalent H atoms of a benzene molecule is removed, the resulting species is called a **phenyl** group. Two phenyl groups may bond together, as in biphenyl, or phenyl groups may be substituents in other molecules, as in phenylhydrazine, used in the detection of sugars.

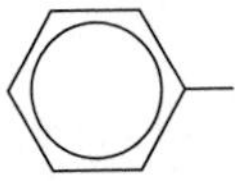

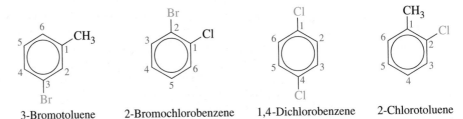

Phenyl group 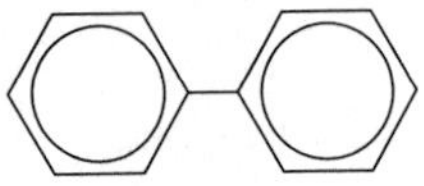   Biphenyl 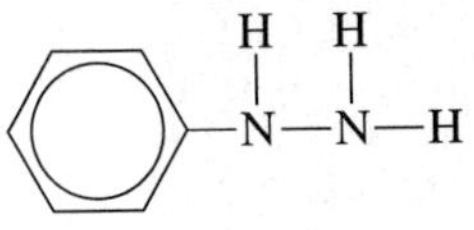   Phenylhydrazine

Naming Aromatic Hydrocarbons

Other atoms or groups may be substituted for H atoms on the benzene molecule, and to name these compounds, we use a numbering system for the C atoms in the ring. If the name of an aromatic compound is based on a common name other than benzene (such as toluene), the characteristic substituent group (for example, the —CH_3 in toluene) is assigned position "1" on the benzene ring. Otherwise, the substituents are listed alphabetically, as in 2-bromochlorobenzene shown below.

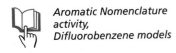

*Aromatic Nomenclature activity,
Difluorobenzene models*

3-Bromotoluene
(*m*-bromotoluene)

2-Bromochlorobenzene
(*o*-bromochlorobenzene)

1,4-Dichlorobenzene
(*p*-dichlorobenzene)

2-Chlorotoluene
(*o*-chlorotoluene)

We can also use the terms *ortho, meta,* and *para* (*o-, m-, p-*) when there are two substituents on the benzene ring. **Ortho** refers to substituents on adjacent carbon atoms, **meta** to substituents with one carbon atom between them, and **para** to substituents opposite one another on the ring.

Uses of Aromatic Hydrocarbons

Current annual production of benzene in the United States is about 16 billion pounds. Over 90% of this is produced from petroleum. The process involves dehydrogenation and cyclization of hexane to the aromatic hydrocarbon. The most important use of the petroleum-produced benzene is in manufacturing ethylbenzene for the production of styrene plastics. Other applications include the manufacture of phenol, the synthesis of dodecylbenzene (for detergents), and as an octane enhancer in gasoline. The production of aromatic compounds by dehydrogenation (removal of hydrogen) of alkanes yields large amounts of hydrogen gas, which is an important reactant in the synthesis of ammonia (page 271).

Aromatic Substitution Reactions

Unlike alkenes and alkynes, aromatic molecules react by *substitution,* not by addition of atoms across a double bond. This difference in reactivity is due to the special stability of the electrons in the aromatic ring. In aromatic rings, certain electrons are delocalized into a π electron cloud. In alkene and alkyne molecules, the regions of high electron density associated with multiple bonds are localized between specific C atoms. These localized π bonds are much more reactive than the delocalized π bonds of the aromatic ring.

The substitution of a single atom or group, X, for a H atom in C_6H_6, can occur at any one of the six positions of the benzene ring. We say that the six positions are equivalent. If a group Y is substituted for a H atom in C_6H_5X, this question arises: To which of the remaining five positions does the Y group go? If all the sites on the benzene ring were equally preferred, the distribution of the products would be a purely statistical one. That is, there are five possible positions where Y can be substituted, and we should get 20% of each one. Since there are two possibilities that lead to an *ortho* isomer and two that lead to a *meta* isomer, however, we should expect the distribution of products to be 40% ortho, 40% meta, and 20% para.

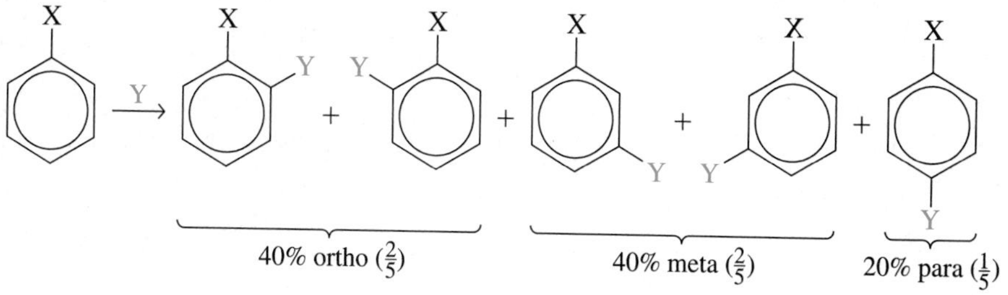

40% ortho ($\frac{2}{5}$)   40% meta ($\frac{2}{5}$)   20% para ($\frac{1}{5}$)

The following scheme describes the products resulting from nitration followed by chlorination (reaction 27.12) and chlorination followed by nitration (reaction 27.13). It shows that the substitution is *not random*. The $-NO_2$ group directs Cl to a meta position. Almost no ortho or para isomer is formed in reaction (27.12). The Cl group, on the other hand, is an ortho, para director. Essentially no meta isomer is produced in reaction (27.13).

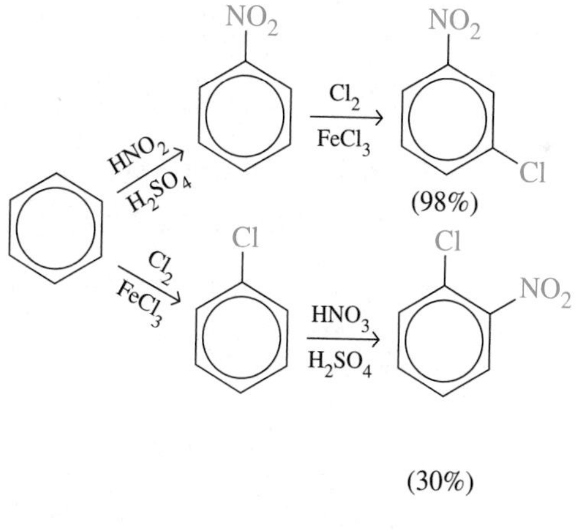

(27.12)

(27.13)

Whether a group is an ortho, para, or a meta, director depends on how the presence of one substituent alters the electron distribution in the benzene ring. As a result, attack by a second group is more likely at one type of position than another. Examination of many reactions leads to the following order.

Ortho, para directors: $-NH_2$, $-OR$, $-OH$, $-OCOR$, $-R$, $-X$ (X = halogen)
(from strongest to weakest)

Meta directors: $-NO_2$, $-CN$, $-SO_3H$, $-CHO$, $-COR$, $-COOH$, $-COOR$
(from strongest to weakest)

When two groups of the same type (both o-, p- or both m-directing) are present, the stronger director wins out. When two groups of different type (one o-, p- and one m-directing) then it has been found that the o-, p-directing group guides the reaction.

EXAMPLE 27-4

Predicting the Products of an Aromatic Substitution Reaction. Predict the products of the mononitration of

OH
NO$_2$

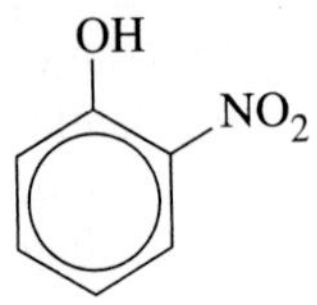

Solution

—OH is an ortho, para director, and we should expect the products

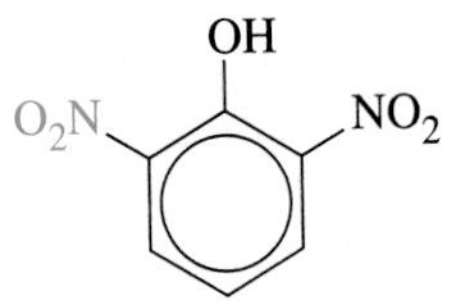

OH
O$_2$N   NO$_2$

2,6-Dinitrophenol

OH
NO$_2$
NO$_2$

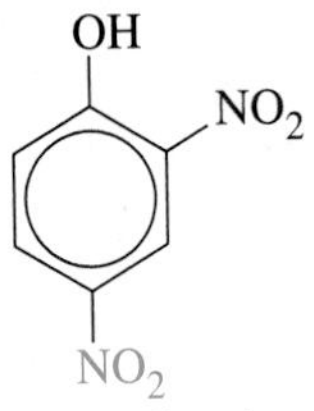

2,4-Dinitrophenol

We consider only the o-, p-directing group in this case.

Practice Example A: Predict the major product(s) of the mononitration of benzaldehyde, C$_6$H$_5$CHO.

Practice Example B: Predict the major product(s) of the mononitration of 1,3-dichlorobenzene.

27-5 Alcohols, Phenols, and Ethers

Alcohols and **phenols** are characterized by the **hydroxyl** group, —OH. In alcohols, the hydroxyl group is bound to an sp^3 hybridized (aliphatic) carbon atom. If this C atom also has *one* R group (and two H atoms) bonded to it, the alcohol is a *primary* alcohol. If the C atom has *two* R groups (and one H atom), the alcohol is a *secondary* alcohol. Finally, if there are *three* R groups on the C atom (and no H atoms), the alcohol is a *tertiary* alcohol.

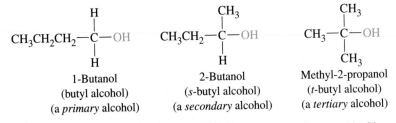

Primary, Secondary, Tertiary Alcohol models

1-Butanol
(butyl alcohol)
(a *primary* alcohol)

2-Butanol
(*s*-butyl alcohol)
(a *secondary* alcohol)

Methyl-2-propanol
(*t*-butyl alcohol)
(a *tertiary* alcohol)

The systematic naming of alcohols uses the suffix *-ol* and was discussed in Chapter 3. In phenols, the hydroxyl group is attached to a benzene ring.

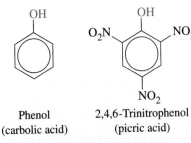

Phenol
(carbolic acid)

2,4,6-Trinitrophenol
(picric acid)

▲ **Giant soap bubble**
A soap bubble consists of an air pocket enclosed in a thin film of soap in water. As the water evaporates, the film breaks and the bubble bursts. Glycerol added to the soap–water mixture forms hydrogen bonds to both the soap and water molecules. This slows the rate of evaporation of the water and increases the strength of the film, allowing the production of very large bubbles.

A molecule may have more than one —OH group. Those with two —OH groups are known as *diols* (or glycols), and those with more than two —OH groups are called *polyols*. Ethylene glycol, a diol, is used in automobile antifreeze solutions; glycerol, a polyol, is an important biological molecule used as part of the body's mechanism for fat storage.

$$\begin{array}{cc} CH_2\!\!-\!\!CH_2 & CH_2\!\!-\!\!CH\!\!-\!\!CH_2 \\ \;|\qquad\; | & \;|\qquad\; |\qquad\; | \\ OH\quad OH & OH\quad OH\quad OH \end{array}$$

$$\begin{array}{cc} \text{1,2-Ethanediol} & \text{1,2,3-Propanetriol} \\ \text{(ethylene glycol)} & \text{(glycerol)} \end{array}$$

Physical properties of aliphatic alcohols are strongly influenced by hydrogen bonding. As the chain length increases, however, the influence of the polar hydroxyl group on the properties of the molecule diminishes. The molecule becomes less like water and more like a hydrocarbon. As a consequence, low-molecular mass alcohols tend to be water soluble, while high-molecular mass alcohols are not. The boiling points and solubilities of the phenols vary widely, depending on the nature of the other substituents on the benzene ring.

Preparation and Uses of Alcohols

Two methods by which alcohols can be synthesized are by the hydration of alkenes and the hydrolysis of alkyl halides.

$$CH_3CH\!\!=\!\!CH_2 + H_2O \xrightarrow{H_2SO_4} \begin{matrix} OH \\ | \\ CH_3CHCH_3 \end{matrix} \qquad (27.14)$$

$$\begin{matrix} \text{propene} & \text{2-propanol} \\ \text{(propylene)} & \text{(isopropyl alcohol)} \end{matrix}$$

$$CH_3CH_2CH_2Br + OH^- \longrightarrow CH_3CH_2CH_2OH + Br^- \qquad (27.15)$$

$$\begin{matrix} \text{1-bromopropane} & \text{1-propanol} \end{matrix}$$

Methanol (wood alcohol) is a highly toxic substance and can lead to blindness or death if ingested. Most methanol is manufactured from carbon monoxide and hydrogen.

$$CO(g) + 2\,H_2(g) \xrightarrow[\substack{200\text{ atm} \\ ZnO,\ Cr_2O_3}]{350\ °C} CH_3OH(g)$$

Methanol is the most extensively produced alcohol. It ranks about twenty-first among all industrial chemicals. It is used to synthesize other organic chemicals and as a solvent, but potentially its most important use may be as a motor fuel (recall Section 7-9).

Ethanol, CH_3CH_2OH, is grain alcohol, which is found in alcoholic beverages. It is easily produced by the fermentation of sugarcane juice or from materials containing natural sugars. The industrial method involves the hydration of ethylene with sulfuric acid catalyst (similar to reaction 27.14).

Ethylene glycol, $HOCH_2CH_2OH$, is water soluble and has a higher boiling point (197 °C) than water. These properties make it an excellent permanent, nonvolatile antifreeze for automobile radiators. It is also used in the manufacture of solvents, paint removers, and plasticizers (softeners).

Glycerol (glycerin), $HOCH_2CH(OH)CH_2OH$, is obtained commercially as a by-product in the manufacture of soap. It is a sweet, syrupy liquid that is miscible with water in all proportions. Because it takes up moisture from the air, glycerol can be used to keep skin moist and soft and is found in lotions and cosmetics.

An interesting derivative of an alcohol is the alkoxide ion, formed by the reaction of sodium or potassium metal with the alcohol.

$$CH_3OH + Na \longrightarrow CH_3O^- + Na^+ + \tfrac{1}{2}H_2(g)$$

The alkoxide ion can react with a haloalkane as follows:

$$CH_3O^- + CH_3CH_2Cl \longrightarrow CH_3CH_2OCH_3 + Cl^-$$

The compound formed is an ether, which we describe next.

Ethers

An **ether** is a compound with the general formula R—O—R. Structurally, ethers can be pure aliphatic, pure aromatic, or mixed.

$$CH_3-O-CH_3$$

Dimethyl ether   Diphenyl ether   Methyl phenyl ether (anisole)

KEEP IN MIND ▶
that although dimethyl ether has the same formula as ethanol (C_2H_6O), the two substances have quite different physical and chemical properties. This is because each has a different functional group (see Table 27.2). The two are skeletal isomers.

Symmetrical ethers, such as diethyl ether, can be prepared by the elimination of a water molecule from between two alcohol molecules with a strong dehydrating agent, such as concentrated H_2SO_4.

$$CH_3CH_2OH + HOCH_2CH_3 \xrightarrow[140\ °C]{concd\ H_2SO_4} CH_3CH_2OCH_2CH_3 + H_2O \quad (27.16)$$

Chemically, ethers are comparatively unreactive. The ether linkage is stable to most oxidizing and reducing agents and to action by dilute acids and alkalis.

Diethyl ether has been used extensively as a general anesthetic. It is easy to administer and produces excellent relaxation of the muscles. Also, it affects the pulse rate, rate of respiration, and blood pressure only slightly. However, it is somewhat irritating to the respiratory passages and produces nausea. Methyl propyl ether (neothyl) is also used as an anesthetic and is less irritating to the respiratory passages. Dimethyl ether, a gas at room temperatures, is used as a propellant for aerosol sprays. Higher molecular mass ethers are used as solvents for varnishes and lacquers. Methyl *t*-butyl ether, an unsymmetrical ether, marketed under the name MTBE, has been used as an octane enhancer in gasoline. However, because of its relatively high solubility in water ($\approx 5\ g/100\ g\ H_2O$), in some localities it has caused rather extensive groundwater pollution through leakage from underground storage tanks; it is currently being phased out of use.

▶ This molecule should be named *t*-butyl methyl ether (TBME), but has been referred to by industrial chemists as MTBE for so long that methyl *t*-butyl ether has become the commonly used name for this compound.

Methyl *t*-butyl ether (MTBE)

27-6 Aldehydes and Ketones

Aldehydes and ketones contain the **carbonyl** group.

$$\underset{R}{\overset{R}{\diagdown}}C=O$$

If one of the R groups is a H atom, the compound is an **aldehyde**. If both R groups are alkyl or aromatic (aryl) groups, the compound is a **ketone**.

Methanal (formaldehyde)   3-Chlorobutanal (β-chlorobutyraldehyde)   Benzaldehyde

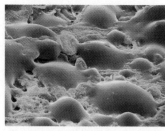

▲ False-color scanning electron micrograph of the sticky surface of a 3M Post-it note. The bubbles are 15–40 μm in diameter and consist of a urea–formaldehyde adhesive. Each time the note is pressed to a surface, fresh adhesive is released. The note can be reattached to a surface as long as some of the bubbles remain.

KEEP IN MIND ▶
that the CHO functional group can be only at the terminus of a carbon chain, whereas the CO functional group cannot be at the end of a carbon chain.

▶ PCC is formed as yellow-orange crystals by the reaction between CrO_3 and HCl in pyridine.

$$\text{Propanone (acetone)} \qquad \text{3-Pentanone (diethyl ketone)} \qquad \text{1-Phenylethanone (acetophenone)}$$

The IUPAC naming system for aldehydes uses the suffix *-al*. The parent chain is the longest chain containing the aldehyde group. Accordingly, the four-carbon aldehyde is called butanal because the name is derived from the alkane (butane) with the *-e* replaced by *-al*. The numbering of the chain starts at the carbon of the aldehyde group; it is always at position one and need not be specified. Thus 3-chloro-2-methylbutanal is

$$CH_3-CH-CH-C-H \quad (\text{Cl, CH}_3, \text{O})$$

The IUPAC naming system for ketones uses the suffix *-one*. The parent chain must include the carbonyl group and be numbered from the end of the chain that reaches the carbonyl group first; this ensures, as required by IUPAC rules, that the number for the carbonyl group is as low as possible. For example, the molecule

$$CH_3-CH-CH-C-CH_3 \quad (\text{Cl, CH}_3, \text{O})$$

is 4-chloro-3-methyl-2-pentanone.

Preparation and Uses

Aldehydes can be produced by the oxidation of a *primary* alcohol with an oxidizing agent such as dichromate ion in acidic solution. However, the aldehyde so formed is readily oxidized further to a carboxylic acid (Section 27-7).

$$CH_3CH_2OH \xrightarrow[H^+]{Cr_2O_7^{2-}} CH_3CHO \xrightarrow[H^+]{Cr_2O_7^{2-}} CH_3CO_2H \qquad (27.17)$$

ethanol (a primary alcohol)   acetaldehyde (an aldehyde)   acetic acid (an acid)

A milder oxidizing agent in a nonaqueous medium is required to stop the oxidation at the aldehyde. This partial oxidation can be done with the reagent pyridinium chlorochromate (PCC) in an organic solvent such as dichloromethane. For example,

$$CH_3(CH_2)_8CH_2OH \xrightarrow{PCC, CH_2Cl_2} CH_3(CH_2)_8CHO$$

Oxidation of a *secondary* alcohol produces a ketone.

$$CH_3CH(OH)CH_3 \xrightarrow[H^+]{Cr_2O_7^{2-}} CH_3CCH_3 \qquad (27.18)$$

2-propanol (a secondary alcohol)   acetone (a ketone)

Formaldehyde ($H_2C{=}O$), a colorless gas that dissolves readily in water, is the simplest aldehyde. Formaldehyde is used in the manufacture of synthetic resins. A polymer of formaldehyde called paraformaldehyde is used as an antiseptic and an insecticide.

$ClCrO_3^-$

Aldehydes and ketones occur widely in nature. Typical natural sources are

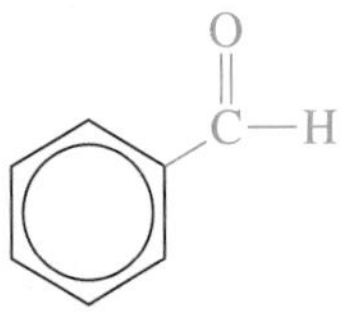

Benzaldehyde (almonds)	Cinnamaldehyde (cinnamon)	Camphor (obtained from camphor tree)

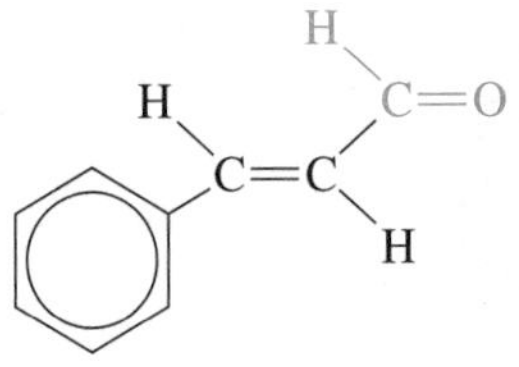

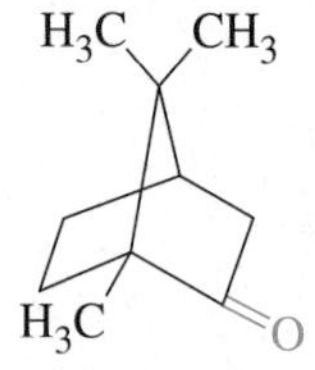

Acetone is the most important of the ketones. It is a volatile liquid (boiling point, 56 °C) and highly flammable. Acetone is a good solvent for a variety of organic compounds and is widely used in solvents for varnishes, lacquers, and plastics. Unlike many common organic solvents, acetone is miscible with water in all proportions.

Addition Reactions of the Carbonyl Group in Ketones and Aldehydes

The double bond of the carbonyl group can undergo addition reactions like the double bond in alkenes. For example, a ketone can be reduced to a secondary alcohol using a reagent such as sodium borohydride, $NaBH_4$ in aqueous solution.

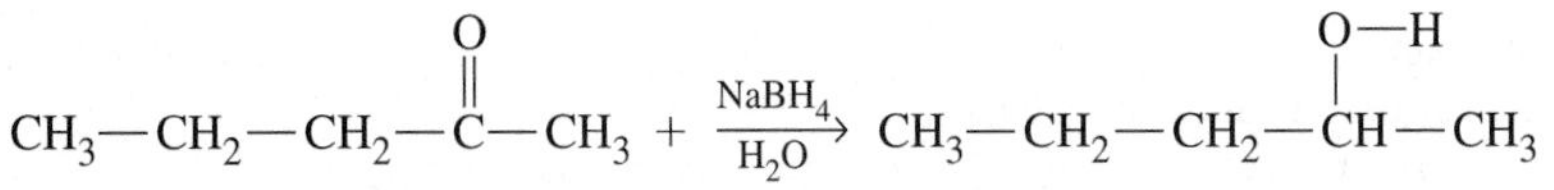

This reaction can be formally thought of as addition of a dihydrogen molecule across the carbonyl double bond, but the actual mechanism is the addition of H^- followed by the addition of a proton.

The difference between the double bond in an alkene and a carbonyl group is the polarity of the carbonyl group, so that when asymmetric reagents such as HCN are added, the negatively charged CN^- adds to the region of partial positive charge on the carbon atom; for example,

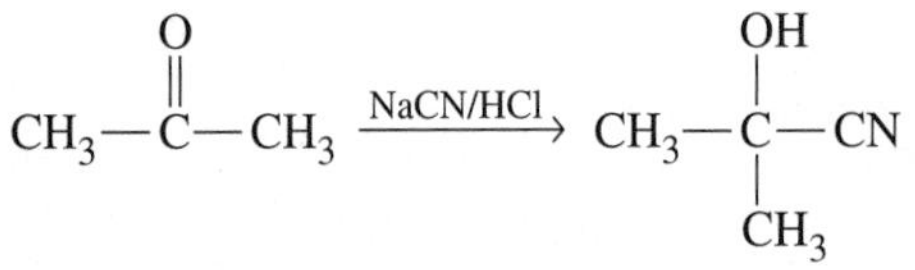

This molecule is called cyanohydrin, and it is the first member of a class of molecules called *cyanohydrins*. The important synthetic consequence of this reaction is that the carbon content of the molecule is increased by one carbon atom.

27-7 Carboxylic Acids and Their Derivatives

▶ The carboxyl group is also represented as —COOH and —CO₂H.

As we learned in Chapter 3, **carboxylic acids** contain the **carboxyl** group (*carbonyl* and hydr*oxyl*).

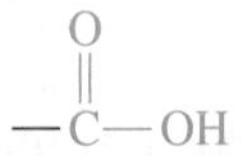

These acids have the general formula R—COOH. In many compounds, R is an aliphatic residue. Such compounds are called fatty acids because high-molecular weight compounds of this type are readily available from naturally occurring fats and

oils. If two carboxyl groups are found on the same molecule, the acid is called a dicarboxylic acid. The carboxyl group can also be found attached to the benzene ring.

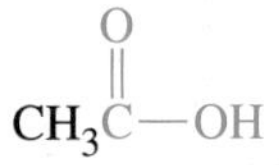

Acetic acid	Benzoic acid	Oxalic acid	o-Phthalic acid
(an aliphatic acid)	(an aromatic acid)	(an aliphatic dicarboxylic acid)	(an aromatic dicarboxylic acid)

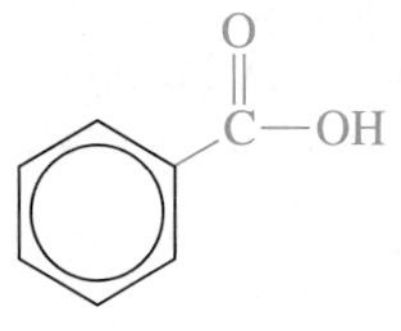

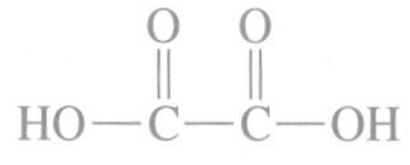

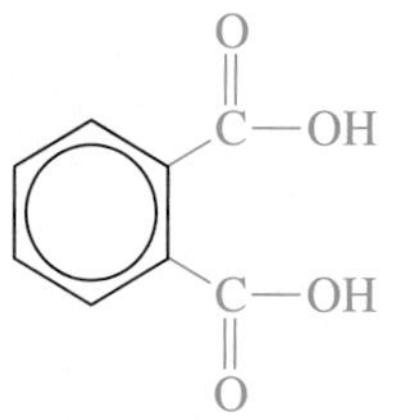

Oxidation of a primary alcohol or an aldehyde can produce a carboxylic acid. For this purpose, the oxidizing agent is generally $KMnO_4(aq)$ in an alkaline medium. Because the medium is alkaline, the product is the potassium salt, but the free carboxylic acid can be regenerated by making the medium acidic.

$$CH_3CH_2OH \xrightarrow[\text{OH}^-, \text{ heat}]{KMnO_4} CH_3COO^-K^+ \xrightarrow{H^+} CH_3COOH + K^+ \quad (27.19)$$

$$CH_3CH_2COH \xrightarrow[\text{OH}^-, \text{ heat}]{KMnO_4} CH_3CH_2COO^-K^+ \xrightarrow{H^+} CH_3CH_2COOH + K^+ \quad (27.20)$$

Substituted aliphatic acids can be named either by their IUPAC names or by using Greek letters in conjunction with common names. Aromatic acids are named as derivatives of benzoic acid.

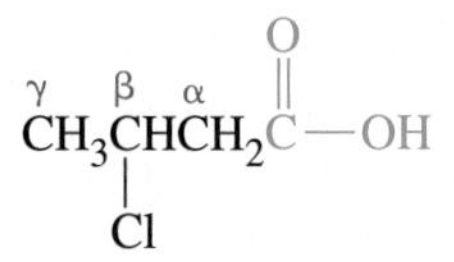

3-Chlorobutanoic acid	3-Methylbenzoic acid	2-Hydroxybenzoic acid
β-chlorobutyric acid	m-methylbenzoic acid	o-hydroxybenzoic acid
	(also toluic acid)	(also salicylic acid)

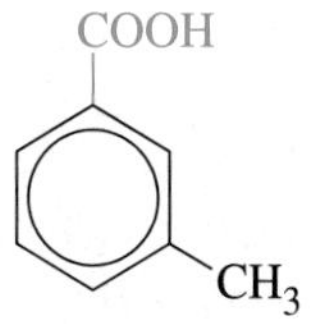

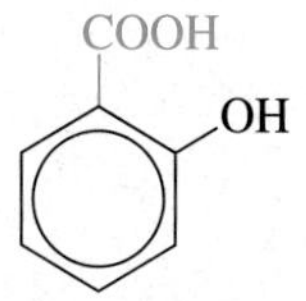

Because many derivatives of the carboxylic acids involve simple replacement of the hydroxyl groups, special names have been developed for the remaining portion of the molecule, that is,

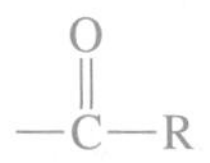

For example, the group

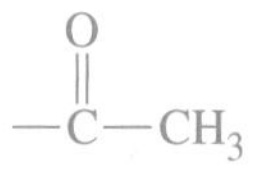

is called the **acetyl** group. The acetyl derivative of o-hydroxybenzoic (salicylic) acid is ordinary aspirin.

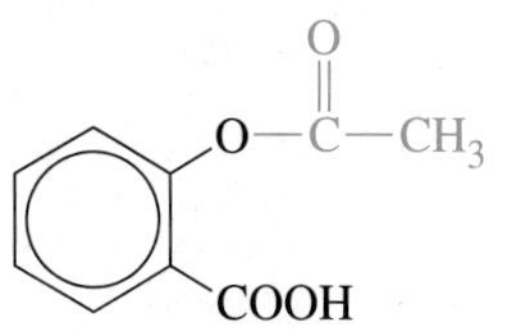

Acetylsalicylic acid
(aspirin)

Carboxyl groups display the chemistry of both carbonyl and hydroxyl groups. Donation of a proton to a base leads to salt formation. The sodium and potassium salts of long-chain fatty acids are known as *soaps*. The structure of the soap sodium palmitate was presented on page 880, and soap formation (saponification) is discussed further in the next chapter. The reaction between an alcohol and a carboxylic acid leads, as we see next, to a different class of compounds.

Esters

The product of the reaction of a carboxylic acid and an alcohol is called an **ester**. Esters are formed by the elimination of H_2O (HOH) between the two molecules. The mechanism of this reaction is such that the —OH comes from the *acid* and the —H from the *alcohol*.

▶ IUPAC names appear below the structural formulas with the more common names in parentheses.

$$CH_3CH_2O-H \ + \ HO-\overset{\overset{\textstyle O}{\|}}{C}CH_3 \ \xrightarrow[\text{heat}]{H^+} \ H_2O \ + \ CH_3CH_2O-\overset{\overset{\textstyle O}{\|}}{C}CH_3 \quad (27.21)$$

ethanol   ethanoic acid   ethyl ethanoate
(ethyl alcohol)   (acetic acid)   (ethyl acetate)

This type of reaction is a **condensation reaction**, a reaction in which two molecules are combined by the elimination of a small molecule such as water.

Esters have two-part names. The first part is the alkyl designation of the alcohol; the second part is the carboxylic acid name with the ending changed to *–oate*. For example, the combination of acetic acid and ethanol is *ethyl ethanoate*.

The reaction of a carboxylic acid and an alcohol is reversible. An ester is hydrolyzed by water in the presence of hydrogen ion to give the alcohol and the carboxylic acid. The hydrolysis goes to completion when it is carried out in the presence of the hydroxide ion; and in this case, the products are the alcohol and a salt of the carboxylic acid.

$$CH_3CH_2CO_2CH_2CH_3 \ \xrightarrow{\ OH^-\ } \ CH_3CH_2COO^- \ + \ CH_3CH_2OH$$

Unlike the pungent odors of the carboxylic acids from which they are derived, esters have very pleasant aromas. The characteristic fragrances of many flowers and fruits can be traced to the esters they contain. Esters are used in perfumes and in the manufacture of flavoring agents for the confectionery and soft drink industries. Most esters are colorless liquids that are insoluble in water. Their melting points and boiling points are generally lower than those of alcohols and acids of comparable carbon content. This is because of the absence of hydrogen bonding in esters.

▲ The distinctive aroma and flavor of oranges are due in part to the ester octyl acetate, $CH_3(CH_2)_6CH_2OOCCH_3$.

Are You Wondering...

If reaction (27.21) is a reversible reaction, how can we ensure a good yield of the ester?

In order to push the reaction toward the ester, we can apply Le Châtelier's principle and either add an excess of alcohol or remove the water as it is formed (there are experimental techniques to do this).

Amides

Heating the ammonium salt of a carboxylic acid results in the elimination of water and the formation of an **amide**. Further loss of water occurs if the amide is heated with a strong dehydrating agent such as P_4O_{10}. The final product, which contains the group $-C\equiv N$, is called a *nitrile*. These reactions can be reversed by stepwise treatment with water.

$$\underset{\text{ammonium acetate}}{CH_3-\overset{\displaystyle O}{\overset{\|}{C}}-O^-NH_4^+} \xrightarrow{\text{heat}} \underset{\text{acetamide}}{CH_3-\overset{\displaystyle O}{\overset{\|}{C}}-NH_2} + H_2O \qquad (27.22)$$

$$\Big\downarrow \text{heat} \bigg| P_4O_{10}$$

$$CH_3C\equiv N + H_2O \qquad (27.23)$$

Amides, like esters, can be hydrolyzed. When simple amides such as acetamide are heated with water in the presence of an acid catalyst, they form the parent carboxylic acid and ammonia.

Despite the presence of the NH_2 group in simple amides, they are not Brønsted bases like amines or ammonia because the carbonyl group promotes the resonance structures shown below.

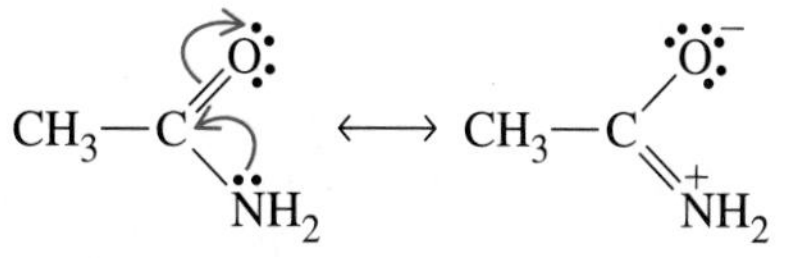

▶
Resonance in an amide.

The lone pair of electrons on the nitrogen atom is delocalized over the carbonyl group toward the more electronegative oxygen atom. In fact, as seen from the resonance structures, it is the carbonyl group that is most likely to be protonated in acidic conditions. The protonation of the oxygen in a carbonyl group is often an important first step in reactions of esters and amides under acidic conditions.

Reduction of Carboxylic Acids, Esters, and Amides

The reduction of carboxylic acids, esters, and amides requires the strong reducing agent lithium aluminum hydride, $LiAlH_4$. An example is the reduction of methyl propionate to give propanol and methanol.

$$CH_3-CH_2-\overset{\displaystyle O}{\overset{\|}{C}}-OCH_3 \xrightarrow{LiAlH_4} CH_3-CH_2-\overset{\displaystyle O-H}{\overset{|}{C}H_2} + HO-CH_3$$

The reduction of ethanamide with $LiAlH_4$ gives the primary amine ethanamine (ethyl amine).

$$CH_3-\overset{\displaystyle O}{\overset{\|}{C}}-NH_2 \xrightarrow{LiAlH_4} CH_3-CH_2-NH_2 + H_2O$$

This reaction provides an alternative synthesis of amines, the class of molecules we discuss next.

27-8   Amines

Amines are organic derivatives of ammonia, NH_3, in which one or more organic groups (R) are substituted for H atoms. Their classification is based on the number of R groups bonded to the nitrogen atom—one for primary amines, two for secondary, and three for tertiary.

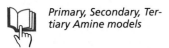
Primary, Secondary, Tertiary Amine models

$$CH_3CH_3NH_2$$

Ethylamine
(a primary amine)

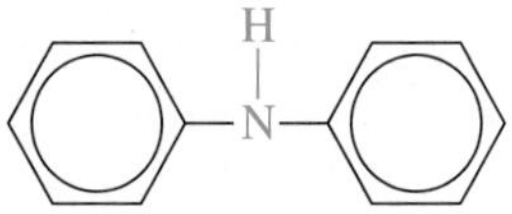

Diphenylamine
(a secondary amine)

$$CH_3NCH_2CH_2CH_3$$
(with CH_2CH_3 group on N)

Ethylmethylpropylamine
(a tertiary amine)

One of the chief methods of preparing an amine is by the reduction of a nitro compound.

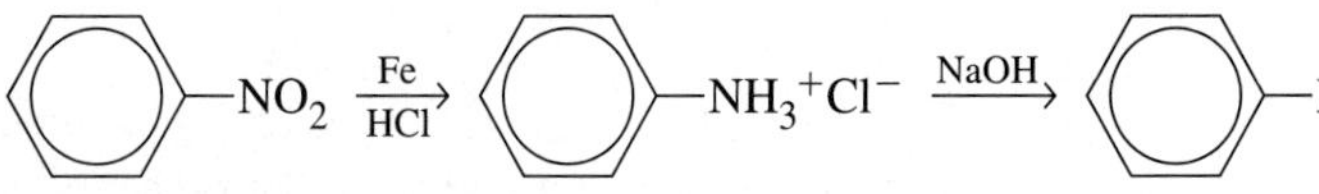

$$\text{—}NO_2 \xrightarrow[\text{HCl}]{\text{Fe}} \text{—}NH_3^+Cl^- \xrightarrow{\text{NaOH}} \text{—}NH_2 \quad (27.24)$$

An alternative method of preparing an amine involves the reaction of ammonia with an alkyl halide.

$$NH_3 + CH_3Br \longrightarrow CH_3NH_3^+ \ Br^-$$

The methylammonium bromide formed reacts with additional ammonia to give the amine.

$$CH_3NH_3^+ \ Br^- + NH_3 \longrightarrow CH_3NH_2 + NH_4^+ \ Br^-$$

The formation of pure primary amines is difficult by this method because when the primary amine is formed, it can take part in a further substitution reaction with the alkyl halide to form a secondary amine, for example,

$$CH_3NH_2 + CH_3Br \longrightarrow CH_3NH \ (\text{with } CH_3 \text{ group on N}) + HBr \quad (27.25)$$

The compound formed here is dimethylamine. Further reaction with bromomethane produces the tertiary amine, trimethylamine.

Amines of low molecular mass are gases that are readily soluble in water, yielding basic solutions. The volatile members have odors similar to ammonia, but more fishlike. Primary and secondary amines form hydrogen bonds, but these bonds are weaker than those in water because nitrogen is less electronegative than oxygen. Like ammonia, amines have trigonal pyramidal structures with a lone pair of electrons on the N atom. Also like ammonia, amines owe their basicity to these lone-pair electrons. In aromatic amines, because of unsaturation in the benzene ring, electrons are drawn into the ring, which reduces the electron charge density on the nitrogen atom. As a result, aromatic amines are *weaker* bases than ammonia. Aliphatic amines are somewhat stronger bases than ammonia.

$$NH_3 + H_2O \rightleftharpoons NH_4^+ + OH^- \qquad K_b = 1.8 \times 10^{-5}$$
$$CH_3NH_2 + H_2O \rightleftharpoons CH_3NH_3^+ + OH^- \qquad K_b = 42 \times 10^{-5}$$
methylamine

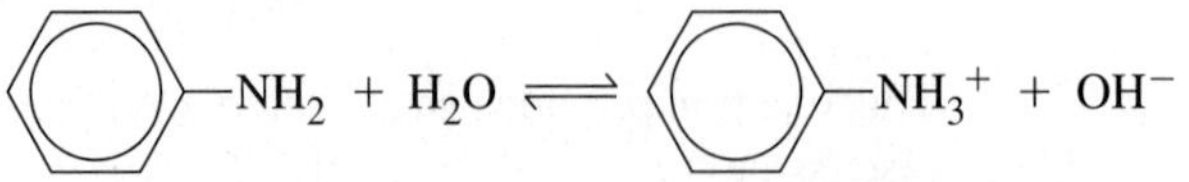

$$\text{—}NH_2 + H_2O \rightleftharpoons \text{—}NH_3^+ + OH^- \qquad K_b = 7.4 \times 10^{-10}$$

aniline

Amines can also be used to make amides. This is done by reaction of an amine and a carboxylic acid. For example, the reaction between methylamine and acetic acid first gives the ammonium carboxylate salt, which on heating gives the N-substituted amide.

$$CH_3-\overset{\overset{\displaystyle O}{\|}}{C}-OH \; + \; CH_3NH_2 \longrightarrow CH_3-\overset{\overset{\displaystyle O}{\|}}{C}-O^-NH_3CH_3{}^+ \;\xrightarrow{\text{heat}}\; CH_3-\overset{\overset{\displaystyle O}{\|}}{C}-NHCH_3 \; + \; H_2O$$

an ammonium   N-methyl acetamide
carboxylate salt   or N-methyl ethanamide

An example of a N-substituted amide is the pain reliever Tylenol.

Amides can be hydrolyzed under both acidic and basic conditions. Under acid conditions, the products are the carboxylic acid and the alkyl ammonium salt of the amine. Under basic conditions, the products are the salt of the carboxylic acid and the amine. The —CO—NH— linkage present in N-substituted amides is called a *peptide bond*, and it is a key bond type in proteins, as we will discover in Chapter 28.

Dimethylamine is an accelerator for the removal of hair from hides in the processing of leather. Butyl- and pentylamines are used as antioxidants, corrosion inhibitors, and in the manufacture of oil-soluble soaps. Dimethylamine and trimethylamine are used in the manufacture of ion-exchange resins. Additional applications are found in the manufacture of disinfectants, insecticides, herbicides, drugs, dyes, fungicides, soaps, cosmetics, and photographic developers.

Important derivatives of ammonia are the *tetraalkyl ammonium salts*. One such salt can be produced by the reaction of ammonia with an excess of iodomethane. The ammonium salt formed, known as a *quaternary ammonium salt*, has four organic groups attached to the nitrogen atom. Tetramethylammonium iodide can be viewed as forming in the reaction of one mole of trimethylamine with an additional mole of iodomethane.

$$CH_3-\overset{\overset{\displaystyle CH_3}{|}}{\underset{\underset{\displaystyle CH_3}{|}}{N}}\!: \; + \; CH_3I \longrightarrow \left[CH_3-\overset{\overset{\displaystyle CH_3}{|}}{\underset{\underset{\displaystyle CH_3}{|}}{N}}-CH_3\right]^+ \; + \; I^- \qquad (27.26)$$

The quaternary ammonium ion is often incorporated into systems involved in transmission of nerve impulses in the human body. One example is acetylcholine. Many poisons that affect the central nervous system also contain a quaternary ammonium functional group.

Tylenol

Acetylcholine

▲ Acetylcholine transmits nerve impulses.

27-9 Heterocyclic Compounds

In the ring structures considered to this point, all the ring atoms have been carbon; these structures are said to be carbocyclic. There are many compounds, however, both natural and synthetic, in which one or more of the atoms in a ring structure is not carbon. These ring structures are said to be **heterocyclic**. The heterocyclic systems most commonly encountered contain N, O, and S atoms, and the rings are of various sizes.

Pyridine is a nitrogen analogue of benzene (Figure 27-8). Unlike benzene, it is water soluble and has basic properties (the unshared pair of electrons on the N atom is not part of the π electron cloud of the ring system). Pyridine, a liquid with a disagreeable odor, was once obtained exclusively from coal tar, but it is now used so extensively that several synthetic methods have been developed for its production. It is used in the production of pharmaceuticals such as sulfa drugs and antihistamines,

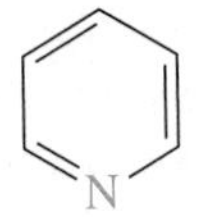

(a)

(b)

▲ **FIGURE 27-8**
Pyridine
In the pyridine molecule, a nitrogen atom replaces one of the CH units of benzene. Its formula is C_5H_5N. **(a)** Structural formula. **(b)** Space-filling model.

as a denaturant for ethyl alcohol, as a solvent for organic chemicals, and in the preparation of waterproofing agents for textiles. We will consider a number of other examples of heterocyclic compounds in Chapter 28.

27-10   Nomenclature of Stereoisomers in Organic Compounds

We have already discussed one form of stereoisomerism in organic compounds, namely cis–trans isomerism. We will look at another form of stereoisomerism in this section—optical isomerism. In addition, we will describe one of the systems of nomenclature used in modern organic chemistry to name stereoisomers.

Chirality

The phenomenon of **optical isomerism** was introduced in Chapter 25 in conjunction with transition metal complexes (coordination compounds). A solution of an optically active compound can rotate the plane of polarized light. The requirement for optical activity is that the molecule be asymmetric, so that its mirror image cannot be superimposed on the original molecule. This situation can arise at a tetrahedral C atom when all four groups attached to the C atom are different. Consider the molecule 3-methylhexane shown in Figure 27-9. The illustration shows that there are two non-superimposable isomers of 3-methylhexane, related as mirror images. The two isomers are said to be *enantiomers*, or optical isomers. A molecule that is not superimposable on its mirror image is said to be **chiral**.

In contrast with chiral molecules, such as 3-methylhexane, compounds possessing structures that are superimposable on their mirror images are **achiral**. Examples of chiral and achiral molecules are shown here,

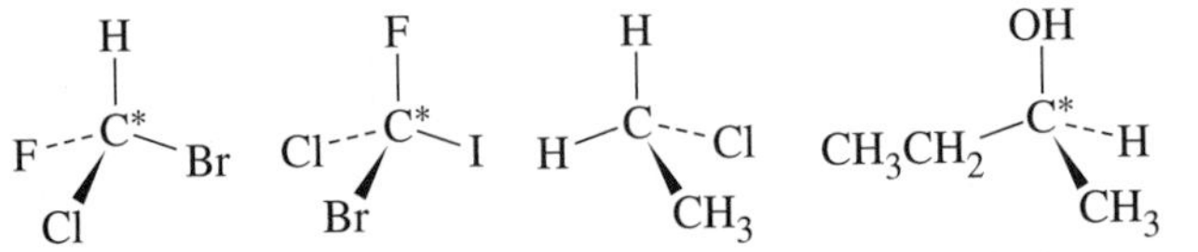

where we have employed the dashed-wedged line notation.

All the chiral molecules shown contain an atom that is connected to four different substituent groups. The C atom to which the four different groups are attached is said to be **asymmetric**, or a *stereocenter*. Centers of this type are often denoted by an asterisk. Molecules with one stereocenter are always chiral. As we will see in

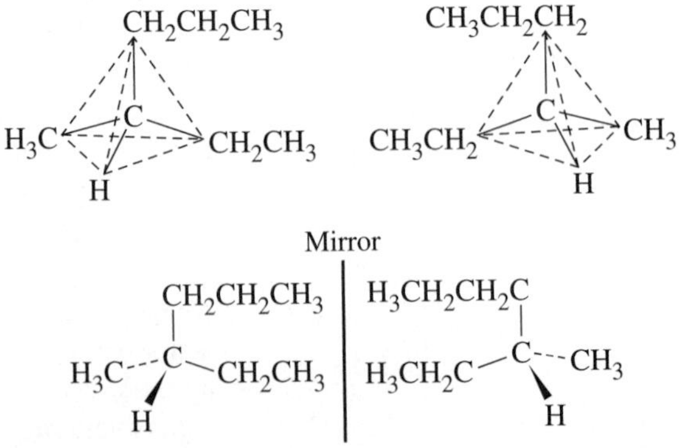

▶ **FIGURE 27-9**
Nonsuperimposable mirror images of 3-methylhexane
Notice in the diagram that we have adopted the convention used by organic chemists in drawing chiral centers: The groups attached to the central carbon in the plane of the paper are connected with solid lines, the group in front of the plane of the paper is attached by a wedge, and the group behind the plane of the paper is indicated by a dashed line.

that a solution of one enantiomer rotates the plane of polarized light in one direction; a solution of the other enantiomer rotates the light in the opposite direction. If we have a 50:50 mixture of both enantiomers—a racemic mixture—no rotation of the plane of polarized light is observed.

Chapter 28, molecules incorporating more than one stereocenter need not be chiral. As we will also see in Chapter 28, many chiral molecules occur in nature, and, as described later in this chapter, the existence of chirality can also play an important role in establishing some reaction mechanisms.

EXAMPLE 27-5

Identifying a Chiral Molecule. Predict whether either 2-chloropentane or 3-chloropentane is chiral.

Solution

To decide whether a molecule is chiral, we look for a C atom that has four different groups attached. The two compounds are depicted below.

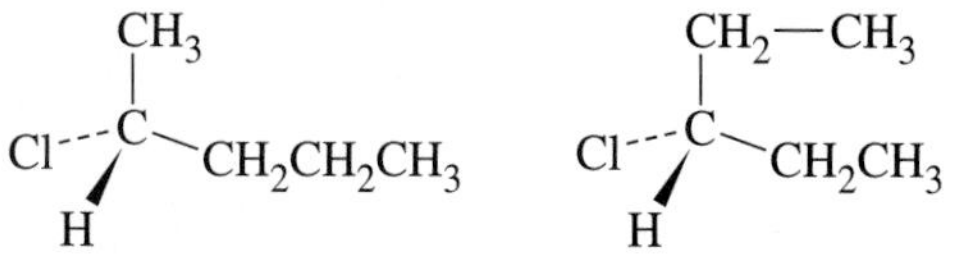

We see that 2-chloropentane contains a C atom that has four different groups attached; hence, it is chiral. However, 3-chloropentane does not have such a C atom; its structure is identical to its mirror image and the molecule is therefore achiral.

Practice Example A: Which of the following chlorofluorohydrocarbons is chiral: **(a)** $CF_3CH_2CCl_3$; **(b)** $CF_2HCHFCCl_3$; **(c)** $CClFHCHHCCl_2F$?

Practice Example B: Which of the following chloroalcohols is chiral: **(a)** $CH_2ClCH_2CH_2OH$; **(b)** $CH_2ClCH(OH)CH_3$; **(c)** $CH(OH)ClCH_2CH_3$?

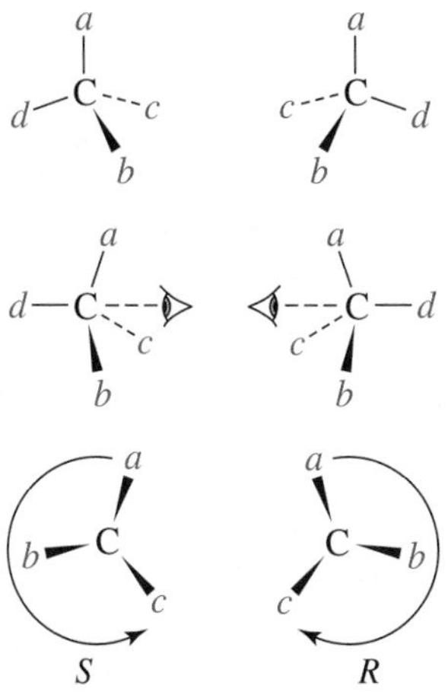

▲ **FIGURE 27-10**
Assignment of *R* and *S* configuration at a tetrahedral stereocenter
The group of lowest priority is placed as far away from the viewer as possible.

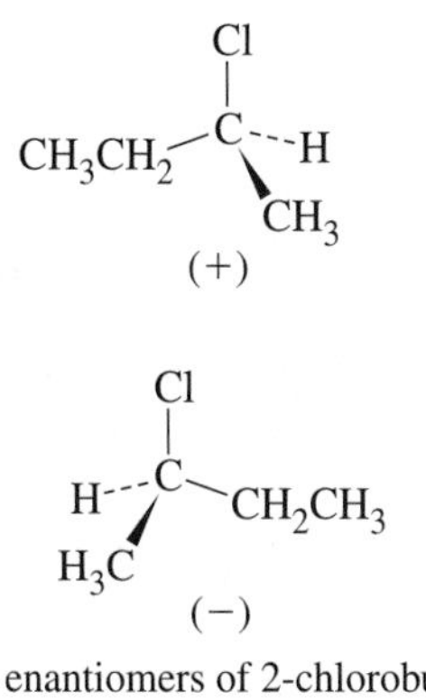

The enantiomers of 2-chlorobutane

Naming Enantiomers: The *R*, *S* System of Nomenclature

The compound 2-chlorobutane has two enantiomers. We need a way to name the individual stereoisomers so that we can distinguish them. That is, we need a system of nomenclature that indicates the arrangement or configuration of the four groups about the chiral center. Such a system was developed by three chemists, R. S. Cahn, C. Ingold, and V. Prelog. The first step is to rank all four substituents in the order of decreasing priority. We will describe the rules for assigning priorities shortly. Suppose that in the hypothetical compound *Cabcd*, substituent *a* has the highest priority, *b* the second highest, *c* the third highest, and *d* the lowest. Now position the molecule (mentally, on paper, or with a model set) so that the lowest priority substituent is as far away from the viewer as possible (Figure 27-10). This procedure results in two possible arrangements of the remaining substituents (one for each enantiomer). In the **R, S system**, if the sequence from *a* to *b* to *c* as viewed toward the substituent of lowest priority is clockwise, the configuration of the stereocenter is named *R* (*rectus*, Latin, meaning right). Conversely, if the sequence is counterclockwise, the stereocenter is named *S* (*sinister, Latin*, meaning left). The symbol (*R*) or (*S*) is added as a prefix to the name of the chiral compound, as in (*R*)-2-chlorobutane and (*S*)-2-chlorobutane. A racemic mixture of the enantiomers is designated (*R, S*) as in (*R, S*)-2-chlorobutane.

Rules for Assigning Priorities to Substituents In order to apply the *R, S* nomenclature to a stereocenter, we must first describe how the priorities are assigned to substituents. When looking at the atoms attached directly to the stereocenter the rules devised by Cahn, Ingold, and Prelog are as follows:

Rule 1.   **A substituent atom of higher atomic number takes precedence over one of lower atomic number.**   Consider the enantiomer of 1-chloro-1-iodoethane shown below.

R, S Enantiomer models

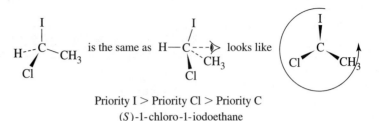

Priority I > Priority Cl > Priority C
(*S*)-1-chloro-1-iodoethane

Based on the atomic numbers of the atoms attached to the stereocenter, we have

Priority I > Priority Cl > Priority C > Priority H

The order of priority as viewed toward the H atom (which has the lowest priority) is counterclockwise. Therefore the enantiomer is (*S*)-1-chloro-1-iodoethane.

Rule 2.   **If two substituent atoms attached to the stereocenter have the same priority, proceed along the two substituent chains until a point of difference is reached. The atom of higher atomic number at this point establishes the priority.**   Thus, an ethyl group takes priority over a methyl group for the following reason. Although at the point of attachment to the stereocenter each substituent has a C atom, equal in priority, beyond these C atoms the methyl group has a H atom and the ethyl group has a higher-priority C atom.

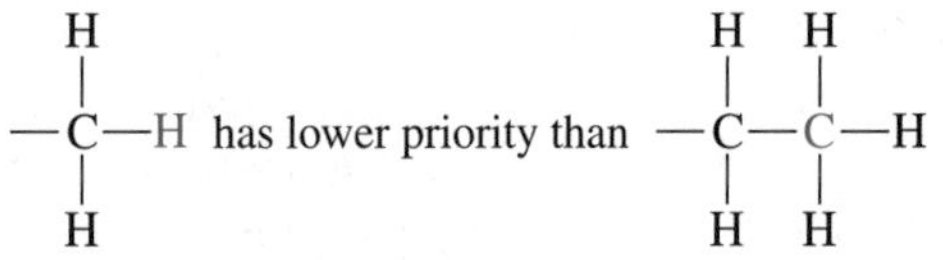

$$
\begin{array}{cc}
\overset{\displaystyle H}{\underset{\displaystyle H}{-C-H}} \quad \text{has lower priority than} & \overset{\displaystyle H\ \ H}{\underset{\displaystyle H\ \ H}{-C-C-H}}
\end{array}
$$

It is important to understand that the decision on priority is made at the *first* point of difference along otherwise similar substituent chains. When that point has been reached, the constitution of the remainder of the chain is immaterial.

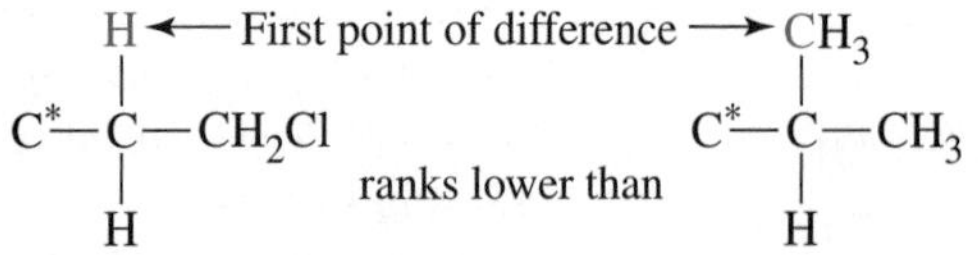

H ← First point of difference → CH_3

$$
C^*-\overset{\displaystyle H}{\underset{\displaystyle H}{C}}-CH_2Cl \qquad \underset{\text{ranks lower than}}{} \qquad C^*-\overset{\displaystyle CH_3}{\underset{\displaystyle H}{C}}-CH_3
$$

Rule 3.   **Double bonds and triple bonds are treated as if they were single, and the atoms in them are duplicated or triplicated at each end by the particular atoms at the *other end of the multiple bond*.**   For example,

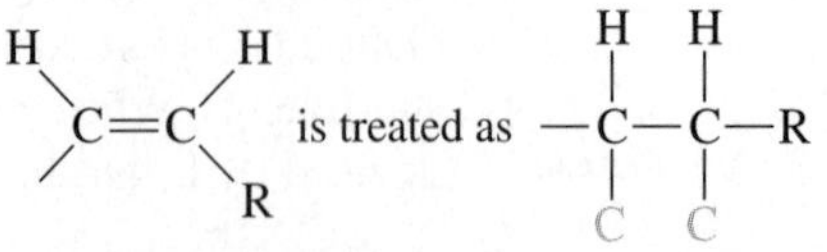

$$
\underset{/}{\overset{H}{\diagdown}}C=C\overset{H}{\underset{R}{\diagup}} \quad \text{is treated as} \quad -\overset{\displaystyle H\ \ H}{\underset{\displaystyle C\ \ C}{C-C}}-R
$$

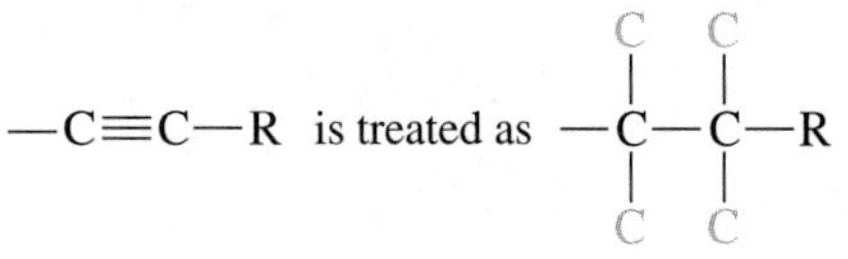

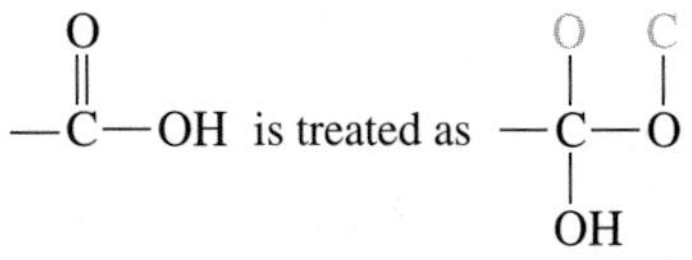

$$
-C{\equiv}C-R \quad \text{is treated as} \quad -\underset{\underset{C}{|}}{\overset{\overset{C}{|}}{C}}-\underset{\underset{C}{|}}{\overset{\overset{C}{|}}{C}}-R
$$

$$
-\overset{\overset{O}{\|}}{C}-OH \quad \text{is treated as} \quad -\underset{\underset{OH}{|}}{\overset{\overset{O\quad C}{|\ \ |}}{C}}-O
$$

Note that the atoms shown in red are added simply for the purpose of assigning priority to the groups containing a multiple bond; they are *not* really there! To illustrate, the —CH₂OH group has lower priority than —CHO.

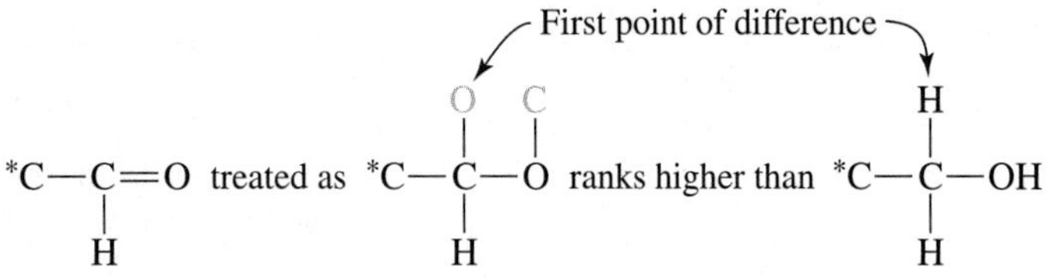

First point of difference

$$
{}^{*}C-\underset{\underset{H}{|}}{C}{=}O \quad \text{treated as} \quad {}^{*}C-\underset{\underset{H}{|}}{\overset{\overset{O\quad C}{|\ \ |}}{C}}-O \quad \text{ranks higher than} \quad {}^{*}C-\underset{\underset{H}{|}}{\overset{\overset{H}{|}}{C}}-OH
$$

The assignment of priorities and configuration of a chiral center is illustrated in Example 27-6.

EXAMPLE 27-6

Assignment of Priorities and Configuration of a Chiral Center. Name each of the following compounds, including the assignment of configuration.

(a) 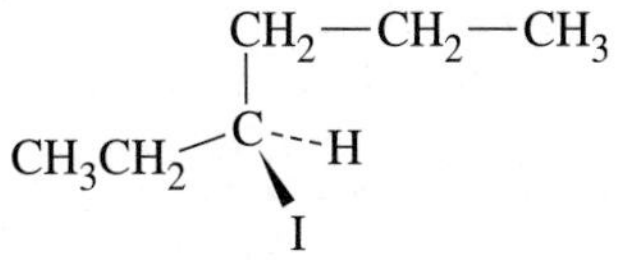 CH₃CH₂—$\overset{\overset{\displaystyle CH_2-CH_2-CH_3}{|}}{\underset{\underset{\displaystyle I}{\big\backslash}}{C}}$---H

(b) 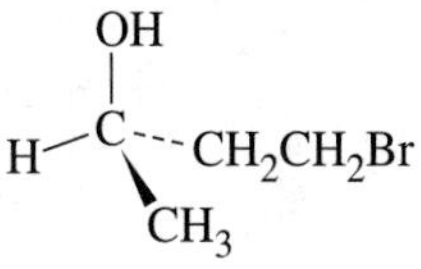 H—$\overset{\overset{\displaystyle OH}{|}}{\underset{\underset{\displaystyle CH_3}{\big\backslash}}{C}}$---CH₂CH₂Br

Solution

To assign the configuration at the stereocenter, we must first assign the priorities of the substituents. Then we determine the *R* or *S* configuration by viewing the molecule toward the atom of lowest priority.

(a) This molecule is 3-iodohexane with C-3 as the chiral center. The order of priorities of atoms attached to C-3 are

$$I > C \text{ (ethyl)} = C \text{ (propyl)} > H$$

In order to decide the ranking of the ethyl group relative to the propyl group, we go to the first point of difference in the chains of these substituents.

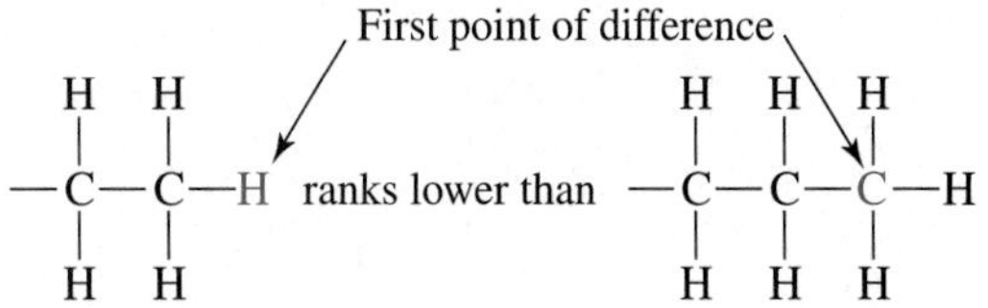

First point of difference

$$
-\underset{\underset{H}{|}}{\overset{\overset{H}{|}}{C}}-\underset{\underset{H}{|}}{\overset{\overset{H}{|}}{C}}-H \quad \text{ranks lower than} \quad -\underset{\underset{H}{|}}{\overset{\overset{H}{|}}{C}}-\underset{\underset{H}{|}}{\overset{\overset{H}{|}}{C}}-\underset{\underset{H}{|}}{\overset{\overset{H}{|}}{C}}-H
$$

We see that the propyl group ranks higher than the ethyl group, so the order of priorities is

$$I > C \text{ (propyl)} > C \text{ (ethyl)} > H$$

Viewing the molecule toward the lowest priority H atom

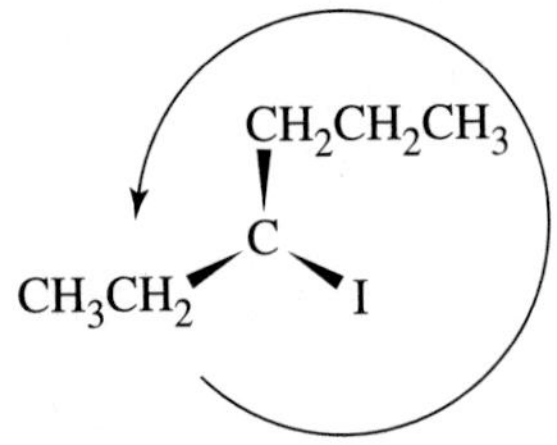

We see that the priorities decrease in a counterclockwise manner, so the configuration at the stereo center is *S*. The complete name of the molecule is

$$(S)\text{-3-iodohexane}$$

(b) This molecule is 4-bromo-2-butanol, with C-2 as the chiral center. The order of priorities of atoms attached to C-2 are

$$O > C \text{ (bromoethyl)} = C \text{ (methyl)} > H$$

In order to decide the ranking of the bromoethyl group relative to the methyl group, we go to the first point of difference in the chains of these substituents.

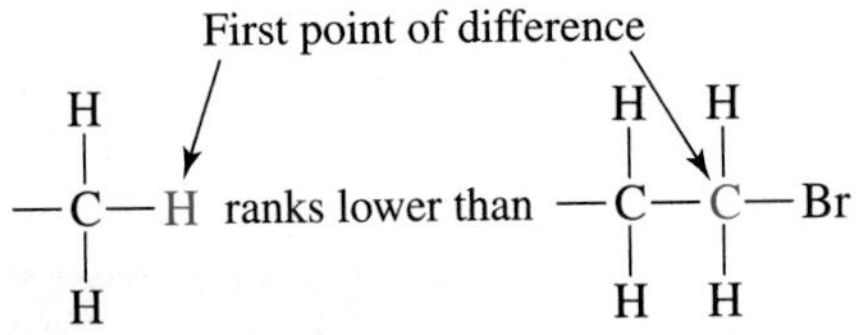

We see that the bromoethyl group ranks higher than the methyl group, so the order of priorities is

$$I > C \text{ (bromoethyl)} > C \text{ (methyl)} > H$$

Viewing the molecule toward the lowest priority H atom is not quite as straightforward as in the previous example. This is because the H atom is in the plane of the paper and thus not as far away from the viewers as possible. We can tackle this problem in one of two ways. The first requires some three-dimensional "vision," or "stereoperception," in that we imagine picking up the molecule by the methyl group and bromoethyl group and reorienting the molecule so that the H atom points away from us, to give

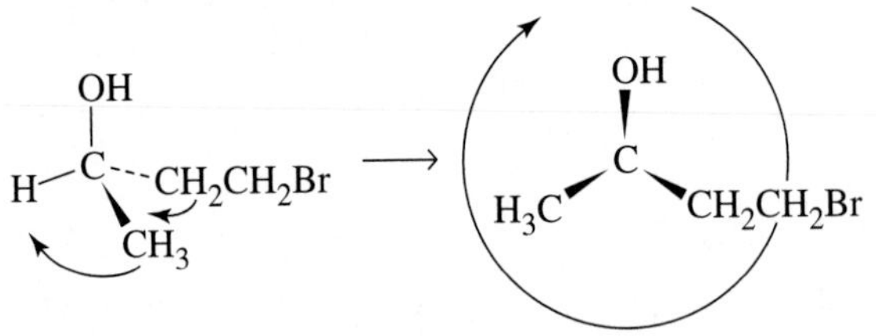

where we can now discern the sequence of priorities as clockwise, so the compound is (R)-4-bromo-2-butanol.

Alternatively, we can switch a pair of groups so that the group of lowest priority is bonded by a dashed line.

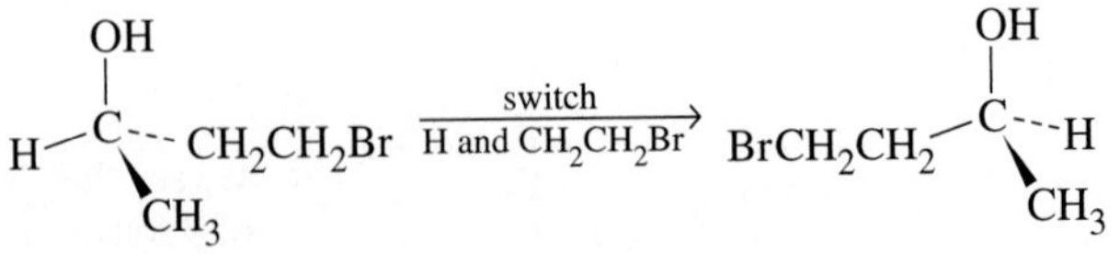

Now draw the view of the molecule toward the group of lowest priority.

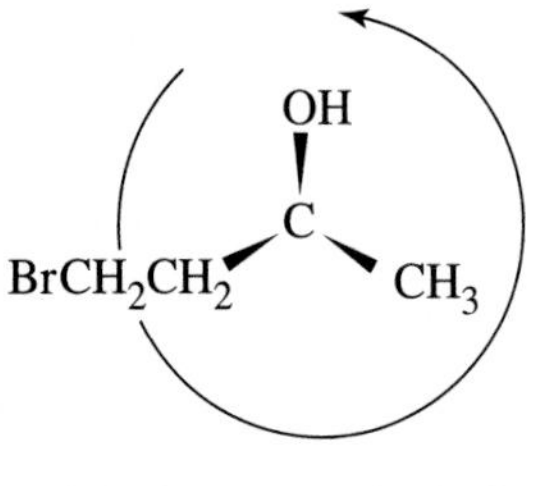

Here the sequence of priorities is counterclockwise. Because we switched two groups, we created the enantiomer of the molecule whose configuration we desire, thus, although in the switched molecule the order of priorities corresponds to an *S* configuration, the original molecule corresponds to the *R* configuration. Therefore, as determined previously, the molecule is (*R*)-4-bromo-2-butanol.

Practice Example A: Indicate whether each of the following structures has the *R* configuration or the *S* configuration.

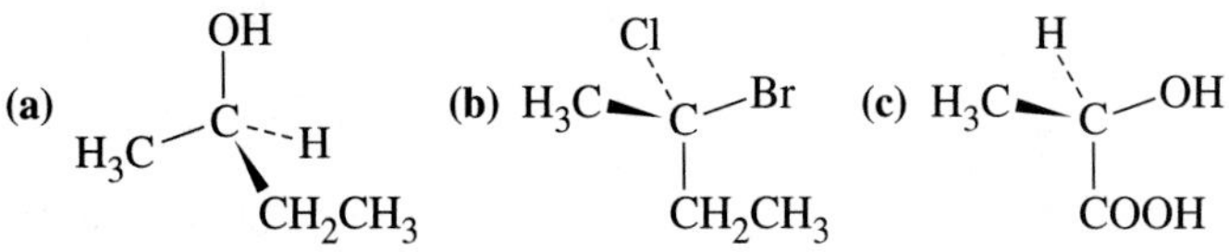

Practice Example B: Do the structures in each of the following pairs represent identical molecules or a pair of enantiomers?

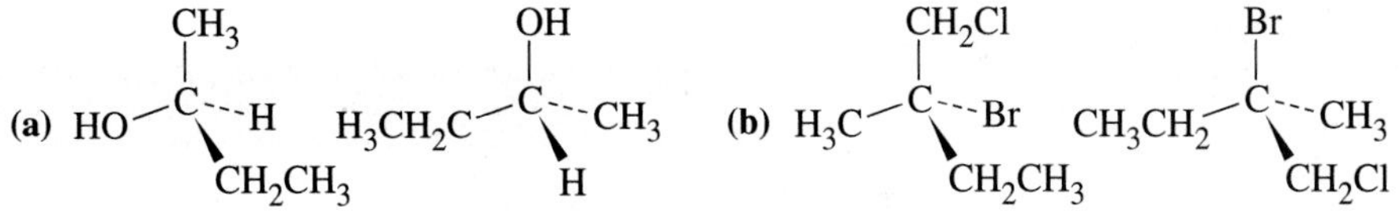

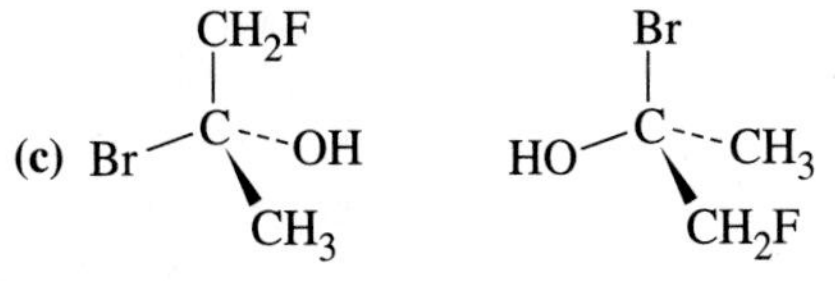

Naming the Stereoisomers of Highly Substituted Alkenes: The *E, Z* System of Nomenclature

As long as each of the sp^2 carbon atoms of an alkene is bonded to only one substituent, we can use the terms *cis* and *trans* to designate the geometric isomers of the alkene. But how can we decide which is the cis or trans isomers for a compound such as 1-chloro-1,2-difluoroethene, which has *three* substituents bonded to the sp^2 carbons?

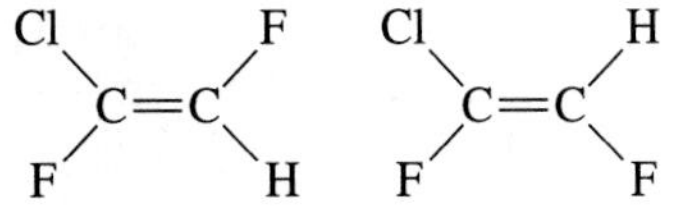

Which is cis? Which is trans?

An alternative system for naming such alkenes has been adopted by IUPAC; the *E*, *Z* **system**. The Cahn-Ingold-Prelog rules, discussed previously, are used systematically to assign priorities to the substituents on the sp^2 carbon atoms of the double bond. The stereochemistry about the double bond is assigned *Z* (from the German word *zusammen*, meaning together) if the two groups of higher priority *at each end of the double bond* are on the same side of the molecule. If the two groups of higher priority are on opposite sides of the double bond, the configuration is denoted by an *E* (from the German word *entgegen*, meaning opposite).

E, Z Isomer models

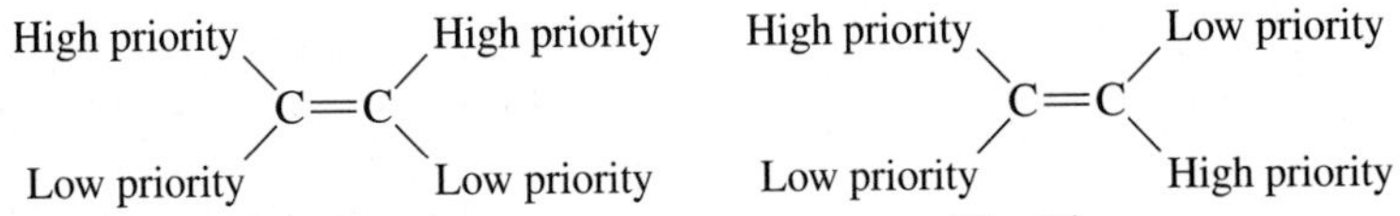

The Z isomer   The *E* isomer

The two stereoisomers of 1-chloro-1,2-difluoroethene are

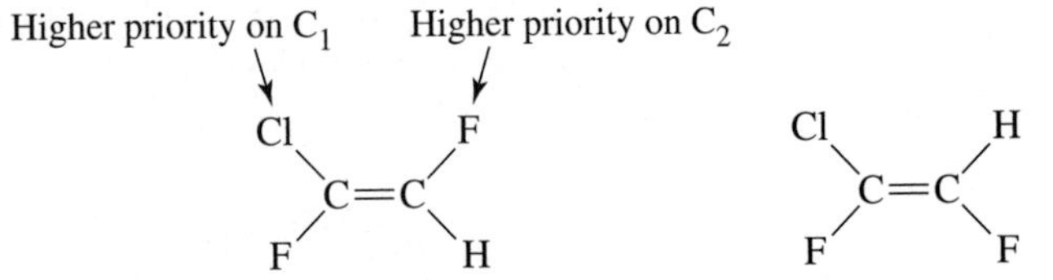

(*Z*)-1-Chloro-1,2-difluorothene   (*E*)-1-Chloro-1,2-difluorothene

Note that, as in *R, S* nomenclature, the *E* or *Z* is placed in parentheses. The *E, Z* system of nomenclature can be used for all alkene stereoisomers; consequently the IUPAC recommends that this system be used exclusively. However many chemists continue to use the cis and trans designations for simple alkenes.

EXAMPLE 27-7

Assignment of Configurations of Alkenes. Name each of the following compounds, including the assignment of configuration.

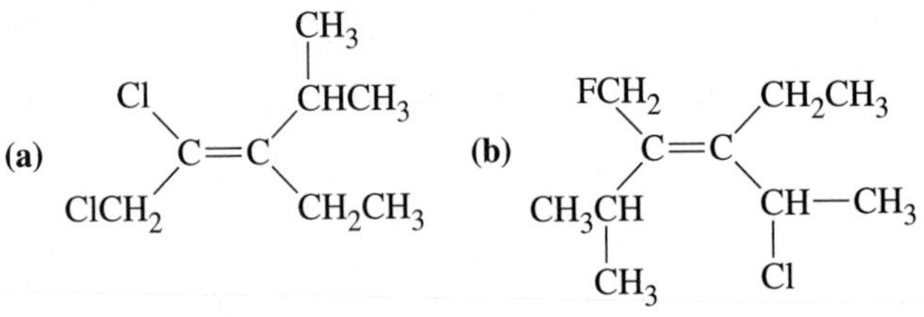

Solution

To assign the configuration of the alkene, we must first assign the priorities to the substituents attached to each carbon in the double bond. Once this is done we can assign the *E* or *Z* designation.

(a) One of the sp^2 carbon atoms is bonded to a Cl atom and a C atom, thus the chlorine atom has highest priority. The other sp^2 carbon is bonded to an ethyl group and a isopropyl group. The carbon of the isopropyl group is bonded to C, C, and H, and the carbon of the ethyl group to C, H, and H. A carbon can be canceled in each group. Of the remaining atoms, the carbon has the highest priority. Thus, the isopropyl group has precedence. The groups of highest priority are on the same side of the double bond. The complete name of the molecule is (*Z*)-1,2-dichloro-3-ethyl-4-methyl-2-pentene.

(b) One of the sp^2 carbon atoms is bonded to a fluoromethyl group and a isopropyl group. The carbon of the fluoromethyl group is bonded to F, H, and

H, and the carbon of the isopropyl group to C, C, and H. Of these six atoms, the fluorine has the highest priority. Thus, the fluoromethyl group has precedence. The other sp^2 carbon atom is bonded to an ethyl group and a chloroethyl group. Because the first point of difference on these substituent chains is a Cl atom, the chloroethyl group has precedence. The groups of highest priority are on opposite sides of the double bond. The complete name of the molecule is (E)-2-chloro-3-ethyl-4-fluoromethyl-5-methyl-3-hexene.

Practice Example A: Assign a configuration to each of the following alkenes.

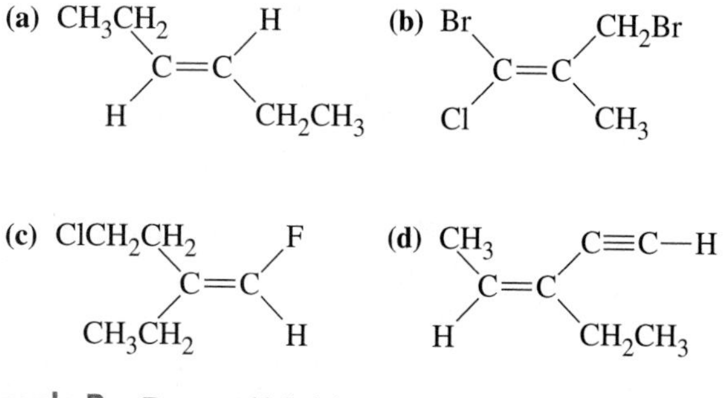

(a)
$$CH_3CH_2 \quad \quad H$$
$$\diagdown C = C \diagup$$
$$H \quad \quad CH_2CH_3$$

(b)
$$Br \quad \quad CH_2Br$$
$$\diagdown C = C \diagup$$
$$Cl \quad \quad CH_3$$

(c)
$$ClCH_2CH_2 \quad \quad F$$
$$\diagdown C = C \diagup$$
$$CH_3CH_2 \quad \quad H$$

(d)
$$CH_3 \quad \quad C \equiv C - H$$
$$\diagdown C = C \diagup$$
$$H \quad \quad CH_2CH_3$$

Practice Example B: Draw and label the E and Z isomers of the following compounds.

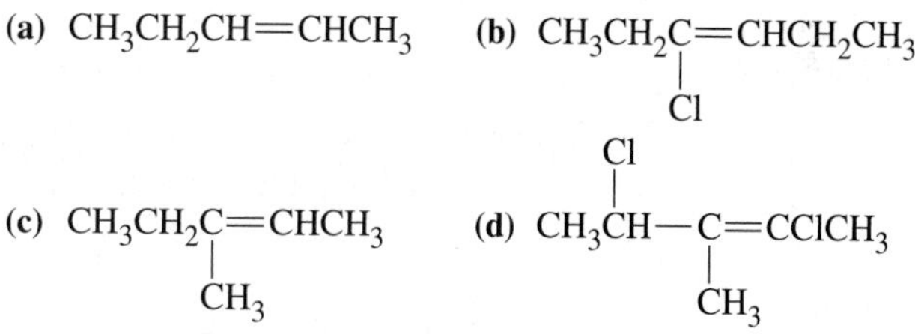

(a) $CH_3CH_2CH = CHCH_3$

(b)
$$CH_3CH_2C = CHCH_2CH_3$$
$$\quad\quad\quad\; |$$
$$\quad\quad\quad Cl$$

(c)
$$CH_3CH_2C = CHCH_3$$
$$\quad\quad\quad |$$
$$\quad\quad\quad CH_3$$

(d)
$$\quad\quad\quad\; Cl$$
$$\quad\quad\quad\; |$$
$$CH_3CH - C = CClCH_3$$
$$\quad\quad\quad\quad |$$
$$\quad\quad\quad\quad CH_3$$

27-11 An Introduction to Substitution Reactions at sp^3 Hybridized Carbon Atoms

We have seen that a H atom in an alkane can be replaced by a halogen. It is also possible to carry out a substitution reaction in which a different atom or group of atoms replaces the halogen. An example of such a reaction is the replacement of a Cl atom in chloromethane by a hydroxyl group to give methanol.

$$CH_3Cl + OH^- \longrightarrow CH_3OH + Cl^- \qquad (27.27)$$

This type of chemical reaction is called a **nucleophilic substitution reaction** because the hydroxide ion is a **nucleophile**—a reactant that seeks centers of positive charge in a molecule. We learned in Chapter 12 that the carbon to chlorine bond is polar, so the C atom is a region of slight positive charge and is said to be **electrophilic** (electron attracting). In this substitution reaction, the hydroxide ion attacks at the slightly positive C atom, pushing out the chloride ion, which is called the **leaving group**.

Nucleophiles, often denoted by the abbreviation Nu, may be negatively charged or neutral, but every nucleophile contains at least one pair of unshared electrons. Nucleophiles are, in fact, Lewis bases. The nucleophilic substitution of a haloalkane is described by either of two general equations.

$$\overset{}{Nu:^-} + \overset{\delta+ \;\; \delta-}{R - \ddot{X}:} \longrightarrow Ru - Nu + :\ddot{X}:^-$$

Nucleophile   Leaving group
                 Electrophile

$$\text{Nu:} + \overset{\delta+}{\underset{\underset{\text{Electrophile}}{\underset{\uparrow}{\text{Nucleophile}}}}{\text{R}}}-\overset{\delta-}{\underset{\text{··}}{\overset{\text{··}}{\text{X}}}}: \longrightarrow [\text{R}-\text{Nu}]^+ + :\overset{\text{··}}{\underset{\text{··}}{\text{X}}}:^{-}$$

As examples, consider these reactions that we have encountered previously in this chapter. In the reaction between sodium methylacetylide and bromoethane

$$\text{Na}^+ \text{ }^-\text{C}\equiv\text{C}-\text{CH}_3 + \text{CH}_3\text{CH}_2\text{Br} \longrightarrow \text{CH}_3\text{CH}_2\text{C}\equiv\text{CCH}_3 + \text{Na}^+\text{Br}^- \quad (27.9)$$

the nucleophile is the methylacetylide anion and the Br^- anion is the leaving group. Similarly, in the reaction between sodium ethoxide and iodomethane

$$\text{Na}^+ \text{ }^-\text{OCH}_2\text{CH}_3 + \text{CH}_3\text{I} \longrightarrow \text{CH}_3\text{CH}_2\text{OCH}_3 + \text{Na}^+\text{I}^- \quad (27.28)$$

the nucleophile is the ethoxide anion. Finally, in the formation of alkyl ammonium salts

$$\text{CH}_3\text{CH}_2\text{Br} + \text{NH}_3 \longrightarrow \text{CH}_3\text{CH}_2\overset{\overset{\displaystyle \text{H}}{|}}{\underset{\underset{\displaystyle \text{H}}{|}}{\text{N}}}{}^+\!-\text{H} + \text{Br}^- \quad (27.29)$$

the nucleophile is the ammonia molecule.

In studying these reactions, you may have noticed the similarity of the reaction

$$\text{Nu:}^- + \text{RX} \longrightarrow \text{RNu} + \text{X}^-$$

to the Brønsted–Lowry acid–base reaction

$$\text{B:}^- + \text{HX} \longrightarrow \text{HB} + \text{X}^-$$

In a Brønsted–Lowry acid–base reaction, the stronger base and the stronger acid react to form the weaker acid and the weaker base. Similarly, in a nucleophilic substitution reaction, a stronger base (conjugate base of a weak acid) displaces a weaker base (conjugate base of a strong acid). This idea enables us to decide on the likelihood of a reaction occurring. Thus, for example, we would not expect the reaction

$$\text{HO}^- + \text{H}_3\text{C}-\text{H} \overset{/}{\longrightarrow} \text{H}_3\text{COH} + \text{H}^-$$

to occur because the hydride ion is a much stronger base than the hydroxide ion. Recall that NaH reacts with water to produce hydrogen gas and aqueous sodium hydroxide.

In a nucleophilic substitution reaction, both the nucleophile and the leaving group are bases. If the difference in base strength between the two is not very large, the equilibrium constant for the resulting reaction will be close to unity. The substitution reaction will reach equilibrium rather than going to completion, as it would if the equilibrium constant were large. For example, the iodide and chloride ions are both weak bases (HI and HCl are both strong acids, but HI is stronger; see page 671) so that the reaction

$$\text{Cl}^- + \text{CH}_3\text{CH}_2\text{CH}_2-\text{I} \rightleftharpoons \text{I}^- + \text{CH}_3\text{CH}_2\text{CH}_2-\text{Cl} \quad (27.30)$$

is shifted toward the chloroalkane because I^- is a weaker base than Cl^-. Such equilibria can be shifted to the right or left by the removal of one of the products. For example, if the reaction is carried out in acetone, the product is iodopropane because sodium chloride is insoluble in acetone, whereas sodium iodide is soluble.

EXAMPLE 27-8

Identifying the Electrophile and Nucleophile. Decide which of the following compounds and ions are electrophiles and which ones are nucleophiles: **(a)** CH_3O^-; **(b)** methanal; **(c)** CH_3CH_2Cl; **(d)** CH_3NH_2.

Solution

To decide whether a molecule is an electrophile or a nucleophile, it is helpful to write a Lewis structure.

(a) The Lewis structure of the methoxide ion is

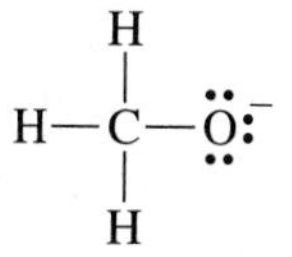

$$H-\overset{\overset{\displaystyle H}{|}}{\underset{\underset{\displaystyle H}{|}}{C}}-\overset{\cdot\cdot}{\underset{\cdot\cdot}{O}}\!:^{-}$$

The methoxide ion is seen to be a nucleophile because it has a pair of electrons to share.

(b) The Lewis structure of methanal is

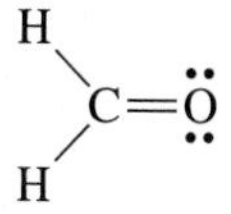

$$\overset{\displaystyle H}{\underset{\displaystyle H}{}}\!\!\diagdown\!\!\diagup\!\! C=\overset{\cdot\cdot}{\underset{\cdot\cdot}{O}}$$

The carbonyl group is polar since the more electronegative oxygen atom has a greater share of the electrons of the double bond. The C atom in the carbonyl group is electrophilic. Methanal is an electrophile.

(c) The Lewis structure of chloroethane is

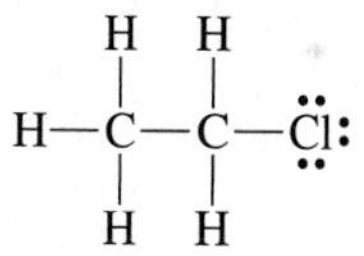

$$H-\overset{\overset{\displaystyle H}{|}}{\underset{\underset{\displaystyle H}{|}}{C}}-\overset{\overset{\displaystyle H}{|}}{\underset{\underset{\displaystyle H}{|}}{C}}-\overset{\cdot\cdot}{\underset{\cdot\cdot}{Cl}}\!:$$

The polar character of the carbon-to-chlorine bond renders this molecule an electrophile.

(d) The Lewis structure of methanamine is

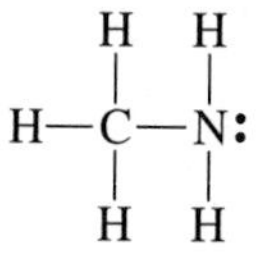

$$H-\overset{\overset{\displaystyle H}{|}}{\underset{\underset{\displaystyle H}{|}}{C}}-\overset{\overset{\displaystyle H}{|}}{\underset{\underset{\displaystyle H}{|}}{N}}\!:$$

The methanamine molecule is seen to be a nucleophile because it has a pair of electrons to share.

Practice Example A: Which of the following are electrophiles, and which are nucleophiles?

$$H^- \qquad AlCl_3 \qquad \overset{\displaystyle H}{\underset{\displaystyle H}{}}\!\!\diagdown\!\!\diagup\!\! C=C\!\!\diagup\!\!\diagdown\!\!\overset{\displaystyle H}{\underset{\displaystyle H}{}} \qquad CH_3S^-$$

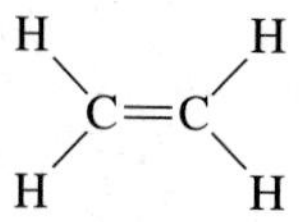

Practice Example B: Identify the electrophiles and the nucleophiles and the likely products in the following reactions.

(a) $CH_3I + NH_3 \longrightarrow$

(b) $CH_3CH_2Cl + CH_3S^- \longrightarrow$

(c) $CN^- + CH_3CH_2CHClCH_3 \longrightarrow$

The S_N1 and S_N2 Mechanisms of Nucleophilic Substitution Reactions

Chemical research, starting in the 1890s, has shown that nucleophilic substitution reactions can involve two types of mechanisms. The reaction between chloromethane and the hydroxide ion (27.27) has a rate law that is first order in both the nucleophile and the electrophile (molecule containing the polar bond). That is,

$$\text{rate} = k[OH^-][CH_3Cl]$$

The mechanism for this reaction involves a bimolecular rate-determining step in which the nucleophilic group approaches the electrophilic carbon atom, and simultaneously, the leaving group departs. The entering and leaving groups act at the same time in what is known as a *concerted step*, as depicted in this schematic way.

$$HO\!:^- + CH_3 \!-\! \ddot{C}l\!: \longrightarrow CH_3 \!-\! \ddot{O}H + :\!\ddot{C}l\!:^-$$

Bimolecular Reaction animation

The pair of curved arrows summarizes our visualization of how electrons flow to form and break bonds. Thus, the arrow starts from an electron-rich center (the electrons on the hydroxyl group in this case) to an electron-poor region (the C atom in the polar carbon-to-chlorine bond).

This mechanism is designated S_N2, with the S indicating substitution, the N indicating nucleophilic, and the 2 indicating that the rate-determining step is bimolecular. The reaction profile for an S_N2 reaction is shown in Figure 27-11, in which the transition state formed by the HO^- ion and the chloromethane is shown. The hydroxide attacks on the side opposite the Cl atom, and the C—Cl bond starts to break as the C—O simultaneously starts to form, producing the transition state shown. In the S_N2 mechanism, the nucleophile donates a pair of electrons to the electrophile to form a covalent bond.

KEEP IN MIND ▶

that the electrophile accepts the pair of electrons. Thus, the nucleophile acts as a Lewis base, and the electrophile as a Lewis acid.

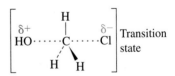

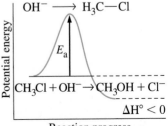

▶ **FIGURE 27-11**
Reaction profile for an S_N2 reaction
The bonds that are forming and breaking in the transition state are shown by dotted lines.

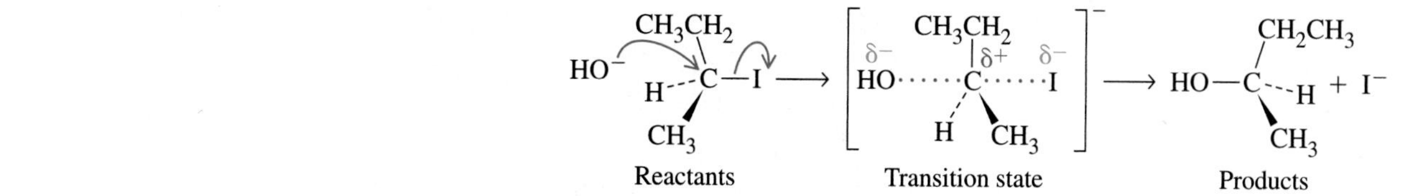

▲ **FIGURE 27-12** **Inversion of configuration in the S$_N$2 mechanism**

Is there any experimental evidence to support this mechanism and postulated transition state? First, the rate law is suggestive of a bimolecular step; second, an observation by Paul Walden in 1893 confirmed the formation of the transition state as shown in Figure 27-12. Walden discovered that if the C atom bonded to the halogen is stereogenic, or chiral, the configuration at the chiral carbon is inverted—that is, the molecular structure is inverted and an enantiomer of the opposite configuration is formed. Thus, when (S)-2-iodobutane undergoes nucleophilic substitution by the hydroxide group, the compound (R)-2-butanol is formed (Figure 27-12). This *inversion of configuration* is taken as confirmation of the proposed bimolecular mechanism and the five-coordinate transition state.

The other mechanism of nucleophilic substitution at haloalkanes has a rate law that is first order in the concentration of haloalkane only. The reaction between 2-bromo-2-methylpropane and water, in which H_2O acts as the nucleophile,

$$(CH_3)_3CBr + H_2O \longrightarrow (CH_3)_3COH + HBr$$

has the rate law

$$\text{rate} = k\,[C(CH_3)_3Br]$$

This rate law suggests that the rate-determining step is unimolecular. The unimolecular step is the dissociation of the haloalkane into ions—a bromide ion and a *carbocation*.

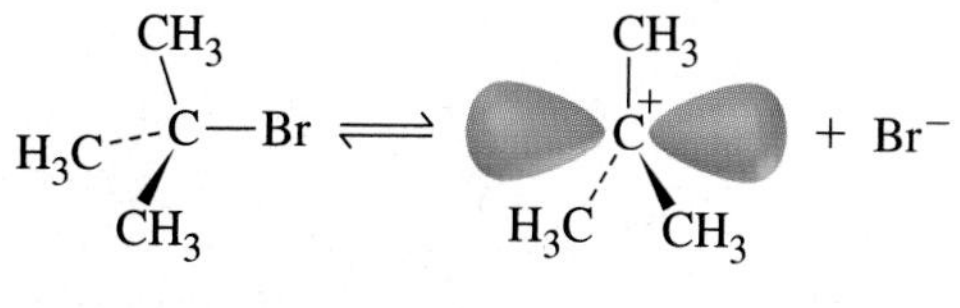

The planar carbocation immediately reacts with the nucleophile, water in this case.

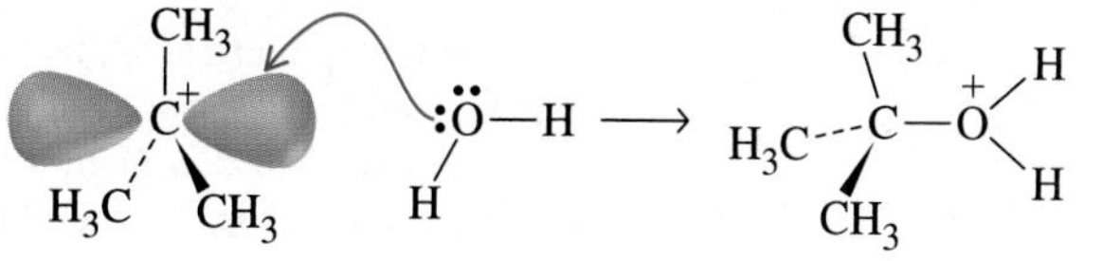

The conjugate acid of an alcohol is formed, which immediately dissociates in the presence of excess water, forming the neutral alcohol and a hydronium ion.

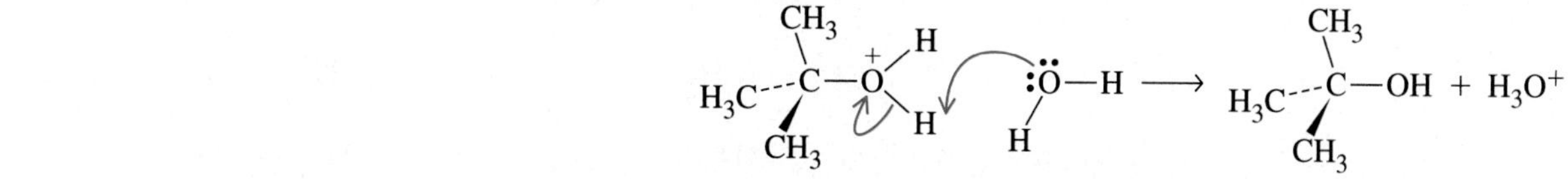

This mechanism is designated **S$_N$1**. Again, the S indicates substitution, and the N nucleophilic; in this case, the 1 indicates that the rate-determining step is unimolecular. The reaction profile for an S$_N$1 reaction is shown in Figure 27-13, which includes three transition states and two intermediates.

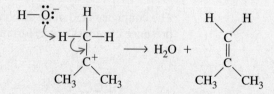

▶ **FIGURE 27-13**
Reaction profile for the S_N1 reaction between *t*-butyl bromide and water
The carbocation formed in the rate-determining step is planar. The central C atom in the carbocation intermediate is sp^2 hybridized. The remaining *p* orbital is used in the formation of the new chemical bond. The central C atom of the 2, 2-dimethylethyloxonium ion (the second intermediate) becomes sp^3 hybridized subsequently.

Apart from the rate law, what other evidence is there for the planar carbocation? Again, we can use a chiral haloalkane and investigate the chirality of the product. When one of the enantiomers of 3-bromo-3-methylhexane reacts with water, the product is a racemic mixture of the enantiomers of 3-methyl-3-hexanol. The carbocation intermediate formed in the rate-determining step is planar. The water molecule (nucleophile) can form a new bond on either face of the carbocation intermediate. This results in a 50:50 mixture of the product enantiomers (Figure 27-14). The formation of a racemic mixture provides confirmation of the unimolecular rate-determining step in the S_N1 mechanism.

When is a mechanism S_N1, and when is it S_N2? To answer this question, we need to briefly discuss some kinetic studies carried out by Christopher Ingold and

Elimination animation

Are You Wondering ...

Why we did not use the hydroxide ion in the hydrolysis of 2-bromo-2-methylpropane?

It is important to recognize that the carbocation is also a Brønsted–Lowry acid, which can react with a strong base such as the hydroxide ion. Loss of a proton in an acid–base reaction results in the formation of an alkene. This type of reaction is called an *elimination* reaction and is illustrated below.

In organic reactions, it is quite common for more than one reaction to take place with the same starting materials. In fact, when 2-bromo-2-methylpropane reacts with hydroxide ion in an aqueous solution, both the alkene and the alcohol are formed. Organic chemists can manipulate the reaction conditions (for example, by changing the solvent) to make one product dominant over the other.

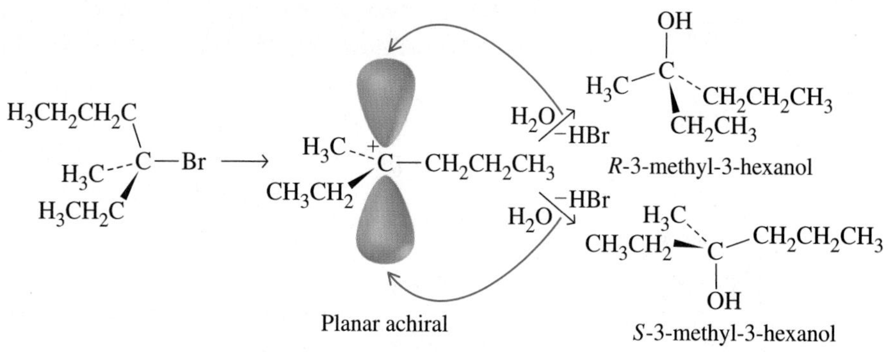

$$H_3CH_2CH_2C \overset{\displaystyle}{\underset{\displaystyle H_3CH_2C}{\overset{\displaystyle H_3C}{}}} C - Br \longrightarrow$$

Planar achiral

$H_2O HBr$ → R-3-methyl-3-hexanol

$H_2O HBr$ → S-3-methyl-3-hexanol

▲ FIGURE 27-14   **Formation of a racemic mixture in an S$_N$1 reaction**

Edward Hughes in 1937. Measurement of rates of reaction under different sets of experimental conditions can provide insight into reaction mechanisms. For example, consider the rate of the following second-order reaction between a bromoalkane and the nucleophile Cl⁻.

$$R-Br + Cl^- \longrightarrow R-Cl + Br^-$$

The rate of reaction decreases with increasing length and branching of the R group, in the following order.

$$CH_3Br > CH_3CH_2Br > CH_3CH_2CH_2Br > CH_3\overset{\displaystyle CH_3}{\underset{\displaystyle CH_3}{C}}HBr > CH_3\overset{\displaystyle CH_3}{\underset{\displaystyle CH_3}{C}}Br$$

Relative rate   1200   40   16   1   Too slow to measure

▶ In the same manner that alcohols are classified as primary, secondary, and tertiary, this principle is also used for haloalkanes. Thus, 2-bromopropane is a secondary haloalkane.

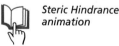

Steric Hindrance animation

In the S$_N$2 mechanism, the nucleophile attacks the electrophilic center on the side *opposite* the leaving group, often called *backside attack*. Because the nucleophile attacks the backside of the carbon that is bonded to the halogen, bulky substituents attached to this carbon will make it harder for the nucleophile to get to the backside and thus will slow down the reaction (Figure 27-15). The S$_N$2 mechanism can therefore provide an explanation for the observed decrease in rate of reaction when bulky groups are substituted for hydrogen atoms attached to the electrophilic C atom. The obstruction of the nucleophile from interacting with the center of partial positive charge is called *steric hindrance*.

With these ideas in mind, we can now appreciate why the reaction of 1,1-dimethyl-1-bromoethane (*t*-butyl bromide) with water proceeds by a different mechanism. The nucleophile (water) cannot attack the carbon in a concerted step (S$_N$2 mechanism) because of steric hindrance. The reaction does not proceed until the carbocation is formed by dissociation of the bromide ion in the rate-determining step. The water molecule then undergoes nucleophilic attack at the positively charged C atom, forming the protonated alcohol (Figure 27-13), which subsequently deprotonates in the aqueous solvent. Thus, in conclusion, we expect haloalkanes that are highly substituted at the C atom to which the halogen is attached to undergo nucleophilic substitution by an S$_N$1 mechanism.

In this brief description of the mechanisms of nucleophilic substitution reactions, we have touched on only one of the factors involved in determining which mechanism prevails. There are a number of others, such as solvent effects and nucleophilicity (how good a nucleophile the nucleophile is), discussions of which are found in organic chemistry texts.

(a)

(b)

(c)

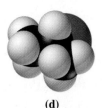

(d)

(e)

▲ **FIGURE 27-15**
Steric hindrance illustrated
Space-filling views of the molecules **(a)** bromomethane,
(b) bromoethane, **(c)** bromopropane, **(d)** 1-methylbromoethane, and
(e) 1,1-dimethylbromoethane viewed from attacking nucleophile's vantage point. The bulk of the substituents decreases access to the center of partial positive charge on the C atom attached to the halogen and therefore decreases the rate of an S_N2 reaction.

EXAMPLE 27-9

Recognizing S_N1 and S_N2 Reactions and Predicting the Products.   Predict whether the following reactions will take place, and suggest the likely mechanism.

  (a) $CN^- + CH_3CH_2CH_2Cl \longrightarrow$
  (b) $Br^- + CH_3CH_2OH \longrightarrow$
  (c) $CH_3OH + (CH_3)_3CCl \longrightarrow$ (in methanol)

Solution
To decide whether a reaction of this type will take place, first identify the electrophile, nucleophile, and the leaving group.

  (a) The nucleophile is the CN^- ion, the electrophile is the chloropropane, and the leaving group is the Cl^- ion. The CN^- is a much stronger base than the Cl^- ion, so the equilibrium constant for the reaction should be large. The chloropropane is a primary haloalkane, so the likely mechanism is S_N2 and the product will be butanonitrile (or 1-cyanopropane).

  (b) The nucleophile is the Br^- ion, the electrophile is the ethanol, and the potential leaving group is the OH^- ion. The OH^- is a much stronger base than the Br^- ion, so the equilibrium constant for the reaction is much *less* than one. No reaction is expected.

  (c) The nucleophile is the CH_3OH molecule, the electrophile is the *t*-butyl chloride, and the leaving group is the Cl^- ion. In this case we have to know the relative basicities of CH_3OH and Cl^- in order to decide in which direction the reaction will proceed. If we assume that the basicities of methanol and chloride ion are about the same (neither likes to accept a proton), we expect an equilibrium to be established. However, the fact that we are using a large excess of methanol (methanol is the solvent) would shift the equilibrium toward the product (Le Châtelier's principle). The product is *t*-butyl alcohol, and the likely mechanism is S_N1 since the electrophile is a tertiary haloalkane.

Practice Example A: Predict whether the following reactions will take place, and suggest the likely mechanism.

  (a) $CH_3CC^- + CH_3Br \longrightarrow$
  (b) $Cl^- + CH_3CH_2CN \longrightarrow$
  (c) $CH_3NH_2 + (CH_3)_3CCl \longrightarrow$

Practice Example B: From the following information, identify and provide chemical equations for the mechanism of the reaction.

  (a) The following reaction is carried out in acetone (propanone).

$$\text{(structure with Cl, H)} + CN^- \longrightarrow \text{(structure with H, CN)} + Cl^-$$

  (b) An optically pure sample of 2-iodobutane is dissolved in methanol. The resultant solution of *sec*-butyl methyl ether is optically inactive.

27-12   Synthesis of Organic Compounds

Originally, all organic compounds were isolated from natural sources. However, as chemists developed an understanding of the chemical behavior of organic compounds, they began to devise methods of synthesizing compounds from simple starting materials. Moreover, some of the compounds synthesized by the newly developed methods had never been observed to occur naturally.

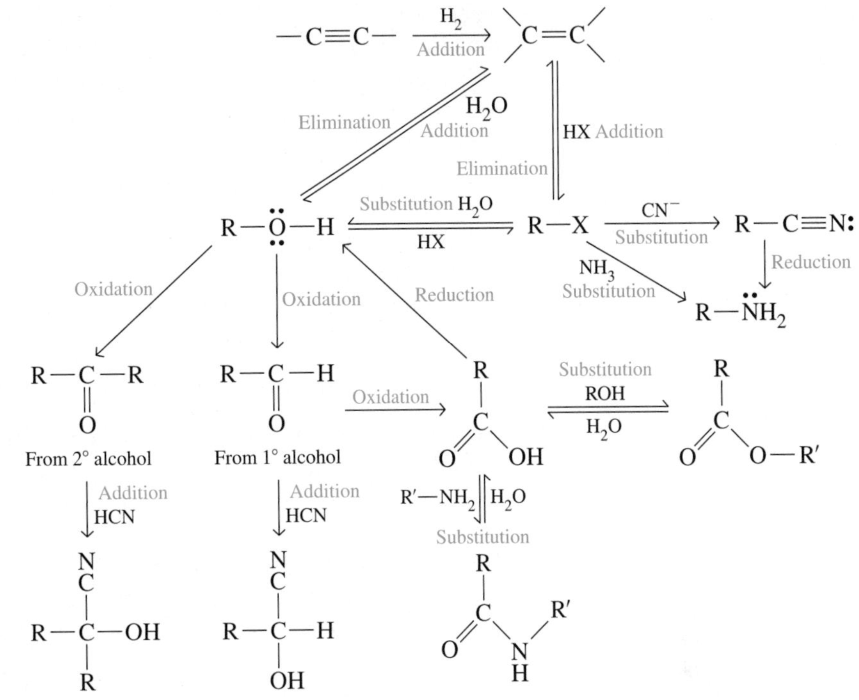

▲ FIGURE 27-16 **Functional group transformations**
A summary of functional group transformations, each of which requires a specific reagent as described in the text. When using this diagram, remember that changes in the R group are not indicated because this depends on the particular group involved. The symbols 1°, 2°, and 3° stand for primary, secondary, and tertiary, respectively.

In organic synthesis, chemists attempt to transform simple, readily available compounds into more complex molecules with desirable physical and chemical properties. Some syntheses are designed to make biologically active compounds that are otherwise available only from natural sources, and in only small quantities at high cost. Other synthetic approaches are designed to make new compounds, similar to naturally occurring ones but even more powerful in biological activity, such as medications to fight disease.

The approach that a synthetic organic chemist takes is to apply a knowledge of a wide variety of reaction types and reaction mechanisms to devise a synthetic scheme for assembling simple molecules into more complex structures. Figure 27-16 summarizes the chemical transformations between functional groups considered in this chapter. This summary stresses that each type of reaction has three components—the starting materials, the products, and the required reagents. For any reaction, we can complete a chemical equation if we have information about two of the three components of the reaction and if we know the type of reaction that converts the starting materials to the products. For example, in the following nucleophilic substitution reaction,

$$CH_3Br + ? \longrightarrow CH_3CN + ?$$

we see that CN has been substituted for Br as a functional group. In order to achieve this transformation, the nucleophile CN^- is required and we complete the equation.

$$CH_3Br + CN^- \longrightarrow CH_3CN + Br^-$$

A somewhat different, and more common, situation is that represented by

$$? + Br^- \longrightarrow CH_3Br + ?$$

where we need to work backwards to decide on the electrophile in a nucleophilic substitution reaction to get the desired product, CH_3Br. The electrophile will need to be connected to the leaving group (for example, another halogen atom) in the same manner and place as the Br atom will be in the product. An appropriate complete equation is

$$CH_3I + Br^- \longrightarrow CH_3Br + I^-$$

Let's apply the approach outlined above to the synthesis of ethyl ethanoate, starting only from inorganic substances and carbon (coke). First, we break the desired compound down into its constituent molecules, namely ethanol and ethanoic acid, which can be combined to give the ester.

$$CH_3COOH + HOCH_2CH_3 \longrightarrow H_2O + CH_3COOCH_2CH_3$$

We therefore need to be able to produce ethanol, which can, in turn, be converted into ethanoic acid.

$$HOCH_2CH_3 \xrightarrow{?} H_3CCOOH$$

The reagent required to oxidize ethanol is $K_2Cr_2O_7$ in acid solution. We now need a route to the synthesis of ethanol from carbon. One inorganic reagent that we have at our disposal that contains a OH group is water. We therefore write the equation

$$? + H_2O \longrightarrow HOCH_2CH_3$$

A consideration of the reactions that we have discussed in this chapter (Figure 27-16) indicates that one possibility is the addition of water to the double bond in ethene in the presence of sulfuric acid as a catalyst. Thus, we write

$$H_2C{=}CH_2 + H_2O \xrightarrow{H_2SO_4} HOCH_2CH_3$$

To continue, the ethene molecule has to be generated. Ethene is related to ethyne by a hydrogenation reaction.

$$H{-}C{\equiv}C{-}H + H_2 \xrightarrow[\text{heat and pressure}]{Pt} H_2C{=}CH_2$$

Our source of carbon is the element carbon and we need a way of producing a molecule with a carbon chain length of two. The production of calcium carbide (CaC_2) and its subsequent hydrolysis to give ethyne is described on page 1072. We have now completed the description of the desired synthesis. The underlying strategy is to break the desired molecule down into its constituent molecules and work backwards, using the chemical transformations available. This approach is known as *retrosynthesis* and is the method of choice in modern synthetic organic chemistry.

 ## Are You Wondering...

Why we did not use the nucleophilic substitution by hydroxide ion in chloroethane to produce ethanol?

This is a perfectly acceptable approach to the synthesis of ethanol. It would require the substitution of a Cl atom for a H atom in ethane which, in turn, can be produced by the hydrogenation of ethyne, which is synthesized as described. Often, in the design of a synthetic pathway, one or more routes are possible. The one chosen is usually the one that produces the greatest yield of the desired product.

27-13 Polymerization Reactions

Synthesis of Nylon 610 movie

The majority of chemists and chemical engineers work with macromolecular substances, or polymers. Of particular interest to them are the reactions that can be used to make polymers. We will briefly survey the main types of polymerization reactions in this final section of the chapter.

Chain-Reaction Polymerization

▶ Molecules of a *poly*mer are formed by the joining together of many simple molecules called *mono*mers.

Monomers with carbon-to-carbon double bonds typically undergo **chain-reaction polymerization**. The net result is that the double bonds open up and monomer units add to growing chains. As with other chain reactions, the mechanism involves three characteristic steps: initiation, propagation, and termination. Let us illustrate this mechanism for the formation of the polymer polyethylene from the monomer ethylene (ethene).

The key to the polymerization reaction is the free-radical initiator. In reaction (27.31), an organic peroxide dissociates into two peroxy radicals. The radicals add to the C$=$C double bonds of ethylene molecules to form radical intermediates that attack more ethylene molecules and form new intermediates of longer and longer length (27.32). The chains terminate as a result of reactions such as (27.33).

Initiation: $R-O:O-R \longrightarrow 2\,R-O\cdot$ (27.31)

an organic peroxide

followed by

$$CH_2=CH_2 + RO\cdot \longrightarrow R-O-CH_2-CH_2\cdot$$

Propagation: $ROCH_2CH_2\cdot + CH_2=CH_2 \longrightarrow ROCH_2CH_2CH_2CH_2\cdot$ (27.32)

$$RO(CH_2)_3CH_2\cdot + CH_2=CH_2 \longrightarrow RO(CH_2)_5CH_2\cdot$$

Termination: $RO(CH_2)_xCH_2\cdot + RO\cdot \longrightarrow RO(CH_2)_xCH_2OR$ (27.33)

or

$$RO(CH_2)_xCH_2\cdot + RO(CH_2)_yCH_2\cdot \longrightarrow RO(CH_2)_xCH_2CH_2(CH_2)_yOR$$

Several polymers formed by chain-reaction polymerization are listed in Table 27.5.

▲ Extruding polyethylene film.

TABLE 27.5 Some Polymers Produced by Chain-Reaction Polymerization

Name	Monomer	Polymer	Uses
Polyethylene	$CH_2=CH_2$	$-(CH_2-CH_2)_n-$	Bags, bottles, tubing, packaging film
Polypropylene	$CH_2=CHCH_3$	$\left(CH_2-\underset{\underset{CH_3}{\mid}}{CH}\right)_n$	Laboratory and household ware, artificial turf, surgical casts, toys
Poly(vinyl chloride) PVC	$CH_2=CHCl$	$\left(CH_2-\underset{\underset{Cl}{\mid}}{CH}\right)_n$	Bottles, floor tile, food wrap, piping, hoses
Poly(tetrafluoroethylene), Teflon	$CF_2=CF_2$	$-(CF_2-CF_2)_n-$	Bearings, insulation, nonstick surfaces, gaskets, industrial ware
Polystyrene	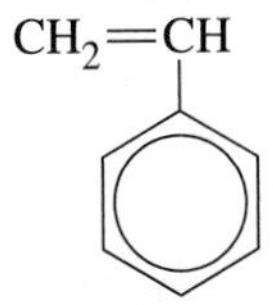 $CH_2=CH$ (with benzene ring)	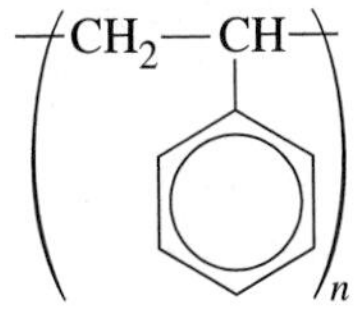 $\left(CH_2-CH\right)_n$ (with benzene ring)	Packaging, refrigerator doors, cups, ice buckets, and coolers (as foam)

> ## Are You Wondering...
>
> ### What effect the —OR groups have on polymers such as polyethylene?
>
> Essentially, they have no effect on the properties of the polymer. This is because there are hundreds, possibly thousands, of monomer units in the polymer, and the —OR groups are only at the ends of each polymer strand. The long chain of monomer units gives the polymer its particular properties, and the —OR group serves only to initiate the reaction and to terminate the polymer chains.

Step-Reaction Polymerization

In **step-reaction polymerization**, also called *condensation polymerization*, the monomers typically have two or more functional groups that react to join the two molecules together. Usually, this involves the elimination of a small molecule, such as H_2O. In chain-reaction polymerization, the reaction of a monomer can occur only at the end of a growing polymer chain, but in step-reaction polymerization, any pair of monomers is free to join into a *dimer*; the dimer can join with a monomer to form a *trimer*; two dimers can join to form a *tetramer*; and so on. Step-reaction polymerization tends to occur slowly and produces polymers of only moderately high molecular masses (less than 10^5 u). The formation of polyethylene glycol terephthalate, Dacron, is illustrated in the following reaction, and several other polymers formed by this method are listed in Table 27.6.

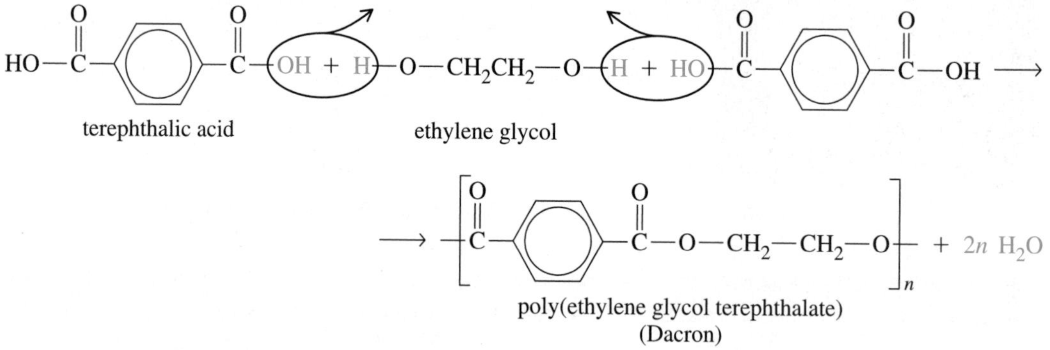

terephthalic acid   ethylene glycol

poly(ethylene glycol terephthalate)
(Dacron)

Stereospecific Polymers

The physical properties of a polymer are determined by a number of factors, such as the average length (average molecular weight) of the polymer chains and the strength of intermolecular forces between chains. Another important factor is whether the polymer chains display any crystallinity—that is, an ordered geometry and spacing of the atoms between polymer chains. In general, amorphous polymers are glasslike or rubbery. A high-strength fiber, on the other hand, must possess some crystallinity. Many polymers have both crystalline and amorphous regions. The relative amount of each type of region affects the physical properties of the polymer.

The usual designation of a polymer, when applied to polypropylene, is not very revealing about the structure of the polymer.

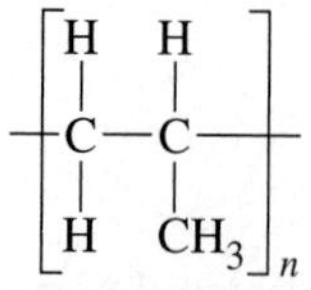

TABLE 27.6 Some Polymers Produced by Step-Reaction Polymerization

Name	Monomer	Polymer	Uses
Poly(ethylene glycol tere-phthalate) (Dacron) and	$HOCH_2CH_2OH$ and 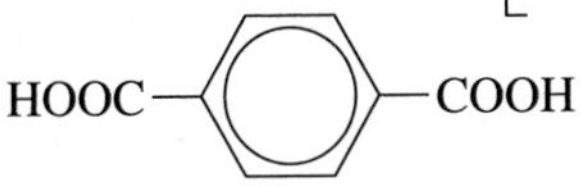 $HOOC\!-\!\!\bigcirc\!\!-\!COOH$	$\left[\!-\!\overset{O}{\overset{\|}{C}}\!-\!\bigcirc\!-\!\overset{O}{\overset{\|}{C}}\!-\!O\!-\!CH_2\!-\!CH_2\!-\!O\!-\!\right]_n$	Textile fabrics, twine and rope, fire hoses, plastic containers
Poly(hexamethyl-eneadipamide) nylon 66	$H_2N(CH_2)_6NH_2$ and $HOOC(CH_2)_4COOH$	$\left[\!-\!\overset{O}{\overset{\|}{C}}\!-\!(CH_2)_4\!-\!\overset{O}{\overset{\|}{C}}\!-\!NH\!-\!(CH_2)_6\!-\!NH\!-\!\right]_n$	Hosiery, rope, tire cord, parachutes, artificial blood vessels
Polyurethane	$HO(CH_2)_4OH$ and $OCN(CH_2)_6NCO$	$\left[\!-\!O\!-\!(CH_2)_4O\!-\!\overset{H}{\overset{\|}{\underset{\overset{\|}{O}}{C}}}\!-\!\overset{H}{\overset{\|}{N}}\!-\!(CH_2)_6\overset{H}{\overset{\|}{N}}\!-\!\overset{}{\underset{\overset{\|}{O}}{C}}\!-\!\right]_n$	Spandex fibers, bristles for brushes, cushions and mattresses (as foam)

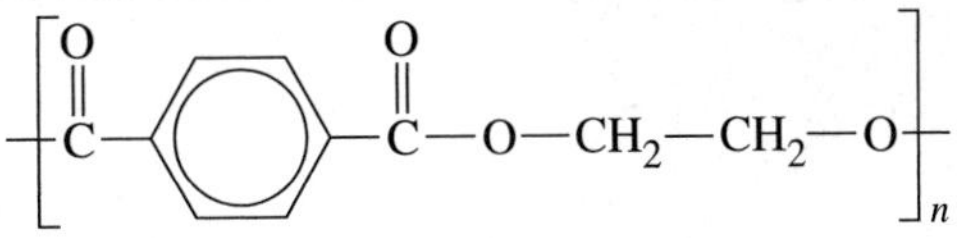

It does not indicate the orientation of the $-CH_3$ groups along the polymer chain.

If propylene ($CH_3-CH\!=\!CH_2$) is polymerized by the method shown for ethylene on page 1105, the orientation of the $-CH_3$ groups is random (Figure 27-17). A polymer of this type is called *atactic*. Because there is no regularity to the

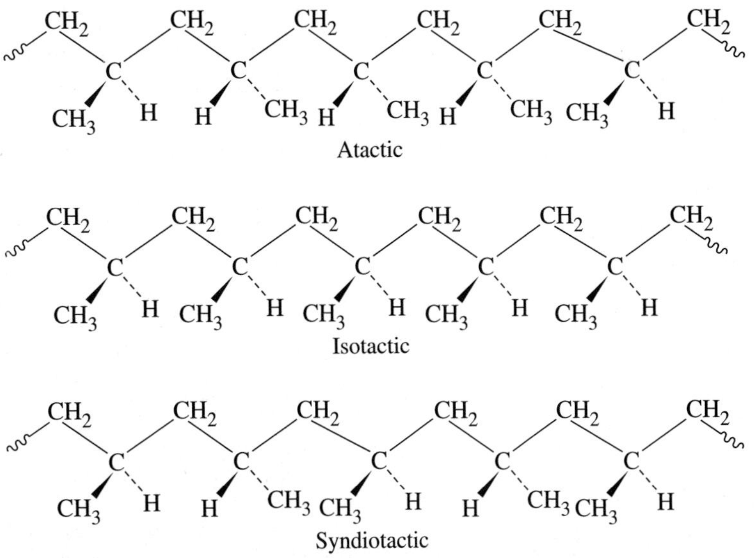

▲ **FIGURE 27-17   Three representations of a polypropylene chain**
In the *atactic* polymer, the $-CH_3$ groups are randomly distributed on the chain. In the *isotactic* polymer, all $-CH_3$ groups are shown coming out of the plane of the page. In the *syndiotactic* polymer, $-CH_3$ groups alternate along the chain—one in front of the plane of the page, the next one behind the plane, and so on.

▲ The synthesis from coal tar chemicals of the purple dye mauveine by William Perkin marked the beginning of the synthetic dye industry.

The production of dyes has been a human activity since the art of weaving was first developed. In the past, dyes were made from animal, vegetable, and mineral materials. One of the most famous historical dyes was *Tyrian purple*, a dye produced from

shells of the marine mollusk *Murex brandaris*. This dye was expensive, and in ancient Rome, it was reserved for the ruling class (hence the term "royal purple").

Tyrian purple

Although it is obtained from a plant rather than a mollusk, *indigo*, the familiar dye providing the blue in blue jeans, has a molecular structure similar to that of Tyrian purple.

Indigo

structure, atactic polymers are amorphous. In an *isotactic* polymer, all $-CH_3$ groups have the same orientation, and in a *syndiotactic* polymer, the $-CH_3$ groups have an orientation that alternates back and forth along the chain. Because of their structural regularity, isotactic and syndiotactic polymers possess crystallinity, which makes them stronger and more resistant to chemical attack than an atactic polymer.

In the 1950s, Karl Ziegler and Giulio Natta developed procedures for controlling the spatial orientation of substituent groups on a polymer chain by using special catalysts [such as $(CH_3CH_2)_3Al + TiCl_4$]. This discovery, recognized through the award of a Nobel Prize in 1963, revolutionized polymer chemistry. Through stereospecific polymerization, it is possible to literally tailor-make large molecules.

Summary

Organic chemistry deals with compounds of carbon. Simplest among these are carbon–hydrogen compounds, hydrocarbons. In hydrocarbons, the C atoms are bonded to one another in straight or branched chains or in rings. Some hydrocarbon molecules contain only single bonds (alkanes), others have some double bonds (alkenes), and still others, triple bonds (alkynes). Yet another class of hydrocarbons—aromatic hydrocarbons—is based on the benzene molecule, C_6H_6.

Functional groups are certain atoms or groupings of atoms that can be introduced into hydrocarbon structures through chemical reactions. With alkanes and aromatic hydrocarbons, these reactions are based on *substitution*: A functional group replaces a

H atom in the hydrocarbon. With alkenes and alkynes, chemical reaction occurs by *addition*: Functional group atoms join to the C atoms at points of unsaturation (double or triple bonds).

Isomerism is frequently found among organic compounds. Some isomers have the same total number of C atoms but different branching of the carbon chain. In other isomers, the position of functional groups on a hydrocarbon chain or ring are different. Still other isomers (such as cis and trans) arise from different orientations of substituent groups in space. Organic compounds also exhibit optical isomerism when four different groups are attached to a carbon atom (chirality). The configuration of the groups around the stereogenic center is described by the *R, S* system of nomenclature.

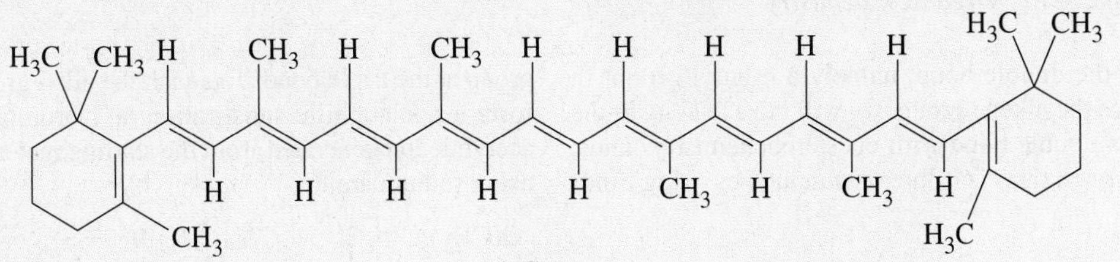

β-Carotene

Tyrian purple and indigo have a feature common to organic dyes: a series of double bonds separated by single bonds, a pattern described as a *conjugated double bond system*. Many organic compounds with conjugated double bonds exhibit color. The orange color of carrots, for example, is due to the presence of β-carotene, whose structure is shown above.

In the last half of the nineteenth century, the dye industry underwent an unexpected and fundamental change. Chemical, or *synthetic*, routes were devised for producing dyes. The first synthetic dye was discovered in 1856 by William Perkin, an 18-year-old student at the Royal College of Chemistry in London. Perkin was studying the reactions of substances obtained from coal tar. When he allowed a mixture of substances containing aniline to react with potassium dichromate, he obtained a tarry product from which he extracted a bright purple, water-soluble substance—the dye *mauveine*. Following this discovery, Perkin went into business as the first manufacturer of synthetic dyes.

By 1900, the BASF corporation had introduced synthetic indigo. This synthetic version of the dye could be produced more cheaply than the natural product, and the result was a dramatic shift from the natural material to a synthetic one.

Plants and animals on Earth have always been an important source of chemicals. A typical pattern has been that a substance with desirable medicinal or other commercial uses has first been discovered in a particular organism. Then chemists have attempted to develop a synthetic route leading to the same substance from cheap raw materials, such as petroleum. To date, only a small fraction of Earth's organisms have been studied for the potential beneficial chemicals they may contain. To permit an expanded search for useful natural products is an important, if not necessarily the most compelling, reason for preserving biological diversity on Earth.

Nucleophilic substitution reactions in haloalkanes occur by two mechanisms, S_N1 and S_N2. In the S_N2 mechanism, the configuration at the C atom is inverted. In the S_N1 mechanism, a racemic mixture is obtained.

Compounds of the general formula ROH are alcohols (phenols if R is a phenyl or substituted phenyl group). One method of preparing alcohols is by *hydration* of an alkene (addition of HOH) or by the *hydrolysis* of an alkyl halide. Ethers (R′OR) result from the elimination of HOH from between two alcohol molecules. Aldehydes (RCHO) and ketones (RCOR′) contain the carbonyl group.

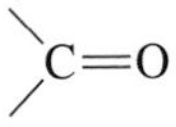

They can be prepared by the controlled oxidation of primary and secondary alcohols, respectively. Carboxylic acids are weak acids having the general formula RCOOH and contain the carboxyl group,

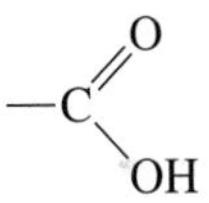

In addition to their typical acid–base reactions, carboxylic acids react with alcohols to form esters. Carboxylic acids can be prepared by the oxidation of a primary alcohol or an aldehyde. Amines are organic derivatives of ammonia, and like ammonia, they have basic properties. Amines and carboxylic acids can be combined to form amides, which do not have basic properties involving the nitrogen lone pair.

The substitution of other atoms (such as N, O, or S) for C atoms in ring structures yields heterocyclic compounds, which are widely encountered among molecules in living organisms (see Chapter 28).

Integrative Example

The female of a species of worm produces the sex attractant *spodoptol*, which has the structure shown.

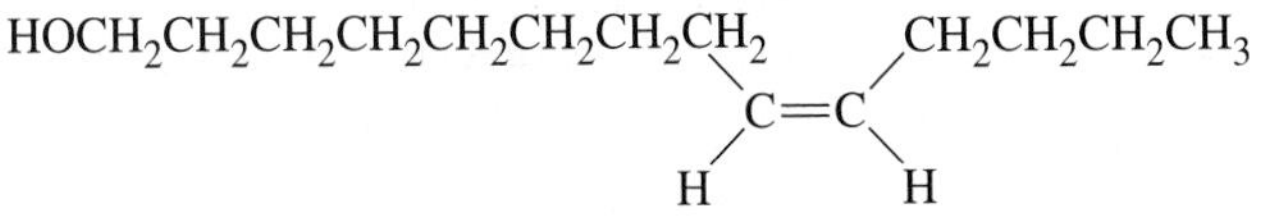

Spodoptol

Spodoptol, a pheromone, attracts the male worms of the species and can be used to control the population of worms by putting synthetic spodoptol in traps. We are asked to devise a synthesis of spodoptol starting from the alcohol shown.

$$HOCH_2CH_2CH_2CH_2CH_2CH_2CH_2CH_2C{\equiv}CH$$

In designing the synthesis, we break the molecule down into the starting material and the necessary alkyl group that has to be attached to the double bond, namely a *n*-butyl group. In order to produce the alkene group, we will have to convert the triple bond to a double bond with cis stereochemistry. Thus, *working backwards*, this cis isomer is obtained by using Lindlar's catalyst.

$$HOCH_2(CH_2)_6CH_2C\equiv C(CH_2)_3CH_3 \xrightarrow[\substack{\text{Lindlar's}\\\text{catalyst}}]{H_2}$$

$$HO(CH_2)_7CH_2 \qquad (CH_2)_3CH_3$$
$$\diagdown C=C \diagup$$
$$\diagup \qquad \diagdown$$
$$H \qquad\qquad H$$

We now need to make the alkyne in order to carry out the above hydrogenation. The starting material provided lacks the butyl group at the triple bond. To attach the alkyl group, we can perform a nucleophilic substitution at 1-bromobutane with the acetylide ion generated from the starting material provided, by using sodium amide.

$$^-O(CH_2)_8C\equiv C{:}^- + CH_3(CH_2)_3Br \longrightarrow$$
$$^-O(CH_2)_8C\equiv C(CH_2)_3CH_3 + Br^-$$

Because the alcohol proton of the acetylide is also acidic, we need to add an *excess* of sodium amide to give the combined alkoxide and acetylide. (Ammonia is also produced in the reaction.)

$$HO(CH_2)_8C\equiv CH \xrightarrow[\text{NaNH}_2]{\text{excess}} {}^-O(CH_2)_8C\equiv C{:}^-$$

We can now combine these steps into a complete synthesis where we have converted the alkoxide to the alcohol in the next to last step by adding ethanol. The ethanol gives up a proton to the alkoxide we have synthesized (the ethoxide ion is the weaker base).

$$HO(CH_2)_8C\equiv CH \xrightarrow[\text{NaNH}_2]{\text{excess}} {}^-O(CH_2)_8C\equiv C{:}^- \xrightarrow{CH_3(CH_2)_3Br}$$

$$^-O(CH_2)_8C\equiv C(CH_2)_3CH_3 \xrightarrow{C_2H_5OH} HO(CH_2)_8C\equiv C(CH_2)_3CH_3$$

$$H_2 \Big\downarrow \substack{\text{Lindlar's}\\\text{catalyst}}$$

$$HO(CH_2)_8 \qquad\qquad (CH_2)_3CH_3$$
$$\diagdown C=C \diagup$$
$$\diagup \qquad \diagdown$$
$$H \qquad\qquad H$$

Key Terms

acetyl (27-7)	chiral (27-10)	nucleophilic substitution reaction (27-11)
achiral (27-10)	condensation reaction (27-7)	optical isomerism (27-10)
addition reaction (27-3)	conformations (27-2)	ortho (*o*) (27-4)
alcohol (27-5)	dashed-wedged line notation (27-1)	para (*p*) (27-4)
aldehyde (27-6)	electrophilic (27-11)	phenol (27-5)
alicyclic (27-2)	elimination reaction (27-3)	phenyl (27-4)
aliphatic (27-2)	ester (27-7)	positional isomer (27-1)
alkane (27-2)	ether (27-5)	*R, S* system (27-10)
alkene (olefin) (27-3)	*E, Z* system (27-10)	saturated hydrocarbon (27-1)
alkyl (27-1)	functional group (27-1)	skeletal isomerism (27-1)
alkyne (27-3)	geometric isomerism (27-3)	S_N1 (27-11)
amide (27-7)	heterocyclic (27-9)	S_N2 (27-11)
amine (27-8)	homologous series (27-2)	step-reaction polymerization (27-12)
aromatic (27-4)	hydrocarbon (27-1)	stereoisomerism (27-3)
asymmetric (27-10)	hydroxyl (27-5)	substitution reaction (27-2, 27-4, 27-11)
carbonyl (27-6)	ketone (27-6)	unsaturated hydrocarbon (27-3)
carboxyl (27-7)	leaving group (27-11)	
carboxylic acid (27-7)	meta (*m*) (27-4)	
chain-reaction polymerization (27-12)	nucleophilic (27-11)	

Review Questions

1. In your own words, define the following terms or symbols: **(a)** *t*; **(b)** R—; **(c)** ⬡ ; **(d)** carbonyl group; **(e)** primary amine.

2. Briefly describe each of the following ideas, phenomena, or methods: **(a)** substitution reaction; **(b)** octane rating of gasoline; **(c)** stereoisomerism; **(d)** ortho, para director; **(e)** step-reaction (condensation) polymerization.

3. Explain the important distinctions between each pair of terms: **(a)** alkane and alkene; **(b)** aliphatic and aromatic compound; **(c)** alcohol and phenol; **(d)** ether and ester; **(e)** amine and ammonia.

4. Describe the characteristics of each of the following types of isomers: **(a)** structural; **(b)** positional; **(c)** cis; **(d)** ortho.

5. Which hydrocarbon has the greater number of isomers, C_4H_8 or C_4H_{10}? Explain your choice.

6. Which of these compounds: C_4H_{10}, $CH_3CH=CHCH_3$, $CH_3C\equiv CCH_3$, or C_6H_6 has the same carbon-to-hydrogen ratio as cyclobutane? Explain your choice.

7. Identify the functional group in each compound (i.e., alcohol, amine, etc.)
   (a) $CH_3CHBrCH_2CH_3$ **(b)** CH_3CH_2COOH
   (c) $C_6H_5CH_2CHO$ **(d)** $(CH_3)_2CHCH_2OCH_3$
   (e) $CH_3COCH_2CH_3$ **(f)** $CH_3CH(NH_2)CH_2CH_3$
   (g) $C_6H_4(OH)_2$ **(h)** CH_3COOCH_3

8. Draw Lewis structures of the following simple organic molecules: **(a)** $CH_3CHClCH_3$; **(b)** $HOCH_2CH_2OH$; **(c)** CH_3CHO.

9. Draw suitable structural formulas to show that there are *four* structural isomers of $C_3H_6Cl_2$.

10. Which of the following pairs of molecules are isomers and which are not? Explain.
    (a) $CH_3CH_2CH_2CH_3$ and $CH_3CH=CHCH_3$
    (b) $CH_3(CH_2)_5CH(CH_3)_2$ and
    $$CH_3(CH_2)_4CH(CH_3)CH_2CH_3$$
    (c) $CH_3CHClCH_2CH_3$ and $CH_3CH_2CH_2CH_2CH_2Cl$

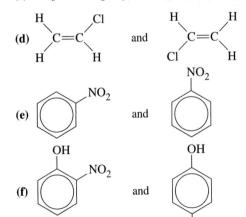

11. Give an acceptable name for each of the following.

    (a) $CH_3CH_2CH_2\overset{\overset{\displaystyle CH_3}{|}}{C}HCH_2\overset{\overset{\displaystyle CH_3}{|}}{C}HCH_2CH_3$

    (b) $CH_3-\overset{\overset{\displaystyle CH_3}{|}}{\underset{\underset{\displaystyle CH_3}{|}}{C}}-CH_3$

    (c) $CH_3CH_2CHCH_2-\overset{\overset{\displaystyle Cl}{|}}{\underset{\underset{\displaystyle Cl}{|}}{C}}-CH_2CH_3$ with C_2H_5

12. Give an acceptable name for each of the following.

    (a) [benzene ring with Cl at position 1 and Cl at position 3]
    (b) [benzene ring with CH_3 and NO_2]
    (c) NH_2—[benzene ring]—$COOH$

13. Draw a structural formula for each of the following compounds.
    (a) 3-isopropyloctane
    (b) 2-chloro-3-methylpentane
    (c) 2-pentene
    (d) dipropyl ether
    (e) *p*-bromophenol

14. Write a condensed formula for each of the following.
    (a) isopropyl alcohol (rubbing alcohol)
    (b) 1,1,1-chlorodifluoroethane (a refrigerant)
    (c) 2-methyl-1,3-butadiene (used to make elastomers)

15. Supply a structural formula for each of the following.
    (a) 1,3,5-trimethylbenzene
    (b) *p*-nitrophenol
    (c) 3-amino-2,5-dichlorobenzoic acid (a plant growth regulator)

16. To prepare methyl ethyl ketone, which of these compounds would you oxidize: 2-propanol, 1-butanol, 2-butanol, or *t*-butyl alcohol? Explain.

17. Indicate the principal product(s) you would expect in
    (a) treating $CH_3CH_2CH=CH_2$ with *dilute* $H_2SO_4(aq)$
    (b) exposing a mixture of chlorine and propane gases to ultraviolet light
    (c) heating a mixture of isopropyl alcohol and benzoic acid
    (d) oxidizing *s*-butyl alcohol with $Cr_2O_7^{2-}$ in acidic solution

18. For each of the following pairs, indicate which substance has
    (a) the higher boiling point, C_6H_{12} or C_6H_6
    (b) the greater solubility in water, C_3H_7OH or $C_7H_{15}OH$
    (c) the greater acidity in aqueous solution, C_6H_5CHO or C_6H_5COOH

19. Identify the following compounds as electrophiles, nucleophiles, or neither:
    (a) NH_3;   **(b)** CH_3Cl;   **(c)** Br^-;   **(d)** CH_3OH;
    (e) $CH_3NH_3^+$.

20. Identify the following compounds as electrophiles, nucleophiles, or neither:
    (a) N_3^-;   **(b)** NH_4^+;   **(c)** $CH_3CHClCH_3$;
    (d) OH^-;   **(e)** $H_2C=O$.

21. Identify the following types of reactions.

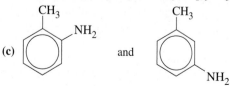

22. Identify the following types of reactions.

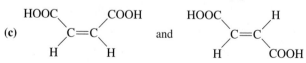

Note the symbol [O] is sometimes used to designate oxidation.

Exercises

Organic Structures

23. Write structural formulas corresponding to these condensed formulas.
    (a) $CH_3CH_2CHBrCHBrCH_3$
    (b) $(CH_3)_3CCH_2C(CH_3)_2CH_2CH_2CH_3$
    (c) $(C_2H_5)_2CHCH=CHCH_2CH_3$

24. Write structural formulas corresponding to these condensed formulas.
    (a) $(CH_3)_3CCH_2CH(CH_3)CH_2CH_2CH_3$
    (b) $(CH_3)_2CHCH_2C(CH_3)_2CH_2Br$
    (c) $Cl_3CCH_2CH(CH_3)CH_2Cl$

25. With appropriate sketches, represent chemical bonding in terms of the overlap of hybridized and unhybridized atomic orbitals in the following molecules.
    (a) C_4H_{10}
    (b) $H_2C=CHCl$
    (c) $CH_3C\equiv CH$

26. With appropriate sketches, represent chemical bonding in terms of the overlap of hybridized and unhybridized atomic orbitals in the following molecules.

    (a) $CH_3\overset{\displaystyle O}{\overset{\|}{C}}CH_3$ **(b)** $CH_3\overset{\displaystyle O}{\overset{\|}{C}}-OH$ **(c)** $H_2C=C=CH_2$

Isomers

27. Indicate the difference in these three types of isomers: structural, positional, and geometric. What term best describes each of the following pairs of isomers?
    (a) $CH_3CH_2CH_2Cl$ and $CH_3CHClCH_3$
    (b) $CH_3CH(CH_3)CH_2CH_3$ and $CH_3(CH_2)_3CH_3$

28. What term—structural, positional, or geometric—best describes each of the following pairs of isomers?
    (a) $CHCl=CHCl$ and $CH_2=CCl_2$
    (b) $CH_2=CHCH_2CH_3$ and $CH_2=C(CH_3)_2$

29. Draw structural formulas for all the isomers of C_7H_{16}.

30. Draw and name all the isomers of **(a)** C_6H_{14}; **(b)** C_4H_8; **(c)** C_4H_6.
    (*Hint:* Do not forget double bonds, rings, and combinations of these.)

31. Identify the chiral carbon atoms, if any, in the following compounds.

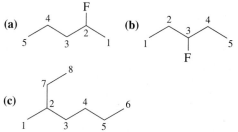

32. Identify the chiral carbon atoms, if any, in the following compounds.

   (a) 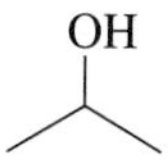 OH structure   (b) 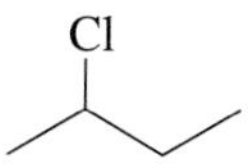 Cl structure

   (c) 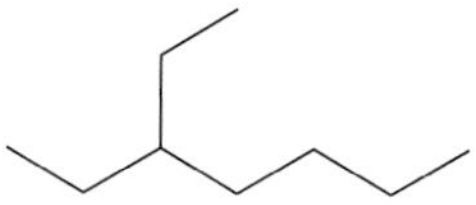 structure

33. Identify the chiral carbon atoms, if any, in the following compounds.

   (a) 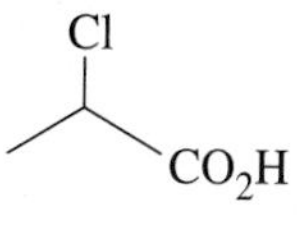 Cl / CO_2H structure   (b) 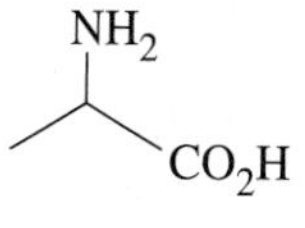 NH_2 / CO_2H structure

   (c) 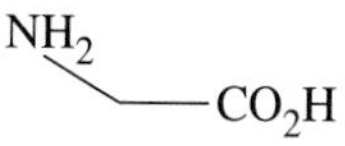 NH_2 / $-CO_2H$ structure

34. Identify the chiral carbon atoms, if any, in the following compounds.

   (a) 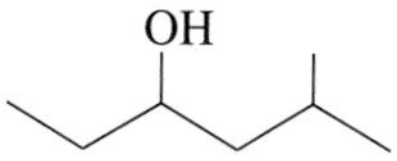 OH structure   (b) 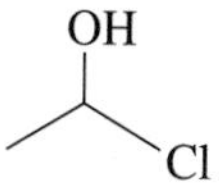 OH / Cl structure

   (c) 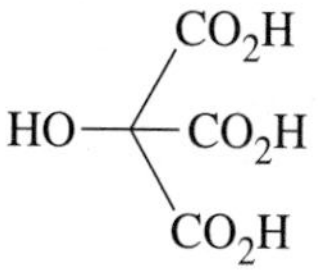 HO—$\overset{CO_2H}{\underset{CO_2H}{C}}$—$CO_2H$

Functional Groups

35. The functional groups in each of the following pairs have certain features in common, but what is the essential difference between them?
    (a) carbonyl and carboxyl
    (b) aldehyde and ketone
    (c) acetic acid and acetyl group
36. Give one example of each of the following types of compounds: (a) aromatic nitro compound; (b) aliphatic amine; (c) chlorophenol; (d) aliphatic diol; (e) unsaturated aliphatic alcohol; (f) alicyclic ketone; (g) halogenated alkane; (h) aromatic dicarboxylic acid.
37. Identify and name the functional groups in each of the following.

    (a) 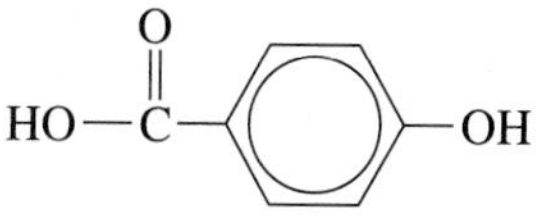 HO—$\overset{O}{\overset{\|}{C}}$—⬡—OH

    (b) CH_3—$\overset{O}{\overset{\|}{C}}$—$OCH_2CH_3$

    (c) CH_3—$\overset{O}{\overset{\|}{C}}$—$CH_2CH_2$—$CO_2H$

    (d) 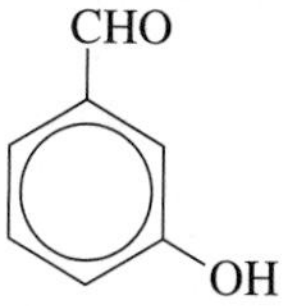 CHO / ⬡ / OH structure

38. Identify and name the functional groups in each of the following.

    (a) CH_3—$\overset{O}{\overset{\|}{C}}$—$CH_2$—$CH_2$—$\overset{O}{\overset{\|}{C}}$—$CH_3$

    (b) CH_3—CH_2—$CONHCH_3$

    (c) 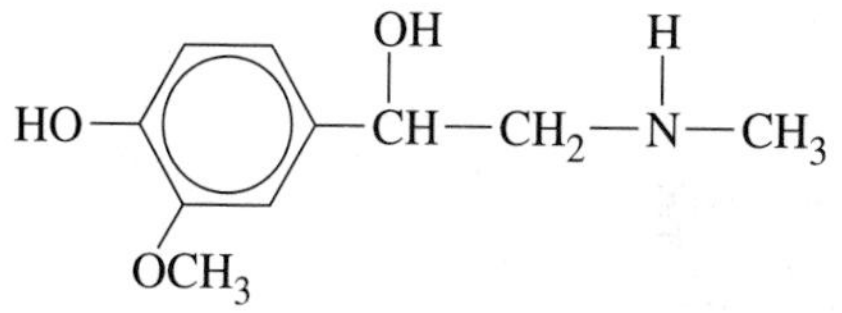 HO—⬡—$\overset{\overset{OH}{|}}{CH}$—$CH_2$—$\overset{\overset{H}{|}}{N}$—$CH_3$ with OCH_3 on ring

    (d) 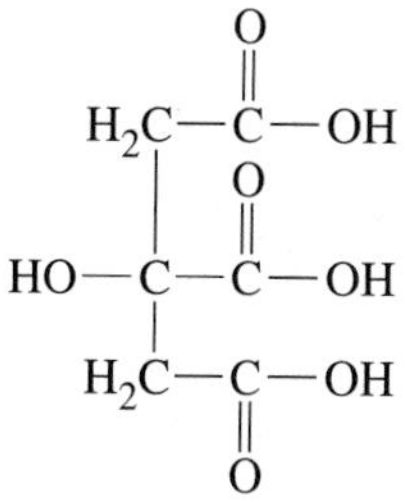
    H_2C—$\overset{O}{\overset{\|}{C}}$—OH
    |
    HO—$\overset{}{C}$—$\overset{O}{\overset{\|}{C}}$—OH
    |
    H_2C—$\overset{O}{\underset{\|}{C}}$—OH

39. Give the isomers of $C_4H_{10}O$ that correspond to an ether.
40. Give the isomers of $C_5H_{12}O$ that correspond to an ether.
41. Give the isomers of the carboxylic acid with molecular formula $C_5H_{10}O_2$.
42. Give the isomers of the carboxylic acid with the molecular formula $C_4H_8O_2$.
43. Give the isomers of the esters that correspond to the molecular formula $C_5H_{10}O_2$.
44. Give the isomers of the esters that correspond to the molecular formula $C_4H_8O_2$.
45. Give the non-cyclic isomers with molecular formula $C_4H_8O_2$ that contain more than one functional group.
46. Give the isomers with molecular formula $C_5H_{10}O_2$ that contain more than one functional group.

Nomenclature and Formulas

47. Give an acceptable name for each of the following structures.
 (a) $CH_3CH_2C(CH_3)_3$
 (b) $(CH_3)_2C{=}CH_2$
 (c) 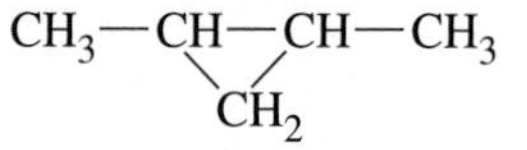 $CH_3{-}CH{-}CH{-}CH_3$
                $\backslash$ $/$
                 CH_2
 (d) $CH_3C{\equiv}CCH(CH_3)_2$
 (e) $CH_3CH(C_2H_5)CH(CH_3)CH_2CH_3$
 (f) $CH_3CH(CH_3)CH(CH_3)C(C_3H_7){=}CH_2$

48. Draw a condensed structure to correspond to each of the following names.
 (a) isopentane
 (b) cyclohexene
 (c) 2-methyl-3-hexyne
 (d) 2-butanol
 (e) isopropyl ethyl ether
 (f) propionaldehyde

49. Does each of the following names convey sufficient information to suggest a specific structure? Explain.
 (a) pentene
 (b) butanone
 (c) butyl alcohol
 (d) methylaniline
 (e) methylcyclopentane
 (f) dibromobenzene

50. Indicate why each of these names is incorrect, and give a correct name.
 (a) 3-pentene
 (b) pentadiene
 (c) 1-propanone
 (d) bromopropane
 (e) 2,6-dichlorobenzene
 (f) 2-methyl-3-pentyne

51. Supply condensed structural formulas for the following substances.
 (a) 2,4,6-trinitrotoluene (TNT—an explosive)
 (b) methyl salicylate (oil of wintergreen)
 [*Hint:* Salicylic acid is *o*-hydroxybenzoic acid.]

 (c) 2-hydroxy-1,2,3-propanetricarboxylic acid (citric acid, $C_6H_8O_7$)

52. Supply condensed structural formulas for the following substances.
 (a) *o-t*-butylphenol (an antioxidant in aviation gasoline)
 (b) 1-phenyl-2-aminopropane (benzedrine—an amphetamine, ingredient in "pep pills")
 (d) 2-methylheptadecane (a sex pheromone of tiger moths—a chemical used for communication among members of the species)
 [*Hint:* Heptadeca means 17.]

53. Name the following amines
 (a)

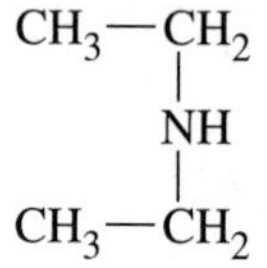

 (b)

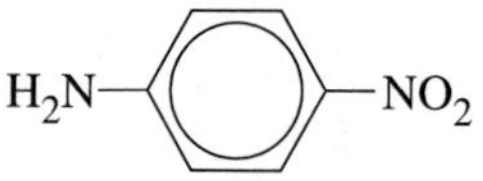

 (c)

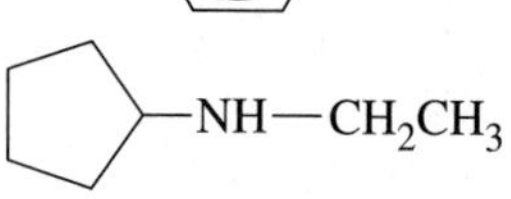

 (d) 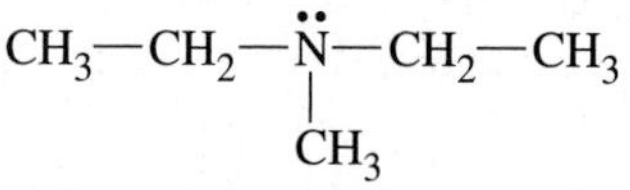 $CH_3{-}CH_2{-}\overset{\cdot\cdot}{N}{-}CH_2{-}CH_3$
                    $|$
                   CH_3

54. Name the following amines.
 (a) $CH_3{-}CH_2{-}NH_2$
 (b)

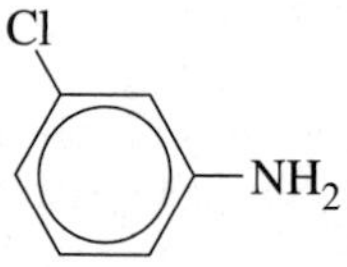

 (c)

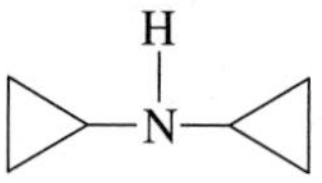

 (d) $Cl{-}CH_2{-}CH_2{-}NH_2$

Alkanes

55. Write the structure of each alkane.
 (a) molecular mass 72 u, forms four monochlorination products
 (b) molecular mass 72 u, forms a single monochlorination product

56. Write the structure of each alkane.
 (a) molecular mass = 44 u; forms two different monochlorination products
 (b) molecular mass = 58 u; forms two different monobromination products

Alkenes

57. Why is it not necessary to refer to ethene and propene as 1-ethene and 1-propene? Can the same be said for butene?

58. Alkenes (olefins) and cyclic alkanes (alicyclics) each have the generic formula C_nH_{2n}. In what important ways do these types of compounds differ structurally?

59. Draw the structures of the products of each of the following reactions.
 (a) propene + hydrogen (Pt, heat)
 (b) 2-butanol + heat (in the presence of sulfuric acid)

60. Use Markovnikov's rule to predict the product of the reaction of
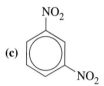
 (a) HCl with $CH_3C{=}CH_2$
 (b) HCN with $CH_3C{\equiv}CH$
 (c) HCl with $CH_3CH{=}C(CH_3)_2$

Aromatic Compounds

61. Supply a name or structural formula for each of the following.
 (a) phenylacetylene
 (b) *m*-dichlorobenzene
 (c)
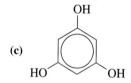

62. Supply a name or structural formula for each of the following.
 (a) *p*-phenylphenol
 (b) 2-hydroxy-4-isopropyltoluene (thymol—flavor constituent of the herb thyme)
 (c)
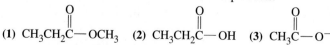

Organic Reactions

63. Describe what is meant by each of the following reaction types, and illustrate with an example from the text: **(a)** aliphatic substitution reaction; **(b)** aromatic substitution reaction; **(c)** addition reaction; **(d)** elimination reaction.

64. Starting with acetylene as the only source of carbon, together with any inorganic reagents required, devise a method to synthesize **(a)** 1,1,2,2-tetrabromoethane; **(b)** acetaldehyde.

65. Draw a structure to represent the principal product of each of the following reactions.
 (a) 1-pentanol + excess dichromate ion (acid solution)
 (b) butyric acid + ethanol (acid solution)
 (c) $CH_3CH_2C{=}CH_2 + H_2O$ (in the presence of H_2SO_4)
 with CH_3 substituent

66. Predict the products of the monobromination of **(a)** *m*-dinitrobenzene; **(b)** aniline; **(c)** *p*-bromoanisole.

67. Write the formulas of the products formed from the reaction of propene with each of the following substances
 (a) H_2
 (b) Cl_2
 (c) HCN
 (d) HCl
 (e) H_2O (in acid)

68. Write the formulas of the products formed from the reaction of 2-butene with each of the following substances.
 (a) H_2
 (b) Cl_2
 (c) HCN
 (d) HCl
 (e) H_2O (in acid)

69. Write the formulas of the products formed in each of the following reactions. If no reaction occurs, write N.R.
 (a) $CH_3CH_2CO_2^-(aq) + HCl(aq) \longrightarrow$
 (b) $CH_3CH_2CO_2CH_3 + H_2O \xrightarrow{H^+}$
 (c) $CH_3CH_2CO_2CH_2CH_3 + NaOH(aq) \longrightarrow$

70. Write the formulas of the products formed in each of the following reactions. If no reaction occurs, write N.R.
 (a) $CH_3CH_2CO_2CH_3 \xrightarrow{LiAlH_4}$
 (b) $CH_3CH_2CO_2CH_3 \xrightarrow{NaBH_4}$
 (c) $HCO_2CH_2CH_3(aq) \xrightarrow{H^+}$

71. Write the formulas of the products formed in each of the following reactions. If no reaction occurs, write N.R.
 (a) $CH_3CO_2H + NH_3 \xrightarrow{heat}$
 (b) $CH_3CH_2CONH_2 \xrightarrow{heat}$
 (c) $CH_3CH_2CONH_2(aq) + NaOH(aq) \longrightarrow$

72. Write the formulas of the products formed in each of the following reactions. If no reaction occurs, write N.R.
 (a) $CH_3CO_2H + CH_3NH_2 \xrightarrow{heat}$
 (b) $CH_3CH_2CONHCH_3 \xrightarrow[H_2O]{LiAlH_4}$
 (c) $CH_3CH_2CONHCH_3 + HCl(aq) \longrightarrow$

73. Which of the following species gives the reaction indicated? Write the structures of the reaction products.
 (1) $CH_3CH_2\overset{O}{\overset{\|}{C}}{-}OCH_3$ **(2)** $CH_3CH_2\overset{O}{\overset{\|}{C}}{-}OH$ **(3)** $CH_3\overset{O}{\overset{\|}{C}}{-}O^-$
 (a) neutralizes dilute HCl
 (b) hydrolyzes
 (c) neutralizes dilute NaOH

74. Which of the following species gives the reaction indicated? Write the structures of the reaction products.

(1) $CH_3CH_2\overset{\overset{O}{\|}}{C}-NH_2$

(2) $CH_3CH_2NH_2$

(3) $CH_3CH_2CH_2NH_3{}^+Cl^-$

(a) neutralizes dilute HCl
(b) hydrolyzes
(c) neutralizes dilute NaOH

75. Write the equilibrium constant expression for the equilibrium present when butanamine is dissolved in water.

76. Write the equilibrium constant expression for the equilibrium present when *N*-methylbutanamine (methylbutyl amine) is dissolved in water.

77. Write the formulas of the products expected to form in the following situations. If no reaction occurs, write N.R.
(a) $CH_3CH_2NH_2(aq) + HCl(aq) \longrightarrow$
(b) $(CH_3)_3N(aq) + HBr(aq) \longrightarrow$
(c) $CH_3CH_2NH_3{}^+(aq) + H_3O^+(aq) \longrightarrow$
(d) $CH_3CH_2NH_3{}^+(aq) + OH^-(aq) \longrightarrow$

78. Write the formulas of the products expected to form in the following situations. If no reaction occurs, write N.R.

(a) 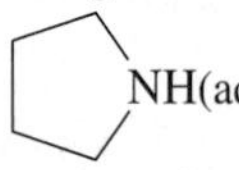 $NH(aq) + HCl(aq) \longrightarrow$

(b) $(CH_3)_4N^+(aq) + HCl(aq) \longrightarrow$
(c) $CH_3CH_2NH_2(aq) + OH^-(aq) \longrightarrow$
(d) $(CH_3)_3NH^+ + OH^- \longrightarrow$

79. Match the following compounds with the chemical properties given below. Write the structure of the products of the reactions described in (a)–(e).

(1)  OH (2)

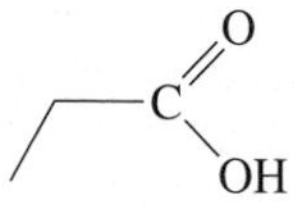

(3)

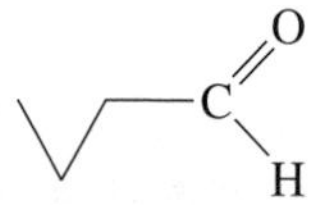

(a) is easily oxidized
(b) neutralizes NaOH(aq)
(c) forms an ester with ethanol
(d) can be oxidized to a carboxylic acid
(e) can be dehydrated to an alcohol.

80. Match the following compounds with the chemical properties given below. Write the structure of the products of the reactions described in (a)–(e).

(1)  NH_2 (2) Cl (3)

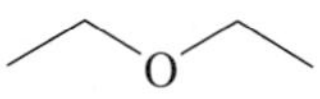

(a) neutralizes HCl(aq)
(b) neutralizes NaOH(aq)
(c) forms an amide with ethanoic acid
(d) reacts with ammonia
(e) reacts with CN^-(aq).

Organic stereochemistry

81. For each pair of structures shown below, indicate whether the two species are identical molecules, enantiomers, or isomers of some other sort.

(a) 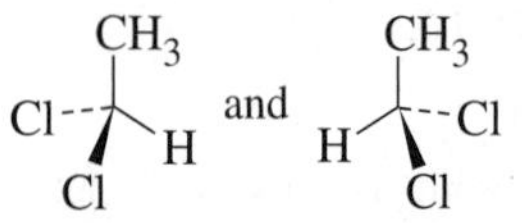
(b) 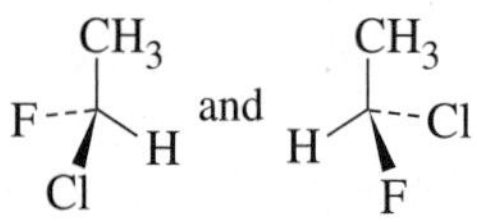
(c) 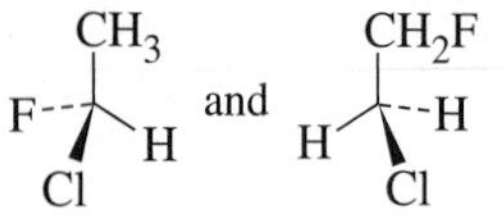
(d) 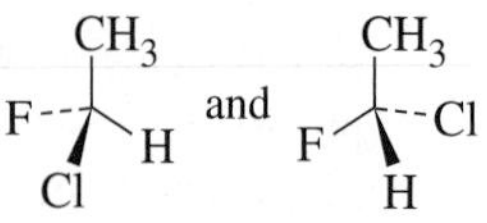
(e) 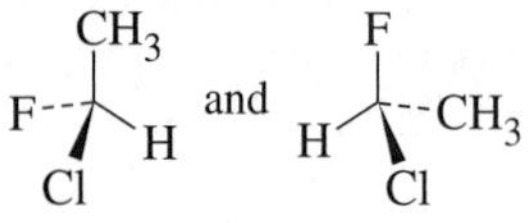
(f)

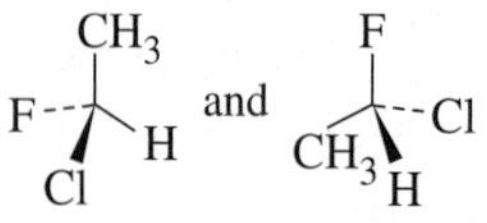

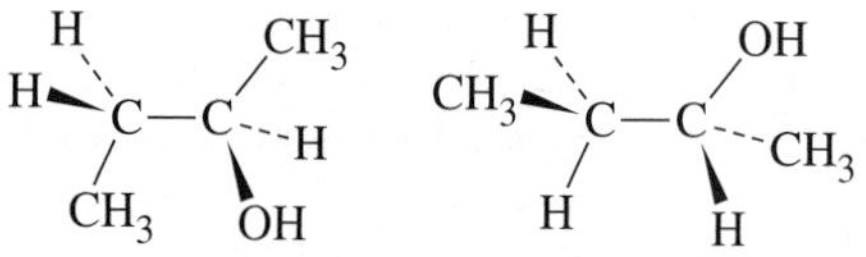

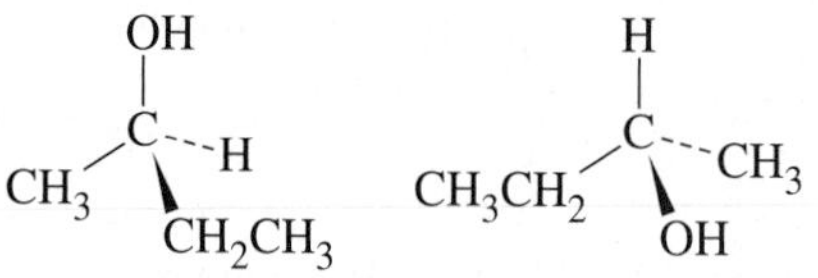

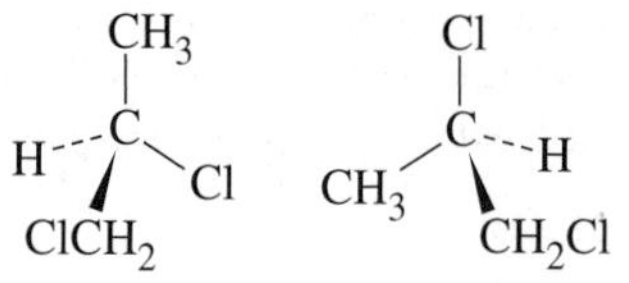

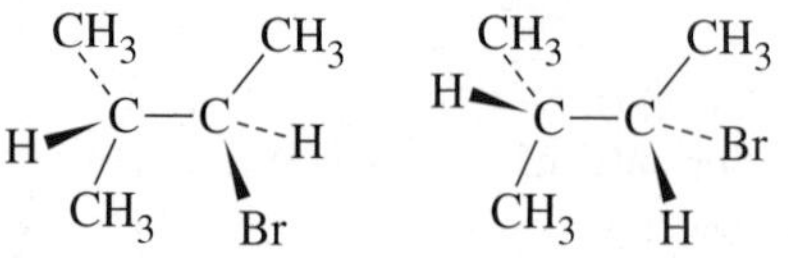

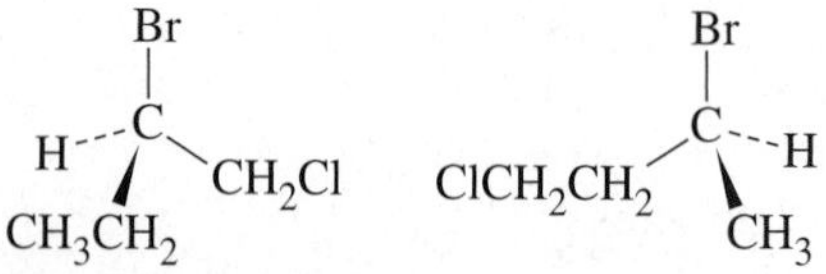

(b)
(c)
(d)
(e)
(f)

82. For each pair of structures shown below, indicate whether the two species are identical molecules, enantiomers, or isomers of some other sort.

(a)

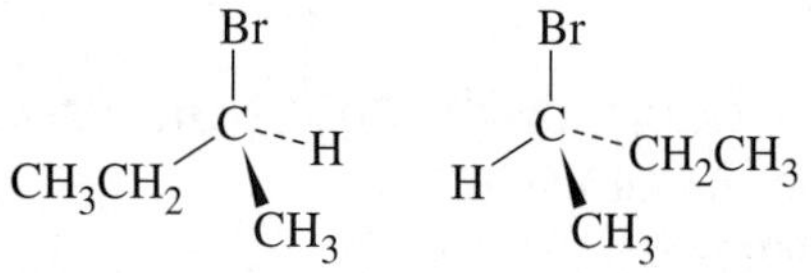

83. Name the following molecules with the appropriate stereochemical designation.

(a)

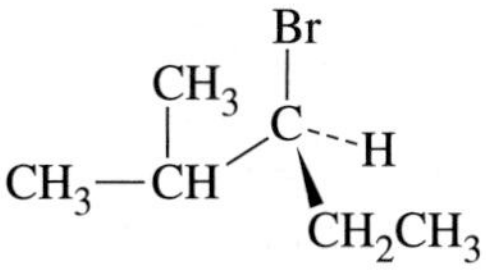

(b)

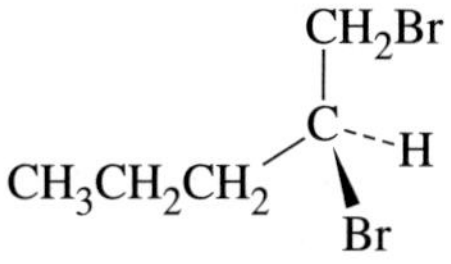

(c)

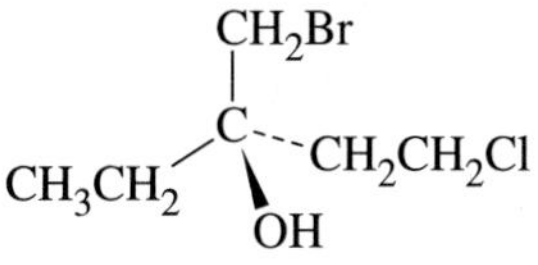

(d)

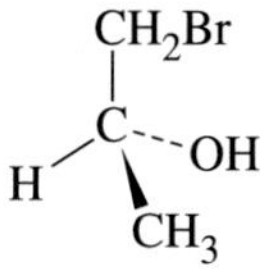

84. Name the following molecules with the appropriate stereochemical designation.

(a)

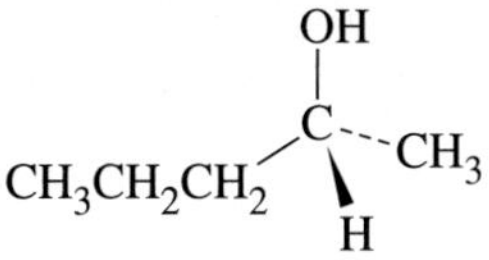

(b)

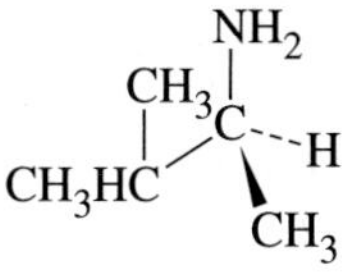

(c)

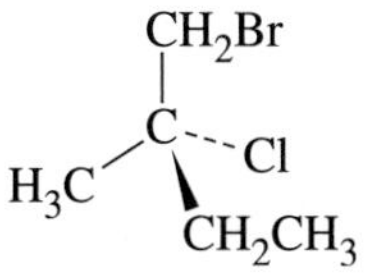

(d)

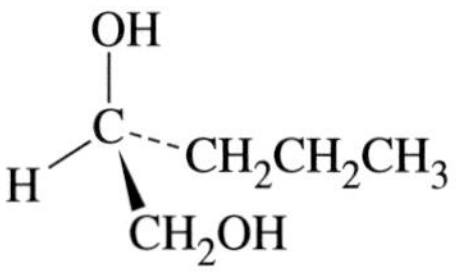

85. Name the following molecules with the appropriate stereochemical designation.

(a)

(b)

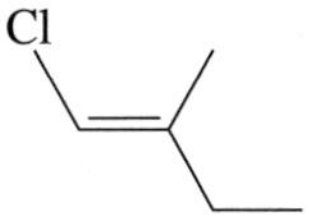

(c)

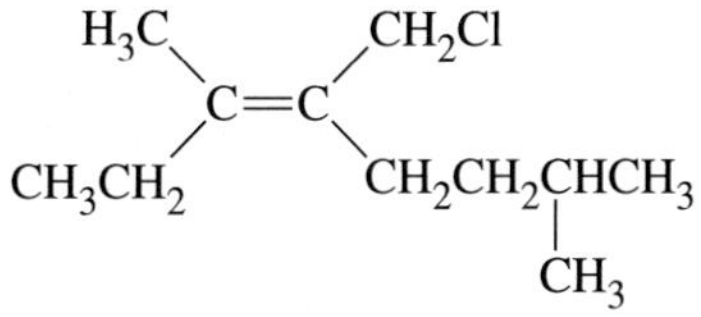

(d)

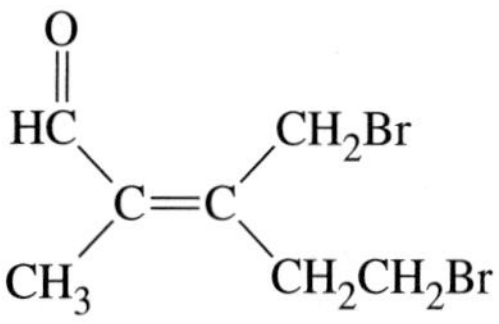

86. Name the following molecules with the appropriate stereochemical designation.

(a)

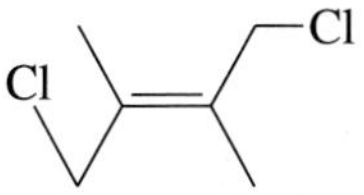

(b)

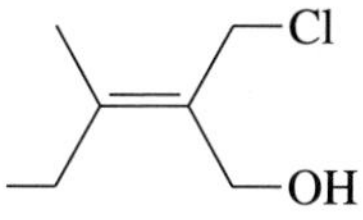

(c)

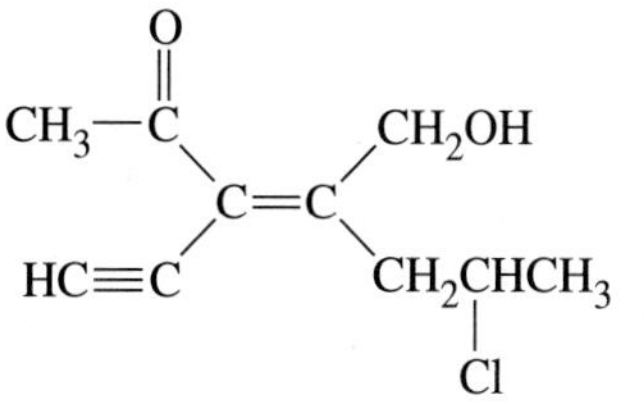

(d)

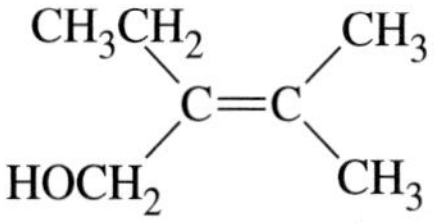

87. Draw the structure for each of the following.
   (a) (Z)-1,3,5-tribromo-2-pentene
   (b) (E)-1,2-dibromo-3-methyl-2-hexene
   (c) (S)-1-bromo-1-chlorobutane
   (d) (R)-1,3-dibromohexane
   (e) (S)-1-chloro-2-propanol
88. Draw the structure for each of the following.
   (a) (R)-1-bromo-1-chloroethane
   (b) (E)-2-bromo-2-pentene
   (c) (Z)-1-chloro-3-ethyl-3-heptene
   (d) (R)-2-hydroxypropanoic acid
   (e) (S)-2-aminopropanoate anion

Nucleophilic Substitution Reactions

89. Answer the following questions for this S_N2 reaction.
$$CH_3CH_2CH_2CH_2Br + NaOH \longrightarrow$$
$$CH_3CH_2CH_2CH_2OH + NaBr$$
   (a) What is the rate expression for the reaction?
   (b) Draw the reaction profile for the reaction. Label all parts. Assume that the products are lower in energy than the reactants.
   (c) What is the effect on the rate of the reaction of doubling the concentration of *n*-butyl bromide?
   (d) What is the effect on the rate of the reaction of halving the concentration of sodium hydroxide?

90. Answer the following questions for this S_N1 reaction.

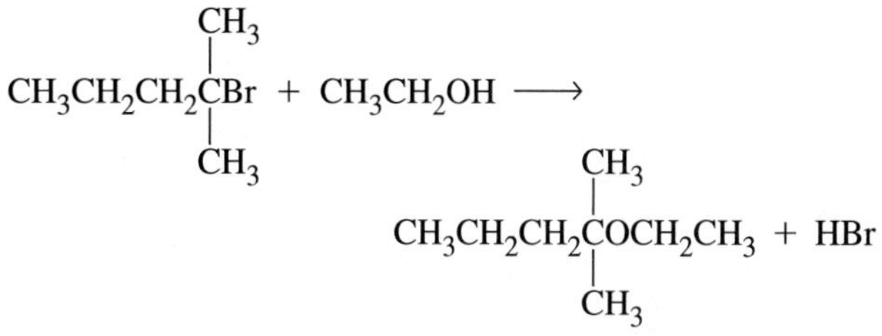

   (a) What is the rate expression for the reaction?

(b) Draw the reaction profile for the reaction. Label all parts. Assume that the products are lower in energy than the reactants.

(c) What is the effect on the rate of the reaction of doubling the concentration of 1-bromo-1-methylpentane?

(d) The solvent for the reaction is ethanol. What is the effect on the rate of the reaction of adding more ethanol?

91. Write equations for the substitution reaction of *n*-bromobutane, a typical primary haloalkane, with the following reagents.
  (a) NaOH;   **(b)** NH_3;
  (c) NaCN;   **(d)** CH_3CH_2ONa.

92. Write equations for the substitution reaction of *n*-bromopentane, a typical primary haloalkane with the following reagents.
  (a) NaN_3;   **(b)** $N(CH_3)_3$;
  (c) $CH_3CH_2 C{\equiv}CNa$;   **(d)** CH_3CH_2SNa.

93. An optically pure sample of $CH_3CH_2CH(CH_3)Cl$ is hydrolyzed by water, and the resulting solution is optically inactive. **(a)** Write the formula of the product. **(b)** By which nucleophilic substitution reaction mechanism does this reaction occur?

94. An optically pure sample of $CH_3CH_2CH(CH_3)Cl$ reacts with CH_3O^- in dimethyl sulfoxide [$(CH_3)_2SO$, a convenient solvent for organic reactions], and the resulting solution is optically active. **(a)** Write the formula of the product. **(b)** By which nucleophilic substitution reaction mechanism does this reaction occur?

95. An optically pure sample of $CH_3CH(Cl)CH_2CH_3$ is dissolved in ethanol, and the resulting solution is optically inactive. **(a)** Write the formula of the product. **(b)** By which nucleophilic substitution reaction mechanism does this reaction occur?

96. An optically pure sample of $CH_3CH(Cl)CH_2CH_3$ reacts with CH_3S^- in dimethyl sulfoxide, and the resulting solution is optically active. **(a)** Write the formula of the product. **(b)** By which nucleophilic substitution reaction mechanism does this reaction occur?

Polymerization Reactions

97. In referring to the molecular mass of a polymer, we can speak only of the average molecular mass. Explain why the molecular mass of a polymer is not a unique quantity, as it is for a substance like benzene.

98. Explain why Dacron is called a polyester. What is the percent oxygen, by mass, in Dacron?

99. Nylon 66 is produced by the reaction of 1,6-hexanediamine with adipic acid. A different nylon polymer is obtained if sebacyl chloride is substituted for the adipic acid. What is the basic repeating unit of this nylon structure?

$$\underset{\displaystyle ClC(CH_2)_8CCl}{\overset{\displaystyle O \qquad\quad O}{\overset{\displaystyle \|\qquad\quad \|}{}}}$$

100. Would you expect a polymer to be formed by the reaction of terephthalic acid with ethyl alcohol in place of ethylene glycol? With glycerol in place of ethylene glycol? Explain.

Integrative and Advanced Exercises

101. In the chlorination of CH_4, some CH_3CH_2Cl is obtained as a product. Explain why this should be so.

102. Supply condensed or structural formulas for the following substances.
  (a) 1,5-cyclooctadiene (an intermediate in the manufacture of resins)
  (b) 3,7,11-trimethyl-2,6,10-dodecatriene-1-ol (farnesol—odor of lily of the valley)
  [*Hint:* Dodeca means 12.]
  (c) 2,6-dimethyl-5-hepten-1-al (used in the manufacture of perfume)

103. Draw structural formulas for all the isomers listed in Table 27.3, and show that, indeed, the substances with more compact structures have lower boiling points.

104. By drawing suitable structural formulas, establish that there are 17 isomers of $C_6H_{13}Cl$.
  [*Hint:* Refer to Example 27-1].

105. Write the structures of the isomers you would expect to obtain in the mononitration of *m*-methoxybenzaldehyde.

What does the fact that no 3-methoxy-5-nitrobenzaldehyde forms imply about the strength of meta and ortho, para directors?

106. The symbol

which is often used to represent benzene, is also the structural formula of cyclohexatriene. Are benzene and cyclohexatriene the same substance? Explain.

107. Use the half-reaction method to balance the following redox equations.

(a) $C_6H_5NO_2 + Fe + H^+ \longrightarrow$
$$C_6H_5NH_3^+ + Fe^{3+} + H_2O$$
(b) $C_6H_5CH_2OH + Cr_2O_7^{2-} + H^+ \longrightarrow$
$$C_6H_5CO_2H + Cr^{3+} + H_2O$$
(c) $CH_3CH{=}CH_2 + MnO_4^- + H_2O \longrightarrow$
$$CH_3CHOHCH_2OH + MnO_2 + OH^-$$

108. A 10.6-g sample of benzaldehyde was allowed to react with 5.9 g $KMnO_4$ in an excess of KOH(aq). After filtration of the $MnO_2(s)$ and acidification of the solution, 6.1 g of benzoic acid was isolated. What was the percent yield of the reaction?
(*Hint:* Write half-equations for the oxidation and reduction half-reactions.)

109. Combustion of a 0.1908-g sample of a compound gave 0.2895 g CO_2 and 0.1192 g H_2O. Combustion of a second sample weighing 0.1825 g yielded 40.2 mL of $N_2(g)$, collected over 50% KOH(aq) (vapor pressure = 9 mmHg) at 25 °C and 735 mmHg barometric pressure. When 1.082 g of compound was dissolved in 26.00 g benzene (mp 5.50 °C, K_f = 5.12 °C m^{-1}), the solution had a freezing point of 3.66 °C. What is the molecular formula of this compound?

110. Explain why a polymer formed by chain-reaction polymerization generally has a higher molecular mass than a corresponding polymer formed by step-reaction polymerization.

111. The three isomeric tribromobenzenes, I, II, and III, when nitrated, form three, two, and one mononitrotribromobenzenes, respectively. Assign correct structures to I, II, and III.

112. Write the name and structure of each aromatic hydrocarbon.
(a) Formula: C_8H_{10}; forms three ring monochlorination products
(b) Formula: C_9H_{12}; forms one ring mononitration product
(c) Formula: C_9H_{12}; forms four ring mononitration products

113. In the molecule 2-methylbutane, the organic chemist distinguishes the different types of H and C atoms as being primary (1°), secondary (2°), and tertiary (3°). For the monochlorination of hydrocarbons, the following ratio of reactivities has been found: 3°/2°/1° 4.3 : 3 : 1. How many different monochloro derivatives of 2-methylbutane are possible, and what percent of each would you expect to find?

114. A particular colorless organic liquid is known to be one of the following compounds: 1-butanol, diethyl ether, methyl propyl ether, butyraldehyde, or propionic acid. Can you identify which it is, based on the following tests? If not, what additional test would you perform? (1) A 2.50-g sample dissolved in 100.0 g water has a freezing point of −0.7 °C. (2) An aqueous solution of the liquid does not change the color of blue litmus paper. (3) When alkaline $KMnO_4(aq)$ is added to the liquid and the mixture is heated, the purple color of the MnO_4^- disappears.

115. The azide anion N_3^- is a nucleophile and, when attached to a carbon atom, undergoes reduction to the amino group and free nitrogen. Suggest a method of preparation of the primary amine propanamine.

116. The cyanide anion is a nucleophile and, when attached to a carbon atom, is reduced to a primary amine. Suggest a method of preparing propanamine from chloroethane.

117. The cyanide anion is a nucleophile and, when attached to a carbon atom, undergoes hydrolysis under basic conditions to the carboxylate anion. Suggest a method of preparing sodium butanoate from chloropropane. How can sodium butanoate be converted into butanoic acid?

118. The iodide ion cannot displace the —OH group in ethanol, but excess HI will react to produce ethyl iodide. Suggest a reason why.

119. Starting with the compounds, chloromethane, chloroethane, sodium azide, sodium cyanide, and a reducing agent, suggest how the following compounds could be synthesized.
(a) *N*-methylpropanamide
(b) ethylethanoate
(c) methylethylamine
(d) tetramethylammonium chloride

120. Write structural formulas for the following.
(a) 2,4-dimethyl-1,4-pentadiene
(b) 2,3-dimethylpentane
(c) 1,2,4-tribromobenzene
(d) methylethanoate
(e) 2-butanone

121. Give the systematic names, including any stereochomical designations, for each of the following.

(a)

(b) $CH_3{-}CH_2{-}\underset{\underset{Cl}{|}}{CH}{-}\underset{\underset{NH_2}{|}}{CH}{-}CO_2H$

(c) $CH_3OCH_2CH_3$

(d) $\underset{\underset{Cl}{|}}{CH_2}{-}CH_2{-}CH_2{-}NH_2$

(e)

122. Write structural formulas for all the isomers of C_4H_7Cl. Indicate any geometrical or optical isomers that occur.

123. Compound A is an alcohol of formula $C_5H_{12}O$ that can be resolved into optical isomers.
 (a) Draw *three* possible structures of A.
 (b) Treatment of A with CrO_3/pyridine gives compound B, which also exhibits optical activity. What are the structural formulas of A and of B? Name and draw the enantiomers of A and B.

124. Levomethadyl acetate (shown below) is used in the treatment of narcotic addiction.

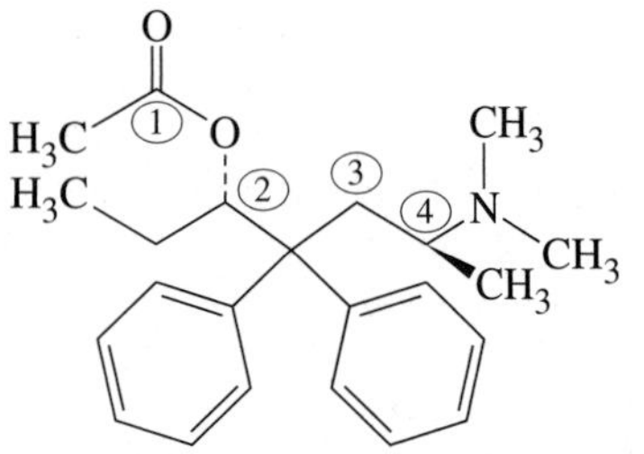

 (a) Name the functional groups in levomethadyl acetate.
 (b) What is the hybridization of the numbered carbon atoms and the nitrogen atom? (c) Which, if any, of the numbered carbon atoms are chiral?

125. Thiamphenicol (shown below) is an antibacterial agent.

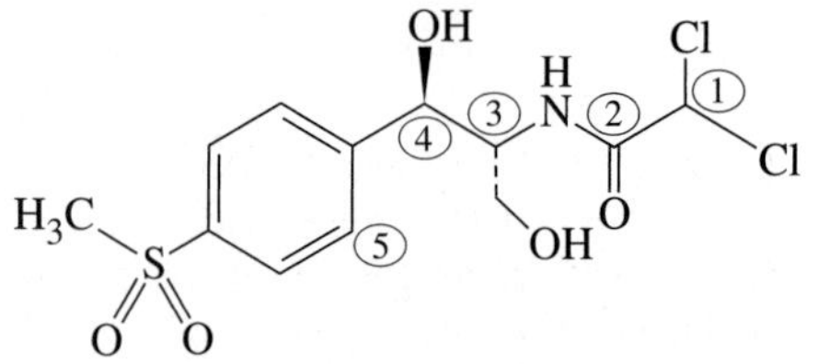

 (a) Name the functional groups of thiamphenicol.
 (b) What is the hybridization of the numbered carbon atoms and the nitrogen atom? (c) Which, if any, of the numbered carbon atoms are chiral?

126. Ephedrine (shown below) is used as a decongestant in cold remedies.

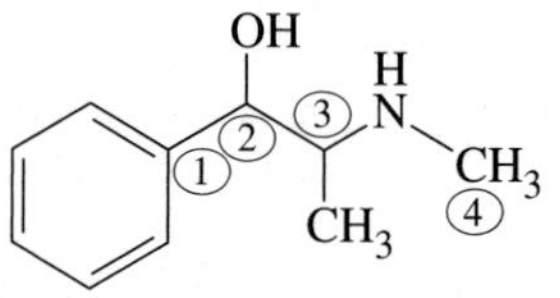

 (a) Name the functional groups of ephedrine. (b) What is the hybridization of the numbered carbon atoms and the nitrogen atom? (c) Which, if any, of the numbered carbon atoms are chiral? (d) The pH of a solution of 1 g of ephedrine in 200 g of water is 10.8. What is the pK_b of ephedrine?

127. How would you synthesize (E)- and (Z)-3-heptene from acetylene and any other chemicals?

128. How would you synthesize (R)-2-butanamine from (S)-2-butanol?

Feature Problem

129. The reduction of aldehydes and ketones with a suitable hydride-containing reducing agent is a good way of synthesizing alcohols. This approach would be even more effective if, instead of a hydride, we could use a source of nucleophilic carbon. Attack by a carbon atom on a carbonyl group would give an alcohol and simultaneously form a carbon to carbon bond. How can we make a C atom in an alkane nucleophilic? This was achieved by Victor Grignard, who created the organometallic reagent

R—MgBr, with the following reaction in diethyl ether.

$$R—Br + Mg \longrightarrow R—MgBr$$

The Grignard reagent (R—MgBr) is rarely isolated. It is formed in solution and used immediately in the desired reaction. The alkylmetal bond is highly polar, with the partial negative charge on the C atom, which makes the C atom highly nucleophilic. The Grignard reagent can attack a carbonyl group in an aldehyde or ketone as follows.

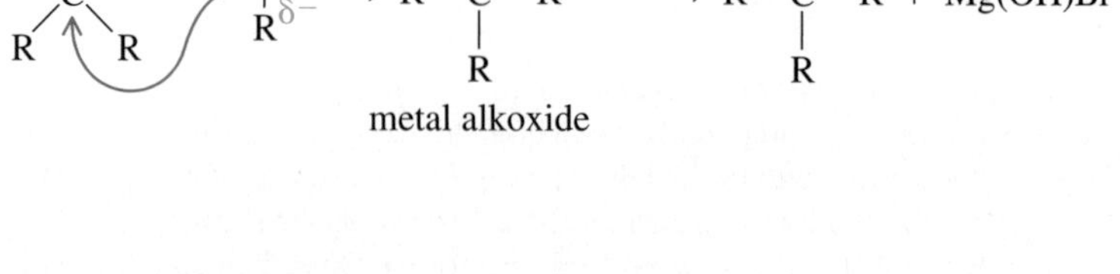

metal alkoxide

Addition of dilute aqueous acid solution to the metal alkoxide furnishes the alcohol. The important synthetic consequence of this procedure is that we have prepared a product with more carbon atoms than present in the starting material. A simple starting material can be transformed into a more complex molecule.

(a) What is the product of the reaction between methanal and the Grignard reagent formed from 1-bromobutane after the addition of dilute acid?

(b) Using a Grignard reagent, devise a synthesis for 2-hexanol.

(c) Using a Grignard reagent, devise a synthesis for 2-methyl-2-hexanol.

(d) Grignard reagents can also be formed with aryl halides such as chlorobenzene. What would be the product of the reaction between the Grignard reagent of chlorobenzene and propanone? Can you think of an alternative synthesis of this product, again using a Grignard reagent?

(e) The basicity of the C atom bound to the magnesium in the Grignard reagent can be used to make Grignard reagents of terminal alkynes. Write the equation of the reaction between ethylmagnesium bromide and 1-hexyne.
[*Hint:* Ethane is evolved.]

(f) Using a Grignard reagent, suggest a synthesis for 2-heptyn-1-ol.

eMedia Exercises

130. View the organic molecules models displayed in *eChapter 27-1*. By right-clicking (Windows) or click-holding (Macintosh) on each molecule, adjust the Display representation (wireframe, sticks, ball-and-stick, space-filling) of the different molecules. What information is emphasized by each of the different types of displays?

131. Compare the physical properties of the entire series of alkanes and alcohols found in the **Boling Points of Organic Molecules** activity *(eChapter 27-12)*. **(a)** How do the trends in boiling point vary between the two series of molecules? **(b)** What physical property is most closely correlated with the boiling points of alkanes? **(c)** What other physical property is also observed to influence the boiling point of alcohols?

132. The nature of addition reactions with alkenes allows one method of detecting their presence, shown in the **Testing for Unsaturated Hydrocarbons with Bromine** movie *(eChapter 27-3)*. Predict the products of the illustrated reaction and explain the observed color difference between the two tests shown.

133. The **Synthesis of Nylon 610** movie *(eChapter 27-10)* illustrates a polymerization reaction. **(a)** Draw an extended structure of the polymer structure formed. **(b)** What properties of the polymer allow it to be removed continuously from the interface of the two reactants? **(c)** Suggest a chemical method of altering the physical properties of polymers formed through condensation reactions.

28 Chemistry of the Living State

Contents

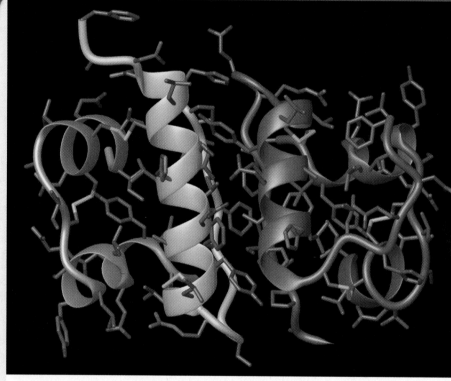

Human insulin protein shown as a ribbon molecular model. A wire frame model of the protein is superimposed on the ribbon model. The ribbon model clearly shows the α-helix portions of the protein.

The planet Earth teems with life—from the tiny amoeba to the enormous whale. In spite of extraordinary differences in outward appearance, all living organisms share similar needs and chemical structures. Raw materials (food) are required for building cells and providing the energy of metabolism and the information of heredity. All forms of life on Earth use the same types of complicated structures to perform specialized functions. In this chapter, we will stress the structures of the macromolecular substances common to all organisms. Our discussion will emphasize how fundamental principles of chemistry acquired throughout the text can contribute to a knowledge of the living state.

28-1 Chemical Structure of Living Matter: An Overview

From one-celled organisms to humans, the living state includes some highly complex forms of matter. Of the known elements, about 50 occur in measurable concentrations in living matter. Of these, about 25 have functions that are definitely known. Four elements together—oxygen, carbon, hydrogen, and nitrogen—account for 96% of human body mass. Figure 28-1 identifies the elements found in living matter. As shown in Figure 28-2, the complex structures of living matter are synthesized in several stages from a few simple environmental precursors—N_2, H_2O, and CO_2.

In addition to water, which is the most abundant compound in the majority of living organisms, other important constituents are lipids, carbohydrates, proteins, and nucleic acids. These four types of macromolecular substances are the principal topics of this chapter.

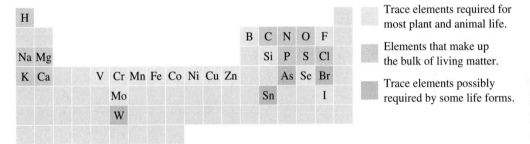

▲ FIGURE 28-1 **Elements in living matter**
For the most part, the elements essential to living matter are also among the more abundant elements in Earth's crust and in seawater. It is likely that life forms developed from the elements available to them.

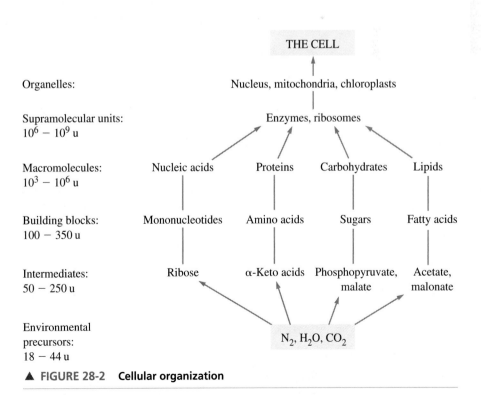

▲ FIGURE 28-2 **Cellular organization**

The **cell** is the fundamental unit of all life. Cells contain a variety of substructures, such as a nucleus, mitochondria, and chloroplasts (plant cells). Cells combine to form tissues; tissues may be grouped into organs; organs combine into organisms.

28-2 Lipids

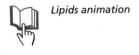

Lipids animation

Lipids are best described through their physical properties rather than in precise structural terms. **Lipids** are those constituents of plant and animal tissue that are soluble in low-polarity solvents, such as chloroform, carbon tetrachloride, diethyl ether, and benzene. Many compounds fit this description, but we will discuss only a few of them.

Triglycerides are esters of glycerol (1,2,3-propanetriol) and long-chain monocarboxylic acids (fatty acids). Some common fatty acids are listed in Table 28.1. *Triglyceride* is a common name; the systematic name of the triglycerides is triacylglycerols. *Glycerol* provides the three-carbon backbone, and the fatty acids provide *acyl* groups (in blue below).

$$\begin{array}{ccccc} & H & OH & H \\ & | & | & | \\ HO- & C- & C- & C- & OH \\ & | & | & | \\ & H & H & H \end{array} \qquad \begin{array}{c} O \\ \| \\ HO-C-R \end{array}$$

Glycerol   Fatty acid

KEEP IN MIND ▶

that in forming an ester, a carboxylic acid loses its —OH group. The group that remains,

$$\begin{array}{c} O \\ \| \\ RC- \end{array}$$

is the *acyl* group.

TABLE 28.1	Some Common Fatty Acids	
Common Name	**IUPAC Name**	**Formula**
Saturated acids		
Lauric acid	Dodecanoic acid	$C_{11}H_{23}CO_2H$
Myristic acid	Tetradecanoic acid	$C_{13}H_{27}CO_2H$
Palmitic acid	Hexadecanoic acid	$C_{15}H_{31}CO_2H$
Stearic acid	Octadecanoic acid	$C_{17}H_{35}CO_2H$
Unsaturated acids		
Oleic acid	9-Octadecenoic acid	$C_{17}H_{33}CO_2H$
Linoleic acid	9,12-Octadecadienoic acid	$C_{17}H_{31}CO_2H$
Linolenic acid	9,12,15-Octadecatrienoic acid	$C_{17}H_{29}CO_2H$
Eleostearic acid	9,11,13-Octadecatrienoic acid	$C_{17}H_{29}CO_2H$

If all acyl groups are the same, the triglyceride is a *simple* triglyceride; otherwise it is a *mixed* triglyceride. In naming a triglyceride, the name *glyceryl* is written first, followed by a compound name for the three acyl groups. The acyl groups are named in the order in which they are attached to the glyceryl backbone. The first two names are given an *o* ending and the third, an *ate* ending. If all acyl groups are the same, only the ending *ate* is used, together with the prefix *tri*.

$$\begin{array}{l} CH_2OH \\ | \\ CHOH \\ | \\ CH_2OH \end{array}$$

Glycerol

$$\begin{array}{l} O \\ \| \\ CH_2OC(CH_2)_{14}CH_3 \\ \quad\quad O \\ \quad\quad \| \\ CHOC(CH_2)_{14}CH_3 \\ \quad\quad O \\ \quad\quad \| \\ CH_2OC(CH_2)_{14}CH_3 \end{array}$$

Glyceryl tripalmitate
Tripalmitin
(a simple triglyceride; a fat)

$$\begin{array}{l} O \\ \| \\ CH_2OC(CH_2)_7CH=CH(CH_2)_7CH_3 \\ \quad\quad O \\ \quad\quad \| \\ CHOC(CH_2)_7CH=CH(CH_2)_7CH_3 \\ \quad\quad O \\ \quad\quad \| \\ CH_2OC(CH_2)_7CH=CH(CH_2)_7CH_3 \end{array}$$

Glyceryl trioleate
Triolein
(a simple triglyceride; an oil)

$$\begin{array}{l} O \\ \| \\ CH_2OC(CH_2)_{10}CH_3 \\ \quad\quad O \\ \quad\quad \| \\ CHOC(CH_2)_{14}CH_3 \\ \quad\quad O \\ \quad\quad \| \\ CH_2OC(CH_2)_{16}CH_3 \end{array}$$

Glyceryl lauropalmitostearate
(a mixed glyceride)

Triglycerides can be *hydrolyzed* in alkaline solution to produce glycerol and salts of the fatty acids. The hydrolysis process is called **saponification,** and the salts are commonly known as **soaps.** For example, the hydrolysis of tristearin with aqueous KOH gives glycerol and the soap potassium stearate

$$
\begin{array}{c}
\text{CH}_2\text{OC(CH}_2)_{16}\text{CH}_3 \\
\text{CHOC(CH}_2)_{16}\text{CH}_3 + 3\ \text{KOH} \longrightarrow \\
\text{CH}_2\text{OC(CH}_2)_{16}\text{CH}_3 \\
\text{tristearin}
\end{array}
\qquad
\begin{array}{c}
\text{CH}_2\text{OH} \\
\text{CHOH} + 3\ \text{CH}_3(\text{CH}_2)_{16}\overset{\text{O}}{\overset{\|}{\text{C}}}\text{O}^-\text{K}^+ \quad (28.1) \\
\text{CH}_2\text{OH} \\
\text{glycerol} \qquad \text{potassium stearate} \\
\text{(a soap)}
\end{array}
$$

The cleansing action of soaps was described in Section 22-1.

Fats and oils are both triglycerides (*glyceryl esters*) but differ from each other by the nature of the acid components in the triglycerides. **Fats** are glyceryl esters in which *saturated* fatty acid components predominate; they are solids at room temperature. **Oils** have a predominance of *unsaturated* fatty acids and are liquids at room temperature. The compositions of fats and oils are variable and depend not only on the particular plant or animal species involved but also on dietary and climatic factors. Some common fats and oils are listed in Table 28.2.

When pure, fats and oils are colorless, odorless, and tasteless. The characteristic colors, odors, and flavors commonly associated with fats and oils come from other organic substances present as impurities. The yellow color of butter is that of β-carotene (a yellow pigment also found in carrots and marigolds). The taste of butter is attributed to 3-hydroxy-2-butanone and diacetyl.

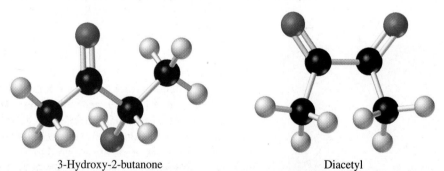

3-Hydroxy-2-butanone   Diacetyl

TABLE 28.2	Some Common Fats and Oils					
	Component Acids, % by Mass					
	Saturated			**Unsaturated**		
Lipid	**Myristic**	**Palmitic**	**Stearic**	**Oleic**	**Linoleic**	**Linolenic**
Fats						
Butter	7–10	24–26	10–13	28–31	1–3	0.2–0.5
Lard	1–2	28–30	12–18	40–50	7–13	0–1
Edible oils						
Corn	1–2	8–12	2–5	19–49	34–62	—
Safflower	—	6–7	2–3	12–14	75–80	0.5–1.5

The formulas of the individual acids are listed in Table 28.1.

Are You Wondering...

What "calorie-free" fats are?

In olestra, an ester of fatty acids and sucrose, the sucrose molecule takes the place of glycerol as the backbone of the ester. When olestra is burned in a bomb calorimeter, the liberated heat is comparable to that obtained in the combustion of other oils and fats. So, in this sense, olestra is *not* calorie free. However, the sucrose backbone of the olestra molecule can bond with six, seven, or eight acyl groups instead of the three acyl groups in a triglyceride, as shown below. Human enzymes cannot break down this bulky molecule, and so it is not digested. It is in this sense that it is calorie free.

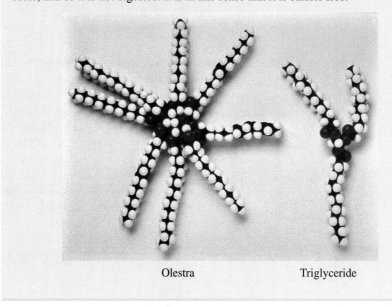

Olestra   Triglyceride

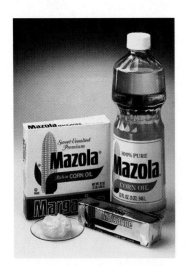

▲ Corn oil has the approximate composition listed in Table 28.2. Hydrogenation converts some of the unsaturated to saturated fatty acid components—the primary change required in making margarine from vegetable oils.

Unsaturated fats and oils can be converted to saturated ones by the catalytic addition of hydrogen (hydrogenation). Thus, oils or low-melting fats are changed to higher-melting fats. These higher-melting fats, when mixed with skim milk, fortified with vitamin A, and artificially colored, are known as *margarines*. Unsaturation in a fat or oil is also removed when the fat or oil decomposes. Edible fats and oils both hydrolyze and cleave at the double bonds by oxidation on exposure to heat, air, and light. When this happens, the fat becomes rancid. The low-molecular-mass fatty acids produced by this cleavage have offensive odors, as exhibited by butyric acid in rancid butter. Antioxidants are commonly added to oils used in the high-temperature cooking of potato chips and other foods to retard this oxidative rancidity.

Medical evidence suggests a relationship between a high intake of saturated fats and the incidence of coronary heart disease. For this reason, many diets call for the substitution of unsaturated for saturated fatty acids in foods. In general, most mammal fats are saturated, whereas those derived from vegetables, seafood, and poultry are unsaturated.

Phospholipids

Phospholipids (phosphatides) occur in all animal cells and are especially prevalent in nerve tissue. They are derived from glycerol, fatty acids, phosphoric acid, and a nitrogen-containing base—ethanolamine in *cephalins* and choline in *lecithins*. In the following structures, R and R′ are long-chain alkyl groups.

$$\begin{array}{c} O \\ \parallel \\ CH_2OCR' \end{array}$$

$$\begin{array}{c} O \\ \parallel \\ RCOCH \end{array}$$

$$CH_2OPOCH_2CH_2\overset{+}{N}H_3$$

Phosphatidylethanolamine
(a cephalin)

$$\begin{array}{c} O \\ \parallel \\ CH_2OCR' \end{array}$$

$$\begin{array}{c} O \\ \parallel \\ RCOCH \end{array}$$

$$CH_2OPOCH_2CH_2\overset{+}{N}(CH_3)_3$$

Phosphatidylcholine
(a lecithin)

Like soap molecules, the phospholipids have a hydrophilic head (the phosphate-ethanolamine or phosphate-choline portions shown in blue above) and hydrophobic tails (the two alkyl chains). This enables phospholipids to solubilize and transport fats and oils in aqueous medium, whether this occurs in transporting lipids in the bloodstream or in emulsifying fats and oils in salad dressings.

Cell membranes (the outer boundary of all living cells) consist of a bilayer of phospholipids having their hydrophilic heads in an aqueous medium and their hydrophobic portions turned inward into a medium of cholesterol and proteins.

28-3 Carbohydrates

The literal meaning of "carbohydrate" is hydrate of carbon: $C_x(H_2O)_y$. Thus, sucrose, or cane sugar, $C_{12}H_{22}O_{11}$, is equivalent to $C_{12}(H_2O)_{11}$. A more useful definition is that **carbohydrates** are polyhydroxy aldehydes, polyhydroxy ketones, their derivatives, and substances that yield them on hydrolysis. Carbohydrates that are aldehydes are called aldoses; those that are ketones are called ketoses. A five-carbon carbohydrate is a pentose, a six-carbon one, a hexose, and so on. The structures in the margin are those of two familiar hexoses—glucose and fructose, an aldose and a ketose, respectively.

The simplest carbohydrates are the **monosaccharides**. **Oligosaccharides** contain from two to ten monosaccharide units bonded together. Names can be assigned to reflect the actual number of such units present, such as *di*saccharide and *tri*saccharide. Mono- and oligosaccharides are also called **sugars**. **Polysaccharides** contain more than ten monosaccharide units. The general term for all carbohydrates is glycoses. In summary,

Glycoses
{
Monosaccharides
  aldoses (aldotriose, aldotetrose, …)
  ketoses (ketotriose, ketotetrose, …)
Oligosaccharides (from two to ten monosaccharide units)
  disaccharides (e.g., sucrose)
  trisaccharides (e.g., raffinose)
  and so on.
Polysaccharides (more than ten monosaccharide units)
  (e.g., starch and cellulose)
}

The simplest glycose is 2,3-dihydroxypropanal (glyceraldehyde), an *aldotriose*. As Figure 28-3 illustrates, the central C atom in glyceraldehyde has four different groups attached to it and is therefore chiral. As we have seen in Chapters 25 and 27, such molecules exhibit an interesting form of stereoisomerism—*optical isomerism*.

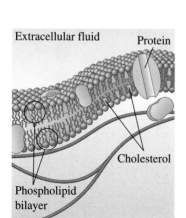

Extracellular fluid
Protein
Cholesterol
Phospholipid bilayer

▲ A cell membrane is a bilayer (double layer) of phospholipid molecules with the polar head groups oriented toward the aqueous phase. Other lipid molecules, such as cholesterol, can be embedded in the bilayer. Some proteins are also found in the bilayer; membrane-bound proteins often act as ion pumps by providing a channel for the ions to pass through.

$$\begin{array}{cc}
HC\!=\!O & CH_2OH \\
\mid & \mid \\
CHOH & C\!=\!O \\
\mid & \mid \\
CHOH & CHOH \\
\mid & \mid \\
CHOH & CHOH \\
\mid & \mid \\
CHOH & CHOH \\
\mid & \mid \\
CH_2OH & CH_2OH \\
\text{Glucose} & \text{Fructose}
\end{array}$$

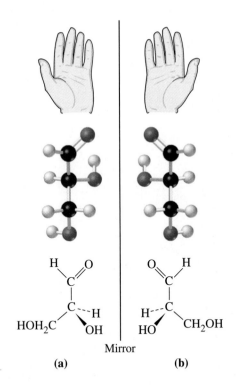

▶ **FIGURE 28-3**
Optical isomers of glyceraldehyde
The structure in (a) is not superimposable
on (b), just as a right hand and a left hand
are not superimposable.

There are *two* nonsuperimposable structures for glyceraldehyde. Such structures are related to each other like a right and a left hand, or like an object and its non-superimposable mirror image; they are called *enantiomers* (see Section 25-4 and Section 27-11).

As we also learned, optically active molecules affect plane-polarized light (Figure 28-4). Interactions between a beam of polarized light and the electrons in an enantiomer cause a rotation of the plane of the polarized light. One enantiomer rotates the plane of polarized light to the right (clockwise) and is said to be **dextrorotatory** (designated +). The other enantiomer rotates the plane of polarized light to the same extent, but to the left (counterclockwise). It is said to be **levorotatory** (−). Because of their ability to rotate the plane of polarized light, isomers of these types are said to be *optically active*, and they are called *optical isomers*. Almost all molecules exhibiting optical isomerism possess at least one asymmetric, or chiral, C atom.

▶ The prefixes *dextro* and *levo* are derived from the Latin words *dexter*, meaning right, and *laevus*, meaning left.

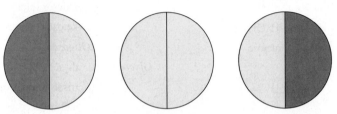

▲ **FIGURE 28-4** **View through the analyzer prism of a polarimeter (see also Figure 25-9)**
The field of view of polarized light from a sodium vapor lamp is split in half. The analyzer prism in the eyepiece has been rotated in the correct direction—clockwise (to the right) or counterclockwise (to the left)—and through the appropriate angle, α, when the two halves transmit light of equal intensity (center). For incorrect angles of rotation, one semicircle is darker than the other.

The arrangement of groups at an asymmetric C atom is called the **absolute configuration**. We have already described the *R, S* system of nomenclature for describing the absolute configuration of chiral centers. The configuration of glyceraldehyde shown in Figure 28-3a is (*R*)-glyceraldehyde. The (*S*)-glyceraldehyde is shown in Figure 28-3b. But which configuration rotates the plane of polarized light in a dextrorotatory (+) sense? Optical rotation studies alone cannot tell us which of the enantiomers rotates the plane of plane polarized light in a positive sense or a negative sense, because there is no way to assign the *R* or *S* configuration to a particular one of a pair of enantiomers. That is, there is no relationship between the absolute configuration and the sign of the rotation; both must be determined experimentally by independent methods. The assignment of which configuration, *R* or *S*, is the dextrorotary (+) isomer has been achieved using X-ray studies. These methods show that the enantiomer that corresponds to the dextrorotatory (+) isomer has the *R* configuration and is named (*R*)-(+)-glyceraldehyde (Figure 28-3a). Correspondingly, the levorotatory (−) isomer is found to correspond to the *S* configuration and is named (*S*)-(−)-glyceraldehydes (Figure 28-3b). It is purely fortuitous that the positive rotation corresponds to the *R* configuration.

Emil Fischer studied carbohydrates in the late nineteenth century when techniques for determining the configuration of compounds were not available. Fischer arbitrarily assigned the configuration shown in Figure 28-3a to the dextrorotatory isomer of glyceraldehyde. Fischer designated this configuration D and named the isomer D-(+)-glyceraldehyde; D-(+)-glyceraldehyde is (*R*)-(+)-glyceraldehyde, and L-(−)-glyceraldehyde is (*S*)-(−)-glyceraldehyde. This system, extended to other chiral structures, is called the D, L convention.

We have already seen how to use a dashed-wedged line diagram to represent the three-dimensional structure of a molecule such as glyceraldehyde. Drawing these diagrams is not too difficult for compounds containing one or two stereocenters, but becomes increasingly cumbersome for compounds having several chiral centers, which carbohydrates often do. The German chemist Emil Fischer, in addition to assigning the configuration to (+)-glyceraldehyde, introduced a convention for representing three-dimensional structures in two-dimensional drawings, now called **Fischer projection formulas**.

The Fischer projections have two aspects to them: The first is how the stereochemistry at a chiral carbon atom is represented in two dimensions, and the second is how the carbon-chain backbone is arranged on the page. First, the bonds between each C atom and its four substituents are drawn in the form of a cross, the central carbon being at the point of intersection. The horizontal lines signify bonds directed toward the viewer; the vertical lines point away. Dashed-wedged line structures have to be arranged in this way to allow their conversion into Fischer projections. Consider (*R*)-(+)-glyceraldehyde.

In order to view the molecule with two groups pointing toward us, we imagine the molecule's being picked up by the hydroxyl group and the hydrogen atom and the molecule turned to bring these groups toward us as shown. The CHO and CH_2OH groups naturally take up positions away from us. The Fischer projection is then drawn with the H and OH groups connected by a horizontal line and the CHO and CH_2OH groups joined by a vertical line. Note that the central C atom is not drawn in, but is implicitly at the point of intersection of the lines; including this C

▲ **Emil Fischer (1852–1919)**
Fischer was awarded the Nobel Prize in 1902 for his research on the structures of sugars. Later he also elucidated how amino acid molecules join to form proteins.

KEEP IN MIND ▶

that D and L are like *R* and *S* in that they indicate the configuration of a chiral carbon atom, but they do not indicate whether the compound rotates plane-polarized light to the right $(+)$ or to the left $(-)$.

atom would make the Fischer projection indistinguishable from a Lewis structure, which contains no stereochemical information.

In the second aspect of the Fischer notation, the structural formula is drawn so that the backbone of the molecule is arranged from top to bottom, with the most oxidized portion of the molecule (—CHO) at the top and the least oxidized (—CH$_2$OH) at the bottom. Attached groups (—H and —OH) are written to the sides. The end groups of the backbone are considered to extend *behind* the plane of the page, *away* from the viewer. The glyceraldehyde enantiomers are written as

$$(S)\text{-}(-)\ \text{Glyceraldehyde} \qquad (R)\text{-}(+)\text{-Glyceraldehyde}$$
$$(\text{L})\text{-}(-)\ \text{Glyceraldehyde} \qquad (\text{D})\text{-}(+)\text{-Glyceraldehyde}$$

and are used to establish the D, L configuration for other sugars: The —H and —OH groups on the next-to-last (penultimate) C atom extend in *front* of the page, *toward* the viewer. The penultimate C atom and groups are shaded with blue. If the —OH group on this penultimate C atom is to the *right*, the configuration is D. If the —OH is to the left, the configuration is L. This convention is applied below to the four-carbon aldoses. All D sugars have the same configuration at this penultimate carbon. Figure 28-5 may help you to picture the relationship between a three-dimensional structure and its two-dimensional representation.

D-(−)-Erythrose   L-(+)-Erythrose   D-(−)-Threose   L-(+)-Threose

D-Erythrose and L-erythrose are enantiomers, as are D-threose and L-threose. If we compare the configurations of D-erythrose and D-threose, we note that these two molecules are *not* mirror images. On the other hand, they are isomers of each other and both are optically active. Optical isomers that are *not* mirror images of each other are called **diastereomers**.

D-Erythrose   D-Threose
Not mirror images

Enantiomers differ only in the *direction*, not the extent of their rotation of plane-polarized light. Diastereomers do differ in the extent to which they rotate plane-polarized light. They also differ in physical and chemical properties; for example, they have different solubilities in a particular solvent and react with chemical reagents at different rates.

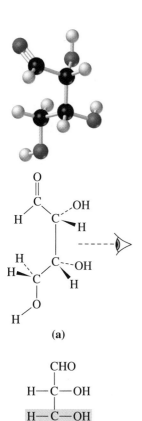

(a)

CHO
|
H—C—OH
|
H—C—OH
|
CH₂OH

(b)

▲ **FIGURE 28-5**
The structure of
D-(−)-erythrose
The three-dimensional structure
(a) is represented in two dimensions by **(b)**.

▲ A test for a reducing sugar, producing a silver mirror from an ammoniacal solution containing silver ion.

Are You Wondering...

How the R, S nomenclature is used in compounds with more than one chiral center?

Each chiral center in the carbon chain is numbered according to the IUPAC convention, and the name is constructed with the prefix R or S, together with the position in the chain. To illustrate this, consider L-threose with two chiral centers, at positions 2 and 3 (numbering starts from the aldehyde group). Remembering that the groups attached to the horizontal lines point toward the viewer, we see that, considering each center in turn, both chiral centers are S. The chiral centers are designated ($2S,3S$), and the complete systematic name is ($2S,3S$)-2,3,4-trihydoxy-1-butanal.

A mixture of equal amounts of the D and L configurations of a particular substance, called a **racemic mixture**, does not rotate the plane of polarized light either to the left or to the right. The designation DL-erythrose, for example, signifies a racemic mixture. Often, when molecules with chiral centers are synthesized, the product is a racemic mixture. This is because the creation of these centers is a random process, like flipping a coin (an equal probability for heads or tails). If optically pure isomers are desired, the racemic mixture must be separated into the component enantiomers by a process called *resolution*. Sometimes this is carried out using an enzyme that reacts with one enantiomer but not the other.

Monosaccharides

Of the 16 possible aldohexoses, only three occur widely in nature: D-glucose, D-mannose, and D-galactose. These three sugar molecules exist in a straight-chain form only to a very small extent (less than 0.5% for glucose). The predominant form for each is *cyclic*. In this ring formation, the —OH group of the fifth C atom (C-5) adds to the carbonyl of the C-1 atom and produces a ring composed of five C atoms and one O atom, as illustrated in Figure 28-6. The conformation of the six-member ring is of the chair type (recall Figure 27-3).

When the chain form of a sugar is converted to the ring form, a new chiral (asymmetrical) center is produced at the C-1 atom. There are two possible orientations at this center. In the α form, the —OH at C-1 is *axial* (directed down); in the β form, it is *equatorial* (extends out from the ring). The α and β forms of glucose are pictured in Figure 28-7.

The naming of monosaccharides in the ring form is complicated, but each item in a name conveys precise information. Thus, D-(+)-glucose refers to the straight-chain form of glucose in the D configuration; the (+) indicates this form is dextrorotatory. The name α-D-(+)-glucose denotes the ring form derived from D-glucose with the α configuration at the C-1 atom.

With many sugars, known as **reducing sugars**, a sufficient amount of the straight-chain form is in equilibrium with the cyclic form that the sugar engages in an oxidation–reduction reaction with Cu^{2+}(aq). The Cu^{2+}(aq) is reduced to insoluble red Cu_2O, and the aldehyde portion of the sugar is oxidized (to an acid). The test for a reducing sugar is conducted with alkaline copper ion complexed with tartrate (Fehling's solution) or citrate ion (Benedict's solution).

$$\text{Certain cyclic sugars} \rightleftharpoons \underset{\text{straight-chain form}}{\overset{\text{CHO}}{\underset{|}{\overset{|}{\text{CHOH}}}}} \xrightarrow{Cu^{2+}} \underset{\text{red ppt}}{\overset{\text{COOH}}{\underset{|}{\overset{|}{\text{CHOH}}}}} + Cu_2O(s) \qquad (28.2)$$

Glucose models

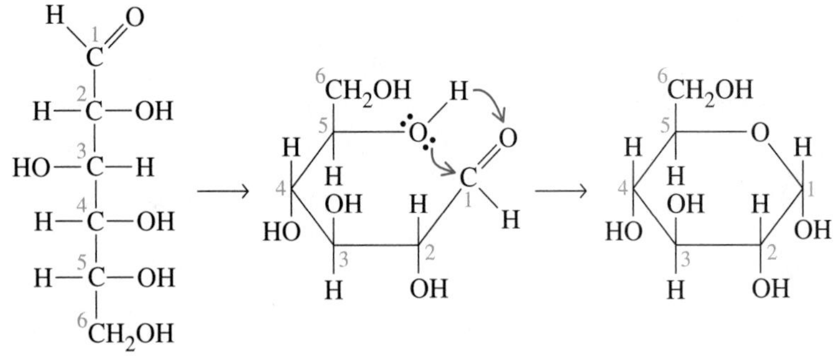

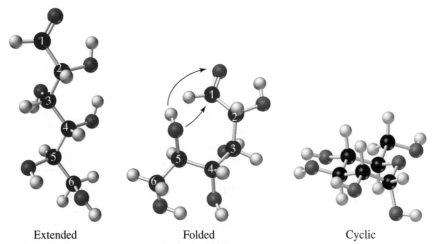

D-(+)-Glucose

α-D-(+)-Glucose

Extended   Folded   Cyclic

▲ **FIGURE 28-6   Representation of ring closure in glucose molecule**
The straight chain D-(+)-glucose molecule folds back on itself to bring the —OH group
on the C-5 atom close to the aldehyde group of the C-1 atom. A proton is transferred, and
this is followed by the formation of a six-member ring with an O atom joining the C-1
and C-5 atoms.

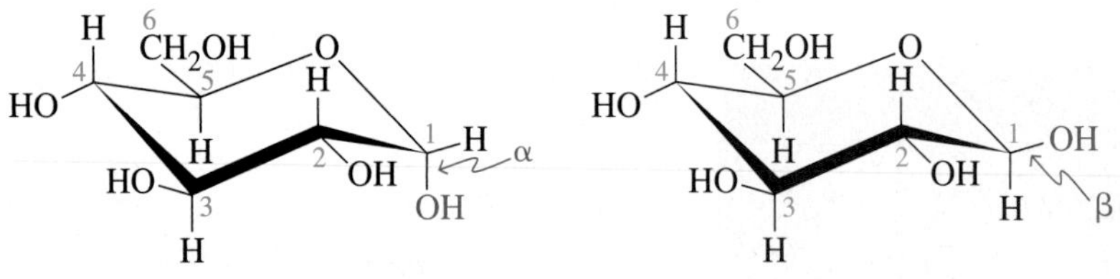

α-D-(+)-Glucose   β-D-(+)-Glucose

▲ **FIGURE 28-7   α and β forms of D-glucose**
In the α form, the —OH group at C-1 (in red) is axial, extending below the chairlike
ring. In the β form, the —OH group is equatorial and extends out from the ring.

Disaccharides

Two monosaccharides can join together by eliminating a H_2O molecule between
them—a condensation reaction. This combination of monosaccharides is called a
disaccharide. In describing a disaccharide, we must consider the identity of the
component monosaccharides and whether the configuration of the linkage between
the monosaccharides is α or β. The important, naturally occurring disaccharides—
maltose, cellobiose, lactose, and sucrose—are presented in Figure 28-8.

Maltose (α form)

Lactose (β form)

Cellobiose

Sucrose

(Glucose unit)

(Fructose unit)

☐ Glucose   ☐ Galactose   ☐ Fructose

▲ **FIGURE 28-8   Some common disaccharides**

Disaccharide Structure activity

▲ **Sucrose**

In *maltose*, a H atom on the C-1 hydroxyl group of one glucose unit reacts with the hydroxyl group on the C-4 atom of a second glucose unit. The two units are linked in the α manner. Equilibrium is possible between the cyclic and straight-chain forms of maltose, so it is a reducing sugar. Maltose is produced by the action of malt enzyme on starch. In the presence of yeast, maltose undergoes fermentation—first to glucose and then to ethanol and $CO_2(g)$.

Cellobiose can be obtained by the careful hydrolysis of cellulose. It is a glucose–glucose disaccharide with β linkages. *Lactose*, or milk sugar, is naturally present in milk, where its concentration may range from 0 to 7%. It is a galactose–glucose disaccharide with β linkages. *Sucrose* is ordinary table sugar (cane or beet sugar). It is a glucose–fructose disaccharide linked 1α, 2β. Neither of the two cyclic sugar units can open up into a chain form, and so sucrose is *not* a reducing sugar.

Polysaccharides

Polysaccharides are composed of monosaccharide units joined into long chains by oxygen linkages. *Starch*, with a molecular mass between 20,000 and 1,000,000 u, is the reserve carbohydrate of many plants and is the bulk constituent of cereals, rice, corn, and potatoes. Its structural features are brought out in Figure 28-9. *Glycogen* is the reserve carbohydrate of animals; it is stored in the liver and in muscle tissue. It has a higher molecular mass than starch, and the polysaccharide chains are more branched. *Cellulose* is the main structural material of plants. It is the chief component of wood pulp, cotton, and straw. Complete hydrolysis of cellulose produces glucose. Cellulose has a molecular mass between 300,000 and 500,000 u, corresponding to 1800–3000 glucose units. Most animals, including human beings, do not possess the necessary enzymes to hydrolyze β linkages. As a result, they cannot digest cellulose. Certain bacteria in ruminants (cows, deer, camels) and termites can hydrolyze cellulose, allowing them to use it as a food. Termites, as we know, subsist on a diet of wood.

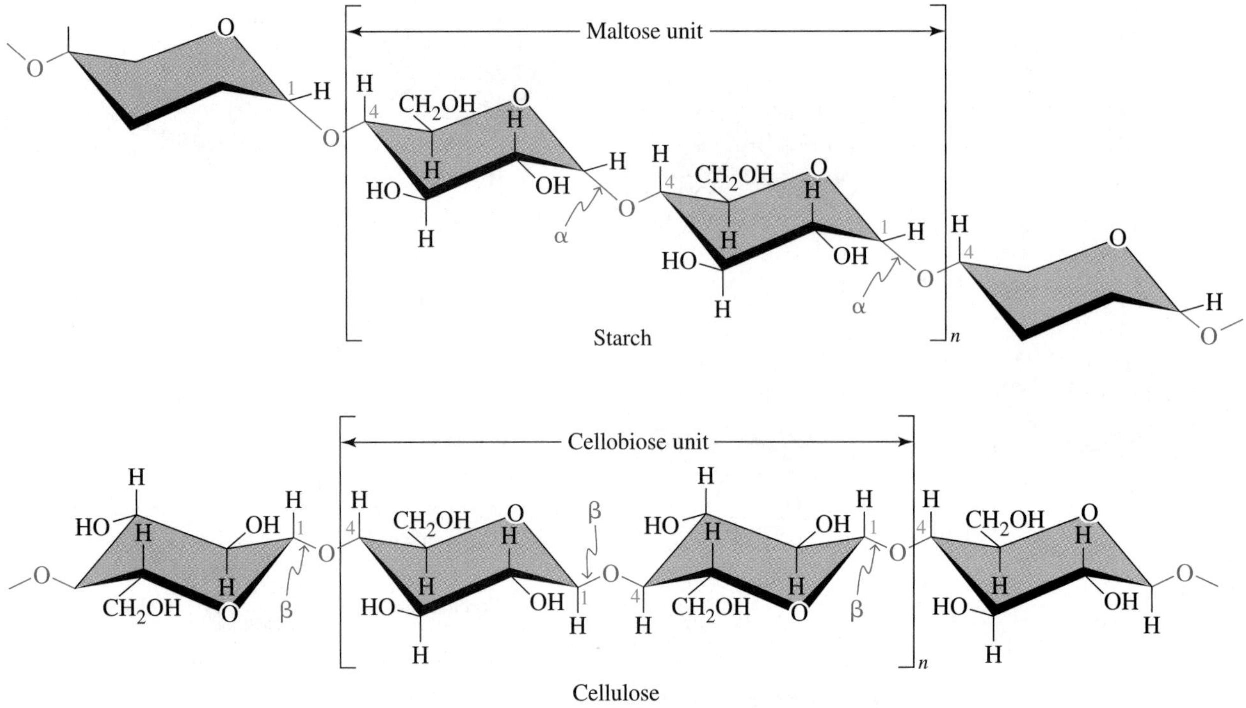

▲ FIGURE 28-9 **Two common polysaccharides**

Photosynthesis

As we have noted on previous occasions (Sections 7-9 and 25-10), the process of photosynthesis involves the conversion by plants of carbon dioxide and water into carbohydrates. Photosynthesis requires the catalyst chlorophyll (see Figure 25-23) and sunlight.

$$n\,CO_2 + n\,H_2O \xrightarrow[\text{chlorophyll}]{\text{sunlight}} C_n(H_2O)_n + n\,O_2 \tag{28.3}$$

Equation (28.3) is greatly oversimplified. The currently accepted mechanism, proposed by Melvin Calvin (Nobel Prize, 1961), involves as many as 100 sequential steps for the conversion of six moles of carbon dioxide to one mole of glucose. The elucidation of this mechanism was greatly aided by the use of carbon-14 as a radioactive tracer. For simplicity, the overall photosynthetic process is divided into two phases: (1) the conversion of solar energy to chemical energy—the light reactions; and (2) the synthesis, promoted by enzymes, of carbohydrates. This latter phase can occur in the absence of light and is called the dark reactions.

Biomass

Biomass is any living matter. An important component of biomass is the organic material produced by photosynthesis, that is, plants or their principal constituents—cellulose, starch, sugars. Plant biomass can be used directly as a fuel, or it may be converted to other gaseous, liquid, or solid materials, which can be used as fuels or chemical raw materials.

Perhaps the best known and most widely used conversion method is the fermentation of sugars to ethanol. Fermentation involves the decomposition of organic matter in the absence of air and through the action of a microorganism.

$$\text{hexose sugar} \xrightarrow[\text{in yeast}]{\text{microorganism}} 2\,CH_3CH_2OH + 2\,CO_2$$

▲ A converter for the pyrolysis of biomass.

In the United States, the main raw material for the production of ethanol by fermentation is corn.

The conversion of plant material to fossil fuels requires geologic processes and geologic time scales, thereby limiting the future availability of these fuels as energy sources and as raw materials. In principle, most of the compounds now being produced from fossil fuels could be made directly from cellulose. Methanol (wood alcohol) is formed by the destructive distillation (pyrolysis) of wood. Cellulose can be hydrolyzed to glucose and then converted to ethanol by fermentation. Fermentation processes can also be used to produce a series of oxygenated compounds—alcohols and ketones—which can be converted to hydrocarbons. Thus, the entire spectrum of organic chemicals could be produced from the simple molecules CO_2 and H_2O. The required energy would be mostly solar.

28-4 Proteins

When a protein is hydrolyzed by dilute acids, bases, or hydrolytic enzymes, the result is a mixture of α-amino acids. An *amino acid* is a carboxylic acid that also contains an amine group, $-NH_2$; an **α-amino acid** has the amino group on the α carbon atom—the C atom next to the carboxyl group. Thus, proteins are high-molecular-mass polymers composed of α-amino acids. Of the known α-amino acids, about 20 have been identified as the building blocks of plant and animal proteins. Some of these amino acids are listed in Table 28.3.

Proteins are the basis of protoplasm and are found in all living organisms. In animals, proteins—as muscle, skin, hair, and other tissue—make up the bulk of the body's nonskeletal structure. As enzymes, proteins catalyze biochemical reactions; as hormones, they regulate metabolic processes; and as antibodies, they counteract the effect of invading organisms.

▶ The name "protein" is derived from the Greek word *proteios*, meaning "of first importance" (similar to the derivation of "proton").

Other than glycine ($H_2NCH_2CO_2H$), naturally occurring amino acids are optically active, mostly with an L configuration.

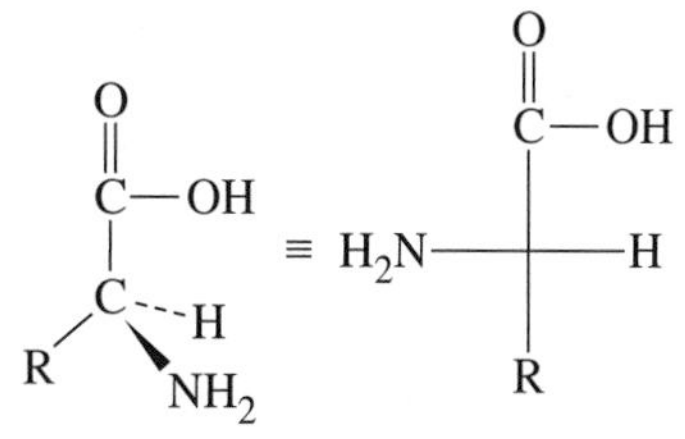

$$
\begin{array}{ccc}
& O & \\
& \| & \\
& C-OH & \\
& | & \\
R & C\cdots H & \equiv \quad H_2N \longrightarrow H \\
& NH_2 & R
\end{array}
$$

An L-amino acid

📖 *Proteins and Amino Acids animation*

The reference structure for establishing the absolute configurations of amino acids is again glyceraldehyde, with the $-NH_2$ group substituting for $-OH$ and $-CO_2H$, for $-CHO$. The molecule shown above has an L configuration because the $-NH_2$ group appears on the *left*.

Certain amino acids that are required for proper health and growth in human beings cannot be synthesized by the body. These amino acids, which are called the *essential amino acids*, must be ingested as food. Eight amino acids are known to be essential; the case of three others is less certain (see Table 28.3).

Amino acids are colorless, crystalline, high-melting-point solids that are moderately soluble in water. In a strongly acidic solution (low pH), the amino acid exists as a cation: A proton from solution attaches itself to the unshared pair of electrons on the nitrogen atom in the $-NH_2$ group. In a strongly basic solution (high pH), an anion forms through the loss of protons by the $-CO_2H$ and $-NH_3^+$ groups. At an intermediate point, a proton is lost from the $-CO_2H$ but retained by the $-NH_3^+$ group. The product is a dipolar ion, or a *zwitterion*.

TABLE 28.3	Some Common Amino Acids		
Name	**Symbol**	**Formula**	**pI^a**
Neutral amino acids			
Glycine	Gly	$HCH(NH_2)CO_2H$	6.03
Alanine	Ala	$CH_3CH(NH_2)CO_2H$	6.10
Valine[b]	Val	$(CH_3)_2CHCH(NH_2)CO_2H$	6.04
Leucine[b]	Leu	$(CH_3)_2CHCH_2CH(NH_2)CO_2H$	6.04
Isoleucine[b]	Ileu or Ile	$CH_3CH_2CH(CH_3)CH(NH_2)CO_2H$	6.04
Serine	Ser	$HOCH_2CH(NH_2)CO_2H$	5.70
Threonine[b]	Thr	$CH_3CHOHCH(NH_2)CO_2H$	5.6
Phenylalanine[b]	Phe	$C_6H_5CH_2CH(NH_2)CO_2H$	5.74
Methionine[b]	Met	$CH_3SCH_2CH_2CH(NH_2)CO_2H$	5.71
Cysteine	Cys	$HSCH_2CH(NH_2)CO_2H$	5.05
Cystine	$(Cys)_2$	$[SCH_2CH(NH_2)CO_2H]_2$	5.1
Tyrosine	Tyr	$4-HOC_6H_4CH_2CH(NH_2)CO_2H$	5.70
Tryptophan[b]	Trp	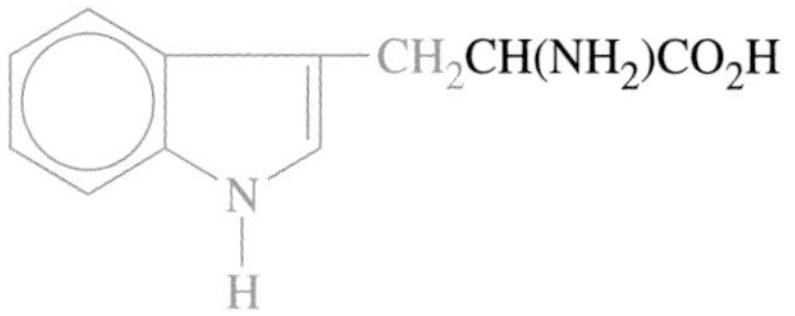	5.89

Proline[c]	Pro	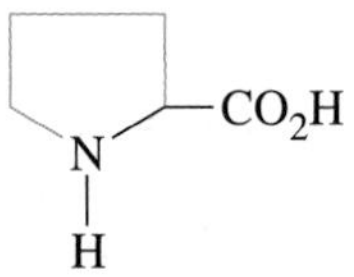	6.21

Acidic amino acids			
Aspartic acid	Asp	$HO_2CCH_2CH(NH_2)CO_2H$	2.96
Glutamic acid	Glu	$HO_2CCH_2CH_2CH(NH_2)CO_2H$	3.22
Basic amino acids			
Lysine[b]	Lys	$H_2N(CH_2)_4CH(NH_2)CO_2H$	9.74
Arginine	Arg	$H_2NC(=NH)NH(CH_2)_3CH(NH_2)CO_2H$	10.73
Histidine	His	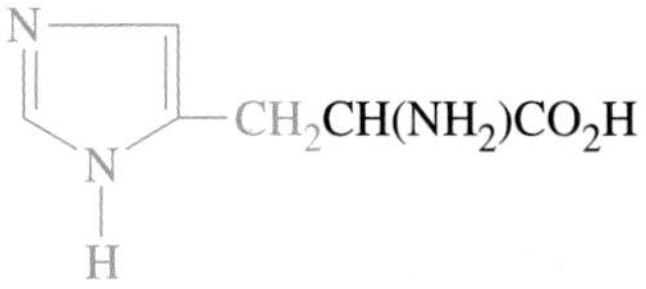	7.58

[a] pH of *isoelectric point*.
[b] Essential amino acids. In addition, arginine and glycine are required by the chick, arginine by the rat, and histidine by human infants.
[c] The secondary amino group makes proline an *α-imino* acid. Nevertheless, it is commonly listed with amino acids.

$$NH_3^+-CH-CO_2H \underset{H^+}{\overset{OH^-}{\rightleftharpoons}} NH_3^+-CH-CO_2^- \underset{H^+}{\overset{OH^-}{\rightleftharpoons}} NH_2-CH-CO_2^- \qquad (28.4)$$

acidic soln   isoelectric point   basic soln

Amino acids are amphoteric. In most amino acids, the acidity of the $-NH_3^+$ group is slightly greater than the basicity of the $-CO_2^-$ group. The largest group of amino acids are close to pH neutral. The pH at which the dipolar structure predominates is called the **isoelectric point**, or pI. At this pH, the molecule does not migrate in an electric field. At a pH above the p*I*, the molecule migrates to the anode (positive electrode); and below the p*I*, to the cathode (negative electrode). Most basic amino acids have a p*I* well above 7, acidic ones well below 7, and most neutral ones slightly less than 7 (5.7–6.1).

Peptides   Two amino acid molecules can be joined by the elimination of a water molecule between them. The amino acids thus joined form a dipeptide. The bond between the two amino acid units is called a **peptide bond**.

Peptide Bond animation

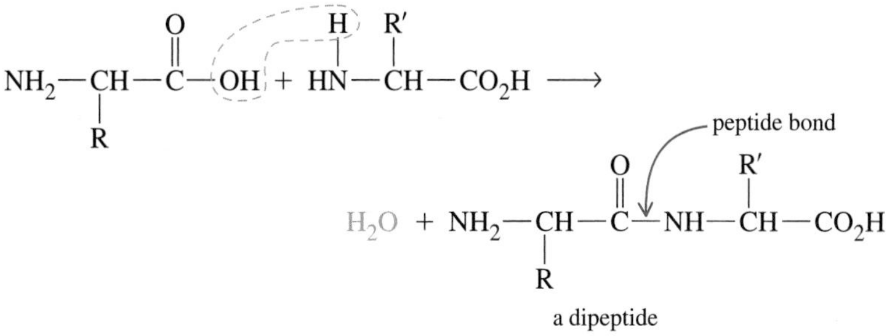

$$NH_2-CH-\underset{R}{\overset{\overset{\displaystyle O}{\|}}{C}}-OH + HN-\underset{R'}{\overset{\displaystyle H}{CH}}-CO_2H \longrightarrow$$

$$H_2O + NH_2-\underset{R}{CH}-\overset{\overset{\displaystyle O}{\|}}{C}-NH-\underset{R'}{CH}-CO_2H \qquad (28.5)$$

peptide bond

a dipeptide

A tripeptide has three amino acid residues and two peptide linkages. A large number of amino acid units may join to form a **polypeptide**.

The amino acid unit present at one end of a polypeptide chain has a free $-NH_2$ group; this is the N-terminal end. The other end of the chain has a free $-CO_2H$ group; this is the C-terminal end. The polypeptide structure is written with the N-terminal end to the left and the C-terminal end to the right. The base name of the polypeptide is that of the C-terminal amino acid. The other amino acid units in the chain are named as substituents of this acid. Their names change from an *ine* to the *yl* ending. Abbreviations are also commonly used in writing polypeptide names, as illustrated in Example 28-1.

EXAMPLE 28-1

Naming a Polypeptide.   What is the name of the polypeptide whose structure is shown below?

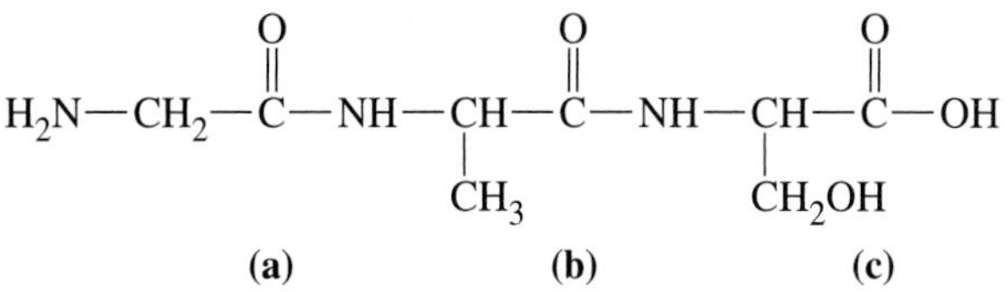

$$H_2N-CH_2-\overset{\overset{\displaystyle O}{\|}}{C}-NH-\underset{CH_3}{CH}-\overset{\overset{\displaystyle O}{\|}}{C}-NH-\underset{CH_2OH}{CH}-\overset{\overset{\displaystyle O}{\|}}{C}-OH$$

               (a)   **(b)**   **(c)**

Solution

We can identify the three amino acids in this tripeptide using Table 28.3. (a) = glycine; (b) = alanine; (c) = serine. The C-terminal amino acid is serine. The name is glycylalanylserine (Gly-Ala-Ser).

Practice Example A: What is the name of the polypeptide shown below?

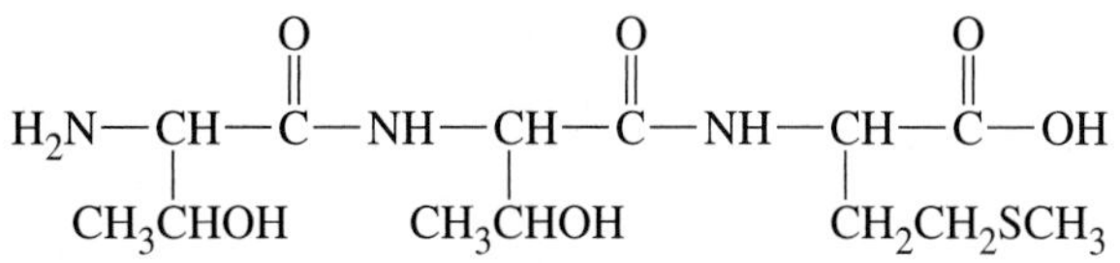

$$H_2N-\underset{CH_3CHOH}{CH}-\overset{\overset{\displaystyle O}{\|}}{C}-NH-\underset{CH_3CHOH}{CH}-\overset{\overset{\displaystyle O}{\|}}{C}-NH-\underset{CH_2CH_2SCH_3}{CH}-\overset{\overset{\displaystyle O}{\|}}{C}-OH$$

Practice Example B: Write the structural formula of the polypeptide serylglycylvaline.

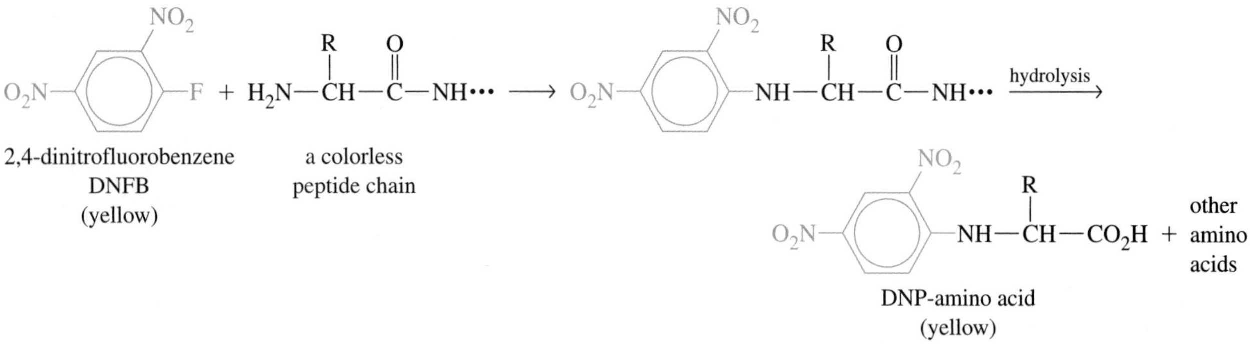

▲ **FIGURE 28-10 Experimental determination of amino acid sequence**
In the reaction between DNFB and a polypeptide, the N-terminal amino acid ends up with the yellow marker (a dinitrophenyl group, DNP) attached to it. By gentle hydrolysis and repeated use of the marker, a polypeptide chain can be broken down and the sequence of the individual units determined.

Sequencing Suppose that a tripeptide is known to consist of the three amino acids: A, B, and C. What is the correct structure: ABC, BAC, ... ? Can you see that there are six possibilities? For longer chains, of course, the number of possibilities is enormous. Determining the sequence of amino acids in a polypeptide chain is one of the most significant problems in all of biochemistry. The method employed is outlined in Figure 28-10, and the structure of a typical polypeptide, beef insulin, is shown in Figure 28-11.

EXAMPLE 28-2

Determining the Sequence of Amino Acids in a Polypeptide. A polypeptide, on complete hydrolysis, yielded the amino acids A, B, C, D, and E. Partial hydrolysis and sequence proof gave single amino acids, together with the following larger fragments: AD, CD, DCB, BE, and BC. What must be the sequence of amino acids in the polypeptide?

Solution

By arranging the fragments in the following manner,

$$
\begin{array}{l}
\text{AD} \\
\ \text{DC} \\
\ \text{DCB} \\
\ \ \ \text{BE} \\
\ \ \text{CB}
\end{array}
$$

we see that the sequence ADCBE is consistent with the fragments observed. However, the sequence EBCDA is also possible because we do not know whether A or E is the N-terminal end of the polypeptide.

Practice Example A: On complete hydrolysis, a pentapeptide yields the amino acids Val, Phe, Gly, Cys, and Tyr. Partial hydrolysis yields the fragments Val-Phe, Gly-Cys, Cys-Val-Phe, and Phe-Tyr. Glycine (Gly) is the N-terminal acid. What is the sequence of amino acids in the polypeptide?

Practice Example B: On complete hydrolysis, a hexapeptide yields the amino acids Ala, Gly, Ser, Trp, and Val. Partial hydrolysis yields the fragments Val-Trp, Gly-Gly-Ala, Ser-Gly-Gly, and Ala-Val-Trp. Serine (Ser) is the N-terminal acid. What is the sequence of amino acids in the polypeptide?

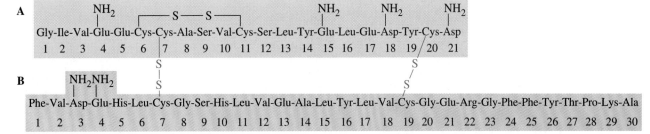

▲ FIGURE 28-11   Amino acid sequence in beef insulin—primary structure of a protein

There are two polypeptide chains joined by disulfide (—S—S—) linkages. One chain has 21 amino acids, and the other 30. In chain A, the Gly at the left end is N-terminal and the Asp is C-terminal. In chain B, Phe is N-terminal and Ala is C-terminal.

The distinction between large polypeptides and proteins is arbitrary. It is generally accepted that if the molecular mass is over 10,000 u (roughly 50–75 amino acid units), the substance is a **protein**. Proteins possess characteristic isoelectric points, and their acidity or basicity depends on their amino acid composition. When proteins are heated, treated with salts, or exposed to UV light, profound and complex changes called **denaturation** occur. Denaturation usually brings about a lowering of solubility and loss of biological activity. The frying or boiling of an egg involves the denaturation (coagulation) of the egg albumin, a protein. The beauty shop permanent wave takes advantage of a denaturation process that is reversible. The proteins found in hair (e.g., keratin) contain disulfide linkages (—S—S—). When hair is treated with a reducing agent, these linkages break—a denaturation process. Following this step, the hair is set into the desired style. Next, the hair is treated with a mild oxidizing agent. The disulfide linkages are reestablished, and the hair remains in the style in which it was set.

The Structure of Proteins

The **primary structure** of a protein refers to the exact sequence of amino acids in the polypeptide chains that make up the protein. But what are the shapes of the long polymeric chains themselves? Are they simply limp and entangled like a plate of spaghetti or is there some order within chains and among chains? The structure or shape of an entire protein chain is referred to as **secondary structure**. In 1951, based on X-ray-diffraction studies on polylysine, a synthetic polypeptide, Linus Pauling and R. B. Corey proposed that the orientation of this polypeptide and thus of the protein chain is *helical*. A spiral, helical, or springlike shape can be either left- or right-handed, but because proteins are composed of L amino acids, their helical structure is right-handed (Figure 28-12).

β-Keratin model

Other types of orientations are also possible. For example, β-keratin and silk fibroin are arranged in pleated sheets. In these proteins, the side chains extend above and below the pleated sheets, and there is hydrogen bonding between *different* molecules (interpeptide bonding) lying next to each other and about 0.47 nm apart in the same sheet. These sheets are stacked on top of one another about 1.0 nm apart, rather like a pile of sheets of corrugated roofing (Figure 28-13). Some proteins, such as gamma globulin, are amorphous: They do not have a definite secondary structure.

Many proteins possess additional structural features. For example, rather than being elongated, the coils may be twisted, knotted, and so forth. The final statement regarding the shape of a protein molecule lies in a description of its **tertiary structure**. Because the internal hydrogen bonding between atoms in successive turns of

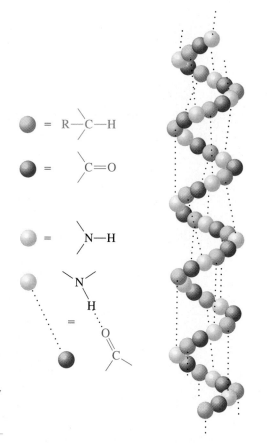

▶ **FIGURE 28-12**
Secondary structure of a protein—an
α helix
The helical structure is stabilized by
hydrogen bonds between

$$\overset{O}{\underset{\parallel}{}}$$

$$\overset{}{\underset{}{}}—C—$$ groups in one turn and —N--- groups in the next turn above. The bulky R groups are directed outward from the atoms in the spiral.

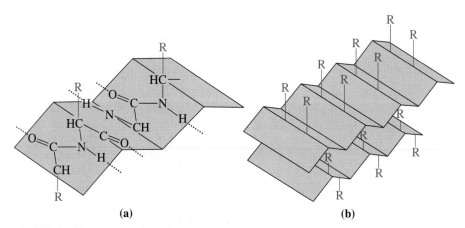

(a)   (b)

▲ **FIGURE 28-13   Pleated-sheet model of β-keratin**
(a) A polypeptide chain showing the direction of interpolypeptide hydrogen bonds (other polypeptide chains lie to the left and to the right of the chain shown). Bulky R groups extend above and below the pleated sheet. (b) The stacking of pleated sheets.

a protein helix is weak, these hydrogen bonds ought to be easily broken. In particular, we should expect them to be replaced by hydrogen bonds to water molecules when the protein is placed in water. That is, the α helix should open up and become a randomized structure when placed in water. But experimental evidence indicates that this does not happen. We are led to the conclusion that other forces

$$\begin{matrix} \overset{\displaystyle O}{\underset{\displaystyle \|}{}} \\ \{\text{—CH}_2\text{C O}^- \quad \text{H}_3\overset{+}{\text{N}}\text{—(CH}_2)_4\text{—}\} \\ \text{Aspartic} \qquad \text{Lysine} \\ \text{acid} \end{matrix} \qquad \begin{matrix} \overset{\displaystyle OH}{\underset{\displaystyle |}{}} \\ \{\text{—CH}_2\text{—C}\text{=}\text{O}\text{---}\text{H}\text{—OCH}_2\text{—}\} \\ \text{Aspartic} \qquad \text{Serine} \\ \text{acid} \end{matrix}$$

(a) **(b)**

$$\{\text{—CH}_2\text{—SH} + \text{HS—CH}_2\text{—}\} \underset{[\text{H}]}{\overset{[\text{O}]}{\rightleftharpoons}} \{\text{—CH}_2\text{—S—S—CH}_2\text{—}\}$$

(c)

▲ **FIGURE 28-14** **Linkages contributing to the tertiary structure of proteins**
(a) Salt linkages. Acid–base interactions between different coils. Here, the carboxyl group of an aspartic acid unit on one coil donates a proton to the free amine group of a lysine unit on another. **(b)** Hydrogen bonding. Interactions between side chains of certain amino acids, for example, aspartic acid and serine. **(c)** Disulfide linkages. Oxidation of the highly reactive thioalcohol group (–SH) of cysteine to a disulfide ($-$S$-$S$-$) can occur (as in beef insulin).

The folding of a polypeptide chain into a tertiary structure is influenced by an additional factor. The hydrophobic hydrocarbon portions of the chains (R groups) tend to pull away from the aqueous medium and retreat to the interior of the structure, leaving ionic groups at the exterior.

▲ **FIGURE 28-15**
Representation of the tertiary structure of myoglobin
The primary structure is that of a peptide of 153 units in a single chain. Secondary structure involves coiling of 70% of the chain into an α helix.

 Hemoglobin model

must be involved in entwining the long α-helical chains into definite geometric shapes. Three types of linkages involved in tertiary structures are described in Figure 28-14. The tertiary structure of myoglobin is shown in Figure 28-15.

The hemoglobin molecule consists of four separate polypeptide chains or subunits. The arrangement of these four subunits constitutes a still higher order of structure referred to as the **quaternary structure**. The levels of protein structure in hemoglobin, from primary to quaternary, are depicted in Figure 28-16.

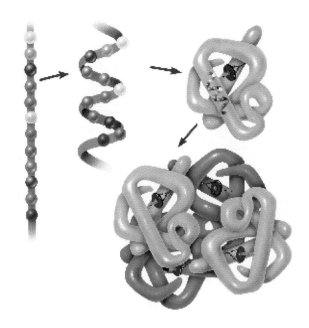

▶ **FIGURE 28-16**
The four levels of protein structure
Hemoglobin is the oxygen-carrying protein in red blood cells. The iron-containing heme units are shown as red discs. The primary structure of the protein is determined by its amino acid sequence. The secondary structure (α helix) is stabilized through formation of hydrogen bonds, as illustrated in Figure 28-12. The tertiary structure is determined by interactions between the R groups and their surroundings, causing the polypeptide chains to fold in a particular manner. Finally, the quaternary structure is an aggregation of two or more folded chains—four chains in the case of hemoglobin. Not all proteins have a quaternary structure.

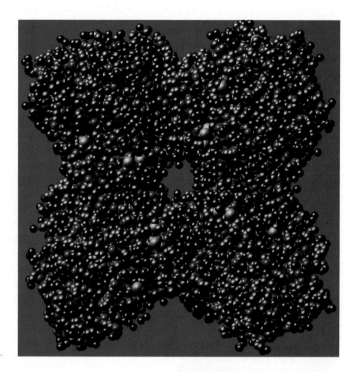

▶ A computer image of the three-dimensional structure of ribulose-1,5-biphosphate carboxylase-oxygenase (RuBisCo), which consists of 37,792 atoms. This structure was elucidated in 1988, after an 18-year effort. RuBisCo is the most abundant protein on Earth. Its estimated annual worldwide production by plants is 4×10^{13} g (40 million tons). RuBisCo is the enzyme that initiates the process of photosynthesis.

Even minor changes in the structure of a protein can have profound effects. Hemoglobin contains four polypeptide chains, each with 146 amino acid units. The substitution of valine for glutamic acid at one site in two of these chains gives rise to the sometimes fatal blood disease known as sickle-cell anemia. Apparently, the altered hemoglobin has a reduced ability to transport oxygen through the blood, although it does seem to provide some defense against malaria.

28-5   Aspects of Metabolism

Although living organisms differ markedly in appearance, there is a striking similarity in the chemical reactions of their life processes. We refer to the totality of these reactions as **metabolism**. Metabolic reactions in which substances are broken down are referred to as *catabolism*. Metabolic reactions in which more complex substances are synthesized from simpler ones are called *anabolism*. Reactions having $\Delta G > 0$ are said to be *endergonic*, and those with $\Delta G < 0$ are *exergonic*. The substances involved in metabolism are called *metabolites*. Metabolism is a complex subject that we can outline only briefly. The discussion that follows centers on the summary of metabolism given in Figure 28-17.

Carbohydrate Metabolism

In glucose-6-phosphate, a phosphate group replaces the —OH group on the C-6 atom of the cyclic glucose molecule (see Figure 28-7).

$$\underset{|}{\overset{|}{H_2C}}{-}O{-}\underset{\underset{O}{|}}{\overset{\overset{O^-}{|}}{P}}{-}OH$$

▶

Starch is the principal source of energy for humans and other animals. Digestion of starch begins in the mouth with the action of salivary enzymes, the amylases. Starch is converted to maltose and polysaccharides known as dextrins. This process continues briefly in the stomach (until the enzymes are denatured by the acid present), and the maltose and polysaccharides pass into the small intestine. Here, amylase from the pancreas completes the conversion of polysaccharides to maltose, and the enzyme *maltase* converts the maltose to glucose. Glucose is absorbed through the wall of the small intestine into the bloodstream and distributed to other organs.

Glucose is ultimately oxidized to carbon dioxide and water, with the liberation of energy. The principal intermediate in this process is glucose-6-phosphate

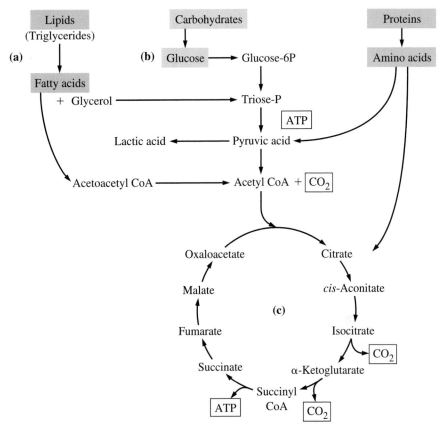

▲ **FIGURE 28-17   Metabolism outline**
(a) *Fatty acid section.* Fatty acids are degraded two C atoms at a time. Acetyl units enter into the citric acid cycle (c) as acetyl CoA. (b) *Glycolysis section (Embden–Meyerhof pathway).* These reactions are anaerobic (no oxygen required). Carbohydrates are degraded to the six-carbon sugar glucose, and then to the three-carbon triose-P (glyceraldehyde-3-phosphate). Next, the three-carbon acid pyruvic acid is formed from triose-P. Pyruvic acid loses a molecule of CO_2, yielding the two-carbon acetyl unit, which combines with coenzyme A (CoA) to form acetyl CoA. (c) *Citric acid cycle (Krebs cycle).* A two-carbon acetyl unit from acetyl CoA joins with the four-carbon oxalacetate unit to produce the six-carbon tricarboxylic acid citric acid (designated here as citrate). A two-step conversion to isocitrate occurs, followed by the loss of a molecule of CO_2 and formation of the five-carbon α-ketoglutarate. Another CO_2 molecule is lost in the formation of succinyl CoA. The remainder of the cycle involves a succession of four-carbon acids leading to oxaloacetate. The oxaloacetate regenerated at the end of the cycle now joins with another acetyl unit, and the cycle is repeated. The overall change occurring in the cycle is that a two-carbon acetyl unit enters the cycle and two molecules of CO_2 leave.

(glucose-6P). The pancreatic hormone *insulin* controls its formation. Once formed, glucose-6P may be converted to glycogen (a polysaccharide stored in the liver), revert back to glucose, or be metabolized. The major metabolic route (Figure 28-17) involves the anaerobic (absence of air) glycolysis pathway, followed by an aerobic cycle, the citric acid cycle.

Lipid Metabolism

The digestion of fats and oils occurs in the small intestine through the action of a combination of lipase enzymes. The products of this enzyme hydrolysis are glycerol, mixtures of mono- and diglycerides, and fatty acids. These are absorbed into

▶ The structure of glyceralde-hyde-3-phosphate is

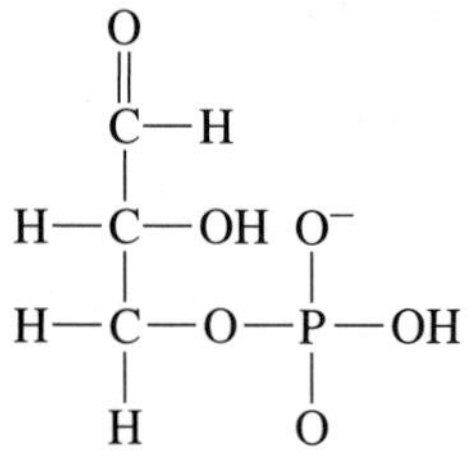

$$
\begin{array}{c}
\text{O} \\
\parallel \\
\text{C}\!-\!\text{H} \\
\mid \\
\text{H}\!-\!\text{C}\!-\!\text{OH} \quad \text{O}^- \\
\mid \qquad\qquad \mid \\
\text{H}\!-\!\text{C}\!-\!\text{O}\!-\!\text{P}\!-\!\text{OH} \\
\mid \qquad\qquad \mid \\
\text{H} \qquad\qquad \text{O}
\end{array}
$$

the bloodstream through the wall of the intestine. Glycerol is converted to glycer-aldehyde-3-phosphate (triose phosphate) and joins into the glucose metabolic route previously described. Fatty acids are oxidized to carbon dioxide and water, with the release of energy, in a series of reactions known as β oxidation. In this process, oxidation occurs at the β carbon atom of a fatty acid, followed by cleavage. Thus, two-carbon pieces (acetic acid) are split off. The process also requires coenzyme A (CoA). For example, with palmitic acid, $C_{15}H_{31}COOH$, the process is repeated seven times, with the formation of eight molecules of acetyl CoA, which enter the Krebs cycle (Figure 28-17).

Protein Metabolism

In the stomach, HCl(aq) and the enzyme *pepsin* hydrolyze about 10% of the amide linkages in proteins and produce polypeptides in the molecular mass range of 500 to several thousand atomic mass units. In the small intestine, peptidases such as trypsin and chymotrypsin (from the pancreas) cleave the polypeptides into very small fragments. These fragments are then acted on by aminopeptidase and carboxypeptidase. The resulting free amino acids pass through the wall of the intestine, into the bloodstream, and to all the cells of the body, where they are the building blocks of proteins. Protein synthesis is directed by the nucleic acids DNA and RNA (page 1149).

Energy Relationships in Metabolism

The fundamental agents involved in energy transfers from exergonic to endergonic reactions are adenosine *di*phosphate (**ADP**) and adenosine *tri*phosphate (**ATP**). The following equation represents the conversion of one mole of ATP to one mole of ADP.

▶ Additional thermodynamic aspects of ADP/ATP conversions were presented in Chapter 20 (page 812).

$$ATP^{4-} + H_2O \longrightarrow ADP^{3-} + HPO_4^{2-} + H^+ \qquad \Delta G^{\circ\prime} = -32.4 \text{ kJ} \qquad (28.6)$$

Figure 28-18 represents the reverse of reaction (28.6).

ADP, ATP models

▶ Based on the discussion of the expanded valence shell (page 408), we represent all the P-to-O bonds in Figure 28-18 as single bonds. Sometimes these structures are written with one P-to-O double bond in each phosphate unit.

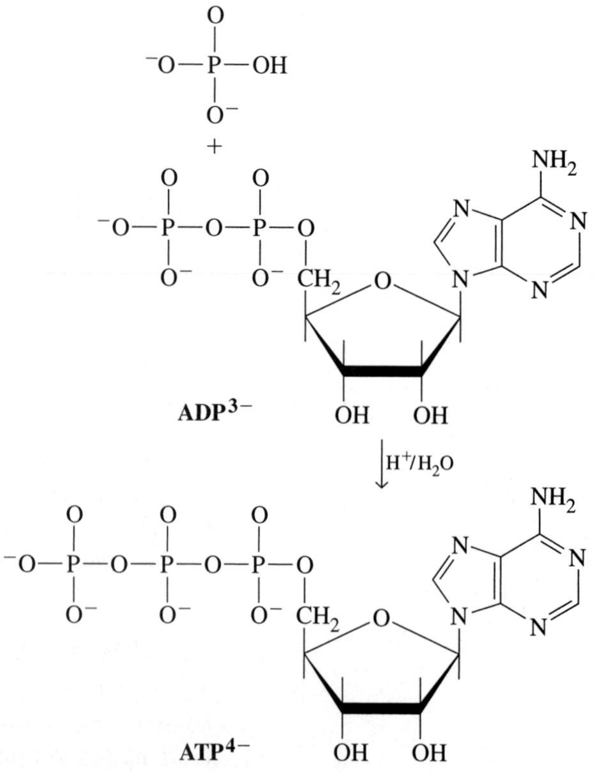

▶ **FIGURE 28-18**
Conversion of ADP to ATP
The H^+ ion entering into the reaction and the H_2O molecule produced are shown beside the main reaction arrow.

The energy released in the oxidation of foods is picked up by ADP, which is converted to ATP. Enzymes catalyze each step in the overall conversion. The oxidation of one mole of glucose to CO_2 and H_2O is accompanied by the conversion of 38 mol ADP to 38 mol ATP. These two processes are represented below.

$$C_6H_{12}O_6 + 6 O_2 \longrightarrow 6 CO_2 + 6 H_2O \quad \Delta G° = -2880 \text{ kJ}$$

$$38 \times \{ADP^{3-} + HPO_4^{2-} + H^+ \longrightarrow ATP^{4-} + H_2O\}$$
$$\Delta G° = 38 \times 32.4 \text{ kJ} = 1230 \text{ kJ}$$

Thus, of the 2880 kJ of energy released in the oxidation of one mole of glucose, 1230 kJ of energy is stored in the high-energy bonds of ATP. The efficiency of this energy storage is $(1230/2880) \times 100\% = 43\%$. That is, nearly half of the energy released in carbohydrate metabolism is stored in the human body for later use. This is a much more efficient use of energy than in an internal combustion engine, for example. If the metabolism of glucose occurred in a single step, with just one mole of ADP converted to one mole of ATP, the efficiency would drop to $(32.4/2880) \times 100\% = 1.1\%$. So we can see why the metabolism of sucrose is such a complex, multistep process.

Enzymes

An **enzyme** is a biological catalyst that contains protein. Enzymes are specific for each biological transformation and catalyze a reaction without requiring a change in temperature or pH. Originally, enzymes were assigned common or trivial names, such as pepsin and catalase. Present practice, however, is to name them after the processes they catalyze, usually employing an *ase* ending.

We introduced a model for enzyme action in our discussion of enzyme kinetics in Section 15-11. According to this model, an enzyme can exert its catalytic activity only after combining with the reacting substance, the *substrate*, to form a complex. The site on the enzyme where the substrate bonds is called the *active site*; some enzymes have more than one active site. Reaction of the substrate (S) with the enzyme (E) to form a complex (ES) permits the reaction to proceed via a path of lower activation energy than the noncatalyzed path. When the complex decomposes, products (P) are formed and the enzyme is regenerated.

$$E + S \rightleftharpoons ES \longrightarrow E + P$$

One example is the hydrolysis of sucrose. Recall from Figure 28-8 that sucrose is a disaccharide of glucose and fructose.

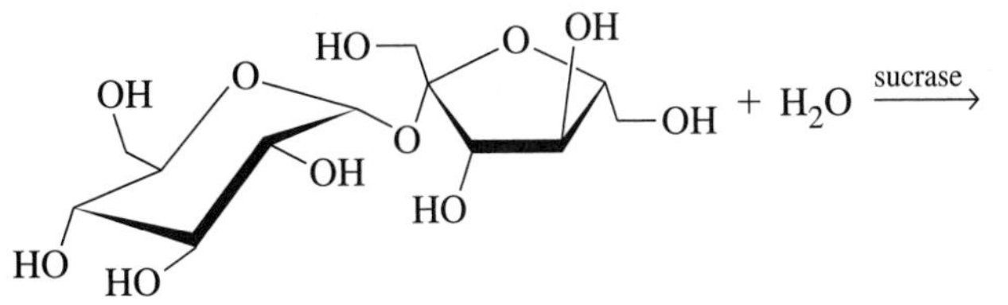

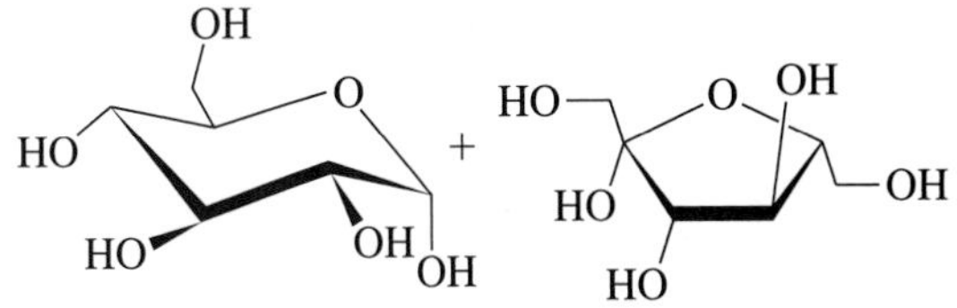

The first step is the binding of sucrose to the active site of the enzyme sucrase, as illustrated in Figure 28-19. When the sucrose is in the active site of the enzyme, a

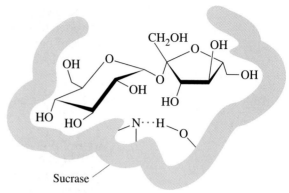

▶ FIGURE 28-19
Sucrase binding a sucrose molecule
The figure illustrates the "pocket" in the enzyme sucrase that the sucrose fits into. The globular protein sucrase (in blue) is much larger than sucrose.

number of chemical processes can take place. For example, a proton can be transferred from the sucrase to the glucose, which causes the active site of the sucrase to open up, causing the C—O bond joining fructose and glucose to weaken. The transfer of a proton from an amino acid side chain of the sucrase molecule may destroy hydrogen bonding within the enzyme and cause it to open up; this process stretches the sucrose molecule by pulling at each end, as suggested in Figure 28-20. In this distorted form, the C—O of the disaccharide is highly susceptible to attack by a water molecule. Figure 28-21 shows that after this reaction, glucose and fructose are formed and the proton transferred back to the sucrase, which causes the active site to return to its original state. The smaller fructose and glucose molecules leave the active site, and the enzyme is ready to go again.

Enzymes such as sucrase do not contain metal ions, but almost a third of known enzymes do. An enzyme with a metal ion in the active site is carboxypeptidase, a metalloenzyme active in the digestion of proteins into individual amino acids. In the process, each peptide bond between amino acids is cleaved. A water molecule is broken apart, the H and OH are added to the ends of the polypeptide chain, and an amino acid is released. Hydrolysis reactions of this type are the reverse of condensation reactions.

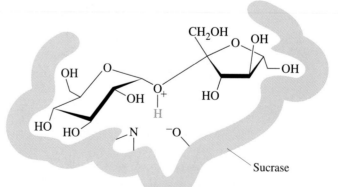

▲ FIGURE 28-20   **Change in enzyme conformation after binding of substrate**
The transfer of a proton (in orange) from the catalase to the sucrose is thought to open up the enzyme structure, leading to a weakening of the C—O bond in sucrose.

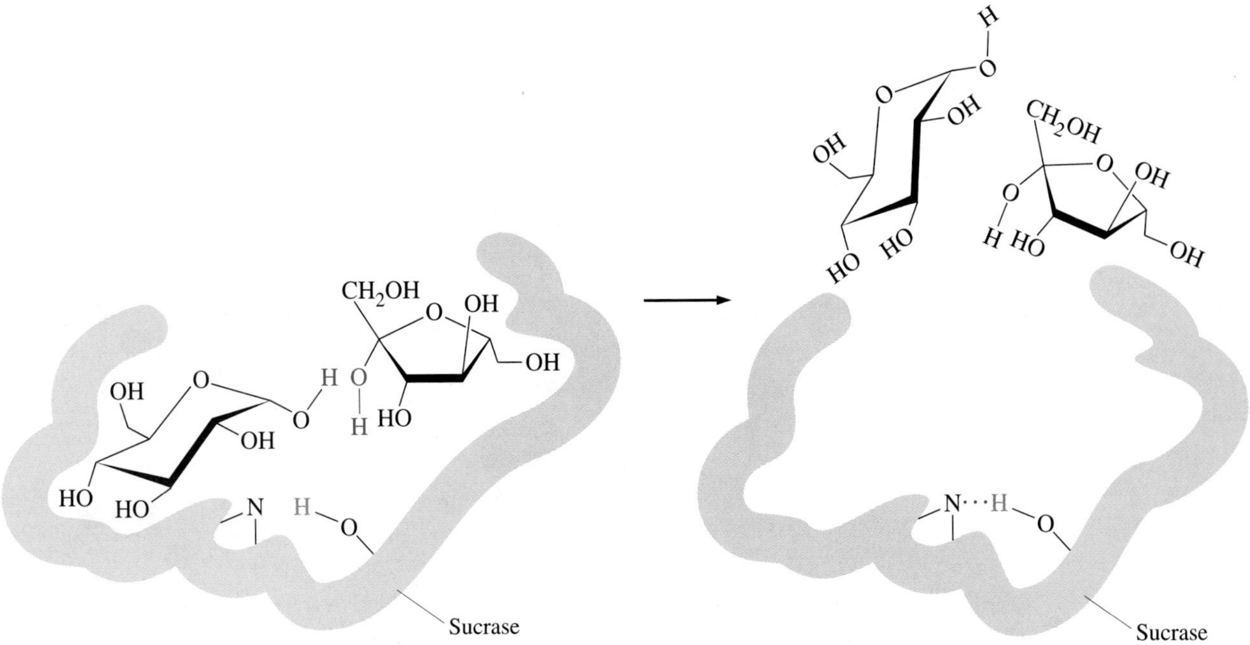

▲ FIGURE 28-21 Completion of the enzyme-catalyzed reaction
The weakened C—O bond in sucrose is susceptible to nucleophilic attack by a water molecule (in red) to produce glucose and fructose. These smaller molecules escape the pocket designed to fit sucrose. And the sucrase returns to its original state with the proton (orange) returned to the hydoxyl group.

Enzyme Catalysis animation

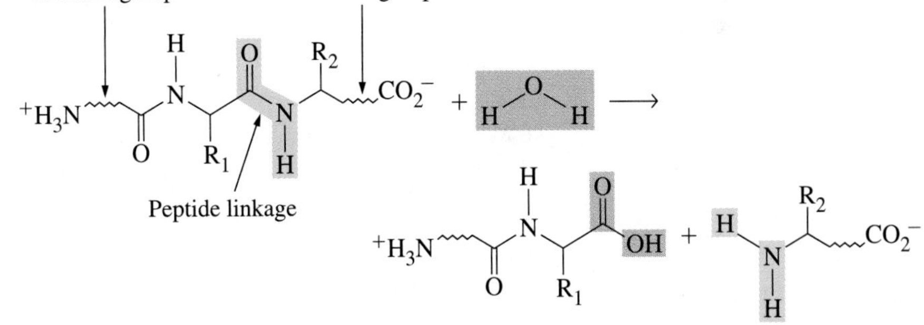

Carboxypeptidase cleaves amino acids one at a time, starting from the carboxyl end of the substrate protein. The enzyme consists of a single polypeptide chain containing 307 amino acids and a single Zn(II) ion. As shown in Figure 28-22, the metal ion is located in a cleft in the molecule, held in place by coordination to three amino acids—two histidines and one glutamic acid. The Zn(II) ion adopts an approximate tetrahedral geometry with the fourth coordination site occupied by a water molecule.

During the hydrolysis of a protein by carboxypeptidase, the protein substrate binds in the active site of the enzyme, as suggested by Figure 28-23. The substrate is held near the zinc center by hydrogen bonding to several amino acids in the cleft. The details of the mechanism have not been conclusively determined, except that it is known that it is the water coordinated to the Zn(II) that is used in the cleavage

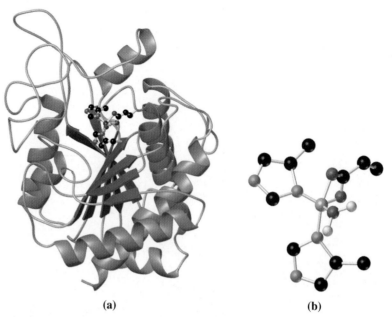

▲ FIGURE 28-22 **Structure of carboxypeptidase**
(a) In this illustration, the secondary structure of the helices (blue), pleated sheets (purple), and nonspecific secondary structure (orange) combine to give a visualization of the overall tertiary structure of carboxypeptidase. (b) Coordination of the Zn(II) ion in the active site.

▶ FIGURE 28-23
Binding of protein to the active site of carboxypeptidase
A schematic of the C-terminus of a protein bound at the active site of carboxypeptidase. Hydrogen bonding to enzyme side chains holds the protein in place to facilitate nucleophilic attack by the water molecule.

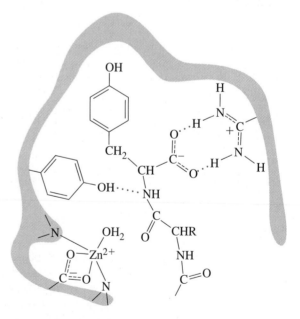

of the peptide bond. The proximity of the peptide linkage to the highly polarized water molecule coordinated to the Zn(II) ion permits the cleavage of the peptide bond and release of an amino acid.

The optimum temperature for enzyme activity is about 37 °C (98 °F) for enzymes present in warm-blooded animals. Above this temperature, enzyme activity declines as the secondary and tertiary structures are disrupted and the active site is distorted.

The two enzymes described here provide only a glimpse of the fascinating chemistry involved in the action of enzymes, which is currently a very active area of research.

28-6   Nucleic Acids

Lipids, carbohydrates, and proteins together with water constitute about 99% of most living organisms. The remaining 1% includes compounds of vital importance to the existence of life. Among these are the **nucleic acids**, which carry the information that directs the metabolic activity of cells. The two nucleic acids are **DNA**, or **deoxyribonucleic acid**, and **RNA**, or **ribonucleic acid**.

DNA is found in the cell nucleus on rodlike structures called chromosomes, which also contain protein. DNA contains the hereditary information that is passed from generation to generation. At specific locations along the chromosomes are *genes*, which are DNA sequences that control particular traits.

Figure 28-24 shows the chemical constituents of DNA and RNA. DNA is made up of the pentose sugar 2-deoxyribose, the purine bases adenine and guanine, and the pyrimidine bases cytosine and thymine. RNA differs in that it contains ribose instead of deoxyribose, and the base uracil instead of thymine. Both DNA and RNA contain phosphate groups. The combination of a pentose sugar and a purine or pyrimidine base is called a *nucleoside*. The combination of a nucleoside and a phosphate group is called a *nucleotide*. A DNA molecule is made up of two strands of nucleotides held together by hydrogen bonds and twisted into a helix. RNA is generally single stranded. Figure 28-25 represents a portion of a nucleic acid chain and shows how the constituents are joined by phosphate groups. If the sugar is 2-deoxyribose and the bases are adenine (A), guanine (G), thymine (T), and cytosine (C), the nucleic acid is DNA. If the sugar is ribose and the bases are adenine (A), guanine (G), uracil (U), and cytosine (C), the nucleic acid is RNA or ribonucleic acid.

The usual *double helix* form of DNA is shown in Figure 28-26. The postulation of this structure by Francis Crick and James Watson in 1953 was one of the great scientific breakthroughs of modern times. Their work was critically dependent on precise X-ray-diffraction studies of DNA done by Maurice Wilkins and Rosalind Franklin and on a set of regularities regarding the purine and pyrimidine bases in nucleic acids discovered by Erwin Chargaff. These regularities, known as the *base-pairing rules*, require the following:

1. The amount of adenine is equal to the amount of thymine (A = T).
2. The amount of guanine is equal to the amount of cytosine (G = C).
3. The total amount of purine bases is equal to the total amount of pyrimidine bases (G + A = C + T).

To maintain the structure of a double helix, hydrogen bonding must occur between the two single strands. The necessary conditions for hydrogen bonding exist only if an A on one strand appears opposite a T on the other, or if a G is opposite a C. C cannot be paired with T because the relatively small molecules (single ring) would not approach each other closely enough. The combination of G and A cannot occur because the molecules are too large (double rings).

DNA molecules have a unique ability to replicate—that is, to make exact copies of themselves. The critical step in DNA replication requires the molecule to unwind into single strands. As the unwinding occurs, free nucleotides present in the nucleus become attached to the exposed portions of the two single strands, converting each to a new double helix of DNA consisting of one old strand and one new strand (Figure 28-27).

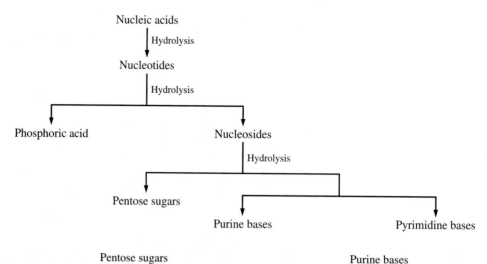

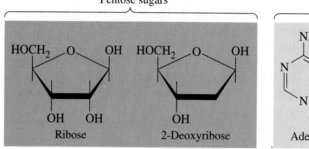

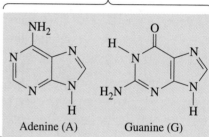

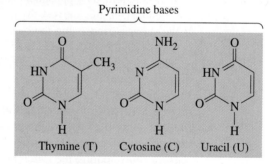

▲ FIGURE 28-24 **Constituents of nucleic acids**
Tracing the hydrolysis reactions in the reverse direction, the combination of a pentose sugar and a purine or a pyrimidine base yields a nucleoside. A nucleoside in combination with phosphoric acid yields a nucleotide. A nucleic acid is a polymer of nucleotides. If the sugar is 2-deoxyribose and the bases are A, G, T, and C, the nucleic acid is DNA. If the sugar is ribose and the bases are A, G, U, and C, the nucleic acid is RNA. (The term "2-deoxy" means without an O atom on the second C atom.)

There are two pieces of evidence, each very convincing, that the process just outlined does indeed occur. First, electron micrographs of the DNA molecule have been obtained that capture DNA in the act of replication. Another elegant experiment involves growing bacteria in a medium containing ^{15}N atoms so that all the N atoms of the bases of the DNA molecules are ^{15}N. The bacteria are then transferred to a nutrient with nucleotides containing normal ^{14}N. Here, the bacteria are allowed to divide and reproduce. The DNA of the offspring cells are then analyzed. Those of the first generation, for example, consist of DNA molecules with one

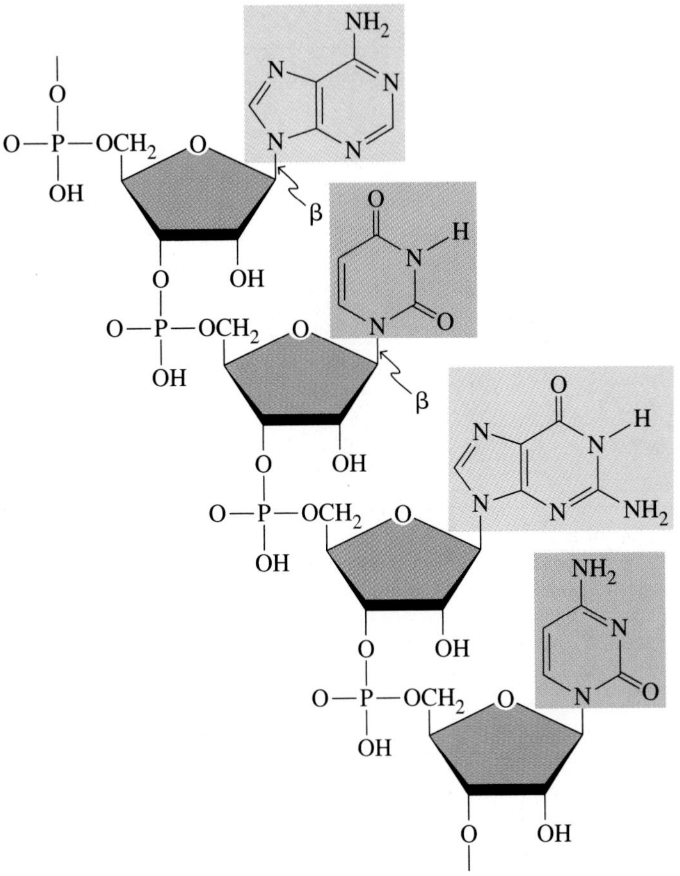

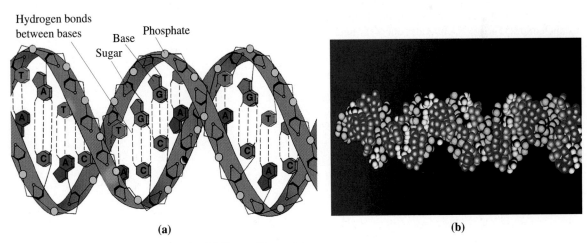

▲ FIGURE 28-25   **A portion of a nucleic acid chain**

Hydrogen bonds
between bases
Base   Phosphate
Sugar

(a)   (b)

▲ FIGURE 28-26   **DNA model**
(a) Schematic representation. (b) Space-filling model.

 DNA Segment model

strand having ^{15}N and the other ^{14}N atoms. This is exactly the result to be expected if replication occurs by the unzipping process described in Figure 28-27. The central importance of the sequence of bases in DNA is that they act as a chemical
template for the formation of an organisms proteins; that is, they act as a gene.
This, together with a discussion of RNA, is outlined in the Focus On feature of this
chapter.

The Genetic Code

Second base

	U	C	A	G
U	UUU UUC } Phe UUA UUG } Leu	UCU UCC UCA UCG } Ser	UAU UAC } Tyr UAA Stop UAG Stop	UGU UGC } Cys UGA Stop UGG Trp
C	CUU CUC CUA CUG } Leu	CCU CCC CCA CCG } Pro	CAU CAC } His CAA CAG } Gln	CGU CGC CGA CGG } Arg
A	AUU AUC AUA } Ile AUG Met	ACU ACC ACA ACG } Thr	AAU AAC } Asn AAA AAG } Lys	AGU AGC } Ser AGA AGG } Arg
G	GUU GUC GUA GUG } Val	GCU GCC GCA GCG } Ala	GAU GAC } Asp GAA GAG } Glu	GGU GGC GGA GGG } Gly

(First base — left column; Second base — top row)

▲ The genetic code

Trinucleotide sequences (*codons*) are used by ribosomes to identify the amino acids needed in protein synthesis. Note that the codons UAA, UAG, and UGA are signal codons. They signal the ribosome to stop protein synthesis.

The puzzle of protein synthesis is that it occurs outside the nucleus of the cell but is directed by DNA, which is found only inside the nucleus. How is the necessary information transmitted? Here is where RNA plays a key role. The following discussion relates to the process outlined in Figure 28-28.

In addition to unwinding for its own replication, DNA unwinds for the synthesis of a kind of RNA molecule called *messenger* RNA, or mRNA. The new strand of mRNA carries the genetic information from DNA coded in the sequence of its nucleotides. The mRNA migrates out of the nucleus into the cell cytoplasm, where it binds to cell components called ribo-somes. The combination of the mRNA and its ribosomes is called a polysome.

Another type of RNA called *transfer* RNA, or tRNA, comes in 20 different forms, each of which can bond with only one specific kind of amino acid. The function of tRNA is to bring a specific amino acid to a site on the ribosome where the amino acid can form a polypeptide bond and become part of a growing polypeptide chain. The ribosome moves along the mRNA chain, picking up different tRNA molecules and incorporating their amino acids into the polypeptide chain. When the ribosome reaches the end of the mRNA chain, it separates from the chain and releases the protein molecule that has been synthesized. The entire process is like the stringing of beads.

The code (set of directions) that determines the exact sequence of amino acids in the synthesis of a protein is incorporated in the chromosomal DNA. It is found in the particular pattern of base molecules on the double helix. Since there are only four different bases possible in an mRNA molecule—U, A, G, and C—but 20 different amino acids, it is clear that the code cannot correspond to individual base molecules. There are $4^2 = 16$ combinations of base molecules taken two at a time (i.e., UU, UA, UG, UC, etc.). But these could account for only 16 amino acids. When the base molecules are taken three at a time, there are $4^3 = 64$ possible combinations. A group of three base molecules in a DNA strand, called a triplet, causes a complementary set of base molecules to appear in the mRNA formed on it. In turn, this triplet, or *codon*, on the mRNA must be matched by a complementary triplet, called an *anticodon*, in a tRNA molecule. The particular tRNA with this anticodon carries a specific amino acid to the site of protein synthesis.

Following are the significant features of the genetic code shown in the table.

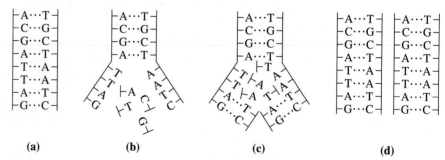

(a)   (b)   (c)   (d)

▲ FIGURE 28-27   Replication of a DNA molecule

As the unzipping process occurs from right to left, hydrogen bonds between the old DNA strands are broken. The new strands grow in the direction of the arrows by attaching nucleotides, which then form hydrogen bonds to the old DNA strands.

- There is more than one triplet code for most amino acids.
- The first two letters of the codon are most significant. There is considerable variation in the third.
- There are three "stop" codons that do not correspond to any amino acids; they end the synthesis of a polypeptide chain. There is also an "initiation" codon that begins the synthesis of a polypeptide chain.
- The various codons direct the incorporation of the same amino acids, whether in bacteria, plants, lower animals, or humans.

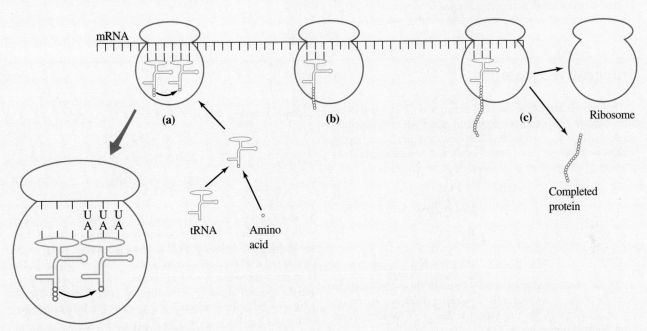

▲ FIGURE 28-28   **Protein synthesis**
(a) Through the action of an enzyme, a tRNA molecule brings a single amino acid to a site on a ribosome. The anticodon of the tRNA (AAA for the example shown) must be complementary to the codon of the mRNA (UUU). (The amino acid carried by this tRNA is phenylalanine.) The amino acid is added to the chain in the manner shown; the chain moves from the tRNA on the left to the one on the right. **(b)** As the ribosome moves along the mRNA strand, more and more amino acid units are added through the proper matching of tRNA molecules with the code on the mRNA. **(c)** When the ribosome reaches the end of the mRNA strand, it and the completed protein are released. The ribosome is free to repeat the process.

Summary

Four categories of substances found in living organisms are lipids, carbohydrates, proteins, and nucleic acids.

One familiar group of lipids is the triglycerides, esters of glycerol with long-chain monocarboxylic (fatty) acids. If saturated fatty acids predominate, the triglyceride is a fat. If unsaturation (in the form of double bonds) occurs in some of the fatty acid components, the triglyceride is an oil. Hydrolysis of a triglyceride with a strong base (saponification) yields glycerol and salts of the fatty acids—soaps.

The simplest carbohydrate molecules are five- and six-carbon-chain polyhydroxy aldehydes and ketones. These molecules, however, convert readily into cyclic structures with five- and six-membered rings. The subject is further complicated by the existence of chiral carbon atoms in these molecules, leading to optical isomers—molecules of identical composition that differ in their spatial orientations and optical activity. Monosaccharides, or simple sugars, can be bound together to form oligosaccharides. The most common of these are the disaccharides (such as sucrose) and trisaccharides. Polysaccharides contain from a few to a few thousand monosaccharide units. The most common of these are starch, glycogen, and cellulose.

The basic building blocks of proteins are some 20 different α-amino acids. Two amino acid molecules may join by eliminating a H_2O molecule between them. The result is a peptide bond. Long polypeptide chains—protein molecules—are formed as this process is repeated. Additional structural features include the twisting of a polypeptide chain into a helical coil and the twisting and folding of coils.

1153

Lipids, carbohydrates, and proteins are complex molecules; in metabolic processes, they are broken down into their simplest units. Carbohydrates are degraded into monosaccharides, proteins into amino acids, and lipids into glycerol and fatty acids. Ultimately, these degradation products are decomposed into still smaller molecules, such as CO_2, H_2O, NH_3, and urea. Energy released in catabolic processes is stored in adenosine triphosphate (ATP), which can then supply the energy needs of other processes.

Nucleic acids are made up of pentose sugars, purine and pyrimidine bases, and phosphate groups. If the sugar is ribose, the nucleic acid is RNA; if the sugar is 2-deoxyribose, the nucleic acid is DNA. Nucleic acids carry the information that directs the metabolic activity of cells and the replication of cells from one generation to another.

Integrative Example

L-Threonine is one of the essential amino acids. It is found in animal protein (for example, eggs and milk) but is deficient in some grains, such as rice. It is now common practice to fortify grains with the essential amino acids that would normally be missing. The structure of L-threonine is shown below.

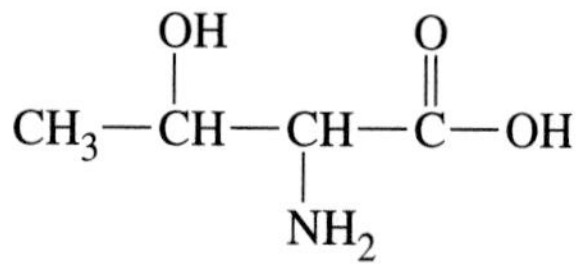

$$CH_3-\overset{\overset{\displaystyle OH}{|}}{CH}-\overset{\overset{\displaystyle }{|}}{\underset{\underset{\displaystyle NH_2}{|}}{CH}}-\overset{\overset{\displaystyle O}{||}}{C}-OH$$

$$pK_{a_1} = 2.15 \quad pK_{a_2} = 9.12$$

1. *Draw condensed structural formulas for L-threonine in strongly acidic solutions, at the isoelectric point, and in strongly basic solutions. The structures parallel those for the general case illustrated in equation (28.4).*

Acidic solution:

$$CH_3CH(OH)\underset{\underset{\displaystyle NH_3^+}{|}}{CH}COOH$$

Isoelectric point:

$$CH_3CH(OH)\underset{\underset{\displaystyle NH_3^+}{|}}{CH}COO^-$$

Basic solution:

$$CH_3CH(OH)\underset{\underset{\displaystyle NH_2}{|}}{CH}COO^-$$

2. *Write chemical equations for the ionization of L-threonine as an acid.* From the two pK values, we know that two equations are required. Note from equation (28.4) that when OH^- is added to an amino acid in acidic solution a proton is extracted from the $-COOH$ group before one is extracted from the $-NH_3^+$ group (that is, $-COOH$ is a stronger acid than $-NH_3^+$). Thus,

First ionization: $CH_3CH(OH)\underset{\underset{\displaystyle NH_3^+}{|}}{CH}COOH + H_2O \rightleftharpoons$

$$H_3O^+ + CH_3CH(OH)\underset{\underset{\displaystyle NH_3^+}{|}}{CH}COO^- \quad pK_{a_1} = 2.15$$

Second ionization: $CH_3CH(OH)\underset{\underset{\displaystyle NH_3^+}{|}}{CH}COO^- + H_2O \rightleftharpoons$

$$H_3O^+ + CH_3CH(OH)\underset{\underset{\displaystyle NH_2}{|}}{CH}COO^- \quad pK_{a_2} = 9.12$$

3. *Show that the isoelectric point of 5.6 in Table 28.3 is about what you would expect for L-threonine.* We showed in Chapter 18 (page 735) that the pH at the equivalence point in the titration of a weak diprotic acid is pH = $(pK_{a_1} + pK_{a_2})/2$. The titration of threonine is essentially that of a weak diprotic acid, and the product of the first neutralization step is the zwitterion. Thus, the pH at the isoelectric point, where the zwitterions predominate, is pH = $(2.15 + 9.12)/2 = 5.64$.

4. *Determine for L-threonine in a gel that is 0.25 M in NaH_2PO_4 and 0.50 M in Na_2HPO_4, the direction of migration when direct electric current is passed.* The key to this question is to recognize that the NaH_2PO_4/Na_2HPO_4 mixture is a buffer that establishes the pH of the gel. We know that L-threonine exists primarily as electrically neutral zwitterions at the isoelectric point. As the pH drops below pI, cations become increasingly important; and as the pH rises above pI, anions begin to predominate. We can calculate the pH of the gel using the Henderson–Hasselbalch equation and a value of pK_{a_2} from Table 17.4.

$$pH = pK_{a_2} + \log([HPO_4^{2-}]/[H_2PO_4^-]) =$$
$$7.20 + \log(0.50/0.25) = 7.50$$

The pH of the gel (7.50) exceeds pI for L-threonine (5.6) by nearly two units. We expect the anionic form, $CH_3CH(OH)CH(NH_2)COO^-$, to be the predominant charged species in the gel and to migrate to the *positive* electrode—the anode.

Key Terms

absolute configuration (28-3)
ADP (28-5)
α-amino acid (28-4)
ATP (28-5)
carbohydrate (28-3)
cell (28-1)
denaturation (28-4)
deoxyribonucleic acid (DNA) (28-6)
dextrorotatory (28-3)
diastereomer (28-3)
enzyme (28-5)
fat (28-2)

Fischer projection formula (28-3)
isoelectric point (28-4)
levorotatory (28-3)
lipid (28-2)
metabolism (28-5)
monosaccharide (28-3)
nucleic acid (28-6)
oil (28-2)
oligosaccharide (28-3)
peptide bond (28-4)
polypeptide (28-4)
polysaccharide (28-3)

primary structure (28-4)
protein (28-4)
quaternary structure (28-4)
racemic mixture (28-3)
reducing sugar (28-3)
ribonucleic acid (RNA) (28-6)
saponification (28-2)
secondary structure (28-4)
soap (28-2)
sugar (28-3)
tertiary structure (28-4)
triglyceride (28-2)

Review Questions

1. In your own words, define the following terms or symbols: (a) (+); (b) L; (c) sugar; (d) α-amino acid; (e) isoelectric point.
2. Briefly describe each of the following ideas, phenomena, or methods: (a) saponification; (b) chiral carbon atom; (c) racemic mixture; (d) denaturation of a protein.
3. Explain the important distinctions between each pair of terms: (a) fat and oil; (b) enantiomer and diastereomer; (c) primary and secondary structure of a protein; (d) DNA and RNA; (e) ADP and ATP.
4. Name the following compounds.

(a)
$$H_2CO-\overset{\displaystyle O}{\overset{\displaystyle \|}{C}}-C_{15}H_{31}$$
$$HCO-\overset{\displaystyle O}{\overset{\displaystyle \|}{C}}-C_{17}H_{29}$$
$$H_2CO-\overset{\displaystyle O}{\overset{\displaystyle \|}{C}}-C_{11}H_{23}$$

(b)
$$H_2CO-\overset{\displaystyle O}{\overset{\displaystyle \|}{C}}-C_{17}H_{33}$$
$$HCO-\overset{\displaystyle O}{\overset{\displaystyle \|}{C}}-C_{17}H_{33}$$
$$H_2CO-\overset{\displaystyle O}{\overset{\displaystyle \|}{C}}-C_{17}H_{33}$$

(c) $C_{13}H_{27}CO_2^- \, Na^+$

5. Write structural formulas for the following.
   (a) glyceryl palmitolauroeleostearate
   (b) tripalmitin
   (c) potassium myristate
   (d) butyl oleate

6. Which of the following statements best describes the way a sugar sample labeled DL-erythrose rotates the plane of polarized light: (1) to the left; (2) to the right; (3) first to the right and then to the left; (4) neither to the left nor the right? Explain.
7. From the given structure of L-(+)-arabinose, derive the structure of (a) D-(−)-arabinose; (b) a diastereomer of L-(+)-arabinose.

$$
\begin{array}{c}
\text{CHO} \\
\text{H}\!-\!\!-\!\!-\!\!-\text{OH} \\
\text{HO}\!-\!\!-\!\!-\!\!-\text{H} \\
\text{HO}\!-\!\!-\!\!-\!\!-\text{H} \\
\text{CH}_2\text{OH}
\end{array}
$$
L-(+)-Arabinose

8. How do the structures represented by the following three names resemble and differ from each other? (a) β-D-(+)-glucose, (b) D-(−)-arabinose, (c) D-(+)-glucose. (See also Review Question 7.)
9. Write the formulas of the species expected if the amino acid phenylalanine is maintained in (a) 1.0 M HCl; (b) 1.0 M NaOH; (c) a buffer solution with pH = 5.7.
10. Write the structures of (a) alanylcysteine; (b) threonylvalylglycine.
11. For the polypeptide Met-Val-Thr-Cys, (a) write the structural formula; (b) name the polypeptide.
    (*Hint:* Which is the N-terminal, and which is the C-terminal amino acid?)
12. With reference to Figure 28-25, identify the purine bases, the pyrimidine bases, the pentose sugars, and the phosphate groups. Is this a chain of DNA or RNA? Explain.

Exercises

Structure and Composition of the Cell

Exercises 13 to 16 refer to a typical Escherichia coli *bacterium. This is a cylindrical cell about 2 μm long and 1 μm in diameter, weighing about 2×10^{-12} g and containing about 80% water by volume.*

13. The intracellular pH is 6.4 and $[K^+] = 1.5 \times 10^{-4}$ M. Determine the number of **(a)** H_3O^+ ions and **(b)** K^+ ions in a typical cell.

14. Calculate the number of lipid molecules present, assuming their average molecular mass to be 700 u and the lipid content to be 2% by mass.

15. The cell is about 15% protein, by mass, with 90% of this protein in the cytoplasm. Assuming an average molecular mass of 3×10^4 u, how many protein molecules are present in the cytoplasm?

16. A single chromosomal DNA molecule contains about 4.5 million nucleotide units. If this molecule were extended so that the nucleotide units were 450 pm apart, what would be the length of the molecule? How does this compare with the length of the cell itself? What does this result suggest about the shape of the DNA molecule?

Lipids

17. Describe the similarities and differences between **(a)** trilaurin and trilinolein, **(b)** a soap and a phospholipid such as a lecithin.

18. Write a structural formula for a generic phosphatidic acid—an acid that produces a cephalin when esterified with ethanolamine or a lecithin when esterified with choline. Mono- and diglycerides are found in many processed foods. Write structural formulas for a generic monoglyceride and a generic diglyceride.

19. Oleic acid is a moderately unsaturated fatty acid. Linoleic acid is *polyunsaturated.* What structural feature characterizes polyunsaturated fatty acids? Is stearic acid polyunsaturated? Is eleostearic acid? Why do you suppose safflower oil is so highly recommended in dietary programs?

20. Corn oil and safflower oil are popular cooking oils containing mostly unsaturated fatty acids. They can be converted to solid fats by hydrogenation. Which would consume the greater amount of $H_2(g)$ in its hydrogenation to a solid fat, 1 kg of corn oil or 1 kg of safflower oil? Explain.

21. Write structural formulas to represent the products of the saponification of tripalmitin with NaOH(aq).

22. Calculate the maximum mass of the sodium soap that can be prepared from 105 g of glyceryl trimyristate.

Carbohydrates

23. Characterize the two sugars represented below, using the terminology from page 1129. Further indicate whether the sugar has a D or L configuration.

24. Write the structure for the straight-chain form of L-glucose. Does the structure determine if this isomer is levorotatory? Explain.

25. The following terms are all related to stereoisomers and their optical activity. Explain the meaning of each: **(a)** dextrorotatory; **(b)** levorotatory; **(c)** racemic mixture; **(d)** (*R*).

26. The following terms are all related to optical isomers. Explain the meaning of each: **(a)** diastereomers; **(b)** enantiomers; **(c)** (−); **(d)** D configuration.

27. Explain the meaning of the term *reducing sugar.* What structural feature characterizes a reducing sugar, and what is the test for a reducing sugar? What mass of $Cu_2O(s)$ should be produced when 0.500 g glucose is subjected to this test?

28. There are eight aldopentoses. Draw their structures, and indicate which are enantiomers.

29. The pure α and β forms of D-glucose rotate the plane of polarized light to the right by 112° and 18.7°, respectively (denoted as +112 and +18.7). Are these two forms of glucose enantiomers or diastereomers?

30. When a mixture of the pure α and β forms of D-glucose is allowed to reach equilibrium in solution, the rotation changes to +52.7 (a phenomenon known as mutarotation). What are the percentages of the α and β forms in the equilibrium mixture?
(*Hint:* Refer to Exercise 29.]

Fischer Projections and *R, S* Nomenclature

31. For each pair of Fischer projections, decide whether they represent the same compound, enantiomers, or diastereomers. Check your conclusion by designating the configuration at each stereocenter *R* or *S*.

(a)

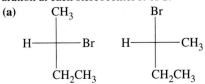
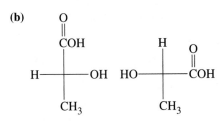

```
    CH₃                    Br
     |                      |
H────┼────Br        H──────┼──────CH₃
     |                      |
   CH₂CH₃                CH₂CH₃
```

(b)

```
     O
     ‖
     COH                   H     O
     |                     |     ‖
H────┼────OH        HO─────┼─────COH
     |                     |
    CH₃                   CH₃
```

(c)

```
    CH₃                   CH₃
     |                     |
H────┼────Br        H──────┼─────Br
     |                     |
H────┼────Br        H──────┼─────CH₃
     |                     |
    CH₃                   Br
```

(d)

```
     O                      H     O
     ‖                      |     ‖
     COH             HO─────┼─────C──OH
     |                      |
H────┼────OH         HO─────┼─────COH
     |                      |     ‖
HO───┼────H                 H     O
     |
     COH
     ‖
     O
```

32. For each pair of Fischer projections, decide whether they represent the same compound, enantiomers, or diastereomers. Check your conclusion by designating the configuration at each stereocenter *R* or *S*.

(a)

```
       CH₃                              NH₂
        |                                |
H───────┼───────CH₂CH₃   CH₃CH₂─────────┼─────────H
        |                                |
       NH₂                              CH₃
```

(b)

```
     O
     ‖
     COH                      NH₂
     |                         |
H────┼────NH₂       HOC───────┼───────H
     |                ‖        |
    CH₃               O       CH₃
```

(c)

```
    CH₃                   NH₂
     |                     |
H────┼────NH₂       H──────┼─────CH₃
     |                     |
H₂N──┼────H         H──────┼─────NH₂
     |                     |
    CH₃                   CH₃
```

(d)

```
    CH₂OH                          H
     |                             |      O
     C═O                           |      ‖
     |                      HO─────┼──────C──CH₂OH
HO───┼────H                        |
     |                      HO─────┼──────CH₂OH
H────┼────OH                       |
     |                             H
    CH₂OH
```

33. Redraw each of the following molecules as a Fischer projection; then assign *R* or *S* to each stereocenter.

(a)
```
H₃C        Cl
   \      /
    C────C
   /      \
  H        H
  |        |
  Cl       CH₃
```

(b)
```
      CH₃      CO₂H
OHC    \      /
   ◀────C────C
       /      ◀
     HO       OH
              CH₃
```

(c)
```
H₂N         OH
   \       /
    C────C──H
   /       ◀
 H₃C       CO₂H
  ◀
  H
```

(d)
```
CH₃         Br
   \       /
    C────C──H
   /       \
  H         CH₃
  |
  Cl
```

34. Redraw each of the following molecules as a Fischer projection; then assign *R* or *S* to each stereocenter.

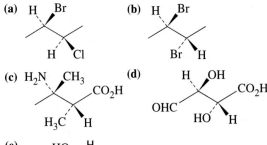

(a)
```
H   Br
 \  /
  \/
  /\
 H  Cl
```

(b)
```
H   Br
 \  /
  \/
  /\
 Br  H
```

(c)
```
H₂N   CH₃
   \  /
    \/     CO₂H
    /\    /
  H₃C  H
```

(d)
```
   H   OH
    \  /
     \/    CO₂H
     /\   /
  OHC  HO  H
```

(e)
```
   HO   H
    \  /
     \/    CO₂H
     /\   /
 HO₂C  HO  H
```

Amino Acids, Polypeptides, and Proteins

35. Describe what is meant by each of the following terms, using specific examples where appropriate: **(a)** α-amino acid; **(b)** zwitterion; **(c)** isoelectric point; **(d)** peptide bond; **(e)** tertiary structure.
36. Describe what is meant by each of the following terms, using specific examples where appropriate: **(a)** polypeptide; **(b)** protein; **(c)** N-terminal amino acid; **(d)** α helix; **(e)** denaturation.
37. A mixture of the amino acids lysine, proline, and aspartic acid is placed in a gel buffered at pH 6.3. An electric current is applied between an anode and a cathode immersed in the gel. Toward which electrode will each amino acid migrate?
38. A mixture of the amino acids histidine, glutamic acid, and phenylalanine is placed in a gel buffered at pH 5.7. An electric current is applied between an anode and a cathode immersed in the gel. Toward which electrode will each amino acid migrate?
39. Draw condensed structural formulas showing what form you would expect for the essential amino acid threonine **(a)** in strongly acidic solutions, **(b)** at the isoelectric point, and **(c)** in strongly basic solutions.
40. Draw condensed structural formulas for the following amino acids buffered at pH 6.0: **(a)** aspartic acid; **(b)** lysine; and **(c)** alanine.
41. Write the structures of **(a)** the different tripeptides that can be obtained from a combination of alanine, serine, and lysine; **(b)** the tetrapeptides containing two serine and two alanine amino acid units.
42. Write the structures of the different tetrapeptides that can be obtained from a combination of alanine, lysine, serine,

and phenylalanine. Give the abbreviated formula of each (such as Ala-Lys-Ser-Phe), starting at the N-terminal end.

43. After undergoing complete hydrolysis, a polypeptide yields the following amino acids: Gly, Leu, Ala, Val, Ser, Thr. Partial hydrolysis yields the following fragments: Ser-Gly-Val, Thr-Val, Ala-Ser, Leu-Thr-Val, Gly-Val-Thr. An experiment using a marker establishes that Ala is the N-terminal amino acid.
    (a) Establish the amino acid sequence in this polypeptide.
    (b) What is the name of the polypeptide?
44. After undergoing complete hydrolysis, a polypeptide yields the following amino acids: Ala, Gly, Lys, Ser, Phe. Partial hydrolysis yields the following fragments: Ala-Lys-Ser, Gly-Phe-Gly, Ser-Gly, Gly-Phe, Lys-Ser-Gly. An experiment using a marker establishes that Ala is the N-terminal amino acid.
    (a) Establish the amino acid sequence in this polypeptide.
    (b) What is the name of the polypeptide?
45. Describe what is meant by the primary, secondary, and tertiary structure of a protein. What is the quaternary structure? Do all proteins have a quaternary structure? Explain.
46. Sickle-cell anemia is sometimes referred to as a "molecular" disease. Comment on the appropriateness of this term.
47. The amino acid (R)-alanine is found in insect larvae. Draw the Fischer projection of this amino acid.
48. The amino acid (R)-serine is found in earthworms. Draw the Fischer projection of this amino acid.
49. Draw the dashed-wedged line structure for (S)-alanine and (S)-phenylalanine.
50. Draw the dashed-wedged line structure for (R)-proline and (S)-valine.

Nucleic Acids

51. What are the two major types of nucleic acids? List their principal components.
52. DNA has been called the "thread of life." Comment on the appropriateness of this expression.
53. If one strand of a DNA molecule has the sequence of bases AGC, what must be the sequence on the opposite strand?

Draw a structure of this portion of the double helix, showing all hydrogen bonds.

54. If one strand of a DNA molecule has the sequence of bases TCT, what must be the sequence on the opposite strand? Draw a structure of this portion of the double helix, showing all hydrogen bonds.

Integrative and Advanced Exercises

55. The protein molecule hemoglobin contains four iron atoms. The mass percent iron in hemoglobin is 0.34%. What is the molecular mass of hemoglobin?
56. A 1.00-mL solution containing 1.00 mg of an enzyme was deactivated by the addition of 0.346 μmol $AgNO_3$. What is the *minimum* molecular mass of the enzyme? Why does this calculation yield only a minimum value?
57. What minimum volume of natural gas that is 92.0% CH_4 and 8.0% C_2H_6 by volume, and measured at 25.5 °C and 756 mmHg pressure, is needed to produce the hydrogen consumed in converting 15.5 kg of the oil glyceryl tri-

oleate (triolein) to the fat glyceryl tristearate (tristearin)? The hydrogen is produced by steam re-forming the natural gas. Assume that natural gas and steam are the only reactants and that carbon dioxide is the only product other than hydrogen.

58. The text states that there are 16 possible aldohexoses. Draw their structures, and indicate which are enantiomers.
59. The term "epimer" is used to describe diastereomers that differ in the configuration about a *single* carbon atom. Which pairs of the eight possible aldopentoses are epimers?

60. The amino acid ornithine, not normally found in proteins, has the following structure at a pH of 1.0.

$$H_3\overset{+}{N}-CH_2-CH_2-CH_2-\underset{\underset{+NH_3}{|}}{CH}-\overset{\overset{O}{\|}}{C}-OH$$

$$pK_{a_1} = 1.94; \quad pK_{a_2} = 8.65; \quad pK_{a_3} = 10.76$$

What is the pI value of this amino acid?

61. In the experiment described on page 1150, the first-generation offspring of DNA molecules each contained one strand with ^{15}N atoms and one with ^{14}N. If the experiment were carried through a second, third, and fourth generation, what fractions of the DNA molecules would still have strands with ^{15}N atoms?

62. Bradykinin is a nonapeptide obtained by the partial hydrolysis of blood serum protein. It causes a lowering of blood pressure and an increase in capillary permeability. Complete hydrolysis of bradykinin yields three proline (Pro), two arginine (Arg), two phenylalanine (Phe), one glycine (Gly), and one serine (Ser) amino acid units. The N-terminal and C-terminal units are both arginine (Arg). In a hypothetical experiment, partial hydrolysis and sequence proof reveals the following fragments: Gly-Phe-Ser-Pro; Pro-Phe-Arg; Ser-Pro-Phe; Pro-Pro-Gly; Pro-Gly-Phe; Arg-Pro-Pro; Phe-Arg. Deduce the sequence of amino acid units in bradykinin.

63. With the aid of the table on page 1152, propose a plausible sequence on a DNA strand that would code for the synthesis of the polypeptide Ser-Gly-Val-Ala. Why is there more than one possible sequence for the DNA strand?

64. Refer to the Integrative Example. A 1.00-g sample of threonine is dissolved in 10.0 mL of 1.00 M HCl, and the solution is titrated with 1.00 M NaOH. Sketch a titration curve for the titration, indicating the approximate pH at representative points on the curve [such as the initial pH, the half-neutralization point(s) and the equivalence point(s)].

65. Projection formulas impose certain limitations and care must be taken in drawing and manipulating them. Show that, using (R)-(+)-glyceraldehyde as an example,
(a) Rotating the Fischer projection by 90° in the plane of the paper produces the opposite absolute configuration.

What is the effect of rotation by 180° and 270° in the plane of the paper?
(b) Interchanging two substituents converts one enantiomer into the other. What is the effect of two such interchanges?

66. A pentapeptide was isolated from a cell extract and purified. A portion of the compound was treated with 2,4-dinitrofluorobenzene (DNFB), and the resulting material hydrolyzed (recall Figure 28-10). Analysis of the hydrolysis products revealed one mole of DNP-methionine, two moles of methionine, and one mole each of serine and glycine. A second portion of the original compound was partially hydrolyzed and separated into four products. Separately, the four products were hydrolyzed further, giving the following four sets of compounds: (1) one mole of DNP-methionine, one mole of methionine, and one mole of glycine; (2) one mole of DNP-methionine and one mole of methionine; (3) one mole of DNP-serine and one mole of methionine; (4) one mole of DNP-methionine, one mole of methionine, and one mole of serine. What is the amino acid sequence of the pentapeptide?

67. Eighteen of the nineteen l-amino acids have the S configuration at the α carbon (the first carbon after the carboxyl carbon). Cysteine is the only l-amino acid that has an R configuration. Explain.

68. The amino acid threonine, $(2S,3R)$-2-amino-3-hydroxybutanoic acid, has two stereocenters. Draw the Fischer projection of threonine. How many other possible stereoisomers are there of threonine?

69. Coupling ATP hydrolysis to a thermodynamically unfavorable reaction can shift the equilibrium of a reaction (see Chapter 20).
(a) Calculate K for the hypothetical reaction A $\longrightarrow$ B when $\Delta G^{\circ\prime}$ is 23 kJ mol^{-1} at 25°C.
(b) Calculate K for the same reaction when it is coupled to the hydrolysis of ATP ($\Delta G^{\circ\prime} = -30$ kJ mol^{-1}). Compare with the value obtained in part (a).
(c) Many cells maintain [ATP] to [ADP] ratios of 400 or more. Calculate the ratio of [B] to [A] when [ATP]/[ADP] = 400 and [P$_i$] = 5 mM. Compare this ratio to that for the uncoupled reaction.

Feature Problems

70. Useful information about the structures of triglycerides can be obtained through saponification reactions. *The saponification value of a triglyceride is the number of milligrams of KOH required to saponify 1.00 g of the triglyceride.* If there is unsaturation in the fatty acids in triglycerides, this can be discovered by studying reactions in which I_2 is added across double bonds. *The iodine number is the number of grams of I_2 that reacts with 100 g of a triglyceride.*
(a) Determine the saponification value of glyceryl tristearate and the iodine number of glyceryl trioleate.
(b) Castor oil is a mixture of triglycerides with about 90% of its fatty acid content in ricinoleic acid,

$CH_3(CH_2)_5CHOHCH_2CH=CH(CH_2)_7COOH$. Estimate the saponification value and the iodine number of castor oil.
(c) Based on the composition given in Table 28.2, estimate the range of values for the saponification value and iodine number of safflower oil.

71. The so-called Ruff degradation is a chain-shortening reaction in which an aldose chain is shortened by one C atom, hexoses, for example, being converted into pentoses. In the Ruff degradation, the calcium salt of an aldonic acid (the corresponding carboxylic acid of an aldose) is oxidized with hydrogen peroxide. Ferric ion catalyzes the reaction. The calcium salt of the aldonic acid necessary for the Ruff degradation is obtained by oxidizing an aldose with an aqueous solution of bromine and then adding calcium hydroxide. The reaction scheme is as follows

$$RCHOHCHO \xrightarrow[\text{2. Ca (OH}_2)]{\text{1. Br}_2, \text{H}_2\text{O}} RCHOHCO_2^- (Ca^{2+})_{\frac{1}{2}}$$

$$\downarrow^{H_2O_2}_{Fe^{3+}(aq)}$$

$$RCHO + CO_2$$

where R represents the rest of the chain of the aldose.
(a) Show that D-glucose can be degraded into D-arabinose. Which other aldose can be degraded into D-arabinose?
(b) Which two monosaccharides can be degraded into D-glyceraldehyde by employing the Ruff degradation only once?

72. If D-(+) glyceraldehyde is treated with HCN in aqueous solution under basic conditions for three days at room temperature, cyanohydrins are formed (see Chapter 27). The cyanohydrins are not isolated, but are hydrolyzed to hydroxyacids in the same reaction mixture using dilute sulfuric acid. In this process, a new stereocenter was formed in the molecule. The products are diastereomers, formed in unequal amounts, and separable from each other by recrystallization because of their different physical properties, including solubilities. The trihydroxybutanoic acids were separated and then oxidized to tartaric

acid with dilute nitric acid, which oxidizes only the primary alcohol group.
(a) Ignoring stereochemistry, draw the reaction sequence for the transformations described above and hence deduce the structure of tartaric acid.
(b) Starting from the Fischer projection of D-(+) glyceraldehyde and using the reaction scheme from part (a), draw Fischer projections of the two trihydroxybutanoic acids formed and designate the chiral centers as *R* or *S*.
(c) Starting from the Fischer projection of D-(+) glyceraldehyde and using the reaction scheme from part (a), draw Fischer projections of the two forms of tartaric formed and designate the chiral centers as *R* or *S*.
(d) One form of tartaric acid obtained is optically active, rotating the plane of polarized light in a negative sense(−). The other isomer formed, called *meso*-tartaric acid, is not optically active. Explain why the other isomer is not optically active. Draw the dashed-wedged line structure that corresponds to the Fischer projection of *meso*-tartaric acid, Can you describe how the two halves of the molecule are related? Using Fischer projections, write equations for the conversion of L-(−)-glyceraldehyde to tartaric acid. Show clearly the stereochemistry of the tartaric acids that are formed, and indicate whether you expect them to be optically active.

---

eMedia Exercises

73. The formation of common table sugar (sucrose) is shown in the **Disaccharide Structure** activity (*eChapter 28-3*).
(a) Write the molecular formula of sucrose in the form describing this molecule as a carbohydrate $C_x(H_2O)_y$. **(b)** Calculate the number of molecules of sucrose found in 1.0 gram of sugar.

74. The characteristic structure of polypeptides is illustrated in the **Proteins and Amino Acids** animation (*eChapter 28-4*). **(a)** Write out the entire molecular structure for the tripeptide Lys-Ala-Phe. **(b)** What aspects of polypeptide structures allow them to be abbreviated in the form Lys-Ala-Phe? **(c)** How many different tripeptides could be formed from these three amino acids?

75. (a) From the **Enzyme Catalysis** animation (*eChapter 28-5)*, describe the role of primary and secondary structure

in the selectivity of the catalyst. **(b)** In this animation, in what sense is the reaction catalytic?

76. View the **DNA Segment** model found in *eChapter 28-6*. There are a number of options for adjusting the image of this segment under the "Display" menu (backbone, ribbon, strands, cartoon). You can also modify the representation of only a portion of the molecule by first using the Select > Nucleic submenu to pick out a particular structure. **(a)** Why are multiple depictions of the DNA structure useful? **(b)** Adjust the display to show the backbone of the segment in ball-and-stick view, and the bases in wireframe view, and adjust the Options to display the hydrogen bonds. What do you learn about the tertiary structure of DNA from this view?

Mathematical Operations

A-1 Exponential Arithmetic

Measured quantities in this text range from very small to very large. For example, the mass of an individual hydrogen atom is 0.00000000000000000000000167 g, and the number of molecules in 18.0153 g of the substance water is 602,214,000,000,000,000,000,000. These numbers are difficult to write in conventional form and are even more cumbersome to handle in numerical calculations. We can greatly simplify them by expressing them in exponential form. The *exponential form* of a number consists of a coefficient (a number with value between 1 and 10) multiplied by a power of 10.

The number 10^n is the *nth power* of 10. If n is a *positive* quantity, 10^n is *greater than 1*. If n is a *negative* quantity, 10^n is *between 0 and 1*. The value of $10^0 = 1$.

Positive powers   *Negative powers*

$$10^0 = 1 \qquad\qquad\qquad 10^0 = 1$$

$$10^1 = 10 \qquad\qquad\qquad 10^{-1} = \frac{1}{10} = 0.1$$

$$10^2 = 10 \times 10 = 100 \qquad\qquad 10^{-2} = \frac{1}{10 \times 10} = \frac{1}{10^2} = 0.01$$

$$10^3 = 10 \times 10 \times 10 = 1000 \qquad 10^{-3} = \frac{1}{10 \times 10 \times 10} = \frac{1}{10^3} = 0.001$$

To express the number 3170 in exponential form, we write

$$3170 = 3.17 \times 1000 = 3.17 \times 10^3$$

For the number 0.00046 we write

$$0.00046 = 4.6 \times 0.0001 = 4.6 \times 10^{-4}$$

A simpler method of converting a number to exponential form that avoids intermediate steps is illustrated below.

$$\underset{3\ 2\ 1}{3\,1\,7\,0} = 3.17 \times 10^3$$

$$\underset{1\ 2\ 3\ 4}{0.0\,0\,0\,4\,6} = 4.6 \times 10^{-4}$$

That is, to convert a number to exponential form,

- Move the decimal point to obtain a coefficient with value between 1 and 10.
- The exponent (power) of 10 is equal to the number of places the decimal point is moved.
- If the decimal point is moved *to the left,* the exponent of 10 is *positive.*
- If the decimal point is moved *to the right,* the exponent of 10 is *negative.*

To convert a number from exponential form to conventional form, move the decimal point the number of places indicated by the power of 10. That is,

$$6.1 \times 10^6 = 6.1\underset{1\ 2\ 3\ 4\ 5\ 6}{00000} = 6,100,000$$

$$8.2 \times 10^{-5} = 0.\underset{5\ 4\ 3\ 2\ 1}{00008}2 = 0.000082$$

Electronic calculators designed for scientific and engineering work easily accommodate exponential numbers. A typical procedure is to key in the number, followed by the key "EXP" or "EE." Thus, the keystrokes required for the number 6.57×10^3 are

$$\boxed{6}\ \boxed{.}\ \boxed{5}\ \boxed{7}\ \boxed{\text{EXP}}\ \boxed{3}$$

and the result displayed is $\boxed{6.57^{03}}$

For the number 6.25×10^{-4}, the keystrokes are

$$\boxed{6}\ \boxed{.}\ \boxed{2}\ \boxed{5}\ \boxed{\text{EXP}}\ \boxed{4}\ \boxed{\pm}$$

and the result displayed is $\boxed{6.25^{-04}}$

Some calculators have a mode setting that automatically converts all numbers and calculated results to the exponential form, regardless of the form in which numbers are entered. In this mode setting you can generally also set the number of significant figures to be carried in displayed results.

▶ The instructions given here are for a typical electronic calculator. The keystrokes required with your calculator may be somewhat different. Look for specific instructions in the instruction manual supplied with the calculator.

Addition and Subtraction.   To add or subtract numbers written in exponential form, first express each quantity as *the same power of 10*. Then add and/or subtract the coefficients as indicated. That is, treat the power of 10 as you would a unit common to the terms being added and/or subtracted. In the example that follows, convert 3.8×10^{-3} to 0.38×10^{-2} and use 10^{-2} as the common power of 10.

$$\left(5.60 \times 10^{-2}\right) + \left(3.8 \times 10^{-3}\right) - \left(1.52 \times 10^{-2}\right) = (5.60 + 0.38 - 1.52) \times 10^{-2}$$

$$= 4.46 \times 10^{-2}$$

Multiplication.   Consider the numbers $a \times 10^y$ and $b \times 10^z$. Their product is $a \times b \times 10^{(y+z)}$. *Coefficients are multiplied, and exponents are added.*

$$0.0220 \times 0.0040 \times 750 = \left(2.20 \times 10^{-2}\right)\left(4.0 \times 10^{-3}\right)\left(7.5 \times 10^2\right)$$

$$= (2.20 \times 4.0 \times 7.5) \times 10^{(-2-3+2)} = 66 \times 10^{-3}$$

$$= 6.6 \times 10^1 \times 10^{-3} = 6.6 \times 10^{-2}$$

Division.   Consider the numbers $a \times 10^y$ and $b \times 10^z$. Their quotient is

$$\frac{a \times 10^y}{b \times 10^z} = (a/b) \times 10^{(y-z)}$$

Coefficients are divided, and the exponent of the denominator is subtracted from the exponent of the numerator.

$$\frac{20.0 \times 636 \times 0.150}{0.0400 \times 1.80} = \frac{\left(2.00 \times 10^1\right)\left(6.36 \times 10^2\right)\left(1.50 \times 10^{-1}\right)}{\left(4.00 \times 10^{-2}\right) \times 1.80}$$

$$= \frac{2.00 \times 6.36 \times 1.50 \times 10^{(1+2-1)}}{(4.00 \times 1.80) \times 10^{-2}} = \frac{19.1 \times 10^2}{7.20 \times 10^{-2}}$$

$$= 2.65 \times 10^{(2-(-2))} = 2.65 \times 10^4$$

Raising a Number to a Power.   To "square" the number $a \times 10^y$ means to determine the value $(a \times 10^y)^2$, or the product $(a \times 10^y)(a \times 10^y)$. According to the rule for multiplication, this product is $(a \times a) \times 10^{(y+y)} = a^2 \times 10^{2y}$. When an exponential number is raised to a power, *the coefficient is raised to that power and the exponent is multiplied by the power.* For example,

$$(0.0034)^3 = (3.4 \times 10^{-3})^3 = (3.4)^3 \times 10^{3\times(-3)} = 39 \times 10^{-9} = 3.9 \times 10^{-8}$$

Extracting the Root of an Exponential Number.   To extract the root of a number is the same as raising the number to a fractional power. This means that the square root of a number is the number to the one-half power; the cube root is the number to the one-third power; and so on. Thus,

$$\sqrt{a \times 10^y} = (a \times 10^y)^{1/2} = a^{1/2} \times 10^{y/2}$$

$$\sqrt{156} = \sqrt{1.56 \times 10^2} = (1.56)^{1/2} \times 10^{2/2} = 1.25 \times 10^1 = 12.5$$

In the following example, where the cube root is sought, the exponent (-5) is not divisible by 3; the number is rewritten with an exponent (-6) that is divisible by 3.

$$(3.52 \times 10^{-5})^{1/3} = (35.2 \times 10^{-6})^{1/3} = (35.2)^{1/3} \times 10^{-6/3} = 3.28 \times 10^{-2}$$

A-2   Logarithms

The *common* logarithm (log) of a number (N) is the exponent (x) to which the base 10 must be raised to yield the number N. That is, $\log N = x$ means that $N = 10^x = 10^{\log N}$. For simple powers of ten, for example,

$$\log 1 = \log 10^0 = 0$$

$$\log 10 = \log 10^1 = 1 \qquad \log 0.10 = \log 10^{-1} = -1$$

$$\log 100 = \log 10^2 = 2 \qquad \log 0.01 = \log 10^{-2} = -2$$

Most of the numbers that result from measurements and appear in calculations are not simple powers of 10, but it is not difficult to obtain logarithms of these numbers with an electronic calculator. To find the logarithm of a number, enter the number, followed by the "LOG" key.

$$\log 734 = 2.866$$

$$\log 0.0150 = -1.824$$

Another common example requires us to find the number having a certain logarithm. This number is often called the *antilogarithm* or the *inverse* logarithm. For example, if $\log N = 4.350$, what is N? N, the antilogarithm, is simply $10^{4.350}$, and to find its value we enter 4.350, followed by the key "10^x." Depending on the calculator used, it is usually necessary to press the key "INV" or "2nd F" before the log key.

$$\log N = 4.350$$

$$N = 10^{4.350}$$

$$N = 2.24 \times 10^4$$

If the task is to find the antilogarithm of -4.350, we again note that $N = 10^{-4.350}$, and $N = 4.47 \times 10^{-5}$. The required keystrokes on a typical electronic calculator are

$$\boxed{4}\ \boxed{.}\ \boxed{3}\ \boxed{5}\ \boxed{0}\ \boxed{\pm}\ \boxed{\text{INV}}\ \boxed{\log}$$

and the display, to three significant figures, is

$$\boxed{4.47^{-05}}$$

Some Useful Relationships. From the definition of a logarithm we can write $M = 10^{\log M}$, $N = 10^{\log N}$, and $M \times N = 10^{\log(M \times N)}$. This means that

$$\log(M \times N) = \log M + \log N$$

Similarly, it is not difficult to show that

$$\log \frac{M}{N} = \log M - \log N$$

Finally, because $N^2 = N \times N$, $10^{\log N^2} = 10^{\log N} \times 10^{\log N}$, and

$$\log N^2 = \log N + \log N = 2 \log N$$

Or, in more general terms,

$$\log N^a = a \log N$$

This expression is especially useful for extracting the roots of numbers. Thus, to determine $(2.5 \times 10^{-8})^{1/5}$, we write

$$\log(2.5 \times 10^{-8})^{1/5} = \tfrac{1}{5}\log(2.5 \times 10^{-8}) = \tfrac{1}{5}(-7.60) = -1.52$$

$$(2.5 \times 10^{-8})^{1/5} = 10^{-1.52} = 0.030$$

Significant Figures in Logarithms. To establish the number of significant figures to use in a logarithm or antilogarithm, use this fundamental rule: All digits to the *right* of the decimal point in a logarithm are significant. Digits to the *left* are used to establish the power of 10. Thus, the logarithm -2.08 is expressed to *two* significant figures. The antilogarithm of -2.08 should also be expressed to *two* significant figures; it is 8.3×10^{-3}. To help settle this point, take the antilogarithms of $-2.07, -2.08$, and -2.09. You will find these antilogs to be $8.5 \times 10^{-3}, 8.3 \times 10^{-3}$, and 8.1×10^{-3}, respectively. Only *two* significant figures are justified.

Natural Logarithms. Logarithms can be expressed to a base other than 10. For instance, because $2^3 = 8$, $\log_2 8 = 3$ (read as "the logarithm of 8 to the base 2 is equal to 3"). Similarly, $\log_2 10 = 3.322$. Several equations in this text are derived by the methods of calculus and involve logarithms. These equations require that the logarithm be a "natural" one. A *natural* logarithm has the base $e = 2.71828\ldots$. A logarithm to the base "e" is usually denoted as *ln*.

The relationship between a "natural" and "common" logarithm simply involves the factor $\log_e 10 = 2.303$. That is, for the number N, $\ln N = 2.303 \log N$. The methods and relationships described for logarithms and antilogarithms to the base 10 all apply to the base e as well, except that the relevant keys on an electronic calculator are "ln" and "e^x" rather than "LOG" and "10^x."

A-3 Algebraic Operations

An algebraic equation is solved when one of the quantities, the unknown, is expressed in terms of all the other quantities in the equation. This effect is achieved when the unknown is present, *alone,* on one side of the equation, and the rest of the terms are on the other side. To solve an equation, a rearrangement of terms may be necessary. The basic principle governing these rearrangements is quite simple. *Whatever is done to one side of the equation must be done to the other as well.*

$$3x^2 + 6 = 33 \qquad \textit{Solve for x.}$$

$$3x^2 + 6 - 6 = 33 - 6 \qquad \text{(1) Subtract 6 from each side.}$$

$$3x^2 = 27$$

$$\frac{3x^2}{3} = \frac{27}{3} \qquad \text{(2) Divide each side by 3.}$$

$$x^2 = 9$$

$$\sqrt{x^2} = \sqrt{9} \qquad \text{(3) Extract the square root of each side.}$$

$$x = 3 \qquad \text{(4) Simplify. The square root of 9 is 3.}$$

Quadratic Equations. A quadratic equation has the form $ax^2 + bx + c = 0$, where a, b, and c are constants (a cannot be equal to 0). A number of calculations in the text require us to solve a quadratic equation. At times, quadratic equations are of the form

$$(x + n)^2 = m^2$$

Such equations can be solved by extracting the square root of each side.

$$x + n = \pm m \qquad \text{and} \qquad x = m - n \qquad \text{or} \qquad x = -m - n$$

More likely, however, the *quadratic formula* will be needed.

$$x = \frac{-b \pm \sqrt{b^2 - 4ac}}{2a}$$

In Example 16-13 on page 653, the following equation must be solved.

$$\frac{(0.300 - x)}{(0.200 + x)(0.100 + x)} = 2.98$$

This is a quadratic equation, but before the quadratic formula can be applied, the equation must be rearranged to the standard form: $ax^2 + bx + c = 0$. This is accomplished in the steps that follow.

$$(0.300 - x) = 2.98(0.200 + x)(0.100 + x)$$

$$0.300 - x = 2.98\left(0.0200 + 0.300x + x^2\right)$$

$$0.300 - x = 0.0596 + 0.894x + 2.98x^2$$

$$2.98x^2 + 1.894x - 0.240 = 0 \qquad\qquad \text{(A.1)}$$

Now we can apply the quadratic formula.

$$x = \frac{-1.894 \pm \sqrt{(1.894)^2 + (4 \times 2.98 \times 0.240)}}{2 \times 2.98}$$

$$= \frac{-1.894 \pm \sqrt{3.587 + 2.86}}{2 \times 2.98}$$

$$= \frac{-1.894 \pm \sqrt{6.45}}{2 \times 2.98} = \frac{-1.894 \pm 2.54}{2 \times 2.98}$$

$$= \frac{-1.894 + 2.54}{2 \times 2.98} = \frac{0.65}{5.96} = 0.11$$

Note that only the $(+)$ value of the $(\pm)$ sign was used in solving for x. If the $(-)$ value had been used, a negative value of x would have resulted. However, for the given situation a negative value of x is meaningless.

The Method of Successive Approximations.   The quadratic equation that was just solved using the quadratic formula can be solved by an alternative method that can be extended to equations of higher order, such as the cubic, quartic, and quintic equations often encountered in solving equilibrium problems. To illustrate the method suppose we wish to solve (A.1) without recourse to the quadratic formula. We can rewrite the equation as follows

$$x = \frac{2.98x^2 - 0.240}{-1.894}$$

and make a guess at the value of x, which we substitute into the right hand side of the equation to calculate a new value of x. If we guess 0.15, which is reasonable given the starting concentrations involved in Example 16-13, we calculate

$$x = \frac{2.98 \times (0.15)^2 - 0.240}{-1.894} = 0.091$$

We can now use this value of x to calculate a new one.

$$x = \frac{2.98 \times (0.091)^2 - 0.240}{-1.894} = 0.114$$

Repeating this procedure one more time, we get

$$x = \frac{2.98 \times (0.114)^2 - 0.240}{-1.894} = 0.106$$

One more attempt gives a value of 0.11, which is in agreement with the answer previously obtained. The method that we have just used is called the method of successive approximations.

Let us now apply the method of successive approximations to the equation obtained in the Integrative Example of Chapter 16, namely

$$256x^5 - 0.74\left[(1.00 - x)(1.00 - 2x)^2\right] = 0 \qquad \text{(A.2)}$$

The approach we can take here is to guess a value of x; evaluate the expression to see how close to zero it comes; and then adjust the value of x accordingly. Let us start with a guess of 0.40. The result is

$$256(0.40)^5 - 0.74\left[(1.00 - 0.40)(1.00 - 2 \times 0.40)^2\right] = 2.60$$

Clearly the value of 0.40 is too large. If we now try 0.10 we obtain a value of −0.42. We have overshot the value of x. We can now try a value of 0.25 (halfway between our two previous guesses) and obtain 0.11. We realize now that we have to reduce the guessed value of x slightly to get closer to zero. If we try 0.20 we obtain −0.13, and if we next try the value $x = 0.225$ we obtain −0.03. We are now very close to our goal of finding the value of x that satisfies the expression. One final guess of 0.23 gives a value of −0.001, a very satisfactory result.

An alternative approach is to rewrite expression (A.2) as

$$x^5 = \frac{0.74\left[(1.00 - x)(1.00 - 2x)^2\right]}{256}$$

and evaluate x as the fifth root of this new expression. If we substitute a value of $x = 0.40$ on the right side, we obtain $x = 0.15$ on the left side. Now using this value on the right we calculate a new value of $x = 0.26$ on the left. Using this value we obtain $x = 0.22$ on the left, and finally with this last value we obtain $x = 0.23$, in agreement with our previous procedure. Which method we use is a matter of convenience, but the second method provides a new value of x whereas the first method may require more trial and error. When using the method of successive approximations it is often a useful strategy to take the average of two results in order to speed up the convergence. The method of successive approximations can be very useful, but sometimes, depending how the equation is set up, the convergence to the correct answer may be slow or the process may even diverge. In such circumstances we can graph the expression as a function of x to ascertain where the solutions occur. In any event, we must always make sure that any answer we obtain is reasonable from a chemical or physical point of view.

A-4   Graphs

Suppose the following sets of numbers are obtained for two quantities x and y by laboratory measurement.

$$x = 0, 1, 2, 3, 4, \ldots$$

$$y = 2, 4, 6, 8, 10, \ldots$$

The relationship between these sets of numbers is not difficult to establish.

$$y = 2x + 2$$

Ideally, the results of experimental measurements are best expressed through a mathematical equation. Sometimes, however, an exact equation cannot be written, or its form is not clear from the experimental data. The graphing of data is very useful in such cases. In Figure A-1 the points listed above are located on a coordinate grid in which x values are placed along the horizontal axis (abscissa) and y values along the vertical axis (ordinate). For each point the x and y values are indicated in parentheses.

The data points are seen to define a straight line. A mathematical equation of a straight line has the form

$$y = mx + b$$

Values of m, the *slope* of the line, and b, the *intercept,* can be obtained from the straight-line graph.

$$\text{Slope} = \frac{\Delta y}{\Delta x} = \frac{10 - 2}{4 - 0} = \frac{8}{4} = 2$$

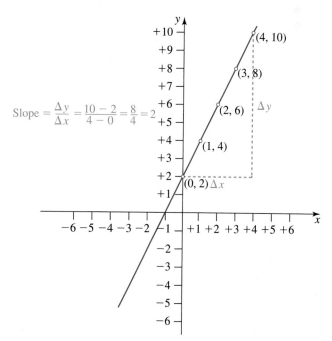

▲ **FIGURE A-1   A straight-line graph: *y + mx + b*.**

When $x = 0$, $y = b$. The intercept is the point where the straight line intersects the y-axis. The slope can be obtained from two points on the graph.

$$y_2 = mx_2 + b \quad \text{and} \quad y_1 = mx_1 + b$$

$$y_2 - y_1 = m(x_2 - x_1) + b - b$$

$$m = \frac{y_2 - y_1}{x_2 - x_1}$$

From the straight line in Figure A-1, can you establish that $m = b = 2$?

The technique used above to eliminate the constant b is applied to logarithmic functions in several places in the text. For example, the expression written below is from page 488. In this expression P is a pressure, T is a Kelvin temperature, and A and B are constants. The equation is that of a straight line.

$$\underbrace{\ln P}_{} = \underbrace{-A}_{} \underbrace{\left(\frac{1}{T}\right)}_{} + \underbrace{B}_{}$$

equation of straight line: $\quad y \quad = \quad m \cdot x \quad + \quad b$

We can write this equation twice, for the point (P_1, T_1) and the point (P_2, T_2).

$$\ln P_1 = -A\left(\frac{1}{T_1}\right) + B \quad \text{and} \quad \ln P_2 = -A\left(\frac{1}{T_2}\right) + B$$

The difference between these equations is

$$\ln P_2 - \ln P_1 = -A\left(\frac{1}{T_2}\right) + B + A\left(\frac{1}{T_1}\right) - B$$

$$\ln \frac{P_2}{P_1} = A\left(\frac{1}{T_1} - \frac{1}{T_2}\right)$$

A-5 Using Conversion Factors (Dimensional Analysis)

Some calculations in general chemistry require that a quantity measured in one set of units be converted to another set of units. Consider this fact.

$$1 \text{ m} = 100 \text{ cm}$$

Divide each side of the equation by 1 m.

$$\frac{1 \text{ m}}{1 \text{ m}} = \frac{100 \text{ cm}}{1 \text{ m}}$$

On the left side of the equation, the numerator and denominator are identical; they cancel.

$$1 = \frac{100 \text{ cm}}{1 \text{ m}} \tag{A.3}$$

On the right side they are not identical, but they are equal because they do represent the *same length*. The ratio 100 cm/1 m, when multiplied by a length in meters, converts that length to centimeters. The ratio is called a *conversion factor.*

Consider the question, "how many centimeters are there in 6.22 m?" The measured quantity is 6.22 m, and multiplying this quantity by 1 does not change its value.

$$6.22 \text{ m} \times 1 = 6.22 \text{ m}$$

Now replace the factor "1" by its equivalent—the conversion factor (A.3). Cancel the unit, m, and carry out the multiplication.

$$6.22 \ m \times \underbrace{\frac{100 \ cm}{1 \ m}}_{\substack{\text{this factor} \\ \text{converts} \\ \text{m to cm}}} = 622 \ cm$$

Next consider the question, "how many meters are there in 576 cm?" If we use the same factor (A.3) as before, the result is nonsensical.

$$576 \ cm \times \frac{100 \ cm}{1 \ m} = 5.76 \times 10^4 \ cm^2/m$$

Factor (A.3) must be rearranged to 1 m/100 cm.

$$576 \ cm \times \underbrace{\frac{1 \ m}{100 \ cm}}_{\substack{\text{this factor} \\ \text{converts} \\ \text{cm to m}}} = 5.76 \ m$$

▶ Because of the importance of the cancellation of units, this problem-solving method is often called *unit analysis* or *dimensional analysis*.

This second example emphasizes two points.

1. There are two ways to write a conversion factor—in one form or its reciprocal (inverse). Because a conversion factor is equal to 1, its value is not changed by the inversion, but

2. A conversion factor must be used in such a way as to produce the necessary cancellaton of units.

Calculations based on conversion factors are always of the form

information sought = information given + conversion factor(s)   (A.4)

Often we must make several conversions in sequence in order to get to the desired result. For example, if we want to know how many yards (yd) there are in 576 cm, we find that there is no direct cm → yd conversion factor available. From the inside back cover of the text, however, we do find a conversion factor for cm → in. Thus, we can develop a *conversion pathway,* that is, a series of conversion factors that will take us from centimeters to yards:

$$\text{cm} \quad \rightarrow \quad \text{in.} \quad \rightarrow \quad \text{ft} \quad \rightarrow \quad \text{yd}$$

$$? \ yards = 576 \ cm \times \frac{1 \ in.}{2.54 \ cm} \times \frac{1 \ ft}{12 \ in.} \times \frac{1 \ yd}{3 \ ft}$$

$$= 6.30 \ yd$$

We can use the same idea of a conversion pathway to deal with the somewhat more challenging situation faced when the units are squared (or cubed). Consider the question, "how many square feet (ft^2) correspond to an area of 1.00 square meter (m^2), given that 1 m = 39.37 in. and 12 in. = 1 ft?" Here, it may be helpful to begin by drawing a sketch or outline of the situation. Figure A-2 represents a 1.00-m^2 area. Think of it as a square with sides 1 m long. Figure A-2 also represents the length 1 ft and an area of 1.00 ft^2. Do you see that there is somewhat more than 9 ft^2 in 1 m^2?

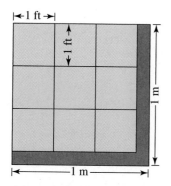

▲ FIGURE A-2
Comparison of one square foot and one square meter. One meter is slightly longer than 3 ft; 1 m² is somewhat larger than 9 ft².

We can write expression (A.4) as follows:

$$? \text{ ft}^2 = 1.00 \text{ m}^2 \times \underbrace{\left(\frac{39.37 \text{ in.}}{1 \text{ m}}\right)\left(\frac{39.37 \text{ in.}}{1 \text{ m}}\right)}_{\substack{\text{to convert} \\ \text{m}^2 \text{ to in.}^2}} \times \underbrace{\left(\frac{1 \text{ ft}}{12 \text{ in.}}\right)\left(\frac{1 \text{ ft}}{12 \text{ in.}}\right)}_{\substack{\text{to convert} \\ \text{in.}^2 \text{ to ft}^2}}$$

This is the same as writing

$$? \text{ ft}^2 = 1.00 \text{ m}^2 \times \frac{(39.37)^2 \text{ in.}^2}{1 \text{ m}^2} \times \frac{1 \text{ ft}^2}{(12)^2 \text{ in.}^2} = 10.8 \text{ ft}^2$$

Another way to look at the problem is to convert the length 1.00 m to feet,

$$? \text{ ft} = 1.00 \text{ m} \times \frac{39.37 \text{ in.}}{1 \text{ m}} \times \frac{1 \text{ ft}}{12 \text{ in.}} = 3.28 \text{ ft}$$

and square the result

$$? \text{ ft}^2 = 3.28 \text{ ft} \times 3.28 \text{ ft} = 10.8 \text{ ft}^2$$

Our last example incorporates several ideas discussed above. Here we will examine the situation in which the units in both the numerator and denominator must be converted. Consider the question, "how many meters per second (m/s) correspond to a speed of 63 mph, given that 1 mi = 5280 ft?"

We need to convert from miles to meters in the numerator and from hours to seconds in the denominator. We will need to use conversion factors from elsewhere in Appendix A-5 in addition to the given value. Also, we must be careful that our conversion factors produce the correct cancellation of units.

$$? \frac{\text{m}}{\text{s}} = \frac{63 \text{ mi}}{1 \text{ h}} \times \frac{1 \text{ h}}{60 \text{ min}} \times \frac{1 \text{ min}}{60 \text{ s}} \times \frac{5280 \text{ ft}}{1 \text{ mi}} \times \frac{12 \text{ in.}}{1 \text{ ft}} \times \frac{1 \text{ m}}{39.37 \text{ in.}}$$

$$= 28 \frac{\text{m}}{\text{s}}$$

In an alternative approach we break down the problem into three steps: (1) Convert 63 miles to a distance in meters; (2) convert 1 hour to a time in seconds; and (3) express the speed as a ratio of distance over time.

Step 1

$$\text{distance} = 63 \text{ mi} \times \frac{5280 \text{ ft}}{1 \text{ mi}} \times \frac{12 \text{ in.}}{1 \text{ ft}} \times \frac{1 \text{ m}}{39.37 \text{ in.}} = 1.0 \times 10^5 \text{ m}$$

Step 2

$$\text{time} = 1 \text{ h} \times \frac{60 \text{ min}}{1 \text{ h}} \times \frac{60 \text{ s}}{1 \text{ min}} = 3.6 \times 10^3 \text{ s}$$

Step 3

$$\text{speed} = \frac{\text{distance}}{\text{time}} = \frac{1.0 \times 10^5 \text{ m}}{3.6 \times 10^3 \text{ s}} = 28 \frac{\text{m}}{\text{s}}$$

In conclusion, we have shown (1) how to make a conversion factor; (2) that a conversion factor may be inverted; (3) that a series of conversion factors may be used to make a conversion pathway; (4) that conversion factors may be raised to powers, if necessary; and (5) that conversions of values with units in both the numerator and the denominator (such as miles per hour or pounds per square inch) can be performed in one step or in several steps.

B Some Basic Physical Concepts

B-1 Velocity and Acceleration

Time elapses as an object moves from one point to another. The *velocity* of the object is defined as the distance traveled per unit of time. An automobile that travels a distance of 60.0 km in exactly one hour has a velocity of 60.0 km/h (or 16.7 m/s).

Table B-1 contains data on the velocity of a free-falling body. For this motion velocity is not constant—it increases with time. The falling body "speeds up" continuously. The rate of change of velocity with time is called *acceleration.* Acceleration has the units of distance per unit time per unit time. With the methods of calculus, mathematical equations can be derived for the velocity (u) and distance (d) traveled in a time (t) by an object that has a constant acceleration (a).

$$u = at \tag{B.1}$$

$$d = \tfrac{1}{2}at^2 \tag{B.2}$$

For a free-falling body, the constant acceleration, called the *acceleration due to gravity,* is $a = g = 9.8$ m/s^2. Equations (B.1) and (B.2) can be used to calculate the velocity and distance traveled by a free-falling body.

TABLE B-1 Velocity and Acceleration of a Free-Falling Body

Time Elapsed, s	Total Distance, m	Velocity, m/s	Acceleration, m/s^2
0	0		
		4.9	
1	4.9		9.8
		14.7	
2	19.6		9.8
		24.5	
3	44.1		9.8
		34.3	
4	78.4		

B-2 Force and Work

Newton's *first law* of motion states that an object at rest remains at rest, and that an object in motion remains in uniform motion, unless acted upon by an external force. The tendency for an object to remain at rest or in uniform motion is called *inertia;* a *force* is what is required to overcome inertia. Since the application of a force either gives an object motion or changes its motion, the actual effect of a force is to change the velocity of an object. Change in velocity is an *acceleration,* so force is what provides an object with acceleration.

Newton's *second law* of motion describes the force F required to produce an acceleration a in an object of mass m.

$$F = ma \tag{B.3}$$

The basic unit of force in the SI system is the *newton* (N). It is the force required to provide a one-kilogram mass with an acceleration of one meter per second per second.

$$1\,N = 1\,kg \times 1\,m\,s^{-2} \tag{B.4}$$

The force of gravity on an object (its weight) is the product of the mass of the object and the acceleration due to gravity, g.

$$F = mg \tag{B.5}$$

Work (w) is performed when a force acts through a distance.

$$\text{work } (w) = \text{force } (F) \times \text{distance } (d) \tag{B.6}$$

The *joule* (J) is the amount of work associated with a force of one newton (N) acting through a distance of one meter.

$$1\,J = 1\,N \times 1\,m \tag{B.7}$$

From the definition of the newton in expression (B.4), we can also write

$$1\,J = 1\,kg \times 1\,m\,s^{-2} \times 1\,m = 1\,kg\,m^2\,s^{-2} \tag{B.8}$$

B-3 Energy

Energy is defined as the capacity to do work, but there are other useful descriptions of energy as well. For example, a moving object possesses a kind of energy known as *kinetic energy.* We can obtain a useful equation for kinetic energy by combining some of the other simple equations in this appendix. Thus, because work is the product of a force and distance (equation B.6), and force is the product of a mass and acceleration (equation B.3), we can write

$$w \text{ (work)} = m \times a \times d \tag{B.9}$$

Now, if we substitute equation (B.2) relating acceleration (a), distance (d), and time (t), into equation (B.9), we obtain

$$w \text{ (work)} = m \times a \times \frac{1}{2}at^2 \tag{B.10}$$

Finally, let's substitute expression (B.1) relating acceleration (a) and velocity (u) into (B.10). That is, because $a = u/t$,

$$w \text{ (work)} = \frac{1}{2}m\left(\frac{u}{t}\right)^2 t^2 \tag{B.11}$$

Think of the work in (B.11) as the amount of work necessary to produce a velocity of u in an object of mass m. This amount of work is the energy that appears in the object as kinetic energy (e_k).

$$e_k \text{ (kinetic energy)} = \frac{1}{2} mu^2 \tag{B.12}$$

An object at rest may also have the capacity to do work by changing its position. The energy it possesses, which can be transformed into actual work, is called *potential energy*. Think of potential energy as energy "stored" within an object. Equations can be written for potential energy, but the exact forms of these equations depend on the manner in which the energy is "stored."

B-4   Magnetism

Attractive and repulsive forces associated with a magnet are centered at regions called *poles*. A magnet has a north and a south pole. If two magnets are aligned such that the north pole of one is directed toward the south pole of the second, an attractive force develops. If the alignment brings like poles into proximity, either both north or both south, a repulsive force develops. *Unlike poles attract; like poles repel.*

A *magnetic field* exists in that region surrounding a magnet in which the influence of the magnet can be felt. Internal changes produced within an iron object by a magnetic field, not produced in a field-free region, are responsible for the attractive force that the object experiences.

B-5   Static Electricity

Another property with which certain objects may be endowed is *electric charge*. Analogous to the case of magnetism, *unlike charges attract, and like charges repel* (recall Figure 2-2). In Coulomb's law, stated below, a *positive* force between electrically charged objects is *repulsive;* a *negative* force is *attractive.*

$$F = \frac{Q_1 Q_2}{\varepsilon r^2} \tag{B.13}$$

where Q_1 is the magnitude of the charge on object 1,

           Q_2 is the magnitude of the charge on object 2,

           r is the distance between the objects, and

           ε is a proportionality constant called the *dielectric constant,* whose numerical value reflects the effect that the medium separating two charged objects has on the force existing between them. For vacuum, $\varepsilon = 1$; for other media, ε is greater than 1 (for example, for water $\varepsilon = 78.5$).

An *electric field* exists in that region surrounding an electrically charged object in which the influence of the electric charge is felt. If an uncharged object is brought into the field of a charged object, the uncharged object may undergo internal changes that it would not experience in a field-free region. These changes may lead to the production of electric charges in the formerly uncharged object, a phenomenon called *induction* (illustrated in Figure B-1).

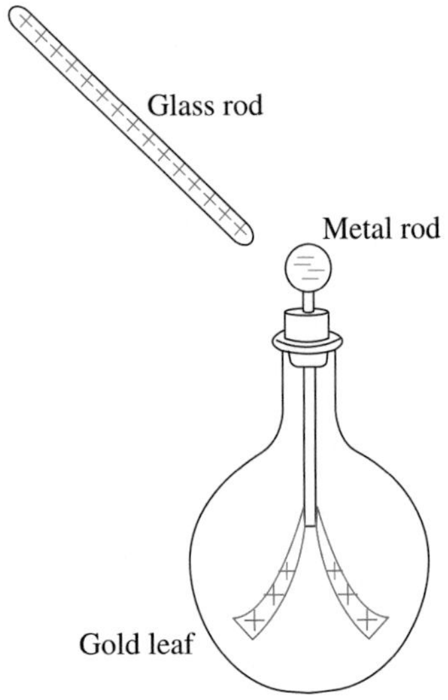

▶ FIGURE B-1
Production of electric charges by induction in a gold leaf electroscope.
The glass rod acquires a positive electric charge by being rubbed with a silk cloth. As the rod is brought near the electroscope, a separation of charge occurs in the electroscope. The leaves become positively charged and repel one another. Negative charge is attracted to the spherical terminal at the end of the metal rod. If the glass rod is removed, the charges on the electroscope redistribute themselves, and the leaves collapse. If the spherical ball is touched by an electric conductor before the glass rod is removed, negative charge is removed from the ball. The electroscope retains a net positive charge, and the leaves remain outstretched.

B-6 Current Electricity

Current electricity is a flow of electrically charged particles. In electric currents in metallic conductors, the charged particles are electrons; in molten salts or in aqueous solutions, the particles are both negatively and positively charged ions.

The unit of electric charge is called a *coulomb* (C). The unit of electric current known as the *ampere* (A) is defined as a flow of one coulomb per second through an electrical conductor. Two variables determine the magnitude of the electric current I flowing through a conductor. These are the potential difference, or voltage drop, E, along the conductor and the electrical resistance of the conductor, R. The units of voltage and resistance are the *volt* (V) and *ohm,* respectively. The relationship of electric current, voltage, and resistance is given by Ohm's law.

$$I = \frac{E}{R} \tag{B.14}$$

One joule of energy is associated with the passage of one coulomb of electric charge through a potential difference (voltage) of one volt. That is, one joule = one volt-coulomb. Electric *power* refers to the rate of production (or consumption) of electric energy. It has the unit *watt* (W).

$$1\,W = 1\,J\,s^{-1} = 1\,V\,C\,s^{-1}$$

Since one coulomb per second is a current of one ampere,

$$1\,W = 1\,V \times 1\,A \tag{B.15}$$

Thus, a 100-watt light bulb operating at 110 V draws a current of $100\,W/110\,V = 0.91\,A$.

B-7 Electromagnetism

The relationship between electricity and magnetism is an intimate one. Interactions of electric and magnetic fields result in (1) magnetic fields associated with the flow of electric current (as in electromagnets), (2) forces experienced by current-carrying conductors when placed in a magnetic field (as in electric motors), and (3) electric current being induced when an electric conductor is moved through a magnetic field (as in electric generators). Several observations described in this text can be understood in terms of electromagnetic phenomena.

Appendix

C

SI Units

The system of units that will in time be used universally for expressing all measured quantities is Le Système International d'Unités (The International System of Units), adopted in 1960 by the Conference Générale des Poids et Measures (General Conference of Weights and Measures). A summary of some of the provisions of the SI convention is provided here.

C-1   SI Base Units

A single unit has been established for each of the basic quantities involved in measurement. These are as follows:

Physical Quantity	Unit	Abbreviation
Length	Meter	m
Mass	Kilogram	kg
Time	Second	s
Electric current	Ampere	A
Temperature	Kelvin	K
Luminous intensity	Candela	cd
Amount of substance	Mole	mol
Plane angle	Radian	rad
Solid angle	Steradian	sr

C-2   SI Prefixes

Distinctive prefixes are attached to the base unit to express quantities that are *multiples* (greater than) or *submultiples* (less than) of the base unit. The multiples and submultiples are obtained by multiplying the base unit by powers of ten.

Multiple	Prefix	Abbreviation	Submultiple	Prefix	Abbreviation
10^{12}	tera	T	10^{-1}	deci	d
10^{9}	giga	G	10^{-2}	centi	c
10^{6}	mega	M	10^{-3}	milli	m
10^{3}	kilo	k	10^{-6}	micro	μ
10^{2}	hecto	h	10^{-9}	nano	n
10^{1}	deka	da	10^{-12}	pico	p
			10^{-15}	femto	f
			10^{-18}	atto	a

C-3   Derived SI Units

A number of quantities must be derived from measured values of the SI base quantities [for example, volume has the unit (length)^3]. Two sets of derived units are given, those whose names follow directly from the base units and those that are given special names. Notice that the units used in the text differ in some respects from those in the table. For example, the text expresses density as g cm^{-3}, molar mass as g mol^{-1}, molar volume as mL mol^{-1} or L mol^{-1}, and molar concentration (molarity) as mol L^{-1} or M.

▶ Two other SI conventions are illustrated through this table:
(a) Units are written in singular form—meter or m, *not* meters or ms; (b) negative exponents are preferred to the shilling bar or solidus (/), that is, m s^{-1} and m s^{-2}, *not* m/s and m/s/s.

Physical Quantity	Unit	Abbreviation
Area	Square meter	m^2
Volume	Cubic meter	m^3
Velocity	Meter per second	m s^{-1}
Acceleration	Meter per second squared	m s^{-2}
Density	Kilogram per cubic meter	kg m^{-3}
Molar mass	Kilogram per mole	kg mol^{-1}
Molar volume	Cubic meter per mole	m^3 mol^{-1}
Molar concentration	Mole per cubic meter	mol m^{-3}

Physical Quantity	Unit	Abbreviation	In Terms of SI Units
Frequency	hertz	Hz	s^{-1}
Force	newton	N	kg m s^{-2}
Pressure	pascal	Pa	N m^{-2}
Energy	joule	J	kg m^2 s^{-2}
Power	watt	W	J s^{-1}
Electric charge	coulomb	C	As
Electric potential difference	volt	V	J A^{-1} s^{-1}
Electric resistance	ohm	Ω	V A^{-1}

C-4   Units to Be Discouraged or Abandoned

There are several commonly used units whose use is to be discouraged and ultimately abandoned. Their gradual disappearance is to be expected, though each is used in this text. A few such units are listed.

▶ Another SI convention is implied here. No commas are used in expressing large numbers. Instead, spaces are left between groupings of three digits, that is, 101 325 rather than 101,235. Decimal points are written either as periods or commas.

Physical Quantity	Unit	Abbreviation	Definition in SI Units
Length	ångstrom	Å	1×10^{-10} m
Force	dyne	dyn	1×10^{-5} N
Energy	erg	erg	1×10^{-7} J
Energy	calorie	cal	4.184 J
Pressure	atmosphere	atm	101 325 Pa
Pressure	millimeter of mercury	mmHg	133.322 Pa
Pressure	torr	Torr	133.322 Pa

Appendix
D
Data Tables

TABLE D-1   Ground-State Electron Configurations

Z	Element	Configuration	Z	Element	Configuration	Z	Element	Configuration
1	H	$1s^1$	37	Rb	[Kr] $5s^1$	72	Hf	[Xe] $4f^{14}5d^26s^2$
2	He	$1s^2$	38	Sr	[Kr] $5s^2$	73	Ta	[Xe] $4f^{14}5d^36s^2$
3	Li	[He] $2s^1$	39	Y	[Kr] $4d^15s^2$	74	W	[Xe] $4f^{14}5d^46s^2$
4	Be	[He] $2s^2$	40	Zr	[Kr] $4d^25s^2$	75	Re	[Xe] $4f^{14}5d^56s^2$
5	B	[He] $2s^22p^1$	41	Nb	[Kr] $4d^45s^1$	76	Os	[Xe] $4f^{14}5d^66s^2$
6	C	[He] $2s^22p^2$	42	Mo	[Kr] $4d^55s^1$	77	Ir	[Xe] $4f^{14}5d^76s^2$
7	N	[He] $2s^22p^3$	43	Tc	[Kr] $4d^55s^2$	78	Pt	[Xe] $4f^{14}5d^96s^1$
8	O	[He] $2s^22p^4$	44	Ru	[Kr] $4d^75s^1$	79	Au	[Xe] $4f^{14}5d^{10}6s^1$
9	F	[He] $2s^22p^5$	45	Rh	[Kr] $4d^85s^1$	80	Hg	[Xe] $4f^{14}5d^{10}6s^2$
10	Ne	[He] $2s^22p^6$	46	Pd	[Kr] $4d^{10}$	81	Tl	[Xe] $4f^{14}5d^{10}6s^26p^1$
11	Na	[Ne] $3s^1$	47	Ag	[Kr] $4d^{10}5s^1$	82	Pb	[Xe] $4f^{14}5d^{10}6s^26p^2$
12	Mg	[Ne] $3s^2$	48	Cd	[Kr] $4d^{10}5s^2$	83	Bi	[Xe] $4f^{14}5d^{10}6s^26p^3$
13	Al	[Ne] $3s^23p^1$	49	In	[Kr] $4d^{10}5s^25p^1$	84	Po	[Xe] $4f^{14}5d^{10}6s^26p^4$
14	Si	[Ne] $3s^23p^2$	50	Sn	[Kr] $4d^{10}5s^25p^2$	85	At	[Xe] $4f^{14}5d^{10}6s^26p^5$
15	P	[Ne] $3s^23p^3$	51	Sb	[Kr] $4d^{10}5s^25p^3$	86	Rn	[Xe] $4f^{14}5d^{10}6s^26p^6$
16	S	[Ne] $3s^23p^4$	52	Te	[Kr] $4d^{10}5s^25p^4$	87	Fr	[Rn] $7s^1$
17	Cl	[Ne] $3s^23p^5$	53	I	[Kr] $4d^{10}5s^25p^5$	88	Ra	[Rn] $7s^2$
18	Ar	[Ne] $3s^23p^6$	54	Xe	[Kr] $4d^{10}5s^25p^6$	89	Ac	[Rn] $6d^17s^2$
19	K	[Ar] $4s^1$	55	Cs	[Xe] $6s^1$	90	Th	[Rn] $6d^27s^2$
20	Ca	[Ar] $4s^2$	56	Ba	[Xe] $6s^2$	91	Pa	[Rn] $5f^26d^17s^2$
21	Sc	[Ar] $3d^14s^2$	57	La	[Xe] $5d^16s^2$	92	U	[Rn] $5f^36d^17s^2$
22	Ti	[Ar] $3d^24s^2$	58	Ce	[Xe] $4f^26s^2$	93	Np	[Rn] $5f^46d^17s^2$
23	V	[Ar] $3d^34s^2$	59	Pr	[Xe] $4f^36s^2$	94	Pu	[Rn] $5f^67s^2$
24	Cr	[Ar] $3d^54s^1$	60	Nd	[Xe] $4f^46s^2$	95	Am	[Rn] $5f^77s^2$
25	Mn	[Ar] $3d^54s^2$	61	Pm	[Xe] $4f^56s^2$	96	Cm	[Rn] $5f^76d^17s^2$
26	Fe	[Ar] $3d^64s^2$	62	Sm	[Xe] $4f^66s^2$	97	Bk	[Rn] $5f^97s^2$
27	Co	[Ar] $3d^74s^2$	63	Eu	[Xe] $4f^76s^2$	98	Cf	[Rn] $5f^{10}7s^2$
28	Ni	[Ar] $3d^84s^2$	64	Gd	[Xe] $4f^75d^16s^2$	99	Es	[Rn] $5f^{11}7s^2$
29	Cu	[Ar] $3d^{10}4s^1$	65	Tb	[Xe] $4f^96s^2$	100	Fm	[Rn] $5f^{12}7s^2$
30	Zn	[Ar] $3d^{10}4s^2$	66	Dy	[Xe] $4f^{10}6s^2$	101	Md	[Rn] $5f^{13}7s^2$
31	Ga	[Ar] $3d^{10}4s^24p^1$	67	Ho	[Xe] $4f^{11}6s^2$	102	No	[Rn] $5f^{14}7s^2$
32	Ge	[Ar] $3d^{10}4s^24p^2$	68	Er	[Xe] $4f^{12}6s^2$	103	Lr	[Rn] $5f^{14}6d^17s^2$
33	As	[Ar] $3d^{10}4s^24p^3$	69	Tm	[Xe] $4f^{13}6s^2$	104	Rf	[Rn] $5f^{14}6d^27s^2$
34	Se	[Ar] $3d^{10}4s^24p^4$	70	Yb	[Xe] $4f^{14}6s^2$	105	Db	[Rn] $5f^{14}6d^37s^2$
35	Br	[Ar] $3d^{10}4s^24p^5$	71	Lu	[Xe] $4f^{14}5d^16s^2$	106	Sg	[Rn] $5f^{14}6d^47s^2$
36	Kr	[Ar] $3d^{10}4s^24p^6$						

The electron configurations printed in blue are those of the noble gases. Each noble gas configuration serves as the core of the electron configurations of the elements that follow it, until the next noble gas is reached. Thus, [He] represents the core configuration of the 2nd period elements; [Ne], the third period; [Ar], the fourth period; [Kr], the fifth period; [Xe], the sixth period; and [Rn], the seventh period.

TABLE D-2 Thermodynamic Properties of Substances at 298.15 K[*]
Substances are at 1 bar pressure. For aqueous solutions, solutes are at unit activity (roughly 1 M). Data for ions in aqueous solution are relative to values of *zero* for ΔH_f°, ΔG_f°, and S° for H^+.

Inorganic Substances

	ΔH_f°, kJ mol^{-1}	ΔG_f°, kJ mol^{-1}	S°, Jmol^{-1}K^{-1}
Aluminum			
Al(s)	0	0	28.33
Al^{3+}(aq)	−531	−485	−321.7
AlCl$_3$(s)	−704.2	−628.8	110.7
Al$_2$Cl$_6$(g)	−1291	−1220.	490.
AlF$_3$(s)	−1504	−1425	66.44
Al$_2$O$_3$(α solid)	−1676	−1582	50.92
Al(OH)$_3$(s)	−1276	—	—
Al$_2$(SO$_4$)$_3$(s)	−3441	−3100.	239
Barium			
Ba(s)	0	0	62.8
Ba^{2+}(aq)	−537.6	−560.8	9.6
BaCO$_3$(s)	−1216	−1138	112.1
BaCl$_2$(s)	−858.6	−810.4	123.7
BaF$_2$(s)	−1207	−1157	96.36
BaO(s)	−553.5	−525.1	70.42
Ba(OH)$_2$(s)	−944.7	—	—
Ba(OH)$_2 \cdot$8 H$_2$O(s)	−3342	−2793	427
BaSO$_4$(s)	−1473	−1362	132.2
Beryllium			
Be(s)	0	0	9.50
BeCl$_2$(α solid)	−490.4	−445.6	82.68
BeF$_2$(α solid)	−1027	−979.4	53.35
BeO(s)	−609.6	−580.3	14.14
Bismuth			
Bi(s)	0	0	56.74
BiCl$_3$(s)	−379.1	−315.0	177.0
Bi$_2$O$_3$(s)	−573.9	−493.7	151.5
Boron			
B(s)	0	0	5.86
BCl$_3$(l)	−427.2	−387.4	206.3
BF$_3$(g)	−1137	−1120.	254.1
B$_2$H$_6$(g)	35.6	86.7	232.1
B$_2$O$_3$(s)	−1273	−1194	53.97
Bromine			
Br(g)	111.9	82.40	175.0
Br$^-$(aq)	−121.6	−104.0	82.4
Br$_2$(g)	30.91	3.11	245.5
Br$_2$(l)	0	0	152.2
BrCl(g)	14.64	−0.98	240.1
BrF$_3$(g)	−255.6	−229.4	292.5
BrF$_3$(l)	−300.8	−240.5	178.2

[*]Data for inorganic substances and for organic compounds with up to two carbon atoms per molecule are adapted from D. D. Wagman, *et. al.*, *The NBS Tables of Chemical Thermodynamic Properties: Selected Values for Inorganic and C_1 and C_2 Organic Substances in SI Units*, *Journal of Physical and Chemical Reference Data*, Volume 11, 1982, Supplement 2. Data for other organic compounds are from J. A. Dean, *Lange's Handbook of Chemistry*, 15/e, McGraw-Hill, Inc., 1999 and other sources.

Inorganic Substances

	ΔH_f°, kJ mol^{-1}	ΔG_f°, kJ mol^{-1}	S°, Jmol^{-1}K^{-1}
Cadmium			
$Cd(s)$	0	0	51.76
$Cd^{2+}(aq)$	-75.90	-77.61	-73.2
$CdCl_2(s)$	-391.5	-343.9	115.3
$CdO(s)$	-258.2	-228.4	54.8
Calcium			
$Ca(s)$	0	0	41.42
$Ca^{2+}(aq)$	-542.8	-553.6	-53.1
$CaCO_3(s)$	-1207	-1129	92.9
$CaCl_2(s)$	-795.8	-748.1	104.6
$CaF_2(s)$	$-1220.$	-1167	68.87
$CaH_2(s)$	-186.2	-147.2	42
$Ca(NO_3)_2(s)$	-938.4	-743.1	193.3
$CaO(s)$	-635.1	-604.0	39.75
$Ca(OH)_2(s)$	-986.1	-898.5	83.39
$Ca_3(PO_4)_2(s)$	-4121	-3885	236.0
$CaSO_4(s)$	-1434	-1322	106.7
Carbon (See also the table of organic substances.)			
$C(g)$	716.7	671.3	158.0
$C(diamond)$	1.90	2.90	2.38
$C(graphite)$	0	0	5.74
$CCl_4(g)$	-102.9	-60.59	309.9
$CCl_4(l)$	-135.4	-65.21	216.4
$C_2N_2(g)$	309.0	297.4	241.9
$CO(g)$	-110.5	-137.2	197.7
$CO_2(g)$	-393.5	-394.4	213.7
$CO_3^{2-}(aq)$	-677.1	-527.8	-56.9
$C_3O_2(g)$	-93.72	-109.8	276.5
$C_3O_2(l)$	-117.3	-105.0	181.1
$COCl_2(g)$	-218.8	-204.6	283.5
$COS(g)$	-142.1	-169.3	231.6
$CS_2(l)$	89.70	65.27	151.3
Chlorine			
$Cl(g)$	121.7	105.7	165.2
$Cl^-(aq)$	-167.2	-131.2	56.5
$Cl_2(g)$	0	0	223.1
$ClF_3(g)$	-163.2	-123.0	281.6
$ClO_2(g)$	102.5	120.5	256.8
$Cl_2O(g)$	80.3	97.9	266.2
Chromium			
$Cr(s)$	0	0	23.77
$[Cr(H_2O)6]^{3+}(aq)$	-1999	—	—
$Cr_2O_3(s)$	$-1140.$	-1058	81.2
$CrO_4^{2-}(aq)$	-881.2	-727.8	50.21
$Cr_2O_7^{2-}(aq)$	$-1490.$	-1301	261.9
Cobalt			
$Co(s)$	0	0	30.04
$CoO(s)$	-237.9	-214.2	52.97
$Co(OH)_2(pink solid)$	-539.7	-454.3	79
Copper			
$Cu(s)$	0	0	33.15
$Cu^{2+}(aq)$	64.77	65.49	-99.6
$CuCO_3 \cdot Cu(OH)_2(s)$	-1051	-893.6	186.2

Inorganic Substances

	ΔH_f°, kJ mol^{-1}	ΔG_f°, kJ mol^{-1}	S°, Jmol^{-1}K^{-1}
CuO(s)	−157.3	−129.7	42.63
Cu(OH)$_2$(s)	−449.8	—	—
CuSO$_4$·5 H$_2$O(s)	−2280.	−1880.	300.4
Fluorine			
F(g)	78.99	61.91	158.8
F$^-$(aq)	−332.6	−278.8	−13.8
F$_2$(g)	0	0	202.8
Helium			
He(g)	0	0	126.2
Hydrogen			
H(g)	218.0	203.2	114.7
H$^+$(aq)	0	0	0
H$_2$(g)	0	0	130.7
HBr(g)	−36.40	−53.45	198.7
HCl(g)	−92.31	−95.30	186.9
HCl(aq)	−167.2	−131.2	56.5
HClO$_2$(aq)	−51.9	5.9	188.3
HCN(g)	135.1	124.7	201.8
HF(g)	−271.1	−273.2	173.8
HI(g)	26.48	1.70	206.6
HNO$_3$(l)	−174.1	−80.71	155.6
HNO$_3$(aq)	−207.4	−111.3	146.4
H$_2$O(g)	−241.8	−228.6	188.8
H$_2$O(l)	−285.8	−237.1	69.91
H$_2$O$_2$(g)	−136.3	−105.6	232.7
H$_2$O$_2$(l)	−187.8	−120.4	109.6
H$_2$S(g)	−20.63	−33.56	205.8
H$_2$SO$_4$(l)	−814.0	−690.0	156.9
H$_2$SO$_4$(aq)	−909.3	−744.5	20.1
Iodine			
I(g)	106.8	70.25	180.8
I$^-$(aq)	−55.19	−51.57	111.3
I$_2$(g)	62.44	19.33	260.7
I$_2$(s)	0	0	116.1
IBr(g)	40.84	3.69	258.8
ICl(g)	17.78	−5.46	247.6
ICl(l)	−23.89	−13.58	135.1
Iron			
Fe(s)	0	0	27.28
Fe^{2+}(aq)	−89.1	−78.90	−137.7
Fe^{3+}(aq)	−48.5	−4.7	−315.9
FeCO$_3$(s)	−740.6	−666.7	92.9
FeCl$_3$(s)	−399.5	−334.0	−142.3
FeO(s)	−272.0	—	—
Fe$_2$O$_3$(s)	−824.2	−742.2	87.40
Fe$_3$O$_4$(s)	−1118	−1015	146.4
Fe(OH)$_3$(s)	−823.0	−696.5	106.7
Lead			
Pb(s)	0	0	64.81
Pb^{2+}(aq)	−1.7	−24.43	10.5
PbI$_2$(s)	−175.5	−173.6	174.9
PbO$_2$(s)	−277.4	−217.3	68.6
PbSO$_4$(s)	−919.9	−813.1	148.6

Inorganic Substances

	ΔH_f°, kJ mol^{-1}	ΔG_f°, kJ mol^{-1}	S°, Jmol^{-1}K^{-1}
Lithium			
Li(g)	159.4	126.7	138.8
Li(s)	0	0	29.12
Li$^+$(aq)	−278.5	−293.3	13.4
LiCl(s)	−408.6	−384.4	59.33
LiOH(s)	−484.9	−439.0	42.80
LiNO$_3$(s)	−483.1	−381.1	90.0
Magnesium			
Mg(s)	0	0	32.68
Mg^{2+}(aq)	−466.9	−454.8	−138.1
MgCl$_2$(s)	−641.3	−591.8	89.62
MgCO$_3$(s)	−1096	−1012	65.7
MgF$_2$(s)	−1123	−1070	57.24
MgO(s)	−601.7	−569.4	26.94
Mg(OH)$_2$(s)	−924.5	−833.5	63.18
MgS(s)	−346.0	−341.8	50.33
MgSO$_4$(s)	−1285	−1171	91.6
Manganese			
Mn(s)	0	0	32.01
Mn^{2+}(aq)	−220.8	−228.1	−73.6
MnO$_2$(s)	−520.0	−465.1	53.05
MnO$_4^-$(aq)	−541.4	−447.2	191.2
Mercury			
Hg(g)	61.32	31.82	175.0
Hg(l)	0	0	76.02
HgO(s)	−90.83	−58.54	70.29
Nitrogen			
N(g)	472.7	455.6	153.3
N$_2$(g)	0	0	191.6
NF$_3$(g)	−124.7	−83.2	260.7
NH$_3$(g)	−46.11	−16.45	192.5
NH$_3$(aq)	−80.29	−26.50	111.3
NH$_4^+$(aq)	−132.5	−79.31	113.4
NH$_4$Br(s)	−270.8	−175.2	113
NH$_4$Cl(s)	−314.4	−202.9	94.6
NH$_4$F(s)	−464.0	−348.7	71.96
NH$_4$HCO$_3$(s)	−849.4	−665.9	120.9
NH$_4$I(s)	−201.4	−112.5	117
NH$_4$NO$_3$(s)	−365.6	−183.9	151.1
NH$_4$NO$_3$(aq)	−339.9	−190.6	259.8
−NH$_4$)$_2$SO$_4$(s)	−1181	−901.7	220.1
N$_2$H$_4$(g)	95.40	159.4	238.5
N$_2$H$_4$(l)	50.63	149.3	121.2
NO(g)	90.25	86.55	210.8
N$_2$O(g)	82.05	104.2	219.9
NO$_2$(g)	33.18	51.31	240.1
N$_2$O$_4$(g)	9.16	97.89	304.3
N$_2$O$_4$(l)	−19.50	97.54	209.2
N$_2$O$_5$(g)	11.3	115.1	355.7
NO$_3^-$(aq)	−205.0	−108.7	146.4
NOBr(g)	82.17	82.42	273.7
NOCl(g)	51.71	66.08	261.7

Inorganic Substances

	ΔH_f°, kJ mol^{-1}	ΔG_f°, kJ mol^{-1}	S°, Jmol^{-1}K^{-1}
Oxygen			
O(g)	249.2	231.7	161.1
O_2(g)	0	0	205.1
O_3(g)	142.7	163.2	238.9
OH$^-$(aq)	−230.0	−157.2	−10.75
OF_2(g)	24.7	41.9	247.4
Phosphorus			
P(α white)	0	0	41.09
P(red)	−17.6	−12.1	22.80
P_4(g)	58.91	24.44	280.0
PCl_3(g)	−287.0	−267.8	311.8
PCl_5(g)	−374.9	−305.0	364.6
PH_3(g)	5.4	13.4	210.2
P_4O_{10}(s)	−2984	−2698	228.9
PO_4^{3-}(aq)	−1277	−1019	−222
Potassium			
K(g)	89.24	60.59	160.3
K(s)	0	0	64.18
K$^+$(aq)	−252.4	−283.3	102.5
KBr(s)	−393.8	−380.7	95.90
KCN(s)	−113.0	−101.9	128.5
KCl(s)	−436.7	−409.1	82.59
$KClO_3$(s)	−397.7	−296.3	143.1
$KClO_4$(s)	−432.8	−303.1	151.0
KF(s)	−567.3	−537.8	66.57
KI(s)	−327.9	−324.9	106.3
KNO_3(s)	−494.6	−394.9	133.1
KOH(s)	−424.8	−379.1	78.9
KOH(aq)	−482.4	−440.5	91.6
K_2SO_4(s)	−1438	−1321	175.6
Silicon			
Si(s)	0	0	18.83
SiH_4(g)	34.3	56.9	204.6
Si_2H_6(g)	80.3	127.3	272.7
SiO_2(quartz)	−910.9	−856.6	41.84
Silver			
Ag(s)	0	0	42.55
Ag$^+$(aq)	105.6	77.11	72.68
AgBr(s)	−100.4	−96.90	107.1
AgCl(s)	−127.1	−109.8	96.2
AgI(s)	−61.84	−66.19	115.5
$AgNO_3$(s)	−124.4	−33.41	140.9
Ag_2O(s)	−31.05	−11.20	121.3
Ag_2SO_4(s)	−715.9	−618.4	200.4
Sodium			
Na(g)	107.3	76.76	153.7
Na(s)	0	0	51.21
Na$^+$(aq)	−240.1	−261.9	59.0
Na_2(g)	142.1	103.9	230.2
NaBr(s)	−361.1	−349.0	86.82
Na_2CO_3(s)	−1131	−1044	135.0
$NaHCO_3$(s)	−950.8	−851.0	101.7

Inorganic Substances

	ΔH_f°, kJ mol^{-1}	ΔG_f°, kJ mol^{-1}	S°, Jmol^{-1}K^{-1}
NaCl(s)	−411.2	−384.1	72.13
NaCl(aq)	−407.3	−393.1	115.5
NaClO$_3$(s)	−365.8	−262.3	123.4
NaClO$_4$(s)	−383.3	−254.9	142.3
NaF(s)	−573.6	−543.5	51.46
NaH(s)	−56.28	−33.46	40.02
NaI(s)	−287.8	−286.1	98.53
NaNO$_3$(s)	−467.9	−367.0	116.5
NaNO$_3$(aq)	−447.5	−373.2	205.4
Na$_2$O$_2$(s)	−510.9	−447.7	95.0
NaOH(s)	−425.6	−379.5	64.46
NaOH(aq)	−470.1	−419.2	48.1
NaH$_2$PO$_4$(s)	−1537	−1386	127.5
Na$_2$HPO$_4$(s)	−1748	−1608	150.5
Na$_3$PO$_4$(s)	−1917	−1789	173.8
NaHSO$_4$(s)	−1126	−992.8	113.0
Na$_2$SO$_4$(s)	−1387	−1270	149.6
Na$_2$SO$_4$(aq)	−1390.	−1268	138.1
Na$_2$SO$_4 \cdot$10 H$_2$O(s)	−4327	−3647	592.0
Na$_2$S$_2$O$_3$(s)	−1123	−1028	155
Sulfur			
S(g)	278.8	238.3	167.8
S(rhombic)	0	0	31.80
S$_8$(g)	102.3	49.63	431.0
S$_2$Cl$_2$(g)	−18.4	−31.8	331.5
SF$_6$(g)	−1209	−1105	291.8
SO$_2$(g)	−296.8	−300.2	248.2
SO$_3$(g)	−395.7	−371.1	256.8
SO$_4^{2-}$(aq)	−909.3	−744.5	20.1
S$_2$O$_3^{2-}$(aq)	−648.5	−522.5	67
SO$_2$Cl$_2$(g)	−364.0	−320.0	311.9
SO$_2$Cl$_2$(l)	−394.1	—	—
Tin			
Sn(white)	0	0	51.55
Sn(gray)	−2.09	0.13	44.14
SnCl$_4$(l)	−511.3	−440.1	258.6
SnO(s)	−285.8	−256.9	56.5
SnO$_2$(s)	−580.7	−519.6	52.3
Titanium			
Ti(s)	0	0	30.63
TiCl$_4$(g)	−763.2	−726.7	354.9
TiCl$_4$(l)	−804.2	−737.2	252.3
TiO$_2$(s)	−944.7	−889.5	50.33
Uranium			
U(s)	0	0	50.21
UF$_6$(g)	−2147	−2064	377.9
UF$_6$(s)	−2197	−2069	227.6
UO$_2$(s)	−1085	−1032	77.03
Zinc			
Zn(s)	0	0	41.63
Zn^{2+}(aq)	−153.9	−147.1	112.1
ZnO(s)	−348.3	−318.3	43.64

Organic Substances

	Name	ΔH_f°, kJ mol^{-1}	ΔG_f°, kJ mol^{-1}	S°, Jmol^{-1} K^{-1}
$CH_4(g)$	Methane(g)	−74.81	−50.72	186.3
$C_2H_2(g)$	Acetylene(g)	226.7	209.2	200.9
$C_2H_4(g)$	Ethylene(g)	52.26	68.15	219.6
$C_2H_6(g)$	Ethane(g)	−84.68	−32.82	229.6
$C_3H_8(g)$	Propane(g)	−103.8	−23.3	270.3
$C_4H_{10}(g)$	Butane(g)	−125.6	−17.1	310.2
$C_6H_6(g)$	Benzene(g)	82.6	129.8	269.3
$C_6H_6(l)$	Benzene(l)	49.0	124.5	173.4
$C_6H_{12}(g)$	Cyclohexane(g)	−123.4	32.0	298.4
$C_6H_{12}(l)$	Cyclohexane(l)	−156.4	26.9	204.4
$C_{10}H_8(g)$	Naphthalene(g)	150.6	224.2	333.2
$C_{10}H_8(s)$	Naphthalene(s)	77.9	201.7	167.5
$CH_2O(g)$	Formaldehyde(g)	−108.6	−102.5	218.8
$CH_3CHO(g)$	Acetaldehyde(g)	−166.2	−128.9	250.3
$CH_3CHO(l)$	Acetaldehyde(l)	−192.3	−128.1	160.2
$CH_3OH(g)$	Methanol(g)	−200.7	−162.0	239.8
$CH_3OH(l)$	Methanol(l)	−238.7	−166.3	126.8
$CH_3CH_2OH(g)$	Ethanol(g)	−235.1	−168.5	282.7
$CH_3CH_2OH(l)$	Ethanol(l)	−277.7	−174.8	160.7
$C_6H_5OH(s)$	Phenol(s)	−165.1	−50.4	144.0
$(CH_3)_2CO(g)$	Acetone(g)	−216.6	−153.0	295.0
$(CH_3)_2CO(l)$	Acetone(l)	−247.6	−155.6	200.5
$CH_3COOH(g)$	Acetic acid(g)	−432.3	−374.0	282.5
$CH_3COOH(l)$	Acetic acid(l)	−484.5	−389.9	159.8
$CH_3COOH(aq)$	Acetic acid(aq)	−485.8	−396.5	178.7
$C_6H_5COOH(s)$	Benzoic acid(s)	−385.2	−245.3	167.6
$CH_3NH_2(g)$	Methylamine(g)	−22.97	32.16	243.4
$C_6H_5NH_2(g)$	Aniline(g)	86.86	166.8	319.3
$C_6H_5NH_2(l)$	Aniline(l)	31.6	149.2	191.3

TABLE D-3 Equilibrium Constants

A. Ionization Constants of Weak Acids at 25 °C

Name of acid	Formula	K_a	Name of acid	Formula	K_a
Acetic	$HC_2H_3O_2$	1.8×10^{-5}	Hyponitrous	$HON{=}NOH$	8.9×10^{-8}
Acrylic	$HC_3H_3O_2$	5.5×10^{-5}		$HON{=}NO^-$	4×10^{-12}
Arsenic	H_3AsO_4	6.0×10^{-3}	Iodic	HIO_3	1.6×10^{-1}
	$H_2AsO_4^-$	1.0×10^{-7}	Iodoacetic	$HC_2H_2IO_2$	6.7×10^{-4}
	$HAsO_4^{2-}$	3.2×10^{-12}	Malonic	$H_2C_3H_2O_4$	1.5×10^{-3}
Arsenous	H_3AsO_3	6.6×10^{-10}		$HC_3H_2O_4^-$	2.0×10^{-6}
Benzoic	$HC_7H_5O_2$	6.3×10^{-5}	Nitrous	HNO_2	7.2×10^{-4}
Bromoacetic	$HC_2H_2BrO_2$	1.3×10^{-3}	Oxalic	$H_2C_2O_4$	5.4×10^{-2}
Butyric	$HC_4H_7O_2$	1.5×10^{-5}		$HC_2O_4^-$	5.3×10^{-5}
Carbonic	H_2CO_3	4.4×10^{-7}	Phenol	HOC_6H_5	1.0×10^{-10}
	HCO_3^-	4.7×10^{-11}	Phenylacetic	$HC_8H_7O_2$	4.9×10^{-5}
Chloroacetic	$HC_2H_2ClO_2$	1.4×10^{-3}	Phosphoric	H_3PO_4	7.1×10^{-3}
Chlorous	$HClO_2$	1.1×10^{-2}		$H_2PO_4^-$	6.3×10^{-8}
Citric	$H_3C_6H_5O_7$	7.4×10^{-4}		HPO_4^{2-}	4.2×10^{-13}
	$H_2C_6H_5O_7^-$	1.7×10^{-5}	Phosphorous	H_3PO_3	3.7×10^{-2}
	$HC_6H_5O_7^{2-}$	4.0×10^{-7}		$H_2PO_3^-$	2.1×10^{-7}
Cyanic	$HOCN$	3.5×10^{-4}	Propionic	$HC_3H_5O_2$	1.3×10^{-5}
Dichloroacetic	$HC_2HCl_2O_2$	5.5×10^{-2}	Pyrophosphoric	$H_4P_2O_7$	3.0×10^{-2}
Fluoroacetic	$HC_2H_2FO_2$	2.6×10^{-3}		$H_3P_2O_7^-$	4.4×10^{-3}
Formic	$HCHO_2$	1.8×10^{-4}		$H_2P_2O_7^{2-}$	2.5×10^{-7}
Hydrazoic	HN_3	1.9×10^{-5}		$HP_2O_7^{3-}$	5.6×10^{-10}
Hydrocyanic	HCN	6.2×10^{-10}	Selenic	H_2SeO_4	strong acid
Hydrofluoric	HF	6.6×10^{-4}		$HSeO_4^-$	2.2×10^{-2}
Hydrogen peroxide	H_2O_2	2.2×10^{-12}	Selenous	H_2SeO_3	2.3×10^{-3}
Hydroselenic	H_2Se	1.3×10^{-4}		$HSeO_3^-$	5.4×10^{-9}
	HSe^-	1×10^{-11}	Succinic	$H_2C_4H_4O_4$	6.2×10^{-5}
Hydrosulfuric	H_2S	1.0×10^{-7}		$HC_4H_4O_4^-$	2.3×10^{-6}
	HS^-	1×10^{-19}	Sulfuric	H_2SO_4	strong acid
Hydrotelluric	H_2Te	2.3×10^{-3}		HSO_4^-	1.1×10^{-2}
	HTe^-	1.6×10^{-11}	Sulfurous	H_2SO_3	1.3×10^{-2}
Hypobromous	$HOBr$	2.5×10^{-9}		HSO_3^-	6.2×10^{-8}
Hypochlorous	$HOCl$	2.9×10^{-8}	Thiophenol	HSC_6H_5	3.2×10^{-7}
Hypoiodous	HOI	2.3×10^{-11}	Trichloroacetic	$HC_2Cl_3O_2$	3.0×10^{-1}

B. Ionization Constants of Weak Bases at 25 °C

Name of base	Formula	K_b	Name of base	Formula	K_b
Ammonia	NH_3	1.8×10^{-5}	Isoquinoline	C_9H_7N	2.5×10^{-9}
Aniline	$C_6H_5NH_2$	7.4×10^{-10}	Methylamine	CH_3NH_2	4.2×10^{-4}
Codeine	$C_{18}H_{21}O_3N$	8.9×10^{-7}	Morphine	$C_{17}H_{19}O_3N$	7.4×10^{-7}
Diethylamine	$(C_2H_5)_2NH$	6.9×10^{-4}	Piperdine	$C_5H_{11}N$	1.3×10^{-3}
Dimethylamine	$(CH_3)_2NH$	5.9×10^{-4}	Pyridine	C_5H_5N	1.5×10^{-9}
Ethylamine	$C_2H_5NH_2$	4.3×10^{-4}	Quinoline	C_9H_7N	6.3×10^{-10}
Hydrazine	NH_2NH_2	8.5×10^{-7}	Triethanolamine	$C_6H_{15}O_3N$	5.8×10^{-7}
	$NH_2NH_3^+$	8.9×10^{-16}	Triethylamine	$(C_2H_5)_3N$	5.2×10^{-4}
Hydroxylamine	NH_2OH	9.1×10^{-9}	Trimethylamine	$(CH_3)_3N$	6.3×10^{-5}

C. Solubility Product Constants[a]

Name of solute	Formula	K_{sp}	Name of solute	Formula	K_{sp}
Aluminum hydroxide	$Al(OH)_3$	1.3×10^{-33}	Lead(II) hydroxide	$Pb(OH)_2$	1.2×10^{-15}
Aluminum phosphate	$AlPO_4$	6.3×10^{-19}	Lead(II) iodide	PbI_2	7.1×10^{-9}
Barium carbonate	$BaCO_3$	5.1×10^{-9}	Lead(II) sulfate	$PbSO_4$	1.6×10^{-8}
Barium chromate	$BaCrO_4$	1.2×10^{-10}	Lead(II) sulfide[b]	PbS	3×10^{-28}
Barium fluoride	BaF_2	1.0×10^{-6}	Lithium carbonate	Li_2CO_3	2.5×10^{-2}
Barium hydroxide	$Ba(OH)_2$	5×10^{-3}	Lithium fluoride	LiF	3.8×10^{-3}
Barium sulfate	$BaSO_4$	1.1×10^{-10}	Lithium phosphate	Li_3PO_4	3.2×10^{-9}
Barium sulfite	$BaSO_3$	8×10^{-7}	Magnesium	$MgNH_4PO_4$	2.5×10^{-13}
Barium thiosulfate	BaS_2O_3	1.6×10^{-5}	ammonium phosphate		
Bismuthyl chloride	$BiOCl$	1.8×10^{-31}	Magnesium carbonate	$MgCO_3$	3.5×10^{-8}
Bismuthyl hydroxide	$BiOOH$	4×10^{-10}	Magnesium fluoride	MgF_2	3.7×10^{-8}
Cadmium carbonate	$CdCO_3$	5.2×10^{-12}	Magnesium hydroxide	$Mg(OH)_2$	1.8×10^{-11}
Cadmium hydroxide	$Cd(OH)_2$	2.5×10^{-14}	Magnesium phosphate	$Mg_3(PO_4)_2$	1×10^{-25}
Cadmium sulfide[b]	CdS	8×10^{-28}	Manganese(II) carbonate	$MnCO_3$	1.8×10^{-11}
Calcium carbonate	$CaCO_3$	2.8×10^{-9}	Manganese(II) hydroxide	$Mn(OH)_2$	1.9×10^{-13}
Calcium chromate	$CaCrO_4$	7.1×10^{-4}	Manganese(II) sulfide[b]	MnS	3×10^{-14}
Calcium fluoride	CaF_2	5.3×10^{-9}	Mercury(I) bromide	Hg_2Br_2	5.6×10^{-23}
Calcium hydroxide	$Ca(OH)_2$	5.5×10^{-6}	Mercury(I) chloride	Hg_2Cl_2	1.3×10^{-18}
Calcium hydrogen phosphate	$CaHPO_4$	1×10^{-7}	Mercury(I) iodide	Hg_2I_2	4.5×10^{-29}
			Mercury(II) sulfide[b]	HgS	2×10^{-53}
Calcium oxalate	CaC_2O_4	4×10^{-9}	Nickel(II) carbonate	$NiCO_3$	6.6×10^{-9}
Calcium phosphate	$Ca_3(PO_4)_2$	2.0×10^{-29}	Nickel(II) hydroxide	$Ni(OH)_2$	2.0×10^{-15}
Calcium sulfate	$CaSO_4$	9.1×10^{-6}	Scandium fluoride	ScF_3	4.2×10^{-18}
Calcium sulfite	$CaSO_3$	6.8×10^{-8}	Scandium hydroxide	$Sc(OH)_3$	8.0×10^{-31}
Chromium(II) hydroxide	$Cr(OH)_2$	2×10^{-16}	Silver arsenate	Ag_3AsO_4	1.0×10^{-22}
Chromium(III) hydroxide	$Cr(OH)_3$	6.3×10^{-31}	Silver azide	AgN_3	2.8×10^{-9}
Cobalt(II) carbonate	$CoCO_3$	1.4×10^{-13}	Silver bromide	$AgBr$	5.0×10^{-13}
			Silver carbonate	Ag_2CO_3	8.5×10^{-12}
Cobalt(II) hydroxide	$Co(OH)_2$	1.6×10^{-15}	Silver chloride	$AgCl$	1.8×10^{-10}
Cobalt(III) hydroxide	$Co(OH)_3$	1.6×10^{-44}	Silver chromate	Ag_2CrO_4	1.1×10^{-12}
Copper(I) chloride	$CuCl$	1.2×10^{-6}	Silver cyanide	$AgCN$	1.2×10^{-16}
Copper(I) cyanide	$CuCN$	3.2×10^{-20}	Silver iodate	$AgIO_3$	3.0×10^{-8}
Copper(I) iodide	CuI	1.1×10^{-12}	Silver iodide	AgI	8.5×10^{-17}
Copper(II) arsenate	$Cu_3(AsO_4)_2$	7.6×10^{-36}	Silver nitrite	$AgNO_2$	6.0×10^{-4}
Copper(II) carbonate	$CuCO_3$	1.4×10^{-10}	Silver sulfate	Ag_2SO_4	1.4×10^{-5}
Copper(II) chromate	$CuCrO_4$	3.6×10^{-6}	Silver sulfide[b]	Ag_2S	6×10^{-51}
Copper(II) ferrocyanide	$Cu_2[Fe(CN)_6]$	1.3×10^{-16}	Silver sulfite	Ag_2SO_3	1.5×10^{-14}
Copper(II) hydroxide	$Cu(OH)_2$	2.2×10^{-20}	Silver thiocyanate	$AgSCN$	1.0×10^{-12}
Copper(II) sulfide[b]	CuS	6×10^{-37}	Strontium carbonate	$SrCO_3$	1.1×10^{-10}
Iron(II) carbonate	$FeCO_3$	3.2×10^{-11}	Strontium chromate	$SrCrO_4$	2.2×10^{-5}
Iron(II) hydroxide	$Fe(OH)_2$	8.0×10^{-16}	Strontium fluoride	SrF_2	2.5×10^{-9}
Iron(II) sulfide[b]	FeS	6×10^{-19}	Strontium sulfate	$SrSO_4$	3.2×10^{-7}
Iron(III) arsenate	$FeAsO_4$	5.7×10^{-21}	Thallium(I) bromide	$TlBr$	3.4×10^{-6}
Iron(III) ferrocyanide	$Fe_4[Fe(CN)_6]_3$	3.3×10^{-41}	Thallium(I) chloride	$TlCl$	1.7×10^{-4}
Iron(III) hydroxide	$Fe(OH)_3$	4×10^{-38}	Thallium(I) iodide	TlI	6.5×10^{-8}
Iron(III) phosphate	$FePO_4$	1.3×10^{-22}	Thallium(III) hydroxide	$Tl(OH)_3$	6.3×10^{-46}
Lead(II) arsenate	$Pb_3(AsO_4)_2$	4.0×10^{-36}	Tin(II) hydroxide	$Sn(OH)_2$	1.4×10^{-28}
Lead(II) azide	$Pb(N_3)_2$	2.5×10^{-9}	Tin(II) sulfide[b]	SnS	1×10^{-26}
Lead(II) bromide	$PbBr_2$	4.0×10^{-5}	Zinc carbonate	$ZnCO_3$	1.4×10^{-11}
Lead(II) carbonate	$PbCO_3$	7.4×10^{-14}	Zinc hydroxide	$Zn(OH)_2$	1.2×10^{-17}
Lead(II) chloride	$PbCl_2$	1.6×10^{-5}	Zinc oxalate	ZnC_2O_4	2.7×10^{-8}
Lead(II) chromate	$PbCrO_4$	2.8×10^{-13}	Zinc phosphate	$Zn_3(PO_4)_2$	9.0×10^{-33}
Lead(II) fluoride	PbF_2	2.7×10^{-8}	Zinc sulfide[b]	ZnS	2×10^{-25}

[a] Data are at various temperatures around "room" temperature, from 18 to 25 °C.
[b] For a solubility equilibrium of the type $MS(s) + H_2O \longleftrightarrow M^{2+}(aq) + HS^-(aq) + OH^-(aq)$.

D. Complex-Ion Formation Constants[c,d]

Formula	K_f	Formula	K_f	Formula	K_f
$[Ag(CN)_2]^-$	5.6×10^{18}	$[Co(ox)_3]^{3-}$	10^{20}	$[HgI_4]^{2-}$	6.8×10^{29}
$[Ag(EDTA)]^{3-}$	2.1×10^7	$[Cr(EDTA)]^-$	10^{23}	$[Hg(ox)_2]^{2-}$	9.5×10^6
$[Ag(en)_2]^+$	5.0×10^7	$[Cr(OH)_4]^-$	8×10^{29}	$[Ni(CN)_4]^{2-}$	2×10^{31}
$[Ag(NH_3)_2]^+$	1.6×10^7	$[CuCl_3]^{2-}$	5×10^5	$[Ni(EDTA)]^{2-}$	3.6×10^{18}
$[Ag(SCN)_4]^{3-}$	1.2×10^{10}	$[Cu(CN)_4]^{3-}$	2.0×10^{30}	$[Ni(en)_3]^{2+}$	2.1×10^{18}
$[Ag(S_2O_3)_2]^{3-}$	1.7×10^{13}	$[Cu(EDTA)]^{2-}$	5×10^{18}	$[Ni(NH_3)_6]^{2+}$	5.5×10^8
$[Al(EDTA)]^-$	1.3×10^{16}	$[Cu(en)_2]^{2+}$	1×10^{20}	$[Ni(ox)_3]^{4-}$	3×10^8
$[Al(OH)_4]^-$	1.1×10^{33}	$[Cu(NH_3)_4]^{2+}$	1.1×10^{13}	$[PbCl_3]^-$	2.4×10^1
$[Al(ox)_3]^{3-}$	2×10^{16}	$[Cu(ox)_2]^{2-}$	3×10^8	$[Pb(EDTA)]^{2-}$	2×10^{18}
$[CdCl_4]^{2-}$	6.3×10^2	$[Fe(CN)_6]^{4-}$	10^{37}	$[PbI_4]^{2-}$	3.0×10^4
$[Cd(CN)_4]^{2-}$	6.0×10^{18}	$[Fe(EDTA)]^{2-}$	2.1×10^{14}	$[Pb(OH)_3]^-$	3.8×10^{14}
$[Cd(en)_3]^{2+}$	1.2×10^{12}	$[Fe(en)_3]^{2+}$	5.0×10^9	$[Pb(ox)_2]^{2-}$	3.5×10^6
$[Cd(NH_3)_4]^{2+}$	1.3×10^7	$[Fe(ox)_3]^{4-}$	1.7×10^5	$[Pb(S_2O_3)_3]^{4-}$	2.2×10^6
$[Co(EDTA)]^{2-}$	2.0×10^{16}	$[Fe(CN)_6]^{3-}$	10^{42}	$[PtCl_4]^{2-}$	1×10^{16}
$[Co(en)_3]^{2+}$	8.7×10^{13}	$[Fe(EDTA)]^-$	1.7×10^{24}	$[Pt(NH_3)_6]^{2+}$	2×10^{35}
$[Co(NH_3)_6]^{2+}$	1.3×10^5	$[Fe(ox)_3]^{3-}$	2×10^{20}	$[Zn(CN)_4]^{2-}$	1×10^{18}
$[Co(ox)_3]^{4-}$	5×10^9	$[Fe(SCN)]^{2+}$	8.9×10^2	$[Zn(EDTA)]^{2-}$	3×10^{16}
$[Co(SCN)_4]^{2-}$	1.0×10^3	$[HgCl_4]^{2-}$	1.2×10^{15}	$[Zn(en)_3]^{2+}$	1.3×10^{14}
$[Co(EDTA)]^-$	10^{36}	$[Hg(CN)_4]^{2-}$	3×10^{41}	$[Zn(NH_3)_4]^{2+}$	4.1×10^8
$[Co(en)_3]^{3+}$	4.9×10^{48}	$[Hg(EDTA)]^{2-}$	6.3×10^{21}	$[Zn(OH)_4]^{2-}$	4.6×10^{17}
$[Co(NH_3)_6]^{3+}$	4.5×10^{33}	$[Hg(en)_2]^{2+}$	2×10^{23}	$[Zn(ox)_3]^{4-}$	1.4×10^8

[c] The ligands referred to in this table are monodentate: Cl^-, CN^-, I^-, NH_3, OH^-, SCN^-, $S_2O_3^{2-}$; bidentate: ethylenediamine (en), oxalate ion (ox); tetradentate: ethylenediaminetetraacetato ion, $EDTA^{4-}$.

[d] The K_f values are cumulative or overall formation constants (see page 1008).

TABLE D-4   Standard Electrode (Reduction) Potentials at 25 °C

Reduction half-reaction	$E°$, V
$F_2(g) + 2\,e^- \longrightarrow 2\,F^-(aq)$	+2.866
$OF_2(g) + 2\,H^+(aq) + 4\,e^- \longrightarrow H_2O(l) + 2\,F^-(aq)$	+2.1
$O_3(g) + 2\,H^+(aq) + 2\,e^- \longrightarrow O_2(g) + H_2O(l)$	+2.075
$S_2O_8^{2-}(aq) + 2\,e^- \longrightarrow 2\,SO_4^{2-}(aq)$	+2.01
$Ag^{2+}(aq) + e^- \longrightarrow Ag^+(aq)$	+1.98
$H_2O_2(aq) + 2\,H^+(aq) + 2\,e^- \longrightarrow 2\,H_2O(l)$	+1.763
$MnO_4^-(aq) + 4\,H^+(aq) + 3\,e^- \longrightarrow MnO_2(s) + 2\,H_2O(l)$	+1.70
$PbO_2(s) + SO_4^{2-}(aq) + 4\,H^+(aq) + 2\,e^- \longrightarrow PbSO_4(s) + 2\,H_2O(l)$	+1.69
$Au^{3+}(aq) + 3\,e^- \longrightarrow Au(s)$	+1.52
$MnO_4^-(aq) + 8\,H^+(aq) + 5\,e^- \longrightarrow Mn^{2+}(aq) + 4\,H_2O(l)$	+1.51
$2\,BrO_3^-(aq) + 12\,H^+(aq) + 10\,e^- \longrightarrow Br_2(l) + 6\,H_2O(l)$	+1.478
$PbO_2(s) + 4\,H^+(aq) + 2\,e^- \longrightarrow Pb^{2+}(aq) + 2\,H_2O(l)$	+1.455
$ClO_3^-(aq) + 6\,H^+(aq) + 6\,e^- \longrightarrow Cl^-(aq) + 3\,H_2O(l)$	+1.450
$Au^{3+}(aq) + 2\,e^- \longrightarrow Au^+(aq)$	+1.36
$Cl_2(g) + 2\,e^- \longrightarrow 2\,Cl^-(aq)$	+1.358
$Cr_2O_7^{2-}(aq) + 14\,H^+(aq) + 6\,e^- \longrightarrow 2\,Cr^{3+}(aq) + 7\,H_2O(l)$	+1.33
$MnO_2(s) + 4\,H^+(aq) + 2\,e^- \longrightarrow Mn^{2+}(aq) + 2\,H_2O(l)$	+1.23
$O_2(g) + 4\,H^+(aq) + 4\,e^- \longrightarrow 2\,H_2O(l)$	+1.229
$2\,IO_3^-(aq) + 12\,H^+(aq) + 10\,e^- \longrightarrow I_2(s) + 6\,H_2O(l)$	+1.20
$ClO_4^-(aq) + 2\,H^+(aq) + 2\,e^- \longrightarrow ClO_3^-(aq) + H_2O(l)$	+1.189
$ClO_3^-(aq) + 2\,H^+(aq) + e^- \longrightarrow ClO_2(g) + H_2O(l)$	+1.175
$NO_2(g) + H^+(aq) + e^- \longrightarrow HNO_2(aq)$	+1.07
$Br_2(l) + 2\,e^- \longrightarrow 2\,Br^-(aq)$	+1.065
$NO_2(g) + 2\,H^+(aq) + 2\,e^- \longrightarrow NO(g) + H_2O(l)$	+1.03
$[AuCl_4]^-(aq) + 3\,e^- \longrightarrow Au(s) + 4\,Cl^-(aq)$	+1.002
$VO_2^+(aq) + 2\,H^+(aq) + e^- \longrightarrow VO^{2+}(aq) + H_2O(l)$	+1.000
$NO_3^-(aq) + 4\,H^+(aq) + 3\,e^- \longrightarrow NO(g) + 2\,H_2O(l)$	+0.956
$Cu^{2+}(aq) + I^-(aq) + e^- \longrightarrow CuI(s)$	+0.86
$Hg^{2+}(aq) + 2\,e^- \longrightarrow Hg(l)$	+0.854
$Ag^+(aq) + e^- \longrightarrow Ag(s)$	+0.800
$Fe^{3+}(aq) + e^- \longrightarrow Fe^{2+}(aq)$	+0.771
$O_2(g) + 2\,H^+(aq) + 2\,e^- \longrightarrow H_2O_2(aq)$	+0.695
$2\,HgCl_2(aq) + 2\,e^- \longrightarrow Hg_2Cl_2(s) + 2\,Cl^-(aq)$	+0.63
$MnO_4^-(aq) + e^- \longrightarrow MnO_4^{2-}(aq)$	+0.56
$I_2(s) + 2\,e^- \longrightarrow 2\,I^-(aq)$	+0.535
$Cu^+(aq) + e^- \longrightarrow Cu(s)$	+0.520
$H_2SO_3(aq) + 4\,H^+(aq) + 4\,e^- \longrightarrow S(s) + 3\,H_2O(l)$	+0.449
$C_2N_2(g) + 2\,H^+(aq) + 2\,e^- \longrightarrow 2\,HCN(aq)$	+0.37
$[Fe(CN)_6]^{3-}(aq) + e^- \longrightarrow [Fe(CN)_6]^{4-}(aq)$	+0.361
$VO^{2+}(aq) + 2\,H^+(aq) + e^- \longrightarrow V^{3+}(aq) + H_2O(l)$	+0.337
$Cu^{2+}(aq) + 2\,e^- \longrightarrow Cu(s)$	+0.340
$PbO_2(s) + 2\,H^+(aq) + 2\,e^- \longrightarrow PbO(s) + H_2O(l)$	+0.28
$Hg_2Cl_2(s) + 2\,e^- \longrightarrow 2\,Hg(l) + 2\,Cl^-(aq)$	+0.2676
$HAsO_2(aq) + 3\,H^+(aq) + 3\,e^- \longrightarrow As(s) + 2\,H_2O(l)$	+0.240
$AgCl(s) + e^- \longrightarrow Ag(s) + Cl^-(aq)$	+0.2223
$SO_4^{2-}(aq) + 4\,H^+(aq) + 2\,e^- \longrightarrow 2\,H_2O(l) + SO_2(g)$	+0.17
$Cu^{2+}(aq) + e^- \longrightarrow Cu^+(aq)$	+0.159
$Sn^{4+}(aq) + 2\,e^- \longrightarrow Sn^{2+}(aq)$	+0.154
$S(s) + 2\,H^+(aq) + 2\,e^- \longrightarrow H_2S(g)$	+0.144
$AgBr(s) + e^- \longrightarrow Ag(s) + Br^-(aq)$	+0.071
$2\,H^+(aq) + 2\,e^- \longrightarrow H_2(g)$	0
$Pb^{2+}(aq) + 2\,e^- \longrightarrow Pb(s)$	−0.125
$Sn^{2+}(aq) + 2\,e^- \longrightarrow Sn(s)$	−0.137
$AgI(s) + e^- \longrightarrow Ag(s) + I^-(aq)$	−0.152

Reduction half-reaction	$E°$, V
$V^{3+}(aq) + e^- \longrightarrow V^{2+}(aq)$	-0.255
$Ni^{2+}(aq) + 2\,e^- \longrightarrow Ni(s)$	-0.257
$H_3PO_4(aq) + 2\,H^+(aq) + 2\,e^- \longrightarrow H_3PO_3(aq) + H_2O(l)$	-0.276
$Co^{2+}(aq) + 2\,e^- \longrightarrow Co(s)$	-0.277
$In^{3+}(aq) + 3\,e^- \longrightarrow In(s)$	-0.338
$PbSO_4(s) + 2\,e^- \longrightarrow Pb(s) + SO_4^{2-}(aq)$	-0.356
$Cd^{2+}(aq) + 2\,e^- \longrightarrow Cd(s)$	-0.403
$Cr^{3+}(aq) + e^- \longrightarrow Cr^{2+}(aq)$	-0.424
$Fe^{2+}(aq) + 2\,e^- \longrightarrow Fe(s)$	-0.440
$2CO_2(g) + 2\,H^+(aq) + 2\,e^- \longrightarrow H_2C_2O_4(aq)$	-0.49
$Zn^{2+}(aq) + 2\,e^- \longrightarrow Zn(s)$	-0.763
$Cr^{2+}(aq) + 2\,e^- \longrightarrow Cr(s)$	-0.90
$Mn^{2+}(aq) + 2\,e^- \longrightarrow Mn(s)$	-1.18
$Ti^{2+}(aq) + 2\,e^- \longrightarrow Ti(s)$	-1.63
$U^{3+}(aq) + 3\,e^- \longrightarrow U(s)$	-1.66
$Al^{3+}(aq) + 3\,e^- \longrightarrow Al(s)$	-1.676
$Mg^{2+}(aq) + 2\,e^- \longrightarrow Mg(s)$	-2.356
$La^{3+}(aq) + 3\,e^- \longrightarrow La(s)$	-2.38
$Na^+(aq) + e^- \longrightarrow Na(s)$	-2.713
$Ca^{2+}(aq) + 2\,e^- \longrightarrow Ca(s)$	-2.84
$Sr^{2+}(aq) + 2\,e^- \longrightarrow Sr(s)$	-2.89
$Ba^{2+}(aq) + 2\,e^- \longrightarrow Ba(s)$	-2.92
$Cs^+(aq) + e^- \longrightarrow Cs(s)$	-2.923
$K^+(aq) + e^- \longrightarrow K(s)$	-2.924
$Rb^+(aq) + e^- \longrightarrow Rb(s)$	-2.924
$Li^+(aq) + e^- \longrightarrow Li(s)$	-3.040

Basic Solution

Reduction half-reaction	$E°$, V
$O_3(g) + H_2O(l) + 2\,e^- \longrightarrow O_2(g) + 2\,OH^-(aq)$	$+1.246$
$ClO^-(aq) + H_2O(l) + 2\,e^- \longrightarrow Cl^-(aq) + 2\,OH^-(aq)$	$+0.890$
$H_2O_2(aq) + 2\,e^- \longrightarrow 2\,OH^-(aq)$	$+0.88$
$BrO^-(aq) + H_2O(l) + 2\,e^- \longrightarrow Br^-(aq) + 2\,OH^-(aq)$	$+0.766$
$ClO_3^-(aq) + 3\,H_2O(l) + 6\,e^- \longrightarrow Cl^-(aq) + 6\,OH^-(aq)$	$+0.622$
$2\,AgO(s) + H_2O(l) + 2\,e^- \longrightarrow Ag_2O(s) + 2\,OH^-(aq)$	$+0.604$
$MnO_4^-(aq) + 2\,H_2O(l) + 3\,e^- \longrightarrow MnO_2(s) + 4\,OH^-(aq)$	$+0.60$
$BrO_3^-(aq) + 3\,H_2O(l) + 6\,e^- \longrightarrow Br^-(aq) + 6\,OH^-(aq)$	$+0.584$
$2\,BrO^-(aq) + 2\,H_2O(l) + 2\,e^- \longrightarrow Br_2(l) + 4\,OH^-(aq)$	$+0.455$
$2\,IO^-(aq) + 2\,H_2O(l) + 2\,e^- \longrightarrow I_2(s) + 4\,OH^-(aq)$	$+0.42$
$O_2(g) + 2\,H_2O(l) + 4\,e^- \longrightarrow 4\,OH^-(aq)$	$+0.401$
$Ag_2O(s) + H_2O(l) + 2\,e^- \longrightarrow 2\,Ag(s) + 2\,OH^-(aq)$	$+0.342$
$Co(OH)_3(s) + e^- \longrightarrow Co(OH)_2(s) + OH^-(aq)$	$+0.17$
$2\,MnO_2(s) + H_2O(l) + 2\,e^- \longrightarrow Mn_2O_3(s) + 2\,OH^-(aq)$	$+0.118$
$NO_3^-(aq) + H_2O(l) + 2\,e^- \longrightarrow NO_2^-(aq) + 2\,OH^-(aq)$	$+0.01$
$CrO_4^{2-}(aq) + 4\,H_2O(l) + 3\,e^- \longrightarrow Cr(OH)_3(s) + 5\,OH^-(aq)$	-0.11
$HPbO_2^-(aq) + H_2O(l) + 2\,e^- \longrightarrow Pb(s) + 3\,OH^-(aq)$	-0.54
$HCHO(aq) + 2\,H_2O(l) + 2\,e^- \longrightarrow CH_3OH(aq) + 2\,OH^-(aq)$	-0.59
$SO_3^{2-}(aq) + 3\,H_2O(l) + 4\,e^- \longrightarrow S(s) + 6\,OH^-(aq)$	-0.66
$AsO_4^{3-}(aq) + 2\,H_2O(l) + 2\,e^- \longrightarrow AsO_2^-(aq) + 4\,OH^-(aq)$	-0.67
$AsO_2^-(aq) + 2\,H_2O(l) + 3\,e^- \longrightarrow As(s) + 4\,OH^-(aq)$	-0.68
$Cd(OH)_2(s) + 2\,e^- \longrightarrow Cd(s) + 2\,OH^-(aq)$	-0.824
$2\,H_2O(l) + 2\,e^- \longrightarrow H_2(g) + 2\,OH^-(aq)$	-0.828
$OCN^-(aq) + H_2O(l) + 2\,e^- \longrightarrow CN^-(aq) + 2\,OH^-(aq)$	-0.97
$As(s) + 3\,H_2O(l) + 3\,e^- \longrightarrow AsH_3(g) + 3\,OH^-(aq)$	-1.21
$Zn(OH)_2(s) + 2\,e^- \longrightarrow Zn(s) + 2\,OH^-(aq)$	-1.246
$Sb(s) + 3\,H_2O(l) + 3\,e^- \longrightarrow SbH_3(g) + 3\,OH^-(aq)$	-1.338
$Al(OH)_4^-(aq) + 3\,e^- \longrightarrow Al(s) + 4\,OH^-(aq)$	-2.310
$Mg(OH)_2(s) + 2\,e^- \longrightarrow Mg(s) + 2\,OH^-(aq)$	-2.687

Appendix

E Glossary

Absolute configuration refers to the spatial arrangement of the groups attached to a chiral carbon atom. The two possibilities are D and L.

The **absolute zero of temperature** is the temperature at which molecular motion is hypothesized to cease: $0\ K = -273.15\ °C$.

Accuracy is the "closeness" of a measured value to the true or accepted value of a quantity.

Acetyl (See **acyl**.)

An **achiral** molecule has a structure that is superimposable on its mirror image. (See also **chiral**.)

An **acid** is (1) a hydrogen-containing compound that can produce hydrogen ions, H^+ (Arrhenius theory); (2) a proton donor (Brønsted–Lowry theory); (3) an atom, ion, or molecule that can accept a pair of electrons to form a covalent bond (Lewis theory).

An **acid anhydride** is an oxide that reacts with water to form a ternary acid as the sole product.

An **acid–base indicator** is a substance used to measure the pH of a solution or to signal the equivalence point in an acid–base titration. The nonionized weak acid form has one color and the anionic form, a different color.

An **acid ionization constant**, K_a, is the equilibrium constant for the ionization reaction of a weak acid.

An **acid salt** contains an anion that can act as an acid (proton donor); examples are $NaHSO_4$ and NaH_2PO_4.

The **actinides** are a series of radioactive elements ($Z = 90 - 103$) characterized by partially filled $5f$ orbitals in their atoms.

An **activated complex** is an intermediate in a chemical reaction formed through collisions between energetic molecules. Once formed, it dissociates either into the products or back to the reactants.

Activation energy is the minimum total kinetic energy that molecules must bring to their collisions for a chemical reaction to occur.

Active sites are the locations at which catalysis occurs, whether on the surface of a heterogeneous catalyst or an enzyme.

The **actual yield** is the *measured* quantity of a product obtained in a chemical reaction. (See also **theoretical yield** and **percent yield**.)

The **acyl group** is $-\overset{\overset{\displaystyle O}{\|}}{C}-R$. If R = H, this is called the **formyl** group; $R = CH_3$, **acetyl**; and $R = C_6H_5$, **benzoyl**.

In an **addition reaction** functional group atoms are joined to the carbon atoms at points of unsaturation in alkene and alkyne hydrocarbon molecules.

An **adduct** is a compound formed by joining together two simpler molecules through a coordinate covalent bond, such as the adduct of $AlCl_3$ and $(C_2H_5)_2O$ pictured on page 893.

Adenosine diphosphate (ADP) and **adenosine triphosphate (ATP)** are agents involved in energy transfers during metabolism. The hydrolysis of ATP produces ADP, the ion HPO_4^{2-}, and a release of energy.

Adhesive forces are intermolecular forces between unlike molecules, such as molecules of a liquid and of a surface with which it is in contact.

Adsorption refers to the attachment of ions or molecules to the surface of a material.

ADP (See **adenosine diphosphate**.)

Alcohols contain the functional group $-OH$ and have the general formula ROH.

Aldehydes have the general formula $R-\overset{\overset{\displaystyle O}{\|}}{C}-H$.

Alicyclic hydrocarbon molecules have their carbon atom skeletons arranged in rings and resemble aliphatic (rather than aromatic) hydrocarbons.

Aliphatic hydrocarbon molecules have their carbon atom skeletons arranged in straight or branched chains.

Alkali metals is the family name for the group 1 elements of the periodic table.

Alkaline earth metals is the family name for the group 2 elements of the periodic table.

Alkane hydrocarbon molecules have only single covalent bonds between carbon atoms. In their chain structures alkanes have the general formula C_nH_{2n+2}.

Alkene hydrocarbons have one or more carbon-to-carbon double bonds in their molecules. The simple alkenes have the general formula C_nH_{2n}.

Alkyl groups are alkane hydrocarbon molecules from which one hydrogen atom has been extracted. For example, the group $-CH_3$ is the **methyl** group; $-CH_2CH_3$ is the **ethyl** group. (See also Table 27.1.)

Alkyne hydrocarbons have one or more carbon-to-carbon triple bonds in their molecules. The simple alkynes have the general formula C_nH_{2n-2}.

Allotropy refers to the existence of an *element* in two or more different *molecular* forms, such as O_2 and O_3 or red and white phosphorus.

An **alloy** is a mixture of two or more metals. Some alloys are solid solutions, some are heterogeneous mixtures, and some are intermetallic compounds.

An **alpha (α) particle** is a combination of two protons and two neutrons identical to the helium ion, that is, $^4He^{2+}$. Alpha particles are emitted in some radioactive decay processes.

Alums are sulfates of the general formula, $M(I)(MIII)(SO_4)_2 \cdot 12\ H_2O$. M(I) is most commonly an alkali metal or ammonium ion, and M(III) is most commonly Al^{3+}, Fe^{3+}, or Cr^{3+}.

Amalgams are metal alloys containing mercury. Depending on their compositions, some are liquid and some are solid.

An **amide** is derived from the ammonium salt of a carboxylic acid and has the general

formula $R\overset{\overset{\displaystyle O}{\|}}{-C}-NH_2$.

An **amine** is an organic base having the formula RNH_2 (primary), R_2NH (secondary), or R_3N (tertiary), depending on the number of hydrogen atoms of an NH_3 molecule that are replaced by R groups.

An **α-amino acid** is a carboxylic acid that has an amino group ($-NH_2$) attached to the carbon atom adjacent to the **carboxyl group** ($-COOH$).

Ampoteric is the term used to describe the ability of certain oxides and hydroxo compounds to act as either acids or bases.

An **angular wave function,** $Y(\theta, \phi)$, is the part of a wave function that depends on the angles θ and ϕ when the Schrödinger wave equation is expressed in spherical polar coordinates. (See also **radial wave function**)

Anhydride is a term meaning "without water." An acid anhydride is an element oxide that reacts with water to form an acid, and a base anhydride, to form a base.

An **anion** is a negatively charged ion. An anion migrates toward the anode in an electrochemical cell.

The **anode** is the electrode in an electrochemical cell at which an oxidation half-reaction occurs.

An **antibonding molecular orbital** describes regions in a molecule in which there is a low electron probability or charge density between two bonded atoms.

Aromatic compounds are organic substances whose carbon atom skeletons are arranged in hexagonal rings, based on benzene, C_6H_6.

In the **Arrhenius acid–base theory** an acid produces H^+ and a base produces OH^- in aqueous solutions.

The **Arrhenius equation** relates the rate constant of a reaction to the temperature and activation energy.

Asymmetric is the term used to describe a C atom with four different substituent groups. A molecule with such a C atom is chiral.

One **atmosphere** (standard atmosphere) is the pressure exerted, under carefully specified conditions, by a column of mercury 760 mm high.

The **atmosphere** is the mixture of gases (nitrogen, oxygen, argon, and traces of others) that lies above the solid crust and oceans of Earth.

The **atom** is the basic building block of matter. The number of different atoms currently known is 115. A chemical element consists of a single type of atom and a chemical compound, of two or more different kinds of atoms.

The **atomic mass (weight)** of an element is the average of the isotopic masses weighted according to the naturally occurring abundances of the isotopes of the element and relative to the value of exactly 12 u for a carbon-12 atom.

An **atomic mass unit, u,** is used to express the masses of individual atoms. One u is 1/12 the mass of a carbon-12 atom.

The **atomic number, Z,** is the number of protons in the nucleus of an atom. It is also the number of electrons outside the nucleus of an electrically neutral atom.

Atomic (line) spectra are produced by dispersing light emitted by excited gaseous atoms. Only a discrete set of wavelength components (seen as colored lines) is present in a line spectrum.

ATP (See **adenosine triphosphate.**)

The **aufbau process** is a method of writing electron configurations. Each element is described as differing from the preceding one in terms of the orbital to which the one additional electron is assigned.

An **average bond energy** is the average of bond-dissociation energies for a number of different species containing a particular covalent bond. (See also **bond dissociation energy.**)

The **Avogadro constant,** N_A, has a value of $6.02214 \times 10^{23} \text{ mol}^{-1}$. It is the number of elementary units in one mole.

Avogadro's law (hypothesis) states that at a fixed temperature and pressure, the volume of a gas is directly proportional to the amount of gas and that equal volumes of different gases, compared under identical conditions of temperature and pressure, contain equal numbers of molecules.

An **azeotrope** is a solution that boils at a constant temperature, producing vapor of the same composition as the liquid. In some cases, the the azeotrope boils at a lower temperature than the solution components, in other cases, at a higher temperature.

A **balanced equation** has the same number of atoms of each type on both sides. (See also **chemical equation.**)

Band theory is a form of molecular orbital theory to describe bonding in metals and semiconductors.

A **barometer** is a device used to measure the pressure of the atmosphere.

Barometric pressure is the prevailing pressure of the atmosphere as indicated by a barometer.

A **base** is (1) a compound that produces hydroxide ions, OH^-, in water solution (Arrhenius theory); (2) a proton acceptor (Brønsted–Lowry theory); (3) an atom, ion, or molecule that can donate a pair of electrons to form a covalent bond (Lewis theory).

A **base anhydride** is an oxide that reacts with water to form a base.

A **base ionization constant,** K_b, is the equilibrium constant for the ionization reaction of a weak base.

The **basic oxygen process** is the principal process used to convert impure iron (pig iron) into steel.

A **battery** is a voltaic cell [or a group of voltaic cells connected in series (+ to −)] used to produce electricity from chemical change.

bcc (See **body-centered cubic.**)

A **beta (β^-) particle** is an electron emitted as a result of the conversion of a neutron to a proton in certain atomic nuclei undergoing radioactive decay.

A **bidentate** ligand attaches itself to the central atom of a complex at two points in the coordination sphere.

A **bimolecular process** is an elementary process involving the collision of two molecules.

Binary compounds are compounds composed of *two* elements.

A **body-centered cubic (bcc)** crystal structure is one in which the unit cell has structural units at each corner and one in the center of the cube.

Boiling is a process in which vaporization occurs throughout a liquid. It occurs when the vapor pressure of a liquid is equal to barometric pressure.

A **bomb calorimeter** is a device used to measure the heat of a combustion reaction. The quantity measured is the heat of reaction at constant volume, $q_V = \Delta U$.

A **bond angle** is the angle between two covalent bonds. It is the angle between hypothetical lines joining the nuclei of two atoms to the nucleus of a third atom to which they are covalently bonded.

Bond dissociation energy, *D*, is the quantity of energy required to break one mole of covalent bonds in a gaseous species, usually expressed in kJ/mol. (See also **average bond energy.**)

Bond length (bond distance) is the distance between the centers of two atoms joined by a covalent bond.

Bond order is one-half the difference between the numbers of electrons in bonding and in antibonding molecular orbitals in a covalent bond. A single bond has a bond order of 1; a double bond, 2; and a triple bond, 3.

A **bond pair** is a pair of electrons involved in covalent bond formation.

A **bonding molecular orbital** describes regions of high electron probability or charge density in the internuclear region between two bonded atoms.

The **Born–Fajans–Haber cycle** relates lattice energies of crystalline ionic solids to ionization energies, electron affinities, and enthalpies of sublimation, dissociation, and formation.

Boyle's law states that the volume of a fixed amount of gas at a constant temperature is inversely proportional to the gas pressure.

A **breeder reactor** is a nuclear reactor that creates more nuclear fuel than it consumes, for example, by converting ^{238}U to ^{239}Pu.

The **Brønsted–Lowry theory** describes acids as proton donors and bases as proton acceptors. An acid–base reaction involves the transfer of protons from an acid to a base.

Buffer capacity refers to the amount of acid and/or base that a buffer solution can neutralize while maintaining an essentially constant pH.

Buffer range is the range of pH values over which a buffer solution can maintain a fairly constant pH.

A **buffer solution** resists a change in its pH. It contains components capable of neutralizing small added amounts of acids and base.

By-products are substances produced along with the principal product in a chemical process, either through the main reaction or a side reaction.

Calcination refers to the decomposition of a solid by heating at temperatures below its melting point, such as the decomposition of calcium carbonate to calcium oxide and $CO_2(g)$.

The **calorie (cal)** is the quantity of heat required to change the temperature of one gram of water by one degree Celsius.

A **calorimeter** is a device (of which there are numerous types) used to measure a quantity of heat.

A **carbohydrate** is a polyhydroxy aldehyde, a polyhydroxy ketone, a derivative of these, or a substance that yields them upon hydrolysis. Carbohydrates can be viewed as "hydrates" of carbon, in the sense that their general formulas are $C_x(H_2O)_y$.

Carbon black is a finely divided amorphous form of carbon prepared by the incomplete combustion of hydrocarbons.

The **carbonyl group,** is found in aldehydes, ketones, and carboxylic acids

$$\diagdown C = O$$

The **carboxyl group** is $-\overset{\overset{\displaystyle O}{\|}}{C}-OH$.

A **carboxylic acid** has one or more carboxyl groups attached to a hydrocarbon chain or ring structure.

A **catalyst** provides an alternative mechanism of lower activation energy for a chemical reaction. The reaction is speeded up, and the catalyst is regenerated.

The **cathode** is the electrode of an electrochemical cell where a reduction half-reaction occurs.

Cathode rays are negatively charged particles (electrons) emitted at the negative electrode (cathode) in the passage of electricity through gases at very low pressures.

Cathodic protection is a method of corrosion control in which the metal to be protected is joined to a more active metal that corrodes instead. The protected metal acts as the cathode of a voltaic cell.

A **cation** is a positively charged ion. A cation migrates toward the cathode in an electrochemical cell.

The **cell** is the fundamental unit of living organisms.

A **cell diagram** is a symbolic representation of an electrochemical cell that indicates the substances entering into the cell reaction, electrode materials, solution concentrations, etc.

The **cell potential, E_{cell},** is the potential difference (voltage) between the two electrodes of an electrochemical cell.

A **central atom** in a structure is an atom that is bonded to two or more other atoms.

The **Celsius** temperature scale is based on a value of 0 °C for the normal melting point of ice and 100 °C for the normal boiling point of water.

In **chain reaction polymerization,** a reaction is initiated by "opening up" a carbon-to-carbon double bond. Monomer units add to free-radical intermediates to produce a long-chain polymer.

Charles's law states that the volume of a fixed amount of gas at a constant pressure is directly proportional to the Kelvin (absolute) temperature.

A **chelate** results from the attachment of polydentate ligands to the central atom of a complex ion. Chelates are five- or six-membered rings that include the central atom and atoms of the ligands.

A **chelating agent** is a polydentate ligand. It simultaneously attaches to two or more positions in the coordination sphere of the central atom of a complex ion.

The **chelation effect** refers to an exceptional stability conferred to a complex ion when polydentate ligands are present.

Chemical change (See **chemical reaction.**)

Chemical energy is the energy associated with chemical bonds and intermolecular forces.

A **chemical equation** is a symbolic representation of a chemical reaction. Symbols and formulas are used to represent reactants and products, and stoichiometric coefficients are used to balance the equation. (See also **balanced equation.**)

A **chemical formula** represents the relative numbers of atoms of each kind in a substance through symbols and numerical subscripts.

A **chemical property** is the ability (or inability) of a sample of matter to undergo a particular chemical reaction.

A **chemical reaction** is a process in which one set of substances (reactants) is transformed into a new set of substances (products).

Chemical symbols are abbreviations of the names of the elements consisting of one or two letters (e.g., N = nitrogen and Ne = neon).

Chiral refers to a molecule with a structure that is not superimposable on its mirror image. (See also **enantiomers.**)

A **chlor-alkali process** involves the electrolysis of NaCl(aq) to produce NaOH(aq), $Cl_2(g)$, and $H_2(g)$.

Chromatography is a separation technique in which the components in one phase (gas or liquid) are separated from one another as they pass over or through another phase (solid or liquid).

The term **cis** describes geometric isomers in which two groups are attached on the same side of a double bond in an organic molecule, or along the same edge of a square in a square-planar complex, or at two adjacent vertices of an octahedral complex. (See also **geometric isomerism.**)

A **closed system** is one that can exchange energy but not matter with its surroundings.

Cohesive forces are intermolecular forces between like molecules, such as within a drop of liquid.

Coke is a relatively pure form of carbon produced by heating coal out of contact with air (destructive distillation).

Colligative properties—vapor pressure lowering, freezing point depression, boiling point elevation, osmotic pressure—have values that depend on the number of solute particles in a solution but not on their identity.

Collision frequency is the number of collisions occurring between molecules in a unit of time.

Collision theory describes reactions in terms of molecular collisions—the frequency of collisions, the fraction of activated molecules, and the probability that collisions will be effective.

A **colloidal mixture** contains particles that are intermediate in size to those of a true solution and an ordinary heterogeneous mixture.

The **common ion effect** describes the effect on an equilibrium by a substance that furnishes ions that can participate in the equilibrium.

A **complementary color** is a secondary color that mixes with the opposite primary color on the color wheel (Figure 25-16) to produce white light in additive color mixing or black in subtractive color mixing.

A **complex** is a polyatomic cation, anion, or neutral molecule in which groups (molecules or ions) called ligands are bonded to a central metal atom or ion.

A **complex ion** is a complex having a net electrical charge.

Composition refers to the components and their relative proportions in a sample of matter.

A **compound** is a substance made up of two or more elements. It does not change its identity in physical changes, but it can be broken down into its constituent elements by chemical changes.

Concentration (1) refers to the composition of a solution. (2) (see **extractive metallurgy.**)

In a **concentration cell** identical electrodes are immersed in solutions of different concentrations. The voltage (emf) of the cell is a function simply of the concentrations of the two solutions.

Condensation is the passage of molecules from the gaseous state to the liquid state.

A **condensation reaction** is one in which two molecules are combined by eliminating a small molecule (such as H_2O) between them.

A **condensed formula** is a simplified representation of a structural formula.

Conformations refer to the different spatial arrangements possible in a molecule. Examples are the "boat" and "chair" forms of cyclohexane.

A **conjugate acid** is formed when a Brønsted–Lowry base gains a proton. Every base has a conjugate acid.

A **conjugate base** remains after a Brønsted–Lowry acid has lost a proton. Every acid has a conjugate base.

Consecutive reactions are two or more reactions carried out in sequence. A product of each reaction becomes a reactant in a following reaction, until a final product is formed.

A **continuous spectrum** is one in which all wavelength components of the visible portion of the electromagnetic spectrum are present.

The **contact process** is a process for the manufacture of sulfuric acid having as its key reaction the oxidation of $SO_2(g)$ to $SO_3(g)$ in contact with a catalyst.

Control rods are neutron-absorbing metal rods (e.g., Cd) that are used to control the neutron flux in a nuclear reactor and thereby control the rate of the fission reaction.

A **conversion factor** is a relationship between quantities expressed in different units. It is written as a *ratio* of two quantities that are either equal or otherwise equivalent to one another.

In a **coordinate covalent bond** electrons shared between two atoms are contributed by just one of the atoms. As a result, the bonded atoms exhibit formal charges.

Coordination compounds are neutral complexes or compounds containing complex ions.

Coordination number is the number of positions around a central atom where ligands can be attached in the formation of a complex. Applied to a crystalline solid, coordination number signifies the number of nearest neighboring atoms (or ions of opposite charge) to any given atom (or ion) in a crystal.

Corrosion is the oxidation of a metal through the action of water, air, and/or salt solutions. A corrosion reaction consists of an oxidation and a reduction half-reaction.

Coupled reactions are sets of chemical reactions that occur together. One (or more) of the reactions taken alone is (are) *nonspontaneous* and other(s), *spontaneous*. The overall reaction is *spontaneous*.

A **covalent bond** is formed when electrons are shared between a pair of atoms. In valence bond theory, the sharing of the electrons is said to occur in the region in which atomic orbitals overlap.

Covalent radius is one-half the distance between the centers of two atoms that are bonded covalently. It is the atomic radius associated with an element in its covalent compounds.

The **critical point** refers to the temperature and pressure at which a liquid and its vapor become identical. It is the highest temperature point on the vapor pressure curve.

Crystal field theory describes bonding in complexes in terms of electrostatic attractions between ligands and the nucleus of the central metal. Particular attention is focused on the splitting of the *d* energy level of the central metal.

Cubic closest packed is one of the two ways in which spheres can be packed to minimize the amount of free space or voids among them. (See Figures 13-39 and 13-40.)

Dalton's law of partial pressures states that in a mixture of gases the total pressure is the sum of the partial pressures of the gases present. (See also **partial pressure.**)

The **dashed-wedged line notation** is a method of conveying a three-dimensional perspective to a structure plotted in a plane (see Figure 27-1).

The *d* **block** refers to that section of the periodic table in which the process of orbital filling (aufbau process) involves a *d* subshell.

A **decay constant** is a first-order rate constant describing radioactive decay.

Degree of ionization refers to the extent to which molecules of a weak acid or weak base ionize. The degree of ionization increases as the weak electrolyte solution is diluted. (See also **percent ionization.**)

Deionized water is water that has been freed of most of its impurity ions by being passed first through a cation and then an anion exchange resin.

Degenerate orbitals are orbitals that are at the same energy level.

Deliquescence is a process in which atmospheric water vapor condenses on a highly soluble solid and the solid dissolves in the water to produce an aqueous solution.

A **delocalized molecular orbital** describes a region of high electron probability or charge density that extends over three or more atoms.

Denaturation refers to the loss of biological activity of a protein brought about by changes in its secondary and tertiary structures.

Density is a physical property obtained by dividing the mass of a material or object by its volume (i.e., mass per unit volume).

Deposition is the passage of molecules from the gaseous to the solid state.

Deoxyribonucleic acid (DNA) is the substance that makes up the genes of the chromosomes in the nuclei of cells.

Detergents are cleansing agents that act by emulsifying oils. Most common among synthetic detergents are the salts of organic sulfonic acids, $RSO_3^-Na^+$.

Dextrorotatory means the ability to rotate the plane of polarized light to the *right,* designated (+).

Diagonal relationships refer to similarities that exist between certain pairs of elements in different groups and periods of the periodic table, such as Li and Mg, Be and Al, and B and Si.

Dialysis is a process, similar to osmosis, in which ions or molecules in solution pass through a semipermeable membrane but colloidal particles do not.

A **diamagnetic** substance has all its electrons paired and is slightly repelled by a magnetic field.

Diastereomers are optically active isomers of a compound, but their structures are *not* mirror images (as are enantiomers).

Diborane is the first in a series of electron-deficient boron hydrides. Its formula is B_2H_6.

Diffraction is the dispersion of light into its different components as a result of the interference produced by the reflection of light from a grooved surface.

Diffusion refers to the spreading of a substance (usually a gas or liquid) into a region where it is not originally present as a result of random molecular motion.

Dilution is the process of reducing the concentration of a solution by adding more solvent.

A **dimer** is a molecule comprised of two simpler formula units, such as Al_2Cl_6, which is a dimer of $AlCl_3$.

Dipole moment, μ, is a measure of the extent to which a separation exists between the centers of positive and negative charge within a molecule. The unit used to measure dipole moment is the **debye,** 3.34×10^{-30} C m.

A **disaccharide** is a sugar molecule formed when two monosaccharide molecules join together by eliminating a H_2O molecule between them.

Dispersion (London) forces are intermolecular forces associated with instantaneous and induced dipoles.

In a **disproportionation reaction** the same substance is both oxidized and reduced.

Distillation is a procedure for separating a liquid from a mixture by vaporizing the liquid and condensing the vapor. In **simple** distillation a volatile liquid is separated from nonvolatile solutes. In **fractional** distillation volatile components of a solution are separated based on their different volatilities.

In a **double covalent bond** *two pairs* of electrons are shared between bonded atoms. The bond is represented by a double-dash sign (=).

Effective nuclear charge, Z_{eff}, is the positive charge acting on a particular electron in an atom. Its value is the charge on the nucleus, reduced to the extent that other electrons screen the particular electron from the nucleus.

Effusion is the escape of a gas through a tiny hole in its container.

An **electrochemical cell** is a device in which the electrons transferred in an oxidation–reduction reaction are made to pass through an electrical circuit. (See also **electrolytic cell** and **voltaic cell.**)

An **electrode** is a metal surface on which an oxidation–reduction equilibrium is established between the metal and substances in solution.

An **electrode potential** is the electric potential developed on a metal electrode when an oxidation and a reduction half-reaction have reached equilibrium on the surface of the electrode.

Electrolysis is the decomposition of a substance, either in the molten state or in an electrolyte solution, by means of electric current.

An **electrolytic cell** is an electrochemical cell in which a nonspontaneous reaction is carried out by electrolysis.

Electromagnetic radiation is a form of energy propagated as mutually perpendicular electric and magnetic fields. It includes visible light, infrared, ultraviolet, X-ray, and radio waves.

Electromotive force (emf) is the potential difference between two electrodes in a voltaic cell, expressed in volts.

Electron affinity is the energy change associated with the gain of an electron by a neutral gaseous atom.

Electron capture (E.C.) is a form of radioactive decay in which an electron from an inner electronic shell is absorbed by a nucleus. In the nucleus the electron is used to convert a proton to a neutron.

An **electron configuration** is a designation of how electrons are distributed among various orbitals in an atom.

Electronegativity is a measure of the electron-attracting power of a bonded atom; metals have low electronegativities, and nonmetals have high electronegativities.

The **electronegativity difference** between two atoms that are bonded together is used to assess the degree of polarity in the bond.

Electron group geometry refers to the geometrical distribution about a central atom of the electron pairs in its valence shell.

Electron spin is a characteristic of electrons giving rise to the magnetic properties of atoms. The two possibilities for electron spin are $+\frac{1}{2}$ and $-\frac{1}{2}$.

Electrons are particles carrying the fundamental unit of negative electric charge and found outside the nuclei of all atoms.

An **electrophilic** center in a molecule is an electron attracting region of positive charge.

An **element** is a substance composed of a single type of atom. It cannot be broken down into simpler substances by chemical reactions.

An **elementary process** is an event that significantly alters a molecule's energy or geometry or produces a new molecule(s). It represents a single step in a reaction mechanism.

An **elimination reaction** is one in which atoms are removed from adjacent positions on a hydrocarbon chain to produce a small molecule (e.g., H_2O) and an additional bond between carbon atoms.

An **empirical formula** is the simplest chemical formula that can be written for a compound, that is, having the smallest integral subscripts possible.

Enantiomers (optical isomers) are molecules whose structures are nonsuperimposable mirror images. The molecules are optically active, that is, able to rotate the plane of polarized light.

An **endothermic** process results in a lowering of the temperature of an isolated system or, the absorption of heat by a system that interacts with its surroundings.

The **end point** is the point in a titration where the indicator used changes color. A properly chosen indicator has its end point coming as closely as possible to the equivalence point of the titration.

Energy is the capacity to do work. (See also **work**.)

An **energy-level diagram** is a representation of the allowed energy states for the electrons in atoms. The simplest energy-level diagram—that of the hydrogen atom—is shown in Figure 9-14.

The **English system** of measurement has the yard as its unit of length, the pound as

its unit of mass, and the second as its unit of time.

Enthalpy, H, is a thermodynamic function used to describe constant-pressure processes: $H = E + PV$, and at constant pressure, $\Delta H = \Delta U + P\Delta V$.

Enthalpy change, ΔH, is the difference in enthalpy between two states of a system. For a chemical reaction carried out at constant temperature and pressure and with work limited to pressure–volume work, the enthalpy change is called the *heat of reaction at constant pressure.*

An **enthalpy diagram** is a diagrammatic representation of the enthalpy changes in a process.

Enthalpy (heat) of formation (See **standard enthalpy of formation.**)

Entropy, S, is a thermodynamic property related to the number of energy levels among which the energy of a system is spread. The greater the number of energy levels for a given total energy, the greater the entropy.

Entropy change, ΔS, is the difference in entropy between two states of a system.

An **enzyme** is a high molar mass protein that catalyzes biological reactions.

An **equation of state** is a mathematical expression relating the amount, volume, temperature, and pressure of a substance (usually applied to gases).

Equilibrium refers to a condition where forward and reverse processes proceed at equal rates and no further net change occurs. For example, amounts of reactants and products in a reversible reaction remain constant over time.

An **equilibrium constant expression** describes the relationship among the concentrations (or partial pressures) of the substances present in a system at equilibrium

The **equivalence point** of a titration is the condition in which the reactants are in stoichiometric proportions. They consume each other, and neither reactant is in excess.

An **ester** is the product of the elimination of H_2O from between an acid and an alcohol molecule. Esters have the general

$$\text{formula } R-\overset{\overset{\displaystyle O}{\|}}{C}-O-R'.$$

An **ether** has the general formula $R-O-R'$.

Eutrophication is the deterioration of a freshwater body caused by nutrients such

as nitrates and phosphates, which stimulate the growth of algae, oxygen depletion, and fish kills.

In an **excited state** of an atom, one or more electrons are promoted to a higher energy level than in the ground state. (See also **ground state**.)

An **exothermic** process produces an increase in temperature in an isolated system or, for a system that interacts with its surroundings, the evolution of heat.

Expanded valence shell is a term used to describe Lewis structures in which certain atoms in the third or higher period of the periodic table appear to require 10 or 12 electrons in their valence shells.

An **extensive property** is one, like mass or volume, whose value depends on the quantity of matter observed.

Extractive metallurgy refers to the process of extracting a metal from its ores. Generally this occurs in four steps. **Concentration** separates the ore from waste rock (gangue). **Roasting** converts the ore to the metal oxide. **Reduction** (usually with carbon) converts the oxide to the metal. **Refining** removes impurities from the metal.

The **E, Z system** is a system of nomenclature used to describe the manner in which substituent groups are attached at a carbon-to-carbon double bond

A **face-centered cubic (fcc)** crystal structure is one in which the unit cell has structural units at the eight corners and in the center of each face of the unit cell. It is derived from the cubic closest packed arrangement of spheres (see Figure 13-40).

The **Fahrenheit** temperature scale is based on a value of 32 °F as the melting point of ice and 212 °F as the boiling point of water.

A **family** of elements is a numbered group from the periodic table, sometimes carrying a distinctive name. For example, group 17 is the halogen family.

The **Faraday constant, F,** is the charge associated with one mole of electrons, 96,485 C/mol e⁻.

Fats are triglycerides in which saturated fatty acid components predominate.

The *f* **block** is that portion of the periodic table where the process of filling of electron orbitals (aufbau process) involves f subshells. These are the lanthanide and actinide elements.

fcc (See **face-centered cubic.**)

Ferromagnetism is a property that permits certain materials (notably Fe, Co, and Ni) to be made into permanent magnets. The magnetic moments of individual atoms are aligned into domains. In the presence of a magnetic field, these domains orient themselves to produce a permanent magnetic moment.

In **filtration** a solid-liquid mixture is passed through a filtering device that retains the solid as the liquid passes through.

The **first law of thermodynamics,** expressed as $\Delta U = q + w$, is an alternative statement of the law of conservation of energy. (See also **law of conservation of energy.**)

A **first-order** reaction is one for which the sum of the concentration-term exponents in the rate equation is 1.

Fission (See **nuclear fission.**)

A **flow battery** is a battery in which materials (reactants, products, electrolyte) pass continuously through the battery. The battery is simply a converter of chemical to electrical energy.

Formal charge is the number of outer-shell (valence) electrons in an isolated atom minus the number of electrons assigned to that atom in a Lewis structure.

The **formation constant, K_f,** describes equilibrium among a complex ion, the free metal ion, and ligands.

Formula mass is the mass of a formula unit of a compound, relative to a mass of exactly 12 u for carbon-12.

A **formula unit** is the smallest collection of atoms or ions from which the empirical formula of a compound can be established.

Fractional crystallization (recrystallization) is a method of purifying a substance by crystallizing the pure solid from a saturated solution while impurities remain in solution.

Fractional distillation (See **distillation.**)

Fractional precipitation is a technique in which two or more ions in solution, each capable of being precipitated by the same reagent, are separated by the use of that reagent.

The **Frasch process** is a method of extracting sulfur from underwater deposits. It is based on the use of superheated water to melt the sulfur.

Free energy, G, is a thermodynamic function designed to produce a criterion for spontaneous change. It is defined through the equation $G = H - TS$.

Free energy change, ΔG, is the change in free energy that accompanies a process and can be used to indicate the direction of spontaneous change. For a spontaneous process at constant temperature and pressure, $\Delta G < 0$. (See also **standard free energy change.**)

Free radicals are highly reactive molecular fragments containing unpaired electrons.

Freezing is the conversion of a liquid to a solid that occurs at a fixed temperature known as the **freezing point.**

The **frequency** of a wave motion is the number of wave crests or troughs that pass through a given point in a unit of time. It is expressed by the unit time^{-1} (e.g., s^{-1}, also called a hertz, Hz).

A **fuel cell** is a voltaic cell in which the cell reaction is the equivalent of the combustion of a fuel. Chemical energy of the fuel is converted to electricity.

A **function of state (state function)** is a property that assumes a unique value when the state or present condition of a system is defined. This value is *independent* of how the state is attained.

A **functional group** is an atom or grouping of atoms attached to a hydrocarbon residue, R. The functional group often confers specific properties to an organic molecule.

Fusion (See **nuclear fusion.**)

Galvanic cell (See **voltaic cell.**)

Galvanizing is the name given to any process in which an iron surface is coated with zinc to give it corrosion protection.

Gamma (γ) rays are a form of electromagnetic radiation of high penetrating power emitted by certain radioactive nuclei.

In a **gas,** atoms or molecules are generally much more widely separated than in liquids and solids. A gas assumes the shape of its container and expands to fill the container, thus having neither definite shape nor volume.

The **gas constant, R,** is the numerical constant appearing in the ideal gas equation ($PV = nRT$) and in several other equations as well.

The **general gas equation** is an expression based on the ideal gas equation and written in the form $P_1V_1/n_1T_1 = P_2V_2/n_2T_2$.

The **genetic code** relates to the sequences of bases in DNA molecules that determine complementary sequences in mRNA and, ultimately, sequences of amino acids in proteins.

Geometric isomerism in organic compounds refers to the existence of nonequivalent structures (cis and trans) that differ in the positioning of substituent groups relative to a double bond. In complexes the nonequivalent structures are based on the positions at which ligands are attached to the metal center.

Glass is a transparent, amorphous solid consisting of Na$^+$ and Ca^{2+} ions in a network of SiO$_4^{4-}$ anions. It is made by fusing together a mixture of sodium and calcium carbonates with sand.

Global warming refers to the warming of Earth that results from an accumulation in the atmosphere of gases such as CO$_2$ that absorb infrared radiation radiated from Earth's surface.

Graham's law states that the rates of effusion or diffusion of two different gases are inversely proportional to the square roots of their molar masses.

The **ground state** is the lowest energy state for the electrons in an atom or molecule.

A **group** is a vertical column of elements in the periodic table. Members of a group have similar properties.

A **half-cell** is a combination of an electrode and a solution. An oxidation–reduction equilibrium is established on the electrode. An electrochemical cell is a combination of two half-cells.

The **half-life** of a reaction is the time required for one-half of a reactant to be consumed. In a nuclear decay process, it is the time required for one-half of the atoms present in a sample to undergo radioactive decay.

A **half-reaction** describes one portion of an overall oxidation–reduction reaction, either the oxidation or the reduction.

Hard water contains dissolved minerals in significant concentrations. If the hardness is primarily due to HCO$_3^-$ and associated cations, the water has **temporary hardness.** If the hardness is due to anions other than HCO$_3^-$ (e.g., SO$_4^{2-}$), the water has **permanent hardness.**

hcp (See **hexagonal closest packed.**)

Heat is a transfer of thermal energy as a result of a temperature difference.

Heat capacity is the quantity of heat required to change the temperature of an object or substance by one degree, usually expressed as $J\,°C^{-1}$ or $cal\,°C^{-1}$. **Specific heat capacity** is the heat capacity per gram of substance, i.e., $J\,°C^{-1}\,g^{-1}$, and **molar heat capacity** is the heat capacity per mole, i.e., $J\,°C^{-1}\,mol^{-1}$.

A **heat of reaction** is energy converted from chemical to thermal (or vice versa) in a reaction. In an isolated system, this energy conversion causes a temperature change, and in a system that interacts with its surroundings, heat (q) is either evolved to or absorbed from the surroundings.

The **Heisenberg uncertainty principle** states that, when measuring the position and momentum of fundamental particles of matter, uncertainties in measurement are inevitable.

The **Henderson-Hasselbalch equation** has the form, $pH = pK_a + \log$ [conjugate base]/[acid], in which stoichiometric concentrations of the weak acid and its conjugate base are used in place of the equilibrium concentrations. There are limitations on its validity.

Henry's law relates the solubility of a gas to the gas pressure maintained above a solution of the gaseous solute. The solubility is directly proportional to the pressure of the gas above the solution.

The **hertz (Hz)** is the SI unit of frequency, equal to s^{-1}.

Hess's law states that the enthalpy change for an overall or net process is the sum of enthalpy changes for individual steps in the process.

Heterocyclic compounds are based on hydrocarbon ring structures in which one or more C atoms is replaced by atoms such as N, O, or S.

In a **heterogeneous mixture,** components separate into physically distinct regions of differing properties and often differing composition.

Hexagonal closest packed is one of the two ways in which spheres can be packed to minimize the amount of free space or voids among them. (See Figures 13-39 and 13-41.) The crystal structure based on this type of packing is referred to as **hcp.**

In a **high-spin complex,** weak crystal field splitting leads to a maximum number of unpaired electrons in the d subshell of the central metal atom or ion.

A **homogeneous mixture (solution)** is a mixture of elements and/or compounds that has a uniform composition and properties within a given sample. However, the composition and properties may vary from one sample to another.

A **homologous series** is a group of compounds that differ in composition by some constant unit ($-CH_2$ in the case of alkanes).

Hund's rule (rule of maximum multiplicity) states that whenever orbitals of equal energy are available, electrons occupy these orbitals singly before any pairing of electrons occurs.

A **hybrid orbital** is one of a set of identical orbitals reformulated from pure atomic orbitals and used to describe certain covalent bonds.

Hybridization refers to combining pure atomic orbitals to generate hybrid orbitals in the valence bond approach to covalent bonding.

A **hydrate** is a compound in which a fixed number of water molecules is associated with each formula unit, such as $CuSO_4 \cdot 5\,H_2O$.

Hydration energy is the energy associated with dissolving gaseous ions in water, usually expressed per mole of ions.

Hydrides are compounds of hydrogen, usually divided into the categories of covalent (e.g., H_2O and HCl), ionic (e.g., LiH and CaH_2), and metallic (mostly non-stoichiometric compounds with the transition metals).

A **hydrocarbon** is a compound containing the two elements carbon and hydrogen. The C atoms are arranged in straight or branched chains or ring structures.

A **hydrogen bond** is an intermolecular force of attraction in which an H atom covalently bonded to one atom is attracted simultaneously to another highly non-metallic atom of the same or a nearby molecule.

The **hydrogen economy** refers to a possible future in which hydrogen may become the principal fuel and metallurgical reducing agent, supplanting fossil fuels.

In a **hydrogenation reaction** H atoms are added to multiple bonds between carbon atoms, converting carbon-to-carbon double bonds to single bonds and carbon-to-carbon triple bonds to double or single bonds. It is a reaction, for example, that converts an unsaturated to a saturated fatty acid.

Hydrolysis is a special name given to acid–base reactions in which ions act as acids or bases. As a result of hydrolysis, many salt solutions are not pH neutral, that is, $pH \neq 7$.

Hydrometallurgy refers to metallurgical procedures where water and aqueous solutions are used to extract metals from their ores. In the first step, **leaching,** the target metal is obtained in soluble form in aqueous solution. Other steps include purifying the leached solution and depositing the metal from solution.

Hydronium ion, H_3O^+, is the form in which protons are found in aqueous solution. The terms "hydrogen ion" and "hydronium ion" are often used synonymously.

The **hydroxyl** group is $-OH$ and is usually found attached to a straight or branched hydrocarbon chain (an alcohol) or a ring structure (a phenol).

A **hypothesis** is a tentative explanation of a series of observations or of a natural law.

An **ICE table** is a format for organizing the data in an equilibrium calculation. It is based on the initial concentrations of reactants and products, changes in concentrations to attain equilibrium, and equilibrium concentrations.

An **ideal (perfect) gas** is one whose behavior can be predicted by the ideal gas equation.

Ideal gas constant (See **gas constant**.)

The **ideal gas equation** relates the pressure, volume, temperature, and number of moles of ideal gas (n) through the expression $PV = nRT$.

An **ideal solution** has $\Delta H_{soln} = 0$ and certain properties (notably vapor pressure) that are predictable from the properties of the solution components.

Incomplete octet is a term used to describe situations in which an atom fails to acquire eight outer-shell electrons in a Lewis structure.

An **indicator** is an added substance that changes color at the equivalence point in a titration.

An **induced dipole** is an atom or molecule in which a separation of charge is produced by a neighboring dipole.

Industrial smog is air pollution in which the chief pollutants are $SO_2(g)$, $SO_3(g)$, H_2SO_4 mist, and smoke.

Inert complex is the term used to describe a complex ion in which the exchange of ligands occurs very slowly.

The **inert-pair effect** refers to the effects on the properties of certain post-transition elements that result from the presence of a pair of electrons in the *s* orbital of the valence shells of their atoms.

The **initial rate of a reaction** is the rate of a reaction immediately after the reactants are brought together.

Inner-transition elements are those of the *f*-block series, that is, the lanthanide and actinide elements.

An **inorganic compound** is any combination of elements that does not fit the category of organic compound. (See also **organic compound**.)

An **instantaneous dipole** is an atom or molecule with a separation of charge produced by a momentary displacement of electrons from their normal distribution.

An **instantaneous rate of reaction** is the exact rate of a reaction at some precise point in the reaction. It is obtained from the slope of a tangent line to a concentration–time graph.

An **integrated rate law (equation)** is derived from a rate law (equation) by the calculus technique of integration. It relates the concentration of a reactant (or product) to elapsed time from the start of a reaction. The equation has different forms depending on the order of the reaction.

An **intensive property** is *independent* of the quantity of matter involved in the observation. Density and temperature are examples of intensive properties.

An **interhalogen compound** is a covalent compound between two or more halogen elements, such as ICl and BrF_3.

An **intermediate** is the product of one reaction that is consumed in a following reaction in a process that proceeds through several steps.

An **intermolecular force** is an attraction *between* molecules.

The **internal energy**, U, of a system is the total energy attributed to the particles of matter and their interactions within a system.

An **ion** is a charged species consisting of a single atom or a group of atoms. It is formed when a neutral atom or a covalently bonded group of atoms either gains or loses electrons.

Ion exchange is a process in which ions held to the surface of an ion exchange material are exchanged for other ions in solution. For example, Na^+ may be exchanged for Ca^{2+} and Mg^{2+}, or OH^- may be exchanged for SO_4^{2-}.

An **ion pair** is an association of a cation and an anion in solution. Such combinations, when they occur, can have a significant effect on solution equilibria.

An **ion product**, Q_{sp}, is formulated in the same manner as a solubility product constant, K_{sp}, but with *nonequilibrium* concentration terms. A comparison of Q_{sp} and K_{sp} provides a criterion for precipitation from solution.

The **ion product of water**, K_w, is the product of $[H_3O^+]$ and $[OH^-]$ in pure water or in an aqueous solution. This product has a unique value that depends only on temperature. At 25°C, $K_w = 1.0 \times 10^{-14}$.

An **ionic bond** results from the transfer of electrons between metal and nonmetal atoms. Positive and negative ions are formed and held together by electrostatic attractions.

An **ionic compound** is a compound consisting of positive and negative ions that are held together by electrostatic forces of attraction.

Ionic radius is the radius of a spherical ion. It is the atomic radius associated with an element in its ionic compounds.

The first **ionization energy**, I_1, is the energy required to remove the most loosely held electron from a *gaseous* atom. The second ionization energy, I_2, is the energy required to remove an electron from a gaseous unipositive ion, and so on.

The **isoelectric point, pI**, of an amino acid is the pH at which the dipolar structure or "zwitterion" predominates.

Isoelectronic species have the same number of electrons (usually in the same configuration). Na^+ and Ne are isoelectronic, as are CO and N_2.

An **isolated system** is one that exchanges neither energy nor matter with its surroundings.

Isomers are two or more compounds having the same formula but different structures and therefore different properties.

Isotopes of an element are atoms with different numbers of neutrons in their nuclei. That is, isotopes of an element have the same atomic numbers but different mass numbers.

Isotopic mass (See **nuclidic mass**.)

IUPAC (or IUC) refers to the International Union of Pure and Applied Chemistry.

The **joule, J,** is the basic SI unit of energy. It is the quantity of work done when a force of one newton acts through a distance of one meter.

K_c is the relationship among the *concentrations* of the reactants and products in a reversible reaction at equilibrium. Concentrations are expressed as molarities.

K_p, the partial pressure equilibrium constant, is the relationship that exists among the partial pressures of gaseous reactants and products in a reversible reaction at equilibrium. Partial pressures are expressed in atm.

The **Kelvin** temperature is an *absolute* temperature. That is, the lowest attainable temperature is 0 K = −273.15 °C (the temperature at which molecular motion ceases). Kelvin and Celsius temperatures are related through the expression $T(K) = t(°C) + 273.15$.

A **ketone** has the general formula

$$R-\overset{\displaystyle O}{\overset{\displaystyle \|}{C}}-R'.$$

A **kilopascal (kPa)** is a unit of pressure equal to 1000 pascals (Pa) or 1000 N/m^2. The standard atmosphere of pressure is 101.325 kPa.

Kinetic energy is energy of motion. The kinetic energy of an object with mass m and velocity u is K.E. $= \frac{1}{2}mu^2$.

The **kinetic molecular theory of gases** is a model for describing gas behavior. It is based on a set of assumptions and yields equations from which various properties of gases can be deduced.

The **Kroll process** is an industrial process for the manufacture of titanium metal from TiO_2. Key steps involve conversion of TiO_2 to $TiCl_4$ and reduction of $TiCl_4$ to Ti with Mg as the reducing agent.

Labile complex is the term used to describe a complex ion in which a rapid exchange of ligands occurs.

The **lanthanide contraction** refers to the decrease in atomic size in a series of elements in which an *f* subshell fills with electrons (an inner transition series). It results from the ineffectiveness of *f* electrons in shielding outer-shell electrons from the nuclear charge of an atom.

The **lanthanides** are the elements ($Z = 58 - 71$) characterized by a partially filled $4f$ subshell in their atoms. Because lanthanum resembles them, La

($Z = 57$) is generally considered together with them.

Lattice energy is the quantity of energy released in the formation of one mole of a crystalline ionic solid from its separated gaseous ions.

Gay-Lussac's **law of combining volumes** states that, when compared at the same temperature and pressure, the volumes of *gases* involved in a reaction are in the ratio of small whole numbers.

The **law of conservation of energy** states that energy can neither be created nor destroyed in ordinary processes.

The **law of conservation of mass** states that the total mass of the products of a chemical reaction is the same as the total mass of the reactants entering into the reaction.

The **law of constant composition (definite proportions)** states that all samples of a compound have the same composition, that is, the same proportions by mass of the constituent elements.

The **law of multiple proportions** states that if two elements form two or more compounds, the masses of one element combined with a fixed mass of the second are in the ratio of small whole numbers when the different compounds are compared.

The **leaving group** is the species expelled from an electrophilic molecule following attack by a nucleophile.

Le Châtelier's principle states that an action that tends to change the temperature, pressure, or concentrations of reactants in a system at equilibrium stimulates a response that partially offsets the change while a new equilibrium condition is established.

Levorotatory means the ability to rotate the plane of polarized light to the *left*, designated ($-$).

Lewis acid (See **acid**.)

Lewis base (See **base**.)

A **Lewis structure** is a combination of Lewis symbols that depicts the transfer or sharing of electrons in a chemical bond.

In the **Lewis symbol** of an element, valence electrons are represented by dots placed around the chemical symbol of the element.

The **Lewis theory** refers to a description of chemical bonding through Lewis symbols and Lewis structures in accordance with a particular set of rules.

Ligands are the groups that are coordinated (bonded) to the central atom in a complex.

The **limiting reactant (reagent)** in a reaction is the reactant that is consumed completely. The quantity of product(s) formed depends on the quantity of the limiting reactant.

Lipids include a variety of naturally occurring substances (e.g., fats and oils) sharing the property of solubility in solvents of low polarity [such as in $CHCl_3$, CCl_4, C_6H_6, and $(C_2H_5)_2O$].

In a **liquid,** atoms or molecules are in close proximity (although generally not as close as in a solid). A liquid occupies a definite volume but has the ability to flow and assume the shape of its container.

Liquid crystals are a form of matter with some of the properties of a liquid and some, of a crystalline solid.

London forces (See **dispersion forces**.)

A **lone pair** is a pair of electrons found in the valence shell of an atom and *not* involved in bond formation.

In a **low-spin complex** strong crystal field splitting leads to a minimum number of unpaired electrons in the d subshell of the central metal atom or ion.

Magic numbers is a term used to describe numbers of protons and neutrons that confer a special stability to an atomic nucleus.

The **main-group elements** are those in which s or p subshells are being filled in the aufbau process. They are also referred to as the s-block and p-block elements. They are found in groups 1, 2, and 13-18 in the periodic table (the A groups).

A **manometer** is a device used to measure the pressure of a gas, usually by comparing the gas pressure with barometric pressure.

Mass describes the quantity of matter in an object.

The **mass number, A,** is the total of the number of protons and neutrons in the nucleus of an atom.

A **mass spectrometer (mass spectrograph)** is a device used to separate and to measure the quantities and masses of different ions in a beam of positively charged gaseous ions.

Matter is anything that occupies space, has the property known as mass, and displays inertia.

Melting is the transition of a solid to a liquid and occurs at the **melting point.** The melting point and freezing point of a substance are identical.

A **meta (m-) isomer** has two substituents on a benzene ring separated by one C atom.

Metabolism refers to the totality of the chemical reactions occurring in living organisms.

A **metal** is an element whose atoms have small numbers of electrons in the outermost electronic shell. Removal of an electron(s) from a metal atom occurs without great difficulty, producing a positive ion (cation). Metals generally have a lustrous appearance, are malleable and ductile, and are able to conduct heat and electricity.

Metal carbonyls are complexes with d-block metals as central atoms and CO molecules as ligands, e.g., $Ni(CO)_4$.

Metallic radius is one-half the distance between the centers of adjacent atoms in a solid metal.

A **metalloid** is an element that may display both metallic and nonmetallic properties under the appropriate conditions.

The **method of initial rates** can be used to establish the order of a reaction from a series of measurements of the initial rate of the reaction. The initial concentration of one reactant is varied while the initial concentrations of all the other reactants are held constant.

The **metric system** is a system of measurement in which the unit of length is the meter, the unit of mass is the kilogram ($1\ kg = 1000\ g$), and the unit of time is the second.

A **millimeter of mercury (mmHg)** is a unit of pressure, usually applied to gases. For example, standard atmospheric pressure is equal to the pressure exerted by 760-mm column of mercury.

A **millimole** is one-thousandth of a mole (0.001 mol). It is especially useful in titration calculations.

A **mixture** is any sample of matter that is not pure, that is, not an element or compound. The composition of a mixture, unlike that of a substance, can be varied. Mixtures are either *homogeneous* or *heterogeneous*.

A **moderator** slows down energetic neutrons from a fission process so that they are able to induce additional fission.

Molality, m, is a solution concentration expressed as the amount of solute, in moles, divided by the mass of solvent, in kg.

Molar mass, *M*, is the mass of one mole of atoms, formula units, or molecules of a substance.

Molar solubility is the molarity of solute (mol L^{-1}) in a saturated solution.

Molarity, M, is the concentration of a solution expressed as the number of moles of solute per liter of solution.

A **mole** is an amount of substance containing Avogadro's number (6.02214×10^{23}) of atoms, formula units, or molecules.

Mole fraction describes a mixture in terms of the fraction of all the molecules that are of a particular type. It is the amount of one component, in moles, divided by the total amount of all the substances in the mixture.

A **mole percent** is a mole fraction expressed on a percentage basis, that is, mole fraction $\times 100\%$.

A **molecular compound** is a compound comprised of discrete molecules.

A **molecular formula** denotes the numbers of the different atoms present in a molecule. In some cases the molecular formula is the same as the empirical formula; in others it is an integral multiple of that formula.

Molecular geometry refers to the geometric shape of a molecule or polyatomic ion. In a species in which all electron pairs are bond pairs, the molecular geometry is the same as the electron-group geometry. In other cases the two properties are related but not the same.

Molecular mass is the mass of a molecule relative to a mass of exactly 12 u for carbon-12.

Molecular orbital theory describes the covalent bonds in a molecule by replacing atomic orbitals of the component atoms by molecular orbitals belonging to the molecule as a whole. A set of rules is used to assign electrons to these molecular orbitals, thereby yielding the electronic structure of the molecule.

A **molecule** is a group of bonded atoms held together by covalent bonds and existing as a separate entity. A molecule is the smallest entity having the characteristic proportions of the constituent atoms present in a substance.

A **monodentate ligand** is a ligand that is able to attach to a metal center in a complex at only one position and using just one lone pair of electrons.

A **monomer** is a simple molecule that is capable of joining with others to form a complex long-chain molecule called a **polymer.**

A **monosaccharide** is a single, simple molecule having the structural features of a carbohydrate. It can also be called a simple sugar.

A **multiple covalent bond** is a bond in which more than two electrons are shared between the bonded atoms.

A **natural law** is a concise statement, often in mathematical terms, that summarizes observations of certain natural phenomena.

The **Nernst equation** is used to relate E_{cell}, E_{cell}°, and the activites of the reactants and products in a cell reaction.

A **net ionic equation** represents a reaction between ions in solution in such a way that all nonparticipant (spectator) ions are eliminated from the equation. The equation must be balanced both atomically and for net electric charge.

A **network covalent solid** is a substance in which covalent bonds extend throughout the crystal, making the covalent bond both an *intra*molecular and an *inter*molecular force.

In a **neutralization** reaction an acid and a base react in stoichiometric proportions, so that there is no excess of either acid or base in the final solution. The products are water and a salt.

Neutrons are electrically neutral fundamental particles of matter found in all atomic nuclei except that of the simple hydrogen atom, protium, ^1H.

The **neutron number** is the number of neutrons in the nucleus of an atom. It is equal to the mass number (A) minus the atomic number (Z).

The **nitrogen cycle** is a series of processes by which atmospheric N_2 is fixed, enters into the food chain of animals, and eventually is returned to the atmosphere by bacteria.

Noble gases are elements whose atoms have the electron configuration ns^2np^6 in the electronic shell of highest principal quantum number. (The noble gas helium has the configuration $1s^2$.)

A **nonbonding molecular orbital** is a molecular orbital that neither contributes to nor detracts from bond formation in a molecule. (It does not affect the bond order.)

A **nonelectrolyte** is a substance that is essentially non-ionized, both in the pure state and in solution.

A **nonideal gas** departs from the behavior predicted by the ideal gas equation. Its behavior can only be predicted by other equations of state (e.g., the van der Waals equation).

A **nonmetal** is an element whose atoms tend to gain small numbers of electrons to form negative ions (anions) with the electron configuration of a noble gas. Nonmetal atoms may also alter their electron configurations by sharing electrons. Nonmetals are mostly gases, liquid (bromine), or low melting point solids and are very poor conductors of heat and electricity.

In a **nonpolar molecule** the centers of positive and negative charge coincide. That is, there is no net separation of charge within the molecule.

A **nonspontaneous** process is one that will not occur naturally. A nonspontaneous process can be brought about only by intervention from outside the system, as in the use of electricity to decompose a chemical compound (electrolysis).

The **normal boiling point** is the temperature at which the vapor pressure of a liquid is 1 atm. It is the temperature at which the liquid boils in a container open to the atmosphere at a pressure of 1 atm.

The **normal melting point** is the temperature at which the melting of a solid occurs at 1 atm pressure. This is the same temperature as the **normal freezing point.**

Nuclear binding energy is the energy released when nucleons (protons and neutrons) are fused into an atomic nucleus. This energy replaces an equivalent quantity of matter.

A **nuclear equation** represents the changes that occur during a nuclear process. The target nucleus and bombarding particle are represented on the left side of the equation, and the product nucleus and ejected particle on the right side.

Nuclear fission is a radioactive decay process in which a heavy nucleus breaks up into two lighter nuclei and several neutrons, accompanied by the release of energy.

In **nuclear fusion** small atomic nuclei are fused into larger ones, with some of their mass being converted to energy.

A **nuclear reactor** is a device in which nuclear fission is carried out as a controlled chain reaction. That is, neutrons produced in one fission event trigger the fission of other nuclei, and so on.

Nucleic acids are cell components comprised of purine and pyrimidine bases, pentose sugars, and phosphoric acid.

A **nucleophile** is a reactant that seeks out a center of positive charge as a point of attack in a chemical reaction.

A **nucleophilic substitution reaction** is a reaction between a nucleophile and an electrophile. The nucleophile attacks at a positive center on the electrophile, and the leaving group is ejected from another point.

Nuclide is a term used to designate an atom with a specific atomic number and mass number. It is represented by the symbolism $_Z^A E$.

Nuclidic mass is the mass, in atomic mass units, of an individual atom, relative to an arbitrarily assigned value of exactly 12 u as the nuclidic mass of carbon-12.

An **octet** refers to *eight* electrons in the outermost (valence) electronic shell of an atom in a Lewis structure.

The **octet rule** states that the number of electrons associated with bond pairs and lone pairs of electrons for each of the Lewis symbols (except H) in a Lewis structure will be eight (an **octet**).

An **odd-electron species** is one in which the total number of valence electrons is an *odd* number. At least one *unpaired* electron is present in such a species.

Oils are triglycerides in which unsaturated fatty acid components predominate.

Oligosaccharides are carbohydrates consisting of two to ten monosaccharide units. (See also **sugar**.)

An **open system** is one that can exchange both matter and energy with its surroundings.

Optical isomerism results from the presence of a chiral atom in a structure, leading to a pair of optical isomers that differ only in the direction that they rotate the plane of polarized light. (See also **enantiomers**.)

Optical isomers, also called enantiomers (nonsuperimposable mirror images), are isomers that differ only in the direction they rotate the plane of polarized light.

An **orbital** is a mathematical function used to describe regions in an atom where the electron charge density or the probability of finding an electron is high. The several kinds of orbitals ($s, p, d, f, \ldots$)

differ from one another in the shapes of the regions of high electron charge density they describe.

An **orbital diagram** is a representation of an electron configuration in which the most probable orbital designation and spin of each electron in the atom are indicated.

The **order of a reaction** relates to the exponents of the concentration terms in the rate law for a chemical reaction. The order can be stated with respect to a particular reactant (first order in A, second order in B, ...) or, more commonly, as the overall order. The overall order is the sum of the concentration-term exponents.

An **organic compound** is made up of carbon and hydrogen or carbon, hydrogen and a small number of other elements, such as oxygen, nitrogen, and sulfur.

An **ortho** (*o-*) **isomer** has two substituents attached to adjacent C atoms in a benzene ring.

Osmosis is the net flow of solvent molecules through a semipermeable membrane, from a more dilute solution (or from the pure solvent) into a more concentrated solution.

Osmotic pressure is the pressure that would have to be applied to a solution to stop the passage through a semipermeable membrane of solvent molecules from the pure solvent.

An **overall reaction** is the overall or net change that occurs when a process is carried out in more than one step.

An **overpotential** is the voltage in excess of the theoretical value required to produce a particular electrode reaction in electrolysis.

Oxidation is a process in which electrons are "lost" and the oxidation state of some atom increases. (Oxidation can occur only in combination with reduction.)

In an **oxidation–reduction** (**redox**) reaction certain atoms undergo changes in oxidation state. The substance containing atoms whose oxidation states *increase* is **oxidized.** The substance containing atoms whose oxidation states *decrease* is **reduced.**

An **oxidation state** relates to the number of electrons an atom loses, gains, or shares in combining with other atoms to form molecules or polyatomic ions.

An **oxidizing agent** (**oxidant**) makes possible an oxidation process by itself being *reduced.*

An **oxoacid** is an acid in which an ionizable hydrogen atom(s) is bonded through an oxygen atom to a central atom, that is, E—O—H. Other groups bonded to the central atom are either additional —OH groups or O atoms (or in a few cases H atoms).

An **oxoanion** is a polyatomic anion containing a nonmetal, such as Cl, N, P, or S, in combination with some number of oxygen atoms.

The **ozone layer** refers to a region of the stratosphere from about 25 to 35 km above Earth's surface having a concentration of ozone, O_3, considerably greater than that at Earth's surface.

Pairing energy is the energy requirement to force an electron into an orbital that is already occupied by one electron.

A **para** (*p-*) **isomer** has two substituents located opposite to one another on a benzene ring.

A **paramagnetic** substance has one or more unpaired electrons in its atoms or molecules. It is attracted into a magnetic field.

A **partial pressure** is the pressure exerted by an individual gas in a mixture, independently of other gases. Each gas in the mixture expands to fill the container and exerts its own partial pressure.

A **pascal** (**pa**) is a pressure of one N/m^2.

The **Pauli exclusion principle** states that no two electrons may have all four quantum numbers alike. This limits occupancy of an orbital to two electrons with opposing spins.

The **p block** is that portion of the periodic table in which the filling of electron orbitals (aufbau process) involves *p* subshells.

A **peptide bond** is formed by the elimination of a water molecule from between two amino acid molecules. The H atom comes from the —NH_2 group of one amino acid and the —OH group, from the —COOH group of the other acid.

Percent is the number of parts of a constituent in 100 parts of the whole.

The **percent ionization** of a weak acid or a weak base is the percent of its molecules that ionize in an aqueous solution.

Percent natural abundances refer to the relative proportions, expressed as percentages by number, in which the isotopes of an element are found in natural sources.

Percent yield is the percent of the theoretical yield of product that is actually obtained in a chemical reaction. (See also **actual yield** and **theoretical yield**.)

A **period** is a horizontal row of the periodic table. All members of a period have atoms with the same highest principal quantum number.

The **periodic law** refers to the periodic recurrence of certain physical and chemical properties when the elements are considered in terms of increasing atomic number.

The **periodic table** is an arrangement of the elements, by atomic number, in which elements with similar physical and chemical properties are grouped together in vertical columns.

Permanent hard water (See **hard water**.)

The **peroxide** ion has the structure

$$\left[:\ddot{O}-\ddot{O}: \right]^{2-}.$$

pH is a shorthand designation for $[H_3O^+]$ in a solution. It is defined as $pH = -\log[H_3O^+]$.

A **phase diagram** is a graphical representation of the conditions of temperature and pressure at which solids, liquids, and gases (vapors) exist, either as single phases or states of matter or as two or more phases in equilibrium.

A **phenol** has the functional group —OH as part of an aromatic hydrocarbon structure.

A **phenyl group** is a benzene ring from which one H atom has been removed: —C_6H_5.

Photochemical smog is air pollution resulting from reactions involving sunlight, oxides of nitrogen, ozone, and hydrocarbons.

The **photoelectric effect** is the emission of electrons by certain materials when their surfaces are struck by electromagnetic radiation of the appropriate frequency.

A **photon** is a "particle" of light. The energy of a beam of light is concentrated into these photons.

In a **physical change** one or more physical properties of a sample of matter change, but the composition remains unchanged.

A **physical property** is a characteristic that a substance can display without undergoing a change in its composition.

A **pi (π) bond** results from the side-to-side overlap of p orbitals, producing a high electron charge density above and below the line joining the bonded atoms.

Pig iron is an impure form of iron (about 95% Fe and 3–4% C, together with small quantities of Mn, Si, and P) produced in a blast furnace.

pK is a shorthand designation for an ionization constant: $pK = -\log K$. pK values are useful when comparing the relative strengths of acids or bases.

Planck's constant, h, is the proportionality constant that relates the energy of a photon of light to its frequency. Its value is 6.626×10^{-34} J s.

Plaster of paris, $CaSO_4 \cdot \frac{1}{2} H_2O$, a hemihydrate of calcium sulfate, is obtained by heating gypsum, $CaSO_4 \cdot 2 H_2O$. It is a widely used material in the construction industry.

pOH is a shorthand designation for $[OH^-]$ in a solution: $pOH = -\log[OH^-]$.

In a **polar covalent bond** a separation exists between the centers of positive and negative charge in the bond.

In a **polar molecule,** the presence of one or more polar covalent bonds leads to a separation of the positive and negative charge centers for the molecule as a whole. A polar molecule has a resultant dipole moment.

Polarizability describes the ease with which the electron cloud in an atom or molecule can be distorted in an electric field, that is, the ease with which a dipole can be induced.

A **polyatomic ion** is a combination of two or more covalently bonded atoms that exists as an ion.

A **polydentate ligand** is capable of donating more than a single electron pair to the metal center of a complex, from different atoms in the ligand and to different sites in the geometric structure.

In a **polyhalide ion** two or more halogen atoms are covalently bonded into a polyatomic anion, e.g., I_3^-.

A **polymer** is a complex, long-chain molecule made up of many (hundreds, thousands) smaller units called *monomers.*

Polymerization is the process of producing a giant molecule (polymer) from simpler molecular units (monomers).

Polymorphism refers to the existence of a solid substance in more than one crystalline form.

In a **polypeptide,** a large number of amino acid units join together through peptide bonds.

A **polyprotic acid** is capable of losing more than a single proton per molecule in acid–base reactions. Protons are lost in a stepwise fashion, with the first proton being the most readily lost.

A **polysaccharide** is a carbohydrate (such as starch or cellulose) consisting of more than ten monosaccharide units.

Positional isomers differ in the position on a hydrocarbon chain or ring where a functional group(s) is attached.

A **positron (β^+)** is a *positive* electron emitted as a result of the conversion of a proton to a neutron in a radioactive nucleus.

Potential energy is energy due to position or arrangement. It is the energy associated with forces of attraction and repulsion between objects.

The term **ppm** (parts per million) refers to the number of parts of a component to one million parts of the medium in which it is found.

The term **ppb** (parts per billion) refers to the number of parts of a component to one billion parts of the medium in which it is found.

The term **ppt** (parts per trillion) refers to the number of parts of a component to one trillion parts of the medium in which it is found.

A **precipitate** is an insoluble solid that deposits from a solution as a result of a chemical reaction.

Precision is the degree of reproducibility of a measured quantity—the closeness of agreement among repeated measurements.

Pressure is a force per unit area. Applied to gases, pressure is most easily understood in terms of the height of a liquid column that can be maintained by the gas.

Pressure–volume work is work associated with the expansion or compression of gases.

A **primary battery** produces electricity from a chemical reaction that cannot be reversed. As a result the battery cannot be recharged.

A **primary color** is one of a set of colors that when added together as light produce white light. Subtractive mixing leads to an absence of color (black). Red, yellow, and blue are a set of primary colors.

Primary structure refers to the sequence of amino acids in the polypeptide chains that make up a protein.

A **principal electronic shell (level)** refers to the collection of all orbitals having the same value of the principal quantum number, n. For example, the $3s$, $3p$, and $3d$ orbitals comprise the third principal shell ($n = 3$).

The **products** are the substances formed in a chemical reaction.

Properties are qualities or attributes that can be used to distinguish one sample of matter from others.

A **protein** is a large polypeptide, that is, having a molecular mass of 10,000 u or more.

A **proton acceptor** is a base in the Brønsted–Lowry acid–base theory.

A **proton donor** is an acid in the Brønsted–Lowry acid–base theory.

Protons are fundamental particles carrying the basic unit of positive electric charge and found in the nuclei of all atoms.

Proton number (See **atomic number.**)

A **pseudo-first-order** reaction is a reaction of higher order (e.g., second-order) that is made to behave like a first-order reaction by using large initial concentrations of all the reactants save one.

Pyrometallurgy is the traditional approach to extractive metallurgy that uses dry solid materials heated to high temperatures. (See also **extractive metallurgy** and **hydrometallurgy.**)

Qualitative cation analysis is a laboratory method, based on a variety of solution equilibrium concepts, for determining the presence or absence of certain cations in a sample.

Quantitative analysis refers to the analysis of substances or mixtures to determine the *quantities* of the various components rather than their mere presence or absence.

Quantum numbers are integral numbers whose values must be specified in order to solve the equations of wave mechanics. Three different quantum numbers are required: the *principal quantum number, n*; the *orbital angular momentum quantum number, l*; and the *magnetic quantum number, m_l*. The permitted values of these numbers are interrelated.

The **quantum theory** is based on the proposition that energy transfers occur in the form of tiny, discrete units called quanta. Whenever an energy transfer occurs, it must involve an entire **quantum.**

Quaternary structure is the highest order structure that is found in some proteins. It describes how separate polypeptide chains may be assembled into a larger, more complex structure.

Quicklime is a common name for calcium oxide, CaO.

A **racemic mixture** is a mixture containing equal amounts of the enantiomers of an optically active substance.

A **rad** is a quantity of radiation able to deposit 1×10^{-2} J of energy per kilogram of matter.

A **radial wave function, $R(r)$**, is the part of a wave function that depends only on the distance r when the Schrödinger wave equation is expressed in spherical polar coordinates. (See also **angular wave function.**)

Radical (See **free radical.**)

The **radioactive decay law** states that the rate of decay of a radioactive material—the activity, A—is directly proportional to the number of atoms present.

A **radioactive decay series** is a succession of individual steps whereby an initial radioactive isotope (e.g., ^{238}U) is ultimately converted to a stable isotope (e.g., ^{206}Pb).

Radioactivity is a phenomenon in which small particles of matter (α or β particles) and/or electromagnetic radiation (γ rays) are emitted by unstable atomic nuclei.

Radiocarbon dating is a method of determining the age of a carbon-containing material based on the rate of decay of radioactive carbon-14.

A **random error** is an error made by the experimenter in performing an experimental technique or measurement, such as the error in estimating a temperature reading on a thermometer.

Raoult's law states that the vapor pressure of a solution component is equal to the product of the vapor pressure of the pure liquid and its mole fraction in solution: $P_A = x_A P_A^\circ$.

The **rate constant, k,** is the proportionality constant in a rate law that permits the rate of a reaction to be related to the concentrations of the reactants.

A **rate-determining step** in a reaction mechanism is an elementary process that is instrumental in establishing the rate of the overall reaction, usually because it is the slowest step in the mechanism.

The **rate law (rate equation)** for a reaction relates the reaction rate to the concentrations of the reactants. It has the form: rate $= k[A]^m[B]^n \ldots$.

The **rate of a chemical reaction** describes how fast reactants are consumed and products are formed, usually expressed as change of concentration per unit time.

Reactants are the substances that enter into a chemical reaction. This term is often applied to *all* the substances involved in a reversible reaction, but it can also be limited to the substances that appear on the *left* side of a chemical equation—the starting substances. (Substances on the *right* side of the equation are usually called products.)

A **reaction intermediate** is a species formed in one elementary reaction in a reaction mechanism and consumed in a subsequent one. As a result, the species does not appear in the equation for the overall reaction.

A **reaction mechanism** is a set of elementary steps or processes by which a reaction is proposed to occur. The mechanism must be consistent with the stoichiometry and rate law of the overall reaction.

A **reaction profile** is a graphical representation of a chemical reaction in terms of the energies of the reactants, activated complex(es), and products.

The **reaction quotient, Q,** is a ratio of concentration terms (or partial pressures) having the same form as an equilibrium constant expression, but usually applied to *nonequilibrium* conditions.

A **reducing agent (reductant)** makes possible a reduction process by itself becoming *oxidized*.

A **reducing sugar** is one that is able to reduce Cu^{2+}(aq) to red, insoluble Cu_2O. The sugar must have available an aldehyde group, which is oxidized to an acid.

A **reduction** process is one in which electrons are "gained" and the oxidation state of some atom decreases. (Reduction can only occur in combination with oxidation.) (See also **extractive metallurgy**.)

Refining (See **extractive metallurgy**.)

The **relative humidity** is the ratio of the partial pressure of water vapor in air to the vapor pressure of water at the same temperature, expressed on a percent basis.

A **rem** is a unit of radiation related to the rad, but taking into account the varying effects on biological matter of different types of radiation of the same energy.

Representative elements (See **main-group elements.**)

Resonance occurs when two or more plausible Lewis structures can be written for a species. The true structure is a composite or *hybrid* of these different contributing structures.

Rest mass is the mass of a particle (e.g., an electron) when it is essentially at rest. As its velocity approaches the speed of light, the mass of a particle increases.

Reverse osmosis is the passage through a semipermeable membrane of solvent molecules *from a solution into a pure solvent.* It can be achieved by applying to the solution a pressure in excess of its osmotic pressure.

A **reversible process** is one that can be made to reverse direction by just an infinitesimal change in a system property.

Ribonucleic acid (RNA), through its **messenger RNA (mRNA)** and **transfer RNA (tRNA)** forms, is involved in the synthesis of proteins.

Roasting (See **extractive metallurgy.**)

The **root-mean-square speed** is the square root of the average of the squares of the speeds of all the gas molecules in a gaseous sample.

The **R, S system** is used to indicate the arrangement of the four groups bonded to a chiral center and to provide names that distinguish between optical isomers.

A **salt bridge** is a device (a U-tube filled with a salt solution) used to join two half-cells in an electrochemical cell. The salt bridge permits the flow of ions between the two half-cells.

The **salt effect** is that of ions *different* from those directly involved in a solution equilibrium. The salt effect is also known as the diverse or "uncommon" ion effect.

Salts are ionic compounds in which hydrogen atoms of acids are replaced by metal ions. Salts are produced by the neutralization of acids with bases.

Saponification is the hydrolysis of a triglyceride by a strong base. The products are glycerol and a soap.

Saturated hydrocarbon molecules contain only single bonds between carbon atoms.

A **saturated solution** is one that contains the maximum quantity of solute that is normally possible at the given temperature.

The ***s* block** refers to the portion of the periodic table in which the filling of electron orbitals (aufbau process) involves the *s* subshell of the electronic shell of highest principal quantum number.

The **Schrödinger equation** describes the electron in a hydrogen atom as a matter wave. Solutions to the Schrödinger equation are called wave functions.

The **scientific method** refers to the general sequence of activities—observation, experimentation, and the formulation of hypotheses, laws, and theories—that lead to the advancement of scientific knowledge.

The **second law of thermodynamics** relates to the direction of spontaneous change. One statement of the law is that all spontaneous processes produce an increase in the entropy of the universe.

A **secondary battery** produces electricity from a reversible chemical reaction. When electricity is passed through the battery in the reverse direction the battery is recharged.

A **secondary color** is the complement of a **primary color.** When light of a primary color and its complement (secondary) color are added, the result is white light. When they are subtracted, the result is an absence of color (black).

The **secondary structure** of a protein describes the structure or shape of a polypeptide chain, for example, a coiled helix.

A **second-order** reaction is one for which the sum of the concentration-term exponents in the rate equation is 2.

Self-ionization is an acid–base reaction in which one molecule acts as an acid and donates a proton to another molecule of the same kind acting as a base.

A **semiconductor** is characterized by a small energy gap between a filled valence band and an empty conduction band.

A **semipermeable membrane** permits the passage of some solution species but restricts the flow of others. It is a film of material containing submicroscopic holes.

The **shielding effect** refers to the effect of inner-shell electrons in shielding or screening outer-shell electrons from the full effects of the nuclear charge. In effect the inner electrons partially reduce the nuclear charge. (See also **effective nuclear charge.**)

SI units are the units of expressing measured quantities preferred by various international scientific agencies. (See Appendix C.)

A **sigma (σ) bond** results from the end-to-end overlap of simple or hybridized atomic orbitals along the straight line joining the nuclei of the bonded atoms.

Significant figures are those digits in an experimentally measured quantity that establish the precision with which the quantity is known.

Silica is a term used to describe various solid forms of silicon dioxide, SiO_2.

A **silicone** is an organosilicon polymer containing $O-Si-O$ bonds.

Simultaneous reactions are two or more reactions that occur at the same time.

A **single covalent** bond results from the sharing of *one pair* of electrons between bonded atoms. It is represented by a single dash sign ($-$).

Skeletal isomerism results from differences in the skeletal structures of molecules having the same composition.

A **skeletal structure** is an arrangement of atoms in a Lewis structure to correspond to the actual arrangement found by experiment.

Slaked lime is a common name for calcium hydroxide, $Ca(OH)_2$.

Smog is the general term used to refer to a condition in which polluted air reduces visibility, causes stinging eyes and breathing difficulties, and produces additional minor and major health problems. (See also **industrial smog** and **photochemical smog.**)

S_N1 is the designation for a nucleophilic substitution reaction in which the rate-determining step is unimolecular.

S_N2 is the designation for a nucleophilic substitution reaction in which the rate-determining step is bimolecular.

Soaps are the salts of fatty acids, e.g., $RCOO^-Na^+$, where the R group is a hydrocarbon chain containing from 3 to 21 C atoms. Sodium and potassium soaps are the common soaps used as cleansing agents.

Solders are low-melting alloys used for joining wires or pieces of metal. They usually contain metals such as Sn, Pb, Bi, and Cd.

In a **solid,** atoms or molecules are in close contact, often in a highly organized arrangement. A solid has a definite shape and occupies a definite volume. (See also **crystal.**)

The **solubility** of a substance is the concentration of its saturated solution.

The **solubility product constant, K_{sp},** is the equilibrium constant that describes the formation of a saturated solution of a slightly soluble ionic compound. It is the product of ionic concentration terms, with each term raised to an appropriate power.

A **solute** is a solution component that is dissolved in a solvent. A solution may have several solutes, with the solutes generally present in lesser amounts than is the solvent.

Solution (See **homogeneous mixture.**)

The **solvent** is the solution component in which one or more solutes are dissolved. Usually the solvent is present in greater amount than are the solutes and determines the state of matter in which the solution exists.

An *sp* **hybrid orbital** is one of the pair of orbitals formed by the hybridization of one *s* and one *p* orbital. The angle between the two orbitals is 180°.

An *sp²* **hybrid orbital** is one of the three orbitals formed by the hybridization of one *s* and two *p* orbitals. The angle between any two of the orbitals is 120°.

An *sp³* **hybrid orbital** is one of the four orbitals formed by the hybridization of one *s* and three *p* orbitals. The angle between any two of the orbitals is the tetrahedral angle—109.5°.

An *sp³d* **hybrid orbital** is one of the five orbitals formed by the hybridization of one *s*, three *p*, and one *d* orbital. The five orbitals are directed to the corners of a trigonal bipyramid.

An *sp³d²* **hybrid orbital** is one of the six orbitals formed by the hybridization of one *s*, three *p*, and two *d* orbitals. The six orbitals are directed to the corners of a regular octahedron.

spdf **notation** is a method of describing electron configurations in which the numbers of electrons assigned to each orbital are denoted as superscripts. For example, the electron configuration of Cl is $1s^2 2s^2 2p^6 3s^2 3p^5$.

The **specific heat** of a substance is the quantity of heat required to change the temperature of one gram of the substance by one degree Celsius.

Spectator ions are ionic species that are present in a reaction mixture but do not take part in the reaction. They are usually eliminated from a chemical equation.

The **spectrochemical series** is a ranking of ligand abilities to produce a splitting of the *d* energy level of a central metal ion in a complex ion.

A **spontaneous (natural) process** is one that is able to take place in a system left to itself. No external action is required to make the process go, although in some cases the process may take a very long time.

Stalactites and **stalagmites** are limestone ($CaCO_3$) formations in limestone caves produced by the slow decomposition of $Ca(HCO_3)_2(aq)$.

The **standard atmosphere (atm)** is the pressure exerted by a column of mercury exactly 760 mm high when the density of mercury is 13.5951 g cm^{-3} and the acceleration due to gravity is $g = 9.80665$ m s^{-2}.

A **standard cell potential, E°_{cell},** is the voltage of an electrochemical cell in which all species are in their standard states. (See also **cell potential**.)

Standard conditions of temperature and pressure (STP) refers to a gas maintained at a temperature of exactly 0 °C (273.15 K) and 760 mmHg (1 atm).

A **standard electrode potential, E°,** is the electric potential that develops on an electrode when the oxidized and reduced forms of some substance are in their *standard* states. Tabulated data are expressed in terms of the reduction process, that is, standard electrode potentials are standard reduction potentials.

The **standard enthalpy of formation, ΔH°_f,** of a substance is the enthalpy change that occurs in the formation of 1 mol of the substance in its standard state from the reference forms of its elements in their standard states. The reference forms of the elements are their most stable forms at the given temperature and 1 bar pressure.

The **standard enthalpy of reaction, ΔH°,** is the enthalpy change of a reaction in which all reactants and products are in their standard states.

Standard free energy change, ΔG°, is the free energy change of a process when the reactants and products are all in their standard states. The equation relating standard free energy change to the equilibrium constant is $\Delta G^{\circ} = -RT \ln K_{eq}$.

The **standard free energy of formation, ΔG°_f,** is the standard free energy change associated with the formation of 1 mol of compound from its elements in their most stable forms at 1 bar pressure.

The **standard hydrogen electrode (SHE)** is an electrode at which equilibrium is established between H_3O^+ ($a = 1$) and H_2(g, 1 bar) on an inert (Pt) surface. The standard hydrogen electrode is *arbitrarily* assigned an electrode potential of exactly 0 V.

The **standard molar entropy** is the absolute entropy evaluated when one mole of a substance is in its standard state at a particular temperature.

The **standard state** of a substance refers to that substance when it is maintained at 1 bar pressure and at the temperature of interest. For a gas it is the (hypothetical) pure gas behaving as an ideal gas at 1 bar pressure and the temperature of interest.

Standardization of a solution refers to establishing the exact concentration of the solution, usually through a titration.

A **standing wave** is a wave motion that reflects back on itself in such a way that the wave contains a certain number of points (nodes) that undergo no motion. A common example is the vibration of a plucked guitar string, and a related example is the description of electrons as matter waves.

Steel is a term used to describe iron alloys containing from 0 to 1.5%C together with other key elements, such as V, Cr, Mn, Ni, W, and Mo.

Step-reaction polymerization is a type of polymerization reaction in which monomers are joined together by the elimination of small molecules between them. For example, a H_2O molecule might be eliminated by the reaction of a H atom from one monomer with an —OH group from another.

In **stereoisomers** the number and types of atoms and bonds in molecules are the same, but certain atoms are oriented differently in space. Cis and trans isomerism is one type of stereoisomerism; optical isomerism is another.

Stoichiometric coefficients are the coefficients used to balance an equation.

A **stoichiometric factor** is a conversion factor relating molar amounts of two species in a chemical reaction (i.e., a reactant to a product, one reactant to another, etc.). The numbers used in formulating the factor are stoichiometric coefficients.

Stoichiometric proportions refer to relative amounts of reactants that are in the same mole ratio as implied by the balanced equation for a chemical reaction. For example, a mixture of 2 mol H_2 and 1 mol O_2 is in stoichiometric proportions, and a mixture of 1 mol H_2 and 1 mol O_2 is not, for the reaction
$$2 H_2 + O_2 \longrightarrow 2 H_2O.$$

Stoichiometry refers to quantitative measurements and relationships involving substances and mixtures of chemical interest.

The **stratosphere** is the region of the atmosphere that extends from approximately 12 to 55 km above Earth's surface.

A **strong acid** is an acid that is completely ionized in aqueous solution.

A **strong base** is a base that is completely ionized in aqueous solution.

A **strong electrolyte** is a substance that is completely ionized in solution.

A **structural formula** for a compound indicates which atoms in a molecule are bonded together, and whether by single, double, or triple bonds.

Structural isomers have the same number and kinds of atoms but they differ in their structural formulas.

Sublimation is the passage of molecules from the solid to the gaseous state.

A **subshell** refers to a collection of orbitals of the same type. For example, the three $2p$ orbitals constitute the $2p$ subshell.

A **substance** has a constant composition and properties throughout a given sample and from one sample to another. All substances are either elements or compounds.

Substitution reactions are typical of those involving alkane and aromatic hydrocarbons. In such a reaction a functional group replaces an H atom on a chain or ring.

A **substrate** is the substance that is acted upon by an enzyme in an enzyme-catalyzed reaction. The substrate is converted to products, and the enzyme is regenerated.

A **sugar** is a monosaccharide (simple sugar), a disaccharide, or an oligosaccharide containing up to ten monosaccharide units.

The **superoxide** ion has the structure

$$[\overset{..}{\underset{..}{O}} - \overset{..}{\underset{..}{O}}]^-.$$

Superphosphate is a mixture of $Ca(H_2PO_4)_2$ and $CaSO_4$ produced by the action of H_2SO_4 on phosphate rock.

A **supersaturated solution** contains more solute than normally expected for a saturated solution, usually prepared from a solution that is saturated at one temperature by changing its temperature to one where supersaturation can occur.

Surface tension is the energy or work required to extend the surface of a liquid.

The **surroundings** represent that portion of the universe with which a system interacts.

Synthesis gas is a mixture of $CO(g)$ and $H_2(g)$, generally made from coal or natural gas, that can be used as a fuel or in the synthesis of organic compounds.

A **system** is the portion of the universe selected for a thermodynamic study. (See also **open, closed,** and **isolated** systems.)

A **systematic error** is one that recurs regularly in a series of measurements because of an inherent error in the measuring system (e.g., through faulty calibration of a measuring device).

Temperature is a measure of the average molecular kinetic energy of a substance (translational kinetic energy for gases and liquids, and vibrational kinetic energy for solids).

Temporary hard water (See **hard water**.)

A **terminal atom** is any atom that is bonded to only one other atom in a molecule or polyatomic ion.

A **termolecular process** is an elementary process in a reaction mechanism in which three atoms or molecules must collide simultaneously.

A **ternary compound** is comprised of *three* elements.

The **tertiary structure** of a protein refers to its three-dimensional structure—for example, the twisting and folding of coils.

The **theoretical yield** is the quantity of product *calculated* to result from a chemical reaction. (See also **actual yield** and **percent yield**.)

A **theory** is a model or conceptual framework with which one is able to explain and make further predictions about natural phenomena.

Thermal energy is energy associated with random molecular motion.

The **thermite reaction** is an oxidation–reduction reaction that uses powdered aluminum metal as a reducing agent to reduce a metal oxide, such as Fe_2O_3, to the free metal.

The **thermodynamic equilibrium constant,** K_{eq}, is an equilibrium constant expression based on activities. In dilute solutions activities can be replaced by molarities and in ideal gases, by partial pressures in atm. The activities of pure solids and liquids are 1.

A **thio** compound is one in which an S atom replaces an O atom. For example, replacement of an O by S converts SO_4^{2-} to $S_2O_3^{2-}$ (thiosulfate ion).

The **third law of thermodynamics** states that the entropy of a pure perfect crystal is *zero* at the absolute zero of temperature, 0 K.

The **titrant** is the solution that is added in a controlled fashion through a buret in a titration reaction. (See also **titration**.)

Titration is a procedure for carrying out a chemical reaction between two solutions by the controlled addition (from a buret) of one solution to the other. In a titration a means must be found, as by the use of an indicator, to locate the equivalence point.

A **titration curve** is a graph of solution pH versus volume of titrant. It outlines how pH changes during an acid–base titration, and it can be used to establish such features as the equivalence point of the titration.

A **torr** is a unit of pressure equal to the unit millimeter of mercury.

The term **trans** is used to describe geometric isomers in which two groups are attached on opposite sides of a double bond in an organic molecule, or at opposite corners of a square in a square-planar complex, or at positions above and below the central plane of an octahedral complex. (See also **geometric isomerism**.)

Transition elements are those elements whose atoms feature the filling of a d or f subshell of an inner electronic shell. If the filling of an f subshell occurs, the elements are sometimes referred to as *inner* transition elements.

The **transition state** in a chemical reaction is an intermediate state between the reactants and products. (See also **activated complex** and **reaction profile**.)

Transmutation is the process in which one element is converted to another as a result of a change affecting the nuclei of atoms (such as in radioactive decay).

A **transuranium element** is one with an atomic number $Z > 92$.

Triglycerides are esters of glycerol (1,2,3-propanetriol) with long-chain monocarboxylic (fatty) acids.

In a **triple covalent bond** *three pairs* of electrons are shared between the bonded atoms. It is represented by a triple-dash sign ($\equiv$).

A **triple point** is a condition of temperature and pressure at which three phases of a substance (usually solid, liquid, and vapor) coexist at equilibrium.

The **troposphere** is the region of the atmosphere that extends from Earth's surface to a height of about 12 km.

Trouton's rule states that at their normal boiling points the entropies of vaporization of many liquids have about the same value: 87 J mol^{-1} K^{-1}.

A **unimolecular process** is an elementary process in a reaction mechanism in which a single molecule, when sufficiently energetic, dissociates.

A **unit cell** is a small collection of atoms, ions, or molecules occupying positions in a crystalline lattice. An entire crystal can be generated by straight-line displacements of the unit cell in the three perpendicular directions.

Unsaturated hydrocarbon molecules contain one or more carbon-to-carbon multiple bonds.

An **unsaturated solution** contains less solute than the solvent is capable of dissolving under the given conditions.

The **valence bond method** treats a covalent bond in terms of the overlap of pure or hybridized atomic orbitals. Electron probability (or electron charge density) is concentrated in the region of overlap.

Valence electrons are electrons in the electronic shell of highest principal quantum number, that is, electrons in the outermost shell.

The **valence-shell electron-pair repulsion (VSEPR) theory** is a theory used to predict probable shapes of molecules and polyatomic ions based on the mutual repulsions of electron pairs found in the valence shell of the central atom in the structure.

The **van der Waals equation** is an equation of state for nonideal gases. It includes correction terms to account for intermolecular forces of attraction and for the volume occupied by the gas molecules themselves.

van der Waals forces is a term used to describe, collectively, intermolecular forces of the London type and interactions between permanent dipoles.

Vaporization is the passage of molecules from the liquid to the gaseous state.

Vapor pressure is the pressure exerted by a vapor when it is in dynamic equilibrium with its liquid at a fixed temperature.

A **vapor-pressure curve** is a graph of vapor pressure as a function of temperature.

Viscosity refers to a liquid's resistance to flow. Its magnitude depends on intermolecular forces of attraction and, in some cases, on molecular sizes and shapes.

A **voltaic (galvanic) cell** is an electrochemical cell in which a *spontaneous* chemical reaction produces electricity.

Water gas is a mixture of CO(g) and H_2(g), together with some of the noncombustible gases CO_2 and N_2, produced by passing steam [H_2O(g)] over heated coke.

A **wave** is a disturbance that transmits energy through a medium.

The **wavelength** is the distance between successive crests or troughs of a wave motion.

Wave mechanics is a form of quantum theory based on the concepts of wave–particle duality, the Heisenberg uncertainty principle, and the treatment of electrons as matter waves. Mathematical solutions of the equations of wave mechanics are known as **wave functions** (ψ).

A **weak acid** is an acid that is only partially ionized in aqueous solution in a reversible reaction.

A **weak base** is a base that it only partially ionized in aqueous solution in a reversible reaction.

A **weak electrolyte** is a substance that is only partially ionized in solution in a reversible reaction.

Weight refers to the force exerted on an object when it is placed in a gravitational field (the "force of gravity"). The terms *weight* and *mass* are often used interchangeably.

Work is a form of energy transfer between a system and its surroundings that can be expressed as a force acting through a distance.

X-ray diffraction is a method of crystal structure determination based on the interaction of a crystal with X rays.

A **zero-order** reaction proceeds at a rate that is *independent* of reactant concentrations. The sum of the concentration-term exponent(s) in the rate equation is equal to *zero*.

The **zero-point energy** is the lowest possible energy in a quantum mechanical system, such as the "particle-in-a-box" energy corresponding to $n = 1$ (page 322).

Zone refining is a purification process in which a rod of material is subjected to successive melting and freezing cycles. Impurities are swept by a moving molten zone to the end of the rod, which is cut off.

Appendix
F

Answers to Selected Exercises

Note: Your answers may differ slightly from those given here, depending on the number of steps used to solve a problem and whether any intermediate results were rounded off.

CHAPTER 1

Practice Examples **1A.** higher **1B.** no: -15 °F $= -26.1$ °C **2A.** 1.46 g/mL **2B.** 2.76 g/mL **3A.** 135 g soln **3B.** 63.4 L ethanol **4A.** 1.8 kg ethanol **4B.** 0.856 g/mL **5A.** 21.3 **5B.** 1.1×10^6 **6A.** 26.7 **6B.** 15.6

Review Questions **4.** (a) 1.55×10^3 g (b) 0.642 kg (c) 289.6 cm (d) 0.86 mm **5.** (a) 127 mL (b) 0.0158 L (c) 0.981 L (d) 2.65×10^6 cm^3 **6.** (a) 174 cm (b) 29 m (c) 644 g (d) 112 kg (e) 7.00 L (f) 3.52×10^3 mL **7.** (a) 1.00×10^6 m^2 (b) 1.00×10^4 cm^2 (c) 2.59×10^6 m^2 **8.** 102 °C. **9.** 90.0 mL of carbon disulfide **10.** 0.958 g/mL **11.** 13.6 g/mL **12.** (a) 502 g (b) 20.6 kg (c) 58.6 mL (d) 21.5 L **13.** 0.629 kg **14.** 24 g **15.** 7.92×10^3 g **16.** 3.69 kg **17.** (a) 8.950×10^3 (b) 1.0700×10^4 (c) 2.40×10^{-2} (d) 4.7×10^{-3} (e) 9.383×10^2 (f) 2.75482×10^5 **18.** (a) 0.0321 (b) 0.000508 (c) 0.001219 (d) 0.162 **19.** (a) two or three (b) three (c) two (d) five (e) four (f) one to four (g) three (h) two to eight **20.** (a) 3985 (b) 422.0 (c) 1.860×10^5 (d) 3.390×10^4 (e) 6.321×10^4 (f) 5.047×10^{-4} **21.** (a) 9.3×10^{-4} (b) 2.2×10^{-1} (c) 5.4×10^{-1} (d) 3.058×10^1 **22.** (a) 1.0×10^5 (b) 1.0 (c) 4.46×10^{-1} (d) 5.79×10^{-1} **23.** 1.04×10^4 g **24.** 1.58×10^3 g

Exercises **25.** No point at which the law is ever verified with certainty. **27.** Cause-and-effect relationships may be difficult to establish ("God is subtle"); they do exist ("he is not malicious"). **29.** Need: controlled situation, careful observations, altering parameters, preferably one at a time. Results must be reproducible resulting in patterns from which a natural law is formulated. **31.** (a) & (d) physical (b) & (c) chemical **33.** (a) heterogeneous mixture (b) homogeneous mixture (c) substance: assumes no gases dissolved (d) heterogeneous mixture **35.** (a), (c), (d) physical (b) chemical **37.** (a) 3.4×10^4 cm/s (b) 6.378×10^3 km (c) 7.4×10^{-11} m (d) 4.6×10^5 **39.** (a) exact (b) measured (c) measured (d) measured **41.** (a) 2.44×10^4 (b) 1.5×10^3 (c) 40.0 (d) 2.131×10^3 (e) 4.8×10^{-3} **43.** (a) 115.76 mi/h (b) 2.80 mi/lb **45.** 2172 μg **47.** 1.5 m **49.** (a) 10. s (b) 9.8 m/s (c) 2.5 min **51.** 2.5 acres **53.** 2.25×10^4 kg/m^2 **55.** high: 48 °C, low: -8.3 °C **57.** No. 240 °F $= 116$ °C **59.** 35.1 °M and -59.2 °M **61.** 0.790 g/mL **63.** iron bar < aluminum foil < water **65.** 0.04 g/shot **67.** 0.9 L **69.** 12% A, 28% B, 47% C, 11% D, and 3% F **71.** 8.61×10^2 g

Integrative and Advanced Exercises **76.** 5.5×10^{16} tons **77.** 38.8 m **79.** (a) 4.9 mg m^{-2} h^{-1} (b) 46.6 years or ~50 years (1 sig fig) **81.** (a)

-40 °C (b) 160. °C (c) -19.1 °C (d) 335 °C **84.** 11 g mL^{-1} **85.** 2.0×10^4 kg

CHAPTER 2

Practice Examples **1A.** 0.529 g **1B.** 6.85 g **2A.** 0.301 g **2B.** 1.20 g magnesium, 0.80 g oxygen **3A.** $^{108}_{47}$Ag **3B.** 16p, 18e, 19n **4A.** 15.0001 u **4B.** mass ratio 16.8308848 **5A.** Boron-11 **5B.** 7.5% lithium-6 and 92.5% lithium-7 **6A.** Li$^+$, S^{2-}, Ra^{2+}, F$^-$, I$^-$, Al^{3+} **6B.** main-group metals: Na, Mg, Al; main-group nonmetals: S, Kr (noble gas), B, I, H; main-group metalloids: As, Si; transition metals U, Re **7A.** 3.05×10^{21} **7B.** 1.20×10^{21} **8A.** 248 g **8B.** 3.40×10^{24} **9A.** 3.46×10^{21} **9B.** 62.5%

Review Questions **4.** 0.268 g oxygen **5.** 4.720 g potassium chloride **6.** product of combustion is a gas, sulfur dioxide **7.** 39.34% Na **8.** (a) 0.066 g oxygen/0.166 g magnesium oxide; (b) 60.2% Mg **9.** (a) yes, 27.3% C; (b) 72.7% O **10.** O-to-S ratios in the compounds are in the ratio 3 : 2. **11.** Cl-to-P ratios in the compounds are in the ratio 3 : 5. **12.** No, need to know the proportions in which the elements react. **14.** (a) $^{40}_{18}$Ar $< ^{39}_{19}$K $< ^{58}_{27}$Co $< ^{59}_{29}$Cu $< ^{120}_{48}$Cd $< ^{112}_{50}$Sn $< ^{122}_{52}$Te (b) $^{39}_{19}$K $< ^{40}_{18}$Ar $< ^{59}_{29}$Cu $< ^{58}_{27}$Co $< ^{112}_{50}$Sn $< ^{122}_{52}$Te $< ^{120}_{48}$Cd (c) $^{39}_{19}$K $< ^{40}_{18}$Ar $< ^{58}_{27}$Co $< ^{59}_{29}$Cu $< ^{112}_{50}$Sn $< ^{120}_{48}$Cd $< ^{122}_{52}$Te **15.** (a) $^{60}_{27}$Co (b) $^{32}_{15}$P (c) $^{131}_{53}$I (d) $^{35}_{16}$S **16.** 59% neutrons **17.** ^{193}Ir **18.** argon **19.** (a) 2.914071 (b) 2.165216 (c) 18.50146 **20.** ^{81}Br $= 80.917$ u **21.** 39.948 u **22.** (a) In (b) like S: O, Se, unlike S: Na, Ba (c) Cs (d) I (e) Xe (f) 15 (g) 2 **23.** noble gas 118, alkali metal 119 **24.** (a) 7.65×10^{21} (b) 2.17×10^{21} (c) 1.1×10^{12} **25.** (a) 362 mol Fe (b) 646 g Kr (c) 20.1 mg Au (d) 9.49×10^{24} Fe atoms **26.** 25.0 g N **27.** 8.7×10^{18} atoms ^{204}Pb **28.** 1.5×10^3 g alloy

Exercises **29.** No: iron and oxygen from air form solid iron oxide, whereas carbon dioxide gas escapes from burning match. **31.** 21.625 g before and after reaction. **33.** Both samples have 39.4% Na $\pm$ 0.1%. **35.** 0.422 g **37.** (a) 3 H (comp. A): 1 H (comp. B): 2 H (comp. C) (b) Comp. B is N$_2$H$_2$ (given). A might be N$_2$H$_6$ (or NH$_3$) and C might be N$_2$H$_4$. **39.** 11% O **41.** Cathode rays have identical properties, no matter their source. **43.** Values are all multiples of e. **45.** (a) Ratio of 1: 1.8×10^3 (b) for proton: 1.044×10^{-5} g/C for electron: 5.686×10^{-9} g/C **47.** (a) 46p, 46e, 62n (b) 8.9919908. **49.** 106.906 u **51.** $^{24}_{12}$Mg^{2+} (b) $^{47}_{24}$Cr (c) $^{226}_{90}$Th **53.** Unlikely because masses of p, n, and e are not integral. **55.** 24.31 u **57.** 108.9 u **59.** 40.962 u **61.** (b) 72.6 u (2 sig fig) **63.** (a) 1.954 mol Rb (b) 3.916×10^{27} Fe atoms (c) 1.8×10^{-10} g Ag (d) 3.15476×10^{-23} g **65.** 2.00×10^{23} **67.** (a) 1.4×10^{-6} mol Pb/L (b) 8.4×10^{14} Pb atoms/mL **69.** 4.38×10^{22} atoms

Integrative and Advanced Exercises **74.** 3×10^{15} g cm^{-3} **77.** ^{27}Al^{3+} **79.** ^{210}Po **82.** 200.6 u

86. 7.0×10^{19} atoms ^{30}Si **89.** 50.1% Sn, 32.0% Pb, 17.9% Cd

Feature Problems **90.** Within 1 mg (similar to a good laboratory balance) **93.** 2.43×10^5 km^3 **94.** 159 ppm Rb

CHAPTER 3

Practice Examples **1A.** 1.5×10^{22} ions **1B.** 4.0×10^1 g MgCl$_2$ **2A.** 3.69×10^{21} **2B.** 9×10^{-3} μmol/m^3 (i.e. > than the detectable limit) **3A.** 17.1 g C **3B.** 132.0 mL **4A.** 40.001% C; 6.714% H; 53.284% O **4B.** 23.681% C; 3.180% H; 13.81% N; 18.32% P; 41.008% O **5A.** C$_6$H$_{10}$O$_3$ and C$_{12}$H$_{20}$O$_6$ **5B.** C$_3$H$_7$O$_3$ and C$_6$H$_{14}$O$_6$ **6A.** C$_7$H$_{14}$O$_2$ **6B.** C$_4$H$_4$S **7A.** 0 for S; +6 for Cr; +1 for Cl; $-1/2$ for O **7B.** +2 for S; +1 for Hg; +7 for Mn; 0 for C **8A.** Li$_2$O; SnF$_2$; Li$_3$N **8B.** Al$_2$S$_3$; Mg$_3$N$_2$; V$_2$O$_3$ **9A.** cesium iodide; calcium fluoride; iron(II) oxide; chromium (III) chloride **9B.** calcium hydride; copper(I) chloride; silver sulfide; mercury(I) chloride **10A.** sulfur hexafluoride; nitrous acid; calcium hydrogen carbonate; iron(II) sulfate **10B.** ammonium nitrate; phosphorus trichloride; hypobromous acid; silver perchlorate; iron(III) sulfate **11A.** BF$_3$; K$_2$Cr$_2$O$_7$; H$_2$SO$_4$; CaCl$_2$ **11B.** Al(NO$_3$)$_3$; P$_4$O$_{10}$; Cr(OH)$_3$; HIO$_3$ **12A.** (a) not isomers (b) isomers **12B.** (a) isomers (b) not isomers **13A.** (a) alkane (b) chloroalkane (c) carboxylic acid. (d) alkene **13B.** (a) alcohol (b) carboxylic acid and an alcohol (c) chlorocarboxylic acid (d) bromoalkene **14A.** (a) 2-propanol (b) 1-iodopropane (c) 3-methylbutanoic acid (d) propene **14B.** (a) 2-chloropropane (b) 1,4-dichlorobutane (c) 2-methyl propanoic acid **15A.** (a) CH$_3$(CH$_2$)$_3$CH$_3$; (b) CH$_3$CO$_2$H (c) ICH$_2$(CH$_2$)$_6$CH$_3$ (d) CH$_2$(OH)(CH$_2$)$_3$CH$_3$ **15B.** (a) CH$_3$CHCH$_2$ (b) CH$_2$(OH)(CH$_2$)$_5$CH$_3$ (c) CH$_2$ClCO$_2$H (d) CH$_3$(CH$_2$)$_4$CO$_2$H

Review Questions **5.** (a) 20 atoms (b) 4.91×10^{21} atoms (c) 2.195×10^{23} atoms **6.** (a) 675 g (b) 168 g (c) 4.64×10^3 g (d) 431 g **7.** (a) 0.134 mol Br$_2$ (b) 1.80 mol Br$_2$ (c) 70.7 mol Br$_2$ (d) 51.4 mol Br$_2$ **8.** (a) 149.2 u (b) 11 moles of H atoms (c) 60.055 g C (d) 2.73×10^{25} C atoms **9.** 15.59% H **10.** 36.18% O **11.** 40.53% H$_2$O **12.** (a) 64.07% Pb (b) 45.50% Fe (c) 2.72% Mg **13.** CH$_3$CH$_2$SH **14.** C$_4$H$_{10}$O **15.** NaC$_5$H$_8$NO$_4$ **16.** C$_8$H$_6$O$_4$ **17.** (a) 75.71% C, 8.795% H, 15.50% O (b) C$_{13}$H$_{18}$O$_2$. **18.** CrO$_3$, **19.** SO$_3$ and S$_2$O **20.** (a) Pb^{2+}; (b) cobalt(III) ion; (c) barium ion; (d) Cr^{2+}; (e) IO$_4^-$ (f) chlorite ion; (g) Au^{3+}; (h) bisulfite *or* hydrogen sulfite ion; (i) HCO$_3^-$; (j) CN$^-$ **21.** (a) potassium bromide; (b) strontium chloride; (c) chlorine trifluoride; (d) dinitrogen tetroxide; (e) phosphorus pentachloride **22.** (a) potassium cyanide; (b) hypochlorous acid; (c) ammonium sulfate (d) potassium iodate **23.** (a) Zn = 0

(b) S = −2 **(c)** N = +4 **(d)** N = +3 **(e)** V = +4 **(f)** P = +5 **24.** **(a)** $MgBr_2$ **(b)** BaO **(c)** $Hg(C_2H_3O_2)_2$ **(d)** $Fe_2(C_2O_4)_3$ **(e)** $Sr(ClO_4)_2$ **(f)** $KHSO_4$ **(g)** NCl_3 **(h)** BrF_5 **25.** **(a)** chlorous acid; **(b)** sulfurous acid; **(c)** hydroselenic acid; **(d)** nitrous acid **26.** **(a)** HI(aq); **(b)** HNO_3; **(c)** H_3PO_4 **(d)** H_2SO_4 **27.** **(b)** 2-butanol **28.** **(c)** butanoic acid

Exercises **29.** **(a)** H_2O_2 **(b)** CH_3CH_2Cl **(c)** P_4O_{10} **(d)** $CH_3CH(OH)CH_3$ **(e)** HCO_2H **31.** **(b)** CH_3CH_2Cl **(d)** $CH_3CH(OH)CH_3$ **(e)** HCO_2H **33.** 65 g SO_2 **35.** **(a)** 7.74 g **(b)** 1.90×10^3 g **(c)** 84.1 g **37.** **(a)** 4.58×10^{-6} mol S_8 **(b)** 2.21×10^{19} S atoms **39.** **(a)** false; **(b)** true; **(c)** false; **(d)** false **41.** **(a)** 8 atoms **(b)** 3 F/2 C **(c)** 1.402 Br/g F **(d)** bromine **(e)** 3.45 g **43.** 75.998% C, 12.755% H, 11.248% O **45.** **(a)** 49.765% Zr **(b)** 5.03004% Be **(c)** 33.659% Fe **(d)** 6.6635% S **47.** In order of increasing %Cr: $CrO_3 < CrO_2 < Cr_2O_3 < CrO$ **49.** $C_{10}H_8$ **51.** **(a)** $C_{19}H_{16}O_4$ **(b)** $C_4H_8Cl_2S$ **53.** $C_{16}H_{10}N_2O_2$ **55.** 31 g/mol X most likely X is P **57.** 894 u **59.** **(a)** 90.51% C and 9.491% H **(b)** C_4H_5 **(c)** C_8H_{10} **61.** CH_4N. **63.** $C_{10}H_8$ **65.** 2.247 g H_2O **67.** **(a)** C = −4 **(b)** S = +4 **(c)** O = −1 **(d)** C = 0 **(e)** Fe = +6 **69.** Cr_2O_3, CrO_2 and CrO_3 **71.** **(a)** strontium oxide; **(b)** zinc sulfide; **(c)** potassium chromate; **(d)** cesium sulfate; **(e)** chromium(III) oxide; **(f)** iron(III) sulfate; **(g)** magnesium hydrogen carbonate; **(h)** ammonium hydrogen phosphate **(i)** calcium hydrogen sulfite; **(j)** copper(II) hydroxide; **(k)** nitric acid; **(l)** potassium perchlorate; **(m)** bromic acid; **(n)** phosphorous acid **73.** **(a)** carbon disulfide; **(b)** silicon tetrafluoride; **(c)** chlorine pentafluoride; **(d)** dinitrogen pentoxide; **(e)** sulfur hexafluoride; **(f)** diiodine hexachloride **75.** **(a)** $Al_2(SO_4)_3$ **(b)** $(NH_4)_2Cr_2O_7$ **(c)** SiF_4; **(d)** Fe_2O_3 **(e)** C_3S_2 **(f)** $Co(NO_3)_2$ **(g)** $Sr(NO_2)_2$ **(h)** HBr(aq) **(i)** HIO_3 **(j)** PCl_2F_3 **77.** **(a)** $TiCl_4$ **(b)** $Fe_2(SO_4)_3$ **(c)** Cl_2O_7 **(d)** $S_2O_8^{2-}$ **79.** $MgCl_2 \cdot 6\,H_2O$ **81.** 15 g $CuSO_4$ **83.** $CuSiF_6 \cdot 6H_2O$ **85.** identical: (a), (c); isomers: (a), (b), (d) **87.** **(a)** $CH_3(CH_2)_4CH_3$ **(b)** HCO_2H **(c)** $CH_3CH_2CH\,(CH_3)CH_2OH$ **(d)** $ClCH_2CH_3$ **89.** **(a)** methanol; CH_3OH; 32.04 u **(b)** 2-chlorohexane; $CH_3(CH_2)_3\,CHClCH_3$; 120.6 u **(c)** pentanoic acid; $CH_3(CH_2)_3CO_2H$; 102.1 u **(d)** 2-methyl-1-propanol; $CH_3CH\,(CH_3)CH_2OH$; 74.12 u

Integrative and Advanced Exercises **91.** 1.24×10^{23} atoms ^6Li **95.** CH_4 **98.** 26.9 u **102.** $ZnSO_4 \cdot 7\,H_2O$ **105.** $C_{12}H_8Cl_6O$

Feature Problems **108.** **(a)** 5% N, 4.4% P, 4.2% K **(b)** (1) 60.6% P_2O_5 (2) 53.7% P_2O_5 **110.** **(a)** 4.7×10^3 m^2 **(b)** 2.5 nm **(c)** 5.8×10^{23} molecules

CHAPTER 4

Practice Examples **1A.** **(a)** $2\,H_3PO_4 + 3\,CaO \longrightarrow Ca_3(PO_4)_2 + 3\,H_2O$ **(b)** $C_3H_8 + 5\,O_2 \longrightarrow 3\,CO_2 + 4\,H_2O$ **1B.** **(a)** $4\,NH_3 + 7\,O_2 \longrightarrow 4\,NO_2 + 6\,H_2O$ **(b)** $6\,NO_2 + 8\,NH_3 \longrightarrow 7\,N_2 + 12\,H_2O$
2A. $4\,HgS + 4\,CaO \longrightarrow 3\,CaS + CaSO_4 + 4\,Hg$
2B. $2\,C_7H_6O_2S + 17\,O_2 \longrightarrow 14\,CO_2 + 6\,H_2O + 2\,SO_2$
3A. 2.64 mol O_2 **3B.** 8.63 mol Ag **4A.** 5.29 g Mg_3N_2 **4B.** 126 g H_2 **5A.** 0.126 g H_2 **5B.** 3.50 g O_2 **6A.** 3.34 cm^3 **6B.** 0.79 g **7A.** 0.4 mg H_2(g) **7B.** 0.15 g CO_2 **8A.** 0.307 M **8B.** 0.524 M **9A.** 115 g **9B.** 50.9 g **10A.** 0.0675 M **10B.** 0.122 M **11A.** 18.1 mL **11B.** 2.00×10^2 mL; 4.96 g **12A.** 936 g PCl_3 **12B.** 1.86 kg $POCl_3$ **13A.** 3.8 g P_4 **13B.** 57 g O_2 **14A.** **(a)** 30.0 g CH_2O **(b)** 25.7 g CH_2O **(c)** 85.6%. **14B.** 93.7% yield **15A.** 41.8 g CO_2 **15B.** 69.0 g **16A.** 2.47×10^3 g HNO_3 **16B.** 0.0734 g H_2

Review Questions
4. **(a)** $Na_2SO_4(s) + 4\,C(s) \longrightarrow Na_2S(s) + 4\,CO(g)$
(b) $4\,HCl(g) + O_2(g) \longrightarrow 2\,H_2O(l) + 2\,Cl_2(g)$
(c) $PCl_5(l) + 4\,H_2O(l) \longrightarrow H_3PO_4(aq) + 5\,HCl(aq)$
(d) $3\,PbO(s) + 2\,NH_3(g) \longrightarrow 3\,Pb(s) + N_2(g) + 3\,H_2O(l)$
(e) $Mg_3N_2(s) + 6\,H_2O(l) \longrightarrow 3\,Mg(OH)_2(s) + 2\,NH_3(g)$
5. **(a)** $2\,Mg + O_2 \longrightarrow 2\,MgO$
(b) $2\,NO + O_2 \longrightarrow 2\,NO_2$
(c) $2\,C_2H_6 + 7\,O_2 \longrightarrow 4\,CO_2 + 6\,H_2O$
(d) $Ag_2SO_4(aq) + BaI_2(aq) \longrightarrow BaSO_4(s) + 2\,AgI(s)$
6. **(a)** $C_7H_{16} + 11\,O_2 \longrightarrow 7\,CO_2 + 8\,H_2O$
(b) $C_4H_9OH + 6\,O_2 \longrightarrow 4\,CO_2 + 5\,H_2O$
(c) $2\,HI(aq) + Na_2CO_3(aq) \longrightarrow 2\,NaI(aq) + H_2O(l) + CO_2(g)$
(d) $3\,NaOH(aq) + FeCl_3(aq) \longrightarrow Fe(OH)_3(s) + 3\,NaCl(aq)$
7. **(a)** and **(b)** atomic O is not a product; **(c)** KCl is the product, not KClO. **8.** 3 and 4 are true. **9.** 4.84 mol $FeCl_3$ **10.** 17.0 g O_2 **11.** 35.8 g Cl_2 and 10.4 g P_4 **12.** **(a)** 5.32 mol O_2 **(b)** 323.1 g KO_2 **(c)** 6.35×10^{18} O_2 molecules **13.** **(a)** 0.408 M **(b)** 0.154 M **(c)** 1.53 M **(d)** 0.675 M **14.** **(a)** 319 mol **(b)** 20.0 g **(c)** 16.4 mg **15.** 373 g KCl in 5.00 L **16.** 169 mL **17.** 0.242 M **18.** 14.1 g $CuCO_3$ **19.** 14.9 mL **20.** 1.52 mol **21.** HNO_3 is limiting, hence, some Cu does not react. **22.** **(a)** 1.80 mole CCl_2F_2 **(b)** 1.55 mol CCl_2F_2 **(c)** 86.1% yield **23.** **(a)** 82.01 g C_6H_{10} **(b)** 78.0% yield **(c)** 156 g **24.** 68.0% $CaCO_3$

Exercises **25.**
(a) $Cr_2O_3(s) + 2\,Al(s) \xrightarrow{\Delta} Al_2O_3(s) + 2\,Cr(l)$
(b) $CaC_2(s) + 2\,H_2O(l) \longrightarrow Ca(OH)_2(s) + C_2H_2(g)$
(c) $3\,H_2(g) + Fe_2O_3(s) \xrightarrow{\Delta} 2\,Fe(l) + 3\,H_2O(g)$
(d) $NCl_3(g) + 3\,H_2O(l) \longrightarrow NH_3(g) + 3\,HOCl(aq)$
27. **(a)** $2\,C_4H_{10} + 13\,O_2 \longrightarrow 8\,CO_2 + 10\,H_2O$
(b) $2\,C_3H_7OH + 9\,O_2 \longrightarrow 6\,CO_2 + 8\,H_2O$
(c) $HC_3H_5O_3 + 3\,O_2 \longrightarrow 3\,CO_2 + 3\,H_2O$
29. **(a)** $NH_4NO_3(s) \xrightarrow{\Delta} N_2O(g) + 2\,H_2O(g)$
(b) $Na_2CO_3(aq) + 2\,HCl(aq) \longrightarrow 2\,NaCl(aq) + H_2O(l) + CO_2(g)$
(c) $2\,CH_4(g) + 2\,NH_3(g) + 3\,O_2(g) \longrightarrow 2\,HCN(g) + 6\,H_2O(g)$
31. $2\,N_2H_4 + N_2O_4 \longrightarrow 4\,H_2O + 3\,N_2$
33. **(a)** 0.401 mol O_2 **(b)** 128 g $KClO_3$ **(c)** 44.0 g KCl **35.** 96.0 g Ag_2CO_3 **37.** 79.7% Fe_2O_3 **39.** 1.03 g H_2 **41.** NH_4NO_3 (reaction 1) **43.** **(a)** 1.753 M **(b)** 0.320 M **(c)** 0.206 M **45.** **(a)** 4.73 g **(b)** 44.1 mL **47.** 46% by mass sucrose **49.** 0.0820 M **51.** 20 : 1 dilution (e.g., 100.0-mL flask and a 5.00-mL pipet) **53.** **(a)** 0.177 g **(b)** 0.562 g **55.** **(a)** 0.0693 mol **(b)** 2.91 M **57.** 59.4 mL **59.** 0.624 g **61.** 0.2649 M **63.** 24.0 g **65.** 143 g **67.** 10.5 g NH_3, 10.1 g excess $Ca(OH)_2$ **69.** **(a)** 27.7 g **(b)** 17.1 g **(c)** 61.7% yield **71.** 76 g **73.** In synthesis, yield may be sacrificed for cost. In an analysis, it is essential that no product is lost (100% yield). **75.** 474 g **77.** 74.4 mol **79.** 1.34×10^3 g

Integrative and Advanced Exercises
84. $3\,FeS(s) + 5\,O_2(g) \longrightarrow Fe_3O_4(s) + 3\,SO_2(g)$ **87.** 0.2 cm^2 **89.** 118 mL **90.** 9.1×10^5 L **92.** 7.25 M **94.** 24% Mg **99.** **(a)** $H_2(g) + \frac{1}{2}O_2(g) \longrightarrow H_2O(g)$ **(b)** (100% reaction, O_2 limiting) yields 0.106 mol H_2, 2.275 mol H_2O. **(c)** 41.20 g initial and final **101.** (a), (b) **103.** 345 mL 1.29M + 29 mL 0.775 M = 374 mL **108.** **(a)** $2\,C_3H_6 + 2\,NH_3 + 3\,O_2 \longrightarrow 2\,C_3H_3N + 6\,H_2O$ **(b)** 503 kg

Feature Problems **109.** 360 or ~ 4×10^2 fish

CHAPTER 5

Practice Examples **1A.** 0.540 M Cl^- **1B.** **(a)** 7.9×10^{-5} M F^- **(b)** 3.1 kg CaF_2 **2A.** **(a)** $Al^{3+}(aq) + 3\,OH^-(aq) \longrightarrow Al(OH)_3(s)$ **(b)** no reaction **(c)** $Pb^{2+}(aq) + 2\,I^-(aq) \longrightarrow PbI_2(s)$ **2B.** **(a)** $Al^{3+}(aq) + PO_4^{3-}(aq) \longrightarrow AlPO_4(s)$ **(b)** $Ba^{2+}(aq) + SO_4^{2-}(aq) \longrightarrow BaSO_4(s)$ **(c)** $Pb^{2+}(aq) + CO_3^{2-}(aq) \longrightarrow PbCO_3(s)$
3A. $NH_3(aq) + HC_3H_5O_2(aq) \longrightarrow NH_4^+(aq) + C_3H_5O_2^-(aq)$
3B. $CaCO_3(s) + 2\,HC_2H_3O_2(aq) \longrightarrow CO_2(g) + H_2O(l) + Ca^{2+}(aq) + 2\,C_2H_3O_2^-(aq)$
4A. **(a)** no **(b)** yes **4B.** VO^{2+} oxidized, MnO_4^- reduced **5A.** ox: $Al \longrightarrow Al^{3+} + 3e^-$; red: $2\,H^+ + 2e^- \longrightarrow H_2$; overall : $2\,Al(s) + 6\,H^+(aq) \longrightarrow 2\,Al^{3+}(aq) + 3\,H_2(g)$ **5B.** ox: $2\,Br^- \longrightarrow Br_2 + 2e^-$; red: $Cl_2 + 2e^- \longrightarrow 2\,Cl^-$; overall: $2\,Br^-(aq) + Cl_2(g) \longrightarrow Br_2(l) + 2\,Cl^-(aq)$
6A. $MnO_4^-(aq) + 8\,H^+(aq) + 5\,Fe^{2+}(aq) \longrightarrow Mn^{2+}(aq) + 4\,H_2O(l) + 5\,Fe^{3+}(aq)$
6B. $3\,UO^{2+}(aq) + Cr_2O_7^{2-}(aq) + 8\,H^+(aq) \longrightarrow 3\,UO_2^{2+}(aq) + 2\,Cr^{3+}(aq) + 4\,H_2O(l)$
7A. $S(s) + 2\,OH^-(aq) + 2\,OCl^-(aq) \longrightarrow SO_3^{2-}(aq) + H_2O(l) + 2\,Cl^-(aq)$
7B. $2\,MnO_4^-(aq) + 3\,SO_3^{2-}(aq) + H_2O(l) \longrightarrow 2\,MnO_2(s) + 3\,SO_4^{2-}(aq) + 2\,OH^-(aq)$
8A. a reducing agent. **8B.** Au-reducing agent, O_2(g)-oxidizing agent **9A.** 0.1019 M **9B.** 0.130 M **10A.** 65.4% Fe **10B.** 0.03129 M

Review Questions **4.** **(a)** 0.10 M NaCl, an ionic compound (strong electrolyte) **(b)** 0.10 M C_2H_5OH, a molecular compound (nonelectrolyte) **5.** strong acid: (e); weak acids: (d), (f); strong bases: (b), (i); weak base (g); salts: (a), (c), (h) **6.** $Al_2(SO_4)_3$ solution has the highest $[SO_4^{2-}]$, 0.24 M. **7.** **(a)** 0.238 M K^+ **(b)** 0.334 M NO_3^- **(c)** 0.17 M Al^{3+} **(d)** 0.627 M Na^+ **8.** 200.0 mL of 0.035 M NaCl **9.** 3.04×10^{-3} M OH^- **10.** 0.126 M K^+, 0.148 M Mg^{2+}, 0.422 M Cl^- **11.** 4.3×10^2 mg **12.** $BaSO_4$(s) and $Al(OH)_3$(s) **13.** $Ca(s) + 2\,HCl(aq) \longrightarrow CaCl_2(aq) + H_2(g)$ $KHSO_3(s) + HCl(aq) \longrightarrow KCl(aq) + H_2O(l) + SO_2(g)$
14. **(a)** $Pb^{2+}(aq) + 2\,Br^-(aq) \longrightarrow PbBr_2(s)$ **(b)** no reaction **(c)** $Fe^{3+}(aq) + 3\,OH^-(aq) \longrightarrow Fe(OH)_3(s)$ **(d)** $Ca^{2+}(aq) + CO_3^{2-}(aq) \longrightarrow CaCO_3(s)$ **(e)** $Ba^{2+}(aq) + SO_4^{2-}(aq) \longrightarrow BaSO_4(s)$ **(f)** no reaction **15.** **(a)** $OH^-(aq) + HC_2H_3O_2(aq) \longrightarrow H_2O(l) + C_2H_3O_2^-(aq)$ **(b)** no reaction **(c)** $FeS(s) + 2\,H^+(aq) \longrightarrow H_2S(g) + Fe^{2+}(aq)$ **(d)** $HCO_3^-(aq) + H^+(aq) \longrightarrow H_2O(l) + CO_2(g)$ **(e)** $Mg(s) + 2\,H^+(aq) \longrightarrow Mg^{2+}(aq) + H_2(g)$ **(f)** no reaction **16.** Use NH_3(aq) in the reaction: $Mg^{2+}(aq) + 2\,NH_3(aq) + 2\,H_2O(l) \longrightarrow Mg(OH)_2(s) + 2\,NH_4^+(aq)$ **17.** 13.3 mL **18.** 0.08683 M NaOH **19.** KOH is the limiting reactant; HCl is in excess and the solution is acidic. **20.** (a) **21.** **(a)** not redox equation; O.S. +2 for Mg, +4 for C, −2 for O, and +1 for H on both sides; **(b)** redox equation; O.S. for Cl, 0 ⟶ −1, for Br, −1 ⟶ 0; **(c)** redox equation; O.S. for Ag, 0 ⟶ +1, for N, +5 ⟶ +4; **(d)** not redox equation; O.S. +1 for Ag, +6 for Cr, and −2 for O on both sides. **22.** **(a)** oxid, agent: NO (O.S. of N, +2 ⟶ −3); red, agent: H_2 (O.S. of H, 0 ⟶ +1); unchanged (O.S. of O, −2 ⟶ −2); **(b)** oxid, agent: NO_3^- (O.S. of N, +5 ⟶ +2); red, agent: Cu (O.S. of Cu, 0 ⟶ +2); unchanged (O.S. of O, −2 ⟶ −2; O.S. of H, +1 ⟶ +1); **(c)** oxid, agent: Cl_2 (O.S. of Cl, 0 ⟶ −1); red, agent: Cl_2 (O.S. of Cl, 0 ⟶ +5); unchanged (O.S. of O,

$-2 \longrightarrow -2$; O.S. of H, $+1 \longrightarrow +1$)
23. (a) Reduction:
$$2\,SO_3^{2-}(aq) + 6\,H^+(aq) + 4\,e^- \longrightarrow$$
$$S_2O_3^{2-}(aq) + 3\,H_2O(l)$$
(b) Reduction:
$$2\,NO_3^-(aq) + 10\,H^+(aq) + 8\,e^- \longrightarrow$$
$$N_2O(g) + 5\,H_2O(l)$$
(c) Oxidation: $I^-(aq) + 3\,H_2O(l) \longrightarrow$
$$IO_3^-(aq) + 6\,H^+(aq) + 6\,e^-$$
(d) Oxidation: $Al(s) + 4\,OH^-(aq) \longrightarrow$
$$Al(OH)_4^-(aq) + 3\,e^-$$
24. (a) $3\,Zn(s) + 2\,NO_3^-(aq) + 8\,H^+(aq) \longrightarrow$
$$3\,Zn^{2+}(aq) + 2\,NO(g) + 4\,H_2O(l)$$
(b) $4\,Zn(s) + NO_3^-(aq) + 10\,H^+(aq) \longrightarrow$
$$4\,Zn^{2+}(aq) + NH_4^+(aq) + 3\,H_2O(l)$$
(c) $Cr_2O_7^{2-}(aq) + 14\,H^+(aq) + 6\,Fe^{2+}(aq) \longrightarrow$
$$6\,Fe^{3+}(aq) + 2\,Cr^{3+}(aq) + 7\,H_2O(l)$$
(d) $2\,MnO_4^-(aq) + 6\,H^+(aq) + 5\,H_2O_2(aq) \longrightarrow$
$$2\,Mn^{2+}(aq) + 8\,H_2O(l) + 5\,O_2(g)$$
25. (a) $2\,MnO_2(s) + ClO_3^-(aq) + 2\,OH^-(aq) \longrightarrow$
$$2\,MnO_4^-(aq) + Cl^-(aq) + H_2O(l)$$
(b) $2\,Fe(OH)_3(s) + 3\,OCl^-(aq) + 4\,OH^-(aq) \longrightarrow$
$$2\,FeO_4^{2-}(aq) + 3\,Cl^-(aq) + 5\,H_2O(l)$$
(c) $6\,ClO_2(aq) + 6\,OH^-(aq) \longrightarrow$
$$5\,ClO_3^-(aq) + Cl^-(aq) + 3\,H_2O$$
26. 0.03421 M $KMnO_4$

Exercises 27. (a) weak electrolyte; **(b)** strong electrolyte; **(c)** strong electrolyte; **(d)** nonelectrolyte; **(e)** strong electrolyte; **29. (a)** barium bromide-strong electrolyte **(b)** propionic acid-weak electrolyte **(c)** ammonia-weak electrolyte **31. (a)** 8.73×10^{-4} M Ca^{2+} **(b)** 6.55×10^{-3} M K^+ **(c)** 2.57×10^{-3} M Zn^{2+} **33.** 14.7 mg Na^+/mL **35.** 0.732 M **37. (a)** no reaction **(b)** $Cu^{2+}(aq) + CO_3^{2-}(aq) \longrightarrow CuCO_3(s)$ **(c)** $3\,Cu^{2+}(aq) + 2\,PO_4^{3-}(aq) \longrightarrow Cu_3(PO_4)_2(s)$ **39. (a)** Use a soluble sulfate (e.g., K_2SO_4) to form $BaSO_4(s)$. **(b)** Use water (Na_2CO_3 is soluble). **(c)** Use a soluble chloride (e.g., KCl) to form $AgCl(s)$. **41. (a)** Add $K_2SO_4(aq)$ to $Sr(NO_3)(aq)$: $Sr^{2+}(aq) + SO_4^{2-}(aq) \longrightarrow SrSO_4(s)$. **(b)** Add $NaOH(aq)$ to $Mg(NO_3)_2(aq)$: $Mg^{2+}(aq) + 2\,OH^-(aq) \longrightarrow Mg(OH)_2$ (s). **(c)** Add $K_2SO_4(aq)$ to $BaCl_2(aq)$: $Ba^{2+}(aq) + SO_4^{2-}(aq) \longrightarrow BaSO_4(s)$; filter solid from the $KCl(aq)$.
43. (a) $NaHCO_3(s) + H^+(aq) \longrightarrow$
$$Na^+(aq) + H_2O(l) + CO_2(g)$$
(b) $CaCO_3(s) + 2\,H^+(aq) \longrightarrow$
$$Ca^{2+}(aq) + H_2O(l) + CO_2(g)$$
(c) $Mg(OH)_2(s) + 2\,H^+(aq) \longrightarrow$
$$Mg^{2+}(aq) + 2\,H_2O(l)$$
(d) reaction (c) and $Al(OH)_3(s) + 3\,H^+(aq) \longrightarrow$
$$Al^{3+}(aq) + 3\,H_2O(l)$$
(e) $NaAl(OH)_2CO_3(s) + 4\,H^+(aq) \longrightarrow$
$$Al^{3+}(aq) + Na^+(aq) + 3\,H_2O(l) + CO_2(g)$$
45. As a salt: $NaHSO_4(aq) \longrightarrow$
$$Na^+(aq) + HSO_4^-(aq);$$
As an acid: $HSO_4^-(aq) + OH^-(aq) \longrightarrow$
$$H_2O(l) + SO_4^{2-}(aq)$$
47. (a) Two reductions and no oxidation; an impossibility. **(b)** Two oxidations and no reduction; also an impossibility.
49. (a) $CH_4(g) + 4\,NO(g) \longrightarrow$
$$2\,N_2(g) + CO_2(g) + 2\,H_2O(g)$$
(b) $16\,H_2S(g) + 8\,SO_2(g) \longrightarrow$
$$3\,S_8(s) + 16\,H_2O(g)$$
(c) $10\,NH_3(g) + 3\,Cl_2O(g) \longrightarrow$
$$6\,NH_4Cl(s) + 2\,N_2(g) + 3\,H_2O(l)$$
51. (a) $10\,I^-(aq) + 2\,MnO_4^-(aq) + 16\,H^+(aq) \longrightarrow$
$$5\,I_2(s) + 2\,Mn^{2+}(aq) + 8\,H_2O(l)$$
(b) $3\,N_2H_4(l) + 2\,BrO_3^-(aq) \longrightarrow$

$$3\,N_2(g) + 2\,Br^-(aq) + 6\,H_2O(l)$$
(c) $Fe^{2+}(aq) + VO_4^{3-}(aq) + 6\,H^+(aq) \longrightarrow$
$$Fe^{3+}(aq) + VO^{2+}(aq) + 3\,H_2O(l)$$
(d) $3\,UO^{2+}(aq) + 2\,NO_3^-(aq) + 2\,H^+(aq) \longrightarrow$
$$3\,UO_2^{2+}(aq) + 2\,NO(g) + H_2O(l)$$
53. (a) $3\,CN^-(aq) + 2\,MnO_4^-(aq) + H_2O(l) \longrightarrow$
$$3\,CNO^-(aq) + 2\,MnO_2(s) + 2\,OH^-(aq)$$
(b) $4[Fe(CN)_6]^{3-}(aq) + N_2H_4(l) + 4\,OH^-(aq) \longrightarrow$
$$4[Fe(CN)_6]^{4-}(aq) + N_2(g) + 4\,H_2O(l)$$
(c) $4\,Fe(OH)_2(s) + O_2(g) + 2\,H_2O(l) \longrightarrow$
$$4\,Fe(OH)_3(s)$$
(d) $3\,C_2H_5OH(aq) + 4\,MnO_4^-(aq) \longrightarrow$
$$3\,C_2H_3O_2^-(aq) + 4\,MnO_2(s) + OH^-(aq) + 4\,H_2O(l)$$
55. (a) $S_2O_3^{2-}(aq) + 5\,H_2O(l) + 4\,Cl_2(g) \longrightarrow$
$$2\,SO_4^{2-}(aq) + 8\,Cl^-(aq) + 10\,H^+(aq)$$
(b) $Cr_2O_7^{2-}(aq) + 14\,H^+(aq) + 3\,Sn^{2+}(aq) \longrightarrow$
$$3\,Sn^{4+}(aq) + 2\,Cr^{3+}(aq) + 7\,H_2O(l)$$
(c) $S_8(s) + 12\,OH^-(aq) \longrightarrow$
$$4\,S^{2-}(aq) + 2\,S_2O_3^{2-}(aq) + 6\,H_2O(l)$$
(d) $As_2S_3(s) + 12\,OH^-(aq) + 14\,H_2O_2(aq) \longrightarrow$
$$2\,AsO_4^{3-}(aq) + 3\,SO_4^{2-}(aq) + 20\,H_2O(l)$$
57. (a) $SO_3^{2-}(aq)$: reducing agent, MnO_4^-: oxidizing agent; **(b)** $H_2(g)$: reducing agent, $NO_2(g)$: oxidizing agent; **(c)** $[Fe(CN)_6]^{4-}(aq)$: reducing agent, $H_2O_2(aq)$: oxidizing agent. **59.** 0.1230 M NaOH **61.** 3.546 mL **63.** 0.077 M NaOH **65.** More acid present than base, therefore acidic. **67.** 34 mL base **69. (d) 71.** 0.01968 M $KMnO_4$ **73.** 53.23% Fe. **75.** 3.70 g $Na_2C_2O_4$
Integrative and Advanced Exercises 79. $3\,Ca^{2+}(aq) + 2\,HPO_4^{2-}(aq) \longrightarrow Ca_3(PO_4)_2(s) + 2\,H^+(aq)$ **82.** 108 ppm Mg **83.** 0.0874 L
88. (a) $2\,FeS_2(s) + 2\,H_2O(l) + 7\,O_2(g) \longrightarrow$
$$2\,Fe^{2+}(aq) + 4\,SO_4^{2-}(aq) + 4\,H^+(aq)$$
(b) $14\,Fe^{3+}(aq) + FeS_2(s) + 8\,H_2O(l) \longrightarrow$
$$15\,Fe^{2+}(aq) + 2\,SO_4^{2-}(aq) + 16\,H^+(aq)$$
91. 44.6 g $Cl_2(g)$ **93.** 5.0×10^2 g ClO_2 **95.** 45.8 g $NaHSO_3$ in step 1; 1.0 L solution in step 2
Feature Problems 96. $x = 1.07$ **97.** 91.0% MnO_2

CHAPTER 6

Practice Examples 1A. 760. mmHg **1B.** 1.39 g/cm³ **2A.** 756.0 mmHg **2B.** 93 mm **3A.** 139 torr **3B.** 346 g (cross-sectional area doesn't affect pressure) **4A.** 4.40 L **4B.** 317 mm Hg **5A.** 1.97×10^3 K **5B.** 79 °C **6A.** 59.0 g **6B.** 65.2 L **7A.** 24.4 L **7B.** 464 K **8A.** 2.11 mol **8B.** 5.59×10^{14} molecules **9A.** 2.11 mL **9B.** 0.383 g **10A.** 86.5 g/mol **10B.** NO **11A.** 0.162 g/L **11B.** 382 K **12A.** 35.6 g Na_3N **12B.** 0.619 g $Na(l)$ **13A.** 1.25 L **13B.** 150. L **14A.** 13 atm **14B.** 5.5 atm **15A.** 0.0348 atm $H_2O(g)$, 2.47 atm $CO_2(g)$ **15B.** 584 mmHg N_2, 157 mmHg O_2, 0.27 mmHg CO_2, 7.0 mmHg Ar **16A.** 0.00278 mol **16B.** 0.395 L **17A.** $NH_3(g)$: 661 m/s **17B.** 76.75 K **18A.** 2.1×10^{-4} mol **18B.** 28.2 s **19A.** 1.90×10^2 g/mol **19B.** 52.3 s **20A.** $Cl_2(g)$ **20B.** $Cl_2(g)$
Review Questions 4. (a) 0.968 atm; **(b)** 0.766 atm; **(c)** 1.17 atm; **(d)** 2.22 atm. **5. (a)** 748 mmHg; **(b)** 928 mmHg; **(c)** 10.4 ft Hg **6.** 753 mmHg **7. (a)** 52.8 L; **(b)** 7.27 L **8. (a)** 1.10×10^3 mL; **(b)** 7.50×10^2 mL **9.** 576 °C **10.** 42.7 L C_2H_4 **11.** 79.03 L Cl_2 **12.** PF_3 **13.** 5.32×10^4 mL **14.** 2.74 atm **15.** 103 g/mol **16.** 1.75 g/L **17.** 1.25 L $H_2(g)$ **18.** 519 L $CO_2(g)$ **19.** 5.59 L **20.** 6.1 L **21. (a)** 737 mmHg; **(b)** 97.5%; **(c)** 0.115 g O_2 **22.** (1) **23.** 18.6 s **24.** (3)
Exercises 25. 11.4 m **27.** 976 mmHg **29.** 1.03 kg cm^{-2} **31.** 50.6 atm **33. (a)** $1° \longrightarrow 2$ °C: increase 0.37%; $10 \longrightarrow 20°C$: increase 3.5%. Volume doubles when temperature in Kelvin doubles. **35. (a)**

41.8 mg PH_3 **(b)** 7.41×10^{20} molecules **37.** Air in the bag expands under the lower pressures at higher altitudes. **39.** 4.30 L **41.** 4.89 g **43.** 2.29 atm **45.** 1.41×10^3 °C **47.** C_3H_6 **49. (a)** 55.8 g mol⁻¹ **(b)** C_4H_8 **51.** 1.22×10^3 mmHg **53. (a)** 1.18 g L⁻¹ **(b)** 1.80 g L⁻¹, Balloon filled with CO_2 has greater density than balloon filled with air at 25 °C, hence CO_2 filled balloon will not rise. **55.** P_4 **57.** 378 L O_2 **59.** 3.3×10^7 L **61.** 109 mL **63.** 2.06×10^3 g **65. (d) 67. (a)** 842 mmHg **(b)** $P_{benzene} = 89.5$ mmHg; $P_{Ar} = 752$ mmHg **69. (b) 71.** 2.37 L **73.** 751 mmHg **75.** 326 m/s **77.** 7.83 u **79.** 1.51×10^3 K **81.** 0.00473 mol **83. (a)** 1.07; **(b)** 1.05; **(c)** 0.978; **(d)** 1.004 **85. (a)** $P_{ideal} = 15.3$ atm, $P_{vdw} = 14.1$ atm; **(b)** $P_{ideal} = 19.4$ atm, $P_{vdw} = 18.3$ atm; **(c)** $P_{ideal} = 27.6$ atm, $P_{vdw} = 26.8$ atm
Integrative and Advanced Exercises 90. C_4H_6 **91.** 0.364 atm **94.** 2.24 atm **95. (a)** 12.5 atm **(b)** 0.373 atm **97.** 153 mmHg **98.** 19.9% He **100.** 22.0 L **104.** 1.95% H_2O by mass **106.** V = 7.39 L **109.** 30.1% Mg, 69.9% Al
Feature Problem 112. X = oxygen(16 u); nitryl fluoride = 2 atoms of O; nitrosyl fluoride = 1 atom of O; thionyl fluoride = 1 atom of O; sulfuryl fluoride = 2 atoms of O.

CHAPTER 7

Practice Examples 1A. 32.7 kJ **1B.** 4.89 kJ **2A.** 3.0×10^2 g **2B.** 37.9 °C **3A.** -3.83×10^3 kJ/mol **3B.** 6.26 kJ/°C **4A.** -65.2 kJ/mol **4B.** 33.4 °C **5A.** 114 J of work is done by the system. **5B.** 14.6 kJ of work is done on the system. **6A.** $+1.70 \times 10^2$ J **6B.** 179 J of work done by system. **7A.** 60.6 g **7B.** 0.15 kJ of heat evolved. **8A.** 3.31 kJ absorbed **8B.** 1.72 kg $H_2O(s)$ **9A.** -124 kJ **9B.** -2006 kJ **10A.** 6 C(graphite) + 13/2 $H_2(g)$ + $O_2(g)$ + 1/2 $N_2(g) \longrightarrow C_6H_{13}O_2N_2(s)$ **10B.** twice the negative of the value in Table 7.2: $+92.22$ kJ **11A.** -1367 kJ/mol **11B.** -2.5×10^3 kJ/mol fuel **12A.** -1273 kJ/mol **12B.** -184 kJ/mol **13A.** -112.3 kJ/mol AgI **13B.** -505.8 kJ/mol
Review Questions 4. (a) $+68.5$ kcal; **(b)** -177 kJ **5. (a)** 39.5 °C; **(b)** 57 °C **6. (a)** 1.7 J g⁻¹ °C⁻¹; **(b)** 21.9 °C **7.** $+49.1$ kJ **8.** 0.450 J g⁻¹ °C⁻¹ **9.** 46 °C **10. (a)** 0 J; **(b)** -236 J; **(d)** $+133$ J; **(d)** -416 J **11. (a)** -2.22×10^3 kJ/mol; **(b)** -5.90×10^3 kJ/mol; **(c)** -1.82×10^3 kJ/mol **12.** 4.98 kJ/°C. **13. (a)** 23.82 °C; **(b)** 29.60 °C. **14.** $C_5H_{10}O_5(g) + 5\,O_2(g) \longrightarrow 5\,CO_2(g) + 5\,H_2O(l)$ $\Delta H = -2.34 \times 10^3$ kJ/mol **15.** -284 kJ/mol Zn **16. (a)** endothermic; **(b)** $+18$ kJ/mol **17.** 26.0 °C **18. (a)** $N_2(g) + \frac{1}{2} O_2(g) \longrightarrow N_2O(g)$; **(b)** $S(rhombic) + O_2(g) + Cl_2(g) \longrightarrow SO_2Cl_2(l)$; **(c)** $CH_3CH_2COOH(l) + \frac{7}{2} O_2(g) \longrightarrow 3\,CO_2(g) + 3\,H_2O(l)$ **19. (a)** 65.59 kJ; **(b)** 3.65×10^3 kJ; **(c)** 1.45×10^3 kJ **20.** $+30.74$ kJ. **21.** -282.97 kJ. **22.** $-290.$ kJ **23.** $\Delta H_{rxn} = 2\,\Delta H_1 + 2\,\Delta H_2 - 3\,\Delta H_3$ **24. (a)** -55.7 kJ; **(b)** -1124 kJ **25.** $+30.6$ kJ **26.** -206.0 kJ/mol
Exercises 27. (a) Zn:0.385 J g⁻¹ °C⁻¹; **(b)** Pt:0.13 J g⁻¹ °C⁻¹; **(c)** Al:0.905 J g⁻¹ °C⁻¹ **29.** 5.4×10^2 °C **31.** 24.0 °C **33.** 2.3×10^2 J mol⁻¹ °C⁻¹ **35.** -2.49×10^5 kJ **37. (a)** 504 kg; **(b)** -6.21×10^5 kJ; **(c)** 2.90×10^3 L **39.** 1.4×10^3 kJ **41. (a)** -5×10^1 kJ/mol; **(b)** Use more solute and less water to increase ΔT **43.** 4.2×10^2 g **45.** -56 kJ/mol **47.** 39.6 g **49.** 2 g **51.** 7.72 kJ/°C **53.** 5.23 °C **55. (a)** 3.4 L atm; **(b)** 3.5×10^2 J; **(c)** 83 cal **57.** No work is done; nothing has been moved. **59. (a)** No pressure-volume work done. **(b)** $\Delta n_{gas} = -1$ mol: work is done on system by the surroundings. **(c)** $\Delta n_{gas} = +1$ mol: work is done on the surroundings by the system. **61.**

(a) 0 J (b) −562 J (c) 0.08 kJ **63. (a)** yes, gas does work (w = negative). **(b)** yes, gas absorbs heat energy so that q = −w **(c)** temperature of the gas stays the same **(d)** $\Delta U = 0$ (temperature is constant). **65.** impossible: The isothermal expansion of an ideal gas has $\Delta U = 0$ **67.** (c) **69. (a)** −2008 kJ/mol (b) $\Delta H = -2012$ kJ/mol **71.** −217.5 kJ **73.** −747.5 kJ. **75.** −35.8 kJ. **77.** Some compounds are more stable and some, less stable, than their elements. $\Delta H_f^\circ = 0$ is unlikely for a compound because the compound would be exactly as stable as the elements from which it is made. **79.** +202.4 kJ **81.** −1367 kJ **83.** −102.9 kJ **85.** −55 kJ **87.** 2.40×10^6 kJ

Integrative and Advanced Exercises 90. 3.5°C **91.** -1.95×10^3 kJ/mol $C_6H_8O_7$ **97.** Sewage gas produces more heat per liter at STP than coal gas **100.** 69.1% efficient **101.** 303 g $C_6H_{12}O_6$ **103.** 87.0% CH_4, 13.0% C_2H_6 (by volume) **106.** $C_2H_6O + 3 O_2 \longrightarrow$ $2 CO_2 + 3 H_2O$, $\Delta H = -1451$ kJ/mol **107.** 69 s **108.** −12 J

Feature Problems 111. (a) sp ht = 0.0113 + (23.9/atomic mass); **(b)** 110 u **3.** 62 u, likely Cu **112. (a)** 45.0 mL of 1.00 M NaOH(aq) and 15.0 mL of 1.00 M citric acid; **(b)** maximum heat is evolved at the stoichiometric point; **(c)** $H_3C_6H_5O_7(aq) +$ $3 OH^-(aq) \longrightarrow 3 H_2O(l) + C_6H_5O_7^{3-}(aq)$ **115. (a)** about −148 J; **(b)** 11 rectangles under the curve at 0.10 atm intervals from 2.40 atm (1.02 L) to 1.30 atm (1.88 L); total work done is the sum of the areas of the 11 rectangles; **(c)** maximum work = −152 J; **(d)** maximum work of compression is for a one-step compression at 2.40 atm, i.e., 209 J; minimum work is the negative of (c), +152 J; **(e)** $\Delta U = 0$; value of q is the negative of the work found in (d); **(f)** The value of q/T for the process in (c) is ΔS, the change in the state function, entropy (S), introduced in Chapter 20.

CHAPTER 8

Practice Examples 1A. 535 L **1B.** 7.5×10^3 L **2A.** 2 $PbO_2(s) + 4 HNO_3(aq) \longrightarrow 2 Pb(NO_3)_2(aq) +$ $O_2(g) + 2 H_2O(l)$ **2B.** 4 $Zn(s) + 10 HNO_3(aq)$ $\longrightarrow 4 Zn(NO_3)_2(aq) + NH_4NO_3(aq) + 3 H_2O(l)$ **3A.** Cathode: $2 H_2O(l) + 2 e^- \longrightarrow H_2(g) + 2 OH^-$ (aq), Anode: $4 OH^-(aq) \longrightarrow O_2(g) + 2 H_2O + 4 e^-$ **3B.** 2 $Li_2O_2(s) + 2 CO_2(g) \longrightarrow 2 Li_2CO_3(s) +$ $O_2(g)$.

Review Questions 4. (a) ozone; **(b)** N_2O; **(c)** KO_2; **(d)** calcium hydride; **(e)** magnesium nitride; **(f)** K_2CO_3; **(g)** $(NH_4)H_2PO_4$ **5. (a)** $C(s) + H_2O(g)$ $\longrightarrow CO(g) + H_2(g)$ and $CO(g) + H_2O(g) \longrightarrow$ $CO_2(g) + H_2(g)$; **(b)** almost pure C; **(c)** $CO(NH_2)_2$; **(d)** principally $CaCO_3$; **(e)** a mixture of $CO(g)$ and $H_2(g)$, produced by steam reforming of a hydrocarbon. **6. (a)** NO_2; **(b)** KO_2; **(c)** N_2O; **(d)** CaH_2; **(e)** CO **7. (a)** $2 KClO_3(s) \xrightarrow{\text{heat/MnO}_2} 2 KCl(s) + 3 O_2(g)$; **(b)** $NH_4NO_3(s) \xrightarrow{200\text{-}260\,°C} N_2O(g) + 2 H_2O(g)$; **(c)** $Zn(s) + 2 HCl(aq) \longrightarrow ZnCl_2(aq) + H_2(g)$; **(d)** $CaCO_3(s) + 2 HCl(aq) \longrightarrow CaCl_2(aq) + H_2O(l) +$ $CO_2(g)$ **8. (b)** NO_2; N is in the +4 oxidation state. **9. (a)** $LiH(s) + H_2O(l) \longrightarrow Li^+(aq) + OH^-(aq) +$ $H_2(g)$; **(b)** $C(s) + H_2O(g) \xrightarrow{\text{heat}} CO(g) + H_2(g)$; **(c)** $3 NO_2(g) + H_2O(l) \longrightarrow 2 HNO_3(aq) + NO(g)$ **10. (a)** $Mg(s) + 2 HCl(aq) \longrightarrow MgCl_2(aq) +$ $H_2(g)$; **(b)** $NH_3(g) + HNO_3(aq) \longrightarrow NH_4NO_3(aq)$; **(c)** $MgCO_3(s) + 2 HCl(aq) \longrightarrow MgCl_2(aq) +$ $H_2O(l) + CO_2(g)$; **(d)** $NaHCO_3(s) + HC_2H_3O_2(aq)$ $\longrightarrow NaC_2H_3O_2(aq) + H_2O(l) + CO_2(g)$ **11.** H_2SO_4

$(aq) + 2 NH_3(aq) \longrightarrow (NH_4)_2SO_4(aq)$ **12.** (8.4) $NH_3(g)$ reducing agent, $O_2(g)$ oxidizing agent; (8.8) $NO_2(g)$ reducing agent and also the oxidizing agent; (8.16) $C_8H_{18}(l)$ reducing agent, $O_2(g)$ oxidizing agent. (8.21) $CaH_2(s)$ reducing agent, $H_2O(l)$ oxidizing agent **13.** $3 NO_2(g) + H_2O(l) \longrightarrow 2 HNO_3(aq) + NO(g)$ **14.** $Fe_2O_3(s) + 3 H_2(g) \longrightarrow 2 Fe(s) + 3 H_2O(g)$ **15.** $3 Cu(s) + 2 NO_3^-(aq) + 8 H^+(aq) \longrightarrow 3 Cu^{2+}$ $(aq) + 2 NO(g) + 4 H_2O(l)$. **16.** 2 $KClO_3(s)$ $\xrightarrow{\text{heat/MnO}_2} 2 KCl(s) + 3 O_2(g)$; molecular oxygen gas is produced, $CaCO_3(s) \xrightarrow{\text{heat}} CaO(s) + CO_2(g)$; carbon dioxide gas is formed, $NH_4NO_3(s)$ $\xrightarrow{200\text{-}260\,°C} 2 H_2O(g) + N_2O(g)$; dinitrogen monoxide gas is formed. **17. (a)** NO_x compounds produced by internal combustion engines. **(b)** NO and chlorofluorocarbons react with O_3 in upper atmosphere. **(c)** Result of increased levels of atmospheric $CO_2(g)$ arising primarily from burning carbon containing fuels (and deforestation). **18.** Catalytic converters are smog control devices: they contain an oxidation catalyst that converts CO and hydrocarbons to CO_2 and water and a reduction catalyst that reduces NO to N_2. **20.** Advantages: H_2 is a pollution free and more efficient fuel; saves precious hydrocarbons for plastics/lubricants. Disadvantages: H_2 is expensive and difficult to store and must be kept away from oxidizing agents.

Exercises 21. Because of the ideal gas law $(V = nRT/P)$, volumes measured at the same T and P are proportional to moles. **23. (a)** $N_2(g) +$ $3 H_2(g) \rightleftharpoons 2 NH_3(g)$; **(b)** $4 NH_3(g) +$ $5 O_2(g) \xrightarrow{850°C/\text{Pt}} 4 NO(g) + 6 H_2O(g)$; **(c)** $2 NO(g) + O_2(g) \longrightarrow 2 NO_2(g)$ and $3 NO_2(g) +$ $H_2O(l) \longrightarrow 2 HNO_3(aq) + NO(g)$. **25.** 4 $HNO_3(l)$ $\longrightarrow 2 N_2O_4(g) + 2 H_2O(l) + O_2(g)$ **27.** 3.32 × 10^{10} kg N_2 **29.** 6×10^9 kg NO_x released. **31. (a)** $2 HgO(s) \xrightarrow{\text{heat}} 2 Hg(l) + O_2(g)$; **(b)** $2 KClO_4(s)$ $\xrightarrow{\text{heat}} 2 KClO_3(s) + O_2(g)$. **33.** CO_2 **35.** 3×10^{-5} mmHg **37.** 2 moles of $H_2(g)$ are produced for each mole of $O_2(g)$. **39.** 7.8×10^5 L air **41.** 0.40 g/L **43. (a)** $2 C_6H_{14}(l) + 19 O_2(g) \longrightarrow 12 CO_2(g) + 14 H_2O$ (l); **(b)** $PbO(s) + CO(g) \xrightarrow{\text{heat}} Pb(s) + CO_2(g)$; **(c)** $2 KOH(aq) + CO_2(g) \longrightarrow K_2CO_3(aq) + H_2O(l)$; **(d)** $MgCO_3(s) + 2 HCl(aq) \longrightarrow MgCl_2(aq) +$ $H_2O(l) + CO_2(g)$ **45.** $CO(g)$ will not form because this would require a reduction in the oxidation state of carbon and no strong reducing agent has been added. **47.** 1.27×10^5 kJ of heat evolved **49.** $\Delta H_{comb}^\circ CH_4(g) =$ -890.3 kJ/mol; $\Delta H_{comb}^\circ C_2H_6(g) = -1559.7$ kJ/mol; ΔH_{comb}° $C_3H_8(g) = -2219.9$ kJ/mol; ΔH_{comb}° $C_4H_{10}(g) = -2877.4$ kJ/mol **51. (a)** $2 Al(s) + 6 HCl$ $(aq) \longrightarrow 2 AlCl_3(aq) + 3 H_2(g)$; **(b)** $C_3H_8(g) +$ $3 H_2O(g) \longrightarrow 3 CO(g) + 7 H_2(g)$; **(c)** $MnO_2(s) +$ $2 H_2(g) \xrightarrow{\text{heat}} Mn(s) + 2 H_2O(g)$ **53. (a)** $CaH_2(s)$; **(b)** $CaH_2(s)$ **55.** $CH_4(g)$

Integrative and Advanced Exercises 58. 75.516% N_2, 23.139% O_2, 1.29% Ar, 0.055% CO_2 (all mass percent) **60.** ~19 °C **62.** CH_4 (−890.3 kJ/mol CO_2) is better than C_8H_{18}(−683.8 kJ/mol CO_2) **64.** carbon suboxide; C_3O_2, malonic acid $C_3O_4H_4$ **66.** $[NO_3^-] = 0.1716$ M **69.** 3.0×10^{-20} J/atom **Feature Problems 72. (a)** N_2 from liquid air contains some Ar. **(b)** N_2 from liquid air; dinitrogen from liquid air contains some argon. **(c)** Mg does not react with Ar. **(d)** 0.50% **73.** First, balance each equation.

Double the first equation and add the result to the second equation and three times the third equation to give the net reaction $2 H_2O(g) \longrightarrow 2$ $H_2(g) + O_2(g)$

CHAPTER 9

Practice Examples 1A. 4.34×10^{14} Hz **1B.** 3.28 m **2A.** 400 to 520 kJ/mol **2B.** 4.612×10^{14} Hz (orange), 6.662×10^{14} Hz (indigo). We see the reflected color, principally green. **3A.** E_9 for $n = 9$ **3B.** no **4A.** 486.2 nm **4B.** 121.6 nm or 1216 Å **5A.** 80.13 nm **5B.** Be^{3+} **6A.** 1.21×10^{-43} m **6B.** 39.6 km/s **7A.** 6.4×10^{-43} m **7B.** 1.3 m/s. **8A.** yes **8B.** $\ell = 2$ or 1 **9A.** 3p **9B.** $n = 3$; $\ell = 0, 1, 2$; $m_\ell = -2, -1, 0, 1, 2$ **10A.** Ti **10B.** $1s^2 2s^2 2p^6$ $3s^2 3p^6 3d^{10} 4s^2 4p^6 4d^{10} 5s^2 5p^5$; ten 3d electrons; one unpaired electron.

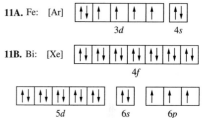

11A. Fe: [Ar] — 3d, 4s

11B. Bi: [Xe] — 4f, 5d, 6s, 6p

12A. (a) 5; **(b)** 6; **(c)** 0; **(d)** 2 **12B. (a)** 10; **(b)** 2; **(c)** 1 **Review Questions 4. (a)** 162.5 nm; **(b)** 0.388 μm; **(c)** 7.27×10^6 nm; **(d)** 5.46×10^{-7} m; **(e)** 1.12×10^7 nm; **(f)** 2.6×10^{-4} cm **5. (a)** 4.4×10^{-5} m, infrared; **(b)** 3.1×10^{-8} m, ultraviolet; **(c)** 11.8 m, radio; **(d)** 2.80 m, radio **6.** 4.68 nm **7. (a)** 6.41×10^{16} Hz; **(b)** 4.25×10^{-17} J **8. (a)** independent; **(b)** inversely; **(c)** directly **9. (a)** 6.9050×10^{14} s^{-1}; **(b)** 397.11 nm; **(c)** $n = 10$ **10.** $v = 3.2881 \times 10^{15}$ s^{-1} $\left[(1/3^2) - (1/n^2)\right]$ where $n = 4, 5, ...$ **11. (a)** 5.71×10^{-18} J/photon; **(b)** 61.1 kJ/mol **12. (a)** 6.31×10^{12} Hz; **(b)** 557 nm **13.** $\Delta E = -1.550 \times 10^{-19}$ J; 2.339×10^{14} s^{-1} **14. (a)** Particle with smallest mass has longest wavelength. **15. (a)** $m_\ell = 0, \pm1$; **(b)** $\ell = 1, 2, 3$; **(c)** $n = \geq 2$ **16. (a)** $n = 4, \ell = 0$; **(b)** $n = 3, \ell = 1$; **(c)** $n = 5, \ell = 3$; **(d)** $n = 3, \ell = 2$ **17. (a)** permitted **(b)** not permitted, $\ell > n$; **(c)** not permitted, $|m_\ell| > \ell$; **(d)** permitted; **(e)** not permitted, $\ell = n$. **(f)** permitted **18. (a)** one; **(b)** no such subshell; **(c)** three; **(d)** five; **(e)** seven; **(f)** three **19. (a)** $1s^2 2s^2 2p^6 3s^2 3p^6 3d^{10} 4s^2 4p^5$; **(b)** $1s^2 2s^2 2p^6 3s^2 3p^4$; **(c)** $1s^2 2s^2 2p^6 3s^2 3p^6 3d^{10} 4s^2 4p^6 4d^{10} 5s^2 5p^3$; **(d)** $1s^2 2s^2 2p^6 3s^2 3p^2$ **20. (a)** 0; **(b)** 1; **(c)** 2; **(d)** 1 **21. (a)** those in groups 1(1A) and 2(2A), along with Al, Ga, In, Tl, Sn, Pb, and Bi; **(b)** H, F, Cl, Br, I, O, S, Se, N, P, C, and B; **(c)** He, Ne, Ar, Kr, Xe, and Rn; **(d)** those in groups 3 – 12 (or 1B – 8B); **(e)** those with atomic numbers from $Z = 58$ (Ce) through $Z = 71$ (Lu) and those from $Z = 90$ (Th) through $Z = 103$ (Lr). **22. (a)** In [Kr] $4d^{10} 5s^2 5p^1$; **(b)** Cd [Kr] $4d^{10} 5s^2$; **(c)** Sb[Kr] $4d^{10}$ $5s^2 5p^3$; **(d)** Au[Xe] $4f^{14} 5d^{10} 6s^1$ **23. (a)** 1; **(b)** 5; **(c)** 10; **(d)** 6; **(e)** 14; **(f)** 8 **24.** Reading left to right: 1. (i), (f), (l); 2. (j), (g); 3. (b), (c); 4. (a), (h), (k)

Exercises 25. (a) and **(d)** are true. **27.** (c). **29.** 8.3 min **31.** 656.46 nm, 486.26 nm, 434.16 nm, 410.28 nm **33.** $n = 8$ **35. (a)** 4.19×10^{-19} J/photon; **(b)** 252 kJ/mol. **37. (b)** has the greatest energy and **(d)** has the least energy. **39.** 3.03×10^{16} s^{-1}, ultraviolet. **41. (a)** 6.60×10^{-19} J/photon; **(b)** ultraviolet yes, not infrared **43. (a)** 1.9 nm; **(b)** -6.053×10^{-20} J **45. (a)** 1.384×10^{14} s^{-1}; **(b)** 2166 nm; **(c)** infrared **47. (a)** 8.5×10^{-9} m; **(b)** no orbit at 4.00 Å; **(c)** $n = 8$, energy is -3.405×10^{-20} J (relative to $E_\infty = 0$ J); **(d)** No -2.50×10^{-17} J not an allowed energy level. **49.** $n = 7$ to $n = 4$ **51.** electrons **53.** 9.79×10^{-35} m **55.** The electron's position and velocity are both exactly known in the Bohr theory. This is not allowed by the Heisenberg uncertainty principle. **57.** $\sim 1 \times 10^{-13}$ m **59.** 1.4×10^7 m s^{-1} **61.** 17 cm **63.** 16 inches **65.** Discrete pathways versus regions of high probability. Both theories give the same value of the most probable distance of the electron from the nucleus. **67. (c) 69. (a)** $5p$; **(b)** $4d$; **(c)** $2s$ **71. (a)** one; **(b)** two; **(c)** ten; **(d)** eighteen; **(e)** five **73.** 106 pm **75.** For $2p_y$ wavefunction is $Y(\theta, \phi)_{py} = (0.75/\pi)^{1/2} \sin\theta \sin\phi$, for all points in the xz plane, $\phi = 0$, and since $\sin 0° = 0$, entire xy plane is a node(zero probability of finding the p_y electron in xz plane).

77.

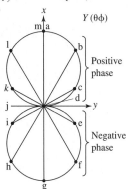

$Y(\theta\phi)$

Positive phase

Negative phase

79.

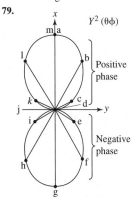

$Y^2(\theta\phi)$

Positive phase

Negative phase

81. A plot of radial probability distribution vs r/a_0 for a $1s$ orbital gives a curve with a maximum at 1.0 $(r = a_0)$; maximum probability for electron in H atom: 0.53 pm. **83.** Configuration (b) is correct, (a) violates the Pauli exclusion principle and (c) and (d) violate Hund's rule **85. (a)** 3; **(b)** 10; **(c)** 2; **(d)** 2; **(e)** 14 **87. (a)** three; **(b)** one; **(c)** zero; **(d)** fourteen; **(e)** two; **(f)** five; **(g)** thirty two **89. (a)** Pb:[Xe] $4f^{14} 5d^{10}$ $6s^2 6p^2$; **(b)** 114:[Rn] $5f^{14} 6d^{10} 7s^2 7p^2$
Integrative and Advanced Exercises 93. (a) 275 nm **(b)** 2.02×10^{-19} J **(c)** 6.66×10^5 m/s **94.** 1.0×10^{20} photons/sec **95.** $\lambda = 525$ nm, green **98.** $m = 3$, $n = 4$ **101.** $n = 138$ **104.** 4.8 mJ **105.** 7×10^2 photons/sec **108.** 7.29×10^5 m/sec, 2.43×10^{14} sec^{-1}
Feature Problems 113. A plot of frequency versus n^{-2} gives a straight line plot with a slope $(-3.291 \times 10^{15}$ Hz) and y-intercept $(8.224 \times 10^{14}$ Hz), virtually identical to that predicted by the Balmer equation.
115. (a) Two lines would be observed in the absorption spectrum $(n = 1 \longrightarrow n = 2$ at 121.6 nm and $n = 1 \longrightarrow n = 3$ at 102.6 nm) **(b)** Six emission lines should be seen for the transitions $n = 4$ to $n = 3$ (1875 nm), $n = 4$ to $n = 2$ (486.2 nm),

$n = 4$ to $n = 1$ (97.24 nm); from $n = 3$ to $n = 2$ (656.4 nm), $n = 3$ to $n = 1$ (102.6 nm); and from $n = 2$ to $n = 1$ (121.6 nm). **(c)** Two lines are common to the absorption and emission spectra.

CHAPTER 10

Practice Examples 1A. S **1B.** Ca **2A.** $V^{3+} < Ti^{2+} < Ca^{2+} < Sr^{2+} < Br^-$ **2B.** As **3A.** $K < Mg < S < Cl$ **3B.** Sb **4A.** Cl and Al are paramagnetic. **4B.** Cr^{2+} **5A.** 280 K **5B.** 570 K
Review Questions 4. Ar-K, Te-I, Co-Ni; order is based on atomic number, not atomic mass. **5. (a)** 50; **(b)** 69; **(c)** 10; **(d)** 2; **(e)** 2; **(f)** 4 **6.** Fe^{2+} and Co^{3+}, Sc^{3+} and Ca^{2+}, F^- and Al^{3+}, Zn^{2+} and Cu^+ **7.** Two neutral atoms may not be isoelectronic, but the other combinations may be. **8. (a)** fluorine; **(b)** scandium; **(c)** silicon **9. (a)** Te; **(b)** K; **(c)** Cs; **(d)** N; **(e)** P; **(f)** Au **10.** smallest: F; largest: I^- **11.** $Cs < Sr < As < S < F$ **12. (a)** C; **(b)** Rb; **(c)** At **13.** 21.6 J **14. (a)** Ba; **(b)** S; **(c)** Bi **15.** $Rb > Ca > Sc > Fe > Te > Br > O > F$ **16.** only Cr^{3+}, Co^{3+}, and Br are paramagnetic **17. (a)** 6; **(b)** 8; **(c)** 5; **(d)** 1; **(e)** 2; **(f)** 4 **18.** 210 K
Exercises 19. 14 to 16 g/cm^3 **21.** Plot of density vs. atomic number has similar shape for both periods. **23.** Moseley permitted only integral atomic numbers. **25. (a)** 118; **(b)** 119; **(c)** about 297 u and 300 u respectively **27.** Electrons are being added to the same subshell in some instances but to a new subshell in others. **29. (a)** B; **(b)** Te **31.** yes, like Cu^+ and Zn^{2+} **33.** $Li^+ < Br < Se < I^-$ **35.** No, the second electron is being removed from a positive ion. **37.** 9952 kJ **39.** −456 kJ, exothermic. **41.** Similar answer to Exercise 35. **43.** Fe^{2+} **45.** All atoms with odd Z are paramagnetic; some with even Z are also paramagnetic, for example, C. **47. (a)** Metals that should exhibit the photoelectric effect: Cs, Rb, and K (small first ionization energy); Zn, Cd, Hg should not (relatively large first ionization energies); **(b)** Rn; **(c)** +600 kJ/mol; **(d)** 5.7 g/cm^3 **49. (a)** 5.6 g/cm^3; **(b)** 74% Ga, Ga_2O_3 **51. (a)** Ba; **(b)** Sr; **(c)** Cl; **(d)** N.
Integrative and Advanced Exercises 55. (b) and **(e)** are best. **58.** Data yield an atomic mass of 75.7 μ, indicating Gruppe V or Gruppe VI. **60.** Al, same or larger, smaller;In, larger, smaller; Se, smaller, larger **62.** BrCl gas, ICl liquid **65.** Yes **68.** 5251 kJ mol **69.** I_1 for F(1681) $\sim$ I_2 for Ba(965) $<$ I_3 for Sc(2389) $<$ I_2 for Na(4562) $<$ I_3 for Mg(7733)
Feature Problems 73. A plot of $\nu^{1/2}$ versus Z provides $A = 2.482 \times 10^{15}$ Hz and $b = 0.969$; value for A is close to R_H (equivalent term in the Moseley equation). The constant b represents the screening afforded by the remaining electron in the K shell. **74. (a)** For $[Ne]3p^1$, the first ionization energy is 293 kJ. For $[Ne]4s^1$, the first ionization energy is 188 kJ. For $[Ne]3d^1$, the first ionization energy is 147 kJ. for $[Ne]4p^1$, the first ionization energy is 134 kJ. **(b)** For $[Ne]3p^1$, $Z_{eff} = 1.42$, for $[Ne]4s^1$, $Z_{eff} = 1.51$, for $[Ne]4p^1$, $Z_{eff} = 1.28$, for $[Ne]3d^1$, $Z_{eff} = 1.00$ **(c)** For $[Ne]3p^1$, $\bar{r}_{3p} = 4.7 \times 10^2$ pm, for $[Ne]4s^1$, $\bar{r}_{4s} = 8.4 \times 10^2$ pm, for $[Ne]4p^1$, $\bar{r}_{4p}$ 9.5×10^2 pm, for $[Ne]3d^1$, $\bar{r}_{3d} = 5.6 \times 10^2$ pm. **(d)** Calculations show that as expected, the more deeply the orbital penetrates, the greater is its Z_{eff} and hence, the closer it is to the nucleus on average.

CHAPTER 11

Practice Examples 1A. $\cdot Mg \cdot$ $\cdot \ddot{Ge} \cdot$ $K \cdot$ $:\ddot{Ne}:$
1B. $\cdot \ddot{Sn} \cdot$ $[:\ddot{Br}:]^-$ $[\cdot Tl \cdot]^+$ $[:\ddot{S}:]^{2-}$

2A. (a) $[Na]^+ [:\ddot{\underset{..}{S}}:]^{2-} [Na]^+$

(b) $[Mg]^{2+}[:\ddot{N}:]^{3-}[Mg]^{2+}[:\ddot{N}:]^{3-}[Mg]^{2+}$

2B. (a) $[:\ddot{I}:]^-[Ca]^{2+}[:\ddot{I}:]^-$ **(b)** $[Ba]^{2+}[:\ddot{S}:]^{2-}$

(c) $[Li]^+[:\ddot{\underset{..}{O}}:]^{2-}[Li]^+$

3A.

$$H - \underset{\displaystyle |}{\overset{\displaystyle |}{C}} - H$$

$:\ddot{Br} - \ddot{Br}:$ $H - \ddot{\underset{..}{O}} - \ddot{\underset{..}{Cl}}:$

3B. $:\ddot{I} - \underset{\displaystyle :\ddot{I}:}{\overset{\displaystyle ..}{N}} - \ddot{I}:$ $H - \underset{\displaystyle |}{\overset{\displaystyle |}{N}} - \underset{\displaystyle |}{\overset{\displaystyle |}{N}} - H$ $H - \underset{\displaystyle |}{\overset{\displaystyle |}{C}} - \underset{\displaystyle |}{\overset{\displaystyle |}{C}} - H$

4A. P—Cl or N—H **4B.** P—O

5A. (a) $\ddot{S} = C = \ddot{S}$ **(b)** $H - C \equiv N:$

(c) $:\ddot{Cl} - \underset{\displaystyle :O:}{\overset{\displaystyle ..}{C}} - \ddot{Cl}:$

5B. (a) $H - \underset{\displaystyle :O:}{\overset{\displaystyle ||}{C}} - \ddot{O} - H$ **(b)** $H - \underset{\displaystyle |}{\overset{\displaystyle :O:}{C}} - C - H$

6A. (a) $[:N \equiv O:]^+$ **(b)** $\left[H - \underset{\displaystyle |}{\overset{\displaystyle |}{N}} - N: \right]^+$ **(c)** $[:\ddot{\underset{..}{O}}:]^{2-}$

6B. (a) $\left[:\ddot{F} - \underset{\displaystyle :\ddot{F}:}{\overset{\displaystyle :\ddot{F}:}{B}} - \ddot{F}: \right]^-$ **(b)** $\left[H - \underset{\displaystyle |}{\overset{\displaystyle |}{N}} - \ddot{O} - H \right]^+$

(c) $[:N \equiv C - \ddot{O}:]^-$ or $[:\ddot{N} = C = \ddot{O}:]^-$

7A. $:\ddot{N} = \underset{\displaystyle \ominus}{\overset{\displaystyle}{O}} - \underset{\displaystyle \oplus}{\overset{\displaystyle}{\ddot{Cl}}}:$

Structure is less plausible because it has a positive formal charge on oxygen, the most electronegative atom in the molecule.

7B. $H - \underset{\displaystyle |}{\overset{\displaystyle H}{N}} - C \equiv N:$ $H - \underset{\displaystyle |}{\overset{\displaystyle H}{N}} = C = \overset{\displaystyle \oplus}{\ddot{N}}:$

8A. $:\underset{\displaystyle \ominus}{\ddot{O}} - \underset{\displaystyle \oplus}{\overset{\displaystyle}{S}} = \ddot{O}: \longleftrightarrow :\ddot{O} = \underset{\displaystyle \oplus}{\overset{\displaystyle}{S}} - \underset{\displaystyle \ominus}{\ddot{O}}:$

8B. $:\underset{\displaystyle \ominus}{\ddot{O}} - \underset{\displaystyle}{\overset{\displaystyle :O:}{N}} - \ddot{O}:^{\ominus} \longleftrightarrow :\ddot{O} = \underset{\displaystyle}{\overset{\displaystyle :O:}{N}} - \ddot{O}:^{\ominus}$

$\longleftrightarrow :\underset{\displaystyle \ominus}{\ddot{O}} - \underset{\displaystyle}{\overset{\displaystyle :\ddot{O}:^{\ominus}}{N}} = \ddot{O}:$

9A. trigonal-pyramidal **9B.** tetrahedral **10A.** linear **10B.** linear

11A.

$$H \cdots \underset{\displaystyle H}{\overset{\displaystyle H}{C}} \diagdown \underset{\displaystyle \ddot{O}:}{\diagup} \diagdown H$$

The H—C—H and H—C—O bond angles are $\sim 109.5°$; C—O—H bond angle is slightly less than $109.5°$.

11B.

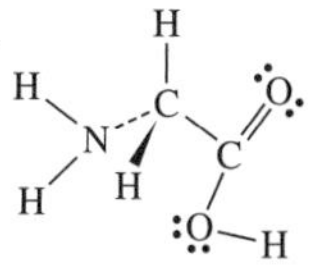

H H
 \ |
H—N—C—C=$\ddot{\text{O}}$
     | |
     H $\ddot{\text{O}}$—H

The H—C—N, H—C—H and H—C—C angles are ~109.5°. C—O—H bond angle is slightly less than 109.5° The O—C—O and O—C—C bond angles are ~120°.

12A. H_2O_2: The individual bond moments do not cancel out. **12B.** PCl_5: The individual bond moments cancel out. **13A.** C—H bond: ~110 pm, C—Br bond: ~191 pm **13B.** $\ddot{\text{O}}$=C=$\ddot{\text{O}}$ **14A.** Enthalpy change = −486 kJ **14B.** ΔH_f = −4 × 10¹ kJ/mol NH_3 **15A.** exothermic **15B.** The reaction is endothermic.

Review Questions

4. (a) [H:]⁻ **(b)** :$\ddot{\text{Kr}}$: **(c)** [·Sn·]²⁺ **(d)** [K]⁺

(e) [:$\ddot{\text{Br}}$:]⁻ **(f)** ·$\dot{\text{Ge}}$· **(g)** :$\dot{\text{N}}$· **(h)** ·Ca·

(i) [:$\ddot{\text{Se}}$:]²⁻ **(j)** [Sc]³⁺

5. (a) [:$\ddot{\text{Cl}}$:]⁻ [Ca]²⁺ [:$\ddot{\text{Cl}}$:]⁻ **(b)** [Ba]²⁺ [:$\ddot{\text{S}}$:]²⁻

(c) [Li]⁺ [:$\ddot{\text{O}}$:]²⁻ [Li]⁺ **(d)** [Na]⁺ [:$\ddot{\text{F}}$:]⁻

(e) [Mg]²⁺ [:$\ddot{\text{N}}$:]³⁻ [Mg]²⁺ [:$\ddot{\text{N}}$:]³⁻ [Mg]²⁺

6. (a) :$\ddot{\text{I}}$—$\ddot{\text{Cl}}$: **(b)** :$\ddot{\text{Br}}$—$\ddot{\text{Br}}$: **(c)** :$\ddot{\text{F}}$—$\ddot{\text{O}}$—$\ddot{\text{F}}$:

(d) :$\ddot{\text{I}}$—$\dot{\text{N}}$—$\ddot{\text{I}}$: **(e)** H—$\ddot{\text{Se}}$—H
          |
        :$\ddot{\text{I}}$:

7. (a) $\ddot{\text{S}}$=C=$\ddot{\text{S}}$ **(b)**
H   :O: H
 | || |
H—C—C—C—H
 | |
 H   H

(c)
      :O:
       ||
:$\ddot{\text{Cl}}$—C—$\ddot{\text{Cl}}$: **(d)** :$\ddot{\text{F}}$—$\dot{\text{N}}$=$\ddot{\text{O}}$:

8. (a) H—H—$\dot{\text{N}}$—$\ddot{\text{O}}$—H has two bonds to (four electrons around) the second hydrogen, and only three bonds to (six electrons around) the nitrogen.

(b) :$\ddot{\text{O}}$—$\ddot{\text{Cl}}$—$\ddot{\text{O}}$: has 20 valence electrons, whereas the molecule ClO_2 has 19 valence electrons.

(c) [·$\dot{\text{C}}$=$\dot{\text{N}}$:]⁻ has only six electrons around the C atom

(d) Ca—$\ddot{\text{O}}$: is improperly written as a covalent Lewis structure.

9. (a) [:$\ddot{\text{O}}$—H]⁻ Ca²⁺ [:$\ddot{\text{O}}$—H]⁻

(b)
$\left[\begin{array}{c} H \\ | \\ H-N-H \\ | \\ H \end{array} \right]^{+}$ [:$\ddot{\text{Br}}$:]⁻

(c) [:$\ddot{\text{O}}$—$\ddot{\text{Cl}}$:]⁻ Ca²⁺ [:$\ddot{\text{O}}$—$\ddot{\text{Cl}}$:]⁻

10.

(a) formal charge
H—	—C≡	≡C
0	0	−1

(b) formal charge
=O	—O	central C
0	−1	0

(c) formal charge
—H	side C	central C
0	0	+1

(d) The formal charge on each I is 0

(e) formal charge
=O	—O	=S—
0	−1	+1

(f) formal charge
=O	—O	= N—
0	−1	+1

11. (a) OH⁻ diamagnetic **(b)** OH paramagnetic **(c)** NO_3 paramagnetic **(d)** SO_3 diamagnetic **(e)** SO_3^{2-} diamagnetic **(f)** HO_2 paramagnetic

12. (a) :$\ddot{\text{Cl}}$—$\ddot{\text{O}}$—$\ddot{\text{Cl}}$:

(b) :$\ddot{\text{F}}$—P—$\ddot{\text{F}}$:
          |
        :$\ddot{\text{F}}$:

(c)
$\left[:\ddot{\text{O}}—C=\ddot{\text{O}}: \right]^{2-}$
          |
        :$\ddot{\text{O}}$:

(d)
        :F: :F:
          \   /
   :$\ddot{\text{F}}$—Br—$\ddot{\text{F}}$:
          |
        :$\ddot{\text{F}}$:

13. (a) bent **(b)** planar **(c)** linear **(d)** octahedral **(e)** tetrahedral **14. (a)** bent **(b)** trigonal pyramidal **(c)** tetrahedral **15. (a)** linear **(b)** tetrahedral **(c)** trigonal pyramidal **(d)** T-shaped **(e)** trigonal bipyramidal **(f)** bent **(g)** octahedral **16.** H—C 110 pm, 414 kJ/mol, C=O 120 pm, 736 kJ/mol; C—C 154 pm, 347 kJ/mol; C—Cl 178 pm, 339 kJ/mol. **17. (c) 18.** endothermic **19. (a)** endothermic; **(b)** exothermic **20.** Bi **21.** C—H < Br—H < F—H < Na—Cl < K—F **22.** polar: **(b), (d), (e), (g)**; bond dipoles do not cancel out in polar molecules.

Exercises 23. H_2, ICl_3, BF_3

25. (a) Cs⁺ [:$\ddot{\text{Br}}$:]⁻ **(b)** H—$\ddot{\text{Sb}}$—H
                                         |
                                         H

(c) :$\ddot{\text{Cl}}$—B—$\ddot{\text{Cl}}$: **(d)** Cs⁺ [:$\ddot{\text{Cl}}$:]⁻
          |
        :$\ddot{\text{Cl}}$:

(e) Li⁺ [:$\ddot{\text{O}}$:]²⁻ Li⁺ **(f)** :$\ddot{\text{I}}$—$\ddot{\text{Cl}}$:

27. (c) is correct. **(a)** C does not have an octet, **(b)** neither C has an octet, **(d)** number of valence electrons is incorrect.

29. (a) Li⁺ [:$\ddot{\text{S}}$:]²⁻ Li⁺ **(b)** Na⁺ [:$\ddot{\text{F}}$:]⁻

(c) [:$\ddot{\text{I}}$:]⁻ Ca²⁺ [:$\ddot{\text{I}}$:]⁻

(d) [:$\ddot{\text{Cl}}$:]⁻ Sc³⁺ [:$\ddot{\text{Cl}}$:]⁻
                    [:$\ddot{\text{Cl}}$:]⁻

31. Oxidation state is fixed, while formal charge is variable. Formal charges are used to decide between alternative Lewis structures, while oxidation state is used in balancing equations and naming compounds. The oxidation state of an element in its compounds is usually not zero, while its formal charge often is. **33. (a)** +1; **(b)** −1; **(c)** +1; **(d)** 0; **(e)** −1. **35.** No, based on formal charge rules alone, structures **(a)** and **(b)** are equally plausible:

(a) $\ddot{\text{O}}$=C=$\overset{\oplus}{\ddot{\text{O}}}$—H **(b)** :O≡C—$\overset{\oplus}{\ddot{\text{O}}}$—H

37. (a) H—$\dot{\text{N}}$—$\ddot{\text{O}}$—H **(b)** :$\ddot{\text{O}}$—$\ddot{\text{Cl}}$—$\ddot{\text{O}}$—H
          | |
          H   :$\ddot{\text{O}}$:

(c) H—$\ddot{\text{O}}$—$\dot{\text{N}}$=$\ddot{\text{O}}$ **(d)**
                              :$\ddot{\text{O}}$:
                               |
                  :$\ddot{\text{Cl}}$—S—$\ddot{\text{Cl}}$:
                               |
                              :$\ddot{\text{O}}$:

39.
(a) $\left[:\ddot{\text{O}}—S—\ddot{\text{O}}: \right]^{2-}$
                |
              :$\ddot{\text{O}}$:

(b) [:$\ddot{\text{O}}$—$\dot{\text{N}}$=$\ddot{\text{O}}$]⁻ ⟷ [:O=$\dot{\text{N}}$—$\ddot{\text{O}}$:]⁻

(c) $\left[:\ddot{\text{O}}—C=\ddot{\text{O}} \right]^{2-}$ ⟷ $\left[\ddot{\text{O}}=C—\ddot{\text{O}}: \right]^{2-}$
             | |
           :$\ddot{\text{O}}$:   :$\ddot{\text{O}}$:

⟷ $\left[:\ddot{\text{O}}—C—\ddot{\text{O}}: \right]^{2-}$
                 ||
               :$\ddot{\text{O}}$:

(d) [H—$\ddot{\text{O}}$—$\ddot{\text{O}}$:]⁻ **41.**
  H H H :O:
   | | | ||
H—C—C=C—C—H
   |
   H

43. (a)
     :O:
      ||
   H—C—H

(b)
  H :$\ddot{\text{Cl}}$: H
   | | |
H—C—C—C—$\ddot{\text{O}}$—H
   | | |
   H   H   H

45. (a)
          H :O:
           | ||
:$\ddot{\text{Cl}}$—C—C—$\ddot{\text{O}}$—H
           |
           H

(b)
      H H :O:
       | | ||
H—$\ddot{\text{O}}$—C—C—C—$\ddot{\text{O}}$—H
       | |
       H H

47. (a) ≈4%; **(b)** ≈5%; **(c)** ≈60%; **(d)** ≈33%

49. [:$\ddot{\text{O}}$—$\dot{\text{N}}$=$\ddot{\text{O}}$]⁻ ⟷ [:O=$\dot{\text{N}}$—$\ddot{\text{O}}$:]⁻

51. The molecule seems best represented as a resonance hybrid of (1) and (2); (3) contains a very long N—N bond, while (4) contains no N—N bonds.

53. (a) H—$\dot{\text{C}}$—H **(b)** :$\ddot{\text{O}}$—$\ddot{\text{Cl}}$—$\ddot{\text{O}}$:
            |
            H

(c) ·$\ddot{\text{O}}$—N=$\ddot{\text{O}}$
          |
        :$\ddot{\text{O}}$:

55. ICl_3 and SF_4

57. (a) linear; **(b)** linear; **(c)** tetrahedral; **(d)** trigonal; **(e)** bent shape **59.** CO_3^{2-} VSEPR class: AX₃

61. (a) $\ddot{\text{O}}$=C=$\ddot{\text{O}}$ **(b)** $\ddot{\text{O}}$=C—$\ddot{\text{Cl}}$:
       Linear |
                                  :$\ddot{\text{Cl}}$:

                            Trigonal planar

(c) 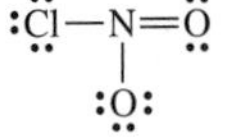 :C̈l—N=Ö̈
      |
     :Ö:

Trigonal planar

63. (a) tetrahedral; **(b)** tetrahedral; **(c)** octahedral; **(d)** linear **65.** tetrahedral
67.

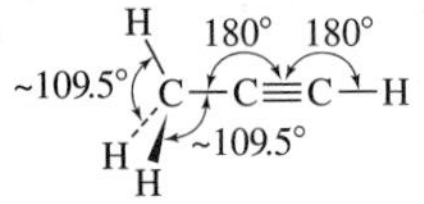

H   180° 180°
~109.5° \\C=C≡C—H   A maximum of
   H/ | ~109.5° 5 atoms can be
       H   in the same plane.

69.

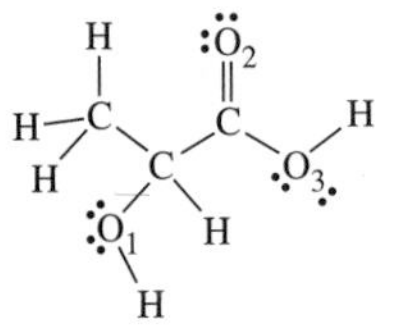

All angles ~109°
except:
H—O_1—C <109°
H—O_3—C <109° A maximum of
O_2—C—C ~120° 7 atoms can be
O_3—C—C ~120° in the same plane.
O_2—C—O_3 ~120°

71. (a) bent, polar; **(b)** trigonal pyramidal, polar; **(c)** bent, polar; **(d)** planar, nonpolar; **(e)** octahedral, non-polar; **(f)** tetrahedral, polar **73.** H_2O_2 cannot be linear(dipole moments cancel in a linear structure). A skewed, chain structure would produce a non-zero molecular dipole moment. **75. (a)** I—Cl bond: 233 pm; **(b)** O—Cl bond: 172 pm; **(c)** C—F bond: 149 pm; **(d)** C—Br bond: 191 pm **77.** N—F bond length = 144 pm **79.** ΔH = −113 kJ/mol **81. (a)** ΔH_f° = 3 kJ/mol or 39 kJ/mol depending on value used for O—H bond energy **(b)** ΔH_f° = 99kJ/mol **83.** ΔH_{rxn}° = −166 kJ

Integrative and Advanced Exercises
85. ΔH = −744 kJ/mol **88.** molar mass = 42.0 g/mol, $H_3CHC=CH_2$ **90.** $H_2C=C=CH_2$ and $H_3C—C≡CH$ **94.** structure: H—N=N=N, shape has 3 Ns in a line, with 120° H—N—N bond angle. **97.** bent **100.** ΔH (1) = +717 kJ, ΔH (2) = −614 kJ, ΔH(net) = +103 kJ **101.** Bond energy [O—O] = 142 kJ **103.** 307 kJ/mol
Feature Problems 106. (a) 91 kJ/mol; **(b)** 0.97; **(c)** ~23% ionic **107. (a)** 1.49 D; **(b)** 92°; **(c)** 97.5°

CHAPTER 12

Practice Examples 1A. trigonal-pyramidal **1B.** trigonal-pyramidal. VSEPR has 109.5° bond angles, valence bond has 90° angles. **2A.** Bent molecular geometry; sp^3-hybridized central Cl atom **2B.** Seesaw molecular geometry; dsp^3-hybridized Br atom **3A.** All central atoms have sp^3 hybridization, approximately 109.5° bond angles. **3B.** The C=O carbon is sp^2-hybridized with 120° bond angles; all other central atoms are sp^3-hybridized with 109.5° bond angles. **4A.** The triple-bonded C is sp-hybridized with 180° bond angles; the other C is sp^3-hybridized with 109.5° bond angles. **4B.**

:N≡N—Ö̈: ⟷ :N̈=N=Ö: the central N is sp-hybridized, with 180° bond angles. **5A.** 53 kJ/mol. **5B.** yes, bond order = 0.5. **6A.** bond orders: **(a)** 2.5; **(b)** 0.5; **(c)** 3.0. **6B.** bond orders: **(a)** 2.5; **(b)** 2.0; **(c)** 1.5; **(d)** 1.0. Bond length increases with decreasing bond order. **7A.**

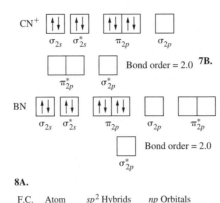

CN⁺   Bond order = 2.0 **7B.**

BN   Bond order = 2.0

8A.

F.C. Atom Label sp^2 Hybrids np Orbitals
+2 S   (3p)
0   O   (2p)
−1 O   (2p)
−1 O   (2p)

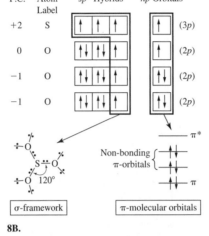

σ-framework   π-molecular orbitals

8B.

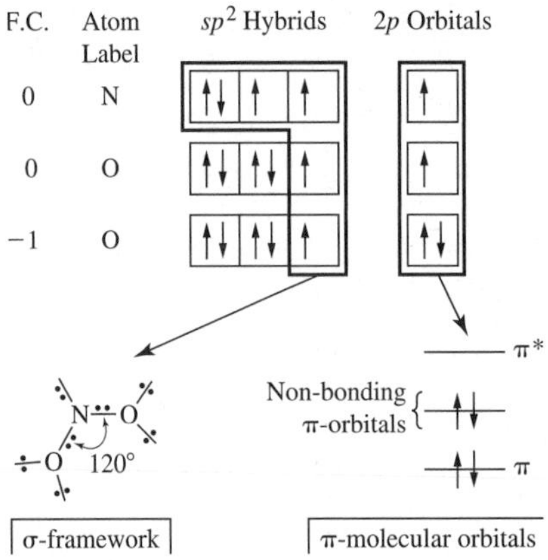

F.C. Atom Label sp^2 Hybrids $2p$ Orbitals
0 N
0 O
−1 O

σ-framework   π-molecular orbitals

Review Questions 4. 1s on H and 4p on Se **5.** CO_3^{2-}, SO_2, and NO_2^- **6. (b)** is true **7.** overlaps are: **(a)** H(1s)-I(5p); **(b)** Br(4p)-Cl(3p); **(c)** Se(4p)-H(1s); **(d)** O(2p)-Cl(3p) **8.** Experimental 120° bond angles contradict predicted 90° p-orbital angles. **9. (a)** C(sp^3) overlaps with H(1s) or Cl(3p), tetrahedral molecule; **(b)** sp hybrid C forms two σ and two π bonds, linear molecule; **(c)** B(sp^2) overlaps with F(2p), trigonal-planar molecule **10. (a)** sp^3d^2; **(b)** sp; **(c)** sp^3; **(d)** sp^2; **(e)** sp^3d **11. (a)** sp on C; **(b)** sp^3 on C and O; **(c)** sp^2 on central C, sp^3 on other C atoms; **(d)** sp^2 on C, sp^3 on N and O **12. (a)** planar, sp^2 on each C; **(b)** linear, sp on each C and N; **(c)** sp^3 on F-surrounded C, sp on other C and N; **(d)** linear, sp on C and N. **13.** A single bond is a σ bond, a double bond is a σ and a π bond, a triple bond is a σ and two π bonds.

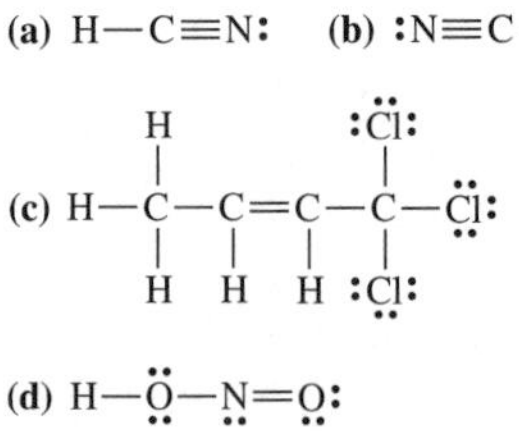

(a) H—C≡N: **(b)** :N≡C—C≡N:

(c) H—C—C=C—C—C̈l: (with H's and :C̈l: lone pairs)

(d) H—Ö̈—N=Ö:

14. (a) :Ö=C=Ö: **(b)** C is sp-hybridized, and to each O is a σ bond [C(sp)-O(2p)] and one π bond [C(2p)-O(2p)] **15. (a)** 6σ; **(b)** 2π **16. (a)** σ_{1s}: bonding MO always lower in energy than the associated antibonding MO; **(b)** σ_{2s}: 2s atomic orbital is lower in energy than the 2p atomic orbital; **(c)** σ_{1s}^*: 1s atomic orbital is lower in energy than the 2s atomic orbital; **(d)** σ_{2p}: a bonding orbital is expected to be more stable than an antibonding orbital. **17.** C_2(bond order = 2.0). **18.** C_2^+ $\sigma_{2s}^2 \sigma_{2s}^{*2} \pi_{2p}^3 \sigma_{2p} \pi_{2p}^* \sigma_{2p}^*$, 1.5 bond order, paramagnetic; O_2^- $\sigma_{2s}^2 \sigma_{2s}^{*2} \sigma_{2p}^2 \pi_{2p}^4 \pi_{2p}^{*3}$, 1.5 bond order, paramagnetic; F_2^+ $\sigma_{2s}^2 \sigma_{2s}^{*2} \sigma_{2p}^2 \pi_{2p}^4 \pi_{2p}^{*3} \sigma_{2p}^*$, 1.5 bond order, paramagnetic **19. (a)** electrical conductor; **(b)** insulator; **(c)** insulator; **(d)** semiconductor; **(e)** electrical conductor; **(f)** insulator; **20.** insulator
Exercises 21. VB method distinguishes between σ and π bonds, predicts molecular shape, explains hindered rotation around double bonds. **23. (a)** no prediction; **(b)** 90° bond angles; **(c)** and **(d)** 109.5° bond angles. **25. (a)** sp on C; **(b)** sp^2 on N, sp^3 on O; **(c)** sp^3 on Cl; **(d)** sp^3 on B **27.** sp^3d on Cl holds two lone pair and forms three overlaps with 2p on F **29. (a)** four C(sp^3)—Cl(3p) σ bonds; **(b)** σ: O($2p_y$)—N(sp^2), σ: N(sp^2)—Cl($3p_z$), π: O($2p_z$)—N($2p_z$); **(c)** σ: H(1s)—O_a(sp^3), σ: O_a(sp^3)—N(sp^2), σ: N(sp^2)—O_b($2p_y$), π: N($2p_z$)—O_b($2p_z$); **(d)** $2x\sigma$. Cl(3p)—C(sp^2), σ. O(2p)—C(sp^2), π. O(2p)—C(2p).
31.

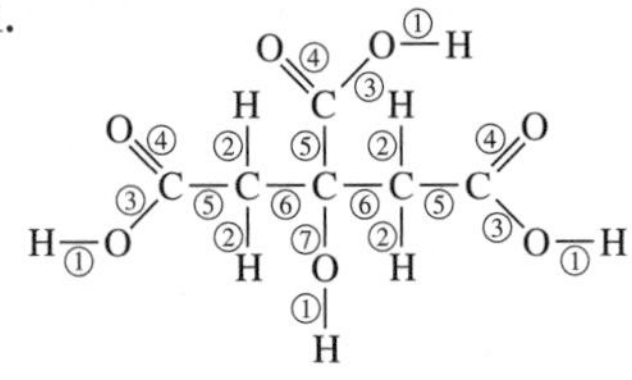

① σ[O(sp^3) − H(1s)]
② σ[C(sp^3) − H(1s)]
③ σ[C(sp^2) − O(sp^3)]
④ σ[C(sp^2) − O(sp^2)] + π[(C(2p) − O(2p)]
⑤ σ[C(sp^3) − C(sp^2)]
⑥ σ[C(sp^3) − C(sp^3)]
⑦ σ[C(sp^3) − O(sp^3)]

33.

	Hybridization	Lewis Diagram	Hybridization/ Orbital Overlap
(a)	Central S: sp^2 hybridized		σ[S(sp^2) − S(3p)] π[S(3p) − S(3p)] σ[S(sp^2) − O(2p)]
(b)	Br atom is sp^3d hybridized		σ[Br(sp^3d) − F(2p)]

35. H_a—C_a≡C_b—C_c=$\ddot{\text{O}}$
                        |
                       H_b

σ: $C_a(sp)$—$H_a(1s)$ σ: $C_a(sp)$—$C_b(sp)$
σ: $C_b(sp)$—$C_c(sp^2)$ σ: $C_c(sp^2)$—$H_b(1s)$
σ: $C_c(sp^2)$—$O(2p_y)$ π: $C_a(2p_y)$—$C_b(2p_y)$
π: $C_a(2p_z)$—$C_b(2p_z)$ π: $C_c(2p_z)^1$—$O(2p_z)$

37. Valence bond theory describes bonds through orbital overlap of atomic orbitals on separated atoms. Molecular orbital theory describes bonds through the electron occupancies of orbitals belonging to a molecule as a whole. **39.** N_2^-, stable: bond order = 2.5, N_2^{2-}, stable: bond order = 2. **41.** Four consecutive bonding molecular orbitals with increasing energies and no intervening antibonding orbitals would be required, and this does not occur. **43.** (a) $\sigma_{2s}^2 \sigma_{2s}^{*2} \pi_{2p}^4 \sigma_{2p}^2 \pi_{2p}^{*1} \sigma_{2p}^*$; (b) $\sigma_{2s}^2 \sigma_{2s}^{*2} \pi_{2p}^4 \sigma_{2p}^2 \pi_{2p}^*$, (c) $\sigma_{2s}^2 \sigma_{2s}^{*2} \pi_{2p}^4 \sigma_{2p}^2 \pi_{2p}^*$ σ_{2p}^*, (d) $\sigma_{2s}^2 \sigma_{2s}^{*2} \pi_{2p}^4 \sigma_{2p}^1 \pi_{2p}^* \sigma_{2p}^*$, (e) $\sigma_{2s}^2 \sigma_{2s}^{*2} \pi_{2p}^4$ $\sigma_{2p}^2 \pi_{2p}^* \sigma_{2p}^*$; (f) and (g) $\sigma_{2s}^2 \sigma_{2s}^{*2} \pi_{2p}^4 \sigma_{2p} \pi_{2p}^* \sigma_{2p}^*$ **45.** NO^+ bond order = 3, diamagnetic, shorter bond; N_2^+ bond order = 2.5, paramagnetic, longer bond **47.** to account for the resonance hybrid required by valence bond theory **49.** (b) is the only one with resonance forms and thus delocalized orbitals. **51.** (c) and (d). **53.** 7.02×10^{20} energy levels and electrons present. **55.** Only (d) and (f) will produce an electron-poor, p-type semiconductor (Ga and B have fewer electrons than Si). **57.** Silicon contaminated with As has extra electrons that can be promoted into the conduction band by thermal excitation; these additional electrons can carry a current. Ultrapure silicon has no extra electrons in the lattice that can be used to carry a current. **59.** The minimum wavelength of light required is 1090 nm (IR radiation) **Integrative and Advanced Exercises 62.** MO theory: First and second shells filled followed by σ_{3s} (↑↓); VB theory: overlap of two 3s orbitals; Lewis theory: No, the structure would have incomplete octets. **64.** (a) O_2^- has a bond order of 1.5, O_2^{2-} has a bond order of 1.0 **68.** (structures) bond angles about N are 125°, so N is sp^2-hybridized, ∠ N—O—F bond angles are 105°, so hybridization is sp^3; overlaps are σ; F$(2p)$—$O_a(sp^3)$, σ: $O_a(sp^3)$—$N(sp^2)$, σ: $N(sp^2)$—$O(2p)$, π: $N(2p)$—$O(2p)$ **70.** ClF_3 is T-shaped, AsF_6^- is octahedral. **72.** He_2^+ $\sigma_{1s}(\uparrow\downarrow)\sigma_{1s}^*(\uparrow)$ $\sigma_{2s}(\uparrow)\sigma_{2s}^*(\ )$ has a bond order of 1.0 **74.** The sp^2 electrons are involved in σ bonding. For N the lone pair is in one sp^2 orbital; the remaining two half-filled sp^2 orbitals overlap with orbitals on adjacent C atoms. The $2p$ electrons are involved in π bonding. **76.** Cl in F_2Cl^- is sp^3d-hybridized; Cl in F_2Cl^+ is sp^3-hybridized. **78.** $C_5H_7NO_2$, NCCH$_2$CO$_2$C$_2$H$_5$ C—H: σ H$(1s)$—C(sp^3) C=O: σ C(sp^2)—O$(2p)$, π: C$(2p)$—O$(2p)$ C—O: σ C(sp^2)—O$(2p)$ C—O: σ C(sp^3)—O$(2p)$ C≡N: σ C(sp) —N(sp), π[2]: C$(2p)$—O$(2p)$ C—C: σ C(sp^3)—C(sp) C—C: σ C(sp^2)—C(sp^3) **80.** C$_4$H$_8$N$_2$O$_2$, H$_3$C—C(=NOH)—C(=NOH)— CH$_3$ C—H: σ H$(1s)$—C(sp^3) C=N: σ C(sp^2)— N(sp^2), π: C$(2p)$—N$(2p)$ N—O: σ N(sp^2)— O(sp^3) C—C: σ C(sp^2)—C(sp^2) C—C: σ C(sp^3)—C(sp^2) O—H: σ H$(1s)$—O(sp^3) **Feature Problems 84.** (a) -205.4 kJ/mol benzene (b) -117.9 kJ/mol cyclohexene (c) Resonance energy for benzene is $\sim$148 kJ (d) $\sim$168 kJ

CHAPTER 13

Practice Examples 1A. 151 torr **1B.** 1.70 g L^{-1} **2A.**

vapor only **2B.** 0.0435 g H$_2$O vapor, 0.089 g liquid water **3A.** 121 mmHg **3B.** 42.9 mmHg **4A.** (a) R $\longrightarrow$ P, H$_2$O(g) liquefies, then solidifies as the temperature is lowered at constant pressure. Then from P $\longrightarrow$ Q, H$_2$O is converted from solid to liquid by increasing the pressure at constant temperature. **4B.** No, the volume decreases from 53.1 L to 15.3 L **5A.** CH$_3$CN (polar) **5B.** (CH$_3$)$_3$CH < CH$_3$CH$_2$CH$_2$CH$_3$ < S O $_3$ < C $_8$ H $_{18}$ < C $_6$ H $_5$ CHO **6A.** He < Ne < O$_2$ < O$_3$ < Cl$_2$ < (CH$_3$)$_2$CO **6B.** Higher molar masses produce larger London forces; polar CH$_3$NO$_2$ has additional intermolecular forces. **7A.** lower: RbI or CsI; higher: MgO **7B.** NaI **8A.** 524 pm **8B.** 6.628×10^7 pm^3 **9A.** 0.903 g/cm^3 **9B.** 6.035×10^{23}/mol. **10A.** 402 pm. **10B.** 2.21 g/cm^3 **11A.** -669.2 kJ/mol. **11B.** -764 kJ/mol.

Review Questions 5. (a) and (b) not necessarily, CO$_2$ has neither; (c) yes **6.** (d): The enthalpy of sublimation is the sum of the enthalpies of fusion and vaporization. **7.** Intermolecular forces (stronger ones produce lower vapor pressure) and temperature (vapor pressure is higher at high temperatures). **8.** (a) 26.8 g; (b) 29.5 kJ/mol; (c) 4.84 kJ **9.** (a) 45 mmHg; (b) 110 °C **10.** (a) 459 K; (b) 518 mmHg **11.** 226 mmHg **12.** 49.7 kJ/mol **13.** Normal boiling point is 339 K **14.** 40 kJ **15.** (a) 776 kJ; (b) 2.5×10^4 kJ **16.** liquid and vapor [assuming that for CO$_2$(l), $d > 0.16$ g/mL] **17.** (a) C$_{10}$H$_{22}$; (b) CH$_3$OCH$_3$; (c) CH$_3$CH$_2$OH **18.** N$_2$ < F$_2$ < Ar < O$_3$ < Cl$_2$, based on increasing molar mass **19.** Ne < C$_3$H$_8$ < CH$_3$CH$_2$OH < CH$_2$OHCHOHCH$_2$OH < KI < K$_2$SO$_4$ < MgO **20.** Yes, it boils at a high temperature compared to HCl, for example. **21.** (a) network covalent solid; (b) molecular solid; (c) ionic solid; (d) metallic solid; (e) molecular solid **22.** (c) is correct **23.** Number of Cl$^-$ ions = $(8 \times \frac{1}{8})$ = 1 Cl$^-$ ion plus 1 Cs$^+$ ion **24.** (a) 362 pm; (b) 4.74×10^7 pm^3; (c) 4 Cu atoms; (d) 74.1%; (e) 4.221×10^{-22} g Cu; (f) 8.91 g/cm^3.

Exercises 25. Nonpolar silicone adheres to cloth or leather and repels polar, hydrogen-bonded water. **27.** Yes, molasses is a very viscous liquid. **29.** Evaporation is endothermic. Uninsulated evaporation involves removal of energy from the surroundings; evaporation from an insulated container involves removal of energy from the liquid that is evaporating. **31.** 8.88 L **33.** 40.5 kJ **35.** 221 L **37.** (a) exothermic condensation; (b) 100 °C. **39.** 120.7 °C **41.** 0.907 mmHg **43.** 6.5 atm **45.** 33 °C **47.** 0.099 atm **49.** CO$_2$, HCl, NH$_3$, SO$_2$, and H$_2$O. **51.** (a) melts at 0 °C, vaporizes at 93.5 °C; (b) liquefies at slightly above 91 mmHg and probably solidifies by 50 atm; (c) liquefies at about 20 °C, solidifies at about -56 °C. **53.** (a) 0.0418 atm; (b) 0.110 atm; (c) 0.117 atm **55.** SO$_2$(g) only. **57.** (a) 0.22 kg; (b) 27 g of steam **59.** Supercooled liquid solidifies. **61.** (a) weak London forces and strong dipole—dipole forces; (b) moderate London forces; (c) moderate London forces, and moderate dipole—dipole forces; (d) strong hydrogen bonding and weak London forces; (e) All forces are weak. **63.** (c) < (b) < (d) < (a) **65.** methanol, due to hydrogen bonding **67.** diamond **69.** (a) diamond structure with alternating silicon and carbon atoms; (b) sp^3 hybridization for B and N; full $2p$ orbitals on N overlap with empty $2p$ orbitals on B to form π bond framework. **71.** Yes, smaller ions produce larger interionic forces. **73.** There are two different ways of placing the third layer in the packing-of-spheres arrangement of Figure 13-39. **75.** (a) A diamond-shaped cell consisting of nine symbols, with a colored diamond at each corner and a gray square at the

center; other unit cells are possible; (b) 1 gray square, 2 open circles, 1 colored diamond **77.** 18.5 g/cm^3 **79.** (a) 335 pm; (b) 9.23 g cm^{-3}; (c) 15.5° **81.** $(8 \times \frac{1}{8}) + (6 \times \frac{1}{2})$ = four Ca^{2+} ions and eight F$^-$ ions; $(8 \times \frac{1}{8}) + 1$ = two Ti^{4+} ions and $(4 \times \frac{1}{2}) + 2$ = four O^{2-} ions. **83.** (a) 6 for each; (b) 4; (c) 7.62×10^{-23} cm^3; (d) 3.51 g/cm^3 **85.** (a), (b) and fcc array of anions with cations occupying octahedral holes. **87.** LiCl(s) (most negative) < NaCl(s) < KCl(s) < RbCl(s) < CsCl(s) (least negative)(lattice energy increases with decreasing cation radius). **89.** -645 kJ/mol; MgCl$_2$ is more stable than MgCl solid because Mg^{2+} ion more strongly attracts anions than the Mg$^+$ ion. **Integrative and Advanced Exercises 93.** Isooctane has the higher vapor pressure (43 vs 23.8 mmHg), so is the more volatile. **96.** ΔH_{vap} decreases as temperature increases. **99.** At 1.00 bar, bp = 99.65 °C **102.** (a) 31 atm; (b) -0.25 °C **103.** 89 °C **106.** 54.3% dimer, decreasing with increasing temperature **107.** 1.84 kg N$_2$ **110.** $\theta = 16.2°$

Feature Problems 120. 21 g/s^2 = 2.1×10^{-2} J/m^2

121. (a) $\ln\left(\dfrac{P_2}{P_1}\right) = \dfrac{15{,}971}{R}\left(\dfrac{1}{T_1} - \dfrac{1}{T_2}\right)$

$$+ \dfrac{14.55}{R}\ln\dfrac{T_2}{T_1} - \dfrac{0.160}{R}(T_2 - T_1)$$

(b) 169 K

CHAPTER 14

Practice Examples 1A. 16.2% **1B.** (a) x_{H2O} = 0.927; (b) 3.73 M; (c) 4.34 m **2A.** 0.03593 **2B.** (a) 0.3038 M; (b) 0.3426 m; (c) 0.005815 **3A.** oxalic acid **3B.** CCl$_4$ **4A.** (1) 31 g NH$_4$Cl; (2) 46 g NH$_4$Cl crystallized if both suggestions (1) and (2) are followed. **4B.** KNO$_3$ > KClO$_4$ > K$_2$SO$_4$ **5A.** 0.378 atm **5B.** 6.33 atm **6A.** 112 mmHg hexane, 239 mmHg total **6B.** 13.0 mmHg toluene, 64.5 mmHg total **7A.** x_{hexane} = 0.469 **7B.** $x_{toluene}$ = 0.202 **8A.** 0.857 atm **8B.** 8.24 mg **9A.** 8.60×10^4g/mol **9B.** 0.0105 atm **10A.** (a) 0.122 m; (b) 377 g/mol; (c) C$_{17}$H$_{20}$O$_6$N$_4$ **10B.** lower than 760.0 mmHg **11A.** 3.9 atm **11B.** 0.56 mL **Review Questions 4.** 53.7% (m/m) **5.** (a) 17.1% (V/V); (b) 13.9% (m/m); (c) 13.7% (m/V) **6.** 1.83×10^4 L **7.** 0.683 g **8.** 0.410 m **9.** 1.85 M **10.** x_{C7} = 0.187, 18.7%, x_{C8} = 0.427, 42.7%, x_{C9} = 0.386, 38.6% **11.** (a) 7.80 M; (b) 24.4 M; (c) 34.0 m; (d) $x_{glycerol}$ = 0.242; (e) 75.8 mol % H$_2$O **12.** unsaturated **13.** (d), Both are nonpolar. **14.** NH$_2$OH (s), because of hydrogen bonding. **15.** (a) The hydrocarbon chain promotes the solubility of C$_4$H$_9$OH in benzene, and hydrogen bonding promotes its solubility in water. **16.** 4.48×10^{-3} M **17.** 40.6 mmHg benzene, 56.9 mmHg total **18.** $x_{benzene}$ = 0.714 **19.** 0.050 m C$_2$H$_5$OH **20.** C$_2$H$_5$OH > HC$_2$H$_3$O$_2$ > NaCl > MgBr$_2$ > Al$_2$(SO$_4$)$_3$ **21.** (a) 0.16 M Na$^+$; (b) 0.32 M; (c) 8.1 atm; (d) -0.60 °C **22.** 3.2×10^4 g/mol **23.** 1.2×10^2 g/mol **24.** 100.02 °C **Exercises 25.** salicyl alcohol **27.** (c) formic acid and (f) propylene glycol are soluble in water; (b) benzoic acid and (d) butyl alcohol are only slightly soluble; (a) iodoform and (e) chlorobenzene are insoluble. **29.** solubilities: CaF$_2$ < MgF$_2$ < NaF < KF **31.** Only true in cases when the other component's density (for example, H$_2$O) is greater than ethanol's. **33.** 21.6 g **35.** 4.80×10^{-4} M **37.** 113 mL **39.** 54.8 g **41.** 4.19 M, 5.26 m **43.** (a) $x_{ethanol}$ = 0.0979; (b) x_{urea} = 0.0122 **45.** 4.93 g **47.** 8.66 m **49.** (a) unsaturated; (b) about 3 g **51.** 4×10^2 g **53.** 1.40×10^{-5} M

55. $c = kP_{gas}$; gas solubilities are low and thus volume doesn't change appreciably. Consequently, the mass of dissolved gas is approximately proportional to its pressure. **57.** 23.2 mmHg **59.** $x_{styrene} = 0.32$ **61.** 1.29×10^3 mmHg **63.** Water is drawn out by osmosis. **65.** 22.4 L of solvent; osmotic pressure equation resembles ideal gas equation. **67.** 2.8×10^5 g/mol **69.** 21 atm **71. (a)** 20. °C/*m*; **(b)** cyclohexane, because of larger K_f **73.** $C_6H_4N_2O_4$ **75.** C_4H_4S **77.** 1.2×10^2 **79. (a)** −0.19 °C; **(b)** −0.37 °C; **(c)** −0.37 °C; **(d)** −0.56 °C; **(e)** −0.37 °C; **(f)** −0.19 °C; **(g)** < −0.19 °C **81.** $NH_3(aq) + HC_2H_3O_2(aq) \longrightarrow NH_4^+(aq) + C_2H_3O_2^-(aq)$ (salt: strong electrolyte).
Integrative and Advanced Exercises 84. 29.3% **85.** 682 g H_2O **88.** 4.5 kg H_2O **91.** ~10% palmitic acid **94.** mole fraction water = 0.99839 **97. (a)** $1.56 \times 10^3\ m^2$ particle area; **(b)** $8.34 \times 10^{-7}\ m^2$ surface area of Au cube **102.** 34.62% O_2 **104.** 1.3×10^2 g
Feature Problems 106. (a) Vapor has $x_{HCl} > 0.50$; **(b)** Because the vapor composition is different from that of the solution, the composition and boiling point of the remaining solution must change as vapor escapes **(c)** $x_{HCl} = 0.12$ and 110 °C; **(d)** $x_{HCl} = 0.111$ **107 (a)** 90.7%; **(b)** yes; **(c)** high water solubility; this will produce a solution with a low x_{H_2O} **108. (a)** 0.92% mass/volume has $t_f = -0.60$ °C; **(b)** $t_f = -0.60$ °C

CHAPTER 15

Practice Examples 1A. $8.93 \times 10^{-5}\ M\ s^{-1}$ **1B.** $1.62 \times 10^{-4}\ M\ s^{-1}$ **2A. (a)** $3.3 \times 10^{-4}\ M\ s^{-1}$; **(b)** 0.37 M **2B.** 2.17 M **3A.** first-order in N_2O_5 **3B.** $3.9 \times 10^{-7}\ M\ min^{-1}$ **4A.** $4.4 \times 10^{-2}\ M^{-2}\ s^{-1}$ **4B.** $2.4 \times 10^{-7}\ M\ s^{-1}$ **5A.** 1.0 M **5B.** first-order, with $k = 7.3 \times 10^{-4}\ s^{-1}$ **6A.** 64.2% **6B.** 25.2 min **7A.** 272 mmHg **7B. (a)** 1.9×10^{-7} mmHg; **(b)** 1.56×10^3 mmHg **8A.** first-order (constant half-life) with $k = 6.93 \times 10^{-3}\ s^{-1}$ **8B.** zero-order from a plot of [A] vs. *t*, with $k = 9.28 \times 10^{-3}$ M/min **9A.** 43 s. at 75 °C **9B.** $T_2 = 311$ K **10A.** $2\ NO_2 \longrightarrow NO + NO_3$; rate $= k_1\ [NO_2]^2$ **10B.** Rate $= (k_1 \times k_3 \times k_2^{-1})[NO_2]\ [F_2]$, derived rate law agrees with the experimental rate law.
Review Questions 4. (a) $2.3 \times 10^{-3}\ M\ min^{-1}$; **(b)** $3.8 \times 10^{-5}\ M\ sec^{-1}$ **5. (a)** $3.1 \times 10^{-4}\ mol\ L^{-1}\ s^{-1}$; **(b)** 3.1×10^{-4} mol B $L^{-1}\ s^{-1}$; **(c)** 9.3×10^{-4} mol D $L^{-1}\ s^{-1}$ **6. (a)** At $t = 800$ s: reaction rate $= 9.4 \times 10^{-4}\ M\ s^{-1}$; **(b)** For $[H_2O_2] = 0.50$ M, $T \approx 2150$ s **7. (b)** is correct **8.** 1.39×10^{-4} M/min **9.** 0.0221 M min^{-1} **10. (a)** 60.0 min; **(b)** 45 min **11. (a)** 42.0 min; **(b)** 0.0165 min^{-1} **12. (a)** 19 min; **(b)** 0.18 g **13. (a)** 0.0160 min^{-1}; **(b)** 43.3 min; **(c)** 77.8 min; **(d)** 0.074 M **14. (a)** first order in A, second order in B; **(b)** third order overall; **(c)** 0.102 $M^{-2}\ s^{-1}$ **15.** Second order: non-constant half-life, doubling of half-life for each half-life period. **16.** (d) **17. (a)** 1.6×10^2 kJ/mol; **(b)** $T = 6.5 \times 10^2$ K **18. (a)** II; **(b)** I; **(c)** III **19.** 0.0100 M/s **20.** slightly less than 75 s **21.** 0.0080 M s^{-1} **22. (a)** 0.010 M s^{-1}; **(b)** 0.0048 M s^{-1}; **(c)** 0.0034 M s^{-1} **23. (a)** 0.00 M; **(b)** 0.32 M; **(c)** 0.47 M **24. (a)** rate $= k_{slow}[A][B]$: equivalent to observed rate law, **(b)** rate $= (k_1 k_3 / k_2)\ [A][B]^2$: differs from empirical rate law
Exercises 25. $1.0 \times 10^{-3}\ M\ s^{-1}$ **27. (a)** 0.575 M; **(b)** 5.4 min **29. (a)** 3.52×10^{-5} M/s; **(b)** 0.357 M; **(c)** 4.5×10^2 s **31. (a)** 3000 mmHg; **(b)** 1400 mmHg **33.** Rate $= (5.8 \times 10^{-3}\ M^{-1}\ min^{-1})[A]^1[B]^1$ **35. (a)** First order in A; second order in B; order of −1 for C; **(b)** $R_5 = \frac{1}{4} R_1$ **37. (a)** true; **(b)** false, half-life is independent of [A] **39. (a)** 3.13%; **(b)** 0.00193 M/s **41.** 20.6 min **43. (a)** 218 min; **(b)** 2.3 L CO_2 **45. (a)** First order, constant value for *k*; **(b)** $1.86 \times 10^{-3}\ s^{-1}$; **(c)** 0.148 M **47.** Rate $= k[HI]^2$, second order in HI and overall with $k = 0.00118\ M^{-1}\ s^{-1}$ **49. (a)** Zeroth order in A; **(b)** Half-life = 72 s **51. (a)** First experiment initial rate = 0.022 M min^{-1}; second experiment: initial rate = 0.089 M min^{-1}; **(b)** Reaction is second order in A **53.** Reaction rate $= k[A]^2$; $k = 0.020\ M^{-1}\ min^{-1}$ **55.** Zeroth order: $t_{1/2} = \dfrac{[A]_0}{2k}$; Second order: $t_{1/2} = \dfrac{[1]}{k[A]_0}$; half-life for zeroth-order reaction is proportional to $[A]_0$ while half-life for a second-order reaction is inversely proportional to $[A]_0$. Consequently, half-life for zeroth order increases with increasing [A] while the half-life for a second-order reaction decreases with increasing [A]. **57. (a)** Also needs fraction of successful collisions; **(b)** Fraction with sufficient energy to react increases dramatically, whereas average kinetic energy increases more gradually; **(c)** Catalyst provides an alternate pathway, with a lower E_a and often a different value for A. **59. (a)** 63 kJ/mol; **(b)** plot of potential energy versus progress of reaction shows $E_{products} > E_{reactants}$ by 21 kJ/mol and $E_{activated\ complex} - E_{reactants} = 84$ kJ/mol **61. (a)** There are two intermediates (B and C); **(b)** There are three transition states; **(c)** Fastest step has the lowest activation energy, hence step 3 is the fastest step in the reaction with step 2 a close second; **(d)** Reactant A(step 1) is the reactant in the rate-limiting step **(e)** endothermic, need energy to go from A $\longrightarrow$ B; **(f)** exothermic, energy released when going from A $\longrightarrow$ D **63. (a)** straight line plot with slope $= -E_a/R$; **(b)** 113 kJ/mol; **(c)** 2.2×10^2 s **65. (a)** 34.8 kJ/mol; **(b)** 61 °C **67. (a)** 51 kJ/mol; **(b)** No, it does not. **69. (a)** A catalyst reacts but is recovered unchanged. **(b)** A catalyst lowers E_a by *changing* the mechanism of the reaction. **71.** Platinum, unlike an enzyme, is rather nonspecific in the reactions it catalyzes. **73.** There must be an excess of substrate. **75.** Molecularity and reaction order correspond only for elementary processes. **77.** first step(fast) $N_2O_2 \rightleftharpoons 2\ NO$, second step(slow) $H_2 + N_2O_2 \longrightarrow H_2O + N_2O$ third step(fast) $N_2 + H_2O \longrightarrow N_2O + H_2$ overall: $2\ NO + 2\ H_2 \longrightarrow N_2 + 2\ H_2O$ **79.** First step: $Cl_2(g) \rightleftharpoons 2\ Cl(g)$, second step: $2\ Cl(g) + 2\ NO(g) \longrightarrow 2\ NOCl(g)$, overall: $Cl_2(g) + 2NO(g) \longrightarrow 2\ NOCl(g)$
Integrative and Advanced Exercises 81. (a) first-order; **(b)** $k = 0.29$/min; **(c)** 0.10 M/min; **(d)** 0.064 M/min; **(e)** 0.29 M/min **82.** 1.86 M **85.** $M^{1-o}\ sec^{-1}$ **90.** 1.9×10^2 minutes **95.** rate $= (k_1[Cl_2]/k_{-1})^{1/2} \times k_2 \times [CHCl_3]$, $k = 0.015$ **98.** $k = 1.0 \times 10^{-6}\ M^{-1}\ sec^{-1}$ **99.** $k = 0.027\ sec^{-1}$ **103.** Mechanism is not consistent with reaction stoichiometry; sum of elementary steps indicates HOBr is a product and this was not observed.
Feature Problems 104. (d) $1.1 \times 10^{-3}\ M\ min^{-1}$; **(e)** $5.5 \times 10^{-3}\ M\ min^{-1}$; **(f)** $k_{avg} = 0.071\ min^{-1}$ and reaction rate $= k[C_6H_5N_2Cl]$; **(g)** 9.8 min; **(h)** about 21 min; **(j)** 0.0661 min^{-1} **105. (a)** first-order in each reactant, second-order overall; **(b)** $3.7 \times 10^{-5}\ M\ s^{-1}$; **(c)** $6.2 \times 10^{-3}\ M^{-1}\ s^{-1}$; **(d)** $k_3 = 0.0014\ M^{-1}\ s^{-1}$, $k_{13} = 0.0030\ M^{-1}s^{-1}$, $k_{24} = 0.0062\ M^{-1}s^{-1}$, $k_{33} = 0.012\ M^{-1}\ s^{-1}$; **(e)** 51 kJ/mol; **(f)** First step is slowest because it involves two anions which will repel one another.

CHAPTER 16

Practice Examples 1A. 0.263 M **1B.** 0.25 M **2A.** 7.1×10^2 **2B.** 1.7×10^{-6} **3A.** 1.7×10^{-6} **3B.** $K_c = 95$ **4A.** $K_c = [Ca^{2+}]^5[HPO_4^{2-}]^3/[H^+]^4$ **4B.** $K_c = [H_2]^4/[H_2O]^4$; $K_p = (P_{H_2})^4/(P_{H_2O})^4$; $K_p = K_c$ **5A.** to the right, forming more products. **5B.** At equilibrium there will be more moles of reactants, and fewer moles of products than there were initially. **6A.** Equilibrium shifts to the right. **6B. (a)** no effect; **(b)** shifts to the left; **(c)** no effect. **7A.** $[N_2O_4]$ will increase and $[NO_2]$ will decrease. **7B.** no effect **8A.** greater at higher temperatures **8B.** $[NH_3(g)]$ will be greater at −100 °C **9A.** 2.3×10^{-4} **9B.** 25.2 g **10A.** $K_c = 1.5 \times 10^{-4}$, $K_p = 3.7 \times 10^{-3}$ **10B.** $K_p = 0.022$ **11A.** 0.481 atm **11B.** 0.716 atm **12A.** 0.262 mol **12B. (a)** decrease **(b)** 0.0118 mol **13A.** $[Ag^+] = [Fe^{2+}] = 0.489$ M; $[Fe^{3+}] = 0.71$ M **13B.** $[V^{3+}] = [Cr^{2+}] = 0.0057$ M; $[V^{2+}] = [Cr^{3+}] = 0.154$ M
Review Questions 4. (a) $[NO_2]^2/[NO]^2[O_2]$; **(b)** $[Zn^{2+}]/[Ag^+]^2$; **(c)** $[OH^-]^2/[CO_3^{2-}]$ **5. (a)** $P\{CH_4\}P\{H_2S\}^2/(P\{CS_2\}P\{H_2\}^4)$; **(b)** $P\{O_2\}^{1/2}$; **(c)** $P\{CO_2\}P\{H_2O\}$ **6. (a)** $[N_2O]/([N_2][O_2]^{1/2})$; **(b)** $[HBr]/([H_2]^{1/2}[Br_2]^{1/2})$; **(c)** $[NH_3]/([N_2]^{1/2}[H_2]^{3/2})$; **(d)** $[ClF_3]/([Cl_2]^{1/2}[F_2]^{3/2})$; **(e)** $[NOF_2]/([N_2]^{1/2}[O_2]^{1/2}[F_2])$ **7. (a)** 0.0431; **(b)** 3.1×10^3; **(c)** 66 **8.** 9.7×10^{-16} **9. (d)**: 1 mol I_2 is the limiting reactant **10.** 1.9 **11.** 2.0×10^{-19} **12. (a)** 0.0578 mol O_2; **(b)** 0.232 mol O_2 **13. (a)** 0.0431; **(b)** 42; **(c)** 6.1 **14.** 1.2×10^2 **15. (a)** $K_c = (n_{CO})(n_{H_2O})/(n_{co})(n_{H_2})$ Since each species has a coefficient of one, therefore, K_c does not depend on the container volume. **(b)** $K_p = K_c = 0.659$ **16. (a)** 26.3; **(b)** 0.613 **17. (a)** 0.011; **(b)** 1.3 **18. (a)** $Q_c \neq K_c$, therefore not at equilibrium; **(b)** to the right **19. (a)** shift right; **(b)** no shift; **(c)** shift right **20. (b)** and **(d)** shift right with an increase in temperature **21. (a)** increase; **(b)** decrease; **(c)** decrease; **(d)** no effect; **(e)** decrease **22.** 0.234 mol HI, 0.033 mol each of H_2 and I_2 **23.** 0.0165 mol I_2 **24.** 1.154 atm
Exercises 25. (a) $2\ COF_2(g) \rightleftharpoons CO_2(g) + CF_4(g)$, $K_c = [CO_2][CF_4]/[COF_2]^2$; **(b)** $Cu(s) + 2\ Ag^+(aq) \rightleftharpoons Cu^{2+}(aq) + 2\ Ag(s)$, $K_c = [Cu^{2+}]/[Ag^+]^2$; **(c)** $S_2O_8^{2-}(aq) + 2\ Fe^{2+}(aq) \rightleftharpoons 2\ SO_4^{2-}(aq) + 2\ Fe^{3+}(aq)$, $K_c = [SO_4^{2-}]^2\ [Fe^{3+}]^2/([S_2O_8^{2-}]\ [Fe^{2+}]^2)$ **27. (a)** 0.0012; **(b)** 5.55×10^5; **(c)** 0.429 **29.** $K_p = 0.0313$, $K_c = 1.28 \times 10^{-3}$ **31.** 1.8×10^6 **33.** 0.106 **35.** 0.516 **37. (a)** $NH_3(g) + 7/4\ O_2(g) \rightleftharpoons NO_2(g) + 3/2\ H_2O(g)$; **(b)** 4.03×10^{19} **39.** no, $Q_c < K_c$ **41.** 0.116 mol $SbCl_5$, 0.17 mol $SbCl_3$, 0.0440 mol Cl_2 **43.** 0.949 g **45. (a)** no, $Q_c \neq K_c$; **(b)** to the right, $Q_c < K_c$; **(c)** 0.112 mol COF_2, 0.279 mol CO_2, 0.0905 mol CF_4 **47.** mole % = 78.0% HI **49.** 5.58 atm **51.** $[Cd^{2+}] = 0.257$ M, $[Cr^{2+}] = 0.514$ M, $[Cr^{3+}] = 0.486$ M; minimum mass of Cd = 10.1 g **53. (b) 55.** 0.529 atm O_2, 0.601 atm total **57.** 30.1 atm $COCl_2$, 32.4 atm total **59.** Constant removal of product(s) makes $Q < K$, the reaction continuously shift right as a result. **61. (a)** decrease; **(b)** increase; **(c)** no effect; **(d)** no effect **63. (a)** increased [NO]; **(b)** Increases the rate of reaction. **65.** Compared to the "perturbed state" achieved through pressurization, P_{N_2} will have decreased when equilibrium is re-established.
Integrative and Advanced Exercises 69. $[Fe^{2+}] = 0.111$ M, $[Fe^{3+}] = 0.049$ M, $[Ag^+] = 0.15$ M **70.** 2.79 atm **72.** $K_p = 2.1 \times 10^{-3}$ **74.** 12.1 **77.** 0.177 **78.** 88.1 g/mol **81.** $K_p = 3.36 \times 10^{-4}$ **83.** 0.085 mol CH_4, 0.070 mol H_2O, 0.12 mol CO_2, 0.16 mol H_2 **87.** 24.6 atm
Feature Problems 88. 4.0 **89.** $K_c \approx 0.015$ at 623 K **90.** $K_P = 4.8 \times 10^{-8}$

CHAPTER 17

Practice Examples 1A. acids: (a) HF and H_3O^+; (b) HSO_4^- and NH_4^+; (c) HCl and $HC_2H_3O_2$; the other species are bases **1B.** $HNO_2(aq) + H_2O(l) \rightleftharpoons NO_2^-(aq) + H_3O^+(aq)$, $HCO_3^-(aq) + H_2O(l) \rightleftharpoons CO_3^{2-}(aq) + H_3O^+(aq)$, $PO_4^{3-}(aq) + H_2O(l) \rightleftharpoons HPO_4^{2-}(aq) + OH^-(aq)$, $HCO_3^-(aq) + H_2O(l) \rightleftharpoons H_2CO_3(aq) + OH^-(aq)$ **2A.** $[H_3O^+] = 1.4 \times 10^{-3}$ M, $[OH^-] = 7.1 \times 10^{-12}$ M **2B.** 3.1 L **3A.** $[I^-] = 0.0025$ M, $[OH^-] = 4.0 \times 10^{-12}$ M, pH = 2.60 **3B.** 1.466 **4A.** 11.52 **4B.** 13.74 **5A.** 2.9×10^{-8} **5B.** 2.6×10^{-6} **6A.** 1.80 **6B.** 2.66 **7A.** 2.29 **7B.** 11.28 **8A.** 5.5% in 0.20 M, 17% in 0.020 M **8B.** 1.4×10^{-4} **9A.** 3.7×10^{-2} M = $[H_3O^+]$ = $[HOOCCH_2COO^-]$, $[^-OOCCH_2COO^-] = 2.0 \times 10^{-6}$ M **9B.** $K_{a_2} = 5.3 \times 10^{-5}$, $K_{a_1} = 5.3 \times 10^{-2}$ **10A.** 0.011 M = $[SO_4^{2-}]$, $[H_3O^+] = 0.21$ M, $[HSO_4^-] = 0.19$ M **10B.** 0.0060 M = $[SO_4^{2-}]$, $[HSO_4^-] = 0.014$ M, $[H_3O^+] = 0.026$ M **11A.** (a) acidic; (b) neutral; (c) basic **11B.** $H_2PO_4^-(aq) + H_2O(l) \rightleftharpoons H_3O^+(aq) + HPO_4^{2-}(aq), H_2PO_4^-(aq) + H_2O(l) \rightleftharpoons OH^-(aq) + H_3PO_4(aq)$; acid ionization occurs to the greater extent. **12A.** codeine hydrochloride **12B.** basic **13A.** 8.08 **13B.** 3.6×10^{-3} M **14A.** $HClO_4$; CH_3FCOOH **14B.** H_2SO_3; CCl_3FCH_2COOH **15A.** (a) BF_3 is a Lewis acid, NH_3 is a Lewis base; (b) H_2O is a Lewis base, Cr^{3+} is a Lewis acid. **15B.** Hydroxide ion and chloride ion are the Lewis bases; $Al(OH)_3$ and $SnCl_4$ are the Lewis acids.

Review Questions 4. (a) acid; (b) base; (c) base; (d) acid; (e) acid **5.** (a) IO_3^-; (b) $C_6H_5COO^-$; (c) PO_4^{3-}; (d) $C_2H_5NH_2$ **6.** $[H_3O^+]$ = (a) 0.00165 M; (b) 1.1×10^{-12} M; (c) 2.3×10^{-12} M; (d) 5.8×10^{-4} M; $[OH^-]$ = (a) 6.1×10^{-12} M; (b) 0.0087 M; (c) 0.00426 M; (d) 1.7×10^{-11} M **7.** (a) 2.35; (b) 3.212; (c) 11.83; (d) 11.98 **8.** $[H_3O^+] = 0.11$ M **9.** 2.7 **10.** answer 4 is correct, pH < 13 If it were a strong base, its $[OH^-] = 0.10$ M **11.** 1.30×10^{-5} **12.** (a) 0.0030 M; (b) 2.62 **13.** 1.4 g **14.** 0.028 M **15.** (a) 1.4×10^{-4} M; (b) 1.4×10^{-4} M; (c) 4.7×10^{-11} M **16.** Just products are written: (a) $NH_3(aq) + NO_3^-(aq) + H_3O^+(aq)$; (b) $Na^+(aq) + HNO_2(aq) + OH^-(aq)$; (c) $K^+(aq) + HC_7H_5O_2(aq) + OH^-(aq)$; (d) no reaction; (e) $C_6H_5NH_2(aq) + Cl^-(aq) + H_3O^+(aq)$ **17.** 4.51 **18.** (a) 6.7×10^{-6}; (b) 5.6×10^{-11}; (c) 1.0×10^{-4} **19.** 8.58 **20.** Electronegative Cl atoms withdraw electron density from the O—H bond. **21.** (a) HI; (b) HOClO; (c) $H_3CCH_2CCl_2COOH$ **22.** (a) base; (b) acid; (c) base **23.** Acids are: (a) SO_3; (b) $Zn(OH)_2$ **24.** decreasing pH: (b) > (c) > (d) > (a) **Exercises 25.** Bases are: (a) H_2O, OBr^-; (b) H_2O, SO_4^{2-}; (c) HS^-, OH^-; (d) OH^-, $C_6H_5NH_2$ **27.** NH_3 **29.** (a) forward; (b) reverse; (c) reverse **31.** 4.0×10^{-14} M, 13.40 **33.** 1.96×10^{-3} M **35.** 8.5 mL **37.** 0.0539 L **39.** 10.20 **41.** 9.8×10^{-3} M; pH = 2.01 **43.** 2.7×10^{-3} M = $[H_3O^+]$, $[OH^-] = 1.4 \times 10^{-13}$ M, pH = 1.14, pOH = 12.86 **47.** 11.0 **49.** 0.063 g **51.** (a) pH = 11.23; (b) 34 mg NaOH **53.** (b) **55.** (a) 0.0053; (b) 0.53% **57.** 0.0098 M **59.** No, the degree of ionization is based on the assumption that $[HC_2H_3O_2]_{initial} \approx [HC_2H_3O_2]_{initial} - [HC_2H_3O_2]_{equilibrium}$, which is invalid at 13% ionization. **61.** There is little PO_4^{3-} produced in the third ionization. **63.** $[S^{2-}] = K_2 = 1 \times 10^{-19}$ M, (a) 8.7×10^{-5} M = $[H_3O^+] = [HS^-]$; (b) 2.2×10^{-5} M = $[H_3O^+] = [HS^-]$; (c) 9.5×10^{-7} M = $[H_3O^+] = [HS^-]$ **65.** 2.0×10^{-3} M = $[H_3O^+] = [HOOC(CH_2)_4COO^-]$, $[OH^-] = 5.0 \times 10^{-12}$ M, $[HOOC(CH_2)_4COOH] = 0.10$ M,

$[^-OOC(CH_2)_4COO^-] = 3.9 \times 10^{-6}$ M **67.** (a) neutral; (b) basic; (c) neutral; (d) basic; (e) acidic **69.** 10.23 **71.** 9.18 **73.** (a) $HSO_3^- + H_2O(l) \rightleftharpoons H_3O^+(aq) + SO_3^{2-}(aq), HSO_3^-(aq) + H_2O(l) \rightleftharpoons OH^-(aq) + H_2SO_3(aq)$, acidic; (b) $HS^-(aq) + H_2O(l) \rightleftharpoons H_3O^+(aq) + S^{2-}(aq),$ $HS^-(aq) + H_2O(l) \rightleftharpoons OH^-(aq) + H_2S(g)$, basic; (c) $HPO_4^{2-}(aq) + H_2O(l) \rightleftharpoons H_3O^+(aq) + PO_4^{3-}(aq),$ $HPO_4^{2-}(aq) + H_2O(l) \rightleftharpoons OH^-(aq) + H_2PO_4^-(aq)$, basic **75.** (a) $HClO_3$; (b) HNO_2; (c) H_3PO_4 **77.** Largest K_b propylamine (hydrocarbon chains have low electronegativity) and smallest K_b is ortho-chloroaniline(electron withdrawing group decreases basicity) **79.** Lewis bases: (a) OH^-; (b) N_2H_4; (c) $(C_2H_5)_2O$ **81.** $I_2(aq)$(Lewis acid) + $I^-(aq)$(Lewis base) $\rightleftharpoons I_3^-$(aq) **83.**

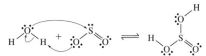

Integrative and Advanced Exercises 85. (a) base; (b) base; (c) acid or base; (d) acid **86.** 2.1 mM **87.** (a) no match; (b) match at pH $\approx$ 5; (c) no match; (d) match at pH $\approx$ 11; (e) match at pH $\approx$ 11; (f) no match; (g) match at pH $\approx$ 5; (h) match at pH $\approx$ 2; (i) no match **91.** 5×10^{-5} **92.** Impractical:(a), (b), or (d). Use (c) 0.096 g NH_4Cl **95.** (b) pH = 2.89, 11.11, 8.88 **96.** (b) 5.43%; (c) $K_a = 1.3 \times 10^{-5}$ **100.** 9.73 g **101.** pH = 0.16

Feature Problems 106. (b) pH = 1.16 **107.** (a) pH = 2.15; (b) pH = 11.95 (c) pH = 9.23

CHAPTER 18

Practice Examples 1A. 0.500 M HF: $[H_3O^+]$ = 0.018 M, [HF] = 0.482 M; 0.500 M HF/0.100 M HCl: $[H_3O^+] = 0.103$ M; [HF] = 0.497 M **1B.** 30 drops **2A.** $[H_3O^+] = 1.2 \times 10^{-4}$ M, $[CHO_2^-] = 0.150$ M **2B.** 15 g **3A.** A reaction occurs producing acetic acid, forming an acetic acid–acetate ion buffer. **3B.** A reaction occurs producing an ammonia–ammonium chloride buffer. **4A.** 3.96 **4B.** 4.51 **5A.** 21 g **5B.** The calculated pH is 5.09. **6A.** (a) 3.54; (b) 3.53; (c) 3.55 **6B.** 1.3 mL **7A.** (a) 0.824; (b) 1.238; (c) 7.00; (d) 11.98 **7B.** (a) 12.21; (b) 11.79; (c) 7.00 **8A.** (a) 2.02; (b) 2.70; (c) 3.18; (d) 8.08 **8B.** (a) 11.14; (b) 9.74; (c) 9.26; (d) 5.20 **9A.** 12.18 **9B.** 10.45

Review Questions 4. (a) 0.0892 M; (b) 1.1×10^{-13} M; (c) 4.0×10^{-5} M; (d) 0.0892 M **5.** (a) 2.9×10^{-5} M; (b) 0.102 M; (c) 0.102 M; (d) 3.4×10^{-10} M **6.** (a) $CHO_2^-(aq) + H_3O^+(aq) \rightleftharpoons HCHO_2(aq) + H_2O(l), HCHO_2(aq) + OH^-(aq) \rightleftharpoons CHO_2^-(aq) + H_2O(l);$ (b) $C_6H_5NH_2(aq) + H_3O^+(aq) \rightleftharpoons C_6H_5NH_3^+(aq) + H_2O(l), C_6H_5NH_3^+(aq) + OH^-(aq) \rightleftharpoons C_6H_5NH_2(aq) + H_2O(l);$ (c) $HPO_4^{2-}(aq) + H_3O^+(aq) \rightleftharpoons H_2PO_4^-(aq) + H_2O(l), H_2PO_4^-(aq) + OH^-(aq) \rightleftharpoons HPO_4^{2-}(aq) + H_2O(l)$ **7.** (a) 4.64; (b) 9.68 **8.** 0.77 M **9.** 0.53 M **10.** 0.60 mol $NaC_2H_3O_2$ **11.** 1.4×10^{-4} g **12.** (a) 9.5 g; (b) 3.86 **13.** (a) neutral: bromothymol blue, basic: thymolphthalein, remainder in acidic solution; (b) 4.7, 6.0 **14.** (a) colorless; (b) red; (c) blue; (d) blue **15.** (a) 36.3 mL; (b) 25.8 mL **16.** equivalence point (a) pH = 7.00, bromothymol blue; (b) pH = 4.62, methyl red; (c) pH = 9.38, phenolphthalein; (d) pH = 10.6, alizarin yellow R; neutralization to $PO_4^{3-}(aq)$ is not possible. **17.** (a) 1.346; (b) 2.034 **18.** (a) 11.85; (b) 1.426 **19.** (a) 2.87; (b) 3.51 **20.** (a) 8.79; (b) 1.66 **21.** (a) 3.10; (b) 4.20; (c) 8.00; (d) 11.11 **22.** $NaHSO_4$.

Exercises 23. (a) +1.05 units; (b) pH unaffected

25. (a) 0.035 M; (b) 4.0×10^{-4} M; (c) 3.9×10^{-5} M **27.** Buffer solutions are: (c), (e), and (f). **29.** 8.88 **31.** (a) 4.8 g; (b) 0.40 g $(NH_4)_2SO_4$ **33.** 9.16 **35.** 100. mL 0.100 M $HCHO_2$, 58 mL 0.100 M $NaCHO_2$ **37.** (a) pH = 3.89 to pH = 5.89; (b) 0.100 mol of acid or base per liter **39.** (a) 3.48; (b) 3.52; (c) 1.23 **41.** (a) 9.01; (b) Yes for dilution to 1.00 L but not to 1000 L, because the buffer components are too dilute; (c) 8.99; (d) 1.1 mL **43.** Each indicator only serves to fix the pH over a quite small region. (b) An indicator is a weak acid and consumes some base in an acid–base titration. **45.** (a) red; (b) yellow; (c) yellow; (d) yellow; (e) red; (f) yellow **47.** $pK_{Hin} = 3.84$; 2,4-dinitrophenol is a relatively good indicator (within 1% of equivalence point volume). For more accurate work, a better indicator is needed. **49.** 0.1508 M **51.** 11.63 **53.** In each case, moles of acid is the same, hence moles of titrant and endpoint volume is the same. At equivalence point, pH differs due to the differences in the strength of the conjugate bases formed. **55.** (a) initial pH = 13.00, equivalence point pH = 7.00, 12.5 mL HI soln, phenolphthalein; (b) initial pH = 11.62, equivalence point pH = 4.96, 40.0 mL HCl solution, methyl red **57.** (a) 11.9 mL; (b) 17.4 mL; (c) 17.5 mL **59.** (a) initial pH = 1.59, equiv pH = 7.43, bromothymol blue; (b) initial pH = 2.26, equiv pH = 8.11, thymol blue; (c) initial pH = 4.35, equiv pH = 10.20, alizarin yellow R **61.** Initial pH = 1.00; after 24.00 mL of HCl, pOH = 2.69; at the equivalence point (25.00 mL HCl added), pOH = 7.00; after addition of 33.00 mL HCl, pOH = 12.14 **63.** basic **65.** (a) $H_3PO_4(aq) + CO_3^{2-}(aq) \rightleftharpoons H_2PO_4^-(aq) + HCO_3^-(aq), H_2PO_4^-(aq) + CO_3^{2-}(aq) \rightleftharpoons HPO_4^{2-}(aq) + HCO_3^-(aq), HPO_4^{2-}(aq) + OH^-(aq) \rightleftharpoons PO_4^{3-}(aq) + H_2O(l)$; (b) Na_2CO_3 is not a strong enough base to remove the third proton of H_3PO_4. **67.** $K_{a_1} = 1.6 \times 10^{-3}$, $K_{a_2} = 1.0 \times 10^{-8}$ **69.** (a) 0.0038 M; (b) 0.49 M **71.** (a) not possible; (b) Solution ends up with an appreciable $[HC_2H_3O_2]$; (c) yes; (d) We end up with $BaCl_2(aq)$, $NH_3(aq)$, unreacted $NH_4Cl(aq)$; (e) yes; (f) not possible, solution would be basic.

Integrative and Advanced Exercises 73. (b) 2.8% NaCl, (c) bromthymol blue or phenol red **74.** (a) 27.8 ml; (b) 3.3 ml; (c) 2.1×10^2 ml **76.** (a) no sudden change; (b) % HCl yet to be neutralized = 18% **79.** (a) pH = pK_a + $\log[f/(1-f)]$; (b) 9.57 **80.** (a) pH = 7.40, $[HPO_4^{2-}]/[H_2PO_4^-]$ = 1.6; (b) 26 g $Na_2HPO_4 \cdot 12 H_2O$ **81.** 3 drops **83.** (a) reaction 1: 1×10^{-14}, reaction 2: 1.8×10^9; (b) the large size of the equilibrium constant **87.** (a) initial pH = 8.35, final pH = 3.33; (b) initial pH = 12.15, 90% titrated: pH = 9.38; (c) 119 ml 0.100 M HCl; (d) 189 ml 0.100 M HCl; (e) 8.3% Na_2CO_3 **91.** (a) no significant change; (b) pH = 4.40; (c) pH = 3.82 **92.** pH at equivalence = 7.9, bromthymol blue or phenol red, phosphate buffer, $[HPO_4^{2-}]/[H_2PO_4^-]$ = 5 **93.** $pK_a = 12.2$ **94.** pH = 8.23

Feature Problems 98. (a) The curves cross at the half-equivalence point in the titration, where pH = pK_a = 4.74. (b) There are three carbonate containing species: H_2CO_3, HCO_3^- and CO_3^{2-}. Intersections occur at the half-equivalence points in each stepwise titration at pH = pK_{a_1} = 6.36 and pH = pK_{a_2} = 10.33; (c) There are four phosphate containing species: H_3PO_4, $H_2PO_4^-$, HPO_4^{2-}, and PO_4^{3-}. Points of intersection should occur at pH = pK_{a_1} = 2.15, pH = pK_{a_2} = 7.20 and pH = pK_{a_3} = 12.38

CHAPTER 19

Practice Examples 1A. $[Mg^{2+}][CO_3^{2-}]$, $[Ag^+]^3[PO_4^{3-}]$ **1B.** $CaHPO_4(s) \rightleftharpoons Ca^{2+}(aq) + HPO_4^{2-}(aq)$, $K_{sp} = [Ca^{2+}][HPO_4^{2-}]$ **2A.** 3×10^{-7} **2B.** 1.9×10^{-9} **3A.** 2×10^{-10} M **3B.** 0.55 mg **4A.** 1.3×10^{-4} M **4B.** 1×10^{-20} M **5A.** Precipitation should occur. **5B.** 10 drops **6A.** Precipitation is not complete. **6B.** 0.02 M **7A.** 0.12% **7B.** Sodium chromate is the best precipitating agent; $[Ba^{2+}] = 5.5 \times 10^{-7}$ M **8A.** Precipitation will not occur. **8B.** Precipitation should occur. **9A.** 0.074 M **9B.** 8.9×10^{-3} M **10A.** (a) $Cu^{2+}(aq) + 2\,OH^-(aq) \longrightarrow Cu(OH)_2(s)$; (b) $Cu(OH)_2(s) + 4\,NH_3(aq) \rightleftharpoons [Cu(NH_3)_4]^{2+}(aq) + 2\,OH^-(aq)$; (c) $[Cu(NH_3)_4]^{2+}(aq) + 4\,H_3O^+(aq) \rightleftharpoons [Cu(H_2O)_4]^{2+}(aq) + 4\ NH_4^+(aq)$ **10B.** (a) $Zn^{2+}(aq) + 4\,NH_3(aq) \rightleftharpoons [Zn(NH_3)_4]^{2+}(aq)$; (b) $[Zn(NH_3)_4]^{2+}(aq) + 4\,H_3O^+(aq) \rightleftharpoons [Zn(H_2O)_4]^{2+}(aq) + 4\,NH_4^+(aq)$; (c) $[Zn(H_2O)_4]^{2+}\ (aq) + 2\ OH^-(aq) \rightleftharpoons Zn(OH)_2(s) + 4\ H_2O(l)$; (d) $Zn(OH)_2(s) + 2\ OH^-(aq) \rightleftharpoons [Zn(OH)_4]^{2-}(aq)$ **11A.** Precipitation should occur. **11B.** Precipitation will not occur. **12A.** 0.84 M **12B.** 0.20 M **13A.** 4×10^{-6} M **13B.** Solubility decreases, as does K_{sp}. **14A.** For FeS Q_{spa} is less than K_{spa}, precipitation of FeS(s) will not occur; For Ag_2S, Q_{spa} is greater than K_{spa}, precipitation of $Ag_2S(s)$ will occur. **14B.** pH = 2.7

Review Questions 4. (a) $[Ag^+]^2[SO_4^{2-}]$; (b) $[Ra^{2+}][IO_3^-]^2$; (c) $[Ni^{2+}]^3[PO_4^{3-}]^2$; (d) $[PuO_2^{2+}]\,[CO_3^{2-}]$ **5.** (a) $Fe(OH)_3(s) \rightleftharpoons Fe^{3+}(aq) + 3\ OH^-(aq)$; (b) $BiOOH(s) \rightleftharpoons BiO^+(aq) + OH^-(aq)$; (c) $Hg_2I_2(s) \rightleftharpoons Hg_2^{2+}(aq) + 2\ I^-(aq)$; (d) $Pb_3(AsO_4)_2(s) \rightleftharpoons 3\ Pb^{2+}(aq) + 2\ AsO_4^{3-}(aq)$ **6.** (a) $[Cr^{3+}][F^-]^3$; (b) $[Au^{3+}]^2[C_2O_4^{2-}]^3$; (c) $[Cd^{2+}]^3[PO_4^{3-}]^2$; (d) $[Sr^{2+}][F^-]^2$ **7.** (a) $s = 1.1 \times 10^{-5}$ M; (b) $s = 2.2 \times 10^{-2}$ M; (c) $s = 7 \times 10^{-5}$ M; (d) $s = 4.5 \times 10^{-5}$ M **8.** (a) 1.4×10^{-5}; (b) 8.8×10^{-8}; (c) 2.7×10^{-11} **9.** (d) is correct **10.** (a) $s = 1.7 \times 10^{-4}$ M; (b) $s = 7.3 \times 10^{-6}$ M; (c) $s = 1.4 \times 10^{-8}$ M **11.** (a) lowers solubility; (b) enhances solubility; (c) enhances solubility **12.** (a) Precipitation should occur; (b) Precipitation should not occur; (c) Precipitation should not occur. **13.** 0.014% unprecipitated **14.** (a) AgI; (b) $[I^-] = 2.7 \times 10^{-4}$ M; (c) $[Ag^+] = 3.1 \times 10^{-13}$ M; (d) Ag^+ and Pb^{2+} can be separated. **15.** $NH_4Cl(aq)$ **16.** (a) $Ag^+(aq) + NO_3^-(aq) + Na^+(aq) + Br^-(aq) \longrightarrow AgBr(s) + Na^+(aq) + NO_3^-(aq)$; (b) no reaction; (c) no reaction; (d) $Cu(OH)_2(s) + 4\,NH_3(aq) \longrightarrow [Cu(NH_3)_4]^{2+}(aq) + 2\ OH^-(aq)$; (e) $Fe^{3+}(aq) + 3\ NH_3(aq) + 3\ H_2O \longrightarrow Fe(OH)_3(s) + 3\ NH_4^+(aq)$; (f) $Ag_2SO_4(s) + 4\,NH_3(aq) \longrightarrow 2\,[Ag(NH_3)_2]^+(aq) + SO_4^{2-}(aq)$; (g) $CaSO_3(s) + 2\ H_3O^+(aq) \longrightarrow Ca^{2+}(aq) + 3\ H_2O(l) + SO_2(g)$ **17.** $Cu(OH)_2(s) + 2\ H_3O^+\ (aq) \longrightarrow Cu^{2+}(aq) + 4\ H_2O(l)$ (with both HCl and HNO_3); $Cu(OH)_2(s) + 4\ NH_3(aq) \longrightarrow [Cu(NH_3)_4]^{2+}(aq) + 2\ OH^-(aq)$ **18.** HCl(aq) **19.** $NH_3(aq)$. **20.** Precipitation should occur. **21.** Precipitation should not occur. **22.** 2.8×10^{-3} M.

Exercises 23. $AgI < AgCN < AgIO_3 < Ag_2SO_4 < AgNO_2$ **25.** yes **27.** 11 mg **29.** 4.5×10^{-3} M **31.** 5.30×10^{-5} **33.** Both KI and $AgNO_3$ reduce the solubility through the common ion effect. **35.** 1.5×10^{-5} **37.** 0.079 M **39.** not with CrO_4^{2-}; yes with Ag^+ **41.** 27 ppm **43.** Precipitation should occur. **45.** 8.32 **47.** (a) Precipitation of AgCl(s) should occur; (b) Precipitation of AgBr(s) should occur; (c) Precipitation of $Mg(OH)_2(s)$ will not occur. **49.** CaC_2O_4 should precipitate from this solution. **51.** 0.052% **53.** 2.5%; 0.16 M **55.** The concentration of the $AgNO_3$ establishes the volume required but does not otherwise affect the separation. **57.** (a) CrO_4^{2-}; (b) 1.6×10^{-6} M; (c) The two anions are effectively separated. **59.** $MgCO_3$, FeS, $Ca(OH)_2$ **61.** 9.60 **63.** (a) 3.41; (b) 0.96 g **65.** $[PbCl_3]^-(aq)$ forms in HCl(aq) **67.** 2.0×10^{30} **69.** Precipitation of AgI(s) should occur because $Q_{sp} > K_{sp}$. **71.** 1.4×10^{-6} g **73.** With $[H_3O^+]$ a bit higher than 1.8×10^{-5} M, separation is essentially complete. **75.** (a) Q_{spa} $(1.7 \times 10^7) < K_{spa}$, therefore, precipitation of MnS (s) will not occur. (b) Raise $[C_2H_3O_2^-]$ to 0.21 M **77.** Pb^{2+} would be missed and might interfere with the test for Ag^+. **79.** (a) and (c) are the valid conclusions.

Integrative and Advanced Exercises 81. 68% **83.** 38% unprecipitated **85.** 0.012 M **88.** 2 g MnS/L **91.** (a) yes; (b) no; (c) yes **94.** 1.2×10^{-16} **96.** 1.6×10^{-3} M **97.** 6.5×10^{-4} M **99.** (a) $Ag_2SO_4(s) + Ba^{2+}(aq) + 2\ Cl^-(aq) \longrightarrow BaSO_4(s) + 2\ AgCl(s)$; (b) 0.63 g $Ag_2SO_4(s)$

Feature Problems 100. 3.0×10^{-4} **102.** (a) 1.7×10^{-4} M; (b) 1.7×10^{-4} M; (c) 2.8×10^{-5} M; (d) $[Mg^{2+}] = 0.059$ M(precipitation barely occurs); (e) 1.4×10^{-9} M (precipitation occurs)

CHAPTER 20

Practice Examples 1A. (a) decreases (b) increases **1B.** (a) uncertain (b) increases **2A.** 83.0 J mol^{-1} K^{-1} **2B.** 402 J/mol **3A.** per mol $NH_3(g)$: -99.4 J K^{-1} **3B.** 312.4 J mol^{-1} K^{-1} **4A.** (a) case 2; (b) case 4 **4B.** (a) high temperatures; (b) low temperatures **5A.** -1484 kJ **5B.** -70.48 kJ **6A.** (a) $K_{eq} = P\{SiCl_4\}/P\{Cl_2\}^2 = K_p$; (b) $K_{eq} = [HOCl][H^+][Cl^-]/P\{Cl_2\}$ **6B.** $K = [Pb^{2+}]^3\ P\{NO\}^2/\{[NO_3^-]^2[H^+]^8\}$ **7A.** $\Delta G^\circ_{rxn} = +4.73$ kJ; reaction as written is nonspontaneous at 298 K. **7B.** NO_2 will spontaneously convert into N_2O_4 (Q > K). **8 A.** 8.5×10^{-17} **8B.** $K_{eq} = 4 \times 10^{-5}$, no appreciable reaction **9A.** 607 K **9B.** (a) 1.5×10^7; (b) 1.7×10^5 **10A.** 1240 K **10B.** 5×10^9.

Review Questions 4. (a) increase; (b) decrease; (c) uncertain; (d) decrease **5.** (a) $H_2O(g)$; (b) 50.0 g Fe; (c) $Br_2(l)$; (d) $SO_2(g)$ **6.** (a) 3; (b) 2; (c) 1; (d) 4 **7.** (b) is correct **8.** (d) is correct **9.** (d) is correct **10.** -285 J mol^{-1} K^{-1} **11.** (a) -242.0 kJ; (b) $+141.8$ kJ; (c) $+101$ kJ; (d) $-970.$ kJ **12.** (a) 114 °C; (b) $\Delta G^\circ = 0$ **13.** (a) 85.9 J mol^{-1} K^{-1}; (b) 7.014 J mol^{-1} K^{-1} **14.** (a) $K_{eq} = P\{NO_2(g)\}^2/P\{NO(g)\}^2\,P\{O_2\}] = K_p$; (b) $K_{eq} = P\{SO_2(g)\} = K_p$; (c) $K_{eq} = [H_3O^+][C_2H_3O_2^-]/[HC_2H_3O_2] = K_c$; (d) $K_{eq} = P\{H_2O(g)\}\,P\{CO_2(g)\} = Kp$; (e) $K_{eq} = [Mn^{2+}(aq)]\,P\{Cl_2(g)\}/\{[H^+]^4\,[Cl^-]^2\}$; this is neither K_p nor K_c. **15.** 127 kJ/mol **16.** 8×10^{-13} **17.** (a) -76.0 kJ, 2.2×10^{13}; (b) -32 kJ, $K_{eq} = 4 \times 10^5$; (c) -28.1 kJ, 8×10^4 **18.** -42.1 J mol^{-1} K^{-1} **19.** (a) does not occur to any measurable extent; (b) occurs only to a very small extent; (c) occurs to some small extent. **20.** 1.36×10^3 K **21.** (a) 34.4 kJ; (b) 9×10^{-7} **22.** (a) 215 J K^{-1}; (b) +91 kJ; (c) 27 kJ; (d) 2×10^{-5} **23.** (a) $\Delta S^\circ = -227.4$ J K^{-1}, $\Delta H^\circ = -165.7$ kJ, $\Delta G^\circ = -97.9$ kJ; (b) low temperatures, high pressures; (c) 0.5 **24.** 3×10^{16}

Exercises 25. (a) decrease; (b) increase; (c) increase **27.** First law states energy is neither created nor destroyed; second law states that total entropy increases in all spontaneous processes. **29.** (a) negative; (b) positive; (c) positive; (d) uncertain; (e) negative **31.** (a) $\Delta H^\circ_{vap} = +44.0$ kJ/mol, $\Delta S^\circ_{vap} = 118.9$ J mol^{-1} K^{-1} (b) Hydrogen bonding in the liquid weakens as the temperature is raised, causing ΔS_{vap} to increase and ΔH_{vap} to decrease. **33.** $C_6H_5CH_3$ **35.** 354 K (81 °C) **37.** (a) 2; (b) unable to predict; (c) 3 **39.** $\Delta G < 0$, $\Delta H = 0$, $\Delta S > 0$ (for non-reacting gases) **41.** (a) it may be that $\Delta S^\circ < 0$; (b) the process is endothermic. **43.** (a) -506 kJ; (b) forward **45.** (a) -1.6 kJ, equilibrium; (b) -474.3 kJ, goes to completion; (c) -574.2 kJ, goes to completion. **47.** $C_6H_6(l) + 15/2\ O_2(g) \longrightarrow 6\ CO_2(g) + 3\ H_2O(l\ or\ g)$; (a) -3202 kJ; (b) -3177 kJ; (c) 25.5 kJ **49.** (a) -54.5 J K^{-1}; (b) -616 kJ (actual value is -636.6 kJ); (c) $-600.$ kJ (reaction is feasible and is favored at low temperatures) **51.** (a) $K_{eq} = K_p = K_c(RT)^{-1}$; (b) $K_{eq} = K_p = K_c$; (c) $K_{eq} = K_p = K_c(RT)^2$ **53.** $\Delta G_{nonstd} = -104$ kJ, therefore, the reaction is spontaneous in the forward direction. **55.** $\Delta G = -1.46$ kJ, spontaneous in the forward direction. **57.** (b) **59.** (a) 0.659; (b) 3.47 kJ; (c) proceeds to the right **61.** (a) $K_{eq} = K_p, \Delta G^\circ = -23.4$ kJ; (b) $K_{eq} = K_p$, $\Delta G^\circ = +68.9$ kJ; (c) $K_{eq} = K_p$, $\Delta G^\circ = +5.40$ kJ; (d) $K_{eq} = K_c, \Delta G^\circ = +28.8$ kJ **63.** -204.6 kJ mol^{-1} (calculated value in excellent agreement with value in appendix D) **65.** (a) $P_{O_2} = 3.0 \times 10^{-21}$ atm; (b) high temperature, oxygen was continuously removed. **67.** 6.0×10^{-4} **69.** ~650 K **71.** (a) 0.014; (b) 329 K **73.** -2.1×10^2 kJ, -206.1 kJ (assumption is reasonable) **75.** (a) $\Delta G^\circ = +152.1$ kJ; (b) $\Delta G^\circ = -362.3$ kJ (consequently, the coupled reaction is spontaneous)

Integrative and Advanced Exercises 79. (a) and (d) are true **80.** 0.208 mol Br_2, 0.208 mol Cl_2, 0.584 mol BrCl **84.** 2.01×10^3 K **86.** 168 J/K **89.** 3.70 atm **91.** 297 K **92.** stable at 25 °C, not stable at 200 °C **94.** 647 K **95.** 4 mg AgBr/L

Feature Problems 98. (a) $\Delta G^\circ = 8.557$ kJ/mol (ΔG°_f values) or $\Delta G^\circ = 8.559$ kJ/mol (using $\Delta H^\circ - T\Delta S^\circ = 44.012$ kJ/mol, $\Delta S^\circ = 118.92$ J K^{-1} mol^{-1}; (b) 0.0317 bar; (c) 23.8 torr; (d) 23.8 torr **99.** (a) The graph for the formation of MgO(s) is below that of ZnO(s) at all temperatures. (b) The graph for CO(g) must be below that for ZnO(s), that is, at temperatures above 1000 °C. (c) Yes, but only above 1850 °C. (d) No, $\Delta G^\circ > 0$ at all temperatures for this reaction. (e) $\Delta S^\circ > 0$ for one reaction, $\Delta S^\circ < 0$ for another, and $\Delta S^\circ \approx 0$ for a third; (f) The C-CO line has a negative slope and, at some temperature, intersects all the metal-metal oxide lines, which have positive slopes. **100.** (a) 4.9×10^2 K; (b) to allow for loss of efficiency beyond the thermodynamic limitations (e.g., compensate for frictional losses). (c) 30. atm; (d) For greater than 100% efficiency, T_l would have to be a negative quantity—an impossibility; for 100% efficiency, $T_l = 0$, which is also unattainable.

CHAPTER 21

Practice Examples 1A. $Sc(s) + 3\ Ag^+(aq) \longrightarrow Sc^{3+}(aq) + 3\ Ag(s)$ **1B.** $Al(s) + 3\ Ag^+(aq) \longrightarrow Al^{3+}(aq) + 3\ Ag(s)$, $Al|Al^{3+}(aq)||Ag^+(aq)|Ag$ **2A.** +0.587 V **2B.** +0.74 V **3A.** -0.48 V **3B.** -0.424 V **4A.** -1587 kJ **4B.** +1.229 V **5A.** Copper will displace $Ag^+(aq)$; $E^\circ_{cell} = +0.460$ V **5B.** Na(s) reacts with H_2O to produce $H_2(g)$ ($E^\circ_{cell} = +1.885$ V) rather than Mg(s) ($E^\circ_{cell} = +0.357$ V). **6A.** $E^\circ_{cell} = -0.78$ V,not a feasible method. **6B.** (1) Oxidation of Sn^{2+} to Sn^{4+} by $O_2(g)$ ($E^\circ_{cell} = +1.075$ V); (2) oxidation of Sn(s) to Sn^{2+} ($E^\circ_{cell} = +1.366$ V), and also reaction of Sn(s) with Sn^{4+} to form Sn^{2+} ($E^\circ_{cell} = +0.291$ V). **7A.** $K_{eq} = 10^{204}$; reaction will go to completion. **7B.** $K_{eq} = 2.5$; reaction does not go to completion. **8A.** +1.815 V **8B.** +0.017 V **9A.** +0.467 V, spontaneous as written **9B.** 2.6 **10A.** +0.23 V **10B.** 6.9×10^{-9} **11A.** $I_2(s)$, KOH(aq), and

$H_2(g)$ **11B.** Ag(s) plates out at the cathode while Ag(s) is oxidized to $Ag^+(aq)$ at the anode. **12A.** 1.89 A **12B.** 5.23 h

Review Questions 4. (d) is correct **5.** (a) Fe(s) + $Cu^{2+}(aq) \longrightarrow Fe^{2+}(aq) + Cu(s)$; (b) 2 $Br^-(aq)$ + $Cl_2(aq) \longrightarrow Br_2(aq) + 2$ $Cl^-(aq)$; (c) Al(s) + 3 $Fe^{3+}(aq) \longrightarrow Al^{3+}(aq) + 3 Fe^{2+}(aq)$; (d) 5 $Cl^-(aq)$ + 6 $MnO_4^-(aq) + 18 H^+(aq) \longrightarrow 5 ClO_3^-(aq)$ + 6 $Mn^{2+}(aq) + 9 H_2O(l)$; (e) $S^{2-}(aq) + 2 O_2(g) \longrightarrow$ $SO_4^{2-}(aq)$ **6.** (a) −1.180 V; (b) +0.37 V; (c) −1.63 V; (d) −1.13 V **7.** (a) 2 Al(s) + 3 $Sn^{2+}(aq) \longrightarrow$ 2 $Al^{3+}(aq) + 3$ Sn(s), +1.539 V; (b) $Fe^{2+}(aq)$ + $Ag^+(aq) \longrightarrow Fe^{3+}(aq) + Ag(s)$, +0.029 V **8.** (a) 0.097 V; (b) −2.03 V; (c) +0.159 V **9.** (a) spontaneous; (b) nonspontaneous; (c) nonspontaneous; (d) spontaneous **10.** 1 M HCl: no; 1 M HNO_3: yes **11.** (a) significant; (b) significant; (c) barely; (d) significant; (e) will not **12.** Reaction stops well short of completion, $K_{eq} = 7$. **13.** (a) and (b) will occur. **14.** (a) -1.165×10^3 kJ; (b) −268 kJ; (c) -3.1×10^2 kJ **15.** (a) $1 \times 10^{-22} = [Sn^{2+}][Ag^+]^2/[Sn^{4+}]$; (b) $4 \times 10^{-5} =$ $[Mn^{2+}]$ $P\{Cl_2(g)\}/\{[Cl^-]^2[H^+]^4\}$; (c) $1 \times 10^{33} =$ $[Cl^-]^2$ $P\{O_2(g)\}/[OCl^-]^2$ **16.** (a) Fe(s) + $Cu^{2+}(aq)$ $\longrightarrow Fe^{2+}(aq) + Cu(s)$, +0.780 V, electron flow from B to A; (b) $Sn^{2+}(aq) + 2 Ag^+(aq) \longrightarrow Sn^{4+}(aq) + 2$ Ag(s), +0.646 V, electron flow from A to B; (c) Zn(s) + $Fe^{2+}(aq) \longrightarrow Zn^{2+}(aq) + Fe(s)$, +0.264 V, electron flow from A to B. **17.** 5.0×10^{-6} M **18.** (a) 0.311 V; (b) 0.118 V; (c) 0.164 V **19.** (a) Anode:$Cl_2(g)$, cathode:Cu(s); (b) Anode:$O_2(g)$, cathode: $H_2(g)$ and $OH^-(aq)$; (c) Anode:$Cl_2(g)$, cathode:Ba(l); (d) Anode: $O_2(g)$, cathode: $H_2(g)$ and $OH^-(aq)$ **20.** $H_2(g)$ and $OH^-(aq)$ are produced at the cathode instead of Mg(s). **21.** (a) 3.3 g; (b) 0.90 g; (c) 11 g; (d) 2.9 g **22.** 1.27 L **Exercises 23.** (a) $0.340 V < E° < 0.800 V$; (b) $-0.440 V < E° < 0.000$ V **25.** +0.755 V **27.** −2.31 V **29.** Sodium reacts with water, producing $H_2(g)$ rather than Zn(s). **31.** (a) 3 Ag(s) + $NO_3^-(aq)$ +4 $H^+(aq) \longrightarrow 3 Ag^+(aq) + NO(g) + 2 H_2O(l)$, $E_{cell} = 0.156$ V; (b) Zn(s) + 2 $H^+(aq) \longrightarrow$ $Zn^{2+}(aq) + H_2(g)$, $E_{cell}° = 0.763$ V; (c) Au(s) does not react with 1.00 M $HNO_3(aq)$ **33.** (a) Cu(s) + 2 $Fe^{3+}(aq) \longrightarrow Cu^{2+}(aq) + 2 Fe^{2+}(aq)$, +0.431 V; (b) 2 Al(s) + 3 $Pb^{2+}(aq) \longrightarrow 2$ $Al^{3+}(aq)$+3 Pb(s), +1.551 V; (c) 2 $H_2O(l) + 2 Cl_2(g) \longrightarrow O_2(g) + 4$ $H^+(aq) + 4$ $Cl^-(aq)$, +0.129 V; (d) 3 Zn(s) + 2 $NO_3^-(aq) + 8 H^+(aq) \longrightarrow 3 Zn^{2+}(aq) + 2 NO(g)$ + 4 $H_2O(l)$, +1.719 V **35.** (a) +0.25 V; (b) -2.4×10^2 kJ; (c) 1×10^{42}; (d) Goes to completion, K_{eq} very large **37.** (a) not possible; (b) a net reaction to the left will occur. **39.** 1.591 V **41.** 0.923 V **43.** (a) 1.249 V; (b) 0.638 V **45.** (a) E depends on pH only if the reduction half-reaction involves either $H^+(aq)$ or $OH^-(aq)$. (b) Reduction of oxoanions is accompanied by the loss of oxygen atoms, requiring H^+ ions on the left. (c) Switch from acidic to basic solution by adding OH^- to both sides. On the left side H^+ and OH^- combine to form H_2O; OH^- remains on the right. **47.** (a) 6×10^{-38} M; (b) goes to completion **49.** (a) +0.818 V; (b) −0.828 V; difference due to reaction reversal and standard (1 M) rather than nonstandard concentration (0.65 M). **51.** 0.177 V **53.** (a) 0.039 V; (b) decrease; (c) 0.025 V; (d) 0.237 M; (e) $[Sn^{2+}] = 0.50$ M, $[Pb^{2+}] = 0.18$ M **55.** (a) 14 $H^+(aq) + Cr_2O_7^{2-}(aq) + 6 Cl^-(aq) \longrightarrow 2 Cr^{3+}(aq)$ + 7 $H_2O(l) + 3 Cl_2(g)$, $E_{cell}° = -0.03$ V; since cell voltage is negative, oxidation of $Cl^-(aq)$ to $Cl_2(g)$ by $Cr_2O_7^{2-}(aq)$ under standard-state conditions won't occur spontaneously. (b) We can drive the reaction in the forward direction by using high $[H^+]$ and $[Cl^-]$

and removing the $Cl_2(g)$ produced. **57.** (a) Cr(s) $|Cr^{2+}(aq), Cr^{3+}(aq)||Fe^{2+}(aq), Fe^{3+}(aq)|$ Fe(s); (b) +1.195 V **59.** (a) 1.229 V; (b) +1.992 V; (c) +2.891 V **61.** The Al-air battery has the greatest quantity of charge transferred per unit mass of metal oxidized because it has the smallest molar mass of the three metals and forms the most highly charged cation of the three metals. **63.** (a) Oxidation of iron metal should be enhanced in the body of the nail; (b) blue precipitate in the vicinity of the scratch; (c) a faint white, but no blue precipitate. **65.** Hooking up an inert electrode to the metal pipe, along with the application of a small voltage across the two metals, converts the metal pipe into a cathode, thereby making it resistant to oxidation. **67.** (a) electrolysis, at >1.229 V; (b) spontaneous; (c) electrolysis, at > 0.236 V; (d) electrolysis, at > 0.183 V **69.** (a) $H_2(g)$, $O_2(g)$; (b) 2 $H_2O(l) \longrightarrow 2 H_2(g) + O_2(g)$, $E_{cell}° = -1.229$ V **71.** (a) 1.62 g; (b) 20.2 min **73.** (a) 1079 C; (b) 0.7642 A.

Integrative and Advanced Exercises 75. −0.255 V **77.** (a) 9.38×10^6 kJ; (b) 2.61×10^3 kWh **80.** 6.9×10^{-4} M **82.** +127.3 kJ **85.** (a) 1.1501 V; (b) 0.079 V **87.** 1.143 V **90.** 0.051% Ag **92.** $[Ag^+] = 1.8 \times 10^{-10}$M, $E = 0.223$ V, SHE is anode **96.** (1) metal must have reduction potential < -0.691 V, be readily available; (2) net: 2 Al(s) + 3 $Ag_2S(s) \longrightarrow 6$ Ag(s) + 3 $S^{2-}(aq)$ + 2 $Al^{3+}(aq)$; (3) $NaHCO_3$ serves as an electrolyte; (4) chemicals are used, Al is consumed

Feature Problems 99. (a) (1) Na(s) $\longrightarrow$ Na(amalg, 0.206%), (2) Na(amalg, 0.206%) + 2 $H^+(1 M) \longrightarrow 2 Na^+(1 M) + H_2(g, 1$ atm); (b) (1) −81.56 kJ, (2) −360.33 kJ; (c) 2 Na(s) + 2 $H^+(aq, 1 M) \longrightarrow 2 Na^+(1 M) + H_2(g, 1$ atm) $\Delta G° = -523.45$ kJ; (d) $E_{cell}° = +2.713$ V,so $E_{Na^+/Na}° = -2.713$ V **101.** (a) 2.66×10^{-13} F; (b) 2.26×10^{-14} C; (c) 1.41×10^5 K^+ ions; (d) 0.93×10^{12} K^+ ions; (e) Only about 1.5×10^{-5}% of the K^+ ions are transferred.

CHAPTER 22

Practice Examples 1A. electrolysis of NaCl(aq) to form NaOH(aq), followed by addition of $NO_2(g)$ to NaOH(aq) **1B.** electrolysis of NaCl(aq) to produce NaOH(aq), the reaction of $SO_2(g)$ with NaOH(aq), and the addition of S to the boiling solution **2A.** pH = 2.394 **2B.** 9.23×10^{-3} M

Review Questions 4. (a) lead(II) oxide; (b) tin(II) fluoride; (c) calcium sulfate hemihydrate; (d) Li_3N; (e) calcium hydroxide; (f) KO_2; (g) $Mg(HCO_3)_2$ **5.** (a) $Li_2CO_3(s) \xrightarrow{heat} Li_2O(s) + CO_2(g)$; (b) $CaCO_3(s) + 2 HCl(aq) \longrightarrow CaCl_2(aq) + H_2O(l)$ + $CO_2(g)$; (c) 2 Al(s) + 2 $Na^+(aq)$ +2 $OH^-(aq)$ + 6 $H_2O(l) \longrightarrow 2$ $Na^+(aq) + 2$ $[Al(OH)_4]^-(aq) + 3$ $H_2(g)$; (d) BaO(s) + $H_2O(l) \longrightarrow Ba(OH)_2(s)$; (e) 2 $Na_2O_2(s) + 2 CO_2(g) \longrightarrow 2 Na_2CO_3(s) + O_2(g)$ **6.** (a) $MgCO_3(s) + 2 HCl(aq) \longrightarrow MgCl_2(aq) +$ $CO_2(g) + H_2O(l)$; (b) 2 Na(s) + 2 $H_2O(l) \longrightarrow$ 2 NaOH(aq) + $H_2(g)$, followed by 2 Al(s) + 2 NaOH(aq) + 6 $H_2O(l) \longrightarrow 2$ $Na[Al(OH)_4](aq)$ + 3 $H_2(g)$; (c) 2 NaCl(s) + H_2SO_4(concd aq) $\longrightarrow$ 2 HCl(g) + $Na_2SO_4(s)$ **7.** (a) $K_2CO_3(aq)$ + Ba(OH)$_2$(aq) $\longrightarrow BaCO_3(s) + 2$ KOH(aq); (b) $Mg(HCO_3)_2(aq) \xrightarrow{heat} MgCO_3(s) + CO_2(g) +$ $H_2O(l)$; (c) SnO(s) + C(s) $\xrightarrow{heat}$ Sn(l) + CO(g); (d) $CaF_2(s) + H_2SO_4$(concd aq) $\longrightarrow 2$ HF(g) + $CaSO_4(s)$; (e) $NaHCO_3(s) + HCl(aq) \longrightarrow$

NaCl(aq) + $H_2O(l) + CO_2(g)$; (f) $PbO_2(s) + 4$ HBr(aq) $\longrightarrow PbBr_2(s) + Br_2(l) + 2$ $H_2O(l)$ **8.** $CaSO_4 \cdot 2H_2O(s) + (NH_4)_2CO_3(aq) \longrightarrow (NH_4)_2S$ $O_4(aq) + CaCO_3(s) + 2 H_2O(l)$ **9.** NH_4Cl (because it lacks basic properties) **10.** (a) $SrCO_3(s) \xrightarrow{heat}$ $SrO(s) + CO_2(g)$; (b) $Al_2O_3(s) \xrightarrow{heat}$ no reaction; (c) $Li_2CO_3(s) \xrightarrow{heat} Li_2O(s) + CO_2(g)$ **11.** (a) $Pb(NO_3)_2(aq) + 2 NaHCO_3(aq) \longrightarrow PbCO_3(s) +$ $H_2O(l) + CO_2(g) + 2 NaNO_3(aq)$; (b) $Li_2O(s) +$ $(NH_4)_2CO_3(aq) \longrightarrow Li_2CO_3(aq) + 2$ $NH_3(aq) +$ $H_2O(l)$; (c) $H_2SO_4(aq) + BaO_2(s) \longrightarrow H_2O_2(aq) +$ $BaSO_4(s)$; (d) PbO(s) + $OCl^-(aq) \longrightarrow PbO_2(s) +$ $Cl^-(aq)$ **12.** Al, because it is protected by a water-insoluble oxide coating. **13.** (a) both yield $H_2(g)$ **14.** (a) $CaCO_3(s)$; (b) $CaSO_4 \cdot 2H_2O(s)$; (c) $Ca[CH_3(CH_2)_{14}COO]_2(s)$; (d) $BaSO_4(s)$; (e) Al_2O_3 with Fe^{3+} and Ti^{4+}

Exercises 15. (a) 2 Cs(s) + $Cl_2(g) \longrightarrow 2$ CsCl(s); (b) 2 Na(s) + $O_2(g) \longrightarrow Na_2O_2(s)$; (c) $Li_2CO_3(s) \xrightarrow{heat} Li_2O(s) + CO_2(g)$; (d) $Na_2SO_4(s) +$ 4 C(s) $\longrightarrow Na_2S(s) + 4$ CO(g); (e) K(s) + $O_2(g) \longrightarrow KO_2(s)$ **17.** flame test: LiCl (red), KCl (violet) **19.** $O^{2-}(aq) + H_2O(l) \longrightarrow 2$ $OH^-(aq)$; 2 $O_2^{2-}(aq)$ + 2 $H_2O(l) \longrightarrow 4$ $OH^-(aq) + O_2(g)$; 4 O_2^- + 2 $H_2O(l) \longrightarrow 4$ $OH^-(aq) + 3 O_2(g)$ **21.** (a) 11.16; (b) NaCl is in excess, solution pH is determined by the number of electrons that pass through the cell (this is assuming that the solution volume stays relatively constant). **23.** (a) 71.5%; (b) NH_3 in intermediate steps is recycled. **25.** $Ca(OH)_2(s) +$ 2 HCl(aq) $\longrightarrow CaCl_2(aq) + 2 H_2O(l)$; $CaCl_2(l) \xrightarrow{\text{molten salt electrolysis}} Ca(l) + Cl_2(g)$; $Ca(OH)_2(s)$ + $CO_2(g) \longrightarrow CaCO_3(s) + H_2O(g)$; $CaCO_3(s) \xrightarrow{heat} CaO(s) + CO_2(g)$; $Ca(OH)_2(s) + H_2SO_4(aq) \longrightarrow CaSO_4(s) + 2 H_2O(l)$; $Ca(OH)_2(s) + H_3PO_4$ (aq) $\longrightarrow CaHPO_4(aq) + 2 H_2O(l)$ **27.** $Mg^{2+}(aq) +$ 2 $Cl^-(aq) \longrightarrow$ Mg(s) + $Cl_2(g)$; thus the process does not violate the principle of conservation of charge. **29.** (a) $BeF_2(s) + Mg(s) \xrightarrow{heat}$ Be(s) + $MgF_2(s)$; (b) Ba(s) + $Br_2(l) \longrightarrow BaBr_2(s)$; (c)$UO_2(s) +$ 2 Ca(s) $\xrightarrow{heat}$ U(s) + 2 CaO(s); (d) $MgCO_3 \cdot$ $CaCO_3(s) \xrightarrow{heat} MgO(s) + CaO(s) + 2 CO_2(g)$; (e) 2 $H_3PO_4(aq) + 3$ CaO(s) $\longrightarrow Ca_3(PO_4)_2(s)$ + 3 $H_2O(l)$ **31.** (a) slightly to the left; (b) to the left; (c) far to the right **33.** 2 $HCO_3^-(aq) +$ $Ca(OH)_2(aq) \longrightarrow CaCO_3(s) + H_2O(l) +$ $CO_2(g) + 2$ $OH^-(aq)$ **35.** (a) 60.76 ppm; (b) 60.76 mg Ca^{2+}/L; (c) 112.3 g **37.** 85 ppm **39.** 40.2 g **41.** (a) 2 Al(s) + 6 HCl(aq) $\longrightarrow 2$ $AlCl_3(aq) + 3$ $H_2(g)$; (b) 2 NaOH(aq) + 2 Al(s) + 6 $H_2O(l) \longrightarrow 2$ $Na^+(aq) + 2$ $[Al(OH)_4]^-(aq) + 3$ $H_2(g)$; (c) 2 Al(s) + 3 $SO_4^{2-}(aq) + 12 H^+(aq) \longrightarrow 2 Al^{3+}(aq)$ + 3 $SO_2(g) + 6 H_2O(l)$ **43.** $Al^{3+}(aq) + 3 HCO_3^-(aq)$ $\longrightarrow Al(OH)_3(s) + 3 CO_2(g)$ **45.** Either acid or base can attack aluminum, but in neutral solutions the metal is protected by $Al_2O_3(s)$. **47.** $[Al(OH)_4]^-(aq) +$ $CO_2(aq) \longrightarrow Al(OH)_3(s) + HCO_3^-(aq)$ **49.** 2 KOH (aq) + 2 Al(s) + 6 $H_2O(l) \longrightarrow 2$ $K[Al(OH)_4](aq)$ + 3 $H_2(g)$; 2 $K[Al(OH)_4](aq) + 4$ $H_2SO_4(aq)$ $\longrightarrow K_2SO_4(aq) + Al_2(SO_4)_3(aq) + 8$ $H_2O(l)$ $\xrightarrow{\text{crystallize}} 2$ $KAl(SO_4)_2$ **51.** (a) PbO(s) + 2 $HNO_3(aq) \longrightarrow Pb(NO_3)_2(s) + H_2O(l)$; (b) $SnCO_3(s) \xrightarrow{heat} SnO(s) + CO_2(g)$; (c) PbO(s) + C(s) $\xrightarrow{heat}$ Pb(l) + CO(g); (d) 2 $Fe^{3+}(aq) +$

$Sn^{2+}(aq) \longrightarrow 2 Fe^{2+}(aq) + Sn^{4+}(aq)$; **(e)** $2 PbS(s) + 3 O_2(g) \xrightarrow{heat} 2 PbO(s) + 2 SO_2(g)$; $2 SO_2(g) + O_2(g) \longrightarrow 2 SO_3(g)$; $SO_3 + PbO \longrightarrow PbSO_4(s)$ **53. (a)** yes; **(b)** no; **(c)** no **Integrative and Advanced Exercises 56. (a)** 0.332 g; **(c)** 69% **60.** $Mg(OH)_2$ should precipitate **61. (a)** 81.4 ppm Ca^{2+}; **(b)** 48.7 g CaO; **(c)** $CaCO_3(s)$ can be precipitated if $[CO_3^{2-}] > 0.0014$ M; **(d)** 2.2×10^2 g **62. (a)** 268 kg Al; **(b)** 1.3 metric tons of coal **64.** 19 g $Pb(NO_3)_2$ **67. (a)** 7.4×10^{-3} M; **(b)** 6.4×10^{-4} M; **(c)** 2.0×10^{-3} M **70.** lattice energies: MgO: -3789 kJ, MgS: -3215 kJ **Feature Problems 71. (a)** 2.87 V **(b)** 3.03 V (3.040 V in Appendix D)

CHAPTER 23

Practice Examples 1A. 0.622 V **1B.** 1.453 V **2A.** $Kp = 2 \times 10^{-48}$ **2B.** $Kp = 4 \times 10^{-34}$, 4×10^{-15}% **3A.** $Na_2B_4O_7 \cdot 10H_2O(l) + H_2SO_4(l) \longrightarrow 4 B(OH)_3(s) + Na_2SO_4(s) + 5 H_2O(l)$, $2 B(OH)_3(s) \xrightarrow{heat} B_2O_3(s) + 3 H_2O(l)$, $2 B_2O_3(s) + 3 C(s) + 6 Cl_2(g) \xrightarrow{heat} 4 BCl_3(l) + 3 CO_2(g)$, $4 BCl_3(l) + 3 LiAlH_4(s) \longrightarrow 2 B_2H_6(g) + 3 LiCl(s) + 3 AlCl_3(s)$ **3B.** $2 ZnS(s) + 3 O_2(g) \longrightarrow 2 ZnO(s) + 2 SO_2(g)$, $2 SO_2(g) + O_2(g) \longrightarrow 2 SO_3(g)$ [V_2O_5 catalyst]; $H_2SO_4(l) + SO_3(g) \longrightarrow H_2S_2O_7(l)$, $H_2S_2O_7(l) + H_2O(l) \longrightarrow 2 H_2SO_4(l)$ **Review Questions 4. (a)** potassium bromate; **(b)** triiodide ion; **(c)** NaClO; **(d)** sodium dihydrogen phosphate; **(e)** lead(II) azide; **(f)** BaS_2O_3 **5. (a)** $CaCl_2(s) + H_2SO_4(concd\ aq) \xrightarrow{heat} CaSO_4(s) + 2 HCl(g)$; **(b)** no reaction; **(c)** $NH_3(aq) + HClO_4(aq) \longrightarrow NH_4ClO_4(aq)$ **6. (a)** $Cl_2(g) + 2 NaOH(aq) \longrightarrow NaCl(aq) + NaOCl(aq) + H_2O(l)$; **(b)** $2 I^-(aq) + SO_4^{2-}(aq) + 4 H^+(aq) \longrightarrow I_2(aq) + SO_2(g) + 2 H_2O(l)$; **(c)** $Cl_2(g) + 2 Br^-(aq) \longrightarrow 2 Cl^-(aq) + Br_2(l)$ **7. (a)** $KI(s) + H_3PO_4(l) \longrightarrow HI(g) + KH_2PO_4(s)$; **(b)** $Na_2SiF_6(s) + 4 Na(l) \xrightarrow{heat} Si(s) + 6 NaF(s)$ **8. (a)** $As_4O_6(s) + 6 CO(g) \longrightarrow 4 As(s) + 6 CO_2(g)$; **(b)** $NH_3(g) + H_3PO_4(aq) \longrightarrow NH_4H_2PO_4(aq)$, $2 NH_3(g) + H_3PO_4(aq) \longrightarrow (NH_4)_2HPO_4(aq)$ **9.** 91% **10.** $3 Cl_2(g) + I^-(aq) + 3 H_2O(l) \longrightarrow 6 Cl^-(aq) + IO_3^-(aq) + 6 H^+(aq)$, $Cl_2(g) + 2 Br^-(aq) \longrightarrow 2 Cl^-(aq) + Br_2(l)$ **11.** 15 km³ **12. (a)** $I_2(s)$; **(b)** $H_2O(l)$ **13. (a)** $KI(s) + H_3PO_4(concd\ aq) \xrightarrow{heat} KH_2PO_4(aq) + HI(g)$; **(b)** $K_2O(s) + H_2O(l) \longrightarrow 2KOH(aq)$; **(c)** $I_2(s) + KI(aq) \longrightarrow KI_3(aq)$ **(d)** $2 MnO_4^-(aq) + 5 HSO_3^-(aq) + H^+(aq) \longrightarrow 2 Mn^{2+}(aq) + 5 SO_4^{2-}(aq) + 3 H_2O(l)$ **14. (a)** sodium perxenate; **(b)** barium peroxide; **(c)** mercury(II) thiocyanate; **(d)** barium nitride; **(e)** silver thiosulfate **15. (a)** PCl_2F_3; **(b)** KNCO; **(c)** $FePO_4$; **(d)** $Ba(N_3)_2$; **(e)** $Mg_2(P_2O_7)$ **16. (a)** AgAt; **(b)** sodium perxenate; **(c)** MgPo; **(d)** tellurous acid; **(e)** K_2SeSO_3; **(f)** potassium perastatate

Exercises 17. (a) trigonal-pyramidal; **(b)** tetrahedral; **(c)** square-pyramidal **19.** $3 XeF_4(s) + 6 H_2O(l) \longrightarrow 2Xe(g) + 3/2 O_2(g) + 12 HF(g) + XeO_3(s)$ **21.** I^- ion is slowly oxidized to I_2 by O_2 in air: $4 I^-(aq) + O_2(g) + 4 H^+(aq) \longrightarrow 2 I_2(aq) + 2 H_2O(l)$ possibly followed by $I_2(aq) + I^-(aq) \longrightarrow I_3^-(aq)$ **23.** Each halogen is able to displace the members of the group below it, but not those above it; only F_2 can displace O_2 from water; none of the halogens can displace H_2 from water. **25. (a)** 1×10^6 kg F_2; **(b)** No, common oxidizing agents cannot oxidize F^- to F_2. **27.** 6 $Cl_2(g) + 6 H_2O(l) \longrightarrow 2 ClO_3^-(aq) + 12 H^+ +$

10 $Cl^-(aq)$; $E^\circ_{cell} = -0.095$ V; since the cell voltage is negative, the disproportionation reaction will not occur under standard state conditions. **29. (a)** T-shaped; **(b)** square pyramidal; **(c)** square planar **31.** pH = 5.89 **33.** 302 kJ/mol **35. (a)** H_2S forms very weak hydrogen bonds compared to $H_2O(l)$; **(b)** All electrons are paired. **37. (a)** $K_{eq} = 3.3 \times 10^{41}$, reaction goes essentially to completion; **(b)** $K_{eq} = 2 \times 10^{-65}$, essentially no reaction in the forward direction; **(c)** $K_{eq} = 9 \times 10^{20}$, reaction goes essentially to completion; **(d)** $K_{eq} = 5.6 \times 10^{-7}$, only slight reaction in the forward direction. **39. (a)** zinc sulfide; **(b)** potassium hydrogen sulfite; **(c)** potassium thiosulfate; **(d)** sulfur tetrafluoride **41. (a)** $FeS(s) + 2 HCl(aq) \longrightarrow FeCl_2(aq) + H_2S(aq)$; **(b)** $CaSO_3(s) + 2 HCl(aq) \longrightarrow CaCl_2(aq) + H_2O(l) + SO_2(g)$; **(c)** $SO_2(g) + MnO_2(s) \longrightarrow Mn^{2+}(aq) + SO_4^{2-}(aq)$; **(d)** $S_2O_3^{2-}(aq) + 2 H^+(aq) \longrightarrow S(s) + SO_2(g) + H_2O(l)$ **43.** $S_2O_3^{2-}(aq) + 2 H^+ (aq) \longrightarrow S(s) + SO_2(g) + H_2O(l)$; no reaction with SO_4^{2-} **45.** 1.21 **47.** 7.002 % Cu by mass **49. (a)** 2 $NO_2(g) \rightleftharpoons N_2O_4(g)$; **(b) (i)** $HNO_2(aq) + N_2H_5^+(aq) \longrightarrow HN_3(aq) + 2 H_2O(l) + H^+(aq)$; **(ii)** $HN_3(aq) + HNO_2(aq) \longrightarrow N_2(g) + H_2O(l) + N_2O$ **(c)** $H_3PO_4(aq) + 2 NH_3(aq) \longrightarrow (NH_4)_2HPO_4(aq)$

51. (a)

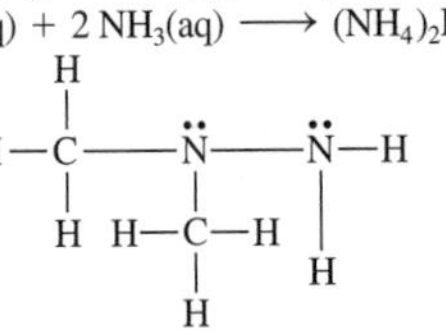

(b)

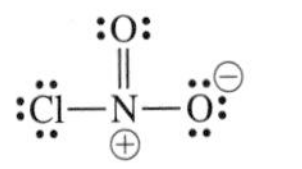

(c)

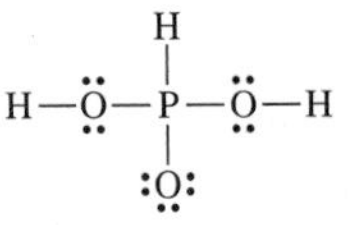

53. (a) hydrogen phosphate ion; **(b)** calcium pyrophosphate; **(c)** tetraphosphoric acid **55.** 1.031 V **57.** Graphite is the most stable form of carbon at room temperature and room pressure, but diamonds convert to graphite imperceptibly slowly under these conditions. **59. (a)** $3 SiO_2(s) + 4 Al(s) \xrightarrow{heat} 2 Al_2O_3(s) + 3 Si(s)$; **(b)** $K_2CO_3(s) + SiO_2(s) \longrightarrow CO_2(g) + K_2SiO_3(s)$; **(c)** $Al_4C_3(s) + 12 H_2O(l) \longrightarrow 3 CH_4(g) + 4Al(OH)_3(s)$ **61.** Silanes have the general formula Si_nH_{2n+2}, silanols substitute O—H for one or more of the H atoms (used to produce silicones). **63.** $2 CH_4(g) + S_8(s) \longrightarrow 2 CS_2(g) + 4 H_2S(g)$; $CS_2(g) + 3 Cl_2(g) \longrightarrow CCl_4(l) + S_2Cl_2(l)$; $4 CS_2(g) + 8 S_2Cl_2(g) \longrightarrow 4 CCl_4(l) + 3 S_8(s)$ **65.** $KAl_3(Si_3O_{10})(OH)_2$; Oxidation states are: K and H = +1, O = -2, Si = +4, Al = +3. Overall charge on compound = 0 = sum of the oxidation numbers. **67. (a)** Ten B—H bonds require 20 of the 22 valence electrons, leaving just two electrons to bond four B atoms together; **(b)** 4; **(c)** C_4H_{10} has a normal Lewis structure **69. (a)** $2 BBr_3 + 3 H_2 \longrightarrow 2 B + 6 HBr$; **(b) (i)** $B_2O_3(s) + 3 C(s) \xrightarrow{heat} 3 CO(g) + 2 B(s)$ **(ii)** $2 B(s) + 3 F_2(g) \xrightarrow{heat} 2 BF_3(g)$; **(c)** $2 B(s) + 3 N_2O(g) \xrightarrow{heat} 3 N_2(g) + B_2O_3(s)$

Integrative and Advanced Exercises 72. 2 $NH_4ClO_4(s) \longrightarrow N_2(g) + 4 H_2O(g) + Cl_2(g) + 2 O_2(g)$ **75.** 9.23 g/cm³ **77. (a)** 339 kJ/mol; **(b)** 353 nm (near UV) **78. (a)** 2.185 g $Ca_3(PO_4)_2$; **(b)** larger %P than $Ca_3(PO_4)_2$; **(c)** 92.4% BPL **82.** 15 g

H_2SO_4 and 37 g Cl_2 **83.** non-spontaneous at pH 4.1 and above **85.** $3 XeF_4(g) + 6 H_2O(l) \longrightarrow 2 Xe(g) + XeO_3(g) + 12 HF(aq) + 3/2 O_2(g)$ **87.** $[Cl_2] = 0.060$ M; $[HOCl] = [H_3O^+] = [Cl^-] = 0.030$ M; **88.** Based on $E^\circ = -3.09$ V for the reduction of N_2 to HN_3, and with data from Figure 23-23, E° for the reduction of HN_3 to NH_4^+ is 0.696 V. **Feature Problems 90.** E° ($ClO_2/HClO_2$) = 1.187 V, $E^\circ(ClO_3^-/ClO_2)$ = 1.175 V

CHAPTER 24

Practice Examples 1A. (a) $2 Cu_2S(s) + 3 O_2(g) \longrightarrow 2 Cu_2O(s) + 2 SO_2(g)$; **(b)** $WO_3(s) + 3 H_2(g) \longrightarrow W(s) + 3 H_2O(g)$; **(c)** $2 HgO(s) \xrightarrow{heat} 2 Hg(l) + O_2(g)$ **1B. (a)** $3 Si(s) + 2 Cr_2O_3(s) \xrightarrow{heat} 3 SiO_2(s) + 4 Cr(l)$; **(b)** $2 Co(OH)_3(s) \xrightarrow{heat} Co_2O_3(s) + 3 H_2O(g)$; **(c)** $Mn^{2+}(aq) + 2 H_2O(l) \longrightarrow MnO_2(s) + 4 H^+(aq) + 2 e^-$ **2A.** $NO_3^-(aq) + 3 V^{3+}(aq) + H_2O(l) \longrightarrow NO(g) + 3 VO^{2+}(aq) + 2 H^+(aq)$, $E^\circ_{cell} = +0.619$ V **2B.** possibilities: $-E^\circ[Cr^{2+}(aq)|Cr^{3+}(aq)] = +0.424$ V, $-E^\circ[Fe(s)|Fe^{2+}(aq)] = +0.440$ V, $-E^\circ[Zn(s)|Zn^{2+}(aq)] = +0.763$ V.

Review Questions 4. (a) scandium hydroxide; **(b)** copper(I) oxide; **(c)** titanium(IV) chloride; **(d)** vanadium(V) oxide; **(e)** potassium chromate; **(f)** potassium manganate **5. (a)** CrO_3; **(b)** $FeSiO_3$; **(c)** $BaCr_2O_7$; **(d)** CuCN; **(e)** $CoCl_2 \cdot 6H_2O$ **6. (a)** about 95% Fe, with 3–4% C and other impurities; **(b)** alloy of iron and manganese; **(c)** $Fe(CrO_2)_2$; **(d)** copper and zinc; **(e)** $HCl(aq)$ and $HNO_3(aq)$; **(f)** impure copper; **(g)** iron, and chromium and nickel **7. (a)** $TiCl_4(g) + 4 Na(l) \xrightarrow{heat} Ti(s) + 4 NaCl(s)$; **(b)** $Cr_2O_3(s) + 2 Al(s) \xrightarrow{heat} 2 Cr(l) + Al_2O_3(s)$; **(c)** no reaction; **(d)** $K_2Cr_2O_7(aq) + 2 KOH(aq) \longrightarrow 2 K_2CrO_4(aq) + H_2O(l)$; **(e)** $MnO_2(s) + 2 C(s) \xrightarrow{heat} Mn(l) + 2 CO(g)$ **8. (a)** $2 Fe_2S_3(s) + 3 O_2(g) + 6 H_2O(l) \longrightarrow 4 Fe(OH)_3(s) + 6 S(s)$; **(b)** $2 Mn^{2+}(aq) + 8 H_2O(l) + 5 S_2O_8^{2-}(aq) \longrightarrow 2 MnO_4^-(aq) + 16 H^+(aq) + 10 SO_4^{2-}(aq)$; **(c)** $4 Ag(s) + 8 CN^-(aq) + O_2(g) + 2 H_2O(l) \longrightarrow 4 [Ag(CN)_2]^-(aq) + 4 OH^-(aq)$ **9. (a)** $Cr(s) + 2 HCl(aq) \longrightarrow CrCl_2(aq) + H_2(g)$; **(b)** $Cr_2O_3(s) + 2 OH^-(aq) + 3 H_2O(l) \longrightarrow 2 Cr(OH)_4^-(aq)$; **(c)** $2 La(s) + 6 HCl(aq) \longrightarrow 2 LaCl_3(aq) + 3 H_2(g)$ **10. (a)** $[Ar]3d^24s^2$; **(b)** $[Ar]3d^2$; **(c)** $[Ar]3d^4$; **(d)** $[Ar]3d^3$; **(e)** $[Ar]3d^5$; **(f)** $[Ar]3d^5$ **11.** $Cr > Fe > Mn^{4+} > Ti^{2+} > Cu^{2+} > Sc^{3+}$ **12. (c)** and **(e)**. The metals tend to have high melting points, low ionization energies, and negative E° values. **13.** Sc^{3+} **14.** Ti **15.** Ag^+ **16.** Fe^{3+} has a d^5 electron configuration.

Exercises 17. Unlike the main group metals, transition metals usually display several oxidation states, readily form complexes and colored compounds, are mostly paramagnetic, and some are ferromagnetic. **19.** Main group elements have electrons added to the outermost subshell. **21.** Mn **23.** The lanthanide atoms are larger and more easily ionized. **25. (a)** $Sc(OH)_3(s) + 3 H^+(aq) \longrightarrow Sc^{3+}(aq) + 3 H_2O(l)$; **(b)** $3 Fe^{2+}(aq) + MnO_4^-(aq) + 2 H_2O(l) \longrightarrow 3 Fe^{3+}(aq) + MnO_2(s) + 4 OH^-(aq)$; **(c)** $2 KOH(l) + TiO_2(s) \xrightarrow{heat} K_2TiO_3(s) + H_2O(g)$; **(d)** $Cu(s) + 2 H_2SO_4(concd\ aq) \longrightarrow CuSO_4(aq) + SO_2(g) + 2 H_2O(l)$ **27. (a)** $FeS(s) + 2 HCl(aq) \longrightarrow FeCl_2(aq) + H_2S(g)$, $4 Fe^{2+}(aq) + O_2(g) + 4 H^+(aq) \longrightarrow 4 Fe^{3+}(aq) + 2 H_2O(l)$, $Fe^{3+}(aq) + 3 OH^-(aq) \longrightarrow Fe(OH)_3(s)$; **(b)** $BaCO_3(s) +$

$2\ HCl(aq) \longrightarrow BaCl_2(aq) + H_2O(l) + CO_2(g)$, $2\ BaCl_2(aq) + K_2Cr_2O_7(aq) + 2\ NaOH(aq) \longrightarrow 2\ BaCrO_4(s) + 2\ KCl(aq) + 2\ NaCl(aq) + H_2O(l)$ **29.** $HgS(s) + O_2(g) \xrightarrow{heat} Hg(l) + SO_2(g)$, $4\ HgS(s) + 4\ CaO(s) \xrightarrow{heat} 4\ Hg(l) + 3\ CaS(s) + CaSO_4(s)$ **31.** For the reaction $2\ Ca(s) + O_2(g) \longrightarrow 2\ CaO(s)$. $\Delta S^\circ_{rxn} = -208.3$ J/K and $\Delta H^\circ_{rxn} = -1270$ kJ. A plot of ΔG° versus temperature will consist of three linear segments of increasing positive slope, the first line is joined to the second at the melting point for Ca(s), while the second is joined to the third at the boiling point for Ca(l). **33.** **(a)** $VO^{2+}(aq) + 2\ H^+(aq) + e^- \longrightarrow V^{3+}(aq) + H_2O(l)$; **(b)** $Cr^{2+}(aq) \longrightarrow Cr^{3+}(aq) + e^-$ **35.** **(a)** no; **(b)** yes; **(c)** yes **37.** Possible reducing agents include H_2, Sn(s), H_2S Pb(s) **39.** $E^\circ[Cr_2O_7^{2-}|Cr^{3+}] = 1.33$ V, E° $[Cr_2O_7^{2-}|Cr^{2+}] = 0.892$ V, $E^\circ[Cr^{3+}|Cr^{2+}] = -0.424$ V, $E^\circ[Cr^{3+}|Cr] = -0.74$ V, $E^\circ[Cr^{2+}|Cr] = -0.90$ V **41.** $Cr_2O_7^{2-}(aq) + H_2O(l) \rightleftharpoons 2\ CrO_4^{2-}(aq) + 2\ H^+(aq)$, $Pb^{2+}(aq) + CrO_4^{2-}(aq) \rightleftharpoons PbCrO_4(s)$ **43.** $3\ Zn(s) + Cr_2O_7^{2-}$(aq, orange) $+ 14\ H^+(aq) \longrightarrow 3\ Zn^{2+}(aq) + 2\ Cr^{3+}$(aq, green) $+ 7\ H_2O(l)$, $Zn(s) + 2\ Cr^{3+}(aq) \longrightarrow Zn^{2+}(aq) + 2\ Cr^{2+}$(aq, blue); $4\ Cr^{2+}$(blue, aq) $+ O_2(g) + 4\ H^+(aq) \longrightarrow 4\ Cr^{3+}$ (green, aq) $+ 2\ H_2O(l)$ **45.** **(a)** 0.74 M; **(b)** 2.6×10^{-5} M **47.** 1.10 g **49.** $E^\circ[Cr_2O_7^{2-}|Cr^{3+}] = +1.33$ V, making $Cr_2O_7^{2-}$ a good oxidizing agent; E° $[CrO_4^{2-}|Cr(OH)_3] = -0.11$ V, making CrO_4^- a poor oxidizing agent. However, many metal chromates are insoluble and can be precipitated from aqueous solution. **51.** $E = 0.24$ V, hence spontaneous under these conditions. **53.** $Fe^{3+}(aq) + K_4[Fe^{(II)}(CN)_6](aq) \longrightarrow KFe^{(III)}[Fe^{(II)}(CN)_6](s) + 3\ K^+$ (aq); alternate formulation: $4\ Fe^{3+}(aq) + 3\ [Fe(CN)_6]^{4-}(aq) \longrightarrow Fe_4[Fe(CN)_6]_3(s)$; **55.** **(a)** $Cu^{2+}(aq) + H_2(g) \longrightarrow Cu(s) + 2\ H^+(aq)$; **(b)** $Au^+(aq) + Fe^{2+}(aq) \longrightarrow Au(s) + Fe^{3+}(aq)$; **(c)** $2\ Cu^{2+}(aq) + SO_2(g) + 2\ H_2O(l) \longrightarrow 2\ Cu^+(aq) + SO_4^{2-}(aq) + 4\ H^+(aq)$ **57.** **(a)** This is an impossibly high concentration for Cu^+(aq), **(b)** This is an entirely reasonable concentration for Cu^+(aq). **59.** $E^\circ_{cell} = 0.912$ V **61.** **(a)** $\Delta G^\circ = 35$ kJ, $K_p = 0.02$; **(b)** $P_{co} = 0.02$ atm **63.** ZnO, $\lambda = 413$ nm(violet light); CdS, $\lambda = 479$ nm(blue light)

Integrative and Advanced Exercises **65.** Ag^+ forms an insoluble chloride, while gold does not; gold forms a stable chloro complex while silver does not. **69.** **(a)** $Mo(CO)_6$; **(b)** $Os(CO)_5$; **(c)** $Re_2(CO)_{10}$; **(d)** Weak intermolecular attractions in the low molecular mass, symmetrical nickel and iron carbonyls result in the liquid state at room temperature. Because of the two metal atoms in the higher molecular mass, less symmetrical cobalt carbonyl molecules, stronger intermolecular attractions lead to the solid state; **(e)** ionic salt containing Na^+ and $V(CO)_6^-$ ions **73.** **(a)** Net: $2\ H_2O + 4\ Ag^{2+}(aq) \longrightarrow O_2(g) + 4\ H^+(aq) + 4\ Ag^+(aq)$; **(b)** $E^\circ_{cell} = +0.75$ V **76.** 82.3% MnO_2 **77.** **(a)** more of the 0.100 M MnO_4^- solution would be required; **(b)** 29.40 ml of the 0.100 M MnO_4^- solution **79.** 28.2% Cr, 3.286% Mn **80.** $NiC_8H_{14}O_4N_4$, empirical molar mass 288.91 g/mol, 3.52% Ni

Feature Problems **82.** **(a)** If $\Delta n_{gas} = 0$, then $\Delta S^\circ \sim 0$ and ΔG° is essentially independent of temperature(reaction b). If $\Delta n_{gas} > 0$, then $\Delta S^\circ > 0$, and ΔG° decreases with temperature (reaction a). If $\Delta n_{gas} < 0$, then $\Delta S^\circ < 0$, and ΔG° increases with temperature (reaction c); **(b)** $P_{co} = 3$ atm

CHAPTER 25

Practice Examples **1A.** Coord # = 5, OS = +2 **1B.** $[Fe(CN)_6]^{3-}$ **2A.** $K_2[PtCl_6]$ **2B.** pentaamminethiocyanato-*S*-cobalt(III) chloride **3A.** The oxalate ion occupies *cis* positions. One isomer has *trans* NH_3 ligands, one has *cis* NH_3 ligands, and one has an NH_3 and a Cl^- in *cis* positions **3B.** There are five: three with the two pyridines *cis*, and two with them *trans*. **4A.** 3 **4B.** 3 in each case **5A.** tetrahedral **5B.** Paramagnetic, but the structure cannot be determined from the magnetic properties. **6A.** greater *d* orbital energy splitting for $[Co(H_2O)_6]^{2+}$ than for $[CoCl_4]^{2-}$ **6B.** $[Fe(CN)_6]^{4-}$ absorbs light of *shorter* wavelength than does $[Fe(H_2O)_6]^{2+}$; $K_4[Fe(CN)_6]$ is yellow.

Review Questions **4.** **(a)** $[CrCl_4(NH_3)_2]^-$, diamminetetrachlorochromate(III) ion; **(b)** $[Fe(CN)_6]^{3-}$, hexacyanoferrate(III) ion; **(c)** $[Cr(en)_3]_2\ [Ni(CN)_4]_3$, tris(ethylenediamine)chromium(III) tetracyanonickelate(II) **5.** **(a)** Coord. #. = 6, O.S. = +2; **(b)** Coord. #. = 6, O.S. = +3; **(c)** Coord. #. = 4, O.S. = +2; **(d)** Coord. #. = 6, O.S. = +3; **(e)** Coord. #. = 6, O.S. = +2; **(f)** Coord. #. = 2, O.S. = +1 **6.** **(a)** diiodoargentate(I) ion; **(b)** pentaaquahydroxoaluminum(III) ion; **(c)** tetracyanozincate(II) ion; **(d)** bis(ethylenediamine)platinum(II) ion; **(e)** tetraamminechloronitrito-*N*-cobaltate(III) ion **7.** **(a)** pentaamminebromocobalt(III) sulfate; **(b)** pentaamminesulfatocobalt (III) bromide; **(c)** hexaamminechromium(III) hexacyanocobaltate(III); **(d)** sodium hexanitrito-*N*-cobaltate(III); **(e)** tris(ethylenediamine) cobalt(III) chloride **8.** **(a)** $[Ag(CN)_2]^-$; **(b)** $[Pt(NO_2)(NH_3)_3]^+$; **(c)** $[CoCl(en)_2(H_2O)]^{2+}$; **(d)** $K_4[Cr(CN)_6]$

9.

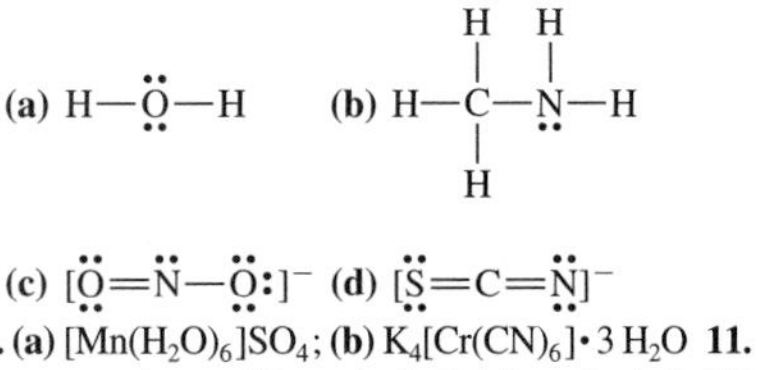

10. **(a)** $[Mn(H_2O)_6]SO_4$; **(b)** $K_4[Cr(CN)_6]\cdot 3\ H_2O$ **11.** **(a)** square-planar; **(b)** octahedral; **(c)** octahedral, Cl^- are *cis*; **(d)** octahedral, Cl^- *trans* to OH^- **12.** **(a)** one; **(b)** two; **(c)** two; **(d)** two **13.** **(a)** coordination; **(b)** linkage; **(c)** none; **(d)** geometric; **(e)** geometric **14.** $[Co(en)_3]^{3+}$ is yellow, $[Co(H_2O)_6]^{3+}$ is blue.

Exercises **15.** **(a)** amminetetraaquahydroxocobalt(III) ion; **(b)** triamminetrinitrito-*O*-cobalt(III); **(c)** tetraaquaplatinum(II) hexachloroplatinate(IV); **(d)** diaquadioxalatoferrate(III) ion; **(e)** silver(I) tetraiodomercurate(II) **17.** **(a)** square-planar; **(b)** octahedral; *fac* isomer has three NH_3 on the same face; **(c)** octahedral **19.** **(a)** octahedral; **(b)** octahedral, three bidentate ligands; **(c)** octahedral, three isomers **21.** **(a)** not possible; all positions in coordination sphere equivalent; **(b)** possible; **(c)** not possible [same reason as in (a)] **23.** **(a)** three isomers; **(b)** yes **25.** The cis-isomer of the ion is chiral and hence optically active; the *trans*-isomer is achiral and hence, not optically active. **27.** Energy differences between the *d* orbitals of the central metal in the field of the ligands correspond to light of different colors. **29.** Because of the larger Δ in the strong-field complex, an additional two *d* electrons pair, leaving only two unpaired. **31.** **(a)** $[Co(en)_3]^{3+}$ is diamagnetic; **(b)** three **33.** No, can only confirm or eliminate square-planar geometry. **35.** **(a)** $Zn(OH)_2(s) + 4\ NH_3(aq) \rightleftharpoons [Zn(NH_3)_4]^{2+}(aq) + 2\ OH^-(aq)$; **(b)** $Cu^{2+}(aq) + 2\ OH^-(aq) \rightleftharpoons Cu(OH)_2(s)$, $Cu(OH)_2(s) + 4\ NH_3(aq) \rightleftharpoons [Cu(NH_3)_4]^{2+}$(aq, dark blue) $+ 2\ OH^-(aq)$, $[Cu(NH_3)_4]^{2+}(aq) + 4\ H_3O^+(aq) \rightleftharpoons [Cu(H_2O)_4]^{2+}(aq) + 4\ NH_4^+(aq)$ **37.** $[Co(en)_3]^{3+}$, chelate effect **39.** $K_f = K_1K_2K_3 = 5.0 \times 10^9$ **41.** **(a)** Al^{3+} forms a hydroxo complex but not an ammine complex. **(b)** $ZnCO_3(s)$ produces a high enough $[Zn^{2+}]$ to form $[Zn(NH_3)_4]^{2+}$, but ZnS(s) is too insoluble. **(c)** At low $[Cl^-]$ the solubility of AgCl is suppressed by the common ion effect, but at higher $[Cl^-]$ the solubility increases due to the formation of $[AgCl_2]^-$. **43.** $[Al(H_2O)_6]^{3+}(aq)$; this ion releases $H^+(aq)$ in water **45.** **(a)** $K = 8.5$; **(b)** K for the dissolution of AgBr(s) in NH_3(aq) is only 8.0×10^{-6}.

Integrative and Advanced Exercises **47.** **(a)** tetraamminecopper(II) ion; **(b)** tetraamminedichlorocobalt(III) chloride; **(c)** hexachloroplatinate(IV) ion; **(d)** sodium tetrachlorocuprate(II); **(e)** potassium pentachloroantimonate(III) **48.** $[Pt(NH_3)_4]\ [PtCl_4]$, tetraammineplatinum(II) tetrachloroplatinate(IV) **53.** The complex has a positive charge, so it is easier electrostatically for the proton to leave. **55.** **(a)** pH = 2.04; **(b)** $[[FeOH](H_2O_5)^{2+}] = 9 \times 10^{-4}$ M; **(c)** No, this would require $[H_3O]^+ = 90$ M, which is not possible. **59.** 0.52 M **60.** **(b)** $[Co^{3+}] = 2.2 \times 10^{-28}$ M; **(c)** $E = -0.142$ V **62.** 0.99 V **64.** $[PtCl(NH_3)_5]Cl_3$ **65.** I. *trans*-chlorobis(ethylenediamine)nitrito-*N*-cobalt(III) nitrite; II. *trans*-bis(ethylenediamine)dinitrito-*N*-cobalt(III) chloride; III. *cis*-bis(ethylenediamine)dinitrito-*N*-cobalt(III) chloride **70.** face-centered cubic, K^+: tetrahedral holes, $[PtCl_6]^{2-}$: octahedral holes.

Feature Problems **71.** **(a)** No, this structure predicts three isomers, which is one more than the actual number of isomers for this complex ion. **(b)** No, the only possible structure with optical activity requires one of the en ligands to span the diagonal distance across the face of the prism, which is not possible.

CHAPTER 26

Practice Examples **1A.** ${}^{241}_{94}Pu \longrightarrow {}^{241}_{95}Am + {}^{0}_{-1}\beta$ **1B.** ${}^{58}_{29}Cu \longrightarrow {}^{58}_{28}Ni + {}^{0}_{+1}\beta$ **2A.** ${}^{139}_{57}La + {}^{12}_{6}C \longrightarrow {}^{147}_{63}Eu + 4\ {}^{1}_{0}n$ **2B.** ${}^{121}_{51}Sb + {}^{4}_{2}He \longrightarrow {}^{124}_{53}I + {}^{1}_{0}n$, ${}^{124}_{53}I \longrightarrow {}^{0}_{+1}\beta + {}^{124}_{52}Te$ **3A.** **(a)** $9.98 \times 10^{-7}\ s^{-1}$; **(b)** 9.40×10^{12} dis s^{-1}; **(c)** 25%; **(d)** 2.36×10^{12} dis s^{-1} **3B.** 75.8 d **4A.** 4.69×10^3 y **4B.** 13 dis $min^{-1}\ g^{-1}$ C **5A.** 2.544 MeV **5B.** 0.006001 u **6A.** **(a)** ${}^{88}Sr$ stable; **(b)** ${}^{118}Cs$ radioactive; **(c)** ${}^{30}S$ radioactive **6B.** positron emission by ${}^{17}F$; $\beta-$ emission by ${}^{22}F$

Review Questions **4.** **(a)** γ rays; **(b)** α particles; **(c)** β particles **5.** **(a)** ${}^{160}_{74}W \longrightarrow {}^{156}_{72}Hf + {}^{4}_{2}He$; **(b)** ${}^{38}_{17}Cl \longrightarrow {}^{38}_{18}Ar + {}^{0}_{-1}\beta$; **(c)** ${}^{214}_{83}Bi \longrightarrow {}^{214}_{84}Po + {}^{0}_{-1}\beta$; **(d)** ${}^{32}_{17}Cl \longrightarrow {}^{32}_{16}S + {}^{0}_{+1}\beta$ **6.** **(a)** ${}^{23}_{11}Na + {}^{2}_{1}H \longrightarrow {}^{24}_{11}Na + {}^{1}_{1}H$; **(b)** ${}^{59}_{27}Co + {}^{1}_{0}n \longrightarrow {}^{56}_{25}Mn + {}^{4}_{2}He$; **(c)** ${}^{238}_{92}U + {}^{2}_{1}H \longrightarrow {}^{240}_{94}Pu + {}^{0}_{-1}\beta$; **(d)** ${}^{246}_{96}Cm + {}^{13}_{6}C \longrightarrow {}^{254}_{102}No + 5\ {}^{1}_{0}n$; **(e)** ${}^{238}_{92}U + {}^{14}_{7}N \longrightarrow {}^{246}_{99}Es + 6\ {}^{1}_{0}n$ **7.** **(a)** ${}^{214}_{88}Ra \longrightarrow {}^{210}_{86}Rn + {}^{4}_{2}He$; **(b)** ${}^{205}_{85}At \longrightarrow {}^{205}_{84}Po + {}^{0}_{+1}\beta$; **(c)** ${}^{212}_{87}Fr + e^- \longrightarrow {}^{212}_{86}Rn$; **(d)** ${}^{2}_{1}H + {}^{2}_{1}H \longrightarrow {}^{3}_{2}He + {}^{1}_{0}n$; **(e)** ${}^{241}_{95}Am + {}^{4}_{2}He \longrightarrow {}^{243}_{97}Bk + 2{}^{1}_{0}n$ **8.** **(a)** ${}^{214}_{84}Po$; **(b)** ${}^{32}_{15}P$; **(c)** ${}^{13}_{8}O$, ${}^{28}_{12}Mg$, ${}^{80}_{35}Br$, ${}^{214}_{84}Po$, and ${}^{222}_{86}Rn$ **9.** 56.3 h **10.** **(a)** 174 d; **(b)** 287 d; **(c)** 375 d **11.** **(a)** 5.42×10^{-9} J; **(b)** 3727 MeV **12.** 8.58 MeV/nucleon **13.** **(c)** ${}^{80}_{35}Br$; **(d)** ${}^{132}_{55}Cs$ **14.** **(a)** Intermediate half-life means continuous moderate activity. **(b)** Those that give off γ or high energy β are hazardous at a distance; α-emitters are dangerous when inside the body. **(c)** Potassium-40 decays by electron capture to argon-40. **(d)** Francium is produced in a radioactive decay series and is found in radioactive minerals rather than in group 1 minerals. **(e)** Fusion involves joining positively charged nuclei and requires extreme thermal energies to initiate it.

Exercises **15.** **(a)** ${}^{230}_{92}U$; **(b)** ${}^{248}_{98}Cf$; **(c)** ${}^{196}_{80}Hg$ **17.** ${}^{14}_{6}C \longrightarrow {}^{14}_{7}N + {}^{0}_{-1}\beta$ **19.** ${}^{232}_{90}Th \longrightarrow {}^{228}_{88}Ra \longrightarrow {}^{228}_{89}Ac \longrightarrow {}^{228}_{90}Th \longrightarrow {}^{224}_{88}Ra \longrightarrow {}^{220}_{86}Rn \longrightarrow {}^{216}_{84}Po$;

first branch, first fork [$^{216}_{84}$Po $\longrightarrow$ $^{212}_{82}$Pb $\longrightarrow$ $^{212}_{83}$Bi], second fork [$^{216}_{84}$Po $\longrightarrow$ $^{216}_{85}$At $\longrightarrow$ $^{212}_{83}$Bi]; second branch, first fork [$^{212}_{83}$Bi $\longrightarrow$ $^{208}_{81}$Tl $\longrightarrow$ $^{208}_{82}$Pb], 2nd fork [$^{212}_{83}$Bi $\longrightarrow$ $^{212}_{84}$Po $\longrightarrow$ $^{208}_{82}$Pb] **21.** "$4n + 2$" is only consistent with the mass numbers: 206, 210, 214, 218, 222, 226, 230, 234, and 238, as seen in Figure 26-2. **23. (a)** $^7_3$Li + $^1_1$H $\longrightarrow$ $^8_4$Be + γ; **(b)** $^9_4$Be + $^2_1$H $\longrightarrow$ $^{10}_5$B + $^1_0$n; **(c)** $^{14}_7$N + $^1_0$n $\longrightarrow$ $^{14}_6$C + $^1_1$H **25.** $^{209}_{83}$Bi + $^{64}_{28}$Ni $\longrightarrow$ $^{272}_{111}$E + $^1_0$n, $^{272}_{111}$E $\longrightarrow$ 5 $^4_2$He + $^{252}_{101}$Md **27.** 4.4×10^8 $^{60}_{27}$Co atoms **29.** 87.5 h **31.** 142 d **33.** 3.35×10^3 y, object probably not from the pyramid era (~3000 BC). **35.** 0.12 g $^{208}_{82}$Pb/1 g $^{232}_{90}$Th **37.** 7.805 MeV/nucleon **39.** 4.06 MeV **41.** 7.25×10^3 neutrons **43. (a)** $^{20}_{10}$Ne; **(b)** $^{18}_8$O; **(c)** $^7_3$Li **45.** $\beta-$ emission: $^{33}_{15}$P, $^{134}_{53}$I; $\beta+$ emission: $^{29}_{15}$P, $^{120}_{53}$I **47.** Doubly magic nuclei have magical atomic numbers and atomic masses: $^4_2$He, $^{16}_8$O, $^{40}_{20}$Ca, $^{56}_{28}$Ni, $^{208}_{82}$Pb **49.** 1.37 mg **51.** Takes into account the quantity of biological damage. **53.** $^{90}_{38}$Sr is in the same periodic family as Ca, and thus can become concentrated in bones. **55.** Mix in a small amount of tritium with the H_2(g) and detect where its radioactivity appears. **57.** Yes, a small amount of radioactive sodium ion will end up in the crystallized NaNO$_3$.

Integrative and Advanced Exercises 61. 2.9×10^3 metric tons **64.** 7.0 µg ^{90}Sr **65.** 29 years **69.** 200 Ci **70.** 0.14 **71.** 1.2×10^7 kJ

Feature Problems 75. The plot of packing fraction versus mass number and the plot in Fig. 26-6 are essentially mirror images of one another (maximum of one is the minimum of the other). Packing fraction is proportional to the negative of the mass defect per nucleon. **76. (a)** Decay rate depends on the mathematical product of half-life and the number of radioactive atoms, large number of atoms/long half-life has greater rate of decay number of atoms/short half-life. **(b)** $dD/dt = \lambda_p(P) - \lambda_d(D) = \lambda_p(P_o e^{-\lambda_p(t)}) - \lambda_d(D)$; **(c)** 1 year (actually, the maximum comes after ~2 month, however, the difference between 1 year and 2 months is negligible). **78. (a)** Average atomic mass of Sr is 87.4 u; **(b)** 266.3 ppm; **(c)** 1.38%; **(d)** 9.65×10^8 y

CHAPTER 27

Practice Examples 1A. $CH_3CH_2CH_2CH_2CH_2CH_3$, $(CH_3)_2CHCH_2CH_2CH_3$, $CH_3CH_2CH(CH_3)CH_2CH_3$, $(CH_3)_2CHCH(CH_3)_2$, $(CH_3)_3CCH_2CH_3$ **1B.** $CH_3CH_2CH_2CH_2CH_2CH_2CH_3$, $(CH_3)_2CHCH_2CH_2CH_2CH_3$, $CH_3CH_2CH(CH_3)CH_2CH_2CH_3$, $(CH_3)_2CHCH_2CH_2CH(CH_3)_2$, $(CH_3)_2CHCH_2CH_2CH_3$, $(CH_3)_3CCH_2CH_2CH_3$, $CH(CH_2CH_3)_3$, $(CH_3)_3CCH(CH_3)_2$, $CH_3CH_2C(CH_3)_2CH_2CH_3$ **2A.** 3,6,6-trimethylnonane **2B.** 3,6-dimethyloctane.

3A. $CH_3CH(CH_3)CH(CH_2CH_3)CH_2$ $CH_2CH(CH_3)_2$ **3B.** $CH_3CH(CH_3)CH[CH(CH_3)_2]CH_2CH_3$ **4A.** 3-nitrobenzaldehyde **4B.** 1,3-dichloro-2-nitrobenzene, 2,4-dichloro-1-nitrobenzene **5A. (b)** and **(c)** are chiral **5B. (b)** and **(c)** are chiral. **6A.** All have *R*-configuration. **6B. (a)**, **(b)** and **(c)** are enantiomeric pairs. **7A. (a)** *E*; **(b)** *Z*; **(c)** *Z*; **(d)** *Z* **7B.**

(a) Z: / E: (structures with H$_3$CH$_2$C, CH$_3$, H groups)

(b) Z: / E: (structures with Cl, CH$_2$CH$_3$, H$_3$CH$_2$C groups)

(c) Z: / E: (structures with H$_3$CH$_2$C, CH$_3$, H$_3$C groups)

(d) Z: / E: (structures with H$_3$CHClC, Cl, CH$_3$, H$_3$C groups)

8A. H$^-$: nucleophile, AlCl$_3$: electrophile, C$_2$H$_4$ nucleophile, CH$_3$S$^-$ nucleophile **8B. (a)** CH$_3$I(electrophile) + NH$_3$(nucleophile) $\longrightarrow$ H$_3$CNH$_3^+$I$^-$; **(b)** CH$_3$CH$_2$Cl(electrophile) + CH$_3$S$^-$(nucleophile) $\longrightarrow$ H$_3$CCH$_2$SCH$_3$ + Cl$^-$; **(c)** CH$_3$CH$_2$CHClCH$_3$ (electrophile) + CN$^-$(nucleophile) $\longrightarrow$ CH$_3$CH$_2$CH (CN)CH$_3$ + Cl$^-$ **9A. (a)** CH$_3$C$\equiv$C$^-$ + CH$_3$Br — S$_N$2 $\longrightarrow$ CH$_3$C$\equiv$CCH$_3$ + Br$^-$; **(b)** No reaction; **(c)** CH$_3$NH$_2$ + (CH$_3$)$_3$CCl — S$_N$1 $\longrightarrow$ CH$_3$NH$_2$ C(CH$_3$)$_3^+$Cl$^-$ **9B. (a)** S$_N$2 mechanism, organic product is (*S*)-2-cyanobutane **(b)** S$_N$1 mechanism, product is a racemic mixture (50% (*R*), 50% (*S*)) of *sec*-butyl methylether.

Review Questions 5. C$_4$H$_8$ has five isomers, C$_4$H$_{10}$ has only two isomers. **6.** CH$_3$CH=CHCH$_3$ **7. (a)** alkyl halide; **(b)** carboxylic acid; **(c)** aldehyde; **(d)** ether; **(e)** ketone; **(f)** amine; **(g)** alcohol groups; **(h)** ester

8. (a) (structure) **(b)** (structure) **(c)** (structure)

9. Only carbon and chlorine atoms are shown

10. (a) different formulas; **(b)** structural isomers; **(c)** different formulas; **(d)** identical; **(e)** identical; **(f)** ortho-para isomers **11. (a)** 3,5-dimethyloctane; **(b)** 2,2-dimethylpropane; **(c)** 3,3-dichloro-5-ethylheptane **12. (a)** *m*-dichlorobenzene; **(b)** *m*-nitrotoluene; **(c)** *p*-aminobenzoic acid

13.

(a) C—C—C—C—C—C—C (structure) **(b)** (structure)
(c) C—C=C—C—C (structure) **(d)** C—C—C—O—C—C—C
(e) Br—(ring)—OH

H atoms attached to carbon are omitted for clarity.
14. (a) CH$_3$CH(OH)CH$_3$; **(b)** CClF$_2$CH$_3$; **(c)** CH$_2$=C(CH$_3$)CH=CH$_2$

15. (a) (structure: ring with CH$_3$, H$_3$C, CH$_3$) **(b)** O$_2$N—(ring)—OH

(c) Cl—(ring with COOH, Cl, NH$_2$)

16. 2-butanol **17. (a)** CH$_3$CH$_2$CH(OH)CH$_3$; **(b)** CH$_3$CH$_2$CH$_2$Cl + CH$_3$CHClCH$_3$ + other chlorinated propanes; **(c)** C$_6$H$_5$COOCH(CH$_3$)$_2$; **(d)** CH$_3$COCH$_2$CH$_3$ **18. (a)** C$_6$H$_6$; **(b)** C$_3$H$_7$OH; **(c)** C$_6$H$_5$COOH **19. (a)** nucleophile; **(b)** electrophile; **(c)** nucleophile; **(d)** nucleophile; **(e)** neither **20. (a)** nucleophile; **(b)** neither; **(c)** electrophile; **(d)** nucleophile; **(e)** both **21. (a)** S$_N$2; **(b)** hydrogenation; **(c)** esterification **22. (a)** Substitution reaction(formation of a secondary amine); **(b)** Substitution reaction(formation of an ether); **(c)** Oxidation of an alcohol to an aldehyde)

Exercises

23. (a) C—C—C—C—C (with Br, Br); **(b)** C—C—C—C—C—C—C (with C, C, C); **(c)** C—C—C—C=C—C—C (with C)

25. (a) sp^3 hybridized carbon, all bonds are σ-bonds; C—H bonds: C_{sp^3} – H_{1s}, C—C bonds: C_{sp^3} – C_{sp^3}; **(b)** sp^2 hybridized carbon, one π-bond and $5 \times \sigma$-bonds; C—H bonds: C_{sp^2} – H_{1s}, C—C σ-bond C_{sp^2} – C_{sp^2}; C—Cl bond: C_{sp^2} – Cl_{3p}, and the C—C π-bond: C_{2p} – C_{2p}; **(c)** sp and sp^3 hybridized carbon atoms, two π-bond and $6 \times \sigma$-bonds; C—H bonds: C_{sp^3} – H_{1s}, and C_{sp} – H_{1s}, C—C σ-bonds: C_{sp^3} – C_{sp} and C_{sp} – C_{sp}, and the two C—C π-bonds: C_{2p} – C_{2p} **27. (a)** positional; **(b)** structural; **(c)** positional **29.** nine possible structures **31. (a)** carbon 2; **(b)** no chiral carbon atoms; **(c)** carbon 2 **33. (a)** Carbon atom bonded to Cl, CH$_3$, H and CO$_2$H; **(b)** Carbon atom bonded to NH$_2$, CH$_3$, H and CO$_2$H; **(c)** No chiral carbon atoms **35. (a)** Carboxyl contains an OH group while a carbonyl does not. **(b)** Aldehyde carbonyl group is bonded to a carbon and a hydrogen, while a ketone carbonyl group is bonded to two carbons. **(c)** The acetyl group, CH$_3$CO—, results by removing the OH group from the acetic acid molecule, CH$_3$COOH. **37. (a)** carboxyl group, hydroxyl groups and a disubstituted aromatic ring; **(b)** ester group; **(c)** ketone and carboxyl groups; **(d)** hydroxyl, aldehyde groups and a disubstituted aromatic ring **39.** Three isomers: diethyl ether, 2-methoxypropane and methyl propyl ether **41.** There are four isomers: pentanoic acid, 3-methylbutanoic acid, 2-methylbutanoic acid, 2,2-dimethylpropanoic acid **43.** There are nine isomers *n*-propyl acetate, ethyl propanoate, isopropyl acetate, *n*-butylformate, *sec*-butylformate, *tert*-butylformate, isobutylformate, methyl-*n*-butanoate, methyl 2-methylpropanoate **45.** There are many isomers, a few are listed below: 1-hydroxy-1-methoxy propene, 1-hydroxy-2-ethoxyethene, 1-hydroxy-1-ethoxyethene, 2-hydroxy *n*-butanal, 3-hydroxy *n*-butanal, 4-hydroxy *n*-butanal, 4-hydroxy *n*-butanone, 1-hydroxy butanone, 3-hydroxy butanone, vinyl oxyethanol, 1-methoxypropanone, etc. **47. (a)** 2,2-dimethylbutane; **(b)** 2-methylpropene; **(c)** 1,2-dimethylcyclopropane; **(d)** 4-methyl-2-pentyne; **(e)** 2-ethyl-3-methylpentane; **(f)** 3,4-dimethyl-2-propyl-1-pentene **49. (a)** 1-pentene or 2-pentene; **(b)** sufficient; **(c)** 1-butanol, 2-butanol, isobutanol, and *t*-butanol; **(d)** need position on ring; **(e)** sufficient; **(f)** need positions on ring **51. (a)** $(NO_2)_3C_6H_2(CH_3)$; **(b)** $(CO_2CH_3)_2C_6H_4(OH)$; **(c)** $HOOCCH_2C(OH)(COOH)CH_2COOH$ **53. (a)** *N,N*-diethylamine; **(b)** *p*-aminonitrobenzene; **(c)** *N*-ethyl-*N*-cyclopentylamine; **(d)** diethylmethylamine

55. (a) 2-methylbutane **(b)** 2,2-dimethylpropane

57. There is only one position for the double bond in ethene and propene but two in butene. **59. (a)** $CH_3CH_2CH_3$; **(b)** $CH_3CH=CHCH_3$ **61. (a)** $C_6H_5—C≡CH$; **(b)** **(c)** 3,5-dihydroxyphenol **63. (a)** H on a carbon chain is displaced by another group; **(b)** H on an aromatic ring is replaced by another group; **(c)** Two fragments of a small molecule add across a double bond; **(d)** Two groups from within the same molecule join to produce a small molecule, such as H and OH to form H_2O. **65. (a)** $CH_3CH_2CH_2COOH$; **(b)** $CH_3CH_2C(O)OCH_2CH_3$; **(c)** $CH_3CH_2C(CH_3)_2OH$ **67. (a)** $H_3CCH_2CH_3$; **(b)** $H_3CCHClCH_2Cl$; **(c)** $H_3CCH(CN)CH_3$; **(d)** $H_3CCHClCH_3$; **(e)** $H_3CCH(OH)CH_3$ **69. (a)** $H_3CCH_2CO_2H(aq) + Cl^-(aq)$; **(b)** $H_3CCH_2CO_2H(aq) + H_2O(aq)$; **(c)** $H_3CCH_2CO_2Na(aq) + HOCH_2CH_3(aq)$ **71. (a)** $H_3CC(O)NH_2 + H_2O$; **(b)** $CH_3CH_2CN + H_2O$; **(c)** $CH_3CH_2C(O)ONa(aq) + NH_3(aq)$ **73. (a)** (3) $CH_3COOH(aq) + Cl^-(aq)$; **(b)** (1), $CH_3C(O)OH(aq) + CH_3OH(aq)$; **(c)** (2) $CH_3CH_2C(O)ONa(aq) + H_2O(l)$ **75.** $\dfrac{[CH_3CH_2CH_2CH_2NH_3^+][OH^-]}{[CH_3CH_2CH_2CH_2NH_2]} = K_b$

77. (a) $CH_3CH_2NH_3^+ Cl^-(aq)$; **(b)** $(CH_3)_3NH^+Br^-$; **(c)** no reaction; **(d)** $CH_3CH_2NH_2(aq) + H_2O(l)$ **79. (a)** (1) **(b)** (2) **(c)** (2); **(d)** (1) and (3); **(e)** none of the compounds **81. (a)** identical molecules, both achiral; **(b)** identical molecules, both *R*-enantiomer; **(c)** structural isomers; **(d)** enantiomers, *R/S*-optical isomers; **(e)** identical molecules, both *R*-enantiomer; **(f)** identical

molecules, both *R*-enantiomer **83. (a)** (*S*)-3-bromo-2-methylpentane; **(b)** (*S*)-1,2-dibromopentane; **(c)** (*R*)-3-(bromomethyl)-5-chloropentan-3-ol; **(d)** (*S*)-1-bromopropan-2-ol **85. (a)** (*Z*)-2-pentene; **(b)** (*E*)-1-chloro-2-methyl-1-butene; **(c)** (*E*)-4-chloromethyl-3,7-dimethyl-3-octene; **(d)** (*Z*)-5-bromo-3-bromomethyl-2-methylpent-2-enal **87.**

89. (a) Reaction rate $= k_{obs}[CH_3CH_2 CH_2CH_2Br][OH^-]$; **(b)** A graph of potential energy versus the progress of the reaction constitutes the reaction profile; plot contains reactants on the left-hand side and products on the right-hand side of an arching line, whose peak is the transition state for the reaction. The energy difference between the reactants and the products is the ΔH_{rxn} (which is negative in this case). The energy difference between the reactants and the transition state is the activation energy for the reaction. **(c)** The reaction rate would increase by a factor of two. **(d)** The reaction rate would decrease by a factor of two. **91. (a)** $CH_3CH_2CH_2CH_2Br(aq) + NaOH(aq) \longrightarrow CH_3CH_2CH_2CH_2OH(aq) + NaBr(aq)$; **(b)** $CH_3CH_2CH_2CH_2Br(aq) + NH_3(aq) \longrightarrow CH_3CH_2CH_2CH_2NH_3^+(aq) + Br^-(aq)$; **(c)** $CH_3CH_2CH_2CH_2Br(aq) + NaCN(aq) \longrightarrow CH_3CH_2CH_2CH_2CN(aq) + NaBr(aq)$; **(d)** $CH_3CH_2CH_2CH_2Br(sol) + CH_3CH_2ONa(sol) \longrightarrow CH_3CH_2CH_2CH_2OCH_2CH_3(sol) + NaBr(sol)$ **93. (a)** $CH_3CH_2CH(CH_3)OH$ (both *R*- and *S*- forms); **(b)** S_N1 **95. (a)** $CH_3CH(OCH_2CH_3)CH_2CH_3$ (both *R*- and *S*- forms); **(b)** S_N1 **97.** Not all polymer chains have the same number of monomer units, and thus individual polymer molecules differ in mass. Hence we can only discuss an average molecular mass (weight average of the molecular masses). **99.** Polymer repeating unit: $—[—C(O)—(CH_2)_8—C(O)—NH—(CH_2)_6—NH—]_x—$ **Integrative and Advanced Exercises 107. (a)** $C_6H_5NO_2 + 7H^+ + 2Fe \longrightarrow C_6H_5NH_3^+ + 2H_2O + 2Fe^{3+}$; **(b)** $3 C_6H_5CH_2OH + 2 Cr_2O_7^{2-} + 16 H^+ \longrightarrow 3 C_6H_5COOH + 4 Cr^{3+} + 11 H_2O$; **(c)** $3 CH_3CH=CH_2 + 2 MnO_4^- + 4 H_2O \longrightarrow 3 CH_2CHOHCH_2OH + 2 MnO_2 + 2 OH^-$ **109.** $C_4H_8N_2O_2$ **114.** (1) Freezing point depression data are not precise enough to distinguish between compounds. (2) Propionic acid in aqueous solution turns litmus red, so unknown is not propionic acid. (3) Both alcohols and aldehydes are oxidized by aqueous permanganate, while ethers are unreactive. The unknown must be 1-butanol or butyraldehyde. Reaction of alcohol with carboxylic acid would produce a pleasant smelling ester; an aldehyde would not react. **115.** $n\text{-}C_3H_7Cl + N_3^- \longrightarrow n\text{-}C_3H_7N_3 +$

Cl^- $n\text{-}C_3H_7N_3$ {reduce} $\longrightarrow n\text{-}C_3H_7NH_2$ **118.** OH is protonated by HI to form $—H_2O^+$, which is a better leaving group than hydroxide **124. (a)** ester, amine, arene; **(b)** (1) sp^2; (2) sp^3; (3) sp^3; (4) sp^3, N:sp^3; **(c)** carbons 2, 4 are chiral.

CHAPTER 28

Practice Examples 1A. dithreonylmethionine **1B.**

2A. pentapeptide sequence: Gly-Cys-Val-Phe-Tyr **2B.** hexapeptide sequence: Ser-Gly-Gly-Ala-Val-Trp **Review Questions 4. (a)** glyceryl palmitolinolenolaurate or glyceryl palmitoeleosterolaurate; **(b)** glyceryl trioleate or triolein; **(c)** sodium myristate

5.

(a)

(b)

(c)

(d)

6. (4) **7. (a)** D-(−)-arabinose is the optical isomer of L-(+)-arabinose. structure (a) below. **(b)** A diastereomer of L-(+)-arabinose is a molecule that is its optical isomer, but not its mirror image. There are several such diastereomers, e.g. (b) below.

(a)
H—C=O
HO—C—H
H—C—OH
H—C—OH
CH₂OH

(b)
H—C=O
H—C—OH
H—C—OH
HO—C—H
CH₂OH

8. All three molecules are optically active monosaccharides, β-D-(+)-glucose is a cyclic molecule. D-(+)-glucose and D-(−)-arabinose are open chain forms. Arabinose is a pentose, while glucose is a hexose.

9. (a)

CH₂—CHCOOH, NH₃⁺ Cl⁻

(b)

CH₂—CHCOO⁻ Na⁺, NH₂

(c)

CH₂—CHCOO⁻, NH₃⁺

10. (a)

O=C
CH—CH₂—SH
NH
O=C
CH—CH₃
NH₂

(b)

O=C
CH₂
NH
O=C
CH—CH—CH₃
NH CH₃
O=C
CH—CHOH—CH₃
NH₂

11. (a)

OH
O=C
CH—CH₂—SH
NH
O=C
CH—CHOH—CH₃
NH
O=C
CH—CH—CH₃
NH CH₃
O=C
CH—(CH₂)₂—S
NH₂ CH₃

(b) methionylvalylthreonylcysteine

12. The pentose sugars in the structure are ribose sugars, and hence this is a chain of RNA.

Exercises 13. (a) 3×10^2 H_3O^+ ions; **(b)** 1.2×10^5 K^+ ions **15.** 5×10^6 protein molecules **17. (a)** Trilaurin is a saturated triglyceride (i.e. a fat), while trilinolein is an unsaturated triglyceride (i.e. an oil). **(b)** Soaps are salts of fatty acids (from saponification of triglycerides), while phospholipids are derived from glycerols, fatty acids, phosphoric acid, and a nitrogen-containing base. Both have hydrophilic heads and hydrophobic tails. **19.** Polyunsaturated fatty acids have two or more carbon-to-carbon double bonds. Stearic is fully saturated, while eleostearic acid has three C=C bonds and hence is unsaturated Safflower oil has the highest percentage of unsaturated fatty acids. **21.** CH₂OHCHOHCH₂OH (glycerol) and Na⁺ ⁻OOC(CH₂)₁₄CH₃ (sodium palmitate)

23.

O H
C
HO—C—H
H—C—OH
HO—C—H
HO—C—H
CH₂OH
L-Glucose
aldohexose

CH₂OH
C=O
H—C—OH
CH₂OH
D-Erythrulose
ketotetrose

25. (a) A dextrorotatory compound rotates the plane of polarized light to the right(clockwise). **(b)** A levorotaory compound rotates the plane of polarized light to the left(counterclockwise). **(c)** An equal mixture of R- and S-enantiomers; **(d)** (R) is the designation given to a chiral carbon atom when the group of lowest priority is directed away from the viewer and the remaining groups are arranged such that they proceed from highest to lowest priority in a clockwise direction. **27.** A reducing sugar will reduce Cu^{2+}(aq) to insoluble, red Cu_2O(s). **29.** They are not enantiomers because the rotational angles are different, and moreover they are of the same sign **31. (a)** Enantiomers: S-config. (left structure), R-config. (right structure); **(b)** Same molecule: both R-configuration; **(c)** Diasteriomers: left structure S,R-configuration with (S) top and (R) bottom; right structure S,S-configuration; **(d)** Diasteriomers: left structure R,R-configuration; right structure R,S-configuration with (R) top and (S) bottom.

33. (a)

H
Cl—Ⓢ—CH₃
H₃C—Ⓢ—Cl
H

(b)

CO₂H
HO—Ⓢ—CH₃
OHC—Ⓡ—OH
CH₃

(c)

CO₂H
HO—Ⓢ—H
H₃C—Ⓢ—NH₂
H

(d)

CH₃
Br—Ⓡ—H
H—Ⓢ—CH₃
Cl

35. (a) An α-amino acid is a carboxylic acid that has an amine group and the carboxyl group on the same carbon atom. **(b)** A zwitterion is a form of an amino acid where the amine group is protonated and the carboxyl group is deprotonated. The zwitterionic form of glycine is ⁺H₃NCH₂COO⁻. **(c)** The pH at which the

zwitterionic form of an amino acid predominates in solution is known as the isoelectric point. The isoelectric point of glycine is pI = 6.03. **(d)** A peptide bond is the bond that forms between the carbonyl group of one amino acid and the amine group of another (with the elimination of H_2O). **(e)** Tertiary structure is the three-dimensional folding of the polypeptide chain in a protein. **37.** Proline will not migrate; lysine toward the cathode; aspartic acid toward the anode.

39. (a) ⁺H₃NCH(CHOHCH₃)COOH;
(b) ⁺H₂NCH(CHOHCH₃)COO⁻;
(c) H₂NCH(CHOHCH₃)COO⁻ **41. (a)** Lys-Ser-Ala, Lys-Ala-Ser, Ser-Lys-Ala, Ser-Ala-Lys, Ala-Ser-Lys, Ala-Lys-Ser; **(b)** Ala-Ser-Ala-Ser, Ala-Ala-Ser-Ser, Ala-Ser-Ser-Ala, Ser-Ser-Ala-Ala, Ser-Ala-Ser-Ala, Ser-Ala-Ala-Ser **43. (a)** Ala-Ser-Gly-Val-Thr-Leu; **(b)** alanylserylglycylvalylthreonylleucine **45.** Primary structure: sequence of amino acids in the chain of the polypeptide. Secondary structure: folding, coiling, or convolutions of the protein chain. Tertiary structure: three dimensional folding of the secondary structure of the polypeptide chain in a protein. Quaternary structure: packing together of two or more protein molecules into a larger protein complex. Not all proteins have a quaternary structure since many proteins have but one polypeptide chain.

47.

CH₃
H₂N—Ⓡ—H
CO₂H

49.

CH₃
H—Ⓢ—NH₂
CO₂H
S-Alanine

CH₃
H···Ⓢ CO₂H
H₂N
S-Alanine

CH₂C₆H₅
H—Ⓢ—NH₂
CO₂H
S-Phenylalanine

CH₂C₆H₅
H···Ⓢ CO₂H
H₂N
S-Phenylalanine

51. DNA and RNA: both contain sugars (deoxyribose and ribose), phosphate groups, and purine and pyrimidine bases. **53.** The complementary sequence is TCG.

Integrative and Advanced Exercises 56. 2.9×10^3 g/mol; more than one Ag^+ per protein might be needed. **60.** $pK_1 = 9.71$ **62.** nonapeptide: Arg Pro Pro Gly Phe Ser Pro Phe Arg **63.** AGA CCA CAA CGA **66.** Met is N-terminal. **(a)** Met Met Gly or Met Gly Met; **(b)** Met Met; **(c)** Ser Met; **(d)** Met Met Ser or Met Ser Met **69. (a)** $K = 9.3 \times 10^{-5}$; **(b)** $K = 1.8 \times 10^5$; **(c)** [B]/[A] = 2.25; ratio coupled/uncoupled = 2.4×10^4

Feature Problems 70. (a) Saponification value = 189 for glyceryl tristearate, iodine number = 86.0 for glyceryl trioleate; **(b)** Castor oil: iodine number = 81.6, saponification value = 180; **(c)** For safflower oil: The iodine number can range from 150 to 144 and the saponification value can range from 211 to 179.

Photo Credits

CHAPTER 14 p. 534, Richard Megna/Fundamental Photographs; **14-9,** Richard Megna/Fundamental Photographs; **14-10,** Carey Van Loon; **p. 548** (top), Charles D. Winter/Photo Researchers, Inc.; **p. 548** (bottom), David Mechlin/Phototake NYC; **p. 553,** Ed Degginger/Color-Pic, Inc.; **14-16b,** Richard Megna/Fundamental Photographs; **p. 557,** Dave M. Phillips/Visuals Unlimited; **p. 558,** © Proctor & Gamble Company. Used by Permission; **p. 560** (center), J. Pat Carter/AP/Wide World Photos; **p. 558** (top left), Wayne Eastep/Stone; **p. 558** (bottom left), Mark Joseph/Stone; **p. 562,** Library of Congress; **14-21,** Stephen Frisch/Stock Boston; **14-23,** Carey Van Loon; **p. 566,** Hank Morgan/Science Source/Photo Researchers, Inc.; **p. 568,** Hewlett Packard/Fundamental Photographs

CHAPTER 15 p. 578, Carey Van Loon; **p. 602,** E. R. Degginger/Color-Pic, Inc.; **15-13,** James Blank/FPG International LLC; **p. 611,** Oxford Molecular Biophysics Laboratory/Science Photo Library/Photo Researchers, Inc.; **p. 614,** NASA/Phototake NYC; **p. 624,** Richard Megna/Fundamental Photographs

CHAPTER 16 p. 626, Steve McCutcheon/Visuals Unlimited; **16-1,** Carey Van Loon; **p. 627** (bottom), Alexandra Winkler/Reuters NewMedia Inc./Corbis; **p. 646,** N. P. Alexander/Visuals Unlimited; **16-8a & b,** Richard Megna/Fundamental Photographs; **p. 654,** Grant Heilman/Grant Heilman Photography, Inc.; **p. 655,** Stone

CHAPTER 17 p. 665, Index Stock Imagery, Inc.; **p. 666,** Carey Van Loon; **17-7,** Tom Pantages; **p. 684,** Carey Van Loon; **p. 687,** Carey Van Loon; **17-9,** Carey Van Loon; **p. 700** (top left), NYC Parks Photo Archive/Fundamental Photographs; **p. 700** (top right), Kristen Brochmann/Fundamental Photographs; **17-12,** Tom Pantages

CHAPTER 18 p. 710, Richard Megna/Fundamental Photographs; **18-1,** Carey Van Loon; **18-2,** Carey Van Loon; **18-3,** Carey Van Loon; **18-4,** Carey Van Loon; **p. 724,** Tom Pantages; **18-8,** Carey Van Loon; **p. 726,** Elena Rooraid/PhotoEdit; **p. 738,** John W. Finley

CHAPTER 19 p. 749, John Heseltine/Science Photo Library/Photo Researchers, Inc.; **p. 751,** Tom Pantages; **p. 752,** Sovereign/ISM/Phototake NYC; **19-1,** Carey Van Loon; **19-3a & b,** Carey Van Loon; **p. 761,** Richard Megna/Fundamental Photographs; **19-5a & b,** Tom Pantages; **19-6,** Richard Megna/Fundamental Photographs; **p. 769,** Carey Van Loon; **19-8,** Carey Van Loon; **p. 772** (top), Jane Burton/Bruce Coleman Inc.; **19-9,** Richard Megna/Fundamental Photographs; **p. 779,** Walter S. Clark/Photo Researchers, Inc.

CHAPTER 20 p. 782, Richard A. Cooke, III/Stone; **p. 783** (top), PhotoDisc, Inc.; **p. 783** (bottom), Michael Dalton/Fundamental Photographs; **p. 785,** Central Library for Physics, Vienna, Austria; **p. 794,** Science Photo Library/Photo Researchers, Inc.; **p. 805,** Joseph Sohm/Stone; **p. 812,** P. Motta & T. Naguro/Science Photo Library/Photo Researchers, Inc.

CHAPTER 21 p. 823, Thomas Kienzle/AP/Wide World Photos; **21-1,** Carey Van Loon; **p. 833,** Corbis;

21-7, Diane Schiumo/Fundamental Photographs; **p. 839,** Stamp from the private collection of Professor C. M. Lang, photography by Gary J. Shulfer, University of Wisconsin, Stevens Point. "1980, Sweden (Scott #1344);" Scott Standard Postage Stamp Catalogue, Scott Pub. Co., Sidney, Ohio; **21-12,** Tom Pantages; **p. 845** (top), Carey Van Loon; **21-15** (left), SuperStock, Inc.; **p. 847,** Lester V. Bergman/Corbis; **p. 848,** Courtesy of Toyota Motor Sales, U.S.A., Inc.; **21-19,** Carey Van Loon; **p. 850,** Robert Erlbacher/Missouri Dry Rock & Repair Co., Inc.; **p. 853,** Runk/Schoenberger/Grant Heilman Photography, Inc.; **p. 856,** Charles E. Rotkin/Corbis; **p. 857,** Richard Pasley/Stock Boston; **p. 860,** Kenneth Eward/BioGrafx; **p. 864,** Carey Van Loon; **p. 866,** New York Convention & Visitors Bureau

CHAPTER 22 p. 872, Tony Perrottet/Omni-Photo Communications, Inc.; **p. 873** (top), Joel E. Arem; **p. 873** (bottom), Ed Degginger/Color-Pic, Inc.; **p. 876,** Department of Clinical Radiology, Salisbury District Hospital/Science Photo Library/Photo Researchers, Inc.; **p. 877,** Kevin Schafer/Corbis; **p. 878,** Ed Degginger/Color-Pic, Inc.; **p. 880,** Jeffrey A. Scovil; **p. 885** (top), William E. Ferguson; **22-9,** Bjorn Bolstad/Peter Arnold, Inc.; **p. 886,** SuperStock, Inc.; **p. 887** (top), Sheila Terry/Science Photo Library/Photo Researchers, Inc.; **p. 887** (bottom), Carey Van Loon; **22-11a-c,** Carey Van Loon; **22-13** (right), Kenneth Eward/Bio Grafx; **p. 894,** Ed Degginger/Color-Pic, Inc.; **p. 895,** Tom Bochsler/Pearson Education/PH College; **p. 898,** Bruce Frisch/Photo Researchers; **p. 904,** Rolf Hicker/Stone

CHAPTER 23 p. 906, Albermarle Corporation; **p. 908,** Argonne National Laboratory; **p. 911** (top), Richard Dunoff/Corbis/Stock Market; **p. 911** (bottom), Tom Pantages; **p. 912,** Richard Megna/Fundamental Photographs; **p. 915,** Carey Van Loon; **23-6a,** Jeffrey A. Scovil; **23-6b,** Tom Bochsler/Pearson Education/PH College; **23-6c,** Carey Van Loon; **23-6d,** Stephen Frisch/Stock Boston; **p. 919,** Bill Pierce/Rainbow; **p. 920,** Tom Pantages; **p. 922** (a & b), Ken Lax/Pearson Education/PH College; **23-12,** Energy Technology Visuals Collection/U.S. Department of Energy; **p. 926,** Ed Degginger/Color-Pic, Inc.; **p. 928,** NASA Headquarters; **p. 929,** Tom Pantages; **p. 932,** Charlie Ott/Photo Researchers, Inc.; **p. 933,** Bruce Frisch/Photo Researchers, Inc.; **p. 934,** Gem Media/Gemological Institute of America; **p. 938** (top), Photo courtesy of Kyocera Industrial Ceramics Corp./Vancouver, WA; **p. 938** (bottom), Frank La Bua/Pearson Education/PH College; **p. 940,** Jan Halaska/Photo Researchers, Inc.; **p. 942,** Carey Van Loon

CHAPTER 24 p. 949, Michael Dalton/Fundamental Photographs; **p. 953,** Reprinted with permission from M. Seul, L. R. Monar, L. O'Gorman and R. Wolfe, Morphology and local structure in labyrinthine stripe domain phase, Science 254:1616, 1991. Copyright 1991 American Association for the Advancement of Science; **24-4b,** Photo courtesy of Peter Bourke, Outokumpu, Perth, Australia, reproduced by permission of Outokumpu OYJ, Riihitontuntie, Finland; **24-5,** Sol Mednick Gallery; **p. 959,** Wah Chang; **p. 960,** Day Williams/Photo Researchers, Inc.; **p. 961,** Robert Semeniuk/Corbis/Stock Market; **p. 963,** Steven Weinberg/Stone; **p. 964** (top), PhotoDisc, Inc.;

p. 964 (bottom), Rich Chisholm/Corbis/Stock Market; **24-10,** Richard Megna/Fundamental Photographs; **p. 966,** Phil Degginger/Color-Pic, Inc.; **24-11,** Carey Van Loon; **24-12,** Carey Van Loon; **p. 970,** Tom Pantages; **p. 973,** Richard Megna/Fundamental Photographs; **p. 974,** NASA/Tom Pantages; **p. 975,** © National Maritime Museum Picture Library, London, England; **p. 978,** David Parker/IMI/University of Birmingham High TC Consortium/Science Photo Library/Photo Researchers, Inc.

CHAPTER 25 p. 985, Karl Hartmann/Traudel Sachs/Phototake NYC; **p. 986,** Science Photo Library/Photo Researchers, Inc.; **25-1,** Carey Van Loon; **25-16a,** Fritz Goro/TimePix; **25-18,** Carey Van Loon; **25-20,** Carey Van Loon; **25-21,** Carey Van Loon; **25-22,** Carey Van Loon; **p. 1014,** Tom Pantages; **p. 1016,** Ed Degginger/Color-Pic, Inc.; **25-26b,** Richard Megna/Fundamental Photographs

CHAPTER 26 p. 1024, NASA/Grant Heilman Photography, Inc.; **p. 1025,** Science Photo Library/Photo Researchers, Inc.; **p. 1026,** Lawrence Berkeley/Science Photo Library/Photo Researchers, Inc.; **p. 1029,** Corbis; **p. 1030,** UPI/Corbis; **p. 1031,** CERN Photo; **p. 1034,** Corbis/Sygma; **p. 1035,** Science VU/NASA-JMP/Visuals Unlimited; **p. 1042,** Yann Arthus-Bertrand/Photo Researchers, Inc.; **26-9b,** David M. Doody/Tom Stack & Associates; **p. 1043** (a), Phil Degginger/Stone; **p. 1043** (b), V. Ivelva/Magnum Photos, Inc.; **p. 1043** ©, Esbin/Anderson/Omni-Photo Communications, Inc.; **p. 1044,** Jerry Mason/Science Photo Library/Photo Researchers, Inc.; **p. 1045,** Yoav Levy/Phototake NYC; **p. 1046,** Ed Degginger/Color-Pic, Inc.; **p. 1048** (a & b), Perrin/CNRI/Phototake NYC; **p. 1049,** (top) Gemaldegalerie, Dahlem-Berlin, Germany/A.K.G., Berlin/SuperStock; **p. 1049** (bottom), Kip Peticolas/Richard Megna/Fundamental Photographs; **p. 1050,** U.S. Department of Energy/Science Source/Photo Researchers, Inc.

CHAPTER 27 p. 1058, Kristen Brochmann/Fundamental Photographs; **p. 1061,** Government of Switzerland; **p. 1070,** Ed Degginger/Color-Pic, Inc.; **p. 1073,** Ed Degginger/Color-Pic, Inc.; **p. 1075,** Stamp from the private collection of Professor C. M. Lang, photography by Gary J. Shulfer, University of Wisconsin, Stevens Point. "Belgium," Scott Standard Postage Stamp Catalogue, Scott Pub. Co., Sidney, Ohio; **p. 1079,** Pete Saloutos/Corbis/Stock Market; **p. 1081,** Science Photo Library/Photo Researchers, Inc.; **p. 1084,** Bill Ross/Corbis; **p. 1105,** Christopher Springmann/Corbis/Stock Market; **p. 1108,** Michael Holford/Michael Holford Photographs

CHAPTER 28 p. 1126, (top) The Proctor & Gamble Company; **p. 1126,** (bottom) Richard Megna/Fundamental Photographs; **p. 1130,** Corbis; **p. 1131,** Jerry Mason/Science Photo Library/Photo Researchers, Inc.; **p. 1135,** National Exposure Research Laboratory, EPA/Tom Pantages; **28-15,** Arthur Lesk/Science Photo Library/Photo Researchers, Inc.; **p. 1142,** Tom Pantages

Index

Selected Physical Constants

Acceleration due to gravity	g	$9.80665 \ \mathrm{m \cdot s^{-2}}$
Speed of light (in vacuum)	c	$2.99792458 \times 10^8 \ \mathrm{m \cdot s^{-1}}$
Gas constant	R	$0.0820574 \ \mathrm{L \cdot atm \cdot mol^{-1} \cdot K^{-1}}$
		$8.314472 \ \mathrm{J \cdot mol^{-1} \cdot K^{-1}}$
Electron charge	e^-	$-1.602176462 \times 10^{-19} \ \mathrm{C}$
Electron rest mass	m_e	$9.10938188 \times 10^{-31} \ \mathrm{kg}$
Planck's constant	h	$6.62606876 \times 10^{-34} \ \mathrm{J \cdot s}$
Faraday constant	F	$9.64853415 \times 10^4 \ \mathrm{C \cdot mol^{-1}}$
Avogadro constant	N_A	$6.02214199 \times 10^{23} \ \mathrm{mol^{-1}}$

Some Common Conversion Factors

Length

1 meter (m) = 39.37007874 inches (in.)

1 in. = 2.54 centimeters (cm) (exact)

Mass

1 kilogram (kg) = 2.2046226 pounds (lb)

1 lb. = 453.59237 grams (g)

Volume

1 liter (L) = 1000 mL = 1000 cm³ (exact)

1 L = 1.056688 quart (qt)

1 gallon (gal) = 3.785412 L

Force

1 newton (N) = $1 \ \mathrm{kg \cdot m \cdot s^{-2}}$

Energy

1 joule (J) = $1 \ \mathrm{N \cdot m} = 1 \ \mathrm{kg \cdot m^2 \cdot s^{-2}}$

1 calorie (cal) = 4.184 J (exact)

1 electronvolt (eV) = $1.602176462 \times 10^{-19}$ J

1 eV/atom = $96.485 \ \mathrm{kJ \cdot mol^{-1}}$

1 kilowatt hour (kWh) = 3600 kJ (exact)

Mass-energy equivalence:

   1 unified atomic mass unit (u)

   = $1.66053873 \times 10^{-27}$ kg

   = 931.4866 MeV

Some Useful Geometric Formulas

Perimeter of a rectangle = $2l + 2w$

Circumference of a circle = $2\pi r$

Area of a rectangle = $l \times w$

Area of a triangle = $\frac{1}{2}$ (base × height)

Area of a circle = πr^2

Area of a sphere = $4\pi r^2$

Volume of a parallelepiped = $l \times w \times h$

Volume of a sphere = $\frac{4}{3} \pi r^3$

Volume of a cylinder or prism = (area of base) × height

$\pi \approx 3.14159$